Intermediate Algebra

Andrea Hendricks
Georgia Perimeter College

Oiyin Pauline Chow
Harrisburg Area Community College

The McGraw-Hill Companies

INTERMEDIATE ALGEBRA

Published by McGraw-Hill, a business unit of The McGraw-Hill Companies, Inc., 1221 Avenue of the Americas, New York, NY 10020.

This book is printed on acid-free paper.

1 2 3 4 5 6 7 8 9 0 DOW/DOW 1 0 9 8 7 6 5 4 3 2

ISBN 978–0–07–338426–9
MHID 0–07–338426–7

ISBN 978–0–07–336097–3 (Annotated Instructor's Edition)
MHID 0–07–336097–X

Vice President, Editor-in-Chief: *Marty Lange*
Vice President, EDP: *Kimberly Meriwether David*
Senior Director of Development: *Kristine Tibbetts*
Editorial Director: *Stewart K. Mattson*
Executive Editor: *Dawn R. Bercier*
Sponsoring Editor: *Mary Ellen Rahn*
Director of Digital Content Development: *Emilie J. Berglund / Nicole Lloyd*
Developmental Editor: *Emily Williams*
Marketing Manager: *Peter A. Vanaria*
Senior Project Manager: *Vicki Krug*
Senior Buyer: *Sherry L. Kane*
Senior Media Project Manager: *Sandra M. Schnee*
Senior Designer: *Laurie B. Janssen*
Cover Illustration: *Imagineering Media Services Inc.*
Senior Photo Research Coordinator: *Lori Hancock*
Photo Research: *Danny Meldung/Photo Affairs, Inc*
Compositor: *Cenveo Publisher Services*
Typeface: *10.5 Times LT Std*
Printer: *R. R. Donnelley*

All credits appearing on page or at the end of the book are considered to be an extension of the copyright page.

Library of Congress Cataloging-in-Publication Data

Hendricks, Andrea.
Intermediate algebra / Andrea Hendricks, Oiyin Pauline Chow.
p. cm.
Includes index.
ISBN 978-0-07-338426-9—ISBN 0-07-338426-7 (hard copy : alk. paper) 1. Algebra. I. Chow, Oiyin Pauline. II. Title.
QA152.3.H3355 2013
512—dc23 2011031865

www.mhhe.com

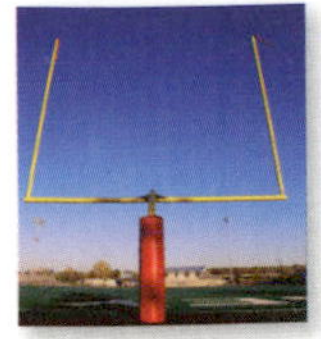

CHAPTER 6 **Exponents, Polynomials, and Polynomial Function 365**

Goal Setting You need to make sure that your goals are attainable. It is important that you review these goals often so that you stay focused on what you are working toward.

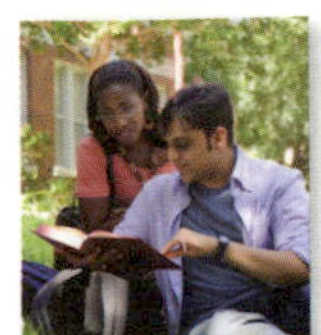

CHAPTER 7 **Rational Expressions, Functions, and Equations 467**

Learning Strategies Some strategies for improving learning includes finding out the kind of learner we are, reviewing class notes a few hours after class and reviewing the notes repeatedly for better retention of course concepts.

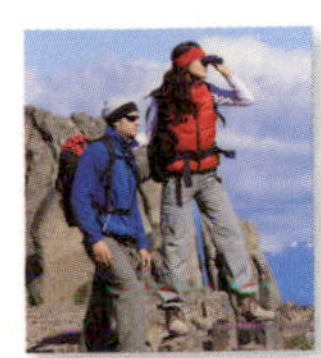

CHAPTER 8 **Rational Exponents, Radicals, and Complex Numbers 551**

Refocus Take a moment to remember why you are enrolled in college and why you are enrolled in this class. Keep sight of your short term and long term goals. It will be worth it!

CHAPTER 9 **Quadratic Equations and Functions and Nonlinear Inequalities 635**

Reflection Think about what you walk away with as you complete this course. Certainly that should include some math, but we hope it might also include a deeper understanding of what it takes to be successful in a college math course.

CHAPTER 10 **Exponential and Logarithmic Functions 725**

Greatness Do you have the motivation to become great? You do not need to have some special gift or talent. You are the only thing you need to make yourself great!

CHAPTER 11 **Conic Sections and Nonlinear Systems 807**

Responsibility As a student, you must be responsible for your education. Remember, college is a privilege and a magnificent way for you to grow yourself while at the same time making yourself extremely marketable and competitive in today's workplace.

CHAPTER 12 **Sequences, Series, and the Binomial Theorem 855**

Persistence Persistence is the key ingredient in getting your degree and achieving success in life. Completing one course at a time is the way to earn your degree. Courses become semesters, semesters become years, and years become degrees.

About the Cover

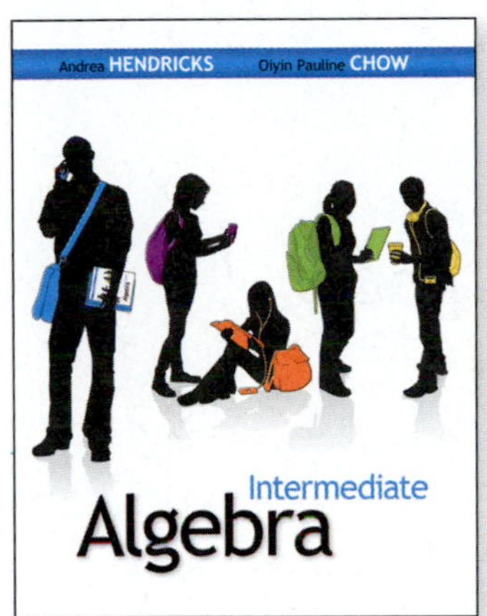

In developmental math, the focus needs to be about the "whole student" and providing students with "more than just the math." As Andrea and Pauline say, we want students to know that we care about their success. Therefore, we chose to include students on the covers to show how interested we are in their pursuit to succeed and persist through their math courses. We also wanted to visually represent today's students in their current environments. Our authors have made this their focus too by providing an entire chapter of robust resources for teaching and learning success strategies beyond just the math. And, no matter the course format, Andrea and Pauline have provided purposeful examples and exercises, current and relevant applications, and critical thinking exercises to reach today's whole student. Our hope is that students will want to envision themselves as the successful and confident math students on these covers.

Dedications

This text is dedicated to my wonderful family. Thank you, Todd, for your support and encouragement through this process and my wonderful boys, Andy, Charlie, and Cory, for understanding all of those times that Mommy had to work on the computer.

—*Andrea Hendricks*

Chow Cheung—my father, who recognized the importance of education and sent all seven of his children abroad to pursue a college education in the 70s. Wong Shing—my mother, who constantly provides us with love and nurture. Michael, Amy, and Andrew—my family, pride and joy.

—*Oiyin Pauline Chow*

Hendricks & Chow

Developmental Math Hardcover Series

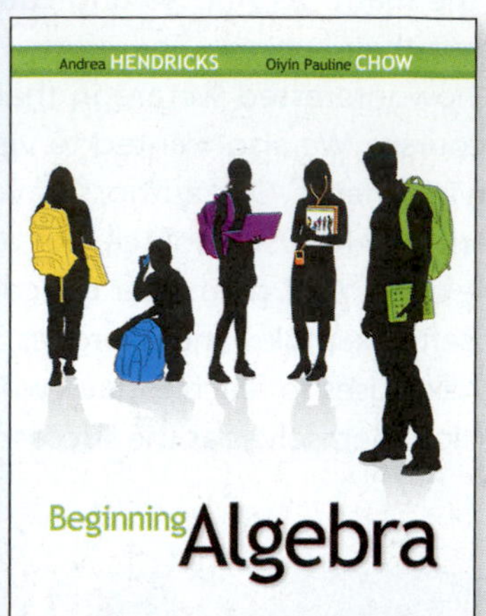

Beginning Algebra

Andrea Hendricks and Oiyin Pauline Chow, ©2013

ISBN: 978-0-07-338427-6
MHID: 0-07-338427-5

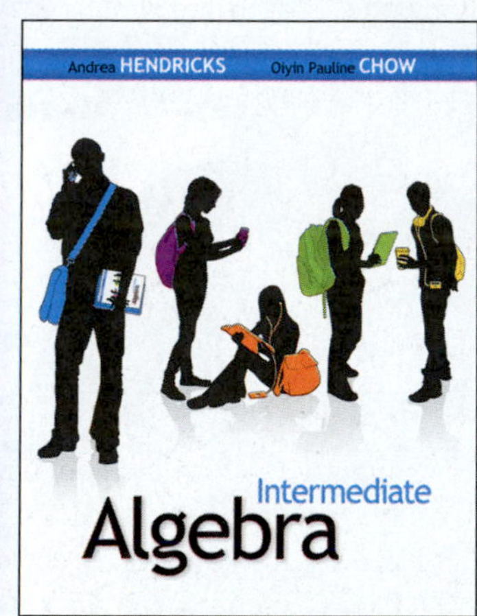

Intermediate Algebra

Andrea Hendricks and Oiyin Pauline Chow, ©2013

ISBN: 978-0-07-338426-9
MHID: 0-07-338426-7

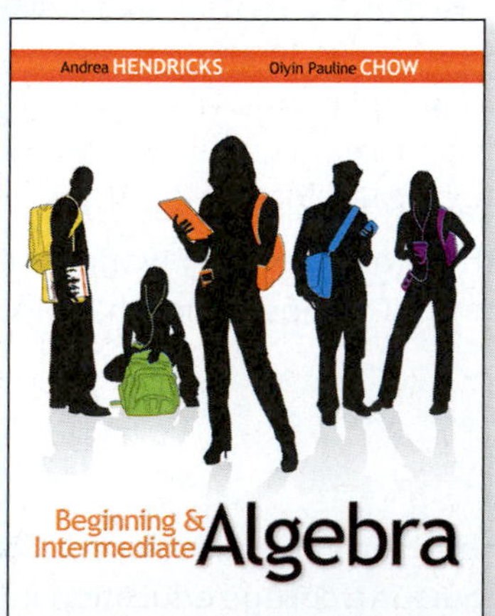

Beginning & Intermediate Algebra

Andrea Hendricks and Oiyin Pauline Chow, ©2013

ISBN: 978-0-07-338453-5
MHID: 0-07-338453-4

About the Authors

Andrea Hendricks I am an Associate Professor of Mathematics at Georgia Perimeter College (GPC), Online Campus. I have been teaching at the college level since 1992 and have taught the full range of math courses. In 2008, I joined the online campus of GPC and teach exclusively online. Prior to joining the online campus, I taught traditional face-to-face classes, hybrid classes, and online classes. During my tenure at GPC, I have served as Assistant Department Chair for one of the ground campuses and am currently Assistant Department Chair for the online math department, which gives me responsibility for the part-time faculty members and for managing and overseeing the developmental math courses. In addition, I have chaired and served on various curriculum committees, the Faculty Senate, Peer Review Committees, Promotion and Tenure Panels, and the Math Conference Committee. I have received the NISOD Teaching Excellence Award, the Faculty Teaching and Service Award, and the GPC Collegiality Award. I particularly enjoy teaching developmental math classes so that I can help students overcome their fear and anxiety toward learning math. I look forward to sharing my strategies and teaching moments on a larger scale within this Developmental Math series. My husband, Todd, is also a professor of mathematics at GPC, and we have three growing boys under the age of 13: Andy, Charlie, and Cory. In additon to teaching, authoring, and being a mom, I enjoy golf, playing the piano, and am involved with my church through teaching a 4-year-old Sunday school class and also AWANA classes.

Oiyin Pauline Chow As a Senior Professor and Chair of the Mathematics and Computer Science Department at Central Pennsylvania's Community College (HACC), I have approximately 30 years of teaching experience, and I have taught a full curriculum—from Developmental Math to Calculus, Linear Algebra, and Differential Equations in both traditional and online formats. Currently, I serve as Mathematics Department Chair at HACC's five regional campuses and their virtual campus. I also participate in many active roles on campus to support the Faculty Council and various developmental education and department committees. My other interests include serving on various state and national math organizations: the executive boards of AMATYC (secretary) and three Pennsylvania math organizations, PCTM (president), PADE (member at large), and PSMATYC (president). I have received the NISOD Teaching Excellent Award, The PADE Exemplary Teaching in Developmental Education award, and the PCTM Outstanding Contribution to Mathematics Education Award. The greatest joy in this profession is being able to work with students, empower them to take ownership in learning, and help them see their own accomplishments. In particular, at the Developmental Math level, I take pride if I can change students' outlook on mathematics, help them achieve an appreciation of math that they never had, and gain a better understanding of math. My husband Mike and I are experiencing an official empty nest, but we enjoy visiting and keeping tabs on our son, Andrew, at his software developer job in Palo Alto California and our daughter, Amy, in New York City at a digital marketing firm.

> "Students want to know that someone is interested in their success."
>
> —Andrea Hendricks and Pauline Chow

Our Development Story

From our years of teaching developmental math students, we have observed that more and more students need a review of study skills and how these can be applied to equal success in college. So, when we teach, we teach more than just the math—we teach life skills. Accordingly, more and more schools are integrating study skills into their developmental math classes to help retain students and improve their success rates. Students want to know that someone is interested in their success and can make the math attainable without sacrificing the integrity of the course. Therefore, we felt that it is our role to provide a new, purposeful developmental math text series. We have thoroughly enjoyed partnering in this quest of authoring and have formed a great team with Andrea creating the organization and framework for the text and writing the narrative and explanations/examples while Pauline manages the exercise sets and oversees the digital content. Together, we have focused on the following three key course needs to provide an accessible, relevant, and motivating series that will help drive students to succeed.

Success Strategies For Today's Students

Since student success is critical, we incorporated a separate chapter on student success strategies including topics like time management, note taking, and preparing for tests. We integrated these topics throughout the text by highlighting each chapter opener with a characteristic of a successful student and a motivational quote. Therefore, students will find new resources needed to be successful at their fingertips while studying math.

Student-Centered Examples and Exercise Sets

Today, we must address and serve a diverse population of developmental math students with various learning styles. When students realize the relevance of math in their lives, they are motivated to learn the material and are prepared to move into various academic disciplines. We have worked diligently to ensure that our clear, concise explanations, exercises, and applications mean something to today's students. Our numerous worked examples are organized by objectives with detailed, step-by-step procedures wherever possible. We also have worked to provide multiple ways for students to learn—troubleshooting their own mistakes, writing about the math, advancing through various levels of problems, identifying other's errors, and critically taking ownership of the material that they are learning.

Digital Solutions For Every Teaching Environment

Because of the growth of online learning, hybrid classes, and course redesigns, it has become even more important to create materials that help students grow into successful, independent learners. This book is designed not only to motivate and inspire students but students can also learn from it in any type of classroom environment. From our experience teaching online, it is clear to us that students depend on the textbook more so than in a traditional classroom. We also worked to take an active role in the creation of the digital content, and while developing the organization and progression of our exercise sets, we considered how a student works through online homework for additional practice. Each and every exercise has a purpose both in the text and online to support the objectives. We were involved in the process of ensuring that the guided solutions in the online homework represent the same author voice and narrative as the printed book to eliminate any inconsistencies.

Andrea M. Hendricks and [signature]

Contents

Preface vi
Application index xxxii

CHAPTER S **Strategies to Succeed in Math** S-1

S.1 Time Management and Goal Setting S-2
S.2 Learning Styles S-6
S.3 Study Skills S-11
S.4 Test Taking S-19
S.5 Blended and Online Classes S-25
Chapter S REVIEW S-28

CHAPTER 1 **Real Numbers and Algebraic Expressions** 1

1.1 Sets and the Real Numbers 2
1.2 Operations with Real Numbers and Algebraic Expressions 15
1.3 Properties of Real Numbers and Simplifying Algebraic Expressions 33
Group Activity A Magical Mathematical Birthday Card 44
Chapter 1 REVIEW 45
Summary 46
Review Exercises 47
Test 49

CHAPTER 2 **Linear Equations and Inequalities in One Variable** 51

2.1 Solving Linear Equations 52
2.2 Introduction to Applications 65
2.3 Formulas and Applications 72
2.4 Linear Inequalities and Applications 85
Piece it Together 2.1–2.4 100
2.5 Compound Inequalities 101
2.6 Absolute Value Equations 116
2.7 Absolute Value Inequalities 125
Group Activity The Mathematics of Controlling Waste 137
Chapter 2 REVIEW 138
Summary 139
Review Exercises 140
Test 143
Cumulative Review Exercises 145

CHAPTER 3 Graphs, Relations, and Functions 147

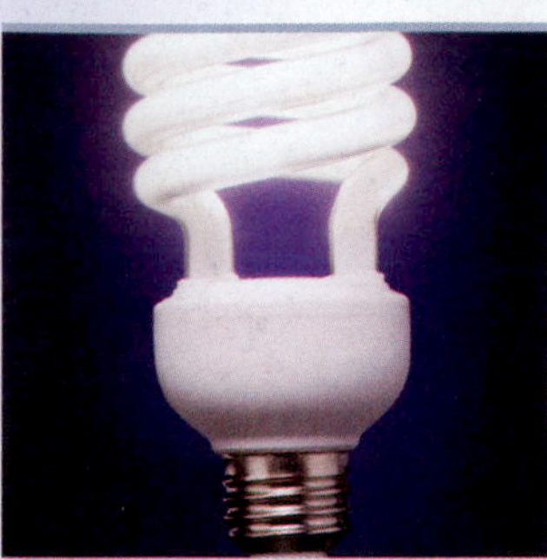

3.1 The Coordinate System, Graphing Equations, and the Midpoint Formula 148
3.2 Relations 165
Piece it Together 3.1–3.2 177
3.3 Functions 178
3.4 The Domain and Range of Functions 190
Group Activity The Mathematics of Professional Sports 200
Chapter 3 REVIEW 201
Summary 201
Review Exercises 202
Test 206
Cumulative Review Exercises 207

CHAPTER 4 Linear Functions and Linear Inequalities in Two Variables 211

4.1 Linear Functions and Linear Equations in Two Variables 212
4.2 Graphing Linear Equations and Linear Functions 224
4.3 The Slope of a Line 238
Piece it Together 4.1–4.3 254
4.4 Writing Equations of Lines 255
4.5 Linear Inequalities in Two Variables 269
Group Activity The Wave 278
Chapter 4 REVIEW 279
Summary 279
Review Exercises 280
Test 283
Cumulative Review Exercises 284

CHAPTER 5 Systems of Linear Equations and Inequalities 289

5.1 Solving Systems of Linear Equations in Two Variables Graphically 290
5.2 Solving Systems of Linear Equations in Two Variables Algebraically 305
Piece it Together 5.1–5.2 319
5.3 Applications of Linear Systems in Two Variables 319
5.4 Solving Linear Systems in Three Variables and Their Applications 335
5.5 Solving Systems of Linear Inequalities and Their Applications 347
Group Activity Linear Programming—Maximize Revenue 356
Chapter 5 REVIEW 357
Summary 357
Review Exercises 358

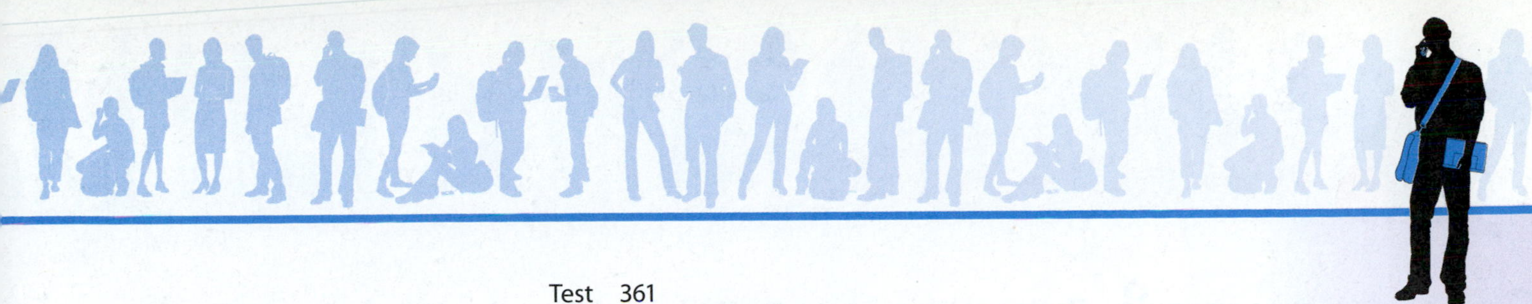

Test 361
Cumulative Review Exercises 362

CHAPTER 6 Exponents, Polynomials, and Polynomial Functions 365

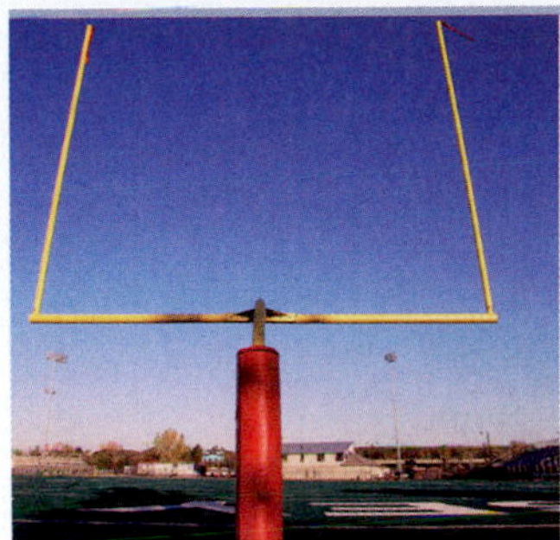

6.1 Rules of Exponents, Zero and Negative Exponents 366
6.2 More Rules of Exponents and Scientific Notation 379
Piece it Together 6.1–6.2 391
6.3 Polynomials, Polynomial Functions, and Their Basic Graphs 391
6.4 Adding, Subtracting, and Multiplying Polynomials and Polynomial Functions 402
Piece it Together 6.3–6.4 417
6.5 Factoring Using the Greatest Common Factor and Grouping 418
6.6 Factoring Trinomials 425
Piece it Together 6.5–6.6 436
6.7 Factoring Binomials and a Factoring Review 437
6.8 Solving Polynomial Equations and Their Applications 445
Group Activity Creating an Amortization Schedule 456
Chapter 6 REVIEW 457
Summary 458
Review Exercises 459
Test 462
Cumulative Review Exercises 463

CHAPTER 7 Rational Expressions, Functions, and Equations 467

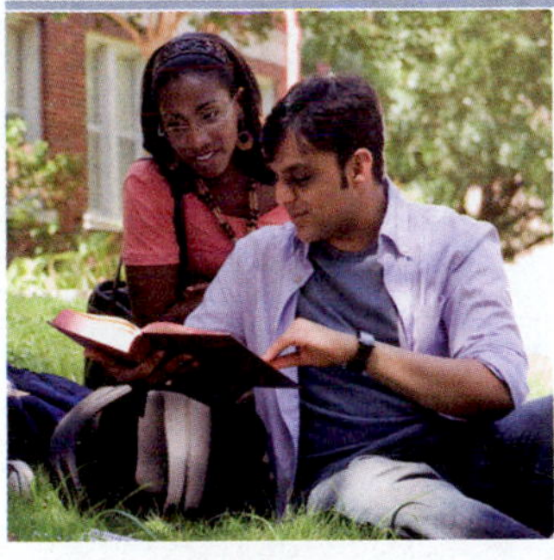

7.1 Rational Functions; Multiplying and Dividing Rational Expressions 468
7.2 More Division of Polynomials: Long Division and Synthetic Division 483
7.3 Adding and Subtracting Rational Expressions 495
7.4 Simplifying Complex Fractions 505
Piece it Together 7.1–7.4 515
7.5 Solving Rational Equations 516
7.6 Applications of Rational Equations 523
7.7 Variation and Applications 530
Group Activity Mathematics of Operating a Vehicle 541
Chapter 7 REVIEW 542
Summary 542
Review Exercises 543
Test 546
Cumulative Review Exercises 547

CHAPTER 8 Rational Exponents, Radicals, and Complex Numbers 551

8.1 Radicals and Radical Functions 552
8.2 Rational Exponents 567
8.3 Simplifying Radical Expressions and the Distance Formula 578
8.4 Adding, Subtracting, and Multiplying Radical Expressions 589
Piece it Together 8.1–8.4 597
8.5 Dividing Radical Expressions and Rationalizing 598
8.6 Radical Equations and Their Applications 606
8.7 Complex Numbers 615
Group Activity Solutions of Quadratic Equations 625
Chapter 8 REVIEW 626
Summary 627
Review Exercises 628
Test 630
Cumulative Review Exercises 631

CHAPTER 9 Quadratic Equations and Functions and Nonlinear Inequalities 635

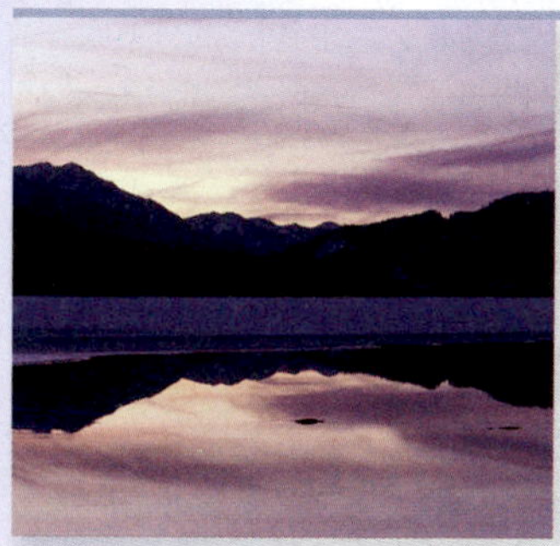

9.1 Quadratic Functions and their Graphs 636
9.2 Solving Quadratic Equations Using the Square Root Property and Completing the Square 651
9.3 Solving Quadratic Equations Using the Quadratic Formula 665
Piece it Together 9.1–9.3 676
9.4 Solving Equations Using Quadratic Methods 667
9.5 More on Graphing Quadratic Functions 690
9.6 Solving Polynomial and Rational Inequalities in One Variable 700
Group Activity The Mathematics of Fatal Crashes 714
Chapter 9 REVIEW 716
Summary 716
Review Exercises 718
Test 720
Cumulative Review Exercises 721

CHAPTER 10 Exponential and Logarithmic Functions 725

10.1 Operations and Composition of Functions 726
10.2 Inverse Functions 739
10.3 Exponential Functions 754
10.4 Logarithmic Functions 762
Piece it Together 10.1–10.4 769
10.5 Properties of Logarithms 769

10.6 The Common Log, Natural Log, and Change-of-Base Formula 778
10.7 Exponential and Logarithmic Equations and Applications 787
Group Activity The Mathematics of Financing a College Education 797
Chapter 10 REVIEW 798
Summary 798
Review Exercises 799
Test 803
Cumulative Review Exercises 803

CHAPTER 11 Conic Sections and Nonlinear Systems 807

11.1 The Parabola and the Circle 808
11.2 The Ellipse and the Hyperbola 817
Piece it Together 11.1–11.2 824
11.3 Solving Nonlinear Systems of Equations 824
11.4 Solving Nonlinear Inequalities and Systems of Inequalities 831
Group Activity The Mathematics of Orbits 839
Chapter 11 REVIEW 840
Summary 840
Review Exercises 841
Test 842
Cumulative Review Exercises 842

CHAPTER 12 Sequences, Series, and the Binomial Theorem 855 (Website)

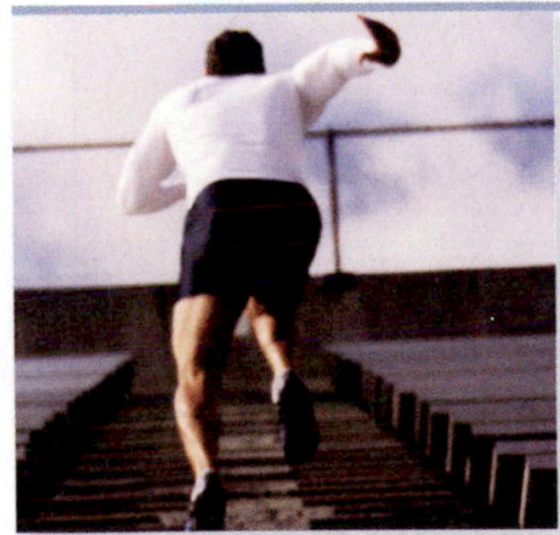

12.1 Sequences 856
12.2 Arithmetic Sequences and Series 863
Piece it Together 12.1–12.2 875
12.3 Geometric Sequences and Series 876
12.4 The Binomial Theorem 887
Group Activity The Mathematics of Accumulated Wealth 894
Chapter 12 REVIEW 895
Summary 895
Review Exercises 896
Test 897
Cumulative Review Exercises 898

Appendix A Gaussian Elimination and Cramer's Rule A-1 (Website)
(Chapter 12 and Appendix A can be found at www.mhhe.com/hendrickschow)

Student Answer Appendix (SE only) SA-1
Instructor Answer Appendix (AIE only) IA-1
Credits C-1
Index I-1

Success Strategies for Today's Students

Today's students need more guidance than ever before on how to succeed in both their current math course and future courses.

- Do your students come to class prepared to study, take good notes, make time for homework, and study for tests?
- How do you cover all of the math curriculum and teach these success strategies?

The Hendricks/Chow series provides materials dedicated to student success strategies that are integrated within both the text and the instructor and student supplements.

How do you incorporate study skills or success strategies into your course?

CHAPTER S

Strategies to Succeed in Math

Success

It is with great delight that we welcome you to this course and the materials we have provided you. You are embarking on perhaps the most rewarding experience of your life. College is a series of challenges that will prepare you for a lifetime of learning and accomplishments. At the end of your college education, you will be awarded a degree—a degree that establishes your ability to learn and to persevere. It is a statement that you can work to accomplish a goal, no matter what the obstacles.

At this point, you must evaluate the reason you are here. Are you here to fulfill your dreams or the dreams of someone else? To successfully complete this journey, you must be here to fulfill your own desires, not those of a friend or family member. It will be most difficult to withstand the trials of college if you do not have a personal desire to see it through.

A semester or quarter can be very overwhelming. Focus on one day at a time and not on everything that you must learn throughout the entire course. Before you know it, the course will be over. We know that it is very easy to get distracted from your goals. Stay committed and motivated by remembering your ultimate reason for attending school.

We wish you success in this course and in your future educational endeavors. It is our hope that you are successful, not only this semester or quarter, but every term until your ultimate goal is achieved. This course will pave the way for that success. It is the door to achieving your dreams.

Mrs. Andrea Hendricks and Mrs. Pauline Chow

Question for Thought: What do you dream about doing? How does attending college make that dream possible? How does this course help you meet your goals? What is your biggest obstacle in being successful in this course? What can you do to overcome that obstacle? What is your plan for succeeding in this course?

Chapter S will introduce some important strategies that can enhance your performance in this course.

Chapter Outline

Section S.1 Time Management and Goal Setting
Section S.2 Learning Styles
Section S.3 Study Skills

An entire chapter, entitled **Strategies to Succeed in Math**, is dedicated to study skills, such as time management, test taking, note taking, and tips to succeed in online/hybrid courses.

A **Learning Style Inventory Quiz** will help students identify if they are a visual, auditory, or kinesthetic learner and then provides specific math strategies to help support the various learning styles.

ACTIVITY 1 Complete the following survey to determine your math learning style.

Math Learning Styles Survey

Answer each question with the number "3" if you agree most of the time, "2" if you agree sometimes, and "1" if you agree rarely or do not agree.

_____ 1. When there is talking or noise in class, I get easily distracted.
_____ 2. If a problem is written on the board, I have difficulty following the steps unless the teacher verbally explains the steps.
_____ 3. I find it easier to have someone explain something to me than to read it in my math book.
_____ 4. If I know how the math is used in real life, it is easier for me to learn.
_____ 5. To remember formulas and definitions, I need to write them down.
_____ 6. I prefer listening to the lecture rather than taking notes.
_____ 7. Using manipulatives, hands-on activities, or games helps me learn math concepts.
_____ 8. When solving a problem, I try to picture working it out in my mind first.

SECTION S.3 Study Skills

OBJECTIVES

As a result of completing this section, you will be able to

1. Identify habits that good math students employ.
2. Read the textbook more effectively and efficiently.
3. Take quality notes.
4. Organize materials in a notebook or portfolio.
5. Complete homework in ways that maximize retention.
6. Establish an action plan.
7. Troubleshoot common errors.

Study skills are the skills students need to improve their learning capacity and to acquire new knowledge. No two students are going to employ the same set of study techniques, though there are some that every student should utilize when studying math. This section will address some of the skills that will enhance the success of all students.

Habits of Successful Math Students

Objective 1 Identify habits that good math students employ.

Good math students study with purpose. When you underline something, it should be because it is important. If you write out a note card, it should be because it is something you want to remember long term. When you work a homework problem, it should lead to a better understanding of the process used to solve that problem. If you can't answer "Why am I doing this?" then what you are doing may not be a productive use of your time. You should always have goals to achieve when you sit down to study and everything you do during your study time should work toward those goals. The following suggestions should help you approach your math class with purpose.

1. *Successful students are responsible.*
   - Attend every class meeting. Try to be a few minutes early so that your materials are out and you are ready for class to begin.
   - Find an accountability partner in class. Encourage one another and use each other as a resource if one of you is absent from class.
   - Adhere to all deadlines for assignments.
2. *Successful students make the most of their time in class.*
   - When you are in class, "be in class." Try not to worry about other things going on in your life. Focus your attention on the topics being discussed. (No texting, surfing, or doing other homework.)
   - Either take notes in class or record the lecture so that you can refer to it later. Some instructors even post their class notes on their website.
   - Actively participate in class. Follow along with the instructor. Ask questions

Instructors and students will benefit from several helpful strategies to use while in these courses.

▶ Are your students coming to your class prepared to practice appropriate study skills?

Each chapter also contains additional materials on success strategies, such as motivational quotes and chapter openers dedicated to a characteristic of a successful student.

CHAPTER 1

Real Numbers and Algebraic Expressions

Time Management

As you begin this chapter, we want you to focus on how you are managing your time. Your success in this course depends on your ability to devote the appropriate time to learning and practicing the material.

- The general rule of thumb is that you should spend 2 hours studying for each hour you are in class.
- Prepare a calendar of your week and schedule an appointment with yourself for study time.
- Review how you are using your time through the week and make an honest assessment of whether you have the time to make this class a priority.
- See the Preface for additional resources concerning time management.

“Imagination is more important than knowledge. For while knowledge defines all we currently know and understand, imagination points to all we might yet discover and create.”

—Albert Einstein (Mathematician and scientist)

Section 1.2 Fractions Review 13

Section 1.3 The Order of Operations, Algebraic Expressions,

Coming Up...

In Section 1.3, we will learn that the number of viewers (in millions) for the season premiere of Fox Network's *American Idol* for Seasons 1 to 8 can be approximated by the expression $0.1x^3 + 18.4x - 2.54x^2 - 4.6$, where x is the number of seasons aired. We will learn how to use this expression to estimate the number of viewers for future seasons.

1

“The key is not to prioritize what's on your schedule, but to schedule your priorities.”

—Stephen Covey

Resources Available in Success Strategies Manual

- Overcoming Math Anxiety Strategies
- Note Taking and Homework Strategies
- Test Preparation Worksheets
- Time Management Activities
- Goal Management
- How to Navigate a Math Textbook
- Organizing a Portfolio
- Printable Math Study Sheets
- Error Analysis Activities
- Math Term Glossaries

In addition to ***Sucess Strategies*** and the chapter opener materials, more resources can be found in our robust ***Success Strategies Manual,*** with both Student and Instructor versions available. The manual will provide additional materials geared toward success strategies, including worksheets, tips, handouts and templates.

“This is a GREAT chapter. All math classes should start with this information!”

—Elise Price, *Tarrant County College*

Relevant Examples and Exercises for Today's Students

Every example and exercise has a purpose! The Hendricks/Chow series offers unique types of examples and exercises to capture students' interest and help them build different skill sets as they move throughout the chapter.

▶ How Relevant are the applications in your text?

SECTION 4.3 The Slope of a Line

▶ OBJECTIVES

As a result of completing this section, you will be able to

1. Find the slope and y-intercept of a line from an equation.
2. Graph a line given its slope and y-intercept.
3. Use the slope formula to determine the slope of a line.
4. Determine if two lines are parallel or perpendicular.
5. Apply the concept of slope to real-world applications.
6. Troubleshoot common errors.

The grade of a highway is the measure of the incline or steepness of the road. The grade is generally expressed as a percentage. One of the steepest roads in the world is Baldwin Street in Dunedin, New Zealand, with a 35% grade. What is the slope of the road and what does it mean? (Source: http://www.geographylists.com/list17y.html)

To answer this question, we must understand the concept of the slope of a line. In this section, we will explore linear equations of the form $y = mx + b$ and show how these equations relate to the slope of a line.

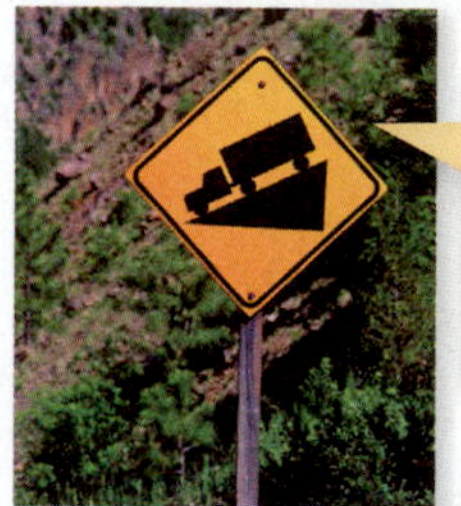

Each section opens with a relevant application that is then revisited in the worked examples.

The Slope-Intercept Form of a Line

From our experience with graphing lines, we know that not all lines have steepness or *slope* as seen in the following graphs. Notice that the graph of

93. The table shows the U.S. unemployment rates between 2001 and 2010. (Source: Bureau of Labor Statistics)

Years After 2001	0	1	2	3	4	5	6	7	8	9
Unemployment Rate (percent)	4.7	5.8	6.0	5.5	5.1	4.6	4.6	5.8	9.3	9.6

a. Write the ordered pairs (x, y) that correspond to the data in the table, where x is the years after 2001 and y is the unemployment rate.

b. Interpret the meaning of the first and last ordered pairs in the context of the problem.

c. In what year was the unemployment rate the highest? The lowest?

scatter plot of the data.

lists the approximate salary of Peyton quarterback for the Indianapolis Colts, for 9. (Source: http://content.usatoday.com/sportsdata/salaries/player/Peyton-Manning)

03	0	1	2	3	4	5	6
ns)	\$11.3	\$35	\$0.7	\$10	\$11	\$11.5	\$14

he pairs that ond to the the table, is years 03 and y is Manning's salary (in millions of dollars).

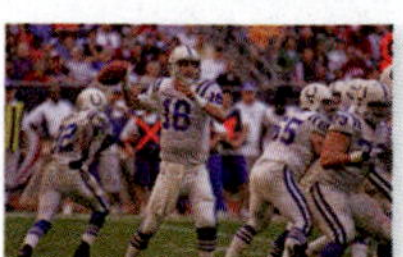

b. Interpret the meaning of first and last ordered pairs in the context of the problem.

c. In what year was his salary the highest? The lowest?

d. Make a scatter plot of the data.

29. The top 10 U.S. Internet search providers processed a total of approximately 9,200,000,000 search requests, during August 2010. Google processed about 6,000,000,000 and Yahoo! processed about 1,210,000,000 search requests. Write fractions that represent the portion of Google search requests

84. The total number of individual songs purchased digitally in 2008 and 2009 was 2.231 billion. The total number of individual songs purchased digitally in 2008 was 0.089 billion less than the total number in 2009. Let x = total number of individual songs purchased digitally in 2008 (in billions) and let y = total number in 2009 (in billions). (Source: http://www.ritholtz.com)

a. Write an equation using x and y that represents the combined total number of individual songs purchased digitally in 2008 and 2009.

b. Write an equation that relates the total number of songs purchased digitally in 2008 to the total number of songs purchased digitally in 2009.

How often do you find the applications in your current text irrelevant and stale? In order to promote active learning, the authors have worked to provide current and relevant applications that are interesting to students and contain subjects that are not always in math books, such as social media, smart phones, consumer topics, and current events.

> "I like the application problems included here—very appropriate for this level student: connected to their real-world future occupations. Kudos to you!"
>
> —Vicki Schell, *Pensacola State College*

How do your current resources help your students progress to become critical thinkers?

Objective 7 Troubleshoot common errors.

Troubleshooting Common Errors

Some common errors associated with the product and quotient negative exponents are shown.

Objective 7 Examples A problem and an incorrect solution are given. Provide the co an explanation of the error.

7a. $y^2 \cdot y^6$

Incorrect Solution	Correct Solution an
$y^2 \cdot y^6 = y^{12}$	The error is that the expone The product of like bases ru we multiply exponential ex bases, we *add* exponents. S is $y^2 \cdot y^6 = y^{2+6} = y^8$.

Troubleshooting Common Errors walks students through a visual example of key exercises where students often struggle. Students will see a problem worked out incorrectly and then correctly with an explanation of how to fix the mistake. This feature will appeal to visual learners and trains students how to identify and overcome common errors when solving a problem. Troubleshooting Common Errors is a consistent feature that appears as the last objective for each section.

The left side shows common mistakes when solving the particular problem.

The right side guides a student step-by-step through the proper solution of the problem.

You Be the Teacher! exercises reinforce and build critical thinking skills by teaching students to constantly reevaluate their work. In these exercises, students are asked to check a student's work and make necessary corrections to ensure the problem is answered correctly. Like Troubleshooting Common Errors, this feature emphasizes Error Analysis and trains students to take ownership and identify their own mistakes.

You Be the Teacher!

Correct each student's errors, if any.

73. Evaluate $g(-4)$ when $g(x) = -x^2 - 7x + 6$.

Chase's work:

$$\begin{aligned} g(-4) &= -(-4)^2 - 7(-4) + 6 \\ &= 16 + 28 + 6 \\ &= 50 \end{aligned}$$

74. Evaluate $P(2)$ when $P(x) = -4x^2 + 12x - 5$.

Josh's work:

$$\begin{aligned} P(2) &= -4(2)^2 + 12(2) - 5 \\ &= 16 + 24 - 5 \\ &= 35 \end{aligned}$$

"The Troubleshooting Common Errors examples make this text stand out among other books. I use this concept in my classes now and would continue to reinforce the material with this feature. Critical thinking is so important for students and this idea helps students to think more about process rather than obtaining the 'right' answer."

—Carol Ann Poore, *Hinds Community College, Ramond Campus*

Chapter Walkthrough & Organization

When developing the organization and framework for this series, Andrea considered her teaching methodologies. **She organized each section with a five-step process, just as she presents her lectures: (1) lead-in, (2) objectives, (3) lesson, (4) check for understanding, and (5) summary.** This five-step process has served as the framework of each section of the textbook. Most every section opens with a real-life application that will be worked through later in the section. The section's objectives, stated in clear, measurable language, follows with many worked examples and student checks for understanding. Finally each section closes with a summary of key concepts.

Section 4.5 Linear Inequalities in Two Variables 269

SECTION 4.5 Linear Inequalities in Two Variables

OBJECTIVES

As a result of completing this section, you will be able to

1. Determine if an ordered ... is a solution of a linear ...uality in two variables.
... ...e solution set of ...ear inequality in two ...bles.
... ...e applications of linear ...ualities in two variables.
... ...bleshoot common ...s.

Each section is organized by **Learning Objectives.** Then, multiple examples are provided to illustrate each objective.

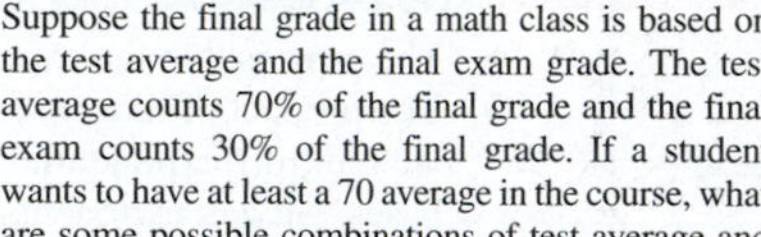

Suppose the final grade in a math class is based on the test average and the final exam grade. The test average counts 70% of the final grade and the final exam counts 30% of the final grade. If a student wants to have at least a 70 average in the course, what are some possible combinations of test average and final exam grades to produce the desired result?

To solve this problem, we must know how to solve linear inequalities in two variables.

Objective 1 Determine if an ordered pair is a solution of a linear inequality in two variables.

Solutions of Linear Inequalities in Two Variables

At this point, we have studied linear equations in two variables and their solutions. We will now investigate *linear inequalities in two variables* and their solutions.

Definition: A **linear inequality in two variables** is an inequality that can be written in one of the following ways, where A, B, C, m, and b are real numbers with A and B not both zero.

$$Ax + By > C \qquad Ax + By \geq C \qquad Ax + By < C \qquad Ax + By \leq C$$

$$y > mx + b \qquad y \geq mx + b \qquad y < mx + b \qquad y \leq mx \text{ ...}$$

Like solutions of linear equations in two variables, **solutions of linear in... two variables** are ordered pairs (x, y) that make the inequality true.

Procedure boxes are included whenever possible to help verbalize the math steps and processes.

Procedure: Determining if an Ordered Pair is a Solution of an Ine...

Step 1: Replace the variables with the corresponding values of x ...
Step 2: Simplify the resulting inequality.
Step 3: If the inequality is true, then the ordered pair is a solution. If the inequality is false, then the ordered pair is not a solution.

Objective 1 Examples Determine if the ordered pair is a solution of the given inequality.

1a. $(-4, 2)$; $x - y < 8$ **1b.** $\left(\frac{3}{2}, -5\right)$; $y > -4x + 10$

1c. $(7, 8)$; $y \leq 3$ **1d.** $(-1, 6)$; $x < 2$

Each objective is followed by thorough and **multiple examples,** numbered exactly as the objective, to reinforce the concepts and show the stepped out solutions.

Solutions

1a.
$x - y < 8$
$-4 - 2 < 8$ Replace x with -4 and y with 2.
$-6 < 8$ True.

Since the resulting statement is true, $(-4, 2)$ is a solution of $x - y < 8$.

1b.
$y > -4x + 10$
$-5 > -4\left(\frac{3}{2}\right) + 10$ Replace x with $\frac{3}{2}$ and y ...
$-5 > -6 + 10$ Multiply -4 and $\frac{3}{2}$.
$-5 > 4$ False.

Since the resulting statement is false, $\left(\frac{3}{2}, -5\right)$ is *not* a ...

The examples provide a high level of **step-by-step** detail so that students do not lose track of the various steps.

After each set of examples for an objective, there is a **Student Check** feature that provides additional exercises for students to try on their own, similar to the examples shown.

Student Check 2 Write the equation of the line that has the given slope and passes through the given point. Write the answer in both slope-intercept form and standard form.

a. $m = -5$; (3, 1) **b.** $m = -\frac{2}{7}$; (4, 5)

c. $m = 0$; (−6, 9) **d.** undefined slope; (8, 1)

Note: *Using method 1 or 2 produces the same linear equation. When using the slope-intercept form, we must substitute the values of m and b into $y = mx + b$ to obtain the equation. When using the point-slope form, the equation is obtained through the process.*

Notes from the author appear throughout the text to give students valuable tips and information.

The end of each section contains the **Answers to the Student Checks**, along with a **Summary of Key Concepts** and most include **Graphing Calculator Skills** for those instructors who incorporate calculator use.

ANSWERS TO STUDENT CHECKS

Student Check 1 **a.** $y = -2x + 9$ **b.** $y = \frac{7}{3}x - 1$ **c.** $y = \frac{2}{3}$ **d.** $y = \frac{3}{2}x + 1$

Student Check 2 **a.** $y = -5x + 16$ **b.** $y = -\frac{2}{7}x + \frac{43}{7}$ **c.** $y = 9$ **d.** $x = 8$

Student Check 3 **a.** $y = -5x + 8$ **b.** $y = -\frac{4}{3}x + \frac{2}{3}$ **c.** $y = -1$ **d.** $x = -8$

Student Check 4 **a.** $y = \frac{1}{2}x - 4$ **b.** $y = -2x + 1$ **c.** $x = 2$ **d.** $y = -3$

Student Check 5 **a. i.** $y = -3000x + 30{,}000$ **ii.** \$18,000 **iii.** 7 yr **b. i.** $y = 0.54x + 15.3$ **ii.** 21.78 million **iii.** 2024

SUMMARY OF KEY CONCEPTS

1. If the slope and y-intercept of an equation are known, writing the equation that satisfies this information is immediate. The value of m and b are substituted into the slope-intercept form $y = mx + b$.
2. There are three other situations that provide enough information for an equation of a line to be written. They are:
   - a point and a slope
   - two points
   - a point and a line parallel or perpendicular

   In each of these situations, the slope must be determined. If it is not given, use the slope formula or the relationship to a given line to find it. After the slope is found, use it with one of the points in either the point-slope form or the slope-intercept form to write the equation of the line.
3. If a line is described as vertical or with undefined slope, the equation of the line will be of the form $x = h$, where h is the x-coordinate of the given point.
4. If a line is described as horizontal or with zero slope, the equation of the line will be of the form $y = k$, where k is the y-coordinate of the given point.
5. In application problems, we will either know the slope (how the values change) and y-intercept (initial value) from the problem or we will be given two points that enable us to write the equation.

GRAPHING CALCULATOR SKILLS

The graphing calculator has the ability to calculate the equation of the line if provided enough information. At this point, it is more beneficial to use the calculator to check our work instead of allowing it to do the work for us.

Example: Use the calculator to verify that $y = \frac{3}{2}x + 6$ is the equation of the line that goes through the points (−4, 0) and (4, 12).

"I think the Hendricks/Chow series will improve the way I teach. The authors offer techniques I've thought about but wasn't sure how to implement. The resources and tips are very helpful and will enhance my teaching!"

—Edward Ennels, *Baltimore City Community College*

Exercise Sets for Today's Whole Student

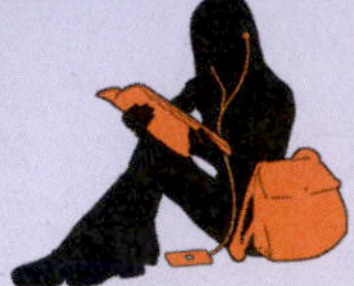

The Exercise Sets contain a wide quantity and variety of exercise types that purposefully help students persist from basic skills through differentiation of topics, to practicing critical thinking and evaluating.

Write About It! exercises help students practice their vocabulary and verbal skills.

SECTION 4.3 EXERCISE SET

Write About It!

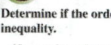

Use complete sentences in your answer to each exercise.

1. Describe what is meant by the slope of a line.
2. How can you find the slope and y-intercept from the equation of a line?
3. Explain how to use the x- and y-intercepts to find the slope of the line passing through the intercepts.
4. Explain how to use the slope and y-intercept to graph a

Practice Makes Perfect!

Find the slope and y-intercept of each line from its equation. Write the y-intercept as an ordered pair. (*See Objective 1.*)

9. $y = x - 5$
10. $y = x + 3$
11. $y = -x + 2$
12. $y = -x - 1$
13. $y = 3x + 9$
14. $y = 7x - 2$
15. $y = \frac{x}{3} + 8$
16. $y = \frac{x}{4} - 7$

Mix 'Em Up! is another layer of practice that mirrors what students might see on a test and also mixes together exercises from various objectives when possible.

Mix 'Em Up!

Determine if the ordered pair is a solution of the given inequality.

43. $x + 3y < 6$; $(0, 0)$
44. $x - 7y < 1$; $(10, 3)$
45. $y > \frac{1}{2}x + 8$; $(8, -5)$
46. $y < -\frac{1}{3}x - 11$; $(-9, 0)$
47. $x > 12y$; $(1, 0)$
48. $y > 6x$; $(0, -1)$
49. $y - 11 < 4$; $(-3, 25)$
50. $x + 5 \geq 1$; $(-2, 4)$
51. $6x - y \leq 1$; $(0.2, -0.5)$
52. $x + 8y > 3$; $(-1.4, 0.5)$

Graph the soluti... variables.

53. $x + 3y <$ ...
55. $5x - 2y >$ ...

at least \$350 per week.

64. If Brasil wants to make at least \$450 per week, (a) write a linear inequality in two variables that represents this situation, (b) graph the linear inequality, and (c) give three possible combinations of hours that Brasil could work at each job to make at least \$450 per week.

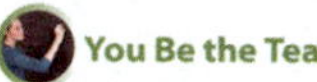

You Be the Teacher!

Correct each student's error, if any.

65. Graph $4x + y < 8$.
Jamie's work: The boundary line is $4x + y = 8$ and my test point is $(0, 0)$.
$4(0) + 0 < 8$
$0 < 8$

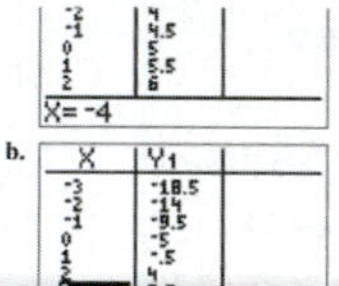

You Be the Teacher! exercises have students correct another students' work and help students analyze common errors.

Calculate It!

109. Each table shows the points on the graph of a given line. Determine which graph contains the points $(0, 2)$ and $(-1, -3)$.

a.

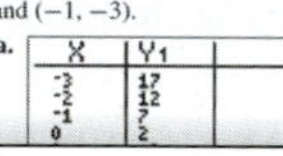

X	Y1
-3	17
-2	12
-1	7
0	2

b.

X	Y1
-3	-18.5
-2	-14
-1	-9.5
0	-5
1	-.5
2	4
	8.5

Think About It! exercises ask students to think critically about conceptual problems.

Think About It!

73. Give an example of a linear inequality for which the point $(0, 0)$ cannot be used as the test point to determine which half-plane to shade.
74. Is it possible for a linear inequality in two variables to have one point as a solution? Explain.
75. Is it possible for a linear inequality in two variables to have only its boundary line as a solution? Explain.
76. Graph the inequalities in parts a–d and use your results to answer parts e–g.

g. How does the graph of $y \geq mx + b$ differ from the graph of $y > mx + b$?

77. Graph the inequalities in parts a–d and use your results to answer parts e–g.
a. $y < -2x + 4$
b. $y < \frac{1}{3}x - 1$
c. $y < 4x$
d. $y < 1$
e. Which half-plane was shaded in each of these inequalities?
f. What can you conclude about the graph of the solution of $y < mx + b$?
g. How does the graph of $y \leq mx + b$ differ from the graph of $y < mx + b$?

PIECE IT TOGETHER SECTIONS 4.1–4.3

Graph each function and state its domain and range. (*Section 4.2, Objective 1*)

1. $f(x) = \frac{4}{5}x + 3$
2. $f(x) = 6$

Graph each line using the x- and y-intercepts. (*Section 4.2, Objectives 2 and 3*)

3. $5x - y = 10$
4. $8x + 5y = 40$
5. $x - 4y = 0$
6. $3x - 5 = 0$

Graph each line using the slope and y-intercept. Label at least two points on the graph. (*Section 4.3, Objective 1*)

7. $4x - 7y = 28$
8. $f(x) = -x + 3$

Use the slope formula to determine the slope of the line containing the points. (*Section 4.3, Objective 3*)

9. $(-7, 5)$ and $(1, 3)$
10. $(4, -5)$, $(8, -5)$

At an appropriate mid-point in the chapter, a review, **Piece It Together,** allows students to review and show mastery of the concepts presented thus far in the chapter.

Group Activity features provide multistep projects for groups of students to complete. This feature appears at the end of each chapter.

GROUP ACTIVITY The Wave

Objective: To use linear functions to model a real-life situation

Resources: Stopwatch, graph paper

1. Assign a person to be the timekeeper. The timer records the time it takes to complete the wave.
2. Begin with a group of five students. At the word GO, students make a wave by standing up, raising their arms, and sitting down in sequence. The last person says STOP as they sit down.
3. Repeat the wave with groups of 10, 15, and 20 students.
4. Record your findings in a table where x represents the number of students and y represents the time to
5. Plot the points on a graph.
6. Write a linear function that describes the relationship.
7. What is the slope of the linear function and what does it mean in the context of the problem?
8. What is the y-intercept of the linear function and what does it mean in the context of the problem?
9. Use your model to predict the time it would take for a group of 100 students to complete the wave in sequence.

Review Material

The Review Material has been designed purposefully for student interaction. Students are provided a list of key terms, formulas, and properties from the chapter, as well as a Chapter Summary that students can quickly complete to review the main concepts that were presented. This section ends with Chapter Review Exercises, a Chapter Test, and Cumulative Review Exercises.

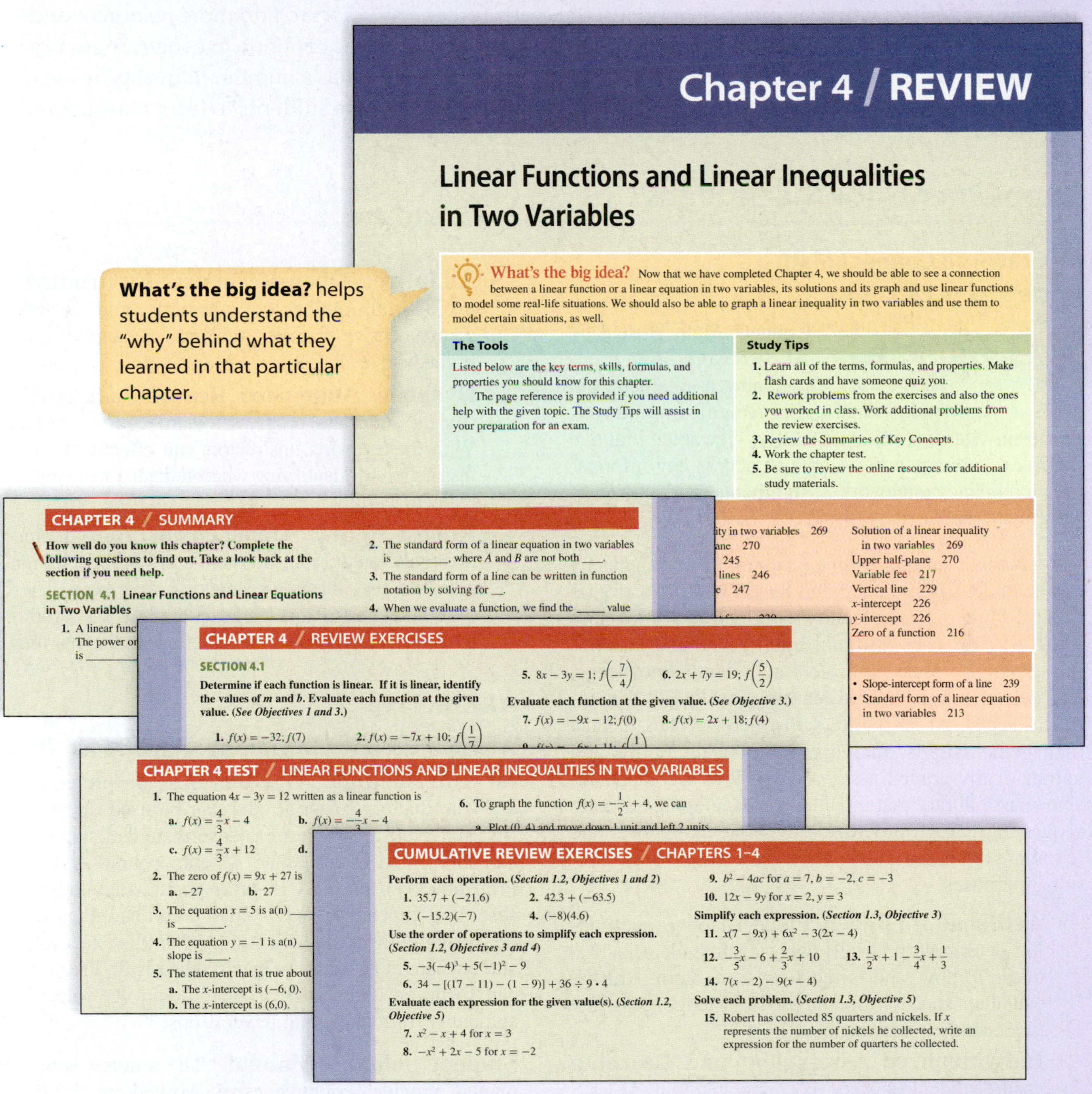

Chapter 4 / REVIEW

Linear Functions and Linear Inequalities in Two Variables

What's the big idea? Now that we have completed Chapter 4, we should be able to see a connection between a linear function or a linear equation in two variables, its solutions and its graph and use linear functions to model some real-life situations. We should also be able to graph a linear inequality in two variables and use them to model certain situations, as well.

The Tools

Listed below are the key terms, skills, formulas, and properties you should know for this chapter.

The page reference is provided if you need additional help with the given topic. The Study Tips will assist in your preparation for an exam.

Study Tips

1. Learn all of the terms, formulas, and properties. Make flash cards and have someone quiz you.
2. Rework problems from the exercises and also the ones you worked in class. Work additional problems from the review exercises.
3. Review the Summaries of Key Concepts.
4. Work the chapter test.
5. Be sure to review the online resources for additional study materials.

ity in two variables 269
ane 270
245
lines 246
e 247

Solution of a linear inequality in two variables 269
Upper half-plane 270
Variable fee 217
Vertical line 229
x-intercept 226
y-intercept 226
Zero of a function 216

- Slope-intercept form of a line 239
- Standard form of a linear equation in two variables 213

What's the big idea? helps students understand the "why" behind what they learned in that particular chapter.

CHAPTER 4 / SUMMARY

How well do you know this chapter? Complete the following questions to find out. Take a look back at the section if you need help.

SECTION 4.1 Linear Functions and Linear Equations in Two Variables

1. A linear func... The power o... is ______

2. The standard form of a linear equation in two variables is ______, where A and B are not both ____.
3. The standard form of a line can be written in function notation by solving for ___.
4. When we evaluate a function, we find the _____ value

CHAPTER 4 / REVIEW EXERCISES

SECTION 4.1

Determine if each function is linear. If it is linear, identify the values of m and b. Evaluate each function at the given value. (*See Objectives 1 and 3.*)

1. $f(x) = -32; f(7)$
2. $f(x) = -7x + 10; f\left(\frac{1}{7}\right)$

5. $8x - 3y = 1; f\left(-\frac{7}{4}\right)$
6. $2x + 7y = 19; f\left(\frac{5}{2}\right)$

Evaluate each function at the given value. (*See Objective 3.*)

7. $f(x) = -9x - 12; f(0)$
8. $f(x) = 2x + 18; f(4)$

CHAPTER 4 TEST / LINEAR FUNCTIONS AND LINEAR INEQUALITIES IN TWO VARIABLES

1. The equation $4x - 3y = 12$ written as a linear function is
   a. $f(x) = \frac{4}{3}x - 4$
   b. $f(x) = -\frac{4}{3}x - 4$
   c. $f(x) = \frac{4}{3}x + 12$
   d.
2. The zero of $f(x) = 9x + 27$ is
   a. -27
   b. 27
3. The equation $x = 5$ is a(n) ______ is ________.
4. The equation $y = -1$ is a(n) ______ slope is _____.
5. The statement that is true about
   a. The x-intercept is $(-6, 0)$.
   b. The x-intercept is $(6, 0)$.

6. To graph the function $f(x) = -\frac{1}{2}x + 4$, we can
   a. Plot (0, 4) and move down 1 unit and left 2 units

CUMULATIVE REVIEW EXERCISES / CHAPTERS 1–4

Perform each operation. (*Section 1.2, Objectives 1 and 2*)

1. $35.7 + (-21.6)$
2. $42.3 + (-63.5)$
3. $(-15.2)(-7)$
4. $(-8)(4.6)$

Use the order of operations to simplify each expression. (*Section 1.2, Objectives 3 and 4*)

5. $-3(-4)^3 + 5(-1)^2 - 9$
6. $34 - [(17 - 11) - (1 - 9)] + 36 \div 9 \cdot 4$

Evaluate each expression for the given value(s). (*Section 1.2, Objective 5*)

7. $x^2 - x + 4$ for $x = 3$
8. $-x^2 + 2x - 5$ for $x = -2$
9. $b^2 - 4ac$ for $a = 7, b = -2, c = -3$
10. $12x - 9y$ for $x = 2, y = 3$

Simplify each expression. (*Section 1.3, Objective 3*)

11. $x(7 - 9x) + 6x^2 - 3(2x - 4)$
12. $-\frac{3}{5}x - 6 + \frac{2}{3}x + 10$
13. $\frac{1}{2}x + 1 - \frac{3}{4}x + \frac{1}{3}$
14. $7(x - 2) - 9(x - 4)$

Solve each problem. (*Section 1.3, Objective 5*)

15. Robert has collected 85 quarters and nickels. If x represents the number of nickels he collected, write an expression for the number of quarters he collected.

"This is a very good book and I would like to use it to teach my courses as soon as possible!"

—Zakia Ibaroudene, *Northeast Lakeview College*

Supplements

Comprehensive Resources for Every Teaching and Learning Environment

Teaching to today's whole student, beyond just the math, forces instructors to find more resources and tools to implement both inside and outside of class. As virtual classrooms continue to evolve, these types of resources become even more necessary. The Hendricks/Chow series has a number of quality, relevant supplements designed to save instructors preparation time and motivate students to learn and succeed.

Supplements for the Student

McGraw-Hill Connect Math Hosted by ALEKS Corp.

Connect Math Hosted by ALEKS Corp. is an exciting, new assignment and assessment ehomework platform. Starting with an easily viewable, intuitive interface, students will be able to access key information, complete homework assignments, and utilize an integrated, media-rich eBook.

ALEKS is a unique, online program that dramatically raises student proficiency and success rates in mathematics, while reducing faculty workload and office-hour lines. ALEKS uses artificial intelligence and adaptive questioning to assess precisely a student's knowledge, and deliver individualized learning tailored to the student's needs. ALEKS offers instructors robust course management tools, including automated reports and automatically-graded assignments. With a comprehensive course library that includes the developmental math sequence, ALEKS provides a dynamic assessment and learning system that can be used for a variety of instructional purposes.

- **Artificial Intelligence** and adaptive questioning determine precisely what each student knows, doesn't know, and is most ready to learn. ALEKS can then successfully target knowledge gaps and guide student learning.
- **Individualized Assessment and Learning** ensure student mastery of course material. ALEKS delivers highly individualized instruction on the exact topics each student is most **ready to learn**. The student is periodically reassessed to fill knowledge gaps and ensure long-term retention.
- **Adaptive, Open-Response Environment** avoids multiple-choice questions and includes comprehensive practice problems, explanations, and immediate feedback.
- **Dynamic, Automated Reports** track detailed student and class progress toward course mastery. With these reports, instructors can effectively direct instruction by identifying what students know and are ready to learn.
- **Robust Course Management Tools** include textbook integration, automatically-graded assignments, a customizable gradebook, and more. These tools allow instructors to spend less time on administrative tasks and more time directing student learning.

ALEKS Prep

ALEKS Prep for Beginning Algebra and **Prep for Intermediate Algebra** focus on prerequisite and introductory material, and can be used during the first six weeks of the term to ensure student success in Beginning and Intermediate Algebra courses. ALEKS Prep quickly fills gaps in prerequisite knowledge by assessing precisely each student's preparedness and delivering individualized instruction on the exact topics students are most **ready to learn**. As a result, instructors can focus on core course concepts and see improved student performance with fewer drops.

Student Solution Manual The student solution manual provides comprehensive, worked-out solutions to the odd-numbered exercises in the section exercises, review exercises, piece-it-together exercises, chapter tests, and the cumulative review. The steps shown in the solutions match the style of solved examples in the textbook.

Lecture and Exercise Videos Online lecture and exercise videos will feature Andrea Hendricks and other instructors guiding students through the learning objectives, examples, and also exercises using the same methodology from the text. Additionally, Andrea Hendricks developed a subset of **Troubleshooting Common Errors videos** that show students how to learn from and overcome common mistakes. These videos support the last objective of each section. All videos are available online as part of Connect Math Hosted by ALEKS Corp. or within ALEKS 360. Other supplemental videos include eProfessor videos, which are animations based on examples in the book. The videos are closed-captioned for the hearing impaired, and meet the Americans with Disabilities Act Standards for Accessible Design.

Student Success Strategies Manual

To support the Chapter S materials in the text, Kelly Jackson from Camden County College, has developed a practical manual of activities, worksheets, tips, and strategies to support Chapter S (*Success Strategies*). This manual is available online as a resource for Connect Math Hosted by ALEKS Corp.

Guided Student Workbook

Developmental Math students often struggle with taking quality notes while in class or listening to a lecture. To support the Hendricks & Chow text, this guided workbook provides a template of a lecture for the students to fill-in the important topics, terms, and procedures so that they spend less time creating notes and more time engaged in class. Also, in addition to the notes sections, students can then practice with additional student check exercises, additional problems similar to those in the text, and extra You Be the Teacher! problems. The Guided Student Workbook is an excellent companion that instructors and students can use to support the main text. This workbook is fully editable and available online as part of Connect Math Hosted by ALEKS Corp. or also available for custom packages. Students can download and print as needed.

Supplements for the Instructor

McGraw-Hill Connect® Math Hosted by ALEKS Corp.

Connect Math Hosted by ALEKS Corp. is an exciting, new assignment and assessment ehomework platform. Instructors can assign an AI-driven ALEKS Assessment to identify the strengths and weaknesses of each student at the beginning of the term rather than after the first exam. Assignment creation and navigation is efficient and intuitive. The grade, based on instructor feedback, has a straightforward design and allows flexibility to import and export additional grades.

Instructor Success Strategies Manual This manual is an excellent resource of additional materials for full- and part-time instructors to incorporate into their courses. This robust manual includes teaching strategies, extra classroom activities, concept reviews, and activities that support the materials in Chapter S to help teach study skills including note taking, test preparation, and studying, etc. This manual is downloadable at Connect Math Hosted by ALEKS Corp.

Annotated Instructor's Edition In the Annotated Instructor's Edition (AIE), answers to exercises, review, and tests appear adjacent to each exercise set, in a color used only for annotations. Instructors will also find helpful hints and notes within the margins to consider while teaching.

Instructor's Solution Manual The instructor's solution manual provides comprehensive, worked-out solutions to all exercises in the section exercises, review exercises, piece-it-together exercises, chapter tests, and the cumulative review. The steps shown in the solutions match the style and methodology of solved examples in the textbook.

Instructor's Testing and Resource Online Among the supplements is a computerized test bank utilizing Brownstone Diploma algorithm-based testing software to create customized exams quickly. This user-friendly program enables instructors to search for questions by topic, format, or difficulty level; to edit existing questions, or to add new ones; and to scramble questions and answer keys for multiple versions of a single test. Hundreds of text-specific, open-ended, and multiple-choice questions are included in the question bank. Sample chapter tests are also provided. CDs are available upon request.

ALEKS 360: A Total Course Solution

A cost-effective total course solution: fully integrated, interactive eBook combined with ALEKS individualized assessment and learning.

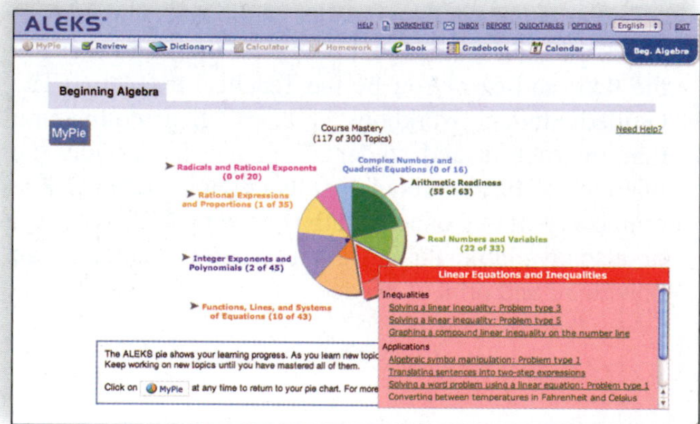

Individualized Learning

- The ALEKS Pie summarizes a student's current knowledge and provides individualized learning on the exact topics the student is **ready to learn**
- Artificial intelligence successfully targets gaps by assessing precisely a student's knowledge and periodically reassessing for long-term retention
- Adaptive, open-response environment avoids multiple-choice and includes problems, explanations, and realistic answer input tools

Interactive eBook

- eBook access provides worked examples, videos, and additional support
- Robust virtual features include highlighting, bookmarking, and note-taking capabilities
- Students can easily access the eBook, multimedia resources, and their notes from within their ALEKS Student Accounts

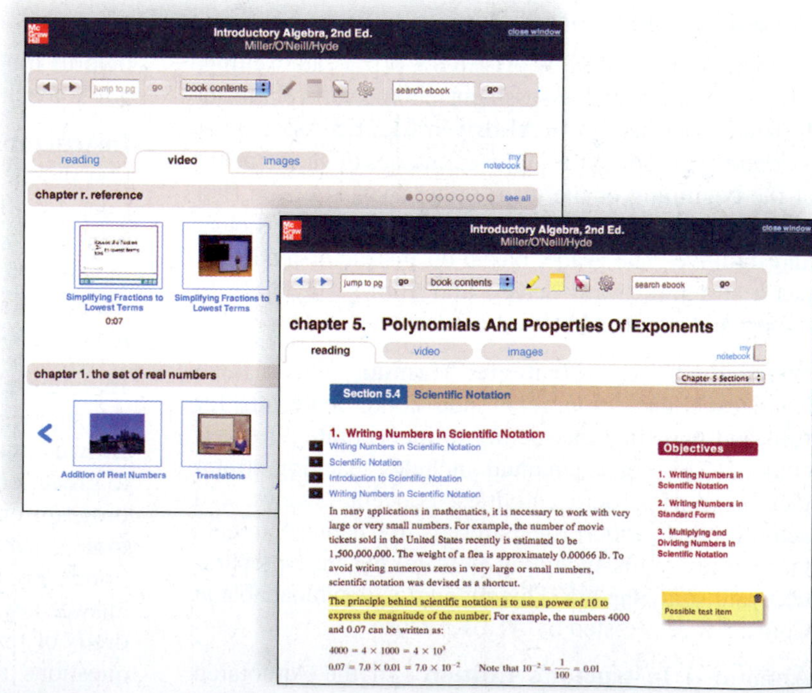

Learn More: www.aleks.com/highered/math/aleks360

ALEKS Course Management Tools

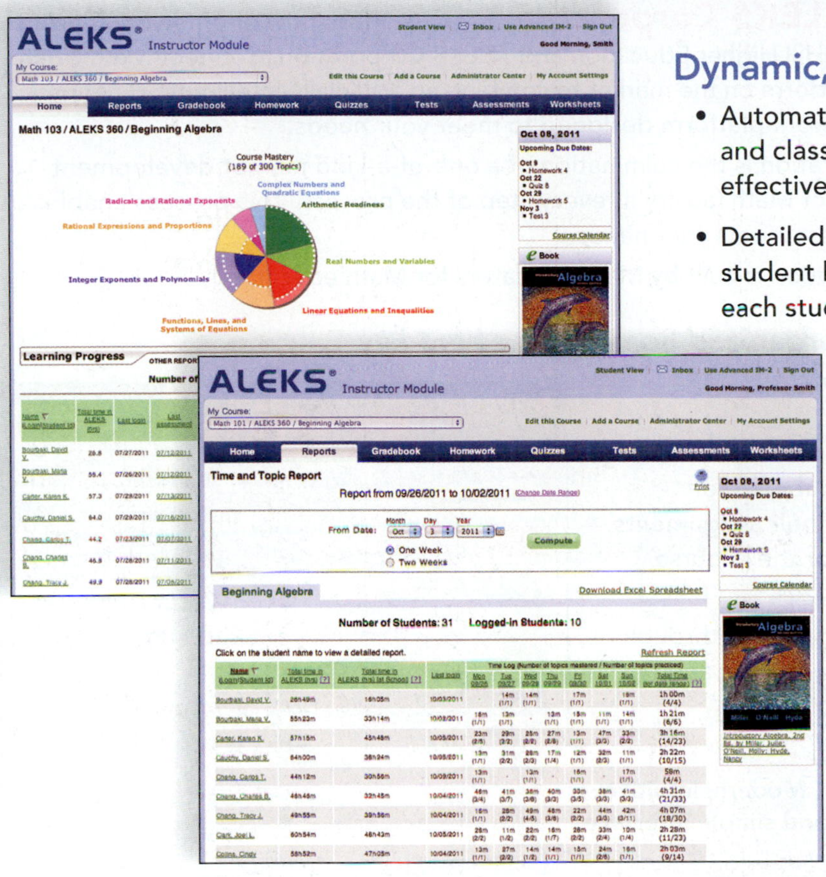

Dynamic, Automated Reporting

- Automated reports dynamically track student and class learning progress so instructors can effectively direct classroom instruction
- Detailed reports identify precisely what each student knows, and more importantly, what each student is ready to learn next
- Time and Topic Report offers up-to-the-minute daily progress, including time logged, topics attempted, and topics mastered

Course Control and Customization

- Align ALEKS topics with a textbook or course syllabus
- Create and customize course objectives and modules
- Set due dates for course objectives to pace student progress
- Assign automatically-graded homework, quizzes, and tests
- Seamlessly track and adjust student scores with the customizable gradebook

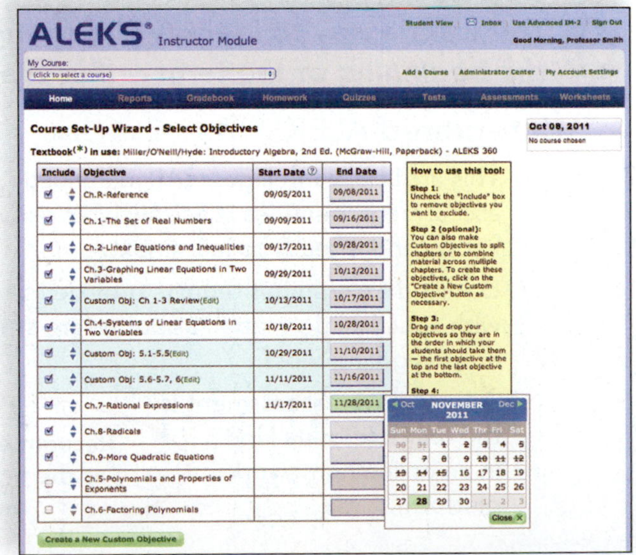

For more information, contact your McGraw-Hill sales representative or visit www.aleks.com

McGraw Hill connect® |MATH

Hosted by ALEKS Corp.

Connect Math Hosted by ALEKS Corporation is an exciting, new ehomework platform combining the strengths of McGraw-Hill Higher Education and ALEKS Corporation. Connect Math Hosted by ALEKS Corporation is the first platform on the market to combine an artificially-intelligent, diagnostic assessment with an intuitive ehomework platform designed to meet your needs.

Connect Math Hosted by ALEKS Corporation is the culmination of a one-of-a-kind market development process involving full-time and adjunct Math faculty at every step of the process. This process enables us to provide you with a solution that best meets your needs.

Connect Math Hosted by ALEKS Corporation is built by Math educators for Math educators!

1 *Your students want a well-organized homepage where key information is easily viewable.*

Modern Student Homepage

- This homepage provides a dashboard for students to immediately view their assignments, grades, and announcements for their course. (Assignments include HW, quizzes, and tests.)
- Students can access their assignments through the course Calendar to stay up-to-date and organized for their class.

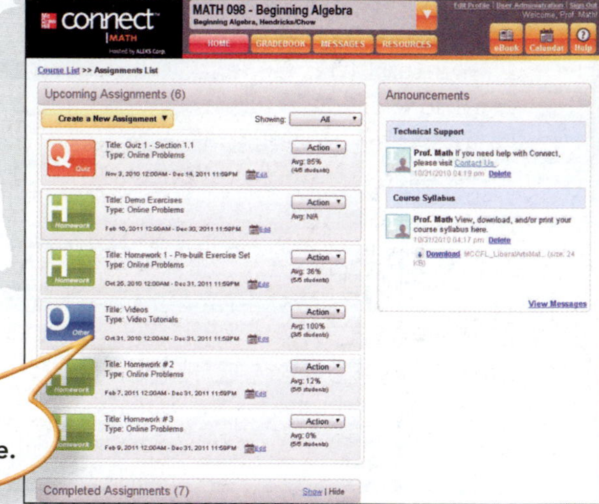

2 *You want a way to identify the strengths and weaknesses of your class at the beginning of the term rather than after the first exam.*

Integrated ALEKS® Assessment

- This artificially-intelligent (AI), diagnostic assessment identifies precisely what a student knows and is ready to learn next.
- Detailed assessment reports provide instructors with specific information about where students are struggling most.
- This AI-driven assessment is the only one of its kind in an online homework platform.

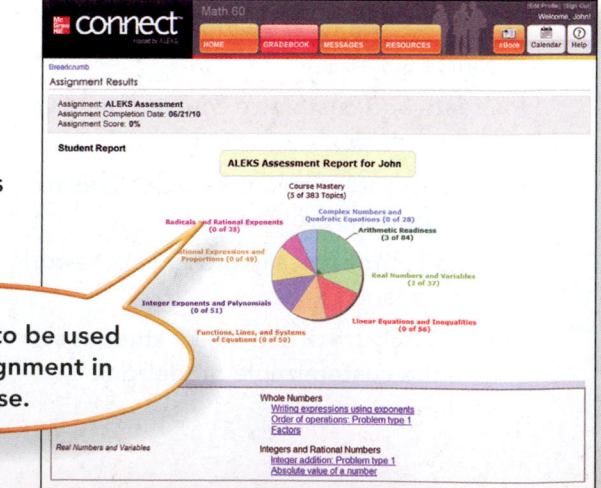

Recommended to be used as the first assignment in any course.

ALEKS is a registered trademark of ALEKS Corporation.

Resources for Online Homework

3 *Your students want an assignment page that is easy to use and includes lots of extra help resources.*

Efficient Assignment Navigation

- Students have access to immediate feedback and help while working through assignments.
- Students have direct access to a media-rich eBook for easy referencing.
- Students can view detailed, step-by-step solutions written by instructors who teach the course, providing a unique solution to each and every exercise.

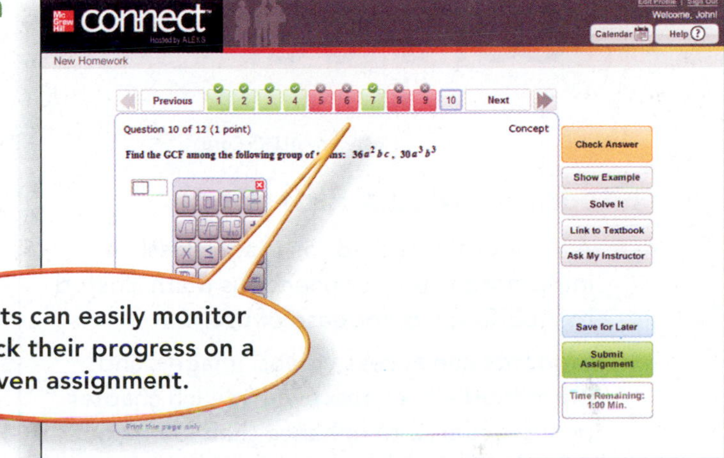

4 *You want a more intuitive and efficient assignment creation process because of your busy schedule.*

Assignment Creation Process

- Instructors can select textbook-specific questions organized by chapter, section, and objective.
- Drag-and-drop functionality makes creating an assignment quick and easy.
- Instructors can preview their assignments for efficient editing.

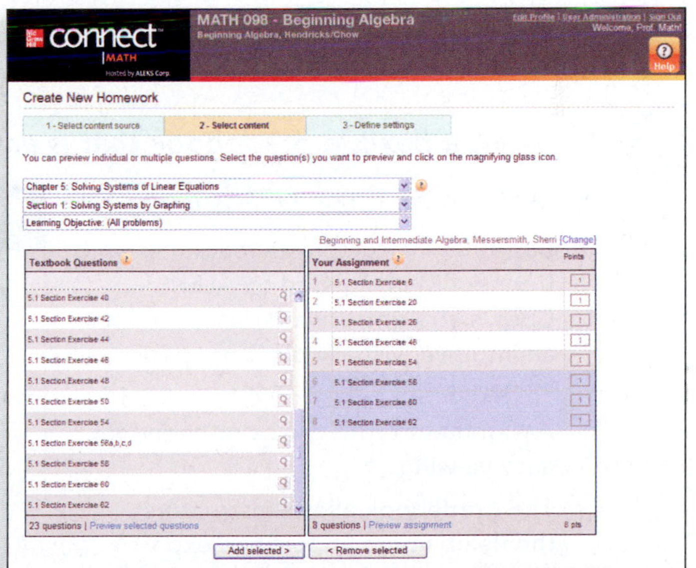

www.connectmath.com

McGraw Hill connect®
MATH
Hosted by ALEKS Corp.

Your students want an interactive eBook with rich functionality integrated into the product.

Hosted by ALEKS Corp.

Integrated Media-Rich eBook

- A Web-optimized eBook is seamlessly integrated within ConnectPlus Math Hosted by ALEKS Corp. for ease of use.
- Students can access videos, images, and other media in context within each chapter or subject area to enhance their learning experience.
- Students can highlight, take notes, or even access shared instructor highlights/notes to learn the course material.
- The integrated eBook provides students with a cost-saving alternative to traditional textbooks.

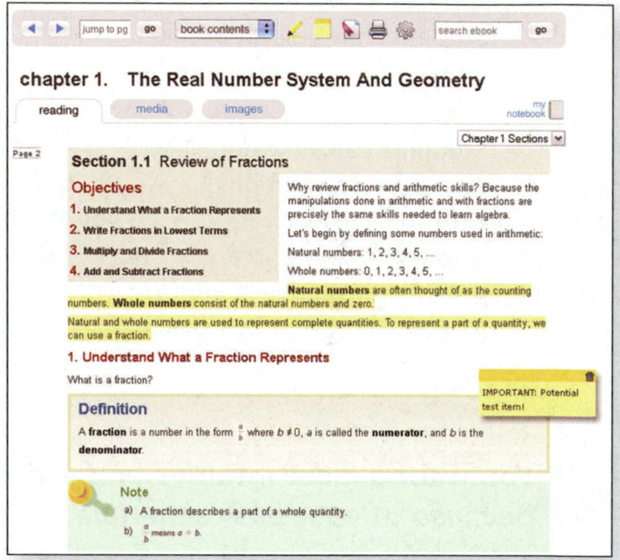

6 You want a flexible gradebook that is easy to use.

Flexible Instructor Gradebook

- Based on instructor feedback, Connect Math Hosted by ALEKS Corp.'s straightforward design creates an intuitive, visually pleasing grade management environment.
- Assignment types are color-coded for easy viewing.
- The gradebook allows instructors the flexibility to import and export additional grades.

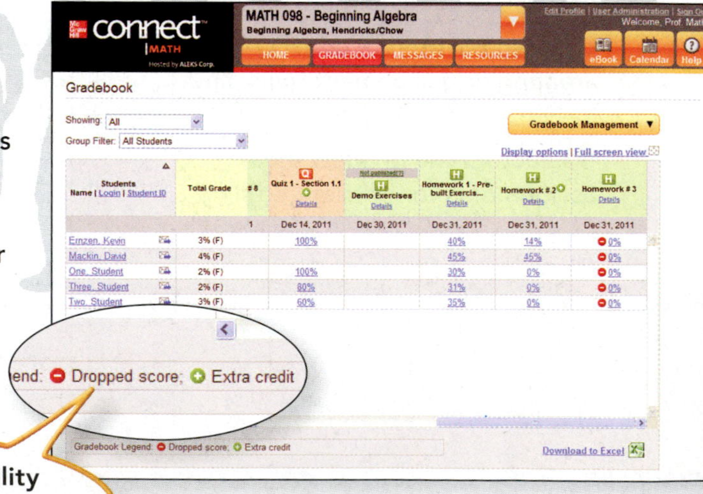

Instructors have the ability to drop grades as well as assign extra credit.

Built by Math Educators for Math Educators

You want algorithmic content that was developed by math faculty to ensure the content is pedagogically sound and accurate.

Digital Content Development Story

As the usage of online homework progresses and evolves, McGraw-Hill understands the need to have author involvement and author approval of the digital content to ensure that what students see in the online homework system is consistent with what they see in their textbooks. For this new developmental math series, co-author Pauline Chow has not only been closely involved with writing exercises for the text but also has overseen and led the creation of the digital content to ensure a seamless transition from print to digital offerings.

The development of McGraw-Hill's Connect Math Hosted by ALEKS Corporation content involved collaboration between McGraw-Hill, our authors, experienced instructors, and ALEKS Corporation, a company known for its high-quality digital content. The result of this process, outlined below, is accurate content created with your students in mind. It is available in a simple-to-use interface with all the functionality tools needed to manage your course.

1. McGraw-Hill partnered with author Pauline Chow to lead and oversee the digital content development.
2. Pauline Chow selected the textbook exercises to be included in the algorithmic content to ensure appropriate coverage of the textbook content.
3. McGraw-Hill auditioned and selected experienced instructors to work as digital contributors and represent the author's voice.
4. These digital contributors created detailed solutions for use in the Guided Solution and Solve It features, matching the voice of authors Andrea Hendricks and Pauline Chow.
5. Pauline and the digital contributors provided detailed instructions for authoring the algorithm specific to each exercise to maintain the original intent and integrity of each unique exercise.
6. Each algorithm was reviewed by Pauline and the contributors, then went through a detailed quality control process by ALEKS Corporation before being copyedited and posted live.

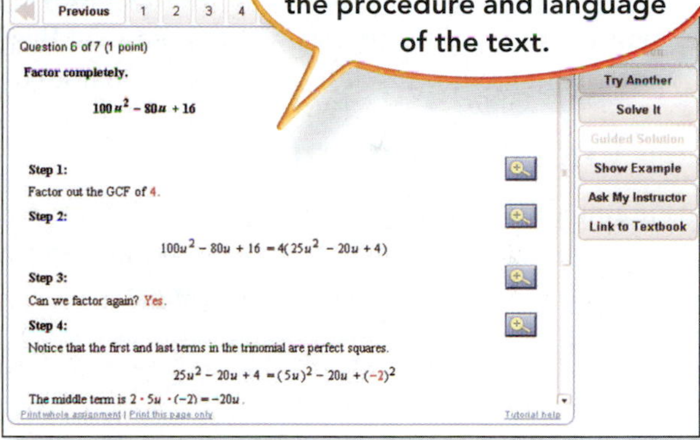

RESULT = Truly Vetted, Consistent Digital Content That Is Approved by the Authors and Supported by ALEKS Corporation.

Author and Lead Digital Contributor, O. Pauline Chow,
Harrisburg Area Community College

Lead Digital Contributor, Amy Naughten

Digital Contributors
Dihema Ferguson, *Georgia Perimeter College*
Marianne Rosato, *Massasoit Community College*
Chris Yarrish, *Harrisburg Area Community College*
Allison Williams, *Georgia Perimeter College*
Eric Bennett, *Lansing Community College*
Katy Cryer

Mc Graw Hill Connect Learn Succeed™

www.connectmath.com

Market Development

Our Commitment to Market Development and Accuracy

McGraw-Hill's Development Process is an ongoing, never-ending, market-oriented approach to building accurate and innovative print and digital products. We begin developing a series by partnering with authors that desire to make an impact within their discipline to help students succeed. Next, we share these ideas and manuscript with instructors for review for feedback and to ensure that the authors' ideas represent the needs within that discipline. Throughout multiple drafts, we help our authors adapt to incorporate ideas and suggestions from reviewers to ensure that the series carries the same pulse as today's classrooms. With any new series, we commit to accuracy across the series and its supplements. In addition to involving instructors as we develop our content, we also utilize accuracy checks through our various stages of development and production. The following is a summary of our commitment to market development and accuracy:

1. 3 drafts of author manuscript
2. 5 rounds of manuscript review
3. 2 focus groups
4. 1 consultative, expert review
5. 3 accuracy checks
6. 3 rounds of proofreading and copyediting
7. Towards the final stages of production, we are able to incorporate additional rounds of quality assurance from instructors as they help contribute towards our digital content and print supplements

This process then will start again immediately upon publication in anticipation of the next edition. With our commitment to this process, we are confident that our series has the most developed content the industry has to offer, thus pushing our desire for quality and accurate content that meets the needs of today's students and instructors.

Acknowledgements

Paramount to the development of *Intermediate Algebra* was the invaluable feedback provided by the instructors from around the country that reviewed the manuscript or attended a market development event over the course of the several years the text was in development.

A Special Thanks To All of The Event Attendees Who Helped Shape Intermediate Algebra.

Focus groups and symposia were conducted with instructors from around the country to provide feedback to editors and the authors and ensure the direction of the text was meeting the needs of students and instructors.

Mihaela Blanariu, *Columbia College Chicago*
Eddie Ennels, *Baltimore City Community College*
Dihema Ferguson, *Georgia Perimeter College–Decatur Campus*
Stephanie Fernandez, *Lewis and Clark College*
Cathy Hoffmaster, *Thomas Nelson Community College*
Joe Howe, *St. Charles Community College*
Kelly Jackson, *Camden County College*
Jason King, *Moraine Valley Community College*
Rob King, *Harrisburg Area Community College*
Michael Kirby, *Tidewater Community College*
Viktoriya Lanier, *Middle Georgia College*
Cindy Light, *Indiana University–Southeast*

Catherine Moushon, *Elgin Community College*
Sandi Nieto, *Santa Rosa Junior College*
Toni Parise, *Southern Maine Community College*
Mari Peddycoart, *Lone Star College–Kingwood*
David Price, *Tarrant County College*
Elise Price, *Tarrant County College*

Amber Rust, *University of Maryland*
Mark Schwartz, *Southern Maine Community College*
Andrew Stephan, *St. Charles Community College*
Brad Stetson, *Schoolcraft College*
Richard Watkins, *Tidewater Community College*
Karen Watson, *Cypress College*
Carol White, *Tarrant County College–Southeast Campus*
Joanna Wilson, *Georgia Perimeter College*

Manuscript Review Panels

Over 200 instructors reviewed the various drafts of manuscript to give feedback on content, design, pedagogy, and organization. Their reviews were used to guide the direction of the text.

Ricki Alexander, *Harrisburg Area Community College*
Marie Aratari, *Oakland Community College*
Dr. Eric Aurand, *Mohave Community College–Lake Havasu Campus*
Chris Barker, *San Joaquin Delta College*
Scott Barnett, *Henry Ford Community College*
Disa Beaty, *Rose State College*
David Behrman, *Somerset Community College*
Sandra Belcher, *Midwestern State University*
Monika Bender, *Central Texas College*
Teresa Betkowski, *Gordon College*
John Beyers, *University of Maryland*
Katrina Bishop, *Craven Community College*
Bret Black, *Oxnard College*
Gregory Bloxom, *Pensacola State College*
Stacey Boggs, *Allegany College of Maryland*
Karen Bond, *Pearl River Community College*
Anthony Bottone, *Arizona Western College*
Cynthia Box, *Georgia Perimeter College*
Susan P. Bradley, *Angelina College*
Chandra Breaux, *Georgia Perimeter College*
Kirby Bunas, *Santa Rosa Junior College*
Julia Burch, *Central Michigan University*
Rebecca Burkala, *Rose State College*
David Busekist, *Southeastern Louisiana University*
Susan Byars, *Gordon College*
Nick Bykov, *Delta College*
Lynn Cade, *Pensacola State College*
Yungchen Cheng, *Missouri State University*
Kim Clark, *Wayne Community College*
Adam Cloutier, *Henry Ford Community College*
Delaine Cochran, *Indiana University Southeast*
David Cooper, *Wake Technical Community College*
Wendy Davidson, *Georgia Perimeter College*
Carlos de la Lama, *San Diego City College*
Marlene Dean, *Oxnard College*
Robert Diaz, *Fullerton College*
Paul Diehl, *Indiana University Southeast*
David Dillard, *Patrick Henry Community College*
Michael Dubrowsky, *Wayne Community College*
Scott M. Dunn, *Central Michigan University*
John Edwards, *University of Oklahoma*
Cheryl Eichenseer, *St. Charles Community College*
Mark Ellis, *Central Piedmont Community College*
Marcos Enriquez, *Moorpark College*
Paul Farnham, *Fullerton College*
Dale Felkins, *Arkansas Tech University*
Dihema Ferguson, *Georgia Perimeter College*
Jacqui Fields, *Wake Technical Community College*
Rhoderick Fleming, *Wake Technical Community College*
Cynthia Fletcher, *Pulaski Technical College*
Donna Flint, *South Dakota State University*
Dorothy French, *Community College of Philadelphia*
John Fulk, *Georgia Perimeter College*
Jenine Galka, *Moraine Valley Community College*
Angela Gallant, *Inver Hills Community College*
Sunshine Gibbons, *Southeast Missouri State University*
Sharon L. Giles, *Grossmont Community College*
Suzette Goss, *Lone Star College–Kingwood*
Kathleen Grigsby, *Moraine Valley Community College*
Kathryn Gunderson, *Three Rivers Community College*
Jin Ha, *Northeast Lakeview College*
Shawna Haider, *Salt Lake Community College*
Mark Harbison, *Sacramento City College*
Jennifer Hastings, *Northeast Mississippi Community College*
Dr. Annette Hawkins, *Wayne Community College*
Alan Hayashi, *Oxnard College*
Kristy Hill, *Hinds Community College–Rankin Campus*
Irene Hollman, *Southwestern College*
Teresa Houston, *East Mississippi Community College*
Heidi Howard, *Florida State College at Jacksonville*
Steven Howard, *Rose State College*
Joe Howe, *St. Charles Community College*
Susan Howell, *University of Southern Mississippi*
Denise Hum, *Canada College*
Zakia Ibaroudene, *Northeast Lakeview College*
Sally Jackman, *Madisonville Community College*
Tina Johnson, *Midwestern State University*
Nancy Johnson, *State College of Florida–Manatee, Sarasota*
Linda Jones, *Vincennes University*
Dynechia Jones, *Baton Rouge Community College*
Paul Jones, *University of Cincinnati*
Diane Joyner, *Wayne Community College*
Laura Kalbaugh, *Wake Technical Community College*
Edward Kavanaugh, *Schoolcraft College*
Pallavi Ketkar, *Arkansas Tech University*

Rob King, *Harrisburg Area Community College*
Jason King, *Moraine Valley Community College*
Jeff Koleno, *Lorain County Community College*
Jacek Kostyrko, *San Joaquin Delta College*
Eugene Kramer, *University of Cincinnati; Raymond Walters College*
Jason Lachowicz, *Patrick Henry Community College*
Marsha Lake, *Brevard Community College*
Debra Landre, *San Joaquin Delta College*
Carol Lanfear, *Central Michigan University*
Betty J. Larson, *South Dakota State University*
Lonnie Larson, *Sacramento City College*
Sungwook Lee, *University of Southern Mississippi*
Lisa Lindloff, *McLennan Community College*
Barbara Little, *Central Texas College*
Wanda J. Long, *St. Charles Community College*
Francine Long, *Edgecombe Community College*
Mike Long, *Shippensburg University of Pennsylvania*
Yixia Lu, *South Suburban College*
Amy Marolt, *Northeast Mississippi Community College*
Dorothy S. Marshall, *Edison College*
Abbas Masum, *Houston Community College & Alvin Community College*
Barabara Maurice, *Three Rivers Community College*
Julie Mays, *Angelina College*
Toni McCall, *Angelina College*
Roger McCoach, *County College of Morris*
Michael McComas, *Marshall Community & Technical College*
Mikal McDowell, *Cedar Valley College*
Bridget Middleton, *Santa Fe Community College*
Edward Migliore, *University of California–Santa Cruz*
Shahnaz Milani, *Blinn College*
Bronte Miller, *Patrick Henry Community College*
Phillip Miller, *Indiana University Southeast*
Jon David Miller, *Lone Star College–CyFair*
Dennis Monbrod, *South Suburban College*
Roya Namavar, *Rogers State University*
Martha Nega, *Georgia Perimeter College*
Cao Nguyen, *Central Piedmont Community College*
Kevin Olwell, *San Joaquin Delta College*
Priti Patel, *Tarrant County College*
Curtis Paul, *Moorpark College*
Mari Peddycoart, *Lone Star College, Kingwood*
Karen Pender, *Chaffey College*
Vic Perera, *Kent State University–Trumbull*
Michele Poast, *Dixie State College of Utah*
Carol Ann Poore, *Hinds Community College–Rankin Campus*
David Price, *Tarrant County College*
Elise Price, *Tarrant County College*
Cynthia Reed, *Moorpark College*
Pamelyn Reed, *Lone Star College, CyFair*
Lynn Rickabaugh, *Aiken Technical College*
Dianne Robinson, *Ivy Tech Community College*
Cosmin Roman, *The Ohio State University*
Jody Rooney, *Jackson Community College*
Elaine Russel, *Angelina College*
Amber Rust, *University of Maryland*
Kristina Sampson, *Lone Star College–CyFair*
Vicki Schell, *Pensacola State College*
Laura Schoppmann, *Seton Hall University*
Mark Schwartz, *Southern Maine Community College*
Daniel Seaton, *University of Maryland Eastern Shore*
Jerry Shawyer, *Florida State College at Jacksonville*
Jenny Shotwell, *Central Texas College*
Jean Shutters, *Harrisburg Area Community College*
Craig Slocum, *Moraine Valley Community College*
Jennifer Smeal, *Wake Technical Community College*
Brad Stetson, *Schoolcraft College*
Mark Stigge, *Baton Rouge Community College*
Daniela Stoevska-Kojouharov, *Tarrant County College*
Panyada Sullivan, *Yakima Valley Community College*
Sharon L. Sweet, *Brevard Community College*
Marcia Swope, *Santa Fe Community College*
Nader Taha, *Kent State University*
M. Kaye Tanner, *Linn Benton Community College*
Linda Tansil, *Southeast Missouri State University*
Carolyn Thomas, *San Diego City College*
Lee Topham, *Lone Star College–Kingwood*
Scott Travis, *Lone Star College–Tomball*
Barbara Jo Tucker, *Tarrant County College–SE Campus*
Laura Tucker, *Central Piedmont Community College*
Chris Turner, *Pensacola State College*
Jewell Valrie, *Wake Technical Community College*
Terry R. Varvil Jr., *Hillsborough Community College–Ybor Campus*
Mansoor Vejdani, *University of Cincinnati*
Mildred Vernia, *Indiana University Southeast*
Carol Walker, *Hinds Community College*
Jimmy Walker, *Hill College*
Jane Wampler, *Housatonic Community College*
Michelle Watts, *Lone Star College, Tomball*
Gail Whitaker, *Wharton County Junior College*
Robert White, *Allan Hancock College*
Suzanne Williams, *Central Piedmont Community College*
Olga Cynthia Wilson Harrison, *Baton Rouge Community College*
Jackie Wing, *Angelina College*
Rick Woodmansee, *Sacramento City College*
Grethe Wygant, *Moorpark College*
Tzu-Yi Alan Yang, *Columbus State Community College*
Mina Yavari, *Allan Hancock College*
Loris Zucca, *Lone Star College–Kingwood*

Acknowledgments

This first edition would not have been possible without the encouragement and support of our families. There really are not words that convey our thanks and appreciation for their understanding as we worked seemingly night and day on the manuscript these last few years. We love you Todd, Andy, Charlie, and Cory Hendricks and Michael, Amy, and Andrew Ko.

Just as it takes a village to raise children, it takes a village to write a book. There are some important members of this village we would like to personally thank. We first extend our thanks to our dear friend and first sponsoring editor at McGraw-Hill, David Millage, who had the foresight to bring us together for this exciting journey. We also thank our Sponsoring Editor, Mary Ellen Rahn, for her support in this project. The resources, guidance, and insight you have provided through this project have been invaluable. We extend our gratitude to our Developmental Editors, Adam Fischer and most recently Emily Williams, for keeping us on task and coordinating all of the helpful feedback we have received through this process. We also truly appreciate the many hours that the production team has spent with us to make this series come together. We thank Vicki Krug for teaching us the world of production and we thank Laurie Janssen for the beautiful, current and creative design. Many thanks also goes to Emilie Berglund and Nicole Lloyd, Directors of Digital Content for overseeing the process of making the book come alive through Connect Math Hosted by ALEKS. Thanks for your countless hours and dedication to this book. We would like to thank every other member of the McGraw-Hill team for their excitement and enthusiasm in this project.

Some of the key contributors that we would like to thank are Calandra Davis, Kelly Jackson, Andrew Stephan, Joe Howe, Lisa Collette, Pat Steele, and Bea Sussman. Thanks, Calandra, for being there and offering your support to get this project going. Thanks, Kelly, for your wonderful insight and enhancements for Chapter S. You really helped our vision for this chapter become a reality. Thanks, Andy and Joe, for your contribution of Chapter 12 for Intermediate Algebra. Your work provided a great addition to the text. Thanks, Lisa, for your priceless input and suggestions for the final manuscript of this text. Your detailed comments were very helpful. Thanks, Pat and Bea, for your keen eyes and thorough review of the final manuscript. Your suggestions have provided the polishing touch to the book for which we are most appreciative.

We also thank all of the supplements and digital content contributors that helped complete this series. Thanks, Emily Whaley, for your dedicated and detailed efforts on the Solutions Manuals. Also, we must thank all of the several instructors that have reviewed this series throughout various stages of manuscript. We appreciate your guidance, feedback and support to help this series come to life. We look forward to you seeing the completed project to see how your suggestions have shaped this series.

Lastly, we extend our thanks to all of the students we have taught in our collective 45 years of teaching for making us the teachers we are today.

Application Index

Business

airline domestic market share, 191
automobile sales, 162
average movie ticket prices, 460
average price per DVD, 460
break-even point, 299, 302, 670, 672, 674, 675, 676, 718
budget for movies, 203
Christmas party budget, 98
consumer's surplus, 355
cost of conference reception, 803
cost of fundraiser, 737, 738, 800
cost of professional organization membership, 199
depreciated value of equipment, 236
depreciated value of limousine, 286
depreciated value of truck, 231–232
equilibrium point, 359
equilibrium point for toys, 297–298
equilibrium point for videos, 302
gaming units sold, 313–314
hybrid vehicle sales, 156–157
library late fees, 175
maximum profit, 696
maximum profit on calculators, 698
maximum profit on cameras, 699, 719
maximum revenue on computers, 698
maximum revenue on laptops, 698, 699
median salary and experience, 184
motel double- and king-sized rooms, 352, 355–356, 360
movie admissions, 44
movie box office grosses, 68, 69
movie market share, 460
net sales of toy manufacturers, 317
pay by job title, 14
price of cookies, 199
price of wedding cakes, 199
producer's surplus, 355
profit, 713, 719
profit from book sales, 733, 737, 738
profit from bottled water, 460
profit from cameras, 710
profit from fundraiser, 412, 415, 416
profit from hot dog sales, 411
profit from pretzels, 460–461
profit from shirts, 713
profit per day, 733
rent collected per month, 412, 415, 416, 461
revenue from airline tickets, 411
revenue from cookie sales, 174
revenue from donut sales, 194–195
revenue from hot dog sales, 174
revenue from lemonade, 416
revenue from pretzels, 174
revenue from video game sales, 400
revenues of Target and Amazon, 303
salary and years at company, 253, 267
salary before raise, 69
salary with bonus, 267
salary with commission, 262
sales at Amazon.com, 390
sales of awards ceremony tickets, 342
sales of concert tickets, 345, 346, 364, 843
sales of movie tickets, 320–321, 331, 332, 364, 460, 843
stock price at beginning of year, 174
stock price at end of year, 173, 842, 843
stock price fall, 40
stock prices at closing, 25–27, 31

Chemistry and Mixtures

acid solution, 326, 333, 334, 364, 631
alcohol solution, 332–333, 334, 359, 632
iodine solution, 331, 334, 359
salt solution, 332, 334
sugar solution, 332

Construction and Work

area of garden and border, 412, 416
area of pool and border, 415
area of shower floor, 411–412
circumference of Ferris wheel, 76, 77, 83
diagonal of room, 593–594
diameter of clock face, 664
dimensions of pen, 672, 674, 675, 718, 722
dimensions of playground, 670–671
distance from ladder to house, 450, 453, 454
distance from tent to stake, 611, 614, 629
grade of highway, 248
grade of road, 248, 253
height of ladder, 448–449, 455, 462, 466
length of city block, 659, 663, 718
length of garden, 84
length of Great Pyramid side, 657–658
load capacity of column, 537, 540
perimeter of maze, 83
perimeter of park, 141
perimeter of parking lot, 83
pitch of roof, 248
slope of stairs, 253
slope of wheelchair ramp, 247–248, 253
time to clean home, 526, 529, 547, 685, 688
time to complete job, 529
time to file, 529, 550
time to mow lawn, 684–685, 688, 689, 719
time to paint house, 529
time for print jobs, 689
time to rake and bag leaves, 526, 529
time to trim bushes, 529, 550
tree-trimmer rates, 142
vehicular incline, 282
well-supplied water, 529
width of garden walkway, 454
width of park, 84
width of pond border, 450, 455, 462
width of pool walkway, 455
width of walkway, 449

Consumer Applications

annual food expenditures, 317
average price of coffee, 71
average price of gas, 31, 395
average price of movie tickets, 44, 164, 263
baby shower budget, 114
birthday party budget, 114
calling plan minutes, 94
carpet cleaning charges, 281
car repair costs, 222
cell phone minutes, 98, 114, 281
cost of amusement park tickets, 333, 359
cost of bowling games, 203
cost of cell phone plan, 195, 217–218, 222
cost of gas for trip, 722
cost of gym membership, 199
cost of laundering shirts, 175
cost of lawn service, 199, 206
cost of manicures, 199, 206
cost at skating rink, 203
cost of Skype calls, 167
cost of text messages, 166
electrician rates, 115, 222
electricity rates, 400
family reunion budget, 142
favorite gaming consoles, 204
favorite movies, 175, 203
favorite sports programs, 204
fitness club membership, 219, 222, 267, 281, 478, 481, 482, 483, 544
maid service rates, 142
movies released per year, 164
new and used car sales, 317
original price of DVD player, 68–69
phone services expenditures, 298–299
postage for first-class mail, 268
price of buffet meals, 332
price of gasoline, 173
price of iced mocha, 174
price of two evening dresses, 43
prices of movie tickets, 48
prices of studio admissions, 332
professional organizer rates, 115, 142
taxicab charges, 218–219, 222, 281
value of car, 253, 267, 282, 283
value of SUV, 263
value of truck, 253, 282
wedding reception costs, 95, 98, 99

Economics

average home sales price, 39
home foreclosures, 71
homes sold in state, 267
median sale price of homes, 287, 548
national debt per capita, 385, 389
unemployment rate, 158, 162, 400
wages from motion picture and television industry, 48

Education

associate degrees conferred, 464
cost of reunion, 477, 481, 482, 544
cost of 2-year institutions, 281
declared majors, 167
degree-granting institutions in U.S., 191
earnings of high school graduates, 39
elementary and secondary school expenditures, 281
elementary and secondary school teachers, 364
enrollment in degree-granting institutions, 197
final exam score and test average, 273, 276–277
final exam score needed, 94, 95, 98, 99, 110, 111, 114, 141, 142
high school dropouts who are Hispanic, 262–263
hours per student, 166
in-state tuition and fees, 39
midterm and final exam grades, 274
number of postsecondary teachers, 282
nursing program enrollments, 43
nursing school enrollments, 32
PTA fundraiser items, 351–352, 354, 360
public school teacher salaries, 363, 464
quiz score needed, 98, 114, 115, 142
SAT scores, 162
STEM in, 136
study group hours, 173
task proficiency, 783, 786, 802
test score needed, 93, 95, 115, 142
tuition and credit hours, 267
tuition at community college, 283
tution and enrollment, 4, 13

Environment

average monthly temperature, 199
cost of removing pollutants, 477, 481, 482
daily change in temperature, 32
earthquake moment magnitude, 783, 786
Fahrenheit to Celsius conversion, 83, 84, 85, 141
temperature as function of date, 205
temperature as function of time, 205
wind chill temperature, 574, 577, 578
wind speed, 578

Finance and Investment

compound interest, 78, 79, 84
cost and size of house, 531, 538, 539, 546
currency conversion, 525, 526, 528, 529, 546
earnings of celebrities, 40
earnings of high school graduates, 39
hourly wage, 173
income from lawn mowing, 277
income from tailoring, 277
incomes of men with bachelor's degrees, 231, 247
incomes of men with master's degrees, 236
incomes of women with bachelor's degrees, 231–232, 248
incomes of women with master's degrees, 236
income tax and hours worked, 733, 737, 738, 800
income tax rates, 219, 222, 281, 286
interest on investment, 141
investment amount with compound interest, 83, 377, 573–574, 577, 628, 757–758, 760, 761, 793, 796, 801, 802, 803, 806
investment amount with continuous compound interest, 792, 793, 795, 796, 802
investment amount with interest, 84
investment amount in three accounts, 341–342, 345, 360, 364
investment amount in two accounts, 38, 40, 67–68, 69, 71, 141, 324–325, 330, 332, 333, 334, 359, 549, 632
investment amount in two stocks, 43
investment rate needed, 658, 659, 663, 664, 718
monthly housing payment, 40
monthly loan payments, 373, 374, 377
net worth of family, 25–27, 30, 31–32, 48
wages at two jobs, 277, 283, 322–323, 332, 359, 364, 548

Food

cost of drinks and popcorn, 323, 334
drinks and hotdogs for student fundraiser, 355
milk fat solution, 325–326, 332, 333, 334
price of hotdogs and fish sandwiches, 359
steel for soup can, 372–373, 376
volume of sugar cone, 374

Geometry

area of circle, 83, 84, 141, 546
area of rectangle, 76, 77, 194–195
area of square, 77
area of trapezoid, 536
area of triangle, 377, 593, 594
base of parallelogram, 491, 493, 494, 544
circumference of circle, 76, 77, 83, 84, 141
complementary angles, 83, 84, 141, 285, 330, 333, 334, 360, 465, 804
height of triangle, 77, 83
length of rectangle, 77, 83
length of triangle base, 85, 141
length of triangle sides, 594, 597, 610, 613, 614, 629
measure of angles, 141
perimeter of rectangle, 76, 77, 415
perimeter of trapezoid, 412, 415, 461
perimeter of triangle, 410–411, 415, 461, 593, 594
radius of sphere, 565
supplementary angles, 83, 85, 141, 285, 329–330, 333, 334, 360, 465, 804
surface area of cylinder, 537
volume of box, 206
volume of cone, 374, 540
volume of cylinder, 32, 48, 537, 546
volume of sphere, 31, 540
width of rectangle, 77, 83, 373, 374, 376, 490–491, 493, 494, 544

Health and Life Sciences

absolute error of scale, 123, 842
AIDS diagnoses, 189, 205
body mass index (BMI), 84, 85, 110, 111
body surface area (BSA), 610–611, 614, 630
calories burned, 71, 253
calories in cheeseburger, 71
calories in chicken sandwich, 141
calories in hamburger, 140–141
calories in ice cream cone, 70
concentration of drug in bloodstream, 509
concentration of drug over time, 761, 796, 802
deaths by lightning, 202
deaths from cancer, 283
diameter of blood platelet, 385
diameter of red blood cell, 389
gym hours per week, 173
health expenditures in U.S., 98–99
health weight among adults, 236
human memory model, 803
length of dog, 121
life expectancy, 188, 286
live births in U.S., 283
number of registered nurses, 282
obesity among Americans, 167
salary of registered nurse, 282
smokers among adults, 189, 205, 236
tobacco use among students, 188
weight loss maintenance, 136
width of hair, 386

Internet and Technology

cell phone subscribers, 47–48, 185
e-book reader owners, 43
Facebook users, 26–27
Internet users in America, 191
Internet users in Asia, 389
Internet users in Europe, 386
Internet users in Japan, 389
iPad sales, 363
number of Facebook accounts, 758
storage capacity for human words, 386
teens who prefer texting, 30–31
type of computers owned, 187
unique visitors to social networks, 299, 303
unique visitors to web sites, 362
visitors to retail websites, 390
width of memory card, 120–121

Politics

absolute error of polling, 123, 124, 143
electoral votes per candidate, 71
governors' party affiliations, 190–191
margin of error for polls, 130, 131, 136, 143
number of representatives in colonies, 203
presidential campaign spending, 71
presidential election votes, 186
president's annual salary, 141
salaries of members of Congress, 631

Science

absolute error of measurement, 124, 143
absolute error of scale, 123, 842
acceleration, 537
ball dropped from roof, 454
ball dropped from tower, 718
Celsius-to-Fahrenheit conversion, 79
coin dropped from building, 395, 400, 450, 455, 462, 466, 659, 663, 676
current in electrical conductor, 539, 540
diameter of hydrogen atom, 389
elasticity of spring, 532, 539, 540
half-life of radioactive substance, 377–378, 761
height of cannonball, 713
height of projectile, 713
height of rocket, 710, 719
kinetic energy, 535–536
law of universal gravitation, 536
period of pendulum, 565, 614
rock dropped from bridge, 454
rock dropped from building, 462
rock dropped from skywalk, 449–450
rocket launch intensity, 786
surface area of can, 465, 632
time for satellite orbit, 536–537
volume of gas, 535, 539, 540, 546, 550, 632
whisper intensity, 786
width of Milky Way, 386

Sociology and Demographics

age of moviegoers, 44
children in foster care, 163
children with siblings, 163–164
disability insurance recipients, 548
employed and unemployed in the U.S., 317
hours of television watched, 186
population by land mass, 3
population increase of county, 267
population of Asia, 389
population of Australia, 795–796
population of Brazil, 802
population of China, 761, 793, 801
population of France, 761
population of Georgia, 283

population of India, 793
population of Japan, 796
population of Mexico, 769, 795
population of Russia, 758
population of United Kingdom, 761
population of United States, 389, 761, 796
resident status of persons from Mexico, 395
Social Security beneficiaries, 267, 287
students who think friendship is "very important," 197
students who think money is "very important," 197
top rated TV programs, 175, 203
twin births, 188

Sports and Hobbies

baseball ticket sales, 345, 346, 360
basketball championship ad revenue, 166–167
cost of golf outing, 737, 738
diameter of stadium, 664
dimensions of basketball court, 329, 360
dimensions of squash court, 333, 334
dimensions of tennis court, 330, 333, 334, 360
dimensions of tennis table, 333
distance around Busch Stadium, 26–27
football ticket sales, 323, 332, 334, 345, 346, 359
golf players, 268
height of ball, 672, 674, 695–696, 718
height of basketball, 669–670
maximum height of ball, 698, 699, 719
speed during biathlon, 683–684
time before BASE jumper lands, 664
time for bungee jump, 658

Travel

airline fuel consumption, 400
airline passengers enplaned, 157–158
airplane travel classes, 340–341, 344–345, 346, 360
distance between bikes, 671–672
distance between cars, 672, 675, 718
distance between cities, 584–585, 588
drivers in fatal crashes, 714–715
miles driven, 175
miles per gallon by vehicle, 197
running speeds, 529
safe distance from chemical spill, 130, 131, 136, 143
speed of airplane and speed of wind, 327, 328, 333, 334, 359, 721
speed of original trip, 688, 689, 719
speed on return journey, 688
speed of skidding car, 566, 614–615
speeds of runner and walker, 547
speeds of two bikes, 546
speeds of two cars, 527, 685
time before ATV and car meet, 364
time before bike and car meet, 328, 333, 334, 359
time before cars meet, 328, 333, 334, 360
time before scooter and car meet, 334
types of cars driven, 186

CHAPTER S

Strategies to Succeed in Math

Success

It is with great delight that we welcome you to this course and the materials we have provided you. You are embarking on perhaps the most rewarding experience of your life. College is a series of challenges that will prepare you for a lifetime of learning and accomplishments. At the end of your college education, you will be awarded a degree—a degree that establishes your ability to learn and to persevere. It is a statement that you can work to accomplish a goal, no matter what the obstacles.

At this point, you must evaluate the reason you are here. Are you here to fulfill your dreams or the dreams of someone else? To successfully complete this journey, you must be here to fulfill your own desires, not those of a friend or family member. It will be very difficult to withstand the trials of college if you do not have a personal desire to see it through.

A semester or quarter can be very overwhelming. Focus on one day at a time and not on everything that you must learn throughout the entire course. Before you know it, the course will be over. We know that it is very easy to get distracted from your goals. Stay committed and motivated by remembering your ultimate reason for attending school.

We wish you success in this course and in your future educational endeavors. It is our hope that you are successful, not only this semester or quarter, but every term until your ultimate goal is achieved. This course will pave the way for that success. It is the door to achieving your dreams.

Mrs. Andrea Hendricks *and* *Mrs. Pauline Chow*

Andrea M. Hendricks

Question for Thought: What do you dream about doing? How does attending college make that dream possible? How does this course help you meet your goals? What is your biggest obstacle in being successful in this course? What can you do to overcome that obstacle? What is your plan for succeeding in this course?

Chapter S will introduce some important strategies that can enhance your performance in this course.

Chapter Outline

Section S.1 Time Management and Goal Setting S-2

Section S.2 Learning Styles S-6

Section S.3 Study Skills S-11

Section S.4 Test Taking S-19

Section S.5 Blended and Online Classes S-25

"Continuous effort—not strength or intelligence—is the key to unlocking our potential."

Winston Churchill

SECTION S.1 Time Management and Goal Setting

OBJECTIVES

As a result of completing this section, you will be able to

1. Set realistic grade and attendance goals.
2. Establish a schedule for studying math.
3. Establish an action plan.
4. Troubleshoot common errors.

A student might have great study skills but without good utilization of time, those skills are a wasted treasure. Time is like money; we should know how every moment is spent so that we do not waste it, and we should spend it wisely. The great news is that everyone has exactly the same amount of time. The bad news is that once the time has passed, it can never be recaptured. In this section, we will learn some helpful ways to manage our time and to set appropriate goals that make the best use of our time.

Objective 1

Set realistic grade and attendance goals.

Setting Realistic Goals

As we begin a new course, we have a "clean slate." No grades are in the grade book, no absences have been recorded, and no assignments are late or outstanding. We can control our behavior in such a way that we maximize our chance of success in this course. The beginning of a semester is an optimal time to set realistic goals that we can achieve by the end of the semester. Setting goals provides us with a sure focus and clarifies the direction in which we are headed. Activity 1 will provide us a chance to think about the goals we have for this course and for college in general.

INSTRUCTOR NOTE:
Tell students the importance of writing down their goals. Writing down goals makes them more tangible to us and it can also help motivate us.

ACTIVITY 1 Answer each question.

1. What was the last math course you took? When did you take it? Was it a successful experience? Why or why not? ____________________
2. What grade would you like to earn in this course? How many hours a week do you think you will need to devote outside of class to meet this goal? ____________________
3. How many absences do you think could put your goal at risk? What other activities could put your goal at risk? ____________________
4. What is your short-term goal? What do you hope to accomplish this semester? ____________________
5. What is your long-term goal? Why are you in college? ____________________
6. What are some things you can do that will enable you to reach your goals? ____________________
7. Are there any other activities that you need to avoid or limit that would prevent you from reaching your goals? ____________________

Note: *Keep in mind that being late or leaving class early can be just as harmful as missing class completely. Attending class regularly, punctually, and for the entire time is a goal that will lead us to success in math.*

Plan Study Time

Objective 2 Establish a schedule for studying math.

Before we can establish study time, we need to know how we currently use and manage our time. One way to do this is by recording our activities in a calendar or planner. The key is to record all of our activities, including time for driving, sleeping, and eating, as shown in Figure S.1. When we do not take the time to record our activities, we are more apt to waste time and to spend time on things that are not aligned with our goals. Once we understand and realize how we use our time, we can be more deliberate in planning a schedule that is beneficial to our goals.

INSTRUCTOR NOTE: Inform students that there are free online planners that can be downloaded.

Figure S.1

Time	Monday, 9/22
7:00	Shower/get dressed
7:30	Breakfast
8:00	Leave for school
8:30	
9:00	Math 1001 (9–10:15)
9:30	
10:00	
10:30	Engl 1101 (10:30–11:30)
11:00	
11:30	
12:00	Lunch
12:30	
1:00	Drive to work
1:30	Work (1:30–6)
2:00	
2:30	
3:00	
3:30	
4:00	
4:30	
5:00	
5:30	
6:00	Drive home
6:30	Dinner
7:00	Watch TV
7:30	
8:00	Study
8:30	
9:00	Watch TV
9:30	
10:00	Exercise
10:30	
11:00	Go to bed

Following are some suggested guidelines for planning study time.

1. Make a daily appointment to study for this class.
   a. A general guideline is to spend 2 hr studying for each hour in class. For example, a 3-credit-hour class will require an average of 6 hr of study time each week.

   b. Devote some time each day to studying math. Do not cram the suggested hours into one day.
   c. The best time to study is the hour immediately following class time. If this is not doable, study time should be as soon after class as possible.
2. Multitask when feasible.
   a. When waiting in line or waiting for an appointment, review your class notes.
   b. When riding a bus, listen to a lecture or review note cards.
3. Prioritize tasks. Differentiate between the things that must be done and the things that you just want to do.
4. Maintain a balanced schedule by not overcommitting your time. Say no when you don't have the time to devote to a task.
5. Make a habit of writing things down in a calendar or a to-do list.

Note: *Doing something right the first time takes less time in the long run than doing it poorly and having to redo it later.*

ACTIVITY 2 Answer each question.

1. Without using a calendar or planner, estimate how much time you spend sleeping, eating, watching TV, playing on the computer, working, socializing with friends, sitting in classes, studying, taking care of children, running errands, traveling from one place to another, exercising, and so on.
   a. What did you discover about yourself and how you spend your time? ______
   b. Is there anything that you want to spend more time doing? ______
   c. Is there anything you want to spend less time doing? ______
2. Create a planned time chart for next week using one similar to Figure S.1.
   a. First record large blocks of time commitments (class time, work time, driving time, church activities, and so on). ______
   b. Schedule necessary activities like sleeping and eating. ______
   c. Schedule regular activities such as grocery shopping, going to the gym, cleaning house, putting kids to bed, and the like. ______
   d. Based on suggested guidelines, you should study 2 hr for each hour you are in class. If you are enrolled in 12-credit hours, you should have 24 hr of study time. Distribute this time over the week in reasonable blocks of time. Do not set yourself up for cram sessions. ______
   e. Schedule yourself some fun time and time to decompress. ______
   f. Allow flexibility in your schedule. Things will come up, so you shouldn't have every moment scheduled. ______
3. Are there things in your life that need to be removed to improve your chances of success? ______

Action Plan

Objective 3 Establish an action plan.

1. Utilize the print and digital resources provided to you that accompany your textbook. Work with your instructor to access the available Success Strategies Manual. You will find things such as
   - a time tracker.
   - a tip sheet for getting to know your teacher.
   - a shopping list for math success that includes a list of supplies you might need.
   - positive affirmations that you can recite to help motivate you.
   - other resources to get organized and to help you have a productive semester.
2. Set some specific goals for this course.
   - I will get a grade of ________ in this course.
   - Three things I can do to ensure that I meet this goal are
     1. ________________________________
     2. ________________________________
     3. ________________________________
   - I plan to spend _______ hours per week outside of class for this course.
   - Three things I can do to be sure that I have enough time to devote to math are
     1. ________________________________
     2. ________________________________
     3. ________________________________
   - At most I will miss _______classes this semester.
   - Three things I can do to be sure I attend each class meeting are
     1. ________________________________
     2. ________________________________
     3. ________________________________
   - What could interfere with my ability to meet these goals? ________________
   - What can I do to overcome or prevent this challenge?________________

INSTRUCTOR NOTE: Encourage students to set high goals for themselves. If, for example, a student has a goal to make a B but falls short, he will most likely pass with a C. If, however, a student has a goal to make a C but falls short, he will not pass the class.

Troubleshooting Common Errors

Objective 4 Troubleshoot common errors.

Some of the common errors associated with time management are shown in the following table along with a more appropriate behavior.

Poor Time Management Behaviors	Better Time Management Behaviors
▶ Student waits until Sunday afternoon to do his homework for the week.	▶ Student sets aside 1 hr per night for math and gets a little bit done each day.
▶ Student arrives late to class each day.	▶ Student arrives on time and is ready for class to start.
▶ Student has no idea when the next test is scheduled.	▶ Student uses a planner to record due dates and exam dates.
▶ Student plans a vacation starting a week before the semester ends.	▶ Student checks the Academic Calendar for the college before making travel plans.

“You can't control how your instructor teaches, but you can control what you do to learn the material. Use strategies that emphasize your strengths and de-emphasize your weaknesses.”

Kelly Jackson, *Camden County College*

SECTION S.2 Learning Styles

OBJECTIVES

As a result of completing this section, you will be able to

1. Identify your dominant learning style(s).
2. Implement strategies to maximize success in math based on your learning style(s).
3. Establish an action plan.
4. Troubleshoot common errors.

People learn in many different ways. The way a person best gathers, processes, organizes, and remembers information refers to their **learning style.** There are three basic learning styles—auditory, visual, and kinesthetic. If you do not know how you best learn, it will be helpful for you to complete a learning styles inventory. One is included in this section, but there are many inventories available online.

Objective 1

Identify your dominant learning style(s).

Learning Styles

The following survey has been created to identify a student's learning style as it pertains to mathematics. Following the survey is a tally sheet for your responses. The category with the highest percentage represents your dominant math learning style.

ACTIVITY 1 Complete the survey.

Math Learning Styles Survey

Answer each question with the number "3" if you agree most of the time, "2" if you agree sometimes, and "1" if you agree rarely or do not agree.

_______ 1. When there is talking or noise in class, I get easily distracted.

_______ 2. If a problem is written on the board, I have difficulty following the steps unless the teacher verbally explains the steps.

_______ 3. I find it easier to have someone explain something to me than to read it in my math book.

_______ 4. If I know how the math is used in real life, it is easier for me to learn.

_______ 5. To remember formulas and definitions, I need to write them down.

_______ 6. I prefer listening to the lecture rather than taking notes.

_______ 7. Using manipulatives, hands-on activities, or games helps me learn math concepts.

_______ 8. When solving a problem, I try to picture working it out in my mind first.

_______ 9. Quiet places are the best places to study math.

_______ 10. Talking myself through a math problem helps me solve it.

_______ 11. When I write things down I remember them better.

_______ 12. I have to do a math problem myself to learn it. I can't really know it by watching someone else do it.

_______ **13.** If someone lectures about math without writing things down, I find it difficult to follow.

_______ **14.** When I take a math test, I recall more of what was said to me than what I read in my notes or in my math book.

_______ **15.** When I take a math test I can picture my notes or problems on the board in my head.

_______ **16.** I can do math problems sometimes, but can't verbally explain what I did.

_______ **17.** Reading math makes my eyes feel tired or strained.

_______ **18.** When I study math I need to take a lot of breaks.

_______ **19.** I am good at using my intuition to know how to solve a math problem.

_______ **20.** I can learn math fast when someone explains it to me.

_______ **21.** Puzzles and games are a good way to practice math.

_______ **22.** When I work a math problem I say the numbers I am working with to myself.

_______ **23.** I try to write down everything and take a lot of notes in math class.

_______ **24.** I do best with math if I just roll up my sleeves and work on problems.

Developed by Kelly Jackson, Camden County College

Math Learning Styles Tally Sheet

Carefully copy your responses (1, 2, or 3) to each question on the survey into the spaces provided.

Auditory:	___ +	___ +	___ +	___ +	___ +	___ +	___ +	___ =		_____
	2	3	6	10	14	17	20	22		A-Total
Visual:	___ +	___ +	___ +	___ +	___ +	___ +	___ +	___ =		_____
	1	5	8	9	11	13	15	23		V-Total
Kinesthetic:	___ +	___ +	___ +	___ +	___ +	___ +	___ +	___ =		_____
	4	7	12	16	18	19	21	24		K-Total

Overall Total = A-Total + V-Total + K-Total = _____

Auditory Percentage = (A-Total ÷ Overall Total) × 100 _____

Visual Percentage = (V-Total ÷ Overall Total) × 100 _____

Kinesthetic Percentage = (K-Total ÷ Overall Total) × 100 _____

Note: *Very few people learn math only one way. We hear things, see things, and do things that help us understand math better. One style may dominate the others though.*

Strategies to Enhance Your Learning

Objective 2

Implement strategies to maximize success in math that are specific to your learning style(s).

Now that you have identified your math learning style, some suggestions for things you can do to increase your success in math class based on which style(s) suits you best are provided. A summary of each type of learner and strategies to apply are shown next. General success strategies that apply to all students will be presented in a later section of this chapter.

Definition: An **auditory learner** learns best through listening to lectures, having discussions, talking things through, and listening to what others have to say. These learners often benefit from reading the text aloud and using a recording device.

Strategies to Assist Auditory Learners

- ✔ Sit near the front of the class so you can hear your instructor.
- ✔ Sit away from auditory distractions (air conditioner, door, window, and so on).
- ✔ Participate in class discussions. Ask and answer questions.
- ✔ Listen carefully to what the teacher says, their tone of voice, inflection, and volume will give cues about what is important.
- ✔ Record lectures to aid in taking notes.
- ✔ Read the text out loud.
- ✔ Use computer tutorials, online sites, or videos with an audio track.
- ✔ Use songs, rhymes, and other auditory memory devices.
- ✔ Discuss your ideas verbally.
- ✔ Dictate to someone while they write down your thoughts.
- ✔ Study with a classmate, which allows you to talk about and hear the information.
- ✔ Recite out loud the information you want to remember several times.
- ✔ Make recordings of important points you want to remember and listen to them repeatedly.
- ✔ Ask your teacher to repeat or restate something, as needed.
- ✔ Verbalize your goals for completing your assignments and say your goals out loud each time you begin work on a particular assignment.

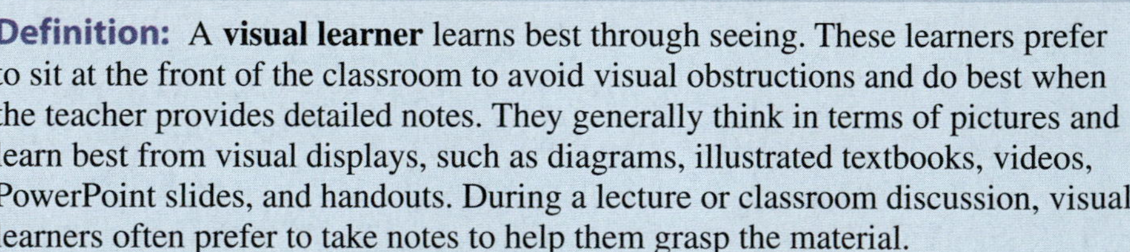

Definition: A **visual learner** learns best through seeing. These learners prefer to sit at the front of the classroom to avoid visual obstructions and do best when the teacher provides detailed notes. They generally think in terms of pictures and learn best from visual displays, such as diagrams, illustrated textbooks, videos, PowerPoint slides, and handouts. During a lecture or classroom discussion, visual learners often prefer to take notes to help them grasp the material.

Strategies to Assist Visual Learners

- ✔ Use visual materials such as pictures, charts, graphs, and the like.
- ✔ Have a clear view of your teacher when she is speaking so you can see body language and facial expressions.
- ✔ Use colored markers, Post-it notes, or highlighters to identify important points in the text.
- ✔ Illustrate your ideas as a picture or brainstorming bubble before writing them down.
- ✔ Use multimedia (e.g., computers and videos) learning tools.
- ✔ Use different colors and pictures in your notes, exercise books, and the like.

- ✔ Study in a quiet place away from auditory disturbances.
- ✔ Visualize information as a picture to aid memorization.
- ✔ Write down things that you want to remember.
- ✔ Take many notes and write down lots of details.
- ✔ Learn new material by writing out notes, cover your notes then rewrite them.
- ✔ Write your goals down and read them as you complete your assignments.
- ✔ Take advantage of study guides, handouts, and workbooks that include worked out solutions.
- ✔ Choose a seat away from visual distractions and close to the front of the class.
- ✔ Prepare flashcards and review them often.

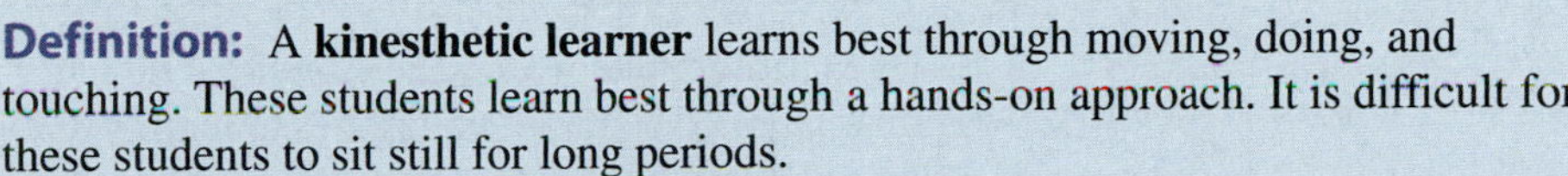

Definition: A **kinesthetic learner** learns best through moving, doing, and touching. These students learn best through a hands-on approach. It is difficult for these students to sit still for long periods.

Strategies to Assist Tactile/Kinesthetic Learners

- ✔ Practice, practice, practice. Because you prefer a hands-on approach, you need to work as many problems as possible.
- ✔ Take a 3- to 5-min study break every 15–25 min.
- ✔ Use manipulatives that can help you investigate the topic you are studying (algebra tiles, base 10 blocks, and the like).
- ✔ Work in a standing position.
- ✔ Chew gum while studying.
- ✔ Use bright colors to highlight what you are reading.
- ✔ If you wish, listen to music while you study (be sure it is not distracting though).
- ✔ Make or use a model.
- ✔ Pace or walk around while reciting to yourself or using flashcards or notes.
- ✔ If the opportunity arises, volunteer to go to the board to work problems.
- ✔ Study while sitting in a comfortable lounge chair or on cushions or a beanbag.
- ✔ Cover your desk with your favorite colored construction paper or even decorate your area to help you focus.
- ✔ Memorize information by closing your eyes and writing the information in the air, try to picture and hear the words in your head as you are doing this.
- ✔ Make flashcards, card games, floor games, and the like to help you process information.
- ✔ When working with someone, after they show you a problem, ask if you can try one.
- ✔ Make a graphic organizer showing the connections between topics.
- ✔ Get a study group together in a classroom where you can write on the board, walk around, talk about the math, but do problems.
- ✔ Use applets or computer tutorials with guided solutions in which you have to enter information throughout.

Note: *There is no right or wrong learning style. Determine what you have a tendency toward and use strategies to enhance your performance in class. You cannot control the way your instructor delivers information, but you can control what you do to best receive the information.*

Action Plan

Objective 3 Establish an action plan.

1. My dominant learning style(s) is(are) dominant ______________________________ ______________________________.
2. Five strategies that I will implement immediately to improve my chances of success in math based on my learning style(s) are
   a. ______________________________.
   b. ______________________________.
   c. ______________________________.
   d. ______________________________.
   e. ______________________________.
3. Does it seem like my teacher's style of delivery matches my style or is it a mismatch? (Example: you prefer to learn visually but your teacher does not write on the board a lot)______________________________ ______________________________.

Troubleshooting Common Errors

Objective 4 Troubleshoot common errors.

One common error is how students deal with the situation in which their teacher's teaching style does not match their learning style. Students can control their own behavior but not those of others. Following are some suggestions on how to deal with this situation.

Learning Style/Teaching Style Mismatch	What Can You Do?
Visual learner with a teacher who talks a lot but doesn't write down a lot.	**1.** Record the lesson. Later go back and listen to the lesson and fill in your notes with anything you missed the first time. **2.** Read the section in the book prior to class so you know the topic that will be discussed and write down some of the vocabulary and steps. **3.** Visit your instructor or a tutor in a setting where you can ask questions and have time to write things down.
Kinesthetic learner with a teacher who does not use activities, does not have you work problems in class, and does not use manipulatives.	**1.** Rework the problems that the teacher did in class, check your work with your instructors, and then try some similar problems on your own. **2.** Use the Explain or Show me features in your online homework system that will ask you to enter each step. **3.** Meet with your instructor or a tutor. Each time they show you a problem ask, "Can I try one now?"
Auditory learner in a blended or online class, with limited access to "lectures."	**1.** Take advantage of lecture and exercise videos that accompany your textbook or from popular Internet sites that have great educational videos. **2.** Make audio notes for yourself, reading important rules, definitions, and processes into a recorder. Listen to your recordings as often as possible. **3.** Have a study group or tutoring session where you can discuss the math concepts with someone.

"Be committed in everything you do. The number of classes you attend, the number of hours you spend studying, the number of homework problems you work are all indications of how committed you are to the course and your ultimate reason for being in college."

Andrea Hendricks, *Georgia Perimeter College*

SECTION S.3 Study Skills

OBJECTIVES

As a result of completing this section, you will be able to

1. Identify habits that good math students employ.
2. Read the textbook more effectively and efficiently.
3. Take quality notes.
4. Organize materials in a notebook or portfolio.
5. Complete homework in ways that maximize retention.
6. Establish an action plan.
7. Troubleshoot common errors.

Study skills are the skills students need to improve their learning capacity and to acquire new knowledge. No two students are going to employ the same set of study techniques, though there are some that every student should utilize when studying math. This section will address some of the skills that will enhance the success of all students.

Habits of Successful Math Students

Objective 1 Identify habits that good math students employ.

Good math students study with purpose. When you underline something, it should be because it is important. If you write out a note card, it should be because it is something you want to remember long term. When you work a homework problem, it should lead to a better understanding of the process used to solve that problem. If you can't answer "Why am I doing this?" then what you are doing may not be a productive use of your time. You should always have goals to achieve when you sit down to study and everything you do during your study time should work toward those goals. The following suggestions will help you approach your math class with purpose.

1. ***Successful students are responsible.***
   - Attend every class meeting. Try to be a few minutes early so that your materials are out and you are ready for class to begin.
   - Find an accountability partner in class. Encourage one another and use each other as a resource if one of you is absent from class.
   - Adhere to all deadlines for assignments.
2. ***Successful students make the most of their time in class.***
   - When you are in class, "be in class." Try not to worry about other things going on in your life. Focus your attention on the topics being discussed. (No texting, surfing, or doing other homework.)
   - Either take notes in class or record the lecture so that you can refer to it later. Some instructors even post their class notes on their website.
   - Actively participate in class. Follow along with the instructor. Ask questions when you don't understand and be prepared to answer questions that you know.
3. ***Successful students dedicate an appropriate amount of time to math outside of class.***
   - Immediately after class, take a few minutes to write down a summary of the key concepts that you remember. (It has been shown that students who write down what they know retain 1½ times as much as those who don't, 6 weeks later.)
   - Review your class notes as soon after class as possible.
   - Begin your homework assignment as soon as you can.
   - Review notes from previous classes on a regular basis so that you do not forget material covered earlier in the course.
   - Devote some time every day to your math class.

4. ***Successful students aren't afraid to ask for help.***
   - Math does not need to be a solo activity. Ask your instructor, a tutor, or a classmate about problems you do not understand. Form a study group.
   - Take advantage of your college's tutoring center and your instructor's office hours.
   - Use self-help books, other texts, online websites, computer programs, DVDs, or any other outside sources you can find to supplement your course materials.
5. ***Successful students are persistent.***
   - If you get "stuck" on a problem, refer to your notes, the book, or other help resources. Rework the problem until you can do it without referring to these things.
   - Be patient with yourself in learning the material. Don't get frustrated that other students may seem to learn more quickly than you. Every student comes to class with a different mathematical background. Some classmates just graduated from high school, while others may be returning after several years.
   - Be willing to try new approaches to solving a problem. If your first attempt fails, continue trying other possible strategies. Often there is more than one way to get to the solution.

Study skills consist of the things you do in class as well as the things you do outside of class. We will take a closer look at four of these topics: reading a math textbook, taking notes, organizing your notebook/portfolio, and completing homework.

Reading a Math Textbook

Objective 2 ▶

Read the textbook more effectively and efficiently.

The textbook is a great resource to aid in the mastery of material presented in class. Many students pay a lot of money for a textbook but do not take the time to read it and often have never been shown how to read a math textbook. Before you attempt homework problems, it is important that you carefully read the relevant sections of your math textbook. The examples should be studied and definitions, properties, and formulas should be learned.

Reading a math textbook is very different than reading a psychology or history text and especially different from reading a novel or a newspaper. Math textbooks generally do not have a whole lot of prose. Textbooks are typically organized around new definitions, new procedures, and worked examples that illustrate these concepts. Graphs, tables, diagrams, equations, formulas, and notations are used to visualize some of the concepts, but these are no more and no less important that the worded passages themselves. Together, they explain the complete math concept you are reading about. When reading a page in your math textbook, you will be reading from left to right, right to left, up and down (think tables), diagonally (think graphs), and from the inside out (think equation solutions).

Some guidelines for reading a math textbook are

1. Skim a new section briefly to identify new vocabulary or processes. Identify the objectives to be presented in the section, which are located in the headings and subheadings.
2. Read the section a second time, using more time and concentration. Try to connect your prior knowledge with the new material you are reading. You will need to read slowly, reread sections, and constantly ask yourself if you understand.
3. When you get to the worked examples, use your own paper to work the problems in detail, making sure that you understand how each step follows from the previous one.
4. Make a list of concepts, formulas, and vocabulary you do *not* understand and seek help.
5. Make a list of formulas and vocabulary that you *do* know and understand.

6. Make sure you understand and use the mathematically correct definition for each vocabulary word. Often words are "borrowed" from everyday language and the meaning can be the same or different in mathematics.
7. You must be able to decode and understand the math notation along with the vocabulary.
8. Learn the formal definitions and properties. It is a good idea to be able to paraphrase into your own words but you don't want to create "new" math rules that may not always work.
9. Get help if you do not understand the reading material.
   - Go back and review the previous section to see if it might be related.
   - Review your instructor's notes on the material.
   - Use the resources that came with your book (videos, CDs, animations, etc.).
   - Ask a classmate, a tutor, or your instructor. Be specific with what you do not understand.

ACTIVITY 1 Get to know the features of this book.

1. Where are the answers located and which answers are available?__________
   ____
2. Where is the chapter review? What is included in the review?__________
   ____
3. What do the different icons used in the book mean?__________
   ____
4. What colors are used to set off definitions, formulas, tips, and other important information?__________
   ____

Taking Notes

Take quality notes.

Effective note taking begins when you enter the classroom. After you get to your seat, prepare to take notes by taking out your paper, pencil, and book. Taking notes enables you to record how your instructor explains processes, to record examples that are worked, and to identify important class information, such as homework assignments and test dates. Note taking also helps you listen more attentively and to be actively engaged in the learning process. Studies show that people may forget 50% of a lecture within 24 hr, 80% in 2 weeks, and 95% within 1 month if they do not take notes.

Some guidelines for taking notes are as follows.

1. To be an effective note taker, it is important that you attend class.

   Don't be late.

   Don't leave early.
2. To be an effective note taker, you must be a good listener.
   - **a.** Be actively involved in the lecture so that you stay focused on what is being presented.
   - **b.** Sit near the front of the class so that distractions are minimized.
   - **c.** Try to relate the new topics to the material that you have already learned.
   - **d.** Ask your instructor for clarification, if you do not understand. For example, "I don't understand how you got from step 2 to step 3…", "I don't know what the symbol ____ means.", and "What is the difference between ____ and ____?"
   - **e.** Listen for words that signal important information.
   - **f.** Notice how your teacher uses formal math vocabulary when speaking.

3. To be an effective note taker, you must actually take notes.
4. Bring pencils and paper to class and take math notes in pencil, not pen.
   a. Each day start a new set of notes on a new page. Date and label them with the chapter (and section) that is being discussed.
   b. Copy down *everything* that the instructor writes on the board. If the instructor takes the time to write something, it is important.
   c. Take notes, even though your understanding may not be complete.
   d. Leave a space where you may have missed steps or have questions so that you can get these filled in after class. Then actually get them filled in by reading your book, going to a tutor, visiting your instructor's office, or asking a study buddy.
   e. Develop a good note-taking system. Ideas for note-taking systems can be found on the Web or also available with the optional Success Strategies Manual.
5. To be an effective note taker, you should review your notes.
   a. Review and reorganize your notes as soon as possible after class. Fill in any steps you missed.
   b. Write clearly and legibly so you can understand what you have written later.
   c. Rewrite ideas in your own words.
   d. Highlight important ideas, examples and issues with colored pens, pencils, or highlighters.
   e. Review your class notes before the next class period.
   f. Ask questions during office hours or the next class period if there are items that are unclear.
   g. Review all of your notes at least once each week to get a perspective on the course.

The following is an example of a page of effective notes.

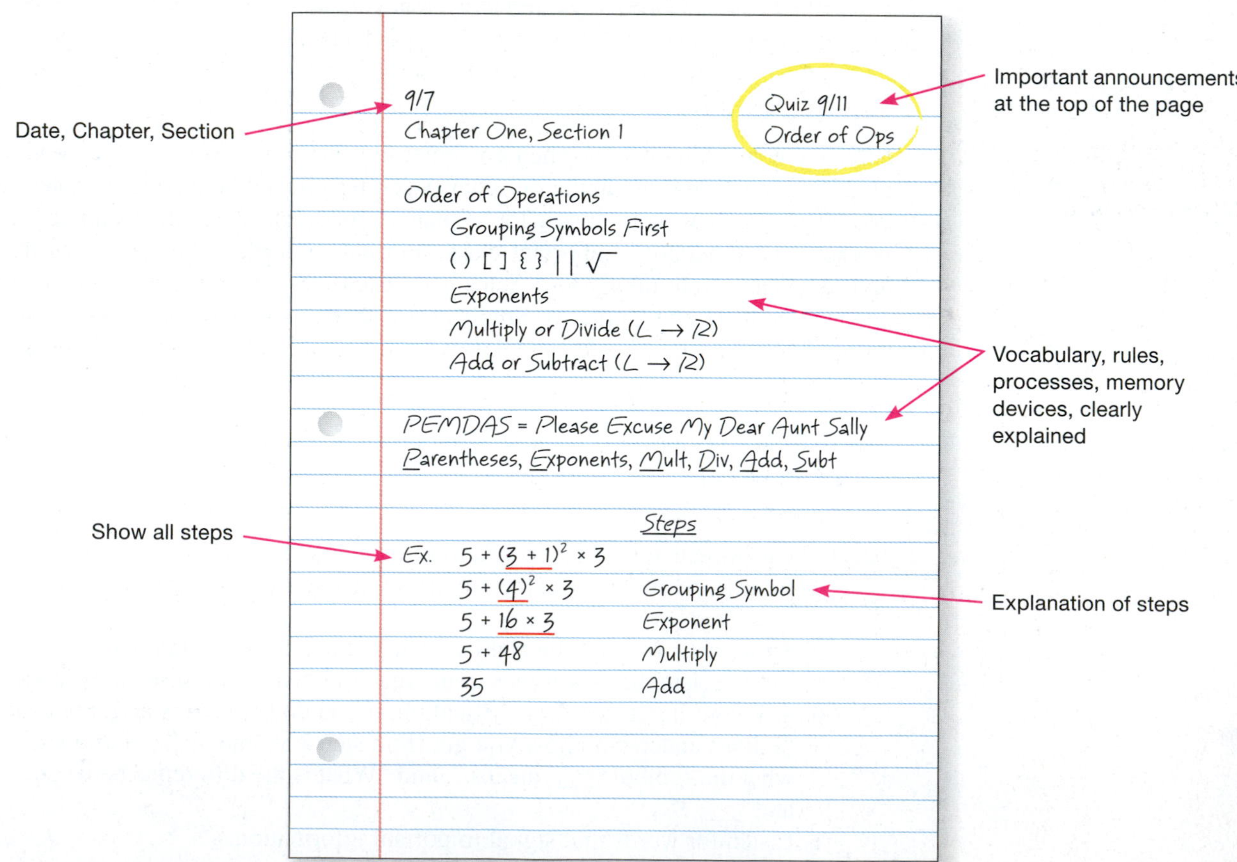

Organizing a Notebook

Objective 4 Organize materials in a notebook or portfolio.

You attend class, take notes, get handouts, complete homework assignments, and so on. All of this information can be overwhelming if it is not kept in an orderly fashion. The purpose of organizing your course materials is so that you can use them as tools to prepare for tests and other assignments. There are many different ways that you can organize your materials. The main thing is that you keep them all together and in a logical manner.

Some organization suggestions are presented next. Use what seems useful to your situation and adapt them as necessary. The key is to have a system that works for you.

1. Keep materials in a three-ring binder with pockets.
2. At the front of the binder, have a calendar and a copy of your syllabus, a list of homework assignments for the term, and other handouts your instructor gives you on the first day of class.
3. In the front pocket, keep a to-do list of what needs to be done in class.
4. Use dividers with tabs to divide the materials into different sections. Have a section for class notes and handouts, homework, study guides and test reviews, and tests and quizzes.
   - Class notes should be ordered by date. If your instructor provides a handout with additional notes or instructions for a section, keep this with your class notes for the day. (Use a three-hole punch on handouts so they fit in the binder.)
   - Homework should be written out neatly and brought to class each day.
   - If your instructor provides a study guide or test reviews, keep these together.
   - Finally, be sure to keep all of your tests and any other graded assignments to assist you in preparing for the final exam.

Completing Homework

Objective 5 ▶

Complete homework in ways that maximize retention.

Have you heard the saying, "Math is not a spectator sport?" What this means is that you cannot expect to watch someone else do it and master the material. A colleague used the following example with her class. One summer, she attended 37 Atlanta Braves baseball games. By the end of the summer, her ability to watch and enjoy baseball improved greatly. However, her ability to perform baseball did not improve at all, because she did not play baseball; she only observed it.

While math is not a baseball game, the same general rule applies. Attending class every day is definitely helpful (and fun!), but it is not the only thing needed to successfully master the subject. You must practice. This is where homework comes in. Homework is not just a necessary "evil." This is your opportunity to apply the new definitions, formulas, and procedures to gain deeper understanding.

Some guidelines for completing your homework are as follows.

1. Start your homework as soon after class as possible. The longer the time between class and homework, the more difficult it will be to remember everything you learned in class.
2. Complete your homework neatly and in order. It might look like this.

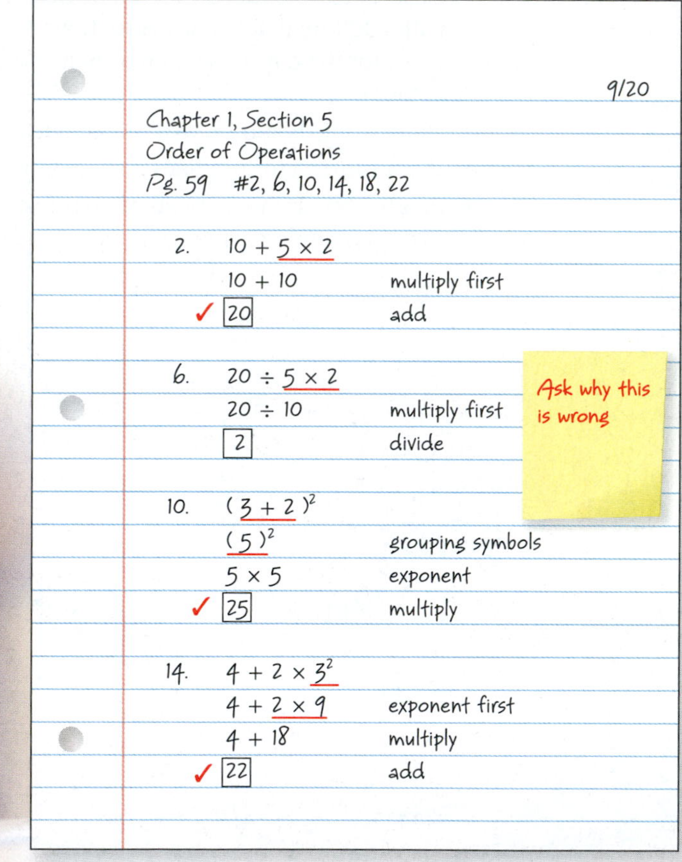

3. If you cannot complete a problem after 10 minutes, skip the problem and continue with your assignment. If, after returning to the problem, you still cannot work it, get help from someone.
4. If you cannot even get started on the homework assignment, review your class notes and rework the problems the instructor worked in class.

5. At the top of the first page of your homework exercises, list the problems that you have questions on. At the next class meeting or during your instructor's office hours, ask for help with these exercises.
6. After completing your homework assignment, compare your answers with those in the back of the book. Most books contain answers to the odd-numbered problems. Rework any problems that you missed.
7. After completing your homework, answer the following questions.
   a. What do I understand clearly?
   b. What do I not understand?
   c. What new definitions, rules, or formulas did I apply?
   d. What types of mistakes did I make on problems I worked incorrectly?
      i. Secretarial: miscopied, misread handwriting, misaligned
      ii. Computational: arithmetic mistake (like $8 \times 7 = 54$)
      iii. Procedural: missed a step, steps out of order, stopped too soon
      iv. Conceptual: no idea how to start, wrong method or formula used
8. Review your homework each day until class meets again so you do not forget your newly acquired skills.

Note: *If your instructor requires you to complete online homework, print a copy of the homework and work it offline. Most systems will allow you to log back in and enter your answers. There really is no substitute for working problems by hand. As your brain directs the motor skills involved in writing the exercises out, a connection is made that impacts your ability to remember the academic concept.*

Action Plan

Objective 6
Establish an action plan.

Check out additional homework strategies in the supplemental Success Strategies Manual for examples of a "Homework Cover Sheet," a "Chapter Preview" note-taking tool for use when you read your textbook, a tip sheet for "Making Note Cards," "Creating Memory Devices and Mnemonics," "Getting the Most Out of Tutoring," and several other study aids.

List one new strategy under each heading that you will implement immediately to improve your chances of success in your math course:

1. I will make it a habit to ______________________________
   ____
2. From now on when I read my book I will ______________________________
   ____
3. My notes would be better if I ______________________________
   ____
4. I will make my notebook more organized by ______________________________
   ____
5. When I do my homework I will ______________________________
   ____

Troubleshooting Common Errors

Objective 7

Troubleshoot common errors.

Think about the behaviors on the left compared with those on the right. Which habits seem more likely to lead to success in your course?

Poor Habits	Better Habits
A student shows up 10 minutes late for class with a latte in hand but no pen or paper.	A student shows up 10 minutes early for class and has his supplies out and has some questions ready for the instructor.
A student texts friends during class and surfs the net.	A student records the lesson, noting in her notebook the counter number that goes with each problem she is working on.
A student emails her instructor with a message that looks like a text… "i missed ur class pls send HW"	Good Morning Dr. Jones, I missed the 10:00 Algebra class this morning. I got the homework and notes from a classmate. Would it be possible for you to email me a copy of the handout that you distributed in class?
A student comes to class and asks, "I wasn't here for the last class; what did I miss?"	A student misses class and contacts a classmate prior to the next class meeting to get the notes, HW assignment, and reads the book to try to learn what she missed.
A student starts his homework and feels stuck. He closes the book and decides to ask his teacher about it next class.	A student starts his homework and feels stuck. He looks to his notebook for a similar problem; he checks the textbook examples to see if he can find a similar problem. He checks the tutoring schedule to see when he can get in for some help.

“Study purposefully and frequently. Everything you do outside of class should prepare you for the “test.” Tests determine our true level of commitment.”

Andrea Hendricks, *Georgia Perimeter College*

SECTION S.4 Test Taking

OBJECTIVES

As a result of completing this section, you will be able to

1. Prepare for tests and quizzes.
2. Maximize success during the test-taking process.
3. Analyze mistakes on a test.
4. Establish an action plan.
5. Troubleshoot common errors.

Tests are used to demonstrate mastery or knowledge. Tests can show that we can apply what we have learned. With adequate preparation, tests should simply be an extension of homework. Test taking is only successful if you have employed successful study strategies prior to the test. As the saying goes, “practice makes perfect.”

Before the Test

Objective 1 Prepare for tests and quizzes.

1. Complete your homework assignments regularly.
2. Review completed sections on a regular basis. Use the weekend to review the week’s sections.
3. Create a practice test from the problems your teacher worked in class. Work the problems without referring to your notes or books so that you simulate a real test environment. Include problems at all levels of difficulty on the practice test. Check the answers by reviewing your notes or have a classmate check your work.
4. Know which formulas will be provided and which ones must be memorized.
5. Understand all formulas and definitions you need to know for the test. Flashcards are very helpful with this task. Use the front of an index card for the name of the formula or definition and write the formula or definition on the back of the card. Be sure you know what each symbol in a formula represents.
6. Begin preparing a week before an exam. Use your weekly planner to assist with study time. Record the date of the exam on your calendar and then schedule test preparation time beginning a week earlier.
7. Find out as much information about the test as you can:
   - **a.** How many questions will be on the test?
   - **b.** What types of questions (multiple choice, free response) will be included?
   - **c.** How long will you be given to complete the test?
   - **d.** What materials are you allowed to use?
   - **e.** Will there be bonus or extra credit problems?
   - **f.** What chapter/sections are covered?

During the Test

Objective 2 Maximize success during the test-taking process.

1. Arrive at class early with all the necessary supplies and any aids you are permitted to use, and be ready to begin.
2. When you get your exam, write down all of the formulas on the top or sides of the first page. If your test is computerized, use scratch paper to write down the formulas.
3. Read the instructions carefully.
4. Review the test and complete the problems you know how to do first. This will build your confidence and bring to mind other things you have learned.
5. Keep an eye on the time. Do not spend too much time on one problem (especially if it is a low-points question).

6. Check your answers to make sure they are reasonable. For example, if you are solving for a length, a negative answer would not make sense.
7. Show all of your work neatly and clearly.
8. If you have time, review your work. Don't change any answers unless you have good reason. Often times, first instincts are correct. Double check your signs and arithmetic.
9. Use all of the allotted time to take your test. There is no prize for finishing first nor is there a penalty for being the last one to turn in the exam.
10. Mark up the test paper, if you are allowed, with information that will help you save time.
    - a. Circle what you are looking for.
    - b. With multiple-choice questions, if you know an answer is impossible, cross it out.
    - c. If you have checked an answer, put a check mark next to the problem.
    - d. If you skipped a problem but want to come back to it, put a plus sign next to it.
    - e. If you see a problem that you don't recognize at all, put a minus sign next to it and come back to it last.
    - f. If you have eliminated some options in a multiple-choice question, put down how many options are left. When running out of time, go back to the ones with the fewest options first.

Here is what your test might look like after the first time through.

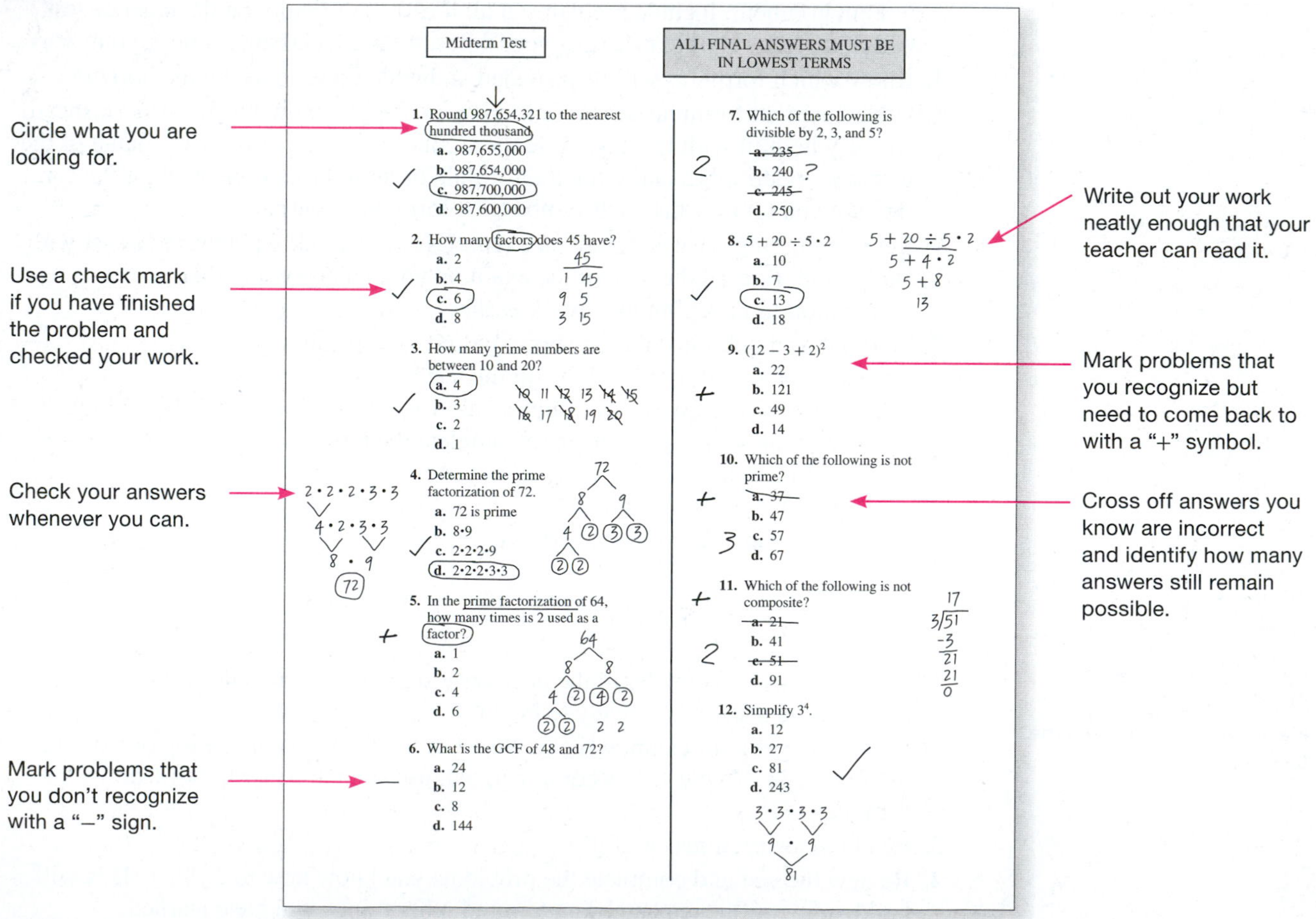

Circle what you are looking for.

Use a check mark if you have finished the problem and checked your work.

Check your answers whenever you can.

Mark problems that you don't recognize with a "–" sign.

Write out your work neatly enough that your teacher can read it.

Mark problems that you recognize but need to come back to with a "+" symbol.

Cross off answers you know are incorrect and identify how many answers still remain possible.

Midterm Test

ALL FINAL ANSWERS MUST BE IN LOWEST TERMS

1. Round 987,654,321 to the nearest hundred thousand.
   a. 987,655,000
   b. 987,654,000
   c. 987,700,000
   d. 987,600,000

2. How many factors does 45 have?
   a. 2
   b. 4
   c. 6
   d. 8

3. How many prime numbers are between 10 and 20?
   a. 4
   b. 3
   c. 2
   d. 1

4. Determine the prime factorization of 72.
   a. 72 is prime
   b. 8•9
   c. 2•2•2•9
   d. 2•2•2•3•3

5. In the prime factorization of 64, how many times is 2 used as a factor?
   a. 1
   b. 2
   c. 4
   d. 6

6. What is the GCF of 48 and 72?
   a. 24
   b. 12
   c. 8
   d. 144

7. Which of the following is divisible by 2, 3, and 5?
   a. 235
   b. 240
   c. 245
   d. 250

8. $5 + 20 \div 5 \cdot 2$
   a. 10
   b. 7
   c. 13
   d. 18

9. $(12 - 3 + 2)^2$
   a. 22
   b. 121
   c. 49
   d. 14

10. Which of the following is not prime?
    a. 37
    b. 47
    c. 57
    d. 67

11. Which of the following is not composite?
    a. 21
    b. 41
    c. 51
    d. 91

12. Simplify 3^4.
    a. 12
    b. 27
    c. 81
    d. 243

After the Test

Objective 3 Analyze mistakes on a test.

1. Look through the test for the problems you missed. Rework these problems until you get them correct.
2. Keep your test corrections with the exam (if it is returned to you) for future studying.
3. Make an appointment to meet with your instructor to review any questions you cannot figure out.
4. Most likely, the material on your test will be covered on your final, so it is important that you keep the exam if your instructor allows you to do so. File it in your notebook so you can review it later in the course.
5. Here is a strategy for test corrections:
   - **a.** Divide a piece of paper in thirds.
   - **b.** On the left side, write out the original problem with your original work.
   - **c.** In the middle, write out the correct answer to the problem.
   - **d.** On the right side, work a similar problem to ensure that you can do this type of problem correctly.

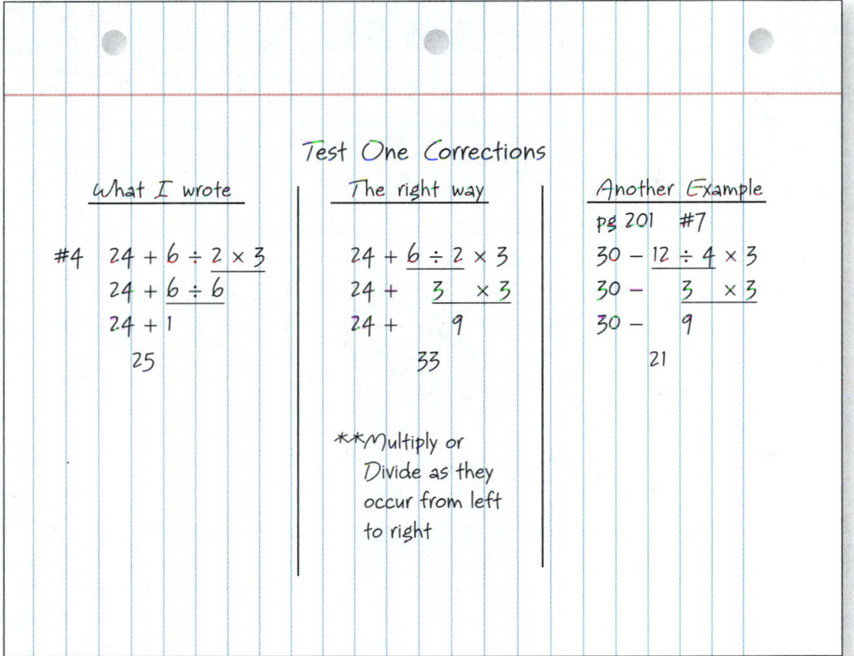

Action Plan

Objective 4 Establish an action plan.

Check out additional homework strategies in the supplemental Success Strategies Manual for a "Test Preparation Worksheet," a "Post-Test Debriefing" form, suggestions for "How to Reduce Test Anxiety," and tips for "Problem-Solving Strategies."

1. Two things that I will commit to doing before a test to improve my chances of success:

____

____

2. Two strategies that I will employ during a test to keep my anxiety down:

____

____

3. After a test I will do the following to learn from my mistakes:

____

____

Troubleshooting Common Errors

Objective 5 ▶

Troubleshoot common errors.

After you check your homework or take a quiz or test, you should review it to determine the types of mistakes you made and think about how you can fix them. It is important to understand whether the errors you make are conceptual, procedural, computational, or secretarial.

A **conceptual error** shows little or no understanding of the underlying concept or procedure.

In a **procedural error**, the correct process is used but there is a mistake made with the steps.

In a **computational error,** an arithmetic mistake is made.

In a **secretarial error**, a miscopy of some kind is made.

Evaluate

$24 + 6 \div 2 \times 3$

$\underline{24 + 6} \div 2 \times 3$ Add

$\underline{30 \div 2} \times 3$ Divide

15×3 Multiply

45

The student shows no understanding of the concept of "order of operations." He just works straight through from left to right. This is an example of a conceptual error.

Evaluate

$24 + 6 \div 2 \times 3$ PEMDAS

$24 + 6 \div \underline{2 \times 3}$ Multiply

$24 + \underline{6 \div 6}$ Divide

$24 + 1$ Add

25

The student clearly identifies that the order of operations is being used as she uses the acronym PEMDAS. She knows the right process. However, she thinks that she must multiply then divide, rather than the correct rule: Multiplication and division are done as they occur, from left to right. This is an example of a procedural error.

Evaluate

$24 + 6 \div 2 \times 3$ PEMDAS

$24 + \underline{6 \div 2} \times 3$ Divide

$24 + \underline{3 \times 3}$ Multiply

$24 + 9$ Add

32

The student clearly identifies that "order of operations" is being used as he uses the acronym PEMDAS. He does all of the right steps, but then adds incorrectly. This is an example of a computational error.

A fourth type of mistake is *simply secretarial*. This type of mistake includes miscopying any part of the problem, misreading your handwriting, misaligning the problem, or any other non-math-related issue with how you write out your work.

How do I correct these errors?

A conceptual error needs to be corrected before the test begins. This mistake is typically about lack of preparation for the test.

Note: *Remember P^5*

Proper Preparation Prevents Poor Performance

- If you have not been doing any homework, do some. If you have been doing some, do more. If you have been doing all of the homework and are still struggling with the concepts, you may need some tutoring.
- Sometimes the issue is problem recognition. Perhaps you can do all of the problems when they are organized into separate sections in the book, but not when they are jumbled together on a test.
  - A great way to mimic this situation is to mix up your notecards.
  - Don't study everything in the same order each time.
  - Mix problems from different sections and chapters together.
  - Work any cumulative reviews in the text.
  - Review old homework each time you complete a new homework assignment to compare how the new assignment is the same or different from the old topics.

A procedural mistake usually requires only a minor fix. You understand the concept; you know what the steps are, but have confused them in some way. This mistake is also about preparation; the steps should be natural by the time you get to a test. If you have repeated a process enough times, you will not even think about the steps any more, you will just know what to do. Whether it is playing music, throwing a baseball, making a pie, or doing a math problem, practice makes perfect!

Computational and secretarial mistakes both have similar fixes. Check your work and slow down. These mistakes are typically made on problems that you are very confident with and are working through quickly. You do not want quickly to mean *carelessly*!

- For many problem types, you can check your answers. When this is possible you should always do so. Check the solution of a problem with the original problem. In this way even a miscopy can be found.
- Estimate your answer before you begin your work. If your answer is not near the estimate, investigate why.
- Think about the reasonableness of your answer. If you made a secretarial error with a decimal point and have an answer where a car is driving 500 mph, it should be clear to you that this type of answer is not reasonable.
- If you are permitted to use a calculator, check hand calculations with the calculator. Also, check the calculator output with your reasonableness criteria. If you press the wrong buttons, a calculator can still give you a wrong answer.

- Use grid paper instead of lined paper to help with misalignment issues. You can also turn lined paper horizontally to create columns.

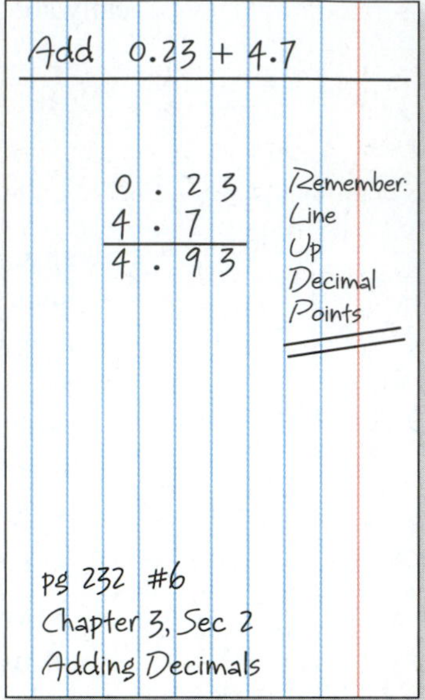

Note: *If you can't read it, you can't review or study it. If your teacher can't read it, you may get it marked wrong.*

Be Neat!

Procedure: Determining the Type of Mistake on a Test or Quiz

Step 1: Conceptual: Left it blank, wrong process, mis-memorized rule
Step 2: Procedural: Performed steps out of order, missed a step, or couldn't finish all steps
Step 3: Computational: Added, subtracted, multiplied, or divided wrong
Step 4: Secretarial: Miscopy, omission, misalignment, misread handwriting

After a test, review each problem that you got wrong and evaluate what went wrong. What types of mistakes did you make? Is there a trend in which type of mistakes you make? What will you do to eliminate these mistakes in the future? Also, reflect about whether the test is helping you meet your grade goal or harming you in making your grade goal. After each test, you should evaluate your progress toward meeting your grade goal.

"Organization and Commitment go hand in hand. For each section, organize the study plan and make a commitment to follow the schedule."

Pauline Chow, *Harrisburg Area Community College*

SECTION S.5 Blended and Online Classes

OBJECTIVES

As a result of completing this section, you will be able to:

1. Differentiate between blended, hybrid, and online classes.
2. Use techniques to maximize your success in blended and online settings.
3. Establish an action plan.
4. Troubleshoot common errors.

The desire for flexibility and convenience in scheduling classes has led to a continued increase in blended and online classes.

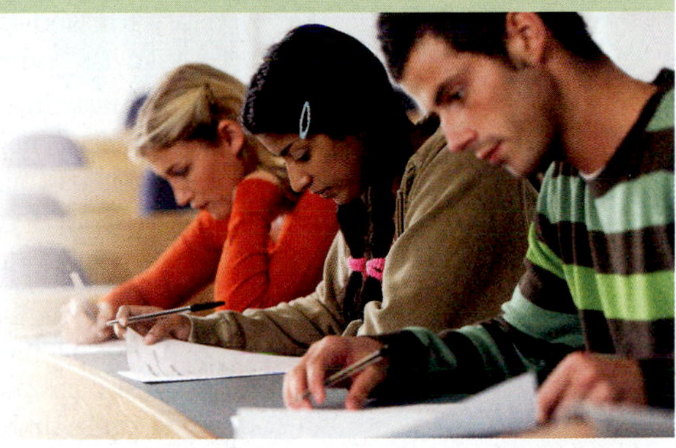

Identifying Different Delivery Methods

Objective 1 Differentiate between blended, hybrid, and online classes.

Blended classes, or **hybrid** classes, incorporate traditional classroom experiences with online learning activities. Students in these types of classes are required to attend a specific number of on-campus class meetings. The remaining portion of the course is delivered online, thereby reducing the amount of time a student spends in class. In summary, blended classes strive to combine the best elements of traditional face-to-face instruction with the best aspects of learning at a distance.

Online classes, unlike blended classes, deliver entire courses at a distance. Some may have required meetings for exams, but typically all aspects of the course are provided online.

With a portion of a course or an entire course delivered at a distance, the roles of the teacher and student change. The teacher becomes more of a facilitator and the student becomes more like a teacher in these environments. While in a traditional classroom, the student's learning is his or her responsibility, this responsibility becomes even more important in blended and online courses. The information provided in the previous sections certainly applies to students in any type of class, but there are some additional tips for students in blended and online classes.

Tips for Blended and Online Classes

Objective 2 Use techniques to maximize your success in blended and online settings.

1. The first day of class is extremely important. If you are in a blended class, be sure to attend the first on-campus meeting. If you are in an online class, be sure to log into the course the first day the course is available.
2. Review the syllabus and course policies to determine what is expected of you.
   - **a.** How many required on-campus meetings are there?
   - **b.** Are there proctored exams?
   - **c.** Upon what activities is your grade based (e.g., online homework, discussion board postings, online quizzes and/or tests, group projects, and so on)?
3. Familiarize yourself with the course resources.
   - **a.** Does your instructor post notes or videos, solutions for quizzes/tests, or other material?
   - **b.** Does your instructor hold online office hours through chat, email, or a live classroom? If so, when?
   - **c.** What learning materials (videos, practice problems, ebook, and the like) are available with your textbook?

4. Make a calendar for the semester (one may be available by your instructor) and write down due dates for homework, quizzes, tests, discussion posts, and other assignments.
5. Determine a time each week to have class with yourself.
   a. If the face-to-face version of the class meets, for example, 2 days a week for 1 hr and 45 min, then you should set aside this same amount of time to "have class" on your own.
   b. Class time is time for you to watch videos, to read the textbook, to do what is necessary to learn the material, and to complete quizzes, tests, and discussions. As you watch videos or read the textbook, take notes as you would if you were attending a traditional class.
6. Determine a time each week to have study time. In addition to "class," you need study time.
   a. For each credit or contact hour for the course, you should have three hours of class plus study time (3 credit class = 9 hr, 3 hr for "class" and 6 hr of study time).
   b. Study time is time for you to complete homework. So that you do not get dependent on the help resources online, we suggest that you print the homework assignment, work it offline, and then log back in to enter your answers. For the problems that you answered incorrectly, use the help features to assist you.
7. Remember that you are not alone in the online classroom. Interact with your instructor and other classmates through discussion boards, email, and other features that are available in your class.
8. Ask your instructor for help when you do not understand the material or the technology.
9. Maintain your notes, homework, printed copies of assignments, and any other printables in your notebook.
10. Check in frequently, daily if possible, for announcements, changes to the calendar or assignments, and for any updates you need to know.

Action Plan

Objective 3 ▶
Establish an action plan.

Set some goals for how you will be successful in a class with a nontraditional delivery method.

1. I will "have class with myself" _________ times a week for _________ minutes each time.
2. If I find I can't do it alone, I will get help by ______________________________

   ____.
3. Five things that I am committed to do to ensure that I maximize my success are

   ____.

   ____

   ____

   ____

   ____

Troubleshooting Common Errors

Objective 4 Troubleshoot common errors.

Blended and online classes present some different challenges than face-to-face classes. Here are some suggestions to overcome those challenges from students who have taken blended and/or online classes.

Study Skills	Suggestions
Homework	"Plan ahead and don't procrastinate in completing your homework. And try to do the take home assignments without looking at notes . . . it will help a lot at test time!" "Follow the study guide, do all the online homework and take-home work. Try the practice tests to help with confidence if you get nervous taking tests."
Textbook	"Read the book first, then do book questions, then do all online homework until a score of 100 percent is reached. Review all materials again before an exam. You have to be disciplined in getting into a routine just like a classroom based course."
Communication	"I would recommend keeping in constant contact with the instructor. Monitor your progress and make sure you are constantly evaluating your study habits. Reviewing feedback is another great way to improve your understanding."
Time Management	"You will need a lot of time to succeed in this course. Just because it's an online class doesn't mean you can sign on for 5 minutes do some work and fly by. Time is essential to succeeding in this course." "This course requires a good bit of time, and dedication. Had I not spend much time on the course at the beginning, I would have not been able to perform well for the remainder of the course." "It is very flexible but students have to be prepared to discipline themselves enough to do the work on time and not wait until the last minute. Don't try to do all the problems at one time, it is very overwhelming. It's not always possible but try to treat it like a normal class and set aside an hour a day or every other day, to work on it. Sometimes you have a busy week and that's what's nice about having an online class, the flexibility to work on the material when you have time." "In order to succeed you have to put the time into this class as if it were on campus. I feel that if individuals apply themselves to this course they will succeed!!!!"

Chapter S / REVIEW

Strategies to Succeed in Math

What's the big idea? Chapter S provides many strategies for maximizing your success in math class. No one will implement all of these ideas, but using even some will improve your chances of success. The more you implement, the better your chance of success.

The Tools

Additional student resources can be found in the supplemental Success Strategies Manual which is available in a workbook or also through your online homework system.

You can learn math!

Summary of Success Strategies

- Set goals.
- Manage your time.
- Adopt habits of successful math students.
- Read your book with purpose.
- Take meaningful notes.
- Keep an organized notebook/portfolio.
- Do homework to ensure deep understanding.
- Adopt good test-taking skills before, during, and after an exam or quiz.
- If you are in a blended or online class, adapt your strategies.

What's Next?

Throughout this text we will continue to support your efforts to succeed. With each chapter we will revisit one of the characteristics of success discussed in this opening chapter. Some chapter openers will focus on skills and others attitudes. When you need a little extra motivation, scan through this chapter or through the chapter openers to find suggestions.

The topics for each chapter opener are as follows.

Chapter 1 Time Management
Chapter 2 Organization
Chapter 3 Commitment and Perseverance
Chapter 4 Study Skills
Chapter 5 Goal Setting
Chapter 6 Motivation
Chapter 7 Learning Strategies
Chapter 8 Refocus
Chapter 9 Reflection

CHAPTER 1

Real Numbers and Algebraic Expressions

Time Management

As you begin this chapter, we want you to focus on how you are managing your time. Your success in this course depends on your ability to devote the appropriate time to learning and practicing the material.

- The general rule of thumb is that you should spend 2 hours studying for each hour you are in class. Some students will take more time; some will take less time.
- It is recommended that after 50 minutes of study, you take a 10-minute break.
- Prepare a calendar of your week and schedule an appointment with yourself for study time.
- Review how you are using your time through the week and make an honest assessment of whether you have the time to make this class a priority.
- See the Preface for additional resources concerning time management.

Question For Thought: Of the activities you participate in, which are essential? Are there any of these activities that you can give up to provide more time for your studies?

Chapter Outline

Section 1.1 Sets and the Real Numbers 2

Section 1.2 Operations with Real Numbers and Algebraic Expressions 15

Section 1.3 Properties of Real Numbers and Simplifying Algebraic Expressions 33

Coming Up...

In Section 1.3, we will learn how to represent the fact that people with a bachelor's degree earn nearly twice as much as those with only a high school diploma. (Source: U.S. Census Bureau)

“Time is the coin of your life. It is the only coin you have, and only you can determine how it will be spent. Be careful lest you let other people spend it for you.”

—Carl Sandburg (Poet and Author)

SECTION 1.1 Sets and the Real Numbers

In **Chapter 1** we will discuss the concepts of sets, the set of real numbers, operations with real numbers, numerical and algebraic expressions.

OBJECTIVES

As a result of completing this section, you will be able to

1. Understand terminology related to sets and perform operations on sets.
2. Classify a number as a natural number, whole number, integer, rational number, or irrational number.
3. Graph real numbers on a real number line.
4. Find the opposite of a real number.
5. Find the absolute value of a real number.
6. Troubleshoot common errors.

How would you classify the following numbers?

5 hr: the approximate time the average American watches TV each day

$-273°$C: the temperature of absolute zero

$6\frac{1}{3}$ ft: the height of the tallest U.S. President, Abraham Lincoln

16π ft^2: the area of a circle with radius 4 ft

The numbers 5, -273, $6\frac{1}{3}$, and 16π are examples of real numbers. We encounter real numbers on a daily basis. In this section, we will discuss the set of real numbers in detail.

Objective 1

Understand terminology related to sets and perform operations on sets.

Sets

A **set** is a collection of objects. Each object in a set is called a **member** or an **element**. A set is written in braces, { }, and is usually denoted with a capital letter.

Sets can either be **finite** or **infinite**. A finite set has a specific number of elements. An example of a finite set is $A = \{1, 2, 3, 4, 5\}$. An infinite set has infinitely many elements. An example of an infinite set is $B = \{1, 3, 5, 7, 9, \ldots\}$. The three dots at the end are called an *ellipsis* and indicate that the set continues in the same manner.

In preceding sets A and B, the elements of the sets are listed explicitly. When we use this method to represent a set, we are using the **roster method**. Another method to represent elements in a set is called **set-builder notation**. In this method, we state the conditions the elements must satisfy; set-builder notation describes the members of a set but does not list them. Set-builder notation is written in the form:

$$\{x \mid \text{condition } x \text{ must satisfy}\}$$

This is read as “the set of x, such that, ______.” In the blank, we insert the condition that x must satisfy. The letter x is a **variable** and represents some unknown number. We will discuss variables in the next section. An example is $C = \{x \mid x \text{ is a positive odd number}\}$. The condition stated tells us that x can be 1, 3, 5, 7, 9, ... This is the same as the previous set B.

A set that contains no elements is called the **empty set** and is denoted by { } or $\varnothing$.

As we talk about sets, there is some terminology and notation that we should learn. To illustrate the notation, let $A = \{3, 4, 5\}$ and $B = \{3, 4, 5, 6, 7\}$.

Symbol	Meaning	Example	Verbal Statement
$\in$	Element	$4 \in A$	“4 is an element of A.” or “4 is a member of A.”
$\notin$	Is not an element of	$6 \notin A$	“6 is not an element of A.” or “6 is not a member of A.”
$\subset$	Subset	$A \subset B$	“Set A is contained in set B.” This means that every element in A is also contained in B.

Symbol	Meaning	Example	Verbal Statement
$\cup$	Union	$A \cup B = \{3, 4, 5, 6, 7\}$	"A union B" The union is the set of elements that are in A or in B or in both. Elements do not need to be repeated if they are in both sets.
$\cap$	Intersection	$A \cap B = \{3, 4, 5\}$	"A intersect B" The intersection is the set of elements that are in both A and B.

Objective 1 Examples **Determine the requested information for the given sets.**

1a. Use the roster method to write the set $A = \{x | x \text{ is a positive even number}\}$ in another way.

Solution **1a.** The roster method lists the numbers in the set explicitly. So, the set $A = \{2, 4, 6, 8, 10, \ldots\}$.

1b. Let $A = \{3, 6, 9, 12, 15\}$ and $B = \{6, 12, 18, 24\}$. Find $A \cup B$ and $A \cap B$.

Solution **1b.** The set $A \cup B$ means to join all elements of A and all elements of B into one set. Therefore, if $A = \{3, 6, 9, 12, 15\}$ and $B = \{6, 12, 18, 24\}$, then

$$A \cup B = \{3, 6, 9, 12, 15, 18, 24\}$$

The set $A \cap B$ means to find the elements that are in both A and B. Therefore,

$$A \cap B = \{6, 12\}$$

INSTRUCTOR NOTE:
Use this activity to help students understand this concept. Let A = {students born in the state} and B = {students less than 25 yr old}. Have students stand up if they belong to set A, B, $A \cup B$, or $A \cap B$.

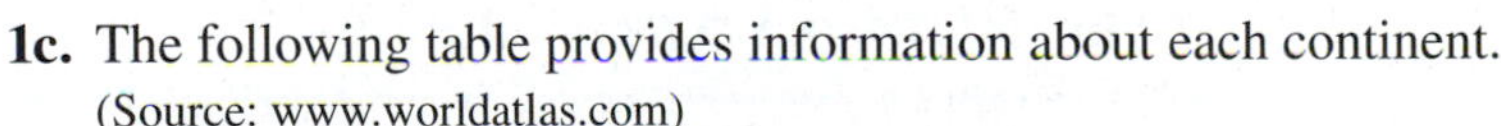

1c. The following table provides information about each continent. (Source: www.worldatlas.com)

Continent	Population in 2009	Land Mass
Africa	1,000,050,000	30,065,000 km^2
Antarctica	0	13,209,000 km^2
Asia	3,879,000,000	44,579,000 km^2
Australia/Oceania	32,000,000	8,112,000 km^2
Europe	731,000,000	9,938,000 km^2
North America	528,720,588	24,474,000 km^2
South America	379,500,000	17,819,000 km^2

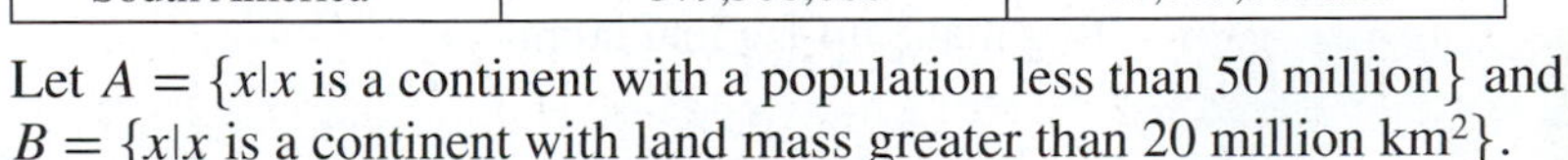

Let $A = \{x|x \text{ is a continent with a population less than 50 million}\}$ and $B = \{x|x \text{ is a continent with land mass greater than 20 million km}^2\}$.

i. State set A using the roster method.

ii. State set B using the roster method.

iii. Is Asia $\in A \cap B$?

Solution **1c.** **i.** To be a member of set A, the continent must have a population less than 50 million. So, $A = \{\text{Antarctica, Australia/Oceania}\}$.

ii. To be a member of set B, the continent must have a land mass greater than 20 million km^2. So, $B = \{\text{Africa, Asia, North America}\}$.

iii. $A \cap B$ is the set of elements that are members of both A and B. There are no continents listed in both sets, so $A \cap B = \varnothing$. Therefore, Asia $\notin A \cap B$.

Student Check 1 Determine the requested information for the given sets.

a. Use the roster method to write the set $A = \{x | x \text{ is a counting number greater than } 3\}$.

b. Let $A = \{5, 10, 15, 20, 25, 30\}$ and $B = \{15, 30, 45, 60, 75\}$. Find $A \cup B$ and $A \cap B$.

c. The following table provides information about the top public colleges according to the Best Colleges Rankings in 2012. (Source: http://colleges.usnews.rankingsandreviews.com/best-colleges/rankings/national-universities/top-public)

College	Undergraduate Enrollment	2011–2012 Tuition (In State)
University of California, Berkeley (UC Berkeley)	25,540	\$11,767
University of California, Los Angeles (UCLA)	26,162	\$11,604
University of Virginia (UVA)	15,595	\$11,576
University of Michigan (U-M)	27,027	\$12,590
University of North Carolina, Chapel Hill (UNC)	18,579	\$ 7,008
College of William and Mary (W&M)	5,898	\$13,132
Georgia Institute of Technology (GT)	13,750	\$ 9,652
University of California, San Diego (UCSD)	23,663	\$12,128
University of California, Davis (UC Davis)	24,737	\$12,794
University of California, Santa Barbara (UCSB)	19,186	\$12,508

Let $A = \{x | x \text{ is a college with enrollment less than } 20{,}000\}$ and $B = \{x | x \text{ is a college with annual cost greater than } \$12{,}500\}$.

i. State set A using the roster method.

ii. State set B using the roster method.

iii. Is UVA $\in A \cap B$?

The Real Numbers

Objective 2

Classify a number as a natural number, whole number, integer, rational number, or irrational number.

Numbers are such a critical component of our lives. We use them to represent so many things, such as, the amount of money in our bank account, the hours in a day, the temperature outside, our current weight, our height, the credit hours we are enrolled in, our GPA, our salary, measurements, and so on. Furthermore, computers store all of their data as numbers. The world as we know it would cease to exist without numbers.

Numbers belong to certain sets. The following table defines the sets of numbers that make up the real numbers.

Natural numbers	$\mathbb{N} = \{1, 2, 3, 4, 5, \ldots\}$
Whole numbers	$\mathbb{W} = \{0, 1, 2, 3, 4, 5, \ldots\}$
Integers	$\mathbb{Z} = \{\ldots, -5, -4, -3, -2, -1, 0, 1, 2, 3, 4, 5, \ldots\}$
Rational numbers	$\mathbb{Q} = \left\{\frac{p}{q} \middle| p, q \in \mathbb{Z} \text{ with } q \neq 0\right\}$
Irrational numbers	$\mathbb{I} = \{\text{numbers that are not rational}\}$
Real numbers	$\mathbb{R} = \{\text{rational or irrational numbers}\}$

The rational and irrational numbers are discussed in more detail next.

The set of *rational numbers* is given in set-builder notation in the previous chart. We read this as "the set of numbers p over q, such that p and q are integers with q not equal to 0." That is, the values of p and q can be replaced with any integer except that q cannot be replaced with zero, because division by zero is undefined.

INSTRUCTOR NOTE:
You may want to show how we generate all of the rational numbers

$\frac{1}{1}, \frac{1}{2}, \frac{1}{3}, \frac{1}{4}, \ldots$

$\frac{2}{1}, \frac{2}{2}, \frac{2}{3}, \frac{2}{4}, \ldots$

$\frac{3}{1}, \frac{3}{2}, \frac{3}{3}, \frac{3}{4}, \ldots$

And so on.

Definition: A **rational number** is a number that can be written as the quotient, or ratio, of two integers, such that, the integer in the denominator is nonzero.

When we divide one integer by another nonzero integer, the result will be either an integer, a terminating decimal, or a repeating decimal. As there are infinitely many rational numbers, it is impossible to list all of them, but an example of each of these is shown.

$10 = \frac{10}{1}$	**Integer**
$-\frac{3}{5} = -0.6$	**Terminating decimal**
$5\frac{2}{7} = \frac{37}{7} = 5.2857142857142\ldots$ $= 5.\overline{285714}$	**Repeating decimal**

The set of *irrational numbers* arose out of the study of geometry, specifically triangles and circles. A simple definition of an irrational number is a number that is not rational. This means that the number cannot be written as a quotient of two integers. Therefore, an irrational number is not an integer, is not a proper fraction, is not an improper fraction, is not a mixed number, and is not a decimal that repeats in a pattern.

INSTRUCTOR NOTE:
Make the connection that irrational numbers are "irrational" in their behavior, while rational numbers behave "rationally."
Also use the number *e* as an example.

Definition: An **irrational number** is a number whose decimal value neither terminates nor repeats in a pattern.

The number π (pi) may be the most commonly known irrational number. We often use the value 3.14 for π but this is only a decimal approximation. The exact value of π is 3.1415926535.... This decimal value does not terminate and continues with no repeating pattern.

Other examples of irrational numbers involve some square roots. The **square root** of a number is the number that must be multiplied by itself to get the original number. For instance, $\sqrt{9} = 3$ since $3 \cdot 3 = 9$. The square root of 9 is a rational number since its value is equivalent to 3. However, not all square roots are rational.

We can use the calculator to approximate square roots and to determine if the number is rational or irrational. If the value on the calculator is a decimal that repeats or terminates, then the number is rational. If the value on the calculator is a decimal that doesn't end or repeat, then the number is irrational. Some irrational numbers are

INSTRUCTOR NOTE:
Roots are covered extensively in Chapter 8.

$$\sqrt{3} = 1.7320508075\ldots$$
$$\sqrt{5} = 2.236067977\ldots$$

The set of rational numbers combined with the set of irrational numbers make up the set of **real numbers**. So, a real number is either rational or irrational; it cannot be both. The following diagram illustrates how the sets of numbers relate to one another.

INSTRUCTOR NOTE:
Have students give an example of a rational number that is not an integer, an integer that is not a whole number, and a whole number that is not a natural number.

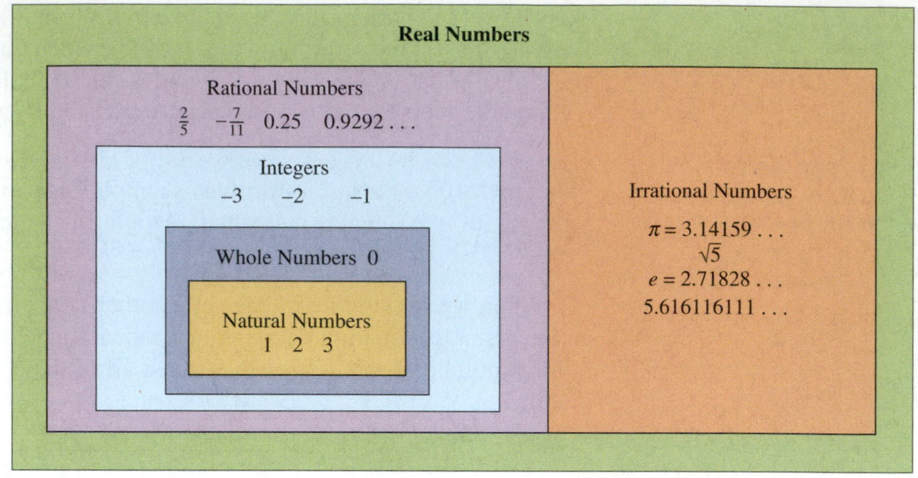

This diagram illustrates the following facts.

- The set of natural numbers is a subset of the set of whole numbers, the set of integers, and the set of rational numbers—that is, $\mathbb{N} \subset \mathbb{W}$, $\mathbb{N} \subset \mathbb{Z}$, $\mathbb{N} \subset \mathbb{Q}$.
- The set of whole numbers is a subset of the set of integers and the set of rational numbers—that is, $\mathbb{W} \subset \mathbb{Z}$, $\mathbb{W} \subset \mathbb{Q}$.
- The set of integers is a subset of the set of rational numbers—that is, $\mathbb{Z} \subset \mathbb{Q}$.
- The set of rational numbers together with the irrational numbers make up the real numbers—that is, $\mathbb{Q} \cup \mathbb{I} = \mathbb{R}$.
- A rational number cannot be an irrational number—that is, $\mathbb{Q} \cap \mathbb{I} = \varnothing$.

In Chapter 8, another set of numbers, called the *complex numbers*, is introduced. This set actually contains all of the real numbers.

Procedure: Classifying a Real Number as a Natural Number, Whole Number, Integer, Rational Number, or Irrational Number

Step 1: If the number is 1, 2, 3, . . . , then the number is a natural number.
Step 2: If the number is 0, 1, 2, 3, . . . , then the number is a whole number.
Step 3: If the number is . . . , −3, −2, −1, 0, 1, 2, 3, . . . , then the number is an integer.
Step 4: Determine if the number is rational or irrational.

a. If the number is equivalent to a decimal that terminates or repeats, then it is rational.

b. If the number is equivalent to a decimal that doesn't terminate or repeat, then it is irrational.

Objective 2 Examples **Classify each number in the set as a natural number, whole number, integer, rational number, irrational number, and/or real number. If the number is a rational number, write it in the form of a fraction. If the number is irrational, approximate its value to two decimal places.**

2a. $\left\{3.2, -4\frac{1}{2}, \sqrt{9}, \sqrt{15}, 0, -8, \frac{3}{4}\right\}$

2b. 5 hr: the time the average American watches TV each day

$-273°C$: the temperature of absolute zero

$6\frac{1}{3}$ ft: the height of the tallest U.S. president, Abraham Lincoln

16π ft^2: the area of a circle with radius 4 ft

Solutions

2a.

Number	Natural	Whole	Integer	Rational	Irrational	Real
$3.2 = \frac{32}{10}$				X		X
$-4\frac{1}{2} = -\frac{9}{2}$				X		X
$\sqrt{9} = 3 = \frac{3}{1}$	X	X	X	X		X
$\sqrt{15} \approx 3.87$					X	X
$0 = \frac{0}{1}$		X	X	X		X
$-8 = -\frac{8}{1}$			X	X		X
$\frac{3}{4}$				X		X

2b.

Number	Natural	Whole	Integer	Rational	Irrational	Real
$5 = \frac{5}{1}$	X	X	X	X		X
$-273 = -\frac{273}{1}$			X	X		X
$6\frac{1}{3} = \frac{19}{3}$				X		X
$16\pi \approx 50.27$					X	X

Student Check 2 Classify each number in the set as a natural number, whole number, integer, rational number, irrational number, and/or real number. If the number is a rational number, write it in the form of a fraction. If the number is irrational, approximate its value to two decimal places.

$$\left\{7.5, 3\frac{4}{5}, \sqrt{30}, \sqrt{25}, -20, 4\pi, \frac{1}{2}\right\}$$

The Real Number Line

Objective 3 Graph real numbers on a real number line.

Graphing or plotting real numbers on a number line is an important skill that will be used in later sections. The **real number line** is a horizontal line drawn with arrows on both ends to indicate that the real numbers are infinite. Tick marks are used to divide the number line into equal segments. Positive numbers are located to the right of 0 and negative numbers are located to the left of 0. Zero is called the **origin** and is neither positive nor negative.

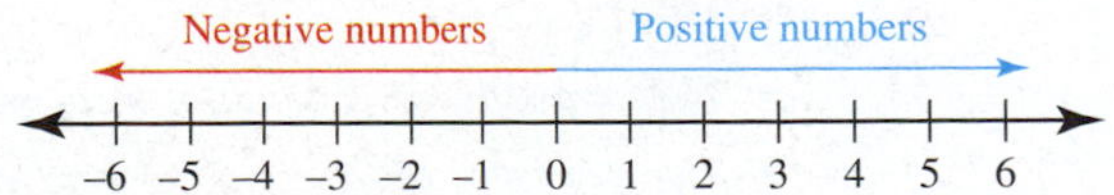

Procedure: Graphing a Point on a Number Line

Step 1: Locate the number on the number line.
  a. If the number is a fraction, convert it to a decimal value.
  b. If the number is irrational, approximate its value to two decimal places.
Step 2: Draw a dot on its position on the number line.

Objective 3 Examples **Graph the numbers -3.2, $-4\frac{1}{2}$, $\sqrt{9}$, $\sqrt{15}$, 0, -6, and $\frac{3}{4}$ on a number line.**

Solution If the number is not an integer, round it to the nearest hundredth to approximate its value and to find its location on the number line.

$$-4\frac{1}{2} = -4.50,\ \sqrt{15} \approx 3.87,\ \frac{3}{4} = 0.75$$

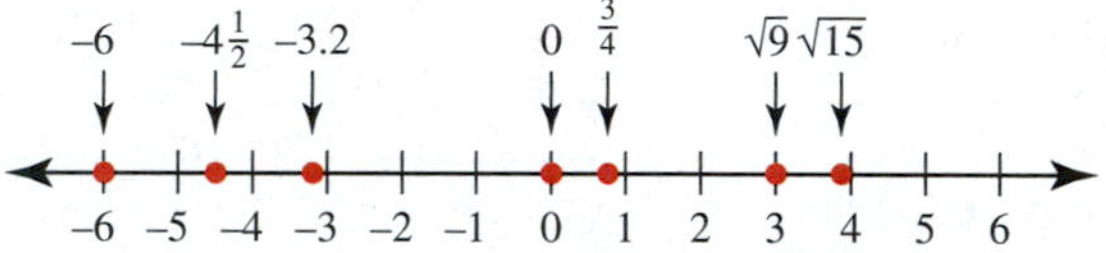

Student Check 3 Graph the numbers -5.4, $-2\frac{1}{4}$, $\sqrt{16}$, $\sqrt{3}$, -3, and $\frac{2}{5}$ on a number line.

Opposites

Objective 4 Find the opposite of a real number.

Two numbers that lie equal distances from zero are *opposites* of one another.

Definition: The **opposite** of a real number a is the number that has the same distance from 0 on a number line but lies on the opposite side of 0 from a. The opposite of a real number a is denoted as $-a$.

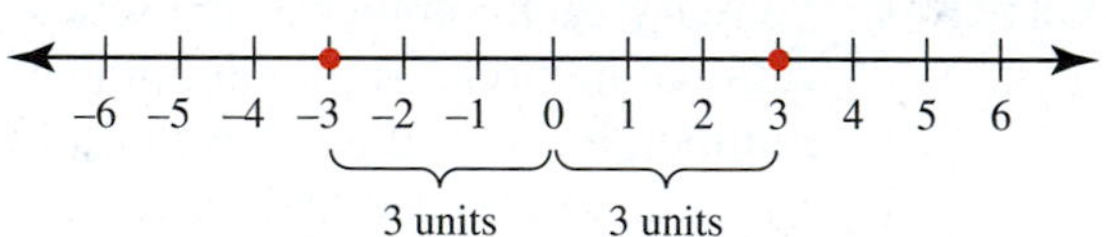

The number line illustrates that -3 and 3 are opposites of one another since they are the same distance from zero and on opposite sides of zero. These numbers are also referred to as **additive inverses**.

Verbal Statement	Mathematical Statement
The opposite of 3 is -3.	$-(3) = -3$
The opposite of -3 is 3.	$-(-3) = 3$

Note:
- *The opposite of a positive real number is a negative number.*
- *The opposite of a negative real number is a positive number.*

Objective 4 Examples Find the opposite of each number.

Problems	Solutions
4a. -4	$-(-4) = 4$
4b. 20	$-(20) = -20$
4c. $\frac{7}{8}$	$-\left(\frac{7}{8}\right) = -\frac{7}{8}$
4d. $-5\frac{1}{3}$	$-\left(-5\frac{1}{3}\right) = 5\frac{1}{3}$
4e. $\sqrt{7}$	$-(\sqrt{7}) = -\sqrt{7}$
4f. $-\pi$	$-(-\pi) = \pi$

Student Check 4 Find the opposite of each number.

a. 10 **b.** -14 **c.** $-\frac{1}{2}$ **d.** $\frac{25}{6}$ **e.** $-\sqrt{6}$ **f.** 8.2

Absolute Value

Objective 5

Find the absolute value of a real number.

The *absolute value* of a number refers to the number's distance from zero on a real number line.

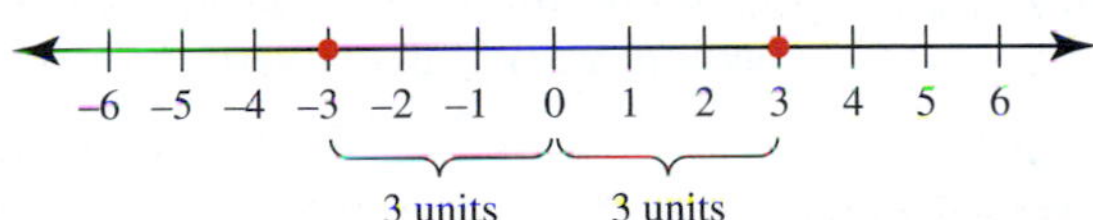

The numbers -3 and 3 are both 3 units from 0 on the real number line, so the absolute value of these numbers is 3. We use vertical bars to denote the absolute value of a number a, $|a|$. Since absolute value refers to distance, the absolute value of a number is always greater than or equal to zero.

Verbal Statement	Mathematical Statement
The absolute value of 3 is 3.	$\|3\| = 3$
The absolute value of -3 is 3.	$\|-3\| = 3$

Definition: The **absolute value** of a real number a, denoted $|a|$, is the distance between a and 0 on the real number line.

$$\text{If } a \geq 0, |a| = a.$$

$$\text{If } a < 0, |a| = -a.$$

Note: *The absolute value of a number that is positive or zero is that number. The absolute value of a negative number is its opposite.*

Objective 5 Examples **Simplify each absolute value expression.**

Problems	Solutions
5a. $\lvert 4 \rvert$	Because 4 is 4 units from 0 on a number line, $\lvert 4 \rvert = 4$.
5b. $\lvert 0 \rvert$	Because 0 is 0 units from 0 on a number line, $\lvert 0 \rvert = 0$.
5c. $\left\lvert -\frac{2}{5} \right\rvert$	Because $-\frac{2}{5}$ is $\frac{2}{5}$ units from 0 on a number line, $\left\lvert -\frac{2}{5} \right\rvert = \frac{2}{5}$.
5d. $-\lvert 8 \rvert$	We need to find the opposite of the absolute value of 8. The absolute value of 8 is 8. Its opposite is -8. So, $$-\lvert 8 \rvert = -(8) = -8$$
5e. $-\lvert -5 \rvert$	We need to find the opposite of the absolute value of -5. The absolute value of -5 is 5. Its opposite is -5. So, $$-\lvert -5 \rvert = -(5) = -5$$

Student Check 5 Simplify each absolute value expression.

a. $\lvert 12 \rvert$ **b.** $\lvert \sqrt{3} \rvert$ **c.** $\left\lvert -\frac{3}{5} \right\rvert$ **d.** $-\lvert 1 \rvert$ **e.** $-\lvert -1 \rvert$

Objective 6

Troubleshoot common errors.

Troubleshooting Common Errors

Some common errors associated with the classification of real numbers and finding absolute values are shown.

Objective 6 Examples **A problem and an incorrect solution are given. Provide the correct solution and an explanation of the error.**

6a. Determine if $\sqrt{81}$ is a natural number, whole number, integer, rational number or irrational number.

Incorrect Solution	Correct Solution and Explanation
$\sqrt{81}$ is an irrational number since it involves a square root symbol.	$\sqrt{81} = 9$ since $9 \cdot 9 = 81$. Therefore, $\sqrt{81}$ is a natural number, whole number, integer, and rational number.

6b. Simplify $-\lvert -3 \rvert$.

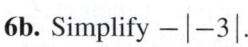

Incorrect Solution	Correct Solution and Explanation
$-\lvert -3 \rvert = 3$	This expression means to find the opposite of the absolute value of -3. Since $\lvert -3 \rvert = 3$, its opposite is -3. $$-\lvert -3 \rvert = -(3) = -3$$

ANSWERS TO STUDENT CHECKS

Student Check 1 **a.** $A = \{4, 5, 6, 7, 8, \ldots\}$ **b.** $A \cup B = \{5, 10, 15, 20, 25, 30, 45, 60, 75\}$; $A \cap B = \{15, 30\}$
c. **i.** $A = \{\text{UVA, UNC, W\&M, GT, UCSB}\}$
ii. $B = \{\text{U-M, W\&M, UC Davis, UCSB}\}$
iii. $A \cap B = \{\text{W\&M}\}$, so UVA $\notin A \cap B$.

Student Check 2 $7.5 = \frac{15}{2}$, rational, real; $3\frac{4}{5} = \frac{19}{5}$, rational, real; $\sqrt{30} \approx 5.48$, irrational, real; $\sqrt{25} = 5 = \frac{5}{1}$, natural, whole, integer, rational, real; $-20 = -\frac{20}{1}$, integer, rational, real; $4\pi \approx 12.57$, irrational, real; $\frac{1}{2}$, rational, real

Student Check 3

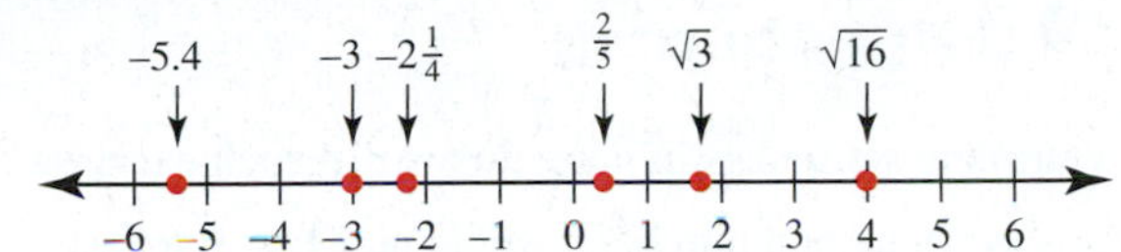

Student Check 4 **a.** -10 **b.** 14 **c.** $\frac{1}{2}$ **d.** $-\frac{25}{6}$ **e.** $\sqrt{6}$ **f.** -8.2

5. a. 12 **b.** $\sqrt{3}$ **c.** $\frac{3}{5}$ **d.** -1 **e.** -1

SUMMARY OF KEY CONCEPTS

1. A set is a collection of objects. The union of two sets is a collection of all elements that are in either of the sets. The intersection of two sets is a collection of the elements that are in each of the sets.
2. The real numbers are comprised of rational numbers and irrational numbers. The set of rational numbers includes all integers. The set of integers includes all whole numbers. The set of whole numbers includes all natural numbers.
3. Every real number can be graphed on a real number line. Irrational numbers must be approximated to be graphed.
4. The numbers a and $-a$ are opposites of one another. These numbers have the same distance from zero and are on opposite sides of zero.
5. The absolute value of a number measures the number's distance from zero on the real number line. The absolute value of a number is either zero or positive. Vertical bars denote absolute value.

GRAPHING CALCULATOR SKILLS

We can use a calculator to approximate irrational numbers, find the opposite of a number, and find the absolute value of a number.

Example 1: Approximate the value of $\sqrt{3}$ and π.

Solution: Enter the square root of 3 using the second function and the squaring function. Enter π using second function and the carat symbol.

```
√(3)
        1.732050808
π
        3.141592654
```

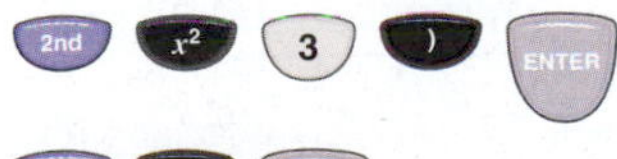

So, $\sqrt{3} \approx 1.73$ and $\pi \approx 3.14$.

Example 2: Find the opposite of -6 and 0.25.

Solution: To find the opposite of a number, we use the symbol (–) not the subtraction sign.

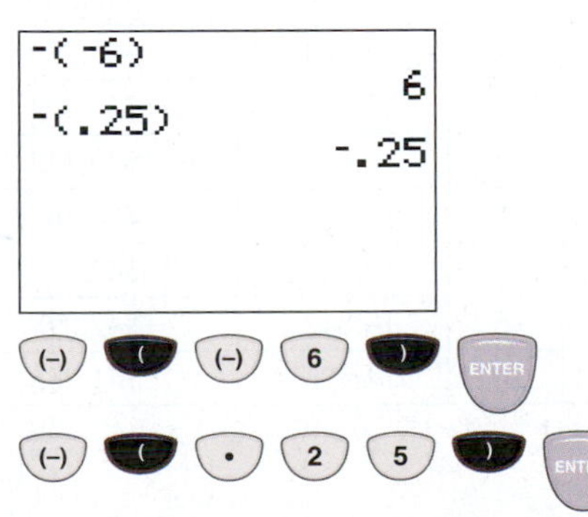

So, $-(-6) = 6$ and $-(0.25) = -0.25$.

Example 3: Find $|-5|$, $|10|$, and $-|-7|$.

Solution: Enter the absolute value function by pressing the MATH menu and then choosing the NUM menu.

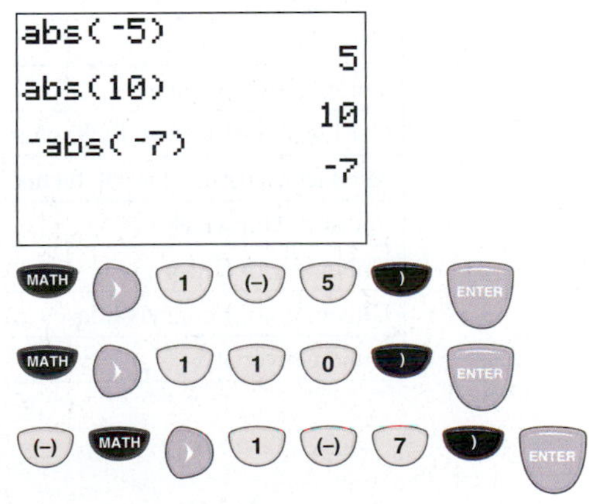

So, $|-5| = 5$, $|10| = 10$, and $-|-7| = -7$.

SECTION 1.1 EXERCISE SET

Write About It!

Use complete sentences in your answer to each exercise.

1. Is every natural number a whole number? Explain.
2. Is every whole number a natural number? Explain.
3. Describe the set of rational numbers.
4. Describe the set of irrational numbers.
5. What does it mean for two numbers to be opposites?
6. What relationship exists between the absolute values of opposite numbers?

Practice Makes Perfect!

Use the roster method to write each set defined in Exercises 7–12. (*See Objective 1.*)

7. $A = \{x | x \text{ is a whole number less than } 5\}$ $\{0, 1, 2, 3, 4\}$
8. $A = \{x | x \text{ is a natural number less than } 8\}$ $\{1, 2, 3, 4, 5, 6, 7\}$
9. $B = \{x | x \text{ is an integer strictly between } -3 \text{ and } 4\}$ $\{-2, -1, 0, 1, 2, 3\}$
10. $B = \{x | x \text{ is an integer strictly between } -5 \text{ and } 2\}$ $\{-4, -3, -2, -1, 0, 1\}$
11. $A = \{x | x \text{ is a natural number less than } \pi\}$ $\{1, 2, 3\}$
12. $B = \{x | x \text{ is an integer greater than } \pi\}$ $\{4, 5, 6, \ldots\}$

For Exercises 13–16, find $A \cup B$ and $A \cap B$ for the given sets. (*See Objective 1.*)

13. Let $A = \{-6, -4, -2, 0, 2, 4, 6\}$ and $B = \{1, 2, 3, \ldots\}$. $\{-6, -4, -2, 0, 1, 2, 3, \ldots\}$; $\{2, 4, 6\}$
14. Let $A = \{\ldots, -3, -1, 1, 3, \ldots\}$ and $B = \{-4, -2, 0, 2, 4, \ldots\}$. $\{\ldots, -7, -5, -4, -3, -2, -1, 0, 1, 2, 3, \ldots\}$; $\varnothing$
15. Let $A = \{3, 6, 9, 12, \ldots\}$ and $B = \{2, 4, 6, 8, \ldots\}$. $\{2, 3, 4, 6, 8, 9, 10, 12, 14, 15, 16, \ldots\}$; $\{6, 12, 18, \ldots\}$
16. Let $A = \{6, 12, 18, 24, 30, 36, 42, 48, 60\}$ and $B = \{10, 20, 30, 40, 50, 60\}$. $\{6, 10, 12, 18, 20, 24, 30, 36, 40, 42, 48, 50, 60\}$; $\{30, 60\}$

Use the following sets and table to complete Exercises 17 through 24. (*See Objective 1.*)

Let $A = \{x | x \text{ is a university with enrollment less than } 10{,}000\}$

$B = \{x | x \text{ is a university with tuition and fees less than } \$38{,}000\}$

$C = \{x | x \text{ is a university with acceptance rate less than or equal to } 10\%\}$

$D = \{x | x \text{ is a university with tuition and fees more than } \$45{,}000\}$

$E = \{x | x \text{ is a university with enrollment greater than } 19{,}000\}$

$F = \{x | x \text{ is a university with ranking less than } 4\}$

The table provides information about the Top 10 National Universities according to the 2012 edition of the *U.S. News & World Report's* annual *Best Colleges* issue. (Source: http://colleges.usnews.rankingsandreviews.com/best-colleges/rankings/national-universities)

Universities	Rank	Total Enrollment	2011–2012 Tuition and Fees	Fall 2010 Acceptance Rate
Harvard University	1	19,627	\$39,849	7%
Princeton University	1	7,802	\$37,000	9%
Yale University	3	11,701	\$40,500	8%
Columbia University	4	22,283	\$45,290	10%
California Institute of Technology	5	2,175	\$37,704	13%
Massachusetts Institute of Technology	5	10,566	\$40,732	10%
Stanford University	5	19,535	\$40,569	7%
University of Chicago	5	12,781	\$42,783	19%
University of Pennsylvania	5	19,842	\$42,098	14%
Duke University	10	14,983	\$41,958	16%

Additional answers can be found in the Instructor Answer Appendix.

17. State the set A using the roster method. {Princeton University, California Institute of Technology}

18. State the set D using the roster method. {Columbia University}

19. State the set $A \cap B$ using the roster method. {Princeton University, California Institute of Technology}

20. State the set $C \cap D$ using the roster method. {Columbia University}

21. State the set $D \cup E$ using the roster method. {Harvard University, Columbia University, Stanford University, University of Pennsylvania}

22. State the set $B \cup F$ using the roster method. {Harvard University, Princeton University, Yale University, California Institute of Technology}

23. Is Columbia University $\in C \cap E$? Yes

24. Is Princeton University $\in A \cap F$? Yes

Classify each real number as natural, whole, integer, rational, or irrational. List each classification that applies. (*See Objective 2.*)

25. -7 integer, rational

26. -5 integer, rational

27. $\frac{2}{5}$ rational

28. $\frac{3}{4}$ rational

29. $\sqrt{3}$ irrational

30. $-\sqrt{10}$ irrational

31. 1.5 rational

32. 3.6 rational

33. $-\frac{7}{4}$ rational

34. $-\frac{8}{3}$ rational

35. 16.375 rational

36. 13.1289 rational

37. $5\frac{1}{2}$ rational

38. $7\frac{2}{3}$ rational

39. -0.001 rational

40. -0.0025 rational

41. $\sqrt{49}$ natural, whole, integer, rational

42. $\sqrt{16}$ natural, whole, integer, rational

43. $-\frac{163}{113}$ rational

44. $-\frac{269}{534}$ rational

45. 3π irrational

46. 4π irrational

47. $0.\overline{3}$ rational

48. $1.\overline{9}$ rational

Graph each number on a real number line. (*See Objective 3.*)

49. -5

50. -1

51. $-\frac{1}{2}$

52. $\frac{3}{4}$

53. $\sqrt{10}$

54. $\sqrt{5}$

55. 3.5

56. 6.2

57. $\frac{5}{2}$

58. $\frac{4}{3}$

59. -0.7

60. -0.1

61. $\sqrt{9}$

62. $\sqrt{100}$

63. $-2\frac{1}{6}$

64. $5\frac{1}{3}$

65. $\frac{25}{2}$

66. $\frac{45}{4}$

Find the opposite of each number. Assume *a* and *b* represent positive real numbers. (*See Objective 4.*)

67. 11 -11

68. 15 -15

69. $-\frac{3}{7}$ $\frac{3}{7}$

70. $-\frac{2}{3}$ $\frac{2}{3}$

71. 4π -4π

72. 7π -7π

73. $-\sqrt{10}$ $\sqrt{10}$

74. $-\sqrt{7}$ $\sqrt{7}$

75. 1001 -1001

76. 121 -121

77. -1.35 1.35

78. -4.67 4.67

79. $2\frac{1}{8}$ $-2\frac{1}{8}$

80. $3\frac{1}{3}$ $-3\frac{1}{3}$

81. $\frac{10}{3}$ $-\frac{10}{3}$

82. $\frac{11}{2}$ $-\frac{11}{2}$

83. $-a$ a

84. b $-b$

85. $-\frac{9}{2}$ $\frac{9}{2}$

86. $-\frac{b}{3}$ $\frac{b}{3}$

87. $3a$ $-3a$

88. $-5b$ $5b$

Simplify each absolute value expression. Assume *m* represents a positive real number. (*See Objective 5.*)

89. $|2|$ 2

90. $|6|$ 6

91. $|-10|$ 10

92. $|-13|$ 13

93. $\left|\frac{10}{3}\right|$ $\frac{10}{3}$

94. $\left|\frac{1}{6}\right|$ $\frac{1}{6}$

95. $\left|-\frac{3}{5}\right|$ $\frac{3}{5}$

96. $\left|-\frac{5}{6}\right|$ $\frac{5}{6}$

97. $-|15|$ -15

98. $-|16|$ -16

99. $-|-4|$ -4

100. $-|-9|$ -9

101. $-|\sqrt{3}|$ $-\sqrt{3}$

102. $-|\sqrt{2}|$ $-\sqrt{2}$

103. $|-\sqrt{6}|$ $\sqrt{6}$

104. $|-\sqrt{18}|$ $\sqrt{18}$

105. $|m|$ m

106. $|-m|$ m

Mix 'Em Up!

Solve each problem.

107. Use the roster method to write $A = \{x|x$ is a whole number greater than or equal to $10\}$. $\{10, 11, 12, \ldots\}$

108. Use the roster method to write $A = \{x|x$ is a natural number less than $\sqrt{15}\}$. $\{1, 2, 3\}$

109. Let $A = \{x|x$ is a positive integer$\}$ and $B = \{x|x$ is a whole number less than $5\}$. Find $A \cup B$ and $A \cap B$. $\{0, 1, 2, 3, \ldots\}$; $\{1, 2, 3, 4\}$

110. Let $A = \{x|x$ is an even integer$\}$ and $B = \{x|x$ is an odd integer$\}$. Find $A \cup B$ and $A \cap B$. $\{x|x$ is an integer$\}$; $\varnothing$

The following table shows the median pay and job growth for the top 15 Best Jobs in America according to CNNMoney.com's *Money Magazine* Best Jobs List in 2010. Use the information for Exercises 111–118. (Source: http://money.cnn.com/magazines/moneymag/bestjobs/2010/full_list/index.html)

Rank	Job Title	Median Pay	Job Growth 10-yr Forecast
1	Software Architect	\$119,000	34%
2	Physician Assistant	\$ 92,000	39%
3	Management Consultant	\$117,000	24%
4	Physical Therapist	\$ 75,000	30%
5	Environmental Engineer	\$ 81,000	31%
6	Civil Engineer	\$ 80,000	24%
7	Database Administrator	\$ 68,000	20%
8	Sales Director	\$142,000	15%
9	Certified Public Accountant	\$ 56,000	22%
10	Biomedical Engineer	\$ 76,000	72%
11	Actuary	\$133,000	21%
12	Dentist	\$142,000	15%
13	Nurse Anesthetist	\$156,000	13%
14	Risk Management Manager	\$107,000	24%
15	Product Management Director	\$148,000	12%

Let $A = \{x | x$ is a job with a median pay less than \$100,000$\}$

$B = \{x | x$ is a job with a median pay greater than \$140,000$\}$

$C = \{x | x$ is a job with a job growth greater than 30%$\}$

$D = \{x | x$ is a job with a job growth less than 15%$\}$

111. State the set B using the roster method. {sales director, dentist, nurse anesthetist, product management director}

112. State the set C using the roster method. {software architect, physician assistant, environmental engineer, biomedical engineer}

113. State the set $A \cap C$ using the roster method. {physician assistant, environmental engineer, biomedical engineer}

114. State the set $B \cap D$ using the roster method. {nurse anesthetist, product management director}

115. State the set $A \cap D$ using the roster method. $\varnothing$

116. State the set $B \cap C$ using the roster method. $\varnothing$

117. Is physician assistant $\in A \cup C$? Yes

118. Is actuary $\in B \cup D$? No

Classify each number as a rational number or irrational number.

119. $\sqrt{7}$ irrational

120. $\sqrt{20}$ irrational

121. 0.6 rational

122. 2.3 rational

123. $\frac{6}{11}$ rational

124. $\frac{9}{14}$ rational

125. $\frac{6}{\pi}$ irrational

126. $\frac{\pi}{3}$ irrational

Find the opposite of each number. Assume *m* is a positive real number.

127. -50.345 50.345

128. -71.43 71.43

129. $\frac{4}{5}$ $-\frac{4}{5}$

130. $\frac{3}{7}$ $-\frac{3}{7}$

131. $-\frac{1}{8}$ $\frac{1}{8}$

132. $-\frac{5}{6}$ $\frac{5}{6}$

133. $7m$ $-7m$

134. $21m$ $-21m$

Simplify each expression. Assume *m* is a positive real number.

135. $-\left|\frac{10}{17}\right|$ $-\frac{10}{17}$

136. $-\left|\frac{2}{5}\right|$ $-\frac{2}{5}$

137. $-|-4\pi|$ -4π

138. $-|-3\pi|$ -3π

139. $\left|-\frac{1}{4}\right|$ $\frac{1}{4}$

140. $\left|-\frac{8}{15}\right|$ $\frac{8}{15}$

141. $-|-m|$ $-m$

142. $\left|\frac{m}{2}\right|$ $\frac{m}{2}$

Graph each number on a real number line.

143. $5\frac{1}{2}$

144. $10\frac{1}{4}$

145. 15.3

146. 12.2

147. $\sqrt{49}$

148. $\sqrt{81}$

You Be the Teacher!

Answer each student's question.

149. Jalen: If the variable x represents a negative number, why does $|x| = -x$?

150. Eddie: Is every integer a rational number? Why or why not?

151. Wally: Is every rational number an integer? Why or why not?

152. Theodore: Is there an easy way to graph large numbers on the real number line?

Calculate It!

Use a graphing calculator to determine each value. Approximate to two decimal places, if needed.

153. $-\sqrt{35}$ -5.92

154. $-\sqrt{24}$ -4.90

155. 7π 21.99

156. 26π 81.68

157. $-(-\sqrt{14})$ 3.74

158. $-(-\sqrt{19})$ 4.36

159. $|-29|$ 29

160. $-|-34|$ -34

SECTION 1.2 Operations with Real Numbers and Algebraic Expressions

▶ OBJECTIVES

As a result of completing this section, you will be able to

1. Add and subtract real numbers.
2. Multiply and divide real numbers.
3. Simplify exponential expressions.
4. Use the order of operations to simplify an expression.
5. Evaluate algebraic expressions.
6. Solve applications involving operations on real numbers and algebraic expressions.
7. Troubleshoot common errors.

From 2004 to 2010, the number of active users (in millions) of Facebook can be modeled by the expression

$$4.3x^4 - 32.5x^3 + 80.9x^2 - 55.5x + 2.2$$

where x is the number of years after 2004. According to the model, how many visitors were there in 2009? (Source: http://www.facebook.com/press/info.php?timeline)

To answer this question, we must know how to perform operations with real numbers and how to evaluate algebraic expressions. In this section, we will review these skills.

Objective 1 ▶

Add and subtract real numbers.

Adding Real Numbers

Adding and subtracting real numbers is a skill that we encounter on a daily basis. Situations in which we must perform these operations involve balancing a checkbook, tracking changes in the stock market, calculating federal taxes, calculating net worth, and understanding changes in temperature, sea level, and altitude. The list could continue. The rules for addition are illustrated next.

Signs Are the Same

Both signs are **positive**: Jan enrolls in 6 semester credit hours and then adds 4 more credit hours. So, Jan enrolls in a total of 10 credit hours. We represent this numerically and graphically as shown.

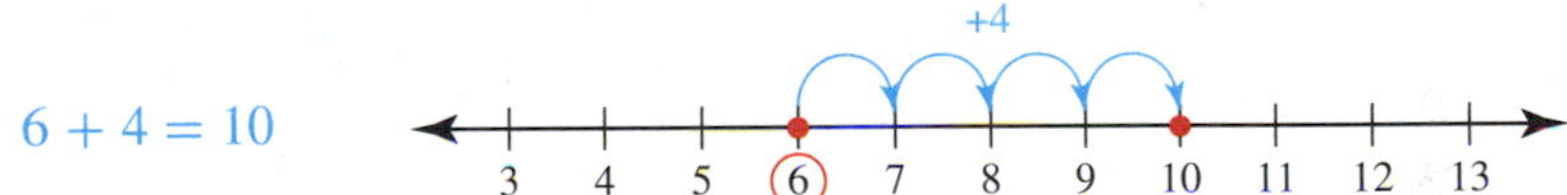

Both signs are **negative**: Jason owes a credit card company \$100 and then makes another charge of \$50. So, Jason owes the credit card company a total of \$150. We represent this numerically and graphically as shown.

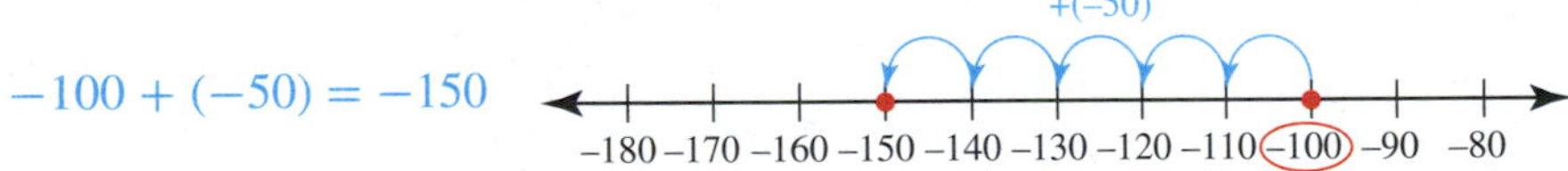

Signs Are Different

Number with the larger absolute value is negative: Maria owes a credit card company \$100 and makes a payment of \$80. After the payment, she owes the company \$20. We represent this numerically and graphically as shown.

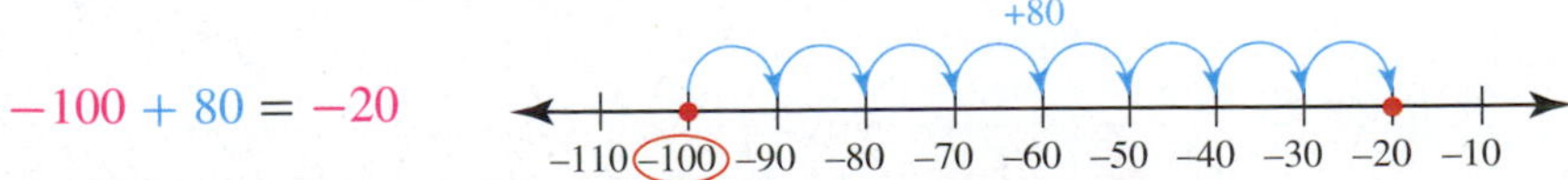

Number with the larger absolute value is positive: The temperature in Boston, Massachusetts at 7 A.M. is −5°F. The temperature rises 20°F in 6 hr. So, the temperature at 1 P.M. is 15°F. We represent this numerically and graphically as shown.

$-5 + 20 = 15$

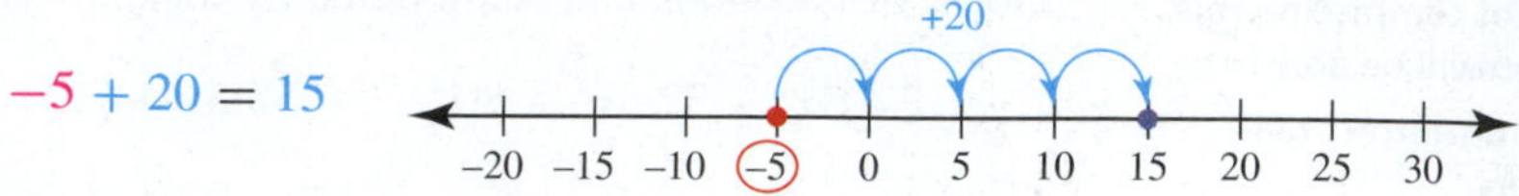

So, there are four different cases of adding real numbers based on the signs of the numbers.

Procedure: Adding Real Numbers

Step 1: Determine the signs of the numbers being added.
Step 2: Add the numbers according to the following rules.
- **a.** If the signs are the same, add the absolute values of the numbers and keep the sign.
- **b.** If the signs are different, subtract the absolute values. The sign of the result is the sign of the number with the larger absolute value.

Subtracting Real Numbers

To determine how to subtract real numbers, we rely on two facts that we already know.

$$6 - 4 = 2 \quad \text{and} \quad 6 + (-4) = 2$$

Since these two statements produce the same result, it must be true that

$$6 - 4 = 6 + (-4)$$

These expressions indicate that subtraction is a form of addition. Subtraction is equivalent to adding the opposite of a number.

Property: Subtracting Real Numbers

For real numbers a and b,

$$a - b = a + (-b)$$

Procedure: Subtracting Real Numbers

Step 1: Change the subtraction sign to addition and write the opposite of the number being subtracted.
Step 2: Apply the rules for adding real numbers to obtain the result.

Note: *If three or more numbers are added or subtracted, add or subtract from left to right.*

Objective 1 Examples **Perform the indicated operation.**

1a. $-\frac{2}{5}+\left(-\frac{1}{5}\right)$ **1b.** $-9+4$ **1c.** $\frac{7}{4}+\left(-\frac{3}{2}\right)$

1d. $0.5+(-0.75)$ **1e.** $-6-(-5)$ **1f.** $\frac{2}{3}-\frac{3}{4}$

1g. $-4-3+(-6)$

Solutions

1a. The numbers being added have the same sign. So, we add their absolute values and keep the sign.

$$-\frac{2}{5}+\left(-\frac{1}{5}\right)=-\frac{3}{5}$$

1b. The numbers being added have opposite signs. $|-9|=9$ and $|4|=4$, the result is negative as -9 has the larger absolute value.

$$-9+4=-5$$

1c. The numbers being added have opposite signs. We first find the LCD, which is 4.

$$\frac{7}{4}+\left(-\frac{3}{2}\right)=\frac{7}{4}+\left(-\frac{6}{4}\right) \quad \text{Convert to a common denominator.}$$

$$=\frac{1}{4} \quad \text{Add opposite signs.}$$

1d. The numbers being added have opposite signs.

$$0.5+(-0.75)=0.50+(-0.75) \quad \text{Rewrite 0.5 as 0.50.}$$

$$=-0.25 \quad \text{Add the decimals.}$$

1e.

$$-6-(-5)=-6+(5) \quad \text{Rewrite as adding the opposite.}$$

$$=-1 \quad \text{Add.}$$

1f.

$$\frac{2}{3}-\frac{3}{4}=\frac{8}{12}-\frac{9}{12} \quad \text{Convert each fraction to an equivalent fraction with a denominator of 12.}$$

$$=\frac{8}{12}+\left(-\frac{9}{12}\right) \quad \text{Rewrite as adding the opposite of } \frac{9}{12}.$$

$$=-\frac{1}{12} \quad \text{Add opposite signs.}$$

1g.

$$-4-3+(-6)=-4+(-3)+(-6) \quad \text{Rewrite as adding the opposite.}$$

$$=-7+(-6) \quad \text{Add like signs.}$$

$$=-13 \quad \text{Add like signs.}$$

Student Check 1 Perform the indicated operation.

a. $-\frac{8}{9}+\left(-\frac{1}{9}\right)$ **b.** $-12+6$ **c.** $\frac{3}{7}+\left(-\frac{1}{7}\right)$

d. $-1.2+(-2.3)$ **e.** $-1-(-9)$ **f.** $\frac{3}{5}-\frac{5}{7}$ **g.** $7-(-3)+(-5)$

Multiplying and Dividing Real Numbers

Objective 2

Multiply and divide real numbers.

Multiplication and division of real numbers also arise in everyday situations, though the occurrences may not be as obvious as adding and subtracting real numbers. Some examples are shown.

Both signs are **positive**: Todd wins \$50 for each of **three** hands of poker. So, he wins a total of \$150.

$$\binom{\text{number of}}{\text{hands played}}\binom{\text{money won}}{\text{each hand}} = \binom{\text{total}}{\text{winnings}}$$

$$3(\$50) = \$150$$

Both signs are **negative**: Nadine typically bets \$10 for a hand of poker. For two hands, she gets lousy cards and decides not to make a bet. By not playing, she saves a total of \$20.

$$\binom{\text{number of hands}}{\text{not played}}\binom{\text{money that}}{\text{would be lost}} = \binom{\text{money}}{\text{saved}}$$

$$(-2)(-\$10) = \$20$$

Signs are **different**: Anthony loses \$25 for each of four hands of poker. So, he loses a total of \$100.

$$\binom{\text{number of}}{\text{hands played}}\binom{\text{money lost}}{\text{each hand}} = \binom{\text{total money}}{\text{lost}}$$

$$(4)(-\$25) = -\$100$$

Note: *Recall multiplication represents repeated addition. So, we can think of this product as*

$$\begin{aligned} 4(-25) &= (-25) + (-25) + (-25) + (-25) \\ &= -100 \end{aligned}$$

These rules can be summarized as follows.

Procedure: Multiplying Real Numbers

Step 1: Determine the signs of the numbers being multiplied.

Step 2: Multiply using the following rules.

- **a.** If the signs of the nonzero numbers are the same, their product is *positive*.
- **b.** If the signs of the nonzero numbers are different, their product is *negative*.

The rules stated previously apply to nonzero numbers. When one of the numbers being multiplied is zero, the result is zero. We state the property as follows.

Property: Product Property of Zero

The product of any number a and zero is zero.

$$a \cdot 0 = 0$$

To define the rules for dividing real numbers, we rely on facts that we already know.

$$\frac{6}{3} = 2 \quad \text{and} \quad 6 \cdot \frac{1}{3} = 2$$

So, it follows that

$$\frac{6}{3} = 6 \div 3 = 6 \cdot \frac{1}{3}$$

Dividing by a number is equivalent to multiplying by the reciprocal of the number. Two nonzero numbers b and $\frac{1}{b}$ are **reciprocals**.

Property: Dividing Real Numbers

For real numbers a and b ($b \neq 0$),

$$\frac{a}{b} = a \div b = a \cdot \frac{1}{b}$$

Because division is another form of multiplication, the sign rules for multiplication also apply to division.

Procedure: Dividing Real Numbers

Step 1: Determine the signs of the numbers being divided.

Step 2: Divide using the following rules.

- **a.** If the signs of the nonzero numbers are the same, their quotient is *positive*.
- **b.** If the signs of the nonzero numbers are different, their quotient is *negative*.

When zero is involved in a quotient, one of the two following cases apply.

Property: Division Properties with Zero

1. Zero divided by a nonzero number is 0: $\frac{0}{b} = 0$ for $b \neq 0$.

2. A nonzero number divided by zero is undefined: $\frac{a}{0}$ is undefined for $a \neq 0$.

Note: *When multiplying or dividing real numbers,*

1. A product or quotient of two real numbers with the same sign is positive.

2. A product or quotient of two real numbers with different signs is negative.

3. A product of a number and zero is zero.

4. A quotient of zero and a nonzero number is zero.

5. A quotient of a nonzero number and zero is undefined.

If three or more numbers are multiplied or divided, then we multiply or divide from left to right.

Objective 2 Examples

Perform the indicated operation.

Problems	Solutions
2a. $(-4)(-8)$	$(-4)(-8) = 32$
2b. $\left(-\frac{5}{2}\right)\left(\frac{2}{5}\right)$	$\left(-\frac{5}{2}\right)\left(\frac{2}{5}\right) = -\frac{10}{10} = -1$
2c. $(-2.3)(-10)$	$(-2.3)(-10) = 23$

Problems	Solutions
2d. $\frac{12}{-3}$	$\frac{12}{-3} = -4$
2e. $\frac{-7}{-2}$	$\frac{-7}{-2} = \frac{7}{2}$
2f. $4 \div \left(-\frac{1}{2}\right)$	$4 \div \left(-\frac{1}{2}\right) = 4 \cdot -2 = -8$
2g. $\frac{0}{-5}$	$\frac{0}{-5} = 0$
2h. $-\frac{3}{0}$	$-\frac{3}{0}$ is undefined
2i. $-6(-2) \div (-4)$	$-6(-2) \div (-4) = 12 \div (-4)$ Multiply. $= -3$ Divide.

Student Check 2 Perform the indicated operation.

a. $(-9)(-2)$ **b.** $\left(-\frac{4}{3}\right)\left(\frac{3}{4}\right)$ **c.** $(-4.12)(-10)$

d. $\frac{32}{-2}$ **e.** $\frac{-5}{-3}$ **f.** $9 \div \left(-\frac{1}{3}\right)$

g. $\frac{-2}{0}$ **h.** $\frac{0}{-10}$ **i.** $3\left(-\frac{2}{3}\right) \div \left(-\frac{1}{2}\right)$

Exponential Expressions

Objective 3

Simplify exponential expressions.

Exponents arise in many different formulas. Exponents indicate repeated multiplication of the same factor. The repeated factor is called the **base** and the number of times it is used as a factor is denoted by the **exponent**.

INSTRUCTOR NOTE: Remind students that when two numbers are multiplied, they are called factors. For example, in $6 \cdot 2 = 12$, 6 and 2 are factors.

Property: Exponential Notation For b a real number and n a natural number, b raised to the nth, b^n, is the product of n factors of b.

exponent

$$b^n = \underbrace{b \cdot b \cdot b \cdots b}_{n \text{ times}}$$

base

Some examples of exponents are shown.

Verbal Phrase	Mathematical Expression
3 squared	$3^2 = 3 \cdot 3 = 9$
5 cubed	$5^3 = 5 \cdot 5 \cdot 5 = 125$
6 to the 4th	$6^4 = 6 \cdot 6 \cdot 6 \cdot 6 = 1296$

Note: *We usually do not write an exponent of* 1. *For example,* 6 *is assumed to be* 6^1.

The following numbers appear often in algebra. It is beneficial to commit these to memory.

Squares			Cubes	Fourths
$1^2 = 1$	$6^2 = 36$	$11^2 = 121$	$1^3 = 1$	$1^4 = 1$
$2^2 = 4$	$7^2 = 49$	$12^2 = 144$	$2^3 = 8$	$2^4 = 16$
$3^2 = 9$	$8^2 = 64$	$13^2 = 169$	$3^3 = 27$	$3^4 = 81$
$4^2 = 16$	$9^2 = 81$	$14^2 = 196$	$4^3 = 64$	$4^4 = 256$
$5^2 = 25$	$10^2 = 100$	$15^2 = 225$	$5^3 = 125$	$5^4 = 625$
			$6^3 = 216$	

Important Facts About Exponents

1. A negative base raised to an even exponent (2, 4, 6, . . .) is *positive*.

$$(-5)^4 = (-5)(-5)(-5)(-5) = 625$$

2. A negative base raised to an odd exponent (1, 3, 5, . . .) is *negative*.

$$(-5)^3 = (-5)(-5)(-5) = -125$$

3. If there are parentheses around a negative number raised to an exponent, then the exponent is applied to the negative number.

$$(-b)^n = \underbrace{(-b)(-b)\cdots(-b)}_{(-b)\text{ is a factor } n \text{ times}}$$

4. If a negative sign is in front of an exponential expression, the negative sign indicates that we take the opposite of the value of the exponential expression. In other words, the negative sign is not part of the base of the exponent.

$$-b^n = -\underbrace{b \cdot b \cdot b \cdots b}_{b \text{ is a factor } n \text{ times}}$$

Objective 3 Examples Simplify each exponential expression.

Problems	Solutions
3a. 2^5	$2^5 = 2 \cdot 2 \cdot 2 \cdot 2 \cdot 2 = 32$
3b. $\left(\frac{3}{4}\right)^2$	$\left(\frac{3}{4}\right)^2 = \left(\frac{3}{4}\right)\left(\frac{3}{4}\right) = \frac{9}{16}$
3c. -3^4	$-3^4 = -(3 \cdot 3 \cdot 3 \cdot 3) = -81$
3d. $(-3)^4$	$(-3)^4 = (-3)(-3)(-3)(-3) = 81$
3e. -4^3	$-4^3 = -(4 \cdot 4 \cdot 4) = -64$
3f. $(-4)^3$	$(-4)^3 = (-4)(-4)(-4) = -64$

Student Check 3 Simplify each exponential expression.

a. 6^3 **b.** $\left(\frac{5}{2}\right)^3$ **c.** -2^6

d. $(-2)^6$ **e.** -7^3 **f.** $(-7)^3$

Order of Operations

Objective 4

Use the order of operations to simplify an expression.

Without a standard order to perform operations in numerical expressions, we could possibly get different values for the same numerical expression. Consider the expression $4 + 5(6)$. There are two possible methods for simplifying this expression.

Method 1		Method 2	
$4 + 5(6) = 4 + 30$	Multiply.	$4 + 5(6) = 9(6)$	Add.
$= 34$	Add.	$= 54$	Multiply.

Both of these values cannot be correct. The *first method is the correct* method to simplify the expression. We must use a special order to simplify numerical expressions. The accepted order is called the *order of operations*. It is stated as follows.

Procedure: Order of Operations

When simplifying a numerical expression, perform the operations in the following order.

Step 1: Simplify expressions inside grouping symbols first. Grouping symbols include parentheses, brackets, absolute value symbols, square root symbols, and fraction bars.
Step 2: Simplify any exponential expressions.
Step 3: Perform multiplication or division in order from left to right. (Since division is multiplying by the reciprocal, one operation doesn't take precedence over the other.)
Step 4: Perform addition or subtraction in order from left to right. (Since subtraction is adding the opposite, one operation doesn't take precedence over the other.)

INSTRUCTOR NOTE: A common mnemonic for the order of operations is "Please excuse my dear Aunt Sally," but students often associate the P with parentheses and not grouping symbols. So, we suggest, "Get every million dollars and save."

Objective 4 Examples **Use the order of operations to simplify each expression.**

INSTRUCTOR NOTE: Formulas for slope (a) and distance (c) have been used to illustrate this concept.

4a. $\dfrac{-2-8}{4-(-1)}$ **4b.** $5(-4)^2 + 2(-4) - 6$

4c. $\sqrt{(7-4)^2 + (1-5)^2}$ **4d.** $2 - 5[9 - 3(4 - 6)] \div 3 \cdot 8$

4e. $\dfrac{2 - 3|5-9|^2}{-4 - \sqrt{4+12}}$

Solutions

4a.
$$\frac{-2-8}{4-(-1)} = \frac{-2+(-8)}{4+1}$$ Rewrite the numerator and denominator as addition.
$$= \frac{-10}{5}$$ Add.
$$= -2$$ Simplify.

4b.
$$5(-4)^2 + 2(-4) - 6 = 5(16) + 2(-4) - 6$$ Simplify the exponent.
$$= 80 + (-8) - 6$$ Multiply from left to right.
$$= 72 - 6$$ Add the first two numbers.
$$= 66$$ Subtract.

4c.
$$\sqrt{(7-4)^2 + (1-5)^2} = \sqrt{(3)^2 + (-4)^2}$$ Simplify within parentheses.
$$= \sqrt{9 + 16}$$ Simplify the exponent.
$$= \sqrt{25}$$ Add
$$= 5$$ Apply the square root.

4d. $2 - 5[9 - 3(4 - 6)] \div 3 \cdot 8$

$= 2 - 5[9 - 3(-2)] \div 3 \cdot 8$	Simplify within the innermost grouping.
$= 2 - 5[9 + 6] \div 3 \cdot 8$	Multiply −3 and −2.
$= 2 - 5[15] \div 3 \cdot 8$	Add 9 and 6.
$= 2 - 75 \div 3 \cdot 8$	Multiply −5 and 15.
$= 2 - 25 \cdot 8$	Divide −75 by 3.
$= 2 - 200$	Multiply −25 by 8.
$= -198$	Subtract.

4e. The fraction bar is a grouping symbol, so we apply the order of operations in the numerator and the denominator

$\dfrac{2 - 3\|5 - 9\|^2}{-4 - \sqrt{4 + 12}} = \dfrac{2 - 3\|-4\|^2}{-4 - \sqrt{16}}$	Simplify within the absolute value symbol and simplify within the square root symbol.
$= \dfrac{2 - 3(4)^2}{-4 - 4}$	Simplify the absolute value of −4 and simplify the square root of 16.
$= \dfrac{2 - 3(16)}{-8}$	Simplify the square of 4 and subtract the numbers in the denominator.
$= \dfrac{2 - 48}{-8}$	Multiply 3 and 16.
$= \dfrac{-46}{-8}$	Subtract the numbers in the numerator.
$= \dfrac{23}{4}$	Simplify the fraction.

Student Check 4 Use the order of operations to simplify each expression.

a. $\dfrac{6 - (-2)}{-3 - 1}$ **b.** $2(-1)^2 - 8(-1) + 5$ **c.** $\sqrt{[1 - (-7)]^2 + (-4 - 2)^2}$

d. $7 + 2|(9 - 8) - (5 + 2)| - 4 \cdot 6 \div 8$ **e.** $\dfrac{3 - 6\sqrt{2 + 7}}{-3|1 - 6|^2}$

Evaluating Algebraic Expressions

Objective 5

Evaluate algebraic expressions.

Thus far we have examined numerical expressions; we will now turn our attention to algebraic expressions. **Algebraic expressions** are expressions joining numbers and letters by mathematical operations. The letters are called *variables* and represent some unknown number. Examples of variables are x, y, z, a, and b. Examples of algebraic expressions are

$$2x \quad 4y - 5 \quad b^2 - 4ac \quad \frac{y_2 - y_1}{x_2 - x_1}$$

The numbers 1 and 2 are called **subscripts**. They are used to differentiate two different, but related, variables.

Algebraic expressions are fundamental to the study of algebra. Algebraic expressions are used in formulas, mathematical models, and equations. Our ability to work with these expressions enables us to manipulate the expressions to obtain what is needed.

The first skill we need is the ability to evaluate algebraic expressions. To **evaluate an algebraic expression** means to find the value of an expression for a specific value of the variable.

Procedure: Evaluating an Algebraic Expression

Step 1: Replace every occurrence of the variable(s) with its given value.
Step 2: Use the order of operations to simplify the resulting expression.

Objective 5 Examples

Evaluate each expression for the given value(s).

5a. $\frac{1}{2}x - 5$ for $x = -4$

5b. $4y^2 - y + 1$ for $y = 5$

5c. $b^2 - 4ac$ for $a = -3$, $b = 2$, and $c = 5$

5d. $\frac{y_2 - y_1}{x_2 - x_1}$ for $x_1 = -2$, $x_2 = 4$, $y_1 = -1$, and $y_2 = -8$

Solutions

5a. $\frac{1}{2}x - 5 = \frac{1}{2}(-4) - 5$ — Replace x with -4.

$= -2 - 5$ — Multiply $\frac{1}{2}$ and -4.

$= -7$ — Subtract.

5b. $4y^2 - y + 1 = 4(5)^2 - (5) + 1$ — Replace y with 5.

$= 4(25) - 5 + 1$ — Simplify the exponent.

$= 100 - 5 + 1$ — Multiply 4 and 25.

$= 95 + 1$ — Subtract 5 from 100.

$= 96$ — Add.

5c. $b^2 - 4ac = (2)^2 - 4(-3)(5)$ — Replace a with -3, b with 2, and c with 5.

$= 4 - 4(-3)(5)$ — Simplify the exponent.

$= 4 + 60$ — Multiply -4, -3, and 5.

$= 64$ — Add.

5d. $\frac{y_2 - y_1}{x_2 - x_1} = \frac{-8 - (-1)}{4 - (-2)}$ — Replace x_1 with -2, x_2 with 4, y_1 with -1, and y_2 with -8.

$= \frac{-8 + 1}{4 + 2}$ — Write the numerator and denominator using addition.

$= \frac{-7}{6}$ or $-\frac{7}{6}$ — Add the numbers in the numerator and in the denominator.

Student Check 5

Evaluate each expression for the given value(s).

a. $\frac{1}{3}x - 8$ for $x = 6$

b. $2a^2 - a + 4$ for $a = -4$

c. $\frac{-b + \sqrt{b^2 - 4ac}}{2a}$ for $a = 1$, $b = -2$, and $c = -3$

d. $\frac{1}{2}h(b_1 + b_2)$ for $h = 4$, $b_1 = 3$, and $b_2 = 5$

Objective 6

Solve applications involving operations on real numbers and algebraic expressions.

Applications

Example 6 will illustrate how we can use operations with real numbers and algebraic expressions to solve real-world applications.

Objective 6 Examples **Solve each problem.**

6a. A person's **net worth** is calculated by subtracting total assets and total liabilities. If the Dobson family has \$250,000 in assets and \$285,000 in total liabilities, find their net worth.

Solution **6a.**

$$\begin{aligned} \text{Net worth} &= \text{total assets} - \text{total liabilities} \\ &= 250{,}000 - 285{,}000 \\ &= 250{,}000 + (-285{,}000) \\ &= -35{,}000 \end{aligned}$$

The Dobson family has a net worth of −\$35,000.

6b. The following chart shows the closing stock price for Walmart for a week in June 2011. (Source: http://investors.walmartstores.com/phoenix.zhtml?c=112761&p=irol-stocklookup)

Date	Closing Price	Daily Change in Price
6/20/2011	\$53.04	n/a
6/21/2011	\$53.29	
6/22/2011	\$53.01	
6/23/2011	\$53.29	
6/24/2011	\$52.41	

i. Complete the chart to show how the price of the stock changed from the previous day to the next.

ii. After completing the chart, find the average daily change in price.

Solution **6b.** **i.** The change in price is found by subtracting the previous day's price from the next day's price.

Date	Closing Price	Daily Change in Price
6/20/2011	\$53.04	n/a
6/21/2011	\$53.29	53.29 − 53.04 = 0.25
6/22/2011	\$53.01	53.01 − 53.29 = −0.28
6/23/2011	\$53.29	53.29 − 53.01 = 0.28
6/24/2011	\$52.41	52.41 − 53.29 = −0.88

ii. To find the average daily change in price, we add the daily changes and divide by the number of changes.

$$\frac{0.25 + (-0.28) + (0.28) + (-0.88)}{4} = \frac{-0.63}{4} = -0.1575$$

So, the price of Walmart stock from June 20 to June 24 decreased by an average of \$0.1575 per day or 15.75 cents per day.

6c. From 2004 to 2010, the number of active users (in millions) of Facebook can be modeled by the expression

$$4.3x^4 - 32.5x^3 + 80.9x^2 - 55.5x + 2.2$$

where x is the number of years after 2004. According to the model, how many active users were there in 2009? (Source: http://www.facebook.com/press/info.php?timeline)

Solution **6c.** The variable x represents the number of years after 2004. Since $2009 - 2004 = 5$, we replace x with 5 to obtain the requested information.

$4.3x^4 - 32.5x^3 + 80.9x^2 - 55.5x + 2.2$	Begin with given model.
$= 4.3(5)^4 - 32.5(5)^3 + 80.9(5)^2 - 55.5(5) + 2.2$	Replace x with 5.
$= 4.3(625) - 32.5(125) + 80.9(25) - 55.5(5) + 2.2$	Simplify each exponent.
$= 2687.5 - 4062.5 + 2022.5 - 277.5 + 2.2$	Multiply from left to right.
$= 372.2$	Add or subtract from left to right.

So, in 2009, there were about 372.2 million active users of Facebook.

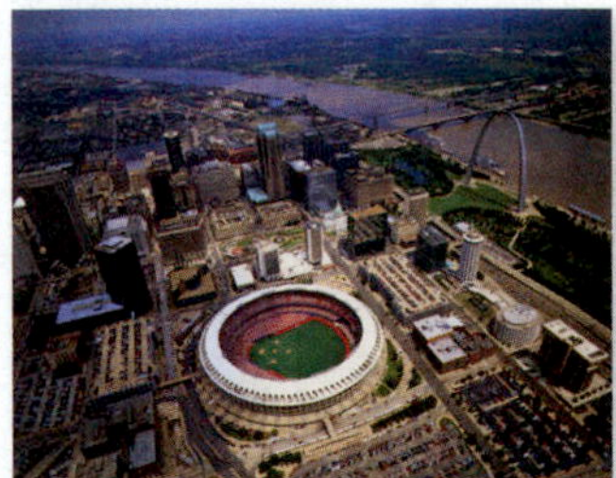

6d. Busch Stadium in St. Louis, Missouri, is home to the St. Louis Cardinals. The stadium is almost a perfect circle with a diameter of 800 ft. The distance around a circle is called its **circumference**. The circumference is represented by the expression πd, where d is the circle's diameter. Find the distance around Busch Stadium. Approximate the answer to two decimal places. (Source: http://www.baseball-statistics.com/Ballparks/StL/index.htm)

Solution **6d.** The stadium's diameter is 800 ft.

$C = \pi d$	State the circumference formula.
$C = \pi(800)$	Replace d with 800.
$C \approx 2513.27$	Multiply and round to two decimal places.

The distance around the stadium is approximately 2513.27 ft. That is nearly half of a mile!

Student Check 6 Solve each problem.

a. What is the net worth of the Bowden family if their assets are \$300,000 and their liabilities are \$450,000?

b. The chart shows the closing price of Target's stock for a week in June 2011. (Source: http://finance.yahoo.com)

Date	Closing Price	Daily Change in Price
6/06/2011	\$47.36	n/a
6/07/2011	\$47.06	
6/08/2011	\$46.86	
6/09/2011	\$47.16	
6/10/2011	\$46.70	

i. Complete the chart to show how the price of the stock changed from the previous day to the next.

ii. After completing the chart, find the average daily change in price.

c. From January 2002 to January 2011, the average retail price (in dollars per gal) of regular gas in California can be approximated by the expression $0.0012x^5 - 0.0265x^4 + 0.1862x^3 - 0.4490x^2 + 0.3783x + 1.5576$, where x is the number of years after 2002. According to the expression, what is the average retail price of regular gas in January 2013? (Source: www.eia.doe.gov)

d. The volume of a sphere is given by $\frac{4}{3}\pi r^3$, where r is the radius of the sphere. Find the volume of a basketball that has a radius of 4.7 in. Round answer to the nearest hundredth.

Objective 7 ▶ Troubleshooting Common Errors

Troubleshoot common errors.

Some common errors associated with exponents and order of operations are shown.

Objective 7 Examples

A problem and an incorrect solution are given. Provide the correct solution and an explanation of the error.

7a. Simplify -10^2.

Incorrect Solution	Correct Solution and Explanation
$-10^2 = (-10)(-10)$ $= 100$	The base of the exponent is 10 not −10 since the negative sign is not in parentheses. So, we find the opposite of 10 squared. $-10^2 = -10 \cdot 10$ $= -100$

7b. Simplify $4 + 5(3)^2$.

Incorrect Solution	Correct Solution and Explanation
$4 + 5(3)^2 = 4 + 15^2$ $= 4 + 225$ $= 229$	Exponents should be applied before multiplication. $4 + 5(3)^2 = 4 + 5(9)$ $= 4 + 45$ $= 49$

ANSWERS TO STUDENT CHECKS

Student Check 1 **a.** −1 **b.** −6 **c.** $\frac{2}{7}$ **d.** −3.5 **e.** 8 **f.** $-\frac{4}{35}$ **g.** 5

Student Check 2 **a.** 18 **b.** −1 **c.** 41.2 **d.** −16 **e.** $\frac{5}{3}$ **f.** −27 **g.** undefined **h.** 0 **i.** 4

Student Check 3 **a.** 216 **b.** $\frac{125}{8}$ **c.** −64 **d.** 64 **e.** −343 **f.** −343

Student Check 4 **a.** −2 **b.** 15 **c.** 10 **d.** −8 **e.** $\frac{1}{5}$

Student Check 5 **a.** −6 **b.** 40 **c.** 3 **d.** 16

Student Check 6 **a.** −$150,000
b. **i.**

Date	Closing Price	Daily Change in Price
6/06/2011	$47.36	n/a
6/07/2011	$47.06	−0.30
6/08/2011	$46.86	−0.20
6/09/2011	$47.16	0.30
6/10/2011	$46.70	−0.46

ii. The average daily change was a decrease of $0.165 per day.

c. about $4.50 per gallon **d.** 434.89 in.^{3}

SUMMARY OF KEY CONCEPTS

1. To add numbers with the same sign, add their absolute values and keep the sign.
2. To add numbers with different signs, subtract the smaller absolute value of the numbers from the larger absolute value. The sign of the answer is the sign of the number with the larger absolute value.
3. Subtracting a real number is equivalent to adding the opposite of the number.
4. The product of two nonzero real numbers with the same sign is a positive real number. The product of two nonzero real numbers with opposite signs is a negative real number.
5. Division by zero is undefined but 0 divided by a nonzero number is 0. The quotient of two nonzero real numbers with the same sign is a positive real number. The quotient of two nonzero real numbers with opposite signs is a negative real number.
6. An exponent indicates how many times to multiply the base by itself.
   - A negative base will be in parentheses and will, therefore, be repeated in the multiplication.
   - If there is a negative sign but no parentheses, then take the opposite of the value of the exponential expression. The negative sign is not repeated.
7. Order of operations is used to simplify an expression containing more than one operation. Remember GEMDAS (grouping, exponents, multiplication and division from left to right, addition and subtraction from left to right).
8. Evaluate an algebraic expression by replacing the variable with its given value and use the order of operations to simplify.

GRAPHING CALCULATOR SKILLS

In this section, the calculator will help us confirm our work and simplify expressions. We should be able to simplify expressions, work with exponents, and evaluate algebraic expressions. When entering expressions into the calculator, be sure to enter parentheses when necessary.

Example 1: Simplify $\frac{5-(-3)}{6-2}$.

Solution:

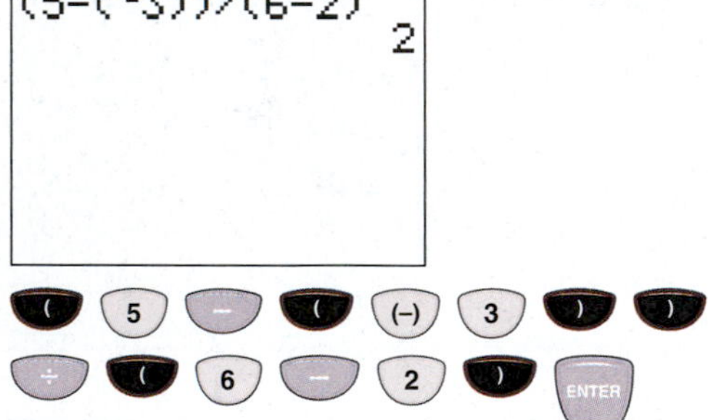

Example 2: Simplify $\sqrt{(5-2)^2+(4-8)^2}$.

Solution:

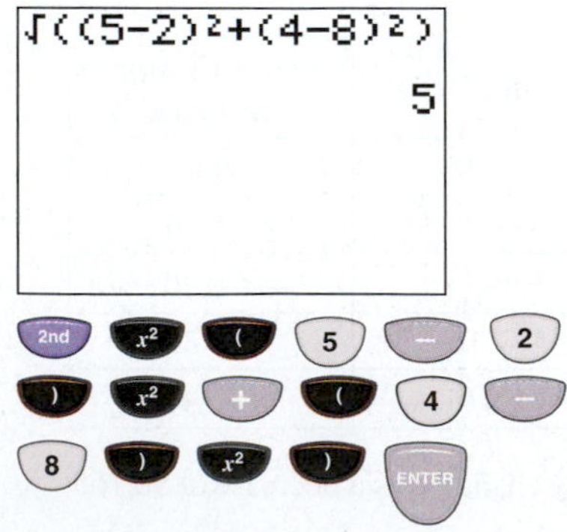

Example 3: Simplify $(-6)^2$, -6^2, and $(-4)^3$.

Solution: The carat symbol ^ is used with exponents. To enter the exponent of 2, we can either press the x^2 key or use ^ and 2. If parentheses are given in the problem, we must enter them on the calculator.

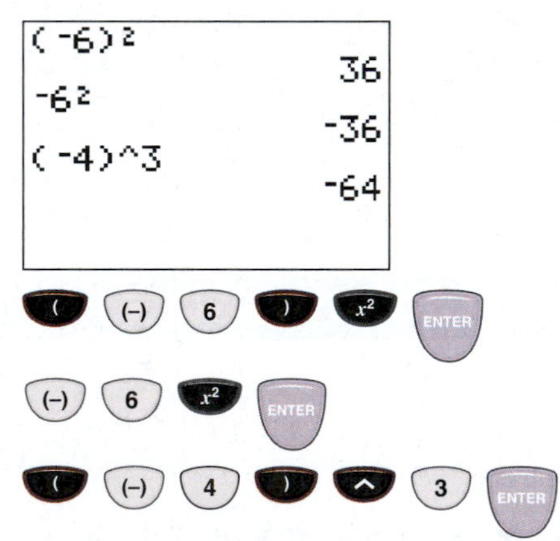

Example 4: Find the value of $x^2 - 3x + 4$ when $x = -2, -1$, and 0.

Solution: Two methods are illustrated for evaluating an algebraic expression.

Method 1: Enter the numerical expression that results after replacing the variable with its given value.

```
(-2)²-3(-2)+4
              14
(-1)²-3(-1)+4
               8
(0)²-3(0)+4
               4
```

Method 2: To evaluate an expression for multiple values, we can use the equation editor and table feature. We enter the expression to be evaluated in the equation editor. In the table, the number in the Y_1 column represents the value of the expression for the corresponding x-value.

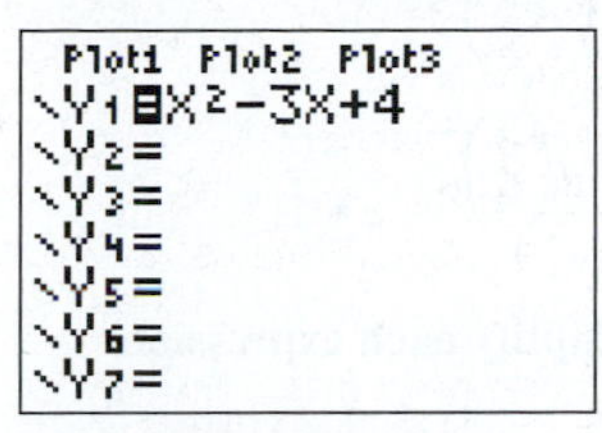

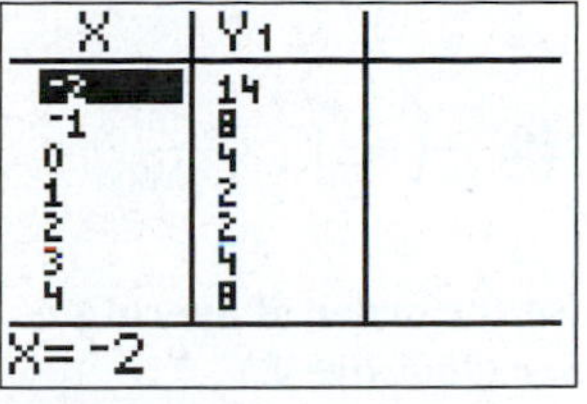

SECTION 1.2 EXERCISE SET

Write About It!

Use complete sentences in your answer to each exercise.

1. Explain how to add two real numbers with the same sign.
2. Explain how to add real numbers with different signs.
3. Suppose a number is raised to the nth. What does this mean?
4. Explain the order of operations needed to simplify an expression with multiplication, division, addition, and subtraction.
5. Write two phrases for the mathematical expression, 7^3.
6. Write two phrases for the mathematical expression, 12^2.
7. If m represents Alan's monthly income and x represents Alan's monthly expenditures, then
   a. What expression represents Alan's balance at the end of the month? $m - x$
   b. What must be true about the values of m and x if Alan is in debt at the end of the month? $m < x$
8. In Exercise 7, what must be true about the values of m and x if Alan has a surplus at the end of the month? $m > x$

Practice Makes Perfect!

Perform the indicated operation. (*See Objective 1.*)

9. $-12 + (-3)$ -15
10. $-5 + (-10)$ -15
11. $-7 + (-2)$ -9
12. $-9 + (-7)$ -16
13. $-\frac{1}{3} + \left(-\frac{2}{3}\right)$ -1
14. $-\frac{1}{5} + \left(-\frac{3}{5}\right)$ $-\frac{4}{5}$
15. $6 + (-10)$ -4
16. $13 + (-14)$ -1
17. $-\frac{1}{3} + \frac{1}{4}$ $-\frac{1}{12}$
18. $-\frac{2}{7} + \frac{4}{5}$ $\frac{18}{35}$

Additional answers can be found in the Instructor Answer Appendix.

19. $-4 - (-3)$ -1
20. $-6 - (-11)$ 5
21. $-2.3 - 6.2$ -8.5
22. $-3.6 - 7.3$ -10.9
23. $-\frac{7}{5} - \left(-\frac{3}{2}\right)$ $\frac{1}{10}$
24. $-\frac{4}{3} - \left(-\frac{5}{6}\right)$ $-\frac{1}{2}$

Perform the indicated operation. (*See Objective 2.*)

25. $4(-3)$ -12
26. $8(-2)$ -16
27. $(-5)(-6)$ 30
28. $(-7)(-10)$ 70
29. $\left(\frac{3}{2}\right)\left(-\frac{2}{5}\right)$ $-\frac{3}{5}$
30. $\left(\frac{2}{3}\right)\left(-\frac{3}{7}\right)$ $-\frac{2}{7}$
31. $\left(-\frac{5}{4}\right)\left(-\frac{3}{2}\right)$ $\frac{15}{8}$
32. $\left(-\frac{4}{3}\right)\left(-\frac{2}{7}\right)$ $\frac{8}{21}$
33. $(-6.1)(-10)$ 61
34. $(4.1)(-2)$ -8.2
35. $\frac{-14}{7}$ -2
36. $\frac{18}{-6}$ -3
37. $\frac{-18}{-9}$ 2
38. $\frac{-24}{-6}$ 4
39. $10 \div (-4)$ -2.5
40. $20 \div (-6)$ $-\frac{10}{3}$
41. $6 \div \left(-\frac{1}{3}\right)$ -18
42. $-16 \div \left(-\frac{1}{4}\right)$ 64
43. $\frac{-10}{0}$ undefined
44. $\frac{12}{0}$ undefined

Simplify each exponential expression. (*See Objective 3.*)

45. 4^3 64
46. 3^4 81
47. $\left(\frac{5}{4}\right)^2$ $\frac{25}{16}$
48. $\left(\frac{2}{3}\right)^3$ $\frac{8}{27}$
49. -11^2 -121
50. -12^2 -144
51. $(-11)^2$ 121
52. $(-12)^2$ 144
53. -5^3 -125
54. -10^3 -1000

55. $(-5)^3$ -125

56. $(-10)^3$ -1000

57. $\left(-\frac{4}{3}\right)^2$ $\frac{16}{9}$

58. $\left(-\frac{1}{6}\right)^2$ $\frac{1}{36}$

59. $-\left(\frac{1}{10}\right)^4$ $-\frac{1}{10{,}000}$

60. $-\left(\frac{3}{5}\right)^4$ $-\frac{81}{625}$

Use the order of operations to simplify each expression. (*See Objective 4.*)

61. $\frac{-6+3}{5-(-4)}$ $-\frac{1}{3}$

62. $\frac{-2+6}{10-(-2)}$ $\frac{1}{3}$

63. $\frac{-1-(-5)}{6-10}$ -1

64. $\frac{-3-(-8)}{1-11}$ $-\frac{1}{2}$

65. $(-2)^3 - 3(4)(-1)$ 4

66. $(-3)^2 - 2(5)(-6)$ 69

67. $12(-10^2) - 3(-7)$ -1179

68. $5(-8^2) - 6(-3)$ -302

69. $-3(1-4)^2 + 7$ -20

70. $-2(3-5)^3 - 10$ 6

71. $\sqrt{[2-(-2)]^2 + (-1-2)^2}$ 5

72. $\sqrt{[1-(-4)]^2 + (-10-2)^2}$ 13

73. $11 - 3[(7-2) - (6+2)] - 24 \div 8 \cdot 3$ 11

74. $-9 - 2[(8-1) + (4+3)] - 30 \cdot 5 \div 6$ -62

75. $\frac{7+\sqrt{27-2}}{-3|8-12|^2}$ $-\frac{1}{4}$

76. $\frac{6+\sqrt{5-1}}{-1|5-1|^2}$ $-\frac{1}{2}$

Evaluate each expression for the given values. (*See Objective 5.*)

77. $\frac{1}{3}x + 2$ for $x = -6$ 0

78. $\frac{1}{2}x - 3$ for $x = -8$ -7

79. $4x^2 + 2x - 5$ for $x = 1$ 1

80. $3x^2 - x + 10$ for $x = 5$ 80

81. $-2x^3 + 6x - 1$ for $x = -2$ 3

82. $-4x^3 + 2x - 3$ for $x = -1$ -1

83. $b^2 - 4ac$ for $a = 3, b = 1, c = -2$ 25

84. $b^2 - 4ac$ for $a = -6, b = 5, c = -3$ -47

85. $\frac{-b - \sqrt{b^2 - 4ac}}{2a}$ for $a = 1, b = -5, c = 4$ 1

86. $\frac{-b - \sqrt{b^2 - 4ac}}{2a}$ for $a = 2, b = -1, c = -3$ -1

Solve each problem. (*See Objective 6.*)

87. What is the net worth of the Gingrich family if their assets are \$860,000 and their liabilities are \$248,000? \$612,000

88. What is the net worth of the Owens family if their assets are \$500,000 and their liabilities are \$310,000? \$190,000

89. What is the net worth of the Lucas family if their assets are \$75,000 and their liabilities are \$92,000? −\$17,000

90. What is the net worth of the Kerns family if their assets are \$93,000 and their liabilities are \$103,500? −\$10,500

As of April 2010, the expression $1.2917x^4 - 74.231x^3 + 1590.403x^2 - 15{,}047x + 53{,}063$ approximates the percent of 12- to 17-year-olds who prefer to contact their friends by text messaging, where x is the age. (Source: http://pewresearch.org/pubs/1572/teens-cell-phones-text-messages)

91. What percent of 12-year-olds prefer to use text messaging to contact their friends? about 30.56

92. What percent of 15-year-olds prefer to use text messaging to contact their friends? about 61.36

93. What percent of 16-year-olds prefer to use text messaging to contact their friends? about 56.84

94. What percent of 17-year-olds prefer to use text messaging to contact their friends? about 77.64

95. The chart shows the closing price of Microsoft's stock for a week in 2011. (Source: http://www.dailyfinance.com/quote/nasdaq/microsoft-corp/msft)

Date	Closing Price of Microsoft Stock	Daily Change in Price
6/20/2011	\$24.47	n/a
6/21/2011	\$24.76	\$0.29
6/22/2011	\$24.65	−\$0.11
6/23/2011	\$24.63	−\$0.02
6/24/2011	\$24.30	−\$0.33

a. Complete the chart to show how the price of the stock changed from one day to the next.

b. After completing the chart, find the average daily change in price. \$0.0425 per day

96. The chart shows the closing price of BP's stock for a week in 2011. (Source: http://www.dailyfinance.com/quote/nyse/bp-plc-adr/bp)

Date	Closing Price of of BP Stock	Daily Change in Price
10/17/2011	\$40.17	n/a
10/18/2011	\$41.11	\$0.94
10/19/2011	\$40.78	−\$0.33
10/20/2011	\$41.32	\$0.54
10/21/2011	\$42.35	\$1.03

a. Complete the chart to show how the price of the stock changed from one day to the next.

b. After completing the chart, find the average daily change in price. −\$0.545 per day

97. The volume of a sphere is given by $\frac{4}{3}\pi r^3$, where r is the radius of the sphere. Find the volume of a sphere that has a radius of 8.5 in. Round answer to the nearest hundredth. 2572.44 in.^{3}

98. The volume of a sphere is given by $\frac{4}{3}\pi r^3$, where r is the radius of the sphere. Find the volume of a sphere that has a radius of 12.4 in. Round answer to the nearest hundredth. 7986.45 in.^{3}

Mix 'Em Up!

Simplify each expression.

99. $-\frac{1}{2}+\frac{1}{6}$ $-\frac{1}{3}$

100. $-\frac{1}{5}+\frac{2}{3}$ $\frac{7}{15}$

101. $\left(-\frac{7}{8}\right)^2$ $\frac{49}{64}$

102. $\left(-\frac{10}{13}\right)^2$ $\frac{100}{169}$

103. $-3+\left(-\frac{1}{4}\right)$ $-\frac{13}{4}$

104. $-5+\left(-\frac{3}{5}\right)$ $-\frac{28}{5}$

105. $\frac{98}{0}$ undefined

106. $-\frac{32}{0}$ undefined

107. $-\left(\frac{6}{5}\right)^3$ $-\frac{216}{125}$

108. $-\left(\frac{9}{7}\right)^3$ $-\frac{729}{343}$

109. $-\left(\frac{1}{2}\right)\left(\frac{4}{5}\right)^2 \div \frac{1}{50}+2$ -14

110. $\left(-\frac{1}{6}\right)\left(\frac{3}{4}\right)^2 \div \frac{1}{96}+3$ -6

111. $\frac{-108}{-3}$ 36

112. $\frac{216}{-4}$ -54

113. $\frac{13}{15}\div\left(-\frac{26}{25}\right)$ $-\frac{5}{6}$

114. $-\frac{12}{11}\div\frac{48}{55}$ $-\frac{5}{4}$

115. $\frac{0}{-3}$ 0

116. $\frac{0}{14}$ 0

117. $\frac{-10+9}{3-(-1)}$ $-\frac{1}{4}$

118. $\frac{-4+6}{11-(-3)}$ $\frac{1}{7}$

119. $2(-3.5)^2-1.6(2.1)$ 21.14

120. $-3(-6.1)^2+4.5(-2.4)$ -122.43

121. $\frac{1+\sqrt{13-4}}{-2|2-5|^2}$ $-\frac{2}{9}$

122. $\frac{1-\sqrt{11+5}}{-1|7-4|^2}$ $\frac{1}{3}$

Evaluate each expression for the given values.

123. $\frac{1}{2}bh$ for $b=6$ and $h=\frac{1}{12}$ $\frac{1}{4}$

124. $\frac{1}{2}bh$ for $b=\frac{4}{7}$ and $h=14$ 4

125. $-\frac{2}{3}x^3+3x^2-10$ for $x=-6$ 242

126. $-\frac{1}{4}x^4+5x^2-3$ for $x=-2$ 13

Solve each problem. Round to two decimal places when appropriate.

127. From June 1995 to June 2009, the average retail price (in cents per gallon) of regular gas in California can be approximated by $0.005x^4+0.08x^3-0.34x^2+2.64x+133.02$, where x is the number of years after 1995. According to this model, what was the average retail price of regular gas in 1995? about $1.33 per gallon

128. According to the model in Exercise 127, what was the average retail price of regular gas in 2007? about $3.58 per gallon

129. What is the net worth of a family with $120,000 in assets and $150,000 in liabilities? −$30,000

130. What is the net worth of a family with $85,000 in assets and $210,000 in liabilities? −$125,000

131. The chart shows the high temperatures in Philadelphia for a week in June.

Date	Temperature	Daily Change in Temperature
Monday	88°F	n/a
Tuesday	89°F	1°F
Wednesday	86°F	−3°F
Thursday	86°F	0°F
Friday	93°F	7°F

a. Complete the chart to show how the high temperature changed from one day to the next.

b. After completing the chart, find the average daily change in the high temperature. 1.25°F per day

132. The chart below shows the 5-day weather forecast of Melbourne, Australia, for a week in June.

Date	Temperature	Daily Change in Temperature
Monday	54°F	n/a
Tuesday	58°F	4°F
Wednesday	59°F	1°F
Thursday	61°F	2°F
Friday	57°F	−4°F

a. Complete the chart to show how the high temperature changed from one day to the next.

b. After completing the chart, find the average daily change in the high temperature. 0.75°F per day

133. The volume of a cylinder is given by $\pi r^2 h$, where r is the radius and h is the height of the cylinder. Find the volume of a cylinder that has a radius of 5.6 in. and a height of 9.5 in. Round answer to the nearest hundredth. 935.94 in.^{3}

134. The volume of a cylinder is given by $\pi r^2 h$, where r is the radius and h is the height of the cylinder. Find the volume of a cylinder that has a radius of 7.5 in. and a height of 5.4 in. Round answer to the nearest hundredth. 954.26 in.^{3}

135. The percentage increase in nursing school enrollments from the previous year can be modeled by the expression $0.24x^3 - 4.02x^2 + 18.57x - 12.03$, where x is the number of years after 2000. Find the percent increase in enrollments in nursing schools in (a) 2001 and (b) 2004. (Source: American Association of Colleges of Nursing) (a) about 2.76% (b) about 13.29%

136. Use the model in Exercise 135 to determine the percentage increase in nursing school enrollments in (a) 2005 and (b) 2010. (a) about 10.32% (b) about 11.67%

You Be the Teacher!

Answer each student's question or correct the error, if any.

137. Clarice: Can you give me a real-life example that would require you to add two real numbers with different signs?

138. Jocelyn: Can you give me a real-life example that would require you to add two negative real numbers?

139. Simplify -16^2.

Zia's work:

$-16^2 = 216$ $\quad -16^2 = -256$

140. Simplify $(-4)(-3)^2 + 6(-1)$.

Diane's work:

$(-4)(-3)^2 + 6(-1)$	$(-4)(-3)^2 + 6(-1)$
$= 12^2 + (-6)$	$= (-4)9 + (-6)$
$= 144 + (-6)$	$= -36 + (-6)$
$= 138$	$= -42$

141. Brent: What punctuation is needed if I want the first step, in simplifying the following expression, to be to subtract $\frac{3}{2}$ and 1?

$$\frac{1}{2}\left(-\frac{5}{2}\right)^2 + 6 \div \frac{3}{2} - 1 \qquad \frac{1}{2}\left(-\frac{5}{2}\right)^2 + 6 \div \left(\frac{3}{2} - 1\right)$$

142. Evaluate $x^2 - 2x$ for $x = -10$.

AJ's work:

$-10^2 - 2(-10)$	$(-10)^2 - 2(-10)$
$= -100 - 2(-10)$	$= 100 - 2(-10)$
$= -102(-10)$	$= 100 + 20$
$= -1020$	$= 120$

Calculate It!

Use a calculator to determine if each expression is simplified correctly. If not, provide the correct value.

143. $\dfrac{(-116)(3) - 14(7)}{13(-9) - (-84)} = \dfrac{446}{33}$ Yes

144. $\dfrac{(-11)(12) - 9(-4)}{11(-4) - (-16)} = \dfrac{24}{7}$ Yes

145. $\left(-\frac{1}{3}\right)^2\left(-\frac{1}{2}\right) + 2\left(\frac{4}{9}\right) + 7 = \dfrac{143}{18}$ No, $\dfrac{47}{6}$

146. $\left(\frac{5}{7}\right)\left(-\frac{1}{3}\right)^3 + 3\left(-\frac{1}{6}\right) + 19 = \dfrac{7003}{378}$ No, $\dfrac{6983}{378}$

SECTION 1.3 Properties of Real Numbers and Simplifying Algebraic Expressions

OBJECTIVES

As a result of completing this section, you will be able to

1. Apply the identity and inverse properties.
2. Apply the commutative, associative, and distributive properties.
3. Simplify algebraic expressions.
4. Translate phrases or statements into algebraic expressions, equations, or inequalities.
5. Translate phrases or statements related to applications.
6. Troubleshoot common errors.

According to the U.S. Census Bureau statistics, people with a bachelor's degree earn nearly twice as much as those with only a high school diploma. If x represents the earnings of a high school graduate, write an expression that represents the earnings of a person with a bachelor's degree.

To answer this question, we need to know how to translate phrases into algebraic expressions. We will learn that and more in this section.

Objective 1

Apply the identity and inverse properties.

The Identity and Inverse Properties

The set of real numbers contains some very interesting properties. The first one that we will discuss is the *identity property*. An **identity element** is a number which leaves another number unchanged when an operation is performed on it.

- When we add numbers, the only number that can be added to another number without changing its value is *zero*. Zero is called the *additive identity*.
- When we multiply numbers, the only number by which another number can be multiplied without changing its value is *one*. One is called the *multiplicative identity*.

	Identity Properties	Identity Elements	Examples ($a = 6$)
Addition	For all real numbers a, $a + 0 = 0 + a = a$	Zero is the **additive identity**.	$6 + 0 = 0 + 6 = 6$
Multiplication	For all real numbers a, $a \cdot 1 = 1 \cdot a = a$	One is the **multiplicative identity**.	$6 \cdot 1 = 1 \cdot 6 = 6$

A related property is the *inverse property*. An **inverse** is a number which produces the identity element when an operation is performed on it.

- The *additive inverse* of a number is its opposite since adding a number and its opposite results in 0.
- The *multiplicative inverse* of a number is its reciprocal since multiplying a number by its reciprocal is 1.

	Inverse Properties	Inverse Elements	Examples ($a = 5$)
Addition	For all real numbers a, $a + (-a) = (-a) + a = 0$	$-a$ is the **additive inverse** (or opposite) of a.	The opposite of 5 is -5 and $5 + (-5) = (-5) + 5 = 0$
Multiplication	For all real numbers $a \neq 0$, $a \cdot \frac{1}{a} = \frac{1}{a} \cdot a = 1$	$\frac{1}{a}$ is the **multiplicative inverse** (or reciprocal) of a, $a \neq 0$	The reciprocal of 5 is $\frac{1}{5}$ and $5 \cdot \frac{1}{5} = \frac{1}{5} \cdot 5 = 1$

Objective 1 Examples **Find the additive inverse (or opposite) and multiplicative inverse (or reciprocal) of each number. Assume any variables are nonzero.**

Problems	Additive Inverse	Multiplicative Inverse
1a. -6	$-(-6) = 6$	$\frac{1}{-6} = -\frac{1}{6}$
1b. $\frac{3}{4}$	$-\left(\frac{3}{4}\right) = -\frac{3}{4}$	$\frac{1}{\frac{3}{4}} = 1 \cdot \frac{4}{3} = \frac{4}{3}$
1c. $2x$	$-(2x) = -2x$	$\frac{1}{2x} = \frac{1}{2x}$
1d. $-3y$	$-(-3y) = 3y$	$\frac{1}{-3y} = -\frac{1}{3y}$
1e. $\frac{x}{7}$	$-\left(\frac{x}{7}\right) = -\frac{x}{7}$	$\frac{1}{\frac{x}{7}} = 1 \cdot \frac{7}{x} = \frac{7}{x}$

Student Check 1 Find the additive inverse (or opposite) and multiplicative inverse (or reciprocal) of each number. Assume any variables are nonzero.

a. -10 **b.** $\frac{7}{8}$ **c.** $4y$ **d.** $-9b$ **e.** $\frac{a}{3}$

The Commutative, Associative, and Distributive Properties

Objective 2

Apply the commutative, associative, and distributive properties.

Additional properties of the real numbers are ones that relate to how we add and multiply them. These properties form the foundation of how we work with algebraic expressions.

- The **commutative property** of the real numbers states that the order in which we add real numbers or multiply real numbers doesn't change the result.
- The **associative property** of the real numbers states that the way numbers are grouped when they are added or multiplied doesn't change the outcome.

	Commutative Properties	Associative Properties	Examples ($a = 2, b = 3, c = 4$)
Addition	For all real numbers a and b, $a + b = b + a$	For all real numbers a, b, and c, $a + (b + c) = (a + b) + c$	$2 + 3 = 3 + 2 = 5$ $2 + (3 + 4) = (2 + 3) + 4$ $2 + (7) = (5) + 4$ $9 = 9$
Multiplication	For all real numbers a and b, $a \cdot b = b \cdot a$	For all real numbers a, b, and c, $(a \cdot b) \cdot c = a \cdot (b \cdot c)$	$2 \cdot 3 = 3 \cdot 2 = 6$ $(2 \cdot 3) \cdot 4 = 2 \cdot (3 \cdot 4)$ $(6) \cdot 4 = 2 \cdot (12)$ $24 = 24$

The **distributive property** illustrates that a factor can be distributed over a sum of numbers.

Distributive Property	Example ($a = 2, b = 3, c = 4$)
For all real numbers a, b, and c, $a(b + c) = ab + ac$	$2(3 + 4) = 2(3) + 2(4)$ $2(7) = 6 + 8$ $14 = 14$

Objective 2 Examples **Apply the commutative, associative, or distributive properties to rewrite each expression as an equivalent expression and then simplify the equivalent expression.**

2a. $2 + y + 7$ **2b.** $2(y)(7)$ **2c.** $(x + 6) + 4$ **2d.** $4(6x)$

2e. $4(x + 6)$ **2f.** $-2(3a + 5)$ **2g.** $-(6x - 9)$

Solutions

2a. $2 + y + 7 = y + 2 + 7$ Apply the commutative property of addition.
$= y + 9$ Add the numbers.

2b. $2(y)(7) = 2(7)y$ Apply the commutative property of multiplication.
$= 14y$ Multiply the numbers.

2c. $(x + 6) + 4 = x + (6 + 4)$ Apply the associative property of addition.
$= x + 10$ Add the numbers.

2d. $4(6x) = (4 \cdot 6)x$ Apply the associative property of multiplication.
$= 24x$ Multiply the numbers.

2e. $4(x + 6) = 4(x) + 4(6)$ Apply the distributive property.
$= 4x + 24$ Simplify each product.

2f. $-2(3a + 5) = -2(3a) + (-2)(5)$ Apply the distributive property.
$= -6a - 10$ Simplify each product.

2g. $-(6x - 9) = -1(6x - 9)$ Write as a product of -1 and $6x - 9$.
$= -1(6x) + (-1)(-9)$ Apply the distributive property.
$= -6x + 9$ Simplify each product.

Student Check 2 Apply the commutative, associative, or distributive properties to rewrite each expression as an equivalent expression and then simplify the equivalent expression.

a. $3 + x + 5$ **b.** $3(x)(5)$ **c.** $(b + 2) + 9$ **d.** $2(9b)$

e. $8(y + 3)$ **f.** $-7(6a + 4)$ **g.** $-(2y - 1)$

Simplifying Algebraic Expressions

Objective 3 Simplify algebraic expressions.

The properties just discussed provide the framework for simplifying algebraic expressions. To **simplify an expression**, we may have to clear parentheses by applying the distributive property and then we combine any like terms.

The **terms** of an expression are the addends of the expression. For example, in the expression $6x + 2$, the terms are $6x$ and 2. Recall **like terms** are terms with identical variables raised to the same exponents.

To combine like terms, we combine their numerical coefficients and keep the variable the same. The **coefficient** of a term is the number that is multiplied by the variable. Recall

$$4x + 2x = \underbrace{x + x + x + x}_{4x} + \underbrace{x + x}_{2x} = 6x$$

So,

$$4x + 2x = (4 + 2)x = 6x$$

Note that combining like terms is based on the distributive property since $ab + ac = a(b + c)$ or $(b + c)a$.

Procedure: Simplifying Algebraic Expressions

Step 1: Remove any parentheses by applying the distributive property.
Step 2: Apply the commutative property to group like terms together.
Step 3: Combine any like terms.

Objective 3 Examples Simplify each expression.

3a. $4y - 9 + 2y$

3b. $x^2 - x - x + 1$

3c. $a^3 + 2a^2 + 4a - 2a^2 - 4a - 8$

3d. $0.05x + 0.03(10{,}000 - x)$

3e. $3 - 5(y - 2)$

3f. $12\left(\frac{2b-7}{6}\right) - 12\left(\frac{b-5}{4}\right)$

Solutions

3a. $4y - 9 + 2y = 4y + 2y - 9$ Apply the commutative property of addition.

$= (4 + 2)y - 9$ Apply the distributive property to add like terms.

$= 6y - 9$ Simplify.

3b. $x^2 - x - x + 1 = x^2 + -1x - 1x + 1$ Recall $-x = -1x$.

$= x^2 + (-1 - 1)x + 1$ Apply the distributive property to add like terms.

$= x^2 - 2x + 1$ Simplify.

3c. $a^3 + 2a^2 + 4a - 2a^2 - 4a - 8$

$= a^3 + 2a^2 - 2a^2 + 4a - 4a - 8$ Apply the commutative property of addition.

$= a^3 + (2 - 2)a^2 + (4 - 4)a - 8$ Apply the distributive property to add like terms.

$= a^3 + 0a^2 + 0a - 8$ Simplify.

$= a^3 - 8$ Simplify.

3d.

$0.05x + 0.03(10{,}000 - x)$ Apply the distributive property.

$= 0.05x + 300 - 0.03x$ Apply the commutative property of addition.

$= (0.05 - 0.03)x + 300$ Apply the distributive property to add like terms.

$= 0.02x + 300$ Simplify.

3e.

$3 - 5(y - 2) = 3 - 5y + 10$ Apply the distributive property.

$= -5y + 3 + 10$ Apply the commutative property of addition.

$= -5y + 13$ Simplify.

3f. Because we are multiplying fractions, we first simplify by dividing out the common factors. Note that 12 divided by 6 is 2 and 12 divided by 4 is 3.

$$12\left(\frac{2b-7}{6}\right) - 12\left(\frac{b-5}{4}\right)$$

$= 2(2b - 7) - 3(b - 5)$ Simplify the products.

$= 4b - 14 - 3b + 15$ Apply the distributive property.

$= 4b - 3b - 14 + 15$ Apply the commutative property of addition.

$= (4 - 3)b + 1$ Apply the distributive property and combine like terms.

$= 1b + 1$ Simplify.

$= b + 1$ Recall $1b = b$.

Student Check 3 Simplify each expression.

a. $7h + 3 - 9h$

b. $y^2 - 6y - 6y + 36$

c. $b^3 - 3b^2 + 9b + 3b^2 - 9b + 27$

d. $0.04n + 0.06(8000 - n)$

e. $8 - 2(x - 6)$

f. $8\left(\frac{5y-1}{4}\right) - 8\left(\frac{y+3}{2}\right)$

Translating into Algebraic Expressions

Objective 4

Translate phrases or statements into algebraic expressions, equations, or inequalities.

Translating phrases into algebraic expressions is an important skill for solving problems. Here is a chart showing some of the common phrases and their translations.

Addition	Subtraction	Multiplication	Division
$a + b$	$a - b$	ab	$\frac{a}{b}$
sum of a and b	difference of a and b	product of a and b	a divided by b
a increased by b	b subtracted from a	a times b	quotient of a and b
b more than a	b less than a	twice b, $(2b)$	ratio of a to b
a added to b	a minus b	a of b	b into a
a plus b	a decreased by b		
total of a and b	from a, subtract b		

There are also key statements that translate into equations or inequalities. Recall that a statement in which two expressions are equal is an **equation**. A statement in which two expressions are not equal is an **inequality**. For instance, to compare the numbers −2 and 5 using an inequality, we could write any of the following.

$$-2 < 5,\ -2 \le 5,\ 5 > -2, \text{ or } 5 \ge -2$$

Note that the inequality symbol always points to the smaller number.

INSTRUCTOR NOTE: Point out that a phrase translates into an expression, but a statement translates into an equation or inequality. In particular, show the difference between "a less than b" and "a is less than b."

Equals	Greater Than or Greater Than or Equal To	Less Than or Less Than or Equal To
$a = b$	$a > b$ or $a \ge b$	$a < b$ or $a \le b$
a is equal to b.	a is greater than b.	a is less than b.
a is the same as b.	a is greater than or equal to b.	a is less than or equal to b.
a results in b.	a is at least b.	a is at most b.
a equals b.	a is not less than b.	a is not greater than b.
a is b.	a is more than b.	a is not more than b.
a yields b.		

Objective 4 Examples **Translate each phrase or statement into an algebraic expression, equation, or inequality. Use the variable x to represent the unknown number.**

Problems	Solutions
4a. Two less than a number	$x - 2$
4b. The sum of twice a number and 3	$2x + 3$
4c. The quotient of three times a number and 8	$\frac{3x}{8}$
4d. The difference of four times a number and 5 is equal to 7.	$4x - 5 = 7$
4e. Twice the sum of a number and 7 is 1 more than the number.	$2(x + 7) = x + 1$
4f. One-third of the difference between a number and 4 yields the sum of the number and 6.	$\frac{1}{3}(x - 4) = x + 6$
4g. The product of a number and 9 is less than 18.	$9x < 18$
4h. Eight plus a number is at least three times the number.	$8 + x \geq 3x$

Student Check 4 Translate each phrase into an algebraic expression, equation, or inequality. Use the variable x to represent the unknown number.

a. A number decreased by 2
b. The sum of three times a number and 5
c. The quotient of a five times a number and 2
d. The difference of twice a number and 9
e. Three times the sum of a number and 1 equals 5.
f. Five less than one-half of a number results in two more than the number.
g. Twice a number is less than 8.
h. The total of a number and 4 is at most 6.

Objective 5

Translate phrases or statements related to applications.

Express Real-Life Situations Algebraically

In Example 5 the translations relate to a specific problem. These translations will prepare us for word problems that we encounter in later sections.

Objective 5 Examples **Write an expression that represents the unknown quantity and/or write an equation that represents the situation.**

5a. Nadia invests \$5000 between two different accounts. If x represents the amount she invests in one account, write an expression that represents the amount invested in the other account.

Solution **5a.** Let's consider some specific numbers to help solve this problem as shown in the table.

Amount Invested in One Account	Amount Invested in the Other Account
\$2000	\$5000 − \$2000 = \$3000
\$4000	\$5000 − \$4000 = \$1000
\$ 500	\$5000 − \$500 = \$4500

From the table, we see that the second amount is the first amount subtracted from \$5000. So, if we let x represent the amount Nadia invests in one account, the other amount is represented as $5000 - x$.

5b. According to the U.S. Census Bureau statistics, people with a bachelor's degree earn nearly twice as much as those with only a high school diploma. If x represents the earnings of a high school graduate, write an expression that represents the earnings of a person with a bachelor's degree.

Solution **5b.** Let x represent the earnings of a high school graduate.

The earnings of a college graduate are twice as much as the earnings of a high school graduate, or $2x$.

5c. According to www.collegeboard.com, the average published tuition and fee charges for in-state students at *four-year public colleges* in 2010–2011 is \$555 more than the average published tuition and fee charges in 2009–2010. If x represents the average tuition and fee charges in 2009–2010, write an expression that represents the average tuition and fee charges in 2010–2011.

Solution **5c.** Let x represent the average tuition and fees for 2009–2010.

The average fees for 2010–2011 are \$555 more than the average fees for 2009–2010, so the expression for the average tuition and fees for 2010–2011 is

$$x + 555$$

5d. In August 2011, the national median home price fell approximately 5.1% from the national median home price in 2010. The 2010 median home price was about \$177,344. Write an equation that represents this situation if x is the national median home price in August 2011. (Source: http://www.realestateabc.com/outlook.htm)

Solution **5d.** Let x represent the national median home price in August 2011. The median home price in August 2011 is equal to the median home price in 2010 minus the amount of decrease. The decrease amount is 5.1% of the price in 2010. So, the equation is

Price in 2010 − decrease amount is the median price in August 2011.

$$x - 0.051x = 177{,}344$$

5e. Shawn has a collection of dimes and quarters. He has seven fewer dimes than quarters.

i. If x represents the number of quarters, write an expression for the number of dimes he has collected.

ii. Write expressions for the value of the quarters and the value of the dimes in dollars.

iii. If Shawn's collection is worth \$6.30, write an equation that represents the total value of the collection.

Solution **5e.** **i.** Let x represent the number of quarters. Since there are seven fewer dimes than quarters, the number of dimes can be represented by $x - 7$.

ii. The value of one quarter is \$0.25, the value of two is $0.25(2) = \$0.50$. So, the value of x quarters is $0.25x$. The value of one dime is \$0.10, the value of two is $0.10(2) = 0.20$. So, the value of $x - 7$ dimes is $0.10(x - 7)$.

iii. The value of the collection is the value of the quarters plus the value of the dimes. So, an equation that represents the total value is

$$0.25x + 0.10(x - 7) = 6.30$$

Student Check 5 Write an expression that represents the unknown quantity and/or write an equation that represents the situation.

a. Juan invests \$2000 in two different accounts. If x represents the amount invests in the first account, write an expression that represents the amount invested in the second account.

b. The maximum amount most financial advisors recommend spending on housing (mortage/rent, repairs, taxes, utilities, and insurance) is $\frac{1}{3}$ of your monthly income. If x represents your monthly income, write an expression for the amount that should be spent on housing.

c. The top earning celebrities in 2010 were Oprah Winfrey and U2. Oprah earned \$95 million more than U2. If x represents U2's earnings in millions, write an expression for Oprah's earnings. (Source: www.forbes.com)

d. In October 2007, the closing price of Google stock was at an all time high. The price decreased by 17% by October 2011. The price of the stock in October 2011 was \$586.31. Write an equation that represents this situation if x represents the price of the Google stock in October 2007. (Source: www.moneycentral.msn.com)

e. Priscilla has a collection of nickels and dimes. She has 20 fewer dimes than nickels.

i. If x represents the number of nickels, write an expression for the number of dimes she has collected.

ii. Write expressions for the value of the nickels and the value of dimes in dollars.

iii. If Priscilla's collection is worth \$2.50, write an equation that represents the total value of the collection.

Objective 6

Troubleshoot common errors.

Troubleshooting Common Errors

Some common errors associated with algebraic expressions are shown.

Objective 6 Examples **A problem and an incorrect solution are given. Provide the correct solution and an explanation of the error.**

6a. Simplify $7 + 3(2x - 5)$.

Incorrect Solution	Correct Solution and Explanation
$7 + 3(2x - 5) = 10(2x - 5)$ $= 20x - 50$	The order of operation tells us that multiplication comes before addition. $7 + 3(2x - 5) = 7 + 6x - 15$ $= 6x - 8$

6b. Simplify $a^2 + 2a + 3 + a^2 - 7a - 4$.

Incorrect Solution	Correct Solution and Explanation
$a^2 + 2a + 3 + a^2 - 7a - 4$ $= 2a^4 - 5a - 1$	The error is that the exponents of the like terms were added. When we add like terms, the variable expression is unchanged. $a^2 + 2a + 3 + a^2 - 7a - 4$ $= 2a^2 - 5a - 1$

6c. Translate the phrase "4 less than a number."

Incorrect Solution	Correct Solution and Explanation
$4 < x$	The phrase *less than* indicates subtraction. For the inequality sign to be used, the statement would be "4 is less than a number." Therefore, the translation is $x - 4$

ANSWERS TO STUDENT CHECKS

Student Check 1 **a.** $10; -\frac{1}{10}$ **b.** $-\frac{7}{8}; \frac{8}{7}$ **c.** $-4y; \frac{1}{4y}$
d. $9b; -\frac{1}{9b}$ **e.** $-\frac{a}{3}; \frac{3}{a}$

Student Check 2 **a.** $x + 8$ **b.** $15x$ **c.** $b + 11$
d. $18b$ **e.** $8y + 24$ **f.** $-42a - 28$
g. $-2y + 1$

Student Check 3 **a.** $-2h + 3$ **b.** $y^2 - 12y + 36$
c. $b^3 + 27$ **d.** $-0.02n + 480$ **e.** $-2x + 20$
f. $6y - 14$

Student Check 4 **a.** $x - 2$ **b.** $3x + 5$ **c.** $\frac{5x}{2}$
d. $2x - 9$ **e.** $3(x + 1) = 5$
f. $\frac{1}{2}x - 5 = x + 2$ **g.** $2x < 8$ **h.** $x + 4 \leq 6$

Student Check 5 **a.** $2000 - x$ **b.** $\frac{1}{3}x$ **c.** $x + 95$
d. $x - 0.17x = 586.31$
e. **i.** $x - 20$
ii. The nickels are worth $0.05x$ and the dimes are worth $0.10(x - 20)$.
iii. $0.05x + 0.10(x - 20) = 2.50$

SUMMARY OF KEY CONCEPTS

1. Apply the identity and inverse properties.
   - The additive identity is the number 0. Adding zero doesn't change the value of the number to which it is added. The additive inverse of a is $-a$. Adding a number and its inverse produces the additive identity, 0.
   - The multiplicative identity is the number 1. Multiplying by 1 doesn't change the value of the number with which it is multiplied. The multiplicative inverse of a is $\frac{1}{a}$. Multiplying a number and its inverse produces the multiplicative identity, 1.
2. Apply the commutative, associative, and distributive properties to rewrite expressions.
   - The commutative property enables us to change the order of terms being added or factors being multiplied without changing the result.
   - The associative property enables us to change the grouping of terms being added or factors being multiplied without changing the result.
   - The distributive property enables us to multiply a number by a sum or difference. The outer number distributes to all terms inside parentheses.
3. Simplify algebraic expressions.
   - To simplify an algebraic expression, clear any parentheses and then combine like terms.
   - To combine like terms, combine the coefficients of the like terms and combine any constant terms.
4. Translate phrases or statements into algebraic expressions, equations, or inequalities. Familiarize yourself with the key phrases for addition, subtraction, multiplication, division, equality, and inequality. These phrases will be used in problems throughout the text.

SECTION 1.3 EXERCISE SET

Write About It!

Use complete sentences in your answer to each exercise.

1. Explain the meaning of an identity.
2. Explain the meaning of an inverse.
3. Explain the commutative property of addition and multiplication.
4. Explain the associative property of addition and multiplication.

Practice Makes Perfect!

Find both the additive inverse and multiplicative inverse of each number. Assume all variables represent nonzero numbers. (*See Objective 1.*)

5. 5 $-5, \frac{1}{5}$

6. 8 $-8, \frac{1}{8}$

7. -20 $20, -\frac{1}{20}$

8. -41 $41, -\frac{1}{41}$

9. $\frac{3}{4}$ $-\frac{3}{4}, \frac{4}{3}$

10. $\frac{5}{2}$ $-\frac{5}{2}, \frac{2}{5}$

11. $-\frac{6}{7}$ $\frac{6}{7}, -\frac{7}{6}$

12. $-\frac{8}{9}$ $\frac{8}{9}, -\frac{9}{8}$

13. $2x$ $-2x, \frac{1}{2x}$

14. $5x$ $-5x, \frac{1}{5x}$

15. $\frac{3}{x}$ $-\frac{3}{x}, \frac{x}{3}$

16. $\frac{5}{x}$ $-\frac{5}{x}, \frac{x}{5}$

Apply the commutative property of addition or the commutative property of multiplication to rewrite each expression as an equivalent expression and then simplify the expression. (*See Objective 2.*)

17. $3 + x + 5$ $x + 3 + 5; x + 8$

18. $5 + x + 2$ $x + 5 + 2; x + 7$

19. $10 + a - 7$ $a + 10 - 7; a + 3$

20. $13 + a - 5$ $a + 13 - 5; a + 8$

21. $-2 + y - 7$ $y + (-2) - 7; y - 9$

22. $-6 + y - 1$ $y + (-6) - 1; y - 7$

23. $3(x)(5)$ $3(5)x; 15x$

24. $4(x)(6)$ $4(6)x; 24x$

25. $-8(a)(-7)$ $(-8)(-7)a; 56a$

26. $-9(a)(-2)$ $(-9)(-2)a; 18a$

Apply the associative property of addition or the associative property of multiplication to rewrite each expression as an equivalent expression and then simplify the expression. (*See Objective 2.*)

27. $(4x + 3) + 5$ $4x + (3 + 5); 4x + 8$

28. $(7x + 9) + 6$ $7x + (9 + 6); 7x + 15$

29. $(x - 2) + 1$ $x + (-2 + 1); x - 1$

30. $(x - 5) + 3$ $x + (-5 + 3); x - 2$

31. $5(2b)$ $(5 \cdot 2)b; 10b$

32. $3(6b)$ $(3 \cdot 6)b; 18b$

33. $-7(-2y)$ $(-7 \cdot -2)y; 14y$

34. $-9(-3y)$ $(-9 \cdot -3)y; 27y$

Apply the distributive property to rewrite each expression and then simplify the expression. (*See Objective 2.*)

35. $3(a + b)$ $3(a) + 3(b); 3a + 3b$

36. $5(q + r)$ $5(q) + 5(r); 5q + 5r$

37. $10(t + 7)$ $10(t) + 10(7); 10t + 70$

38. $9(l + 3)$ $9(l) + 9(3); 9l + 27$

39. $-4(9x + 4)$ $-4(9x) + (-4)(4); -36x - 16$

40. $-12(6x + 5)$ $-12(6x) + (-12)(5); -72x - 60$

41. $-(2x - 5)$ $-1(2x) + (-1)(-5); -2x + 5$

42. $-(3x - 1)$ $-1(3x) + (-1)(-1); -3x + 1$

43. $2(4x + 7y)$ $2(4x) + 2(7y); 8x + 14y$

44. $4(5f + 6g)$ $4(5f) + 4(6g); 20f + 24g$

45. $-6(5x + 2y + 1)$ $-6(5x) + (-6)(2y) + (-6)(1); -30x - 12y - 6$

46. $-2(-3x + 4y - 2)$ $-2(-3x) + (-2)(4y) + (-2)(-2); 6x - 8y + 4$

Simplify each expression. (*See Objective 3.*)

47. $14d - 13 + 12d - 15$ $26d - 28$

48. $16a - 11 + 18a + 12$ $34a + 1$

49. $x^2 + 5x + 3x + 15$ $x^2 + 8x + 15$

50. $a^2 + 3a + 4a + 12$ $a^2 + 7a + 12$

51. $6x^2 - 8x - 3x + 4$ $6x^2 - 11x + 4$

52. $9y^2 - 15y - 15y + 25$ $9y^2 - 30y + 25$

53. $y^3 - 3y^2 + 9y + 3y^2 - 9y + 27$ $y^3 + 27$

54. $y^3 + 5y^2 + 25y - 5y^2 - 25y - 125$ $y^3 - 125$

55. $8b^3 + 24b^2 + 18b + 12b^2 + 36b + 27$ $8b^3 + 36b^2 + 54b + 27$

56. $64x^3 - 160x^2 + 100x - 80x^2 + 200x - 125$ $64x^3 - 240x^2 + 300x - 125$

57. $0.03x + 0.05(1000 - x)$ $-0.02x + 50$

58. $0.04x + 0.01(3000 - x)$ $0.03x + 30$

59. $0.25x + 0.60(40 - x)$ $-0.35x + 24$

60. $0.10x + 0.30(600 - x)$ $-0.20x + 180$

61. $4 - 5(x - 2)$ $-5x + 14$

62. $7 - 4(x + 3)$ $-4x - 5$

63. $-8 + 2(4x - 1)$ $8x - 10$

64. $-9 + 3(10x + 6)$ $30x + 9$

65. $6\left(\frac{2x - 5}{3}\right) - 4\left(\frac{x + 1}{2}\right)$ $2x - 12$

66. $10\left(\frac{1 - 6x}{2}\right) - 6\left(\frac{3x + 2}{3}\right)$ $1 - 36x$

67. $12\left(\frac{x - 7}{4}\right) + 6\left(\frac{3 - 5x}{2}\right)$ $-12x - 12$

68. $-4\left(\frac{3x - 1}{2}\right) + 15\left(\frac{4x - 1}{5}\right)$ $6x - 1$

Additional answers can be found in the Instructor Answer Appendix.

Translate each phrase or sentence into an algebraic expression, equation, or inequality. Use the variable *x* to represent the unknown number. (*See Objective 4.*)

69. Three less than a number $x - 3$
70. Twelve less than a number $x - 12$
71. Six more than twice a number $2x + 6$
72. Seven more than three times a number $3x + 7$
73. A number divided by 6 $\frac{x}{6}$
74. The ratio of a number and 4 $\frac{x}{4}$
75. Half the difference between a number and 11 $\frac{1}{2}(x - 11)$
76. One-third the sum of a number and 5 $\frac{1}{3}(x + 5)$
77. Thirteen is ten more than twice a number. $13 = 2x + 10$
78. Two is three less than four times a number. $2 = 4x - 3$
79. Five more than twice a number is greater than the number plus 3. $2x + 5 > x + 3$
80. Six less than four times a number is less than two more than twice the number. $4x - 6 < 2x + 2$
81. The total of eight and a number yields half the sum of the number and 5. $8 + x = \frac{1}{2}(x + 5)$
82. The total of nine and twice a number results in one-half the difference between the number and 10. $9 + 2x = \frac{1}{2}(x - 10)$

Solve each problem. (*See Objective 5.*)

83. Marty invests \$6000 in two different stocks. If x represents the amount invested in one stock, write an expression for the amount invested in the other stock. $6000 - x$
84. Emma spends \$534 on two evening dresses. If she spends x dollars on the first dress, how much did she spend on the second dress? $534 - x$
85. Ricky has been collecting pennies and nickels in a jar and he has 20 fewer nickels than pennies.
    a. If x represents the number of pennies Ricky has collected, write an expression for the number of nickels he has collected. $x - 20$
    b. Write an expression for the value, in cents, of the nickels Ricky has collected. $5(x - 20)$
    c. Write an expression for the value, in cents, of the pennies Ricky has collected. x
    d. If Ricky's collection is worth \$1.70, write an equation that represents this total value, in cents. $x + 5(x - 20) = 170$
86. Diana has been saving \$2 bills and silver dollars. She has 10 fewer \$2 bills than silver dollars.
    a. If x represents the number of silver dollars Diana has saved, write an expression for the number of \$2 bills she has saved. $x - 10$
    b. Write an expression for the value, in dollars, of Diana's silver dollars. x
    c. Write an expression for the value, in dollars, of Diana's \$2 bills. $2(x - 10)$
    d. If Diana has \$40 total, write an equation that represents the total value of her collection. $x + 2(x - 10) = 40$
87. The American Association of Colleges of Nursing found that enrollment in entry-level baccalaureate nursing programs increased by 5.7% from 2009 to 2010. If x represents the enrollment in baccalaureate nursing programs in 2009, write an expression for the enrollment in 2010. (Source: http://www.aacn.nche.edu/media/factsheets/nursingshortage.htm) $x + 0.057x$ or $1.057x$
88. The number of adults in the United States who own an e-book reader in May 2011 increased by 100% over the number who owned one in November 2010. If x represents the number who owned an e-book reader in November 2010, write an expression for the number who owned one in May 2011. (Source: http://pewresearch.org/pubs/?Year=2011) $x + 1x = 2x$

Mix 'Em Up!

Find the additive inverse and multiplicative inverse for each expression. Assume all denominators are nonzero.

89. $x - 5$ $-x + 5; \frac{1}{x - 5}$
90. $x + 2$ $-x - 2; \frac{1}{x + 2}$
91. $\frac{4}{a}$ $-\frac{4}{a}; \frac{a}{4}$
92. $\frac{b}{3}$ $-\frac{b}{3}; \frac{3}{b}$
93. $\frac{3}{x + 1}$ $-\frac{3}{x + 1}; \frac{x + 1}{3}$
94. $\frac{x + 3}{2}$ $-\frac{x + 3}{2}; \frac{2}{x + 3}$

Identify the property that is used to rewrite each expression.

95. $(5x - 29) + 70y = 5x + (-29 + 70y)$ associative property of addition
96. $(3x - 10) + 6y = 3x + (-10 + 6y)$ associative property of addition
97. $[6(v - u)] \cdot 9 = 6[(v - u) \cdot 9]$ associative property of multiplication
98. $[7(g + h)] \cdot 4 = 7[(g + h) \cdot 4]$ associative property of multiplication
99. $(x + y)(m + n) = (x + y)m + (x + y)n$ distributive property
100. $(x + y)(m + n) = (m + n)(x + y)$ commutative property of multiplication

Simplify each expression.

101. $a^3 - 5a^2 + 12a + 6a^2 - 8a + 17$ $a^3 + a^2 + 4a + 17$
102. $2b^3 - 7b^2 - 15b + 10 - 3b^2 + 9b - 21$ $2b^3 - 10b^2 - 6b - 11$
103. $\frac{1}{3}(6x - 3) + \frac{3}{4}[8(x - 1)] + 10$ $8x + 3$
104. $\frac{3}{4}(4x - 12) - \frac{5}{2}[4(3 - x)] - 3$ $13x - 42$
105. $0.035m + 0.042(3000 - m)$ $126 - 0.007m$
106. $0.064x + 0.032(4500 - x)$ $0.032x + 144$
107. $4(m - 3) + 2(m^2 - 5)$ $2m^2 + 4m - 22$
108. $12(2b + 4) - 4(b^2 + 3)$ $-4b^2 + 24b + 36$
109. $\frac{3}{5}x - \frac{2}{5} - \frac{1}{5}x + \frac{7}{5}$ $\frac{2}{5}x + 1$
110. $\frac{3}{4}x - \frac{7}{5} - \frac{5}{4}x + \frac{1}{5}$ $-\frac{1}{2}x - \frac{6}{5}$

Translate each expression. Use x for the unknown quantity.

111. Half the sum of a number and 9 equals the difference between the number of 15. $\frac{1}{2}(x + 9) = x - 15$

112. Triple the difference of a number and 10 equals the sum of half the number and 5. $3(x - 10) = \frac{1}{2}x + 5$

113. A number decreased by 3 is greater than 10 less than twice the number. $x - 3 > 2x - 10$

114. A number decreased by 12 is greater than half the sum of the number and 15. $x - 12 > \frac{1}{2}(x + 15)$

115. Persia is 9 years older than Obie. If x represents Obie's age, write an expression for Persia's age. $x + 9$

116. Brent is 3 years younger than AJ. If x represents AJ's age, write an expression for Brent's age. $x - 3$

117. There was a 5% decrease in U.S./Canada movie admissions from 2009 to 2010. If x represents the number of movie admissions in 2009, write an expression for the number of movie admissions in 2010. (Source: http://www.mpaa.org) $x - 0.05x$ or $0.95x$

118. The average price of a movie ticket in 2010 was \$2.23 more than the average price of a movie ticket in 2001. If x represents the average price of a movie ticket in 2001, write an expression for the price of a movie ticket in 2010. (Source: http://www.mpaa.org) $x + 2.23$

119. Of the moviegoers in 2010, 23% of them are aged 25 to 39. If x represents the number of moviegoers, write an algebraic expression for the number of moviegoers who are aged 25 to 39. (Source: http://www.mpaa.org) $0.23x$

120. Referring to Exercise 119, write an algebraic expression for the number of moviegoers who are not aged 25 to 39. $x - 0.23x$ or $0.77x$

You Be the Teacher!

Answer each student's question or correct the student's error, if any.

121. There was a 4.9% increase in the number of people who patronized a restaurant from 2010 to 2011. If x represents the number of people who patronized that restaurant in 2010, write an algebraic expression for the number of people who patronized the restaurant in 2011.

Knox's work: $x + 0.49x$ $x + 0.049x$ or $1.049x$

122. There was a 12% decrease in the number of memberships to a gym from 2009 to 2010. If x represents the number of people who had membership to the gym in 2009, write an expression for the number of people who had a gym membership in 2010.

Vivienne's work: $0.12x$ $x - 0.12x$ or $0.88x$

123. Sandra: How do you find the additive inverse when there are variables? Multiply the expression by -1.

124. Dave: How do you find the multiplicative inverse when there is a variable involved? Find the reciprocal of the expression.

GROUP ACTIVITY A Magical Mathematical Birthday Card

For Part 1, begin with your birth month and follow Steps 1–12.

For Part 2, use the variables as described to explain how the trick works.

Part 1

1. Write the number of the month you were born, or enter it in your calculator.
2. Multiply this number by 4.
3. Add 13 to your result.
4. Multiply this number by 25.
5. Subtract 200 from your result.
6. Add the day of the month on which you were born to the previous result.
7. Multiply this number by 2.
8. Subtract 40 from your result.
9. Multiply this number by 50.
10. Add the last two digits of your birth year.
11. Subtract 10,500 from your result.
12. Put slashes in your final result, so that the digits are in groups of two, starting from the right. You should see your birth date displayed!

Part 2

1. Let m = month, d = day, and y = last two digits of year you were born.
2. Write the algebraic statements that correspond to Steps 1–12.
3. Explain how this trick works.

Chapter 1 / REVIEW

Real Numbers and Algebraic Expressions

What's the big idea? Chapter 1 provides us with the skills to simplify numerical expressions by applying the rules for signed numbers, the properties of real numbers, and the order of operations. These rules and properties also provide the framework for us to simplify and evaluate algebraic expressions. Translating phrases to their mathematical form was also shown. These skills form the foundation for algebra and will be used in every section of this text. Success in this course depends on mastering these skills.

The Tools

Listed below are the key terms, skills, and formulas, and properties you should know for this chapter.

The page reference is provided if you need additional help with the given topic. The Study Tips will assist in your preparation for an exam.

Study Tips

1. Learn all of the terms, formulas, and properties. Make flash cards and have someone quiz you.
2. Rework problems from the exercises and also the ones you worked in class. Work additional problems from the review exercises.
3. Review the summaries of key concepts.
4. Work the chapter test.
5. Be sure to review the online resources for additional study materials.

Terms

Absolute value 9
Additive identity 33
Additive inverse 8, 33
Algebraic expressions 23
Associative property 34
Base 20
Circumference 26
Coefficients 35
Commutative property 34
Empty set 2
Equation 37
Evaluate an algebraic expression 24
Exponent 20
Finite set 2
Identity element 33
Inequality 37
Infinite set 2
Integers 4
Intersection of sets 3
Inverse 33
Irrational numbers 5
Like terms 35
Member or element 2
Multiplicative identity 33
Multiplicative inverse 33
Natural numbers 4
Net worth 25
Opposite 8
Origin 7
Rational numbers 5
Real number line 7
Real numbers 5
Reciprocal 19
Roster method 2
Set 2
Set-builder notation 2
Simplify an expression 35
Square root 5
Subscripts 24
Subset 2
Term 35
Union of sets 3
Variable 2
Whole numbers 4

Formulas and Properties

- Associative properties 34
- Commutative properties 34
- Distributive property 34
- Division properties with zero 19
- Dividing rule 19
- Exponential notation 20
- Identity properties 33
- Inverse properties 33
- Order of operations 22
- Product property of zero 18
- Subtraction rule 16

CHAPTER 1 / SUMMARY

How well do you know this chapter? Complete the following questions to find out. Take a look back at the section if you need help.

SECTION 1.1 Sets and the Real Numbers

1. A(n) set is a collection of objects. Each object in the set is called a(n) member or a(n) element.
2. Sets can be represented in two ways: the roster method or set-builder notation.
3. The symbol ∪ denotes the union of two sets. The symbol ∩ denotes the intersection of two sets.
4. The set of natural numbers is $\{1, 2, 3, \ldots\}$.
5. The set of whole numbers is $\{0, 1, 2, 3, \ldots\}$.
6. The set of integers is $\{\ldots, -3, -2, -1, 0, 1, 2, 3, \ldots\}$.
7. A(n) rational number is a number that can be written as the quotient of integers.
8. A(n) irrational number is a number whose decimal form continues indefinitely without a repeating pattern.
9. We can graph or plot real numbers on a(n) real number line. Numbers to the left of zero are negative numbers. Numbers to the right of zero are positive numbers. Zero is called the origin and is neither positive nor negative.
10. The opposite of a number is a number with the same distance from zero but lies on the other side of zero on the number line.
11. The absolute value of a number is the distance the number is from zero on the real number line.

SECTION 1.2 Operations with Real Numbers and Algebraic Expressions

12. To add numbers with the same sign, add their absolute values and keep the sign the same.
13. To add numbers with different signs, subtract their absolute values. The sign of the answer has the same sign as the number with the larger absolute value.
14. Subtracting real numbers is the same as adding the opposite of a number. In symbols, $a - b = a + (-b)$.
15. The product of real numbers with the same signs is positive. The product of real numbers with opposite signs is negative.
16. To divide real numbers is the same as multiplying by the reciprocal, $a \div b = a \cdot \frac{1}{b}$.
17. Zero divided by a nonzero number is zero. A nonzero number divided by zero is undefined.
18. A(n) exponent indicates repeated multiplication.
19. In the expression b^n, b is called the base and n is called the exponent.
20. A negative base raised to an even exponent is positive.
21. A negative base raised to an odd exponent is negative.
22. The order of operations provides us a way to simplify numerical expressions. First, simplify what is in grouping symbols, then simplify exponents, multiply or divide from left to right, and finally add or subtract from left to right.
23. A(n) variable represents an unknown number and is represented by a(n) letter.
24. A(n) algebraic expression is an expression that involves variables and/or numbers.
25. To evaluate an algebraic expression, replace the variable with the given value and simplify.

SECTION 1.3 Properties of Real Numbers and Simplifying Algebraic Expressions

26. A(n) identity element is a number which leaves another number unchanged when an operation is performed on it. Zero is the additive identity. One is the multiplicative identity. In symbols, $a + 0 = a$ and $a \cdot 1 = a$.
27. A(n) inverse is a number which produces the identity element when an operation is performed on it. The additive inverse of number a is $-a$. The multiplicative inverse of a number a, $a \neq 0$ is $\frac{1}{a}$. In symbols, $a + -a = 0$ and $a \cdot \frac{1}{a} = 1$.
28. The commutative property of real numbers states that the order in which we add or multiply real numbers doesn't change the result. In symbols, $a + b = b + a$ and $ab = ba$.
29. The associative property of real numbers states that the grouping of the things being added or multiplied doesn't change the result. In symbols, $a + (b + c) = (a + b) + c$ and $a(bc) = (ab)c$.
30. The distributive property enables us to multiply a number by a sum or difference. $a(b + c) = ab + ac$.
31. The coefficient of a term is the number multiplied by the variable.
32. Terms that have identical variables raised to the same exponents are like terms.
33. To combine like terms, add their coefficients and keep the variable the same.

CHAPTER 1 / REVIEW EXERCISES

SECTION 1.1

Solve each problem. (*See Objectives 1–4.*)

1. Use the roster method to write the set: $A = \{x|x$ is a natural number between $\sqrt{5}$ and $\sqrt{38}\}$ $\{3, 4, 5, 6\}$
2. Use the roster method to write the set: $B = \{x|x$ is an odd integer between -5 and $7\}$ $\{-3, -1, 1, 3, 5\}$
3. Let $A = \{x|x$ is an even integer less than $11\}$ and $B = \{x|x$ is a whole number greater $2\}$. Find $A \cup B$ and $A \cap B$. $\{\ldots, -4, -2, 0, 2, 3, 4, 5, \ldots\}$; $\{4, 6, 8, 10\}$
4. Let $A = \{x|x$ is a real number greater than $1.9\}$ and $B = \{x|x$ is a positive integer less than $6.2\}$. Find $A \cup B$ and $A \cap B$.
$\{x|x$ is a real number greater than 1.9 and less than $6.2\}$; $\{2, 3, 4, 5, 6\}$

The following table shows information for the Top 10 Liberal Art Colleges. Use the information for Exercises 5–10. (Source: http://colleges.usnews.rankingsandreviews.com/bestcolleges/liberal-arts-rankings)

College	2009 Total Enrollment	2010–2011 Tuition and Fees
Williams College	2067	$41,434
Amherst College	1744	$40,862
Swarthmore College	1525	$39,600
Middlebury College	2482	$52,500
Wellesley College	2324	$39,666
Bowdoin College	1777	$41,565
Pomona College	1550	$38,394
Carleton College	2009	$41,304
Davidson College	1774	$36,683
Haverford College	1190	$40,624

Let $A = \{x|x$ is a college with tuition and fees more than $\$40{,}000\}$

$B = \{x|x$ is a college with enrollment less than $1550\}$

$C = \{x|x$ is a college with tuition and fees less than $\$40{,}000\}$

5. State the set A using the roster method.
6. State the set B using the roster method.
{Swarthmore College, Haverford College}
7. State the set $B \cap C$ using the roster method.
{Swarthmore College}
8. State the set $A \cap B$ using the roster method.
{Haverford College}
9. Is Haverford College $\in A \cap B$? Yes
10. Is Bowdoin College $\in B \cap C$? No
11. Classify the real number: $\sqrt{48}$. irrational
12. Classify the real number: $\sqrt{\pi}$. irrational
13. Find the opposite of -12.98. 12.98
14. Find the opposite of $\frac{4}{5}$. $-\frac{4}{5}$
15. Simplify the expression: $-\left|-\frac{20}{9}\right|$. $-\frac{20}{9}$
16. Simplify the expression: $-|45|$. -45
17. Graph on a real number line: $21\frac{1}{2}$.
18. Graph on a real number line: 8.25.

SECTION 1.2

Simplify each expression. (*See Objectives 1–4.*)

19. $-\frac{3}{4} + \frac{7}{8}$ $\frac{1}{8}$
20. $-\frac{9}{14} - \frac{5}{6}$ $-\frac{31}{21}$
21. $\left(-\frac{4}{5}\right)^2$ $\frac{16}{25}$
22. $\left(-\frac{11}{9}\right)^2$ $\frac{121}{81}$
23. $-\left(\frac{7}{3}\right)^3$ $-\frac{343}{27}$
24. $-\left(-\frac{5}{4}\right)^3$ $\frac{125}{64}$
25. $-\left(\frac{1}{3}\right)\left(\frac{6}{7}\right)^2 \div \frac{1}{28} + 4$ $-\frac{20}{7}$
26. $\left(-\frac{1}{2}\right)\left(\frac{1}{6}\right)^2 \div \frac{1}{54} + \frac{1}{4}$ $-\frac{1}{2}$
27. $\frac{18}{0}$ undefined
28. $\frac{0}{-14}$ 0
29. $\frac{-21 + 5}{2 - (-1)}$ $-\frac{16}{3}$
30. $\frac{-9 + 4}{11 - (-4)}$ $-\frac{1}{3}$
31. $3(-2.4)^2 - 1.5(3.2)$ 12.48
32. $-2(-4.5)^2 + 1.5(-6.4)$ -50.1
33. $\frac{6 + \sqrt{13 + 3}}{-5|2 - 1|^2}$ -2
34. $\frac{9 - \sqrt{27 - 2}}{-2|1 - 2|^2}$ -2

Evaluate each expression for the given values. (*See Objective 5.*)

35 $\frac{1}{2}bh$ for $b = 8$ and $h = \frac{3}{4}$ 3

36. $\frac{1}{2}bh$ for $b = \frac{3}{5}$ and $h = 20$ 6
37. $-\frac{1}{2}x^2 + 5x - 6$ for $x = -4$ -34
38. $-\frac{1}{4}x^3 + 6x^2 - 1$ for $x = -2$ 25
39. $\frac{-b + \sqrt{b^2 - 4ac}}{2a}$ for $a = 6$, $b = -7$, and $c = -5$ $\frac{5}{3}$
40. $\frac{-b - \sqrt{b^2 - 4ac}}{2a}$ for $a = 2$, $b = 17$, and $c = 35$ -5

Solve each problem. (*See Objective 6.*)

41. From 1995 to 2008, the number of cell phone subscribers (in millions) in the United States can be modeled by $0.0147x^4 - 0.3337x^3 + 2.9827x^2 + 5.9939x + 34.077$, where x is the number of years after 1995. According to this model, what was the number of cell phone subscribers (in millions) in 2000? (Source: http://www.infoplease.com) about 106.09 million

42. According to the model in Exercise 41, what was the number of cell phone subscribers (in millions) in 2007? about 263.70 million

43. What is the net worth of a family with \$2,250,000 in assets and \$125,000 in liabilities? \$2,125,000

44. What is the net worth of a family with \$76,500 in assets and \$234,500 in liabilities? –\$158,000

45. The volume of a cylinder is given by $\pi r^2 h$, where r is the radius and h is the height of the cylinder. Find the volume of a cylinder that has a radius of 7.2 in. and a height of 3.5 in. Round answer to the nearest hundredth. 570.01 in.^{3}

46. The volume of a cylinder is given by $\pi r^2 h$, where r is the radius and h is the height of the cylinder. Find the volume of a cylinder that has a radius of 2.4 in. and a height of 10.5 in. Round answer to the nearest hundredth. 190 in.^{3}

SECTION 1.3

Find the additive and multiplicative inverse of each expression. Assume that all denominators are nonzero. (*See Objective 1.*)

47. $-\frac{23}{12}$ $\frac{23}{12}, -\frac{12}{23}$

48. -17 $17, -\frac{1}{17}$

49. -12 $12, -\frac{1}{12}$

50. 25 $-25, \frac{1}{25}$

51. $\frac{5}{x+3}$ $-\frac{5}{x+3}, \frac{x+3}{5}$

52. $\frac{x-5}{2}$ $-\frac{x-5}{2}, \frac{2}{x-5}$

Identify the property that was used to write each equation. (*See Objective 2.*)

53. $(4a - 19) + 45b = 4a + (-19 + 45b)$
associative property of addition

54. $(5x - 8) + 12y = 5x + (-8 + 12y)$
associative property of addition

55. $[3(v - u)] \cdot 12 = 3[(v - u) \cdot 12]$
associative property of multiplication

56. $[23(g + h)] \cdot 47 = 23[(g + h) \cdot 47]$
associative property of multiplication

57. $(a + b)(c + d) = (a + b)c + (a + b)d$
distributive property

58. $(a + b)(c + d) = (c + d)(a + b)$
commutative property of multiplication

Simplify each expression. (*See Objectives 2 and 3.*)

59. $6m(m - 2) + 3(m^2 - 7)$ $9m^2 - 12m - 21$

60. $14b(3b + 2) - 9(b^2 + 3)$ $33b^2 + 28b - 27$

61. $\frac{2}{3}x - \frac{1}{6} - \frac{5}{6}x + \frac{7}{4}$ $\frac{19}{12} - \frac{x}{6}$

62. $\frac{2}{5}x - \frac{7}{3} - \frac{3}{10}x + \frac{1}{2}$ $\frac{x}{10} - \frac{11}{6}$

63. $a^3 - 15a^2 + 2a + 8a^2 - 5a + 11$ $a^3 - 7a^2 - 3a + 11$

64. $4b^3 - 10b^2 - 8b + 13 - 9b^2 + 11b - 23$ $4b^3 - 19b^2 + 3b - 10$

65. $-6\left(\frac{1 - 2x}{3}\right) + 8\left(\frac{3x - 5}{4}\right)$ $10x - 12$

66. $4\left(\frac{2x - 7}{2}\right) - 12\left(\frac{1 - x}{3}\right)$ $8x - 18$

67. $\frac{1}{4}(12x - 5) + \frac{3}{5}[10(x - 1) + 12]$ $9x - \frac{1}{20}$

68. $\frac{3}{5}(15x - 8) - \frac{1}{3}[6(3 - 2x)] - 6$ $13x - \frac{84}{5}$

69. $0.024m + 0.036(4300 - m)$ $154.8 - 0.012m$

70. $0.049x + 0.018(5600 - x)$ $0.031x + 100.8$

Translate each statement into an algebraic equation, using x to represent the unknown. (*See Objective 4.*)

71. Half the sum of a number and 6 equals the difference between the number and 20. $\frac{1}{2}(x + 6) = x - 20$

72. Double the difference of a number and 4 equals the sum of half the number and 19. $2(x - 4) = \frac{1}{2}x + 19$

Write an algebraic expression that represents each situation. (*See Objective 5.*)

73. Wesley has collected 120 quarters and dimes. If x represents the number of dimes he collected, write an expression for the number of quarters he collected. $120 - x$

74. Larry has collected 210 nickels and pennies. If x represents the number of nickels he collected, write an expression for the number of pennies he collected. $210 - x$

75. Ming is 10 years older than Fung. If x represents Fung's age, write an expression for Ming's age. $x + 10$

76. Owen is 5 years younger than CJ. If x represents CJ's age, write an expression for Owen's age. $x - 5$

77. The price of a movie ticket for all shows before 6 P.M. is \$2.50 less than the price of a movie ticket after 6 P.M. If x represents the price of a movie ticket after 6 P.M., write an expression for the price of a movie ticket before 6 P.M. $x - 2.5$

78. The motion picture and television industry provides tremendous economic growth to cities and towns across America. One day of on-site shooting brings about \$225,000 to the local economy. In 2010, the motion picture and television industry brought California \$0.6 billion more in wages than twice the amount it brought New York. If x represents the amount in wages in billions of dollars brought to New York, write an expression for the amount in wages brought to California. (Source: http://www.mpaa.org/policy/state-by-state) $2x + 0.6$

CHAPTER 1 TEST / REAL NUMBERS AND ALGEBRAIC EXPRESSIONS

1. Let $A = \{2, 4, 8, 16, 32, 64\}$ and $B = \{4, 16, 64, 256\}$. The number that is not an element of $A \cap B$ is b.

a. 4 **b.** 8
c. 16 **d.** 64

2. The number that is not a natural number is d.

a. $(-2)^2$ **b.** $\sqrt{25}$
c. $\frac{9-3}{4-2}$ **d.** $3 - 4(2)$

3. Provide an example of a rational number that is also an integer.
Answers vary but some examples are $\frac{-3}{1} = -3$ and $\frac{16}{2} = 8$.

4. Simplify each number and then classify it as a natural number, whole number, integer, rational, or irrational number. Then state the opposite and absolute value of each number.

a. $\sqrt{6^2 + 8^2}$ **b.** $\frac{4\pi}{2}$
c. $\frac{4 + 5(-2)}{3^2}$ **d.** $\frac{0}{-4}$

5. The following table shows the top 10 professional golfers for 2009. (Source: http://sports.espn.go.com/golf/statistics?tour=pga)

Rank	Player	Age	Earnings
1	Tiger Woods	34	$10,508,163.00
2	Steve Stricker	43	$ 6,332,636.00
3	Phil Mickelson	39	$ 5,332,754.50
4	Zach Johnson	34	$ 4,714,812.50
5	Kenny Perry	49	$ 4,445,562.00
6	Sean O'Hair	27	$ 4,316,493.00
7	Jim Furyk	39	$ 3,946,515.00
8	Geoff Ogilvy	32	$ 3,866,270.00
9	Lucas Glover	30	$ 3,693,353.30
10	Y.E. Yang	38	$ 3,489,515.80

a. Let $A = \{x|x$ is a player younger than $35\}$. Write set A using the roster method. $A = \{$Woods, Johnson, O'Hair, Ogilvy, Glover$\}$

b. Let $B = \{x|x$ is a player earning \$4 million or more$\}$. Write set B using the roster method. $B = \{$Woods, Stricker, Mickelson, Johnson, Perry, O'Hair$\}$

c. Find $A \cap B$. What conditions do these players satisfy? $A \cap B = \{$Woods, Johnson, O'Hair$\}$; These are players who are younger than 35 that earned more than \$4 million.

6. Simplify each expression.

a. $-\left(-\frac{7}{2}\right)$ $\frac{7}{2}$ **b.** $-(2 - 5)$ 3
c. $|-64|$ 64 **d.** $-|-9|$ -9
e. $(-3)^4$ 81 **f.** -3^4 -81

7. Perform the indicated operation. Write answers in lowest terms, if appropriate.

a. $\frac{12}{25} \cdot -\frac{5}{3}$ $-\frac{4}{5}$ **b.** $32 \div \frac{1}{4}$ 128
c. $-\frac{5}{9} + \left(-\frac{4}{9}\right)$ -1 **d.** $\frac{1}{4} - \frac{2}{7}$ $-\frac{1}{28}$
e. $\frac{7}{15} + \left(-\frac{1}{2}\right)\left(\frac{2}{5}\right) \div 3$ $\frac{2}{5}$
f. $(-4) + (-5)(2)$ -14
g. $(9 - 12) - (7 - 9)$ -1
h. $(-2)^3 - 4 \div 2 \cdot 6$ -20
i. $\sqrt{[1 - (-4)]^2 + (-4 - 8)^2}$ 13
j. $\frac{6 - 3(-2)}{4 - 2^2}$ undefined
k. $[1 + 4(5 - 6)] - 12 \div 2 + 5 \cdot 2$ 1

8. Identify the property that is illustrated.

a. $5(6 + 2) = (6 + 2)5$ commutative property of multiplication
b. $4 \cdot \frac{1}{4} = 1$ inverse property of multiplication
c. $3 + (4 + 7) = (3 + 4) + 7$ associative property of addition
d. $\frac{1}{5} + \left(-\frac{1}{5}\right) = 0$ inverse property of addition
e. $3 \cdot \frac{6}{7} \cdot \frac{7}{6} = 3$ identity property of multiplication
f. $-9 + 0 = -9$ identity property of addition
g. $4 + (-3) = (-3) + 4$ commutative property of addition
h. $3(x + 2) = 3x + 6$ distributive property

9. Evaluate the expression $4x^2 - 2x + 1$ for $x = -3$. 43

10. Simplify each expression.

a. $4(3x + 7)$ $12x + 28$
b. $8(6x - 5)$ $48x - 40$
c. $-6\left(\frac{1}{3}x + \frac{1}{2}\right)$ $-2x - 3$
d. $x + 2 + 5x - 4$ $6x - 2$
e. $2(5x - 3) - (7x - 9)$ $3x + 3$
f. $2x^2 - 8x + 1 + x^2 - x - 3$ $3x^2 - 9x - 2$
g. $x + \frac{3}{2}x$ $\frac{5}{2}x$
h. $5 + 4(3x + 2)$ $12x + 13$

11. Translate each phrase or sentence into an algebraic expression, equation, or inequality. Use the variable x to represent the unknown number.

a. Two more than a number $x + 2$
b. The difference of a number and 5 $x - 5$
c. The sum of three times a number and 4 $3x + 4$

d. Twice the sum of 7 less than a number $2(x - 7)$

e. The ratio of four times a number and 3 $\frac{4x}{3}$

f. Six less than one-fourth of a number $\frac{1}{4}x - 6$

g. One less than four times a number is 9. $4x - 1 = 9$

h. A number plus 1 is less than the product of 6 and the number. $x + 1 < 6x$

12. Write an expression that represents the unknown quantity and/or write an equation that represents the situation.

a. On average, a person with a master's degree earns \$31,900 more per year than a high school graduate. If x represents the average yearly income of a high school graduate, write an expression that represents the average yearly income of a person with a master's degree. (Source: http://www.earnmydegree.com/onlineeducation/learning-center/education-value.html) $x + 31{,}900$

b. Nikki invests \$10,000 in two different accounts. If x represents the amount she invested in one account, write an expression that represents the amount invested in the second account. $10{,}000 - x$

c. Dexter has a collection of nickels and dimes. He has 6 more dimes than nickels.

i. If x represents the number of nickels, write an expression for the number of dimes he has collected. $x + 6$

ii. Write expressions for the value of the nickels and the value of dimes in dollars.
nickels: $0.05x$; dimes: $0.10(x + 6)$

iii. If Dexter's collection is worth \$10.20, write an equation that represents the total value of the collection. $0.05x + 0.10(x + 6) = 10.20$

13. Solve each problem.

a. There are 20 countries in the world that have either negative or zero natural population growth. A negative or zero natural population growth means that a country has more deaths than births or an even number of deaths and births. As of March 2011, Japan is the only non-European country on the list. It has a 0% natural birth increase and is expected to lose 21% of its population by 2050 (shrinking from 127.8 million to 100.6 million in 2050). What is the difference in the current population and the population in 2050? (Source: http://geography.about.com/od/populationgeography/a/zero.htm) 27.2 million

b. If \$1,000,000 is invested in a trust fund account that earns 4.5% annual interest for 17 yr, the amount in the account at the end of the 17 yr is given by $1{,}000{,}000(1.045)^{17}$. Simplify this expression to determine the amount of money that would be in the account after 17 yr. Round to the nearest hundredth. \$2,113,376.81

c. Based on a Federal Reserve Statistical Release, the amount of consumer credit in billions of dollars x years after 2005 can be modeled by $-16.14x^3 + 60.32x^2 + 56.25x + 2290$. Use this model to determine the amount of consumer credit in 2012. (Source: http://www.federalreserve.gov/releases/g19/Current/) about \$103.41 billion

Linear Equations and Inequalities in One Variable

CHAPTER

2

Organization

Organization is critical for your success in college. Good organizational skills will assist you in

- meeting assignment deadlines.
- memory skills.
- test preparation and test taking.
- keeping you focused on your tasks.
- note taking.
- writing.

Organization is needed for your personal study space and your course information. Your personal study space should be organized so that it is

- free from clutter.
- easy to access class materials.
- free from distractions.

Organize your course information with a

- binder with dividers for notes, exams, homework, handouts, and course information.
- course calendar at the front of each binder to quickly determine assignments and important dates.

There are many websites that contain helpful information on how to get organized. Take a few moments to review these if you need assistance.

Question For Thought: Do you consider yourself organized? If not, what is one area that could use improvement?

Chapter Outline

Section 2.1 Solving Linear Equations 52

Section 2.2 Introduction to Applications 65

Section 2.3 Formulas and Applications 72

Section 2.4 Linear Inequalities and Applications 85

Section 2.5 Compound Inequalities 101

Section 2.6 Absolute Value Equations 116

Section 2.7 Absolute Value Inequalities 125

Coming Up...

In Section 2.4, we will learn how to use linear inequalities to solve the following problem. Jamaal's final grade in his math class is based upon a weighted average. The weights are as follows: homework (10%), quizzes (15%), tests (50%), and final exam (25%). If Jamaal's homework average is 85, quiz average is 80, and test average is 75, what does he need on his final exam to have at least a 70 average in the class?

"By failing to prepare you are preparing to fail."

—Benjamin Franklin
(Inventor, Scientist, Politician)

SECTION 2.1 Solving Linear Equations

In **Chapter 1,** we learned how to evaluate algebraic expressions and to simplify algebraic expressions. We will now turn our attention to a fundamental skill of algebra, solving equations. The first few sections of this chapter will focus on linear equations. We will learn how to solve equations and inequalities using the addition and multiplication properties. Real-life applications of these equations and inequalities, along with absolute value equations and inequalities, will also be examined.

OBJECTIVES

As a result of completing this section, you will be able to

1. **Define a linear equation.**
2. **Use the addition property of equality.**
3. **Use the multiplication property of equality.**
4. **Solve linear equations.**
5. **Identify linear equations with no solution.**
6. **Identify linear equations with infinitely many solutions.**
7. **Troubleshoot common errors.**

In Section 2.3, we will determine how much money needs to be invested at 5% annual interest to have \$4000 in an account after 3 yr. To find this amount of money requires us to solve the equation, $4000 = P(1 + 0.05)^3$. This is an example of a *linear equation.* This section will provide the skills we need to solve such an equation.

Linear Equations

Objective 1 Define a linear equation.

Now that we know how to work with algebraic expressions, we will learn how to solve algebraic equations. Specifically, our focus will be on linear equations. Some examples of linear equations are shown.

$$x + 5 = 9 \qquad 2m - 3 = 4(m + 2) \qquad \frac{3}{2}y - 6 = \frac{2}{3}y + 1$$

Note that in each of these equations, the variable has an exponent of 1. Recall that $x = x^1$. We can define a linear equation as follows.

Definition: A **linear equation in one variable** is an equation that can be written in the form $ax + b = c$, where a, b, and c are real numbers and $a \neq 0$. The exponent of the variable in a linear equation is 1. Linear equations are also called **first-degree equations**.

Some examples of equations that are not linear are $x^2 - x = 6$ and $4a^3 + 1 = 9$. Note that the largest exponent of the variable in these equations is greater than 1, which contradicts the definition of a linear equation.

Our ultimate goal is to solve a linear equation. A **solution of a linear equation in one variable** is the value of the variable that satisfies the equation. That is, when the variable is replaced by this value, the resulting statement is true. The **solution set** of an equation is the set of all numbers that make the equation true. For example, if 3 is a solution of a linear equation, then the solution set is written as $\{3\}$.

Once we obtain a proposed solution of an equation, it is important to verify that this value is, in fact, the solution of the equation. For example, we know that 10 is a solution of the linear equation, $5x - 5 = 45$, since this value makes the equation true as shown.

$$\begin{aligned} 5x - 5 &= 45 \\ 5(10) - 5 &= 45 && \text{Replace } x \text{ with 10.} \\ 50 - 5 &= 45 && \text{Simplify the left side.} \\ 45 &= 45 && \text{True} \end{aligned}$$

However, -10 is not a solution of the linear equation $5x - 5 = 45$ since the resulting equation is not true as shown.

$$5x - 5 = 45$$
$$5(-10) - 5 = 45 \quad \text{Replace } x \text{ with } -10.$$
$$-50 - 5 = 45 \quad \text{Simplify the left side.}$$
$$-55 = 45 \quad \text{False}$$

Procedure: Determining if a Value Is a Solution of an Equation

Step 1: Replace the variable with the given value.
Step 2: Simplify each side of the resulting equation.
Step 3: If the simplified equation is true, the value is a solution. Otherwise, it is not a solution.

Objective 1 Examples

Determine if the equation is a linear equation. If the equation is linear, determine if –3 is a solution of the equation.

1a. $4x - 10 = 2$ **1b.** $\frac{3x + 7}{2} = -1$ **1c.** $2x^2 + 3x = 11$

Solutions

1a. The equation $4x - 10 = 2$ is a linear equation because the exponent of the variable is 1. To determine if -3 is a solution, we replace the variable with -3.

$$4x - 10 = 2 \quad \text{Begin with the equation.}$$
$$4(-3) - 10 = 2 \quad \text{Replace } x \text{ with } -3.$$
$$-12 - 10 = 2 \quad \text{Multiply.}$$
$$-22 = 2 \quad \text{Simplify.}$$

Since -3 makes the equation *false*, it is *not* a solution of $4x - 10 = 2$.

1b. The equation $\frac{3x + 7}{2} = -1$ is equivalent to $\frac{3}{2}x + \frac{7}{2} = -1$. It is linear because the exponent of the variable is 1. To determine if -3 is a solution, we replace the variable with -3.

$$\frac{3x + 7}{2} = -1 \quad \text{Begin with the equation.}$$
$$\frac{3(-3) + 7}{2} = -1 \quad \text{Replace } x \text{ with } -3.$$
$$\frac{-9 + 7}{2} = -1 \quad \text{Multiply.}$$
$$\frac{-2}{2} = -1 \quad \text{Add.}$$
$$-1 = -1 \quad \text{Simplify.}$$

Since -3 makes the equation *true*, it is a solution of $\frac{3x + 7}{2} = -1$.

1c. The equation is not linear because the largest exponent on the variable is 2.

Student Check 1 Determine if the equation is a linear equation. If the equation is linear, determine if $-\frac{1}{2}$ is a solution of the equation.

a. $8x - 3 = -7$ **b.** $6(x + 1) + 5 = -2$ **c.** $\frac{4x^2 - 2x}{2} = 5$

The Addition Property of Equality

Objective 2 ▶

Use the addition property of equality.

Now that we know what a linear equation is and how to determine if a value is a solution of a linear equation, we will learn the process for solving a linear equation. When we solve a linear equation in one variable, our goal is to find the solution set of the equation by isolating the variable to one side of the equation. We do this by producing a series of *equivalent equations* until we reach an equation of the form

$$x = \textit{some number} \quad \text{or} \quad \textit{some number} = x$$

Equivalent equations are equations with the same solution set.

Since two sides of an equation are equal, whatever we do to one side of an equation must be done to the other side to maintain this equality. For instance, we can add the same number to each side of an equation or subtract the same number from each side of an equation. Consider the equation $x - 3 = 2$. When we add 3 to each side of the equation, we get

$$x - 3 = 2$$
$$x - 3 + 3 = 2 + 3$$
$$x = 5$$

Adding 3 to each side of the equation $x - 3 = 2$ produces an equivalent equation of the form $x = 5$, which provides the solution of the equation. This operation is an example of the addition property of equality.

Property: Addition Property of Equality

For real numbers a, b, and c,

$a = b$ is equivalent to $a + c = b + c$.

INSTRUCTOR NOTE:
Use the concept of a balance scale to explain this idea to students.

When the same value is added to two equal quantities (or each side of an equation), the two resulting expressions are also equal. Because we define subtraction in terms of addition, this property also guarantees that when we *subtract* the same number from each side of an equation, an equivalent equation is obtained.

Consider the equation $x + 5 = -1$. We can either add -5 to each side of the equation or subtract 5 from each side of the equation to obtain an equivalent equation in which x is isolated on one side.

$x + 5 = -1$	$x + 5 = -1$
$x + 5 + (-5) = -1 + (-5)$	$x + 5 - 5 = -1 - 5$
$x = -6$	$x = -6$

Note that in either case, the same solution is obtained.

Procedure: Using the Addition Property of Equality to Solve an Equation

Step 1: Determine the operation that will isolate the variable on one side of the equation. Perform this operation on each side of the equation. Remember the inverse property for addition: $a + (-a) = -a + a = a - a = 0$.
Step 2: Simplify each side of the equation, as necessary. The result should be of the form $x =$ some number or some number $= x$.
Step 3: Check the solution by substituting the value into the original equation.
Step 4: Write the solution in set notation.

Objective 2 Examples **Use the addition property of equality to solve each linear equation and check the solution.**

2a. $12.5 = d - 2.5$ **2b.** $12 + 2g = 3g + 15.4$

Solutions **2a.**

$$12.5 = d - 2.5$$

$$12.5 + 2.5 = d - 2.5 + 2.5 \quad \text{Add 2.5 to each side.}$$

$$15 = d \quad \text{Simplify.}$$

Check:

$$12.5 = d - 2.5 \quad \text{Begin with the original equation.}$$

$$12.5 = 15 - 2.5 \quad \text{Replace } d \text{ with 15.}$$

$$12.5 = 12.5 \quad \text{Simplify.}$$

The value 15 makes the equation true, so the solution set is $\{15\}$.

2b. We must isolate the variable terms on one side of the equation and the constant terms on the other side.

$$12 + 2g = 3g + 15.4$$

$$12 + 2g - 2g = 3g - 2g + 15.4 \quad \text{Subtract } 2g \text{ from each side.}$$

$$12 = g + 15.4 \quad \text{Simplify.}$$

$$12 - 15.4 = g + 15.4 - 15.4 \quad \text{Subtract 15.4 from each side.}$$

$$-3.4 = g \quad \text{Simplify.}$$

Check:

$$12 + 2g = 3g + 15.4 \quad \text{Begin with the original equation.}$$

$$12 + 2(-3.4) = 3(-3.4) + 15.4 \quad \text{Replace } g \text{ with } -3.4.$$

$$12 - 6.8 = -10.2 + 15.4 \quad \text{Simplify each side.}$$

$$5.2 = 5.2 \quad \text{Simplify.}$$

The value -3.4 makes the equation true, so the solution set is $\{-3.4\}$.

Student Check 2 Use the addition property of equality to solve each linear equation and check the solution.

a. $4.3 = f + 9.1$ **b.** $2 + 6y = -10 + 5y$

The Multiplication Property of Equality

Objective 3 Use the multiplication property of equality.

The addition property of equality can be used to eliminate a value that is added to or subtracted from a variable term. To eliminate a value that multiplies or divides a variable term, we must use the multiplication property of equality.

Property: Multiplication Property of Equality
For a, b, and $c\,(c \neq 0)$ real numbers,

$$a = b \text{ is equivalent to } ac = bc$$

So, when two equal quantities are multiplied by the same nonzero number, the resulting expressions are equal. Because division is defined in terms of multiplication, we can also divide each side of an equation by the same nonzero number and obtain an equivalent equation.

Objective 3 Examples **Use the multiplication property of equality to solve each linear equation and check the answer.**

3a. $\frac{x}{12} = -4$ **3b.** $8 - 7a = -6$

Solutions **3a.** Since the variable x is divided by 12, we can multiply each side by 12 to isolate the variable.

$$\frac{x}{12} = -4$$

$$12\left(\frac{x}{12}\right) = 12(-4) \quad \text{Multiply each side by 12.}$$

$$x = -48 \quad \text{Simplify.}$$

Check:

$$\frac{x}{12} = -4 \quad \text{Begin with the original equation.}$$

$$\frac{-48}{12} = -4 \quad \text{Replace } x \text{ with } -48.$$

$$-4 = -4 \quad \text{Simplify.}$$

The value -48 makes the equation true, so the solution set is $\{-48\}$.

3b. We first isolate the variable term on one side of the equation.

$$8 - 7a = -6$$

$$8 - 7a - 8 = -6 - 8 \quad \text{Subtract 8 from each side.}$$

$$-7a = -14 \quad \text{Simplify.}$$

$$\frac{-7a}{-7} = \frac{-14}{-7} \quad \text{Divide each side by } -7.$$

$$a = 2 \quad \text{Simplify.}$$

Check:

$$8 - 7a = -6 \quad \text{Begin with the original equation.}$$

$$8 - 7(2) = -6 \quad \text{Replace } a \text{ with 2.}$$

$$8 - 14 = -6 \quad \text{Simplify.}$$

$$-6 = -6 \quad \text{Simplify.}$$

The value 2 makes the equation true, so the solution set is $\{2\}$.

Student Check 3 Use the multiplication property of equality to solve each linear equation and check the solution.

a. $\frac{y}{5} = -2$ **b.** $-2 + 3y = 7$

Solving Linear Equations

Objective 4 Solve linear equations.

Most equations we encounter cannot be solved by using only the addition property of equality or only the multiplication property of equality. Most will require us to apply both properties. The following strategy provides the steps for solving any linear equation.

Procedure: General Strategy for Solving Linear Equations

Step 1: Write an equivalent equation that doesn't contain fractions or decimals, if necessary.
- **a.** Clear fractions from the equation by multiplying each side by the LCD of the fractions in the equation.
- **b.** Clear decimals from the equation by multiplying by an appropriate power of 10.

Step 2: Clear any parentheses by applying the distributive property.
Step 3: Apply the addition property of equality to isolate variable terms on one side of the equation and constant terms on the other side.
Step 4: Apply the multiplication property of equality to obtain a coefficient of 1 on the variable.
Step 5: Check by substituting the proposed solution in the original equation.

Objective 4 Examples

Solve each linear equation.

4a. $4a - 9 + 5a = 2a - 7 + 3$ **4b.** $-2(3x + 4) = 6 - 4(x - 2)$

4c. $\frac{2m}{3} + \frac{3}{2} = 2m$ **4d.** $0.05(x - 2) + 0.10x = 3.65$

Solutions **4a.**

$4a - 9 + 5a = 2a - 7 + 3$ Combine like terms on each side of the equation.

$9a - 9 = 2a - 4$

$9a - 9 - 2a = 2a - 4 - 2a$ Subtract $2a$ from each side.

Simplify.

$7a - 9 = -4$

Add 9 to each side.

$7a - 9 + 9 = -4 + 9$

Simplify.

$7a = 5$

Divide each side by 7.

$\frac{7a}{7} = \frac{5}{7}$

Simplify.

$a = \frac{5}{7}$

Check:

$4a - 9 + 5a = 2a - 7 + 3$ Begin with the original equation.

$4\left(\frac{5}{7}\right) - 9 + 5\left(\frac{5}{7}\right) = 2\left(\frac{5}{7}\right) - 7 + 3$ Replace a with $\frac{5}{7}$.

$\frac{20}{7} - 9 + \frac{25}{7} = \frac{10}{7} - 4$ Simplify each product.

$\frac{20}{7} - \frac{63}{7} + \frac{25}{7} = \frac{10}{7} - \frac{28}{7}$ Convert each number to a fraction with an LCD of 7.

$-\frac{18}{7} = -\frac{18}{7}$ Simplify each side.

The value $\frac{5}{7}$ makes the equation true, so the solution set is $\left\{\frac{5}{7}\right\}$.

4b.

$-2(3x+4) = 6 - 4(x-2)$	
$-6x - 8 = 6 - 4x + 8$	Apply the distributive property.
$-6x - 8 = -4x + 14$	Combine like terms.
$-6x - 8 + 4x = -4x + 14 + 4x$	Add $4x$ to each side.
$-2x - 8 = 14$	Simplify.
$-2x - 8 + 8 = 14 + 8$	Add 8 to each side.
$-2x = 22$	Simplify.
$\dfrac{-2x}{-2} = \dfrac{22}{-2}$	Divide each side by -2.
$x = -11$	Simplify.

The value -11 makes the equation true, so the solution set is $\{-11\}$. The check is left for the reader.

INSTRUCTOR NOTE: Remind students that clearing fractions is not required. It just makes the terms in the equation easier to manipulate.

4c.

$\dfrac{2m}{3} + \dfrac{3}{2} = 2m$	
$6\left(\dfrac{2m}{3} + \dfrac{3}{2}\right) = 6(2m)$	Multiply each side by the LCD, 6.
$6\left(\dfrac{2m}{3}\right) + 6\left(\dfrac{3}{2}\right) = 12m$	Apply the distributive property on the left and simplify the right side.
$4m + 9 = 12m$	Simplify the left side.
$4m + 9 - 4m = 12m - 4m$	Subtract $4m$ from each side.
$9 = 8m$	Simplify.
$\dfrac{9}{8} = \dfrac{8m}{8}$	Divide each side by 8.
$\dfrac{9}{8} = m$	Simplify.

Check:

$\dfrac{2m}{3} + \dfrac{3}{2} = 2m$	Begin with the original equation.
$\dfrac{2\left(\frac{9}{8}\right)}{3} + \dfrac{3}{2} = 2\left(\dfrac{9}{8}\right)$	Replace m with $\dfrac{9}{8}$.
$\dfrac{\frac{9}{4}}{3} + \dfrac{3}{2} = \dfrac{9}{4}$	Simplify each product.
$\dfrac{9}{4} \cdot \dfrac{1}{3} + \dfrac{3}{2} = \dfrac{9}{4}$	Multiply $\dfrac{9}{4}$ by the reciprocal of 3, $\dfrac{1}{3}$.
$\dfrac{3}{4} + \dfrac{6}{4} = \dfrac{9}{4}$	Simplify the product and write $\dfrac{3}{2}$ as $\dfrac{6}{4}$.
$\dfrac{9}{4} = \dfrac{9}{4}$	Simplify.

The value $\dfrac{9}{8}$ makes the equation true, so the solution set is $\left\{\dfrac{9}{8}\right\}$.

INSTRUCTOR NOTE: Illustrate how multiplying by powers of 10 eliminates the decimal point from a number.

4d.

$$0.05(x-2)+0.10x=3.65$$

$$100[0.05(x-2)+0.10x]=100(3.65) \quad \text{Multiply each side by 100.}$$

$$100[0.05(x-2)]+100(0.10x)=365 \quad \text{Apply the distributive property.}$$

$$5(x-2)+10x=365 \quad \text{Simplify.}$$

$$5x-10+10x=365 \quad \text{Apply the distributive property.}$$

$$15x-10=365 \quad \text{Combine like terms.}$$

$$15x-10+10=365+10 \quad \text{Add 10 to each side.}$$

$$15x=375 \quad \text{Simplify.}$$

$$\frac{15x}{15}=\frac{375}{15} \quad \text{Divide each side by 15.}$$

$$x=25 \quad \text{Simplify.}$$

INSTRUCTOR NOTE: Point out that clearing decimals is not required.

It is important to note that clearing decimals is not required in this example.

$$0.05(x-2)+0.10x=3.65$$

$$0.05x-0.1+0.10x=3.65 \quad \text{Apply the distributive property.}$$

$$0.15x-0.1=3.65 \quad \text{Combine like terms.}$$

$$0.15x-0.1+0.1=3.65+0.1 \quad \text{Add 0.1 to each side.}$$

$$0.15x=3.75 \quad \text{Simplify.}$$

$$\frac{0.15x}{0.15}=\frac{3.75}{0.15} \quad \text{Divide each side by 0.15.}$$

$$x=25 \quad \text{Simplify.}$$

The value 25 makes the equation true, so the solution set is $\{25\}$. The check is left for the reader.

Student Check 4 Solve each linear equation.

a. $9x-3+2x=7x-2-13$

b. $4(3-2x)=6-(3x-1)$

c. $\dfrac{3y}{4}-\dfrac{y}{3}=\dfrac{1}{12}$

d. $0.25(x+5)+0.05x=4.85$

Linear Equations with No Solution

Objective 5 Identify linear equations with no solution.

INSTRUCTOR NOTE: Explain that { } and {0} are not the same. The solution set {0} means zero is a solution. The solution set { } means there is *no* solution.

Thus far, all of the linear equations we have solved had one real number as a solution. These equations are examples of *conditional* equations. A **conditional equation** is an equation that is true for some values of the variable and not true for other values.

Not every linear equation, however, has a solution. A linear equation with no solution is called a *contradiction*. A **contradiction** is an equation that is *never* true. Some simple examples of contradictions are $3=5$, $-19=9$, and $0=1$.

If an equation is a contradiction, there is no value of the variable that produces a true statement when it is substituted into the equation. When an equation has no solution, we write the solution set as the empty set, $\varnothing$, or $\{\ \}$.

Although we may not be able to determine if a linear equation is a contradiction by examination, we can use the addition and multiplication properties of equality to go through the process of solving the equation. If a false equation results, then the linear equation is a contradiction and has no solution.

Objective 5 Example **Solve the linear equation $4x - 3x + 2 = 3 - (5 - x)$.**

Solution

$$4x - 3x + 2 = 3 - (5 - x)$$

$$4x - 3x + 2 = 3 - 5 + x \quad \text{Apply the distributive property.}$$

$$x + 2 = -2 + x \quad \text{Combine like terms on each side.}$$

$$x + 2 - x = -2 + x - x \quad \text{Subtract } x \text{ from each side.}$$

$$2 = -2 \quad \text{Simplify.}$$

The resulting equation $2 = -2$ is false; it is a contradiction which means this equation has no solution. We write the solution set as $\varnothing$.

Student Check 5 Solve the linear equation $2x - 3 - x = 2 + x + 1$.

Linear Equations with Infinitely Many Solutions

Identify linear equations with infinitely many solutions.

Just as there are some linear equations with no solution, there are some linear equations that have all real numbers as solutions. This is a linear equation with infinitely many solutions, and it is called an *identity*. An **identity** is an equation that is *always* true. An identity occurs when both sides of the equation are equal for all values substituted in place of the variable. Some simple examples of identities are $5 = 5$, $-2 = -2$, $0 = 0$, and $x = x$.

If an equation is an identity, then every value of the variable produces a true statement when it is substituted into the equation. We write the solution set as $\{x \mid x \text{ is a real number}\}$ (read "the set of x such that x is a real number") or $\mathbb{R}$.

Again, we may not be able to determine if a linear equation is an identity by examination, but we can use the addition and multiplication properties of equality to solve the equation. If a true statement results, then the linear equation is an identity and has all real numbers as its solution set.

Objective 6 Example **Solve the linear equation $4(x + 3) + 5 = 17 + 4x$.**

Solution

$$4(x + 3) + 5 = 17 + 4x$$

$$4x + 12 + 5 = 17 + 4x \quad \text{Apply the distributive property.}$$

$$4x + 17 = 17 + 4x \quad \text{Combine like terms on the left side.}$$

$$4x + 17 - 4x = 17 + 4x - 4x \quad \text{Subtract } 4x \text{ from each side.}$$

$$17 = 17 \quad \text{Simplify.}$$

The equation $17 = 17$ is true; it is an identity, which means there are infinitely many solutions. We write the solution set as $\{x \mid x \text{ is a real number}\}$ or $\mathbb{R}$.

Student Check 6 Solve the linear equation $6(2 - x) + 4 - 3x = 12 - (9x - 4)$.

Troubleshooting Common Errors

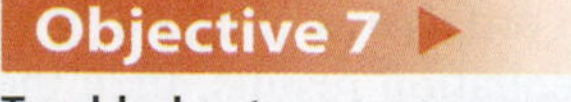

Troubleshoot common errors.

Some common errors associated with solving linear equations using the addition and multiplication properties are shown.

Objective 7 Examples

A problem and an incorrect solution are given. Provide the correct solution and an explanation of the error.

7a. Solve $-3m + 5 = -7$.

Incorrect Solution

$$\begin{aligned} -3m + 5 &= -7 \\ -3m + 5 - 5 &= -7 - 5 \\ -3m &= -12 \\ -3m + 3 &= -12 + 3 \\ m &= -9 \end{aligned}$$

The solution set is $\{-9\}$.

Correct Solution and Explanation

To make the coefficient of m 1, we must divide by -3, not add -3. Keep in mind that $-3m + 3$ can't be simplified but $\frac{-3m}{-3} = m$.

$$\begin{aligned} -3m + 5 &= -7 \\ -3m + 5 - 5 &= -7 - 5 \\ -3m &= -12 \\ \frac{-3m}{-3} &= \frac{-12}{-3} \\ m &= 4 \end{aligned}$$

The solution set is $\{-4\}$.

7b. Solve $\frac{x}{4} + \frac{5}{2} = 3$.

Incorrect Solution

$$\begin{aligned} \frac{x}{4} + \frac{5}{2} &= 3 \\ 8\left(\frac{x}{4} + \frac{5}{2}\right) &= 3 \\ 8\left(\frac{x}{4}\right) + 8\left(\frac{5}{2}\right) &= 3 \\ 2x + 20 &= 3 \\ 2x + 20 - 20 &= 3 - 20 \\ 2x &= -17 \\ x &= -\frac{17}{2} \end{aligned}$$

The solution set is $\left\{-\frac{17}{2}\right\}$.

Correct Solution and Explanation

The error was made in not multiplying each side by the LCD.

$$\begin{aligned} \frac{x}{4} + \frac{5}{2} &= 3 \\ 8\left(\frac{x}{4} + \frac{5}{2}\right) &= 8(3) \\ 8\left(\frac{x}{4}\right) + 8\left(\frac{5}{2}\right) &= 24 \\ 2x + 20 &= 24 \\ 2x + 20 - 20 &= 24 - 20 \\ 2x &= 4 \\ x &= 2 \end{aligned}$$

The solution set is $\{2\}$.

7c. Solve $4 - 3(2x - 4) = 4(x + 2) + 8$.

Incorrect Solution

$$\begin{aligned} 4 - 3(2x - 4) &= 4(x + 2) + 8 \\ 4 - 6x + 12 &= 4x + 8 + 8 \\ -6x + 16 &= 4x + 16 \\ 16 &= 10x + 16 \\ 0 &= 10x \end{aligned}$$

The solution set is $\varnothing$.

Correct Solution and Explanation

In the last step, we should divide each side by 10 to make the coefficient of x 1.

$$\begin{aligned} 4 - 3(2x - 4) &= 4(x + 2) + 8 \\ 4 - 6x + 12 &= 4x + 8 + 8 \\ -6x + 16 &= 4x + 16 \\ 16 &= 10x + 16 \\ 0 &= 10x \\ \frac{0}{10} &= \frac{10x}{10} \\ 0 &= x \end{aligned}$$

The solution set is $\{0\}$.

ANSWERS TO STUDENT CHECKS

Student Check 1 **a.** linear; yes **b.** linear; no **c.** not linear

Student Check 2 **a.** $\{-4.8\}$ **b.** $\{-12\}$

Student Check 3 **a.** $\{-10\}$ **b.** $\{3\}$

Student Check 4 **a.** $\{-3\}$ **b.** $\{1\}$ **c.** $\left\{\frac{1}{5}\right\}$ **d.** $\{12\}$

Student Check 5 $\varnothing$

Student Check 6 $\mathbb{R}$

SUMMARY OF KEY CONCEPTS

1. A linear equation is an equation in which the exponent of the variable is 1. The standard form of the equation is $ax + b = c$.
2. The addition property of equality enables us to add the same number to each side of an equation or subtract the same number from each side of an equation and obtain an equivalent equation.
3. The multiplication property of equality enables us to multiply or divide each side of an equation by a nonzero number and obtain an equivalent equation.
4. To solve a linear equation, we must isolate the variable on one side of the equation. Clear fractions, decimals, and any parentheses. Then isolate the variable on one side of the equation and the constant on the other by applying the addition property. Finally, divide each side by the coefficient of the variable by applying the multiplication property.
5. A linear equation with no solution is a contradiction. This type of equation results when each side of the equation has the same variable term but different constant terms. The variables are eliminated from the equation and the resulting statement is false.
6. A linear equation with infinitely many solutions is an identity. All real numbers satisfy the equation. This type of equation results when each side of the equation has the same variable term and the same constant term. The variables are eliminated from the equation and the resulting statement is true.

GRAPHING CALCULATOR SKILLS

The graphing calculator can be used to determine if a number is a solution of an equation.

Example: Determine if -2 is a solution of $-4x + 2 = 2x - 1$.

Method 1: Use the calculator to evaluate the left and right side of the equation.

```
-4(-2)+2
                10
2(-2)-1
                -5
```

Since the left side, 10, does not equal the right side, -5, the value -2 is *not* a solution of the equation.

Method 2: Enter the left side and right side of the equation in the equation editor and use the table to determine the value of the each side.

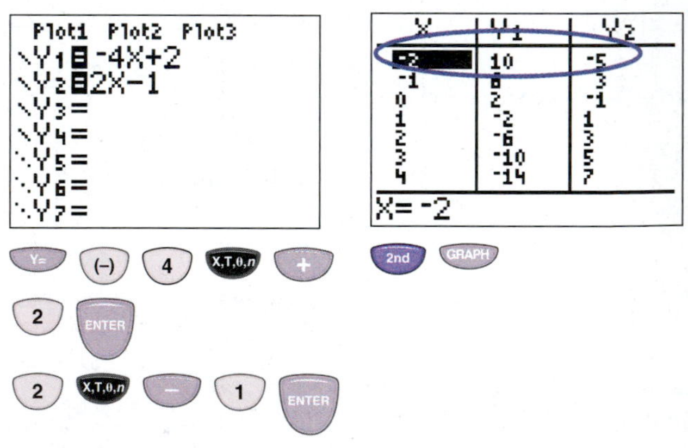

When $x = -2$, the values in the columns for Y_1 and Y_2 are not equal. This shows that -2 is not a solution.

SECTION 2.1 EXERCISE SET

Write About It!

Use complete sentences in your answer to each exercise.

1. What is a linear equation?
2. What does it mean for an equation to be conditional? Give an example.
3. What does it mean for an equation to be a contradiction? Give an example.
4. What does it mean for an equation to be an identity? Give an example.
5. Explain why the addition property of equality covers subtraction. Subtraction is defined in terms of addition.
6. Explain why the multiplication property of equality covers division. Division is defined in terms of multiplication.
7. How can you determine if a linear equation is a contradiction?
8. How can you determine if a linear equation is an identity?

Practice Makes Perfect!

Determine if each equation is a linear equation. If it is a linear equation, determine if -1 is a solution of the equation. (*See Objective 1.*)

9. $5x + 1 = -4$ linear; It is a solution.
10. $2x + 7 = 5$ linear; It is a solution.
11. $4x - 3 = 7$ linear; It is not a solution.
12. $6x - 5 = -1$ linear; It is not a solution.
13. $\dfrac{3x-2}{5} = -1$ linear; It is a solution.
14. $\dfrac{4x+3}{4} = 3$ linear; It is not a solution.
15. $2m^2 + 3 = 5$ not linear
16. $5m^2 - 6 = -1$ not linear
17. $2(x - 1) + 3 = 4$ linear; It is not a solution.
18. $5(x - 2) + 1 = 3$ linear; It is not a solution.
19. $1 - 3x = 2(x + 3)$ linear; It is a solution.
20. $5 - 4x = 3(x - 1)$ linear; It is not a solution.
21. $\dfrac{10x^2 + 5x - 1}{x} = 7$ not linear
22. $\dfrac{6x^2 - 10x + 4}{x} = 12$ not linear

Use the addition property of equality to solve each linear equation and check the answer. (*See Objective 2.*)

23. $x + \dfrac{1}{2} = 3$ $\left\{\dfrac{5}{2}\right\}$
24. $x + \dfrac{1}{3} = 1$ $\left\{\dfrac{2}{3}\right\}$
25. $y - 1 = \dfrac{1}{4}$ $\left\{\dfrac{5}{4}\right\}$
26. $y - 4 = \dfrac{2}{3}$ $\left\{\dfrac{14}{3}\right\}$
27. $\dfrac{3x+9}{3} = 2$ $\{-1\}$
28. $\dfrac{4x+8}{4} = 3$ $\{1\}$
29. $\dfrac{6x-12}{6} = 10$ $\{12\}$
30. $\dfrac{5x-15}{5} = 9$ $\{12\}$
31. $6.5 = a + 14.5$ $\{-8\}$
32. $12.3 = a + 10.3$ $\{2\}$
33. $14 = x - 23$ $\{37\}$
34. $10 = x - 11$ $\{21\}$
35. $16 - d = 13.5$ $\{2.5\}$
36. $19 - d = 12.2$ $\{6.8\}$
37. $5 - d = 6.5$ $\{-1.5\}$
38. $7 - d = 14.9$ $\{-7.9\}$

Use the multiplication property of equality to solve each linear equation and check the answer. (*See Objective 3.*)

39. $\dfrac{x}{3} = -2$ $\{-6\}$
40. $\dfrac{x}{6} = -5$ $\{-30\}$
41. $\dfrac{a}{11} = 6$ $\{66\}$
42. $\dfrac{a}{9} = 12$ $\{108\}$
43. $\dfrac{3x-8}{4} = 4$ $\{8\}$
44. $\dfrac{4x-9}{3} = 2$ $\left\{\dfrac{15}{4}\right\}$
45. $\dfrac{2y-7}{7} = 1$ $\{7\}$
46. $\dfrac{5y-8}{8} = 2$ $\left\{\dfrac{24}{5}\right\}$
47. $6x + 13 = 9$ $\left\{-\dfrac{2}{3}\right\}$
48. $4x - 9 = 11$ $\{5\}$
49. $2x - 6 = 7$ $\left\{\dfrac{13}{2}\right\}$
50. $3x - 10 = 5$ $\{5\}$
51. $1 - 2x = 9$ $\{-4\}$
52. $2 - 3x = 8$ $\{-2\}$
53. $10 - 4g = 2$ $\{2\}$
54. $9 - 5g = 4$ $\{1\}$

Solve each linear equation and check the answer. If there is no solution, then write $\varnothing$ for the answer. (*See Objectives 4 and 5.*)

55. $6x - 7x + 1 = 4 - x$ $\varnothing$
56. $2x - 3x - 5 = 5 - x$ $\varnothing$
57. $5x + 3 = 5(1 + x)$ $\varnothing$
58. $7x - 2 = 7(3 + x)$ $\varnothing$
59. $2(1 - 3a) = 4(a - 1)$ $\left\{\dfrac{3}{5}\right\}$
60. $3(4 - 2a) = 5(a + 3)$ $\left\{-\dfrac{3}{11}\right\}$
61. $9 - 3x = 3(x - 7)$ $\{5\}$
62. $12 - 4x = 4(x + 1)$ $\{1\}$
63. $\dfrac{4x-12}{2} = 3x + 6$ $\{-12\}$
64. $\dfrac{15x-20}{5} = 2x + 4$ $\{8\}$
65. $5(2y - 1) = 4 - (6y + 2)$ $\left\{\dfrac{7}{16}\right\}$
66. $6(3y - 4) = 7 - (10y - 5)$ $\left\{\dfrac{9}{7}\right\}$
67. $\dfrac{4m}{3} - \dfrac{1}{2} = \dfrac{5}{2}$ $\left\{\dfrac{9}{4}\right\}$
68. $\dfrac{5m}{2} - \dfrac{1}{3} = \dfrac{2}{3}$ $\left\{\dfrac{2}{5}\right\}$
69. $\dfrac{3m}{2} - \dfrac{1}{4} = -\dfrac{5}{4}$ $\left\{-\dfrac{2}{3}\right\}$
70. $\dfrac{3m}{5} - \dfrac{1}{5} = -\dfrac{2}{5}$ $\left\{-\dfrac{1}{3}\right\}$
71. $\dfrac{5x}{3} - \dfrac{3x}{2} = \dfrac{1}{6}$ $\{1\}$
72. $\dfrac{7x}{4} - \dfrac{x}{2} = \dfrac{5}{8}$ $\left\{\dfrac{1}{2}\right\}$

Additional answers can be found in the Instructor Answer Appendix.

73. $0.03(x - 6) + 0.02(x - 4) = 5.24$ $\{110\}$

74. $0.38x + 0.42(x + 7.2) = 15.6$ $\{15.72\}$

Solve each linear equation and check the answer. If there are infinitely many solutions, then write $\mathbb{R}$ for the answer. (*See Objectives 4 and 6.*)

75. $3(x + 1) + 10 = 13 + 3x$ $\mathbb{R}$

76. $4(x + 5) + 2 = 22 + 4x$ $\mathbb{R}$

77. $6(a - 2) - 5a = a - 12$ $\mathbb{R}$

78. $5(a - 3) - 4a = a - 15$ $\mathbb{R}$

79. $(3 - x) + 9 - 5x = 17 - (10x - 7)$ $\{3\}$

80. $5(6 - x) + 1 - 3x = 12 - 2(x - 8)$ $\left\{\frac{1}{2}\right\}$

81. $0.25(8y - 6) = 2(y - 0.75)$ $\mathbb{R}$

82. $0.24(5y - 15) = 0.2(6y - 18)$ $\mathbb{R}$

Mix 'Em Up!

Determine if the given value is a solution of the equation.

83. $3x - 2(x + 5) = x - 4;\ 3$ It is not a solution.

84. $5x - 3(x - 1) = x + 6;\ 5$ It is not a solution.

85. $4x + 3 - 8x = 3 - 4x;\ \frac{1}{2}$ It is a solution.

86. $6x - 5 - 10x = -5 - 4x;\ \frac{1}{2}$ It is a solution.

87. $\frac{4x - 5}{7} = 1;\ 3$ It is a solution.

88. $\frac{6x - 1}{10} = \frac{1}{2};\ 1$ It is a solution.

Solve each equation and check the answer.

89. $2x + 7 = 5$ $\{-1\}$

90. $5x + 15 = -10$ $\{-5\}$

91. $\frac{6x - 15}{3} = -2$ $\left\{\frac{3}{2}\right\}$

92. $\frac{12x - 15}{3} = -4$ $\left\{\frac{1}{4}\right\}$

93. $12 - 3y = 5$ $\left\{\frac{7}{3}\right\}$

94. $2 - 10y = 6$ $\left\{-\frac{2}{5}\right\}$

95. $6(x - 1) - 3(1 - x) = 9x - 7$ $\varnothing$

96. $2(x - 9) - 3(2 - x) = 5x + 18$ $\varnothing$

97. $\frac{n}{4} = -5$ $\{-20\}$

98. $\frac{t}{5} = -10$ $\{-50\}$

99. $6(2x + 10) - 3(x - 1) = 2x + 7$ $\{-8\}$

100. $4(3x - 5) - 2(x + 5) = 4x - 6$ $\{4\}$

101. $\frac{1 - 3x}{5} = 12$ $\left\{-\frac{59}{3}\right\}$

102. $\frac{2 - 4x}{3} = 7$ $\left\{-\frac{19}{4}\right\}$

103. $4x + 5(1 - 2x) = 5 - 6x$ $\mathbb{R}$

104. $6x + 3(2 - 4x) = 6 - 6x$ $\mathbb{R}$

105. $23 - d = 15.4$ $\{7.6\}$

106. $30 - d = 12.1$ $\{17.9\}$

107. $2.5 = x + 8.6$ $\{-6.1\}$

108. $4.5 = x - 11.7$ $\{16.2\}$

109. $m + \frac{2}{3} = 3$ $\left\{\frac{7}{3}\right\}$

110. $m - \frac{3}{4} = 1$ $\left\{\frac{7}{4}\right\}$

111. $\frac{3y}{2} + \frac{1}{3} = \frac{5}{2}$ $\left\{\frac{13}{9}\right\}$

112. $\frac{4y}{3} - \frac{1}{4} = \frac{5}{3}$ $\left\{\frac{23}{16}\right\}$

113. $\frac{6x - 1}{6} = x - 1$ $\varnothing$

114. $\frac{5x - 3}{5} = x - 3$ $\varnothing$

115. $12 - 4x = 3(7 - 2x) + 2x - 9$ $\mathbb{R}$

116. $1 - 3x = 5(6 - 3x) + 12x - 29$ $\mathbb{R}$

You Be the Teacher!

Correct each student's errors, if any.

117. Solve $\frac{4d}{3} - 4 = \frac{2}{5}$.

Brandi's work:

$$\frac{4d}{3} - 4 = \frac{2}{5}$$
$$15\left(\frac{4d}{3}\right) - 4 = 15\left(\frac{2}{5}\right)$$
$$20d - 4 = 6$$
$$20d = 10$$
$$d = \frac{1}{2}$$

118. Solve $\frac{5m}{3} - 1 = \frac{4}{7}$.

Naomi's work:

$$\frac{5m}{3} - 1 = \frac{4}{7}$$
$$35m - 1 = 12$$
$$35m = 13$$
$$m = \frac{13}{35}$$

119. Solve $0.75(2x - 4) + 0.3x = 7.8$.

Ruthann's work:

$$0.75(2x - 4) + 0.3x = 7.8$$
$$100(0.75(2x - 4)) + 100(0.3x) = 100(7.8)$$
$$75(200x - 400) + 30x = 78$$
$$15{,}000x - 30{,}000 + 30x = 78$$
$$15{,}030x = 30{,}078$$
$$x = \frac{30{,}078}{15{,}030}$$
$$x \approx 2$$

120. Solve $0.45(6 - 2x) - 0.2x = 1249$.

Benjamin's work:

$$0.45(6 - 2x) - 0.2x = 1249$$
$$100(0.45(6 - 2x) - 100(0.2x) = 1249$$
$$45(6 - 2x) - 20x = 1249$$
$$270 - 90x - 20x = 1249$$
$$-110x = 979$$
$$x = -\frac{979}{110}$$
$$x = -8.9$$

121. Solve $3x - 2(x + 4) = x + 5$.

David's work:

$$3x - 2(x + 4) = x + 5$$
$$3x - 2x - 8 = x + 5$$
$$x - 8 = x + 5$$
$$x = 13$$

122. Solve $8 + 5(x - 1) = 3(2x + 1) - x$.

Amber's work:

$$8 + 5(x - 1) = 3(2x + 1) - x$$
$$8 + 5x - 5 = 6x + 3 - x$$
$$5x + 3 = 5x + 3$$
$$0 = 0$$

So, the solution set is $\{0\}$.

Calculate It!

Use a graphing calculator to determine if -3, $\frac{1}{2}$, or 0 is a solution of the linear equation.

123. $6(1 - 3x) - 4x = 10(4 - x)$ None are solutions of the equation.

124. $3(20 - x) - 5x = 12(1 - 2x)$ -3 is the solution of the equation.

125. $\frac{10x - 9}{5} = -\frac{9}{5}$ 0 is the solution of the equation.

126. $\frac{9x - 8}{3} = -\frac{8}{3}$ 0 is the solution of the equation.

Think About It!

127. What value of c is required for -2 to be a solution of $4x + 3 = c$? -5

128. What value of c is required for $\frac{1}{2}$ to be a solution of $-6x + 9 = c$? 6

129. What value of b is required for -4 to be a solution of $3x + b = 10$? 22

130. What value of b is required for $\frac{2}{3}$ to be a solution of $-6x + b = 1$? 5

131. Complete the equation $3x - 2 =$ ______ so that its solution set is $\varnothing$. Answers vary; $3x - 2 = 3x$

132. Complete the equation $-2x + 6 =$ ______ so that its solution set is $\varnothing$. Answers vary; $-2x + 6 = 1 - 2x$

133. Complete the equation $8x + 3 =$ ______ so that its solution set is $\mathbb{R}$. Answers vary; $8x + 3 = 4(2x + 1) - 1$

134. Complete the equation $7x - 5 =$ ______ so that its solution set is $\mathbb{R}$. Answers vary; $7x - 5 = 3(x - 3) + 4(x + 1)$

SECTION 2.2 Introduction to Applications

OBJECTIVES

As a result of completing this section, you will be able to

1. Translate applications into linear equations and solve.
2. Troubleshoot common errors.

Iron Man 2 ranks tenth in the list of all-time opening weeks at the box office. It earned approximately \$36.1 million more than *Iron Man* in its opening week. Together, they earned \$282.3 million in their opening weeks. How much did *Iron Man* and *Iron Man 2* each earn in their opening week? (Source: http://boxofficemojo.com/)

To solve this problem, we need to know how to translate the given information into an equation. We will then use the process from Section 2.1 to solve the equation.

Objective 1

Translate applications into linear equations and solve.

Translating Statements into Linear Equations

In Chapter 1, we discussed how to translate phrases into mathematical expressions. We can now combine this information with the properties that we learned in Section 2.1 to solve applications that can be translated into linear equations.

Throughout this textbook, we will encounter applications that relate the mathematics that we are learning to real-life situations. Often the most difficult part of solving application problems is setting up the equation. So, some guidelines are provided.

Procedure: Setting up Equations for Application Problems

Step 1: Read the problem carefully and determine the unknown and assign a variable to it. Other unknowns in the problem should be represented in terms of the same variable initially chosen.

Step 2: Determine what is given. Use this information as well as your knowledge of translating phrases into mathematical expressions to write the equation.

Step 3: Solve the equation.

Step 4: Check the proposed solution and state the result.

One of the first applications we will solve deals with consecutive integers. Consecutive means "successive" or "following one after another without interruption." So, **consecutive integers** are integers that follow one another. Some examples are listed in the table.

	Specific Example	Variable Representation
Consecutive integers	4 (+1) 5 (+1) 6	$x, x + 1, x + 2, \ldots$
Consecutive even integers	10 (+2) 12 (+2) 14	$x, x + 2, x + 4, \ldots$
Consecutive odd integers	11 (+2) 13 (+2) 15	$x, x + 2, x + 4, \ldots$

Note: *Consecutive odd and consecutive even integers are represented in the same way since the difference between these types of integers is 2 units. The value of x determines the numbers that will be generated from the expressions* $x + 2,\ x + 4, \ldots.$

Objective 1 Examples

Translate each problem into a linear equation and solve the problem.

1a. The sum of two consecutive integers is −9. Find the numbers.

Solution

1a. What is unknown? Two consecutive integers are unknown.
Let x represent one integer. The next integer is $x + 1$.

What is given? The sum of the consecutive integers is 9.

First integer plus second integer is −9.

$$x + (x + 1) = -9 \quad \text{Express the relationship.}$$
$$2x + 1 = -9 \quad \text{Combine like terms.}$$
$$2x + 1 - 1 = -9 - 1 \quad \text{Subtract 1 from each side.}$$
$$2x = -10 \quad \text{Simplify.}$$
$$\frac{2x}{2} = \frac{-10}{2} \quad \text{Divide each side by 2.}$$
$$x = -5 \quad \text{Simplify.}$$

One integer is −5. The other integer is $x + 1 = -5 + 1 = -4$. So, the numbers −4 and −5 are consecutive integers whose sum is −9.

1b. When the quotient of a number and 2 is subtracted from the number, the result is 6. Find the number.

Solution

1b. What is unknown? A number is unknown. Let x represent the number.
What is given? The quotient of a number and 2, subtracted from the number, is 6.

A number less the quotient of the number and 2 is 6.

$$x - \frac{x}{2} = 6 \quad \text{Express the relationship.}$$
$$2\left(x - \frac{x}{2}\right) = 2(6) \quad \text{Multiply each side by 2.}$$
$$2x - x = 12 \quad \text{Apply the distributive property and simplify.}$$
$$x = 12 \quad \text{Combine like terms.}$$

Since 12 checks in the equation, the number is 12.

1c. Aidan saves nickels and dimes in a jar. He has 20 more nickels than dimes and his collection is worth \$8.50. Find the number of nickels and dimes in his collection.

Solution **1c.** What is unknown? The number of nickels and dimes is unknown. Let d represent the number of dimes. Then $\underbrace{d + 20}$ represents the number of nickels.

20 more nickels than dimes

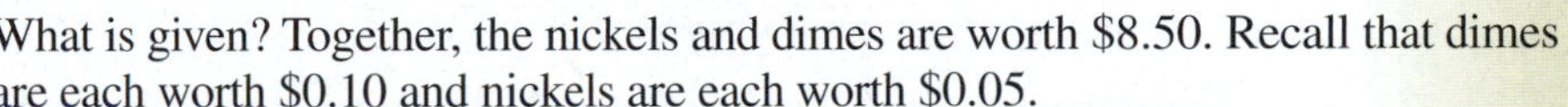

What is given? Together, the nickels and dimes are worth \$8.50. Recall that dimes are each worth \$0.10 and nickels are each worth \$0.05.

Value of dimes plus value of nickels is \$8.50.

$$0.10d + 0.05(d + 20) = 8.50 \quad \text{Express the relationship.}$$
$$0.10d + 0.05d + 1 = 8.50 \quad \text{Apply the distributive property.}$$
$$0.15d + 1 = 8.50 \quad \text{Combine like terms.}$$
$$0.15d + 1 - 1 = 8.50 - 1 \quad \text{Subtract 1 from each side.}$$
$$0.15d = 7.50 \quad \text{Simplify.}$$
$$\frac{0.15d}{0.15} = \frac{7.50}{0.15} \quad \text{Divide each side by 0.15.}$$
$$d = 50 \quad \text{Simplify.}$$

Check:

$$0.10d + 0.05(d + 20) = 8.50 \quad \text{Begin with the original equation.}$$
$$0.10(50) + 0.05(50 + 20) = 8.50 \quad \text{Replace } d \text{ with 50.}$$
$$5 + 0.05(70) = 8.50 \quad \text{Multiply and add inside parentheses.}$$
$$5 + 3.5 = 8.50 \quad \text{Simplify.}$$
$$8.50 = 8.50 \quad \text{True}$$

So, 50 is the solution of the equation and $d + 20 = 50 + 20 = 70$. Therefore, Aidan has 70 nickels and 50 dimes in his collection.

1d. Carrie invests \$6000 in two accounts. If the amount she invests in the second account is \$300 more than twice the amount invested in the first account, how much does Carrie invest in each account.

Solution **1d.** What is unknown? The amount of money invested in each account is unknown. Let f represent the amount invested in the first account. Then, $\underbrace{2f + 300}$ represents the amount in the second account.

Twice the amount of the first account (↑ $2f$) ; 300 more (↑ 300)

What is given? Carrie invested \$6000 in the two accounts.

Amount invested in first account plus amount invested in second account is \$6000.

$$f + 2f + 300 = 6000 \quad \text{Express the relationship.}$$
$$3f + 300 = 6000 \quad \text{Combine like terms.}$$
$$3f + 300 - 300 = 6000 - 300 \quad \text{Subtract 300 from each side.}$$
$$3f = 5700 \quad \text{Simplify.}$$
$$\frac{3f}{3} = \frac{5700}{3} \quad \text{Divide each side by 3.}$$
$$f = 1900 \quad \text{Simplify.}$$

Since 1900 checks in the equation, it is the solution and $2f + 300 = 2(1900) + 300 = 4100$. So, Carrie invests \$1900 in the first account and \$4100 in the second account.

1e. *Iron Man 2* ranks tenth in the list of all-time opening weeks at the box office. It earned approximately \$36.1 million more than *Iron Man* in its opening week. Together, they earned \$282.3 million in their opening weeks. How much did *Iron Man* and *Iron Man 2* each earn in their opening weeks? (Source: http://boxofficemojo.com/)

Solution **1e.** What is unknown? The opening week earnings of *Iron Man* and *Iron Man 2* are unknown. Let x represent the earnings of *Iron Man* in millions of dollars. Then $\underbrace{x + 36.1}_{\text{36.1 more than Iron Man}}$ represents the earnings of *Iron Man 2* in millions of dollars.

What is given? Together, the two movies earned \$282.3 million.

Iron Man earnings + *Iron Man 2* earnings is 282.3.

$$x + x + 36.1 = 282.3 \quad \text{Express the relationship.}$$

$$2x + 36.1 = 282.3 \quad \text{Combine like terms.}$$

$$2x + 36.1 - 36.1 = 282.3 - 36.1 \quad \text{Subtract 36.1 from each side.}$$

$$2x = 246.2 \quad \text{Simplify.}$$

$$\frac{2x}{2} = \frac{246.2}{2} \quad \text{Divide each side by 2.}$$

$$x = 123.1 \quad \text{Simplify.}$$

Since 123.1checks, it is the solution and $x + 36.1 = 123.1 + 36.1 = 159.2$. So, *Iron Man* earned \$123.1 million and *Iron Man 2* earned \$159.2 million in their opening weeks.

1f. A DVD player now sells for \$30. This is a 90% decrease in price from the original selling price of the DVD player. Find the original selling price of the DVD player.

Solution **1f.** What is unknown? The original selling price of the DVD player is unknown. Let x represent the original selling price.

What is given? The DVD player now sells for \$30. This is a 90% decrease from the original selling price, or $0.90x$. We also know that the original selling price minus the decrease in price is the current selling price.

original selling price − decrease in price is current selling price

$$x - 0.90x = 30 \quad \text{Express the relationship.}$$

$$0.10x = 30 \quad \text{Combine like terms. Recall } x - 0.90x = 1x - 0.90x = 0.10x.$$

$$\frac{0.10x}{0.10} = \frac{30}{0.10} \quad \text{Divide each side by 0.10.}$$

$$x = 300 \quad \text{Simplify.}$$

Check:

$$x - 0.90x = 30 \quad \text{Begin with the original equation.}$$

$$300 - 0.90(300) = 30 \quad \text{Replace } x \text{ with 300.}$$

$$300 - 270 = 30 \quad \text{Simplify.}$$

$$30 = 30 \quad \text{True}$$

So, the original selling price of the DVD player was \$300.

Student Check 1 Translate each problem into a linear equation and solve the problem.

a. Five less than the sum of two consecutive integers yields 14. Find the integers.

b. The quotient of the sum of two consecutive integers and three is 13. Find the integers.

c. A jar of Susan B. Anthony dollars and half-dollars totals \$90.50. If the number of half-dollars is 20 more than five times the number of Susan B. Anthony dollars, find the number of Susan B. Anthony dollars in the jar.

d. Miranda invests \$2000 in two accounts. The amount she invests in the second account is \$400 less than twice the amount invested in the first account. How much does Miranda invest in the first account?

e. *Harry Potter and the Deathly Hallows Part 2* grossed \$56 million more in its opening week (July 2011) than *Harry Potter and the Deathly Hallows Part 1* in its opening week (November 2010). If the two movies grossed a total of \$396 million in their opening weeks, how much did each movie gross? (Source: www.boxofficemojo.com)

f. Jon receives a 4% raise over his previous year's salary. If his current salary is \$75,000, what was his salary last year?

Objective 2 ▶ Troubleshoot common errors.

Troubleshooting Common Errors

A common error associated with word problems is shown next.

Objective 2 Example **A problem and an incorrect solution are given. Provide the correct solution and an explanation of the error.**

A collection of quarters and dimes is worth \$12.50. If there are 20 fewer quarters than dimes, how many of each type of coin is in the collection?

Incorrect Solution

$$\begin{aligned} 25(d-20)+10d &= 12.50 \\ 25d-500+10d &= 12.50 \\ 35d-500 &= 12.50 \\ 35d &= 512.50 \\ d &= 14.6 \end{aligned}$$

Correct Solution and Explanation

The error is that one side of the equation is in terms of cents and the other side is in terms of dollars. We must be consistent and use the same units on each side of the equation.

$$\begin{aligned} 0.25(d-20)+0.10d &= 12.50 \\ 0.25d-5+0.10d &= 12.50 \\ 0.35d-5 &= 12.50 \\ 0.35d &= 17.50 \\ d &= 50 \end{aligned}$$

There are 50 dimes and $50 - 20 = 30$ quarters in the collection.

ANSWERS TO STUDENT CHECKS

Student Check 1 **a.** The integers are 9 and 10.
b. The integers are 19 and 20.
c. There are 23 Susan B. Anthony dollars in the collection.
d. Miranda invested \$800 in the first account.
e. *Harry Potter and the Deathly Hallows Part 1* grossed \$170 million and *Harry Potter and the Deathly Hallows Part 2* grossed \$226 million.
f. Jon's salary last year was \$72,115.38.

SUMMARY OF KEY CONCEPTS

Linear equations are used to model many different situations. The key to solving word problems is in determining what is known and what is unknown. Then use the given relationship to write an equation.

SECTION 2.2 EXERCISE SET

Write About It!

Use complete sentences in your answer to each exercise.

1. What does it mean for two integers to be consecutive and how can we represent these numbers symbolically?
2. What does it mean for two numbers to be consecutive even integers and how can we represent these numbers symbolically?
3. Explain the steps to set up a mathematical equation from an application.
4. When solving an equation based on an application, will the solution of the equation always be the answer to the problem? Explain.

Practice Makes Perfect!

Translate each problem into a linear equation and solve the equation. (*See Objective 1.*)

5. The sum of two consecutive integers is 15. Find the numbers. 7 and 8
6. The sum of two consecutive integers is 35. Find the numbers. 17 and 18
7. The product of a number and one-half is 18. Find the number. 36
8. The product of a number and one-third is 45. Find the number. 135
9. The quotient of a number and two is four. Find the number. 8
10. The quotient of a number and five is 25. Find the number. 125
11. The quotient of the sum of two consecutive numbers and five is 15. Find the numbers. 37 and 38
12. The quotient of the sum of two consecutive odd numbers and four is 25. Find the numbers. 49 and 51
13. Five less than three times a number yields 15. Find the number. $\frac{20}{3}$
14. Six less than four times a number yields 14. Find the number. 5
15. Twice the sum of two consecutive even numbers is 68. Find the numbers. 16 and 18
16. Twice the sum of two consecutive odd numbers is 48. Find the numbers. 11 and 13

Additional answers can be found in the Instructor Answer Appendix.

17. Five decreased by triple a number is 20. Find the number. -5
18. Twelve decreased by triple a number is 12. Find the number. 0
19. Lindsay saves quarters and dimes in a piggy bank. If she has 15 more quarters than dimes and her collection is worth $12.50, how many quarters has Lindsay saved? 25 dimes and 40 quarters
20. Colin saves half-dollars and quarters in a piggy bank. If he has 30 more half-dollars than quarters and his collection is worth $63, how many half-dollars and quarters has Colin saved? 94 half-dollars and 64 quarters
21. Austin collects dimes and nickels. If he has 44 more nickels than dimes and his collection is worth $29.5, how many nickels and dimes has he collected? 226 nickels and 182 dimes
22. Jamie saves half-dollars and silver dollars. If she has 43 more half-dollars than silver dollars and her collection is worth $128, how many half-dollars has Jamie saved? 114 half-dollars and 71 silver dollars
23. The number of calories in a McDonald's reduced-fat vanilla ice cream cone is 150 Cal. This is 15 calories less than half the number calories in a McDonald's hot fudge sundae. How many calories are in a McDonald's hot fudge sundae? (Source: http://nutrition.mcdonalds.com) 330 Cal
24. The number of calories in a McDonald's cheeseburger is 300 calories. This value is 45 calories more than half the number of calories in a McDonald's Quarter Pounder with cheese. How many calories are in a McDonald's Quarter Pounder with cheese? (Source: http://nutrition.mcdonalds.com) 510 Cal
25. In 2011 the number of foreclosure filings in the United States was about 2.2 million less than twice the number of foreclosure filings in 2009. If there were about 3.5 million foreclosure filings in 2011, how many were there in 2009? (Source: http://www.msnbc.msn.com) about 2.85 million
26. According to Forbes.com, the average price of a cup of coffee in New York is $3.75. This is $1.35 less than half the average price of coffee in Moscow. How much is an average cup of coffee in Moscow? $10.20
27. Corrine invests a total of $1000 in Secure Bank and Lender's Bank. The amount she invests in Secure Bank is $200 less than twice the amount she invests in Lender's Bank. How much does Corinne invest in each bank? $400 at Lender's Bank, $600 at Secure Bank

28. Marcus invests a total of \$10,000 in Trust Bank and Towne Bank. The amount he invests in Trust Bank is \$600 less than a third of the amount he invests in Town Bank. How much does Marcus invest in each bank? \$7950 at Towne Bank, \$2050 at Trust Bank

29. Emily invests a total of \$5400 in PNC Bank and Wachovia Bank. The amount she invests in PNC Bank is \$300 more than half of the amount she invests in Wachovia Bank. How much does Emily invest in each bank? \$3400 at Wachovia Bank and \$2000 at PNC Bank

30. Charles invests a total of \$15,000 in People's Bank and Community Bank. The amount he invests in People's Bank is \$1200 less than twice the amount he invests in Community Bank. How much does Charles invest in each bank? \$9600 at Wachovia Bank and \$5400 at Community Bank

31. In the 2008 U.S. presidential election, Senator John McCain earned approximately 20 electoral votes less than half the number of electoral votes earned by President Barack Obama. If Senator McCain earned 162 electoral votes, how many electoral votes were earned by President Obama? (Source: *The Washington Post*) 364 votes

32. According to the 2008 U.S. presidential election campaign finance report, the amount spent on Obama's campaign was \$57 million less than three times the amount spent on McCain's campaign. If the amount spent on Obama's campaign was \$594 million, how much was spent on McCain's campaign? (Source: http://elections.nytimes.com/2008/president/campaign-finance) \$217 million

33. The number of calories burned while watching television for an hour is 8 calories less than one-fourth the number of calories burned while mowing a lawn for an hour. If 72 calories are burned while watching television for an hour, find the number of calories burned from mowing a lawn for an hour. (Source: www.cancer.org) 320 calories

34. The number of calories burned while sleeping for an hour is 54 calories less than one-third the number of calories burned when walking briskly for an hour. If 45 calories are burned while sleeping for an hour, find the number of calories burned from walking briskly for an hour. (Source: www.cancer.org) 297 calories

You Be the Teacher!

Answer each student's question.

35. Ian: I am trying to write the mathematical equation for the sentence, "Five less than a number is 15." I think the answer should be $n - 5 = 15$, but it seems like $5 - n = 15$ could be correct, too. Which is right, and why?

36. Terrell: My teacher said that the sentence, "Ten more than a number is 20," can be expressed as $x + 10 = 20$ or $10 + x = 20$. Does this mean that the sentence, "Ten less than a number is 20" can be expressed as $x - 10 = 20$ or $10 - x = 20$?
No, $x - 10 = 20$; Subtraction is not commutative so the order is important.

Think About It!

37. Suppose you have 15 quarters and 20 dimes. Write a word problem that would provide these values as its solution. (See Example 1c.)

38. Suppose you have 30 nickels and 10 quarters. Write a word problem that would provide these values as its solution. (See Example 1c.)
Answers vary. Wanda saves nickels and quarters in a box. She has 20 more nickels than quarters and her collection is worth \$4. Find the number of nickels and quarters in her collection.

SECTION 2.3 Formulas and Applications

OBJECTIVES

As a result of completing this section, you will be able to

1. Solve problems involving angles.
2. Use perimeter, area, and circumference formulas.
3. Use other formulas.
4. Solve formulas for specified variables.
5. Troubleshoot common errors.

In February 2008, the Singapore Flyer became the largest operating Ferris wheel in the world. It has a diameter of 150 m (approximately 492 ft). What is the distance of one revolution of the Ferris wheel? (Source: www.singaporeflyer.com)

To answer this question, we must find the wheel's circumference. This formula and others will be discussed in this section.

Objective 1 Solve problems involving angles.

Problems Involving Angles

We will turn our attention to problems that involve special pairs of angles, namely, complementary angles, supplementary angles, and vertical angles.

First recall that a **right angle** is an angle whose measure is 90° and a **straight angle** is an angle whose measure is 180°.

Definition: Complementary angles are two angles whose measures add to 90°.

Complementary angles form a right angle when joined together. Some examples of complementary angles are shown in the table. In the following tables, a represents the measure of angle A and b represents the measure of angle B.

a	b	$a + b$
20°	70°	20° + 70° = 90°
30°	60°	30° + 60° = 90°
45°	45°	45° + 45° = 90°
50°	40°	50° + 40° = 90°
$x°$	$(90 - x)°$	$x° + (90 - x)° = 90°$

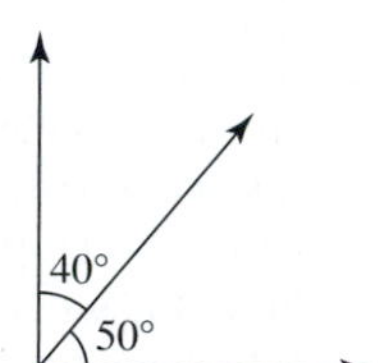

Definition: Supplementary angles are two angles whose measures add to 180°.

Supplementary angles form a straight angle when joined. Some examples of supplementary angles are shown in the table.

a	b	$a + b$
20°	160°	20° + 160° = 180°
30°	150°	30° + 150° = 180°
45°	135°	45° + 135° = 180°
50°	130°	50° + 130° = 180°
$x°$	$(180 - x)°$	$x° + (180 - x)° = 180°$

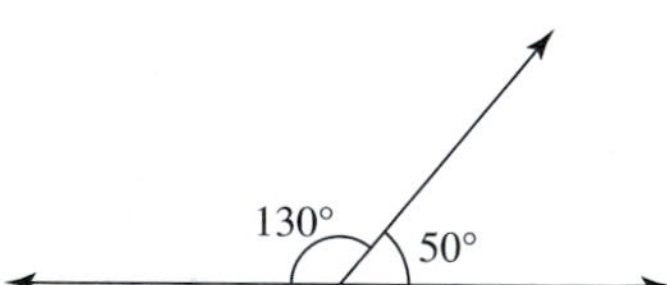

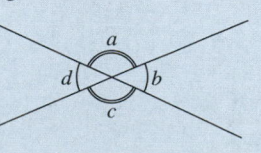

When two lines intersect one another, four angles are created. Angles opposite from one another are *vertical angles*. These angles have equal measure.

Definition: Vertical Angles

Angles A and C and angles B and D are vertical angles. The measures of these angles are a, c, b, and d, respectively. So,

$$a = c \quad \text{and} \quad b = d$$

We use the angle relationships involving complementary, supplementary, and vertical angles to solve the problems in Example 1. We write a corresponding linear equation that represents the situation and then we solve it.

Objective 1 Examples **Find the measure of each unknown angle.**

1a. Find the measure of an angle whose complement is 15° less than twice the measure of the angle.

Solution **1a.** What is unknown? The measures of an angle and its complement are unknown.
Let a represent the measure of the angle.
Then $90 - a$ represents the measure of the complement.

To write the equation, we use the following statement.

Complement is 15 less than twice the measure of the angle.

$$
\begin{aligned}
90 - a &= 2a - 15 && \text{Express the relationship.}\\
90 - a + a &= 2a - 15 + a && \text{Add } a \text{ to each side.}\\
90 &= 3a - 15 && \text{Simplify.}\\
90 + 15 &= 3a - 15 + 15 && \text{Add 15 to each side.}\\
105 &= 3a && \text{Simplify.}\\
\frac{105}{3} &= \frac{3a}{3} && \text{Divide each side by 3.}\\
35 &= a && \text{Simplify.}
\end{aligned}
$$

So, the measure of the angle is 35°. The check is left for the reader.

1b. Find the measure of an angle whose supplement is 40° more than the measure of the angle.

Solution **1b.** What is unknown? The measures of an angle and its supplement are unknown.
Let a represent the measure of the angle.
Then $180 - a$ represents the measure of the supplement.

To write the equation, use the following statement.

Supplement is 40° more than the measure of the angle.

$$
\begin{aligned}
180 - a &= a + 40 && \text{Express the relationship.}\\
180 - a + a &= a + 40 + a && \text{Add } a \text{ to each side.}\\
180 &= 2a + 40 && \text{Simplify.}\\
180 - 40 &= 2a + 40 - 40 && \text{Subtract 40 from each side.}\\
140 &= 2a && \text{Simplify.}\\
\frac{140}{2} &= \frac{2a}{2} && \text{Divide each side by 2.}\\
70 &= a && \text{Simplify.}
\end{aligned}
$$

So, the measure of the angle is 70°. The check is left for the reader.

1c. The following angles make a straight angle. Find the measure of each angle.

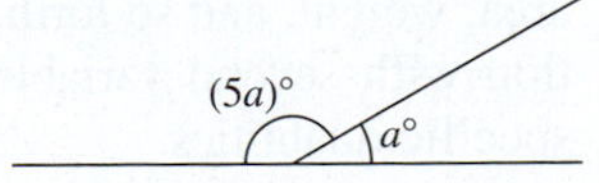

Solution **1c.** The two angles form a straight angle and, therefore, have a sum of 180°.

$$5a + a = 180 \quad \text{Express the relationship.}$$
$$6a = 180 \quad \text{Combine like terms.}$$
$$\frac{6a}{6} = \frac{180}{6} \quad \text{Divide each side by 6.}$$
$$a = 30 \quad \text{Simplify.}$$

So, one angle is 30° and the other angle is 5(30) = 150°. The check is left for the reader.

1d. The diagram illustrates two vertical angles. Find the measure of each angle.

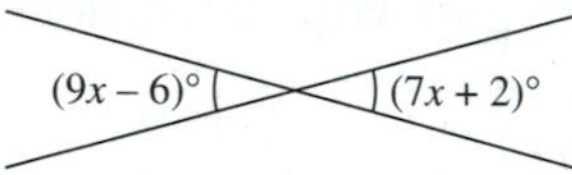

Solution **1d.** Since the angles are vertical angles, their measures are equal.

$$9x - 6 = 7x + 2 \quad \text{Express the relationship.}$$
$$9x - 6 - 7x = 7x + 2 - 7x \quad \text{Subtract } 7x \text{ from each side.}$$
$$2x - 6 = 2 \quad \text{Simplify.}$$
$$2x - 6 + 6 = 2 + 6 \quad \text{Add 6 to each side.}$$
$$2x = 8 \quad \text{Simplify.}$$
$$\frac{2x}{2} = \frac{8}{2} \quad \text{Divide each side by 2.}$$
$$x = 4 \quad \text{Simplify.}$$

Check:

$$9x - 6 = 7x + 2 \quad \text{Begin with the original equation.}$$
$$9(4) - 6 = 7(4) + 2 \quad \text{Replace } x \text{ with 4.}$$
$$36 - 6 = 28 + 2 \quad \text{Simplify.}$$
$$30 = 30 \quad \text{True}$$

The measure of each angle is 30° because 9(4) – 6 = 30 and 7(4) + 2 = 30.

Student Check 1 Find the measure of each unknown angle.

a. Find the measure of an angle whose complement is 10° more than three times the angle.

b. Find the measure of an angle whose supplement is 60° more than the angle.

c. The angles in the diagram make a straight angle. Find the measure of each angle.

d. The diagram illustrates two vertical angles. Find the measure of each angle.

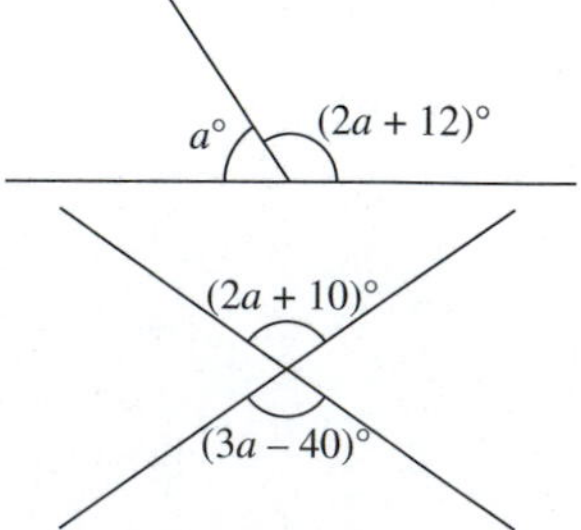

Perimeter, Area, and Circumference Formulas

Objective 2 ▶ Use perimeter, area, and circumference formulas.

An equation that describes a known relationship between quantities, such as distance, area, weight, and so forth, is called a **formula**. We can think of a formula as an equation with several variables. Formulas enable us to find valuable information about specific quantities.

In Objectives 2 and 3, we will solve problems that are modeled by known formulas. First we will work with formulas that deal with various geometric shapes and circles, namely, perimeter, area, and circumference.

> **Definition:** The **perimeter** of a polygon is the distance around the outside of the figure. In other words, it is the sum of the lengths of the sides of the polygon. The distance around a circle is called the **circumference** of the circle.

> **Definition:** The **area** of a figure is the number of square units it takes to cover the inside of the figure. (A square unit is a square whose length and width are both 1 unit long.)

The perimeter and area formulas for some common shapes are as follows.

Triangle	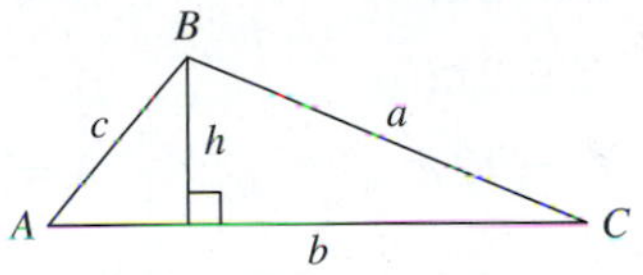	$P = a + b + c$ $A = \frac{1}{2}bh$
Rectangle	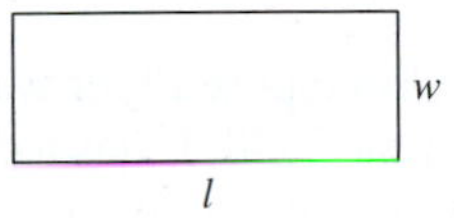	$P = 2l + 2w$ $A = lw$
Square	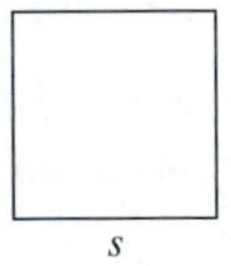	$P = 4s$ $A = s^2$
Circle	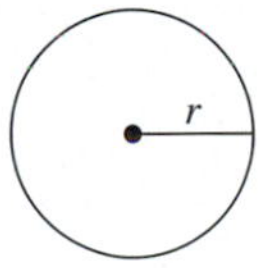	$C = 2\pi r$ or $C = \pi d$ $A = \pi r^2$

The perimeter, area, and circumference formulas will be used in two ways. In Example 2 parts (a)–(c), the dimensions of a geometric shape are known. We will calculate the area, perimeter, and/or circumference by substituting the given dimensions into the appropriate formula.

In Example 2 parts (d)–(e), the perimeter or area is known as is one of the shape's dimensions. The known values must be substituted into the appropriate formula. It will be necessary to solve the resulting equation to find the unknown dimension.

Objective 2 Examples **Use the perimeter, area, or circumference formulas to solve each problem.**

2a. What is the perimeter and area of a rectangle whose length is 5 ft and whose width is 12 ft?

Solution **2a.**

5 ft

12 ft

Perimeter	**Area**	
$P = 2l + 2w$	$A = lw$	Begin with the appropriate formula.
$P = 2(5) + 2(12)$	$A = (5)(12)$	Replace l with 5 and w with 12.
$P = 10 + 24$	$A = 60$	Simplify.
$P = 34$		

The perimeter of the rectangle is 34 ft and the area of the rectangle is 60 ft^2.

2b. What is the circumference and area of a circle whose radius is 6 cm?

Solution **2b.** Substitute the value of the radius in the circumference and area formulas.

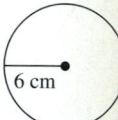

6 cm

Circumference	**Area**	
$C = 2\pi r$	$A = \pi r^2$	Begin with the appropriate formula.
$C = 2\pi(6)$	$A = \pi(6)^2$	Replace r with 6.
$C = 12\pi$	$A = 36\pi$	Simplify.

The circumference of the circle is 12π cm and the area is 36π cm^2.

2c. The Singapore Flyer was the largest operating Ferris wheel in the world in the year 2008. It has a diameter of 150 m (approximately 492 ft). What is the distance of one revolution of the Ferris wheel? Find the distance in meters and feet. Use the value 3.14 for π.

Solution **2c.** To find the distance of one revolution, we must find the circumference of a circle whose diameter is 150 m, or 492 ft.

$$C = \pi d = (3.14)(150) = 471$$

$$C = \pi d = (3.14)(492) = 1544.88$$

The distance traveled one time around the Singapore Flyer is approximately 471 m or 1544.88 ft.

2d. What is the height of a triangle whose area is 16 ft^2 and whose base is 8 ft?

Solution **2d.** We know the area, $A = 16$, and the base, $b = 8$, of the triangle, so we substitute these into the area formula and solve for h.

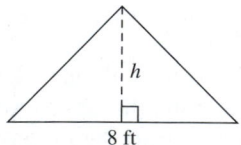

h

8 ft

$A = \frac{1}{2}bh$	Begin with the area formula.
$16 = \frac{1}{2}(8)h$	Substitute the values of A and b.
$16 = 4h$	Multiply on the right side.
$\frac{16}{4} = \frac{4h}{4}$	Divide each side by 4.
$4 = h$	Simplify.

So, the height of the triangle is 4 ft.

Check: We can verify the solution by substituting $h = 4$ in the area formula. We get $A = \frac{1}{2}(8)(4) = \frac{1}{2}(32) = 16.$

2e. The length of a rectangle is 5 ft less than three times its width. If the perimeter is 150 ft, find the length and width of the rectangle.

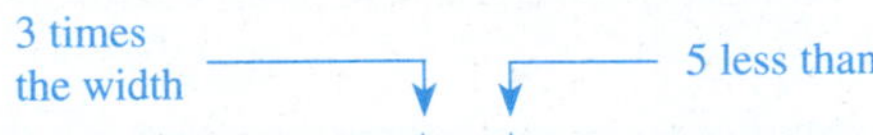

Solution

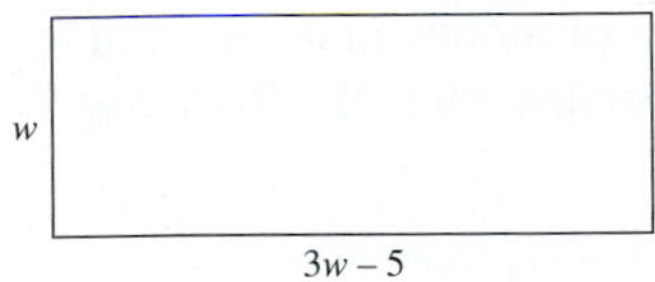

2e. Let w represent the width. Then $3w - 5$ represents the length.

	$P = 2l + 2w$	Begin with the perimeter formula.
	$150 = 2(3w - 5) + 2w$	Replace l with $3w - 5$.
	$150 = 6w - 10 + 2w$	Apply the distributive property.
	$150 = 8w - 10$	Combine like terms.
	$150 + 10 = 8w - 10 + 10$	Add 10 to each side.
	$160 = 8w$	Simplify.
	$\frac{160}{8} = \frac{8w}{8}$	Divide each side by 8.
	$20 = w$	Simplify.
Check:	$150 = 2(3w - 5) + 2w$	Begin with the original equation.
	$150 = 2[3(20) - 5] + 2(20)$	Replace w with 20.
	$150 = 2(60 - 5) + 40$	Multiply.
	$150 = 2(55) + 40$	Subtract inside the parentheses.
	$150 = 110 + 40$	Multiply.
	$150 = 150$	True

So, the width is 20 ft and the length is $l = 3w - 5 = 3(20) - 5 = 60 - 5 = 55$ ft.

Use the perimeter, area, or circumference formulas to solve each problem.

a. Find the area and perimeter of a rectangle whose length is 18 in. and whose width is 12 in.

b. Find the area and circumference of a circle whose radius is 4 in.

c. Find the area of a square whose perimeter is 12 ft.

d. The London Eye in London, England, was the tallest Ferris wheel in the world until the Singapore Flyer was constructed. It has a diameter of 122 m (approximately 400 ft). What is the distance (in meters and feet) of one revolution around the London Eye? Use the value 3.14 for π.

e. The width of a rectangle is 4 ft more than five times its length. If the perimeter of the rectangle is 428 ft, find the length and width of the rectangle.

Other Formulas

Objective 3 Use other formulas.

In Objective 2, we used formulas to find the area and perimeter of certain geometric shapes. Other important formulas deal with converting degrees Fahrenheit to degrees Celsius, calculating the amount of money in a savings account, and many more. Some of the formulas are shown.

Compound Interest Formula (compounded annually)	$A = P(1 + r)^t$	P: principal or amount invested r: annual interest rate (as a decimal) t: time of investment (in years) A: total amount saved
Celsius-to-Fahrenheit conversion	$F = \frac{9}{5}C + 32$	F: degrees Fahrenheit C: degrees Celsius

Note: *To use a given formula, substitute the known values in place of the appropriate variables to find the unknown value.*

Objective 3 Examples

Use the appropriate formula to find the requested information.

3a. Use the compound interest formula to find the amount of money in an account if $10,000 is invested for 5 yr at 4.5% interest compounded annually. Round to the nearest cent.

Solution 3a.

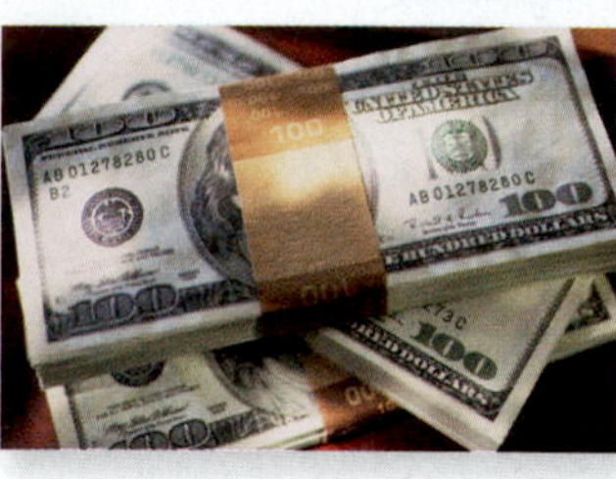

$A = P(1 + r)^t$ Begin with the appropriate formula.

$A = 10{,}000(1 + 0.045)^5$ Substitute the given values of P, r, and t.

$A = 10{,}000(1.045)^5$ Add inside the parentheses.

$A \approx 10{,}000(1.2461819)$ Approximate the exponent.

$A \approx 12{,}461.82$ Multiply.

So, the amount saved in the account would be $12,461.82 if $10,000 is invested for 5 yr at 4.5% interest compounded annually.

3b. Use the compound interest formula to find how much money must be invested at 5% interest compounded annually to have $4000 in an account after 3 yr. Round to the nearest cent.

Solution 3b.

$A = P(1 + r)^t$ Begin with the appropriate formula.

$4000 = P(1 + 0.05)^3$ Substitute the given values of A, r, and t.

$4000 = P(1.05)^3$ Add inside the parentheses.

$4000 \approx P(1.157625)$ Approximate the exponent.

$\dfrac{4000}{1.157625} \approx \dfrac{P(1.157625)}{1.157625}$ Divide each side by 1.157625.

$3455.35 \approx P$ Simplify.

So, an investment of $3455.35 will yield $4000 if invested at 5% interest compounded annually for 3 yr.

3c. The temperature of boiling water is 100°C. Use the Celsius-to-Fahrenheit conversion formula to find this temperature in degrees Fahrenheit.

Solution 3c.

$F = \dfrac{9}{5}C + 32$ Begin with the appropriate formula.

$F = \dfrac{9}{5}(100) + 32$ Substitute the given value of C.

$F = 180 + 32$ Multiply.

$F = 212$ Add.

So, the temperature of 100°C is equivalent to 212°F.

3d. Use the Celsius-to-Fahrenheit conversion formula to find the value of C if $F = 59°$.

Solution 3d.

$F = \dfrac{9}{5}C + 32$ Begin with the appropriate formula.

$59 = \dfrac{9}{5}C + 32$ Substitute the given value of F.

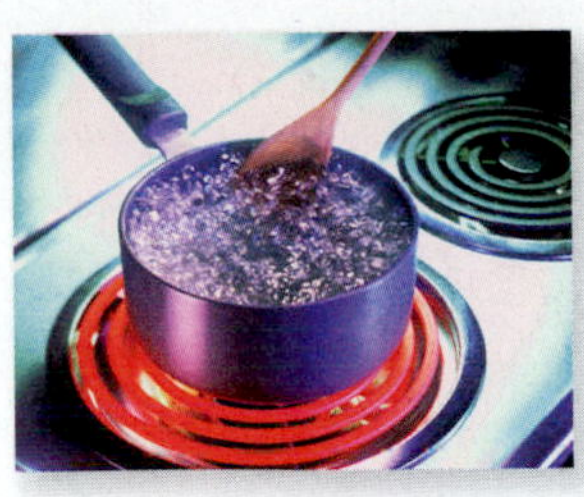

$$5(59) = 5\left(\frac{9}{5}C\right) + 5(32)$$ Multiply each side by the LCD, 5.

$$295 = 9C + 160$$ Simplify.

$$295 - 160 = 9C + 160 - 160$$ Subtract 160 from each side.

$$135 = 9C$$ Simplify.

$$\frac{135}{9} = \frac{9C}{9}$$ Divide each side by 9.

$$15 = C$$ Simplify.

So, the temperature of 59°F is equivalent to 15°C.

Student Check 3 Use the appropriate formula to find the requested information.

a. Use the compound interest formula to determine how much money is in an account if \$20,000 is invested for 10 yr at 3.5% interest compounded annually. Round to the nearest cent.

b. Use the compound interest formula to find how much money must be invested at 6% interest compounded annually to have \$100,000 in an account after 5 yr. Round to the nearest cent.

c. Use the Celsius-to-Fahrenheit conversion formula to find the value of F if C is 20°.

d. Use the Celsius-to-Fahrenheit conversion formula to find the value of C if $F = 104°$.

Solve Formulas for a Specified Variable

Objective 4

Solve formulas for specified variables.

At times, it is helpful to rewrite formulas so that a different variable is isolated. When asked to solve a formula for a specified variable, the goal is to isolate that variable on one side of the equation. The result will be another version of the given formula.

We will use these skills in Chapter 3 when we rewrite linear equations in two variables in different forms. Example 4 parts (d) and (e) each illustrate this skill.

Procedure: Solving a Formula for a Specific Variable

Step 1: Clear fractions from the equation, if necessary.

Step 2: Use the addition property of equality to get all terms with the appropriate variable on one side of an equation and the other terms on the other side. Simplify.

Step 3: Use the multiplication property of equality to get a coefficient of 1 on the appropriate variable.

Step 4: Simplify.

Objective 4 Examples **Solve each formula for the specified variable.**

4a. $d = rt$ is the distance formula. Solve this formula for r.

4b. $P = 2l + 2w$ is the perimeter formula for a rectangle. Solve this formula for w.

4c. $A = \frac{1}{2}bh$ is the area formula for a triangle. Solve this formula for b.

4d. $2x + y = 4$; Solve this formula for y.

4e. $3x - 4y = 12$; Solve this formula for y.

Solutions

INSTRUCTOR NOTE: When illustrating this concept, it is helpful to highlight the variable to be isolated.

4a.

$d = rt$

$\frac{d}{t} = \frac{rt}{t}$ Divide each side by t.

$\frac{d}{t} = r$ Simplify.

4b.

$P = 2l + 2w$

$P - 2l = 2l + 2w - 2l$ Subtract $2l$ from each side.

$P - 2l = 2w$ Simplify.

$\frac{P - 2l}{2} = \frac{2w}{2}$ Divide each side by the coefficient of w, 2.

$\frac{P - 2l}{2} = w$ Simplify.

Note: $\frac{P - 2l}{2}$ is also equivalent to $\frac{P}{2} - \frac{2l}{2}$ or $\frac{P}{2} - l$. It is not equivalent to $P - l$.

4c.

$A = \frac{1}{2}bh$

$2(A) = 2\left(\frac{1}{2}bh\right)$ Multiply each side by the LCD, 2.

$2A = bh$ Simplify.

$\frac{2A}{h} = \frac{bh}{h}$ Divide each side by the coefficient of b, h.

$\frac{2A}{h} = b$ Simplify.

4d.

$2x + y = 4$

$2x + y - 2x = 4 - 2x$ Subtract $2x$ from each side.

$y = -2x + 4$ Simplify.

4e.

$3x - 4y = 12$

$3x - 4y - 3x = 12 - 3x$ Subtract $3x$ from each side.

$-4y = -3x + 12$ Simplify.

$\frac{-4y}{-4} = \frac{-3x + 12}{-4}$ Divide each side by the coefficient of y, -4.

$y = \frac{-3x}{-4} + \frac{12}{-4}$ Divide each term by -4.

$y = \frac{3}{4}x - 3$ Simplify.

Student Check 4 Solve each formula for the specified variable.

a. $A = lw$ is the formula for the area of a rectangle. Solve for w.

b. $P = a + b + c$ is the formula for the perimeter of a triangle. Solve for a.

c. $V = \frac{1}{3}Bh$ is the formula for the volume of a regular triangular pyramid. Solve for h.

d. $6x - y = 6$ for y.

e. $2x + 7y = -14$ for y.

Objective 5 ▶

Troubleshoot common errors.

Troubleshooting Common Errors

Some common errors associated with formulas are shown.

Objective 5 Examples

A problem and an incorrect solution are given. Provide the correct solution and an explanation of the error.

5a. The supplement of an angle is 40° less than three times the measure of the angle. Find the measure of the angle.

Incorrect Solution	Correct Solution and Explanation
$x - 180 = 3x - 40$ $-140 = 2x$ $-70 = x$ There is no solution since we can't have a negative angle.	The error was made in the representation of the supplement of an angle. The supplement is $180 - x$. $180 - x = 3x - 40$ $220 = 4x$ $55 = x$ The measure of the angle is 55°.

5b. Solve the equation for y: $3x - 4y = -12$.

Incorrect Solution	Correct Solution and Explanation
$3x - 4y = -12$ $4y = -3x - 12$ $y = \frac{-3x - 12}{4}$ $y = -3x - 3$	When $3x$ is subtracted from both sides, the coefficient of y should remain negative. Also, in the last step the denominator of 4 must divide into each term. $3x - 4y = -12$ $-4y = -3x - 12$ $y = \frac{-3x - 12}{-4}$ $y = \frac{3}{4}x + 3$

ANSWERS TO STUDENT CHECKS

Student Check 1 **a.** 20° **b.** 60° **c.** 56° and 124° **d.** The measure of each angle is 110°.

Student Check 2 **a.** $A = 216$ in.^{2}, $P = 60$ in. **b.** $A = 16\pi$ in.^{2}, $C = 8\pi$ in. **c.** $A = 9$ ft^2 **d.** The distance is 383.1 m or 1256 ft. **e.** The length is 35 ft and the width is 179 ft.

Student Check 3 **a.** $A \approx \$28{,}211.98$ **b.** $P \approx \$74{,}725.82$ **c.** 68°F **d.** 40°C

Student Check 4 **a.** $w = \frac{A}{l}$ **b.** $a = P - b - c$ **c.** $h = \frac{3V}{B}$ **d.** $y = 6x - 6$ **e.** $y = -\frac{2}{7}x - 2$

SUMMARY OF KEY CONCEPTS

1. The following are important facts about angles.
   - Right angles measure 90° and straight angles measure 180°.
   - Complementary angles and supplementary angles are two angles whose measures sum to 90° and 180°, respectively. If a is the measure of an angle, then $90 - a$ represents the measure of its complement and $180 - a$ represents the measure of its supplement.
   - Vertical angles are two nonadjacent angles formed by two intersecting lines. Vertical angles are equal in measure.
2. A formula is an equation that describes the relationship between quantities. It is important to be able to use these formulas to find the perimeter, area, or circumference. These formulas can also be used to find unknown dimensions of a figure if given its perimeter, area, or circumference.
3. Substitute known values into a formula to find missing values.
4. To solve a formula for a specified variable requires us to rewrite the equation so that the stated variable is isolated on one side of the equation. No substitutions are made in this process. We apply the addition and multiplication properties of equality to solve for a different variable.

GRAPHING CALCULATOR SKILLS

The graphing calculator can aid us in evaluating formulas.

Example: Use $F = \frac{9}{5}C + 32$ to find the value of F if C is 100°.

Method 1: Enter the expression in the main window of the calculator.

```
(9/5)(100)+32
             212
```

Method 2: Enter the formula in the equation editor. Use the table to enter the known value.

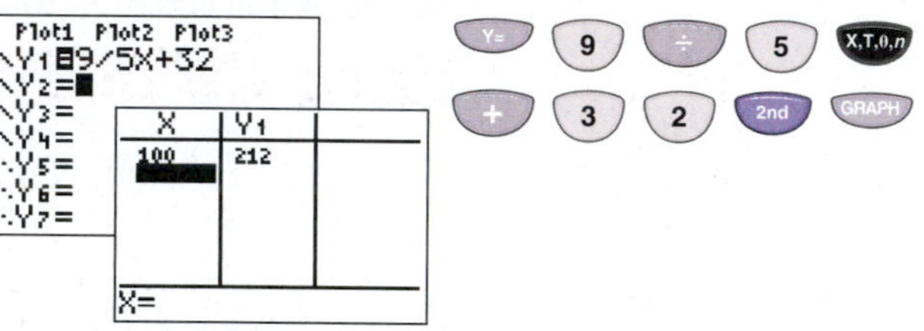

SECTION 2.3 EXERCISE SET

Write About It!

Use complete sentences in your answer to each exercise.

1. What are complementary angles?
2. What are supplementary angles?
3. What are vertical angles?
4. What relationship exists between vertical angles?
5. Explain the meaning of the perimeter of a figure.
6. Explain the meaning of the area of a figure.
7. Explain how to apply perimeter and area formulas to solve for missing dimensions of a figure.
8. Explain how to solve a formula for a specific variable.

Practice Makes Perfect!

Find the measure of each angle described or pictured. (*See Objective 1.*)

9. Find the measure of an angle whose complement is 18° more than twice the measure of the angle. 24°

Additional answers can be found in the Instructor Answer Appendix.

10. Find the measure of an angle whose complement is 30° more than triple the measure of the angle. 15°
11. Find the measure of an angle whose complement is 25° less than the measure of the angle. 57.5°
12. Find the measure of an angle whose complement is 40° less than the measure of the angle. 65°
13. Find the measure of an angle whose supplement is 90° more than triple the measure of the angle. 22.5°
14. Find the measure of an angle whose supplement is 99° more than twice the measure of the angle. 27°
15. Find the measure of an angle whose supplement is 120° less than the angle. 150°
16. Find the measure of an angle whose supplement is 110° less than the angle. 145°

17. 30°, 60° 18. 22.5°, 67.5°

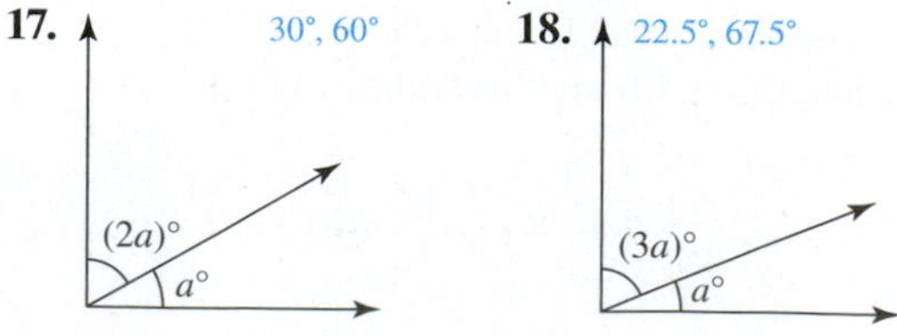

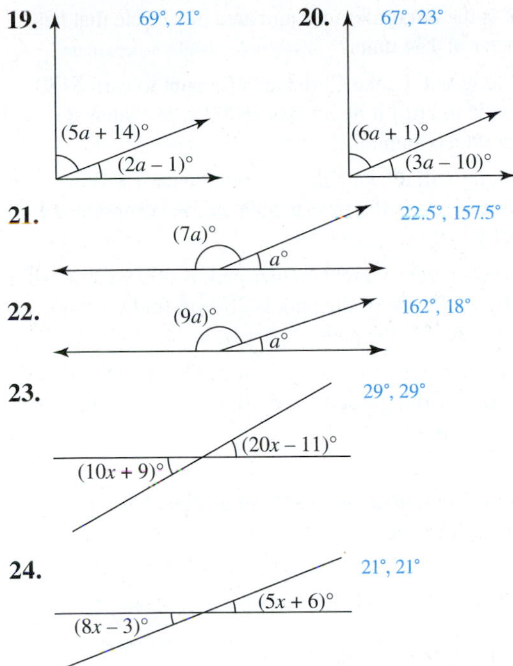

Find the perimeter of each triangle. (*See Objective 2.*)

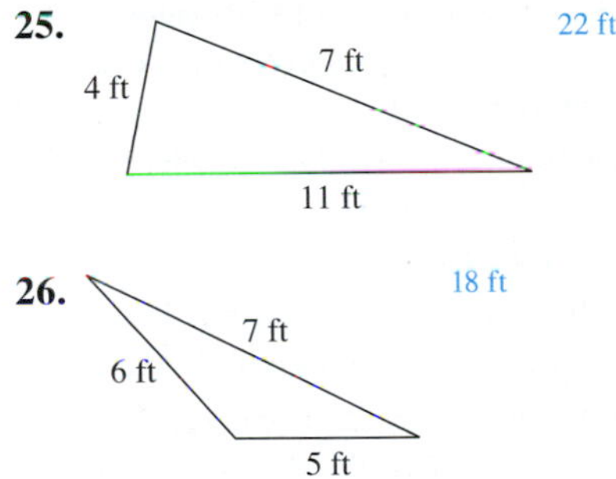

Solve each problem. (*See Objective 2.*)

27. The dimensions of a parking lot are 1400 ft long and 1000 ft wide. Find the perimeter of the parking lot. 4800 ft

28. The dimensions of a maze are 200 ft long by 300 ft wide. Find the perimeter of the maze. 1000 ft

29. What is the circumference and area of a circle with radius 10 in.? 20π in., 100π in.^{2}

30. What is the circumference and area of a circle with radius 15 in.? 30π in., 225π in.^{2}

31. As of 2009, the Beijing Great Wheel was the largest Ferris wheel in the world. If its diameter is 682 ft, what is the distance of one revolution of the Ferris wheel? (Source: www.popularmechanics.com) 682π ft

32. Canada's Niagara Sky Wheel has a diameter of approximately 174 ft. What is the distance of one revolution of this Ferris wheel? (Source: www.cliftonhill.com) 174π ft

33. Find the height of a triangle whose area is 25 ft^2 and whose base is 5 ft. 10 ft

34. Find the height of a triangle whose area is 54 ft^2 and whose base is 6 ft. 18 ft

35. The length of a rectangle is 2 yd less than three times its width. If the perimeter is 180 yd, find the length and width of the rectangle. The length is 67 yd and the width is 23 yd.

36. The length of a rectangle is 5 m less than six times its width. If the perimeter is 200 m, find the length and width of the rectangle. The length is 85 m and the width is 15 m.

The formula $A = P(1 + r)^t$ gives the amount of money A in an account after t yr if P dollars is invested at an interest rate r (in decimal form) compounded annually. Use this formula to answer each question. Round to the nearest cent. (*See Objective 3.*)

37. Find the amount of money Deborah must invest at 5% interest compounded annually to have \$5000 in an account after 2 yr. \$4535.15

38. Find the amount of money Sheree must invest at 4% interest compounded annually to have \$1000 in an account after 3 yr. \$889.00

The formula $F = \frac{9}{5}C + 32$ is used to convert degrees Celsius to degrees Fahrenheit. Use this formula to answer each question. Round to two decimal places. (*See Objective 3.*)

39. In 1983, the coldest temperature on earth was recorded at Vostok Station located on the East Antarctica ice sheet. The temperature was −89.2°C. Convert this temperature to degrees Fahrenheit. (Source: World Meteorological Organization) −128.56°F

40. The hottest temperature on earth was recorded in 1922 in El Azizia, Libya. The temperature was 57.8°C. Convert this temperature to degrees Fahrenheit. (Source: World Meteorological Organization) 136.04°F

41. Juneau, Alaska, has an average December temperature of 28.5°F. Convert this temperature to degrees Celsius. (Source: www.weather.com) −1.94°C

42. Newburgh, New York has an average July temperature of 74.5°F. Convert this temperature to degrees Celsius. (Source: www.weather.com) 23.61°C

The formula $I = Prt$ gives the amount of simple interest an account earns after t yr if P dollars is invested at an annual interest rate r (in decimal form). Use this formula to answer each question. (*See Objective 3.*)

43. How much interest is earned after 10 yr if Sebastian invests \$500 in an account that earns 3.5% annual interest? \$175

44. How much interest is earned after 3 yr if Padma invests \$700 in an account that earns 4.5% annual interest? \$94.50

45. If Stacy wants to earn \$600 in 3 yr from an account that earns 5% annual interest, how much must she invest in the account? \$4000

46. If Richard wants to earn \$1200 in 5 yr from an account that earns 3% annual interest, how much must he invest in the account? \$8000

The body mass index (BMI) is an indicator used to determine if a person is at a healthy weight. A normal BMI is in the interval 18.5 – 24.9. If w is a person's weight (in pounds) and h is a person's height (in inches), then the person's BMI $= \dfrac{703w}{h^2}$. Use this formula to answer each question. Round to one decimal place. (*See Objective 3.*)

47. Carlos has a BMI of 39. If his height is 6 ft, determine his weight. 287.6 lb
48. Kate has a BMI of 15. If her height is 5 ft, determine her weight. 76.8 lb
49. Gretchen is striving for a BMI of 20. If she is 5 ft 3 in. tall, how much should she weigh? 112.9 lb
50. Marcus is striving for a BMI of 25. If he is 6 ft 2 in. tall, how much should he weigh? 194.7 lb

Solve each formula for the specified variable. (*See Objective 4.*)

51. Distance: $d = rt$ for t $\quad t = \dfrac{d}{r}$
52. Area of rectangle: $A = lw$ for l $\quad l = \dfrac{A}{w}$
53. Perimeter of rectangle: $P = 2l + 2w$ for l $\quad l = \dfrac{P - 2w}{2}$
54. Area of triangle: $A = \dfrac{1}{2}bh$ for h $\quad h = \dfrac{2A}{b}$
55. Volume of cylinder: $V = \pi r^2 h$ for h $\quad h = \dfrac{V}{\pi r^2}$
56. Surface area of cylinder: $S = 2\pi r^2 + 2\pi rh$ for h $\quad h = \dfrac{S - 2\pi r^2}{2\pi r}$
57. $y = mx + b$ for m $\quad m = \dfrac{y - b}{x}$
58. $y = mx + b$ for x $\quad x = \dfrac{y - b}{m}$
59. $4x + y = 8$ for y $\quad y = -4x + 8$
60. $5x + y = 10$ for y $\quad y = -5x + 10$
61. $x - y = 3$ for y $\quad y = x - 3$
62. $2x - y = -6$ for y $\quad y = 2x + 6$
63. $8x + 3y = 9$ for y $\quad y = -\dfrac{8}{3}x + 3$
64. $6x + 5y = 30$ for y $\quad y = -\dfrac{6}{5}x + 6$
65. $x - 7y = -14$ for y $\quad y = \dfrac{1}{7}x + 2$
66. $x - 4y = -4$ for y $\quad y = \dfrac{1}{4}x + 1$

Mix 'Em Up!

Solve each problem.

67. Find the measure of an angle whose complement is 20° more than the angle. 35°
68. Find the measure of an angle whose complement is 10° more than the angle. 40°
69. What is the circumference and area of a circle that has a diameter of 60 units? 60π units, 900π square units
70. What is the circumference and area of a circle that has a diameter of 134 units? 134π units, 4489π square units
71. How long will it take Christian's account to earn \$180 in simple interest if he invests \$6000 at 6% interest compounded annually? 0.5 yr or 6 months
72. How long will it take Kiki's account to earn \$50 in simple interest if she invests \$500 at 5% compounded annually? 2 yr
73. If the perimeter of a rectangular park is 600 yd and the length of one side of the park is 200 yd, find the width of the other side of the park. 100 yd
74. If the perimeter of a rectangular garden is 300 yd and the width of the garden is 50 yd, find the length of the garden. 100 yd

The diagram illustrates two vertical angles. Find the measure of each angle.

75. 76.

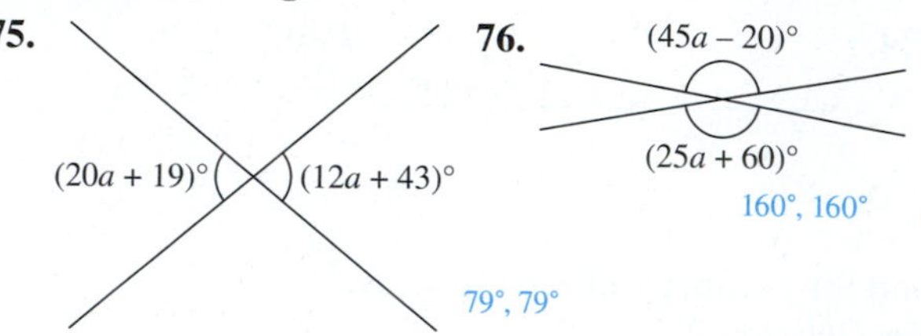

77. The average July high temperature in Death Valley, California, is 115°F. Convert this temperature to degrees Celsius. (Source: www.weather.com) 46.11°C
78. The average January high temperature in North Pole, Alaska, is −2°F. Convert this temperature to degrees Celsius. (Source: www.weather.com) −18.89°C
79. Find the length of the base of a triangle whose area is 49 square units and whose height is 14 units. 7 units
80. Find the length of the base of a triangle whose area is 100 square units and whose height is 32 units. 6.25 units
81. Michelle is striving for a BMI of 25. If she is 5 ft tall, how much should she weigh? 128.0 lb
82. Sasha is striving for a BMI of 20. If her height is 5 ft 5 in., how much should she weigh? 120.2 lb
83. Find the measure of an angle whose supplement is 21° more than twice the measure of the angle. 53°
84. Find the measure of an angle whose supplement is 24° less than five times the measure of the angle. 34°

Solve each formula for the specified variable.

85. $B = \dfrac{703w}{h^2}$ for w $\quad w = \dfrac{Bh^2}{703}$
86. $p = \dfrac{i}{s}$ for s $\quad s = \dfrac{i}{p}$
87. $F = s(1 + c)^n$ for s $\quad s = \dfrac{F}{(1 + c)^n}$
88. $D = Pz - a$ for z $\quad z = \dfrac{D + a}{P}$
89. $x = -\dfrac{b}{2a}$ for b $\quad b = -2ax$
90. $x = -\dfrac{b}{2a}$ for a $\quad a = -\dfrac{b}{2x}$

91. $y - y_1 = m(x - x_1)$ for x $x = \dfrac{y - y_1 + mx_1}{m}$

92. $A = \dfrac{h(b_1 + b_2)}{2}$ for b_1 $b_1 = \dfrac{2A - hb_2}{h}$

You Be the Teacher!

Correct each student's errors, if any.

93. Agnes invests \$4500 in two accounts. The amount she invests in the second account is \$300 less than twice the amount invested in the first account. How much does Agnes invest in the first account?

Jada's work: Let x be the amount invested in the first account and $2x$ be the amount invested in the second account.

$$x + 2x = 4500$$
$$3x = 4500$$
$$x = 1500$$

94. Solve the formula $S = \dfrac{a}{1 - r}$ for the variable r.

Maureen's work:

$$S = \frac{a}{1 - r}$$
$$S(1 - r) = a$$
$$S - r = a$$
$$-r = a - S$$
$$r = a + S$$

Calculate It!

The formula $A = 100(1.01)^{4t}$ specifies the amount A in an account after t yr when \$100 is invested at 4% annual interest rate compounded quarterly. Use a graphing calculator to determine the number of years it will take for the account to grow to the given amounts. Round to the nearest year.

95. \$150 10 yr

96. \$200 17 yr

97. \$700 49 yr

98. \$1000 58 yr

SECTION 2.4 Linear Inequalities and Applications

OBJECTIVES

As a result of completing this section, you will be able to

1. Graph the solution set of an inequality and write its solution set in interval notation and set-builder notation.
2. Solve a linear inequality using the addition and multiplication properties of inequalities.
3. Solve applications of linear inequalities.
4. Troubleshoot common errors.

Susan uses a phone service provided by a high-speed Internet connection. The service offers a basic residential calling plan for \$14.99 for the first 500 min of local and long distance calling plus 3.9¢ for each additional minute. The monthly cost of Susan's phone bill is represented by $14.99 + 0.039x$, where x is the number of additional minutes. How many minutes of calling can Susan use for her cost to be at most \$20.00 per month?

To answer this question, we must solve a special type of inequality. In this section, we will learn how to solve linear inequalities and their applications.

Objective 1

Graph the solution set of an inequality and write its solution set in interval notation and set-builder notation.

Solutions of Inequalities

Until now, we have solved only linear equations in one variable. We now turn our focus to solving linear inequalities in one variable. An *inequality* is a statement that two quantities are not equal. The following symbols are used to denote inequalities.

$p < q$	p is less than q.
$p \leq q$	p is less than or equal to q.
$p > q$	p is greater than q.
$p \geq q$	p is greater than or equal to q.

A *linear inequality* is the same as a linear equation except the equality sign is replaced with an inequality symbol.

Definition: A **linear inequality in one variable** is an inequality of the form $ax + b < c$, where a, b, and c are real numbers and $a \neq 0$.

Note: *The definition of a linear inequality in one variable also applies to inequalities containing the symbols $\leq$, $>$, or $\geq$.*

Some examples of linear inequalities in one variable are

$$x > 3 \qquad 2x + 3 \le -7 \qquad 7x - 2 \ge 6x + 4 \qquad 4(3y - 1) < -2(y - 9)$$

A **solution of a linear inequality** is a value that makes the inequality true. The *solution set* of an inequality is the set of all the solutions.

Consider the inequality $x > 3$. Some solutions of this inequality are 3.05, 3.9, 4, 4.2, 10, and 250. There are infinitely many solutions of this inequality since there are infinitely many numbers larger than 3. It is impossible to list all of the solutions, so we visualize the solution set by graphing the solution set on a number line. We also use a special notation to represent the solution set.

The solution set of $x > 3$ includes all the numbers greater than 3 but not equal to 3. So, the graph of the solution set of $x > 3$ looks like the following.

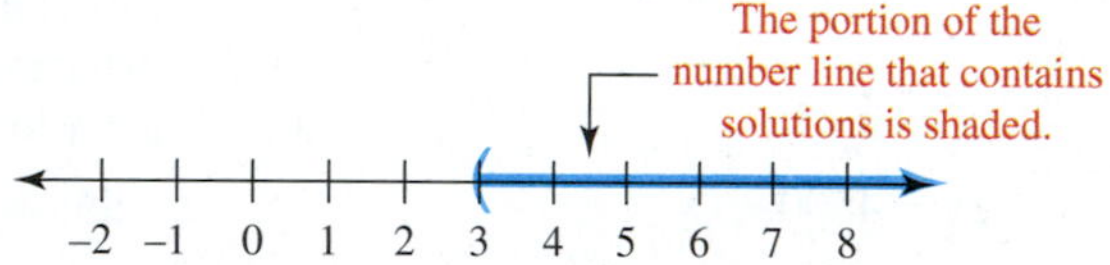

Notice that a parenthesis is used when the endpoint is not included in the solution set. A bracket is used when the endpoint is included in the solution set.

We can also write the solution set of an inequality using two special notations, interval notation and set-builder notation. **Interval notation** is a concise way to represent the solution set of an inequality. In interval notation, we represent the smallest and largest values in the solution set.

- When the graph of an inequality extends to the right indefinitely, we say that the numbers in the set approach ∞ (infinity). This means that the numbers in the solution set of the inequality increase indefinitely.
- When the graph of an inequality extends to the left indefinitely, we say that the numbers in the set approach $-\infty$ (negative infinity). This means that the numbers in the solution set of the inequality decrease indefinitely.
- Interval notation makes use of parentheses and brackets, like graphing solution sets of inequalities.
  - A bracket, [or], denotes that the endpoint of the solution set is included.
  - A parenthesis, (or), denotes that the endpoint of the solution set is not included.
  - Parentheses are always used on positive and negative infinity.

So, the solution set of $x > 3$, written in interval notation, is $(3, \infty)$.

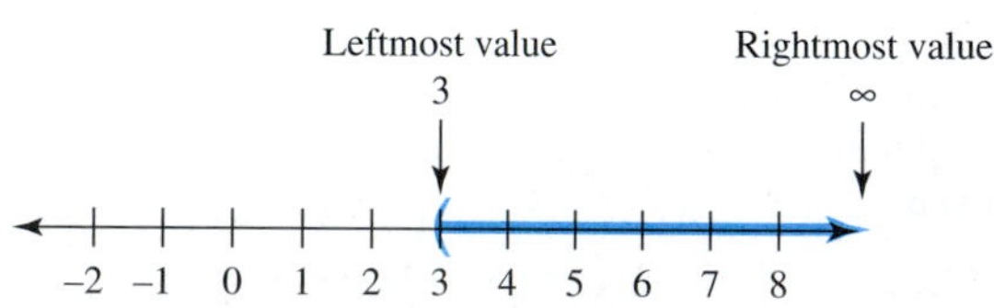

Set-builder notation was discussed in Chapter 1. Recall that **set-builder notation** is used to state the conditions the solutions must meet to be included in the set and is written in the form $\{x | \text{condition } x \text{ must satisfy}\}$.

So, the solution set of $x > 3$, written in set-builder notation, is $\{x | x > 3\}$.

Mathematical Notation	Verbal Expression
$(3, \infty)$	The interval from 3 to infinity, not including 3
$\{x \| x > 3\}$	The set of x such that x is greater than 3

The following table shows some inequalities and their equivalent interval and set-builder notations for real numbers a and b.

Inequality	Graph	Interval Notation	Set-Builder Notation
$x > a$	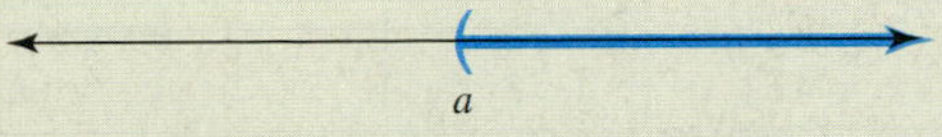	(a, ∞)	$\{x \mid x > a\}$
$x \geq a$	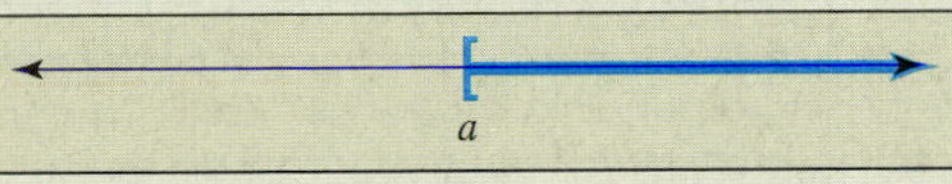	$[a, \infty)$	$\{x \mid x \geq a\}$
$x < a$	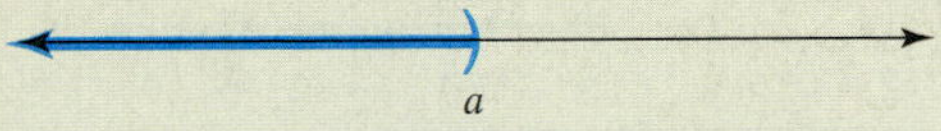	$(-\infty, a)$	$\{x \mid x < a\}$
$x \leq a$	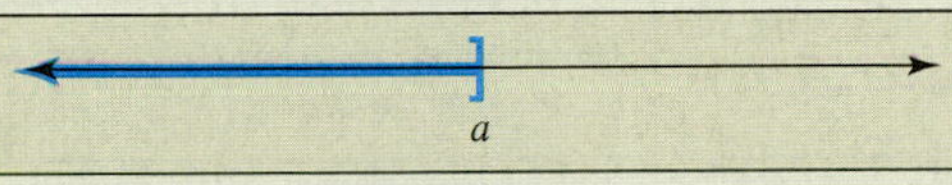	$(-\infty, a]$	$\{x \mid x \leq a\}$
$a < x < b$		(a, b)	$\{x \mid a < x < b\}$
$a \leq x \leq b$		$[a, b]$	$\{x \mid a \leq x \leq b\}$
$a < x \leq b$		$(a, b]$	$\{x \mid a < x \leq b\}$
$a \leq x < b$	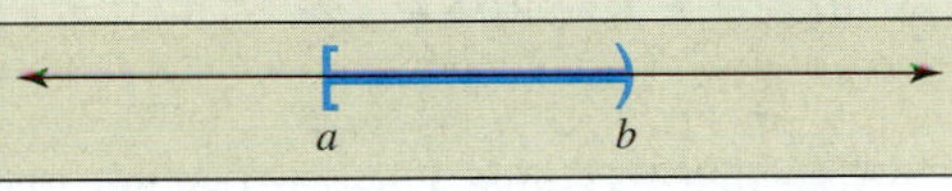	$[a, b)$	$\{x \mid a \leq x < b\}$

Procedure: Graphing the Solution Set of an Inequality

Step 1: Shade the portion of the number line that contains numbers that satisfy the inequality.

Step 2: Determine the appropriate symbol on the endpoint of the solution set.

a. If the endpoint of the solution set is a solution of the inequality, place a bracket on this number (a closed-circle can also be used to represent this).

b. If the endpoint of the solution set is *not* a solution of the inequality, place a parenthesis on this number (an open-circle can also be used to represent this).

Procedure: Writing the Solution Set of an Inequality in Interval Notation or Set-Builder Notation

Interval Notation

Step 1: Begin the interval with the smallest endpoint in the set. Use a parenthesis if this endpoint is *not* included or a bracket if it is included.

Step 2: End the interval with the largest endpoint in the set. Use a parenthesis if this endpoint is *not* included or a bracket if it is included.

Set-Builder Notation

Write the set as $\{x \mid \text{put the inequality here}\}$.

Objective 1 Examples For each inequality, graph the solution set and write the solution set in interval notation and set-builder notation.

Inequality	Graph	Interval Notation	Set-Builder Notation
1a. $x > 4$	−1 0 1 2 3 4 5 6 7 8 9	$(4, \infty)$	$\{x \mid x > 4\}$
1b. $x \geq -2$	−7 −6 −5 −4 −3 −2 −1 0 1 2 3	$[-2, \infty)$	$\{x \mid x \geq -2\}$
1c. $\frac{1}{2} > x$	−5 −4 −3 −2 −1 0 1 2 3 4 5	$\left(-\infty, \frac{1}{2}\right)$	$\left\{x \mid x < \frac{1}{2}\right\}$
1d. $x \leq 0$	−5 −4 −3 −2 −1 0 1 2 3 4 5	$(-\infty, 0]$	$\{x \mid x \leq 0\}$
1e. $-1 < x < 3$	−4 −3 −2 −1 0 1 2 3 4 5 6	$(-1, 3)$	$\{x \mid -1 < x < 3\}$

Student Check 1 For each inequality, graph the solution set and write the solution set in interval notation and set-builder notation.

a. $x > -1$ **b.** $x \geq 3$ **c.** $x < -1$ **d.** $x \leq -3$ **e.** $2 < x < 7$

Note that Example 1 part (c) shows that the inequality $\frac{1}{2} > x$ is equivalent to $x < \frac{1}{2}$. Example 1 parts (a)–(d) are illustrations of **simple inequalities**. The solutions of a simple inequality have to satisfy only one inequality. Example 1 part (e) is an illustration of a *compound inequality*. The solutions of a compound inequality must satisfy two inequalities, not just one. For instance,

$$-1 < x < 3 \text{ means that "} -1 < x \text{ and } x < 3\text{" or "} x > -1 \text{ and } x < 3\text{"}$$

Compound inequalities will be studied more in Section 2.5.

Addition and Multiplication Properties of Inequalities

Objective 2

Solve a linear inequality using the addition and multiplication properties of inequalities.

Solving linear inequalities is very similar to solving linear equations. The goal is the same, to isolate the variable on one side of the inequality. The properties that enable us to do this are the addition and multiplication properties of inequality. While only one inequality symbol is used in the statement of the following properties, they work with all inequality symbols.

Property: Addition Property of Inequality

If a, b, and c are real numbers, then

$a < b$ and $a + c < b + c$ and $a - c < b - c$ are equivalent inequalities.

The property states that adding the same number to each side of an inequality or subtracting the same number from each side of an inequality produces an equivalent inequality. For an illustration of this property, let $a = -5$, $b = 2$, and $c = 4$.

$a < b$	$a + c < b + c$	$a - c < b - c$
$-5 < 2$	$-5 + 4 < 2 + 4$	$-5 - 4 < 2 - 4$
True	$-1 < 6$	$-9 < -2$
	True	True

So, adding the same number to each side of an inequality or subtracting the same number from each side of an inequality maintains the inequality relationship.

Property: Multiplication Property of Inequality

If a, b, and c are real numbers and

c is *positive*, then $a < b$ and $ac < bc$ and $\frac{a}{c} < \frac{b}{c}$ are equivalent inequalities.

c is *negative*, then $a < b$ and $ac > bc$ and $\frac{a}{c} > \frac{b}{c}$ are equivalent inequalities.

This property states that

- Multiplying or dividing each side of an inequality by a *positive* number produces an equivalent inequality.
- Multiplying or dividing each side of an inequality by a *negative* number produces an equivalent inequality only if the inequality symbol is *reversed*.

For an illustration of the property, let $a = -6$ and $b = 12$.

Let $c = 3$.	$a < b$	$ac < bc$	$\frac{a}{c} < \frac{b}{c}$
	$-6 < 12$	$-6(3) < 12(3)$	$\frac{-6}{3} < \frac{12}{3}$
	True	$-18 < 36$	$-2 < 4$
		True	True

Multiplying each side of an inequality by a positive number maintains the inequality relationship.

Let $c = -1$.	$a < b$	$ac > bc$	$\frac{a}{c} > \frac{b}{c}$
	$-6 < 12$	$-6(-1) > 12(-1)$	$\frac{-6}{-1} > \frac{12}{-1}$
	True	$6 > -12$	$6 > -12$
		True	True

Multiplying each side of an inequality by a negative number requires us to *reverse* the inequality symbol to maintain the inequality relationship.

Procedure: Solving a Linear Inequality

Step 1: **a.** Clear any parentheses from the equation by applying the distributive property.
b. Remove any fractions by multiplying each side by the LCD.

Step 2: Use the addition property of inequality to collect all variable terms on one side and all constant terms on the other side.

Step 3: Use the multiplication property of inequality to get a coefficient of 1 on the variable. Remember that if we multiply or divide by a negative number, we must *reverse* the inequality symbol.

Step 4: Graph the solution set and write the solution set in interval notation and set-builder notation.

Objective 2 Examples **Solve each inequality. Graph the solution set and write the solution set in interval notation and set-builder notation.**

2a. $x + 4 < -1$ **2b.** $-2.5y \le 50$ **2c.** $-2a - 4 \ge 10$

2d. $3(2y - 4) - (y + 3) \ge 5(3y + 1)$ **2e.** $\frac{3}{2}(x - 5) > \frac{1}{4}x + 4$

2f. $4x - 5(x - 2) < 6(x + 1) - 7x$ **2g.** $x + 5(x - 2) < 3(2x + 1) + 7$

Solutions

INSTRUCTOR NOTE:
After the solution set is found, ask students for some specific solutions to each inequality. Ask students to check to make sure their specific solution checks in the original inequality.

2a.

$$x + 4 < -1$$
$$x + 4 - 4 < -1 - 4 \quad \text{Subtract 4 from each side.}$$
$$x < -5 \quad \text{Simplify.}$$

We graph the solution set and write the solution set in interval notation and set-builder notation.

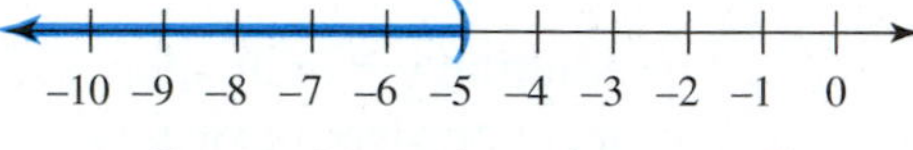

Interval notation: $(-\infty, -5)$

Set-builder notation: $\{x|x < -5\}$

Check: It is impossible to check every solution but we can check two values to make sure our answer is reasonable. A value in the shaded region should make the original inequality true, while other values will not satisfy the inequality.

$x = -6$ (in solution set):	$x = 1$ (not in solution set):
$x + 4 < -1$	$x + 4 < -1$
$-6 + 4 < -1$	$1 + 4 < -1$
$-2 < -1$ True	$5 < -1$ False

2b.

$$-2.5y \le 50$$
$$\frac{-2.5y}{-2.5} \ge \frac{50}{-2.5} \quad \text{Divide each side by } -2.5 \text{ and reverse the inequality symbol.}$$
$$y \ge -20 \quad \text{Simplify.}$$

We graph the solution set and write the solution set in interval notation and set-builder notation.

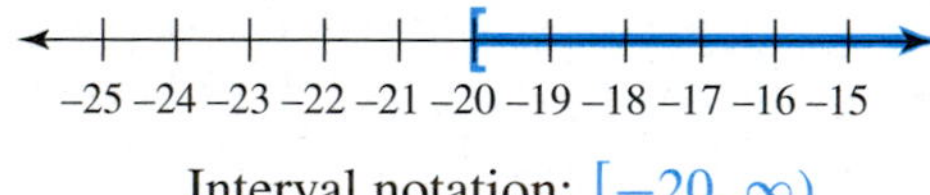

Interval notation: $[-20, \infty)$

Set-builder notation: $\{y|y \ge -20\}$

Check: We will check two values to make sure our answer is reasonable.

$y = -10$ (in solution set):	$y = -30$ (not in solution set):
$-2.5y \le 50$	$-2.5y \le 50$
$-2.5(-10) \le 50$	$-2.5(-30) \le 50$
$25 \le 50$ True	$75 \le 50$ False

2c.

$$-2a - 4 \ge 10$$
$$-2a - 4 + 4 \ge 10 + 4 \quad \text{Add 4 to each side.}$$
$$-2a \ge 14 \quad \text{Simplify.}$$
$$\frac{-2a}{-2} \le \frac{14}{-2} \quad \text{Divide each side by } -2 \text{ and reverse the inequality symbol.}$$
$$a \le -7 \quad \text{Simplify.}$$

We graph the solution set and write the solution set in interval notation and set-builder notation.

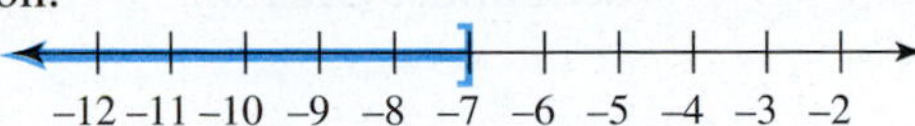

Interval notation: $(-\infty, -7]$
Set-builder notation: $\{a \mid a \leq -7\}$

Check: We will check two values to make sure our answer is reasonable.

$a = -8$ (in solution set):	$a = 1$ (not in solution set):
$-2a - 4 \geq 10$	$-2a - 4 \geq 10$
$-2(-8) - 4 \geq 10$	$-2(1) - 4 \geq 10$
$16 - 4 \geq 10$	$-2 - 4 \geq 10$
$12 \geq 10$ True	$-6 \geq 10$ False

2d. $3(2y - 4) - (y + 3) \geq 5(3y + 1)$

$6y - 12 - y - 3 \geq 15y + 5$ Apply the distributive property.
$5y - 15 \geq 15y + 5$ Combine like terms.
$5y - 15 + 15 \geq 15y + 5 + 15$ Add 15 to each side.
$5y \geq 15y + 20$ Simplify.
$5y - 15y \geq 15y + 20 - 15y$ Subtract $15y$ from each side.
$-10y \geq 20$ Simplify.
$\dfrac{-10y}{-10} \leq \dfrac{20}{-10}$ Divide each side by -10 and reverse the inequality symbol.
$y \leq -2$ Simplify.

We graph the solution set and write the solution set in interval notation and set-builder notation.

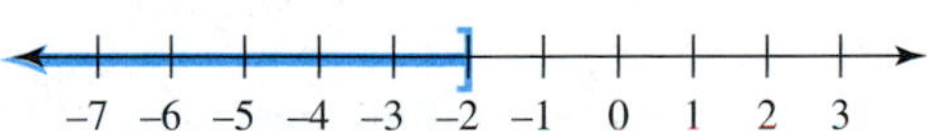

Interval notation: $(-\infty, -2]$
Set-builder notation: $\{y \mid y \leq -2\}$

2e. $\dfrac{3}{2}(x - 5) > \dfrac{1}{4}x + 4$

$4\left[\dfrac{3}{2}(x - 5)\right] > 4\left(\dfrac{1}{4}x + 4\right)$ Multiply each side by the LCD, 4.
$6(x - 5) > x + 16$ Simplify on the left: $4\left(\dfrac{3}{2}\right) = \dfrac{12}{2} = 6$ and apply the distributive property on the right.
$6x - 30 > x + 16$ Apply the distributive property.
$6x - 30 + 30 > x + 16 + 30$ Add 30 to each side.
$6x > x + 46$ Simplify.
$6x - x > x + 46 - x$ Subtract x from each side.
$5x > 46$ Simplify.
$\dfrac{5x}{5} > \dfrac{46}{5}$ Divide each side by 5.
$x > \dfrac{46}{5}$ Simplify.

We graph the solution set and write the solution set in interval notation and set-builder notation.

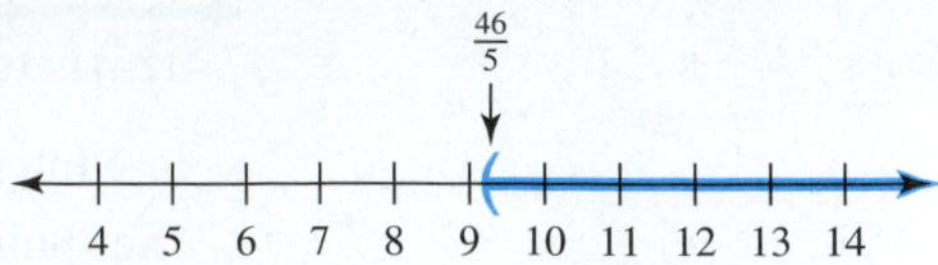

Interval notation: $\left(\frac{46}{5}, \infty\right)$

Set-builder notation: $\left\{x \middle| x > \frac{46}{5}\right\}$

2f. $4x - 5(x - 2) < 6(x + 1) - 7x$

$4x - 5x + 10 < 6x + 6 - 7x$ Apply the distributive property.

$-x + 10 < -x + 6$ Combine like terms.

$-x + 10 + x < -x + 6 + x$ Add x to each side.

$10 < 6$ Simplify.

The resulting inequality, $10 < 6$, is an inequality that is always false. Therefore, the oroginal inequality has no solution. The solution set is the empty set, or $\varnothing$. The graph of the solution set is an unshaded number line as shown.

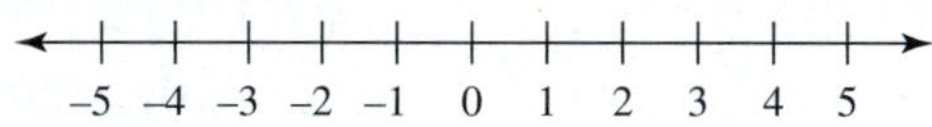

2g. $x + 5(x - 2) < 3(2x + 1) + 7$

$x + 5x - 10 < 6x + 3 + 7$ Apply the distributive property.

$6x - 10 < 6x + 10$ Combine like terms.

$6x - 10 - 6x < 6x + 10 - 6x$ Subtract $6x$ from each side.

$-10 < 10$ Simplify.

The resulting inequality, $-10 < 10$, is always true. Therefore, any real number is a solution of the original inequality. So, the solution set is all real numbers, or $\mathbb{R}$.

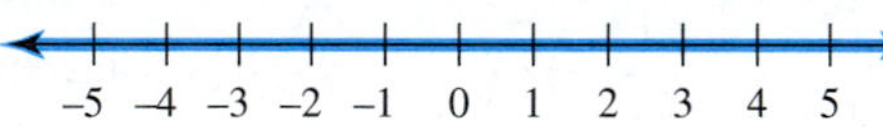

Interval notation: $(-\infty, \infty)$

Set-builder notation: $\{x \mid x \text{ is a real number}\}$

Student Check 2 Solve each inequality. Graph the solution set and write the solution set in interval notation and set-builder notation.

a. $y + 3 < -2$

b. $-x \leq 2$

c. $7y - 1 > 6$

d. $3(a + 2) - 7 \geq -4a + 10$

e. $\frac{1}{3}y - 2\left(y + \frac{1}{6}\right) > \frac{2}{3}y + \frac{5}{6}$

f. $4(x - 3) + 1 \geq 5(x + 2) - x$

g. $7x - 2(4x + 3) < 3(x + 5) - 4x$

Objective 3 ▶

Solve applications of linear inequalities.

Applications

There are a few key phrases that we need to learn before more applications are introduced.

INSTRUCTOR NOTE:
To help students learn the inequality symbol that corresponds to the statement, use the concept of money. Say, "If I have at most $10 in my pocket, how much money could I have?" or "If I have at least $50, how much could I have?"

Verbal Statement	Mathematical Statement
a is less than b.	$a < b$
a is less than or equal to b. a is no more than b. a is at most b.	$a \le b$
a is greater than b.	$a > b$
a is greater than or equal to b. a is no less than b. a is at least b.	$a \ge b$

Procedure: Solving Word Problems with Inequalities

Step 1: Read the problem and determine the unknown. Assign a variable to the unknown value.
Step 2: Read the problem and determine the given information.
Step 3: Find the statement in the problem that states the inequality relationship, looking for key phrases that are listed in the preceding chart.
Step 4: Use the statement in step 3 to write the inequality.
Step 5: Apply the addition and multiplication properties of inequalities to solve the inequality.
Step 6: Answer the question with a complete sentence.

Objective 3 Examples

Write an inequality that models each situation. Solve the inequality and answer the problem in a complete sentence.

3a. Sonia has math test scores of 73, 85, and 80. What score does she need to make on her fourth math test to have a test average of at least 80?

Solution

3a. What is unknown? The score on the fourth test is unknown. Let x represent the fourth test score.

What is given? The first three scores, 73, 85, and 80, are given.

To obtain the inequality, we use the following statement.

Average of the four tests is at least 80.

$$\frac{73 + 85 + 80 + x}{4} \ge 80$$ Recall average is $\frac{\text{the sum of all items}}{\text{the number of items}}$.

$$\frac{238 + x}{4} \ge 80$$ Simplify the numerator of the fraction.

$$4\left(\frac{238 + x}{4}\right) \ge 4(80)$$ Multiply each side by 4.

$$238 + x \ge 320$$ Simplify.

$$238 + x - 238 \ge 320 - 238$$ Subtract 238 from each side.

$$x \ge 82$$ Simplify.

Sonia needs to make 82 or higher on her fourth test to have at least an 80 test average. We can check by substituting a number greater than or equal to 82 into the original inequality to see that the average is greater than or equal to 80.

3b. Jamaal's final grade in his math class is based upon a weighted average. The weights are as follows.

Homework (10%)
Quizzes (15%)
Tests (50%)
Final exam (25%)

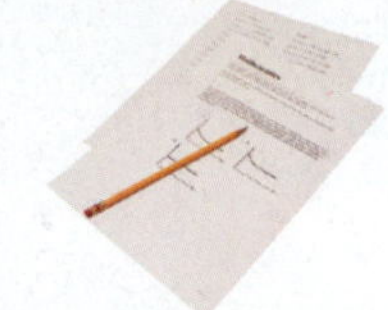

If Jamaal's homework average is 85, quiz average is 80, and test average is 75, what does he need on his final exam to have at least a 70 average in the class?

Solution **3b.** What is unknown? The final exam grade is unknown. Let f represent the final exam grade.

What is given? The final grade is calculated as follows.

$$10\%(\text{homework average}) + 15\%(\text{quiz average}) + 50\%(\text{test average}) + 25\%(\text{final exam}) = 0.10(85) + 0.15(80) + 0.50(75) + 0.25f$$

To obtain the inequality, we use the following statement.

Final grade is at least 70.

$0.10(85) + 0.15(80) + 0.50(75) + 0.25f \geq 70$	Translate the statement.
$8.5 + 12 + 37.5 + 0.25f \geq 70$	Multiply on the left.
$58 + 0.25f \geq 70$	Combine like terms.
$58 + 0.25f - 58 \geq 70 - 58$	Subtract 58 from each side.
$0.25f \geq 12$	Simplify.
$\dfrac{0.25f}{0.25} \geq \dfrac{12}{0.25}$	Divide each side by 0.25.
$f \geq 48$	Simplify.

Jamaal needs at least a 48 on the final exam to have at least a 70 final grade. We can check by substituting a number greater than or equal to 48 into the original inequality to see that the final grade is at least 70.

3c. Susan uses a phone service provided by a high-speed Internet connection. The basic residential calling plan costs $14.99 for the first 500 min of local and long distance calling plus 3.9¢ for each additional minute. How many minutes can she use for her cost to be at most $20.00 per month?

Solution **3c.** What is unknown? The number of additional minutes over 500 is unknown. Let m represent the number of additional minutes.

What is given? The monthly cost is represented by $14.99 + 0.039m$.

To obtain the inequality, we use the following statement.

Monthly cost is at most 20.

$14.99 + 0.039m \leq 20$	Translate to obtain the inequality.
$14.99 + 0.039m - 14.99 \leq 20 - 14.99$	Subtract 14.99 from each side.
$0.039m \leq 5.01$	Simplify.
$\dfrac{0.039m}{0.039} \leq \dfrac{5.01}{0.039}$	Divide each side by 0.039.
$m \leq 128.46$	Simplify.

Susan can talk for 128 additional minutes, or for a total of 628 min, for her cost to be at most $20.

Student Check 3 Write an inequality that models each situation. Solve the inequality and answer the problem in a complete sentence.

a. Bryan has math test scores of 68, 83, and 75. What score does he need to make on his fourth math test to have a test average of at least 70?

b. Tomekia's final grade in her math class is based on a weighted average. The weights are as follows: homework (5%), quizzes (20%), tests (45%), and final exam (30%). If Tomekia's homework average is 100, quiz average is 84, and test average is 80, what does she need on her final exam to have at least an 80 average in the class?

c. Jared and Angela are planning their wedding reception. A reception hall charges \$1500 plus \$20 per person. How many people can they invite to their reception to have a cost of at most \$4000?

Objective 4 Troubleshoot common errors.

Troubleshooting Common Errors

Some common errors associated with linear inequalities are shown.

Objective 4 Examples **A problem and an incorrect solution are given. Provide the correct solution and an explanation of the error.**

4a. Write the set $\{x|x < -5\}$ in interval notation.

Incorrect Solution	Correct Solution and Explanation
The interval notation is $(-5, -\infty)$.	We always begin the interval with the smallest number and end the interval with the largest number in the set. Therefore, it should be $(-\infty, -5)$.

4b. Draw the graph that represents the set given by $\{x|x < -5\}$.

Incorrect Solution

The graph is:

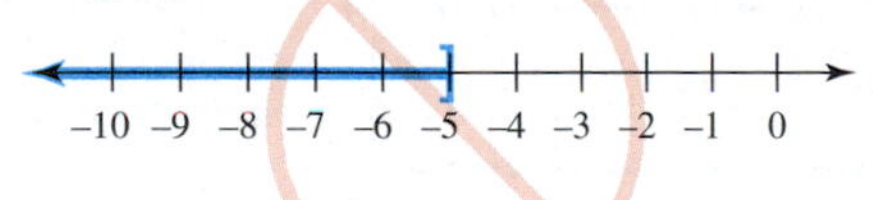

Correct Solution and Explanation

The number -5 is not included in the solution set. So, we should have a parenthesis on -5.

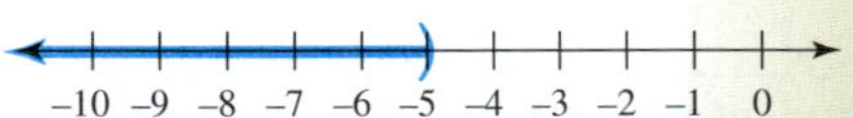

4c. Solve the inequality $5 - 3x \le -4$.

Incorrect Solution

$$5 - 3x \le -4$$
$$5 - 3x - 5 \le -4 - 5$$
$$-3x \le -9$$
$$x \le 3$$

So, the solution set is $(-\infty, 3]$.

Correct Solution and Explanation

When each side of an inequality is divided by a negative number, we must reverse the inequality symbol.

$$5 - 3x \le -4$$
$$5 - 3x - 5 \le -4 - 5$$
$$-3x \le -9$$
$$x \ge 3$$

So, the solution set is $[3, \infty)$.

ANSWERS TO STUDENT CHECKS

		Graph	Interval Notation	Set-Builder Notation
Student Check 1	**a.**	−6 −5 −4 −3 −2 −1 0 1 2 3 4	$(-1, \infty)$	$\{x \mid x > -1\}$
	b.	−2 −1 0 1 2 3 4 5 6 7 8	$[3, \infty)$	$\{x \mid x \geq 3\}$
	c.	−6 −5 −4 −3 −2 −1 0 1 2 3 4	$(-\infty, -1)$	$\{x \mid x < -1\}$
	d.	−8 −7 −6 −5 −4 −3 −2 −1 0 1 2	$[-\infty, -3]$	$\{x \mid x \leq -3\}$
	e.	−1 0 1 2 3 4 5 6 7 8 9	$(2, 7)$	$\{x \mid 2 < x < 7\}$
Student Check 2	**a.**	−10 −9 −8 −7 −6 −5 −4 −3 −2 −1 0	$(-\infty, -5)$	$\{y \mid y < -5\}$
	b.	−7 −6 −5 −4 −3 −2 −1 0 1 2 3	$[-2, \infty)$	$\{x \mid x \geq -2\}$
	c.	−4 −3 −2 −1 0 1 2 3 4 5 6	$(1, \infty)$	$\{y \mid y > 1\}$
	d.	$\frac{11}{7}$ ↓ −4 −3 −2 −1 0 1 2 3 4 5 6	$\left[\frac{11}{7}, \infty\right)$	$\left\{a \,\middle\vert\, a \geq \frac{11}{7}\right\}$
	e.	$-\frac{1}{2}$ ↓ −5 −4 −3 −2 −1 0 1 2 3 4 5	$\left(-\infty, -\frac{1}{2}\right)$	$\left\{y \,\middle\vert\, y < -\frac{1}{2}\right\}$
	f.	−5 −4 −3 −2 −1 0 1 2 3 4 5		$\varnothing$
	g.	−5 −4 −3 −2 −1 0 1 2 3 4 5	$(-\infty, \infty)$	$\{x \mid x \text{ is a real number}\}$

Student Check 3 **a.** Bryan needs at least 54 on his fourth test to have at least a 70 test average.

b. Tomekia needs at least 74 on the final exam to have a final average of at least 80.

c. Jared and Angela can invite at most 125 people to have a cost of no more than \$4000.

SUMMARY OF KEY CONCEPTS

1. The graph of the solution set of an inequality is a picture of all real numbers that make the inequality a true statement. A parenthesis (used with $<$ or $>$) on a number indicates the number is not included in the solution set. A bracket (used with $\leq$ or $\geq$) on a number indicates the number is included in the solution set. We can use interval notation or set-builder notation to express the solution set.
   a. Interval notation begins with the left bound of the solution set and ends with the right bound of the solution set. When the numbers in a set continue indefinitely to the right, the right bound is represented by ∞. When the numbers in a set continue indefinitely to the left, the left bound is represented by $-\infty$. A parenthesis is always used with ∞ or $-\infty$.
   b. Set-builder notation is written using { }. We write {variable | final inequality}. Example: $\{y \mid y < 5\}$.
2. Inequalities are solved using the addition and multiplication properties of inequalities. The most important thing to remember is that when we multiply or divide by a *negative* number, we must also *reverse* the inequality symbol.

3. Applications are solved by translating the given statement into an appropriate inequality. Key phrases are "is at least" and "is at most." A good way to remember this is to think of money. If we have at least \$10, we would have \$10 or more. If we have at most \$10, we would have \$10 or less.

GRAPHING CALCULATOR SKILLS

The graphing calculator can be used as a reference to check the work we do by hand. We can do this using the Test menu and the Store feature.

Example: $x - 5 > 2$

Solution: The solution of this inequality is $x > 7$. To check the work on the calculator, we should determine if the inequality is true for a number larger than 7, is false for a number less than 7, and determine what happens at 7.

Let $x = 8$ (a value larger than 7).

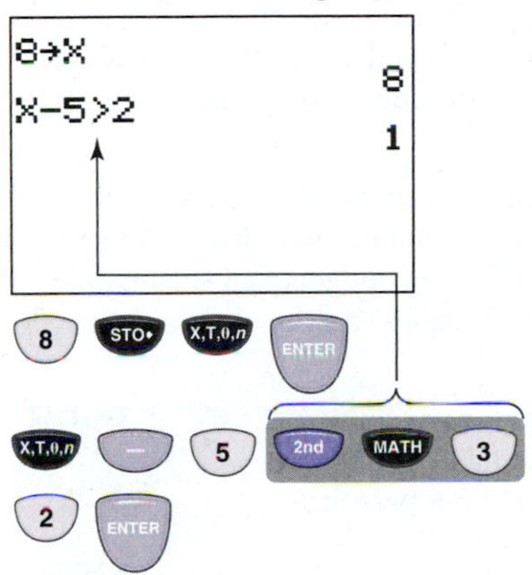

The result of 1 confirms that 8 is a solution of the inequality. The shaded portion of the graph should contain 8 (to the right of 7).

Let $x = 6$ (a value less than 7).

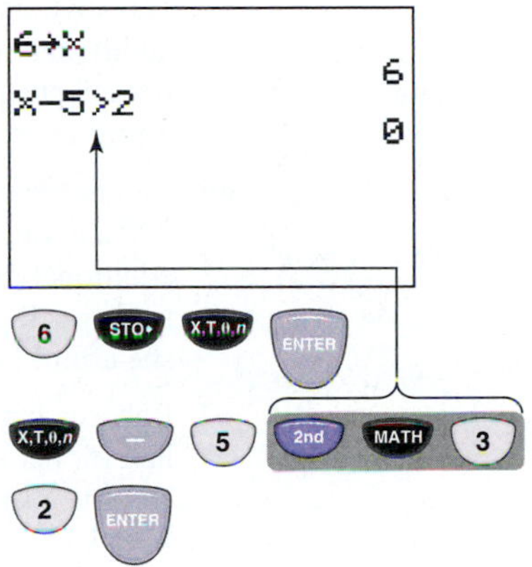

The result of 0 means that 6 is *not* a solution of the inequality. The shaded portion of the graph should not include the side with 6.

Let $x = 7$.

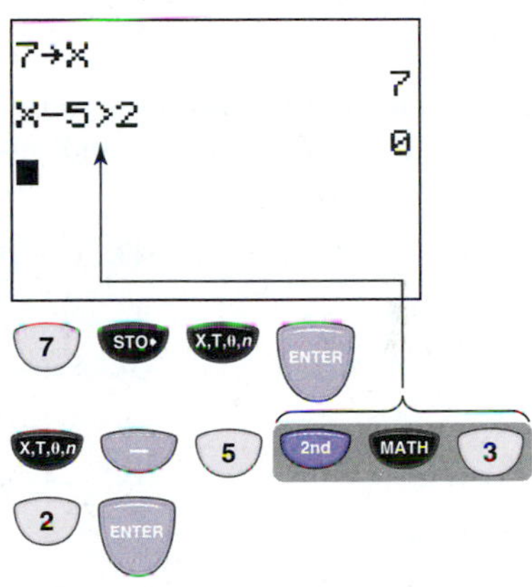

The result is 0 which means that 7 is *not* a solution of the inequality. Therefore, we should use a parenthesis on the value 7 on the graph.

SECTION 2.4 EXERCISE SET

Write About It!

Use complete sentences in your answer to each exercise.

1. If a boundary number of an inequality is not included in the solution set of the inequality, how is this represented when graphing the solution set on a number line? Place a parenthesis on this number.
2. If a boundary number of an inequality is included in the solution set of the inequality, how is this represented when graphing the solution set on a number line? Place a bracket on this number.
3. Explain how to express an inequality in interval notation.
4. What similarities exist between the solution sets of inequalities on a number line and the solution sets of inequalities represented in interval notation?
5. Should parentheses or brackets be used with ∞ in interval notation? Why?
6. When solving an inequality, what operations cause the direction of the inequality to change?

Practice Makes Perfect!

Graph the solution set of each inequality on a number line and express the solution set in interval notation and set-builder notation. (*See Objective 1.*)

7. $x > 5$
8. $x > 7$
9. $x < -4$
10. $x < -8$
11. $x \geq 10$
12. $x \geq -12$
13. $x \leq \frac{1}{2}$
14. $x \leq -\frac{3}{4}$

Additional answers can be found in the Instructor Answer Appendix.

15. $-9 < x \leq -4$
16. $0 < x \leq 2$
17. $-4 \leq x \leq 0$
18. $5 \leq x \leq 6$
19. $3 < x$
20. $4 \leq x$
21. $-2 > x \geq -10$
22. $3 \geq x > -6$
23. $-10 \geq x$
24. $-9 \geq x$
25. $x > 0$
26. $x > 14$
27. $x \geq \frac{1}{3}$
28. $x \geq -\frac{1}{4}$
29. $x < -5$
30. $x < -16$
31. $-15 < x < -4$
32. $-1 \leq x < 45$
33. $-13 > x$
34. $-26 > x$
35. $11 < x$
36. $5 \leq x$
37. $1 \geq x \geq 0$
38. $-7 \geq x > -20$

Solve each inequality. Graph the solution set and write the answer in interval notation and set-builder notation. (*See Objective 2.*)

39. $y + 9 > -3$
40. $y - 10 > 6$
41. $6a \leq 12$
42. $3a \leq 24$
43. $2x - 10 < 0$
44. $5x - 15 > 0$
45. $6x + 11 \geq -7$
46. $7x - 1 \geq 8$
47. $-3x < 14$
48. $-4x \leq 15$
49. $b + 14 \geq 6b - 9$
50. $2b - 8 < 4b + 1$
51. $16 \leq -12a$
52. $-24 \leq -6a$
53. $4a + 15 < 13a - 9$
54. $a - 16 \leq 6a + 10$
55. $5(x - 1) > 4(2x + 1)$
56. $6(3x + 2) \geq 3(7x - 1)$
57. $\frac{2}{3}(x + 3) < -x + 5$
58. $\frac{5}{2}(x - 4) \geq -2x + 8$
59. $\frac{1}{4}(5x - 2) \geq \frac{1}{3}(2x + 9)$
60. $\frac{1}{2}(3x + 8) \leq \frac{2}{3}(4x - 1)$

Write an inequality that represents each situation. Solve the inequality and explain the answer using a complete sentence. (*See Objective 3.*)

61. If Lucinda's quiz scores are 85, 78, 100, and 87, what must she earn on her fifth quiz to have an average of at least 90?

62. If Aaron's quiz scores are 66, 79, 60, and 65, what must he earn on his fifth quiz to have an average of at least 70?

63. Aleksandr's course grade is based on a weighted average. The weights are as follows: homework (15%), quizzes (25%), tests (40%), and final exam (20%). If Aleksandr's homework average is 80, quiz average is 60, and test average is 68, what must he score on the final exam to have an average of at least 70?

64. Christopher's course grade is based on a weighted average. The weights are as follows: homework (15%), quizzes (25%), tests (40%), and final exam (20%). If Christopher's homework average is 100, quiz average is 75, test average is 80, what must he score on the final exam for his course average to be at least 80?

65. Connect2Me charges a monthly fee of \$39.99 for 450 anytime cell phone minutes, and then \$0.45 per additional minute. If Joyce budgets \$60 each month for her cell phone, how many additional minutes can she afford to use?

66. MobileMania charges a monthly fee of \$29.99 for 200 anytime cell phone minutes, and then \$0.30 per additional minute. If Sandra budgets \$45 each month for her cell phone, how many additional minutes can she afford to use?

67. Rebecca pays \$1500 for a reception hall for her son's graduation party, and she is charged \$15 per person for a buffet by the caterer. If Rebecca budgets \$3000 for the reception hall and the caterer, how many people can Rebecca invite to her son's party?

68. Ilene is planning her office Christmas party. She finds a banquet hall at a local hotel that charges \$700, and the hotel will also prepare dinner for \$30 per person. If Ilene has a budget of \$2000, how many people can attend the Christmas party?

The total U.S. national health expenditures, in trillions of dollars, can be approximated by the equation $E = 0.118x + 2.03$, where x is the number of years after 2005. (Source: http://www.cms.gov)

69. In what year will the total national health expenditures exceed \$3 trillion? 2014

70. In what year will the total national health expenditures exceed \$3.2 trillion? 2015

The total national health expenditures for the Department of Veterans' Affairs, in billions of dollars, can be approximated by the equation, $V = 3.154x + 28.94$, where x is the number of years after 2005. (Source: http://www.cms.gov)

71. In what year will the total national health expenditures for the Department of Veterans' Affairs exceed \$51 billion? 2012

72. In what year will the total national health expenditures for the Department of Veterans' Affairs exceed \$60 billion? 2015

Mix 'Em Up!

Solve each inequality. Graph the solution set and write it in interval notation and set-builder notation.

73. $13 < -5a$

74. $8 \geq -4a$

75. $5 + 2(3 - x) < 5(x - 2)$

76. $12 - 3(x - 4) \geq 6(x + 1)$

77. $4(2x - 1) \geq 9x + 1$

78. $5(3x + 4) < 20x - 10$

79. $1.3 + 0.2(8 - x) < 0.7(2x - 5)$

80. $3.5 - 0.3(6 + x) > 0.8(7 - 2x)$

81. $6y - 9 < 14y + 3$

82. $3y + 2 < 10y - 6$

83. $64 \geq -8x$

84. $-54 > -18x$

85. $\frac{1}{2}(3x - 6) \geq 2(x - 4)$

86. $\frac{2}{3}(6x + 3) > 4(3x - 5)$

87. $4(8 - 7x) < 6(1 + 3x)$

88. $-4(1 - 6x) \geq 2(4x - 3)$

89. $0.36(5x - 3) \leq 0.6(7 - x)$

90. $0.25(8x - 6) \geq 1.8(1 - 5x)$

91. Hope's history course grade is based on a weighted average. The weights are as follows: tests (50%), term paper (30%), and final exam (20%). If Hope's test average is 80 and she scores 89 on her term paper, what must she score on the final exam to maintain at least an 80 course average? 66.5

92. Fiona's English course grade is based on a weighted average. The weights are as follows: midterm (30%), papers (45%), and final exam (25%). If Fiona earns 70 on her midterm and 75 on her papers, what must she score on her final exam to maintain at least a 70 course average? 61

93. Mandy rents a small reception hall for her wedding for \$500. She hires a caterer who charges \$25 per person. If she has \$2500 to spend on her reception and caterer, how many people can she have at her reception? 80 people

94. Bette rents a banquet hall for her wedding reception for \$5000 and they provide dinner at \$40 per person. If Bette's budget is \$15,000, how many people can she have at her reception? 250 people

The following students are taking algebra from Dr. Cheng. In Dr. Cheng's class, the course grade is determined by tests (55%), quizzes (20%), and a final exam (25%). Answer each student's question.

95. Jada: My test average is 76, and my quiz average is 71. I need to earn a B in the class to keep my 3.0 GPA. If an 80–89 is a B, what do I need to earn on the final exam to earn at least B in the class? 96

96. Fredericka: My test average is 95 and my quiz average is 85. What do I have to get on the final exam to earn an A in the class, if 90–100 is an A? 83

The total of all Low Income Home Energy Assistance Program (LIHEAP) funds allocated to states, in billions of dollars can be approximated by the equation $L = 1.19x + 3.017$, where x is the number of years after 2008.
(Source: http://www.acf.hhs.gov)

97. In what year will the total of all LIHEAP funds exceed \$6.5 billion? 2011

98. In what year will the total of all LIHEAP funds exceed \$9 billion? 2014

You Be the Teacher!

Correct each student's errors, if any.

99. $2x + 5 \geq 8x - 7$

Ada's work:	
$2x + 5 \geq 8x - 7$	$2x + 5 \geq 8x - 7$
$12 \geq 6x$	$-6x \geq -12$
$x \geq \frac{1}{2}$	$x \leq 2$
$\left(\frac{1}{2}, \infty\right)$	$(-\infty, 2]$
$\left\{x \middle\vert x \geq \frac{1}{2}\right\}$	$\{x \vert x \leq 2\}$

100. $\frac{1}{3}(2x + 10) \leq -2(x - 1)$

Wanda's work:	$\frac{1}{3}(2x + 10) \leq -2(x - 1)$
$\frac{1}{3}(2x + 10) \leq -2(x - 1)$	$2x + 10 \leq -6(x - 1)$
$2x + 10 \leq -6x - 1$	$2x + 10 \leq -6x + 6$
$8x \leq -11$	$8x \leq -4$
$x \geq -\frac{11}{8}$	$x \leq -\frac{1}{2}$
$\left[-\frac{11}{8}, \infty\right)$	$\left(-\infty, -\frac{1}{2}\right]$
$\left\{x \middle\vert x \geq -\frac{11}{8}\right\}$	$\left\{x \middle\vert x \leq -\frac{1}{2}\right\}$

Calculate It!

Use a graphing calculator to determine if the given number is in the solution set of the inequality.

101. $12x - 15 \leq 16 - 25x$; 13 no

102. $17x - 3 \geq 10 + 6x$; -2 no

103. $\frac{3}{5}(4x - 7) > \frac{5}{3}(2x + 5)$; -15 yes

104. $\frac{1}{6}(8 - 9x) < \frac{3}{4}(7x + 13)$; 15 yes

PIECE IT TOGETHER SECTIONS 2.1–2.4

Solve each equation and check the answer. (*See Section 2.1, Objectives 2–6.*)

1. $\frac{7x}{3} - \frac{x}{2} = \frac{5}{6}$ $\left\{\frac{5}{11}\right\}$
2. $12 - (1 - 4x) = 4(x + 1)$ $\varnothing$
3. $6(x - 1) - 3x = 2x - 17$ $\{-11\}$
4. $7(x - 2) - 3x = 2(2x - 7)$ $\mathbb{R}$

Translate each problem into a linear equation and solve. (*See Section 2.2, Objective 1, and Section 2.3, Objectives 1–3.*)

5. Eight less than twice a number yields 50. Find the number. 29
6. Twice the sum of two consecutive even integers is 52. Find the integers. 12 and 14
7. Bao invests a total of \$5000 in two banks. The amount she invests in the first bank is \$800 more than half of the amount she invests in the second bank. How much does Boa invest in each bank? \$2800 at the first bank and \$2200 at the second bank
8. Find the measure of an angle whose complement is 30° more than four times the measure of the angle. 12°
9. Find the measure of an angle whose supplement is 63° more than twice the measure of the angle. 39°
10. Find the height of a triangle whose area is 108 ft^2 and whose base is 9 ft. 24 ft
11. The length of a rectangle is 2 ft more than three times its width. If the perimeter is 200 ft, find the length and width of the rectangle. The length is 75.5 ft and the width is 24.5 ft.

Solve each formula for the specified variable. (*See Section 2.3, Objective 4.*)

12. $\frac{2}{3}a = b - 2cd$ for c $c = \frac{3b - 2a}{6d}$
13. $A = P + Prt$ for r $r = \frac{A - P}{Pt}$

Solve each inequality. Graph the solution set and write the solution set in interval notation and set-builder notation. (*See Section 2.4, Objectives 1 and 2.*)

14. $\frac{1}{2}(x - 4) < -x + 5$
15. $4(8 - 7x) < 6(1 + 3x)$

SECTION 2.5 Compound Inequalities

OBJECTIVES

As a result of completing this section, you will be able to

1. Determine the intersection of two sets.
2. Determine the union of two sets.
3. Solve compound inequalities involving "and."
4. Solve compound inequalities involving "or."
5. Solve applications of compound inequalities.
6. Troubleshoot common errors.

Body mass index (BMI) is a measure of how much body fat a person has. A normal BMI is between 18.5 and 24.9. The formula to calculate BMI is $\text{BMI} = \frac{703w}{h^2}$, where w is weight (in pounds) and h is height (in inches). Find the weight necessary for a person with a height of 5 ft 4 in. to have a normal BMI. To solve this problem, we need to find the solution of the inequality

$$18.5 \leq \frac{703w}{64^2} \leq 24.9$$

This is an example of a *compound inequality*, which we will discuss in this section.

Objective 1

Determine the intersection of two sets.

The Intersection of Sets

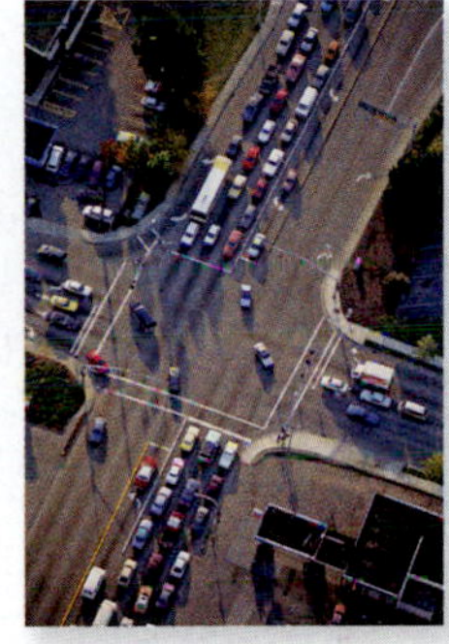

In Chapter 1, we discussed the concept of sets and their intersection and union. We will explore this concept further as we deal with sets which are intervals of the real numbers. Knowing how to find the intersection and union of sets will enable us to solve special types of inequalities, namely, compound inequalities.

We will first investigate the *intersection* of two sets. To do this, think about the intersection of two streets. The intersection is where the streets meet one another or where they cross. The intersection is the part of the road that the two streets have in common. This idea extends to sets, as discussed in Section 1.1.

Recall, the intersection of two sets is the set of elements that the two sets have in common.

INSTRUCTOR NOTE:
To help illustrate this concept, define two sets *A* and *B*. For example, let *A* be the set of students 20–25 yr old and let *B* be the set of students native to your state. Have students in each set stand up. Then have students in the intersection remain standing.

Definition: The Intersection of Two Sets
Let A and B be sets. The **intersection** of A and B is denoted by $A \cap B$ and is the set of elements that are members of both sets A **and** B.

In the figure, the intersection is represented by the purple piece.

A $A \cap B$ B

Procedure: Finding the Intersection of Two Intervals A and B

Step 1: Draw the graph of each interval.
Step 2: Find the interval of the real number line that contains members from both sets, that is, where the two sets overlap.

Objective 1 Examples

Find the intersection of the sets. Draw the graph of the intersection and write each solution set in interval notation and set-builder notation.

1a. $A = (-\infty, 7]$ and $B = [1, \infty)$
1b. $A = (3, \infty)$ and $B = (0, \infty)$
1c. $A = [4, \infty)$ and $B = (-\infty, 2]$

Solutions

INSTRUCTOR NOTE: Draw a number line on the floor or the board and have students represent numbers in the two sets. Have them visualize the overlap.

1a.

Set	Graph	Interval
A	−1 0 1 2 3 4 5 6 7 8 9	$(-\infty, 7]$ $\{x \mid x \le 7\}$
B	−1 0 1 2 3 4 5 6 7 8 9	$[1, \infty)$ $\{x \mid x \ge 1\}$
$A \cap B$	Overlaps here −1 0 1 2 3 4 5 6 7 8 9 −1 0 1 2 3 4 5 6 7 8 9	$[1, 7]$ $\{x \mid 1 \le x \le 7\}$

So, $A \cap B = [1, 7]$.

1b.

Set	Graph	Interval
A	−3 −2 −1 0 1 2 3 4 5 6 7	$(3, \infty)$ $\{x \mid x > 3\}$
B	−3 −2 −1 0 1 2 3 4 5 6 7	$(0, \infty)$ $\{x \mid x > 0\}$
$A \cap B$	Overlaps here −3 −2 −1 0 1 2 3 4 5 6 7 −3 −2 −1 0 1 2 3 4 5 6 7	$(3, \infty)$ $\{x \mid x > 3\}$

So, $A \cap B = (3, \infty)$.

1c.

Set	Graph	Interval
A	−2 −1 0 1 2 3 4 5 6 7 8	$[4, \infty)$ $\{x \mid x \ge 4\}$
B	−2 −1 0 1 2 3 4 5 6 7 8	$(-\infty, 2]$ $\{x \mid x \le 2\}$
$A \cap B$	No overlap −2 −1 0 1 2 3 4 5 6 7 8 −2 −1 0 1 2 3 4 5 6 7 8	{ } $\varnothing$

Notice that the two sets do not overlap; therefore, there is no intersection. So, $A \cap B$ is the empty set, or $\varnothing$.

Student Check 1 Find the intersection of the sets. Draw the graph of the intersection and write each solution set in interval notation and set-builder notation.

a. $A = [5, \infty)$ and $B = (-\infty, 6)$ **b.** $A = (-\infty, -1)$ and $B = (-\infty, -4)$

c. $A = (-\infty, -3)$ and $B = (5, \infty)$

The Union of Sets

Objective 2 Determine the union of two sets.

We will now examine another operation on sets called the *union*. The union of sets is the joining together of the elements from each set.

The word "union" conjures up images of joining together people or things for a common purpose. For instance, groups of players come together to form a team, and workers often join unions to have someone look out for their interests. Furthermore, when two people get married, they form a union of sorts.

INSTRUCTOR NOTE:
To help illustrate this concept, use the previous example but this time have students in the union remain standing.

Definition: The Union of Two Sets

Let A and B be sets. The **union** of A and B, denoted $A \cup B$, is the set of elements that belong to either set A **or** B.

In the diagram, $A \cup B$ consists of everything in set A and everything in set B.

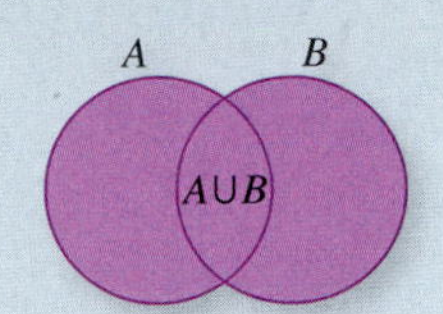

Procedure: Finding the Union of Two Intervals A and B

Step 1: Draw the graph of each interval.
Step 2: Combine each interval on one number line to form the union of the two sets.

Objective 2 Examples

Find the union of the sets. Draw the graph of each union and write the solution set in interval notation and set-builder notation.

2a. $A = (-\infty, 3)$ and $B = (6, \infty)$ **2b.** $A = (0, \infty)$ and $B = (3, \infty)$
2c. $A = (-\infty, 5]$ and $B = [-2, \infty)$

Solutions **2a.**

Set	Graph	Interval
A	–1 0 1 2 3 4 5 6 7 8 9	$(-\infty, 3)$ $\{x \mid x < 3\}$
B	–1 0 1 2 3 4 5 6 7 8 9	$(6, \infty)$ $\{x \mid x > 6\}$
$A \cup B$	–1 0 1 2 3 4 5 6 7 8 9 Include all members of set A and all members of set B	$(-\infty, 3) \cup (6, \infty)$ $\{x \mid x < 3 \text{ or } x > 6\}$

So, $A \cup B = (-\infty, 3) \cup (6, \infty)$.

2b.

Set	Graph	Interval
A	–5 –4 –3 –2 –1 0 1 2 3 4 5	$(0, \infty)$ $\{x \mid x > 0\}$
B	–5 –4 –3 –2 –1 0 1 2 3 4 5	$(3, \infty)$ $\{x \mid x > 3\}$
$A \cup B$	–5 –4 –3 –2 –1 0 1 2 3 4 5 Include all members of set A and all members of set B. Note all members of set B are included in set A, so we do not need to list these again.	$(0, \infty)$ $\{x \mid x > 0\}$

So, $A \cup B = (0, \infty)$.

2c.

Set	Graph	Interval
A	–5 –4 –3 –2 –1 0 1 2 3 4 5	$(-\infty, 5]$ $\{x \mid x \leq 5\}$
B	–5 –4 –3 –2 –1 0 1 2 3 4 5	$[-2, \infty)$ $\{x \mid x \geq -2\}$
$A \cup B$	–5 –4 –3 –2 –1 0 1 2 3 4 5 Include all members of set A and all members of set B. The two sets together cover the entire number line.	$(-\infty, \infty)$ $\{x \mid x \in \mathbb{R}\}$

Notice that the union of the two sets covers the entire set of real numbers. So, $A \cup B = (-\infty, \infty)$.

Student Check 2 Find the union of the sets. Draw the graph of each union and write the solution set in interval notation and set-builder notation.

a. $A = [4, \infty)$ and $B = (-\infty, 0]$ **b.** $A = (-\infty, -2)$ and $B = (-\infty, 3)$

c. $A = [-1, \infty)$ and $B = (-\infty, 5]$

Compound Inequalities Joined by "And"

Objective 3 Solve compound inequalities involving "and."

Compound inequalities are mathematical compound sentences. Recall that a compound sentence is formed by joining two simple sentences by a coordinating conjunction, such as "and" or "or." A compound inequality is defined similarly.

Definition: A **compound inequality** consists of two inequalities joined by the terms "and" or "or."

We will first focus on compound inequalities joined by the term "and." Recall that a compound sentence joined by the term "and" is a statement in which two conditions must be met. Suppose an advertisement for a job states that "Candidates must have a Bachelor's degree *and* candidates must have 5 years of experience." Both conditions must be met for a person to be eligible to apply for the job.

Suppose we want to solve the inequality: $x \geq -1$ and $x \leq 5$. Solutions of this compound inequality are numbers that satisfy both parts of the inequality. Consider the possible solutions.

Value	$x \geq -1$ and $x \leq 5$	Is the value a solution?
$x = 3$	$3 \geq -1$ and $3 \leq 5$	Because 3 makes both inequalities true, the compound inequality is *true*. So, 3 is a solution of the compound inequality.
$x = -2$	$-2 \geq -1$ and $-2 \leq 5$	Because -2 makes only the second inequality true, the compound inequality is *false*. So, -2 is not a solution of the compound inequality.

So, solutions of the compound inequality, $x \geq -1$ and $x \leq 5$, are values that are both greater than or equal to -1 and also less than or equal to 5. The solutions lie between and include -1 and 5. This can be written in a more compact way as $-1 \leq x \leq 5$. Thus, the solution set is $[-1, 5]$ or $\{x \mid -1 \leq x \leq 5\}$.

The values of the variable that make both inequalities true are solutions of a compound inequality joined by "and." So, solutions must lie in the *intersection* of the solution sets of each individual inequality in the compound inequality.

Procedure: Solving a Compound Inequality Involving "And"

Step 1: Find the solution set of inequality 1.
Step 2: Find the solution set of inequality 2.
Step 3: Find the intersection of the solution sets of inequalities 1 and 2.
Step 4: Write the final solution set in interval notation and provide its graph.

Objective 3 Examples Solve each compound inequality. Write each solution set in interval notation and graph the solution set.

3a. $x + 5 \geq 8$ and $-2x \geq -10$ **3b.** $3y + 1 < 5$ and $2(y - 3) < 2$

3c. $4 - 2x > 0$ and $5x \geq 15$ **3d.** $2x - 1 \leq 5$ and $2x - 1 \geq -5$

3e. $-7 \leq 4x + 3 \leq 10$

Solutions **3a.**

Subtract 5 from each side.	$x + 5 \geq 8$ $x + 5 - 5 \geq 8 - 5$	and	$-2x \geq -10$ $\dfrac{-2x}{-2} \leq \dfrac{-10}{-2}$	Divide each side by −2 and reverse the inequality symbol.
Simplify.	$x \geq 3$		$x \leq 5$	

Now we find the intersection of the solution sets of the two inequalities.

Inequality 1	$x \geq 3$ $[3, \infty)$	−1 0 1 2 3 4 5 6 7 8 9
Inequality 2	$x \leq 5$ $(-\infty, 5]$	−1 0 1 2 3 4 5 6 7 8 9
Intersection of the inequalities	$[3, 5]$	−1 0 1 2 3 4 5 6 7 8 9

Check: The numbers in the intersection of the two inequalities should make the compound inequality true. Numbers not in the intersection of the two inequalities will make the compound inequality false.

$$
\begin{aligned}
x = 4:\quad x + 5 &\geq 8 \quad \text{and} \quad -2x \geq -10 \\
4 + 5 &\geq 8 \quad \text{and} \quad -2(4) \geq -10 \\
9 &\geq 8 \quad \text{and} \quad -8 \geq -10 \qquad \text{True}
\end{aligned}
$$

$$
\begin{aligned}
x = 0:\quad x + 5 &\geq 8 \quad \text{and} \quad -2x \geq -10 \\
0 + 5 &\geq 8 \quad \text{and} \quad -2(0) \geq -10 \\
5 &\geq 8 \quad \text{and} \quad 0 \geq -10 \qquad \text{False}
\end{aligned}
$$

Note that 4 makes both inequalities true and so is a solution of the compound inequality. The value 0 makes one of the inequalities false and is, therefore, not a solution of the compound inequality. So, the solution set is [3, 5].

3b.

	$3y + 1 < 5$	and	$2(y - 3) < 2$	
Subtract 1 from each side.	$3y + 1 - 1 < 5 - 1$		$2y - 6 < 2$	Distribute.
Simplify.	$3y < 4$		$2y - 6 + 6 < 2 + 6$	Add 6 to each side.
Divide each side by 3.	$y < \dfrac{4}{3}$		$2y < 8$	Simplify.
			$y < 4$	Divide each side by 2.

Now we find the intersection of the solution sets of the two inequalities.

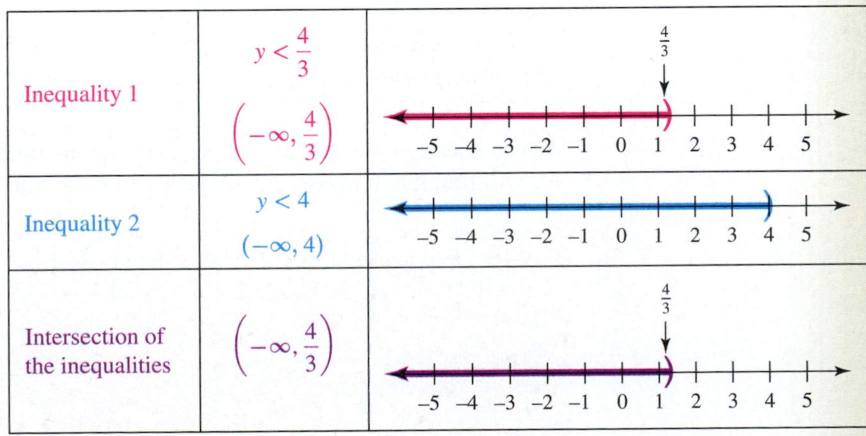

Inequality 1	$y < \dfrac{4}{3}$ $\left(-\infty, \dfrac{4}{3}\right)$	$\frac{4}{3}$ −5 −4 −3 −2 −1 0 1 2 3 4 5
Inequality 2	$y < 4$ $(-\infty, 4)$	−5 −4 −3 −2 −1 0 1 2 3 4 5
Intersection of the inequalities	$\left(-\infty, \dfrac{4}{3}\right)$	$\frac{4}{3}$ −5 −4 −3 −2 −1 0 1 2 3 4 5

We can check by substituting a value from the intersection of the sets and will find that this value makes both inequalities true. So, the solution set is $\left(-\infty, \frac{4}{3}\right)$.

3c.

	$4 - 2x > 0$	and	$5x \geq 15$	
Subtract 4 from each side.	$4 - 2x - 4 > 0 - 4$		$\frac{5x}{5} \geq \frac{15}{5}$	Divide each side by 5.
Simplify.	$-2x > -4$			
Divide each side by −2 and reverse the inequality sign.	$\frac{-2x}{-2} < \frac{-4}{-2}$		$x \geq 3$	Simplify.
	$x < 2$			

Now we find the intersection of the solution sets of the two inequalities.

Inequality 1	$x < 2$ $(-\infty, 2)$	−5 −4 −3 −2 −1 0 1 2 3 4 5
Inequality 2	$x \geq 3$ $[3, \infty)$	−5 −4 −3 −2 −1 0 1 2 3 4 5
Intersection of the inequalities	$\varnothing$	−5 −4 −3 −2 −1 0 1 2 3 4 5

We can check by substituting values into the compound inequality. We will find that there are no values that make the inequality true. So, the solution set is the empty set, or $\varnothing$.

3d.

	$2x - 1 \leq 5$	and	$2x - 1 \geq -5$	
Add 1 to each side.	$2x - 1 + 1 \leq 5 + 1$		$2x - 1 + 1 \geq -5 + 1$	Add 1 to each side.
Divide each side by 2.	$2x \leq 6$		$2x \geq -4$	Divide each side by 2.
	$x \leq 3$		$x \geq -2$	

Now we find the intersection of the solution sets of the two inequalities.

Inequality 1	$x \leq 3$ $(-\infty, 3]$	−4 −3 −2 −1 0 1 2 3 4 5 6
Inequality 2	$x \geq -2$ $[-2, \infty)$	−4 −3 −2 −1 0 1 2 3 4 5 6
Intersection of the inequalities	$[-2, 3]$	−4 −3 −2 −1 0 1 2 3 4 5 6

We can check by substituting a value from the intersection into the compound inequality. We will find that it makes the inequality true. So, the solution set is [−2, 3].

Note: *This problem can also be worked using a compact form. The inequality* $2x - 1 \geq -5$ *is the same as* $-5 \leq 2x - 1$. *So, we have that*

$$-5 \leq 2x - 1 \quad \text{and} \quad 2x - 1 \leq 5$$

This is equivalent to

$$-5 \leq 2x - 1 \leq 5$$

This three-part inequality can be solved by applying the same operation to each part. The goal is to isolate the variable in the middle.

$$\begin{array}{rcccl l} -5 & \le & 2x-1 & \le & 5 & \\ -5+1 & \le & 2x-1+1 & \le & 5+1 & \text{Add 1 to each part.} \\ -4 & \le & 2x & \le & 6 & \text{Simplify.} \\ \frac{-4}{2} & \le & \frac{2x}{2} & \le & \frac{6}{2} & \text{Divide each part by 2.} \\ -2 & \le & x & \le & 3 & \text{Simplify.} \end{array}$$

3e.

$$\begin{array}{rcccl l} -7 & \le & 4x+3 & \le & 10 & \\ -7-3 & \le & 4x+3-3 & \le & 10-3 & \text{Subtract 3 from each part.} \\ -10 & \le & 4x & \le & 7 & \text{Simplify.} \\ \frac{-10}{4} & \le & \frac{4x}{4} & \le & \frac{7}{4} & \text{Divide each part by 4.} \\ -\frac{5}{2} & \le & x & \le & \frac{7}{4} & \text{Simplify.} \end{array}$$

So, the graph of the solution set is

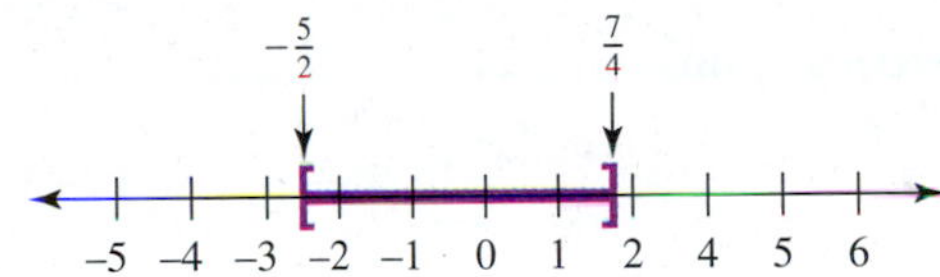

The solution set is $\left[-\frac{5}{2}, \frac{7}{4}\right]$.

Student Check 3 Solve each compound inequality. Write each solution set in interval notation and graph the solution set. Solve (d) also using the compact form.

a. $x + 4 \ge 1$ and $-3x \ge -6$

b. $2y - 3 > 7$ and $4(y + 2) > 4$

c. $8 + 5x > -2$ and $7x \le -21$

d. $1 - 9x < 8$ and $1 - 9x > -8$

e. $-6 \le 7x + 5 \le 19$

Compound Inequalities Joined by "Or"

Objective 4 Solve compound inequalities involving "or."

We will now focus on compound inequalities joined by the term "or." Recall that a compound sentence joined by the term "or" is a statement in which one of two conditions must be met. Suppose the advertisement for a job states that "Candidates must have a Bachelor's degree *or* candidates must have 5 years of experience." In this case, only one of the conditions is necessary for a person to apply for the job. Note a person may also meet both conditions and be eligible to apply.

Suppose we want to solve the inequality: $x < -1$or $x > 5$. Solutions of this compound inequality are numbers that satisfy at least one part of the inequality. Consider the possible solutions.

Value	$x < -1$ or $x > 5$	Is the value a solution?
$x = 3$	$3 < -1$ or $3 > 5$	Because 3 makes both inequalities false, the compound inequality is *false*. So, 3 is a not a solution of the compound inequality.
$x =- 2$	$-2 < -1$ or $-2 > 5$	Because −2 makes the first inequality true, the compound inequality is *true*. So, −2 is a solution of the compound inequality.
$x = 7$	$7 < -1$ or $7 > 5$	Because 7 makes the second inequality true, the compound inequality is *true*. So, 7 is a solution of the compound inequality.

The values of the variable that make at least one inequality true are solutions of a compound inequality joined by "or." Thus, values that lie in the *union* of the solution sets of each inequality are solutions of the compound inequality.

Procedure: Solving a Compound Inequality Involving "Or"

Step 1: Find the solution set of inequality 1.
Step 2: Find the solution set of inequality 2.
Step 3: Find the union of the solution sets of inequalities 1 and 2.
Step 4: Write the final solution set in interval notation and provide its graph.

Objective 4 Examples

Solve each compound inequality. Write each solution set in interval notation and graph the solution set.

4a. $x + 3 < -1$ or $2x > -4$

4b. $\frac{2}{3}x - 4 \geq 2$ or $-3(x + 1) \geq 5$

4c. $2x - 4 > -8$ or $7x - 5 < 2$

Solutions

4a. $x + 3 < -1$ or $2x > -4$

Subtract 3 from each side.	$x + 3 - 3 < -1 - 3$		$\frac{2x}{2} > \frac{-4}{2}$	Divide each side by 2.
Simplify.	$x < -4$		$x > -2$	Simplify.

Now we find the union of the two solution sets.

Inequality 1	$x < -4$ $(-\infty, -4)$	−8 −7 −6 −5 −4 −3 −2 −1 0 1 2
Inequality 2	$x > -2$ $(-2, \infty)$	−8 −7 −6 −5 −4 −3 −2 −1 0 1 2
Union of two sets	$(-\infty, -4) \cup (-2, \infty)$	−8 −7 −6 −5 −4 −3 −2 −1 0 1 2

We can check by substituting values from the union in the original compound inequality and will find that these values make at least one of the inequalities true. So, the solution set is $(-\infty, -4) \cup (-2, \infty)$.

4b. $\frac{2}{3}x - 4 \geq 2$ or $-3(x + 1) \geq 5$

		$-3x - 3 \geq 5$	Distribute.
Multiply each side by 3.	$3\left(\frac{2}{3}x - 4\right) \geq 3(2)$	$-3x - 3 + 3 \geq 5 + 3$	Add 3 to each side.
Simplify.	$2x - 12 \geq 6$	$-3x \geq 8$	Simplify.
Add 12 to each side.	$2x - 12 + 12 \geq 6 + 12$	$\frac{-3x}{-3} \leq \frac{8}{-3}$	Divide each side by −3 and reverse the inequality symbol.
Simplify.	$2x \geq 18$	$x \leq -\frac{8}{3}$	
Divide each side by 2.	$x \geq 9$		

Now we find the union of the two solution sets.

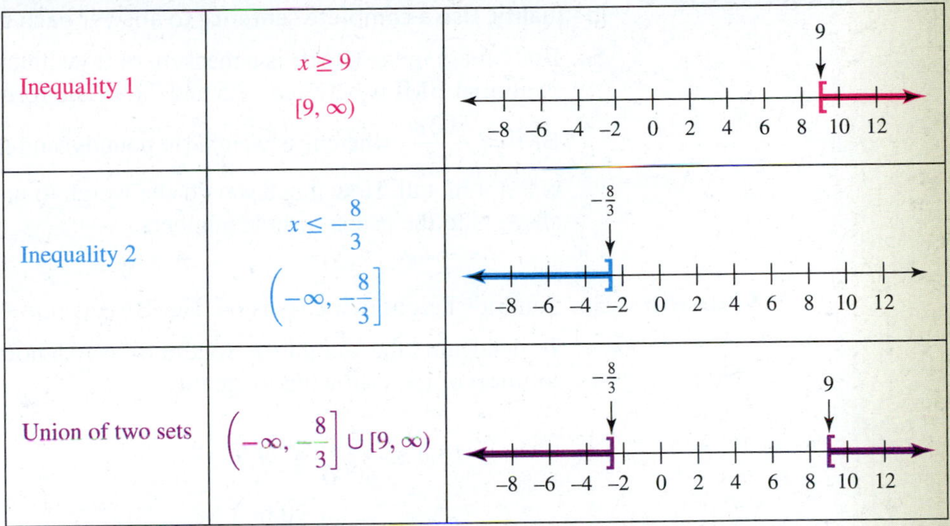

Inequality 1	$x \geq 9$ $[9, \infty)$	−8 −6 −4 −2 0 2 4 6 8 10 12
Inequality 2	$x \leq -\frac{8}{3}$ $\left(-\infty, -\frac{8}{3}\right]$	−8 −6 −4 −2 0 2 4 6 8 10 12
Union of two sets	$\left(-\infty, -\frac{8}{3}\right] \cup [9, \infty)$	−8 −6 −4 −2 0 2 4 6 8 10 12

We can check by substituting values from the union of the inequalities and will find that these values make at least one of the inequalities true. So, the solution set is $\left(-\infty, -\frac{8}{3}\right] \cup [9, \infty)$.

4c.

	$2x - 4 > -8$	or	$7x - 5 < 2$	
Add 4 to each side.	$2x - 4 + 4 > -8 + 4$		$7x - 5 + 5 < 2 + 5$	Add 5 to each side.
Simplify.	$2x > -4$		$7x < 7$	Simplify.
Divide each side by 2.	$x > -2$		$x < 1$	Divide each side by 7.

Now we find the union of the two solution sets.

Inequality 1	$x > -2$ $(-2, \infty)$	−5 −4 −3 −2 −1 0 1 2 3 4 5
Inequality 2	$x < 1$ $(-\infty, 1)$	−5 −4 −3 −2 −1 0 1 2 3 4 5
Union of two sets	$(-\infty, \infty)$	−5 −4 −3 −2 −1 0 1 2 3 4 5

We can check by substituting any number into the original inequality and find that it makes the compound inequality true. So, the solution set is all real numbers, $\mathbb{R}$, or $(-\infty, \infty)$.

Student Check 4 Solve each compound inequality. Write each solution set in interval notation and graph the solution set.

a. $x - 1 > 4$ or $-2x > -10$

b. $\frac{9}{4}x - 5 \geq -3$ or $6(x + 2) < 7$

c. $-4x - 9 > -1$ or $2 - 3x < 11$

Objective 5 ▶

Solve applications of compound inequalities.

Applications

Applications of compound inequalities arise when there are two conditions that define a situation. When both conditions must be satisfied, an "and" statement is used. When only one of the conditions must be satisfied, an "or" statement is used.

Objective 5 Examples **Write a compound inequality that represents each situation and solve the inequality. Use a complete sentence to answer each question.**

5a. Body mass index (BMI) is a measure of how much body fat a person has. A normal BMI is between 18.5 and 24.9. The formula to calculate BMI is $\text{BMI} = \frac{703w}{h^2}$, where w is weight (in pounds) and h is height (in inches). Sondra is 5 ft 4 in. tall. How much should she weigh to have a normal BMI? Round answers to the nearest whole numbers.

Solution **5a.** Sondra's height in inches is 64. Her BMI is represented by $\frac{703w}{64^2} = \frac{703w}{4096}$. To determine the weight for Sondra to have a normal BMI, we need to find the solution of the following inequality.

$$18.5 \le \frac{703w}{4096} \le 24.9$$

$$4096(18.5) \le 4096\left(\frac{703w}{4096}\right) \le 4096(24.9) \quad \text{Multiply each part by 4096.}$$

$$75{,}776 \le 703w \le 101{,}990.4 \quad \text{Simplify.}$$

$$\frac{75{,}776}{703} \le \frac{703w}{703} \le \frac{101{,}990.4}{703} \quad \text{Divide each part by 703.}$$

$$107.8 \le w \le 145.1 \quad \text{Approximate.}$$

Sondra should weigh between 108 and 145 lb to have a normal BMI.

5b. Neal needs to earn a B in his math class to keep his 3.0 GPA. His final grade is calculated using the following percentages: tests (45%), quizzes (20%), projects (10%), and final (25%). Neal has a 73 test average, 85 quiz average, and 82 project average. What grades could he earn on the final exam for his average to be between 80 and 89? (The maximum grade on the final exam is 100.)

Solution **5b.** The unknown is Neal's final exam grade. So, we let f represent the final exam grade. Neal's final average is computed as shown.

$$\begin{aligned}\text{Final average} &= 0.45(73) + 0.20(85) + 0.10(82) + 0.25f \\ &= 32.85 + 17 + 8.2 + 0.25f \\ &= 58.05 + 0.25f\end{aligned}$$

For Neal's final average to be between 80 and 89, we solve the following inequality.

$$80 \le 58.05 + 0.25f \le 89$$

$$80 - 58.05 \le 58.05 + 0.25f - 58.05 \le 89 - 58.05 \quad \text{Subtract 58.05 from each part.}$$

$$21.95 \le 0.25f \le 30.95 \quad \text{Simplify.}$$

$$\frac{21.95}{0.25} \le \frac{0.25f}{0.25} \le \frac{30.95}{0.25} \quad \text{Divide each part by 0.25.}$$

$$87.5 \le f \le 123.8 \quad \text{Simplify.}$$

To get a B in the course, Neal needs to score between 88 and 100 since 100 is the highest grade possible. This also tells us that Neal cannot earn a grade higher than a B, since a grade of 123.8 would be needed to earn an 89 average.

Student Check 5 Write a compound inequality that represents each situation and solve the inequality. Use a complete sentence to answer each question.

a. Bryan is 6 ft tall. How much should he weigh to have a normal BMI? Use the formula in Example 5(a).

b. Ameena is attempting to earn a C in her chemistry class. Her final grade is computed as follows: 60% (test average), 15% (quizzes), and 25% (final exam). Ameena has a test average of 62 and a quiz average of 78. What grades could she earn on her final to have a final grade between 70 and 79?

Objective 6 Troubleshoot common errors.

Troubleshooting Common Errors

Some common errors related to solving compound inequalities are shown.

Objective 6 Examples A problem and an incorrect solution are given. Provide the correct solution and an explanation of the error.

6a. Solve $5x - 5 < -5$ or $-2x < 0$.

Incorrect Solution	Correct Solution and Explanation
$5x - 5 < -5$ or $-2x < 0$ $5x < 0$ or $x > 0$ $x < 0$ The solution set is all real numbers, or $\mathbb{R}$.	Since neither inequality includes zero as a solution, the union can't include zero. So, the solution set is $(-\infty, 0) \cup (0, \infty)$.

6b. Solve $5x - 5 \le -5$ or $-2x \le 0$.

Incorrect Solution	Correct Solution and Explanation
$5x - 5 \le -5$ or $-2x \le 0$ $5x \le 0$ or $x \ge 0$ $x \le 0$ The solution set is the empty set, or $\varnothing$.	Since both inequalities include zero as a solution, the intersection will include zero. So, the solution set is $\{0\}$.

ANSWERS TO STUDENT CHECKS

Student Check 1

a.

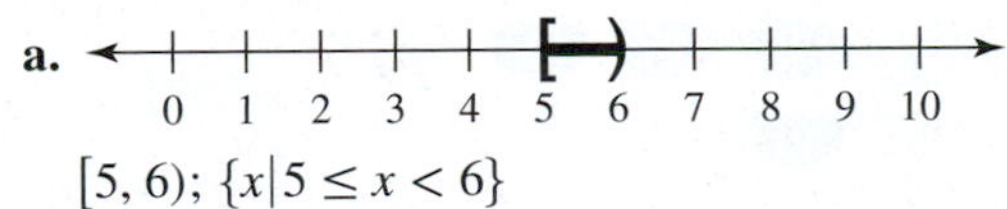

$[5, 6)$; $\{x \mid 5 \le x < 6\}$

b.

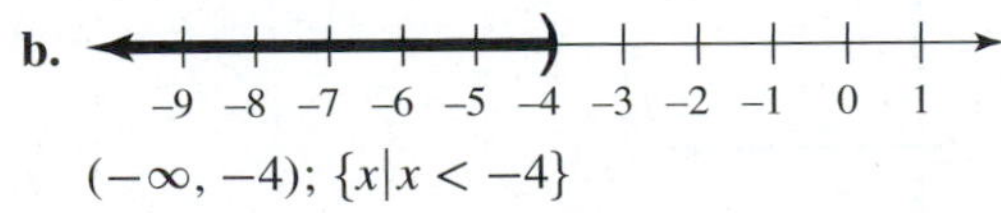

$(-\infty, -4)$; $\{x \mid x < -4\}$

c.

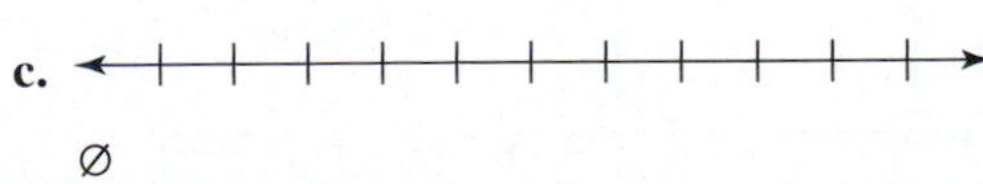

$\varnothing$

Student Check 2 **a.** $(-\infty, 0] \cup (4, \infty)$ **b.** $(-\infty, 3)$ **c.** $(-\infty, \infty)$

Student Check 3 **a.** $[-3, 2]$

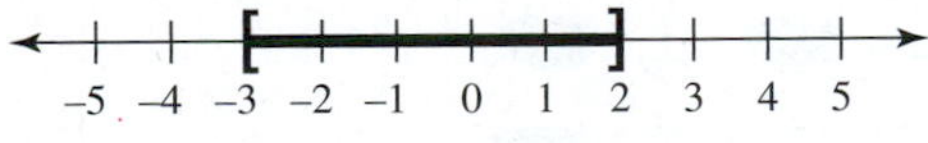

b. $(5, \infty)$

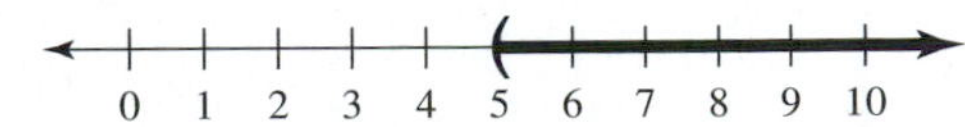

c. $\varnothing$

d. $\left(-\frac{7}{9}, 1\right)$

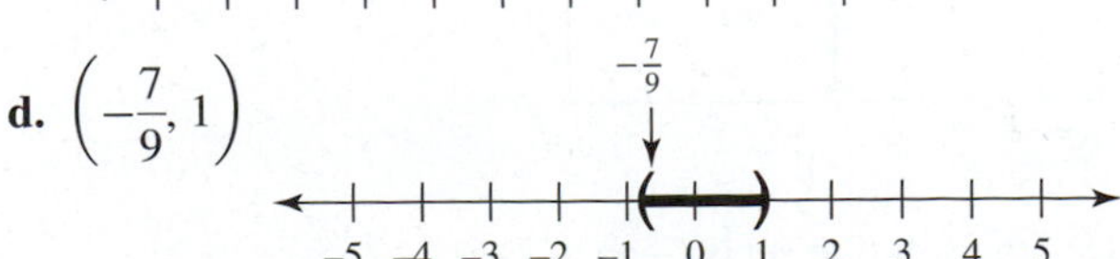

e. $\left[-\frac{11}{7}, 2\right]$

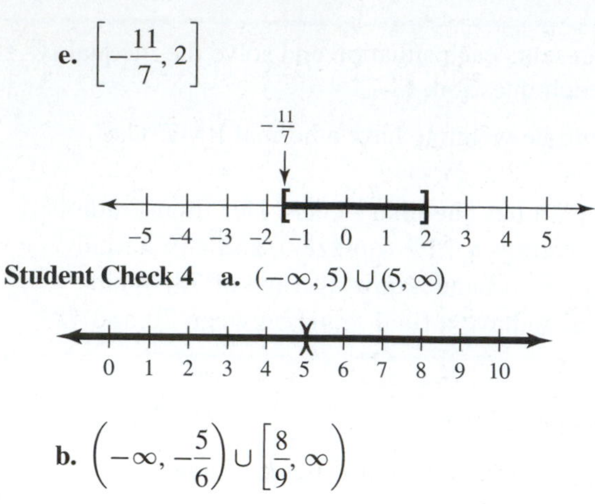

Student Check 4 **a.** $(-\infty, 5) \cup (5, \infty)$

b. $\left(-\infty, -\frac{5}{6}\right) \cup \left[\frac{8}{9}, \infty\right)$

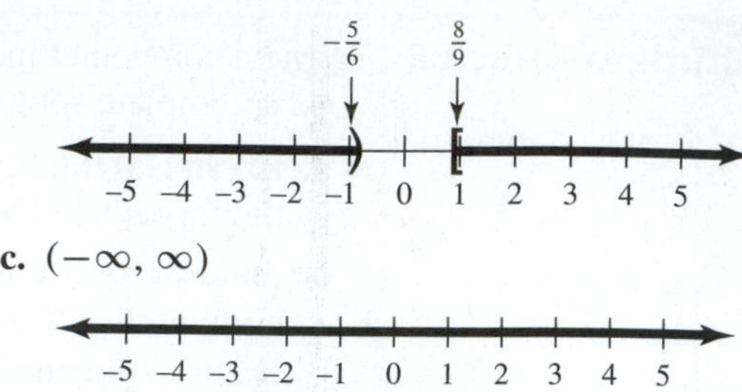

c. $(-\infty, \infty)$

Student Check 5 **a.** Bryan should weigh between 136 and 184 lb. **b.** Ameena must score between 85 and 100 on her final to make a 70–79 in chemistry.

SUMMARY OF KEY CONCEPTS

1. Elements that two sets have in common form the intersection of the two sets.
2. The elements that are in either set form the union of the two sets.
3. To solve a compound inequality involving "and" means to find the intersection of the solution sets of each inequality.
4. To solve a compound inequality involving "or" means to find the union of the solution sets of each inequality.

GRAPHING CALCULATOR SKILLS

We can use the graphing calculator to graph the solution set of a compound inequality. It is best to solve the inequality by hand and then use the calculator to confirm the solution.

Example 1: Solve $x + 6 < -1$ or $2x > -4$.

Solution: Enter the compound inequality into Y_1. Insert parentheses around each inequality. Use the Test menu (2nd Math) to access the inequality symbols and the coordinating conjunction "or" (2nd Math, Logic). Then graph the solution. The interval that contains the solutions will be highlighted above the x-axis.

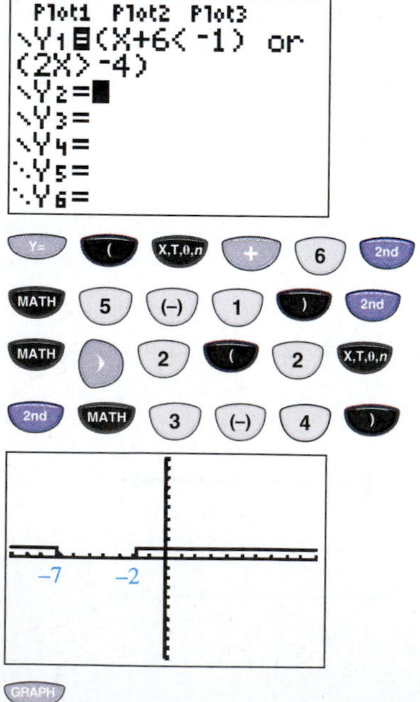

So, the solution set is $(-\infty, -7) \cup (-2, \infty)$.

Example 2: $x + 5 \ge 8$ and $-2x \ge -10$.

Solution: Enter the compound inequality into Y_1. Use the Test menu (2nd Math) to access the inequality symbols and the coordinating conjunction "and" (2nd Math, Logic). Then graph the solution. The interval that contains the solutions will be highlighted above the x-axis.

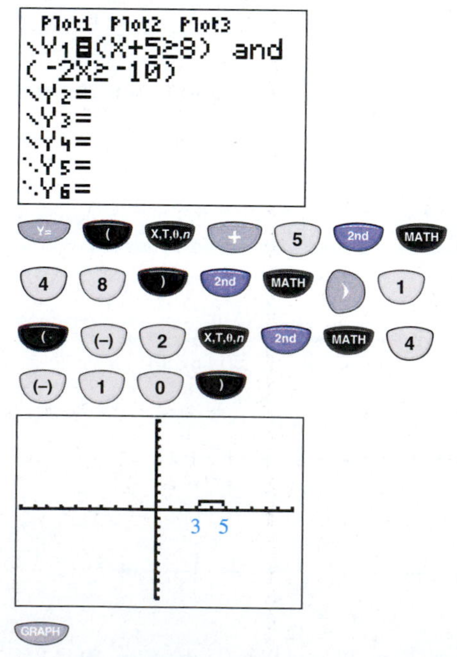

So, the solution set is $[3, 5]$.

SECTION 2.5 EXERCISE SET

Write About It!

Use complete sentences in your answer to each exercise.

1. Explain the meaning of the intersection of sets.
2. Explain the meaning of the union of sets.
3. Is it possible for the intersection of two sets to be the empty set? Explain.
4. Is it possible for the union of two sets to be the empty set? Explain.
5. What is a compound inequality?
6. Explain the difference of the graphs of compound inequalities $-a < x < a$ and $x < -a$ or $x > a$, where $a > 0$.
7. How do you find the solutions of a compound inequality involving "and"?
8. How do you find the solutions of a compound inequality involving "or"?

Practice Makes Perfect!

Find the intersection and union of the following sets. Write each solution set in interval notation. (*See Objectives 1 and 2.*)

9. $A = (-10, 6)$, $B = [-4, 2]$ $A \cap B = [-4, 2], A \cup B = (-10, 6)$
10. $A = (-8, 3)$, $B = [-7, 1]$ $A \cap B = [-7, 1], A \cup B = (-8, 3)$
11. $A = [-10, -3]$, $B = [-9, -3]$ $A \cap B = [-9, -3], A \cup B = [-10, -3]$
12. $A = [-13, -5]$, $B = [-12, -5]$ $A \cap B = [-12, -5], A \cup B = [-13, -5]$
13. $A = [6, 20]$, $B = (6, 20)$ $A \cap B = (6, 20), A \cup B = [6, 20]$
14. $A = [-7, 5]$, $B = (-7, 5)$ $A \cap B = (-7, 5), A \cup B = [-7, 5]$
15. $A = (-\infty, 7)$, $B = [-1, 7]$ $A \cap B = [-1, 7), A \cup B = (-\infty, 7]$
16. $A = (-\infty, -2]$, $B = [-4, -2)$ $A \cap B = [-4, -2), A \cup B = (-\infty, -2]$
17. $A = (-11, \infty)$, $B = (-\infty, 11)$ $A \cap B = (-11, 11), A \cup B = (-\infty, \infty)$
18. $A = (-\infty, 53)$, $B = (-53, \infty)$ $A \cap B = (-53, 53), A \cup B = (-\infty, \infty)$
19. $A = [-12, 5]$, $B = (-10, 8)$ $A \cap B = (-10, 5], A \cup B = [-12, 8)$
20. $A = [-2, 10]$, $B = (-4, 6)$ $A \cap B = [-2, 6), A \cup B = (-4, 10]$

Solve each compound inequality. Write each solution set in interval notation. (*See Objective 3.*)

21. $x \le 5$ and $x \ge -1$ $[-1, 5]$
22. $x \le 1$ and $x \ge -10$ $[-10, 1]$
23. $x > 3$ and $x < 2$ $\varnothing$
24. $x \le 8$ and $x > 10$ $\varnothing$
25. $2x + 4 > 8$ and $3x < 9$ $(2, 3)$
26. $9x + 3 > 12$ and $10x \le 40$ $(1, 4]$
27. $3x + 6 \ge 3$ and $7x - 13 < -20$ $\varnothing$
28. $6x - 1 > 17$ and $4x - 5 \le 29$ $\left(3, \frac{17}{2}\right]$
29. $6 - 4x > 2$ and $1 - 3x < 5$ $\left(-\frac{4}{3}, 1\right)$
30. $7 - 8x > -1$ and $2 - 5x < 2$ $(0, 1)$
31. $9x + 11 \ge 2$ and $4x - 3 \le -7$ $\{-1\}$
32. $12x + 1 \ge 13$ and $5x - 6 \le -1$ $\{1\}$
33. $5(3x - 2) \le 15$ and $2(4x + 1) \ge 10$ $\left[1, \frac{5}{3}\right]$
34. $4(2x - 3) \le 6$ and $3(5x + 10) \ge 0$ $\left[-2, \frac{9}{4}\right]$
35. $x > 7$ and $x > 5$ $(7, \infty)$
36. $x \le -3$ and $x \le -7$ $(-\infty, -7]$
37. $6x - 5 > 1$ and $7x + 6 > 20$ $(2, \infty)$
38. $4x - 3 > 5$ and $5x - 1 \ge 9$ $(2, \infty)$
39. $3x - 2 \ge 10$ and $1 - 3x < 40$ $[4, \infty)$
40. $2x - 7 > 0$ and $1 - 2x < 15$ $\left(\frac{7}{2}, \infty\right)$
41. $6 - 3x > 4$ and $2 - 2x > 10$ $(-\infty, -4)$
42. $3 - 10x \le 23$ and $10 - 11x \le 43$ $[-2, \infty)$
43. $3x + 9 \ge 4$ and $9x > -15$ $\left(-\frac{5}{3}, \infty\right)$
44. $4x - 1 \le 11$ and $3x < 9$ $(-\infty, 3)$
45. $2(4x - 3) < 4$ and $6(x - 5) < 12$ $\left(-\infty, \frac{5}{4}\right)$
46. $4(2x + 1) > 5$ and $8(x - 2) > 3$ $\left(\frac{19}{8}, \infty\right)$

Solve each compound inequality. Write each solution set in interval notation. (*See Objective 4.*)

47. $x < -3$ or $x > 1$ $(-\infty, -3) \cup (1, \infty)$
48. $x < -4$ or $x > 0$ $(-\infty, -4) \cup (0, \infty)$
49. $x \ge -7$ or $x < 3$ $(-\infty, \infty)$
50. $x > 2$ or $x \le 10$ $(-\infty, \infty)$
51. $x \le 0$ or $x \le 6$ $(-\infty, 6]$
52. $x \ge 10$ or $x \ge 15$ $[10, \infty)$
53. $11x - 3 \ge 52$ or $6x \le 42$ $(-\infty, \infty)$
54. $13x < -52$ or $3 - 5x > 17$ $\left(-\infty, -\frac{14}{5}\right)$

Additional answers can be found in the Instructor Answer Appendix.

55. $12x \le -60$ or $7 - 9x > 3$ $\left(-\infty, \frac{4}{9}\right)$

56. $13x \ge 65$ or $11 - 5x > 6$ $(-\infty, 1) \cup [5, \infty)$

57. $5x + 3 > 8$ or $7x - 4 \le -3$ $\left(-\infty, \frac{1}{7}\right] \cup (1, \infty)$

58. $2x + 5 > 6$ or $3x - 1 \le -4$ $(-\infty, -1] \cup \left(\frac{1}{2}, \infty\right)$

59. $5x + 1 \le 7$ or $6x - 2 < 1$ $\left(-\infty, \frac{6}{5}\right]$

60. $4x - 7 < 9$ or $2x + 3 < 4$ $(-\infty, 4)$

61. $10(2 - 9x) \ge 20$ or $2(1 + 3x) > 6$ $(-\infty, 0] \cup \left(\frac{2}{3}, \infty\right)$

62. $10(3 - x) \le 15$ or $3(2 - 5x) > 13$ $\left(-\infty, -\frac{7}{15}\right) \cup \left[\frac{3}{2}, \infty\right)$

Write a compound inequality that represents each situation and solve the inequality. Use a complete sentence to answer each question. (*See Objective 5.*)

63. If Lawrence's quiz scores are 90, 82, 75, and 78, what must he earn on his fifth quiz to have an average between 80 and 89? Lawrence must earn between 75 and 100 on his fifth quiz.

64. If Emma's quiz scores are 99, 89, 80, and 70, what must she earn on her fifth quiz to have an average between 80 and 89? Emma must earn between 62 and 100 on her fifth quiz.

65. Sakinah's course grade is based on a weighted average. The weights are as follows: homework (15%), quizzes (25%), tests (40%), and final exam (20%). If Sakinah's homework average is 100, quiz average is 90, and test average is 86, what must she score on the final exam to have an average between 90 and 100? Sakinah must score between 90.5 and 100 on the final exam.

66. Su Ming's course grade is based on a weighted average. The weights are as follows: homework (15%), quizzes (25%), tests (40%), and final exam (20%). If Su Ming's homework average is 75, quiz average is 70, test average is 83, what must she score on the final exam for her course average to be between 80 and 89? Su Ming must earn between 90.25 and 100 on the final exam.

67. A cell phone company charges a monthly fee of \$29.99 for 300 anytime cell phone minutes and \$0.40 per additional minute. How many additional minutes can a subscriber use if he budgets between \$40 and \$50 for his phone bill? The subscriber can use between 25 and 50 additional minutes.

68. A cell phone company charges a monthly fee of \$49.99 for 800 anytime cell phone minutes and \$0.25 per additional minute. If Callie budgets a range of \$70 to \$100 each month for her phone bill, how many additional minutes can she afford to use? Callie can use between 80 and 200 additional minutes.

69. Miguel pays \$3000 to rent a hall for his fiftieth birthday party. The buffet costs \$18 per person. If Miguel has between \$4000 and \$5000 to spend on his party, how many people can he invite to his birthday party? Miguel can invite between 55 and 111 people to his birthday party.

70. Marianne is planning a graduation party for her daughter. She finds a banquet hall that charges \$850 and a caterer who charges \$20 per person. If Marianne can spend between \$2000 and \$3000 on the party, how many people can attend the party? Between 57 and 107 people can attend the party.

Mix 'Em Up!

Solve each problem. Write each answer in interval notation when applicable.

71. $[-7, 8] \cup (-3, 12)$ $[-7, 12)$

72. $(-20, 5) \cup [-5, 9]$ $(-20, 9]$

73. $[-20, 9] \cap (-20, 9]$ $(-20, 9]$

74. $[-1, 2] \cap (-1, 2]$ $(-1, 2]$

75. $(-\infty, 6] \cup (6, \infty)$ $(-\infty, \infty)$

76. $(-\infty, 3) \cup [3, \infty)$ $(-\infty, \infty)$

77. $(-\infty, 3) \cap (6, \infty)$ $\emptyset$

78. $(-\infty, 5] \cap [10, \infty)$ $\emptyset$

79. $(-\infty, 6.5) \cup [-3.5, 12.2)$ $(-\infty, 12.2)$

80. $(-\infty, 6.5) \cap [-3.5, 12.2)$ $[-3.5, 6.5)$

81. $\emptyset \cap [2, 7]$ $\emptyset$

82. $\emptyset \cap (-9, 7)$ $\emptyset$

83. $\emptyset \cup (-10, \infty)$ $(-10, \infty)$

84. $\emptyset \cup [-1, \infty)$ $[-1, \infty)$

85. $7x + 6 < 11$ or $5x - 5 > 12$ $\left(-\infty, \frac{5}{7}\right) \cup \left(\frac{17}{5}, \infty\right)$

86. $10x - 5 \ge 2$ and $4x + 3 < 10$ $\left[\frac{7}{10}, \frac{7}{4}\right)$

87. $2(1 - 4x) > 10$ and $4x - 5 < 11$ $(-\infty, -1)$

88. $3(2 - x) > -6$ and $6x + 1 < 7$ $(-\infty, 1)$

89. $\frac{2}{3}x - \frac{7}{6} > \frac{5}{6}$ or $\frac{3}{5}x + \frac{1}{2} < \frac{7}{10}$ $\left(-\infty, \frac{1}{3}\right) \cup (3, \infty)$

90. $\frac{3}{4}x - \frac{1}{8} > 1$ or $\frac{5}{6}x + 2 < \frac{2}{3}$ $\left(-\infty, -\frac{8}{5}\right) \cup \left(\frac{3}{2}, \infty\right)$

91. $1 - \frac{3}{7}x < \frac{1}{2}$ and $1 + \frac{1}{2}x < \frac{3}{5}$ $\emptyset$

92. $2 - \frac{5}{2}x > \frac{13}{2}$ and $\frac{7}{3}x - 1 > \frac{2}{5}$ $\emptyset$

93. $3 - 4x \le 0$ or $6x + 1 > 3$ $\left(\frac{1}{3}, \infty\right)$

94. $2 - 5x \le 1$ or $4x - 3 > 5$ $\left[\frac{1}{5}, \infty\right)$

95. $8x + 9 > 3$ and $7x - 2 > -3$ $\left(-\frac{1}{7}, \infty\right)$

96. $7x - 1 \geq 5$ and $3x + 4 > -2$ $\left[\frac{6}{7}, \infty\right)$

97. $3x - 5 > 7$ or $2x + 6 < 1$ $\left(-\infty, -\frac{5}{2}\right) \cup (4, \infty)$

98. $12x - 4 \leq 8$ and $3x + 2 > 2$ $(0, 1]$

99. $0.3x + 0.12 < 3.27$ and $0.4x - 1.36 > 0.54$ $(4.75, 10.5)$

100. $1.56x - 3.2 \geq 2.26$ and $0.8x + 5.6 \leq 2.4$ $\varnothing$

101. $2x + 6 < 8$ or $4x - 9 < 2$ $\left(-\infty, \frac{11}{4}\right)$

102. $8x - 3 \leq 3$ or $5x - 1 < 7$ $\left(-\infty, \frac{8}{5}\right)$

Write a compound inequality that represents each situation and solve the inequality. Use a complete sentence to answer each question.

103. Frederick needs the work of an electrician. He finds an electrician that charges \$50 to make a service call plus \$30 per hour. If Frederick plans to spend between \$110 and \$500, how many hours can he afford to hire the electrician? Frederick can afford to hire the electrician for 2 to 15 hr.

104. Roberto hires a professional organizer to organize his home office. The professional organizer charges a flat fee of \$75 plus \$25 per hour. If Roberto budgets between \$600 and \$800 to get his office organized, how many hours can he afford to hire the professional organizer? Roberto can afford to hire the professional organizer for 21 to 29 hr.

105. Maddy wants to maintain a test average between an 80 and 89. If Maddy's test scores are 85, 96, 89, and 78, what must she score on her fifth test? Maddy can score between 52 and 97 on her fifth test.

106. Isaac wants to maintain a quiz average between 70 and 79. If Isaac's quiz scores are 84, 69, 73, 81, and 70, what must he score on his sixth quiz? Isaac must score between 43 and 97 on his sixth quiz.

You Be the Teacher!

Answer each student's question.

107. Jacquesha: On a test, my answer to a question was $\{0\}$. My teacher marked it wrong and said that the answer was the empty set. What is the difference between $\{0\}$ and $\varnothing$? $\{0\}$ has a number 0 as an answer and $\varnothing$ is for no answer at all.

108. Lori: I do not understand the exercises that require me to find the intersection of the empty set and another set. Please explain how to do this type of problem.

Correct each student's errors, if any.

109. Solve $10 - 3x < 4$ and $5x + 23 > 3$.

Josh's work:

$$\begin{array}{rcr} 10 - 3x < 4 & \text{and} & 5x + 23 > 3 \\ -3x < -6 & \text{and} & 5x > -20 \\ x < 2 & \text{and} & x > -4 \end{array}$$

The solution set is $(-4, 2)$.

110. Solve $28 - 4x > 4$ or $28 - 4x < -4$.

Mark's work:

$$\begin{array}{rcr} 28 - 4x > 4 & \text{or} & 28 - 4x < -4 \\ -4x > -24 & \text{or} & -4x < -32 \\ x > 6 & \text{or} & x < 8 \end{array}$$

The solution set is $(-\infty, \infty)$.

Calculate It!

Solve each compound inequality and use a graphing calculator to confirm each solution.

111. $6x + 7 > 1$ and $5x - 2 < 13$ $(-1, 3)$

112. $3x - 2 > 10$ and $4x - 5 < 23$ $(4, 7)$

113. $5x - 7 \leq 33$ or $7x > 14$ $(-\infty, \infty)$

114. $2x \geq 18$ or $3x - 4 < 23$ $(-\infty, \infty)$

SECTION 2.6 Absolute Value Equations

OBJECTIVES

As a result of completing this section, you will be able to

1. Solve absolute value equations of the form $|X| = k$, where k is a real number.
2. Solve absolute value equations of the form $|X| = |Y|$.
3. Solve applications of absolute value equations.
4. Troubleshoot common errors.

A memory card is measured with a ruler with centimeter gradations. The width is reported to be 2.5 cm. Because of the inaccuracy of measuring, the actual width is 2.5 ± 0.1 cm. This states that the distance between the exact width and the measured width is 0.1 cm. What is the actual width of the memory card? To solve this problem, we must solve the equation $|w - 2.5| = 0.1$, where w represents the exact width of the memory card.

In this section, we will learn how to solve equations containing absolute values.

Objective 1

Solve absolute value equations of the form $|X| = k$, where k is a real number.

Absolute Value Equations

In Section 1.1, the absolute value of a number was presented. The absolute value of a number is the number's distance from zero on a real number line. We can visualize this on a number line as shown. Note that both 5 and -5 are 5 units from zero. So, their absolute values are the same.

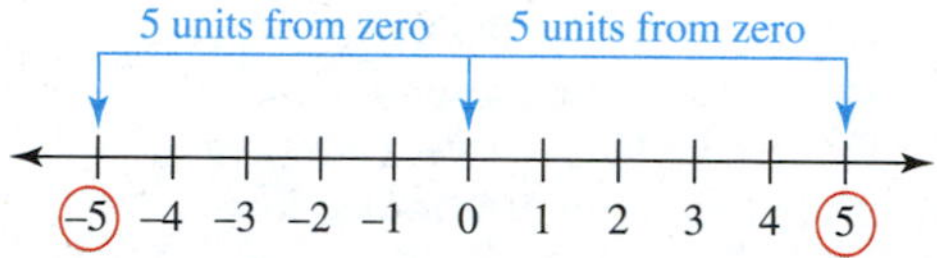

In fact, every real number and its opposite have the same absolute value, since they are the same distance from zero on a number line. Some examples are shown.

$|-5| = 5$ and $|5| = 5$ $|-6| = 6$ and $|6| = 6$ $|-7| = 7$ and $|7| = 7$

$|-4| = 4$ and $|4| = 4$ $|0| = 0$

Important Facts About Absolute Value

- The absolute value of any real number is nonnegative.
- There are two numbers (a number and its opposite) that have the same absolute value. For instance, both -5 and 5 have the same absolute value, which is 5.
- The number zero has an absolute value of zero.

In this section, we will solve **absolute value equations**. Some examples are

$$|x| = 5 \qquad |y - 4| = 7 \qquad |2x - 1| - 3 = 2$$

We will use the following property to solve absolute value equations.

INSTRUCTOR NOTE: Tell students that X can represent a single variable or a variable expression.

Property: Property 1 for Absolute Value Equations

Let k be a real number.

1. If $|X| = k$ and $k > 0$, then $X = k$ or $X = -k$.
2. If $|X| = 0$, then $X = 0$.
3. If $|X| = k$ and $k < 0$, then there are no solutions.

Note: *The expression inside the absolute value can be a single variable or a variable expression.*

Procedure: Solving an Absolute Value Equation

Step 1: Isolate the absolute value on one side of the equation and the constant on the other side.
Step 2: Set the expression inside the absolute value equal to the number(s) whose absolute value is the constant.

a. If the constant is positive, there are two solutions.
b. If the constant is zero, there is one solution.
c. If the constant is negative, there are no solutions.

Step 3: Check the solutions by substituting them in the original equation.

Objective 1 Examples

Solve each equation.

1a. $|x| = 5$ **1b.** $|a| = -3$ **1c.** $|y - 4| = 7$ **1d.** $|2x - 1| - 3 = 2$

1e. $\left|\frac{4x + 6}{5}\right| + 1 = 1$

Solutions

1a. We set the variable x equal to the numbers whose absolute value is 5, namely 5 and -5.

$$|x| = 5$$

$$x = 5 \quad \text{or} \quad x = -5$$

So, the solution set is $\{-5, 5\}$ or $\{\pm 5\}$. We can check by replacing x with 5 and -5 in the original equation.

1b. The equation asks us to find all real numbers a whose absolute value, or distance from zero, is -3. There is no real number whose absolute value is a negative number. Therefore, the solution set of this equation is the empty set, or $\varnothing$.

1c. We set the expression $y - 4$ equal to the numbers whose absolute value is 7, namely 7 and -7.

$$|y - 4| = 7$$

$y - 4 = 7$	or	$y - 4 = -7$	Apply property 1.
$y - 4 + 4 = 7 + 4$		$y - 4 + 4 = -7 + 4$	Add 4 to each side.
$y = 11$		$y = -3$	Simplify.

Check: $y = 11$:

$$|y - 4| = 7$$
$$|11 - 4| = 7$$
$$|7| = 7$$
$$7 = 7 \quad \text{True}$$

$y = -3$:

$$|y - 4| = 7$$
$$|-3 - 4| = 7$$
$$|-7| = 7$$
$$7 = 7 \quad \text{True}$$

So, the solution set is $\{-3, 11\}$.

1d. We must first isolate the absolute value expression.

$|2x - 1| - 3 = 2$	
$|2x - 1| - 3 + 3 = 2 + 3$	Add 3 to each side.
$|2x - 1| = 5$	Simplify.

$2x - 1 = 5$	or	$2x - 1 = -5$	Apply property 1.
$2x - 1 + 1 = 5 + 1$		$2x - 1 + 1 = -5 + 1$	Add 1 to each side.
$2x = 6$		$2x = -4$	Simplify.
$\frac{2x}{2} = \frac{6}{2}$		$\frac{2x}{2} = \frac{-4}{2}$	Divide each side by 2.
$x = 3$		$x = -2$	Simplify.

So, the solution set is $\{-2, 3\}$. We can check by replacing x with -2 and 3 in the original equation.

1e. We must first isolate the absolute value expression.

$$\left|\frac{4x+6}{5}\right| + 1 = 1$$

$$\left|\frac{4x+6}{5}\right| + 1 - 1 = 1 - 1 \quad \text{Subtract 1 from each side.}$$

$$\left|\frac{4x+6}{5}\right| = 0 \quad \text{Simplify.}$$

$$\frac{4x+6}{5} = 0 \quad \text{Apply property 1.}$$

$$5\left(\frac{4x+6}{5}\right) = 5(0) \quad \text{Multiply each side by 5.}$$

$$4x + 6 = 0 \quad \text{Simplify.}$$

$$4x + 6 - 6 = 0 - 6 \quad \text{Subtract 6 from each side.}$$

$$4x = -6 \quad \text{Simplify.}$$

$$\frac{4x}{4} = \frac{-6}{4} \quad \text{Divide each side by 4.}$$

$$x = -\frac{3}{2} \quad \text{Simplify.}$$

So, the solution set is $\left\{-\frac{3}{2}\right\}$. We can check by replacing x with 0 in the original equation.

Student Check 1 Solve each equation.

a. $|x| = 7$ **b.** $|a| = -10$ **c.** $|y + 1| = 6$ **d.** $|3x + 2| - 4 = 8$

e. $\left|\frac{5x-9}{4}\right| + 3 = 3$

Absolute Value Equations, $|X| = |Y|$

Objective 2 Solve absolute value equations of the form $|X| = |Y|$.

To solve equations containing two absolute values equal to one another, we need to understand when the absolute value of two expressions is equal. This occurs when the expressions have the same distance from zero. So, the expressions must either be the *same* or *opposite*. The following are examples of absolute value expressions that are equal.

$|5| = |5|$ $\quad |-5| = |-5|$ The expressions inside the absolute values are the same.

$|5| = |-5|$ The expressions inside the absolute values are opposites.

We will use the following property to solve equations containing two absolute values.

Property: Property 2 for Absolute Value Equations

If $|X| = |Y|$, then $X = Y$ or $X = -Y$.

Procedure: Solving an Equation Containing Two Absolute Value Expressions

Step 1: Set the expressions inside the absolute values equal to one another.
Step 2: Set the expression inside the first absolute value equal to the opposite of the expression inside the second absolute value.
Step 3: Solve the resulting equations.
Step 4: Check the solutions by substituting them in the original equation.

Objective 2 Examples Solve each equation.

2a. $|x-2| = |3x-4|$ **2b.** $\left|\frac{y-1}{5}\right| = \left|\frac{y}{2}\right|$

Solutions **2a.**

$$|x-2| = |3x-4|$$

$$\begin{aligned} x-2 &= 3x-4 \\ x-2-x &= 3x-4-x \\ -2 &= 2x-4 \\ -2+4 &= 2x-4+4 \\ 2 &= 2x \\ \frac{2}{2} &= \frac{2x}{2} \\ 1 &= x \end{aligned} \quad \text{or} \quad \begin{aligned} x-2 &= -(3x-4) \\ x-2 &= -3x+4 \\ x-2+3x &= -3x+4+3x \\ 4x-2 &= 4 \\ 4x-2+2 &= 4+2 \\ 4x &= 6 \\ \frac{4x}{4} &= \frac{6}{4} \\ x &= \frac{3}{2} \end{aligned}$$

Check:

$x = 1$:

$$\begin{aligned} |x-2| &= |3x-4| \\ |1-2| &= |3(1)-4| \\ |-1| &= |3-4| \\ 1 &= |-1| \\ 1 &= 1 \quad \text{True} \end{aligned}$$

$x = \frac{3}{2}$:

$$\begin{aligned} |x-2| &= |3x-4| \\ \left|\frac{3}{2}-2\right| &= \left|3\left(\frac{3}{2}\right)-4\right| \\ \left|\frac{3}{2}-\frac{4}{2}\right| &= \left|\frac{9}{2}-\frac{8}{2}\right| \\ \left|-\frac{1}{2}\right| &= \left|\frac{1}{2}\right| \\ \frac{1}{2} &= \frac{1}{2} \quad \text{True} \end{aligned}$$

So, the solution set is $\left\{1, \frac{3}{2}\right\}$.

2b. $\left|\frac{y-1}{5}\right| = \left|\frac{y}{2}\right|$

$$\frac{y-1}{5} = \frac{y}{2} \quad \text{or} \quad \frac{y-1}{5} = -\frac{y}{2}$$

$$10\left(\frac{y-1}{5}\right) = 10\left(\frac{y}{2}\right) \qquad 10\left(\frac{y-1}{5}\right) = 10\left(-\frac{y}{2}\right)$$

$$2(y-1) = 5y \qquad 2(y-1) = -5y$$

$$2y - 2 = 5y \qquad 2y - 2 = -5y$$

$$2y - 2 - 2y = 5y - 2y \qquad 2y - 2 - 2y = -5y - 2y$$

$$-2 = 3y \qquad -2 = -7y$$

$$\frac{-2}{3} = \frac{3y}{3} \qquad \frac{-2}{-7} = \frac{-7y}{-7}$$

$$-\frac{2}{3} = y \qquad \frac{2}{7} = y$$

So, the solution set is $\left\{-\frac{2}{3}, \frac{2}{7}\right\}$. We can check by replacing y with $-\frac{2}{3}$ and $\frac{2}{7}$.

Student Check 2 Solve each equation.

a. $|5z - 7| = |z + 1|$ **b.** $\left|\frac{x+3}{2}\right| = \left|\frac{2x}{5}\right|$

Applications

Objective 3 Solve applications of absolute value equations.

Since the absolute value of a number represents that number's distance from zero on a number line, distance between objects is a key application of absolute value equations. In measurements, this distance is often referred to as absolute error. In science-related fields, accuracy in measurement is vital. However, there will almost always be error in measurement due to the precision of the measuring instruments and human error.

INSTRUCTOR NOTE: Explain to students that the absolute error gives the maximum and minimum for the true value of a measurement.

Definition: Absolute error is the absolute value (sometimes called magnitude) of the difference between the measured value and the exact value. We can use the following formula to represent the absolute error.

$E_{abs} = |x - a|$, where a is the approximated value and x is the exact measure.

Objective 3 Examples Solve each problem.

3a. A memory card is measured with a ruler with centimeter gradations. The width is reported to be 2.5 cm. The absolute error is 0.1 cm. Find the possible values for the exact width of the memory card.

Solution **3a.** Let w represent the exact width of the memory card. The approximated value is 2.5 cm and the absolute error is 0.1 cm. To find the exact width, we must solve the following equation.

$$|w - 2.5| = 0.1$$

$w - 2.5 = 0.1$	or	$w - 2.5 = -0.1$	Apply property 1.
$w - 2.5 + 2.5 = 0.1 + 2.5$		$w - 2.5 + 2.5 = -0.1 + 2.5$	Add 2.5 to each side.
$w = 2.6$		$w = 2.4$	Simplify.

So, the exact width of the memory card is at most 2.6 cm and at least 2.4 cm.

3b. Find the absolute error when the value of 3.14 is used for π. Round the answer to four decimal places.

Solution **3b.** We can find the absolute error using the given formula since we know the approximated value is 3.14 and the exact value is π. So, $a = 3.14$ and $x = \pi$.

$$E_{\text{abs}} = |x - a|$$
$$E_{\text{abs}} = |\pi - 3.14|$$
$$E_{\text{abs}} = 0.0016$$

So, the absolute error is 0.0016.

Student Check 3 Solve each problem.

a. Suppose a dog measures 80 cm long with a 3 cm absolute error. Find the possible values for the exact length of the dog.

b. Find the absolute error when the value of 2.7 is used for the number e, where $e = 2.71828\ldots$. Round the answer to four decimal places.

Objective 4 ▶

Troubleshoot common errors.

Troubleshooting Common Errors

Some common errors associated with solving absolute value equations are shown.

Objective 4 Examples

A problem and an incorrect solution are given. Provide the correct solution and an explanation of the error.

4a. Solve: $|2x - 3| = 7$.

Incorrect Solution	Correct Solution and Explanation
$|2x - 3| = 7$ $2x - 3 = 7$ $2x = 10$ $x = 5$ The solution set is $\{5\}$.	There are two solutions of this equation since there are two numbers whose absolute value is 7. $|2x - 3| = 7$ $2x - 3 = 7$ or $2x - 3 = -7$ $2x = 10$ $\quad$ $2x = -4$ $x = 5$ $\quad$ $x = -2$ The solution set is $\{-2, 5\}$.

4b. Solve: $|y + 2| - 1 = 4$.

Incorrect Solution	Correct Solution and Explanation
$|y + 2| - 1 = 4$ $y + 2 - 1 = 4$ or $y + 2 - 1 = -4$ $y + 1 = 4$ $\quad$ $y + 1 = -4$ $y = 3$ $\quad$ $y = -5$ The solution set is $\{-5, 3\}$.	The absolute value expression must be isolated prior to solving. We must first add 1 each side. $|y + 2| - 1 = 4$ $|y + 2| = 5$ $y + 2 = 5$ or $y + 2 = -5$ $y = 3$ $\quad$ $y = -7$ The solution set is $\{-7, 3\}$.

4c. Solve: $|3a + 2| = -7$.

Incorrect Solution	Correct Solution and Explanation
$|3a + 2| = -7$ $3a + 2 = -7$ or $3a + 2 = 4$ $3a = -9$ $\quad 3a = 2$ $a = -3$ $\quad a = \frac{2}{3}$ The solution set is $\left\{-3, \frac{2}{3}\right\}$.	Because the absolute value expression is equal to a negative number, this equation has no solution. The absolute value of any real number is always greater than or equal to zero. So, the solution set is the empty set, or $\varnothing$.

ANSWERS TO STUDENT CHECKS

Student Check 1 **a.** $\{\pm 7\}$ **b.** $\varnothing$ **c.** $\{-7, 5\}$ **d.** $\left\{-\frac{14}{3}, \frac{10}{3}\right\}$ **e.** $\left\{\frac{9}{5}\right\}$

Student Check 2 **a.** $\{1, 2\}$ **b.** $\left\{-15, -\frac{5}{3}\right\}$

Student Check 3 **a.** The length of the dog is at most 83 cm and at least 77 cm. **b.** The absolute error is 0.0183.

SUMMARY OF KEY CONCEPTS

1. To solve an absolute value equation, we must write the equation so that it is in the form $|X| = c$, where c is a real number. If the constant c is positive, there are two solutions. If the constant c is zero, there is one solution. If the constant c is negative, there are no solutions.
2. To solve an equation in which the absolute value of one expression is equal to the absolute value of another expression, we must form two equations. One equation is formed by setting the two expressions equal to one another. The other equation is formed by setting one expression equal to the opposite of the other expression.
3. A key application of absolute value involves the error in measurement. The absolute error is calculated as the absolute value of the difference of the exact value and the approximated value.

GRAPHING CALCULATOR SKILLS

We can use the graphing calculator to verify solutions of absolute value equations.

Example: Verify that the solutions of the equation $|2x - 1| - 3 = 2$ are -2 and 3.

Solution:

Method 1: Enter the solutions as the stored value for the variable x. Then compute the value of the left side of the equation. The result should be the right side of the equation.

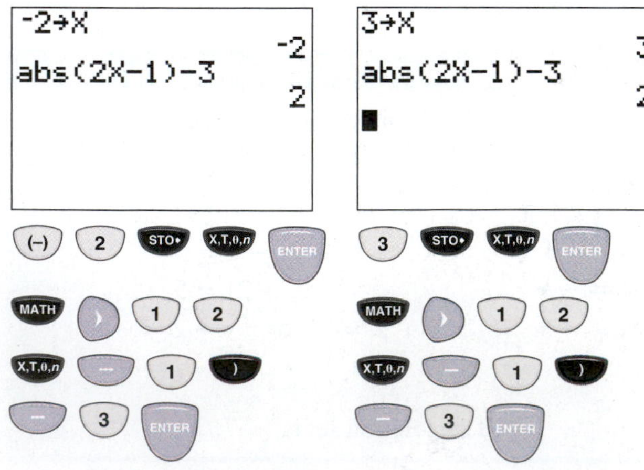

Method 2: Enter the left side of the equation in the equation editor. Verify that the table shows that the corresponding y-value of $x = -2$ and $x = 3$ is $y = 2$.

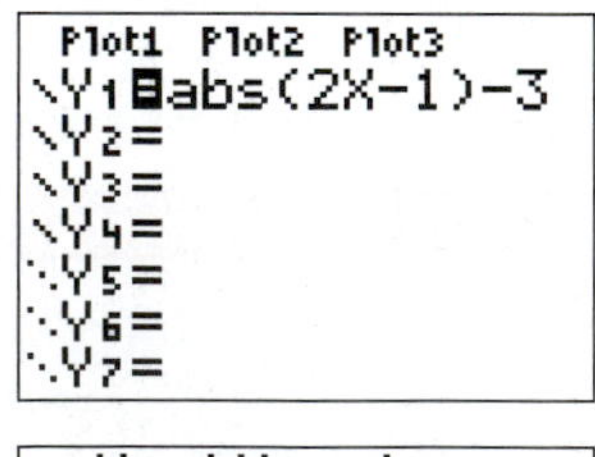

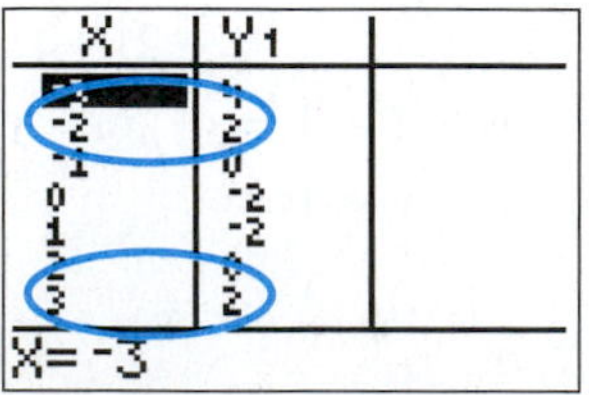

Since $Y_1 = 2$ for both $x = -2$ and $x = 3$, we know they are solutions of the equation.

SECTION 2.6 EXERCISE SET

Write About It!

Use complete sentences in your answer to each exercise.

1. Can the absolute value of an expression equal zero? Explain. The number zero has an absolute value of zero.
2. Can the absolute value of an expression equal a negative number? Explain. The absolute value of any real number is nonnegative.
3. Explain the steps for solving an equation with one absolute value expression.
4. Explain the steps for solving an equation with two absolute value expressions.

Practice Makes Perfect!

Solve each equation. (*See Objective 1.*)

5. $|x| = 10$ $\{-10, 10\}$
6. $|x| = 1$ $\{-1, 1\}$
7. $|2x + 3| = 2$ $\left\{-\frac{5}{2}, -\frac{1}{2}\right\}$
8. $|3x - 1| = 5$ $\left\{-\frac{4}{3}, 2\right\}$
9. $|4 - y| = 6$ $\{-2, 10\}$
10. $|6 - y| = 3$ $\{3, 9\}$
11. $|a| = -9$ $\varnothing$
12. $|a| = -4$ $\varnothing$
13. $|4x + 3| = -6$ $\varnothing$
14. $|5x - 1| = -2$ $\varnothing$
15. $|4x + 3| - 6 = 8$
16. $|6x - 5| + 1 = 3$ $\left\{\frac{1}{2}, \frac{7}{6}\right\}$
17. $|4 - 7x| + 8 = 2$ $\varnothing$
18. $|1 - 3x| + 5 = 4$ $\varnothing$
19. $\left|\frac{1 - 4x}{5}\right| + 5 = 5$ $\left\{\frac{1}{4}\right\}$
20. $\left|\frac{2 - 3x}{6}\right| + 1 = 1$ $\left\{\frac{2}{3}\right\}$
21. $\left|\frac{5x + 3}{2}\right| + 4 = 9$
22. $\left|\frac{6x + 5}{3}\right| + 2 = 3$

Solve each equation. (*See Objective 2.*)

23. $|4x + 3| = |5x - 1|$
24. $|6x - 3| = |7x + 4|$
25. $|5 - x| = |3x + 2|$
26. $|5 - 2x| = |x + 5|$ $\{0, 10\}$
27. $|x - 1| = |6x - 6|$ $\{1\}$
28. $|4x + 8| = |3x - 9|$ $\left\{-17, \frac{1}{7}\right\}$
29. $|3x + 7| = |2x - 8|$
30. $|7x - 10| = |10x - 5|$
31. $\left|\frac{y - 3}{2}\right| = |y + 2|$
32. $\left|\frac{y + 1}{3}\right| = |2y - 1|$ $\left\{\frac{2}{7}, \frac{4}{5}\right\}$
33. $\left|\frac{a + 2}{5}\right| = |3a + 3|$
34. $\left|\frac{a + 4}{6}\right| = |a - 4|$ $\left\{\frac{20}{7}, \frac{28}{5}\right\}$
35. $\left|\frac{x - 6}{3}\right| = \left|\frac{3x}{4}\right|$
36. $\left|\frac{x - 4}{4}\right| = \left|\frac{2x}{7}\right|$ $\left\{-28, \frac{28}{15}\right\}$

Solve each problem. (*See Objective 3.*)

37. Estimate the absolute error when the value $\frac{22}{7}$ is used for π. Round to five decimal places. 0.00126
38. Estimate the absolute error when the value 3.141593 is used for π. Round to 10 decimal places. 0.0000003464

Additional answers can be found in the Instructor Answer Appendix.

39. Estimate the absolute error when the value 2.72 is used for e, where e is 2.71828182846. Round to five decimal places. 0.00172
40. Estimate the absolute error when the value 2.71828 is used for e, where e is 2.71828182846. Round to 10 decimal places. 0.0000018285
41. Suppose a digital scale reflects a weight (in pounds) with an absolute error of 0.4 lb. If someone weighs 184.3 lb according to the scale, find the possible values for the exact weight of the person.
42. Suppose a countertop food scale reflects a weight (in ounces) with an absolute error of 0.1 oz. If a piece of chicken weighs 3 oz according to the scale, find the possible values for the exact weight of the piece of chicken.
43. Jana Lipsey and Linda Easley are running for the same congressional seat. Currently, 51% of those polled will reportedly vote for Lipsey and 45% of those polled will reportedly vote for Easley. If the pollsters report an absolute error of 3 points, find the possible values for the actual percentage of those polled who will reportedly vote for Lipsey and Easley.
44. Alexis Johnson and Steve Hunter are running for the same office. Currently, 42% of those polled will vote for Johnson and 48% of those polled will vote for Hunter. If the pollsters report an absolute error of 2 points, find the possible values for the actual percentage of those polled who will reportedly vote for Johnson and Hunter.

Mix 'Em Up!

Solve each problem.

45. $|10x - 3| = 15$ $\left\{-\frac{6}{5}, \frac{9}{5}\right\}$
46. $|8x - 11| = 12$ $\left\{-\frac{1}{8}, \frac{23}{8}\right\}$
47. $|4 - 5x| = 4$ $\left\{0, \frac{8}{5}\right\}$
48. $|6 - 2x| = 3$ $\left\{\frac{3}{2}, \frac{9}{2}\right\}$
49. $\left|\frac{7x + 5}{3}\right| = \left|\frac{2x}{5}\right|$
50. $\left|\frac{4x - 9}{6}\right| = \left|\frac{3x}{5}\right|$ $\left\{\frac{45}{38}, \frac{45}{2}\right\}$
51. $\left|\frac{5x - 6}{2}\right| - 1 = 12$
52. $\left|\frac{3x + 10}{4}\right| + 2 = 5$ $\left\{-\frac{22}{3}, \frac{2}{3}\right\}$
53. $|10 - 3x| + 8 = 3$ $\varnothing$
54. $|2 - 5x| + 3 = 1$ $\varnothing$
55. $|5x + 2| = |6x - 7|$
56. $|8x - 9| = |2 - 3x|$ $\left\{1, \frac{7}{5}\right\}$
57. $\left|\frac{9x}{5}\right| = 3$ $\left\{-\frac{5}{3}, \frac{5}{3}\right\}$
58. $\left|\frac{10x}{3}\right| = 2$ $\left\{-\frac{3}{5}, \frac{3}{5}\right\}$
59. $|2x + 1| = 0$ $\left\{-\frac{1}{2}\right\}$
60. $|7x - 8| = 0$ $\left\{\frac{8}{7}\right\}$
61. $|0.4x - 1.5| = 6.3$ $\{-12, 19.5\}$
62. $|0.3x + 4.2| = 12.6$ $\{-56, 28\}$
63. $\left|\frac{9x + 10}{5}\right| - 3 = 12$
64. $\left|\frac{8x - 3}{4}\right| + 4 = 9$ $\left\{-\frac{17}{8}, \frac{23}{8}\right\}$

65. $|2.6 - 0.2x| = 14.6$ $\{-60, 86\}$ **66.** $|5.4 - 0.6x| = 13.8$ $\{-14, 32\}$

67. Suppose a machine with an absolute error reading of 1.3 units reports the measurement of an object as 10.98 units. What are the possible values for the actual measurement of the object? The actual measurement of the object is between 9.68 units and 12.28 units.

68. Suppose a machine with an absolute error reading of 5.65 units reports the measurement of an object as 2564.42 units. What are the possible values for the actual measurement of the object?

69. Suppose the actual length of an object is 4.569 m. If this length is estimated by 4.6 m, what the absolute error? The absolute error is 0.031 m.

70. Suppose the actual length of an object is 10.32 in. If this length is estimated by 9 in., what is the absolute error? The absolute error is 1.32 in.

71. A poll shows that candidate A will receive 30% of the vote and candidate B will receive 39% of the vote. If this poll has an absolute error of 5%, what are the possible values for how much of the vote each candidate will receive? Candidate A will receive between 25% and 35% of the vote and candidate B will receive between 34% and 44% of the vote.

72. A poll shows that candidate C will receive 60% of the vote and candidate D will receive 29% of the vote. If this poll has an absolute error of 7%, what are the possible values for how much of the vote each candidate will receive? Candidate C will receive between 53% and 67% of the vote and candidate D will receive between 22% and 36% of the vote.

You Be the Teacher!

Correct each student's errors, if any.

73. Solve $|4x - 5| = 11$.

Kelly's work:

$|4x - 5| = 11$

$4x - 5 = 11$

$4x = 16$

$x = 4$

So, the solution set is $\{4\}$.

Correction:

$|4x - 5| = 11$

$4x - 5 = 11$ or $4x - 5 = -11$

$4x = 16$ $\quad$ $4x = -6$

$x = 4$ $\quad$ $x = -\frac{3}{2}$

So, the solution set is $\left\{-\frac{3}{2}, 4\right\}$.

74. Solve $|9x + 1| - 4 = -2$.

Kyle's work:

The solution set is ∅ since the right side of the equation is negative.

Correction:

$|9x + 1| - 4 = -2$

$|9x + 1| = 2$

$9x + 1 = 2$ or $9x + 1 = -2$

$9x = 1$ $\quad$ $9x = -3$

$x = \frac{1}{9}$ $\quad$ $x = -\frac{1}{3}$

So, the solution set is $\left\{-\frac{1}{3}, \frac{1}{9}\right\}$.

75. Solve $|x + 5| = -3$.

Mark's work:

$|x + 5| = -3$

$x + 5 = 3$ or $x + 5 = -3$

$x = -2$ $\quad$ $x = -8$

So, the solution set is $\{-8, -2\}$.

The solution set is ∅ since the absolute value expression is equal to a negative number.

76. Solve $|2x - 3| + 5 = 8$.

Ashley's work:

$|2x - 3| + 5 = 8$

$2x - 3 + 5 = 8$ or $2x - 3 + 5 = -8$

$2x + 2 = 8$ $\quad$ $2x + 2 = -8$

$2x = 6$ $\quad$ $2x = -10$

$x = 3$ $\quad$ $x = -5$

So, the solution set is $\{-5, 3\}$.

Calculate It!

Solve each absolute value equation. Use a graphing calculator to check each answer.

77. $|17x - 5| = 33$ **78.** $|13x + 15| = 21$

79. $\left|\frac{6x + 11}{4}\right| - 2 = 5$ **80.** $\left|\frac{5x - 12}{4}\right| + 5 = 8$

Think About It!

81. For what values of x and a does the absolute value equation $|x - a| = x$ have solutions?
$x \geq 0$ and $a = 0$ or $x \geq 0$ and $a = 2x$

82. For what values of x and a does the absolute value equation $|x - a| = -x$ have solutions?
$x \leq 0$ and $a = 2x$ or $x \leq 0$ and $a = 0$

83. Write an example of an absolute value equation that has one solution.
Answers vary. $|2x - 6| = 0$

84. Write an example of an absolute value equation that has no solution.
Answers vary. $|2x + 5| = -6$

SECTION 2.7 Absolute Value Inequalities

OBJECTIVES

As a result of completing this section, you will be able to

1. Solve absolute value inequalities involving $<$ or $\leq$.
2. Solve absolute value inequalities involving $>$ or $\geq$.
3. Solve special cases of absolute value inequalities.
4. Solve applications of absolute value inequalities.
5. Solve absolute value inequalities using test points.
6. Troubleshoot common errors.

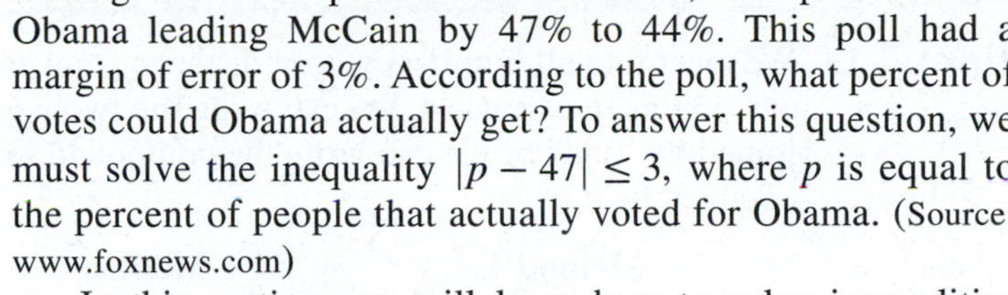

During the 2008 presidential election, one poll showed Obama leading McCain by 47% to 44%. This poll had a margin of error of 3%. According to the poll, what percent of votes could Obama actually get? To answer this question, we must solve the inequality $|p - 47| \leq 3$, where p is equal to the percent of people that actually voted for Obama. (Source: www.foxnews.com)

In this section, we will learn how to solve inequalities involving absolute values.

Objective 1 Solve absolute value inequalities involving $<$ or $\leq$.

Absolute Value Inequalities with $<$ or $\leq$

When we solve an absolute value equation, we are looking for values that have a specific distance from 0 on the number line. But when we solve absolute value inequalities, we are looking for numbers whose distance from 0 may be less than or greater than a given number.

Suppose we want to solve $|x| \leq 2$. This means that we need to find all real numbers whose distance from zero is less than or equal to 2. Consider the following number line.

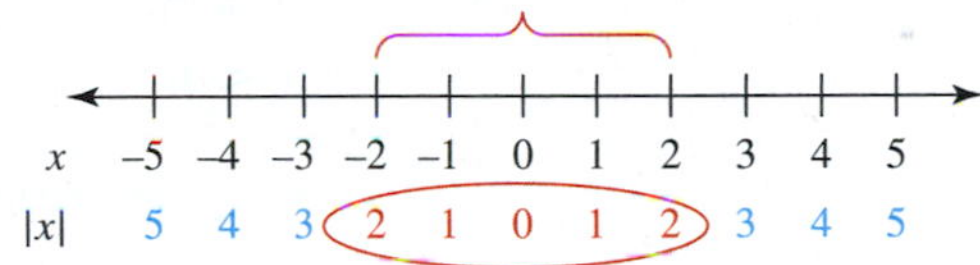

So, we see from the number line that numbers between, and including, -2 and 2 have an absolute value less than or equal to 2. So, the solution set is $[-2, 2]$.

This leads to a property that will enable us to solve absolute value inequalities of the form $|X| < k$ or $|X| \leq k$.

Property: Property 1 for Absolute Value Inequalities

Let k be a positive real number.

If $|X| < k$, then $X < k$ and $X > -k$.
If $|X| \leq k$, then $X \leq k$ and $X \geq -k$.

Note: *These compound inequalities can also be written as*

$$-k < X < k \quad \text{and} \quad -k \leq X \leq k$$

Procedure: Solving an Inequality of the Form $|X| < k$ or $|X| \leq k$

Step 1: Isolate the absolute value expression on one side of the inequality.
Step 2: Apply property 1 to remove the absolute value sign.
Step 3: Solve the resulting compound inequality.
Step 4: Graph the solution set and write the answer in interval notation.

Objective 1 Examples Solve each absolute value inequality.

1a. $|x| < 5$ **1b.** $|y + 3| \le 4$ **1c.** $|2x - 1| - 5 \le 4$

Solutions **1a.** We must find all numbers whose distance from zero is less than 5. We can do this in two ways. We can write the two appropriate inequalities joined by "and" or we can write the compound inequality using the compact form.

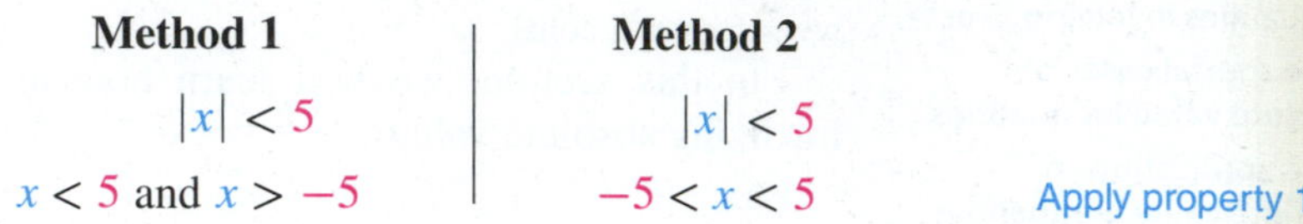

Method 1	**Method 2**	
$|x| < 5$	$|x| < 5$	
$x < 5$ and $x > -5$	$-5 < x < 5$	Apply property 1.

Graph:

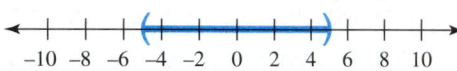

Interval: $(-5, 5)$

1b. We will solve the inequality using two separate inequalities.

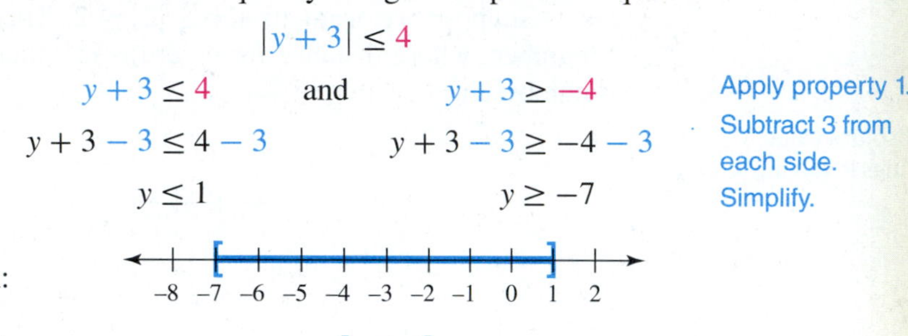

$$|y + 3| \le 4$$

$y + 3 \le 4$	and	$y + 3 \ge -4$	Apply property 1.
$y + 3 - 3 \le 4 - 3$		$y + 3 - 3 \ge -4 - 3$	Subtract 3 from each side.
$y \le 1$		$y \ge -7$	Simplify.

Graph:

Interval: $[-7, 1]$

1c. We first isolate the absolute value expression and then apply property 1. We will use the compact form to solve the inequality.

$$|2x - 1| - 5 \le 4$$

$$|2x - 1| - 5 + 5 \le 4 + 5 \quad \text{Add 5 to each side.}$$

$$|2x - 1| \le 9 \quad \text{Simplify.}$$

$$-9 \le 2x - 1 \le 9 \quad \text{Apply property 1.}$$

$$-9 + 1 \le 2x - 1 + 1 \le 9 + 1 \quad \text{Add 1 to each part.}$$

$$-8 \le 2x \le 10 \quad \text{Simplify.}$$

$$\frac{-8}{2} \le \frac{2x}{2} \le \frac{10}{2} \quad \text{Divide each part by 2.}$$

$$-4 \le x \le 5 \quad \text{Simplify.}$$

Graph:

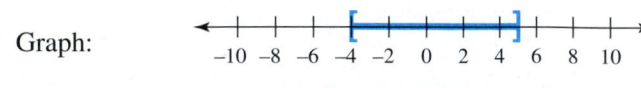

Interval: $[-4, 5]$

Student Check 1 Solve each absolute value inequality.

a. $|x| < 10$ **b.** $|y - 4| \le 5$ **c.** $|4x + 3| - 1 \le 6$

Absolute Value Inequalities with > or ≥

Objective 2 ▶

Solve absolute value inequalities involving > or ≥.

The graph from Objective 1 will help us solve the equation $|x| \geq 2$.

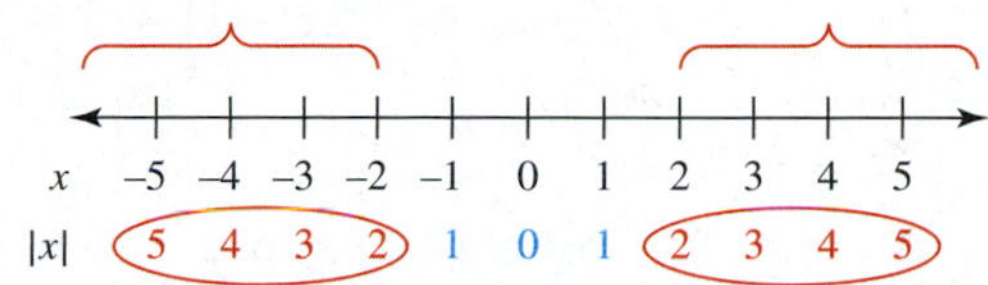

The numbers larger than and including 2 or less than and including −2 have an absolute value greater than or equal to 2. So, the solution set is $(-\infty, -2] \cup [2, \infty)$.

This leads to a property that enables us to solve absolute value inequalities of the form $|X| > k$ or $|X| \geq k$.

Property: Property 2 for Absolute Value Inequalities

Let k be a positive real number.

If $|X| > k$, then $X < -k$ or $X > k$.

If $|X| \geq k$, then $X \leq -k$ or $X \geq k$.

Procedure: Solving an Inequality of the Form $|X| > k$ or $|X| \geq k$

Step 1: Isolate the absolute value expression on one side of the inequality.
Step 2: Apply property 2 to remove the absolute value sign.
Step 3: Solve the resulting compound inequality.
Step 4: Graph the solution set and write the answer in interval notation.

Objective 2 Examples Solve each absolute value inequality.

2a. $|x| > 5$ **2b.** $|y + 3| \geq 4$ **2c.** $|2x - 1| - 5 > 4$

Solutions

2a.

$$|x| > 5$$

$$x > 5 \quad \text{or} \quad x < -5$$ Apply property 2.

Graph: −10 −8 −6 −4 −2 0 2 4 6 8 10

Interval: $(-\infty, -5) \cup (5, \infty)$

2b.

$$|y + 3| \geq 4$$

$$y + 3 \geq 4 \quad \text{or} \quad y + 3 \leq -4$$ Apply property 2.

$$y + 3 - 3 \geq 4 - 3 \qquad y + 3 - 3 \leq -4 - 3$$ Subtract 3 from each side.

$$y \geq 1 \qquad y \leq -7$$ Simplify.

Graph:

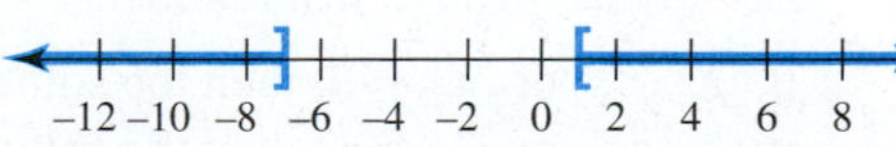

Interval: $(-\infty, -7] \cup [1, \infty)$

2c. We must first isolate the absolute value expression and then apply property 2.

$$|2x - 1| - 5 > 4$$

$$|2x - 1| - 5 + 5 > 4 + 5 \quad \text{Add 5 to each side.}$$

$$|2x - 1| > 9 \quad \text{Simplify.}$$

$$2x - 1 > 9 \quad \text{or} \quad 2x - 1 < -9 \quad \text{Apply property 2.}$$

$$2x - 1 + 1 > 9 + 1 \qquad 2x - 1 + 1 < -9 + 1 \quad \text{Add 1 to each side.}$$

$$2x > 10 \qquad 2x < -8 \quad \text{Simplify.}$$

$$\frac{2x}{2} > \frac{10}{2} \qquad \frac{2x}{2} < \frac{-8}{2} \quad \text{Divide each side by 2.}$$

$$x > 5 \qquad x < -4 \quad \text{Simplify.}$$

Graph: 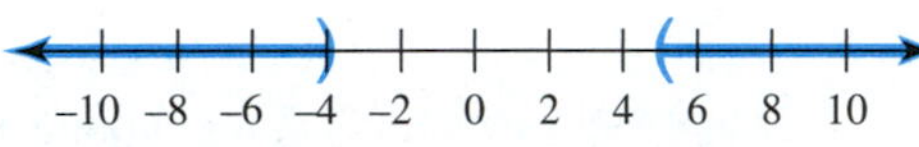

Interval: $(-\infty, -4) \cup (5, \infty)$

Student Check 2 Solve each absolute value inequality.

a. $|x| > 10$ **b.** $|y - 4| \geq 5$ **c.** $|4x + 3| - 1 > 6$

Special Cases of Absolute Value Inequalities

Objective 3 Solve special cases of absolute value inequalities.

We will now investigate how to solve absolute value inequalities in which the number opposite the absolute value expression is negative. We will use the number line to solve $|x| < -3$ and $|x| > -3$.

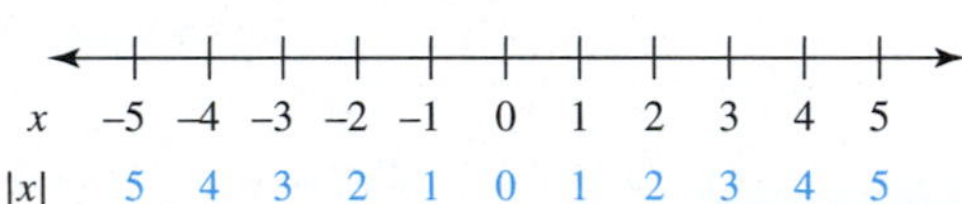

$\|x\| < -3$	There are no real numbers whose absolute value is less than −3. So, the solution set of this inequality is the empty set, ∅.
$\|x\| > -3$	Every real number has an absolute value that is greater than −3. So, the solution set of this inequality is all real numbers, ℝ, or $(-\infty, \infty)$.

Note that the absolute value of any real number is nonnegative. That is, the absolute value of any real number is always greater than or equal to zero. This tells us two important facts:

- The absolute value of a number will never be less than a negative number.
- The absolute value of a number will always be greater than a negative number.

These facts yield the following property.

Property: Property 3 for Absolute Value Inequalities

Let k be a negative real number.

If $|X| < k$ or $|X| \leq k$, then the solution set is ∅.
If $|X| > k$ or $|X| \geq k$, then the solution set is $(-\infty, \infty)$ or all real numbers, ℝ.

So, inequalities of the form $|X| < k$, $|X| \leq k$, $|X| > k$, or $|X| \geq k$, where $k < 0$, will either have no solution or all real numbers as solutions.

The last property deals with absolute value inequalities in which the number opposite the absolute value expression is zero.

Property: Property 4 for Absolute Value Inequalities

a. $|X| < 0$ This inequality has no solution since the absolute value of any real number is nonnegative.

b. $|X| \le 0$ The only solution that satisfies this inequality is $X = 0$.

c. $|X| > 0$ This inequality is solved using property 2.
$X > 0$ or $X < 0$. The only point not included is $X = 0$.

d. $|X| \ge 0$ This inequality has all real numbers as its solution since the absolute value of any real number is always positive or zero.

Objective 3 Examples

Solve each absolute value inequality.

3a. $|x| < -5$ **3b.** $|y + 1| \le -7$ **3c.** $|a| > -5$

3d. $|3b + 5| + 4 \ge 1$ **3e.** $\left|\dfrac{4x + 1}{3}\right| > 0$

Solutions

3a. The solution set is the empty set, or $\varnothing$, since the absolute value of a number is never less than -5.

3b. Since the absolute value of a number is never less than or equal to -7, the solution set is the empty set, or $\varnothing$.

3c. Since the absolute value of a number is always greater than -5, the solution set is all real numbers, or $(-\infty, \infty)$. The graph of the solution set is

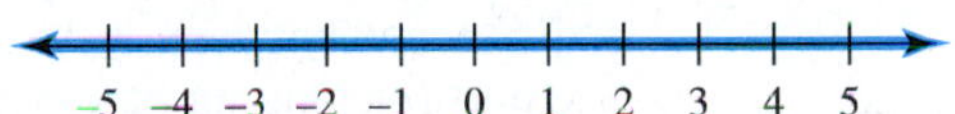

3d. We must first isolate the absolute value by subtracting 4 from each side.

$$|3b + 5| + 4 \ge 1$$

$$|3b + 5| + 4 - 4 \ge 1 - 4 \quad \text{Subtract 4 from each side.}$$

$$|3b + 5| \ge -3 \quad \text{Simplify.}$$

Since the absolute value of a number is always greater than or equal to -3, the solution set is all real numbers, or $(-\infty, \infty)$. The graph is

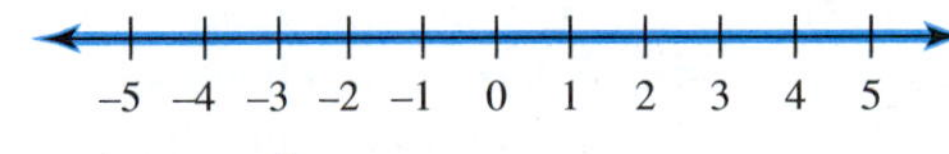

3e.

$$\left|\frac{4x + 1}{3}\right| > 0$$

$\dfrac{4x + 1}{3} > 0$	or	$\dfrac{4x + 1}{3} < 0$	Apply property 4.
$3\left(\dfrac{4x + 1}{3}\right) > 3(0)$		$3\left(\dfrac{4x + 1}{3}\right) < 3(0)$	Multiply each side by 3.
$4x + 1 > 0$		$4x + 1 < 0$	Simplify.
$4x + 1 - 1 > 0 - 1$		$4x + 1 - 1 < 0 - 1$	Subtract 1 from each side.
$4x > -1$		$4x < -1$	Simplify.
$\dfrac{4x}{4} > \dfrac{-1}{4}$		$\dfrac{4x}{4} < \dfrac{-1}{4}$	Divide each side by 4.
$x > -\dfrac{1}{4}$		$x < -\dfrac{1}{4}$	Simplify.

So, the solution set is $\left(-\infty, -\frac{1}{4}\right) \cup \left(-\frac{1}{4}, \infty\right)$. The graph is

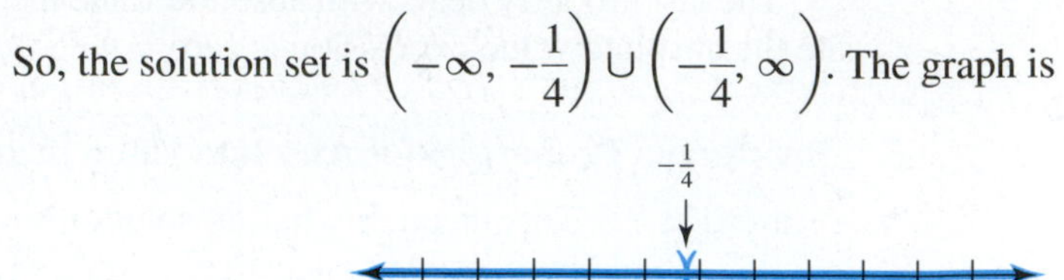

Student Check 3 Solve each absolute value inequality.

a. $|x| < -4$ **b.** $|y - 8| \leq -1$ **c.** $|a| > -3$

d. $|2b - 7| + 6 \geq 3$ **e.** $\left|\frac{3x - 5}{4}\right| > 0$

Objective 4 Solve applications of absolute value inequalities.

Applications

Applications of absolute value inequalities arise when a certain quantity must have a certain distance from another quantity. Some examples are shown.

Objective 4 Examples **Solve each problem using an absolute value inequality.**

4a. During the 2008 presidential election, one poll showed Obama leading McCain by 47% to 44%. This poll had a margin of error of 3%. According to the poll, what percent of votes would Obama actually get? (Source: www.foxnews.com)

Solution **4a.** Let p represent the percent of people who said they would vote for Obama. Since the margin of error is 3%, the absolute value of the difference between the percent surveyed and the actual percentage that voted for Obama is less than or equal to 3. So, we solve the following inequality.

$$|p - 47| \leq 3$$
$$-3 \leq p - 47 \leq 3 \quad \text{Apply property 1.}$$
$$-3 + 47 \leq p - 47 + 47 \leq 3 + 47 \quad \text{Add 47 to each part.}$$
$$44 \leq p \leq 50 \quad \text{Simplify.}$$

Based on the poll, Obama would have received between 44% and 50% of the votes.

4b. A hazardous chemical spill of titanium tetrachloride occurred at mile marker 331 on Interstate 40 in Tennessee. Emergency management personnel stated that motorists should be a minimum of 12 mi from the site of the spill. What are safe locations on Interstate 40 for motorists?

Solution **4b.** Let x represent the mile marker of a motorist. The distance from the site of the spill to the motorist should be at least 12 mi. Since motorists can be on either side of the spill, we must solve the following inequality.

$$|x - 331| \geq 12$$
$$x - 331 \geq 12 \quad \text{or} \quad x - 331 \leq -12 \quad \text{Apply property 2.}$$
$$x - 331 + 331 \geq 12 + 331 \qquad x - 331 + 331 \leq -12 + 331 \quad \text{Add 331 to each side.}$$
$$x \geq 343 \qquad x \leq 319 \quad \text{Simplify.}$$

Motorists should be located at or below mile marker 319 or located at or above mile marker 343 to be a safe distance from the spill.

Student Check 4 Solve each word problem with an absolute value inequality.

a. A Gallup Poll, taken in August 2008, revealed that 48% of U.S. workers are completely satisfied with their job. The poll had a margin of error of 5%. What is the range of workers who are completely satisfied with their job? (Source: www.gallup.com)

b. A tractor trailer accident at mile marker 51 on I-68 in West Virginia caused a hazardous chemical spill. Emergency management personnel advised that motorists should be at least 8 mi from the site of the spill. What are safe locations on I-68 for motorists?

Using Test Points to Solve Absolute Value Inequalities

Objective 5 Solve absolute value inequalities using test points.

In the first two objectives, compound inequalities were used to solve absolute value inequalities. Absolute value inequalities can be solved using an alternate method involving test points. In this method, the graph of the solution set is first constructed and, from this, the solution set is obtained.

We will examine the graph of the solution set of the inequality we solved in Example 2 part (b). The inequality $|y + 3| \geq 4$ was shown to have the following solution.

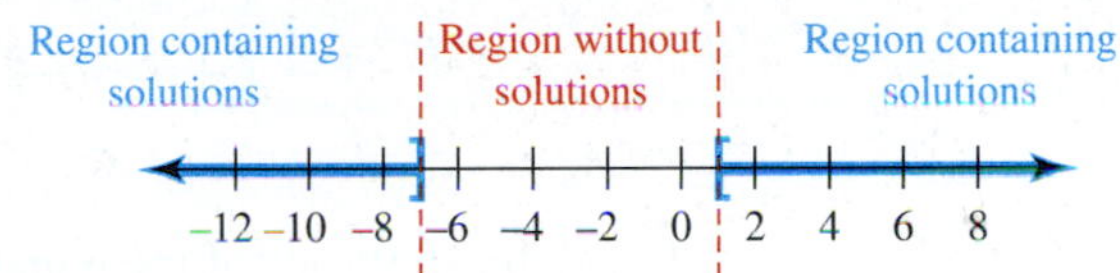

Notice that the number line is separated into regions separated by the numbers -7 and 1. The significance of the numbers -7 and 1 is that they are the solutions of the equation $|y + 3| = 4$.

$$|y + 3| = 4$$

$$\begin{aligned} y + 3 &= -4 \quad &\text{or}\quad y + 3 &= 4 \\ y + 3 - 3 &= -4 - 3 & y + 3 - 3 &= 4 - 3 \\ y &= -7 & y &= 1 \end{aligned}$$

The parts of the number line that are shaded include solutions of the inequality. If numbers from these regions are substituted into the inequality, a true statement will result. The values -10 and 3 are included in the shaded regions. When we substitute these numbers in for y, we get

$$|-10 + 3| \geq 4 \rightarrow |-7| \geq 4 \rightarrow 7 \geq 4 \quad \text{True}$$

$$|3 + 3| \geq 4 \rightarrow |6| \geq 4 \rightarrow 6 \geq 4 \quad \text{True}$$

The part of the number line that is *not* shaded includes numbers that are *not* solutions of the inequality. If numbers from this region are substituted into the inequality, a false statement will result. The value -4 is in the region that is not shaded. When we substitute this value for y, we get

$$|-4 + 3| \geq 4 \rightarrow |-1| \geq 4 \rightarrow 1 \geq 4 \quad \text{False}$$

So, this example illustrates that the solutions of the associated absolute value equation separate the number line into intervals of numbers that are or are not solutions of the absolute value inequality.

Procedure: Solving an Absolute Value Inequality Using Test Points

Step 1: Solve the associated absolute value equation. This equation comes from replacing the inequality symbol with an equals sign.

Step 2: Place the solutions of the equation on a number line.

Step 3: Select a number (a **test point**) from each of the resulting intervals and substitute it into the original inequality to determine if it is a solution.
Step 4: If the test point is a solution of the inequality, shade the interval of the number line that includes this point. If the test point is not a solution of the inequality, do not shade this interval of the number line.
Step 5: Put the appropriate symbol on the endpoints of the intervals.
a. If the absolute value inequality contains $\le$ or $\ge$, the endpoints are solutions of the inequality. Brackets are used to represent this.
b. If the absolute value inequality contains $<$ or $>$, the endpoints are *not* solutions of the inequality. Parentheses are used to represent this.
Step 6: Write the solution set in interval notation.

Objective 5 Examples Use test points to solve each absolute value inequality.

5a. $|x - 2| < 5$ **5b.** $|2y - 7| \ge -1$

Solutions **5a.** Solve the equation $|x - 2| = 5$.

$$x - 2 = -5 \quad \text{or} \quad x - 2 = 5$$
$$x - 2 + 2 = -5 + 2 \quad \text{or} \quad x - 2 + 2 = 5 + 2$$
$$x = -3 \quad \text{or} \quad x = 7$$

Place the solutions of the equation on a number line to form intervals.

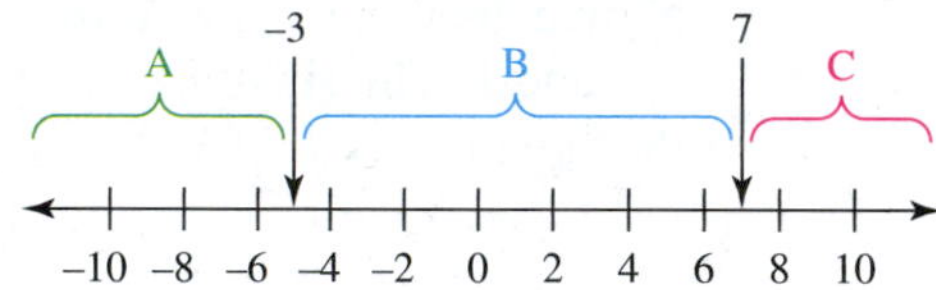

Test a point from intervals A, B, and C.

Interval	Test Point	$\lvert x - 2\rvert < 5$	True/False
A: $(-\infty, -3)$	-6	$\lvert -6 - 2\rvert < 5$ $\lvert -8\rvert < 5$ $8 < 5$	False
B: $(-3, 7)$	0	$\lvert 0 - 2\rvert < 5$ $\lvert -2\rvert < 5$ $2 < 5$	True
C: $(7, \infty)$	9	$\lvert 9 - 2\rvert < 5$ $\lvert 7\rvert < 5$ $7 < 5$	False

We shade interval B since its test point is a solution of the inequality. The endpoints are not included since the original inequality is $<$.

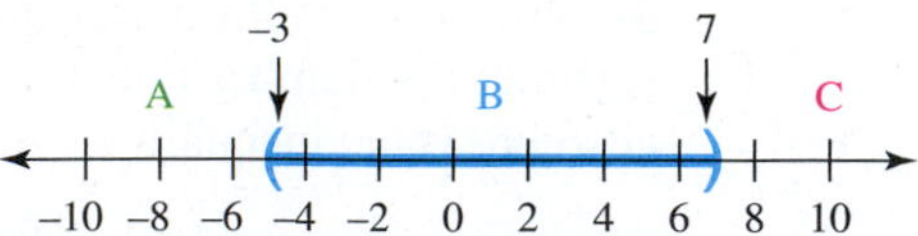

So, the interval notation for the solution set is $(-3, 7)$.

5b. Solve the equation $|2y - 7| = -1$. This equation has no solution since the constant on the right side is negative.

Since there are no solutions of the equation, the number line is not separated into any additional intervals. The entire number line is the interval we need to test.

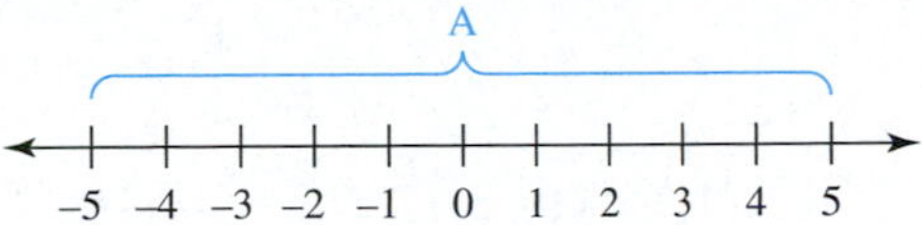

Test a point from interval A.

Interval	Test Point	$\lvert 2y - 7 \rvert \geq -1$	True/False
A: $(-\infty, \infty)$	0	$\lvert 2(0) - 7 \rvert \geq -1$ $\lvert -7 \rvert \geq -1$ $7 \geq -1$	True

Since the test point is a solution of the inequality, we shade interval A.

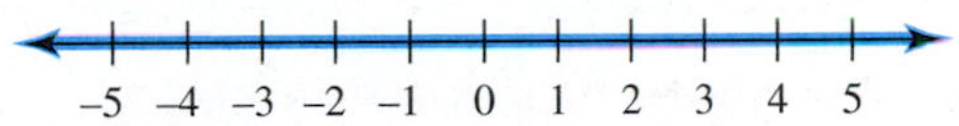

The solution set is $(-\infty, \infty)$ or all real numbers.

Student Check 5 Use test points to solve each absolute value inequality.

a. $\lvert x + 5 \rvert \geq 4$ **b.** $\lvert 6a - 1 \rvert < -2$

Objective 6 Troubleshoot common errors.

Troubleshooting Common Errors

Some common errors associated with absolute value inequalities are shown.

Objective 6 Examples **A problem and an incorrect solution are given. Provide the correct solution and an explanation of the error.**

6a. Solve $\lvert x + 2 \rvert < 5$.

Incorrect Solution	Correct Solution and Explanation
$\lvert x + 2 \rvert < 5$ $\lvert x + 2 \rvert < 5$ and $\lvert x + 2 \rvert > -5$ $\lvert x \rvert < 3$ and $\lvert x \rvert > -7$ So, the solution set is $(-7, 3)$.	While the solution set is correct, the notation used to solve the problem is wrong. When we write the compound inequality that satisfies the absolute value inequality, we no longer include the absolute value bars. $\lvert x + 2 \rvert < 5$ $x + 2 < 5$ and $x + 2 > -5$ $x < 3$ and $x > -7$

6b. Solve $\lvert y - 5 \rvert + 3 < 2$.

Incorrect Solution	Correct Solution and Explanation
$\lvert y - 5 \rvert + 3 < 2$ $y - 5 + 3 < 2$ or $y - 5 + 3 > -2$ $y - 2 < 2$ or $y - 2 > -2$ $y < 4$ or $y > 0$ Solution: $(-\infty, \infty)$	We must first isolate the absolute value expression. We should subtract 3 from each side. $\lvert y - 5 \rvert + 3 < 2$ $\lvert y - 5 \rvert + 3 - 3 < 2 - 3$ $\lvert y - 5 \rvert < -1$ The absolute value of a real number is never less than a negative number. So, the solution set is $\varnothing$.

ANSWERS TO STUDENT CHECKS

Student Check 1 **a.** $(-10, 10)$ **b.** $[-1, 9]$

c. $\left[-\frac{5}{2}, 1\right]$

Student Check 2 **a.** $(-\infty, -10) \cup (10, \infty)$

b. $(-\infty, -1] \cup [9, \infty)$ **c.** $\left(-\infty, -\frac{5}{2}\right) \cup (1, \infty)$

Student Check 3 **a.** $\varnothing$ **b.** $\varnothing$ **c.** $(-\infty, \infty)$

d. $(-\infty, \infty)$ **e.** $\left(-\infty, \frac{5}{3}\right) \cup \left(\frac{5}{3}, \infty\right)$

Student Check 4 **a.** According to the poll, 43% to 53% of U.S. workers are completely satisfied with their job.

b. Motorists should be at or below mile marker 43 or at or above mile marker 59.

Student Check 5 **a.** $(-\infty, -9] \cup [-1, \infty)$ **b.** $\varnothing$

SUMMARY OF KEY CONCEPTS

1. Solving an absolute value inequality involves solving a compound inequality. If the absolute value of an expression is less than a positive number, the compound inequality involves *and*. $|X| < k$, is equivalent to $X < k$ and $X > -k$, or we can write this as $-k < X < k$, where $k > 0$.
2. If the absolute value of an expression is greater than a positive number, the compound inequality involves *or*. $|X| > k$ is equivalent to $X < -k$ or $X > k$, where $k > 0$.
3. Special cases of absolute value inequalities arise when the constant $k < 0$ or when $k = 0$.
   - If the absolute value of an expression is less than a negative number, the inequality has no solution since the absolute value will never be less than a negative number.
   - If the absolute value of an expression is greater than a negative number, then the solution is all real numbers. This is because the absolute value of any number is positive and will always be greater than a negative number.
   - The inequality $|X| < 0$ has no solution. The inequality $|X| \leq 0$ has only one solution, $X = 0$. The inequality $|X| \geq 0$ has all real numbers as solutions. To solve $|X| > 0$, we must solve it by solving $X > 0$ or $X < 0$.
4. To solve applications of absolute values, the expression inside the absolute value represents the distance between two quantities. The number that the absolute value is greater than or less than is the given distance.
5. Test points can be used to solve absolute value inequalities. The solutions of the associated equation must be found first. The solutions partition the number line into regions that must be tested. A test point from each region is substituted into the original inequality to determine whether it is a solution. The regions that test true are shaded. Brackets or parentheses should be placed on the end points of the regions.

GRAPHING CALCULATOR SKILLS

The graphing calculator has the capacity to graph the solution set of an absolute value inequality. At this point, make sure that you know how to solve these by hand and use the calculator only to check your work.

Example: Solve $|y + 3| \leq 4$.

Solution: Enter the inequality in the equation editor and graph.

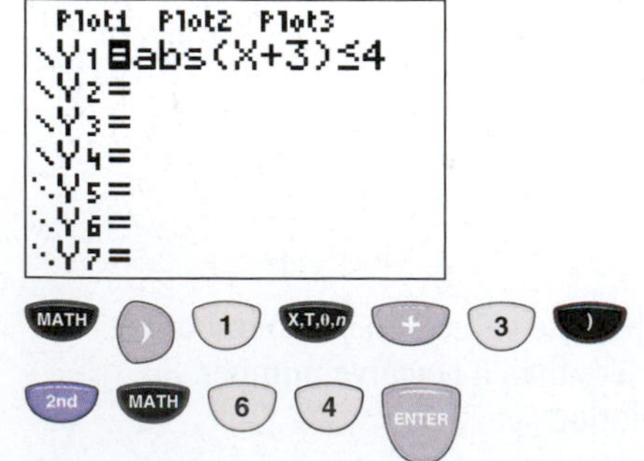

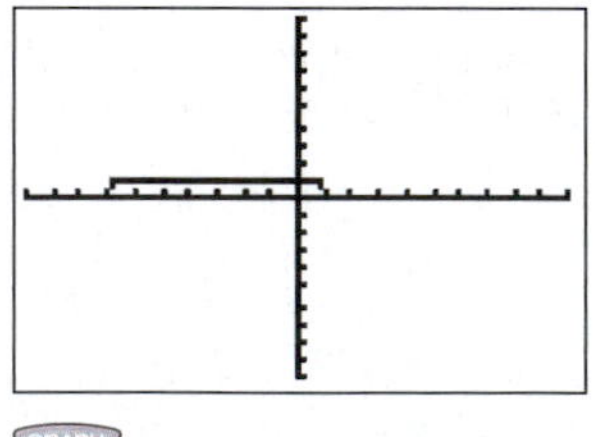

We TRACE along the graph to the endpoints to determine whether they should be included or not. If $y = 1$, then the endpoint is included. If $y = 0$, the endpoint is not included. Notice both $x = -7$ and $x = 1$ correspond to a y-value of 1, so they are included in the solution set. So, the solution set is $[-7, 1]$.

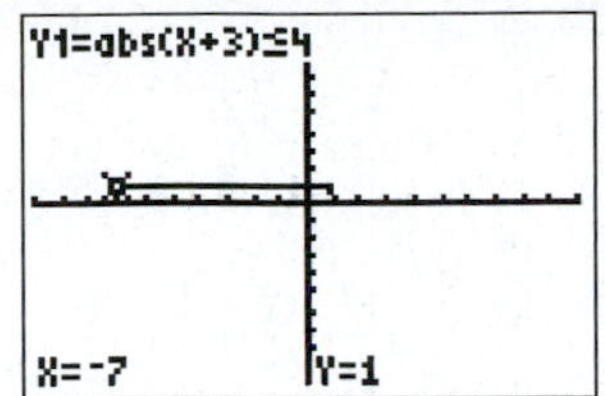

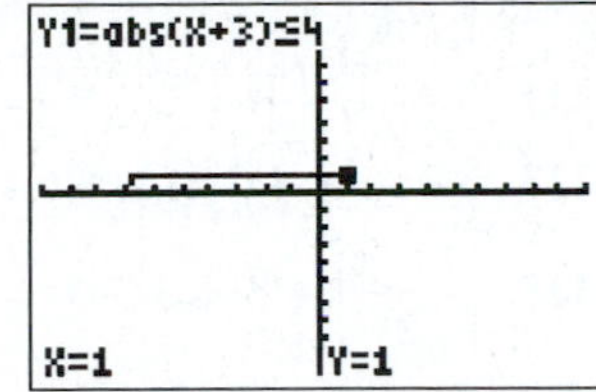

The TABLE also shows a picture of the solution set. The x-values whose y-value is 1 are part of the solution of the inequality. The x-values whose y-value is 0 are *not* solutions of the inequality. The table confirms the solution $[-7, 1]$.

X	Y1
-9	0
-8	0
-7	1
-6	1
-5	1
-4	1
-3	1

X=-9

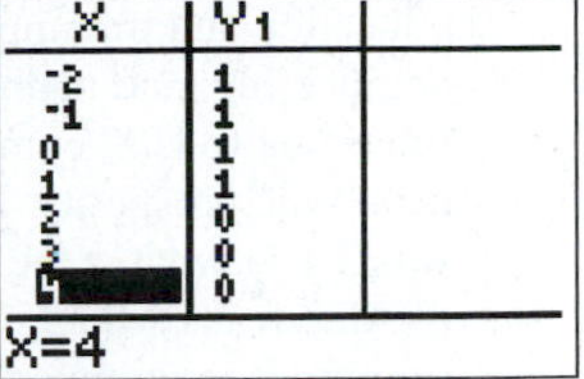

X
Y1
-2 1
-1 1
0 1
1 1
2 0
3 0
4 0
X=4

SECTION 2.7 EXERCISE SET

Write About It!

Use complete sentences in your answer to each exercise.

1. Explain how to solve an absolute value inequality involving a less than symbol.
2. Explain how to solve an absolute value inequality involving a greater than symbol.
3. Describe the special cases that may arise when solving an absolute value inequality.
4. Use an example to explain how to use test points to solve an absolute value inequality, $|x - a| < b$, $b > 0$. Answers vary.
5. Use an example to explain how to use test points to solve an absolute value inequality, $|x - a| > b$, $b > 0$. Answers vary.
6. Use an example to explain how to solve an absolute value inequality, $|x - a| \le 0$. Answers vary.

Practice Makes Perfect!

Solve each inequality. Write each answer in interval notation. (*See Objective 1.*)

7. $|x| \le 6$ $[-6, 6]$
8. $|x| \le 3$ $[-3, 3]$
9. $|y - 4| < 19$ $(-15, 23)$
10. $|y + 6| < 20$ $(-26, 14)$
11. $|2x + 3| < 4$ $\left(-\frac{7}{2}, \frac{1}{2}\right)$
12. $|4x - 3| < 5$ $\left(-\frac{1}{2}, 2\right)$
13. $|3x - 1| \le 6$ $\left[-\frac{5}{3}, \frac{7}{3}\right]$
14. $|6x - 7| \le 12$ $\left[-\frac{5}{6}, \frac{19}{6}\right]$
15. $|2 - 5x| < 1$ $\left(\frac{1}{5}, \frac{3}{5}\right)$
16. $|6 - 3x| < 7$ $\left(-\frac{1}{3}, \frac{13}{3}\right)$
17. $|1 - 4x| \le 5$ $\left[-1, \frac{3}{2}\right]$
18. $|3 - 9x| \le 15$ $\left[-\frac{4}{3}, 2\right]$
19. $|3x - 1| + 5 < 12$
20. $|5x + 3| + 8 < 10$
21. $|7 + 4x| - 1 < 4$
22. $|6 - 2x| - 8 < 3$
23. $6 + |8x - 5| \le 16$ $\left[\frac{-5}{8}, \frac{15}{8}\right]$
24. $5 + |3x + 2| < 15$ $\left(-4, \frac{8}{3}\right)$

Additional answers can be found in the Instructor Answer Appendix.

Solve each inequality. Write each answer in interval notation. (*See Objective 2.*)

25. $|y| > 4$ $(-\infty, -4) \cup (4, \infty)$
26. $|y| > 9$ $(-\infty, -9) \cup (9, \infty)$
27. $|x - 3| \ge 2$ $(-\infty, 1] \cup [5, \infty)$
28. $|x + 7| \ge 12$ $(-\infty, -19] \cup [5, \infty)$
29. $|5x + 5| > 17$
30. $|6x + 7| > 8$
31. $|10x - 4| > 16$
32. $|8x - 3| > 13$
33. $|8 - 4x| > 6$
34. $|7 - 3x| > 5$
35. $|2x + 7| - 3 \ge 6$
36. $|3x - 5| - 9 \ge 1$
37. $|6 - 2x| + 3 \ge 11$
38. $|7 - 4x| + 6 \ge 13$
39. $4 + |6x - 1| > 9$
40. $7 + |2x + 9| > 15$

Solve each inequality. Write each answer in interval notation. (*See Objective 3.*)

41. $|a| \le -1$ $\varnothing$
42. $|a| \le -5$ $\varnothing$
43. $|x + 5| < 0$ $\varnothing$
44. $|3x - 6| < 0$ $\varnothing$
45. $\left|\frac{x + 1}{2}\right| < 0$ $\varnothing$
46. $\left|\frac{2x - 3}{5}\right| < 0$ $\varnothing$
47. $|2x + 5| + 6 \le 3$ $\varnothing$
48. $|5x - 1| + 8 \le 2$ $\varnothing$
49. $|x| > -8$ $\mathbb{R}$
50. $|x| > -2$ $\mathbb{R}$
51. $|4x + 10| > 0$
52. $|5x - 6| > 0$
53. $|3x - 18| \le 0$ $\{6\}$
54. $|15 - 5x| \le 0$ $\{3\}$
55. $\left|\frac{4x - 3}{2}\right| > 0$
56. $\left|\frac{x + 3}{4}\right| > 0$
57. $|4 - 3x| + 6 \le 3$ $\varnothing$
58. $|2 - 7x| + 9 \le 6$ $\varnothing$
59. $|7 - x| \le 0$ $\{7\}$
60. $|6 - 2x| \le 0$ $\{3\}$
61. $5 + |4x + 6| > 5$
62. $1 + |2x + 5| > 1$
63. $\left|\frac{3x - 1}{5}\right| + 4 > 6$
64. $\left|\frac{2x + 5}{3}\right| - 1 \ge 3$
65. $\left|\frac{5x + 3}{2}\right| - 3 \le 1$
66. $\left|\frac{x - 4}{6}\right| + 3 < 8$ $(-26, 34)$

Use an absolute value inequality to solve each problem. (*See Objective 4.*)

67. A 2011 poll revealed that 74% of Americans think education and training in science, technology, engineering, and mathematics (STEM) is very important to U.S. competitiveness and our future economic prosperity. If the margin of error of the poll was 3.1%, what is the possible range of percentages of Americans who think that STEM education and training is very important to U.S. competitiveness and prosperity? (Source: www.researchamerica.org)

68. A 2010 poll revealed that 88% of Americans with private health insurance thought that the quality of their health care was excellent or good. If the margin of error of the poll was 4%, what is the possible range of percentages of Americans who were satisfied with the quality of their health care? (Source: www.gallup.com)

69. Rita read in a magazine that the best way to maintain one's weight loss is making sure one's weight stays within 5 lb of his or her goal weight. If Rita's goal weight is 145 lb, how much can she weigh while maintaining her weight loss? Rita can weigh between 140 and 150 lb.

70. Randal wants to stay within 5 lb of his ideal weight of 185 lb. How much can Randal weigh while meeting his goal? Randal can weigh between 180 and 190 lb.

Use test points to solve each absolute value inequality. (*See Objective 5.*)

71. $|x| > 16$ $(-\infty, -16) \cup (16, \infty)$
72. $|x| > 25$ $(-\infty, -25) \cup (25, \infty)$
73. $|2x - 5| < 11$ $(-3, 8)$
74. $|6x - 3| < 21$ $(-3, 4)$
75. $|1 - 4x| \geq 5$
76. $|3 - 6x| \geq 3$
77. $|y - 4| \leq 0$ $\{4\}$
78. $|y + 3| \leq 0$ $\{-3\}$
79. $|5x + 4| - 6 \geq 10$
80. $|7x + 3| - 8 \geq 2$
81. $|7 - 7x| < -3$ $\emptyset$
82. $|5 - 3x| < -4$ $\emptyset$
83. $|x| > -1$ $\mathbb{R}$
84. $|x| > -7$ $\mathbb{R}$
85. $|x - 6| \geq 0$ $\mathbb{R}$
86. $|x + 3| \geq 0$ $\mathbb{R}$
87. $|6x - 3| > -3$ $\mathbb{R}$
88. $|3x - 2| > -5$ $\mathbb{R}$
89. $|2x - 10| > 0$ $(-\infty, 5) \cup (5, \infty)$
90. $|0.4x - 1.2| > 0$ $(-\infty, 3) \cup (3, \infty)$
91. $|2x - 5| + 9 \leq 21$
92. $|3x - 1| + 7 \leq 18$
93. $|2 - x| - 5 < 16$
94. $|3 - 2x| - 7 < 23$

Mix 'Em Up!

Solve each inequality. Write each answer in interval notation.

95. $|6x + 9| \leq 17$
96. $|4x + 10| \leq 2$
97. $|a| > -9$ $\mathbb{R}$
98. $|a| \geq -\frac{1}{2}$ $\mathbb{R}$
99. $|2y - 1| - 8 < -2$
100. $|4y - 3| - 9 < -1$
101. $-|x| < 6$ $\mathbb{R}$
102. $-|x| < -10$
103. $|5x + 6| - 2 \geq 7$
104. $|8x - 3| - 4 \geq 5$
105. $|2x - 12| > 0$
106. $|5x + 15| > 0$
107. $|8 - 4x| \leq 0$ $\{2\}$
108. $|24 - 6x| \leq 0$ $\{4\}$
109. $\left|\frac{2x - 6}{5}\right| + 1 \leq 2$
110. $\left|\frac{2 - 3x}{4}\right| - 2 \geq 1$
111. $3 - |x + 4| < 5$ $\mathbb{R}$
112. $1 - |x + 3| < 6$ $\mathbb{R}$
113. $3.2 - |x - 2.1| \geq 8.6$ $\emptyset$
114. $12.5 + |x + 1.3| \geq 3.5$ $\mathbb{R}$
115. $\left|\frac{1 - x}{2}\right| + 3 < 3$ $\emptyset$
116. $\left|\frac{2 + x}{3}\right| - 2 \geq -2$ $\mathbb{R}$
117. $2.8 + |x - 0.9| \leq 6.3$ $[-2.6, 4.4]$
118. $4.5 + |x - 2.4| \leq 7.2$ $[-0.3, 5.1]$
119. Use test points to solve: $|x - 3| \leq -3$ $\emptyset$
120. Use test points to solve: $|x + 9| \leq 0.$ $\{-9\}$
121. Use test points to solve: $|x - 8| < 9$ $(-1, 17)$
122. Use test points to solve: $|x + 3| < 11$ $(-14, 8)$
123. Use test points to solve: $|6x + 3| - 1 \geq 4$
124. Use test points to solve: $|5x + 8| + 3 \geq 6$

125. A 2008 *USA Today*/Gallup Poll found that 53% of Americans described themselves as angry about the country's financial crisis. If the margin of error was 3%, what are the possible percentages of Americans polled who described themselves as angry about the country's financial crisis? (Source: www.gallup.com) According to the poll, 50% to 56% of Americans described themselves as angry about the country's financial crisis.

126. A 2007/2008 Gallup Poll found that approximately 28% of people polled in Latin America with a college degree desired to move to another country. If the margin of error was 4%, what are the possible percentages of people in Latin America with a college degree who desired to move to another country. (Source: www.gallup.com) According to the poll, 24% to 32% of people in Latin America with a college degree desired to move to another country.

127. A truck overturn results in a hazardous chemical spill. Emergency management personnel advise that motorists should be restricted from being within a 10-mi radius of the spill. If the spill happens at mile marker 35 on Interstate 20 in South Carolina, what are the safe locations on Interstate 20 for motorists? The safe locations on Interstate 20 for motorists will be outside markers 25 and 45.

128. A truck overturn on the New Jersey Turnpike results in a hazardous chemical spill. Emergency management personnel advise that motorists should be restricted from being within a 25-mi radius of the spill. If the spill happens at mile marker 40 on the New Jersey Turnpike, what are the safe locations on the turnpike for motorists? The safe locations on the turnpike for motorists will be outside markers 15 and 65.

You Be the Teacher!

Answer each student's questions.

129. Lee: Will there always be two disjoint intervals in the solution set when I solve an absolute value inequality of the form $|X| > k$? Please explain. There will always be two disjoint intervals in the solution set if $k > 0$.

130. Brielle: When solving an absolute value inequality of the form $|X| < k$, will the answer always be a single interval? Please explain. There will always be a single interval if $k > 0$.

Correct each student's errors, if any.

131. Solve $|2x - 5| < 13$.

Ed's work:

$$|2x - 5| < 13$$
$$2x - 5 < 13 \quad \text{or} \quad 2x - 5 > -13$$
$$2x < 18 \quad \text{or} \quad 2x > -8$$
$$x < 9 \quad \text{or} \quad x > -4$$

$|2x - 5| < 13$
$-13 < 2x - 5 < 13$
$-8 < 2x < 18$
$-4 < x < 9$
The solution set is $(-4, 9)$.

132. Solve $|4x + 1| + 2 \geq 15$.

Monica's work:

$$|4x + 1| + 2 \geq 15$$
$$4x + 1 + 2 \geq 15 \quad \text{or} \quad 4x + 1 + 2 \leq -15$$
$$4x \geq 12 \quad \text{or} \quad 4x \leq -18$$
$$x \geq 3 \quad \text{or} \quad x \leq -4.5$$

Calculate It!

Find the test points of the inequalities algebraically and then use a graphing calculator to solve the inequality.

133. $|8x + 4| < 12$ $(-2, 1)$

134. $|7x + 3| < 10$ $\left(-\frac{13}{7}, 1\right)$

135. $|5x - 14| \geq 14$

136. $|4x - 9| \geq 9$

Think About It!

137. Solve the inequality: $|x + 5| > |x - 7|$ $(1, \infty)$

138. Solve the inequality: $|2x - 6| \leq |x + 4|$ $\left[\frac{2}{3}, 10\right]$

139. Use test points to solve: $|x - 3| > |x + 2|$ $\left(-\infty, \frac{1}{2}\right)$

140. Use test points to solve: $|2x - 3| \geq |x + 1|$ $\left(-\infty, \frac{2}{3}\right] \cup [4, \infty)$

GROUP ACTIVITY The Mathematics of Controlling Waste

Your employer has assigned you with the task of controlling waste in the cutting room for a wrapping paper manufacturer. The company wants to lose less than 1% of wrapping paper for each run of 2000 sheets of paper. The reported width of each piece of square wrapping paper is 1 yd. The absolute error of the cutting machines is 0.25 in. in all directions. Determine if the current machines are within the desired standards. (Recall from Section 2.6 that $E_{abs} = |x - a|$, where x is the exact measure and a is the approximated value.)

1. Let x be the actual length of the cut wrapping paper. Write an absolute value inequality that shows that the absolute error of one side of the paper is within 0.25 in. $|x - 36| \leq 0.25$
2. Solve the inequality from part 1. Interpret the meaning in the context of the problem. $35.75 \leq x \leq 36.25$; The length of each piece of wrapping paper can range from 37.75 in. to 36.25 in.
3. What are the maximum and minimum values for the area of a cut sheet of wrapping paper? Since the wrapping paper is a square, each sheet of wrapping paper can be from 1278.0625 in.^{2} to 1314.0625 in.^{2}.
4. What is the maximum error of the area of the cut paper? The area of the desired piece of wrapping paper is $36^2 = 1296$ in.^{2}, so the maximum error is $1314.0625 - 1296 = 18.0625$ in.^{2}.
5. If the cutter wastes the maximum amount of paper (as found in part 4) for each piece of paper in the run, how much total waste would there be? $18.0625(2000) = 36{,}125$ in.^{2}
6. How many sheets of wrapping paper does the cutter waste for a run of 2000 sheets of paper? $\frac{36{,}125}{1296} = 27.87$ pieces of paper per run of 2000 sheets
7. What percentage of waste does this represent? approximately 0.0139 or 1.39% waste
8. What would you report to your employer? The cutter is potentially wasting more than 1% of the paper, so it is time to get a new cutting machine.

Chapter 2 / REVIEW

Linear Equations and Inequalities in One Variable

What's the big idea? Chapter 2 provides us with the skills to solve equations and inequalities in one variable. Knowing how to solve these equations and inequalities gives us the foundation we need to solve some word problems. The properties that enable us to solve these types of problems also provide the framework for us to solve more difficult equations, which we will encounter in later chapters.

The Tools

Listed below are the key terms, skills, formulas, and properties you should know for this chapter.

The page reference is provided if you need additional help with the given topic. The Study Tips will assist in your preparation for an exam.

Study Tips

1. Learn all of the terms, formulas, and properties. Make flash cards and have someone quiz you.
2. Rework problems from the exercises and also the ones you worked in class. Work additional problems from the review exercises.
3. Review the summaries of key concepts.
4. Work the chapter test.
5. Be sure to review the online resources for additional study materials.

Terms

Absolute error 120
Absolute value equation 116
Area 75
Circumference 75
Complementary angles 72
Compound inequality 104
Conditional equation 59
Consecutive integers 66
Contradiction 59
Equivalent equations 54
First-degree equation 52
Formula 75
Identity 60
Intersection 101
Interval notation 86
Linear equation in one variable 52
Linear inequality in one variable 86
Perimeter 75
Right angle 72
Set-builder notation 87
Simple inequality 88
Solution of a linear inequality in one variable 86
Solution of a linear equation in one variable 52
Solution set 52
Straight angle 72
Supplementary angles 72
Test point 132
Union 103
Vertical angles 72

Formulas and Properties

- Absolute value equations (Property 1) 116
- Absolute value equations (Property 2) 119
- Absolute value inequalities (Property 1) 125
- Absolute value inequalities (Property 2) 127
- Absolute value inequalities (Property 3) 128
- Absolute value inequalities (Property 4) 129
- Addition property of equality 54
- Addition property of inequality 88
- Area of triangle, rectangle, square, and circle 75
- Circumference of a circle 75
- Multiplication property of equality 55
- Multiplication property of inequality 89
- Perimeter of triangle, rectangle, and square 75

CHAPTER 2 / SUMMARY

How well do you know this chapter? Complete the following questions to find out. Take a look back at the section if you need help.

SECTION 2.1 Solving Linear Equations

1. A(n) linear equation in one variable is an equation of the form $ax + b = c$. Linear equations are also called first-degree equations.
2. A(n) solution is a value of the variable that makes the equation true.
3. To solve a linear equation in one variable, the goal is to produce an equation of the form $x =$ constant or constant $= x$.
4. Equivalent equations are equations with the same solution set.
5. The addition property of equality states that we can add or subtract the same number from an equation and not change the solution set.
6. The multiplication property of equality enables us to multiply or divide each side of an equation by the same non-zero number and not change the solution set.
7. To solve equations with fractions, we can multiply each side of the equation by the LCD to obtain an equivalent equation without fractions.
8. To solve equations with decimals, we can multiply each side of the equation by the power of 10 that eliminates the decimals.
9. A(n) conditional equation is an equation that is true for some values of the variable but not true for others.
10. A(n) contradiction is an equation that is not true for any values of the variable. The solution set for these equations is $\varnothing$.
11. A(n) identity is an equation that is true for all values of the variable. The solution set for these equations is $\mathbb{R}$.

SECTION 2.2 Introduction to Applications

12. To solve an application problem, the first thing to do is read the problem.
13. Determine the unknown and assign a variable to it. Other unknowns in the problem should be expressed in terms of this variable.
14. Use the known information to translate phrases into mathematical expressions to write the equation. Solve the equation and answer the question.

SECTION 2.3 Formulas and Applications

15. Complementary angles are angles whose sum is 90°. If one angle has a measure of $x°$, its complement has measure $(90 - x)^\circ$.
16. Supplementary angles are angles whose sum is 180°. If one angle has a measure of $x°$, its complement has measure $(180 - x)^\circ$.
17. Vertical angles, or opposite angles, are formed by intersecting lines. These angles are equal in measure.
18. The perimeter of a polygon is the distance around the figure.
19. The distance around a circle is its circumference.
20. The area of a figure is the number of square units its takes to cover the inside of the figure.
21. A mathematical formula is an equation that expresses the relationship between two or more variables.

SECTION 2.4 Linear Inequalities and Applications

22. A linear inequality in one variable is an inequality of the form $ax + b < c$.
23. A(n) solution of a linear inequality is a value that makes the inequality true.
24. The picture of the solution set of an inequality is the graph of the solution set.
25. If an endpoint of a solution set of an inequality is included in the solution, a(n) bracket is used. If an endpoint of a solution set is not included in the solution set a(n) parenthesis is used.
26. A concise way to express the solution set of an inequality is interval notation.
27. The symbol ∞ indicates that the solutions of an inequality continue to the right indefinitely. The symbol $-\infty$ indicates that the solutions of an inequality continue indefinitely to the left. A(n) parenthesis is always used with these notations.
28. When solving linear inequalities, we can add or subtract the same number from each side of an inequality and not change the relationship between the two expressions.
29. When solving linear inequalities, we can multiply or divide each side by a(n) positive number and not change the relationship between the two expressions.
30. When solving linear inequalities, we can multiply or divide each side by a(n) negative number but we must also reverse the inequality symbol to maintain the relationship between the two expressions.
31. The phrase "is at most" can be translated by the symbol $\leq$. The phrase "is at least" can be translated by the symbol $\geq$.

SECTION 2.5 Compound Inequalities

32. The intersection of two sets consists of the elements the sets have in common.
33. The union of two sets consists of the elements that are in either set.
34. A(n) compound inequality consists of two inequalities joined by "and" or "or."
35. To solve a compound inequality using "and" requires us to find the intersection of the solution sets.
36. To solve a compound inequality using "or" requires us to find the union of the solution sets.

SECTION 2.6 Absolute Value Equations

37. The absolute value of a number is its distance from zero on the number line.
38. The absolute value of any real number is nonnegative.
39. If $|X| = k$, $k > 0$, then $X = k$ or $X = -k$.
40. If $|X| = 0$, then $X = 0$.
41. If $|X| = k$, $k < 0$, then there is no solution.
42. To solve an absolute value equation, we must first isolate the absolute value on one side of the equation.
43. If $|X| = |Y|$, then $X = Y$ or $X = -Y$. That is, the expressions inside the absolute value must be the same or opposite.
44. The absolute error is the absolute value of the difference between the measured value and the exact value.

SECTION 2.7 Absolute Value Inequalities

45. Solving $|X| < k$, $k > 0$ is equivalent to solving the compound inequality $X < k$ and $X > -k$. This can be written in a compact form, $-k < X < k$.
46. Solving $|X| > k$, $k > 0$ is equivalent to solving the compound inequality $X > k$ or $X < -k$.
47. If $|X| < k$, $k < 0$, then there is no solution.
48. If $|X| > k$, $k < 0$, then the solution set is all real numbers.
49. Another method to solve absolute value inequalities involves using test points.

CHAPTER 2 / REVIEW EXERCISES

SECTION 2.1

Determine if each equation is a linear equation. If so, determine if the given number is a solution of the equation. (*See Objective 1.*)

1. $3x - 2(x + 5) = x - 4;\ x = 8$ yes; no
2. $4x + 3 - 8x = 3 - 4x;\ x = -\dfrac{3}{2}$ yes; yes
3. $\dfrac{4x + 3}{7} = 1;\ x = 1$ yes; yes
4. $\dfrac{2x - 3}{10} = -\dfrac{1}{2};\ x = -1$ yes; yes

Solve each equation. (*See Objectives 2–4.*)

5. $4x + 1 = 17$ $\{4\}$
6. $5 - 3x = -10$ $\{5\}$
7. $\dfrac{n}{6} = -2$ $\{-12\}$
8. $\dfrac{3 - t}{4} = -5$ $\{23\}$
9. $12.6 - d = -5.4$ $\{18\}$
10. $26.8 - 2d = 12.4$ $\{7.2\}$
11. $-2.5 = 3x + 6.8$ $\{-3.1\}$
12. $3.6 = 2x - 9.4$ $\{6.5\}$
13. $\dfrac{5y}{3} + \dfrac{1}{6} = \dfrac{3}{2}$ $\left\{\dfrac{4}{5}\right\}$
14. $\dfrac{5y}{6} - \dfrac{2}{3} = \dfrac{1}{2}$ $\left\{\dfrac{7}{5}\right\}$
15. $3(2x - 11) - 4(x - 7) = 5x + 10$ $\{-5\}$
16. $8(x - 2) - 2(1 - 3x) = 2(7x - 5)$ $\varnothing$
17. $2(x - 9) - 3(2 - x) = 5x - 24$ $\mathbb{R}$
18. $7x + 5(3 - 2x) = -3(x - 5)$ $\mathbb{R}$

SECTION 2.2

Translate each problem into a linear equation and find the solution. (*See Objective 1.*)

19. The sum of two consecutive even integers is 66. Find the integers. 32, 34
20. The sum of two consecutive integers is 43. Find the integers. 21, 22
21. Seven decreased by triple a number is -17. Find the number. 8
22. Twenty decreased by twice a number is -10. Find the number. 15
23. The quotient of a number and 4 is 7. Find the number. 28
24. The quotient of a number and 6 is 12. Find the number. 72
25. The product of a number and $\frac{1}{3}$ is 48. Find the number. 144
26. The product of a number and $\frac{1}{5}$ is 45. Find the number. 225
27. Nine less than twice times a number yields 21. Find the number. 15
28. Thirteen less than three times a number yields 17. Find the number. 10
29. The quotient of a number and 4 subtracted from the number is 9. Find the number. $\frac{9}{2}$
30. The quotient of a number and 3 subtracted from the number is 20. Find the number. $\frac{60}{19}$
31. Three times the sum of a number and 8 is 15. Find the number. -3
32. Twice the sum of a number and 4 is 24. Find the number. 8
33. Wesley saves quarters and dimes in a jar. If he has 18 more quarters than dimes and his collection is worth \$16.05, how many quarters has Wesley saved? 51 quarters
34. George saves half-dollars and silver dollars. If he has 41 more half-dollars than silver dollars and his collection is worth \$148, how many half-dollars has George saved? 126 half-dollars
35. The number of calories in a Burger King Whopper is 670 Cal. This is 220 Cal more than half the number

calories in a Burger King double Whopper. How many calories are in a Burger King double Whopper? (Source: http://fastfood.com/nutrition/burger_king.html) 900 Cal

36. The number of calories in a Wendy's crispy chicken sandwich is 350 Cal. This is 70 Cal more than half the number of calories in a Wendy's apple pecan chicken salad. How many calories are in a Wendy's apple pecan chicken salad? (Source: http://wendys.com) 560 Cal

37. Lois invests a total of \$1500 in Community Bank and Fulton Bank. The amount she invests in Community Bank is \$140 less than the amount she invests in Fulton Bank. How much does Lois invest in each bank? Lois invests \$680 in Community Bank and \$820 in Fulton Bank.

38. In 2011, the annual salary of Vice President Biden was \$230,700. The annual salary of President Obama was \$61,400 less than twice the salary of Vice President Biden. What was the annual salary for President Obama? (Source: http://usgovinfo.about.com) \$400,000

SECTION 2.3

Find the measure of each unknown angle. (*See Objective 1.*)

39. Find the measure of an angle whose complement is 34° more than the angle. 28°

40. Find the measure of an angle whose complement is 12° more than the angle. 39°

41. Find the measure of an angle whose supplement is 42° more than twice the measure of the angle. 46°

42. Find the measure of an angle whose supplement is 28° less than five times the measure of the angle. Round answer to two decimal places. 34.67°

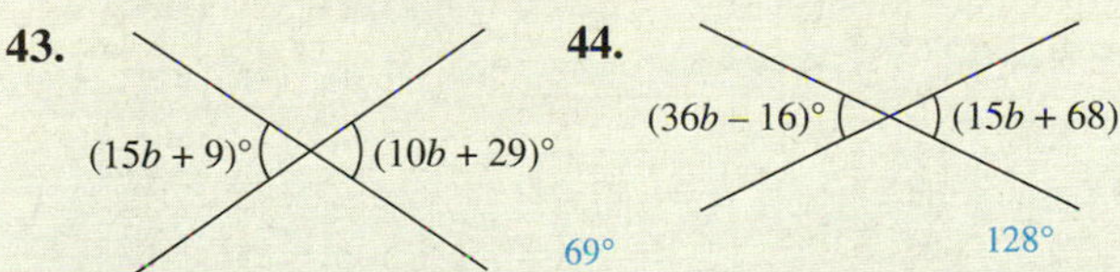

Use the perimeter, area, or circumference formulas to solve each problem. (*See Objective 2.*)

45. Find the length of the base of a triangle whose area is 60 square units and whose height is 15 units. 8 units

46. Find the length of the base of a triangle whose area is 75 square units and whose height is 25 units. 6 units

47. Find the circumference and area of a circle that has a diameter of 42 units. 42π units, 441π square units

48. If the perimeter of a rectangular park is 720 yd and the length of the park is 224 yd, find the width of the park. 136 yd

Use an appropriate formula to solve each problem. (*See Objective 3.*)

49. How long will it take Ricki's account to earn \$288 in simple interest if she invests \$2400 at 3% annual interest? 4 yr

50. How long will it take Bernadette's account to earn \$636 in simple interest if she invests \$3180 at 4% annual interest? 5 yr

51. As of 2011, the coldest temperature recorded on Earth is −129°F which occurred at Vostok, Antarctica, on July 21, 1983. Convert this temperature to degrees Celsius. Round answer to two decimal places. (Source: http://hypertextbook.com/facts/2000/YongLiLiang.shtml) −89.44°C

52. As of 2011, the highest temperature officially recorded in the United States is 134°F, which occurred at Greenland Ranch, California, on July 10, 1913. Convert this temperature to degrees Celsius. Round answer to two decimal places. (Source: http://hypertextbook.com/facts/2000/MichaelLevin.shtml) 56.67°C

Solve each formula for the specified variable. (*See Objective 4.*)

53. $y = mx + b$ for m $\quad m = \dfrac{y - b}{x}$

54. $Ax + By = C$ for A $\quad A = \dfrac{C - By}{x}$

55. $A = P(1 + r)$ for r $\quad r = \dfrac{A}{P} - 1$

56. $V = 4\pi r^2 h$ for h $\quad h = \dfrac{V}{4\pi r^2}$

SECTION 2.4

Graph the solution set of each inequality. Write the solution set in interval notation and set-builder notation. (*See Objectives 1.*)

57. $x \geq -6$

58. $x \geq -\dfrac{4}{9}$

59. $x < 10$

60. $-2 < x < 0$

61. $-20 > x > -25$

62. $-3 < x \leq 9$

Solve each inequality. Write each answer in interval notation. (*See Objective 2.*)

63. $30 < -5a$ $\quad (-\infty, -6)$

64. $12 \geq -4a$ $\quad [-3, \infty)$

65. $0.16(3x - 1) \leq 0.8(4 - x)$ $\quad (-\infty, 2.625]$

66. $0.24(6x - 3) \geq 0.28(1 - 2x)$ $\quad [0.5, \infty)$

67. $5 + 2(3 - x) < 5(x - 2)$ $\quad (3, \infty)$

68. $12 - 3(x - 4) \geq 6(x + 1)$ $\quad (-\infty, 2]$

69. $4(3x - 5) \geq 9x + 1$ $\quad [7, \infty)$

70. $5(2x + 1) < 18x - 19$ $\quad (3, \infty)$

71. $2.4 + 1.2(6 - x) < 0.9(2x + 7)$ $\quad (1.1, \infty)$

72. $6.3 - 3.2y < 6.8y - 13.7$ $\quad (2, \infty)$

73. $\dfrac{2}{3}(6x - 3) > 4(2x - 7)$ $\quad \left(-\infty, \dfrac{13}{2}\right)$

74. $\dfrac{1}{2}(2x - 8) \geq 2(x - 3)$ $\quad (-\infty, 2]$

Write an inequality that models each situation. Solve the inequality and answer the question in a complete sentence. (*See Objective 3.*)

75. Ruben's English course grade is based on a weighted average. The weights are as follows: tests (50%), term paper (30%), and final exam (20%). If Ruben's test average is 78 and he scores 85 on his term paper, what must he score on the final exam to maintain at least an 80 course average? Ruben must score at least a 77.5 on his final to maintain at least an 80 course average.

76. Hong's Psychology course grade is based on a weighted average. The weights are as follows: midterm (35%), papers (40%), and final exam (25%). If Hong earns a 75 on his midterm and 78 on his papers, what must he score on his final exam to maintain at least an 80 course average? Hong must score at least a 90.2 on his final exam to maintain at least an 80 course average.

77. Amy rents a small fire hall for her family reunion for $400. She hires a caterer who charges $20 per person. If she has $2000 to spend on the reunion, how many family members and friends can attend? Amy can have at most 80 people at her family reunion.

78. Nam hires a maid service to clean her house biweekly. The company charges a one-time fee of $129 and $65 per cleaning visit. If Nam budgets $1800 for the maid service for the year, how many cleaning visits can she have? Nam can have at most 25 cleaning visits.

SECTION 2.5

Find the intersection or union of the two sets. Write each answer in interval notation when applicable. (*See Objectives 1 and 2.*)

79. $(-\infty, 0) \cap (-4, \infty)$ $(-4, 0)$

80. $(-\infty, 10] \cap (10, \infty)$ $\varnothing$

81. $(-\infty, 3.2) \cap [-1.8, 16.4)$ $[-1.8, 3.2)$

82. $\varnothing \cap [4, 12]$ $\varnothing$ **83.** $\varnothing \cap (-2, 1)$ $\varnothing$

84. $[-10, 3] \cup (-4, 6)$ $[-10, 6)$

85. $(-\infty, 4) \cup [4, \infty)$ $(-\infty, \infty)$

86. $(-\infty, 8.6) \cup [-2.9, 13.8)$ $(-\infty, 13.8)$

87. $\varnothing \cup (-4, \infty)$ $(-4, \infty)$ **88.** $\varnothing \cup [8, \infty)$ $[8, \infty)$

Solve each compound inequality. Write each answer in interval notation when applicable. (*See Objectives 3 and 4.*)

89. $6x - 7 \geq 5$ and $4x - 3 < 9$ $[2, 3)$

90. $2(1 - 5x) > 12$ and $3x - 4 < 14$ $(-\infty, -1)$

91. $6(2 - x) < -4$ and $4x + 3 < 10$ $\varnothing$

92. $3x - 5 > 7$ and $2x + 6 < 1$ $\varnothing$

93. $0.7x + 0.34 < 4.274$ and $0.4x - 2.56 > -1.7$ $(2.15, 5.62)$

94. $1.24x - 4.35 \geq -2.49$ and $0.8x + 2.6 \leq 2.4$ $\varnothing$

95. $8x + 3 < 19$ or $3x - 1 > 14$ $(-\infty, 2) \cup (5, \infty)$

96. $4 - 3x \leq 0$ or $8x + 3 > -5$ $(-1, \infty)$

97. $2 - 5x \leq 1$ or $4x - 3 > 5$ $\left[\frac{1}{5}, \infty\right)$

98. $12x - 4 \leq 8$ or $3x + 2 > 8$ $(-\infty, 1] \cup (2, \infty)$

Solve each problem using compound inequalities. (*See Objective 5.*)

99. Margie wants to maintain a test average between an 80 and 89. If Margie's test scores thus far are 81, 70, 85, and 76, what must she score on her fifth test to make the desired grade? (Assume the highest grade possible is 100.) Margie must score between 88 and 100 on her fifth test.

100. Grace wants to make a quiz average between 70 and 79. If Grace's quiz scores are 65, 72, 65, 71, and 68, then what must she score on her sixth quiz to get the desired grade? (Assume the highest grade possible is 100.) Grace must score between 79 and 100 on her sixth quiz.

101. Michael needs someone to trim two-thirds of his bamboo forest. He finds a company that charges $35 per hour plus $90 to haul the trees away. If Michael plans to spend between $195 and $500, how many hours can he afford to hire the tree-trimmer? Michael can afford to hire the tree-trimmer for at least 3 hr and at most 11 hr.

102. Patty hires a professional organizer to organize her home office. The professional organizer charges a flat fee of $80 plus $32 per hour. If Patty budgets $720 to $1000 to get her office organized, how many hours can she afford to hire the professional organizer? Patty can afford to hire the professional organizer at least 20 hr and at most 28 hr.

SECTION 2.6

Solve each equation. (*See Objective 1.*)

103. $|12x - 5| = 25$ $\left\{-\frac{5}{3}, \frac{5}{2}\right\}$

104. $|5x - 16| = 14$ $\left\{\frac{2}{5}, 6\right\}$

105. $|7 - 2x| = 18$ $\left\{-\frac{11}{2}, \frac{25}{2}\right\}$

106. $|9 - 4x| = 3$ $\left\{\frac{3}{2}, 3\right\}$

107. $\left|\frac{8x - 4}{2}\right| - 1 = 6$ $\left\{-\frac{5}{4}, \frac{9}{4}\right\}$

108. $\left|\frac{2x + 9}{5}\right| + 1 = 4$ $\{-12, 3\}$

109. $\left|\frac{2x}{3}\right| = -12$ $\varnothing$ **110.** $\left|\frac{15x}{4}\right| = 0$ $\{0\}$

111. $|3x + 2| = 0$ $\left\{-\frac{2}{3}\right\}$ **112.** $|6x - 9| = 0$ $\left\{\frac{3}{2}\right\}$

113. $|0.6x - 2.5| = 7.1$ $\left\{-\frac{23}{3}, 16\right\}$

114. $|0.4x + 4.1| = 11.5$ $\{-39, 18.5\}$

Solve each equation. (*See Objective 2.*)

115. $\left|\frac{5x + 1}{3}\right| = \left|\frac{x}{6}\right|$ $\left\{-\frac{2}{9}, -\frac{2}{11}\right\}$

116. $\left|\frac{3x - 1}{4}\right| = \left|\frac{x}{4}\right|$ $\left\{\frac{1}{4}, \frac{1}{2}\right\}$

117. $|2x + 10| = |4x - 8|$ $\left\{-\frac{1}{3}, 9\right\}$

118. $|6x - 11| = |5 - 4x|$ $\left\{\frac{8}{5}, 3\right\}$

Solve each problem using an absolute value equation. (*See Objective 3.*)

119. Suppose the actual length of an object is 6.245 m. If the length is approximated to be 6.512 m, what is the absolute error? 0.267 m

120. Suppose the actual length of an object is 20.35 in. If the length is approximated to be 21 in., what is the absolute error? 0.65 in.

121. Suppose a machine with an absolute error reading of 1.15 units reports the measurement of an object as 12.98 units. What are the possible values for the actual measurement of the object? The actual measurement of the object is between 11.83 units and 14.13 units.

122. Suppose a machine with an absolute error reading of 8.25 units reports the measurement of an object as 3163.78 units. What is the actual measurement of the object? The actual measurement of the object is between 3155.53 units and 3172.03 units.

123. A poll shows that Candidate A will receive 32% of the votes and Candidate B will receive 36% of the votes. If the poll has an absolute error of 4%, what are the possible values for how much of the vote each candidate will receive? Candidate A will receive between 28% and 36% of the vote and candidate B will receive between 32% and 40% of the vote.

124. A poll shows that Candidate C will receive 58% of the votes and Candidate D will receive 32% of the votes. If the poll has an absolute error of 5%, what are the possible values for how much of the vote each candidate will receive? Candidate C will receive between 53% and 63% of the vote and candidate D will receive between 27% and 37% of the vote.

SECTION 2.7

Solve each absolute value inequality. Write the answer in interval notation. (*See Objectives 1–3.*)

125. $|3x + 2| \le 16$ $\left[-6, \frac{14}{3}\right]$

126. $|12 - 4x| \ge 8$ $(-\infty, 1] \cup [5, \infty)$

127. $-|x| \le -2$ $(-\infty, -2] \cup [2, \infty)$

128. $14 - |x + 3| > 8$ $(-9, 3)$

129. $10 - |5 - 2x| < 7$ $(-\infty, 1) \cup (4, \infty)$

130. $3.2 + |x - 0.6| \ge 5.4$ $(-\infty, -1.6] \cup [2.8, \infty)$

131. $-|x| > 7$ $\varnothing$

132. $|4 - x| > 0$ $(-\infty, 4) \cup (4, \infty)$

133. $|2x + 10| \le 0$ $\{-5\}$

134. $|2 - 9x| < 0$ $\varnothing$

135. $|8 - 3x| \ge 0$ $(-\infty, \infty)$

136. $4.6 - |x - 2.3| \le 9.6$ $(-\infty, \infty)$

137. $|b| > -12$ $(-\infty, \infty)$ **138.** $|b| \ge -\frac{4}{5}$ $(-\infty, \infty)$

139. $|5y - 9| - 2 < -12$ $\varnothing$ **140.** $|3y - 2| - 1 < -14$ $\varnothing$

Use an absolute value inequality to solve each problem. (*See Objective 4.*)

141. A November 2009 Gallup Poll found that 38% of Americans rating healthcare coverage in this country as excellent or good. If the margin of error was 4%, what are the possible percentages of Americans rating healthcare coverage as excellent or good? (Source: www.gallup.com) According to the poll, 34% to 42% of Americans rate healthcare coverage as excellent or good.

142. A June 2010 Gallup Poll found that approximately 54% of Americans said they were energized Americans who were well-rested and did not experience ailments such as physical pain, headache, cold or flu. If the margin of error was 3%, what are the possible percentages of Americans who were energized? According to the poll, 51% to 57% of Americans were energized.

143. A truck overturn resulted in a hazardous chemical spill. Emergency management personnel advised that motorists should be restricted from being within a 12-mi radius of the spill. If the spill happened at mile marker 226 on Pennsylvania Turnpike 76, what are the safe locations on the turnpike for motorists? The safe locations on the turnpike for motorists are outside markers 214 and 238.

144. A truck overturn on the Pennsylvania Turnpike resulted in a hazardous chemical spill. Emergency management personnel advised that motorists should be restricted from being within a 20-mi radius of the spill. If the spill happened at mile marker 74 on the turnpike, what are the safe locations on the turnpike for motorists? The safe locations on the turnpike for motorists are outside markers 54 and 94.

Use test points to solve each inequality. (*See Objective 5.*)

145. $|x - 12| < 19$ $(-7, 31)$

146. $|x + 7| > 21$ $(-\infty, -28) \cup (14, \infty)$

147. $|1 - 10x| - 3 \ge 18$ $(-\infty, -2] \cup \left[\frac{11}{5}, \infty\right)$

148. $|3 - 2x| + 5 \ge 20$ $(-\infty, -6] \cup [9, \infty)$

CHAPTER 2 TEST / LINEAR EQUATIONS AND INEQUALITIES IN ONE VARIABLE

1. The solution set for $10y + 9 = 19$ is

a. $\left\{\frac{14}{5}\right\}$ **(b.)** $\{1\}$ **c.** $\{0\}$ **d.** $\left\{-\frac{71}{10}\right\}$

2. The solution set for $6x - 4 - 4x = 2x - 4$ is

a. $\{2\}$ **b.** $\left\{\frac{2}{3}\right\}$

c. $\varnothing$ **(d.)** $\{x \mid x \text{ is a real number}\}$

3. The solution set for $-y - 2(2y - 1) = 5(1 - y)$ is

a. $\{-2\}$ **(b.)** $\varnothing$

c. $\{2\}$ **d.** $\{x \mid x \text{ is a real number}\}$

4. If $2x - 3y = 6$ is solved for y, then $y =$

a. $-\frac{3}{2}x - 3$ **b.** $\frac{3}{2}x + 3$

(c.) $\frac{2}{3}x - 2$ **d.** $-\frac{2}{3}x + 2$

5. The graph of $6 - 3x \le -3$ is

a.

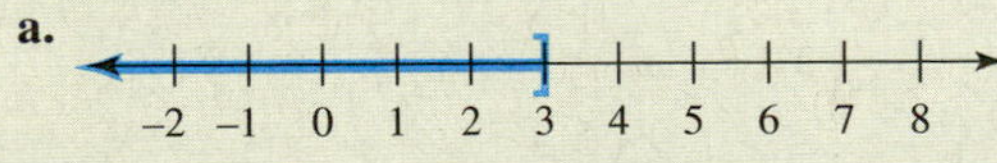

(b.)

-2 -1 0 1 2 3 4 5 6 7 8

c. 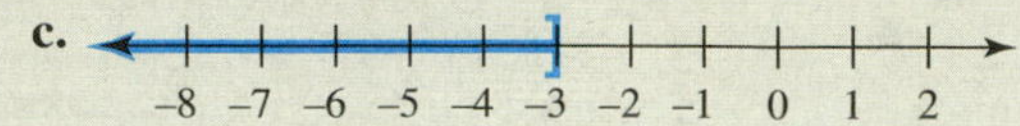

d. –8 –7 –6 –5 –4 –3 –2 –1 0 1 2

6. The solution for $4x + 1 > 9x - 4$ in interval notation is
   a. $(-\infty, 1)$ b. $(-\infty, -1)$
   c. $(1, \infty)$ d. $(-1, \infty)$

7. Use the figure to determine the measure of the smaller angle.
   a. $5.6°$
   b. $12°$
   c. $30°$
   d. $38°$

   $(12x + 6)°$ $(2x + 6)°$

8. The interval that represents $(2, \infty) \cup (4, \infty)$ is
   a. $\varnothing$ b. $(2, 4)$
   c. $(4, \infty)$ d. $(2, \infty)$

9. The solution set of $|3x - 5| - 4 = -1$ is
   a. $\left\{\frac{2}{3}, \frac{8}{3}\right\}$ b. $\left\{\frac{8}{3}, \frac{10}{3}\right\}$
   c. $\left\{\frac{8}{3}\right\}$ d. $\varnothing$

10. The solution set for $|x + 3| > -2$ is
    a. $(-5, \infty)$ b. $\varnothing$
    c. $(-\infty, \infty)$ d. $(-5, -1)$

Solve each equation. Write the answer in a solution set.

11. $\frac{5}{3}(2y + 3) - \frac{1}{6}y = \frac{1}{2}(y - 4)$ $\left\{-\frac{21}{8}\right\}$

12. $0.05x + 0.10(30 - x) = 2.4$ $\{12\}$

13. $|7a + 4| - 5 = 8$ $\left\{-\frac{17}{7}, \frac{9}{7}\right\}$

14. $\left|\frac{1}{2}b - \frac{2}{3}\right| + 5 = 1$ $\varnothing$

15. $|3 - 4x| = \left|\frac{1}{5}x + 2\right|$ $\left\{\frac{5}{21}, \frac{25}{19}\right\}$

Solve each inequality. Graph the solution set and write the solution set in interval and set-builder notation.

16. $6 - 2(3x - 1) > 4(x + 2) - 10$

17. $\frac{9}{8}m - \frac{m + 2}{6} \geq \frac{m}{2}$

18. $-0.25x + 4 > -1$ and $\frac{1}{2}x - \frac{2}{3} < x$

19. $7(4 - x) + 3 < -2(x + 5)$ or $6(x + 1) - 4(2x - 3) > 7x$

20. $|8x + 3| \leq 5$

21. $|3x - 2| - 1 \geq 6$

Solve each problem by solving an appropriate equation or inequality. Use complete sentences to state the answer.

22. A cashier has a total of 28 bills made up of tens and twenties. The total value of the money is \$400. How many of each bill does she have?

23. The width of a rectangle is five ft more than four times the length of the rectangle. If the perimeter is 210 ft, find the length and width of the rectangle.

24. A resting metabolic rate (RMR) is the rate at which a person burns energy or calories at rest. It is given by the formula $\text{RMR} = 4.541w + 15.875h - 4.92a + 166g - 161$, where w is weight in pounds, h is height in inches, a is age in years, and g is gender ($g = 0$ for females and $g = 1$ for males). Find Jan's RMR if she weighs 145 lb, is 65 in. tall, and is 35 yr old.

25. The Smartphone Company sells its smartphones for \$599.00 each. So, the company's revenue is represented by $R = 599x$, where x is the number of phones sold. The company's monthly cost of producing x phones is $C = 194.05x + 2500$. The company has a profit when its revenue is greater than its cost. How many phones does the company need to sell in one month to make a profit?

26. Juan's final grade in his math class is based on a weighted average as follows.

Test average	50%
Homework	10%
Quizzes	15%
Final exam	25%

If Juan has a 79 test average, 95 homework average, and 83 quiz average, what scores can he make on his final exam to have a final grade of a C (70 to 79)?

27. A utility company installs a power line 100 ft from the entrance of a subdivision. The utility company has an easement that is 40 ft on either side of the power line. Write an absolute value equation that will determine where the easement begins and ends, where x represents the location of the easement, and solve it.

28. On a portion of an interstate highway, the speed limit is 65 mph. The highway patrol officers stop vehicles which travel at a speed that is more than 15 mph from the speed limit. Write an absolute value inequality that models this situation, where s is the speed of the vehicle, and solve the inequality.

CUMULATIVE REVIEW EXERCISES / CHAPTERS 1 AND 2

Determine the requested information about the given sets. (*Section 1.1, Objective 1*)

1. Use the roster method to write the set: $B = \{x \mid x \text{ is an integer strictly between } -6 \text{ and } 5\}$ $\{-5, -4, -3, -2, -1, 0, 1, 2, 3, 4\}$
2. Use the roster method to write the set: $A = \{x \mid x \text{ is a natural number less than } 2\pi\}$ $\{1, 2, 3, 4, 5, 6\}$

Classify each real number as natural, whole, integer, rational, or irrational. List each classification that applies. (*Section 1.1, Objective 2*)

3. $12\frac{2}{3}$ rational
4. $-\sqrt{17}$ irrational
5. -8.901 rational
6. $\sqrt{16}$ natural, whole, integer, rational

Graph each real number on the real number line. (*Section 1.1, Objective 3*)

7. $\sqrt{20}$
8. 4.3

Find the opposite of the given number. Assume *a* and *b* represent positive real numbers. (*Section 1.1, Objective 4*)

9. $5a$ $-5a$
10. $-24b$ $24b$

Simplify each absolute value expression. (*Section 1.1, Objective 5*)

11. $-|-43|$ -43
12. $|-62|$ 62

Perform the indicated operation. (*Section 1.2, Objectives 1 and 2*)

13. $5.1 + (-8.6)$ -3.5
14. $6.3 + (-14.9)$ -8.6
15. $\frac{1}{6} + \left(-\frac{5}{6}\right)$ $-\frac{2}{3}$
16. $\frac{9}{5} + \left(-\frac{2}{5}\right)$ $\frac{7}{5}$
17. $24 \div \left(-\frac{1}{4}\right)$ -96
18. $\frac{12}{0}$ undefined
19. $(-13.5)(-10)$ 135
20. $(-42)(5)$ -210

Use the order of operations to simplify each expression. (*Section 1.2, Objectives 3 and 4*)

21. $-6(-1)^3 - 10(-2)^2 + 18$ -16
22. $6 - [(15 - 7) - (1 - 3)] - 32 \div 8 \cdot 2$ -12
23. $\dfrac{19 - 2\sqrt{27 - 2}}{2|2 - 3|^2}$ $\frac{9}{2}$
24. $\dfrac{-23 + 35}{10 - (-8)}$ $\frac{2}{3}$
25. $-4 - \left(-\frac{7}{5}\right)$ $-\frac{13}{5}$

Evaluate each expression for the given values. (*Section 1.2, Objective 5*)

26. $2x^2 - 5x + 1$ for $x = -3$ 34
27. $-x^3 + 4x - 7$ for $x = 2$ -7
28. $b^2 - 4ac$ for $a = 6, b = -3, c = 1$ -15
29. $15x - 10y$ for $x = 4, y = 0$ 60

Find both the additive inverse and multiplicative inverses of each number. (*Section 1.3, Objective 1*)

30. 12 $-12, \frac{1}{12}$
31. -4 $4, -\frac{1}{4}$

Apply the commutative, associative, or distributive properties to rewrite the given expression as an equivalent expression. (*Section 1.3, Objective 2*)

32. $14a - 8$ Answers vary. $-8 + 14a$
33. $(6 - x) + 2y$ Answers vary. $6 - x + 2y$
34. $23(m)n$ Answers vary. $23(mn)$
35. $-2(15x - 3y + 2)$ Answers vary. $-30x + 6y - 4$

Simplify each expression. (*Section 1.3, Objective 3*)

36. $-3x(10 + 2x) + 12x^2 - 2(5x - 3)$ $6x^2 - 40x + 6$
37. $\frac{2}{3}x + 7 - \frac{1}{3}x + 17$ $\frac{1}{3}x + 24$
38. $\frac{4}{3}x + 1 - \frac{1}{5}x + \frac{1}{2}$ $\frac{17}{15}x + \frac{3}{2}$
39. $5(2x + 6) - 2(x - 14)$ $8x + 58$

Translate each phrase or sentence into an algebraic expression, equation, or inequality. Use the variable *x* to represent the unknown number. (*Section 1.3, Objective 4*)

40. One-third the sum of a number and 13 $\frac{1}{3}(x + 13)$
41. Six more than twice a number is less than the number plus ten $6 + 2x < x + 10$

Solve each problem. (*Section 1.3, Objective 5*)

42. Jacob has collected 120 quarters and dimes. If *x* represents the number of dimes he collected, write an expression for the number of quarters he collected. $120 - x$
43. Referring to Exercise 42, if Jacob's coins total \$24, write an algebraic expression that represents the total value of his coins. $0.1x + 0.25(120 - x) = 24$

Determine if the equation is a linear equation. If it is a linear equation, determine if $x = -2$ is a solution of the equation. (*Section 2.1, Objective 1*)

44. $-4x - 3 = 5$ linear; It is a solution.
45. $2m^2 + 7 = 15$ not linear

Use the addition property of equality to solve each linear equation and check your answer. (*Section 2.1, Objective 2*)

46. $19 - d = 12.2$ $\{6.8\}$
47. $x - 5 = 7$ $\{12\}$

Use the multiplication property of equality to solve each linear equation and check your answer. (*Section 2.1, Objective 3*)

48. $\frac{a}{13} = 3$ $\{39\}$
49. $3x = 24$ $\{8\}$

Solve each linear equation and check your answer. If there is no solution, then write ∅ for your answer. (*Section 2.1, Objectives 4 and 5*)

50. $12 + 4x = 4(1 - x)$ $\{-1\}$

51. $\frac{5x}{3} - \frac{3x}{2} = \frac{1}{6}$ $\{1\}$

52. $5(x + 3) - 4x = x - 5$ ∅

Solve the linear equation and check your answer. If there is no solution, then write ℝ for your answer. (*Section 2.1, Objectives 4 and 6*)

53. $6(x - 2) - 5x = x - 12$ ℝ

54. $8(2x + 1) - 6x = x - 10$ $\{-2\}$

55. $7(x - 4) - x = 6(x - 5) + 2$ ℝ

Translate each problem into a linear equation and solve the problem. (*Section 2.2, Objective 1*)

56. The product of a number and one-third is 60. Find the number. 180

57. The quotient of a number and four is eight. Find the number. 32

58. Five greater than three times a number yields 17. Find the number. 4

59. Twice the sum of two consecutive odd numbers is 88. Find the numbers. 21, 23

Find the measure of each unknown angle. (*Section 2.3, Objective 1*)

60. Find the measure of an angle whose complement is 15° more than twice the measure of the angle. 25°

61. Find the measure of an angle whose supplement is 33° less than twice the measure of the angle. 71°

62. 68°, 22° **63.** 31°, 31°

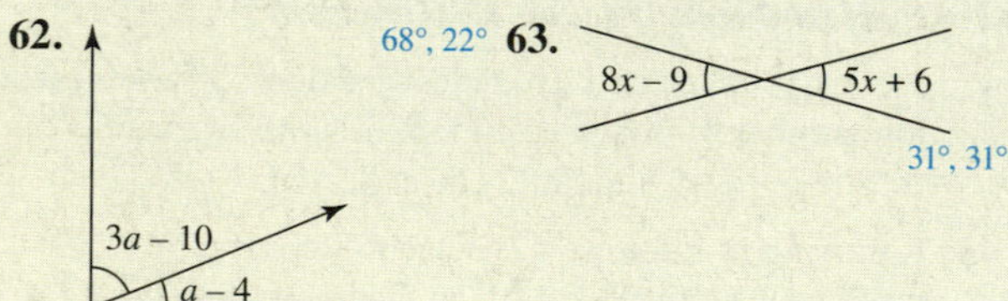

64. Find the height of a triangle whose area is 108 ft^2 and whose base is 9 ft. 24 ft

65. The length of a rectangle is 2 ft less than three times its width. If the perimeter is 280 ft, find the length and width of the rectangle. The length is 104.5 ft and the width is 35.5 ft.

Solve each formula for the specified variable. (*Section 2.3, Objective 4*)

66. $a = \frac{2}{5}b - cd$ for b $b = \frac{5a + 5cd}{2}$

67. $x = \frac{1}{2}y + 3zw$ for z $z = \frac{2x - y}{6w}$

Graph the solution set of each inequality on a number line and express the solution set in interval notation and set-builder notation. (*Section 2.4, Objective 1*)

68. $3 < x$

69. $-10 \geq x$

Solve each inequality. Graph the solution set and write your answer in interval notation and set-builder notation. (*Section 2.4, Objective 2*)

70. $-24 \leq -6a$

71. $\frac{2}{3}(x + 3) < -x + 5$

The following students are taking algebra from Professor Fowser. In Professor Fowser's class, the course grade is determined by tests (40%), homework (10%), quizzes (20%), and a final exam (30%). Answer each student's question. (*Section 2.4, Objective 3*)

72. Paulos: This semester, my test average is 85, homework average is 100, and quiz average is 65. What do I have to get on the final exam to earn a B in the class, if a B is 80–89?

73. Jennifer: My test average is 74, and my homework average is 85, and my quiz average is 63. I just want to earn a C in the class, so what do I have to get on the final exam, if a C is 70–79?

Find the intersection and union of the sets. Write the solution in interval notation. (*Section 2.5, Objectives 1 and 2*)

74. $A = [-13, -1], B = [-12, -1]$ $A \cap B = [-12, -1]$ $A \cup B = [-13, -1]$

75. $A = [4, 20], B = (4, 20)$ $A \cap B = (4, 20)$ $A \cup B = [4, 20]$

Solve each compound inequality. Write the solution in interval notation. (*Section 2.5, Objective 3*)

76. $x > 7$ and $x > 5$ $(7, \infty)$

77. $x \leq -3$ and $x \leq -7$ $(-\infty, -7]$

Solve each compound inequality. Write the solution in interval notation. (*Section 2.5, Objective 4*)

78. $x \geq 10$ or $x \geq 15$ $[10, \infty)$

79. $5x + 3 > 8$ or $7x - 4 \leq -3$ $\left(-\infty, \frac{1}{7}\right] \cup (1, \infty)$

Solve each equation. (*Section 2.6, Objectives 1 and 2*)

80. $|5 - x| = 12$ $\{-7, 17\}$

81. $|8 - 2x| = |3x + 6|$ $\left\{-14, \frac{2}{5}\right\}$

82. $|3x + 6| = 0$ $\{-2\}$

83. $|16 - 2x| = 0$ $\{8\}$

Solve each problem. (*Section 2.6, Objective 3*)

84. Suppose a digital scale reflects a weight (in pounds) with an absolute error of 0.4 lb. If someone weighs 184.3 lb according to the scale, find the possible values for the exact weight of the person. The exact weight of the person is between 183.9 lb and 184.7 lb.

85. Suppose a countertop food scale reflects a weight (in ounces) with an absolute error of 0.1 oz. If a piece of chicken weighs 3 oz according to the scale, find the possible values for the exact weight of the piece of chicken. The exact weight of the piece of chicken is between 2.9 oz and 3.1 oz.

CHAPTER 3

Graphs, Relations, and Functions

Creativity

Success in a math class depends largely on the ability to be creative. The willingness to think "outside of the box" is necessary for critical thinking and problem solving to take place. It is not enough to learn the rules and definitions; you must be able to apply them correctly. While you are not going to create new mathematics in this class, you will create new associations with the concepts. So, keep an open mind and allow yourself the time to truly think about what you are learning and how it applies to the things you know mathematically as well as how it applies in your everyday life.

Question For Thought: What are some things that stifle your creativity when it comes to learning math? Do you feel you are more creative in other areas of your life? The site http://www.virtualsalt.com/crebook1.htm has a great exposition on creative thinking. Review this and find one thing that you can apply to your life.

Chapter Outline

Section 3.1 The Coordinate System, Graphing Equations, and the Midpoint Formula 148

Section 3.2 Relations 165

Section 3.3 Functions 178

Section 3.4 The Domain and Range of Functions 190

Coming Up...

In Section 3.3, we will learn how to use functions to determine the median salary for the fastest-growing occupation for the years 2006–2016, a network system and data communications analyst.

"There is no doubt that creativity is the most important human resource of all. Without creativity, there would be no progress, and we would be forever repeating the same patterns."

—Edward de Bono (Authority in the Field of Creative Thinking)

SECTION 3.1 The Coordinate System, Graphing Equations, and the Midpoint Formula

In **Chapter 2**, we explored linear equations and inequalities in one variable. In Chapter 4, we will examine linear equations and inequalities in two variables. This chapter will provide a foundation for studying the equations in Chapter 4. We will learn how to visualize solutions of equations in two variables. The difference between a relation and a function and some important properties of functions are also presented.

OBJECTIVES

As a result of completing this section, you will be able to

1. Plot points and identify their location on a rectangular coordinate system.
2. Determine algebraically if an ordered pair is a solution of an equation.
3. Graph an equation in two variables.
4. Determine graphically if an ordered pair is a solution of an equation.
5. Determine the midpoint of a line segment.
6. Solve application problems.
7. Troubleshoot common errors.

This chart shows the number of hybrid electric vehicles (HEVs) sold for the years 1999 through 2010. Each year corresponds to a specific number of cars. This pairing of numbers provides points that we can graph on a coordinate system. Graphing these points enables us to visualize data so that trends and patterns in these data can be observed. In Example 6, we will graph these data and use the graph to observe trends in the sales of HEVs. (Source: http://www.afdc.energy.gov/afdc/data/vehicles.html and http://www.hybridcars.com)

In this section, we will learn how to graph points and solutions of equations and to apply these skills to data from the real world.

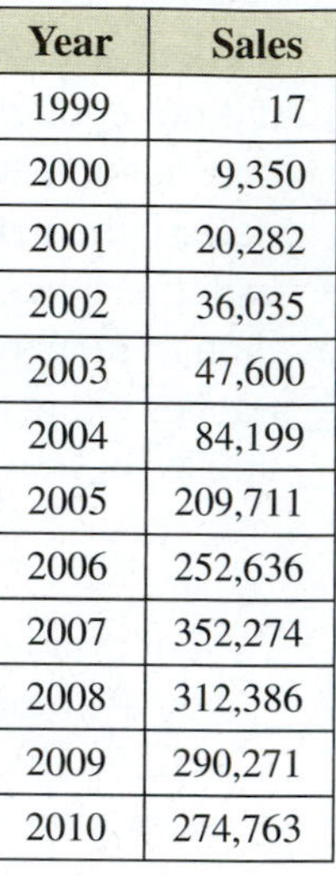

Year	Sales
1999	17
2000	9,350
2001	20,282
2002	36,035
2003	47,600
2004	84,199
2005	209,711
2006	252,636
2007	352,274
2008	312,386
2009	290,271
2010	274,763

Objective 1 ▶

Plot points and identify their location on a rectangular coordinate system.

INSTRUCTOR NOTE: Introduce this concept by talking about the Battleship game and the grid used to name a point on the board. You can also use a map grid to introduce this concept.

The Rectangular Coordinate System

Each pair of numbers in the table showing the total sales for HEVs can be expressed as an ordered pair. For example, in 2010 the total number of HEVs sold was 274,763. So, we say the year 2010 corresponds to sales of 274,763, which we express as (2010, 274,763). This pairing of numbers is called an **ordered pair** and is denoted by (x, y).

An important skill involving ordered pairs is the ability to plot them so we can visualize them. To plot ordered pairs, we use a **rectangular coordinate system** or the **Cartesian coordinate system**. René Descartes, a seventeenth-century French mathematician, is credited with inventing the Cartesian coordinate system.

A rectangular coordinate system consists of two real number lines intersecting at right angles. The horizontal number line is usually referred to as the ***x*-axis** and the vertical number line is usually referred to as the ***y*-axis**. The point where the two number lines intersect is called the **origin**.

Every point on the coordinate system is associated with an ordered pair, (x, y).

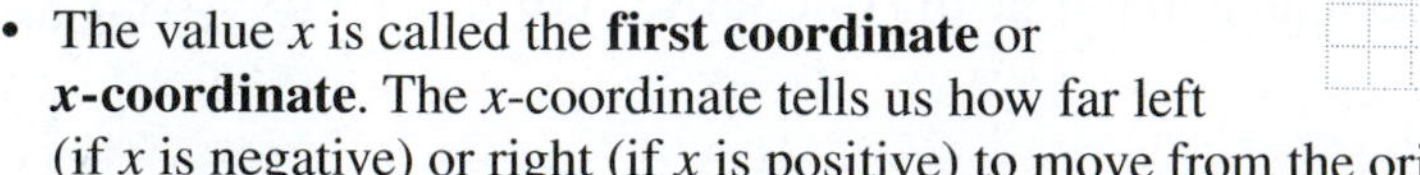

- The value x is called the **first coordinate** or ***x*-coordinate**. The x-coordinate tells us how far left (if x is negative) or right (if x is positive) to move from the origin.
- The value y is called the **second coordinate** or ***y*-coordinate**. The y-coordinate tells us how far up (if y is positive) or down (if y is negative) to move from the x-axis.

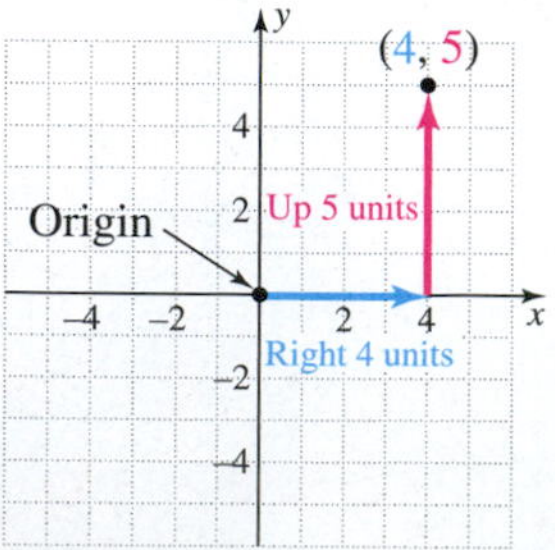

For example, in the ordered pair (4, 5), 4 is the x-coordinate, which tells us to move 4 units right from the origin. The number 5 is the y-coordinate, which tells us to move 5 units up from the x-axis.

Procedure: Plotting an Ordered Pair (x, y)

Step 1: From the origin, move left or right to the x-value given in the ordered pair.
Step 2: From this x-value, move up or down to the y-value given in the ordered pair.
Step 3: Plot a point at this location on the coordinate system.

The two axes divide the plane into four regions called **quadrants**. Quadrants are labeled with Roman numerals I, II, III, and IV, beginning in the upper right quadrant and rotating counterclockwise.

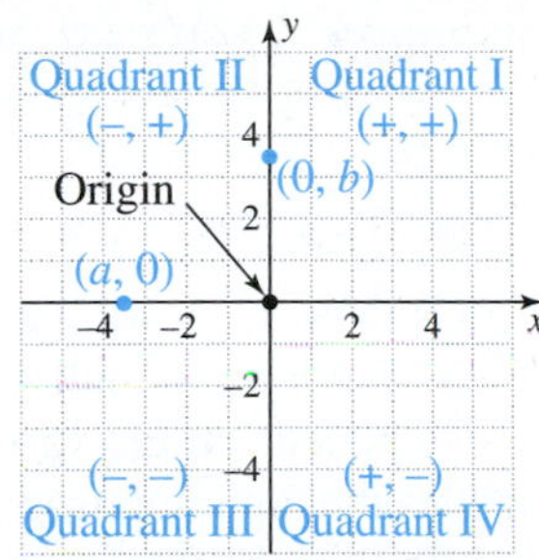

Procedure: Identifying a Point's Location on the Coordinate System

Step 1: Convert the coordinates to decimal form, if necessary.
Step 2: Identify the point's location.
- **a.** If both x and y are positive, the point is in Quadrant I.
- **b.** If x is negative and y is positive, the point is in Quadrant II.
- **c.** If both x and y are negative, the point is in Quadrant III.
- **d.** If x is positive and y is negative, the point is located in Quadrant IV.
- **e.** If $x = 0$, the point is on the y-axis.
- **f.** If $y = 0$, the point is on the x-axis.

Note: *Points that lie on the axes do not lie in any of the quadrants.*

Objective 1 Examples

Plot each ordered pair on a rectangular coordinate system and identify its location.

1a. $(2, -3)$ **1b.** $(-3, 2)$ **1c.** $(0, -2)$ **1d.** $(4, 0)$

1e. $\left(\frac{1}{2}, \frac{9}{2}\right)$ **1f.** $\left(-5, -\frac{4}{3}\right)$

Solutions See Figure 3.1.

1a. $(2, -3)$; Move 2 units right from the origin and 3 units down; Quadrant IV

1b. $(-3, 2)$; Move 3 units left from the origin and 2 units up; Quadrant II

1c. $(0, -2)$; Move 0 units right or left from the origin and 2 units down; y-axis

1d. $(4, 0)$; Move 4 units right from the origin and 0 units up or down; x-axis

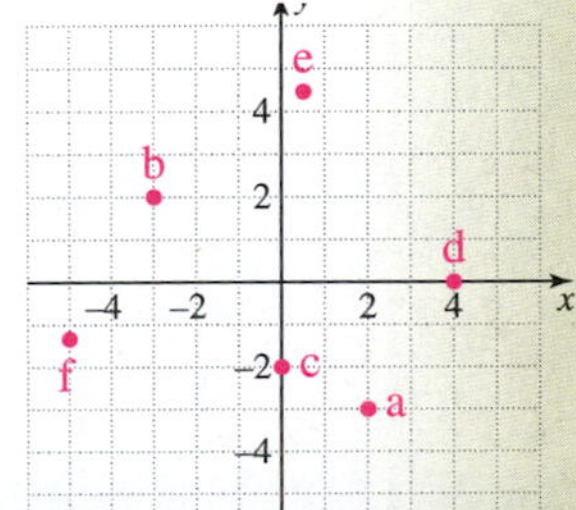

Figure 3.1

1e. $\left(\frac{1}{2}, \frac{9}{2}\right) = (0.5, 4.5)$; Move 0.5 units right from the origin and 4.5 units up; Quadrant I

1f. $\left(-5, -\frac{4}{3}\right) = (-5, -1.3)$; Move 5 units left from the origin and 1.3 units down; Quadrant III

 **Student Check 1** Plot each ordered pair on a Cartesian coordinate system and identify its location.

a. $(-5, -1)$ **b.** $(4, -1)$ **c.** $(-1, 4)$ **d.** $(0, 3)$

e. $(-3, 0)$ **f.** $\left(\frac{5}{2}, \frac{5}{4}\right)$

Note: *Changing the order of the numbers in an ordered pair changes the location of the point. Compare the locations of* $(3, -2)$ *and* $(-2, 3)$ *from Example 1 parts (a) and (b).*

Solutions of Linear Equations in Two Variables

Objective 2 ▶

Determine algebraically if an ordered pair is a solution of an equation.

INSTRUCTOR NOTE: Linear equations in two variables will be covered extensively in Chapter 4.

In Chapter 2, we studied linear equations with one variable and learned how to find their solutions. Recall that a linear equation in one variable is an equation in which the exponent on the variable is 1. A solution of this type of equation is a number that makes the equation true.

In this chapter, equations in *two* variables are introduced. One type of equation in two variables is a *linear equation in two variables.*

Definition: A **linear equation in two variables** is an equation of the form $Ax + By = C$, where A, B, and C are real numbers with A and B not both zero.

Examples of linear equations in two variables include $2x + 3y = -12$ and $y = 3x - 1$. The main characteristic of a linear equation in two variables is that the exponent on both x and y is 1.

Not all equations in two variables are linear equations. In upcoming Examples 2c and 2d, we will see illustrations of an *absolute value equation in two variables* and a *quadratic equation in two variables*, respectively. These types of equations will be studied in more detail in later chapters.

A **solution of an equation in two variables** is an ordered pair (x, y) that makes the equation true. For example, the ordered pair $(3, -6)$ is a solution of $2x + 3y = -12$ because when x is replaced with 3 and y is replaced with -6, we obtain a true equation.

$$2x + 3y = -12$$

$$2(3) + 3(-6) = -12 \quad \text{Replace } x \text{ with 3 and } y \text{ with } -6.$$

$$6 - 18 = -12 \quad \text{Simplify.}$$

$$-12 = -12 \quad \text{True}$$

Procedure: Determining if an Ordered Pair Is a Solution of an Equation in Two Variables

Step 1: Replace the values of x and y with the numbers given in the ordered pair.
Step 2: Simplify each side of the equation.
Step 3: If the resulting equation is true, then the ordered pair is a solution of the equation. If it is not true, then the ordered pair is not a solution.

Objective 2 Examples **Determine if each ordered pair is a solution of the given equation.**

2a. $3x - y = 6$; $(1, -3)$, $(0, 6)$, $(2, 0)$

2b. $y = -4x + 5$; $\left(-\frac{5}{4}, 0\right)$, $\left(\frac{3}{4}, 8\right)$, $\left(\frac{1}{4}, 4\right)$

2c. $y = |x - 2|$; $(-5, 7)$, $(-3, 5)$, $(4, 2)$

2d. $y = x^2 - x + 1$; $(2, 3)$, $(-1, 3)$, $(1, 2)$

Solutions

2a. Replace the variables with their corresponding values.

$(1, -3)$	$(0, 6)$	$(2, 0)$
$3x - y = 6$	$3x - y = 6$	$3x - y = 6$
$3(1) - (-3) = 6$	$3(0) - (6) = 6$	$3(2) - (0) = 6$
$3 + 3 = 6$	$0 - 6 = 6$	$6 - 0 = 6$
$6 = 6$	$-6 = 6$	$6 = 6$
True	False	True

The ordered pairs $(1, -3)$ and $(2, 0)$ are solutions of $3x - y = 6$ since they make the equation a true statement. The ordered pair $(0, 6)$ is not a solution of $3x - y = 6$ since it doesn't make the equation true.

2b. Replace the variables with their corresponding values.

$\left(-\frac{5}{4}, 0\right)$	$\left(\frac{3}{4}, 8\right)$	$\left(\frac{1}{4}, 4\right)$
$y = -4x + 5$	$y = -4x + 5$	$y = -4x + 5$
$0 = -4\left(-\frac{5}{4}\right) + 5$	$8 = -4\left(\frac{3}{4}\right) + 5$	$4 = -4\left(\frac{1}{4}\right) + 5$
$0 = 5 + 5$	$8 = -3 + 5$	$4 = -1 + 5$
$0 = 10$	$8 = 2$	$4 = 4$
False	False	True

The ordered pair $\left(\frac{1}{4}, 4\right)$ is a solution of $y = -4x + 5$ since it makes the equation a true statement. The ordered pairs $\left(-\frac{5}{4}, 0\right)$ and $\left(\frac{3}{4}, 8\right)$ are not solutions of $y = -4x + 5$ since they do not make the equation true.

2c. Replace the variables with their corresponding values.

$(-5, 7)$	$(-3, 5)$	$(4, 2)$
$y = \lvert x - 2\rvert$	$y = \lvert x - 2\rvert$	$y = \lvert x - 2\rvert$
$7 = \lvert -5 - 2\rvert$	$5 = \lvert -3 - 2\rvert$	$2 = \lvert 4 - 2\rvert$
$7 = \lvert -7\rvert$	$5 = \lvert -5\rvert$	$2 = \lvert 2\rvert$
$7 = 7$	$5 = 5$	$2 = 2$
True	True	True

All three ordered pairs make the equation true, so all are solutions of $y = |x - 2|$.

2d. Replace the variables with their corresponding values.

$(2, 3)$	$(-1, 3)$	$(1, 2)$
$y = x^2 - x + 1$	$y = x^2 - x + 1$	$y = x^2 - x + 1$
$3 = (2)^2 - (2) + 1$	$3 = (-1)^2 - (-1) + 1$	$2 = (1)^2 - (1) + 1$
$3 = 4 - 2 + 1$	$3 = 1 + 1 + 1$	$2 = 1 - 1 + 1$
$3 = 3$	$3 = 3$	$2 = 1$
True	True	False

The ordered pairs $(2, 3)$ and $(-1, 3)$ are solutions of the equation $y = x^2 - x + 1$ since they make the equation true. The ordered pair $(1, 2)$ is not a solution of $y = x^2 - x + 1$ since it doesn't make the equation true.

Student Check 2 Determine if each ordered pair is a solution of the given equation.

a. $3x - y = 0$; $(-2, 6)$, $(2, 6)$, $\left(\frac{1}{3}, 1\right)$

b. $y = 2x - 1$; $\left(\frac{1}{2}, 0\right)$, $(0, -1)$, $\left(\frac{11}{2}, 10\right)$

c. $y = |x + 3|$; $(-3, 0)$, $(-4, 1)$, $(-7, -4)$

d. $y = 3x^2 - 4x + 2$; $(0, -2)$, $(-1, 9)$, $(1, 3)$

Graphing Equations in Two Variables

Objective 3 Graph an equation in two variables.

Example 2 shows that an equation in two variables has more than one solution. In fact, an equation in two variables can have infinitely many ordered pairs that satisfy it. It is impossible to state every solution of this type of equation. Our goal then is to visualize the solution set of an equation in two variables by graphing the equation.

To *graph an equation in two variables* means to draw a picture of all the solutions of the equation. We plot several ordered pairs that satisfy the equation and use these points to determine the shape of the graph. Some of the graphs we will obtain are shown as follows.

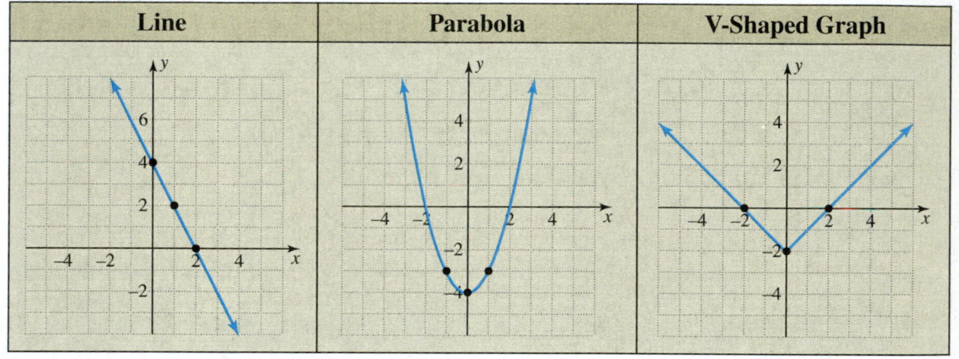

Each type of graph comes from a special type of equation. A line comes from a linear equation in two variables. A parabola comes from a quadratic equation in two variables. A V-shaped graph comes from an absolute value equation in two variables. These special equations will be studied in more detail in later chapters. At this point, our goal is to construct a graph of an equation in two variables by plotting points.

Procedure: Graphing an Equation in Two Variables

Step 1: Determine at least three solutions of the equation.

- **a.** Substitute a numerical value for x. Choose values that are negative, positive, and zero so that we get a complete picture of the graph.
- **b.** Simplify the resulting expression or solve the resulting equation to find y.
- **c.** The ordered pair (x, y) is a solution of the equation.
- **d.** Repeat this process as many times as needed.
- **e.** Organize the information in a chart.

Step 2: Plot the solutions found in step 1 on a coordinate system.

Step 3: Connect the points and use their pattern to sketch the graph.

Objective 3 Examples Graph each equation by plotting points.

3a. $y = x - 2$ **3b.** $y = x^2 - 4$ **3c.** $y = |x + 1|$

Solutions **3a.**

x	$y = x - 2$	(x, y)
-2	$y = -2 - 2 = -4$	$(-2, -4)$
-1	$y = -1 - 2 = -3$	$(-1, -3)$
0	$y = 0 - 2 = -2$	$(0, -2)$
1	$y = 1 - 2 = -1$	$(1, -1)$
2	$y = 2 - 2 = 0$	$(2, 0)$

Plot the resulting ordered pairs. Note that the points lie on a line. Connect the points to form the line. Arrows are used on the ends of the line to indicate the graph continues indefinitely in both directions.

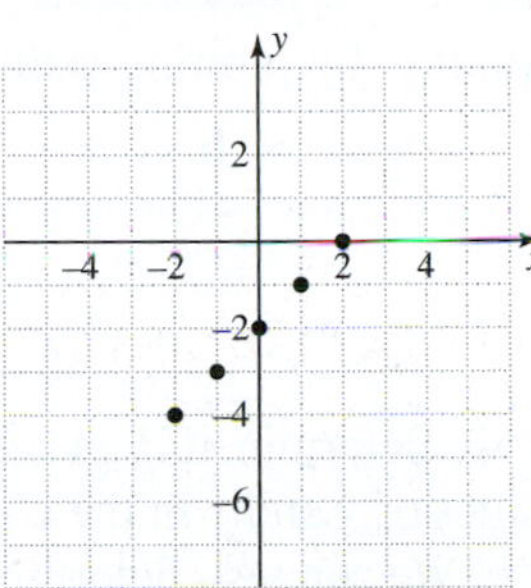

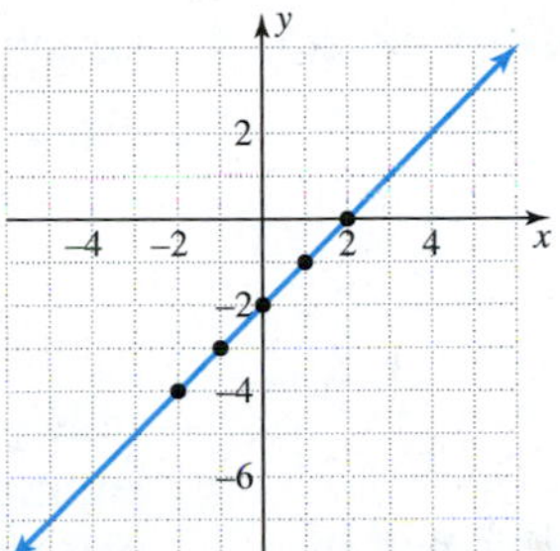

The graph of the line represents the solution set of the equation $y = x - 2$.

3b.

x	$y = x^2 - 4$	(x, y)
-2	$y = (-2)^2 - 4 = 4 - 4 = 0$	$(-2, 0)$
-1	$y = (-1)^2 - 4 = 1 - 4 = -3$	$(-1, -3)$
0	$y = (0)^2 - 4 = 0 - 4 = -4$	$(0, -4)$
1	$y = (1)^2 - 4 = 1 - 4 = -3$	$(1, -3)$
2	$y = (2)^2 - 4 = 4 - 4 = 0$	$(2, 0)$

Plot the resulting ordered pairs. Note that the points lie in a U-shaped pattern, called a *parabola*. Connect the points using a smooth curve to form the parabola. Arrows are used on both ends of the graph to indicate the graph continues indefinitely.

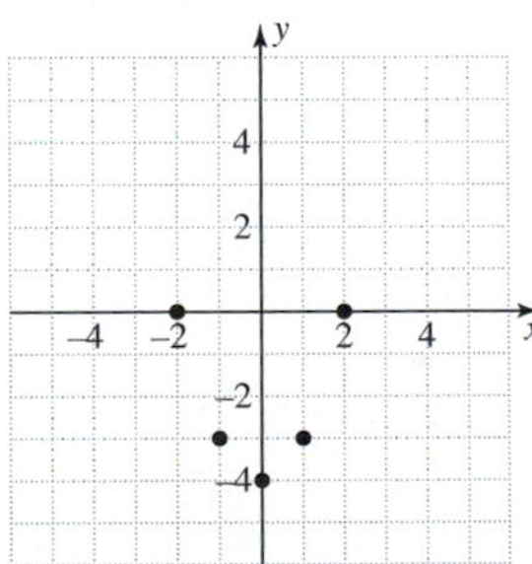

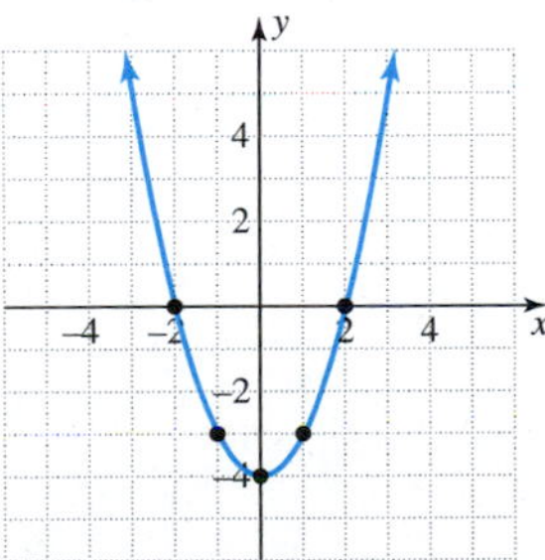

The graph of the parabola represents the solution set of $y = x^2 - 4$.

3c.

x	$y = \|x + 1\|$	(x, y)
-2	$y = \|-2 + 1\| = \|-1\| = 1$	$(-2, 1)$
-1	$y = \|-1 + 1\| = \|0\| = 0$	$(-1, 0)$
0	$y = \|0 + 1\| = \|1\| = 1$	$(0, 1)$
1	$y = \|1 + 1\| = \|2\| = 2$	$(1, 2)$
2	$y = \|2 + 1\| = \|3\| = 3$	$(2, 3)$

Plot the points. Note that the points lie in a V-shaped pattern. Connect the points to form the V-shape. Arrows are used to indicate the graph continues indefinitely in each direction.

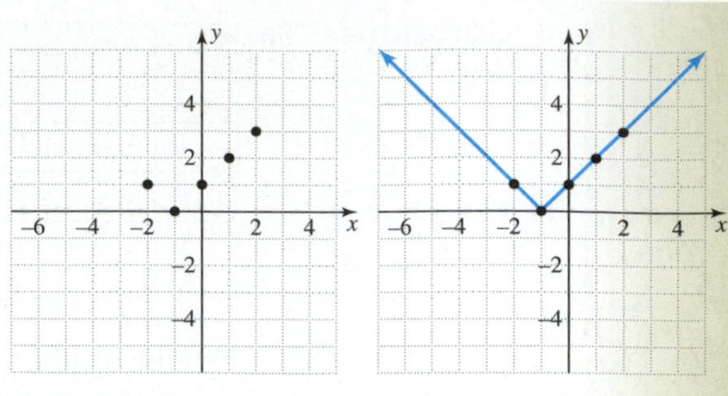

The V-shaped graph represents the solution set of $y = |x + 1|$.

Student Check 3 Graph each equation by plotting points.

a. $y = x + 3$ **b.** $y = x^2 + 1$ **c.** $y = |x|$

Using a Graph to Determine Solutions

Objective 4 ▶ Determine graphically if an ordered pair is a solution of an equation.

In Objective 2, we determined if an ordered pair was a solution of an equation by substituting the values of x and y in the equation and determining if the resulting equation was true or false. We can also determine if an ordered pair is a solution of an equation by examining its graph.

Procedure: Determining Graphically if an Ordered Pair Is a Solution of an Equation

Step 1: Graph the equation, if needed.

Step 2: Plot the point.

a. If the point lies on a graph, then it is a solution of the graphed equation.

b. If the point does not lie on a graph, then it is not a solution of the graphed equation.

Objective 4 Examples

The graph of $y = x^2 - 4$ is provided. Use the graph to determine if the given ordered pair is a solution of $y = x^2 - 4$.

4a. $(-4, 0)$ **4b.** $(0, -4)$ **4c.** $(2, 0)$

4d. $(0, 2)$ **4e.** $(-3, 5)$

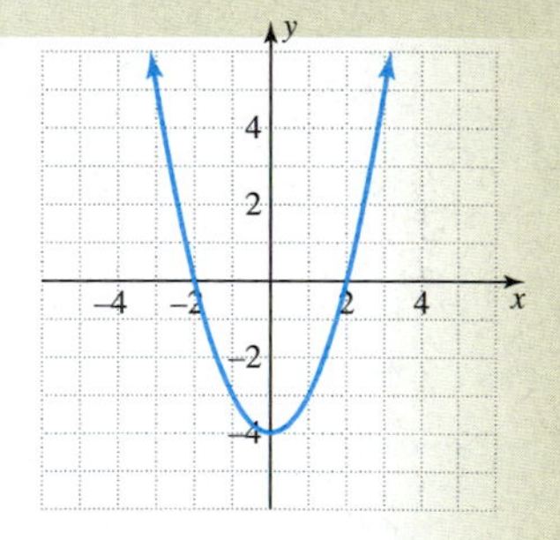

Solution The points $(0, -4)$, $(2, 0)$, and $(-3, 5)$ lie on the graph. So, these points are solutions of the equation. The points $(-4, 0)$ and $(0, 2)$ do not lie on the graph. So, these points are not solutions of the equation.

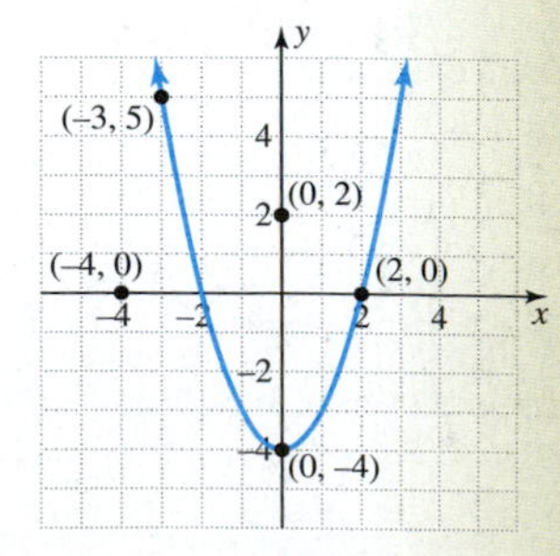

Student Check 4 The graph of $y = |x| - 3$ is provided. Use the graph to determine if the given ordered pair is a solution of $y = |x| - 3$.

a. (3, 0) **b.** (0, 3) **c.** (0, −3)

d. (4, −1) **e.** (−4, 1)

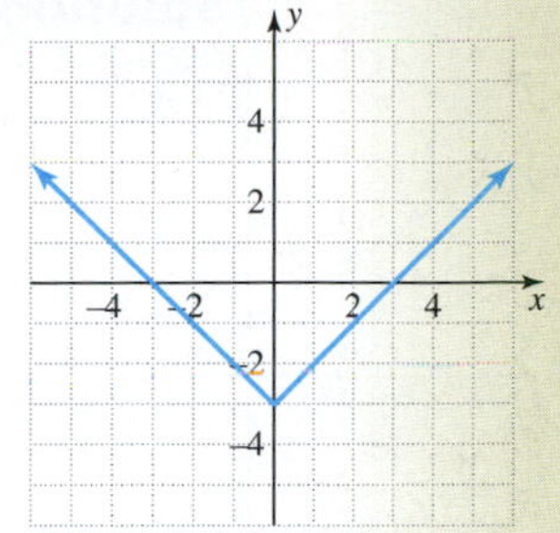

The Midpoint Formula

Objective 5

Determine the midpoint of a line segment.

The **midpoint** of a line segment is the ordered pair that lies exactly in the middle of the line segment. For example, the midpoint of 4 and 6 is 5 since 5 lies exactly in the middle of 4 and 6. We can also obtain the midpoint of 4 and 6 with the following calculation. Recall the midpoint of two numbers on a real number line is the average of the two numbers.

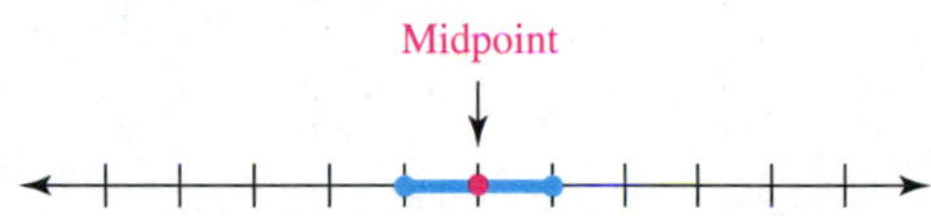

$$\frac{4+6}{2} = \frac{10}{2} = 5$$

We can extend this concept to finding the midpoint of a line segment formed by two ordered pairs. We find the average of the x-coordinates and the average of the y-coordinates to determine the midpoint.

Consider the line segment formed by (0, 4) and (2, 0).

The average of the x-values is $\frac{0+2}{2} = \frac{2}{2} = 1$.

The average of the y-values is $\frac{4+0}{2} = \frac{4}{2} = 2$.

So, the midpoint of the line segment is (1, 2).
We can state the midpoint formula as follows.

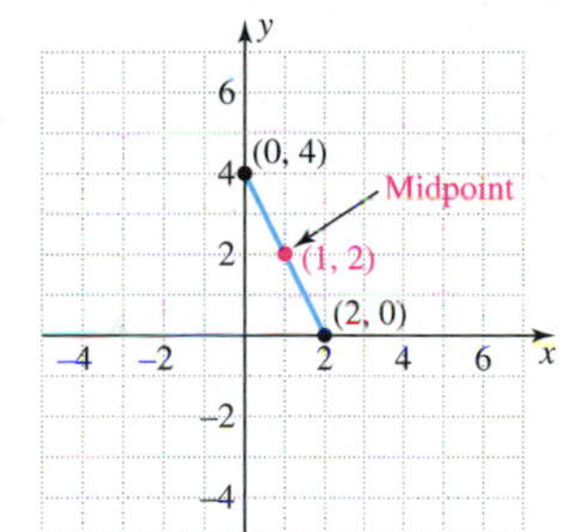

Property: Midpoint Formula

If (x_1, y_1) and (x_2, y_2) are ordered pairs, then the *midpoint* of the line segment formed by these ordered pairs is given by

$$\text{Midpoint} = \left(\frac{x_1 + x_2}{2}, \frac{y_1 + y_2}{2}\right)$$

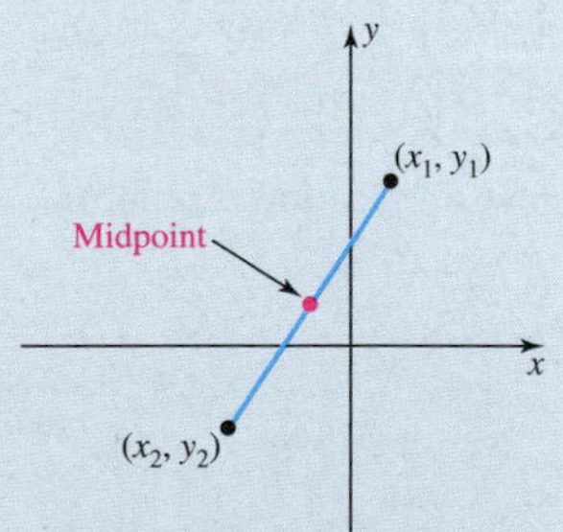

Procedure: Finding the Midpoint of a Line Segment Given Two Ordered Pairs

Step 1: Label the ordered pairs as (x_1, y_1) and (x_2, y_2).
Step 2: Substitute the appropriate values into the midpoint formula and simplify.

Objective 5 Examples Find the midpoint of the line segment formed by the ordered pairs.

5a. (4, −1) and (−6, 5) **5b.** $\left(\frac{1}{2}, -3\right)$ and $\left(\frac{7}{3}, 4\right)$

Solutions

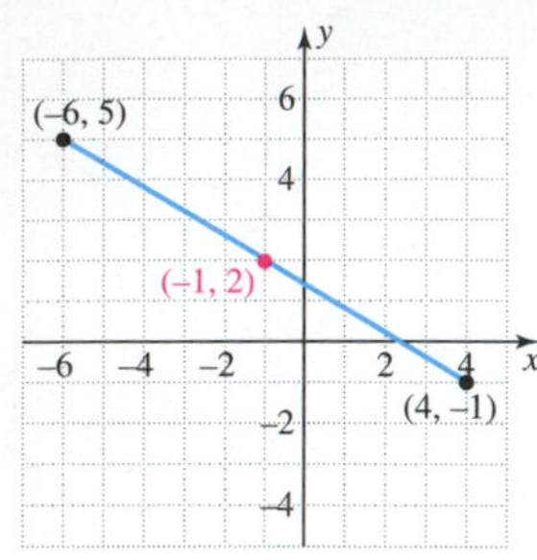

5a. Let $(x_1, y_1) = (4, -1)$ and $(x_2, y_2) = (-6, 5)$.

$$\left(\frac{x_1 + x_2}{2}, \frac{y_1 + y_2}{2}\right) = \left(\frac{4 + (-6)}{2}, \frac{-1 + 5}{2}\right)$$ Substitute the appropriate values in the midpoint formula.

$$= \left(\frac{-2}{2}, \frac{4}{2}\right)$$ Add the numerators of each fraction.

$$= (-1, 2)$$ Simplify.

So, the midpoint of the line segment formed by $(4, -1)$ and $(-6, 5)$ is $(-1, 2)$.

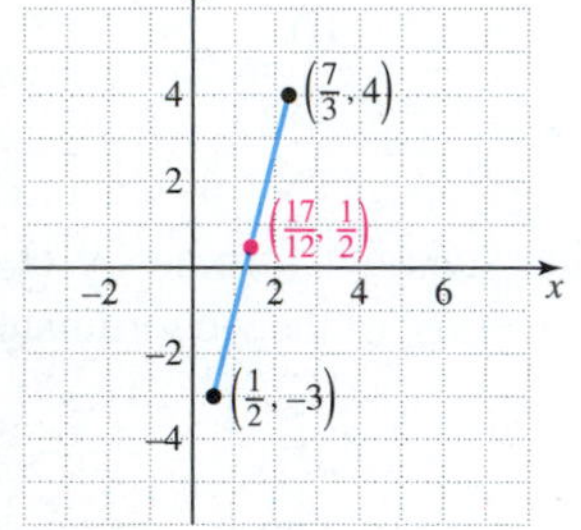

5b. Let $(x_1, y_1) = \left(\frac{1}{2}, -3\right)$ and $(x_2, y_2) = \left(\frac{7}{3}, 4\right)$.

$$\left(\frac{x_1 + x_2}{2}, \frac{y_1 + y_2}{2}\right) = \left(\frac{\frac{1}{2} + \frac{7}{3}}{2}, \frac{-3 + 4}{2}\right)$$ Substitute the appropriate values in the midpoint formula.

$$= \left(\frac{\frac{3}{6} + \frac{14}{6}}{2}, \frac{1}{2}\right)$$ Write equivalent fractions in the x-coordinate and simplify the y-coordinate.

$$= \left(\frac{17}{6} \cdot \frac{1}{2}, \frac{1}{2}\right)$$ Simplify the x-coordinate.

$$= \left(\frac{17}{12}, \frac{1}{2}\right)$$ Simplify.

So, the midpoint of the line segment formed by $\left(\frac{1}{2}, -3\right)$ and $\left(\frac{7}{3}, 4\right)$ is $\left(\frac{17}{12}, \frac{1}{2}\right)$.

Student Check 5 Find the midpoint of the line segment formed by the ordered pairs.

a. $(-5, -2)$ and $(7, 0)$ **b.** $\left(-\frac{1}{4}, 1\right)$ and $\left(\frac{2}{3}, -5\right)$

Applications

Objective 6 ▶ Solve application problems.

There are many situations where a relationship exists between two quantities. For example, a person's weight on a specific day can be written as an ordered pair in the form (day, weight). The population of a country for a specific year can be written as an ordered pair in the form (year, population). A tip applied to a meal can be written as an ordered pair in the form (cost of meal, amount of tip). Data that can be represented as an ordered pair is called **paired data**.

When data is provided in tables, bar graphs, or line graphs, we can convert the data to ordered pairs and use them to plot the data. The plot of this data is called a **scatter plot** or *scatter diagram*. Scatter plots can be used to look for patterns or trends that occur in paired data.

Note: When plotting data points, it is often necessary to determine an appropriate scale for the axes and to use appropriate minimum and maximum values on the axes.

Objective 6 Examples **Solve each problem using the given information.**

6a. The table shows the number of hybrid electric vehicles (HEVs) sold for the years 1999 through 2010. Use the table to answer the questions. (Source: http://www.afdc.energy.gov/afdc/data/vehicles.html and http://www.hybridcars.com)

Year	1999	2000	2001	2002	2003	2004	2005	2006	2007	2008	2009	2010
Sales	17	9350	20,282	36,035	47,600	84,199	209,711	252,636	352,274	312,386	290,271	274,763

i. Write the ordered pairs that represent the given data, where x is the number of years after 1999 and y is the number of HEVs sold that year.

ii. Create a scatter plot of the data.

iii. What can you conclude about HEV sales?

Solution **6a.** **i.** Since x is the number of years after 1999, we must subtract 1999 from each of the years to determine the x-coordinates of the ordered pairs. The ordered pairs are. (0, 17), (1, 9350), (2, 20,282), (3, 36,035), (4, 47,600), (5, 84,199), (6, 209,711), (7, 252,636), (8, 352,274), (9, 312,386), (10, 290,271), and (11, 274,763).

INSTRUCTOR NOTE:
Use the real-life problems as an opportunity to discuss with students the reasons that the sales in HEVs increased.

ii. Because the x- and y-values are positive, all of the ordered pairs lie in Quadrant I. We use a scale of 1 for the x-axis and a scale of 50,000 for the y-axis. The scatter plot of the paired data is as follows.

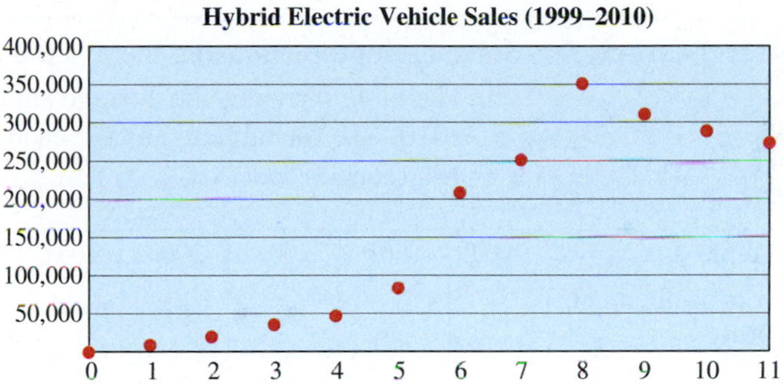

iii. The graph shows that the number of HEVs sold increased steadily through 2007 but is now on the decline.

6b. The graph shows the total number of passengers enplaned (in thousands) for domestic airline flights for 1996 to 2009, where x is the number of years after 1996. Use the scatter plot to answer the following questions. (Source: http://www.bts.gov)

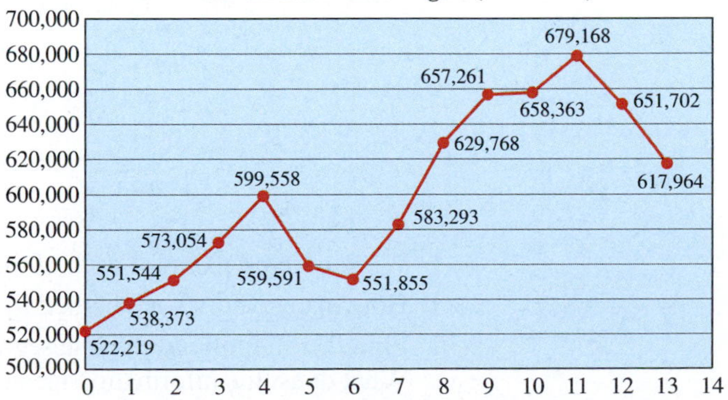

i. Write ordered pairs for each of the points on the graph.

ii. How many domestic airline passengers were enplaned in 2001? In 2005?

iii. What year had the greatest number of enplaned passengers? The least? How many enplaned passengers were there in each of these cases?

iv. Use the graph to describe the number of enplaned passengers.

Solution **6b.** **i.** The ordered pairs are (0, 522,219), (1, 538,373), (2, 551,544), (3, 573,054), (4, 599,558), (5, 559,591), (6, 551,855), (7, 583,293), (8, 629,768), (9, 657,261), (10, 658,363), (11, 679,168), (12, 651,702), and (13, 617,964).

ii. The year 2001 is 5 yr after 1996. The ordered pair (5, 559,591) lies on the graph, so there were 559,591 thousand domestic airline passengers enplaned in 2001. The year 2005 is 9 yr after 1996. The ordered pair (9, 657,261) lies on the graph, so there were 657,261 thousand domestic airline passengers enplaned in 2005.

iii. The greatest number of enplaned passengers was 679,168 thousand. This occurred 11 yr after 1996, or in 2007. The least number of enplaned passengers was 522,219 thousand. This occurred 0 yr after 1996, or in 1996.

iv. The graph shows that the number of enplaned passengers increased between 1996 and 2000, decreased between 2000 to 2002, increased between 2002 and 2007, and decreased from 2007 to 2009.

Student Check 6 Solve each problem using the given information.

a. The table provides the average annual unemployment rate for 2000 to 2010. Use the table to answer each question. (Source: http://www.bls.gov/cps/demographics.htm)

Year	2000	2001	2002	2003	2004	2005	2006	2007	2008	2009	2010
Unemployment Rate	4.0	4.7	5.8	6.0	5.5	5.1	4.6	4.6	5.8	9.3	9.6

i. Write ordered pairs for the given data, where x is the number of years after 2000 and y is the unemployment rate.

ii. Create a scatter plot of the data.

iii. Use the graph to describe the unemployment rate since 2000.

b. The graph shows the calories burned from a 30-min brisk (3.5 mph) walk by a person who weighs x pounds. Use the graph to answer each question. (Source: http://www.buzzle.com/articles/calories-burned-in-brisk-walking.html)

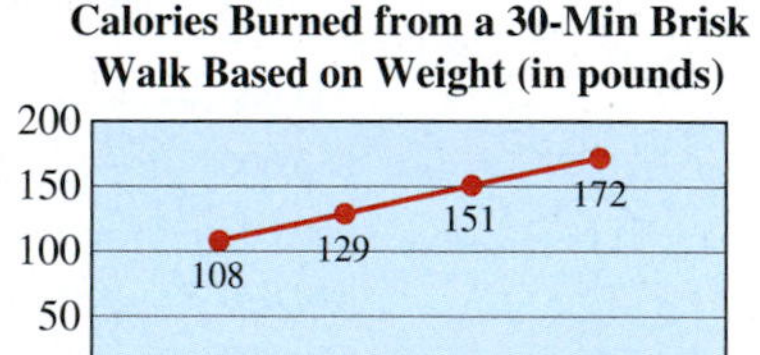

i. Write ordered pairs for each of the points on the graph.

ii. How many calories are burned by someone who weighs 150 lb?

iii. Find the midpoint of the line segment formed by $x = 125$ and $x = 150$. What does this information mean in the context of the problem?

Objective 7 ▶ Troubleshoot common errors.

Troubleshooting Common Errors

Some common errors associated with plotting points and determining if a point is a solution of an equation in two variables are shown next.

Objective 7 Examples A problem and an incorrect solution are given. Provide the correct solution and an explanation of the error.

Solutions **7a.** Plot the point $(0, -3)$.

Incorrect Solution	Correct Solution and Explanation
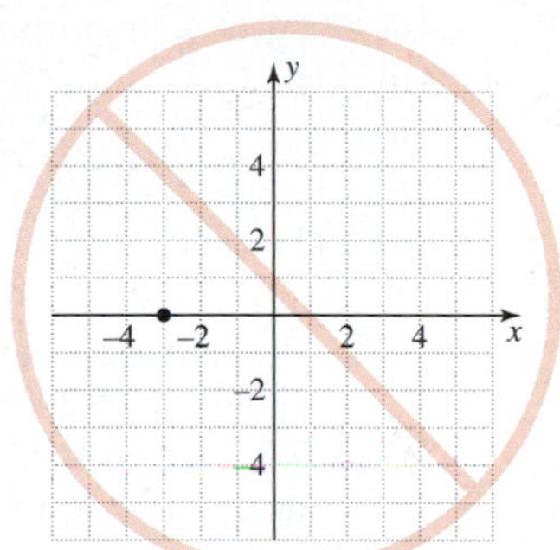	To plot $(0, -3)$, we must not move left or right from the origin but move down 3 units. 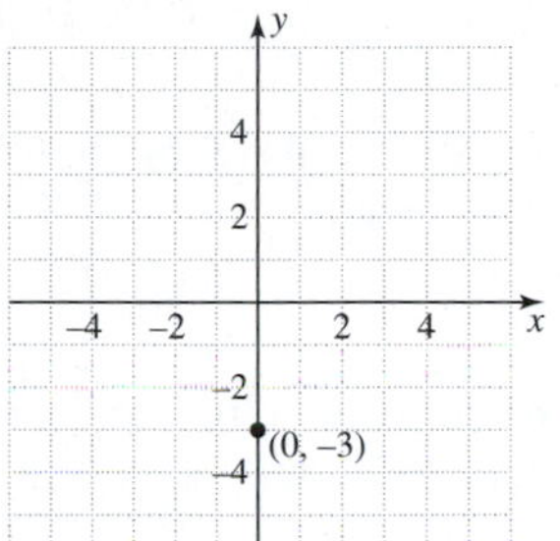

7b. Determine if $\left(-\frac{2}{3}, -1\right)$ is a solution of $y = -6x - 5$.

Incorrect Solution	Correct Solution and Explanation
$y = -6x - 5$ $-1 = -6\left(-\frac{2}{3}\right) - 5$ $-1 = -4 - 5$ $-1 = -9$ Since the statement is false, the ordered pair is not a solution of the equation.	The error was made in multiplying the values -6 and $-\frac{2}{3}$. Their product is 4 not -4. $y = -6x - 5$ $-1 = -6\left(-\frac{2}{3}\right) - 5$ $-1 = 4 - 5$ $-1 = -1$ Since the statement is true, the ordered pair is a solution of the equation.

ANSWERS TO STUDENT CHECKS

Student Check 1

a. Quadrant III
b. Quadrant IV
c. Quadrant II
d. y-axis
e. x-axis
f. Quadrant I

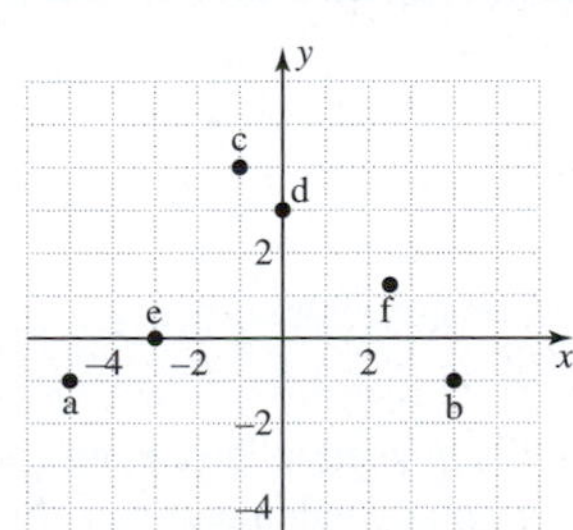

Student Check 2 **a.** $(2, 6)$ and $\left(\frac{1}{3}, 1\right)$ are solutions.

b. $\left(\frac{1}{2}, 0\right)$, $(0, -1)$, and $\left(\frac{11}{2}, 10\right)$ are solutions.

c. $(-3, 0)$ and $(-4, 1)$ are solutions.

d. $(-1, 9)$ is a solution.

Student Check 3

a. $y = x + 3$

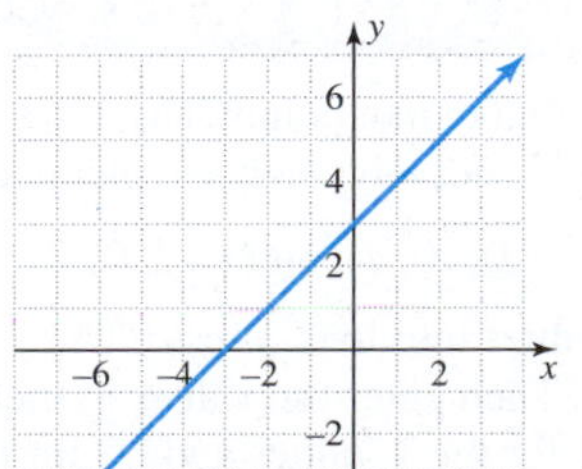

b. $y = x^2 + 1$

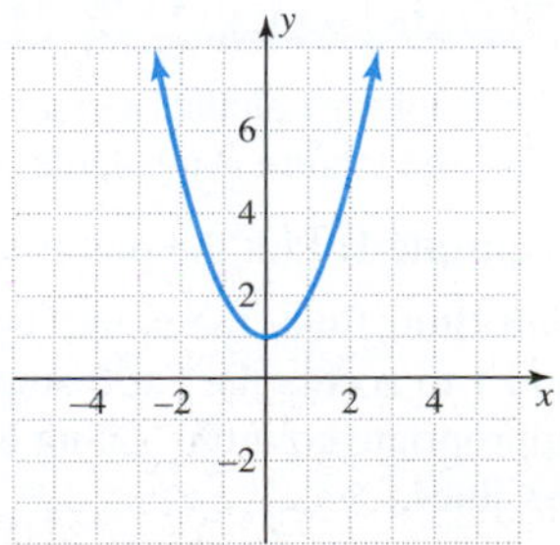

c. $y = |x|$

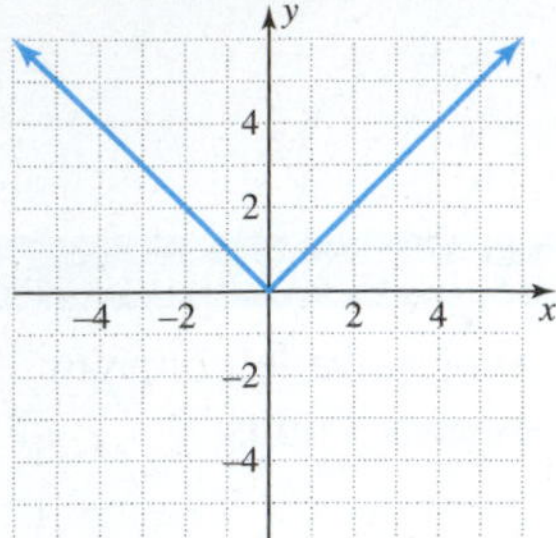

Student Check 4 The points (3, 0), (0, −3), and (−4, 1) are solutions of the equation.

Student Check 5 **a.** (1, −1) **b.** $\left(\frac{5}{24}, -2\right)$

Student Check 6 **a. i.** (0, 4.0), (1, 4.7), (2, 5.8), (3, 6.0), (4, 5.5), (5, 5.1), (6, 4.6), (7, 4.6), (8, 5.8), (9, 9.3), (10, 9.6)

ii.

Average Annual Unemployment Rate (2000–2010)

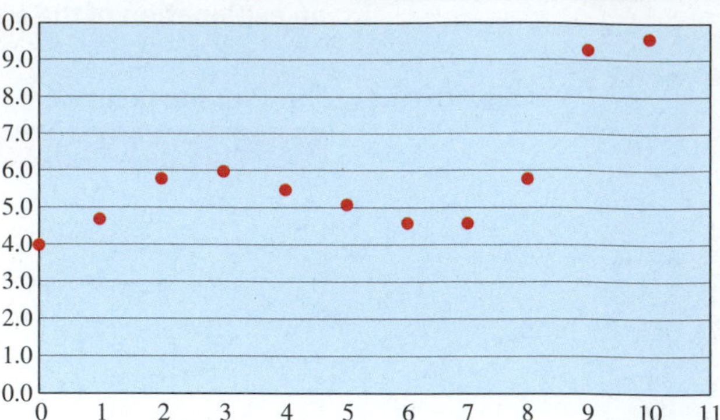

iii. The annual average unemployment rate increased slightly from 2000 to 2003 and then decreased until 2006. It remained unchanged through 2007 and then increased sharply until 2010.

b. i. (125, 108), (150, 129), (175, 151), (200, 172)
ii. 129 calories **iii.** (137.5, 118.5); A person who weighs 137.5 lb will burn 118.5 calories from a 30-min brisk walk.

SUMMARY OF KEY CONCEPTS

1. Ordered pairs (x, y) are plotted on a Cartesian coordinate system.
   - The x-value determines the movement from the origin on the x-axis.
   - The y-value determines how far up or down the point is located from the x-axis.
   - Quadrants are the four regions formed by the axes. Quadrant I is in the upper right, Quadrant II is in the upper left, Quadrant III is in the lower left, and Quadrant IV is in the lower right.
   - All points will lie in either one of the four quadrants or on one of the axes.
2. A solution of an equation in two variables is an ordered pair that makes the equation a true statement.
3. To graph solutions of an equation, find several solutions of the equation by completing a table. Plot the solutions to obtain the graph. There can be infinitely many solutions of an equation in two variables. The graph is a picture of the solution set.
4. An ordered pair is a solution of an equation if it lies on the graph of the equation.
5. The midpoint of a line segment formed by two ordered pairs is found by averaging the x-coordinates and averaging the y-coordinates.
6. There are two main types of applications—using tables and using graphs. For each table given, write the paired data and then plot these points to create a scatter plot. For each given graph, write the ordered pairs that correspond to the points and answer questions about them.

GRAPHING CALCULATOR SKILLS

The graphing calculator can be used to plot points. This is a skill you will most likely need in later math courses. At this point, just know the feature is available but do not rely on this to plot points by hand. The graphing calculator can also be used to graph equations.

Example 1: Plot the points (3, −2), (0, 4), and (−2, 0).

Solution: Enter the x- and y-values into lists. Press STAT and 1 to access the list feature. Then enter the values in the appropriate columns, using L1 for the x-values and L2 for the y-values.

L1	L2	L3 2
3	-2	------
0	4	
-2	0	

L2(4) =

Once the points are entered, turn the STAT PLOT feature ON and graph.

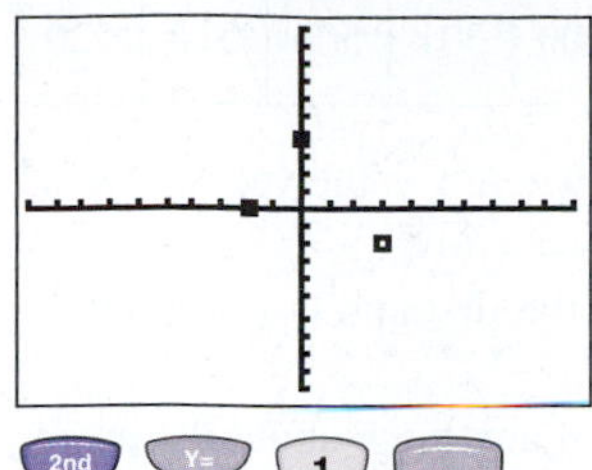

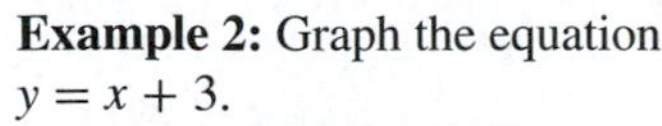

Example 2: Graph the equation $y = x + 3$.

Solution: Enter the equation into the equation editor by pressing Y =.

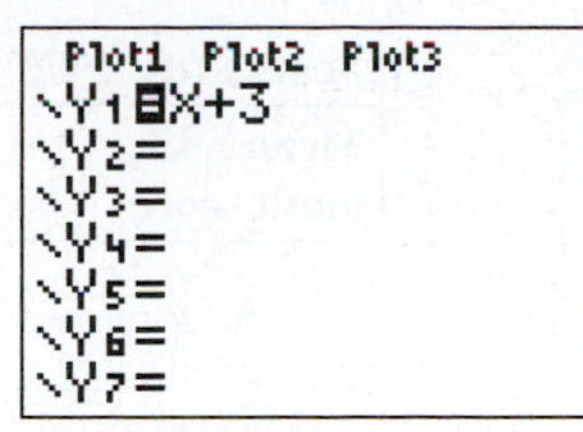

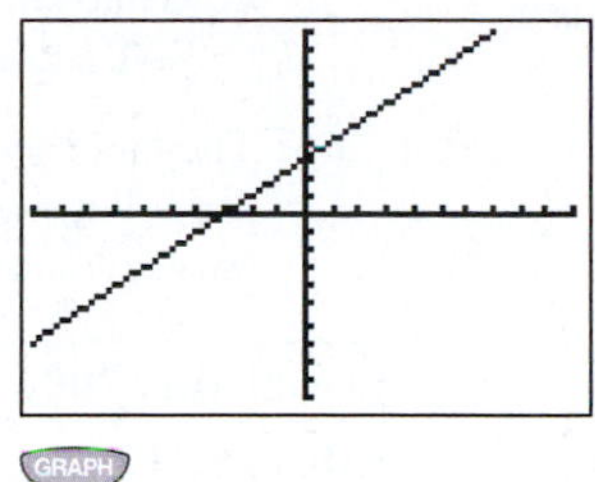

To view a table of specific solutions, access the TABLE.

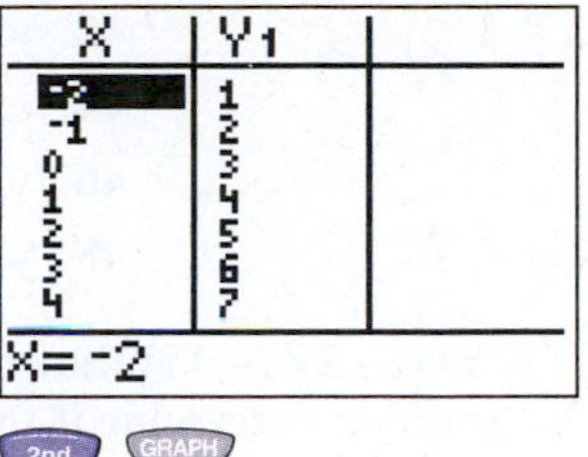

SECTION 3.1 EXERCISE SET

Write About It!

Use complete sentences in your answer to each exercise.

1. What is an ordered pair?
2. What is the rectangular coordinate system?
3. Explain the quadrants of a rectangular coordinate system.
4. How can we determine algebraically if an ordered pair is a solution of an equation?
5. Explain how to graph an equation.
6. How can we determine graphically if an ordered pair is a solution of an equation?
7. How does the concept of average pertain to the midpoint of a line segment?
8. Where is the midpoint of a line segment if the input values of both points are the same?

Practice Makes Perfect!

Plot each ordered pair on a rectangular coordinate system and identify the quadrant in which it is located. (*See Objective 1.*)

9. $(5, 7)$
10. $(3, 6)$
11. $(-1, 4)$
12. $(-2, 10)$
13. $(0, -13)$
14. $(0, 17)$
15. $\left(\frac{1}{2}, -\frac{1}{4}\right)$
16. $\left(-\frac{1}{3}, -\frac{3}{4}\right)$
17. $(7.3, 0)$
18. $(9.6, 0)$

Additional answers can be found in the Instructor Answer Appendix.

Choose an appropriate scale to plot the three ordered pairs on the same rectangular coordinate system. (*See Objective 1.*)

19. $(10, -7), (4, 3), (-6, 0)$
20. $(-8, 14), (2, 10), (-6, -2)$
21. $(20, 48), (14, 23), (12, 35)$
22. $(32, 16), (18, 52), (7, 21)$
23. $(-15, 3), (6, -7), (-3, 12)$
24. $(-4, -15), (-13, 6), (7, 21)$

Determine if each ordered pair is a solution of the equation. (*See Objective 2.*)

25. $4x + y = 3$; $(1, -1)$ a solution
26. $5x + y = 6$; $(2, -4)$ a solution
27. $2x - 3y = 4$; $(-1, -1)$ not a solution
28. $9x - 5y = 1$; $(-2, 3)$ not a solution
29. $y = -3x + 8$; $\left(\frac{2}{3}, -5\right)$ not a solution
30. $y = -7x + 14$; $\left(\frac{3}{7}, 11\right)$ a solution
31. $y = \frac{1}{2}x - 10$; $(-6, -3)$ not a solution
32. $y = \frac{3}{4}x - 3$; $(-4, 7)$ not a solution
33. $y = |x + 3|$; $(6, -9)$ not a solution
34. $y = |x - 12|$; $(5, 7)$ a solution
35. $y = |x + 1| - 3$; $(-4, -6)$ not a solution
36. $y = |x - 8| - 5$; $(-2, 5)$ a solution
37. $y = x^2 - 2$; $(-4, 14)$ a solution
38. $y = x^2 + 3$; $(-6, 39)$ a solution

Graph each equation. (*See Objective 3.*)

39. $y = x - 5$
40. $y = x + 1$
41. $y = 2x - 3$
42. $y = 3x + 4$
43. $y = x^2 - 5$
44. $y = -x^2 + 2$
45. $y = |x + 2|$
46. $y = |x + 5|$
47. $y = -|x| + 1$
48. $y = -|x - 4|$

The graph of $y = (x - 2)^2 - 4$ is shown. In Exercises 49–52, use the graph to determine if the ordered pair is a solution of the equation. (*See Objective 4.*)

49. $(0, -1)$ not a solution
50. $(4, 0)$ a solution
51. $(3, -3)$ a solution
52. $(5, 6)$ not a solution

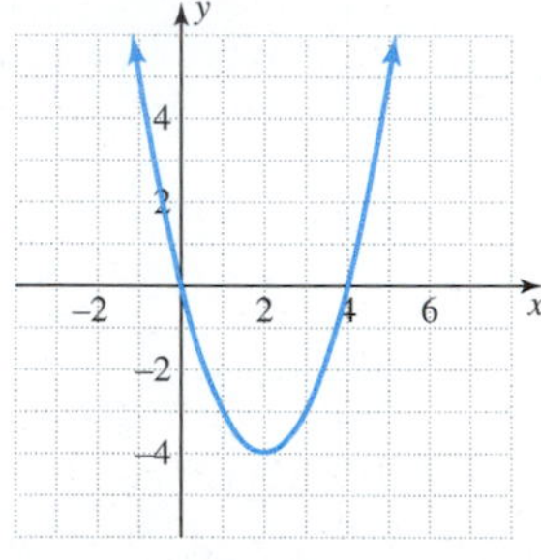

The graph of $y = -2|x + 5| + 2$ is shown. In Exercises 53–56, use the graph to determine if the ordered pair is a solution of the equation. (*See Objective 4.*)

53. $(-5, 2)$ a solution
54. $(0, -4)$ not a solution
55. $(-2, -3)$ not a solution
56. $(-7, -2)$ a solution

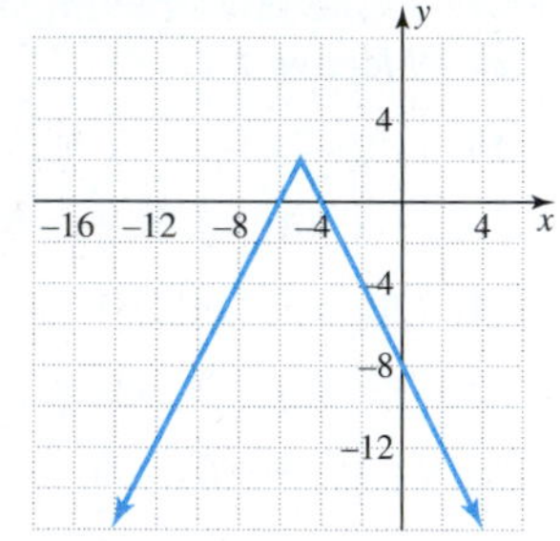

Find the midpoint of each line segment formed by the given ordered pairs. (*See Objective 5.*)

57. $(-7, -2)$ and $(3, 6)$ $(-2, 2)$
58. $(-4, 3)$ and $(0, 5)$ $(-2, 4)$
59. $(1, -2)$ and $(4, 9)$ $(2.5, 3.5)$
60. $(2, -5)$ and $(3, 8)$ $\left(\frac{5}{2}, \frac{3}{2}\right)$
61. $\left(\frac{3}{5}, -4\right)$ and $\left(-\frac{1}{2}, 6\right)$ $\left(\frac{1}{20}, 1\right)$
62. $\left(-\frac{2}{3}, 1\right)$ and $\left(\frac{1}{4}, -3\right)$ $\left(-\frac{5}{24}, -1\right)$
63. $\left(\frac{5}{6}, -1\right)$ and $\left(-\frac{1}{2}, 5\right)$ $\left(\frac{1}{6}, 2\right)$
64. $\left(-\frac{7}{8}, -4\right)$ and $\left(-\frac{3}{4}, 2\right)$ $\left(-\frac{13}{16}, -1\right)$

Solve each problem. (*See Objective 6.*)

65. The table shows the mean SAT mathematics score for college-bound Georgia seniors for the given years. (Source: The College Board)

Years after 1975	0	5	10	15	20	25	30
Mean SAT math score	498	492	500	501	506	514	520

a. Plot each pair of points, where x is the year and y is the mean SAT mathematics score.

b. Interpret the meaning of the first and last ordered pairs in the table.

c. What can you conclude, if anything, about the mean SAT score?

66. The table shows the mean SAT critical reading score for college-bound Georgia seniors for the given years. (Source: The College Board)

Years after 2002	0	1	2	3	4	5	6
Mean SAT critical reading score	504	507	508	508	503	502	502

a. Plot each pair of points, where x is the year and y is the mean SAT critical reading score.

b. Interpret the meaning of the first and last ordered pairs in the table.

c. What can you conclude, if anything, about the mean SAT score?

67. The table shows the total number (in thousands) of Honda automobiles sold for the years 2006 to 2010. (Source: http://world.honda.com/investors/library/annual_report/)

Years after 2006	0	1	2	3	4
Thousands of Honda automobiles sold	3391	3652	3925	3517	3392

a. Write the ordered pairs that represent the given data, where x is the number of years after 2006 and y is the total number of units of Honda automobiles sold.

b. Create a scatter plot of the data.

c. What can you conclude about Honda automobile sales?

68. The table shows the percent of the labor force that is unemployed (not seasonally adjusted) in California for the years 2004 to 2011. (Source: U.S. Bureau of Labor Statistics)

Years after 2004	0	1	2	3	4	5	6	7
Percent unemployed	7	6.2	5.3	5.4	6.4	10.3	13.2	12.4

a. Write the ordered pairs that represent the given data, where x is the number of years after 2004 and y is percent of the labor force that is unemployed in California.

b. Create a scatter plot of the data.

c. What can you conclude about the percent of the labor force that is unemployed in California?

Mix 'Em Up!

Choose an appropriate scale to plot the three ordered pairs on the same rectangular coordinate system. Then specify the quadrant where each point is located.

69. $(-10.4, 12.5), (-7, 3), (2.5, -1.8)$

70. $(2.4, -6.6), (8, 13), (-3.6, 8.5)$

71. $\left(-1, -\frac{1}{5}\right), \left(2, -\frac{3}{5}\right), \left(-6, -\frac{5}{13}\right)$

72. $\left(-10, \frac{13}{15}\right), \left(\frac{54}{17}, -\frac{33}{23}\right), \left(-\frac{23}{4}, -\frac{19}{22}\right)$

73. $\left(-\frac{1}{2}, -\frac{1}{4}\right), (1, -2), \left(3, \frac{5}{2}\right)$

74. $\left(\frac{3}{4}, -\frac{7}{2}\right), (-4, 3), \left(2, \frac{1}{2}\right)$

Determine if each ordered pair is a solution of the equation.

75. $y = 5x - x^2; (1, 4)$ a solution

76. $y = x - 2x^2 + 1; (2, -5)$ a solution

77. $10x - 3y = 4; \left(0, \frac{4}{3}\right)$ not a solution

78. $6x + 5y = 9; \left(\frac{3}{2}, 0\right)$ a solution

79. $y = |2x - 1| - 3; (-3, -4)$ not a solution

80. $y = |4 - 3x| + 2; (2, 0)$ not a solution

81. $y = -3|x + 9| - 10; (-10, -13)$ a solution

82. $y = -6|x + 10| - 12; (-8, 0)$ not a solution

Graph each equation.

83. $y = 4 - 3x$

84. $y = 7 - 2x$

85. $y = (x + 1)^2 - 2$

86. $y = (x + 3)^2 - 4$

87. $y = |x - 3| - 1$

88. $y = |x - 1| + 2$

89. $y = (x - 1)^2$

90. $y = (x + 3)^2$

The graph of $y = -(x + 2)(x - 1)(x - 2)$ is shown. Use the graph to determine if the ordered pair is a solution of the equation.

91. $(-2, 0)$ a solution

92. $(0, -4)$ a solution

93. $(2, 0)$ a solution

94. $(1, 1)$ not a solution

95. $(-1.5, -3)$ not a solution

96. $(0.5, -2)$ a solution

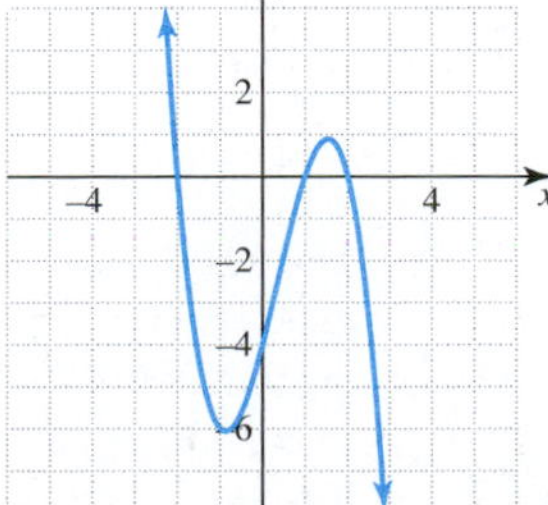

Find the midpoint of each line segment formed by the given ordered pairs.

97. $(-10, 1)$ and $(19, -5)$ $\left(\frac{9}{2}, -2\right)$

98. $(4, 12)$ and $(-3, -25)$ $\left(\frac{1}{2}, -\frac{13}{2}\right)$

99. $(5.2, -2.8)$ and $(-3.6, -0.2)$ $(0.8, -1.5)$

100. $(-7.6, 2.5)$ and $(4.3, 16.4)$ $(-1.65, 9.45)$

101. $\left(2\frac{3}{5}, 0\right)$ and $\left(-4\frac{1}{2}, -6\right)$ $(-0.95, -3)$

102. $\left(-2\frac{1}{4}, 2\right)$ and $\left(4\frac{1}{2}, 9\right)$ $\left(\frac{9}{8}, \frac{11}{2}\right)$

Solve each problem.

103. During 2010, 254,375 children entered foster care in the United States. The table shows the approximate percentage of these children who were at a particular age. (Source: http://www.acf.hhs.gov/programs/cb/stats_research/afcars/tar/report18.htm)

Age	1	3	7	10	11	13	16	18
Percentage who entered foster care at this age	8	6	4	3	3	4	7	0

a. Plot each pair of points, where x is the age and y is the approximate percentage of those children who entered foster care at this age.

b. Interpret the meaning of the first and last ordered pairs in the table.

c. What can you conclude, if anything, about the age of the children who entered foster care in 2010?

104. The table shows the percentage of children in the United States living with the given number of siblings in 2009. (Source: http://www.census.gov/prod/2011pubs/p70–126.pdf)

Number of siblings	0	1	2	3	4 or more
Percentage of children	22.1	38	24.1	10.5	5.4

a. Plot each pair of points, where x is the number of siblings and y is the percentage of children living with x siblings.

b. Interpret the meaning of the first and last ordered pairs in the table.

c. What can you conclude, if anything, about the percentage of children living with siblings in 2009?

105. The table shows the number of domestic movies released in a given year. (Source: www.boxofficemojo.com)

Number of years after 1980	0	5	10	15	20	25
Number of domestic movies released	161	470	410	411	478	547

a. Plot each pair of points, where x is the number of years after 1980 and y is the number of domestic movies released.

b. Interpret the meaning of the first and last ordered pairs in the table.

c. What can you conclude, if anything, about the number of domestic movies released?

106. The table shows the average price of a U.S. movie ticket in a given year. (Source: www.boxofficemojo.com)

Year	2001	2002	2003	2004	2005	2006	2007	2008
Average price (in dollars)	5.66	5.81	6.03	6.21	6.40	6.58	6.88	7.08

a. Plot each pair of points, where x is the year and y is the average price of a U.S. movie ticket.

b. Interpret the meaning of the first and last ordered pairs in the table.

c. What can you conclude, if anything, about the average price of a U.S. movie ticket?

You Be the Teacher!

Answer each student's questions.

107. Gunnar: Is there an easy way to graph ordered pairs with large numbers like $(-16, 23)$ without drawing too many tick marks?

108. Sherry: Is there an easy way to determine if a point is a solution of an equation other than looking at the graph?

Correct each student's errors, if any.

109. Determine if the point $(-25, -33)$ is a solution of $y = |x + 14| - 12$.

Lisa's work:

$$y = |x + 14| - 12$$
$$-33 = |-25 + 14| - 12$$
$$-33 = |-11| - 12$$
$$-33 = -11 - 12$$
$$-33 = -33 \quad \text{True}$$

Since the statement is true, $(-25, -33)$ is a solution of the equation.

110. Determine if the point $\left(\frac{3}{2}, 12\right)$ is a solution of the equation $y = -8x^2 + 4x$.

Harrison's work:

$$y = -8x^2 + 4x.$$
$$12 = -8\left(\frac{3}{2}\right)^2 + 4\left(\frac{3}{2}\right)$$
$$12 = -8\left(\frac{9}{4}\right) + 2 \cdot 3$$
$$12 = -2 \cdot 9 + 6$$
$$12 = -18 + 6$$
$$12 = -12 \quad \text{True}$$

Since the statement is true, $\left(\frac{3}{2}, 12\right)$ is a solution of the equation.

Calculate It!

Use the TABLE feature of a graphing calculator to find five solutions of each equation.

111. $y = 43x^2 - 12x + 5$

112. $y = -5x^3 + 9x^2 + 13$

113. $y = \dfrac{3x - 7}{2}$

114. $y = \dfrac{6 - 4x}{5}$

Think About It!

115. The ordered pair $(3, -1)$ is a solution of $y = \frac{2}{3}x + b$. What value of b makes this possible? $b = -3$

116. The ordered pair $\left(\frac{1}{2}, 5\right)$ is a solution of $y = -4x + b$. What value of b makes this possible? $b = 7$

117. The ordered pair $(a, -2)$ is a solution of $y = |x + 1| - 3$. What value(s) of a makes this possible? $a = -2$ or $a = 0$

118. The ordered pair $(a, 6)$ is a solution of $y = |2x + 3| - 5$. What value(s) of a makes this possible. $a = -7$ or $a = 4$

119. Suppose the midpoint between two points $(x, -6)$ and $(4, y)$ is $(-3, 1)$. Find the values of x and y. $x = -10$ and $y = 8$

120. Suppose the midpoint between two points $(x, 9)$ and $(-2, y)$ is $(-5, 6)$. Find the values of x and y. $x = -8$ and $y = 3$

SECTION 3.2 Relations

OBJECTIVES

As a result of completing this section, you will be able to

1. Express a relation in various forms and determine its domain and range.
2. Determine the domain and range from the graph of a relation.
3. Determine input and output values for a relation.
4. Troubleshoot common errors.

Mathematicians, scientists, economists, and others are intrigued by the relations that exist between two quantities. They might study questions such as these. How does drinking alcohol affect a person's response time? How does cholesterol intake affect the likelihood of a heart attack? How does study time affect a grade in a course? How do years of education affect income?

In this section, we will study this concept of relations in more detail.

Relations

Objective 1 Express a relation in various forms and determine its domain and range.

When we look up the definition of a relation, we find that almost every description involves the words connection and/or association. This idea extends to the definition of a mathematical relation. A *mathematical relation* describes the connection or association between two sets of information. This description is denoted by a set of ordered pairs.

We use relations every day. When we checkout at the grocery store, each item scanned corresponds to a specific price. When a topic in Google is entered, search results are displayed. When we search a class schedule via the Web, we enter a specific course and are then given a list of all the classes for that course.

INSTRUCTOR NOTE: Ask students what they think of when they hear the word relation. They should come up with something about a connection between two people, two quantities, and the like. Use this to develop the term mathematically.

In each of these cases, we pair together a piece of information from one set of values to a piece of information from another set of values. The first piece of information is the **input value**, or x-value. The second piece of information is the **output value**, or y-value. This set of ordered pairs is called a *relation*.

> **Definition:** A **relation** is a set of ordered pairs, in which the first coordinate of the ordered pairs comes from a set called the **domain** and the second coordinate of the ordered pairs comes from a set called the **range**. The domain is the set of x-values, or input values, and the range is the set of y-values, or output values.

Relations can be expressed in various forms: a table, a mapping, a graph, an equation, and a set of ordered pairs.

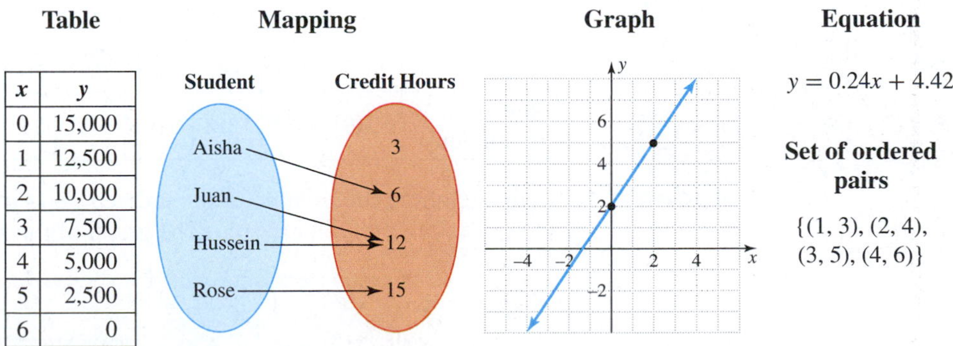

Table

x	y
0	15,000
1	12,500
2	10,000
3	7,500
4	5,000
5	2,500
6	0

Recall that an equation represents a relation because an equation defines a set of ordered pair solutions.

When we state the domain or range of a relation, it is not necessary to list values more than once.

Objective 1 Examples **Express each relation in the requested form. Identify the domain and range of each relation.**

1a. A class is surveyed to find out how many hours each student is taking. The results are shown in the mapping. Write this relation as a set of ordered pairs.

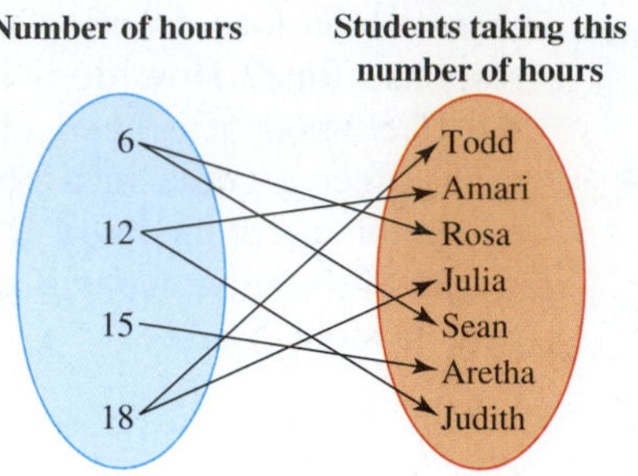

Solution **1a.** In this relation, the x-value of the ordered pairs represents the number of credit hours and the y-value represents the student.

The set of ordered pairs for this relation is {(6, Rosa), (6, Sean), (12, Amari), (12, Judith), (15, Aretha), (18, Todd), (18, Julia)}.

The domain is the set of x-values and is {6, 12, 15, 18}. The range is the set of y-values and is {Rosa, Sean, Amari, Judith, Aretha, Todd, Julia}.

1b. According to T-Mobile.com, each text message sent or received in the United States costs \$0.20 (if you do not have a messaging plan). Express the relation between the number of text messages sent or received and the cost associated with them as an equation.

Solution **1b.** In this relation, the x-value of the ordered pairs represents the number of text messages and the y-value represents the cost.

The equation for the relation is $y = 0.20x$.

The domain is the set of x-values or input values. Since it only makes sense to have 0, 1, 2, 3, and so on text messages, the domain is $\{0, 1, 2, 3, \ldots\}$. The range is the set of y-values or output values. Since each input is multiplied by 0.20, the range is $\{0, 0.20, 0.40, 0.60, 0.80, \ldots\}$.

1c. The network TV ad revenue (in millions of dollars) for the NCAA Division I Men's Basketball Championship is given in the table. Express this relation in a graph. (Source: http://www.internetadsales.com/march-madness-advertising-trends-report)

Year	2000	2001	2002	2003	2004	2005	2006	2007	2008	2009
Revenue (in millions)	\$319	\$318	\$358	\$380	\$451	\$475	\$500	\$520	\$643	\$589

Solution **1c.** In this relation, the x-value of the ordered pairs represents the year and the y-value represents the revenue. The graph of the relation is shown.

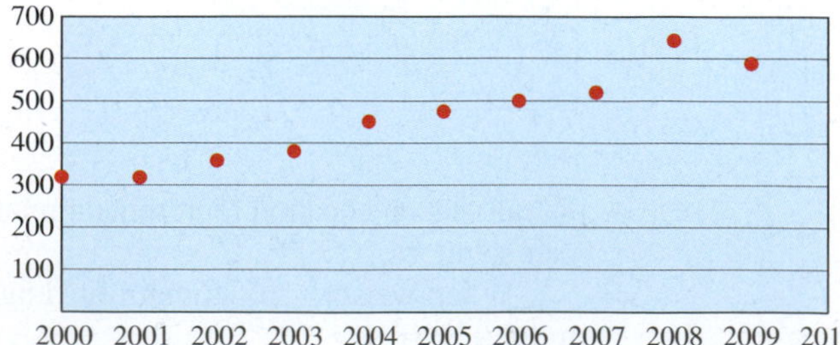

The domain is the set of x-values and is {2000, 2001, 2002, 2003, 2004, 2005, 2006, 2007, 2008, 2009}.

The range is the set of y-values and is {318, 319, 358, 380, 451, 475, 500, 520, 589, 643}.

Student Check 1 Express each relation in the requested form. State the domain and range of each relation. For part (b), use only whole numbers for the input.

a. A class is surveyed to find each student's declared major. The results are shown in the mapping. Write this relation as a set of ordered pairs.

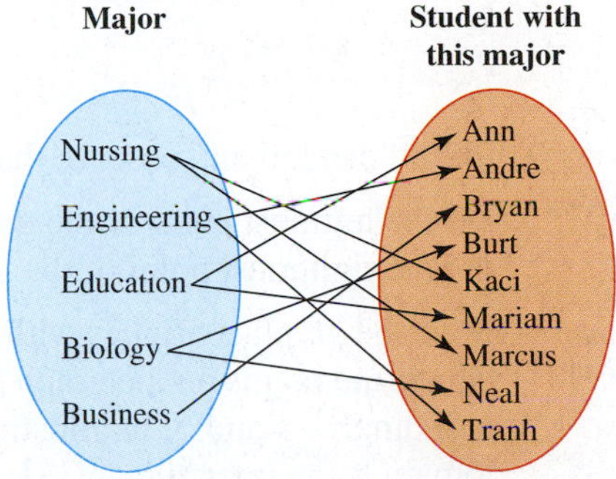

b. Skype allows calls from computer-to-computer or computer-to-land line or mobile phone. Calls from computer-to-computer are free but calls made from a computer to phones have an associated cost. Calls from a computer in the United States to a phone in Iraq cost approximately \$0.39 per minute. Express the relation between the number of minutes and the associated cost of using Skype to call Iraq from the United States as an equation. (Source: http://www.skype.com/intl/en-us/prices/payg-rates/)

c. The percentage of Americans who are obese (BMI is greater than or equal to 30) is given in the table. Express this relation in a graph. (Source: http://www.cdc.gov and http://www.data360.org/index.aspx)

Year	1995	1996	1997	1998	1999	2000	2001	2002	2003	2004	2005	2006	2007	2008	2009
Percentage	15.8	16.8	16.6	18.3	19.8	20.1	21.1	22.2	22.8	23.2	24.4	25.1	26.3	26.6	27.2

The Domain and Range of a Relation

Objective 2 Determine the domain and range from the graph of a relation.

The most common type of relation that we work with in math classes is an algebraic equation. From an equation, we can graph the relation and then use its graph to determine the domain and range of the relation.

In this objective, we will examine graphs of some basic relations and learn how to read a graph to determine its domain and range. When writing the domain and range of a relation from a graph, we use interval notation to denote the set of x-values represented by the graph and the set of y-values represented by the graph. Recall a graph represents the solution set of an equation in two variables and can have infinitely many points on it. Hence, interval notation is a concise way to represent the domain and range of a graph.

To read the domain from a graph, we examine the graph to find the smallest and largest values of x in the set of ordered pairs that lie on the graph. Since x-values are on the horizontal axis, the smallest value of x corresponds to the leftmost point on the graph and the largest value of x corresponds to the rightmost point on the graph.

To read the range from a graph, we examine the graph to find the smallest and largest values of y in the set of ordered pairs that lie on the graph. Since y-values are on the vertical axis, the smallest value of y corresponds to the lowest point on the graph and the largest value of y corresponds to the highest point on the graph.

Consider the relation given by the graph as shown.

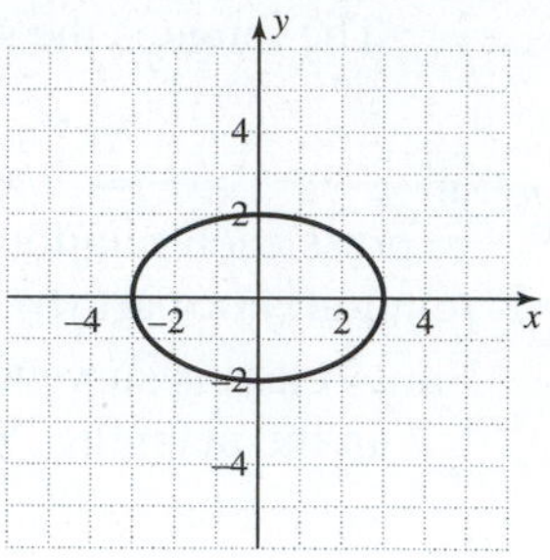

INSTRUCTOR NOTE:
To explain the concept of domain, draw a copy of the x-axis below the graph and extend a line from each point on the graph to its corresponding x-value. The points that are "touched" are in the domain.

To find the domain, observe that

- the leftmost point of the graph is $(-3, 0)$.
- the rightmost point of the graph is $(3, 0)$.

Since the graph continues without stopping between these values, the points on the graph have x-values between, and including, -3 and 3, as illustrated by Figure 3.2. So, the domain is the interval $[-3, 3]$.

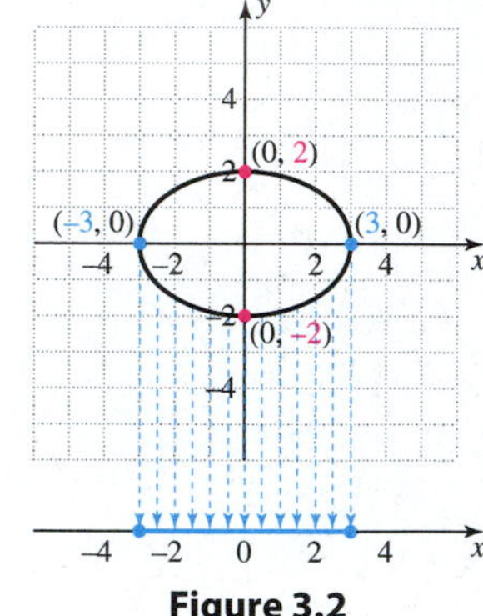

Figure 3.2

INSTRUCTOR NOTE:
To explain the concept of range, draw a copy of the y-axis to the right of the graph and extend a line from each point on the graph to its corresponding y-value. The points that are "touched" are in the range.

To find the range, observe that

- the lowest point of the graph is $(0, -2)$.
- the highest point of the graph is $(0, 2)$.

Since the graph continues without stopping between these values, the points on the graph have y-values between, and including, -2 and 2, as illustrated by Figure 3.3. So, the range is the interval $[-2, 2]$.

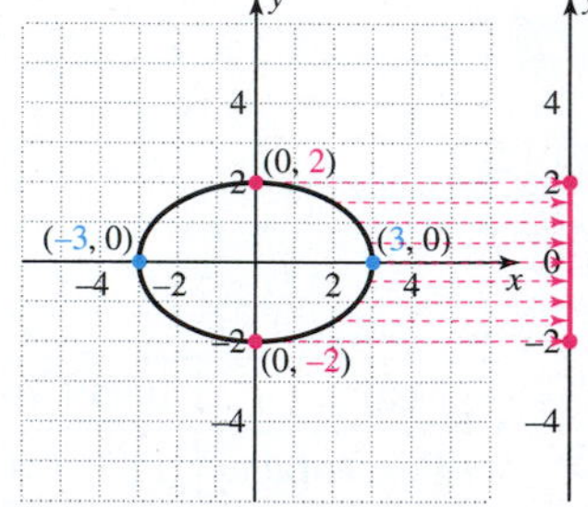

Figure 3.3

Recall that brackets are used to denote that a number is included in a set and parentheses are used to denote that a number is not included in a set. Brackets are used with the endpoints -3 and 3 since points with these x-values are included on the graph. Brackets are also used on -2 and 2 since points with these y-values are included on the graph.

We will examine one more relation. In this case, the graph continues indefinitely in both directions.

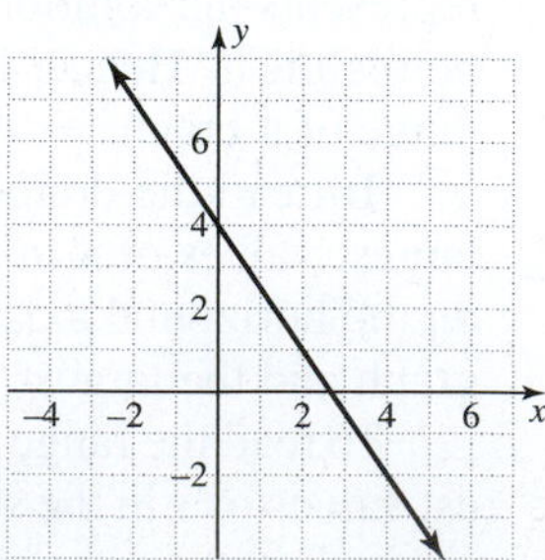

To find the domain, observe that there is no leftmost point or rightmost point of the graph. The arrows indicate that the graph continues left and right indefinitely, that is, the graph extends left to $-\infty$ and right to ∞.

Since the graph continues without stopping between these values, the points on the graph have x-values between $-\infty$ and ∞, as illustrated by Figure 3.4. So, the domain is $(-\infty, \infty)$.

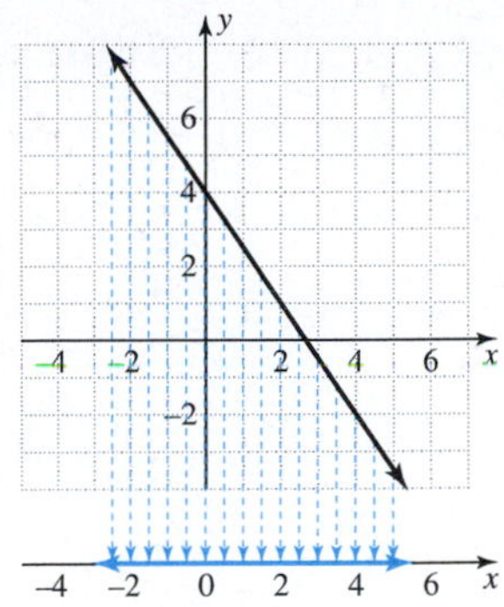

Figure 3.4

To find the range, observe that the graph does not have a lowest or highest point. The arrows indicate that the graph continues upward and downward indefinitely, that is, the graph extends down to $-\infty$ and up to ∞.

Since the graph continues without stopping between these values, the points on the graph have y-values between $-\infty$ and ∞, as illustrated by Figure 3.5. So, the range is $(-\infty, \infty)$.

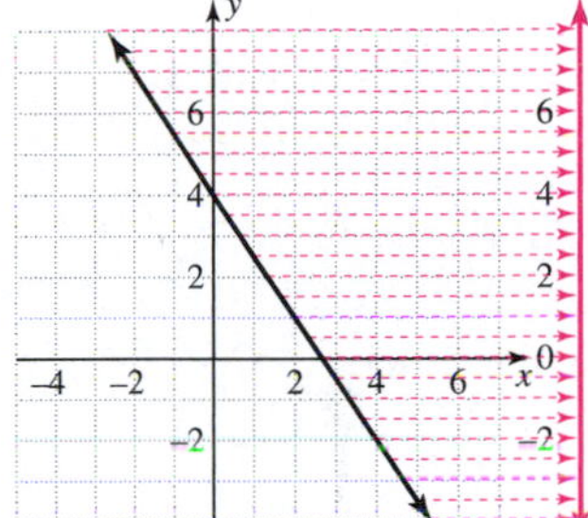

Figure 3.5

Procedure: Finding the Domain by Reading the Graph from Left to Right

Step 1: What is the x-value of the leftmost point on the graph?
Step 2: What is the x-value of the rightmost point on the graph?
Step 3: Ask yourself, "Between what values of x is the graph contained?"
Step 4: If the graph extends indefinitely from the left to the right, the domain is all real numbers, denoted by $(-\infty, \infty)$.

Procedure: Finding the Range by Reading the Graph from Bottom to Top

Step 1: What is the y-value of the lowest point on the graph?
Step 2: What is the y-value of the highest point on the graph?
Step 3: Ask yourself, "Between what values of y is the graph contained?"
Step 4: If the graph extends indefinitely from the bottom to the top, the range is all real numbers, denoted by $(-\infty, \infty)$.

Objective 2 Examples

Determine the domain and range of each relation from its graph.

INSTRUCTOR NOTE: Remind students that they should be able to construct the graphs of these equations using the skills of Section 3.1.

2a. $y = x - 2$

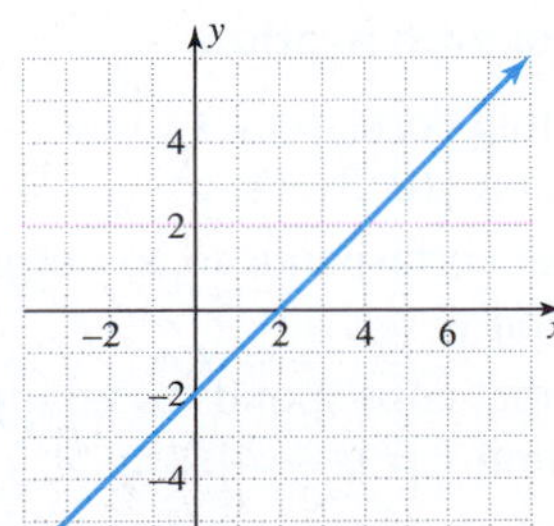

2b. $y = x^2 - 4$

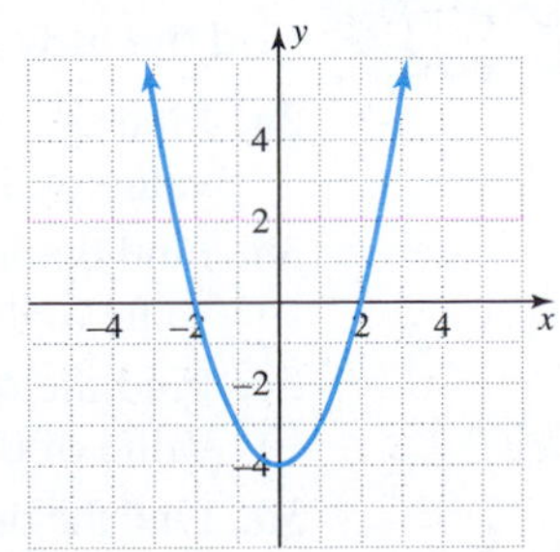

Solutions

2a. There is no leftmost or rightmost point on the graph as indicated by the arrows. The graph extends indefinitely left to $-\infty$ and right to ∞. So, the domain is all real numbers, or $(-\infty, \infty)$.

There is no lowest or highest point on the graph as indicated by the arrows. The graph extends indefinitely downward to $-\infty$ and upward to ∞. So, the range is all real numbers, or $(-\infty, \infty)$.

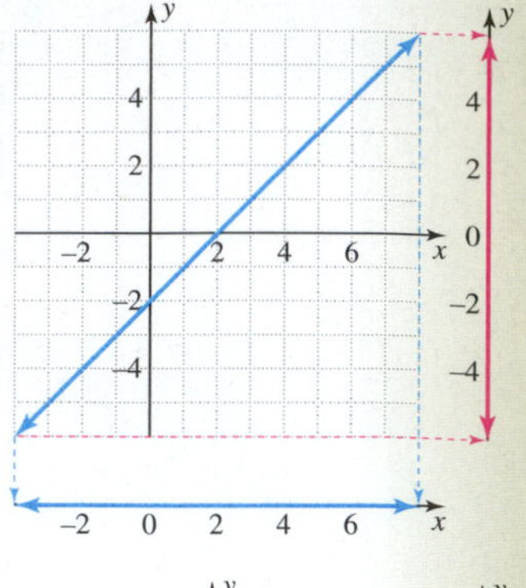

2b. There is no leftmost or rightmost point on the graph as indicated by the arrows. The graph extends indefinitely left to $-\infty$ and right to ∞. So, the domain is all real numbers, or $(-\infty, \infty)$.

The lowest point on the graph is $(0, -4)$. There is no highest point on the graph as indicated by the arrows. The graph extends up to ∞. So, the range is $[-4, \infty)$.

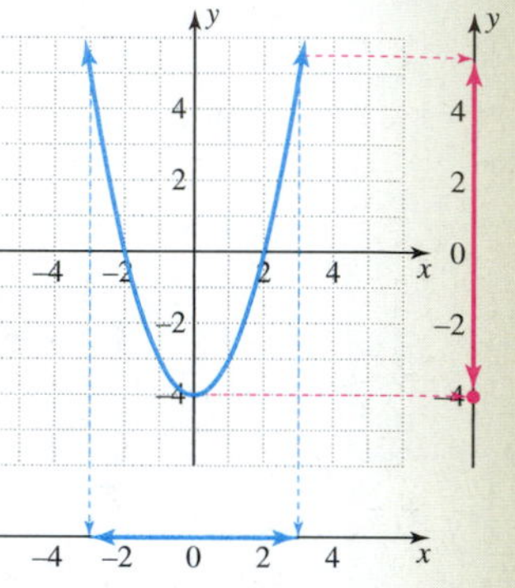

Student Check 2 Determine the domain and range of each relation from its graph.

a. $y = x - 3$

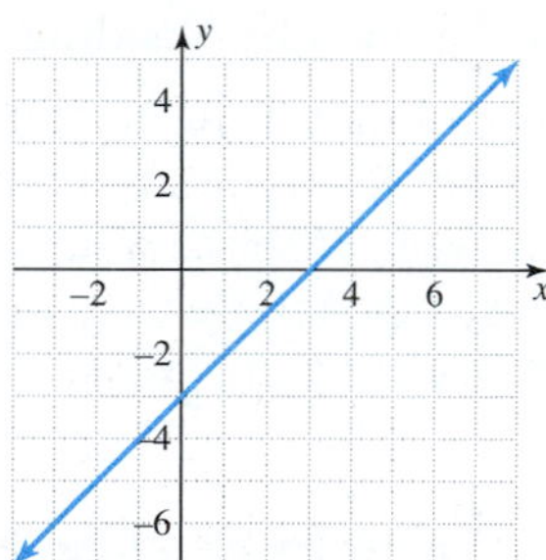

b. $y = |x + 1|$

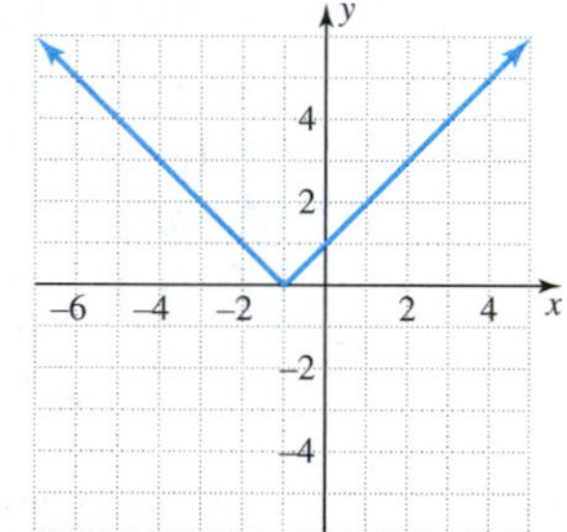

Determining Input and Output Values

Objective 3 Determine input and output values for a relation.

Knowing the relationship between two sets of information enables us to find unknown information. We often have to find an output value for a given input value or find an input value for a given output value. This task is performed nearly every day by each of us. How often have we had to find the tip on a meal, find the sales price of an item, or find directions to a specific location? In each of these cases, we know one value and must find another value that corresponds to it.

Objective 3 Examples **Find the indicated value for each relation.**

3a. Find the output value that corresponds to the input value of 3 for the relation $\{(0, 3), (1, 5), (2, 7), (3, 9)\}$

3b. Find the input value that corresponds to an output value of 4 for the relation $y^2 = x$.

3c. Find the output value that corresponds to an input value of 0 for the relation $y = |x - 4|$.

3d. Find the output value that corresponds to an input value of 1 for the relation given by the graph in Figure 3.6.

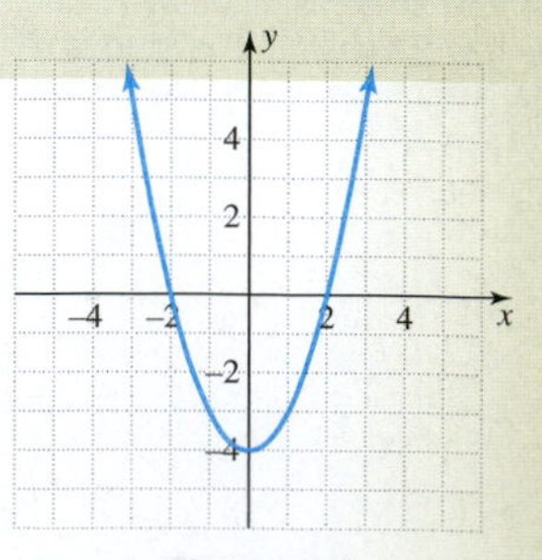

Figure 3.6

x	3	0	−1	0	3
y	−2	−1	0	1	2

3e. Find the output value that corresponds to an input value of 0 for the relation given by the table.

Solutions

3a. The input value is the x-value. The ordered pair in the relation with an x-value of 3 is (3, 9). Therefore, the output value that corresponds to $x = 3$ is $y = 9$.

3b.

$y^2 = x$	Given relation
$(4)^2 = x$	Replace the output value, y, with 4.
$16 = x$	Simplify the exponent.

The input value that corresponds to $y = 4$ is $x = 16$.

3c.

$y = \lvert x - 4\rvert$	Given relation
$y = \lvert 0 - 4\rvert$	Replace the input value, x, with 0.
$y = \lvert -4\rvert$	Simplify.
$y = 4$	Simplify the absolute value.

The output value that corresponds to $x = 0$ is $y = 4$.

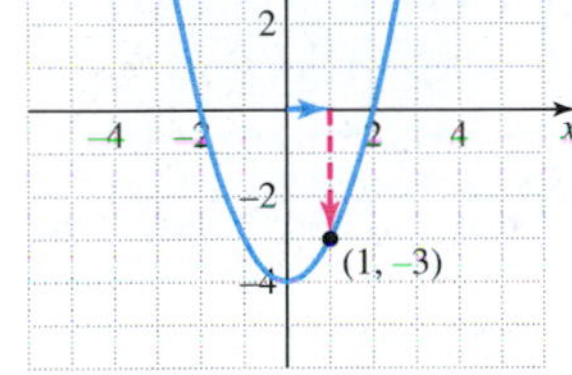

3d. The input value is 1, so $x = 1$. We need to find the point on the graph which has an x-value of 1. Since (1, −3) lies on the graph, the output value that corresponds to $x = 1$ is $y = -3$.

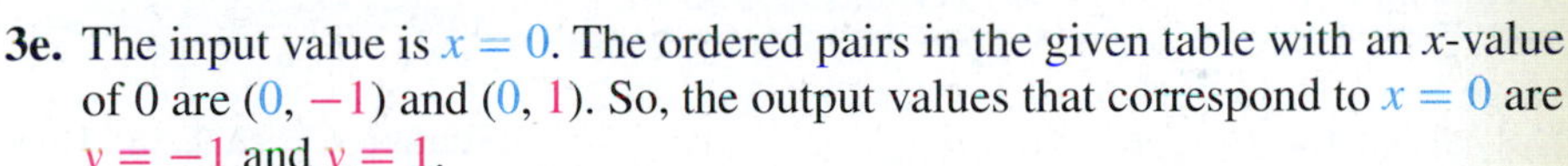

3e. The input value is $x = 0$. The ordered pairs in the given table with an x-value of 0 are (0, −1) and (0, 1). So, the output values that correspond to $x = 0$ are $y = -1$ and $y = 1$.

Student Check 3

Find the indicated value for each relation.

a. Find the output value that corresponds to the input value of 1 for the relation {(0, 3), (1, 5), (2, 7), (3, 9)}.

b. Find the input value that corresponds to an output value of 5 for the relation $y = 4x + 1$.

c. Find the output value that corresponds to an input value of −2 for the relation $y = |x - 4|$.

d. Find the output value that corresponds to an input value of 0 for the relation given by the graph in Figure 3.7.

e. Find the output value that corresponds to an input value of −3 for the relation given by the table.

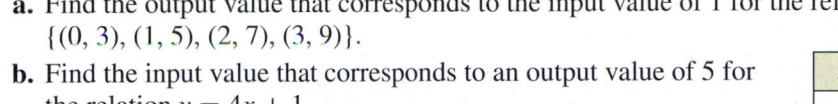

x	y
−4	3
−3	4
−4	−3
−3	−4
−5	0
5	0

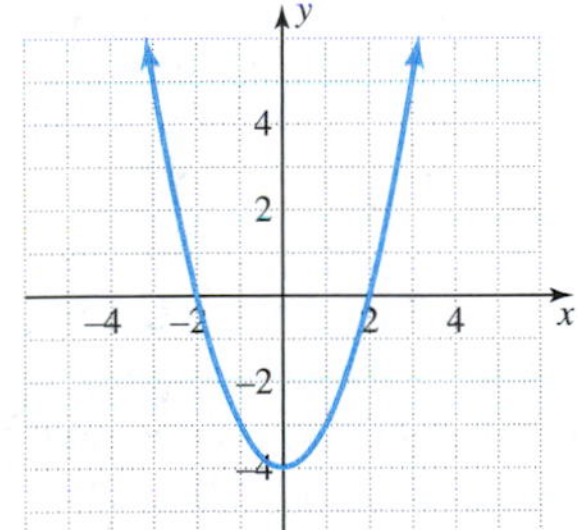

Figure 3.7

Objective 4

Troubleshoot common errors.

Troubleshooting Common Errors

A common error associated with relations is shown.

Objective 4 Example

A problem and an incorrect solution are given. Provide the correct solution and an explanation of the error.

Determine the domain and range of the relation represented by the graph.

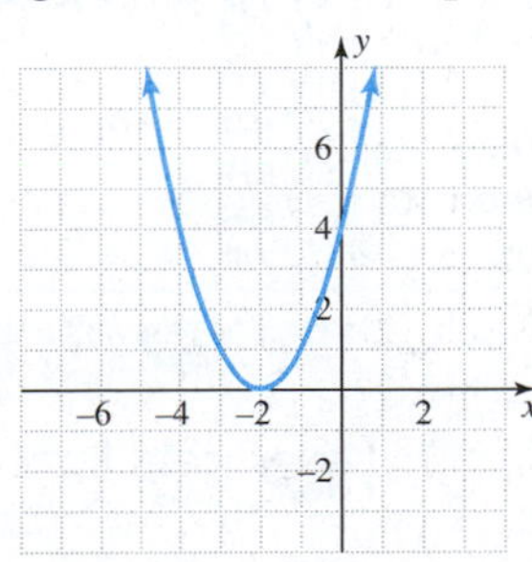

Incorrect Solution	Correct Solution and Explanation
The domain is $[-2, \infty)$ and the range is $[-2, \infty)$.	The domain is read from left to right. Since the graph continues indefinitely in the left and right directions, the domain is $(-\infty, \infty)$. The range is read from bottom to top and represents the y-values. The bottom-most point is $(-2, 0)$. So, the range is $[0, \infty)$.

ANSWERS TO STUDENT CHECKS

Student Check 1 **a.** {(Nursing, Kaci), (Nursing, Marcus), (Engineering, Andre), (Engineering, Tranh), (Education, Ann), (Education, Mariam), (Biology, Burt), (Biology, Neal), (Business, Bryan)}
Domain = {Nursing, Engineering, Education, Biology, Business} Range = {Ann, Andre, Bryan, Burt, Kaci, Mariam, Marcus, Neal, Tranh}

b. $y = 0.39x$, where x is the number of minutes; Domain = $\{0, 1, 2, 3, 4, 5, \ldots\}$, Range = $\{0, 0.39, 0.78, 1.17, \ldots\}$

c. Domain = {1995, 1996, 1997, 1998, 1999, 2000, 2001, 2002, 2003, 2004, 2005, 2006, 2007, 2008, 2009} Range = {15.9, 16.8, 16.6, 18.3, 19.8, 20.1, 21.1, 22.2, 22.8, 23.2, 24.4, 25.1, 26.3, 26.6, 27.2}

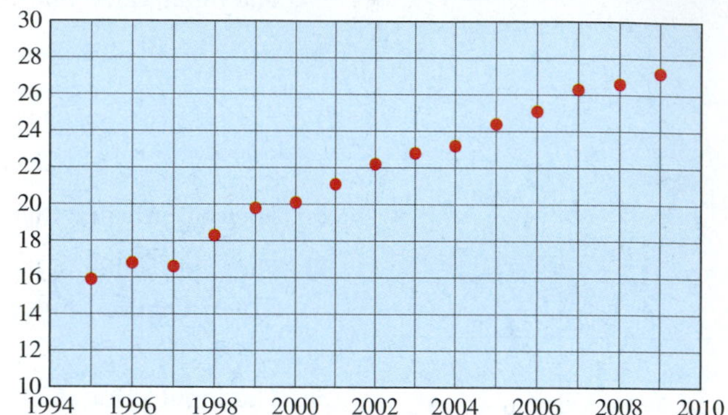

Student Check 2 **a.** Domain = $(-\infty, \infty)$, Range = $(-\infty, \infty)$ **b.** Domain = $(-\infty, \infty)$, Range = $[0, \infty)$

Student Check 3 **a.** $y = 5$ **b.** $x = 1$ **c.** $y = 6$ **d.** $y = -4$ **e.** $y = 4, y = -4$

SUMMARY OF KEY CONCEPTS

1. A relation is a set of ordered pairs. A relation can be expressed as a set, a mapping, a table, an equation, or a graph. The set of x-values (input values) for the relation is the domain and the set of y-values (output values) is called the range.
2. To determine the domain from a graph, find the interval of x that contains the graph. To determine the range from a graph, find the interval of y that contains the graph.
3. To determine output values for a given input value, find the ordered pair for the relation with the given value as its x-value. The corresponding y-value is the needed output value. If the output value is known, find the ordered pair for the relation with the given value as its y-value. The corresponding x-value is the needed input.

GRAPHING CALCULATOR SKILLS

The graphing calculator can assist us in finding input or output values for a given relation. If the relation is given in terms of an equation solved for y, we can input the equation in the calculator to find the requested information.

Example: Find the output value that corresponds to an input value of 0 for the relation $y = |x - 4|$.

Solution: We can do this by using the TABLE feature or the TRACE feature. Input the equation into the calculator.

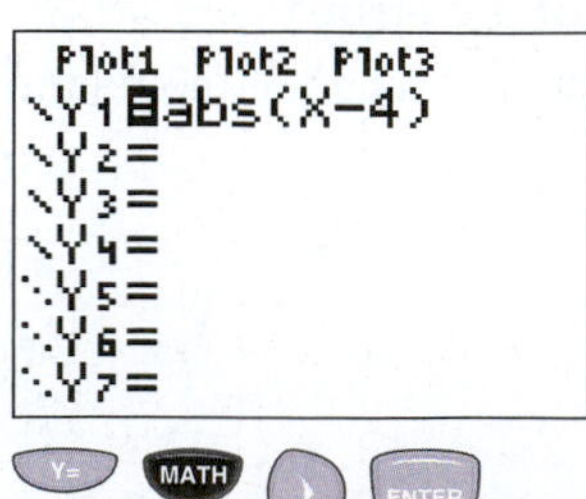

Method 1: Use the TABLE feature to find the y-value when $x = 0$.

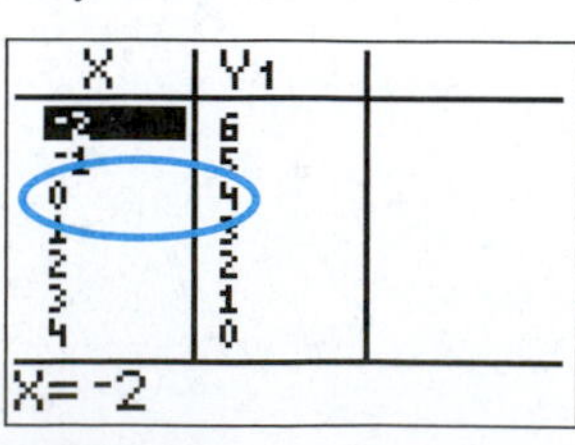

The table shows us that when the input value is 0 ($x = 0$), the output value is 4 ($y = 4$).

Method 2: Use the TRACE feature to find the point that has an x-value of 0. We first graph the equation and then access the TRACE function. The starting point when we press the TRACE key is $x = 0$. (Note: If we need to find the output for another input value, enter the value and press enter.)

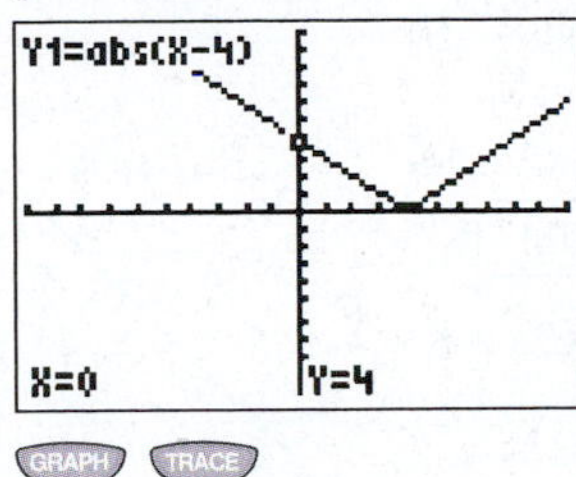

GRAPH TRACE

The display at the bottom of the calculator window shows the y-value that corresponds to the x-value. So, we see that the input value of 0 corresponds to an output value of 4.

SECTION 3.2 EXERCISE SET

Write About It!

Use complete sentences in your answer to each exercise.

1. Explain the different ways that a relation can be expressed. A relation can be expressed as a set, a mapping, a table, an equation, or a graph.
2. Define a relation. A relation is a set of ordered pairs.
3. What is the domain of a relation?
4. What is the range of a relation?
5. Explain how to determine the domain of a relation expressed as a graph. Determine the interval of x that contains the graph.
6. Explain how to determine the range of a relation expressed as a graph. Determine the interval of y that contains the graph.

Practice Makes Perfect!

Express each relation in the specified form. Identify its domain and range. (*See Objective 1.*)

7. Annie's algebra study group shares the number of hours they work each week. The results are shown in the mapping. Write this relation as a set of ordered pairs.

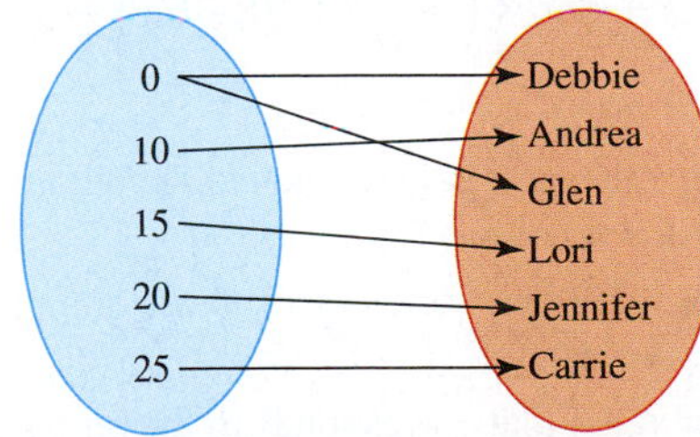

{(0, Debbie), (0, Glen), (10, Andrea), (15, Lori), (20, Jennifer), (25, Carrie)}; Domain = {0, 10, 15, 20, 25}, Range = {Debbie, Andrea, Glen, Lori, Jennifer, Carrie}

8. A group of friends share the number of times they worked at the gym each week. The results are expressed in a mapping. Write this relation as a set of ordered pairs.

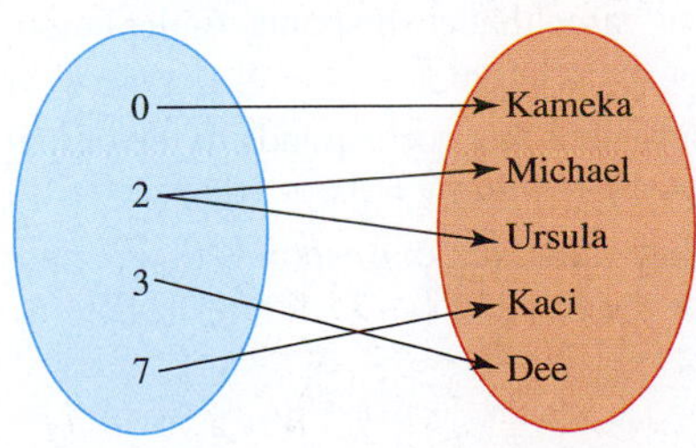

{(0, Kameka), (2, Michael), (2, Ursula), (3, Dee), (7, Kaci)}; Domain = {0, 2, 3, 7}, Range = {Kameka, Michael, Ursula, Dee, Kaci}

9. Express this relation as a set of ordered pairs.

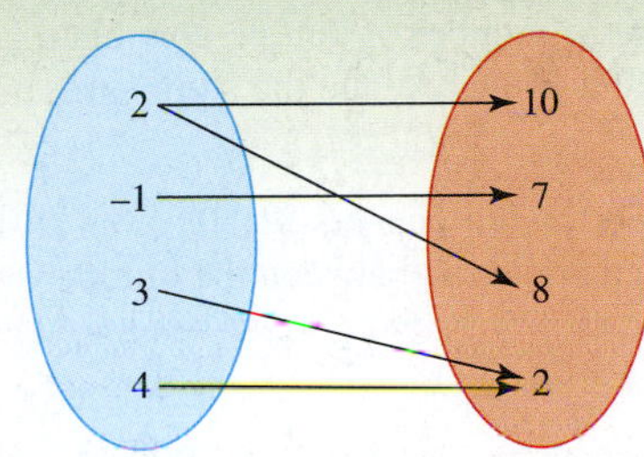

{(2, 10), (2, 8), (−1, 7), (3, 2), (4, 2)}; Domain = {−1, 2, 3, 4}, Range = {2, 7, 8, 10}

10. Express this relation as a set of ordered pairs.

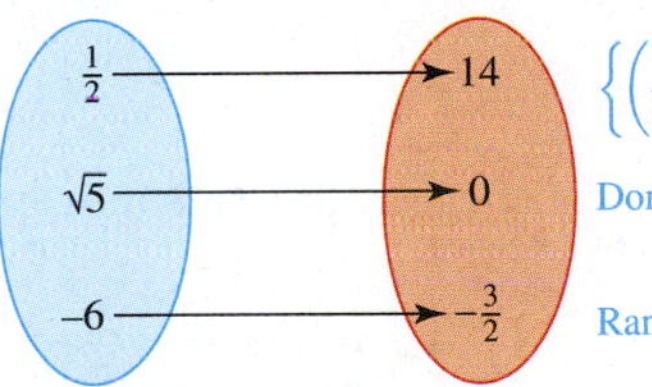

$\left\{\left(\frac{1}{2}, 14\right), (\sqrt{5}, 0)\left(-6, -\frac{3}{2}\right)\right\}$; Domain $= \left\{-6, \frac{1}{2}, \sqrt{5}\right\}$, Range $= \left\{-\frac{3}{2}, 0, 14\right\}$

11. Manny earns \$8.10 per hour at his coffee house job. Express the relation between the number of hours Manny works and the amount of money he earns as an equation.
12. Calista fills up her SUV with premium unleaded gas at a price of \$4.50 per gallon. Express the relation between the number of gallons Calista puts in her car and the cost of the sale as an equation.
13. Express this relation as a set of ordered pairs.

x	−4	−2	0	2
y	10	6	−4	3

{(−4, 10), (−2, 6), (0, −4), (2, 3)}; Domain = {−4, −2, 0, 2}, Range = {−4, 3, 6, 10}

14. Express this relation as a set of ordered pairs.

x	−12	−9	−5	−2
y	−7	−8	−4	−9

{(−12, −7), (−9, −8), (−5, −4), (−2, −9}; Domain = {−12, −9, −5, −2}, Range = {−9, −8, −7, −4}

15. The opening price of Apple Inc. stock in the beginning of January of the specified year is given in the table. Express this relation as a set of ordered pairs. (Source: http://www.dailyfinance.com)

Year	2007	2008	2009	2010	2011
Price (in dollars)	84.05	199.27	85.84	213.43	325.65

16. The opening price of Microsoft Corp. stock in the beginning of January of the specified year is given in the table. Express this relation as a set of ordered pairs. (Source: http://www.dailyfinance.com)

Year	2007	2008	2009	2010	2011
Price (in dollars)	30.19	32.20	17.10	28.18	27.75

17. A popular coffee shop charges $3.10 for a tall iced mocha. Express the relation between the number of iced mochas sold and the revenue the coffee shop earns as an equation.

18. A cookie store in the mall charges $2.50 for a large double-chocolate chip cookie. Express the relation between the number of double-chocolate chip cookies sold and the revenue the store earns as an equation.

19. A New York street vendor charges $1.50 for an all-beef hotdog. Express the relation between the number of hotdogs sold and the revenue the street vendor earns as an equation.

20. A Philadelphia street vendor charges $2.00 for a salted pretzel with mustard. Express the relation between the number of pretzels sold and the revenue the street vendor earns as an equation.

Find the domain and range of each relation whose graph is given. (*See Objective 2.*)

21. Domain = $[-3, 2]$, Range = $[-4, 3]$

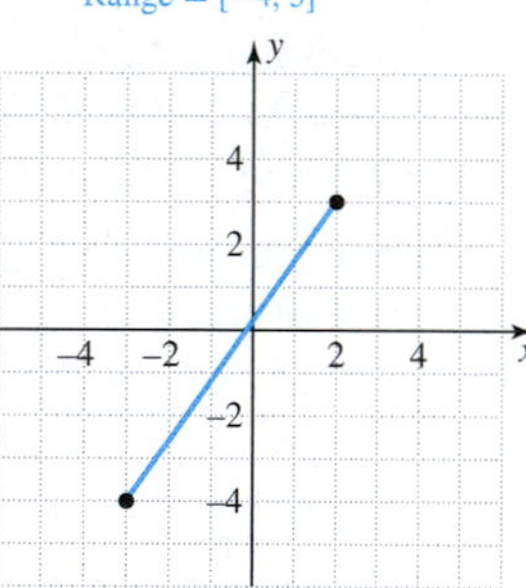

22. Domain = $[-5, 4]$, Range = $[-5, 5]$

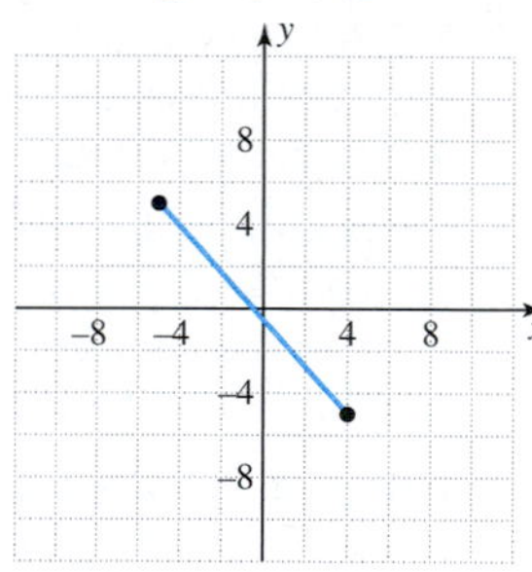

23. Domain = $(-\infty, \infty)$, Range = $[-4, \infty)$

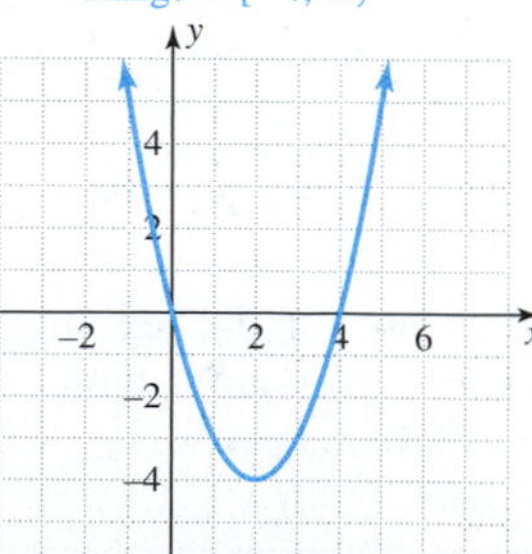

24. Domain = $(-\infty, \infty)$, Range = $(-\infty, 4]$

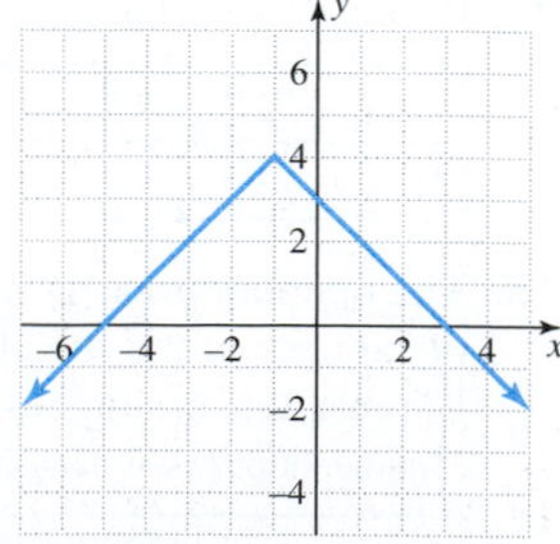

25. Domain = $(-\infty, \infty)$, Range = $(-\infty, \infty)$

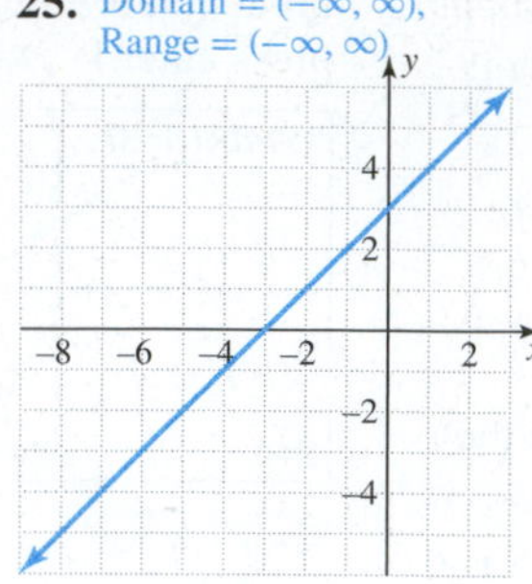

26. Domain = $(-\infty, 3]$, Range = $(-\infty, \infty)$

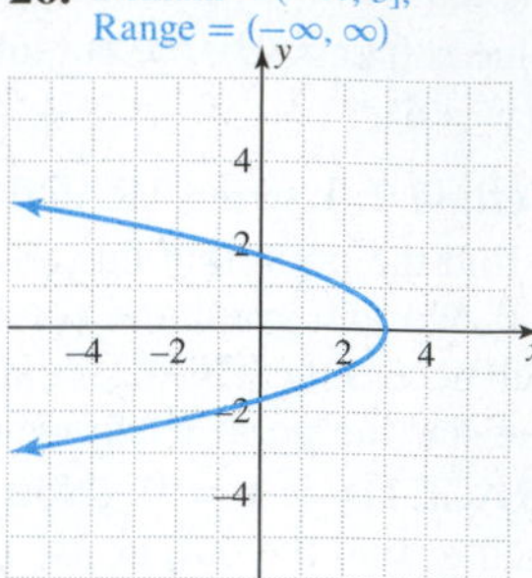

27. Domain = $[0, \infty)$, Range = $[0, \infty)$

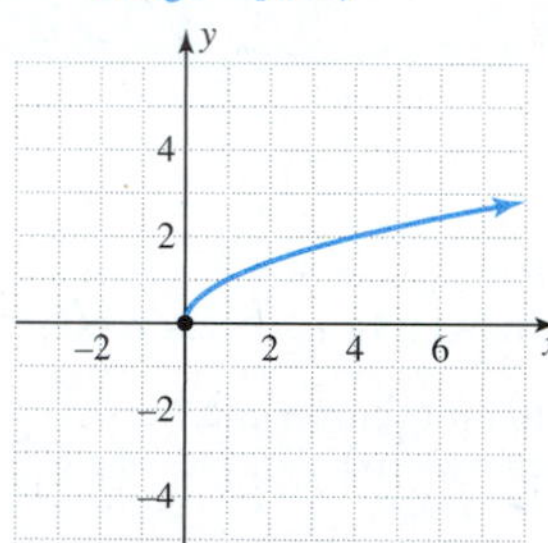

28. Domain = $[-8, \infty)$, Range = $(-\infty, 4]$

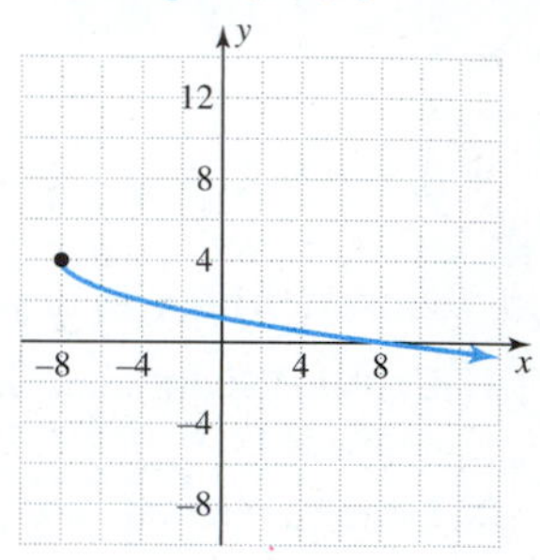

Find the indicated value for each relation. (*See Objective 3.*)

29. Find the output value that corresponds to the input value of -1 for the relation $\{(-2, -1), (-1, 3), (0, 5), (6, -4)\}$. $y = 3$

30. Find the output value that corresponds to the input value of 5 for the relation $\{(0, 10), (2, -16), (5, 13), (-16, 5)\}$. $y = 13$

31. Find the output value that corresponds to the input value of 3 for the relation $\left\{(-4, -1), \left(-\frac{1}{2}, 2\right), (3, 4), (5, 2)\right\}$. $y = 4$

32. Find the output value that corresponds to the input value of $-\frac{3}{4}$ for the relation $\left\{\left(-\frac{3}{4}, 7\right), (0, -6), \left(\frac{1}{2}, -\frac{3}{4}\right), (4, -10)\right\}$. $y = 7$

33. Find the input value that corresponds to the output value of 9 for the relation $y^2 = x$. $x = 81$

34. Find the input value that corresponds to the output value of 1 for the relation $y^2 = x$. $x = 1$

35. Find the input value that corresponds to the output value of 3 for the relation $2x + y = 5$. $x = 1$

36. Find the input value that corresponds to the output value of -2 for the relation $3x + y = 7$. $x = 3$

37. Find the input value that corresponds to the output value of 6 for the relation $y = |x - 5|$. $x = -1, x = 11$

38. Find the input value that corresponds to the output value of 1 for the relation $y = |x - 10|$. $x = 9, x = 11$

39. Find the output value that corresponds to the input value of 3 for the relation $\{(-11, 7), (-9, 12), (3, -5), (0, -10), (3, 14)\}$. $y = -5, y = 14$

40. Find the output value that corresponds to the input value of -2 for the relation $\{(-5, -2), (-3, 10), (-2, -8), (2, 0), (-2, 0)\}$. $y = -8, y = 0$

Mix 'Em Up!

Express each given relation in the specified form and then state the domain and range.

41. A group of friends share their all-time favorite movies. The results are shown in the mapping. Write this relation as a set of ordered pairs.

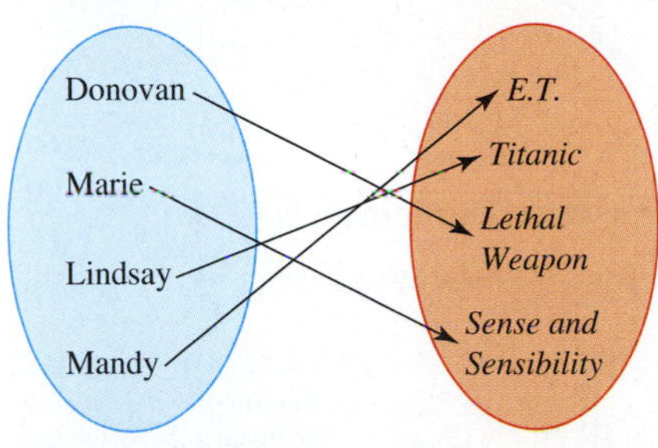

{(Donovan, *Lethal Weapon*), (Marie, *Sense and Sensibility*), (Lindsay, *Titanic*), (Mandy, *E.T.*)}; Domain = {Donovan, Marie, Lindsay, Mandy}, Range = {*E.T.*, *Titanic, Lethal Weapon, Sense and Sensibility*}

42. A group of friends share the number of miles they drive to campus. The results are shown in the mapping. Write this relation as a set of ordered pairs.

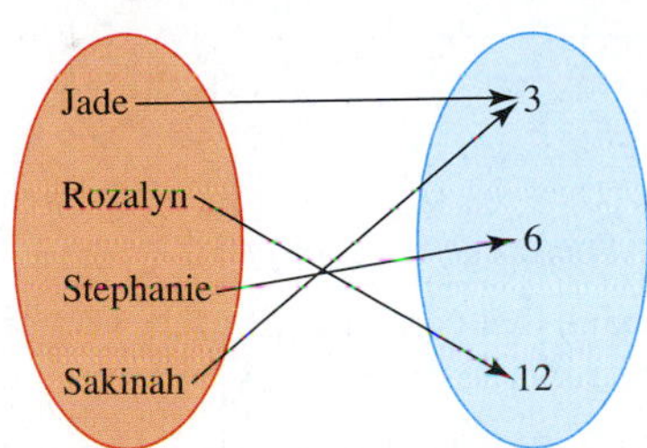

{(Jade, 3), (Rozalyn, 12), (Stephanie, 6), (Sakinah, 3)}; Domain = {Jade, Rozalyn, Stephanie, Sakinah}, Range = {3, 6, 12}

43. A library charges \$0.50 for each day that a book is turned in late. Express the relation between the number of days late and the late fee.

44. A local cleaners charges \$1.25 to wash and iron cotton shirts. Express the relation between the number of shirts washed and the total cost.

45. Express the relation as a set of ordered pairs.

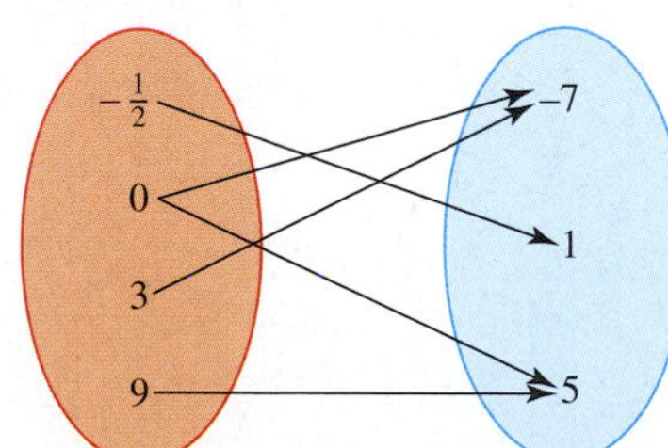

$\left\{\left(-\frac{1}{2}, 1\right), (0, -7), (0, 5), (3, -7), (9, 5)\right\}$; Domain = $\left\{-\frac{1}{2}, 0, 3, 9\right\}$, Range = $\{-7, 1, 5\}$

46. Express the relation as a set of ordered pairs.

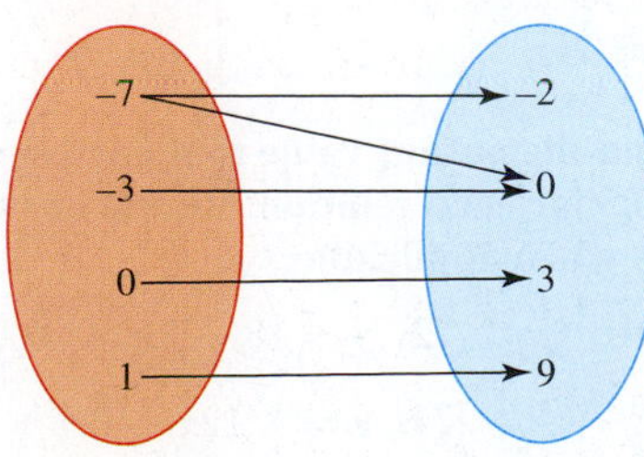

$\{(-7, -2), (-7, 0), (-3, 0), (0, 3), (1, 9)\}$; Domain = $\{-7, -3, 0, 1\}$, Range = $\{-2, 0, 3, 9\}$

47. The number of viewers of the top five rated broadcast TV programs for the week of August 15, 2011, is given in the table. Express this relation as a graph. (Source: Nielsen Media Research)

Rating	1	2	3	4	5
Viewers (in millions)	10.7	10.4	8.7	9.7	8.6

48. The number of viewers of the top five rated cable TV programs for the week of August 15, 2011, is given in the table. Express this relation as a graph. (Source: Nielsen Media Research)

Rating	1	2	3	4	5
Viewers (in millions)	7.8	6.7	6.7	5.4	5.5

Find the domain and range of each relation.

49. $\{(-90, 4), (-71, 91), (-15, 9), (20, 34), (53, -34)\}$

50. $\{(-100, 23), (-93, 83), (33, 42), (52, 83)\}$

51. $\{(-17, 18), (-16, 4), (0, 18), (18, -17)\}$

52. $\{(-9, 7), (-6, 24), (1, 7), (24, 3)\}$

53.

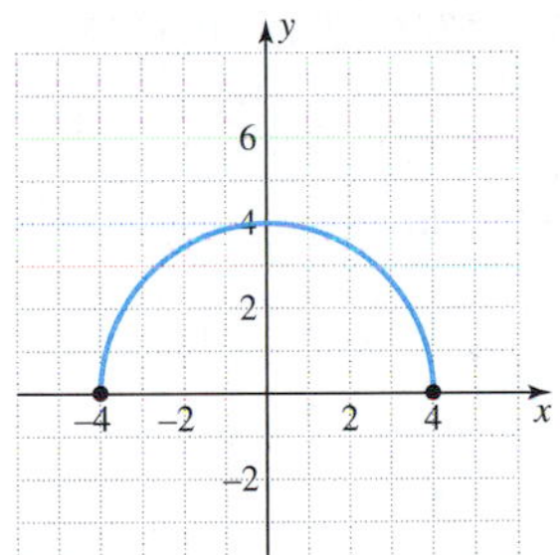

Domain = $[-4, 4]$, Range = $[0, 4]$

54.

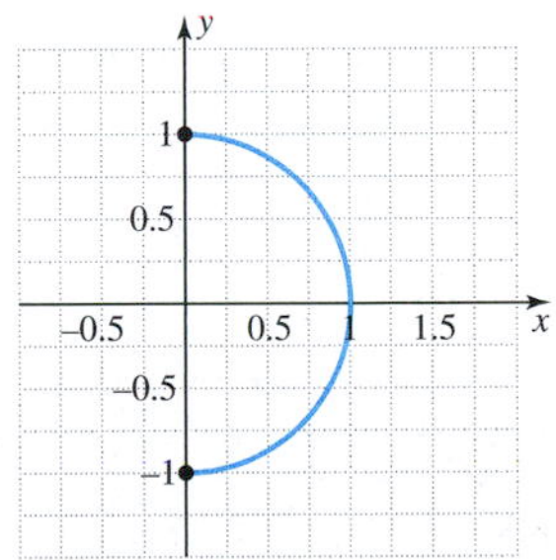

Domain = $[0, 1]$, Range = $[-1, 1]$

55.

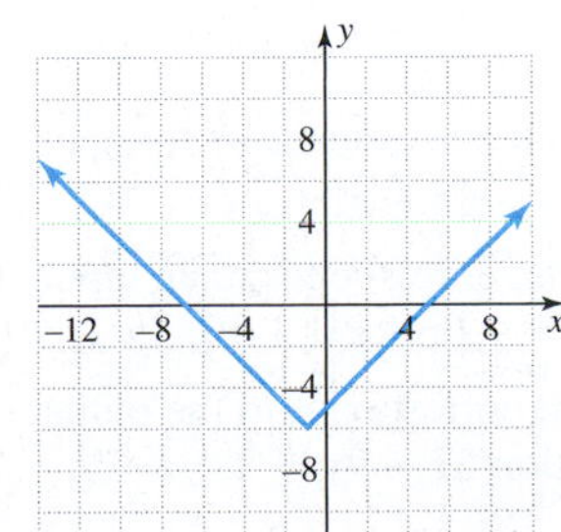

Domain = $(-\infty, \infty)$, Range = $[-6, \infty)$

56.

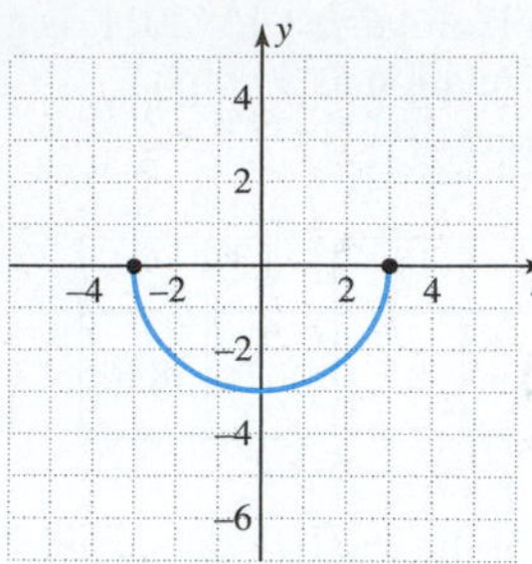

Domain = [−3, 3],
Range = [−3, 0]

57.

x	−22	−11	−8	−7
y	18	9	12	15

Domain = {−22, −11, −8, −7},
Range = {9, 12, 15, 18}

58.

x	3	12	14	25
y	−5	−12	10	0

Domain = {3, 12, 14, 15},
Range = {−12, −5, 0, 10}

Find the indicated value for each given relation.

59. Find the output value that corresponds to the input value of $\frac{1}{4}$ for the relation $y = 16x^2$. $y = 1$

60. Find the output value that corresponds to the input value of $\frac{1}{3}$ for the relation $y = -9x^2$. $y = -1$

61. Find the output values that correspond to the input value of 0 for the relation whose graph is shown.

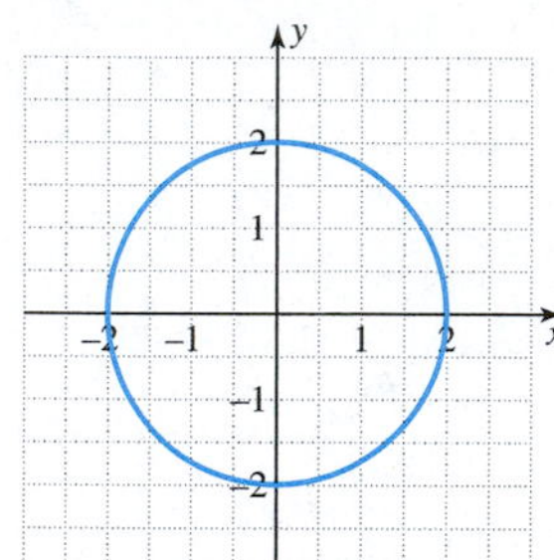

$y = -2, y = 2$

62. Find the input values that correspond to the output value of 0 for the relation whose graph is shown.

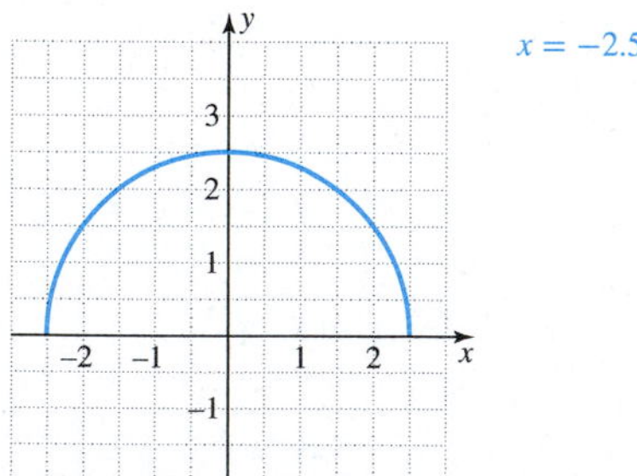

$x = -2.5, x = 2.5$

63. Find the output value that corresponds to an input value of 11 for the relation $y = -|3 - x| + 11$. $y = 3$

64. Find the output value that corresponds to an input value of 23 for the relation $y = -|6 - 2x| + 13$. $y = -27$

65. Find the input value that corresponds to the output value of 14 for the relation $5x - 7y = 2$. $x = 20$

66. Find the input value that corresponds to the output value of 9 for the relation $4x - 5y = 3$. $x = 12$

You Be the Teacher!

Correct each student's errors, if any.

67. Find the domain of the relation given by the graph.

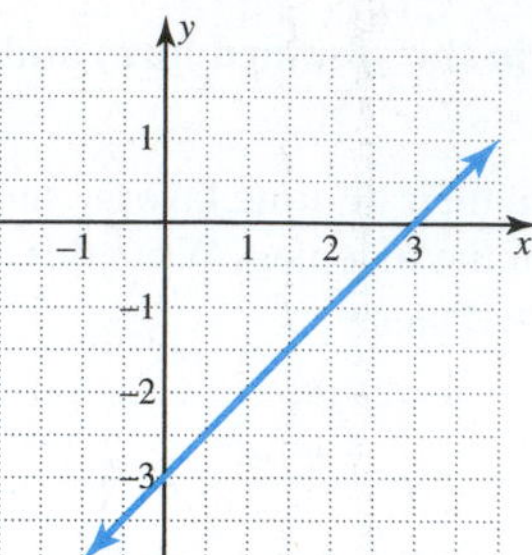

Because the graph extends left and right indefinitely, the domain is $(-\infty, \infty)$.

Stephanie's work: The domain is $[3, \infty)$.

68. Find the range of the relation given by the graph.

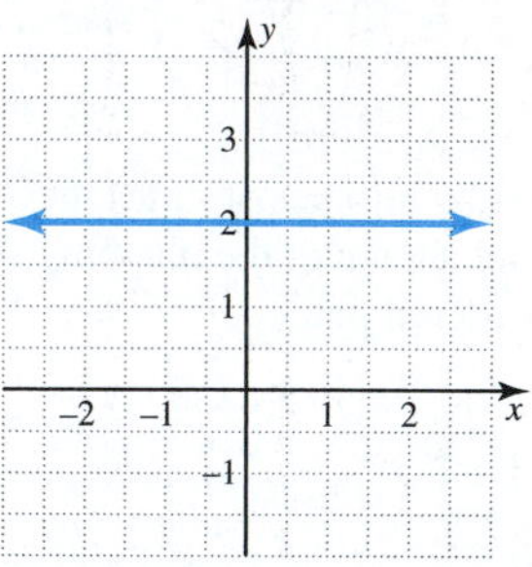

The range is the set of y-values. The smallest and largest y-value on the graph is 2. So, the range is {2}.

Monica's work: The range is $[2, \infty)$.

69. Find the input value that corresponds to the output value of 2 for the relation $2x - y = 4$.

Brandon's work:

$2(2) - y = 4$

$4 - y = 4$

$-y = 4$

$y = 0$

So, the input value of 0 corresponds to the output value of 2.

The output value is y. So, replace y with 2 and solve for x.
$2x - 2 = 4$
$2x = 6$
$x = 3$
So, the input value of 3 corresponds to the output value of 2.

70. Find the input value that corresponds to the output value of 3 for the relation $y = |x - 1| + 2$.

Marcus's work:

$|x - 1| + 2 = 3$

$|x - 1| = 1$

$x - 1 = 1$

$x = 2$

So, the input value of 2 corresponds to an output value of 3.

The absolute value equation was solved incorrectly.
$|x - 1| + 2 = 3$
$|x - 1| = 1$
$x - 1 = 1$ or $x - 1 = -1$
$x = 2$ $\quad$ $x = 0$
So, the input values of 0 and 2 correspond to an output value of 3.

Calculate It!

Use a calculator to find the output value that corresponds to an input value of 3.5 for each relation. Round answers to two decimal places when applicable.

71. $y = x^2 - 6x - 10$ $\quad$ $y = -18.75$

72. $y = 18 - 3x^2 + 4x$ $\quad$ $y = -4.75$

73. $y = \sqrt{10x - 20}$ $\quad$ $y = 3.87$

74. $y = \sqrt{12x - 3}$ $\quad$ $y = 6.24$

Think About It!

75. Write a real-life example of a relation in which there is more than one output for a single input.

76. Write a real-life example of a relation in which for every input, there is only one output.

77. What must be true about a graph whose domain can be represented by a finite set of numbers?

78. What must be true about a graph whose domain can be represented with interval notation?

PIECE IT TOGETHER SECTIONS 3.1 AND 3.2

Choose an appropriate scale to plot the following three ordered pairs on the same rectangular coordinate system. Identify the quadrants in which each point is located. (*See Section 3.1, Objective 1.*)

1. $(-20, 25), (0, -15), (30, 40)$

Determine if the ordered pair is a solution of the equation. (*See Section 3.1, Objective 2.*)

2. $5x + y = 17; (2, -7)$ not a solution

3. $y = |x + 1| - 2; (-4, 1)$ a solution

4. $y = 2.5x^2 + 1.6; (-1.2, 5.52)$ not a solution

Graph each equation. (*See Section 3.1, Objective 3.*)

5. $y = x^2 - 9$

6. $y = 1 - 3x$

7. $y = |2x + 6|$

Find the midpoint of the line segment formed by the ordered pairs. (*See Section 3.1, Objective 5.*)

8. $(7, -15)$ and $(19, 18)$ $\left(13, \frac{3}{2}\right)$

9. $\left(-6, \frac{7}{2}\right)$ and $\left(11, -\frac{5}{3}\right)$ $\left(\frac{5}{2}, \frac{11}{12}\right)$

10. A New York street vendor charges $12.50 for a 100% wool scarf. (*See Section 3.2, Objective 1.*)
    a. Express the relation between the number of 100% wool scarves sold and the revenue the street vendor earns as an equation. $y = 12.50x$, where x is the number of 100% wool scarves sold and y is the revenue earned.
    b. Find the domain and range of the relation. Domain = {0, 1, 2, 3, . . .}, Range = {0, 12.50, 25.00, 37.50, . . .}

Find the domain and range of the relation whose graph is shown. (*See Section 3.2, Objective 2.*)

11.

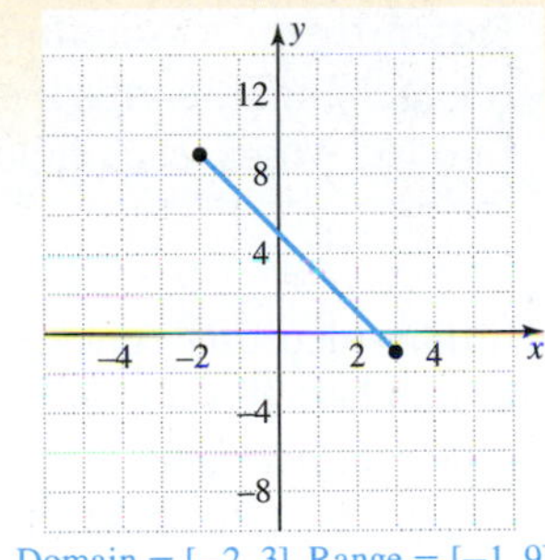

Domain = [−2, 3], Range = [−1, 9]

12.

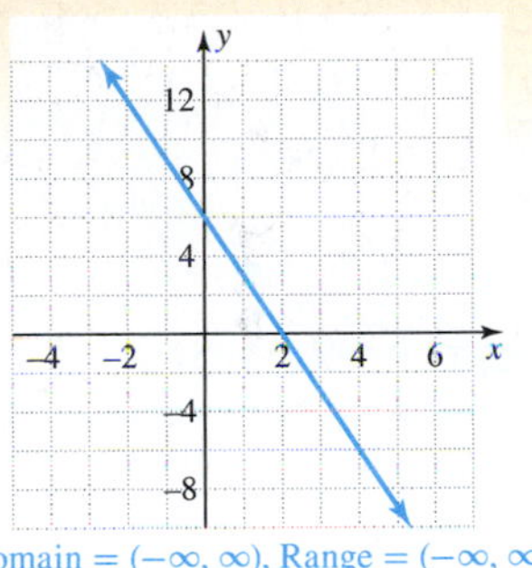

Domain = (−∞, ∞), Range = (−∞, ∞)

13.

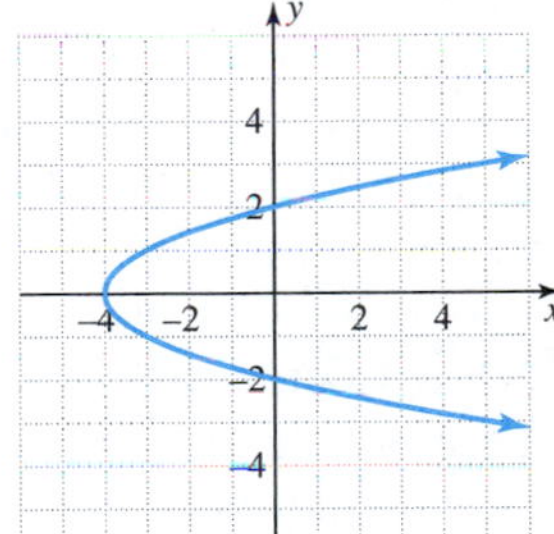

Domain = [−4, ∞), Range = (−∞, ∞)

Find the indicated value for each relation. (*See Section 3.2, Objective 3.*)

14. Find the output value that corresponds to the input value of -2 for the relation $\{(-2, -1), (-1, 3), (0, 5), (6, -4)\}$. $y = -1$

15. Find the input value that corresponds to the output value of 5 for the relation $\{(0, 10), (2, -16), (5, 13), (-16, 5)\}$. $x = -16$

SECTION 3.3 Functions

OBJECTIVES

As a result of completing this section, you will be able to

1. Determine if a relation is a function.
2. Use the vertical line test.
3. Use function notation.
4. Apply functions to real life.
5. Troubleshoot common errors.

According to the U.S. Department of Labor, the fastest-growing occupation for the years 2006–2016 is a network system and data communications analyst. The function $f(x) = 1343x + 45{,}657$ approximates the median salary for x years of experience. (Source: http://www.bls.gov and http://www.payscale.com)

In this section, we will learn the definition of a function and how to use functions to obtain certain information.

Functions

Objective 1

Determine if a relation is a function.

In Section 3.2, we defined a relation as a set of ordered pairs, a correspondence between a set of input values and output values. A *function* is a special type of relation.

Definition: A **function** is a relation in which each member of the domain (or each input) corresponds to *exactly one* member of the range (or an output). In other words, each input, or x-value, can have only one output, or y-value.

INSTRUCTOR NOTE: Introduce the concept of a function using a relation that works "properly." The concept of a cell phone and a number assigned to it is referenced in the text. Have students come up with other real-life examples.

INSTRUCTOR NOTE: Remind students that a function is a special type of relation. All functions are relations but not all relations are functions.

To introduce this concept, consider cell phone numbers and the person assigned to the number. If we think of the phone number as the *input* and the person assigned to the number as the *output*, we have a relation with ordered pairs of the form (number, person). This relation satisfies the definition of a function since each input value corresponds to only one output value. That is, each x-value corresponds to only one y-value. There would be complete chaos if a cell phone number was assigned to more than one person!

Since a function is a special relation, functions can also be expressed as a set of ordered pairs, a mapping, an equation, a table, or a graph. Consider the set of ordered pairs $\{(1, 3), (2, 4), (3, 4), (4, 6)\}$. We can use a mapping to visualize that each x-value corresponds to exactly one y-value. Therefore, the mapping is a function.

Now consider the following table. The input value of 12 corresponds to two different output values—Aisha and Hussein. Therefore, this relation is *not* a function.

Credit Hours	Student
12	Aisha
6	Juan
12	Hussein
15	Rose

Finally, we will consider a relation defined by the equation, $y = 5x + 2$. Note that each input value is multiplied by 5 and then added to 2, so there is only one possible output value. So, this equation is a function.

Note: *If there is only one corresponding y-value for each x-value, then the relation is a function. If at least one x-value corresponds to more than one y-value, then the relation is not a function.*

Objective 1 Examples Determine if each relation is a function. If not, explain why.

1a. $\{(-4, 0), (-3, 2), (-3, -2), (0, 4), (0, -4)\}$

1b. $\{(-2, -3), (-1, -3), (0, -3), (1, -3), (2, -3), (3, 3)\}$

1c. Let the relation be defined by the given table, where x is the math grade and y is the student earning the grade.

Grade	Student
A	Dylan, Hunter
B	Angela, Edward
C	Aretha, Ginger
D	Jake
F	Natalia

1d. $x = y^2$

Solutions

1a.

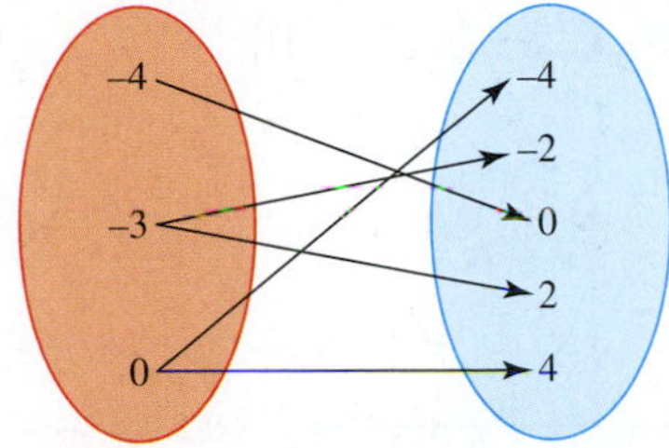

Write the relation as a mapping.

The x-value of -3 corresponds to 2 y-values, 2 and -2.

The x-value of 0 correspond to 2 y-values, -4 and 4.

So, the relation is not a function.

1b.

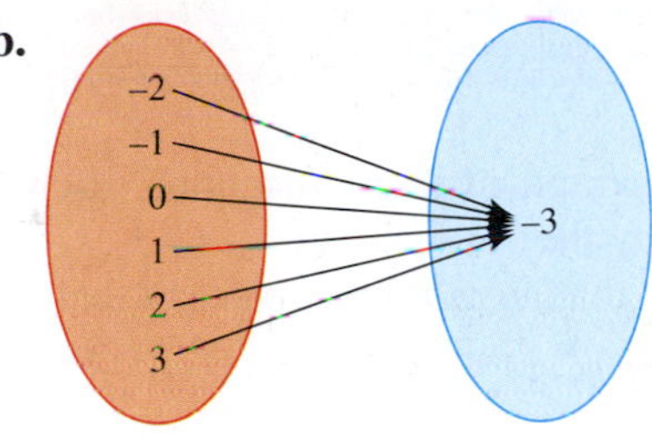

Write the relation as a mapping.

Each x-value corresponds to only 1 y-value

So, the relation is a function.

1c. The relation is *not* a function since the grades of A, B, and C each correspond to more than 1 student. For example, the grade of A corresponds to Dylan and Hunter. The grade of B corresponds to Angela and Edward. The grade of C corresponds to Aretha and Ginger.

1d. We must determine how many output values correspond to a given input.

$$x = y^2$$

$$4 = y^2 \qquad \text{Let } x = 4 \text{, for example.}$$

$$y = 2 \quad \text{or} \quad y = -2 \qquad \text{Note that } 4 = (2)^2 \text{ and } 4 = (-2)^2.$$

Since $(4, 2)$ and $(4, -2)$ are solutions of this equation, there is an input value that corresponds to 2 output values. So, this equation is *not* a function.

Student Check 1 Determine if each relation is a function. If not, explain why.

a. $\{(5, 1), (5, -2), (5, 6), (5, 3), (5, 0)\}$

b. $\{(1, 5), (-2, 5), (6, 5), (3, 5), (0, 5)\}$

c. Let the relation be defined by the given mapping, where x is the student and y is the grade earned in math class.

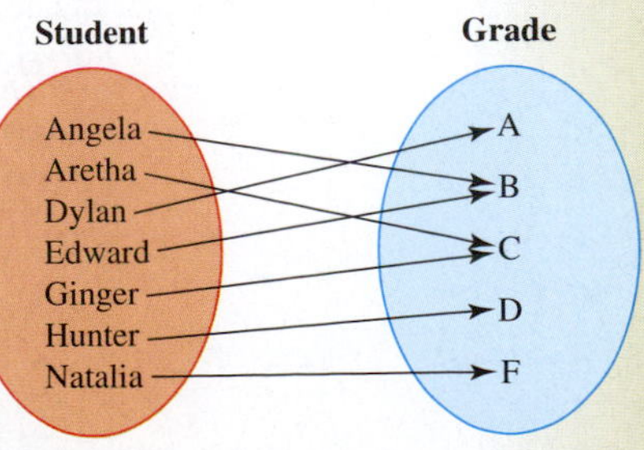

d. $y = x^2$

Note: *Example 1 illustrates two important facts.*
1. *When the x-values of a relation repeat, the relation is not a function.*
2. *The y-values of a function can repeat.*

The Vertical Line Test

Objective 2 ▶
Use the vertical line test.

Now we will examine graphs of relations and learn how we can determine if they represent a function. Some examples of functions and their corresponding graphs are shown.

$\{(-2, -3), (-1, -3), (0, -3), (1, -3), (2, -3), (3, -3)\}$ $\qquad$ $y = |x|$

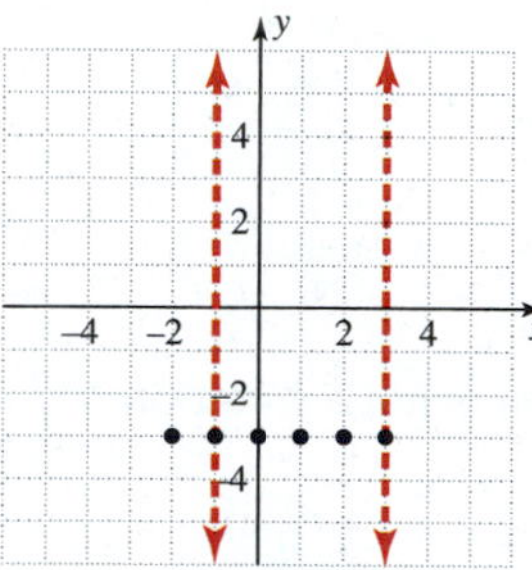

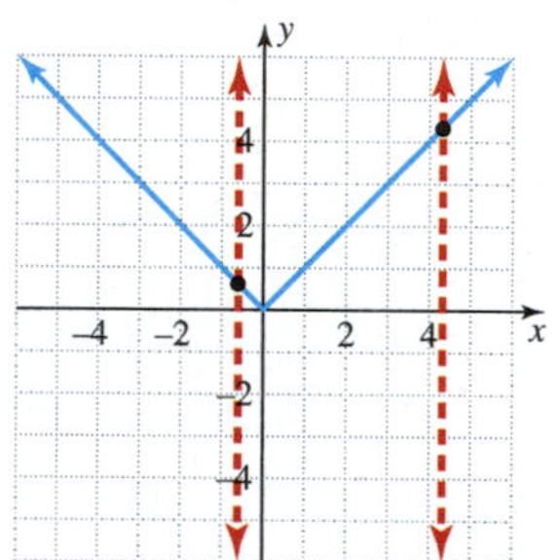

Notice that in the graph of these functions, each x-value corresponds to only one y-value, as shown by the vertical lines.

Some relations that are *not* functions and their corresponding graphs are shown.

$\{(-4, 0), (-3, 2), (-3, -2), (0, 4), (0, -4)\}$ $\qquad$ $x = y^2$

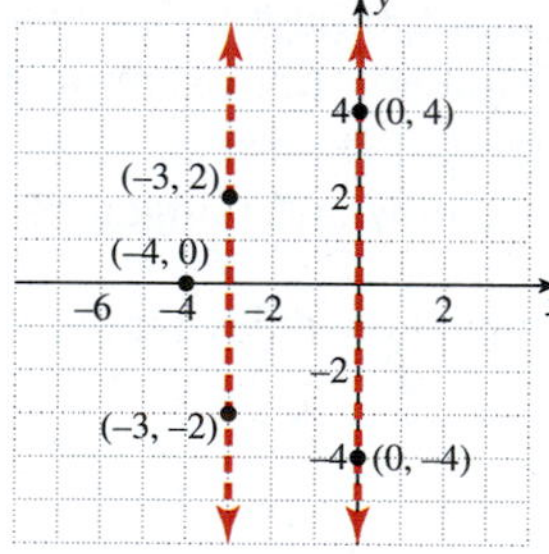

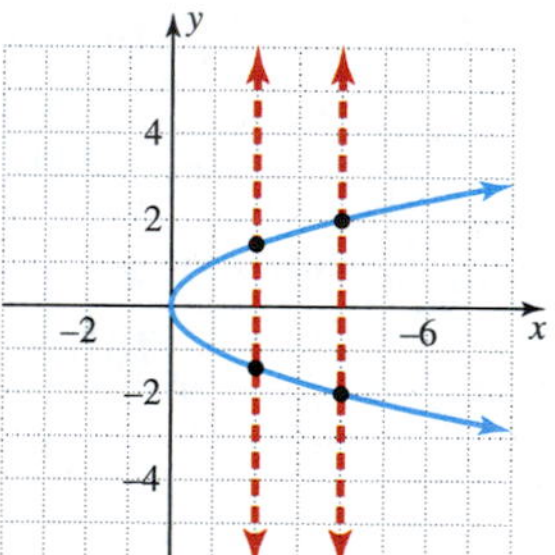

Notice that in these graphs that are not functions, there are values of x that have more than one corresponding y-value, as shown by the vertical lines.

We can use the facts shown in these illustrations to determine if a relation is a function by examining its graph. We can determine if a graph represents a function by performing the **vertical line test**.

Procedure: Using the Vertical Line Test

Step 1: Draw vertical lines through the graph of the relation.

Step 2: Determine how many times each vertical line intersects the graph of the relation.

- **a.** If each vertical line intersects the graph in at most one point, then the graph is a function.
- **b.** If at least one vertical line intersects the graph in more than one point, then the graph is *not* a function.

Objective 2 Examples **Use the vertical line test to determine if each relation is a function.**

2a. 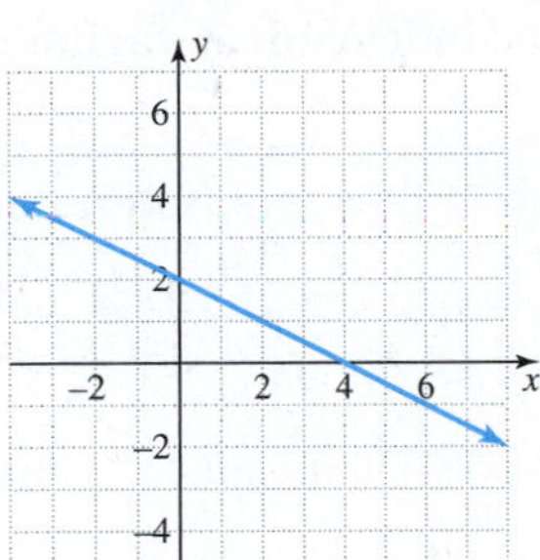

2b. 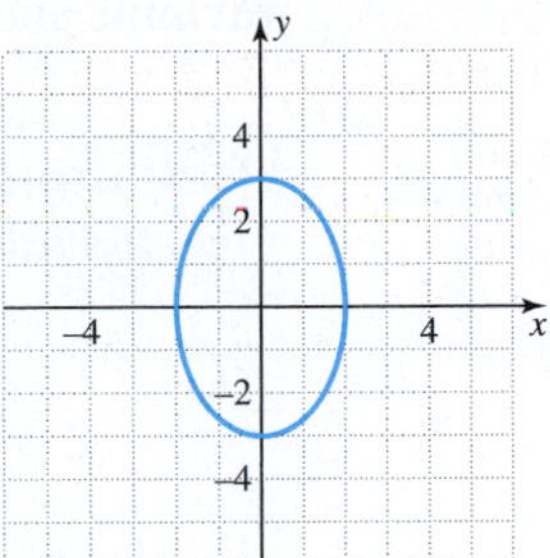

Solutions **2a.**

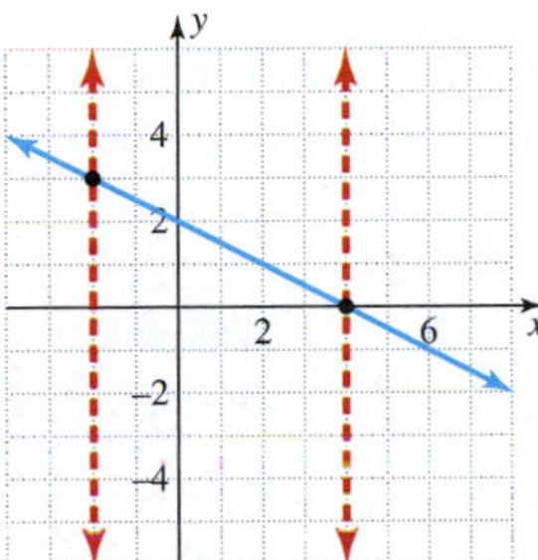

Draw vertical lines through the graph.

Each vertical line intersects the graph in at most one point.

So, the graph is a function

2b.

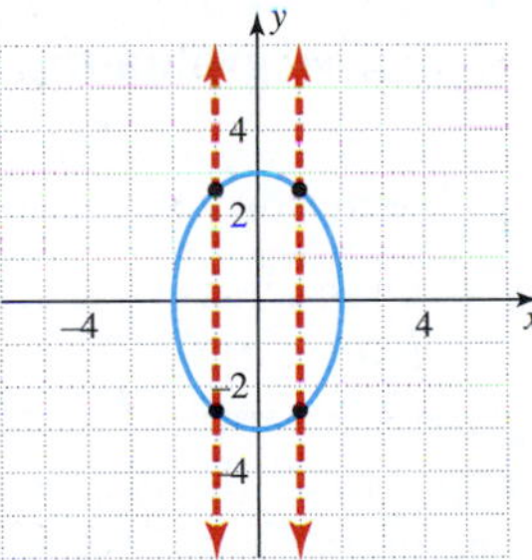

Draw vertical lines through the graph.

There is at least one vertical line that intersects the graph at more than one point.

So, the graph is not a function

Student Check 2 Use the vertical line test to determine if each relation is a function.

a. 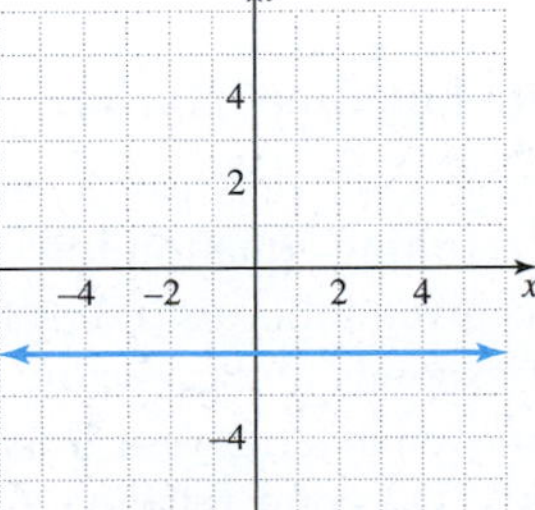

b. 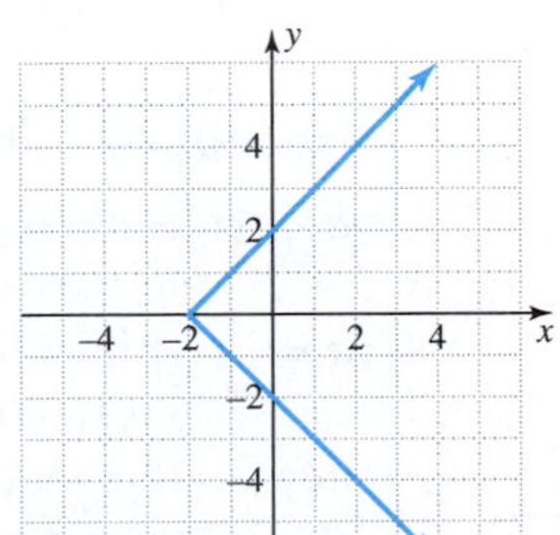

Function Notation

Objective 3

Use function notation.

In Objective 1 of this section, we learned that $y = 5x + 2$ is a function since each input value, x, corresponds to only one output value, y. We can use a special notation, called *function notation*, to indicate that an equation is a function. In function notation,

$$y = 5x + 2 \text{ is written as } f(x) = 5x + 2$$

In mathematics, we use letters such as f, g, and h to name functions. The symbol $f(x)$ is read "f of x" and means "the function of x." Function notation means that y is a function of x. In other words, y depends on the value of x. As a result, y is called the **dependent variable** and x is called the **independent variable**.

INSTRUCTOR NOTE: Point out that $f(x)$ is read "f of x" and is not f times x. Also, be sure to state that $f(x)$ and y are interchangeable for functions.

Definition: The notation $f(x)$ is **function notation** and denotes that y is a function of x.

$$y = f(x)$$

Name of function

Output value

Input value

In this notation, x is the input value and $f(x)$, or y, is the output value.

We can also use function notation to make the process of finding input and output values easier. For instance, if we want to find the value of $y = 5x + 2$ when $x = 3$, we do the following.

$f(x) = 5x + 2$	Write the equation with function notation.
$f(3) = 5(3) + 2$	Replace x with 3.
$f(3) = 15 + 2$	Multiply.
$f(3) = 17$	Add.

So, $f(3) = 17$ means that when $x = 3$, $y = 17$. It also means that the ordered pair $(3, 17)$ is a solution of the equation $y = 5x + 2$. The process of finding an output value, y, for a given input value, x, is called **evaluating a function**.

Procedure: Evaluating a Function $f(x)$ at the Value k

Step 1: If the function is given in terms of an equation, replace the variable with the given number k and simplify.

Step 2: If the function is given in terms of a set, a mapping, or a table, find the y-value that corresponds to the x-value of k.

Step 3: If the function is given in terms of a graph, find the ordered pair on the graph whose x-value is k. The corresponding y-value is $f(k)$.

We can also use function notation to solve an equation of the form $f(x) = h$, where h is a real number. For example, if $f(x) = 5x + 2$, to solve $f(x) = 7$ means to find the solution of $5x + 2 = 7$. Solving this equation gives us $x = 1$. So, $f(x) = 7$ when $x = 1$.

Procedure: Solving the Equation $f(x) = h$, Where h Is a Real Number

Step 1: If the function is given in terms of an equation, replace $f(x)$ with the value h and solve the resulting equation for x.

Step 2: If the function is given in terms of a set, a mapping, or a table, find the x-value that corresponds to the y-value of h.

Step 3: If the function is given in terms of a graph, find the ordered pair on the graph whose y-value is h. The corresponding x-value is the solution of $f(x) = h$.

Note that there may be more than 1 solution of $f(x) = h$.

Objective 3 Examples **Find the requested information.**

3a. Find $g(-3)$ if $g(x) = x^2 + 1$.

3b. Let $f(x) = 3x - 12$. Find $f(0)$ and find x such that $f(x) = 0$.

3c. Let $c(x)$ be given by Y_1. Find $c(4)$ and find x such that $c(x) = -6$.

3d. Let $f(x)$ be given by the graph. Find $f(2)$ and find x such that $f(x) = 1$.

X	Y1
-1	4.5
0	0
1	-3.5
2	-6
3	-7.5
4	-8
5	-7.5

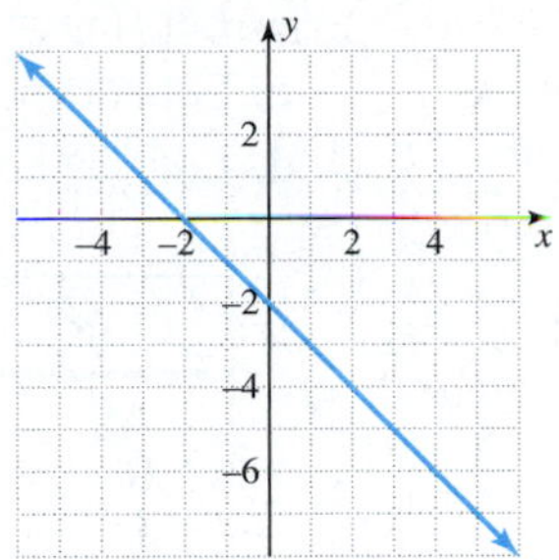

Solutions

INSTRUCTOR NOTE:
Remind students that $y = f(x)$. Stress the importance of using parentheses in place of the variable first and then inserting the given value.

3a.

$g(x) = x^2 + 1$

$g(-3) = (-3)^2 + 1$ — Replace x with -3.

$g(-3) = 9 + 1$ — Simplify the exponent.

$g(-3) = 10$ — Add.

So, $g(-3) = 10$.

3b.

$f(x) = 3x - 12$

$f(0) = 3(0) - 12$ — Replace x with 0.

$f(0) = 0 - 12$ — Multiply.

$f(0) = -12$ — Subtract.

So, $f(0) = -12$.

Now to solve $f(x) = 0$, replace $f(x)$ with 0.

$f(x) = 3x - 12$

$0 = 3x - 12$ — Replace $f(x)$ with 0.

$0 + 12 = 3x - 12 + 12$ — Add 12 to each side.

$12 = 3x$ — Simplify.

$\frac{12}{3} = \frac{3x}{3}$ — Divide each side by 3.

$4 = x$ — Simplify.

So, $f(x) = 0$ when $x = 4$.

3c. To find $c(4)$, we must find the point in the table whose x-value is 4. The point $(4, -8)$ is one of the ordered pairs in the table. So, $c(4) = -8$.

To solve $c(x) = -6$, we must find the point(s) in the table whose Y_1-value is -6. The point $(2, -6)$ is in the table. So, $c(x) = -6$ when $x = 2$.

3d. To find $f(2)$, we need to find the point on the graph whose x-value is 2. The point $(2, -4)$ lies on the graph of the function. So, $f(2) = -4$.

To solve $f(x) = 1$, we need to find the point on the graph whose y-value is 1. The point $(-3, 1)$ lies on the graph. So, $f(x) = 1$ when $x = -3$.

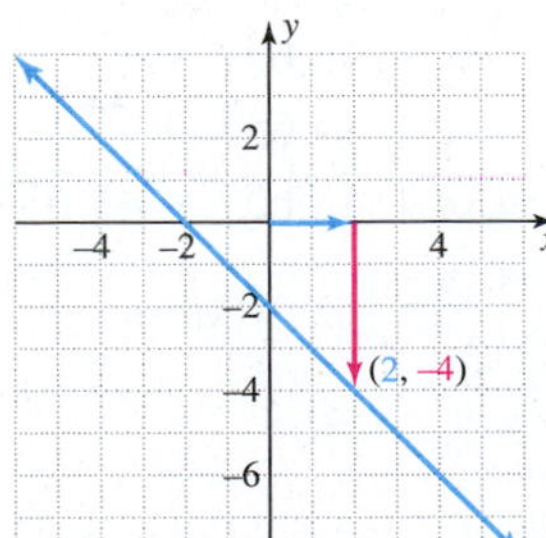

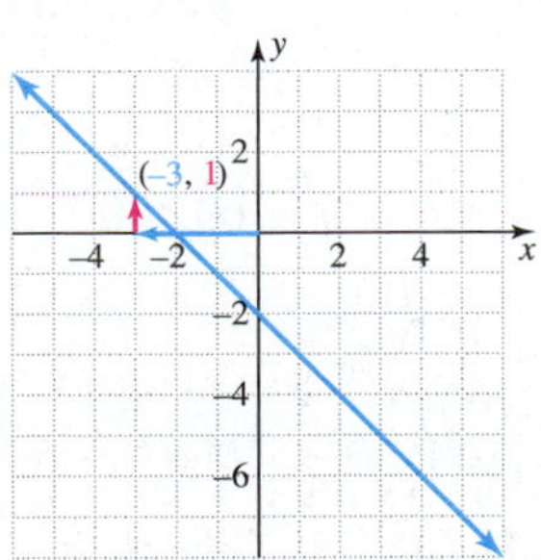

Student Check 3 Find the requested information.

a. Find $g(-5)$ if $g(x) = -2x^2 - 4x + 1$.

b. Let $f(x) = 9x + 4$. Find $f(0)$ and find x such that $f(x) = 0$.

c. Let $c(x)$ be given by Y_1. Find $c(2)$ and find x such that $c(x) = 4.5$.

d. Let $f(x)$ be given by the graph. Find $f(-2)$ and find x such that $f(x) = -3$.

X	Y1
-1	4.5
0	0
1	-3.5
2	-6
3	-7.5
4	-8
5	-7.5

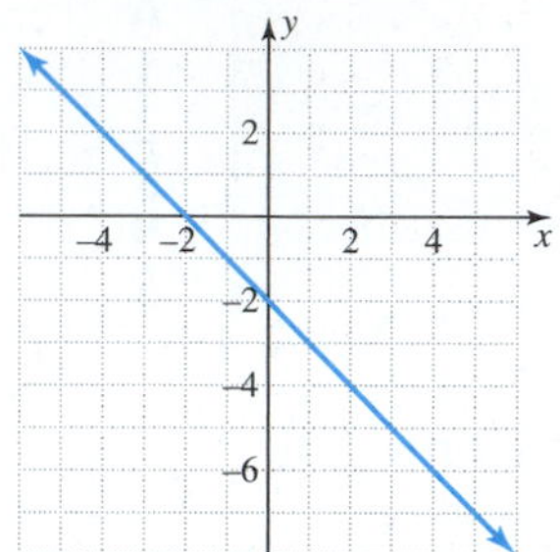

Applications

Objective 4 Apply functions to real life.

We encounter functions every day of our lives. Every time we go to the grocery store and an item is scanned, a price appears on the register. This function pairs together an item with its price. Have you ever registered to use a password-protected software program? Have you ever entered a login name only to receive the message that the name you entered has already been used? This is because a login name must be assigned to only one person. This function pairs together login name to the person. There are many other examples of functions from the real world. In Example 4, we will see a function that relates to a real-world situation.

Objective 4 Examples According to the U.S. Department of Labor, the fastest-growing occupation for the years 2006–2016 is a network system and data communications analyst. The function $f(x) = 1343x + 45{,}657$ models the median salary in the United States for x years of experience. (Source: http://www.bls.gov and http://www.payscale.com)

4a. Find the median salary with 5 yr of experience.

4b. How many years of experience is required for the median salary to be $60,430?

Solutions

4a. The input, x, represents the number of years of experience and $f(x)$, or y, represents the median salary. To find the salary with 5 yr of experience, we let $x = 5$.

$$f(x) = 1343x + 45{,}657$$

$$f(5) = 1343(5) + 45{,}657 \quad \text{Replace } x \text{ with 5.}$$

$$f(5) = 6715 + 45{,}657 \quad \text{Multiply.}$$

$$f(5) = 52{,}372 \quad \text{Add.}$$

So, the median salary is about $52,372 with 5 yr of experience.

4b. To find how many years of experience is required for the salary to be $60,430, we set $f(x) = 60{,}430$ and solve for x.

$$f(x) = 1343x + 45{,}657$$

$$60{,}430 = 1343x + 45{,}657 \quad \text{Replace } f(x) \text{ with 60,430.}$$

$$60{,}430 - 45{,}657 = 1343x + 45{,}657 - 45{,}657 \quad \text{Subtract 45,657 from each side.}$$

$$14{,}773 = 1343x \quad \text{Simplify.}$$

$$\frac{14{,}773}{1343} = \frac{1343x}{1343} \quad \text{Divide each side by 1343.}$$

$$11 = x \quad \text{Simplify.}$$

So, the median salary is about $60,430 with 11 yr of experience.

Student Check 4 The number of cell phone subscribers in the United States, in millions, between 1985–2008 can be approximated by $f(x) = 0.6501x^2 - 3.0122x + 2.258$, where x is the number of years after 1985. Use the function to determine the number of cell phone subscribers in the United States in the year 2008. (Source: http://www.infoplease.com)

Objective 5 ▶

Troubleshoot common errors.

Troubleshooting Common Errors

Some common errors associated with functions are shown.

Objective 5 Examples

A problem and an incorrect solution are given. Provide the correct solution and an explanation of the error.

5a. Determine if the relation $\{(-2, 4), (-1, 1), (0, 0), (1, 1), (2, 4)\}$ is a function.

Incorrect Solution	Correct Solution and Explanation
The relation is not a function because -2 and 2 correspond to the same y-value.	In a function, two x-values can correspond to the same y-value as long as each x-value corresponds to only 1 y-value. So, this relation is a function.

5b. Find $f(3)$ if $f(x) = 4x + 5$.

Incorrect Solution	Correct Solution and Explanation
$f(3) = (4x + 5)(3)$ $f(3) = 12x + 15$	$f(3)$ does not mean multiplication. It means to find the output value for $x = 3$. $f(3) = 4(3) + 5$ $f(3) = 12 + 5$ $f(3) = 17$

ANSWERS TO STUDENT CHECKS

Student Check 1 **a.** not a function **b.** function **c.** function **d.** function

Student Check 2 **a.** function **b.** not a function

Student Check 3 **a.** $f(0) = 4$ **b.** $g(-5) = -29$ **c.** $c(2) = -6$ **d.** $f(-2) = 0$ **e.** $x = -\frac{4}{9}$ **f.** $x = -1$ **g.** $x = 1$

Student Check 4 about 276.88 million

SUMMARY OF KEY CONCEPTS

1. A function is a relation in which each input can have only one output, that is, each x-value corresponds to exactly one y-value. The x-values of the function cannot occur more than once with different y-values if the relation is to be called a function.
2. The vertical line test can be used to determine if a graph represents a function. If any vertical line intersects a graph in more than one point, the graph does *not* represent a function.
3. Function notation, $f(x)$, is another name for the output value y. The notation $f(a) = b$ corresponds to the ordered pair (a, b). If the value of $f(a)$ is unknown, we can evaluate the function at $x = a$ to determine it. In an equation, we do this by substituting the number a for the variable in the function f. In a table or graph, we find the ordered pair in the table or on the graph whose x-value is a. The y-value of this ordered pair is $f(a)$.
4. Functions occur in many real-world situations. It is important to understand from the problem what the input and output of the function represent. This enables us to find the requested information.

GRAPHING CALCULATOR SKILLS

The graphing calculator can assist us in evaluating functions when the function is given in terms of an equation.

Example: Find $f(0)$ if $f(x) = |x - 4|$.

Solution: Input the equation into the equation editor.

```
Plot1 Plot2 Plot3
\Y1=abs(X-4)
\Y2=
\Y3=
\Y4=
\Y5=
\Y6=
\Y7=
```

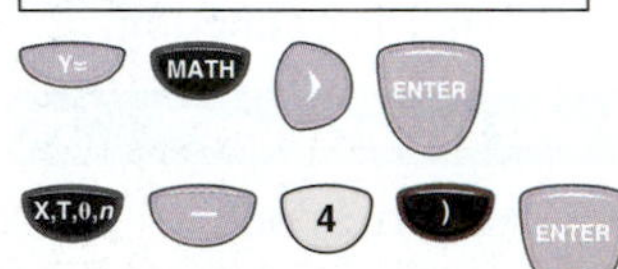

Method 1: Use the TABLE feature to find the y-value.

X	Y1
-2	6
-1	5
0	4
1	3
2	2
3	1
4	0

X=-2

From the table, we see that $Y_1 = 4$ when $x = 0$, so $f(0) = 4$.

Method 2: Graph the equation. Press TRACE.

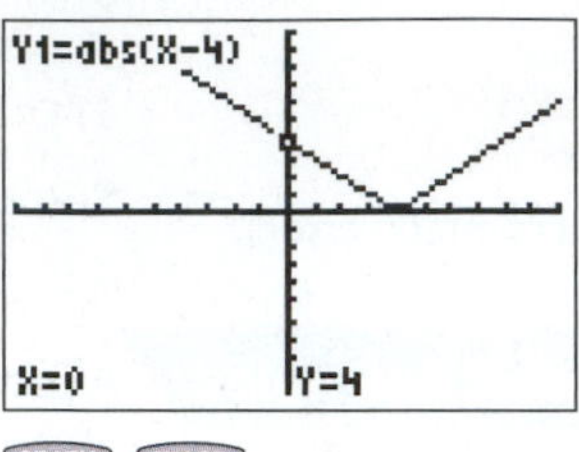

GRAPH TRACE

The display at the bottom shows $y = 4$ at $x = 0$. So, $f(0) = 4$.

Note: *The initial value displayed for* x *when we press TRACE is* $x = 0$. *If we need to evaluate a function at a different value, then we enter this number and press ENTER.*

SECTION 3.3 EXERCISE SET

Write About It!

Use complete sentences in your answer to each exercise.

1. Is every relation a function? Explain.
2. Is every function a relation? Explain.
3. Explain the vertical line test.
4. What is function notation?

Practice Makes Perfect!

Determine if each relation is a function. If not, explain why. (*See Objective 1.*)

5. $\{(3, 1), (5, 2), (9, 7), (10, 7)\}$ function
6. $\{(-1, 6), (-1, 3), (4, -3), (5, -10)\}$ not a function
7. $\{(-7, 12), (-6, -9), (-6, 10), (1, 5), (2, 3)\}$ not a function
8. $\{(-8, 4), (-3.6, 7), (-2.5, 7), (0, 5)\}$ function
9. Cala polls her study group to find out the average number of hours of television watched by each member daily. The results are given in the mapping. function

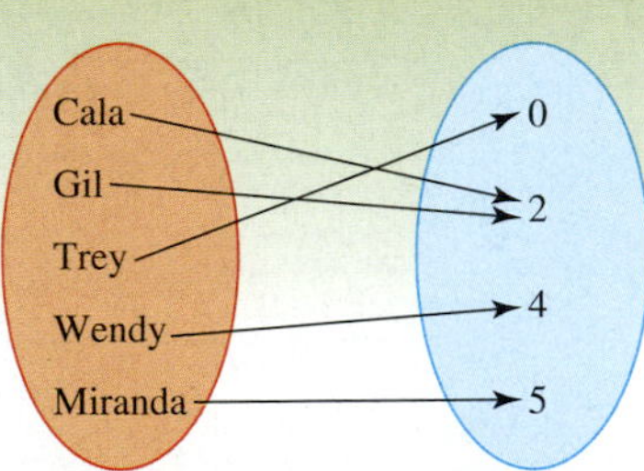

10. Christian's study group shares the types of cars they drive. The results are given in the mapping.

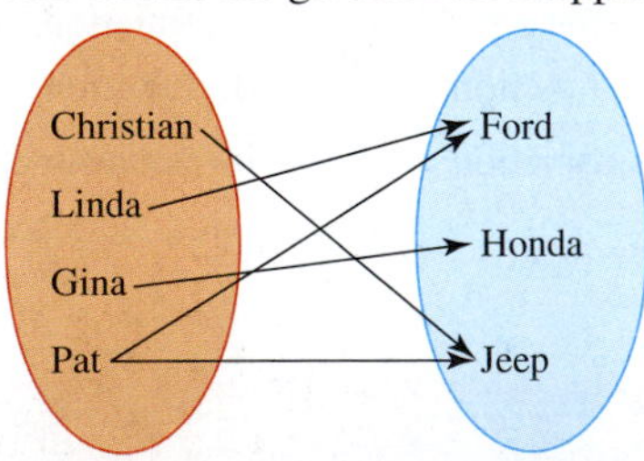

not a function; Pat corresponds to both Ford and Jeep.

11. Dominique and her friends share their 2008 U.S. Presidential election votes. The results are given in the mapping.

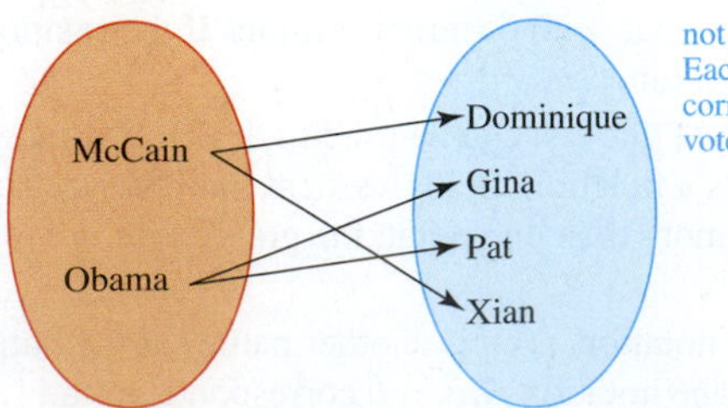

not a function; Each candidate corresponds to two voters.

12. Several students complete a poll about the type of computer they own. The results are given in the mapping.

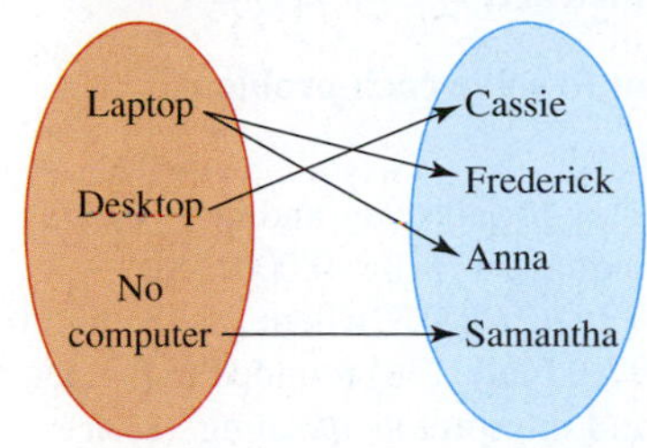

not a function; Laptop corresponds to Frederick and Anna.

13. $y = x + 3$ function
14. $y = 2x + 1$ function
15. $y = 1 - x^2$ function
16. $y = 3 + x^2$ function
17. $x = y^2 + 1$ not a function
18. $x = y^2 - 1$ not a function
19. $x = |y + 3|$ not a function
20. $x = |y - 2|$ not a function
21. $x = 2y$ function
22. $x = y + 5$ function
23. $y = (x - 9)^2$ function
24. $y = (x + 4)^2$ function
25. $y = |x + 6|$ function
26. $y = |x - 4|$ function

Use the vertical line test to determine if each relation is a function. (*See Objective 2.*)

27. function

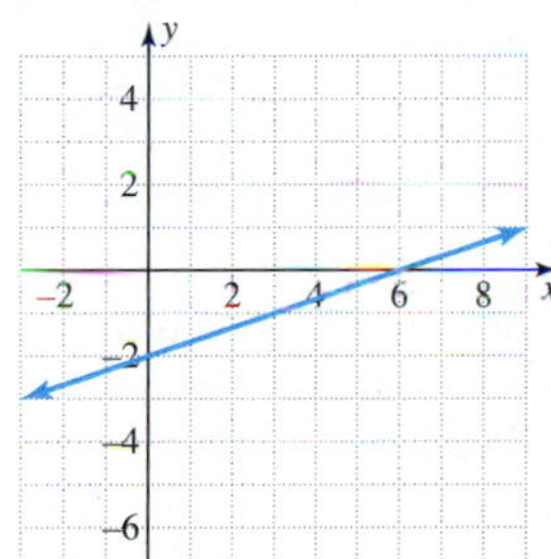

28. function

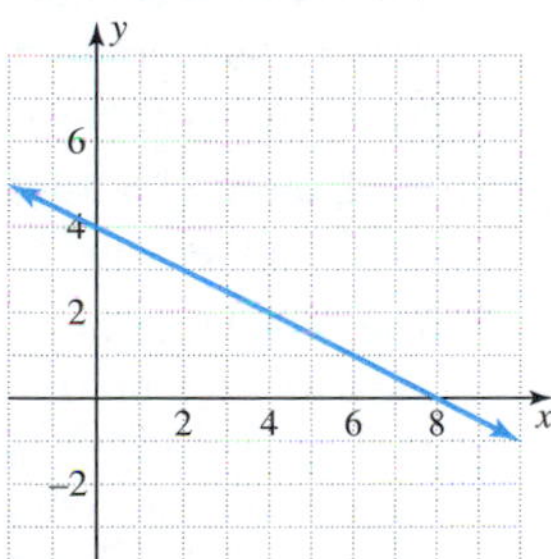

29. function

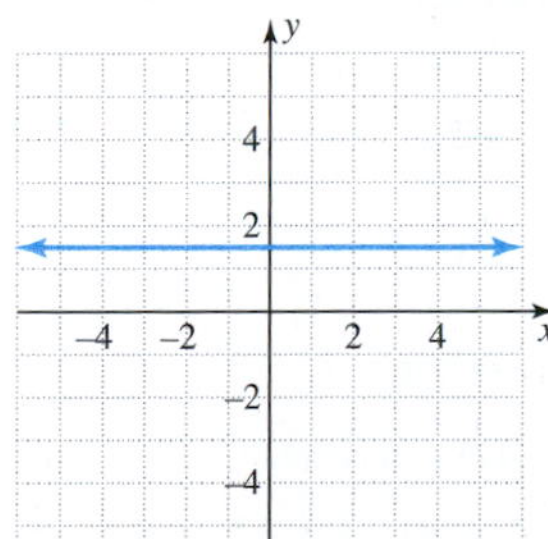

30. not a function

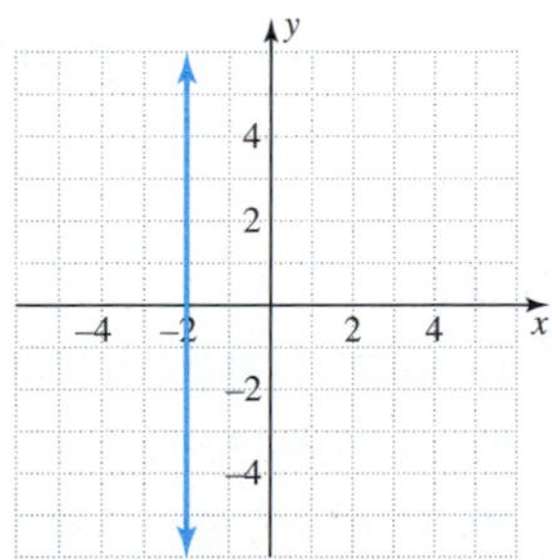

31. not a function

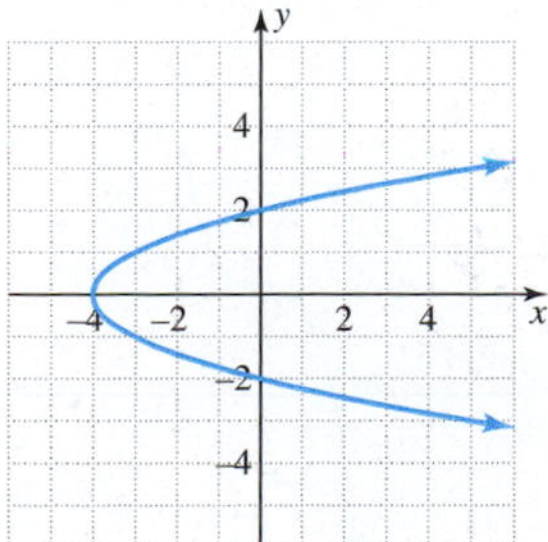

32. function

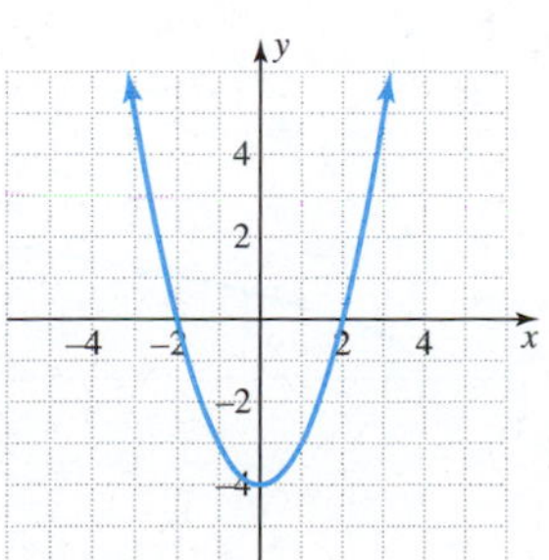

33. function

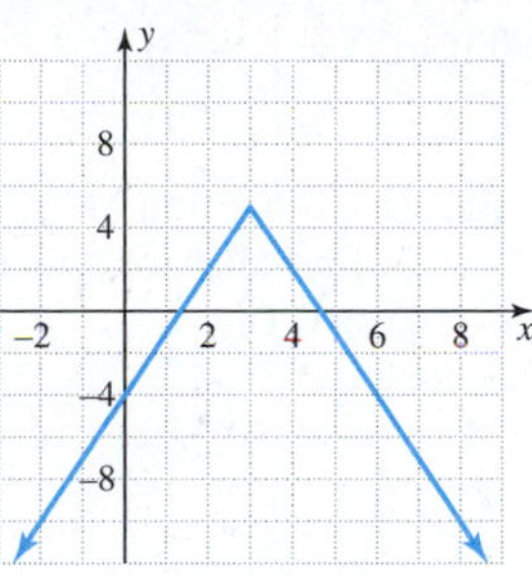

34. function

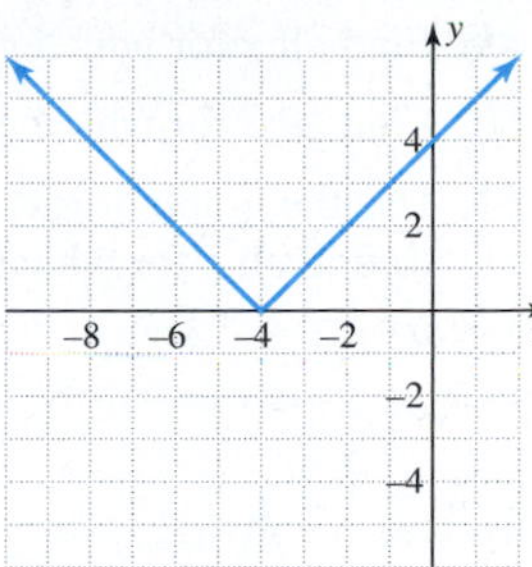

35. not a function

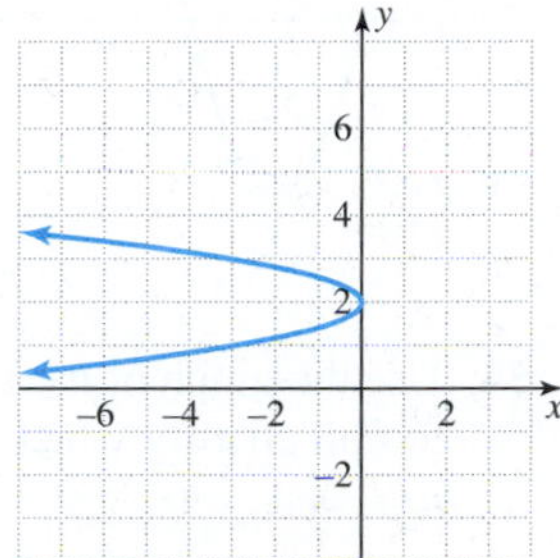

36. function

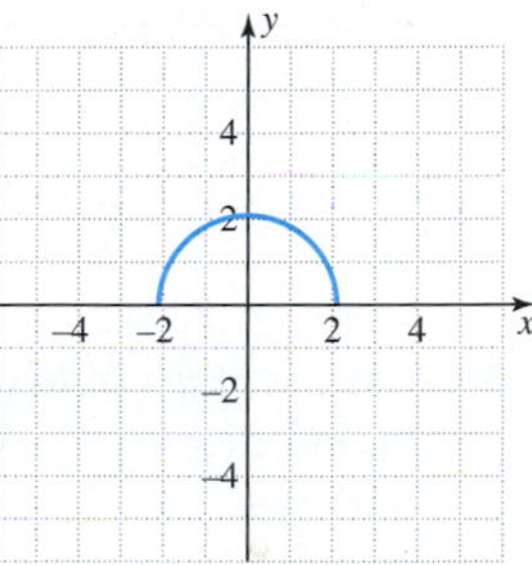

Find the requested information. (*See Objective 3.*)

37. Find $f(-1)$ if $f(x) = 5x - 2$. -7
38. Find $f(-5)$ if $f(x) = 6x - 1$. -31
39. Find $h(-7)$ if $h(x) = -x^2$. -49
40. Find $h(3)$ if $h(x) = 2x^2 + 9$. 27
41. Find $h(-10)$ if $h(x) = |x - 10|$. 20
42. Find $h(-9)$ if $h(x) = |3x + 20|$. 7
43. Find $f(3)$ from the graph of $f(x)$. 5

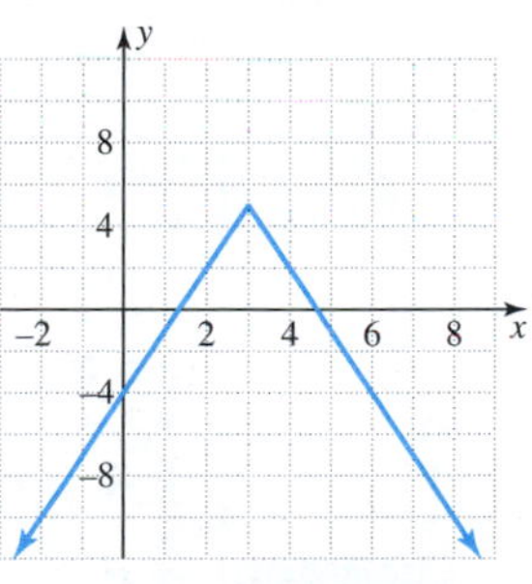

44. Find $g(6)$ from the graph of $g(x)$. 0

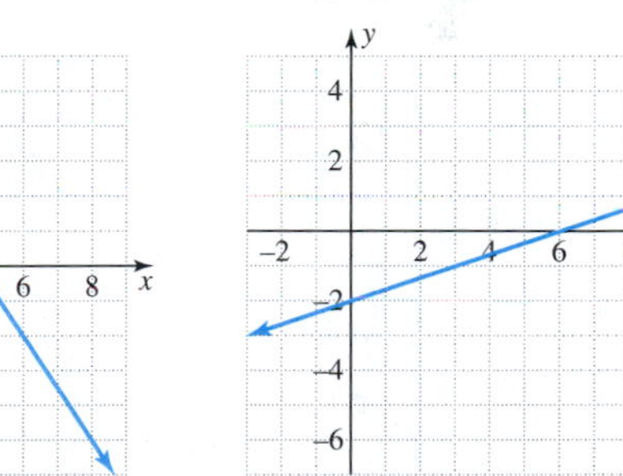

45. Find $f(-2)$ from the graph of $f(x)$. 0

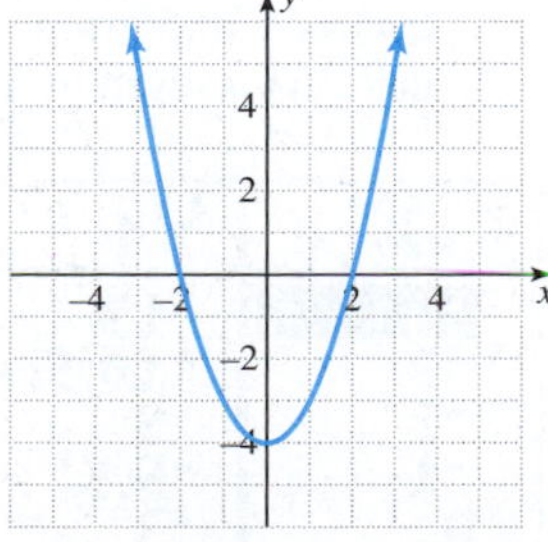

46. Find $g(0)$ from the graph of $g(x)$. 4

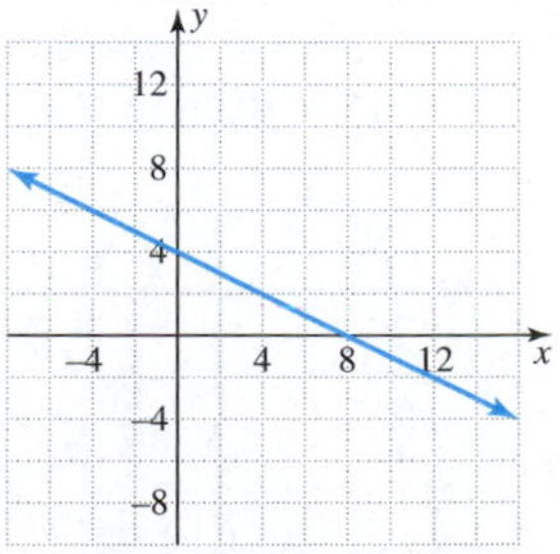

47. Find all x for which $f(x) = 3$ if $f(x) = 4x + 7$. -1

48. Find all x for which $f(x) = -2$ if $f(x) = -12x + 5$. $\frac{7}{12}$

49. Find all x for which $g(x) = 9$ if $g(x) = |x + 3|$. 6, −12

50. Find all x for which $g(x) = 8$ if $g(x) = |x + 1|$. 7, −9

51. Use the graph of $f(x)$ to find all x for which $f(x) = 2$. 2, 4

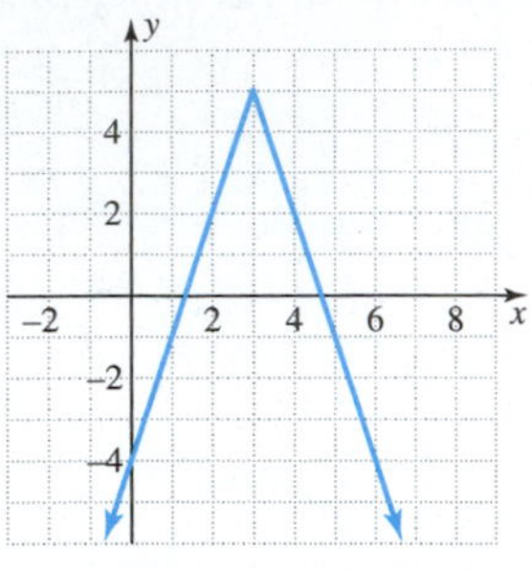

52. Use the graph of $f(x)$ to find all x for which $f(x) = 0$. −2, 2

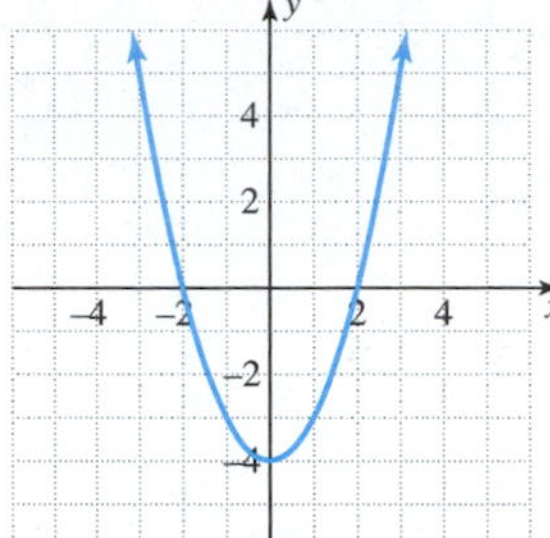

53. Use the graph of $h(x)$ to find all x for which $h(x) = 4$. 0

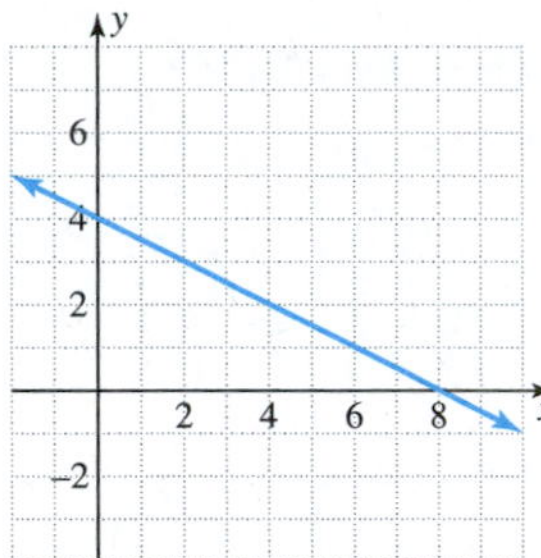

54. Use the graph of $h(x)$ to find all x for which $h(x) = 1$. 9

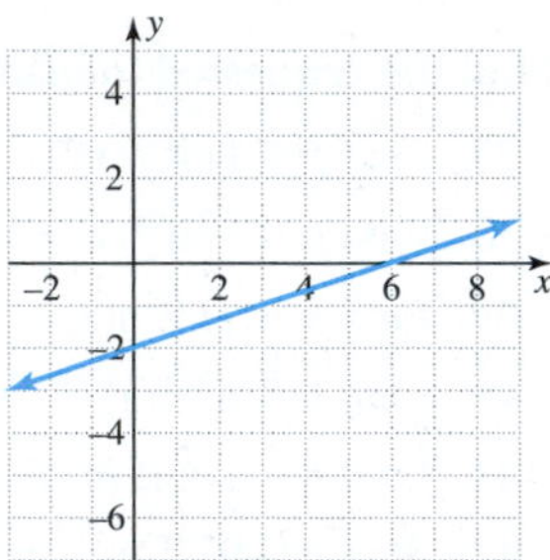

55. Find $f(-1)$ if the function $f(x)$ is given by Y_1. 1.5

X	Y1
-3	10.5
-2	5
-1	1.5
0	0
1	.5
2	3
3	7.5

56. Find $f(-2)$ if the function $f(x)$ is given by Y_1. −9

X	Y1
-3	-12.5
-2	-9
-1	-5.5
0	-2
1	1.5
2	5
3	8.5

57. Find all x for which $f(x) = -2$ given that $f(x)$ is given by Y_1. 0

X	Y1
-3	-12.5
-2	-9
-1	-5.5
0	-2
1	1.5
2	5
3	8.5

58. Find all x for which $c(x) = -10.5$ given that $c(x)$ is given by Y_1. −3

X	Y1
-4	-13
-3	-10.5
-2	-8
-1	-5.5
0	-3
1	-.5
2	2

59. Find all x for which $k(x) = 16$ if $k(x) = x^2$. −4, 4

60. Find all x for which $k(x) = 25$ if $k(x) = x^2$. −5, 5

Use the given function to solve each problem. (*See Objective 4.*)

61. The life expectancy for all sexes and races can be modeled by the equation $f(x) = 0.000158x^3 - 0.0104x^2 + 0.372x + 70.813$, where x is the number of years after 1970. Find $f(30)$ rounded to the nearest whole number and interpret its meaning. (Source: National Center for Health Statistics)

62. For Exercise 61, find $f(15)$ rounded to the nearest whole number and interpret its meaning.

63. The number of twin births (in thousands) in the United States can be modeled by the equation $f(x) = 0.0204x^2 + 1.906x + 67.8596$, where x is the number of years after 1980. Find $f(39)$ rounded to the nearest whole number and interpret its meaning. (Source: http://www.cdc.gov)

64. For Exercise 63, find $f(20)$ rounded to the nearest whole number and interpret its meaning.

65. The percentage of middle school students who currently use a tobacco product can be modeled by the equation, $m(x) = -0.92x + 15.17$, where x is the number of years after 2000. Find $m(10)$ rounded to the nearest whole number and interpret its meaning. (Source: National Youth Tobacco Survey)

66. For Exercise 65, find $m(12)$ rounded to the nearest whole number and interpret its meaning.

67. The percentage of high school students who currently use a tobacco product can be modeled by the equation $h(x) = -0.14x^3 + 1.5x^2 - 5.6x + 34.5$, where x is the number of years after 2000. Find $h(4)$ rounded to the nearest whole number and interpret its meaning. (Source: National Youth Tobacco Survey)

68. For Exercise 67, find $h(8)$ rounded to the nearest whole number and interpret its meaning.

Mix 'Em Up!

Determine if each relation is a function.

69. $\{(-9, 4), (-5, 4), (0, 4), (1, 4)\}$ function

70. $\{(-2, 15), (0, -15), (2, 9), (3, -8), (3, 9)\}$ not a function

71. function

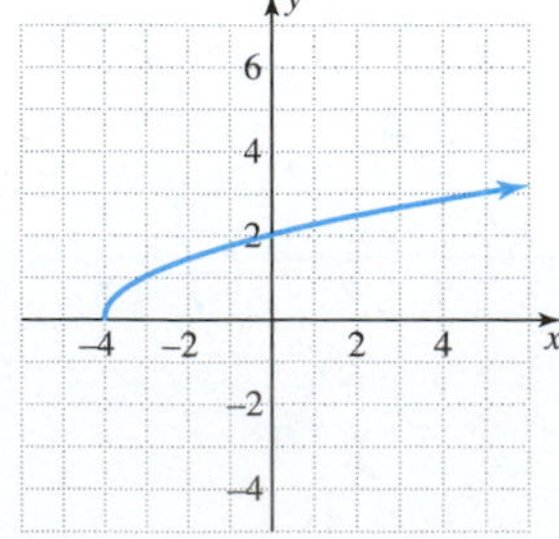

72. function

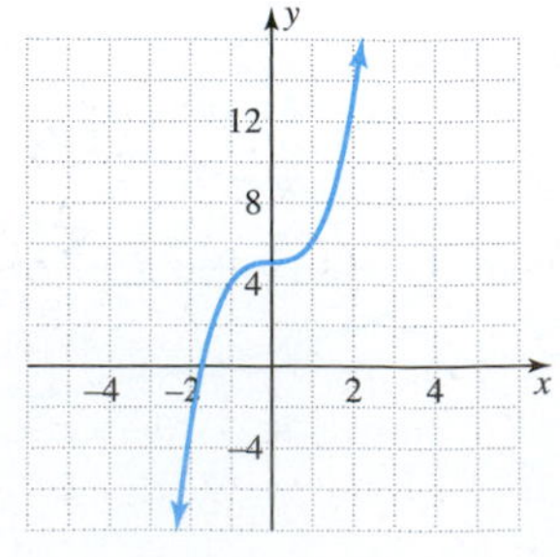

73. not a function

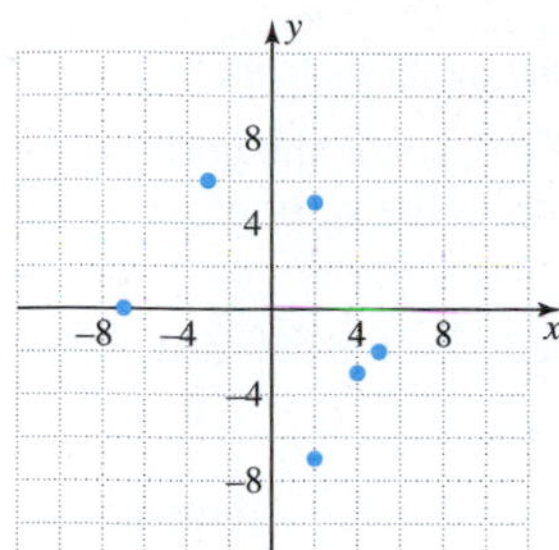

74. function

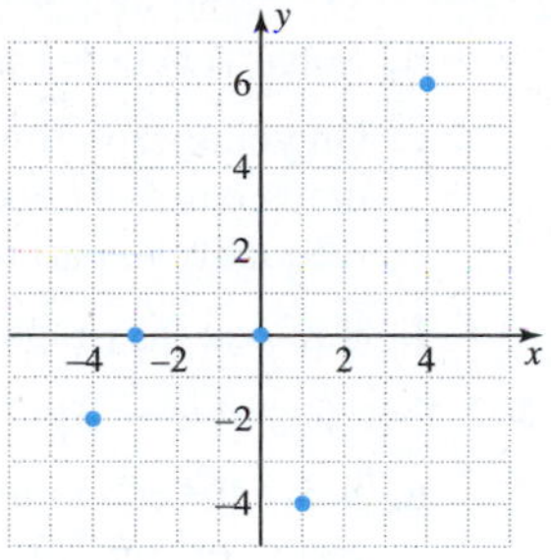

75. $x = 3y + 1$ function

76. $x = y - 6$ function

77. $y = |x + 3| + 1$ function

78. $y = |5 - 2x|$ function

79. $x = (y - 4)^2$ not a function

80. $x = (y + 7)^2$ not a function

81. not a function

x	y
0	3
1	4
2	6
0	5

82. function

x	y
−7	3
0	4
2	4
5	5

Given the graph of each function, find the indicated values.

83. Find $g(-3)$ for $g(x)$. 5

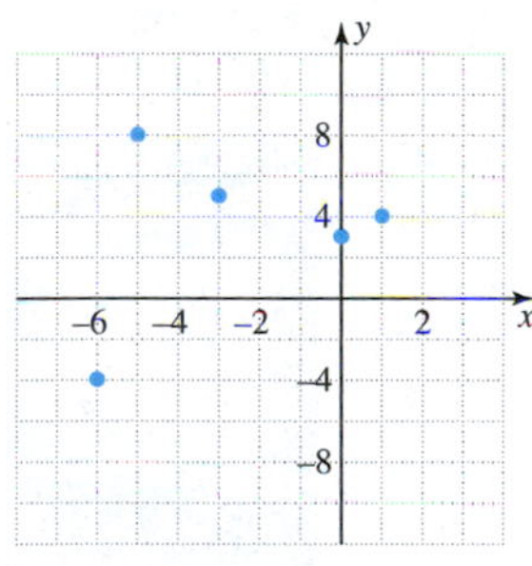

84. Find $g(8)$ for $g(x)$. 0

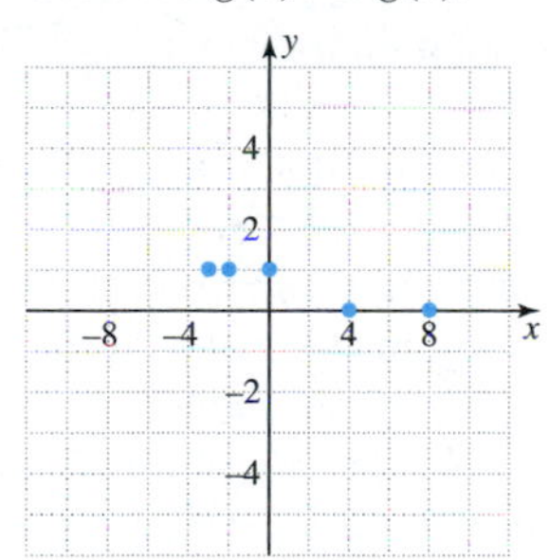

85. Find $f(-3)$ for $f(x)$. 1

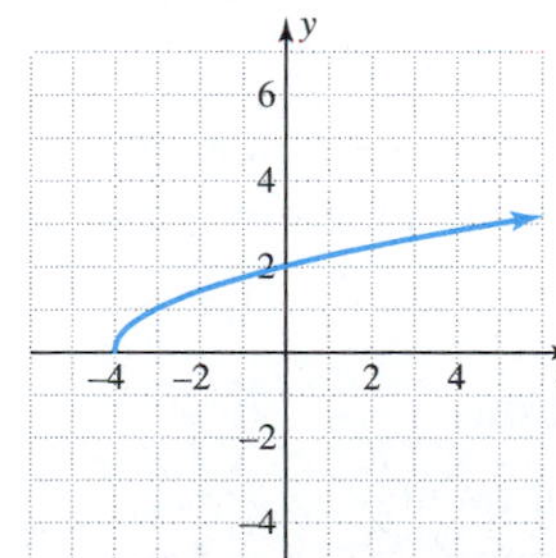

86. Find $f(1)$ for $f(x)$. 6

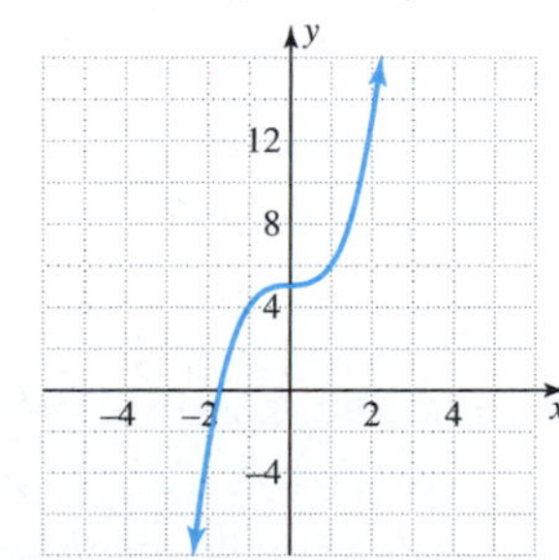

Use the given function to solve each problem.

87. Find $f(0)$ if $f(x) = 3x^2 - 4x + 5$. 5

88. Find $f(-2)$ if $f(x) = \frac{1}{2}x^5 - 5x^4 + 3$. −93

89. The estimated number of AIDS diagnoses can be modeled by the equation $d(x) = 222x^3 - 2712x^2 + 9812x + 27{,}565$, where x is the number of years after 2000. Find $d(10)$ rounded to the nearest whole number and interpret. (Source: http://www.cdc.gov)

90. Use the function given in Exercise 89 to find $d(13)$ rounded to the nearest whole number and interpret. (Source: http://www.cdc.gov)

91. The percentage of smokers among adults age 18 and over is modeled by the equation $k(x) = -0.44x + 23.11$, where x is the number of years after 2000. Find all x rounded to the nearest whole number such that $k(x) = 20$ and interpret. (Source: National Center for Health Statistics)

92. Use the function given in Exercise 91 to find all x rounded to the nearest whole number such that $k(x) = 16$ and interpret. (Source: National Center for Health Statistics)

You Be the Teacher!

Correct each student's errors, if any.

93. Evaluate $g(3)$ for $g(x) = 7 - x^2$.

Debbie's work:

$g(x) = 7 - x^2$

$g(3) = (7 - x^2)(3) = 21 - 3x^2$

94. Evaluate $g(-4)$ for $g(x) = 2x^2 + 5$.

Evan's work:

$g(x) = 2x^2 + 5$

$g(-4) = 2(-4)^2 + 5 = 64 + 5 = 69$

Calculate It!

Use the TABLE feature in a calculator to evaluate each function at the specified value. Round each answer to two decimal places, when applicable.

95. $f(x) = -4x^2 + 6x - 3, f(2.5)$ −13

96. $f(x) = x^3 - 3x^2 + 5, f(-1.5)$ −5.13

97. $g(x) = |7x + 5|, g(15)$ 110

98. $g(x) = |19x - 21|, g(-19)$ 382

Think About It!

99. Let $f(x) = |x + 2|$.

a. Solve the equation $f(x) = 4$ using the methods from Section 2.6.

b. Graph $f(x)$ and draw a horizontal line through $y = 4$ on the same coordinate system.

c. What are the points, if any, where the graph of $f(x)$ and the horizontal line intersect? How do the points of intersection relate to the solutions of the equation?

d. Solve $f(x) < 4$ using the methods from Section 2.7.

e. On the graph from part (b), shade the portion of the x-axis that corresponds to the solution set of the inequality in part (d).

f. How can the graph be used to solve the inequality?

100. Let $f(x) = |x - 1|$.

a. Solve the equation $f(x) = 2$ using the methods from Section 2.6.

b. Graph $f(x)$ and draw a horizontal line through $y = 2$ on the same coordinate system.

c. What are the points, if any, where the graph of $f(x)$ and the horizontal line intersect? How do the points of intersection relate to the solutions of the equation?

d. Solve $f(x) > 2$ using the methods from Section 2.7.

e. On your graph from part (b), shade the portion of the x-axis that corresponds to the solution set of the inequality in part (d).

f. How can the graph be used to solve the inequality?

101. Let $f(x) = |x + 4|$.

a. Solve the equation $f(x) = -1$ using the methods from Section 2.6.

b. Graph $f(x)$ and draw a horizontal line through $y = -1$ on the same coordinate system.

c. What are the points, if any, where the graph of $f(x)$ and the horizontal line intersect? How do the points of intersection relate to the solutions of the equation?

d. Solve $f(x) > -1$ using the methods from Section 2.7.

e. On your graph from part (b), shade the portion of the x-axis that corresponds to the solution set of the inequality in part (d).

f. How can the graph be used to solve the inequality?

102. Let $f(x) = |x - 3|$.

a. Solve the equation $f(x) = -2$ using the methods from Section 2.6.

b. Graph $f(x)$ and draw a horizontal line through $y = -2$ on the same coordinate system.

c. What are the points, if any, where the graph of $f(x)$ and the horizontal line intersect? How do the points of intersection relate to the solutions of the equation?

d. Solve $f(x) < -2$ using the methods from Section 2.7.

e. On your graph from part (b), shade the portion of the x-axis that corresponds to the solution set of the inequality in part (d).

f. How can the graph be used to solve the inequality?

SECTION 3.4 The Domain and Range of Functions

OBJECTIVES

As a result of completing this section, you will be able to

1. Find the domain and range of a function given a set, a mapping, or a table.
2. Find the domain and range of a function given a graph.
3. Find the domain of a function given an equation.
4. Apply the concept of domain to real-world situations.
5. Troubleshoot common errors.

Many websites require us to enter a login name and a password to access their site. The correspondence of a login name and password form a function since each login name can have only one password. Have you ever entered your login name incorrectly and received the message that your login name is invalid? This is because the incorrect login name is not in the *domain* of this function.

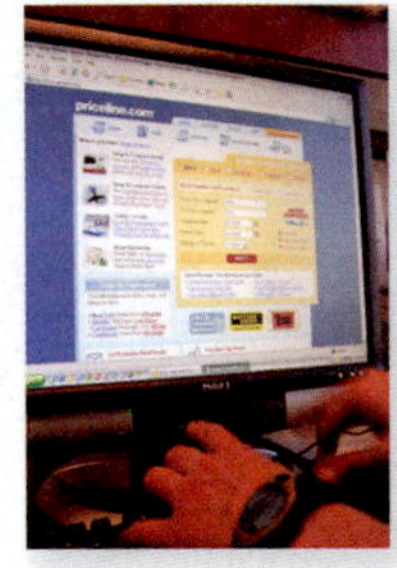

In this section, we will define and determine the domain and range of a function.

Domain and Range of a Function

Objective 1 Find the domain and range of a function given a set, a mapping, or a table.

In Section 3.2, we learned how to find the domain and range of a relation. Because every function is a relation, the steps for finding the domain and range of a function are the same.

Procedure: Finding the Domain and Range of a Function

Step 1: The domain is the set of x-values in the given set of points.
Step 2: The range is the set of y-values in the given set of points.

Objective 1 Examples Find the domain and range of each function.

1a. $\{(-2, 4), (-1, 1), (0, 0), (1, 1), (2, 4)\}$

1b. The table shows the five most populated states along with the party of their Governor in 2011. (Source: en.wikipedia.org)

State	California	Texas	New York	Florida	Illinois
Party	Democrat	Republican	Democrat	Republican	Democrat

1c. According to a September 2010 survey, 79% of American adults use the Internet. The table shows the percentage of Internet users that have ever engaged in the given activities. (Source: http://www.pewinternet.org/Static-Pages/Trend-Data/Online-Activites-Total.aspx)

Activity	Percent
Sends/reads e-mail	94
Uses a search engine	87
Gets driving directions	86
Checks the weather	81
Takes online classes	12
Uses online dating sites	8
Uses social networking sites	61
Uses online banking	58

Solutions

1a. The domain is the set of x-values and is $\{-2, -1, 0, 1, 2\}$.

The range is the set of y-values and is $\{0, 1, 4\}$.

Recall the values of 1 and 4 do not need to be repeated since the range is just a set of the different y-values. Also, note that we list the elements in ascending order when the set is numerical.

1b. The domain is the set of x-values and is {California, Texas, New York, Florida, Illinois}.

The range is the set of y-values and is {Democrat, Republican}.

1c. The domain is the set of x-values and is {sends/reads e-mail, uses a search engine, gets driving directions, checks the weather, takes online classes, uses online dating sites, uses social networking sites, uses online banking}.

The range is the set of y-values and is $\{8, 12, 58, 61, 81, 86, 87, 94\}$.

Student Check 1 Find the domain and range of each function.

a. $\{(-3, -27), (-2, -8), (-1, -1), (0, 0), (1, 1)\}$

b. The table shows the total number of degree-granting institutions in the United States in the fall of each year. (Source: National Center for Education Statistics)

Year	1869	1919	1969	1989	1999	2005	2006	2007
Number of institutions	563	1041	2525	3535	4084	4276	4314	4352

c. The table shows the domestic market share (in percent) for the given airline for August 2009–July 2010. (Source: Bureau of Transportation Statistics)

Airline	AirTran	Alaska	American	Continental	Delta	JetBlue	Northwest	Southwest	United	US Airways
Market share	3.4	3.2	13.8	7.5	14.2	4.3	2.5	13.9	10.3	7.9

Reading a Graph to Find the Domain and Range

Objective 2 Find the domain and range of a function given a graph.

Recall that functions can also be represented by graphs. Graphs that pass the vertical line test represent functions. We can use the graph to determine the domain and range of a function. As shown in Section 3.2, we use interval notation to represent the domain and range of a function when a graph is a smooth, continuous curve. If a graph consists of a finite set of points, we list the domain and range explicitly.

Between what values of x is the graph contained?

Procedure: Finding the Domain of a Function from Its Graph

Step 1: Determine the x-value of the leftmost point on the graph.
Step 2: Determine the x-value of the rightmost point on the graph.
Step 3: The domain is the interval between the x-values of the points found in steps 1 and 2.

Note: *If the graph extends indefinitely from the left to the right, the domain is all real numbers, denoted by* $(-\infty, \infty)$.

Between what values of y is the graph contained?

Procedure: Finding the Range of a Function from Its Graph

Step 1: Determine the y-value of the lowest point on the graph.
Step 2: Determine the y-value of the highest point on the graph.
Step 3: The range is the interval between the y-values of the points found in steps 1 and 2

Note: *If the graph extends indefinitely from the bottom to the top, the range is all real numbers, denoted by* $(-\infty, \infty)$.

Objective 2 Examples Find the domain and range of each function.

2a.

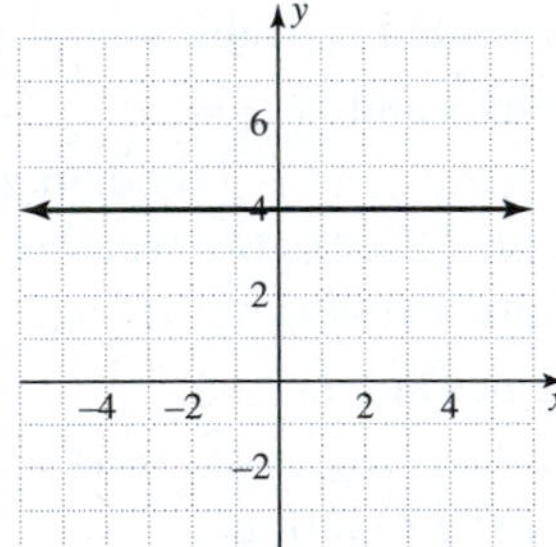

2b.

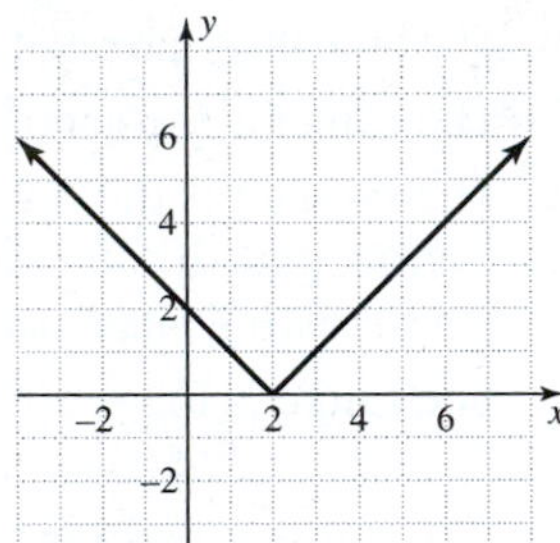

2c.

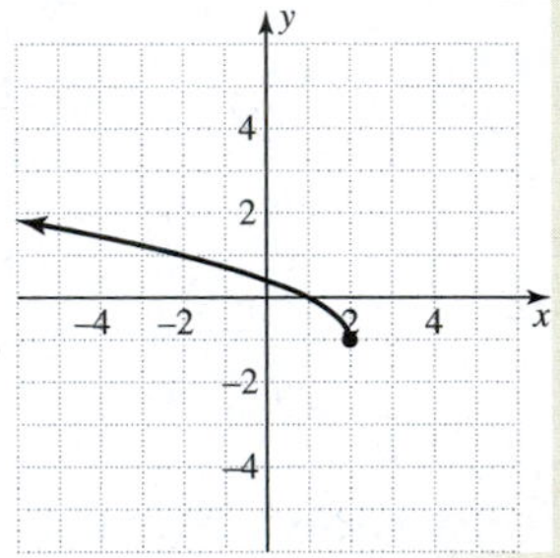

Solutions

2a. The graph extends indefinitely to the left and right, so the domain is $(-\infty, \infty)$. The lowest and highest points on the graph are both on the line $y = 4$ since every point on the line has a y-value of 4. So, the range is $\{4\}$. (See Figure 3.4.1.)

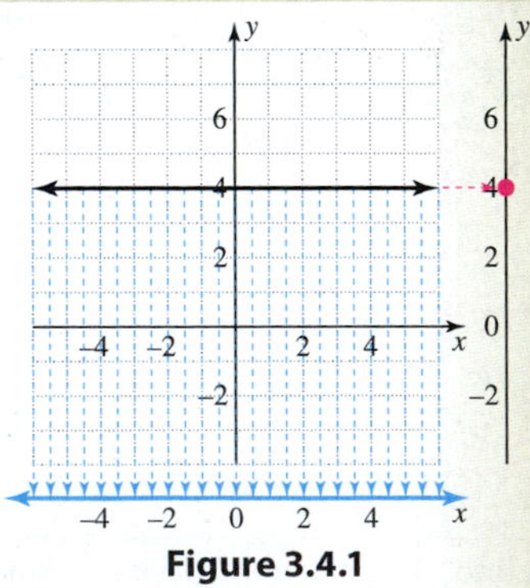

Figure 3.4.1

2b. The graph extends indefinitely to the left and right, so the domain is $(-\infty, \infty)$. The lowest point on the graph is $(2, 0)$. The graph doesn't have a highest point since it extends indefinitely upward. So, the range is $[0, \infty)$. (See Figure 3.4.2.)

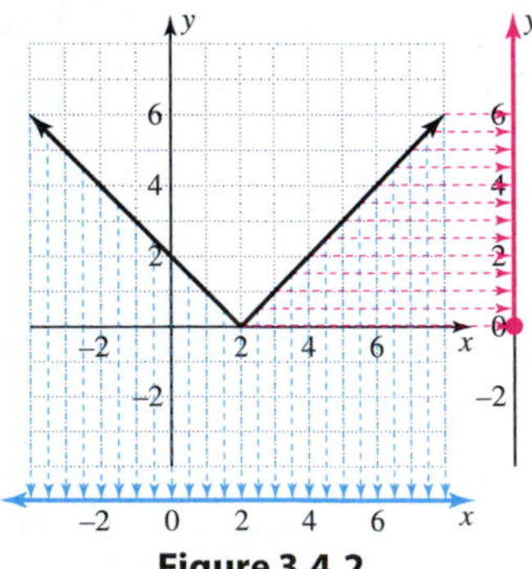

Figure 3.4.2

2c. The graph extends indefinitely to the left. The rightmost point is $(2, -1)$. So, the domain is $(-\infty, 2]$. The lowest point on the graph is $(2, -1)$. The graph extends indefinitely upward. So, the range is $[-1, \infty)$. (See Figure 3.4.3.)

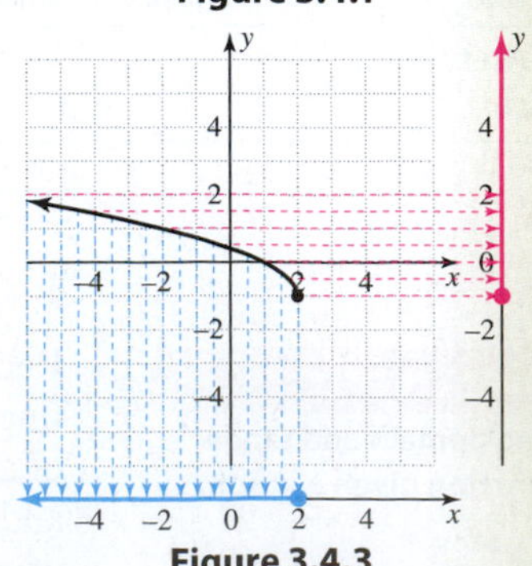

Figure 3.4.3

Student Check 2 Find the domain and range of each function.

a.

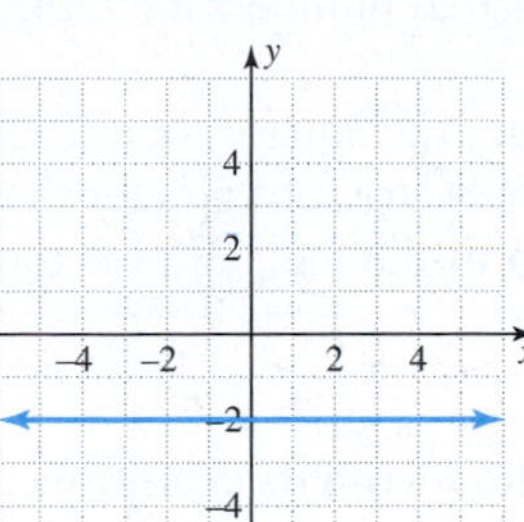

b.

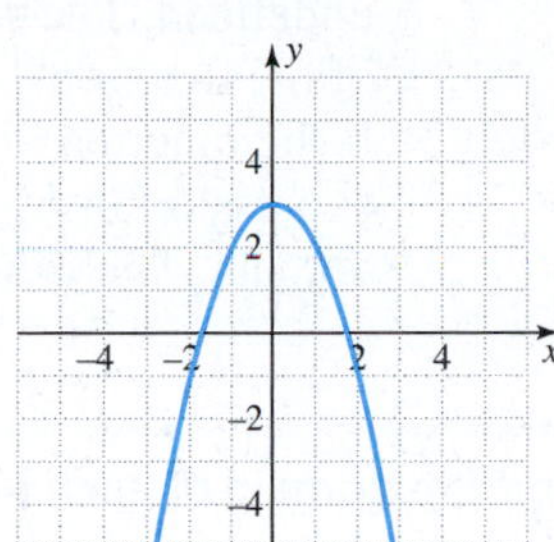

c.

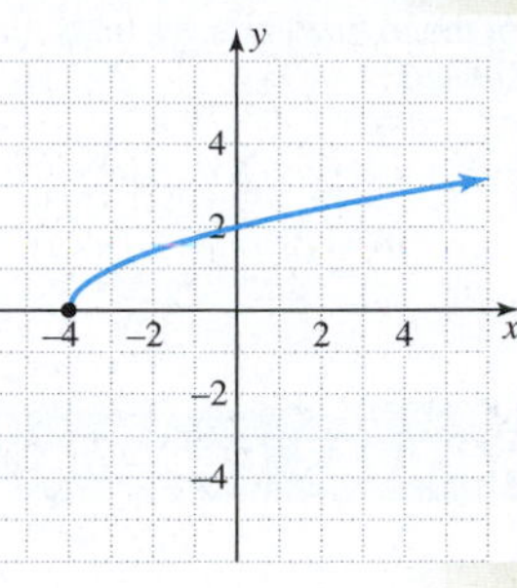

Using an Equation to Find Its Domain

Objective 3 Find the domain of a function given an equation.

To find the domain of a function given its equation, we can graph the equation and read the graph to determine its domain. This, however, is not necessary as we can determine the domain of a function from its equation.

When examining an equation to find its domain, the goal is to determine what values of x make the function *defined*. A function is defined at a value of x if the corresponding y-value is a real number. If there are any values of x that make the function *undefined* or *not a real number*, they must be excluded from the domain of the function.

Definition: The **domain of a function** represented algebraically is the set of real numbers, excluding any number that makes the equation undefined or *not a real number*.

Recall that division by zero is undefined and that the square root of a negative number is not a real number. If a function involves a fraction with a variable in the denominator or involves the square root of an algebraic expression, the function may have values that make it undefined or not real. Examples of these types of functions are

$$f(x) = \frac{1}{x+5} \qquad g(x) = \sqrt{x+3}$$

The function $f(x)$ is undefined when the denominator is equal to zero, or when

$$x + 5 = 0$$
$$x = -5$$

So, the domain of $f(x)$ is all real numbers except $x = -5$. We can write this as $(-\infty, -5) \cup (-5, \infty)$.

The function $g(x)$ is not a real number when the expression inside the square root is negative. So, as long as the expression inside the square root is positive or zero, the function is defined. We must solve the following inequality to determine the domain of $g(x)$.

$$x + 3 \geq 0$$
$$x \geq -3$$

So, the domain of $g(x)$ is $[-3, \infty)$.

Procedure: Determining the Domain of a Function from Its Equation

Step 1: If the function does not contain a fraction with a variable in the denominator or the square root of an expression, the domain is all real numbers.

INSTRUCTOR NOTE: These types of functions will be discussed in much more detail in later chapters. At that point, the range of these functions will also be discussed.

Step 2: If the function has a fraction with a variable in the denominator, set the denominator equal to zero to find the value(s) that make(s) the function undefined. The set of real numbers excluding these values is the domain of the function.

Step 3: If the function involves the square root of an expression, set the expression inside the square root greater than or equal to zero to find its domain. The solution of the inequality is the domain of the function.

Objective 3 Examples Find the domain of each function from its equation.

3a. $f(x) = 3x - 5$ **3b.** $f(x) = \dfrac{2}{x - 3}$ **3c.** $f(x) = \sqrt{x - 4}$

Solutions

3a. The expression $3x - 5$ is defined for all real numbers. So, the domain is $(-\infty, \infty)$.

3b. The expression $\dfrac{2}{x - 3}$ is undefined when $x - 3 = 0$ or when $x = 3$. The domain is all real numbers except $x = 3$. We write this as $(-\infty, 3) \cup (3, \infty)$.

3c. The expression $\sqrt{x - 4}$ is defined only when $x - 4 \geq 0$ or $x \geq 4$. We write the domain as $[4, \infty)$.

Student Check 3 Find the domain of each function from its equation.

a. $f(x) = 2x + 1$ **b.** $f(x) = \dfrac{4}{x + 7}$ **c.** $f(x) = \sqrt{x - 2}$

Applications

Objective 4

Apply the concept of domain to real-world situations.

We have already established that functions occur in many real-world situations. When a student uses a student ID number to register for classes, the number is paired with the student name so that the student gets registered for the appropriate classes. The domain of this function is the set of student ID numbers and the range is the set of students assigned to an ID.

Another example of a function involves e-mail. When a person e-mails someone, an e-mail address is paired with a particular recipient. The domain in this function is the e-mail address and the range is the recipient. The list could go on with these types of examples.

In each of these situations, the domain and range of these functions has to be defined. Someone has to set up restrictions on what is going to be a reasonable value for the domain (input) and the range (output) of the function.

In Example 4, we are in charge of determining an appropriate domain of a particular function. These examples come from problems that will be discussed later in the book.

Objective 4 Examples Determine an appropriate domain for each situation.

4a. The revenue R for selling d donuts a month for a bakery is given by

$$R(d) = 0.65d - 4500$$

4b. The area of the rectangle can be represented by the function $A(x) = x(10 - 2x)$.

x

$10 - 2x$

Solutions

4a. It doesn't make sense to sell a negative number of donuts or a partial donut. Only whole donuts can be sold. So, the reasonable domain for this situation is $d = 0, 1, 2, 3, \ldots$, or the domain is the set of whole numbers, denoted by $\mathbb{W} = \{0, 1, 2, 3, \ldots\}$.

Note: *It is not practical for a bakery to make infinitely many donuts in a month. There is an upper limit to the domain but we need more information to determine what that is.*

4b. Since the expressions x and $10 - 2x$ represent the lengths of the sides of a rectangle, it is appropriate that these values be positive. So, to find the domain we have to solve the compound inequality, $x > 0$ and $10 - 2x > 0$.

$x > 0$	and	$10 - 2x > 0$	Set each side greater than zero.
		$10 - 2x - 10 > 0 - 10$	Subtract 10 from each side.
		$-2x > -10$	Simplify.
		$\dfrac{-2x}{-2} < \dfrac{-10}{-2}$	Divide each side by -2 and reverse the inequality symbol.
		$x < 5$	Simplify.

The intersection of the sets $x > 0$ and $x < 5$ is the interval $(0, 5)$. So, the domain of $A(x)$ is $(0, 5)$.

Student Check 4 Determine an appropriate domain for each situation.

a. The monthly cost of a cell phone plan is $C(x) = 69.99 + 0.45x$, where x is the number of minutes over 4000.

b. The area of the rectangle is $A(x) = x(18 - 2x)$.

x

$18 - 2x$

Objective 5 Troubleshoot common errors.

Troubleshooting Common Errors

A common error associated with the domain and range of functions is shown.

Objective 5 Example **A problem and an incorrect solution are given. Provide the correct solution and an explanation of the error.**

Find the domain and range of the given function.

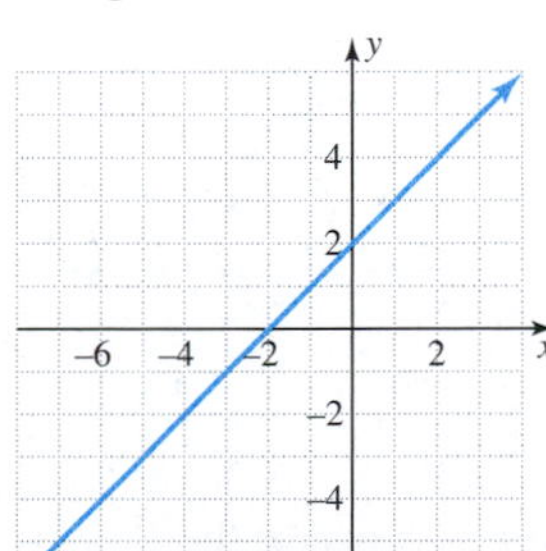

Incorrect Solution	Correct Solution and Explanation
The domain is $[-2, \infty)$ and the range is $[2, \infty)$.	The graph of the function extends indefinitely to the left of $x = -2$. So, the domain is $(-\infty, \infty)$. The graph also extends indefinitely below $y = 2$. So, the range is $(-\infty, \infty)$.

ANSWERS TO STUDENT CHECKS

Student Check 1 **a.** Domain = {−3, −2, −1, 0, 1} Range = {−27, −8, −1, 0, 1}

b. Domain = {1869, 1919, 1969, 1989, 1999, 2005, 2006, 2007}

Range = {563, 1041, 2525, 3535, 4084, 4276, 4314, 4352}

c. Domain = {AirTran, Alaska, American, Continental, Delta, JetBlue, Northwest, Southwest, United, US Airways}, Range = {2.5, 3.2, 3.4, 4.3, 7.5, 7.9, 10.3, 13.8, 13.9, 14.2}

Student Check 2 **a.** Domain = $(-\infty, \infty)$ and Range = {−2}
b. Domain = $(-\infty, \infty)$ and Range = $(-\infty, 3]$
c. Domain = $[-4, \infty)$ and Range = $[0, \infty)$

Student Check 3 **a.** Domain = $(-\infty, \infty)$ **b.** Domain = $(-\infty, -7) \cup (-7, \infty)$ **c.** Domain = $[2, \infty)$

Student Check 4 **a.** $\mathbb{W} = \{0, 1, 2, 3, \ldots\}$ **b.** (0, 9)

SUMMARY OF KEY CONCEPTS

1. The domain of a function is the set of x-values for which the function is defined. The range is the set of corresponding y-values. In a set, table, or mapping, the x- and y-values are listed explicitly. To write the domain and range, list the appropriate values in a set.
2. To find the domain and range from a graph, state the intervals of x and y, respectively, that contain the graph. The domain is found by reading the graph horizontally from left to right and the range is found by reading the graph vertically from the bottom up.
3. The domain of a function that is represented algebraically is the set of real numbers that make the function defined.
   a. If the function contains a fraction with variables in the denominator, exclude the values that make the denominator zero.
   b. If the function contains a square root, exclude the values that make the expression inside the square root negative.
4. When dealing with real-world applications, the domain consists not only of the values that make the function defined, but also the values that are reasonable. For example, a negative number may make the function defined but it may not make sense in the context of the application.

GRAPHING CALCULATOR SKILLS

The graphing calculator can assist us in determining values for the domain of a function that is represented algebraically. We input the function into the calculator and then graph it or examine the table of ordered pairs to determine input values that make the function defined or undefined.

Example 1: Find the domain of $f(x) = \dfrac{2}{x-3}$.

Solution:

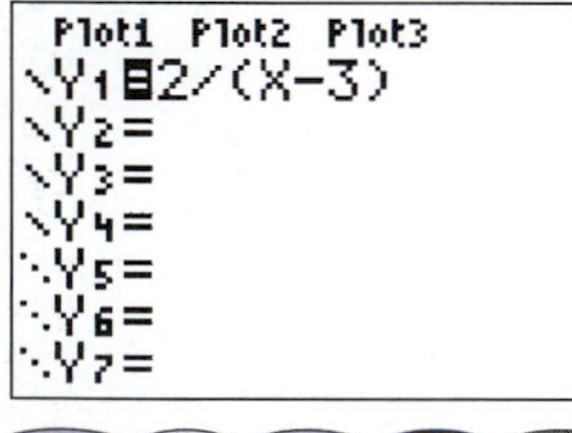

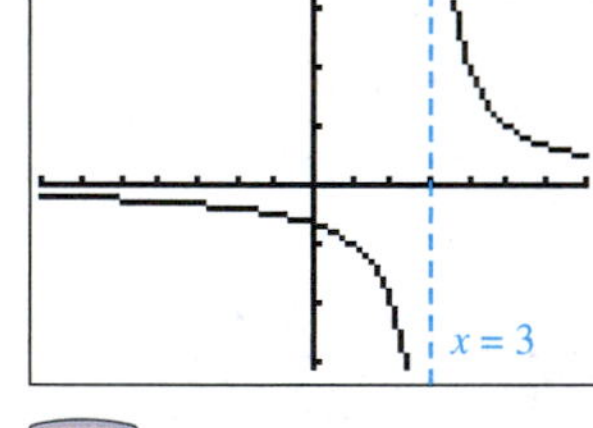

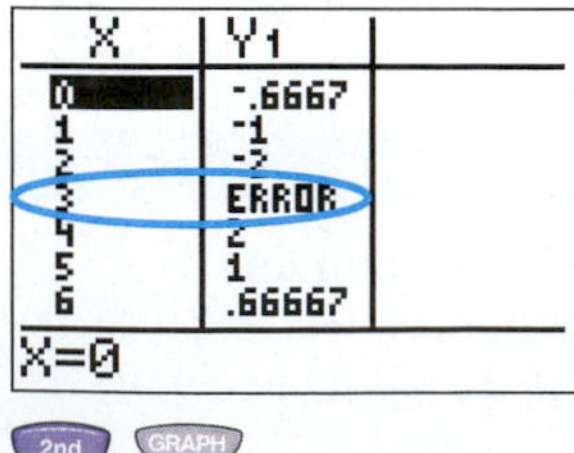

Note $x = 3$ makes the function undefined as evidenced by the error message in the Y_1-column of the table and also the gap in the graph. So, the graph and table show us that the domain is $(-\infty, 3) \cup (3, \infty)$.

Example 2: Find the domain of $f(x) = \sqrt{x-4}$.

Solution:

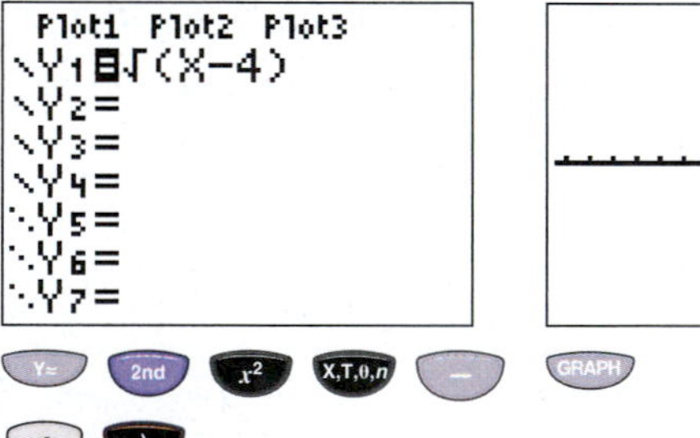
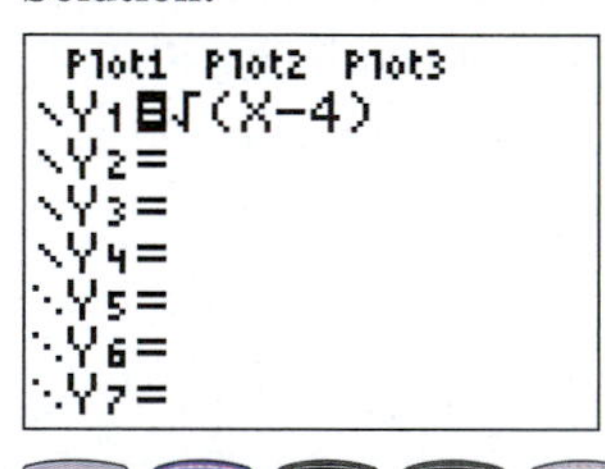

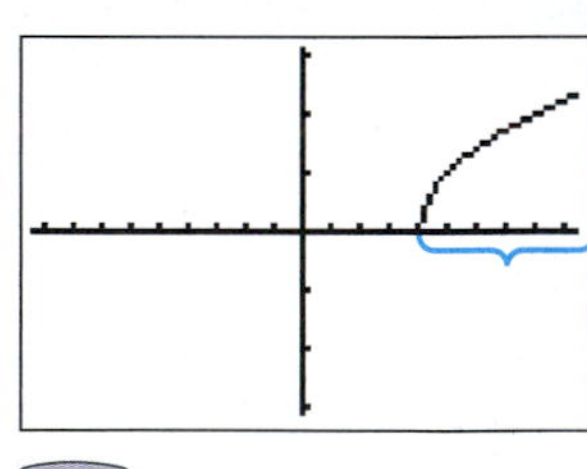

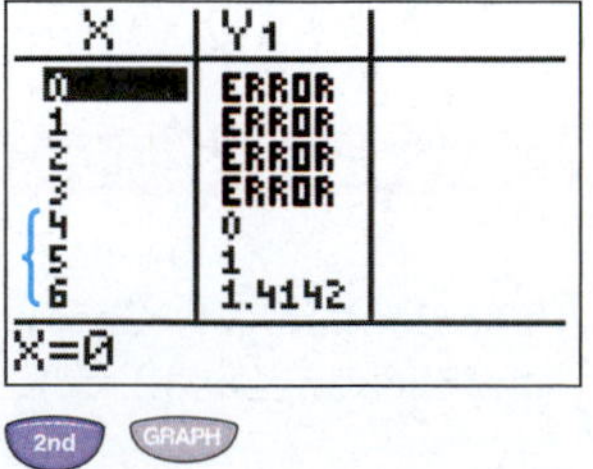

Note that the graph shows the leftmost x-value is 4 and it extends indefinitely to the right. The table shows an error message for all values less than 4, which means these numbers are not included in the domain. So, the table and graph show us that the domain is $[4, \infty)$.

SECTION 3.4 EXERCISE SET

Write About It!

Use complete sentences in your answer to each exercise.

1. Explain how to find the domain of a function that is given as a set of ordered pairs.
2. Explain how to find the domain of a function from its graph.
3. Explain how to find the range of a function that is given as a set of ordered pairs.
4. Explain how to find the range of a function from its graph.
5. Explain how to find the domain of an algebraic function with a fraction.
6. Explain how to find the domain of an algebraic function with a square root.

Practice Makes Perfect!

Find the domain and range of each function. (*See Objective 1.*)

7. $\{(-3, 1), (-2, -4), (-1, 0), (0, 5)\}$
8. $\{(-1, 6), (0, 4), (1, -10), (2, -7)\}$
9. $\{(-10, 4), (0, 3), (2, 4), (5, -7)\}$
10. $\{(-9, -11), (-8, 11), (6, 11), (7, -9)\}$
11. The number of highway miles per gallon is a function of the vehicle model. (Source: http:www.fueleconomy.gov)

Vehicle Model	Highway mpg
2011 Ford Edge AWD	26
2011 Honda CR-V 4WD	27
2011 Subaru Forester AWD	27
2011 Toyota RAV4 4WD	27

12. The number of city miles per gallon is a function of the vehicle model. (Source: http:www.fueleconomy.gov)

Vehicle Model	City mpg
2011 Ford Edge AWD	18
2011 Honda CR-V 4WD	21
2011 Subaru Forester AWD	21
2011 Toyota RAV4 4WD	21

13. The total fall enrollment in U.S. degree-granting institutions is a function of the year. (Source: http://nces.ed.gov/)

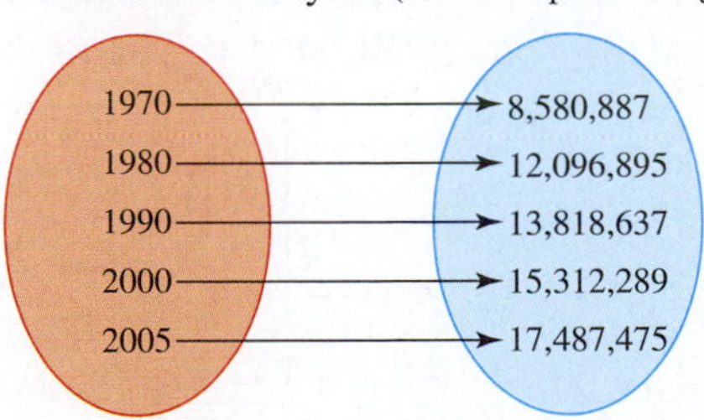

Domain = {1970, 1980, 1990, 2000, 2005}, Range = {8,580,887, 12,096,895, 13,818,637, 15,312,289, 17,487,475}

14. The percentage of male high school seniors who felt that having lots of money was "very important" is a function of the year. (Source: http://nces.ed.gov/)

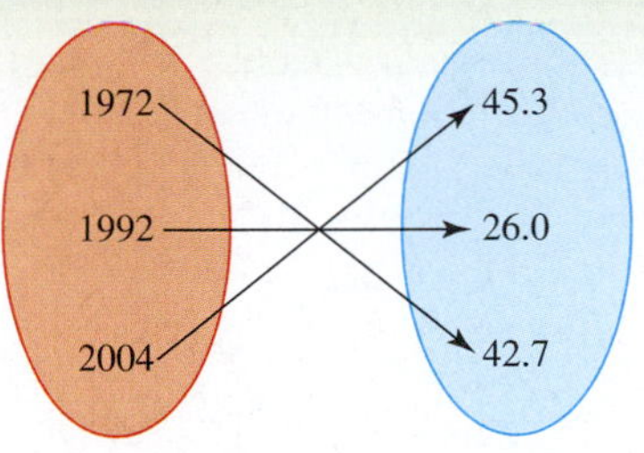

Domain = {1972, 1992, 2004}, Range = {26.0, 42.7, 45.3}

15. The percentage of female high school seniors who felt that having lots of money was "very important" is a function of the year. (Source: http://nces.ed.gov/)

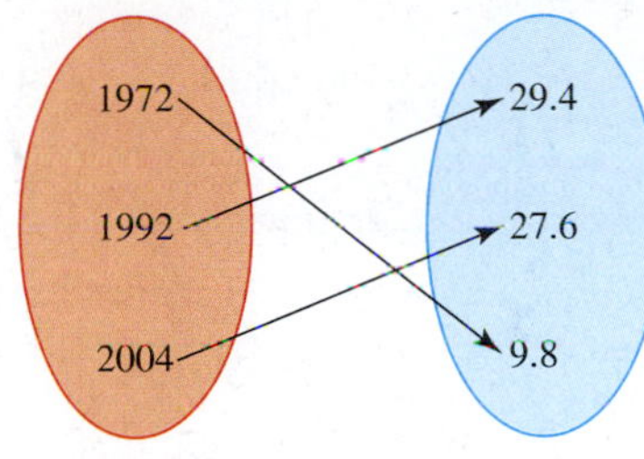

Domain = {1972, 1992, 2004}, Range = {9.8, 27.6, 29.4}

16. Percentage of male high school seniors who felt that having strong friendships was "very important" is a function of the year. (Source: http://nces.ed.gov/)

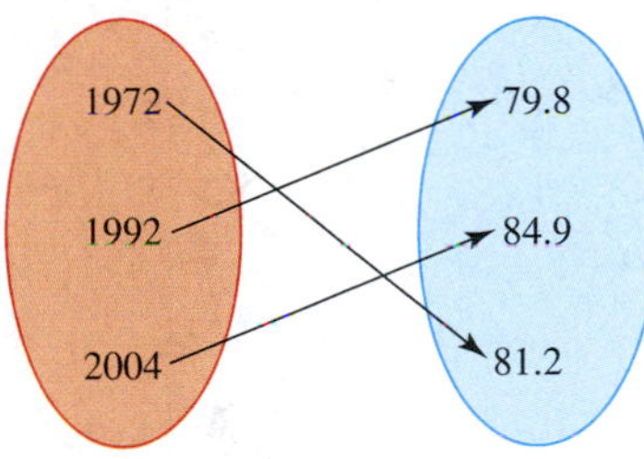

Domain = {1972, 1992, 2004}, Range = {79.8, 81.2, 84.9}

Find the domain and range of each function. (*See Objective 2.*)

17.

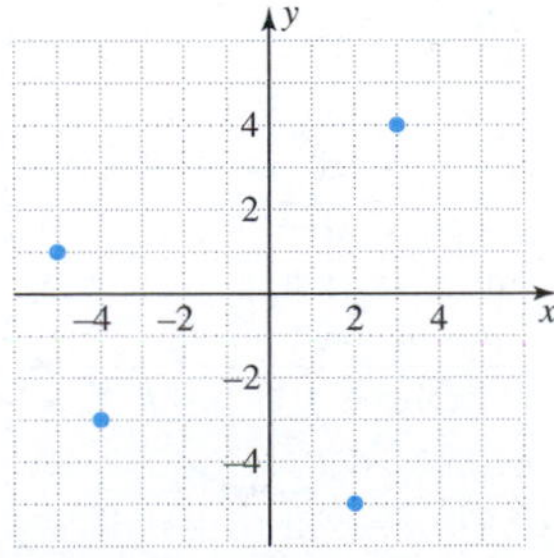

Domain = {−5, −4, 2, 3}, Range = {−5, −3, 1, 4}

18.

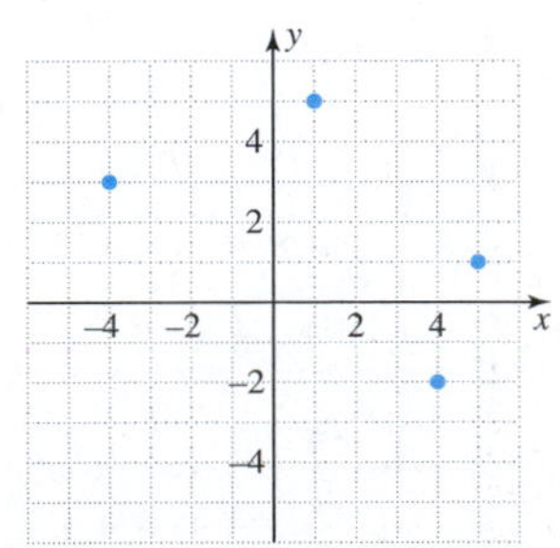

Domain = {−4, 1, 4, 5}, Range = {−2, 1, 3, 5}

19.

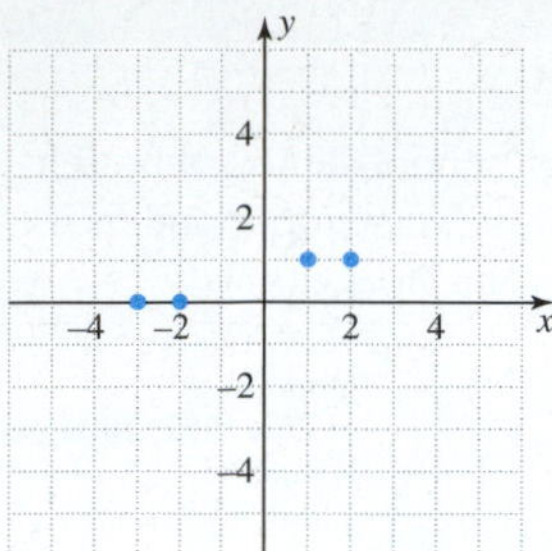

20.

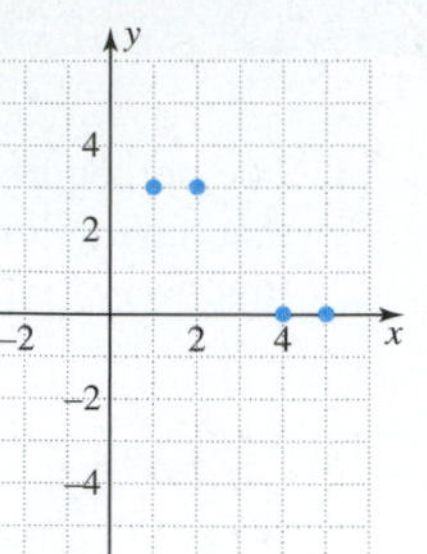

21.

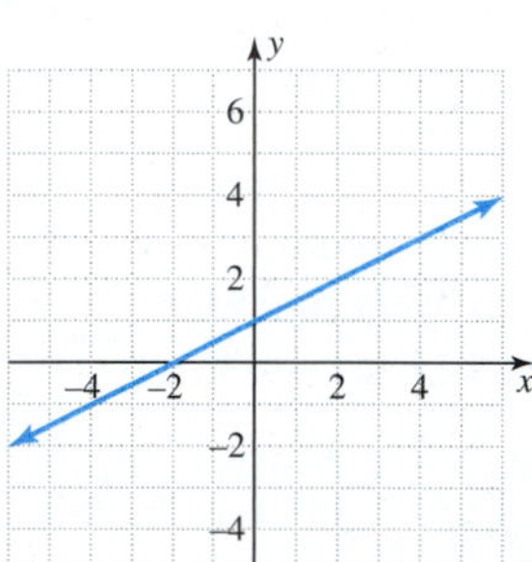

22.

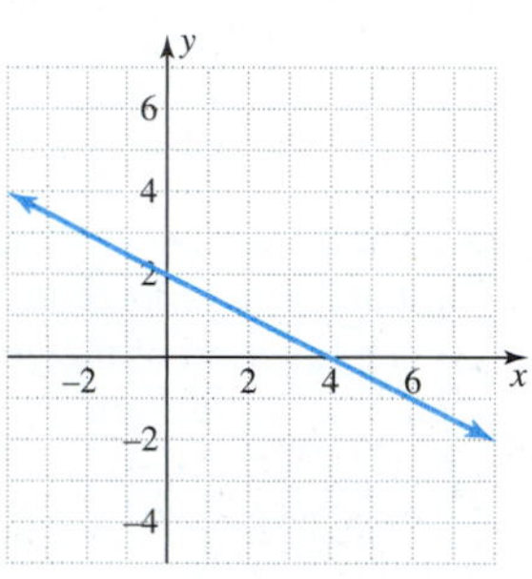

23.

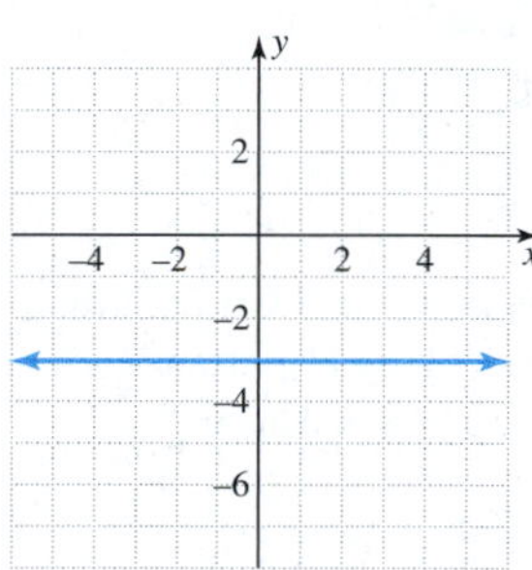

24.

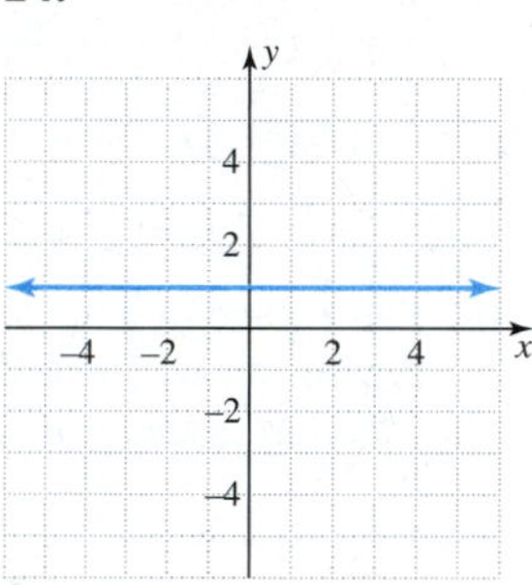

25.

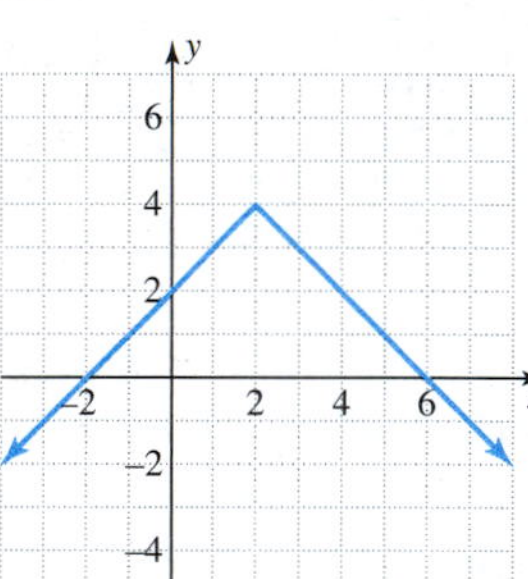

26.

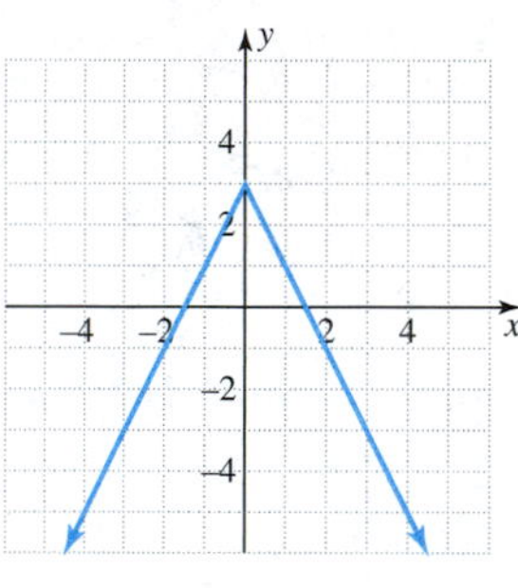

27.

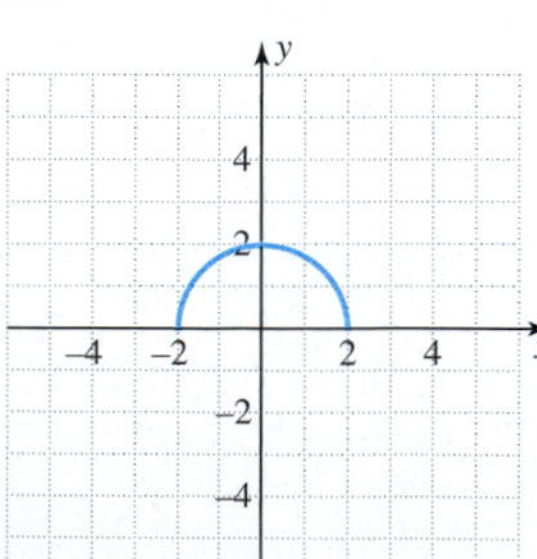

28.

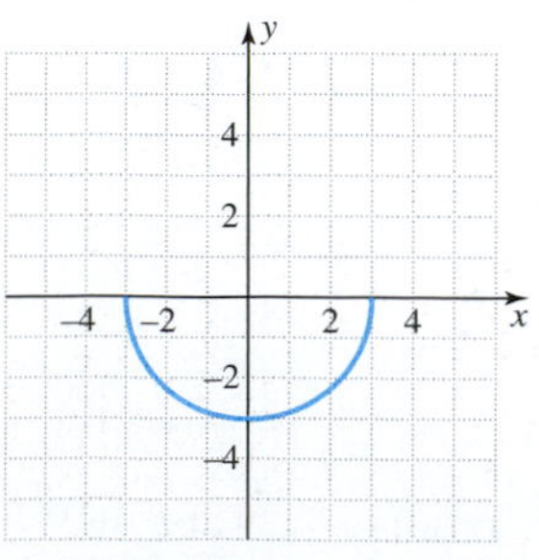

29.

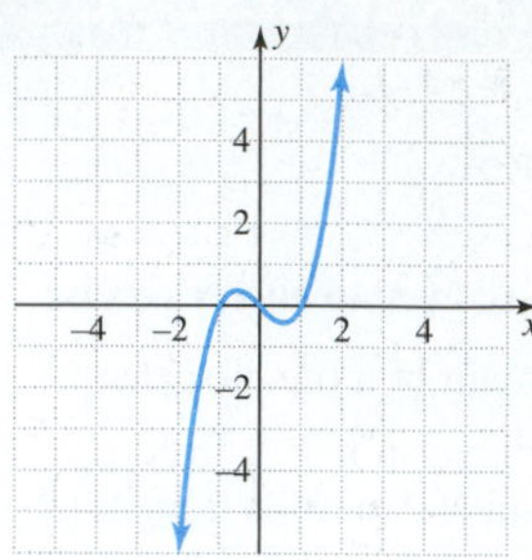

30.

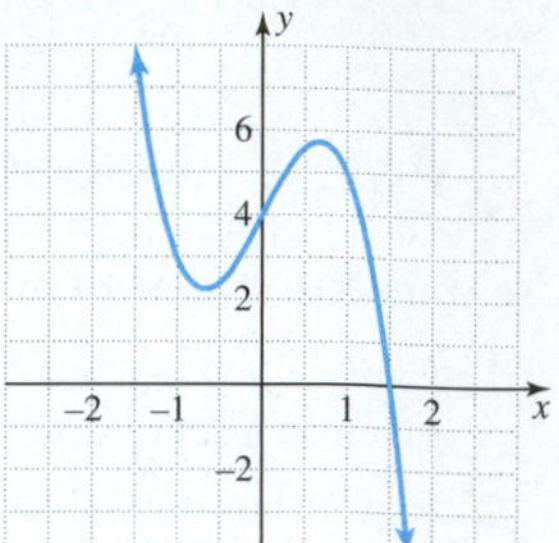

31.

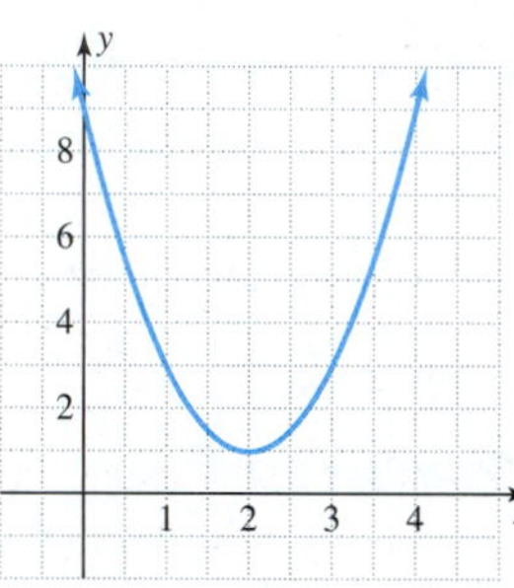

32.

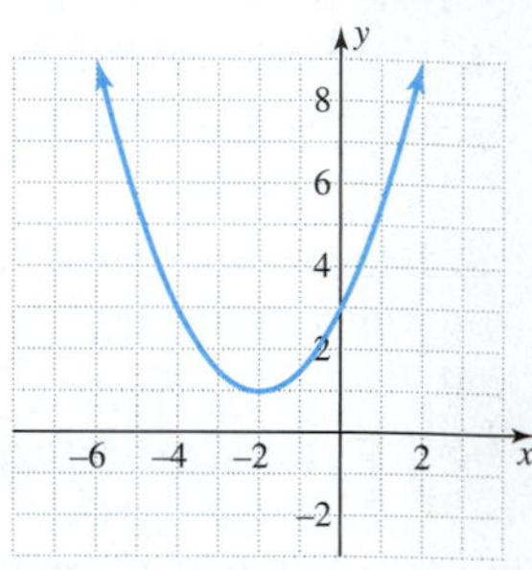

Find the domain of each function. (*See Objective 3.*)

33. $f(x) = 4x - 7$ $(-\infty, \infty)$

34. $f(x) = 2x + 1$ $(-\infty, \infty)$

35. $f(x) = \frac{3}{5}x + 2$ $(-\infty, \infty)$

36. $f(x) = -\frac{9}{2}x - 5$ $(-\infty, \infty)$

37. $g(x) = \frac{1}{x}$ $(-\infty, 0) \cup (0, \infty)$

38. $g(x) = \frac{2}{x}$ $(-\infty, 0) \cup (0, \infty)$

39. $h(x) = \frac{x}{x + 5}$

40. $h(x) = \frac{x}{x - 7}$

41. $f(x) = \sqrt{x + 6}$ $[-6, \infty)$

42. $f(x) = \sqrt{5x + 10}$

43. $f(x) = \sqrt{6 - 8x}$

44. $f(x) = \sqrt{5 - 3x}$

Find the domain of each function described. (*See Objective 4.*)

45. $A(x) = x(8 - x)$ $(0, 8)$

x

8 – 2x

46. $A(x) = x(2 - x)$ $(0, 2)$

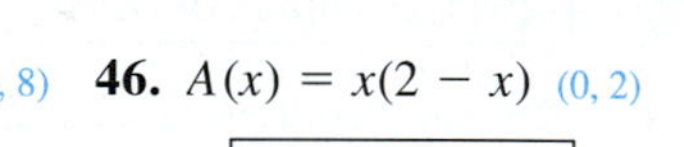

47. $A(x) = (4 - 2x)(5 - x)$

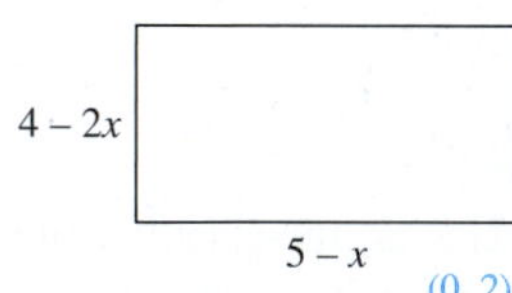

$(0, 2)$

48. $A(x) = (9 - 3x)(7 - x)$

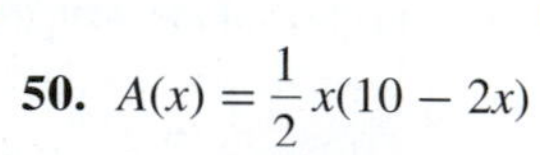

$(0, 3)$

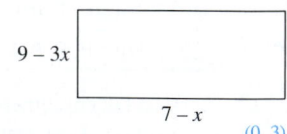

49. $A(x) = \frac{1}{2}x(18 - 2x)$ $(0, 9)$

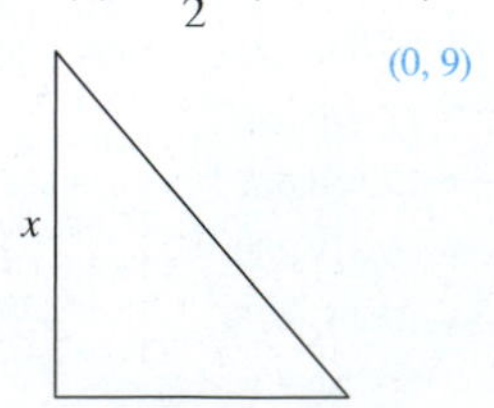

50. $A(x) = \frac{1}{2}x(10 - 2x)$ $(0, 5)$

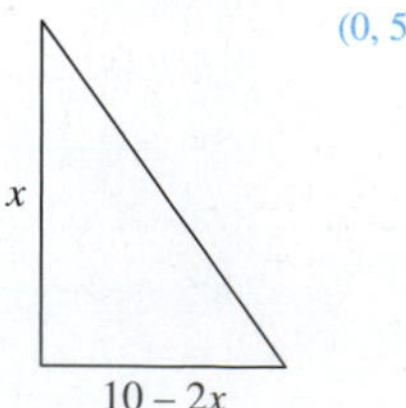

51. Vivienne owns a wedding cake business. The price she charges per wedding cake when x cakes are scheduled to be made in a month is given by the equation $p = 700 - 50x$. $\{1, 2, \ldots, 13\}$

52. Janelle sells gourmet praline cookies through the Internet. The price she charges per order depends on the dozens of cookies ordered. If x dozen cookies are ordered, the price of the order is $p = 150 - 6x$. $\{1, 2, \ldots, 24\}$

53. The total cost of joining a gym for x months can be modeled by the equation $C = 75 + 40x$. $\{0, 1, 2, \ldots\}$

54. The total cost of joining a professional organization for x years is $C = 75x$. $\{0, 1, 2, \ldots\}$

Mix 'Em Up!

Find the domain and range of each function.

55. The average temperature in Stone Mountain, Georgia, is a function of the month. (Source: http://www.accuweather.com)

Month	Average Temperature
January 2011	39°F
February 2011	49°F
March 2011	55°F
April 2011	63°F

Domain = {January 2011, February 2011, March 2011, April 2011}, Range = {39°F, 49°F, 55°F, 63°F}

56. The average high temperature for Amarillo, Texas, is a function of the date. (Source: http://www.accuweather.com)

Date	Average High Temperature
September 4	85°F
September 11	83°F
September 18	81°F
September 25	79°F

Domain = {Sept. 4, Sept. 11, Sept. 18, Sept. 25}, Range = {79°F, 81°F, 83°F, 85°F}

57. $\{(0, -2), (4, -3), (13, 44), (20, -3)\}$.

58. $\{(-4, 21), (-2, 13), (0, -14), (4, 21)\}$.

59. Domain = $\{-5, -4, -2, 1, 2\}$, Range = $\{0, 3, 4, 5\}$

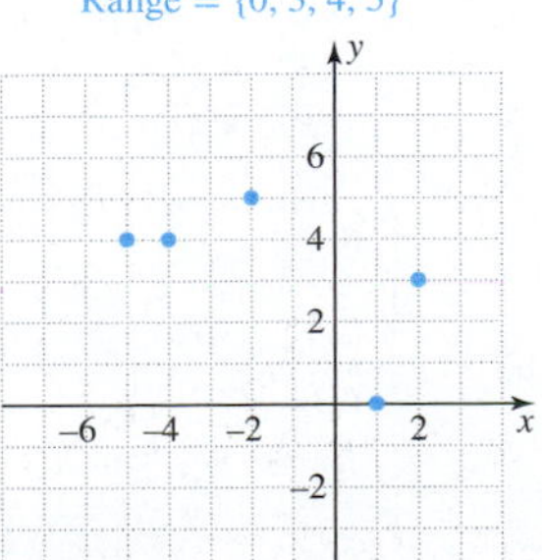

60. Domain = $\{-2, -1, 0, 2, 3\}$, Range = $\{-5, -2, 1, 3, 4\}$

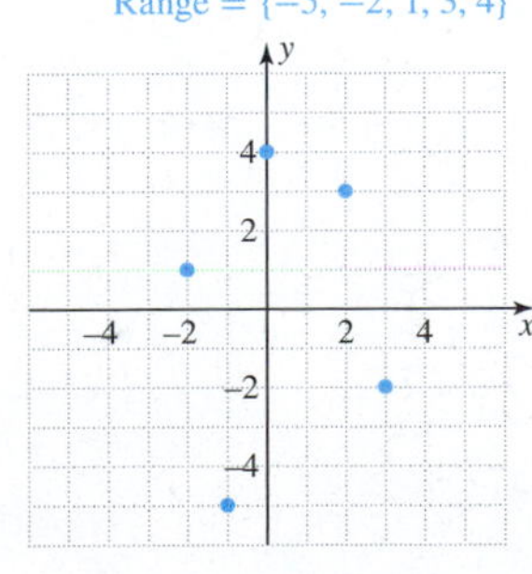

61. Domain = $(-\infty, 7]$, Range = $[0, \infty)$

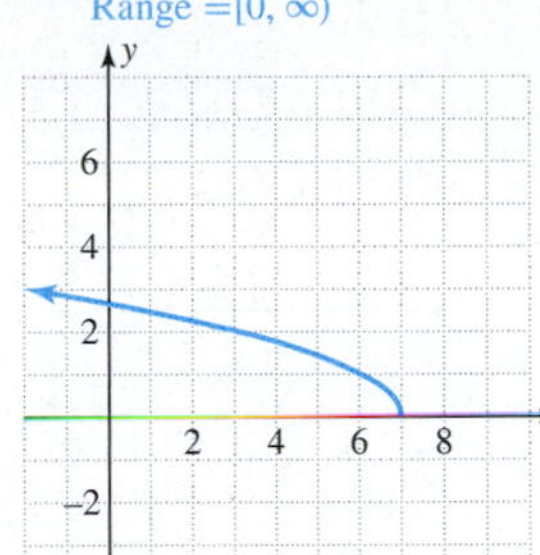

62. Domain = $[2, \infty)$, Range = $(-\infty, 0]$

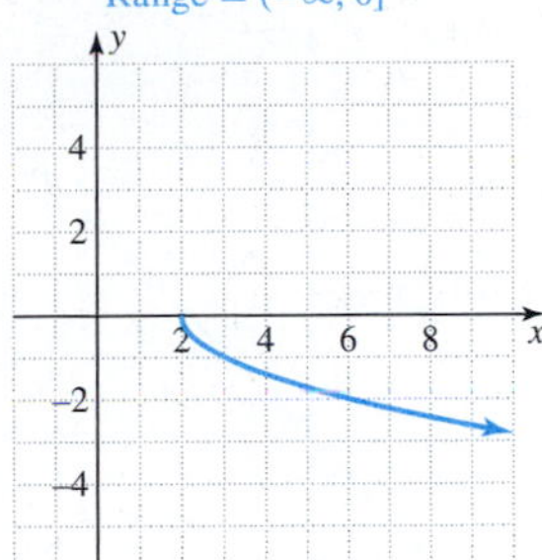

Find the domain of each function.

63. $h(x) = \sqrt{3x - 7}$. $\left[\frac{7}{3}, \infty\right)$

64. $h(x) = \sqrt{7x + 8}$. $\left[-\frac{8}{7}, \infty\right)$

65. $g(x) = x + 4$ Domain = $(-\infty, \infty)$

66. $g(x) = \frac{1}{2}x - 3$ Domain = $(-\infty, \infty)$

67. $f(x) = \frac{x + 10}{x - 19}$ Domain = $(-\infty, 19) \cup (19, \infty)$

68. $f(x) = \frac{x - 8}{x + 12}$ Domain = $(-\infty, -12) \cup (-12, \infty)$

69. $V(x) = x(8 - 2x)(14 - 2x)$ Domain = $(0, 4)$

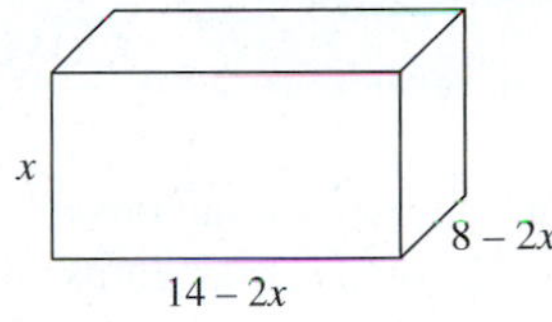

70. $V(x) = 3(24 - 3x)(6 - x)$ Domain = $(0, 6)$

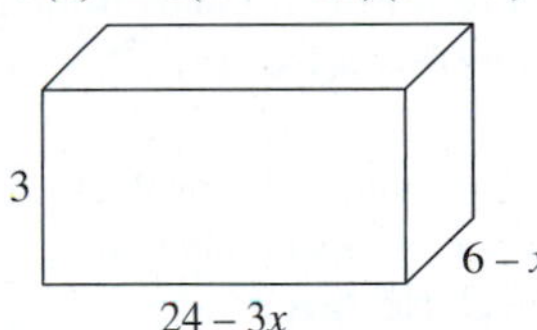

71. The cost that Boris pays for x months of lawn service is modeled by the equation $C = 125x$. Domain = $\{0, 1, 2, 3, \ldots\}$

72. Olivia regularly pays for manicures. When she makes x visits to the nail salon in a month, Olivia's monthly manicure cost can be modeled by the equation $C = 15x$. Domain = $\{0, 1, 2, 3, \ldots\}$

You Be the Teacher!

Correct each student's errors, if any.

73. Find the domain of $f(x) = 4x - 1$.

Petra's work:

$$4x - 1 \geq 0$$
$$4x \geq 1$$
$$x \geq \frac{1}{4}$$

The domain is $\left[\frac{1}{4}, \infty\right)$.

If the function was $f(x) = \sqrt{4x - 1}$, then Petra's work would be correct. The function $f(x) = 4x - 1$ is defined for all real numbers. So, the domain is $(-\infty, \infty)$.

74. Find the domain of $f(x) = \frac{x + 2}{x - 1}$.

Rob's work: The domain is $(-\infty, -2) \cup (-2, \infty)$.

75. Find the domain of the graph.

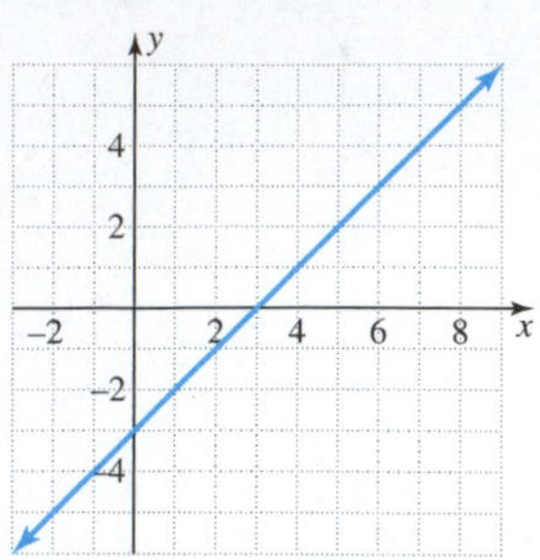

The graph extends left and right indefinitely, so the domain is $(-\infty, \infty)$.

Stephanie's work: The domain is $[3, \infty)$.

Calculate It!

Determine which of the given values are in the domain of the function and sketch a graph of each function.

76. $f(x) = \sqrt{9 - x^2}$; $x = -4, x = -3, x = 0, x = 2, x = 5$

77. $f(x) = \sqrt{x^2 - 9}$; $x = -4, x = -3, x = 0, x = 2, x = 5$

78. $g(x) = \sqrt{x^2 - 16}$; $x = -4, x = -3, x = 0, x = 2, x = 5$

79. $g(x) = \sqrt{16 - x^2}$; $x = -4, x = -3, x = 0, x = 2, x = 5$

Think About It!

Draw an example of a graph of a function that has the given domain and range. Answers vary.

80. Domain $(-\infty, \infty)$ and range $(-\infty, \infty)$

81. Domain $(-\infty, \infty)$ and range $[0, \infty)$

82. Domain $(-\infty, \infty)$ and range $(-\infty, 2]$

83. Domain $(-\infty, \infty)$ and range $\{3\}$

84. Domain $[0, \infty)$ and range $[0, \infty)$

85. Domain $[4, \infty)$ and range $[3, \infty)$

86. Domain $[-3, 3]$ and range $\{-2\}$

87. Domain $[-2, 2]$ and range $[-4, 4]$

GROUP ACTIVITY The Mathematics of Professional Sports

We will explore salaries of professional sports teams and players and present the findings as a relation and determine if it represents a function.

1. Collect data for the average salary of a professional sports team using http://content.usatoday.com/sportsdata/baseball/mlb/salaries/team.

a. Click the link of a professional sports league (MLB, NFL, NBA, or NHL) of your choosing.

b. Select a professional sports team in the league you have chosen and record its average salary for the years provided in the database. The most recent year is displayed on the initial page. To obtain data for other years, click the arrow to the right of year, select a year, and click Go. Repeat this process to obtain the data for all years. (Note: The MLB has data available from 1988 until the present. If you choose this organization, record data for every other year.)

2. Present your findings in a table and a scatter plot. Choose an appropriate scale for the axes on your graph.

3. What makes the data a relation? State the domain and range of the relation. Is this relation a function? Why or why not?

4. Use the scatter plot to describe the average salary of your professional sports team. What year had the highest average salary? What year had the lowest average salary?

Chapter 3 / REVIEW

Graphs, Relations, and Functions

What's the big idea? Now that you have completed Chapter 3, you should be able to graph an equation in two variables and determine from that graph if the equation represents a function. You should also be able to determine if relations defined by sets, tables, or mappings are functions. For relations and functions, you should be able to state the domain and range and find output and input values.

The Tools

Listed below are the key terms, skills, formulas, and properties you should know for this chapter.

The page reference is provided if you need additional help with the given topic. The Study Tips will assist in your preparation for an exam.

Study Tips

1. Learn all of the terms, formulas, and properties. Make flash cards and have someone quiz you.
2. Rework problems from the exercises and also the ones you worked in class. Work additional problems from the review exercises.
3. Review the summaries of key concepts.
4. Work the chapter test.
5. Be sure to review the online resources for additional study materials.

Terms

Cartesian coordinate system 148
Dependent variable 182
Domain 165
Domain of a function 193
Evaluating a function 182
First coordinate or x-coordinate 148
Function 178
Function notation 182
Independent variable 182
Input value 165
Linear equation in two variables 150
Midpoint 155
Origin 148
Ordered pair 148
Paired data 156
Output value 165
Quadrant 149
Range 165
Rectangular coordinate system 148
Relation 165
Scatter plot 156
Second coordinate or y-coordinate 148
Solution of an equation in two variables 150
Vertical line test 180
x-axis 148
y-axis 148

Formulas and Properties

- Midpoint formula 155

CHAPTER 3 / SUMMARY

How well do you know this chapter? Complete the following questions to find out! Take a look back at the section if you need help.

SECTION 3.1 The Coordinate System, Graphing Equations, and the Midpoint Formula

1. The rectangular coordinate system consists of two real number lines intersecting at right angles. The horizontal number line is referred to as the x-axis and the vertical number line is referred to as the y-axis. The point where the two cross is called the origin and has the ordered pair (0, 0).
2. The two axes divide the plane into four regions called quadrants. Quadrants are labeled as I, II, III, IV. Quadrant I is the upper-right quadrant. From here we rotate counterclockwise to label the other quadrants.
3. For a point to be in quadrant I, x is positive and y is positive. For a point to be in quadrant II, x is negative and y is positive. For a point to be in quadrant III, x is negative and y is negative. For a point to be in quadrant IV, x is positive and y is negative.
4. A point not in a quadrant lies on one of the axes. For a point to be on the x-axis, x is a real number and y is zero. For a point to be on the y-axis, x is zero and y is a real number.
5. A solution of an equation in two variables is a(n) ordered pair that makes the equation true.

6. The graph of an equation is a(n) picture of all the solutions of the equation.
7. If a point lies on the graph of an equation, it is a(n) solution of the equation.
8. If a point does not lie on the graph of an equation, it is not a solution of the equation.
9. When we plot data points, a(n) scatter plot is formed.
10. The midpoint of a line segment is the point that lies exactly in the middle of the endpoints of the line segment. It is found by averaging the x-values and averaging the y-values.

SECTION 3.2 Relations

11. A(n) relation is a set of ordered pairs.
12. The domain of a relation is the set of all x-values, the set of all input values, the set of all first coordinates, or the set of all starting values.
13. The range of a relation is the set of all y-values, the set of all output values, the set of all second coordinates, or the set of all ending values.
14. To find the domain of a relation from a graph, we read it from left to right to find the starting and ending x-values.
15. To find the range of a relation from a graph, we read it from bottom to top to find the starting and ending y-values.
16. To determine output values for a given input value, find the ordered pair for the relation with the given value as its x-value. The corresponding y-value is the output value.

SECTION 3.3 Functions

17. A function is a relation in which each member of the domain corresponds to exactly one member of the range.
18. To determine if a graph represents a function, we can use the vertical line test. If each line intersects the graph in at most one point, then the graph is a function. If at least one line intersects the graph in more than one point, then the graph is not a function.
19. Function notation is denoted by $f(x)$ and is read f of x. The input of the function is x. The output of the function is $f(x)$. The name of the function is f.
20. If $f(5) = -7$, then the input value is 5 and the output value is -7. The point $(5, -7)$ lies on the graph of the function.

SECTION 3.4 The Domain and Range of Functions

21. The domain of the function is the set of x-values.
22. The range of the function is the set of y-values.
23. To find the domain of a function from a graph, read it from left to right.
24. To find the range of a function from a graph, read it from bottom to top.
25. To determine the domain of an algebraic function, we must determine the values of x for which the function is defined.
26. If the function involves a fraction with a variable in the denominator, we must exclude from the domain values that make the denominator zero since division by zero is undefined.
27. If the function involves the square root of an algebraic expression, we must find the values that make the expression inside the square root greater than or equal to zero to find its domain.

CHAPTER 3 / REVIEW EXERCISES

SECTION 3.1

Plot the given points on the same coordinate system. (*See Objective 1.*)

1. $(-6, 3), (4, 0), (-2, -5)$
2. $(-6, 10), (12, -4), (0, 8)$

Identify the quadrant where each point is located. (*See Objective 1.*)

3. $\left(-2, -\frac{5}{2}\right)$ III
4 $(-3.5, 1.5)$ II

Determine if the ordered pair is a solution of the equation. (*See Objective 2.*)

5. $y = 3x - x^2 + 4; (-2, -6)$ yes
6. $y = -2|3 - x| + 11; (-3, -1)$ yes

Graph each equation. (*See Objective 3.*)

7. $y = (x + 2)^2 - 3$
8. $y = -|x - 2| + 1$

Find the midpoint of the line segment formed by the ordered pairs. (*See Objective 5.*)

9. $(4.6, 2.1)$ and $(-8.2, 11.3)$ $(-1.8, 6.7)$
10. $(-12.5, 7.2)$ and $(-0.9, -4.9)$ $(-6.7, 1.15)$

Solve each problem. (*See Objective 6.*)

11. The table shows the number of lightning deaths per state for the top ten states between 1959 and 2010. (Source: http://www.lightningsafety.noaa.gov/stats/59-10_fatalities_rates.pdf)

State	Rank	Number of Deaths
Florida	1	461
Texas	2	212
North Carolina	3	192
Ohio	4	144
Colorado	5	140
Tennessee	6	140
Louisiana	7	138
New York	8	138
Pennsylvania	9	129
Maryland	10	126

a. Plot each pair of points, where x is the rank and y is the number of lightning deaths.
b. Interpret the meaning of the first and last ordered pairs in the table.

12. The table shows the number of U.S. Representatives and the number of counties in the thirteen original colonies. (Source: www.factmonster.com)

State	Number of U.S. Representatives	Number of Counties in the State
DE	1	3
CT	5	8
GA	13	159
MD	8	23
NH	2	10
MA	10	14
NJ	13	21
NY	29	62
NC	13	100
PA	19	67
RI	2	5
VA	11	95
SC	6	46

a. Plot each pair of points, where x is the number of U.S. Representatives and y is the number of counties in the state.

b. Interpret the meaning of your first and last ordered pairs in the table.

c. What can you conclude, if anything, about the number of U.S. Representatives and the number of counties in the thirteen original colonies?

SECTION 3.2

Express each relation as a set of ordered pairs. (*See Objective 1.*)

13. A group of friends share their all-time favorite movies. The results are shown in the mapping.

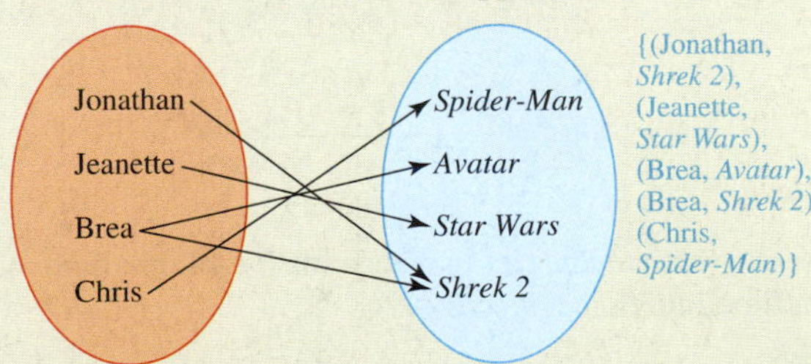

{(Jonathan, *Shrek 2*), (Jeanette, *Star Wars*), (Brea, *Avatar*), (Brea, *Shrek 2*), (Chris, *Spider-Man*)}

14. The budgets (in millions) for selected movies are shown in the mapping. (Source: http://www.the-numbers.com/movies/records/allbudgets.php)

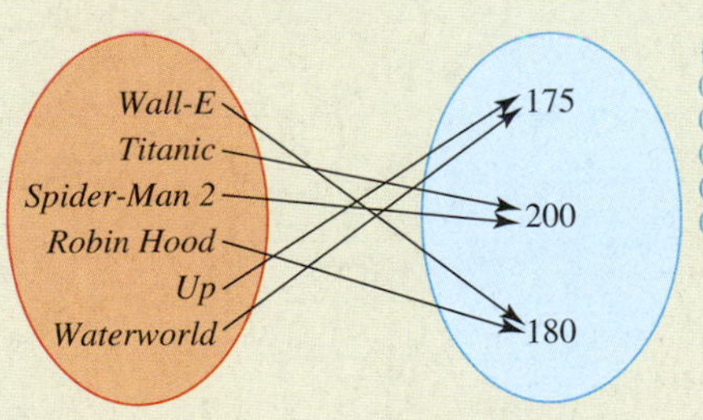

{(*Wall-E*, 180), (*Titanic*, 200), (*Spider-Man 2*, 200), (*Robin Hood*, 180), (*Up*, 175), (*Waterworld*, 175)}

Express each relation as a graph. (*See Objective 1.*)

15.

x	−4	−5	6	0
y	15	1	0	4

16.

x	3	4	5	9
y	−2	−4	0	0

17. The number of viewers of the top five rated syndicated programs for the week of June 7, 2010, is given in the table. Ratings are the percentage of TV homes in the United States tuned into television. (Source: Nielsen Media Research)

Rating	5.8	5.0	4.5	4.3	4.1
Number of viewers (in millions)	8.9	7.3	6.3	6.7	5.8

18. The number of viewers of the top five rated cable TV programs during for week of June 14, 2010, is given in the table. Ratings are the percentage of TV homes in the United States tuned into television. (Source: Nielsen Media Research)

Rating	3.7	3.4	3.3	3.3	3.2
Number of viewers (in millions)	5.9	5.5	5.3	5.3	4.8

Express each relation as an equation and find its domain and range. (*See Objective 1.*)

19. A bowling alley charges $5.25 per person per game. Express the relation between the number of games and the cost as an equation.

20. A skating rink charges $4 per person. Express the relation between the number of skaters and the total cost as an equation.

$y = 4x$, where x is the number of skaters; Domain = {0, 1, 2, . . .}, Range = {0, 4, 8, . . .}

Determine the domain and range of each relation. (*See Objectives 1 and 2.*)

21. {(−27, 10), (−6, 5), (0, 7), (8, −6)}

Domain = {−27, −6, 0, 8}, Range = {−6, 5, 7, 10}

22. {(−1, 5), (−3, 14), (6, 9), (14, −3)}

Domain = {−3, −1, 6, 14}, Range = {−3, 5, 9, 14}

23.

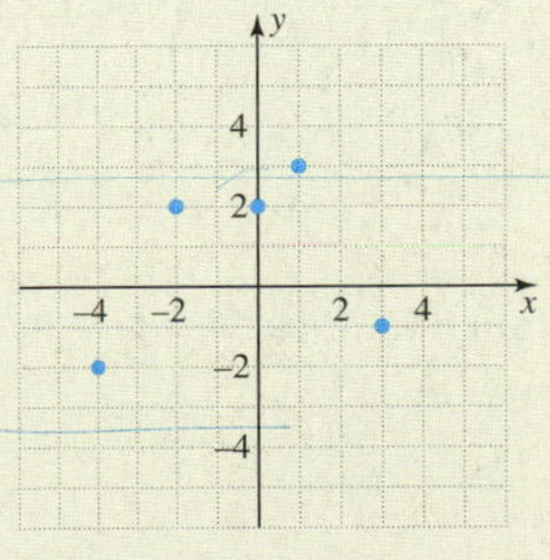

Domain = {−4, −2, 0, 1, 3}, Range = {−2, −1, 2, 3}

24.

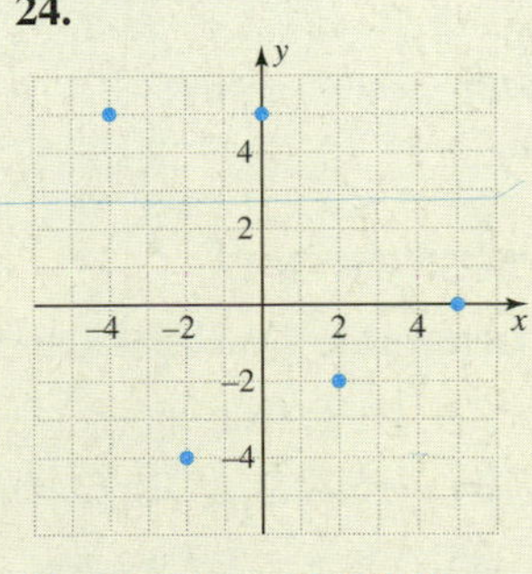

Domain = {−4, −2, 0, 2, 5}, Range = {−4, −2, 0, 5}

25.

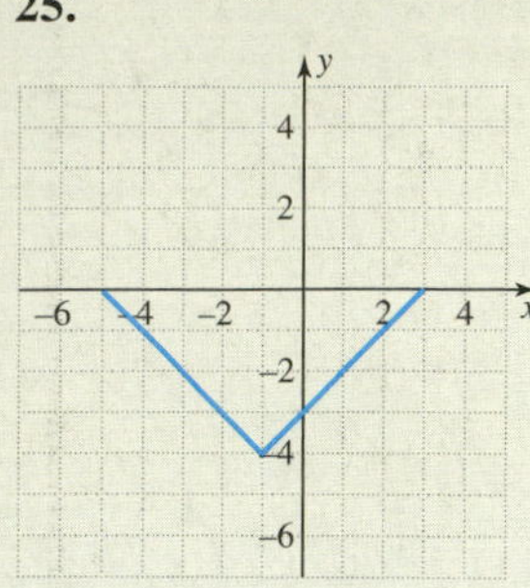

Domain = [−5, 3], Range = [−4, 0]

26.

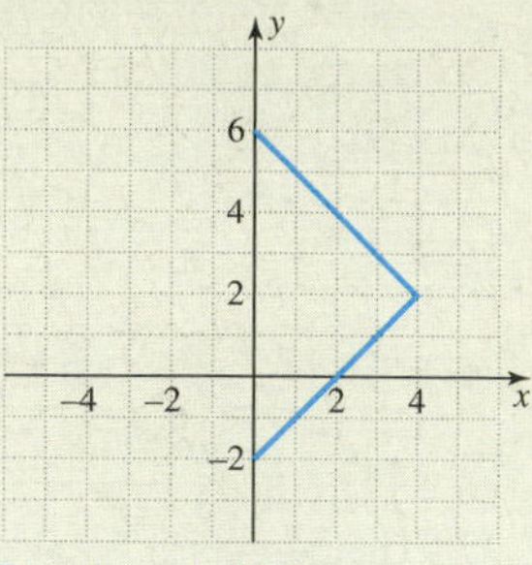

Domain = [0, 4], Range = [−2, 6]

Find the requested information for the given relation. (*See Objective 3.*)

27. Find the output value that corresponds to the input value of −1 for the relation $y = x^2 - 2x$. $y = 3$

28. Find the output value that corresponds to the input value of $\frac{1}{3}$ for the relation $y = 3x - 9x^2$. $y = 0$

29. Find the input value that corresponds to the output value of −4 for the relation $4x - 3y = 20$. $x = 2$

30. Find the input value that corresponds to the output value of 7 for the relation $6x + 3y = -9$. $x = -5$

31. Find the output value that corresponds to the input value of 0 for the relation whose graph is shown. $y = -2, y = 2$

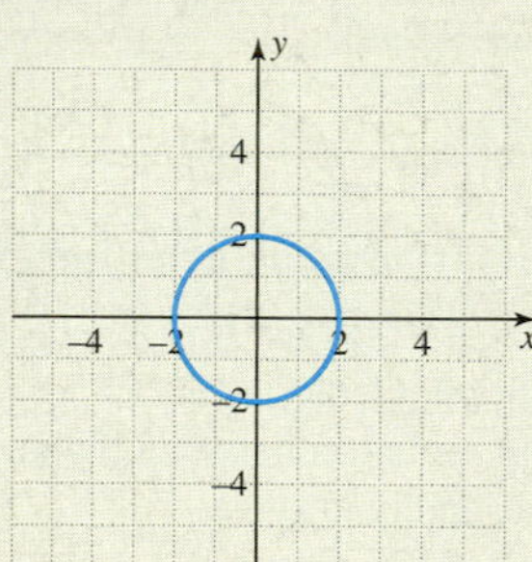

32. Find the input value that corresponds to the output value of 2 for the relation whose graph is shown. $x = -1.5, x = 1.5$

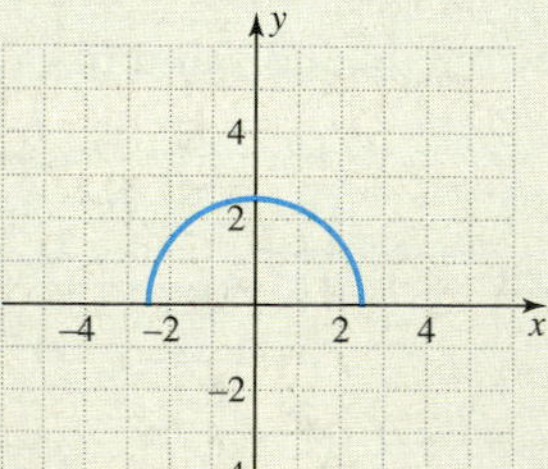

SECTION 3.3

Determine if each relation is a function. (*See Objective 1.*)

33. $\{(-6, 0), (-6, -7), (-6, 1), (-6, 4)\}$ not a function

34. $\{(5, -3), (1, 2), (3, 4), (6, -2)\}$ function

35. $x = 4 - 3y$ function

36. $x = 2y + 5$ function

37. William and his friends share their favorite gamine consoles as shown in the mapping. not a function

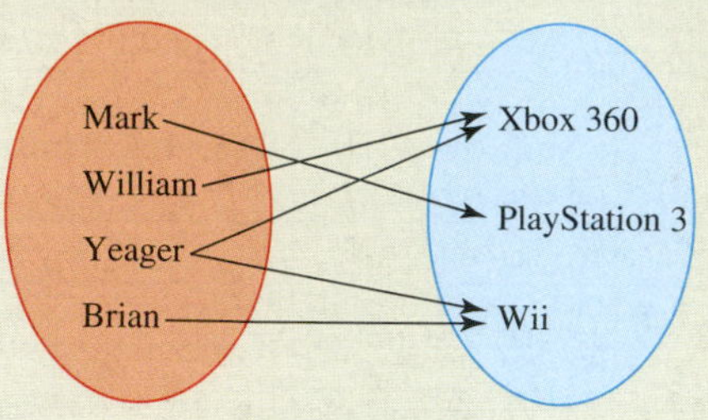

38. David and his friends share their favorite sport programs as shown in the mapping. not a function

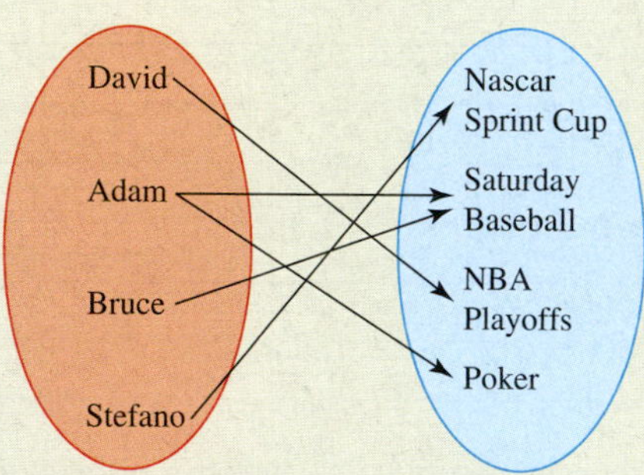

39.

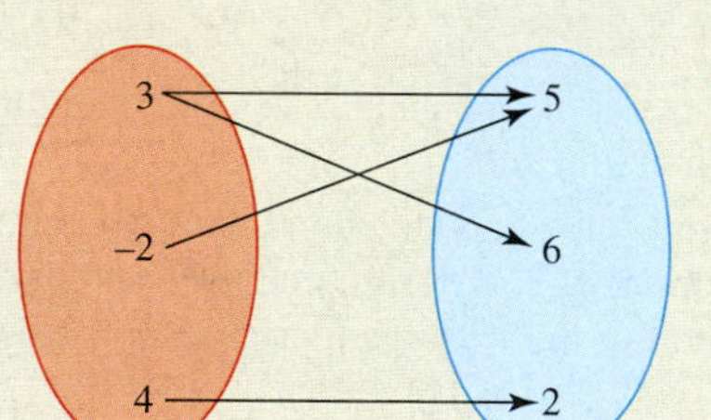

not a function

40.

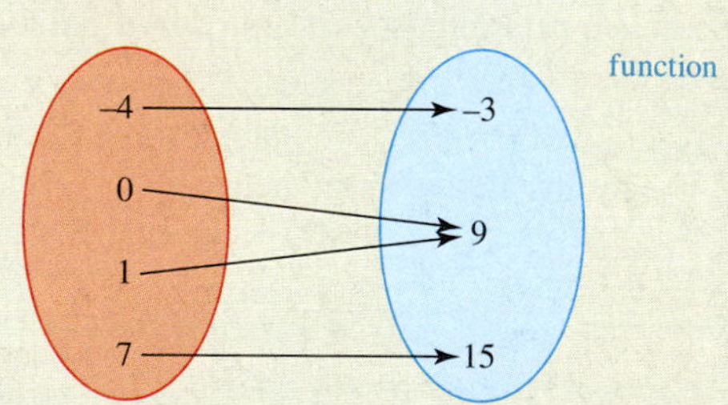

function

Use the vertical line test to determine if each relation is a function. (*See Objective 2.*)

41.

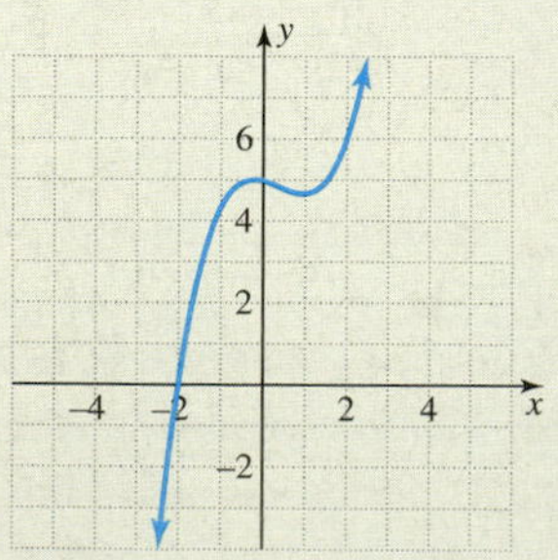

function

42.

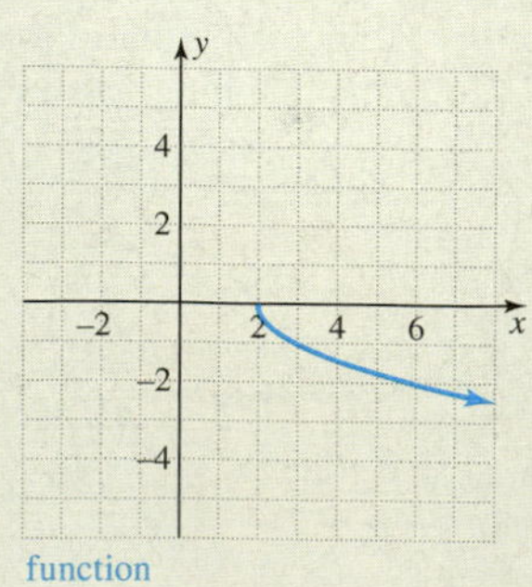

function

43. 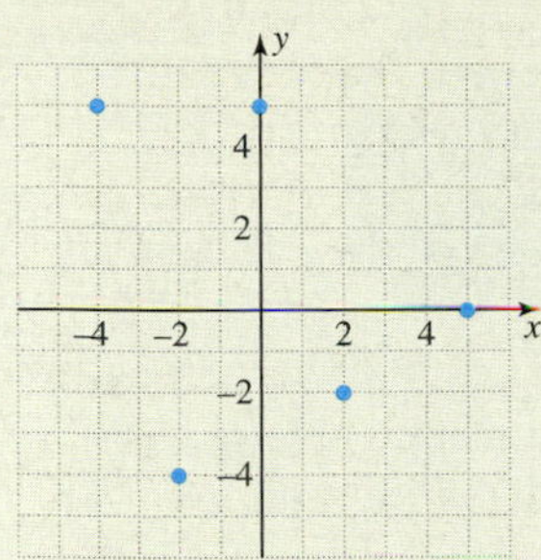

function

44. 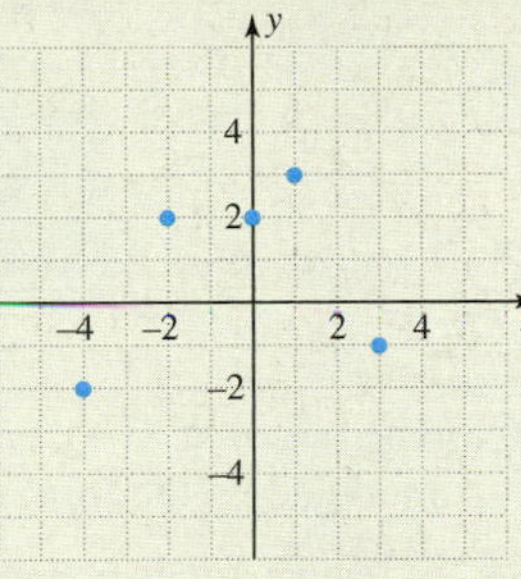

function

Find the requested information. (*See Objective 3.*)

45. Use the graph of $f(x)$ to find $f(3)$. 2

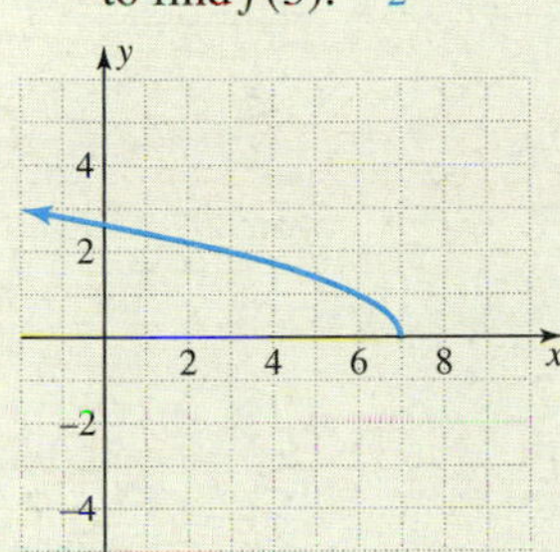

46. Use the graph of $f(x)$ to find $f(-4)$. 0

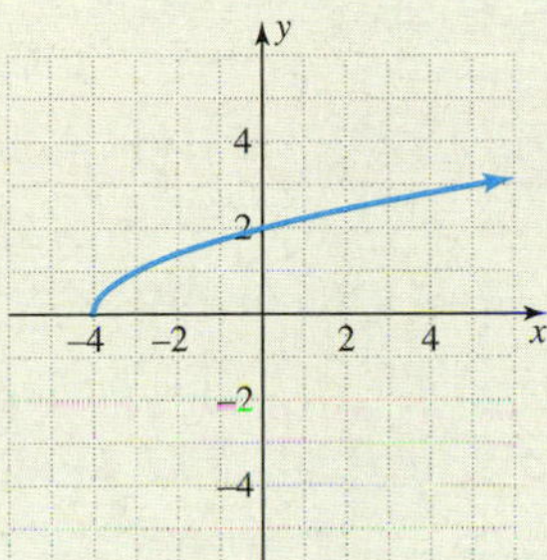

47. Use the graph of $g(x)$ to find $g(-3)$. -5

48. Use the graph of $g(x)$ to find $g(3)$. 4

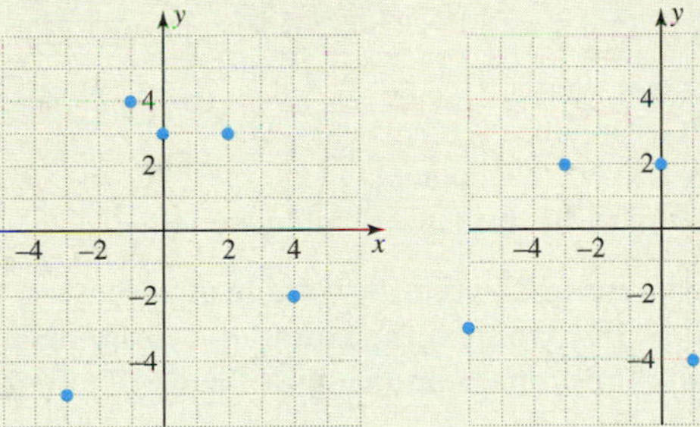

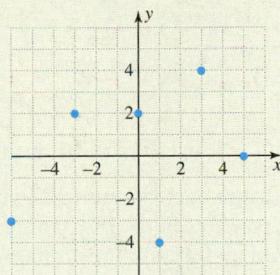

49. Find all x for which $c(x) = 0.75$ given that $c(x)$ is given by Y_1.

X	Y1
-1.5	-5.25
-1	-3
-.5	-1.25
0	0
.5	.75
1	1
1.5	.75

$x = 0.5, x = 1.5$

50. Find all x for which $c(x) = 4$ given that $c(x)$ is given by Y_1.

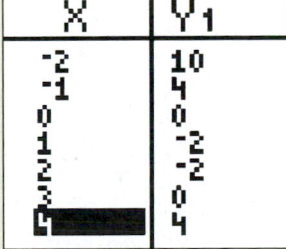

X	Y1
-2	10
-1	4
0	0
1	-2
2	-2
3	0
4	4

$x = -1, x = 4$

51. Find $f(-4)$ if $f(x) = 2x^2 - x + 1$. 37

52. Find $f(0)$ if $f(x) = 3 - 2x^2$. 3

Solve each problem. (*See Objective 4.*)

53. The estimated number of AIDS cases diagnosed can be modeled by the equation $d(x) = 222x^3 - 2712x^2 + 9812x + 27{,}565$, where x is the number of years after 2000. Find $d(8)$ and interpret the result. (Source: http://www.cdc.gov) $d(8) = 46{,}147$; In 2008, there were about 46,147 AIDS cases diagnosed.

54. The percentage of smokers among adults age 18 and over is modeled by the equation $k(x) = -0.44x + 23.11$, where x is the number of years after 2000. Find all x for $k(x) = 17.83$ and interpret the result. (Source: National Center for Health Statistics) $x = 12$; In 2012, about 17.83% of adults age 18 and over were smokers.

SECTION 3.4

Find the domain and range of each function. (*See Objective 1.*)

55. The temperature (in degrees Fahrenheit) in New York, on June 26, 2010, is a function of time of day. (Source: www.accuweather.com)

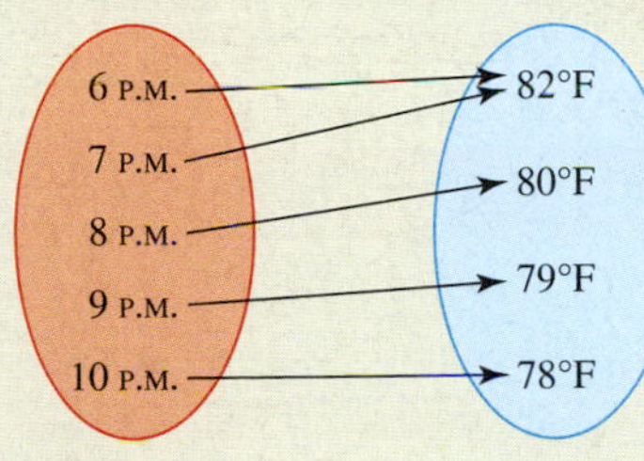

Domain = {6 P.M., 7 P.M., 8 P.M., 9 P.M., 10 P.M.}, Range = {78°F, 79°F, 80°F, 82°F}

56. The temperature (in degrees Fahrenheit) for Chicago, Illinois, is a function of the date. (Source: www.weather.com)

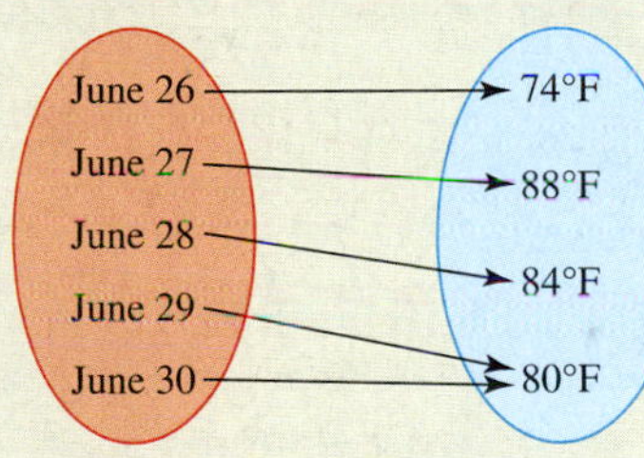

Domain = {June 26, June 27, June 28, June 29, June 30}, Range = {74°F, 80°F, 84°F, 88°F}

57. $\{(-5, -4), (-1, -2), (0, 11), (2, 13)\}$
Domain = $\{-5, -1, 0, 2\}$, Range = $\{-4, -2, 11, 13\}$

58. $\{(-4, 1), (-3, 3), (-2, 0), (4, 8)\}$
Domain = $\{-4, -3, -2, 4\}$, Range = $\{0, 1, 3, 8\}$

59. 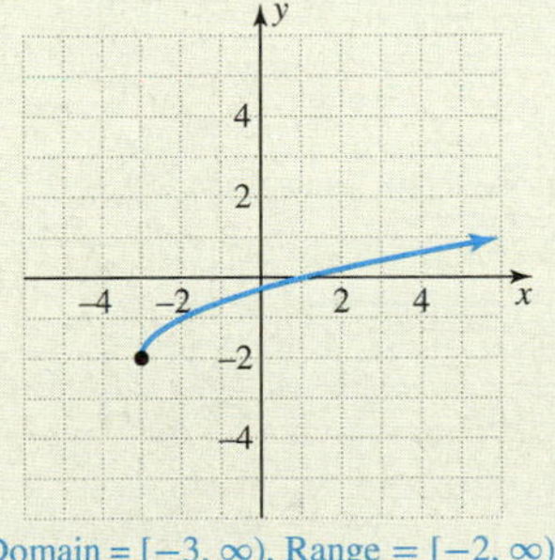

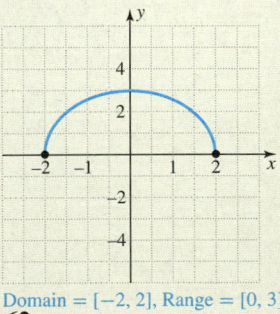

Domain = $[-3, \infty)$, Range = $[-2, \infty)$

60.

Domain = $[-2, 2]$, Range = $[0, 3]$

61. 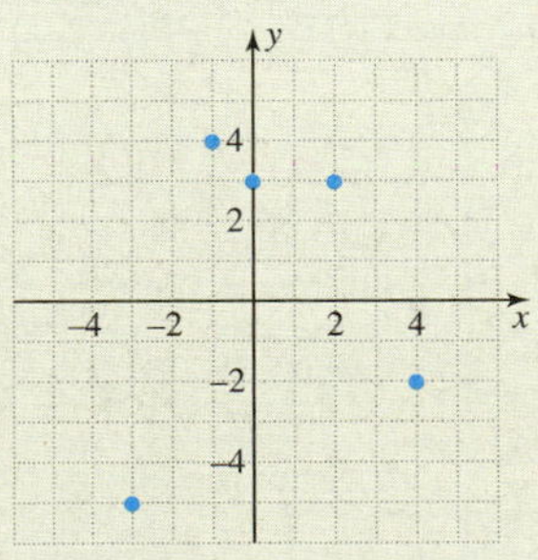

Domain = $\{-3, -1, 0, 2, 4\}$, Range = $\{-5, -2, 3, 4\}$

62. 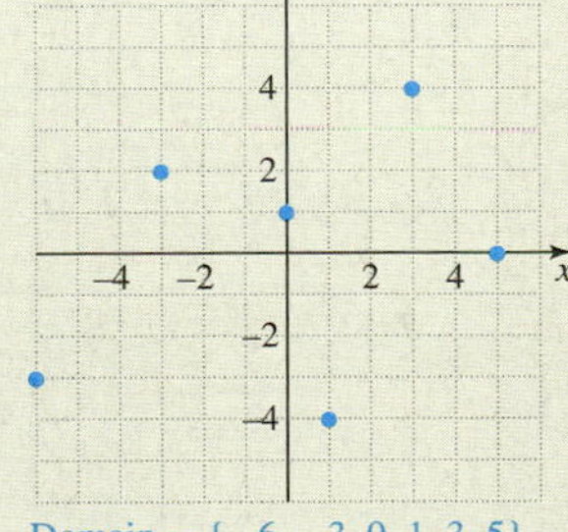

Domain = $\{-6, -3, 0, 1, 3, 5\}$, Range = $\{-4, -3, 0, 1, 2, 4\}$

Find the domain of each function. (*See Objective 3.*)

63. $h(x) = \sqrt{2x - 8}$ $[4, \infty)$

64. $h(x) = \sqrt{3x + 6}$ $[-2, \infty)$

65. $f(x) = \dfrac{x + 6}{x - 1}$ $(-\infty, 1) \cup (1, \infty)$

66. $f(x) = \dfrac{x}{x - 2}$ $(-\infty, 2) \cup (2, \infty)$

Find the domain of each function described. (*See Objective 4.*)

67. The volume of the box is represented by the function $V(x) = x(12 - 2x)(24 - 2x)$. $(0, 6)$

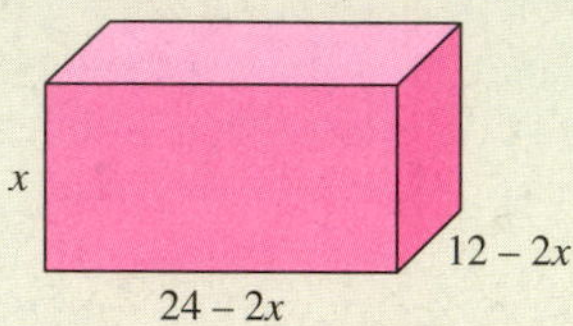

68. The volume of the box is represented by the function $V(x) = 4(18 - 3x)(10 - x)$. $(0, 6)$

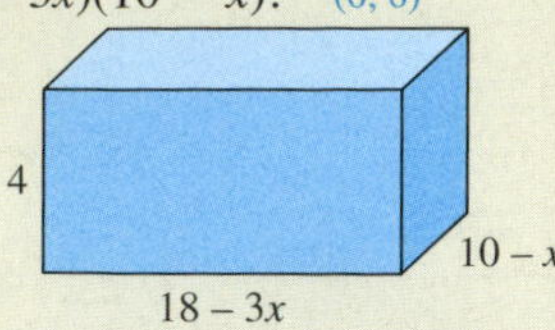

69. The cost that Kos pays for x months of lawn service is modeled by the function $C(x) = 40x$. $\{0, 1, 2, \ldots\}$

70. When Bea makes x visits to the nail salon in a month, her monthly manicure cost can be modeled by the function $C(x) = 18x$. $\{0, 1, 2, \ldots\}$

CHAPTER 3 TEST / GRAPHS, RELATIONS, AND FUNCTIONS

1. If $x > 0$ and $y < 0$, the point $\left(-3x, \frac{1}{2}y\right)$ lies in Quadrant

   a. I b. II (c.) III d. IV

2. The ordered pair that is a solution of $y = -\frac{1}{2}x - 3$ is

   a. $(0, 3)$ (b.) $(-6, 0)$ c. $(2, -2)$ d. $(-4, 1)$

3. The midpoint of the line segment formed by $(-3, 7)$ and $(5, -3)$ is

   a. $(2, 4)$ b. $(-2, 10)$ (c.) $(1, 2)$ d. $(-1, 5)$

4. The domain of the relation $\{(-2, 5), (-1, 2), (0, 1), (1, 2), (2, 5)\}$ is

   (a.) $\{-2, -1, 0, 1, 2\}$ b. $\{1, 2, 5\}$
   c. $\{-2, -1, 0, 1, 2, 5\}$

5. The relation that is *not* a function is

   a. $\{(-2, 4), (-1, 1), (0, 0), (1, 1), (2, 4)\}$

   b. $\{(x, y) | x$ is a person and y is the person's Social Security Number$\}$

   c.

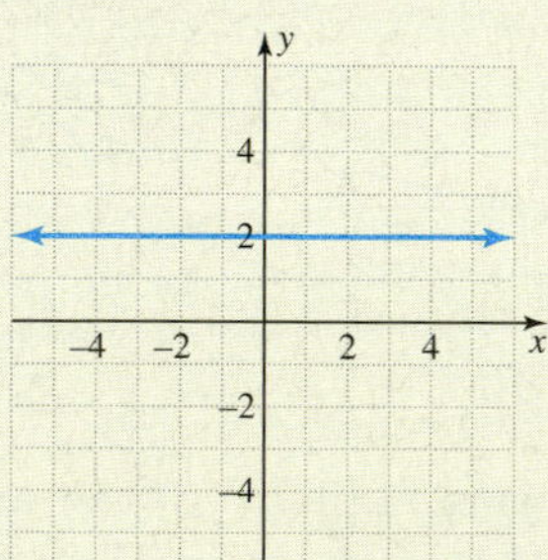

   (d.)

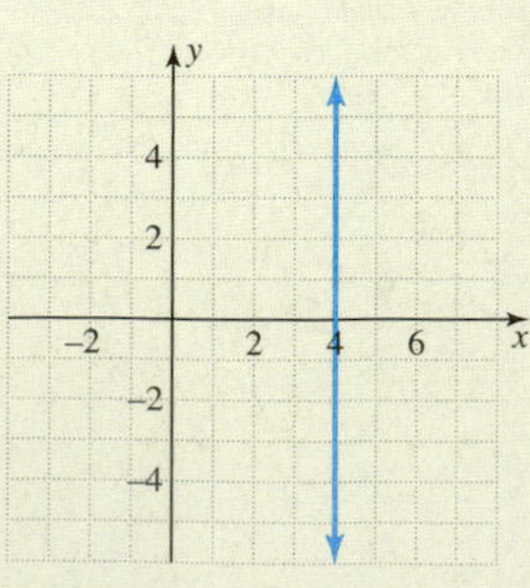

6. If $f(x) = 2x^2 - 5x + 1$, then $f(-4)$ is

   a. -11 (b.) 53 c. 37 d. 5

7. If $f(x)$ is represented by the graph, then the solution(s) of $f(x) = 0$ is/are

   a. 1 b. 3 (c.) $-1, 3$
   d. $-1, 1, 3$

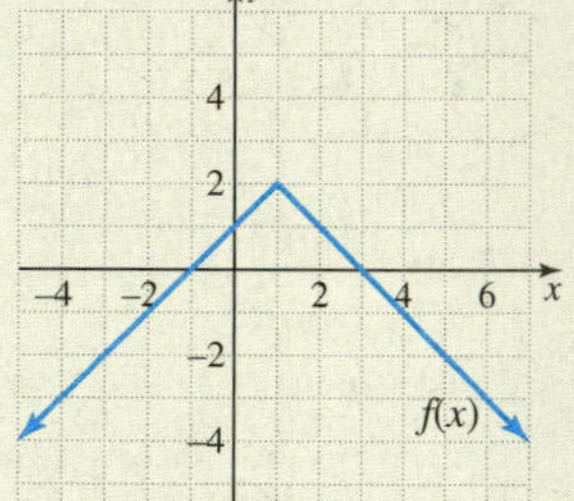

8. In your own words, explain what it means for a relation to be a function. Provide a real-life example of a function.

Graph each equation by creating a table of at least three solutions. Determine if the equation represents a function. State the domain and range of each relation.

9. $y = \frac{2}{3}x - 3$

10. $y = x^2 - 2$

11. $y = |x + 1|$

12. Explain how to use the vertical line test.

13. Determine if the relation is a function. State the domain and range of the relation.

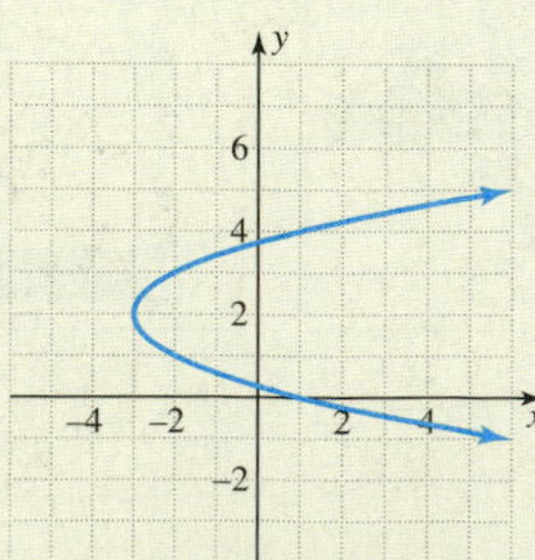

not a function; Domain: $[-3, \infty)$, Range: $(-\infty, \infty)$

Find the domain of each function. Express the domain in interval notation.

14. $f(x) = \dfrac{4x}{x - 9}$ $(-\infty, 9) \cup (9, \infty)$

15. $g(x) = \sqrt{3 - x}$ $(-\infty, 3]$

16. $y = 7x + 2$ $(-\infty, \infty)$

Use the function to find the requested information.

17. Let $f(x) = 9x + 18$. Find $f(0)$ and solve $f(x) = 0$. $f(0) = 18$; $x = -2$

18. The graph of $h(x)$ is given. Find $h(0)$ and solve $h(x) = 5$.

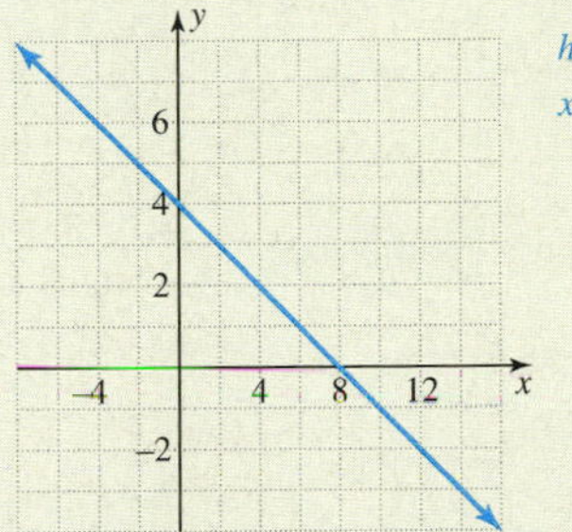

$h(0) = 4$; $x = -2$

For Exercises 19–22, use the following information to answer each question.

The average hours of daylight in Barrow, Alaska, for the first day of each month is given in the table. Let x be the month and y be the number of hours of daylight. (Source: http://www.absak.com/library/average-annual-insolation-alaska)

Month	Hours	Month	Hours
Jan.	0.00	July	24.00
Feb.	4.08	Aug.	24.00
Mar.	9.33	Sept.	14.75
Apr.	14.22	Oct.	11.05
May	19.73	Nov.	5.87
June	24.00	Dec.	0.00

19. Are hours of daylight a function of the month, that is, is y a function of x? Why or why not?

20. What are the domain and range of the relation?

21. Write ordered pairs for each piece of data in the table and create a scatter plot. Let $x = 1$ represent January, $x = 2$ represent February, and so on.

22. What months have the largest average hours of daylight and how much do they have? What months have the lowest average hours of daylight and how much do they have?

CUMULATIVE REVIEW EXERCISES / CHAPTERS 1–3

Determine the requested information about each set. (*Section 1.1, Objective 1*)

1. Use the roster method to write the set: $B = \{x \mid x \text{ is an integer between } -2 \text{ and } 5\}$. $\{-1, 0, 1, 2, 3, 4\}$

2. Use the roster method to write the set: $A = \{x \mid x \text{ is a whole number less than } 2\pi\}$. $\{0, 1, 2, 3, 4, 5, 6\}$

Classify each real number as natural, whole, integer, rational, or irrational. List each classification that applies. (*Section 1.1, Objective 2*)

3. π irrational

4. $-\sqrt{25}$ rational, integer

5. $-4.\overline{6}6$ rational

6. $\sqrt{16}$ natural, whole, integer, rational

Graph each real number on a real number line. (*Section 1.1, Objective 3*)

7. $\sqrt{20}$

8. 4.3

Find the opposite of each number. Assume *a* and *b* represent positive real numbers. (*Section 1.1, Objective 4*)

9. $15a$ $-15a$

10. $-29b$ $29b$

Simplify each absolute value expression. (*Section 1.1, Objective 5*)

11. $-|-39|$ -39

12. $|-61|$ 61

Perform each indicated operation. (*Section 1.2, Objectives 1 and 2*)

13. $15.1 + (-18.2)$ -3.1

14. $26.4 + (-33.9)$ -7.5

15. $\frac{1}{6} + \left(-\frac{11}{6}\right)$ $-\frac{5}{3}$

16. $-\frac{9}{5} + \left(-\frac{2}{5}\right)$ $-\frac{11}{5}$

17. $48 \div \left(-\frac{1}{4}\right)$ -192

18. $\frac{10}{0}$ undefined

19. $(-13.5)(-9)$ 121.5

20. $(-26)(5)$ -130

Use the order of operations to simplify each expression. (*Section 1.2, Objectives 3 and 4*)

21. $-9(-2)^3 - 5(-2)^2 + 1$ 53

22. $13 - [(18 - 6) - (1 - 3)] - 24 \div 12 \cdot 2$ -5

23. $\dfrac{29 - 3\sqrt{27 - 2}}{2|2 - 3|^2}$ 7

24. $\dfrac{-23 + 38}{10 - (-8)}$ $\frac{5}{6}$

25. $-4 + \left(-\frac{7}{5}\right)$ $-\frac{27}{5}$

26. $\dfrac{14 + (-14)}{2 - (-2)}$ 0

Evaluate each expression for the given values. (*Section 1.2, Objective 5*)

27. $2x^2 - 7x + 3$ for $x = -3$ 42

28. $-x^3 + 4x - 7$ for $x = 1$ -4

29. $b^2 - 4ac$ for $a = 6, b = -3, c = 0$ 9

30. $15x + 10y$ for $x = 0, y = 4$ 40

Find both the additive inverse and multiplicative inverse of each number. (*Section 1.3, Objective 1*)

31. 17 $-17, \frac{1}{17}$

32. -6 $6, -\frac{1}{6}$

Apply the commutative, associative, and distributive properties to rewrite each given expression as an equivalent expression. (*Section 1.3, Objective 2*)

33. $24a - 18$ $-18 + 24a$

34. $(11 - x) + 2y$ $11 - x + 2y$

35. $(38r)s$ $38(rs)$

36. $-4(5x - 2y + 6)$ $-20x + 8y - 24$

Simplify each expression. (*Section 1.3, Objective 3*)

37. $3x(10 - 2x) - 27x^2 - 2(5x - 6)$ $-33x^2 + 20x + 12$

38. $-\frac{2}{3}x - 17 + \frac{1}{3}x + 8$ $-\frac{1}{3}x - 9$

39. $\frac{4}{3}x + 1 - \frac{3}{5}x + \frac{3}{2}$ $\frac{11}{15}x + \frac{5}{2}$

40. $5(2x + 7) - 2(x - 12)$ $8x + 59$

Translate each phrase or sentence into an algebraic expression, equation, or inequality. Use the variable *x* to represent the unknown number. (*Section 1.3, Objective 4*)

41. One-fourth the difference of a number and 12 $\frac{1}{4}(x - 12)$

42. Six less than three times a number is more than the number plus 9. $3x - 6 > x + 9$

Solve each problem. (*Section 1.3, Objective 5*)

43. Jacob has collected 120 dimes and nickels. If *x* represents the number of nickels he collected, write an expression for the number of dimes he collected. $120 - x$

44. Referring to Exercise 43, if Jacob's coins total \$9.60, write an algebraic expression that represents the total value of his coins. $0.05x + 0.10(120 - x) = 9.6$

Determine if each equation is a linear equation. If it is a linear equation, determine if $x = -2$ is a solution of the equation. (*Section 2.1, Objective 1*)

45. $-4x - 13 = -5$ linear; It is a solution.

46. $2x^2 + 7 = 14$ not linear

Use the addition and multiplication properties of equality to solve each linear equation. (*Section 2.1, Objectives 2–4*)

47. $17.8 - d = 12.1$ $\{5.7\}$

48. $\frac{5x - 12}{6} = 3$ $\{6\}$

49. $\frac{a - 4}{5} = 2$ $\{14\}$

50. $3x - 13 = 8$ $\{7\}$

Solve each linear equation. If there is no solution, then write ∅ for the answer. If there are infinitely many solutions, then write ℝ for the answer. (*Section 2.1, Objectives 4–6*)

51. $29 + 8x = 4(2x - 3)$ ∅

52. $\frac{7x}{6} - \frac{3x}{2} = \frac{2}{3}$ $\{-2\}$

53. $4(x - 2) - 5x = 2x - 3x - 8$ ℝ

54. $7(x + 3) - 4x = x - 5$ $\{-13\}$

Translate each statement into a linear equation and solve the problem. (*Section 2.2, Objective 1*)

55. The product of a number and one-third is 12. Find the number. 36

56. The quotient of a number and four is 80. Find the number. 320

57. Four less than three times a number yields 17. Find the number. 7

58. Twice the sum of two consecutive even numbers is 76. Find the numbers. 18, 20

Find the measure of each angle described or pictured. (*Section 2.3, Objective 1*)

59. Find the measure of an angle whose complement is 15° less than twice the measure of the angle. 35°

60. Find the measure of an angle whose supplement is 42° more than twice the measure of the angle. 46°

61. 62°, 28°

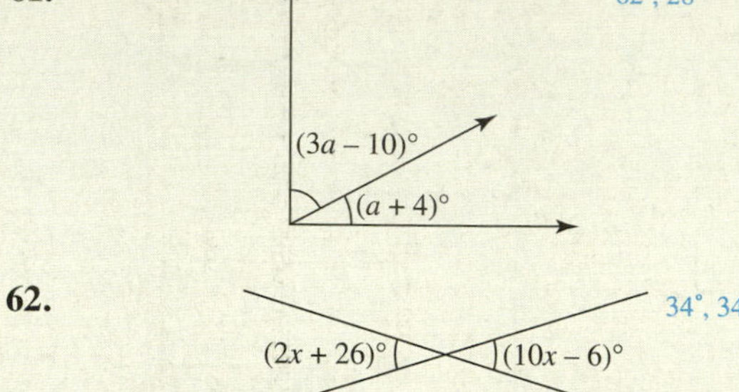

62. 34°, 34°

Use the appropriate formula to solve each problem. (*Section 2.3, Objective 2*)

63. Find the height of a triangle whose area is 168 ft² and whose base is 12 ft. 28 ft

64. The length of a rectangle is 6 ft less than three times its width. If the perimeter is 132 ft, find the length and width of the rectangle. length is 48 ft and width is 18 ft

Solve each formula for the specified variable. (*Section 2.3, Objective 4*)

65. $a = b - 3cd$ for c $c = \frac{b - a}{3d}$

66. $x = \frac{1}{2}y + 3zw$ for y $y = 2x - 6zw$

Graph the solution set of each inequality on a number line and express the solution set in interval notation and set-builder notation. (*Section 2.4, Objective 1*)

67. $7 < x$

68. $-12 \geq x$

Solve each inequality. Graph the solution set and write each answer in interval notation and set-builder notation. (*Section 2.4, Objective 2*)

69. $2(a + 7) - 8 \leq a + 10$

70. $\frac{1}{3}(x + 15) < -\frac{4}{3}(x - 6)$

In Professor Long's algebra class, the course grade is determined by tests (40%), homework (10%), quizzes (20%), and a final exam (30%). (*Section 2.4, Objective 3*)

71. Ashlee's test average is 82, homework average is 96, and quiz average is 71. What must she get on the final exam to earn a *B* in the class, if a *B* is 80–89? between 78 and 100

72. Jennifer's test average is 68, her homework average is 80, and her quiz average is 63. What must she get on the final exam to earn a *C* in the class, if a *C* is 70–79? between 74 and 100

Find the intersection and union of the sets. Write the solution in interval notation. (*Section 2.5, Objectives 1 and 2*)

73. $A = [-13, 1], B = [-12, 4]$ $A \cap B = [-12, 1]$ $A \cup B = [-13, 4]$

74. $A = (4, 20], B = [3, 18)$ $A \cap B = (4, 18)$ $A \cup B = [3, 20]$

Solve each compound inequality. Write the solution in interval notation. (*Section 2.5, Objective 3*)

75. $x > -7$ and $x > -5$ $(-5, \infty)$

76. $x \leq 3$ and $x \leq 7$ $(-\infty, 3]$

Solve each compound inequality. Write the solution in interval notation. (*Section 2.5, Objective 4*)

77. $x \geq 8$ or $x \geq 15$ $[8, \infty)$

78. $5x + 13 > 18$ or $7x - 14 \leq -13$ $\left(-\infty, \frac{1}{7}\right] \cup (1, \infty)$

Solve each equation. (*Section 2.6, Objectives 1–3*)

79. $|11 - x| = 23$ $\{-12, 34\}$

80. $|7 - 2x| = |3x + 16|$ $\left\{-23, -\frac{9}{5}\right\}$

81. $|2x + 10| = 0$ $\{-5\}$

82. $|26 - 2x| = 0$ $\{13\}$

Solve each problem. (*Section 2.6, Objective 3*)

83. Suppose a digital scale reflects a weight (in pounds) with an absolute error of 0.2 1b. If someone weighs 110.1 according to the scale, find the exact weight of the person. The exact weight of the person is between 109.9 lb and 110.3 1b.

84. Suppose a countertop food scale reflects a weight (in ounces) with an absolute error of 0.1 oz. If a piece of meat weighs 4 oz according to the scale, find the exact weight of the piece of meat. The exact weight of the piece of meat is between 3.9 oz and 4.1 oz.

85. Plot the three ordered pairs, $(-12, 25)$, $(10, -5)$, and $(24, 40)$, on the same rectangular coordinate system. Then specify the quadrant where each point is located. (*Section 3.1, Objective 1*)

86. Graph $y = -(x - 2)^2 + 4$. (*Section 3.1, Objective 3*)

87. Find the midpoint of the line segment formed by $(4.3, -16.2)$ and $(-18.5, -21.4)$. (*Section 3.1, Objective 5*) $(-7.1, -18.8)$

88. The table shows the number of Honda motorcycles sold (in thousands) for the years 2006 to 2010. (Source: http://world.honda.com/investors/financial_data/segment/) (*Section 3.1, Objective 6*)

Years after 2006	0	1	2	3	4
Number of Honda motorcycles sold	10,271	10,369	9320	10,114	9639

a. Write the ordered pairs that represent the given data, where x is the number of years after 2006 and y is the number of units of Honda motorcycles sold.

b. Create a scatter plot of the data.

c. What can you conclude about Honda motorcycle sales?

Express each relation as a graph. (*Section 3.2, Objective 1*)

89. The price of Walt Disney Company stock in the beginning of January of the specified year is given in the table. (Source: http://finance.yahoo.com/q/hp?s=DIS)

Year	2006	2007	2008	2009	2010	2011
Price (in dollars)	25.7	35.03	29.83	21.46	30.60	39.29

90. The percentage of active reach of the top five global Web companies in 2010 is given in the table. Use rating as x and percentage of active reach as y. (Source: http://www.nielsen.com)

Rating	1	2	3	4	5
Company	Google	Yahoo!	Facebook	Bing	YouTube
Percentage of active reach	76.0	65.0	61.9	56.2	47.6

91.

x	5	9	16	28
y	−8	−12	21	15

Write each relation in the specified form and state its domain and range. (*Section 3.2, Objective 1*)

92. According to the Nielsen Ratings, the total number of viewers (in millions) for the four major networks in January of a recent year is shown in the mapping. Write this relation as a set of ordered pairs. (Source: http://tvbythenumbers.zap2it.com)

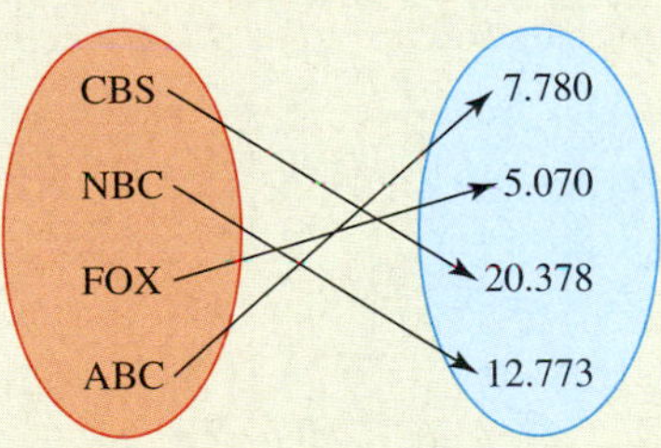

{(CBS, 20.378), (NBC, 12.773), (FOX, 5.070), (ABC, 7.780)}; Domain = {CBS, NBC, FOX, ABC}, Range = {5.070, 7.780, 12.773, 20.378}

93. A bus company runs a route between New York City and Philadelphia and charges \$12.00 per person for a one-way ticket. Express the relation between the number of one-way tickets sold and the revenue the bus company earns as an equation.

94. Find the domain and range of the relation given by the graph. (*Section 3.2, Objective 2*)

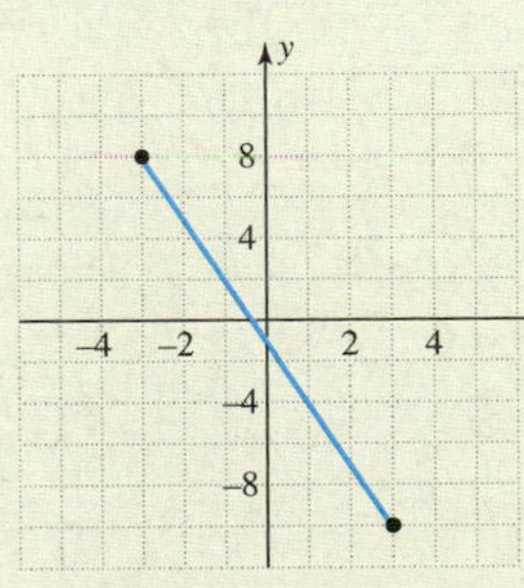

Domain = [−3, 3], Range = [−10, 8]

Find the requested information. (*Section 3.2, Objective 3*)

95. Find the output value that corresponds to the input value of 3 for the relation $\{(-5, -2), (-3, 0), (3, 8), (3, -5)\}$. $y = 8, y = -5$

96. Find the input values that correspond to the output value of 0 for the relation whose graph is shown.

$x = -2, x = 0, x = 3$

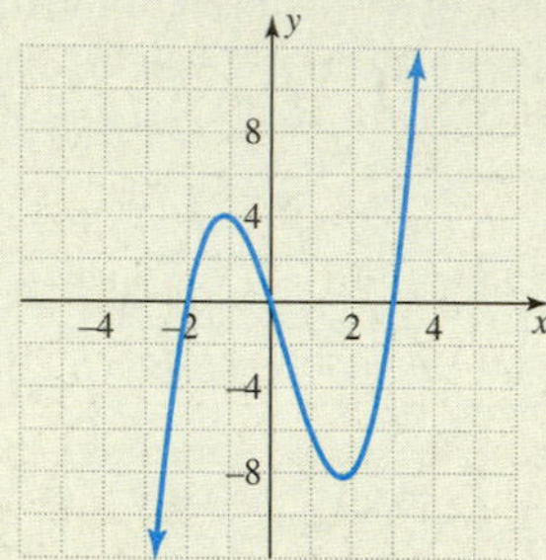

97. Mary and her friends share their 2008 U.S. Presidential election votes as shown in the mapping. Determine if the relation is a function. If not, explain why. (*Section 3.3, Objective 1*)

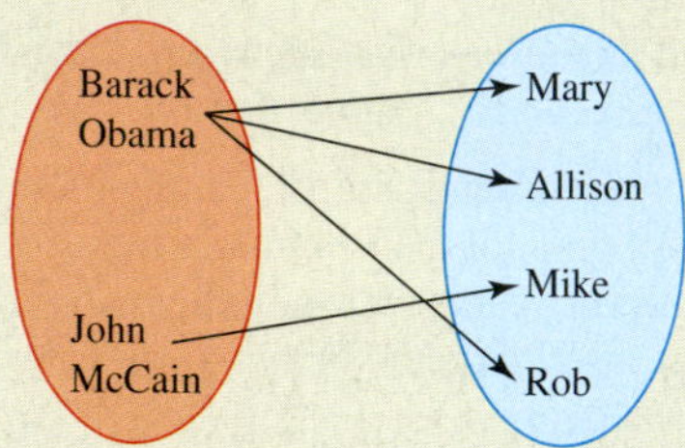

not a function; The x-value Obama corresponds to more than one y-value.

Find the requested information. (*Section 3.3, Objective 3*)

98. Find $f(-1)$ if $f(x) = 6x - 14$. -20

99. Find $f(-4)$ if $f(x) = 12x + 23$. -25

100. Use the graph of $h(x)$ to find all x for which $h(x) = 2$. $x = 4$

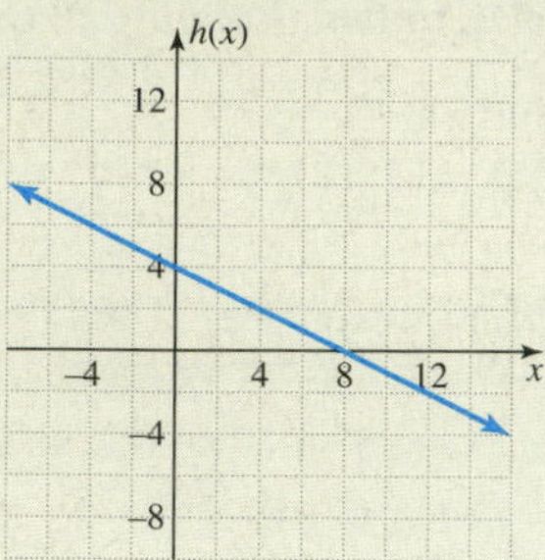

101. Find $f(2)$ if $f(x) = -x^2 + 3x$. 2

102. Find $f(3)$ if the function $f(x)$ is given by Y_1. 7.5

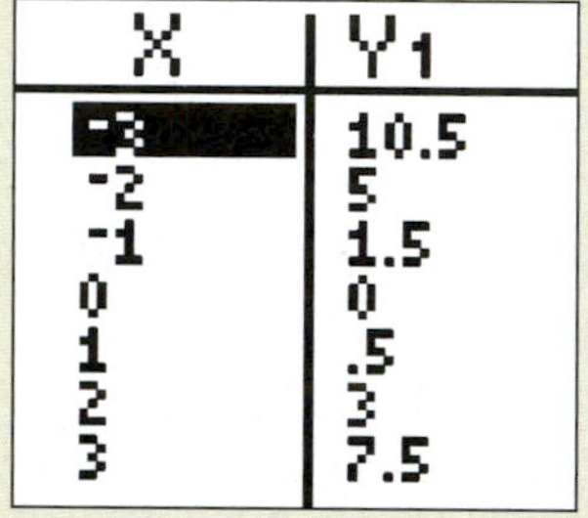

X	Y1
-3	10.5
-2	5
-1	1.5
0	0
1	.5
2	3
3	7.5

103. The life expectancy for all sexes and races can be modeled by the equation $f(x) = 0.000158x^3 - 0.0104x^2 + 0.372x + 70.813$, where x is the number of years after 1970. Find $f(40)$ and interpret its meaning. Round the answer to the nearest whole number. (Source: National Center for Health Statistics) (*Section 3.3, Objective 4*) $f(40) = 79$; In 2010, the life expectancy for all sexes and races was about 79 yr.

104. Find the domain and range of the function represented by the set $\{(-12, 7), (-8, -20), (15, 0), (0, 19)\}$. (*Section 3.4, Objective 1*)
Domain = $\{-12, -8, 0, 15\}$, Range = $\{-20, 0, 7, 19\}$

Find the domain of each function. (*Section 3.4, Objective 3*)

105. $f(x) = 4x^2 - 1$ $(-\infty, \infty)$

106. $f(x) = \sqrt{6 + 2x}$ $[-3, \infty)$

Linear Functions and Linear Inequalities in Two Variables

CHAPTER 4

Commitment and Perseverance

Commitment and perseverance are necessary to be successful in any college classroom. Some courses may require more commitment and perseverance than others, especially if you find the material somewhat difficult to conquer. We believe that, given the right resources and the right amount of time, you can pass this class. You cannot compare yourself to other students in the classroom since college brings together students of all backgrounds—some may have seen the material last year, some five years ago, and some may never have seen the material. So, do not get frustrated when others seem to learn the concepts faster than you. This does not mean that you will not learn them; it just means you need to have a little more perseverance until you do.

If you believe you can do something and believe that what you are doing is important, you will be committed to doing it. Know that what you are doing in the classroom is important. It is one step in achieving your college education. Visualize the big picture as you take these small steps to reach your goal. Find a person that is supportive of your goals; allow them to hold you to your commitment of being successful.

Question For Thought: How would you evaluate your ability to see things through to the end? Do you typically give up on your commitments or do you stick with them? Think about a time you have given up and a time when you have stuck it out . . . what made the difference?

Chapter Outline

Section 4.1 Linear Functions and Linear Equations in Two Variables 212

Section 4.2 Graphing Linear Equations and Linear Functions 224

Section 4.3 The Slope of a Line 238

Section 4.4 Writing Equations of Lines 255

Section 4.5 Linear Inequalities in Two Variables 269

Coming Up...

In Section 4.4, we will write a linear function that represents the percentage of Hispanics who are high school dropouts and use that function to estimate the percentage of Hispanics who are high school dropouts in 2015.

“What this power is, I cannot say. All I know is that it exists . . . and it becomes available only when you are in that state of mind in which you know exactly what you want . . . and are fully determined not to quit until you get it.”

—Alexander Graham Bell (Scientist, Inventor)

SECTION 4.1 Linear Functions and Linear Equations in Two Variables

In Chapter 3, the concept of a function was introduced. In this chapter, we will focus on a specific type of function, linear functions. We will learn how to identify and graph linear functions and to determine important properties of their graphs. In addition, we will study real-life applications of these functions and learn how to write a function that represents given situations. Lastly, linear inequalities in two variables will be presented.

▶ OBJECTIVES

As a result of completing this section, you will be able to

1. Define and recognize a linear function.
2. Rewrite a linear equation in two variables using function notation.
3. Evaluate a linear function.
4. Find solutions of $f(x) = c$, where c is a real number.
5. Solve applications of linear functions.
6. Troubleshoot common errors.

Samuel purchased a voice and data plan for his smartphone. The plan he selected allows 1350 monthly voice minutes for \$79.99 per month with a charge of \$0.35 for each minute over 1350 min. The data plan is an additional \$30 each month. How much does Samuel owe if he talks 1500 min in a month?

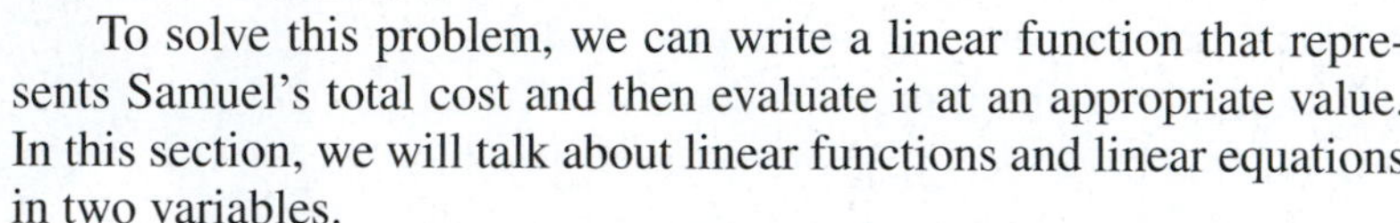

To solve this problem, we can write a linear function that represents Samuel's total cost and then evaluate it at an appropriate value. In this section, we will talk about linear functions and linear equations in two variables.

Linear Functions

Objective 1 ▶ Define and recognize a linear function.

INSTRUCTOR NOTE: Inform students that the form $mx + b$ has a specific meaning, which will be addressed in Section 4.3.

INSTRUCTOR NOTE: Remind students that the independent variable is x in these examples.

In Chapter 3, we defined the concept of a function. Recall that a function is a relation in which each input corresponds to exactly one output. In Section 3.3, we worked with a function of the form $f(x) = 4x - 5$. This function is an example of a *linear function.*

Definition: A **linear function** is a function of the form

$$y = f(x) = mx + b, \quad \text{where } m \text{ and } b \text{ are real numbers.}$$

Some examples of linear functions are

$$y = 2x - 5, \quad y = -\frac{4}{3}x + 7, \quad y = 2, \quad \text{and} \quad f(x) = 3x + 1$$

A function is linear if the exponent of the independent variable is 1.

- The coefficient of the independent variable is the value m.
- The constant term is the value b.

Objective 1 Examples Determine if each function is linear. If it is linear, identify the values of m and b.

	Solutions			
Problems	**Is it linear?**	**Why or why not?**	**Value of m**	**Value of b**
1a. $f(x) = 2x - 6$	Yes	$f(x) = 2x^1 - 6$ The largest exponent of the variable is 1.	$m = 2$	$b = -6$
1b. $f(x) = 3x^2 - 2x + 1$	No	The largest exponent of the variable is 2.	n/a	n/a

1c. $f(x) = \frac{x}{2}$	Yes	$f(x) = \frac{x}{2} = \frac{1}{2}x^1 + 0$ The largest exponent of the variable is 1.	$m = \frac{1}{2}$	$b = 0$
1d. $f(x) = \frac{2}{x}$	No	$f(x) = \frac{2}{x} = 2 \cdot \frac{1}{x}$ The variable is in the denominator. $x^1 = \frac{x^1}{1}$ not $\frac{1}{x^1}$	n/a	n/a
1e. $f(x) = 7$	Yes	$f(x) = 7 = 0x^1 + 7$ The largest exponent of the variable is 1.	$m = 0$	$b = 7$

Student Check 1 Determine if each function is linear. If it is linear, identify the values of m and b.

a. $f(x) = -4$ **b.** $f(x) = \frac{2}{3x}$ **c.** $f(x) = x^2 + 5x - 6$

d. $f(x) = \frac{2x}{3}$ **e.** $f(x) = -\frac{1}{3}x + 1$

Linear Equations in Two Variables in Function Notation

Objective 2 Rewrite a linear equation in two variables using function notation.

If both variables in an equation are raised to an exponent of 1, the equation is a **linear equation in two variables**. Often we encounter equations in two variables that are linear but are not in function notation. Some examples are

$$3x - 4y = 12 \quad \text{and} \quad x + 2y = 8$$

Equations of this form are in the *standard form of a linear equation*.

Definition: The **standard form of a linear equation** in two variables is an equation of the form

$$Ax + By = C, \quad \text{where } A, B, \text{ and } C \text{ are real numbers.}$$

It is important to be able to rewrite a linear equation in standard form in function notation. We will use this skill in later sections of this chapter as well as in Chapter 5.

Procedure: Rewriting a Linear Equation in Two Variables in Function Notation

Step 1: Solve the equation for y.
Step 2: Replace y with $f(x)$.

Objective 2 Examples Rewrite each linear equation using function notation, if possible. Identify m and b.

2a. $y = 3x + 1$ **2b.** $4x - y = 10$ **2c.** $x - 7y = 8$

2d. $y - 3 = 0$ **2e.** $x - 5 = 0$

Solutions

2a.

$$y = 3x + 1 \quad \text{The equation is solved for } y.$$
$$f(x) = 3x + 1 \quad \text{Replace } y \text{ with } f(x).$$

So, $y = 3x + 1$, in function notation, is $f(x) = 3x + 1$ with $m = 3$ and $b = 1$.

INSTRUCTOR NOTE: Remind students that function notation for the standard form of a line is equivalent to the function form of a line.

2b.

$$4x - y = 10$$
$$4x - y - 4x = 10 - 4x \quad \text{Subtract } 4x \text{ from each side.}$$
$$-y = -4x + 10 \quad \text{Simplify.}$$
$$-1(-y) = -1(-4x + 10) \quad \text{Multiply each side by } -1.$$
$$y = 4x - 10 \quad \text{Simplify.}$$
$$f(x) = 4x - 10 \quad \text{Replace } y \text{ with } f(x).$$

So, $4x - y = 10$, in function notation, is $f(x) = 4x - 10$ with $m = 4$ and $b = -10$.

2c.

$$x - 7y = 8$$
$$x - 7y - x = 8 - x \quad \text{Subtract } x \text{ from each side.}$$
$$-7y = -x + 8 \quad \text{Simplify.}$$
$$\frac{-7y}{-7} = \frac{-x + 8}{-7} \quad \text{Divide each side by } -7.$$
$$y = \frac{1}{7}x - \frac{8}{7} \quad \text{Simplify.}$$
$$f(x) = \frac{1}{7}x - \frac{8}{7} \quad \text{Replace } y \text{ with } f(x).$$

So, $x - 7y = 8$, in function notation, is $f(x) = \frac{1}{7}x - \frac{8}{7}$ with $m = \frac{1}{7}$ and $b = -\frac{8}{7}$.

2d.

$$y - 3 = 0$$
$$y - 3 + 3 = 0 + 3 \quad \text{Add 3 to each side.}$$
$$y = 3 \quad \text{Simplify.}$$
$$f(x) = 3 \quad \text{Replace } y \text{ with } f(x).$$
$$f(x) = 0x + 3 \quad \text{Note that the coefficient of } x \text{ is 0.}$$

So, $y - 3 = 0$, in function notation, is $f(x) = 3$ with $m = 0$ and $b = 3$.

2e. This equation cannot be solved for y, so it cannot be written in function notation.

$$x - 5 = 0$$
$$x - 5 + 5 = 0 + 5 \quad \text{Add 5 to each side.}$$
$$x = 5 \quad \text{There is no } y\text{-variable to replace with } f(x).$$

Student Check 2 Rewrite each linear equation using function notation, if possible. Identify m and b.

a. $y = \frac{6}{5}x + 3$ **b.** $7x - y = 1$ **c.** $x - 2y = 0$

d. $y - 4 = 0$ **e.** $x + 2 = 0$

Evaluating Linear Functions

Objective 3 Evaluate a linear function.

In Section 3.3, we learned how to evaluate a function. The process of *evaluating a linear function* is no different. Our goal is to find the output value that corresponds to a given input value.

Procedure: Evaluating a Linear Function

Step 1: Replace the independent variable x with its assigned value.
Step 2: Simplify the result to find the output value.

Objective 3 Examples

Evaluate each function at the given values.

3a. $f(x) = -3x + 2$ at $x = 0$ and $x = \frac{1}{3}$ **3b.** $f(x) = \frac{2}{3}x - 1$ at $x = 0$ and $x = -6$
3c. $f(x) = 7$ at $x = 0$ and $x = \pi$

Solutions **3a.**

x	$f(x) = -3x + 2$	
0	$f(0) = -3(0) + 2$ $f(0) = 0 + 2$ $f(0) = 2$	Replace x with 0. Multiply -3 and 0. Add.
$\frac{1}{3}$	$f\left(\frac{1}{3}\right) = -3\left(\frac{1}{3}\right) + 2$ $f\left(\frac{1}{3}\right) = -1 + 2$ $f\left(\frac{1}{3}\right) = 1$	Replace x with $\frac{1}{3}$. Multiply -3 and $\frac{1}{3}$. Add.

So, $f(0) = 2$ and $f\left(\frac{1}{3}\right) = 1$. Recall this means that the ordered pairs (0, 2) and $\left(\frac{1}{3}, 1\right)$ are solutions of the linear equation $f(x) = -3x + 2$.

3b.

x	$f(x) = \frac{2}{3}x - 1$	
0	$f(0) = \frac{2}{3}(0) - 1$ $f(0) = 0 - 1$ $f(0) = -1$	Replace x with 0. Multiply $\frac{2}{3}$ and 0. Add.
-6	$f(-6) = \frac{2}{3}(-6) - 1$ $f(-6) = -4 - 1$ $f(-6) = -5$	Replace x with -6. Multiply $\frac{2}{3}$ and -6. Add.

So, $f(0) = -1$ and $f(-6) = -5$. Recall this means that the ordered pairs $(0, -1)$ and $(-6, -5)$ are solutions of the linear equation $f(x) = \frac{2}{3}x - 1$.

3c. The function $f(x) = 7$ is equivalent to $f(x) = 0x + 7$. When we replace the variable x with any value, the result is always 7.

INSTRUCTOR NOTE:
Point out that this type of linear function is called a constant function since its y-value never changes.

x	$f(x) = 0x + 7$	
0	$f(0) = 0(0) + 7$ $f(0) = 0 + 7$ $f(0) = 7$	Replace x with 0. Multiply 0 and 0. Add.
π	$f(\pi) = 0(\pi) + 7$ $f(\pi) = 0 + 7$ $f(\pi) = 7$	Replace x with π. Multiply 0 and π. Add.

So, $f(0) = 7$ and $f(\pi) = 7$. Recall this means that the ordered pairs (0, 7) and $(\pi, 7)$ are solutions of the linear equation $f(x) = 7$.

Student Check 3 Evaluate each function at the given values.

a. $f(x) = 5x - 3$ at $x = 0$ and $x = \frac{4}{5}$ **b.** $f(x) = \frac{1}{4}x + 6$ at $x = 0$ and $x = -8$

c. $f(x) = -2$ at $x = 0$ and $x = 2.5$

Solving $f(x) = c$

Objective 4 Find solutions of $f(x) = c$, where c is a real number.

In Objective 3, we found the y-value that corresponds to a given x-value. Now we will find the x-value that corresponds to a given y-value, just as we did in Section 3.3.

When the value $c = 0$, we are solving the equation $f(x) = 0$. The solutions of this equation are called the *zeros* of the function. An illustration of this is shown in Examples 4a and 4b that follow.

Definition: Zero of a Function

The number a is a **zero** of $f(x)$ if $f(a) = 0$.

Procedure: Solving an Equation of the Form $f(x) = c$, Where c Is a Real Number

Step 1: Replace the notation $f(x)$ with the given value of c.
Step 2: Solve the resulting equation for x.

Objective 4 Examples Find the value of x that solves the given equation for each function.

4a. $f(x) = 4x + 8$; Solve $f(x) = 0$. **4b.** $f(x) = -\frac{3}{2}x + 9$; Solve $f(x) = 0$.

4c. $f(x) = \frac{1}{4}x - 5$; Solve $f(x) = -3$.

Solutions **4a.**

$$f(x) = 4x + 8$$

$$0 = 4x + 8 \quad \text{Replace } f(x) \text{ with } 0.$$

$$0 - 8 = 4x + 8 - 8 \quad \text{Subtract 8 from each side.}$$

$$-8 = 4x \quad \text{Simplify.}$$

$$\frac{-8}{4} = \frac{4x}{4} \quad \text{Divide each side by 4.}$$

$$-2 = x \quad \text{Simplify.}$$

So, $f(x) = 0$ when $x = -2$. Recall this means that the point $(-2, 0)$ is a solution of $f(x)$. Also, note that $x = -2$ is a zero of $f(x)$.

4b.

$$f(x) = -\frac{3}{2}x + 9$$

$$0 = -\frac{3}{2}x + 9 \quad \text{Replace } f(x) \text{ with } 0.$$

$$2(0) = 2\left(-\frac{3}{2}x + 9\right) \quad \text{Multiply each side by 2.}$$

$$0 = -3x + 18 \quad \text{Simplify.}$$

$$0 - 18 = -3x + 18 - 18 \quad \text{Subtract 18 from each side.}$$

$$-18 = -3x \quad \text{Simplify.}$$

$$\frac{-18}{-3} = \frac{-3x}{-3} \quad \text{Divide each side by } -3.$$

$$6 = x \quad \text{Simplify.}$$

So, $f(x) = 0$ when $x = 6$. Recall this means that the point $(6, 0)$ is a solution of $f(x)$. Also, note that $x = 6$ is a zero of $f(x)$.

4c.
$$f(x) = \frac{1}{4}x - 5$$
$$-3 = \frac{1}{4}x - 5 \qquad \text{Replace } f(x) \text{ with } -3.$$
$$4(-3) = 4\left(\frac{1}{4}x - 5\right) \qquad \text{Multiply each side by 4.}$$
$$-12 = x - 20 \qquad \text{Simplify.}$$
$$-12 + 20 = x - 20 + 20 \qquad \text{Add 20 to each side.}$$
$$8 = x \qquad \text{Simplify.}$$

So, $f(x) = -3$ when $x = 8$. Recall this means that the point $(8, -3)$ is a solution of $f(x)$.

Student Check 4 Find the value of x that solves the given equation for each function.

a. $f(x) = 2x - 5$; Solve $f(x) = 0$.

b. $f(x) = -\frac{4}{5}x + 8$; Solve $f(x) = 0$.

c. $f(x) = \frac{1}{3}x + 7$; Solve $f(x) = 4$.

Note: *In Examples 3 and 4, it is important to be able to distinguish between the following notations.*

- $f(0) \rightarrow$ *The input value, or* x, *is* 0.
- $f(x) = 0 \rightarrow$ *The output value, or* y, *is* 0.

Applications of Linear Functions

Objective 5 Solve applications of linear functions.

Linear functions arise in many everyday situations. In this section, we will specifically work with examples of linear functions that represent a *base fee* plus a *variable fee*.

- A **base fee** is an amount charged for the use of a service.
- A **variable fee** is a fee that is charged for the length (minutes, miles, and so on) of the service.

Procedure: Solving an Application

Step 1: Write a linear function that represents the situation.
Step 2: Evaluate the function if given an input value or solve an appropriate equation if given an output value.

Objective 5 Examples Solve each problem using a linear function.

5a. Samuel purchased a voice and data plan for his smartphone. The plan he selected allows 1350 monthly voice minutes for \$79.99 per month with a charge of \$0.35 for each minute over 1350 min. The data plan is an additional \$30 each month.

i. Write a linear function $f(x)$ that represents the monthly cost, where x is the number of additional minutes in a month.
ii. Find $f(150)$ and interpret the result.
iii. How many minutes did Samuel talk if his bill was \$284.99?

Solution **5a.** **i.** The total cost of Samuel's plan is the sum of the base fees and the variable fees. The base fees are the charges per month, which are \$79.99 for the voice plan and \$30 for the data plan. The variable fee is the fee for the additional minutes. If x is the number of additional minutes over 1350, the variable fee is $0.35x$. So, a linear function that represents the monthly cost of Samuel's smartphone, where x is the number of additional monthly minutes, is

$$f(x) = \overbrace{79.99 + 30 + 0.35x}^{\text{Voice plan cost + data plan cost + cost for additional minutes}}$$
$$f(x) = 109.99 + 0.35x$$

ii.

$$f(x) = 109.99 + 0.35x \qquad \text{Begin with the function.}$$
$$f(150) = 109.99 + 0.35(150) \qquad \text{Replace } x \text{ with 150.}$$
$$f(150) = 109.99 + 52.50 \qquad \text{Multiply 0.35 and 150.}$$
$$f(150) = 162.49 \qquad \text{Add.}$$

Since $f(150) = 162.49$, Samuel's cost for talking 150 additional minutes in a month, or a total of $1350 + 150 = 1500$ minutes, is \$162.49.

iii.

$$f(x) = 109.99 + 0.35x \qquad \text{Begin with the function.}$$
$$284.99 = 109.99 + 0.35x \qquad \text{Replace } f(x) \text{ with 284.99.}$$
$$284.99 - 109.99 = 109.99 + 0.35x - 109.99 \qquad \text{Subtract 109.99 from each side.}$$
$$175 = 0.35x \qquad \text{Simplify.}$$
$$\frac{175}{0.35} = \frac{0.35x}{0.35} \qquad \text{Divide each side by 0.35.}$$
$$500 = x \qquad \text{Simplify.}$$

So, $f(x) = 284.99$ for $x = 500$. This means that Samuel talked 500 additional minutes in a month, or $1350 + 500 = 1850$ min, for his bill to be \$284.99.

5b. New York City taxi cabs charge an initial fee of \$2.50 plus \$0.40 for each one-fifth mile for a cab ride.

i. Write a linear function $f(x)$ that represents the cab fare, where x is the number of one-fifth miles traveled.

ii. Find the cab fare for a 2-mi ride.

iii. Solve $f(x) = 12.50$ and interpret the result.

Solution **5b.** **i.** The total cab fare is the base fee plus the variable fee. The base fee is the initial fee, or \$2.50. The variable fee is the fee charged for each one-fifth mile traveled, which is $0.40x$. So, a linear function that represents the cab fare, where x is the number of one-fifth miles traveled is

$$f(x) = \overbrace{2.50 + 0.40x}^{\text{Initial fee + cost for each one-fifth mile}}$$
$$f(x) = 0.40x + 2.50$$

ii. To find the cab fare for a 2-mi cab ride, we must determine how many one-fifth miles equal 2 mi. Because 2 mi $= \frac{10}{5}$ mi, we must evaluate the function at $x = 10$, to determine the fare.

$$f(x) = 0.40x + 2.50 \qquad \text{Begin with the function.}$$
$$f(10) = 0.40(10) + 2.50 \qquad \text{Replace } x \text{ with 10.}$$
$$f(10) = 4 + 2.50 \qquad \text{Multiply 0.4 and 10.}$$
$$f(10) = 6.50 \qquad \text{Add.}$$

Since $f(10) = 6.50$, the cab fare for 10 one-fifth miles, or 2 mi, is \$6.50.

iii.

$f(x) = 0.40x + 2.50$	Begin with the function.
$12.50 = 0.40x + 2.50$	Replace $f(x)$ with 12.50.
$12.50 - 2.50 = 0.40x + 2.50 - 2.50$	Subtract 2.50 from each side.
$10 = 0.40x$	Simplify.
$\frac{10}{0.40} = \frac{0.40x}{0.40}$	Divide each side by 0.40.
$25 = x$	Simplify.

So, $f(x) = 12.50$ for $x = 25$. This means that the cab fare for 25 one-fifth miles traveled, or $\frac{25}{5} = 5$ mi, is \$12.50.

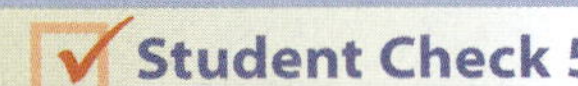

Student Check 5

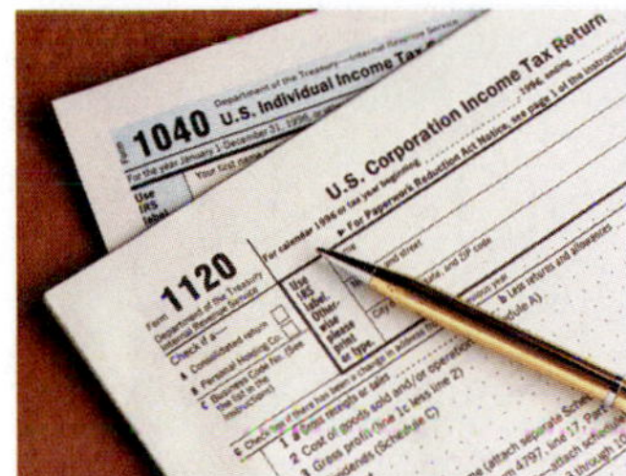

Solve each problem using a linear function.

a. The 2011 U.S. federal income tax for a person filing single with an income over \$8500, but not over \$34,500, is \$850 plus 15% of the amount over \$8500.

i. Write a linear function $f(x)$ that represents the tax owed, where x is the amount of income over \$8500.

ii. Find $f(11{,}800)$ and interpret the result.

iii. If a single person pays \$4075 in federal income tax, what is his annual income?

b. A fitness club charges an initiation fee of \$129 to join plus \$39 per month.

i. Write a linear function $f(x)$ that represents the cost of joining a fitness club for x months.

ii. What is the cost of joining a fitness club for 1 yr?

iii. Solve $f(x) = 1533$ and interpret the results.

Objective 6

Troubleshoot common errors.

Troubleshooting Common Errors

Some common errors associated with linear functions are shown.

Objective 6 Examples

A problem and an incorrect solution are given. Provide the correct solution and an explanation of the error.

6a. Determine if $f(x) = -2$ is a linear function.

Incorrect Solution	Correct Solution and Explanation
This is not a linear function since it doesn't have a variable x with an exponent of 1.	The function $f(x) = -2$ is equivalent to $f(x) = 0x - 2$. Because the variable has an exponent of 1, $f(x) = -2$ is a linear function.

6b. Rewrite $4x - 5y = 10$ using function notation.

Incorrect Solution	Correct Solution and Explanation
$4x - 5y = 10$ $5y = -4x + 10$ $y = \frac{-4x + 10}{5}$ $y = -4x + 2$ $f(x) = -4x + 2$	When $4x$ is subtracted from each side, the left side should be $-5y$ not $5y$. Also, the final result should be obtained by dividing each term by 5. $4x - 5y = 10$ $-5y = -4x + 10$ $y = \frac{-4x + 10}{-5}$ $y = \frac{-4x}{-5} + \frac{10}{-5}$ $f(x) = \frac{4}{5}x - 2$

6c. Let $f(x) = 7x - 3$. Find the value of x for which $f(x) = 0$.

Incorrect Solution	Correct Solution and Explanation
$f(0) = 7(0) - 3$ $f(0) = 0 - 3$ $f(0) = -3$	To solve $f(x) = 0$ means to set the function equal to zero and solve for x. $f(x) = 7x - 3$ $0 = 7x - 3$ $3 = 7x$ $\frac{3}{7} = x$ So, $f(x) = 0$ when $x = \frac{3}{7}$.

ANSWERS TO STUDENT CHECKS

Student Check 1 **a.** Linear, $m = 0, b = -4$ **b.** not linear **c.** not linear **d.** linear, $m = \frac{2}{3}, b = 0$ **e.** linear, $m = -\frac{1}{3}, b = 1$

Student Check 2 **a.** $f(x) = \frac{6}{5}x + 3, m = \frac{6}{5}, b = 3$ **b.** $f(x) = 7x - 1, m = 7, b = -1$ **c.** $f(x) = \frac{1}{2}x, m = \frac{1}{2}, b = 0$ **d.** $f(x) = 4, m = 0, b = 4$ **e.** can't be written in function notation

Student Check 3 **a.** $f(0) = -3, f\left(\frac{4}{5}\right) = 1$ **b.** $f(0) = 6, f(-8) = 4$ **c.** $f(0) = -2, f(2.5) = -2$

Student Check 4 **a.** $x = \frac{5}{2}$ **b.** $x = 10$ **c.** $x = -9$

Student Check 5 **a. i.** $f(x) = 850 + 0.15x$ **ii.** $f(11{,}800) = 2620$; A single person making \$8500 + \$11,800 = \$20,300 must pay \$2620 in federal income tax. **iii.** A person paying \$4075 in taxes earns an annual income of \$30,000. **b. i.** $f(x) = 129 + 39x$ **ii.** $f(12) = 597$; A 1-yr membership costs \$597. **iii.** $x = 36$; It costs \$1533 to be a member of the fitness club for 36 months or 3 yr.

SUMMARY OF KEY CONCEPTS

1. A linear function is a function of the form $f(x) = mx + b$, where m and b are real numbers.
2. To rewrite a linear equation in two variables using function notation, solve the equation for y and replace y with $f(x)$. Equations of the form $x = a$, where a is a real number cannot be written in function notation.
3. To evaluate a linear function, substitute the given value into the function in place of the variable and simplify.
4. To solve an equation of the form $f(x) = c$, where c is a real number, set the function equal to c and solve for x.
5. A linear function $f(x) = mx + b$ can be used to represent real-life situations in which b is a base fee and m is a variable fee.

GRAPHING CALCULATOR SKILLS

The graphing calculator can be used to evaluate functions.

Example: Evaluate $f(x) = \frac{2}{3}x - 1$ at $x = 0$ and $x = -6$.

Method 1: Input the expression in the main window of the calculator.

```
2/3(0)-1
                 -1
2/3(-6)-1
                 -5
```

So, $f(0) = -1$ and $f(-6) = -5$.

Method 2: Input the function into the equation editor and use the TABLE feature.

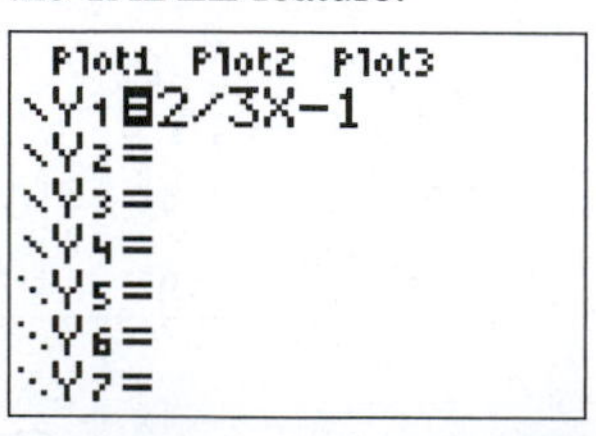

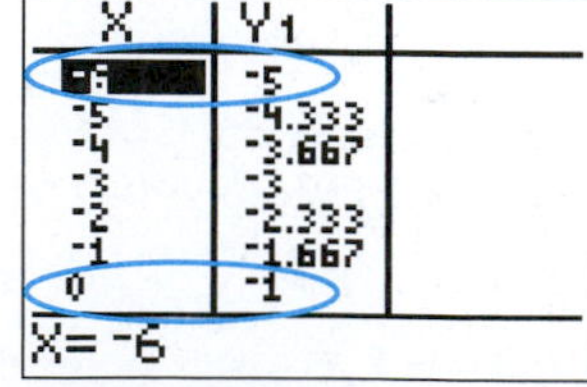

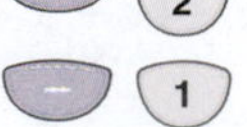

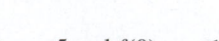

The value in $Y_1 = f(x)$. So, $f(-6) = -5$ and $f(0) = -1$.

Method 3: Input the function into the equation editor, graph, and use the TRACE feature.

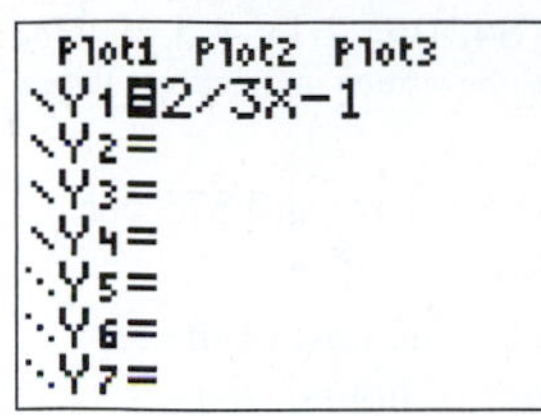

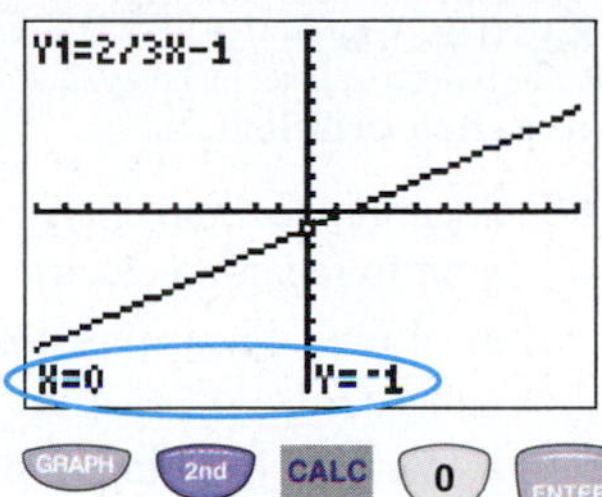

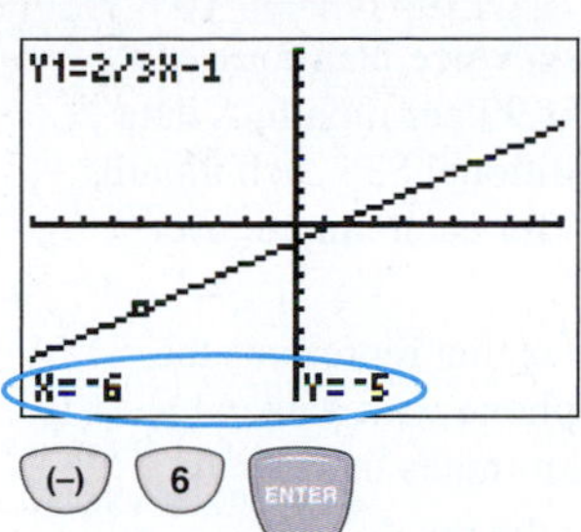

GRAPH 2nd CALC 0 ENTER (–) 6 ENTER

SECTION 4.1 EXERCISE SET

Write About It!

Use complete sentences in your answer to each exercise.

1. What is a linear function? A linear function is a function of the form $f(x) = mx + b$, where m and b are real numbers.
2. How do you evaluate a linear function?
3. How do you solve the equation $f(x) = c$, where c is a real number? You set the function equal to the given number c and solve for x.
4. What is the difference between $f(0)$ and $f(x) = 0$?
5. How do you write a linear equation in two variables in function notation? Can all linear equations in two variables be written in function notation? Explain.
6. Do all linear equations in two variables represent functions? Explain. No, not all linear equations in two variables represent functions. An equation of the form $x = a$, where a is a real number, does not represent a function.

Practice Makes Perfect!

Determine if each function is linear. If it is linear, identify the values of m and b. (*See Objective 1.*)

7. $f(x) = 3x + 4$ linear, $m = 3, b = 4$
8. $f(x) = 5x + 1$ linear, $m = 5, b = 1$
9. $f(x) = \dfrac{1}{2x}$ not linear
10. $f(x) = \dfrac{3}{x}$ not linear
11. $f(x) = -x - 6$ linear, $m = -1, b = -6$
12. $f(x) = -x - 3$ linear, $m = -1, b = -3$
13. $f(x) = x^2 - 2x$ not linear
14. $f(x) = x^2 + 9x$ not linear
15. $f(x) = \dfrac{3}{2}x$
16. $f(x) = \dfrac{2}{5}x$
17. $f(x) = -\dfrac{x}{4} - 8$
18. $f(x) = -\dfrac{x}{6} + 12$
19. $f(x) = 5$ linear, $m = 0, b = 5$
20. $f(x) = -3$ linear, $m = 0, b = -3$

Rewrite each equation using function notation, if possible. Identify m and b. (*See Objective 2.*)

21. $y = 5x + 7$ $f(x) = 5x + 7, m = 5, b = 7$
22. $y = 9x - 1$ $f(x) = 9x - 1, m = 9, b = -1$
23. $x + y = 10$ $f(x) = -x + 10, m = -1, b = 10$
24. $x + y = 15$ $f(x) = -x + 15, m = -1, b = 15$
25. $x - y = 6$ $f(x) = x - 6, m = 1, b = -6$
26. $x - y = 2$ $f(x) = x - 2, m = 1, b = -2$
27. $y = \dfrac{2}{3}x - 5$
28. $y = \dfrac{5}{4}x + 1$
29. $4x + y = -8$ $f(x) = -4x - 8, m = -4, b = -8$
30. $7x + y = -14$ $f(x) = -7x - 14, m = -7, b = -14$
31. $5x - y = 5$ $f(x) = 5x - 5, m = 5, b = -5$
32. $3x - y = 12$ $f(x) = 3x - 12, m = 3, b = -12$
33. $x + 6y = -18$
34. $x + 9y = -9$
35. $4x + 3y = 12$
36. $6x + 5y = 30$
37. $9x - 2y = -72$
38. $8x - 3y = -24$
39. $x - 7y = 0$
40. $x + 4y = 0$
41. $2x - 7y = 0$
42. $3x - 8y = 0$
43. $y + 7 = 0$ $f(x) = -7, m = 0, b = -7$
44. $y + 1 = 0$ $f(x) = -1, m = 0, b = -1$
45. $x - 5 = 0$ can't be written in function notation
46. $x - 3 = 0$ can't be written in function notation

Evaluate each function at the given value. (*See Objective 3.*)

47. $f(x) = 2x + 3; x = 4$ 11
48. $f(x) = 3x + 1; x = 6$ 19
49. $f(x) = -5x + 7; x = \dfrac{1}{5}$ 6
50. $f(x) = -4x - 2; x = \dfrac{3}{4}$ −5
51. $f(x) = \dfrac{4}{7}x + 8; x = -14$ 0
52. $f(x) = \dfrac{2}{9}x + 5; x = -9$ 3
53. $f(x) = 10x - 6; x = 0$ −6
54. $f(x) = -7x + 3; x = 0$ 3
55. $f(x) = -\dfrac{2}{9}x - \dfrac{1}{9}; x = 0$ $-\dfrac{1}{9}$
56. $f(x) = -\dfrac{3}{7}x + \dfrac{2}{7}; x = 0$ $\dfrac{2}{7}$
57. $f(x) = 11; x = 1$ 11
58. $f(x) = 6; x = 5$ 6

Find the value of x for which $f(x) = c$. (*See Objective 4.*)

59. $f(x) = 2x + 4; f(x) = 0$ $x = -2$
60. $f(x) = 4x + 12; f(x) = 0$ $x = -3$
61. $f(x) = -5x + 20; f(x) = 5$ $x = 3$
62. $f(x) = -9x + 72; f(x) = 9$ $x = 7$
63. $f(x) = 5x - 3; f(x) = 0$
64. $f(x) = 7x - 1; f(x) = 0$
65. $f(x) = \dfrac{3}{2}x + 7; f(x) = 0$
66. $f(x) = \dfrac{5}{6}x + 2; x = 0$ $x = -\dfrac{12}{5}$
67. $f(x) = -\dfrac{2}{3}x + 6; f(x) = 4$ $x = 3$
68. $f(x) = -\dfrac{4}{9}x + 9; f(x) = 5$ $x = 9$

Additional answers can be found in the Instructor Answer Appendix.

Solve each problem. (*See Objective 5.*)

69. A cell phone provider offers a voice plan for a cell phone with 1000 min for \$64.99 per month. A data plan can be added for an additional \$35 each month. The company charges \$0.25 for each minute over 1000 min in a month.

a. Write a linear function $f(x)$ that represents the monthly cost of the cell phone with data, where x is the number of additional minutes in a month. $f(x) = 99.99 + 0.25x$

b. Find $f(300)$ and interpret the result.

c. Solve $f(x) = 124.99$ and interpret the result.

70. The Verizon Wireless Nationwide 900 plan costs \$59.99 per month for 900 anytime minutes. Additional anytime minutes cost \$0.40 each. There is also a one-time \$35 activation fee. (Source: www.letstalk.com)

a. Write a linear function $f(x)$ that represents the first month's cost of this cell phone plan as a function of x, the number of additional anytime minutes.

b. Find $f(100)$ and interpret the result.

c. Solve $f(x) = 214.99$ and interpret the result.

71. San Francisco City taxi cab rates are \$3.10 for the first one-fifth mile and \$0.45 for each additional one-fifth mile. (Source: http://www.sfgov.org/site/taxicommission_index.asp?id=8125)

a. Write a linear function $f(x)$ for the cost of a cab ride, where x is the number of additional one-fifth miles traveled. $f(x) = 3.10 + 0.45x$

b. Find $f(19)$ and interpret the result.

c. Solve $f(x) = 16.15$ and interpret the result.

72. A general guideline for taxi cab rates in Seattle, Washington, is an initial fee of \$2.50 plus \$2.00 per mile. (Source: http://www.taxigrab.com/Washington/seattle-taxi-service.html)

a. Write a linear function $f(x)$ for the cost of a cab ride, where x is the number of miles traveled. $f(x) = 2.50 + 2x$

b. Find $f(5)$ and interpret the result.

c. Solve $f(x) = 6.50$ and interpret the result. $x = 2$; The cost of a 2-mi cab ride is \$6.50.

Mix 'Em Up!

Rewrite each equation in function notation, if possible. Identify the values of m and b. Then find $f(0)$ and solve $f(x) = 0$ for x.

73. $5x - 2y = -10$

74. $7x + 3y = 21$

75. $y = -1.1x + 4.4$

76. $y = -3.7x - 11.1$

77. $2y = 6x - 10$

78. $-3y = 4x - 15$

79. $3.6 = 4.8x - y$

80. $2.1 = 0.3x - y$

81. $3y + 6 = 0$

82. $-2y + 8 = 0$

83. $0 = 3.4x + 7.1$ can't be written in function notation

84. $0 = 2.1x - 5.2$ can't be written in function notation

Solve each problem.

85. A car repair shop charges \$250 for parts and \$75 per hour to repair an electric window.

a. Write a linear function $f(x)$ for the cost of the repair as a function of x, the number of hours. $f(x) = 75x + 250$

b. Find $f(5)$ and interpret the result.

c. Solve $f(x) = 400$ and interpret the result.

86. An electrician charges \$50 for a house call plus \$45 per hour for his services.

a. Write a linear function $f(x)$ for the cost of a house call as a function of x hours.

b. Find $f(3)$ and interpret the result.

c. Solve $f(x) = 117.50$ and interpret the result.

87. A fitness club is advertising a special one-time offer. This offer includes a \$19 membership fee plus a \$9 monthly charge.

a. Write a linear function $f(x)$ for the cost of joining the fitness club as a function of x months. $f(x) = 9x + 19$

b. Find $f(24)$ and interpret the result.

c. Solve $f(x) = 343$ and interpret the result.

88. A fitness club charges a one-time enrollment fee of \$99 plus \$19.95 per month to be a member.

a. Write a linear function $f(x)$ for the cost of joining the fitness club as a function of x months.

b. Find $f(12)$ and interpret the result.

c. Solve $f(x) = 1296$ and interpret the result.

89. The 2011 U.S. federal income tax for a person filing single with an income over \$34,500 but not over \$83,600 is \$4,750 plus 25% of the amount over \$34,500.

a. Write a linear function $f(x)$ that represents the tax owed, where x is the amount of income over \$34,500. $f(x) = 0.25x + 4750$

b. Find $f(41{,}000)$ and interpret the result.

c. Solve $f(x) = 9000$ and interpret the result.

90. The U.S. 2011 federal income tax for a married person filing separate with an income over \$69,675 but not over \$106,150 is \$13,543.75 plus 28% of the amount over \$69,675.

a. Write a linear function $f(x)$ that represents the tax owed, where x is the amount of income over \$69,675. $f(x) = 0.28x + 13{,}543.75$

b. Find $f(1475)$ and interpret the result.

c. Solve $f(x) = 20{,}543.75$ and interpret the result. $x = 25{,}000$; A married person filing separate who makes \$69,675 + \$25,000 = \$94,675 must pay \$20,543.75 in federal income tax.

You Be the Teacher!

Correct each student's errors, if any.

91. Write the equation $7x - y = 9$ in function notation.

Isabella's work:

$$7x - y = 9$$
$$-7x + 7x - y = 9 - 7x$$
$$y = -7x + 9$$
$$f(x) = -7x + 9$$

$$7x - y = 9$$
$$-7x + 7x - y = 9 - 7x$$
$$-y = -7x + 9$$
$$-1(-y) = -1(-7x + 9)$$
$$y = 7x - 9$$
$$f(x) = 7x - 9$$

92. Write the equation $4x - 3y = 12$ in function notation.

George's work:

$$4x - 3y = 12$$
$$-4x + 4x - 3y = 12 - 4x$$
$$-3y = -4x + 12$$
$$\frac{-3y}{-3} = -4x + \frac{12}{-3}$$
$$y = -4x - 4$$
$$f(x) = -4x - 4$$

$$4x - 3y = 12$$
$$-4x + 4x - 3y = 12 - 4x$$
$$-3y = -4x + 12$$
$$\frac{-3y}{-3} = \frac{-4x}{-3} + \frac{12}{-3}$$
$$y = \frac{4}{3}x - 4$$
$$f(x) = \frac{4}{3}x - 4$$

93. Find $f(-3)$ for $f(x) = 5x + 2$.

April's work:

$$f(x) = 5x + 2$$
$$f(-3) = (5x + 2)(-3)$$
$$= -15x - 6$$

$$f(x) = 5x + 2$$
$$f(-3) = 5(-3) + 2$$
$$= -15 + 2$$
$$= -13$$

94. Find $f(4)$ for $f(x) = 11x - 1$.

Tyrone's work:

$$f(x) = 11x - 1$$
$$f(4) = (11x - 1)(4)$$
$$= 44x - 4$$

$$f(x) = 11x - 1$$
$$f(4) = 11(4) - 1$$
$$= 44 - 1$$
$$= 43$$

95. Find $f(0.5)$ for $f(x) = 25$.

Pat's work: $f(0.5) = 25(0.5) = 12.5$

$$f(x) = 0x + 25$$
$$f(0.5) = 0(0.5) + 25 = 25$$

96. Find $f(-2.1)$ for $f(x) = -6$.

Adam's work: $f(-2.1) = -6(-2.1) = 12.6$

$$f(x) = 0x - 6$$
$$f(-2.1) = 0(-2.1) - 6 = -6$$

97. Find the solution of $f(x) = 0$ for $f(x) = 9x - 2$.

Allyssa's work:

$$f(x) = 9x - 2$$
$$f(0) = 9(0) - 2$$
$$= -2$$

98. Find the solution of $f(x) = 0$ for $f(x) = -4x + 8$.

Peter's work:

$$f(x) = -4x + 8$$
$$f(0) = -4(0) + 8$$
$$= 8$$

$$f(x) = -4x + 8$$
$$f(x) = 0$$
$$-4x + 8 = 0$$
$$-4x + 8 - 8 = 0 - 8$$
$$-4x = -8$$
$$\frac{-4x}{-4} = \frac{-8}{-4}$$
$$x = 2$$

Calculate It!

Use the main window of a graphing calculator to evaluate each function at the given values.

99. $f(x) = \frac{5}{6}x - 1$ at $x = 0$ and $x = -12$

100. $f(x) = -3x + 4$ at $x = 0$ and $x = -5$

Use the TABLE feature of a graphing calculator to evaluate each function at the given values.

101. $f(x) = -\frac{1}{2}x + 4$ at $x = 0$ and $x = 6$

102. $f(x) = -\frac{1}{3}x - 2$ at $x = 0$ and $x = -3$

Use the given table to find $f(0)$ and the value of x such that $f(x) = 0$ for each function given that $Y_1 = f(x)$.

103.

X	Y1
-5	6
-4	0
-3	-6
-2	-12
-1	-18
0	-24
1	-30

X= -5

$f(0) = -24, x = -4$

104.

X	Y1
-1	-2.5
0	-2
1	-1.5
2	-1
3	-.5
4	0
5	.5

X=5

$f(0) = -2, x = 4$

Think About It!

Solve each problem.

105. Find the value of m for $f(x) = mx + 4$ if $f(-2) = 6$.
$m = -1$

106. Find the value of m for $f(x) = mx - 3$ if $f\left(-\frac{5}{2}\right) = -8$.
$m = 2$

107. Find the value of m for $f(x) = mx + 1$ if $f(7) = 1$.
$m = 0$

108. Find the value of m for $f(x) = mx - 6$ if $f(13) = -6$.
$m = 0$

109. Find the value of b for $f(x) = 2x + b$ if $f(0) = 4$. $b = 4$

110. Find the value of b for $f(x) = -\frac{1}{3}x + b$ if $f(9) = 0$.
$b = 3$

SECTION 4.2 Graphing Linear Equations and Linear Functions

OBJECTIVES

As a result of completing this section, you will be able to

1. Plot points to graph a linear equation in two variables or a linear function. Determine its domain and range.
2. Use intercepts to graph linear equations in two variables.
3. Recognize and graph vertical and horizontal lines.
4. Solve application problems.
5. Troubleshoot common errors.

The median annual income for men with a bachelor's degree is given by the linear function $f(x) = 1443x + 38{,}843$, where x is the number of years after 1990. What is the y-intercept and what does it mean? (Source:www.infoplease.com)

To answer this question, we need to understand the y-intercept of a graph and how we can use a function to determine this piece of information. In this section, we will learn how to graph linear functions by plotting points and finding intercepts.

Objective 1

Plot points to graph a linear equation in two variables or a linear function. Determine its domain and range.

Graph Linear Equations and Find Their Domain and Range

In this section, we will relate together some former concepts that we have learned. Recall the following facts.

- A linear function is a function of the form, $f(x) = mx + b$, or $y = mx + b$, where m and b are real numbers. (See Section 4.1.)
- A linear equation in standard form, $Ax + By = C$, can be written in function form except when $B = 0$. (See Section 4.1.)
- An equation in two variables can be graphed by plotting several ordered pairs that satisfy the equation. (See Section 3.1.)

So, to graph a linear equation in two variables or a linear function, we need to plot several solutions of the equation.

Consider the linear equation $y = x - 3$. Note this can be written as $f(x) = x - 3$. The solutions of this equation are found by substituting values for x and finding the corresponding y-value. Some of the solutions are shown in the table.

x	$y = x - 3$	(x, y)
-2	$y = -2 - 3$ $y = -5$	$(-2, -5)$
-1	$y = -1 - 3$ $y = -4$	$(-1, -4)$
0	$y = 0 - 3$ $y = -3$	$(0, -3)$
1	$y = 1 - 3$ $y = -2$	$(1, -2)$
2	$y = 2 - 3$ $y = -1$	$(2, -1)$
3	$y = 3 - 3$ $y = 0$	$(3, 0)$
4	$y = 4 - 3$ $y = 1$	$(4, 1)$
5.5	$y = 5.5 - 3$ $y = 2.5$	$(5.5, 2.5)$

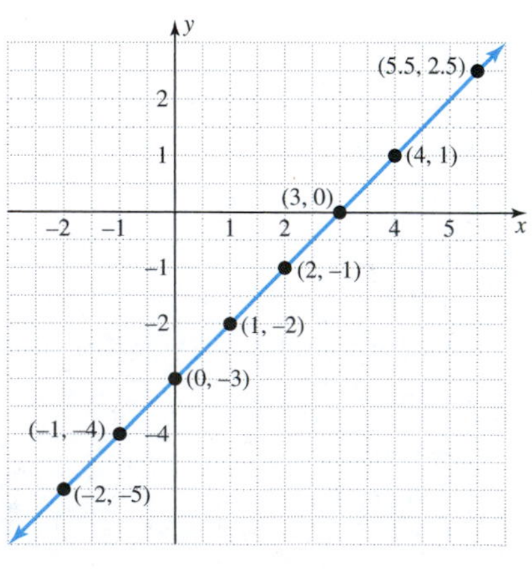

The solutions of this equation lie in a line. We draw a straight line that passes through all the points to obtain the complete graph of $y = x - 3$. Every point on the line is a solution of the equation. Though we found only eight solutions in the table, there are infinitely many solutions of the equation, $y = x - 3$, as indicated by the arrow on each end of the line.

Note: *The graph of any linear equation in two variables or linear function is a line.*

To graph a line, it is sufficient to know two points on the line. For the purposes of accuracy, it is good practice to find at least three points on the line. If one of the points does not lie on the line, an error was made.

INSTRUCTOR NOTE: Inform students that they can also choose values for y to solve for x.

Procedure: Graphing a Linear Equation in Two Variables or a Linear Function

Step 1: Replace $f(x)$ with y, if needed.

Step 2: Make a table of at least three ordered pairs that satisfy the equation by choosing values for x, substituting them in the equation, and solving for y.

Step 3: Plot the ordered pairs found in step 2.

Step 4: Draw the line through the points. The line should extend beyond the points and have arrows at both ends to indicate that there are infinitely many solutions of the equation.

After the graph is obtained, we will use it to determine its domain and range. Recall from Chapter 3 that we read the graph from left to right to find its domain, and we read the graph from bottom to top to find its range.

Objective 1 Examples

Plot points to graph each linear equation or function. State the domain and range.

1a. $y = -x - 1$ **1b.** $f(x) = \frac{1}{4}x - 2$ **1c.** $x + 2y = 4$

Solutions

INSTRUCTOR NOTE: Let students know that the same line will be obtained even if they find different solutions on the line.

1a.

x	$y = -x - 1$	y	(x, y)
-3	$y = -(-3) - 1$ $= 3 - 1$ $= 2$	2	$(-3, 2)$
0	$y = 0 - 1$ $= -1$	-1	$(0, -1)$
3	$y = -(3) - 1$ $= -4$	-4	$(3, -4)$

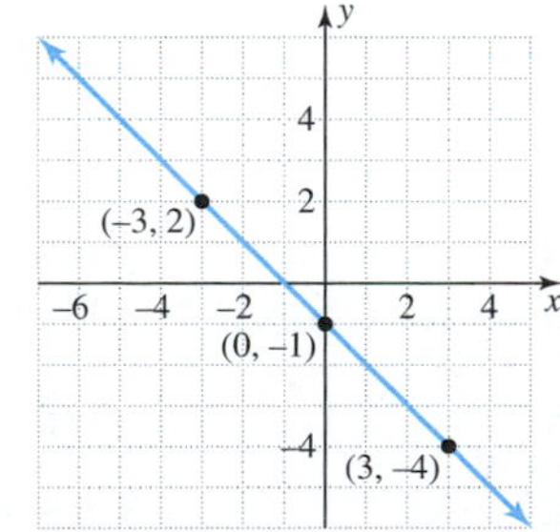

Because the graph extends indefinitely to the left and right, the domain of the function is all real numbers, or $(-\infty, \infty)$. The graph also extends indefinitely from bottom to top, so the range is all real numbers, or $(-\infty, \infty)$.

INSTRUCTOR NOTE: Remind students that the graphs of these linear equations pass the vertical line test and are, therefore, functions.

1b. The function is equivalent to $y = \frac{1}{4}x - 2$. We find solutions of the equation by assigning values to the variable x. Since the variable x is multiplied by $\frac{1}{4}$, we choose values that are divisible by 4, so the calculations are easier.

x	$y = \frac{1}{4}x - 2$	y	(x, y)
-4	$y = \frac{1}{4}(-4) - 2$ $= -1 - 2$ $= -3$	-3	$(-4, -3)$
0	$y = \frac{1}{4}(0) - 2$ $= 0 - 2$ $= -2$	-2	$(0, -2)$
4	$y = \frac{1}{4}(4) - 2$ $= 1 - 2$ $= -1$	-1	$(4, -1)$

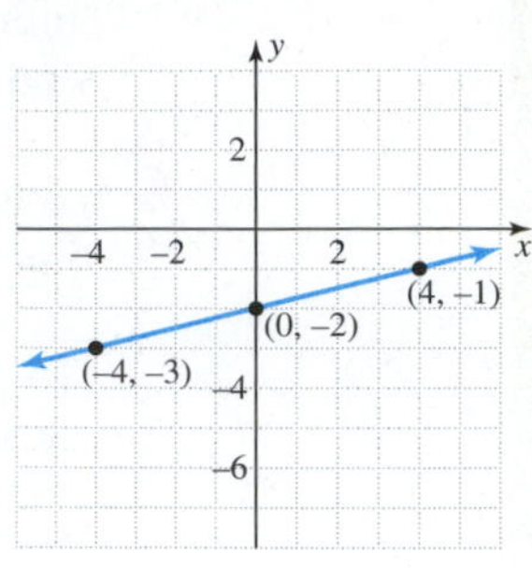

Because the graph extends indefinitely to the left and right, the domain of the function is all real numbers, or $(-\infty, \infty)$. The graph also extends indefinitely from bottom to top, so the range is all real numbers, or $(-\infty, \infty)$.

1c.

x	$x + 2y = 4$	y	(x, y)
-2	$-2 + 2y = 4$ $2y = 6$ $y = 3$	3	$(-2, 3)$
0	$0 + 2y = 4$ $2y = 4$ $y = 2$	2	$(0, 2)$
2	$2 + 2y = 4$ $2y = 2$ $y = 1$	1	$(2, 1)$

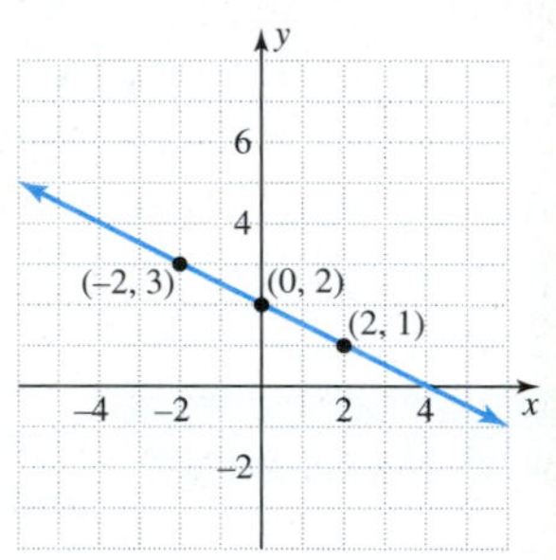

Because the graph extends indefinitely to the left and right, the domain of the function is all real numbers, or $(-\infty, \infty)$. The graph also extends indefinitely from bottom to top, so the range is all real numbers, or $(-\infty, \infty)$.

Recall $x + 2y = 4$ can be written in function notation as $f(x) = -\frac{1}{2}x + 2$. We could have used this form to obtain solutions as well.

Student Check 1 Plot points to graph each linear equation or function. State the domain and range.

a. $y = -2x + 1$ **b.** $f(x) = \frac{2}{3}x - 3$ **c.** $x - 3y = 6$

Graph Linear Equations Using Intercepts

Objective 2 Use intercepts to graph linear equations in two variables.

Another method of graphing a linear equation in two variables involves finding two important points that lie on the graph of the line. These points are called the *intercepts* of the graph and are the points where the graph crosses the axes.

Definition: Intercepts of a Graph

- The **x-intercept** is the point on the graph where the graph intersects the x-axis.
- The **y-intercept** is the point on the graph where the graph intersects the y-axis.

In the graph from Example 1c, the x-intercept is (4, 0) and the y-intercept is (0, 2).

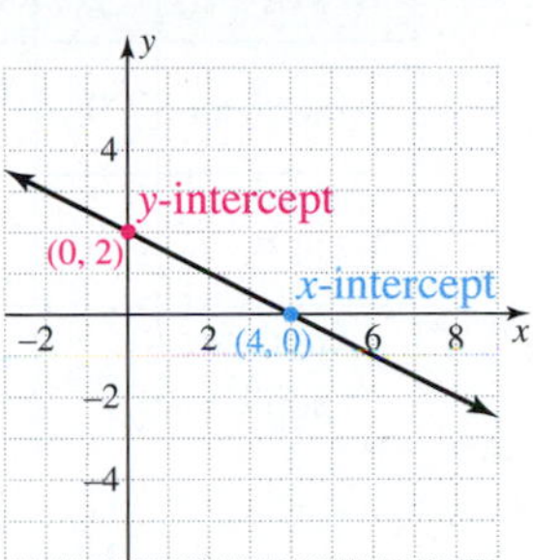

Notice that the intercepts are the points in which one of the coordinates is *zero*.

Procedure: Finding the Intercepts of a Graph from Its Equation

Step 1: To find the x-intercept, replace y with 0 and solve for x. The point is of the form $(x, 0)$.

Step 2: To find the y-intercept, replace x with 0 and solve for y. The point is of the form $(0, y)$.

Objective 2 Examples

Find the x- and y-intercepts and use them to graph each linear equation.

2a. $x - 3y = -6$ **2b.** $f(x) = -\frac{5}{2}x + 3$ **2c.** $y = 2x$

Solutions

2a.

	x	$x - 3y = -6$	y	(x, y)
x-intercept	-6	$x - 3(0) = -6$ $x = -6$	0	$(-6, 0)$
y-intercept	0	$0 - 3y = -6$ $-3y = -6$ $y = 2$	2	$(0, 2)$
Checkpoint	2	$2 - 3y = -6$ $-3y = -8$ $y = \frac{8}{3}$	$\frac{8}{3}$	$\left(2, \frac{8}{3}\right)$

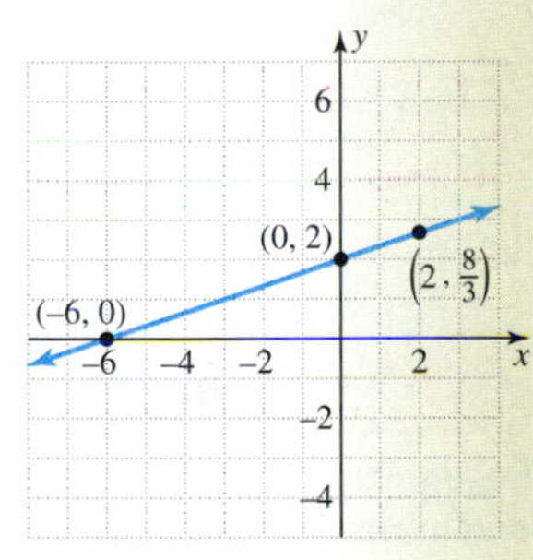

2b.

	x	$y = -\frac{5}{2}x + 3$	y	(x, y)
x-intercept	$\frac{6}{5}$	$0 = -\frac{5}{2}x + 3$ $2(0) = 2\left(-\frac{5}{2}x + 3\right)$ $0 = -5x + 6$ $5x = 6$ $x = \frac{6}{5}$	0	$\left(\frac{6}{5}, 0\right)$
y-intercept	0	$y = -\frac{5}{2}(0) + 3$ $y = 3$	3	$(0, 3)$
Checkpoint	4	$y = -\frac{5}{2}(4) + 3$ $y = -10 + 3$ $y = -7$	-7	$(4, -7)$

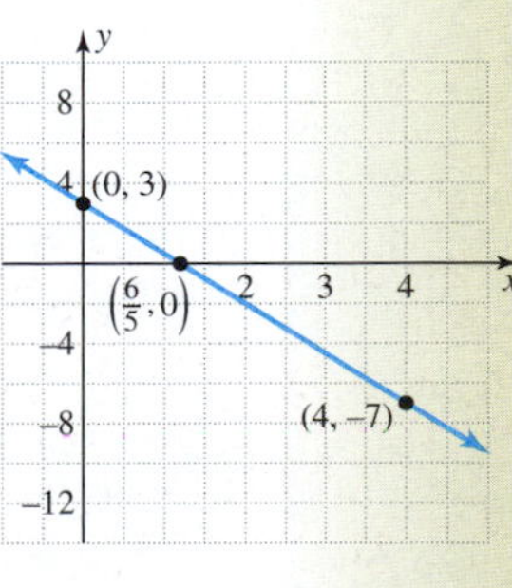

2c.

	x	$y = 2x$	y	(x, y)
x-intercept	0	$0 = 2x$ $0 = x$	0	(0, 0)
y-intercept	0	$y = 2(0)$ $y = 0$	0	(0, 0)
Another point	1	$y = 2(1)$ $y = 2$	2	(1, 2)

Note that the x- and y-intercepts for the graph of this equation are the same point, the origin. So, we must find another solution of the equation to have enough information to graph the line. Choose any value for x and solve for y.

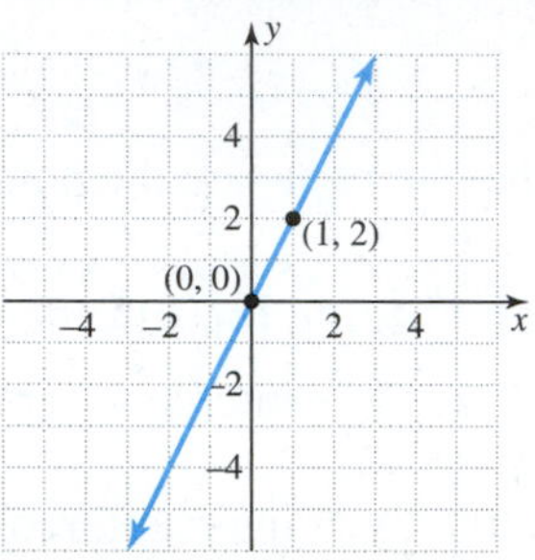

Note: *The equation* $y = 2x$ *can also be written as* $-2x + y = 0$ *or* $2x - y = 0$. *So, when the constant of a linear equation in standard form is zero, the graph goes through the origin.*

Student Check 2 Find the x- and y-intercepts and use them to graph each linear equation.

a. $4x - y = -4$ **b.** $f(x) = -\frac{2}{3}x + 5$ **c.** $y = -3x$

Vertical and Horizontal Lines

Objective 3 Recognize and graph vertical and horizontal lines.

In Objectives 1 and 2, the lines that we graphed were oblique, that is, slanted. We will now examine two special cases of linear equations in two variables whose graphs are vertical or horizontal lines. Consider the following graphs of the vertical and horizontal lines.

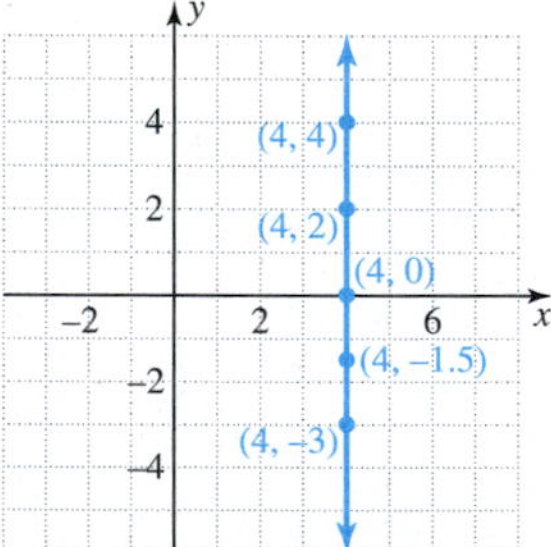

The graph of this *vertical* line goes through the points (4, 0), (4, 2), (4, 4), (4, −1.5), (4, −3), and so on. The only points that lie on the graph of this line are points whose x-value is 4. So, the points that lie on this graph satisfy the equation $x = 4$.

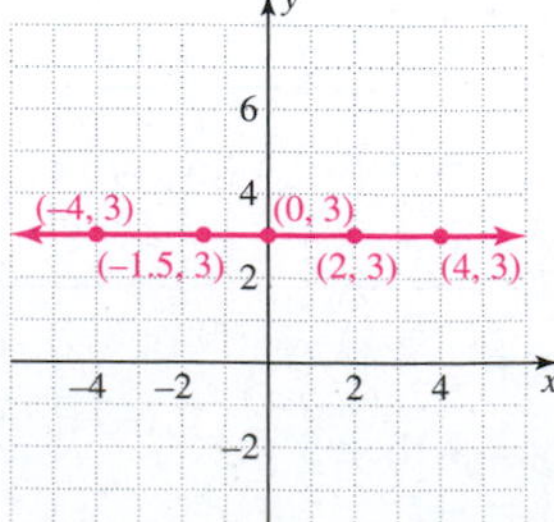

The graph of this *horizontal* line goes through the points (−4, 3), (−1.5, 3), (0, 3), (2, 3), (4, 3), and so on. The only points that lie on the graph of this line are points whose y-value is 3. So, the points that lie on this graph satisfy the equation $y = 3$.

So, linear equations of the form $x = h$ and $y = k$, where h and k are real numbers, are equations of vertical and horizontal lines, respectively.

Definition: The graph of an equation of the form $x = h$, where h is a real number, is a **vertical line** through h on the x-axis. The point $(h, 0)$ is the x-intercept of the graph.

The equation $x = h$ can be written as $x + 0y = h$. So, no matter what value is substituted in place of y, the x-value will always be h.

Recall that the graph of a vertical line does not pass the vertical line test, which means that an equation of the form $x = h$ does not represent a function. Its domain is $\{h\}$ since h is the only x-value of the points on the graph. The range is $(-\infty, \infty)$ since the graph extends indefinitely from the bottom to the top.

Procedure: Graphing Equations of the Form $x = h$

Step 1: Isolate x on one side of the equation, if necessary. The constant on the other side of the equation is the value h.
Step 2: Plot the point $(h, 0)$.
Step 3: Additional points whose x-value is h can be plotted to obtain other points on the graph.
Step 4: Draw a vertical line through the point $(h, 0)$ to complete the graph.

Definition: The graph of an equation of the form $y = k$, where k is a real number, is a **horizontal line** through k on the y-axis. The point $(0, k)$ is the y-intercept of the graph.

The equation $y = k$ can be written as $0x + y = k$. So, no matter what value is substituted in place of x, the y-value will always be k.

Recall that the graph of a horizontal line passes the vertical line test, which means that an equation of the form $y = k$ represents a function and can be written as $f(x) = k$. Its domain is $(-\infty, \infty)$ since the graph extends indefinitely from the left to the right. Its range is $\{k\}$ since k is the only y-value of the points on the graph.

Procedure: Graphing Equations of the Form $y = k$

Step 1: Isolate y on one side of the equation, if necessary. The constant on the other side of the equation is the value k.
Step 2: Plot the point $(0, k)$.
Step 3: Additional points whose y-value is k can be plotted to obtain other points on the graph.
Step 4: Draw a horizontal line through the point $(0, k)$ to complete the graph.

Objective 3 Examples Graph each linear equation and state its domain and range.

3a. $x = -2$ **3b.** $3x - 2 = 1$ **3c.** $y = \frac{5}{2}$ **3d.** $-2y + 3 = 11$

Solutions **3a.** This equation represents a vertical line through $x = -2$. The x-intercept is $(-2, 0)$. Some other points on the line are $(-2, 3)$ and $(-2, -4)$.

This equation does not represent a function since the graph fails the vertical line test. The domain of the relation is $\{-2\}$, and the range is $(-\infty, \infty)$.

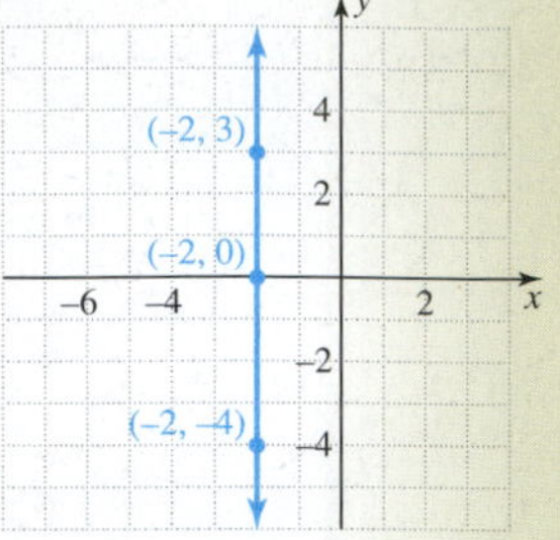

3b. Isolate the variable x to determine the graph of the equation.

$$3x - 2 = 1$$
$$3x - 2 + 2 = 1 + 2 \quad \text{Add 2 to each side.}$$
$$3x = 3 \quad \text{Simplify.}$$
$$x = 1 \quad \text{Divide each side by 3.}$$

This equation represents a vertical line through $x = 1$. The x-intercept is $(1, 0)$. Some other points on the line are $(1, 2)$ and $(1, -3)$. The domain of the relation is $\{1\}$, and the range is $(-\infty, \infty)$.

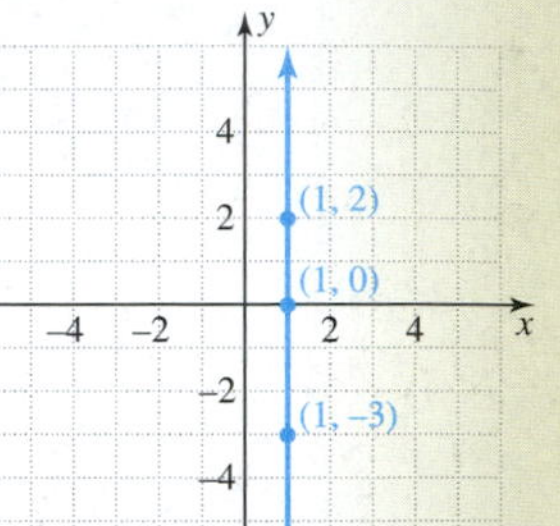

3c. This equation represents a horizontal line through $y = \frac{5}{2}$. The y-intercept is $\left(0, \frac{5}{2}\right)$. Some other points on the line are $\left(2, \frac{5}{2}\right)$ and $\left(-3, \frac{5}{2}\right)$.

The equation represents a function since its graph passes the vertical line test. The domain of the function is $(-\infty, \infty)$, and the range is $\left\{\frac{5}{2}\right\}$.

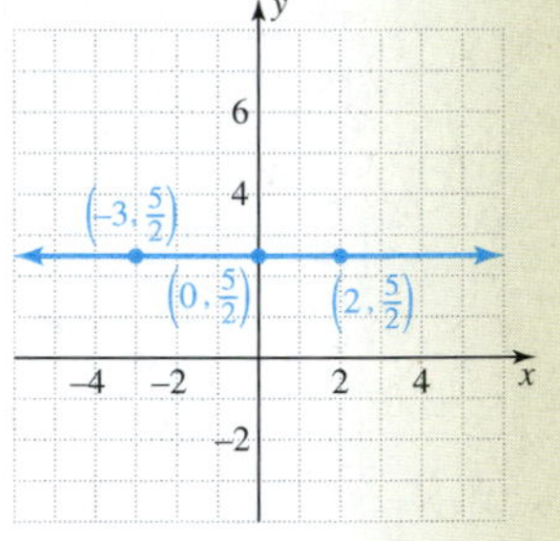

3d. Isolate the variable y to determine the graph of the equation.

$$-2y + 3 = 11$$
$$-2y + 3 - 3 = 11 - 3 \quad \text{Subtract 3 from each side.}$$
$$-2y = 8 \quad \text{Simplify.}$$
$$y = -4 \quad \text{Divide each side by } -2.$$

The equation represents a horizontal line through $y = -4$. The y-intercept is $(0, -4)$. Some other points on the line are $(-3, -4)$ and $(5, -4)$. The domain of the function is $(-\infty, \infty)$, and the range is $\{-4\}$.

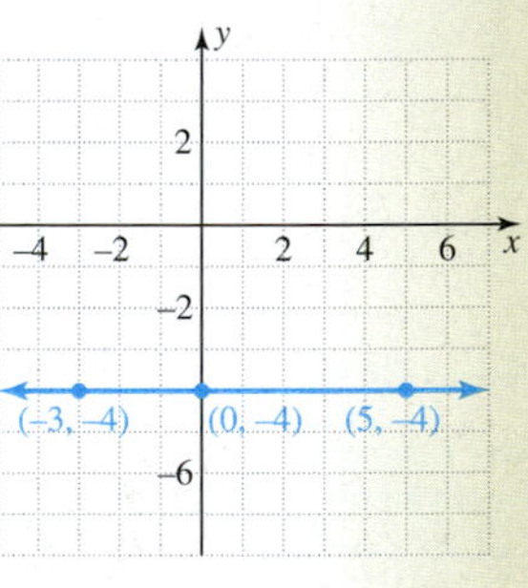

Student Check 3 Graph each linear equation and state its domain and range.

a. $x = -2$ **b.** $2x - 5 = 0$ **c.** $y = -2$ **d.** $6y + 3 = 0$

Applications

Objective 4 Solve application problems.

In the application problems encountered in this section, a linear equation or function that models a real-world situation will be given. We will obtain information from the equation and also construct the graph of the function.

Objective 4 Examples **Use the equation or function to solve each problem.**

4a. The median annual income for men with a bachelor's degree can be modeled by the linear function $f(x) = 1443x + 38{,}843$, where x is the number of years after 1990. (Source: www.infoplease.com)

i. Find the y-intercept of the equation and interpret its meaning.

ii. Find the median annual income for the years 2010 and 2015.

Solution **4a.** **i.** The y-intercept of $f(x) = 1443x + 38{,}843$ is found by replacing x with 0, or by finding $f(0)$.

$$f(x) = 1443x + 38{,}843$$
$$f(0) = 1443(0) + 38{,}843$$
$$f(0) = 38{,}843$$

So, the y-intercept is (0, 38,843). So, the median annual income for men with a bachelor's degree in 0 years after 1990, or 1990, was \$38,843.

ii. The median annual income for 2010 is found by replacing x with 20 $(2010 - 1990 = 20)$ and the median annual income for 2015 is found by replacing x with 25 $(2015 - 1990 = 25)$.

Let $x = 20$.

$$f(x) = 1443x + 38{,}843$$
$$f(20) = 1443(20) + 38{,}843$$
$$f(20) = 28{,}860 + 38{,}843$$
$$f(20) = 67{,}703$$

Let $x = 25$.

$$f(x) = 1443x + 38{,}843$$
$$f(25) = 1443(25) + 38{,}843$$
$$f(25) = 36{,}075 + 38{,}843$$
$$f(25) = 74{,}918$$

So, the median annual income for men with a bachelor's degree in 2010 was about \$67,703. In 2015 the median annual income is about \$74,918.

4b. Straight-line depreciation is the most common method of depreciating business assets. A surveyor company purchases a truck for their business. On the company's tax records, the owner must report the depreciated value of the truck. The truck's value is given by $y = -4000x + 32{,}000$, where x is the age of the truck in years.

i. Find the x- and y-intercepts and interpret the results.

ii. Find the value of the truck after 2 yr and after 4 yr.

iii. Use the ordered pairs from parts (i) and (ii) to draw the graph of the equation.

Solution **4b.** **i.** x-intercept (let $y = 0$):

$$y = -4000x + 32{,}000$$
$$0 = -4000x + 32{,}000$$
$$0 + 4000x = -4000x + 32{,}000 + 4000x$$
$$4000x = 32{,}000$$
$$\frac{4000x}{8} = \frac{32{,}000}{8}$$
$$x = 8$$

(8, 0)

y-intercept (let $x = 0$):

$$y = -4000x + 32{,}000$$
$$y = -4000(0) + 32{,}000$$
$$y = 32{,}000$$

(0, 32,000)

The x-intercept is (8, 0). So, when the truck is 8 yr old, its value will be \$0.

The y-intercept is (0, 32,000). So, the truck's initial value is \$32,000.

ii. To find the value of the truck after 2 yr and after 4 yr, we replace the variable x with these numbers.

Let $x = 2$.	Let $x = 4$.
$y = -4000x + 32{,}000$	$y = -4000x + 32{,}000$
$y = -4000(2) + 32{,}000$	$y = -4000(4) + 32{,}000$
$y = -8000 + 32{,}000$	$y = -16{,}000 + 32{,}000$
$y = 24{,}000$	$y = 16{,}000$

So, after 2 yr, the truck is worth \$24,000. After 4 yr, the truck is worth \$16,000.

iii. The graph of the equation is found by plotting the points (8, 0), (0, 32,000), (2, 24,000), and (4, 16,000). Only the first quadrant is shown since it doesn't make sense to have a negative depreciated value for negative years. Since the largest y-value is 32,000, we let each tick mark on the y-axis represent 4000 units.

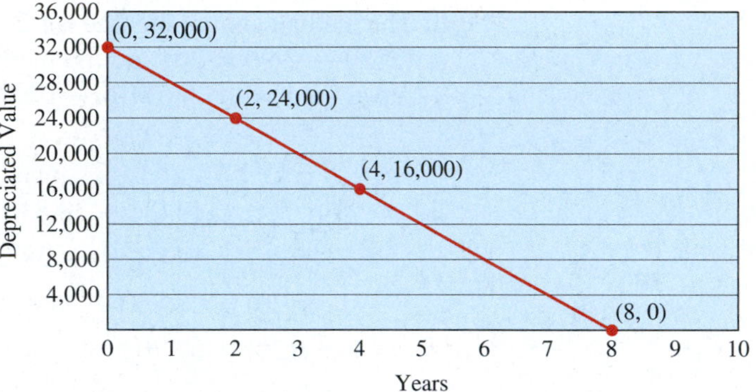

Student Check 4 Use the equation to solve each problem. The median annual income for women with a bachelor's degree is given by the equation, $y = 1100x + 27{,}645$, where x is the number of years after 1990. (Source: www.infoplease.com)

a. Find the y-intercept of the equation and interpret its meaning.

b. Find the median annual income for the years 2012 and 2025.

c. Graph the equation.

Objective 5

Troubleshoot common errors.

Troubleshooting Common Errors

Some common errors associated with graphing linear equations and linear functions are shown.

Objective 5 Examples

A problem and an incorrect solution are given. Provide the correct solution and an explanation of the error.

5a. Graph the equation $y = \frac{1}{2}x$ using the intercepts.

Incorrect Solution	Correct Solution and Explanation
The intercepts of the graph of this equation are (0, 0) and (0, 0).	The x- and y-intercepts are the same point, so we need another point to graph the line accurately. If we substitute $x = 2$, we get $y = \frac{1}{2}(2) = 1$. So, another point is (2, 1).

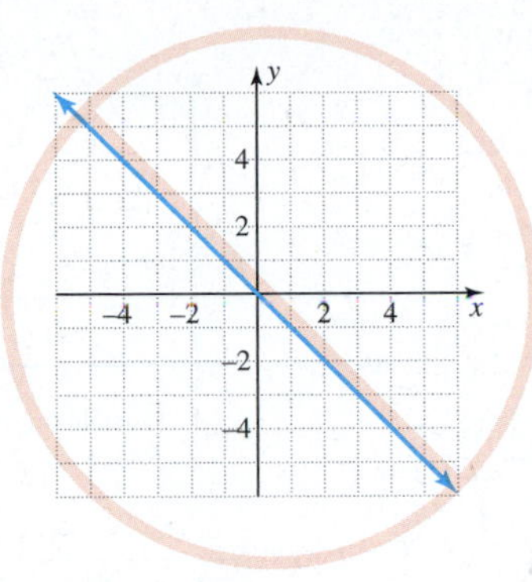

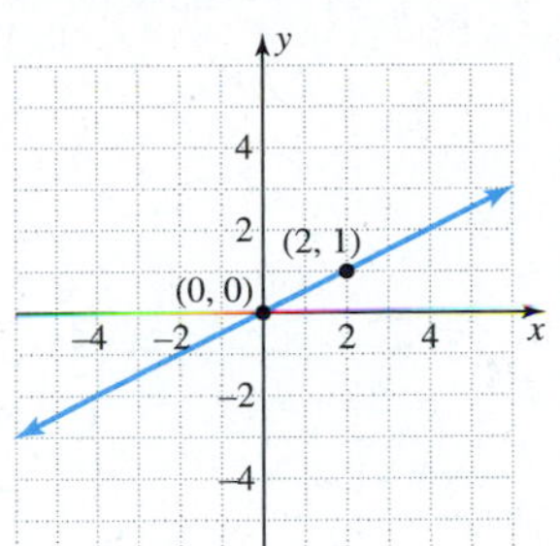

5b. Graph the equation $x = 3$.

Incorrect Solution	Correct Solution and Explanation
The equation is a horizontal line and the graph is	An equation of the form $x = h$ is a vertical line, not a horizontal line. Every point on the line has an x-value of 3.

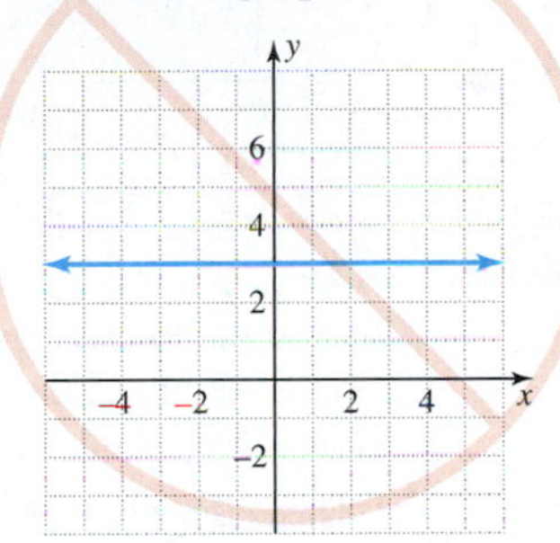

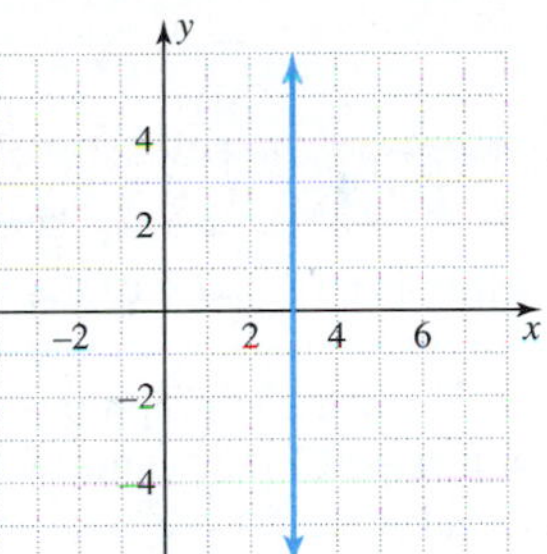

ANSWERS TO STUDENT CHECKS

Student Check 1

a.

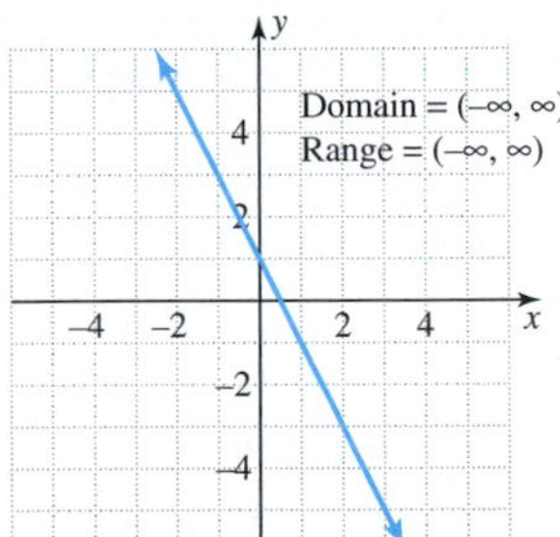

b.

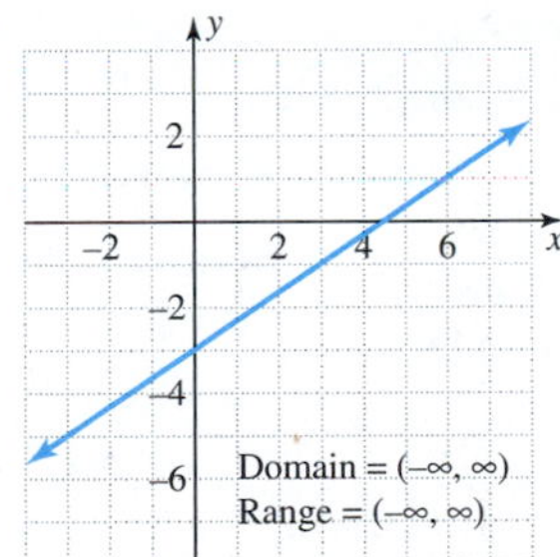

c.

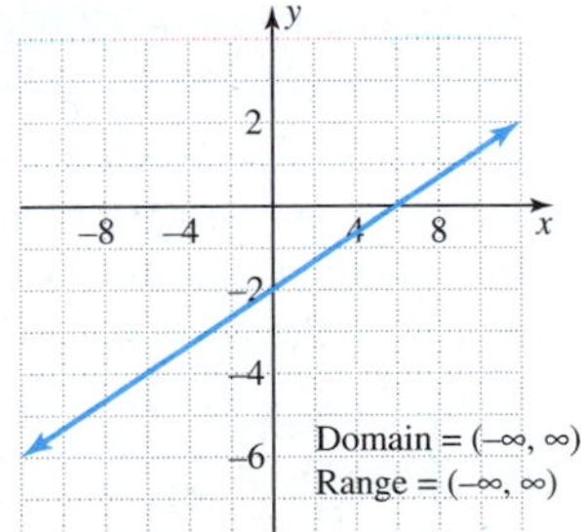

Student Check 2

a.

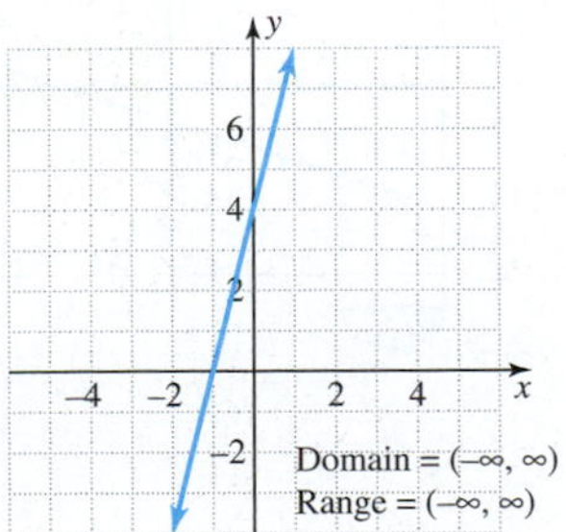

b.

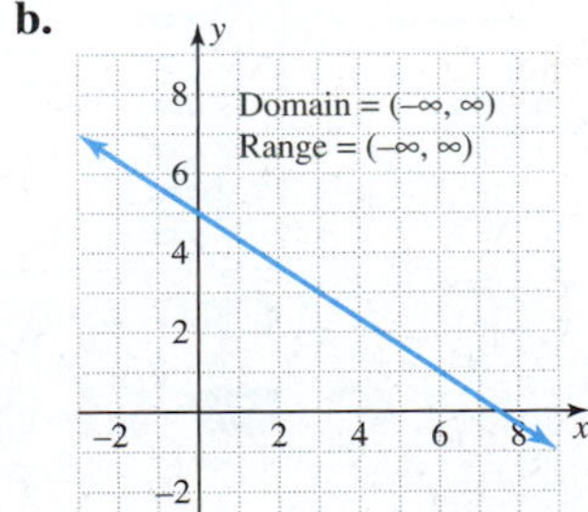

c.

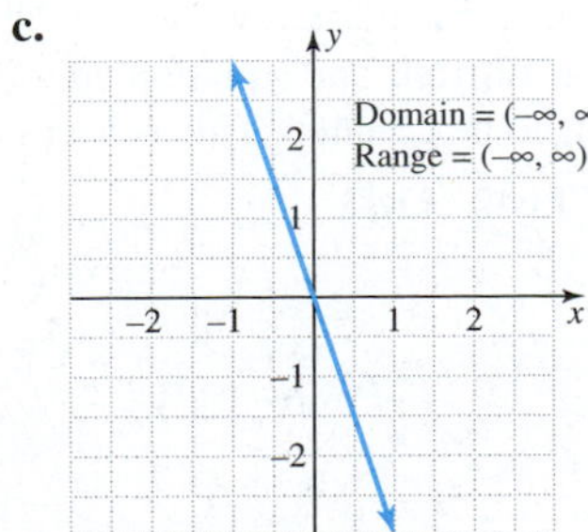

Student Check 3

a.

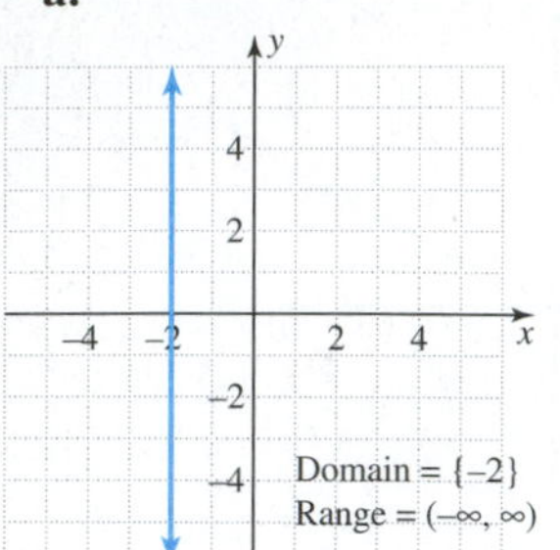

b.

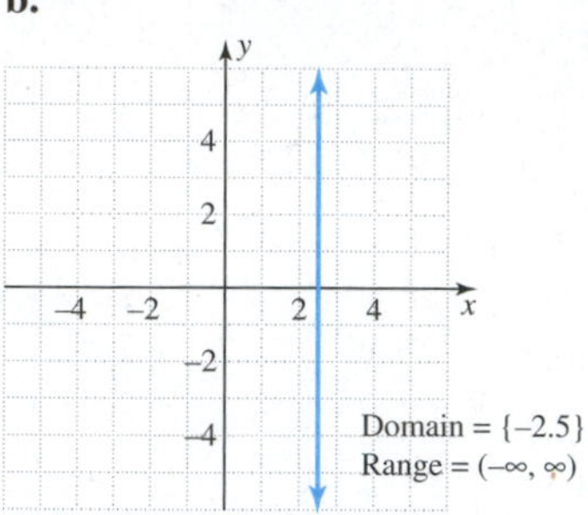

c.

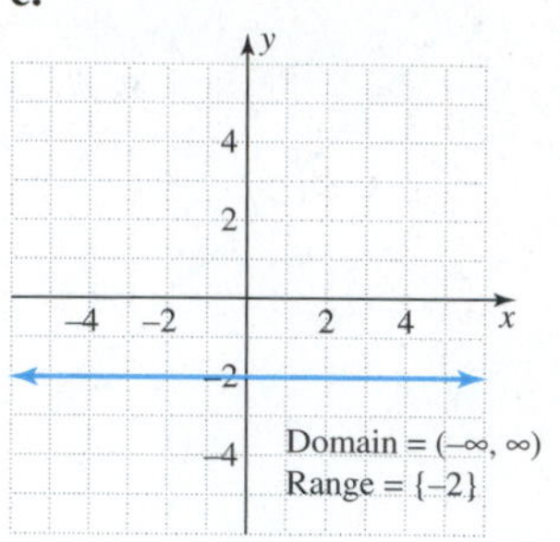

d.

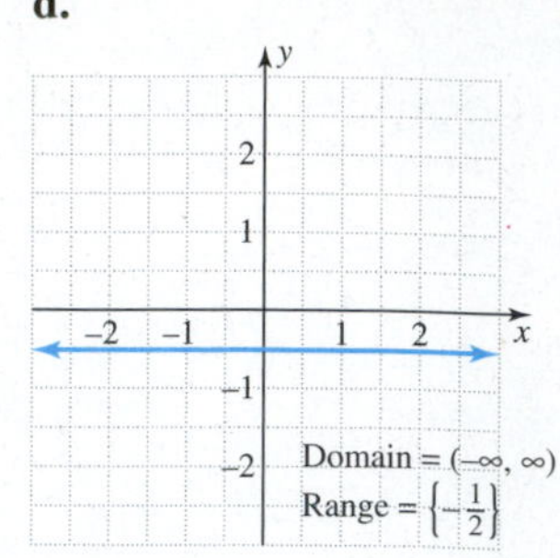

Student Check 4 **a.** The y-intercept is (0, 27,645). In 0 years after 1990, or in 1990, the median annual income for women with a bachelor's degree is \$27,645.

b. The median annual income in 2012 is \$51,845. The median annual income in 2025 is \$66,145.

c.

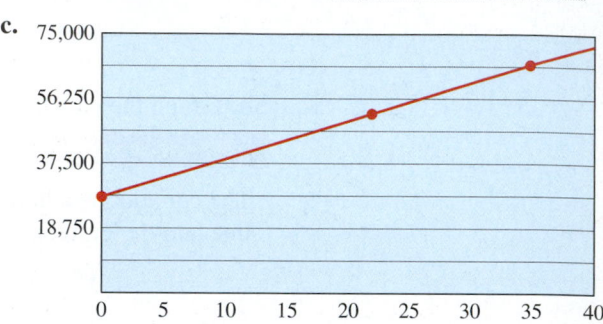

SUMMARY OF KEY CONCEPTS

1. A linear function is a function of the form $f(x) = mx + b$, and a linear equation in two variables is an equation of the form $Ax + By = C$. To graph this equation/function, select at least three different values for x, substitute into the equation, and solve for the corresponding y-value. The points should lie in a line. If they do not, a mistake was made. The domain and range of a nonvertical line is $(-\infty, \infty)$.
2. A linear equation can also be graphed by finding the x- and y-intercepts. To find the x-intercept, replace y with zero and solve for x. This is a point of the form $(x, 0)$. To find the y-intercept, replace x with zero and solve for y. This is a point of the form $(0, y)$.
3. Vertical lines have an equation of the form $x = h$.
   - These lines have an x-intercept of $(h, 0)$.
   - Vertical lines are not functions.
   - The domain of $x = h$ is $\{h\}$ and the range is $(-\infty, \infty)$.
4. Horizontal lines have an equation of the form $y = k$.
   - These lines have a y-intercept of $(0, k)$.
   - Horizontal lines are functions.
   - The domain of $y = k$ is $(-\infty, \infty)$ and the range is $\{k\}$.

GRAPHING CALCULATOR SKILLS

Graphing Window: The graphing window on a calculator comes with a standard setting. Press GRAPH to access this window. The x- and y-values range from -10 to 10, with each tick mark representing 1 unit. The window is viewed by pressing WINDOW.

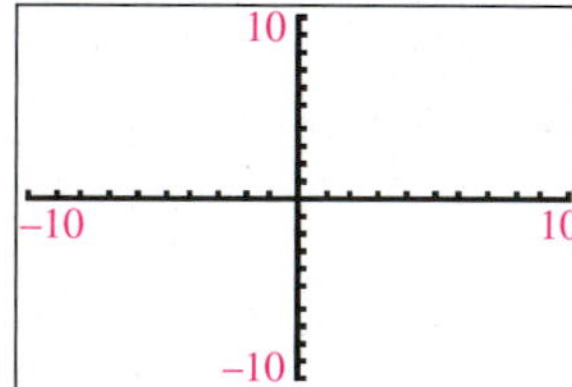


```
WINDOW
 Xmin=-10
 Xmax=10
 Xscl=1
 Ymin=-10
 Ymax=10
 Yscl=1
 Xres=1
```

Equation Editor: The equation editor is where we enter equations to be graphed. This is accessed by pressing Y=.

```
Plot1 Plot2 Plot3
\Y1=
\Y2=
\Y3=
\Y4=
\Y5=
\Y6=
\Y7=
```

Example 1: Graph $x - 2y = 4$ on the calculator.

Solution: We solve the equation for y. Then we enter it into the equation editor and graph.

$$x - 2y = 4$$
$$-2y = -x + 4$$
$$y = \frac{1}{2}x - 2$$

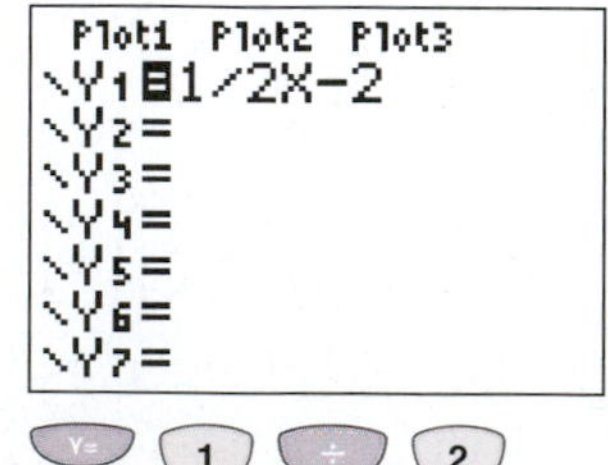

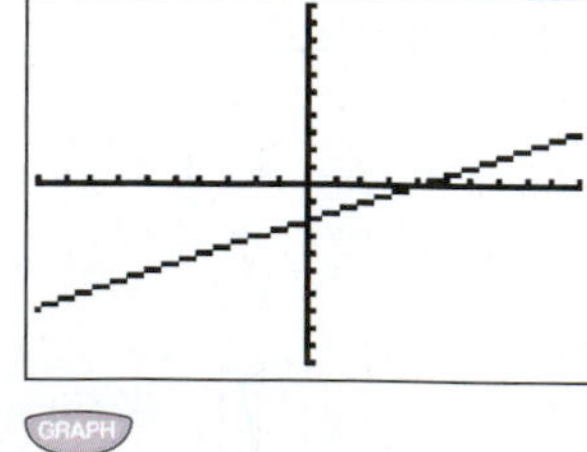

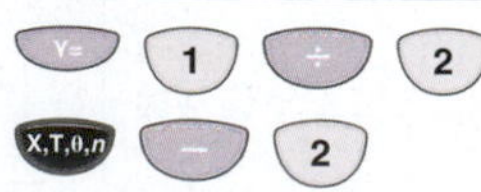

GRAPH

The TABLE feature enables us to view ordered pairs that satisfy the equation.

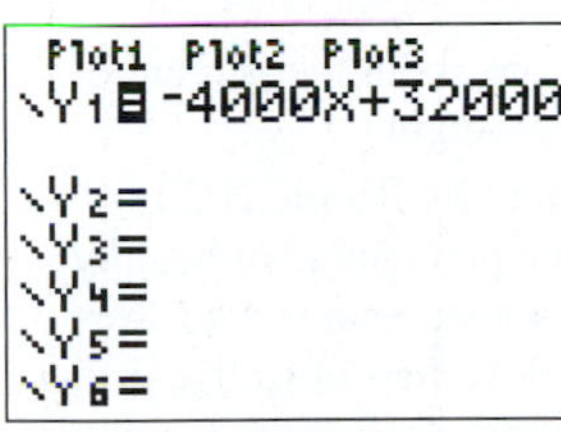

Example 2: Graph $y = -4000x + 32{,}000$.

Solution: This equation is solved for y, so we enter it into the equation editor.

Press GRAPH. The graph looks like a vertical line (which is not the case).

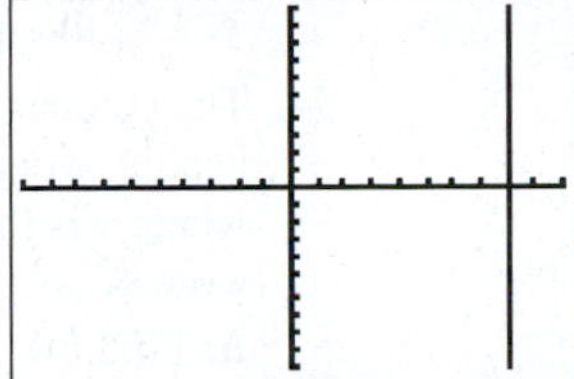

We must change the window settings to view the complete graph. From Example 4 part b, the x-intercept is (8, 0) and the y-intercept is (0, 32,000). We also know that negative x- and y-values do not make sense in this problem.

The graph crosses the x-axis at (0, 8), so we use an Xmax = 10. The graph crosses the y-axis at (0, 32,000), so we use a Ymax = 35,000. The yscl value represents how many units each tick mark on the y-axis represents. So, we will use Yscl = 5000.

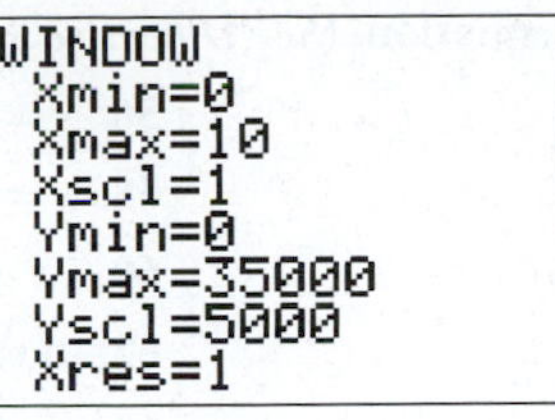

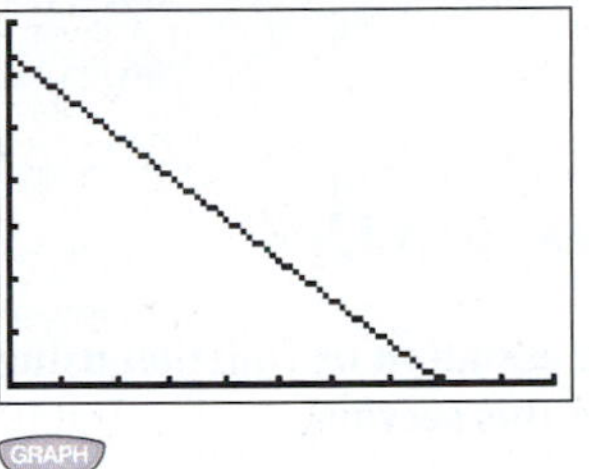

GRAPH

SECTION 4.2 EXERCISE SET

Write About It!

Use complete sentences in your answer to each exercise.

1. Describe how you know that an equation is a linear equation in two variables.
2. How can you graph a linear equation by plotting points?
3. How do you find the x-intercept of a linear equation?
4. How do you find the y-intercept of a linear equation?
5. If the origin is both the x-intercept and the y-intercept, what must you do to graph the line?
6. Describe how you can recognize if an equation represents a vertical or horizontal line.

Practice Makes Perfect!

Plot points to graph each linear equation and then state the domain and range. (*See Objective 1.*)

7. $x + y = 6$
8. $x + y = 8$
9. $2x - y = 5$
10. $x - 2y = 3$
11. $2x + y = -4$
12. $3x + y = -6$
13. $x + 2y = 2$
14. $x + 5y = 5$
15. $x - 4y = -8$
16. $x - 6y = 6$
17. $7x + 3y = -7$
18. $8x + 5y = -5$
19. $9x + 2y = 0$
20. $3x + 4y = 0$
21. $f(x) = -x + 2$
22. $f(x) = -x + 3$
23. $f(x) = 2x + 4$
24. $f(x) = -3x + 6$
25. $f(x) = 4x - 8$
26. $f(x) = 5x - 5$
27. $f(x) = \frac{1}{2}x - 2$
28. $f(x) = \frac{1}{3}x - 6$
29. $f(x) = -\frac{1}{5}x + 5$
30. $f(x) = -\frac{1}{4}x + 1$
31. $f(x) = \frac{2}{5}x - 1$
32. $f(x) = \frac{3}{4}x - 2$
33. $f(x) = 3x$
34. $f(x) = -4x$
35. $f(x) = \frac{1}{2}x$
36. $f(x) = -\frac{1}{3}x$

Find the x- and y-intercepts of each line and use them to draw the graph. (*See Objective 2.*)

37. $3x + y = 6$
38. $2x + y = 8$
39. $x - 2y = 0$
40. $x + 5y = 0$
41. $f(x) = 9x + 9$
42. $f(x) = 8x - 16$
43. $3x - 2y = 12$
44. $3x - 5y = 15$
45. $f(x) = -x + 7$
46. $f(x) = -x - 4$
47. $f(x) = \frac{1}{2}x - 3$
48. $f(x) = \frac{1}{3}x + 4$
49. $f(x) = -\frac{2}{3}x + 5$
50. $f(x) = -\frac{5}{2}x - 3$
51. $f(x) = \frac{2}{7}x$
52. $f(x) = -\frac{4}{9}x$
53. $-\frac{1}{4}x + \frac{2}{3}y = 1$
54. $\frac{2}{3}x + \frac{1}{2}y = 1$

Additional answers can be found in the Instructor Answer Appendix.

Graph each equation. (*See Objective 3.*)

55. $x = 6$

56. $x = 4$

57. $x = -7$

58. $x = -5$

59. $2x + 10 = 0$

60. $7x + 9 = 0$

61. $f(x) = 2$

62. $f(x) = 5$

63. $f(x) = -1$

64. $f(x) = -3$

65. $2y + 5 = 0$

66. $5y - 8 = 0$

Mix 'Em Up!

Graph each equation or function using any of the methods described in this section.

67. $x = -5$

68. $f(x) = 3.5$

69. $2x - 7y = -14$

70. $f(x) = \frac{7}{9}x$

71. $x + 4y = 0$

72. $3y + 4.5 = 0$

73. $f(x) = -x + 2$

74. $f(x) = \frac{8}{3}x - 1$

75. $6x - 2 = 0$

76. $f(x) = -\frac{2}{3}x + 4$

77. $3x + 6y = 6$

78. $x - 4y = -4$

79. $2x - 3y = 0$

80. $4x + 5y = 0$

81. $0.7x + 0.2y = 2.8$

82. $f(x) = -\frac{8}{3}x$

83. $0.8x - 0.2y = 0$

84. $3.2 + 2x = 0$

85. $f(x) = -6.5$

86. $0.5x - 0.6y = 4.5$

87. $f(x) = 3.5$

88. $f(x) = -\frac{5}{2}x$

89. $x - 4y = 80$

90. $5x + 2y = 120$

Solve each problem.

91. The median annual income for men with a master's degree is given by the equation $y = 1540x + 94{,}921$, where x is the number of years after 2005. (Source: www .census.gov)

a. Find the y-intercept of the equation and interpret its meaning. (0, 94,921); In 0 yr after 2005, or 2005, the median annual income for men with a master's degree was \$94,921.

b. Find the median annual income for men in 2011. \$104,161

c. Find the median annual income for men in 2013. \$107,241

92. The median annual income for women with a master's degree is given by the equation $y = 2660x + 59{,}536$, where x is the number of years after 2005. (Source: www.census.gov)

a. Find the y-intercept of the equation and interpret its meaning. (0, 59,536); In 0 yr after 2005, the median annual income for women with a master's degree was \$59,536.

b. Find the median annual income for women in 2011. \$75,496

c. Find the expected median annual income for women in 2013. \$80,816

93. The percent of U.S. adults who are smokers can be modeled by the equation $y = -0.27x + 22.5$, where x is the number of years after 2002. (Source: media.cleveland.com)

a. Find the x-intercept of the equation and interpret its meaning. (83, 0); In 83 yr after 2002, or 2085, about 0% of U.S. adults will smoke.

b. Find the y-intercept of the equation and interpret its meaning. (0, 22.5); In 2002, about 22.5% of U.S. adults were smokers.

c. What percent of U.S. adults are smokers in 2011? about 20.07%

d. What percent of U.S. adults are smokers in 2014? about 19.26%

e. Use the intercepts to graph the equation.

94. The percent of adults aged 20 and over who are at a healthy weight can be modeled by $y = -0.5x + 55.43$, where x is the number of years after 1960. (Source: http://www.cdc.gov/nchs/data/hus/hus10.pdf)

a. Find the y-intercept of the equation and interpret its meaning. (0, 55.43); In 1960, about 55.43% of adults aged 20 and over were at a healthy weight.

b. What percent of adults were at a healthy weight in 2010? In 2010, about 30.43% of adults aged 20 and over were at a healthy weight.

c. What percent of adults will be at a healthy weight in 2020 if this trend continues?

d. Use the information obtained in parts (a)–(c) to graph the equation.

e. What does the graph indicate about the percent of adults who are at a healthy weight?

f. One of the objectives of Healthy People 2020 Campaign is to increase the percentage of healthy weight adults by 10% of the percentage who were at a healthy weight in 2010. If their objective is met, what percent of adults will be at a healthy weight in 2020? (Source: http://www.healthypeople.gov)

95. A manufacturing company purchases a piece of equipment for \$50,000. Its depreciated value can be represented by $y = -5000x + 50{,}000$, where x is the age of the equipment in years.

a. Find the x- and y-intercepts and interpret the results.

b. Find the value of the piece of equipment after 4 yr.

c. Use the ordered pairs from parts (a) and (b) to draw the graph of the equation.

96. A company purchases a computer for \$2400. Its depreciated value can be represented by $y = -800x + 2400$, where x is the age of the computer in years.

a. Find the x- and y-intercepts and interpret the results.

b. Find the value of the computer after 2 yr.

c. Use the ordered pairs from parts (a) and (b) to draw the graph of the equation.

You Be the Teacher!

Correct each student's errors, if any.

97. Graph $x + 2 = 6$.

Derek's work: $x + 2 = 6$ has intercepts of (0, 3) and (6, 0).

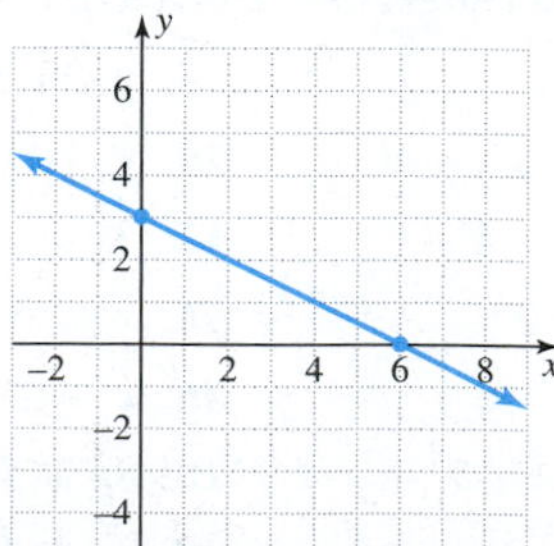

98. Graph by $4x - 3y = -12$ finding the intercepts.

Denny's work: The intercepts are -3 and 4.

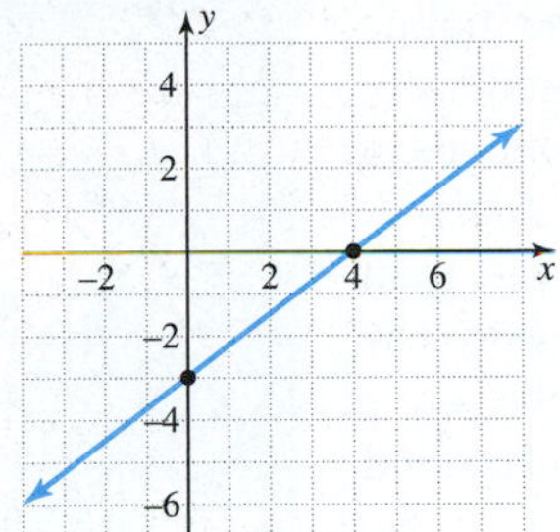

99. Graph $f(x) = 5x$.

Anna's work:

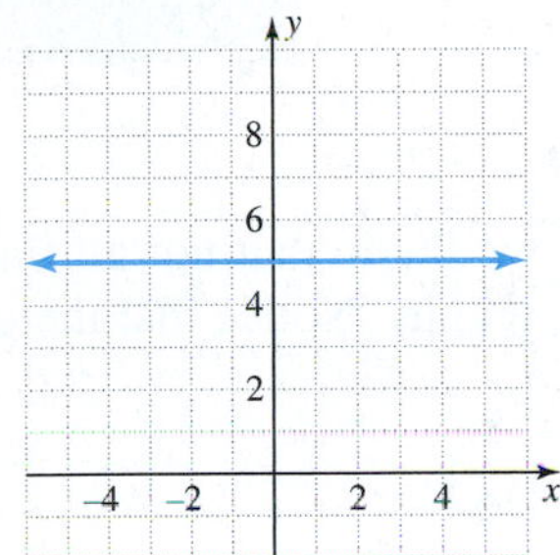

100. Find the intercepts of $6x - 4y = 18$.

Greg's work:

$$6(0) - 4y = 18 \qquad 6x - 4(0) = 18$$
$$4y = 18 \qquad 6x = 18$$
$$y = \frac{9}{2} \qquad x = 3$$
$$\left(0, \frac{9}{2}\right) \qquad (3, 0)$$

The mistake was made in finding the y-intercept. The second step should be $-4y = 18$, so $y = -\frac{9}{2}$.

Calculate It!

101. Draw, by hand, the coordinate system that is defined by these window settings. Label the x-axis and y-axis with the appropriate tick marks.

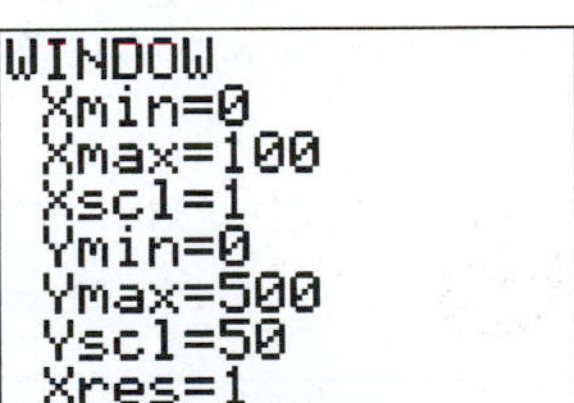

102. Draw, by hand, the coordinate system that is defined by these window settings. Label the x-axis and y-axis with appropriate tick marks.

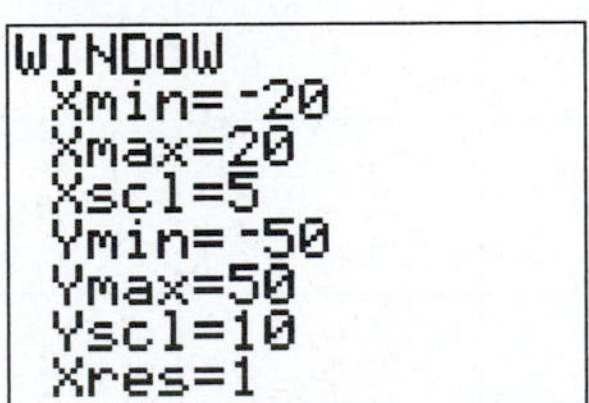

103. Determine the ordered pairs for the x-intercept and y-intercept of the graph if each tick mark represents one unit. (2, 0) and (0, 6)

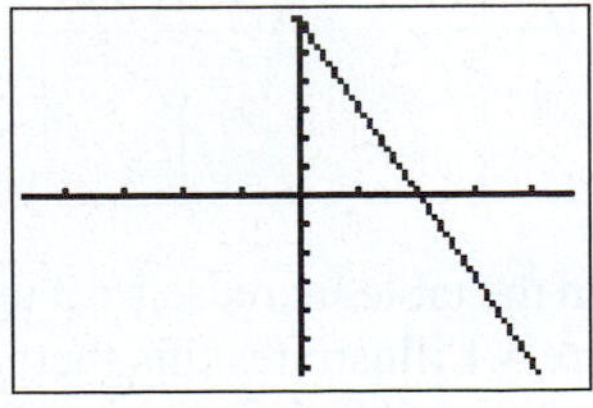

104. Determine the ordered pairs for the x-intercept and y-intercept of the graph if each tick mark on the x-axis represents one unit and each tick mark on the y-axis represents five units. (5, 0) and (0, −15)

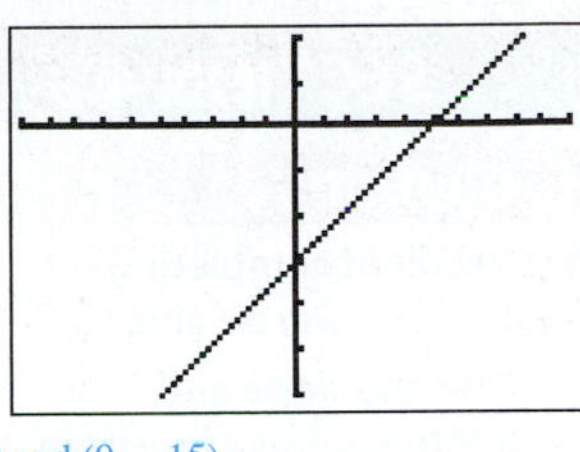

105. Use a graphing calculator to graph $y = -2x + 20$ in the standard viewing window.

a. Sketch the graph shown on the calculator.

b. Which intercept(s) can be seen on the calculator? The x-intercept (10, 0)

c. Find the x-intercept and y-intercept by hand. (10, 0) and (0, 20)

d. What viewing window will show the complete graph that contains both intercepts? State the values that you would use for Xmin, Xmax, Xscl, Ymin, Ymax, and Yscl. Answers vary.

e. Sketch the graph shown on the calculator using the window settings from part d. Answers vary.

106. Use a graphing calculator to graph $y = 0.5x + 40$ in the standard viewing window.

a. Sketch the graph that is shown on the calculator.

b. Which intercept(s) can be seen on the calculator? none

c. Find the x- and y-intercepts by hand. (−80, 0) and (0, 40)

d. What viewing window will show the complete graph that contains both intercepts? State the values you would use for Xmin, Xmax, Xscl, Ymin, Ymax, and Yscl. Answers vary.

e. Sketch the graph shown on the calculator using the window settings from part d. Answers vary.

Think About It!

Solve each problem.

107. Are the graphs of $f(x) = 0.1x + 2$ and $g(x) = 2$ the same? If not, explain. No. The graph of $f(x) = 0.1x + 2$ has $m = 0.1$ and $b = 2$. The graph of $g(x) = 2$ is a horizontal line.

108. Are the graphs of $f(x) = 100x$ and $x = 0$ the same? If not, explain. No. The graph of $f(x) = 100x$ has $m = 100$ and $b = 0$. The graph of $x = 0$ is the y-axis.

109. Determine if the graphs of $2x - 3y = 6$ and $0.002x - 0.003y = 0.006$ are the same. Yes, the graphs are the same.

110. Determine if the graphs of $x + 2y = 1$ and $100x + 200y = 100$ are the same. Yes, the graphs are the same.

SECTION 4.3 The Slope of a Line

OBJECTIVES

As a result of completing this section, you will be able to

1. Find the slope and y-intercept of a line from an equation.
2. Graph a line given its slope and y-intercept.
3. Use the slope formula to determine the slope of a line.
4. Determine if two lines are parallel or perpendicular.
5. Apply the concept of slope to real-world applications.
6. Troubleshoot common errors.

The grade of a highway is the measure of the incline or steepness of the road. The grade is generally expressed as a percentage. One of the steepest roads in the world is Baldwin Street in Dunedin, New Zealand, with a 35% grade. What is the slope of the road and what does it mean? (Source: http://www.geography-lists.com/list17y.html)

To answer this question, we must understand the concept of the slope of a line. In this section, we will explore linear equations of the form $y = mx + b$ and show how these equations relate to the slope of a line.

Objective 1 Find the slope and y-intercept of a line from an equation.

The Slope-Intercept Form of a Line

From our experience with graphing lines, we know that not all lines have the same steepness or *slope* as seen in the following graphs. Notice that the graph of $y = 2x$ is steeper than the graph of $y = \frac{1}{2}x$.

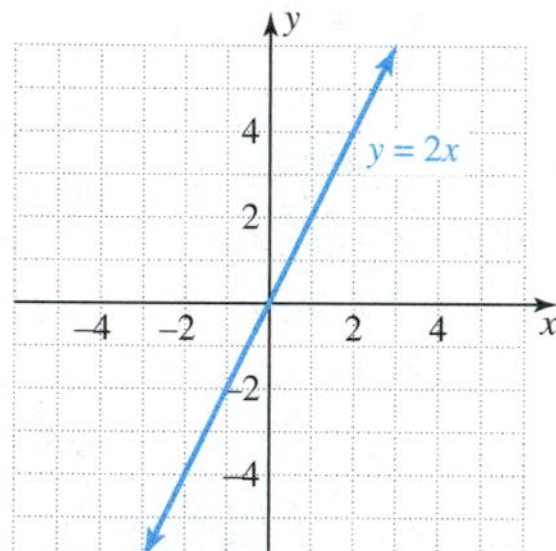

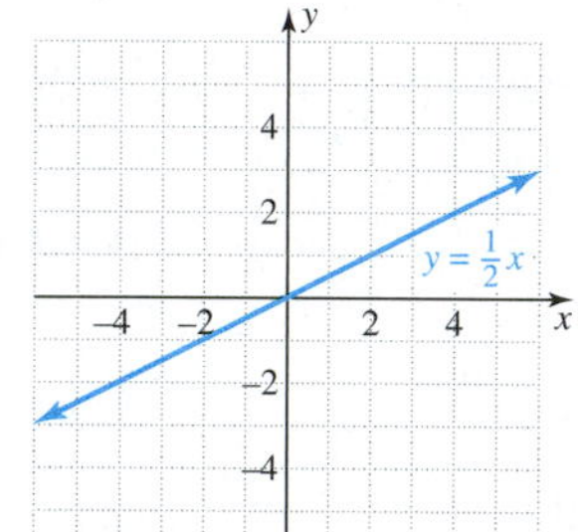

Mathematically speaking, the slope of a line is a measure of the steepness of the line. Consider the linear equation $y = \frac{3}{2}x - 1$. Several solutions of this equation are shown in the following table. We will examine these points and determine how they relate to the graph.

x	y	Change in x	Change in y	$\frac{\text{Change in } y}{\text{Change in } x}$
0	$\frac{3}{2}(0) - 1 = -1$			
2	$\frac{3}{2}(2) - 1 = 2$	$2 - 0 = 2$	$2 - (-1) = 3$	$\frac{3}{2}$
4	$\frac{3}{2}(4) - 1 = 5$	$4 - 2 = 2$	$5 - 2 = 3$	$\frac{3}{2}$
6	$\frac{3}{2}(6) - 1 = 8$	$6 - 4 = 2$	$8 - 5 = 3$	$\frac{3}{2}$

The y-values of the solutions in the table increase by 3 units as the x-values of the solutions increase by 2 units. Figure 4.1 illustrates this fact; as we move from one point on the line to another point on the line, we rise 3 units and move right 2 units.

INSTRUCTOR NOTE: Point out that the slope of the line is the same regardless of which two points are used. Illustrate this with the points (0, −1) and (6, 8). The change in x is $6 - 0 = 6$ and the change in y is $8 - (-1) = 9$. The ratio of these changes is $\frac{9}{6} = \frac{3}{2}$.

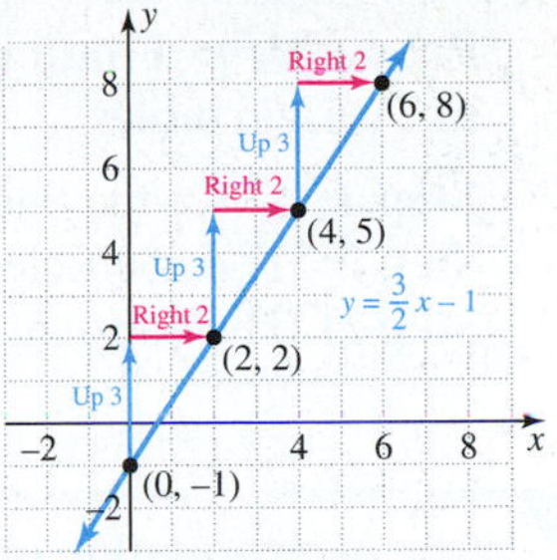

Figure 4.1

So, the value 3 describes how the y-values change (vertical change) and the value 2 describes how the x-values change (horizontal change) as we move from one point on the line to another point.

The ratio of these changes, $\frac{\text{change in } y \text{ (vertical change)}}{\text{change in } x \text{ (horizontal change)}}$, is called the slope of the line.

So, in this example, the slope of the line is

$$\text{slope} = \frac{\text{change in } y}{\text{change in } x} = \frac{3}{2}$$

Note that the slope of the line is the same for every pair of points on the line. If the slope is not the same, then the graph is not a line.

Definition: Slope of a Line

The **slope** of a line is the ratio of the change in y to the change in x between two points on a line. The slope is denoted by the letter m. It is often called "vertical change over horizontal change" or "rise over run."

$$m = \frac{\text{change in } y}{\text{change in } x} = \frac{\text{vertical change}}{\text{horizontal change}} = \frac{\text{rise}}{\text{run}}$$

So, from the table and graph of the line, $y = \frac{3}{2}x - 1$, illustrated in Figure 4.1, we see that the slope $m = \frac{3}{2}$ and the y-intercept of the line is $(0, -1)$. Note that both the slope and the y-coordinate of the y-intercept are provided in the given equation, $y = \frac{3}{2}x - 1$.

- The coefficient of x, $\frac{3}{2}$, is the slope of the line.
- The constant term, -1, is the y-coordinate of the y-intercept.

Because this information is provided in this form of a line, it is referred to as the *slope-intercept form* of a line.

Property: The Slope-Intercept Form of a Line

An equation or function in the form of $y = mx + b$ or $f(x) = mx + b$ is the **slope-intercept form** of a line.

- The coefficient of x, denoted by m, is the slope of the line.
- The constant term, denoted by b, is the y-coordinate of the y-intercept. The point $(0, b)$ is the y-intercept of the graph of the line.

$$y = mx + b$$

slope (points to m) $(0, b)$ is the y-intercept (points to b)

INSTRUCTOR NOTE:
Remind students the coefficient is the number in front of the variable. The variable is not included in the slope.

Procedure: Finding the Slope and y-Intercept of a Linear Equation in Two Variables

Step 1: Solve the linear equation for y, if necessary.
Step 2: The coefficient of x is the slope of the line.
Step 3: The constant term is the y-coordinate of the y-intercept of the line.

Objective 1 Examples Write each equation in slope-intercept form and find its slope and y-intercept.

Problems	Slope-intercept Form	Value of m	Value of b	Slope	y-intercept
1a. $y = \frac{1}{2}x + 3$	$y = \frac{1}{2}x + 3$	$m = \frac{1}{2}$	$b = 3$	$\frac{1}{2}$	$(0, 3)$
1b. $f(x) = -x - 4$	$y = -1x - 4$	$m = -1$	$b = -4$	-1	$(0, -4)$
1c. $2x - 3y = 6$	$2x - 3y = 6$ $2x - 3y - 2x = 6 - 2x$ $-3y = -2x + 6$ $\frac{-3y}{-3} = \frac{-2x}{-3} + \frac{6}{-3}$ $y = \frac{2}{3}x - 2$	$m = \frac{2}{3}$	$b = -2$	$\frac{2}{3}$	$(0, -2)$
1d. $5x + y = 0$	$5x + y = 0$ $y = -5x$ $y = -5x + 0$	$m = -5$	$b = 0$	-5	$(0, 0)$
1e. $y = 6$	$y = 0x + 6$	$m = 0$	$b = 6$	0	$(0, 6)$

Student Check 1 Write each equation in slope-intercept form and find its slope and y-intercept.

a. $y = \frac{5}{4}x - 3$ **b.** $f(x) = x + 7$ **c.** $8x - y = 6$

d. $x - 3y = 0$ **e.** $y = -5$

Graph Using Slope and a y-Intercept

Objective 2

Graph a line given its slope and y-intercept

In Section 4.2, we graphed lines by finding two points that lie on the line. We can also graph a line if we know its slope and y-intercept.

- The y-intercept $(0, b)$ is one point on the line.
- The slope, m, determines how to move from the y-intercept to another point on the line.

Since slope is the ratio of the change in y to the change in x, it tells us how to move from the y-intercept to another point on the line. Recall that the numerator of the slope corresponds to the change in y, or the vertical movement between the two points. The denominator of the slope corresponds to the change in x, or the horizontal movement between the two points.

The following table illustrates different values of slope and the movement between points that each represents. If the slope is not given as a fraction, we must convert it to an equivalent fractional form to determine the movement between points on the line. Any equivalent fraction can be used since the slope is a ratio.

Value of m	Movement Indicated by the Numerator	Movement Indicated by the Denominator
$\frac{1}{2}$	Vertical change is 1 → move up 1 unit	Horizontal change is 2 → move right 2 units
$2 = \frac{2}{1}$	Vertical change is 2 → move up 2 units	Horizontal change is 1 → move right 1 unit
$0.7 = \frac{7}{10}$	Vertical change is 7 → move up 7 units	Horizontal change is 10 → move right 10 units
$-\frac{3}{4} = \frac{-3}{4}$	Vertical change is −3 → move down 3 units	Horizontal change is 4 → move right 4 units
$-\frac{3}{4} = \frac{3}{-4}$	Vertical change is 3 → move up 3 units	Horizontal change is −4 → move left 4 units

The last two rows in the table show that if the slope is a negative value, the negative sign can be applied to the numerator or to the denominator but not both. Recall the following relationship.

$$-\frac{A}{B} = \frac{-A}{B} = \frac{A}{-B}$$

Procedure: Graphing a Line Using Its Slope and y-Intercept

Step 1: Plot the y-intercept $(0, b)$.

Step 2: From the y-intercept, move to another point based on the information from the slope.

a. The numerator of the slope indicates how many units to go up (if positive) or to go down (if negative) from the y-intercept.

b. The denominator indicates how many units to move right (if positive) or to move left (if negative).

Step 3: Draw the graph through these two points.

Note: *When graphing $y = mx + b$, think of b as begin and m as move.*

Objective 2 Examples

Graph each line using its slope and y-intercept. Label the y-intercept and one other point.

2a. $y = \frac{2}{3}x - 2$

2b. $f(x) = -x + 4$

2c. $y = 6$

2d. $3x - y = -2$

Solutions

2a. The equation is in slope-intercept form, so we identify the slope and y-intercept.

$$y = \frac{2}{3}x - 2$$

$$m = \frac{2}{3}, \text{ } y\text{-intercept is } (0, -2)$$

Step 1: Plot $(0, -2)$.

Step 2: $m = \frac{2}{3} \rightarrow$ move up 2 units and right 3 units

This movement places us at the point $(3, 0)$.
The graph is the line through these two points.

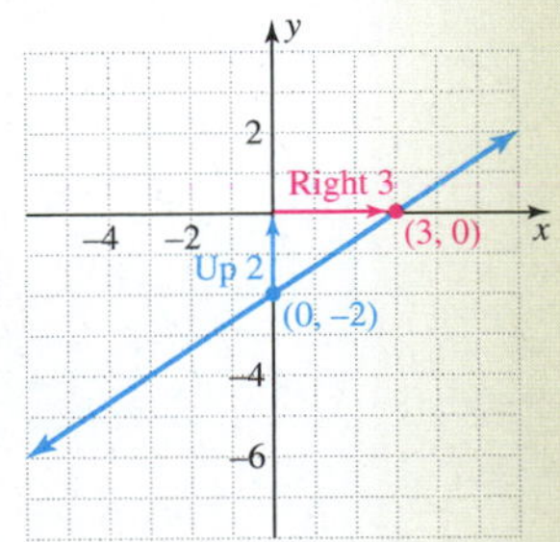

INSTRUCTOR NOTE:
Show students the calculation to determine the 2nd point.

$$\begin{array}{r} (0, -2) \\ +3 \quad +2 \\ \hline (3, 0) \end{array}$$

2b. The linear function is in slope-intercept form, so we identify the slope and y-intercept.

$$f(x) = -1x + 4$$
$$m = -1,\ y\text{-intercept is } (0, 4)$$

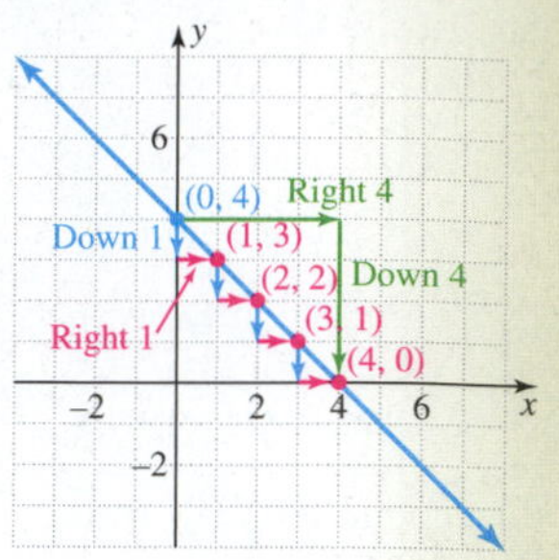

Step 1: Plot (0, 4).

Step 2: $m = -1 = \frac{-1}{1} \rightarrow$ move down 1 unit and right 1 unit

This movement places us at the point (1, 3). The graph is the line through these two points. We can use the slope repeatedly to obtain more points on the graph to enable us to draw the line accurately.

Note that any equivalent form of the slope will provide the same line. For instance, $m = -1$ is equivalent to $m = \frac{-4}{4}$. So, we could move down 4 units and right 4 units to obtain the same graph.

2c. The equation is in slope-intercept form, so we identify the slope and y-intercept.

$$y = 0x + 6$$
$$m = 0,\ y\text{-intercept is } (0, 6)$$

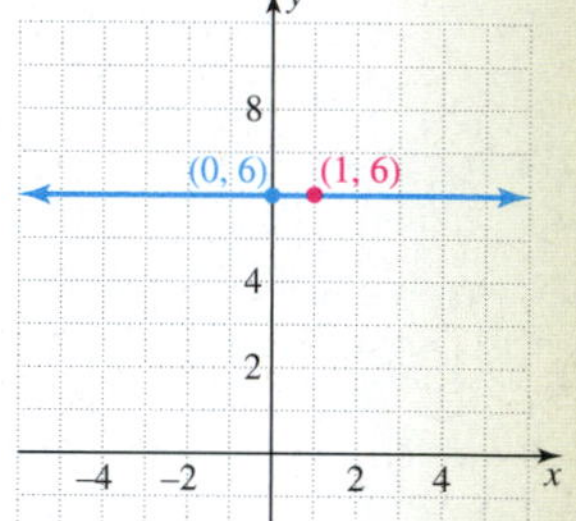

Step 1: Plot (0, 6).

Step 2: $m = 0 = \frac{0}{1} \rightarrow$ move up 0 units and right 1 unit

This places us at the point (1, 6). The graph is the line through these two points.

2d. We must write the equation in slope-intercept form so that we can identify the slope and y-intercept.

$$3x - y = 2$$
$$3x - y - 3x = 2 - 3x$$
$$-y = -3x + 2$$
$$y = 3x - 2$$
$$m = 3,\ y\text{-intercept is } (0, -2)$$

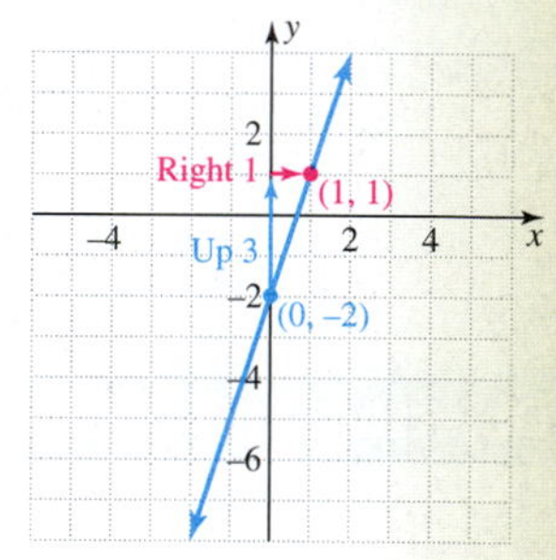

Step 1: Plot (0, −2).

Step 2: $m = 3 = \frac{3}{1} \rightarrow$ move up 3 units and right 1 unit

This places us at the point (1, 1). The graph is the line through these two points.

Student Check 2 Graph each line using its slope and y-intercept. Label the y-intercept and one other point.

a. $y = -\frac{3}{5}x + 3$ **b.** $f(x) = x + 1$ **c.** $y = 2$ **d.** $6x + 5y = -15$

Note: *Example 2 illustrates two important facts about slope.*

1. *As the absolute value of the slope increases, the line becomes steeper. Note that $m = 3$ is greater than $m = \frac{3}{2}$ and the line in Example 2d is steeper than the line in Example 2a.*

2. *Slope not only measures the steepness of a line but its direction, as well.*

- *The graph of a line with positive slope is rising or increasing from left to right (Example 2a).*
- *The graph of a line with negative slope is falling or decreasing from left to right (Example 2b).*
- *The graph of a line with zero slope is horizontal (Example 2c).*

The Slope Formula

Objective 3 Use the slope formula to determine the slope of a line.

So far, we have found the slope of a line from its equation. The slope can also be found using a formula since it is defined as $\frac{\text{change in } y}{\text{change in } x}$.

Property: The Slope Formula

The slope of a line is the ratio of the change in y, denoted Δy, to the change in x, denoted Δx. If (x_1, y_1) and (x_2, y_2) are two points that lie on a line such that $x_1 \neq x_2$, then the slope of the line is

$$m = \frac{\text{change in } y}{\text{change in } x} = \frac{\Delta y}{\Delta x} = \frac{y_2 - y_1}{x_2 - x_1}$$

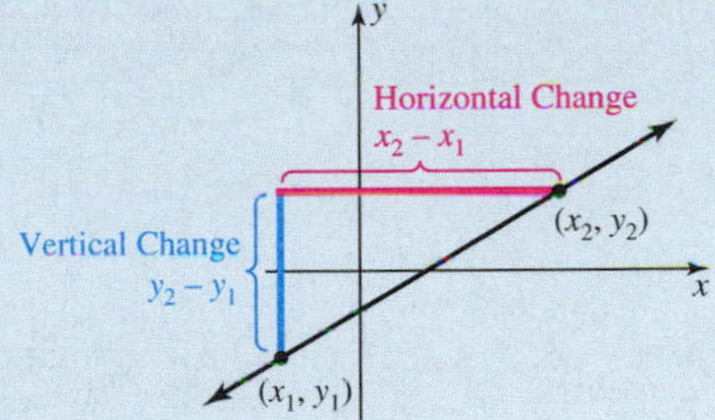

If $x_1 = x_2$, then the line through the points is vertical and the slope is undefined.

The triangle symbol Δ is the Greek letter delta. In mathematics and the sciences, this symbol usually represents a change in the indicated quantity.

Procedure: Finding the Slope of a Line Using the Slope Formula

Step 1: Label one point as (x_1, y_1) and the other point as (x_2, y_2).

Step 2: Substitute the values into the formula $m = \frac{y_2 - y_1}{x_2 - x_1}$.

Step 3: Simplify the numerator and denominator and simplify the resulting fraction, if necessary.

Objective 3 Examples Find the slope of the line through the given points. Graph the line and observe the direction of the line.

3a. $(3, -1)$ and $(-2, 5)$

3b.

x	y
−3	−2
−2	1
−1	4
0	7
1	10

3c. $(2, -3)$ and $(-4, -3)$

3d. $(5, 6)$ and $(5, -1)$

Solutions **3a.** $(x_1, y_1) = (3, -1)$ and $(x_2, y_2) = (-2, 5)$

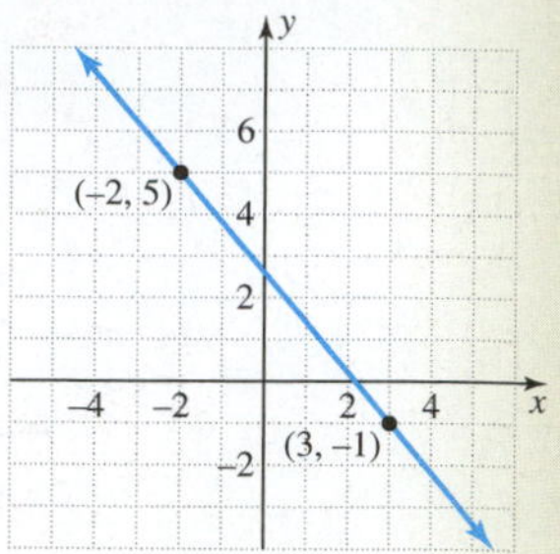

$$m = \frac{y_2 - y_1}{x_2 - x_1} = \frac{5 - (-1)}{-2 - 3} = \frac{6}{-5} = -\frac{6}{5}$$

The line is *decreasing* from left to right and its slope is *negative*.

Note: It doesn't matter which point we label as (x_1, y_1) and (x_2, y_2). The slope of the line will still be the same. If $(x_1, y_1) = (-2, 5)$ and $(x_2, y_2) = (3, -1)$, then the slope is $m = \frac{y_2 - y_1}{x_2 - x_1} = \frac{-1 - (5)}{3 - (-2)} = \frac{-6}{5} = -\frac{6}{5}$, which is exactly what we found when the points were interchanged.

3b. We will use two different pairs of points to find the slope to show that it does not matter which points we choose.

First pair: $(x_1, y_1) = (-3, -2)$ and $(x_2, y_2) = (-1, 4)$

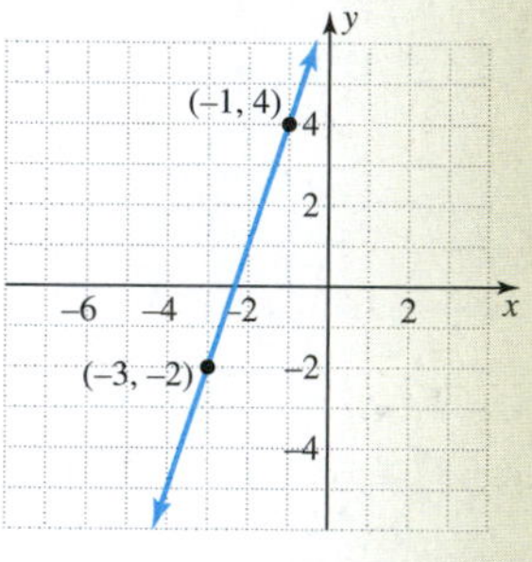

$$m = \frac{y_2 - y_1}{x_2 - x_1} = \frac{4 - (-2)}{-1 - (-3)} = \frac{6}{2} = 3$$

Second pair: $(x_1, y_1) = (-2, 1)$ and $(x_2, y_2) = (1, 10)$

$$m = \frac{y_2 - y_1}{x_2 - x_1} = \frac{10 - (1)}{1 - (-2)} = \frac{9}{3} = 3$$

The line is *increasing* from left to right and its slope is *positive*.

3c. $(x_1, y_1) = (2, -3)$ and $(x_2, y_2) = (-4, -3)$

$$m = \frac{y_2 - y_1}{x_2 - x_1} = \frac{-3 - (-3)}{-4 - (2)} = \frac{0}{-6} = 0$$

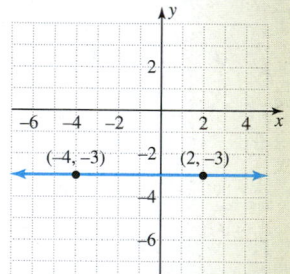

The line is *horizontal* and its slope is *zero*. We also say that the line is constant since it is neither increasing nor decreasing from left to right.

INSTRUCTOR NOTE: Remind students that "zero" slope is not the same as "no slope." No slope refers to an undefined slope. Use zero speed limit and no speed limit for an illustration of the difference between zero and no.

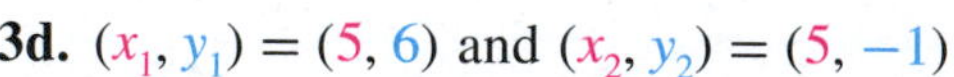

3d. $(x_1, y_1) = (5, 6)$ and $(x_2, y_2) = (5, -1)$

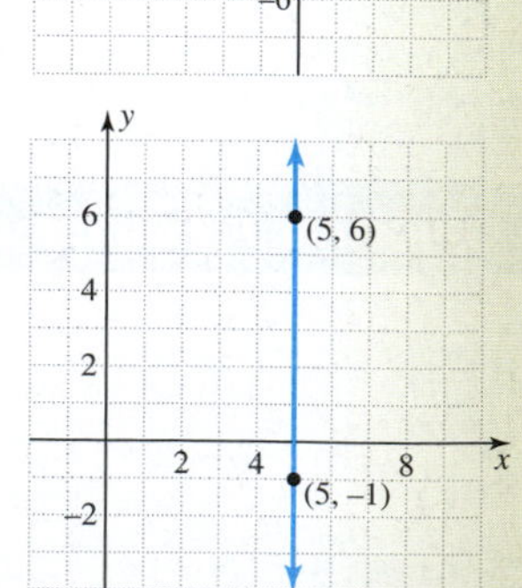

$$m = \frac{y_2 - y_1}{x_2 - x_1} = \frac{-1 - (6)}{5 - (5)} = \frac{-7}{0} = \text{undefined}$$

The line is *vertical* and its slope is *undefined*.

Student Check 3 Find the slope of the line through the given points.

a. (1, −6) and (7, −6) **b.** (−2, 5) and (3, 2)

c. (4, 7) and (2, −1) **d.** (−4, 2) and (−4, −1)

Note: Here are some hints to remember the slopes of horizontal and vertical lines:

If we can "walk" on the line, then the line has a slope. If we can't "walk" on the line, then the slope is undefined. Since we cannot walk on a vertical line, it has slope that is undefined.

Another way to remember the slopes is to think "uv" (undefined goes with vertical) and both zero and horizontal have a "z" in them.

INSTRUCTOR NOTE:
Remind students that we must look at the line from left to right to determine if the line "goes up" or "goes down."

Summary of Slopes

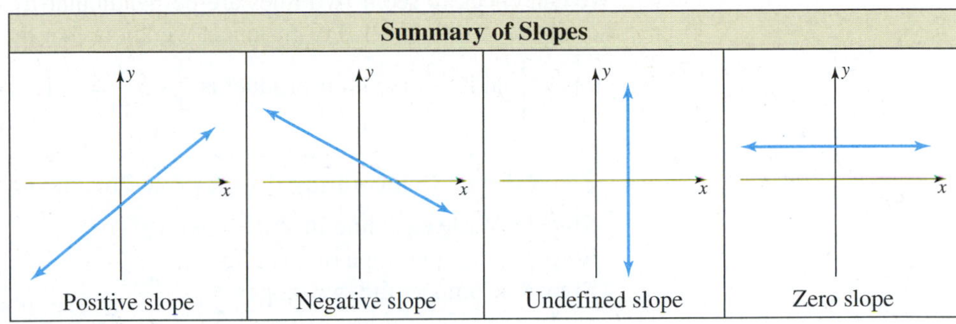

Parallel and Perpendicular Lines

Objective 4

Determine if two lines are parallel or perpendicular.

We will now discuss special ways in which two lines can relate to one another, parallel lines or perpendicular lines.

Two lines, in the same plane, are *parallel* if they never intersect. For lines to be parallel, they must have the same slope but different y-intercepts.

The graphs of the equations $y = \frac{3}{4}x - 1$ and $y = \frac{3}{4}x + 3$ are shown. Both lines have a slope of $\frac{3}{4}$ but have different y-intercepts. The lines do not intersect one another and are, therefore, parallel.

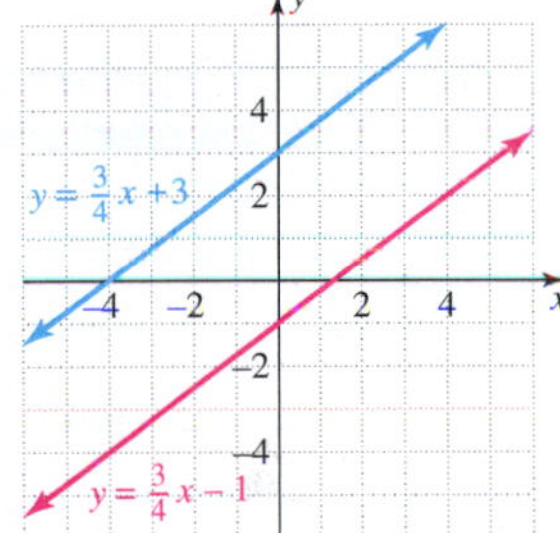

Definition: Parallel Lines

Two nonvertical lines that lie in the same plane are **parallel** if they have the same slope. Vertical lines are parallel to each other.

Two nonvertical lines are *perpendicular* to each other if the lines form a right angle (90°) at their point of intersection. For this to happen, the graph of one line must be increasing (have positive slope) and the other must be decreasing (have negative slope). Also, the rise of one line must be the run for the other line and vice versa. Mathematically, this means that the slopes of the lines are *negative reciprocals* of each other.

The given graph contains lines that intersect at a right angle. On one line, we rise 3 units and run 4 units to reach another point. Therefore, the slope is $\frac{3}{4}$. On the other line, we go down 4 units and run 3 units to reach another point. Therefore, the slope is $-\frac{4}{3}$. These lines have slopes that are negative reciprocals and the lines are perpendicular to one another. The graphs of $y = \frac{3}{4}x + 3$ and $y = -\frac{4}{3}x + \frac{2}{3}$ are shown.

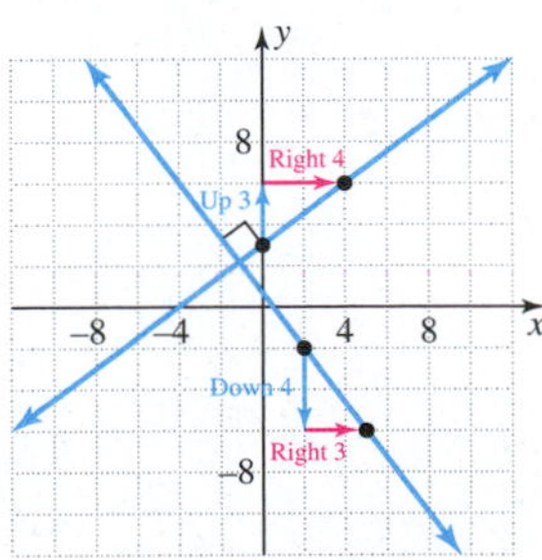

Definition: Perpendicular Lines

Two nonvertical lines $y_1 = m_1x + b_1$ and $y_2 = m_2x + b_2$ are **perpendicular** if $m_1 = -\frac{1}{m_2}$; that is, the slopes of the lines are negative reciprocals of one another. The lines may or may not have the same y-intercept. Vertical and horizontal lines are perpendicular.

We can check to see if two lines are perpendicular by determining whether the product of the slopes is -1. For instance, we know that the lines previously graphed have slopes $\frac{3}{4}$ and $-\frac{4}{3}$ and their product is $\frac{3}{4} \cdot -\frac{4}{3} = -1$.

Procedure: Determining if Lines Are Parallel or Perpendicular

Step 1: Write each line in slope-intercept form.
Step 2: Find the slope of each line.
Step 3: Compare the two slopes.
- **a.** If the slopes are the same (and the y-intercepts are different), the lines are parallel.
- **b.** If the slopes are negative reciprocals, then the lines are perpendicular.
- **c.** If the slopes are neither the same nor negative reciprocals, then the lines are neither parallel nor perpendicular.

Objective 4 Examples Determine if the lines are parallel, perpendicular, or neither.

4a. $y = 2x - 4$ and $4x - 2y = -6$ **4b.** $3x - y = 3$ and $x + 3y = 9$

4c. $7x + 2y = 14$ and $6x + 21y = 0$

Solutions **4a.**

$y = 2x - 4$

$m = 2$, y-intercept $= (0, -4)$

$$4x - 2y = -6$$
$$-2y = -4x - 6$$
$$\frac{-2y}{-2} = \frac{-4x}{-2} - \frac{6}{-2}$$
$$y = 2x + 3$$

$m = 2$, y-intercept $= (0, 3)$

The slope of each line is $m = 2$. The two lines have different y-intercepts. Therefore, the lines are parallel.

4b.
$$3x - y = 3$$
$$-y = -3x + 3$$
$$\frac{-y}{-1} = \frac{-3x}{-1} + \frac{3}{-1}$$
$$y = 3x - 3$$

$m = 3$, y-intercept $= (0, -3)$

$$x + 3y = 9$$
$$3y = -x + 9$$
$$\frac{3y}{3} = \frac{-x}{3} + \frac{9}{3}$$
$$y = -\frac{1}{3}x + 3$$

$m = -\frac{1}{3}$, y-intercept $= (0, 3)$

The slope of the first equation is $m = 3$ and the slope of the second equation is $m = -\frac{1}{3}$. These slopes are negative reciprocals of one another. So, the lines are perpendicular.

4c. $7x + 2y = 14$

$$2y = -7x + 14$$

$$\frac{2y}{2} = \frac{-7x}{2} + \frac{14}{2}$$

$$y = -\frac{7}{2}x + 7$$

$m = -\frac{7}{2}$, y-intercept $= (0, 7)$

$6x + 21y = 0$

$$21y = -6x$$

$$\frac{21y}{21} = \frac{-6x}{21}$$

$$y = -\frac{2}{7}x$$

$m = -\frac{2}{7}$, y-intercept $= (0, 0)$

The slope of the first equation is $m = -\frac{7}{2}$ and the slope of the second equation is $m = -\frac{2}{7}$. The slopes are reciprocals of one another but are not negative reciprocals. The lines are neither parallel nor perpendicular.

Student Check 4 Determine if the lines are parallel, perpendicular, or neither.

a. $y = 3x - 4$ and $3x + y = 5$

b. $2x + 3y = 6$ and $3x - 2y = 12$

c. $2x - 5y = 10$ and $4x - 10y = 6$

Applications of Slope

Objective 5 Apply the concept of slope to real-world applications.

When a linear equation in two variables represents a real-world situation, the slope and y-intercept have practical meanings that relate to the situation. In the equation $y = mx + b$, the point $(0, b)$ is the y-intercept. The y-intercept indicates the beginning or **initial value**. The slope m represents the **rate of change**, or how the y-values change as the x-values change.

Slope has many meaningful applications in real life. Slope is used, for example, to construct stairs, wheelchair ramps, highways, and roofs. In these situations, the slope is often described in terms of rise and run.

In Example 5, we will express the slope of the given situation and also interpret the meaning of the slope and y-intercept.

Objective 5 Examples **Solve each problem.**

5a. The median annual income for men with a bachelor's degree is given by the linear equation, $y = 1443x + 38{,}843$, where x is the number of years after 1990. Interpret the meaning of the slope and y-intercept in the context of this problem. (Source: www.infoplease.com)

Solution **5a.** The slope of the equation is $m = 1443 = \frac{1443}{1}$. This means that the median annual income for men with a bachelor's degree increases by $1443 each year after 1990.

The y-intercept is $(0, 38{,}843)$. This represents the initial median annual income for men with a bachelor's degree. That is, the median annual income in 1990 was $38,843.

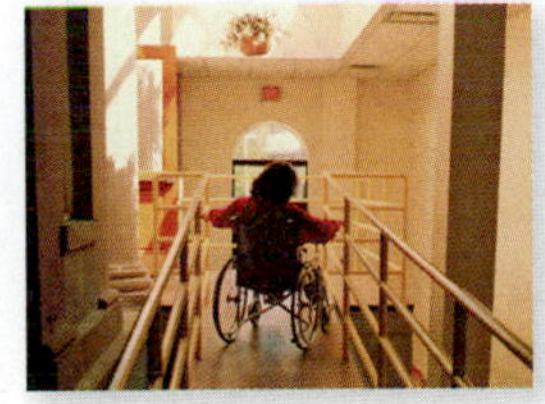

5b. The guideline for constructing a wheelchair ramp is that the maximum slope should be 1 in. of rise for each 12 in. of run. Express the slope of the wheelchair ramp.

Solution **5b.** The slope of the wheelchair ramp is

$$\frac{\text{rise}}{\text{run}} = \frac{1}{12}$$

5c. The grade of a highway is the measure of the incline or steepness of the road. The grade is generally expressed as a percentage. One of the steepest roads in the world is Baldwin Street in Dunedin, New Zealand, with a 35% grade. What is the slope of the road and what does it mean? (Use meters as the units.) (Source: http://www.geographylists.com/list17y.html)

Solution **5c.** The slope is

$$35\% = \frac{35}{100} = \frac{7}{20}$$

This slope means that the road rises 7 m for every 20 m traveled.

Student Check 5 Solve each problem.

a. The median annual income for women with a bachelor's degree is given by the equation, $y = 1100x + 27{,}645$, where x is the number of years after 1990. Interpret the slope and y-intercept in the context of this problem.

b. The pitch of a roof is defined as the number of inches it rises vertically for each 12 in. it extends horizontally. If a roof rises 8 in. for each 12 in. it extends horizontally, what is its pitch?

c. The steepest road in the United States is Canton Avenue in Pittsburgh, Pennsylvania, with a grade of 37%. What is the slope of the road and what does it mean? (Use feet as the units.) (Source: http://www.geographylists.com/list17y.html)

Objective 6 ▶

Troubleshoot common errors.

Troubleshooting Common Errors

Some common errors associated with the slope of a line are shown.

Objective 6 Examples

A problem and an incorrect solution are given. Provide the correct solution and an explanation of the error.

6a. Find the slope of the line $4x + 3y = 12$.

Incorrect Solution	Correct Solution and Explanation
The slope of the line is the coefficient of x and is $m = 4$.	The equation must be written in slope-intercept form, $y = mx + b$, before we can identify the slope. $4x + 3y = 12$; $3y = -4x + 12$; $y = -\frac{4}{3}x + \frac{12}{3}$; $y = -\frac{4}{3}x + 4$. So, the slope of the line is $m = -\frac{4}{3}$.

6b. Find the slope of the line through (4, −2) and (−1, 7).

Incorrect Solution	Correct Solution and Explanation
The slope of the line is $m = \frac{-1-4}{7-(-2)}$ $m = \frac{-5}{9}$ $m = -\frac{5}{9}$	The formula for slope is $\frac{\text{change in } y}{\text{change in } x}$. $m = \frac{y_2 - y_1}{x_2 - x_1}$ $m = \frac{7-(-2)}{-1-4}$ $m = \frac{9}{-5}$ $m = -\frac{9}{5}$

6c. Explain how to use the slope and y-intercept to graph the line $y = -\frac{3}{2}x + 5$.

Incorrect Solution	Correct Solution and Explanation
The slope $m = -\frac{3}{2}$ and the y-intercept is (0, 5). I would plot (0, 5) and then move down 3 units and left 2 units to get to the point (−2, 2).	The value $-\frac{3}{2} = \frac{-3}{2}$ or $\frac{3}{-2}$, so this means we would plot (0, 5) and then move down 3 units and right 2 units to the point (2, 2) or we would move up 3 units and left 2 units to the point (−2, 8).

ANSWERS TO STUDENT CHECKS

Student Check 1 **a.** $m = \frac{5}{4}$, y-intercept = (0, −3) **b.** $m = 1$, y-intercept = (0, 7) **c.** $m = 8$, y-intercept = (0, −6) **d.** $m = \frac{1}{3}$, y-intercept = (0, 0) **e.** $m = 0$, y-intercept = (0, −5)

Student Check 2

a.

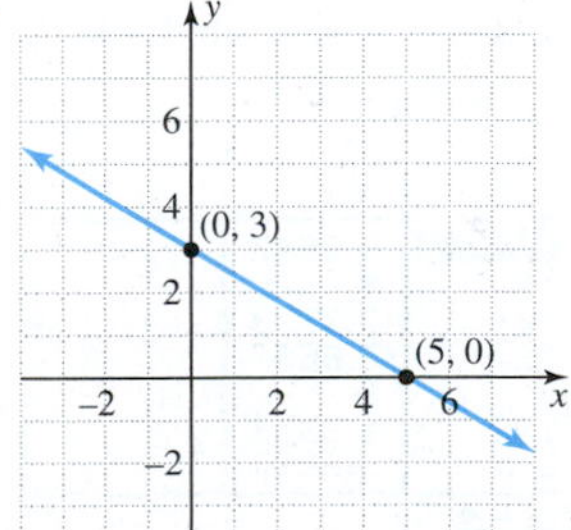

b.

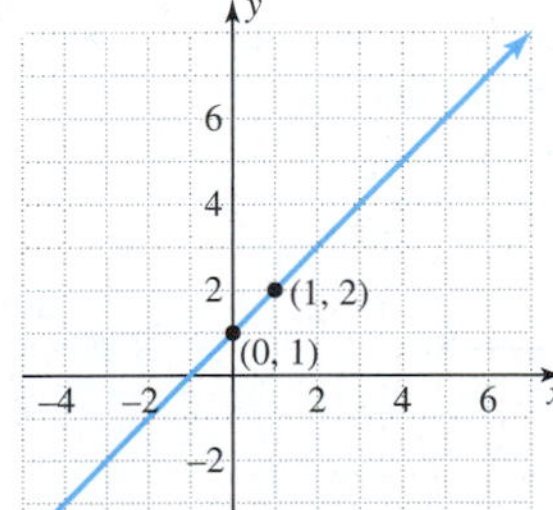

c.

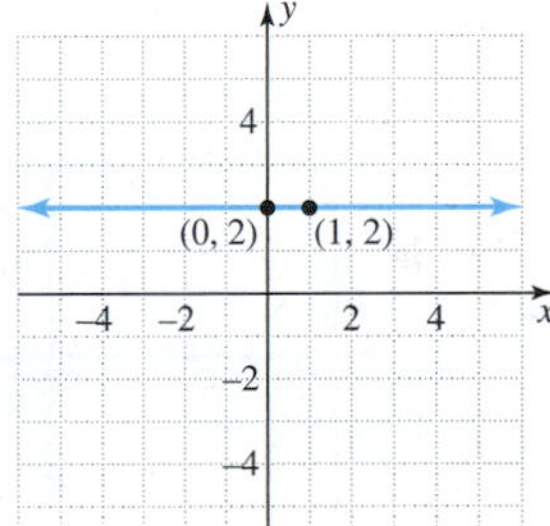

d.

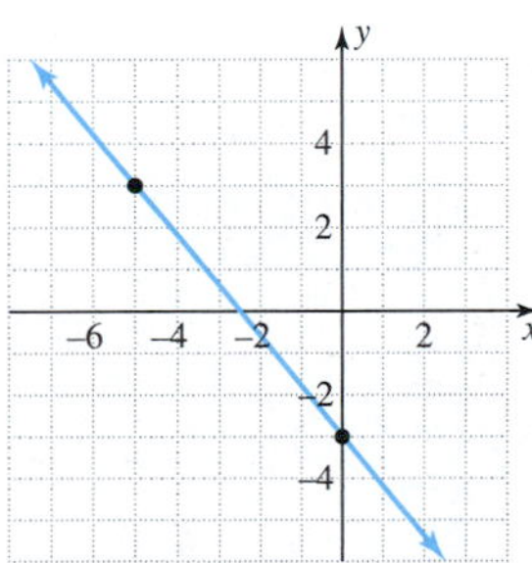

Student Check 3 **a.** $m = 0$ **b.** $m = -\frac{3}{5}$ **c.** $m = 4$ **d.** undefined slope

Student Check 4 **a.** neither **b.** perpendicular **c.** parallel

Student Check 5 **a.** The slope means that the median annual income for women with a bachelor's degree increases by \$1100 each year after 1990. The median annual income in 1990 was \$27,645. **b.** The pitch is the slope which is $\frac{8}{12} = \frac{2}{3}$. **c.** The slope is $37\% = \frac{37}{100}$. This means that the road rises by 37 ft for every 100 ft traveled.

SUMMARY OF KEY CONCEPTS

1. The slope of a line is the ratio of the vertical change between two points on the line to the horizontal change between the points. It tells us the direction of the line and its steepness.

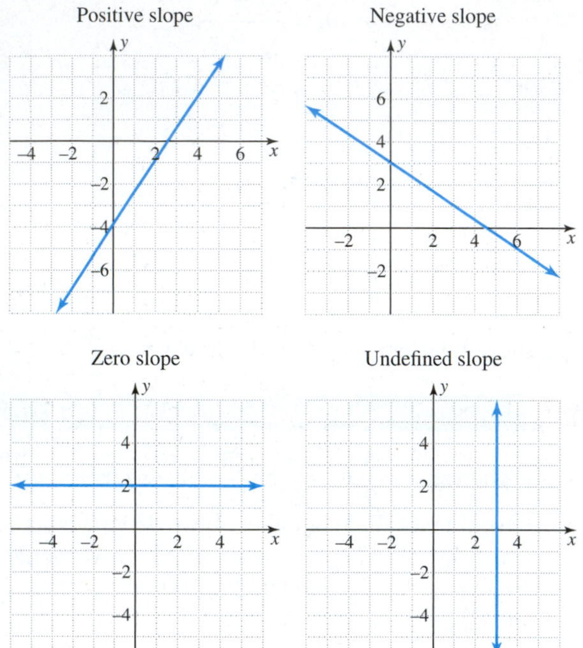

2. The slope of a line can be found using its equation, a table of solutions, its graph, or the slope formula.
   - Equation—The equation must be written in the form $y = mx + b$, if possible. The coefficient of x is the slope.
   - Table of solutions—Find the change in the y-values and the change in the x-values between two ordered pairs in the table. The ratio of these changes is the slope.
   - Graph—Determine the rise and run between two points on the graph. The ratio of the rise to the run is the slope.
   - Slope Formula—Substitute the x and y coordinates of the two points into the formula and simplify.
3. A line can be graphed using the slope intercept form. From an equation, identify the y-intercept and the slope. Plot the y-intercept and use the slope to determine how to move to the next point.
4. Lines that have the same slope and different y-intercepts are parallel to one another. Lines that have slopes that are negative reciprocals are perpendicular to one another. The negative reciprocal means that the signs must be different and the fractions must be reciprocals of each other. Also, the product of the slopes of perpendicular lines is -1.
5. The slope and y-intercept have real meaning when the linear equation represents a real-life situation. The slope indicates how the y-values are changing as the x-values change and the y-intercept represents the initial value. Slope has many real-life applications pertaining to construction of roads, wheelchair ramps, and roofs.

GRAPHING CALCULATOR SKILLS

The graphing calculator can be used to verify the slope of a line by examining the table of solutions and the graph. In addition, the calculator can help to determine if lines are parallel or perpendicular.

Example 1: Graph $y = -\frac{2}{3}x + 2$ using its slope and y-intercept.

Solution: Enter the equation in the equation editor.

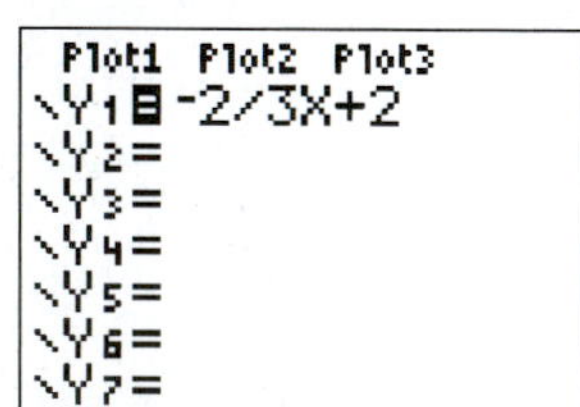

Graph the equation in the ZDecimal window. The ZDecimal view makes each tick mark 0.1 units. The ZDecimal or ZSquare format does not distort the graphs and therefore, enables, us to get a more accurate picture especially when we need to compare two graphs.

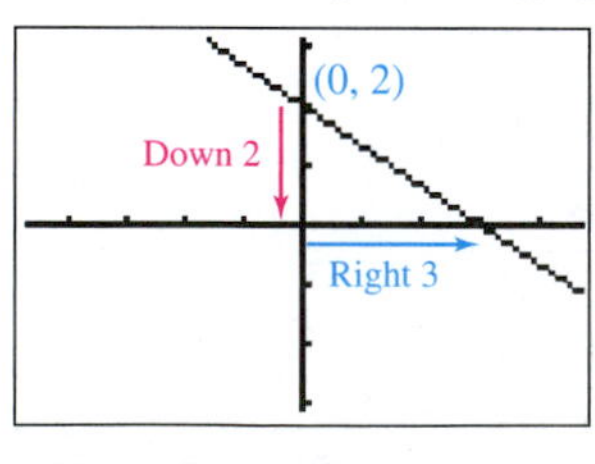

ZOOM 4 GRAPH

From the graph, we can verify that the slope is $-\frac{2}{3}$ since the movement between points is down 2 and right 3.

We can also examine the table to verify the slope of the line.

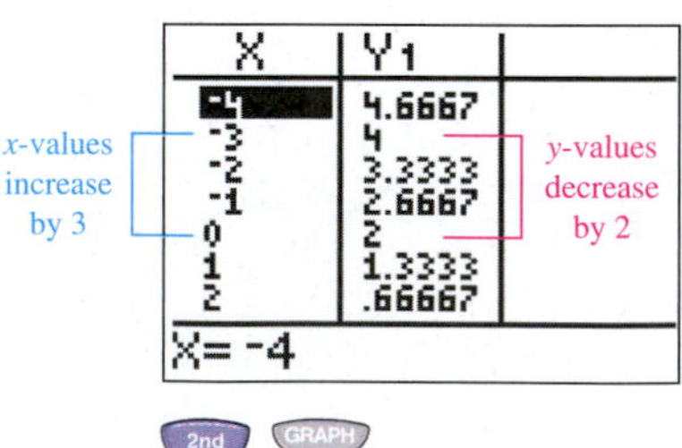

2nd GRAPH

Example 2: Are $y = 2x - 3$ and $y = \frac{1}{2}x + 1$ parallel, perpendicular, or neither?

Solution: Enter each equation in the equation editor and graph.

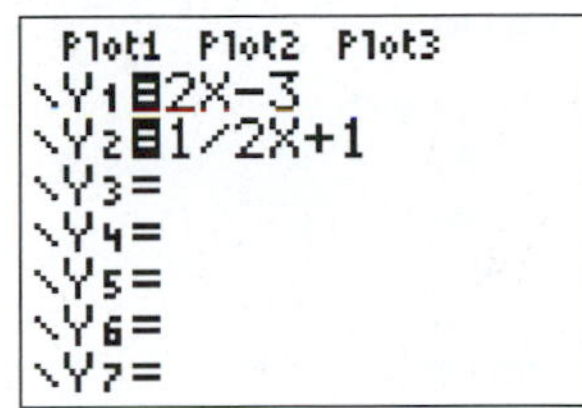

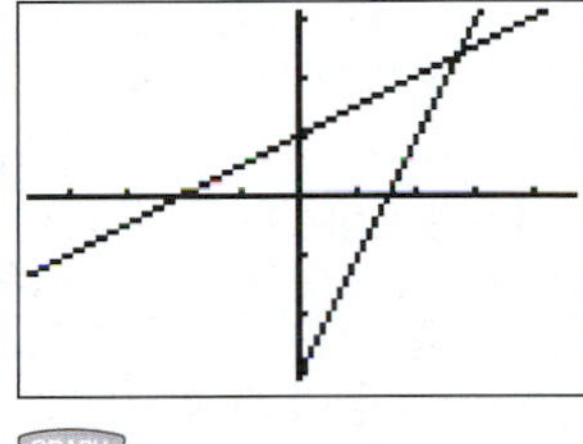

GRAPH

By graphing the lines on the calculator, we can verify that the lines are neither parallel nor perpendicular.

Example 3: Are $y = 2x - 3$ and $y = -\frac{1}{2}x + 1$ parallel, perpendicular, or neither?

Solution: Enter each equation in the equation editor and graph.

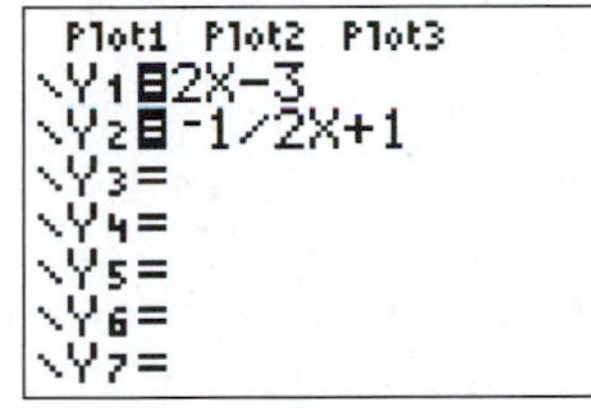

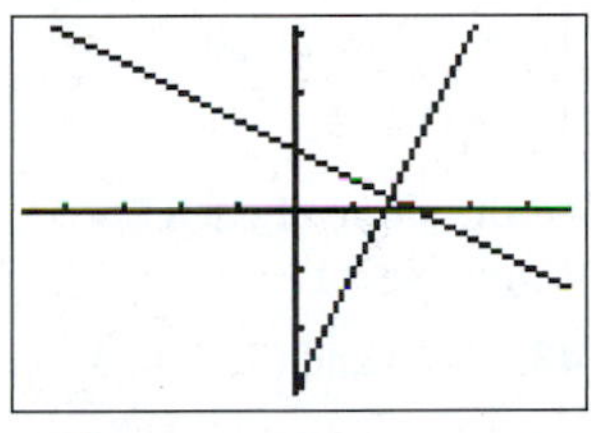

The graph, in the ZDecimal view, shows that the lines are perpendicular to one another. This is not always the case in the standard viewing window.

Note: *Looking at the graphs isn't a completely accurate method of determining if the lines are parallel or perpendicular. We should always check algebraically.*

SECTION 4.3 EXERCISE SET

Write About It!

Use complete sentences in your answer to each exercise.

1. Describe what is meant by the slope of a line.
2. How can you find the slope and y-intercept from the equation of a line?
3. Explain how to use the x- and y-intercepts to find the slope of the line passing through the intercepts.
4. Explain how to use the slope and y-intercept to graph a line.
5. How do you determine if lines are parallel, perpendicular, or neither?
6. How can you find the slope of a line if given two points on the line?
7. If the slope of a line is $-\frac{3}{5}$, explain two ways that you could move from one point on the line to another point.
8. Explain the four different types of slope a line may possess. Provide a graphical illustration of each.

Additional answers can be found in the Instructor Answer Appendix.

Practice Makes Perfect!

Find the slope and y-intercept of each line from its equation. Write the y-intercept as an ordered pair. (*See Objective 1.*)

9. $y = x - 5$ $m = 1, (0, -5)$

10. $y = x + 3$ $m = 1, (0, 3)$

11. $y = -x + 2$ $m = -1, (0, 2)$

12. $y = -x - 1$ $m = -1, (0, -1)$

13. $y = 3x + 9$ $m = 3, (0, 9)$

14. $y = 7x - 2$ $m = 7, (0, -2)$

15. $y = \frac{x}{2} + 8$ $m = \frac{1}{2}, (0, 8)$

16. $y = \frac{x}{4} - 7$ $m = \frac{1}{4}, (0, -7)$

17. $y = -\frac{2}{3}x - 2$ $m = -\frac{2}{3}, (0, -2)$

18. $y = -\frac{5}{4}x + 5$

19. $f(x) = \frac{7}{5}x$ $m = \frac{7}{5}, (0, 0)$

20. $f(x) = \frac{2}{9}x$ $m = \frac{2}{9}, (0, 0)$

21. $4x - 5y = 3$ $m = \frac{4}{5}, \left(0, -\frac{3}{5}\right)$

22. $7x - 3y = 4$

23. $6x + 2y = 0$ $m = -3, (0, 0)$

24. $9x - 3y = 0$ $m = 3, (0, 0)$

25. $8x + 5y = -40$

26. $2x + 9y = -18$

27. $y = 4$ $m = 0, (0, 4)$

28. $y = -9$ $m = 0, (0, -9)$

29. $y - 4 = 3$ $m = 0, (0, 7)$

30. $y - 10 = 1$ $m = 0, (0, 11)$

Graph each line using its slope and y-intercept. Label at least two points on the graph. (*See Objective 2.*)

31. $y = 3x - 4$

32. $y = 6x - 3$

33. $y = -2x + 2$

34. $y = -5x + 5$

35. $y = \frac{2}{3}x - 2$

36. $y = \frac{4}{5}x - 4$

37. $f(x) = -\frac{1}{2}x + 3$

38. $f(x) = -\frac{5}{7}x + 5$

39. $f(x) = 4$

40. $f(x) = -3$

41. $4x + y = 8$

42. $2x + y = 6$

43. $x - 3y = -9$

44. $x - 5y = -20$

45. $f(x) = -1$

46. $f(x) = 2$

Use the slope formula to determine the slope of each line containing the given points. (*See Objective 3.*)

47. (0, 4) and (5, 14) 2

48. (0, 2) and (3, 11) 3

49. (1, −3) and (2, 7) 10

50. (2, −1) and (3, 5) 6

51 (−1, −8) and (4, −3) 1

52. (−3, −6) and (6, −1) $\frac{5}{9}$

53. (−2, 0) and (0, −5) $-\frac{5}{2}$

54. (−6, 0) and (0, −3) $-\frac{1}{2}$

55. (5, −1) and (−4, −1) 0

56. (−6, 7) and (3, 7) 0

57. (9, −2) and (9, 3) undefined

58. (−1, 5) and (−1, −4) undefined

59. (−3, 5.2) and (−3, 6) undefined

60. (−5, 7.1) and (3, 7.1) 0

61.

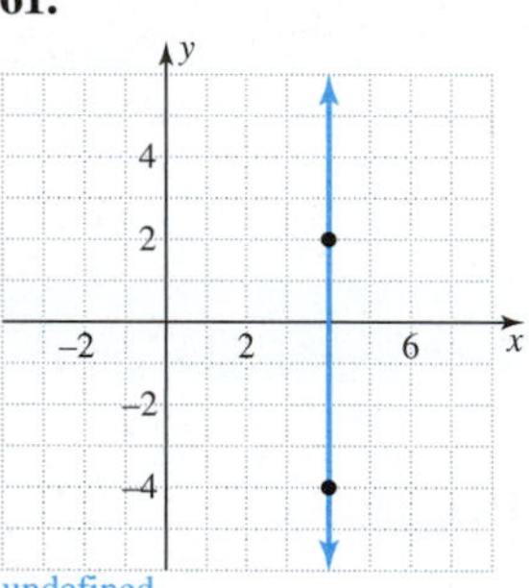

undefined

62.

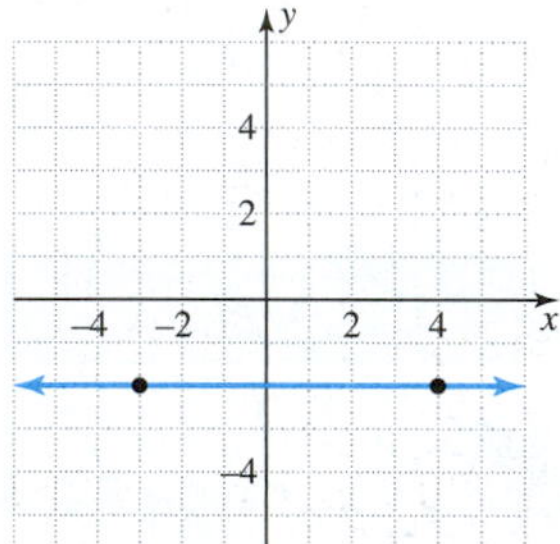

0

63.

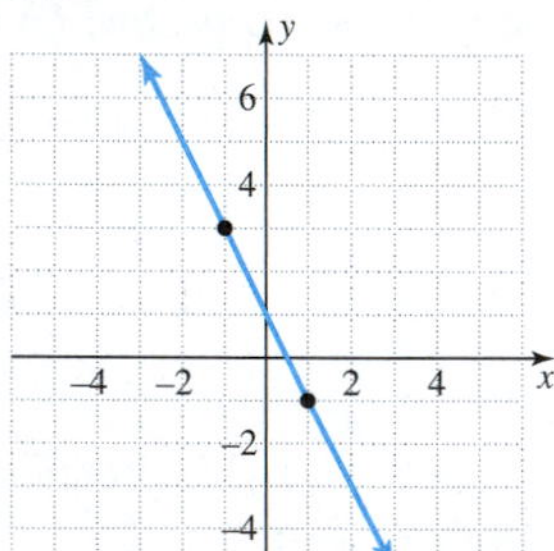

−2

64.

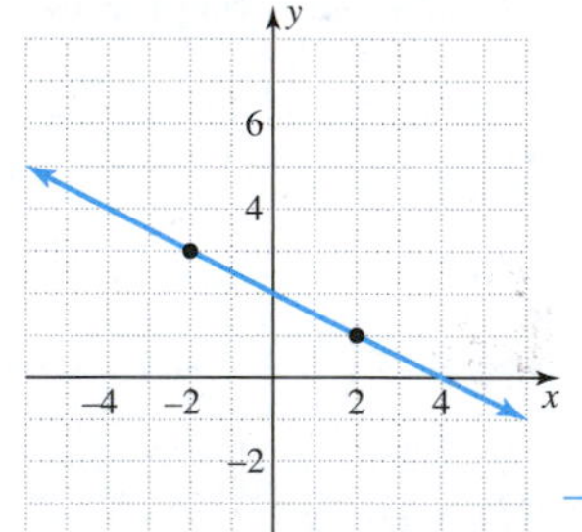

$-\frac{1}{2}$

65.

X	Y1
-3	-2
-2	-1.667
-1	-1.333
0	-1
1	-.6667
2	-.3333
3	0

X=3

$\frac{1}{3}$

66.

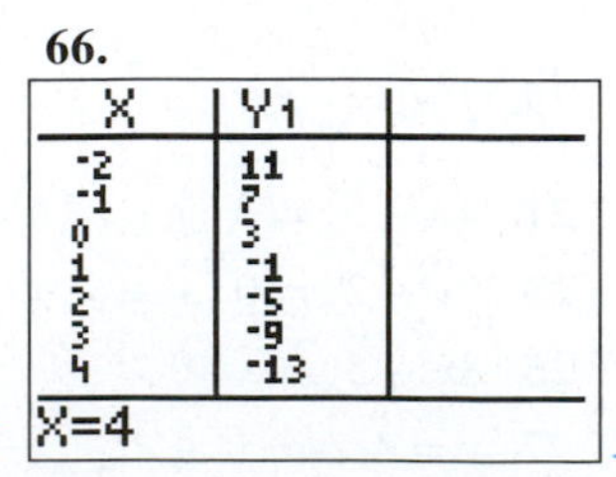

X	Y1
-2	11
-1	7
0	3
1	-1
2	-5
3	-9
4	-13

X=4

−4

Determine if the two lines are parallel, perpendicular, or neither. (*See Objective 4.*)

67. $y = 7x - 5$ and $y = -7x - 5$ neither

68. $y = -2x + 1$ and $y = 2x - 3$ neither

69. $y = 4x + 2$ and $y = 4x - 1$ parallel

70. $y = -3x - 8$ and $y = -3x + 4$ parallel

71. $y = \frac{1}{2}x - 7$ and $y = 2x + 3$ neither

72. $y = \frac{1}{4}x + 5$ and $y = 4x - 9$ neither

73. $y = \frac{1}{7}x - 2$ and $y = -7x + 3$ perpendicular

74. $y = -\frac{1}{5}x + 8$ and $y = 5x + 10$ perpendicular

75. $y = -2$ and $y = 7$ parallel

76. $y = 3$ and $y = -4$ parallel

77. $x = 1$ and $y = 2$ perpendicular

78. $x = -3$ and $y = -5$ perpendicular

79. $3x - 2y = 6$ and $2x + 3y = 9$ perpendicular

80. $6x - 2y = -2$ and $15x - 5y = 6$ parallel

Mix 'Em Up!

Find the slope and y-intercept of each line from its equation. Write the y-intercept as an ordered pair. Graph the line using the slope and y-intercept. Label at least two points on the graph.

81. $x + 6y = 6$

82. $5x - y = 5$

83. $2y - 5 = 0$

84. $4x + 8 = 0$

85. $y = 7x - 5$

86. $x - 4y = -8$

87. $f(x) = -2$

88. $f(x) = \frac{2}{3}x - 4$

89. $y = -\frac{1}{2}x$

90. $3x - y = -9$

Use the slope formula to determine the slope of the line containing the given points.

91. −3

X	Y1
-3	14
-2	11
-1	8
0	5
1	2
2	-1
3	-4

X= -3

92. $\frac{5}{4}$

X	Y1
-8	-14
-4	-9
0	-4
4	1
8	6
12	11
16	16

X= -8

93. (1, −2) and (−3, 4) $-\frac{3}{2}$

94. (−4, −6) and (3, 8) 2

95.

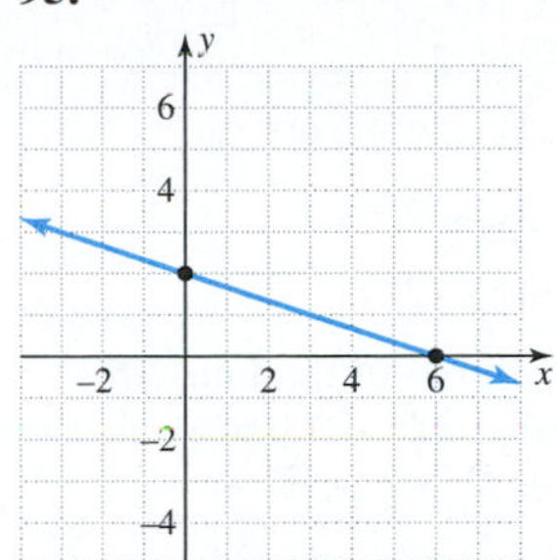

$-\frac{1}{3}$

96.

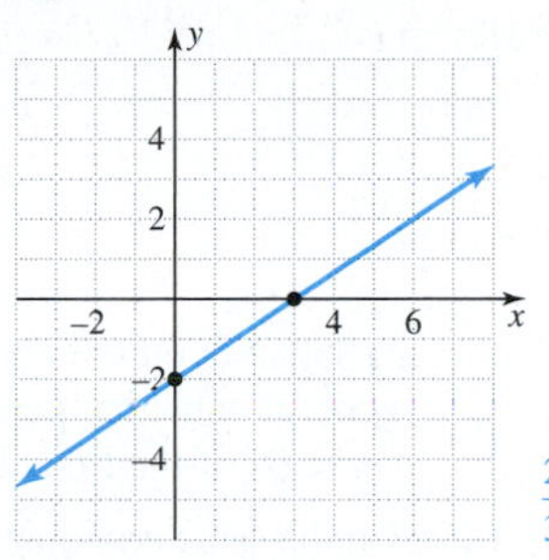

$\frac{2}{3}$

Determine if the two lines are parallel, perpendicular, or neither.

97. $y = 2x + 1$ and $x + 2y = 10$ perpendicular

98. $y = -3x + 5$ and $x + 3y = 6$ neither

99. $-2x + 3y = 21$ and $2x - 3y = 3$ parallel

100. $y = 0.2x - 3.5$ and $5x - y = -1.5$ neither

101. $0.4x + y = 2.1$ and $2.5x + y = -7.5$ neither

102. $1.6x + 2.4y = -1.8$ and $0.6x + 0.9y = 0$ parallel

Solve each problem.

103. A company pays a business and financial analyst according to the salary given by $y = 72{,}000 + 5000x$, where x is the number of years the employee has been with the company. Interpret the meaning of the slope and y-intercept in the context of the problem.

104. A company pays an audiologist according to the salary given by $f(x) = 61{,}500 + 6000x$, where x is the number of years the employee has been with the company. Interpret the meaning of the slope and y-intercept in the context of the problem.

105. A person walking at a rate of 3 mph will burn a total of $y = 0.5x + 53$ calories per mile, where x is the number of pounds over 100. Interpret the slope and y-intercept in the context of the problem.

106. A person walking at a rate of 5 mph will burn a total of $y = 0.7x + 73$ calories per mile, where x is the number of pounds over 100. Interpret the slope and y-intercept in the context of the problem.

107. A car's value is given by $f(x) = -2000x + 25{,}000$, where x is the age of the car in years. Interpret the slope and y-intercept in the context of the problem.

108. A truck's value is given by $f(x) = -3500x + 38{,}000$, where x is the age of the truck in years. Interpret the slope and y-intercept in the context of the problem.

109. A staircase is shown in the diagram. What is the slope of the stairs?

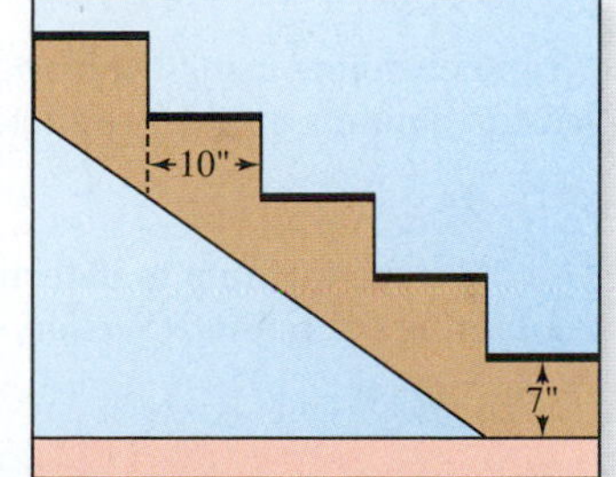

110. Suggested guidelines for constructing wheelchair ramps for the elderly state that there should be no more than 1 ft of rise for every 18 ft in length. What is the slope of a wheelchair ramp that meets these suggested guidelines? Would a ramp with a 3 ft rise and a length of 12 ft satisfy the guidelines? Why or why not?

111. For a residential road to have proper drainage during a rainstorm, it must have $\frac{1}{10}$ of a foot rise for a run of 10 ft from the curb to the middle of the road. What is the slope of the road from the curb to the center of the road? Convert the slope to a percent.

112. One of the steepest roads in Hawaii is the Waipio Road, an access road from the Waipio Overlook at the western end of Honokaa. The road has an average grade of 25%. The road is restricted to four by fours and hikers. What is the slope of the road and what does it mean? (Source: http://www.hawaiihighways.com/photos-Waipio-Valley.htm)

You Be the Teacher!

Correct each student's errors, if any.

113. Find the slope of the line $8x + y = -16$.

Tara's work:

$$\begin{aligned} 8x + y &= -16 \\ -8x + 8x + y &= 8x - 16 \\ y &= 8x - 16 \\ m &= 8 \end{aligned}$$

$$\begin{aligned} 8x + y &= -16 \\ -8x + 8x + y &= -8x - 16 \\ y &= -8x - 16 \\ m &= -8 \end{aligned}$$

114. Graph the line $y = -\frac{3}{5}x + 3$.

Diego's work: The slope is $m = -\frac{3}{5}$ and the y-intercept is (0, 3). I plot (0, 3) and then go down 3 and left 5 to get to the point $(-5, 0)$.

115. Find the slope between $(-5, 2)$ and $(3, 7)$.

Angeline's work:

$$m = \frac{-5 - 3}{2 - 7} = \frac{-8}{-5} = \frac{8}{5} \qquad m = \frac{2 - 7}{-5 - 3} = \frac{-5}{-8} = \frac{5}{8}$$

116. Find the slope of a line that is perpendicular to $y = -\frac{4}{9}x - 6$.

Tyrone's work:

The slope of the given line is $-\frac{4}{9}$. So, the slope of a line perpendicular to this is $-\frac{9}{4}$.

Calculate It!

Determine the slope of each line using the given information.

117. 0.8

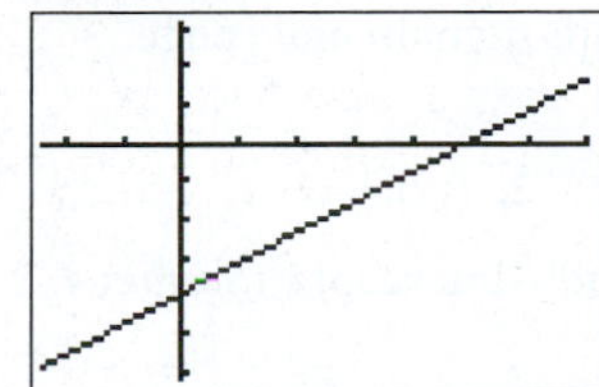

X	Y_1
-1	-4.8
0	-4
1	-3.2
2	-2.4
3	-1.6
4	-.8
5	0

X=-1

118.

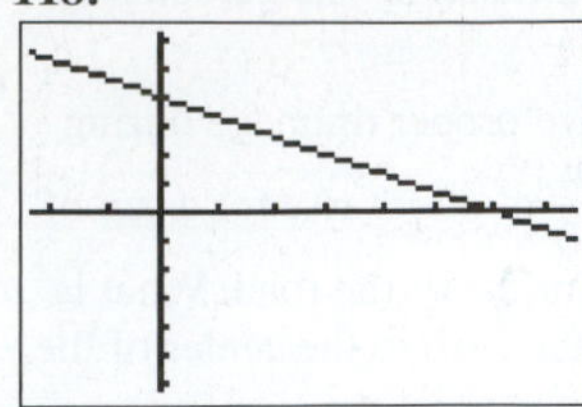

X	Y_1
0	4
1	3.3333
2	2.6667
3	2
4	1.3333
5	.66667
6	0

X=6

$-\frac{2}{3}$

Graph each pair of equations in the ZSquare or ZDecimal format to determine if the lines are parallel, perpendicular, or neither.

119. $y = 4x - 2$ and $y = \frac{1}{4}x + 2$ neither

120. $y = -\frac{3}{4}x + 1$ and $y = \frac{4}{3}x - 3$ perpendicular

121. $y = \frac{1}{5}x + 4$ and $y = -5x$ perpendicular

122. $y = \frac{1}{2}x - 4$ and $y = -\frac{1}{2}x + 5$ neither

Think About It!

123. On the same coordinate system, graph the following lines.

$$y = x - 2$$
$$y = 2x - 2$$
$$y = 3x - 2$$
$$y = 4x - 2$$

What can you tell about the slopes of the lines? In general, what happens to the graph of a line as the slope ($m > 0$) of the line increases?

124. On the same coordinate system, graph the following lines.

$$y = 2x + 3$$
$$y = x + 3$$
$$y = \frac{3}{4}x + 3$$
$$y = \frac{1}{2}x + 3$$
$$y = 0.3x + 3$$

What can you tell about the slopes of the lines? In general, what happens to the graph of a line as the slope ($m > 0$) of the line decreases?

125. On the same coordinate system, graph the following lines.

$$y = -x - 2$$
$$y = -2x - 2$$
$$y = -3x - 2$$
$$y = -4x - 2$$

What can you tell about the slopes of the lines? In general, what happens to the graph of a line as the slope ($m < 0$) of the line decreases?

126. On the same coordinate system, graph the following lines.

$$y = -2x + 3$$
$$y = -x + 3$$
$$y = -\frac{3}{4}x + 3$$
$$y = -\frac{1}{2}x + 3$$
$$y = -0.3x + 3$$

What can you tell about the slopes of the lines? In general, what happens to the graph of a line as the slope ($m < 0$) of the line increases?

127. If a door is 15 in. above the ground, how long must the wheelchair ramp be for its slope to be $\frac{1}{12}$? Express your answer in inches and feet. 15 ft

Draw a possible graph of a line with the following characteristics, where *m* is the slope of the line and *b* is the *y*-intercept of the line.

128. $m > 0$ and $b < 0$ Answers vary.

129. $m > 0$ and $b = 0$ Answers vary.

130. $m < 0$ and $b < 0$ Answers vary.

131. $m < 0$ and $b = 0$ Answers vary.

132. $m < 0$ and $b > 0$ Answers vary.

133. Slope is undefined. Answers vary.

134. Slope is zero. Answers vary.

PIECE IT TOGETHER SECTIONS 4.1–4.3

Graph each function and state its domain and range. (*Section 4.2, Objective 1*)

1. $f(x) = \frac{4}{5}x + 3$

2. $f(x) = 6$

Graph each line using the *x*- and *y*-intercepts. (*Section 4.2, Objectives 2 and 3*)

3. $5x - y = 10$

4. $8x + 5y = 40$

5. $x - 4y = 0$

6. $3x - 5 = 0$

Graph each line using the slope and *y*-intercept. Label at least two points on the graph. (*Section 4.3, Objective 1*)

7. $4x - 7y = 28$

8. $f(x) = -x + 3$

Use the slope formula to determine the slope of the line containing the points. (*Section 4.3, Objective 3*)

9. $(-7, 5)$ and $(1, 3)$ $-\frac{1}{4}$

10. $(4, -5)$, $(8, -5)$ 0

11. $(-4, 10), (-4, 8)$ undefined

12. $\frac{7}{2}$

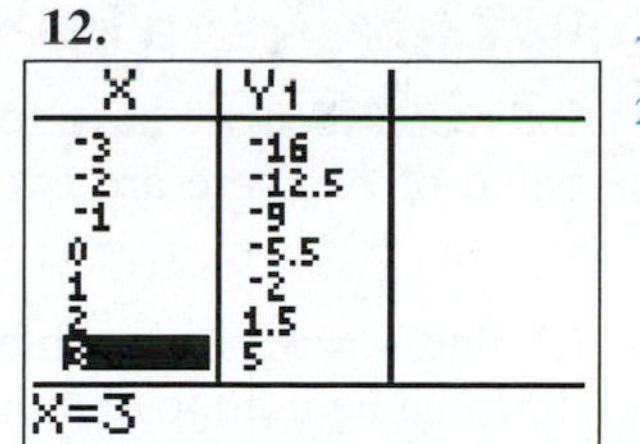

Determine if the two lines are parallel, perpendicular, or neither. (*Section 4.3, Objective 4*)

13. $y = -\frac{5}{2}x - 1$ and $y = -\frac{2}{5}x + 4$ neither

14. $y = -\frac{3}{7}x + 2$ and $y = \frac{7}{3}x$ perpendicular

15. $2x - 6y = 4$ and $x - 3y = 5$ parallel

SECTION 4.4 Writing Equations of Lines

OBJECTIVES

As a result of completing this section, you will be able to

1. Write the equation of a line given its slope and y-intercept.
2. Write the equation of a line given a point and its slope.
3. Write the equation of a line given two points.
4. Write the equation of parallel and perpendicular lines.
5. Solve application problems.
6. Troubleshoot common errors.

The National Center for Education Statistics states that 35.2% of high school dropouts were Hispanic in 1980. In 2008, 18.3% of high school dropouts were Hispanic. Write a linear function that represents the percentage of high school dropouts who were Hispanic in terms of the years after 1980 and use the function to estimate the percentage of high school dropouts who are Hispanic in 2015. (Source: http://nces.ed.gov/fastfacts/display.asp?id=16)

To address this problem, we need to know how to write the equation of a line given certain criteria.

Objective 1 Write the equation of a line given its slope and y-intercept.

Writing the Equation of a Line Given a Slope and y-Intercept

In the previous sections, we learned that linear equations of the form $y = mx + b$ have slope m and y-intercept $(0, b)$. So, if we know the slope and y-intercept of a line, we can write the equation of the line that satisfies this information.

Procedure: Writing the Equation of a Line Given Its Slope and y-Intercept

Step 1: State the slope-intercept form of a line, $y = mx + b$.
Step 2: Substitute the slope for the value m and substitute the y-coordinate of the y-intercept for the value b.

Objective 1 Examples Write the equation of each line, in slope-intercept form, that satisfies the given information.

1a. $m = -3$ and y-intercept $= (0, 4)$ **1b.** $m = \frac{2}{5}$ and y-intercept $= (0, -7)$
1c. $m = 0$ and y-intercept $= (0, 1)$

Solutions

INSTRUCTOR NOTE: Remind students that the y-value of the y-intercept is b.

1a. $y = mx + b$ State the slope-intercept form.
$y = -3x + 4$ Replace m with -3 and b with 4.

1b. $y = mx + b$ State the slope-intercept form.
$y = \frac{2}{5}x - 7$ Replace m with $\frac{2}{5}$ and b with -7.

1c. $y = mx + b$ State the slope-intercept form.
$y = 0x + 1$ Replace m with 0 and b with 1.
$y = 1$ Simplify.

Student Check 1 Write the equation of each line, in slope-intercept form, that satisfies the given information.

a. $m = -6$ and y-intercept $= (0, -2)$ **b.** $m = \frac{1}{3}$ and y-intercept $= (0, 5)$
c. $m = 0$ and y-intercept $= (0, -3)$

Writing the Equation of a Line Given a Point and a Slope

Objective 2 ▶

Write the equation of a line given a point and its slope.

If the point given is not the y-intercept, then we have to perform some work to determine the y-intercept and the value of b. There are two methods that we can use to determine the value of b.

- The slope-intercept form of a line, $y = mx + b$, can be used to directly solve for b.
- The *point-slope form* of a line can be used to obtain the equation of the line.

Suppose we need to write the equation of the line that passes through the point $(1, -3)$ with slope $m = 2$. We can use $y = mx + b$ to solve for b since we know the values of x, y, and m.

$y = mx + b$	State the slope-intercept form of a line.
$-3 = 2(1) + b$	Replace x with 1, y with -3, and m with 2.
$-3 = 2 + b$	Multiply 2 and 1.
$-3 - 2 = 2 + b - 2$	Subtract 2 from each side.
$-5 = b$	Simplify.

When we use the slope-intercept form, note that we obtain the value of b, not the equation. We can write the equation of the line that satisfies the given conditions by replacing the value of m and b with their known values in $y = mx + b$. In this case, we replace m with 2 and b with -5 to obtain the equation. So, the equation of the line that passes through $(1, -3)$ with $m = 2$ is

$$y = 2x - 5$$

The point-slope form is a formula that can be used to obtain the equation of the line with the given conditions. It is derived from the slope formula using (x_1, y_1) as the given point and (x, y) as another point on the line.

$m = \dfrac{y_2 - y_1}{x_2 - x_1}$	State the slope formula.
$m = \dfrac{y - y_1}{x - x_1}$	Let (x_1, y_1) be the given point and $(x_2, y_2) = (x, y)$.
$(x - x_1)m = \dfrac{y - y_1}{x - x_1}(x - x_1)$	Clear the fraction by multiplying each side by $(x - x_1)$.
$y - y_1 = m(x - x_1)$	Simplify. The result is the point-slope form.

Property: The Point-Slope Form of a Line

Let (x_1, y_1) be a point on a line with slope m, then the **point-slope form** of the line is

$$y - y_1 = m(x - x_1)$$

Slope (m); Point (x_1, y_1)

So, to write the equation of the line that passes through the point $(1, -3)$ with slope $m = 2$ using the point-slope form, we do the following.

$y - y_1 = m(x - x_1)$	State the point-slope form.
$y - (-3) = 2(x - 1)$	Replace x_1 with 1, y_1 with -3, and m with 2.
$y + 3 = 2x - 2$	Simplify on the left and distribute on the right.
$y + 3 - 3 = 2x - 2 - 3$	Subtract 3 from each side.
$y = 2x - 5$	Simplify.

When we use the point-slope form, note that the equation is obtained through this process. Also note that the two methods are very similar in that we make the same replacements in each form. So, to write the equation of a line given a point and a slope, we can use either method described.

Procedure: Writing the Equation of a Line Given a Point and a Slope

Method 1: Use the slope-intercept form of a line.

Step 1: Substitute the given slope m and point (x, y) in $y = mx + b$.
Step 2: Solve the resulting equation for b.
Step 3: Write the equation in the form $y = mx + b$ by replacing the values of m and b with their known values.

Method 2: Use the point-slope form of a line.

Step 1: Substitute the given slope m and point (x_1, y_1) in the equation $y - y_1 = m(x - x_1)$.
Step 2: Simplify each side of the equation.
Step 3: Solve for y.

Note: *If the slope of a line is undefined, the line is vertical and has the form $x = h$, where h is the x-value of the given point. These methods cannot be used to find the equation. This is a special case we must remember.*

Objective 2 Examples **Use methods 1 and 2 to write the equation of each line, in slope-intercept form, that has the given slope and passes through the given point.**

2a. $m = -3$; $(4, -2)$ **2b.** $m = \frac{2}{3}$; $(-5, 1)$

2c. $m = 0$; $(7, -6)$ **2d.** $m =$ undefined; $(-2, 5)$

Solutions **2a.**

Slope-Intercept Form	Point-Slope Form
$m = -3, (x, y) = (4, -2)$	$m = -3, (x_1, y_1) = (4, -2)$
$y = mx + b$	$y - y_1 = m(x - x_1)$
$-2 = -3(4) + b$	$y - (-2) = -3(x - 4)$
$-2 = -12 + b$	$y + 2 = -3x + 12$
$-2 + 12 = -12 + b + 12$	$y + 2 - 2 = -3x + 12 - 2$
$10 = b$	$y = -3x + 10$
The equation is $y = -3x + 10$.	

Notice that we obtained the same equation with both methods. We can check our equation by showing that $(4, -2)$ is a solution of $y = -3x + 10$.

2b.

Slope-Intercept Form	Point-Slope Form
$m = \frac{2}{3}, (x, y) = (-5, 1)$	$m = \frac{2}{3}, (x_1, y_1) = (-5, 1)$
$y = mx + b$	$y - y_1 = m(x - x_1)$
$1 = \frac{2}{3}(-5) + b$	$y - 1 = \frac{2}{3}[x - (-5)]$
$1 = -\frac{10}{3} + b$	$y - 1 = \frac{2}{3}(x + 5)$
$3(1) = 3\left(-\frac{10}{3} + b\right)$	$y - 1 = \frac{2}{3}x + \frac{10}{3}$
$3 = -10 + 3b$	$3(y - 1) = 3\left(\frac{2}{3}x + \frac{10}{3}\right)$
$3 + 10 = -10 + 3b + 10$	$3y - 3 = 2x + 10$
$13 = 3b$	$3y - 3 + 3 = 2x + 10 + 3$
$\frac{13}{3} = b$	$3y = 2x + 13$
	$\frac{3y}{3} = \frac{2x + 13}{3}$
The equation is $y = \frac{2}{3}x + \frac{13}{3}$.	$y = \frac{2}{3}x + \frac{13}{3}$

Notice we obtained the same equation with both methods. We can check our equation by showing that $(-5, 1)$ is a solution of $y = \frac{2}{3}x + \frac{13}{3}$.

2c.

Slope-Intercept Form	Point-Slope Form
$m = 0, (x, y) = (7, -6)$	$m = 0, (x_1, y_1) = (7, -6)$
$y = mx + b$	$y - y_1 = m(x - x_1)$
$-6 = 0(7) + b$	$y - (-6) = 0(x - 7)$
$-6 = 0 + b$	$y + 6 = 0$
$-6 = b$	$y + 6 - 6 = 0 - 6$
The equation is $y = 0x - 6$ or $y = -6$.	$y = -6$

2d. An undefined slope corresponds to a vertical line, which has an equation of the form $x = h$. Since the x-value of the given point is -2, the equation of the line is $x = -2$.

Student Check 2 Write the equation of each line that has the given slope and passes through the given point.

a. $m = -5$; $(3, -2)$ **b.** $m = \frac{1}{5}$; $(-9, 2)$

c. $m = 0$; $(-4, 1)$ **d.** $m =$ undefined; $\left(\frac{1}{2}, \frac{2}{3}\right)$

Note: *While answers in Example 2 parts (a) and (b) were written in slope-intercept form, it is acceptable to write the equations in standard form, as well. For instance,*

$$y = -3x + 10 \textit{ is equivalent to } 3x + y = 10.$$

$$y = \frac{2}{3}x + \frac{13}{3} \textit{ is equivalent to } 2x - 3y = -13.$$

Writing the Equation of a Line Given Two Points

Objective 3 Write the equation of a line given two points.

The previous methods described are also used to write the equation of a line if two points on the line are known. The only difference is that the slope of the line is not provided. Writing the equation of a line given two points requires an additional step of calculating the slope using the slope formula. Once the slope is known, method 1 or method 2 can be used to write the equation of the line that contains the given points.

Procedure: Writing the Equation of a Line Given Two Points

Step 1: Find the slope of the line using the slope formula, $m = \frac{y_2 - y_1}{x_2 - x_1}$.

Step 2: Use the slope and one of the given points to find the equation of the line using either the slope-intercept form of a line or the point-slope form of a line.

Objective 3 Examples Write the equation of the line that passes through the given points.

3a. $(1, -4)$ and $(3, 6)$ **3b.** $(8, -1)$ and $(12, 4)$

3c. $(2, 6)$ and $(-5, 6)$ **3d.** $(-3, 9)$ and $(-3, -7)$

Solutions

INSTRUCTOR NOTE:
Explain that the same line will be obtained with either of the points. However, the point with only positive values is often easier to use.

3a.

Step 1: Find the slope: $\overset{(x_1, y_1)}{(1, -4)}$ and $\overset{(x_2, y_2)}{(3, 6)}$

$$m = \frac{y_2 - y_1}{x_2 - x_1} = \frac{6 - (-4)}{3 - 1} = \frac{10}{2} = 5$$

Step 2: Use method 1 with the slope $m = 5$ and one of the given points, (3, 6).

$y = mx + b$	State the slope-intercept form.
$6 = 5(3) + b$	Replace m with 5, x with 3 and y with 6.
$6 = 15 + b$	Multiply 5 and 3.
$6 - 15 = 15 + b - 15$	Subtract 15 from each side.
$-9 = b$	Simplify.

Since $m = 5$ and the y-intercept is $(0, -9)$, the equation is $y = 5x - 9$.

We check the equation by showing that $(1, -4)$ and $(3, 6)$ are solutions of $y = 5x - 9$.

$(1, -4)$:	$(3, 6)$:
$y = 5x - 9$	$y = 5x - 9$
$-4 = 5(1) - 9$	$6 = 5(3) - 9$
$-4 = 5 - 9$	$6 = 15 - 9$
$-4 = -4$	$6 = 6$

3b.

Step 1: Find the slope: $\overset{(x_1, y_1)}{(8, -1)}$ and $\overset{(x_2, y_2)}{(12, 4)}$

$$m = \frac{y_2 - y_1}{x_2 - x_1} = \frac{4 - (-1)}{12 - 8} = \frac{5}{4}$$

Step 2: Use method 2 with the slope $m = \frac{5}{4}$ and one of the given points, (12, 4).

$y - y_1 = m(x - x_1)$	State the point-slope form.
$y - 4 = \frac{5}{4}(x - 12)$	Replace m with 0, x_1 with 12 and y_1 with 4.
$y - 4 = \frac{5}{4}x - 15$	Apply the distributive property.
$y - 4 + 4 = \frac{5}{4}x - 15 + 4$	Add 4 to each side.
$y = \frac{5}{4}x - 11$	Simplify.

We check the equation by showing that $(8, -1)$ and $(12, 4)$ are solutions of $y = \frac{5}{4}x - 11$.

$(8, -1)$:	$(12, 4)$:
$y = \frac{5}{4}x - 11$	$y = \frac{5}{4}x - 11$
$-1 = \frac{5}{4}(8) - 11$	$4 = \frac{5}{4}(12) - 11$
$-1 = 10 - 11$	$4 = 15 - 11$
$-1 = -1$	$4 = 4$

3c.

Step 1: Find the slope: $\overset{(x_1, y_1)}{(2, 6)}$ and $\overset{(x_2, y_2)}{(-5, 6)}$

$$m = \frac{y_2 - y_1}{x_2 - x_1} = \frac{6 - (6)}{-5 - 2} = \frac{0}{-7} = 0$$

Step 2: Use method 2 with the slope $m = 0$ and one of the given points, (2, 6).

$$y - y_1 = m(x - x_1) \quad \text{State the point-slope form.}$$
$$y - 6 = 0(x - 2) \quad \text{Replace } m \text{ with 0, } x_1 \text{ with 2 and } y_1 \text{ with 6.}$$
$$y - 6 = 0 \quad \text{Simplify the right side.}$$
$$y - 6 + 6 = 0 + 6 \quad \text{Add 6 to each side.}$$
$$y = 6 \quad \text{Simplify.}$$

Note: *The slope* $m = 0$ *tells us that the line is horizontal. If we recognize this and know the form of a horizontal line is* $y = k$, *then there is no need to perform step 2.*

3d. (x_1, y_1) (x_2, y_2)

Step 1: Find the slope: (−3, 9) and (−3, −7)

$$m = \frac{y_2 - y_1}{x_2 - x_1} = \frac{-7 - (9)}{-3 - (-3)} = \frac{-16}{0}$$

The slope is undefined, which means the line is vertical. So, the equation of the line is $x = -3$.

Student Check 3 Write the equation of the line that passes through the given points.

a. (−1, 4) and (1, −8) **b.** (2, −11) and (−6, −9)

c. (4, −5) and (−3, −5) **d.** (9, −7) and (9, −4)

Writing the Equations of Parallel and Perpendicular Lines

Objective 4 Write the equation of parallel or perpendicular lines.

The last case of writing equations of lines involves parallel and perpendicular lines. In these types of problems, a point and an equation of a parallel or perpendicular line will be given. The given line enables us to find the slope of the unknown line. Recall that if two lines are parallel, their slopes are the same. If two lines are perpendicular, their slopes are negative reciprocals. Once the slope of the line is determined, the previous methods in this section apply.

Procedure: Writing the Equations of Parallel or Perpendicular Lines

Step 1: Determine the slope of the given line (write equation in slope-intercept form, if necessary).

Step 2: Determine the slope of the parallel or perpendicular line.

Step 3: Use the slope from step 2 and the given point to write the equation of the line using either method 1 or method 2.

Objective 4 Examples **Write the equation of the line that passes through the given point and is parallel *and* perpendicular to the given line.**

4a. (6, −4), $x - 2y = 10$ **4b.** (−8, 4), $y = -3$

Solutions **4a.** Write the given equation in slope-intercept form to identify its slope.

$$x - 2y = 10$$
$$-2y = -x + 10 \quad \text{Subtract } x \text{ from each side.}$$
$$\frac{-2y}{-2} = \frac{-x}{-2} + \frac{10}{-2} \quad \text{Divide each side by } -2.$$
$$y = \frac{1}{2}x + 5 \quad \text{Simplify.}$$

Equation of Parallel Line	Equation of Perpendicular Line
Step 1: Slope of given line is $m = \frac{1}{2}$. **Step 2:** Slope of a parallel line is $m = \frac{1}{2}$. **Step 3:** Use the point-slope form with (6, −4) and $m = \frac{1}{2}$. $y - y_1 = m(x - x_1)$ $y - (-4) = \frac{1}{2}(x - 6)$ $y + 4 = \frac{1}{2}x - 3$ $y + 4 - 4 = \frac{1}{2}x - 3 - 4$ $y = \frac{1}{2}x - 7$	**Step 1:** Slope of given line is $m = \frac{1}{2}$. **Step 2:** Slope of a perpendicular line is $m = -\left(\frac{2}{1}\right) = -2$ **Step 3:** Use the point-slope form with (6, −4) and $m = -2$. $y - y_1 = m(x - x_1)$ $y - (-4) = -2(x - 6)$ $y + 4 = -2x + 12$ $y + 4 - 4 = -2x + 12 - 4$ $y = -2x + 8$

4b. Recall that $y = -3$ is a horizontal line whose slope is $m = 0$.

Equation of Parallel Line	Equation of Perpendicular Line
Step 1: Slope of the given line is $m = 0$. **Step 2:** Slope of a parallel line is $m = 0$. **Step 3:** The line through (−8, 4) with $m = 0$ is a horizontal line. So, its equation is $y = 0x + 4$ or $y = 4$.	**Step 1:** Slope of the given line is $m = 0$. **Step 2:** Slope of a perpendicular line is undefined. **Step 3:** The line through (−8, 4) with undefined slope is a vertical line. So, its equation is $x = -8$.

Student Check 4 Write the equation of the line that passes through the given point and is parallel *and* perpendicular to the given line.

a. $(1, -6)$; $x + 3y = 9$ **b.** $(-3, 10)$; $x = 1$

Applications

Objective 5 Solve application problems.

There are two types of application problems that we will examine in this section. One type involves knowing an initial value and information about how that value changes. The other type involves knowing two data points that describe a particular situation.

Procedure: Solving Applications Involving an Initial Value and How That Value Changes

Step 1: Identify the initial (or beginning) value of the given quantity. This value is b in the slope-intercept form of a line.

Step 2: Identify how the initial value changes. This information is used to find the slope, m.

Step 3: Write the equation in slope intercept form, $y = mx + b$, by substituting the values of m and b into the equation.

Procedure: Solving Applications Given Two Data Points

Step 1: Find the slope of the line containing the two points.
Step 2: Use one data point and the slope to determine the equation by using either the slope-intercept form or the point-slope form of a line, as previously described.

Objective 5 Examples

Solve each problem.

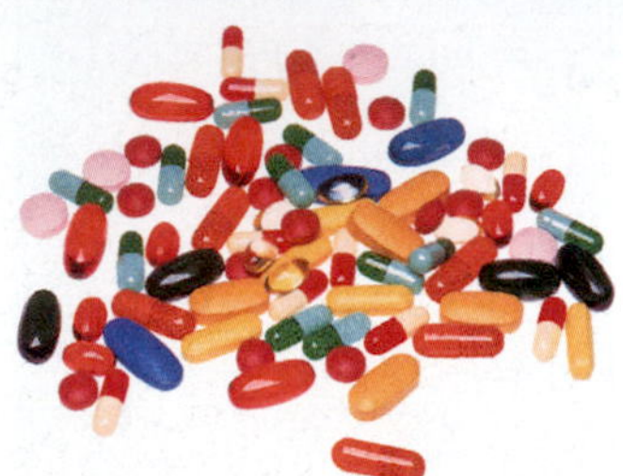

5a. David works as a sales representative for a pharmaceutical company. He gets a base salary of \$30,000 a year plus 5% commission on his total annual sales.

i. Write a linear equation that represents his income, where x is his annual sales.

ii. Use the equation to determine David's income if he sells \$150,000 in products.

iii. Use the equation to determine the sales he needs to earn an income of \$70,000.

Solution **5a.** **i.** Let x represent David's annual sales. The base salary, \$30,000, is the initial value b. This is the income if no sales are made. The commission rate, 5% = 0.05, is the value m. The commission rate tells us that David gets an additional \$0.05 for each dollar in sales. So, the equation that represents his income is

$$y = 0.05x + 30{,}000$$

ii. If David sells \$150,000 worth of products, his income is found by letting $x = 150{,}000$.

$y = 0.05x + 30{,}000$	Begin with the model.
$y = 0.05(150{,}000) + 30{,}000$	Replace x with 150,000.
$y = 7500 + 30{,}000$	Multiply.
$y = 37{,}500$	Add.

David's income will be \$37,500 if he sells \$150,000 worth of products.

iii. If David's income is \$70,000, the amount of product (in dollars) that he sold can be found by letting $y = 70{,}000$.

$y = 0.05x + 30{,}000$	Begin with the model.
$70{,}000 = 0.05x + 30{,}000$	Replace y with 70,000.
$70{,}000 - 30{,}000 = 0.05x + 30{,}000 - 30{,}000$	Subtract 30,000 from each side.
$40{,}000 = 0.05x$	Simplify.
$\frac{40{,}000}{0.05} = \frac{0.05x}{0.05}$	Divide each side by 0.05.
$800{,}000 = x$	Simplify.

For David to have an income of \$70,000, he must sell \$800,000 in products.

5b. The National Center for Education Statistics states that 35.2% of high school dropouts were Hispanic in 1980. In 2008, 18.3% of high school dropouts were Hispanic. (Source: http://nces.ed.gov/fastfacts/display.asp?id=16)

i. Write a linear function that represents the percentage of high school dropouts who were Hispanic in terms of the years after 1980.

ii. Use the function to estimate the percentage of high school dropouts who will be Hispanic in 2015.

iii. When will only 10% of high school dropouts be Hispanic?

Solution 5b. i. Since x is the years after 1980, the year 1980 corresponds to $x = 1980 - 1980 = 0$. The year 2008 corresponds to $x = 2008 - 1980 = 28$. The given information corresponds to the ordered pairs (0, 35.2) and (28, 18.3). To find the linear function, we must first find the slope.

$$m = \frac{18.3 - 35.2}{28 - 0} = \frac{-16.9}{28} \approx -0.60$$

The percentage of Hispanic dropouts in 1980 represents the initial value, so $b = 35.2$. So, the linear equation that represents the percentage of high school dropouts who were Hispanic x years after 1980 is $y = -0.60x + 35.2$. In function notation, we write this as $f(x) = -0.60x + 35.2$.

ii.

$f(x) = -0.60x + 35.2$ Begin with the model.

$f(35) = -0.60(35) + 35.2$ Replace x with 35 (2015 − 1980 = 35).

$f(35) = 14.2$ Simplify.

So, approximately 14.2% of high school dropouts will be Hispanic in 2015.

iii.

$f(x) = -0.60x + 35.2$ Begin with the model.

$10 = -0.60x + 35.2$ Replace $f(x)$ with 10.

$10 - 35.2 = -0.60x + 35.2 - 35.2$ Subtract 34.3 from each side.

$-25.2 = -0.60x$ Simplify.

$\frac{-25.2}{-0.60} = \frac{-0.60x}{-0.60}$ Divide each side by −0.41.

$42 = x$ Simplify.

According to the model, 10% of high school dropouts will be Hispanic 42 yr after 1980, or in 2022.

Student Check 5 Solve each problem.

a. Ryan bought an SUV for \$30,000. The value of the SUV decreases by \$3000 each year.

i. Write a linear equation that represents the value of the SUV, where x is its age in years.

ii. Use the equation to find the value of the SUV after 4 yr.

iii. When will the value of the SUV be \$9000?

b. The average price of a movie ticket in 1996 was \$4.42. The average price of a movie ticket in 2010 was \$7.89. (Source: http://www.natoonline.org/statisticstickets.htm)

i. Write a linear equation that represents the average price of a movie ticket, where x is the years after 1996.

ii. Use equation to predict the average price of a movie ticket in 2015.

iii. In what year will the average price of a movie ticket be \$10.67?

Objective 6 ▶ Troubleshoot common errors.

Troubleshooting Common Errors

Some common errors associated with writing equations of lines are shown.

Objective 6 Examples **A problem and an incorrect solution are given. Provide the correct solution and an explanation of the error.**

6a. Write the equation of the line through (3, 0) with slope $m = \frac{2}{3}$.

Incorrect Solution	Correct Solution and Explanation
Since the point (3, 0) is given, the value of $b = 3$. So, the equation is $y = \frac{2}{3}x + 3$	The point (3, 0) is the x-intercept, *not* the y-intercept. So, we use the slope-intercept form to find the value of b. $y = mx + b$; $0 = \frac{2}{3}(3) + b$; $0 = 2 + b$; $-2 = b$. So, the equation is $y = \frac{2}{3}x - 2$.

6b. Write the equation of the line through (2, −1) and (5, −2).

Incorrect Solution	Correct Solution and Explanation
$m = \frac{5-2}{-2-(-1)} = \frac{3}{-1} = -3$ So, the equation is $y - y_1 = m(x - x_1)$; $y - (-2) = -3(x - 5)$; $y + 2 = -3x + 15$; $y = -3x + 13$	The slope should be $m = \frac{\text{change in } y}{\text{change in } x} = \frac{-2-(-1)}{5-2} = -\frac{1}{3}$ We use the point-slope form to find the equation. $y - y_1 = m(x - x_1)$; $y - (-2) = -\frac{1}{3}(x - 5)$; $y + 2 = -\frac{1}{3}x + \frac{5}{3}$; $y + 2 - 2 = -\frac{1}{3}x + \frac{5}{3} - 2$; $y = -\frac{1}{3}x + \frac{5}{3} - \frac{6}{3}$; $y = -\frac{1}{3}x - \frac{1}{3}$

6c. Write the equation of the line through (4, −1) that is perpendicular to $y = 2x + 4$.

Incorrect Solution	Correct Solution and Explanation
The slope of the given line is $m = 2$. The slope of a perpendicular line is $m = -\frac{1}{2}$. The value of $b = 4$. So, the equation is $y = mx + b$; $y = -\frac{1}{2}x + 4$	The y-intercept of the given line is (0, 4), but this is not necessarily the y-intercept of the perpendicular line. We use $m = -\frac{1}{2}$ and (4, −1) to get $y = mx + b$; $-1 = -\frac{1}{2}(4) + b$; $-1 = -2 + b$; $1 = b$. So, the equation is $y = -\frac{1}{2}x + 1$.

ANSWERS TO STUDENT CHECKS

Student Check 1 **a.** $y = -6x - 2$ **b.** $y = \frac{1}{3}x + 5$ **c.** $y = -3$

Student Check 2 **a.** $y = -5x + 13$ **b.** $y = \frac{1}{5}x + \frac{19}{5}$ **c.** $y = 1$ **d.** $x = \frac{1}{2}$

Student Check 3 **a.** $y = -6x - 2$ **b.** $y = -\frac{1}{4}x - \frac{21}{2}$ **c.** $y = -5$ **d.** $x = 9$

Student Check 4 **a.** parallel: $y = -\frac{1}{3}x - \frac{17}{3}$; perpendicular: $y = 3x - 9$ **b.** parallel: $x = -3$; perpendicular: $y = 10$

Student Check 5 **a. i.** $y = -3000x + 30{,}000$ **ii.** \$18,000 **iii.** in 7 yr **b. i.** $y = 0.25x + 4.42$ **ii.** \$9.17 **iii.** in the year 2021

SUMMARY OF KEY CONCEPTS

1. If the slope and y-intercept of an equation are known, writing the equation that satisfies this information is immediate. The value of m and b are substituted into the slope-intercept form $y = mx + b$.
2. There are three other situations that provide enough information for an equation of a line to be written. They are

   a. a point and a slope.
   b. two points.
   c. a point and a line parallel or perpendicular.

   In each of these situations, the slope must be determined. If it is not given, use the slope formula or the relationship to a given line to find it. After the slope is found, use the slope with one of the given points in either the point-slope form or the slope-intercept form to write the equation of the line.
3. If a line is described as vertical or with undefined slope, the equation of the line is $x = h$, where h is the x-coordinate of the given point.
4. If a line is described as horizontal or with zero slope, the equation of the line is $y = k$, where k is the y-coordinate of the given point.
5. In application problems, we will either know the slope (how the values change) and y-intercept (initial value) from the problem or we will be given two points that enable us to write the equation.

GRAPHING CALCULATOR SKILLS

The graphing calculator can calculate the equation of the line if provided enough information. At this point, it is more beneficial to you if you use the calculator to check your work instead of allowing it to do the work for you.

Example: Verify that $y = \frac{3}{2}x + 6$ is the equation of the line that passes through the points $(-4, 0)$ and $(4, 12)$.

Solution: Enter the equation in the equation editor.

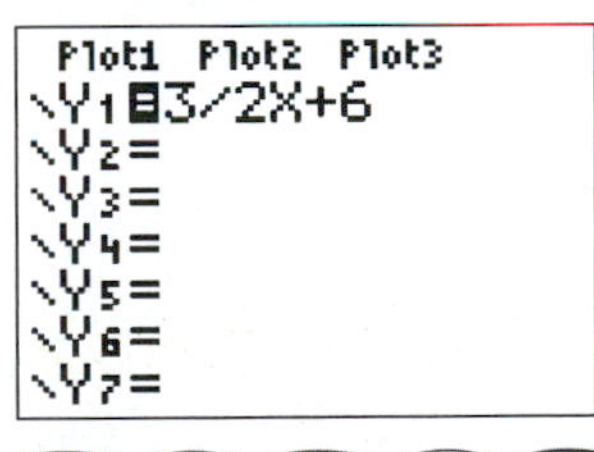

Graph the line, press TRACE, enter the x-value of the first point and press ENTER. Notice how this takes us to the point $(-4, 0)$ on the graph.

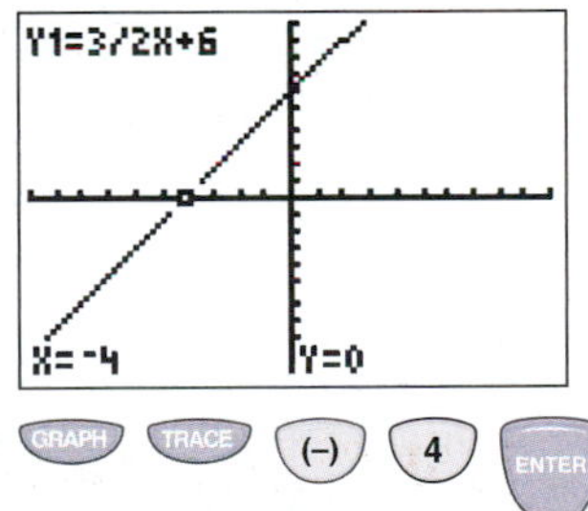

Now enter the x-value of the second point. Notice the display is $x = 4$ and $y = 12$. The point is not shown since it is outside of the standard viewing window.

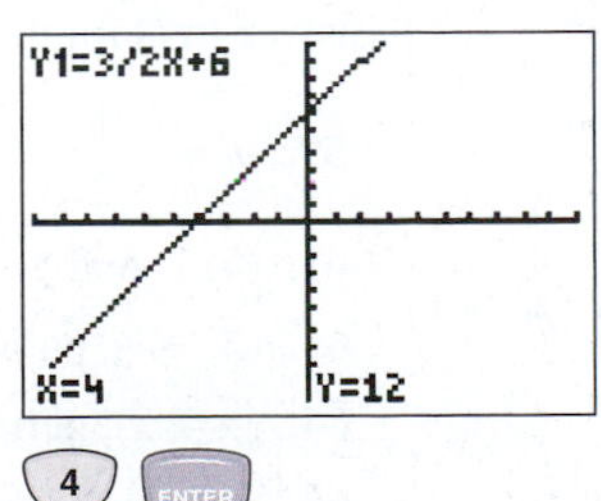

4 ENTER

Another way to check to see if the equation contains the points as solutions is to use the TABLE feature. Press 2nd GRAPH to verify that the points (4, 12) and $(-4, 0)$ are in the table.

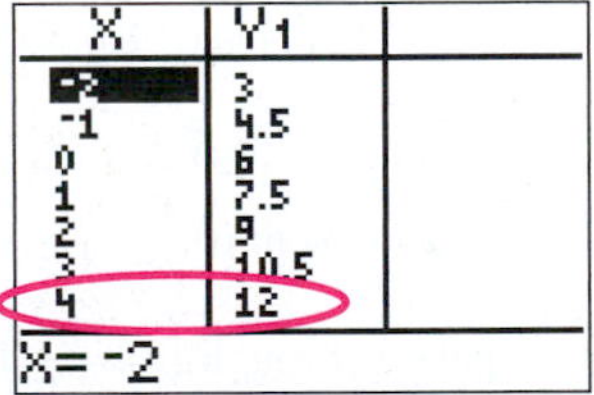

X	Y1
-2	3
-1	4.5
0	6
1	7.5
2	9
3	10.5
4	12

X= -2

X	Y1
-4	0
-3	1.5
-2	3
-1	4.5
0	6
1	7.5
2	9

X= -4

SECTION 4.4 EXERCISE SET

Write About It!

Use complete sentences in your answer to each exercise.

1. How can you determine the equation of a line if you know its slope and y-intercept?
2. How can you determine the equation of a line if you know two points on the line?
3. Which method do you prefer to find the equation of a line—using the slope-intercept form or using the point-slope form? Why?

Determine if each statement is true or false. If a statement is false, explain why it is false.

4. The equation of the line with slope 7 that passes through the point (5, 0) is $y = 7x + 5$.
5. The equation of the line that passes through the points (0, 4) and (−3, 0) is $y = -3x + 4$.
6. The equation of the line that passes through the point (2, 6) and is perpendicular to $y = 3x + 4$ is $y = -\frac{1}{3}x + 6$.

Practice Makes Perfect!

Write the equation of each line, in slope-intercept form, that satisfies the given information. (*See Objective 1.*)

7. $m = 4, b = 5$ $y = 4x + 5$
8. $m = -2, b = 1$ $y = -2x + 5$
9. $m = 3$, passes through $(0, -4)$ $y = 3x - 4$
10. $m = \frac{1}{2}$, passes through $(0, 7)$ $y = \frac{1}{2}x + 7$
11. $m = -\frac{2}{3}$ with y-intercept $(0, -10)$ $y = -\frac{2}{3}x - 10$
12. $m = \frac{4}{5}$ with y-intercept $(0, 20)$ $y = \frac{4}{5}x + 20$
13. $m = 0$, passes through $(0, 7)$ $y = 7$
14. $m = 0$, passes through $(0, -14)$ $y = -14$
15. $m = -\frac{1}{9}$, passes through $\left(0, -\frac{4}{9}\right)$ $y = -\frac{1}{9}x - \frac{4}{9}$
16. $m = \frac{3}{7}$, passes through $\left(0, \frac{2}{7}\right)$ $y = \frac{3}{7}x + \frac{2}{7}$

Write the equation of each line, in slope-intercept form, that has the given slope and passes through the given point. (*See Objective 2.*)

17. $m = 4$; $(1, 5)$ $y = 4x + 1$
18. $m = 2$; $(4, 3)$ $y = 2x - 5$
19. $m = -5$; $(-2, -1)$ $y = -5x - 11$
20. $m = -7$; $(1, -6)$ $y = -7x + 1$
21. $m = 8$; $(9, -3)$ $y = 8x - 75$
22. $m = -12$; $(-5, -1)$ $y = -12x - 61$

Additional answers can be found in the Instructor Answer Appendix.

23. $m = 6$; $(0, 0)$ $y = 6x$
24. $m = -1$; $(0, 0)$ $y = -x$
25. $m = 0$; $(2, -5)$ $y = -5$
26. $m = 0$; $(2, 8)$ $y = 8$
27. $m = \frac{2}{3}$; $(6, -8)$
28. $m = -\frac{1}{4}$; $(6, -5)$
29. $m =$ undefined; $(9, 3)$ $x = 9$
30. $m =$ undefined; $(-5, 4)$ $x = -5$

Write the equation of each line, in slope-intercept form, that contains the given points. (*See Objective 3.*)

31. $(-4, 1)$ and $(2, 7)$ $y = x + 5$
32. $(-2, -5)$ and $(2, -1)$ $y = x - 3$
33. $(-4, -8)$ and $(2, 4)$ $y = 2x$
34. $(-5, 15)$ and $(1, -3)$ $y = -3x$
35. $(-6, 21)$ and $(3, -15)$ $y = -4x - 3$
36. $(-1, 11)$ and $(4, -19)$ $y = -6x + 5$
37. $(4, -4)$ and $(12, 0)$
38. $(8, -3)$ and $(-8, -15)$
39. $(-2, 5)$ and $(4, 5)$ $y = 5$
40. $(1, -7)$ and $(-4, -7)$ $y = -7$
41. $(2, -1)$ and $(2, 6)$ $x = 2$
42. $(-3, 5)$ and $(-3, 2)$ $x = -3$

Write the equation of each line, in slope-intercept form, that passes through the given point and is either parallel or perpendicular to the given line. (*See Objective 4.*)

43. $(0, -4)$, parallel to $y = -3x + 6$ $y = -3x - 4$
44. $(0, 3)$, parallel to $y = 7x - 3$ $y = 7x + 3$
45. $(-7, 8)$, parallel to $2x + y = 4$ $y = -2x - 6$
46. $(5, -2)$, parallel to $3x - y = 9$ $y = 3x - 17$
47. $(3, -5)$, parallel to $4x + 3y = -12$ $y = -\frac{4}{3}x - 1$
48. $(-6, -1)$, parallel to $5x - 6y = -30$ $y = \frac{5}{6}x + 4$
49. $(-1, 4)$, parallel to $y = 3$ $y = 4$
50. $(9, -3)$, parallel to $x = -4$ $x = 9$
51. $(0, -4)$, perpendicular to $y = -3x + 6$ $y = \frac{1}{3}x - 4$
52. $(0, 3)$, perpendicular to $y = 7x - 3$ $y = -\frac{1}{7}x + 3$
53. $(-7, 8)$, perpendicular to $2x + y = 4$ $y = \frac{1}{2}x + \frac{23}{2}$
54. $(5, -2)$, perpendicular to $3x - y = 9$ $y = -\frac{1}{3}x - \frac{1}{3}$
55. $(3, -5)$, perpendicular to $3x - 4y = -12$ $y = -\frac{4}{3}x - 1$
56. $(-6, -1)$, perpendicular to $5x - 6y = -30$ $y = -\frac{6}{5}x - \frac{41}{5}$
57. $(-5, 4)$, perpendicular to $y = 7$ $x = -5$
58. $(-3, 6)$, perpendicular to $x = -12$ $y = 6$

Mix 'Em Up!

Write the equation of each line described. Express each answer in slope-intercept form.

59. $m = -\frac{1}{2}$; $(-2, 4)$
60. $m = -\frac{2}{3}$; $(3, -3)$
61. $(4, 1)$ and $(-4, -1)$ $y = \frac{1}{4}x$
62. $(-5, 0)$ and $(6, -10)$
63. $(3, -4)$, parallel to $y + 2 = 0$ $y = -4$
64. $(-4, 3)$, perpendicular to $y - 3 = 0$ $x = -4$
65. $m =$ undefined; $(-10, 12)$ $x = -10$

66. $m = 0$; $(-15, 20)$ $y = 20$

67. $(4, -5)$, parallel to $y = -6x + 3$ $y = -6x + 19$

68. $(-1, 3)$, perpendicular to $y = -\frac{1}{10}x + 15$ $y = 10x + 13$

69. $(5, 4)$ and $(5, -6)$ $x = 5$

70. $(-1, -3)$ and $(2, -3)$ $y = -3$

71. $(6, -5)$, perpendicular to $3x - 2y = 0$ $y = -\frac{2}{3}x - 1$

72. $(-4, -2)$, parallel to $x + 4y = 0$ $y = -\frac{1}{4}x - 3$

73. $m = 4$; $(1, -2)$ $y = 4x - 6$

74. $m = -2$; $(-4, 3)$ $y = -2x - 5$

75. $m = 0$; $(5, 2)$ $y = 2$

76. $m =$ undefined; $(-6, -1)$ $x = -6$

77. $m = 5$; $(0, -5)$ $y = 5x - 5$

78. $m = -6$; $(0, 12)$ $y = -6x + 12$

79. $\left(\frac{7}{2}, -\frac{5}{3}\right)$ and $\left(\frac{9}{4}, \frac{1}{3}\right)$ $y = -\frac{8}{5}x + \frac{59}{15}$

80. $\left(-\frac{3}{4}, -\frac{2}{5}\right)$ and $\left(\frac{1}{2}, \frac{8}{5}\right)$ $y = \frac{8}{5}x + \frac{4}{5}$

81. $(0, 5)$, parallel to $3x - 5y = 4$ $y = \frac{3}{5}x + 5$

82. $\left(0, \frac{3}{5}\right)$, perpendicular to $6x - 3y = 12$ $y = -\frac{1}{2}x + \frac{3}{5}$

83. $(-5, 4)$, perpendicular to $x = -5$ $y = 4$

84. $(-4, 0)$, parallel to $x = 10$ $x = -4$

85. $(4.5, -2.5)$, perpendicular to $0.5x - 0.1y = 12$ $y = -0.2x - 1.6$

86. $(-1.2, 6.4)$, parallel to $0.1x + 0.4y = 8$ $y = -0.25x + 6.1$

87. $(5.6, -9.2)$, $(1.6, 3.2)$ $y = -3.1x + 8.16$

88. $(1.4, 5.3)$, $(-4.6, 2.3)$ $y = 0.5x + 4.6$

Solve each problem.

89. Pedro has a new job as a computer programmer. His starting salary is \$45,000. He will receive a raise of \$2000 each year he works for the company.

a. Write a linear equation that represents Pedro's salary, where x is the years he has worked for the company. $y = 2000x + 45{,}000$

b. What is Pedro's salary after 5 yr of working for the company? \$55,000

c. How long does he have to work for the company to earn a salary of \$71,000? 13 yr

90. Sue has a new job as a pharmaceutical sales representative. Her starting salary is \$50,000. She earns a bonus based on her sales. Her bonus is 25% of her total sales for the year.

a. Write a linear equation that represents Sue's yearly income, where x is her total sales for the year. $y = 0.25x + 50{,}000$

b. If Sue's sales total \$30,000 for the year, what is her income? \$57,500

c. How much would Sue need to sell to have an income of \$65,000? \$60,000

91. Shanika registers for her first semester in college. Her tuition includes fees of \$400 plus \$150 per credit hour.

a. Write a linear equation that represents Shanika's total tuition, where x is the number of credit hours she takes. $y = 150x + 400$

b. What is Shanika's tuition if she registers for 12 credit hours? \$2200

c. How many credit hours can Shanika register for if she has \$1300 for tuition? 6 credit hours

92. Heath enrolls in a fitness club. There is a one-time membership fee of \$300 plus a monthly charge of \$35.

a. Write a linear equation that represents the total cost of enrolling in the fitness club, where x is the number of months Heath is a member. $y = 35x + 300$

b. How much money will Heath spend for his fitness club membership if he is a member for 1 yr? \$720

c. How long can Heath be a member of the club if he pays the fitness club a total of \$1560? 36 months

93. Abdul purchases a new car for \$20,000. The car's value decreases by \$1500 each year.

a. Write a linear equation that represents the value of the car, where x is the age of the car in years. $y = -1500x + 20{,}000$

b. What is the car's value after 6 yr? \$11,000

c. What will be the age of the car when its value is \$5000? 10 yr

94. One of the fastest growing counties in the United States is Flagler County, Florida. In 2004, its population was approximately 69,000. In 2008, the population was approximately 91,000. (Source: U.S. Census Bureau)

a. Assuming linear growth, write an equation that approximates the population of Flagler County, where x is the number of years after 2004. $y = 5500x + 69{,}000$

b. What is the estimated population of Flagler county in 2020? 157,000

c. When will the population reach 135,000? 2016

95. The number of existing single-family homes sold in Florida was approximately 124,168 in 2008 and 170,848 in 2010. (Source: media.living.net/statistics/2010)

a. Assuming linear growth, write an equation that approximates the number of existing single-family homes sold in Florida, where x is the number of years after 2008. $y = 23{,}340x + 124{,}168$

b. If this trend continues, how many existing single-family homes will be sold in 2015? 287,548 homes

96. There were approximately 50.9 million Social Security beneficiaries in 2008, and approximately 54.0 million beneficiaries in 2010. (Source: www.ssa.gov)

a. Write a linear equation that approximates the number of Social Security beneficiaries (in millions), where x is the number of years after 2008. $y = 1.55x + 50.9$

b. If this growth continues, how many beneficiaries will there be in 2014? 60.2 million

97. The graph shows the number of people (ages 7 and older) who played golf at least once during the year from 1999 to 2009. (Source: http://www.nsga.org)

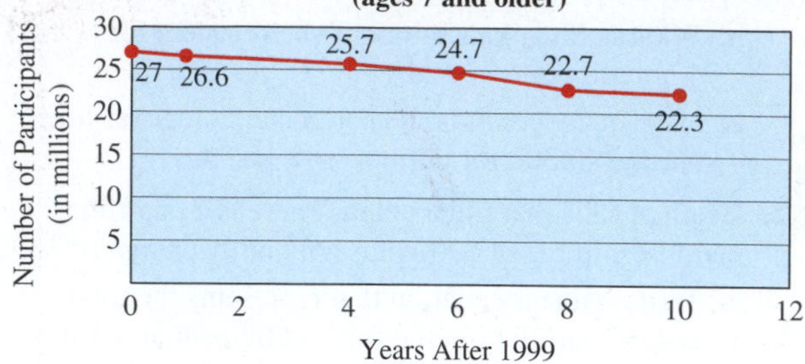

a. Use the points (0, 27) and (10, 22.3) to write a linear equation that models the number of people who have played golf at least once during the year, x years after 1999. $y = -0.47x + 27$

b. If this trend continues, how many people will play golf at least once during 2014? 19.95 million

98. The graph shows the postage for first-class mail. (Source: https://www.usps.com/send/first-class.htm)

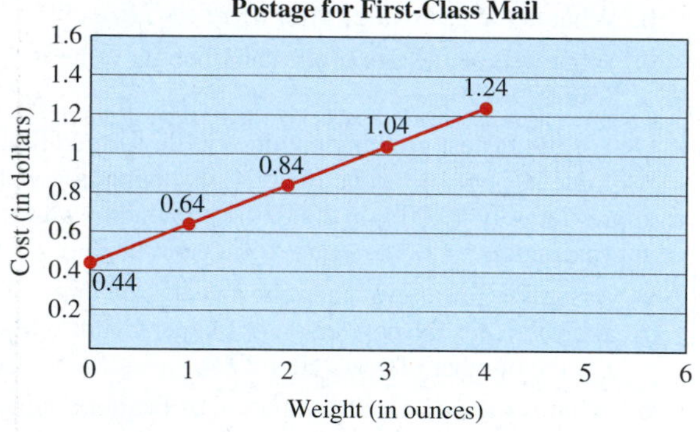

a. Use the points (1, 0.64) and (4, 1.24) to write a linear equation that models the postage for first-class mail, where x is the number of ounces over 1 oz the package weighs. $y = 0.20x + 0.44$

b. If this trend continues, how much is the postage for a first-class package that weighs 10 oz? $2.24

You Be the Teacher!

Correct each student's errors, if any.

99. Find the equation of the line that that is parallel to $4x - 3y = 6$ that passes through the point (8, 3).

Vivian's work:

$$4x - 3y = 6$$
$$\underline{-4x \qquad\quad -4x}$$
$$\frac{-3y}{-3} = \frac{-4x}{-3} + \frac{6}{-3}$$
$$y = \frac{4}{3}x - 2$$

So, the slope is $\frac{4}{3}$. Since it passes through the point (8, 3), the equation of the line is $y = \frac{4}{3}x + 3$.

100. Find the equation of the line that passes through the points $(-3, 4)$ and $\left(0, \frac{1}{2}\right)$.

William's work:

$$m = \frac{\frac{1}{2} - 4}{0 + 3} = \frac{\frac{1}{2} - \frac{8}{2}}{3}$$
$$= \frac{-\frac{7}{2}}{3} = -\frac{7}{6}$$

The equation should be $y = -\frac{7}{6}x + 4$ or $y = -\frac{7}{6}x + \frac{1}{2}$.

There can only be one equation that satisfies the given information. The slope $m = -\frac{7}{6}$ is correct. The point $\left(0, \frac{1}{2}\right)$ means that $b = \frac{1}{2}$. So, $y = -\frac{7}{6}x + \frac{1}{2}$.

101. Use the slope-intercept form to find the equation of the line with slope $m = -2$ that passes through the point $(5, -1)$.

Desi's work:

$$y = mx + b$$
$$y = -2x - 1$$

Calculate It!

Use the TABLE feature of a calculator to determine if the points (0, 2) and (−1, −3) lie on the given line.

102. $y = 4x + 1$

103. $y = -x + 2$

104. $y = -6x - 9$

105. $y = 5x + 2$

Think About It!

Let $Ax + By = C$ be the equation of a line in standard form.

106. What is the slope of the line? $-\frac{A}{B}$

107. What is the slope of a line perpendicular to the given line? $\frac{B}{A}$

108. What is the value of C if the line passes through the origin? 0

109. If the line has a positive slope, is $\frac{A}{B}$ positive or negative? negative

SECTION 4.5 Linear Inequalities in Two Variables

OBJECTIVES

As a result of completing this section, you will be able to

1. Determine if an ordered pair is a solution of a linear inequality in two variables.
2. Graph the solution set of a linear inequality in two variables.
3. Solve applications of linear inequalities in two variables.
4. Troubleshoot common errors.

Suppose the final grade in a math class is based on the test average and the final exam grade. The test average counts 70% of the final grade and the final exam counts 30% of the final grade. If a student wants to have at least a 70 average in the course, what are some possible combinations of test average and final exam grades to produce the desired result?

To solve this problem, we must know how to solve linear inequalities in two variables.

Objective 1 Determine if an ordered pair is a solution of a linear inequality in two variables.

Solutions of Linear Inequalities in Two Variables

At this point, we have studied linear equations in two variables and their solutions. We will now investigate *linear inequalities in two variables* and their solutions.

> **Definition:** A **linear inequality in two variables** is an inequality that can be written in one of the following ways, where A, B, C, m, and b are real numbers with A and B not both zero.
>
> $Ax + By > C$ $\quad$ $Ax + By \geq C$ $\quad$ $Ax + By < C$ $\quad$ $Ax + By \leq C$
>
> $y > mx + b$ $\quad$ $y \geq mx + b$ $\quad$ $y < mx + b$ $\quad$ $y \leq mx + b$

Like solutions of linear equations in two variables, **solutions of linear inequalities in two variables** are ordered pairs (x, y) that make the inequality true.

> **Procedure: Determining if an Ordered Pair Is a Solution of an Inequality**
>
> **Step 1:** Replace the variables with the corresponding values of x and y.
> **Step 2:** Simplify the resulting inequality.
> **Step 3:** If the inequality is true, then the ordered pair is a solution. If the inequality is false, then the ordered pair is not a solution.

Objective 1 Examples Determine if the ordered pair is a solution of the given inequality.

1a. $x - y < 8; (-4, 2)$ $\quad$ **1b.** $y > -4x + 10; \left(\frac{3}{2}, -5\right)$

1c. $y \leq 3; (7, 8)$ $\quad$ **1d.** $x < 2; (-1, 6)$

Solutions

1a.

$x - y < 8$

$-4 - 2 < 8$ $\quad$ Replace x with -4 and y with 2.

$-6 < 8$ $\quad$ Simplify.

Since the resulting statement is true, $(-4, 2)$ is a solution of $x - y < 8$.

1b.

$y > -4x + 10$

$-5 > -4\left(\frac{3}{2}\right) + 10$ $\quad$ Replace x with $\frac{3}{2}$ and y with -5.

$-5 > -6 + 10$ $\quad$ Multiply -4 and $\frac{3}{2}$.

$-5 > 4$ $\quad$ Simplify.

Since the resulting statement is false, $\left(\frac{3}{2}, -5\right)$ is *not* a solution of $y > -4x + 10$.

1c. $y \le 3$

$8 \le 3$ Replace y with 8.

Since the resulting statement is false, (7, 8) is *not* a solution of $y \le 3$.

1d. $x < 2$

$-1 < 2$ Replace x with -1.

Since the resulting statement is true, $(-1, 6)$ is a solution of $x < 2$.

Student Check 1 Determine if the ordered pair is a solution of the given inequality.

a. $2x - y \ge 15$; $(6, -3)$ **b.** $y < -\frac{2}{5}x - 4$; (0, 0)

c. $y > -3$; $(2, -1)$ **d.** $x \le 5$; $(-7, 2)$

Graphing Linear Inequalities in Two Variables

Objective 2 Graph the solution set of a linear inequality in two variables.

INSTRUCTOR NOTE: Point out that if the boundary line is a vertical line, the half-planes are the right and left half-plane.

Now that we know what it means for an ordered pair to be a solution of a linear inequality in two variables, we will graph the solution set of a linear inequality in two variables. Just as there are infinitely many solutions of linear inequalities in one variable, there are infinitely many solutions of linear inequalities in two variables.

To graph a linear inequality in two variables, we first graph the linear equation in two variables that is formed by replacing the inequality symbol with an equals sign. This line, called the **boundary line**, divides the plane into two regions called **half-planes**. The region above the line is the **upper half-plane** and the region below the line is the **lower half-plane**. (See Figure 4.2.) We will see that one region of the plane contains solutions of the linear inequality in two variables while the other does not. The boundary line may or may not be included in the solution set.

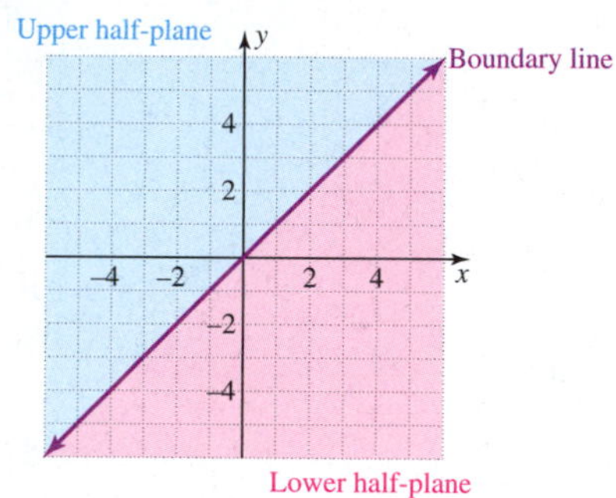

Figure 4.2

Consider the inequality $y \ge 2x - 2$. To solve the inequality, we first draw the graph of the linear equation associated with this inequality, $y = 2x - 2$. Then we identify some ordered pairs in each of the half-planes. These ordered pairs will either satisfy $y > 2x - 2$ or $y < 2x - 2$. That is, they will either make the inequality we are attempting to solve true or false. We can determine if the points are solutions of $y \ge 2x - 2$ by substituting them into the inequality. We also need to determine if points on the boundary line are solutions of the inequality.

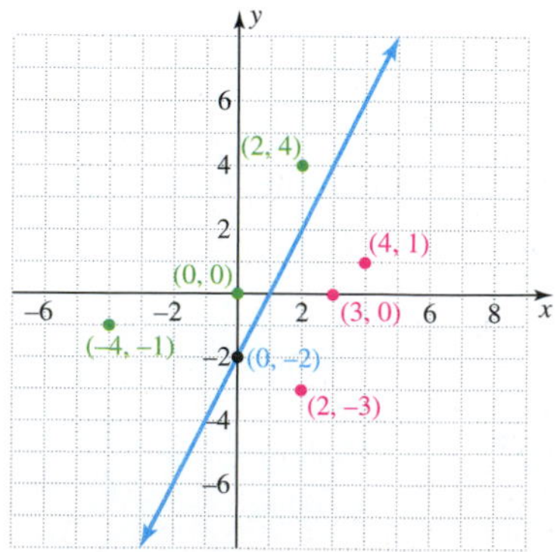

(x, y)	$y \ge 2x - 2$	True?	Solution?
(2, 4)	$4 \ge 2(2) - 2$ $4 \ge 2$	Yes	Yes
(0, 0)	$0 \ge 2(0) - 2$ $0 \ge -2$	Yes	Yes
$(-4, -1)$	$-1 \ge 2(-4) - 2$ $-1 \ge -10$	Yes	Yes
(4, 1)	$1 \ge 2(4) - 2$ $1 \ge 6$	No	No
(3, 0)	$0 \ge 2(3) - 2$ $0 \ge 4$	No	No
$(2, -3)$	$-3 \ge 2(2) - 2$ $-3 \ge 2$	No	No
$(0, -2)$	$-2 \ge 2(0) - 2$ $-2 \ge -2$	Yes	Yes

From the table, we observe that solutions of the inequality, $y \geq 2x - 2$, are located in the upper half-plane and on the boundary line. In fact, any point in the upper-half plane or on the boundary line is a solution of the linear inequality. Note that the points located in the lower half-plane are not solutions of the inequality.

So, the solutions of $y \geq 2x - 2$ consist of the solutions of the boundary line, $y = 2x - 2$, and all points in the upper half-plane, that is, the points that satisfy $y \geq 2x - 2$. We indicate this by shading the region of the plane that contains the solutions and making the boundary line solid. The graph of the solution set of $y \geq 2x - 2$ is shown.

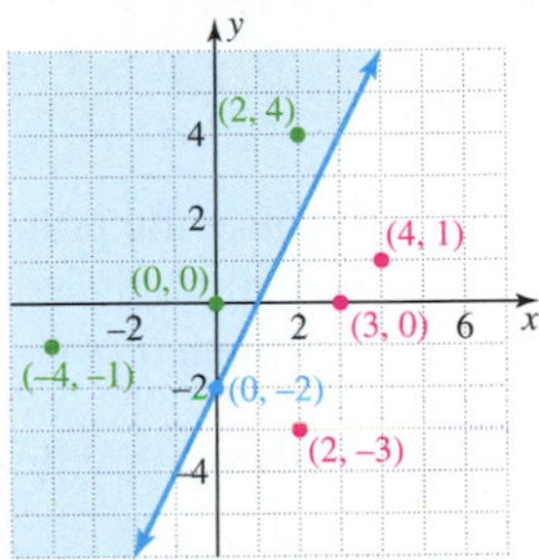

If points on the boundary line do not satisfy the inequality, we make the boundary line a dashed line.

Definition: The **graph of the solution set of a linear inequality** in two variables has three parts:

- The half-plane that contains solutions of the inequality.
- The half-plane that does not contain solutions of the inequality.
- The boundary line that divides the coordinate system into these two half-planes.

Our job is to determine which of the half-planes contain solutions of the inequality. We will use the following facts.

- If one point in a half-plane is a solution of an inequality, then all points in that half-plane are also solutions.
- If one point in a half-plane is not a solution of an inequality, then none of the points in that half-plane are solutions.

These facts enable us to use a *single test point* to determine the graph of the solution set of a linear inequality in two variables. The only requirement of our test point is that it cannot be on the boundary line; it must be located in one of the half-planes.

Procedure: Graphing the Solution Set of a Linear Inequality in Two Variables

Step 1: Graph the associated linear equation in two variables by replacing the inequality symbol with an equals sign.
  a. The line is solid if the inequality symbol is $\leq$ or $\geq$.
  b. The line is dashed if the inequality symbol is $<$ or $>$.

Step 2: Identify a point in either one of the half-planes formed by the boundary line and test it in the original inequality.
  a. If the point makes the original inequality *true*, then shade the half-plane that contains the test point.
  b. If the point makes the original inequality *false*, then shade the half-plane that does not contain the test point.

Note: *The point* $(0, 0)$ *is the easiest point to test as long as it is not on the boundary line. Any other point in the plane also works as long as the point is not on the line.*

Objective 2 Examples **Graph the solution set of each linear inequality in two variables.**

2a. $x - y > 4$ **2b.** $y < -\frac{2}{3}x + 2$ **2c.** $x \geq -1$ **2d.** $y \leq -3$

Solutions **2a.** The boundary line is $x - y = 4$.

The boundary line is dashed since the inequality is >.

Test the point (0, 0) in the original inequality.

$$x - y > 4$$
$$0 - 0 > 4$$
$$0 > 4 \quad \text{False}$$

Because the point (0, 0) is *not* a solution of the inequality, the solutions of the inequality lie in the half-plane *not* containing (0, 0).

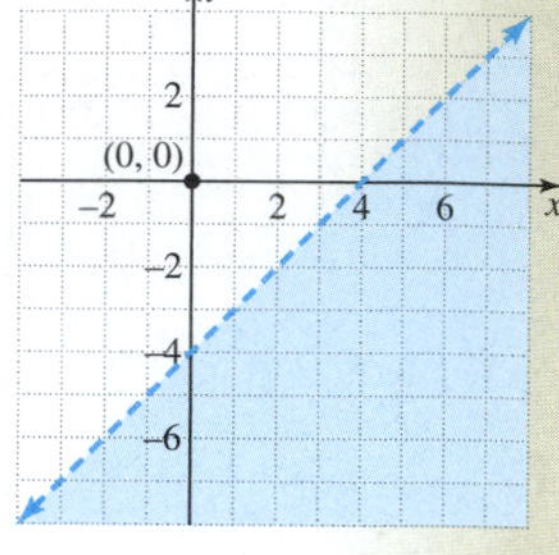

2b. The boundary line is $y = -\frac{2}{3}x + 2$.

The boundary line is dashed since the inequality is <.

Test the point (0, 0) in the original inequality.

$$y < -\frac{2}{3}x + 2$$
$$0 < -\frac{2}{3}(0) + 2$$
$$0 < 2 \quad \text{True}$$

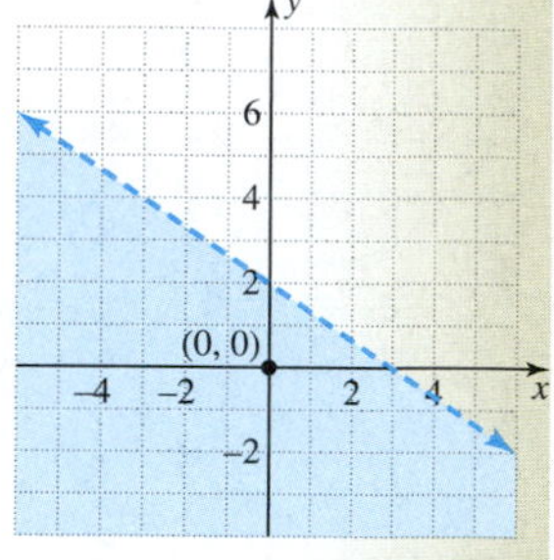

Because the point (0, 0) is a solution of the inequality, the solutions of the inequality lie in the half-plane containing (0, 0).

2c. The boundary line is $x = -1$.

The boundary line is solid since the inequality is ≥.

Test the point (0, 0) in the original inequality.

$$x \geq -1$$
$$0 \geq -1 \quad \text{True}$$

Because the point (0, 0) is a solution of the inequality, the solutions of the inequality lie in the half-plane containing (0, 0).

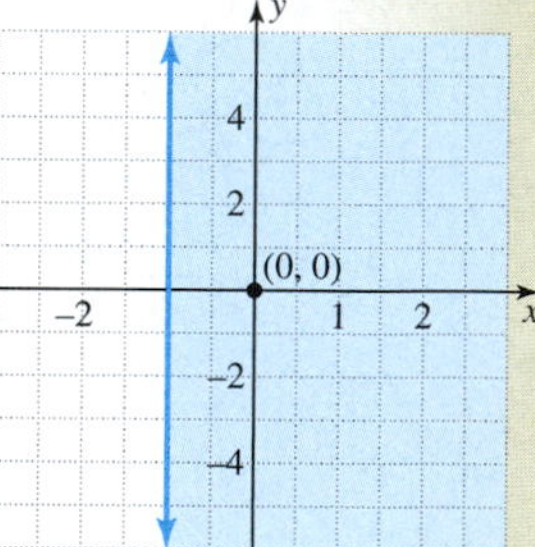

2d. The boundary line is $y = -3$.

The boundary line is solid since the inequality is ≤.

Test the point (0, 0) in the original inequality.

$$y \leq -3$$
$$0 \leq -3 \quad \text{False}$$

Because the point (0, 0) is a *not* a solution of the inequality, the solutions of the inequality lie in the half-plane *not* containing (0, 0).

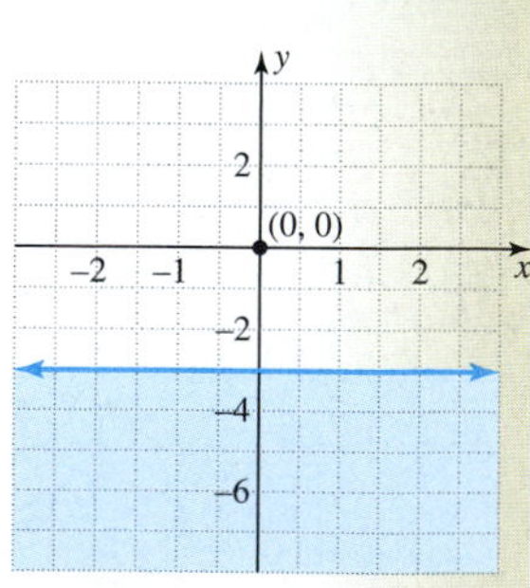

Student Check 2 Graph the solution set of each linear inequality in two variables.

a. $2x + y > 2$ **b.** $y < \frac{2}{5}x - 2$ **c.** $x \geq 2$ **d.** $y \leq 4$

Objective 3 ▶

Solve applications of linear inequalities in two variables.

Applications of Linear Inequalities in Two Variables

Applications of linear inequalities in two variables are very important for a branch of mathematics called *linear programming*. This field of mathematics is used to determine how to maximize or minimize different values. While this topic is beyond the scope of this class, we will explore other applications of linear inequalities.

The key to solving applications of linear inequalities lies in the ability to identify the inequality that pertains to the problem. Once the inequality is determined, we will solve it using the previous method described.

Objective 3 Examples

The final grade in Lucy's math class is based on the test average and final exam grade. The test average counts 70% of the final grade and the final exam counts 30% of the final grade. Lucy wants to earn at least a 70 in the course.

3a. Write a linear inequality in two variables that models this situation and graph it.

3b. State some possible combinations of test average and final exam grades that will produce the desired result. (The test average and final exam grade cannot be higher than 100.)

Solutions

3a. The unknowns are the test average and final exam grade. So, let $x =$ the test average and let $y =$ the final exam grade.

The final grade is based on the fact that the test average counts 70% and the final exam grade counts 30%. So,

$$0.70x + 0.30y = \text{final grade}$$

For the final grade to be *at least* a 70 average, we must solve the following inequality.

$$0.70x + 0.30y \geq 70$$

We solve the inequality by graphing the "solid" boundary line $0.70x + 0.30y = 70$. The line is solid because the inequality symbol is $\geq$. The line passes through the points (100, 0) and (70, 70).

Test the point (0, 0) since it is not on the boundary line.

$$0.70x + 0.30y \geq 70$$
$$0.70(0) + 0.30(0) \geq 70$$
$$0 \geq 70 \quad \text{False}$$

Because the point (0, 0) makes the inequality false, the solutions lie in the half-plane that does not contain the origin.

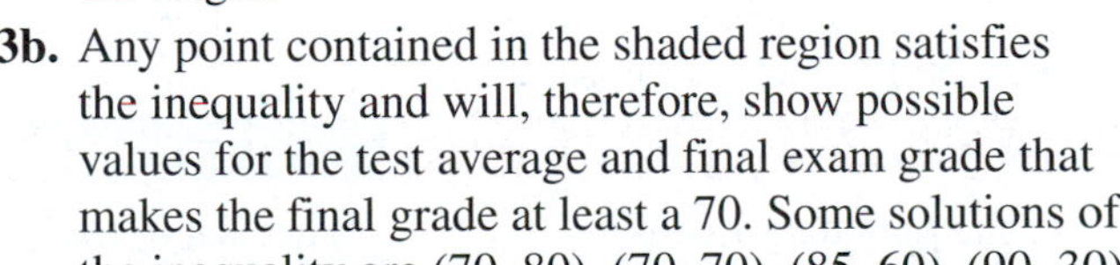

3b. Any point contained in the shaded region satisfies the inequality and will, therefore, show possible values for the test average and final exam grade that makes the final grade at least a 70. Some solutions of the inequality are (70, 80), (70, 70), (85, 60), (90, 30), and (100, 0). The table confirms that the final average would be at least a 70 for these values.

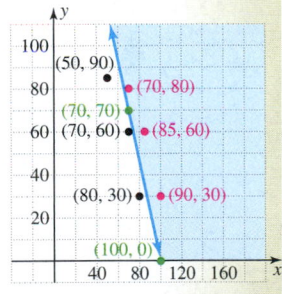

Test Average, x	Final Exam Grade, y	Final Average $= 0.70x + 0.30y$
70	80	$0.70(70) + 0.30(80) = 73$
70	70	$0.70(70) + 0.30(70) = 70$
85	60	$0.70(85) + 0.30(60) = 77.5$
90	30	$0.70(90) + 0.30(30) = 72$
100	0	$0.70(100) + 0.30(0) = 70$

There are many other combinations of test averages and final exam grades that would provide the desired course grade. The points labeled in black (50, 90), (70, 60), and (80, 30) are ordered pairs that do not satisfy the inequality and would, therefore, make the final grade less than 70.

Student Check 3 The final grade in a history class is based on the midterm exam grade and the final exam grade. The midterm exam counts 40% of the final grade and the final exam counts 60% of the final grade. Juan wants to earn at least an 80 in the course.

a. Write a linear inequality in two variables that models the situation and graph it.

b. State some possible grades on the midterm and final exam that produce the desired grade.

Objective 4

Troubleshoot common errors.

Troubleshooting Common Errors

Some common errors associated with linear inequalities in two variables are shown.

Objective 4 Examples

A problem and an incorrect solution are given. Provide the correct solution and an explanation of the error.

4a. Graph the solution set of $y > -x + 3$.

Incorrect Solution	Correct Solution and Explanation
The boundary line is $y = -x + 3$. The point (0, 0) makes the inequality false. 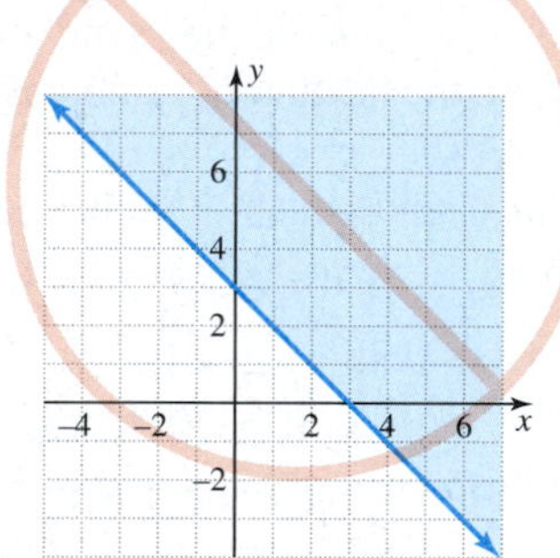	Everything is correct except for the boundary line. It should be dashed since the inequality is $>$. 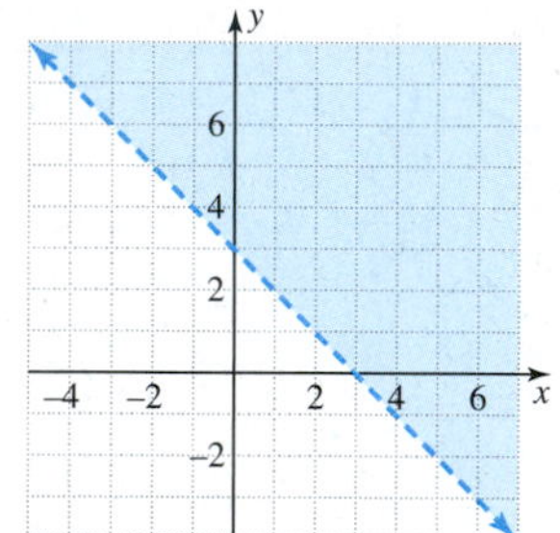

4b. Graph the solution set of $x - 5y < 10$.

Incorrect Solution	Correct Solution and Explanation
The boundary line is $x - 5y = 10$. The point (0, 0) makes the inequality true. 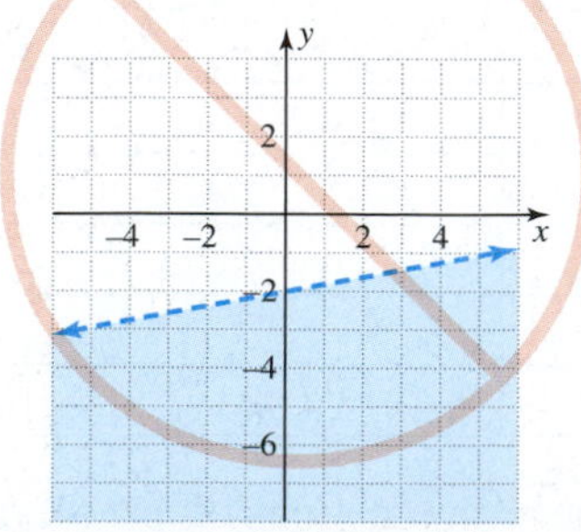	Since (0, 0) makes the inequality true, the solution must include this point. We shade the half-plane that contains the origin. 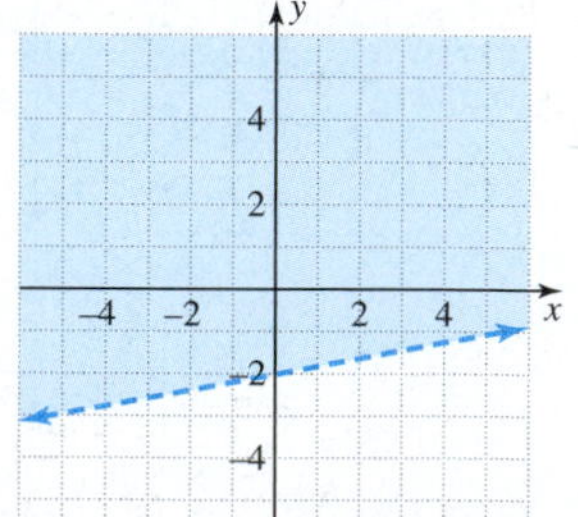

ANSWERS TO STUDENT CHECKS

Student Check 1 **a.** yes **b.** no **c.** yes **d.** yes

Student Check 2

a.

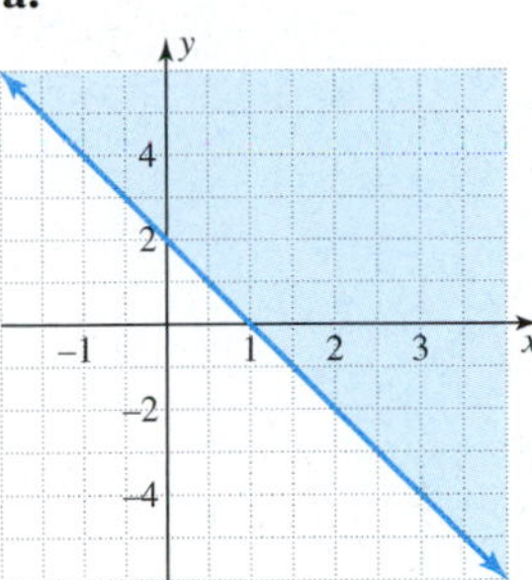

b.

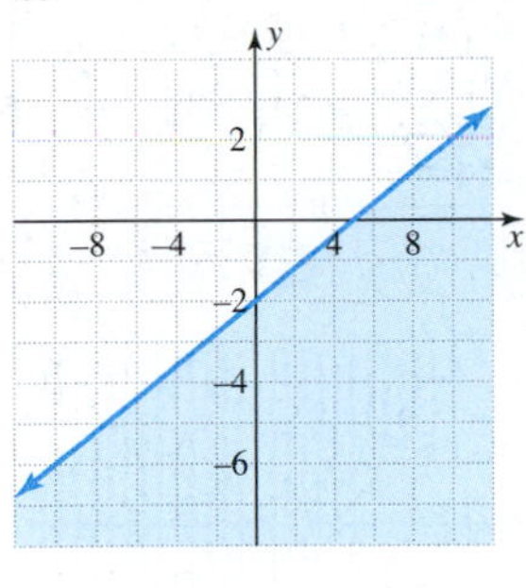

c.

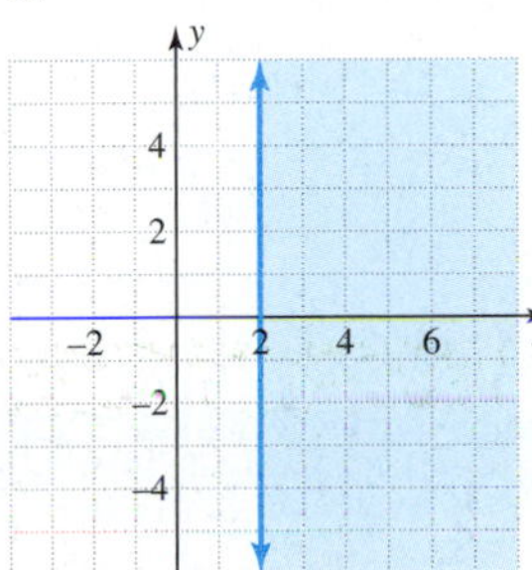

d.

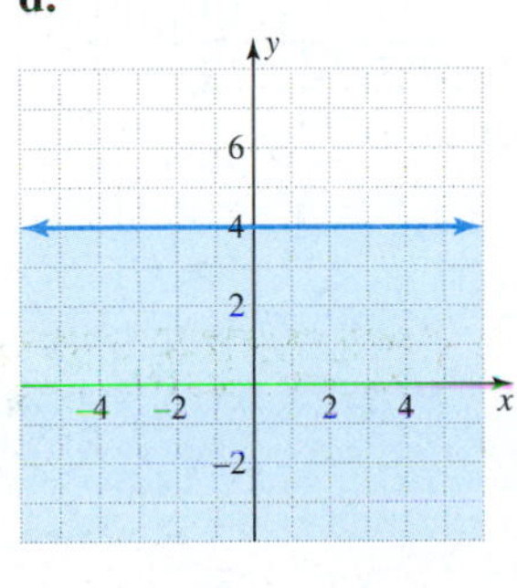

Student Check 3 **a.** $0.4x + 0.6y \geq 80$

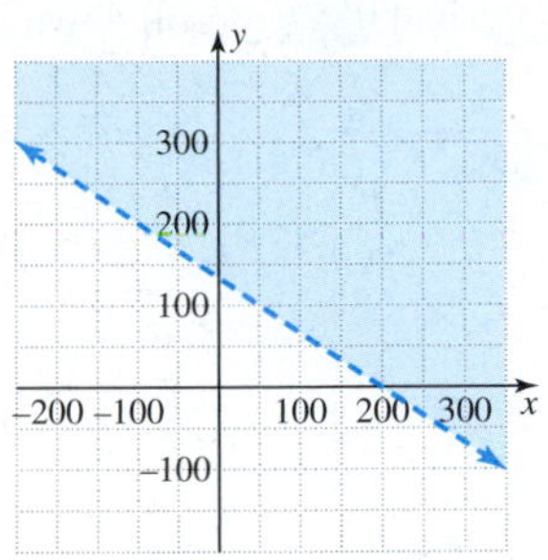

b. Answers vary; (80, 80), (60, 100), and (70, 90)

SUMMARY OF KEY CONCEPTS

1. Solutions of linear inequalities in two variables are ordered pairs that satisfy the linear inequality. There are infinitely many solutions of linear inequalities in two variables.
2. Solution sets of linear inequalities can be illustrated by a graph. The graph of the associated linear equation forms a boundary line between two half-planes. The solutions of the linear inequality lie in one of the half-planes. A single test point can be used to determine which half-plane contains the solutions.
3. To use a linear inequality in two variables to solve an applied problem, graph the solution of the inequality as shown in this section. Any ordered pair in the shaded region is a solution of the problem.

GRAPHING CALCULATOR SKILLS

To solve a linear inequality in two variables using the graphing calculator, we must first solve the inequality for y. If the resulting inequality is of the form $y < mx + b$, shade below the boundary line. If the resulting inequality is of the form $y > mx + b$, shade above the boundary line. If the inequality is < or >, the boundary line will be dashed. If the inequality is < or >, then the boundary line is solid.

Example 1: Solve $x - 2y > 4$.

Solution: Solve the inequality for y. Recall that when we multiply or divide an inequality by a negative number, the inequality symbol reverses.

$$-2y > -x + 4$$

$$\frac{-2y}{-2} < \frac{-x + 4}{-2}$$

$$y < \frac{1}{2}x - 2$$

Now enter the associated equation, $y = \frac{1}{2}x - 2$, in the equation editor. For the calculator to graph the appropriate region, we need to "tell it" which region to shade. Move the cursor to the left of Y_1. Press ENTER until we reach the symbol ◣ (shade below).

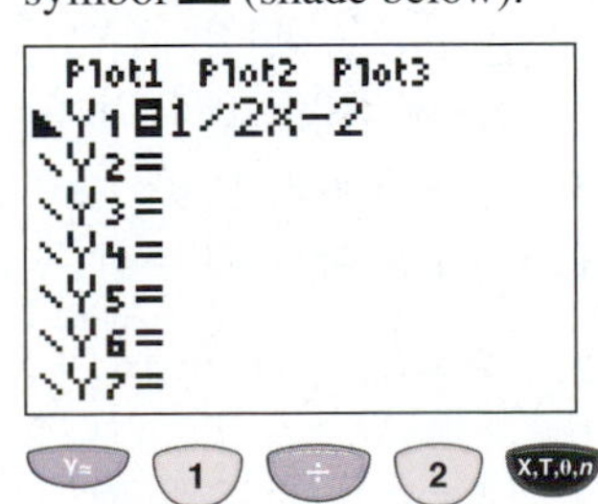

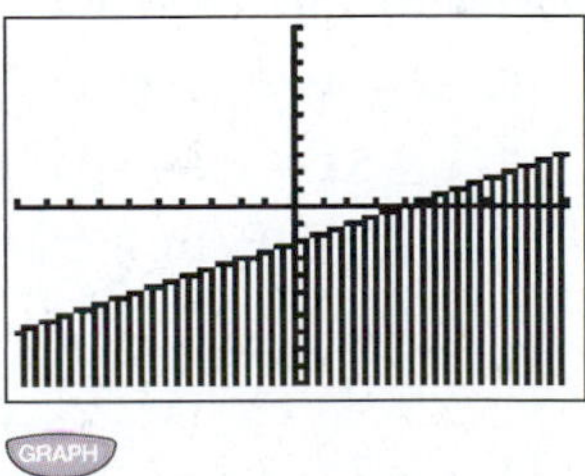

Alternate method for TI-84 users: Use the APPS feature to graph solutions of inequalities.

Example 2: Solve $y > -2x - 4$.

Solution: Press Y= and APPS. Scroll down in the menu and select Inequalz.

Press any key to continue.

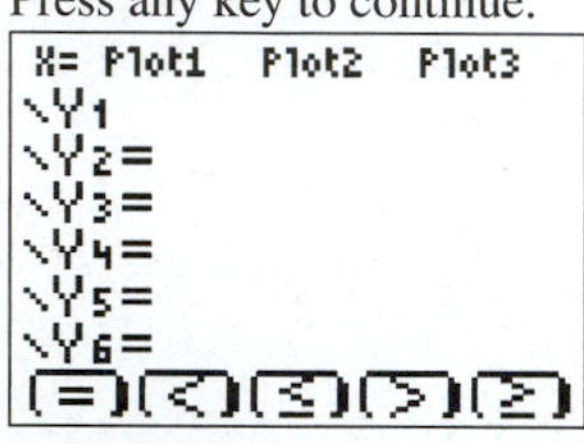

Select the > symbol and then enter the remaining part of the inequality.

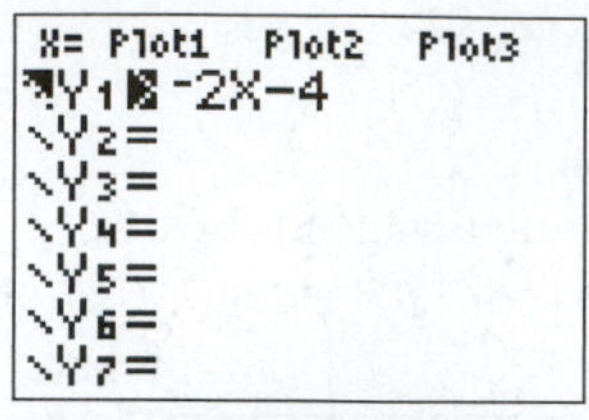

(-) 2 X,T,θ,n — 4

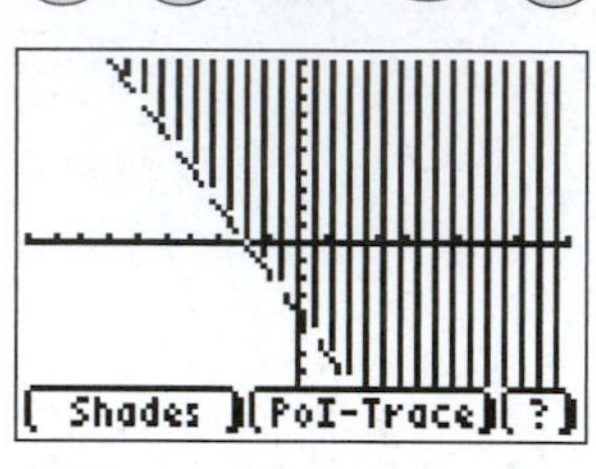

GRAPH

SECTION 4.5 EXERCISE SET

Write About It!

Use complete sentences in your answer to each exercise.

1. How do you determine if an ordered pair (x, y) is a solution of a linear inequality?
2. How do you determine the boundary line of a linear inequality? The boundary line is the graph of the linear equation associated with the linear inequality.
3. How will you know whether the boundary line of the graph of a linear inequality should be a dashed line or a solid line?
4. How many points in a half-plane must be tested to determine if the half-plane includes solutions of a linear inequality?

Practice Makes Perfect!

Determine if the ordered pair is a solution of the given inequality. (*See Objective 1.*)

5. $3x - 4y < 12$; $(1, 1)$ yes
6. $2x + y < -3$; $(0, 1)$ no
7. $y \geq 5x - 1$; $(4, -1)$ no
8. $y \leq -2x + 3$; $(0, 0)$ yes
9. $x + y < 9$; $(10, 3)$ no
10. $x - y > -5$; $(-2, -4)$ yes
11. $y > \frac{1}{2}x + 8$; $(6, -3)$ no
12. $y < -\frac{1}{3}x - 11$; $(9, 0)$ no
13. $y - 7 > 2$; $(-3, 5)$ no
14. $y + 2 < 0$; $(8, -1)$ no
15. $x - 3 < 0$; $(-4, 11)$ yes
16. $x + 15 > 1$; $(13, -2)$ yes
17. $4x - 2y > 1$; $\left(\frac{1}{2}, \frac{3}{4}\right)$ no
18. $x - 2y < 3$; $\left(\frac{2}{3}, -\frac{1}{6}\right)$ yes

Additional answers can be found in the Instructor Answer Appendix.

Graph the solution set of each linear inequality in two variables. (*See Objective 2.*)

19. $2x + y > 4$
20. $3x + y < 5$
21. $y \leq 4x - 8$
22. $y \geq 3x - 6$
23. $y < -2x + 10$
24. $y > -3x - 9$
25. $y \leq 5x - 15$
26. $y \geq 10x - 20$
27. $x - y < 2$
28. $y > 2x$
29. $y < 3x$
30. $y - 7x \leq 0$
31. $3x - 3y \geq 0$
32. $4x - y > -3$
33. $x > 8$
34. $x \leq -1$
35. $y \leq -4$
36. $y \geq 3$

Use a linear inequality in two variables to solve each problem. (*See Objective 3.*)

37. Todd's algebra course grade is based on his test average and his final exam grade. The test average counts 60% of the course grade and the final exam counts 40% of the course grade. If Todd wants to have at least an 80 average in the course,
    a. write a linear inequality in two variables that represents this situation. $0.6x + 0.4y \geq 80$
    b. Graph the linear inequality.
    c. Give three possible combinations of test averages and final exam grades that produce the desired grade.
38. Tyrone's chemistry course grade is based on his test average and his final exam grade. The test average counts 65% of the course grade and the final exam

counts 35% of the course grade. If Tyrone wants to have at least an 80 average in the course,

a. write a linear inequality in two variables that represents this situation. $0.65x + 0.35y \geq 80$

b. Graph the linear inequality.

c. Give three possible combinations of test averages and final exam grades that produce the desired grade.

39. Jenna works two jobs to put herself through college. She works at a bookstore for $8 per hour, and she works at the school library for $10.50 per hour. If Jenna wants to make at least $300 per week,

a. write a linear inequality in two variables that represents this situation. $8x + 10.50y \geq 300$

b. Graph the linear inequality.

c. Give three possible combinations of hours that Jenna could work at each job to earn her desired income.

40. Courtney works two jobs to put herself through college. She works at a food factory for $8.25 per hour, and she works at a company as a part-time secretary for $10.50 per hour. If Courtney wants to make at least $231 per week,

a. write a linear inequality in two variables that represents this situation. $8.25x + 10.50y \geq 231$

b. Graph the linear inequality.

c. Give three possible combinations of hours that Courtney could work to earn her desired income.

41. Shenita recently started a tailoring business. She charges $15 to hem a pair of pants and she charges $25 to tailor a suit jacket. If Shenita wants to make at least $500 in one week from tailoring,

a. write a linear inequality in two variables that represents this situation. $15x + 25y \geq 500$

b. Graph the linear inequality.

c. Give three possible combinations of the number of pants she must hem and the number of suit jackets she must tailor.

42. David started a lawn mowing business. He charges $35 to mow a lawn that is less than half an acre and $50 for a lawn that is between half an acre to an acre. If David wants to make at least $1500 per week,

a. write a linear inequality in two variables that represents this situation. $35x + 50y \geq 1500$

b. Graph the linear inequality.

c. Give three possible combinations of mowing jobs that David needs to make at least $1500 per week.

You Be the Teacher!

Correct each student's errors, if any.

43. Graph $x - y < 5$.

Adrianna's work: $x - y = 5$ has intercepts (0, 5) and (5, 0).

$0 - 0 < 5$

$0 < 5$ True

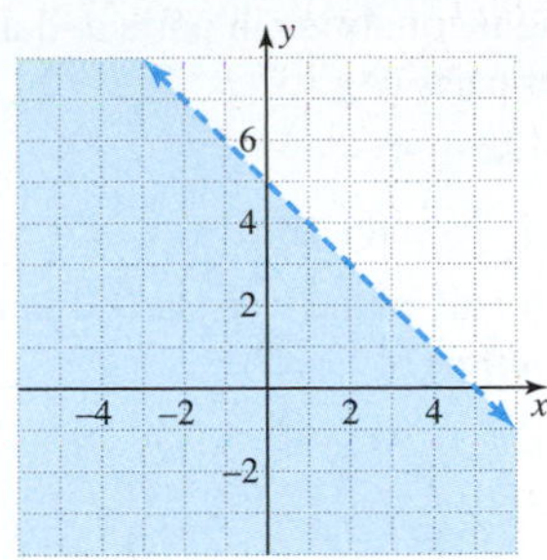

44. Graph $3x + y > 0$.

Scott's work: The boundary line goes through $(1, -3)$ and $(0, 0)$.

$3(0) + (0) > 0$

$0 > 0$ False

So, there is no solution.

45. Graph $4x - 3y \leq 12$.

Lily's work: The intercepts of $4x - 3y = 12$ are $(0, -4)$ and $(3, 0)$.

$4(0) - 3(0) \leq 12$

$0 \leq 12$ True

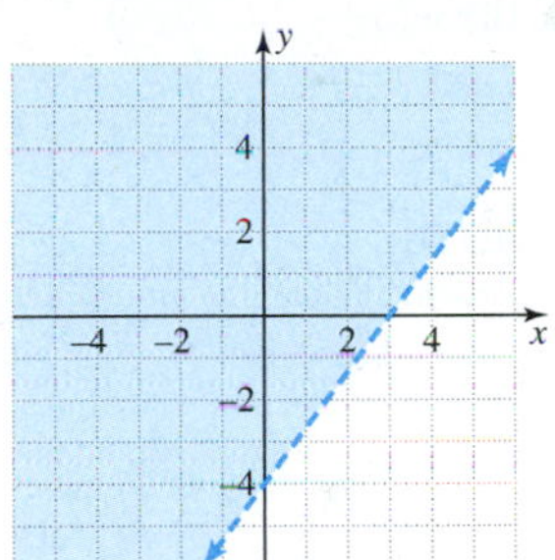

Calculate It!

Use a graphing calculator to graph each linear inequality in two variables.

46. $3x + y \geq 1$

47. $6x + y \leq 2$

48. $x + 3y < 2$

49. $x - 5y < 4$

Think About It!

50. Graph the inequalities in parts a–d and use your results to answer parts e–g.

a. $y > -2x + 4$

b. $y > \frac{1}{3}x - 1$

c. $y > 4x$

d. $y > 1$

e. Which half-plane was shaded in each of these inequalities? upper half-plane

f. What can you conclude about the graph of the solution of $y > mx + b$? The solution set will always be the upper half-plane.

g. How does the graph of $y \geq mx + b$ differ from the graph of $y > mx + b$?

51. Graph the inequalities in parts a–d and use your results to answer parts e–g.

a. $y < -2x + 4$

b. $y < \frac{1}{3}x - 1$

c. $y < 4x$

d. $y < 1$

e. Which half-plane was shaded in each of these inequalities? lower half-plane

f. What can you conclude about the graph of the solution of $y < mx + b$? The solution set will always be the lower half-plane.

g. How does the graph of $y \leq mx + b$ differ from the graph of $y < mx + b$?

GROUP ACTIVITY The Wave

1. Assign a person to be the timekeeper. The timer records the time it takes to complete the wave.
2. Begin with a group of five students. At the word GO, students make a wave by standing up, raising their arms, and sitting down in sequence. The last person says STOP as they sit down.
3. Repeat the wave with groups of 10, 15, and 20 students.
4. Record your findings in a table where x represents the number of students and y represents the time to complete the wave.
5. Plot the points on a graph.
6. Write a linear function that describes the relationship.
7. What is the slope of the linear function and what does it mean in the context of the problem?
8. What is the y-intercept of the linear function and what does it mean in the context of the problem?
9. Use your model to predict the time it would take for a group of 100 students to complete the wave in sequence.

INSTRUCTOR NOTE:
Objective: To use linear functions to model a real-life situation
Resources: Stopwatch, graph paper

Linear Functions and Linear Inequalities in Two Variables

What's the big idea? Now that we have completed Chapter 4, we should be able to see a connection between a linear function or a linear equation in two variables, its solutions and its graph, and use linear functions to model some real-life situations. We should also be able to graph a linear inequality in two variables and use them to model certain situations, as well.

The Tools

Listed below are the key terms, skills, formulas, and properties you should know for this chapter.

The page reference is provided if you need additional help with the given topic. The Study Tips will assist in your preparation for an exam.

Study Tips

1. Learn all of the terms, formulas, and properties. Make flash cards and have someone quiz you.
2. Rework problems from the exercises and also the ones you worked in class. Work additional problems from the review exercises.
3. Review the summaries of key concepts.
4. Work the chapter test.
5. Be sure to review the online resources for additional study materials.

Terms

Base fee 217
Boundary line 270
Half-plane 270
Horizontal line 229
Initial value 247
Linear function 212
Linear equation in two variables 213
Linear inequality in two variables 269
Lower half-plane 270
Parallel lines 245
Perpendicular lines 246
Rate of change 247
Slope 239
Slope-intercept form 239
Solution of a linear inequality in two variables 269
Upper half-plane 270
Variable fee 217
Vertical line 229
x-intercept 226
y-intercept 226
Zero of a function 216

Formulas and Properties

- Equation of a horizontal line 229
- Equation of a vertical line 229
- Point-slope form of a line 256
- Slope formula 243
- Slope-intercept form of a line 239
- Standard form of a linear equation in two variables 213

CHAPTER 4 / SUMMARY

How well do you know this chapter? Complete the following questions to find out. Take a look back at the section if you need help.

SECTION 4.1 Linear Functions and Linear Equations in Two Variables

1. A linear function is a function of the form $f(x) = mx + b$. The exponent on the variable is one. An example is $f(x) = 3x - 4$ (Answers vary.).
2. The standard form of a linear equation in two variables is $Ax + By = C$, where A and B are not both zero.
3. The standard form of a line can be written in function notation by solving for y.
4. When we evaluate a function, we find the output value that corresponds to a given input value.
5. To solve $f(x) = c$ means to replace $f(x)$ with c and solve for x.

6. The solutions of $f(x) = 0$ are called the zeros of a function.

7. The graph of a linear function is a(n) line.

SECTION 4.2 Graphing Linear Equations and Linear Functions

8. To graph a line by plotting points, two points are required. However, a total of three points is recommended so that one of the points serves as a check.

9. The point where a graph crosses the x-axis is called the x-intercept. To find this point, set y = 0 and solve the resulting equation. The point where a graph crosses the y-axis is called the y-intercept. To find this point, set x = 0 and solve the resulting equation.

10. If the value of the constant in a linear equation equals zero, then the graph of the equation will pass through the point (0, 0). To graph such an equation, find another point in addition to the intercepts.

11. The equation of a horizontal line has the form $y = k$.

12. The equation of a vertical line has the form $x = h$.

13. All linear equations in two variables are functions except for equations of the form $x = h$.

14. The domain of a linear function is $(-\infty, \infty)$ and the range is $(-\infty, \infty)$ unless the line is horizontal. The range of $f(x) = c$, where c is a real number is $\{c\}$.

SECTION 4.3 The Slope of a Line

15. The slope of a line measures the line's steepness.

16. The slope is defined as the ratio of the change in y to the change in x. It is also called rise over run.

17. The slope-intercept form of a line is $y = mx + b$. The m represents the slope and b represents the y-coordinate of the y-intercept.

18. To graph a line using the slope and y-intercept, plot $(0, b)$ first. From this point, use the slope to move to another point on the line. For instance, a slope of $\frac{4}{3}$ would mean to move up 4 units and move right 3 units.

19. The slope formula is $m = \frac{y_2 - y_1}{x_2 - x_1}$.

20. A positive slope corresponds to a line that is increasing from left to right. A negative slope corresponds to a line that is decreasing from left to right. A slope of zero corresponds to a(n) horizontal line. An undefined slope corresponds to a(n) vertical line.

21. Two lines are parallel if they have the same slope.

22. Two lines are perpendicular if their slopes are negative reciprocals.

23. In application problems the value of b represents the initial value. The value m represents the rate of change.

SECTION 4.4 Writing Equations of Lines

24. To write the equation of a line, we must know the slope and at least one point. The slope-intercept form or the point-slope form may be used to find the equation.

SECTION 4.5 Linear Inequalities in Two Variables

25. A linear inequality in two variables is an inequality of the form $y < mx + b$ or $Ax + By < C$.

26. Solutions of linear inequalities in two variables are ordered pairs that make the inequality true.

27. To graph a linear inequality in two variables, we must first graph the boundary line, which is formed by replacing the inequality symbol with a(n) equals sign. If the inequality symbol is $\leq$ or $\geq$, the boundary line is solid. If the inequality symbol is $<$ or $>$, the boundary line is dashed.

28. The boundary line divides the plane into two half-planes. One half-plane contains solutions of the inequality and the other does not.

29. After the boundary line is drawn, we must use a(n) test point in the original inequality to determine which half-plane contains solutions of the inequality. The ideal test point is (0, 0) as long as the boundary line does not go through the origin.

CHAPTER 4 / REVIEW EXERCISES

SECTION 4.1

Determine if each function is linear. If it is linear, identify the values of m and b. Evaluate each function at the given value. (*See Objectives 1 and 3.*)

1. $f(x) = -32; f(7)$
yes, $m = 0$, $b = -32$, $f(7) = -32$

2. $f(x) = -7x + 10; f\left(\frac{1}{7}\right)$

Rewrite each equation using function notation, if possible. Evaluate each function at the given value. (*See Objectives 2 and 3.*)

3. $6x - 5y = -15; f(-5)$
$f(x) = \frac{6}{5}x + 3; f(-5) = -3$

4. $y + 8 = 0; f(1)$
$f(x) = -8;\ f(1) = -8$

5. $8x - 3y = 1;\ f\left(-\frac{7}{4}\right)$

6. $2x + 7y = 19;\ f\left(\frac{5}{2}\right)$

Evaluate each function at the given value. (*See Objective 3.*)

7. $f(x) = -9x - 12; f(0)$ -12

8. $f(x) = 2x + 18; f(4)$ 26

9. $f(x) = -6x + 11;\ f\left(\frac{1}{3}\right)$ 9

Find the value of x for which $f(x) = c$. (*See Objective 4.*)

10. $f(x) = -10x - 20; f(x) = 0$ $x = -2$

11. $f(x) = 1.8x - 4.8; f(x) = 16.8$ $x = 12$

12. $f(x) = \frac{3}{7}x - 2; f(x) = -2$ $x = 0$

Solve each problem. (*See Objective 5.*)

13. Sprint offers a voice plan for a cell phone with 450 minutes for \$69.99 per month. The phone service charges \$0.45 for each minute over 450 minutes. Write a linear function $f(x)$ that represents the monthly cost, where x is the number of additional minutes in a month. Find $f(100)$ and explain what it means. Solve $f(x) = 182.49$ and explain what it means. (www.shop.sprint.com)

14. Cab fares in Kansas City, Missouri, are \$2.50 with an additional \$2 per mile and \$1 for any time spent waiting in traffic. Assuming there is no traffic, write a linear function $f(x)$ for the cost of a cab ride, where x is the number of additional miles traveled. Find $f(8)$ and explain what it means. Solve $f(x) = 34.5$ and explain what it means. (Source: http://www.visitkc.com/getting-around/transportation/taxi--car-service/index.aspx)

15. A carpet cleaning service charges \$89 to deep clean the carpet of two rooms in a house plus \$30 for each additional room up to 8 rooms. Write a linear function $f(x)$ for the cost of carpet cleaning, where x is the number of additional rooms. Find $f(2)$ and explain what it means. Solve $f(x) = 239$ and explain what it means.

16. A fitness club is advertising a special one-time offer. This offer includes a \$1 membership fee, a one-time \$19.99 administrative fee, plus a \$29 monthly charge. Write a linear function $f(x)$ for the cost of joining the fitness club as a function of x months. Find $f(12)$ and explain what it means. Solve $f(x) = 542.99$ and explain what it means.

17. The 2009 federal income tax for a person filing single with an income over \$82,250 but not over \$171,550 is \$16,750 plus 28% of the amount over \$82,250. Write a linear function $f(x)$ that represents the tax owed, where x is the amount of income over \$82,250. Find $f(85{,}400)$ and explain what it means. Solve $f(x) = 24{,}128$ and explain what it means.

Find the requested information given that $Y_1 = f(x)$. (*See Objectives 3 and 4.*)

18. $f(0)$ $f(0) = 32$

X	Y1
-3	77
-2	62
-1	47
0	32
1	17
2	2
3	-13

X= -3

19. Find the value of x such that $f(x) = 0$. $x = 3$

X	Y1
-2	-60
-1	-48
0	-36
1	-24
2	-12
3	0
4	12

X= -2

20. Find the value of x such that $f(x) = -4.5$. $x = 3$

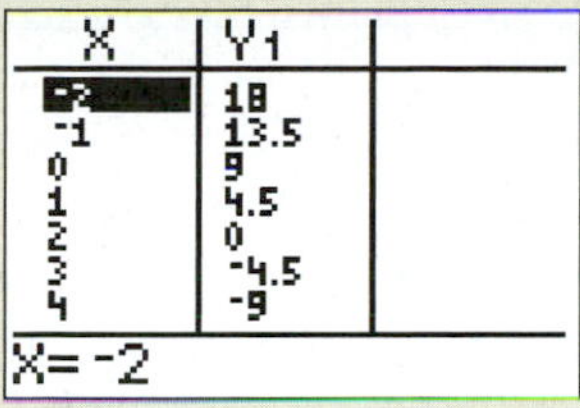

X	Y1
-2	18
-1	13.5
0	9
1	4.5
2	0
3	-4.5
4	-9

X= -2

SECTION 4.2

Graph each equation or function. (*See Objectives 1–3.*)

21. $7x - 2y = 35$

22. $f(x) = -2x + 6$

23. $f(x) = -\frac{1}{2}x$

24. $f(x) = 6x - 12$

25. $y = -\frac{5}{4}x - 2$

26. $0.09x + 0.01y = 2.7$

27. $5x - y = 0$

28. $0.5x - 0.8y = 0$

29. $x = -8$

30. $2x - 7 = 0$

31. $2y + 9 = 0$

32. $f(x) = 6.5$

Solve each problem. (*See Objective 4.*)

33. The cost of higher education at 2-year public institutions is modeled by the equation $y = 335.6x + 4889.5$, where x is the number of years after 2000. (Source: www.infoplease.com)

a. What is the y-intercept of the equation and what does it mean in the context of the problem?

b. Find the cost of higher education at 2-year public institution in 2005.

c. Find the expected cost of higher education at 2-year public institution in 2015.

34. The expenditure per pupil in public elementary and secondary schools is modeled by the equation $y = 370.4x + 5969.8$ where x is the number of years after 1994. (Source: www.infoplease.com)

a. What is the y-intercept of the equation and what does it mean in the context of the problem?

b. Find the expenditure per pupil in public elementary and secondary schools in 2001.

c. Find the expected expenditure per pupil in public elementary and secondary schools in 2014.

SECTION 4.3

Find the slope and y-intercept of each line from its equation. Write the y-intercept as an ordered pair. (*See Objective 1.*)

35. $x + 9y = 12$

36. $x - 3y = -11$

Find the slope of a line that satisfies the given condition. (*See Objective 4.*)

37. perpendicular to $3x - y = -6$ $m = -\frac{1}{3}$

38. parallel to $7x - 14y = 21$ $m = \frac{1}{2}$

Graph each line using its slope and y-intercept. Label at least two points on the graph. (*See Objective 2.*)

39. $y = -0.25x + 2.25$

40. $4x - y = 16$

Use the slope formula to determine the slope of each line containing the given points. (*See Objective 3.*)

41.

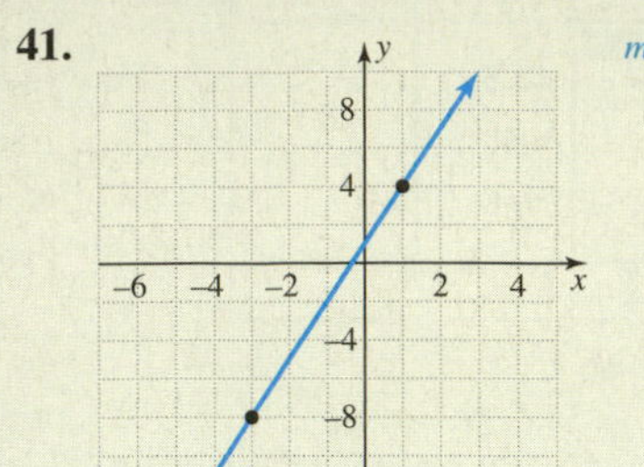

$m = 3$

42. $(6, -7)$ and $(-2, 5)$ $m = -\frac{3}{2}$

43.

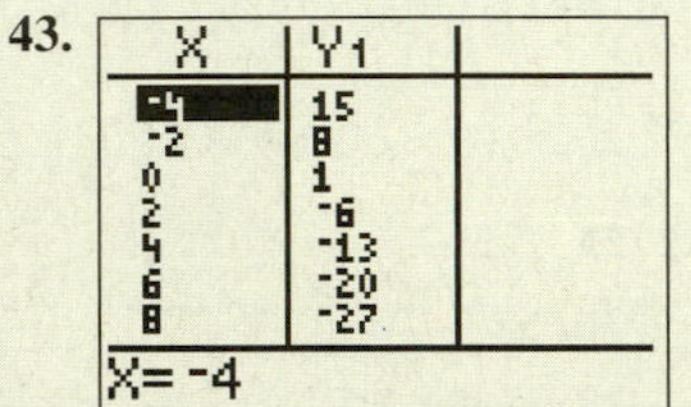

X	Y1
-4	15
-2	8
0	1
2	-6
4	-13
6	-20
8	-27

X= -4

$m = -\frac{7}{2}$

44.

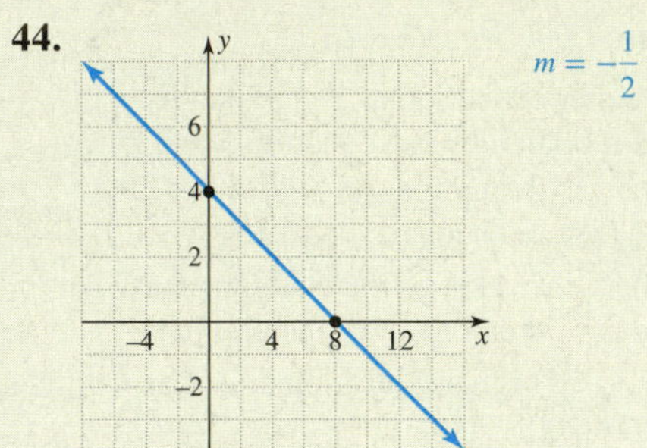

$m = -\frac{1}{2}$

Determine if the two lines are parallel, perpendicular, or neither. (*See Objective 4.*)

45. $3y = 2x + 3$ and $3x + 2y = 7$ perpendicular

46. $y = 0.5x - 4.5$ and $2x - y = -3.5$ neither

Interpret the meaning of the slope and *y*-intercept in the context of each problem. (*See Objective 5.*)

47. The number of registered nurses in 2004 was 2394 thousand. The number of registered nurses is predicted to be 3096 thousand in 2014. Find a linear function to represent the number of registered nurses, where x is the number of years after 2004. (Source: www.bls.gov)

48. The number of postsecondary teachers in 2004 was 1628 thousand. The number of postsecondary teachers is predicted to be 2153 thousand in 2014. Find a linear function to represent the number of postsecondary teachers, where x is the number of years after 2004. (Source: www.bls.gov)

49. A car's value is given by $f(x) = -2035x + 18{,}500$, where x is the age of the car in years.

50. A truck's value is given by $f(x) = -3960x + 36{,}000$, where x is the age of the truck in years.

Find the slope of each incline and explain what it means in terms of rise over run. (*See Objective 5.*)

51. The steepest vehicular incline in the world is the Johnstown Inclined Plane, which was built to provide transportation to a new hilltop community in 1891 and has a grade of 70.9%. (Source: http://www.inclinedplane.com/History.html)

52. The Duquesne Heights Incline has a beautiful view of Pittsburgh's downtown Golden Triangle and has a grade of 58.5%. (Source: http://incline.pghfree.net/openhours.htm)

SECTION 4.4

Write the equation of each line described. Express each answer in slope-intercept form. (*See Objectives 1–4.*)

53. $m = -\frac{3}{5}$; passes through $(-5, 4)$ $y = -\frac{3}{5}x + 1$

54. $m = 0$; passes through $(1, -6)$ $y = -6$

55. $m = -6$; passes through $(-5, 0)$ $y = -6x - 30$

56 $m = \frac{2}{3}$; passes through $(0, 8)$ $y = \frac{2}{3}x + 8$

57. $m =$ undefined; passes through $(4, -9)$ $x = 4$

58. $m =$ undefined; passes through $(-5, 3)$ $x = -5$

59. passes through $\left(\frac{1}{2}, \frac{2}{3}\right)$ and $\left(\frac{5}{4}, \frac{1}{6}\right)$ $y = -\frac{2}{3}x + 1$

60. passes through $(-0.25, -1.2)$ and $(1.75, 1.8)$

61. passes through $(0, -4.2)$ and $(1.4, 0)$ $y = 3x - 4.2$

62. passes through $(-1.6, 5.2)$ and $(-4.6, 8.2)$ $y = -x + 3.6$

63. passes through $(-3, 4)$, parallel to $y = -\frac{5}{6}x + 1$

64. passes through $(2, 0)$, perpendicular to $y - 7 = 0$ $x = 2$

65. passes through $(-6, -5)$, perpendicular to a line with slope $m = 4$ $y = -\frac{1}{4}x - \frac{13}{2}$

66. passes through $(-1, 2)$, parallel to a line with slope $m = -3$ $y = -3x - 1$

67. passes through $(-2, 12)$, perpendicular to $y = -5$ $x = -2$

68. passes through $(0, -4)$, parallel to $y = 3$ $y = -4$

69. passes through $(4.8, 0)$, perpendicular to $0.1x - 0.5y = 10$ $y = -5x + 24$

70. passes through $(-1.2, 0)$, parallel to $0.2x + 0.3y = 6$

Solve each problem. (*See Objective 5.*)

71. Myra has a new job as a registered nurse. Her starting salary is \$48,000. She will receive a raise of \$2,500 each year she works with the acute care department of the hospital.

a. Write a linear equation that represents Myra's salary, where x is the number of years she has worked with the department. $f(x) = 2500x + 48{,}000$

b. What is Myra's salary after 4 years of working for the hospital? \$58,000

c. How long does she have to work for the hospital to earn a salary of \$65,500? 7 yr

72. Loretta registers for her first semester in a local community college. Her tuition is \$199.50 per credit hour.

a. Write a linear equation that represents Loretta's total tuition, where x is the number of credit hours for which Loretta registers. $f(x) = 199.5x$

b. What is Loretta's tuition if she registers for 12 credit hours? \$2394

c. How many hours did Loretta enroll in if her tuition is \$2,992.50? 15 credit hours

73. David purchases a new car for \$18,500. The car's value decreases by \$1500 each year.

a. Write a linear equation that represents the value of the car, where x is the age of the car in years. $f(x) = -1500x + 18,500$

b. What is the car's value after 5 yr? \$11,000

c. What will be the age of the car when its value is \$6500? 8 yr

74. In 2004, the population in Georgia was approximately 8,829,383. In 2008, the population was approximately 9,685,744. (Source: U.S. Census Bureau)

a. Assuming this growth is linear, write an equation that approximates the population of Georgia, where x is the number of years after 2004. $f(x) = 214,090.25x + 8,829,383$

b. What was the estimated population of Georgia in 2010 to the nearest integer? 10,113,925

c. When will the population reach 12,250,000? 2020

75. The number of live births in the United States was 4059 (in thousands) in 2000 and 4116 in 2004.

a. Assuming linear growth, write an equation that approximates the number of live births in the United States where x is the number of years after 2000. $f(x) = 14.25x + 4059$

b. How many live births were there in the United States in 2005? (Source: http://www.infoplease.com/ipa/A0922289.html) 4130.25 thousand

76. The number of deaths caused by cancer in the United States was approximately 208.7 (per 100,000) in 1996 and 180.7 (per 100,000) in 2006.

a. Assuming linear decline, write a linear equation that approximates the number of deaths caused by cancer (per 100,000), where x is the number of years after 1996. $f(x) = -2.8x + 208.7$

b. How many deaths were caused by cancer (per 100,000) in 2013? (Source: http://www.infoplease.com/ipa/A0922289.html) 161.1 per 100,000

SECTION 4.5

Determine if the ordered pair is a solution of the given inequality. (*See Objective 1.*)

77. $3x + 7y > 0; (-1, -6)$ no

78. $x - 2y < 0; (5, -2)$ no

79. $4.8x + 0.8y < 2.8; (-3.2, 6.1)$ yes

80. $-12x + 5y \geq 4; \left(-\frac{1}{6}, -\frac{2}{5}\right)$ no

Graph the solution set of each linear inequality in two variables. (*See Objective 2.*)

81. $x - 5y \geq 15$

82. $4x + y \leq -8$

83. $2x - y \geq 0$

84. $x + 1 < 0$

85. Sidney works two jobs to put herself through college. She works as a manager at a fast-food restaurant for \$15.25 per hour and at a bank as a part-time teller for \$10.50 per hour. If Sidney wants to make at least \$400 per week, (*See Section 4.5, Objective 3.*)

a. write a linear inequality in two variables that represents this situation.

b. Graph the linear inequality.

c. Give three possible combinations of hours that Sidney could work at each job to earn her desired income.

CHAPTER 4 TEST / LINEAR FUNCTIONS AND LINEAR INEQUALITIES IN TWO VARIABLES

1. The equation $4x - 3y = 12$ written as a linear function is

a. $f(x) = \frac{4}{3}x - 4$ **(b.)** $f(x) = -\frac{4}{3}x - 4$

c. $f(x) = \frac{4}{3}x + 12$ **d.** $f(x) = -\frac{4}{3}x + 12$

2. The zero of $f(x) = 9x + 27$ is

a. -27 **b.** 27 **c.** 3 **(d.)** -3

3. The equation $x = 5$ is a(n) vertical line and its slope is undefined.

4. The equation $y = -1$ is a(n) horizontal line and its slope is zero.

5. The statement that is true about $4x - y = 6$ is

a. The x-intercept is $(-6, 0)$.

b. The x-intercept is $(6,0)$.

(c.) The y-intercept is $(0, -6)$.

d. The y-intercept is $(0, 6)$.

6. To graph the function $f(x) = -\frac{1}{2}x + 4$, we can

a. Plot (0, 4) and move down 1 unit and left 2 units.

(b.) Plot (0, 4) and move down 1 unit and right 2 units.

c. Plot (4, 0) and move down 1 unit and left 2 units.

d. Plot (4, 0) and move down 1 unit and right 2 units.

7. The slope of a line perpendicular to $y = -3x + 4$ is

a. $m = -3$ **b.** $m = 3$

(c.) $m = \frac{1}{3}$ **d.** $m = -\frac{1}{3}$

8. The statement that is true about the graph of the inequality $x + 4y < -8$ is

a. The boundary line is solid and the lower half-plane is shaded.

(b.) The boundary line is dashed and the lower half-plane is shaded.

c. The boundary line is solid and the upper half-plane is shaded.

d. The boundary line is dashed and the upper half-plane is shaded.

9. Find the x- and y-intercepts and graph each line.

a. $2x - y = 6$ b. $y = \frac{1}{5}x + 1$

c. $x + 3y = 0$

10. Suppose an airplane descends at a rate of 500 ft/min from an altitude of 8000 ft above the ground.

a. Write a linear equation that represents the plane's altitude, y, after x min. $y = -500x + 8000$

b. Find the x-intercept and interpret its meaning.

c. Find the y-intercept and interpret its meaning.

11. State the slope of each line.

a. $y = \frac{4}{5}x + 3$ b. $3x + 5y = 10$

c. Through $(4, -1)$ and $(6, 5)$ $m = 3$

d. Parallel to $7x - y = 4$ $m = 7$

e. $m = 1$

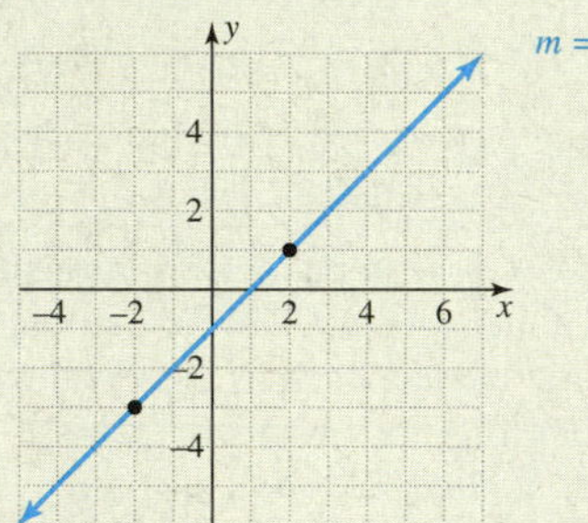

12. Graph the line $y = -\frac{3}{2}x + 4$ using the slope and y-intercept. Label the y-intercept and at least one other point.

13. Determine if the lines $5x + 2y = 10$ and $4x - 10y = -4$ are parallel, perpendicular, or neither. perpendicular

14. The equation $y = 138.41x + 2961.8$ models the average undergraduate cost of attending a public 2-yr college x yr after 1986. (Source: http://nces.ed.gov/fastfacts)

a. What is the slope and y-intercept?

b. What do the slope and y-intercept mean in the context of this problem?

c. Use this model to find the cost of attending a 2-yr college in 2016. \$7114.10

15. Write the equation of the line that satisfies each condition.

a. $m = \frac{3}{4}$ and $(0, -6)$ b. $m = -2$ and $(-7, 4)$ $y = -2x - 10$

c. $(4, -1)$ and $(6, 5)$ $y = 3x - 13$

d. through $(2, -8)$ and parallel to $y = 4x - 1$ $y = 4x - 16$

e. through $(3, 0)$ and perpendicular to $y = -3x + 9$

f. through $(5, -1)$ with undefined slope $x = 5$

g. through $(-9, -2)$ with $m = 0$ $y = -2$

16. In January 2009, one share of Google stock was worth \$321.32. In January 2011, one share of Google stock was worth \$616.44. If x is the number of years after 2009 and y is the price of the stock, write a linear equation that models the price of a share of Google's stock x years after 2009. Use the model to predict the price of a share of Google stock in 2015. (Source: http://finance.yahoo.com/q?s=GOOG) $y = 147.56x + 321.32$; about \$1206.68

17. Write the linear equation $4x - 7y = 14$ in function notation. Then find $f(-14)$. Write the corresponding ordered pair. $f(x) = \frac{4}{7}x - 2$; $f(-14) = -10$; $(-14, -10)$

18. Graph the solution set of each linear inequality.

a. $x - 4y \geq -8$ b. $y > -\frac{3}{5}x$

19. Tickets to a high school football game cost \$6 for students and \$12 for general admission. If the school wants to raise at least \$12,000,

a. write a linear inequality that represents this situation,

b. graph the inequality, and Answers vary.

c. state three combinations of student tickets and general admission tickets that must be sold to accomplish this. Answers vary.

CUMULATIVE REVIEW EXERCISES / CHAPTERS 1–4

Perform each operation. (*Section 1.2, Objectives 1 and 2*)

1. $35.7 + (-21.6)$ 14.1 **2.** $42.3 + (-63.5)$ −21.2

3. $(-15.2)(-7)$ 106.4 **4.** $(-8)(4.6)$ −36.8

Use the order of operations to simplify each expression. (*Section 1.2, Objectives 3 and 4*)

5. $-3(-4)^3 + 5(-1)^2 - 9$ 188

6. $34 - [(17 - 11) - (1 - 9)] + 36 \div 9 \cdot 4$ 36

Evaluate each expression for the given value(s). (*Section 1.2, Objective 5*)

7. $x^2 - x + 4$ for $x = 3$ 10

8. $-x^2 + 2x - 5$ for $x = -2$ −13

9. $b^2 - 4ac$ for $a = 7, b = -2, c = -3$ 88

10. $12x - 9y$ for $x = 2, y = 3$ −3

Simplify each expression. (*Section 1.3, Objective 3*)

11. $x(7 - 9x) + 6x^2 - 3(2x - 4)$ $-3x^2 + x + 12$

12. $-\frac{3}{5}x - 6 + \frac{2}{3}x + 10$ **13.** $\frac{1}{2}x + 1 - \frac{3}{4}x + \frac{1}{3}$

14. $7(x - 2) - 9(x - 4)$ $-2x + 22$

Solve each problem. (*Section 1.3, Objective 5*)

15. Robert has collected 85 quarters and nickels. If x represents the number of nickels he collected, write an expression for the number of quarters he collected. $85 - x$

16. Referring to Exercise 15, if Robert's coins total \$8.80, write an algebraic expression that represents the total value of his coins. $0.05x + 0.25(85 - x) = 8.8$

Solve each linear equation. (*Section 2.1, Objectives 2–6*)

17. $23.2 - a = 15.9$ $\{7.3\}$ **18.** $\dfrac{2x - 21}{3} = 5$ $\{18\}$

19. $3(18 + 4x) = 6(2x + 9)$ $\mathbb{R}$

20. $5(x - 1) - 8x = x - 4x - 1$ $\varnothing$

Translate each statement into a linear equation and solve the problem. (*Section 2.2, Objective 1*)

21. The product of a number and two-thirds is 10. Find the number. 15

22. The quotient of a number and five is 20. Find the number. 100

23. Four more than three times a number yields 19. Find the number. 5

24. Twice the sum of two consecutive odd numbers is 72. Find the numbers. $\{17, 19\}$

Find the measure of each angle described or pictured. (*Section 2.3, Objective 1*)

25. Find the measure of an angle whose complement is 10° more than three times the measure of the angle. 20°

26. Find the measure of an angle whose supplement is 42° less than twice the measure of the angle. 74°

27. 142.5°, 37.5°

28. 50°, 50°

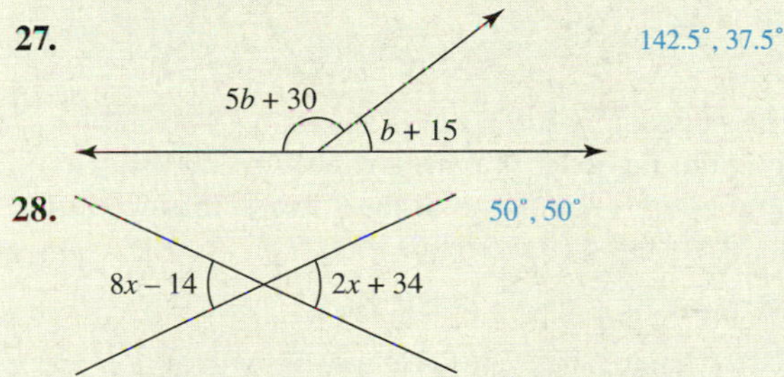

Solve each formula for the specified variable. (*Section 2.3, Objective 4*)

29. $a = 3b + 8cd$ for d **30.** $x = \dfrac{3}{5}y - 2zw$ for z

Solve each inequality. Graph the solution set and write each answer in interval notation and set-builder notation. (*Section 2.4, Objectives 1 and 2*)

31. $3(x + 5) < 5x - 7$ **32.** $4(y - 2) - 12 \le 2y + 26$

Find the intersection and union of the following sets. Write each solution in interval notation. (*Section 2.5, Objectives 1 and 2*)

33. $A = (-11, \infty), B = (-\infty, 4)$ $A \cap B = (-11, 4)$; $A \cup B = (-\infty, \infty)$

34. $A = [2, 12], B = (4, 18)$ $A \cap B = (4, 12]$; $A \cup B = [2, 18)$

Solve each compound inequality. Write the solution in interval notation. (*Section 2.5, Objectives 3 and 4*)

35. $x > -10$ and $x > -2$ $(-2, \infty)$ **36.** $x \le 1$ and $x \le 12$ $(-\infty, 1]$

37. $2x - 7 \ge 3$ or $-2x \ge 16$ $(-\infty, -8] \cup [5, \infty)$

38. $4x + 13 > 17$ or $5x - 14 \le -12$ $\left(-\infty, \dfrac{2}{5}\right] \cup (1, \infty)$

Solve each equation. (*Section 2.6, Objectives 1 and 2*)

39. $|15 - x| = 18$ $\{-3, 33\}$ **40.** $|27 - 4x| = |3x + 11|$

41. $|2x + 18| = 0$ $\{-9\}$ **42.** $|34 - 2x| = 0$ $\{17\}$

Graph each equation. (*Section 3.1, Objective 3*)

43. $y = -(x - 2)^2 + 9$ **44.** $y = (x + 2)^2 - 9$

Find the midpoint of the line segment formed by the ordered pairs. (*Section 3.1, Objective 5*)

45. $(4.3, -16.2)$ and $(-18.5, -21.4)$ $(-7.1, -18.8)$

46. $(-16, -12)$ and $(-20, 8)$ $(-18, -2)$

47. Find the domain and range of the relation represented by the graph. (*Section 3.2, Objective 2*)
Domain $= [-2, 5]$, Range $= [-12, 8]$

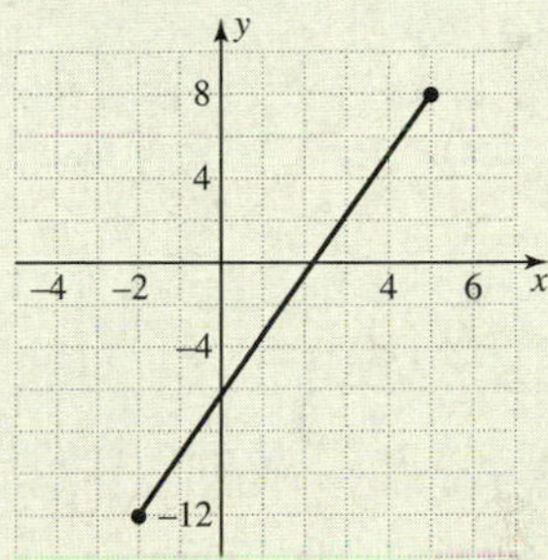

48. Represent the relation shown in the table as a graph. (*Section 3.2, Objective 1*)

x	-2	10	18	21
y	-8	-4	6	12

49. Find the output value(s) corresponding to the input value of 4 for the relation. (*Section 3.2, Objective 3*)
$\{(-4, -1), (-3, 6), (4, 8), (4, -2)\}$ $y = 8$ and $y = -2$

50. Find the input value(s) corresponding to the output value of 0 for the relation whose graph is shown. (*Section 3.2, Objective 3*) $x = -3, 0, 2$

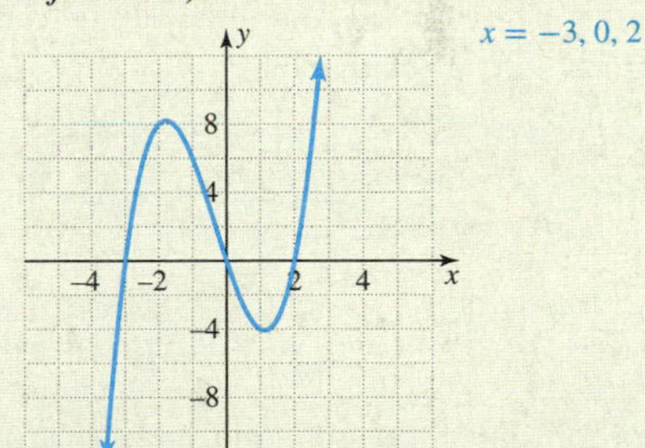

Find the requested information. (*Section 3.3, Objective 3*)

51. Find $f(-4)$ if $f(x) = 14x + 11$. -45

52. Use the graph of $h(x)$ to find all x for which $h(x) = 1$. $x = 4$

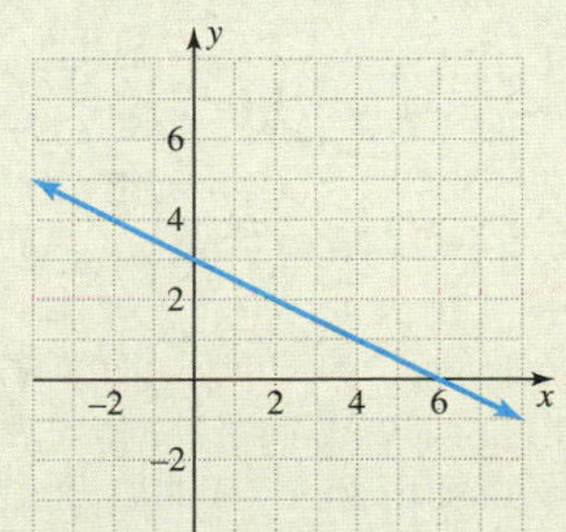

53. Find $f(-2)$ if $f(x) = -x^2 + 5x$. -14

54. Find $f(-3)$ if the function $f(x)$ is given by Y_1. 10.5

X	Y_1
-3	10.5
-2	5
-1	1.5
0	0
1	.5
2	3
3	7.5

55. The life expectancy for a person of any sex and race can be approximated by the equation $f(x) = 0.000158x^3 - 0.0104x^2 + 0.372x + 70.813$, where x is the number of years after 1970. Find $f(45)$ and interpret its meaning. (Source: National Center for Health Statistics) (*Section 3.3, Objective 4*)

56. Find the domain and range of the function. (*Section 3.4, Objective 1*)
$\{(-16, 7), (-1, -20), (10, 0), (0, 12)\}$
Domain = $\{-16, -1, 0, 10\}$, Range = $\{-20, 0, 7, 12\}$

Find the domain of each function. (*Section 3.4, Objective 3*)

57. $f(x) = x^2 - 9$ $(-\infty, \infty)$

58. $f(x) = \sqrt{8 - 2x}$ $(-\infty, 4]$

Determine if each function is linear. If it is linear, identify the values of *m* and *b*. (*Section 4.1, Objective 1*)

59. $f(x) = 5x - 12$ linear, $m = 5$, $b = -12$

60. $f(x) = -\dfrac{15}{x} + 1$ not linear

Rewrite each equation using function notation, if possible. Identify the values of *m* and *b*. (*Section 4.1, Objective 2*)

61. $3x + y = -18$ $f(x) = -3x - 18$, $m = -3$, $b = -18$

62. $7x + 2y = 0$ $f(x) = -\dfrac{7}{2}x$, $m = -\dfrac{7}{2}$, $b = 0$

63. $5x - 15 = 0$ can't be written in function notation

64. $2y + 6 = 0$ $f(x) = -3$, $m = 0$, $b = -3$

Evaluate each linear function at the given value. (*Section 4.1, Objective 3*)

65. $f(x) = 5x + 6$; $x = -4$ -14

66. $f(x) = -\dfrac{5}{6}x - \dfrac{1}{3}$; $x = -2$ $\dfrac{4}{3}$

67. $f(x) = -14$; $x = 3$ -14

68. $f(x) = -\dfrac{3}{10}x + \dfrac{1}{10}$, $f(x) = 0$ $\dfrac{1}{10}$

Find solutions of each equation of the form $f(x) = c$ for the given function. (*Section 4.1, Objective 4*)

69. $f(x) = -2x + 11$; $f(x) = 15$ $x = -2$

70. $f(x) = \dfrac{2}{5}x + 4$; $f(x) = 4$ $x = 0$

71. $f(x) = -\dfrac{5}{2}x + \dfrac{15}{2}$, $f(x) = 0$ 3

72. $f(x) = -\dfrac{1}{3}x + 2$; $f(x) = 6$ $x = -12$

Solve each problem. (*Section 4.1, Objective 5*)

73. The owner of a limousine rental company purchases a stretch limousine for \$110,000. For tax purposes, the owner uses straight-line depreciation for reporting the value of the limo. The depreciated value of the limo is given by $y = -22{,}000x + 110{,}000$, where x is the age of the limo in years.

a. What is the x-intercept and what does it mean in the context of the problem? (5, 0); The depreciated value of the limo is \$0 in 5 yr.

b. What is the y-intercept and what does it mean in the context of the problem? (0, 110,000); The initial value of the limo is \$110,000.

c. When will the depreciated value of the limo be \$66,000? The depreciated value of the limo will be \$66,000 in 2 yr.

d. What will be the depreciated value of the limo when it is 3 yr old? The depreciated value of the limo will be \$44,000 in 3 yr.

e. Use the information obtained in parts (a) and (b) to graph the equation.

74. The 2009 U.S. federal income tax for a person filing head of household with an income over \$45,500, but not over \$117,450, is \$6227.25 plus 25% of the amount over \$45,500.

a. Write a linear function $f(x)$ that represents the tax owed, where x is the amount of income over \$45,500

b. Find $f(76{,}400)$ and explain what it means.

c. If a single person pays \$12,977.25 in federal income tax, what income does he earn? \$72,500

Graph each linear function and state its domain and range. (*Section 4.2, Objective 1*)

75. $f(x) = \dfrac{3}{2}x - 1$

76. $f(x) = -2$

77. $f(x) = 2x$

78. $f(x) = -3x + 12$

Find the slope and *y*-intercept of each line. Write the *y*-intercept as an ordered pair. Graph the line using its slope and *y*-intercept. Label at least two points on your graph. (*Section 4.3, Objectives 1 and 2*)

79. $f(x) = -9x + 6$

80. $f(x) = -2x - 5$

81. $3x - 2y = 12$

82. $x - 5y = 0$

Use the slope formula to determine the slope of the line containing the given points. (*Section 4.3, Objective 3*)

83. $(-4, 6)$ and $(2, 4)$ $-\dfrac{1}{3}$

84. $(-7, 5)$ and $(1, 3)$ $-\dfrac{1}{4}$

85. $(-2.5, 0)$ and $(-2.5, 1)$ undefined

86. $(-4, -6.3)$ and $(2, -6.3)$ 0

Determine if the two lines are parallel, perpendicular, or neither. (*Section 4.3, Objective 4*)

87. $y = -\dfrac{3}{4}x + 2$ and $y = -\dfrac{4}{3}x$ neither

88. $y = -\dfrac{5}{2}x + 2$ and $y = \dfrac{2}{5}x$ perpendicular

89. $4x + 7y = -14$ and $8x + 14y = 8$ parallel

90. $4x + 7y = -14$ and $x + 4y = 8$ neither

Write the equation of each line described. Express each answer in slope-intercept form. (*Section 4.4, Objectives 2–4*)

91. $m = 0$; passes through $(-1, 7)$ $y = 7$

92. $m = \dfrac{1}{9}$; passes through $(-9, 2)$ $y = \dfrac{1}{9}x + 3$

93. passes through $(-2, 7)$ and $(-2, 1)$ $x = -2$

94. passes through $(12, -3)$, perpendicular to $y = -1$ $x = 12$

95. passes through $(2, -4)$, parallel to $y = \frac{4}{5}x + 6$ $y = \frac{4}{5}x - \frac{28}{5}$

96. passes through $(7, -9)$, perpendicular to $y = -\frac{1}{6}x + 5$ $y = 6x - 51$

Solve each problem. (*See Section 4.4, Objective 5*)

97. The median sale price of existing single-family homes sold in Florida was approximately $142,500 in 2009 and $136,500 in 2010.

a. Assuming linear decline, write an equation that approximates the median sale price of existing single-family homes sold in Florida, where x is the number of years after 2009. $y = -6000x + 142{,}500$

b. If this trend continues, what will be the median sale price of existing single-family homes in 2012? (Source: media.living.net/statistics/2010) $124,500

98. There were approximately 9.3 million Social Security beneficiaries receiving disability insurance in 2008. There were approximately 10.2 million beneficiaries receiving disability insurance in 2010.

a. Write a linear equation that approximates the number of Social Security beneficiaries (in millions) receiving disability insurance where x is the number of years after 2008. $y = 0.45x + 9.3$

b. If this growth continues, how many beneficiaries receiving disability insurance will there be in 2014? (Source: www.ssa.gov) 12 million

Graph the solution set of each linear inequality in two variables. (*Section 4.5, Objective 2*)

99. $x - 4y > 12$

100. $2x - 6y < 6$

101. $3x - 12y \geq 1.5$

102. $4x + 5y \leq 100$

CHAPTER 5

Systems of Linear Equations and Inequalities

Study Skills

Studying math is very much like studying a foreign language. The way to learn a foreign language is to learn the alphabet, learn words, learn the rules of grammar, and so on. Learning math takes a similar approach. You must learn the notation and symbols, definitions, mathematical rules and processes, and so on. If you do not have a firm foundation in the basics of math, then it will be very difficult to build upon this foundation.

To learn mathematics, you must employ certain study skills. Some of these include

- Take good notes.
- Review your notes.
- Study immediately after class.
- Seek help on topics you do not understand before the next class meeting.
- Don't compare yourself with classmates since everyone comes from a different mathematical background.

Question For Thought: What study techniques have been successful for you in other courses? Can they work for you in this class? What is your biggest obstacle in studying?

Chapter Outline

Section 5.1 Solving Systems of Linear Equations in Two Variables Graphically 290

Section 5.2 Solving Systems of Linear Equations in Two Variables Algebraically 305

Section 5.3 Applications of Linear Systems in Two Variables 319

Section 5.4 Solving Linear Systems in Three Variables and Their Applications 335

Section 5.5 Solving Systems of Linear Inequalities and Their Applications 347

Coming Up...

In Section 5.1, we will learn how to use a graph to approximate when annual expenditures for residential phone services equal annual expenditures for cell phone services.

“I will study and prepare and perhaps my chance will come.”
—Abraham Lincoln (U.S. President)

SECTION 5.1 Solving Systems of Linear Equations in Two Variables Graphically

In **this chapter,** we will apply our skills of solving equations and graphing equations in two variables to solve systems of linear equations and inequalities. Linear systems with both two and three variables will be solved. Applications of linear systems will also be discussed.

OBJECTIVES

As a result of completing this section, you will be able to

1. Determine if an ordered pair is a solution of a system of linear equations.
2. Solve systems of linear equations graphically.
3. Solve special cases of systems of linear equations graphically.
4. Determine how the lines relate, the number of solutions, and the type of system without graphing.
5. Solve applications of systems of linear equations.
6. Troubleshoot common errors.

The following graph shows consumer spending for residential telephone services and cell phone services for the years 2001–2008. In what year does consumer spending for residential telephone services equal consumer spending for cell phone services? (Source: http://www.census.gov/compendia/statab/2011/tables/11s1147.pdf)

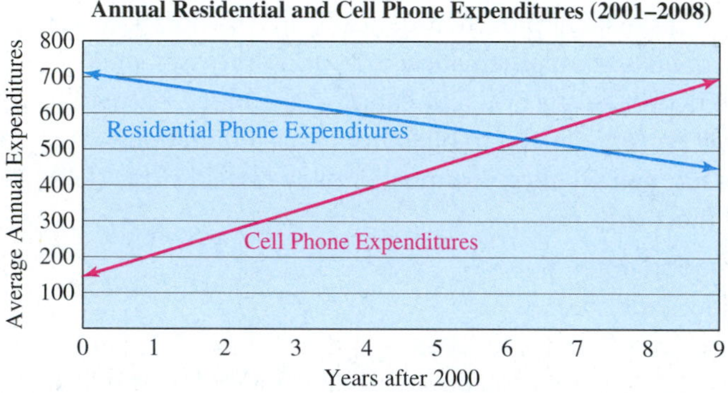

To answer this question, we must know how to solve a system of equations graphically. We will use our skills of graphing linear equations in two variables to accomplish this.

Solutions of Systems of Linear Equations

Objective 1 Determine if an ordered pair is a solution of a system of linear equations.

When an ordered pair satisfies more than one equation, we are solving a *system of equations*. We will focus on systems involving linear equations.

> **Definition:** A **system of linear equations** is a set of two or more linear equations that must be solved simultaneously.

An example of a system of linear equations in two variables is

$$\begin{cases} x + 6y = -11 \\ 2x - y = 4 \end{cases}$$

Equations in a system are often grouped together with a brace to indicate that they belong together.

> **Definition:** A **solution of a system of linear equations in two variables** is an ordered pair that satisfies both equations in the system.

> **Procedure: Determining if an Ordered Pair Is a Solution of a Linear System**
>
> **Step 1:** Substitute the given ordered pair into each equation in the system.
> **Step 2:** Simplify the resulting equations.
> **Step 3:** If the ordered pair makes both equations true, it is a solution of the system. If the ordered pair makes one or both equations false, then it is not a solution of the system.

Objective 1 Examples **Determine if the given ordered pair is a solution of the system.**

1a. $\left(-4, \frac{1}{2}\right), \begin{cases} x - 2y = -5 \\ y = \frac{3}{4}x + \frac{7}{2} \end{cases}$ **1b.** $(5, -3), \begin{cases} y = -3x + 5 \\ y = \frac{1}{5}x - 4 \end{cases}$

Solutions **1a.** Replace x with -4 and y with $\frac{1}{2}$ in each of the equations.

$x - 2y = -5$	$y = \frac{3}{4}x + \frac{7}{2}$
$-4 - 2\left(\frac{1}{2}\right) = -5$	$\frac{1}{2} = \frac{3}{4}(-4) + \frac{7}{2}$
$-4 - 1 = -5$	$\frac{1}{2} = -3 + \frac{7}{2}$
$-5 = -5$	$\frac{1}{2} = -\frac{6}{2} + \frac{7}{2}$
True	$\frac{1}{2} = \frac{1}{2}$
	True

Since $\left(-4, \frac{1}{2}\right)$ makes each equation true, it is a solution of the system.

1b. Replace x with 5 and y with -3 in each of the equations.

$y = -3x + 5$	$y = \frac{1}{5}x - 4$
$-3 = -3(5) + 5$	$-3 = \frac{1}{5}(5) - 4$
$-3 = -15 + 5$	$-3 = 1 - 4$
$-3 = -10$	$-3 = -3$
False	True

Since $(5, -3)$ makes one of the equations false, it is *not* a solution of the system.

Student Check 1 Determine if the given ordered pair is a solution of the system.

a. $(4, 7), \begin{cases} y = -2x + 15 \\ x - 4y = -18 \end{cases}$ **b.** $\left(\frac{3}{4}, -2\right), \begin{cases} y = -\frac{4}{3}x - 1 \\ y = 12x - 11 \end{cases}$

Solving Systems Graphically

Objective 2 ▶

Solve systems of linear equations graphically.

As stated in Objective 1, a solution of a system of linear equations in two variables is an ordered pair that satisfies both equations in the system. The graph of the system containing the equations $x - 3y = 6$ and $x + y = 2$ is shown next. Each of these linear equations has infinitely many solutions. When we graph the lines on the same coordinate system, we find that they share a common point. This common point is the solution of the system.

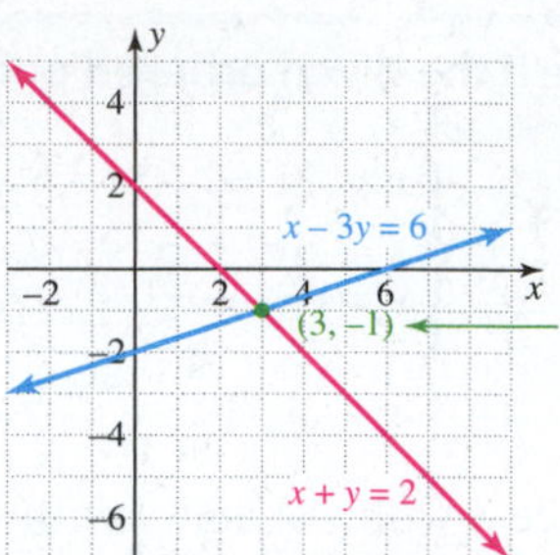

The point (3, −1) is a solution of each equation since it is a point on each line. Therefore, the point (3, −1) is the solution of the system containing these two linear equations.

Graphically, a solution of a system of two linear equations is an ordered pair, (x, y), where the two lines intersect. This ordered pair is called the **point of intersection**.

Procedure: Solving a System of Linear Equations Graphically

Step 1: Graph each equation on the same coordinate system. Recall that a linear equation can be graphed by plotting points, plotting the x- and y-intercepts, or by using the slope and y-intercept.

Step 2: Identify the ordered pair where the two lines intersect. This is the solution of the system.

Step 3: Check the ordered pair in each equation to verify the solution.

Note: *Use graph paper and a straight edge for precision.*

Objective 2 Examples

Solve each system of linear equations graphically.

2a. $\begin{cases} x + y = 5 \\ x - 3y = -3 \end{cases}$ **2b.** $\begin{cases} x - 2y = 2 \\ y = -3 \end{cases}$

Solutions

2a. Graph each equation by finding the x- and y-intercepts.

$x + y = 5$		
x	y	(x, y)
0	$0 + y = 5$ $y = 5$	(0, 5)
$x + 0 = 5$ $x = 5$	0	(5, 0)

$x - 3y = -3$		
x	y	(x, y)
0	$0 - 3y = -3$ $-3y = -3$ $y = 1$	(0, 1)
$x - 3(0) = -3$ $x = -3$	0	(−3, 0)

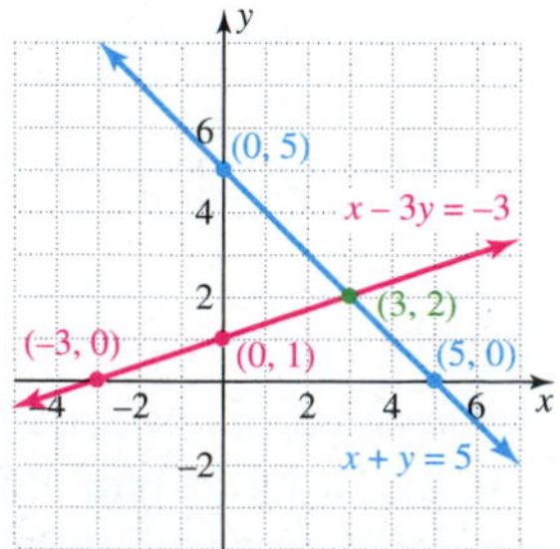

The point of intersection of the lines is (3, 2). So, the solution set of the system is $\{(3, 2)\}$. We can check by substituting the point into each equation.

Check:

$x + y = 5$	$x - 3y = -3$
$3 + 2 = 5$	$3 - 3(2) = -3$
$5 = 5$	$3 - 6 = -3$
True	$-3 = -3$
	True

2b. Graph each equation using its slope and y-intercept.

$$x - 2y = 2$$

$$-2y = -x + 2$$

$$\frac{-2y}{-2} = \frac{-x}{-2} + \frac{2}{-2}$$

$$y = \frac{1}{2}x - 1$$

$m = \frac{1}{2}$, y-intercept $= (0, -1)$

Plot $(0, -1)$ and move up 1 unit and right 2 units.

$$y = -3$$

This is the equation of a horizontal line through $(0, -3)$.

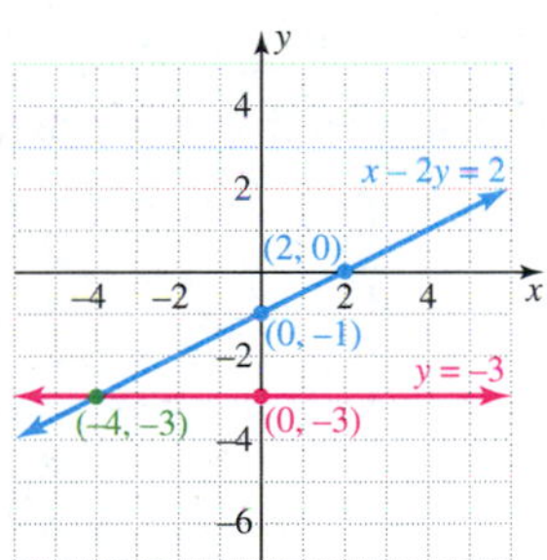

The point of intersection of the lines is $(-4, -3)$. So, the solution set of the system is $\{(-4, -3)\}$. We can check by substituting the point into each equation.

Student Check 2 Solve each system of linear equations graphically.

a. $\begin{cases} 3x - y = 6 \\ 5x + 2y = 10 \end{cases}$ **b.** $\begin{cases} x + y = -3 \\ x = -1 \end{cases}$

Note: *Examples 2a and 2b each have a point of intersection with integer coordinates. If the point of intersection involves a coordinate that is a fraction, graphing is not the best method to use as we cannot determine this solution with accuracy. Section 5.2 demonstrates algebraic methods that can be used to solve systems with precision.*

Solving Special Cases of Systems Graphically

Objective 3 ▶

Solve special cases of systems of linear equations graphically.

In Examples 2a and 2b, the graphs of the lines in the system of linear equations intersected in one point, which yielded one solution of the system of linear equations. There are two additional possibilities for the graphs of the lines in a system of linear equations. The lines in the system can be parallel to one another or they can be coinciding lines (the same line) as shown.

Parallel Lines	Same Lines
The two lines do not share any common points. So, there is *no solution* of the system. The solution of the system is the empty set, ∅.	Every point on the line is a point of intersection. So, there are *infinitely many solutions* of the system. The solution of the system is written with set-builder notation to denote that every point on the line is a solution, $\{(x, y) \mid \text{write equation from system here}\}$.

Objective 3 Examples Solve each system of equations graphically.

3a. $\begin{cases} y = -\frac{4}{5}x + 2 \\ \frac{2}{5}x + \frac{1}{2}y = -2 \end{cases}$ **3b.** $\begin{cases} 3x - 6y = -12 \\ 2x - 4y = -8 \end{cases}$

Solutions

3a. Graph each line using the slope-intercept form.

$$y = -\frac{4}{5}x + 2$$

$$m = -\frac{4}{5}, \text{ } y\text{-intercept} = (0, 2)$$

Plot (0, 2) and move down 4 units and right 5 units.

$$\frac{2}{5}x + \frac{1}{2}y = -2$$

$$10\left(\frac{2}{5}x + \frac{1}{2}y\right) = 10(-2)$$

$$4x + 5y = -20$$

$$5y = -4x - 20$$

$$\frac{5y}{5} = \frac{-4x}{5} - \frac{20}{5}$$

$$y = -\frac{4}{5}x - 4$$

$$m = -\frac{4}{5}, \text{ } y\text{-intercept} = (0, -4)$$

Plot (0, −4) and move down 4 units and right 5 units.

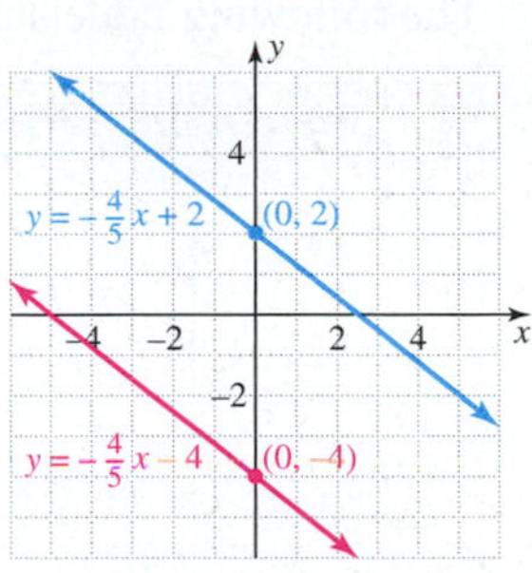

The slopes of the lines are the same but the y-intercepts are different. So, the lines are parallel. The system has no solution. We write the solution set as the empty set, ∅.

3b. Graph each line using the slope-intercept form.

$3x - 6y = -12$	$2x - 4y = -8$
$-6y = -3x - 12$	$-4y = -2x - 8$
$\frac{-6y}{-6} = \frac{-3x}{-6} - \frac{12}{-6}$	$\frac{-4y}{-4} = \frac{-2x}{-4} - \frac{8}{-4}$
$y = \frac{1}{2}x + 2$	$y = \frac{1}{2}x + 2$
$m = \frac{1}{2}$, y-intercept $= (0, 2)$	$m = \frac{1}{2}$, y-intercept $= (0, 2)$

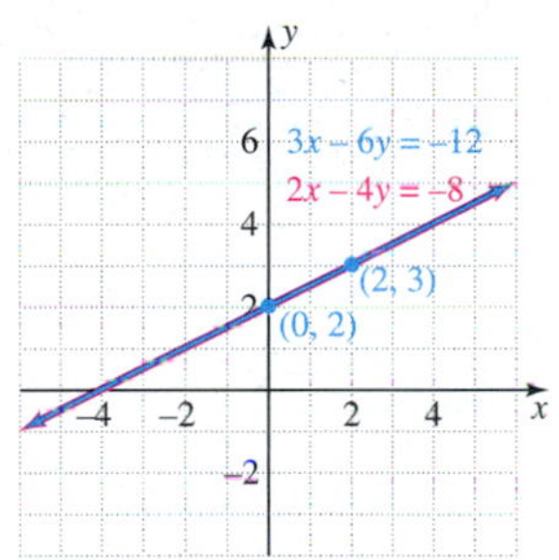

The slopes and y-intercepts of the lines are the same. So, the lines are coinciding. Every point on the line is a solution of the system. The system has infinitely many solutions. We write the solution set as

$$\{(x, y) | 3x - 6y = -12\}, \{(x, y) | 2x - 4y = -8\}, \text{ or } \left\{(x, y) \middle| y = \frac{1}{2}x + 2\right\}.$$

Student Check 3 Solve each system of equations graphically.

a. $\begin{cases} x + 5y = -5 \\ \frac{x}{15} + \frac{y}{3} = -\frac{1}{3} \end{cases}$

b. $\begin{cases} 6x - 8y = 24 \\ y = \frac{3}{4}x + 3 \end{cases}$

Understand the Types of Solutions of Linear Systems

Objective 4 ▶

Determine how the lines relate, the number of solutions, and the type of system without graphing.

As we have seen in Objectives 2 and 3, there are three possibilities for the solutions of systems of linear equations in two variables. The system can have one solution, no solution, or infinitely many solutions. The number of solutions is determined by the slopes and y-intercepts of the lines in the system.

- A system that contains parallel lines and has no solution is called an **inconsistent system**.
- A system that contains the same lines with infinitely many solutions is called a **consistent system with dependent equations**.
- A system that contains intersecting lines with one solution is called a **consistent system with independent equations**.

The following table summarizes the three cases.

Intersecting Lines	Parallel Lines	Same Lines
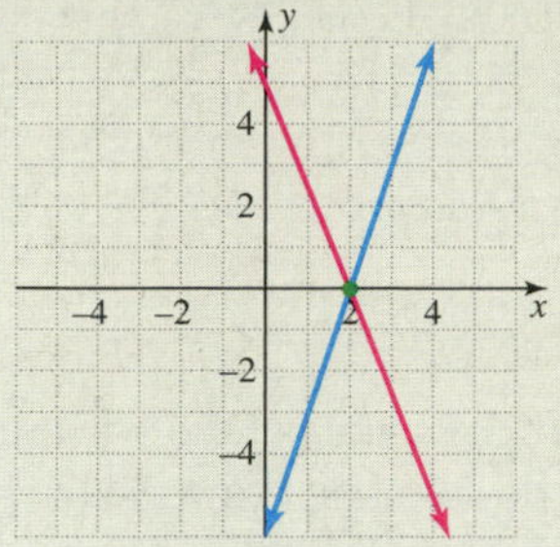	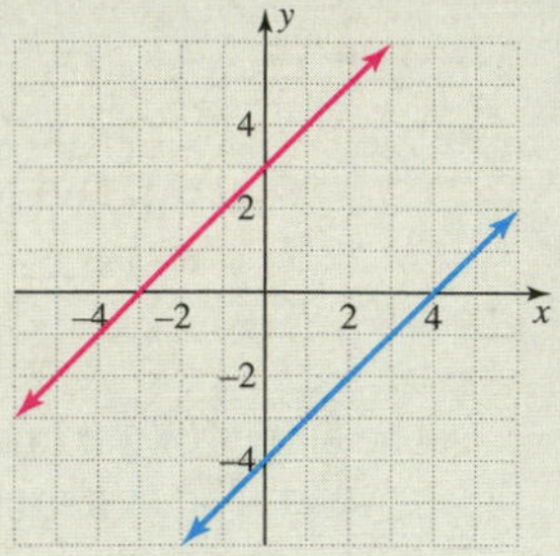	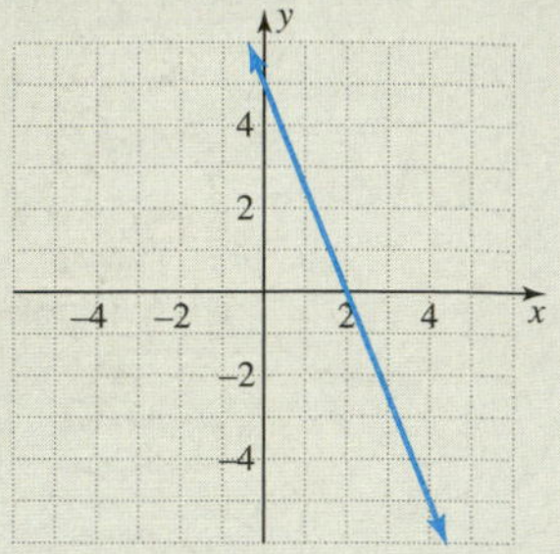
One solution	No solution	Infinitely many solutions
Different slopes; same or different y-intercepts	Same slopes; different y-intercepts	Same slopes; same y-intercepts
Consistent system with independent equations	Inconsistent system	Consistent system with dependent equations

Objective 4 Examples

Determine how the two lines relate, the number of solutions of the system, and the type of system without graphing.

4a. $\begin{cases} 7x + 3y = -6 \\ 2x - 5y = 15 \end{cases}$ **4b.** $\begin{cases} x = \frac{1}{4}y + 3 \\ y = 4x - 3 \end{cases}$ **4c.** $\begin{cases} \frac{3}{2}x + \frac{5}{3}y = 1 \\ \frac{9}{5}x + 2y = \frac{6}{5} \end{cases}$

Solutions

4a.

Slope-intercept form	$7x + 3y = -6$ $3y = -7x - 6$ $\frac{3y}{3} = -\frac{7x}{3} - \frac{6}{3}$ $y = -\frac{7}{3}x - 2$	$2x - 5y = 15$ $-5y = -2x + 15$ $\frac{-5y}{-5} = \frac{-2x}{-5} + \frac{15}{-5}$ $y = \frac{2}{5}x - 3$
Slopes	$m = -\frac{7}{3}$	$m = \frac{2}{5}$
y-intercepts	$(0, -2)$	$(0, -3)$
How do the lines relate?	Because the slopes are different, the two lines intersect.	
Number of solutions?	One solution	
Type of system?	The system is consistent with independent equations.	

4b.

Slope-intercept form	$x = \frac{1}{4}y + 3$ $4(x) = 4\left(\frac{1}{4}y + 3\right)$ $4x = y + 12$ $4x - 12 = y$	$y = 4x - 3$
Slopes	$m = 4$	$m = 4$
y-intercepts	$(0, -12)$	$(0, -3)$
How do the lines relate?	Because the slopes are the same and the y-intercepts are different, the two lines are parallel.	
Number of solutions?	No solution	
Type of system?	The system is inconsistent.	

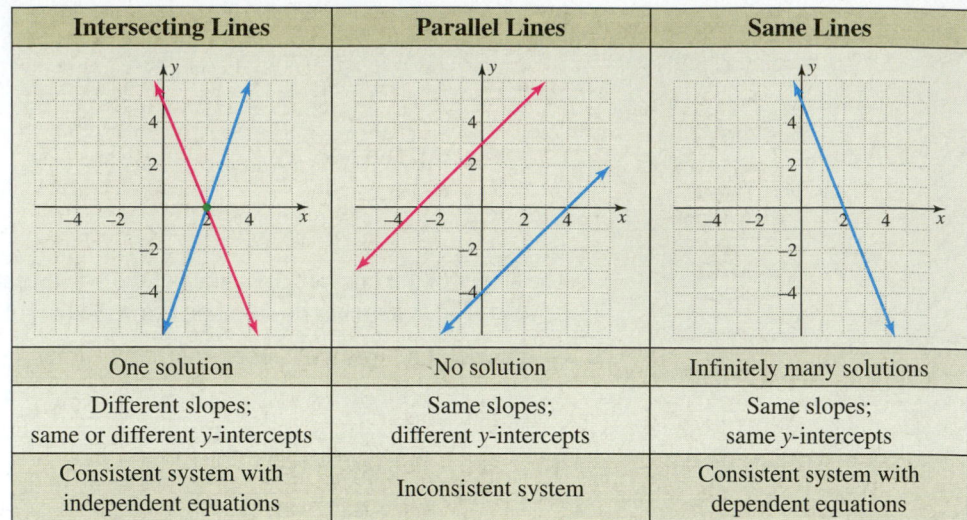

4c.

Slope-intercept form	$\frac{3}{2}x + \frac{5}{3}y = 1$ $6\left(\frac{3}{2}x + \frac{5}{3}y\right) = 6(1)$ $9x + 10y = 6$ $10y = -9x + 6$ $\frac{10y}{10} = -\frac{9}{10}x + \frac{6}{10}$ $y = -\frac{9}{10}x + \frac{3}{5}$	$\frac{9}{5}x + 2y = \frac{6}{5}$ $5\left(\frac{9}{5}x + 2y\right) = 5\left(\frac{6}{5}\right)$ $9x + 10y = 6$ $10y = -9x + 6$ $\frac{10y}{10} = -\frac{9}{10}x + \frac{6}{10}$ $y = -\frac{9}{10}x + \frac{3}{5}$
Slopes	$m = -\frac{9}{10}$	$m = -\frac{9}{10}$
y-intercepts	$\left(0, \frac{3}{5}\right)$	$\left(0, \frac{3}{5}\right)$
How do the lines relate?	Because the slopes and the y-intercepts are the same, the two lines are the same.	
Number of solutions?	Infinitely many solutions	
Type of system?	The system is consistent with dependent equations.	

Student Check 4 Determine how the lines relate, the number of solutions of the system, and the type of system without graphing.

a. $\begin{cases} 6x + 2y = 10 \\ 15x + 5y = 25 \end{cases}$ **b.** $\begin{cases} x = 2y + 3 \\ y = \frac{1}{2}x - 7 \end{cases}$ **c.** $\begin{cases} \frac{3}{4}x - 2y = 8 \\ x + 6y = 12 \end{cases}$

Applications of Systems

Objective 5 ▶ Solve applications of systems of linear equations.

The skills learned in this section apply to many real-life situations. We can use graphs and equations to solve systems related to real-world problems. One of the following examples involves concepts studied in an Economics class—supply and demand. The amount of merchandise that a supplier has available to sell is called *supply*. The amount of merchandise that consumers wish to buy is called *demand*. Supply can be affected by storage space, time for production, and other things. In general, as prices increase, the supply of the product increases. The price of a product affects demand in the opposite manner—the higher the price, the lower the demand for the product. The **equilibrium point** is the price at which supply equals demand.

Another example from Economics is related to the break-even point. The **break-even point** occurs where cost equals revenue. The *cost* is the manufacturer's cost to produce a certain quantity of their product. The *revenue* is the money made from selling a certain quantity of this product. The break-even point is where the manufacturer doesn't make any profit since the money paid out (cost) equals the money coming in (revenue).

Objective 5 Examples **Solve each problem by using a given graph or by solving an appropriate system by graphing.**

5a. A certain manufacturer produces a toy train. The supply for the train can be represented by $S(x) = 5x - 10$, where x is the price of the toy and y is the number of trains (in hundreds) supplied. The demand for this toy can be represented by $D(x) = -5x + 30$, where x is the price of the toy and y is the number of trains (in hundreds) demanded at this price. What is the equilibrium point for this product? Interpret the result.

Solution **5a.** To find the equilibrium point, we solve the system $\begin{cases} y = 5x - 10 \\ y = -5x + 30 \end{cases}$ by graphing.

Since x and y refer to the price of an item and the quantity of the item, we need to graph only in Quadrant I. We can draw the graph of $y = 5x - 10$ by plotting the points (2, 0), (3, 5), and (4, 10). We can draw the graph of $y = -5x + 30$ by plotting the points (6, 0), (5, 5), and (4, 10).

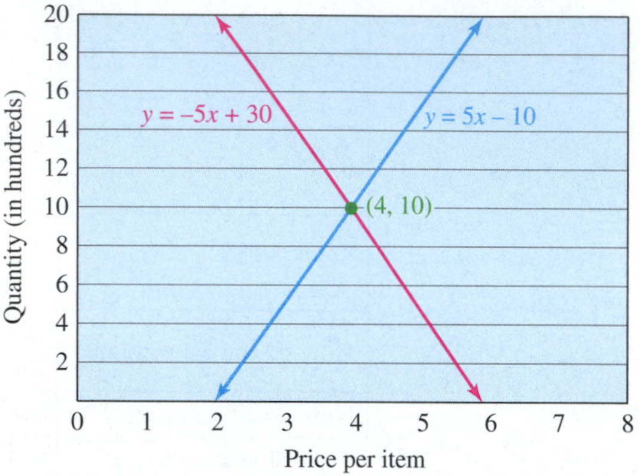

The solution of the system occurs at the point of intersection, which is (4, 10). When the price is \$4, the number of toy trains supplied and demanded is 10 hundred or 1000.

5b. A scatter plot for annual consumer expenditures for residential phone services and cell phone services for 2001–2008 is shown. The graph also includes the line that best models each data set. Use the graphs of the lines to approximate when annual expenditures for residential phone services equal annual expenditures for cell phone services. (Source: http://www.census.gov/compendia/statab/2011/tables/11s1147.pdf)

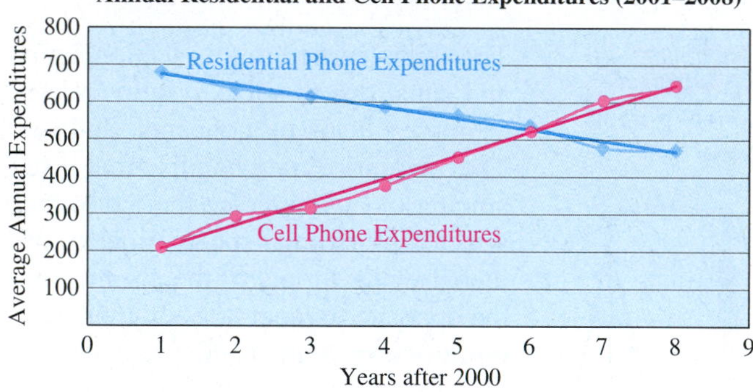

Solution **5b.** Annual expenditures for residential phone services is equal to the annual expenditures for cell phone services at the point of intersection of the two lines. We approximate the point of intersection to be (6, 525) as shown on the graph.

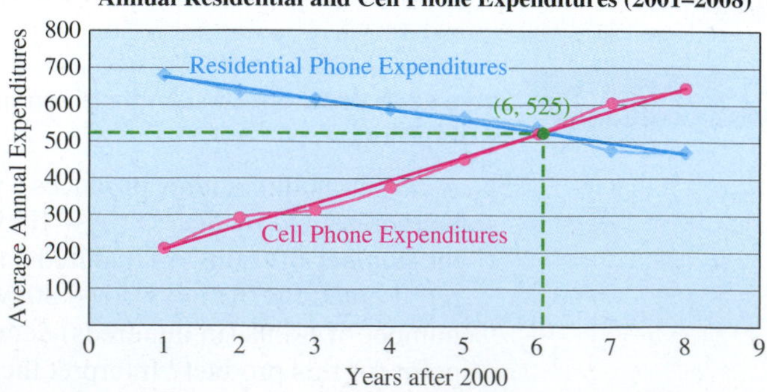

This ordered pair means that in 6 years after 2000, or in 2006, annual expenditures for residential phone services and cell phone services were both \$525.

Student Check 5 Solve each problem by using the given graph or by solving an appropriate system by graphing.

a. The graph models the approximate number of unique visitors, in millions, to MySpace and to Facebook, where x is the years after March 2008. Approximate the point of intersection of the graphs and interpret its meaning in the context of the problem. Compare the number of visitors to each site before and after the point of intersection. (Source: http://blog.nielsen.com/nielsenwire/global/facebook-and-twitter-post-large-year-over-year-gains-in-unique-users/)

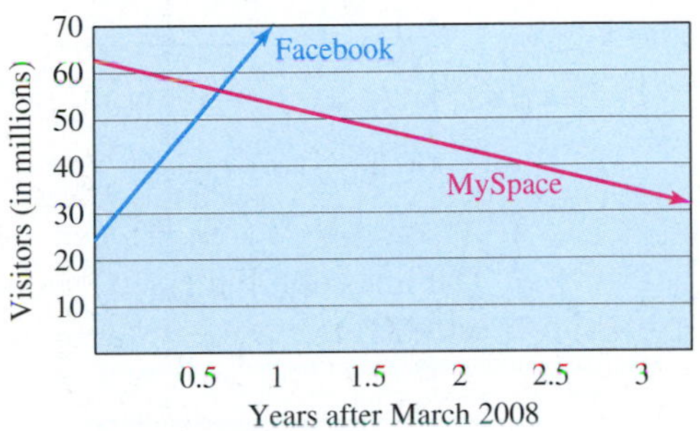

b. A crafter knits scarves and sells them at local craft shows. The cost for making x scarves can be represented by the equation $y = 3x + 40$. Her revenue from selling x scarves can be represented by the equation $y = 5x$. Find the break-even point by graphing and interpret the result.

Objective 6 Troubleshoot common errors.

Troubleshooting Common Errors

Some common errors associated with solving systems by graphing are shown.

Objective 6 Examples **A problem and an incorrect solution are given. Provide the correct solution and an explanation of the error.**

6a. The graph of a system of equations is given. What is its solution?

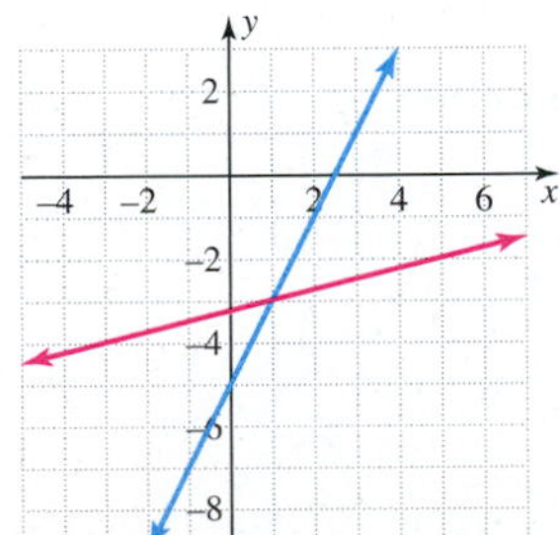

Incorrect Solution	Correct Solution and Explanation
The solution is the ordered pair $(-3, 1)$.	The solution should state the x-value of the ordered pair first. So, the solution is $(1, -3)$.

6b. Solve the system $\begin{cases} y = \dfrac{1}{2}x - 4 \\ y = \dfrac{1}{3}x - 6 \end{cases}$ graphically.

Incorrect Solution

The graph of the system is

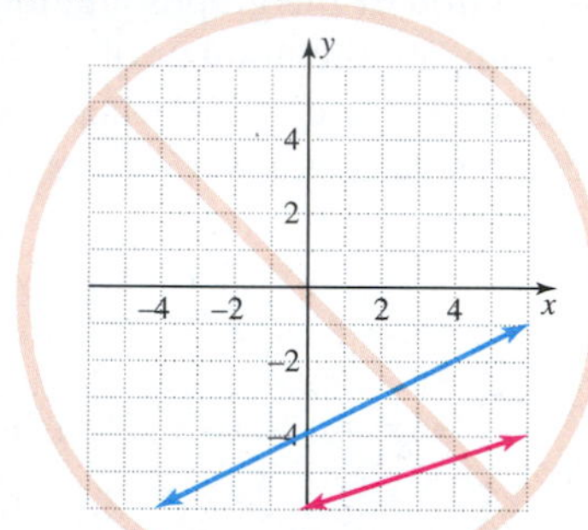

The lines don't intersect, so the solution set is ∅.

Correct Solution and Explanation

The lines have different slopes so they will intersect. If we extend the graphs, we can find the point of intersection.

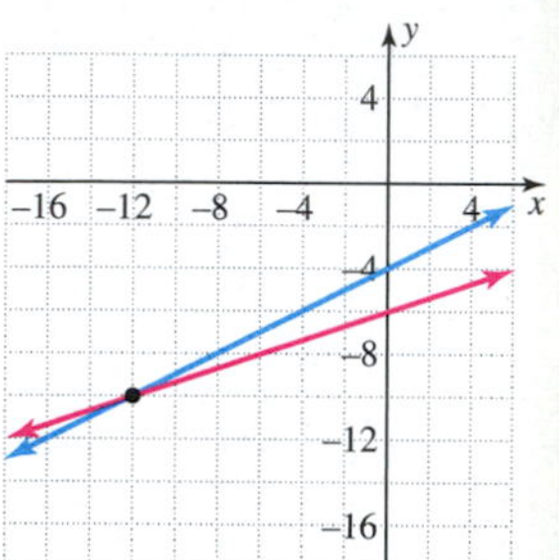

The point of intersection is approximately (−12, −10).

ANSWERS TO STUDENT CHECKS

Student Check 1 **a.** no **b.** yes

Student Check 2 **a.** $\{(2, 0)\}$ **b.** $\{(-1, -2)\}$

Student Check 3 **a.** $\{(x, y) | x + 5y = -5\}$ **b.** ∅

Student Check 4 **a.** same line, infinitely many solutions, consistent system with dependent equations. **b.** parallel lines, no solution, inconsistent system. **c.** intersecting lines, one solution, consistent system with independent equations.

Student Check 5 **a.** The point of intersection is approximately (0.7, 56), which means that in 0.7 yr after March 2008, or in November 2008, there were 56 million unique visitors to MySpace and to Facebook. Before this time, MySpace had more visitors than Facebook. After this time, Facebook had more visitors than MySpace **b.** The break-even point is (20, 100). This means that the crafter needs to sell 20 scarves to break even. When she sells 20 scarves, her cost and revenue are both $100.

SUMMARY OF KEY CONCEPTS

1. A solution of a system of linear equations in two variables is an ordered pair that makes both equations in the system true. Graphically, a solution is the point where the graphs intersect.
2. There are three possible solutions of a system of linear equations in two variables as shown.

One Ordered-Pair Solution	No Solution	Infinitely Many Solutions
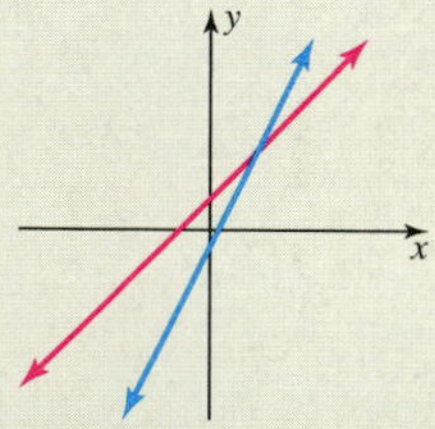	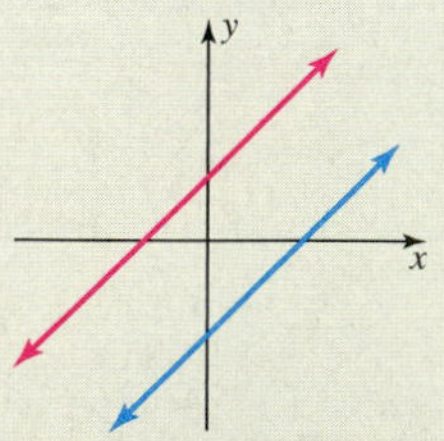	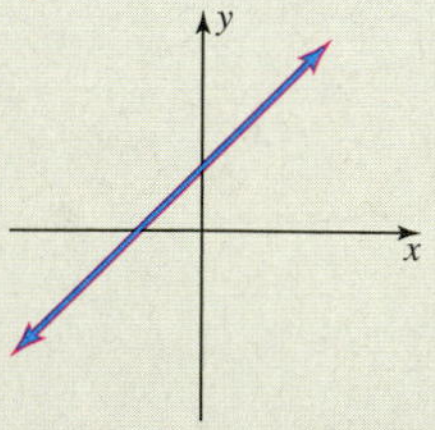
Solution set: $\{(x, y)\}$	Solution set: ∅	Solution set: $\{(x, y) \mid \text{equation}\}$
Lines have different slopes.	Lines have same slopes but different y-intercepts.	Lines have same slopes and same y-intercepts.
Consistent system; independent equations	Inconsistent system	Consistent system; dependent equations

3. Applications of systems of linear equations in two variables arise from finding the equilibrium point (when supply equals demand) and from finding the break-even point (when revenue equals cost). We can also read graphs of real data to determine their point of intersection.

GRAPHING CALCULATOR SKILLS

The graphing calculator can help us solve a system of equations. To solve a system on the calculator, we must

1. Rewrite each equation in slope-intercept form ($y = mx + b$).
2. Enter both equations in the equation editor.
3. Graph the equations.
4. Execute the Intersect command.

Example: Solve the system

$$\begin{cases} x + y = 5 \\ x - 3y = -3 \end{cases}$$

Solution: Solve each equation for y and enter into the equation editor.

$$y = -x + 5 \quad \text{and} \quad y = \frac{1}{3}x + 1$$

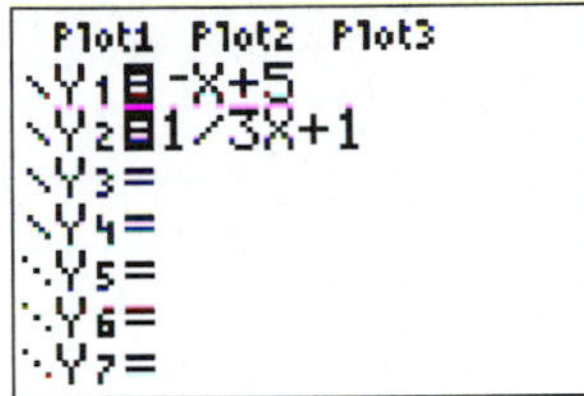

Graph the equations.

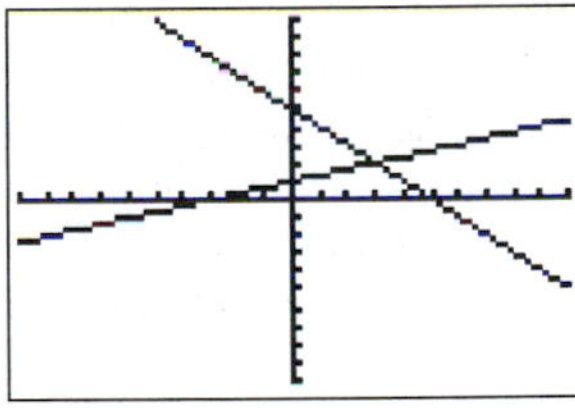

Go to the CALC menu to select the Intersect command.

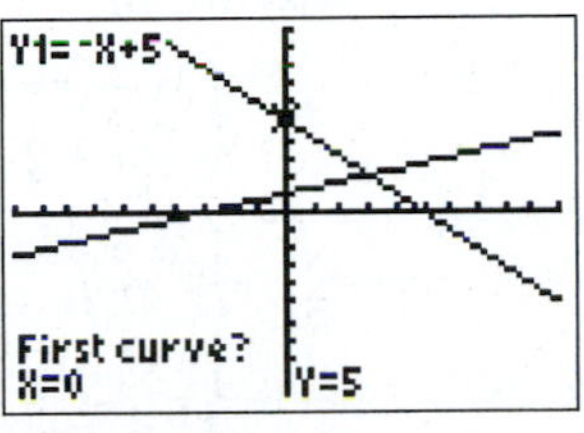

Press ENTER to accept First curve. Note the first equation is displayed in the upper left corner.

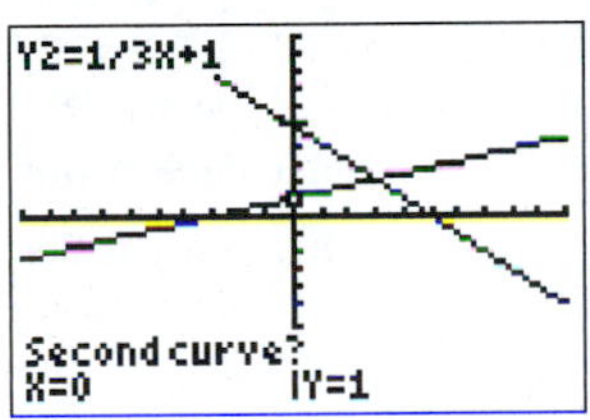

Press ENTER to accept Second curve. Note the second equation is displayed in the upper left corner.

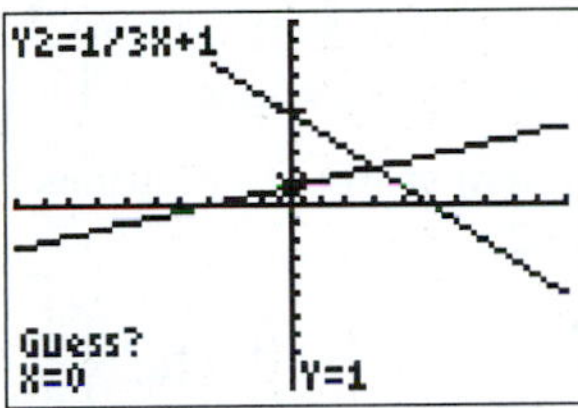

To input a Guess, move the cursor left or right to the point of intersection and press ENTER.

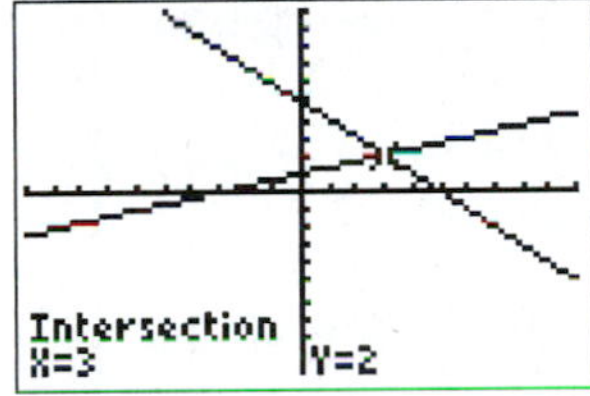

The point of intersection is displayed at the bottom of the screen. So, the point of intersection is (3, 2).

SECTION 5.1 EXERCISE SET

Write About It!

Use complete sentences in your answer to each exercise.

1. What is a system of equations?
2. How do you solve a system of equations graphically?
3. What does a system of dependent equations look like graphically?
4. What does an inconsistent system look like graphically?
5. Define demand and supply.
6. What is the break-even point from a business perspective? Explain.

Additional answers can be found in the Instructor Answer Appendix.

Practice Makes Perfect!

Determine if the ordered pair is the solution of the system of linear equations. (*See Objective 1.*)

7. $\begin{cases} x - 5y = 3 \\ y = \frac{1}{5}x - \frac{3}{5} \end{cases}; \left(6, \frac{3}{5}\right)$ is a solution

8. $\begin{cases} 6x - 5y = 2 \\ y = \frac{1}{2}x + \frac{3}{10} \end{cases}; \left(1, \frac{4}{5}\right)$ is a solution

9. $\begin{cases} 4x - 6y = -25 \\ y = \frac{5}{6}x + 5 \end{cases}$; $\left(-5, \frac{5}{6}\right)$ is a solution

10. $\begin{cases} 6x - 2y = -43 \\ y = \frac{3}{4}x + \frac{23}{4} \end{cases}$; $\left(-7, \frac{1}{2}\right)$ is a solution

11. $\begin{cases} y = \frac{1}{5}x + 3 \\ y = x - 1 \end{cases}$; (5, 4) is a solution

12. $\begin{cases} y = \frac{8}{9}x - 10 \\ y = 2x - 16 \end{cases}$; (9, 2) not a solution

13. $\begin{cases} y = \frac{4}{5}x + 7 \\ y = x + 2 \end{cases}$; (−5, 3) not a solution

14. $\begin{cases} y = \frac{1}{7}x - 4 \\ y = -x + 4 \end{cases}$; (7, −3) is a solution

Solve each system of linear equations graphically. (*See Objective 2.*)

15. $\begin{cases} x - y = -8 \\ x + y = -10 \end{cases}$ $\{(-9, -1)\}$

16. $\begin{cases} -x + y = -11 \\ x + y = 3 \end{cases}$ $\{(7, -4)\}$

17. $\begin{cases} -x + y = 6 \\ x - 2y = -9 \end{cases}$ $\{(-3, 3)\}$

18. $\begin{cases} 2x - 3y = -15 \\ x - 2y = -8 \end{cases}$ $\{(-6, 1)\}$

19. $\begin{cases} 2x + 2y = 7 \\ x - 2y = 11 \end{cases}$ $\{(6, -2.5)\}$

20. $\begin{cases} 2x - 3y = -18 \\ -x + 3y = 6 \end{cases}$ $\{(-12, -2)\}$

21. $\begin{cases} 3x + 4y = 6 \\ 5x + 2y = 10 \end{cases}$ $\{(2, 0)\}$

22. $\begin{cases} x - 4y = 2 \\ 2x + y = -5 \end{cases}$ $\{(-2, -1)\}$

23. $\begin{cases} 2x + 3y = 6 \\ y = 4 \end{cases}$ $\{(-3, 4)\}$

24. $\begin{cases} 2x - 3y = 15 \\ y = -5 \end{cases}$ $\{(0, -5)\}$

25. $\begin{cases} 4x - y = 8 \\ y = 2 \end{cases}$ $\{(2.5, 2)\}$

26. $\begin{cases} 2x - 9y = 18 \\ y = -1 \end{cases}$ $\{(4.5, -1)\}$

27. $\begin{cases} 3x - y = -6 \\ y = 3 \end{cases}$ $\{(-1, 3)\}$

28. $\begin{cases} 2x - 5y = 10 \\ y = 1 \end{cases}$ $\{(7.5, 1)\}$

29. $\begin{cases} -4x + 7y = 28 \\ y = 8 \end{cases}$ $\{(7, 8)\}$

30. $\begin{cases} 3x - 10y = 15 \\ y = -3 \end{cases}$ $\{(-5, -3)\}$

31. $\begin{cases} 2x - y = 6 \\ y = -x \end{cases}$ $\{(2, -2)\}$

32. $\begin{cases} x + 2y = 4 \\ -x + 2y = 0 \end{cases}$ $\{(2, 1)\}$

Solve each system graphically. (*See Objective 3.*)

33. $\begin{cases} 5x - 2y = 10 \\ -5x + 2y = -3 \end{cases}$ $\varnothing$

34. $\begin{cases} 2x - y = -8 \\ -4x + 2y = 6 \end{cases}$ $\varnothing$

35. $\begin{cases} -x - 5y = 10 \\ x + 5y = 15 \end{cases}$ $\varnothing$

36. $\begin{cases} 2x - 3y = 6 \\ -6x + 9y = -18 \end{cases}$

37. $\begin{cases} 3x + y = -9 \\ -6x - 2y = 18 \end{cases}$

38. $\begin{cases} -x - 2y = 14 \\ 2x + 4y = 28 \end{cases}$ $\varnothing$

39. $\begin{cases} y = 2x + 5 \\ x - \frac{1}{2}y = -\frac{5}{2} \end{cases}$

40. $\begin{cases} y = \frac{3}{4}x + 1 \\ \frac{1}{4}x - \frac{1}{3}y = -\frac{1}{3} \end{cases}$

41. $\begin{cases} y = \frac{5}{4}x - 2 \\ \frac{1}{4}x - \frac{1}{5}y = \frac{2}{5} \end{cases}$

42. $\begin{cases} y = \frac{3}{4}x - 4 \\ \frac{1}{4}x - \frac{1}{3}y = \frac{4}{3} \end{cases}$

43. $\begin{cases} y = \frac{5}{2}x - 1 \\ \frac{1}{2}x - \frac{1}{5}y = \frac{2}{5} \end{cases}$ $\varnothing$

44. $\begin{cases} y = -\frac{6}{5}x - 3 \\ -\frac{1}{5}x - \frac{1}{6}y = \frac{3}{5} \end{cases}$ $\varnothing$

45. $\begin{cases} y = \frac{3}{5}x + 2 \\ \frac{1}{5}x - \frac{1}{3}y = -\frac{2}{3} \end{cases}$

46. $\begin{cases} y = -\frac{6}{5}x - 4 \\ -\frac{1}{5}x - \frac{1}{6}y = \frac{1}{3} \end{cases}$ $\varnothing$

47. $\begin{cases} x - 2y = 8 \\ -x + 2y = 0 \end{cases}$ $\varnothing$

48. $\begin{cases} 3x = 2y \\ -6x + 4y = 0 \end{cases}$ $\{(x, y) \mid 3x = 2y\}$

Determine how the lines relate, the number of solutions of the system, and the type of system without graphing. (*See Objective 4.*)

49. $\begin{cases} 3x - y = 4 \\ -6x + 2y = -8 \end{cases}$

50. $\begin{cases} x + 4y = 1 \\ 4x + 16y = 4 \end{cases}$

51. $\begin{cases} 5x - y = 0 \\ x = \frac{1}{5}y \end{cases}$

52. $\begin{cases} x = -5y + 3 \\ y = -\frac{1}{5}x - 4 \end{cases}$

53. $\begin{cases} x = \frac{3}{7}y + \frac{10}{7} \\ 3y = 7x + 5 \end{cases}$

54. $\begin{cases} x = -6y \\ y = -\frac{1}{6}x + \frac{1}{2} \end{cases}$

55. $\begin{cases} \frac{1}{2}x - y = 4 \\ 3x - 4y = 16 \end{cases}$

56. $\begin{cases} 2x + y = 2 \\ 3x - 6y = 5 \end{cases}$

Solve each problem. (*See Objective 5.*)

57. Sally makes T-shirts and sells them at local craft shows. The cost for making x T-shirts can be represented by the equation $y = 3x + 36$. Her revenue from selling x T-shirts can be represented by the equation $y = 12x$. Find the break-even point by graphing and interpret the results. The break-even point is reached when she produces four T-shirts. The cost and revenue are both $48.

58. Tammy makes custom jewelry and sells the pieces at local craft shows. The cost for making x pairs of earrings can be represented by the equation $y = 15x + 120$. Her revenue from selling x pairs of earrings can be represented by the equation $y = 30x$. Find the break-even point by graphing and interpret the results.

59. The supply of videos to a video store for rental can be represented by $S(x) = 10x$, where x is the price of each video and y is the number of videos (in hundreds) supplied. The demand for videos can be represented by $D(x) = -10x + 60$, where x is the price of each video and y is the number of videos (in hundreds) demanded at this price. Find the equilibrium point for this product and interpret the results.

60. The supply of videos to a video store for rental can be represented by $S(x) = 12x$, where x is the price of each video and y is the number of videos (in hundreds) supplied. The demand for videos can be represented by $D(x) = -16x + 126$, where x is the price of each video and y is the number of videos (in hundreds) demanded at this price. Find the equilibrium point for this product and interpret the results.

61. The graph models the number of unique visitors, in millions, to MySpace and to Twitter, where x is the years after March 2008. Approximate the point of intersection of the graphs and interpret its meaning in the context of the problem. Compare the number of visitors to each site before and after the point of intersection. (Source: http://blog.nielsen.com/nielsenwire/global/facebook-and-twitter-post-large-year-over-year-gains-in-unique-users/.)

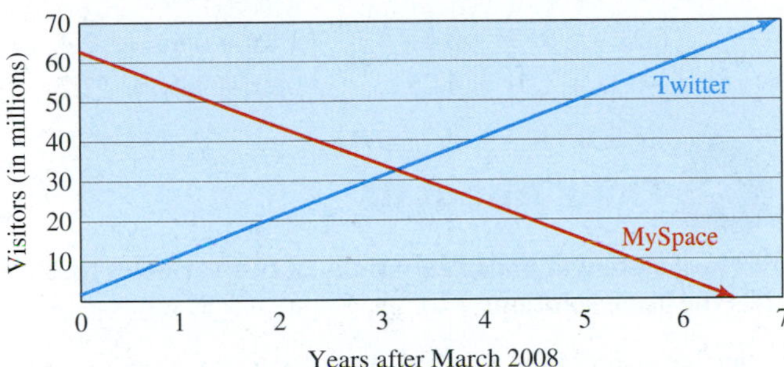

62. The graph models the revenue, in billions, for Target Corporation and Amazon, where x is the years after 2006. Approximate the point of intersection of the graphs and interpret its meaning in the context of the problem. Compare the revenue of each company before and after the point of intersection. (Sources: http://www.target.com and http://www.amazon.com)

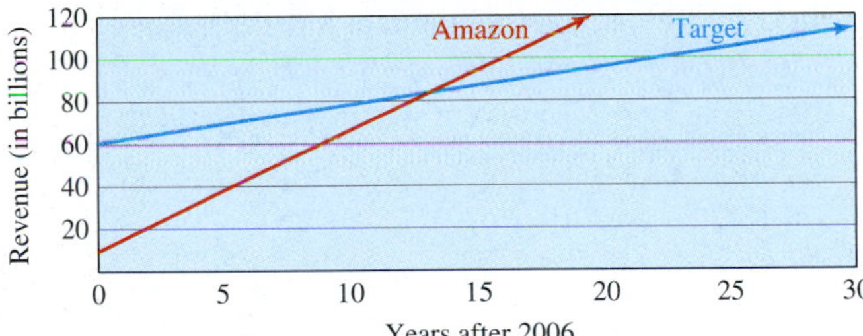

Mix 'Em Up!

Determine if the ordered pair is a solution of the system of linear equations.

63. $\begin{cases} 6x - y = -23 \\ y = \frac{2}{3}x + 7 \end{cases}$; $(-3, 5)$ is a solution

64. $\begin{cases} 4x + y = 5 \\ y = -\frac{1}{2}x + 2 \end{cases}$; $(-2, 3)$ not a solution

65. $\begin{cases} y = -\frac{4}{7}x + 5 \\ y = -3x + 20 \end{cases}$; $(7, -1)$ not a solution

66. $\begin{cases} y = -\frac{1}{9}x + 4 \\ y = -2x + 15 \end{cases}$; $(9, -3)$ not a solution

67. $\begin{cases} 0.2x + 0.3y = 2.48 \\ -0.3x + 0.1y = -1.08 \end{cases}$; $(5.2, 4.8)$ is a solution

68. $\begin{cases} 0.4x - 0.5y = -1.71 \\ 0.6x + 0.2y = -1.14 \end{cases}$; $(-2.4, 1.5)$ is a solution

69. $\begin{cases} 0.7x - 1.2y = -4.48 \\ -0.2x + 2.5y = 6.32 \end{cases}$; $(-1.6, 2.8)$ not a solution

70. $\begin{cases} 1.8x + 0.6y = -3 \\ 0.4x - 3.2y = 6.4 \end{cases}$; $(-0.8, -2.1)$ not a solution

Solve each system of linear equations graphically.

71. $\begin{cases} x + 4y = 8 \\ -2x + y = 2 \end{cases}$ $\{(0, 2)\}$

72. $\begin{cases} 4x + y = -12 \\ -x + 2y = 3 \end{cases}$ $\{(-3, 0)\}$

73. $\begin{cases} x - 3y = 12 \\ y = \frac{1}{3}x \end{cases}$ $\varnothing$

74. $\begin{cases} 5x + 2y = 10 \\ y = -\frac{5}{2}x \end{cases}$ $\varnothing$

75. $\begin{cases} x + 3y = 6 \\ y = -\frac{1}{2}x + 4 \end{cases}$ $\{(12, -2)\}$

76. $\begin{cases} -4x + 3y = -6 \\ y = -\frac{1}{3}x + 3 \end{cases}$ $\{(3, 2)\}$

77. $\begin{cases} -4x + y = 2 \\ y = 4x + 3 \end{cases}$ $\varnothing$

78. $\begin{cases} -x + 5y = -5 \\ y = \frac{1}{5}x - 1 \end{cases}$ $\{(x, y) \mid x - 5y = 5\}$

79. $\begin{cases} x + 2y = 6 \\ y = -\frac{1}{2}x + 3 \end{cases}$ $\{(x, y) \mid x + 2y = 6\}$

80. $\begin{cases} 3x - y = 9 \\ y = 3x + 1 \end{cases}$ $\varnothing$

Determine how the lines relate, the number of solutions of the system, and the type of system without graphing.

81. $\begin{cases} 2x - 5y = -9 \\ y = \frac{2}{5}x + 1 \end{cases}$

82. $\begin{cases} y = \frac{1}{3}x - 5 \\ -2x + y = 1 \end{cases}$

83. $\begin{cases} y = \frac{7}{3}x - 2 \\ y = x \end{cases}$

84. $\begin{cases} 3x + y = 5 \\ x = -\frac{1}{3}y + \frac{5}{3} \end{cases}$

85. $\begin{cases} y = -\frac{3}{2}x + \frac{1}{2} \\ x = -\frac{2}{3}y + \frac{1}{3} \end{cases}$

86. $\begin{cases} y = -2x + 9 \\ 4x + 2y = 9 \end{cases}$

The supply of a certain product can be represented by $S(x)$, where x is the price of each item and y is the number of items (in hundreds) supplied. The demand for the same product can be represented by $D(x)$, where x is the price of each item and y is the number of items (in hundreds) demanded at this price. Find the equilibrium point for this product and interpret the results.

87. $S(x) = 2x + 56$; $D(x) = 9x$

88. $S(x) = 4x + 16$; $D(x) = 8x$

89. $S(x) = 7x + 42$; $D(x) = 21x$

90. $S(x) = 5x + 45$; $D(x) = 20x$ The equilibrium point is when the price of each item is \$3. The number of items demanded and supplied is 60 hundred or 6000.

You Be the Teacher!

Correct each student's errors, if any.

91. Solve the system $\begin{cases} x - 2y = -8 \\ y = -\frac{1}{2}x + 2 \end{cases}$ graphically.

Warren's work: I solve the first equation for y.

$x - 2y = -8$

$2y = -x - 8$

$y = -\frac{1}{2}x - 4$

The second equation, $y = -\frac{1}{2}x + 2$, has the same slope as the first equation. So, the lines are parallel and the solution is the empty set, $\varnothing$.

92. Solve the system $\begin{cases} x - y = 3 \\ y = -3x + 1 \end{cases}$ graphically.

Mary's work:

$x - y = 3$ $\quad$ $y = -3x + 1$

Intercepts: (3, 0) and (0, 3) $\quad$ Points: (0, 1) and (1, 4)

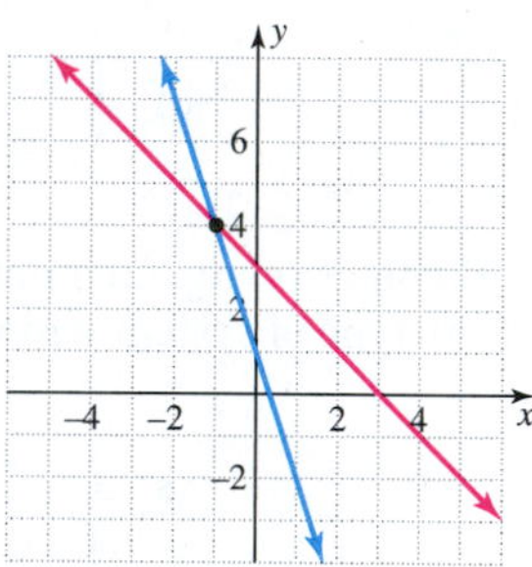

The solution set is $\{(-1, 4)\}$.

Calculate It!

Determine the solution set of the system given the scale of each axis.

93. Xscl = 2 and Yscl = 10

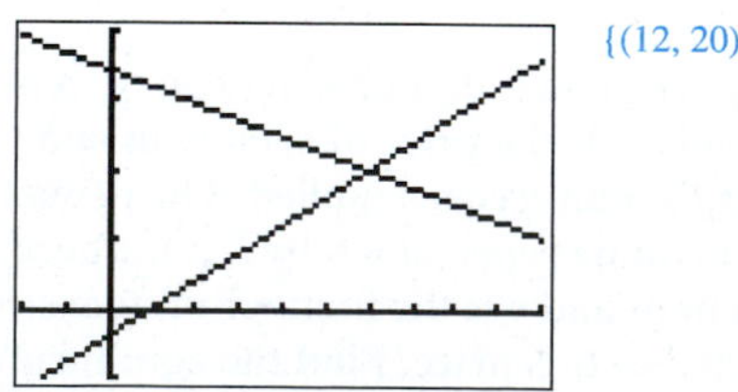

{(12, 20)}

94. Xscl = 10 and Yscl = 4

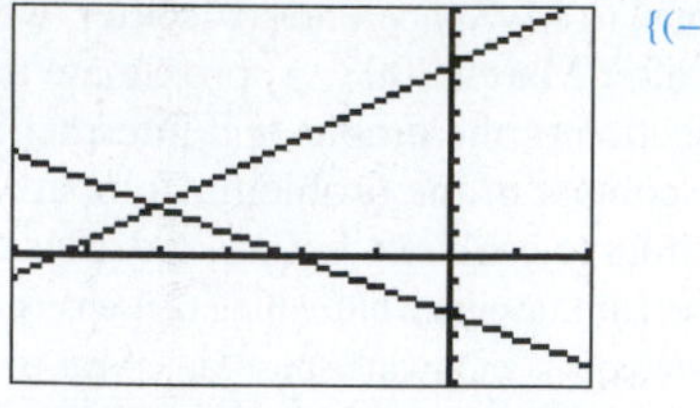

{(−40, 8)}

Use a graphing calculator to solve each system.

95. $\begin{cases} 0.2x - 0.3y = -0.4 \\ -0.1x + 1.5y = 4.25 \end{cases}$

96. $\begin{cases} 1.8x + 0.6y = -2.4 \\ 0.4x - 3.2y = -2.2 \end{cases}$

Think About It!

Write a system of linear equations in two variables that has the given solution.

97. (3, −2) **98.** (−4, 1)

99. (0, 0) **100.** (1, 2)

A system of linear equations in two variables contains the given equation. Give an example of the other equation in the system that will satisfy the given condition.

101. $x - 4y = 8$, inconsistent system

102. $3x + 5y = 2$, inconsistent system

103. $x - 4y = 8$, consistent system with dependent equations Answers vary; $2x - 8y = 16$.

104. $3x + 5y = 2$, consistent system with dependent equations Answers vary; $9x + 15y = 6$.

SECTION 5.2 Solving Systems of Linear Equations in Two Variables Algebraically

OBJECTIVES

As a result of completing this section, you will be able to

1. Solve a system of linear equations by substitution.
2. Solve a system of linear equations by elimination.
3. Solve special systems of linear equations by substitution or elimination.
4. Solve applications of systems.
5. Troubleshoot common errors.

In November 2009, Nintendo DS and Nintendo Wii were the two top selling gaming units. A total of 2,960,000 gaming units were sold. If there were 440,000 more Nintendo DS units sold than Nintendo Wii units, how many of each gaming unit was sold? (Source: http://www.1up.com/do/newsStory?cId=3177273)

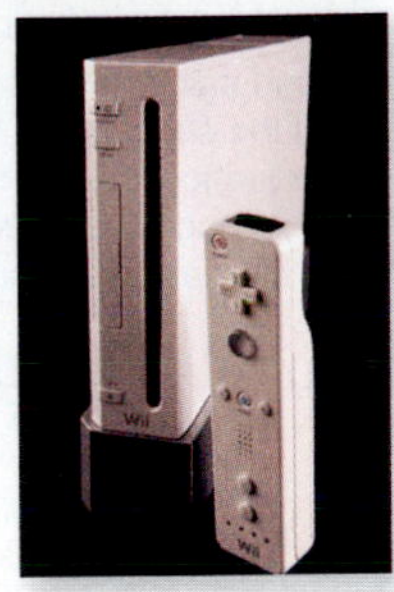

To solve this problem, we need to create a system of equations that represents the situation. Because the numbers are quite large, we need an algebraic method to solve the system.

Objective 1 Solve a system of linear equations by substitution.

Solve Linear Systems by Substitution

In Section 5.1, we solved systems by graphing. Graphing by hand, however, is not the most precise method to solve a system. If the point of intersection does not occur at integer values, it is very difficult to know the exact solution. In this section, we will learn two algebraic methods to solve systems.

The algebraic methods provide a precise way to solve a system of linear equations. The first method shown is **substitution**. Our goal is to create one equation with one unknown from the two equations in the system.

Consider the system $\begin{cases} x + 2y = 4 \\ y = x - 1 \end{cases}$. The second equation has the variable y expressed in terms of x. If we replace the variable y in the first equation with the expression from the second equation, we obtain an equation with one variable that we know how to solve.

$x + 2y = 4$	Begin with the first equation.
$x + 2(x - 1) = 4$	Replace y with $x - 1$.
$x + 2x - 2 = 4$	Apply the distributive property.
$3x - 2 = 4$	Combine like terms.
$3x - 2 + 2 = 4 + 2$	Add 2 to each side.
$3x = 6$	Simplify.
$x = \frac{6}{3}$	Divide each side by 3.
$x = 2$	Simplify.

Now we can find the corresponding y-value by substituting the known x-value into one of the equations in the system.

$y = x - 1$	Begin with the second equation.
$y = 2 - 1$	Replace x with 2.
$y = 1$	Simplify.

So, the solution is (2, 1). We can check the solution by substituting 2 for x and 1 for y into each of the original equations.

The steps that were used to solve the previous system can be generalized to solve any system of linear equations in two variables.

INSTRUCTOR NOTE: Point out that if one of the variables has a coefficient of 1 or −1, it is best to isolate this variable in order to avoid working with fractions.

Procedure: Solving a System of Linear Equations by Substitution

Step 1: Solve one of the equations in the system for one of the variables.
Step 2: Substitute the expression found in step 1 into the other equation and solve the resulting equation.
Step 3: Substitute the value from step 3 into one of the original or equivalent equations to find the value of the other coordinate.
Step 4: Check the proposed solution in both of the original equations.
Step 5: The solution set of the system consists of the ordered pair $\{(x, y)\}$.

Objective 1 Examples

Solve each system by substitution.

INSTRUCTOR NOTE: Remind students that when using the substitution method, one equation is used to isolate a variable and the other equation is used for making the substitution.

1a. $\begin{cases} x = 1 - 5y \\ 3x - 10y = -7 \end{cases}$ **1b.** $\begin{cases} 4x - y = 11 \\ 5x + 3y = 1 \end{cases}$ **1c.** $\begin{cases} \frac{x}{3} + \frac{2y}{3} = 1 \\ x - \frac{y}{2} = -\frac{1}{8} \end{cases}$

Solutions

1a. The variable x is isolated in the first equation, $x = 1 - 5y$.

Substitute and solve.

$3x - 10y = -7$	Begin with the second equation.
$3(1 - 5y) - 10y = -7$	Substitute $1 - 5y$ in place of x.
$3 - 15y - 10y = -7$	Apply the distributive property.
$3 - 25y = -7$	Combine like terms.
$3 - 25y - 3 = -7 - 3$	Subtract 3 from each side.
$-25y = -10$	Simplify.
$\frac{-25y}{-25} = \frac{-10}{-25}$	Divide each side by -25.
$y = \frac{2}{5}$	Simplify.

Now substitute $\frac{2}{5}$ for y in either of the equations to find the value of x. Both equations are shown but only one is necessary.

$x = 1 - 5y$	First equation	$3x - 10y = -7$	Second equation
$x = 1 - 5\left(\frac{2}{5}\right)$	Substitute $\frac{2}{5}$ for y.	$3x - 10\left(\frac{2}{5}\right) = -7$	Substitute $\frac{2}{5}$ for y.
$x = 1 - 2$	Multiply 5 and $\frac{2}{5}$.	$3x - 4 = -7$	Multiply 10 and $\frac{2}{5}$.
$x = -1$	Simplify.	$3x - 4 + 4 = -7 + 4$	Add 4 to each side.
		$3x = -3$	Simplify.
		$x = -1$	Divide each side by 3.

Check the solution $\left(-1, \frac{2}{5}\right)$ in each of the equations in the system.

$x = 1 - 5y$	$3x - 10y = -7$
$-1 = 1 - 5\left(\frac{2}{5}\right)$	$3(-1) - 10\left(\frac{2}{5}\right) = -7$
$-1 = 1 - 2$	$-3 - 4 = -7$
$-1 = -1$	$-7 = -7$
True	True

Since $\left(-1, \frac{2}{5}\right)$ makes each equation true, the solution set is $\left\{\left(-1, \frac{2}{5}\right)\right\}$. Because there is one solution, this system contains lines that intersect at this point.

1b. We solve the first equation for y since its coefficient is -1.

$4x - y = 11$	Begin with the first equation.
$4x - y - 4x = 11 - 4x$	Subtract $4x$ from each side.
$-y = -4x + 11$	Simplify.
$y = 4x - 11$	Multiply each side by -1.

Substitute and solve.

$5x + 3y = 1$	Begin with the second equation.
$5x + 3(4x - 11) = 1$	Substitute $4x - 11$ in place of y.
$5x + 12x - 33 = 1$	Apply the distributive property.
$17x - 33 = 1$	Simplify.
$17x - 33 + 33 = 1 + 33$	Add 33 to each side.
$17x = 34$	Simplify.
$x = 2$	Divide each side by 17.

Now substitute 2 for x in either of the equations to find the value of y.

$4x - y = 11$	Begin with the first equation.
$4(2) - y = 11$	Substitute 2 for y.
$8 - y = 11$	Multiply 4 and 2.
$8 - y - 8 = 11 - 8$	Subtract 8 from each side.
$-y = 3$	Simplify.
$y = -3$	Multiply each side by -1.

Check the solution $(2, -3)$ in each of the equations in the system.

$4x - y = 11$	$5x + 3y = 1$
$4(2) - (-3) = 11$	$5(2) + 3(-3) = 1$
$8 + 3 = 11$	$10 - 9 = 1$
$11 = 11$	$1 = 1$
True	True

Since $(2, -3)$ makes each equation true, the solution set is $\{(2, 3)\}$. Because there is one solution, this system contains lines that intersect at this point.

1c. Before we isolate a variable, we rewrite the system so that the fractions are eliminated by multiplying each equation by its LCD.

$$\begin{cases} 3\left(\dfrac{x}{3} + \dfrac{2y}{3} = 1\right) \\ 8\left(x - \dfrac{y}{2} = -\dfrac{1}{8}\right) \end{cases} \rightarrow \begin{cases} x + 2y = 3 \\ 8x - 4y = -1 \end{cases}$$

Now, isolate the variable x in the first equation since its coefficient is 1.

$x + 2y = 3$	Begin with the first equation.
$x + 2y - 2y = 3 - 2y$	Subtract $2y$ from each side.
$x = 3 - 2y$	Simplify.

Substitute and solve.

$8x - 4y = -1$	Begin with the second equation.
$8(3 - 2y) - 4y = -1$	Substitute $3 - 2y$ in place of x.
$24 - 16y - 4y = -1$	Apply the distributive property.
$24 - 20y = -1$	Simplify.
$24 - 20y - 24 = -1 - 24$	Subtract 24 from each side.
$-20y = -25$	Simplify.
$\dfrac{-20y}{-20} = \dfrac{-25}{-20}$	Divide each side by -20.
$y = \dfrac{5}{4}$	Simplify.

INSTRUCTOR NOTE: Point out that the original equations contain fractions so we use the equivalent form without fractions to find the value of x.

Now substitute $\frac{5}{4}$ for y in either of the equations to find the value of x.

$x + 2y = 3$ Begin with the first equation.

$x + 2\left(\frac{5}{4}\right) = 3$ Substitute $\frac{5}{4}$ for y.

$x + \frac{5}{2} = 3$ Multiply 2 and $\frac{5}{4}$.

$2\left(x + \frac{5}{2}\right) = 2(3)$ Multiply each side by the LCD, 2.

$2x + 5 = 6$ Simplify.

$2x + 5 - 5 = 6 - 5$ Subtract 5 from each side.

$2x = 1$ simplify.

$x = \frac{1}{2}$ Divide each side by 2.

Checking confirms that $\left(\frac{1}{2}, \frac{5}{4}\right)$ makes each equation true. So, the solution set is $\left\{\left(\frac{1}{2}, \frac{5}{4}\right)\right\}$.

Student Check 1 Solve each system by substitution.

a. $\begin{cases} 2x - 5y = -4 \\ y = 4x - 1 \end{cases}$ **b.** $\begin{cases} 3x - y = -11 \\ 7x + 6y = -34 \end{cases}$ **c.** $\begin{cases} \frac{x}{2} - y = \frac{7}{2} \\ x + \frac{y}{3} = \frac{28}{9} \end{cases}$

Solve Linear Systems by Elimination

Objective 2 Solve a system of linear equations by elimination.

The method of substitution will always yield an exact solution for a system of linear equations. However, this method can be tedious to use, especially if no variable in the system has a coefficient of 1 or −1.

Another process for solving a system of linear equations that produces an exact answer is the method of **elimination** or the addition method. The goal of this method is to add the equations in the system or to add equivalent forms of the equations in the system so that one of the variables is eliminated.

Consider the system of equations $\begin{cases} 5x + 2y = 16 \\ 3x - 2y = -8 \end{cases}$. If we add the two equations in the system together, we get an equation in one variable that we can solve.

$$\begin{cases} 5x + 2y = 16 \\ 3x - 2y = -8 \end{cases}$$

$8x = 8$ Add the equations.

$\frac{8x}{8} = \frac{8}{8}$ Divide each side by 8.

$x = 1$ Simplify.

Now we can solve for y by replacing x with 1 in one of the original equations in the system.

$5x + 2y = 16$ Begin with the first equation.

$5(1) + 2y = 16$ Replace x with 1.

$5 + 2y = 16$ Multiply 5 and 1.

$$5 + 2y - 5 = 16 - 5 \quad \text{Subtract 5 from each side.}$$
$$2y = 11 \quad \text{Simplify.}$$
$$y = \frac{11}{2} \quad \text{Divide each side by 2.}$$

So, the solution set of the system is $\left\{\left(1, \frac{11}{2}\right)\right\}$.

This process is based on the addition property of equality which enables us to add the same quantity to each side of equation. Since $3x - 2y$ is equal to -8, we obtain an equivalent equation when they are added to each side of the first equation, that is, $5x + 2y + 3x - 2y = 16 + (-8)$. In this system, the coefficients of the y-variables are opposites, so their sum is zero.

Property: The Addition Property of Equality

If $A = B$ and $C = D$, then $A + C = B + D$.

Procedure: Solving a System of Linear Equations by Elimination

Step 1: Write each equation in the system in standard form, $Ax + By = C$.
Step 2: Create a new system with opposites as coefficients on one of the variables, if necessary, by multiplying one or both equations by an appropriate nonzero number.
Step 3: Add the equations in the system together.
Step 4: Solve the resulting equation.
Step 5: Substitute the solution from Step 4 in one of the original equations to find the value of the other coordinate.
Step 6: Check the proposed solution in both of the original equations.
Step 7: Write the solution set.

Objective 2 Examples Solve each system by elimination.

2a. $\begin{cases} x + 3y = 7 \\ 4x - 3y = 3 \end{cases}$ **2b.** $\begin{cases} 6x - 7y = 9 \\ 3x - 5y = 9 \end{cases}$ **2c.** $\begin{cases} x + y = 50 \\ 0.05x + 0.10y = 4 \end{cases}$

Solutions **2a.** The coefficients of y are opposite, so we add the equations in the system.

$$x + 3y = 7 \quad \text{First equation}$$
$$4x - 3y = 3 \quad \text{Second equation}$$
$$5x = 10 \quad \text{Add.}$$
$$x = 2 \quad \text{Divide each side by 5.}$$

Substitute 2 for x in one of the original equations to solve for the value of y.

$$x + 3y = 7 \quad \text{Begin with the first equation.}$$
$$2 + 3y = 7 \quad \text{Substitute 2 for } x.$$
$$2 + 3y - 2 = 7 - 2 \quad \text{Subtract 2 from each side.}$$
$$3y = 5 \quad \text{Simplify.}$$
$$y = \frac{5}{3} \quad \text{Divide each side by 3.}$$

Now check the solution in each of the original equations.

$x + 3y = 7$	$4x - 3y = 3$
$2 + 3\left(\frac{5}{3}\right) = 7$	$4(2) - 3\left(\frac{5}{3}\right) = 3$
$2 + 5 = 7$	$8 - 5 = 3$
$7 = 7$ True	$3 = 3$ True

Since $\left(2, \frac{5}{3}\right)$ makes each equation in the system true, the solution set is $\left\{\left(2, \frac{5}{3}\right)\right\}$.

2b. We choose to eliminate x from the system. A number that both 6 and 3 divide into is 6.

$$\begin{cases} 6x - 7y = 9 \\ 3x - 5y = 9 \end{cases}$$

The coefficients of x must be opposite, 6 and -6. The first equation has the needed coefficient, 6. The second equation must have -6 for the coefficient of x. So, divide the coefficient we need, -6, by the given coefficient, 3, to determine the number we distribute to this equation, that is $\frac{-6}{3} = -2$.

INSTRUCTOR NOTE:
Remind students to multiply both sides of the equation by the number that makes the coefficients opposite.

$$\begin{cases} 6x - 7y = 9 \\ -2(3x - 5y = 9) \end{cases} \quad \text{simplifies to} \quad \begin{cases} 6x - 7y = 9 \\ -6x + 10y = -18 \end{cases}$$

Add the equations in the system and solve the resulting equation.

$6x - 7y = 9$	First equation
$-6x + 10y = -18$	Second equation
$3y = -9$	Add.
$y = -3$	Divide each side by 3.

Substitute -3 for y in one of the original equations in the system to solve for the value of x.

$6x - 7y = 9$	Begin with the first equation.
$6x - 7(-3) = 9$	Substitute -3 for x.
$6x + 21 = 9$	Multiply -7 and -3.
$6x + 21 - 21 = 9 - 21$	Subtract 21 from each side.
$6x = -12$	Simplify.
$\frac{6x}{6} = \frac{-12}{6}$	Divide each side by 6.
$x = -2$	Simplify.

Now check the solution in each of the original equations.

$6x - 7y = 9$	$3x - 5y = 9$
$6(-2) - 7(-3) = 9$	$3(-2) - 5(-3) = 9$
$-12 + 21 = 9$	$-6 + 15 = 9$
$9 = 9$	$9 = 9$

Since $(-2, -3)$ makes both equations in the system true, the solution set is $\{(-2, -3)\}$.

2c. Multiply the second equation by 100 to clear decimals.

$$\begin{cases} x + y = 50 \\ 100(0.05x + 0.10y = 4) \end{cases} \text{ simplifies to } \begin{cases} x + y = 50 \\ 5x + 10y = 400 \end{cases}$$

We choose to eliminate x from the system. The number that both 1 and 5 divide into is 5. We multiply the first equation by $\frac{-5}{1} = -5$.

$$\begin{cases} -5(x + y = 50) \\ 5x + 10y = 400 \end{cases} \text{ simplifies to } \begin{cases} -5x - 5y = -250 \\ 5x + 10y = 400 \end{cases}$$

Add the equations and solve the resulting equation.

$-5x - 5y = -250$	First equation
$5x + 10y = 400$	Second equation
$5y = 150$	Add.
$y = 30$	Divide each side by 5.

Now substitute 30 for y in one of the original equations to solve for x.

$x + y = 50$	Begin with the first equation.
$x + 30 = 50$	Substitute 30 for y.
$x + 30 - 30 = 50 - 30$	Subtract 30 from each side.
$x = 20$	Simplify.

Checking confirms that (20, 30) makes each equation true. So, the solution set is $\{(20, 30)\}$.

Student Check 2 Solve each system by elimination.

a. $\begin{cases} 2x - y = 4 \\ 3x + y = 1 \end{cases}$ **b.** $\begin{cases} 4x + 3y = 12 \\ 2x - 5y = 6 \end{cases}$ **c.** $\begin{cases} x + y = 80 \\ 0.05x + 0.25y = 8 \end{cases}$

Specials Cases of Linear Systems

Objective 3 Solve special systems of linear equations by substitution or elimination.

From Section 5.1, we know that there are two special cases of systems—systems with no solution and systems with infinitely many solutions. Recall that a system with no solution consists of lines whose graphs are parallel; a system with infinitely many solutions consists of two lines that are the same.

Procedure: Solving Special Systems by Substitution or Elimination

Step 1: Solve the system by substitution or elimination.

Step 2: **a.** If the resulting equation is a false statement (i.e., a contradiction), then there is no solution of the system.

b. If the resulting equation is a true statement (i.e., an identity), then there are infinitely many solutions.

Objective 3 Examples Solve each system by either substitution or elimination.

3a. $\begin{cases} x + 5y = -1 \\ 2x + 10y = 5 \end{cases}$ **3b.** $\begin{cases} 3x - 6y = -3 \\ 5x - 10y = -5 \end{cases}$

Solutions **3a.** We choose to use substitution to solve the system and will solve the first equation for x since its coefficient is 1.

$$x + 5y = -1 \quad \text{Begin with the first equation.}$$
$$x + 5y - 5y = -1 - 5y \quad \text{Subtract } 5y \text{ from each side.}$$
$$x = -5y - 1 \quad \text{Simplify.}$$

Substitute and solve.

$$2x + 10y = 5 \quad \text{Begin with the second equation.}$$
$$2(-5y - 1) + 10y = 5 \quad \text{Substitute } -5y - 1 \text{ for } x.$$
$$-10y - 2 + 10y = 5 \quad \text{Apply the distributive property.}$$
$$-2 = 5 \quad \text{Combine like terms.}$$

Because we obtained a contradiction, $-2 = 5$, in the substitution process, this system has no solution. This means that the two lines are parallel and that the system is inconsistent. So, the solution set of the system is the empty set, or $\varnothing$.

3b. We choose to use elimination to solve the system and will eliminate y from the system. The number that both 6 and 10 divide into is 30.

$$\begin{cases} 3x - 6y = -3 \\ 5x - 10y = -5 \end{cases}$$

The coefficients of y must be 30 and -30, $\frac{30}{-6} = -5$ and $\frac{-30}{-10} = 3$.

$$\begin{cases} -5(3x - 6y = -3) \\ 3(5x - 10y = -5) \end{cases} \quad \text{simplifies to} \quad \begin{cases} -15x + 30y = 15 \\ 15x - 30y = -15 \end{cases}$$

Now add the equations in the system and solve the resulting equation.

$$-15x + 30y = 15 \quad \text{First equation}$$
$$15x - 30y = -15 \quad \text{Second equation}$$
$$0 = 0 \quad \text{Add.}$$

Because we obtained an identity, $0 = 0$, through the elimination process, this system has infinitely many solutions. The lines are the same and the system is consistent with dependent equations. The solution set consists of all ordered pairs on the line, which is denoted as $\{(x, y) | 3x - 6y = -3\}$.

Student Check 3 Solve each system by substitution or elimination.

a. $\begin{cases} 8x - 12y = -8 \\ 2x - 3y = -2 \end{cases}$ **b.** $\begin{cases} 10x + 5y = -20 \\ 8x + 4y = 10 \end{cases}$

A summary of the different cases for solving systems by substitution or elimination is shown.

Intersecting Lines	Parallel Lines	Same Lines
After substitution or adding the two equations, the result is an equation of the form $x = k$ or $y = k$, for k a real number.	After substitution or adding the equations, a *false* statement or contradiction results, for example, $0 = 8$.	After substitution or adding the equations, a *true* statement or identity results, for example, $0 = 0$.
One solution	No solution	Infinitely many solutions
Consistent system with independent equations	Inconsistent system	Consistent system with dependent equations

Applications

Objective 4 Solve applications of systems.

Many word problems that we solve involve more than one unknown. When this is the case, systems are the best method to solve the problem. A system enables us to assign a different variable for each of the unknown values. There are four general steps we will follow to solve such problems.

Procedure: General Problem-Solving Strategy

Step 1: Read the problem and determine the two unknown values. Assign a different variable to represent each of these values.
Step 2: Analyze the problem and write two equations that define a relationship between the two variables.
Step 3: Solve the system of equations by either substitution or elimination.
Step 4: Write the answer to the problem.

Objective 4 Example In November 2009, Nintendo DS and Nintendo Wii were the two top-selling gaming units. In total, 2,960,000 of these gaming units were sold. If there were 440,000 more Nintendo DS units sold than Nintendo Wii units, how many of each gaming unit was sold? (Source: http://www.1up.com/do/newsStory?cId=3177273)

Solution What are the unknowns? The unknowns are the number of Nintendo DS units and the number of Nintendo Wii units sold.

Let x = number of Nintendo DS units sold.

Let y = number of Nintendo Wii games sold.

What is known? The total number of gaming units sold was 2,960,000. There were 440,000 more DS units sold than Wii units. The system of equations that represents this information is

$$\begin{cases} x + y = 2{,}960{,}000 \\ x = y + 440{,}000 \end{cases}$$

Solve the system using elimination.

$$\begin{cases} x + y = 2{,}960{,}000 \\ x = y + 440{,}000 \end{cases}$$ Begin with the system.

$$\begin{cases} x + y = 2{,}960{,}000 \\ x - y = 440{,}000 \end{cases}$$ Rewrite the second equation in standard form by subtracting y from each side.

$$2x = 3{,}400{,}000$$ Add the equations.

$$\frac{2x}{2} = \frac{3{,}400{,}000}{2}$$ Divide each side by 2.

$$x = 1{,}700{,}000$$ Simplify.

Now solve for y by replacing x with 1,700,000 in either equation.

$x + y = 2{,}960{,}000$	Begin with the first equation.
$1{,}700{,}000 + y = 2{,}960{,}000$	Substitute 1,700,000 for x.
$1{,}700{,}000 + y - 1{,}700{,}000 = 2{,}960{,}000 - 1{,}700{,}000$	Subtract 1,700,000 from each side.
$y = 1{,}260{,}000$	Simplify.

In November 2009, there were 1,700,000 Nintendo DS gaming units sold and 1,260,000 Nintendo Wii gaming units sold.

Student Check 4 In November 2009, the Xbox 360 and the PlayStation 3 (PS3) were the third and fourth top-selling gaming units. In total, 1,529,900 of these gaming units were sold. If there were 109,100 more Xbox 360 units sold than PS3 units, how many of each gaming unit was sold? (Source: http://www.1up.com/do/newsStory?cId=3177273)

Objective 5 Troubleshoot common errors.

Troubleshooting Common Errors

Some common errors associated with substitution and elimination are shown.

Objective 5 Examples **A problem and an incorrect solution are given. Provide the correct solution and an explanation of the error.**

5a. Solve the system $\begin{cases} y = 4x - 9 \\ 3x - y = 6 \end{cases}$.

Incorrect Solution

$$y = 4x - 9$$
$$3x - 4x - 9 = 6$$
$$-x - 9 = 6$$
$$-x = 15$$
$$x = -15$$

Now substitute -15 for x and solve for y.

$$y = 4(-15) - 9$$
$$y = -60 - 9$$
$$y = -69$$

The solution set is $\{(-15, -69)\}$.

Correct Solution and Explanation

The expression $4x - 9$ should be put in parentheses because the negative sign must be applied to it.

$$3x - (4x - 9) = 6$$
$$3x - 4x + 9 = 6$$
$$-x + 9 = 6$$
$$-x = -3$$
$$x = 3$$

Now substitute 3 for x and solve for y.

$$y = 4(3) - 9$$
$$y = 12 - 9$$
$$y = 3$$

The solution set is $\{(3, 3)\}$.

5b. Solve the system $\begin{cases} 6x - 5y = 27 \\ 3x + y = 3 \end{cases}$.

Incorrect Solution

$$\begin{cases} 6x - 5y = 27 \\ 5(3x + y = 3) \end{cases} \rightarrow \begin{cases} 6x - 5y = 27 \\ 15x + 5y = 3 \end{cases}$$
$$21x = 30$$
$$x = \frac{30}{21} = \frac{10}{7}$$

Correct Solution and Explanation

We must distribute 5 to each side of the equation.

$$\begin{cases} 6x - 5y = 27 \\ 15x + 5y = 15 \end{cases}$$
$$21x = 42$$
$$x = 2$$

Substitute $\frac{10}{7}$ for x and solve for y.

$$3\left(\frac{10}{7}\right) + y = 3$$
$$\frac{30}{7} + y = 3$$
$$30 + 7y = 21$$
$$7y = -9$$
$$y = -\frac{9}{7}$$

The solution set is $\left\{\left(\frac{10}{7}, -\frac{9}{7}\right)\right\}$.

Substitute 2 for x and solve for y.

$$3(2) + y = 3$$
$$6 + y = 3$$
$$y = -3$$

The solution set is $\{(2, -3)\}$.

5c. Solve the system $\begin{cases} x + 5y = 20 \\ 9x - y = -4 \end{cases}$.

Incorrect Solution

$$\begin{cases} x + 5y = 20 \\ 5(9x - y = -4) \end{cases} \rightarrow \begin{cases} x + 5y = 20 \\ 45x - 5y = -20 \end{cases}$$
$$46x = 0$$
$$x = 0$$

So, there are infinitely many solutions of the system and the solution set is $\{(x, y) | x + 5y = 20\}$.

Correct Solution and Explanation

For the system to have infinitely many solutions, the resulting equation must be $0 = 0$. Since the variable x is still in the resulting equation, we can solve for y.

Replace x with 0.

$$x + 5y = 20$$
$$0 + 5y = 20$$
$$5y = 20$$
$$y = 4$$

The solution set is $\{(0, 4)\}$.

ANSWERS TO STUDENT CHECKS

Student Check 1 **a.** $\left\{\left(\frac{1}{2}, 1\right)\right\}$ **b.** $\{(-4, -1)\}$ **c.** $\left\{\left(\frac{11}{3}, -\frac{5}{3}\right)\right\}$

Student Check 2 **a.** $\{(1, -2)\}$ **b.** $\{(3, 0)\}$ **c.** $\{(60, 20)\}$

Student Check 3 **a.** $\{(x, y) | 2x - 3y = -2\}$ **b.** $\varnothing$

Student Check 4 **a.** There were 819,500 Xbox 360 units sold and 710,400 PS3 units sold.

SUMMARY OF KEY CONCEPTS

Substitution and elimination are algebraic methods that enable us to find the exact solution of a system of linear equations. Answers can always be checked by substituting the solution in each equation in the system or by graphing.

1. The goal of substitution is to solve one of the equations for one of the variables. This expression is substituted into the other equation to yield one equation in one variable. The solution of this equation is then substituted into one of the original equations in the system to determine the value of the other variable.
2. The goal of elimination is to add the equations so that one variable is eliminated. If the equations in the system do not have opposites for one of the coefficients, we must multiply one or both equations by some nonzero number to produce an equivalent system with opposites as coefficients of one of the variables.
3. If, after applying substitution or elimination, the resulting equation is false or true, then the solution of the system is no solution or infinitely many solutions, respectively.
4. Systems can be used to solve application problems. From the information given, we need to determine the two unknowns and two equations that represent the situation.

GRAPHING CALCULATOR SKILLS

The graphing calculator can be used to check our work. In Section 5.1, the Intersect command was shown. We can also check by substituting the x and y coordinates of the solution into each equation.

Example: Check that $\left(\frac{1}{2}, \frac{4}{3}\right)$ is the solution of $\begin{cases} 2x - 3y = -3 \\ 4x + 9y = 14 \end{cases}$.

Solution: Substitute the coordinates of the ordered pair into the two equations in the system. If the ordered pair makes each equation true, then the solution is correct.

```
2(1/2)-3(4/3)
               -3
4(1/2)+9(4/3)
               14
```

Because we obtain the right side of each of the equations in the system, the solution is correct.

SECTION 5.2 EXERCISE SET

Write About It!

Use complete sentences in your answer to each exercise.

1. When solving a system by substitution, how do you determine which variable to isolate?
2. How are the two equations in a system used when solving by substitution?
3. When solving a system by elimination, why must the coefficients of one of the variables have opposite coefficients? What must be done if neither of the variables has opposites for their coefficients?
4. After you find the solution for one of the variables in a system by either substitution or elimination, what is the next step?
5. If the resulting equation after substitution or elimination is $0 = -5$, what can you conclude about the solution of the system? Answers vary. If the resulting equation is $0 = -5$, then the system has no solution.
6. If the resulting equation after substitution or elimination is $4 = 4$, what can you conclude about the solution of the system? Answers vary. If the resulting equation is $4 = 4$, then the system contains lines that are the same and there are infinitely many solutions of the system.

Practice Makes Perfect!

Solve each system by substitution. (*See Objective 1.*)

7. $\begin{cases} x = 4y + 15 \\ 6x + 5y = -26 \end{cases}$ $\{(-1, -4)\}$
8. $\begin{cases} y = -2x + 20 \\ 5x + 2y = 49 \end{cases}$ $\{(9, 2)\}$
9. $\begin{cases} x = -2y + 10 \\ -7x - 4y = -10 \end{cases}$ $\{(-2, 6)\}$
10. $\begin{cases} y = 6x + 4 \\ 3x - 8y = 58 \end{cases}$ $\{(-2, -8)\}$
11. $\begin{cases} x + 3y = 2 \\ 4x - 5y = -43 \end{cases}$ $\{(-7, 3)\}$
12. $\begin{cases} 4x - y = -38 \\ 3x + 5y = 6 \end{cases}$ $\{(-8, 6)\}$
13. $\begin{cases} x - y = 10 \\ x + 4y = -15 \end{cases}$ $\{(5, -5)\}$
14. $\begin{cases} -3x - 4y = 18 \\ 2x + y = -7 \end{cases}$ $\{(-2, -3)\}$
15. $\begin{cases} 5x - 3y = 41 \\ -2x - 5y = -35 \end{cases}$ $\{(10, 3)\}$
16. $\begin{cases} -3x - 2y = -3 \\ 5x + 4y = 1 \end{cases}$ $\{(5, -6)\}$
17. $\begin{cases} x + 2y = 10 \\ \frac{x}{2} - y = 1 \end{cases}$ $\{(6, 2)\}$
18. $\begin{cases} -3x + y = -9 \\ x + \frac{y}{3} = 1 \end{cases}$ $\{(2, -3)\}$
19. $\begin{cases} x - 4y = 18 \\ 3x + \frac{y}{2} = 4 \end{cases}$ $\{(2, -4)\}$
20. $\begin{cases} -4x + 5y = -60 \\ \frac{2}{5}x + 2y = -4 \end{cases}$ $\{(10, -4)\}$
21. $\begin{cases} 6x - 5y = 32 \\ y = \frac{1}{4}x - \frac{7}{10} \end{cases}$ $\left\{\left(6, \frac{4}{5}\right)\right\}$
22. $\begin{cases} 4x - 8y = 17 \\ y = \frac{1}{5}x - \frac{5}{8} \end{cases}$ $\left\{\left(5, \frac{3}{8}\right)\right\}$
23. $\begin{cases} 4x - 7y = 6 \\ y = \frac{1}{2}x - \frac{5}{7} \end{cases}$ $\left\{\left(2, \frac{2}{7}\right)\right\}$
24. $\begin{cases} 2x - 5y = -13 \\ \frac{3}{4}x - y = -\frac{87}{20} \end{cases}$

Solve each system by elimination. (*See Objective 2.*)

25. $\begin{cases} -5x - 6y = 4 \\ 2x - y = -5 \end{cases}$ $\{(-2, 1)\}$
26. $\begin{cases} 2x - 3y = 16 \\ -3x - y = -2 \end{cases}$ $\{(2, -4)\}$
27. $\begin{cases} -x - 2y = 2 \\ 8x + y = 14 \end{cases}$ $\{(2, -2)\}$
28. $\begin{cases} 5x - y = 53 \\ -4x - 3y = -12 \end{cases}$ $\{(9, -8)\}$
29. $\begin{cases} 7x + 2y = -11 \\ 5x - 4y = -35 \end{cases}$ $\{(-3, 5)\}$
30. $\begin{cases} 7x - 5y = 40 \\ -4x + 3y = -23 \end{cases}$ $\{(5, -1)\}$
31. $\begin{cases} -x + y = -8 \\ -0.3x + 0.1y = -1.8 \end{cases}$ $\{(5, -3)\}$
32. $\begin{cases} 2x - 5y = 21 \\ -0.3x - 0.1y = 1.1 \end{cases}$ $\{(-2, -5)\}$
33. $\begin{cases} 5x + 4y = 2 \\ 0.8x + 0.3y = 2.7 \end{cases}$ $\{(6, -7)\}$
34. $\begin{cases} x - y = 2 \\ 0.3x - 0.5y = 1.2 \end{cases}$ $\{(-1, -3)\}$

Additional answers can be found in the Instructor Answer Appendix.

35. $\begin{cases} -0.5x + 0.1y = 4.5 \\ -0.8x - 0.3y = 4.9 \end{cases}$ $\{(-8, 5)\}$

36. $\begin{cases} 0.2x + 0.5y = -0.3 \\ 3y - x = 7 \end{cases}$ $\{(-4, 1)\}$

37. $\begin{cases} 0.4x - 0.3y = 10.8 \\ \dfrac{1}{5}x + \dfrac{1}{2}y = -5 \end{cases}$ $\{(15, -16)\}$

38. $\begin{cases} 0.1x + 0.2y = -0.5 \\ -\dfrac{1}{3}x + \dfrac{1}{4}y = 9 \end{cases}$ $\{(-21, 8)\}$

39. $\begin{cases} -0.2x + 0.1y = 2.6 \\ \dfrac{1}{6}x - \dfrac{1}{2}y = 2 \end{cases}$ $\{(-18, -10)\}$

40. $\begin{cases} 0.4x + 0.5y = -5.2 \\ -\dfrac{1}{2}x + \dfrac{3}{4}y = -21 \end{cases}$ $\{(12, -20)\}$

Solve each system by substitution or elimination. (*See Objective 3.*)

41. $\begin{cases} \dfrac{4}{7}x - y = 1 \\ 4x - 7y = 3 \end{cases}$ $\varnothing$

42. $\begin{cases} \dfrac{1}{3}x + y = 2 \\ x + 3y = 5 \end{cases}$ $\varnothing$

43. $\begin{cases} -x + 3y = 2 \\ y = \dfrac{1}{3}x + \dfrac{2}{3} \end{cases}$ $\{(x, y) \mid x - 3y = -2\}$

44. $\begin{cases} 5x - y = 1 \\ -10x + 2y = -2 \end{cases}$ $\{(x, y) \mid 5x - y = 1\}$

45. $\begin{cases} -3x + 6y = 4 \\ x - 2y = 0 \end{cases}$ $\varnothing$

46. $\begin{cases} 6x + 2y = 6 \\ 3x + y = 0 \end{cases}$ $\varnothing$

47. $\begin{cases} -3x + \dfrac{1}{2}y = 4 \\ y = 6x + 8 \end{cases}$ $\{(x, y) \mid y = 6x + 8\}$

48. $\begin{cases} 2x - \dfrac{1}{5}y = 1 \\ y = 10x - 5 \end{cases}$ $\{(x, y) \mid y = 10x - 5\}$

49. $\begin{cases} -2x + y = -5 \\ 6x - 3y = 15 \end{cases}$ $\{(x, y) \mid 2x - y = 5\}$

50. $\begin{cases} x - 6y = 3 \\ -3x + 18y = -9 \end{cases}$ $\{(x, y) \mid x - 6y = 3\}$

51. $\begin{cases} -4x + y = 0 \\ 4x - y = 1 \end{cases}$ $\varnothing$

52. $\begin{cases} \dfrac{1}{2}x - y = 3 \\ 2x - 4y = 12 \end{cases}$ $\{(x, y) \mid x - 2y = 6\}$

Solve each problem using a system of linear equations. (*See Objective 4.*)

53. In 2010, the two largest toy manufacturers, Mattel and Hasbro, reported a total net sales of \$9.86 billion. If the reported net sales of Mattel is \$2.14 billion less than twice that of Hasbro, find the net sales of each company. (Source: http://www.ita.doc.gov, http://investor.shareholder.com/mattel/results.cfm; www.zacks.com) Mattel reported net sales of \$5.86 billion and Hasbro reported net sales of \$4 billion.

54. In 2010, there were 48,464 thousand new and used vehicles sold in the United States. The number of used vehicles sold was 2144 thousand more than 3 times the number of new vehicles sold. Find the number of new and used vehicles sold in the United States in 2010. (Source: http://www.bts.gov/publications/national_transportation_statistics/) There were 11,580 thousand new vehicles and 36,884 thousand used vehicles sold in 2010.

55. In April 2011, there were a total of 153,421 thousand employed and unemployed persons in the United States. The number employed was 2204 thousand more than 10 times the number unemployed. How many people were employed and unemployed in April 2011? (Source: http://www.bls.gov/news.release/empsit.t01.htm) There were 139,674 thousand employed and 13,747 thousand unemployed in April 2011.

56. The Consumer Expenditure Survey conducted by the Department of Labor in 2009 showed that each consumer unit spends an average of \$6372 on food each year. This includes food at home and food away from home. The amount spent on food at home was \$1134 more than the amount spent on food away from home. What is the average annual expenditure for food at home and food away from home? (Source: http://www.bls.gov/news.release/cesan.nr0.htm) An average of \$3753 was spent on food at home each year and an average of \$2619 was spent on food away from home.

Mix 'Em Up!

Solve each system by substitution or elimination.

57. $\begin{cases} x - y = -8 \\ 4x - y = -26 \end{cases}$ $\{(-6, 2)\}$

58. $\begin{cases} x - y = 8 \\ 7x + 4y = -21 \end{cases}$ $\{(1, -7)\}$

59. $\begin{cases} x - 5y = -1 \\ y = \dfrac{4}{5}x - 1 \end{cases}$ $\left\{\left(2, \dfrac{3}{5}\right)\right\}$

60. $\begin{cases} x - 3y = -6 \\ y = \dfrac{1}{3}x + 2 \end{cases}$ $\left\{\left(-5, \dfrac{1}{3}\right)\right\}$

61. $\begin{cases} 5x - 4y = 13 \\ 0.3x - 0.2y = 0.9 \end{cases}$ $\{(5, 3)\}$

62. $\begin{cases} 3x + y = 17 \\ -0.08x + 0.05y = -0.99 \end{cases}$ $\{(8, -7)\}$

63. $\begin{cases} 2x + y = 12 \\ -2y - 4x = -26 \end{cases}$ $\varnothing$

64. $\begin{cases} 3x - y = -5 \\ -6x + 2y = 12 \end{cases}$ $\varnothing$

65. $\begin{cases} -4x - 3y = 13 \\ -5x - y = 30 \end{cases}$ $\{(-7, 5)\}$

66. $\begin{cases} -5x + 4y = 42 \\ -3x + y = 28 \end{cases}$ $\{(-10, -2)\}$

67. $\begin{cases} 2x - y = 1 \\ -4x + 2y = -2 \end{cases}$ $\{(x, y) \mid 2x - y = 1\}$

68. $\begin{cases} \dfrac{2}{3}x + y = 2 \\ x + \dfrac{3}{2}y = 3 \end{cases}$ $\{(x, y) \mid 2x + 3y = 6\}$

69. $\begin{cases} \dfrac{3}{8}x - \dfrac{7}{8}y = -1 \\ -\dfrac{5}{6}x - \dfrac{7}{3}y = 1 \end{cases}$ $\left\{\left(-2, \dfrac{2}{7}\right)\right\}$

70. $\begin{cases} \dfrac{6}{7}x - \dfrac{8}{7}y = 1 \\ \dfrac{1}{4}x - \dfrac{1}{2}y = \dfrac{3}{16} \end{cases}$ $\left\{\left(2, \dfrac{5}{8}\right)\right\}$

71. $\begin{cases} 0.5x - 0.2y = 3 \\ -2.5x + y = -15 \end{cases}$ $\{(x, y) \mid 5x - 2y = 30\}$

72. $\begin{cases} 0.3x - 0.7y = 1.4 \\ 15x - 35y = 70 \end{cases}$ $\{(x, y) \mid 3x - 7y = 14\}$

73. $\begin{cases} 0.2x + 0.3y = 3 \\ 0.6x - 0.1y = 5 \end{cases}$ $\{(9, 4)\}$

74. $\begin{cases} 0.1x + 0.1y = 1.3 \\ 0.1x + 0.5y = 3.3 \end{cases}$ $\{(8, 5)\}$

75. $\begin{cases} x = 0.6y + 3.2 \\ 0.7x + 0.4y = 4.7 \end{cases}$ $\{(5, 3)\}$

76. $\begin{cases} x = 3y + 15 \\ 0.6x - 0.5y = 3.8 \end{cases}$ $\{(3 -4)\}$

77. $\begin{cases} -4x - 4 = 8y \\ x + 2y = -3 \end{cases}$ $\varnothing$

78. $\begin{cases} x - 3y = 7 \\ 2x - 10 = 6y \end{cases}$ $\varnothing$

79. $\begin{cases} \dfrac{4}{7}x - \dfrac{6}{7}y = 1 \\ \dfrac{1}{4}x - \dfrac{3}{2}y = -\dfrac{1}{2} \end{cases}$ $\left\{\left(3, \dfrac{5}{6}\right)\right\}$

80. $\begin{cases} \dfrac{3}{13}x - \dfrac{2}{13}y = -1 \\ \dfrac{1}{2}x - \dfrac{3}{2}y = -\dfrac{11}{4} \end{cases}$ $\left\{\left(-4, \dfrac{1}{2}\right)\right\}$

You Be the Teacher!

Answer each student's question.

81. Collin: I am trying to solve the following system by substitution but my resulting equation doesn't have any variables. What does this mean?

$$\begin{cases} 3x - 6y = 8 \\ y = \frac{1}{2}x - 1 \end{cases}$$

$$3x - 6\left(\frac{1}{2}x - 1\right) = 8 \quad \varnothing$$
$$3x - 3x + 6 = 8$$
$$6 = 8$$

82. Jeffrey: I am trying to solve the following system by substitution but my resulting equation doesn't have any variables. What does this mean?

$$\begin{cases} 5x + 2y = 6 \\ y = -\frac{5}{2}x + 3 \end{cases}$$

$$5x + 2\left(-\frac{5}{2}x + 3\right) = 6 \quad \{(x, y)|5x + 2y = 6\}$$
$$5x - 5x + 6 = 6$$
$$6 = 6$$

Correct each student's errors, if any.

83. Solve the system by elimination.

$$\begin{cases} 6x + 5y = 3 \\ 3x + 2y = 0 \end{cases}$$

Zach's work:

$$\begin{cases} 6x + 5y = 3 \\ -2(3x + 2y = 0) \end{cases} \rightarrow \begin{cases} 6x + 5y = 3 \\ -6x - 2y = 0 \end{cases}$$

$$6x + 5y = 3$$
$$\underline{-6x - 2y = 0}$$
$$3y = 3$$
$$y = 1$$
$$6x + 5y = 3$$
$$6x + 5(1) = 3$$
$$6x = -2$$
$$x = -\frac{1}{3}$$

$$\begin{cases} 6x + 5y = 3 \\ -2(3x + 2y = 0) \end{cases} \rightarrow \begin{cases} 6x + 5y = 3 \\ -6x - 4y = 0 \end{cases}$$

$$6x + 5y = 3$$
$$\underline{-6x - 4y = 0}$$
$$y = 3$$
$$6x + 5y = 3$$
$$6x + 5(3) = 3$$
$$6x + 15 = 3$$
$$6x = -12$$
$$x = -2$$

The solution set is $\{(-2, 3)\}$.

84. Solve the system by elimination.

$$\begin{cases} (2x - 7y = 24) \\ (5x + 2y = 21) \end{cases}$$

Maryem's work:

$$\begin{cases} 2(2x - 7y = 24) \\ 7(5x + 2y = 21) \end{cases} \rightarrow \begin{cases} 4x - 14y = 24 \\ 35x + 14y = 21 \end{cases}$$

$$4x - 14y = 24$$
$$\underline{35x + 14y = 21}$$
$$39x = 45$$
$$x = \frac{45}{39} = \frac{15}{13}$$
$$5x + 2y = 21$$
$$5\left(\frac{15}{13}\right) + 2y = 21$$
$$2y = \frac{198}{13}$$
$$y = \frac{99}{13}$$

$$\begin{cases} 2(2x - 7y = 24) \\ 7(5x + 2y = 21) \end{cases} \rightarrow \begin{cases} 4x - 14y = 48 \\ 35x - 14y = 147 \end{cases}$$

$$4x - 14y = 48$$
$$\underline{35x + 14y = 147}$$
$$39x = 195$$
$$x = \frac{195}{39} = 5$$
$$5x + 2y = 21$$
$$5(5) + 2y = 21$$
$$2y = -4$$
$$y = \frac{-4}{2} = -2$$

The solution set is $\{(5, -2)\}$.

Calculate It!

Use a graphing calculator to solve each system.

85. $\begin{cases} x = 0.52y + 172 \\ 0.32x + 0.12y = 312.8 \end{cases}$ $\{(640, 900)\}$

86. $\begin{cases} y = -0.64x - 312 \\ 0.24x - 0.18y = 216 \end{cases}$ $\{(450, -600)\}$

87. $\begin{cases} 0.15x - 0.42y = -345 \\ -0.12x + 0.05y = -24.3 \end{cases}$ $\{(640, 1050)\}$

88. $\begin{cases} 0.25x + 0.36y = 432 \\ -0.56x + 0.84y = -226.8 \end{cases}$ $\{(1080, 450)\}$

89. $\begin{cases} 0.75x - 0.03y = 81 \\ -0.15x + 0.12y = 18 \end{cases}$ $\{(120, 300)\}$

90. $\begin{cases} 0.09x + 0.07y = 47 \\ -0.54x + 0.56y = 61 \end{cases}$ $\{(250, 350)\}$

Think About It!

91. The expression $4x - 2$ is substituted in place of y in the second equation of a system. If the solution of the system is (3, 10), what is the equation of the other line in the system? Answers vary; $2x - y = -4$

92. The expression $\frac{1}{3}y + 1$ is substituted in place of x in the second equation of a system. If the solution of the system is (−1, −6), what is the equation of the other line in the system? Answers vary; $5x + y = -11$

93. The expression $-5x + 3$ is substituted in place of y in the second equation of a system. If the solution of the system is $\varnothing$, what is the equation of the other line in the system? Answers vary; $m = -5$ and $b \neq 3$, one example is $y = -5x - 6$.

94. The expression $7y - 1$ is substituted in place of x in the second equation of a system. If the solution of the system is $\varnothing$, what is the equation of the other line in the system?

95. If one equation in a system is $2x + y = 6$ and the equation that results after elimination is $0 = 5$, what is the equation of the other line in the system? $2x + y = 1$

96. If one equation in a system is $x - 3y = 4$ and the equation that results after elimination is $0 = -1$, what is the equation of the other line in the system? $x - 3y = 5$

97. If one equation in a system is $2x + y = 6$ and the equation that results after elimination is $0 = 0$, what is the equation of the other line in the system?
Answers vary; $4x + 2y = 12$

98. If one equation in a system is $x - 3y = 4$ and the equation that results after elimination is $0 = 0$, what is the equation of the other line in the system?
Answers vary; $3x - 9y = 12$

PIECE IT TOGETHER SECTIONS 5.1–5.2

Determine if the ordered pair is a solution of the system. (*Section 5.1, Objective 1*)

1. $\begin{cases} 3x - 5y = -17 \\ y = \dfrac{3}{4}x + \dfrac{83}{20} \end{cases}; \left(5, \dfrac{2}{5}\right)$ no

2. $\begin{cases} y = \dfrac{7}{9}x - 2 \\ y = 4x - 31 \end{cases}; (9, 5)$ yes

Solve each system graphically. (*Section 5.1, Objectives 2 and 3*)

3. $\begin{cases} 2x + y = 1 \\ x + 2y = -10 \end{cases}$ $\{(4, -7)\}$

4. $\begin{cases} x - 2y = 1 \\ 8x + y = 8 \end{cases}$ $\{(1, 0)\}$

5. $\begin{cases} -7x - y = 14 \\ 7x + y = 14 \end{cases}$ $\varnothing$

6. $\begin{cases} -x + 4y = 8 \\ 3x - 12y = -24 \end{cases}$ $\{(x, y) | x - 4y = -8\}$

Determine how the lines relate, the number of solutions of the system, and the type of system without graphing. (*Section 5.1, Objective 4*)

7. $\begin{cases} x = 2y + 7 \\ y = \dfrac{1}{2}x + 3 \end{cases}$ parallel lines, no solution, inconsistent system

8. $\begin{cases} \dfrac{1}{3}x + y = 2 \\ 6x + 5y = 10 \end{cases}$

9. $\begin{cases} y = -\dfrac{4}{3}x + 2 \\ -\dfrac{1}{3}x - \dfrac{1}{4}y = -\dfrac{1}{2} \end{cases}$

10. $\begin{cases} 3x - y = -3 \\ y = 4x \end{cases}$

Solve each system by substitution or elimination. (*Section 5.2, Objectives 1 and 2*)

11. $\begin{cases} y = -3x - 7 \\ 3y + 5x = -11 \end{cases}$ $\{(-2, -1)\}$

12. $\begin{cases} 2x + 3y = -11 \\ 5x - 2y = -18 \end{cases}$ $\{(-4, -1)\}$

13. $\begin{cases} -2x + y = 9 \\ 4x - 2y = -18 \end{cases}$ $\{(x, y) | 2x - y = -9\}$

14. $\begin{cases} -0.2x + 0.5y = -1.7 \\ 2.1x + 1.5y = -2.4 \end{cases}$ $\{(1, -3)\}$

15. $\begin{cases} 3x + 5y = 23 \\ 6x - y = 35 \end{cases}$ $\{(6, 1)\}$

16. $\begin{cases} 0.1x - 0.3y = 0.8 \\ -0.6x + 1.8y = 6 \end{cases}$ $\varnothing$

17. $\begin{cases} \dfrac{5}{4}x - 2y = 18 \\ 7x - y = 60 \end{cases}$ $\{(8, -4)\}$

18. $\begin{cases} \dfrac{1}{3}x - 2y = \dfrac{14}{3} \\ 2x - \dfrac{1}{2}y = -\dfrac{13}{2} \end{cases}$ $\{(-4, -3)\}$

SECTION 5.3 Applications of Linear Systems in Two Variables

OBJECTIVES

As a result of completing this section, you will be able to

1. Solve money applications.
2. Solve investment applications.
3. Solve mixture applications.
4. Solve distance applications.
5. Solve geometry applications.
6. Troubleshoot common errors.

Sam and Lori flew from Atlanta, Georgia, to San Francisco, California, a distance of approximately 2135 mi. The flight from Atlanta to San Francisco took 5 hr since the plane was flying against the wind. The return trip took 4.5 hr since the plane was flying with the wind. What was the speed of the plane in still air and what was the speed of the wind?

In this section we will discuss how to use a system of linear equations in two variables to solve this problem and other applications.

The applications in this section are similar to the applications covered in Chapter 2. In Chapter 2, there were two unknowns in the problem and we expressed both unknowns in terms of one variable and we used one linear equation to solve the problem. Sometimes, it is easier to assign a different variable to each unknown. This is what we will do in this section. We will assign two different variables to the two unknown values, and then use a system of linear equations in two variables to solve the problem. There are four general steps we will follow to solve such problems.

Procedure: General Problem-Solving Strategy

Step 1: Read the problem and determine the two unknown values. Assign a different variable to represent each of these values.
Step 2: Read the problem and write two equations that define a relationship between the two variables.
Step 3: Solve the system of equations by substitution or elimination.
Step 4: Write the answer to the problem.

Money Applications

Objective 1 Solve money applications.

Some of the applications covered in this objective require us to find the value of a collection of objects. The following table illustrates how to find the total worth of a collection of books, for example, if each book in the collection has the same value.

Number of Books	Value of Each Book	Total Worth of Collection
3	\$10	$3(10) = \$30$
7	\$15	$7(15) = \$105$
10	\$20	$10(20) = \$200$
x	\$30	$x(30) = 30x$

So, if we know the value of an individual item, we can determine the worth of a collection of these items by multiplying the value of the individual item by the total number of items in the collection.

This concept can be extended to money, as we have seen before when dealing with applications of linear equations in one variable. Let's suppose we have two different types of coins in a collection. If we know how many of each coin we have in the collection as well as the value of each coin, we can find the total value of the collection of coins.

Suppose we have a total of 30 coins consisting of dimes and quarters. The total value of a collection of coins is provided in the following table for several cases.

Number of Dimes	Number of Quarters	Value of Dimes + Value of Quarters = Total Worth of Collection
6	$30 - 6 = 24$	$6(0.10) + 24(0.25) = \$0.60 + \$6.00 = \$6.60$
20	$30 - 20 = 10$	$20(0.10) + 10(0.25) = \$2.00 + \$2.50 = \$4.50$
12	$30 - 12 = 18$	$12(0.10) + 18(0.25) = \$1.20 + \$4.50 = \$5.70$
x	$30 - x = y$	$x(0.10) + y(0.25) = 0.10x + 0.25y$

The objects in the collection can extend beyond coins. They might be a collection of tickets or a collection of hours at work, as we will see in Examples 1a through 1c.

Objective 1 Examples Solve each problem using a system of linear equations.

1a. A movie theater sold 2500 tickets. Adult tickets cost \$8.25 each and student tickets cost \$6.00 each. If the movie theater collected a total of \$17,475 for the tickets, how many adult tickets and student tickets were sold?

Solution **1a.** What is unknown? The number of adult tickets and the number of student tickets sold are unknown.

Let $a =$ the number of adult tickets sold.
Let $s =$ the number of student tickets sold.

What is known? The total number of tickets sold is 2500. The total money collected is \$17,475. We can organize this information in a table.

Unknowns	Number of tickets sold ×	Price per ticket =	Total money collected
Adult tickets sold	a	\$8.25	\$8.25a
Student tickets sold	s	\$6.00	\$6.00s
Totals	2500		\$17,475

From the table, we write a system that enables us to solve the problem.

$$\begin{cases} a + s = 2500 \\ 8.25a + 6.00s = 17{,}475 \end{cases}$$

A movie theater sold 2500 tickets.
The movie theater collected \$17,475.

We solve the system by elimination since both equations are in standard form.

$$\begin{cases} -6(a + s = 2500) \\ 8.25a + 6.00s = 17{,}475 \end{cases}$$

$$\begin{cases} -6a - 6s = -15{,}000 \\ 8.25a + 6.00s = 17{,}475 \end{cases}$$ Multiply the first equation by −6. Second equation

$2.25a = 2475$ Add.

$a = 1100$ Divide each side by 2.25.

Now solve for s by substituting 1100 for a in one of the equations in the system.

$a + s = 2500$ Begin with the first equation.

$1100 + s = 2500$ Substitute 1100 for a.

$1100 + s - 1100 = 2500 - 1100$ Subtract 1100 from each side.

$s = 1400$ Simplify.

The movie theater sold 1100 adult tickets and 1400 student tickets.

1b. Monte shows his children a jar of coins that consists of only quarters and nickels. He tells them there are 300 coins in the jar, totaling \$51. How many quarters and nickels are in the jar?

Solution **1b.** What is unknown? The number of quarters and the number of nickels are unknown.

Let q = number of quarters.

Let n = number of nickels.

What is known? There are 300 coins in the jar. The value of the coins is \$51. We can organize the information in a table.

Unknowns	Number of coins ×	Value per coin =	Total value of coins
Quarters	q	\$0.25	\$0.25q
Nickels	n	\$0.05	\$0.05n
Totals	300		\$51

From the table, we write a system that enables us to solve the problem.

$$\begin{cases} q + n = 300 \\ 0.25q + 0.05n = 51 \end{cases}$$

There is a total of 300 coins in the jar.
The total value of the collection is \$51.

Because the coefficients in the first equation are 1, we solve the system by substitution.

$$
\begin{aligned}
q + n &= 300 && \text{Begin with the first equation.}\\
q + n - n &= 300 - n && \text{Subtract } n \text{ from each side.}\\
q &= 300 - n && \text{Simplify.}
\end{aligned}
$$

Now we substitute this expression in place of q in the other equation.

$$
\begin{aligned}
0.25q + 0.05n &= 51 && \text{Begin with the second equation.}\\
0.25(300 - n) + 0.05n &= 51 && \text{Substitute } 300 - n \text{ for } q.\\
75 - 0.25n + 0.05n &= 51 && \text{Apply the distributive property.}\\
75 - 0.20n &= 51 && \text{Combine like terms.}\\
75 - 0.20n - 75 &= 51 - 75 && \text{Subtract 75 from each side.}\\
-0.20n &= -24 && \text{Simplify.}\\
\frac{-0.20n}{-0.20} &= \frac{-24}{-0.20} && \text{Divide each side by } -0.20.\\
n &= 120 && \text{Simplify.}
\end{aligned}
$$

Finally we solve for q by replacing n with 120 in the first equation.

$$
\begin{aligned}
q + n &= 300 && \text{Begin with the first equation.}\\
q + 120 &= 300 && \text{Replace } n \text{ with 120.}\\
q + 120 - 120 &= 300 - 120 && \text{Subtract 120 from each side.}\\
q &= 180 && \text{Simplify.}
\end{aligned}
$$

Monte's jar contains 120 nickels and 180 quarters.

1c. Janna is working two jobs to help pay her way through college. She works as an office assistant and a store clerk. One week she earns \$440 by working 25 hr as an office assistant and 20 hr as a store clerk. Another week she earns \$390 by working 15 hr as an office assistant and 30 hr as a store clerk. What is Janna's hourly wage as an office assistant and as a store clerk?

Solution **1c.** What is unknown? The hourly wages Janna makes as an office assistant and as a store clerk are unknown.

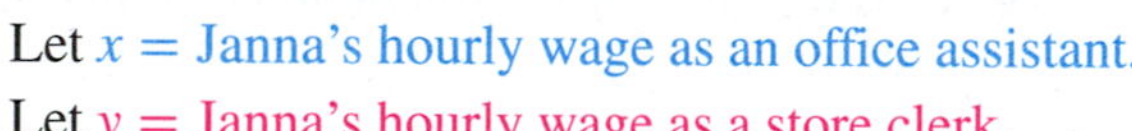

Let x = Janna's hourly wage as an office assistant.

Let y = Janna's hourly wage as a store clerk.

What is known? Janna's earnings for week 1 were \$440 and Janna's earnings for week 2 were \$390. We can organize the information in a table.

Unknowns	Hourly Wage	Hours Worked for Week 1	Total Paid	Hours Worked for Week 2	Total Paid
Office assistant wage	x	25	$25x$	15	$15x$
Store clerk wage	y	20	$20y$	30	$30y$
Totals			440		390

The system that represents these facts is shown.

$$
\begin{cases}
25x + 20y = 440 & \text{Week 1 earnings were \$440.}\\
15x + 30y = 390 & \text{Week 2 earnings were \$390.}
\end{cases}
$$

We solve the system by elimination.

$$\begin{cases} 3(25x + 20y = 440) \\ -2(15x + 30y = 390) \end{cases} \rightarrow \begin{cases} 75x + 60y = 1320 \\ -30x - 60y = -780 \end{cases}$$ Multiply the first equation by 3. Multiply the second equation by −2.

$45x = 540$ Add the equations.

$\frac{45x}{45} = \frac{540}{45}$ Divide each side by 45.

$x = 12$ Simplify.

Now solve for y.

$25x + 20y = 440$ Begin with the first equation.

$25(12) + 20y = 440$ Substitute 12 for x.

$300 + 20y = 440$ Simplify.

$300 + 20y - 300 = 440 - 300$ Subtract 300 from each side.

$20y = 140$ Simplify.

$\frac{20y}{20} = \frac{140}{20}$ Divide each side by 20.

$y = 7$ Simplify.

So, Janna earns \$12 an hour as an office assistant and \$7 an hour as a store clerk.

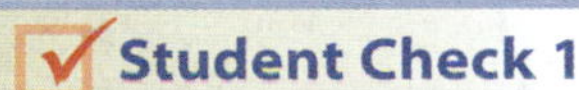

Student Check 1 Solve each problem using a system of linear equations.

a. Admission to a football game is \$10 for students and \$14 for general admission. If 300 tickets are sold and \$3500 is collected, how many student tickets and general admission tickets are sold?

b. Thomas has a collection of coins consisting of nickels and dimes. If he has 500 coins totaling \$40, how many nickels and dimes are in his collection?

c. Two families go to the movies together. One family purchases four drinks and two large popcorns for \$26. The other family purchases two drinks and three large popcorns for \$24. How much does each drink and popcorn cost?

Investment Applications

Objective 2 Solve investment applications.

Investment problems in this course deal primarily with *simple interest*. **Interest** is a fee paid for the use of money. Interest is calculated as a percentage of the amount of money invested. Simple interest is the most basic type of interest that can be earned.

Property: Simple Interest Formula

Let P be the principal or initial money invested, r the annual interest rate (as a decimal), and t the length of the investment in years. The earned interest I is

$$I = Prt$$

Suppose we invest \$5000 in two different savings accounts. One account earns 5% annual interest and the other earns 4% annual interest. The following table shows how much annual interest is earned from the two accounts for two different principal amounts and rates.

	Principal ×	Rate ×	Time =	Interest
Account 1	\$3000	0.05	1	(3000)(0.05)(1) = \$150
Account 2	\$2000	0.04	1	(2000)(0.04)(1) = \$80
Totals	\$5000			\$230

If the amount invested in each account is unknown, we assign a variable to represent those amounts. The amount of interest earned is shown in the table.

	Principal	× Rate	× Time	= Interest
Account 1	x	0.05	1	$(x)(0.05)(1) = 0.05x$
Account 2	y	0.04	1	$(y)(0.04)(1) = 0.04y$
Totals	\$5000			$0.05x + 0.04y$

We can use these ideas to solve word problems involving investments.

Objective 2 Example Sonja invests \$3000 in two different accounts. She invests part of her money in a savings account that earns 4% simple interest and the rest of her money in a money market account that earns 8% simple interest. If she earns a total of \$200 interest in one year, how much did she invest in each account?

Solution What are the unknowns? The amount of money invested at 4% and at 8% is unknown.

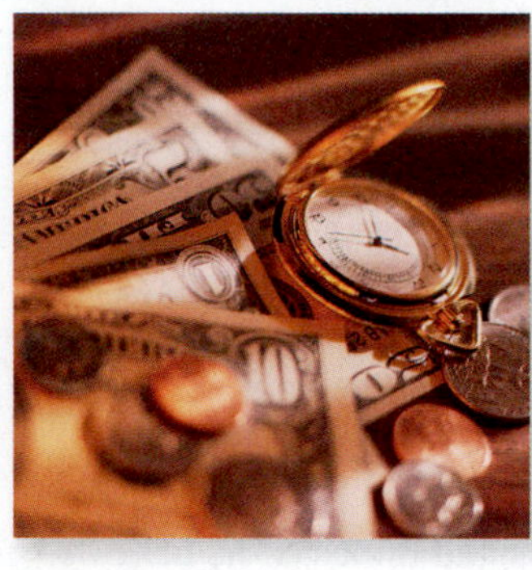

Let $x =$ amount invested at 4%.

Let $y =$ amount invested at 8%.

What is known? The total amount invested is \$3000. The total interest earned is \$200. We can organize the information in a table.

	Principal	× Rate	× Time	= Interest
Savings account	x	0.04	1	$(x)(0.04)(1) = 0.04x$
Money market account	y	0.08	1	$(y)(0.08)(1) = 0.08y$
Totals	\$3000			\$200

From the table, we write a system that enables us to solve the problem.

$$\begin{cases} x + y = 3000 & \text{Total invested is \$3000.} \\ 0.04x + 0.08y = 200 & \text{Total interest earned is \$200.} \end{cases}$$

We solve by elimination. We first clear the decimals from the system by multiplying the second equation by 100.

$$\begin{cases} x + y = 3000 \\ 100(0.04x + 0.08y = 200) \end{cases} \quad \text{simplifies to} \quad \begin{cases} x + y = 3000 \\ 4x + 8y = 20{,}000 \end{cases}$$

Now multiply the first equation by -8 to eliminate the y-terms from the equations.

$$\begin{cases} -8(x + y = 3000) \\ 4x + 8y = 20{,}000 \end{cases} \quad \text{simplifies to} \quad \begin{cases} -8x - 8y = -24{,}000 \\ 4x + 8y = 20{,}000 \end{cases} \quad \text{Multiply the first equation by } -8.$$

$$-4x = -4000 \quad \text{Add the equations.}$$

$$\frac{-4x}{-4} = \frac{-4000}{-4} \quad \text{Divide each side by } -4.$$

$$x = 1000 \quad \text{Simplify.}$$

Now solve for y.

$$x + y = 3000 \quad \text{Begin with the first equation.}$$

$$1000 + y = 3000 \quad \text{Replace } x \text{ with 1000.}$$

$$1000 + y - 1000 = 3000 - 1000 \quad \text{Subtract 1000 from each side.}$$

$$y = 2000 \quad \text{Simplify.}$$

Sonja invested \$1000 in the 4% account and \$2000 in the 8% account.

Student Check 2 Georgia receives an inheritance of \$100,000. She invests part of the money in a certificate of deposit (CD) that earns 8% simple interest. She invests the rest of the money in a money market account that earns 12% simple interest. If Georgia earns \$11,000 in interest for one year, how much did she invest in each account?

Mixture Applications

Objective 3 Solve mixture applications.

Mixture applications involve combining two or more substances to get a new substance. To illustrate this concept, we will examine the percent alcohol in beer, wine, and liquor. Each of the following is equivalent to one standard drink, which have equal effects on the body.

Solution	Amount of Pure Alcohol
1.5 oz of 80 proof liquor (40% alcohol)	$1.5(0.40) = 0.6$ oz of pure alcohol
12 oz of regular beer (5% alcohol)	$12(0.05) = 0.6$ oz of pure alcohol
5 oz of table wine (12% alcohol)	$5(0.12) = 0.6$ oz of pure alcohol

Though these three drinks are very different, they contain the same amount of pure alcohol. So, if someone had these three drinks, they would be drinking $0.6 + 0.6 + 0.6 = 1.8$ oz of pure alcohol, not $1.5 + 12 + 5 = 18.5$ oz of alcohol.

So, when we combine different substances, we can calculate the amount of pure substance by multiplying the strength of the substance by the amount of the substance.

Objective 3 Example A dairy farmer produces two types of milk. How many gallons of a 3.25% milk fat solution and how many gallons of a 1% milk fat solution should be mixed together to obtain 750 gal of a 2% milk fat solution?

Solution What are the unknowns? The gallons of 3.25% milk fat solution and the gallons of 1% milk fat solution are unknown.

Let $x =$ number of gallons of 3.25% milk fat solution.

Let $y =$ number of gallons of 1% milk fat solution.

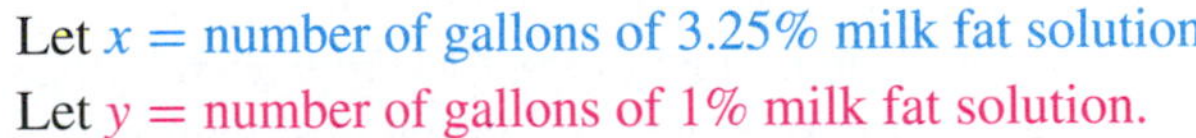

What is known? The total mixture is 750 gal. The final mixture is 2% milk fat. We can organize the information in the table.

Unknowns	Amount of Substance	Strength of Substance	Amount of Milk Fat
Gallons of milk with 3.25% milk fat	x	$3.25\% = 0.0325$	$0.0325x$
Gallons of milk with 1% milk fat	y	$1\% = 0.01$	$0.01y$
Final Mixture	750	$2\% = 0.02$	$0.02(750) = 15$

From the table, we write a system that enables us to solve the problem.

$$\begin{cases} x + y = 750 & \text{Total obtained is 750 gal.} \\ 0.0325x + 0.01y = 15 & \text{Mixture needs to be 2\% milk fat.} \end{cases}$$

We solve the system by elimination. We first clear the decimals from the system by multiplying the second equation by 10,000.

$$\begin{cases} x + y = 750 \\ 10{,}000(0.0325x + 0.01y = 15) \end{cases} \rightarrow \begin{cases} x + y = 750 \\ 325x + 100y = 150{,}000 \end{cases}$$

Now multiply the first equation by -100 to eliminate the y-terms from the equations.

$$\begin{cases} -100(x + y = 750) \\ 325x + 100y = 150{,}000 \end{cases} \rightarrow \begin{cases} -100x - 100y = -75{,}000 \\ 325x + 100y = 150{,}000 \end{cases}$$

$$225x = 75{,}000 \quad \text{Add the equations.}$$

$$\frac{225x}{225} = \frac{75{,}000}{225} \quad \text{Divide each side by 225.}$$

$$x \approx 333.33 \quad \text{Approximate.}$$

Now solve for y.

$$x + y = 750 \quad \text{Begin with first equation.}$$

$$333.33 + y \approx 750 \quad \text{Substitute 333.33 for } x.$$

$$333.33 + y - 333.33 \approx 750 - 333.33 \quad \text{Subtract 333.33 from each side.}$$

$$y \approx 416.67 \quad \text{Simplify.}$$

The farmer needs to mix 333.33 gal of the 3.25% milk fat solution with 416.67 gal of the 1% milk fat solution to obtain 750 gal of a 2% milk fat solution.

Student Check 3 A chemist has a 5% acid solution and 50% acid solution. She needs 30 L of a 15% acid solution. How many liters of the 5% acid solution and how many liters of the 50% acid solution must she mix together to obtain 30 L of a 15% acid solution?

Distance Applications

Objective 4 Solve distance applications.

The total distance traveled equals the rate (or speed) times the time elapsed. Algebraically, we write the **distance formula** as $d = rt$, where d is the distance, r is the rate or speed, and t is the time traveled. We can consider flying to illustrate this concept.

- If a plane is flying *directly with the wind*, the plane will arrive at its destination sooner than it would if flying in still air since the speed of the wind increases the plane's speed relative to the ground.
- If, however, a plane flies *directly against the wind*, the plane will take longer to reach its destination since the speed of the wind decreases the speed of the plane relative to the ground.

The following table provides some illustrations of how the speed of the wind affects the speed of a plane.

Speed of a Plane in Still Air	Speed of the Wind	Speed of Plane Flying with the Wind	Speed of Plane Flying Against the Wind
500 mph	20 mph	$500 + 20 = 520$ mph	$500 - 20 = 480$ mph
600 mph	x mph	$600 + x$ mph	$600 - x$ mph

Note:

- *Flying directly with the wind increases the speed of the plane by the speed of the wind.*
- *Flying directly against the wind decreases the speed of the plane by the speed of the wind.*

Objective 4 Examples

Solve each problem using a system of linear equations.

4a. Sam and Lori flew from Atlanta, Georgia (ATL), to San Francisco, California (SFO), a distance of approximately 2135 mi. The flight from Atlanta to San Francisco took 5 hr since the plane was flying against the wind. The return trip took 4.5 hr since the plane was flying with the wind. What was the speed of the plane in still air and what was the speed of the wind?

Solution

4a. What are the unknowns? The speed of the plane and the speed of the wind are unknown.

Let $x =$ speed of the plane.

Let $y =$ speed of the wind.

What is known? The total distance between the cities is 2135 mi. The trip against the wind was 5 hr. The trip with the wind was 4.5 hr. We can organize this information in a table as shown.

	Rate	• Time	= Distance
From ATL to SFO	$x - y$	5	2135
From SFO to ATL	$x + y$	4.5	2135

The system of linear equations that represents this problem is

$$\begin{cases} 5(x - y) = 2135 \\ 4.5(x + y) = 2135 \end{cases}$$

We choose to solve the system by elimination after first applying the distributive property to clear parentheses. Then we clear decimals by multiplying the second equation by 10.

$$\begin{cases} 5x - 5y = 2135 \\ 4.5x + 4.5y = 2135 \end{cases} \rightarrow \begin{cases} 5x - 5y = 2135 \\ 10(4.5x + 4.5y = 2135) \end{cases} \rightarrow \begin{cases} 5x - 5y = 2135 \\ 45x + 45y = 21{,}350 \end{cases}$$

Now multiply the first equation by 9 to eliminate the y-terms from the system.

$$\begin{cases} 9(5x - 5y = 2135) \\ 45x + 45y = 21{,}350 \end{cases} \rightarrow \begin{cases} 45x - 45y = 19{,}215 \\ 45x + 45y = 21{,}350 \end{cases} \quad \text{Multiply first equation by 9.}$$

$$90x = 40{,}565 \quad \text{Add the equations.}$$

$$\frac{90x}{90} = \frac{40{,}565}{90} \quad \text{Divide each side by 90.}$$

$$x \approx 450.7 \quad \text{Approximate.}$$

Next solve for y.

$$5(x - y) = 2135 \quad \text{Begin with the first equation.}$$

$$5(450.7 - y) \approx 2135 \quad \text{Replace } x \text{ with 450.7.}$$

$$\frac{5(450.7 - y)}{5} \approx \frac{2135}{5} \quad \text{Divide each side by 5.}$$

$$450.7 - y \approx 427 \quad \text{Simplify.}$$

$$450.7 - y - 450. \approx 427 - 450.7 \quad \text{Subtract 450.7 from each side.}$$

$$-y \approx -23.7 \quad \text{Simplify.}$$

$$\frac{-y}{-1} \approx \frac{-23.7}{-1} \quad \text{Divide each side by } -1.$$

$$y \approx 23.7 \quad \text{Simplify.}$$

So, the speed of the plane is 450.7 mph and the speed of the wind is 23.7 mph.

4b. At 9 A.M., David leaves his home traveling south on his bike at an average speed of 10 mph. At 11 A.M., his wife Judy leaves their home to find David to give him an urgent message, since he left his cell phone at home. She travels in her car at an average speed of 65 mph. How long will it take for Judy to reach David?

Solution **4b.** What are the unknowns? The time that David and Judy each travel is unknown.

Let $x =$ time David travels.

Let $y =$ time Judy travels.

What is known? David's speed is 10 mph and Judy's speed is 65 mph. The distance traveled by both David and Judy is the same. Judy's traveling time is 2 hr less than David's since she left 2 hr after David. We can organize this information in a table.

	Rate	• Time =	Distance
David	10	x	$10x$
Judy	65	y	$65y$

The system of linear equations that represents this problem is

$$\begin{cases} 10x = 65y \\ y = x - 2 \end{cases}$$

The distances traveled are the same.
Judy's time is 2 hr less than David's.

Because the second equation is solved for y, we solve the system by substitution.

$10x = 65y$ Begin with the first equation.

$10x = 65(x - 2)$ Replace y with $x - 2$.

$10x = 65x - 130$ Apply the distributive property.

$10x - 65x = 65x - 130 - 65x$ Subtract $65x$ from each side.

$-55x = -130$ Simplify.

$\dfrac{-55x}{-55} = \dfrac{-130}{-55}$ Divide each side by -55.

$x \approx 2.4$ Approximate.

INSTRUCTOR NOTE: Remind students that 0.4 hr is equivalent to $0.4 \cdot 60$ min or 24 min.

Finally, we solve for y.

$y = x - 2$ Begin with the second equation.

$y \approx 2.4 - 2$ Substitute 2.4 for x.

$y \approx 0.4$ Simplify.

It will take Judy about 0.4 hr or 24 min to reach David.

Student Check 4 Solve each problem using a system of linear equations.

a. A plane can travel 1200 mi with the wind in 3.2 hr. The return trip against the wind takes 4 hr. What are the speed of the plane and the speed of the wind?

b. Cherie and her mom live 330 mi apart. They want to meet each other so that Cherie's kids can go home with her mom for a few days. Cherie leaves at 8 A.M. and travels toward her mom at an average speed of 50 mph. Her mom leaves at 8:30 A.M. and travels toward Cherie at an average speed of 65 mph. What time will they meet? (Round solutions to the nearest tenth.)

Geometry Applications

Objective 5 Solve geometry applications.

We will use some topics from Chapter 2 to solve applications involving geometry. Some of the geometry problems we will solve involve perimeter of a rectangle and the relationships between complementary and supplementary angles. Recall these important facts.

- The **perimeter of a rectangle** is $2l + 2w$, where l is the length and w is the width.
- **Complementary angles** are two angles whose measures add to 90°.
- **Supplementary angles** are two angles whose measures add to 180°.

Objective 5 Examples Solve each problem using a system of linear equations.

5a. The perimeter of a high school basketball court is 268 ft. The length of the court is 16 ft less than two times the width of the court. Find the dimensions of the court.

Solution

5a. What are the unknowns? The length and the width of the basketball court are unknown.

Let $l =$ the length of the court.

Let $w =$ the width of the court.

What is known? The perimeter of the court is 268 ft. The length is 16 ft less than two times the width. The system is

$$\begin{cases} 2l + 2w = 268 & \text{The perimeter of the court is 268 ft.} \\ l = 2w - 16 & \text{The length is 16 ft less than two times the width.} \end{cases}$$

Because the second equation is solved for l, we solve the system by substitution.

$2l + 2w = 268$	Begin with the first equation.
$2(2w - 16) + 2w = 268$	Substitute $2w - 16$ for l.
$4w - 32 + 2w = 268$	Apply the distributive property.
$6w - 32 = 268$	Combine like terms.
$6w - 32 + 32 = 268 + 32$	Add 32 to each side.
$6w = 300$	Simplify.
$\frac{6w}{6} = \frac{300}{6}$	Divide each side by 6.
$w = 50$	Simplify.

Next solve for the length.

$l = 2w - 16$	Begin with the second equation.
$l = 2(50) - 16$	Substitute 50 for w.
$l = 100 - 16$	Multiply 2 and 50.
$l = 84$	Subtract.

So, the dimensions of the basketball court are 84 ft by 50 ft.

5b. Two angles are supplementary. The measure of one angle is 6° less than twice the measure of the other angle. Find the measure of each angle.

Solution

5b. What are the unknowns? The measure of each angle is unknown.

Let $x =$ the measure of one angle.

Let $y =$ the measure of its supplement.

What is known? The angles are supplementary. The measure of one angle is 6° less than twice the measure of the other. The system is

$$\begin{cases} x + y = 180 & \text{The angles are supplementary.} \\ x = 2y - 6 & \text{The measure of one angle is 6° less than twice the measure of the other.} \end{cases}$$

Because the second equation is solved for x, we solve the system by substitution.

$$\begin{aligned} x + y &= 180 && \text{Begin with the first equation.} \\ 2y - 6 + y &= 180 && \text{Substitute } 2y - 6 \text{ for } x. \\ 3y - 6 &= 180 && \text{Combine like terms.} \\ 3y - 6 + 6 &= 180 + 6 && \text{Add 6 to each side.} \\ 3y &= 186 && \text{Simplify.} \\ \frac{3y}{3} &= \frac{186}{3} && \text{Divide each side by 3.} \\ y &= 62 && \text{Simplify.} \end{aligned}$$

Now solve for x.

$$\begin{aligned} x &= 2y - 6 && \text{Begin with the second equation.} \\ x &= 2(62) - 6 && \text{Substitute 62 for } y. \\ x &= 124 - 6 && \text{Multiply 2 and 62.} \\ x &= 118 && \text{Simplify.} \end{aligned}$$

So, the measure of the first angle is 118° and the measure of its supplement is 62°.

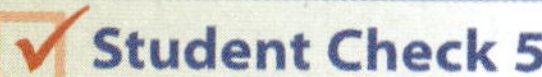

Student Check 5 Solve each problem using a system of equations.

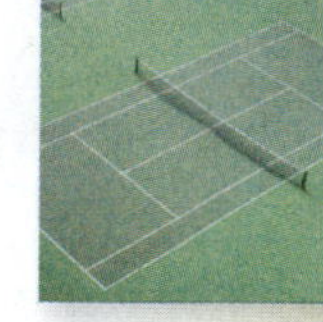

a. A regulation size tennis court has a perimeter of 228 ft. The length of the court is 6 ft more than twice the width. Find the dimensions of the court.

b. Two angles are complementary. The measure of one angle is 10° more than the measure of the other. Find the measure of each angle.

Objective 6 ▶ Troubleshoot common errors.

Troubleshooting Common Errors

Some common errors associated with applications of systems of linear equations are shown next. Most of the errors deal with setting up the equations in the system incorrectly.

Objective 6 Examples **A problem and an incorrect solution are given. Provide the correct solution and an explanation of the error.**

6a. Micah invests some money in two different accounts, a CD that earns 5% annual interest and a money market account that earns 8% annual interest. In the money market account, he invests \$2000 less than 3 times the amount invested in the CD account. If he earns a total of \$3320 in yearly interest, how much was invested in each account? State the system that solves this problem. Do not solve.

Incorrect Solution	Correct Solution and Explanation
Let x = amount invested in the CD acct. Let y = amount invested in the money market account. $\begin{cases} y = 2000 - 3x \\ 0.05x + 0.08y = 3320 \end{cases}$	To represent 2000 less than 3 times an amount is $3x - 2000$. So, the system consists of the following equations. $\begin{cases} y = 3x - 2000 \\ 0.05x + 0.08y = 3320 \end{cases}$

6b. A chemist needs to make a 20% iodine solution. He has a 50% iodine solution and a 10% iodine solution. How much of each should he mix together to obtain 60 mL of a 20% iodine solution? State the system that solves this problem. Do not solve.

Incorrect Solution	Correct Solution and Explanation
Let x = amount of 50% iodine solution. Let y = amount of 10% iodine solution. $\begin{cases} x + y = 60 \\ 0.50x + 0.10y = 60 \end{cases}$	The left side of the equation represents the amount of pure iodine in the two solutions, so the right side should also represent the amount of pure iodine. $\begin{cases} x + y = 60 \\ 0.50x + 0.10y = 0.20(60) \end{cases}$

ANSWERS TO STUDENT CHECKS

Student Check 1 **a.** 175 student tickets and 125 general admission tickets **b.** 200 nickels and 300 dimes **c.** \$3.75 per drink and \$5.50 per popcorn

Student Check 2 \$75,000 invested in the money market account and \$25,000 in the stock market

Student Check 3 $23\frac{1}{3}$ L of the 5% acid solution and $6\frac{2}{3}$ L of the 50% acid solution

Student Check 4 **a.** The plane's speed is 337.5 mph and the wind speed is 37.5 mph. **b.** They will meet at approximately 11:09 A.M.

Student Check 5 **a.** The tennis court is 78 ft long by 36 ft wide. **b.** The angles are 50° and 40°.

SUMMARY OF KEY CONCEPTS

While many different types of applications were solved in this section, there is really only one method that we use to solve these problems. The key to solving an application with a system of two linear equations in two variables is to

1. Determine the two unknowns in the problem. Use two different variables to represent these unknowns.
2. Then write two equations that relate these unknowns to each other.
3. Once the system is set up, use substitution or elimination to solve the system.

SECTION 5.3 EXERCISE SET

Write About It!

Use complete sentences in your answer to each exercise.

1. What is the first step in solving an application problem?
2. If we know the value of an item and the total number of items in the collection, how can we determine the worth of the collection? Answers vary. To determine the worth of a collection of items, multiply the number of items by the value of the item.
3. If we have x dimes and y quarters, explain how to find the total value of the coins. Answers vary. To find the value of x dimes and y quarters, we find the value of $0.10x + 0.25y$.
4. If Warren invests x dollars in a savings account that earns 2.5% simple interest, how much interest he will earn in 1 yr? Answers vary. Warren will earn $0.025x$ dollars in interest if he invests x dollars at 2.5% simple interest.
5. If a plane is flying at a speed of x mph in still air, explain how to define the speed of the plane with and against wind at y mph.

Additional answers can be found in the Instructor Answer Appendix.

6. After setting up a system of two equations with two unknowns, explain how to solve the system.

Practice Makes Perfect!

Solve each problem with a system of linear equations. (*See Objective 1.*)

7. At a movie theater, adult tickets cost \$9.00 each and student tickets cost \$6.00 each. If the movie theater collects a total of \$12,030 for 1500 tickets, how many adult tickets and student tickets are sold? 1010 adult tickets and 490 student tickets

8. At a movie theater, adult tickets cost $8.00 each and student tickets cost $6.50 each. If the movie theater collects a total of $9380 for 1300 tickets, how many adult tickets and student tickets are sold?
620 adult tickets and 680 student tickets

9. Fung has a jar of 470 coins that consists of only quarters and nickels. The total of the coins in the jar is $79.50. How many quarters and nickels are in the jar? 280 quarters and 190 nickels

10. Hong has a jar of 440 coins that consists of only dimes and nickels. The total of the coins in the jar is $30. How many dimes and nickels are in the jar?
160 dimes and 280 nickels

11. Two families go to a buffet restaurant together. One family pays $59 for four adults and two children. Another family pays $98.50 for five adults and seven children. What is the cost for each adult and each child? $12 for an adult and $5.50 for a child

12. Two families go to a grill and buffet restaurant together. One family pays $63.50 for three adults and four children. Another family pays $95.50 for four adults and seven children. What is the cost for each adult and each child? $12.50 for an adult and $6.50 for a child

13. Two families visit Universal Studios Hollywood in Los Angeles. The ticket types are general admission or discounted admission for people who are under 48 in. tall. One family pays $288 for three general admission tickets and one discounted admission ticket. Another family pays $412 for two general admission tickets and four discounted admission tickets. What is the price for each general admission ticket and each discounted admission ticket?
$74 for a general admission ticket and $66 for a discounted ticket

14. Two families take their children to visit a zoo. One family pays $84 for three adults and two children. Another family pays $105 for five adults and one child. What is the cost for each adult and each child to visit the zoo? $18 for an adult and $15 for a child

15. Amie is working two jobs to help pay her way through college. She works as an office assistant and a store clerk. One week she earns $327.50 by working 15 hr as an office assistant and 20 hr as a store clerk. Another week she earns $432.50 by working 25 hr as an office assistant and 20 hr as a store clerk. What is Amie's hourly wage as an office assistant and a store clerk?

16. Bertha is working two jobs to help pay her way through college. She works as a nursing assistant and a teacher assistant. One week she earns $335 by working 20 hr as a nursing assistant and 5 hr as a teacher assistant. Another week she earns $570 by working 25 hr as a nursing assistant and 20 hr as a teacher assistant. What is Bertha's hourly wage as a nursing assistant and a teacher assistant? Bertha's hourly wage as a nursing assistant is $14 and her hourly wage as a teacher assistant is $11.

17. Admission to a football game is $13 for student tickets and $15 for general admission. If 2900 tickets are sold and $42,700 is collected, how many of each type of ticket are sold? 400 student tickets and 2500 general admission tickets

18. Admission to a football game is $10 for student tickets and $24 for general admission. If 2700 tickets are sold and $49,400 is collected, how many of each type of ticket are sold? 1100 student tickets and 1600 general admission tickets

Solve each problem with a system of linear equations. (*See Objective 2.*)

19. Li invests $20,000 in two different accounts. She invests part of her money in a savings account that earns 4.5% simple interest and the rest of her money in a CD that earns 5.25% simple interest. If she earns a total of $1035 interest in 1 yr, how much did she invest in each? $2000 at 4.5% in the savings account and $18,000 at 5.25% in the CD

20. Tanya invests $12,000 in two different accounts. She invests part of her money in a savings account that earns 5% simple interest and the rest of her money in a money market account that earns 6.75% simple interest. If she earns $731.25 interest in 1 yr, how much did she invest in each account? $4500 at 5% in the savings account and $7500 at 6.75% in the money market account

21. Sally receives an inheritance of $88,000. She invests part of the money in a CD that earns 7% simple interest and the rest of the money in a money market account that earns 7.5% simple interest. If she earns $6430 in interest for 1 yr, how much did she invest in each account? $34,000 at 7% in the CD and $54,000 at 7.5% in the money market account

22. Georgia receives an inheritance of $72,000. She invests part of the money in a stock that pays 4.75% in dividends and the rest of the money in a money market account that earns 4.8% simple interest. If she earns $3444 in investment income for 1 yr, how much did she invest in each account? $24,000 at 4.75% in the stock market and $48,000 at 4.8% in the money market account

Solve each problem with a system of linear equations. (*See Objective 3.*)

23. A dairy farmer produces two typesof milk. How many gallons of a 3% milk fat solution and how many gallons of a 1% milk fat solution should he mix together to obtain 600 gal of a 1.5% milk fat solution? 150 gal of 3% milk fat solution and 450 gal of 1% milk fat solution

24. A dairy farmer produces two types of milk. How many gallons of a 3.25% milk fat solution and how many gallons of skim milk (0% milk fat) should he mix together to obtain 650 gal of a 2% milk fat solution?
400 gal of 3.25% milk fat solution and 250 gal of 0% milk fat solution

25. A chemist has a 50% salt solution. She needs 30 L of a 10% salt solution. How many liters of water and how many liters of the 50% salt solution must she mix together to obtain 30 L of a 10% salt solution?
24 L of water and 6 L of 50% salt solution

26. A chemist has a 60% sugar solution. He needs 30 gal of a 20% sugar solution. How many gallons of water and how many gallons of the 60% sugar solution must he mix together to obtain 30 gal of a 20% sugar solution? 20 gal of water and 10 gal of 60% sugar solution

27. A chemist has a 45% alcohol solution and a 25% alcohol solution. He needs 75 gal of a 33.8% alcohol solution. How many gallons of the 45% alcohol solution and how many gallons of the 25% alcohol

solution must he mix together to obtain 75 gal of a 33.8% alcohol solution?
33 gal of 45% alcohol solution and 42 gal of 25% alcohol solution

28. A chemist has a 55% acid solution and a 10% acid solution. He needs 30 gal of a 49% acid solution. How many gallons of the 55% acid solution and how many gallons of the 10% acid solution must he mix together to obtain 30 gal of a 49% acid solution?
26 gal of 55% acid solution and 4 gal of 10% acid solution

Solve each problem with a system of linear equations. (*See Objective 4.*)

29. The distance from San Francisco, California, to Charlotte, North Carolina, is approximately 2450 mi. Flying from San Francisco to Charlotte takes 5.4 hr with the wind. The return flight takes 6.6 hr since the plane is flying against the wind. What is the speed of the plane in still air and what is the speed of the wind? Round to the nearest integer. The speed of the plane in still air is 412 mph and the speed of the wind is 41 mph.

30. The distance from Las Vegas, Nevada, to Washington, D.C., is approximately 2435 mi. Flying from Las Vegas to Washington takes 5.1 hr with the wind. The return flight takes 6.5 hr since the plane is flying against the wind. What is the speed of the plane in still air and what is the speed of the wind? Round to the nearest integer. The speed of the plane in still air is 426 mph and the speed of the wind is 51 mph.

31. At 7 A.M. Lawson leaves his home traveling south on his bike at an average speed of 17 mph. At 8:30 A.M. his wife Jill leaves their home to find Lawson to give him an urgent message, since he left his cell phone at home. She travels south in her car at an average speed of 51 mph. How long will it take Jill to reach Lawson? It will take Jill 0.75 hr to reach him.

32. At 8 A.M. Terry leaves his home traveling south on his bike at an average speed of 18 mph. At 9:45 A.M. his wife Peggy leaves their home to meet Terry for brunch. She travels south in her car at an average speed of 53 mph. How long will it take Peggy to reach Terry?
It will take Peggy 0.9 hr to reach him.

33. A plane travels 1800 mi with the wind in 5 hr. The return trip against the wind takes 6 hr. What is the speed of the plane in still air and what is the speed of the wind?
The speed of the plane is 330 mph and the speed of the wind is 30 mph.

34. A plane travels 1350 mi with the wind in 4.5 hr. The return trip against the wind takes 6 hr. What is the speed of the plane in still air and what is the speed of the wind? The speed of the plane is 262.5 mph and the speed of the wind is 37.5 mph.

Solve each problem with a system of linear equations. (*See Objective 5.*)

35. The perimeter of a table tennis table is 28 ft. The length of the table is 1 ft less than twice the width of the table. Find the dimensions of the table.
The length is 9 ft and the width is 5 ft.

36. The perimeter of a squash court for singles is 34,740 cm. The length of the court is 2130 cm more than the width of the court. Find the dimensions of the court.

The length is 9750 cm and the width is 7620 cm.

37. The perimeter of a racquetball court is 120 ft. The length of the court is 40 ft less than four times the width of the court. Find the dimensions of the court.
The length is 40 ft and the width is 20 ft.

38. The perimeter of a tennis court for singles is 210 ft. The length of the court is 30 ft less than four times the width of the court. Find the dimensions of the court.
The length is 78 ft and the width is 27 ft.

39. Two angles are supplementary. The measure of one angle is 19° less than 4 times the measure of the other angle. Find the measure of each angle. 140.2° and 39.8°

40. Two angles are supplementary. The measure of one angle is 24° less than 2 times the measure of the other angle. Find the measure of each angle. 112° and 68°

41. Two angles are complementary. The measure of one angle is 24° more than 5 times the measure of the other angle. Find the measure of each angle. 11° and 79°

42. Two angles are complementary. The measure of one angle is 22° more than 3 times the measure of the other angle. Find the measure of each angle. 73° and 17°

Mix 'Em Up!

Solve each problem using a linear system.

43. Ashlee invests $13,000 in two different accounts. She invests part of her money in a savings account that earns 4.25% simple interest and the rest of her money in a stock account that pays 7.25% in dividends. If she earns $702.50 in investment income in 1 yr, how much did she invest in each account?
$8000 at 4.25% and $5000 at 7.25%

44. Martha invests $17,000 in two different accounts. She invests part of her money in a savings account that earns 4.5% simple interest and the rest of her money in a CD that earns 5% simple interest. If she earns a total of $825 interest in 1 yr, how much did she invest in each account? $5000 at 4.5% and $12,000 at 5%

45. Two families visit an amusement park together. One family pays $281.70 for four adults and two children. Another family pays $488.45 for six adults and five children. What is the cost for each adult and each child to visit the amusement park?
$53.95 for an adult and $32.95 for a child

46. Two families visit an amusement park together for a 3-day vacation. One family pays $1585 for five adults and three children. The other family pays $2716 for six adults and eight children. What is the cost for each adult and each child to visit the amusement park for 3 days? $206 for an adult and $185 for a child

47. A dairy farmer produces two types of milk. How many gallons of a 3% milk fat solution and how many gallons of skim milk (0% milk fat) should he mix together to obtain 540 gal of a 1% milk fat solution?
180 gal of 3% milk fat solution and 360 gal of skim milk fat solution

48. A dairy farmer produces two types of milk. How many gallons of a 1.5% milk fat solution and how many gallons of a 3% milk fat solution should she mix together to obtain 240 gal of a 2% milk fat solution? 160 gal of 1.5% milk fat solution and 80 gal of 3% milk fat solution

49. Two families go to the movies together. One family purchases six drinks and two large popcorns for $21.50. Another family purchases three drinks and five large popcorns for $26.75. What is the price of each drink and each popcorn? $2.25 for a drink and $4 for a popcorn

50. Two families go to the movies together. One family purchases four drinks and five large popcorns for $25.25. Another family purchases three drinks and four large popcorns for $19.75. What is the price of each drink and each popcorn? $2.25 for a drink and $3.25 for a popcorn

51. Admission to a football game is $10.50 for student tickets and $23 for general admission. If 2600 tickets are sold and $51,050 is collected, how many of each type of ticket was sold? 700 student tickets and 1900 general admission tickets were sold.

52. Admission to a football game is $11.50 for student tickets and $15.50 for general admission. If 2300 tickets are sold and $31,650 is collected, how many of each type of ticket was sold? 1000 student tickets and 1300 general admission tickets were sold.

53. Kyle receives an inheritance of $69,000. He invests part of the money in a stock that pays an average of 8% yearly dividends and the rest of the money in a money market account that earns 6.8% simple interest. If he earns $5100 in investment income for 1 yr, how much did he invest in each account? $34,000 in stock at 8% and $35,000 in money market at 6.8%

54. Tom receives a bonus of $73,000. He invests part of the money in a savings account that earns 3% simple interest and the rest of the money in a money market account that earns 4.1% simple interest. If he earns $2696 in interest for 1 yr, how much did he invest in each account?
$27,000 in savings at 3% and $46,000 in money market at 4.1%

55. A chemist has a 60% salt solution. She needs 90 L of a 10% salt solution. How many liters of water and how many liters of the 60% salt solution must she mix together to obtain 90 L of a 10% salt solution?
75 L of water and 15 L of 60% salt solution

56. A chemist has a 50% alcohol solution. She needs 45 L of a 20% alcohol solution. How many liters of water and how many liters of the 50% alcohol solution must she mix together to obtain 45 L of a 20% alcohol solution? 27 L of water and 18 L of 50% alcohol solution

57. The distance from Los Angeles, California, to Boston, Massachusetts, is approximately 2987 mi. Flying from Los Angeles to Boston takes 5.48 hr with the wind. The return flight takes 6.29 hr since the plane is flying against the wind. What is the speed of the plane in still air and what is the speed of the wind? Round to the nearest integer. The speed of the plane in still air is 510 mph and the speed of the wind is 35 mph.

58. The distance from Phoenix, Arizona, to Washington, D.C., is approximately 2305 mi. Flying from Phoenix to Washington takes 6.88 hr with the wind. The return flight takes 8.7 hr since the plane is flying against the wind. What is the speed of the plane in still air and what is the speed of the wind? Round to the nearest integer. The speed of the plane in still air is 300 mph and the speed of the wind is 35 mph.

59. At 9 A.M. Steve leaves his home traveling north on his bike at an average speed of 10 mph. At 10 A.M. his wife May leaves their home to meet Steve for brunch. She travels north in her car at an average speed of 50 mph. How long will it take May to reach Steve?
It will take May 0.25 hr to reach Steve.

60. At 10 A.M. Andrew leaves his home traveling east on his scooter at an average speed of 14 mph. At 11 A.M. his friend Charlene leaves home to meet Andrew. She travels east in her car at an average speed of 56 mph. How long will it take Charlene to reach Andrew? It will take Charlene $\frac{1}{3}$ hr to reach Andrew.

61. A chemist has a 50% acid solution and a 25% acid solution. He needs 75 gal of a 44% acid solution. How many gallons of the 50% acid solution and how many gallons of the 25% acid solution must he mix together to obtain 75 gal of a 44% acid solution?
57 gal of 50% acid solution and 18 gal of 25% acid solution

62. A chemist has a 41% iodine solution and a 23% iodine solution. He needs 150 L of a 38.6% iodine solution. How many liters of the 41% iodine solution and how many liters of the 23% iodine solution must he mix together to obtain 150 L of a 38.6% iodine solution?
130 L of 41% iodine solution and 20 L of 23% iodine solution

63. A plane travels 1440 mi with the wind in 5 hr. The return trip against the wind takes 6.25 hr. What is the speed of the plane in still air and what is the speed of the wind?
The speed of the plane is 259.2 mph and the speed of the wind is 28.8 mph.

64. A plane travels 725 mi with the wind in 5 hr. The return trip against the wind takes 6.75 hr. What is the speed of the plane in still air and what is the speed of the wind?
The speed of the plane is 126.2 mph and the speed of the wind is 18.8 mph.

65. Two angles are supplementary. The measure of one angle is 35° less than 4 times the measure of the other angle. Find the measure of each angle. 137° and 43°

66. Two angles are complementary. The measure of one angle is 16° more than 4 times the measure of the other angle. Find the measure of each angle. 75.2° and 14.8°

67. Flora and her mom live 170.2 mi apart. They want to meet each other for lunch. Flora leaves at 10 A.M. and travels toward her mom at an average speed of 60 mph. Her mom leaves at 11:15 A.M. and travels toward Flora at an average speed of 59 mph. What time will they meet? Flora travels 2.05 hr after 10 A.M. and they will meet at 12:03 P.M.

68. Helen and her mom live 351 mi apart. They want to meet each other so that Helen's kids can go home with her mom for a few days. Helen leaves at 8 A.M. and travels toward her mom at an average speed of 58 mph. Her mom leaves at 10:30 A.M. and travels toward Helen at an average speed of 45 mph. What time will they meet? Helen travels 4.5 hr after 8 A.M. and they will meet at 12:30 P.M.

69. The perimeter of a tennis court for doubles is 228 ft. The length of the court is 42 ft more than the width of the court. Find the dimensions of the court.
The length is 78 ft and the width of the court is 36 ft.

70. The perimeter of a squash court for doubles is 32,300 cm. The length of the court is 15,850 cm less than four times the width of the court. Find the dimensions of the court. The length is 9750 cm and the width is 6400 cm.

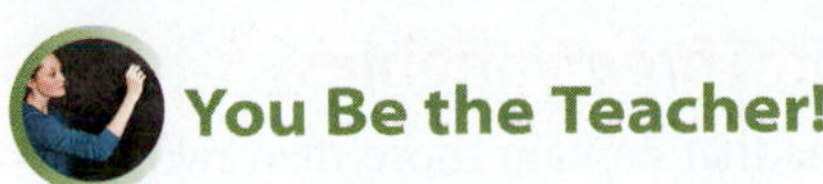

You Be the Teacher!

Correct each student's errors, if any.

71. Two angles are supplementary angles. The measure of one angle is 34° less than 3 times the measure of the other angle. Find the measure of each angle.

Bradley's work:
Let x and y be the two supplementary angles.

$$\begin{aligned} x + y &= 180 \\ x &= 34 - 3y \\ 34 - 3y + y &= 180 \\ 34 - 2y &= 180 \\ -2y &= 146 \\ y &= -73 \end{aligned}$$

72. Wang invests \$9000 in two different accounts, one paying 5.5% simple interest and the other paying 7% simple interest. Wang earns a total of \$615 interest in one year. Find the amount invested in each account.

Ashlee's work: Let x represent the amount in the account paying 5.5% simple interest and y be the amount in the account paying 7% simple interest.

$$\begin{aligned} x + y &= 9000 \\ 5.5x + 7y &= 615 \\ -5.5(x + y &= 9000) \\ 5.5x + 7y &= 615 \\ -5.5x - 5.5y &= -49{,}500 \\ 5.5x + 7y &= 615 \\ \hline 1.5y &= -48{,}885 \\ y &= -32{,}590 \end{aligned}$$

73. Irene and her mom live 339 mi apart. They want to meet each other so that Irene's kids can go home with her mom for a few days. Irene leaves at 10 A.M. and travels toward her mom at an average speed of 60 mph. Her mom leaves at 11:45 P.M. and travels toward Irene at an average speed of 57 mph. When they will meet?

Marybeth's work: Let x hr be the time taken by Irene and y hr be the time taken by her mom.

$$\begin{aligned} 60x + 57y &= 339 \\ x &= y + 1.45 \\ 60x + 57y &= 339 \\ 60(y + 1.45) + 57y &= 339 \\ 60y + 87 + 57y &= 339 \\ 117y &= 252 \\ y &= \frac{252}{117} \\ y &\approx 2.15 \\ x \approx 2.15 + 1.45 &= 3.6 \end{aligned}$$

They will meet at 1:06 P.M.

$$\begin{aligned} 60x + 57y &= 339 \\ x &= y + 1.75 \\ 60x + 57y &= 339 \\ 60(y + 1.75) + 57y &= 339 \\ 60y + 105 + 57y &= 339 \\ 117y &= 234 \\ y &= \frac{234}{117} = 2 \\ x = 2 + 1.75 &= 3.75 \end{aligned}$$

They will meet at 1:45 P.M.

74. A plane travels 1680 mi with the wind in 4 hr. The return trip against the wind takes 6 hr. Find the speed of the plane in still air and the speed of the wind.

Chris's work: Let x mph be the speed of the plane and y mph be the speed of the wind.

$$\begin{aligned} 6(x + y) &= 1680 \\ 4(x - y) &= 1680 \\ x + y &= 280 \\ x - y &= 420 \\ \hline 2x &= 700 \\ x &= 350 \\ x + y &= 280 \\ 350 + y &= 280 \\ y &= -70 \end{aligned}$$

$$\begin{aligned} 4(x + y) &= 1680 \\ 6(x - y) &= 1680 \\ x + y &= 420 \\ x - y &= 280 \\ \hline 2x &= 700 \\ x &= 350 \\ x + y &= 420 \\ 350 + y &= 420 \\ y &= 70 \end{aligned}$$

The speed of the plane in still air is 350 mph and the speed of the wind is 70 mph.

SECTION 5.4 Solving Linear Systems in Three Variables and Their Applications

OBJECTIVES

As a result of completing this section, you will be able to

1. Solve a system of linear equations in three variables by elimination.
2. Solve applications of linear systems in three variables.
3. Troubleshoot common errors.

An international flight from Atlanta, Georgia, to Dubai, United Arab Emirates, consists of first-class, business-class, and coach travelers. A first-class ticket costs \$4700. A business-class ticket costs \$2300 and a coach ticket costs \$750. The airline made \$342,000 in airfare revenue for one trip from Atlanta to Dubai. The flight had a total of 300 paying passengers. If there were four times as many coach travelers as first-class and business-class travelers combined, how many of each type of traveler were on this flight?

Notice that there are three unknowns in this problem. Solving this problem requires us to use a system that consists of three equations and three variables. In this section, we will learn how to set up and solve a system involving three linear equations and three unknowns.

Solving a System of Linear Equations in Three Variables

Objective 1 ▶

Solve a system of linear equations in three variables by elimination.

In business, engineering, and other fields, systems that contain more than two equations and more than two unknowns are needed to represent real-life situations. These systems are solved using computers or calculators. Systems involving more than two unknowns can be difficult to solve by hand, so we will only solve systems of three linear equations in three variables algebraically. We will use elimination, which was covered in Section 5.2. Substitution can also be used to solve these systems but can be tedious to use.

Definition: A **system of linear equations in three variables** can be represented by

$$\begin{cases} Ax + By + Cz = D \\ Ex + Fy + Gz = H \\ Ix + Jy + Kz = L \end{cases}$$

Notice that each variable in a linear system in three variables is raised to an exponent of 1. A **solution of a system of linear equations in three variables** is an *ordered triple* (x, y, z) that makes each equation in the system a true statement.

Just as the graph of a linear equation in two variables is a line in two-dimensional space, the graph of a linear equation in three variables is a plane in three-dimensional space provided the coefficients of all the variables are nonzero. Linear systems in three variables have solution sets similar to linear systems in two variables.

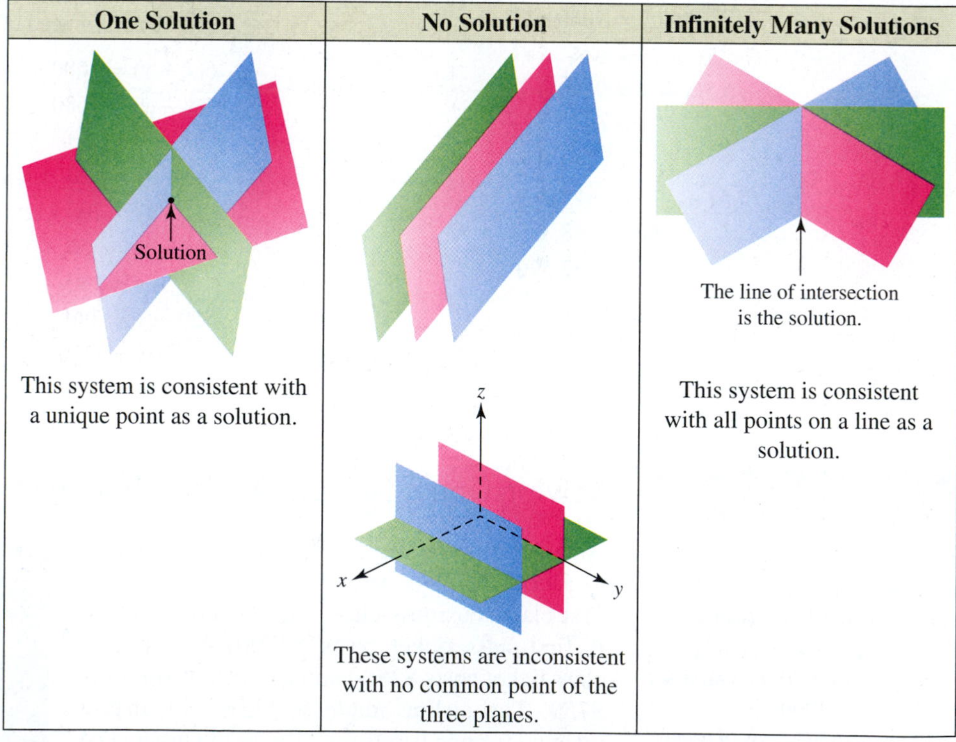

One Solution
Solution
This system is consistent with a unique point as a solution.
No Solution
z
x
y
These systems are inconsistent with no common point of the three planes.
Infinitely Many Solutions
The line of intersection is the solution.
This system is consistent with all points on a line as a solution.

Visualizing solutions of linear systems in three variables can be difficult. It is sufficient to understand that the solution of such a system consists of the points where the three planes intersect. A plane is a flat surface—think of a wall or ceiling that extends indefinitely on all four sides.

To solve a system of three linear equations in three variables, we will create an equivalent system of two linear equations with two variables. To do this, we work with pairs of equations from the original system and eliminate the same variable from each pair of equations obtaining the new system we need.

Procedure: Solving a System of Linear Equations in Three Variables by Elimination

Step 1: Write all equations in standard form.
Step 2: Pair together two of the equations and eliminate one variable from this set of equations.
Step 3: Pair together two other equations and eliminate the variable that was eliminated in step 2.
Step 4: Form a system of the two equations obtained from steps 2 and 3. Solve this system by elimination.
Step 5: Substitute the solution found in step 4 in one of the equations in the system from step 4 to find the value of another variable.
Step 6: Substitute the two known values in one of the original equations to find the remaining value.
Step 7: Check the solution by substituting it in all three equations in the given system.

Note: *If in the process of solving the system, we obtain a*

- *False statement (for example,* $0 = 12$*), then the system has no solution.*
- *True statement (for example,* $12 = 12$*), then the system has infinitely many solutions.*

Objective 1 Examples **Solve each system by elimination. The equations in the system have been numbered for reference.**

1a. $\begin{cases} x + 4y + 2z = -1 & (1) \\ 2x - 4y - z = 6 & (2) \\ 5x + 4y - z = 23 & (3) \end{cases}$

1b. $\begin{cases} x - 2y + z = 6 & (1) \\ 2x + y - z = 0 & (2) \\ 4x - 8y + 4z = -12 & (3) \end{cases}$

1c. $\begin{cases} x + 4y - 5z = 8 & (1) \\ y - 3z = 5 & (2) \\ x + 2y + z = -2 & (3) \end{cases}$

Solutions **1a.** We form pairs of equations and eliminate the variable y from each pair.

Pair together Equations (1) and (2) and add.

$$\begin{cases} x + 4y + 2z = -1 & (1) \\ 2x - 4y - z = 6 & (2) \end{cases}$$
$$2x + z = 5 \quad (4)$$

Pair together Equations (2) and (3) and add.

$$\begin{cases} 2x - 4y - z = 6 & (2) \\ 5x + 4y - z = 23 & (3) \end{cases}$$
$$7x + 2z = 29 \quad (5)$$

Now form a new system using Equations (4) and (5) and solve by elimination.

$$\begin{cases} 3x + z = 5 & (4) \\ 7x - 2z = 29 & (5) \end{cases}$$

We choose to eliminate z from this system, so we multiply the first equation by 2.

$$\begin{cases} 3x + z = 5 \\ 7x - 2z = 29 \end{cases} \rightarrow \begin{cases} 2(3x + z = 5) \\ 7x - 2z = 29 \end{cases} \rightarrow \begin{cases} 6x + 2z = 10 \\ 7x - 2z = 29 \end{cases}$$

Now add the equations in the new system to solve for x.

$$\begin{cases} 6x + 2z = 10 & \text{Equation (4)} \\ 7x - 2z = 29 & \text{Equation (5)} \end{cases}$$

$$13x = 39 \quad \text{Add.}$$

$$x = 3 \quad \text{Divide each side by 13.}$$

Now we can solve for z by replacing x with 3 in one of the new equations, Equation (4) or Equation (5).

$$\begin{aligned} 3x + z &= 5 && \text{Equation (4)} \\ 3(3) + z &= 5 && \text{Substitute 3 for } x. \\ 9 + z &= 5 && \text{Simplify.} \\ z &= -4 && \text{Subtract 9 from each side.} \end{aligned}$$

Solve for y using one of the original equations.

$$\begin{aligned} x + 4y + 2z &= -1 && \text{Equation (1)} \\ 3 + 4y + 2(-4) &= -1 && \text{Substitute 3 for } x \text{ and } -4 \text{ for } z. \\ 3 + 4y - 8 &= -1 && \text{Simplify.} \\ 4y - 5 &= -1 && \text{Combine like terms.} \\ 4y &= 4 && \text{Add 5 to each side.} \\ y &= 1 && \text{Divide each side by 4.} \end{aligned}$$

The solution for the system is $x = 3$, $y = 1$, and $z = -4$, or we write the solution set as $\{(3, 1, -4)\}$. We can check the solution by substituting these values in the three original equations.

$x + 4y + 2z = -1$	$2x - 4y - z = 6$	$5x + 4y - z = 23$
$3 + 4(1) + 2(-4) = -1$	$2(3) - 4(1) - (-4) = 6$	$5(3) + 4(1) - (-4) = 23$
$3 + 4 - 8 = -1$	$6 - 4 + 4 = 6$	$15 + 4 + 4 = 23$
$-1 = -1$	$6 = 6$	$23 = 23$
True	True	True

1b. We form pairs of equations and eliminate the variable z from each pair.

Pair together Equations (1) and (2) and add.

$$\begin{cases} x - 2y + z = 6 & (1) \\ 2x + y - z = 0 & (2) \end{cases}$$

$$3x - y = 6 \quad (4)$$

Pair together Equations (1) and (3). To eliminate z, we must multiply Equation (1) by -4.

$$\begin{cases} -4(x - 2y + z = 6) & (1) \\ 4x - 8y + 4z = -12 & (3) \end{cases}$$

$$\begin{cases} -4x + 8y - 4z = -24 \\ 4x - 8y + 4z = -12 \end{cases}$$

$$0 = -36 \quad (5)$$

Because we obtained a false statement, $0 = -36$, this system is an inconsistent system and has no solution.

1c. Since Equation (2) is missing the variable x, we can use it as one of the equations in the system of two linear equations.

$$y - 3z = 5$$

Now we pair together Equations (1) and (3) and eliminate x. We multiply Equation (3) by -1 and add.

$$\begin{cases} x + 4y - 5z = 8 & (1) \\ -1(x + 2y + z = -2) & (3) \end{cases} \rightarrow \begin{cases} x + 4y - 5z = 8 \\ \underline{-x - 2y - z = 2} \end{cases}$$
$$2y - 6z = 10 \quad (4)$$

Now pair Equations (2) and (4).

$$\begin{cases} y - 3z = 5 & (2) \\ 2y - 6z = 10 & (4) \end{cases}$$

Solve the system by multiplying Equation (2) by -2 to eliminate y and then add.

$$\begin{cases} -2(y - 3z = 5) & (2) \\ 2y - 6z = 10 & (4) \end{cases} \rightarrow \begin{cases} -2y + 6z = -10 \\ \underline{2y - 6z = 10} \end{cases}$$
$$0 = 0$$

Because we obtained a true statement, $0 = 0$, this system has infinitely many solutions. We can represent the solutions in terms of the variable z, a real number.

Equation (2) expresses y in terms of z, $y = 3z + 5$. We substitute $y = 3z + 5$ in Equation (1) to solve for x in terms of z.

$$x + 4y - 5z = 8$$
$$x + 4(3z + 5) - 5z = 8 \qquad \text{Replace } y \text{ with } 3z + 5.$$
$$x + 12z + 20 - 5z = 8 \qquad \text{Apply the distributive property.}$$
$$x + 7z + 20 = 8 \qquad \text{Combine like terms.}$$
$$x = -7z - 12 \qquad \text{Subtract } 7z \text{ and } 20 \text{ from each side.}$$

The solution set is $\{(-7z - 12, 3z + 5, z) | z \text{ is a real number}\}$.

Student Check 1 Solve each system by elimination.

a. $\begin{cases} x + 6y - 5z = -14 \\ 3x - 6y + z = -10 \\ 7x + 6y + z = -26 \end{cases}$ **b.** $\begin{cases} x + 4y - z = -10 \\ 6x + 2y + z = 8 \\ 4x - 6y + 3z = 16 \end{cases}$ **c.** $\begin{cases} x - 3y + 2z = -1 \\ -5x + 16y - 14z = 13 \\ 4x - 10y = 12 \end{cases}$

Applications of Linear Systems in Three Variables

Objective 2 Solve applications of linear systems in three variables.

Many of the word problems we solved that involved systems of two linear equations can be extended to situations with three unknowns—mixture problems, investment problems, geometry problems, and the like. If a problem contains three unknown quantities, then this is our clue that we must use a system of three linear equations and three variables to solve it.

Procedure: Solving an Application of Linear Systems in Three Variables

Step 1: Read the problem carefully to determine the unknown quantities and assign a variable to represent each of them.
Step 2: From the problem determine three equations that define the relationships between the unknowns.
Step 3: Solve the system by elimination.
Step 4: Check the solution by substituting it in the three equations in the system.

Objective 2 Examples **Use a linear system in three variables to solve each problem.**

2a. An international flight from Atlanta, Georgia to Dubai, United Arab Emirates, consists of first-class, business-class, and coach travelers. A first-class ticket costs \$4700. A business-class ticket costs \$2300 and a coach ticket costs \$750. The airline made \$342,000 in airfare revenue for one trip from Atlanta to Dubai. The flight had a total of 300 paying passengers. If there were four times as many coach travelers as first-class and business-class travelers combined, how many of each type of traveler were on this flight?

Solution **2a.** What is unknown? The number of first-class passengers, the number of business-class passengers, and the number of coach passengers are unknown.

Let f = number of first-class passengers.
Let b = number of business-class passengers.
Let c = number of coach passengers.

We use the given facts to write a system that represents the problem.

$$\begin{cases} f + b + c = 300 & \text{There are a total of 300 passengers.} \\ 4700f + 2300b + 750c = 342{,}000 & \text{Total revenue is \$342,000.} \\ c = 4(f + b) & \text{There are 4 times as many coach travelers as the other two combined.} \end{cases}$$

Rewrite the system so that each equation is in standard form.

$$\begin{cases} f + b + c = 300 & (1) \\ 4700f + 2300b + 750c = 342{,}000 & (2) \\ 4f + 4b - c = 0 & (3) \end{cases}$$

To solve the system by elimination, pair together Equations (1) and (2) and Equations (1) and (3) and eliminate c from each pair. To eliminate c from Equations (1) and (2), multiply the first equation by -750 and add.

$$\begin{cases} -750(f + b + c = 300) & (1) \\ 4700f + 2300b + 750c = 342{,}000 & (2) \end{cases} \rightarrow \begin{cases} -750f - 750b - 750c = -225{,}000 & (1) \\ 4700f + 2300b + 750c = 342{,}000 & (2) \end{cases}$$
$$3950f + 1550b = 117{,}000 \quad (4)$$

To eliminate c from Equations (1) and (3), add the equations.

$$\begin{cases} f + b + c = 300 & (1) \\ 4f + 4b - c = 0 & (3) \end{cases}$$
$$5f + 5b = 300 \quad (5)$$

Now solve the system $\begin{cases} 3950f + 1550b = 117{,}000 & (4) \\ 5f + 5b = 300 & (5) \end{cases}$

We can multiply Equation (5) by -310 to eliminate the variable b. Then add.

$$\begin{cases} 3950f + 1550b = 117{,}000 & (4) \\ -310(5f + 5b = 300) & (5) \end{cases} \rightarrow \begin{cases} 3950f + 1550b = 117{,}000 \\ -1550f - 1550b = -93{,}000 \end{cases}$$
$$2400f = 24{,}000$$
$$f = 10$$

Now we solve for b:

$$\begin{aligned} 5f + 5b &= 300 && \text{Equation (5)} \\ 5(10) + 5b &= 300 && \text{Substitute 10 for } f. \\ 50 + 5b &= 300 && \text{Simplify.} \\ 5b &= 250 && \text{Subtract 50 from each side.} \\ b &= 50 && \text{Divide each side by 5.} \end{aligned}$$

Now solve for c:

$$\begin{aligned} f + b + c &= 300 && \text{Equation (1)} \\ 10 + 50 + c &= 300 && \text{Substitute 10 for } f \text{ and 50 for } b. \\ 60 + c &= 300 && \text{Simplify.} \\ c &= 240 && \text{Subtract 60 from each side.} \end{aligned}$$

So, the flight contained 10 first-class passengers, 50 business-class passengers, and 240 coach passengers. We can check our answer by substituting the three values into each equation in the original system.

$$\begin{aligned} f + b + c &= 300 \\ 10 + 50 + 240 &= 300 \\ 300 &= 300 \end{aligned}$$

$$\begin{aligned} 4f + 4b - c &= 0 \\ 4(10) + 4(50) - 240 &= 0 \\ 40 + 200 - 240 &= 0 \\ 0 &= 0 \end{aligned}$$

$$\begin{aligned} 4700f + 2300b + 750c &= 342{,}000 \\ 4700(10) + 2300(50) + 750(240) &= 342{,}000 \\ 47{,}000 + 115{,}000 + 180{,}000 &= 342{,}000 \\ 342{,}000 &= 342{,}000 \end{aligned}$$

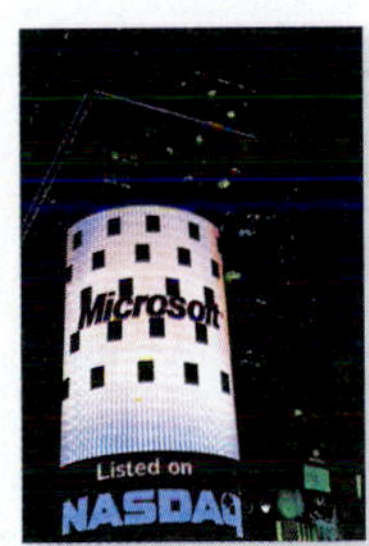

2b. A retired couple made a profit of \$300,000 on the sale of their family business. They invested this money in three areas: a high-risk stock fund with an expected annual rate of return of 15%, a low-risk stock fund with an expected annual rate of return of 10%, and government bonds at an annual return of 8%. To protect their investment, they invested twice as much money in the low-risk stock fund as the high-risk stock fund and used the remainder to buy bonds. How much did the couple allocate per investment if they earned \$32,800 annually on their investment?

Solution **2b.** What is unknown? The amounts invested in the high-risk stock fund, the low-risk stock fund, and government bonds are the unknowns.

Let x = amount invested in the high-risk stock fund.
Let y = amount invested in the low-risk stock fund.
Let z = amount invested in government bonds.

We use the given facts to write a system that represents the problem.

$$\begin{cases} x + y + z = 300{,}000 & \text{Total invested is \$300,000.} \\ 0.15x + 0.10y + 0.08z = 32{,}800 & \text{Total earned is \$32,800.} \\ y = 2x & \text{Twice as much is invested in the low-risk stock fund as the high-risk stock fund.} \end{cases}$$

Rewrite each equation in standard form and clear the decimals by multiplying the second equation by 100.

$$\begin{cases} x + y + z = 300{,}000 \\ 100(0.15x + 0.10y + 0.08z = 32{,}800) \rightarrow \\ 2x - y = 0 \end{cases} \quad \begin{cases} x + y + z = 300{,}000 & (1) \\ 15x + 10y + 8z = 3{,}280{,}000 & (2) \\ 2x - y = 0 & (3) \end{cases}$$

Solve the system by elimination. We use Equation (3) as one of the equations in the new system since it only contains two variables. Then we pair together Equations (1) and (2) and eliminate the variable z.

$$\begin{cases} x + y + z = 300{,}000 & (1) \\ 15x + 10y + 8z = 3{,}280{,}000 & (2) \end{cases}$$

Multiply Equation (1) by -8 to eliminate z and add the equations.

$$\begin{cases} -8(x + y + z = 300{,}000) & (1) \\ 15x + 10y + 8z = 3{,}280{,}000 & (2) \end{cases} \rightarrow \begin{cases} -8x - 8y - 8z = -2{,}400{,}000 & (1) \\ 15x + 10y + 8z = 3{,}280{,}000 & (2) \end{cases}$$
$$7x + 2y = 880{,}000 \quad (4)$$

Now solve the system $\begin{cases} 2x - y = 0 & (3) \\ 7x + 2y = 880{,}000 & (4) \end{cases}$

Multiply Equation (3) by 2 and add the equations.

$$\begin{cases} 2(2x - y = 0) & (3) \\ 7x + 2y = 880{,}000 & (4) \end{cases} \rightarrow \begin{cases} 4x - 2y = 0 & (3) \\ 7x + 2y = 880{,}000 & (4) \end{cases}$$
$$11x = 880{,}000$$
$$x = 80{,}000$$

Solve for y:

$y = 2x$	Equation (3)
$y = 2(80{,}000)$	Substitute 80,000 for x.
$y = 160{,}000$	Simplify.

Solve for z:

$x + y + z = 300{,}000$	Equation (1)
$80{,}000 + 160{,}000 + z = 300{,}000$	Substitute 80,000 for x and 160,000 for y.
$240{,}000 + z = 300{,}000$	Simplify.
$z = 60{,}000$	Subtract 240,000 from each side.

The couple invested \$80,000 in the high-risk stock fund, \$160,000 in the low-risk stock fund, and \$60,000 in government bonds.

Student Check 2 Solve each problem using a linear system with three equations and three variables.

a. A music hall hosted an awards ceremony. There were three types of tickets available to the public—first mezzanine (\$539 each), second mezzanine (\$392 each), and third mezzanine (\$294 each). A total of 2300 tickets were sold totaling \$1,004,500. The number of first mezzanine tickets sold was 500 less than the other two combined. Find the number of each type of ticket that was sold.

b. A young couple has saved six months of living expenses, which is \$30,000. They invested this money in three different accounts—a 1-yr CD that pays 5% annual interest, a high-interest savings account that pays 4.5% annual interest, and a money market account that pays 10% annual interest. To keep their money as liquid as possible, they invested three times as much money in the high-interest savings account as the amount invested in the CD. Find the amount of money invested in each account if they earned \$1925 interest in 1 yr.

Objective 3 ▶ Troubleshoot common errors.

Troubleshooting Common Errors

Some common errors associated with systems of linear equations in three variables are shown.

Objective 3 Example **A problem and an incorrect solution are given. Provide the correct solution and an explanation of the error.**

Solve the system $\begin{cases} 4x - 5y + 3z = 3 \\ 2x + 5y - z = 9 \\ x + 2y + z = 8 \end{cases}$.

Incorrect Solution	Correct Solution and Explanation
Add Equations (1) and (2) to get $6x + 2z = 12$ Add Equations (2) and (3) to get $3x + 7y = 17$ Now solve $\begin{cases} 6x + 2z = 12 \\ 3x + 7y = 17 \end{cases}$. There is no solution since the system doesn't have the same variables.	We must eliminate the same variable from the pairs of equations. So, we can eliminate the variable y from Equations (2) and (3) by multiplying Equation (2) by -2 and Equation (3) by 5. This yields $\begin{cases} -4x - 10y + 2z = -18 \\ 5x + 10y + 5z = 40 \end{cases}$ Adding the equations gives us $x + 7z = 22$ Now we solve the system $\begin{cases} 6x + 2z = 12 \\ x + 7z = 22 \end{cases}$. $\begin{cases} 6x + 2z = 12 \\ -6(x + 7z = 22) \end{cases} \rightarrow \begin{cases} 6x + 2z = 12 \\ -6x - 42z = -132 \end{cases}$ $-40z = -120$ $z = 3$ Solve for x: $x + 7z = 22$; $x + 7(3) = 22$; $x + 21 = 22$; $x = 1$ Solve for y: $x + 2y + z = 8$; $1 + 2y + 3 = 8$; $4 + 2y = 8$; $2y = 4$; $y = 2$ The solution set is $\{(1, 2, 3)\}$.

ANSWERS TO STUDENT CHECKS

Student Check 1 **a.** $\{(-4, 0, 2)\}$ **b.** no solution

c. $\left\{\left(x, \frac{2x-6}{5}, \frac{x-23}{10}\right) \middle| x \text{ is a real number}\right\}$

Student Check 2 **a.** There were 900 first mezzanine, 1100 second mezzanine, and 300 third mezzanine tickets sold. **b.** \$5000 in the CD, \$15,000 in the savings account, and \$10,000 in the money market account

SUMMARY OF KEY CONCEPTS

1. The first step in solving a linear system in three variables is to use elimination to reduce it to a system involving two variables.
   - Pair together two different sets of equations from the linear system in three variables and eliminate the same variable from each pair.
   - The resulting equations form a linear system in two variables. Solve this system by elimination.
   - The solution for a linear system in three variables may be an ordered triple, no solution, or infinitely many solutions.
2. To solve an application with a system in three variables, determine three different equations that represent the given information and solve it.

SECTION 5.4 EXERCISE SET

Write About It!

Use complete sentences in your answer to each exercise.

1. Describe the solution of a system of three linear equations.

2. Explain how to reduce a system of three linear equations to a system of two linear equations.

3. When you obtain a false statement during the process of solving a system of three linear equations, what is the solution of the system?

4. When you obtain a true statement during the process of solving a system of three linear equations, what is the solution of the system?

5. How do you know that you must use a system of three linear equations to solve a word problem?

6. Let A be the total amount you have to invest, and let r, s, and t be the annual rate of return in percents for three investments. Describe how to set up the first two equations in the system if a total of \$I is earned in interest in one year.

Practice Makes Perfect!

Solve each system of linear equations in three variables by elimination. (*See Objective 1.*)

7. $\begin{cases} x + 6y + 2z = 5 \\ 2x + 9y + z = -2 \\ 3x - y - 4z = -16 \end{cases}$ $\{(1, -1, 5)\}$

8. $\begin{cases} -2x + 5y + 7z = -11 \\ 2x - 6y + 3z = -10 \\ 3x + y + 3z = -2 \end{cases}$ $\{(1, 1, -2)\}$

9. $\begin{cases} x + 2y + 4z = -15 \\ 6x - 5y - 2z = -3 \\ -x - 4y - 3z = 9 \end{cases}$ $\{(-1, 1, -4)\}$

10. $\begin{cases} 3x + y - 4x = 3 \\ 2x + 5y + 4x = 0 \\ x + 3y - 4x = -7 \end{cases}$ $\{(3, -2, 1)\}$

11. $\begin{cases} 2x + y - 3z = 0 \\ -5x + 2y + 3z = -9 \\ 4x + 4y - 5z = -1 \end{cases}$ $\{(2, -1, 1)\}$

12. $\begin{cases} 2x - 3y - 6z = -4 \\ -4x + y + z = -4 \\ 2x + 7y + z = -10 \end{cases}$ $\{(1, -2, 2)\}$

13. $\begin{cases} 3x - y + 6z = 2 \\ -2x + 3y + z = 6 \\ 2x + 5y - 4z = -1 \end{cases}$ $\{(-1, 1, 1)\}$

14. $\begin{cases} x + 3y - 2z = 15 \\ x + y - z = 8 \\ 3x + 2y = 8 \end{cases}$ $\{(2, 1, -5)\}$

15. $\begin{cases} x + 5y - 3z = -4 \\ x - y - z = 10 \\ x + 3y = -4 \end{cases}$ $\{(5, -3, -2)\}$

16. $\begin{cases} -x + y + 2z = -3 \\ 4x + 3y - 5z = 14 \\ 5x - 4y = -23 \end{cases}$ $\{(-3, 2, -4)\}$

17. $\begin{cases} x + 5y + 7z = 30 \\ -2x + z = 5 \\ 6x - 6y + 5z = -3 \end{cases}$ $\{(-1, 2, 3)\}$

18. $\begin{cases} 2x + 3y + 4z = 4 \\ 5x + 7z = 22 \\ 9x + y - 2z = 23 \end{cases}$ $\{(3, -2, 1)\}$

19. $\begin{cases} -3x + 4y - 7z = 2 \\ x + 3z = 0 \\ 6x + 8y - 5z = 31 \end{cases}$ $\{(3, 1, -1)\}$

20. $\begin{cases} 2x - 3y - 2z = -3 \\ x - 2z = 2 \\ 5x + 2y - 4z = 0 \end{cases}$ $\{(-2, 1, -2)\}$

Additional answers can be found in the Instructor Answer Appendix.

21. $\begin{cases} 5x + 6y + 6z = 35 \\ y + z = 5 \\ -x + y + z = 4 \end{cases}$ $\{(1, -z + 5, z) \mid z \text{ is a real number}\}$

22. $\begin{cases} 2x + y - 3z = 4 \\ y - z = 0 \\ 3x - y - 2z = 6 \end{cases}$ $\{(z + 2, z, z) \mid z \text{ is a real number}\}$

23. $\begin{cases} -4x + y + 5z = 0 \\ y - 27z = -48 \\ -3x + y - 3z = -12 \end{cases}$ $\{(8z - 12, 27z - 48, z) \mid z \text{ is a real number}\}$

24. $\begin{cases} -x + 5y + 5z = -19 \\ 2y + 3z = -8 \\ x + y + 4z = -5 \end{cases}$

25. $\begin{cases} x - y - 2z = 3 \\ -2y + z = -7 \\ -2x + 4y + 3z = 2 \end{cases}$ no solution

26. $\begin{cases} 5x + 2y + 6z = 0 \\ 2y + 26z = 11 \\ -4x - y + 3z = 15 \end{cases}$ no solution

27. $\begin{cases} 1.5x + y - 2.5z = 4 \\ 0.1x + 0.1y - 0.3z = 0.7 \\ 1.5x - 1.5y + 0.5z = -0.5 \end{cases}$ $\{(-2, -3, -4)\}$

28. $\begin{cases} 0.3x + 0.2y + 0.6z = -1.1 \\ 3x - 2y - 1.5z = 3.5 \\ -x + 0.8y + 0.2z = 0 \end{cases}$ $\{(1, 2, -3)\}$

29. $\begin{cases} 2x - 5y - 3z = -39 \\ 17x - 23z = -7 \\ -3x - y + 4z = -10 \end{cases}$ no solution

30. $\begin{cases} 2x + 5y + 3z = 0 \\ -7x - 13z = 8 \\ x - y + 2z = 5 \end{cases}$ no solution

31. $\begin{cases} x - y - 4z = 11 \\ x - \frac{1}{2}y + \frac{3}{2}z = -6 \\ -\frac{5}{3}x - \frac{2}{3}y + z = 1 \end{cases}$ $\{(-2, -1, -3)\}$

32. $\begin{cases} \frac{3}{2}x - \frac{1}{2}y + 3z = 1 \\ 2x - y - z = -4 \\ x + \frac{5}{2}y - 2z = -\frac{1}{2} \end{cases}$ $\{(-1, 1, 1)\}$

Solve each problem using a linear system. (*See Objective 2.*)

33. An international flight from London to Singapore consists of first-class, business-class, and coach travelers. A first-class ticket costs \$18,000. A business-class ticket costs \$7400 and a coach ticket costs \$1600. The airline made \$858,400 in airfare revenue for one trip from London to Singapore. The flight had a total of 336 paying passengers. If there were six times as many coach travelers as first-class and business-class travelers combined, how many of each type of traveler were on this flight? 4 in first-class, 44 in business-class, and 288 in coach

34. An international flight from Paris to Singapore consists of first-class, business-class, and coach travelers. A first-class ticket costs \$11,200. A business-class ticket costs \$5300 and a coach ticket costs \$1100. The airline made \$431,300 in airfare revenue for one trip from Paris to Singapore. The flight had a total of 240 paying passengers. If there were seven times as many coach travelers as first-class and business-class travelers combined, how many of each type of traveler were on this flight?
7 in first-class, 23 in business-class, and 210 in coach

35. An international flight from Los Angeles to Hong Kong consists of first-class, business-class, and coach travelers. A first-class ticket costs \$15,200. A business-class ticket costs \$5700 and a coach ticket costs \$1160. The airline made \$746,980 in airfare revenue for one trip from Los Angeles to Hong Kong. The flight had a total of 371 paying passengers. If there were six times as many coach travelers as first-class and business-class travelers combined, how many of each type of traveler were on this flight? 8 in first-class, 45 in business-class, and 318 in coach

36. An international flight from Honolulu to Hong Kong consists of first-class, business-class, and coach travelers. A first-class ticket costs \$11,200. A business-class ticket costs \$5300 and a coach ticket costs \$1100. The airline made \$636,300 in airfare revenue for one trip from Honolulu to Hong Kong. The flight had a total of 350 paying passengers. If there were six times as many coach travelers as first-class and business-class travelers combined, how many of each type of traveler were on this flight? 7 in first-class, 43 in business-class, and 300 in coach

37. The Winters have saved \$15,000. They invest this money in three different accounts—a 1-yr CD that pays 4.2% annual interest, a high-interest savings account that pays 5.6% annual interest, and a money market account that pays 4.1% annual interest. They invest twice as much money in the high-interest savings account as the amount invested in the CD. Find the amount invested in each account if they earn \$735.90 interest in 1 yr. \$3900 in CD, \$7800 in high-interest savings account, \$3300 in money market account

38. The Browns invest \$29,500 in three different accounts—a 1-yr CD that pays 4.1% annual interest, a high-interest savings account that pays 7.9% annual interest, and a money market account that pays 3.3% annual interest. They invest twice as much money in the high-interest savings account as the amount invested in the CD. Find the amount invested in each account if they earn \$1681.50 interest in 1 yr. \$7080 in CD, \$14,160 in high-interest savings account, \$8260 in money market account

39. A concert is held at a music hall. There are three types of tickets available to the public—first mezzanine (\$120 each), second mezzanine (\$105 each), and third mezzanine (\$65 each). A total of 10,300 tickets are sold totaling \$1,091,500. The number of first mezzanine tickets sold is 200 less than twice the other two combined. Find the number of tickets sold in each mezzanine. 6800 in first mezzanine, 1,200 in second mezzanine, and 2300 in third mezzanine

40. A concert is held at a music hall. There are three types of tickets available to the public—first mezzanine (\$185 each), second mezzanine (\$120 each), and third mezzanine (\$70 each). A total of 11,450 tickets are sold totaling \$1,701,500. The number of first mezzanine tickets sold is 400 less than twice the other two combined. Find the number of tickets sold in each mezzanine. 7500 in first mezzanine, 750 in second mezzanine, and 3200 in third mezzanine

41. There are three types of tickets sold for a college football game—main (\$115 each), terrace (\$65 each), and grandstand (\$50 each). A total of 3040 tickets are sold for a revenue of \$199,050. The number of the grandstand tickets sold is 100 less than the other two combined. Find the number of tickets sold in each section. 470 in main, 1100 in terrace, 1470 in grandstand

42. There are three types of tickets sold for a college baseball championship game—main (\$110 each), terrace (\$70 each), and grandstand (\$35 each). A total of 3000 tickets are sold for a revenue of \$147,500. The number of the grandstand tickets sold is 300 less than twice the other two combined. Find the number of tickets sold in each section. 100 in main, 1000 in terrace, 1900 in grandstand

Mix 'Em Up!

Solve each system of linear equations.

43. $\begin{cases} x + 3y - 6z = -11 \\ -5x + 4y + 3z = -10 \\ 4x + y - 5z = -3 \end{cases}$ $\{(1, -2, 1)\}$

44. $\begin{cases} 4x + y - 6z = -19 \\ -5x + 4y + z = -18 \\ x + 5y - 3z = -25 \end{cases}$ $\{(2, -3, 4)\}$

45. $\begin{cases} x + 5y - 2z = 28 \\ 3y - z = 16 \\ x + 2y - z = 12 \end{cases}$

46. $\begin{cases} x - 2y + 4z = 3 \\ -5y + 14z = -4 \\ 3x - y - 2z = 13 \end{cases}$

47. $\begin{cases} 0.1x - 0.2y - 0.4z = 1.3 \\ 0.7x + 0.3y + 0.2z = 1.8 \\ 1.5x + 4y + z = 5.5 \end{cases}$ $\{(3, 1, -3)\}$

48. $\begin{cases} 0.5x - y - z = 0 \\ 0.5x + y - z = -2 \\ x + y + 0.4z = 4.2 \end{cases}$ $\{(4, -1, 3)\}$

49. $\begin{cases} 5x + 6y + z = 43 \\ 9x + 11z = -8 \\ -2x + 3y - 5z = 6 \end{cases}$ no solution

50. $\begin{cases} 4x - 5y - 6z = 31 \\ 8x - 7y = 2 \\ -2x + 3y + 5z = -8 \end{cases}$ no solution

51. $\begin{cases} 4x + \frac{1}{2}y + z = -23 \\ 2y + 3z = 31 \\ \frac{1}{4}x + \frac{1}{4}y = 0 \end{cases}$ $\{(-8, 8, 5)\}$

52. $\begin{cases} x + \frac{2}{3}y + \frac{5}{3}z = -1 \\ 3y - z = 17 \\ \frac{8}{5}x + y = \frac{17}{5} \end{cases}$ $\{(-1, 5, -2)\}$

Solve each problem using a linear system.

53. The Shutters invest \$42,500 in three different accounts—a 1-yr CD that pays 4.9% annual interest, a savings account that pays 8% annual interest, and a money market account that pays 3.3% annual interest. They invest three times as much money in the savings account as the amount invested in the CD. Find the amount invested in each account if they earn \$2870.45 interest in 1 yr. \$9350 in CD, \$28,050 in high-interest savings account, \$5100 in money market account

54. The Chans invest \$35,000 in three different accounts—a 1-yr CD that pays 4.2% annual interest, a savings account that pays 6.8% annual interest, and a money market account that pays 2.7% annual interest. They invest three times as much money in the savings account as the amount invested in the CD. Find the amount invested in each account if they earn \$1911 interest in 1 yr. \$7000 in CD, \$21,000 in high-interest savings account, \$7000 in money market account

55. There are three types of tickets sold for a Las Vegas show—category 1 (\$130 each), category 2 (\$115 each), category 3 (\$95 each). A total of 1500 tickets are sold for a revenue of \$168,500. The number of category 3 tickets sold is 500 less than the other two combined. Find the number of tickets sold in each category. 400 in category 1, 600 in category 2, 500 in category 3

56. There are three types of tickets sold for a concert—floor ($300 each), lower level ($160 each), and upper level ($110 each). A total of 5000 tickets are sold for a revenue of $770,000. The number of upper level tickets sold is 1000 less than the other two combined. Find the number of tickets sold in each section.
500 floor level, 2500 lower level, and 2000 upper level

57. An international flight from Los Angeles to Shanghai consists of first-class, business-class, and coach travelers. A first-class ticket costs $5662. A business-class ticket costs $2400 and a coach ticket costs $1000. The airline made $444,234 in airfare revenue for one trip from Los Angeles to Shanghai. The flight had a total of 357 paying passengers. If the number of coach travelers was 35 more than six times the first-class and business-class travelers combined, how many of each type of traveler were on this flight?
7 in first-class, 39 in business-class, and 311 in coach

58. An international flight from Los Angeles to Seoul consists of first-class, business-class, and coach travelers. A first-class ticket costs $7690. A business-class ticket costs $3900 and a coach ticket costs $700. The airline made $402,240 in airfare revenue for one trip from Los Angeles to Seoul. The flight had a total of 341 paying passengers. If the number of coach travelers was 11 less than seven times the first-class and business-class travelers combined, how many of each type of traveler were on this flight?
6 in first-class, 38 in business-class, and 297 in coach

59. There are three types of tickets sold for a college football game—main ($120 each), terrace ($65 each), and grandstand ($45 each). A total of 3990 tickets are sold for a revenue of $251,050. The number of the grandstand tickets sold is 300 less than twice the other two combined. Find the number of tickets sold in each.
780 in main, 650 in terrace, 2560 in grandstand

60. There are three types of tickets sold for a college baseball game—main ($125 each), terrace ($70 each), and grandstand ($30 each). A total of 6970 tickets are sold for a revenue of $341,150. The number of the grandstand tickets sold is 500 less than twice the other two combined. Find the number of tickets sold in each.
590 in main, 1900 in terrace, 4480 in grandstand

You Be the Teacher!

Correct each student's errors, if any.

61. Solve $\begin{cases} 3x - 4y + 6z = 3 \\ x - y - 2z = 8 \\ -2x + 7y - 3z = -17 \end{cases}$.

Marian's work:

$3x - 4y + 6z = 3 \rightarrow 3x - 4y + 6z = 3$

$x - y - 2z = 8 \rightarrow -3(x - y - 2z = 8)$

$$\begin{array}{r} 3x - 4y + 6z = 3 \\ \underline{-3x - 3y - 6z = 24} \\ -7y = 27 \\ y = -\dfrac{27}{7} \end{array}$$

62. Solve $\begin{cases} x + 2y + 3z = -1 \\ 4y + 4z = -5 \\ 3x - 2y + z = 7 \end{cases}$.

Josh's work:

$$\begin{cases} x + 2y + 3z = -1 \\ \underline{3x - 2y + z = 7} \end{cases}$$
$$4x + 4z = 6$$

$$\begin{cases} 4y + 4z = -5 \\ 4x + 4z = 6 \end{cases} \rightarrow \begin{array}{r} 4y + 4z = -5 \\ \underline{-4x - 4z = -6} \\ -4x + 4y = -11 \end{array}$$

No solution

Think About It!

63. Write a system of linear equations in three variables that has $(5, -1, 4)$ as its solution.

64. Write a system of linear equations in three variables that has $(-2, 3, 7)$ as its solution.

65. The equation $y = ax^2 + bx + c$ has solutions of $(1, 6)$, $(-1, 8)$, and $(2, 11)$. Use a system of three linear equations to find the values of a, b, and c. [Hint: Substitute each ordered pair, (x, y), in $y = ax^2 + bx + c$ to obtain three equations with unknowns of a, b, and c.
$a = 2, b = -1, c = 5$

66. The equation $y = ax^2 + bx + c$ has solutions of $(1, 2)$, $(2, 2)$, and $(3, 4)$. Use a system of three linear equations to find the values of a, b, and c. [Hint: Substitute each ordered pair, (x, y), in $y = ax^2 + bx + c$ to obtain three equations with unknowns of a, b, and c.
$a = 1, b = -3, c = 4$

SECTION 5.5 Solving Systems of Linear Inequalities and Their Applications

OBJECTIVES

As a result of completing this section, you will be able to

1. Solve a system of linear inequalities in two variables.
2. Solve applications involving systems of linear inequalities.
3. Troubleshoot common errors.

The Parent Teacher Association (PTA) of a local school is selling rolls of wrapping paper and boxes of chocolate for a fund-raiser. The PTA can order at most 500 items. Each roll of wrapping paper costs \$2 and each box of chocolate costs \$3. The PTA can spend no more than \$1200 on these items. From past experience, the PTA knows that they sell at least twice as many boxes of chocolate as they do rolls of wrapping paper. How many rolls of wrapping paper and boxes of chocolate should the PTA order for their conditions to be satisfied?

To solve this problem, we need to solve a system of linear inequalities.

Systems of Linear Inequalities in Two Variables

Objective 1

Solve a system of linear inequalities in two variables.

Just as we can solve a system of linear equations, we can also solve a system of linear inequalities. A **system of linear inequalities** is a set of two or more linear inequalities that must be solved simultaneously. Systems of linear inequalities are used in an important field of mathematics called linear programming. *Linear programming* is a process used to find the maximum or minimum value of a linear function given certain constraints (that is, inequalities).

Recall that the solution set of a linear inequality is a half-plane. The **solution set of a system of linear inequalities** in two variables is the set of all ordered pairs that satisfies each inequality in the system. Graphically, this is the set of all ordered pairs where the half-planes in the system intersect. When we solve a system of linear inequalities, we will graph each inequality in the system.

Procedure: Solving a System of Linear Inequalities in Two Variables

Step 1: Graph each linear inequality on the same coordinate system using the method presented in Section 4.5.

Step 2: The intersection of the shaded regions is the solution set of the system.

Objective 1 Examples Solve each system of linear inequalities.

1a. $\begin{cases} y \leq -2x + 6 \\ y \geq \dfrac{1}{4}x \end{cases}$ **1b.** $\begin{cases} 2x + 3y > 6 \\ x - 4y < -8 \end{cases}$

1c. $\begin{cases} x - y > -5 \\ x + 2y > 0 \\ x < 2 \end{cases}$ **1d.** $\begin{cases} x \geq 0 \\ y \geq 0 \\ 4x + 3y \geq 12 \end{cases}$

Solutions

1a. Graph each inequality and find their intersection.

$y \leq -2x + 6$	$y \geq \dfrac{1}{4}x$
Boundary line: $y = -2x + 6$ is solid. Plot (0, 6) and move down 2 units, right 1 unit.	Boundary line: $y = \dfrac{1}{4}x$ is solid. Plot (0, 0) and move up 1 unit, right 4 units.

Test point (0, 0):

$$y \le -2x + 6$$
$$0 \le -2(0) + 6$$
$$0 \le 6 \quad \text{True}$$

Shade the half-plane that contains (0, 0).

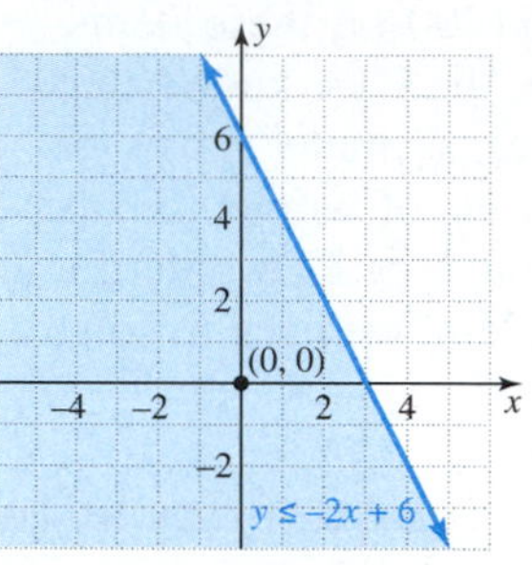

Test point (4, 0): $y \ge \frac{1}{4}x$

$$0 \ge \frac{1}{4}(4)$$
$$0 \ge 1 \quad \text{False}$$

Shade the half-plane that does not contain (4, 0).

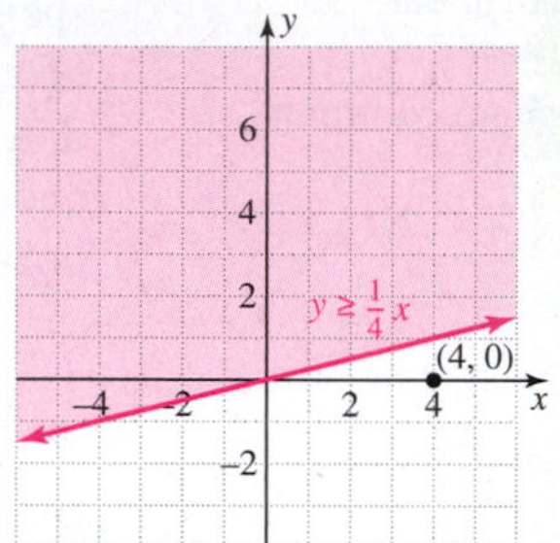

The region where the half-planes intersect is the solution set of the system of inequalities as shown in purple.

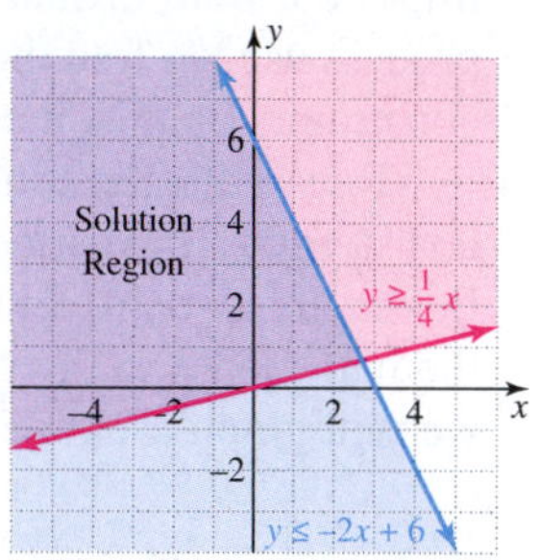

1b. Graph each inequality and find their intersection.

$$2x + 3y > 6$$

Boundary line: $2x + 3y = 6$ is dashed. Plot (0, 2) and (3, 0).

Test point (0, 0):

$$2x + 3y > 6$$
$$2(0) + 3(0) > 6$$
$$0 > 6 \quad \text{False}$$

Shade the half-plane that does not contain (0, 0).

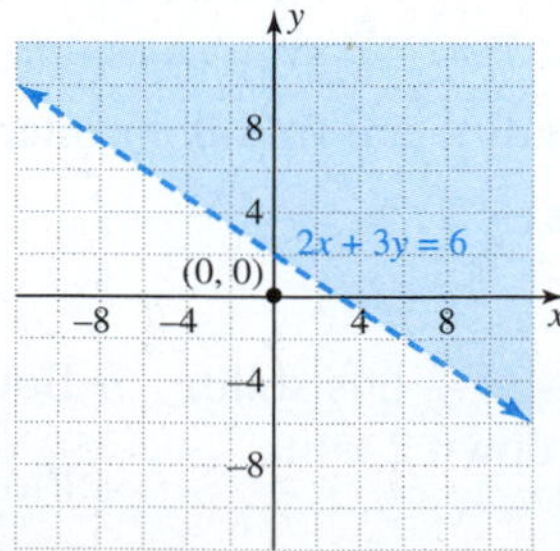

$$x - 4y < -8$$

Boundary line: $x - 4y = -8$ is dashed. Plot (0, 2) and (−8, 0).

Test point (0, 0):

$$x - 4y < -8$$
$$0 - 4(0) < -8$$
$$0 < -8 \quad \text{False}$$

Shade the half-plane that does not contain (0, 0).

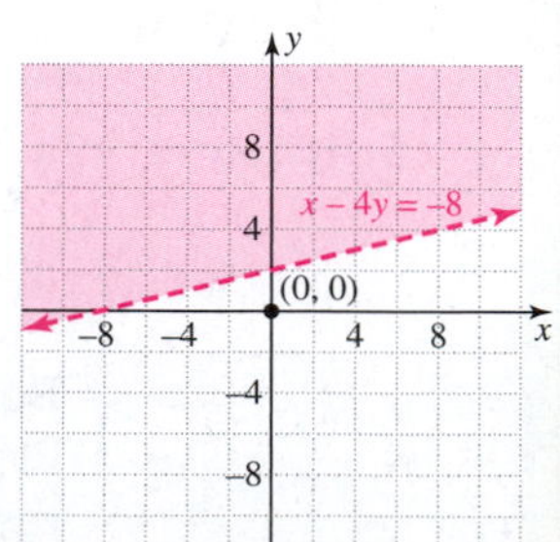

The region where the half-planes intersect is the solution set of the system of inequalities as shown.

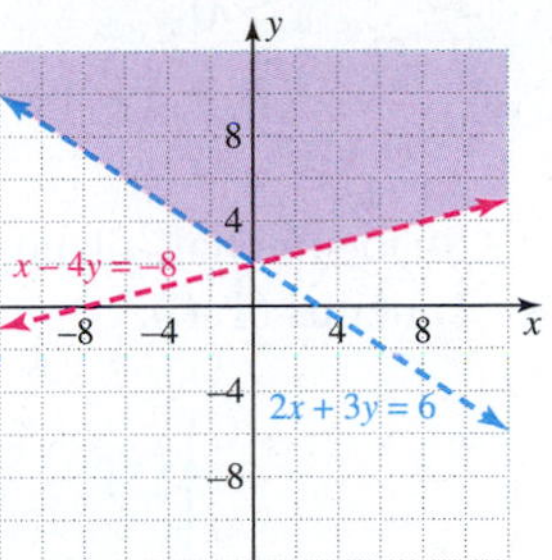

1c. Graph each inequality and find their intersection.

$x - y > -5$

Boundary line: $x - y = -5$ is dashed.

Plot (0, 5) and (−5, 0).

Test point (0, 0):

$$x - y > -5$$
$$0 - 0 > -5$$
$$0 > -5 \quad \text{True}$$

Shade the half-plane that contains (0, 0).

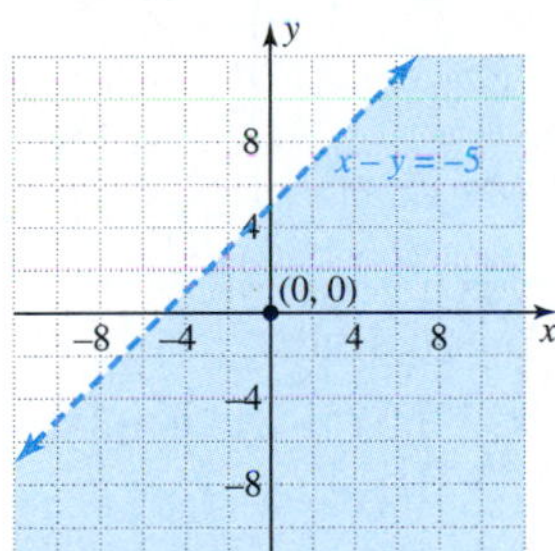

$x + 2y > 0$

Boundary line: $x + 2y = 0$ is dashed.

Plot (0, 0) and (2, −1).

Test point (3, 1):

$$x + 2y > 0$$
$$3 + 2(1) > 0$$
$$5 > 0 \quad \text{True}$$

Shade the half-plane that contains (3, 1).

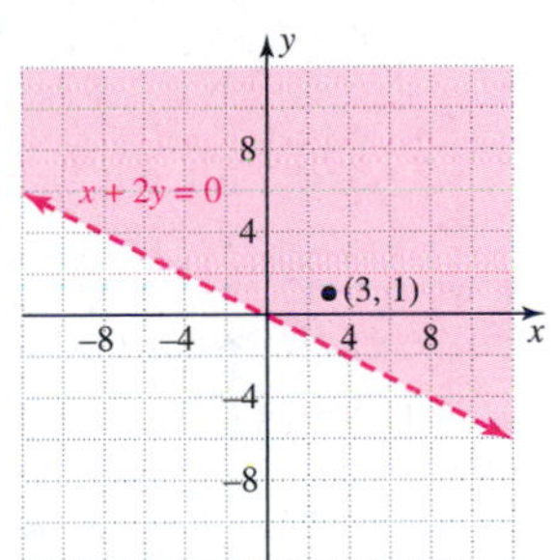

$x < 2$

Boundary line: $x = 2$ is dashed.

Vertical line through $x = 2$.

Test point (0, 0):

$$x < 2$$
$$0 < 2 \quad \text{True}$$

Shade the half-plane that contains (0, 0).

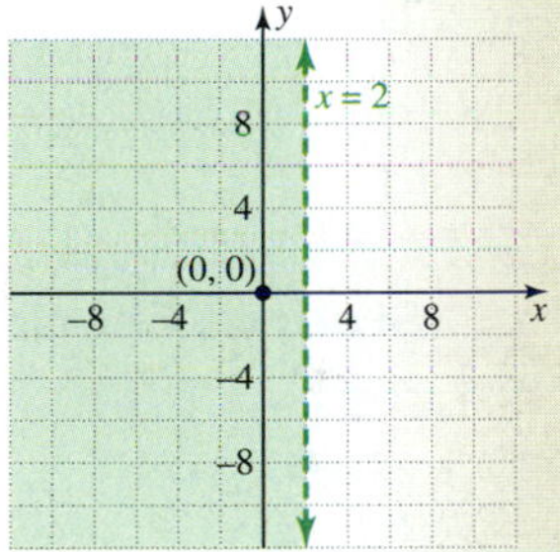

The region where the half-planes intersect is the solution set of the system of inequalities as shown.

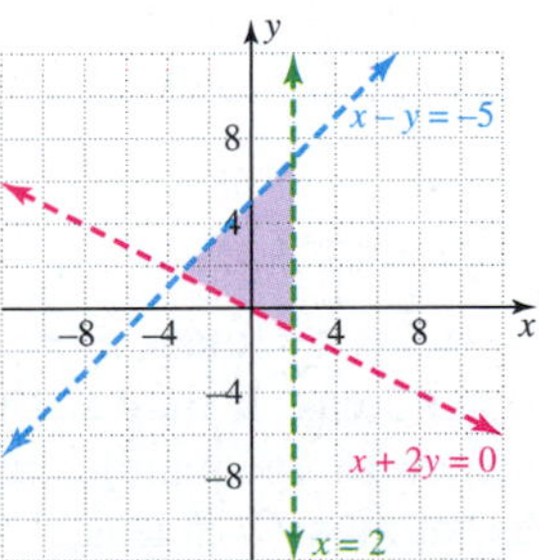

1d. Graph each inequality and find their intersection.

$x \geq 0$

Boundary line: $x = 0$ is solid.

Vertical line through $x = 0$.

$y \geq 0$

Boundary line: $y = 0$ is solid.

Horizontal line through $y = 0$.

$4x + 3y \geq 12$

Boundary line: $4x + 3y = 12$ is solid.

Plot (0, 4) and (3, 0).

Test point (1, 0):	Test point (1, 4):	Test point (0, 0):
$x \geq 0$	$y \geq 0$	$4x + 3y \geq 12$
$1 \geq 0$ True	$4 \geq 0$ True	$4(0) + 3(0) \geq 12$
		$0 \geq 12$ False
Shade the half-plane that contains (1, 0).	Shade the half-plane that contains (1, 4).	Shade the half-plane that does not contain (0, 0).

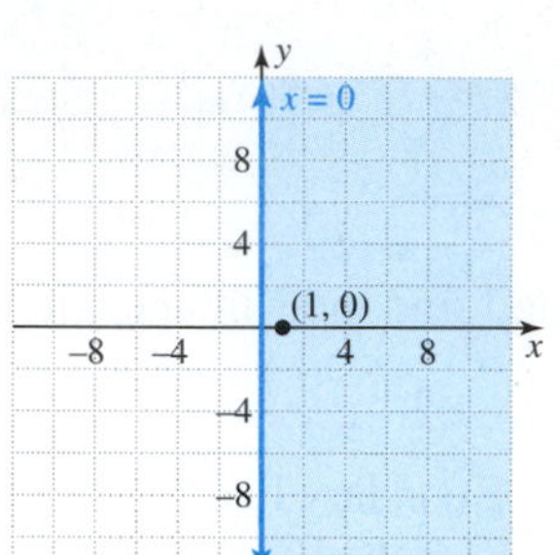

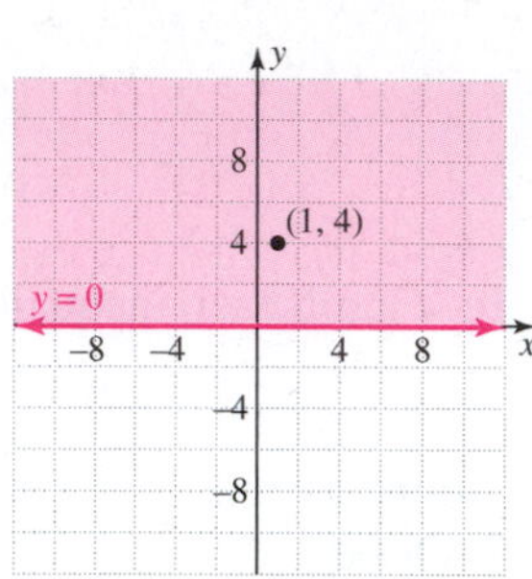

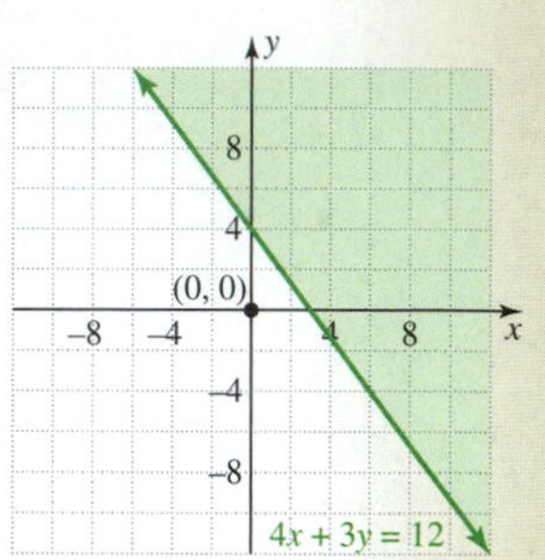

The region where the half-planes intersect is the solution of the system of inequalities as shown.

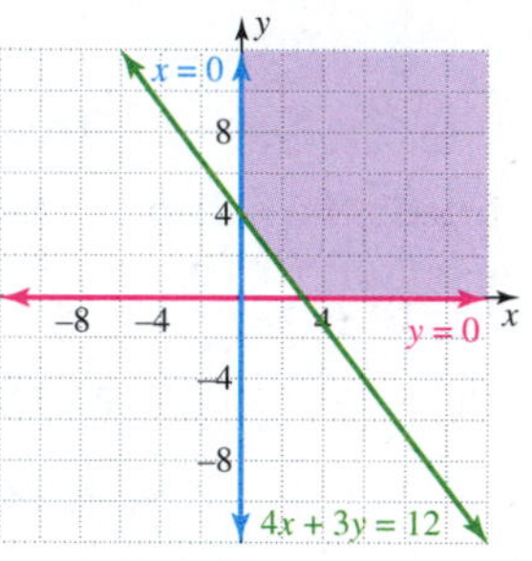

Student Check 1 Solve each system of linear inequalities.

a. $\begin{cases} y \geq x + 5 \\ y \leq -x + 2 \end{cases}$ **b.** $\begin{cases} 5x + 2y > -10 \\ 4x - y < 8 \end{cases}$

c. $\begin{cases} 3x - y < 3 \\ x + 2y > 0 \\ y \geq 1 \end{cases}$ **d.** $\begin{cases} x \geq 0 \\ y \geq 0 \\ 3x + 2y < 6 \end{cases}$

Applications

Objective 2 Solve applications involving systems of linear inequalities.

Systems of linear inequalities can be used to solve problems in business as well as other areas. If the variables in a problem satisfy several constraints that can be represented by inequalities, then a system of linear inequalities is used. Key phrases such as "at most," "at least," "not more than," and "not less than" indicate that an inequality is needed.

Procedure: Solving an Application Involving a System of Linear Inequalities

Step 1: Determine the unknowns and define variables for each of them.
Step 2: Write a system of linear inequalities to represent the constraints given in the problem.
Step 3: Solve the system graphically.
Step 4: Solutions of the word problem are ordered pairs within the solution set.

Objective 2 Example The Parent Teacher Association (PTA) of a local school is selling rolls of wrapping paper and boxes of chocolate for a fund-raiser. The PTA can order at most 500 items. Each roll of wrapping paper costs \$2 and each box of chocolate costs \$3. The PTA can spend no more than \$1200 on these items. From past experience, the PTA knows that they sell at least twice as many rolls of wrapping paper as they do boxes of chocolate. How many rolls of wrapping paper and how many boxes of chocolate should the PTA order for their conditions to be satisfied? Provide three specific pairs of numbers that are in the solution set.

Solution What are the unknowns? The number of rolls of wrapping paper and the number of boxes of chocolate that should be ordered are unknown.

Let w = number of rolls of wrapping paper.

Let c = number of boxes of chocolate.

The system can be defined as

$$\begin{cases} 2w + 3c \le 1200 \\ w + c \le 500 \\ w \ge 2c \\ c \ge 0 \\ w \ge 0 \end{cases}$$

- $2w + 3c \le 1200$: The PTA can spend no more than \$1200.
- $w + c \le 500$: The PTA can order at most 500 items.
- $w \ge 2c$: Sell at least twice as many rolls of wrapping paper as boxed choc.
- $c \ge 0$, $w \ge 0$: The number of boxes of chocolate and rolls of wrapping paper must be greater than or equal to zero.

Now we graph each inequality and find their intersection. The last two inequalities tell us that the graph of the solution will be in Quadrant I. So, we will graph the first three inequalities in Quadrant I only. The following notation BL represents the boundary line.

$2w + 3c \le 1200$

BL: $2w + 3c = 1200$ (solid)

(0, 0) makes the inequality true, so shade the half-plane that contains the origin.

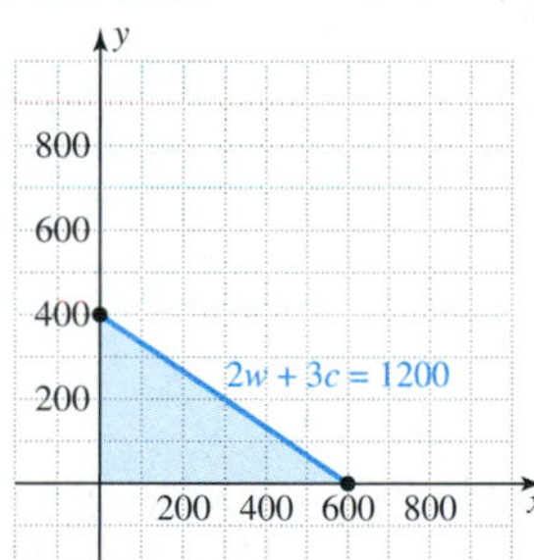

$w + c \le 500$

BL: $w + c = 500$ (solid)

(0, 0) makes the inequality true, so shade the half-plane that contains the origin.

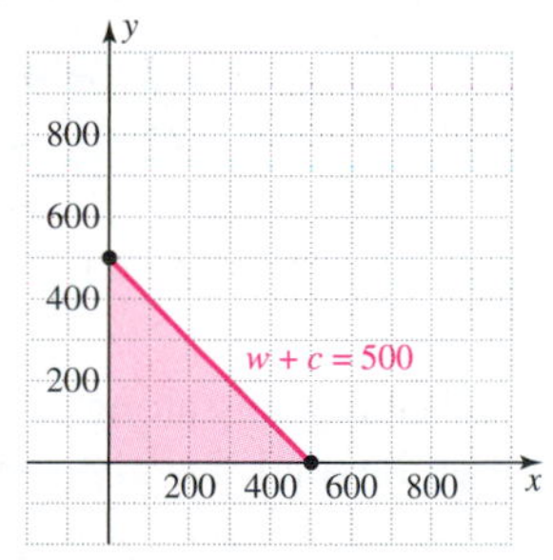

$w \ge 2c$

BL: $w = 2c$ (solid)

(100, 0) makes the inequality true, so shade the half-plane containing (100, 0).

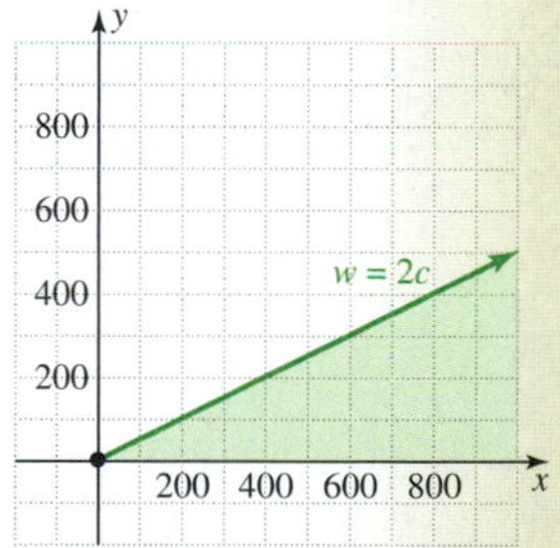

The intersection of these three graphs is the solution set of the system of inequalities as shown.

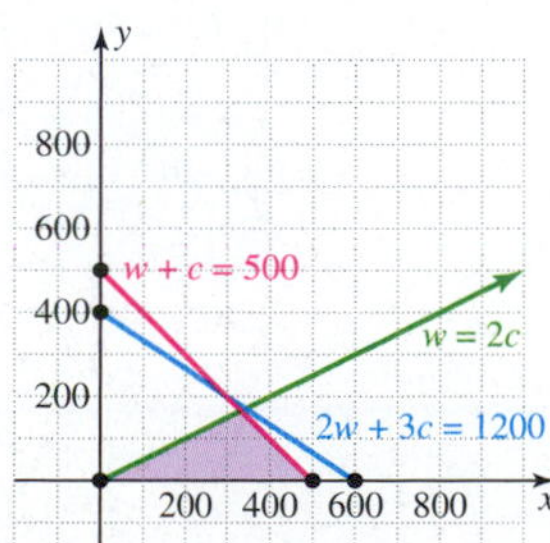

Any ordered pair in the shaded region or on the boundary lines of the shaded region satisfies the system. Some possible combinations that will satisfy the PTA's requirements are

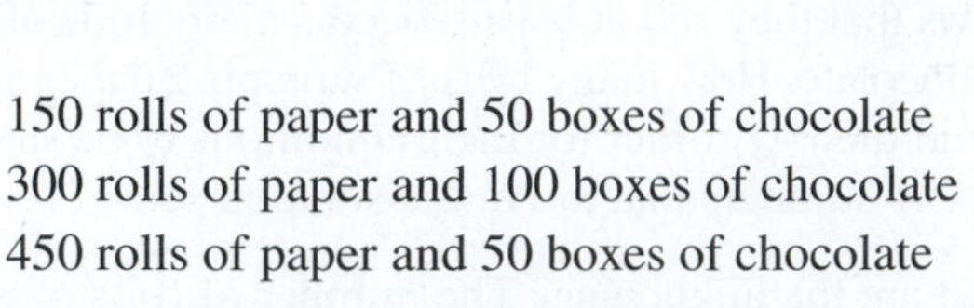

150 rolls of paper and 50 boxes of chocolate
300 rolls of paper and 100 boxes of chocolate
450 rolls of paper and 50 boxes of chocolate

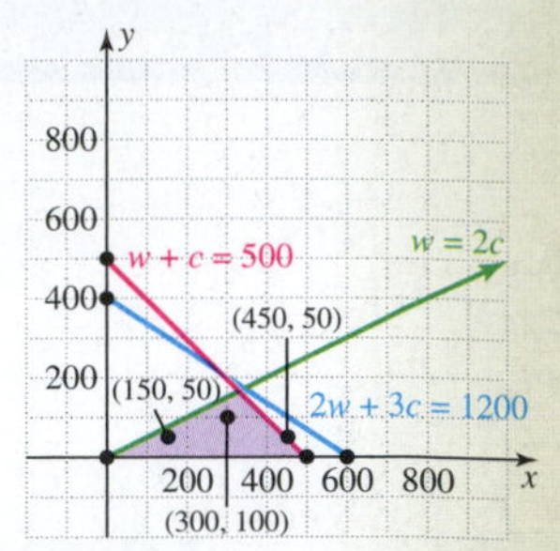

Student Check 2 A motel chain plans to open a new motel. The motel will have a combination of double rooms and king-size rooms. The motel will have at most 200 rooms. Based on consumer trends, the motel knows that they should have at least 125 double rooms. They also know that they need at least twice as many double rooms as king-size rooms. Use a system of linear inequalities to find at least three combinations of double rooms and king-size rooms that will satisfy these conditions. Let $x =$ king-size rooms and $y =$ double rooms.

Objective 3 ▶

Troubleshoot common errors.

Troubleshooting Common Errors

A common error associated with systems of linear inequalities is shown.

Objective 3 Example **A problem and an incorrect solution are given. Provide the correct solution and an explanation of the error.**

Solve the system $\begin{cases} x - 2y \geq 4 \\ x \geq 0 \\ y \geq 0 \end{cases}$.

Incorrect Solution

$$x - 2y \geq 4$$
$$-2y \geq -x + 4$$
$$y \geq \frac{1}{2}x - 2$$

Graphing this along with the other two inequalities gives the solution,

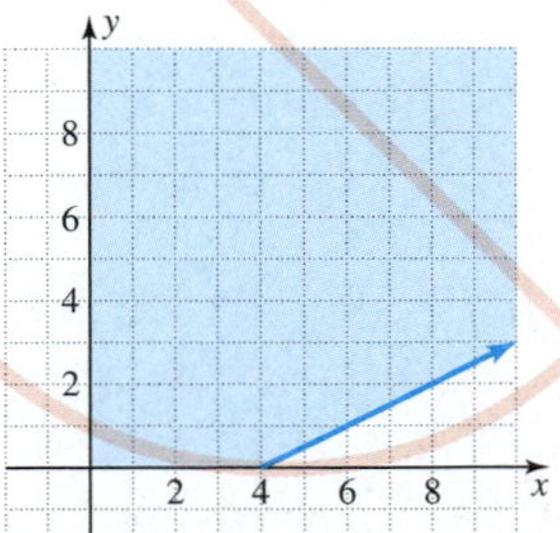

Correct Solution and Explanation

When we divide by a negative sign, the inequality symbol reverses. So, we should get

$$y \leq \frac{1}{2}x - 2$$

So, the solution set of the system is

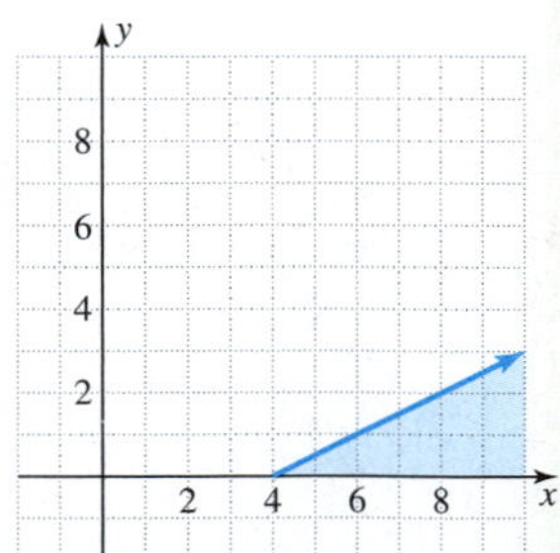

ANSWERS TO STUDENT CHECKS

Student Check 1

a.

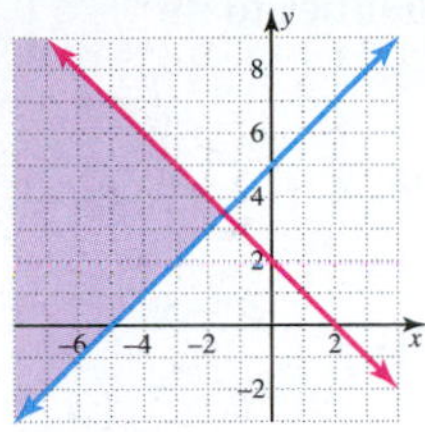

b.

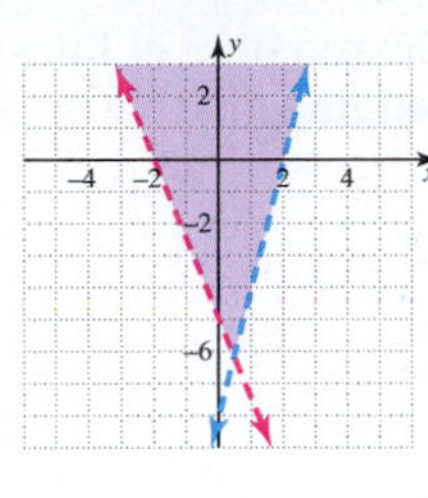

c.

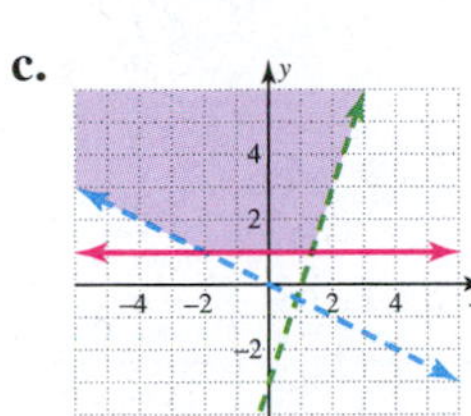

d.

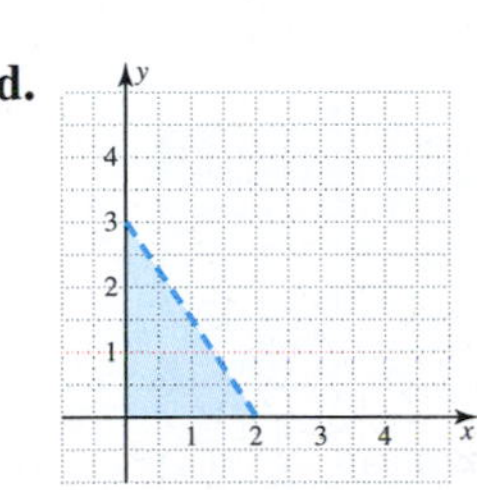

Student Check 2

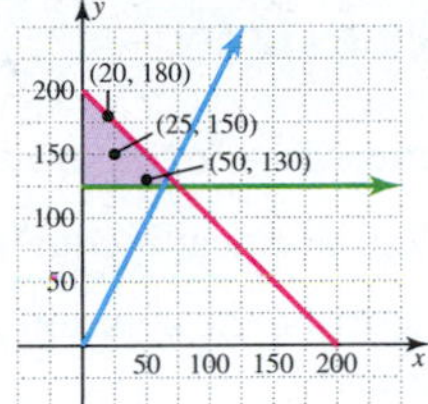

SUMMARY OF KEY CONCEPTS

1. Solving a system of linear inequalities requires us to graph each linear inequality on the same coordinate system. The solution set for the system is the intersection of the half-planes formed by each inequality. Be careful to draw the boundary lines correctly for each inequality. If the inequality symbol is $<$ or $>$, then the boundary line is dashed. If the inequality symbol is $\leq$ or $\geq$, then the boundary line is solid.
2. Applications of systems arise when there are several inequalities that describe a given situation.

GRAPHING CALCULATOR SKILLS

The graphing calculator can provide the solution set of a system of linear inequalities.

Example: Solve the system $\begin{cases} 2x + y < 6 \\ x - y \leq 4 \end{cases}$.

Method 1: Solve each inequality in the system for y. This gives us

$$\begin{cases} y < -2x + 6 \\ y \geq x - 4 \end{cases}.$$

Enter each inequality in the equation editor. Since the first inequality involves a $<$ symbol, select the option to graph below the line. The second inequality involves a $\geq$ symbol, so we select the option to graph above the line.

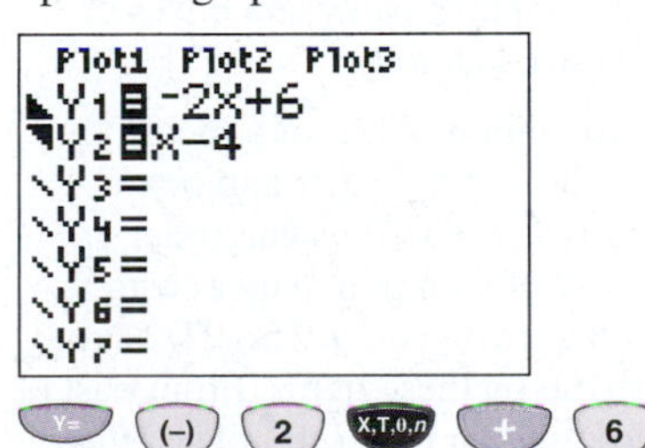

Y= (-) 2 X,T,θ,n + 6 ENTER

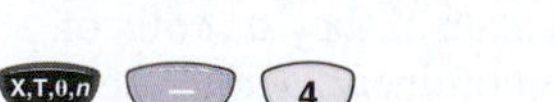

Graph to view the intersection of the two inequalities.

GRAPH

Method 2: Use the APPS menu to graph the inequalities. Press APPS and scroll down to Inequalz, and press any key to enter the program. Select the appropriate inequality symbol by using the ALPHA key and the appropriate function key. Graph to view the solution.

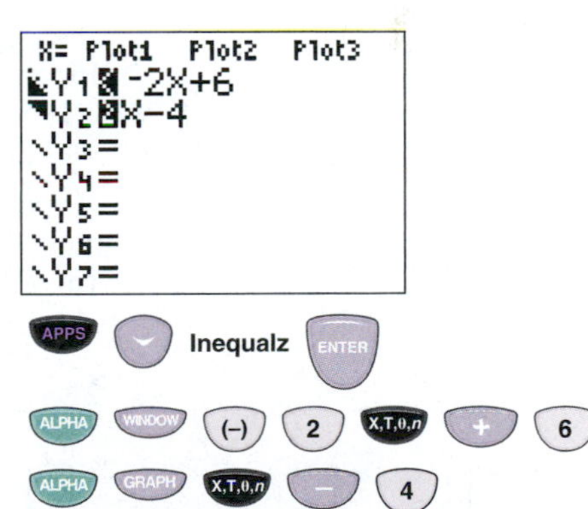

Note: The cursor must be placed on the equals sign for the inequality symbols to display at the bottom of the screen.

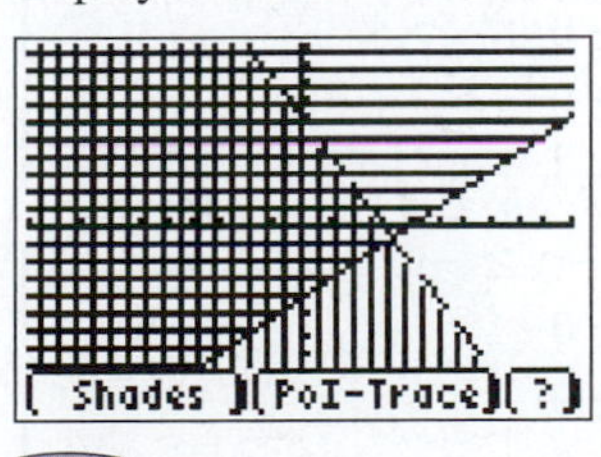

GRAPH

SECTION 5.5 EXERCISE SET

Write About It!

Use complete sentences in your answer to each exercise.

1. What is a system of linear inequalities?
2. How do you describe the solution set of a system of linear inequalities?
3. How can you tell if inequalities will be needed in an application problem?
4. Describe the steps to solve a system of linear inequalities.
5. How do you draw the boundary line for each inequality?

Practice Makes Perfect!

Solve each system of linear inequalities. (*See Objective 1.*)

6. $\begin{cases} y \le 2x - 2 \\ y \ge -\dfrac{2}{3}x + 6 \end{cases}$

7. $\begin{cases} y \le 4x - 8 \\ y \ge \dfrac{7}{4}x + 1 \end{cases}$

8. $\begin{cases} y \le 3x - 1 \\ y \ge -\dfrac{1}{2}x + 6 \end{cases}$

9. $\begin{cases} y \le 3x - 5 \\ y \ge \dfrac{1}{4}x + 6 \end{cases}$

10. $\begin{cases} y > -3x + 5 \\ y < 10x - 8 \end{cases}$

11. $\begin{cases} y > x + 1 \\ y < 4x - 8 \end{cases}$

12. $\begin{cases} y < -5x + 8 \\ y < 10x - 7 \end{cases}$

13. $\begin{cases} y > -x + 5 \\ y > 8x - 4 \end{cases}$

14. $\begin{cases} 15x + y \ge 8 \\ 3x + y \ge -4 \end{cases}$

15. $\begin{cases} 6x + 5y > 20 \\ 3x - 5y < 25 \end{cases}$

16. $\begin{cases} x - 2y \le 6 \\ y \ge 2x \end{cases}$

17. $\begin{cases} -3x + y \le 10 \\ x \le 2y \end{cases}$

18. $\begin{cases} 4x - y < 9 \\ x + 2y > 0 \end{cases}$

19. $\begin{cases} x + 2y > 5 \\ 2x - y < 0 \end{cases}$

20. $\begin{cases} x \le 2 \\ y \ge -2 \end{cases}$

21. $\begin{cases} y - 3 > 0 \\ x + 2 < 0 \end{cases}$

22. $\begin{cases} x + y \le 5 \\ x + 2y \ge 8 \\ x \ge 0 \end{cases}$

23. $\begin{cases} x + y \le 4 \\ -3x + 2y \ge -2 \\ y \ge 0 \end{cases}$

24. $\begin{cases} x - 3y \ge 3 \\ x \ge 0 \\ y \ge 0 \\ y \le 3 \end{cases}$

25. $\begin{cases} 2x + y \le 4 \\ x \ge 0 \\ y \ge 0 \\ y \le 2 \end{cases}$

26. $\begin{cases} 2x - y \ge -4 \\ x \ge 0 \\ y \ge 0 \\ x \le 3 \end{cases}$

27. $\begin{cases} x - 2y \ge -6 \\ x \ge 0 \\ y \ge 0 \\ x \le 4 \end{cases}$

Additional answers can be found in the Instructor Answer Appendix.

Match each system of linear inequalities to its corresponding graph. (*See Objective 1.*)

28. $\begin{cases} x > -2 \\ y < 4 \end{cases}$ D

29. $\begin{cases} x \le 4 \\ y \ge -2 \end{cases}$ A

30. $\begin{cases} x \le -1 \\ y \ge -3 \end{cases}$ B

31. $\begin{cases} y < -3 \\ x > -1 \end{cases}$ C

A. **B.**

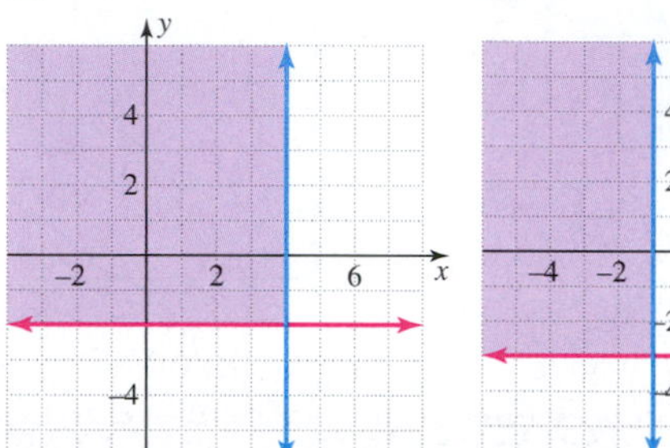

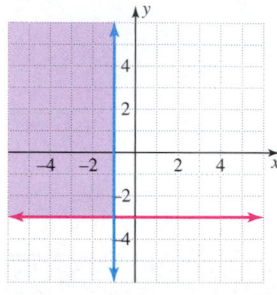

C. **D.**

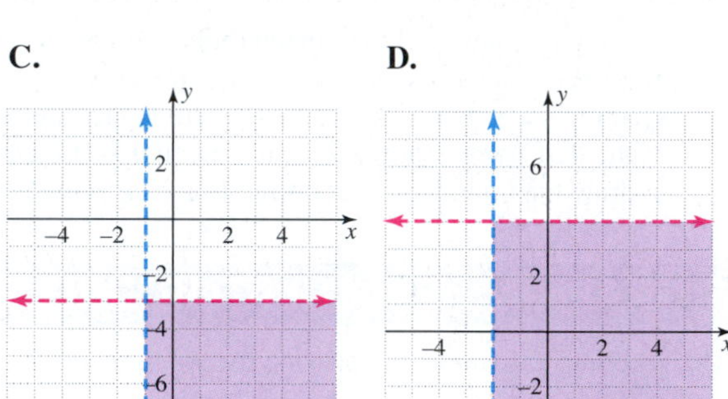

Solve each problem using a system of linear inequalities. (*See Objective 2.*)

32. The Parent Teacher Association (PTA) of a local school is selling rolls of wrapping paper and boxes of chocolate for a fund-raiser. The PTA can order at most 1200 items. Each roll of wrapping paper costs \$1.50 and each box of chocolate costs \$4. The PTA can spend no more than \$4200 on these items. From past experience, the PTA knows that they sell at least two times as many boxes of chocolate as they do rolls of wrapping paper. Use a system of linear inequalities to find at least three combinations of rolls of wrapping paper and boxes of chocolate that satisfy these conditions.

33. The Parent Teacher Association (PTA) of a local school is selling rolls of wrapping paper and boxes of chocolate for a fund-raiser. The PTA can order at most 750 items. Each roll of wrapping paper costs \$2 and each box of chocolate costs \$6.50. The PTA can spend no more than \$5100 on these items. From past experience, the PTA knows that they sell at least three times as many boxes of chocolate as they do rolls of wrapping paper. Use a system of linear inequalities to find at least three combinations of rolls of wrapping paper and boxes of chocolate that satisfy these conditions.

34. The Student Government Association (SGA) of a local college is selling drinks and hot dogs for a fund-raiser. The SGA can order at most 810 items. Each drink costs \$0.75 and each hot dog costs \$1.50. The SGA can spend no more than \$1200 on these items. From past experience, the SGA knows that they sell at least twice as many hot dogs as they do drinks. Use a system of linear inequalities to find at least three combinations of drinks and hot dogs that satisfy these conditions.

35. The Student Government Association (SGA) of a local college is selling drinks and hot dogs for a fund-raiser. The SGA can order at most 720 items. Each drink costs \$1 and each hot dog costs \$2. The SGA can spend no more than \$1200 on these items. From past experience, the SGA knows that they sell at least twice as many hot dogs as they do drinks. Use a system of linear inequalities to find at least three combinations of drinks and hot dogs that satisfy these conditions.

36. A motel chain plans to open a new motel. The motel will have a combination of double rooms and king-size rooms. The motel will have at most 360 rooms. Based on consumer trends, the motel knows that they should have at least 160 double rooms. They also know that they need at least twice as many double rooms as king-size rooms. Use a system of linear inequalities to find at least three combinations of double rooms and king-size rooms that will satisfy these conditions.

37. A motel chain plans to open a new motel. The motel will have a combination of double rooms and king-size rooms. The motel will have at most 500 rooms. Based on consumer trends, the motel knows that they should have at least 120 double rooms. They also know that they need at least three times as many double rooms as king-size rooms. Use a system of linear inequalities to find at least three combinations of double rooms and king-size rooms that will satisfy these conditions.

In the business world, the price per item is determined by the demand equation $y = D(x)$, where x is the number of items sold. The consumer's surplus is the area of the graph of the solution set of the following system of linear inequalities.

$$\begin{cases} y \le D(x) \\ y \ge k \\ x \ge 0 \end{cases}, \text{ where } k \text{ is a fixed price}$$

38. The demand equation for an item is $y = D(x) = 20 - 0.05x$ and $k = 10$.

a. Set up and solve a system of linear inequalities.

b. Find the consumer's surplus. \$1000

39. The demand equation for an item is $y = D(x) = 30 - 0.45x$ and $k = 9$.

a. Set up and solve a system of linear inequalities.

b. Find the consumer's surplus. \$490

The price per item is determined by the supply equation $y = S(x)$, where x is the number of items produced. The producer's surplus is the area of the graph of the solution set of the following system of linear inequalities.

$$\begin{cases} y \ge S(x) \\ y \le k \\ x \ge 0 \end{cases}, \text{ where } k \text{ is a fixed price}$$

40. The supply equation for an item is $y = S(x) = 25 + 0.10x$ and $k = 50$.

a. Set up and solve a system of linear inequalities.

b. Find the producer's surplus. \$3125

41. The supply equation for an item is $y = S(x) = 8 + 0.004x$ and $k = 20$.

a. Set up and solve a system of linear inequalities.

b. Find the producer's surplus. \$18,000

Mix 'Em Up!

Solve each system of linear inequalities in two variables.

42. $\begin{cases} y \le \frac{1}{2}x + 8 \\ y \ge \frac{3}{2}x + 6 \end{cases}$

43. $\begin{cases} y \le \frac{4}{3}x + 2 \\ y \ge -\frac{2}{3}x + 6 \end{cases}$

44. $\begin{cases} x + 2y > -18 \\ x - y < 12 \end{cases}$

45. $\begin{cases} 2x - y > -6 \\ x + y < 3 \end{cases}$

46. $\begin{cases} 4x + 3y < 14 \\ 3x - 2y < 4 \end{cases}$

47. $\begin{cases} 7x + y \ge -6 \\ 5x + 2y \le 6 \end{cases}$

48. $\begin{cases} x + 2y \le -6 \\ y \ge -2x \end{cases}$

49. $\begin{cases} x + 4y < -9 \\ 2x - y > 0 \end{cases}$

50. $\begin{cases} x \le -3 \\ y \ge 4 \end{cases}$

51. $\begin{cases} x \ge 2 \\ y \le -1 \end{cases}$

52. $\begin{cases} x + y \le 6 \\ y \ge 2x \\ x \ge -3 \end{cases}$

53. $\begin{cases} x - y \le -2 \\ 6x + y \le 12 \\ y \ge -1 \end{cases}$

54. $\begin{cases} 3x + y < 6 \\ x \ge 0 \\ y \ge 0 \end{cases}$

55. $\begin{cases} 5x - 2y < 10 \\ x \ge 0 \\ y \ge 0 \end{cases}$

Solve each problem using a system of linear inequalities.

56. The Student Government Association (SGA) of a local college is selling drinks and hot dogs for a fund-raiser. The SGA can order at most 1080 items. Each drink costs \$1.50 and each hot dog costs \$2.50. The SGA can spend no more than \$2400 on these items. From past experience, the SGA knows that they sell at least twice as many hot dogs as they do drinks. Use a system of linear inequalities to find at least three combinations of drinks and hot dogs that satisfy these conditions.

57. A motel chain plans to open a new motel. The motel will have a combination of double rooms and king-size rooms. The motel will have at most 320 rooms. Based on consumer trends, the motel knows that they should have at least 60 double rooms. They also know that they need at least three times as many double rooms as king-size rooms. Use a system of linear inequalities to find at least three combinations of double rooms and king-size rooms that will satisfy these conditions.

58. The demand equation for an item is $y = D(x) = 42 - 0.07x$ and $k = 14$.

a. Set up and solve a system of linear inequalities.

b. Find the consumer's surplus. 5600

59. The supply equation for an item is $y = S(x) = 18 + 0.06x$ and $k = 33$.

a. Set up and solve a system of linear inequalities.

b. Find the producer's surplus. 1875

You Be the Teacher!

Correct each student's errors, if any.

60. Solve the system of linear inequalities.

$$\begin{cases} x - y \le 2 \\ y \ge 0 \\ x \ge 0 \end{cases}$$

Isabella's answer:

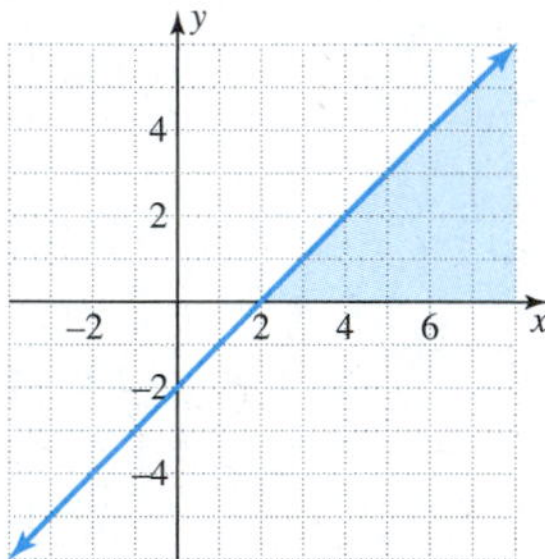

61. Solve the system of linear inequalities.

$$\begin{cases} x - y \le 0 \\ -3x + y \ge -3 \\ x \ge 0 \\ y \ge 0 \end{cases}$$

Nicole's answer:

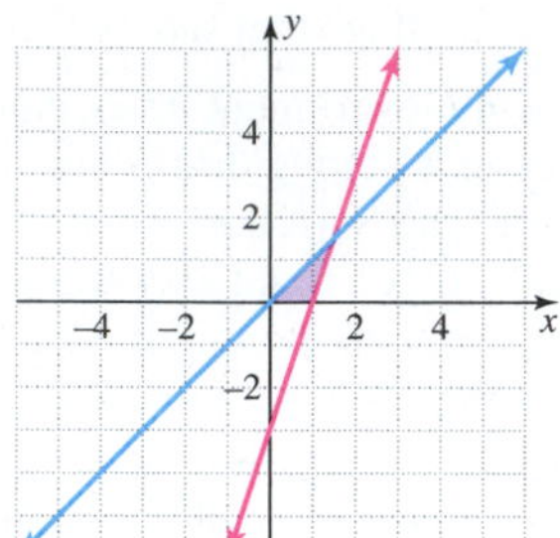

Calculate It!

Use a graphing calculator to solve each system of inequalities.

62. $\begin{cases} 4x + 3y \le 9 \\ x - y \ge -5 \end{cases}$

63. $\begin{cases} 5x - 7y \le 49 \\ x - y \ge 13 \end{cases}$

64. $\begin{cases} 3x - 2y \le 9 \\ y \le -\frac{1}{2}x - 1 \\ y \ge -6 \end{cases}$

65. $\begin{cases} 2x - y \ge 30 \\ y \le -\frac{1}{3}x - 5 \\ y \le -15 \end{cases}$

GROUP ACTIVITY Linear Programming—Maximize Revenue

Linear programming is a method to minimize or maximize certain values while satisfying given constraints. The constraints are inequalities that represent the given situation.

A youth group is planning a retreat. They must rent buses and vans to make the trip. Each bus can transport 30 students, requires 2 chaperones, and costs \$700 to rent. Each van can transport 6 students, requires 1 chaperone, and costs \$100 to rent. There are at least 90 students and at most 9 chaperones going on the trip. How many buses and vans should be rented in order to minimize transportation costs? What is the minimum cost?

1. What are the unknowns? Define a variable to represent each unknown.
2. Write a system of linear inequalities that represents the statements given in the problem.
3. Solve the system of linear inequalities. The solution set of the system of linear inequalities is called the *feasible region*.
4. Find the points where the boundary lines intersect. These points are called *corner points*.
5. What quantity needs to be minimized? What expression represents this quantity?
6. It has been shown that the minimum occurs at one of the corner points. Evaluate the expression that needs to be minimized for each of your corner points. Which point produces the minimum value?
7. Use the result from part 6 to determine how many buses and vans should be rented. What are the transportation costs for this result?

Chapter 5 / REVIEW

Systems of Linear Equations and Inequalities

What's the big idea? Systems can be used to solve applications involving two and three unknowns. Now that we have completed Chapter 5, we should be able to solve these types of systems with two unknowns in three different ways and to identify the number of solutions of a system. We should also be able to solve systems in three variables and their applications using elimination. Graphing systems of linear inequalities in two variables and their applications was also presented.

The Tools

Listed below are the key terms, skills, formulas, and properties you should know for this chapter.

The page reference is provided if you need additional help with the given topic. The Study Tips will assist in your preparation for an exam.

Study Tips

1. Learn all of the terms, formulas, and properties. Make flash cards and have someone quiz you.
2. Rework problems from the exercises and also the ones you worked in class. Work additional problems from the review exercises.
3. Review the summaries of key concepts.
4. Work the chapter test.
5. Be sure to review the online resources for additional study materials.

Terms

Break-even point 297
Consistent systems with dependent equations 295
Consistent systems with independent equations 295
Dependent equations 295
Elimination 308
Equilibrium point 297
Inconsistent system 295
Independent equations 295
Interest 323
Point of intersection 292
Solution of a system of linear equations in three variables 336
Solution of a system of linear equations in two variables 290
Solution set of a system of linear inequalities in two variables 347
Substitution 305
System of linear equations 290
System of linear equations in three variables 336
System of linear equations in two variables 290
System of linear inequalities in two variables 347

Formulas and Properties

- Addition property of equality 309
- Complementary angles 329
- Distance formula 326
- Perimeter of a rectangle 329
- Simple interest formula 323
- Supplementary angles 329

CHAPTER 5 / SUMMARY

How well do you know this chapter? Complete the following questions to find out. Take a look back at the section if you need help.

SECTION 5.1 Solving Systems of Linear Equations in Two Variables Graphically

1. A system of linear equations consists of two or more equations that must be _solved_ _simultaneously_.
2. A solution of a system of linear equations in two variables is a(n) _ordered_ _pair_ that satisfies both equations.
3. To solve a system of linear equations in two variables by graphing, we graph each _equation_ on the same coordinate system and find the _point_ _of_ _intersection_.
4. A system of linear equations in two variables has three possible types of solutions: _unique_ solution, _no_ solution, or _infinitely_ _many_ solutions.

5. The lines in a system may either _intersect_ one another, be _parallel_ to each other, or be the _same_ line.

6. A system that has at least one solution is called a(n) _consistent_ system.

7. A system that has no solution is called a(n) _inconsistent_ system.

8. If the two equations in a consistent system are multiples of one another, then the equations are _dependent_. Otherwise, the equations are _independent_.

SECTION 5.2 Solving Systems of Linear Equations in Two Variables Algebraically

9. To solve a system by substitution, we must first _solve_ one of the equations for one of the _variables_ and _substitute_ this expression into the other equation.

10. If after substitution, the resulting equation is $5 = 3$, for example, then the system is _inconsistent_ and has _no_ solution.

11. If after substitution, the resulting equation is $-3 = -3$, for example, then the system is _consistent_ with _dependent_ equations and has _infinitely_ _many_ solutions.

12. The goal of elimination is to _add_ the equations in the system so that one variable is _eliminated_.

13. For a variable to be eliminated, its coefficients must be _opposites_ of one another. If they are not, we can _multiply_ one or both equations by some nonzero number.

14. If after adding the two equations together, the resulting equation is $0 = 0$, for example, this means that the system contains two lines that are the _same_ and there are _infinitely_ _many_ solutions.

15. If after adding the two equations together, the resulting equation is $0 = 5$, for example, this means that the system contains two lines that are _parallel_ and there is _no_ solution of the system.

SECTION 5.3 Applications of Linear Systems in Two Variables

16. Applications that involve two unknowns may be solved using a(n) _system_ of equations.

17. We assign two _variables_ to the unknowns and write two different _equations_ that relate the two unknowns and then solve using _substitution_ or _elimination_.

SECTION 5.4 Solving Linear Systems in Three Variables and Their Applications

18. To solve a linear system in three variables, we use _elimination_ to reduce it to a system involving _two_ variables.

19. To solve an application with a system of three variables, determine the three _unknowns_ and the three different _equations_ that represent the given information.

SECTION 5.5 Solving Systems of Linear Inequalities and Their Applications

20. The solution set of a system of linear inequalities in two variables consists of the _intersection_ of the half-planes formed from each linear inequality.

21. Draw the boundary lines correctly: if the inequality symbol is $<$ or $>$, the boundary line is _dashed_; if the inequality symbol is $\leq$ or $\geq$, the boundary line is _solid_.

CHAPTER 5 / REVIEW EXERCISES

SECTION 5.1

Determine if the given ordered pair is a solution of the system. (*See Objective 1.*)

1. $(-4, 1)$; $\begin{cases} 6x - y = -21 \\ y = \frac{1}{2}x + 3 \end{cases}$ is not a solution

2. $(-3, 2)$; $\begin{cases} 4x + y = -13 \\ y = -\frac{1}{3}x + 1 \end{cases}$ is not a solution

3. $(-1.5, 2.5)$; $\begin{cases} 0.4x - 0.1y = -0.85 \\ -0.1x + 0.2y = 0.65 \end{cases}$ is a solution

4. $(0.8, -1.2)$; $\begin{cases} 0.3x - 0.2y = 0.48 \\ -0.5x + 0.1y = -0.52 \end{cases}$ is a solution

Solve each system graphically. (*See Objectives 2 and 3.*)

5. $\begin{cases} x - 3y = -10 \\ -2x + y = 0 \end{cases}$ $\{(2, 4)\}$

6. $\begin{cases} 5x - y = -9 \\ x - 2y = 0 \end{cases}$ $\{(-2, -1)\}$

7. $\begin{cases} 3x - y = 3 \\ y = 6 \end{cases}$ $\{(3, 6)\}$

8. $\begin{cases} -2x + 5y = 13 \\ x = -4 \end{cases}$ $\{(-4, 1)\}$

9. $\begin{cases} 2x - 5y = 10 \\ y = \frac{2}{5}x \end{cases}$ no solution

10. $\begin{cases} x = \frac{1}{5}y + \frac{1}{5} \\ y = 5x - 1 \end{cases}$ $\{(x, y) | y = 5x - 1\}$

Determine how the lines relate, the number of solutions of the system, and the type of system without graphing. (*See Objective 4.*)

11. $\begin{cases} y + 5 = 0 \\ x - 2 = 0 \end{cases}$ intersecting lines, one solution, consistent system with independent equations

12. $\begin{cases} x = \frac{5}{2}y + \frac{1}{2} \\ y = \frac{2}{5}x + \frac{1}{5} \end{cases}$ parallel lines, no solution, inconsistent system

13. $\begin{cases} x + 4y = 0 \\ 3x = -12y \end{cases}$ same lines, infinitely many solutions, consistent system with dependent equations

14. $\begin{cases} y = \frac{1}{2}x - 7 \\ y = -3x \end{cases}$ intersecting lines, one solution, consistent system with independent equations

Solve each problem. **(*See Objective 5.*)**

15. The supply of a certain product can be represented by $y = 5x + 30$, where x is the price of each item and y is the number of items (in hundreds) supplied. The demand for the same product can be represented by $y = 8x$, where x is the price of each item and y is the number of items (in hundreds) demanded at this price. Find the equilibrium point for this product and explain what it means. (10, 8000); When the product is priced at \$10, 8000 units will be supplied and demanded.

16. The supply of a certain product can be represented by $y = 3x + 48$, where x is the price of each item and y is the number of items (in hundreds) supplied. The demand for the same product can be represented by $y = 7x$, where x is the price of each item and y is the number of items (in hundreds) demanded at this price. Find the equilibrium point for this product and explain what it means. (12, 8400); When the product is priced at \$12, 8400 units will be supplied and demanded.

SECTION 5.2

Solve each system of linear equations by substitution or elimination. **(*See Objectives 1–3.*)**

17. $\begin{cases} 2x + y = -1 \\ 3x - 2y = -19 \end{cases}$ $\{(-3, 5)\}$

18. $\begin{cases} x - 5y = 7 \\ 3x + 7y = -1 \end{cases}$ $\{(2, -1)\}$

19. $\begin{cases} 4x - 5y = 3.2 \\ 0.2x - 0.7y = 0.34 \end{cases}$ $\{(0.3, -0.4)\}$

20. $\begin{cases} x + 3y = 4.3 \\ -0.6x + 0.8y = 1.32 \end{cases}$ $\{(-0.2, 1.5)\}$

21. $\begin{cases} x - 2y = 1 \\ -2x + 4y = -2 \end{cases}$ $\{(x, y) | x - 2y = 1\}$

22. $\begin{cases} -\frac{3}{4}x + y = 6 \\ x - \frac{4}{3}y = -8 \end{cases}$ $\{(x, y) | -3x + 4y = 24\}$

23. $\begin{cases} \frac{1}{3}x - \frac{1}{2}y = -\frac{1}{4} \\ -\frac{2}{3}x + \frac{1}{6}y = \frac{7}{4} \end{cases}$ $\left\{\left(-3, -\frac{3}{2}\right)\right\}$

24. $\begin{cases} \frac{1}{5}x - \frac{4}{5}y = \frac{1}{2} \\ \frac{1}{2}x - \frac{2}{3}y = \frac{3}{4} \end{cases}$ $\left\{\left(1, -\frac{3}{8}\right)\right\}$

25. $\begin{cases} x + 2y = 9 \\ y = -\frac{1}{2}x + 4 \end{cases}$ $\varnothing$

26. $\begin{cases} 5x - 7y = 1 \\ -10x + 14y = 1 \end{cases}$ $\varnothing$

27. $\begin{cases} x - 6y = -6 \\ y = \frac{5}{3}x + 4 \end{cases}$ $\left\{\left(-2, \frac{2}{3}\right)\right\}$

28. $\begin{cases} 5x - 6y = 24 \\ y = \frac{1}{3}x - 2 \end{cases}$ $\left\{\left(4, -\frac{2}{3}\right)\right\}$

29. $\begin{cases} 3.5x - 2.8y = 0 \\ -15x + 12y = 0 \end{cases}$ $\{(x, y) | -5x + 4y = 0\}$

30. $\begin{cases} 0.7x - 0.3y = 2.8 \\ 35x - 15y = 140 \end{cases}$ $\{(x, y) | 7x - 3y = 28\}$

31. $\begin{cases} 0.7x + 0.1y = 3.9 \\ 0.2x - 0.5y = 2.7 \end{cases}$ $\{(6, -3)\}$

32. $\begin{cases} -0.2x + 0.3y = 1.1 \\ 0.4x + 0.1y = -1.5 \end{cases}$ $\{(-4, 1)\}$

SECTION 5.3

Solve each problem using a system of linear equations. **(*See Objectives 1–5.*)**

33. Two families attend the season home opener game at Citi Field, the new home for the New York Mets in Flushing, New York. One family purchases five Shack-cago Dogs and three fried flounder sandwiches for \$53.50. Another family purchases two Shack-cago Dogs and five fried flounder sandwiches for \$52.75. What is the price of each Shack-cago Dog and each fried flounder sandwich? (Source: http://www.nytimes.com/2009/04/03/dining/03food.html) \$5.75 for a Shack-cago Dog and \$8.25 for a fried flounder sandwich

34. Two families visit LEGOLAND Park in San Diego, California. One family pays \$325 for three adults and two children for admission tickets. Another family pays \$453 for four adults and three children for admission. What is the price of admission for each adult and each child to LEGOLAND Park? (Source: http://california.legoland.com/tickets/admission_tickets/) \$69 for each adult and \$59 for each child

35. Admission to a football game is \$15.50 for student tickets and \$26.25 for general admission. If 3200 tickets are sold and \$74,862.50 is collected, how many of each type of ticket is sold? 850 student tickets and 2350 general admission tickets

36. Hong is working two jobs to help pay his way through college. He works as a laboratory assistant for the Physics Department and as a peer tutor at the math tutoring lab. One week he earns \$183.75 by working 5 hr as a laboratory assistant and 15 hr as a peer tutor. Another week he earns \$216 by working 8 hr as a laboratory assistant and 16 hr as a peer tutor. What is Hong's hourly wage as a laboratory assistant and as a peer tutor? \$7.50 per hour as a laboratory assistant and \$9.75 per hour as a peer tutor

37. Allison invests \$11,000 in two different accounts. She invests part of her money in a savings account that earns 3.75% simple interest and the rest of her money in a CD that earns 5% simple interest. If she earns a total of \$456.25 interest in 1 yr, how much did she invest in each account? \$7500 at 3.75% and \$3500 at 5%

38. Dihema receives an inheritance of \$88,000. She invests part of the inheritance in a money market account that earns 3.75% simple interest. She invests the rest of her inheritance in a stock account that pays 5.8% in dividends. If she earns a total of \$4714.50 investment income in 1 yr, how much did she invest in each account? \$19,000 at 3.75% and \$69,000 at 5.8%

39. How many gallons of a 2.5% iodine solution and how many gallons of a 6% iodine solution should be mixed together to obtain 800 gal of a 3.2% iodine solution? 640 gal of 2.5% iodine solution and 160 gal of 6% iodine

40. A chemist has a 60% alcohol solution. He needs 90 L of a 10% alcohol solution. How many liters of water and how many liters of the 60% alcohol solution must he mix together to obtain 60 L of a 90% alcohol solution? 75 L of water and 15 L of 60% alcohol solution

41. The distance from San Diego, California, to Tampa, Florida, is approximately 2452 mi. Flying from San Diego to Tampa takes approximately 4.58 hr with the wind. The return flight takes approximately 5.27 hr since the plane is flying against the wind. What is the speed of the plane in still air and what is the speed of the wind? Round to the nearest integer. The speed of the plane in still air is 500 mph and the speed of the wind is 35 mph.

42. A plane travels 1968 mi with the wind in 5 hr. The return trip against the wind takes 6 hr. What is the speed of the plane in still air and what is the speed of the wind? The speed of the plane in still air is 360.8 mph and the speed of the wind is 32.8 mph.

43. At 7 A.M. Andrew leaves his home traveling south on his bike at an average speed of 9 mph. At 8 A.M. his wife Beth leaves their home to meet Andrew. She travels south in her car at an average speed of 45 mph. How long will it take Beth to reach Andrew? It will take Beth 0.25 hr to reach Andrew.

44. Amy and her mom live 219.4 mi apart. They plan to meet each other for lunch. Amy leaves at 10 A.M. and travels toward her mom at an average speed of 57 mph. Her mom leaves at 11:30 A.M. and travels toward Amy at an average speed of 46 mph. What time will they meet?
Amy travels 2.8 hr after 10 A.M., so they will meet at 12:48 P.M.

45. The perimeter of a tennis court is 120 ft. The length of the court is 20 ft less than three times the width of the court. Find the dimensions of the court.
The length of the court is 40 ft and the width of the court is 20 ft.

46. The perimeter of a basketball court is 268 ft. The length of the court is 116 ft less than four times the width of the court. Find the dimensions of the court.
The length of the court is 84 ft and the width of the court is 50 ft.

47. Two angles are supplementary. The measure of one angle is 46° more than 4 times the measure of the other angle. Find the measure of each angle. 153.2° and 26.8°

48. Two angles are complementary. The measure of one angle is 26° more than 3 times the measure of the other angle. Find the measure of each angle. 74° and 16°

SECTION 5.4

Solve each system of linear equations. (*See Objective 1.*)

49. $\begin{cases} 2x + y - 2z = -1 \\ -4x - 3y - 3z = 6 \\ -x + 8y - 2z = 29 \end{cases}$ $\{(-3, 3, -1)\}$

50. $\begin{cases} 3x + 3y + 5z = 25 \\ x + y + z = 5 \\ 3x - 3y + 4z = 14 \end{cases}$ $\{(-1, 1, 5)\}$

51. $\begin{cases} 7x + 6y + z = -34 \\ 9x + 2z = -10 \\ x + 9y + z = -34 \end{cases}$ $\{(-2, -4, 4)\}$

52. $\begin{cases} 6x + y = 28 \\ 5x + 4y - 5z = 7 \\ -x - 3y + 3z = 7 \end{cases}$ $\{(5, -2, 2)\}$

53. $\begin{cases} 5x + y + z = 10 \\ 3y + 13z = -45 \\ 2x + y + 3z = -5 \end{cases}$

54. $\begin{cases} -5x + 6y + 3z = -10 \\ y + 8z = -15 \\ -x + y - z = 1 \end{cases}$ $\{(-16 - 9z, -15 - 8z, z) \mid z \text{ is a real number}\}$

Solve each problem using a system of linear equations in three variables. (*See Objective 2.*)

55. Ted and Kristi invest $35,750 in three different accounts—a 1-yr CD that pays 3.5% annual interest, a savings account that pays 4.5% annual interest, and a money market account that pays 2.5% annual interest. They invest twice as much money in the savings account as the amount invested in the CD. Find the amount invested in each account if they earn $1443.75 interest in a year. $11,000 in CD, $22,000 in high-interest savings account, $2750 in money market account

56. A baseball game offers three types of admission tickets. Dugout-level tickets are $120 each, club-level tickets are $75 each, and general admission tickets are $65 each. A total of 7050 tickets are sold totaling $647,500. The number of dugout-level tickets sold was 350 less than the other two combined. How many of each type of ticket were sold? 3350 dugout-level tickets, 500 in club-level tickets, and 3200 general admission tickets

57. An international flight from Los Angeles to Beijing consists of first-class, business-class, and coach travelers. A first-class ticket costs $9500. A business-class ticket costs $4600 and a coach ticket costs $650. The airline made $306,300 in airfare revenue for the trip from Los Angeles to Beijing. The flight had a total of 224 paying passengers. If the number of coach travelers was 6 times as many as first-class and business-class combined, how many of each type of traveler were on the flight?
7 first-class travelers, 25 business-class travelers, and 192 coach travelers

SECTION 5.5

Solve each system of linear inequalities in two variables. (*See Objective 1.*)

58. $\begin{cases} y < -\dfrac{5}{8}x + 8 \\ y > \dfrac{9}{8}x - 6 \end{cases}$

59. $\begin{cases} y \le \dfrac{13}{5}x - 5 \\ y \ge \dfrac{6}{5}x + 2 \end{cases}$

60. $\begin{cases} x - y \ge -3 \\ 3x + y \le 7 \\ y \ge 1 \end{cases}$

61. $\begin{cases} x + y \le 5 \\ y \ge \dfrac{1}{2}x \\ x \ge -2 \end{cases}$

62. $\begin{cases} 2x + 3y \le 6 \\ x \ge 0 \\ y \ge 0 \end{cases}$

63. $\begin{cases} 2x - 5y \ge 5 \\ x \ge 0 \\ y \ge 0 \end{cases}$

64. $\begin{cases} 3x - y > 9 \\ x - 4y < 8 \end{cases}$

65. $\begin{cases} x + 5y > 5 \\ 3x - y > 6 \end{cases}$

66. $\begin{cases} x > 4 \\ y < -2 \end{cases}$

67. $\begin{cases} y \ge -3 \\ x \le 1 \end{cases}$

Solve each problem using a system of linear inequalities. (*See Objective 2.*)

68. The Parent Teacher Association (PTA) of a local school is selling candles and cookbooks for a fund-raiser. The PTA can order at most 1400 items. Each candle costs $2 and each cookbook costs $5.25. The PTA can spend no more than $6640 on these items. From past experience, the PTA knows that they sell at least four times as many cookbooks as they do candles. Use a system of linear inequalities to find at least three combinations of candles and cookbooks that satisfy these conditions.

69. A motel chain plans to open a new motel. The motel will have a combination of double rooms and king rooms. The motel will have at most 790 rooms. Based on consumer trends, the motel knows that they should have at least 250 double rooms. They also know that they need at least three times as many double rooms as king rooms. Use a system of linear inequalities to find at least three combinations of double rooms and king rooms that will satisfy these conditions.

CHAPTER 5 TEST / SYSTEMS OF LINEAR EQUATIONS AND INEQUALITIES

1. The ordered pair that is a solution of the system $\begin{cases} 4x - y = 3 \\ y = 6x - 4 \end{cases}$ is

   a. $(2, 8)$ b. $\left(\frac{1}{2}, -1\right)$

   c. $\left(-\frac{1}{2}, -7\right)$ d. $(-2, -11)$

2. Use graphing to solve the system $\begin{cases} x - y = -2 \\ y = -\frac{2}{3}x - 3 \end{cases}$.

3. Without graphing the system, determine the number of solutions of the system, how the lines in the system relate, the type of system and the solution set. Complete each table.

   a. $\begin{cases} 7x - 4y = 8 \\ \frac{14}{8}x - y = 2 \end{cases}$

Slope-intercept form	$y = \frac{7}{4}x - 2$	$y = \frac{7}{4}x - 2$
Slopes	$m = \frac{7}{4}$	$m = \frac{7}{4}$
y-intercepts	$(0, -2)$	$(0, -2)$
How do lines relate?	The lines are the same.	
Number of solutions?	Infinitely many solutions	
Type of system?	Consistent system with dependent equations	
Solution set?	$\{(x, y) \mid 7x - 4y = 8\}$	

   b. $\begin{cases} y = \frac{1}{5}x + 3 \\ x - 5y = 10 \end{cases}$

Slope-intercept form	$y = \frac{1}{5}x + 3$	$y = \frac{1}{5}x - 2$
Slopes	$m = \frac{1}{5}$	$m = \frac{1}{5}$
y-intercepts	$(0, 3)$	$(0, -2)$
How do lines relate?	The lines are parallel.	
Number of solutions?	No solution	
Type of system?	Inconsistent system	
Solution set?	$\varnothing$	

4. The graph models the percentage of Americans who use the newspaper or Internet as their source for news, where x is the years after 2003. (Source: http://pewresearch.org/pubs/1066/internet-overtakes-newspapers-as-news-outlet)

News Sources for Americans

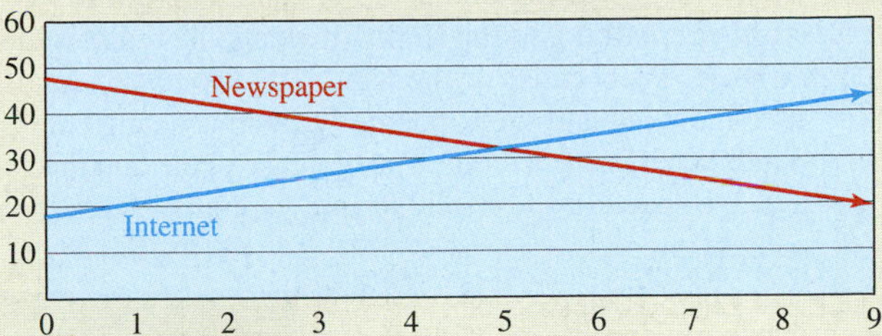

   a. Approximate the point of intersection. $(5, 32)$

   b. Interpret what the point of intersection means in the context of this problem.

5. What are the three methods for solving a system of linear equations in two variables by hand? Which method is the least precise? Why? How do you decide when to use the other two methods?

6. Solve each system by substitution.

   a. $\begin{cases} x - 8y = 13 \\ 5y - 9x = -50 \end{cases}$ $\{(5, -1)\}$ b. $\begin{cases} x - y = 5 \\ 3x + 5y = -41 \end{cases}$ $\{(-2, -7)\}$

7. Solve each system by elimination.

   a. $\begin{cases} 7x + 3y = -22 \\ 6x - 3y = -30 \end{cases}$ $\{(-4, 2)\}$ b. $\begin{cases} 3x - 2y = 19 \\ 2x + 6y = -46 \end{cases}$ $\{(1, -8)\}$

8. Solve each system using any method.

   a. $\begin{cases} 24x - 4y = 24 \\ 6x = y - 6 \end{cases}$ $\varnothing$ b. $\begin{cases} \frac{x}{5} + y = \frac{11}{5} \\ \frac{3x}{5} + \frac{y}{2} = -\frac{17}{5} \end{cases}$ $\{(-9, 4)\}$

   c. $\begin{cases} x + \frac{1}{4}y = y - 4 \\ \frac{1}{3}x + y = x + y \end{cases}$ $\left\{\left(0, \frac{16}{3}\right)\right\}$

Use a system of equations to solve each problem.

9. Ms. Jones invested \$18,000 in two accounts, one yielding 8% interest and the other yielding 9%. She received a total of \$1490 in interest at the end of the year. How much did she invest in each account?
   Ms. Jones invested \$13,000 in the 8% account and \$5000 in the 9% account.

10. Goodyear Ballpark in Goodyear, Arizona, holds 10,000 people. It is the practice field for the Cincinnati Reds and Cleveland Indians. Suppose you are the manager of the stadium and want to host a special event to raise money for the Cystic Fibrosis Foundation. You plan to sell VIP seats for \$100 each and regular admission seats for \$37.50. If you want to raise \$500,000 by selling every seat, how many of your seats should you designate as VIP seats? (Source: http://www.arizona-vacation-planner.com/goodyear-ballpark.html)
    You should designate 2000 seats as VIP seats.

11. Dominick is on a flight from Chicago O'Hare Airport to London Heathrow Airport, a distance of approximately 3950 mi. He notices on his ticket that the travel time from Chicago to London is 7.5 hr and the return trip is 9 hr. After talking to the flight attendant, he realizes the difference is because of the wind. The trip to London is with the wind and the trip from London is against the wind. What is the average speed of the plane and the average speed of the wind? Round your answers to the nearest integer. The average speed of the plane is 483 mph and the average speed of the wind is 44 mph.

12. The Lincoln Memorial Reflecting Pool is the largest reflecting pool in Washington, D.C. It is in the shape of a rectangle whose perimeter is 1338 m. Its length is 6 m more than 12 times its width. Find the length and width of the Reflecting Pool. The length is 618 m and the width is 51 m.

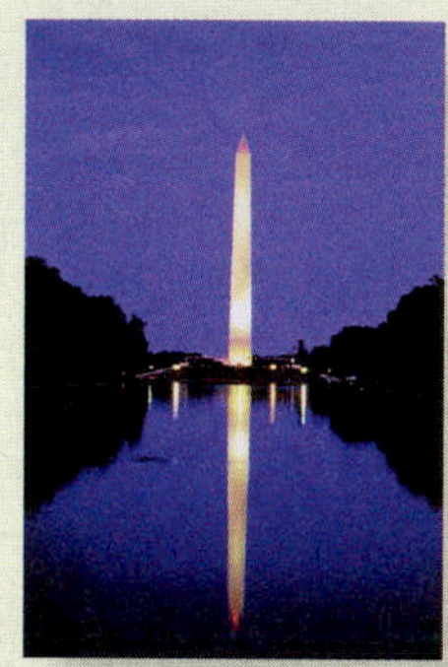

13. Solve the system of linear inequalities.

$$\begin{cases} x + y < 6 \\ y > \frac{1}{2}x - 4 \end{cases}$$

14. Solve the system by elimination.

$$\begin{cases} 4x - y + 3z = -1 \\ 2x + y - z = -9 \\ x + 5y + 2z = 10 \end{cases}$$ $\{(-3, 1, 4)\}$

15. Amelia receives a \$5800 bonus at work, after taxes. She invests this money in the stock market. She purchases three different stocks—Home Health Services (HHS) for \$5 per share, ApplianceMart (AM) for \$20 per share, and SearchWeb (SW) for \$40 per share. She purchases twice as many shares of AM stock as the other two combined because this stock has the highest dividend yield. The annual dividend yield of each stock is as follows: HHS (4%), AM (20%), and SW (15%). If Amelia earns \$45.10 in dividends the first year, how many shares of each stock did she purchase? 40 shares of HHS stock, 180 shares of AM stock, and 50 shares of SW stock

CUMULATIVE REVIEW EXERCISES / CHAPTERS 1–5

Use the order of operations to simplify each expression. (*Section 1.2, Objectives 1–4*)

1. $-2(-3)^2 - 5(-4) + 3$ 5

2. $\dfrac{5 - 2\sqrt{20 - 4}}{2|5 - 2|^2}$ $-\frac{1}{6}$

3. $\dfrac{-1 - (-5)}{6 - 16}$ $-\frac{2}{5}$

4. $-14 - \left(-\frac{2}{9}\right)$ $-13\frac{7}{9}$

Evaluate expression for the given values. (*Section 1.2, Objective 5*)

5. $\dfrac{-b + \sqrt{b^2 - 4ac}}{2a}$ for $a = 4, b = -13, c = 9$ $\frac{9}{4}$

6. $\dfrac{-b - \sqrt{b^2 - 4ac}}{2a}$ for $a = 1, b = -3, c = -4$ -1

From 2006 to 2011, the average number of unique visitors each month, in millions, to YouTube can be modeled by $0.898x^3 - 9.825x^2 + 126.094x + 68.884$, where x is the number of years after 2006. Round each answer to two decimal places. (Source: http://www.trefis.com) (*Section 1.2, Objective 6*)

7. How many unique visitors were there in 2006? about 68.88 million

8. How many unique visitors were there in 2011? about 565.98 million

Translate each sentence into an algebraic inequality. Use the variable x to represent the unknown number. (*Section 1.3, Objective 4*)

9. Five more than twice a number is greater than the number plus 16. $5 + 2x > x + 16$

10. Nine less than four times a number is less than 23 more than twice the number. $4x - 9 < 23 + 2x$

Solve each linear equation. (*Section 2.1, Objectives 2–6*)

11. $\dfrac{7x}{4} - \dfrac{x}{2} = \dfrac{5}{8}$ $\left\{\frac{1}{2}\right\}$

12. $\dfrac{5x}{3} - \dfrac{3x}{2} = \dfrac{1}{6}$ $\{1\}$

13. $6(x - 2) - 5x = x - 12$ $\mathbb{R}$

14. $6(x - 1) - 8x = x - 3x - 1$ $\varnothing$

Find the perimeter of each figure. (*Section 2.3, Objective 2*)

15. 65 cm

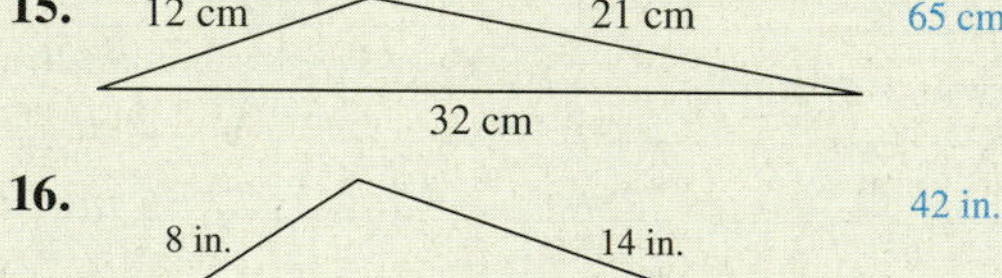

16. 42 in.

Solve each inequality. Graph the solution set and write each answer in interval notation and set-builder notation. (*Section 2.4, Objectives 1 and 2*)

17. $\frac{2}{3}(x - 1) < 4x - 2$

18. $5(y - 1) - 13 \leq 2y + 30$

Find the intersection and union of the sets. Write each solution in interval notation. (*Section 2.5, Objectives 1 and 2*)

19. $\varnothing \cap (-11, 18)$ $\varnothing$

20. $\varnothing \cup (-26, \infty)$ $(-26, \infty)$

Solve each compound inequality. Write the solution in interval notation. (*Section 2.5, Objectives 3 and 4*)

21. $2x - 11 \geq 3$ or $-2x \geq 26$ $(-\infty, -13] \cup [7, \infty)$

22. $4x + 13 > 17$ and $5x - 14 \leq 11$ $(1, 5]$

Solve each equation. (*Section 2.6, Objectives 1–3*)

23. $\left|\dfrac{a - 2}{5}\right| = |2a + 1|$

24. $\left|\dfrac{a + 2}{6}\right| = |a - 4|$

25. $|3x - 8| = |2x + 7|$

26. $|12 - 3x| = 0$ $\{4\}$

27. The table shows the approximate number (in millions) of iPads sold between April 2010 and March 2011. (Source: http://ipod.about.com/od/ipadmodelsandterms/f/ipad-sales-to-date.htm) (*Section 3.1, Objective 6*)

x	April 2010	May 2010	July 2010	September 2010	January 2011	March 2011
y	0.45	2	3.27	7.5	14.8	19

a. Write the ordered pairs that represent the given data, where x is the number of months after April 2010 and y is the total number (in millions) of iPads sold.

b. Create a scatter plot of the data.

c. What can you conclude about iPad sales?

28. Find the input values that correspond to the output value of 0 for the relation whose graph is shown. (*Section 3.2, Objective 3*)

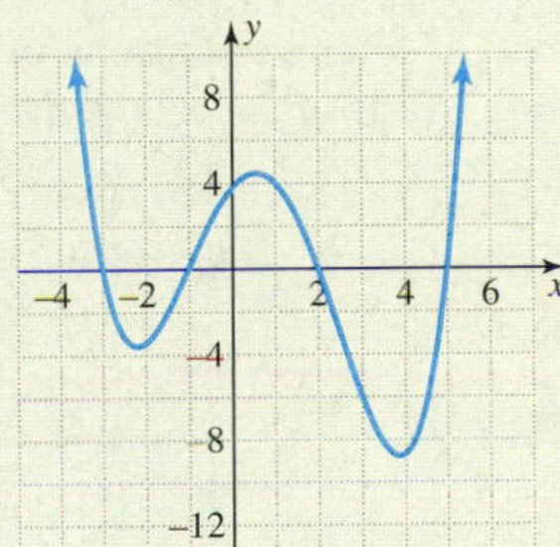

$x = -3, -1, 2, 5$

Find the requested information. (*Section 3.3, Objective 3*)

29. Use the graph of $h(x)$ to find all x for which $h(x) = 3$.

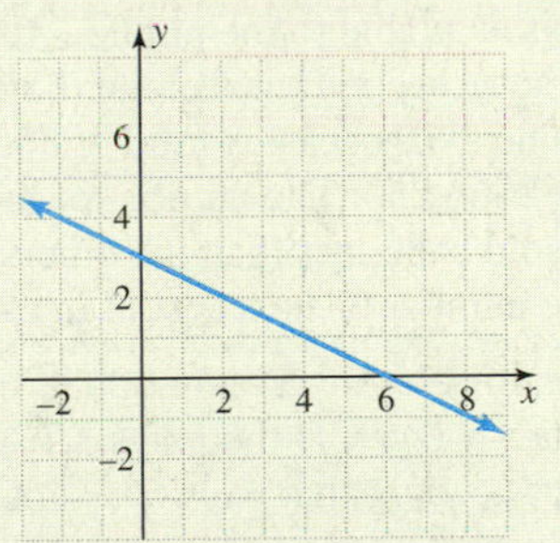

$x = 0$

30. Find $f(2)$ if the function $f(x)$ is given by Y_1.

X	Y1
-3	10.5
-2	5
-1	1.5
0	0
1	.5
2	3
3	7.5

3

Find the domain of each function. (*Section 3.4, Objective 3*)

31. $f(x) = 1 - x^2$ $(-\infty, \infty)$ 32. $f(x) = \sqrt{18 - 3x}$ $(-\infty, 6]$

Rewrite each equation using function notation, if possible. Identify *m* and *b*. (*Section 4.1, Objective 2*)

33. $y = \frac{3}{5}x - 2$

34. $6x + 5y = 0$

35. $2x + 11 = 0$ can't be written in function notation

36. $2x + 4y = 8$

Evaluate each linear function at the given value. (*Section 4.1, Objective 3*)

37. $f(x) = 5x + 2; x = -6$ -28

38. $f(x) = -\frac{5}{6}x + \frac{2}{3}; x = 2$ -1

39. $f(x) = -4; x = 3$ -4

40. $3x + 4y = 2, f(x) = 0$ $\frac{2}{3}$

Find solutions of each equation of the form $f(x) = c$ for the given function. (*Section 4.1, Objective 4*)

41. $f(x) = -\frac{1}{3}x + 2; f(x) = 0$ $x = 6$

42. $f(x) = -\frac{2}{3}x + 2; f(x) = 14$ $x = -18$

Graph each linear function and state its domain and range. (*Section 4.2, Objective 1*)

43. $f(x) = -\frac{1}{4}x + 8$

44. $f(x) = -4$

45. $f(x) = -x$

46. $f(x) = 2x - 18$

Find the slope and *y*-intercept of each line from its equation. Graph the line using its slope and *y*-intercept. (*Section 4.3, Objectives 1 and 2*)

47. $f(x) = -9x + 6$

48. $f(x) = -2x - 5$

49. $3x - 2y = 12$

50. $x - 5y = 0$

Determine if the lines are parallel, perpendicular, or neither. (*Section 4.3, Objective 4*)

51. $y = -\frac{3}{4}x + 2$ and $y = -\frac{4}{3}x$ neither

52. $y = -\frac{5}{2}x + 2$ and $y = \frac{2}{5}x$ perpendicular

53. $4x + 7y = -14$ and $8x + 14y = 8$ parallel

54. $4x + 7y = -14$ and $x + 4y = 8$ neither

Write the equation of the line that satisfies the given conditions. Express each answer in slope-intercept form, if possible. (*Section 4.4, Objective 2*)

55. $m = 0$; passes through $(-1, 7)$ $y = 7$

56. $m = \frac{1}{9}$; passes through $(-9, 2)$ $y = \frac{1}{9}x + 3$

57. passes through $(-2, 7)$ and $(-2, 1)$ $x = -2$

58. passes through $(12, -3)$, perpendicular to $y = -1$ $x = 12$

59. passes through $(2, -4)$, parallel to $y = \frac{4}{5}x + 6$ $y = \frac{4}{5}x - \frac{28}{5}$

60. passes through $(7, -9)$, perpendicular to $y = -\frac{1}{6}x + 5$ $y = 6x - 51$

Write a linear equation that represents the given information. (*Section 4.4, Objective 3*)

61. The average salary for U.S. public elementary and secondary school teachers was \$41,807 in 1999 and \$55,350 in 2009. (Source: http://www.nea.org)

a. Write an equation that models the average salary for public school teachers, where x is the number of years after 1999. $y = 1354.3x + 41{,}807$

b. Use the linear equation to determine the average U.S. salary for public school teachers in 2019. about \$68,893

62. There were approximately 2.89 million U.S. public elementary and secondary classroom teachers in 1999 and approximately 3.23 million classroom teachers in 2009. (Source: www.nea.org)

a. Write a linear equation that models the number of classroom teachers (in millions), where x is the number of years after 1999. $y = 0.034x + 2.89$

b. If this growth continues, how many classroom teachers will there be in 2024? about 3.74 million

Graph each linear inequality. (*Section 4.5, Objective 2*)

63. $x - 4y > 12$ **64.** $2x - 6y < 6$

65. $3x - 12y \geq 1.5$ **66.** $4x + 5y \geq 100$

Determine if the ordered pair is a solution of the system of linear equations. (*Section 5.1, Objective 1*)

67. $\begin{cases} x - 3y = 0 \\ y = \frac{5}{6}x - 1 \end{cases}$; $\left(2, \frac{2}{3}\right)$ yes **68.** $\begin{cases} y = \frac{5}{8}x + 10 \\ y = 2x + 21 \end{cases}$; $(8, -5)$ no

Solve each system graphically. (*Section 5.1, Objectives 2 and 3*)

69. $\begin{cases} x - 3y = 6 \\ y = 1 \end{cases}$ $\{(9, 1)\}$ **70.** $\begin{cases} -4x + y = 24 \\ 8x - 2y = 16 \end{cases}$ $\varnothing$

71. $\begin{cases} x - 6y = -12 \\ -2x + 12y = 24 \end{cases}$ $\{(x, y) | x - 6y = -12\}$ **72.** $\begin{cases} x + 4y = 8 \\ x + 2y = 0 \end{cases}$ $\{(-8, 4)\}$

Determine how the lines relate, the number of solutions of the system, and the type of system without graphing. (*Section 5.1, Objective 4*)

73. $\begin{cases} y = -\frac{2}{3}x \\ 4x + 6y = 0 \end{cases}$ **74.** $\begin{cases} y = -\frac{1}{2}x \\ y = 2x + 3 \end{cases}$

75. $\begin{cases} y = -\frac{3}{4}x + 2 \\ -\frac{1}{4}x - \frac{1}{3}y = -\frac{1}{3} \end{cases}$ **76.** $\begin{cases} 3x - y = -5 \\ y = 4x \end{cases}$

Solve each system of linear equations by substitution or elimination. (*Section 5.2, Objectives 1 and 2*)

77. $\begin{cases} x = 2y + 18 \\ -x - 3y = 2 \end{cases}$ $\{(10, -4)\}$ **78.** $\begin{cases} 5x + 3y = -1 \\ -7x - 5y = 3 \end{cases}$ $\{(1, -2)\}$

79. $\begin{cases} 0.1x - 0.6y = 1.4 \\ 4x - y = -13 \end{cases}$ $\{(-4, -3)\}$ **80.** $\begin{cases} 0.2x + 0.5y = 3.3 \\ -0.3x + 0.2y = 1.7 \end{cases}$ $\{(-1, 7)\}$

81. $\begin{cases} \frac{3}{11}x - \frac{8}{11}y = -1 \\ -\frac{2}{7}x + \frac{4}{7}y = 1 \end{cases}$ $\left\{\left(-3, \frac{1}{4}\right)\right\}$ **82.** $\begin{cases} \frac{1}{9}x - \frac{2}{3}y = 1 \\ \frac{1}{2}x - \frac{3}{4}y = 3 \end{cases}$ $\left\{\left(5, -\frac{2}{3}\right)\right\}$

Solve each problem using a system of linear equations. (*Section 5.3, Objectives 1–4*)

83. Derek is working two jobs to help pay his way through college. He works as a campus tour guide in the college admission office and a night shift manager at a local supermarket. One week he earns $175 by working 5 hr as a tour guide and 10 hr as a night shift manager. Another week he earns $295 by working 20 hr as a tour guide and 10 hr as a night shift manager. What is Derek's hourly wage as a campus tour guide and as a night shift manager?

84. A movie theater sells 1900 tickets. Adult tickets cost $8.75 and student tickets cost $5.50. If the movie theater collects a total of $13,342.50 for the tickets, how many adult tickets and student tickets are sold? 890 adult tickets and 1010 student tickets

85. A chemist has a 45% acid solution. She needs 75 L of a 15% acid solution. How many liters of water and how many liters of the 45% acid solution must she mix together to obtain 75 L of a 15% acid solution? 50 L of water and 25 L of 45% acid solution

86. At 7:30 A.M. Mikail leaves his home traveling north on his ATV at an average speed of 13 mph. At 9 A.M. his wife Nancy leaves their home to meet Mikail. She travels north in her car at an average speed of 43 mph. How long will it take Nancy to reach Mikail? It will take Nancy 0.65 hr to reach Mikail.

Solve each system of linear equations in three variables using elimination. (*Section 5.4, Objective 1*)

87. $\begin{cases} 2x - 3y + 5z = 13 \\ -x + 5y - 3z = 2 \\ 5x - 6y + z = -9 \end{cases}$ $\{(1, 3, 4)\}$ **88.** $\begin{cases} -3x + y + 2z = -9 \\ 9x + 2y + 2z = 10 \\ 6x + 7y = -23 \end{cases}$ $\{(2, -5, 1)\}$

89. $\begin{cases} -x + 6y + 6z = 22 \\ 2y + z = 5 \\ x + 2y - 2z = -2 \end{cases}$ **90.** $\begin{cases} 2x - y - 3z = 9 \\ x + y = 12 \\ -3x + y + 4z = 13 \end{cases}$ $\varnothing$

Solve each problem. (*Section 5.4, Objective 2*)

91. A concert is held at a music hall. There are three types of tickets available to the public—first mezzanine ($190 each), second mezzanine ($85 each), and third mezzanine ($70 each). A total of 5050 tickets are sold totaling $666,250. The number of first mezzanine tickets sold is 50 less than the other two combined. Find the number of tickets sold in each mezzanine. 2500 in first mezzanine, 850 in second mezzanine, and 1700 in third mezzanine

92. The Hendrickses invest $36,000 in three different accounts—a 1-yr CD that pays 3.7% annual interest, a savings account that pays 6% annual interest, and a money market account that pays 2.9% annual interest. They invest three times as much money in the savings account as the amount invested in the CD. Find the amount invested in each account if they earn $1807.56 in interest in 1 yr. $7560 in CD, $22,680 in savings account, $5760 in money market account

Solve each system of linear inequalities in two variables. (*Section 5.5, Objective 1*)

93. $\begin{cases} 11x + 7y < 49 \\ x - 7y < 35 \end{cases}$ **94.** $\begin{cases} 7x + y \geq 6 \\ 5x + 2y \leq -6 \end{cases}$

95. $\begin{cases} -x + 4y > 4 \\ 5x + 4y < 20 \\ x > 0 \end{cases}$

CHAPTER

6

Exponents, Polynomials, and Polynomial Functions

Goal Setting

Having goals is essential to your success in life, in college, and in the classroom. Without goals, it is difficult to know what you are working toward. Take the time to write down the goals you hope to accomplish by being in this class and by being in college. Start with specific and realistic goals. Here are some examples.

- I will review my class notes each day.
- I will spend one hour per day working homework problems from this class.
- I will ask questions if I do not understand.
- I will attend class each day.

Be sure that your goals are attainable. Review these goals often so that you do not lose focus. Defining goals may require you to make changes in other areas of your life. Be willing to do what is necessary to accomplish your dreams.

Question For Thought: What do you hope to accomplish by attending college? How does this class help you reach that goal? What do you hope to accomplish in this class?

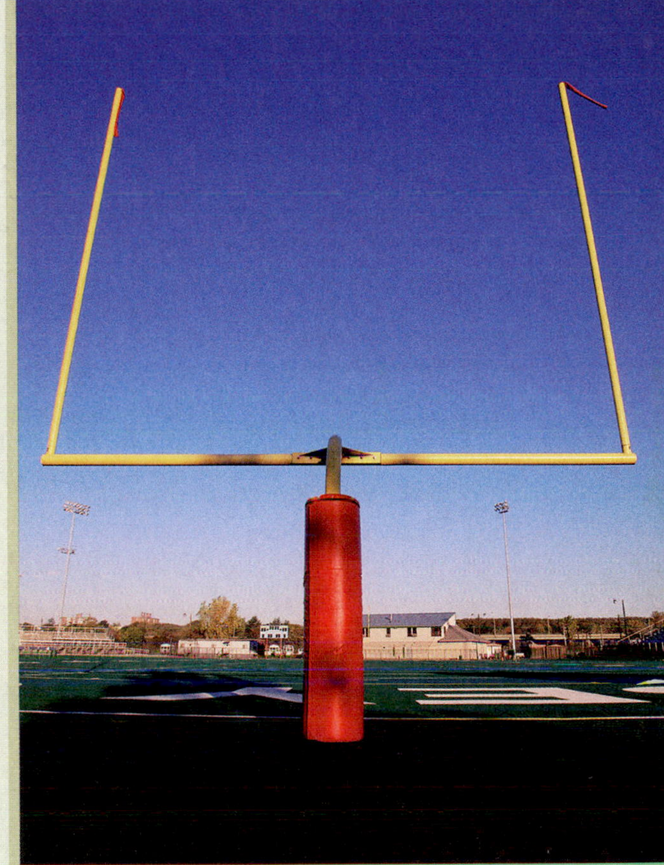

Chapter Outline

Section 6.1 Rules of Exponents, Zero and Negative Exponents 366

Section 6.2 More Rules of Exponents and Scientific Notation 379

Section 6.3 Polynomials, Polynomial Functions, and Their Basic Graphs 391

Section 6.4 Adding, Subtracting, and Multiplying Polynomials and Polynomial Functions 402

Section 6.5 Factoring Using the Greatest Common Factor and Grouping 418

Section 6.6 Factoring Trinomials 425

Section 6.7 Factoring Binomials and a Factoring Review 437

Section 6.8 Solving Polynomial Equations and Their Applications 445

Coming Up...

In Section 6.8, we will use factoring to find the time that it takes for a rock dropped from the Grand Canyon Skywalk to reach the floor of the Grand Canyon.

"Where there is no vision, there is no hope."

—George Washington Carver
(American Scientist and Inventor)

SECTION 6.1 Rules of Exponents, Zero and Negative Exponents

In **Chapter 1,** the concept of an exponent was defined and introduced briefly. In this chapter, we will look more closely at exponents and their properties. We will also examine a new class of functions, polynomial functions. Operations of polynomials and polynomial equations will also be studied.

OBJECTIVES

As a result of completing this section, you will be able to

1. Apply the product of like bases rule for exponents.
2. Apply the quotient of like bases rule for exponents.
3. Simplify expressions with zero as an exponent.
4. Simplify expressions with negative exponents.
5. Apply a combination of properties and definitions.
6. Solve application problems.
7. Troubleshoot common errors.

Negative exponents arise in many formulas. One such formula is the amortization formula, the formula to determine the monthly payment needed to repay a loan. The expression for the monthly payment needed to repay a home loan of \$200,000 at an annual interest rate of 5% in 360 months is

$$\frac{200{,}000\left(\frac{0.05}{12}\right)}{1-\left(1+\frac{0.05}{12}\right)^{-360}}$$

The graphing calculator skills will illustrate how to simplify this expression.

In this section, we will learn some of the basic rules for exponents. These rules are used quite extensively in algebra, so it is very important that you become familiar with them.

Objective 1

Apply the product of like bases rule for exponents.

INSTRUCTOR NOTE: Explain the difference between the terms power and exponent. The expression b^n is called a power and n is the exponent. For example, some powers of 2 are 2^1, 2^2, and 2^3; they are not 1, 2, and 3.

The Product of Like Bases

Recall from Chapter 1 that exponents are used to write repeated factors in a more compact form.

Definition: Exponent

For a real number b and a natural number n,

$$b^n = \underbrace{b \cdot b \cdot b \cdots b}_{b \text{ is a factor } n \text{ times}}$$

Exponent: n; Base: b

The exponent n tells us how many times b is repeated as a factor. Expressions that contain exponents, such as 2^5 and y^4, are called **exponential expressions** or *powers*. Operations such as addition, subtraction, multiplication, division, and raising to an exponent can be performed on exponential expressions. The definition of an exponent can be used to develop rules for simplifying these types of exponential expressions.

The first rule relates to products of exponential expressions with the same base, what we call like bases. In the examples shown in the table, we obtain each product by first using the definition of the exponent. Then we observe a rule that relates the final result to the original problem.

Based on Definition	Observed Rule
$b \cdot b = b^1 \cdot b^1 = b^2$	$b \cdot b = b^{1+1} = b^2$
$b^2 \cdot b = b^2 \cdot b^1 = (b \cdot b) \cdot b = b^3$	$b^2 \cdot b = b^{2+1} = b^3$
$b^2 \cdot b^3 = (b \cdot b) \cdot (b \cdot b \cdot b) = b^5$	$b^2 \cdot b^3 = b^{2+3} = b^5$

So, we observe that when exponential expressions with like bases are multiplied, the exponent of the product is the *sum* of the exponents of the like bases.

Property: The Product of Like Bases Rule for Exponents

For a real number b and natural numbers m and n,

$$b^m \cdot b^n = b^{m+n}$$

The product rule can be applied *only* if the bases of each factor are the same.

Procedure: Multiplying Exponential Expressions with Like Bases

Step 1: Keep the base the same.
Step 2: Add the exponents of the factors.
Step 3: Simplify the exponents, if needed.

Objective 1 Examples Simplify each expression by applying the product of like bases rule.

1a. $t^7 \cdot t^4 \cdot t$ **1b.** $(-5)^3 \cdot (-5)^2$ **1c.** $(-2a^7)(-3a^2)$ **1d.** $(-4xy^2)(-5x^2y^4)$

Solutions

1a. $t^7 \cdot t^4 \cdot t = t^7 \cdot t^4 \cdot t^1$ Recall $t = t^1$.
$= t^{7+4+1}$ Apply the product of like bases rule.
$= t^{12}$ Add the exponents.

INSTRUCTOR NOTE: Discuss how the expression $2^3 \cdot 5^2$ is simplified since the bases are not the same.

1b. $(-5)^3 \cdot (-5)^2 = (-5)^{3+2}$ Apply the product of like bases rule.
$= (-5)^5$ Add the exponents.
$= -3125$ Simplify the exponent.

1c. The commutative property enables us to change the order of the factors, so that the coefficients are together and the like bases are together.

$(-2a^7)(-3a^2) = (-2)(-3)a^7 \cdot a^2$ Apply the commutative property.
$= 6a^{7+2}$ Multiply and apply the product of like bases rule.
$= 6a^9$ Simplify.

1d. $(-4x^1y^2)(-5x^2y^4) = (-4)(-5)x^1x^2y^2y^4$ Apply the commutative property.
$= 20x^{1+2}y^{2+4}$ Multiply and apply the product of like bases rule.
$= 20x^3y^6$ Simplify.

Student Check 1 Simplify each expression by applying the product of like bases rule.

a. $y^6 \cdot y^2 \cdot y^{12}$ **b.** $(-3)^6 \cdot (-3)^{11}$ **c.** $(7b^5)(-4b^7)$ **d.** $(-2x^4y^5)(-8x^3y^7)$

The Quotient of Like Bases

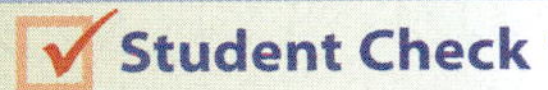

Objective 2 Apply the quotient of like bases rule for exponents.

Now we will find a rule that applies to simplifying quotients of exponential expressions with like bases. The following example illustrates this concept. Assume that the denominator is not zero.

Based on Definition

$\frac{a^6}{a^4} = \frac{a \cdot a \cdot a \cdot a \cdot a \cdot a}{a \cdot a \cdot a \cdot a}$ Six factors of a in the numerator and four factors of a in the denominator

$= \frac{\not{a} \cdot \not{a} \cdot \not{a} \cdot \not{a} \cdot a \cdot a}{\not{a} \cdot \not{a} \cdot \not{a} \cdot \not{a}}$ Divide out the common factors.

$= \frac{a \cdot a}{1}$ Write the remaining factors.

$= a^2$ Express the answer with exponents.

Observed Rule

The same result is obtained by subtracting the exponents.

$$\frac{a^6}{a^4} = a^{6-4}$$
$$= a^2$$

So, we observe that when exponential expressions with like bases are divided, the exponent of the quotient is the *difference* of the exponents of the exponential expressions in the numerator and denominator.

Property: The Quotient of Like Bases Rule for Exponents

For a real number b ($b \neq 0$) and natural numbers m and n,

$$\frac{b^m}{b^n} = b^{m-n}$$

The quotient of like bases rule can be applied *only* if the bases of the exponential expression in the numerator and denominator are the same.

Procedure: Dividing Exponential Expressions with Like Bases

Step 1: Keep the base the same.
Step 2: Subtract the denominator's exponent from the numerator's exponent.
Step 3: Simplify the expression, if needed.

Objective 2 Examples

Simplify each expression by applying the quotient of like bases rule.

2a. $\frac{a^6}{a^2}$ **2b.** $\frac{4^5}{4^3}$ **2c.** $\frac{-10x^4}{2x^3}$ **2d.** $\frac{-18a^8b^{10}}{-9a^4b^2}$

Solutions

2a. $\frac{a^6}{a^2} = a^{6-2}$ Apply the quotient of like bases rule.

$= a^4$ Subtract the exponents.

2b. $\frac{4^5}{4^3} = 4^{5-3}$ Apply the quotient of like bases rule.

$= 4^2$ Subtract the exponents.

$= 16$ Simplify the exponent.

2c. $\frac{-10x^4}{2x^3} = \frac{-10}{2} \cdot \frac{x^4}{x^3}$ Divide the coefficients and the like bases.

$= -5x^{4-3}$ Apply the quotient of like bases rule.

$= -5x^1$ Subtract the exponents.

$= -5x$ Rewrite x^1 as x.

2d. $\frac{-18a^8b^{10}}{-9a^4b^2} = \frac{-18}{-9} \cdot \frac{a^8}{a^4} \cdot \frac{b^{10}}{b^2}$ Divide the coefficients and the like bases.

$= 2a^{8-4}b^{10-2}$ Apply the quotient of like bases rule.

$= 2a^4b^8$ Subtract the exponents.

Student Check 2 Simplify each expression by applying the quotient of like bases rule.

a. $\frac{y^{10}}{y}$ **b.** $\frac{(-1)^{11}}{(-1)^5}$ **c.** $\frac{-21r^8}{-3r^6}$ **d.** $\frac{-16x^9y^{12}}{4x^6y^3}$

The Zero Exponent

Objective 3 Simplify expressions with zero as an exponent.

Examples 1 and 2 used only natural number (that is, positive integer) exponents. We will now investigate how to define an exponent of zero. Consider the expression, $\frac{x^5}{x^5}$. We know that to divide exponential expressions with like bases we subtract the exponents. So,

$$\frac{x^5}{x^5} = x^{5-5} = x^0$$

On the other hand, we also know that any nonzero number divided by itself is 1. So, it is also true that

$$\frac{x^5}{x^5} = 1$$

We have that $\frac{x^5}{x^5} = x^0$ but $\frac{x^5}{x^5} = 1$, so it must follow that $x^0 = 1$. This leads to the following definition.

INSTRUCTOR NOTE: Show students why $b \neq 0$. We know that $0^n = 0$ but the zero exponent rule would imply that $0^0 = 1$, which contradicts the previous statement.

Definition: The Zero Exponent

For a real number b ($b \neq 0$),

$$b^0 = 1$$

So, any nonzero number b raised to an exponent of 0 is 1.

Objective 3 Examples

Use the definition of the zero exponent to simplify each expression. Assume all bases are nonzero real numbers.

3a. 3^0 **3b.** $(-6)^0$ **3c.** $-(-4)^0$ **3d.** $8b^0$ **3e.** $(5x + 4)^0$

Solutions

3a. $3^0 = 1$

3b. $(-6)^0 = 1$

3c. The opposite of -4 to the 0 is $-(-4)^0 = -(1) = -1$.

3d. Eight times b to the 0 is $8b^0 = 8(1) = 8$.

3e. $(5x + 4)^0 = 1$

Student Check 3 Use the definition of the zero exponent to simplify each expression. Assume all bases are nonzero real numbers.

a. 17^0 **b.** $(-9)^0$ **c.** $-(-10)^0$ **d.** $3d^0$ **e.** $(7a - 1)^0$

Negative Exponents

Objective 4 Simplify expressions with negative exponents.

We now know how to work with exponents that are whole numbers {0, 1, 2, . . .}. It is important to understand how to work with exponents that are negative integers as well, such as 2^{-3}. It is impossible to repeat the base 2 a "−3" times, so we need another way to simplify this exponential expression. The following example will help us understand the concept of a negative exponent.

Based on Definition		Based on the Quotient of Like Bases Rule
$\frac{a^3}{a^6} = \frac{a \cdot a \cdot a}{a \cdot a \cdot a \cdot a \cdot a \cdot a}$	Three factors of a in the numerator and six factors of a in the denominator	$\frac{a^3}{a^6} = a^{3-6}$
$= \frac{\cancel{a}^1 \cdot \cancel{a}^1 \cdot \cancel{a}^1}{\cancel{a}_1 \cdot \cancel{a}_1 \cdot \cancel{a}_1 \cdot a \cdot a \cdot a}$	Divide out the common factors.	$= a^{-3}$
$= \frac{1}{a^3}$	Simplify.	

INSTRUCTOR NOTE:
Point out that when the exponent in the denominator is larger than the exponent in the numerator, you will obtain a negative exponent when applying the quotient of like bases rule.

So, we see that $\frac{a^3}{a^6} = \frac{1}{a^3}$ and $\frac{a^3}{a^6} = a^{-3}$. Therefore,

$$a^{-3} = \frac{1}{a^3}$$

An exponential expression with a *negative exponent* is equivalent to the reciprocal of the base of the expression raised to the positive exponent.

Definition: Negative Exponent

For a real number b ($b \neq 0$) and a positive integer n,

$$b^{-n} = \frac{1}{b^n}$$

Procedure: Evaluating an Expression Raised to a Negative Exponent

Step 1: Identify the base of the negative exponent.
Step 2: Rewrite the expression as 1 divided by the base to its positive exponent.
Step 3: Simplify the expression.

Objective 4 Examples

Simplify each expression. Write each answer with positive exponents. Assume all bases are nonzero real numbers.

INSTRUCTOR NOTE:
Point out that negative exponents do not impact the sign of the expression. That is, $b^{-n} \neq -b^n$.

4a. 6^{-2} **4b.** $(-4)^{-3}$ **4c.** -2^{-4} **4d.** $\left(\frac{5}{4}\right)^{-2}$ **4e.** $6x^{-2}$ **4f.** $\frac{x^{-3}}{y^{-4}}$

Solutions

4a. $6^{-2} = \frac{1}{6^2}$ Apply the definition of a negative exponent.

$= \frac{1}{36}$ Simplify the exponent.

4b. $(-4)^{-3} = \frac{1}{(-4)^3}$ Apply the definition of a negative exponent.

$= \frac{1}{-64}$ Simplify the exponent.

$= -\frac{1}{64}$ Simplify.

4c. Without parentheses, the negative exponent only applies to the base 2.

$-2^{-4} = -\frac{1}{2^4}$ Apply the definition of a negative exponent.

$= -\frac{1}{16}$ Simplify the exponent.

4d. $\left(\frac{5}{4}\right)^{-2} = \frac{1}{\left(\frac{5}{4}\right)^2}$ Apply the definition of a negative exponent.

$= \frac{1}{\frac{25}{16}}$ Simplify the exponent.

$= 1 \cdot \frac{16}{25}$ Multiply by the reciprocal of $\frac{25}{16}$, which is $\frac{16}{25}$.

$= \frac{16}{25}$ Simplify.

4e. The base of the negative exponent is x. Rewrite x^{-2} with a positive exponent and multiply by 6.

$6x^{-2} = 6 \cdot \frac{1}{x^2}$ Apply the definition of a negative exponent to x^{-2}.

$= \frac{6}{x^2}$ Multiply.

4f. We must apply the definition of a negative exponent twice.

$\frac{x^{-3}}{y^{-4}} = \frac{\frac{1}{x^3}}{\frac{1}{y^4}}$ Apply the definition of a negative exponent to the numerator and denominator.

$= \frac{1}{x^3} \div \frac{1}{y^4}$ Write the numerator divided by the denominator.

$= \frac{1}{x^3} \cdot \frac{y^4}{1}$ Multiply by the reciprocal of $\frac{1}{y^4}$, or $\frac{y^4}{1}$.

$= \frac{y^4}{x^3}$ Simplify.

So, note that $\frac{x^{-3}}{y^{-4}} = \frac{y^4}{x^3}$.

Student Check 4 Simplify each expression. Write each answer with positive exponents. Assume all bases are nonzero real numbers.

a. 7^{-3} **b.** $(-2)^{-4}$ **c.** -5^{-2} **d.** $\left(\frac{1}{4}\right)^{-1}$ **e.** $3y^{-3}$ **f.** $\frac{a^{-5}}{b^{-2}}$

Example 4 illustrates two important facts about negative exponents.

Fact 1: *When a factor crosses the fraction bar, the sign of its exponent changes.*

$$\frac{1}{b^{-n}} = b^n \quad \text{and} \quad b^{-n} = \frac{1}{b^n}, \quad \text{for } n \text{ an integer and } b \neq 0$$

Fact 2: *A fraction raised to an exponent is equivalent to its reciprocal raised to the opposite of the exponent.*

$$\left(\frac{a}{b}\right)^{-n} = \left(\frac{b}{a}\right)^n$$

Combining Properties and Definitions

Objective 5 Apply a combination of properties and definitions.

Now that we know how to work with all integers as exponents, the product and quotient of like bases rules apply for integer exponents, not just natural numbers. We will now simplify expressions that require us to apply both rules.

Objective 5 Examples **Simplify each expression. Write each answer with positive exponents. Assume all bases are nonzero real numbers.**

5a. $a^{-4} \cdot a^{-6} \cdot a^5$ **5b.** $(-4x^2y^{-3})(2x^{-4}y^6)$ **5c.** $\frac{6b^3}{2b^{-2}}$ **5d.** $\frac{8a^3b^{-5}}{16a^7b^{-2}}$

Solutions

5a. $a^{-4} \cdot a^{-6} \cdot a^{5} = a^{-4+(-6)+5}$ — Apply the product of like bases rule.

$= a^{-5}$ — Add the exponents.

$= \dfrac{1}{a^{5}}$ — Rewrite with positive exponents.

5b. $(-4x^{2}y^{-3})(2x^{-4}y^{6}) = (-4)(2)x^{2}x^{-4}y^{-3}y^{6}$ — Apply the commutative property.

$= -8x^{2+(-4)}y^{-3+6}$ — Multiply coefficients and apply the product of like bases rule.

$= -8x^{-2}y^{3}$ — Add exponents.

$= -8 \cdot \dfrac{1}{x^{2}} \cdot y^{3}$ — Rewrite with positive exponents.

$= -\dfrac{8y^{3}}{x^{2}}$ — Simplify.

INSTRUCTOR NOTE: Remind students that subtracting a negative is the same as adding a positive.

5c. $\dfrac{6b^{3}}{2b^{-2}} = \dfrac{6}{2} \cdot b^{3-(-2)}$ — Divide the coefficients and apply the quotient of like bases rule.

$= 3b^{5}$ — Simplify.

5d. $\dfrac{8a^{3}b^{-5}}{16a^{7}b^{-2}} = \dfrac{8}{16} \cdot a^{3-7}b^{-5-(-2)}$ — Divide the coefficients and apply the quotient of like bases rule.

$= \dfrac{1}{2}a^{-4}b^{-3}$ — Simplify the coefficient and subtract the exponents.

$= \dfrac{1}{2} \cdot \dfrac{1}{a^{4}} \cdot \dfrac{1}{b^{3}}$ — Rewrite with positive exponents.

$= \dfrac{1}{2a^{4}b^{3}}$ — Simplify.

Student Check 5 Simplify each expression. Write each answer with positive exponents. Assume all bases are nonzero real numbers.

a. $b^{-6} \cdot b^{2} \cdot b^{5}$ **b.** $(-5x^{-3}y)(4xy^{-2})$ **c.** $\dfrac{4a^{5}}{10a^{-4}}$ **d.** $\dfrac{9x^{-2}y^{4}}{27x^{-5}y^{6}}$

Applications

Objective 6 Solve application problems.

There are many applications in real life that use exponents. Exponents appear in formulas for area, volume, surface area of geometric figures, population growth, investments, the growth of bacteria, and the half-life of chemical elements. In Example 6 we will simplify some exponential expressions as well as construct expressions that represent the situation.

Objective 6 Examples **Solve each problem by evaluating an expression or by constructing an expression that represents the situation. Round each answer to two decimal places, when needed.**

6a. A soup can has a diameter of 6.7 cm and a height of 10.2 cm. How much steel is needed to produce one soup can? The formula for surface area is $S = 2\pi r^{2} + 2\pi rh$. (Use $\pi \approx 3.14$.)

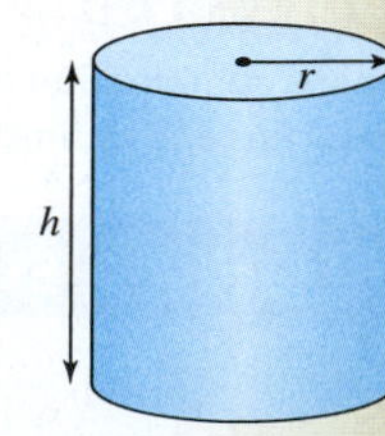

Solution

6a. $S = 2\pi r^{2} + 2\pi rh$ — Begin with the surface area formula.

$S = 2\pi(3.35)^{2} + 2\pi(3.35)(10.2)$ — Replace h with 10.2 and r with $\dfrac{6.7}{2} = 3.35$.

$S = 2\pi(11.2225) + 2\pi(34.17)$ Simplify the exponent and multiply.

$S \approx 70.4773 + 214.5876$ Multiply.

$S \approx 285.0649$ Add.

So, approximately 285.06 cm² of steel are needed to produce one soup can.

6b. Write an expression in terms of the height for the surface area of a can if its radius is one-half its height. Recall that the formula for surface area is $S = 2\pi r^2 + 2\pi rh$.

Solution **6b.** Let h represent the height of the can. Then $r = \frac{h}{2}$.

$S = 2\pi r^2 + 2\pi rh$ Begin with the surface area formula.

$S = 2\pi\left(\frac{h}{2}\right)^2 + 2\pi\left(\frac{h}{2}\right)(h)$ Replace r with $\frac{h}{2}$.

$S = 2\pi\left(\frac{h^2}{4}\right) + \frac{2\pi h^2}{2}$ Simplify the exponent and apply the product of like bases rule.

$S = \frac{\pi h^2}{2} + \frac{2\pi h^2}{2}$ Simplify the first term.

$S = \frac{3\pi h^2}{2}$ Add the fractions.

6c. The area of a rectangle is given by $A = 32x^3y$ and the length is $l = 8x$. What expression represents the width of the rectangle?

Solution **6c.** Since $A = lw$, the width of the rectangle can be found by dividing both sides by l to obtain $w = \frac{A}{l}$.

$w = \frac{A}{l}$ State the formula to find w.

$w = \frac{32x^3y}{8x}$ Replace A with $32x^3y$ and l with $8x$.

$w = \frac{32}{8} \cdot \frac{x^3}{x} \cdot y$ Divide coefficients and the like bases.

$w = 4x^2y$ Simplify the coefficient and subtract exponents.

6d. The formula to determine the monthly payment to repay a loan of P dollars, at an annual interest rate of r (as a decimal), in n months is given by

$$A = \frac{P\left(\frac{r}{12}\right)}{1 - \left(1 + \frac{r}{12}\right)^{-n}}$$

Write the expression to find the monthly payment to repay a home loan of \$200,000 at an annual interest rate of 5% in 360 months and then simplify the expression.

Solution **6d.** $A = \frac{P\left(\frac{r}{12}\right)}{1 - \left(1 + \frac{r}{12}\right)^{-n}}$ State the formula.

$A = \frac{200{,}000\left(\frac{0.05}{12}\right)}{1 - \left(1 + \frac{0.05}{12}\right)^{-360}}$ Replace P with 200,000, r with 0.05, and n with 360.

$A = \$1073.64$ Approximate using a calculator.

So, the monthly payment to repay the \$200,000 loan is \$1073.64.

Student Check 6 Solve each problem by evaluating an expression or by constructing an expression that represents the situation. Round each answer to two decimal places, when needed.

a. A sugar cone has a radius of 2.5 cm and a height of 11.2 cm. Find how much ice cream the cone will hold. The volume of a cone is $V = \frac{1}{3}\pi r^2 h$. (Use $\pi \approx 3.14$.)

b. Suppose the height of a cone is 6 times its radius. Write an expression in terms of the radius that represents the volume of the cone.

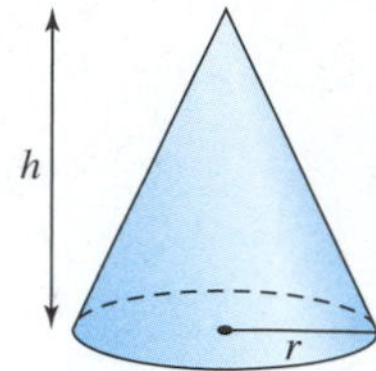

c. The area of a rectangle is $A = 60x^3y^4$. The length of the rectangle is $l = 15x^2y^3$. What expression represents the width of the rectangle?

d. Use the formula in Example 6d to find the monthly payment to repay a loan of \$30,000 at 6% interest in 60 months.

Objective 7 Troubleshoot common errors.

Troubleshooting Common Errors

Some common errors associated with the product and quotient of like bases rules and negative exponents are shown.

Objective 7 Examples **A problem and an incorrect solution are given. Provide the correct solution and an explanation of the error.**

7a. $y^2 \cdot y^6$

Incorrect Solution	Correct Solution and Explanation
$y^2 \cdot y^6 = y^{12}$	The product of like bases rule states that when we multiply exponential expressions with like bases, we *add* exponents not multiply them. So, the correct answer is $y^2 \cdot y^6 = y^{2+6} = y^8$.

7b. $5^2 \cdot 5^3$

Incorrect Solution	Correct Solution and Explanation
$5^2 \cdot 5^3 = 25^5$	The product of like bases rule states that when we multiply exponential expressions with like bases, we must keep the base the same. So, the correct answer is $5^2 \cdot 5^3 = 5^5 = 3125$.

7c. Rewrite 2^{-3} with positive exponents.

Incorrect Solution	Correct Solution and Explanation
$2^{-3} = -8$	A negative exponent doesn't make the sign of the expression negative; it means to take the reciprocal of the base. $2^{-3} = \frac{1}{2^3} = \frac{1}{8}$

7d. Simplify $\frac{x^{-5}}{x^{-6}}$.

Incorrect Solution	Correct Solution and Explanation
$\frac{x^{-5}}{x^{-6}} = x^{-5-6} = x^{-11} = \frac{1}{x^{11}}$	The denominator's exponent is negative, so when we subtract it, we get a positive number. $\frac{x^{-5}}{x^{-6}} = x^{-5-(-6)} = x^{-5+6} = x^1 = x$

ANSWERS TO STUDENT CHECKS

Student Check 1 **a.** y^{20} **b.** -3^{17} **c.** $-28b^{12}$ **d.** $16x^7y^{12}$

Student Check 2 **a.** y^9 **b.** 1 **c.** $7r^2$ **d.** $-4x^3y^9$

Student Check 3 **a.** 1 **b.** 1 **c.** -1 **d.** 3 **e.** 1

Student Check 4 **a.** $\frac{1}{343}$ **b.** $\frac{1}{16}$ **c.** $-\frac{1}{25}$ **d.** 4 **e.** $\frac{3}{y^3}$ **f.** $\frac{b^2}{a^5}$

Student Check 5 **a.** b **b.** $-\frac{20}{x^2y}$ **c.** $\frac{2a^9}{5}$ **d.** $\frac{x^3}{3y^2}$

Student Check 6 **a.** 73.27 cm^3 **b.** $V = 2\pi r^3$ **c.** $w = 4xy$ **d.** \$579.98

SUMMARY OF KEY CONCEPTS

1. The product of like bases rule only applies when the bases are the same. It states that the product of exponential expressions with like bases is the like base raised to the sum of its exponents.
2. The quotient of like bases rule only applies when the bases are the same. The quotient of exponential expressions with like bases is the like base raised to the difference of its exponents. If the resulting exponent is negative, rewrite the expression with a positive exponent.
3. Any nonzero base raised to the zero exponent is 1. Be careful with negative signs. Note the difference in the following expressions.

$$(-b)^0 = 1 \quad \text{but} \quad -b^0 = -1$$

4. Negative exponents involve taking reciprocals. The base with the negative exponent always crosses the fraction bar for the exponent to become positive. The expression b^{-n} is the reciprocal of b^n. That is, $b^{-n} = \frac{1}{b^n}$.

GRAPHING CALCULATOR SKILLS

The calculator skills for this section involve working with negative exponents and evaluating formulas.

Example 1: Simplify 7^{-3} and express the result as a fraction.

Solution:

```
7^-3
         .0029154519
Ans▸Frac
               1/343
```

Keys: 7 ^ (−) 3 ENTER; MATH 1 ENTER

Example 2: Find the monthly payment to repay a home loan of \$200,000 at an annual interest rate of 5% in 360 months. (Refer to Example 6d for the monthly payment formula.)

Solution:

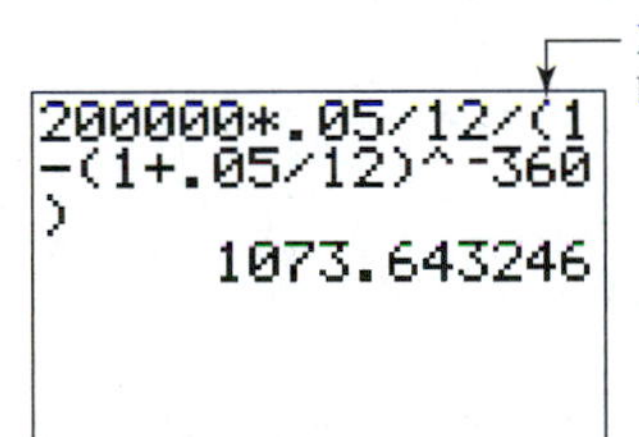

Note that the denominator must be entered in parentheses

SECTION 6.1 EXERCISE SET

Write About It!

Use complete sentences in your answer to each exercise.

1. Explain why a nonzero real number raised to the zero exponent is 1.
2. Explain the difference between $(-5)^4$ and -5^4.
3. Explain why $3^{-2} \neq -3^2$.
4. Use an example to show how to simplify a nonzero real number raised to a negative exponent.

Additional answers can be found in the Instructor Answer Appendix.

5. Can we apply the product of like bases rule of exponents to $(-7)^m(-7)^n$? Explain.
6. Can we apply the product of like bases rule of exponents to $a^m b^n$, $a \neq b$? Explain.
7. Can we apply the quotient of like bases rule of exponents to $\frac{x^{-4}}{x^4}$? Explain. Yes, we can apply the quotient of like bases rule of exponents because the bases are the same.
8. Can we apply the quotient of like bases rule of exponents to $\frac{a^m}{b^n}$, $a \neq b$? Explain. No, we cannot apply the quotient of like bases rule of exponents because the bases are not the same.

Practice Makes Perfect!

Simplify each expression by applying the product of like bases rule. (*See Objective 1.*)

9. $r^3 \cdot r^4$ r^7
10. $s^8 \cdot s^2$ s^{10}
11. $m \cdot m^4$ m^5
12. $x \cdot x^7$ x^8
13. $x^2 \cdot x^3 \cdot x$ x^6
14. $y^3 \cdot y^5 \cdot y$ y^9
15. $m^3 \cdot m^5 \cdot m^4$ m^{12}
16. $n^4 \cdot n^5 \cdot n^6$ n^{15}
17. $3^{11} \cdot 3^4$ 3^{15}
18. $4^2 \cdot 4^{10}$ 4^{12}
19. $(-2)^5 \cdot (-2)^9$ 2^{14}
20. $(-5)^6 \cdot (-5)^8$ 5^{14}
21. $(5x^3) \cdot (-4x^2)$ $-20x^5$
22. $(-3y^8) \cdot (6y^3)$ $-18y^{11}$
23. $(-12a^4) \cdot (-5a^6)$ $60a^{10}$
24. $(15y^7) \cdot (-5y^{12})$ $-75y^{19}$
25. $(-2x^2y^3)(-3x^4y^4)$ $6x^6y^7$
26. $(6x^5y)(-12x^7y^3)$ $-72x^{12}y^4$
27. $(-3a^4b^2)(5a^9b^7)$ $-15a^{13}b^9$
28. $(-8a^2b^4)(-10a^6b^8)$ $80a^8b^{12}$

Simplify each expression by applying the quotient of like bases rule. (*See Objective 2.*)

29. $\dfrac{x^{13}}{x^7}$ x^6
30. $\dfrac{y^{19}}{y^{11}}$ y^8
31. $\dfrac{(-1)^9}{(-1)^2}$ -1
32. $\dfrac{(-1)^{12}}{(-1)^4}$ 1
33. $\dfrac{2^9}{2^5}$ 16
34. $\dfrac{3^{12}}{3^7}$ 243
35. $\dfrac{45p^9}{5p}$ $9p^8$
36. $\dfrac{-18q^7}{-6q^3}$ $3q^4$
37. $\dfrac{-72r^{15}}{12r^4}$ $-6r^{11}$
38. $\dfrac{68s^9}{-4s^3}$ $-17s^6$
39. $\dfrac{-12x^9y^{13}}{6xy^9}$ $-2x^8y^4$
40. $\dfrac{-36a^3b^8}{-9ab^3}$ $4a^2b^5$
41. $\dfrac{112p^{19}q^5}{7p^{13}q}$ $16p^6q^4$
42. $\dfrac{-120r^{10}s^7}{-15r^2s^6}$ $8r^8s$

Simplify each expression. Assume all bases are nonzero real numbers. (*See Objective 3.*)

43. 29^0 1
44. 13^0 1
45. $(-20)^0$ 1
46. $(-15)^0$ 1
47. $-(-18)^0$ -1
48. $-(-25)^0$ -1
49. $(6c)^0$ 1
50. $(9a)^0$ 1
51. $(2b+3)^0$ 1
52. $(3x-4)^0$ 1
53. $2(4z-3)^0$ 2
54. $-2(b-1)^0$ -2

Simplify each expression. Express each answer with positive exponents. Assume all bases are nonzero real numbers. (*See Objective 4.*)

55. 2^{-5} $\dfrac{1}{32}$
56. 3^{-4} $\dfrac{1}{81}$
57. $(-4)^{-2}$ $\dfrac{1}{16}$
58. $(-5)^{-2}$ $\dfrac{1}{25}$
59. $\left(\dfrac{2}{3}\right)^{-3}$ $\dfrac{27}{8}$
60. $\left(\dfrac{2}{5}\right)^{-1}$ $\dfrac{5}{2}$
61. $\left(-\dfrac{1}{5}\right)^{-4}$ 625
62. $\left(-\dfrac{7}{12}\right)^{-2}$ $\dfrac{144}{49}$
63. $-(-3)^{-4}$ $-\dfrac{1}{81}$
64. $-(-4)^{-3}$ $\dfrac{1}{64}$
65. $4v^{-5}$ $\dfrac{4}{v^5}$
66. $-15w^{-8}$ $\dfrac{-15}{w^8}$
67. $\dfrac{y^{-6}}{y^{12}}$ $\dfrac{1}{y^{18}}$
68. $\dfrac{x^{-4}}{x^5}$ $\dfrac{1}{x^9}$
69. $\dfrac{a^{-4}}{b^{-3}}$ $\dfrac{b^3}{a^4}$
70. $\dfrac{x^{-6}}{y^{-2}}$ $\dfrac{y^2}{x^6}$
71. $\dfrac{p^2}{q^{-4}}$ p^2q^4
72. $\dfrac{a^{-1}}{b^3}$ $\dfrac{1}{ab^3}$

Simplify each expression. Write each answer with positive exponents. Assume all bases are nonzero real numbers. (*See Objective 5.*)

73. $q^{-7} \cdot q^{-12}$ $\dfrac{1}{q^{19}}$
74. $p^{-4} \cdot p^{-9}$ $\dfrac{1}{p^{13}}$
75. $x^5 \cdot x^3 \cdot x^{-14}$ $\dfrac{1}{x^6}$
76. $x^3 \cdot x^{-10} \cdot x^4$ $\dfrac{1}{x^3}$
77. $(-4a^7 \cdot b^{-12})(a^{10}b^2)$
78. $(-s^3t^{-7})(-3s^{-4}t^{10})$ $\dfrac{3t^3}{s}$
79. $(-r^4t^{-9})(-6r^{-7}t^{-1})$
80. $(-3p^{-5} \cdot q^{-11})(-p^{-12}q^6)$
81. $(5m^5n^8)(2m^{-2}n^{-7})$ $10m^3n$
82. $(5y^6z^2)(4y^{-5}z^7)$ $20yz^9$
83. $(-5p^{-7} \cdot h^2)(3p^3h^{-3})$
84. $(-3x^5y^4)(-9x^{-2}y^{-8})$
85. $(2m^{-9} \cdot n^{-10})(-m^9n^{-8})$
86. $(4m^{-8} \cdot n^{-6})(-m^2n^6)$
87. $\dfrac{-5u^{-3}v^{10}}{-15u^{-6}v^8}$ $\dfrac{u^3v^2}{3}$
88. $\dfrac{10cd^{-4}}{40c^{-3}d^{-6}}$ $\dfrac{c^4d^2}{4}$
89. $\dfrac{-18u^{-3}v^{-1}}{5u^{11}v^2}$ $-\dfrac{18}{5u^{14}v^3}$
90. $\dfrac{-14m^3n^{-12}}{-24m^{12}n^7}$ $\dfrac{7}{12m^9n^{19}}$
91. $\dfrac{-6e^{-5}t^{10}}{-18e^3t^{-5}}$ $\dfrac{t^{15}}{3e^8}$
92. $\dfrac{-16f^{-5}g^{12}}{-20f^{-3}g^{-6}}$ $\dfrac{4g^{18}}{5f^2}$

Solve each problem. Use $\pi \approx 3.14$ and round each answer to two decimal places where applicable. (*See Objective 6.*)

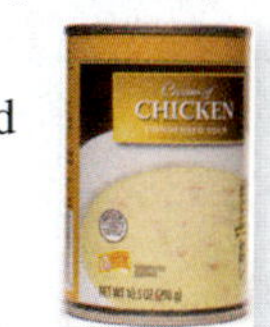

93. A soup can has a diameter of 6.8 cm and a height of 12.4 cm. How much steel is needed to produce one can? Use the formula for surface area $S = 2\pi r^2 + 2\pi rh$. 337.36 cm²
94. A soup can has a diameter of 8.1 cm and a height of 11.1 cm. How much steel is needed to produce one can? Use the formula for surface area $S = 2\pi r^2 + 2\pi rh$. 385.33 cm²
95. Write an expression in terms of the radius for the surface area of a can if the height is three-fourths of its radius. $\dfrac{7}{2}\pi r^2$
96. Write an expression in terms of the radius for the surface area of a can if the height is one-third of its radius. $\dfrac{8}{3}\pi r^2$
97. The area of a rectangle is given by $A = 42x^3y$ and the length is $l = 3x$. What expression represents the width of the rectangle? $14x^2y$
98. The area of a rectangle is given by $A = 18x^4y^5$ and the length is $l = 2x^2y^2$. What expression represents the width of the rectangle? $9x^2y^3$

99. The area of a triangle is given by $A = \frac{1}{2}bh$. If the base of a triangle is $b = 2x^4y$ and the height is $h = 6xy^7$, what expression represents the area of the triangle? $6x^5y^8$

100. The area of a triangle is given by $A = \frac{1}{2}bh$. If the base of a triangle is $b = 4x^4y^2$ and the height is $h = 9x^4y^5$, what expression represents the area of the triangle? $18x^8y^7$

The formula to determine the monthly payment A to repay a loan of P dollars in n months, at an annual interest rate of r (as a decimal), is given by

$$A = \frac{P\left(\frac{r}{12}\right)}{1 - \left(1 + \frac{r}{12}\right)^{-n}}$$

Find the monthly payment needed to repay each loan. Round each answer to the nearest cent.

101. Home loan of \$150,000 at an annual interest rate of 4.3% in 360 months \$742.31

102. Home loan of \$258,000 at an annual interest rate of 5.2% in 360 months \$1416.71

The formula to determine the accumulated amount A for an investment of P dollars at an annual interest rate of r (as a decimal) compounded quarterly for t years is given by

$$A = P\left(1 + \frac{r}{4}\right)^{4t}$$

Find the accumulated amount for each investment. Round each answer to the nearest cent.

103. \$5600 at an annual interest rate of 2.5% compounded quarterly for 2 yr \$5886.20

104. \$12,600 at an annual interest rate of 3.2% compounded quarterly for 4 yr \$14,313.28

Mix 'Em Up!

Simplify each expression and write the result with positive exponents. Assume all bases are nonzero real numbers.

105. $-7(6m)^0$ -7

106. $12(34a)^0$ 12

107. $-19d^{-5}$ $-\frac{19}{d^5}$

108. $5y^{-9}$ $\frac{5}{y^9}$

109. $\frac{11^9}{11^4}$ 11^5 or 161,051

110. $\frac{5^4}{5^6}$ $\frac{1}{25}$

111. $\frac{90p^4}{5p}$ $18p^3$

112. $\frac{-80q^{11}}{-4q^5}$ $20q^6$

113. $\frac{s^{-4}}{b^{-12}}$ $\frac{b^{12}}{s^4}$

114. $\frac{c^{-3}}{d^2}$ $\frac{1}{c^3d^2}$

115. $(-2)^5 \cdot (-2)^{-8}$ $-\frac{1}{8}$

116. $(-5)^9 \cdot (-5)^{12}$ -125

117. $\frac{3u^{-1}v^{-7}}{12u^{-9}v^7}$ $\frac{u^8}{4v^{14}}$

118. $\frac{7x^7y^{-9}}{63x^{-3}y^5}$ $\frac{x^{10}}{9y^{14}}$

119. $(-4d^6r^9)(2d^3r^{-7})$ $-8d^9r^2$

120. $(-2a^7b^{-2})(5a^{-6}b^4)$ $-10ab^2$

121. $-c^0 + 3$ 2

122. $-2c^0 + 5$ 3

123. $\frac{-20a^{-8}n}{-a^{13}n^3}$ $\frac{20}{a^{21}n^2}$

124. $\frac{12u^{-8}v}{-10u^6v^2}$ $-\frac{6}{5u^{14}v}$

125. $(2b + 3)^0$ 1

126. $-2(b - 1)^0$ -2

127. $7^{-3} \cdot 7^3$ 1

128. $9^{-4} \cdot 9^4$ 1

129. $\frac{8t^{-7}z^6}{10t^{-14}z^4}$ $\frac{4t^7z^2}{5}$

130. $\frac{9a^4b^{12}}{-18a^{-10}b^4}$ $-\frac{a^{14}b^8}{2}$

131. $-(-3)^{-6}$ $-\frac{1}{729}$

132. $\left(-\frac{1}{5}\right)^{-4}$ 625

133. $(-c^{-3}p^{-1})(5c^4p^{-5})$ $-\frac{5c}{p^6}$

134. $(-3m^{-1}k^{-6})(m^{-2}k)$ $-\frac{3}{m^3k^5}$

Find the monthly payment needed to repay each loan. Round each answer to the nearest cent.

135. Home loan of \$125,000 at an annual interest rate of 3.9% in 360 months \$589.59

136. Car loan of \$6450 at an annual interest rate of 2.4% in 60 months \$114.19

The formula to determine the accumulated amount for an investment of P dollars at an annual interest rate of r (as a decimal) compounded weekly for t years is given by

$$A = P\left(1 + \frac{r}{52}\right)^{52t}$$

Find the accumulated amount for each investment. Round each answer to the nearest cent.

137. \$2200 at an annual interest rate of 4% compounded weekly for 2 yr \$2383.16

138. \$29,000 at an annual interest rate of 3.6% compounded weekly for 3 yr \$32,306.18

The formula to determine the amount P of radioactive material present at time t, where P_0 is the original amount of the material and m is the half-life of the material, is given by

$$P = P_0 \cdot 2^{-t/m}$$

The half-life of a radioactive substance is the time it takes for half of a certain amount of radioactive substance to decay.

139. The radioactive isotope sodium-24 is used in medical and nonmedical applications. It can be used to detect circulatory problems in patients as well as used to detect leaks in oil pipe lines. The half-life of sodium-24 is 15 hr. If a hospital has 20 g of sodium-24, how many grams will be present after 15 hr? 60 hr? (Sources: http://www.chemistryexplained.com/elements/P-T/Sodium.html and http://www.3rd1000.com/nuclear/halflife.htm) 10 g, 1.25 g

140. The radioactive isotope gallium-67 can be used to locate sites of infection in a patient, tumor imaging, and chemotherapy for pediatric patients. Its half-life is approximately 3 days. If the Imaging Center has 400 mg of gallium-67, how much will be present after 3 days? after 15 days? (Source: http://www.radiochemistry.org/nuclearmedicine/radioisotopes/ex_iso_medicine.htm) 200 mg after 3 days, 12.5 mg after 15 days

You Be the Teacher!

Correct each student's errors, if any.

141. Simplify $-2^4 \cdot x^3 \cdot x^2$.

Isabella's work:

$-2^4 \cdot x^3 \cdot x^2 = 16x^{3+2} = 16x^5$

$-2^4 \cdot x^3 \cdot x^2 = -2 \cdot 2 \cdot 2 \cdot 2x^{3+2} = -16x^5$

142. Simplify $(2^{-3}x^4y)(x^{-5}y^2)$.

Jason's work:

$(2^{-3}x^4y)(x^{-5}y^2) = -6x^{4-5}y^{1+2} = -6x^{-1}y^3 = \dfrac{-6y^3}{x}$

$(2^{-3}x^4y)(x^{-5}y^2) = 2^{-3}x^{4-5}y^{1+2} = 2^{-3}x^{-1}y^3 = \dfrac{y^3}{2^3x} = \dfrac{y^3}{8x}$

143. Simplify $\dfrac{c^4d^7}{c^{-3}d^2}$.

Kyle's work:

$\dfrac{c^4d^7}{c^{-3}d^2} = c^{4-3}d^{7-2} = cd^5$ $\quad \dfrac{c^4d^7}{c^{-3}d^2} = c^{4-(-3)}d^{7-2} = c^7d^5$

144. Simplify $3500\left(1 + \dfrac{0.0225}{12}\right)^{12(4)}$.

Fora's work:

```
3500(1+.0225/12)
^12*4
        14318.26883
```

Calculate It!

Explain and correct errors in each screen shot, if any.

145. Simplify $2^5 \cdot 2^8$.

```
4^13
           67108864
```

146. Simplify $(-3)^5 \cdot (-3)^4$.

```
9^9
          387420489
```

For Exercises 147 and 148, use the formula $A = \dfrac{P\left(\dfrac{r}{12}\right)}{1 - \left(1 + \dfrac{r}{12}\right)^{-n}}$ to determine the monthly payment A needed to repay each loan of P dollars in n months at an annual interest rate r (as a decimal).

147. Home loan of \$165,000 at an annual interest rate of 4.5% in 360 months

```
165000(.045/12)/
1-(1+.045/12)^(-
360)
        618.4901043
```

148. Car loan of \$3200 at an annual interest rate of 2.9% in 60 months

```
3200(.29/12)/(1-
(1+.29/12)^(-60)
)
        101.5739843
```

For Exercises 149 and 150, use the formula $A = P\left(1 + \dfrac{r}{52}\right)^{52t}$ to determine the accumulated amount A for each investment of P dollars at an annual interest rate r (as a decimal) that is compounded weekly for t years.

149. \$10,800 at an annual interest rate of 1.35% compounded weekly for 2 yr

```
10800(1+1.35/52)
^(52*2)
        155258.7512
```

150. \$7600 at an annual interest rate of 2.1% compounded weekly for 3 yr

```
7600(1+.021/52)^
52*3
        23283.76407
```

Think About It!

Use the properties of exponents to simplify each expression.

151. $3^{4x} \cdot 3^{x+1}$ $\quad 3^{5x+1}$

152. $2^{3(a+2)} \cdot 2$ $\quad 2^{3a+7}$

153. $\dfrac{5^{2a}}{5^{a-1}}$ $\quad 5^{a+1}$

154. $\dfrac{4^{x+3}}{4^{x+4}}$ $\quad \dfrac{1}{4}$

SECTION 6.2 More Rules of Exponents and Scientific Notation

OBJECTIVES

As a result of completing this section, you will be able to

1. Apply the power rules for exponents.
2. Simplify exponential expressions using a combination of exponent properties.
3. Convert between standard notation and scientific notation.
4. Perform operations with numbers in scientific notation.
5. Troubleshoot common errors.

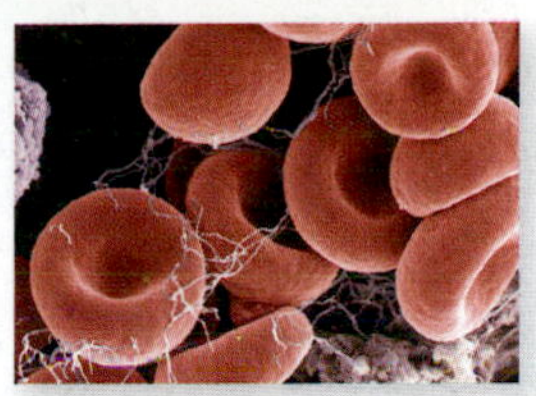

Blood platelets are very small disk-shaped particles that are necessary for clotting blood. The diameter of a single blood platelet is approximately 0.003 mm. A healthy person produces approximately 150 billion platelets each day. If the platelets could be placed side by side, how many miles would they extend? (Source: http://medical-dictionary.thefreedictionary.com/Blood+platelets)

To answer this question, we will represent each number in a special form called scientific notation and then use properties of exponents to find their product.

Objective 1

Apply the power rule for exponents.

Power Rules for Exponents

We will now investigate how to find powers of exponential expressions. The following illustrations show how to use the definition of an exponent to obtain the result as well as a rule that can be applied to the original problem to obtain the result.

Based on Definition

$(x^2)^3 = x^2 \cdot x^2 \cdot x^2$ Three factors of x^2

$= x^{2+2+2}$ Apply the product of like bases rule.

$= x^6$ Add the exponents.

Observed Rule

The same result is obtained by multiplying the exponents.

$(x^2)^3 = x^{2\cdot 3}$

$= x^6$

So, we observe that when an exponential expression is raised to an exponent, the original exponents are multiplied to obtain the exponent in the final result. This brings us to the power of a power rule.

Property: The Power of a Power Rule for Exponents

For a real number b and integers m and n,

$$(b^m)^n = b^{m\cdot n}$$

The base of an exponential expression can be a product, a quotient, a sum, or a difference of expressions. When the base of an exponential expression contains an operation, the base must be written in parentheses when it is expressed in factored form. The entire expression in parentheses repeats when the exponent is applied to it. The following example illustrates how to simplify a power of a product.

Based on Definition

$(3x^4)^2 = (3x^4)(3x^4)$ Two factors of $3x^4$

$= 3 \cdot 3 \cdot x^4 \cdot x^4$ Group coefficients and like bases.

$= 9x^8$ Multiply the coefficients and add the exponents.

Observed Rule

The same result is the obtained by squaring 3 and squaring x^4.

$(3x^4)^2 = (3)^2(x^4)^2$

$= 9x^8$

So, we observe that when a product is raised to an exponent, the result is the same as applying the exponent to each factor in the base. This brings us to another property of exponents.

Property: The Power of a Product Rule for Exponents

For real numbers a and b and an integer n,

$$(ab)^n = a^n b^n$$

A similar result occurs for powers of quotients.

Based on Definition		Observed Rule
$\left(\frac{y}{3}\right)^2 = \left(\frac{y}{3}\right)\left(\frac{y}{3}\right)$	Two factors of $\frac{y}{3}$	The same result is the obtained by squaring y and squaring 3.
$= \frac{y \cdot y}{3 \cdot 3}$	Group coefficients and like bases.	$\left(\frac{y}{3}\right)^2 = \frac{(y)^2}{(3)^2}$
$= \frac{y^2}{9}$	Multiply the coefficients and add the exponents.	$= \frac{y^2}{9}$

So, we observe that when a quotient is raised to an exponent, the exponent is applied to both the numerator and denominator. This leads to another property.

Property: The Power of a Quotient Rule for Exponents

For real numbers a and b ($b \neq 0$) and n an integer,

$$\left(\frac{a}{b}\right)^n = \frac{a^n}{b^n}$$

Procedure: Applying the Power Rules

Step 1: If the base is a product, apply the power of a power rule.
Step 2: If the base is a quotient, apply the power of a quotient rule.
Step 3: Apply the power of a power rule to simplify any exponential expressions.
Step 4: Rewrite all exponents as positive exponents.

Objective 1 Examples **Simplify each expression by applying the power rules. Write each result using positive exponents. Assume all bases are nonzero real numbers.**

1a. $(x^{-5})^2$ **1b.** $[(-3)^4]^5$ **1c.** $(-6x^{-3})^{-1}$

1d. $(-4a^5b^{-2})^3$ **1e.** $\left(\frac{2a}{5b^2}\right)^3$ **1f.** $\left(\frac{4x^2}{3z^{-1}}\right)^{-2}$

Solutions

1a. $(x^{-5})^2 = x^{-5\cdot 2}$ Apply the power of a power rule.

$= x^{-10}$ Multiply the exponents.

$= \frac{1}{x^{10}}$ Rewrite with a positive exponent.

1b. $[(-3)^4]^5 = (-3)^{4\cdot 5}$ Apply the power of a power rule.

$= (-3)^{20}$ Multiply the exponents.

$= 3^{20}$ Recall a negative base to an even exponent is positive.

1c. $(-6x^{-3})^{-1} = (-6)^{-1}(x^{-3})^{-1}$ Apply the power of a product rule.

$= \frac{1}{(-6)^1} \cdot x^3$ Apply the definition of a negative exponent and the power of a power rule.

$= \frac{1}{-6} \cdot x^3$ Simplify the exponent.

$= -\frac{x^3}{6}$ Multiply.

1d. $(-4a^5b^{-2})^3 = (-4)^3(a^5)^3(b^{-2})^3$ Apply the power of a product rule.

$= -64a^{15}b^{-6}$ Simplify the exponent and apply the power of a power rule.

$= -64a^{15} \cdot \dfrac{1}{b^6}$ Rewrite with positive exponents.

$= -\dfrac{64a^{15}}{b^6}$ Simplify.

1e. $\left(\dfrac{2a}{5b^2}\right)^3 = \dfrac{(2a)^3}{(5b^2)^3}$ Apply the power of a quotient rule.

$= \dfrac{(2)^3(a)^3}{(5)^3(b^2)^3}$ Apply the power of a product rule in the numerator and denominator.

$= \dfrac{8a^3}{125b^6}$ Simplify the exponents and apply the power of a power rule.

1f. $\left(\dfrac{4x^2}{3z^{-1}}\right)^{-2} = \dfrac{(4x^2)^{-2}}{(3z^{-1})^{-2}}$ Apply the power of a quotient rule.

$= \dfrac{4^{-2}x^{-4}}{3^{-2}z^2}$ Apply the power of a product rule in the numerator and denominator.

$= \dfrac{3^2}{4^2x^4z^2}$ Rewrite with positive exponents.

$= \dfrac{9}{16x^4z^2}$ Simplify.

Student Check 1 Simplify each expression by applying the power rules. Write each result using positive exponents. Assume all bases are nonzero real numbers.

a. $(y^{-3})^3$ **b.** $[(-2)^7]^3$ **c.** $(-11x^{-4})^2$

d. $(-2m^{-6}n^4)^5$ **e.** $\left(\dfrac{6x^2}{7y}\right)^3$ **f.** $\left(\dfrac{-2a^{-1}}{5b^4}\right)^{-3}$

Note: *Once we become comfortable with the rules, we can omit some of the steps. For instance, in Example 1f, we can directly apply the exponent of -2 to each factor in the numerator and denominator.*

Combine Properties of Exponents

Objective 2

Simplify exponential expressions using a combination of exponent properties.

We will now review all of the exponent properties presented in Sections 6.1 and 6.2 so that we can combine several properties in one problem.

Property: Summary of Exponent Rules

If a and b are real numbers and m and n are integers, then

Product of like bases: $b^m \cdot b^n = b^{m+n}$

Quotient of like bases: $\dfrac{b^m}{b^n} = b^{m-n}, b \neq 0$

Zero exponent: $b^0 = 1, b \neq 0$

Negative exponent: $b^{-n} = \dfrac{1}{b^n}, b \neq 0$

Negative exponent of a fraction: $\left(\dfrac{a}{b}\right)^{-n} = \left(\dfrac{b}{a}\right)^n, a \neq 0, b \neq 0$

Power of a power:	$(b^m)^n = b^{mn}$
Power of a product:	$(ab)^n = a^n b^n$
Power of a quotient:	$\left(\frac{a}{b}\right)^n = \frac{a^n}{b^n}, b \neq 0$

Procedure: Simplifying Exponential Expressions

When problems require us to use multiple properties to simplify, it is generally best to

Step 1: Clear parentheses by applying the power rule for products or quotients.
Step 2: If like bases occur, apply the product or quotient rule.
Step 3: Finally, rewrite any negative exponents and simplify any exponential expressions.

Objective 2 Examples

Simplify each expression and write the result using positive exponents. Assume all bases are nonzero real numbers.

2a. $(-3x^5y^{-2})^3 \cdot x^{-7}y^5$ **2b.** $\left(\frac{5a^{-2}}{4a^3}\right)^{-4}$ **2c.** $\left(\frac{6b^4}{a^2}\right)^{-1}\left(\frac{6a^{-4}}{b^{-2}}\right)^2$

Solutions

2a. $(-3x^5y^{-2})^3 \cdot x^{-7}y^5 = (-3)^3(x^5)^3(y^{-2})^3 \cdot x^{-7}y^5$ Apply the power of a product rule.

$= -27x^{15}y^{-6} \cdot x^{-7}y^5$ Simplify the exponent and apply the power of a power rule.

$= -27x^{15}x^{-7}y^{-6}y^5$ Apply the commutative property.

$= -27x^{15+(-7)}y^{-6+5}$ Apply the product of like bases rule.

$= -27x^8y^{-1}$ Simplify.

$= \frac{-27x^8}{y}$ Rewrite with positive exponents.

2b. $\left(\frac{5a^{-2}}{4a^3}\right)^{-4} = \frac{(5a^{-2})^{-4}}{(4a^3)^{-4}}$ Apply the power of a quotient rule.

$= \frac{5^{-4}a^8}{4^{-4}a^{-12}}$ Apply the power of a product rule and the power of a power rule.

$= \frac{4^4a^8a^{12}}{5^4}$ Rewrite with positive exponents.

$= \frac{256a^{20}}{625}$ Simplify the exponents and apply the product of like bases rule.

2c. $\left(\frac{6b^4}{a^2}\right)^{-1}\left(\frac{6a^{-4}}{b^{-2}}\right)^2 = \frac{6^{-1}b^{-4}}{a^{-2}} \cdot \frac{6^2a^{-8}}{b^{-4}}$ Apply the power of a quotient rule for each fraction. Apply the power of a power rule.

$= \frac{6^{-1}6^2a^{-8}b^{-4}}{a^{-2}b^{-4}}$ Multiply the fractions.

$= 6^{-1+2}a^{-8-(-2)}b^{-4-(-4)}$ Apply the product of like bases rule and the quotient of like bases rule.

$= 6^1a^{-6}b^0$ Simplify.

$= \frac{6(1)}{a^6}$ Rewrite with positive exponents and simplify the exponents.

$= \frac{6}{a^6}$ Simplify.

Student Check 2 Simplify each expression and write the result using positive exponents. Assume all bases are nonzero real numbers.

a. $(-2m^{-4}n^2)^4 \cdot m^{-3}n^5$ **b.** $\left(\dfrac{3x^{-3}}{2x^5}\right)^{-2}$ **c.** $\left(\dfrac{8h^3}{r^4}\right)^{-2}\left(\dfrac{8r^{-3}}{h^{-1}}\right)^3$

Scientific Notation

Objective 3

Convert between standard notation and scientific notation.

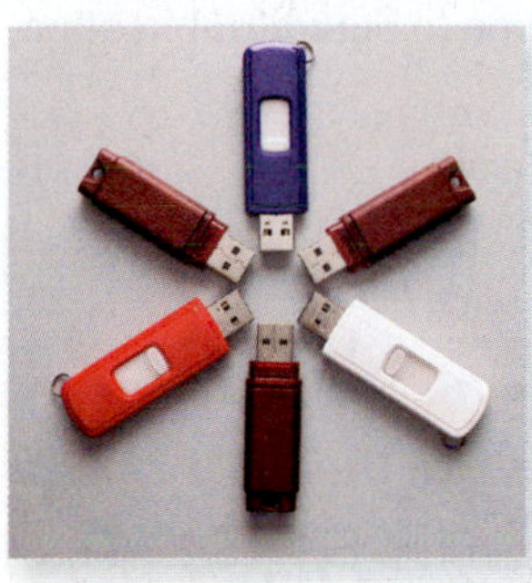

In real life, we deal with numbers that are very large and very small. For example, extremely large numbers are needed to represent the amount of information stored on a computer. Information stored on computers is measured in bytes. Some examples are

- It takes 1 byte (B) to digitally store a single character.
- A typewritten page takes 2000 bytes or 2 kilobytes (kB) to store.
- A movie takes 1 billion bytes or 1 gigabyte (GB) to store.
- An academic research library would take 2 trillion bytes or 2 terabytes (TB) to store.
- It would take 5 quintillion bytes or 5 exabytes (EB) to store all words ever spoken by humans. (Source: http://www2.sims.berkeley.edu/research/projects/how-much-info/datapowers.html)

We can also find some very small numbers when dealing with the body and things in nature. Some examples are

- The thickness of human skin is 0.07 in.
- The diameter of a skin cell is 0.001181 in.
- The diameter of a red blood cell is 0.000315 in.
- The diameter of an influenza virus is 0.00000472 in.
- The diameter of a water molecule is 0.0000000108 in.
(Source: http://learn.genetics.utah.edu/content/begin/cells/scale/)

We will now use exponents to write these very large or very small numbers in *scientific notation*. Because our numbering system is based on the number 10, every digit in the number represents a power of 10. Recall the place value of the digits in a number.

Ten-thousands	Thousands	Hundreds	Tens	Ones	●	Tenths	Hundredths
$10{,}000 = 10^4$	$1000 = 10^3$	$100 = 10^2$	$10 = 10^1$	$1 = 10^0$	Decimal point	$\dfrac{1}{10} = 10^{-1}$	$\dfrac{1}{100} = 10^{-2}$

So, every number in standard notation can be written as a product of a number, called the *coefficient*, and a power of 10. For example,

INSTRUCTOR NOTE: Point out that while 5000 can also be written as 50×10^2 or 500×10^1, these numbers are not in scientific notation since their coefficients are greater than 10.

$$50{,}000 = 5 \times 10{,}000 = 5 \times 10^4$$
$$5000 = 5 \times 1000 = 5 \times 10^3$$
$$500 = 5 \times 100 = 5 \times 10^2$$
$$50 = 5 \times 10 = 5 \times 10^1$$
$$5 = 5 \times 1 = 5 \times 10^0$$
$$0.5 = 5 \times 0.1 = 5 \times \frac{1}{10} = 5 \times 10^{-1}$$
$$0.05 = 5 \times 0.01 = 5 \times \frac{1}{100} = 5 \times 10^{-2}$$

When the exponent of 10 is positive, the number (e.g., 50,000, 5000, 500, and 50) is greater than or equal to 10. When the exponent of 10 is negative, the number (e.g., 0.5 and 0.05) is between 0 and 1. When the exponent of 10 is zero, the number (e.g., 5) is greater than or equal to 1 but less than 10. The power of 10 represents the place value of the coefficient, 5.

Definition: A number written in the form $a \times 10^n$ is a number written in **scientific notation** as long as $1 \le a < 10$ and n is an integer. The number a is called the coefficient.

Procedure: Writing a Number in Scientific Notation

Step 1: Move the original decimal point to the left or right until you reach a number between 1 and 10. This number is the coefficient.

Step 2: The exponent of 10 represents the number of decimal places that you moved in step 1. If the decimal is moved left, the exponent of 10 is positive. If the decimal is moved right, the exponent of 10 is negative.

Procedure: Converting a Number from Scientific Notation to Standard Notation

Step 1: Drop " $\times 10^n$".

Step 2: If $n > 0$, move the decimal point to the right n places.

Step 3: If $n < 0$, move the decimal point to the left $|n|$ places.

Note: *Helpful Method to Remember How to Convert to the Standard Form*

Think of a number line. If a number is positive, we move to the right. If the number is negative, we move to the left. This is the same movement needed to convert numbers from scientific notation to standard form.

Objective 3 Examples Write the given number in either scientific or standard notation.

3a. The storage needed for all words spoken by humans is 5 exabytes or 5 quintillion bytes. Write this number in scientific notation. (Source: http://www2.sims.berkeley.edu/research/projects/how-much-info/datapowers.html)

3b. The diameter of a red blood cell is about 0.000315 in. Write this number in scientific notation. (Source: http://www.wadsworth.org/chemheme/heme/microscope/rbc.htm)

3c. The memory capacity of the brain is 1×10^{12} bytes. Write this number in standard form. (Source: http://www.moah.org/exhibits/archives/brains/technology.html)

3d. The radius of a carbon atom is 1.34×10^{-8} in. Write this number in standard form. (Source: http://learn.genetics.utah.edu/content/begin/cells/scale/)

Solutions

3a. 5 quintillion = 5,000,000,000,000,000,000.0 = 5×10^{18}

Move the decimal point left 18 places

3b. 0.000315 in. = 3.15×10^{-4} in.

Move the decimal point right 4 places

3c. 1×10^{12} = 1,000,000,000,000. bytes

Move the decimal point right 12 places

3d. 1.34×10^{-8} = 0.0000000134

Move the decimal point left 8 places

Student Check 3 Write the given number in either scientific or standard notation.

a. The world population in 2011 was approximately 6,900,000,000. Write this number in scientific notation. (Source: http://www.census.gov/main/www/popclock.html)

b. A hydrogen atom is about 0.00000005 mm in diameter. Write this number in scientific notation. (Source: http://web.jjay.cuny.edu/~acarpi/NSC/3-atoms.htm)

c. The population of the United States in 2011 was approximately 3.1×10^8. Write this number in standard form. (Source: http://www.census.gov/main/www/popclock.html)

d. A carbon atom has a mass of 2×10^{-23}g. Write this number in standard form. (Source: http://www.cavendishscience.org/phys/mole/mole.htm)

Operations with Scientific Notation

Objective 4

Perform operations with numbers in scientific notation.

It is nearly impossible to perform calculations with numbers that are very large or very small in standard notation on a calculator. It is much easier to handle calculations with these types of numbers if they are written in scientific notation. We will use properties of exponents to perform the computations.

Objective 4 Examples **Use scientific notation to solve each problem. Convert each answer to standard notation.**

4a. In 2011, the national debt of the United States was approximately \$14,096,000,000,000 and the population of the United States was approximately 310,000,000. If the debt was evenly distributed among all people in the country, how much would each person owe? (Sources: http://www.usdebtclock.org/ and http://www.census.gov/main/www/poplock.html)

Solution **4a.** We must divide the national debt by the population.

$$\frac{14{,}096{,}000{,}000{,}000}{310{,}000{,}000} = \frac{1.4096 \times 10^{13}}{3.1 \times 10^{8}}$$ Convert each number to scientific notation.

$$= \frac{1.4096}{3.1} \times \frac{10^{13}}{10^{8}}$$ Divide the coefficients and divide the powers of 10.

$$\approx 0.4547 \times 10^{5}$$ Apply the quotient of like bases rule.

$$\approx \$45{,}470$$ Convert to standard notation.

So, each person would owe approximately \$45,470 to pay off the national debt.

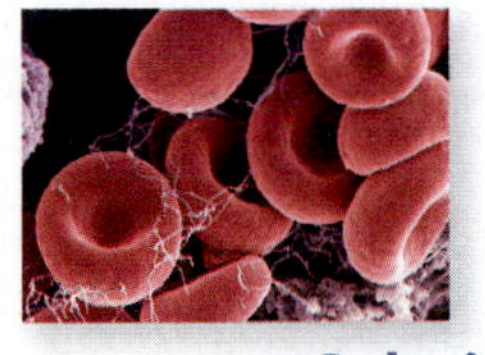

4b. The diameter of a blood platelet is approximately 0.003 mm. A healthy person produces approximately 150 billion platelets each day. If the platelets could be placed side by side, how many miles would they extend? (Note: There are 1,609,344 mm in 1 mi.) (Source: http://medical-dictionary.thefreedictionary.com/Blood+platelets)

Solution **4b.** We must multiply the number of platelets produced by their diameter.

$$150{,}000{,}000{,}000 \times 0.003$$

$$= (1.5 \times 10^{11})(3 \times 10^{-3})$$ Write each number in scientific notation.

$$= (1.5 \times 3) \times (10^{11} \times 10^{-3})$$ Multiply the coefficients and the powers of 10.

$$= 4.5 \times 10^{8} \text{ mm}$$ Apply the product of like bases rule.

$$= 450{,}000{,}000 \text{ mm}$$ Convert to standard notation.

Dividing 450,000,000 mm by 1,609,344 gives us the number of miles the platelets would extend. So, the blood platelets would extend approximately 280 mi if placed side by side.

4c. It would take 5 EB or 5 quintillion bytes to store all words ever spoken by humans. How many 1-GB flash drives would it take to store this amount of information? (Source: http://www2.sims.berkeley.edu/research/projects/how-much-info/datapowers.html)

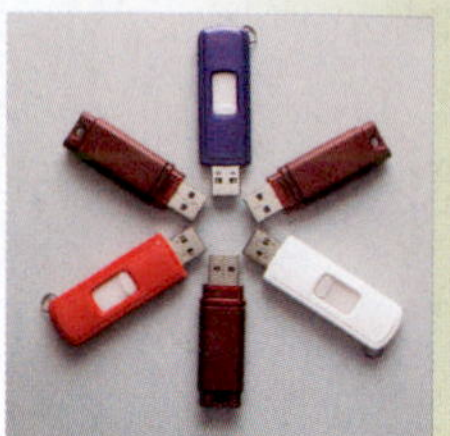

Solution **4c.** We must divide 5 EB by 1 GB. Convert each number to scientific notation. Recall that 5 EB is the same as 5 quintillion bytes and 1 GB is 1 billion bytes.

$$\frac{5 \text{ EB}}{1 \text{ GB}} = \frac{5{,}000{,}000{,}000{,}000{,}000{,}000}{1{,}000{,}000{,}000}$$ Write each number in terms of bytes.

$$= \frac{5 \times 10^{18}}{1 \times 10^{9}}$$ Convert each number to scientific notation.

$$= \frac{5}{1} \times \frac{10^{18}}{10^{9}}$$ Divide the coefficients and the powers of 10.

$$= 5 \times 10^{9}$$ Apply the quotient of like bases rule.

$$= 5{,}000{,}000{,}000$$ Convert to standard notation.

So, it would take 5,000,000,000 or 5 billion 1-GB flash drives to store all the words ever spoken by humans.

Student Check 4 Use scientific notation to solve each problem. Convert each answer to standard notation.

a. As of June 2010, there were approximately 1,970,000,000 worldwide users of the Internet. Europe had approximately 475,000,000 users of the Internet. What percent of the worldwide users were in Europe? (Source: http://www.internetworldstats.com/)

b. Fine hair is approximately 0.002 in. wide. The typical person has 100,000 hairs on their head. If all the hairs could be put side by side, how many inches would be covered? (Sources: http://hypertextbook.com/facts/1999/BrianLey.shtml and http://www.enotes.com)

c. The Milky Way Galaxy is approximately 100,000 light-years in diameter. One light-year is the distance light can travel in a year, or approximately 5.9 trillion mi. How many miles wide is the Milky Way Galaxy? (Source: http://www.hartrao.ac.za/other/howfar/howfar.html)

Objective 5

Troubleshoot common errors.

Troubleshooting Common Errors

Some common errors associated with properties of exponents are shown.

Objective 5 Examples **A problem and an incorrect solution are given. Provide the correct solution and an explanation of the error.**

5a. Simplify $(-4x^3)^2$.

Incorrect Solution	Correct Solution and Explanation
$(-4x^3)^2 = -4x^9$	The exponent was not applied to -4 and the exponents were not multiplied. $(-4x^3)^2 = (-4)^2(x^3)^2 = 16x^6$

5b. Simplify $(2y^{-5})^{-3}$.

Incorrect Solution	Correct Solution and Explanation
$(2y^{-5})^{-3} = -8y^{15}$	The exponent was not applied to the coefficient of 2 correctly. $(2y^{-5})^{-3} = (2)^{-3}(y^{-5})^{-3} = \frac{1}{8}y^{15} = \frac{y^{15}}{8}$

ANSWERS TO STUDENT CHECKS

Student Check 1 **a.** $\frac{1}{y^9}$ **b.** -2^{21} **c.** $\frac{121}{x^8}$ **d.** $-\frac{32n^{20}}{m^{30}}$ **e.** $\frac{216x^6}{343y^3}$ **f.** $-\frac{125a^3b^{12}}{8}$

Student Check 2 **a.** $\frac{16n^{13}}{m^{19}}$ **b.** $\frac{4x^{16}}{9}$ **c.** $\frac{8}{rh^3}$

Student Check 3 **a.** 6.9×10^9 **b.** 5×10^{-8} **c.** 310,000,000 **d.** 0.000 000 000 000 000 000 000 02

Student Check 4 **a.** 24% **b.** 200 in. **c.** 590 quadrillion mi

SUMMARY OF KEY CONCEPTS

1. The power of a power rule applies when an exponential expression is raised to an exponent. In this case, the exponents are multiplied to simplify the expression. It is important that we know when to add and when to multiply the exponents. Note the difference in the following expressions.

$$x^3 \cdot x^2 = x^5 \quad \text{but} \quad (x^3)^2 = x^6$$

2. The power of a product rule applies when the base is a product. We apply the exponent to each factor and simplify. Note the difference in the expressions $3x^4$ and $(3x)^4$.

$3x^4 = 3 \cdot x \cdot x \cdot x \cdot x$ (The base is x.)

$(3x)^4 = (3x)(3x)(3x)(3x) = 3^4x^4 = 81x^4$. (The base is $3x$.)

3. The power of a quotient rule applies when the base is a quotient or a fraction. We apply the exponent to the numerator and denominator. To simplify the resulting expression, we may have to use the power of a power rule and/or the power of a product rule.

4. Scientific notation is a useful way to express very large or very small numbers. A number of the form $a \times 10^n$ is a number written in scientific notation as long as $1 \le a < 10$ and n is an integer.
   - If the exponent in scientific notation is negative, the number has a value between 0 and 1.
   - If the exponent in scientific notation is positive, the number has a value greater than or equal to 10.
   - A number is converted from scientific notation to standard notation by moving the decimal point as indicated by the exponent of 10. A positive exponent tells us to move the decimal to the right. A negative exponent tells us to move the decimal point to the left.

5. Operations with numbers written in scientific notation can be performed by using the properties of exponents.

GRAPHING CALCULATOR SKILLS

The calculator skills for this section include entering exponential expressions and working with scientific notation.

Example 1: Simplify $(-6)^4 \cdot (-6)^8$.

Solution:

```
(-6)^4*(-6)^8
          2176782336
(-6)^12
          2176782336
```

So, $(-6)^4 \cdot (-6)^8 = (-6)^{12} = 6^{12}$.

Example 2: Simplify $[(-3)^4]^5$.

Solution:

```
((-3)^4)^5
          3486784401
(-3)^20
          3486784401
```

So, $[(-3)^4]^5 = (-3)^{20} = 3^{20}$.

Example 3: Enter the number 10,000,000,000,000 on your calculator and interpret the display.

Solution:

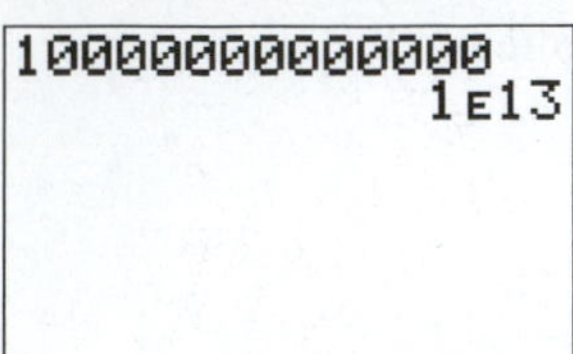

Notice that the display is 1 E 13. This means that the number entered is equivalent to 1×10^{13}. "E" represents "times 10 to the exponent of."

Example 4: Enter the number 7.53×10^{-10} on the calculator. To enter "E" on the calculator, access the EE function (2nd ,) on the calculator.

Solution:

```
7.53E-10
                7.53E-10
7.53*10^-10
                7.53E-10
```

SECTION 6.2 EXERCISE SET

Write About It!

Use complete sentences in your answer to each exercise.

1. Explain how to apply the power of a product rule to $(ab)^m$.
2. Explain how to apply the power of a quotient rule to $\left(\frac{a}{b}\right)^m$.
3. Explain why $(-2x^m)^n \neq 2^n x^{mn}$.
4. Define the scientific notation of a number.
5. Use an example to explain how to convert a number less than one from standard form to scientific notation.
6. Use an example to explain how to convert a number in scientific notation to standard form.
7. Use an example to explain how to convert a number greater than 10 from standard form to scientific notation.
8. Name two advantages of using scientific notation in the calculation of very large or very small numbers.

Practice Makes Perfect!

Simplify each expression by applying the power rules. Assume all bases are nonzero real numbers. Express each answer with positive exponents. (*See Objective 1.*)

9. $(b^3)^5$ b^{15}
10. $(y^7)^2$ y^{14}
11. $(5^{-3})^2$ $\frac{1}{15{,}625}$
12. $(6^9)^{-2}$ $\frac{1}{6^{18}}$
13. $[(-2)^3]^{-5}$ $-\frac{1}{32{,}768}$
14. $[(-3)^2]^{-4}$ $\frac{1}{6561}$
15. $[(-6)^{-7}]^5$ $-\frac{1}{6^{35}}$
16. $[(-5)^{-3}]^6$ $\frac{1}{5^{18}}$
17. $(2a^9b)^5$ $32a^{45}b^5$
18. $(5u^7v^2)^3$ $125u^{21}v^6$
19. $(-4r^9s^2)^4$ $256r^{36}s^8$
20. $(-5u^3v^{11})^2$ $25u^6v^{22}$
21. $(-6x^{-3})^{-2}$ $\frac{x^6}{36}$
22. $(-3v^{-6})^{-1}$ $-\frac{v^6}{3}$
23. $\left(\frac{2x^3}{3y}\right)^4$ $\frac{16x^{12}}{81y^4}$
24. $\left(-\frac{7r^2}{10s^4}\right)^2$ $\frac{49r^4}{100s^8}$
25. $\left(\frac{3u^{-1}}{2v^3}\right)^{-4}$ $\frac{16u^4v^{12}}{81}$
26. $\left(\frac{2a^{-3}}{3b}\right)^{-3}$ $\frac{27a^9b^3}{8}$

Additional answers can be found in the Instructor Answer Appendix.

Simplify each expression using the rules of exponents. Write each answer with positive exponents. Assume all bases are nonzero real numbers (*See Objective 2.*)

27. $(a^7b^{-5})^3 \cdot a^{-17}b^{-3}$
28. $(p^5q^{-7})^6 \cdot p^{-16}q^{-3}$ $\frac{p^{14}}{q^{45}}$
29. $(-2r^3s^{-2})^4 \cdot r^{-10}s^{-5}$
30. $(-4u^9v^{-10})^3 \cdot u^{16}v^{12}$ $-\frac{64u^{43}}{v^{18}}$
31. $\left(-\frac{9x^{-4}}{4x^4}\right)^{-2}$ $\frac{16x^{16}}{81}$
32. $\left(-\frac{2x^{-1}}{3x^2}\right)^{-5}$ $-\frac{243x^{15}}{32}$
33. $\left(\frac{7q^8}{6q^{-10}}\right)^2$ $\frac{49q^{36}}{36}$
34. $\left(\frac{13u^2}{2u^{-11}}\right)^2$ $\frac{169u^{26}}{4}$
35. $\left(\frac{9a^{-4}}{b^{-3}}\right)^4\left(\frac{9b^{-1}}{a^{-6}}\right)^{-4}$ $\frac{b^{16}}{a^{40}}$
36. $\left(\frac{3x^{-5}}{y^{-1}}\right)^3\left(\frac{3y^{-4}}{x^{-7}}\right)^{-3}$ $\frac{y^{15}}{x^{36}}$
37. $\left(\frac{12h^{-7}}{k^{-4}}\right)^3\left(\frac{12k^9}{h^6}\right)^{-2}$ $\frac{12}{h^9k^6}$
38. $\left(\frac{10p^{-4}}{q^{-2}}\right)^{-2}\left(\frac{10q^{-5}}{p^{-3}}\right)^5$ $\frac{1000p^{23}}{q^{29}}$
39. $\frac{(y^{-1}z^8)^4}{2^5y^3z^{15}}$ $\frac{z^{17}}{32y^7}$
40. $\frac{(3^{-1}a^2b^8)^3}{5a^{10}b^{14}}$ $\frac{b^{10}}{135a^4}$
41. $\frac{(2^{-3}s^{-2}t^{12})^{-1}}{s^3t^{12}}$ $\frac{8}{st^{24}}$
42. $\frac{(r^{-5}s^{-8})^4}{2^4r^{-8}s^{10}}$ $\frac{1}{16r^{12}s^{42}}$

Write each number in an alternate form, either in scientific or standard notation. (*See Objective 3.*)

43. 0.000000603 6.03×10^{-7}
44. 0.000000563 5.63×10^{-7}

45. 945,000 9.45×10^5

46. 812,000,000 8.12×10^8

47. 4.65×10^4 46,500

48. 6.68×10^3 6680

49. 4.52×10^{-3} 0.00452

50. 1.89×10^{-5} 0.0000189

Use scientific notation to perform the indicated operation. Express each answer in scientific notation. (*See Objective 4.*)

51. $(7.6 \times 10^{-20})(4.8 \times 10^{-5})$ 3.648×10^{-24}

52. $(1.8 \times 10^3)(5 \times 10^{-8})$ 9×10^{-5}

53. $(4 \times 10^{13})(5.3 \times 10^{-8})$ 2.12×10^6

54. $(4 \times 10^{-18})(1.4 \times 10^{-3})$ 5.6×10^{-21}

55. $(2 \times 10^{-5})^5$ 3.2×10^{-24}

56. $(5 \times 10^{-6})^3$ 1.25×10^{-16}

57. $(0.071)(120{,}000)$ 8.52×10^3

58. $(0.15)(900{,}000)$ 1.35×10^5

59. $(0.00096)(970{,}000{,}000{,}000)$ 9.312×10^8

60. $(0.000000054)(0.0000008)$ 4.32×10^{-14}

61. $\dfrac{1.62 \times 10^{-11}}{4.5 \times 10^3}$ 3.6×10^{-15}

62. $\dfrac{2.496 \times 10^8}{3.9 \times 10^2}$ 6.4×10^5

63. $\dfrac{804{,}000}{0.00012}$ 6.7×10^9

64. $\dfrac{18{,}200}{0.0000013}$ 1.4×10^{10}

65. $\dfrac{(1.29 \times 10^3)(4 \times 10^{12})}{3 \times 10^{-4}}$ 1.72×10^{19}

66. $\dfrac{(6.27 \times 10^5)(9 \times 10^8)}{5.7 \times 10^{-1}}$ 9.9×10^{14}

67. $\dfrac{(6.052 \times 10^7)(7 \times 10^{-12})}{8.9 \times 10^6}$ 4.76×10^{-11}

68. $\dfrac{(6.02 \times 10^{-7})(5 \times 10^{-2})}{3.5 \times 10^{-11}}$ 8.6×10^2

Use scientific notation to solve each problem. Convert each result to standard notation. (*See Objective 4.*)

69. As of June 2010, there were approximately 1,970,000,000 worldwide users of the Internet. Japan had approximately 99,000,000 users of the Internet. What percent of the worldwide users were in Japan? (Source: http://www.internetworldstats.com) 5.03%

70. As of June 2010, there were approximately 1,970,000,000 worldwide users of the Internet. Asia had approximately 825,000,000 users of the Internet. What percent of the worldwide users were in Asia? (Source: http://www.internetworldstats.com) 41.88%

71. As of June 2010, the world population was approximately 6,800,000,000. Asia had a population of approximately 3,800,000,000. What percent of the world population lived in Asia? (Source: http://www.internetworldstats.com) 55.88%

72. As of June 2010, the world population was approximately 6,800,000,000. The United States had a population of approximately 310,000,000. What percent of the world population lived in the United States? (Source: http://www.internetworldstats.com) 4.56%

73. As of April 2011, the public sector net debt of the United Kingdom was £910,000,000,000 and the population of the UK was approximately 60,000,000. If the debt was evenly distributed among all people in the UK, how much debt would each person owe? (Source: http://www.statistics.gov.uk/default.asp) £15,166.67

74. As of April 2011, the national debt of Canada was approximately \$576,000,000,000 USD and the population of Canada was approximately 34,000,000. If the debt was evenly distributed among all people in Canada, how much debt would each person owe? (Sources: http://www.cia.gov/library/publicatons/the-world-factbook/index.html and http://taxpayer.com/federal/national-debt-clock-tour-2011) \$16,941.18

75. One hydrogen atom has a diameter of approximately 5×10^{-8} mm. How many hydrogen atoms would it take to form a line 1 in. long if they were put side by side? Note: 1 in. is equivalent to 25.4 mm or 2.54×10^1 mm. (Source: http://web.jjay.cuny.edu/~acarpi/NSC/3-atoms.htm) 508,000,000 hydrogen atoms

76. The average red blood cell is 3×10^{-4} in. in diameter. The diameter of the head of a pin is approximately 7.87×10^{-2} in. How many red blood cells would it take to stretch the length of the pin's diameter if they were put side by side? (Source: http://www.wadsworth.org/chemheme/heme/microscope/rbc.htm) approximately 262 red blood cells

Mix 'Em Up!

Simplify each expression using the rules of exponents. Express each answer with positive exponents. Assume all bases are nonzero real numbers.

77. $(-5x^6y^{-2})^{-3}$ $-\dfrac{y^6}{125x^{18}}$

78. $(-3u^{-3}v^2)^{-2}$ $\dfrac{u^6}{9v^4}$

79. $(-2a^7b^{-1})^5$ $-\dfrac{32a^{35}}{b^5}$

80. $(-4p^{-5}q^8)^4$ $\dfrac{256q^{32}}{p^{20}}$

81. $(-3v^{-10})^{-4}$ $\dfrac{v^{40}}{81}$

82. $(-2y^5)^{-3}$ $-\dfrac{1}{8y^{15}}$

83. $(7s^{-5})^{-2}$ $\dfrac{s^{10}}{49}$

84. $(5r^8)^{-4}$ $\dfrac{1}{625r^{32}}$

85. $\left(-\dfrac{x^{-2}}{3}\right)^{-3}$ $-27x^6$

86. $\left(-\dfrac{y^{-6}}{5}\right)^{-2}$ $25y^{12}$

87. $\left(\dfrac{q^{-4}}{6}\right)^2$ $\dfrac{1}{36q^8}$

88. $\left(\dfrac{p^{-9}}{2}\right)^5$ $\dfrac{1}{32p^{45}}$

89. $\left(\dfrac{6u^{-5}}{11v^3}\right)^{-2}$ $\dfrac{121u^{10}v^6}{36}$

90. $\left(-\dfrac{5p^{-1}}{2q^2}\right)^{-3}$ $-\dfrac{8p^3q^6}{125}$

91. $\left(-\frac{9x^{-5}}{10y^3}\right)^{-3}$ $-\frac{1000x^{15}y^9}{729}$

92. $\left(\frac{2a^6}{7b^{-1}}\right)^2$ $\frac{4a^{12}b^2}{49}$

93. $(2a^3b^{-11})^4 \cdot a^{-14}b^3$ $\frac{16}{a^2b^{41}}$

94. $(3r^8s^{-14})^2 \cdot r^{-7}s^{12}$ $\frac{9r^9}{s^{16}}$

95. $(-x^2y^{-13})^3 \cdot x^9y^7$ $-\frac{x^{15}}{y^{32}}$

96. $(-a^3b^{-7})^5 \cdot a^6b^4$ $-\frac{a^{21}}{b^{31}}$

97. $\frac{(5^{-3}s^{-8}t^{-10})^{-1}}{4^{-1}s^{-3}t^9}$ $500s^{11}t$

98. $\frac{(r^{-9}s^{-3})^2}{2^4r^{12}s^{-7}}$ $\frac{s}{16r^{30}}$

99. $\frac{(y^{-3}z^{-9})^2}{2y^6z^{-7}}$ $\frac{1}{2y^{12}z^{11}}$

100. $\frac{(2^4r^7s^{10})^{-1}}{5^{-3}r^{-4}s^{-20}}$ $\frac{125s^{10}}{16r^3}$

101. $\left(\frac{3u^{-7}}{5u^{-3}}\right)^{-2}$ $\frac{25u^8}{9}$

102. $\left(\frac{10y^{-1}}{y^{-8}}\right)^2$ $100y^{14}$

103. $\left(-\frac{y^8}{4y^{11}}\right)^4$ $\frac{1}{256y^{12}}$

104. $\left(-\frac{2q^{-8}}{3q^{-7}}\right)^{-5}$ $-\frac{243q^5}{32}$

105. $\left(\frac{12p^{-3}}{q^6}\right)^{-3}\left(\frac{12q^{-1}}{p^{-7}}\right)^5$ $144p^{44}q^{13}$

106. $\left(\frac{5h^{-6}}{k^7}\right)^{-1}\left(\frac{5k^{-7}}{h^6}\right)^4$ $\frac{125}{h^{18}k^{21}}$

107. $\left(\frac{15r}{h^8}\right)\left(\frac{15h^{-9}}{r^{10}}\right)^{-1}$ hr^{11}

108. $\left(\frac{8u^2}{v^{-6}}\right)^{-2}\left(\frac{8v^{-8}}{u^{-7}}\right)^2$ $\frac{u^{10}}{v^{28}}$

Use scientific notation to perform the indicated operation. Express each answer in scientific notation.

109. $(5 \times 10^{-12})(7.7 \times 10^5)$ 3.85×10^{-6}

110. $(5.2 \times 10^{-9})(8.9 \times 10^{23})$ 4.628×10^{15}

111. $\frac{(5.4464 \times 10^{11})(5.5 \times 10^5)}{4.6 \times 10^{12}}$ 6.512×10^4

112. $\frac{(2.4948 \times 10^5)(2.5 \times 10^{-9})}{2.7 \times 10^9}$ 2.31×10^{-13}

113. $(0.000032)(280{,}000{,}000)$ 8.96×10^3

114. $(0.0000042)(12{,}500)$ 5.25×10^{-2}

115. $\frac{7.885 \times 10^7}{9.5 \times 10^{-4}}$ 8.3×10^{10}

116. $\frac{(9.672 \times 10^5)(5.5 \times 10^{-4})}{1.3 \times 10^{-9}}$ 4.092×10^{11}

117. $(1.7 \times 10^6)^2$ 2.89×10^{12}

118. $(3.1 \times 10^{-5})^3$ 2.9791×10^{-14}

119. $\frac{(2.68 \times 10^2)(3 \times 10^{15})}{4 \times 10^{-7}}$ 2.01×10^{24}

120. $\frac{(1.29 \times 10^3)(4 \times 10^{12})}{3 \times 10^{-4}}$ 1.72×10^{19}

Use scientific notation to solve each problem. Convert each answer to standard notation.

121. In 2009 the average number of monthly visits to the top 500 retail websites was 2,580,000,000. The number of Internet users in the United States was approximately 228,000,000. Find the average number of monthly visits per Internet user in the United States. Round to two decimal places. (Source: http://www.internetretailer.com) 11.32 visits per month

122. Amazon was the largest Web retailer in 2010. It had sales of \$34,200,000,000. Their international sales were approximately \$15,000,000,000 in 2010. Find the percent of Amazon's total sales that were international. Round to the nearest hundredth of a percent. (Sources: http://www.internetretailer.com and http://www.amazon.com) 43.86%

You Be the Teacher!

Correct each student's errors, if any.

123. Simplify $(-4a^3b^2)^3$.

Adrian's work:

$(-4a^3b^2)^3 = -12a^9b^6$

124. Simplify $\left(\frac{5x^{-2}}{2x^5}\right)^{-3}$.

Joshua's work:

$$\left(\frac{5x^{-2}}{2x^5}\right)^{-3} = \frac{-125x^6}{-8x^{-15}} = \frac{125x^{-9}}{8} = \frac{125}{8x^9}$$

125. Simplify $\frac{6.3 \times 10^{-21}}{1.4 \times 10^{-6}}$.

Atesa's work:

$$\frac{6.3 \times 10^{-21}}{1.4 \times 10^{-6}} = 4.9 \times 10^{-27}$$

126. Simplify $(4.1 \times 10^8)(2.9 \times 10^{-5})$.

Mark's work:

$(4.1 \times 10^{-8})(2.8 \times 10^{-5}) = 6.9 \times 10^{-3}$

Calculate It!

Correct the error in each screen shot.

127. $\frac{1.512 \times 10^{-12}}{7.2 \times 10^{-4}}$

```
1.512*10^-12/7.2
*10^-4
          2.1E-17
```

128. $\frac{2.47 \times 10^2}{1.3 \times 10^{-7}}$

```
2.47E2-1.3E-7
     246.9999999
```

Think About It!

129. What expression needs to be squared to obtain $49a^2b^{10}$? Check your result.

130. What expression needs to be squared to obtain $144x^4y^6$? Check your result.

131. What expression needs to be cubed to obtain $-\frac{27r^3}{s^6}$? Check your result.

132. What expression needs to be cubed to obtain $\frac{125x^{12}}{y^9}$? Check your result.

133. Use the properties of exponents to simplify $\frac{(x^{3a})^2}{x^{a-6}}$.

134. Use the properties of exponents to simplify $\frac{(2^{4x})^3 \cdot 2^{x-1}}{2^{x+3}}$.

PIECE IT TOGETHER SECTIONS 6.1–6.2

Simplify each expression by applying the rules of exponents. Assume all bases are nonzero real numbers. Express each answer with positive exponents. (*Section 6.1, Objectives 1–5, and Section 6.2, Objectives 1 and 2*)

1. $n^5 \cdot n^3$ n^8
2. $6^4 \cdot 6^{20}$ 6^{24}
3. $(12y^{17})(-7y^{12})$ $-84y^{29}$
4. $(-2a^5b)(8a^8b^{11})$ $-16a^{13}b^{12}$
5. $\frac{10^{21}}{10^{16}}$ 100,000
6. $\frac{54x^{14}}{6x^6}$ $9x^8$
7. $7(-9c)^0$ 7
8. $6x^0 + 2$ 8
9. $(-8)^{-3}$ $-\frac{1}{512}$
10. $\left(\frac{1}{6}\right)^{-2}$ 36
11. $\frac{a^{-11}}{b^{-6}}$ $\frac{b^6}{a^{11}}$
12. $\frac{y^3}{z^{-7}}$ y^3z^7
13. $(3b^{10}p^{-7})(4b^2p^{-5})$ $\frac{12b^{12}}{p^{12}}$
14. $(4j^9 \cdot k^{10})(-7j^{-14}k^5)$
15. $(-2x^7y^8)(-7x^{-5}y^{-18})$
16. $(-7b^{-4})^{-2}$ $\frac{b^8}{49}$
17. $(-2a^3b^7)^5$ $-32a^{15}b^{35}$
18. $(-3a^3b^{-10})^4 \cdot a^{-18}b^{10}$
19. $\left(-\frac{5p^2}{2q^6}\right)^3$ $-\frac{125p^6}{8q^{18}}$
20. $\left(\frac{6u^4}{v^2}\right)^3\left(\frac{6v}{u^{-4}}\right)^{-1}$ $\frac{36u^8}{v^7}$
21. The area of a rectangle is given by $A = 36x^3y^4$ and the length is $l = 6x^2y$. What expression represents the width of the rectangle? (*Section 6.1, Objective 6*) $6xy^3$
22. The area of a triangle is given by $A = \frac{1}{2}bh$. If the base is $b = 2x^4y$ and the height is $h = 3xy^4$. What expression represents the area of the triangle? (*See Section 6.1, Objective 6.*) $3x^5y^5$

Use scientific notation to perform the indicated operation. Express answers in scientific notation. (*Section 6.2, Objectives 3 and 4*)

23. $(0.0003)(0.00000038)$ 1.14×10^{-10}
24. $(1.5 \times 10^6)^4$ 5.0625×10^{24}
25. $\frac{(5.893 \times 10^{-4})(7.5 \times 10^{-3})}{7.1 \times 10^{12}}$ 6.225×10^{-19}

SECTION 6.3 Polynomials, Polynomial Functions, and Their Basic Graphs

OBJECTIVES

As a result of completing this section, you will be able to

1. Identify the coefficient and degree of a term.
2. Classify a polynomial and identify its terms, leading coefficient, and degree.
3. Evaluate polynomial functions.
4. Graph basic polynomial functions.
5. Troubleshoot common errors.

In 2010 the Burj Khalifa, in Dubai, became the tallest skyscraper in the world, reaching 2717 ft. If a penny is dropped from the top of this building, its height above the ground h, in feet, t seconds after it is dropped is given by

$$h(t) = -16t^2 + 2717$$

This function is an example of a *polynomial function*. We can evaluate it at different values to determine the height of the penny using the concepts we will learn from this chapter.

Terms

Objective 1 Identify the coefficient and degree of a term.

A *term* is a number or the product of a number and powers of variables. If a term only contains a number, it is called a **constant term**.

Definition: A **term** is an expression of the form ax^n, where a is a real number and n is a whole number. The real number a is the **coefficient** of the term. The exponent n is the *degree* of the term.

Terms can have one variable or multiple variables. Each term has a degree.

Definition: The **degree of a term** is the sum of all the exponents of the variables in the term.

Some terms and their degrees are shown in the table.

INSTRUCTOR NOTE: Point out that a constant term has degree zero.

Term	Degree
$4x = 4x^1$	1
$-3y^2$	2
$7ab^3 = 7a^1b^3$	$1 + 3 = 4$
$5 = 5x^0$	0

Objective 1 Examples Determine the coefficient and the degree of each term.

Problems	Solutions	Coefficient	Degree
1a. $-3x$	$-3x = -3x^1$	-3	1
1b. $5.3a^2$	$5.3a^2$	5.3	2
1c. $-\frac{y}{2}$	$-\frac{y}{2} = -\frac{1}{2}y^1$	$-\frac{1}{2}$	1
1d. $\frac{b^3}{6}$	$\frac{b^3}{6} = \frac{1}{6}b^3$	$\frac{1}{6}$	3
1e. -10	$-10 = -10x^0$	-10	0
1f. $7x^4y^2$	$7x^4y^2$	7	$4 + 2 = 6$

Student Check 1 Determine the coefficient and the degree of each term.

a. $-y$ **b.** $6.7h^5$ **c.** $-\frac{r}{4}$ **d.** $\frac{x^2}{5}$ **e.** -7 **f.** $4a^3b^8$

Polynomials

Objective 2 Classify a polynomial and identify its terms, leading coefficient, and degree.

INSTRUCTOR NOTE: Point out that the definition of a polynomial can be extended to include more than one variable.

In Section 1.2, we discussed the concept of algebraic expressions. *Polynomials* are special types of algebraic expressions in which the variables have whole numbers as exponents. Polynomials are very important to the study of algebra.

Definition: A **polynomial** is an algebraic expression that consists of a finite sum of terms of the form ax^n, where a is a real number and n is a whole number. The **standard form** is to write the polynomials so that the degrees of the terms are in descending order.

An example of a polynomial is $-3x^4 + 2x^2 - 5x + 7$. This polynomial is written in standard form because the exponents of x are in descending order. The expressions $\frac{2}{x}$, $6x^{-2} + 4x^{-1}$, and $\sqrt{y^2 + 3}$ are not polynomials since a variable occurs in the denominator, has a negative exponent, or occurs inside a square root, respectively. Polynomials can have one, two, three, or more terms. The following table shows examples of each of these cases and provides the special name given to some of these polynomials.

Number of Terms	Name	Examples
One	**Monomial**	$5x, -3y, 7t, \frac{2}{3}ab$
Two	**Binomial**	$x-4, \frac{1}{2}t-1, y+2, -3a-7$
Three	**Trinomial**	$x^2-xy+6y^2, 4y^2+3y-1$
Four or more	**Polynomial**	$-7y^3+9y^2+6y-5$

Like terms, polynomials have a degree as well.

Definition: The **degree of a polynomial** is the largest degree of the terms in the polynomial.

A polynomial can be classified by not only its number of terms but also by its degree.

Degree	Type of Polynomial
0	Constant
1	Linear or first-degree
2	Quadratic or second-degree
3	Cubic or third-degree
4	Quartic or fourth-degree
5	Quintic or fifth-degree

Definition: The **leading coefficient** of a single-variable polynomial is the coefficient of the term with the largest degree.

An example of a polynomial with its degree and leading coefficient is shown.

Degree

$$-5y^2+4y+3$$

Leading coefficient

The degree and leading coefficient of a polynomial are very important when we graph polynomial functions.

The leading coefficient for a multiple-variable polynomial is not defined, so we use not applicable (n/a).

Objective 2 Examples **Classify each polynomial and identify its terms, degree, type, and leading coefficient, if applicable.**

Problems	Solutions: Terms and Classification	Degree and Type	Leading Coefficient
2a. $-x^2+7x-6$	$-1x^2, 7x, -6$ Trinomial	2 Quadratic	-1
2b. $4.2x^3$	$4.2x^3$ Monomial	3 Cubic	4.2

Problems	Terms and Classification	Degree and Type	Leading Coefficient
2c. $\frac{3}{5}x + 2$	$\frac{3}{5}x^1, 2$ Binomial	1 Linear	$\frac{3}{5}$
2d. $3x^3y - 4x^2y^3 + 2xy^2$	$3x^3y^1, -4x^2y^3, 2x^1y^2$ Trinomial	5 Quintic	n/a

Student Check 2 Classify each polynomial and identify its terms, degree, type, and leading coefficient, if applicable.

a. $-4y^4 + y^3 + 3y$ **b.** $x^2 + 16x$ **c.** $9\pi r^3$ **d.** $8r^2s^5 - 7rs^8 + 9rs$

Polynomial Functions

Objective 3 Evaluate polynomial functions.

In Chapter 4, we discussed linear functions. Recall a linear function is a function of the form $f(x) = ax + b$, where a and b are real numbers. Linear functions are *polynomial functions* of degree 1, because the expression $ax + b$ is a first-degree polynomial. Some other examples of polynomial functions are $f(x) = x^2 - 4x + 3$ and $P(x) = x^3 + 5x^2 - 7x + 1$.

Definition: A **polynomial function** is a function of the form

$$P(x) = a_nx^n + a_{n-1}x^{n-1} + \cdots + a_1x + a_0$$

where $a_n, a_{n-1}, \ldots, a_1$, and a_0 are real numbers and n is a whole number.

Function notation provides a convenient way for us to evaluate polynomials. Recall that to evaluate a function is to find the output value that corresponds to a given input value.

Procedure: Evaluating a Polynomial Function

Step 1: Replace the independent variable, x, with the assigned value.
Step 2: Simplify the result to find the output value.

Objective 3 Examples Evaluate each function for the given value.

3a. Find $f(0)$ if $f(x) = 3x^2 - 5x - 8$.

Solution **3a.**

$f(x) = 3x^2 - 5x - 8$
$f(0) = 3(0)^2 - 5(0) - 8$ Replace x with 0.
$f(0) = 3(0) - 0 - 8$ Simplify.
$f(0) = -8$ Add the resulting values.

3b. Find $P(-3)$ if $P(x) = -4x^3 - x^2 + 5x + 2$.

Solution **3b.**

$P(x) = -4x^3 - x^2 + 5x + 2$
$P(-3) = -4(-3)^3 - (-3)^2 + 5(-3) + 2$ Replace x with -3.
$P(-3) = -4(-27) - (9) - 15 + 2$ Simplify the exponents.
$P(-3) = 108 - 9 - 15 + 2$ Multiply.
$P(-3) = 86$ Add the resulting values.

3c. In 2010 the Burj Khalifa in Dubai became the tallest skyscraper in the world reaching 2717 ft. If a penny is dropped from the top of this building, its height above ground h, in feet, t seconds after it is dropped is given by $h(t) = -16t^2 + 2717$. Find $h(5)$ and $h(12)$ and interpret the results. (Source: http://www.emporis.com)

Solution **3c.** We evaluate the function at $t = 5$ and $t = 12$.

$t = 5$	$t = 12$
$h(t) = -16t^2 + 2717$	$h(t) = -16t^2 + 2717$
$h(5) = -16(5)^2 + 2717$	$h(12) = -16(12)^2 + 2717$
$h(5) = -16(25) + 2717$	$h(12) = -16(144) + 2717$
$h(5) = -400 + 2717$	$h(12) = -2304 + 2717$
$h(5) = 2317$	$h(12) = 413$

So, the penny will be 2317 ft above the ground 5 sec after it is dropped and 413 ft above the ground 12 sec after it is dropped.

3d. The number of persons from Mexico obtaining legal permanent resident status in the United States from 1999–2008 can be modeled by the function

$$f(x) = 1028.12x^3 - 15{,}178.23x^2 + 58{,}211.6x + 141{,}625.65$$

where x is the number of years after 1999. Find $f(5)$ and interpret the result. (Source: www.dhs.gov)

Solution **3d.** We evaluate the function at $x = 5$.

$$f(x) = 1028.12x^3 - 15{,}178.23x^2 + 58{,}211.6x + 141{,}625.65$$
$$f(5) = 1028.12(5)^3 - 15{,}178.23(5)^2 + 58{,}211.6(5) + 141{,}625.65$$
$$f(5) \approx 181{,}743$$

So, in 2004 (5 yr after 1999), approximately 181,743 people from Mexico obtained legal permanent resident status in the United States.

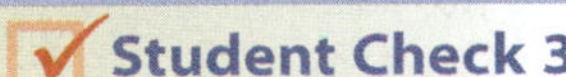

Student Check 3 Evaluate each function for the given value.

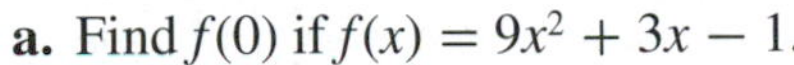

a. Find $f(0)$ if $f(x) = 9x^2 + 3x - 1$.

b. Find $P(-2)$ if $P(x) = -x^3 + 7x^2 - x + 4$.

c. The Eiffel Tower is 1063 ft tall. If a penny is dropped from the top of this tower, its height above ground h, in feet, t seconds after it is dropped is given by $h(t) = -16t^2 + 1063$. Find $h(3)$ and $h(8)$ and interpret the results. (Source: http://www.emporis.com)

d. The average price of a gallon of gasoline in the United States in the first week of January between 2006 and 2011 can be modeled by $p(x) = -0.1251x^5 + 1.5647x^4 - 6.7497x^3 + 11.539x^2 - 6.1691x + 2.236$, where x is the years after 2006. Find $p(5)$ and interpret its meaning. (Source: http://www.eia.gov)

Graphs of Polynomial Functions

Objective 4 Graph basic polynomial functions.

Now that we know the algebraic form of polynomial functions, we will look at the shape of their graphs. We begin by examining the graph of the basic polynomial functions

$$f(x) = x, \quad f(x) = x^2, \quad f(x) = x^3, \quad \text{and} \quad f(x) = x^4$$

Recall from Section 3.1 that we can graph functions by plotting points.

Graph of $f(x) = x$

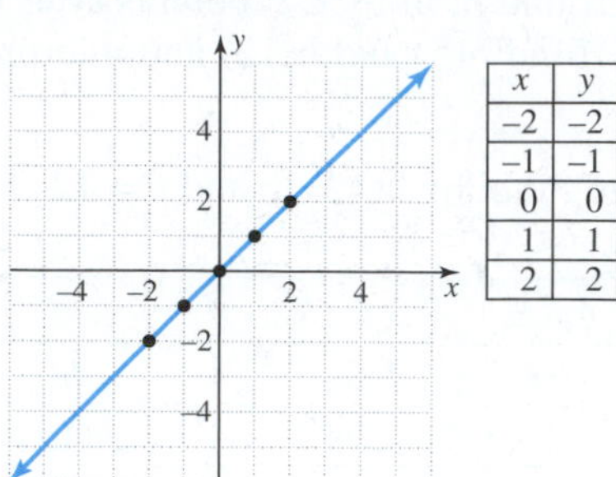

x	y
−2	−2
−1	−1
0	0
1	1
2	2

Graph of $f(x) = x^2$

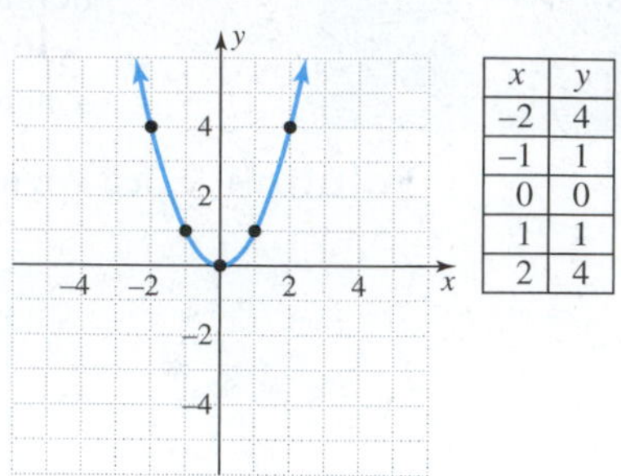

x	y
−2	4
−1	1
0	0
1	1
2	4

Graph of $f(x) = x^3$

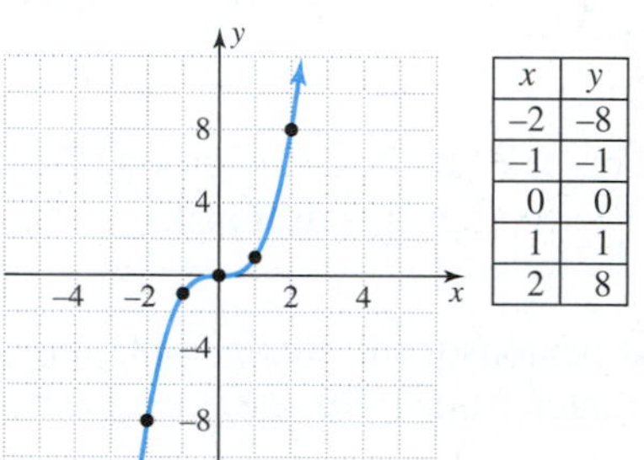

x	y
−2	−8
−1	−1
0	0
1	1
2	8

Graph of $f(x) = x^4$

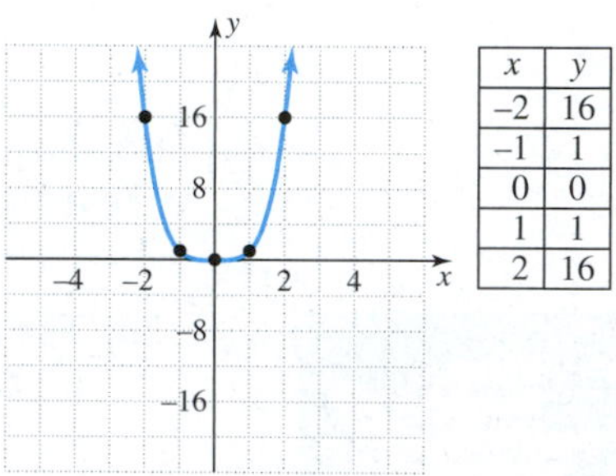

x	y
−2	16
−1	1
0	0
1	1
2	16

From the graphs of these basic polynomial functions, we can make the following observations.

Facts: *Features of Graphs of Polynomial Functions*

1. *The graph of a polynomial function is smooth with no sharp curves. The graph of a polynomial function of degree* 1 *is a line.*
2. *The domain of a polynomial function is* $(-\infty, \infty)$ *since the graphs extend indefinitely to the left and right.*
3. *The graphs of polynomial functions with odd degree have a similar shape. The graphs go in opposite directions at their ends. The graphs of* $y = x$ *and* $y = x^3$ *fall to the left and rise to the right. The range of these graphs is* $(-\infty, \infty)$.
4. *The graphs of polynomial functions with even degree also have a similar shape. The graphs go in the same direction at their ends. The graphs of* $y = x^2$ *and* $y = x^4$ *rise to the left and rise to the right. The range of these graphs is* $[0, \infty)$.
5. *All of these polynomial functions have an x- and y-intercept of* $(0, 0)$.

Now let's examine the graphs of $f(x) = -x$, $f(x) = -x^2$, $f(x) = -x^3$, and $f(x) = -x^4$.

Graph of $f(x) = -x$

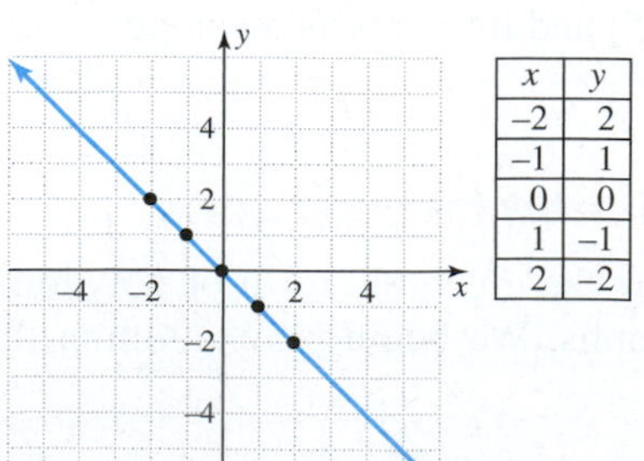

x	y
−2	2
−1	1
0	0
1	−1
2	−2

Graph of $f(x) = -x^2$

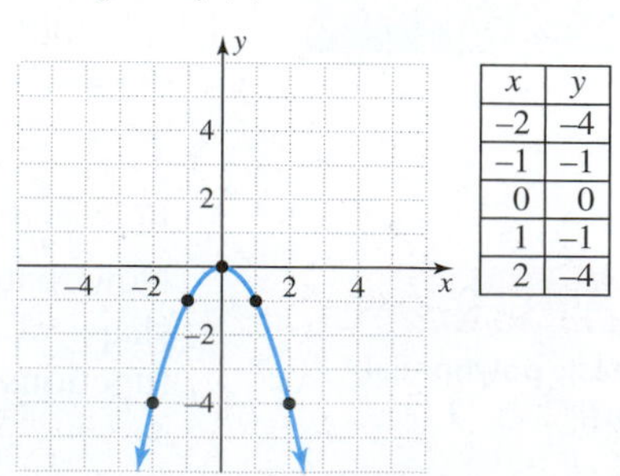

x	y
−2	−4
−1	−1
0	0
1	−1
2	−4

Graph of $f(x) = -x^3$

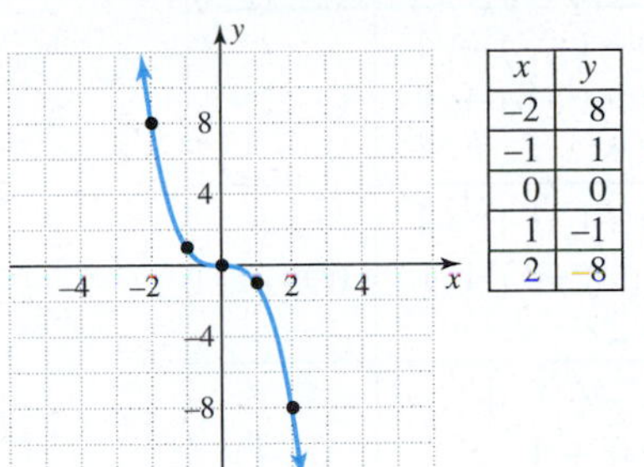

x	y
−2	8
−1	1
0	0
1	−1
2	−8

Graph of $f(x) = -x^4$

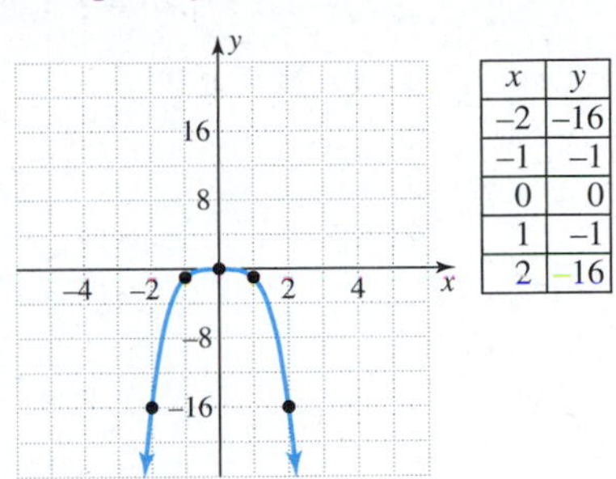

x	y
−2	−16
−1	−1
0	0
1	−1
2	−16

Observe that for polynomial functions of the same degree, the direction of the graph with a negative leading coefficient is opposite that of the graph with a positive leading coefficient.

Procedure: Graphing a Polynomial Function and Finding Its Domain and Range

Step 1: Make a table of at least five points. Include two negative values, zero, and two positive values for x. Find the corresponding y-value.
Step 2: Plot the points and connect the points with a smooth curve.
Step 3: The domain of any polynomial function is all real numbers, or $(-\infty, \infty)$.
Step 4: The range of an odd-degree polynomial function is all real numbers. The range of an even-degree polynomial function must be obtained from its graph.

Objective 4 Examples

Graph each polynomial function by plotting points. State the domain and range of each function.

4a. $f(x) = x^2 - 4$ **4b.** $f(x) = -x^3 + 1$

Solutions **4a.**

x	$y = x^2 - 4$	(x, y)
−2	$y = (-2)^2 - 4$ $= 4 - 4$ $= 0$	$(-2, 0)$
−1	$y = (-1)^2 - 4$ $= 1 - 4$ $= -3$	$(-1, -3)$
0	$y = (0)^2 - 4$ $= 0 - 4$ $= -4$	$(0, -4)$
1	$y = (1)^2 - 4$ $= 1 - 4$ $= -3$	$(1, -3)$
2	$y = (2)^2 - 4$ $= 4 - 4$ $= 0$	$(2, 0)$

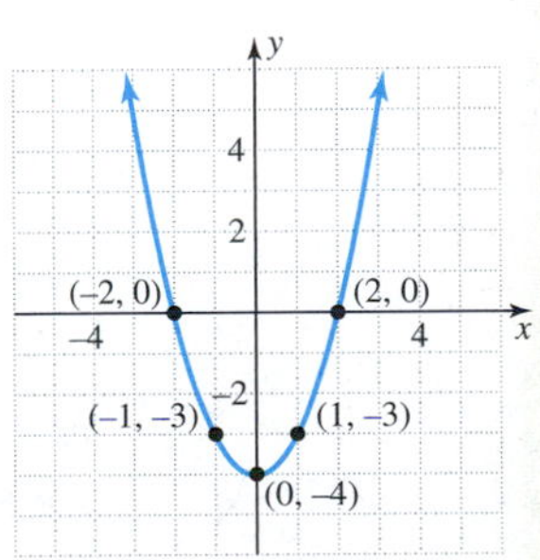

The domain of the function is $(-\infty, \infty)$. The range of the function is $[-4, \infty)$.

4b.

x	$y = -x^3 + 1$	(x, y)
-2	$y = -(-2)^3 + 1$ $= -(-8) + 1$ $= 9$	$(-2, 9)$
-1	$y = -(-1)^3 + 1$ $= -(-1) + 1$ $= 2$	$(-1, 2)$
0	$y = -(0)^3 + 1$ $= 0 + 1$ $= 1$	$(0, 1)$
1	$y = -(1)^3 + 1$ $= -1 + 1$ $= 0$	$(1, 0)$
2	$y = -(2)^3 + 1$ $= -8 + 1$ $= -7$	$(2, -7)$

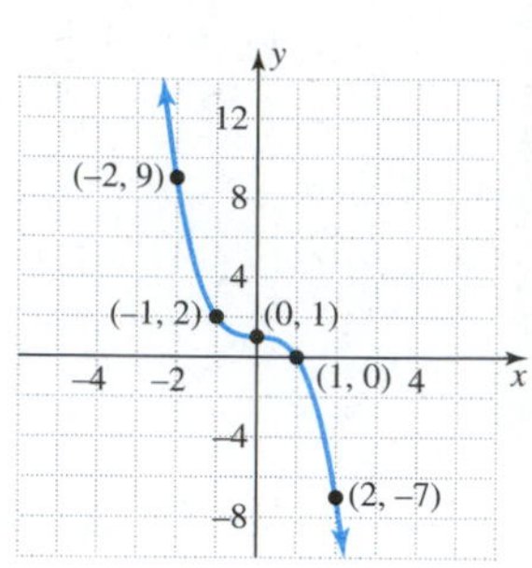

The domain of the function is $(-\infty, \infty)$. The range of the function is $(-\infty, \infty)$.

Student Check 4 Graph each polynomial function by plotting points. State the domain and range of each function.

a. $f(x) = x^4 + 2$ **b.** $f(x) = -x^5 - 3$

Objective 5

Troubleshoot common errors.

Troubleshooting Common Errors

Example 5 illustrates a common error associated with evaluating polynomial functions.

Objective 5 Example **A problem and an incorrect solution are given. Provide the correct solution and an explanation of the error.**

Find $f(-4)$ if $f(x) = -2x^2 + 3x - 5$.

Incorrect Solution	Correct Solution and Explanation
$f(-4) = -2(-4)^2 + 3(-4) - 5$ $f(-4) = 64 - 12 - 5$ $f(-4) = 47$	The mistake was made in evaluating the first term $-2(-4)^2$. Based on the order of operations, we square -4 first and then multiply by -2. $-2(-4)^2 = -2(16) = -32$ $f(-4) = -2(-4)^2 + 3(-4) - 5$ $f(-4) = -32 - 12 - 5$ $f(-4) = -49$

ANSWERS TO STUDENT CHECKS

Student Check 1 **a.** $-1, 1$ **b.** $6.7, 5$ **c.** $-\frac{1}{4}, 1$ **d.** $\frac{1}{5}, 2$ **e.** $-7, 0$ **f.** $4, 11$

Student Check 2 **a.** $-4y^4, y^3, 3y$; trinomial; degree $= 4$, quartic; leading coefficient $= -4$ **b.** $x^2, 16x$; binomial; degree $= 2$, quadratic; leading coefficient $= 1$ **c.** $9\pi r^3$; monomial; degree $= 3$, cubic; leading coefficient $= 9\pi$ **d.** $8r^2s^5, -7rs^8, 9rs$; trinomial; degree $= 9$, ninth-degree; leading coefficient: n/a

Student Check 3 **a.** -1 **b.** 42 **c.** $h(3) = 919$, $h(8) = 39$; After 3 sec the penny will be 919 ft above the ground and, after 8 sec, it will be 39 ft above the ground. **d.** $p(5) \approx 3.15$; The average price of gas was about \$3.15 in the first week of January 2011.

Student Check 4 **a.**

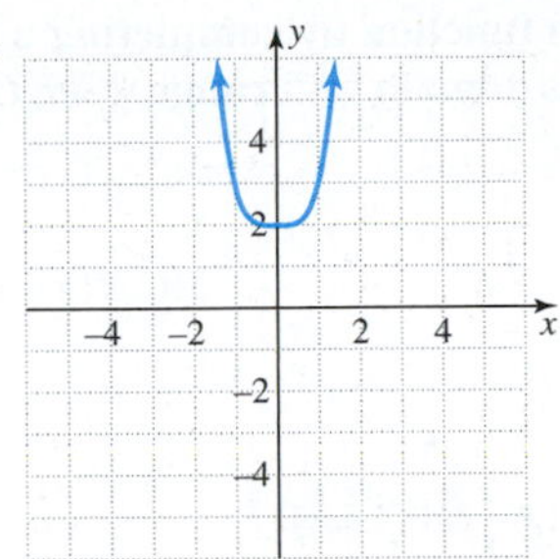

b.

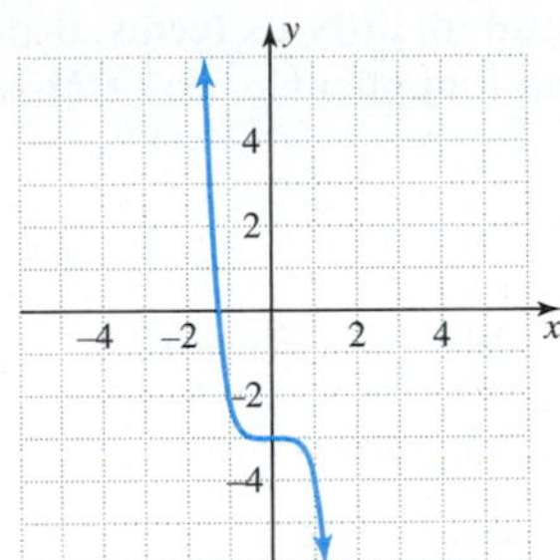

SUMMARY OF KEY CONCEPTS

1. An expression of the form ax^n is a term with coefficient a and degree n.
2. Polynomials consist of a finite sum of terms of the form ax^n.
   - If the polynomial consists of one term, it is a monomial.
   - If it consists of two terms, it is a binomial.
   - If it consists of three terms, it is a trinomial.
   - The degree of a term is the sum of the exponents of the variables.
   - The degree of a polynomial is the largest degree of the terms in the polynomial.
   - The leading coefficient of a single-variable polynomial is the coefficient of the term with the largest degree.
3. To evaluate a polynomial function at a given value, replace the variable with the given value and use the order of operations to simplify the expression.
4. The graph of a polynomial function is a smooth curve. Even degree polynomial functions go in the same direction at their ends. Odd-degree polynomial functions go in opposite directions at their ends.

GRAPHING CALCULATOR SKILLS

The graphing calculator can be used to evaluate and graph polynomial functions.

Example: Find $f(1)$ if $f(t) = -16t^2 + 32t + 100$.

Method 1: Enter the resulting numerical expression on the calculator.

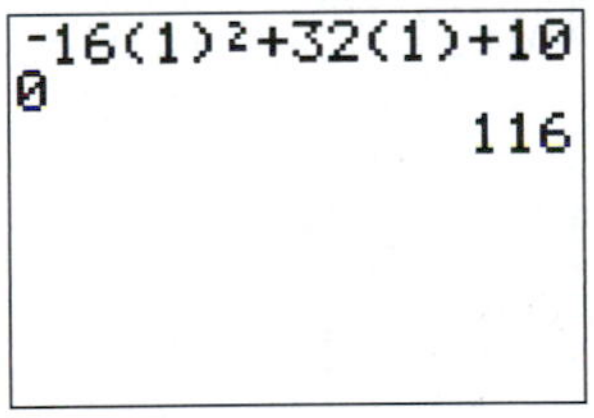

```
-16(1)²+32(1)+10
0
                116
```

So, $f(1) = 116$.

Method 2: Enter the function in the equation editor and use the table to evaluate it.

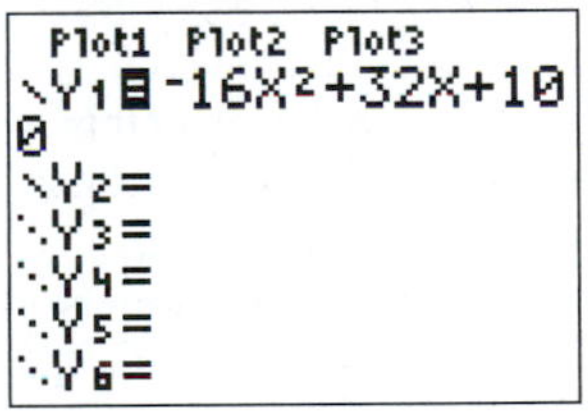

```
Plot1 Plot2 Plot3
\Y1=-16X²+32X+10
0
\Y2=
\Y3=
\Y4=
\Y5=
\Y6=
```

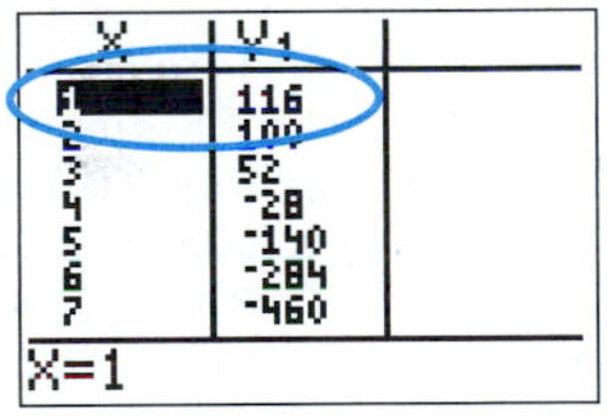

X	Y_1
1	116
2	100
3	52
4	-28
5	-140
6	-284
7	-460

X=1

So, $f(1) = 116$.

SECTION 6.3 EXERCISE SET

Write About It!

Use complete sentences in your answer to each exercise.

1. Define a second-degree polynomial.
2. Explain the difference between the degree of a term and the degree of a polynomial.
3. Explain the difference between the coefficient of a term and the leading coefficient.
4. Explain how to evaluate a polynomial function at a given x-value, i.e., $x = c$.

Additional answers can be found in the Instructor Answer Appendix.

Practice Makes Perfect!

Determine the coefficient and degree of each term. (*See Objective 1.*)

5. $2b^3$ 2, 3

6. $4.7z^2$ 4.7, 2

7. $-c^5$ −1, 5

8. $-w^4$ −1, 4

9. $-\dfrac{x^2}{3}$ $-\dfrac{1}{3}$, 2

10. $\dfrac{y^3}{4}$ $\dfrac{1}{4}$, 3

11. 12 12, 0

12. $\dfrac{1}{2}$ $\dfrac{1}{2}$, 0

13. $3a^5b^3$ 3, 8

14. $-2cd^4$ −2, 5

15. $-7.5u^2v^4w$ −7.5, 7

16. $12.3xy^6z^3$ 12.3, 10

Classify each polynomial and identify its terms, degree, type, and leading coefficient, if applicable. (*See Objective 2.*)

17. $3a^3 + 2a^2 - 5a + 7$

18. $-2p^3 + 5p^2 - 4p + 11$

19. $-11r^4 + 13r + 2$ **20.** $\frac{3}{2}s^2 - 7s + 10$

21. $32\,abc^2$ **22.** $9rs^2t$

23. $7 - 16s$ **24.** $2t - 8$

Evaluate each function. (*See Objective 3.*)

If $f(x) = 4x^2 - 5x + 1$ and $g(x) = -x^4 + 5x^2 + 3x - 10$, find the following.

25. $f(0)$ 1 **26.** $g(0)$ −10

27. $f(3)$ 22 **28.** $f(-2)$ 27

29. $g(-1)$ −9 **30.** $g(3)$ −37

31. $f(2) + g(1)$ 4 **32.** $f(-3) + g(-2)$ 40

33. $f(-1) - g(2)$ 10 **34.** $f(5) - g(-3)$ 131

If $P(x) = 12x^2 + 6x - 3$ and $Q(x) = -12x + 1$, find the following

35. $P\left(\frac{1}{2}\right)$ 3 **36.** $P\left(\frac{1}{3}\right)$ $\frac{1}{3}$

37. $Q\left(-\frac{1}{2}\right)$ 7 **38.** $Q\left(-\frac{1}{6}\right)$ 3

39. $P\left(\frac{1}{4}\right) + Q\left(\frac{1}{3}\right)$ $-\frac{15}{4}$ **40.** $P\left(\frac{1}{3}\right) + Q\left(-\frac{1}{4}\right)$ $\frac{13}{3}$

41. $P\left(\frac{3}{4}\right) - Q\left(\frac{2}{3}\right)$ $\frac{61}{4}$ **42.** $P\left(\frac{2}{3}\right) - Q\left(-\frac{5}{6}\right)$ $-\frac{14}{3}$

43. The Shanghai World Financial Center in China is 1614 ft high. If a penny is dropped from the top of this building, its height above the ground h, in feet, t seconds after it is dropped is given by $h(t) = -16t^2 + 1614$. Find $h(4)$ and $h(9)$ and interpret the results. (Source: http://www.emporis.com)

44. If a dime is dropped from the top of a building that is 509 m high, its height above the ground h, in meters, t seconds after it is dropped is given by $h(t) = -9.8t^2 + 509$. Find $h(5)$ and $h(7)$ and interpret the results.

45. Video game revenue, in billions of dollars, in the United States from 2002 to 2008 can be modeled by the function

$$R(x) = 0.004167x^3 + 0.5482x^2 - 1.6917x + 11.9286$$

where x is the number of years after 2002. Find $R(5)$ and $R(7)$ and interpret the results. (Source: http://www.bls.gov/data/)

46. The unemployment rate in percent for men over 20 yr old from 1999 to 2009 can be modeled by the function

$$f(x) = 0.04227x^3 - 0.5832x^2 + 2.2242x + 2.6042$$

where x is the number of years after 1999. Find $f(8)$ and $f(11)$ and interpret the results. (Source: http://www.bls.gov/data/)

Graph each function by completing a table of values and state its domain and range. (*See Objective 4.*)

47. $f(x) = -2x^2$ **48.** $f(x) = 4x$

49. $f(x) = \frac{1}{2}x^4$ **50.** $f(x) = -\frac{1}{3}x^3$

Mix 'Em Up!

Classify each polynomial and identify its terms, degree, type, and leading coefficient, if applicable.

51. $6.25b^2 - 0.64$ **52.** $\frac{1}{27}z^3 - 8$

53. $5a^3b^2c$ **54.** $-6p^2q^4r$

55. $0.1x^3 + 1.6x^2 - 2.9x + 1.8$

56. $-0.1y^3 - 0.9y^2 + 1.2y - 2.1$

57. $-\frac{x^2}{5}$ **58.** $\frac{v}{7}$

Evaluate each function.

If $f(x) = 2.4x^2 - 1.5x + 1.2$ and $g(x) = -3.6x + 4.8$, find the following.

59. $f(5)$ 53.7 **60.** $g(-5)$ 22.8

61. $f(-2)$ 13.8 **62.** $g(2)$ −2.4

63. $f(0) + g(-1)$ 9.6 **64.** $f(2) + g(0)$ 12.6

65. $f(-1) - g(1)$ 3.9 **66.** $f(6) - g(-6)$ 52.2

67. The average cost, in dollars, of electricity per 500 kilowatt hour (kWh) in a city from 1999 to 2008 can be modeled by the function

$$C(x) = -0.0168x^4 + 0.3194x^3 - 1.6895x^2 + 3.7336x + 45.5973$$

where x is the number of years after 1999. Find $C(7)$ and $C(10)$ and interpret the results. Round to the nearest cent.

68. The average quarterly domestic airline fuel consumption, in millions of gallons, from 2000 to 2009 can be modeled by the function

$$Q(x) = -0.6x^4 + 1.51x^3 + 60.94x^2 - 329.1x + 4561.2$$

where x is the number of years after 2000. Find $Q(6)$ and $Q(9)$ and interpret the results. Round to the nearest million gallons. (Source: http://www.transtats.bts.gov)

Graph each function by completing a table of values and state its domain and range.

69. $f(x) = \frac{1}{2}x^2 + 1$ **70.** $f(x) = -2x + 4$

71. $f(x) = -\frac{1}{4}x^3 + 2$ **72.** $f(x) = \frac{1}{4}x^4 - 4$

You Be the Teacher!

Correct each student's errors, if any.

73. Evaluate $g(-4)$ when $g(x) = -x^2 - 7x + 6$.

Chase's work:

$$\begin{aligned} g(-4) &= -(-4)^2 - 7(-4) + 6 \\ &= 16 + 28 + 6 \\ &= 50 \end{aligned}$$

$$\begin{aligned} g(-4) &= -(-4)^2 - 7(-4) + 6 \\ &= -16 + 28 + 6 \\ &= 18 \end{aligned}$$

74. Evaluate $P(2)$ when $P(x) = -4x^2 + 12x - 5$.

Josh's work:

$$\begin{aligned} P(2) &= -4(2)^2 + 12(2) - 5 \\ &= 16 + 24 - 5 \\ &= 35 \end{aligned}$$

$$\begin{aligned} P(2) &= -4(2)^2 + 12(2) - 5 \\ &= -16 + 24 - 5 \\ &= 3 \end{aligned}$$

Calculate It!

Correct the errors in each screen shot, if any.

75. Evaluate $f(-5)$ for $f(x) = -3x^2 + 10x - 7$.

```
-3*-5²+10*-5-7
            18
```

76. Evaluate $g(-1)$ for $g(x) = -x^2 + 6x - 4$.

```
--1²+6*-1-4
            -9
```

Use the TABLE feature to find each value.

77. $f(-3)$ for $f(x) = -2x^3 + 7x^2 - 5x + 4$ 136

78. $f(8)$ for $f(x) = x^3 - 6x^2 - 19x + 24$ 0

Think About It!

79. Write an example of a trinomial that has degree 3 with leading coefficient of −2. Answers vary; $-2x^3 + 5x - 1$

80. Write an example of a trinomial that has degree 4 with leading coefficient of 7. Answers vary; $7x^4 - 6x^3 + 3$

81. Write an example of a monomial that has degree 0. Answers vary; 6

82. Write an example of a monomial that has degree 1. Answers vary; $-4x$

Use the degree and leading coefficient of the polynomial function to match each function with its graph.

83. $y = x^2 + 2x - 8$ D

84. $y = -x^3 + 2x^2 + 5x - 6$ A

85. $y = -x^4 + 9x^2 + 4x - 12$ E

86. $y = x^3 - x^2 - 2x$ C

87. $y = -2x + 3$ B

88. $y = 2x - 4$ F

A.

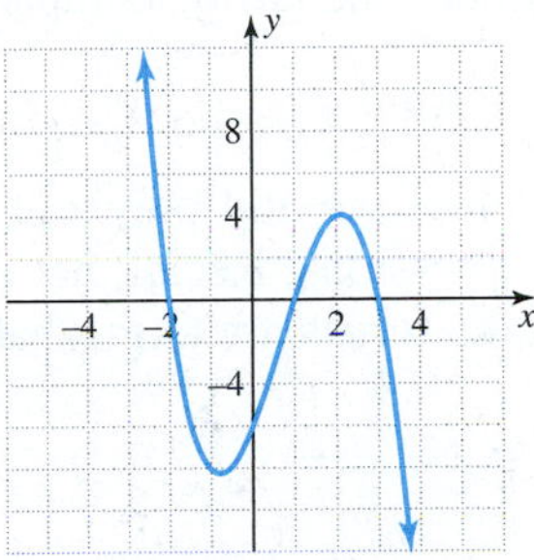

B.

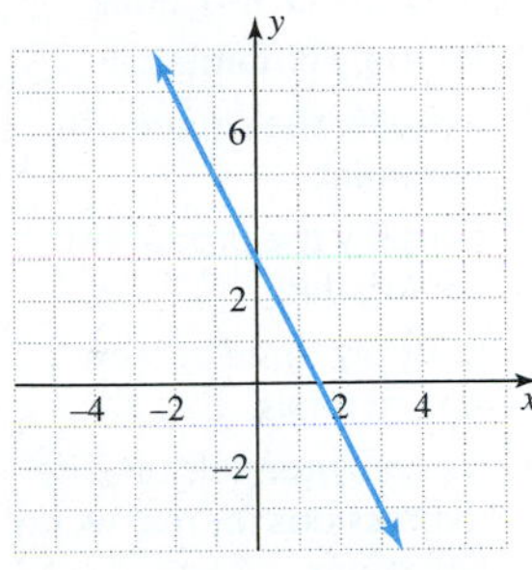

C.

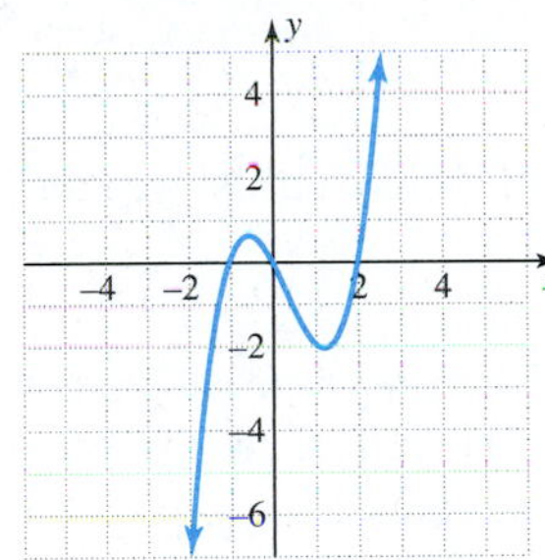

D.

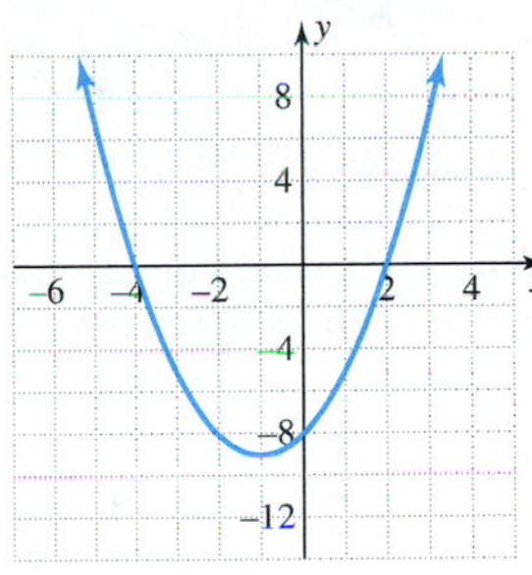

E.

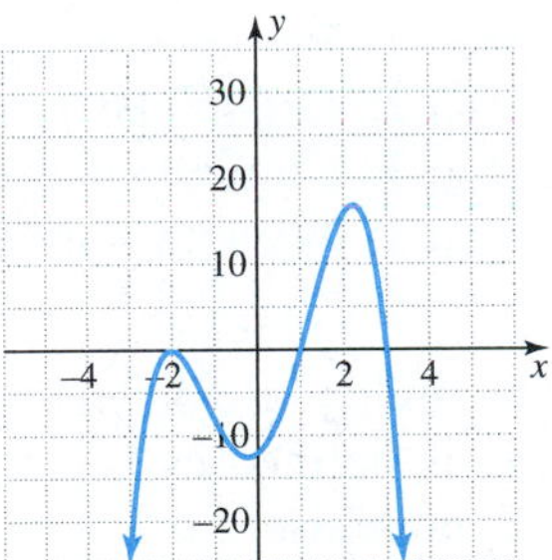

F.

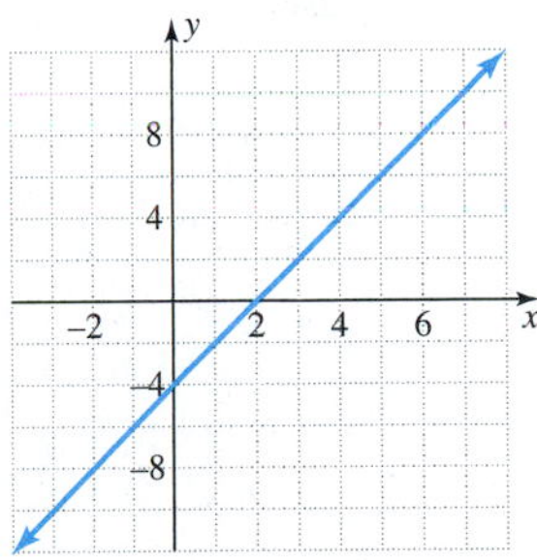

SECTION 6.4

Adding, Subtracting, and Multiplying Polynomials and Polynomial Functions

OBJECTIVES

As a result of completing this section, you will be able to

1. Add and subtract polynomials and polynomial functions.
2. Multiply polynomials and polynomial functions.
3. Multiply binomials.
4. Simplify the square of a binomial.
5. Simplify the product of conjugates.
6. Multiply three or more polynomials.
7. Use polynomials to write expressions for real-world situations.
8. Troubleshoot common errors.

When an airline company charges \$100 per seat, they can sell 120 tickets. For each \$10 increase in price, they sell one less ticket. If x represents the number of increases in price, then $100 + 10x$ is the expression for the ticket price and $120 - x$ is the number of tickets sold. In this section, we will learn how to write an expression for the amount of money the airline company earns for this trip.

Objective 1

Add and subtract polynomials and polynomial functions.

Adding and Subtracting Polynomials

Before we add polynomials, we will review some facts about simplifying algebraic expressions. Recall that only algebraic expressions with *like terms* can be combined. **Like terms** are terms that contain the same variables raised to the same exponents.

Like Terms	Unlike Terms
$-a, \frac{1}{2}a$	$3y^2, 3y$
$\frac{2x^2}{3}, -x^2$	$4x, 7$

To add like terms, we add their coefficients and the variable component remains the same.

$$6x^2 + 3x^2 = (6 + 3)x^2 = 9x^2$$

Since polynomials consist of terms, we add polynomials by combining the like terms of the polynomials. When polynomials are added or subtracted, it is most often the case that each polynomial will be written in a set of parentheses.

Procedure: Adding Polynomials

Step 1: Remove the parentheses.
Step 2: Group like terms together.
Step 3: Combine like terms.
Step 4: Write the answer in standard form.

The ability to add polynomials enables us to subtract them as well, since subtraction is defined as adding the opposite. The rule $a - b = a + (-b)$ also extends to polynomials.

Property: Subtracting Polynomials

If P and Q are polynomials, then

$$P - Q = P + (-Q)$$

To find the **opposite of a polynomial**, we multiply the polynomial by -1.

Opposite of $(x - 4) = -(x - 4) = -1(x - 4) = -x + 4$

Opposite of $(2y^2 - 3y + 5) = -(2y^2 - 3y + 5) = -1(2y^2 - 3y + 5) = -2y^2 + 3y - 5$

Notice that each sign of the polynomial is *changed* when the opposite is found.

Procedure: Subtracting Polynomials

Step 1: Find the opposite of the polynomial that is being subtracted.
Step 2: Combine like terms.
Step 3: Write the answer in standard form.

Note: We can add or subtract polynomial functions to obtain another polynomial function by simply adding or subtracting the two polynomials.

Objective 1 Examples

Perform each indicated operation.

1a. $(0.3x - 5.2) + (x + 9.4)$

1b. $(9y^2 - 3y + 7) + (-y^2 - 6y + 1)$

1c. $\left(\frac{3}{2}a^2 - \frac{1}{4}ab + 2b^2\right) + \left(\frac{5}{2}a^2 - \frac{3}{4}ab + b^2\right)$

1d. $(0.3x - 5.2) - (x + 9.4)$

1e. $(9y^2 - 3y + 7) - (-y^2 - 6y + 1)$

1f. $\left(\frac{3}{2}a^2 - \frac{1}{4}ab + 2b^2\right) - \left(\frac{5}{2}a^2 - \frac{3}{4}ab + b^2\right)$

1g. Let $f(x) = 3x^2 - 4x + 1$ and $g(x) = 2x - 3$. Find $f(x) + g(x)$ and $f(x) - g(x)$.

Solutions

1a. $(0.3x - 5.2) + (x + 9.4)$

$= 0.3x - 5.2 + x + 9.4$ Remove parentheses.
$= 0.3x + x - 5.2 + 9.4$ Group like terms.
$= 1.3x + 4.2$ Combine like terms. Recall $x = 1x$.

1b. $(9y^2 - 3y + 7) + (-y^2 - 6y + 1)$

$= 9y^2 - 3y + 7 - y^2 - 6y + 1$ Remove parentheses.
$= 9y^2 - y^2 - 3y - 6y + 7 + 1$ Group like terms.
$= 8y^2 - 9y + 8$ Combine like terms.

1c. $\left(\frac{3}{2}a^2 - \frac{1}{4}ab + 2b^2\right) + \left(\frac{5}{2}a^2 - \frac{3}{4}ab + b^2\right)$

$= \frac{3}{2}a^2 - \frac{1}{4}ab + 2b^2 + \frac{5}{2}a^2 - \frac{3}{4}ab + b^2$ Remove parentheses.

$= \frac{3}{2}a^2 + \frac{5}{2}a^2 - \frac{1}{4}ab - \frac{3}{4}ab + 2b^2 + b^2$ Group like terms.

$= \frac{8}{2}a^2 - \frac{4}{4}ab + 3b^2$ Combine like terms.

$= 4a^2 - ab + 3b^2$ Simplify each fraction.

1d. $(0.3x - 5.2) - (x + 9.4)$

$= (0.3x - 5.2) - 1(x + 9.4)$ Find the opposite by multiplying by -1.
$= 0.3x - 5.2 - x - 9.4$ Apply the distributive property.
$= 0.3x - x - 5.2 - 9.4$ Group like terms.
$= -0.7x - 14.6$ Combine like terms.

1e. $(9y^2 - 3y + 7) - (-y^2 - 6y + 1)$

$= (9y^2 - 3y + 7) - 1(-y^2 - 6y + 1)$ Find the opposite by multiplying by -1.
$= 9y^2 - 3y + 7 + y^2 + 6y - 1$ Apply the distributive property.
$= 9y^2 + y^2 - 3y + 6y + 7 - 1$ Group like terms.
$= 10y^2 + 3y + 6$ Combine like terms.

1f. $\left(\frac{3}{2}a^2 - \frac{1}{4}ab + 2b^2\right) - \left(\frac{5}{2}a^2 - \frac{3}{4}ab + b^2\right)$

$= \left(\frac{3}{2}a^2 - \frac{1}{4}ab + 2b^2\right) - 1\left(\frac{5}{2}a^2 - \frac{3}{4}ab + b^2\right)$ Find the opposite by multiplying by −1.

$= \frac{3}{2}a^2 - \frac{1}{4}ab + 2b^2 - \frac{5}{2}a^2 + \frac{3}{4}ab - b^2$ Apply the distributive property.

$= \frac{3}{2}a^2 - \frac{5}{2}a^2 - \frac{1}{4}ab + \frac{3}{4}ab + 2b^2 - b^2$ Group like terms.

$= -\frac{2}{2}a^2 + \frac{2}{4}ab + b^2$ Combine like terms.

$= -a^2 + \frac{1}{2}ab + b^2$ Simplify fractions.

1g. Since both $f(x)$ and $g(x)$ are polynomial functions, we add and subtract them by adding and subtracting the polynomials.

$f(x) + g(x)$	$f(x) - g(x)$
$= (3x^2 - 4x + 1) + (2x - 3)$	$= (3x^2 - 4x + 1) - (2x - 3)$
$= 3x^2 - 4x + 1 + 2x - 3$	$= (3x^2 - 4x + 1) - 1(2x - 3)$
$= 3x^2 - 4x + 2x + 1 - 3$	$= 3x^2 - 4x + 1 - 2x + 3$
$= 3x^2 - 2x - 2$	$= 3x^2 - 4x - 2x + 1 + 3$
	$= 3x^2 - 6x + 4$

Student Check 1 Perform each indicated operation.

a. $(2.1p - 3.9) + (7.5p - 5.1)$ **b.** $(y^2 - 9y + 10) + (y^2 - y + 1)$

c. $\left(\frac{6}{7}x^2 + \frac{3}{4}xy + 4y^2\right) + \left(\frac{1}{7}x^2 - \frac{7}{4}xy + 2y^2\right)$

d. $(2.1p - 3.9) - (7.5p - 5.1)$ **e.** $(y^2 - 9y + 10) - (y^2 - y + 1)$

f. $\left(\frac{6}{7}x^2 + \frac{3}{4}xy + 4y^2\right) - \left(\frac{1}{7}x^2 - \frac{7}{4}xy + 2y^2\right)$

g. Let $f(x) = -2x^2 - 6x - 5$ and $g(x) = -5x + 6$. Find $f(x) + g(x)$ and $f(x) - g(x)$.

Note: *We can also add and subtract polynomials vertically by aligning like terms.*

$(9y^2 - 3y + 7) + (-y^2 - 6y + 1)$

$$\begin{array}{r} 9y^2 - 3y + 7 \\ -y^2 - 6y + 1 \\ \hline 8y^2 - 9y + 8 \end{array}$$

$(9y^2 - 3y + 7) - (-y^2 - 6y + 1)$

$$\begin{array}{r} 9y^2 - 3y + 7 \\ -(-y^2 - 6y + 1) \\ \hline 10y^2 + 3y + 6 \end{array}$$

Multiplying Polynomials

Objective 2 ▶ Multiply polynomials and polynomial functions.

To multiply polynomials, we rely on skills we have learned in previous sections. The distributive property and the product of like bases rule for exponents provide the foundation for multiplying polynomials.

Recall the distributive property enables us to multiply an expression by a sum or difference.

$$a(b + c) = ab + ac$$

For example,

$$5(x + 4) = 5(x) + 5(4) = 5x + 20$$

The distributive property can be used when we multiply a binomial by a polynomial. We will, in fact, have to use the distributive property twice. For instance, to multiply $x + 3$ by the polynomial $x^2 + x + 6$, we distribute $x^2 + x + 6$ to each term in the binomial, x and 3. Then we apply the distributive property again to distribute x and 3 to each term in the trinomial.

INSTRUCTOR NOTE: Remind students that we must use the product of like bases rule, $a^m a^n = a^{m+n}$, to multiply exponential expressions with the same base.

$(x + 3)(x^2 + x + 6)$

$= x(x^2 + x + 6) + 3(x^2 + x + 6)$	Apply the distributive property.
$= x(x^2) + x(x) + x(6) + 3(x^2) + 3(x) + 3(6)$	Apply the distributive property.
$= x^3 + x^2 + 6x + 3x^2 + 3x + 18$	Multiply.
$= x^3 + x^2 + 3x^2 + 6x + 3x + 18$	Group like terms.
$= x^3 + 4x^2 + 9x + 18$	Add.

This multiplication results in each term of the binomial being multiplied by each term of the trinomial. So we can perform the multiplication by distributing the first term of the binomial to each term of the polynomial and then distributing the second term of the binomial to each term of the polynomial. This concept can be extended in order to multiply any two polynomials.

Procedure: Multiplying Polynomials

Step 1: Distribute the first term of the first polynomial to each term of the second polynomial.

Step 2: Distribute the second term of the first polynomial to each term of the second polynomial.

Step 3: Continue this process until you have distributed each term of the first polynomial.

Step 4: Remember to add exponents when multiplying like bases but add coefficients when adding like terms.

Objective 2 Examples

Multiply the polynomials and simplify the result.

2a. $-6x(2x - 3)$ **2b.** $3y^2(6y^2 - 2y + 9)$

2c. $-2a^3b^2(4a^2 + 5ab - 2b^2)$ **2d.** $(x + 3)(x + 4)$

2e. $(x - 2)(x^2 + 2x + 4)$

2f. Let $f(x) = x^2 + 2x - 4$ and $g(x) = 8x + 3$. Find $f(x)g(x)$.

Solutions

2a.

$-6x(2x - 3) = -6x(2x) - (-6x)(3)$	Apply the distributive property.
$= -12x^2 + 18x$	Simplify.

2b.

$3y^2(6y^2 - 2y + 9) = 3y^2(6y^2) - 3y^2(2y) + 3y^2(9)$	Apply the distributive property.
$= 3 \cdot 6y^2y^2 - 3 \cdot 2y^2y^1 + 3 \cdot 9y^2$	Multiply coefficients and like bases.
$= 18y^4 - 6y^3 + 27y^2$	Simplify.

2c. Distribute $-2a^3b^2$ to each term in the trinomial and simplify each product.

$$\begin{aligned}-2a^3b^2(4a^2 + 5ab - 2b^2) &= -2a^3b^2(4a^2) + (-2a^3b^2)(5ab) - (-2a^3b^2)(2b^2)\\ &= -2 \cdot 4a^3a^2b^2 - 2 \cdot 5a^3a^1b^2b^1 + 2 \cdot 2a^3b^2b^2\\ &= -8a^5b^2 - 10a^4b^3 + 4a^3b^4\end{aligned}$$

INSTRUCTOR NOTE: Caution students that $x \cdot x = x^2$ but $x + x = 2x$.

2d. Distribute each term in the first binomial to each term in the second binomial. Simplify each product and combine like terms.

$$\begin{aligned}(x + 3)(x + 4) &= x(x) + x(4) + 3(x) + 3(4)\\ &= x^2 + 4x + 3x + 12\\ &= x^2 + 7x + 12\end{aligned}$$

Note: *Distributing* $x + 3$ *to each term of the binomial* $x + 4$ *gives us*

$$\begin{aligned}(x + 3)(x + 4) &= (x + 3)(x) + (x + 3)(4)\\ &= x(x) + 3(x) + x(4) + 3(4)\end{aligned}$$

This is equivalent to distributing each term in the first binomial to each term in the second binomial.

2e. Distribute each term in the binomial to each term in the trinomial. Simplify each product and combine like terms.

$$\begin{aligned}(x - 2)(x^2 + 2x + 4) &= x(x^2) + x(2x) + x(4) - 2(x^2) - 2(2x) - 2(4)\\ &= x^3 + 2x^2 + 4x - 2x^2 - 4x - 8\\ &= x^3 - 8\end{aligned}$$

2f. We find the product of the functions by multiplying the polynomials.

$$\begin{aligned}f(x)g(x) &= (x^2 + 2x - 4)(8x + 3)\\ &= x^2(8x) + x^2(3) + 2x(8x) + 2x(3) - 4(8x) - 4(3)\\ &= 8x^3 + 3x^2 + 16x^2 + 6x - 32x - 12\\ &= 8x^3 + 19x^2 - 26x - 12\end{aligned}$$

Student Check 2 Multiply the polynomials and simplify the result.

a. $-7y(4y - 3)$ **b.** $8x^2(2x^2 - 5x + 1)$ **c.** $6a^2b^2(3a^2 - 2ab + b^2)$

d. $(x + 5)(x + 6)$ **e.** $(a - 4)(a^2 + 4a + 16)$

f. Let $f(x) = y^2 - 5y + 7$ and $g(x) = 2y - 9$. Find $f(x)g(x)$.

Multiplying Binomials

Objective 3 Multiply binomials.

In Example 2d, we multiplied binomials using the distributive property. This type of product repeatedly comes up in algebra, so we observe that there is a special order of multiplying the terms of two binomials. This is called the *FOIL method*. Recall

$$(x + 3)(x + 4) = x(x) + x(4) + 3(x) + 3(4) = x^2 + 7x + 12$$

This product was formed by distributing x to $(x + 4)$ and then 3 to $(x + 4)$. To ensure that we get all of the terms in the product, we can remember the word FOIL.

FOIL = First + Outer + Inner + Last

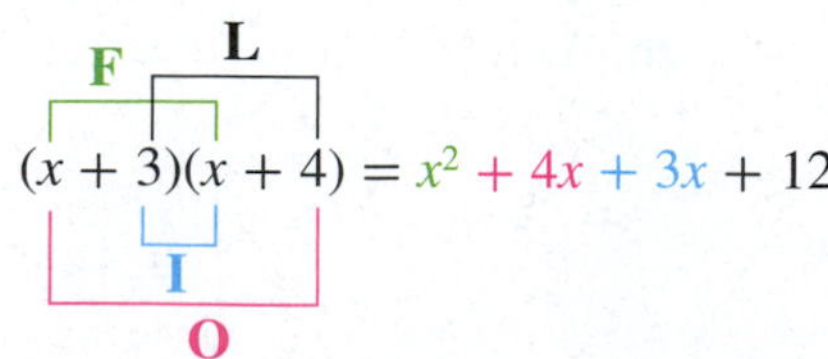

Procedure: Multiplying Two Binomials Using FOIL

Step 1: Find the product of the two first (F) terms of the binomials.
Step 2: Find the product of the two outer (O) terms of the binomials.
Step 3: Find the product of the two inner (I) terms of the binomials.
Step 4: Find the product of the two last (L) terms of the binomials.
Step 5: Add the products and combine like terms.

Note: *This method only applies to the product of two binomials and does not extend to other types of products. Also note that this method is exactly what was covered in the last objective; it is simply that we are giving the process a name, the FOIL method.*

Objective 3 Examples Use the FOIL method to multiply the binomials.

3a. $(y - 5)(y - 7)$ **3b.** $(2x - 3y)(5x + 2y)$ **3c.** $(a^2 + 9)(a^2 - 4)$

Solutions

3a.

$$\begin{aligned} & \quad\ \ \text{F} \qquad\quad \text{O} \qquad\quad \text{I} \qquad\quad \text{L} \\ (y - 5)(y - 7) &= (y)(y) + (y)(-7) - 5(y) - 5(-7) \\ &= y^2 - 7y - 5y + 35 \\ &= y^2 - 12y + 35 \end{aligned}$$

3b.

$$\begin{aligned} & \quad\ \ \text{F} \qquad\quad \text{O} \qquad\quad \text{I} \qquad\quad \text{L} \\ (2x - 3y)(5x + 2y) &= (2x)(5x) + (2x)(2y) - 3y(5x) - 3y(2y) \\ &= 10x^2 + 4xy - 15xy - 6y^2 \\ &= 10x^2 - 11xy - 6y^2 \end{aligned}$$

3c.

$$\begin{aligned} & \quad\ \ \text{F} \qquad\quad \text{O} \qquad\quad \text{I} \qquad\quad \text{L} \\ (a^2 + 9)(a^2 - 4) &= (a^2)(a^2) - (a^2)(4) + 9(a^2) + 9(-4) \\ &= a^4 - 4a^2 + 9a^2 - 36 \\ &= a^4 + 5a^2 - 36 \end{aligned}$$

Student Check 3 Use the FOIL method to multiply the binomials.

a. $(a + 6)(a + 2)$ **b.** $(4x - 7y)(3x + 8y)$ **c.** $(a^2 - 2)(a^2 + 10)$

Squaring a Binomial

Objective 4 ▶

Simplify the square of a binomial.

In Sections 6.1 and 6.2, we learned properties of exponents that enable us to simplify products and quotients raised to exponents. We will now learn how to find the square of a binomial, that is the square of a sum or difference of two terms.

Consider the expression $(x + 4)^2$. To simplify this expression, we must apply the definition of the exponent. The exponent 2 tells us to multiply two factors of the base, $x + 4$. This results in the product of two binomials, which we can simplify by applying the distributive property or the FOIL method.

$$\begin{aligned} & \qquad\qquad\qquad\qquad\qquad\quad \text{F} \qquad\quad \text{O} \qquad\quad \text{I} \qquad\quad \text{L} \\ (x + 4)^2 = (x + 4)(x + 4) &= (x)(x) + x(4) + 4(x) + 4(4) \\ &= x^2 + 4x + 4x + 16 \\ &= x^2 + 8x + 16 \end{aligned}$$

So, $(x + 4)^2 = x^2 + 8x + 16$.

Distribution will always provide the result of squaring a binomial but there is also a rule that can be applied. From the preceding example, we observe the following about squaring a binomial.

1. The square of a binomial is a trinomial.
2. The first term of the trinomial is the square of the first term of the binomial. x^2 is the first term of the trinomial and is the first term of the binomial squared: $(x)^2 = x^2$.
3. The middle term of the trinomial is twice the product of the terms of the binomial. $8x$ is the middle term and is twice the product of the terms in the binomial: $2(x)(4) = 8x$.
4. The last term of the trinomial is the square of the last term of the binomial. 16 is the last term and is the second term of the binomial squared: $(4)^2 = 16$.

We can generalize these observations as follows.

Property: The Square of a Binomial

For a and b real numbers,

$$(a + b)^2 = a^2 + 2ab + b^2$$
$$(a - b)^2 = a^2 - 2ab + b^2$$

Geometrically, we can think of squaring a binomial as calculating the area of a square whose side has a length of $a + b$.

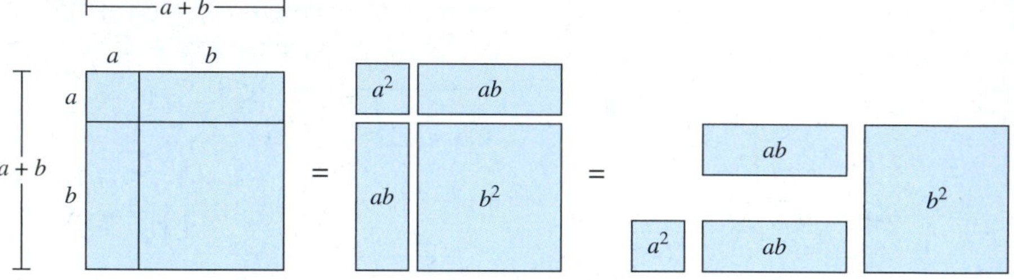

The area of the large square is

$$A = (a + b)(a + b) = (a + b)^2$$

This is also equivalent to the sum of the areas of the four rectangles, that is,

$$A = a^2 + ab + ab + b^2 = a^2 + 2ab + b^2$$

The trinomials that result from squaring a binomial are called **perfect square trinomials**.

Procedure: Squaring a Binomial

Step 1: Square the first term.
Step 2: Find twice the product of the terms in the binomial.
Step 3: Square the last term.
Step 4: Add the products and combine like terms.
Step 5: Verify the answer by multiplying.

Objective 4 Examples Simplify each expression.

Problems	Solutions
4a. $(x + 8)^2$	$(x + 8)^2 = (x)^2 + 2(x)(8) + (8)^2 = x^2 + 16x + 64$
4b. $(9y - 4)^2$	$(9y - 4)^2 = (9y)^2 - 2(9y)(4) + (4)^2 = 81y^2 - 72y + 16$
4c. $(m - 2n)^2$	$(m - 2n)^2 = (m)^2 - 2(m)(2n) + (2n)^2 = m^2 - 4mn + 4n^2$
4d. $(x^2 + 3)^2$	$(x^2 + 3)^2 = (x^2)^2 + 2(x^2)(3) + (3)^2 = x^4 + 6x^2 + 9$

Student Check 4 Simplify each expression.

a. $(y - 8)^2$ **b.** $(4x + 2)^2$ **c.** $(m + 5n)^2$ **d.** $(a^2 - 1)^2$

Multiplying Conjugates

Objective 5 Simplify the product of conjugates.

There is another special case that involves the product of two binomials, the product of conjugates. *Conjugates* are binomials with the same terms connected by different operations (one a sum and one a difference).

Definition: The expressions $a + b$ and $a - b$ are **conjugates**.

Some examples of conjugates are

$$x + 3 \quad \text{and} \quad x - 3$$
$$4b + 5 \quad \text{and} \quad 4b - 5$$

To find the product of conjugates, we multiply the binomials. We can use the FOIL method, as shown.

$$(x + 3)(x - 3) = x^2 - 3x + 3x - 9 = x^2 - 9$$

$$(4b + 5)(4b - 5) = 16b^2 - 20b + 20b - 25 = 16b^2 - 25$$

When we multiply conjugates using the FOIL method, the outer product and inner product are additive inverses and, therefore, add to zero. The product of the first terms minus the product of the last terms remains. Since the first and last terms are the same in conjugates, this is equivalent to the *difference of the squares* of the first term and last term, respectively.

$$(x + 3)(x - 3) = (x)^2 - (3)^2 = x^2 - 9$$
$$(4b + 5)(4b - 5) = (4b)^2 - (5)^2 = 16b^2 - 25$$

Property: The Product of Conjugates

$$(a + b)(a - b) = a^2 - b^2$$

Even though the product of conjugates is a special case, we can multiply the binomials using the distributive property or by the FOIL method rather than memorizing a rule, if preferred.

Procedure: Multiplying Conjugates

Step 1: Find the square of the first term.
Step 2: Find the square of the last term.
Step 3: Subtract the results from step 1 and step 2.

Objective 5 Examples Simplify each product.

Problems	Solutions
5a. $(x + 6)(x - 6)$	$(x + 6)(x - 6) = (x)^2 - (6)^2 = x^2 - 36$
5b. $(5m - 2n)(5m + 2n)$	$(5m - 2n)(5m + 2n) = (5m)^2 - (2n)^2 = 25m^2 - 4n^2$
5c. $(a^2 - 5)(a^2 + 5)$	$(a^2 - 5)(a^2 + 5) = (a^2)^2 - (5)^2 = a^4 - 25$
5d. $\left(2b^3 - \frac{9}{2}\right)\left(2b^3 + \frac{9}{2}\right)$	$\left(2b^3 - \frac{9}{2}\right)\left(2b^3 + \frac{9}{2}\right) = (2b^3)^2 - \left(\frac{9}{2}\right)^2 = 4b^6 - \frac{81}{4}$

Student Check 5 Simplify each product.

a. $(x + 1)(x - 1)$ **b.** $(9m - 7n)(9m + 7n)$

c. $(a^3 - 6)(a^3 + 6)$ **d.** $\left(3b^2 - \frac{4}{5}\right)\left(3b^2 + \frac{4}{5}\right)$

Objective 6 ▸ Multiplying Three or More Polynomials

Multiply three or more polynomials.

To multiply three or more polynomials, multiply two of the polynomials together and then multiply this result by the remaining polynomial or polynomials.

Objective 6 Examples **Simplify each product.**

6a. $(x + 5)(x - 5)(2x + 3)$ **6b.** $(x + 2)^3$

Solutions

6a. $(x + 5)(x - 5)(2x + 3) = (x^2 - 25)(2x + 3)$ Multiply the conjugates.

$= 2x^3 + 3x^2 - 50x - 75$ Apply the FOIL method.

6b. $(x + 2)^3 = (x + 2)^2 (x + 2)$ Rewrite $(x + 2)^3$ as $(x + 2)^2 (x + 2)$.

$= (x^2 + 4x + 4)(x + 2)$ Apply the property for squaring a binomial.

$= x^3 + 2x^2 + 4x^2 + 8x + 4x + 8$ Apply the distributive property.

$= x^3 + 6x^2 + 12x + 8$ Combine like terms.

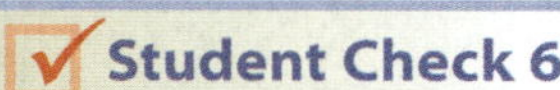

Student Check 6 Simplify each product.

a. $(4x + 7)(x - 2)(x + 2)$ **b.** $(5x + 3)^3$

Objective 7 ▸ Applications

Use polynomials to write expressions for real-world situations.

Polynomials are used to represent real-life situations. Polynomial models occur in geometry, physics, engineering, and business. In Example 7, we write a polynomial expression that represents each situation.

Objective 7 Examples **Write a polynomial that represents each situation and use it to find the requested information.**

7a. One side of a triangle is 2 ft shorter than the cube of the shortest side. The other side is 3 ft longer than the square of the shortest side. Find a polynomial that represents the perimeter of the triangle. What is the perimeter if the shortest side is 2 ft?

Solution

7a. Let x represent the length of the shortest side of the triangle. We translate the statements to obtain an expression for the other two sides of the triangle.

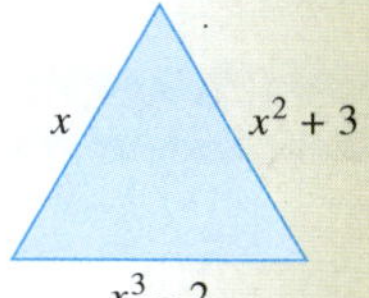

2 ft shorter than the cube of the shortest side, x, is $x^3 - 2$.

3 ft longer than the square of the shortest side, x, is $x^2 + 3$.

The perimeter is the sum of the length of the sides.

$P = x + (x^3 - 2) + (x^2 + 3)$ Find the sum of the three sides.

$P = x + x^3 - 2 + x^2 + 3$ Add the polynomials.

$P = x^3 + x^2 + x + 1$ Combine like terms.

If the shortest side is 2 ft, then the perimeter is

$P = x^3 + x^2 + x + 1$	State the expression for perimeter.
$P = (2)^3 + (2)^2 + (2) + 1$	Substitute 2 for x.
$P = 8 + 4 + 2 + 1$	Simplify the exponents.
$P = 15$ ft	Add.

So, the perimeter of the triangle whose shortest side is 2 ft is 15 ft.

7b. The operator of a hot dog stand can make hot dogs for \$0.25 each. The cost to operate his hot dog stand is \$500 per month. So, the monthly cost of operating the hot dog stand is $C(h) = 0.25h + 500$, where h is the number of hot dogs made in a month. Each hot dog sells for \$0.99. The revenue (in dollars) made from selling h hot dogs in a month is $R(h) = 0.99h$. Write a polynomial function that represents the profit (in dollars) from selling h hot dogs in a month. What is the profit from selling 800 hot dogs in a month? [Note: Profit, $P(h)$, is defined as revenue minus cost, $R(h) - C(h)$.]

Solution **7b.**

$P(h) = R(h) - C(h)$	Write the profit function.
$P(h) = 0.99h - (0.25h + 500)$	Replace $R(h)$ and $C(h)$ with the appropriate functions.
$P(h) = 0.99h - 1(0.25h + 500)$	Subtract by multiplying the second polynomial by -1.
$P(h) = 0.99h - 0.25h - 500$	Apply the distributive property.
$P(h) = 0.74h - 500$	Combine like terms.

The profit from selling 800 hot dogs in a month is represented by $P(800)$.

$P(h) = 0.74h - 500$	State the polynomial function.
$P(800) = 0.74(800) - 500$	Replace h with 800.
$P(800) = 92$	Simplify.

So, the profit from selling 800 hot dogs in a month is \$92.

7c. When an airline company charges \$100 per seat, they can sell 120 tickets. For each \$10 increase in price, they sell one less ticket. If x represents the number of increases in price, then $100 + 10x$ is the expression for the ticket price (in dollars) and $120 - x$ is the number of tickets sold. What expression represents the amount of revenue (in dollars) the airline company earns for the trip? (Note: Revenue, R, is calculated by multiplying the price per ticket by the number of tickets sold.)

Solution **7c.**

$R = \text{price} \times \text{number}$	Write the revenue equation.
$R = (100 + 10x)(120 - x)$	Substitute the expressions for price and number.
$R = 100(120) - 100(x) + 10x(120) - 10x(x)$	Apply the FOIL method.
$R = 120{,}000 - 100x + 1200x - 10x^2$	Simplify each product.
$R = -10x^2 + 1100x + 120{,}000$	Combine like terms.

So, the polynomial $-10x^2 + 1100x + 120{,}000$ represents the amount of revenue the airline company earns for the trip.

7d. A homeowner plans to remodel his master bathroom. He wants to use a more expensive tile to make a uniform border around the floor of the shower stall.

If the shower stall measures 3 ft by 4 ft, what expression represents the area of the shower floor that is not covered by the expensive tile?

Solution

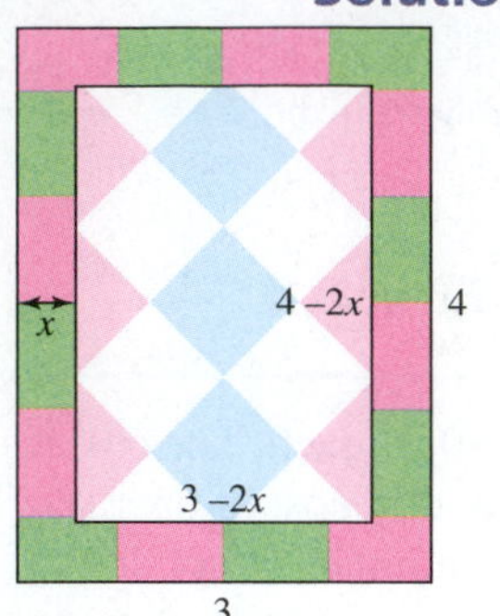

7d. Let x be the width (in feet) of the border with the expensive tile. Then the dimensions of the floor not covered by the expensive tile are $3 - 2x$ and $4 - 2x$.

$A = lw$	State the area formula.
$A = (3 - 2x)(4 - 2x)$	Replace length and width with the appropriate expressions.
$A = 3(4) - 3(2x) - 2x(4) - 2x(-2x)$	Apply the FOIL method.
$A = 12 - 6x - 8x + 4x^2$	Simplify each product.
$A = 4x^2 - 14x + 12$	Combine like terms.

So, the polynomial $4x^2 - 14x + 12$ (in ft^2) represents the area of the floor not covered by the expensive tile.

Student Check 7 Write a polynomial that represents each situation and use it to find the requested information.

a. Write a polynomial function that represents the perimeter of a trapezoid with sides of lengths $y + 2$, $4y^2 + y + 5$, $2y - 3$, and $3y^2 - 2y + 1$ units.

b. A Kiwanis club is organizing a pancake breakfast as a fund-raiser. The cost of preparing the breakfast is \$1.25 per person and the cost to rent the location for the breakfast is \$200. So, the total cost (in dollars) is $C(x) = 1.25x + 200$, where x is the number of people who attend the breakfast. A ticket to attend the breakfast is \$6 per person. So, the revenue (in dollars) is $R(x) = 6x$, where x is the number of tickets sold. Write a polynomial function that represents the profit for the fund-raiser. What is the profit from selling 100 tickets?

c. An apartment manager can rent all 150 units if he charges \$400 per month rent. If he increases the price by \$50, he rents two fewer apartments. If x represents the number of increases in price, then the rent (in dollars) for an apartment is given by $400 + 50x$ and the number of apartments rented is $150 - 2x$. What polynomial represents the total rent (in dollars) collected each month?

d. A gardener has a rectangular garden that measures 5 ft by 7 ft. She wants to put a uniform border of gravel around each side of the garden. If x represents the width of the border (in feet), what polynomial represents the area of the garden and the border?

Objective 8 ▶

Troubleshoot common errors.

Troubleshooting Common Errors

Example 8 illustrates some common errors associated with operations of polynomials.

Objective 8 Examples

A problem and an incorrect solution are given. Provide the correct solution and an explanation of the error.

8a. Simplify $(x^2 + 3x - 4) + (x^2 - 7x + 1)$.

Incorrect Solution	Correct Solution and Explanation
$(x^2 + 3x - 4) + (x^2 - 7x + 1)$ $= x^2 + 3x - 4 + x^2 - 7x + 1$ $= x^2 + x^2 + 3x - 7x - 4 + 1$ $= x^4 - 4x - 3$	The sum of x^2 and x^2 is not x^4 These are like terms so their coefficients should be added to get $1x^2 + 1x^2 = 2x^2$. $(x^2 + 3x - 4) + (x^2 - 7x + 1)$ $= x^2 + 3x - 4 + x^2 - 7x + 1$ $= x^2 + x^2 + 3x - 7x - 4 + 1$ $= 2x^2 - 4x - 3$

8b. Simplify $(x^2 + 3x - 4) - (x^2 - 7x + 1)$.

Incorrect Solution	Correct Solution and Explanation
$(x^2 + 3x - 4) - (x^2 - 7x + 1)$ $= x^2 + 3x - 4 - x^2 - 7x + 1$ $= x^2 - x^2 + 3x - 7x - 4 + 1$ $= -4x - 3$	Each sign of the polynomial being subtracted will change. $(x^2 + 3x - 4) - (x^2 - 7x + 1)$ $= x^2 + 3x - 4 - x^2 + 7x - 1$ $= x^2 - x^2 + 3x + 7x - 4 - 1$ $= 10x - 5$

8c. Simplify $(x + 6)^2$.

Incorrect Solution	Correct Solution and Explanation
$(x + 6)^2 = x^2 + 36$	We cannot simply square each term of the binomial. We must either multiply the two binomials or use the squaring pattern. $(x + 6)^2 = (x + 6)(x + 6)$ $= x^2 + 6x + 6x + 36$ $= x^2 + 12x + 36$

ANSWERS TO STUDENT CHECKS

Student Check 1 **a.** $9.6p - 9$
b. $2y^2 - 10y + 11$ **c.** $x^2 - xy + 6y^2$
d. $-5.4p + 1.2$ **e.** $-8y + 9$
f. $\frac{5}{7}x^2 + \frac{5}{2}xy + 2y^2$
g. $-2x^2 - 11x + 1$; $-2x^2 - x - 11$

Student Check 2 **a.** $-28y^2 + 21y$
b. $16x^4 - 40x^3 + 8x^2$ **c.** $18a^4b^2 - 12a^3b^3 + 6a^2b^4$
d. $x^2 + 11x + 30$ **e.** $a^3 - 64$
f. $2y^3 - 19y^2 + 59y - 63$

Student Check 3 **a.** $a^2 + 8a + 12$
b. $12x^2 + 11xy - 56y^2$ **c.** $a^4 + 8a^2 - 20$

Student Check 4 **a.** $y^2 - 16y + 64$ **b.** $16x^2 + 16x + 4$
c. $m^2 + 10mn + 25n^2$ **d.** $a^4 - 2a^2 + 1$

Student Check 5 **a.** $x^2 - 1$ **b.** $81m^2 - 49n^2$
c. $a^6 - 36$ **d.** $9b^2 - \frac{16}{25}$

Student Check 6 **a.** $4x^3 + 7x^2 - 16x - 28$
b. $125x^3 + 225x^2 + 135x + 27$

Student Check 7 **a.** $P = 7y^2 + 2y + 5$
b. $P(x) = 4.75x - 200$; $P(100) = \$275$
c. $60{,}000 + 6700x - 100x^2$ **d.** $4x^2 + 24x + 35$

SUMMARY OF KEY CONCEPTS

1. Polynomials are added by combining like terms. When we add, the exponent stays the same and the coefficients are added.
2. To subtract two polynomials, we add the first polynomial to the opposite of the second polynomial. The opposite is found by multiplying the polynomial by -1.
3. All of the products discussed in this section can be found by applying the distributive property, but we can also use the patterns shown in this section.
   - Two binomials can be multiplied by using the FOIL method. FOIL stands for First, Outer, Inner, and Last.
   - The square of a binomial is *always* a trinomial. The middle term results from twice the product of the terms in the binomial.
   - Conjugates are expressions of the form $a + b$ and $a - b$. Their product is the difference of two squares.
4. We multiply three or more polynomials by first multiplying two of the polynomials together. Then we multiply the result by the remaining polynomials one at a time.

GRAPHING CALCULATOR SKILLS

The graphing calculator can be used to verify polynomial arithmetic.

Example: Verify that $(x^2 - x + 6) - (x^2 + 4x - 3) = -5x + 9$.

Solution: Enter the given problem in the equation editor for Y_1 and the result in Y_2. Then compare the table entries. If the columns agree, then the polynomial operation has been performed correctly.

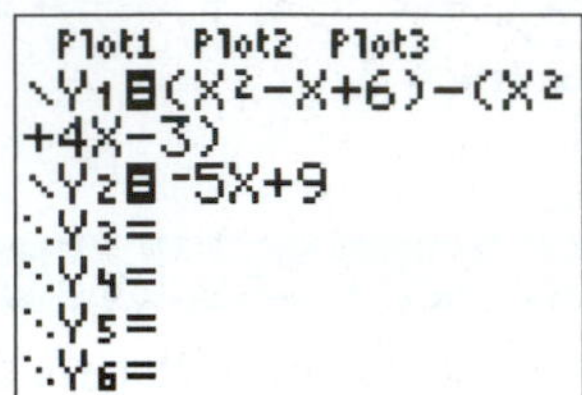

X	Y1	Y2
1	4	4
2	-1	-1
3	-6	-6
4	-11	-11
5	-16	-16
6	-21	-21
7	-26	-26

X=1

2nd GRAPH

Since the Y_1 and Y_2 columns agree for each value of x, we can conclude that the subtraction is correct.

SECTION 6.4 EXERCISE SET

Write About It!

Use complete sentences in your answer to each exercise.

1. Explain how to add two polynomials.
2. Explain how to subtract two polynomials.
3. Explain how to subtract two polynomials vertically.
4. Give an example of a common mistake made by students when subtracting two polynomials.
5. Multiply the polynomials (a) $(2x + y)(4x^2 - 2xy + y^2)$ and (b) $(2x + y)(4x^2 + 2xy - y^2)$ and explain the difference in the answers.
6. Explain why $(2x + 5y)^2 \neq 4x^2 + 25y^2$.
7. Explain why $(a + b)^3 \neq a^3 + b^3$.
8. Explain why $(x + y)^2 \neq (x + y)(x - y)$.

Practice Makes Perfect!

Perform the indicated operation. (*See Objective 1.*)

9. $(11y - 13) + (-20y - 3)$ $-9y - 16$
10. $(-10x - 4) + (23x + 13)$ $13x + 9$
11. $(-2y + 7) - (-15y - 8)$ $13y + 15$
12. $(-3x + 4) - (5x - 15)$ $-8x + 19$
13. $(7.1a - 3.8) + (-4.3a + 1.4)$ $2.8a - 2.4$
14. $(-9.4b - 7.5) + (6.2b - 12.1)$ $-3.2b - 19.6$
15. $(0.6a + 10.3) - (2.1a + 6.1)$ $-1.5a + 4.2$
16. $(9.7b - 0.1) - (-3.1b + 2)$ $12.8b - 2.1$
17. $(3y^2 - 12y + 25) + (-11y^2 + 2y - 5)$ $-8y^2 - 10y + 20$
18. $(11x^2 - 20x - 14) + (18x^2 + 7x - 17)$ $29x^2 - 13x - 31$
19. $(9x^2 + 14x + 25) - (-15x^2 + 28x - 15)$ $24x^2 - 14x + 40$
20. $(-9y^2 - 4y + 4) - (25y^2 - 9y + 14)$ $-34y^2 + 5y - 10$
21. $(8.2q^2 + 7.5q + 7.8) + (-8.8q^2 - 14.6q + 1.8)$ $-0.6q^2 - 7.1q + 9.6$
22. $(0.6p^2 - 6.4p - 7.2) + (-2.3p^2 + 5.2p + 15.5)$ $-1.7p^2 - 1.2p + 8.3$
23. $\left(\frac{10}{7}x^2 + \frac{5}{12}xy + 12y^2\right) - \left(-\frac{4}{7}x^2 - \frac{7}{12}xy + 16y^2\right)$ $2x^2 + xy - 4y^2$
24. $\left(\frac{1}{2}a^2 - \frac{5}{14}ab - 5b^2\right) + \left(\frac{9}{2}a^2 - \frac{23}{14}ab + 18b^2\right)$ $5a^2 - 2ab + 13b^2$
25. Let $f(x) = 5x^2 + 7x - 21$ and $g(x) = -3x^2 + 3x + 13$. Find $f(x) + g(x)$ and $f(x) - g(x)$. $f(x) + g(x) = 2x^2 + 10x - 8$ and $f(x) - g(x) = 8x^2 + 4x - 34$
26. Let $f(x) = 8x^2 + 3x - 10$ and $g(x) = -13x + 2$. Find $f(x) + g(x)$ and $f(x) - g(x)$. $f(x) + g(x) = 8x^2 - 10x - 8$ and $f(x) - g(x) = 8x^2 + 16x - 12$

Multiply the polynomials and simplify the result. (*See Objective 2.*)

27. $-4y(9y - 3)$ $-36y^2 + 12y$
28. $-3x(5x - 7)$ $-15x^2 + 21x$
29. $4a(10a^2 - 6a + 12)$ $40a^3 - 24a^2 + 48a$
30. $3z^3(6z^2 + 4z - 8)$ $18z^5 + 12z^4 - 24z^3$
31. $-p^2q^2(5p^2 + 9pq - q^2)$ $-5p^4q^2 - 9p^3q^3 + p^2q^4$
32. $-4s^2t^2(5s^2 + 4st - 6t^2)$ $-20s^4t^2 - 16s^3t^3 + 24s^2t^4$
33. $(z - 6)(8z + 5)$ $8z^2 - 43z - 30$
34. $(6r + 1)(5r - 1)$ $30r^2 - r - 1$
35. $(6t - 7)(t - 3)$ $6t^2 - 25t + 21$
36. $(x + 1)(3x + 4)$ $3x^2 + 7x + 4$

Additional answers can be found in the Instructor Answer Appendix.

37. $(z - 4)(z^2 + 4z - 16)$
$z^3 - 32z + 64$

38. $(4q + 1)(16q^2 + 4q - 1)$
$64q^3 + 32q^2 - 1$

39. $(a - 5)(a^2 + 5a + 25)$
$a^3 - 125$

40. $(3y - 1)(9y^2 + 3y + 1)$
$27a^3 - 1$

41. $(2x + 3)(4x^2 - 6x + 9)$
$8x^3 + 27$

42. $(4b + 1)(16b^2 - 4b + 1)$
$64b^3 + 1$

43. Let $f(x) = 8x^2 - 7x - 5$ and $g(x) = x - 3$. Find $f(x)g(x)$.
$8x^3 - 31x^2 + 16x + 15$

44. Let $f(x) = 6x^2 - 2x - 5$ and $g(x) = 3x + 10$. Find $f(x)g(x)$. $18x^3 + 54x^2 - 35x - 50$

Use the FOIL method to multiply the binomials. (*See Objective 3.*)

45. $(8x + 5)(7x - 2)$
$56x^2 + 19x - 10$

46. $(5u + 3)(2u + 1)$
$10u^2 + 11u + 3$

47. $(r - 3)(4r + 1)$
$4r^2 - 11r - 3$

48. $(6y - 5)(y - 2)$
$6y^2 - 17y + 10$

49. $(2a^2 - 1)(3a^2 - 11)$
$6a^4 - 25a^2 + 11$

50. $(10b^2 + 3)(b^2 - 2)$
$10b^4 - 17b^2 - 6$

51. $(6b^3 - 1)(5b^3 - 7)$
$30b^6 - 47b^3 + 7$

52. $(x^3 + 3)(7x^3 - 9)$
$7x^6 + 12x^3 - 27$

Simplify each expression. (*See Objective 4.*)

53. $(3x - 5)^2$
$9x^2 - 30x + 25$

54. $(4y + 1)^2$
$16y^2 + 8y + 1$

55. $(5x + 3y)^2$
$25x^2 + 30xy + 9y^2$

56. $(6z - 5w)^2$
$36z^2 - 60wz + 25w^2$

57. $(8t^2 - 1)^2$
$64t^4 - 16t^2 + 1$

58. $(5u^3 + 7)^2$
$25u^6 + 70u^3 + 49$

59. $(2c^3 - 11)^2$
$4c^6 - 44c^3 + 121$

60. $(2s^2 + 3)^2$
$4s^4 + 12s^2 + 9$

Simplify each product. (*See Objective 5.*)

61. $(3x + 7)(3x - 7)$
$9x^2 - 49$

62. $(4y + 3)(4y - 3)$
$16y^2 - 9$

63. $(9c^2 + d^2)(9c^2 - d^2)$
$81c^4 - d^4$

64. $(3a^2 - 5b^2)(3a^2 + 5b^2)$
$9a^4 - 25b^4$

65. $\left(s - \frac{5}{2}t\right)\left(s + \frac{5}{2}t\right)$
$s^2 - \frac{25}{4}t^2$

66. $\left(x - \frac{2}{3}y\right)\left(x + \frac{2}{3}y\right)$
$x^2 - \frac{4}{9}y^2$

67. $\left(2a - \frac{1}{3}\right)\left(2a + \frac{1}{3}\right)$
$4a^2 - \frac{1}{9}$

68. $\left(3b - \frac{7}{5}c\right)\left(3b + \frac{7}{5}c\right)$
$9b^2 - \frac{49}{25}c^2$

Simplify each product. (*See Objective 6.*)

69. $(c + 1)(c - 1)(3c + 5)$
$3c^3 + 5c^2 - 3c - 5$

70. $(d + 2)(d - 2)(6d + 1)$
$6d^3 + d^2 - 24d - 4$

71. $(r + 3)(r - 3)(r + 2)$
$r^3 + 2r^2 - 9r - 18$

72. $(s + 6)(s - 6)(s + 2)$
$s^3 + 2s^2 - 36s - 72$

73. $(a - 2b)^3$
$a^3 - 6a^2b + 12ab^2 - 8b^3$

74. $(5b + 2)^3$
$125b^3 + 150b^2 + 60b + 8$

75. $(3a - 2)^4$
$81a^4 - 216a^3 + 216a^2 - 96a + 16$

76. $(5n + 2)^4$
$625n^4 + 1000n^3 + 600n^2 + 160n + 16$

Write a polynomial that represents each situation and use it to find the requested information. (*See Objective 7.*)

77. One side of a triangle is 2 in. shorter than the cube of the shortest side. The other side is 4 in. longer than the cube of the shortest side. Find a polynomial that represents the perimeter of the triangle. What is the perimeter if the shortest side is 8 in.? $P(x) = 2x^3 + x + 2$; 1034 in.

78. One side of a triangle is 3 in. shorter than the square of shortest side. The other side is 2 in. longer than the shortest side. Find a polynomial that represents the perimeter of the triangle. What is the perimeter if the shortest side is 3 in.? $P(x) = x^2 + 2x - 1$; 14 in.

79. Find the perimeter of a rectangle with sides of lengths $2y^2 + 11y - 2$ ft and $y^2 + 6y - 5$ ft. $6y^2 + 34y - 14$ ft

80. Find the perimeter of a triangle with sides of lengths $6x - 15$ cm, $4x^2 - 4x - 2$ cm, and $5x^2 + 4x - 2$ cm.
$9x^2 + 6x - 19$ cm

81. Find the perimeter of a trapezoid with sides of lengths $4y - 7$ m, $2y + 5$ m, $5y^2 - 11y - 5$ m, and $9y^2 - 5y + 6$ m. $14y^2 - 10y - 1$ m

82. Find the perimeter of a trapezoid with sides of lengths $6x - 8$ ft, $4x - 1$ ft, $9x^2 + 5x + 6$ ft, and $8x^2 - 4x - 14$ ft. $17x^2 + 11x - 17$ ft

83. A Boy Scout troop is organizing a barbeque chicken dinner as a fund-raiser. The cost of preparing the dinner is \$1.45 per person and the cost to rent the location for the dinner is \$310. So, the total cost (in dollars) is $C(x) = 1.45x + 310$, where x is the number of people who attend the dinner. A ticket to attend the dinner is \$7.25 per person. So, the revenue (in dollars) is $R(x) = 7.25x$. (a) Write a polynomial function that represents the profit (in dollars) for the fund-raiser. (b) What is the profit from selling 185 tickets? (a) $P(x) = 5.8x - 310$; (b) $P(185) = \$763$

84. A Kiwanis club is organizing a spaghetti dinner as a fund-raiser. The cost of preparing the dinner is \$1.25 per person and the cost to rent the location for the dinner is \$280. So, the total cost (in dollars) is $C(x) = 1.25x + 280$, where x is the number of people who attend the dinner. A ticket to attend the dinner is \$7.50 per person. So, the revenue (in dollars) is $R(x) = 7.5x$. (a) Write a polynomial function that represents the profit (in dollars) for the fund-raiser. (b) What is the profit from selling 150 tickets?
(a) $P(x) = 6.25x - 280$; (b) $P(150) = \$657.50$

85. An apartment manager can rent all 110 units if he charges \$810 per month. If he increases the price by \$87, he rents four fewer apartments. If x represents the number of increases in price, what polynomial represents the total rent (in dollars) collected each month? $R(x) = -348x^2 + 6330x - 89{,}100$

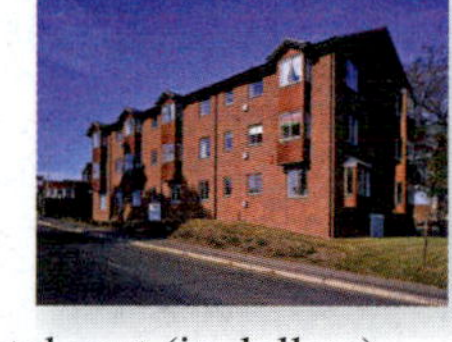

86. An apartment manager can rent all 105 units if he charges \$830 per month. If he increases the price by \$69, he rents three fewer apartments. If x represents the number of increases in price, what polynomial represents the total rent (in dollars) collected each month? $R(x) = -207x^2 + 4755x - 87{,}150$

87. A homeowner installed a rectangular pool that measures 12 ft by 15 ft. She wants to put a uniform border of colored concrete around each side of the pool. If x represents the width of the border (in feet), what polynomial represents the area of the pool and its border? $4x^2 + 54x + 180$ ft²

88. A gardener has a rectangular garden that measures 4 ft by 6 ft. She wants to put a uniform border of gravel around each side of the garden. If x represents the width of the border (in feet), what polynomial represents the area of the garden and the border? $4x^2 + 20x + 24$ ft^2

Mix 'Em Up!

Perform each operation.

89. $(3x - 10) + (18x + 16)$ $21x + 6$

90. $(-5y + 12) + (-14y - 4)$ $-19y + 8$

91. $4x^4(6x^2 - 4x + 2)$ $24x^6 - 16x^5 + 8x^4$

92. $3y^2(y^2 - 4y + 5)$ $3y^4 - 12y^3 + 15y^2$

93. $(5q - 4)(3q - 1)$ $15q^2 - 17q + 4$

94. $(4v + 7)(v - 5)$ $4v^2 - 13v - 35$

95. $(x - 2)(x^2 - 2x - 4)$ $x^3 - 4x^2 + 8$

96. $(2y + 3)(4y^2 + 6y + 9)$ $8y^3 + 24y^2 + 36y + 27$

97. $(2p^2 - 1)(p^2 + 5)$ $2p^4 + 9p^2 - 5$

98. $(3q^2 - 5)(2q^2 + 6)$ $6q^4 + 8q^2 - 30$

99. $(-9r - 11) - (-10r + 3)$ $r - 14$

100. $(4s + 5) - (8s - 13)$ $-4s + 18$

101. $(5a + 6)^2$ $25a^2 + 60a + 36$

102. $(2b - 7)^2$ $4b^2 - 28b + 49$

103. $(-0.2a + 10.4) + (-9.8a + 5.1)$ $-10a + 15.5$

104. $(-2.7b - 8.3) + (-6.9b - 5)$ $-9.6b - 13.3$

105. $-2x^2y^2(4x^2 + 7xy - 12y^2)$ $-8x^4y^2 - 14x^3y^3 + 24x^2y^4$

106. $-5xy(10x^2 - xy - 8y^2)$ $-50x^3y + 5x^2y^2 + 40xy^3$

107. $(5a + 3)(5a - 3)(a + 1)$ $25a^3 + 25a^2 - 9a - 9$

108. $(b + 2)(b - 2)(7b + 5)$ $7b^3 + 5b^2 - 28b - 20$

109. $(w - 6)(5w + 12)$ $5w^2 - 18w - 72$

110. $(9z + 8)(7z + 1)$ $63z^2 + 65z + 8$

111. $(20a^2 - 8a - 10) - (8a^2 + 31a - 16)$ $12a^2 - 39a + 6$

112. $(11b^2 + 23b - 18) - (-15b^2 + 17b + 6)$ $26b^2 + 6b - 24$

113. $(3v^3 + 1)^2$ $9v^6 + 6v^3 + 1$

114. $(2u^3 - 5)^2$ $4u^6 - 20u^3 + 25$

115. $(4r - 1)^3$ $64r^3 - 48r^2 + 12r - 1$

116. $(x + 5y)^3$ $x^3 + 15x^2y + 75xy^2 + 125y^3$

117. $\left(5m^2 + \dfrac{2}{3}\right)\left(5m^2 - \dfrac{2}{3}\right)$

118. $\left(3n^2 + \dfrac{4}{7}\right)\left(3n^2 - \dfrac{4}{7}\right)$

119. $(2a + b)^4$ $16a^4 + 32a^3b + 24a^2b^2 + 8ab^3 + b^4$

120. $(u - 3v)^4$ $u^4 - 12u^3v + 54u^2v^2 - 108uv^3 + 81v^4$

121. $\left(\dfrac{4}{3}p^2 + \dfrac{1}{4}pq - 11q^2\right) + \left(\dfrac{8}{3}p^2 - \dfrac{15}{4}pq + 2q^2\right)$ $4p^2 - 4pq - 9q^2$

122. $\left(\dfrac{5}{2}x^2 + \dfrac{3}{5}xy - 7y^2\right) - \left(\dfrac{1}{2}x^2 - \dfrac{2}{5}xy + y^2\right)$ $2x^2 + xy - 8y^2$

123. $(1.8x^2 + 5x - 0.6) - (-2.6x^2 - 13.5x - 8.6)$ $4.4x^2 + 18.5x + 8$

124. $(-1.5y^2 - 1.7y + 6.3) - (-7.4y^2 - 14.6x + 21.5)$ $5.9y^2 + 12.9y - 15.2$

125. A street vendor sells lemonade. She can make each drink for \$0.45. The cost to operate her stand is \$320 per month. So, the monthly cost (in dollars) of operating the stand is $C(x) = 0.45x + 320$, where x is the number of lemonades sold in a month. If she sells lemonades for \$1.60 each, the revenue (in dollars) made from selling x lemonades in a month is $R(x) = 1.6x$. (a) Write a polynomial function that represents the profit (in dollars) of the lemonade stand in one month. (b) What is the profit from selling 1200 lemonades in a month? (a) $P(x) = 1.15x - 320$; (b) $P(1200) = \$1060$

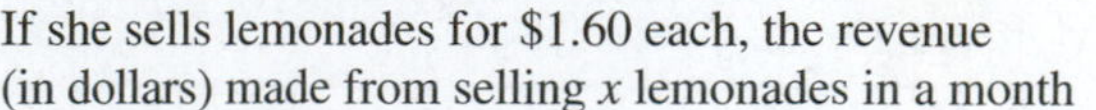

126. A Girl Scout troop is organizing a pancake breakfast as a fund-raiser. The cost of preparing the breakfast is \$1.05 per person and the cost to rent the location for the breakfast is \$330. So, the total cost (in dollars) is $C(x) = 1.05x + 330$, where x is the number of people who attend the breakfast. A ticket to attend the breakfast is \$5.10. So, the revenue (in dollars) is $R(x) = 5.10x$. (a) Write a polynomial function that represents the profit (in dollars) for the fund-raiser. (b) What is the profit from selling 205 tickets? (a) $P(x) = 4.05x - 330$; (b) $P(205) = \$500.25$

127. An apartment manager can rent all 190 units if he charges \$850 per month. If he increases the price by \$100, he rents three fewer apartments. If x represents the number of increases in price, what polynomial represents the total rent (in dollars) collected each month? $R(x) = -300x^2 + 16{,}450x + 161{,}500$

You Be the Teacher!

Correct each student's errors, if any.

128. Simplify $(a^2 + 23a - 18) + (a^2 + 4a - 33)$.

Andrea's work:
$(a^2 + 23a - 18) + (a^2 + 4a - 33) = a^4 + 27a^2 - 51$

129. Simplify $(9b^2 - 15b + 2) - (-2b^2 - 6b + 7)$.

Matt's work:
$(9b^2 - 15b + 2) - (-2b^2 - 6b + 7)$
$= 9b^2 - 15b + 2 + 2b^2 - 6b + 7$
$= 11b^2 - 21b + 9$

130. Simplify $-2y(8y + 5)$.

Alyssa's work:
$-2y(8y + 5) = -16y^2 + 5$

$-2y(8y + 5)$
$= (-2y)(8y) + (-2y)(5)$
$= -16y^2 - 10y$

131. Simplify $(3x - 8y)^2$.

Mark's work:
$(3x - 8y)^2 = 9x^2 - 64y^2$

$(3x - 8y)^2$
$= (3x)^2 - 2(3x)(8y) + (8y)^2$
$= 9x^2 - 48xy + 64y^2$

132. Simplify $(5x - 2y)^3$.

Joan's work:
$(5x - 2y)^3 = 125x^3 - 8y^3$

$(5x - 2y)^3$
$= (5x - 2y)^2(5x - 2y)$
$= (25x^2 - 20xy + 4y^2)(5x - 2y)$
$= 125x^3 - 50x^2y - 100x^2y + 40xy^2 + 20xy^2 - 8y^3$
$= 125x^3 - 150x^2y + 60xy^2 - 8y^3$

133. Simplify $(x - 2y)^4$.

Vincent's work:
$(x - 2y)^4 = x^4 - 2y^4$

$(x - 2y)^4$
$= (x - 2y)^2(x - 2y)^2$
$= (x^2 - 4xy + 4y^2)(x^2 - 4xy + 4y^2)$
$= x^4 - 4x^3y + 4x^2y^2 - 4x^3y + 16x^2y^2 - 16xy^3 + 4x^2y^2 - 16xy^3 + 16y^4$
$= x^4 - 8x^3y + 24x^2y^2 - 32xy^3 + 16y^4$

Calculate It!

Use a graphing calculator to determine if each statement is true.

134. $(-2x^2 + 5x - 4) + (5x^2 - 7x + 3) = 3x^2 - 2x - 1$

135. $(5x - 2)^2 = 25x^2 + 4$

136. $(x-2)^3 = x^3 - 6x^2 + 12x - 8$

137. $(x+2)^4 = x^4 + 16$

Think About It!

138. Write an example of a perfect square trinomial and determine the product that produces it.
Answers vary; $x^2 - 6x + 9 = (x-3)^2$

139. Write an example of a perfect square trinomial and determine the product that produces it.
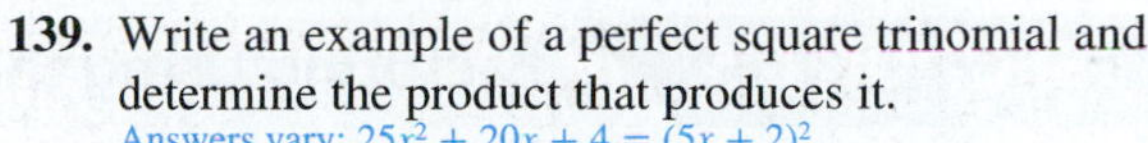
Answers vary; $25x^2 + 20x + 4 = (5x+2)^2$

140. Write an example of a difference of two squares and determine the product that produces it.
Answers vary; $x^2 - 9 = (x+3)(x-3)$

141. Write an example of a difference of two squares and determine the product that produces it.
Answers vary; $36x^2 - 1 = (6x+1)(6x-1)$

PIECE IT TOGETHER SECTIONS 6.3–6.4

Classify each polynomial and identify its terms, degree, type, and leading coefficient, if applicable. (*Section 6.3, Objectives 1 and 2*)

1. $-5.4c^6$

2. $-x + 3.5$

3. $6x^2 - 4.5x - 9.4$

4. $4s^2t$

If $f(x) = 7x^2 - 2x + 3$ and $g(x) = -x^3 + 2x^2 - 12$, find the following. (*Section 6.3, Objective 3*)

5. $g(-2)$ 4

6. $f(2)$ 27

7. $f(3) - g(-1)$ 69

8. $f(-1) + g(2)$ 0

Graph each function by completing a table of values and state its domain and range. (*Section 6.3, Objective 4*)

9. $y = x^3 + 1$

10. $y = 3x - 6$

Perform the indicated operation. (*Section 6.4, Objectives 1–4*)

11. $(14b - 6) + (3b - 10)$ $17a - 16$

12. $(7q + 15) - (18q - 14)$ $-11q + 29$

13. $(9.2b + 6.4) - (-8.1b - 5.5)$ $17.3b + 11.9$

14. $(3.1c - 4.7) - (3.4c + 10.2)$ $-0.3d - 14.9$

15. $-4p^2(5p^2 + 9p - 11)$
$-20p^4 - 36p^3 + 44p^2$

16. $-5q(10q^2 + 8q + 5)$
$-50q^4 - 40q^3 - 25q^2$

17. $(4y - 1)(16y^2 + 4y + 1)$
$64a^3 - 1$

18. $(2x + 5)(4x^2 - 10x + 25)$
$8x^3 + 125$

19. $(a^2 - 5)(2a^2 - 9)$
$2a^4 - 19a^2 + 45$

20. $(10b^2 + 7)(3b^2 - 2)$
$10b^4 + b^2 - 14$

21. $(5c^2 - 6)^2$
$25c^4 - 60c^2 + 36$

22. $(7z - 10w)(7z + 10w)$
$49z^2 - 100w^2$

23. $(x + 2)(x - 2)(x + 5)$
$x^3 + 5x^2 - 4x - 20$

24. $(5a - b)^3$
$125a^3 - 75a^2b + 15ab^2 - b^3$

25. $(2c + d)^4$ $16c^4 + 32c^3d + 24c^2d^2 + 8cd^3 + d^4$

SECTION 6.5 Factoring Using the Greatest Common Factor and Grouping

OBJECTIVES

As a result of completing this section, you will be able to

1. Find the greatest common factor of a set of terms.
2. Use the greatest common factor to factor a polynomial.
3. Use grouping to factor a polynomial.
4. Troubleshoot common errors.

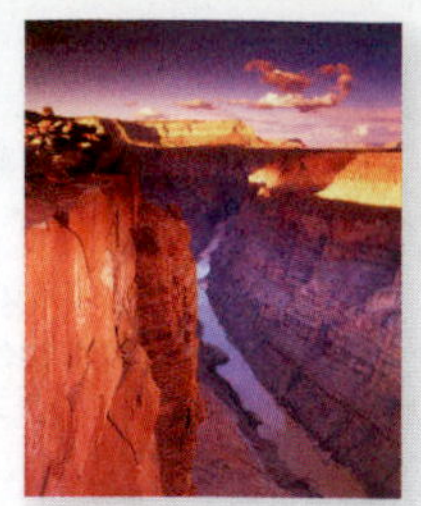

The Grand Canyon Skywalk is a glass walkway that stands about 3600 ft above the floor of the Grand Canyon. The distance an object would be above the floor of the canyon if dropped from the skywalk is given by $s = -16t^2 + 3600$ ft after t seconds. In Section 6.8, we will determine how long it will take for a rock dropped from the skywalk to reach the floor of the canyon. In order to do this, we must be able to factor the binomial, $-16t^2 + 3600$.

In Section 6.4, we learned how to multiply polynomials. In this section, our focus will be on the reverse process. That is, given a polynomial, we will find what must be multiplied together to obtain that polynomial. This process is called *factoring*.

The Greatest Common Factor

Objective 1 Find the greatest common factor of a set of terms.

When factoring a polynomial, the first step is to factor out the *greatest common factor* (GCF) of its terms.

> **Definition:** The **greatest common factor (GCF)** of a set of terms is the largest term that divides evenly into each term in the set.

The GCF of 24 and 30, which can be expressed as GCF(24, 30), is the largest integer that divides evenly into both of these numbers. To find the GCF, we factor each number into its *prime* factors. The GCF is the product of the number's common factors. Recall *prime numbers* are numbers divisible by only 1 and the number itself.

$$24 = 4 \cdot 6 = \textcircled{2} \cdot 2 \cdot 2 \cdot \textcircled{3} \qquad 30 = 6 \cdot 5 = \textcircled{2} \cdot \textcircled{3} \cdot 5$$

From the prime factors, we see that the common factors of 24 and 30 are 2 and 3. Therefore, GCF(24, 30) $= 2 \cdot 3 = 6$.

The greatest common factor of x^3 and x^5, denoted GCF(x^3, x^5), is the largest variable expression that divides both terms. To find the GCF of variable terms, rewrite each term in expanded form and find the product of their common factors.

$$x^3 = \textcircled{x} \cdot \textcircled{x} \cdot \textcircled{x} \qquad x^5 = \textcircled{x} \cdot \textcircled{x} \cdot \textcircled{x} \cdot x \cdot x$$

We see that the common factors of x^3 and x^5 are three factors of x, so GCF(x^3, x^5) $= x^3$.

> **Note:** *The GCF of a set of variable terms with the same base is the base raised to the smallest exponent of the terms.*

> **Procedure: Finding the GCF of a Set of Terms**
>
> **Step 1:** Find the GCF of the coefficients of the terms. Find the prime factors of each coefficient and form the product of those factors that are common to each number.
>
> **Step 2:** Find the GCF of each variable contained in the terms. The GCF is the variable raised to the smallest exponent of that variable.
>
> **Step 3:** The GCF is the product of the GCF of the coefficients and the GCF of each variable.

Objective 1 Examples **Find the GCF of each set of terms.**

1a. $8xy^2$ and $10xy$ **1b.** $-12a^4$ and $48a^3$

1c. $4y^2$, $9y$, and 6 **1d.** $3(x - 2)$ and $y(x - 2)$

Solutions

1a. $8 = 2 \cdot 2 \cdot 2$ Factor 8.

$10 = 2 \cdot 5$ Factor 10.

GCF(8, 10) = 2. The smallest exponent of the common variable x is 1 and the smallest exponent of the common variable y is 1, so GCF(xy^2, xy) = xy. So, GCF($8xy^2$, $10xy$) = $2xy$.

1b. $-12 = -4 \cdot 3 = -2 \cdot 2 \cdot 3$ Factor −12.

$48 = 4 \cdot 12 = 2 \cdot 2 \cdot 2 \cdot 2 \cdot 3$ Factor 48.

GCF(−12, 48) = $2 \cdot 2 \cdot 3 = 12$. The smallest exponent of the common variable a is 3, so GCF(a^4, a^3) = a^3. So, GCF($-12a^4$, $48a^3$) = $2 \cdot 2 \cdot 3 \cdot a^3 = 12a^3$ or $-12a^3$.

Note: *When one of the terms has a negative coefficient, the GCF can be positive or negative.*

1c. $4 = 2 \cdot 2$ Factor 4.

$9 = 3 \cdot 3$ Factor 9.

$6 = 2 \cdot 3$ Factor 6.

The coefficients do not share any common factors, so GCF(4, 9, 6) = 1. The variable is not in all three terms, so it is not common. So, GCF($4y^2$, $9y$, 6) = 1.

1d. The terms are already in factored form. The common factor to each term is the binomial $x - 2$.

$3(x - 2)$

$y(x - 2)$

So, GCF[$3(x - 2)$, $y(x - 2)$] = $x - 2$.

Student Check 1 Find the GCF of each set of terms.

a. $6x^3y$ and $9xy^2$ **b.** $24b^5$ and $32b^4$

c. $8x$, $20x^2$, and 25 **d.** $6(y + 4)$ and $z(y + 4)$

Factor out the Greatest Common Factor

Objective 2 ▶

Use the greatest common factor to factor a polynomial.

To **factor a polynomial** means to write the polynomial as a product of prime factors. To use the greatest common factor when factoring, we apply the distributive property in reverse. Recall the following product.

$$2x(3x + 5) = 2x(3x) + 2x(5) = 6x^2 + 10x$$

When this statement is reversed, we have

$$6x^2 + 10x = \underbrace{2x(3x) + 2x(5)}_{2x \text{ is the GCF of } 6x^2 \text{ and } 10x} = \underbrace{2x(3x + 5)}_{\text{Factored form of } 6x^2 + 10x}$$

So, the polynomial $6x^2 + 10x$ has been written as a product, or in factored form, where one if its factors is the GCF of its terms.

Property: The Distributive Property

For a, b, and c real numbers,

$$ab + ac = a(b + c)$$

where a represents the greatest common factor of the terms in the polynomial.

Procedure: Factoring a Polynomial Using the Greatest Common Factor

Step 1: Find the GCF of the terms in the polynomial.
Step 2: Rewrite each term of the polynomial as a product of the GCF and the remaining factor.
Step 3: Apply the distributive property to factor out the GCF. The GCF is written on the outside of parentheses and the remaining factors are written in parentheses.
Step 4: Check the answer by multiplying.

Objective 2 Examples

Use the GCF to factor each polynomial. Check each answer by multiplying.

2a. $6x^3 - 24x^2$ **2b.** $14xy^3 + 56x^2y^2 - 7x^3y$ **2c.** $-2a^3 + 8a$
2d. $-x^2 - 3xy + 2y^2$ **2e.** $12x^3y^2 + 6x^2y^2$ **2f.** $x(x + 5) + 3(x + 5)$

Solutions

INSTRUCTOR NOTE: Point out that students can divide each term of the polynomial by the GCF to determine the remaining factor.

2a. $\text{GCF}(6x^3, 24x^2) = 6x^2$

$$6x^3 - 24x^2 = 6x^2(x) - 6x^2(4)$$ Rewrite each term as a product of $6x^2$ and another factor.
$$= 6x^2(x - 4)$$ Apply the distributive property.

2b. $\text{GCF}(14xy^3, 56x^2y^2, 7x^3y) = 7xy$

$$14xy^3 + 56x^2y^2 - 7x^3y = 7xy(2y^2) + 7xy(8xy) - 7xy(x^2)$$
$$= 7xy(2y^2 + 8xy - x^2)$$

So, $14xy^3 + 56x^2y^2 - 7x^3y = 7xy(2y^2 + 8xy - x^2)$. We can check our factoring by multiplying.

2c. $\text{GCF}(-2a^3, 8a) = 2a$ or $-2a$. We can factor the polynomial in two ways.

Method 1 $-2a^3 + 8a = 2a(-a^2) + 2a(4)$ Rewrite each term as a product of $2a$ and another factor.
$= 2a(-a^2 + 4)$ Apply the distributive property.

Method 2 $-2a^3 + 8a = -2a(a^2) + 2a(-4)$ Rewrite each term as a product of $-2a$ and another factor.
$= -2a(a^2 - 4)$ Apply the distributive property.

So, $-2a^3 + 8a = 2a(-a^2 + 4)$ or $-2a(a^2 - 4)$. We can check our factoring by multiplying.

Note: *While either method is correct, it is standard to factor out a negative GCF if the first term of the polynomial has a negative coefficient.*

2d. $\text{GCF}(-x^2, 3xy, 2y^2) = -1$

$$-x^2 - 3xy + 2y^2 = -1(x^2) - 1(3xy) - 1(-2y^2)$$ Rewrite each term as a product of -1 and another factor.
$$= -1(x^2 + 3xy - 2y^2)$$ Apply the distributive property.
$$= -(x^2 + 3xy - 2y^2)$$ Multiplying by -1 is the same as the opposite of the polynomial.

So, $-x^2 - 3xy + 2y^2 = -(x^2 + 3xy - 2y^2)$.

Note: *The signs of the polynomial change when a negative GCF is factored out.*

INSTRUCTOR NOTE:
Remind students that when one of the terms is the GCF, we must have a 1 or −1 when we factor it out.

2e. $\text{GCF}(12x^3y^2, 6x^2y^2) = 6x^2y^2$

$12x^3y^2 + 6x^2y^2 = 6x^2y^2(2x) + 6x^2y^2(1)$ — Rewrite each term as a product of $6x^2y^2$ and another factor.

$= 6x^2y^2(2x + 1)$ — Apply the distributive property.

So, $12x^3y^2 + 6x^2y^2 = 6x^2y^2(2x + 1)$.

2f. $\text{GCF}[x(x + 5), 3(x + 5)] = x + 5$. Each term is already written as a product of $x + 5$ and another factor. So, we apply the distributive property to factor.

$$x(x + 5) + 3(x + 5) = (x + 5)(x + 3)$$

So, $x(x + 5) + 3(x + 5) = (x + 5)(x + 3)$.

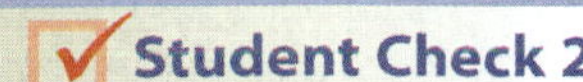

Student Check 2 Use the GCF to factor each polynomial. Check each answer by multiplying.

a. $5x^2 - 20x^3$ **b.** $21y^4 + 63y^3 + 3y^2$ **c.** $-8a^3 + 64a$

d. $-4b^2 + 12b - 20$ **e.** $36r^2s^3 - 9rs$ **f.** $x(x - 9) - 6(x - 9)$

Factoring by Grouping

Objective 3 ▶

Use grouping to factor a polynomial.

The *grouping method of factoring* applies to polynomials with four or more terms in which pairs of terms have a common factor. If after factoring out the GCF from pairs of terms, the remaining factors are the same, the polynomial can be factored. The following example illustrates this concept.

INSTRUCTOR NOTE:
This course will focus only on grouping pairs of terms together. Grouping can be done in other ways, namely grouping three of the four terms together.

$x^3 + x^2 + 2x + 2$	There is no common factor to all four terms.
$(x^3 + x^2) + (2x + 2)$	Group the first two terms and the last two terms.
$x^2(x + 1) + 2(x + 1)$	Factor out the GCF from each pair of terms.
$(x + 1)(x^2 + 2)$	Factor out the common binomial of $x + 1$.

Note: *The polynomial must be written as a product for it to be factored. A common mistake is to stop after factoring out the GCF from the pairs of terms. At this point the polynomial is still expressed as a sum. We must complete the factoring by factoring out the common factor from the resulting binomial.*

In this illustration, note that after the GCF was factored out of each pair of terms, the binomial that remained was the same, namely $x + 1$. If, however, the binomial is not the same after factoring out the GCF from the two pairs of terms, then different terms may need to be grouped together. If no grouping produces a common binomial, then we can conclude that grouping cannot be used to factor the polynomial.

Procedure: Factoring by Grouping

Step 1: Group together the first two terms and the last two terms.
Step 2: Factor out the GCF from each pair of terms.
Step 3: Factor out the common factor from the resulting binomial. This common factor should be a binomial.
Step 4: If there is no common factor in the resulting binomial, rearrange the grouping of terms. Repeat the preceding steps. If the resulting binomial still doesn't have a common factor, then the polynomial cannot be factored by grouping.
Step 5: Check the answer by multiplying.

Objective 3 Examples **Use grouping to factor each polynomial. Check each answer by multiplying.**

3a. $8x^2 - 10x + 12x - 15$ **3b.** $5x^4 - 15x^3 + x - 3$

3c. $3x^3 - 7x^2y - 6x + 14y$

Solutions

3a. $8x^2 - 10x + 12x - 15 = (8x^2 - 10x) + (12x - 15)$ Group pairs of terms.

$= 2x(4x - 5) + 3(4x - 5)$ Factor each binomial.

$= (4x - 5)(2x + 3)$ Factor out the common factor, $4x - 5$.

So, $8x^2 - 10x + 12x - 15 = (4x - 5)(2x + 3)$.

Check: $(4x - 5)(2x + 3) = 8x^2 + 12x - 10x - 15$

3b. $5x^4 - 15x^3 + x - 3 = (5x^4 - 15x^3) + (x - 3)$ Group pairs of terms.

$= 5x^3(x - 3) + 1(x - 3)$ Factor each binomial.

$= (x - 3)(5x^3 + 1)$ Factor out the common factor, $x - 3$.

So, $5x^4 - 15x^3 + x - 3 = (x - 3)(5x^3 + 1)$.

Check: $(x - 3)(5x^3 + 1) = 5x^4 + x - 15x^3 - 3$

3c. Note the negative sign on $6x$. When factoring out the GCF from the second grouping, we must factor out a negative GCF. Remember factoring out a negative GCF changes the signs.

$3x^3 - 7x^2y - 6x + 14y = (3x^3 - 7x^2y) + (-6x + 14y)$ Group pairs of terms.

$= x^2(3x - 7y) - 2(3x - 7y)$ Factor each binomial.

$= (3x - 7y)(x^2 - 2)$ Factor out the common factor, $3x - 7y$.

So, $3x^3 - 7x^2y - 6x + 14y = (3x - 7y)(x^2 - 2)$. We can check by multiplying.

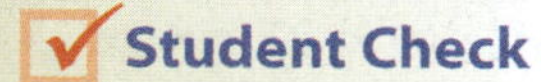

Student Check 3 Use grouping to factor each polynomial. Check each answer by multiplying.

a. $8y^2 - 2y + 20y - 5$ **b.** $6x^4 - 42x^3 + x - 7$ **c.** $2x^3 - 3x^2 - 12x + 18$

Troubleshooting Common Errors

Objective 4 Troubleshoot common errors.

Some common errors that occur when factoring out the GCF are a result of sign errors or not factoring out the *greatest* common factor. Grouping errors occur by not applying the distributive property correctly.

Objective 4 Examples **A problem and an incorrect solution are given. Provide the correct solution and an explanation of the error.**

4a. Factor $6x^3 - 18x^4$ completely.

Incorrect Solution	Correct Solution and Explanation
$6x^3 - 18x^4 = 3x(2x^2) - 3x(6x^3)$ $= 3x(2x^2 - 6x^3)$	The answer shown is a correct factorization; it is just not the complete factorization because $2x^2$ and $6x^3$ still have common factors. Since $\text{GCF}(6x^3, 18x^4) = 6x^3$, $6x^3 - 18x^4 = 6x^3(1) - 6x^3(3x)$ $= 6x^3(1 - 3x)$

4b. Factor $-28x^2y + 42xy^2$ completely.

Incorrect Solution	Correct Solution and Explanation
$-28x^2y + 42xy^2$ $= -14xy(2x) - 14xy(3y)$ $= -14xy(2x + 3y)$	Factoring out a negative number changes the signs of the terms in the polynomial. $-28x^2y + 42xy^2 = -14xy(2x) - 14xy(-3y)$ $= -14xy(2x - 3y)$

4c. Factor $8z - 10zy + 12 - 15y$ completely.

Incorrect Solution	Correct Solution and Explanation
$8z - 10zy + 12 - 15y$ $= (8z - 10zy) + (12 - 15y)$ $= 2z(4 - 5y) + 3(4 - 5y)$ $= (4 - 5y)^2(2z + 3)$	The binomial $4 - 5y$ is a common factor and should be factored out. Since this factor only occurs to the exponent of 1 in each term, it can only be raised to the exponent of 1 in the factored form. $8z - 10zy + 12 - 15y$ $= (8z - 10zy) + (12 - 15y)$ $= 2z(4 - 5y) + 3(4 - 5y)$ $= (4 - 5y)(2z + 3)$

ANSWERS TO STUDENT CHECKS

Student Check 1 **a.** $3xy$ **b.** $8b^4$ **c.** 1 **d.** $y + 4$

Student Check 2 **a.** $5x^2(1 - 4x)$ **b.** $3y^2(7y^2 + 21y + 1)$ **c.** $-8a(a^2 - 8)$ **d.** $-4(b^2 - 3b + 5)$ **e.** $9rs(4rs^2 - 1)$ **f.** $(x - 9)(x - 6)$

Student Check 3 **a.** $(4y - 1)(2y + 5)$ **b.** $(x - 7)(6x^3 + 1)$ **c.** $(2x - 3)(x^2 - 6)$

SUMMARY OF KEY CONCEPTS

1. The greatest common factor (GCF) of a set of terms is the largest term that divides into each of the terms in the set. To find the GCF, express each term as a product of primes and then form the product of those factors that are common to each term. For common variables, the GCF contains the smallest exponent in the set.
2. The distributive property enables us to factor a polynomial by factoring out the GCF. Identify the GCF of the terms in the polynomial. Rewrite each term as a product of the GCF and another factor and then apply the distributive property. Factoring can be checked by multiplying.
3. Grouping is used to factor a polynomial with four or more terms. Pair together terms. If possible, factor out a GCF from each pair of terms. Then, if there is a common binomial as a factor, factor it out and use the distributive property to rewrite the expression as a product of two binomials.

GRAPHING CALCULATOR SKILLS

The graphing calculator can find factors of numbers as well as find the GCF of a set of numbers.

Example 1: Find factors of 42.

Solution: To find factors of 42 by hand, we divide it by 1, 2, 3, and so on. To do this on the calculator, we divide 42 by x and then view the table. When both X and Y_1 are integers, then X and Y_1 are factors of 42.

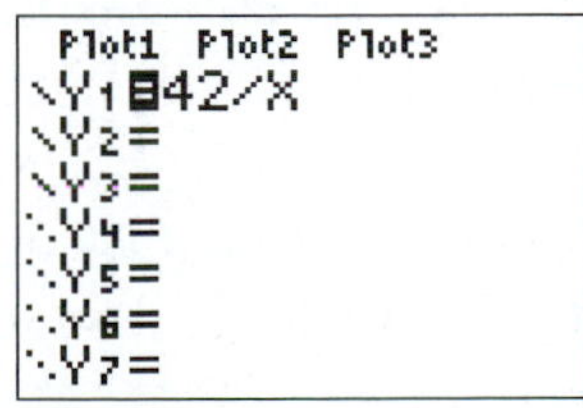

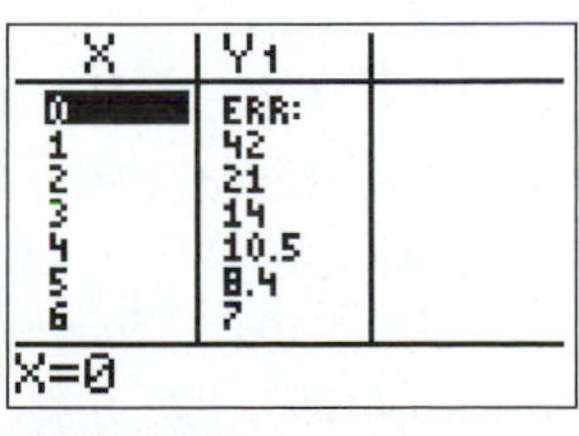

From the table, the factors of 42 are 1, 2, 3, 6, 7, 14, 21, and 42.

Example 2: Find GCF(60, 84).

Solution: The calculator has a built-in command called gcd, which stands for the greatest common divisor, or greatest common factor. To access the command, press MATH and use the NUM menu, option 9.

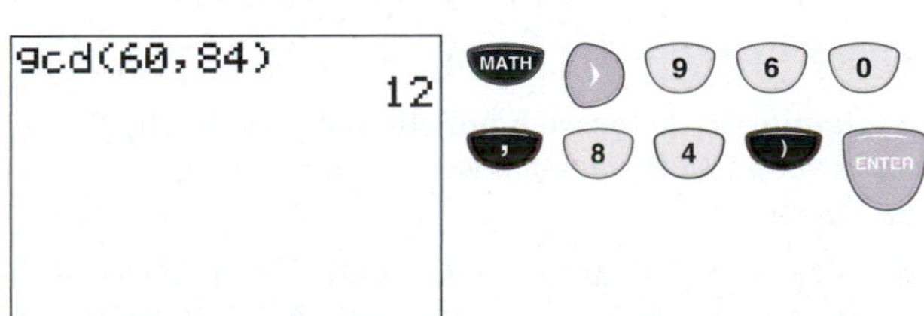

So, GCF(60, 84) = 12.

SECTION 6.5 EXERCISE SET

Write About It!

Use complete sentences in your answer to each exercise.

1. Define the greatest common factor of two integers.
2. Define the greatest common factor of two variable terms with the same base.
3. Explain how to factor a polynomial using the greatest common factor.
4. Explain how to factor a polynomial if the greatest common factor is a negative term.
5. Describe the first step of factoring a polynomial with four terms.
6. If a polynomial with four terms has been changed to a sum of two products with no common factor, explain the next step.

Practice Makes Perfect!

Find the GCF of each set of terms. (*See Objective 1.*)

7. $5x^3y$ and $15x^5y^3$ $5x^3y$
8. $14x^8y$ and $24x^4y^5$ $2x^4y$
9. $30x^2y^4$ and $18xy^6$ $6xy^4$
10. $35x^3y$ and $14x^6y^2$ $7x^3y$
11. $-10a^3b^2$ and $6a^7b^5$ $2a^3b^2$ or $-2a^3b^2$
12. $-26a^2b^8$ and $39a^5b^3$ $13a^2b^3$ or $-13a^2b^3$
13. $18a^4$, $24a$, and $12a^2$ $6a$
14. $27b^2, 18b^5$, and $54b^7$ $9b^2$
15. $30x^8$, $40x^6$, and $25x^3$ $5x^3$
16. $22z^4$, $44z^7$, and $11z^5$ $11z^4$
17. $16y^2(x-5)$ and $72y(x-5)$ $8y(x-5)$
18. $27x^2(y-3)$ and $18x(y-3)$ $9x(y-3)$

Use the GCF to factor each polynomial completely. Check each answer by multiplying. (*See Objective 2.*)

19. $36x^2 + 14x^6$ $2x^2(18 + 7x^4)$
20. $30b^4 - 56b$ $2b(15b^3 - 28)$
21. $-54a^3 - 48a^8$ $-6a^3(9 + 8a^5)$
22. $-40w^3 - 35w^2$ $-5w^2(8w + 7)$
23. $42a^2 - 21a + 7$ $7(6a^2 - 3a + 1)$
24. $56b^2 - 40b + 16$ $8(7b^2 - 5b + 2)$
25. $-64y^2 - 36y + 72$ $-4(16y^2 + 9y - 18)$
26. $-52a^2 + 26a + 13$ $-13(4a^2 - 2a - 1)$
27. $80x^3 + 50x^2 - 60x$ $10x(8x^2 + 5x - 6)$
28. $-14b^3 + 35b^2 - 21b$ $-7b(2b^2 - 5b + 3)$
29. $76x(y+1) + 57(y+1)$ $19(y+1)(4x+3)$
30. $20y(x-4) - 32(x-4)$ $4(x-4)(5y-8)$
31. $2x(y-2) - (y-2)$ $(y-2)(2x-1)$
32. $y(x+2) - (x+2)$ $(x+2)(y-1)$
33. $-6x^2(y-3) + 12x(y-3)$ $-6x(y-3)(x-2)$
34. $-5y^2(x+2) + 15y(x+2)$ $-5y(x+2)(y-3)$

Use grouping to factor each polynomial completely, if applicable. Check each answer by multiplying. (*See Objective 3.*)

35. $12a^3 - 18a^2 + 10a - 15$ $(2a-3)(6a^2+5)$
36. $8b^3 - 7b^2 + 24b - 21$ $(8b-7)(b^2+3)$
37. $15w^3 + 24w^2 - 5w - 8$ $(5w+8)(3w^2-1)$
38. $20w^3 + 35w^2 - 4w - 7$ $(4w+7)(5w^2-1)$

Additional answers can be found in the Instructor Answer Appendix.

39. $10b^3 + 25b^2 - 4b + 10$ cannot be factored by grouping
40. $3x^3 + 6x^2 - 7x + 14$ cannot be factored by grouping
41. $25x^4 - 5x^3 - 10x + 2$ $(5x-1)(5x^3-2)$
42. $30b^4 + 5b^3 - 18b - 3$ $(6b+1)(5b^3-3)$

Mix 'Em Up!

Find the GCF of each set of terms.

43. $36x^3y^{10}$ and $48x^7y^8$ $12x^3y^8$
44. $50x^8y^4$ and $75x^2y^5$ $25x^2y^4$
45. $-12a^4b^3$ and $18a^9b^2$ $6a^4b^2$ or $-6a^4b^2$
46. $-45a^3b^9$ and $36a^6b^7$ $9a^3b^7$ or $-9a^3b^7$
47. $14x^8(y-3)$ and $24x^5(y-3)$ $2x^5(y-3)$
48. $64x^9(y+1)$ and $24x^3(y+1)$ $8x^3(y+1)$
49. $32(x+2)$ and $15y(x-2)$ 1
50. $27x(y-1)$ and $16y(y+1)$ 1

Factor each polynomial completely, if possible. Check each answer by multiplying.

51. $14c^9 + 30c^{12}$ $2c^9(7 + 15c^3)$
52. $-33b^7 - 3b^6$ $-3b^6(11b + 1)$
53. $15y^2 + 16x$ prime
54. $-12a^2 + 5b$ prime
55. $-72a^3 + 24a^2 - 20a$ $-4a(18a^2 - 6a + 5)$
56. $-68b^3 - 28b^2 + 76b$ $-4b(17b^2 + 7b - 19)$
57. $18y^4(x-3) - 54y^2(x-3)$ $18y^2(x-3)(y^2-3)$
58. $70y(x-1) - 30(x-1)$ $10(x-1)(7y-3)$
59. $5x(y+3) - (y+3)$ $(y+3)(5x-1)$
60. $y^2(x-5) - 2(x-5)$ $(x-5)(y^2-2)$
61. $14w^3 - 6w^2 - 7w + 3$ $(7w-3)(2w^2-1)$
62. $42b^3 - 48b^2 + 7b - 8$ $(7b-8)(6b^2+1)$
63. $-52c^4 + 39c^3 - 13c^2$ $-13c^2(4c^2 - 3c + 1)$
64. $-50w^4 - 30w^3 + 10w^2$ $-10w^2(5w^2 + 3w - 1)$
65. $24a^3 + 32a^2 + 6a + 8$ $2(3a+4)(4a^2+1)$
66. $25c^3 - 25c^2 - 10c + 10$ $5(c-1)(5c^2-2)$
67. $-25a^9 - 15a^7$ $-5a^7(5a^2+3)$
68. $-5c^8 + 45c^9$ $-5c^8(1-9c)$
69. $15a^3 + 30a^2 - 3a - 6$ $3(a+2)(5a^2-1)$
70. $8y^3 - 2y^2 + 8y - 2$ $2(4y-1)(y^2+1)$
71. $81a^3 - 45a^2 - 27a$ $9a(9a^2 - 5a - 3)$
72. $36b^3 - 30b^2 + 18$ $6b(6b^2 - 5b + 3)$

You Be the Teacher!

Correct each student's errors, if any.

73. Factor $24a^3 - 16a^2 + 8$.

Fred's work:

$24a^3 - 16a^2 + 8 = 8a(3a^2 - 2a + 1)$

a is not a common factor. $24a^3 - 16a^2 + 8 = 8(3a^3 - 2a^2 + 1)$

74. Factor $-25a^3 + 10a^2 - 5a$.

Joanna's work:

$-25a^3 + 10a^2 - 5a = -5a(5a^2 + 2a + 1)$

$-25a^3 + 10a^2 - 5a = -5a(5a^2 - 2a + 1)$

75. Factor $12b^3 - 8b^2 + 4b$.

Diane's work:

$12b^3 - 8b^2 + 4b = 4b(3b^2 - 2b)$

$12b^3 - 8b^2 + 4b = 4b(3b^2 - 2b + 1)$

76. Factor $8c^3 - 2c^2 + 12c - 3$.

Elisa's work:

$$\begin{aligned}8c^3 - 2c^2 + 12c - 3 &= (8c^3 - 2c^2) + (12c - 3)\\ &= 2c^2(4c - 1) + 3(4c - 1)\\ &= (4c - 1)^2(2c^2 + 3)\end{aligned}$$

77. Factor $3w^3 + 2w^2 - 6w + 4$.

Mark's work:

$$\begin{aligned}3w^3 + 2w^2 - 6w + 4 &= (3w^3 + 2w^2) - (6w + 4)\\ &= w^2(3w + 2) - 2(3w + 2)\\ &= (3w + 2)(w^2 - 2)\end{aligned}$$

78. Factor $30y^3 + 25y^2 - 6y - 5$.

Andrew's work:

$$\begin{aligned}30y^3 + 25y^2 - 6y - 5 &= (30y^3 + 25y^2) - (6y - 5)\\ &= 5y^2(6y + 5) - 1(6y - 5)\end{aligned}$$

Calculate It!

Use the built-in command, gcd, of a graphing calculator to find the greatest common factor for each set of numbers.

79. 3225 and 3975 **80.** 1344 and 936

81. 228, 336, and 552 **82.** 480, 1152, and 2400

Think About It!

Write a polynomial that has the given GCF.

83. $3x^2$ Answers vary; $6x^3 + 15x^2$

84. $5a^3$ Answers vary; $10a^4 - 5a^3$

85. $-2x$ Answers vary; $-4x^2 + 6x$

86. $-7y$ Answers vary; $-21y^3 + 14y$

Find each product. Then use grouping to factor the resulting polynomial. Observe the connection between the factors and the polynomial.

87. $(5b^2 + 4)(2b - 1)$ $10b^3 - 5b^2 + 8b - 4$; $(10b^3 - 5b^2) + (8b - 4) = 5b^2(2b - 1) + 4(2b - 1) = (2b - 1)(5b^2 + 4)$

88. $(6a - 7)(2a^2 + 3)$ $12a^3 + 18a - 14a^2 - 21$; $(12a^3 + 18a) + (-14a^2 - 21) = 6a(2a^2 + 3) - 7(2a^2 + 3) = (2a^2 + 3)(6a - 7)$

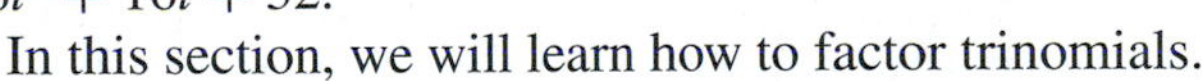

SECTION 6.6 Factoring Trinomials

OBJECTIVES

As a result of completing this section, you will be able to

1. Factor trinomials of the form $x^2 + bx + c$.
2. Factor trinomials of the form $ax^2 + bx + c$ by trial and error.
3. Factor trinomials of the form $ax^2 + bx + c$ by grouping.
4. Factor by substitution.
5. Troubleshoot common errors.

A ball is thrown upward with an initial velocity of 16 ft/sec from a height of 32 ft above the ground. The height above the ground, in feet, of the ball t seconds after it is thrown can be represented by $h = -16t^2 + 16t + 32$. To find when the ball reaches the ground, we must factor the trinomial $-16t^2 + 16t + 32$.

In this section, we will learn how to factor trinomials.

Factoring Trinomials with a Leading Coefficient of 1

Objective 1 Factor trinomials of the form $x^2 + bx + c$.

In Section 6.5, we learned to factor polynomials using the GCF and grouping. We will now learn additional ways to factor trinomials. Our goal is to express a trinomial of the form $x^2 + bx + c$ as a product, if possible. Some examples of this type of trinomial are $x^2 + 10x + 24$ and $y^2 - 5y - 24$. Recall the products that produce these trinomials.

$$\begin{aligned}(x + 4)(x + 6) &= x^2 + 4x + 6x + 24\\ &= x^2 + 10x + 24\end{aligned}$$

$$\begin{aligned}(y - 8)(y + 3) &= y^2 - 8y + 3y - 24\\ &= y^2 - 5y - 24\end{aligned}$$

If we reverse these statements, we begin with a trinomial and end with its factored form. The table shows some important observations of the binomial factors and the trinomial.

Trinomial	Factored Form	Observations
$x^2 + 10x + 24$	$(x + 4)(x + 6)$	4 and 6 are the factors of 24 whose sum is 10.
$y^2 - 5y - 24$	$(y - 8)(y + 3)$	-8 and 3 are the factors of -24 whose sum is -5.

These examples illustrate that when we factor the trinomial $x^2 + bx + c$, the factors of the constant c have a sum equal to b.

Property: Factoring $x^2 + bx + c$

For b and c real numbers,

$$x^2 + bx + c = (x + \#)(x + \#)$$

These numbers have a product of c and a sum of b.

Procedure: Factoring $x^2 + bx + c$

Step 1: Factor out common factors, if possible.
Step 2: List the factors of the last term.
Step 3: Choose the pair of factors whose sum is the middle coefficient. If no such factors exist, then the trinomial is prime over the integers and cannot be factored.
Step 4: Arrange the factors in the binomials.
Step 5: Check the factoring by multiplying.

Note: *A **prime polynomial** is a polynomial that can be factored as only 1 times the polynomial.*

Objective 1 Examples

Factor each trinomial. Check each answer by multiplying.

1a. $b^2 + 19b + 48$ **1b.** $x^2 - 33xy + 32y^2$ **1c.** $y^2 + 3y + 4$
1d. $2y^3 - 20y^2 - 150y$ **1e.** $-3a^2 - 21a + 90$

Solutions

1a. We find the factors of 48 whose sum is 19. Because the last term is positive, the factors must have the same sign. For the sum of the factors to be positive means that the factors are both positive.

Factors of 48	Sum of the Factors
1, 48	$1 + 48 = 49$
2, 24	$2 + 24 = 26$
3, 16	$3 + 16 = 19$
4, 12	$4 + 12 = 16$
6, 8	$6 + 8 = 14$

So, $b^2 + 19b + 48 = (b + 3)(b + 16)$.

Check: $(b + 3)(b + 16) = b^2 + 16b + 3b + 48 = b^2 + 19b + 48$

1b. We find the factors of 32 whose sum is -33. Because the last term is positive, the factors must have the same sign. For the sum of the factors to be negative, the factors are both negative.

Factors of 32	Sum of the Factors
$-1, -32$	$-1 + (-32) = -33$
$-2, -16$	$-2 + (-16) = -18$
$-4, -8$	$-4 + (-8) = -12$

So, $x^2 - 33xy + 32y^2 = (x - 1y)(x - 32y) = (x - y)(x - 32y)$.

Check: $(x - y)(x - 32y) = x^2 - 32xy - xy + 32y^2 = x^2 - 33xy + 32y^2$

1c. We find the factors of 4 whose sum is 3. Because the constant term is positive, the factors must have the same sign. For the sum of the factors to be positive, the factors are both positive.

Factors of 4	Sum of the Factors
1, 4	$1 + 4 = 5$
2, 2	$2 + 2 = 4$

There are not any factors of 4 whose sum is 3. Therefore, the trinomial $y^2 + 3y + 4$ is prime.

1d. We first factor out the greatest common factor of $2y$.

$$2y^3 - 20y^2 - 150y = 2y(y^2 - 10y - 75)$$

Now we factor the trinomial $y^2 - 10y - 75$ by finding the factors of -75 whose sum is -10. Because the last term is negative, the factors will have opposite signs. For the sum of the factors to be negative, the larger factor is negative.

Factors of -75	Sum of the Factors
$1, -75$	$1 + (-75) = -74$
$3, -25$	$3 + (-25) = -22$
$5, -15$	$5 + (-15) = -10$

So, $y^2 - 10y - 75 = (y + 5)(y - 15)$. The complete factorization of the original polynomial is

$$\begin{aligned} 2y^3 - 20y^2 - 150y &= 2y(y^2 - 10y - 75) \\ &= 2y(y + 5)(y - 15) \end{aligned}$$

Check:

$$\begin{aligned} 2y(y + 5)(y - 15) &= 2y(y^2 - 15y + 5y - 75) \\ &= 2y(y^2 - 10y - 75) \\ &= 2y^3 - 20y^2 - 150y \end{aligned}$$

1e. We first factor out the greatest common factor of -3.

$$-3a^2 - 21a + 90 = -3(a^2 + 7a - 30)$$

Now we factor $a^2 + 7a - 30$ by finding the factors of -30 whose sum is 7. Because the constant term is negative, the factors have opposite signs. For the sum of the factors to be positive, the larger factor is positive.

Factors of -30	Sum of the Factors
$-1, 30$	$-1 + 30 = 29$
$-2, 15$	$-2 + 15 = 13$
$-3, 10$	$-3 + 10 = 7$
$-5, 6$	$-5 + 6 = 1$

So, $a^2 + 7a - 30 = (a - 3)(a + 10)$. The complete factorization of the original polynomial is

$$\begin{aligned} -3a^2 - 21a + 90 &= -3(a^2 + 7a - 30) \\ &= -3(a - 3)(a + 10) \end{aligned}$$

Check:

$$\begin{aligned} -3(a - 3)(a + 10) &= -3(a^2 + 10a - 3a - 30) \\ &= -3(a^2 + 7a - 30) \\ &= -3a^2 - 21a + 90 \end{aligned}$$

Student Check 1 Factor each trinomial. Check each answer by multiplying.

a. $x^2 + 12x + 32$ **b.** $a^2 - 38ab + 72b^2$ **c.** $a^2 - 5a + 24$

d. $-y^3 + 4y^2 + 21y$ **e.** $2a^2 + 24a - 90$

Trial and Error

Objective 2

Factor trinomials of the form $ax^2 + bx + c$ by trial and error.

We now take a look at factoring trinomials of the form $ax^2 + bx + c$, where $a \neq 1$. Some examples of these polynomials are $6x^2 + 19x + 10$ and $12c^2 + 4c - 5$. Recall the products that produce these trinomials.

$$(2x + 5)(3x + 2) = 6x^2 + 4x + 15x + 10 = 6x^2 + 19x + 10$$
$$(6c + 5)(2c - 1) = 12c^2 - 6c + 10c - 5 = 12c^2 + 4c - 5$$

If we reverse these statements, we begin with a trinomial and end with its factored form. The table shows some important observations.

Trinomial	Factored Form	Observations
$6x^2 + 19x + 10$	$(2x + 5)(3x + 2)$	$2x$ and $3x$ are factors of $6x^2$. 5 and 2 are factors of 10. The sum of the outer and inner products, $4x$ and $15x$, is $19x$.
$12c^2 + 4c - 5$	$(6c + 5)(2c - 1)$	$6c$ and $2c$ are factors of $12c^2$ 5 and -1 are factors of -5. The sum of the outer and inner products, $-6c$ and $10c$, is $4c$.

These examples illustrate the relationship between the binomial factors and the trinomial $ax^2 + bx + c$.

1. The first terms of the binomials are factors of the first term of the trinomial, ax^2.
2. The last terms of the binomials are factors of the last term of the trinomial, c.
3. The sum of the outer and inner products is the middle term of the trinomial, bx.

We can use these facts as guidelines when factoring trinomials by the *trial and error method*. The process of trial and error involves arranging all possible factors of the first term of the trinomial and factors of the last term of the trinomial until we obtain the correct outer and inner products.

Property: Factoring $ax^2 + bx + c$ by Trial and Error

For a, b, and c real numbers and $a \neq 0$, the factored form of $ax^2 + bx + c$ is

F O+I L

$$ax^2 + bx + c = (\_x + \_)(\_x + \_)$$

F L I O

Procedure: Factoring $ax^2 + bx + c, a \neq 1$ by Trial and Error

Step 1: Factor out common factors, if possible.

Step 2: Determine the signs of the factors.

a. If the last sign of the trinomial is positive, the signs of the factors of the last term are the same as the sign of the middle coefficient of the trinomial.

b. If the last sign of the trinomial is negative, the signs of the factors of the last term are different.

Step 3: List the factors of the first term of the trinomial.

Step 4: List the factors of the last term of the trinomial.
Step 5: Arrange the factors of the first and last terms in two binomials and check if the product of the binomials produces the given trinomial.

When trying different factors, be mindful of the following.

- If the first arrangement of factors doesn't work, obtain another combination by reversing the position of the factors of the last term.
- If the trinomial does not have a common factor, the binomial factors cannot have a common factor. So, we can eliminate any combinations in which one or both of the binomials has a common factor in this case.
- If none of the combinations work, then the trinomial is prime.

Objective 2 Examples

Factor each trinomial by trial and error.

2a. $5x^2 + 17x + 6$ **2b.** $2a^2 - 5a - 12$ **2c.** $12x^3 - 60x^2 + 75x$

Solutions

2a. Since the last sign of the trinomial is positive, the signs of the factors of the last term and hence the binomial factors are the same. The signs are positive since the coefficient of the middle term is positive.

INSTRUCTOR NOTE: Point out that because of the way the factors are selected, the first terms and last terms of the trinomials will always be correct. The difference is that they each produce different middle terms.

Factors of $5x^2$ and 6		Possible Factorizations	
$5x, x$	1, 6	$(5x + 1)(x + 6) = 5x^2 + 30x + x + 6 = 5x^2 + 31x + 6$	Wrong middle term
		$(5x + 6)(x + 1) = 5x^2 + 5x + 6x + 6 = 5x^2 + 11x + 6$	Wrong middle term
	2, 3	$(5x + 2)(x + 3) = 5x^2 + 15x + 2x + 6 = 5x^2 + 17x + 6$	Correct middle term
		$(5x + 3)(x + 2) = 5x^2 + 10x + 3x + 6 = 5x^2 + 13x + 6$	Wrong middle term

So, $5x^2 + 17x + 6 = (5x + 2)(x + 3)$.

2b. Since the last term of the trinomial is negative, the signs of the factors of the last term and hence the binomial factors are different.

Factors of $2a^2$ and -12		Possible Factorizations	
$2a, a$	$1, -12$	$(2a + 1)(a - 12) = 2a^2 - 12a + a - 12 = 2a^2 - 11a - 12$	Wrong middle term
		$(2a - 12)(a + 1)$	Common factor
	$-1, 12$	$(2a - 1)(a + 12) = 2a^2 + 24a - a - 12 = 2a^2 + 23a - 12$	Wrong middle term
		$(2a + 12)(a - 1)$	Common factor
	$2, -6$	$(2a + 2)(a - 6)$	Common factor
		$(2a - 6)(a + 2)$	Common factor
	$-2, 6$	$(2a - 2)(a + 6)$	Common factor
		$(2a + 6)(a - 2)$	Common factor
	$3, -4$	$(2a + 3)(a - 4) = 2a^2 - 8a + 3a - 12 = 2a^2 - 5a - 12$	Correct middle term
		$(2a - 4)(a + 3)$	Common factor
	$-3, 4$	$(2a - 3)(a + 4) = 2a^2 + 8a - 3a - 12 = 2a^2 + 5a - 12$	Wrong middle term
		$(2a + 4)(a - 3)$	Common factor

So, $2a^2 - 5a - 12 = (2a + 3)(a - 4)$.

INSTRUCTOR NOTE:
Point out that students don't have to list every possible factor. Once they find the correct middle term, their work is done.

2c. The terms of the trinomial have a GCF of $3x$. We factor this out and then factor the remaining trinomial.

$$12x^3 - 60x^2 + 75x = 3x(4x^2 - 20x + 25)$$

Now we factor $4x^2 - 20x + 25$. Since the last sign of this trinomial is positive, the signs of the factors of the last term and hence the binomial factors are the same. The signs are negative since the coefficient of the middle term is negative.

Factors of $4x^2$ and 25		Possible Factorizations	
$4x, x$	$-1, -25$	$(x-1)(4x-25) = 4x^2 - 25x - 4x + 25$ $= 4x^2 - 29x + 25$	Wrong middle term
		$(x-25)(4x-1) = 4x^2 - x - 100x + 25$ $= 4x^2 - 101x + 25$	Wrong middle term
	$-5, -5$	$(x-5)(4x-5) = 4x^2 - 5x - 20x + 25$ $= 4x^2 - 25x + 25$	Wrong middle term
$2x, 2x$	$-1, -25$	$(2x-1)(2x-25) = 4x^2 - 50x - 2x + 25$ $= 4x^2 - 52x + 25$	Wrong middle term
	$-5, -5$	$(2x-5)(2x-5) = 4x^2 - 10x - 10x + 25$ $= 4x^2 - 20x + 25$	Correct middle term

So, $12x^3 - 60x^2 + 75x = 3x(4x^2 - 20x + 25) = 3x(2x-5)(2x-5) = 3x(2x-5)^2$.

Note: *The trinomial* $4x^2 - 20x + 25$ *is a* perfect square trinomial *since it can be factored as a binomial times itself.*

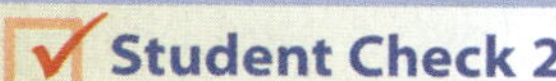

Student Check 2 Factor each trinomial by trial and error.

a. $5x^2 + 21x + 4$ **b.** $7x^2 - 11x - 6$ **c.** $45y^3 - 60y^2 + 20y$

Factoring by Grouping

Objective 3 ▶

Factor trinomials of the form $ax^2 + bx + c$ by grouping.

Factoring by trial and error is a method that works, but it can often be a lengthy process when the first term of the trinomial and/or the last term have a lot of factors. We will now learn a systematic approach to factoring trinomials. This process expands the middle term of the trinomial to two terms, specifically the outer and inner products, that would result from multiplying the two binomial factors together. This expanded polynomial has four terms which we can factor by grouping. This method is basically FOIL in reverse.

We will use the products that we examined in Objective 2 to illustrate this concept and will observe how the outer and inner products relate to the terms in the trinomial.

Trinomial and Its Factored Form	Observations
$(2x+5)(3x+2) = 6x^2 + 4x + 15x + 10$ $= 6x^2 + 19x + 10$	$6(10) = 60$ $4(15) = 60$ and $4 + 15 = 19$
$(6c+5)(2c-1) = 12c^2 - 6c + 10c - 5$ $= 12c^2 + 4c - 5$	$12(-5) = -60$ $(-6)(10) = -60$ and $(-6) + 10 = 4$

From the observations, we conclude that the coefficients of the outer and inner terms have a product equal to $a \cdot c$ and have a sum equal to the middle coefficient of the trinomial, b. We can summarize this as follows.

Property: Factoring $ax^2 + bx + c$ by Grouping

The coefficients of the outer and inner terms are factors of the number $a \cdot c$ and have a sum of b.

$$ax^2 + bx + c = ax^2 + \underbrace{b_1x + b_2x}_{\substack{b_1 \cdot b_2 = a \cdot c \\ b_1 + b_2 = b}} + c = \underbrace{(\_x + \_)(\_x + \_)}_{\text{Obtained by grouping}}$$

Procedure: Factoring $ax^2 + bx + c$ $(a \neq 1)$ by Grouping

Step 1: Factor out common factors, if possible.
Step 2: Find the product of the leading coefficient and the constant term, that is $a \cdot c$.
Step 3: List the factors of $a \cdot c$ and find the factors that have a sum of b, the middle coefficient of the trinomial. If there is no such pair, the trinomial is prime.
Step 4: Rewrite the term bx as a sum using the factors found in step 3.
Step 5: Factor by grouping.
Step 6: Check by multiplying.

Objective 3 Examples

Factor each trinomial by grouping.

3a. $15x^2 + 23x + 4$ **3b.** $7a^2 - 19a - 6$

Solutions

3a. The terms of this trinomial contain no common factors other than 1. The trinomial is in the form $ax^2 + bx + c$, where $a = 15$ and $c = 4$. The product $a \cdot c = 15(4) = 60$. We find the factors of 60 whose sum is $b = 23$, the middle coefficient. Since the product is positive and the sum is positive, both factors are positive.

Factors of $a \cdot c = 60$	Sum of Factors
1, 60	$1 + 60 = 61$
2, 30	$2 + 30 = 32$
3, 20	$3 + 20 = 23$
4, 15	$4 + 15 = 19$
6, 10	$6 + 10 = 16$

The factors whose sum produces the middle coefficient are 3 and 20.

INSTRUCTOR NOTE:
Show students that the order of the terms that replace the middle term of the trinomial does not matter. Either order will produce the correct factorization.

$$\begin{aligned} 15x^2 + 23x + 4 &= 15x^2 + 3x + 20x + 4 && \text{Replace } 23x \text{ with } 3x + 20x. \\ &= (15x^2 + 3x) + (20x + 4) && \text{Group pairs of terms.} \\ &= 3x(5x + 1) + 4(5x + 1) && \text{Factor out the GCF from each pair.} \\ &= (5x + 1)(3x + 4) && \text{Factor out the common binomial, } 5x + 1. \end{aligned}$$

So, $15x^2 + 23x + 4 = (5x + 1)(3x + 4)$.

3b. The terms of this trinomial contain no common factors other than 1. The trinomial is in the form $ax^2 + bx + c$, where $a = 7$ and $c = -6$. The product $a \cdot c = 7(-6) = -42$. We need to find the factors of -42 whose sum is the middle coefficient, $b = -19$. The signs of the factors are different since the product is negative.

Factors of $a \cdot c = -42$	Sum of Factors
$-1, 42$	$-1 + 42 = 41$
$1, -42$	$1 + (-42) = -41$
$-2, 21$	$-2 + 21 = 19$
$2, -21$	$2 + (-21) = -19$
$-3, 14$	$-3 + 14 = 11$
$3, -14$	$3 + (-14) = -11$
$-6, 7$	$-6 + 7 = 1$
$6, -7$	$6 + (-7) = -1$

The factors whose sum produces the middle coefficient are 2 and -21.

$7a^2 - 19a - 6 = 7a^2 + 2a - 21a - 6$ — Replace $-19a$ with $2a - 21a$.

$= (7a^2 + 2a) + (-21a - 6)$ — Group pairs of terms.

$= a(7a + 2) - 3(7a + 2)$ — Factor out the GCF from each pair.

$= (7a + 2)(a - 3)$ — Factor out the common binomial, $7a + 2$.

So, $7a^2 - 19a - 6 = (7a + 2)(a - 3)$.

Note: *The order in which we insert the two terms will not affect the factorization. For instance, we could replace $-19a$ with $-21a + 2a$.*

$7a^2 - 19a - 6 = 7a^2 - 21a + 2a - 6$ — Replace $-19a$ with $-21a + 2a$.

$= (7a^2 - 21a) + (2a - 6)$ — Group pairs of terms.

$= 7a(a - 3) + 2(a - 3)$ — Factor out the GCF from each pair of terms.

$= (a - 3)(7a + 2)$ — Factor out the common binomial, $a - 3$.

Student Check 3 Factor each trinomial by grouping.

a. $4x^2 + 13x + 3$ **b.** $6x^2 - x - 12$

Substitution

Objective 4 Factor by substitution.

Not all polynomials are given to us in the form $ax^2 + bx + c$, but that doesn't mean that we can't apply the methods of this section to factor them. We will make an appropriate substitution to write the given trinomial in the form $ax^2 + bx + c$ and then factor it.

The examples in the table illustrate two different substitutions.

Original Polynomial	Substitution to be Made	New Trinomial
$x^4 + 8x^2 - 9 = (x^2)^2 + 8x^2 - 9$	$u = x^2$	$u^2 + 8u - 9$
$3(y + 1)^2 + 5(y + 1) + 2$	$u = (y + 1)$	$3u^2 + 5u + 2$

Objective 4 Examples Use an appropriate substitution to factor each polynomial.

4a. $x^4 + 8x^2 - 9$ **4b.** $3(y + 1)^2 + 5(y + 1) + 2$

Solutions **4a.** Let $u = x^2$. Rewrite the polynomial in terms of u.

$$\begin{aligned} x^4 + 8x^2 - 9 &= (x^2)^2 + 8x^2 - 9 && \text{Write } x^4 \text{ in terms of } x^2. \\ &= u^2 + 8u - 9 && \text{Substitute } u \text{ for } x^2. \\ &= (u - 1)(u + 9) && \text{Factor } u^2 + 8u - 9. \\ &= (x^2 - 1)(x^2 + 9) && \text{Replace } u \text{ with } x^2. \end{aligned}$$

So, $x^4 + 8x^2 - 9 = (x^2 - 1)(x^2 + 9)$. We can check by multiplying.

Note: *This trinomial can be factored without using substitution. Since* $x^4 = x^2 \cdot x^2$, *we get* $x^4 + 8x^2 - 9 = (x^2 + 9)(x^2 - 1)$.

4b. Let $u = (y + 1)$. Rewrite the polynomial in terms of u.

$$\begin{aligned} &3(y + 1)^2 + 5(y + 1) + 2 \\ &\quad = 3u^2 + 5u + 2 && \text{Substitute } u \text{ for } (y + 1). \\ &\quad = (3u + 2)(u + 1) && \text{Factor } 3u^2 + 5u + 2. \\ &\quad = [3(y + 1) + 2]\,[(y + 1) + 1] && \text{Replace } u \text{ with } (y + 1). \\ &\quad = (3y + 3 + 2)(y + 1 + 1) && \text{Apply the distributive property.} \\ &\quad = (3y + 5)(y + 2) && \text{Simplify each factor.} \end{aligned}$$

So, $3(y + 1)^2 + 5(y + 1) + 2 = (3y + 5)(y + 2)$.

Student Check 4 Use an appropriate substitution to factor each polynomial.

a. $x^4 - 9x^2 + 20$ **b.** $5(h - 2)^2 + 8(h - 2) - 4$

Troubleshooting Common Errors

Objective 5 Troubleshoot common errors.

Common errors with factoring trinomials involve mistakes with signs, failure to factor completely, failure to exhaust all possible factorizations, and confusing the methods of grouping and trial and error.

Objective 5 Examples **A problem and an incorrect solution are given. Provide the correct solution and an explanation of the error.**

5a. Factor $x^2 + 12x - 35$.

Incorrect Solution	Correct Solution and Explanation
$x^2 + 12x - 35 = (x - 7)(x - 5)$	When we check by multiplying, we get $(x - 7)(x - 5) = x^2 - 5x - 7x + 35 = x^2 - 12x + 35$ which is not the given trinomial. To factor $x^2 + 12x - 35$, we find the factors of -35 whose sum is 12. The factors of -35 are $-1, 35$; $1, -35$; $-5, 7$; and $5, -7$. None of these has a sum of 12. Therefore, the polynomial is prime.

5b. Factor $6x^2 + 13x + 6$.

Incorrect Solution	Correct Solution and Explanation
$a \cdot c = 6 \cdot 6 = 36$ The factors of 36 whose sum is 13 are 4 and 9. So, $6x^2 + 13x + 6 = (2x + 9)(3x + 4)$.	The factors of 36 whose sum is 13 are 4 and 9 but these do not go in the binomial factors. These numbers are used to expand the middle term. $6x^2 + 13x + 6 = 6x^2 + 4x + 9x + 6 = 2x(3x + 2) + 3(3x + 2) = (3x + 2)(2x + 3)$

ANSWERS TO STUDENT CHECKS

Student Check 1 **a.** $(x + 4)(x + 8)$ **b.** $(a - 36b)(a - 2b)$ **c.** prime **d.** $-y(y - 7)(y + 3)$ **e.** $2(a + 15)(a - 3)$

Student Check 2 **a.** $(5x + 1)(x + 4)$ **b.** $(7x + 3)(x - 2)$ **c.** $5y(3y - 2)(3y - 2) = 5y(3y - 2)^2$

Student Check 3 **a.** $(4x + 1)(x + 3)$ **b.** $(3x + 4)(2x - 3)$

Student Check 4 **a.** $(x^2 - 4)(x^2 - 5)$ **b.** $h(5h - 12)$

SUMMARY OF KEY CONCEPTS

To factor a trinomial, we must observe the clues that are given to us in the polynomial.

1. To factor a trinomial $x^2 + bx + c$, we find the factors of c whose sum is b.
   a. The sign of the constant term tells us if the factors are going to have the same sign or different signs.
      i. If $c > 0$, the factors will have the same sign.
      ii. If $c < 0$, the factors will have different signs.
   b. The sign of the middle coefficient, b, tells us the specific signs.
      i. If the signs of the factors are the same, they are the same as the sign of the middle coefficient.
      ii. If the signs of the factors are different, the sign of the middle coefficient is the sign of the larger factor.
2. There are two methods that can be used to factor a trinomial whose leading coefficient is not 1. We can use trial and error or grouping.
   a. Trial and error requires us to try every combination of factors of the first term with factors of the constant term until the correct middle term is obtained.
   b. Grouping is a methodical process that requires us to find factors of the product of the leading coefficient and the constant term that adds to the middle coefficient.
3. If the trinomial is not in the form $ax^2 + bx + c$, determine if there is an appropriate substitution that will put it in this form.
4. Every factoring problem can be checked by multiplying.
5. Summary of signs

___+___+___ = (___+___)(___+___)
___−___+___ = (___−___)(___−___)
___+___−___ = (___+___)(___−___)
___−___−___ = (___+___)(___−___)

GRAPHING CALCULATOR SKILLS

The graphing calculator can assist us in finding factors of a number and the sum of the factors. The calculator can also be used to verify our factored form.

Example 1: Find the factors of -75 whose sum is 10.

Solution: In Y_1, enter the expression for finding the factors of -75. In Y_2, enter the expression to add the two factors.

```
Plot1 Plot2 Plot3
\Y1=-75/X
\Y2=X+Y1
\Y3=
\Y4=
\Y5=
\Y6=
\Y7=
```

X	Y1	Y2
-5	15	10
-4	18.75	14.75
-3	25	22
-2	37.5	35.5
-1	75	74
0	ERR:	ERR:
1	-75	-74

X=-5

The first two columns, X and Y_1, are factors of -75. The last column, Y_2, shows the sum of those factors. So, -5 and 15 are the factors we need.

Example 2: Verify that $x^2 - 6x - 16 = (x + 2)(x - 8)$.

Solution: Enter the trinomial in Y_1 and the factored form in Y_2. If the factored form is correct, these two expressions should have the same value when evaluated for different values of X. We can confirm this using the TABLE feature.

```
Plot1 Plot2 Plot3
\Y1=X²-6X-16
\Y2=(X-8)(X+2)
\Y3=
\Y4=
\Y5=
\Y6=
\Y7=
```

X	Y1	Y2
-3	11	11
-2	0	0
-1	-9	-9
0	-16	-16
1	-21	-21
2	-24	-24
3	-25	-25

X=3

Since the values of Y_1 and Y_2 are the same for every value of X, the factorization is correct.

SECTION 6.6 EXERCISE SET

Write About It!

Use complete sentences in your answer to each exercise.

1. Explain how to determine the signs of the factors of c if c is a positive constant when factoring $x^2 + bx + c$.
2. Explain how to determine the signs of the factors of c if c is a negative constant when factoring $x^2 + bx + c$.
3. Explain how to determine the factors of c if c is a positive constant and b is a negative coefficient when factoring $x^2 + bx + c$.
4. Explain how to determine the factors of c if c is a negative constant and b is a positive coefficient when factoring $x^2 + bx + c$.
5. Describe how to factor $5x^2 - 19x + 14$ using the grouping method.
6. Describe how to factor $a^8 - 4a^4 - 21$ using substitution.
7. Describe how to factor $10(y + 5)^2 - 13(y + 5) - 9$ using substitution.
8. If you carry out the multiplication in Exercise 7 and factor, do you expect to have the same solution? Please explain.

Practice Makes Perfect!

Factor each trinomial completely. Check each answer by multiplying. (*See Objective 1.*)

9. $y^2 - 14y + 40$ $(y - 4)(y - 10)$
10. $x^2 - 13x + 12$ $(x - 1)(x - 12)$
11. $p^2 + 7p - 98$ $(p - 7)(p + 14)$
12. $b^2 + 9b - 36$ $(b - 3)(b + 12)$
13. $a^2 + 11a + 28$ $(a + 7)(a + 4)$
14. $s^2 - 5s - 24$ $(s - 8)(s + 3)$
15. $x^2 - 17x + 84$ prime
16. $c^2 - 8c - 36$ prime
17. $x^2 - 13x - 30$ $(x - 15)(x + 2)$
18. $c^2 - 10c - 24$ $(c + 2)(c - 12)$
19. $2x^3 - 2x^2 - 112x$ $2x(x - 8)(x + 7)$
20. $2y^3 + 42y^2 + 220y$ $2y(y + 10)(y + 11)$
21. $-3a^3 - 12a^2 + 180a$ $-3a(a - 6)(a + 10)$
22. $-6x^3 - 48x^2 - 72x$ $-6x(x + 2)(x + 6)$
23. $-6b^3 - 6b^2 + 18b$ $-6b(b^2 + b - 3)$
24. $5v^3 - 30v^2 - 60v$ $5v(v^2 - 6v - 12)$
25. $4b^3 - 68b^2 + 120b$ $4b(b - 15)(b - 2)$
26. $7a^3 + 21a^2 - 126a$ $7a(a - 3)(a + 6)$

Factor each trinomial completely by trial and error. (*See Objective 2.*)

27. $5c^2 + 13c - 28$ $(5c - 7)(c + 4)$
28. $8b^2 - 15b - 27$ $(b - 3)(8b + 9)$
29. $20x^2 + 9x + 1$ $(5x + 1)(4x + 1)$
30. $56y^2 - 29y + 3$ $(7y - 1)(8y - 3)$
31. $49a^2 - 7a - 12$ $(7a - 4)(7a + 3)$
32. $48b^2 + 2b - 63$ $(6b + 7)(8b - 9)$
33. $5x^2 - 18x - 12$ prime
34. $7y^2 + 16y - 18$ prime

Additional answers can be found in the Instructor Answer Appendix.

35. $24p^3 - 81p^2 - 60p$ $3p(8p + 5)(p - 4)$
36. $6q^3 - 46q^2 - 16q$ $2q(3q + 1)(q - 8)$
37. $-18a^3 + 84a^2 - 48a$ $-6a(a - 4)(3a - 2)$
38. $-24b^3 + 6b^2 + 9b$ $-3b(4b - 3)(2b + 1)$
39. $-a^3 + a^2 - 12a$ $-a(a^2 - a + 12)$
40. $4c^3 + 18c^2 - 14c$ $2c(2c^2 + 9c - 7)$
41. $-8x^3 - 84x^2 - 108x$ $-4x(2x + 3)(x + 9)$
42. $-10y^3 + 58y^2 + 84y$ $-2y(5y + 6)(y - 7)$

Factor each trinomial completely by grouping. (*See Objective 3.*)

43. $x^2 - 12x + 32$ $(x - 8)(x - 4)$
44. $y^2 - 19y + 84$ $(y - 12)(y - 7)$
45. $p^2 + 2p - 63$ $(p + 9)(p - 7)$
46. $q^2 - 6q - 40$ $(q + 4)(q - 10)$
47. $3a^2 - 19a - 14$ $(3a + 2)(a - 7)$
48. $2b^2 - 9b + 9$ $(2b - 3)(b - 3)$
49. $35x^2 - 57x + 10$ $(5x - 1)(7x - 10)$
50. $10y^2 + 3y - 27$ $(2x - 3)(5x + 9)$
51. $5c^2 + 11c - 12$ $(5c - 4)(c + 3)$
52. $42a^2 + 25a + 3$ $(6a + 1)(7a + 3)$
53. $-12a^2 + 56a - 60$ $-4(a - 3)(3a - 5)$
54. $-40b^2 + 408b - 80$ $-8(5b - 1)(b - 10)$

Use an appropriate substitution to factor each polynomial. (*See Objective 4.*)

55. $a^4 - 16a^2 + 60$ $(a^2 - 10)(a^2 - 6)$
56. $b^4 + 16b^2 + 48$ $(b^2 + 4)(b^2 + 12)$
57. $x^6 + x^3 - 30$ $(x^3 + 6)(x^3 - 5)$
58. $y^6 - 11y^3 - 60$ $(y^3 - 15)(y^3 + 4)$
59. $(x - 3)^2 - 2(x - 3) - 80$ $(x + 5)(x - 13)$
60. $(v + 1)^2 - 14(v + 1) + 24$ $(v - 11)(v - 1)$
61. $24(x - 1)^2 - 38(x - 1) + 15$ $(6x - 11)(4x - 7)$
62. $9(y + 1)^2 + 23(y + 1) - 12$ $(9y + 5)(y + 4)$
63. $6(a + 5)^2 + 7(a + 5) - 5$ $(2a + 9)(3a + 20)$
64. $5(b - 1)^2 - 34(b - 1) - 7$ $(5b - 4)(b - 8)$

Mix 'Em Up!

Factor each trinomial completely.

65. $v^2 + 18v + 56$ $(v + 4)(v + 14)$
66. $r^2 - 20r + 36$ $(r - 2)(r - 18)$
67. $18x^2 - 51xy + 8y^2$ $(3x - 8y)(6x - y)$
68. $42x^2 + 43xy + 6y^2$ $(7x + 6y)(6x + y)$
69. $a^2 - 23a - 50$ $(a - 25)(a + 2)$
70. $b^2 - 11b - 80$ $(b - 16)(b + 5)$
71. $y^2 - 9y - 70$ $(y - 14)(y + 5)$
72. $x^2 + x - 30$ $(x - 5)(x + 6)$
73. $-5b^3 - 50b^2 + 120b$ $-5b(b - 2)(b + 12)$
74. $-36x^3 - 96x^2y + 80xy^2$ $-4x(3x - 2y)(3x + 10y)$
75. $x^2 - 12xy + 35y^2$ $(x - 7y)(x - 5y)$
76. $a^2 - 7ab - 30b^2$ $(a + 3b)(a - 10b)$
77. $8y^2 - 19y - 52$ $(8y + 13)(y - 4)$
78. $4b^2 + 13b - 35$ $(4b - 7)(b + 5)$
79. $7x^2 - 39x + 20$ $(x - 5)(7x - 4)$
80. $20p^2 - 56p + 15$ $(10p - 3)(2p - 5)$
81. $10x^2 + 7x + 3$ prime
82. $5y^2 - 3y + 8$ prime
83. $40a^3 - 75a^2b - 10ab^2$ $5a(a - 2b)(8a + b)$
84. $140x^3 + 12x^2y - 8xy^2$ $4x(7x + 2y)(5x - y)$

85. $35b^2 - 48b - 27$ $(5b - 9)(7b + 3)$

86. $8a^2 + 2a - 45$ $(4a - 9)(2a + 5)$

87. $12a^2 - 11ab - 15b^2$ $(4a + 3b)(3a - 5b)$

88. $21a^2 - 25ab - 4b^2$ $(3a - 4b)(7a + b)$

89. $6v^3 + 6v^2 - 252v$ $6v(v + 7)(v - 6)$

90. $2p^3 - 30p^2 + 108p$ $2p(p - 9)(p - 6)$

91. $-a^3 + 9a^2 - 18a$ $-a(a - 3)(a - 6)$

92. $-32a^3 + 64a^2b - 24ab^2$ $-8a(2a - 3b)(2a - b)$

93. $x^6 + 21x^3 + 110$ $(x^3 + 11)(x^3 + 10)$

94. $y^6 + 20y^3 + 91$ $(y^3 + 13)(y^3 + 7)$

95. $4(x - 5)^2 - 5(x - 5) - 6$ $(x - 7)(4x - 17)$

96. $6(y + 4)^2 - 17(y + 4) - 3$ $(y + 1)(6y + 25)$

97. $a^4 - 14a^2 - 15$ $(a^2 - 15)(a^2 + 1)$

98. $b^4 - 7b^2 - 30$ $(b^2 - 10)(b^2 + 3)$

99. $40x^4 + x^2y^2 - 15y^4$ $(5x^2 - 3y^2)(8x^2 + 5y^2)$

100. $4x^4 + 13x^2y^2 - 35y^4$ $(4x^2 - 7y^2)(x^2 + 5y^2)$

101. $(a + 6)^2 - 11(a + 6) + 24$ $(a + 3)(a - 2)$

102. $(b - 4)^2 + 2(b - 4) - 15$ $(b + 1)(b - 7)$

103. $6p^8 + 11p^4 - 2$ $(p^4 + 2)(6p^4 - 1)$

104. $14q^8 + 9q^4 - 8$ $(7q^4 + 8)(2q^4 - 1)$

105. $42(a - 5)^2 + 29(a - 5) - 5$ $(6a - 25)(7a - 36)$

106. $20(b - 2)^2 - 9(b - 2) + 1$ $(5b - 11)(4b - 9)$

You Be the Teacher!

Correct each student's errors, if any.

107. Factor $a^2 - 15a - 54$.

Elisa's work:

$a^2 - 15a - 54 = (a - 9)(a - 6)$

$a^2 - 15a - 54 = (a - 18)(a + 3)$

108. Factor $x^2 + 20x + 96$.

Ngoc's work:

$x^2 + 20x + 96 = (x + 24)(x - 4)$

$x^2 + 20x + 96 = (x + 12)(x + 8)$

109. Factor $-4b^2 + 2b - 30$,

Wing's work:

$$\begin{aligned} -4b^2 + 2b - 30 &= -2(2b^2 + b - 15) \\ &= -2(2b^2 + 6b - 5b - 15) \\ &= -2[2b(b + 3) - 5(b + 3)] \\ &= -2(b + 3)(2b - 5) \end{aligned}$$

$-4b^2 + 2b - 30 = -2(2b^2 - b + 15)$

110. Factor $-6a^2 - 30a + 36$.

Shane's work:

$$\begin{aligned} -6a^2 - 30a + 36 &= -6(a^2 - 5a + 6) \\ &= -6(a^2 - 3a - 2a + 6) \\ &= -6[a(a - 3) - 2(a - 3)] \\ &= -6(a - 3)(a - 2) \end{aligned}$$

111. Factor $12(a - 4)^2 + 4(a - 4) - 5$.

Loretta's work:

$$\begin{aligned} 12(a - 4)^2 + 4(a - 4) - 5 &= (6a - 4 + 5)(2a - 4 - 1) \\ &= (6a + 1)(2a - 5) \end{aligned}$$

112. Factor $2(b - 6)^2 + 23(b - 6) + 45$.

Sam's work:

$$\begin{aligned} 2(b - 6)^2 + 23(b - 6) + 45 &= (2b - 6 + 5)(b - 6 + 9) \\ &= (2b - 1)(b + 3) \end{aligned}$$

Calculate It!

Use a graphing calculator to solve each problem.

113. Find the factors of 45 whose sum is -14. -9 and -5

114. Find the factors of 120 whose sum is 22. 12 and 10

115. Find the factors of -156 whose sum is -1. -13 and 12

116. Find the factors of -108 whose sum is -3. -12 and 9

117. Verify $x^2 - 12x + 35 = (x - 5)(x - 7)$.

118. Verify $x^2 + 5x - 104 = (x - 8)(x + 13)$.

119. Verify $x^2 - 9x - 90 = (x - 15)(x + 6)$.

120. Verify $x^2 - 12x - 45 = (x - 15)(x + 3)$.

Think About It!

Determine the integer values of *b* for which the trinomial is factorable.

121. $x^2 + bx + 10$ $-11, -7, 7, 11$

122. $x^2 + bx + 24$ $-25, -14, -11, -10, 10, 11, 14, 25$

123. $x^2 + bx - 10$ $-9, -3, 3, 9$

124. $x^2 + bx - 24$ $-23, -10, -5, -2, 2, 5, 10, 23$

125. $3x^2 + bx + 5$ $-16, -8, 8, 16$

126. $4x^2 + bx - 7$ $-27, -3, 3, 27$

PIECE IT TOGETHER SECTIONS 6.5–6.6

Find the GCF of each set of terms. (*Section 6.5, Objective 1*)

1. $28z^3$, $35z^6$, and $7z^2$ $7z^2$

2. $12c^7$, $15c^2$, and $48c^4$ $3c^2$

3. $42y^2(x - 5)$ and $28y(x - 5)$ $14y(x - 5)$

4. $51x^2(y + 1)$ and $34x(y + 1)$ $17x(y + 1)$

Use the GCF to factor each polynomial completely. Check each answer by multiplying. (*Section 6.5, Objective 2*)

5. $-52b^4 + 39b^2$ $-13b^2(4b^2 - 3)$

6. $42a^2 - 35a + 7$ $7(3a - 1)(2a - 1)$

7. $28y(x + 3) - 32(x + 3)$ $4(x + 3)(7y - 8)$

8. $30x^3(y - 1) - 45x^2(y - 1)$ $15x^2(y - 1)(2x - 3)$

Use grouping to factor each polynomial completely, if applicable. Check each answer by multiplying. (*Section 6.5, Objective 3*)

9. $-24y^5 + 28y^4$ $-4y^4(6y - 7)$

10. $-52b^4 + 39b^2$ $-13b^2(4b^2 - 3)$

11. $28y^4 + 28y^3 - 21y - 21$ $7(y + 1)(4y^3 - 3)$

12. $18x^5 + 12x^4 + 18x + 12$ $6(3x + 2)(x^4 + 1)$

Factor each trinomial. (*Section 6.6, Objectives 1–3*)

13. $x^2 - 11x - 26$ $(x - 13)(x + 2)$

14. $x^2 - 20xy + 99y^2$ $(x - 9y)(x - 11y)$

15. $36v^3 + 63v^2 + 27v$ $9(v + 1)(4v + 3)$

16. $-8x^3 + 56x^2 - 48x$ $-8x(x - 6)(x - 1)$

17. $(y + 2)^2 + 5(y + 2) - 36$ $(y - 2)(y + 11)$

18. $(u + 8)^2 - (u + 8) - 20$ $(u + 12)(u + 3)$

19. $35x^2 - 57x + 10$ $(5x - 1)(7x - 10)$

20. $10y^2 + 3y - 27$ $(2x - 3)(5x + 9)$

SECTION 6.7 Factoring Binomials and a Factoring Review

OBJECTIVES

As a result of completing this section, you will be able to

1. Factor the difference of two squares.
2. Factor the sum or difference of two cubes.
3. Apply factoring techniques.
4. Troubleshoot common errors.

B.A.S.E. jumping is an activity whereby people jump from structures with only a brief period before they must open a parachute to survive. If a person jumps from a 1600-ft structure, his or her height can be represented by $h = -16t^2 + 1600$, where t is the number of seconds after the fall. To determine when the jumper reaches the ground if his chute doesn't open, we can factor $-16t^2 + 1600$.

This section will complete our study of factorization techniques. In this section, we will concentrate on factoring binomials that fit a special pattern.

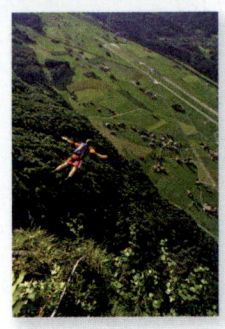

Difference of Two Squares

Objective 1 Factor the difference of two squares.

In Section 6.4, we learned that there are products which result in the difference of two squares. Recall,

$$(x + 3)(x - 3) = x^2 - 9$$
$$(2y + 7)(2y - 7) = 4y^2 - 49$$

So, the product of the sum and difference of two terms results in a binomial which is the difference of two squares. That is, the *product of conjugates is the difference of two squares*. To use this as a factoring method, we reverse the previous statements.

Difference of Two Squares	Factored Form
$x^2 - 9 = (x)^2 - (3)^2$	$(x + 3)(x - 3)$
$4y^2 - 49 = (2y)^2 - (7)^2$	$(2y + 7)(2y - 7)$

Property: The Difference of Two Squares

$$a^2 - b^2 = (a + b)(a - b)$$

It is helpful to remember some perfect squares.

$1^2 = 1$	$6^2 = 36$	$11^2 = 121$	$(x)^2 = x^2$
$2^2 = 4$	$7^2 = 49$	$12^2 = 144$	$(x^2)^2 = x^4$
$3^2 = 9$	$8^2 = 64$	$13^2 = 169$	$(x^3)^2 = x^6$
$4^2 = 16$	$9^2 = 81$	$14^2 = 196$	$(x^4)^2 = x^8$
$5^2 = 25$	$10^2 = 100$	$15^2 = 225$	$(x^5)^2 = x^{10}$

Procedure: Factoring the Difference of Two Squares

Step 1: Factor out any common factor.
Step 2: Rewrite the first term as a^2, where a is the quantity that must be squared to obtain the first term.
Step 3: Rewrite the second term as b^2, where b is the quantity that must be squared to obtain the second term.
Step 4: Write the factorization as $(a - b)(a + b)$.
Step 5: Check by multiplying.

Objective 1 Examples Factor each binomial completely.

1a. $b^2 - \dfrac{4}{9}$ **1b.** $25x^2 - 36y^2$ **1c.** $-3x^3 + 12x$

1d. $(x + 5)^2 - 16$ **1e.** $x^2 + 4$

Solutions

1a. $b^2 - \frac{4}{9} = b^2 - \left(\frac{2}{3}\right)^2$ — Write as $a^2 - b^2$.

$= \left(b - \frac{2}{3}\right)\left(b + \frac{2}{3}\right)$ — Apply the difference of two squares property.

1b. $25x^2 - 36y^2 = (5x)^2 - (6y)^2$ — Write as $a^2 - b^2$.

$= (5x - 6y)(5x + 6y)$ — Apply the difference of two squares property.

1c. $-3x^3 + 12x = -3x(x^2 - 4)$ — Factor out the common factor.

$= -3x(x^2 - 2^2)$ — Write $x^2 - 4$ as $a^2 - b^2$.

$= -3x(x - 2)(x + 2)$ — Apply the difference of two squares property.

1d. $(x + 5)^2 - 16 = (x + 5)^2 - (4)^2$ — Write as $a^2 - b^2$.

$= [(x + 5) - 4][(x + 5) + 4]$ — Apply the difference of two squares property.

$= (x + 5 - 4)(x + 5 + 4)$ — Add or subtract the expressions in the factors.

$= (x + 1)(x + 9)$ — Simplify.

So, $(x + 5)^2 - 16 = (x + 1)(x + 9)$. We could also factor this polynomial using substitution.

$(x + 5)^2 - 16 = u^2 - 16$ — Let $u = (x + 5)$.

$= u^2 - 4^2$ — Write as $a^2 - b^2$.

$= (u - 4)(u + 4)$ — Apply the difference of two squares property.

$= [(x + 5) - 4][(x + 5) + 4]$ — Replace u with $(x + 5)$.

$= (x + 1)(x + 9)$ — Simplify each factor.

1e. Each term is a perfect square, but this binomial is the *sum of squares*, not the difference of squares. This binomial cannot be factored as the product of conjugates.

We can think of $x^2 + 4$ as $x^2 + 0x + 4$. Using the rules from Section 6.6, factoring this trinomial requires us to find the factors of 4 whose sum is 0. No such factors exist. So, $x^2 + 4$ is a *prime* polynomial.

Note: *Except for a possible common factor, the sum of squares $a^2 + b^2$ cannot be factored over the real numbers.*

Student Check 1 Factor each binomial completely.

a. $y^2 - \frac{1}{9}$ **b.** $81a^2 - 64b^2$ **c.** $-5b^3 + 125b$

d. $(x + 4)^2 - 1$ **e.** $25x^2 + 9$

Sum or Difference of Cubes

Objective 2 ▶

Factor the sum or difference of two cubes.

The next type of binomial to factor is the sum or difference of two cubes. This factoring pattern is not as obvious as the difference of two squares. Earlier in this chapter, we multiplied some polynomials that produced the sum or difference of two cubes. For example,

$$(x + 2)(x^2 - 2x + 4) = x^3 - 2x^2 + 4x + 2x^2 - 4x + 8$$
$$= x^3 + 8$$

$$(3y - 4)(9y^2 + 12y + 16) = 27y^3 + 36y^2 + 48y - 36y^2 - 48y - 64$$
$$= 27y^3 - 64$$

Reversing the preceding statements, we obtain the factorization.

Sum/Difference of Cubes	Factored Form	Observations
$x^3 + 8 = (x)^3 + (2)^3$	$(x + 2)(x^2 - 2x + 4)$	$x^2 - 2x + 4 = x^2 - 2(x) + (2)^2$
$27y^3 - 64 = (3y)^3 - (4)^3$	$(3y - 4)(9y^2 + 12y + 16)$	$9y^2 + 12y + 16 = (3y)^2 - 4(3y) + (4)^2$

Note that the factorization of the sum or difference of cubes involves a binomial and a trinomial.

- The binomial consists of the quantities that, when cubed, produce the terms in the given binomial.
- The trinomial is obtained from the binomial. From the observations, we see that the first and last terms of the trinomial factor are the square of the terms in the binomial factor. The middle term of the trinomial factor involves the product of the terms in the binomial factor.

We can write the factoring patterns as shown.

Property: Sum and Difference of Two Cubes

$$a^3 + b^3 = (a + b)(a^2 - ab + b^2)$$
$$a^3 - b^3 = (a - b)(a^2 + ab + b^2)$$

It is helpful to recognize some perfect cubes. Some are shown next.

$1^3 = 1$	$5^3 = 125$	$(x)^3 = x^3$
$2^3 = 8$	$6^3 = 216$	$(x^2)^3 = x^6$
$3^3 = 27$	$7^3 = 343$	$(x^3)^3 = x^9$
$4^3 = 64$	$8^3 = 512$	$(x^4)^3 = x^{12}$

Procedure: Factoring the Sum or Difference of Two Cubes

Step 1: Factor out common factors, if possible.
Step 2: Rewrite the first term as a^3, where a is what must be cubed to obtain the first term.
Step 3: Rewrite the second term as b^3, where b is what must be cubed to obtain the second term.
Step 4: Apply the sum and difference of two cubes property.
Step 5: Check factorization by multiplying

Note: *Helpful hint to remember the signs of the factors is the word SOAP.*

S = same sign
O = opposite sign
AP = always positive

$$x^3 + 8 = (x + 2)(x^2 - 2x + 4)$$

Objective 2 Examples Factor each binomial completely.

2a. $y^3 + 64$ **2b.** $125x^3 - 8y^3$ **2c.** $-2a^4b - 54ab$

Solutions

2a. $y^3 + 64$

$= y^3 + 4^3$	Write as $a^3 + b^3$.
$= (y + 4)[y^2 - (y)(4) + (4)^2]$	Factor with $a = y$ and $b = 4$.
$= (y + 4)(y^2 - 4y + 16)$	Simplify.

Check: $(y + 4)(y^2 - 4y + 16) = y^3 - 4y^2 + 16y + 4y^2 - 16y + 64 = y^3 + 64$

2b. $125x^3 - 8y^3$

$= (5x)^3 - (2y)^3$	Write as $a^3 - b^3$.
$= (5x - 2y)[(5x)^2 + (5x)(2y) + (2y)^2]$	Factor with $a = 5x$ and $b = 2y$.
$= (5x - 2y)(25x^2 + 10xy + 4y^2)$	Simplify.

2c. $-2a^4b - 54ab$

$= -2ab(a^3 + 27)$	Factor out the common factor.
$= -2ab(a^3 + 3^3)$	Write as $a^3 + b^3$.
$= -2ab(a + 3)[a^2 - a(3) + (3)^2]$	Factor as $(a + b)(a^2 - ab + b^2)$.
$= -2ab(a + 3)(a^2 - 3a + 9)$	Simplify.

Student Check 2 Factor each binomial completely.

a. $h^3 + 1$ **b.** $8a^3 - 343y^3$ **c.** $-4x^4y + 32xy$

Factoring Review

Objective 3 Apply factoring techniques.

Now we will review all of the factoring techniques that we have learned and will develop a general factoring strategy. We will then factor polynomials, which may require us to use more than one factoring technique.

The key to factoring lies in the ability to identify the type of polynomial, as there are specific methods that apply to binomials, trinomials, and four-term polynomials.

Procedure: A General Strategy for Factoring Polynomials

Step 1: Factor out any common factors.
Step 2: Identify the polynomial as a binomial, trinomial, or one with four terms and apply the appropriate method.

a. Binomials
- Difference of Squares: $a^2 - b^2 = (a + b)(a - b)$
- Sum of Cubes: $a^3 + b^3 = (a + b)(a^2 - ab + b^2)$
- Difference of Cubes: $a^3 - b^3 = (a - b)(a^2 + ab + b^2)$

b. Trinomials
- Factor by trial and error or the grouping method.
- Perfect square trinomials are special cases that factor as a binomial squared.
- Use substitution as necessary to write as $au^2 + bu + c$.

c. Four terms
- Factor by grouping, if possible.

Step 3: Apply appropriate factoring techniques to any of the factors, if possible.
Step 4: Check by multiplying.

Objective 3 Examples **Factor each polynomial completely.**

3a. $36y^3 + 9y$ **3b.** $x^3 + 2x^2 - 9x - 18$ **3c.** $x^4 - 16$
3d. $24x^4y - 52x^3y - 20x^2y$ **3e.** $y^6 - 27$ **3f.** $x^2 - 6x + 9 - 25y^2$

Solutions **3a.** This binomial isn't a type that we can factor, so we look for a common factor first.

$36y^3 + 9y = 9y(4y^2 + 1)$ Factor out the common factor of $9y$.

The resulting binomial, $4y^2 + 1$, is the sum of squares and can't be factored any further.

3b. This polynomial has four terms, so we try factoring by grouping.

$x^3 + 2x^2 - 9x - 18$
$= (x^3 + 2x^2) + (-9x - 18)$ Group terms.
$= x^2(x + 2) - 9(x + 2)$ Factor out the common factor from each pair of terms.
$= (x + 2)(x^2 - 9)$ Factor out the common binomial, $(x + 2)$.
$= (x + 2)(x - 3)(x + 3)$ Factor the difference of squares.

3c. This binomial is the difference of two squares.

$x^4 - 16$
$= (x^2)^2 - 4^2$ Write as the difference of two squares.
$= (x^2 - 4)(x^2 + 4)$ Factor the difference of two squares.
$= (x + 2)(x - 2)(x^2 + 4)$ Factor the difference of two squares.

Note that $x^2 + 4$ is the sum of squares and can't be factored further.

3d. We factor out the common factor and then factor the resulting trinomial by grouping.

$24x^4y - 52x^3y - 20x^2y$
$= 4x^2y(6x^2 - 13x - 5)$ Factor out the common factor, $4x^2y$.
$= 4x^2y(6x^2 - 15x + 2x - 5)$ Factor the trinomial by grouping.
$= 4x^2y[3x(2x - 5) + 1(2x - 5)]$ Factor out the GCFs from the first two and last two terms.
$= 4x^2y(2x - 5)(3x + 1)$ Factor out the common binomial, $2x - 5$.

3e. This binomial is the difference of two cubes.

$y^6 - 27$
$= (y^2)^3 - 3^3$ Write as the difference of cubes.
$= (y^2 - 3)[(y^2)^2 + (3)y^2 + (3)^2]$ Factor the difference of two cubes.
$= (y^2 - 3)(y^4 + 3y^2 + 9)$ Simplify.

3f. This polynomial has four terms but grouping pairs of terms will not work. So, we group the first three terms together. The trinomial $x^2 - 6x + 9$ is a perfect square trinomial and is $(x - 3)^2$. The resulting binomial is the difference of two squares.

$x^2 - 6x + 9 - 25y^2$

$= (x - 3)(x - 3) - 25y^2$ — Factor $x^2 - 6x + 9$ as $(x - 3)(x - 3)$.

$= (x - 3)^2 - (5y)^2$ — Write as the difference of two squares.

$= [(x - 3) - 5y][(x - 3) + 5y]$ — Factor the difference of two squares.

$= (x - 3 - 5y)(x - 3 + 5y)$ — Simplify each factor.

Check: $(x - 3 - 5y)(x - 3 + 5y) = x^2 - 3x + 5xy - 3x + 9 - 15y - 5xy + 15y - 25y^2$

$= x^2 - 6x + 9 - 25y^2$

Student Check 3 Factor each polynomial completely.

a. $4y^3 + 4y$ **b.** $a^3 + 6a^2 - 4a - 24$ **c.** $b^4 - 81$

d. $20x^4y^2 - 25x^3y^2 - 30x^2y^2$ **e.** $x^6 - 8$ **f.** $y^2 - 8y + 16 - 4x^2$

Objective 4 ▶

Troubleshoot common errors.

Troubleshooting Common Errors

Some common errors related to factoring binomials are shown next.

Objective 4 Examples

A problem and an incorrect solution are given. Provide the correct solution and an explanation of the error.

4a. Factor $x^2 - 64x$.

Incorrect Solution	Correct Solution and Explanation
$x^2 - 64x = (x - 8)(x + 8)$	The binomial is not the difference of two squares since there is a common factor of x. $x^2 - 64x = x(x - 64)$

4b. Factor $x^2 + 64$.

Incorrect Solution	Correct Solution and Explanation
$x^2 + 64 = (x + 8)(x + 8)$	The binomial is the *sum* of two squares, not the difference of two squares. Therefore, it is prime.

4c. Factor $x^3 - 64$.

Incorrect Solution	Correct Solution and Explanation
$x^3 - 64 = (x - 4)(x - 4)(x - 4)$	The binomial is the difference of two cubes, so we must apply the factoring property. $x^3 - 64 = (x - 4)(x^2 + 4x + 16)$

ANSWERS TO STUDENT CHECKS

Student Check 1 **a.** $\left(y - \frac{1}{3}\right)\left(y + \frac{1}{3}\right)$
b. $(9a - 8b)(9a + 8b)$ **c.** $-5b(b - 5)(b + 5)$
d. $(x + 5)(x + 3)$ **e.** prime

Student Check 2 **a.** $(h + 1)(h^2 - h + 1)$
b. $(2a - 7y)(4a^2 + 14ay + 49y^2)$
c. $-4xy(x - 2)(x^2 + 2x + 4)$

Student Check 3 **a.** $4y(y^2 + 1)$
b. $(a + 6)(a + 2)(a - 2)$
c. $(b + 3)(b - 3)(b^2 + 9)$
d. $5x^2y^2(4x + 3)(x - 2)$
e. $(x^2 - 2)(x^4 + 2x^2 + 4)$
f. $(y - 4 - 2x)(y - 4 + 2x)$

SUMMARY OF KEY CONCEPTS

1. The difference of two squares can be factored using conjugates.
$$a^2 - b^2 = (a - b)(a + b)$$
The sum of two squares is prime.
2. The sum and difference of two cubes can be factored as:
$$a^3 + b^3 = (a + b)(a^2 - ab + b^2)$$
$$a^3 - b^3 = (a - b)(a^2 + ab + b^2)$$
3. When factoring a polynomial, apply the general factoring strategy by taking out any common factors and then applying the appropriate method.

GRAPHING CALCULATOR SKILLS

The graphing calculator can be used to verify factorization as described in previous sections. It can also be used to generate a list of perfect squares and perfect cubes that will help in identifying the values of a and b in the properties.

Example 1: List the perfect squares on the calculator.

Solution: Enter x^2 in the equation editor. Then view the table. The values in Y_1 are the perfect squares. The corresponding x-value is the number that must be squared to obtain the value in Y_1.

```
Plot1 Plot2 Plot3
\Y1=X²
\Y2=
\Y3=
\Y4=
\Y5=
\Y6=
\Y7=
```

X	Y1
0	0
1	1
2	4
3	9
4	16
5	25
6	36

X=6

Example 2: List the perfect cubes on the calculator.

Solution: Enter x^3 in the equation editor. Then view the table. The values in Y_1 are the perfect cubes. The corresponding x-value is the number that must be cubed to obtain the value in Y_1.

```
Plot1 Plot2 Plot3
\Y1=X^3
\Y2=
\Y3=
\Y4=
\Y5=
\Y6=
\Y7=
```

X	Y1
0	0
1	1
2	8
3	27
4	64
5	125
6	216

X=0

SECTION 6.7 EXERCISE SET

Write About It!

Use complete sentences in your answer to each exercise.

1. Use an example to explain how to factor a difference of two squares.
2. Explain why the expressions $(2x + y)^3$ and $8x^3 + y^3$ are not the same.
3. Explain why the expressions $8x^3 + y^3$ and $(2x + y)(2x^2 - 2xy + y^2)$ are not the same.
4. Use an example to explain how to apply the sum of two cubes formula.
5. Explain why taking out the greatest common factor should be the first step in factoring the expression $18x^2 - 50y^2$.
6. Explain how to apply the difference of two cubes formula in factoring the expression $w^6 - 64z^6$.
7. Explain how to apply the difference of two squares formula in factoring the expression $w^6 - 64z^6$.
8. Explain why the solutions to Exercises 6 and 7 should be the same.

Additional answers can be found in the Instructor Answer Appendix.

Practice Makes Perfect!

Factor each binomial completely. (*See Objective 1.*)

9. $x^2 - \frac{9}{4}$ $\left(x - \frac{3}{2}\right)\left(x + \frac{3}{2}\right)$
10. $y^2 - \frac{25}{9}$ $\left(y - \frac{5}{3}\right)\left(y + \frac{5}{3}\right)$
11. $a^2 - \frac{4}{49}$ $\left(a - \frac{2}{7}\right)\left(a + \frac{2}{7}\right)$
12. $b^2 - \frac{121}{100}$ $\left(b - \frac{11}{10}\right)\left(b + \frac{11}{10}\right)$
13. $4z^2 - 9w^2$ $(2z + 3w)(2z - 3w)$
14. $64x^2 - 25y^2$ $(8x + 5y)(8x - 5y)$
15. $c^2 - 625d^2$ $(c + 25d)(c - 25d)$
16. $100a^2 - 49b^2$ $(10a + 7b)(10a - 7b)$
17. $25z^2 + 4$ prime
18. $x^2 + 9y^2$ prime
19. $6ac^2 + 54ad^2$ $6a(c^2 + 9d^2)$
20. $-10rs^2 - 40r^3$ $-10r(s^2 + 4r^2)$
21. $-18t^3 + 32t$ $-2t(3t + 4)(3t - 4)$
22. $150x^3 - 6x$ $6x(5x + 1)(5x - 1)$
23. $(6z + 4)^2 - 9$ $(6z + 7)(6z + 1)$
24. $(5w + 2)^2 - 121$ $(5w + 13)(5w - 9)$
25. $(b - 2)^2 + 1$ prime
26. $(a - 3)^2 + 4$ prime
27. $(c + 3)^2 - 1$ $(c + 4)(c + 2)$
28. $(d - 8)^2 - 121$ $(d + 3)(d - 19)$

Factor each binomial completely. (*See Objective 2.*)

29. $a^3 + 27$
$(a + 3)(a^2 - 3a + 9)$

30. $b^3 + 125$
$(b + 5)(b^2 - 5b + 25)$

31. $x^3 - 1000$
$(x - 10)(x^2 + 10x + 100)$

32. $y^3 - 8$
$(y - 2)(y^2 + 2y + 4)$

33. $8r^3 - 125s^3$
$(2r - 5s)(4r^2 + 10rs + 25s^2)$

34. $64u^3 - 27v^3$
$(4u - 3v)(16u^2 + 12uv + 9v^2)$

35. $-8z^3 + 512w^3$
$-8(z - 4w)(z^2 + 4wz + 16w^2)$

36. $-7a^3 - 56b^3$
$-7(a + 2b)(a^2 - 2ab + 4b^2)$

37. $-27s^4t^2 - st^2$
$-st^2(3s + 1)(9s^2 - 3s + 1)$

38. $2s^6t - 128s^3t$
$2s^3t(s - 4)(s^2 + 4s + 16)$

Factor each polynomial completely. (*See Objective 3.*)

39. $-25w^3 - w$
$-w(25w^2 + 1)$

40. $63y^4 + 7y^2$
$7y^2(9y^2 + 1)$

41. $4s^3 + 8s^2 - s - 2$
$(s + 2)(2s + 1)(2s - 1)$

42. $48t^3 - 32t^2 - 3t + 2$
$(3t - 2)(4t + 1)(4t - 1)$

43. $4a^3 + 3a^2 - 4a - 3$
$(4a + 3)(a + 1)(a - 1)$

44. $4b^3 + 20b^2 - b - 5$
$(b + 5)(2b + 1)(2b - 1)$

45. $81x^4 - 256$
$(9x^2 + 16)(3x + 4)(3x - 4)$

46. $625y^4 - 16$
$(25y^2 + 4)(5y + 2)(5y - 2)$

47. $-40s^6t^2 + 68s^5t^2 + 80s^4t^2$ $-4s^4t^2(5s + 4)(2s - 5)$

48. $4x^4y + 34x^3y + 42x^2y$ $2x^2y(2x + 3)(x + 7)$

49. $u^6 - 8$
$(u^2 - 2)(u^4 + 2u^2 + 4)$

50. $v^6 + 27$
$(v^2 + 3)(v^4 - 3v^2 + 9)$

51. $1000x^6 + 1$
$(10x^2 + 1)(100x^4 - 10x^2 + 1)$

52. $216y^6 - 1$
$(6y^2 - 1)(36y^4 + 6y^2 + 1)$

53. $8r^9 + 27$
$(2r^3 + 3)(4r^6 - 6r^3 + 9)$

54. $125s^9 - 8$
$(5s^3 - 2)(25s^6 + 10s^3 + 4)$

55. $v^2 - 12v + 36 - 144u^2$
$(v - 6 - 12u)(v - 6 + 12u)$

56. $s^2 - 16s + 64 - 36t^2$
$(s - 8 - 6t)(s - 8 + 6t)$

57. $a^2 + 10a + 25 - 64b^2$
$(a + 5 - 8b)(a + 5 + 8b)$

58. $c^2 - 6c + 9 - 16d^2$
$(c - 3 - 4d)(c - 3 + 4d)$

59. $4x^2 + 12x + 9 - 25y^2$
$(2x + 3 - 5y)(2x + 3 + 5y)$

60. $9y^2 - 30y + 25 - 4x^2$
$(3y - 5 - 2x)(3y - 5 + 2x)$

Mix 'Em Up!

Factor each polynomial completely.

61. $49x^2 - \dfrac{25}{9}$

62. $25y^2 - \dfrac{1}{64}$

63. $0.81a^2 - b^2$
$(0.9a + b)(0.9a - b)$

64. $c^2 - 0.16d^2$
$(c + 0.4d)(c - 0.4d)$

65. $18s^2 - 242t^2$
$2(3s - 11t)(3s + 11t)$

66. $441u^2 - 81v^2$
$9(7u + 3v)(7u - 3v)$

67. $2x^2 - 98y^2$
$2(x + 7y)(x - 7y)$

68. $4z^2 - 64w^2$
$4(z + 4w)(z - 4w)$

69. $x^2 + 10x + 25 - 9y^2$
$(x + 5 - 3y)(x + 5 + 3y)$

70. $9a^2 - 12a + 4 - 64b^2$
$(3a - 2 + 8b)(3a - 2 - 8b)$

71. $9u^3 - 121u$
$u(3u + 11)(3u - 11)$

72. $-48v^3 + 75v$
$-3v(4v - 5)(4v + 5)$

73. $81a^2 + 4$ prime

74. $25b^2 + 1$ prime

75. $(2a - 1)^2 - 49b^2$
$(2a - 1 + 7b)(2a - 1 - 7b)$

76. $(5c + 2)^2 - 16d^2$
$(5c + 2 + 4d)(5c + 2 - 4d)$

77. $2(x - 1)^2 - 18$
$2(x + 2)(x - 4)$

78. $3(y + 2)^2 - 12$
$3y(y + 4)$

79. $24a^3 + 81b^3$
$3(2a + 3b)(4a^2 - 6ab + 9b^2)$

80. $c^6 + 8d^3$
$(c^2 + 2d)(c^4 - 2c^2d + 4d^2)$

81. $5x^3 - 40y^3$
$5(x - 2y)(x^2 + 2xy + 4y^2)$

82. $125u^3 - v^6$
$(5u - v^2)(25u^2 + 5uv^2 + v^4)$

83. $a^6 + b^6$
$(a^2 + b^2)(a^4 - a^2b^2 + b^4)$

84. $x^6 - y^6$

85. $256c^4 - d^4$
$(16c^2 + d^2)(4c + d)(4c - d)$

86. $x^4 - 625y^4$
$(x^2 + 25y^2)(x + 5y)(x - 5y)$

87. $14xy + 7y + 42x + 21$
$7(2x + 1)(y + 3)$

88. $20ab - 8b + 30a - 12$
$2(5a - 2)(2b + 3)$

89. $12cx^2 + 4x^2 - 27c - 9$
$(2x + 3)(2x - 3)(3c + 1)$

90. $2ay^2 + 5y^2 - 8a - 20$
$(2a + 5)(y - 2)(y + 2)$

91. $a^2 - 2ab + b^2 + 1$
prime

92. $c^2 + 4cd + d^2 + 25$
prime

You Be the Teacher!

Correct each student's errors, if any.

93. Factor $x^2 + 9y^2$.

Amiee's work:

$x^2 + 9y^2 = (x + 3y)^2$ The expression $x^2 + 9y^2$ is prime.

94. Factor $a^3 + 8$.

Brett's work:

$a^3 + 8 = (a + 8)(a^2 - 8a + 64)$

$a^3 + 8 = (a)^3 + (2)^3$
$= (a + 2)[(a)^2 - (a)(2) + (2)^2]$
$= (a + 2)(a^2 - 2a + 4)$

95. Factor $3xy - 21y - 2x + 14$ by grouping.

Roy's work:

$3xy - 21y - 2x + 14 = 3y(x - 7) - 2(x + 7)$

$3xy - 21y - 2x + 14 = 3y(x - 7) - 2(x - 7)$
$= (x - 7)(3y - 2)$

96. Factor $5ab + 3b - 10a - 6$ by grouping.

Jean's work:

$5ab + 3b - 10a - 6 = b(5a + 3) - 2(5a + 3)$
$= (5a + 3)^2(b - 2)$

97. Factor $x^2 + 6x + 9 - z^2$.

Rosa's work:

$x^2 + 6x + 9 - z^2 = (x^2 - z^2) + (6x + 9)$
$= (x - z)(x + z) + 3(2x + 3)$

98. Factor $x^2 - 4x + 4 - z^2$.

Doug's work:

$x^2 - 4x + 4 - z^2 = (x^2 - 4x) + (4 - z^2)$
$= x(x - 4) + (2 + z)(2 - z)$

Calculate It!

Use a graphing calculator to verify if each factorization is correct or not.

99. $x^3 - 1 = (x - 1)(x^2 + x + 1)$

100. $9x^2 - 16 = (3x + 4)(3x - 4)$

101. $16x^4 - 81 = (4x^2 + 9)(2x + 3)(2x - 3)$

102. $x^3 + 27 = (x + 3)(x^2 - 3x + 9)$

Think About It!

103. Write a binomial that is the difference of two squares and factor it. Answers vary; $25x^2 - 144 = (5x - 12)(5x + 12)$

104. Write a binomial that is the sum of two squares and factor it, if possible. Answers vary; $y^2 + 9$ is prime

105. Write a binomial that is the difference of two cubes and factor it. Answers vary; $a^3 - 8 = (a - 2)(a^2 + 2a + 4)$

106. Write a binomial that is the sum of two cubes and factor it. Answers vary; $64x^3 + 1 = (4x + 1)(16x^2 - 4x + 1)$

107. The trinomials $x^2 - 10x + 25$ and $4y^2 + 12y + 9$ are perfect square trinomials. Factor these and determine a special factoring pattern that applies to perfect square trinomials. (Refer to Section 6.4.) $x^2 - 10x + 25 = (x - 5)^2$ and $4y^2 + 12y + 9 = (2y + 3)^2$; $a^2 + 2ab + b^2 = (a + b)^2$

SECTION 6.8 Solving Polynomial Equations and Their Applications

OBJECTIVES

As a result of completing this section, you will be able to

1. Solve polynomial equations by factoring.
2. Solve applications of polynomial equations.
3. Find the zeros of the graphs of polynomial functions.
4. Troubleshoot common errors.

The Grand Canyon Skywalk is a glass walkway that stands about 3600 ft above the floor of the Grand Canyon. The distance s an object would be above the floor of the canyon if dropped from the skywalk is given by

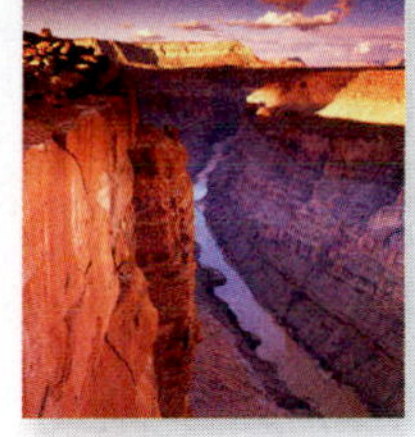

$$s = -16t^2 + 3600$$

feet after t sec. How long will it take for a rock dropped from the skywalk to reach the floor of the canyon? To answer this question, we must solve the equation $-16t^2 + 3600 = 0$.

In this section, we will learn how to use factoring to solve polynomial equations like the one shown.

Solving Polynomial Equations

Objective 1 Solve polynomial equations by factoring.

So far in this chapter, we have learned how to factor different types of polynomials. We will use our knowledge of factoring to solve a new type of equation called a *polynomial equation*.

> **Definition:** A **polynomial equation** is an equation that can be written in the form $P(x) = 0$, where $P(x)$ is a polynomial.

Some examples of polynomial equations are

$$5x^2 - x - 6 = 0 \qquad x^3 = 2x^2 \qquad x^4 - 5x^2 + 4 = 0$$

A polynomial equation is written in *standard form* when one side of the equation is equal to 0. The **degree of a polynomial equation** in standard form is its largest exponent.

Example	Standard Form	Degree of Equation	Type of Equation
$4x - 5 = 3$	$4x^1 - 8 = 0$	1	Linear
$5x^2 - 6 = x$	$5x^2 - x - 6 = 0$	2	Quadratic
$x(x^2 + 2) = 0$	$x^3 + 2x^2 = 0$	3	Cubic
$-5x^2 + x^4 + 4 = 0$	$x^4 - 5x^2 + 4 = 0$	4	Quartic

A *solution* of a polynomial equation is a value of the variable that makes the equation a true statement. In this section, we will concentrate on solving polynomial equations by factoring. In Chapter 9, we will explore other methods for solving polynomial equations that do not factor. A property from the real numbers provides us with the tools to solve polynomial equations. Recall that if two numbers are multiplied together and the answer is zero, then at least one of the numbers has to be zero. For example,

$$6 \cdot 0 = 0; \quad 0 \cdot 6 = 0; \quad -2 \cdot 0 = 0; \quad 0 \cdot -2 = 0$$

The only way for a product to equal zero is if one of the factors is zero. This concept is called the *zero products property*.

> **Property: Zero Products Property**
>
> If A and B are real numbers and $A \cdot B = 0$, then,
>
> $$A = 0 \quad \text{or} \quad B = 0$$

This property also holds for three or more factors.

INSTRUCTOR NOTE:
Remind students that the equation must be in standard form so that the zero products property can be used.

Procedure: Solving Polynomial Equations by Factoring

Step 1: Write the equation in standard form.
Step 2: Factor the resulting polynomial if not already done.
Step 3: Apply the zero products property by setting each factor equal to zero.
Step 4: Solve the resulting equations.
Step 5: Check the solutions in the original equation.

Note *A quadratic equation, written in standard form, is an equation of the form* $ax^2 + bx + c = 0$, *where* a, b, *and* c *are real nonzero numbers.*

Objective 1 Examples

Solve each equation by factoring.

1a. $(x - 3)(x + 4) = 0$ **1b.** $5x^2 - x - 6 = 0$ **1c.** $4y(y - 3) = -5$

1d. $a^2 + (a + 1)^2 = (a + 2)^2$ **1e.** $\frac{3x^2}{4} - \frac{5}{2}x + 2 = 0$ **1f.** $x^3 - 4x = 0$

1g. $x^3 + 2x^2 = 9x + 18$

Solutions

1a.

$(x - 3)(x + 4) = 0$ The equation is factored and equal to zero.

$x - 3 = 0$ or $x + 4 = 0$ Apply the zero products property.

$x = 3$ $\quad x = -4$ Solve each equation.

Check: $x = 3$

$$(x - 3)(x + 4) = 0$$
$$(3 - 3)(3 + 4) = 0$$
$$(0)(7) = 0$$
$$0 = 0$$

Check: $x = -4$

$$(x - 3)(x + 4) = 0$$
$$(-4 - 3)(-4 + 4) = 0$$
$$(-7)(0) = 0$$
$$0 = 0$$

The solutions check in the original equation. So, the solution set is $\{-4, 3\}$.

1b.

$5x^2 - x - 6 = 0$ The equation is in standard form.

$(5x - 6)(x + 1) = 0$ Factor $5x^2 - x - 6$.

$5x - 6 = 0$ or $x + 1 = 0$ Apply the zero products property.

$5x = 6$ $\quad x = -1$ Solve each equation.

$x = \frac{6}{5}$

Check: $x = \frac{6}{5}$

$$5x^2 - x - 6 = 0$$
$$5\left(\frac{6}{5}\right)^2 - \left(\frac{6}{5}\right) - 6 = 0$$
$$5\left(\frac{36}{25}\right) - \frac{6}{5} - 6 = 0$$
$$\frac{36}{5} - \frac{6}{5} - \frac{30}{5} = 0$$
$$0 = 0$$

Check: $x = -1$

$$5x^2 - x - 6 = 0$$
$$5(-1)^2 - (-1) - 6 = 0$$
$$5(1) + 1 - 6 = 0$$
$$5 + 1 - 6 = 0$$
$$0 = 0$$

The solutions check in the original equation. So, the solution set is $\left\{-1, \frac{6}{5}\right\}$.

1c.

$4y(y-3)=-5$	
$4y^2-12y=-5$	Apply the distributive property.
$4y^2-12y+5=-5+5$	Add 5 to each side.
$4y^2-12y+5=0$	Simplify.
$(2y-5)(2y-1)=0$	Factor $4y^2-12y+5$.
$2y-5=0$ or $2y-1=0$	Apply the zero products property.
$2y=5 \qquad 2y=1$	Solve each equation.
$y=\frac{5}{2} \qquad y=\frac{1}{2}$	

The solutions check in the original equation. So, the solution set is $\left\{\frac{5}{2}, \frac{1}{2}\right\}$.

INSTRUCTOR NOTE: Remind students that $(a+2)^2$ does not equal a^2+4.

1d.

$a^2+(a+1)^2=(a+2)^2$	
$a^2+a^2+2a+1=a^2+4a+4$	Square the binomials.
$2a^2+2a+1=a^2+4a+4$	Combine like terms.
$2a^2+2a+1-a^2-4a-4=a^2+4a+4-a^2-4a-4$	Subtract a^2, $4a$, and 4 from each side.
$a^2-2a-3=0$	Simplify.
$(a-3)(a+1)=0$	Factor a^2-2a-3.
$a-3=0$ or $a+1=0$	Apply the zero products property.
$a=3 \qquad a=-1$	Solve each equation.

The solutions check in the original equation. So, the solution set is $\{-1, 3\}$.

1e. The equation is in standard form but it is easier to factor if we clear the fractions.

$\frac{3x^2}{4}-\frac{5}{2}x+2=0$	
$4\left(\frac{3x^2}{4}-\frac{5}{2}x+2\right)=4(0)$	Multiply each side by the LCD, 4.
$3x^2-10x+8=0$	Simplify.
$(3x-4)(x-2)=0$	Factor $3x^2-10x+8$.
$3x-4=0$ or $x-2=0$	Apply the zero products property.
$3x=4 \qquad x=2$	Solve each equation.
$x=\frac{4}{3}$	

The solutions check in the original equation. So, the solution set is $\left\{\frac{4}{3}, 2\right\}$.

1f.

$x^3-4x=0$	The equation is in standard form.
$x(x^2-4)=0$	Factor out the GCF, x.
$x(x-2)(x+2)=0$	Factor the difference of squares, x^2-4.
$x=0$ or $x-2=0$ or $x+2=0$	Apply the zero products property.
$x=2 \qquad x=-2$	Solve each equation.

The solutions check in the original equation. So, the solution set is $\{-2, 0, 2\}$.

1g.

$x^3+2x^2=9x+18$	
$x^3+2x^2-9x-18=0$	Subtract $9x$ and 18 from each side.
$x^2(x+2)-9(x+2)=0$	Factor $x^3+2x^2-9x-18$ by grouping.
$(x+2)(x^2-9)=0$	

$(x + 2)(x + 3)(x - 3) = 0$	Factor the difference of squares, $x^2 - 9$.
$x + 2 = 0$ or $x + 3 = 0$ or $x - 3 = 0$	Apply the zero products property.
$x = -2$ $\quad x = -3$ $\quad x = 3$	Solve each equation.

The solutions check in the original equation. So, the solution set is $\{-3, -2, 3\}$.

Student Check 1 Solve each equation by factoring.

a. $(y - 9)(y + 6) = 0$

b. $8a^2 - 13a - 6 = 0$

c. $9y(y + 2) = 7$

d. $b^2 + (b + 2)^2 = (b + 4)^2$

e. $\frac{4}{5}x^2 - \frac{11}{10}x - 1 = 0$

f. $2a^3 - 18a = 0$

g. $x^3 - 4x^2 - 25x + 100 = 0$

Applications

Objective 2 Solve applications of polynomial equations.

Many real-world situations can be modeled by polynomial equations. Some are illustrated in Example 2.

Some of the problems require us to apply the *Pythagorean theorem*. The Greek mathematician Pythagoras is credited with discovering a special relationship that exists among the lengths of the sides of a *right triangle* (i.e., a triangle with a 90° angle). The **hypotenuse** of a right triangle is the side opposite the right angle and is the longest side. The **legs** are the remaining sides of the triangle.

Property: Pythagorean Theorem

For any right triangle, where a and b are the lengths of the legs and c is the length of the hypotenuse,

$$a^2 + b^2 = c^2$$

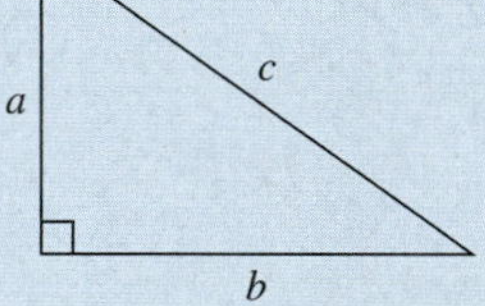

That is, $(\text{leg})^2 + (\text{other leg})^2 = (\text{hypotenuse})^2$.

Procedure: A Problem-Solving Strategy

Step 1: Determine the unknown quantity and assign a variable to it, if necessary.
Step 2: Determine what is known and draw a diagram to visualize the question.
Step 3: Write an equation that models the situation.
Step 4: Use the zero products property to solve the equation.
Step 5: Since the solutions of the equation represent a physical quantity, some may need to be discarded. Check for the reasonableness of an answer.

Objective 2 Examples **For each problem, write a polynomial equation that represents the situation, solve the equation using the zero products property, and state the answer in a sentence.**

2a. A 13-ft ladder leans against the side of a house. If the distance from the bottom of the ladder to the house is 5 ft, how high up the house does the ladder reach?

Solution

2a. What is unknown? The distance that the ladder reaches up the house is unknown. Let x = distance from the ground to the top of the ladder (in feet).

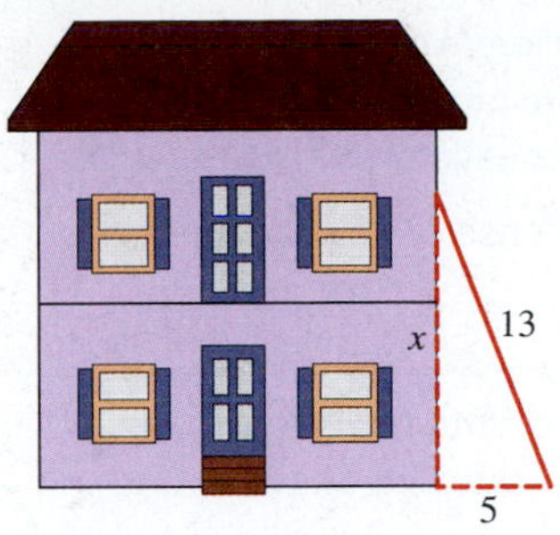

What is known? The ladder forms a right triangle with the house. The legs of the triangle are x and 5. The hypotenuse is 13.

Equation: We use the Pythagorean theorem to write the equation.

$x^2 + 5^2 = 13^2$	Apply the Pythagorean theorem.
$x^2 + 25 = 169$	Simplify each exponent.
$x^2 + 25 - 169 = 169 - 169$	Subtract 169 from each side.
$x^2 - 144 = 0$	Simplify.
$(x - 12)(x + 12) = 0$	Factor the difference of squares.
$x - 12 = 0$ or $x + 12 = 0$	Apply the zero products property.
$x = 12$ $\qquad x = -12$	Solve each equation.

Since x represents a length, the only possible answer is 12. So, the distance from the ground to the top of the ladder is 12 ft.

2b. Jentezen plans to pour concrete in a uniform width around his rectangular pool to form a walkway. If his pool measures 20 ft by 50 ft and he has enough concrete to cover 800 ft^2, how wide can the walkway be?

Solution **2b.** What is unknown? The width of the walkway is unknown. Let $x =$ the width of the walkway.

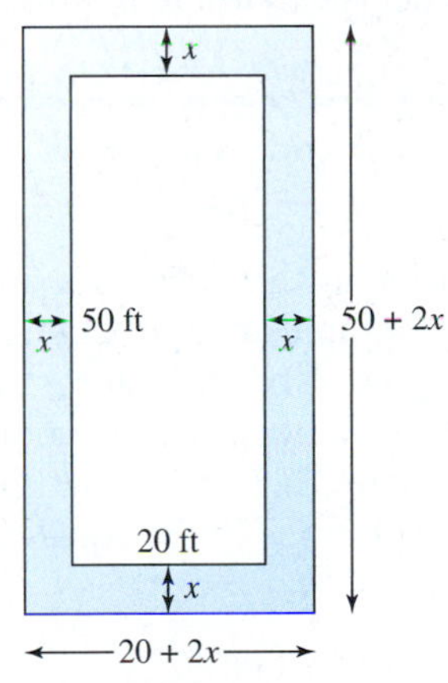

What is known? Jentezen has enough concrete to cover an area of 800 ft^2. The pool together with the border has a width of $20 + 2x$ and a length of $50 + 2x$.

Equation: We need an expression that represents the area of the walkway. So, we need to apply the formula $A = lw$.

area of the pool and walkway − area of the pool = area of the walkway

$(50 + 2x)(20 + 2x) - (50)(20) = 800$	
$1000 + 100x + 40x + 4x^2 - 1000 = 800$	Simplify each product.
$4x^2 + 140x = 800$	Combine like terms.
$4x^2 + 140x - 800 = 0$	Subtract 800 from each side.
$4(x^2 + 35x - 200) = 0$	Factor out the common factor, 4.
$4(x - 5)(x + 40) = 0$	Factor $x^2 + 35x - 200$.
$x - 5 = 0$ or $x + 40 = 0$	Apply the zero products property.
$x = 5$ $\qquad x = -40$	Solve each equation.

Since x represents the width of the walkway, 5 is the answer because we cannot have a negative width. So, the walkway can be 5 ft wide.

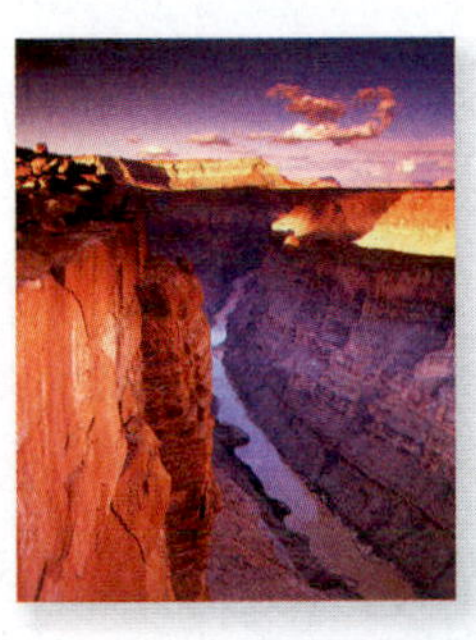

2c. The Grand Canyon Skywalk is a glass walkway that stands about 3600 ft above the floor of the Grand Canyon. The distance s an object would be above the floor of the canyon after t sec if dropped from the skywalk is given by $s = -16t^2 + 3600$ ft. How long will it take a rock dropped from the skywalk to reach the floor of the canyon?

Solution **2c.** What is unknown? The time, t, that it takes for the rock to reach the canyon floor is unknown.

What is known? When the rock reaches the floor of the canyon, its distance s is 0 ft.

Equation:

$s = -16t^2 + 3600$	State the given model.
$0 = -16t^2 + 3600$	Replace s with 0.
$0 = -16(t^2 - 225)$	Factor out the common factor, −16.

$$0 = -16(t - 15)(t + 15) \quad \text{Factor the difference of squares.}$$

$$t - 15 = 0 \quad \text{or} \quad t + 15 = 0 \quad \text{Apply the zero products property.}$$

$$t = 15 \qquad t = -15 \quad \text{Solve each equation.}$$

Since t represents time, the only solution that makes sense is 15. So, the rock will reach the canyon floor 15 sec after it is dropped.

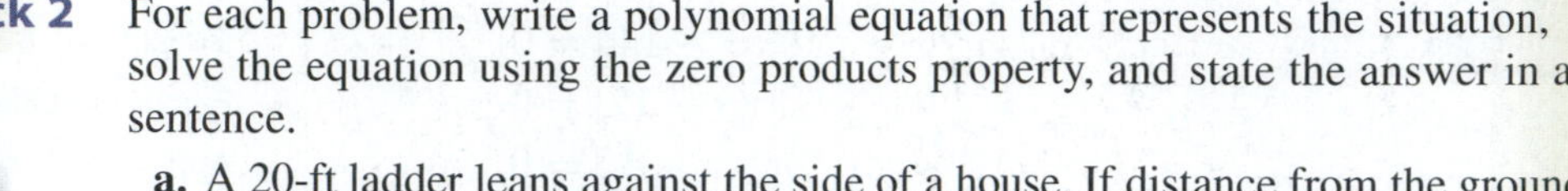

Student Check 2 For each problem, write a polynomial equation that represents the situation, solve the equation using the zero products property, and state the answer in a sentence.

a. A 20-ft ladder leans against the side of a house. If distance from the ground to the top of the ladder is 16 ft, what is the distance from the bottom of the ladder to the house?

b. Kathleen built a rectangular koi pond that measures 5 ft by 4 ft in her back yard. She purchased gravel to spread uniformly around the pond for decoration. If Kathleen has enough gravel to cover 52 ft^2, how wide can she make the gravel border?

c. The Empire State Building in New York City is the third tallest skyscraper in the United States, standing at 1250 ft tall. Suppose a penny was dropped from the top of the Empire State Building. The equation $s = -16t^2 + 1250$ approximates the distance s from the ground (in feet) that the penny will be after t sec. How long will it take the penny to be 850 ft from the ground?

Zeros of Polynomial Functions

Objective 3 Find the zeros of the graphs of polynomial functions.

We briefly examined graphs of polynomial functions earlier in this chapter. Now we will learn how to find the zeros or x-intercepts of the graph of a polynomial function. In Chapter 3, we defined the **zero of a function**, $f(x)$, to be the value c for which $f(c) = 0$. The zero of a function corresponds to the x-intercept of the graph of the function. The graph of $f(x) = x^2 - 4$ is shown in Figure 6.1. Note that the graph crosses the x-axis at 2 and -2. These values are also solutions of the equation $f(x) = 0$, and, therefore, the zeros of $f(x)$.

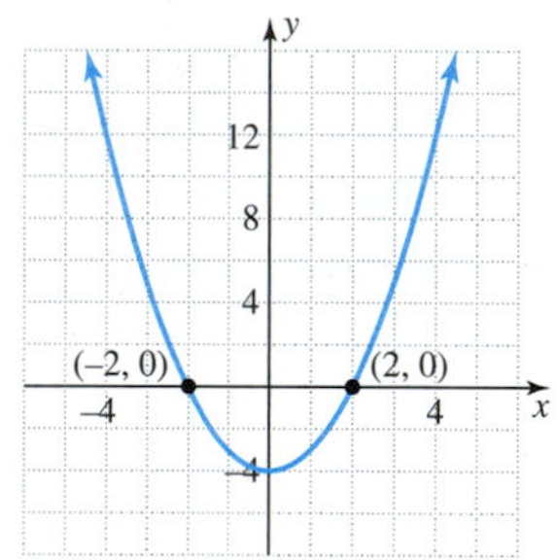

Figure 6.1

$$f(x) = 0$$
$$x^2 - 4 = 0$$
$$(x - 2)(x + 2) = 0$$
$$x - 2 = 0 \quad \text{or} \quad x + 2 = 0$$
$$x = 2 \qquad x = -2$$

The graph of $f(x) = x^3 + x^2 - 12x$ is shown in Figure 6.2. Note that the graph crosses the x-axis at -4, 0, and 3. These values are also solutions of the equation $f(x) = 0$, and, therefore, the zeros of $f(x)$.

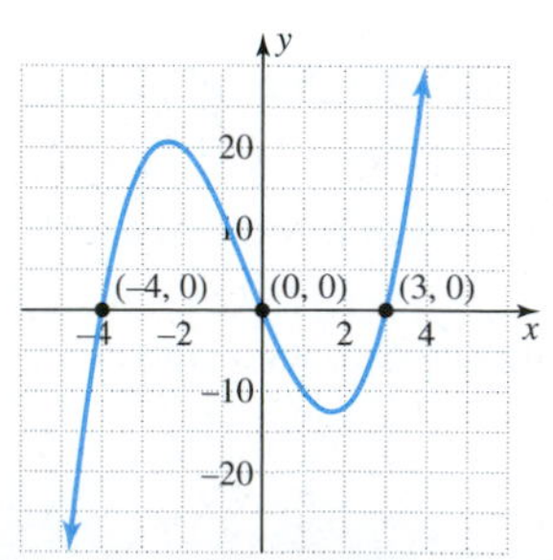

Figure 6.2

$$f(x) = 0$$
$$x^3 + x^2 - 12x = 0$$
$$x(x^2 + x - 12) = 0$$
$$x(x + 4)(x - 3) = 0$$
$$x = 0 \quad \text{or} \quad x + 4 = 0 \quad \text{or} \quad x - 3 = 0$$
$$x = -4 \qquad x = 3$$

So, when we find zeros of a function, we are finding valuable information regarding the graph of the function. The solutions of $f(x) = 0$, or the zeros of $f(x)$, correspond to the x-intercepts of the graph of $f(x)$.

Procedure: Finding Zeros of the Graph of a Polynomial Function

Step 1: Set the function equal to 0.
Step 2: Use factoring to solve the resulting equation.

Objective 3 Examples Determine the zeros of each function and the x-intercepts of the graph.

3a. $f(x) = x^2 - x - 6.$ **3b.** $f(x) = -x^3 + 9x.$

Solutions

3a.

$$x^2 - x - 6 = 0 \quad \text{Set } f(x) = 0.$$
$$(x - 3)(x + 2) = 0 \quad \text{Factor the trinomial.}$$
$$x - 3 = 0 \quad \text{or} \quad x + 2 = 0 \quad \text{Apply the zero products property.}$$
$$x = 3 \qquad x = -2 \quad \text{Solve each equation.}$$

So, the zeros are $x = 3$ and $x = -2$ and the x-intercepts are $(3, 0)$ and $(-2, 0)$.

3b.

$$-x^3 + 9x = 0 \quad \text{Set } f(x) = 0.$$
$$-x(x^2 - 9) = 0 \quad \text{Factor out the common factor, } -x.$$
$$-x(x + 3)(x - 3) = 0 \quad \text{Factor the difference of squares.}$$
$$-x = 0 \quad \text{or} \quad x + 3 = 0 \quad \text{or} \quad x - 3 = 0 \quad \text{Apply the zero products property.}$$
$$x = 0 \qquad x = -3 \qquad x = 3 \quad \text{Solve each equation.}$$

The zeros of the function are $x = 0$, $x = -3$, and $x = 3$. So, the x-intercepts are $(0, 0)$, $(-3, 0)$, and $(3, 0)$.

Student Check 3 Determine the zeros of each function and the x-intercepts of the graph.

a. $f(x) = -x^2 - x + 6$ **b.** $f(x) = x^3 + 5x^2 - x - 5$

Objective 4 ▶
Troubleshoot common errors.

Troubleshooting Common Errors

Many of the errors made in solving equations come from factoring incorrectly, simplifying the equation incorrectly, or from applying the zero products property incorrectly.

Objective 4 Examples A problem and an incorrect solution are given. Provide the correct solution and an explanation of the error.

4a. Solve $x^2 - 2x - 8 = 0.$

Incorrect Solution	Correct Solution and Explanation
	The polynomial was factored incorrectly.
$x^2 - 2x - 8 = 0$	$x^2 - 2x - 8 = 0$
$(x + 4)(x - 2) = 0$	$(x - 4)(x + 2) = 0$
$x + 4 = 0$ or $x - 2 = 0$	$x - 4 = 0$ or $x + 2 = 0$
$x = -4$ $\quad$ $x = 2$	$x = 4$ $\quad$ $x = -2$
	The solution set is $\{-2, 4\}$.

4b. Solve $x(x - 4) = 12$.

Incorrect Solution	Correct Solution and Explanation
$x(x - 4) = 12$ $x = 12$ or $x - 4 = 12$ $x = 16$	While the left side of the equation is factored, the right side is *not* zero. We must write the equation in standard form before we apply the zero products property. $x(x - 4) = 12$ $x^2 - 4x = 12$ $x^2 - 4x - 12 = 0$ $(x - 6)(x + 2) = 0$ $x - 6 = 0$ or $x + 2 = 0$ $x = 6$ $\quad x = -2$ The solution set is $\{-2, 6\}$.

4c. Solve $x^3 - 4x = 0$.

Incorrect Solution	Correct Solution and Explanation
$x^3 - 4x = 0$ $\frac{x^3 - 4x}{x} = \frac{0}{x}$ $x^2 - 4 = 0$ $(x - 2)(x + 2) = 0$ $x - 2 = 0$ or $x + 2 = 0$ $x = 2$ $\quad x = -2$	Since x may be zero, we cannot divide by it. Also dividing by a variable loses a solution of the equation. $x^3 - 4x = 0$ $x(x^2 - 4) = 0$ $x(x + 2)(x - 2) = 0$ $x = 0$ or $x + 2 = 0$ or $x - 2 = 0$ $x = -2$ $\quad x = 2$ The solution set is $\{-2, 0, 2\}$.

ANSWERS TO STUDENT CHECKS

Student Check 1 **a.** $\{-6, 9\}$ **b.** $\left\{-\frac{3}{8}, 2\right\}$ **c.** $\left\{-\frac{7}{3}, \frac{1}{3}\right\}$ **d.** $\{-2, 6\}$ **e.** $\left\{-\frac{5}{8}, 2\right\}$ **f.** $\{-3, 0, 3\}$ **g.** $\{-5, 4, 5\}$

Student Check 2 **a.** The distance is 12 ft. **b.** She can make the border 2 ft wide. **c.** It will take 5 sec.

Student Check 3 **a.** The zeros are $x = -3$ and $x = 2$. So, the x-intercepts are $(-3, 0)$ and $(2, 0)$. **b.** The zeros are $x = -5$, $x = -1$, and $x = 1$. So, the x-intercepts are $(-5, 0)$, $(-1, 0)$, and $(1, 0)$.

SUMMARY OF KEY CONCEPTS

1. To solve a polynomial equation, write the equation so that zero is on one side of the equation. Factor the polynomial. Set each factor equal to zero and solve the resulting equations.
2. When solving applications of polynomial equations, apply the problem-solving strategy.
   a. When solving applications involving area, identify the shape of the object whose area is being examined. Next, write down the area formula for that shape. Drawing a picture will help visualize the problem.
   b. The Pythagorean theorem defines a relationship among the lengths of the sides of a right triangle. The Pythagorean theorem is $(\text{leg})^2 + (\text{other leg})^2 = (\text{hypotenuse})^2$.
3. The zeros of a polynomial function are solutions of the equation $f(x) = 0$. The zeros correspond to the x-intercepts of the graph.

GRAPHING CALCULATOR SKILLS

The graphing calculator can be used to verify that we solved a polynomial equation correctly.

Example: Determine if $x^2 - 8x - 20 = 0$ has the solution set of $\{2, 10\}$.

Solution: Enter one side of the equation as Y_1 and the other side of the equation as Y_2. Then we view the table and determine if the y-values are equal.

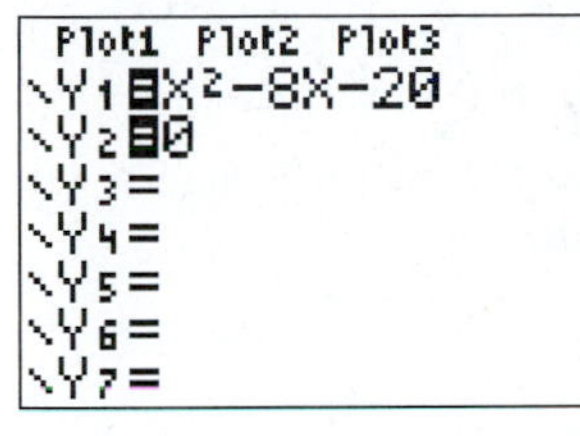

X	Y1	Y2
10	0	0

X=

2nd GRAPH

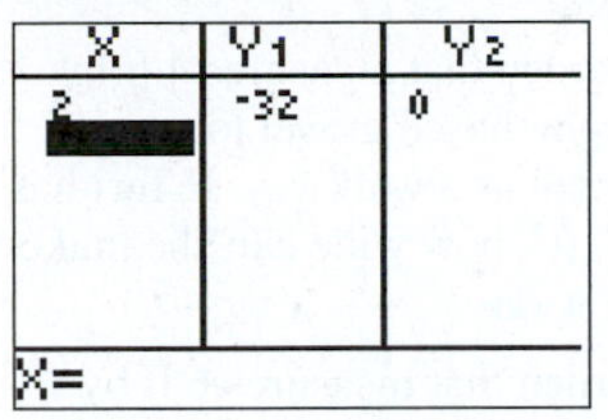

The y-values are equal when $x = 10$, so 10 is a solution of the equation. The y-values do not agree when $x = 2$, so 2 is not a solution of the equation.

SECTION 6.8 EXERCISE SET

Write About It!

Use complete sentences in your answer to each exercise.

1. Use an example to illustrate a third-degree polynomial equation in standard form.
2. Use an example to illustrate the zero products property.
3. Use an example to explain how to solve a quadratic equation by factoring.
4. Explain the difference in the solution sets when solving the two equations $(x - 3)(x + 4) = 0$ and $x(x - 3)(x + 4) = 0$.
5. Explain the difference in solving the two equations $(x - 3)(x + 2) = 0$ and $(x - 3)(x + 2) = 6$.
6. Explain how to apply the zero products property to solve the quadratic equation $x(x - 2) = 15$.
7. What is the graphical connection between the zeros of a polynomial function and the solutions of the corresponding polynomial equation?
8. Explain how to find the zeros of the function $f(x) = x^3 - 3x^2 - 10x$.

Practice Makes Perfect!

Solve each equation by factoring. (*See Objective 1.*)

9. $(x - 7)(2x - 5) = 0$
10. $(5y - 9)(y + 3) = 0$
11. $(6r - 21)(3r - 7) = 0$
12. $(3s - 12)(5s - 7) = 0$
13. $12x^2 - 19x + 4 = 0$
14. $5y^2 + 14y + 8 = 0$
15. $4x^2 + 35x - 9 = 0$
16. $4y^2 - 27y + 44 = 0$
17. $9x(x - 3) = 22$
18. $y(4y + 15) = 25$
19. $2a(20a - 21) = -9$
20. $2b(24b - 11) = 15$
21. $a^2 + (a - 5)^2 = (a - 10)^2$ $\{-15, 5\}$
22. $b^2 + (b - 6)^2 = (b - 12)^2$ $\{-18, 6\}$
23. $\frac{1}{27}x^2 + \frac{7}{18}x + 1 = 0$ $\left\{-6, -\frac{9}{2}\right\}$
24. $\frac{6}{7}x^2 + \frac{17}{7}x + 1 = 0$ $\left\{-\frac{7}{3}, -\frac{1}{2}\right\}$
25. $\frac{2}{23}x^2 + \frac{21}{23}x - 1 = 0$ $\left\{-\frac{23}{2}, 1\right\}$
26. $\frac{4}{5}y^2 - \frac{21}{5}y + 1 = 0$ $\left\{\frac{1}{4}, 5\right\}$
27. $z^3 - 36z = 0$ $\{-6, 0, 6\}$
28. $w^3 - 64w = 0$ $\{-8, 0, 8\}$
29. $-49a^3 + 36a = 0$ $\left\{-\frac{6}{7}, 0, \frac{6}{7}\right\}$
30. $-45b^3 + 80b = 0$ $\left\{-\frac{4}{3}, 0, \frac{4}{3}\right\}$
31. $x^3 - 4x^2 - 4x + 16 = 0$ $\{-2, 2, 4\}$
32. $y^3 + 2y^2 - 25y - 50 = 0$ $\{-5, -2, 5\}$
33. $a^3 + 2a^2 - a - 2 = 0$ $\{-2, -1, 1\}$
34. $b^3 + 9b^2 - 4b - 36 = 0$ $\{-9, -2, 2\}$
35. $r^3 - 3r^2 - 16r + 48 = 0$ $\{-4, 3, 4\}$
36. $s^3 + s^2 - 49s - 49 = 0$ $\{-7, -1, 7\}$

Solve each problem. (*See Objective 2.*)

37. A 10-ft ladder leans against the side of a house. If distance from the ground to the top of the ladder is 6 ft, what is the distance from the bottom of the ladder to the house? 8 ft

Additional answers can be found in the Instructor Answer Appendix.

38. A 26-ft ladder leans against the side of a house. If the distance from the ground to the top of the ladder is 24 ft, what is the distance from the bottom of the ladder to the house? 10 ft

39. Teri built a rectangular garden that measures 4 ft by 5 ft in her backyard. She purchased gravel to spread uniformly around the garden as a walkway. If Teri has enough gravel to cover 70 ft^2, how wide can she make the gravel walkway? 2.5 ft wide

40. Ko built a rectangular garden that measures 6 ft by 7 ft in her backyard. She purchased gravel to spread uniformly around the garden for a walkway. If Ko has enough gravel to cover 30 ft^2, how wide can she make the gravel border? 1 ft wide

41. The deck of the Golden Gate Bridge in San Francisco, California, stands about 220 ft above the water. Suppose a rock is dropped off the deck of the bridge. The equation $s = -16t^2 + 220$ gives the distance, s in feet, above the water the rock will be after t sec. When will the rock be 76 ft above the water? (Source: http://www.goldengatebridge.org/research/factsGGBDesign.php)
The rock will be 76 ft above the water in 3 sec.

42. The roof of a tower stands about 1474 ft above the ground. Suppose a ball is dropped from the roof of the tower. The equation $s = -16t^2 + 1474$ gives the distance, s in feet, above the ground the ball will be after t sec. When will the ball be 450 ft above the ground? The ball will be 450 ft above the ground in 8 sec.

Find the zeros of each function and identify the x-intercepts of the graph of the function. *(See Objective 3.)*

43. $f(x) = x^2 - 8x + 15$ $\quad x = 3, x = 5; (3, 0), (5, 0)$

44. $f(x) = 3x^2 - x - 2$ $\quad x = -\frac{2}{3}, x = 1; \left(-\frac{2}{3}, 0\right), (1, 0)$

45. $f(x) = -8x^2 + 10x + 3$ $\quad x = \frac{3}{2}, x = -\frac{1}{4}; \left(-\frac{1}{4}, 0\right), \left(\frac{3}{2}, 0\right)$

46. $f(x) = -5x^2 + 17x + 12$ $\quad x = 4, x = -\frac{3}{5}; \left(-\frac{3}{5}, 0\right), (4, 0)$

47. $f(x) = 27x^3 - 12x$ **48.** $f(x) = 25x^3 - 81x$

49. $f(x) = -54x^3 + 6x$ **50.** $f(x) = -18x^3 + 50x$

51. $f(x) = -2x^3 + 3x^2 + 9x$

52. $f(x) = -2x^3 + 5x^2 + 25x$

Match each function with its graph. (*See Objective 3.*)

53. **i.** $f(x) = -3x^2 - 5x + 2$ c **ii.** $f(x) = -2x^2 + 5x - 3$ a
iii. $f(x) = 2x^2 - 5x - 3$ d **iv.** $f(x) = 3x^2 + 5x - 2$ b

a.

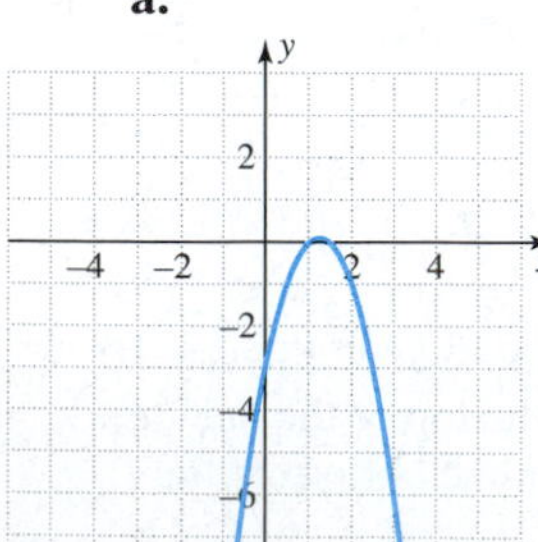

b.

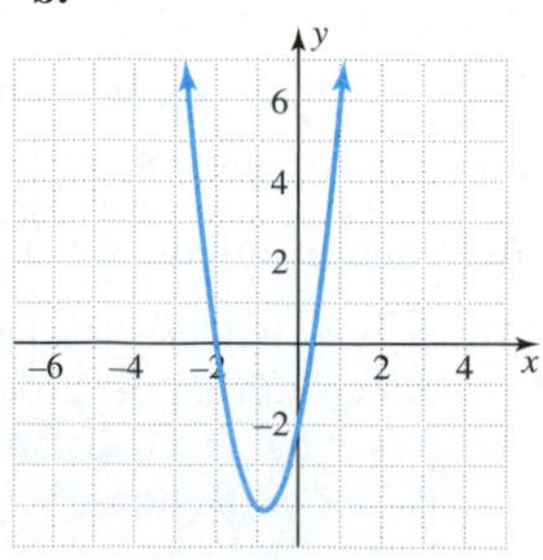

c.

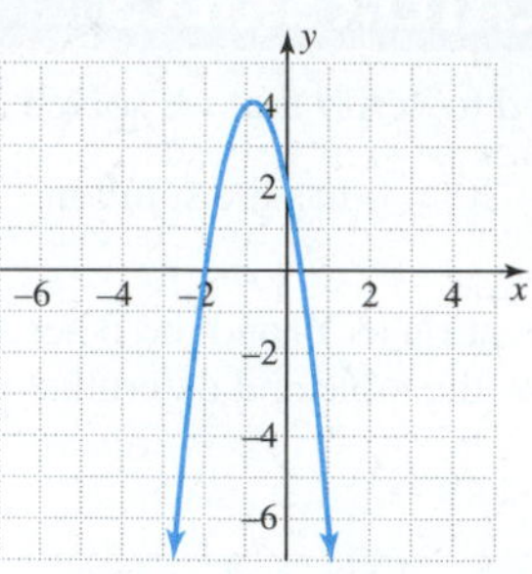

d.

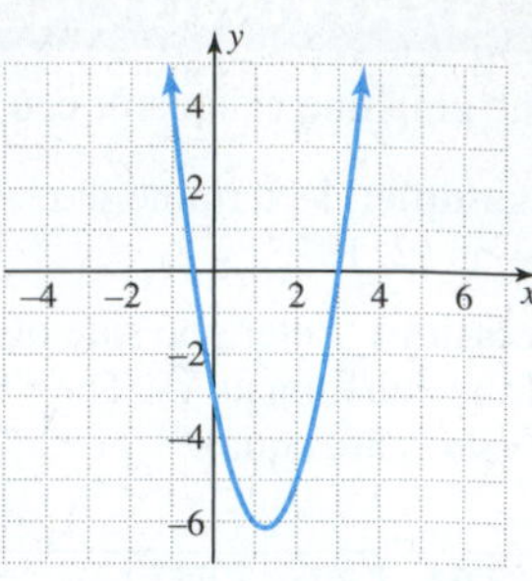

54. **i.** $f(x) = -x^3 + 4x$ d **ii.** $f(x) = x^3 - 4x$ b

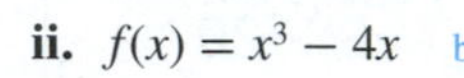

iii. $f(x) = x^2 - 4$ a **iv.** $f(x) = -x^2 + 4$ c

a.

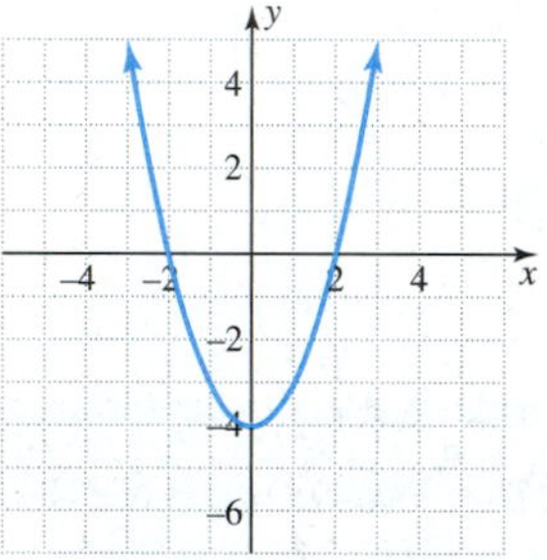

b.

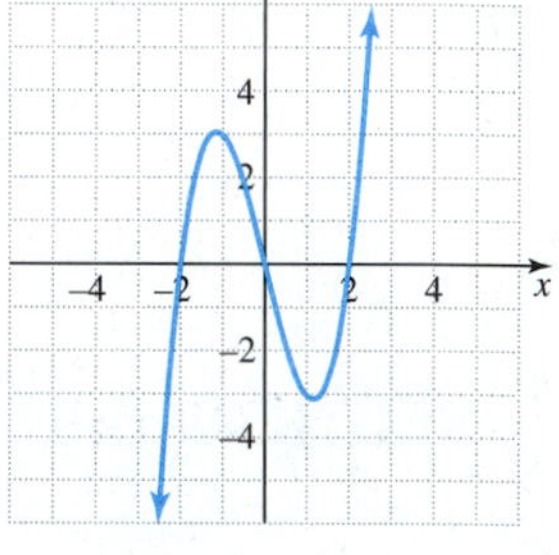

c.

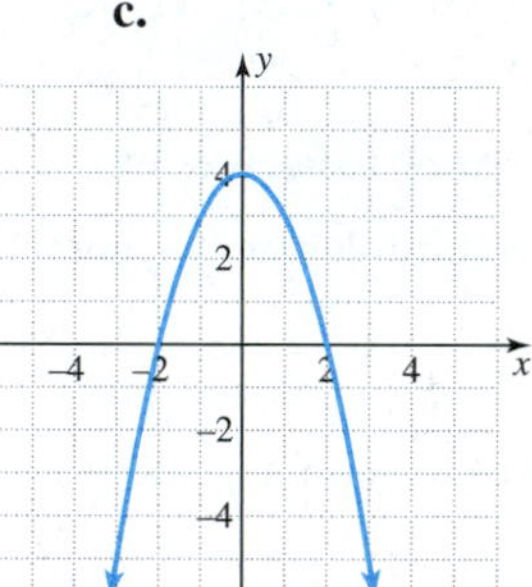

d.

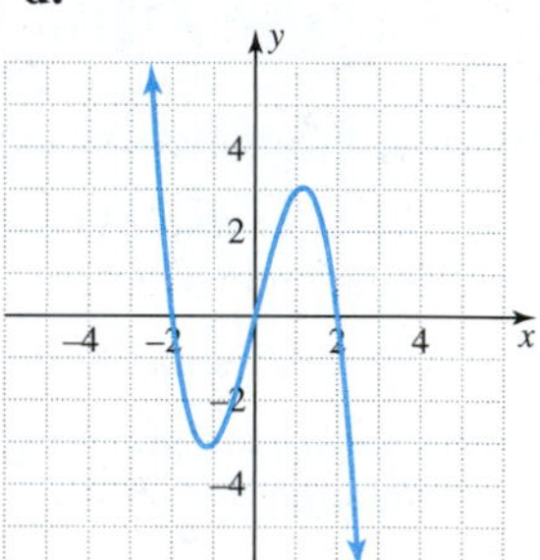

Mix 'Em Up!

Solve each equation.

55. $x(4x + 7)(5x - 9) = 0$ **56.** $y(5y + 13)(y - 2) = 0$

57. $2(a - 13)(2a - 5) = 0$ **58.** $3(t + 15)(3t + 7) = 0$

59. $5x^2 - 11x + 6 = 0$ **60.** $10z^2 + 7z - 12 = 0$

61. $10x^3 - 4x^2 - 6x = 0$ **62.** $20y^3 - 22y^2 + 6y = 0$

63. $x(x + 11) = -30$ **64.** $5y(5y - 9) = 22$

65. $a(3a + 11) = -10$ **66.** $b(9b - 19) = -2$

67. $x^2 + (x + 3)^2 = (x + 6)^2$ $\{-3, 9\}$ **68.** $y^2 + (y - 2)^2 = (y - 4)^2$ $\{-6, 2\}$

69. $r^2 + (r - 4)^2 = (r - 8)^2$ $\{-12, 4\}$ **70.** $s^2 + (s + 5)^2 = (s + 10)^2$ $\{-5, 15\}$

71. $\frac{3}{25}x^2 + \frac{1}{10}x - 1 = 0$ **72.** $\frac{1}{46}x^2 - \frac{31}{92}x + 1 = 0$

73. $\frac{15}{8}a^2 + \frac{7}{4}a - 1 = 0$ **74.** $\frac{2}{15}b^2 - \frac{13}{15}b + 1 = 0$

75. $-36z^3 + 16z = 0$ **76.** $-8w^3 + 50w = 0$

77. $36a^3 - 121a = 0$ **78.** $-45b^3 + 5b = 0$

79. $x^3 + 25x = 0$ $\{0\}$

80. $y^3 + 9x = 0$ $\{0\}$

81. $x^3 - 2x^2 - 36x + 72 = 0$ $\{-6, 2, 6\}$

82. $y^3 - 7y^2 - 4y + 28 = 0$ $\{-2, 2, 7\}$

83. $2w^3 + 5w^2 - 32w - 80 = 0$ $\left\{-4, -\frac{5}{2}, 4\right\}$

84. $2z^3 + z^2 - 8z - 4 = 0$ $\left\{-2, -\frac{1}{2}, 2\right\}$

85. $3x^3 - 2x^2 + 27x - 18 = 0$ $\left\{\frac{2}{3}\right\}$

86. $7y^3 + 2y^2 + 7y + 2 = 0$ $\left\{-\frac{2}{7}\right\}$

Solve each problem.

87. A 10-ft ladder leans against the side of a house. If the distance from the bottom of the ladder to the house is 6 ft, how high up the house does the ladder reach? 8 ft

88. A 25-ft ladder leans against the side of a house. If the distance from the bottom of the ladder to the house is 7 ft, how high up the house does the ladder reach? 24 ft

89. Johnson built a rectangular 12 ft by 24 ft swimming pool in his backyard. He purchased travertine pavers to build a walkway around the pool. If Johnson has enough travertine pavers to cover 160 ft^2, how wide is the walkway? 2 ft wide

90. Wing built a rectangular koi pond that measures 5 ft by 10 ft in her backyard. She purchased gravel to spread uniformly around the pond for decoration. If Wing has enough gravel to cover 76 ft^2, how wide can she make the gravel border? 2 ft wide

91. Suppose a coin was dropped from the top of the Central Plaza in Hong Kong. The function $s = -16t^2 + 1227$ approximates the distance s from the ground (in feet) of the coin after t seconds. When will the coin be 651 ft from the ground? (Source: http://skyscraperpage.com/cities/?buildingID=4637) 6 sec

92. Suppose a coin was dropped from the top of the Q1 Tower in Gold Coast QLD Australia. The function $s = -16t^2 + 1058$ approximates the distance s from the ground (in feet) that the coin will be after t seconds. When will the coin be 274 ft from the ground? (Source: http://skyscraperpage.com/cities/?buildingID=4637) 7 sec

Find the zeros of each function.

93. $f(x) = 6x^2 - 54x$ $x = 0, x = 9$

94. $f(x) = 3x^2 - 75x$ $x = 0, x = 25$

95. $f(x) = 20x^2 + 19x + 3$ $x = -\frac{3}{4}, x = -\frac{1}{5}$

96. $f(x) = 15x^2 - 26x + 8$ $x = \frac{2}{5}, x = \frac{4}{3}$

97. $f(x) = -8x^3 + 50x$ $x = -\frac{5}{2}, x = \frac{5}{2}, x = 0$

98. $f(x) = -25x^3 + 36x$ $x = -\frac{6}{5}, x = \frac{6}{5}, x = 0$

99. $f(x) = 12a^3 + 27a$ $x = 0$

100. $f(x) = -18x^3 - 50x$ $x = 0$

101. $f(x) = 6x^3 - 4x^2 - 32x$ $x = -2, x = \frac{8}{3}, x = 0$

102. $f(x) = 2x^3 - 12x^2 - 80x$ $x = -4, x = 0, x = 10$

Match each function with its graph.

103. **i.** $f(x) = 2x^2 - x - 3$ d **ii.** $f(x) = -2x^2 + x + 3$ c

iii. $f(x) = 2x^2 + x - 3$ a **iv.** $f(x) = -2x^2 + 7x - 3$ b

a.

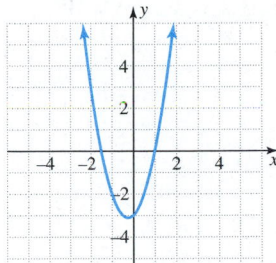

b.

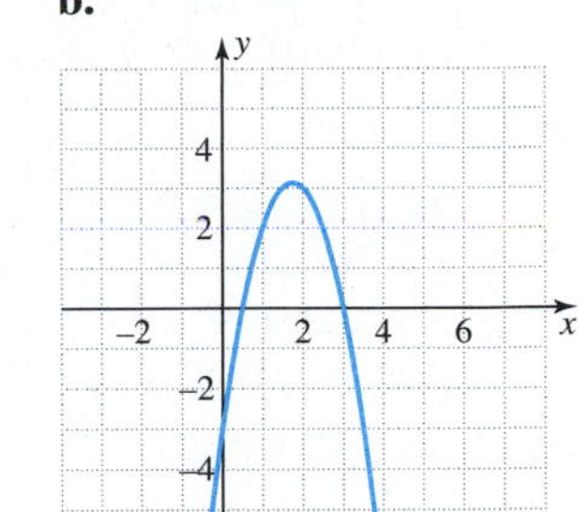

c.

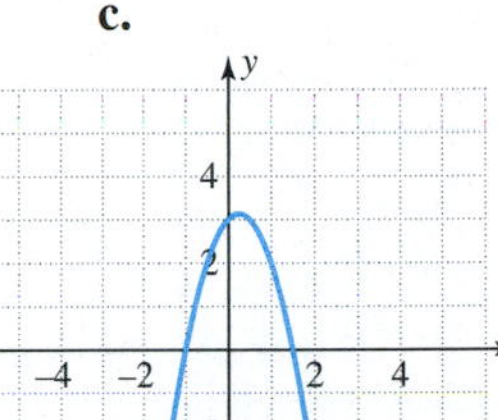

d.

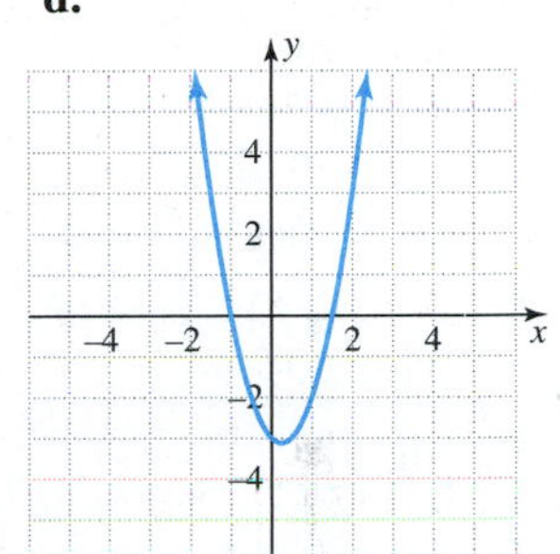

104. **i.** $f(x) = x^3 + x$ c **ii.** $f(x) = -x^3 - x$ b

iii. $f(x) = x^3 - x$ d **iv.** $f(x) = -x^3 + x$ a

a.

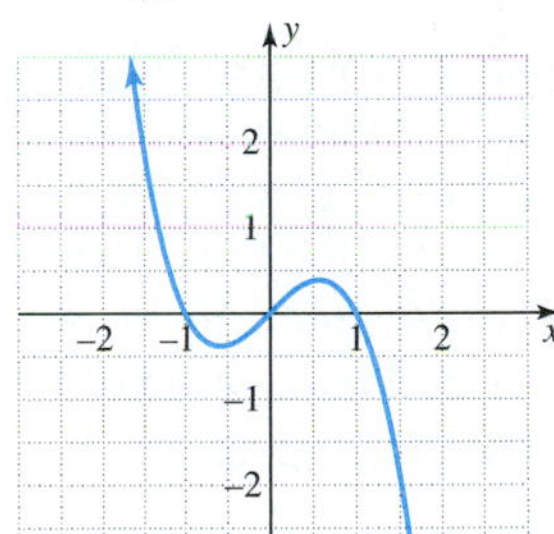

b.

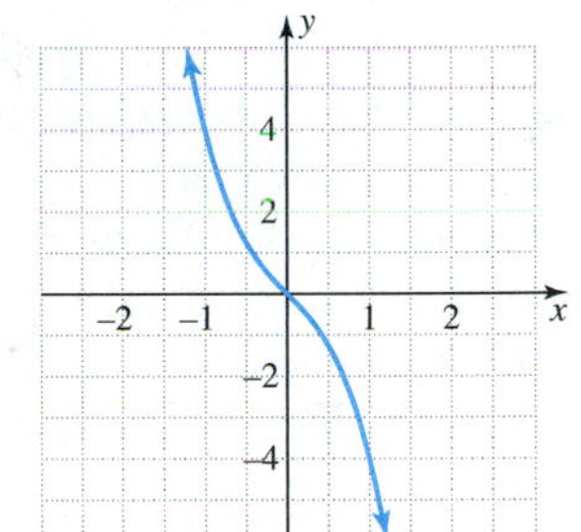

c.

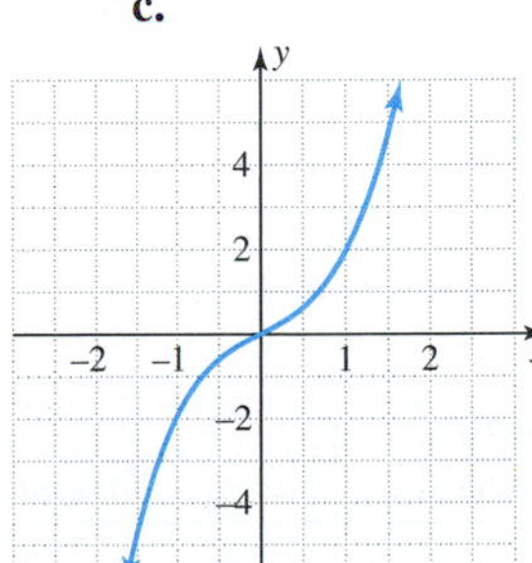

d.

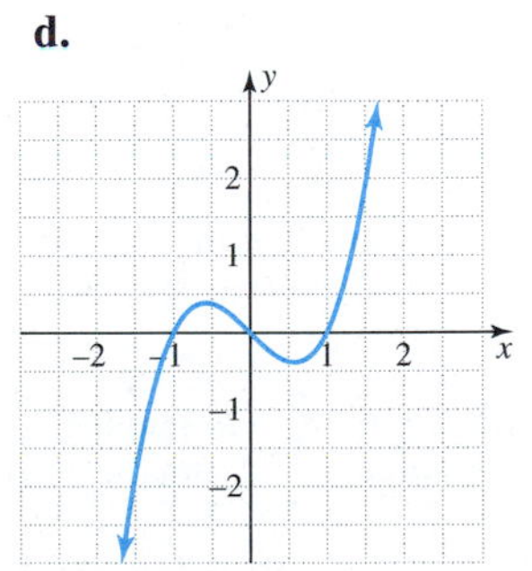

You Be the Teacher!

Correct each student's errors, if any.

105. Solve $2x(x + 3)(x - 1) = 0$.

Jody's work:

$2x(x + 3)(x - 1) = 0$

$x = 2, x = -3, x = 1$

$\{2, -3, 1\}$

$2x(x + 3)(x - 1) = 0$

$2x = 0, x + 3 = 0, x - 1 = 0$

$x = 0, \quad x = -3, \quad x = 1$

The solution set is $\{-3, 0, 1\}$.

106. Solve $-3y(y - 1)(y + 4) = 0$.

Miket's work:

$-3y(y - 1)(y + 4) = 0$

$y = 3, y = 1, y = -4$

$\{3, 1, -4\}$

$-3y(y - 1)(y + 4) = 0$

$-3y = 0, y - 1 = 0, y + 4 = 0$

$y = 0, \quad y = 1, \quad y = -4$

The solution set is $\{-4, 0, 1\}$.

107. Solve $x^3 - 4x = 0$.

Joanna's work:

$x^3 - 4x = 0$

$x^2 - 4 = 0$

$(x + 2)(x - 2) = 0$

$x = -2, x = 2$

$\{-2, 2\}$

$x^3 - 4x = 0$

$x(x^2 - 4) = 0$

$x(x + 2)(x - 2) = 0$

$x = 0, x + 2 = 0, x - 2 = 0$

$x = 0, \quad x = -2, \quad x = 2$

The solution set is $\{-2, 0, 2\}$.

108. Solve $x^3 + 9x = 0$.

Jill's work:

$x^3 + 9x = 0$

$x(x^2 + 9) = 0$

$x(x + 3)(x - 3) = 0$

$x = 0, x = -3, x = 3$

$\{0, -3, 3\}$

$x^2 + 9$ is the sum of squares and does not factor.

$x^3 + 9x = 0$

$x(x^2 + 9) = 0$

$x = 0$

The solution set is $\{0\}$.

Calculate It!

Use a graphing calculator to determine if the equation has the given solution set.

109. $5x^2 - 43x + 24 = 0, \left\{\frac{3}{5}, 8, 1\right\}$

110. $49x^3 - 9x = 0, \left\{0, -\frac{3}{7}, \frac{3}{7}, 3\right\}$

Think About It!

111. Write a quadratic equation that has solutions 5 and -1. Answers vary; $x^2 - 4x - 5 = 0$

112. Write a quadratic equation that has solutions 2 and -2. Answers vary; $x^2 - 4 = 0$

113. Write a polynomial equation that has solutions 3, 0, and 4. Answers vary; $x^3 + 7x^2 + 12x = 0$

114. Write a polynomial equation that has solutions 0, 1, and -3. Answers vary; $x^3 + 2x^2 - 3x = 0$

GROUP ACTIVITY Creating an Amortization Schedule

When we take out a mortgage or car loan, we generally agree to repay the lending institution in a specific number of equal monthly payments. Each payment consists of the principal and interest required to repay the loan. The formula, introduced in Section 6.2, to determine the monthly payment to repay a loan of P dollars, at an annual interest rate of r (as a decimal) in n months is given by

$$A = \frac{P\left(\frac{r}{12}\right)}{1 - \left(1 + \frac{r}{12}\right)^{-n}}$$

1. Show that the payment required to repay a loan of \$100,000 in 360 months at 6% interest is \$599.55.

2. How much will be paid to the lending institution after 360 months? How much of the amount paid was interest?

3. An amortization schedule shows the portion of the payment that goes toward interest and the portion that goes toward repaying the principal. The scheduled payment is the payment amount calculated in step 1. The interest paid each month is the monthly interest, $\frac{r}{12}$, times the beginning balance. The amount paid toward the principal each month is the difference between the scheduled payment and the interest. The ending balance is the beginning balance minus the principal paid each month. The first row is completed. Complete the schedule for the first 12 payments.

Payment Number	Beginning Balance	Scheduled Payment	Interest	Principal	Ending Balance
1	\$100,000	\$599.55	$\frac{0.06}{12}(100{,}000) = \$500$	\$599.55 − 500 = \$99.55	\$100,000 − 99.55 = \$99,900.45
2	\$99,900.45	\$599.55			
3					
⋮					
12					

4. Do the early payments pay more toward the principal amount borrowed or the interest?

5. Suppose that you make an additional payment of \$500 toward principal for each payment. Complete a new amortization schedule to determine the affect of this additional payment.

Chapter 6 / REVIEW

Exponents, Polynomials, and Polynomial Functions

What's the big idea? Polynomials are very important to the study of math, chemistry, science, economics, and so on. Polynomials are used to model real-life phenomena in these fields. Since polynomials are constructed of terms of the form ax^n, the properties of exponents enable us to perform operations on the polynomials. Factoring polynomials provides us with a method for solving polynomial equations.

The Tools

Listed below are the key terms, skills, formulas, and properties you should know for this chapter.

The page reference is provided if you need additional help with the given topic. The Study Tips will assist in your preparation for an exam.

Study Tips

1. Learn all of the terms, formulas, and properties. Make flash cards and have someone quiz you.
2. Rework problems from the exercises and also the ones you worked in class. Work additional problems from the review exercises.
3. Review the summaries of key concepts.
4. Work the chapter test.
5. Be sure to review the online resources for additional study materials.

Terms

Binomial 394
Coefficient 391
Conjugates 409
Constant term 391
Degree of a polynomial 393
Degree of a polynomial equation 445
Degree of a term 392
Exponent 366
Exponential expression 366
Factor a polynomial 419
Greatest common factor 418
Hypotenuse 448
Leading coefficient 393
Legs 448
Like terms 402
Monomial 393
Opposite of a polynomial 402
Perfect square trinomial 408
Polynomial 392
Polynomial equation 445
Polynomial function 394
Prime polynomial 426
Quadratic equation 446
Scientific notation 384
Standard form 392
Term 391
Trinomial 393
Zero of a function 450

Formulas and Properties

- Area of a rectangle 373
- Area of a triangle 376
- Difference of two cubes 439
- Difference of two squares 437
- Distributive property 420
- FOIL 406
- Pythagorean theorem 448
- Square of a binomial 408
- Subtracting polynomials 402
- Sum of two cubes 439
- Negative exponent 370
- Power of a power rule 379
- Power of a product rule 379
- Power of a quotient rule 380
- Product of conjugates 409
- Product of like bases rule 367
- Quotient of like bases rule 368
- Surface area of a cylinder 372
- Zero exponent 369
- Zero products property 445

CHAPTER 6 / SUMMARY

How well do you know this chapter? Complete the following questions to find out. Take a look back at the section if you need help.

SECTION 6.1 Rules for Exponents, Zero and Negative Exponents

1. When exponential expressions with like bases are multiplied, the resulting exponent is the sum of the exponents in the original expression. The base stays the same. This rule can be written as $x^a \cdot x^b = x^{a+b}$.
2. When exponential expressions with like bases are divided, the resulting exponent is the difference of the exponents in the original expression. The base stays the same. This rule can be written as $\frac{x^a}{x^b}$ x^{a-b}.
3. When a nonzero number is raised to the exponent of zero, the result is one.
4. A negative exponent is equivalent to the reciprocal of the base raised to the positive exponent. This can be stated as $b^{-n} = \frac{1}{b^n}$ for $b \neq 0$.

SECTION 6.2 More Rules of Exponents and Scientific Notation

5. When an exponential expression is raised to an exponent, the original exponents are multiplied to obtain the exponent in the final result. This rule can be written as $(x^a)^b = x^{ab}$.
6. When a product is raised to an exponent, the exponent must be applied to each factor in the base. This rule can be written as $(xy)^b$ $x^b y^b$.
7. When a quotient is raised to an exponent, the exponent must be applied to the numerator and denominator of the base. This rule can be written as $\left(\frac{x}{y}\right)^b = \frac{x^b}{y^b}$.
8. A number written in the form $a \times 10^n$ is a number written in scientific notation as long as $1 \leq a < 10$ and n is an integer.
9. When a number greater than 10 is written in scientific notation, the exponent of 10 is positive.
10. When a number between 0 and 1 is written in scientific notation, the exponent of 10 is negative.

SECTION 6.3 Polynomials, Polynomial Functions, and Their Basic Graphs

11. A(n) polynomial consists of a finite sum of terms of the form ax^n, where a is a real number and n is a whole number.
12. Polynomials with one term are called monomials.
13. Polynomials with two terms are called binomials.
14. Polynomials with three terms are called trinomials.
15. When the powers are written in descending order, the polynomial is in standard form.
16. The degree of a polynomial in a single variable is the largest exponent on the terms.
17. The leading coefficient is the coefficient of the term with the largest exponent.
18. The degree of a term with more than one variable is the sum of the exponents of the variables.
19. The degree of a polynomial in more than one variable is the largest degree of the terms in the polynomial.
20. To evaluate a polynomial, substitute the variable with the given number and simplify.
21. The graph of a polynomial function is smooth with no sharp curves. The domain of all polynomial functions is $(-\infty, \infty)$.

SECTION 6.4 Adding, Subtracting, and Multiplying Polynomials and Polynomial Functions

22. To add polynomials, combine like terms.
23. To subtract two polynomials, add the opposite of the second polynomial. When we find the opposite of a polynomial, all of the signs are changed.
24. When a monomial is multiplied by a polynomial, we apply the distributive property and the product of like bases rule for exponents.
25. To multiply polynomials, each term in the first polynomial must be distributed to each term in the second polynomial.
26. Two binomials may be multiplied using the FOIL method. This only works for two binomials.
27. When a binomial is squared, the result is always a(n) trinomial. The rule for squaring a binomial can be written as $(a + b)^2 = a^2 + 2ab + b^2$. The result of squaring a binomial is called a(n) perfect square trinomial.
28. Conjugates are binomials of the form $a + b$ and $a - b$.
29. When $(a + b)$ and $(a - b)$ are multiplied, the result is $a^2 - b^2$. This is called the difference of two squares.
30. When we raise a binomial to an exponent of 3 or higher, we must apply the definition of the exponent.

SECTION 6.5 Factoring Using the Greatest Common Factor and Grouping

31. To factor a polynomial means to write the polynomial as a(n) product.
32. The greatest common factor of integers is the largest factor that is common to each integer.
33. The GCF of a set of variable terms is the common variable raised to the smallest exponent of the terms.
34. The GCF of a set of terms is the product of the GCF of the coefficients and the GCF of the variable expressions.
35. To factor using the GCF, we apply the distributive property in reverse. That is, $ab + ac = a(b + c)$.
36. The grouping method applies to factoring polynomials with four or more terms.
37. All factoring can be checked by multiplying.

SECTION 6.6 Factoring Trinomials

38. To factor a trinomial of the form $x^2 + bx + c$, we must find the factors of c whose sum is b.

39. If the last term of the trinomial is positive, the signs in the binomial factors are the same.

40. If the last term of the trinomial is negative, the signs in the binomial factors are different.

41. Before factoring a trinomial, we should factor out any common factors.

42. There are two methods for factoring trinomials of the form $ax^2 + bx + c$, $a \neq 1$. We may use trial and error or the method of grouping.

43. A(n) perfect square trinomial factors as a binomial squared.

SECTION 6.7 Factoring Binomials and a Factoring Review

44. A binomial of the form $a^2 - b^2$ can be factored as $(a - b)(a + b)$. An example is $x^2 - 16$ (answers vary).

45. A binomial of the form $a^3 - b^3$ can be factored as $(a - b)(a^2 + ab + b^2)$. An example is $x^3 - 8$ (answers vary).

46. A binomial of the form $a^3 - b^3$ can be factored as $(a + b)(a^2 - ab + b^2)$. An example is $x^3 + 27$ (answers vary).

47. Except for a possible common factor, the sum of squares is prime. An example is $x^2 + 4$ (answers vary).

SECTION 6.8 Solving Polynomial Equations and Their Applications

48. A polynomial equation is an equation that can be written in the form of the form $P(x) = 0$, where $P(x)$ is a polynomial.

49. The zero products property states that if a product is zero, then one of the factors must be zero.

50. To solve a polynomial equation by factoring, write the equation in standard form, factor the polynomial, set each factor equal to zero, and solve.

51. A polynomial equation with degree 3 is called a(n) cubic equation. A polynomial equation with degree 4 is called a(n) quartic equation.

52. The degree of the equation determines the maximum number of solutions of the equation.

53. Quadratic equations can be used to solve problems involving the area of rectangles and triangles.

54. The Pythagorean theorem is $a^2 + b^2 = c^2$, where a and b represent the legs of a triangle and c represents the hypotenuse.

CHAPTER 6 / REVIEW EXERCISES

SECTION 6.1

Use the properties of exponents to simplify each expression. Write answers with positive exponents. (*See Objectives 1–5.*)

1. $-4a^{-4}$ $\quad -\frac{4}{a^4}$

2. $\frac{a^{-2}}{b^{-3}}$ $\quad \frac{b^3}{a^2}$

3. $(-p^{-2}q^{-4})(2p^5q^{-7})$ $\quad -\frac{2p^3}{q^{11}}$

4. $(-5m^{-7}n^{-2})(m^{-3}n)$ $\quad -\frac{5}{m^{10}n}$

5. $-a^0 + 6$ $\quad 5$

6. $-2b^0 - 3$ $\quad -5$

7. $\frac{-12a^{-5}n}{6a^9n^4}$ $\quad -\frac{2}{a^{14}n^3}$

8. $\frac{16s^{-2}t^6}{10s^{-4}t^4}$ $\quad \frac{8}{5}s^2t^2$

9. $-(-2)^{-5}$ $\quad \frac{1}{32}$

10. $-(-3)^{-3}$ $\quad \frac{1}{27}$

11. $\frac{5u^{-9}v^{-4}}{20u^{-1}v^{-4}}$ $\quad \frac{1}{4u^8}$

12. $\frac{9x^6y^{-5}}{63x^{-2}y^9}$ $\quad \frac{x^8}{7y^{14}}$

13. $(2a + 1)^0$ $\quad 1$

14. $-2(b - 3)^0$ $\quad -2$

For Exercises 17 and 18, the formula to determine the monthly payment needed to repay a loan of P dollars, at an annual interest rate of r (as a decimal), in n months is given by

$$A = \frac{P\left(\frac{r}{12}\right)}{1 - \left(1 + \frac{r}{12}\right)^{-n}}$$

Find the monthly payment needed to repay each loan. Round each answer to the nearest cent. (*See Objective 6.*)

15. Home loan of \$180,000 at an annual interest rate of 5.9% and compounded in 360 months $\quad$ \$1067.65

16. Car loan of \$25,450 at an annual interest rate of 4.3% and compounded in 60 months $\quad$ \$472.15

SECTION 6.2

Simplify each expression using the rules of exponents. Write each answer with positive exponents. (*See Objectives 1 and 2.*)

17. $(-2x^4y^{-5})^{-3}$ $\quad -\frac{y^{15}}{8x^{12}}$

18. $(-3a^{-6}b)^{-2}$ $\quad \frac{a^{12}}{9b^2}$

19. $(-2c^3d^{-6})^3$ $\quad -\frac{8c^9}{d^{18}}$

20. $(-p^{-2}q^3)^4$ $\quad \frac{q^{12}}{p^8}$

21. $\left(-\frac{x^{-4}}{2}\right)^{-5}$ $\quad -32x^{20}$

22. $\left(-\frac{y^{-2}}{6}\right)^{-2}$ $\quad 36y^4$

23. $\left(\frac{2x^{-2}}{5y^5}\right)^{-3}$ $\quad \frac{125x^6y^{15}}{8}$

24. $\left(\frac{7u^5}{3v^{-2}}\right)^3$ $\quad \frac{343u^{15}v^6}{27}$

25. $(3a^2b^{-9})^2 \cdot a^{-9}b^{12}$ $\quad \frac{9}{a^5b^6}$

26. $(-x^{-3}y^{-2})^3 \cdot x^9y^8$ $\quad -y^2$

27. $\frac{(2^{-4}s^8t^{-5})^{-2}}{3^{-2}s^{-11}t^7}$ $\quad \frac{2304t^3}{s^5}$

28. $\frac{(9r^7s^{10})^{-1}}{3^{-2}r^{-9}s^{-6}}$ $\quad \frac{r^2}{s^4}$

29. $\left(\frac{10p^{-5}}{q^3}\right)^{-2}\left(\frac{20q^{-4}}{p^{-8}}\right)^3$ $\quad \frac{80p^{34}}{q^6}$

30. $\left(\frac{6h^{-9}}{k^5}\right)^{-2}\left(\frac{6k^{-8}}{h^3}\right)^3$ $\quad \frac{6h^9}{k^{14}}$

Use scientific notation to perform each operation. Express each answer in scientific notation. (*See Objectives 3 and 4.*)

31. $(3.5 \times 10^{-10})(7.2 \times 10^{15})$ 2.52×10^{6}

32. $(1.2 \times 10^{-7})(4.5 \times 10^{13})$ 5.4×10^{6}

33. $(0.00018)(25{,}000{,}000)$ 4.5×10^{3}

34. $(0.00026)(15{,}500{,}000)$ 4.03×10^{3}

35. $\dfrac{4.75 \times 10^{9}}{3.8 \times 10^{-5}}$ 1.25×10^{14}

36. $\dfrac{1.221 \times 10^{-4}}{7.4 \times 10^{-12}}$ 1.65×10^{7}

37. $\dfrac{3726}{276{,}000{,}000}$ 1.35×10^{-5}

38. $\dfrac{0.00001598}{425{,}000}$ 3.76×10^{-11}

39. $\dfrac{(3.8 \times 10^{-9})(6.5 \times 10^{-6})}{1.25 \times 10^{-12}}$ 1.976×10^{-2}

40. $\dfrac{(6.75 \times 10^{5})(3.2 \times 10^{-4})}{4.5 \times 10^{-9}}$ 4.8×10^{10}

41. In 2010 the total box office gross for the U.S. movie market was approximately \$10,456,000,000. The number of tickets sold was approximately 1,325,000,000. Find the average price per ticket. Round to two decimal places. (Source: http://www.the-numbers.com) \$7.89

42. In 2010 the total number of DVD units sold for the movie *Avatar* was approximately 10,173,000. The total DVD sales was approximately \$183,638,000. Find the average price per DVD unit. Round to two decimal places. (Source: http://www.the-numbers.com) \$18.05

SECTION 6.3

Determine the coefficient and degree of each term. (*See Objective 1.*)

43. $-120x^5$ $-120, 5$

44. $24y^4$ $24, 4$

45. $-\dfrac{x^3}{12}$ $-\dfrac{1}{12}, 3$

46. $\dfrac{y}{15}$ $\dfrac{1}{15}, 1$

47. -18 $-18, 0$

48. 17 $17, 0$

49. $6s^3t^4$ $6, 7$

50. $-8c^5d$ $-8, 6$

Classify each polynomial and identify its terms, degree, type, and leading coefficient, if applicable. (*See Objective 2.*)

51. $5.76b^3 - 1.24b + 2.1$

52. $\dfrac{1}{8}x^3 - 27$

53. a^2b^4c

54. $-pq^4r^3$

55. $0.4x^2 - 1.2x^3 + 3.1x - 1.6$

56. $-0.5y^2 - 0.6y^3 - 2.8y + 1.1$

If $f(x) = 4x^2 - 5x + 1$ and $g(x) = -6x + 7$, evaluate each expression. (*See Objective 3.*)

57. $f(-2)$ 27

58. $g(-1)$ 13

59. $f(0) + g(-3)$ 26

60. $f(-1) + g(0)$ 17

61. $f(1) - g(-1)$ -13

62. $f(0.1) - g(-0.1)$ -7.06

63. According to the box office history for a movie distributor, the number of ticket sales (in millions) from 2000 to 2009 can be modeled by the function

$$C(x) = -0.051x^3 - 0.667x^2 + 5.987x + 190.325$$

where x is the number of years after 2000. Find $C(2)$ and $C(10)$ and interpret their meaning. Round to two decimal places.

64. According to the box office market share history for a movie distributor, the percent of market share from 2000 to 2009 can be modeled by the function

$$P(x) = -0.0672x^3 + 1.1375x^2 - 4.4618x + 12.041$$

where x is the number of years after 2000. Find $P(3)$ and $P(10)$ and interpret their meaning. Round to two decimal places.

SECTION 6.4

Perform each operation. (*See Objectives 1–6.*)

65. $(5x - 13) + (9x + 26)$ $14x + 13$

66. $(-7y + 18) + (-12y - 14)$ $-19y + 4$

67. $-1.2a - 15.6 + (-3.8a + 7.2)$ $-5a - 8.4$

68. $(-4.5b + 11.3) + (-2.9b - 5.6)$ $-7.4b + 5.7$

69. $(19.2b + 26.4) - (-11.5b - 15.8)$ $30.7b + 42.2$

70. $(3.1c - 4.5) - (6.4c + 17.2)$ $-3.3c - 21.7$

71. $\left(\dfrac{5}{3}x^2 - \dfrac{7}{10}xy - 14y^2\right) + \left(\dfrac{11}{3}x^2 - \dfrac{1}{5}xy + 9y^2\right)$ $\dfrac{16}{3}x^2 - \dfrac{9}{10}xy - 5y^2$

72. $\left(\dfrac{7}{2}a^2 + \dfrac{1}{5}ab - 6b^2\right) - \left(\dfrac{1}{4}a^2 - \dfrac{2}{5}ab + b^2\right)$ $\dfrac{13}{4}a^2 + \dfrac{3}{5}a - 7b^2$

73. $-3x^4(2x^2 - 5x + 1)$

74. $-y^2(2y^2 + 6y - 5)$

75. $-5xy^2(4x^2 - 9x + 2y^2)$

76. $-6x^2y(2x^2 - xy + 4y^2)$

77. $(4q - 5)(q - 3)$

78. $(7v + 1)(4v - 5)$

79. $(5x + 1)(25x^2 - 5x + 1)$ $125x^3 + 1$

80. $(2y - 3)(4y^2 + 6y + 9)$ $8y^3 - 27$

81. $(w - 2)(7w + 15)$

82. $(3z + 16)(2z - 17)$

83. $(5p^2 - 3)(p^2 + 2)$

84. $(3x - 5)^2$ $9x^2 - 30x + 25$

85. $\left(3m^2 + \dfrac{2}{5}\right)\left(3m^2 - \dfrac{2}{5}\right)$

86. $\left(4n^2 - \dfrac{3}{2}\right)\left(4n^2 + \dfrac{3}{2}\right)$

87. $(a + 2)(a - 2)(a + 3)$

88. $(x - 4y)^3$

89. $(3a - b)^3$

90. $(u - 2v)^4$

91. A street vendor selling bottles of water can purchase bottles of water for \$0.35 each. The cost to operator her stand is \$540 per month. So, the monthly cost is $C(x) = 0.35x + 540$, where x is the number of bottles of water sold in a month. If she sells the bottles of water for \$1.80 each, the revenue made from selling x bottles in a month is $R(x) = 1.8x$. Write a polynomial function that represents the monthly profit of the bottled of water stand. What is the profit from selling 1500 bottles of water in a month? $P(x) = 1.45x - 540$; \$1635

92. A street vendor selling pretzels can purchase pretzels for \$0.45 each. The cost to operate his stand is \$380 per month. So, the monthly cost is $C(x) = 0.45x + 380$, where x is the number of pretzels sold in a month. If he sells pretzels for \$2.10 each, the revenue made from selling x pretzels in a month is $R(x) = 2.10x$. Write a polynomial function that represents the monthly profit of the pretzel stand. What is the profit from selling 1500 pretzels in a month? $P(x) = 1.65x - 380$; \$2095

93. An apartment manager can rent 200 units if he charges \$750 per month. If he increases the price by \$50, he rents three fewer apartments. If x represents the number of increases in price, what polynomial represents the total rent collected each month? $R(x) = -150x^2 + 7750x + 150{,}000$

94. An apartment manager can rent all 150 units if he charges \$850 per month. If he increases the price by \$35, he rents two fewer apartments. If x represents the number of increases in price, what polynomial represents the total rent collected each month? $R(x) = -70x^2 + 3550x + 127{,}500$

95. Write a polynomial function that represents the perimeter of a trapezoid with sides of lengths $2y - 9$, $6y + 11$, $5y^2 - 8y + 13$, and $13y^2 - 16y + 1$ units. $18y^2 - 16y + 16$ units

96. Write a polynomial function that represents the perimeter of a triangle with sides of lengths $15x - 11$, $3x^2 - 10x + 5$, and $2x^2 - 5x - 19$ units. $5x^2 - 25$ units

SECTION 6.5

Find the GCF of each set of terms. (*See Objective 1.*)

97. $8x^2y^5$ and $28x^3y^3$ $4x^2y^3$

98. $-25a^3b^6$ and $10a^2b^8$ $5a^2b^6$ or $-5a^2b^6$

99. $18x^4(y + 1)$ and $24x^3(y - 1)$ $6x^3$

100. $42x^2(y - 2)$ and $35x^3(y + 2)$ $7x^2$

Factor each polynomial completely. Check each answer by multiplying. (*See Objectives 2 and 3.*)

101. $-74a^9 - 40a^6$ $-2a^6(37a^3 + 20)$

102. $21b^5 - 12b^3$ $3b^3(7b^2 - 4)$

103. $54a^3 - 50a^2 - 42a$ $2a(27a^2 - 25a - 21)$

104. $-48b^3 - 24b^2 + 8b$ $-8b(6b^2 + 3b - 1)$

105. $15y^6(x - 5) - 45y^3(x - 5)$ $15y^3(x - 5)(y^3 - 3)$

106. $-28y(x + 2) + 35y(x + 2)$ $7y(x + 2)$

107. $7x(y + 4) - (y - 4)$ prime

108. $y^2(x + 3) - 6(x - 3)$ prime

109. $2y^3 + 7y^2 - 6y - 21$ $(2y + 7)(y^2 - 3)$

110. $20b^3 - 25b^2 + 24b - 30$ $(4b - 5)(5b^2 + 6)$

111. $20a^3 + 40a^2 - 16a - 32$ $4(a + 2)(5a^2 - 4)$

112. $12c^3 + 12c^2 - 6c - 6$ $6(c + 1)(2c^2 - 1)$

SECTION 6.6

Factor each trinomial. Use an appropriate substitution if needed. (*See Objectives 1–4.*)

113. $v^2 - 7v - 30$ $(v - 10)(v + 3)$

114. $r^2 - 5r - 150$ $(r - 15)(r + 10)$

115. $a^2 + 16a + 55$ $(a + 5)(a + 11)$

116. $b^2 - 19b + 90$ $(b - 10)(b - 9)$

117. $x^2 - 7xy + 12y^2$ $(x - 3y)(x - 4y)$

118. $x^2 + 6xy - 135y^2$ $(x - 9y)(x + 15y)$

119. $5a^3 + 50a^2 + 105a$ $5a(a + 3)(a + 7)$

120. $-4q^3 + 4q^2 + 24q$ $-4q(q - 3)(q + 2)$

121. $63a^2 - 59a + 10$ $(7a - 5)(9a - 2)$

122. $14q^2 - 17q - 6$ $(2q - 3)(7q + 2)$

123. $-20a^2 + 104ab - 20b^2$ $-4(5a - b)(a - 5b)$

124. $-6x^2 + 62xy - 72y^2$ $-2(3x - 4y)(x - 9y)$

125. $6v^3 - 6v^2 - 336v$ $6v(v - 8)(v + 7)$

126. $-80p^3 - 44p^2 + 12p$ $-4p(5p - 1)(4p + 3)$

127. $40a^3 - 75a^2b - 10ab^2$ $5a(8a + b)(a - 2b)$

128. $140x^3 + 12x^2y - 8xy^2$ $4x(7x + 2y)(5x - y)$

129. $-36x^3 - 96x^2y + 80xy^2$ $-4x(3x + 10y)(3x - 2y)$

130. $-32a^3 + 64a^2b - 24ab^2$ $-8a(2a - b)(2a - 3b)$

131. $x^6 + x^3 - 12$ $(x^3 - 3)(x^3 + 4)$

132. $y^6 + 16y^3 + 48$ $(y^3 + 12)(y^3 + 4)$

133. $a^4 + 7a^2 - 30$ $(a^2 + 10)(a^2 - 3)$

134. $b^4 + 10b^2 - 24$ $(b^2 - 2)(b^2 + 12)$

135. $32p^8 - 8p^4 - 84$ $4(4p^4 - 7)(2p^4 + 3)$

136. $45q^8 - 57q^4 - 24$ $3(5q^4 - 8)(3q^4 + 1)$

137. $(a - 5)^2 - 6(a - 5) - 7$ $(a - 12)(a - 4)$

138. $(b + 2)^2 - 5(b + 2) - 14$ $(b - 5)(b + 4)$

139. $28(x - 1)^2 - 29(x - 1) + 6$ $(4x - 7)(7x - 9)$

140. $8(s + 4)^2 - 61(s + 4) - 24$ $(8s + 35)(s - 4)$

SECTION 6.7

Factor each polynomial completely. (*See Objectives 1–3.*)

141. $9x^2 - \dfrac{1}{64}$

142. $49y^2 - \dfrac{1}{16}$

143. $441u^2 - 144v^2$ $9(7u - 4v)(7u + 4v)$

144. $4r^2 - 36s^2$ $4(r - 3s)(r + 3s)$

145. $16a^2 + 9$ prime

146. $64b^2 + 1$ prime

147. $(2a - 3)^2 - 16b^2$ $(2a - 3 - 4b)(2a - 3 + 4b)$

148. $(7c + 1)^2 - 25d^2$ $(7c + 1 - 5d)(7c + 1 + 5d)$

149. $3(x - 4)^2 - 27$ $3(x - 1)(x - 7)$

150. $4(y + 1)^2 - 16$ $4(y - 1)(y + 3)$

151. $54a^3 - 16b^3$ $2(3a - 2b)(9a^2 + 6ab + 4b^2)$

152. $c^6 + 125d^3$ $(c^2 + 5d)(c^4 - 5c^2d + 25d^2)$

153. $40x^3 + 5y^3$ $5(2x + y)(4x^2 - 2xy + y^2)$

154. $u^3 - 27v^6$ $(u - 3v^2)(u^2 + 3uv^2 + 9v^4)$

155. $a^6 - 8b^6$ $(a^2 - 2b^2)(a^4 + 2a^2b^2 + 4b^4)$

156. $216x^6 + y^6$ $(6x^2 + y^2)(36x^4 - 6x^2y^2 + y^4)$

157. $c^4 + 256d^4$ prime

158. $x^4 + 625y^4$ prime

159. $45v^3 + 36v^2 - 80v - 64$ $(5v + 4)(3v + 4)(3v - 4)$

160. $7z^3 - z^2 - 112z + 16$ $(7z - 1)(z + 4)(z - 4)$

SECTION 6.8

Solve each equation. (*See Objective 1.*)

161. $x(3x + 7)(5x - 13) = 0$ $\left\{-\dfrac{7}{3}, 0, \dfrac{13}{5}\right\}$

162. $y(5y + 11)(y - 14) = 0$ $\left\{-\dfrac{11}{5}, 0, 14\right\}$

163. $12x^2 - 36x + 24 = 0$ $\{1, 2\}$

164. $5z^2 - 44z - 60 = 0$ $\left\{-\dfrac{6}{5}, 10\right\}$

165. $2x(12x + 23) = -7$

166. $y(4y - 7) = -3$

167. $x^2 + (x - 4)^2 = (x - 8)^2$ $\{-12, 4\}$

168. $y^2 + (y - 3)^2 = (y - 6)^2$ $\{-9, 3\}$

169. $\dfrac{1}{39}x^2 - \dfrac{10}{39}x - 1 = 0$

170. $\dfrac{5}{4}x^2 - 2x - 1 = 0$

171. $-9z^3 + 16z = 0$

172. $4w^3 - 144w = 0$

173. $x^3 - 9x^2 - 4x + 36 = 0$ $\{-2, 9, 2\}$

174. $y^3 - 2y^2 - 64y + 128 = 0$ $\{-8, 2, 8\}$

Solve each problem. (*See Objective 2.*)

175. A 41-ft ladder leans against the side of a house. If the distance from the bottom of the ladder to the house is 9 ft, how high up the house does the ladder reach? 40 ft

176. Jill built a rectangular koi pond that measures 3 ft by 12 ft in her backyard. She purchased gravel to spread uniformly around the pond for decoration. If Jill has enough gravel to cover 54 ft^2, how wide can she make the gravel border? 1.5 ft

177. Suppose a rock was dropped from the top of the Empire State Building in New York City, New York. The function $s = -4.9t^2 + 381$ approximates the distance s from the ground (in meters) that the rock will be after t seconds. When will the rock be 258.5 m from the ground? 5 sec

178. Suppose a penny was dropped from the top of the Shanghai World Financial Center in China. The function $s = -16t^2 + 1614$ approximates the distance s from the ground (in feet) that the penny will be after t seconds. How long will it take the penny to be 590 ft from the ground? 8 sec

Find the zeros of each polynomial function. (*See Objective 5.*)

179. $f(x) = 6x^2 - 18x$ $\{0, 3\}$

180. $f(x) = 3x^2 - 45x$ $\{0, 15\}$

181. $f(x) = 10x^2 - 28x - 48$ $\left\{-\frac{6}{5}, 4\right\}$

182. $f(x) = 2x^2 + 11x - 40$ $\left\{-8, \frac{5}{2}\right\}$

183. $f(x) = 48x^3 - 192x$

184. $f(x) = -50x^3 + 2x$

185. $f(x) = 12x^3 - 5x^2 - 3x$ $\left\{-\frac{1}{3}, 0, \frac{3}{4}\right\}$

186. $f(x) = 6x^3 - 14x^2 - 80x$ $\left\{-\frac{8}{3}, 0, 5\right\}$

Match each graph with its function.

187. **i.** $f(x) = x^2 + 2x - 3$ d **ii.** $f(x) = -x^2 - 2x + 3$ c
iii. $f(x) = 2x^2 - x - 3$ a **iv.** $f(x) = -3x^2 - 7x + 2$ b

a.

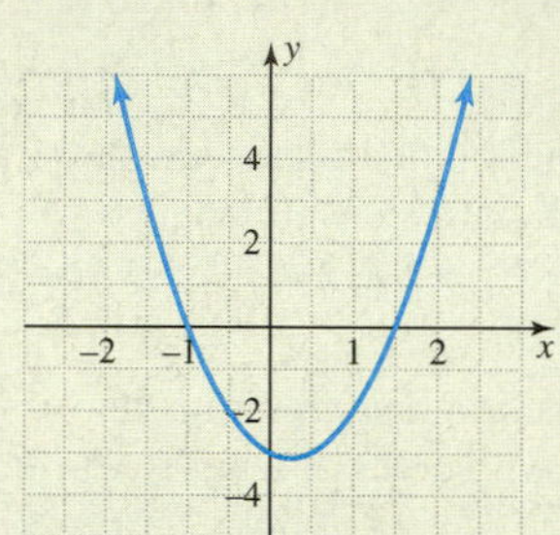

b.

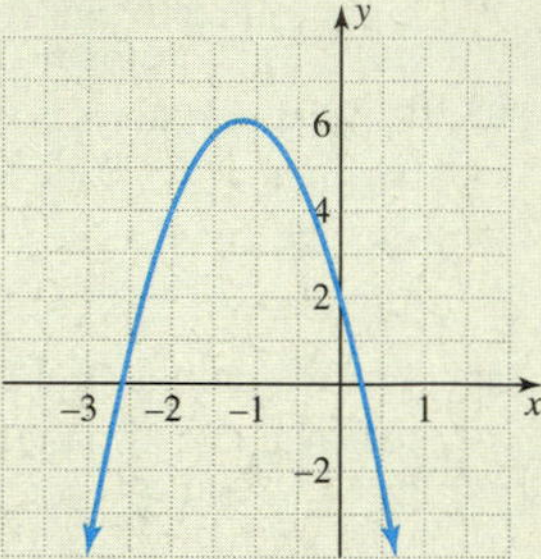

c.

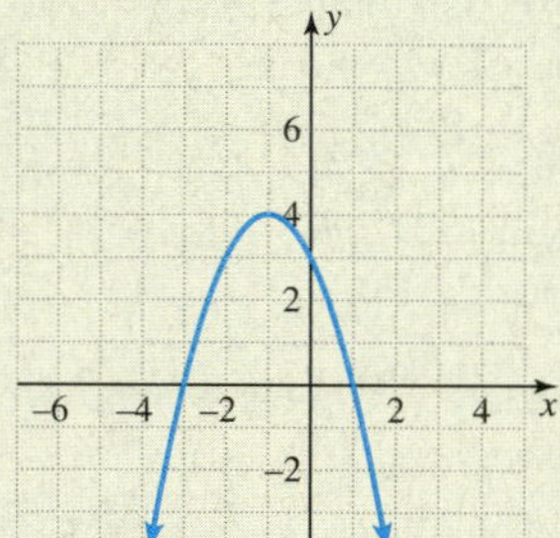

d.

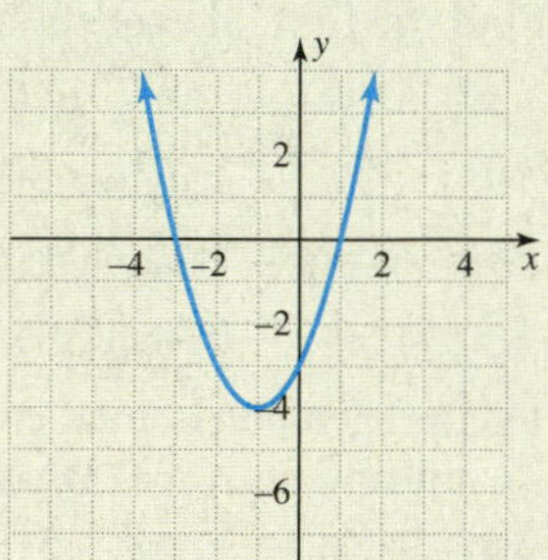

CHAPTER 6 TEST / EXPONENTS, POLYNOMIALS, AND POLYNOMIAL FUNCTIONS

1. When simplified $(4x^5)(-2x^3)^2$ is

a. $16x^{11}$ (circled) **b.** $-8x^{11}$ **c.** $-8x^{10}$ **d.** $16x^{10}$

2. -8^0 is equivalent to

a. -8 **b.** -1 (circled) **c.** 1 **d.** 0

3. When simplified $\frac{a^6}{a^{-2}}$ is equivalent to

a. a^4 **b.** a^8 (circled) **c.** $\frac{1}{a^3}$ **d.** $\frac{1}{a^{12}}$

4. When 0.0000037 is written in scientific notation, it is

a. 37×10^5 **b.** 37×10^{-5}
c. 3.7×10^6 **d.** 3.7×10^{-6} (circled)

5. When simplified $(-4x^2 - 6x + 2) - (3x^2 + 2x - 7)$ is

a. $-7x^2 - 4x + 5$ **b.** $-7x^2 - 8x + 9$ (circled)
c. $x^2 - 4x - 5$ **d.** $x^2 - 8x + 9$

6. When simplified $(x - 5)^2$

a. $x^2 + 25$ **b.** $x^2 - 25$
c. $x^2 + 10x + 25$ **d.** $x^2 - 10x + 25$ (circled)

7. The greatest common factor of $4x^5 - 8x^4 + 12x^3$ is

a. 4 **b.** $4x$ **c.** $4x^3$ (circled) **d.** $4x^5$

8. One of the prime factors of $6t^2 - 19t - 20$ is

a. $(t + 5)$ **b.** $(2t + 5)$ **c.** $(6t + 5)$ (circled) **d.** $(t + 1)$

9. The complete factorization of $16x^2 + 48x + 36$ is

a. $(8x + 12)(2x + 3)$ **b.** $4(2x + 3)^2$ (circled)
c. $4(2x - 3)^2$ **d.** $(4x + 9)^2$

10. The solution set for $6x^2 + x = 2$ is

a. $\left\{-\frac{3}{2}, 2\right\}$ **b.** $\left\{-\frac{1}{2}, \frac{2}{3}\right\}$
c. $\left\{-\frac{3}{2}, \frac{1}{2}\right\}$ **d.** $\left\{-\frac{2}{3}, \frac{1}{2}\right\}$ (circled)

Perform each operation. Write answers using positive exponents.

11. $(-6a^3)(2a^{-4})^2$ $-\frac{24}{a^5}$

12. $\left(-\frac{3x^5y}{2x^{-1}y^3}\right)^{-2}$ $\frac{4y^4}{9x^{12}}$

13. $-4x^0 + (-4)^0$ -3

Write each polynomial in standard form and identify its type, degree, and leading coefficient.

14. $-4y - y^2 + 5$

15. $1 - 27x^3$

16. $9a^6$

17. Simplify: $(6x^4y^3 - 2x^3y^2 + x^2y) + (2x^4y^3 + 3x^3y^2 - 7x^2y)$ $8x^4y^3 + x^3y^2 - 6x^2y$

18. What is the degree of the polynomial that results in Exercise 17? 7

19. Evaluate the resulting polynomial in Exercise 17 when $x = -2$ and $y = 3$. 3312

Perform each operation.

20. $(2x + 5) - (3x - 7)$ $-x + 12$

21. $(2x + 5)(3x - 7)$ $6x^2 + x - 35$

22. $(5a^3 + 2b)^2$ $25a^6 + 20a^3b + 4b^2$

23. $(5a^3 + 2b)(5a^3 - 2b)$ $25a^6 - 4b^2$

24. $(3x - 5)(9x^2 + 15x + 25)$ $27x^3 - 125$

25. $(3x - 5)^3$ $27x^3 - 135x^2 + 225x - 125$

Solve each problem.

26. The distance traveled in 1 light-year is approximately 6 trillion miles. The North Star, Polaris, is approximately 300 light-years away from Earth. Convert each of these numbers to scientific notation and then perform an operation using these numbers that will determine how many miles away the North Star is from Earth. (Sources: http://www.worsleyschool.net/science/files/polaris/thenorthstar.html and http://science.howstuffworks.com/dictionary/astronomy-terms/question94.htm)
6×10^{12} miles; 3×10^{12} light-years; 1.8×10^{15} miles

27. Write a polynomial that represents the perimeter and area of a rectangle whose length is $5x^2 + 3x + 1$ ft and whose width is $6x + 7$ ft.
Perimeter $10x^2 + 18x + 16$ ft; Area $30x^3 + 53x^2 + 27x + 7$ ft^2

28. Allison owns a jewelry business. The cost for making x bracelets in a month can be represented by $C(x) = 2x + 50$ and the revenue for selling x bracelet's in a month is $R(x) = 12x$.
 a. Find a polynomial that represents the profit for Allison's business. Recall profit is revenue minus cost. $10x - 50$
 b. Find the profit for selling 30 bracelets in a month. \$250

Factor each polynomial completely.

29. $5x^2 + 10x$ $5x(x + 2)$

30. $20 - 5y - 4p + yp$ $(4 - y)(5 - p)$

31. $r^3 + r^2 - 4r - 4$ $(r + 1)(r - 2)(r + 2)$

32. $3a^2 - 3a - 18$ $3(a - 3)(a + 2)$

33. $9t^2 + 6t - 8$ $(3t - 2)(3t + 4)$

34. $10b^3 - 35b^2 - 20b$ $5b(2b + 1)(b - 4)$

35. $9x^2 + 30x + 25$ $(3x + 5)^2$

36. $81p^2 - 49$ $(9p - 7)(9p + 7)$

37. $4a^2 + 25$ prime

38. $125x^3 - 27y^3$ $(5x - 3y)(25x^2 + 15xy + 9y^2)$

39. $81t^4 - 16$ $(3t - 2)(3t + 2)(9t^2 + 4)$

40. $64y^6 + 1$ $(4y^2 + 1)(16y^4 - 4y^2 + 1)$

Solve each equation.

41. $4x^2 = 36$ $\{-3, 3\}$

42. $a^2 + a = 56$ $\{-8, 7\}$

43. $6z^2 = 5z + 6$ $\left\{-\frac{2}{3}, \frac{3}{2}\right\}$

44. $(y + 1)(y - 3) = 5$ $\{-2, 4\}$

45. $2x^3 - 2x^2 - 40x = 0$ $\{-4, 0, 5\}$

46. Find the lengths of the legs of a right triangle if the lengths are consecutive even integers and the hypotenuse is 10 units.
The lengths of the legs are 6 units and 8 units.

CUMULATIVE REVIEW EXERCISES / CHAPTERS 1–6

Evaluate each expression for the given value. (*Section 1.2, Objective 5*)

1. $5x^2 + 2x - 6$ for $x = -3$ 33

2. $-x^3 + 7x + 5$ for $x = 2$ 11

Simplify each expression. (*Section 1.3, Objective 3*)

3. $4x(3x - 6) - 15x^2 - 3(9 - 2x)$ $-3x^2 - 18x - 27$

4. $6(5x + 8) - 4(x - 15)$ $26x + 108$

Solve each linear equation and check the answer. (*Section 2.1, Objective 2–6*)

5. $\frac{15x}{7} - \frac{3x}{2} = \frac{18}{7}$ $\{4\}$

6. $1.5x - 0.9x = 4.8$ $\{8\}$

Translate each problem into a linear equation and solve the problem. (*Section 2.2, Objective 1*)

7. Five less than three times a number yields 31. Find the number. The number is 12.

8. The sum of two consecutive even numbers is 54. Find the numbers. The numbers are 26 and 28.

Solve each compound inequality. Write each solution in interval notation. (*Section 2.5, Objectives 3 and 4*)

9. $4x + 6 > 26$ or $4x + 6 < -26$ $(-\infty, -8) \cup (5, \infty)$

10. $2x - 13 > -23$ and $2x - 13 \le 23$ $(-5, 18]$

Solve each equation. (*Section 2.6, Objectives 1–3*)

11. $|x + 2| = |4x - 19|$ $\left\{\frac{17}{5}, 7\right\}$

12. $|3x - 25| = -8$ $\varnothing$

13. $|8 - 2x| = 0$ $\{4\}$

14. $|2x| = |x - 6|$ $\{-6, 2\}$

15. Choose an appropriate scale to plot the three ordered pairs, $(-25, 45)$, $(20, -15)$, $(5, 60)$, on the same rectangular coordinate system. Then specify the quadrant where each point is located. (*Section 3.1, Objective 1*)

16. Find the midpoint of the line segment formed by the ordered pairs $(-14, -8)$ and $(8, -24)$. (*Section 3.1, Objective 5*) $(-3, -16)$

Find the domain and range of each relation. **(*Section 3.2, Objective 2*)**

17.

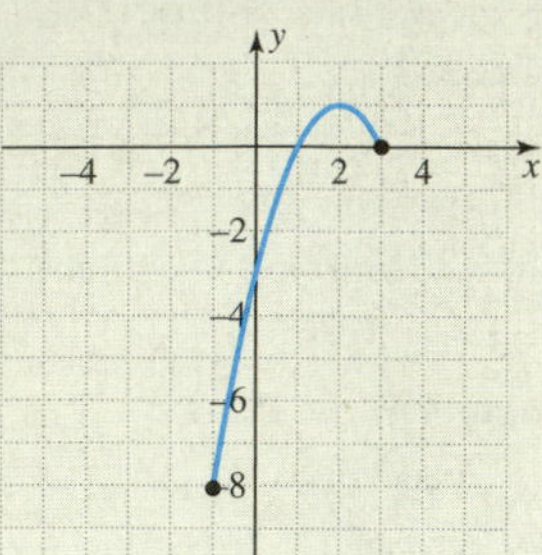

Domain = [−1, 3], Range = [−8, 1]

18.

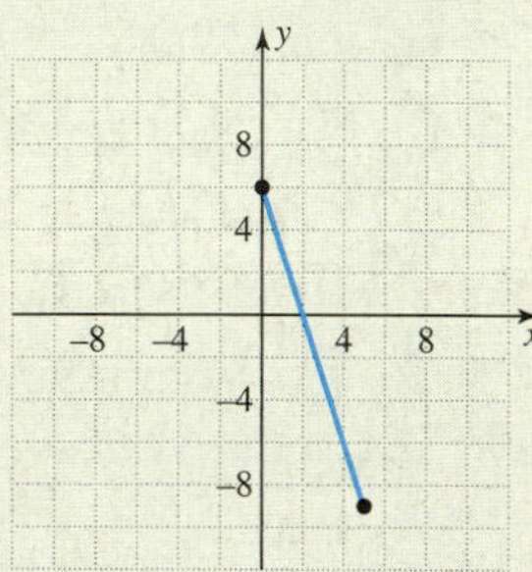

Domain = [0, 5], Range = [−9, 6]

Find the requested information. **(*Section 3.3, Objective 3*)**

19. Use the graph of $h(x)$ to find all x for which $h(x) = 0$. $x = 1, x = 3$

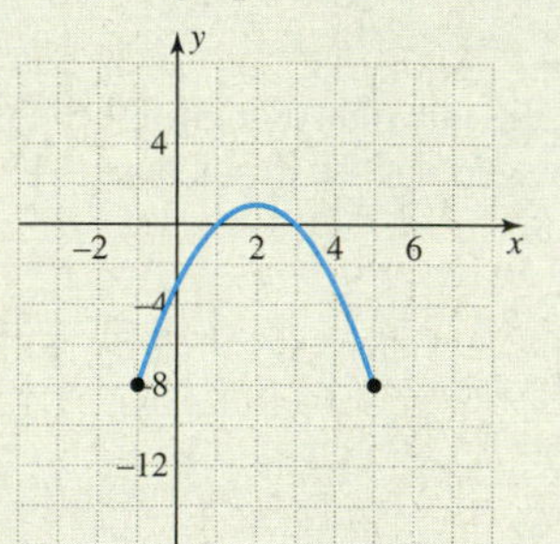

20. Find $f(0)$ if the function $f(x)$ is given by Y_1. 0

X	Y1
-3	-12
-2	0
-1	2
0	0
1	0
2	8
3	30

Rewrite each linear equation using function notation, if possible. Identify m and b. **(*Section 4.1, Objective 2*)**

21. $y = -2$ $f(x) = -2; m = 0; b = -2$

22. $3x - 8y = 12$ $f(x) = \frac{3}{8}x - \frac{3}{2}; m = \frac{3}{8}; b = -\frac{3}{2}$

Evaluate each linear function at the given value. **(*Section 4.1, Objective 3*)**

23. $f(x) = -4x + 7, x = -\frac{5}{2}$ 17

24. $f(x) = \frac{3}{5}x - 2, x = 5$ 1

25. $f(x) = 5, x = 0$ 5

26. $f(x) = -\frac{3}{2}x, x = 4$ −6

Graph each linear function and state its domain and range. **(*Section 4.2, Objective 1*)**

27. $f(x) = \frac{1}{2}x$

28. $f(x) = 2$

Find the slope and y-intercept of each line from its equation. Write the y-intercept as an ordered pair. Graph the line using its slope and y-intercept. Label at least two points on your graph. **(*Section 4.3, Objectives 1 and 2*)**

29. $f(x) = 2x - 3$

30. $3x + y = 6$

Write the equation of the line that has the given slope and passes through the given point. Express each answer in slope-intercept form and standard form. **(*Section 4.4, Objective 2*)**

31. $m = 0$; $(0, -7)$ $y = -7; y + 7 = 0$

32. $m = \frac{5}{2}$; $(0, -6)$ $y = \frac{5}{2}x - 6; 5x - 2y = 12$

33. parallel to $y = -2$; $(-3, 5)$ $y = 5$

34. perpendicular to $x + 6y = 1$; $(-5, 2)$ $6x - y = -32$

35. The number of associate's degrees awarded by colleges and universities in the United States was 738,000 in May 2008 and 813,000 in May 2012. (Source: nces.ed.gov) (*Section 4.4, Objective 5*)

a. Write a linear equation that models the number of associate's degrees awarded by colleges and universities, where x is the number of years after 2008. $y = 18{,}750x + 738{,}000$

b. If this trend continues, how many associate's degrees will be awarded in 2015? about 869,250

36. The average annual salary for public school teachers in the United States was about \$50,758 in May 2007 and \$53,910 in 2009. (Source: nces.ed.gov) (*Section 4.4, Objective 5*)

a. Write a linear equation that models the average annual salary for public school teachers, where x is the number of years after 2007. $y = 1576x + 50{,}758$

b. If this trend continues, what will be the average annual salary for public school teachers in 2015? about \$63,366

Graph each linear inequality in two variables. **(*Section 4.5, Objective 2*)**

37. $4x - y \le 6$

38. $x + 3y < 9$

Solve each system graphically. **(*Section 5.1, Objectives 2 and 3*)**

39. $\begin{cases} 5x - y = 4 \\ x = 2 \end{cases}$ {(2, 6)}

40. $\begin{cases} x + 3y = 3 \\ y = \frac{1}{2}x - 4 \end{cases}$ {(6, −1)}

Solve each system by substitution or elimination. **(*Section 5.2, Objectives 1 and 2*)**

41. $\begin{cases} y = 2x + 6 \\ x + 7y = 27 \end{cases}$ {(−1, 4)}

42. $\begin{cases} 8x - 5y = 11 \\ x - 3y = 18 \end{cases}$ {(−3, −7)}

43. $\begin{cases} 1.2x + 0.8y = 21.2 \\ 1.8x - y = 23 \end{cases}$ {(15, 4)}

44. $\begin{cases} x + y = 45 \\ 0.1x + 0.4y = 11.7 \end{cases}$ {(21, 24)}

45. $\begin{cases} \frac{1}{2}x - y = -4 \\ \frac{3}{5}x + \frac{1}{5}y = -2 \end{cases}$ {(−4, 2)}

46. $\begin{cases} \frac{5}{4}x - \frac{1}{4}y = \frac{21}{2} \\ \frac{1}{2}x + \frac{7}{6}y = -11 \end{cases}$ {(6, −12)}

Solve each problem. (*Section 5.3, Objective 5*)

47. Two angles are supplementary. The measure of one angle is 38° less than three times the measure of the other angle. Find the measure of each angle. 54.5° and 125.5°

48. Two angles are complementary. The measure of one angle is 9° more than twice the measure of the other angle. Find the measure of each angle. 27° and 63°

Solve each system of linear equations in three variables by elimination. (*Section 5.4, Objective 1*)

49. $\begin{cases} x - 2y - 6z = -14 \\ -x + 4y + 8z = 22 \\ 3x - y + 7z = -2 \end{cases}$

50. $\begin{cases} x + 2y = -1 \\ 3y - z = -10 \\ 2x + 5z = 26 \end{cases}$

51. $\begin{cases} x + 3y - 6z = 7 \\ y - 4z = 5 \\ -x - 2y + 2z = -2 \end{cases}$

52. $\begin{cases} 3x - 2y + 4z = 35 \\ 7x + 4y = -9 \\ 5x + y + 2z = 18 \end{cases}$

Solve each system of linear inequalities in two variables. (*Section 5.5, Objective 1*)

53. $\begin{cases} 4x + y < 24 \\ x - 3y < 15 \end{cases}$

54. $\begin{cases} 2x + y \geq 0 \\ x - 4y \leq -8 \end{cases}$

Simplify each expression by applying the product or quotient rule. Assume all bases are nonzero real numbers. (*Section 6.1, Objectives 1 and 2*)

55. $(x)^{10} \cdot (x)^7$ x^{17}

56. $(7)^{13} \cdot (7)^5$ 7^{18}

57. $(-14a^5) \cdot (-2a^{16})$ $28a^{21}$

58. $(-6a^5b^2)(-9a^2b^{12})$ $54a^7b^{14}$

59. $\frac{p^7}{p^6}$ p

60. $\frac{q^{11}}{q^9}$ q^2

61. $\frac{5^{16}}{5^{13}}$ 125

62. $\frac{-42q^9}{-7q^6}$ $6q^3$

Simplify each expression. Write each with positive exponents. Assume all bases are nonzero real numbers. (*Section 6.1, Objectives 3 and 4*)

63. $-8(10b)^0$ -8

64. $-2(-5e)^0$ -2

65. $-2c^0 + 5$ 3

66. $-5x^0 + 11$ 6

67. $\left(-\frac{5}{6}\right)^{-3}$ $-\frac{216}{125}$

68. $-\left(-\frac{10}{3}\right)^{-3}$ $\frac{27}{1000}$

69. $(3c^7q^2)(-6c^{-12}q)$ $-\frac{18q^3}{c^5}$

70. $(n^{10}a^{-12})(-4n^{-6}a^{-7})$ $-\frac{4n^4}{a^{19}}$

71. $(2x^5y^4)(-4x^{-5}y^{-8})$ $-\frac{8}{y^4}$

72. $(5x^3y)(-3x^{-3}y^{-4})$ $-\frac{15}{y^3}$

73. $\frac{-16x^3y^7}{2xy^{10}}$ $-\frac{8x^2}{y^3}$

74. $\frac{-6r^6t^{-10}}{20r^{-11}t^{-5}}$ $-\frac{3r^{17}}{10t^5}$

75. $\frac{-12pq^{-8}}{3p^{-5}q^{-6}}$ $-\frac{4p^6}{q^2}$

76. $\frac{18x^{-5}y^{-6}}{2x^2y^{-13}}$ $\frac{9y^7}{x^7}$

Solve each problem. Use the formula $A = 2\pi r^2 + 2\pi rh$ for the total surface area of a can with height h and radius r. (*Section 6.1, Objective 6*)

77. Write an expression for the surface area of a can if the height is three times its radius. $8\pi r^2$

78. Write an expression for the surface area of a can if the height is half of its radius. $3\pi r^2$

Simplify each expression using the rules of exponents. Write each answer with positive exponents. Assume all bases are nonzero real numbers. (*Section 6.2, Objectives 1 and 2*)

79. $(-3x^4y^{10})^3$ $-27x^{12}y^{30}$

80. $(-2a^{-5})^{-3}$ $-\frac{a^{15}}{8}$

81. $\left(\frac{8p^2}{7q^{-3}}\right)^{-2}$ $\frac{49}{64p^4q^6}$

82. $\left(\frac{4r^3}{3s^{-1}}\right)^{-3}$ $\frac{27}{64r^9s^3}$

83. $\frac{(4y^{10}z^{-9})^{-3}}{y^{-6}z^{-17}}$ $\frac{z^{44}}{64y^{24}}$

84. $\frac{(x^{-5}y^{-1})^{-1}}{6^2xy^{-4}}$ $\frac{x^4y^5}{36}$

Use scientific notation to perform the indicated operation. Write each answer in scientific notation. (*Section 6.2, Objectives 3 and 4*)

85. $(1.2 \times 10^4)^3$ 1.728×10^{12}

86. $\frac{1.476 \times 10^{29}}{4.1 \times 10^{18}}$ 3.6×10^{10}

87. $\frac{0.297}{5{,}500{,}000}$ 5.4×10^{-8}

88. $\frac{(9.36 \times 10^{-3})(7.5 \times 10^{-1})}{4.5 \times 10^7}$ 1.56×10^{-10}

Classify each polynomial and identify its terms, degree, type, and leading coefficient, if applicable. (*Section 6.3, Objectives 1 and 2*)

89. $2x^3 - 5x + 17$

90. $2x^3 - 5x + 17$

91. $-ab^2c$

92. $9 - 5t$

If $f(x) = x^2 - 6x + 2$ and $g(x) = -x^3 + 8x - 1$, find the following. (*Section 6.3, Objective 3*)

93. $f(0) - g(-2)$ 11

94. $f(2) + g(-1)$ -14

Simplify each expression. (*Section 6.4, Objectives 1–5*)

95. $(-19q^2 - 15q - 4) + (23q^2 - 24q - 15)$
$4q^2 - 39q - 19$

96. $(-7b^2 + 9b - 17) - (-22b^2 - 8b + 15)$
$15b^2 + 17b - 32$

97. $\left(\frac{2}{3}r^2 + \frac{4}{3}rs - 8s^2\right) + \left(\frac{4}{3}r^2 - \frac{7}{3}rs + s^2\right)$ $2r^2 - rs - 7s^2$

98. $\left(\frac{19}{10}s^2 + \frac{5}{3}st - 9t^2\right) - \left(\frac{9}{10}s^2 + \frac{8}{3}st - 6t^2\right)$ $s^2 - st - 3t^2$

99. $(1.2w^2 + 8w + 10.6) - (-3.2w^2 + 8.3w - 10.1)$
$4.4w^2 - 0.3w + 20.7$

100. $(-2.7z^2 - 8z - 4.4) - (8.5z^2 - 6.8z + 1.6)$
$-11.2z^2 - 1.2z - 6$

101. $-3ab(9a^2 - 10ab - 12b^2)$ $-27a^3b + 30a^2b^2 + 36ab^3$

102. $5xy(9x^2 - 4xy + 6y^2)$

103. $(w + 3)(w^2 + 3w - 9)$

104. $(3w - 2)(9w^2 + 6w - 4)$ $27w^3 - 24w + 8$

105. $(4r - 3)(r + 1)$ **106.** $(y - 5)(6y + 1)$

107. $(2a + 9)^2$ **108.** $(7b - 3)^2$

Use grouping to factor each polynomial completely, if applicable. Check each answer by multiplying. (*Section 6.5, Objectives 1–3*)

109. $15w^3 + 24w^2 - 5w + 8$ can't factor by grouping

110. $20w^3 + 35w^2 - 4w + 7$ can't factor by grouping

111. $10b^3 + 25b^2 - 4b - 10$ $(2b + 5)(5b^2 - 2)$

112. $3x^3 + 6x^2 - 7x - 14$ $(x + 2)(3x^2 - 7)$

113. $12x^5(y - 2) - 40x^3(y - 2)$ $4x^3(y - 2)(3x^2 - 10)$

114. $51y^3(x + 2) - 34y^2(x + 2)$ $17y^2(x + 2)(3y - 2)$

Factor each trinomial. (*Section 6.6, Objectives 1–4*)

115. $10p^2 + 39p + 14$ **116.** $64q^2 - 64q + 7$

117. $-98x^2 + 91x + 70$ **118.** $-40y^2 - 37y - 4$

119. $(r - 3)^2 - 5(r - 3) - 66$ $(r - 14)(r + 3)$

120. $(s + 4)^2 + 13(s + 4) + 40$ $(s + 12)(s + 9)$

Factor each expression completely. (*Section 6.7, Objectives 1 and 2*)

121. $9r^3 - 64r$ **122.** $-5s^3 + 80s$

123. $(5a - 1)^2 - 25$ **124.** $(4b - 3)^2 - 81$

125. $(b + 1)^2 + 4$ prime **126.** $(a - 1)^2 + 9$

127. $16z^4 - 81$ **128.** $16w^4 - 1$

129. $27a^3 + 1$ **130.** $8b^3 - 125$

Factor each polynomial completely. (*Section 6.7, Objective 3*)

131. $36c^3 - 9c^2 - 64c + 16$ **132.** $4d^3 - d^2 - 36d + 9$

133. $25u^2 - 10u + 1 - 9v^2$ **134.** $9r^2 + 6rs + s^2 - 49$

Solve each equation by factoring. (*Section 6.8, Objective 1*)

135. $(6r - 21)(3r - 7) = 0$ **136.** $(3s - 12)(5s - 7) = 0$

137. $y(4y + 15) = 25$ **138.** $2a(20a - 21) = -9$

139. $98y^3 - 162y = 0$ **140.** $y^3 + 49y = 0$ $\{0\}$

Solve each problem. (*Section 6.8, Objective 2*)

141. Suppose a penny was dropped from the top of the Willis Tower in Chicago. The quadratic equation $s = -9.8t^2 + 442$ approximates the distance from the ground (in meters) that the penny will be after t seconds. How long will it take the penny to fall to 197 m from the ground? 5 sec

142. A 25-ft ladder leans against the side of a house. If the distance from the bottom of the ladder to the house is 7 ft, how high up the house does the ladder reach? 24 ft

Find the zeros of each function. (*Section 6.8, Objective 3*)

143. $f(x) = x^2 + 3x - 10$ **144.** $f(x) = 3x^2 + 5x - 8$

145. $f(x) = 2x^3 - x^2 - 10x$ **146.** $f(x) = 6x^3 - 9x^2 + 3x$

CHAPTER

7

Rational Expressions, Functions, and Equations

Learning Strategies

Albert Einstein is probably the epitome of an incredible learner. Could you ever be like him? What made him have the capacity to acquire all of his knowledge? While you and I are not Albert Einstein, there are things that we can do to improve our own ability to learn new skills.

- Find out what kind of learner you are—visual, auditory, kinesthetic, and so on. The Internet offers many different learning styles inventories that you can complete to determine this information.
- As soon as class ends, take a few moments to write down a summary of the main points that were presented in class.
- It has been proven that most people forget up to 40% of new material within a few hours of being exposed to it. Therefore, you should find the time to review class notes within a couple hours of attending class.
- Review your notes frequently over the next few days so that the concepts get transferred to your long-term memory.
- Distribute your studying over several sessions rather than one long cram session. Some studies have shown that students who distribute their practice over several sessions remember 67% more than those who learn by cramming. [Source: John J. Donovan and David J. Radosevich, "A Meta-Analytic Review of the Distribution of Practice Effect: Now You See It, Now You Don't," *Journal of Applied Psychology*, 84 (1999): 795–805.]

Question For Thought: What type of learner are you? How long after new material is presented do you review it?

Chapter Outline

Section 7.1 Rational Functions; Multiplying and Dividing Rational Expressions 468

Section 7.2 More Division of Polynomials: Long Division and Synthetic Division 483

Section 7.3 Adding and Subtracting Rational Expressions 495

Section 7.4 Simplifying Complex Fractions 505

Section 7.5 Solving Rational Equations 516

Section 7.6 Applications of Rational Equations 523

Section 7.7 Variation and Applications 530

Coming Up...

In Section 7.4, we will evaluate the function $C(t) = \dfrac{300t}{36t^2 + 3.3}$ to calculate the concentration, in micrograms per milliliter, of a drug in the bloodstream t hr after it is administered.

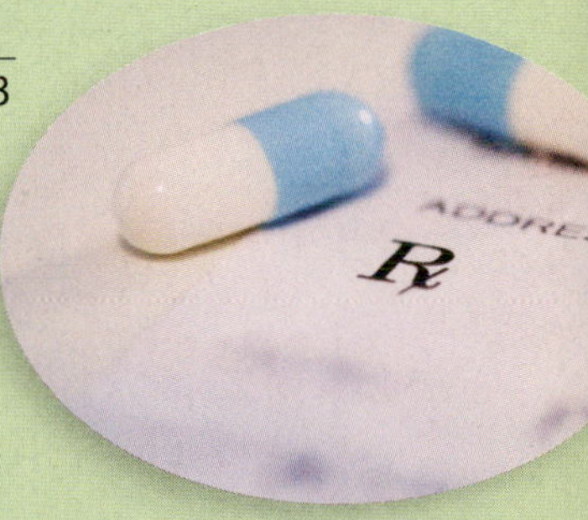

> "When the student is ready, the master appears."
>
> —Buddhist Proverb

SECTION 7.1 Rational Functions; Multiplying and Dividing Rational Expressions

In Chapter 6, we were introduced to polynomials and polynomial operations. We continue that study with quotients of polynomials, or rational expressions. We perform operations on rational expressions, solve equations involving rational expressions, and solve applications of rational expressions. Factoring polynomials plays a key role in working with rational expressions so we must have a good grasp of factoring skills.

OBJECTIVES

As a result of completing this section, you will be able to

1. Evaluate rational expressions and functions.
2. Find the domain of a rational function.
3. Simplify rational expressions.
4. Multiply rational expressions.
5. Divide rational expressions.
6. Solve applications of rational expressions.
7. Troubleshoot common errors.

A high school class is organizing its 25-yr reunion. The venue, DJ, decorations, and other expenses total \$6000 and the dinner costs \$35 per person. The total cost for x attendees is $C(x) = 35x + 6000$. The cost per person is the

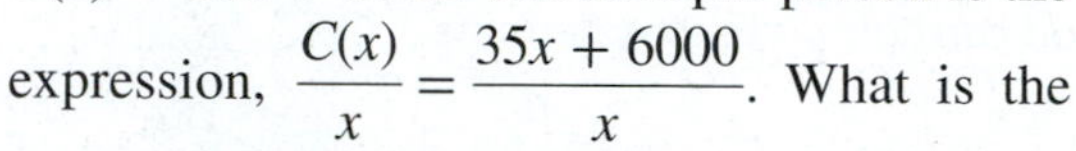

expression, $\dfrac{C(x)}{x} = \dfrac{35x + 6000}{x}$. What is the

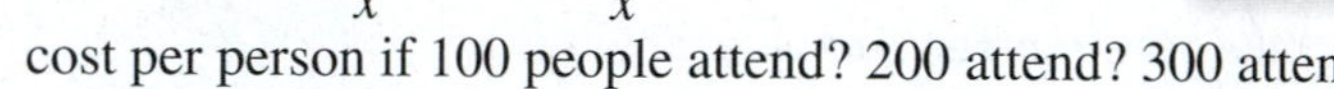

cost per person if 100 people attend? 200 attend? 300 attend?

The expression in the preceding example is a rational expression. In this section, we learn how to evaluate and simplify these expressions as well as multiply and divide them.

Objective 1 Evaluate rational expressions and functions.

Evaluating Rational Expressions

At this point, we have only studied rational numbers. Rational numbers, as you recall, are numbers that can be expressed as a quotient of integers, such as $-\frac{5}{3}, \frac{15}{2}, 0.25 = \frac{1}{4}, -2 = \frac{-2}{1}$. A *rational expression* is also a quotient. It is a quotient of two polynomials.

Definition: The expression $\dfrac{P}{Q}$, $Q \neq 0$, is a **rational expression** if P and Q are both polynomials.

Some examples of rational expressions are

$$\frac{x^2 - x - 6}{x} \qquad -\frac{5}{y + 1} \qquad b^2 - 4 = \frac{b^2 - 4}{1}$$

The fraction $\dfrac{\sqrt{x + 2}}{x - 1}$, for example, is not a rational expression since the numerator is not a polynomial.

When we use function notation to represent quotients of polynomials, we have a *rational function*.

Definition: A **rational function** is a function of the form $f(x) = \dfrac{P(x)}{Q(x)}$, where $P(x)$ and $Q(x)$ are polynomial functions and $Q(x) \neq 0$.

The definitions of a rational expression and a rational function have a restriction that the denominator cannot equal zero. Recall the following facts about fractions.

- When the *numerator* of a fraction is zero, the value of the fraction is zero. For instance, $\dfrac{0}{5} = 0$.

- When the *denominator* of a fraction is zero, the value of the fraction is undefined. For example, $\frac{5}{0}$ is undefined.

So, the requirement that the denominator cannot equal zero is stated to ensure that the expression is defined.

When working with rational expressions and negative signs, recall that when the numerator or denominator of a fraction is negative, we can make the entire fraction negative. For instance,

$$\frac{-2}{3} = -\frac{2}{3} \quad \text{and} \quad \frac{2}{-3} = -\frac{2}{3}$$

We state this in the following property.

Property: Negative Rational Expressions

If P and Q are polynomials, and $Q \neq 0$, then

$$\frac{-P}{Q} = \frac{P}{-Q} = -\frac{P}{Q}$$

Throughout the course, we have evaluated algebraic expressions and functions. Evaluating a rational expression or function is no different.

Procedure: Evaluating a Rational Expression or Rational Function

Step 1: Replace the variable with the given number.
Step 2: Apply the order of operations to simplify the resulting expression.

Objective 1 Examples

Evaluate each expression or function for the given values. Round each answer to two decimal places when necessary.

1a. $\frac{3x^2 - 5x}{x - 2}$; $x = -50, 0, 1.9, 2, 2.001$, and 100

1b. $f(x) = \frac{4x + 5}{x + 1}$. Find $f(-25)$, $f(-1.05)$, $f(-1)$, $f(0)$, $f\left(\frac{1}{4}\right)$, and $f(100)$.

Solutions **1a.**

x	$\frac{3x^2 - 5x}{x - 2}$
-50	$\frac{3(-50)^2 - 5(-50)}{-50 - 2} = \frac{3(2500) + 250}{-52} = \frac{7500 + 250}{-52} = \frac{7750}{-52} \approx -149.04$
0	$\frac{3(0)^2 - 5(0)}{0 - 2} = \frac{0}{-2} = 0$
1.9	$\frac{3(1.9)^2 - 5(1.9)}{1.9 - 2} = \frac{3(3.61) - 9.5}{-0.1} = \frac{10.83 - 9.5}{-0.1} = \frac{1.33}{-0.1} = -13.3$
2	$\frac{3(2)^2 - 5(2)}{2 - 2} = \frac{3(4) - 10}{0} = \frac{12 - 10}{0} = \frac{2}{0}$ undefined
2.001	$\frac{3(2.001)^2 - 5(2.001)}{2.001 - 2} = \frac{3(4.004001) - 10.005}{0.001} = \frac{2.007003}{0.001} = 2007.003$
100	$\frac{3(100)^2 - 5(100)}{100 - 2} = \frac{3(10{,}000) - 500}{98} = \frac{30{,}000 - 500}{98} = \frac{29{,}500}{98} \approx 301.02$

1b.

x	$f(x) = \dfrac{4x+5}{x+1}$
-25	$f(-25) = \dfrac{4(-25)+5}{-25+1} = \dfrac{-100+5}{-24} = \dfrac{-95}{-24} \approx 3.96$
-1.05	$f(-1.05) = \dfrac{4(-1.05)+5}{-1.05+1} = \dfrac{-4.2+5}{-0.05} = \dfrac{0.8}{-0.05} = -16$
-1	$f(-1) = \dfrac{4(-1)+5}{-1+1} = \dfrac{1}{0}$ undefined
0	$f(0) = \dfrac{4(0)+5}{0+1} = \dfrac{5}{1} = 5$
$\dfrac{1}{4}$	$f\left(\dfrac{1}{4}\right) = \dfrac{4\left(\dfrac{1}{4}\right)+5}{\dfrac{1}{4}+1} = \dfrac{1+5}{\dfrac{1}{4}+\dfrac{4}{4}} = \dfrac{6}{\dfrac{5}{4}} = 6 \cdot \dfrac{4}{5} = \dfrac{24}{5} = 4.8$
100	$f(100) = \dfrac{4(100)+5}{100+1} = \dfrac{405}{101} \approx 4.01$

Student Check 1 Evaluate each expression or function for the given values. Round each answer to two decimal places when necessary.

a. $\dfrac{x^2+5x}{x-4}$; $x = -100, 0, 3.98, 4, 4.001$, and 200

b. $f(x) = \dfrac{6x-2}{x+5}$. Find $f(-25)$, $f(-5)$, $f(-4.99)$, $f(0)$, $f\left(\dfrac{1}{2}\right)$, and $f(100)$.

The Domain of a Rational Function

Objective 2 Find the domain of a rational function.

INSTRUCTOR NOTE: Remind students that fractions are undefined when we divide by zero.

In Example 1a and 1b, $x = 2$ makes $\dfrac{3x^2-5x}{x-2}$ undefined and the value $x = -1$ makes $\dfrac{4x+5}{x+1}$ undefined. The expressions are undefined at these values because they make the denominator of the rational expression equal to zero. The values that make the denominator of a rational function zero are the values that must be excluded from the *domain of a rational function*.

Definition: The **domain of a rational function** consists of all real numbers except for the values that make the denominator equal to zero. In other words, the domain is the set of numbers for which the rational expression is defined.

Procedure: Finding the Domain of a Rational Function

Step 1: Set the denominator equal to zero and solve the resulting equation.
Step 2: The domain is all real numbers, excluding the solutions of the equation obtained in step 1.
Step 3: Write the domain in either set-builder notation or interval notation.

Objective 2 Examples Find the domain of each rational function.

2a. $f(x) = \dfrac{2x+3}{4x-8}$ **2b.** $f(x) = \dfrac{x^2+4}{x^2-5x-14}$ **2c.** $f(x) = \dfrac{x-1}{5}$

Solutions **2a.**

$$4x - 8 = 0 \quad \text{Set the denominator equal to zero.}$$
$$4x - 8 + 8 = 0 + 8 \quad \text{Add 8 to each side.}$$
$$4x = 8 \quad \text{Simplify.}$$
$$x = 2 \quad \text{Divide each side by 4.}$$

So, the domain of the function is all real numbers, except 2. We write this as $\{x | x \text{ is a real number, } x \neq 2\}$ or $(-\infty, 2) \cup (2, \infty)$.

2b. Since the denominator is a quadratic expression, we use factoring to solve the resulting equation.

$$x^2 - 5x - 14 = 0 \quad \text{Set the denominator equal to zero.}$$
$$(x-7)(x+2) = 0 \quad \text{Factor the trinomial.}$$
$$x - 7 = 0 \quad \text{or} \quad x + 2 = 0 \quad \text{Apply the zero products property.}$$
$$x = 7 \qquad x = -2 \quad \text{Solve each equation.}$$

The domain of the function is all real numbers, except 7 and -2. We write this as $\{x | x \text{ is a real number and } x \neq 7, x \neq -2\}$ or $(-\infty, -2) \cup (-2, 7) \cup (7, \infty)$.

2c. $5 = 0$ Set the denominator equal to zero.

Since this statement is false, there is no solution of the equation, which indicates that there are no values that need to be restricted from the domain. The domain of the function is all real numbers, or $(-\infty, \infty)$.

 **Student Check 2** Find the domain of each rational function.

a. $f(x) = \dfrac{3x+10}{5x+5}$ **b.** $f(x) = \dfrac{x+4}{x^2-9}$ **c.** $f(x) = \dfrac{7x-4}{3}$

The Fundamental Property of Rational Expressions

Objective 3 Simplify rational expressions.

A fraction is in **lowest terms** if 1 is the only common factor of its numerator and denominator. The fraction $\frac{26}{39}$, for example, is not in lowest terms because the common factor of 26 and 39 is 13. To simplify the fraction $\frac{26}{39}$, we must factor the numerator and denominator and divide out the common factor, 13.

$$\frac{26}{39} = \frac{2 \cdot 13}{3 \cdot 13} = \frac{2}{3} \cdot \frac{13}{13} = \frac{2}{3} \cdot 1 = \frac{2}{3}$$

Likewise, a rational expression is in lowest terms if 1 is the only common factor of its numerator and denominator. Consider the expression, $\dfrac{a^2-9}{a^2-4a+3}$. To simplify this expression, we must factor the numerator and denominator and divide out the common factors.

$$\frac{a^2-9}{a^2-4a+3} = \frac{(a-3)(a+3)}{(a-1)(a-3)} = \frac{a+3}{a-1} \cdot \frac{a-3}{a-3} = \frac{a+3}{a-1} \cdot 1 = \frac{a+3}{a-1}$$

Note that the original rational expression, $\frac{a^2 - 9}{a^2 - 4a + 3}$, is equivalent to the simplified rational expression, $\frac{a + 3}{a - 1}$, except for the values that make either denominator zero.

The process of writing a rational expression in lowest terms is called *simplifying the expression*. To simplify a rational expression, we use the *fundamental property of rational expressions*.

Property: Fundamental Property of Rational Expressions

For a rational expression $\frac{P}{Q}$ and a polynomial R $(Q, R \neq 0)$,

$$\frac{PR}{QR} = \frac{P}{Q} \cdot \frac{R}{R} = \frac{P}{Q} \cdot 1 = \frac{P}{Q}$$

The key in simplifying rational expressions is recognizing quotients that are equivalent to 1 or -1.

- A quotient has a value of 1 when an expression is divided by itself. For example, assuming the denominator is not zero,

$$\frac{3x}{3x} = 1 \qquad \frac{y + 4}{y + 4} = 1 \qquad \frac{2a - 5}{2a - 5} = 1$$

- A quotient has a value of -1 when an expression and its opposite are divided. For example, assuming the denominator is not zero,

$$\frac{5}{-5} = -1 \qquad \frac{-2y}{2y} = -1 \qquad \frac{t - 2}{2 - t} = -1$$

$$\begin{aligned} 2 - t &= -t + 2 \\ &= -1(t - 2) \end{aligned}$$

Property: Dividing Opposites

The expressions $x - y$ and $y - x$ are opposites of one another. As long as $x \neq y$,

$$\frac{x - y}{y - x} = -1$$

Procedure: Simplifying a Rational Expression

Step 1: Rewrite the numerator and denominator in their factored forms.
Step 2: Divide the numerator and denominator by their common factors.

Note: *We will assume that the variables in a rational expression do not represent the values that make the denominator zero.*

Objective 3 Examples

Simplify each rational expression.

3a. $\frac{4y - 2}{2 - 4y}$

3b. $\frac{5x^2 - 10x}{5x^2 - 20}$

3c. $\frac{x^2 - x - 12}{27 + x^3}$

3d. $\frac{6x^2 + 11x - 10}{3x^3 - 2x^2 + 6x - 4}$

Solutions **3a.** $\dfrac{4y-2}{2-4y} = \dfrac{4y-2}{-1(4y-2)}$ Rewrite the denominator as the opposite of $4y - 2$.

$= \dfrac{1}{-1} \cdot \dfrac{4y-2}{4y-2}$ The common factor of the numerator and denominator is $4y - 2$.

$= \dfrac{1}{-1} \cdot 1$ Note that $\dfrac{4y-2}{4y-2} = 1$.

$= -1$ Simplify.

We can also apply the property of dividing opposites to note that

$$\frac{4y-2}{2-4y} = -1$$

3b. $\dfrac{5x^2-10x}{5x^2-20} = \dfrac{5x(x-2)}{5(x-2)(x+2)}$ Factor the numerator and denominator completely.

$= \dfrac{x}{x+2} \cdot \dfrac{5(x-2)}{5(x-2)}$ The common factor of the numerator and denominator is $5(x - 2)$.

$= \dfrac{x}{x+2} \cdot 1$ Note that $\dfrac{5(x-2)}{5(x-2)} = 1$.

$= \dfrac{x}{x+2}$ Simplify.

We can also show the simplification by dividing the numerator and denominator by their common factors.

$$\frac{5x^2-10x}{5x^2-20} = \frac{5x(x-2)}{5(x-2)(x+2)}$$

$$= \frac{\overset{1}{\cancel{5}}x\overset{1}{\cancel{(x-2)}}}{\underset{1}{\cancel{5}}\underset{1}{\cancel{(x-2)}}(x+2)}$$

$$= \frac{x}{x+2}$$

Divide the numerator and denominator by their common factors, 5 and $x - 2$. Write a "1" above and below the common factors to show that $\dfrac{5}{5} = 1$ and $\dfrac{x-2}{x-2} = 1$.

3c. **Method 1:** Factor out the common factor from the numerator and denominator and show that the quotient is 1.

$$\frac{x^2-x-12}{27+x^3} = \frac{(x-4)(x+3)}{(3+x)(9+3x+x^2)}$$

$$= \frac{x-4}{9+3x+x^2} \cdot \frac{x+3}{3+x}$$

$$= \frac{x-4}{9+3x+x^2} \cdot 1$$

$$= \frac{x-4}{x^2+3x+9}$$

Method 2: Divide the numerator and denominator by their common factor.

$$\frac{x^2-x-12}{27+x^3} = \frac{(x-4)\overset{1}{\cancel{(x+3)}}}{\underset{1}{\cancel{(3+x)}}(9+3x+x^2)}$$

$$= \frac{x-4}{x^2+3x+9}$$

3d. $\dfrac{6x^2+11x-10}{3x^3-2x^2+6x-4} = \dfrac{\overset{1}{\cancel{(3x-2)}}(2x+5)}{\underset{1}{\cancel{(3x-2)}}(x^2+2)}$ Factor the numerator and denominator.

$= \dfrac{2x+5}{x^2+2}$ Divide the numerator and denominator by their common factor, $3x - 2$.

Student Check 3 Simplify each rational expression.

a. $\dfrac{7p-5}{5-7p}$ **b.** $\dfrac{3y^2-12y}{3y^2-48}$

c. $\dfrac{x^2+6x+5}{1+x^3}$ **d.** $\dfrac{9a^2-20a-21}{a^3-3a^2+7a-21}$

Multiplying Rational Expressions

Objective 4 Multiply rational expressions.

Multiplying rational expressions is similar to multiplying fractions. Recall that to multiply fractions, we multiply the numerators together, multiply the denominators together, and then express the result in lowest terms. For instance,

$$\frac{2}{3}\cdot\frac{4}{5}=\frac{2\cdot 4}{3\cdot 5}=\frac{8}{15}$$

This process extends to rational expressions, as well.

Property: Multiplying Rational Expressions

If $\dfrac{P}{Q}$ and $\dfrac{R}{S}$ are rational expressions, where Q and S are not zero, then

$$\frac{P}{Q}\cdot\frac{R}{S}=\frac{PR}{QS}$$

Procedure: Multiplying Rational Expressions

Step 1: Factor each numerator and denominator completely.
Step 2: Divide the numerator and denominator by their common factors.
Step 3: Multiply the remaining rational expressions.

Objective 4 Examples Multiply the rational expressions. Write each result in lowest terms.

4a. $\dfrac{5x-15}{3x+9}\cdot\dfrac{4x+12}{6x-18}$ **4b.** $\dfrac{6x^2}{x^2-1}\cdot\dfrac{x^2-x-2}{-2x+4}$

Solutions

4a. $\dfrac{5x-15}{3x+9}\cdot\dfrac{4x+12}{6x-18}=\dfrac{5(x-3)}{3(x+3)}\cdot\dfrac{4(x+3)}{6(x-3)}$ Factor each numerator and denominator completely.

$=\dfrac{5\cancel{(x-3)}^{1}}{3\cancel{(x+3)}_{1}}\cdot\dfrac{\cancel{4}^{2}\cancel{(x+3)}^{1}}{\cancel{6}_{3}\cancel{(x-3)}_{1}}$ Divide out the common factors of $x+3$, $x-3$, and 2.

$=\dfrac{5\cdot 1\cdot 2\cdot 1}{3\cdot 1\cdot 3\cdot 1}$ Multiply the remaining factors in the numerator and denominator.

$=\dfrac{10}{9}$ Simplify.

4b. $\dfrac{6x^2}{x^2-1}\cdot\dfrac{x^2-x-2}{-2x+4}=\dfrac{6x^2}{(x+1)(x-1)}\cdot\dfrac{(x-2)(x+1)}{-2(x-2)}$ Factor each numerator and denominator completely.

$=\dfrac{\overset{3}{\cancel{6}}x^2}{\underset{1}{\cancel{(x+1)}}(x-1)}\cdot\dfrac{\overset{1}{\cancel{(x-2)}}\overset{1}{\cancel{(x+1)}}}{\underset{-1}{\cancel{-2}}\underset{1}{\cancel{(x-2)}}}$ Divide out the common factors of $x+1$, $x-2$, and 2.

$=\dfrac{3x^2\cdot 1\cdot 1}{1\cdot(x-1)\cdot(-1)\cdot 1}$ Multiply the remaining factors in the numerator and denominator.

$=\dfrac{3x^2}{-1(x-1)}$ Simplify.

$=-\dfrac{3x^2}{x-1}$ Apply the property of negative rational expressions.

Student Check 4 Multiply the rational expressions. Write each result in lowest terms.

a. $\dfrac{6x-12}{5x+20}\cdot\dfrac{3x+12}{9x-18}$ **b.** $\dfrac{2x^2}{x^2-25}\cdot\dfrac{x^2-2x-15}{-4x-12}$

Dividing Rational Expressions

Objective 5 Divide rational expressions.

We divide rational expressions just like we divide rational numbers. Recall that we divide fractions by multiplying by the reciprocal. For instance,

$$\frac{2}{3}\div\frac{4}{5}=\frac{2}{3}\cdot\frac{5}{4}=\frac{2\cdot 5}{3\cdot 4}=\frac{10}{12}=\frac{5}{6}$$

We also divide rational expressions by multiplying by the *reciprocal* of the second expression.

Definition: Two numbers are **reciprocals** if their product is 1. The reciprocal of a number a is $\frac{1}{a}$. The reciprocal of a rational expression $\frac{P}{Q}$ is $\frac{Q}{P}$ since $\frac{P}{Q}\cdot\frac{Q}{P}=1$.

Some examples of an expression and its reciprocal are

Expression	Reciprocal
$\dfrac{4x}{x+2}$	$\dfrac{x+2}{4x}$
$3x=\dfrac{3x}{1}$	$\dfrac{1}{3x}$

Property: Dividing Rational Expressions

If $\frac{P}{Q}$ and $\frac{R}{S}$ are rational expressions, where Q, R, and S are not zero, then

$$\frac{P}{Q}\div\frac{R}{S}=\frac{P}{Q}\cdot\frac{S}{R}=\frac{PS}{QR}$$

Procedure: Dividing Rational Expressions

Step 1: Convert the division to multiplication by the reciprocal of the divisor (second fraction).
Step 2: Multiply the resulting rational expressions.
Step 3: Write the result in lowest terms.

Objective 5 Examples Divide the rational expressions. Write each result in lowest terms.

5a. $\dfrac{8m^3-27}{9-4m^2} \div \dfrac{2m+8}{2m^2+11m+12}$ **5b.** $\dfrac{y^2+4y+4}{2y+1} \div \dfrac{(y+2)^4}{2y^2+5y+2}$

Solutions

5a. $\dfrac{8m^3-27}{9-4m^2} \div \dfrac{2m+8}{2m^2+11m+12}$

$= \dfrac{8m^3-27}{9-4m^2} \cdot \dfrac{2m^2+11m+12}{2m+8}$ Multiply by the reciprocal of the divisor.

$= \dfrac{(2m-3)(4m^2+6m+9)}{(3-2m)(3+2m)} \cdot \dfrac{(2m+3)(m+4)}{2(m+4)}$ Factor completely.

$= \dfrac{\overset{-1}{\cancel{(2m-3)}}(4m^2+6m+9)}{\underset{1}{\cancel{(3-2m)}}\underset{1}{\cancel{(3+2m)}}} \cdot \dfrac{\overset{1}{\cancel{(2m+3)}}\overset{1}{\cancel{(m+4)}}}{2\underset{1}{\cancel{(m+4)}}}$ Divide out the common factors. Note that $2m-3$ and $3-2m$ are opposites.

$= \dfrac{-1(4m^2+6m+9)(1)(1)}{1 \cdot 1 \cdot 2 \cdot 1}$ Multiply the remaining factors.

$= -\dfrac{4m^2+6m+9}{2}$ Simplify.

5b. $\dfrac{y^2+4y+4}{2y+1} \div \dfrac{(y+2)^4}{2y^2+5y+2}$

$= \dfrac{y^2+4y+4}{2y+1} \cdot \dfrac{2y^2+5y+2}{(y+2)^4}$ Multiply by the reciprocal of the divisor.

$= \dfrac{(y+2)^2}{2y+1} \cdot \dfrac{(2y+1)(y+2)}{(y+2)^4}$ Factor completely.

$= \dfrac{\overset{1}{\cancel{(y+2)^2}}}{\underset{1}{\cancel{2y+1}}} \cdot \dfrac{\overset{1}{\cancel{(2y+1)}}\overset{1}{\cancel{(y+2)}}}{\cancel{(y+2)^4}\ \cancel{(y+2)^2}\ y+2}$ Divide out the common factors.

$= \dfrac{1 \cdot 1 \cdot 1}{1(y+2)}$ Multiply the remaining factors.

$= \dfrac{1}{y+2}$ Simplify.

Student Check 5 Divide the rational expressions. Write each result in lowest terms.

a. $\dfrac{27x^3-125}{25-9x^2} \div \dfrac{4x-2}{6x^2+7x-5}$ **b.** $\dfrac{y^2+6y+9}{4y+3} \div \dfrac{(y+3)^4}{4y^2+15y+9}$

Objective 6

Solve applications of rational expressions.

Applications

We now look at some rational expressions that describe real-life events and evaluate them to find some specific information.

Objective 6 Examples

Solve each problem.

6a. A high school class is organizing its 25-yr reunion. The venue, DJ, decorations, and other expenses total \$6000 and the dinner costs \$35 per person. The total cost for x attendees is $C(x) = 35x + 6000$. The cost per person is the expression, $\frac{C(x)}{x} = \frac{35x + 6000}{x}$. What is the cost per person if 100 people attend? 200 attend? 300 attend?

Solution **6a.** We evaluate $\frac{C(x)}{x} = \frac{35x + 6000}{x}$ for $x = 100$, $x = 200$, and $x = 300$.

x	$\frac{C(x)}{x} = \frac{35x + 6000}{x}$	Interpretation
100	$\frac{C(100)}{100} = \frac{35(100) + 6000}{100} = \frac{3500 + 6000}{100} = \frac{9500}{100} = 95$	The reunion will cost \$95 per person if 100 people attend.
200	$\frac{C(200)}{200} = \frac{35(200) + 6000}{200} = \frac{7000 + 6000}{200} = \frac{13{,}000}{200} = 65$	The reunion will cost \$65 per person if 200 people attend.
300	$\frac{C(300)}{300} = \frac{35(300) + 6000}{300} = \frac{10{,}500 + 6000}{300} = \frac{16{,}500}{300} = 55$	The reunion will cost \$55 per person if 300 people attend.

6b. A utility company burns coal to generate electricity. The cost C (in dollars) of removing p percent of the smokestack pollutants is given by $C = \frac{70{,}000p}{100 - p}$. What is the cost of removing 75% of the pollutants? What is the cost of removing 95% of the pollutants? What happens to the cost as more and more pollutants are removed?

Solution **6b.** We first evaluate $C = \frac{70{,}000p}{100 - p}$ for $p = 75$ and $p = 95$.

p	$C = \frac{70{,}000p}{100 - p}$	Interpretation
75	$C = \frac{70{,}000(75)}{100 - 75} = \frac{5{,}250{,}000}{25} = 210{,}000$	The cost to remove 75% of pollutants is \$210,000.
95	$C = \frac{70{,}000(95)}{100 - 95} = \frac{6{,}650{,}000}{5} = 1{,}330{,}000$	The cost to remove 95% of pollutants is \$1,330,000.

As the percent of pollutants to be removed increases, the cost of removing the pollutants increases.

Student Check 6 Solve each problem.

a. A popular fitness club charges a one-time membership fee of \$120 when a new member signs a contract. The new member is charged a monthly fee of \$48, and the one-time fee is divided equally among the monthly payments. If x represents the number of months a new member commits to in the contract, the expression $\frac{120 + 48x}{x}$ represents the member's monthly payment. What is the monthly payment if the member signs a contract for 6 months? 12 months? 24 months?

b. Use the formula in Example 6b to find the cost of removing 99% of the pollutants.

Objective 7 Troubleshoot common errors.

Troubleshooting Common Errors

Some of the common mistakes made with rational expressions are shown next.

Objective 7 Examples **A problem and an incorrect solution are given. Provide the correct solution and an explanation of the error.**

7a. Find the domain of $f(x) = \frac{4x - 5}{3x}$.

Incorrect Solution	Correct Solution and Explanation
$3x = 0$ $x = -3$ The domain is all real numbers except -3 or $(-\infty, -3) \cup (-3, \infty)$.	It is correct to set the denominator equal to zero. However, the equation was solved incorrectly. Instead of subtracting 3 from each side, we must divide each side by 3. $3x = 0$ $\frac{3x}{3} = \frac{0}{3}$ $x = 0$ So, the domain is all real numbers except 0 or $(-\infty, 0) \cup (0, \infty)$

7b. Simplify the expression $\frac{6x + 2}{6}$.

Incorrect Solution	Correct Solution and Explanation
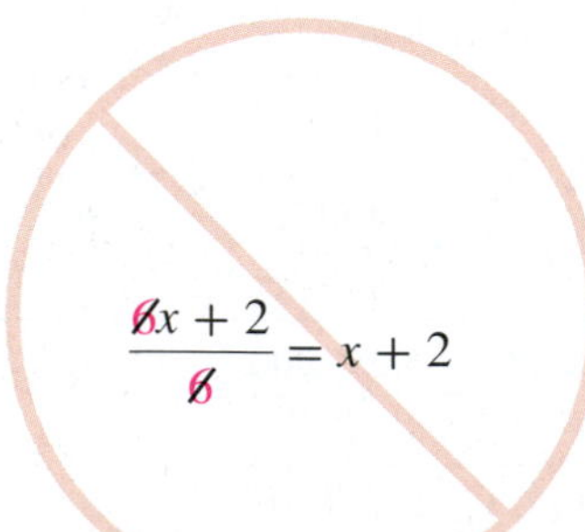	We cannot divide 6 out of the first term and not the second term of the numerator. To simplify the expression, we must factor the numerator. $\frac{6x + 2}{6} = \frac{2(3x + 1)}{2 \cdot 3} = \frac{3x + 1}{3}$ The expression is simplified as far as possible since 3 is not a factor of the numerator. We can rewrite the expression by dividing each term in the numerator by 3. $\frac{3x + 1}{3} = \frac{3x}{3} + \frac{1}{3} = x + \frac{1}{3}$

ANSWERS TO STUDENT CHECKS

Student Check 1 **a.** $-91.35, 0, -1787.02$, undefined, 36,013.001, and 209.18 **b.** $\frac{38}{5} = 7.6$, undefined, -3194, $-\frac{2}{5} = -0.4$, $\frac{2}{11} \approx 0.18$, and $\frac{598}{105} \approx 5.70$

Student Check 2 **a.** $(-\infty, -1) \cup (-1, \infty)$ **b.** $(-\infty, -3) \cup (-3, 3) \cup (3, \infty)$ **c.** $(-\infty, \infty)$

Student Check 3 **a.** -1 **b.** $\frac{y}{y+4}$ **c.** $\frac{x+5}{x^2 - x + 1}$ **d.** $\frac{9a+7}{a^2+7}$

Student Check 4 **a.** $\frac{2}{5}$ **b.** $-\frac{x^2}{2(x+5)}$

Student Check 5 **a.** $-\frac{9x^2 + 15x + 25}{2}$ **b.** $\frac{1}{y+3}$

Student Check 6 **a.** The payment for a 6-month contract would be \$68 per month, for a 12-month contract would be \$58 per month, and for a 24-month contract would be \$53 per month. **b.** The cost to remove 99% of pollutants is \$6,930,000.

SUMMARY OF KEY CONCEPTS

1. A rational expression or a rational function is a quotient of two polynomials, in which the denominator cannot be zero. Rational expressions can be evaluated by substituting the given value in place of the variable.
2. Rational functions are undefined for the values that make the denominator equal to zero. So, the domain of a rational function is all real numbers excluding the values that make the denominator equal to 0.
3. To simplify a rational expression, factor the numerator and denominator completely and divide out their common factors.
4. The key to multiplying rational expressions is factoring. Once the product is formed, rewrite the numerator and denominator in factored form. Divide out the common factors and multiply the remaining factors.
5. Division of rational expressions is just like dividing rational numbers. Convert the division to multiplication by the reciprocal of the divisor and then follow the steps for multiplying rational expressions.

GRAPHING CALCULATOR SKILLS

We can use the Store command in the calculator to evaluate rational expressions. When entering rational expressions in the calculator, parentheses are important to preserve the order of operations. The calculator can be used to verify products or quotients.

Example 1: Use the Store command to evaluate the expression $\frac{x+1}{4x-3}$ for $x = 3$.

Solution: Enter the function in the equation editor.

```
Plot1 Plot2 Plot3
\Y1=(X+1)/(4X-3)
\Y2=
\Y3=
\Y4=
\Y5=
\Y6=
```

Go to the home screen and store the value 3 for the variable x.

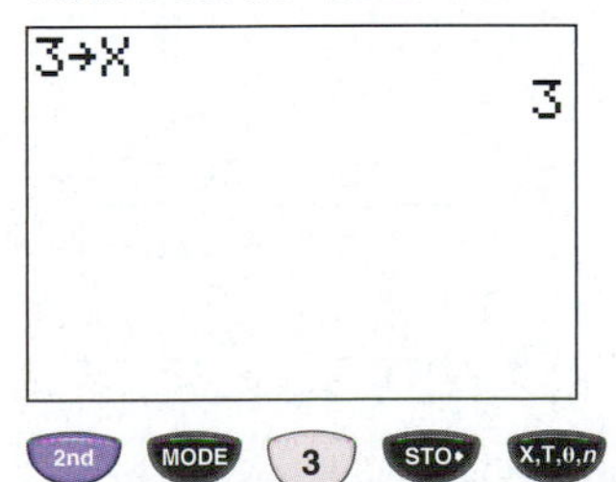

2nd MODE 3 STO▸ X,T,θ,n ENTER

Evaluate the function Y_1. Use the Fraction command to put the answer in Fraction form.

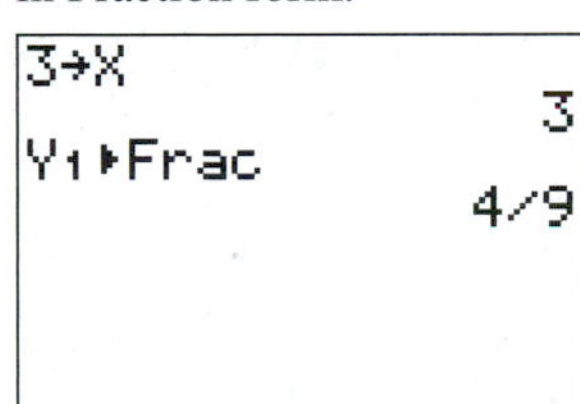

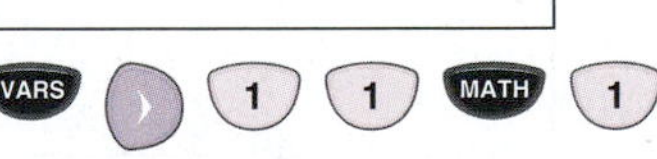

So, the answer is $\frac{4}{9}$.

Example 2: Verify that

$$\frac{x^2 - 36}{x+1} \div \frac{2x+12}{x-1} = \frac{(x-6)(x-1)}{2(x+1)}.$$

Solution: Enter the original problem in Y_1 and the answer in Y_2. If the denominator has a sum or product, the entire expression must be entered in parentheses.

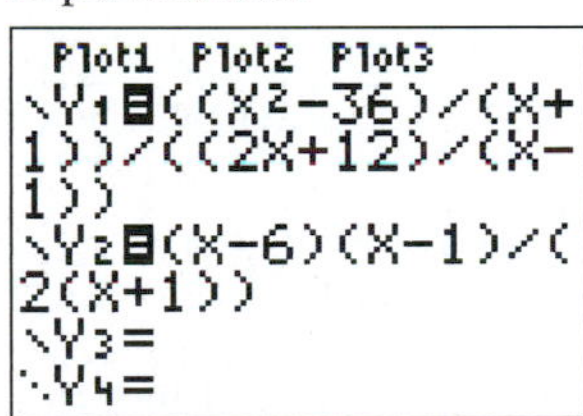

View the Table to compare the y-values. If the y-values agree for each x-value, then the two expressions are equal.

X	Y1	Y2
-3	-9	-9
-2	-12	-12
-1	ERR:	ERR:
0	3	3
1	ERR:	0
2	-.6667	-.6667
3	-.75	-.75

X=3

Note: *The original problem and the final answer may have different values for which the expressions are undefined. The y-value has an ERR: for the values at which it is undefined.*

SECTION 7.1 EXERCISE SET

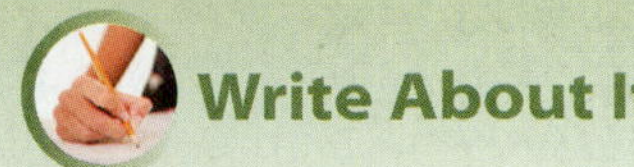

Write About It!

Use complete sentences in your answer to each exercise.

1. Describe the difference between a rational expression and a rational function.
2. Describe how to find the domain of a rational function.
3. Explain how to simplify a rational expression.
4. Give an example of a common mistake made by students when multiplying two rational expressions.
5. Explain how to divide two rational expressions.
6. Explain why $\frac{x^2 + 5x + 6}{x^2 - 3} = 5x - 2$ is incorrect. (The x^2 terms and the 6 and 3 are crossed out, with -2 written above.)
7. Explain why $\frac{x^2 - 4}{x^2 + 1} \cdot \frac{x}{x + 2} = -2$ is incorrect. (The x^2 terms and the x terms are crossed out.)
8. Explain why you have to exclude $x = -3$ from the domain of the function $f(x) = \frac{2x - 5}{x + 3}$.

Practice Makes Perfect!

Evaluate each expression or function for the given values. Round each answer to two decimal places when necessary. (*See Objective 1.*)

9. $\frac{3x + 4}{x - 3}$, $x = -2, 3.001, 2.99, 200, -100$ 0.4, 13,003, −1297, 3.07, 2.87
10. $\frac{5x - 2}{x - 2}$, $x = 3, 2.001, 1.99, 100, -200$ 13, 8005, −795, 5.08, 4.96
11. $\frac{3x^2 + x + 1}{x + 4}$, $x = 1, -3.99, -4.01, 200, -200$ 1, 4477.03, −4523.03, 589.22, −611.23
12. $\frac{x^2 + 2x + 3}{x - 2}$, $x = -4, 2.001, 1.9, 100, -100$ −1.83, 11,006, −104.1, 104.11, −96.11
13. $f(x) = \frac{4x + 3}{x + 2}$, $f(0), f(-1.99), f(-2.01), f(100), f(-100)$ 1.5, −496, 504, 3.95, 4.05
14. $f(x) = \frac{3x + 8}{x - 5}$, $f(0), f(5.01), f(4.9), f(100), f(-100)$ −1.6, 2303, −227, 3.24, 2.78

Find the domain of each rational function. Write each answer in set-builder notation. (*See Objective 2.*)

15. $f(x) = \frac{4x + 11}{x - 3}$
16. $f(x) = \frac{5x - 4}{x + 9}$
17. $f(x) = \frac{3x - 1}{x^2 - 49}$
18. $f(x) = \frac{x - 5}{x^2 - 36}$
19. $f(x) = \frac{x + 3}{x^2 - 2x}$
20. $f(x) = \frac{2x + 1}{x^2 + 3x}$
21. $f(x) = \frac{x^2 - 3x - 4}{2x + 7}$
22. $f(x) = \frac{x^2 + 3x + 2}{5x - 2}$

Find the domain of each rational function. Write each answer in interval notation. (*See Objective 2.*)

23. $f(x) = \frac{2x + 3}{x - 5}$ $(-\infty, 5) \cup (5, \infty)$
24. $f(x) = \frac{6x - 1}{x + 7}$ $(-\infty, -7) \cup (-7, \infty)$
25. $f(x) = \frac{3x}{4x^2 - 1}$
26. $f(x) = \frac{4x}{25x^2 - 4}$
27. $f(x) = \frac{x + 1}{3x^2 - 2x}$
28. $f(x) = \frac{4x - 3}{2x^2 + 5x}$
29. $f(x) = \frac{2x^2 - x - 3}{2x + 1}$
30. $f(x) = \frac{x^2 - 4x - 5}{x}$
31. $f(x) = \frac{3x^2 + 1}{2x - 5}$
32. $f(x) = \frac{x^2 + 4}{3x - 2}$

Simplify each rational expression. (*See Objective 3.*)

33. $\frac{12x^2 - 10x}{18x^2 - 3x - 10}$ $\frac{2x}{3x + 2}$
34. $\frac{3x^2 + 6x}{2x^2 - 3x - 14}$ $\frac{3x}{2x - 7}$
35. $\frac{6x^2 - 5x}{18x^2 - 3x - 10}$ $\frac{x}{3x + 2}$
36. $\frac{2x^2 + 5x}{2x^2 - x - 15}$ $\frac{x}{x - 3}$
37. $\frac{3x^2 + 2x}{9x^2 - 4}$ $\frac{x}{3x - 2}$
38. $\frac{30x^2 + 75x}{12x^2 - 75}$ $\frac{5x}{2x - 5}$
39. $\frac{3x + 2}{2 + 3x}$ 1
40. $\frac{2x + 5}{5 + 2x}$ 1
41. $\frac{x - 2}{2 - x}$ −1
42. $\frac{3 - x}{x - 3}$ −1
43. $\frac{3 + x}{x - 3}$ can't be simplified
44. $\frac{4 + 3x}{3x - 4}$ can't be simplified
45. $\frac{30x^2 + 12x}{50x^2 - 8}$ $\frac{3x}{5x - 2}$
46. $\frac{6x^2 - 8x}{18x^2 - 32}$ $\frac{x}{3x + 4}$
47. $\frac{2x^2 + 17x + 35}{x^3 + 125}$ $\frac{2x + 7}{x^2 - 5x + 25}$
48. $\frac{5x^2 + 11x - 12}{x^3 + 27}$
49. $\frac{6x^2 - 7x - 10}{x^3 - 8}$ $\frac{6x + 5}{x^2 + 2x + 4}$
50. $\frac{9x^2 + 15x - 14}{27x^3 - 8}$
51. $\frac{6x^2 + 11x + 5}{24x^3 + 20x^2 + 18x + 15}$
52. $\frac{21x^2 - 2x - 8}{6x^3 - 4x^2 + 21x - 14}$
53. $\frac{4x^2 + 18x + 8}{8x^3 + 4x^2 + 26x + 13}$
54. $\frac{15x^2 - 36x - 27}{5x^3 + 3x^2 + 20x + 12}$

Multiply the rational expressions. Write each answer in lowest terms. (*See Objective 4.*)

55. $\frac{10x - 14}{21x + 24} \cdot \frac{56x + 64}{35x - 49}$ $\frac{16}{21}$
56. $\frac{7x + 3}{20x - 36} \cdot \frac{10x - 18}{14x + 6}$ $\frac{1}{4}$

Additional answers can be found in the Instructor Answer Appendix.

57. $\dfrac{4x-16}{16x+30}\cdot\dfrac{24x+45}{8x-32}$ $\dfrac{3}{4}$

58. $\dfrac{x-12}{56x+63}\cdot\dfrac{48x+54}{x-12}$ $\dfrac{6}{7}$

59. $\dfrac{5x^2}{49x^2-81}\cdot\dfrac{21x^2+13x-18}{-9x+6}$ $-\dfrac{5x^2}{3(7x-9)}$

60. $\dfrac{9x^2}{9x^2-16}\cdot\dfrac{3x^2-25x+28}{-x+7}$ $-\dfrac{9x^2}{3x+4}$

61. $\dfrac{-6x^2}{16x^2-81}\cdot\dfrac{28x^2+95x+72}{-42x-48}$ $\dfrac{x^2}{4x-9}$

62. $\dfrac{2x^2}{64x^2-1}\cdot\dfrac{56x^2+31x+3}{35x+15}$ $\dfrac{2x^2}{5(8x-1)}$

Divide the rational expressions. Write each answer in lowest terms. (*See Objective 5.*)

63. $\dfrac{x^3+8}{4-x^2}\div\dfrac{30x+35}{6x^2-5x-14}$ $-\dfrac{x^2-2x+4}{5}$

64. $\dfrac{x^3+64}{16-x^2}\div\dfrac{-11x-4}{11x^2-40x-16}$ $x^2-4x+16$

65. $\dfrac{x^3-1}{1-x^2}\div\dfrac{30x+66}{5x^2+16x+11}$ $-\dfrac{x^2+x+1}{6}$

66. $\dfrac{x^3-343}{49-x^2}\div\dfrac{-21x-98}{3x^2+35x+98}$ $\dfrac{x^2+7x+49}{7}$

67. $\dfrac{x^2-16x+64}{7x-2}\div\dfrac{(x-8)^6}{21x^2-48x+12}$ $\dfrac{3(x-2)}{(x-8)^4}$

68. $\dfrac{x^2+2x+1}{9x+2}\div\dfrac{(x+1)^6}{36x^2-28x-8}$ $\dfrac{4(x-1)}{(x+1)^4}$

69. $\dfrac{9x^2-30x+25}{6x-5}\div\dfrac{(3x-5)^3}{18x^2+21x-30}$ $\dfrac{3(x+2)}{3x-5}$

70. $\dfrac{16x^2-24x+9}{2x+7}\div\dfrac{(4x-3)^3}{10x^2+21x-49}$ $\dfrac{5x-7}{4x-3}$

Solve each problem. (*See Objective 6.*)

71. A high school class is organizing its 40-yr reunion. The venue, DJ, decorations, and other expenses total \$4500 and the dinner costs \$15 per person. The total cost for x attendees is $C(x) = 15x + 4500$. The cost per person is the expression $\dfrac{C(x)}{x} = \dfrac{15x+4500}{x}$.

a. What is the cost per person if 50 people attend?

b. 75 attend? \$75 per person if 75 attend

c. 100 attend? \$60 per person if 100 attend

72. A high school class is organizing its 15-yr reunion. The venue, DJ, decorations, and other expenses total \$3000 and the dinner costs \$50 per person. The total cost for x attendees is $C(x) = 50x + 3000$. The cost per person is the expression $\dfrac{C(x)}{x} = \dfrac{50x+3000}{x}$.

a. What is the cost per person if 100 people attend?

b. 125 attend? \$74 per person if 125 attend

c. 150 attend? \$70 per person if 150 attend

73. A utility company burns coal to generate electricity. The cost C (in dollars) of removing p amount (percent) of the smokestack pollutants is given by $C = \dfrac{82{,}500p}{100-p}$.

a. What is the cost of removing 80% of the pollutants?

b. 95% of the pollutants? The cost to remove 95% of the pollutants is \$1,567,500.

74. A utility company burns coal to generate electricity. The cost C (in dollars) of removing p amount (percent) of the smokestack pollutants is given by $C = \dfrac{87{,}500p}{100-p}$.

a. What is the cost of removing 75% of the pollutants?

b. 98% of the pollutants? The cost to remove 98% of the pollutants is \$4,287,500.

75. A fitness club charges a one-time membership fee of \$180 when a new member signs a contract. The new member is charged a monthly fee of \$54, and the one-time fee is divided equally among the monthly payments. If x represents the number of months a new member commits to in the contract, the expression $\dfrac{180+54x}{x}$ represents the member's monthly payment.

a. What is the monthly payment if a member signs a contract for 6 months? The monthly payment is \$84 for 6 months.

b. 12 months? \$69 for 1 month

c. 24 months? \$61.50 for 24 months

76. A fitness club charges a one-time membership fee of \$90 when a new member signs a contract. The new member is charged a monthly fee of \$48, and the one-time fee is divided equally among the monthly payments. If x represents the number of months a new member commits to in the contract, the expression $\dfrac{90+48x}{x}$ represent the member's monthly payment.

a. What is the monthly payment if a member signs a contract for 3 months? The monthly payment is \$78 for 3 months.

b. 6 months? \$63 for 6 months

c. 12 months? \$55.50 for 12 months

Mix 'Em Up!

Evaluate each expression or function for the given values. Round each answer to two decimal places when necessary.

77. $\dfrac{3x^2-4x+5}{x+1}$, $x = 4, -0.99, -1.1, 100, -100$
7.4, 1190.03, −130.3, 293.12, −307.12

78. $\dfrac{4x^2+x-5}{x+3}$, $x = 0, -2.999, -3.01, 100, -100$
$-\dfrac{5}{3}$, 27,977, −2823.04, 389.27, −411.29

79. $f(x) = \dfrac{3x-8}{x+6}$, $f(0), f(-5.99), f(-6.01), f(100), f(-100)$
$-\dfrac{4}{3}$, −2597, 2603, 2.75, 3.28

80. $f(x) = \dfrac{3x+4}{x-4}$, $f(0), f(4.01), f(3.9), f(100), f(-100)$
−1, 1603, −157, 3.17, 2.85

Find the domain of each rational function. Write each answer in both set-builder and interval notation.

81. $f(x) = \dfrac{6x - 5}{3x + 4}$

82. $f(x) = \dfrac{x + 9}{5x - 12}$

83. $f(x) = \dfrac{2x + 7}{4x^2 - 3x}$

84. $f(x) = \dfrac{x + 6}{x^2 + 7x}$

85. $f(x) = \dfrac{2x}{4x^2 - 9}$

86. $f(x) = \dfrac{x^2 - x - 6}{x + 4}$

87. $f(x) = \dfrac{x^2 - 1}{x - 3}$

88. $f(x) = \dfrac{x + 3}{9x^2 - 25}$

Perform the indicated operation. Write each answer in lowest terms.

89. $\dfrac{2x - 10}{3x + 6} \cdot \dfrac{6x + 12}{3x + 15}$ $\dfrac{4(x-5)}{3(x+5)}$

90. $\dfrac{6x - 6}{3x + 12} \cdot \dfrac{2x + 8}{2x + 2}$ $\dfrac{2(x-1)}{x+1}$

91. $\dfrac{3x + 15}{3x - 9} \div \dfrac{6x}{2x - 6}$ $\dfrac{x+5}{3x}$

92. $\dfrac{4x - 10}{5x + 10} \div \dfrac{8x}{10x + 20}$ $\dfrac{2x-5}{2x}$

93. $\dfrac{12x^2 - 14x}{36x^2 + 42x - 98}$ $\dfrac{x}{3x+7}$

94. $\dfrac{4x^2 + 14x}{24x^2 + 64x - 70}$ $\dfrac{x}{6x-5}$

95. $\dfrac{3x^2}{25x^2 - 1} \cdot \dfrac{40x^2 + 33x + 5}{16x + 10}$ $\dfrac{3x^2}{2(5x-1)}$

96. $\dfrac{-x^2}{x^2 - 4} \cdot \dfrac{9x^2 - 10x - 16}{-18x - 16}$ $\dfrac{x^2}{2(x+2)}$

97. $\dfrac{64x^3 - 1}{1 - 16x^2} \div \dfrac{-2x + 26}{4x^2 - 51x - 13}$ $\dfrac{16x^2 + 4x + 1}{2}$

98. $\dfrac{8x^3 - 1}{1 - 4x^2} \div \dfrac{-4x - 15}{8x^2 + 34x + 15}$ $4x^2 + 2x + 1$

99. $\dfrac{9x^2 - 9x + 2}{15x^3 - 10x^2 - 33x + 22}$ $\dfrac{3x-1}{5x^2-11}$

100. $\dfrac{7x^2 - 24x + 9}{2x^3 - 6x^2 - 5x + 15}$ $\dfrac{7x-3}{2x^2-5}$

101. $\dfrac{27x^3 + 64}{9x^2 - 16} \cdot \dfrac{3x^2 - 28x + 32}{3x^2 - 28x + 32}$ $\dfrac{9x^2 - 12x + 16}{3x-4}$

102. $\dfrac{125x^3 + 8}{25x^2 - 4} \div \dfrac{5x^2 + 3x - 2}{3x + 3}$ $\dfrac{3(25x^2 - 10x + 4)}{(5x-2)^2}$

103. $\dfrac{64x^2 + 48x + 9}{x - 5} \div \dfrac{(8x + 3)^2}{x^2 - x - 20}$ $x + 4$

104. $\dfrac{x^2 - 8x + 16}{3x + 5} \div \dfrac{(x - 4)^3}{3x^2 + 2x - 5}$ $\dfrac{x-1}{x-4}$

105. $\dfrac{5x^2 - 20}{3x^2 - 75} \div \dfrac{x^3 + 2x^2}{2x^2 + 10x} \cdot \dfrac{3x^2 - 15x}{x + 1}$ $\dfrac{10(x-2)}{x+1}$

106. $\dfrac{2x^2 - 18}{4x^2 - 4} \div \dfrac{x^3 - 3x^2}{5x^2 - 5x} \cdot \dfrac{2x^2 + 2x}{x - 6}$ $\dfrac{5(x+3)}{x-6}$

107. $\dfrac{3x^2 - 14x - 5}{2x^2 + 5x - 3} \cdot \left(\dfrac{x^2 - x - 12}{x^2 - x - 20} \div \dfrac{3x^2 - 11x - 4}{2x^2 - 7x + 3}\right)$ $\dfrac{x-3}{x+4}$

108. $\dfrac{3x^2 + 2x - 8}{x^2 - 36} \cdot \left(\dfrac{x^2 + 12x + 36}{3x^2 - 7x + 4} \div \dfrac{x^2 - 3x - 10}{x^2 - 7x + 6}\right)$ $\dfrac{x+6}{x-5}$

109. $\dfrac{x^3 + 8}{x^2 - 49} \div \left(\dfrac{2x^2 - 4x + 8}{x^2 + 6x - 7} \cdot \dfrac{3x + 6}{2x - 14}\right)$ $\dfrac{x-1}{3}$

110. $\dfrac{x^3 - 27}{x^2 - 16} \div \left(\dfrac{3x^2 + 9x + 27}{x^2 + x - 20} \cdot \dfrac{2x - 6}{3x + 12}\right)$ $\dfrac{x+5}{2}$

Solve each problem.

111. A high school class is organizing its 10-yr reunion. The venue, DJ, decorations, and other expenses total \$4500 and the dinner costs \$25 per person. The total cost for x attendees is $C(x) = 25x + 4500$. The cost per person is the expression, $\dfrac{C(x)}{x} = \dfrac{25x + 4500}{x}$.

a. What is the cost per person if 100 people attend?

b. 125 attend? \$61 per person if 125 attend

c. 150 attend? \$55 per person if 150 attend

112. A high school class is organizing its 5-yr reunion. The venue, DJ, decorations, and other expenses total \$7500 and the dinner costs \$40 per person. The total cost for x attendees is $C(x) = 40x + 7500$. The cost per person is the expression $\dfrac{C(x)}{x} = \dfrac{40x + 7500}{x}$.

a. What is the cost per person if 75 people attend?

b. 100 attend? \$115 per person if 100 attend

c. 125 attend? \$100 per person if 125 attend

113. A utility company burns coal to generate electricity. The cost C (in dollars) of removing p amount (percent) of the smokestack pollutants is given by $C = \dfrac{67{,}500p}{100 - p}$.

a. What is the cost of removing 85% of the pollutants?

b. 97% of the pollutants? The cost to remove 97% of the pollutants is \$2,182,500.

114. A utility company burns coal to generate electricity. The cost C (in dollars) of removing p amount (percent) of the smokestack pollutants is given by $C = \dfrac{72{,}000p}{100 - p}$.

a. What is the cost of removing 82% of the pollutants?

b. 96% of the pollutants? The cost to remove 96% of the pollutants is \$1,728,000.

115. A fitness club charges a one-time membership fee of \$90 when a new member signs a contract. The new member is charged a monthly fee of \$51, and the one-time fee is divided equally among the monthly payments. If x represents the number of months a new member commits to in the contract, the expression $\dfrac{90 + 51x}{x}$ represents the member's monthly payment.

a. What is the monthly payment if a member signs a contract for 18 months? The monthly payment is \$56 for 18 months.

b. 24 months? \$54.75 for 24 months

c. 30 months? \$54 for 30 months

116. A fitness club charges a one-time membership fee of \$105 when a new member signs a contract. The new member is charged a monthly fee of \$27, and the one-time fee is divided equally among the monthly payments. If x represents the number of months a new member commits to in the contract, the expression $\frac{105 + 27x}{x}$ represents the member's monthly payment.

a. What is the monthly payment if a member signs a contract for 3 months? The monthly payment is \$62 for 3 months.

b. 6 months? \$44.50 for 6 months

c. 12 months? \$35.75 for 12 months

You Be the Teacher!

Correct each student's errors, if any.

117. Find the domain of the function $f(x) = \frac{3x - 1}{2x}$.

Michael's work:

$2x \neq 0$

$x \neq -2$

The domain of the function is $(-\infty, -2) \cup (-2, \infty)$.

118. Find the domain of the function $f(x) = \frac{x - 3}{x + 1}$.

Romona's work:

$x - 3 \neq 0$

$x \neq 3$

The domain of the function is $(-\infty, 3) \cup (3, \infty)$.

119. Simplify the rational expression $\frac{9x^2 - 4}{3x^2 + x - 2}$.

Tom's work:

$$\frac{9x^2 - 4}{3x^2 + x - 2} = \frac{\cancel{9x^2}^{3} - \cancel{4}^{2}}{\cancel{3x^2} + x - \cancel{2}} = \frac{1}{x}$$

120. Simplify the rational expression $\frac{4x^2 - x - 3}{2x^2 + x - 3}$.

Taylor's work:

$$\frac{4x^2 - x - 3}{2x^2 + x - 3} = \frac{\cancel{4x^2}^{2} - \cancel{x}^{1} - \cancel{3}}{\cancel{2x^2} + \cancel{x} - \cancel{3}} = 2$$

Calculate It!

121. Use the Store command to evaluate $f(x) = \frac{5x + 4}{2x - 1}$ for $x = 2$.

122. Use the Store command to evaluate $f(x) = \frac{3x - 1}{x + 2}$ for $x = -6$.

123. Verify that $\frac{4x^2 + 12x + 9}{x - 2} \div \frac{(2x + 3)^4}{x^2 - x - 2} = \frac{x + 1}{(2x + 3)^2}$.

124. Verify that $\frac{x^3 + 1}{x^2 - 1} \div \frac{x + 2}{x^2 + x - 2} = x^2 - x + 1$.

Think About It!

Write a rational function that has the given domain.

125. $(-\infty, 3) \cup (3, \infty)$ Answers vary; $f(x) = \frac{x + 5}{x - 3}$

126. $(-\infty, -2) \cup (-2, \infty)$ Answers vary; $f(x) = \frac{4x}{x + 2}$

127. $(-\infty, 1) \cup (1, 5) \cup (5, \infty)$ Answers vary; $f(x) = \frac{2x}{(x - 1)(x - 5)}$

128. $(-\infty, -2) \cup (-2, 2) \cup (2, \infty)$

Answers vary; $f(x) = \frac{x + 10}{(x + 2)(x - 2)}$

SECTION 7.2 More Division of Polynomials: Long Division and Synthetic Division

OBJECTIVES

As a result of completing this section, you will be able to

1. Divide a polynomial by a monomial.
2. Use long division to divide polynomials.
3. Use synthetic division to divide a polynomial by a binomial of the form $x - c, c \neq 0$.
4. Use the remainder theorem to evaluate polynomials.
5. Solve application problems.
6. Troubleshoot common errors.

The area of this rectangle is $4x^3y + 2x^2y$ square units and its length is $2x^2y$ units. Find an expression for the width of the rectangle.

$2x^2y$

$A = 4x^3y + 2x^2y$

To solve this problem, we divide the area by its length, which requires us to divide polynomials. In this section, we will learn several methods to divide polynomials.

Dividing Polynomials by Monomials

Objective 1 Divide a polynomial by a monomial.

In Chapter 6, we discussed how to add, subtract, and multiply polynomials. In Section 7.1, we defined a rational expression as the quotient of two polynomials. This is nothing more than dividing polynomials. We will first illustrate how to divide a polynomial by a monomial.

To divide a polynomial by a monomial, we use a basic property of fractions. Let's look at an example to review this property.

$$\frac{14}{7} + \frac{21}{7} = \frac{14 + 21}{7} = \frac{35}{7} = 5$$

To use it for a property relating to division, we need to view it in reverse order.

$$\frac{14 + 21}{7} = \frac{14}{7} + \frac{21}{7} = 2 + 3 = 5$$

In this order, we see that if a sum is divided by a single term, each term in the numerator must be divided by the denominator. This leads to the following property.

Property: Property of Division

If A, B, and C are monomials with $C \neq 0$, then $\frac{A + B}{C} = \frac{A}{C} + \frac{B}{C}$.

The properties of exponents are also needed to divide a polynomial by a monomial. Recall the quotient of like bases rule.

$$\frac{b^m}{b^n} = b^{m-n}$$

Procedure: Dividing a Polynomial by a Monomial

Step 1: Rewrite the expression so that each term in the numerator (or dividend) is divided by the monomial in the denominator (or divisor).
Step 2: Simplify each expression using properties of exponents.
Step 3: Check by multiplying.

Objective 1 Examples

Perform each operation and check by multiplying.

1a. Divide $5x^2 - 10x$ by $5x$. **1b.** $\frac{12y^4 - 18y^3 + 6y^2}{-6y^2}$

1c. Divide $9a^3 + 3a^2 - 6a$ by $9a^2$.

Solutions

1a. $\frac{5x^2 - 10x}{5x} = \frac{5x^2}{5x} - \frac{10x}{5x}$ Apply the property of division.

$= x - 2$ Apply the quotient of like bases rule.

We check by multiplying the quotient, $x - 2$, by the divisor, $5x$.

$$5x(x - 2) = 5x(x) - 5x(2) = 5x^2 - 10x$$

1b. $\frac{12y^4 - 18y^3 + 6y^2}{-6y^2} = \frac{12y^4}{-6y^2} - \frac{18y^3}{-6y^2} + \frac{6y^2}{-6y^2}$ Apply the property of division.

$= -2y^2 + 3y - 1$ Apply the quotient of like bases rule.

We check by multiplying the quotient, $-2y^2 + 3y - 1$, by the divisor, $-6y^2$.

$$-6y^2(-2y^2 + 3y - 1) = -6y^2(-2y^2) - 6y^2(3y) - 6y^2(-1)$$
$$= 12y^4 - 18y^3 + 6y^2$$

1c. $\dfrac{9a^3 + 3a^2 - 6a}{9a^2} = \dfrac{9a^3}{9a^2} + \dfrac{3a^2}{9a^2} - \dfrac{6a}{9a^2}$ Apply the property of division.

$= a + \dfrac{1}{3} - \dfrac{2}{3}a^{-1}$ Apply the quotient of like bases rule.

$= a + \dfrac{1}{3} - \dfrac{2}{3a}$ Rewrite with positive exponents.

Student Check 1 Perform each operation and check by multiplying.

a. Divide $8y^2 - 24y$ by $8y$.

b. $\dfrac{14x^4 - 21x^3 + 7x^2}{-7x^2}$

c. Divide $6b^5 + 16b^3 - 24b^2$ by $4b^4$.

Long Division of Polynomials

Objective 2 Use long division to divide polynomials.

When the denominator of a quotient is not a monomial, we must use **long division** to perform the division. Long division of polynomials involves the same steps as long division of real numbers. The following example reviews the steps involved in long division of real numbers.

$$\begin{array}{r} 21 \\ 15\overline{)325} \\ \underline{30} \\ 25 \\ \underline{15} \\ 10 \end{array}$$

Divide: $\dfrac{32}{15} \approx 2$.
Multiply: $2(15) = 30$.
Subtract: $32 - 30 = 2$.
Bring down the next digit, 5.

Divide: $\dfrac{25}{15} \approx 1$.
Multiply: $1(15) = 15$.
Subtract: $25 - 15 = 10$.

The divisor, 15, doesn't go into 10 so we are done. So, $\dfrac{325}{15} = 21$ R 10, which means 21 with a remainder of 10. We can also write this as $\dfrac{325}{15} = 21 + \dfrac{10}{15}$. To check, note that $21(15) + 10 = 325$, which is the dividend.

The same basic steps are applied to long division of polynomials. An example follows.

$$\begin{array}{r} 2x + 3 \\ x + 4\overline{)2x^2 + 11x + 12} \\ -(2x^2 + 8x)\downarrow \\ 3x + 12 \\ -(3x + 12) \\ 0 \end{array}$$

Divide: $\dfrac{2x^2}{x} = 2x$.
Multiply: $2x(x + 4) = 2x^2 + 8x$.
Subtract: $(2x^2 + 11x) - (2x^2 + 8x) = 3x$.
Bring down 12.

Divide: $\dfrac{3x}{x} = 3$.
Multiply: $3(x + 4) = 3x + 12$.
Subtract: $(3x + 12) - (3x + 12) = 0$.

So, $\dfrac{2x^2 + 11x + 12}{x + 4} = 2x + 3$, which means that $(2x + 3)(x + 4) = 2x^2 + 11x + 12$.

Procedure: Using Long Division to Divide Polynomials

Step 1: Write the numerator (dividend) in standard form, inserting 0 for any missing terms.

Step 2: Divide the first term of the polynomial in the denominator (divisor) into the first term of the polynomial in the numerator.

Step 3: Multiply this result by the divisor and line up like terms.
Step 4: Subtract the product found in step 3 from the polynomial in the row above the product. Remember that subtraction changes the signs of each term. Bring down the next term.
Step 5: Repeat this process until the degree of the polynomial that results from subtraction is less than the degree of the divisor.
Step 6: Check by multiplying. If q is the quotient after dividing A by B and r is the remainder, then

$$\frac{A}{B} = q + \frac{r}{B}$$

To check the division, the dividend = quotient × divisor + remainder.

$$A = q \cdot B + r$$

Objective 2 Examples **Perform each operation.**

2a. Divide $x^2 + 2x - 35$ by $x - 5$. **2b.** $\dfrac{x^3 + 8}{x + 2}$ **2c.** $\dfrac{3y^2 + 7y - 6}{3y + 4}$

Solutions **2a.**

$$\begin{array}{r} x + 7 \\ x - 5 \overline{) x^2 + 2x - 35} \\ -(x^2 - 5x) \downarrow \\ \hline 7x - 35 \\ -(7x - 35) \\ \hline 0 \end{array}$$

Divide: $\dfrac{x^2}{x} = x$.
Multiply: $x(x - 5) = x^2 - 5x$.
Subtract: $(x^2 + 2x - 35) - (x^2 - 5x)$.
Divide: $\dfrac{7x}{x} = 7$.
Multiply: $7(x - 5) = 7x - 35$.
Subtract: $(7x - 35) - (7x - 35)$.

So, $\dfrac{x^2 + 2x - 35}{x - 5} = x + 7$.

Quotient • Divisor = Dividend

Check: $(x + 7)(x - 5) = x^2 - 5x + 7x - 35 = x^2 + 2x - 35$

2b. The polynomial in the numerator does not have an x^2-term or an x-term. We insert $0x^2 + 0x$ as a placeholder for these terms.

$$\begin{array}{r} x^2 - 2x + 4 \\ x + 2 \overline{) x^3 + 0x^2 + 0x + 8} \\ -(x^3 + 2x^2) \downarrow \quad \downarrow \\ \hline -2x^2 + 0x + 8 \\ -(-2x^2 - 4x) \downarrow \\ \hline 4x + 8 \\ -(4x + 8) \\ \hline 0 \end{array}$$

Divide: $\dfrac{x^3}{x} = x^2$.
Multiply: $x^2 (x + 2) = x^3 + 2x^2$.
Subtract: $(x^3 + 0x^2 + 0x + 8) - (x^3 + 2x^2)$.
Divide: $\dfrac{-2x^2}{x} = -2x$.
Multiply: $-2x(x + 2) = -2x^2 - 4x$.
Subtract: $(-2x^2 + 0x + 8) - (-2x^2 - 4x)$.
Divide: $\dfrac{4x}{x} = 4$.
Multiply: $4(x + 2) = 4x + 8$.
Subtract: $(4x + 8) - (4x + 8)$.

So, $\dfrac{x^3 + 8}{x + 2} = x^2 - 2x + 4$.

2c.

$$\begin{array}{r} y + 1 \\ 3y + 4\overline{)3y^2 + 7y - 6} \\ -(3y^2 + 4y) \downarrow \\ \hline 3y - 6 \\ -(3y + 4) \\ \hline -10 \end{array}$$

Divide: $\dfrac{3y^2}{3y} = y$.

Multiply: $y(3y + 4) = 3y^2 + 4y$.

Subtract: $(3y^2 + 7y - 6) - (3y^2 + 4y)$.

Divide: $\dfrac{3y}{3y} = 1$.

Multiply: $1(3y + 4) = 3y + 4$.

Subtract: $(3y - 6) - (3y + 4)$.

So, $\dfrac{3y^2 + 7y - 6}{3y + 4} = y + 1 - \dfrac{10}{3y + 4}$.

Student Check 2 Perform each operation.

a. Divide $x^2 - x - 72$ by $x - 9$. **b.** $\dfrac{a^3 + 125}{a + 5}$ **c.** $\dfrac{4y^2 - 6y + 3}{4y - 2}$

Synthetic Division

Objective 3 Use synthetic division to divide a polynomial by a binomial of the form $x - c, c \neq 0$.

Synthetic division is a process that can be used as a shortcut for long division as long as the divisor is a binomial of the form $x - c$. The expression on the left below is an example of long division. The expression on the right is the same long division problem with the variables removed.

$$\begin{array}{r} 2x + 3 \\ x + 4\overline{)2x^2 + 11x + 12} \\ -(2x^2 + 8x) \\ 3x + 12 \\ -(3x + 12) \end{array} \qquad \begin{array}{r} 2 \;+\; 3 \\ 1\;\;4\overline{)2 \;+ 11 \;+ 12} \\ 2 \;+ 8 \quad\quad \\ \hline 3 \;+ 12 \\ 3 \;+ 12 \end{array}$$

Notice that after each multiplication step, the first terms add to zero ($2x^2 - 2x^2 = 0$ and $3x - 3x = 0$), leaving the term that determines the quotient. Also remember that when we subtract, the signs change.

This leads us to the process of synthetic division. In order to use synthetic division, the divisor must be of the form, $x - c$. Since the divisor in the previous example is $x + 4$, or $x - (-4)$, $c = -4$. We use a symbol that looks like an upside down division symbol to perform our work.

$$\begin{array}{r|rrr} -4 & 2 & 11 & 12 \\ \hline & 2 & & \end{array}$$

Bring down the first term.

$$\begin{array}{r|rrr} -4 & 2 & 11 & 12 \\ & & -8 & \\ \hline & 2 & & \end{array}$$

Multiply -4 and 2 and write result in the second column.

$$\begin{array}{r|rrr} -4 & 2 & 11 & 12 \\ & & -8 & \\ \hline & 2 & 3 & \end{array}$$

Add the values in the second column.

$$\begin{array}{r|rrr} -4 & 2 & 11 & 12 \\ & & -8 & -12 \\ \hline & 2 & 3 & \end{array}$$

Multiply -4 and 3 and write result in the third column.

$$\begin{array}{r|rrr} -4 & 2 & 11 & 12 \\ & & -8 & -12 \\ \hline & 2 & 3 & 0 \end{array}$$

Add the values in the third column.

So, the quotient is $2x + 3$ with remainder 0.

Procedure: Dividing a Polynomial by a Binomial of the Form $x - c$ Using Synthetic Division

Step 1: Write down the coefficients of the polynomial in the numerator, inserting zeros for any missing terms.
Step 2: Determine the value of c from the divisor and write it to the left of the inverted division symbol.
Step 3: Bring down the coefficient of the first term and multiply it by the number c. Write the result under the number in the second column.
Step 4: Add the numbers in the second column.
Step 5: Multiply the result from step 4 by the value of c and write this result under the number in the third column.
Step 6: Add the numbers in the third column.
Step 7: Continue this process until there are no more columns.
Step 8: The quotient is on the bottom row. The numbers are the coefficients of the polynomial whose degree is one less than the degree of the dividend. The last number in the row is the remainder.

Objective 3 Examples

Use synthetic division to divide the polynomials.

3a. Divide $5x^2 - 7x + 3$ by $x - 2$.

3b. Divide $4x^3 - 5x^2 + 2x - 9$ by $x + 1$.

Solutions

3a. The coefficients of the dividend are 5, −7, and 3 and the value of c is 2 since the divisor is $x - 2$.

$$\begin{array}{r|rrr} 2 & 5 & -7 & 3 \\ & & & \\ \hline & 5 & & \end{array}$$

Write the coefficients of the dividend and bring down 5 in the first column.

INSTRUCTOR NOTE: Point out that students do not have to show each separate step. All of the work can be done with one symbol.

$$\begin{array}{r|rrr} 2 & 5 & -7 & 3 \\ & & 10 & \\ \hline & 5 & & \end{array}$$

Multiply 2 and 5 and record the result, 10, in the second column.

$$\begin{array}{r|rrr} 2 & 5 & -7 & 3 \\ & & 10 & \\ \hline & 5 & 3 & \end{array}$$

Add the values in the second column, −7 and 10, and write the result, 3, below the line.

$$\begin{array}{r|rrr} 2 & 5 & -7 & 3 \\ & & 10 & 6 \\ \hline & 5 & 3 & \end{array}$$

Multiply 2 and 3 and record the result, 6, in the third column.

$$\begin{array}{r|rrr} 2 & 5 & -7 & 3 \\ & & 10 & 6 \\ \hline & 5 & 3 & 9 \end{array}$$

Add the values in the third column, 3 and 6, and write the result, 9, below the line.

The numbers 5 and 3 are the coefficients of the quotient polynomial and 9 is the remainder. The degree of the quotient polynomial is one less than the degree of the dividend.

So, $\dfrac{5x^2 - 7x + 3}{x - 2} = 5x + 3 + \dfrac{9}{x - 2}$.

3b. The coefficients of the dividend are 4, −5, 2, and −9. The value of c is −1, since the divisor is $x + 1 = x - (-1)$.

$$\begin{array}{r|rrrr} -1 & 4 & -5 & 2 & -9 \\ & & & & \\ \hline & 4 & & & \end{array}$$

Write the coefficients of the dividend and bring down 4 in the first column.

$$\begin{array}{r|rrrr} -1 & 4 & -5 & 2 & -9 \\ & & -4 & & \\ \hline & 4 & & & \end{array}$$

Multiply −1 and 4 and write the result, −4, in the second column.

$$\begin{array}{r|rrrr} -1 & 4 & -5 & 2 & -9 \\ & & -4 & & \\ \hline & 4 & -9 & & \end{array}$$

Add the values in the second column, −5 and −4, and record the result, −9, below the line.

$$\begin{array}{r|rrrr} -1 & 4 & -5 & 2 & -9 \\ & & -4 & 9 & \\ \hline & 4 & -9 & & \end{array}$$

Multiply −1 and −9 and write the result, 9, in the third column.

$$\begin{array}{r|rrrr} -1 & 4 & -5 & 2 & -9 \\ & & -4 & 9 & \\ \hline & 4 & -9 & 11 & \end{array}$$

Add the values in the third column, 2 and 9, and write the result, 11, below the line.

$$\begin{array}{r|rrrr} -1 & 4 & -5 & 2 & -9 \\ & & -4 & 9 & -11 \\ \hline & 4 & -9 & 11 & \end{array}$$

Multiply −1 and 11 and write the result, −11, in the fourth column.

$$\begin{array}{r|rrrr} -1 & 4 & -5 & 2 & -9 \\ & & -4 & 9 & -11 \\ \hline & 4 & -9 & 11 & -20 \end{array}$$

Add the values in the fourth column, −9 and −11, and write the result, −20, below the line.

The numbers 4, −9, and 11 are the coefficients of the quotient polynomial and −20 is the remainder. The degree of the quotient polynomial is one less than the degree of the dividend.

So, $\dfrac{4x^3 - 5x^2 + 2x - 9}{x + 1} = 4x^2 - 9x + 11 - \dfrac{20}{x + 1}$.

Student Check 3 Use synthetic division to divide the polynomials.

a. Divide $6x^2 + 8x - 3$ by $x - 4$. **b.** Divide $2x^3 + 3x^2 - x + 5$ by $x + 2$.

The Remainder Theorem

Objective 4 Use the remainder theorem to evaluate polynomials.

There is an interesting relationship between the remainder of dividing a polynomial $P(x)$ by the quantity $x - c$ and the value of $P(c)$. In Example 3a, we found that the remainder after dividing $P(x) = 5x^2 - 7x + 3$ by $x - 2$ is 9. When we evaluate $P(x)$ for $x = 2$, we get

INSTRUCTOR NOTE:
Show students that
$P(x) = (x - c)Q(c) + R$, so
$P(c) = (c - c)Q(c) + R$
$= 0 \cdot Q(c) + R$
$= R$

$$\begin{aligned} P(2) &= 5(2)^2 - 7(2) + 3 \\ &= 5(4) - 14 + 3 \\ &= 20 - 14 + 3 \\ &= 6 + 3 \\ &= 9 \end{aligned}$$

So, $P(2) = 9$, which is the remainder obtained after dividing $P(x)$ by $x - 2$. This result can be stated as the *remainder theorem.*

Property: The Remainder Theorem

The remainder when dividing $P(x)$ by $x - c$ is $P(c)$.

Objective 4 Example **Use the remainder theorem and synthetic division to find P(3) if**

$$P(x) = 9x^4 - 3x^3 + 7x^2 - 5x + 6$$

Solution The coefficients of the polynomial are 9, −3, 7, −5, and 6. The value of c is 3.

$$\begin{array}{r|rrrrr} 3 & 9 & -3 & 7 & -5 & 6 \\ & & 27 & 72 & 237 & 696 \\ \hline & 9 & 24 & 79 & 232 & 702 \end{array}$$

The remainder after dividing $P(x)$ by $x - 3$ is 702. So, $P(3) = 702$.

Student Check 4 Use the remainder theorem and synthetic division to find $P(-5)$ if

$$P(x) = 3x^4 + 2x^3 - 5x^2 + 4x + 1$$

Applications

Objective 5
Solve application problems.

The problems in Example 5 involve finding an expression that represents the length of a side of a geometric figure.

Procedure: Solving Application Problems

Step 1: Write the formula for the area, volume, or perimeter given.
Step 2: Write a quotient that represents the expression you need to find.
Step 3: Divide using the methods from Objective 1 or 2.

Objective 5 Example **The area of a rectangle is 4x3y + 2x2y square units and its length is 2x2y units as shown in the figure. Find an expression for the width of the rectangle.**

$l = 2x^2y$

w

$A = 4x^3y + 2x^2y$

Solution Since $A = lw$, we can find an expression for width by dividing area by length.

$$w = \frac{A}{l}$$

$$w = \frac{4x^3y + 2x^2y}{2x^2y}$$

$$w = \frac{4x^3y}{2x^2y} + \frac{2x^2y}{2x^2y}$$

$$w = 2x + 1$$

So, the width can be represented by $2x + 1$ units.

Check: $(2x + 1)(2x^2y) = 4x^3y + 2x^2y$

Student Check 5 The area of a parallelogram is $9a^3b^3 - 12a^2b^2$ square units. Its height is $3a^2b$ units, as shown in the figure. Find an expression for the base of the parallelogram.

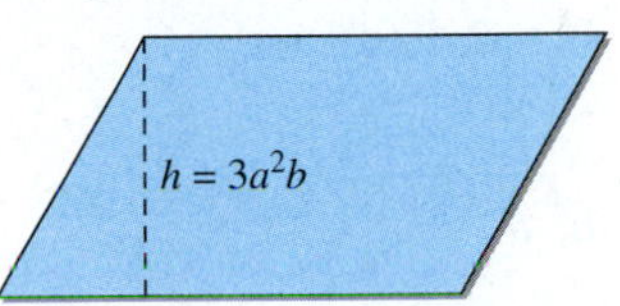

Objective 6 Troubleshoot common errors.

Troubleshooting Common Errors

Some common errors associated with dividing polynomials are illustrated.

Objective 6 Examples **A problem and an incorrect solution are given. Provide the correct solution and an explanation of the error.**

6a. Divide $6x^3 - 9x$ by $3x$.

Incorrect Solution	Correct Solution and Explanation
$\frac{6x^3 - 9x}{3x} = 2x^2 - 9x$	When dividing by a monomial, each term in the numerator must be divided by the monomial. $\frac{6x^3 - 9x}{3x} = \frac{6x^3}{3x} - \frac{9x}{3x} = 2x^2 - 3$

6b. Divide $2x^2 - 5x + 3$ by $2x - 1$ using long division.

Incorrect Solution	Correct Solution and Explanation
$\begin{array}{r} x - 3 \\ 2x - 1 \overline{)\, 2x^2 - 5x + 3} \\ \underline{2x^2 - x} \\ -6x + 3 \\ \underline{-6x + 3} \\ 0 \end{array}$	In long division, subtracting changes the signs: $-5x - (-x) = -5x + x = -4x$. $\begin{array}{r} x - 2 \\ 2x - 1 \overline{)\, 2x^2 - 5x + 3} \\ \underline{2x^2 - x} \\ -4x + 3 \\ \underline{-4x + 2} \\ 1 \end{array}$ The answer is $x - 2 + \frac{1}{2x - 1}$.

6c. Use synthetic division to divide $7x^2 + 8x - 9$ by $x + 1$.

Incorrect Solution	Correct Solution and Explanation
$\begin{array}{r\|rrr} 1 & 7 & 8 & -9 \\ & & 7 & 15 \\ \hline & 7 & 15 & 6 \end{array}$ So, $\dfrac{7x^2 + 8x - 9}{x + 1} = 7x + 15 + \dfrac{6}{x + 1}$.	Synthetic division works with divisors of the form $x - c$. Since $x + 1 = x - (-1)$, $c = -1$. $\begin{array}{r\|rrr} -1 & 7 & 8 & -9 \\ & & -7 & -1 \\ \hline & 7 & 1 & -10 \end{array}$ So, $\dfrac{7x^2 + 8x - 9}{x + 1} = 7x + 1 - \dfrac{10}{x + 1}$.

ANSWERS TO STUDENT CHECKS

Student Check 1 **a.** $y - 3$ **b.** $-2x^2 + 3x - 1$ **c.** $\dfrac{3b}{2} + \dfrac{4}{b} - \dfrac{6}{b^2}$

Student Check 2 **a.** $x + 8$ **b.** $a^2 - 5a + 25$ **c.** $y - 1 + \dfrac{1}{4y - 2}$

Student Check 3 **a.** $6x + 32 + \dfrac{125}{x - 4}$ **b.** $2x^2 - x + 1 + \dfrac{3}{x + 2}$

Student Check 4 $P(-5) = 1481$

Student Check 5 $3ab^2 - 4b$

SUMMARY OF KEY CONCEPTS

1. To divide a polynomial by a monomial, divide each term in the numerator by the monomial. Use rules of exponents to simplify each term. Check by multiplying.
2. Division by a polynomial other than a monomial requires long division. Repeat the steps involved in long division of real numbers—divide, multiply, subtract, and bring down. Continue these steps until the remainder has a degree less than the degree of the divisor. Check by multiplying. The quotient times the divisor plus the remainder should equal the dividend.
3. Synthetic division is a shortcut for dividing a polynomial by a binomial of the form $x - c$. Only the coefficients are used in this process.
4. The remainder theorem tells us that the remainder after dividing a polynomial $P(x)$ by a binomial of the form $x - c$ is equivalent to $P(c)$.
5. Division of polynomials can be used to find expressions that represent the length of a side of a geometric figure if an area, volume, or perimeter is given.

GRAPHING CALCULATOR SKILLS

The calculator can be used to check answers of division problems that involve a single variable.

Example: Verify that $\dfrac{3y^2 + 7y - 6}{3y + 4} = y + 1 - \dfrac{10}{3y + 4}$.

Method 1: Input the original division problem in Y_1 and the answer in Y_2. Then compare the table of values. They should agree everywhere except for the value of x that makes the denominator zero.

```
Plot1 Plot2 Plot3
\Y1■(3X²+7X-6)/(
3X+4)
\Y2■X+1-10/(3X+4
)
\Y3=
\Y4=
\Y5=
```

X	Y1	Y2
-3	0	0
-2	4	4
-1	-10	-10
0	-1.5	-1.5
1	.57143	.57143
2	2	2
3	3.2308	3.2308

X=-3

Method 2: The quotient times the divisor plus the remainder should equal the dividend. So, enter the quotient times the divisor plus the remainder in Y_1 and the dividend in Y_2. Compare the table of values. If they agree, then the answer is correct.

```
Plot1 Plot2 Plot3
\Y1■3X²+7X-6
\Y2■(X+1)(3X+4)-
10
\Y3=
\Y4=
\Y5=
\Y6=
```

X	Y1	Y2
-3	0	0
-2	-8	-8
-1	-10	-10
0	-6	-6
1	4	4
2	20	20
3	42	42

X=-3

Since the tables agree for Y_1 and Y_2, the answer is correct.

SECTION 7.2 EXERCISE SET

Write About It!

Use complete sentences in your answer to each exercise.

1. Explain how to divide a polynomial by a monomial and give an example.
2. In long division, name the four major steps. In long division, the four major steps are divide, multiply, subtract, and bring down.
3. In long division, what is the first step in writing the dividend? In long division, the first step is to write the dividend in standard form, inserting a zero for any missing term.
4. Can you apply synthetic division to any form of divisor? Please explain. No, you can apply synthetic division to only one form of a divisor, $x - c$.
5. In synthetic division, what is the first step in writing the dividend?
6. Explain how to write the quotient and remainder after completing synthetic division.
7. Explain how to use synthetic division to evaluate a polynomial function at a given x-value.
8. Explain why $\dfrac{x^3 + 64}{x + 4} \neq x^2 + 16$.

Practice Makes Perfect!

Divide and check by multiplying. (*See Objective 1.*)

9. $\dfrac{24x^2 - 18x}{6x}$ $4x - 3$
10. $\dfrac{35y^2 + 25y}{5y}$ $7y + 5$
11. $\dfrac{-12a^2 + 6a}{3a}$ $-4a + 2$
12. $\dfrac{-21b^2 - 14b}{7b}$ $-3b - 2$
13. $\dfrac{36x^3 - 16x^2 + 44x}{4x}$ $9x^2 - 4x + 11$
14. $\dfrac{-6x^3 - 48x^2 + 60x}{-6x}$ $x^2 + 8x - 10$
15. $\dfrac{-10x^4 + 45x^3 - 30x^2}{5x^2}$ $-2x^2 + 9x - 6$
16. $\dfrac{40x^4 - 8x^3 - 16x^2}{-4x^2}$ $-10x^2 + 2x + 4$
17. $\dfrac{-9x^4 + 11x^3 + 6x^2}{6x^2}$
18. $\dfrac{-12x^5 + 8x^4 + 9x^3}{2x^3}$ $-6x^2 + 4x + \dfrac{9}{2}$

Divide using long division. (*See Objective 2.*)

19. $\dfrac{18x^2 - 36x + 10}{3x - 5}$ $6x - 2$
20. $\dfrac{9x^2 + 61x - 14}{x + 7}$ $9x - 2$
21. $\dfrac{6x^2 + 22x - 40}{x + 5}$ $6x - 8$
22. $\dfrac{4x^2 + 8x - 5}{2x - 1}$ $2x + 5$
23. $\dfrac{4x^2 + 7x - 1}{x + 2}$
24. $\dfrac{10x^2 + 33x - 25}{x + 4}$ $10x - 7 + \dfrac{3}{x + 4}$
25. $\dfrac{12x^3 - 29x^2 - 3x + 30}{3x - 5}$ $4x^2 - 3x - 6$
26. $\dfrac{30x^3 - 27x^2 - 54x + 24}{5x - 2}$ $6x^2 - 3x - 12$
27. $\dfrac{6x^3 + 13x^2 - 4}{2x - 1}$ $3x^2 + 8x + 4$
28. $\dfrac{21x^3 - 2x^2 + 1}{3x + 1}$ $7x^2 - 3x + 1$
29. $\dfrac{27x^3 - 8}{3x - 2}$ $9x^2 + 6x + 4$
30. $\dfrac{8x^3 - 27}{2x - 3}$ $4x^2 + 6x + 9$
31. $\dfrac{x^3 - 64}{x + 4}$
32. $\dfrac{125x^3 + 8}{5x - 2}$

Divide using synthetic division. (*See Objective 3.*)

33. $\dfrac{3x^2 + 5x - 28}{x + 4}$ $3x - 7$
34. $\dfrac{2x^2 + 3x - 2}{x + 2}$ $2x - 1$
35. $\dfrac{2x^2 - x - 6}{x - 2}$ $2x + 3$
36. $\dfrac{3x^2 - 15x - 42}{x - 7}$ $3x + 6$
37. $(x^2 + 14x + 27) \div (x + 10)$
38. $(2x^2 - x - 11) \div (x + 3)$ $2x - 7 + \dfrac{10}{x + 3}$
39. $\dfrac{5x^3 + 24x^2 - 28x + 48}{x + 6}$ $5x^2 - 6x + 8$
40. $\dfrac{7x^3 - 2x^2 - 21x - 12}{x + 1}$ $7x^2 - 9x - 12$
41. $\dfrac{x^3 - 13x - 12}{x - 4}$ $x^2 + 4x + 3$
42. $\dfrac{x^3 - 2x^2 + 45}{x + 3}$ $x^2 - 5x + 15$
43. $\dfrac{2x^3 - 15x^2 + 25x + 8}{x - 3}$
44. $\dfrac{6x^3 + 47x^2 - 6x - 7}{x + 8}$
45. $\dfrac{x^3 + 27}{x + 3}$ $x^2 - 3x + 9$
46. $\dfrac{x^3 - 64}{x - 4}$ $x^2 - 4x + 16$

Use the remainder theorem and synthetic division or long division to evaluate $P(x)$ for the specified x-value. (*See Objective 4.*)

47. $P(x) = 2x^3 + 3x^2 + 8x - 17, P(1)$ $P(1) = -4$
48. $P(x) = -8x^3 + 2x^2 + x + 10, P(2)$ $P(2) = -44$
49. $P(x) = -8x^3 + 4x^2 + 1, P(2)$ $P(2) = -47$
50. $P(x) = 5x^3 - 12x^2 + 8x - 5, P(3)$ $P(3) = 46$
51. $P(x) = 4x^4 + 10x^3 - 3x^2 + 4x - 2, P(-3)$ $P(-3) = 13$
52. $P(x) = 2x^4 + 8x^3 - 9x^2 + 2x - 12, P(-5)$ $P(-5) = 3$
53. $P(x) = 4x^3 - 12x^2 + 10, P(4)$ $P(4) = 74$
54. $P(x) = 2x^4 - 13x^2 - 39, P(3)$ $P(3) = 6$

Solve each problem. (*See Objective 5.*)

55. The area of a rectangle is $30a^4b^4 + 50a^3b^3$ square units. Its height is $10a^2b$ units. Find an expression for the width of the rectangle. $3a^2b^3 + 5ab^2$ units
56. The area of a rectangle is $10r^2s^2 + 26r^3s^4$ square units. Its height is $2rs$ units. Find an expression for the width of the rectangle. $5rs + 13r^2s^3$ units
57. The area of a parallelogram is $4x^5y^3 - 56x^3y^2$ square units. Its height is $4x^2y^2$ units. Find an expression for the base of the parallelogram. $x^3y - 14x$ units
58. The area of a parallelogram is $33a^3b^2 + 30a^2b$ square units. Its height is $3ab$ units. Find an expression for the base of the parallelogram. $11a^2b + 10a$ units

Additional answers can be found in the Instructor Answer Appendix.

Mix 'Em Up!

Divide and check by multiplying.

59. $\dfrac{126a^3 - 30a^2 + 78a}{6a}$ $21a^2 - 5a + 13$

60. $\dfrac{-56x^3 - 84x^2 + 14x}{-7x}$ $8x^2 + 12x - 2$

61. $\dfrac{6b^4 + 9b^3 - 24b^2}{-3b^2}$ $-2b^2 - 3b + 8$

62. $\dfrac{-10x^4 + 45x^3 - 30x^2}{5x^2}$ $-2x^2 + 9x - 6$

Divide using long division or synthetic division.

63. $\dfrac{7x^2 - 61x - 18}{7x + 2}$ $x - 9$

64. $\dfrac{6x^2 + 4x - 10}{3x + 5}$ $2x - 2$

65. $\dfrac{2x^2 + 14x + 23}{x + 3}$

66. $\dfrac{x^2 + 2x + 14}{x - 1}$

67. $\dfrac{4x^3 + 20x^2 - 27x - 88}{2x + 11}$ $2x^2 - x - 8$

68. $\dfrac{21x^3 + 29x^2 - 25x - 25}{7x + 5}$ $3x^2 + 2x - 5$

69. $\dfrac{x^3 + 512}{x + 8}$ $x^2 - 8x + 64$

70. $\dfrac{x^3 - 1000}{x - 10}$ $x^2 + 10x + 100$

71. $\dfrac{20x^4 + x^2 - 12}{4x^2 - 3}$ $5x^2 + 4$

72. $\dfrac{6x^4 + 13x^2 - 28}{2x^2 + 7}$ $3x^2 - 4$

73. $\dfrac{3x^2 + x - 2}{x + 1}$ $3x - 2$

74. $\dfrac{8x^2 - 30x + 18}{x - 3}$ $8x - 6$

75. $\dfrac{25x^3 - 21x - 4}{5x + 4}$ $5x^2 - 4x - 1$

76. $\dfrac{27x^3 - 75x^2 + 4}{9x + 2}$ $3x^2 - 9x + 2$

77. $\dfrac{x^3 - 16x + 45}{x + 5}$ $x^2 - 5x + 9$

78. $\dfrac{4x^3 + 6x^2 + 13}{x + 2}$

79. $\dfrac{24x^3 - 10x^2 + 3}{3x + 1}$

80. $\dfrac{20x^3 - 33x^2 - 12}{5x - 2}$

81. $\dfrac{x^3 + 1}{x - 1}$ $x^2 + x + 1 + \dfrac{2}{x - 1}$

82. $\dfrac{x^3 - 1}{x + 1}$ $x^2 - x + 1 - \dfrac{2}{x + 1}$

Use the remainder theorem and synthetic division or long division to evaluate $P(x)$ for the specified x-values.

83. $P(x) = 3x^3 + 4x^2 - 5x - 18,\ P(2)$ $P(2) = 12$

84. $P(x) = -3x^3 + 6x^2 - 5,\ P(3)$ $P(3) = -32$

85. $P(x) = 5x^4 + 8x^3 - 6x + 4,\ P(-2)$ $P(-2) = 32$

86. $P(x) = 4x^4 + 10x^3 + 12x - 14,\ P(-3)$ $P(-3) = 4$

87. $P(x) = x^4 + 3x^2 - 10x - 1,\ P(2)$ $P(2) = 7$

88. $P(x) = -x^2 - 2x + 3,\ P(-5)$ $P(-5) = -12$

Solve each problem.

89. The area of a rectangle is $24a^2b^3 - 39a^3b^2$ square units. Its height is $3a^2b^2$ units. Find an expression for the width of the rectangle. $8b - 13a$ units

90. The area of a rectangle is $28x^4y^3 + 42x^2y^2$ square units. Its height is $14xy^2$ units. Find an expression for the width of the rectangle. $2x^3y + 3x$ units

91. The area of a parallelogram is $30a^4b^2 + 28a^2b^3$ square units. Its height is $2ab^2$ units. Find an expression for the base of the parallelogram. $15a^3 + 14ab$ units

92. The area of a parallelogram is $35r^2s^3 + 85r^3s^2$ square units. Its height is $5rs$ units. Find an expression for the base of the parallelogram. $7rs^2 + 17r^2s$ units

You Be the Teacher!

Correct each student's errors, if any.

93. Divide: $\dfrac{8x^3 + 12x^2 + 4x}{4x}$.

Celine's work:

$$\frac{8x^3 + 12x^2 + 4x}{4x} = \frac{\overset{2x^2}{\cancel{8x^3}}}{\cancel{4x}} + \frac{\overset{3x}{\cancel{12x^2}}}{\cancel{4x}} + \frac{\cancel{4x}}{\cancel{4x}} = 2x^2 + 3x$$

94. Divide: $\dfrac{6x^2 - x - 2}{2x + 1}$.

Marie's work:

$$\begin{array}{r} 3x + 1 \\ 2x + 1\overline{)6x^2 - x - 2} \\ \underline{6x^2 + 3x} \\ 2x - 2 \\ \underline{2x + 1} \\ -1 \end{array}$$

So, $\dfrac{6x^2 - x - 2}{2x + 1} = 3x + 1 - \dfrac{1}{2x + 1}$.

$$\begin{array}{r} 3x - 2 \\ 2x + 1\overline{)6x^2 - x - 2} \\ \underline{6x^2 + 3x} \\ -4x - 2 \\ \underline{-4x - 2} \end{array}$$

So, $\dfrac{6x^2 - x - 2}{2x + 1} = 3x - 2$.

95. Divide: $\dfrac{6x^3 + 4x - 16}{2x + 4}$.

Al's work:

$$\begin{array}{r} 3x^2 - 4 \\ 2x + 4\overline{)6x^3 + 4x - 16} \\ \underline{6x^3 + 12x^2} \\ -8x - 16 \\ \underline{-8x - 16} \end{array}$$

So, $\dfrac{6x^3 + 4x - 16}{2x + 4} = 3x^2 - 4$.

96. Divide: $(5x^3 - 2x - 16) \div (x - 2)$.

Bert's work:

$$\begin{array}{r|rrr} 2 & 5 & -2 & -16 \\ & & 10 & 16 \\ \hline & 5 & 8 & 0 \end{array}$$

So, $(5x^3 - 2x - 16) \div (x - 2) = 5x^2 + 8$.

Calculate It!

Use a calculator to verify each division.

97. $\dfrac{4x^2 + 13x + 3}{x + 3} = 4x + 1$

98. $\dfrac{x^3 + 7x^2 + 13x - 5}{x + 2} = x^2 + 5x + 3 - \dfrac{11}{x + 2}$

99. $\dfrac{64x^3 - 125}{4x - 5} = 16x^2 + 20x + 25$

100. $\dfrac{64x^3 + 125}{4x - 5} = 16x^2 + 20x + 25 + \dfrac{250}{4x - 5}$

Think About It!

101. What polynomial must be divided by $x - 4$ to yield $3x + 2$? $3x^2 - 10x - 8$

102. What polynomial must be divided by $x + 5$ to yield $4x - 1$? $4x^2 + 19x - 5$

103. If 5 is a zero of the polynomial $P(x)$, what is the remainder after dividing $P(x)$ by $(x - 5)$? Why?
If 5 is a zero of $P(x)$, then we know that $P(5) = 0$. By the remainder theorem, this is equivalent to the remainder after dividing $P(x)$ by $x - 5$. So, the remainder is 0.

SECTION 7.3 Adding and Subtracting Rational Expressions

OBJECTIVES

As a result of completing this section, you will be able to

1. Add or subtract rational expressions with like denominators.
2. Find the least common denominator of rational expressions.
3. Add or subtract rational expressions with unlike denominators.
4. Troubleshoot common errors.

In this section, we will learn the last of the basic arithmetic operations for rational expressions—addition and subtraction. Adding and subtracting rational expressions is similar to adding and subtracting rational numbers.

Objective 1 Add or subtract rational expressions with like denominators.

Adding or Subtracting Rational Expressions with Like Denominators

Recall that fractions can be added or subtracted only when their denominators are the same.

$$\frac{1}{5} + \frac{2}{5} = \frac{1 + 2}{5} = \frac{3}{5}$$

This is also true with rational expressions.

Property: Adding and Subtracting Rational Expressions

If $\dfrac{P}{Q}$ and $\dfrac{R}{Q}$ are rational expressions with $Q \neq 0$, then

$$\frac{P}{Q} + \frac{R}{Q} = \frac{P + R}{Q} \quad \text{and} \quad \frac{P}{Q} - \frac{R}{Q} = \frac{P - R}{Q}$$

Procedure: Adding or Subtracting Rational Expressions with Like Denominators

Step 1: Add or subtract the numerators.
Step 2: Place the result of step 1 over the denominator.
Step 3: Simplify, if necessary.

Objective 1 Examples Add or subtract the rational expressions. Write each answer in lowest terms.

1a. $\dfrac{4x}{x + 3} + \dfrac{2x}{x + 3}$

1b. $\dfrac{3a}{8a - 6} + \dfrac{7a - 4}{8a - 6}$

1c. $\dfrac{y}{y^2 - 36} - \dfrac{6}{y^2 - 36}$

1d. $\dfrac{4x}{x^2 + x - 12} - \dfrac{3x - 4}{x^2 + x - 12}$

Solutions

1a. $\dfrac{4x}{x+3}+\dfrac{2x}{x+3}=\dfrac{4x+2x}{x+3}=\dfrac{6x}{x+3}$

1b.

$\dfrac{3a}{8a-6}+\dfrac{7a-4}{8a-6}=\dfrac{3a+7a-4}{8a-6}$ Add the numerators and place over the like denominator.

$=\dfrac{10a-4}{8a-6}$ Simplify the numerator.

$=\dfrac{2(5a-2)}{2(4a-3)}$ Factor the numerator and denominator.

$=\dfrac{5a-2}{4a-3}$ Divide out the common factor, 2.

1c.

$\dfrac{y}{y^2-36}-\dfrac{6}{y^2-36}=\dfrac{y-6}{y^2-36}$ Subtract the numerators and place over the like denominator.

$=\dfrac{y-6}{(y-6)(y+6)}$ Factor the difference of squares.

$=\dfrac{1}{y+6}$ Divide out the common factor, $y-6$.

1d. $\dfrac{4x}{x^2+x-12}-\dfrac{3x-4}{x^2+x-12}$

$=\dfrac{4x-(3x-4)}{x^2+x-12}$ Subtract the numerators and place over the like denominator.

$=\dfrac{4x-3x+4}{x^2+x-12}$ Apply the distributive property. Recall $-(3x-4)=-1(3x-4)$.

$=\dfrac{x+4}{x^2+x-12}$ Simplify the numerator.

$=\dfrac{x+4}{(x+4)(x-3)}$ Factor the denominator.

$=\dfrac{1}{x-3}$ Divide out the common factor, $x+4$.

Student Check 1 Add or subtract the rational expressions. Write each answer in lowest terms.

a. $\dfrac{6x}{x+2}+\dfrac{4x}{x+2}$ **b.** $\dfrac{6b}{3b+15}+\dfrac{3b+3}{3b+15}$

c. $\dfrac{a}{a^2-4}-\dfrac{2}{a^2-4}$ **d.** $\dfrac{5m^2+m}{m^2-2m-3}-\dfrac{4m^2+4m}{m^2-2m-3}$

The Least Common Denominator

Objective 2 Find the least common denominator of rational expressions.

We cannot add rational expressions with unlike denominators until we convert them to equivalent fractions with the same denominator.

The following illustration reviews the process of finding the least common denominator of two fractions. Consider the fractions $\dfrac{5}{36}$ and $\dfrac{7}{45}$. The **least common denominator (LCD)** is the smallest number that is divisible by each denominator. We factor each denominator and then "build" the LCD from these factors.

$36=4\cdot9=2\cdot2\cdot3\cdot3$ $\quad$ $45=5\cdot9=5\cdot3\cdot3$

The LCD must contain the factors from each denominator. So, we "build it" by including all of the factors from 36 and any additional factors from 45 that have not already been included.

$$\text{LCD} = \underbrace{2 \cdot 2 \cdot 3 \cdot \overbrace{3 \cdot 5}^{45}}_{36} = 180$$

The LCD consists of the product of each unique factor used the largest number of times it occurs in any of the denominators.

Procedure: Finding the Least Common Denominator (LCD)

Step 1: Factor each denominator completely.
Step 2: Form the product of all factors from the first denominator.
Step 3: Multiply the product from step 2 by the factors from the other denominators that have not been included.
Step 4: The product of the factors is the LCD. Leave the LCD in factored form.

Note: *The LCD is equal to the least common multiple (LCM) of the denominators.*

Objective 2 Examples

Find the least common denominator of the given rational expressions.

2a. $\dfrac{3}{x^3y^2}, \dfrac{4}{xy^4}$ **2b.** $\dfrac{x+3}{x-4}, \dfrac{1}{x}$ **2c.** $\dfrac{3y}{4y^2-9}, \dfrac{y+1}{2y^2-5y-12}, \dfrac{7y-9}{3y-12}$

2d. $\dfrac{4}{x-5}, \dfrac{2}{5-x}$ **2e.** $\dfrac{a+2}{a^2-6a+9}, \dfrac{5}{12-4a}$

Solutions

2a.
$$x^3y^2 = x \cdot x \cdot x \cdot y \cdot y$$
$$xy^4 = x \cdot y \cdot y \cdot y \cdot y$$
$$\longrightarrow \text{LCD} = \underbrace{x \cdot x \cdot \overbrace{x \cdot y \cdot y}}_{x^3y^2}\overbrace{\cdot\, y \cdot y}^{xy^4} = x^3y^4$$

2b. Each denominator is a prime polynomial. Therefore, the LCD is the product of these two factors.

$$x - 4 = (x - 4)$$
$$x = (x)$$
$$\longrightarrow \text{LCD} = (x-4)(x) = x(x-4)$$

2c.
$$4y^2 - 9 = (2y+3)(2y-3)$$
$$2y^2 - 5y - 12 = (2y+3)(y-4)$$
$$3y - 12 = 3(y-4)$$
$$\longrightarrow \text{LCD} = 3(2y+3)(2y-3)(y-4)$$

2d. The denominators $x - 5$ and $5 - x$ are opposites.

$$x - 5 = (x - 5)$$
$$5 - x = -x + 5 = -1(x - 5)$$
$$\longrightarrow \text{LCD} = -1(x-5)$$

It is, however, sufficient to use $x - 5$ as the LCD since the second fraction can be written as

$$\frac{2}{5-x} = \frac{2}{-(x-5)} = -\frac{2}{x-5}$$

So, the LCD is either $x - 5$ or $-(x - 5)$.

2e. $a^2 - 6a + 9 = (a-3)(a-3)$
$12 - 4a = 4(3-a) = -4(a-3)$ $\longrightarrow$ LCD $= -4(a-3)(a-3)$
$= -4(a-3)^2$

While the steps of including the different factors give us a negative LCD, it is standard practice to write the LCD as a positive expression, that is, as $4(a-3)^2$.

Student Check 2 Find the least common denominator of the given rational expressions.

a. $\dfrac{4}{a^2b^3}, \dfrac{5}{a^3b}$ **b.** $\dfrac{x+2}{x}, \dfrac{2x+5}{x+3}$ **c.** $\dfrac{y-4}{y^2-9}, \dfrac{3y}{3y^2+10y+3}, \dfrac{7}{9y+3}$

d. $\dfrac{7}{p-2}, \dfrac{1}{2-p}$ **e.** $\dfrac{x+2}{x^2-10x+25}, \dfrac{3x}{10-2x}$

Note: *In Example 1a, the LCD consists of the factors x^3 and y^4. The largest exponent of the variable x in the given denominators is 3, and the largest exponent of the variable y in the given denominators is 4. So, the LCD will involve a factor raised to the largest exponent that it occurs in the given denominators.*

Adding and Subtracting Rational Expressions with Unlike Denominators

Objective 3 Add or subtract rational expressions with unlike denominators.

Now that we know how to find the LCD, we can add or subtract rational expressions with different denominators. After we find the LCD, we convert each fraction to an equivalent fraction with the LCD as its denominator. We do this by multiplying the fraction by a form of 1. For instance, to write the fraction $\dfrac{2}{x-5}$ as an equivalent fraction with $(x+5)(x-5)$ as its denominator, we must multiply the fraction by a form of 1 that consists of the factor needed to obtain the new denominator. In this case, the needed factor is $x + 5$.

$$\frac{2}{x-5} = \frac{2}{x-5} \cdot \frac{x+5}{x+5} = \frac{2(x+5)}{(x-5)(x+5)} = \frac{2x+10}{(x-5)(x+5)}$$

Procedure: Adding or Subtracting Rational Expressions with Unlike Denominators

Step 1: Determine the LCD of the expressions.
Step 2: Convert each fraction to an equivalent fraction with the LCD as its denominator.
Step 3: Add or subtract the fractions.
Step 4: Simplify to lowest terms, if necessary.

Objective 3 Examples Add or subtract the rational expressions. Write each answer in lowest terms.

3a. $\frac{7}{3x^2y^3} + \frac{4}{xy^4}$ **3b.** $\frac{4}{x-1} + \frac{3}{x}$ **3c.** $\frac{6}{a-2} - \frac{3}{2-a}$

3d. $\frac{y+9}{y^2-4} + \frac{1}{4y-8}$ **3e.** $\frac{x+2}{x^2+x-20} - \frac{3x}{2x^2-7x-4}$

Solutions

3a. The LCD $= 3 \cdot x \cdot x \cdot y \cdot y \cdot y \cdot y = 3x^2y^4$. The first denominator needs a factor of y and the second denominator needs a factor of $3x$ to obtain the LCD.

$$\frac{7}{3x^2y^3} + \frac{4}{xy^4} = \frac{7}{3x^2y^3} \cdot \frac{y}{y} + \frac{4}{xy^4} \cdot \frac{3x}{3x}$$ Write equivalent rational expressions.

$$= \frac{7y}{3x^2y^4} + \frac{12x}{3x^2y^4}$$ Multiply the fractions.

$$= \frac{7y+12x}{3x^2y^4}$$ Add the rational expressions.

3b. The LCD $= x(x-1)$. The first denominator needs a factor of x and the second denominator needs a factor of $x - 1$ to obtain the LCD.

$$\frac{4}{x-1} + \frac{3}{x} = \frac{4}{x-1} \cdot \frac{x}{x} + \frac{3}{x} \cdot \frac{x-1}{x-1}$$ Write equivalent rational expressions.

$$= \frac{4x}{x(x-1)} + \frac{3x-3}{x(x-1)}$$ Multiply the fractions.

$$= \frac{4x+3x-3}{x(x-1)}$$ Add the rational expressions.

$$= \frac{7x-3}{x(x-1)}$$ Simplify.

3c. The denominators are opposites, so the LCD is $a - 2$.

$$\frac{6}{a-2} - \frac{3}{2-a} = \frac{6}{a-2} - \frac{3}{-1(a-2)}$$ Write $2 - a$ as the opposite of $a - 2$.

$$= \frac{6}{a-2} - \frac{-1 \cdot 3}{a-2}$$ Apply the property of negative rational expressions.

$$= \frac{6}{a-2} - \frac{-3}{a-2}$$ Simplify the second numerator.

$$= \frac{6-(-3)}{a-2}$$ Subtract the rational expressions.

$$= \frac{9}{a-2}$$ Simplify.

3d. We first find the LCD.

$y^2 - 4 = (y-2)(y+2)$
$4y - 8 = 4(y-2)$ ⟶ LCD $= 4(y-2)(y+2)$

The first denominator needs a factor of 4 and the second denominator needs a factor of $y + 2$ to obtain the LCD.

$$\frac{y+9}{y^2-4}+\frac{1}{4y-8}=\frac{y+9}{(y-2)(y+2)}\cdot\frac{4}{4}+\frac{1}{4(y-2)}\cdot\frac{y+2}{y+2}$$ Write equivalent fractions.

$$=\frac{4y+36}{4(y-2)(y+2)}+\frac{y+2}{4(y-2)(y+2)}$$ Multiply the fractions.

$$=\frac{4y+36+y+2}{4(y-2)(y+2)}$$ Add the rational expressions.

$$=\frac{5y+38}{4(y-2)(y+2)}$$ Simplify.

3e. We first find the LCD.

$$x^2+x-20=(x+5)(x-4)$$
$$2x^2-7x-4=(2x+1)(x-4) \longrightarrow \text{LCD}=(x+5)(x-4)(2x+1)$$

The first denominator needs a factor of $2x + 1$ and the second denominator needs a factor of $x + 5$ to obtain the LCD.

$$\frac{x+2}{x^2+x-20}-\frac{3x}{2x^2-7x-4}$$

$$=\frac{x+2}{(x+5)(x-4)}\cdot\frac{2x+1}{2x+1}-\frac{3x}{(2x+1)(x-4)}\cdot\frac{x+5}{x+5}$$

$$=\frac{2x^2+5x+2}{(x+5)(x-4)(2x+1)}-\frac{3x^2+15x}{(2x+1)(x-4)(x+5)}$$

$$=\frac{2x^2+5x+2-(3x^2+15x)}{(x+5)(x-4)(2x+1)}$$

$$=\frac{2x^2+5x+2-3x^2-15x}{(x+5)(x-4)(2x+1)}$$

$$=\frac{-x^2-10x+2}{(x+5)(x-4)(2x+1)}$$

Student Check 3 Add or subtract the rational expressions. Write each answer in lowest terms.

a. $\dfrac{2}{7ab^2}+\dfrac{5}{a^2b^3}$ **b.** $\dfrac{8}{x-6}+\dfrac{1}{x}$ **c.** $\dfrac{4}{x-5}-\dfrac{2}{5-x}$

d. $\dfrac{b+3}{b^2-25}+\dfrac{3}{2b+10}$ **e.** $\dfrac{y+1}{y^2-y-6}-\dfrac{2y}{4y^2+5y-6}$

Objective 4 ▶ Troubleshoot common errors.

Troubleshooting Common Errors

Some common errors associated with adding and subtracting rational expressions are shown.

Objective 4 Examples **A problem and an incorrect solution are given. Provide the correct solution and an explanation of the error.**

4a. Subtract: $\dfrac{x+2}{x+4}-\dfrac{x+5}{x+4}$.

Incorrect Solution	Correct Solution and Explanation
$\frac{x+2}{x+4} - \frac{x+5}{x+4} = \frac{x+2-x+5}{x+4}$ $= \frac{7}{x+4}$	The negative sign was not distributed to each term in the numerator of the second fraction. $\frac{x+2}{x+4} - \frac{x+5}{x+4} = \frac{x+2-(x+5)}{x+4}$ $= \frac{x+2-x-5}{x+4}$ $= \frac{-3}{x+4}$ $= -\frac{3}{x+4}$

4b. Find the least common denominator of the fractions $\frac{3}{x-2}$ and $\frac{5}{x}$.

Incorrect Solution	Correct Solution and Explanation
The LCD is $x - 2$.	The LCD must consist of the factors of each denominator. Since the two denominators are prime, the LCD consists of the product of these denominators. $\text{LCD} = x(x-2)$

4c. Add: $\frac{x+3}{x^2-16} + \frac{1}{2x-8}$.

Incorrect Solution	Correct Solution and Explanation
$\frac{x+3}{x^2-16} + \frac{1}{2x-8}$ $= \frac{x+3}{(x+4)(x-4)} + \frac{1}{2(x-4)}$ $= 2(x+4)(x-4)\frac{x+3}{(x+4)(x-4)}$ $+ 2(x+4)(x-4)\frac{1}{2(x-4)}$ $= 2(x+3) + x + 4$ $= 2x + 6 + x + 4$ $= 3x + 10$	This is an expression not an equation. We cannot eliminate the denominators since there are not two sides of an equation to balance the multiplication by the LCD. We must convert each fraction to an equivalent fraction with the LCD as its denominator. $\frac{x+3}{x^2-16} + \frac{1}{2x-8}$ $= \frac{x+3}{(x+4)(x-4)} + \frac{1}{2(x-4)}$ $= \frac{(x+3)\cdot 2}{(x+4)(x-4)\cdot 2} + \frac{1\cdot(x+4)}{2(x-4)\cdot(x+4)}$ $= \frac{2x+6}{2(x+4)(x-4)} + \frac{x+4}{2(x-4)(x+4)}$ $= \frac{2x+6+x+4}{2(x+4)(x-4)}$ $= \frac{3x+10}{2(x+4)(x-4)}$

ANSWERS TO STUDENT CHECKS

Student Check 1 **a.** $\dfrac{10x}{x+2}$ **b.** $\dfrac{3b+1}{b+5}$ **c.** $\dfrac{1}{a+2}$ **d.** $\dfrac{m}{m+1}$

Student Check 2 **a.** a^3b^3 **b.** $x(x+3)$ **c.** $3(y+3)(y-3)(3y+1)$ **d.** $p-2$ **e.** $2(x-5)^2$

Student Check 3 **a.** $\dfrac{2ab+35}{7a^2b^3}$ **b.** $\dfrac{9x-6}{x(x-6)}$ **c.** $\dfrac{6}{x-5}$ **d.** $\dfrac{5b-9}{2(b-5)(b+5)}$ **e.** $\dfrac{2y^2+7y-3}{(y-3)(y+2)(4y-3)}$

SUMMARY OF KEY CONCEPTS

1. If the rational expressions being added or subtracted have like denominators, then add or subtract the numerators and put the sum or difference over the like denominator. Simplify, if possible.
2. The least common denominator is found by first factoring all of the denominators. The LCD consists of the product of all the unique factors used the largest number of times they occur in any one denominator.
3. To add or subtract rational expressions with unlike denominators, convert each fraction to an equivalent fraction with the LCD as its denominator. Then add or subtract the numerators and put the sum or difference over the LCD. Simplify, if possible.

GRAPHING CALCULATOR SKILLS

We can use the graphing calculator to find the least common multiple of two numbers. This can assist us in finding the least common denominator for some rational expressions. The graphing calculator can also help us determine if we worked a problem correctly.

Example 1: Determine the coefficient of the least common denominator of the two fractions $\dfrac{5}{14x^4y^5}$ and $\dfrac{x}{104y^4}$.

Solution: The calculator cannot help us with the variables, but it can help us determine the coefficient of the LCD. We find the least common multiple (lcm) of 14 and 104. The LCM is found by pressing MATH and accessing the NUM menu and selecting option 8.

```
lcm(14,104)
                728
```

The coefficient of the LCD of the rational expressions is 728.

Example 2: Verify that $\dfrac{2x}{x^2-3x} - \dfrac{1}{x-3} = \dfrac{1}{x-3}$.

Solution: Enter the left side of the equation in Y_1 and the right side in Y_2. Compare the y-values of the two expressions by examining the table. If the two columns agree, except for where each expression is undefined, then we have performed the operation correctly.

```
Plot1 Plot2 Plot3
\Y1=(2X)/(X²-3X)
-1/(X-3)
\Y2=1/(X-3)
\Y3=
\Y4=
\Y5=
\Y6=
```

X	Y1	Y2
-3	-.1667	-.1667
-2	-.2	-.2
-1	-.25	-.25
0	ERR:	-.3333
1	-.5	-.5
2	-1	-1
3	ERR:	ERR:

X=-3

Since the two columns agree, our answer is correct.

SECTION 7.3 EXERCISE SET

Write About It!

Use complete sentences in your answer to each exercise.

1. Explain how to find the least common denominator for the rational expressions $\frac{5}{12x^3y}$ and $\frac{1}{18x^2y^5}$.
2. Explain why $\frac{4x}{3x-2} - \frac{x-2}{3x-2} \neq \frac{3x-2}{3x-2}$.
3. Explain how to find the least common denominator and build the numerators when adding the two rational expressions $\frac{3}{2x-5} + \frac{1}{5-2x}$.
4. Explain how to find the least common denominator and build the numerators when adding the two rational expressions $\frac{x}{(x-1)(2x+1)} + \frac{5x}{(x+3)(2x+1)}$.
5. Explain why $\frac{4x+1}{x^2-x-6} - \frac{3x-1}{x^2-2x-8} \neq \frac{x+2}{x+2}$.
6. Explain how to subtract the rational expressions $\frac{4x+1}{x^2-x-6} - \frac{3x-1}{x^2-2x-8}$.
7. Explain how to find the least common denominator for the three rational expressions $\frac{3}{(5x-2)^2}, \frac{1}{30x-12}$, and $\frac{5}{40x-16}$.
8. Explain how to simplify $\frac{3}{(5x-2)^2} + \frac{1}{30x-12} - \frac{5}{40x-16}$.

Practice Makes Perfect!

Add or subtract the rational expressions. (*See Objective 1.*)

9. $\frac{10x}{4x+9} + \frac{2x}{4x+9}$ $\frac{12x}{4x+9}$
10. $\frac{5y}{y+8} + \frac{10y}{y+8}$ $\frac{15y}{y+8}$
11. $\frac{5a}{5a-4} - \frac{7a}{5a-4}$ $-\frac{2a}{5a-4}$
12. $\frac{-18b}{5b-7} - \frac{5b}{5b-7}$ $-\frac{23b}{5b-7}$
13. $\frac{9y+4}{y+3} - \frac{12y-4}{y+3}$ $\frac{-3y+8}{y+3}$
14. $\frac{a-9}{a+10} + \frac{7a+3}{a+10}$ $\frac{8a-6}{a+10}$
15. $\frac{3x}{9x^2-64} - \frac{8}{9x^2-64}$ $\frac{1}{3x+8}$
16. $\frac{x}{x^2-9} - \frac{3}{x^2-9}$ $\frac{1}{x+3}$
17. $\frac{7x}{49x^2-4} + \frac{2}{49x^2-4}$ $\frac{1}{7x-2}$
18. $\frac{5x}{25x^2-36} + \frac{6}{25x^2-36}$ $\frac{1}{5x-6}$
19. $\frac{2x^2+8x}{3x^2+17x+20} + \frac{x^2+4x}{3x^2+17x+20}$ $\frac{3x}{3x+5}$
20. $\frac{6x^2+14x}{3x^2+x-14} + \frac{15x^2+35x}{3x^2+x-14}$ $\frac{7x}{x-2}$
21. $\frac{5x^2+10x}{x^2+5x+6} - \frac{x^2+2x}{x^2+5x+6}$ $\frac{4x}{x+3}$
22. $\frac{-9y^2-18y}{4y^2+5y-6} - \frac{-6y^2-12y}{4y^2+5y-6}$ $-\frac{3y}{4y-3}$

Find the least common denominator for the given rational expressions. (*See Objective 2.*)

23. $\frac{9}{x^5y}, \frac{-3}{x^3y^4}$ x^5y^4
24. $\frac{10}{x^2y^4}, \frac{15}{x^6y^3}$ x^6y^4
25. $\frac{8}{a^3b}, \frac{-6}{a^3b^4}$ a^3b^4
26. $\frac{3}{a^3b^5}, \frac{14}{a^4b^2}$ a^4b^5
27. $\frac{9x+7}{x}, \frac{x-3}{7x-2}$ $x(7x-2)$
28. $\frac{5x+2}{10x}, \frac{3x-1}{x-1}$ $10x(x-1)$
29. $\frac{5x-4}{4x}, \frac{6x+1}{9x-6}$ $12x(3x-2)$
30. $\frac{x-1}{4x}, \frac{6x}{3x+21}$ $12x(x+7)$
31. $\frac{x+6}{x^2-49}, \frac{2x}{x^2+8x+7}, \frac{1}{x+1}$ $(x+7)(x-7)(x+1)$
32. $\frac{8x+7}{x^2-16}, \frac{-9x}{x^2+5x+4}, \frac{11}{x+1}$ $(x+4)(x-4)(x+1)$
33. $\frac{3x+5}{x^2-4}, \frac{-x}{7x^2+17x+6}, \frac{9}{14x+6}$ $2(x+2)(x-2)(7x+3)$
34. $\frac{x+6}{x^2-49}, \frac{2x}{x^2+8x+7}, \frac{1}{x+1}$ $(x+7)(x-7)(x+1)$
35. $\frac{1}{(x-6)^2}, \frac{11}{2x-12}$ $2(x-6)^2$
36. $\frac{5x}{(4x-7)^2}, \frac{-2}{20x-35}$ $5(4x-7)^2$
37. $\frac{x}{(3x-2)^2}, \frac{5x}{12-18x}$ $6(3x-2)^2$
38. $\frac{2x+1}{(x-2)^2}, \frac{3x}{8-4x}$ $4(x-2)^2$

Add or subtract the rational expressions. Write each answer in lowest terms. (*See Objective 3.*)

39. $\frac{4}{x^2y^5} + \frac{17}{2x^4y^6}$ $\frac{8x^2y+17}{2x^4y^6}$
40. $\frac{9}{4x^5y^2} - \frac{3}{x^4y^6}$ $\frac{9y^4-12x}{4x^5y^6}$
41. $\frac{11}{5xy^3} - \frac{3}{10x^3y^2}$ $\frac{22x^2-3y}{10x^3y^3}$
42. $\frac{-5}{6x^2y^2} + \frac{1}{8x^3y}$ $\frac{-20x+3y}{24x^3y^2}$
43. $\frac{-5}{3x-2} + \frac{3}{x}$ $\frac{4x-6}{x(3x-2)}$
44. $\frac{-2}{5x-3} + \frac{1}{2x}$ $\frac{x-3}{2x(5x-3)}$
45. $\frac{-2}{3x-9} - \frac{5}{2x}$ $\frac{-19x+45}{6x(x-3)}$
46. $\frac{5}{4x-2} - \frac{1}{8x}$ $\frac{18x+1}{8x(2x-1)}$
47. $\frac{2}{6x-7} + \frac{14}{7-6x}$ $-\frac{12}{6x-7}$
48. $\frac{8}{4x-1} + \frac{2}{1-4x}$ $\frac{6}{4x-1}$
49. $\frac{4x}{x^2-1} - \frac{3}{x-1}$
50. $\frac{6x}{x^2-4} + \frac{3}{x+2}$
51. $\frac{5x-1}{9x^2-16} + \frac{3}{6x-8}$
52. $\frac{x-2}{4x^2-9} + \frac{5}{6x+9}$

Additional answers can be found in the Instructor Answer Appendix.

53. $\dfrac{3x}{5x^2 - 9x + 4} + \dfrac{x}{2x^2 - 3x + 1}$ $\dfrac{11x^2 - 7x}{(x - 1)(5x - 4)(2x - 1)}$

54. $\dfrac{3x}{5x^2 - 9x - 2} + \dfrac{x}{4x^2 - 7x - 2}$ $\dfrac{17x^2 + 4x}{(x - 2)(5x + 1)(4x + 1)}$

55. $\dfrac{3x + 7}{4x^2 + 25x - 21} - \dfrac{3x + 5}{4x^2 + 17x - 15}$ $\dfrac{-4x}{(4x - 3)(x + 7)(x + 5)}$

56. $\dfrac{x - 3}{3x^2 - 2x - 5} - \dfrac{x - 6}{3x^2 - 11x + 10}$ $\dfrac{12}{(3x - 5)(x + 1)(x - 2)}$

57. $\dfrac{x - 7}{6x^2 + 17x + 12} - \dfrac{x + 3}{12x^2 - 5x - 28}$ $\dfrac{2x^2 - 44x + 40}{(3x + 4)(2x + 3)(4x - 7)}$

58. $\dfrac{x + 3}{6x^2 - 7x - 5} - \dfrac{3x + 1}{9x^2 - 25}$ $\dfrac{-3x^2 + 9x + 14}{(3x - 5)(2x + 1)(3x + 5)}$

Mix 'Em Up!

Perform each operation. Write each answer in lowest terms.

59. $\dfrac{-3x}{3x + 8} + \dfrac{19x}{3x + 8}$ $\dfrac{16x}{3x + 8}$

60. $\dfrac{8x}{x + 7} - \dfrac{9x}{x + 7}$ $-\dfrac{x}{x + 7}$

61. $\dfrac{5x + 3}{2x + 7} - \dfrac{3x - 4}{2x + 7}$ 1

62. $\dfrac{22x - 9}{12x - 5} - \dfrac{10x - 4}{12x - 5}$ 1

63. $\dfrac{3x}{9x^2 - 25} - \dfrac{5}{9x^2 - 25}$

64. $\dfrac{8x}{64x^2 - 9} - \dfrac{3}{64x^2 - 9}$

65. $\dfrac{7x}{49x^2 - 25} + \dfrac{5}{49x^2 - 25}$

66. $\dfrac{x}{x^2 - 100} + \dfrac{10}{x^2 - 100}$

67. $\dfrac{-21x^2 + 35x}{3x^2 + x - 10} + \dfrac{18x^2 - 30x}{3x^2 + x - 10}$ $-\dfrac{x}{x + 2}$

68. $\dfrac{4y^2 - 7y}{4y^2 + 17y - 42} - \dfrac{20y^2 - 35y}{4y^2 + 17y - 42}$ $-\dfrac{4y}{y + 6}$

69. $\dfrac{11}{6x^2y} + \dfrac{15}{2xy^2}$ $\dfrac{11y + 45x}{6x^2y^2}$

70. $\dfrac{4}{3c^3d^2} - \dfrac{3}{2c^2d^4}$ $\dfrac{8d^2 - 9c}{6c^3d^4}$

71. $\dfrac{4}{2x - 3} - \dfrac{1}{6x}$ $\dfrac{22x + 3}{6x(2x - 3)}$

72. $\dfrac{-2}{5x - 2} + \dfrac{3}{7x}$ $\dfrac{x - 6}{7x(5x - 2)}$

73. $\dfrac{10}{2x - 3} - \dfrac{2}{3 - 2x}$ $\dfrac{12}{2x - 3}$

74. $\dfrac{18}{9x - 1} + \dfrac{3}{1 - 9x}$ $\dfrac{15}{9x - 1}$

75. $\dfrac{3x - 5}{2x - 7} + \dfrac{x + 2}{7 - 2x}$ 1

76. $\dfrac{2x - 3}{3x - 4} - \dfrac{x - 1}{4 - 3x}$ 1

77. $\dfrac{3x - 1}{4x^2 - 1} + \dfrac{10}{6x - 3}$

78. $\dfrac{x - 3}{9x^2 - 4} + \dfrac{5}{6x + 4}$

79. $\dfrac{3}{14r^4s^3} - \dfrac{5}{12r^5s}$

80. $\dfrac{-7}{ab^2} - \dfrac{5}{3a^2b}$ $\dfrac{-21a - 5b}{3a^2b^2}$

81. $\dfrac{3x - 8}{9x^2 + 9x + 2} - \dfrac{x + 3}{3x^2 - 11x - 4}$ $\dfrac{-31x + 26}{(3x + 1)(3x + 2)(x - 4)}$

82. $\dfrac{x + 5}{3x^2 + x - 10} - \dfrac{x}{3x^2 + 22x - 45}$ $\dfrac{12x + 45}{(3x - 5)(x + 2)(x + 9)}$

83. $\dfrac{1}{3x + 9} - \dfrac{7}{6x}$ $-\dfrac{5x + 21}{6x(x + 3)}$

84. $\dfrac{9}{6x - 8} - \dfrac{3}{10x}$ $\dfrac{18x + 6}{5x(3x - 4)}$

85. $\dfrac{3}{x^2 - 2x} + \dfrac{1}{x^2 - 4}$

86. $\dfrac{5}{x^2 - 3x} + \dfrac{4}{x^2 - 9}$

87. $\dfrac{2x + 4}{x^2 - 81} + \dfrac{7}{45 - 5x}$

88. $\dfrac{9x - 5}{36x^2 - 49} - \dfrac{1}{35 - 30x}$

89. $\dfrac{5x}{4x^2 + 4x + 1} - \dfrac{2x}{4x^2 - 1}$ $\dfrac{6x^2 - 7x}{(2x + 1)^2(2x - 1)}$

90. $\dfrac{2x}{9x^2 + 12x + 4} - \dfrac{x}{9x^2 - 4}$ $\dfrac{3x^2 - 6x}{(3x + 2)^2(3x - 2)}$

91. $\dfrac{4}{2x^2 - x - 3} + \dfrac{5}{x^2 - 1} - \dfrac{3}{2x^2 - 5x + 3}$

92. $\dfrac{1}{x^2 - 2x - 15} - \dfrac{2}{x^2 - 9} + \dfrac{3}{x^2 - 8x + 15}$

93. $\dfrac{4x}{x^3 + 8} + \dfrac{3x}{x^2 - 4}$

94. $\dfrac{4x}{x^3 - 27} + \dfrac{3x}{x^2 - 9}$

95. $\dfrac{3}{x^3 + 1} - \dfrac{1}{x^2 - 1} + \dfrac{1}{x^2 - x + 1}$ $\dfrac{4x - 5}{(x + 1)(x - 1)(x^2 - x + 1)}$

96. $\dfrac{2}{x^3 - 1} + \dfrac{3}{x^2 - 1} - \dfrac{3}{x^2 + x + 1}$ $\dfrac{5x + 8}{(x + 1)(x - 1)(x^2 + x + 1)}$

You Be the Teacher!

Correct each student's errors, if any.

97. Add: $\dfrac{5}{4x - 3} + \dfrac{3}{3 - 4x}$.

Jane's work:

$$\frac{5}{4x - 3} + \frac{3}{3 - 4x}$$
$$= \frac{5(3 - 4x)}{(4x - 3)(3 - 4x)} + \frac{3(4x - 3)}{(3 - 4x)(4x - 3)}$$
$$= \frac{15 - 20x + 12x - 9}{-16x^2 + 24x - 9}$$
$$= \frac{-8x + 6}{-16x^2 + 24x - 9}$$

98. Add: $\dfrac{3}{x^3 + 64} + \dfrac{2x}{x^2 - 16}$.

Marianne's work:

$$\frac{3}{x^3 + 64} + \frac{2x}{x^2 - 16} = \frac{3}{(x + 4)^3} + \frac{2x}{(x - 4)(x + 4)}$$
$$= \frac{3(x - 4)}{(x + 4)^3(x - 4)} + \frac{2(x + 4)^2}{(x - 4)(x + 4)^3}$$
$$= \frac{2x^2 + 19x + 20}{(x + 4)^3(x - 4)}$$

99. Subtract: $\dfrac{3}{12x^3y^4} - \dfrac{5}{18x^2y^5}$.

Al's work:

$$\frac{3}{12x^3y^4} - \frac{5}{18x^2y^5} = \frac{3x - 5y}{6x^2y^4}$$

$\dfrac{3}{12x^3y^4} - \dfrac{5}{18x^2y^5} = \dfrac{3(3y)}{36x^3y^5} - \dfrac{5(2x)}{36x^3y^5}$

$= \dfrac{9y - 10x}{36x^3y^5}$

100. Combine the expressions:

$$\frac{x+2}{x(x-1)}+\frac{x-2}{x(x+1)}-\frac{x}{x^2-1}.$$

Nick's work:

$$\frac{x+2}{x(x-1)}+\frac{x-2}{x(x+1)}-\frac{x}{x^2-1}$$

$$=\frac{x+2}{(x-1)^2}+\frac{x-2}{(x+1)^2}-\frac{x}{x^2-1}$$

$$=\frac{(x+2)(x+1)^2}{(x-1)^2(x+1)^2}+\frac{(x-2)(x-1)^2}{(x+1)^2(x-1)^2}-\frac{x(x^2-1)}{(x+1)^2(x-1)^2}$$

$$=\frac{x^3+4x^2+5x+2+x^3-4x^2+5x-2-x^3+x}{(x-1)^2(x+1)^2}$$

$$=\frac{x^3+11x}{(x-1)^2(x+1)^2}$$

Calculate It!

Use a calculator to verify each equation.

101. $\frac{7}{2x+6}-\frac{5}{7x}=\frac{39x-30}{14x(x+3)}$

102. $\frac{14}{4x-3}+\frac{13}{3-4x}=\frac{1}{4x-3}$

103. $\frac{6x+2}{9x^2-49}-\frac{5}{6x-14}=\frac{-3x-31}{2(3x+7)(3x-7)}$

104. $\frac{3x-5}{x^2+x-12}+\frac{3x-5}{5x^2-7x-24}=\frac{18x^2+6x-60}{(x-3)(x+4)(5x+8)}$

SECTION 7.4 Simplifying Complex Fractions

OBJECTIVES

As a result of completing this section, you will be able to

1. Simplify complex fractions using division (method 1).
2. Simplify complex fractions using multiplication by the LCD (method 2).
3. Solve problems involving complex fractions.
4. Troubleshoot common errors.

The function $C(t)=\frac{300t}{36t^2+3.3}$ describes the concentration, in micrograms per milliliter, of a drug in the bloodstream t hr after it is administered. Find the concentration of the drug 24 min after it is administered. To determine this information, we must evaluate the function for $t=\frac{24}{60}=\frac{2}{5}$ hr. This gives us

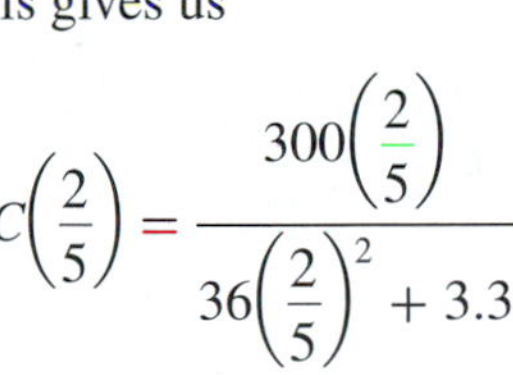

$$C\left(\frac{2}{5}\right)=\frac{300\left(\frac{2}{5}\right)}{36\left(\frac{2}{5}\right)^2+3.3}$$

This expression is a *complex fraction* because the numerator and denominator of the fraction contain a fraction. We will learn two different methods to simplify complex fractions.

Objective 1 Simplify complex fractions using division (method 1).

Simplifying Complex Fractions—Method 1

Definition: A **complex fraction** is a fraction in which the numerator and/or denominator of a fraction contains a fraction or rational expression.

Some examples of complex fractions are

$$\frac{\frac{1}{3}}{\frac{2}{5}} \qquad \frac{2-\frac{1}{4}}{\frac{1}{8}} \qquad \frac{\frac{6}{x}}{x+\frac{1}{x}}$$

Numerator of the complex fraction

Denominator of the complex fraction

Our goal is to simplify a complex fraction by writing it as a rational expression, $\frac{P}{Q}$, where P and Q ($Q \neq 0$) are polynomials with no common factors. In other words, our goal is to remove fractions from the numerator and denominator of the complex fraction.

The first method to simplify a complex fraction requires us to combine the terms in the numerator and the terms in the denominator and write them as a single fraction. Then we can apply the rules for dividing fractions.

Procedure: Simplifying Complex Fractions Using Division (Method 1)

Step 1: Add or subtract the fractions in the numerator and/or denominator of the complex fraction, if necessary. This will result in a fraction divided by a fraction.

Step 2: Divide the resulting fractions by multiplying the fraction in the numerator by the reciprocal of the denominator.

$$\frac{\frac{a}{b}}{\frac{c}{d}} = \frac{a}{b} \div \frac{c}{d} = \frac{a}{b} \cdot \frac{d}{c}$$

Step 3: Simplify the answer by dividing out any common factors.

Objective 1 Examples

Simplify each complex fraction using method 1.

1a. $\dfrac{\frac{4a^2}{5b}}{\frac{10ab}{6b^2}}$ **1b.** $\dfrac{\frac{7x-14}{6x+6}}{\frac{x^2-4}{x+1}}$ **1c.** $\dfrac{x+y}{\frac{1}{x}+\frac{1}{y}}$

Solutions

1a.

$$\frac{\frac{4a^2}{5b}}{\frac{10ab}{6b^2}} = \frac{4a^2}{5b} \div \frac{10ab}{6b^2}$$ Rewrite the complex fraction as a division problem.

$$= \frac{4a^2}{5b} \cdot \frac{6b^2}{10ab}$$ Multiply by the reciprocal of $\frac{10ab}{6b^2}$, which is $\frac{6b^2}{10ab}$.

$$= \frac{4a^2 \cdot 6b^2}{5b \cdot 10ab}$$ Multiply the rational expressions.

$$= \frac{24a^2b^2}{50ab^2}$$ Simplify.

$$= \frac{12a}{25}$$ Divide out the common factor, $2ab^2$.

1b.

$$\frac{\frac{7x-14}{6x+6}}{\frac{x^2-4}{x+1}} = \frac{7x-14}{6x+6} \div \frac{x^2-4}{x+1}$$ Rewrite the complex fraction as a division problem.

$$= \frac{7x-14}{6x+6} \cdot \frac{x+1}{x^2-4}$$ Multiply by the reciprocal of $\frac{x^2-4}{x+1}$, which is $\frac{x+1}{x^2-4}$.

$$= \frac{7(x-2)}{6(x+1)} \cdot \frac{x+1}{(x+2)(x-2)}$$ Factor each numerator and denominator.

$$= \frac{7\cancel{(x-2)}}{6\cancel{(x+1)}} \cdot \frac{\cancel{(x+1)}}{(x+2)\cancel{(x-2)}}$$ Divide out the common factors.

$$= \frac{7}{6(x+2)}$$ Multiply the remaining factors.

1c. $$\frac{x+y}{\frac{1}{x}+\frac{1}{y}} = \frac{x+y}{\frac{1}{x}\cdot\frac{y}{y}+\frac{1}{y}\cdot\frac{x}{x}}$$ Write the fractions in the denominator as equivalent fractions with an LCD of *xy*.

$$= \frac{x+y}{\frac{y}{xy}+\frac{x}{xy}}$$ Simplify the products in the denominator.

$$= \frac{x+y}{\frac{y+x}{xy}}$$ Add the fractions in the denominator.

$$= \frac{x+y}{1} \div \frac{y+x}{xy}$$ Rewrite the complex fraction as a division problem.

$$= \frac{x+y}{1} \cdot \frac{xy}{y+x}$$ Multiply by the reciprocal of $\frac{y+x}{xy}$, which is $\frac{xy}{y+x}$.

$$= \frac{\cancel{x+y}}{1} \cdot \frac{xy}{\cancel{y+x}}$$ Divide out the common factor, $x + y$.

$$= \frac{xy}{1}$$ Multiply the remaining factors.

$$= xy$$ Simplify.

Student Check 1 Simplify each complex fraction using method 1.

a. $\dfrac{\frac{3}{8y^2}}{\frac{15x}{14y}}$ **b.** $\dfrac{\frac{9x-12}{6x+9}}{\frac{9x^2-16}{2x+3}}$ **c.** $\dfrac{\frac{1}{a}+\frac{2}{b}}{2a+b}$

Simplifying Complex Fractions—Method 2

Objective 2 Simplify complex fractions using multiplication by the LCD (method 2).

Another method to simplify complex fractions is to eliminate the fractions from the numerator and denominator rather than combine the fractions in the numerator and denominator. We do this by multiplying the numerator and denominator of the complex fraction by the LCD of all the fractions in the complex fraction.

Procedure: Simplifying a Complex Fraction Using Multiplication (Method 2)

Step 1: Determine the LCD of all the fractions in the complex fraction.
Step 2: Multiply the numerator and denominator of the complex fraction by the LCD.
Step 3: Simplify the resulting fraction.

Objective 2 Examples Simplify each complex fraction using method 2.

2a. $\dfrac{\frac{4a^2}{5b}}{\frac{10ab}{6b^2}}$ **2b.** $\dfrac{x+y}{\frac{1}{x}+\frac{1}{y}}$ **2c.** $\dfrac{\frac{1}{y}-\frac{1}{xy}}{\frac{1}{x}-\frac{1}{x^2}}$

Solutions **2a.** The LCD of the fractions is $30b^2$.

$$\frac{\frac{4a^2}{5b}}{\frac{10ab}{6b^2}} = \frac{\left(\frac{4a^2}{5b}\right)\cdot(30b^2)}{\left(\frac{10ab}{6b^2}\right)\cdot(30b^2)}$$ Multiply the numerator and denominator by the LCD.

$$= \frac{4a^2\cdot 6b}{10ab\cdot 5}$$ Simplify the products $\left(\frac{30b^2}{5b} = 6b \text{ and } \frac{30b^2}{6b^2} = 5\right)$.

$$= \frac{24a^2b}{50ab}$$ Multiply the remaining factors.

$$= \frac{12a}{25}$$ Divide out the common factor, $2ab$.

2b. The LCD of the fractions is xy.

$$\frac{x+y}{\frac{1}{x}+\frac{1}{y}} = \frac{(x+y)\cdot xy}{\left(\frac{1}{x}+\frac{1}{y}\right)\cdot xy}$$ Multiply the numerator and denominator by the LCD.

$$= \frac{xy(x+y)}{\frac{1}{x}\cdot xy + \frac{1}{y}\cdot xy}$$ Apply the distributive property in the denominator.

$$= \frac{xy(x+y)}{y+x}$$ Simplify the products in the denominator.

$$= \frac{xy\cancel{(x+y)}}{\cancel{y+x}}$$ Divide out the common factor, $x + y$.

$$= xy$$ Simplify.

2c. The LCD of the fractions is x^2y.

$$\frac{\frac{1}{y}-\frac{1}{xy}}{\frac{1}{x}-\frac{1}{x^2}} = \frac{\left(\frac{1}{y}-\frac{1}{xy}\right)\cdot x^2y}{\left(\frac{1}{x}-\frac{1}{x^2}\right)\cdot x^2y}$$ Multiply the numerator and denominator by the LCD.

$$= \frac{\left(\frac{1}{y}\right)x^2y - \left(\frac{1}{xy}\right)x^2y}{\left(\frac{1}{x}\right)x^2y - \left(\frac{1}{x^2}\right)x^2y}$$ Apply the distributive property in the numerator and denominator.

$$= \frac{\left(\frac{1}{\cancel{y}}\right)\frac{x^2\cancel{y}}{1} - \left(\frac{1}{\cancel{xy}}\right)\frac{\cancel{x^2y}^{\,x}}{1}}{\left(\frac{1}{\cancel{x}}\right)\frac{\cancel{x^2}^{\,x}y}{1} - \left(\frac{1}{\cancel{x^2}}\right)\frac{\cancel{x^2}y}{1}}$$ Divide out the common factors of each product.

$$= \frac{x^2 - x}{xy - y}$$ Simplify each product.

$$= \frac{x(x-1)}{y(x-1)}$$ Factor the numerator and denominator.

$$= \frac{x}{y}$$ Divide out the common factor, $x - 1$.

Student Check 2 Simplify each complex fraction using method 2.

a. $\dfrac{\dfrac{3}{8y^2}}{\dfrac{15x}{14y}}$ **b.** $\dfrac{\dfrac{1}{a}+\dfrac{2}{b}}{2a+b}$ **c.** $\dfrac{\dfrac{1}{ab}+\dfrac{2}{a^2b}}{\dfrac{1}{b^2}+\dfrac{2}{ab^2}}$

Problems Involving Complex Fractions

Objective 3 Solve problems involving complex fractions.

Complex fractions can occur when we use different formulas, such as slope or midpoint, or when we evaluate functions. Complex fractions can also result from problems involving negative exponents. Recall that $x^{-n} = \dfrac{1}{x^n}$, provided $x \neq 0$.

Objective 3 Examples **Solve each problem.**

3a. The function $C(t) = \dfrac{300t}{36t^2 + 3.3}$ describes the concentration, in micrograms per milliliter (µg/mL), of a drug in the bloodstream t hr after it is administered. Find the concentration of the drug 24 min after it is administered.

Solution **3a.** We evaluate the function for $t = \dfrac{24}{60} = \dfrac{2}{5}$ hr.

$$C\left(\frac{2}{5}\right) = \frac{300\left(\frac{2}{5}\right)}{36\left(\frac{2}{5}\right)^2 + 3.3}$$ Replace t with $\frac{2}{5}$.

$$= \frac{120}{36\left(\frac{4}{25}\right) + 3.3}$$ Simplify the numerator and denominator.

$$= \frac{120}{\frac{144}{25} + \frac{33}{10}}$$ Simplify the product in the denominator and write 3.3 as $\frac{33}{10}$.

$$= \frac{(120)\cdot 50}{\left(\frac{144}{25}\right)\cdot 50 + \left(\frac{33}{10}\right)\cdot 50}$$ Multiply the numerator and denominator by the LCD, 50.

$$= \frac{6000}{288 + 165}$$ Simplify the products.

$$= \frac{6000}{453}$$ Add the denominators.

$$\approx 13.25$$ Approximate.

So, in 24 min, there will be approximately 13.25 µg/mL of the drug in the patient's bloodstream.

3b. Find the slope of the line that passes through the points $\left(\frac{1}{4}, -\frac{2}{3}\right)$ and $\left(-\frac{5}{6}, \frac{1}{12}\right)$.

Solution 3b.

INSTRUCTOR NOTE: Refer students to Chapter 4 if they need to review the concept of slope.

$$m = \frac{y_2 - y_1}{x_2 - x_1}$$ State the slope formula.

$$m = \frac{\frac{1}{12} - \left(-\frac{2}{3}\right)}{-\frac{5}{6} - \frac{1}{4}}$$ Substitute the appropriate values: $(x_1, y_1) = \left(\frac{1}{4}, -\frac{2}{3}\right)$ and $(x_2, y_2) = \left(-\frac{5}{6}, \frac{1}{12}\right)$

$$m = \frac{\left(\frac{1}{12} + \frac{2}{3}\right) \cdot 12}{\left(-\frac{5}{6} - \frac{1}{4}\right) \cdot 12}$$ Multiply the numerator and denominator by the LCD, 12.

$$m = \frac{\frac{1}{12} \cdot 12 + \frac{2}{3} \cdot 12}{-\frac{5}{6} \cdot 12 - \frac{1}{4} \cdot 12}$$ Apply the distributive property in the numerator and denominator.

$$m = \frac{1 + 8}{-10 - 3}$$ Simplify each product.

$$m = \frac{9}{-13}$$ Add.

$$m = -\frac{9}{13}$$ Simplify.

3c. Rewrite with positive exponents and then simplify: $\frac{2^{-2} + 3^{-1}}{6^{-1}}$.

Solution 3c. Recall $2^{-2} = \frac{1}{2^2} = \frac{1}{4}$, $3^{-1} = \frac{1}{3^1} = \frac{1}{3}$, and $6^{-1} = \frac{1}{6^1} = \frac{1}{6}$.

$$\frac{2^{-2} + 3^{-1}}{6^{-1}} = \frac{\frac{1}{4} + \frac{1}{3}}{\frac{1}{6}}$$ Rewrite each negative exponent with positive exponents.

$$= \frac{\frac{3}{12} + \frac{4}{12}}{\frac{1}{6}}$$ Convert the fractions in the numerator to equivalent fractions with the LCD, 12.

$$= \frac{\frac{7}{12}}{\frac{1}{6}}$$ Add the fractions in the numerator.

$$= \frac{7}{12} \cdot \frac{6}{1}$$ Divide the fractions by multiplying by the reciprocal of $\frac{1}{6}$.

$$= \frac{7}{2}$$ Simplify.

Student Check 3 Solve each problem.

a. Let $f(x) = \dfrac{x+3}{2x}$. Find $f\left(\dfrac{1}{3}\right)$.

b. Find the slope of the line that passes through the points $\left(\dfrac{3}{2}, \dfrac{1}{5}\right)$ and $\left(\dfrac{3}{10}, \dfrac{5}{4}\right)$.

c. Rewrite with positive exponents and then simplify: $\dfrac{4^{-2}+2^{-3}}{2^{-1}}$.

Objective 4

Troubleshoot common errors.

Troubleshooting Common Errors

Some common errors associated with complex fractions are shown.

Objective 4 Examples A problem and an incorrect solution are given. Provide the correct solution and an explanation of the error.

4a. Simplify $\dfrac{x+\dfrac{1}{x}}{\dfrac{2}{x^2}}$.

Incorrect Solution	Correct Solution and Explanation
$\dfrac{x+\dfrac{1}{x}}{\dfrac{2}{x^2}} = x + \dfrac{1}{x}\cdot\dfrac{x^2}{2} = x + \dfrac{x}{2}$	We must combine the fractions in the numerator before multiplying by the reciprocal of the fraction in the denominator. $\dfrac{x+\dfrac{1}{x}}{\dfrac{2}{x^2}} = \dfrac{\dfrac{x^2}{x}+\dfrac{1}{x}}{\dfrac{2}{x^2}} = \dfrac{\dfrac{x^2+1}{x}}{\dfrac{2}{x^2}} = \dfrac{x^2+1}{x}\cdot\dfrac{x^2}{2} = \dfrac{x(x^2+1)}{2}$

4b. Simplify $\dfrac{2^{-2}+3^{-1}}{6^{-1}}$.

Incorrect Solution	Correct Solution and Explanation
$\dfrac{2^{-2}+3^{-1}}{6^{-1}} = \dfrac{6^1}{2^2+3^1} = \dfrac{6}{4+3} = \dfrac{6}{7}$	Since the numerator is a sum, we must rewrite the terms as fractions and then simplify the resulting complex fraction. $\dfrac{2^{-2}+3^{-1}}{6^{-1}} = \dfrac{\dfrac{1}{4}+\dfrac{1}{3}}{\dfrac{1}{6}} = \dfrac{\dfrac{1}{4}\cdot 12+\dfrac{1}{3}\cdot 12}{\dfrac{1}{6}\cdot 12} = \dfrac{3+4}{2} = \dfrac{7}{2}$

ANSWERS TO STUDENT CHECKS

Student Check 1 **a.** $\frac{7}{20xy}$ **b.** $\frac{1}{3x+4}$ **c.** $\frac{1}{ab}$

Student Check 2 **a.** $\frac{7}{20xy}$ **b.** $\frac{1}{ab}$ **c.** $\frac{b}{a}$

Student Check 3 **a.** 5 **b.** $-\frac{7}{8}$ **c.** $\frac{3}{8}$

SUMMARY OF KEY CONCEPTS

1. There are two methods to simplify a complex fraction. Method 1 requires us to combine the numerators and/or denominators of the complex fractions into a single fraction so that we have one fraction divided by another fraction. Then we multiply by the reciprocal of the denominator to simplify the result.
2. Method 2 requires us to eliminate the fractions from the numerator and/or denominator of the complex fraction initially. This is done by multiplying the numerator and the denominator of the complex fraction by the LCD of all the fractions within the complex fraction. Combine any like terms and simplify the resulting fraction.

GRAPHING CALCULATOR SKILLS

When entering complex fractions in the calculator, it is important that parentheses are entered around the numerator and denominator of the complex fractions.

Example: Show that $\dfrac{\frac{1}{3}}{\frac{2}{5}} = \frac{5}{6}$.

Solution:

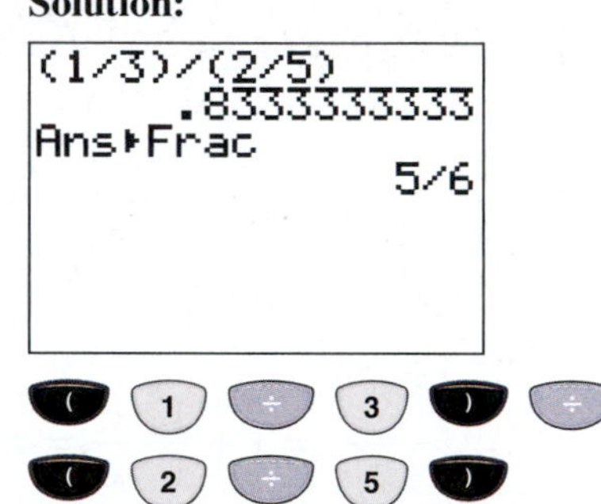

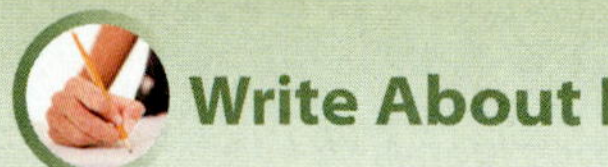

SECTION 7.4 EXERCISE SET

Write About It!

Use complete sentences in your answer to each exercise.

1. Explain how to simplify the complex fraction $\dfrac{\frac{6x^3}{20y^3}}{\frac{36x^4}{15y}}$ using method 1.

Rewrite the complex fraction by multiplying the fraction in the numerator by the reciprocal of the fraction in the denominator. Simplify by dividing out common factors.

2. Explain how to simplify the complex fraction $\dfrac{\frac{6x^3}{20y^3}}{\frac{36x^4}{15y}}$ using method 2.

3. Explain how to simplify the complex fraction $\dfrac{\frac{1}{a}+\frac{5}{b}}{10a+2b}$.

4. Find the mistakes in the work of simplifying the complex fraction $\dfrac{\frac{49}{x^2}-4}{\frac{7}{x}-2}$.

$$\dfrac{\frac{49}{x^2}-4}{\frac{7}{x}-2} = \dfrac{\frac{\cancel{49}^{7}}{\cancel{x^2}_{x}}-\cancel{4}^{2}}{\frac{\cancel{7}}{\cancel{x}}-\cancel{2}} = \frac{7}{x}-4$$

You cannot divide out terms in the complex fraction. You need to multiply the numerator and denominator by the LCD and simplify.

5. Explain why the solution is incorrect.

$$\frac{4x^{-2}-y^{-2}}{2x^{-1}-y^{-1}} \neq \frac{2x-y}{4x^2-y^2} = \frac{1}{4x-y}$$

You need to rewrite each negative exponent as a reciprocal and simplify the resulting complex fraction.

6. Explain why the solution is incorrect.

$$\dfrac{\frac{3}{5x-4}+\frac{3}{4}}{x} = \dfrac{\frac{3}{5x}-\frac{3}{4}+\frac{3}{4}}{x} = \frac{3}{5x^2}$$

You cannot split the first fraction in the numerator into two fractions.

Additional answers can be found in the Instructor Answer Appendix.

7. Explain how to simplify the complex fraction in Exercise 6. You need to multiply the numerator and denominator by the LCD and simplify.

8. Explain how to simplify the expression $\frac{3^{-1}+4^{-1}}{2^{-3}}$.

You need to rewrite each negative exponent as a reciprocal and use method 1 and method 2 to simplify.

Practice Makes Perfect!

Simplify each complex fraction using method 1. (*See Objective 1.*)

9. $\dfrac{\frac{7x}{10y^3}}{\frac{21x^2}{9y}}$ $\frac{3}{10xy^2}$

10. $\dfrac{\frac{3r}{7s^2}}{\frac{28r^3}{8s}}$ $\frac{6}{49r^2s}$

11. $\dfrac{\frac{3a}{10b^4}}{\frac{6a^2}{36b^3}}$ $\frac{9}{5ab}$

12. $\dfrac{\frac{1}{4y^3}}{\frac{16x}{24y^2}}$ $\frac{3}{8xy}$

13. $\dfrac{\frac{15a+12}{a-7}}{\frac{25a^2-16}{5a-35}}$ $\frac{15}{5a-4}$

14. $\dfrac{\frac{24b+16}{6b+8}}{\frac{9b^2-4}{9b+12}}$ $\frac{12}{3b-2}$

15. $\dfrac{\frac{8a}{4a+10}}{\frac{a^2}{6a+15}}$ $\frac{12}{a}$

16. $\dfrac{\frac{7b}{6b+12}}{\frac{2b^2}{3b+6}}$ $\frac{7}{4b}$

17. $\dfrac{\frac{8}{r}+\frac{5}{s}}{5r+8s}$ $\frac{1}{rs}$

18. $\dfrac{\frac{3}{a}+\frac{2}{b}}{2a+3b}$ $\frac{1}{ab}$

19. $\dfrac{\frac{5}{x}-\frac{3}{y}}{9x-15y}$ $-\frac{1}{3xy}$

20. $\dfrac{\frac{3}{r}-\frac{4}{s}}{20r-15s}$ $-\frac{1}{5rs}$

Simplify each complex fraction using method 2. (*See Objective 2.*)

21. $\dfrac{\frac{3x^2}{21y^3}}{\frac{35x^3}{28}}$ $\frac{4}{35xy^3}$

22. $\dfrac{\frac{12r^2}{2s}}{\frac{25r^4}{20}}$ $\frac{24}{5r^2s}$

23. $\dfrac{\frac{10a^3}{18b^3}}{\frac{20a^4}{21b^5}}$ $\frac{7b^2}{12a}$

24. $\dfrac{\frac{4r}{15s^2}}{\frac{36r^4}{40s^3}}$ $\frac{8s}{27r^3}$

25. $\dfrac{\frac{2}{x}-\frac{3}{y}}{9x-6y}$ $-\frac{1}{3xy}$

26. $\dfrac{\frac{1}{a}+\frac{7}{b}}{14a+2b}$ $\frac{1}{2ab}$

27. $\dfrac{\frac{8}{x}+\frac{1}{y}}{5x+40y}$ $\frac{1}{5xy}$

28. $\dfrac{\frac{3}{r}-\frac{7}{s}}{28r-12s}$ $-\frac{1}{4rs}$

29. $\dfrac{\frac{6}{rs}-\frac{3}{r^2s}}{\frac{12}{s^2}-\frac{6}{rs^2}}$ $\frac{s}{2r}$

30. $\dfrac{\frac{10}{ab}+\frac{8}{a^2b}}{\frac{15}{b^2}+\frac{12}{ab^2}}$ $\frac{2b}{3a}$

31. $\dfrac{\frac{7}{xy}-\frac{8}{x^2y}}{\frac{35}{y^2}-\frac{40}{xy^2}}$ $\frac{y}{5x}$

32. $\dfrac{\frac{24}{ab}+\frac{6}{a^2b}}{\frac{4}{b^2}+\frac{1}{ab^2}}$ $\frac{6b}{a}$

33. $\dfrac{\frac{21}{xy}+\frac{12}{x^2}}{\frac{7}{y^2}+\frac{4}{xy}}$ $\frac{3y}{x}$

34. $\dfrac{\frac{14}{ab}-\frac{7}{a^2}}{\frac{16}{b^2}-\frac{8}{ab}}$ $\frac{7b}{8a}$

Evaluate each rational function for the given x-value. (*See Objective 3.*)

35. $f(x)=\frac{3x+4}{7x-2}, f\left(\frac{4}{3}\right)$ $\frac{12}{11}$

36. $f(x)=\frac{7x-9}{3x+2}, f\left(\frac{6}{7}\right)$ $-\frac{21}{32}$

37. $f(x)=\frac{2x+1}{x+6}, f\left(-\frac{2}{7}\right)$ $\frac{3}{40}$

38. $f(x)=\frac{x-5}{3x-2}, f\left(-\frac{3}{4}\right)$ $\frac{23}{17}$

Find the slope of the line that passes through the given points. (*See Objective 3.*)

39. $\left(\frac{1}{3}, \frac{2}{9}\right)$ and $\left(-\frac{1}{5}, \frac{4}{9}\right)$ $-\frac{5}{12}$

40. $\left(\frac{3}{2}, -\frac{6}{7}\right)$ and $\left(\frac{3}{7}, \frac{5}{2}\right)$ $-\frac{47}{15}$

41. $\left(\frac{1}{10}, -\frac{2}{3}\right)$ and $\left(-\frac{2}{5}, -\frac{1}{10}\right)$ $-\frac{17}{15}$

42. $\left(\frac{3}{2}, -\frac{1}{7}\right)$ and $\left(-\frac{3}{10}, -\frac{1}{5}\right)$ $\frac{2}{63}$

Rewrite with positive exponents and then simplify each expression. (*See Objective 3.*)

43. $\frac{3^{-3}-6^{-2}}{4^{-2}}$ $\frac{4}{27}$

44. $\frac{2^{-1}-8^{-2}}{2^{-3}}$ $\frac{31}{8}$

45. $\frac{3^{-3}+3^{-2}}{2^{-1}}$ $\frac{8}{27}$

46. $\frac{6^{-2}-7^{-1}}{6^{-1}}$ $-\frac{29}{42}$

47. $\frac{4^{-2}-2^{-2}}{5^{-1}}$ $-\frac{15}{16}$

48. $\frac{4^{-2}+6^{-1}}{2^{-1}}$ $\frac{11}{24}$

Mix 'Em Up!

Simplify each complex fraction using method 1 or method 2.

49. $\dfrac{\frac{3x}{9y^4}}{\frac{24x^2}{18y^5}}$ $\frac{y}{4x}$

50. $\dfrac{\frac{4a}{12b^2}}{\frac{18a^3}{32b^4}}$ $\frac{16b^2}{27a^2}$

51. $\dfrac{\frac{3a}{4a+8}}{\frac{a^2}{8a+16}}$ $\frac{6}{a}$

52. $\dfrac{\frac{7x}{18x+30}}{\frac{x^2}{6x+10}}$ $\frac{7}{3x}$

53. $\dfrac{\frac{8x+4}{20x-35}}{\frac{4x^2-1}{28x-49}}$ $\frac{28}{10x-5}$

54. $\dfrac{\frac{5a-15}{6a-12}}{\frac{a^2-9}{8a-16}}$ $\frac{20}{3a+9}$

55. $\dfrac{\frac{6}{x}+\frac{5}{y}}{20x+24y}$ $\frac{1}{4xy}$

56. $\dfrac{\frac{3}{r}-\frac{1}{s}}{4r-12s}$ $-\frac{1}{4rs}$

57. $\dfrac{\frac{5}{a}-\frac{7}{b}}{21a-15b}$ $-\frac{1}{3ab}$

58. $\dfrac{\frac{6}{a}+\frac{1}{b}}{2a+12b}$ $\frac{1}{2ab}$

59. $\dfrac{\frac{8}{rs}+\frac{1}{r^2s}}{\frac{32}{s^2}+\frac{4}{rs^2}}$ $\frac{s}{4r}$

60. $\dfrac{\frac{14}{xy}+\frac{49}{x^2y}}{\frac{2}{y^2}+\frac{7}{xy^2}}$ $\frac{7y}{x}$

61. $\dfrac{\frac{2}{xy}-\frac{1}{x^2}}{\frac{12}{y^2}-\frac{6}{xy}}$ $\frac{y}{6x}$

62. $\dfrac{\frac{42}{ab}-\frac{49}{a^2}}{\frac{6}{b^2}-\frac{7}{ab}}$ $\frac{7b}{a}$

63. $\dfrac{\frac{9}{x^2}-\frac{25}{y^2}}{\frac{3}{x}+\frac{5}{y}}$ $\frac{3y-5x}{xy}$

64. $\dfrac{\frac{25}{r^2}-\frac{4}{s^2}}{\frac{5}{r}+\frac{2}{s}}$ $\frac{5s-2r}{rs}$

65. $\dfrac{\frac{16}{a^2}-\frac{1}{b^2}}{\frac{12}{a}-\frac{3}{b}}$ $\frac{4b+a}{3ab}$

66. $\dfrac{\frac{1}{x^2}-\frac{49}{y^2}}{\frac{5}{x}-\frac{35}{y}}$ $\frac{y+7x}{5xy}$

67. $\dfrac{r^{-2}-81s^{-2}}{2r^{-1}+18s^{-1}}$ $\frac{s-9r}{2rs}$

68. $\dfrac{100a^{-2}-b^{-2}}{30a^{-1}+3b^{-1}}$ $\frac{10b-a}{3ab}$

69. $\dfrac{x^{-2}-9y^{-2}}{5x^{-1}-15y^{-1}}$ $\frac{3x+y}{5xy}$

70. $\dfrac{a^{-2}-36b^{-2}}{4a^{-1}-24b^{-1}}$ $\frac{b+6a}{4ab}$

71. $\dfrac{\frac{2}{6x-1}+2}{x}$ $\frac{12}{6x-1}$

72. $\dfrac{\frac{1}{7y+6}-\frac{1}{6}}{y}$ $-\frac{7}{42y+36}$

73. $\dfrac{\frac{4}{2a-7}+\frac{4}{7}}{a}$ $\frac{8}{14a-49}$

74. $\dfrac{\frac{3}{b+2}-\frac{3}{2}}{b}$ $-\frac{3}{2b+4}$

Evaluate each rational function for the given *x*-value.

75. $f(x)=\dfrac{7x+5}{x-2}$, $f\left(\dfrac{1}{4}\right)$ $-\frac{27}{7}$

76. $f(x)=\dfrac{3x-5}{9x-4}$, $f\left(\dfrac{5}{2}\right)$ $\frac{5}{37}$

77. $f(x)=\dfrac{x}{4x+1}$, $f\left(-\dfrac{3}{7}\right)$ $\frac{3}{5}$

78. $f(x)=\dfrac{x}{8x+3}$, $f\left(-\dfrac{6}{7}\right)$ $\frac{2}{9}$

Find the slope of the line that passes through the given points.

79. $\left(\dfrac{1}{6},-\dfrac{5}{3}\right)$ and $\left(-4,\dfrac{2}{3}\right)$ $-\frac{14}{25}$

80. $\left(-\dfrac{5}{4},-6\right)$ and $\left(5,\dfrac{2}{7}\right)$ $\frac{176}{175}$

81. $\left(-\dfrac{1}{6},-\dfrac{3}{4}\right)$ and $\left(5,\dfrac{3}{2}\right)$ $\frac{27}{62}$

82. $\left(2,-\dfrac{7}{6}\right)$ and $\left(-\dfrac{4}{9},-\dfrac{3}{2}\right)$ $\frac{3}{22}$

Simplify each rational expression.

83. $\dfrac{8^{-1}-3^{-3}}{6^{-3}}$ 19

84. $\dfrac{6^{-3}-3^{-3}}{6^{-2}}$ $-\frac{7}{6}$

85. $\dfrac{8^{-2}-6^{-1}}{4^{-3}}$ $-\frac{29}{3}$

86. $\dfrac{9^{-2}+3^{-1}}{3^{-3}}$ $\frac{28}{3}$

You Be the Teacher!

Correct each student's errors, if any.

87. Simplify: $\dfrac{\frac{2}{x}+\frac{5}{y}}{10x+4y}$.

Jen's work:

$$\dfrac{\frac{2}{x}+\frac{5}{y}}{10x+4y}=\dfrac{\frac{2y+5x}{xy}}{2(5x+2y)}=\dfrac{2y+5x}{xy}\cdot\dfrac{2(2y+5x)}{1}$$
$$=\dfrac{(5x+2y)^2}{2xy}$$

88. Simplify: $\dfrac{\frac{3}{x}-\frac{2}{y}}{\frac{1}{x}+\frac{1}{y}}$.

Marie's work:

$$\dfrac{\frac{3}{x}-\frac{2}{y}}{\frac{1}{x}+\frac{1}{y}}=\left(\dfrac{3}{x}-\dfrac{2}{y}\right)(x+y)=\dfrac{3y-2x}{xy}(x+y)$$
$$=\dfrac{(x+y)(3y-2x)}{xy}$$

89. Simplify: $\dfrac{4x^{-2}-y^{-2}}{2x^{-1}-y^{-1}}$.

Al's work:

$$\dfrac{4x^{-2}-y^{-2}}{2x^{-1}-y^{-1}}=\dfrac{2x-y}{4x^2-y^2}=\dfrac{2x-y}{(2x-y)(2y+x)}=\dfrac{1}{2y+x}$$

90. Simplify: $\dfrac{7^{-2}+8^{-1}}{2^{-3}}$.

Sally's work:

$$\dfrac{7^{-2}+8^{-1}}{2^{-3}}=\dfrac{2^3}{7^2+8}=\dfrac{8}{49+8}=\dfrac{8}{57}$$

Use a calculator for each problem.

91. Let $f(x) = \frac{x-2}{4x-3}$. Find $f\left(-\frac{5}{8}\right)$. Find the error in the screen shot and explain why $-\frac{53}{16}$ is incorrect.

```
-5/8-2/4(-5/8)-3
              -3.3125
Ans▸Frac
              -53/16
```

92. Let $f(x) = \frac{6x+1}{x-4}$. Find $f\left(\frac{2}{13}\right)$. Find the error in the screen shot and explain why $-\frac{79}{26}$ is incorrect.

```
6*2/13+1/2/13-4
         -3.038461538
Ans▸Frac
              -79/26
```

93. Find the slope of the line that passes through $\left(-\frac{4}{3}, -\frac{1}{2}\right)$ and $\left(-\frac{5}{3}, 4\right)$.

94. Find the slope of the line that passes through $\left(\frac{1}{3}, \frac{5}{9}\right)$ and $\left(-7, -\frac{5}{2}\right)$.

95. Simplify $\frac{4^{-2}+8^{-2}}{2^{-3}}$.

96. Simplify $\frac{9^{-3}-4^{-1}}{6^{-2}}$.

PIECE IT TOGETHER — SECTIONS 7.1–7.4

Evaluate the expression for the given values. (*Section 7.1, Objective 1*)

1. $\frac{2x+5}{x-1}$, $x = 1, 1.01, 0.99$ undefined; 702; −698

Find the domain of the rational function. Write the answer in both set-builder and interval notation. (*Section 7.1, Objectives 2 and 3*)

2. $f(x) = \frac{4x^2-3}{2x+7}$ $\left\{x \,\middle|\, x \text{ is a real number and } x \neq -\frac{7}{2}\right\}$; $\left(-\infty, -\frac{7}{2}\right) \cup \left(-\frac{7}{2}, \infty\right)$

Simplify each rational expression. (*Section 7.1, Objective 3*)

3. $\frac{7x-1}{-1+7x}$ 1

4. $\frac{2x^2+8x}{7x^2+26x-8}$ $\frac{2x}{7x-2}$

5. $\frac{16x^2-28x-30}{64x^3+27}$ $\frac{2(2x-5)}{16x^2-12x+9}$

Multiply or divide the rational expressions. Write each answer in lowest terms. (*Section 7.1, Objectives 4 and 5*)

6. $\frac{x-12}{56x+63} \cdot \frac{48x+54}{x-12}$ $\frac{6}{7}$

7. $\frac{x^3-125}{25-x^2} \div \frac{3x^2+15x+75}{2x^2+7x-15}$ $-\frac{2x-3}{3}$

Divide and check by multiplying. (*Section 7.2, Objective 1*)

8. $\frac{2x^4-7x^3-4x^2}{6x^2}$ $\frac{1}{3}x^2 - \frac{7}{6}x - \frac{2}{3}$

Use long division or synthetic division to divide. (*Section 7.2, Objectives 2 and 3*)

9. $\frac{x^3-13x-12}{x-4}$ x^2+4x+3

10. $\frac{x^3-3x^2-48x-54}{x-9}$ x^2+6x+6

Use the remainder theorem and synthetic division to evaluate $P(x)$ for the specified x-value. (*Section 7.2, Objective 4*)

11. $P(x) = x^4 - 5x^3 + 8x^2 + 6x - 13$, $P(2)$ $P(2) = 7$

12. $P(x) = x^4 + 7x^2 - 10x - 12$, $P(-1)$ $P(-1) = 6$

Add or subtract the rational expressions. Write each answer in lowest terms. (*Section 7.3, Objectives 1 and 3*)

13. $\frac{7}{4x-2} + \frac{1}{8x}$ $\frac{30x-1}{8x(2x-1)}$

14. $\frac{x}{(3x-2)^2} + \frac{3x}{6x-4}$ $\frac{9x^2-4x}{2(3x-2)^2}$

15. $\frac{x+4}{x^2-25} - \frac{8x}{3x^2-16x+5} + \frac{1}{3x-1}$ $\frac{-4x^2-29x-29}{(x-5)(x+5)(3x-1)}$

Simplify each complex fraction using method 1 or method 2. (*Section 7.4, Objectives 1 and 2*)

16. $\frac{\frac{3a}{10b^4}}{\frac{6a^2}{36b^3}}$ $\frac{9}{5ab}$

17. $\frac{\frac{2}{xy}-\frac{3}{x^2y}}{\frac{10}{y^2}-\frac{15}{xy^2}}$ $\frac{y}{5x}$

Evaluate the rational function for the given x-value. (*Section 7.4, Objective 3*)

18. $f(x) = \frac{3x+4}{x-6}$, $f\left(-\frac{3}{7}\right)$ $-\frac{19}{45}$

Find the slope of the line that passes through the given points. (*Section 7.4, Objective 3*)

19. $\left(-\frac{2}{3}, \frac{1}{4}\right)$ and $\left(-\frac{7}{5}, -\frac{1}{2}\right)$ $\frac{45}{44}$

Simplify the rational expression. (*Section 7.4, Objective 3*)

20. $\frac{2^{-3}-3^{-2}}{3^{-2}}$ $\frac{1}{8}$

SECTION 7.5 Solving Rational Equations

▶ OBJECTIVES

As a result of completing this section, you will be able to

1. Solve rational equations.
2. Troubleshoot common errors.

In Sections 7.6 and 7.7, we will explore some applications of rational expressions which require us to solve equations involving these types of expressions.

Rational Equations

Objective 1 ▶

Solve rational equations.

At this point in the chapter, we have only performed operations with rational expressions. We now turn our attention to solving equations that contain rational expressions. These equations are called *rational equations*.

Definition: A **rational equation** is an equation involving rational expressions.

It is important to distinguish between rational expressions and rational equations. Some examples of rational expressions and rational equations are shown.

Rational Expressions	**Rational Equations**
$3 - \frac{2}{x}$	$3 - \frac{2}{x} = x$
$\frac{4}{x+2} - \frac{3}{x+2}$	$\frac{4}{x+2} - \frac{3}{x+2} = \frac{1}{2}$

Note: A rational equation must contain an equals sign.

We *simplify* rational expressions but we *solve* rational equations. In simplifying rational expressions, we use the LCD to add or subtract rational expressions as well as simplify complex fractions. To solve a rational equation, we use the LCD in a different way. We multiply each side of the equation by the LCD to eliminate fractions from the equation and then solve the resulting equation. This process is the method we also used in Chapter 2 when we solved linear equations containing fractions.

Rational equations are unlike any of the equations we have solved because rational equations contain variables in the denominator of the expression. This means that there can be values that make the equation undefined. These values must be excluded from the solution set of a rational equation.

Because there are values that make rational expressions undefined, we must check the solutions in the original equation. After we multiply each side of a rational equation by the LCD, we have a "new" type of equation—either a linear equation or a quadratic equation. These proposed solutions will satisfy the "new" equation, but they may not satisfy the original rational equation. The proposed solutions that do not satisfy the original equation are **extraneous solutions**.

Procedure: Solving Rational Equations

Step 1: Multiply each side of the equation by the LCD of the rational expressions in the equation.

Step 2: Solve the resulting equation.

Step 3: Check the proposed solutions in the original equation.

a. If the proposed solution makes one of the rational expressions undefined, then it is an extraneous solution and must be excluded.

b. If the proposed solution makes the original equation a true statement, then it is a solution of the equation.

Objective 1 Examples Solve each rational equation.

1a. $\frac{3}{x} + \frac{4}{x} = 2$ **1b.** $\frac{y-6}{3} = \frac{y+4}{2}$ **1c.** $\frac{k}{k-4} - 5 = \frac{4}{k-4}$

1d. $\frac{2}{t^2-4} = \frac{3}{t^2-2t}$ **1e.** $1 - \frac{5}{x} + \frac{6}{x^2} = 0$ **1f.** $\frac{r}{r-3} - \frac{1}{r+3} = \frac{6}{r^2-9}$

Solutions **1a.**

$$\frac{3}{x} + \frac{4}{x} = 2$$

$$x\left(\frac{3}{x} + \frac{4}{x}\right) = x(2)$$ Multiply each side by the LCD, x.

$$x\left(\frac{3}{x}\right) + x\left(\frac{4}{x}\right) = x(2)$$ Apply the distributive property.

$$3 + 4 = 2x$$ Simplify each product.

$$7 = 2x$$ Combine like terms.

$$\frac{7}{2} = x$$ Divide each side by 2.

Check:

$$\frac{3}{x} + \frac{4}{x} = 2$$ Original equation

$$\frac{3}{\frac{7}{2}} + \frac{4}{\frac{7}{2}} = 2$$ Replace x with $\frac{7}{2}$.

$$3 \cdot \frac{2}{7} + 4 \cdot \frac{2}{7} = 2$$ Simplify the complex fraction by multiplying by the reciprocals.

$$\frac{6}{7} + \frac{8}{7} = 2$$ Simplify each product.

$$\frac{14}{7} = 2$$ Add.

$$2 = 2$$ Simplify.

The proposed solution makes the original equation true. So, the solution set is $\left\{\frac{7}{2}\right\}$.

1b.

$$\frac{y-6}{3} = \frac{y+4}{2}$$

$$6\left(\frac{y-6}{3}\right) = 6\left(\frac{y+4}{2}\right)$$ Multiply each side by the LCD, 6.

$$2(y-6) = 3(y+4)$$ Divide out the common factors from each product.

$$2y - 12 = 3y + 12$$ Apply the distributive property.

$$2y - 12 - 2y = 3y + 12 - 2y$$ Subtract $2y$ from each side.

$$-12 = y + 12$$ Simplify.

$$-12 - 12 = y + 12 - 12$$ Subtract 12 from each side.

$$-24 = y$$ Simplify.

Check:

$$\frac{y-6}{3} = \frac{y+4}{2} \quad \text{Original equation}$$

$$\frac{-24-6}{3} = \frac{-24+4}{2} \quad \text{Replace } y \text{ with } -24.$$

$$\frac{-30}{3} = \frac{-20}{2} \quad \text{Add the terms in the numerators.}$$

$$-10 = -10 \quad \text{Simplify each quotient.}$$

The proposed solution makes the original equation true, so the solution set is $\{-24\}$.

1c.

$$\frac{k}{k-4} - 5 = \frac{4}{k-4}$$

$$(k-4)\left(\frac{k}{k-4} - 5\right) = (k-4)\left(\frac{4}{k-4}\right) \quad \text{Multiply each side by the LCD, } k-4.$$

$$(k-4)\left(\frac{k}{k-4}\right) - 5(k-4) = 4 \quad \text{Apply the distributive property.}$$

$$k - 5k + 20 = 4 \quad \text{Simplify each product.}$$

$$-4k + 20 = 4 \quad \text{Combine like terms.}$$

$$-4k + 20 - 20 = 4 - 20 \quad \text{Subtract 20 from each side.}$$

$$-4k = -16 \quad \text{Simplify.}$$

$$k = 4 \quad \text{Divide each side by } -4.$$

Check:

$$\frac{k}{k-4} - 5 = \frac{4}{k-4} \quad \text{Original equation}$$

$$\frac{4}{4-4} - 5 = \frac{4}{4-4} \quad \text{Replace } k \text{ with 4.}$$

$$\frac{4}{0} - 5 = \frac{4}{0} \quad \text{Simplify.}$$

Because 4 makes the rational expressions in the equation *undefined*, it cannot be a solution of the equation. Therefore, the equation has no solution. So, the solution set is the empty set, $\varnothing$.

1d. We first factor the denominators to determine the LCD.

$$\left.\begin{aligned} t^2 - 4 &= (t+2)(t-2) \\ t^2 - 2t &= t(t-2) \end{aligned}\right\} \text{LCD} = t(t-2)(t+2)$$

Now multiply each side of the equation by the LCD.

$$\frac{2}{(t+2)(t-2)} = \frac{3}{t(t-2)}$$

$$t(t-2)(t+2)\frac{2}{(t+2)(t-2)} = t(t-2)(t+2)\frac{3}{t(t-2)} \quad \text{Multiply each side by the LCD.}$$

$$2t = 3(t+2) \quad \text{Simplify each product.}$$

$$2t = 3t + 6 \quad \text{Apply the distributive property.}$$

$$2t - 3t = 3t + 6 - 3t \quad \text{Subtract } 3t \text{ from each side.}$$

$$-t = 6 \quad \text{Simplify.}$$

$$t = -6 \quad \text{Divide each side by } -1.$$

Check:

$\dfrac{2}{t^2-4}=\dfrac{3}{t^2-2t}$	Original equation
$\dfrac{2}{(-6)^2-4}=\dfrac{3}{(-6)^2-2(-6)}$	Replace t with -6.
$\dfrac{2}{36-4}=\dfrac{3}{36+12}$	Simplify each denominator.
$\dfrac{2}{32}=\dfrac{3}{48}$	Add the values in the denominators.
$\dfrac{1}{16}=\dfrac{1}{16}$	Simplify each fraction.

The proposed solution makes the equation true, so the solution set is $\{-6\}$.

1e.

$1-\dfrac{5}{x}+\dfrac{6}{x^2}=0$	
$x^2\left(1-\dfrac{5}{x}+\dfrac{6}{x^2}\right)=x^2(0)$	Multiply each side by the LCD, x^2.
$x^2(1)-x^2\left(\dfrac{5}{x}\right)+x^2\left(\dfrac{6}{x^2}\right)=0$	Apply the distributive property.
$x^2-5x+6=0$	Simplify each product.
$(x-3)(x-2)=0$	Factor the trinomial.
$x-3=0$ or $x-2=0$	Apply the zero products property.
$x=3$ $\quad$ $x=2$	Solve each equation.

Neither of the proposed solutions makes the rational equation undefined, so the solution set is $\{2, 3\}$. We can verify this by substituting these values in the original equation.

1f. Since $r^2-9=(r+3)(r-3)$, the LCD is $(r-3)(r+3)$.

$$\frac{r}{r-3}-\frac{1}{r+3}=\frac{6}{r^2-9}$$

$$(r-3)(r+3)\frac{r}{r-3}-(r-3)(r+3)\frac{1}{r+3}=(r-3)(r+3)\frac{6}{(r-3)(r+3)}$$

$$r(r+3)-1(r-3)=6$$

$$r^2+3r-r+3=6$$

$$r^2+2r+3=6$$

$$r^2+2r+3-6=6-6$$

$$r^2+2r-3=0$$

$$(r+3)(r-1)=0$$

$$r+3=0 \quad \text{or} \quad r-1=0$$

$$r=-3 \qquad\qquad r=1$$

The value -3 makes two of the rational expressions in the equation undefined, so it must be excluded from the solution set. Therefore, the solution set is $\{1\}$.

Student Check 1 Solve each rational equation.

a. $\frac{1}{x-3}+\frac{5}{x-3}=6$ **b.** $\frac{k+2}{4}=\frac{k-1}{3}$ **c.** $\frac{y}{y-6}-2=\frac{6}{y-6}$

d. $\frac{4}{t^2-1}=\frac{-2}{t^2+t}$ **e.** $1+\frac{3}{x}-\frac{4}{x^2}=0$ **f.** $\frac{a}{2a-1}+\frac{1}{a+5}=\frac{-11}{2a^2+9a-5}$

Objective 2 ▶ Troubleshoot common errors.

Troubleshooting Common Errors

The most common errors of solving rational equations occur from not checking solutions and from not distributing correctly when clearing fractions. Some examples are shown.

Objective 2 Examples A problem and an incorrect solution are given. Provide the correct solution and an explanation of the error.

2a. Solve $\frac{y}{y+2}-4=\frac{2}{y+2}$.

Incorrect Solution

$$\frac{y}{y+2}-4=\frac{2}{y+2}$$
$$(y+2)\left(\frac{y}{y+2}-4\right)=(y+2)\left(\frac{2}{y+2}\right)$$
$$y-4=2$$
$$y=6$$

Correct Solution and Explanation

We should multiply 4 by the LCD as well.

$$(y+2)\left(\frac{y}{y+2}\right)-4(y+2)=(y+2)\left(\frac{2}{y+2}\right)$$
$$y-4y-8=2$$
$$-3y-8=2$$
$$-3y=10$$
$$y=-\frac{10}{3}$$

So, the solution set is $\left\{-\frac{10}{3}\right\}$.

2b. Solve $\frac{x}{x+2}+\frac{4}{x-3}=\frac{20}{x^2-x-6}$.

Incorrect Solution

$$\frac{x}{x+2}+\frac{4}{x-3}=\frac{20}{x^2-x-6}$$
$$\frac{x}{x+2}+\frac{4}{x-3}=\frac{20}{(x+2)(x-3)}$$
$$(x+2)(x-3)\frac{x}{x+2}+(x+2)(x-3)\frac{4}{x-3}=(x+2)(x-3)\frac{20}{(x+2)(x-3)}$$
$$x^2-3x+4x+8=20$$
$$x^2+x-12=0$$
$$(x+4)(x-3)=0$$
$$x=-4 \quad \text{or} \quad x=3$$

The solution set is $\{-4, 3\}$.

Correct Solution and Explanation

The problem is solved correctly but the proposed solutions were not checked. The value 3 makes two of the rational expressions in the original equation undefined. So, this value must be excluded from the solution set. The solution set is $\{-4\}$.

ANSWERS TO STUDENT CHECKS

Student Check 1 **a.** $\{4\}$ **b.** $\{10\}$ **c.** $\varnothing$ **d.** $\left\{\frac{1}{3}\right\}$ **e.** $\{-4, 1\}$ **f.** $\{-2\}$

SUMMARY OF KEY CONCEPTS

1. The method to solve a rational equation involves clearing fractions from the equation by multiplying by the LCD. Once the fractions are cleared, we solve the resulting linear or quadratic equation. It is essential that the proposed solutions are checked in the original equation to make sure that extraneous solutions are discarded.

GRAPHING CALCULATOR SKILLS

We can use the TABLE function of a graphing calculator to check proposed solutions of rational equations.

Example: The proposed solutions of the equation $x + \dfrac{6}{x-2} = \dfrac{3x}{x-2} + 1$ are 4 and 2. Determine if these check in the original equation.

Solution: Enter the two sides of the equation into Y_1 and Y_2 as shown.

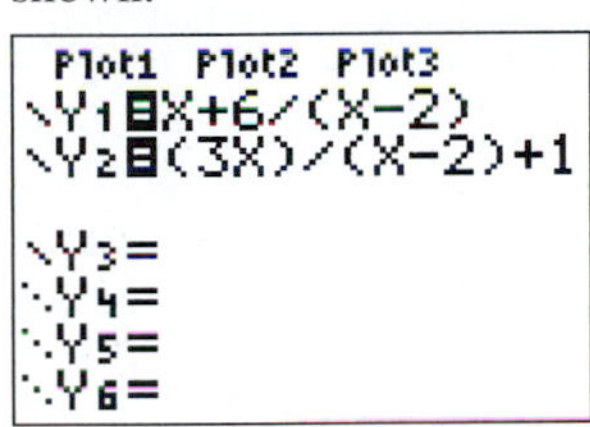

Now view the table at $x = 4$ and at $x = 2$.

X	Y1	Y2
0	-3	1
1	-5	-2
2	ERR:	ERR:
3	9	10
4	7	7
5	7	6
6	7.5	5.5

X=4

Since the two y-values are equal when $x = 4$, this confirms that 4 is a solution. Note that for $x = 2$, the y-values have the ERR: message. This means that 2 makes the rational expressions undefined and, therefore, cannot be a solution of the equation. From the table, we can see that the solution is 4.

SECTION 7.5 EXERCISE SET

Write About It!

Use complete sentences in your answer to each exercise.

1. Explain how to solve a rational equation of the form $\dfrac{c}{ax+b} = \dfrac{e}{cx+d}$. Multiply both sides by the LCD, $(ax + b)(cx + d)$ and solve the resulting equation.
2. Explain how to solve a rational equation of the form $a + \dfrac{b}{x} + \dfrac{c}{x^2} = 0$. Multiply both sides of the equation by the LCD, x^2 and solve the resulting quadratic equation.
3. When solving rational equations, explain the steps of removing the fractions from the equation.
4. When solving rational equations, explain why it is important to check the answers.
5. Explain how to find the least common denominator and solve the rational equation $\dfrac{2}{x^2 - 4x + 4} - \dfrac{1}{x-2} = \dfrac{4}{x^2 - x - 2}$.

Practice Makes Perfect!

Solve each rational equation. (*See Objective 1.*)

6. $\dfrac{2a+3}{3} = \dfrac{6a-5}{5}$ $\left\{\frac{15}{4}\right\}$
7. $\dfrac{b-7}{3} = \dfrac{7b+2}{4}$ $\{-2\}$
8. $\dfrac{10x-3}{5} = \dfrac{8x+1}{2}$ $\left\{-\frac{11}{20}\right\}$
9. $\dfrac{4y-7}{3} = \dfrac{5y+6}{2}$ $\left\{-\frac{32}{7}\right\}$
10. $\dfrac{x}{x-2} - 5 = -\dfrac{10}{x-2}$ $\{5\}$
11. $\dfrac{y}{y+9} + 2 = \dfrac{9}{y+9}$ $\{-3\}$
12. $\dfrac{x}{x+1} + 4 = -\dfrac{1}{x+1}$ $\varnothing$
13. $\dfrac{y}{y+3} + 8 = -\dfrac{3}{y+3}$ $\varnothing$
14. $\dfrac{a}{a+1} - 8 = -\dfrac{7}{a+1}$ $-\frac{1}{7}$
15. $\dfrac{b}{b+3} - 6 = \dfrac{3}{b+3}$ $-\frac{21}{5}$
16. $\dfrac{2}{s} = 10 + \dfrac{8}{s}$ $\left\{-\frac{3}{5}\right\}$
17. $\dfrac{3}{r} = 2 - \dfrac{4}{r}$ $\left\{\frac{7}{2}\right\}$
18. $\dfrac{7}{x} = -4 - \dfrac{6}{x}$ $\left\{-\frac{13}{4}\right\}$
19. $\dfrac{3}{y} = 7 + \dfrac{5}{y}$ $\left\{-\frac{2}{7}\right\}$

Additional answers can be found in the Instructor Answer Appendix.

20. $\frac{r^2}{r+2} = \frac{4}{r+2}$ $\{2\}$

21. $\frac{36s^2}{6s+1} = \frac{1}{6s+1}$ $\left\{\frac{1}{6}\right\}$

22. $\frac{9x^2}{3x-2} = \frac{4}{3x-2}$ $\left\{-\frac{2}{3}\right\}$

23. $\frac{25y^2}{5y-4} = \frac{16}{5y-4}$ $\left\{-\frac{4}{5}\right\}$

24. $3 - \frac{5}{x} - \frac{8}{x^2} = 0$ $\left\{-1, \frac{8}{3}\right\}$

25. $1 + \frac{17}{y} + \frac{70}{y^2} = 0$ $\{-10, -7\}$

26. $28 - \frac{1}{r} - \frac{2}{r^2} = 0$ $\left\{-\frac{1}{4}, \frac{2}{7}\right\}$

27. $5 + \frac{43}{s} - \frac{18}{s^2} = 0$ $\left\{-9, \frac{2}{5}\right\}$

28. $\frac{4}{x} + \frac{3}{x-7} = \frac{21}{x^2-7x}$ $\varnothing$

29. $\frac{2}{y} - \frac{6}{y+4} = \frac{24}{y^2+4y}$ $\varnothing$

30. $\frac{1}{r} - \frac{8}{r+6} = \frac{9}{r^2+6r}$

31. $\frac{3}{s} - \frac{7}{s+3} = \frac{1}{s^2+3s}$ $\{2\}$

32. $\frac{2}{a} + \frac{8}{a-2} = \frac{5}{a^2-2a}$

33. $\frac{1}{b} - \frac{5}{b+7} = \frac{10}{b^2+7b}$

34. $\frac{3}{9r^2-16} = -\frac{10}{3r^2+4r}$

35. $\frac{4}{49s^2-16} = -\frac{6}{7s^2+4s}$

36. $\frac{8x}{x+1} - \frac{7}{x-1} = -\frac{5}{x^2-1}$ $\left\{2, -\frac{1}{8}\right\}$

37. $\frac{3y}{y-5} + \frac{4}{y+5} = -\frac{26}{y^2-25}$ $\left\{-6, -\frac{1}{3}\right\}$

38. $\frac{20r}{r-3} - \frac{19}{r+3} = \frac{37}{r^2-9}$

39. $\frac{24s}{s-1} + \frac{1}{s+1} = \frac{24}{s^2-1}$

40. $\frac{3x}{x-6} + \frac{8}{x+6} = -\frac{96}{x^2-36}$ $\left\{-\frac{8}{3}\right\}$

41. $\frac{9x}{x+2} + \frac{40}{x-2} = \frac{72}{x^2-4}$ $\left\{-\frac{4}{9}\right\}$

42. $\frac{9a}{a-1} - \frac{2}{a+1} = \frac{4}{a^2-1}$ $\left\{\frac{2}{9}\right\}$

43. $\frac{8b}{b-3} - \frac{5}{b+3} = \frac{30}{b^2-9}$ $\left\{\frac{5}{8}\right\}$

44. $\frac{2}{5x+9} + \frac{3}{3x-2} = \frac{2}{15x^2+17x-18}$ $\{-1\}$

45. $\frac{4}{3y-5} + \frac{4}{7y+4} = \frac{1}{21y^2-23y-20}$ $\left\{\frac{1}{8}\right\}$

46. $\frac{5}{2a-7} - \frac{10}{4a+1} = \frac{1}{8a^2-26a-7}$ $\varnothing$

47. $\frac{4}{2b-1} - \frac{6}{3b+5} = \frac{1}{6b^2+7b-5}$ $\varnothing$

48. $\frac{6}{s-2} - \frac{1}{2s+1} = \frac{10}{2s^2-3s-2}$ $\left\{\frac{2}{11}\right\}$

49. $\frac{5}{t-4} + \frac{1}{t+7} = \frac{8}{t^2+3t-28}$ $\left\{-\frac{23}{6}\right\}$

52. $\frac{10x}{x-1} - \frac{27}{x+1} = \frac{33}{x^2-1}$ $\left\{2, -\frac{3}{10}\right\}$

53. $\frac{20r}{r-1} + \frac{23}{r+1} = -\frac{44}{r^2-1}$ $\left\{-\frac{3}{4}, -\frac{7}{5}\right\}$

54. $\frac{x}{x+10} + 3 = -\frac{10}{x+10}$ $\varnothing$

55. $\frac{y}{y-7} + 4 = \frac{7}{y-7}$ $\varnothing$

56. $\frac{9s^2}{3s-8} = \frac{64}{3s-8}$

57. $\frac{16s^2}{4s-3} = \frac{9}{4s-3}$ $\left\{-\frac{3}{4}\right\}$

58. $9 - \frac{18}{x} - \frac{16}{x^2} = 0$

59. $3 - \frac{5}{y} - \frac{2}{y^2} = 0$ $\left\{-\frac{1}{3}, 2\right\}$

60. $\frac{2r+5}{5} = \frac{3r+4}{4}$ $\{0\}$

61. $\frac{10s-3}{3} = \frac{s-4}{4}$ $\{0\}$

62. $\frac{1}{x} - \frac{10}{x+3} = \frac{9}{x^2+3x}$

63. $\frac{2}{y} + \frac{8}{y-2} = \frac{8}{y^2-2y}$ $\left\{\frac{6}{5}\right\}$

64. $\frac{2}{s} + \frac{1}{s+1} = \frac{2}{s^2+s}$ $\varnothing$

65. $\frac{2}{s} - \frac{4}{s+4} = \frac{16}{s^2+4s}$ $\varnothing$

66. $\frac{8}{a} = -4 - \frac{6}{a}$ $\left\{-\frac{7}{2}\right\}$

67. $\frac{9}{y} = 5 - \frac{8}{y}$ $\left\{\frac{17}{5}\right\}$

68. $\frac{y}{y+2} + \frac{11}{y-2} = \frac{8}{y^2-4}$ $\{-7\}$

69. $\frac{8x}{x-5} + \frac{9}{x+5} = -\frac{90}{x^2-25}$ $\left\{-\frac{9}{8}\right\}$

70. $\frac{2a+14}{a^2-3a} - \frac{6}{a-3} = \frac{2}{a^2-3a}$ $\varnothing$

71. $\frac{2b-7}{b^2-5b} - \frac{2}{b-5} = \frac{11}{b^2-5b}$ $\varnothing$

72. $\frac{6}{49x^2-36} = \frac{1}{7x^2-6x}$ $\{-6\}$

73. $\frac{3}{9y^2-4} = \frac{7}{3y^2+2y}$ $\left\{\frac{7}{9}\right\}$

74. $\frac{1}{r-7} - \frac{7}{9r-2} = \frac{5}{9r^2-65r+14}$ $\{-21\}$

75. $\frac{4}{5s-4} + \frac{5}{4s-3} = \frac{9}{20s^2-31s+12}$ $\{1\}$

76. $\frac{5}{x} = 3 + \frac{7}{x}$ $\left\{-\frac{2}{3}\right\}$

77. $\frac{10}{y} = 6 + \frac{5}{y}$ $\left\{\frac{5}{6}\right\}$

78. $\frac{2}{x+10} + \frac{1}{4x-3} = \frac{2}{4x^2+37x-30}$ $\left\{-\frac{2}{9}\right\}$

79. $\frac{5}{10y+9} + \frac{7}{y+2} = \frac{1}{10y^2+29y+18}$ $\left\{-\frac{24}{25}\right\}$

Mix 'Em Up!

Solve each rational equation.

50. $\frac{x}{x+5} - 2 = \frac{3}{x+5}$ $\{-13\}$

51. $\frac{y}{y+1} - 3 = \frac{9}{y+1}$ $\{-6\}$

You Be the Teacher!

Correct each student's errors, if any.

80. Solve $\frac{y}{y-7} - 3 = \frac{9}{y-7}$.

Sue's work:

$$\frac{y}{y-7} - 3 = \frac{9}{y-7}$$
$$y - 3y - 7 = 9$$
$$-2y - 7 = 9$$
$$-2y = 16$$
$$y = -8$$

$$\frac{y}{y-7} - 3 = \frac{9}{y-7}$$
$$y - 3(y-7) = 9$$
$$y - 3y + 21 = 9$$
$$-2y = -12$$
$$y = 6$$

The solution set is $\{6\}$.

81. Solve $5 - \frac{8}{x} - \frac{21}{x^2} = 0$.

Mike's work:

$$5 - \frac{8}{x} - \frac{21}{x^2} = 0$$
$$5x - 8 - 21x = 0$$
$$-16x = 8$$
$$x = -\frac{1}{2}$$

$$5 - \frac{8}{x} - \frac{21}{x^2} = 0$$
$$5x^2 - 8x - 21 = 0$$
$$(5x + 7)(x - 3) = 0$$
$$x = -\frac{7}{5}, x = 3$$

The solution set is $\left\{-\frac{7}{5}, 3\right\}$.

82. Solve $\frac{2}{x} + \frac{1}{x+5} = -\frac{5}{x^2 + 5x}$.

Ed's work:

$$\frac{2}{x} + \frac{1}{x+5} = -\frac{5}{x^2 + 5x}$$
$$2(x + 5) + x = -5$$
$$2x + 10 + x = -5$$
$$3x = -15$$
$$x = -5$$

$$\frac{2}{x} + \frac{1}{x+5} = -\frac{5}{x^2 + 5x}$$
$$\frac{2}{(-5)} + \frac{1}{(-5)+5} = -\frac{5}{(-5)^2 + 5(-5)}$$
$$-\frac{2}{5} + \frac{1}{0} = -\frac{5}{0}$$

It doesn't check. There is no solution: $\varnothing$.

83. Solve $\frac{4x + 28}{x^2 - x} + \frac{5}{x - 1} = \frac{10}{x^2 - x}$.

Kelly's work:

$$\frac{4x + 28}{x^2 - x} + \frac{5}{x - 1} = \frac{10}{x^2 - x}$$
$$(4x + 28)(x - 1) + 5(x^2 - x) = 10(x - 1)$$
$$4x^2 + 24x - 28 + 5x^2 - 5x = 10x - 10$$
$$9x^2 + 9x - 18 = 0$$
$$9(x^2 + x - 2) = 0$$
$$9(x + 2)(x - 1) = 0$$
$$x = 9, x = -2, x = 1$$

Calculate It!

Solve each equation and then use a calculator to check the proposed solutions.

84. $\frac{x}{x-9} - 4 = \frac{6}{x-9}$.

85. $\frac{6}{x} = 2 + \frac{2}{x}$

86. $\frac{4}{x} - \frac{6}{x+9} = \frac{10}{x^2 + 9x}$

87. $10 - \frac{33}{x} - \frac{28}{x^2} = 0$

SECTION 7.6 Applications of Rational Equations

OBJECTIVES

As a result of completing this section, you will be able to

1. Solve rational equations for a specified variable.
2. Solve applications involving number relationships.
3. Solve applications involving proportions.
4. Solve applications involving work.
5. Solve applications involving distance.

When Iason returns to the United States from England, he has 245 euros in his wallet. If the conversion rate is 1 euro = \$1.42, how much money does Iason have in U.S. dollars? To solve this problem, we can set up and solve a rational equation.

Rewriting Rational Equations

Objective 1 Solve rational equations for a specified variable.

It is often necessary to rewrite a mathematical formula in terms of another variable. We will solve formulas that involve rational expressions for a specified variable by using the skills of Section 7.5.

Procedure: Solving a Rational Equation for a Specified Variable

Step 1: Multiply each side of the equation by the LCD to clear fractions from the equation.
Step 2: Clear all grouping symbols by applying the distributive property.
Step 3: Simplify each side of the equation.
Step 4: Use the addition property of equality to get all of the terms containing the specified variable on one side of the equation and all other terms on the opposite side.
Step 5: If there is more than one term containing the specified variable, apply the distributive property to factor this variable out.
Step 6: Apply the multiplication property of equality to solve for the specified variable.

Objective 1 Examples **Solve each rational equation for the specified variable.**

1a. $a = \frac{GM}{r^2}$ for M (acceleration of gravity) **1b.** $\frac{1}{a} + \frac{1}{b} = \frac{1}{c}$ for b

Solutions **1a.**

$$a = \frac{GM}{r^2}$$ Highlight the variable to isolate.

$$r^2(a) = r^2\left(\frac{GM}{r^2}\right)$$ Multiply each side of the equation by the LCD, r^2.

$$ar^2 = GM$$ Simplify each product.

$$\frac{ar^2}{G} = \frac{GM}{G}$$ Divide each side by G.

$$\frac{ar^2}{G} = M$$ Simplify.

1b.

$$\frac{1}{a} + \frac{1}{b} = \frac{1}{c}$$ Highlight the variable to isolate.

$$abc\left(\frac{1}{a} + \frac{1}{b}\right) = abc\left(\frac{1}{c}\right)$$ Multiply each side of the equation by the LCD, abc.

$$abc\left(\frac{1}{a}\right) + abc\left(\frac{1}{b}\right) = abc\left(\frac{1}{c}\right)$$ Apply the distributive property.

$$bc + ac = ab$$ Simplify each product.

$$bc + ac - bc = ab - bc$$ Subtract bc from each side.

$$ac = ab - bc$$ Simplify.

$$ac = b(a - c)$$ Factor b out of the terms on the right side.

$$\frac{ac}{a - c} = \frac{b(a - c)}{a - c}$$ Divide each side by the factor $a - c$.

$$\frac{ac}{a - c} = b$$ Simplify.

Student Check 1 Solve each rational equation for the specified variable.

a. $r^2 = \frac{V}{ph}$ for p **b.** $\frac{1}{r} - \frac{1}{s} = \frac{1}{t}$ for r

Objective 2 ▶

Solve applications involving number relationships.

Applications Involving Number Relationships

Rational equations occur in different types of applications. We will look at examples involving proportions, work hours, distance, and number relationships.

Objective 2 Example **A number plus three times its reciprocal is 4. Find the number.**

Solution Let x be a number. Then $\frac{1}{x}$ is its reciprocal and $3\left(\frac{1}{x}\right) = \frac{3}{x}$ is three times its reciprocal. The sum of these two expressions is 4. The equation is

A number + 3 times its reciprocal is 4.

$$x + \frac{3}{x} = 4$$ Translate the relationship.

$$x\left(x + \frac{3}{x}\right) = x(4)$$ Multiply each side by the LCD, x.

$$x(x) + x\left(\frac{3}{x}\right) = 4x$$ Apply the distributive property.

$$x^2 + 3 = 4x$$ Simplify each product.

$$x^2 + 3 - 4x = 4x - 4x$$ Subtract $4x$ from each side.

$$x^2 - 4x + 3 = 0$$ Simplify.

$$(x - 3)(x - 1) = 0$$ Factor.

$$x - 3 = 0 \quad \text{or} \quad x - 1 = 0$$ Apply the zero products property.

$$x = 3 \qquad x = 1$$ Solve each equation.

So, the numbers 1 and 3 satisfy the given relationship.

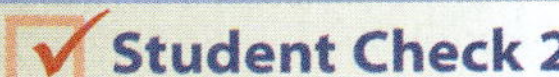

Student Check 2 The sum of a number and five times its reciprocal is 6. Find the number.

Objective 3 Solve applications involving proportions.

Proportions

A **ratio** is a quotient of two quantities, $\frac{A}{B}$, where $B \neq 0$. Rational expressions are ratios since they are quotients of two quantities. *Proportions* come from setting two ratios equal to one another.

Definition: A **proportion** is an equation of the form $\frac{A}{B} = \frac{C}{D}$.

Our goal in using proportions is to set up a known ratio equal to an unknown ratio. Generally, the numerators of the two ratios represent one quantity and the denominator represents another quantity.

Objective 3 Example When Iason returns to the United States from England, he has 245 euros in his wallet. If the conversion rate is 1 euro = \$1.42, how much money does Iason have in U.S. dollars?

Solution Let the ratio be euros to U.S. dollars, $\frac{\text{euros}}{\text{U.S. dollars}}$. Let x represent the amount in U.S. dollars. The known ratio is the currency rate: 1 euro = \$1.42 can be expressed as $\frac{1}{1.42}$. The unknown ratio is $\frac{245 \text{ euros}}{x \text{ dollars}}$.

$$\frac{1}{1.42} = \frac{245}{x}$$ Set the known ratio equal to the unknown ratio.

$$1.42x\left(\frac{1}{1.42}\right) = 1.42x\left(\frac{245}{x}\right)$$ Multiply each side by the LCD, $1.42x$.

$$x = 1.42(245)$$ Simplify each product.

$$x = \$347.90$$ Multiply.

So, Iason has the equivalent of \$347.90 in U.S. dollars.

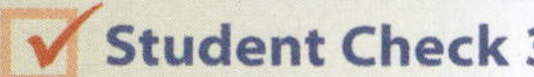

Student Check 3 How many U.S. dollars is equivalent to 3000 Pakistani rupees if the conversion rate is 1 U.S. dollar is equal to 61.2 Pakistan rupees?

Objective 4 Solve applications involving work.

Work Applications

Work applications involve finding the amount of time it takes two individuals to complete a task when working together. If a person can do a job in 5 hr, then in 1 hr, $\frac{1}{5}$ of the job is completed. We add up the portion of the job that can be completed in 1 hr by each person working alone and set it equal to the portion of the job that can be completed in 1 hr by their combined efforts.

Property: Work Equation

Work equations are of the form

$$\frac{1}{a} + \frac{1}{b} = \frac{1}{t}$$

where a is the time for one person to complete the job working alone, b is the time for another person to complete the job working alone, and t is the time for the two people working together to complete the job.

Objective 4 Example Suppose it takes Harold 5 hr to rake and bag the leaves in his yard when he works alone. His son, Carrey, can complete the job by himself in 3 hr. How long will it take Harold and Carrey to complete the job if they work together?

Solution Let t represent the time it takes for Harold and Carrey to complete the job together.

	Harold Alone	Carrey Alone	Harold and Carrey Together
Time to complete the job	5 hr	3 hr	t hr
Portion of job completed in 1 hr	$\frac{1}{5}$	$\frac{1}{3}$	$\frac{1}{t}$

$$\frac{1}{5} + \frac{1}{3} = \frac{1}{t}$$ Work equation

$$15t\left(\frac{1}{5} + \frac{1}{3}\right) = 15t\left(\frac{1}{t}\right)$$ Multiply each side by the LCD, $15t$.

$$15t\left(\frac{1}{5}\right) + 15t\left(\frac{1}{3}\right) = 15$$ Apply the distributive property.

$$3t + 5t = 15$$ Simplify each product.

$$8t = 15$$ Combine like terms.

$$t = \frac{15}{8}$$ Divide each side by 8.

$$t = 1\frac{7}{8}$$ Convert the fraction to a mixed number.

So, Harold and Carrey can complete the job together in approximately 1 hr 53 min.

Student Check 4 Helena thoroughly cleans newly constructed homes for a real estate company. Alone, Helena can clean a 2400-ft^2 home in 4 hr. Her partner Kenneth can clean the same size home in 6 hr. How long will it take them to clean the home if they work together?

Objective 5 ▶

Solve applications involving distance.

Distance Applications

Our last application involves rates. A **rate** is a ratio of two different quantities, such as distance to time. A rate that we are most likely familiar with is the speed of a car, usually given as miles/hour or kilometers/hour, depending on the country where one lives. We can see that rate (speed) is equal to distance (miles) divided by time (hours). This is true no matter what unit is being used for distance or time.

Property: Distance-Rate-Time

If d is the distance a car travels, t is the time, and r is the rate, then $d = r \times t$.

Solving for r, we obtain the formula $r = \frac{d}{t}$.

Solving for t, we obtain the formula $t = \frac{d}{r}$.

Objective 5 Example **Drew drives 200 mi in the same amount of time it takes Nancy to drive 150 mi. If Drew drives 15 mph faster than Nancy, find Drew's speed and Nancy's speed.**

Solution Let r represent Nancy's speed. Complete the chart with the information we know.

	Distance (miles)	Rate (mph)	Time (hours) $t = \frac{d}{r}$
Drew	200	$r + 15$ Drew traveled 15 mph faster than Nancy	$\frac{200}{r + 15}$
Nancy	150	r	$\frac{150}{r}$

Since we know that Drew and Nancy travel for the same amount of time, we set their times equal to obtain the equation.

$$\frac{200}{r + 15} = \frac{150}{r} \quad \text{Set times equal to one another.}$$

$$r(r + 15)\left(\frac{200}{r + 15}\right) = r(r + 15)\left(\frac{150}{r}\right) \quad \text{Multiply each side by the LCD, } r(r + 15).$$

$$200r = 150(r + 15) \quad \text{Simplify each product.}$$

$$200r = 150r + 2250 \quad \text{Apply the distributive property.}$$

$$200r - 150r = 150r + 2250 - 150r \quad \text{Subtract } 150r \text{ from each side.}$$

$$50r = 2250 \quad \text{Simplify.}$$

$$r = 45 \quad \text{Divide each side by 50.}$$

So, Nancy traveled at 45 mph and Drew traveled at $45 + 15 = 60$ mph.

Student Check 5 Courtney runs 3.5 mi in the same amount of time it takes Tammy to run 6 mi. If Courtney ran 2 mi/min slower than Tammy, how fast did each woman run?

ANSWERS TO STUDENT CHECKS

Student Check 1 **a.** $p = \frac{V}{hr^2}$ **b.** $r = \frac{st}{t + s}$

Student Check 2 The numbers are 1 and 5.

Student Check 3 $49.02

Student Check 4 2.4 hr or 2 hr 24 min

Student Check 5 Tammy ran at a rate of 4.8 mph and Courtney ran at a rate of 2.8 mph.

SUMMARY OF KEY CONCEPTS

1. The techniques used to solve rational equations are the same techniques that we use to solve a formula for a specified variable.
2. A proportion is an equation in which two ratios are equal. A ratio is the quotient of two quantities. Proportions are solved by multiplying each side by the LCD.
3. To solve a proportion, set a known ratio equal to an unknown ratio. In these ratios, let the numerators represent one quantity and the denominators represent the other quantity.
4. To solve work problems, we either find how long it will take one person working alone or two people working together to complete a job. Represent the portion of the job that each person can do in 1 hr. Put the equation in the form $\frac{1}{a} + \frac{1}{b} = \frac{1}{t}$, where a is one person's time alone, b is another person's time alone, and t is the time working together.
5. To solve distance problems, use the relationship that $d = rt$. Assign appropriate variables to the unknown quantities and then use the formula to write an equation that represents the situation.

SECTION 7.6 EXERCISE SET

Write About It!

Use complete sentences in your answer to each exercise.

1. When solving a rational equation for a specific variable, explain the steps of how to get rid of the denominators.
2. If David can do a job in x hr, explain the meaning of $\frac{1}{x}$. It means the portion of the job completed by David in 1 hr.
3. Explain the difference between a ratio and a proportion. A ratio is a fraction and a proportion is an equation in which two ratios are equal.
4. If Melissa can complete a job in x hr and Jenny can do the same job in y hr, explain the meaning of $\frac{1}{x} + \frac{1}{y}$.
5. If Tammi drives x mi in t hr, explain how to find her speed. Since rate is $\frac{\text{distance}}{\text{time}}$, Tammi's speed is $\frac{x}{t}$ mph.
6. If Nicole drives x mph and Mark drives y mph slower than Nicole, explain how to find Mark's speed.

Practice Makes Perfect!

Solve each rational equation for the specified variable. (*See Objective 1.*)

7. $R = \frac{K}{st^2}$ for s $s = \frac{K}{Rt^2}$
8. $Z = \frac{p}{wv^2}$ for p $p = v^2wZ$
9. $P = \frac{h}{rq}$ for q $q = \frac{h}{Pr}$
10. $t = \frac{s}{KR}$ for K $K = \frac{s}{Rt}$
11. $\frac{10}{x} = \frac{3}{y} + \frac{5}{z}$ for y
12. $\frac{9}{a} = \frac{2}{c} + \frac{12}{b}$ for b
13. $\frac{5}{r} = \frac{3}{s} - \frac{2}{t}$ for s
14. $\frac{6}{a} = \frac{10}{b} - \frac{7}{c}$ for c
15. $m = \frac{y_2 - y_1}{x_2 - x_1}$ for y_1 $y_1 = y_2 - mx_2 + mx_1$
16. $m = \frac{y_2 - y_1}{x_2 - x_1}$ for x_1
17. $c = \frac{ab}{a - b}$ for a $a = \frac{bc}{c - b}$
18. $c = \frac{ab}{a + b}$ for b $b = \frac{ac}{a - c}$
19. $s = \frac{a - br}{1 - r}$ for r $r = \frac{s - a}{s - b}$
20. $s = \frac{a - br}{1 - r}$ for b $b = \frac{a - s + rs}{r}$
21. $\frac{P_1V_1}{R_1} = \frac{P_2V_2}{R_2}$ for V_1
22. $\frac{P_1V_1}{R_1} = \frac{P_2V_2}{R_2}$ for R_2

Solve each problem. (*See Objective 2.*)

23. A number plus eight times its reciprocal is 6. Find the number. 2 and 4
24. A number plus nine times its reciprocal is 10. Find the number. 1 and 9
25. A number minus seven times its reciprocal is −6. Find the number. 1 and −7
26. A number minus ten times its reciprocal is −3. Find the number. 2 and −5

Solve each problem using a proportion. (*See Objective 3.*)

27. When Seth returns to the United States from Japan, he has 2300 yen. If the conversion rate is 1 U.S. dollar = 94.382 yen, how much money in U.S. dollars does Seth have? \$24.37
28. When Hung returns to the United States from China, he has 2400 yuan. If the conversion rate is 1 U.S. dollar = 6.8313 yuan, how much money in U.S. dollars does Hung have? \$351.32
29. When Amanda returns to the United States from England, she has 100 British pounds. If the conversion rate is 1 U.S. dollar = 0.6058 British pounds, how much money in U.S. dollars does Amanda have? \$165.07
30. When Paul returns to the United States from Canada, he has 300 Canadian dollars. If the conversion rate is 1 U.S. dollar = 1.0812 Canadian dollars, how much money in U.S. dollars does Paul have? \$277.47

Additional answers can be found in the Instructor Answer Appendix.

Solve each work problem. (*See Objective 4.*)

31. It takes Barry 4 hr to paint his house by himself when working alone. Christy can complete the job by herself in 11 hr. How long will it take Barry and Christy to complete the job if they work together? 2 hr 56 min

32. It takes Mary 2 hr to clean a 1200-ft^2 apartment when working alone. John can complete the job by himself in 6 hr. How long will it take Mary and John to complete the job if they work together? 1 hr 30 min

33. It takes Joann 4 hr to trim the bushes around the house by herself when working alone. Tim, her son, can complete the job by himself in 2 hr. How long will it take Joann and Tim to complete the job if they work together? 1 hr 20 min

34. It takes Hong 8 hr to do filing at the office when working alone. Shan can complete the job by herself in 7 hr. How long will it take Hong and Shan to complete the job if they work together? 3 hr 44 min

Solve each distance problem. (*See Objective 5.*)

35. Joan runs 3.5 mi in the same amount of time it takes Barbara to run 6 mi. If Joan ran 3 mph slower than Barbara, how fast did each person run? Barbara at 7.2 mph and Joan at 4.2 mph

36. Marian runs 2 mi in the same amount of time it takes Geri to run 4 mi. If Marian ran 4 mph slower than Geri, how fast did each person run? Geri at 8 mph and Marian at 4 mph

Mix 'Em Up!

Solve each rational equation for the specified variable.

37. $q = \frac{r}{Ph^3}$ for P $\quad P = \frac{r}{h^3 q}$

38. $w = \frac{p}{zV^2}$ for p $\quad p = v^2wZ$

39. $\frac{8}{r} = \frac{14}{s} - \frac{1}{t}$ for s

40. $\frac{4}{r} = \frac{9}{s} - \frac{2}{t}$ for r $\quad r = \frac{4st}{9t - 2s}$

41. $m = \frac{y_2 - y_1}{x_2 - x_1}$ for y_2 $\quad y_2 = y_1 + mx_2 - mx_1$

42. $m = \frac{y_2 - y_1}{x_2 - x_1}$ for x_2

43. $c = \frac{ab}{3a - 2b}$ for a

44. $c = \frac{ab}{4a + 3b}$ for b $\quad b = \frac{4ac}{a - 3c}$

45. $A = \frac{1}{2}h(a + b)$ for b

46. $A = \frac{1}{2}h(a + b)$ for h $\quad h = \frac{2A}{a + b}$

Solve each problem.

47. When Jon returns to the United Stated from Israel, he has 1400 Israeli new sheqalim. If the conversion rate is 1 U.S. dollar = 3.7975 Israeli new sheqalim, how much money in U.S. dollars does Jon have? $368.66

48. When Luis returns to the United States from Mexico, he has 2150 pesos. If the conversion rate is 1 U.S. dollar = 12.833 pesos, how much money in U.S. dollars does Luis have? $167.54

49. A number plus six times its reciprocal is 5. Find the number. 2 and 3

50. A number minus five times its reciprocal is −4. Find the number. 1 and −5

51. Linda runs 2.5 mi in the same amount of time it takes Maria to run 3.5 mi. If Linda ran 1 mph slower than Maria, how fast did each person run? Maria at 3.5 mph and Linda at 2.5 mph

52. Jan runs 5 mi in the same amount of time it takes Gary to run 6 mi. If Jan ran 0.5 mph slower than Gary, how fast did each person run? Gary at 3 mph and Jan at 2.5 mph

53. It takes Hugh 5 hr to rake and bag the leaves in the yard when working alone. Chris, his son, can complete the job by himself in 7 hr. How long would it take Hugh and Chris to complete the job if they work together? 2 hr 55 min

54. It takes Teri 4 hr to do a job when working alone. Joan can complete the job by herself in 1 hr. How long would it take Teri and Joan to complete the job if they work together? 48 min

55. If 6 out of 15 homes in a community have wells for their water supply, how many homes have wells in a community of 18,000 homes? 7200 homes

56. If 16 out of 20 homes in a community have wells for their water supply, how many homes have wells in a community of 15,000 homes? 12,000 homes

You Be the Teacher!

Correct each student's errors, if any.

57. It takes Howard 10 hr to do a job when working alone. Keith can do the same job by himself in 6 hr. Find the time it takes for Howard and Keith to complete the job when working together.

Josh's work:

$$\frac{10 + 6}{2} = \frac{16}{2} = 8$$

58. Nathan runs 3.5 mi in the same amount of time it takes Chris to run 6.5 mi. If Nathan ran 4 mph slower that Chris, how fast did each person run?

Roy's work: Let Nathan's speed be x mph and Chris's speed be $x - 4$ mph.

$$\frac{3.5}{x} = \frac{6.5}{x - 4}$$

$$3.5x - 14 = 6.5x$$

$$-3x = 14$$

$$x = -\frac{14}{3}$$

Let Chris's speed be x mph and Nathan's speed be $x - 4$ mph.

$$\frac{3.5}{x - 4} = \frac{6.5}{x}$$

$$3.5x = 6.5x - 26$$

$$-3x = -26$$

$$x = \frac{26}{3} = 8\frac{2}{3}$$

Chris's speed was $8\frac{2}{3}$ mph and Nathan's speed was $4\frac{2}{3}$ mph.

SECTION 7.7 Variation and Applications

OBJECTIVES

As a result of completing this section, you will be able to

1. Solve problems involving direct variation.
2. Solve problems involving inverse variation.
3. Solve problems involving joint and combined variation.
4. Troubleshoot common errors.

The cost of a house in Florida is directly proportional to the size of the house. If a 2600-ft^2 house costs \$195,000, then what is the cost of a 3500-ft^2 house? We will learn a formula for direct variation to solve this problem.

Objective 1 Solve problems involving direct variation.

Direct Variation

Direct variation may be new terminology, but it is a relationship that we have used many times before. Here are some examples of direct variation and their mathematical models.

Distance traveled is directly proportional to the time traveled.	$d = 60t$
Salary varies directly as hours worked.	$s = 15h$
The circumference of a circle varies directly as the diameter of the circle.	$C = \pi d$
The sales tax on a car is directly proportional to the price of the car.	$t = 0.07p$

Each of the preceding equations is written in a similar form: one quantity is equal to a constant times another quantity.

Definition: Direct Variation

We say *y varies directly as x*, or *y is directly proportional to x*, if

$$y = kx$$

for some nonzero constant k, called the **constant of variation** or **constant of proportionality**.

Another way to think of direct variation is that y is directly proportional to x if $\frac{y}{x} = k$, where k is a nonzero constant. In other words, the quotient of y and x is a constant.

Procedure: Solving Problems Involving Direct Variation

Step 1: If y varies directly as x, then $y = kx$.
Step 2: Solve for the constant of variation by substituting the known values of x and y.

Objective 1 Examples Use direct variation to solve each problem.

1a. Suppose y varies directly as x. If y is 10 when x is 5, find the constant of variation and the direct variation equation.

Solution **1a.** Since y varies directly as x, we can write $y = kx$. To find the constant of variation, let $x = 5$ and $y = 10$ and solve for k.

$y = kx$ State the direct variation equation.

$10 = k(5)$ Replace x with 5 and y with 10.

$\frac{10}{5} = k$ Divide each side by 5.

$2 = k$ Simplify.

So, the constant of variation is 2 and the variation equation is $y = 2x$.

1b. The cost of a house in Florida is directly proportional to the size of the house. If a 2600-ft² house costs \$195,000, then what is the cost of a 3500-ft² house?

Solution **1b.** Let c represent the cost of a house and s represent its size. Since c is directly proportional to s, we know that $c = ks$.

We use the fact that a 2600-ft² house costs \$195,000 to find the constant of variation.

$$c = ks \quad \text{Write the variation equation.}$$

$$195{,}000 = k(2600) \quad \text{Replace } c \text{ with 195,000 and } k \text{ with 2600.}$$

$$\frac{195{,}000}{2600} = k \quad \text{Divide each side by 2600.}$$

$$75 = k \quad \text{Simplify.}$$

So, the variation equation is $c = 75s$.

We use this equation to find the cost of a 3500-ft² house.

$$c = 75s \quad \text{Write the variation equation.}$$

$$c = 75(3500) \quad \text{Replace } s \text{ with 3500.}$$

$$c = 262{,}500 \quad \text{Simplify.}$$

So, the cost of a 3500-ft² house is \$262,500.

1c. Determine if each table shows that y varies directly as x. If the table represents direct variation, find the variation equation.

Table 1

x	y
3	2
6	4
9	6
12	8

Table 2

x	y
1	5
2	10
4	15
8	20

Solution **1c.** For y to be directly proportional to x, $\frac{y}{x}$ must be constant.

Table 1			Table 2		
x	y	$\frac{y}{x}$	x	y	$\frac{y}{x}$
3	2	$\frac{2}{3}$	1	5	$\frac{5}{1} = 5$
6	4	$\frac{4}{6} = \frac{2}{3}$	2	10	$\frac{10}{2} = 5$
9	6	$\frac{6}{9} = \frac{2}{3}$	4	15	$\frac{15}{4}$
12	8	$\frac{8}{12} = \frac{2}{3}$	8	20	$\frac{20}{8} = \frac{5}{2}$

In Table 1, we see that for each ordered pair, $\frac{y}{x} = \frac{2}{3}$, so y varies directly as x. So, the variation equation is $y = \frac{2}{3}x$.

In Table 2, the first two entries have $\frac{y}{x} = 5$ but the last two entries do not satisfy this relationship. Therefore, Table 2 is not an example of direct variation.

1d. Which graph shows y varies directly as x? If the graph represents direct variation, find the variation equation.

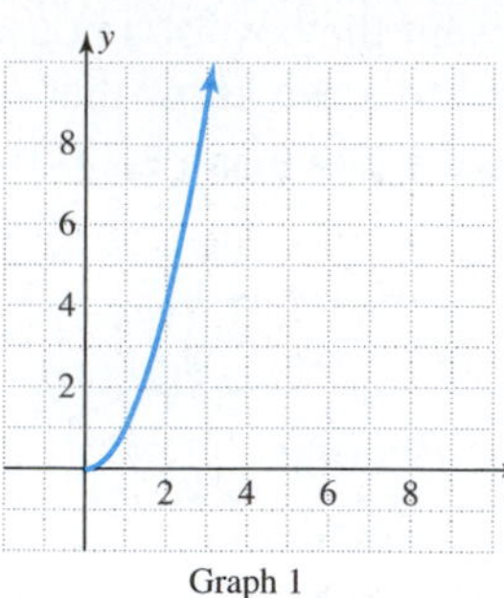
Graph 1

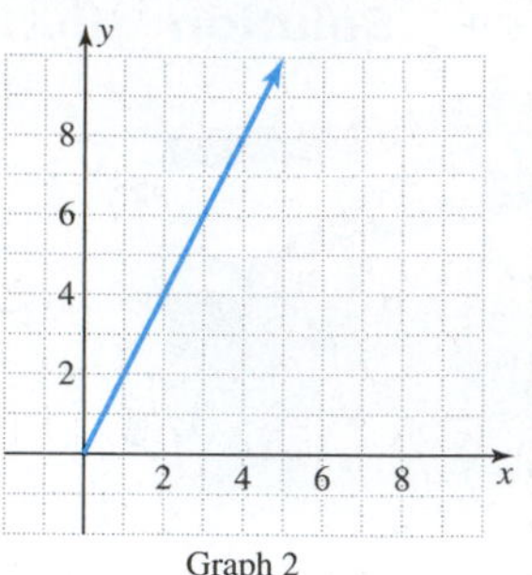
Graph 2

Solution **1d.** Direct variation is an equation of the form $y = kx$. The graph of this type of equation is a straight line with slope $m = k$ and y-intercept $(0, 0)$. In other words, it is a straight line through the origin.

Graph 2 is a line through the origin and so y is directly proportional to x. To find the constant of proportionality, we need to find $\frac{y}{x}$ for points on the line. The points $(1, 2)$, $(2, 4)$, $(3, 6)$, . . . , are on the line, so the constant of proportionality is $\frac{6}{3} = \frac{4}{2} = \frac{2}{1} = 2$. Therefore, the variation equation is $y = 2x$.

Student Check 1 Use direct variation to solve each problem.

a. Suppose y varies directly as x. If $y = -4$ when $x = 1$, find the constant of variation and the variation equation.

b. Hooke's law of elasticity states that the distance a spring stretches vertically is directly proportional to the weight on the end of the spring. If a 10-lb weight attached to the spring stretches the spring 2 in., find the distance that a 35-lb weight stretches the spring.

c. Determine if each table shows that y varies directly as x. If the table represents direct variation, find the variation equation.

Table 1

x	y
1	2
2	4
3	8
4	16

Table 2

x	y
1	−2
2	−4
4	−8
8	−16

d. Which graph shows y varies directly as x? If the graph represents direct variation, find the variation equation.

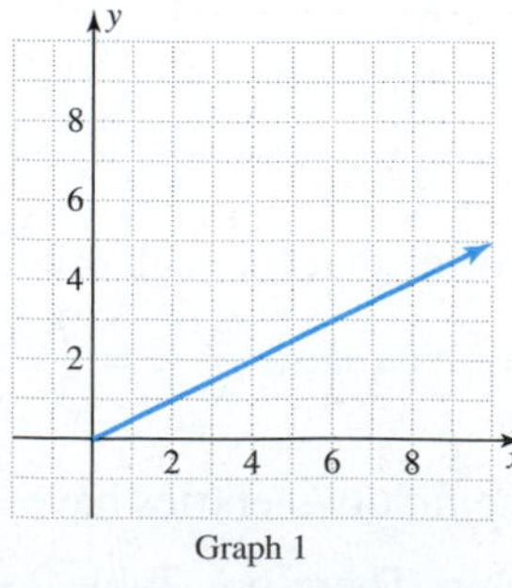
Graph 1

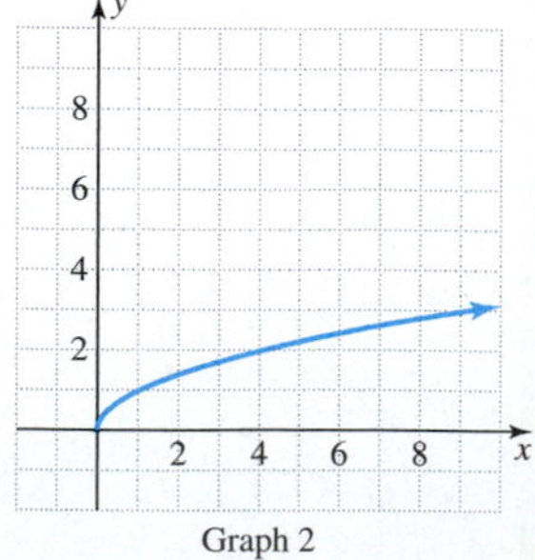
Graph 2

Inverse Variation

Objective 2 Solve problems involving inverse variation.

When the values of two quantities change in an opposite manner, they are inversely related. So, if one quantity increases, the other will decrease. Some examples of this relationship are shown in the following table.

For a fixed distance, the time at which one travels is inversely proportional to the average rate.	$t = \dfrac{100}{r}$
For a fixed area, the length of a rectangle is inversely proportional to its width.	$l = \dfrac{50}{w}$
For a salaried employee, the hourly wage, w, is inversely proportional to the hours worked, h.	$w = \dfrac{1500}{h}$

Two quantities x and y are said to be *inversely proportional* to each other if y is proportional to the *reciprocal* of x.

Definition: Inverse Variation

We say y *varies inversely as* x, or y *is indirectly proportional to* x, if

$$y = \frac{k}{x}$$

for some nonzero constant k, called the *constant of variation* or *constant of proportionality*.

Another way to think of inverse variation is that y is inversely proportional to x if $xy = k$, where k is a nonzero constant. In other words, the product and x and y is a constant.

Procedure: Solving Problems Involving Inverse Variation

Step 1: If y varies inversely as x, then $y = \dfrac{k}{x}$.

Step 2: Solve for the constant of variation by substituting the known values of x and y.

Objective 2 Examples Use indirect variation to solve each problem.

2a. Suppose y varies inversely as x. If $y = 5$ when $x = 30$, find the constant of proportionality and the variation equation.

Solution **2a.** Since y varies inversely as x, $y = \dfrac{k}{x}$. To find the constant of proportionality, let $y = 5$ and $x = 30$.

$$y = \frac{k}{x} \quad \text{Write the variation equation.}$$

$$5 = \frac{k}{30} \quad \text{Replace } y \text{ with 5 and } x \text{ with 30.}$$

$$30(5) = 30\left(\frac{k}{30}\right) \quad \text{Multiply each side by 30.}$$

$$150 = k \quad \text{Simplify.}$$

So, the constant of proportionality is 150 and the variation equation is $y = \dfrac{150}{x}$.

2b. Ohm's law states that the current, I, in an electrical conductor varies inversely as the resistance, R, of the conductor. If the current is 6 amperes (amp) when the resistance is 170 ohms (Ω), what is the current when the resistance is 34 Ω?

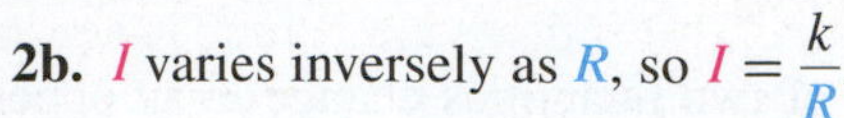

Solution **2b.** I varies inversely as R, so $I = \frac{k}{R}$.

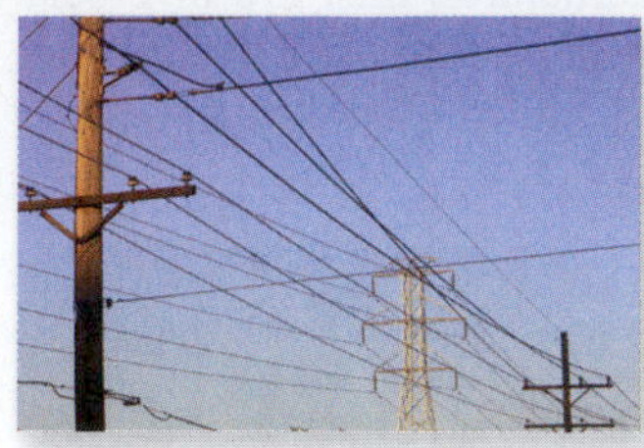

$$I = \frac{k}{R}$$ Write the variation equation.

$$6 = \frac{k}{170}$$ Replace I with 6 and R with 170.

$$170(6) = 170\left(\frac{k}{170}\right)$$ Multiply each side by 170.

$$1020 = k$$ Simplify.

So, the constant of variation is 1020 and the variation equation is $I = \frac{1020}{R}$.

We use this equation to find the current when the resistance is 34 Ω.

$$I = \frac{1020}{R}$$ State the variation equation.

$$I = \frac{1020}{34}$$ Replace R with 34.

$$I = 30$$ Simplify.

So, the current is 30 amp when the resistance is 34 Ω.

2c. Determine if each table shows that y is inversely proportional to x. Find the variation equation for a table that has this relationship.

Table 1

x	y
1	8
2	4
4	2
16	$\frac{1}{2}$

Table 2

x	y
1	16
2	32
4	64
8	128

Solution **2c.** For a set of points to be inversely related, their product must be a constant.

Table 1

x	y	xy
1	8	$1(8) = 8$
2	4	$2(4) = 8$
4	2	$4(2) = 8$
16	$\frac{1}{2}$	$16\left(\frac{1}{2}\right) = 8$

Table 2

x	y	xy
1	16	$1(16) = 16$
2	32	$2(32) = 64$
4	64	$4(64) = 256$
8	128	$8(128) = 1024$

In Table 1, the product $xy = 8$ for each pair of points. So, y is inversely proportional to x and the variation equation is $y = \frac{8}{x}$.

In Table 2, the product of x and y is not the same for the ordered pairs. This table does not represent inverse variation.

Student Check 2 Use inverse variation to solve each problem.

a. Suppose y is inversely proportional to x. If $y = 8$ when $x = 3$, find the constant of proportionality and the variation equation.

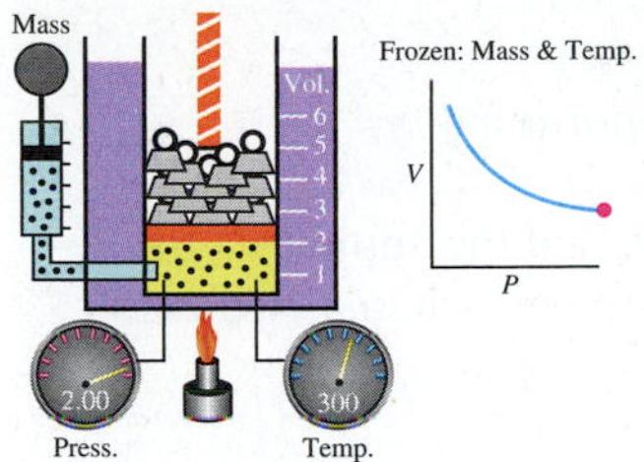

b. Boyle's law states that if the temperature remains the same, the volume V of a gas is inversely proportional to the pressure P applied to it. A given mass of a gas occupies 240 mL at a pressure of 800 mm of mercury (mm Hg). If the temperature remains constant, what volume will the gas occupy if the pressure is increased to 1200 mm Hg?

c. Determine if each table shows that y is inversely proportional to x. Find the variation equation for table that has this relationship.

Table 1

x	y
2	30
3	20
4	15
5	12

Table 2

x	y
2	120
3	180
4	240
5	300

Objective 3 Solve problems involving joint and combined variation.

Joint and Combined Variation

Joint variation occurs when a quantity can be expressed as the product of a constant and two or more variables.

Definition: Joint Variation

We say *z varies jointly as x and y* or *z is jointly proportional to x and y* if

$$z = kxy \quad \text{or} \quad \frac{z}{xy} = k$$

where k is the *constant of proportionality* or the *constant of variation*.

An example of joint variation is the area of a rectangle, $A = lw$. Area is equal to the product of the constant 1 and length and width.

Definition: Combined variation involves combinations of direct, inverse, or joint variation.

Objective 3 Examples Use joint or combined variation to solve each problem.

3a. Suppose z varies jointly as x and y. Find an equation that relates the variables if $z = 2$ when $x = 4$ and $y = -3$.

Solution 3a. Since z varies jointly as x and y, we know $z = kxy$.

$z = kxy$ — Write the variation equation.

$2 = k(4)(-3)$ — Let $z = 2$, $x = 4$, and $y = -3$.

$2 = -12k$ — Simplify.

$\frac{2}{-12} = k$ — Divide each side by -12.

$-\frac{1}{6} = k$ — Simplify.

So, the constant of variation is $-\frac{1}{6}$ and the variation equation is $z = -\frac{1}{6}xy$.

3b. When an object is in motion, it is said to have kinetic energy. Kinetic energy, K, varies jointly as the mass, m, of an object and the square of the object's velocity, v. Write a variation formula that represents this relationship.

Solution **3b.** K varies jointly as m and the square of v. So,

$$K = kmv^2, \text{ where } k \text{ is the constant of proportionality.}$$

3c. The area A of a trapezoid varies jointly as the height h and the sum of its bases b_1 and b_2. Find the equation of joint variation if $A = 48$ cm² when $h = 8$ cm, $b_1 = 5$ cm, and $b_2 = 7$ cm.

Solution **3c.** Since A varies jointly as h and $b_1 + b_2$, we write

$$A = kh(b_1 + b_2)$$

We find k by substituting the given values.

$A = kh(b_1 + b_2)$	State the variation equation.
$48 = k(8)(5 + 7)$	Let $A = 48$, $h = 8$, $b_1 = 5$, $b_2 = 7$.
$48 = k(8)(12)$	Simplify in parentheses.
$48 = 96k$	Multiply 8 and 12.
$\frac{48}{96} = k$	Divide each side by 96.
$\frac{1}{2} = k$	Simplify.

So, $A = \frac{1}{2}h(b_1 + b_2)$.

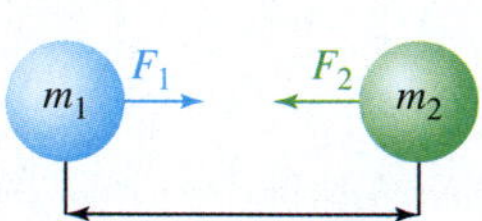

3d. Newton's law of universal gravitation states that every object in the universe attracts every other object with a force, F, directed along the line through the centers of the two objects that is jointly proportional to the objects' masses, m_1 and m_2, and inversely proportional to the square of the distance, r, between the two masses. Write a variation formula that represents this relationship.

Solution **3d.** F varies jointly as m_1 and m_2 and is inversely proportional to the square of the distance, r. So,

$$F = \frac{km_1m_2}{r^2}, \text{ where } k \text{ is the constant of proportionality.}$$

3e. The time T, in hours, required for a satellite to complete a circular orbit around the Earth varies directly as the radius r of the orbit measured from the center of the Earth and inversely as the orbital velocity v in miles per hour. If a satellite traveling at 17,000 mph completes an orbit 4500 miles from the center of the Earth in 100 min, find the constant of variation. Then determine how long it will take a satellite to complete one orbit if it is circling the earth at 4800 mi from the center of the Earth at 16,000 mph.

Solution **3e.** T varies directly as r and inversely as v, so $T = \frac{kr}{v}$. To find k, substitute the given values.

$T = \frac{kr}{v}$	Write the variation equation.
$\frac{5}{3} = \frac{k(4500)}{17{,}000}$	Let $T = 100 \text{ min} = \frac{100}{60} \text{ hr} = \frac{5}{3} \text{ hr}$, $r = 4500$, and $v = 17{,}000$.
$\frac{5}{3} \cdot \frac{17{,}000}{4500} = \frac{k(4500)}{17{,}000} \cdot \frac{17{,}000}{4500}$	Multiply each side by $\frac{17{,}000}{4500}$.
$\frac{85{,}000}{13{,}500} = k$	Simplify.
$6.296 = k$	Divide.

So, the constant of variation is 6.296 and the variation equation is $T = \frac{6.296r}{v}$.

We use the variation equation to find how long it will take a satellite to complete one orbit if it is circling the Earth at 4800 mi from the center of the Earth at 16,000 mph.

$$T = \frac{6.296r}{v} \qquad \text{Write the variation equation.}$$

$$T = \frac{6.296(4800)}{16{,}000} \qquad \text{Let } r = 4800 \text{ and } v = 16{,}000.$$

$$T = 1.89 \text{ hours} \qquad \text{Simplify.}$$

So, it will take approximately 1 hr and 53 min for the satellite to complete one orbit.

Student Check 3 Use joint or combined variation to solve each problem.

a. Suppose z varies jointly as x and y. Find an equation that relates the variables if $z = 6$ when $x = 3$ and $y = -4$.

b. The volume of a cylinder varies jointly as the square of its radius, r, and its height, h. Write a variation formula that represents this relationship.

c. The surface area, A, of a cylinder varies jointly as the radius, r, and the sum of the radius and the height, h. A cylinder with height 8 cm and radius 4 cm has a surface area of 96π cm^2. Find the surface area of a cylinder with radius 3 cm and height 10 cm.

d. Newton's second law states that the acceleration, a, of an object is directly proportional to the force, F, acting on it and inversely proportional to the mass, m, of the object. Write a variation formula that represents this relationship.

e. The maximum load, l, that a cylindrical column with a circular cross section can hold varies directly as the fourth power of the diameter, d, and inversely as the square of the height, h. A 9-m column that is 2 m in diameter can support 64 metric tons. Find the constant of variation and then determine how many metric tons can be supported by a 9-m column that is 3 m in diameter.

Objective 4 ▶ Troubleshoot common errors.

Troubleshoot Common Errors

Some common errors associated with variation are shown.

Objective 4 Examples A problem and an incorrect solution are given. Provide the correct solution and an explanation of the error.

4a. Suppose y varies directly as the cube of x. Find the variation equation if $y = 16$ when $x = 2$.

Incorrect Solution	Correct Solution and Explanation
$y = k(3x)$ $16 = k(3 \cdot 2)$ $16 = 6k$ $\frac{16}{6} = k$ $\frac{8}{3} = k$ $y = \frac{8}{3}(3k) = 8k$	Since y varies directly as the cube of x, we should have $y = kx^3$ $16 = k(2)^3$ $16 = 8k$ $2 = k$ $y = 2x^3$

4b. Suppose y varies inversely as x. Find the constant of variation if $y = 12$ when $x = 3$.

Incorrect Solution	Correct Solution and Explanation
$y = \frac{x}{k}$ $12 = \frac{3}{k}$ $12k = 3$ $k = \frac{3}{12}$ $k = \frac{1}{4}$	The constant of variation goes in the numerator and the variable x is in the denominator. $y = \frac{k}{x}$ $12 = \frac{k}{3}$ $36 = k$

ANSWERS TO STUDENT CHECKS

Student Check 1 **a.** $k = -4, y = -4x$ **b.** $d = 7$ in. **c.** Table 2, $y = -2x$ **d.** Graph 1, $y = \frac{1}{2}x$

Student Check 2 **a.** $k = 24, y = \frac{24}{x}$ **b.** $V = 160$ mL **c.** Table 1, $y = \frac{60}{x}$

Student Check 3 **a.** $z = -\frac{1}{2}xy$ **b.** $V = kr^2h$ **c.** $A = 78\pi$ cm^2 **d.** $a = \frac{kF}{m}$ **e.** 324 metric tons

SUMMARY OF KEY CONCEPTS

1. When two quantities x and y are related by $y = kx$ for some constant k, then we say y varies directly as x.
2. When two quantities x and y are related by $y = \frac{k}{x}$ for some constant k, then we say y varies indirectly or inversely as x.
3. Quantities can be jointly proportional. If $z = kxy$, then z varies jointly as x and y. Combined variation involves combinations of direct, inverse, or joint variation.

SECTION 7.7 EXERCISE SET

Write About It!

Use complete sentences in your answer to each exercise.

1. Define direct variation.
2. Define inverse variation.
3. Define joint variation.

For Exercises 4–8, translate each equation into words using variation.

4. $A = \pi r^2$
5. $V = \frac{1}{3}\pi r^2 h$
6. $V = \frac{4}{3}\pi r^3$
7. $G = \frac{km_1m_2}{r^2}$
8. $y = \frac{k}{\sqrt[3]{x}}$

Practice Makes Perfect!

Suppose y varies directly as x. Find the constant of variation and the equation of variation for each case. (*See Objective 1.*)

9. $y = 35$ when $x = 7$ $k = 5; y = 5x$
10. $y = 24$ when $x = 3$ $k = 8; y = 8x$
11. $y = -4$ when $x = -6$
12. $y = -12$ when $x = -10$
13. $y = 1.2$ when $x = 2$ $k = 0.6; y = 0.6x$
14. $y = 36$ when $x = 15$ $k = 2.4; y = 2.4x$

Solve each problem. (*See Objective 1.*)

15. The cost of a house in California is directly proportional to the size of the house. If a 1200-ft^2 house costs \$180,000, then what is the cost of a 2200-ft^2 house? \$330,000

Additional answers can be found in the Instructor Answer Appendix.

16. The cost of a house in Pennsylvania is directly proportional to the size of the house. If a 1800-ft² house costs \$210,000, then what is the cost of a 2700-ft² house? \$315,000

Hooke's law of elasticity states that the distance a spring stretches vertically is directly proportional to the weight on the end of the spring. (*See Objective 1.*)

17. If a 5-lb weight attached to the spring stretches the spring 6 in., find the distance that a 17-lb weight stretches the spring. 20.4 in.

18. If an 8-lb weight attached to the spring stretches the spring 3 in., find the distance that a 48-lb weight stretches the spring. 18 in.

Determine if each table shows that y varies directly as x. If the table represents direct variation, $y = kx$, find the variation equation. (*See Objective 1.*)

19.

Table 1

x	y
2	1
4	2
6	3
8	4

Table 2 Table 1; $y = \frac{1}{2}x$

x	y
2	−1
4	2
6	−3
8	4

20.

Table 1

x	y
2	2
4	8
6	18
8	32

Table 2 Table 2; $y = \frac{3}{7}x$

x	y
7	3
14	6
21	9
28	12

Determine if each graph shows that y varies directly as x. If the graph represents direct variation, $y = kx$, find the variation equation. (*See Objective 1.*)

21.

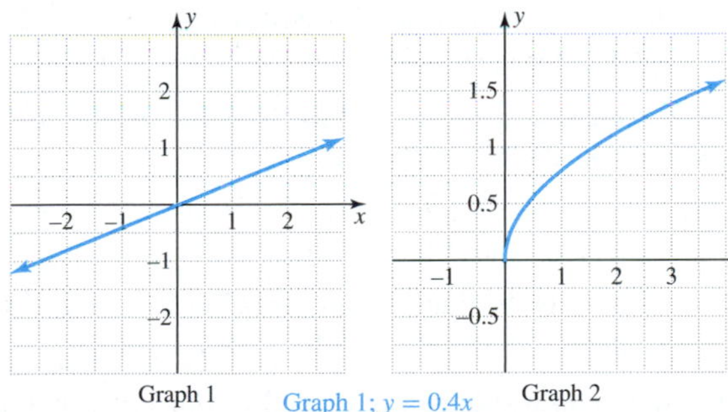

Graph 1 Graph 2

Graph 1; $y = 0.4x$

22.

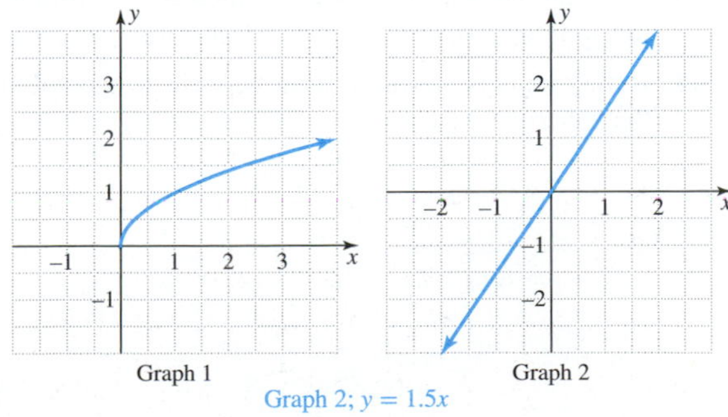

Graph 1 Graph 2

Graph 2; $y = 1.5x$

Suppose y varies inversely as x. Find the constant of variation and the equation of variation for each case. (*See Objective 2.*)

23. $y = \frac{1}{2}$ when $x = 6$ $k = 3$; $y = \frac{3}{x}$

24. $y = 2$ when $x = 2$ $k = 4$; $y = \frac{4}{x}$

25. $y = -3$ when $x = -\frac{1}{2}$

26. $y = -0.3$ when $x = -2$

27. $y = 0.32$ when $x = 5$

28. $y = 3$ when $x = 0.7$

Boyle's law states that if the temperature remains the same, the volume V of a gas is inversely proportional to the pressure P applied to it. (*See Objective 2.*)

29. A given mass of a gas occupies 930 mL at a pressure of 450 mm Hg. If the temperature remains constant, what volume will the gas occupy if the pressure is increased to 1350 mm Hg? 310 mL

30. A given mass of a gas occupies 360 mL at a pressure of 430 mm Hg. If the temperature remains constant, what volume will the gas occupy if the pressure is increased to 1200 mm Hg? 129 mL

Ohm's law states that if the voltage remains the same, the current, I, in an electrical conductor varies inversely as the resistance, R, of the conductor. (*See Objective 2.*)

31. If the current is 41 amp when the resistance is 270 Ω, what is the current when the resistance is 300 Ω? 36.9 amp

32. If the current is 57 amp when the resistance is 320 Ω, what is the current when the resistance is 480 Ω? 38 amp

Determine if each table shows that y varies inversely as x. If the table represents inverse variation, $y = \frac{k}{x}$, find the variation equation. (*See Objective 2.*)

33.

Table 1

x	y
2	1
4	2
6	3
8	4

Table 2

x	y
0.5	5
2	1.25
4	0.625
8	0.3125

Table 2; $y = \frac{2.5}{x}$

34.

Table 1

x	y
0.5	7
1	3.5
2	1.75
3	0.875

Table 2

x	y
1	3
2	6
3	9
4	12

Table 1; $y = \frac{7}{2x}$

Suppose z varies jointly as x and y. Find the constant of variation and the equation of variation for each case. (*See Objective 3.*)

35. $z = 10$ when $x = 2$ and $y = 4$ $k = \frac{5}{4}$; $z = \frac{5}{4}xy$

36. $z = 40$ when $x = 3$ and $y = 10$ $k = \frac{4}{3}$; $z = \frac{4}{3}xy$

37. $z = \frac{1}{3}$ when $x = 2$ and $y = 3$ $k = \frac{1}{18}$; $z = \frac{1}{18}xy$

38. $z = \frac{1}{4}$ when $x = 3$ and $y = \frac{5}{4}$ $k = \frac{1}{15}$; $z = \frac{1}{15}xy$

39. $z = 0.9$ when $x = 3$ and $y = 12$ $k = 0.025$; $z = 0.025xy$

40. $z = 0.36$ when $x = 0.5$ and $y = 9$ $k = 0.08$; $z = 0.08xy$

Find the constant of variation and variation equation for each situation. (*See Objective 3.*)

41. y varies jointly as the square of p and cube of q; $y = 8$ when $p = 3$ and $q = 2$. $k = \frac{1}{9}$; $y = \frac{1}{9}p^2q^3$

42. z varies jointly as x and square of y; $z = 16$ when $x = 5$ and $y = 2$. $k = \frac{4}{5}$; $z = \frac{4}{5}xy^2$

43. c varies jointly as the square root of a and b; $c = 3.6$ when $a = 4$ and $b = 1.2$. $k = 1.5$; $c = 1.5b\sqrt{a}$

44. F varies jointly as the cube root of G and square of H; $F = 4.5$ when $G = 27$ and $H = 0.5$. $k = 6$; $F = 6H^2\sqrt[3]{G}$

45. w varies directly as the square of u and inversely as the cube of v; $w = 1.8$ when $u = 1.5$ and $v = 2$.

46. z varies directly as the square root of x and inversely as the fourth power of y; $z = 2.7$ when $x = 2.25$ and $y = 0.2$.

The maximum load that a cylindrical column with a circular cross section can hold varies directly as the fourth power of the diameter and inversely as the square of the height. (*See Objective 3.*)

47. A 12-m column that is 2.4 m in diameter can support 72 metric tons. Find the constant of variation and then determine how many metric tons can be supported by a 10-m column that is 3.2 m in diameter. $k = 312.5$; 327.68 metric tons

48. A 10-m column that is 2 m in diameter can support 72 metric tons. Find the constant of variation and then determine how many metric tons can be supported by a 9-m column that is 3 m in diameter. $k = 450$; 450 metric tons

Mix 'Em Up!

Find the constant of variation and the variation equation for each situation.

49. y varies directly as the cube root of x; $y = 0.5$ when $x = 125$. $y = \frac{1}{10}\sqrt[3]{x}$

50. b varies directly as square root of a; $b = 5$ when $a = 4$.

51. y varies inversely as the square of x; $y = 2.5$ when $x = 1.2$.

52. F varies inversely as the cube of G; $F = 4.8$ when $G = 1.5$. $F = \frac{16.2}{G^3}$

53. w varies directly as the cube of u and inversely as the square root of v; $w = 24$ when $u = 2$ and $v = 25$. $w = \frac{15u^3}{\sqrt{v}}$

54. c varies directly as a and inversely as the cube root of b; $c = 0.9$ when $a = 1.8$ and $b = 0.064$. $c = \frac{0.2a}{\sqrt[3]{b}}$

Boyle's law states that if the temperature remains the same, the volume V of a gas is inversely proportional to the pressure P applied to it.

55. A given mass of a gas occupies 520 mL at a pressure of 880 mm Hg. If the temperature remains constant, what volume will the gas occupy if the pressure is increased to 1040 mm Hg? 440 mL

56. A given mass of a gas occupies 880 mL at a pressure of 680 mm Hg. If the temperature remains constant, what volume will the gas occupy if the pressure is increased to 1700 mm Hg? 352 mL

Hooke's law of elasticity states that the distance a spring stretches vertically is directly proportional to the weight on the end of the spring.

57. If a 10-lb weight attached to the spring stretches the spring 7 in., find the distance that a 43-lb weight stretches the spring. 30.1 in.

58. If a 24-lb weight attached to the spring stretches the spring 16 in., find the distance that a 27-lb weight stretches the spring. 18 in.

Ohm's law states that if the voltage remains the same, the current, *I*, in an electrical conductor varies inversely as the resistance, *R*, of the conductor.

59. If the current is 59 amp when the resistance is 480 Ω, what is the current when the resistance is 600 Ω? 47.2 amp

60. If the current is 54 amp when the resistance is 320 Ω, what is the current when the resistance is 450 Ω? 38.4 amp

The volume of a cone varies jointly as the height and the square of the radius of the base.

61. If the height is doubled and the radius is halved, what happens to the volume of the cone? Original volume is halved.

62. If the height is halved and the radius is doubled, what happens to the volume of the cone? Original volume is doubled.

The volume of a sphere varies as the cube of the radius.

63. If the radius is halved, what happens to the volume of the sphere? New volume is one-eighth of the original volume.

64. If the radius is doubled, what happens to the volume of the sphere? New volume is eight times the original volume.

You Be the Teacher!

Correct each student's errors, if any.

65. Find the equation of variation if y varies directly as the square root of x and $y = 48$ when $x = 4$.

Joseph's work:

$$y = kx^2$$
$$48 = k(4)^2$$
$$48 = 16k$$
$$k = 3$$
$$y = 3x^2$$

$y = k\sqrt{x}$
$48 = k\sqrt{4}$
$48 = 2k$
$k = 24$
$y = 24\sqrt{x}$

66. Find the equation of variation if z varies inversely as the square of w and $z = 6$ when $w = 9$.

Mike's work:

$$z = \frac{k}{\sqrt{w}}$$
$$6 = \frac{k}{\sqrt{9}}$$
$$18 = k$$
$$z = \frac{18}{\sqrt{w}}$$

$z = \frac{k}{w^2}$
$6 = \frac{k}{9^2}$
$486 = k$
$z = \frac{486}{w^2}$

GROUP ACTIVITY Mathematics of Operating a Vehicle

Rational expressions can be used to determine the cost efficiency of our cars. Because the issue of oil and the economy are so important these days, we will examine the cost of gas per year for a particular car.

1. Select a traditional gas-powered vehicle to examine from http://www.fueleconomy.gov. Determine the average mpg for the selected vehicle. Answers vary.
2. Select a vehicle to examine from the list of the most fuel efficient cars at http://www.thesupercars.org/ford/most-fuel-efficient-cars/. Determine the average mpg for this vehicle. Answers vary.
3. Calculate the average number of miles that you drive in a week. Use this information to set up a proportion to calculate the number of miles that you drive in a year. Answers vary.
4. What is the cost per gallon of gas where you live? Answers vary.
5. Convert the mpg of each vehicle to gallons per mile. Write these values as ratios. Answers vary.
6. Calculate the cost/year of each vehicle using the following equation.
$$\frac{\text{cost}}{\text{year}} = \frac{\text{miles}}{\text{year}} \cdot \frac{\text{cost}}{\text{gallon}} \cdot \frac{\text{gallons}}{\text{mile}}$$
Answers vary.
7. Use the cost/year that was found in step 6 to determine the cost/month and the cost/week. Answers vary.
8. How do the efficiencies of the two vehicles compare to one another? Answers vary.
9. What percent of your weekly or monthly expenses does the cost of gas represent? Answers vary.

Chapter 7 / REVIEW

Rational Expressions, Functions, and Equations

What's the big idea? Now that you have completed Chapter 7, you should understand that the quotient of polynomials is a rational expression. You should be able to simplify rational expressions and to perform operations on the rational expressions. Factoring is a key skill that is used to reduce rational expressions, multiply and divide rational expressions, and find the LCD. The LCD is used in adding and subtracting rational expressions, simplifying complex fractions, and solving rational equations.

The Tools

Listed below are the key terms, skills, formulas, and properties you should know for this chapter.

The page reference is provided if you need additional help with the given topic. The Study Tips will assist in your preparation for an exam.

Study Tips

1. Learn all of the terms, formulas, and properties. Make flash cards and have someone quiz you.
2. Rework problems from the exercises and also the ones you worked in class. Work additional problems from the review exercises.
3. Review the summaries of key concepts.
4. Work the chapter test.
5. Be sure to review the online resources for additional study materials.

Terms

Combined variation 535
Complex fraction 505
Constant of proportionality or constant of variation 530
Domain of a rational function 470
Extraneous solution 516
Least common denominator (LCD) 496
Long division 485
Lowest terms 471
Proportion 525
Rate 527
Ratio 525
Rational equation 516
Rational expression 468
Rational function 468
Reciprocals 475
Synthetic division 487

Formulas and Properties

- Adding and subtracting rational expressions 495
- Direct variation 530
- Distance-rate-time 527
- Dividing opposites 472
- Dividing rational expressions 475
- Fundamental property of rational expressions 472
- Inverse variation 533
- Joint variation 535
- Multiplying rational expressions 474
- Negative rational expressions 469
- Property of division 484
- Remainder theorem 490
- Work equation 526

CHAPTER 7 / SUMMARY

How well do you know this chapter? Complete the following questions to find out. Take a look back at the section if you need help.

SECTION 7.1 Rational Functions; Multiplying and Dividing Rational Expressions

1. A rational expression is a(n) quotient of two polynomials.
2. A rational function is a function of the form $f(x) = \frac{P(x)}{Q(x)}$, where $P(x)$ and $Q(x)$ are polynomials.
3. To evaluate a rational expression or function, substitute the given value in place of the variable and simplify.
4. When the denominator of a rational expression is zero, the expression is undefined. The domain of a rational function is all real numbers except the values that make the expression undefined.
5. To find the values that make a rational expression undefined, set the denominator equal to zero and solve.

6. To simplify a rational expression, factor the numerator and denominator and divide out any common factors.

7. The expressions $x - y$ and $y - x$ are opposites. The quotient of opposites is -1.

8. To multiply rational expressions, factor each numerator and denominator, divide out common factors, and multiply the remaining factors in the numerator and denominator.

9. To divide rational expressions, multiply by the reciprocal of the second fraction. That is, $\frac{A}{B} \div \frac{C}{D} = \frac{A}{B} \cdot \frac{D}{C}$.

SECTION 7.2 More Division of Polynomials: Long Division and Synthetic Division

10. To divide a polynomial by a monomial, divide each term of the numerator by the monomial.

11. To divide polynomials when the denominator is not a monomial, we must use long division.

12. Synthetic division can be used to divide a polynomial by an expression of the form $x - c$, $c \neq 0$.

13. The remainder theorem states that the remainder after dividing $P(x)$ by $x - c$ is $P(c)$

SECTION 7.3 Adding and Subtracting Rational Expressions

14. To add or subtract rational expressions, we must have like denominators.

15. The least common denominator consists of the product of each unique factor used the largest number of times it occurs in any of the denominators.

16. To add or subtract rational expressions with different denominators, find the LCD and convert each fraction to an equivalent fraction with the LCD as its denominator. Then add or subtract the expressions.

SECTION 7.4 Simplifying Complex Fractions

17. A(n) complex fraction is a fraction that contains fractions in its numerator and/or denominator.

18. There are two methods to simplify a complex fraction. We can either use division or multiply the numerator and denominator of the complex fraction by the LCD.

SECTION 7.5 Solving Rational Equations

19. A(n) rational equation is an equation that contains a rational expression.

20. To solve a rational equation, multiply both sides by the LCD and then solve the resulting equation.

21. A solution that makes the LCD equal to zero or the rational expression undefined is a(n) extraneous solution and must be excluded from the solution set.

SECTION 7.6 Applications of Rational Equations

22. A(n) ratio is a quotient of two quantities.

23. A(n) proportion is an equation with two ratios set equal to each other.

24. Work applications involve finding the time it takes for two individuals to complete a task when working together.

25. To solve motion problems, use the formula $d = rt$.

SECTION 7.7 Variation and Applications

26. When y varies directly as x, we write $y = kx$. The quotient of y and x is constant.

27. When y varies indirectly as x, we write $y = \frac{k}{x}$. The product of y and x is constant.

28. When z varies jointly as x and y, we write $z = kxy$.

29. Combined variation involves combinations of direct, inverse, or joint variation.

CHAPTER 7 / REVIEW EXERCISES

SECTION 7.1

Perform each operation. Round each answer to two decimal places when necessary. (*See Objective 1.*)

1. $\frac{2x^2 + x + 3}{x - 4}$, $x = 1, 4.01, 3.99, 100$ $-2, 3917.02, -3883.02, 209.41$

2. $\frac{x^2 - 5x + 1}{x - 2}$, $x = 0, 1.99, 2.01, 100$ $-0.5, 498.99, -500.99, 96.95$

3. $f(x) = \frac{5x + 2}{x + 2}$, $f(0)$, $f(-1.99)$, $f(-2.01)$, $f(100)$ $1, -795, 805, 4.92$

4. $f(x) = \frac{2x - 1}{x + 3}$, $f(0)$, $f(-3.01)$, $f(-2.99)$, $f(100)$ $-0.33, 702, -698, 1.93$

Find the domain of each rational function. Write each answer in both set-builder and interval notation. (*See Objective 2.*)

5. $f(x) = \frac{2x - 11}{x - 10}$

6. $f(x) = \frac{5x + 12}{3x - 2}$

7. $f(x) = \frac{2x - 1}{x^2 - 36}$

8. $f(x) = \frac{2x - 3}{x^2 - 49}$

9. $f(x) = \frac{x - 6}{2x^2 + 4x}$

10. $f(x) = \frac{x + 5}{2x^2 - 6x}$

Simplify each rational expression. (*See Objective 3.*)

11. $\frac{x - 10}{10 + x}$ prime

12. $\frac{x - 10}{10 - x}$ -1

13. $\frac{6x^2 - 2x}{6x^2 + 70x - 24}$ $\frac{x}{x + 12}$

14. $\frac{20x^2 + 25x}{12x^2 - 5x - 25}$ $\frac{5x}{3x - 5}$

15. $\frac{3x^2 - 5x}{9x^2 - 25}$ $\frac{x}{3x + 5}$

16. $\frac{x^2 + 2x}{x^2 - 100}$ prime

17. $\frac{6x^2 - 4x - 2}{27x^3 + 1}$ $\frac{2x - 2}{9x^2 - 3x + 1}$

18. $\frac{7x^2 - 17x + 6}{x^3 - 8}$ $\frac{7x - 3}{x^2 + 2x + 4}$

19. $\dfrac{5x^2 + 37x + 14}{2x^3 + 14x^2 + 15x + 105}$ $\quad \dfrac{5x+2}{2x^2+15}$

20. $\dfrac{3x^2 - x - 2}{3x^3 + 2x^2 + 9x + 6}$

21. $\dfrac{x^3 - 6x^2 - 4x + 24}{x^2 - 8x + 12}$ $\quad x + 2$

22. $\dfrac{x^3 + 8x^2 - 9x - 72}{x^2 + 5x - 24}$ $\quad x + 3$

Perform each operation. Give each answer in simplest form. (*See Objectives 4 and 5.*)

23. $\dfrac{7x + 6}{24x - 21} \cdot \dfrac{64x - 56}{14x + 12}$ $\quad \dfrac{4}{3}$

24. $\dfrac{4x - 3}{3x + 5} \cdot \dfrac{18x + 30}{16x - 12}$ $\quad \dfrac{3}{2}$

25. $\dfrac{40x + 8}{6x - 4} \div \dfrac{15x + 3}{9x - 6}$ $\quad 4$

26. $\dfrac{14x - 4}{x + 2} \div \dfrac{70x - 20}{9x + 18}$ $\quad \dfrac{9}{5}$

27. $\dfrac{7x^2}{x^2 - 100} \cdot \dfrac{3x^2 + 23x - 70}{-3x + 7}$ $\quad -\dfrac{7x^2}{x - 10}$ or $\dfrac{7x^2}{10 - x}$

28 $\dfrac{2x^2}{25x^2 - 16} \cdot \dfrac{30x^2 + 11x - 28}{-18x - 21}$ $\quad -\dfrac{2x^2}{3(5x + 4)}$

29. $\dfrac{1000x^3 + 1}{1 - 100x^2} \div \dfrac{4x + 1}{40x^2 + 6x - 1}$ $\quad -(100x^2 - 10x + 1)$

30. $\dfrac{8x^3 - 27}{9 - 4x^2} \div \dfrac{36x - 32}{18x^2 + 11x - 24}$ $\quad -\dfrac{4x^2 + 6x + 9}{4}$

31. $\dfrac{4x^2 + 12x + 9}{7x + 9} \div \dfrac{(2x + 3)^4}{7x^2 + 2x - 9}$ $\quad \dfrac{x - 1}{(2x + 3)^2}$

32. $\dfrac{x^2 - 18x + 81}{x - 4} \div \dfrac{(x - 9)^5}{3x^2 - 7x - 20}$ $\quad \dfrac{3x + 5}{(x - 9)^3}$

33. $\dfrac{x^2 + x - 6}{2x^2 + 8x} \div \dfrac{2x^2 - 4x}{x^3 + 4x^2} \cdot \dfrac{4x^2 - 8x}{x + 3}$ $\quad x(x - 2)$

34. $\dfrac{x^3 + 1}{2x^2 - 6x} \div \dfrac{3x^2 + 12x}{x^3 - 3x^2} \cdot \dfrac{4x^2 + 16x}{2x^2 - 2x + 2}$ $\quad \dfrac{x(x + 1)}{3}$

35. $\dfrac{x^3 - 8}{2x^3 + x^2 - 8x - 4} \cdot \left(\dfrac{2x + 1}{5x - 2} \div \dfrac{5x^2 + 3x - 2}{x^2 + x}\right)$

36. $\dfrac{x^2 - 25}{x^3 + 125} \div \left(\dfrac{3x^2 + 7x + 2}{2x^2 - 10x + 50} \cdot \dfrac{4x - 20}{3x + 1}\right)$ $\quad \dfrac{1}{2(x + 2)}$

Solve each problem. (*See Objective 6.*)

37. A high school class is organizing its 10-yr reunion. The venue, DJ, decorations, and other expenses total $3600 and the dinner will cost $30 per person. The total cost for x attendees is $C(x) = 30x + 3600$. The cost per person is the expression, $\dfrac{C(x)}{x} = \dfrac{30x + 3600}{x}$. What is the cost per person if 80 people attend? 100 attend? 120 attend? $75, $66, $60

38. A fitness club charges a one-time membership fee of $90 when a new member signs a contract. The new member is charged a monthly fee of $54, and the one-time fee is divided equally among the monthly payments. If x represents the number of months a new member commits to in a contract, the expression $\dfrac{90 + 54x}{x}$ represents the member's monthly payment. What is the member's monthly payment if the member signs a contract for 12 months? 18 months? 24 months? $61.50, $59, $57.75

SECTION 7.2

Perform each operation. (*See Objective 1.*)

39. $\dfrac{-6x^2 + 16x}{2x}$ $\quad -3x + 8$

40. $\dfrac{8y^2 - 20y}{4y}$ $\quad 2y - 5$

41. $\dfrac{-12x^5 - 4x^4 + 56x^3}{4x^3}$ $\quad -3x^2 - x + 14$

42. $\dfrac{42x^4 + 18x^3 - 66x^2}{6x^2}$ $\quad 7x^2 + 3x - 11$

43. $\dfrac{10x^3y + 30x^2y^2 - 75xy^3}{-5xy}$ $\quad -2x^2 - 6xy + 15y^2$

44. $\dfrac{-24x^4y^2 + 40x^3y^3 - 20x^2y^4}{-4x^2y^2}$ $\quad 6x^2 - 10xy + 5y^2$

Perform each operation using long division. (*See Objective 2.*)

45. $\dfrac{8x^2 - 18x - 35}{2x - 7}$ $\quad 4x + 5$

46. $\dfrac{6x^2 + 4x - 10}{3x + 5}$ $\quad 2x - 2$

47. $\dfrac{12x^2 - 4x - 65}{6x + 13}$ $\quad 2x - 5$

48. $\dfrac{2x^2 + 18x + 28}{2x + 14}$ $\quad x + 2$

49. $\dfrac{64x^3 + 729}{4x + 9}$ $\quad 16x^2 - 36x + 81$

50. $\dfrac{125x^3 - 27}{5x - 3}$ $\quad 25x^2 + 15x + 9$

51. $\dfrac{9x^3 + 24x^2 + 49x + 44}{3x + 4}$ $\quad 3x^2 + 4x + 11$

52. $\dfrac{12x^3 - 49x^2 - 6x + 27}{4x - 3}$ $\quad 3x^2 - 10x - 9$

Perform each operation using synthetic division. (*See Objective 3.*)

53. $\dfrac{5x^2 - 12x - 32}{x - 4}$ $\quad 5x + 8$

54. $\dfrac{7x^3 + 43x^2 + 34x - 30}{x + 5}$ $\quad 7x^2 + 8x - 6$

55. $\dfrac{x^3 - 8x^2 - 20}{x - 6}$

56. $\dfrac{x^3 + 4x - 12}{x + 3}$

57. $\dfrac{x^3 + 8}{x - 2}$ $\quad x^2 + 2x + 4 + \dfrac{16}{x - 2}$

58. $\dfrac{x^3 - 27}{x + 3}$ $\quad x^2 - 3x + 9 - \dfrac{54}{x + 3}$

Use the remainder theorem and synthetic division to evaluate $P(x)$ for the specified x-values. (*See Objective 4.*)

59. $P(x) = 5x^3 - 8x^2 + 10x - 4$, $P(-2)$ $\quad -96$

60. $P(x) = -2x^3 + 3x^2 + 2x - 18$, $P(-4)$ $\quad 150$

61. $P(x) = 2x^4 - 5x^3 - 12x^2 - 2$, $P(3)$ $\quad -83$

62. $P(x) = 2x^4 + 12x^3 - x - 3$, $P(-1)$ $\quad -12$

Solve each problem. (*See Objective 5.*)

63. The area of a rectangle is $26a^4b^2 + 40a^2b^4$ square units. Its height is $3ab$ units. Find an expression for the width of the rectangle. $\dfrac{26}{3}a^3b + \dfrac{40}{3}ab^3$ units

64. The area of a parallelogram is $154r^5s^3 + 121r^4s^5$ square units. Its height is $11r^2s^2$ units. Find an expression for the base of the parallelogram. $14r^3s + 11r^2s^3$ units

SECTION 7.3

Find the least common denominator for the rational expressions. (*See Objective 2.*)

65. $\dfrac{1}{15a^3b}, \dfrac{2}{40a^2b^2}$ $\quad 120a^3b^2$

66. $\dfrac{7}{20xy^2}, \dfrac{3}{28x^2y}$ $\quad 140x^2y^2$

67. $\dfrac{2x - 4}{10x}, \dfrac{2x + 1}{8x - 10}$ $\quad 10x(4x - 5)$

68. $\dfrac{x + 10}{4x}, \dfrac{11x}{4x - 5}$ $\quad 4x(4x - 5)$

69. $\dfrac{3x-8}{x^2-9}, \dfrac{-10x}{2x^2+13x+21}, \dfrac{1}{8x+28}$ $\quad 4(2x+7)(x+3)(x-3)$

70. $\dfrac{10x+7}{x^2-1}, \dfrac{-6x}{3x^2+4x+1}, \dfrac{8}{-15x-5}$ $\quad -5(3x+1)(x+1)(x-1)$

71. $\dfrac{5x-1}{(3x+1)^2}, \dfrac{9x}{15x+5}$ $\quad 5(3x+1)(3x+1)$

72. $\dfrac{x-6}{(x+6)^2}, \dfrac{-5x}{4x+24}$ $\quad 4(x+6)(x+6)$

Add or subtract the rational expressions. Write each answer in lowest terms. (*See Objectives 1–3.*)

73. $\dfrac{-10x}{2x+9} - \dfrac{4x}{2x+9}$

74. $\dfrac{2x}{x-4} + \dfrac{6x}{x-4}$ $\quad \dfrac{8x}{x-4}$

75. $\dfrac{7a-8}{5a+7} - \dfrac{5a+12}{5a+7}$

76. $\dfrac{8b+2}{2b-4} + \dfrac{8b-5}{2b-4}$ $\quad \dfrac{16b-3}{2b-4}$

77. $\dfrac{2x}{4x^2-1} - \dfrac{1}{4x^2-1}$ $\quad \dfrac{1}{2x+1}$

78. $\dfrac{-x^2-8x}{5x^2+38x-16} - \dfrac{2x^2+16x}{5x^2+38x-16}$ $\quad -\dfrac{3x}{5x-2}$

79. $\dfrac{1}{4x^2y} + \dfrac{5}{3xy^2}$ $\quad \dfrac{20x+3y}{12x^2y^2}$

80. $\dfrac{3}{2c^2d^3} - \dfrac{1}{3c^3d^2}$ $\quad \dfrac{9c-2d}{6c^3d^3}$

81. $\dfrac{4}{x+6} - \dfrac{3}{10x}$ $\quad \dfrac{37x-18}{10x(x+6)}$

82. $\dfrac{-5}{4x-3} + \dfrac{2}{5x}$ $\quad -\dfrac{17x+6}{5x(4x-3)}$

83. $\dfrac{10x+4}{x^2-36} - \dfrac{11}{5x-30}$

84. $\dfrac{5x-7}{9x^2-4} + \dfrac{3}{3x-2}$

85. $\dfrac{x+7}{x^2+7x-30} - \dfrac{x+3}{x^2-9}$ $\quad -\dfrac{3}{(x+10)(x-3)}$

86. $\dfrac{3x+1}{x^2+5x-36} + \dfrac{x+4}{4x^2-13x-12}$ $\quad \dfrac{13(x^2+2x+3)}{(x-4)(x+9)(4x+3)}$

87. $\dfrac{3}{5x-6} - \dfrac{12}{6-5x}$

88. $\dfrac{4}{6x-1} + \dfrac{2}{1-6x}$ $\quad \dfrac{2}{6x-1}$

89. $\dfrac{2}{3x-9} + \dfrac{10}{9x}$ $\quad \dfrac{16x-30}{9x(x-3)}$

90. $\dfrac{1}{2x-3} - \dfrac{1}{2x}$ $\quad \dfrac{3}{2x(2x-3)}$

91. $\dfrac{4}{x^2-x-2} - \dfrac{2}{x^2-2x} - \dfrac{1}{x^2+x}$ $\quad \dfrac{1}{(x-2)(x+1)}$

92. $\dfrac{5}{x^2+3x-4} - \dfrac{3}{x^2+4x} - \dfrac{1}{x^2-x}$ $\quad \dfrac{1}{x(x+4)}$

93. $\dfrac{3x}{x^3-1} - \dfrac{2}{x^2-1}$

94. $\dfrac{5x}{x^3+8} + \dfrac{2}{x^2-4}$

95. $\dfrac{3}{x^3+1} - \dfrac{3}{x^3-x^2+x} + \dfrac{1}{x^2+x}$ $\quad \dfrac{x-2}{x(x^2-x+1)}$

96. $\dfrac{3x}{x^3-27} - \dfrac{5}{2x^3+6x^2+18x} - \dfrac{1}{x^2-3x}$ $\quad \dfrac{4x+1}{2x(x^2+3x+9)}$

SECTION 7.4

Simplify each complex fraction using method 1. (*See Objective 1.*)

97. $\dfrac{\frac{2x^3}{6y^2}}{\frac{28x^2}{14y^3}}$ $\quad \dfrac{xy}{6}$

98. $\dfrac{\frac{3a}{12b^3}}{\frac{15a^2}{45b^2}}$ $\quad \dfrac{3}{4ab}$

99. $\dfrac{3^{-3}-6^{-3}}{6^{-1}}$ $\quad \dfrac{7}{36}$

100. $\dfrac{5^{-1}-2^{-3}}{4^{-2}}$ $\quad \dfrac{6}{5}$

101. $\dfrac{\frac{5a}{2a+6}}{\frac{2a^2}{8a+24}}$ $\quad \dfrac{10}{a}$

102. $\dfrac{\frac{6x^3}{3x-15}}{\frac{9x^2}{7x-35}}$ $\quad \dfrac{14x}{9}$

103. $\dfrac{\frac{3x}{x^2+3x-10}}{\frac{2x^2-10x}{x^2-25}}$ $\quad \dfrac{3}{2(x-2)}$

104. $\dfrac{\frac{5a}{a^2-4}}{\frac{3a^2+9a}{a^2+a-6}}$ $\quad \dfrac{5}{3(a+2)}$

105. $\dfrac{\frac{2}{x}-\frac{3}{y}}{12x-8y}$ $\quad -\dfrac{1}{4xy}$

106. $\dfrac{\frac{4}{r}+\frac{1}{s}}{8s+2r}$ $\quad \dfrac{1}{2rs}$

Simplify each complex fraction using method 2. (*See Objective 2.*)

107. $\dfrac{\frac{4}{xy}+\frac{16}{x^2y}}{\frac{5}{y^2}+\frac{20}{xy^2}}$ $\quad \dfrac{4y}{5x}$

108. $\dfrac{\frac{2}{ab}-\frac{16}{a^2b}}{\frac{8}{b^2}-\frac{64}{ab^2}}$ $\quad \dfrac{b}{4a}$

109. $\dfrac{\frac{16}{x^2}-\frac{49}{y^2}}{\frac{4}{x}-\frac{7}{y}}$

110. $\dfrac{\frac{16}{r^2}-\frac{1}{s^2}}{\frac{4}{r}+\frac{1}{s}}$ $\quad \dfrac{4s-r}{rs}$

111. $\dfrac{r^{-2}-25s^{-2}}{r^{-1}-5s^{-1}}$ $\quad \dfrac{5r+s}{rs}$

112. $\dfrac{9a^{-2}-4b^{-2}}{3a^{-1}+2b^{-1}}$

113. $\dfrac{\frac{10}{3a+7}-\frac{10}{7}}{x}$ $\quad -\dfrac{30a}{7x(3a+7)}$

114. $\dfrac{\frac{1}{2y-7}+\frac{1}{7}}{y}$ $\quad \dfrac{2}{7(2y-7)}$

Evaluate each rational function for the given x-value. (*See Objective 3.*)

115. $f(x) = \dfrac{3x-2}{x-7}, f\left(-\dfrac{2}{3}\right)$

116. $f(x) = \dfrac{5x-4}{10x+9}, f\left(\dfrac{1}{2}\right)$

117. $f(x) = \dfrac{5x+1}{x}, f\left(\dfrac{2}{5}\right)$

118. $f(x) = \dfrac{2x-7}{x}, f\left(\dfrac{3}{4}\right)$

Find the slope of the line that passes through the given points. (*See Objective 3.*)

119. $\left(\dfrac{3}{10}, \dfrac{1}{8}\right)$ and $\left(-\dfrac{5}{2}, \dfrac{3}{10}\right)$ $\quad -\dfrac{1}{16}$

120. $\left(\dfrac{1}{10}, -\dfrac{7}{5}\right)$ and $\left(-\dfrac{1}{10}, \dfrac{3}{5}\right)$ $\quad -10$

121. $\left(-\dfrac{3}{2}, -\dfrac{1}{4}\right)$ and $\left(\dfrac{7}{6}, -\dfrac{1}{4}\right)$ $\quad 0$

122. $\left(\dfrac{5}{6}, -\dfrac{7}{4}\right)$ and $\left(\dfrac{5}{6}, \dfrac{5}{9}\right)$ $\quad$ undefined

SECTION 7.5

Solve each rational equation. (*See Objective 1.*)

123. $\dfrac{10b-3}{5} = \dfrac{4b-7}{3}$ $\quad \left\{-\dfrac{13}{5}\right\}$

124. $\dfrac{s}{s+4} + 5 = \dfrac{10}{s+4}$ $\quad \left\{-\dfrac{5}{3}\right\}$

125. $\frac{5}{t} = -10 + \frac{2}{t}$ $\left\{-\frac{3}{10}\right\}$

126. $\frac{r^2}{r-3} = \frac{9}{r-3}$ $\{-3\}$

127. $1 - \frac{13}{r} + \frac{36}{r^2} = 0$ $\{4, 9\}$

128. $3 - \frac{2}{s} - \frac{21}{s^2} = 0$ $\left\{-\frac{7}{3}, 3\right\}$

129. $\frac{3}{x} - \frac{6}{2x-3} = \frac{2}{2x^2-3x}$ $\varnothing$

130. $\frac{5}{5s+1} - \frac{2}{2s-3} = \frac{16}{10s^2-13s-3}$ $\varnothing$

131. $\frac{5r}{r-5} - \frac{2}{r+5} = \frac{20}{r^2-25}$ $\left\{\frac{2}{5}\right\}$

132. $\frac{6x}{x-2} - \frac{29}{x+2} = \frac{48}{x^2-4}$ $\left\{\frac{5}{6}\right\}$

133. $\frac{3}{81x^2-16} = -\frac{9}{9x^2-4x}$ $\left\{-\frac{3}{7}\right\}$

134. $\frac{5}{4y^2-1} = \frac{7}{4y^2-4y+1}$ $\{-3\}$

135. $\frac{1}{2x+5} - \frac{1}{3x+2} = \frac{1}{6x^2+19x+10}$ $\{4\}$

136. $\frac{1}{10y+3} + \frac{2}{y-4} = \frac{2}{10y^2-37y-12}$ $\{0\}$

SECTION 7.6

Solve each rational equation for the specified variable. (*See Objective 1.*)

137. $\frac{5}{a} = \frac{4}{b} - \frac{3}{c}$ for c $c = \frac{3ab}{4a-5b}$

138. $\frac{1}{x} = \frac{7}{y} + \frac{4}{z}$ for z $z = \frac{4xy}{y-7x}$

139. $s = \frac{h}{2}(a + 2b + c)$ for b $b = \frac{2s - ah - ch}{2h}$

140. $s = \frac{h}{3}(a + 4b + c)$ for c $c = \frac{3s - ah - 4bh}{h}$

Solve each problem. (*See Objectives 2, 3, and 5.*)

141. Twice a number minus fifteen times its reciprocal is 7. Find the number. $-\frac{3}{2}$ and 5

142. Twice a number plus six times its reciprocal is 13. Find the number. $\frac{1}{2}$ and 6

143. When Beth returns to the United States from Germany, she has 611.6 euros. If the conversion rate is 1 U.S. dollar = 0.695 euros, how much money in U.S. dollars does Beth have? \$880

144. When Steve returns to the United States from Japan, he has 50,930 yen. If the conversion rate is 1 U.S. dollar = 81.488 yen, how much money in U.S. dollars does Wing have? \$625

145. Linda can bike 40 mi in the same amount of time it takes Tobin to bike 30 mi. If Linda's speed is 4 mph faster than Tobin's, how fast does each person bike? Linda at 16 mph and Tobin at 12 mph

146. Charlie can bike 60 mi in the same amount of time it takes Elisa to bike 72 mi. If Charlie's speed is 3 mph slower than Elisa, how fast does each person bike? Charlie at 15 mph and Elisa at 18 mph

SECTION 7.7

Find the constant of variation and the variation equation for each situation. (*See Objectives 1–3.*)

147. r varies directly as the square of s; $r = 8$ when $s = 2$. $k = 2, r = 2s^2$

148. b varies directly as cube of a; $b = 54$ when $a = 3$. $k = 2; b = 2a^3$

149. y varies inversely as the cube root of x; $y = 2$ when $x = 64$. $k = 8; y = \frac{8}{\sqrt[3]{x}}$

150. M varies inversely as the square root of N; $M = 4.8$ when $N = 2.25$. $k = 7.2; M = \frac{7.2}{\sqrt{N}}$

151. w varies directly as the cube root of u and inversely as the square of v; $w = 6$ when $u = 27$ and $v = 5$.

152. z varies directly as x and inversely as the cube of y; $z = 6$ when $x = 24$ and $y = 2$. $k = 2; z = \frac{2x}{y^3}$

Solve each problem. (*See Objectives 1–3.*)

153. Boyle's law states that if the temperature remains the same, the volume V of a gas is inversely proportional to the pressure P applied to it. A given mass of a gas occupies 320 mL at a pressure of 600 mm Hg. If the temperature remains constant, what volume will the gas occupy if the pressure is increased to 1200 mm Hg? 160 mL

154. The area A of a circle varies directly as the square of the radius. A circle with radius 6 cm has an area of 36π cm^2. Find the area of a circle with radius 21 cm. 441π cm^2

155. The volume of a cylinder varies jointly as the square of the radius and the height. If the radius is halved and the height is doubled, what happens to the volume of the cylinder? half of the original volume

156. The cost of a house in Iowa is directly proportional to the size of the house. If a 2000-ft^2 house costs \$210,000, then what is the cost of a 3350-ft^2 house? \$351,750

CHAPTER 7 TEST / RATIONAL EXPRESSIONS, FUNCTIONS, AND EQUATIONS

1. When simplified completely, $\frac{2t-1}{t+2} \cdot \frac{t^2+2t}{2t^2+t-1}$ is equivalent to

(a.) $\frac{t}{t+1}$

b. $\frac{t(2t-1)}{2t^2+t+1}$

c. $\frac{t^2+2t}{(t+2)(t+1)}$

d. $\frac{t}{t-1}$

2. When simplified completely, $\frac{x}{x^2-1} \div \frac{x-1}{x^2-3x-4}$ is equivalent to

a. $\frac{x-1}{x}$

(b.) $\frac{x(x-4)}{(x-1)^2}$

c. $\frac{x}{x-1}$

d. $\frac{x(x-1)}{(x^2-1)(x^2-3x-4)}$

3. When simplified completely, $\frac{x^2}{x+4} + \frac{4x}{x+4}$ is equivalent to

(a.) x b. $\frac{4x^3}{x+4}$

c. $\frac{5x}{x+4}$ d. $\frac{x^2+4x}{x+4}$

4. When simplified completely, $\frac{5-2x}{x+7} - \frac{x+3}{x+7}$ is equivalent to

a. $\frac{-3x+2}{0}$ b. $\frac{-3x+8}{x+7}$

c. $\frac{-x+8}{x+7}$ (d.) $\frac{-3x+2}{x+7}$

5. The solution set for $\frac{6}{3x} - \frac{2}{x} = 4$ is

a. $\{0\}$ b. $\left\{\frac{1}{12}\right\}$

c. $\{12\}$ (d.) $\varnothing$

6. The solution set for $\frac{x}{x-1} = \frac{2x}{x+1}$ is

(a.) $\{0, 3\}$ b. $\{3\}$

c. $\{0\}$ d. $\varnothing$

7. $\frac{x-7}{(x-1)(x+3)}$ is undefined for the x-value(s)

a. 0 b. 7, −3, and 1

c. −1 and 3 (d.) 1 and −3

8. Find the value of $\frac{2x^2-4x+5}{3x+1}$ when $x = -1$. $-\frac{11}{2}$

9. Simplify each expression.

a. $\frac{5-3x}{9x^2-25}$ $-\frac{1}{3x+5}$

b. $\frac{2x^2-5x-7}{x^2+5x+4}$ $\frac{2x-7}{x+4}$

c. $\frac{8a^3-27}{4a^2-9}$ $\frac{4a^2+6a+9}{2a+3}$

10. What numerator will make the statement $\frac{6y}{5y+10} = \frac{?}{5y^2-20}$ true? $6y^2 - 12y$

11. Find the LCD of the rational expressions.

a. $\frac{3}{8x^2}, \frac{5}{10x^3}$ $40x^3$

b. $\frac{4}{b}, -\frac{2}{b+5}$ $b(b+5)$

c. $\frac{x+5}{x^2+6x-7}, \frac{3x-1}{2x^2+3x-5}$ $(x+7)(x-1)(2x+5)$

12. Perform the indicated operation.

a. $\frac{4}{x-2} - \frac{6}{2-x}$ $\frac{10}{x-2}$

b. $\frac{2a}{a+1} - \frac{5}{3a}$ $\frac{6a^2-5a-5}{3a(a+1)}$

c. $\frac{3}{2x+8} - \frac{x}{x^2-16}$ $\frac{x-12}{2(x-4)(x+4)}$

13. Simplify each complex fraction.

a. $\dfrac{\frac{12a^3}{5b^2}}{\frac{21a^4}{10b}}$ $\frac{8}{7ab}$

b. $\dfrac{\frac{1}{a}-\frac{1}{b}}{\frac{1}{a}+\frac{1}{b}}$ $\frac{b-a}{a+b}$

14. Solve each equation.

a. $\frac{5}{x} + \frac{3}{2} = \frac{x+1}{x}$ $\{-8\}$

b. $1 - \frac{4}{x} - \frac{21}{x^2} = 0$ $\{-3, 7\}$

c. $\frac{x}{x+3} + \frac{6}{x+1} = \frac{6}{x^2+4x+3}$ $\{-4\}$

15. Solve each problem.

a. Mary and Sue have a maid service. Mary can clean an average size house in 3 hr when she works alone. Sue can clean an average size house in 4 hr when she works alone. How long will it take the two of them to clean an average size house if they work together? 1 hr and 43 min

b. Todd runs for total of 10 mi in the same time that Andrea walks 8 mi. Todd runs 1 mph faster than Andrea walks. How fast does Todd run? 5 mph

16. Find the domain of $f(x) = \frac{3x}{x^2 - x + 42}$. $(-\infty, -6) \cup (-6, 7) \cup (7, \infty)$

17. Solve each problem involving variation.

a. y varies directly as x. If $y = 10$, when $x = 15$, find the variation equation. $y = \frac{2}{3}x$

b. y varies inversely as x. If $y = 6$, when $x = 4$, find the variation equation. $y = \frac{24}{x}$

CUMULATIVE REVIEW EXERCISES / CHAPTERS 1–7

Translate each sentence into an algebraic inequality. Use the variable x to represent the unknown number. (*Section 1.3, Objective 4*)

1. Five more than twice a number is greater than the number plus 16. $5 + 2x > x + 16$

2. Nine less than four times a number is less than 23 more than twice the number. $4x - 9 < 23 + 2x$

Solve each linear equation. (*Section 2.1, Objectives 2–4 and 6*)

3. $\frac{5x}{3} - \frac{3x}{2} = \frac{1}{6}$ $\{1\}$

4. $6(x-2) - 5x = x - 12$ $\mathbb{R}$

Solve each inequality. Graph the solution set and write each answer in interval notation and set-builder notation. (*Section 2.4, Objectives 1 and 2*)

5. $\frac{5}{4}(x-8) \geq 2x - 7$

6. $6(y+3) - 19 > 4y + 5$

Find the requested information. (*Section 3.2, Objective 3, Section 3.3, Objective 3*)

7. Find the output value that corresponds to the input value of 0 for the relation whose graph is shown. $y = -6$

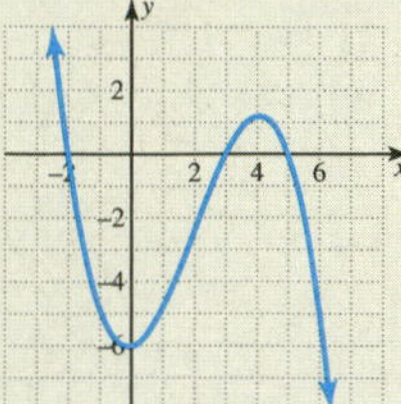

8. Find $f(-3)$ if the function $f(x)$ is given by Y_1. 8

X	Y1
-4	20
-3	8
-2	0
-1	-4
0	-4
1	0
2	8

Rewrite each linear equation using function notation, if possible. Identify *m* and *b*. (*Section 4.1, Objective 2*)

9. $y - 4x = 0$ $f(x) = 4x$; $m = 4, b = 0$ 10. $2y + 8 = 0$ $f(x) = -4$; $m = 0, b = -4$

Graph each linear function and state its domain and range. (*Section 4.2, Objective 1*)

11. $f(x) = -\frac{1}{4}x + 8$ 12. $f(x) = -4$

Find the slope and *y*-intercept of each line from its equation. Write the *y*-intercept as an ordered pair. Graph each line using its slope and *y*-intercept. Label at least two points on the graph. (*Section 4.3, Objectives 1 and 2*)

13. $f(x) = -2x - 5$ 14. $3x - 2y = 12$

Write the equation, in slope-intercept form, of the line that has the given slope and passes through the given point. (*Section 4.4, Objective 2*)

15. $(2, -4)$ and parallel to $y = \frac{4}{5}x + 6$ $y = \frac{4}{5}x - \frac{28}{5}$

16. $(7, -9)$ and perpendicular to $y = -\frac{1}{6}x + 5$ $y = 6x - 51$

Solve each problem. (*Section 4.4, Objective 5*)

17. The median sale price of single-family existing homes sold in Florida was approximately \$142,500 in 2009 and \$136,500 in 2010.
   a. Assuming linear decline, write an equation that approximates the median sale price of single-family existing homes sold in Florida, where x is the number of years after 2009. $y = -6000x + 142{,}500$
   b. If this trend continues, what will be the median sale price of single-family existing homes in 2012? (Source: http://media.living.net/statistics/2010) \$124,500

18. There were approximately 9.3 million Social Security beneficiaries receiving disability insurance in 2008. There were approximately 10.2 million beneficiaries receiving disability insurance in 2010.
   a. Write a linear equation that approximates the number of Social Security beneficiaries (in millions) receiving disability insurance, where x is the number of years after 2008. $y = 0.45x + 9.3$
   b. If this growth continues, how many beneficiaries receiving disability insurance will there be in 2014? (Source: www.ssa.gov) 12 million

Graph each linear inequality in two variables. (*Section 4.5, Objective 2*)

19. $x - 4y > 12$ 20. $4x + 5y < 100$

Solve each system of linear equations graphically. (*Section 5.1, Objectives 2 and 3*)

21. $\begin{cases} x - 3y = 6 \\ y = 1 \end{cases}$ $\{(9, 1)\}$ 22. $\begin{cases} x - 6y = -12 \\ -2x + 12y = 24 \end{cases}$ $\{(x, y) \mid x - 6y = -12\}$

Determine how the lines relate, the number of solutions of each system, and the type of system without graphing. (*Section 5.1, Objective 4*)

23. $\begin{cases} y = -\frac{2}{3}x \\ 4x + 6y = 0 \end{cases}$ 24. $\begin{cases} y = -\frac{1}{2}x \\ y = 2x + 3 \end{cases}$

Solve each system of linear equations using substitution or elimination. (*Section 5.2, Objectives 1 and 2*)

25. $\begin{cases} x = 2y + 18 \\ -x - 3y = 2 \end{cases}$ $\{(10, -4)\}$ 26. $\begin{cases} 0.1x - 0.6y = 1.4 \\ 4x - y = -13 \end{cases}$ $\{(-4, -3)\}$

27. $\begin{cases} 0.2x + 0.5y = 3.3 \\ -0.3x + 0.2y = 1.7 \end{cases}$ $\{(-1, 7)\}$ 28. $\begin{cases} \frac{3}{11}x - \frac{8}{11}y = -1 \\ -\frac{2}{7}x + \frac{4}{7}y = 1 \end{cases}$ $\left\{\left(-3, \frac{1}{4}\right)\right\}$

Solve each problem. (*Section 5.3, Objective 1*)

29. Derek works two jobs to help pay his way through college. He works as a clerk in his college admissions office and a night shift manager at a local supermarket. One week he earns \$175 by working 5 hr as a clerk and 10 hr as a night shift manager. Another week he earns \$295 by working 20 hr as a clerk and 10 hr as a night shift manager. What is Derek's hourly wage as a clerk and as a night shift manager? Derek's hourly wage as a clerk is \$8 and his hourly wage as a night shift manager is \$13.50.

30. Colleen works two jobs to help pay her way through college. She works as a peer tutor in the college math lab and a part-time receptionist at a local doctor's office. One week she earns \$390 by working 15 hr as the part-time receptionist and 25 hr as a peer tutor. Another week she earns \$240 by working 15 hr as the part-time receptionist and 5 hr as a peer tutor. What is Colleen's hourly wage as a part-time receptionist and a peer tutor? Colleen's hourly wage as a part-time receptionist is \$13.50 and her hourly wage as a peer tutor is \$7.50.

Solve each system of linear equations in three variables using elimination. (*Section 5.4, Objective 1*)

31. $\begin{cases} 2x - 3y + 5z = 13 \\ -x + 5y - 3z = 2 \\ 5x - 6y + z = -9 \end{cases}$ (1, 3, 4)

32. $\begin{cases} 3x + y - 3z = -4 \\ -5x + 5y + 4z = 1 \\ 6x - 7y + z = -6 \end{cases}$ (−2, −1, −1)

Solve each problem. (*Section 5.4, Objective 2*)

33. The Seyfrieds have saved $32,700. They invest this money in three different accounts—a 1-yr CD that pays 3.4% annual interest, a savings account that pays 4.6% annual interest, and a money market account that pays 3.2% annual interest. They invest four times as much money in the savings account as the amount invested in the CD. Find the amount invested in each account if they earn $1330.89 interest in a year. $4905 in the CD, $19,620 in the savings account, $8175 in the money market account

34. The Hendrickses saved $36,000. They invest this money in three different accounts—a 1-yr CD that pays 3.7% annual interest, a savings account that pays 6% annual interest, and a money market account that pays 2.9% annual interest. They invest three times as much money in the savings account as the amount invested in the CD. Find the amount invested in each account if they earn $1807.56 interest in a year. $7560 in the CD, $22,680 in the savings account, $5760 in the money market account

Solve each system of linear inequalities in two variables. (*Section 5.5, Objective 1*)

35. $\begin{cases} x \geq -3 \\ y \leq 5 \end{cases}$

36. $\begin{cases} x - 4 < 0 \\ y + 1 > 0 \end{cases}$

Simplify each expression by applying the product or quotient rule of like bases. (*Section 6.1, Objectives 1 and 2*)

37. $(-14a^5) \cdot (-2a^{16})$ $28a^{21}$

38. $(-8a^2b^4)(-10a^6b^8)$ $80a^8b^{12}$

39. $\dfrac{p^7}{p^6}$ p

40. $\dfrac{q^{11}}{q^9}$ q^2

Simplify each expression. Write each answer with positive exponents. (*Section 6.1, Objectives 3 and 4*)

41. $(2x^5y^4)(-4x^{-5}y^{-8})$ $-\dfrac{8}{y^4}$

42. $(5x^3y)(-3x^{-3}y^{-4})$ $-\dfrac{15}{y^3}$

43. $\dfrac{-16x^3y^7}{2xy^{10}}$ $-\dfrac{8x^2}{y^3}$

44. $\dfrac{-6r^6t^{-10}}{20r^{-11}t^{-5}}$ $-\dfrac{3r^{17}}{10t^5}$

Simplify each expression using the rules of exponents. Write each answer with positive exponents. (*Section 6.2, Objectives 1 and 2*)

45. $\dfrac{(4y^{10}z^{-9})^{-3}}{y^{-6}z^{-17}}$ $\dfrac{z^{44}}{64y^{24}}$

46. $\dfrac{(x^{-5}y^{-1})^{-1}}{6^2xy^{-4}}$ $\dfrac{x^4y^5}{36}$

Use scientific notation to perform the indicated operation. Write each answer in scientific notation. (*Section 6.2, Objectives 3 and 4*)

47. $(1.2 \times 10^4)^3$ 1.728×10^{12}

48. $\dfrac{1.476 \times 10^{29}}{4.1 \times 10^{18}}$ 3.6×10^{10}

49. $\dfrac{0.297}{5{,}500{,}000}$ 5.4×10^{-8}

50. $\dfrac{(9.36 \times 10^{-3})(7.5 \times 10^{-1})}{4.5 \times 10^7}$ 1.56×10^{-10}

Classify each polynomial and identify its terms, degree, type, and leading coefficient, if applicable. (*Section 6.3, Objectives 1 and 2*)

51. $\dfrac{3}{2}x^2 - x + 1$ $\dfrac{3}{2}x^2, -x, 1$; Trinomial; 2, Quadratic; $\dfrac{3}{2}$

52. $243a^2bc^2$ $243a^2bc^2$; Monomial; 5, Fifth-degree; n/a

53. $5 - 6y$ $-6y, 5$; Binomial; 1, Linear; -6

54. $2t - 8$ $2t, -8$; Binomial; 1, Linear; 2

Simplify each expression. (*Section 6.4, Objectives 1–5*)

55. $(-9a + 6) + (-12a - 13)$ $-21a - 7$

56. $(5p + 10) - (-17p + 9)$ $22p + 1$

57. $(-19q^2 - 15q - 4) + (23q^2 - 24q - 15)$ $4q^2 - 39q - 19$

58. $(-7b^2 + 9b - 17) - (-22b^2 - 8b + 15)$ $15b^2 + 17b - 32$

59. $(5r - 3)(4r - 1)$ $20r^2 - 17r + 3$

60. $(6y - 5)(y + 2)$ $6y^2 + 7y - 10$

61. $(2a - b)^3$ $8a^3 - 12a^2b + 6ab^2 - b^3$

62. $(b + 5)^3$ $b^3 + 15b^2 + 75b + 125$

Use grouping to factor each polynomial completely. (*Section 6.5, Objectives 1–3*)

63. $10w^4 - 5w^3 + 6w - 3$ $(2w - 1)(5w^3 + 3)$

64. $4w^4 + w^3 - 24w - 6$ $(4w + 1)(w^3 - 6)$

Factor each trinomial. (*Section 6.6, Objectives 1–4*)

65. $10p^2 + 39p + 14$ $(5p + 2)(2p + 7)$

66. $64q^2 - 64q + 7$ $(8q - 1)(8q - 7)$

67. $(r - 3)^2 - 5(r - 3) - 66$ $(r - 14)(r + 3)$

68. $(s + 4)^2 + 13(s + 4) + 40$ $(s + 12)(s + 9)$

Factor completely. (*Section 6.7, Objectives 1–3*)

69. $9r^3 - 64r$ $r(3r + 8)(3r - 8)$

70. $-5s^3 + 80s$ $-5s(s + 4)(s - 4)$

71. $(b - 1)^2 + 4$ prime

72. $(a - 5)^2 + 16$ prime

73. $a^3 + 8b^3$ $(a + 2b)(a^2 - 2ab + 4b^2)$

74. $c^3 - d^3$ $(c - d)(c^2 + cd + d^2)$

75. $36c^3 - 9c^2 - 64c + 16$ $(4c - 1)(3c + 4)(3c - 4)$

76. $4d^3 - d^2 - 36d + 9$ $(4d - 1)(d + 3)(d - 3)$

77. $25u^2 - 10u + 1 - v^2$ $(5u - 1 - v)(5u - 1 + v)$

78. $9r^2 + 6rs + s^2 - 64$ $(3r + s - 8)(3r + s + 8)$

Solving each equation by factoring. (*Section 6.8, Objective 1*)

79. $(6r - 21)(3r - 7) = 0$

80. $(3s - 12)(5s - 7) = 0$

81. $y(4y + 15) = 25$ $\left\{-5, \dfrac{5}{4}\right\}$

82. $2a(20a - 21) = -9$ $\left\{\dfrac{3}{10}, \dfrac{3}{4}\right\}$

Evaluate each expression or function for the given values. Round each answer to two decimal places when necessary. (*Section 7.1, Objective 1*)

83. $\dfrac{x - 1}{x + 2}$, $x = -2, -2.01, -1.99$ undefined; 301; −299

84. $f(x) = \dfrac{2x + 3}{x - 1}$, $x = 1, 1.01, 0.99$ undefined; 502; −498

Find the domain of each rational function. Write each answer in set-builder notation. (*Section 7.1, Objective 2*)

85. $f(x) = \dfrac{5x - 4}{x + 9}$ $\{x \mid x \text{ is a real number and } x \neq -9\}$

86. $f(x) = \dfrac{3x - 1}{x^2 - 9}$ $\{x \mid x \text{ is a real number and } x \neq -3, x \neq 3\}$

Simplify each rational expression. (*Section 7.1, Objective 3*)

87. $\dfrac{4x + 5}{5 + 4x}$ 1

88. $\dfrac{7 - 2x}{2x - 7}$ -1

89. $\dfrac{14x^2 + 3x - 2}{8x^3 + 1}$ $\dfrac{7x - 2}{4x^2 - 2x + 1}$

90. $\dfrac{8x^2 - 32x}{2x^2 - 32}$ $\dfrac{4x}{x + 4}$

91. $\dfrac{30x^2 - 84x + 48}{5x^3 - 4x^2 + 30x - 24}$

92. $\dfrac{6x^2 + 11x + 5}{24x^3 + 20x^2 + 18x + 15}$

Multiply or divide the rational expressions. Write each answer in simplest form. (*Section 7.1, Objectives 4 and 5*)

93. $\dfrac{4x-16}{16x+30} \cdot \dfrac{24x+45}{8x-32}$ $\dfrac{3}{4}$

94. $\dfrac{9x^2}{9x^2-16} \cdot \dfrac{3x^2-25x+28}{-x+7}$ $-\dfrac{9x^2}{3x+4}$

95. $\dfrac{x^3-1}{1-x^2} \div \dfrac{30x+66}{5x^2+16x+11}$ $-\dfrac{x^2+x+1}{6}$

96. $\dfrac{x^3-343}{49-x^2} \div \dfrac{-21x-98}{3x^2+35x+98}$ $\dfrac{x^2+7x+49}{7}$

Perform each operation. (*Section 7.2, Objective 1*)

97. $\dfrac{-5x^5+9x^4-10x^3}{5x^3}$

98. $\dfrac{2x^3-12x^2-9x}{-6x}$

Perform each operation using long division. (*Section 7.2, Objective 2*)

99. $\dfrac{6x^3-17x^2+4x-28}{2x+1}$

100. $\dfrac{15x^3+4x^2+5x-39}{5x-7}$

Perform each operation using synthetic division. (*Section 7.2, Objective 3*)

101. $\dfrac{x^3-216}{x+6}$

102. $\dfrac{2x^3-35x+24}{x-3}$

Use the remainder theorem and synthetic division to evaluate *P*(*x*) for the specified *x*-value. (*Section 7.2, Objective 4*)

103. $P(x) = x^4-5x^3+8x^2+6x-13,\ P(-1)$ $P(-1) = -5$

104. $P(x) = x^4+7x^2-10x-12,\ P(-2)$ $P(-2) = 52$

Add or subtract the rational expressions. Write each answer in lowest terms. (*Section 7.3, Objectives 1–3*)

105. $\dfrac{3x}{3x^2+17x+20} + \dfrac{4x}{x^2-16}$ $\dfrac{15x^2+8x}{(3x+5)(x+4)(x-4)}$

106. $\dfrac{5x}{3x^2-x-14} + \dfrac{2x}{x^2-4}$ $\dfrac{11x^2-24x}{(3x-7)(x-2)(x+2)}$

107. $\dfrac{16}{3x-21} - \dfrac{5}{21-3x}$

108. $\dfrac{15}{2x-6} - \dfrac{11}{6-2x}$ $\dfrac{13}{x-3}$

109. $\dfrac{x}{(x+1)^2} - \dfrac{2}{5x+5}$

110. $\dfrac{3x}{(4x+5)^2} - \dfrac{1}{8x+10}$

Simplify each complex fraction. (*Section 7.4, Objectives 1 and 2*)

111. $\dfrac{\dfrac{15a}{4b^4}}{\dfrac{5a^3}{12b^2}}$ $\dfrac{9}{a^2b^2}$

112. $\dfrac{\dfrac{x}{6y^2}}{\dfrac{5x^2}{18y}}$ $\dfrac{3}{5xy}$

113. $\dfrac{\dfrac{6}{xy}-\dfrac{1}{x^2}}{\dfrac{48}{y^2}-\dfrac{8}{xy}}$ $\dfrac{y}{8x}$

114. $\dfrac{\dfrac{7}{ab}-\dfrac{5}{a^2}}{\dfrac{35}{b^2}-\dfrac{25}{ab}}$ $\dfrac{b}{5a}$

Evaluate each rational function for the given *x*-value. (*Section 7.4, Objective 3*)

115. $f(x) = \dfrac{3x-2}{x+6},\ f\left(-\dfrac{8}{3}\right)$ -3

116. $f(x) = \dfrac{3x+10}{9x+5},\ f\left(-\dfrac{5}{3}\right)$ $-\dfrac{1}{2}$

Find the slope of the line that passes through the given points. (*Section 7.4, Objective 3*)

117. $\left(\dfrac{4}{5}, \dfrac{3}{4}\right)$ and $\left(\dfrac{2}{3}, \dfrac{1}{2}\right)$ $\dfrac{15}{8}$

118. $\left(\dfrac{5}{2}, -\dfrac{5}{6}\right)$ and $\left(\dfrac{2}{5}, \dfrac{1}{3}\right)$ $-\dfrac{5}{9}$

Simplify each rational expression. (*Section 7.4, Objective 3*)

119. $\dfrac{2^{-3}+3^{-1}}{2^{-1}}$ $\dfrac{11}{12}$

120. $\dfrac{5^{-1}+4^{-3}}{2^{-3}}$ $\dfrac{69}{40}$

Solve each rational equation. (*Section 7.5, Objective 1*)

121. $10 + \dfrac{27}{a} + \dfrac{18}{a^2} = 0$

122. $40 + \dfrac{9}{b} - \dfrac{10}{b^2} = 0$

123. $\dfrac{3}{y} + \dfrac{2}{y+6} = \dfrac{3}{y^2+6y}$ $\{-3\}$

124. $\dfrac{1}{x} + \dfrac{2}{x-10} = \dfrac{8}{x^2-10x}$ $\{6\}$

125. $\dfrac{4x+32}{x^2-x} - \dfrac{8}{x-1} = \dfrac{2}{x^2-x}$ $\left\{\dfrac{15}{2}\right\}$

126. $\dfrac{y+9}{y^2+10y} + \dfrac{5}{y+10} = \dfrac{2}{y^2+10y}$ $\left\{-\dfrac{7}{6}\right\}$

Solve each rational equation for the specified variable. (*Section 7.6, Objective 1*)

127. $\dfrac{12}{x} + \dfrac{1}{y} = \dfrac{2}{z}$ for y

128. $\dfrac{1}{a} = \dfrac{4}{c} - \dfrac{3}{b}$ for b

Solve each problem. (*Section 7.6, Objective 4*)

129. It takes Joann 4 hr to trim the bushes around the house by herself when working alone. Tim, her son, can complete the job by himself in 2 hr. How long would it take Joann and Tim to complete the job if they work together? 1 hr 20 min

130. It takes Hong 8 hr to do filing at the office when working alone. Shan can complete the job by herself in 7 hr. How long would it take Hong and Shan to complete the job if they work together? 3 hr 44 min

Boyle's law states that if the temperature remains the same, the volume *V* of a gas is inversely proportional to the pressure *P* applied to it. (*Section 7.7, Objective 2*)

131. A given mass of a gas occupies 450 mL at 610 mm Hg. What volume will the gas occupy if the pressure is increased to 720 mm Hg, temperature remaining constant? 381.25 mL

132. A given mass of a gas occupies 340 mL at 800 mm Hg. What volume will the gas occupy if the pressure is increased to 1600 mm Hg, temperature remaining constant? 170 mL

Find the constant of variation and the equation of variation for each case. (*Section 7.7, Objective 3*)

133. F varies jointly as the cube of G and square root of H; $F = 12$ when $G = 2$ and $H = 0.25$ $F = 3G^3\sqrt{H}$

134. w varies directly as the square of u and inversely as the cube root of v; $w = 7.5$ when $u = 5$ and $v = 64$

$w = \dfrac{1.2u^2}{\sqrt[3]{v}}$

CHAPTER 8

Rational Exponents, Radicals, and Complex Numbers

Refocus

Do you need to refocus your efforts to be successful? Maybe it has been a long term for you and it is a struggle to stay motivated and attentive to your studies. Don't give up. There is light at the end of the tunnel. Take a moment to remember why you enrolled in college and in this class. Keep sight of your short- and long-term goals. We know it will be worth it!

"What comes first, the compass or the clock? Before one can truly manage time (the clock), it is important to know where you are going, what your priorities and goals are, in which direction you are headed (the compass). Where you are headed is more important than how fast you are going. Rather than always *focusing* on what's urgent, learn to focus on what is really important." (Author Unknown)

Question For Thought: What are the things that distract your from accomplishing tasks or drain you of your time and energy? (Examples: Facebook, Web surfing, gaming, talking to co-workers, watching TV, etc.) How can you refocus your time and energy toward your studies?

Chapter Outline

Section 8.1 Radicals and Radical Functions 552

Section 8.2 Rational Exponents 567

Section 8.3 Simplifying Radical Expressions and the Distance Formula 578

Section 8.4 Adding, Subtracting, and Multiplying Radical Expressions 589

Section 8.5 Dividing Radical Expressions and Rationalizing 598

Section 8.6 Radical Equations and Their Applications 606

Section 8.7 Complex Numbers 615

Coming Up...

In Section 8.1, we will use the function $f(d) = 3.13\sqrt{d}$, which determines the wave speed, in meters per second, of surface waves in water that is d meters deep to find the wave speed of a tsunami in the Pacific Ocean. (Source: http://oceanservice.noaa.gov/facts/oceandepth.html)

“The best advice I ever came across on the subject of concentration is: Wherever you are, be there. When you work, work. When you play, play. Don't mix the two.”

—Jim Rohn (Entrepreneur, Author, Motivational Speaker)

SECTION 8.1 Radicals and Radical Functions

Chapter 8 deals with the concept of roots and radicals. We will review square roots and introduce cube roots and higher roots. This chapter will also explore operations with radical expressions and solve equations containing them. We will learn how roots are connected to exponents and how to express some roots as complex numbers.

OBJECTIVES

As a result of completing this section, you will be able to

1. Simplify square roots.
2. Simplify cube roots.
3. Simplify *n*th roots.
4. Approximate roots.
5. Simplify expressions of the forms $(\sqrt[n]{a})^n$ and $\sqrt[n]{a^n}$.
6. Find the domain of radical functions and evaluate radical functions.
7. Graph square root and cube root functions.
8. Troubleshoot common errors.

The function $f(d) = 3.13\sqrt{d}$ determines the wave speed, in meters per second, of surface waves in water that is d meters deep. The fastest surface waves are those caused by tsunamis. The average depth of the Pacific Ocean is approximately 4300 m. Estimate the wave speed of a tsunami in the Pacific Ocean.

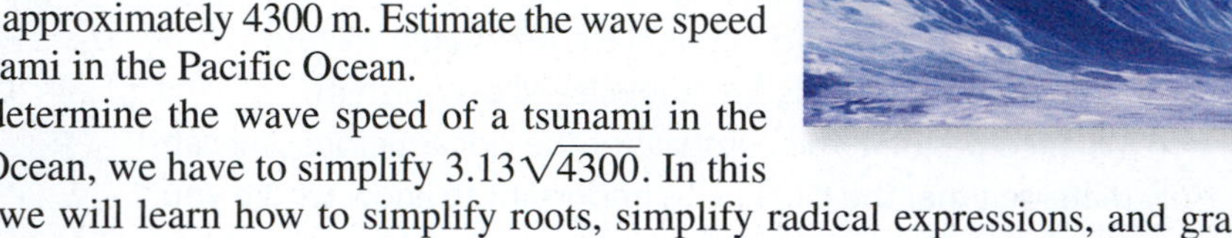

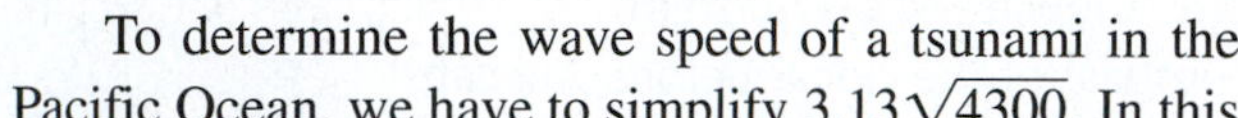

To determine the wave speed of a tsunami in the Pacific Ocean, we have to simplify $3.13\sqrt{4300}$. In this section, we will learn how to simplify roots, simplify radical expressions, and graph radical functions.

Objective 1

Simplify square roots.

Square Roots

Recall that the *square root* of a number, b, is the number that must be squared to obtain b. Since $(-3)^2 = 9$ and $3^2 = 9$, we say that the square roots of 9 are -3 and 3. We write this mathematically as

$$\sqrt{9} = 3 \quad \text{and} \quad -\sqrt{9} = -3$$

The notation $\sqrt{9}$ denotes the *principal square root* of 9, which is the nonnegative square root, or 3. The notation $-\sqrt{9}$ denotes the *negative square root* of 9, which is -3.

Definition: A number a is a **square root** of b if $a^2 = b$, for $b \geq 0$. The notation $\sqrt{b}$ denotes the **principal square root** of b and $-\sqrt{b}$ denotes the **negative square root** of b.

A nonnegative number whose square root is an integer value is a **perfect square**. For example, $\sqrt{9} = 3$, so 9 is a perfect square. It is very helpful to know perfect squares when simplifying square roots. Some of the perfect squares are shown.

$$\begin{array}{llll}
1^2 = 1 & 6^2 = 36 & 11^2 = 121 & (b)^2 = b^2 \\
2^2 = 4 & 7^2 = 49 & 12^2 = 144 & (b^2)^2 = b^4 \\
3^2 = 9 & 8^2 = 64 & 13^2 = 169 & (b^3)^2 = b^6 \\
4^2 = 16 & 9^2 = 81 & 14^2 = 196 & (b^4)^2 = b^8 \\
5^2 = 25 & 10^2 = 100 & 15^2 = 225 & (b^5)^2 = b^{10}
\end{array}$$

The square root of a negative number, however, is *not a real number*. For instance, the notation $\sqrt{-9}$ represents the number whose square is -9. Recall that $(3)^2 = 9$ and $(-3)^2 = 9$, so there is not a real number whose square is -9. This concept will be revisited in Section 8.7.

The expression $\sqrt{9}$ is an example of a **radical**, which is the root of a number. The notation used to denote a radical is defined as follows.

Definition: Radical

The expression $\sqrt{b}$ is called a **radical**, or **radical expression**. The symbol $\sqrt{\ }$ is called a **radical sign** and b is called the **radicand**.

$\sqrt{b}$ ⟵ Radical Sign

↑ Radicand

Procedure: Simplifying Square Roots

Step 1: Find the expression that must be squared to obtain the radicand. If the radicand is negative, the square root is not a real number.

Step 2: Check by squaring the result. The square should yield the radicand.

Objective 1 Examples

Simplify each expression. Assume all variables represent positive real numbers.

1a. $\sqrt{49}$ **1b.** $\sqrt{\frac{4}{9}}$ **1c.** $-\sqrt{4}$ **1d.** $\sqrt{-4}$ **1e.** $\sqrt{0.36}$ **1f.** $\sqrt{y^8}$ **1g.** $\sqrt{16b^6}$

Solutions

1a. $\sqrt{49} = 7$ because $7^2 = 49$. Note that $\sqrt{49} = \sqrt{7^2} = 7$.

1b. $\sqrt{\frac{4}{9}} = \frac{2}{3}$ because $\left(\frac{2}{3}\right)^2 = \frac{4}{9}$. Note that $\sqrt{\frac{4}{9}} = \sqrt{\left(\frac{2}{3}\right)^2} = \frac{2}{3}$.

1c. $-\sqrt{4} = -2$

1d. $\sqrt{-4}$ is not a real number.

1e. $\sqrt{0.36} = 0.6$ because $0.6^2 = 0.36$. Note that $\sqrt{0.36} = \sqrt{(0.6)^2} = 0.6$.

1f. We must find the expression that we square to get y^8: $(?)^2 = y^8$. Recall that when we raise a power to an exponent, we multiply exponents. For instance, $(y^4)^2 = y^8$. So,

$$\sqrt{y^8} = y^4$$

Note that $\sqrt{y^8} = \sqrt{(y^4)^2} = y^4$.

1g. We must find the expression that we square to get $16b^6$: $(?)^2 = 16b^6$. The power of a product rule for exponents tells us that $(4b^3)^2 = (4)^2(b^3)^2 = 16b^6$, so

$$\sqrt{16b^6} = 4b^3$$

Note that $\sqrt{16b^6} = \sqrt{(4b^3)^2} = 4b^3$.

Student Check 1 Simplify each expression. Assume all variables represent positive real numbers.

a. $\sqrt{64}$ **b.** $\sqrt{\frac{25}{36}}$ **c.** $-\sqrt{49}$ **d.** $\sqrt{-49}$ **e.** $\sqrt{0.04}$ **f.** $\sqrt{h^{10}}$ **g.** $\sqrt{9a^{12}}$

Cube Roots

Objective 2 ▶
Simplify cube roots.

The idea of square roots can be extended to other roots. The *cube root* of a number b, denoted $\sqrt[3]{b}$, is the number that must be cubed to obtain b.

Definition: The **cube root** of a real number b, denoted $\sqrt[3]{b}$, is the number that must be cubed to obtain b. That is,

$$\sqrt[3]{b} = a \text{ if and only if } a^3 = b$$

For example,

$$\sqrt[3]{64} = 4 \text{ because } 4^3 = 64$$
$$\sqrt[3]{-64} = -4 \text{ because } (-4)^3 = -64$$

This example illustrates that the cube root of a negative number is a real number, unlike the square root of a negative number.

A real number whose cube root is an integer is a **perfect cube**. For example, the number 64 is a perfect cube since its cube root is 4. It is very helpful to know perfect cubes when simplifying cube roots. Some of the perfect cubes are shown.

$1^3 = 1$	$6^3 = 216$	$(b)^3 = b^3$
$2^3 = 8$	$7^3 = 343$	$(b^2)^3 = b^6$
$3^3 = 27$	$8^3 = 512$	$(b^3)^3 = b^9$
$4^3 = 64$	$9^3 = 729$	$(b^4)^3 = b^{12}$
$5^3 = 125$	$10^3 = 1000$	$(b^5)^3 = b^{15}$

Procedure: Simplifying Cube Roots

Step 1: Find the expression that must be cubed to obtain the radicand.
Step 2: Check by cubing the result. The cube should yield the radicand.

Objective 2 Examples Simplify each expression.

2a. $\sqrt[3]{27}$ **2b.** $\sqrt[3]{-125}$ **2c.** $\sqrt[3]{\frac{1}{8}}$ **2d.** $\sqrt[3]{x^{12}}$ **2e.** $\sqrt[3]{-64y^6}$

Solutions

2a. $\sqrt[3]{27} = 3$ because $3^3 = 27$. Note that $\sqrt[3]{27} = \sqrt[3]{3^3} = 3$.

2b. $\sqrt[3]{-125} = -5$ because $(-5)^3 = -125$. Note that $\sqrt[3]{-125} = \sqrt[3]{(-5)^3} = -5$.

2c. $\sqrt[3]{\frac{1}{8}} = \frac{1}{2}$ because $\left(\frac{1}{2}\right)^3 = \frac{1}{8}$. Note that $\sqrt[3]{\frac{1}{8}} = \sqrt[3]{\left(\frac{1}{2}\right)^3} = \frac{1}{2}$.

2d. We must find the expression we cube to get x^{12}: $(?)^3 = x^{12}$. Recall that $(x^4)^3 = x^{12}$. So,

$$\sqrt[3]{x^{12}} = x^4$$

Note that $\sqrt[3]{x^{12}} = \sqrt[3]{(x^4)^3} = x^4$.

2e. We must find the expression we cube to get $-64y^6$: $(?)^3 = -64y^6$. Recall that $(-4y^2)^3 = (-4)^3(y^2)^3 = -64y^6$. So,

$$\sqrt[3]{-64y^6} = -4y^2$$

Note that $\sqrt[3]{-64y^6} = \sqrt[3]{(-4y^2)^3} = -4y^2$.

Student Check 2 Simplify each expression.

a. $\sqrt[3]{343}$ **b.** $\sqrt[3]{-8}$ **c.** $\sqrt[3]{\frac{27}{64}}$ **d.** $\sqrt[3]{a^3}$ **e.** $\sqrt[3]{-125y^9}$

Objective 3

Simplify *n*th roots.

The *n*th Root

We can generalize this concept of roots and define an *nth root*. The square root is the expression that is squared to obtain the radicand. The cube root is the expression that is cubed to obtain the radicand. The fourth root is the expression that is raised to the fourth to obtain the radicand and so on.

Definition: For a natural number $n \geq 2$, the ***n*th root** of a number b, denoted $\sqrt[n]{b}$, is the value that must be raised to the exponent of n to obtain b.

$$\sqrt[n]{b} = a \text{ if and only if } a^n = b$$

The number n is called the **index**. For square roots, the value of $n = 2$ but this is generally omitted.

The nth root is not a real number for all indices. It depends on whether n is even or odd.

- When the value of n is *even* (2, 4, 6, . . .), the radicand must be a *nonnegative number* for the nth root to be a real number.

  $$\sqrt[4]{16} = 2 \text{ because } 2^4 = 16 \text{ but } \sqrt[4]{-16} \text{ is not a real number}$$

  Recall that when we raise a nonzero number to an *even* exponent, the result is positive.
- When the value of n is *odd* (3, 5, 7, . . .), the radicand can be any real number for the nth root to be a real number.

  $$\sqrt[5]{32} = 2 \text{ because } 2^5 = 32$$
  $$\sqrt[5]{-32} = -2 \text{ because } (-2)^5 = -32$$

It is helpful to know the following powers when simplifying nth roots.

$1^4 = 1$	$1^5 = 1$	$(b)^4 = b^4$	$(b)^5 = b^5$
$2^4 = 16$	$2^5 = 32$	$(b^2)^4 = b^8$	$(b^2)^5 = b^{10}$
$3^4 = 81$	$3^5 = 243$	$(b^3)^4 = b^{12}$	$(b^3)^5 = b^{15}$
$4^4 = 256$	$4^5 = 1024$	$(b^4)^4 = b^{16}$	$(b^4)^5 = b^{20}$

Procedure: Simplifying *n*th Roots

Step 1: Determine the expression that must be raised to the nth to obtain the radicand.

Step 2: Check by raising the result to the exponent of n, $(\text{result})^n = \text{radicand}$.

Objective 3 Examples Simplify each expression. Assume all variables represent positive real numbers.

3a. $\sqrt[4]{256}$ **3b.** $\sqrt[4]{-625}$ **3c.** $\sqrt[5]{1024}$ **3d.** $\sqrt[5]{-243x^{10}}$ **3e.** $\sqrt[6]{x^6y^{12}}$

Solutions

3a. $\sqrt[4]{256} = 4$ because $4^4 = 256$. Note that $\sqrt[4]{256} = \sqrt[4]{4^4} = 4$.

3b. $\sqrt[4]{-625}$ is not a real number because there is no real number that we can raise to the fourth to get -625.

3c. $\sqrt[5]{1024} = 4$ because $4^5 = 1024$. Note that $\sqrt[5]{1024} = \sqrt[5]{4^5} = 4$.

3d. We must find the expression that, when raised to the fifth, is $-243x^{10}$. By the properties of exponents, we know that $(-3x^2)^5 = (-3)^5(x^2)^5 = -243x^{10}$. So,

$$\sqrt[5]{-243x^{10}} = -3x^2$$

Note that $\sqrt[5]{-243x^{10}} = \sqrt[5]{(-3x^2)^5} = -3x^2$.

3e. We must find the expression that, when raised to the sixth, is x^6y^{12}. By the properties of exponents, we know that $(xy^2)^6 = (x)^6(y^2)^6 = x^6y^{12}$. So,

$$\sqrt[6]{x^6y^{12}} = xy^2$$

Note that $\sqrt[6]{x^6y^{12}} = \sqrt[6]{(xy^2)^6} = xy^2$.

Student Check 3 Simplify each expression. Assume all variables represent positive real numbers.

a. $\sqrt[4]{625}$ **b.** $\sqrt[4]{-81}$ **c.** $\sqrt[5]{1}$ **d.** $\sqrt[5]{-32x^5}$ **e.** $\sqrt[7]{x^{14}y^7}$

Approximating Roots

Objective 4 Approximate roots.

In Objectives 1–3, we simplified some roots that were rational numbers. For instance,

$$\sqrt{49} = \sqrt{7^2} = 7$$
$$\sqrt[3]{27} = \sqrt[3]{3^3} = 3$$
$$\sqrt[4]{256} = \sqrt[4]{4^4} = 4$$

INSTRUCTOR NOTE: Remind students that a rational number is a number that can be written as a quotient of two integers.

Each of the radicands can be written as a factor raised to an exponent equal to the index. When the radicand is in this form, the radical simplifies to a *rational number*. If the radicand cannot be written as a factor raised to an exponent equal to the index, the radical is an *irrational number*. For the purpose of this discussion, we will assume that the radicals are real numbers.

Consider $\sqrt{5}$. To simplify this expression, we must find the number whose square is 5. There is no integer whose square is 5, but this does not mean that the square root does not exist. Because $2^2 = 4$ and $3^2 = 9$, $\sqrt{5}$ has a value between 2 and 3.

$$4 < 5 < 9$$
$$\sqrt{4} < \sqrt{5} < \sqrt{9}$$
$$2 < \sqrt{5} < 3$$

Using a calculator, we find that $\sqrt{5} \approx 2.24$. This number is irrational since its decimal form never terminates or repeats. Note that the symbol "$\approx$" means "approximately equal to." Depending on the instructions, we will be asked to provide either the exact value of a radical or an approximation of it.

$\sqrt{5}$ is the *exact value*, but 2.24 is the *approximated value*.

Procedure: Approximating Roots

Step 1: Use a calculator to find the decimal approximation. (Calculator instructions are provided at the end of the section.)

Step 2: Determine if the answer is reasonable. Find the two perfect squares (if the index is 2), perfect cubes (if the index is 3), and so on that the radicand lies between. Take the root of these numbers. The root should lie between these two values.

Step 3: Check the solution by raising the decimal approximation to the index. This number should be very close to the radicand. It will not be exact since the value has been approximated.

Objective 4 Examples Use a calculator to approximate each root. Round each answer to two decimal places. Verify the reasonableness of the result.

4a. $\sqrt{50}$ **4b.** $\sqrt[3]{15}$

7. Explain how to find the domain of the function $f(x) = \sqrt[3]{2x - 1}$.
Since it is a cube root function, the domain is $(-\infty, \infty)$.

8. Explain how to simplify $\sqrt[5]{x^{10}}$.

Practice Makes Perfect!

Simplify each expression. Assume all variables represent positive real numbers. (*See Objective 1.*)

9. $\sqrt{81}$ 9
10. $\sqrt{100}$ 10
11. $-\sqrt{25}$ -5
12. $-\sqrt{49}$ -7
13. $\sqrt{-64}$ not a real number
14. $\sqrt{-9}$ not a real number
15. $\sqrt{\frac{16}{49}}$ $\frac{4}{7}$
16. $\sqrt{\frac{25}{64}}$ $\frac{5}{8}$
17. $-\sqrt{\frac{9}{100}}$ $-\frac{3}{10}$
18. $-\sqrt{\frac{25}{4}}$ $-\frac{5}{2}$
19. $-\sqrt{-16}$ not a real number
20. $-\sqrt{-81}$ not a real number
21. $\sqrt{1.44}$ 1.2
22. $\sqrt{0.81}$ 0.9
23. $\sqrt{s^4}$ s^2
24. $\sqrt{t^8}$ t^4
25. $\sqrt{81a^{10}}$ $9a^5$
26. $\sqrt{121b^{12}}$ $11b^6$
27. $-\sqrt{16h^6}$ $-4h^3$
28. $-\sqrt{36k^{10}}$ $-6k^5$
29. $\sqrt{\frac{4x^{16}}{9}}$ $\frac{2x^8}{3}$
30. $\sqrt{\frac{64y^{14}}{25}}$ $\frac{8y^7}{5}$

Find each cube root. (*See Objective 2.*)

31. $\sqrt[3]{64}$ 4
32. $\sqrt[3]{125}$ 5
33. $\sqrt[3]{-1000}$ -10
34. $\sqrt[3]{-512}$ -8
35. $\sqrt[3]{\frac{64}{125}}$ $\frac{4}{5}$
36. $\sqrt[3]{\frac{27}{8}}$ $\frac{3}{2}$
37. $\sqrt[3]{a^6}$ a^2
38. $\sqrt[3]{b^9}$ b^3
39. $\sqrt[3]{64x^6}$ $4x^2$
40. $\sqrt[3]{27y^{12}}$ $3y^4$
41. $\sqrt[3]{-1000c^3}$ $-10c$
42. $\sqrt[3]{-125d^{15}}$ $-5d^5$
43. $\sqrt[3]{0.064a^6}$ $0.4a^2$
44. $\sqrt[3]{0.216b^{12}}$ $0.6b^4$
45. $\sqrt[3]{-\frac{729x^3}{125}}$ $-\frac{9x}{5}$
46. $\sqrt[3]{-\frac{27y^6}{64}}$ $-\frac{3y^2}{4}$
47. $-\sqrt[3]{-\frac{343a^{12}}{125}}$ $\frac{7a^4}{5}$
48. $-\sqrt[3]{-\frac{27y^9}{1000}}$ $\frac{3y^3}{10}$

Simplify each expression. Assume all variables represent positive real numbers. (*See Objective 3.*)

49. $\sqrt[4]{16}$ 2
50. $\sqrt[5]{243}$ 3
51. $\sqrt[5]{-32}$ -2
52. $\sqrt[4]{-16}$ not a real number
53. $\sqrt[4]{256x^4}$ $4x$
54. $\sqrt[4]{81y^8}$ $3y^2$
55. $\sqrt[5]{-32x^{15}y^{10}}$ $-2x^3y^2$
56. $\sqrt[5]{-243a^5b^{30}}$ $-3ab^6$
57. $\sqrt[7]{128a^{14}b^7}$ $2a^2b$
58. $\sqrt[7]{c^{28}d^{21}}$ c^4d^3
59. $\sqrt[6]{-x^6y^{12}}$ not a real number
60. $\sqrt[6]{-x^{18}y^{30}}$ not a real number
61. $\sqrt[6]{64a^{12}b^{24}}$ $2a^2b^4$
62. $\sqrt[6]{x^{18}y^6}$ x^3y

Use a calculator to approximate each root. Round the answer to two decimal places. Verify that the answer is reasonable. (*See Objective 4.*)

63. $\sqrt{42}$ 6.48
64. $\sqrt{95}$ 9.75
65. $\sqrt[3]{145}$ 5.25
66. $\sqrt[3]{387}$ 7.29

Simplify each expression. Assume all radical expressions with variables are real numbers. (*See Objective 5.*)

67. $(\sqrt{65})^2$ 65
68. $(\sqrt{48})^2$ 48
69. $(\sqrt{(-10)})^2$ not a real number
70. $(\sqrt{(-24)})^2$ not a real number
71. $\sqrt[3]{(-7)^3}$ -7
72. $\sqrt[3]{(-12)^3}$ -12
73. $(\sqrt{x-5})^2$ $x-5$
74. $(\sqrt{x+3})^2$ $x+3$
75. $(\sqrt{2x-3})^2$ $2x-3$
76. $(\sqrt{x+6})^2$ $x+6$
77. $\sqrt{(2x-5)^2}$ $|2x-5|$
78. $\sqrt{(7x+10)^2}$ $|7x+10|$
79. $\sqrt[5]{(6x)^5}$ $6x$
80. $\sqrt[5]{(15y)^5}$ $15y$
81. $\sqrt[7]{(13a)^7}$ $13a$
82. $\sqrt[7]{(3b)^7}$ $3b$

Find the domain of each radical function and write it in interval notation. Evaluate each function for the indicated values. (*See Objective 6.*)

83. $f(x) = \sqrt{x+5};\ x = -6, x = 0, x = 4$
84. $f(x) = \sqrt{x-2};\ x = 0, x = 3, x = 7$
85. $f(x) = \sqrt{6x+1};\ x = -1, x = -\frac{1}{6}, x = 4$
86. $f(x) = \sqrt{5x-3};\ x = 0, x = \frac{3}{5}, x = 2$
87. $f(x) = \sqrt[3]{x-3};\ x = -5, x = 0, x = 3$
88. $f(x) = \sqrt[3]{x+7};\ x = -7, x = 0, x = 1$

89. The function $f(S) = \sqrt{\frac{S}{4\pi}}$ gives the radius of a sphere with surface area S. Find the radius of a sphere whose surface area is 400 ft^2. Round the answer to two decimal places. 5.64 ft

90. Use the function given in Exercise 89 to find the radius of a sphere whose surface area is 650 m^2. Round the answer to two decimal places. 7.19 m

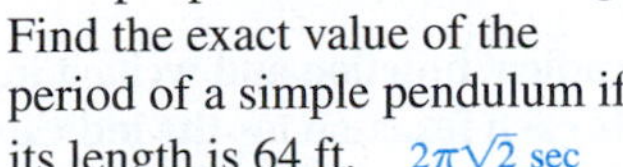

91. The function $f(L) = 2\pi\sqrt{\frac{L}{32}}$ gives the period in seconds of a simple pendulum L feet long. Find the exact value of the period of a simple pendulum if its length is 64 ft. $2\pi\sqrt{2}$ sec

92. Use the function in Exercise 91 to find the exact value of the period of a simple pendulum if its length is 128 ft. 4π sec

Graph each radical function using the accompanying tables. (*See Objective 7.*)

93. $f(x) = \sqrt{x-3}$

x	3	4	7	8
$f(x)$				

94. $f(x) = \sqrt{x+2}$

x	-2	-1	2	6
$f(x)$				

95. $f(x) = \sqrt{2x + 5}$

x	−2.5	−1	0	2
$f(x)$				

96. $f(x) = \sqrt{2x - 3}$

x	1.5	2	4	6
$f(x)$				

97. $f(x) = \sqrt[3]{x - 2}$

x	−6	0	1	2	3	10
$f(x)$						

98. $f(x) = \sqrt[3]{x + 2}$

x	−10	−2	−1	0	6
$f(x)$					

99. $f(x) = \sqrt[3]{x} - 2$

x	−8	−1	0	1	8	12
$f(x)$						

100. $f(x) = \sqrt[3]{x} + 2$

x	−8	−1	0	1	8
$f(x)$					

Mix 'Em Up!

Simplify each radical. Assume all radical expressions with variables are real numbers.

101. $-\sqrt{400x^6}$ $-20|x^3|$

102. $\sqrt{196y^2}$ $14|y|$

103. $\sqrt{225h^2k^8}$ $15|h|k^4$

104. $\sqrt{81m^6n^{10}}$ $9|m^3n^5|$

105. $\sqrt{(5x-9)^2}$ $|5x-9|$

106. $\sqrt{(8x+3)^2}$ $|8x+3|$

107. $\sqrt{0.49a^4}$ $0.7a^2$

108. $\sqrt{1.21b^6}$ $1.1|b^3|$

109. $(\sqrt{5x-6})^2$ for $x \geq \frac{6}{5}$ $5x - 6$

110. $(\sqrt{7x+2})^2$ for $x \geq -\frac{2}{7}$ $7x + 2$

111. $\sqrt[3]{64a^3}$ $4a$

112. $\sqrt[3]{27b^6}$ $3b^2$

113. $\sqrt[5]{-32r^{10}t^5}$ $-2r^2t$

114. $\sqrt[4]{81x^4y^{12}}$ $3|xy^3|$

115. $\sqrt{\frac{25r^{12}}{49}}$ $\frac{5r^6}{7}$

116. $\sqrt{\frac{9t^{10}}{121}}$ $\frac{3|t^5|}{11}$

117. $-\sqrt[4]{625r^8s^4}$ $-5r^2|s|$

118. $\sqrt[5]{-243a^{15}b^{20}}$ $-3a^3b^4$

119. $\sqrt[6]{a^6b^{18}}$ $|ab^3|$

120. $\sqrt[6]{c^{12}d^6}$ $c^2|d|$

121. $\sqrt[7]{x^{21}y^{14}}$ x^3y^2

122. $\sqrt[7]{x^{14}y^7}$ x^2y

Find the domain of each radical function and write it in interval notation. Evaluate each function for the indicated values.

123. $f(x) = \sqrt{3x+4}$; $x = -\frac{4}{3}, x = 0, x = 2$

124. $f(x) = \sqrt{2x-7}$; $x = 0, x = \frac{7}{2}, x = 5$

125. $f(x) = \sqrt{3-x}$; $x = -6, x = 0, x = 4$

126. $f(x) = \sqrt{5-x}$; $x = -4, x = 0, x = 5$

127. $f(x) = \sqrt[3]{x+4}$; $x = -4, x = 0, x = 4$

128. $f(x) = \sqrt[3]{x-5}$; $x = -3, x = 0, x = 6$

For Exercises 129 and 130, the initial speed s, in miles per hour, of a car that skids to a stop can be calculated by measuring the distance d, in feet, of the skid marks. The formula is $s = \sqrt{30kd}$, where k is the coefficient of friction, which depends on the road surface and condition.

Road Surface and Condition	k
Dry tar	1.0
Wet tar	0.5
Dry concrete	0.8
Wet concrete	0.4

129. Find the initial speed of a car if the distance of the skid marks is 145 ft on a wet tar road. 46.6 mph

130. Find the initial speed of a car if the distance of the skid marks is 280 ft on a dry concrete road. 82.0 mph

Graph each radical function using the accompanying tables.

131. $f(x) = \sqrt{3-2x}$

x	−3	0	1	1.5
$f(x)$				

132. $f(x) = \sqrt{1-2x}$

x	−3	−1	0	0.5
$f(x)$				

133. $f(x) = 1 - \sqrt[3]{x}$

x	−8	−1	0	1	8
$f(x)$					

134. $f(x) = 1 + \sqrt[3]{x}$

x	−8	−1	0	1	8
$f(x)$					

You Be the Teacher!

Correct each student's errors, if any.

135. Find the domain of the function $f(x) = \sqrt{x-1}$.

Vincent's work:

$x \neq 1$

$(-\infty, 1) \cup (1, \infty)$

136. Find the domain of the function $f(x) = \sqrt[3]{2x-5}$.

Tom's work:

$2x - 5 \geq 0$

$2x \geq 5$

$x \geq \frac{5}{2}$

$\left[\frac{5}{2}, \infty\right)$

137. Simplify: $-\sqrt{-9x^6y^{12}}$.

Ashley's work:

$-\sqrt{-9x^6y^{12}} = 3x^3y^6$

138. Simplify: $-\sqrt[4]{-81x^4}$.

Kimberly's work:

$-\sqrt[4]{-81x^4} = 3x$

Use a calculator to find each root and round each answer to two decimal places.

139. $\sqrt{1834}$ **140.** $\sqrt[3]{432}$

141. $\sqrt[5]{1492}$ **142.** $\sqrt[7]{654}$

SECTION 8.2 Rational Exponents

OBJECTIVES

As a result of completing this section, you will be able to

1. Simplify expressions of the form $b^{1/n}$.
2. Simplify expressions of the form $b^{m/n}$.
3. Simplify expressions of the form $b^{-m/n}$.
4. Use exponent rules to simplify expressions with rational exponents.
5. Use rational exponents to simplify radical expressions.
6. Solve applications with rational exponents.
7. Troubleshoot common errors.

The compound interest formula is given by $A = P\left(1 + \frac{r}{n}\right)^{nt}$, where A is the amount of money in an account in t years if P dollars is invested at an annual interest rate r, compounded n times a year. How much money will be in an account in 6 months $\left(\frac{1}{2} \text{ of a year}\right)$ if \$3000 is invested at an annual interest rate of 5% compounded annually?

To answer this question, we must simplify the expression

$$A = 3000\left(1 + \frac{0.05}{1}\right)^{1(1/2)} = 3000(1.05)^{1/2}$$

Notice that this expression has an exponent that is a fraction. This is called a *fractional exponent*, or a *rational exponent*. In this section, we will learn how to deal with this type of exponent. We will also discover how rational exponents are related to radicals.

Rational Exponents of the Form $b^{1/n}$

Simplify expressions of the form $b^{1/n}$.

Until this point, we have dealt only with integer exponents. We know the following.

- A positive integer exponent indicates how many times a base is repeated as a factor. For example, $4^3 = 4 \cdot 4 \cdot 4$.
- A nonzero number raised to the exponent of 0 is 1. For example, $4^0 = 1$.
- A negative integer exponent indicates that we take the reciprocal of the base to the positive exponent. For example, $4^{-3} = \frac{1}{4^3}$.

We will now learn how to deal with rational exponents of the form $\frac{1}{2}, \frac{1}{3}, \frac{1}{4}, \frac{1}{5}$, and so on.

Using a calculator, we can determine the value of the exponential expressions in the left column of the following table. The value of each exponential expression is also the same as a value of a radical expression, as shown.

Value Obtained by a Calculator	Radical Expression with the Same Value
$4^{1/2} = 2$	$\sqrt{4} = 2$
$9^{1/2} = 3$	$\sqrt{9} = 3$
$16^{1/2} = 4$	$\sqrt{16} = 4$
$8^{1/3} = 2$	$\sqrt[3]{8} = 2$
$27^{1/3} = 3$	$\sqrt[3]{27} = 3$

INSTRUCTOR NOTE: Inform students that a fractional exponent must be entered in parentheses.

This example shows equivalent expressions that will help us define an exponent of the form $\frac{1}{n}$. Note the following.

$$4^{1/2} = \sqrt{4}$$
$$9^{1/2} = \sqrt{9}$$
$$16^{1/2} = \sqrt{16}$$
$$8^{1/3} = \sqrt[3]{8}$$
$$27^{1/3} = \sqrt[3]{27}$$

So, **rational exponents** are simply another way to write a radical expression. Notice that the denominator of a rational exponent is the index of the equivalent radical expression.

Definition: $b^{1/n}$

For n a positive integer greater than 1 and $\sqrt[n]{b}$ a real number,

$$b^{1/n} = \sqrt[n]{b}$$

Recall that the nth root of a number b is defined as $\sqrt[n]{b} = a$ if and only if $a^n = b$. This definition also holds true for rational exponents.

$$(b^{1/n})^n = b^{(1/n)(n)} = b^1 = b$$

Procedure: Evaluating Expressions of the Form $b^{1/n}$

Step 1: Convert the expression to a radical of the form $\sqrt[n]{b}$.
- **a.** The denominator of the fractional exponent becomes the index of the radical.
- **b.** The base of the fractional exponent becomes the radicand of the radical.

Step 2: Simplify the resulting radical expression, if possible.

Objective 1 Examples Rewrite each expression as a radical expression and simplify, if possible.

1a. $81^{1/2}$ **1b.** $(-8)^{1/3}$ **1c.** $y^{1/4}$ **1d.** $-144^{1/2}$ **1e.** $(-144)^{1/2}$ **1f.** $(32x^{10})^{1/5}$

Solutions

1a. $81^{1/2} = \sqrt{81} = 9$

1b. $(-8)^{1/3} = \sqrt[3]{-8} = -2$

1c. $y^{1/4} = \sqrt[4]{y}$

1d. The base of the exponent is 144. So, we find the opposite of the square root of 144.

$$-144^{1/2} = -\sqrt{144} = -12$$

1e. The base of the exponent is -144. So, we find the square root of -144.

$$(-144)^{1/2} = \sqrt{-144} \text{ not a real number}$$

1f. This problem is equivalent to the fifth root of $32x^{10}$. So, we must find the expression that when raised to the exponent of 5 gives us $32x^{10}$.

$$(30x^{10})^{1/5} = \sqrt[5]{32x^{10}} = 2x^2$$

Student Check 1 Rewrite each expression as a radical expression and simplify, if possible.

a. $25^{1/2}$ **b.** $(-64)^{1/3}$ **c.** $x^{1/5}$ **d.** $-36^{1/2}$ **e.** $(-36)^{1/2}$ **f.** $(81x^{12})^{1/4}$

Rational Exponents of the Form $b^{m/n}$

Objective 2 ▶

Simplify expressions of the form $b^{m/n}$.

To be thorough in our discussion of rational exponents, we need to consider rational exponents whose numerators are different from one. In other words, we will define what it means to raise an expression to exponents such as $\frac{2}{3}$, $\frac{3}{2}$, and $\frac{3}{4}$. The power of a power rule for exponents helps us define this type of rational exponent. Recall that

$$(b^m)^n = b^{mn}$$

To simplify $16^{3/4}$, we think of it as follows.

$$16^{3/4} = 16^{(1/4)(3)} = (16^{1/4})^3 = \left(\sqrt[4]{16}\right)^3 = 2^3 = 8$$

Definition: $b^{m/n}$

If m and n are positive integers greater than 1 and $\sqrt[n]{b}$ a real number, then

$$b^{m/n} = \left(\sqrt[n]{b}\right)^m \quad \text{or} \quad \sqrt[n]{b^m}$$

Note: *The exponent m can be applied after the root is simplified or before the root is taken.*

Procedure: Evaluating an Expression of the Form $b^{m/n}$

Step 1: Convert the expression to radical form $\left(\sqrt[n]{b}\right)^m$.

- **a.** The base of the exponent is the radicand of the radical.
- **b.** The denominator of the fractional exponent is the index of the radical.
- **c.** The numerator of the fractional exponent is the exponent of the radical expression.

Step 2: Simplify the radical expression.

Step 3: Simplify the exponential expression.

Objective 2 Examples **Rewrite each expression as a radical expression and simplify, if possible. Assume all variable bases represent positive real numbers.**

2a. $16^{5/4}$ **2b.** $-25^{3/2}$ **2c.** $(-64)^{4/3}$ **2d.** $\left(\frac{1}{8}\right)^{2/3}$ **2e.** $(x-3)^{5/6}$

Solutions

2a.

$$\begin{aligned} 16^{5/4} &= \left(\sqrt[4]{16}\right)^5 && \text{Apply the definition of } b^{m/n}. \\ &= 2^5 && \text{Simplify the radical expression.} \\ &= 32 && \text{Simplify the exponential expression.} \end{aligned}$$

2b.

$$\begin{aligned} -25^{3/2} &= -\left(\sqrt{25}\right)^3 && \text{Apply the definition of } b^{m/n}. \\ &= -5^3 && \text{Simplify the radical expression.} \\ &= -125 && \text{Simplify the exponential expression.} \end{aligned}$$

Notice the base is 25, not −25, because there are no parentheses in the given problem.

2c.

$$\begin{aligned} (-64)^{4/3} &= \left(\sqrt[3]{-64}\right)^4 && \text{Apply the definition of } b^{m/n}. \\ &= (-4)^4 && \text{Simplify the radical expression.} \\ &= 256 && \text{Simplify the exponential expression.} \end{aligned}$$

Notice the base is −64 since it is contained in parentheses in the given problem.

2d. $\left(\frac{1}{8}\right)^{2/3} = \left(\sqrt[3]{\frac{1}{8}}\right)^2$ Apply the definition of $b^{m/n}$.

$= \left(\frac{1}{2}\right)^2$ Simplify the radical expression.

$= \frac{1}{4}$ Simplify the exponential expression.

2e. $(x-3)^{5/6} = \left(\sqrt[6]{x-3}\right)^5$ Apply the definition of $b^{m/n}$.

$= \sqrt[6]{(x-3)^5}$ Apply the exponent of 5.

Student Check 2 Rewrite each expression as a radical expression and simplify, if possible. Assume all variable bases represent positive real numbers.

a. $9^{3/2}$ **b.** $-81^{3/4}$ **c.** $(-27)^{2/3}$ **d.** $\left(\frac{1}{32}\right)^{3/5}$ **e.** $(x+2)^{6/7}$

Rational Exponents of the Form $b^{-m/n}$

Objective 3 ▶ Simplify expressions of the form $b^{-m/n}$.

Rational exponents can also be negative. In this case, we have to use the definition of the negative exponent to help us rewrite the expression. Recall that for $n > 0$, $x^{-n} = \frac{1}{x^n}$. So, using this definition together with the definition of $b^{m/n}$, we get the following.

Definition: $b^{-m/n}$

For $b \neq 0$,

$$b^{-m/n} = \frac{1}{b^{m/n}}$$

Procedure: Evaluating an Expression of the Form $b^{-m/n}$

Step 1: Rewrite the expression with a positive exponent.
Step 2: Convert the expression to its equivalent radical form $\left(\sqrt[n]{b}\right)^m$.
Step 3: Simplify the radical expression.
Step 4: Simplify the exponential expression, if needed.

Objective 3 Examples Rewrite each expression with a positive exponent and simplify.

3a. $16^{-1/2}$ **3b.** $(-8)^{-4/3}$ **3c.** $-81^{-3/4}$

Solutions

3a. $16^{-1/2} = \frac{1}{16^{1/2}}$ Apply the definition of $b^{-m/n}$.

$= \frac{1}{\sqrt{16}}$ Apply the definition of $b^{1/n}$.

$= \frac{1}{4}$ Simplify the radical expression.

3b. $(-8)^{-4/3} = \frac{1}{(-8)^{4/3}}$ Apply the definition of $b^{-m/n}$.

$= \frac{1}{\left(\sqrt[3]{-8}\right)^4}$ Apply the definition of $b^{m/n}$.

$= \dfrac{1}{(-2)^4}$ Simplify the radical expression.

$= \dfrac{1}{16}$ Simplify the exponential expression.

3c. $-81^{-3/4} = -\dfrac{1}{81^{3/4}}$ Apply the definition of $b^{-m/n}$.

$= -\dfrac{1}{(\sqrt[4]{81})^3}$ Apply the definition of $b^{m/n}$.

$= -\dfrac{1}{(3)^3}$ Simplify the radical expression.

$= -\dfrac{1}{27}$ Simplify the exponential expression.

Student Check 3 Rewrite each expression with a positive exponent and simplify.

a. $32^{-1/5}$ **b.** $(-27)^{-2/3}$ **c.** $-25^{-3/2}$

Simplify Exponential Expressions Involving Rational Exponents

Objective 4 Use exponent rules to simplify expressions with rational exponents.

The exponent rules that we learned previously work not only for integer exponents, but also for rational exponents. A review of the exponent rules follows.

Property: Summary of Exponent Rules

If a and b are real numbers and m and n are integers, then

Product of like bases:	$b^m \cdot b^n = b^{m+n}$
Quotient of like bases:	$\dfrac{b^m}{b^n} = b^{m-n}, b \neq 0$
Power of a power:	$(b^m)^n = b^{mn}$
Power of a product:	$(ab)^n = a^n b^n$
Power of a quotient:	$\left(\dfrac{a}{b}\right)^n = \dfrac{a^n}{b^n}, b \neq 0$
Zero exponent:	$b^0 = 1$
Negative exponent:	$b^{-n} = \dfrac{1}{b^n}, b \neq 0$

Objective 4 Examples Simplify each expression and write each result with positive exponents. Assume all variables represent positive real numbers.

4a. $4^{4/3} \cdot 4^{2/3}$ **4b.** $x^{3/5} \cdot x^{-4/5}$ **4c.** $y^{1/2} \cdot y^{1/4}$ **4d.** $\dfrac{8^{1/5}}{8^{6/5}}$

4e. $(a^4)^{3/2}$ **4f.** $(8x^6)^{2/3}$ **4g.** $\left(\dfrac{3b^{1/2}c^{-1/3}}{b^2}\right)^2$ **4h.** $x^{1/3}(x^{1/3} + 2x^{2/3})$

Solutions

4a. $4^{4/3} \cdot 4^{2/3} = 4^{4/3+2/3}$ Apply the product of like bases rule.

$= 4^{6/3}$ Add the exponents.

$= 4^2$ Simplify the exponent.

$= 16$ Simplify.

4b. $x^{3/5} \cdot x^{-4/5} = x^{3/5+(-4/5)}$ Apply the product of like bases rule.

$= x^{-1/5}$ Add the exponents.

$= \dfrac{1}{x^{1/5}}$ Apply the negative exponent rule.

4c.

$y^{1/2} \cdot y^{1/4} = y^{1/2+1/4}$	Apply the product of like bases rule.
$= y^{2/4+1/4}$	Find a common denominator.
$= y^{3/4}$	Add the exponents.

4d.

$\dfrac{8^{1/5}}{8^{6/5}} = 8^{1/5-6/5}$	Apply the quotient of like bases rule.
$= 8^{-5/5}$	Subtract the exponents.
$= 8^{-1}$	Simplify the exponent.
$= \dfrac{1}{8^1}$	Apply the negative exponent rule.
$= \dfrac{1}{8}$	Simplify.

4e.

$(a^4)^{3/2} = a^{4(3/2)}$	Apply the power of a power rule.
$= a^6$	Multiply exponents.

4f.

$(8x^6)^{2/3} = (8)^{2/3}(x^6)^{2/3}$	Apply the power of a product rule.
$= (\sqrt[3]{8})^2 x^{6(2/3)}$	Apply the definition of $b^{m/n}$ and the power of a power rule.
$= 4x^4$	Simplify the radical expression and multiply exponents.

4g.

$\left(\dfrac{3b^{1/2}c^{-1/3}}{b^2}\right)^2 = \dfrac{(3b^{1/2}c^{-1/3})^2}{(b^2)^2}$	Apply the power of a quotient rule.
$= \dfrac{3^2 b^{(1/2)(2)} c^{(-1/3)(2)}}{b^4}$	Apply the power of a product rule in the numerator and the power of a power rule in the denominator.
$= \dfrac{9b^1 c^{-2/3}}{b^4}$	Simplify 3^2 and multiply exponents.
$= 9b^{1-4}c^{-2/3}$	Apply the quotient of like bases rule.
$= 9b^{-3}c^{-2/3}$	Subtract the exponents.
$= \dfrac{9}{b^3c^{2/3}}$	Apply the negative exponent rule.

4h.

$x^{1/3}(x^{1/3} + 2x^{2/3}) = x^{1/3}(x^{1/3}) + x^{1/3}(2x^{2/3})$	Apply the distributive property.
$= x^{1/3+1/3} + 2x^{1/3+2/3}$	Apply the product of like bases rule.
$= x^{2/3} + 2x^{3/3}$	Add the exponents.
$= x^{2/3} + 2x^1$	Simplify the exponent.
$= x^{2/3} + 2x$	Recall $x^1 = x$.

Student Check 4 Simplify each expression and write each result with positive exponents. Assume all variables represent positive real numbers.

a. $7^{1/6} \cdot 7^{5/6}$ **b.** $y^{3/7} \cdot y^{-6/7}$ **c.** $x^{1/3} \cdot x^{1/6}$ **d.** $\dfrac{3^{2/3}}{3^{5/3}}$

e. $(a^4)^{5/4}$ **f.** $(9x^6)^{3/2}$ **g.** $\left(\dfrac{2x^{2/3}y^{-1/2}}{x^2}\right)^3$ **h.** $x^{1/4}(x^{1/4} - 3x^{3/4})$

Objective 5 Use rational exponents to simplify radical expressions.

Simplifying Radical Expressions

Radical expressions can often be simplified if we write the radical as an exponential expression. Rational exponents also enable us to multiply radicals that have the same radicand but difference indices.

Procedure: Simplifying Radical Expressions

Step 1: Convert the radical expression to an exponential expression.
Step 2: Apply the properties of exponents.
Step 3: Write the result in radical form.

Objective 5 Examples Use rational exponents to simplify each expression. Assume all variables represent positive real numbers.

5a. $\sqrt[3]{x^6}$ **5b.** $\sqrt[6]{x^3}$ **5c.** $\sqrt[4]{9}$ **5d.** $\sqrt[3]{2} \cdot \sqrt{2}$

Solutions

5a.
$\sqrt[3]{x^6} = x^{6/3}$ Convert the radical to a rational exponent.
$= x^2$ Simplify the exponent.

5b.
$\sqrt[6]{x^3} = x^{3/6}$ Convert the radical to a rational exponent.
$= x^{1/2}$ Simplify the exponent.
$= \sqrt{x}$ Convert the rational exponent to a radical.

5c.
$\sqrt[4]{9} = \sqrt[4]{3^2}$ Rewrite 9 as an exponent, $9 = 3^2$.
$= 3^{2/4}$ Convert the radical to a rational exponent.
$= 3^{1/2}$ Simplify the exponent.
$= \sqrt{3}$ Convert the rational exponent to a radical.

5d.
$\sqrt[3]{2} \cdot \sqrt{2} = 2^{1/3} \cdot 2^{1/2}$ Convert each radical to a rational exponent.
$= 2^{1/3 + 1/2}$ Apply the product of like bases rule.
$= 2^{2/6 + 3/6}$ Find a common denominator.
$= 2^{5/6}$ Add the exponents.
$= \sqrt[6]{2^5}$ Convert the rational exponent to radical form.
$= \sqrt[6]{32}$ Simplify the radicand.

Student Check 5 Use rational exponents to simplify each expression. Assume all variables represent positive real numbers.

a. $\sqrt{x^8}$ **b.** $\sqrt[8]{x^2}$ **c.** $\sqrt[4]{4}$ **d.** $\sqrt[3]{5} \cdot \sqrt[4]{5}$

Applications

Objective 6 Solve applications with rational exponents.

Rational exponents are used in many formulas, such as those for compound interest and wind chill. Any formula that contains a radical can be written with rational exponents. Formulas that involve exponents are applications of rational exponents when those exponents are rational numbers. Example 6 illustrates a few of these applications.

Objective 6 Examples Solve each problem.

6a. The compound interest formula is $A = P\left(1 + \frac{r}{n}\right)^{nt}$, where A is the amount of money in an account in t years if P dollars is invested at an annual interest rate r, compounded n times a year. How much money will be in an account in 6 months if \$3000 is invested at an annual interest rate of 5% compounded annually?

Solution **6a.** $P = 3000$, $r = 0.05$, $n = 1$, and $t = \frac{1}{2}$.

$$A = P\left(1 + \frac{r}{n}\right)^{nt}$$ State the compound interest formula.

$$A = 3000\left(1 + \frac{0.05}{1}\right)^{1(1/2)}$$ Let $P = 3000$, $r = 0.05$, $n = 1$, and $t = \frac{1}{2}$.

$$A = 3000(1.05)^{1/2}$$ Simplify the exponent and in parentheses.

$$A \approx 3074.09$$ Approximate the value.

So, \$3074.09 will be in the account after 6 months.

INSTRUCTOR NOTE: Remind students that the power $0.16 = \frac{16}{100} = \frac{4}{25}$. So, this is an application of rational exponents.

6b. The wind chill temperature, WC, in °F, can be determined by the given formula, where T is the air temperature in °F and V is the wind speed in mph.

$$WC = 35.74 + 0.6215T - 35.75V^{0.16} + 0.4275TV^{0.16}$$

Use the formula to determine the wind chill temperature if the air temperature is 45°F and the wind speed is 15 mph. Round answer to the nearest hundredth. (Source: http://www.weather.gov/om/windchill/)

Solution **6b.** Replace T with 45 and V with 15.

$$WC = 35.74 + 0.6215T - 35.75V^{0.16} + 0.4275TV^{0.16}$$

$$WC = 35.74 + 0.6215(45) - 35.75(15)^{0.16} + 0.4275(45)(15)^{0.16}$$

$$WC \approx 35.74 + 27.9675 - 35.75(1.54232) + 0.4275(45)(1.54232)$$

$$WC \approx 35.74 + 27.9675 - 55.13794 + 29.67038$$

$$WC \approx 38.24$$

So, the wind chill temperature in these conditions is approximately 38.24°F.

Student Check 6 Solve each problem.

a. How much money will be in an account after 3 months if \$5000 is invested at an annual interest rate of 4% compounded quarterly ($n = 4$)?

b. Use the wind chill formula to calculate the wind chill temperature when the air temperature is 20°F and the wind speed is 30 mph. Round the answer to the nearest hundredth of a degree.

Objective 7 ▶ Troubleshoot common errors.

Troubleshooting Common Errors

Some common errors associated with rational exponents are shown next.

Objective 7 Examples **A problem and an incorrect solution are given. Provide the correct solution and an explanation of the error.**

7a. Simplify $(-27)^{2/3}$.

Incorrect Solution	Correct Solution and Explanation
$(-27)^{2/3} = (\sqrt[3]{-27})^2 = -3^2 = -9$	The error was made in not applying the exponent of 2 to -3. $(-27)^{2/3} = (\sqrt[3]{-27})^2 = (-3)^2 = 9$

7b. Simplify $5^{1/4} \cdot 5^{3/4}$.

Incorrect Solution	Correct Solution and Explanation
$5^{1/4} \cdot 5^{3/4} = 25^{4/4} = 25^1 = 25$	When we multiply like bases, the bases stay the same. $5^{1/4} \cdot 5^{3/4} = 5^{4/4} = 5^1 = 5$

ANSWERS TO STUDENT CHECKS

Student Check 1 **a.** 5 **b.** -4 **c.** $\sqrt[5]{x}$ **d.** -6 **e.** not a real number **f.** $3x^3$

Student Check 2 **a.** 27 **b.** -27 **c.** 9 **d.** $\frac{1}{8}$ **e.** $\sqrt[7]{(x+2)^6}$

Student Check 3 **a.** $\frac{1}{2}$ **b.** $\frac{1}{9}$ **c.** $-\frac{1}{125}$

Student Check 4 **a.** 7 **b.** $\frac{1}{y^{3/7}}$ **c.** $x^{1/2}$ **d.** $\frac{1}{3}$ **e.** a^5 **f.** $27x^9$ **g.** $\frac{8}{x^4y^{3/2}}$ **h.** $x^{1/2} - 3x$

Student Check 5 **a.** x^4 **b.** $\sqrt[4]{x}$ **c.** $\sqrt{2}$ **d.** $\sqrt[12]{5^7} = \sqrt[12]{78{,}125}$

Student Check 6 **a.** $5050.00 **b.** 1.30 °F

SUMMARY OF KEY CONCEPTS

1. Radical expressions can be converted to exponential form, or vice versa, by using these formulas. If $\sqrt[n]{b}$ is a real number, then

$$b^{1/n} = \sqrt[n]{b}$$

$$b^{m/n} = \left(\sqrt[n]{b}\right)^m \quad \text{or} \quad \sqrt[n]{b^m}$$

$$b^{-m/n} = \left(\sqrt[n]{b}\right)^{-m} = \frac{1}{\left(\sqrt[n]{b}\right)^m}$$

2. All of the exponent rules are defined for rational exponents. If a and b are real numbers and m and n are rational numbers, then

Product of like bases: $b^m \cdot b^n = b^{m+n}$

Quotient of like bases: $\frac{b^m}{b^n} = b^{m-n}, b \neq 0$

Power of a power: $(b^m)^n = b^{mn}$

Power of a product: $(ab)^n = a^nb^n$

Power of a quotient: $\left(\frac{a}{b}\right)^n = \frac{a^n}{b^n}, b \neq 0$

Zero exponent: $b^0 = 1$

Negative exponent: $b^{-n} = \frac{1}{b^n}, b \neq 0$

3. A radical expression can often be simplified if we convert it to exponential form. This allows us to simplify the exponents and, thereby, simplify the expression. After the exponents are simplified, the expression can be converted back to radical form.

GRAPHING CALCULATOR SKILLS

We can use the graphing calculator to simplify expressions with rational exponents. The key is to put parentheses around the entire exponent.

Example: Simplify $(-32)^{6/5}$.

Solution:

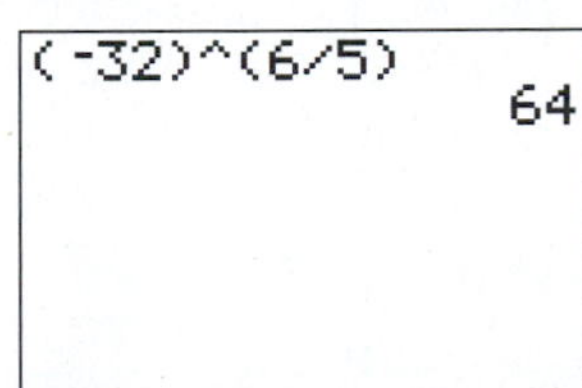

SECTION 8.2 EXERCISE SET

Write About It!

Use complete sentences in your answer to each exercise.

1. Define a rational exponent and give an example.
2. Explain how an expression with a rational exponent is related to a radical.
3. Explain how to apply the rules of exponents to simplify $x^{2/5} \cdot x^{1/3}$.
4. Explain how to apply the rules of exponents to simplify $\frac{x^{1/2}}{x^{1/3}}$.
5. Explain how to simplify $32^{-3/5}$ by converting it to a radical.
6. Explain why $2^{1/3} \cdot 3^{1/2} \neq 6^{5/6}$.

Practice Makes Perfect!

Rewrite each expression as a radical expression and simplify, if possible. Assume all variables represent positive real numbers. (*See Objective 1.*)

7. $49^{1/2}$ 7
8. $25^{1/2}$ 5
9. $121^{1/2}$ 11
10. $169^{1/2}$ 13
11. $(-64)^{1/3}$ -4
12. $(-125)^{1/3}$ -5
13. $-225^{1/2}$ -15
14. $-9^{1/2}$ -3
15. $(-64)^{1/2}$ not a real number
16. $(-225)^{1/2}$ not a real number
17. $x^{1/3}$ $\sqrt[3]{x}$
18. $y^{1/4}$ $\sqrt[4]{y}$
19. $5^{1/4}$ $\sqrt[4]{5}$
20. $20^{1/5}$ $\sqrt[5]{20}$
21. $64^{1/6}$ 2
22. $625^{1/4}$ 5
23. $(8x^6)^{1/3}$ $2x^2$
24. $(125y^9)^{1/3}$ $5y^3$
25. $(243a^{15})^{1/5}$ $3a^3$
26. $(625b^{12})^{1/4}$ $5b^3$

Rewrite each expression as a radical and simplify, if possible. Assume all variables represent positive real numbers. (*See Objective 2.*)

27. $16^{3/4}$ 8
28. $27^{4/3}$ 81
29. $25^{3/2}$ 125
30. $49^{3/2}$ 343
31. $-16^{5/4}$ -32
32. $-625^{3/4}$ -125
33. $-27^{2/3}$ -9
34. $-64^{2/3}$ -16
35. $(-125)^{2/3}$ 25
36. $(-8)^{4/3}$ 16
37. $\left(\frac{1}{27}\right)^{2/3}$ $\frac{1}{9}$
38. $\left(\frac{1}{1000}\right)^{2/3}$ $\frac{1}{100}$
39. $\left(\frac{16}{625}\right)^{3/4}$ $\frac{8}{125}$
40. $\left(\frac{256}{81}\right)^{3/4}$ $\frac{64}{27}$

Additional answers can be found in the Instructor Answer Appendix.

41. $(64x^9y^6)^{2/3}$ $16x^6y^4$
42. $(27x^6y^{12})^{4/3}$ $81x^8y^{16}$
43. $(32x^{15}y^{10})^{4/5}$ $16x^{12}y^8$
44. $(81x^{12}y^8)^{5/4}$ $243x^{15}y^{10}$
45. $\left(\frac{a^6}{64b^3}\right)^{2/3}$ $\frac{a^4}{16b^2}$
46. $\left(\frac{x^{10}}{243y^5}\right)^{3/5}$ $\frac{x^6}{27y^3}$
47. $(x+8)^{4/5}$ $\sqrt[5]{(x+8)^4}$
48. $(2x+1)^{2/5}$ $\sqrt[5]{(2x+1)^2}$
49. $(x+3)^{2/3}$ $\sqrt[3]{(x+3)^2}$
50. $(3x-1)^{2/3}$ $\sqrt[3]{(3x-1)^2}$

Rewrite each expression with positive exponents and simplify. Assume all variables represent positive real numbers. (*See Objective 3.*)

51. $64^{-1/3}$ $\frac{1}{4}$
52. $125^{-1/3}$ $\frac{1}{5}$
53. $-32^{-2/5}$ $-\frac{1}{4}$
54. $-243^{-3/5}$ $-\frac{1}{27}$
55. $(-125)^{-2/3}$ $\frac{1}{25}$
56. $(-216)^{-2/3}$ $\frac{1}{36}$
57. $(32x^{10})^{-2/5}$ $\frac{1}{4x^4}$
58. $(125y^9)^{-2/3}$ $\frac{1}{25y^6}$
59. $(64a^6b^{12})^{-5/6}$ $\frac{1}{32a^5b^{10}}$
60. $(c^{10}d^{25})^{-4/5}$ $\frac{1}{c^8d^{20}}$
61. $(x^{12}y^{-15})^{-2/3}$ $\frac{y^{10}}{x^8}$
62. $(a^{-8}b^{12})^{-5/4}$ $\frac{a^{10}}{b^{15}}$

Simplify each expression. Assume all variables represent positive real numbers. Write each answer with positive exponents. (*See Objective 4.*)

63. $5^{1/6} \cdot 5^{5/6}$ 5
64. $3^{2/5} \cdot 3^{3/5}$ 3
65. $2^{4/3} \cdot 2^{2/3}$ 4
66. $6^{4/5} \cdot 6^{6/5}$ 36
67. $x^{3/5} \cdot x^{1/2}$ $x^{11/10}$
68. $y^{4/7} \cdot y^{1/3}$ $y^{19/21}$
69. $a^{1/6} \cdot a^{-5/6}$ $\frac{1}{a^{2/3}}$
70. $b^{1/2} \cdot b^{-5/6}$ $\frac{1}{b^{1/3}}$
71. $\frac{5^{1/5}}{5^{6/5}}$ $\frac{1}{5}$
72. $\frac{7^{1/3}}{7^{4/3}}$ $\frac{1}{7}$
73. $\frac{x^{1/2}}{x^{2/5}}$ $x^{1/10}$
74. $\frac{y^{3/7}}{y^{1/3}}$ $y^{2/21}$
75. $(x^3y^6)^{4/3}$ x^4y^8
76. $(a^3b^{12})^{5/3}$ a^5b^{20}
77. $(32x^5y^{15})^{6/5}$ $64x^6y^{18}$
78. $(729a^{18}b^6)^{5/6}$ $243a^{15}b^5$
79. $(2x^{2/3}y^{-1/2})^3$ $\frac{8x^2}{y^{3/2}}$
80. $(2a^{-3/5}b^{1/3})^5$ $\frac{32b^{5/3}}{a^3}$
81. $\left(\frac{x^{3/5}y^{1/3}}{x^{1/2}}\right)^5$ $x^{1/2}y^{5/3}$
82. $\left(\frac{a^{3/2}b^{2/3}}{a^{1/3}}\right)^3$ $a^{7/2}b^2$
83. $\frac{(x^{5/6}y^{1/5})^3}{y}$ $\frac{x^{5/2}}{y^{2/5}}$
84. $\frac{(a^{2/7}b^{1/3})^7}{b^3}$ $\frac{a^2}{b^{2/3}}$
85. $x^{1/2}(x^{1/2}+x^{1/3})$ $x + x^{5/6}$
86. $y^{2/5}(y^{3/5}-y^{1/5})$ $y - y^{3/5}$
87. $(2a^{1/2}-3b^{1/3})(a^{1/2}+4b^{1/3})$ $2a + 5a^{1/2}b^{1/3} - 12b^{2/3}$
88. $(7x^{2/5}+2y^{1/2})(x^{2/5}+y^{1/2})$ $7x^{4/5} + 9x^{2/5}y^{1/2} + 2y$
89. $(4r^{1/2}-s^{2/3})(4r^{1/2}+s^{2/3})$ $16r - s^{4/3}$
90. $(a^{3/5}+3b^{2/3})(a^{3/5}-3b^{2/3})$ $a^{6/5} - 9b^{4/3}$

Use rational exponents to simplify each expression. Assume all variables represent positive real numbers. (*See Objective 5.*)

91. $\sqrt[3]{a^{12}}$ a^4

92. $\sqrt[5]{b^{15}}$ b^3

93. $\sqrt[6]{x^2}$ $\sqrt[3]{x}$

94. $\sqrt[15]{y^3}$ $\sqrt[5]{y}$

95. $\sqrt[6]{a^3}$ $\sqrt{a}$

96. $\sqrt[4]{b^2}$ $\sqrt{b}$

97. $\sqrt[6]{8}$ $\sqrt{2}$

98. $\sqrt[6]{125}$ $\sqrt{5}$

99. $\sqrt[4]{49}$ $\sqrt{7}$

100. $\sqrt[4]{25}$ $\sqrt{5}$

101. $\sqrt[6]{27a^3}$ $\sqrt{3a}$

102. $\sqrt[6]{8b^3}$ $\sqrt{2b}$

103. $\sqrt[4]{121x^2}$ $\sqrt{11x}$

104. $\sqrt[4]{49y^2}$ $\sqrt{7y}$

105. $\sqrt[5]{2}\cdot\sqrt{2}$ $\sqrt[10]{128}$

106. $\sqrt[3]{3}\cdot\sqrt{3}$ $\sqrt[6]{243}$

107. $\sqrt[5]{x^3}\cdot\sqrt{x}$ $\sqrt[10]{x^{11}}$

108. $\sqrt[3]{y}\cdot\sqrt{y^3}$ $\sqrt[6]{y^{11}}$

109. $\sqrt[6]{a}\cdot\sqrt[3]{a^2}$ $\sqrt[6]{a^5}$

110. $\sqrt{b}\cdot\sqrt[4]{b^3}$ $\sqrt[4]{b^5}$

For Exercises 111–114, use the compound interest formula $A = P\left(1 + \dfrac{r}{n}\right)^{nt}$, where A is the amount of money in an account in t years if P dollars is invested at an annual interest rate r, compounded n times a year. Round each answer to the nearest cent. (*See Objective 6.*)

111. How much money will be in an account after 2 months if you invest \$1000 at an annual interest rate of 2% compounded annually ($n = 1$)? \$1003.31

112. How much money will be in an account after 4 months if you invest \$1500 at an annual interest rate of 2.5% compounded annually ($n = 1$)? \$1512.40

113. How much money will be in an account after 2 months if you invest \$800 at an annual interest rate of 2% compounded quarterly ($n = 4$)? \$802.66

114. How much money will be in an account after 4 months if you invest \$2400 at an annual interest rate of 1.5% compounded quarterly ($n = 4$)? \$2412.01

For Exercises 115 and 116, use the wind chill formula, $WC = 35.74 + 0.6215T - 35.75V^{0.16} + 0.4275TV^{0.16}$, where T is the air temperature in °F and V is the wind speed in mph. Round each answer to the nearest hundredth of a degree. (*See Objective 6.*)

115. What is the wind chill temperature when the air temperature is 32°F and the wind speed is 20 mph? 19.99°F

116. What is the wind chill temperature when the air temperature is 0°F and the wind speed is 25 mph? −24.09°F

Mix 'Em Up!

Rewrite each expression as a radical expression and simplify, if possible. Assume all variables represent positive real numbers.

117. $(-8x^9)^{1/3}$ $-2x^3$

118. $(-64y^6)^{1/3}$ $-4x^2$

119. $\left(\dfrac{x^3}{8y^6}\right)^{5/3}$ $\dfrac{x^5}{32y^{10}}$

120. $\left(\dfrac{a^8}{81b^4}\right)^{1/4}$ $\dfrac{a^2}{3b}$

121. $(125a^3b^6)^{2/3}$ $25a^2b^4$

122. $(8c^3d^6)^{5/3}$ $32c^5d^{10}$

123. $(243x^{-10}y^5)^{3/5}$ $\dfrac{27y^3}{x^6}$

124. $(625a^8b^{-4})^{3/4}$ $\dfrac{125a^6}{b^3}$

125. $(y + 5)^{4/3}$ $\sqrt[3]{(y+5)^4}$

126. $(2x + 3)^{3/2}$ $\sqrt{(2x+3)^3}$

Simplify each expression and write each answer with positive exponents. Assume all variables represent positive real numbers.

127. $(-64x^6)^{-1/3}$ $-\dfrac{1}{4x^2}$

128. $(125y^{-3})^{-2/3}$ $\dfrac{y^2}{25}$

129. $2^{5/6}\cdot 2^{7/6}$ 4

130. $3^{4/5}\cdot 3^{6/5}$ 9

131. $(16x^8)^{-3/4}$ $\dfrac{1}{8x^6}$

132. $(-27y^6)^{-2/3}$ $\dfrac{1}{9y^4}$

133. $(a^{-6}b^{12})^{-5/3}$ $\dfrac{a^{10}}{b^{20}}$

134. $(-c^5d^{-15})^{-3/5}$ $-\dfrac{d^9}{c^3}$

135. $r^{1/4}\cdot r^{5/3}$ $r^{23/12}$

136. $s^{4/7}\cdot s^{-1/3}$ $s^{5/21}$

137. $(-x^5y^{10})^{2/5}$ x^2y^4

138. $(-a^3b^9)^{2/3}$ a^2b^6

139. $\dfrac{x^{2/7}}{x^{1/2}}$ $\dfrac{1}{x^{3/14}}$

140. $\dfrac{y^{1/5}}{y^{2/3}}$ $\dfrac{1}{y^{7/15}}$

141. $(3x^{2/5}y^{-2/3})^5$ $\dfrac{243x^2}{y^{10/3}}$

142. $(4a^{-1/2}b^{5/6})^3$ $\dfrac{64b^{5/2}}{a^{3/2}}$

143. $\dfrac{(x^{-3/7}y^{1/2})^4}{x^3}$ $\dfrac{y^2}{x^{33/7}}$

144. $\dfrac{(a^{-5/2}b^{1/2})^2}{a^{4/3}}$ $\dfrac{b}{a^{19/3}}$

145. $(3x^{1/2} - y^{1/3})(3x^{1/2} + y^{1/3})$ $9x - y^{2/3}$

146. $(2a^{2/5} + 3b^{1/2})(2a^{2/5} - 3b^{1/2})$ $4a^{4/5} - 9b$

147. $r^{1/2}(5r^{1/2} + r^{2/3})$ $5r + r^{7/6}$

148. $s^{2/3}(s^{2/3} - 4s^{1/2})$ $s^{4/3} - 4s^{7/6}$

149. $(5x^{1/2} - 3y^{1/3})(2x^{1/2} + 4y^{1/3})$ $10x + 14x^{1/2}y^{1/3} - 12y^{2/3}$

150. $(a^{2/5} - 3b^{1/2})(a^{2/5} - 5b^{1/2})$ $a^{4/5} - 8a^{2/5}b^{1/2} + 15b$

Use rational exponents to simplify each expression. Assume all variables represent positive real numbers.

151. $\sqrt[3]{64x^6}$ $4x^2$

152. $\sqrt[5]{32y^{20}}$ $2y^4$

153. $\sqrt[6]{343a^3}$ $\sqrt{7a}$

154. $\sqrt[6]{216b^3}$ $\sqrt{6b}$

155. $\sqrt[4]{\dfrac{49}{x^4}}$ $\dfrac{\sqrt{7}}{x}$

156. $\sqrt[4]{\dfrac{9}{y^4}}$ $\dfrac{\sqrt{3}}{y}$

157. $\sqrt{x}\cdot\sqrt[3]{x^5}$ $\sqrt[6]{x^{13}}$

158. $\sqrt{y^3}\cdot\sqrt[4]{y}$ $\sqrt[4]{y^7}$

Solve each problem.

159. How much money will be in an account after 8 months if you invest \$500 at an annual interest rate of 1.8% compounded quarterly ($n = 4$)? Round to the nearest cent. \$506.02

160. How much money will be in an account after 9 months if you invest \$600 at an annual interest rate of 2.4% compounded quarterly ($n = 4$)? Round to the nearest cent. \$610.86

161. On top of Mount Everest, the wind speed can reach 115 mph and the temperature is $-15°$F. Use the wind chill formula to determine the wind chill temperature in these conditions. (Source: http://www.mounteverest2008.com/) $-63.66°$ F

162. The formula $v = 5.289(T + 4)^{3/2}$ gives the wind speed in miles per hour of a tornado with a Torro intensity value, T. A violent tornado has an intensity of T-10. Estimate the wind speed of this tornado. Round the answer to the nearest hundredth. (Source: http://www.torro.org.uk/TORRO/severeweather/Tscaleorigin.php) 277.05 mph

You Be the Teacher!

Correct each student's errors, if any.

163. Simplify: $5^{1/3} \cdot 5^{2/3}$.

Celine's work: $5^{1/3} \cdot 5^{2/3} = 25^{1/3+2/3} = 25^{3/3} = 25^1 = 25$

164. Simplify: $(-16x^8)^{3/4}$. Assume $x \geq 0$.

Christine's work: $(-16x^8)^{3/4} = -16^{3/4}(x^8)^{3/4} = -8x^6$

165. Simplify: $-(-x^3y^6)^{2/3}$. Assume $x, y \geq 0$.

Tony's work:
$-(-x^3y^6)^{2/3} = (x^3y^6)^{2/3} = (x^3)^{2/3}(y^6)^{2/3} = x^2y^4$

166. Simplify: $-\sqrt[4]{9x^2}$. Assume $x \geq 0$.

Kim's work: $-\sqrt[4]{9x^2} = -3x$

Calculate It!

Use a calculator to simplify each expression. Write the answer to two decimal places. If possible, convert the result to fractional form.

167. $(36)^{5/3}$

168. $(7)^{-2/5}$

169. $(32)^{-4/5}$

170. $(125)^{-4/3}$

SECTION 8.3 Simplifying Radical Expressions and the Distance Formula

OBJECTIVES

As a result of completing this section, you will be able to

1. Use the product rule to simplify radicals.
2. Use the quotient rule to simplify radicals.
3. Use the distance formula.
4. Troubleshoot common errors.

Suppose we want to find the distance between Key West, Florida, and Pensacola, Florida, as shown on the map (Source: www.siteatlas.com). We can impose a grid on the map, label the coordinates of the two cities and then use a distance formula to find the distance between them. In this section, we will learn the distance formula and two basic rules of radicals that enable us to simplify radical expressions.

Simplifying Radical Expressions Using the Product Rule

Objective 1 Use the product rule to simplify radicals.

We know that $\sqrt{25}$ and $\sqrt[3]{8}$ are radical expressions that are not in their *simplest form* since $\sqrt{25} = 5$ and $\sqrt[3]{8} = 2$. When the radicand can be written as a number raised to the index, the radical can be simplified fairly easily. If a radicand contains a factor raised to the index, it can also be simplified. Before this is illustrated, we need to learn the *product rule for radicals*. The following illustration shows some radical expressions that are equivalent.

$$\sqrt{4 \cdot 4} = \sqrt{16} = 4 \quad \Big| \quad \sqrt{4} \cdot \sqrt{4} = 2 \cdot 2 = 4 \longrightarrow \sqrt{4 \cdot 4} = \sqrt{4} \cdot \sqrt{4} = 4$$

$$\sqrt[3]{8 \cdot 27} = \sqrt[3]{216} = 6 \quad \Big| \quad \sqrt[3]{8} \cdot \sqrt[3]{27} = 2 \cdot 3 = 6 \longrightarrow \sqrt[3]{8 \cdot 27} = \sqrt[3]{8} \cdot \sqrt[3]{27} = 6$$

These examples show that the nth root of a product is the same as the product of the nth roots.

Property: Product Rule for Radicals

If $\sqrt[n]{a}$ and $\sqrt[n]{b}$ are real numbers, and n is a positive integer, then

$$\sqrt[n]{a \cdot b} = \sqrt[n]{a} \cdot \sqrt[n]{b}$$

For a radical to be in its simplest form, the radicand cannot contain any factor raised to the index. Some examples of radicals that are not in their simplest form are $\sqrt{12}$, $\sqrt[3]{16}$, $\sqrt{y^7}$, and $\sqrt[3]{x^5}$.

- $\sqrt{12}$ is not in simplest form since $12 = 4 \cdot 3$ and 4 is a perfect square factor of 12.
- $\sqrt[3]{16}$ is not in simplest form since $16 = 8 \cdot 2$ and 8 is a perfect cube factor of 16.
- $\sqrt{y^7}$ is not in simplest form since $y^7 = y^6 \cdot y$ and y^6 is a perfect square factor of y^7.
- $\sqrt[3]{x^5}$ is not in simplest form since $x^5 = x^3 \cdot x^2$ and x^3 is a perfect cube factor of x^5.

Note: *The key to simplifying higher roots lies in our ability to recognize perfect powers. Refer to the chart in Section 8.1 for more perfect powers.*

Perfect squares: $1, 4, 9, 16, 25, 36, \ldots, x^2, x^4, x^6, \ldots$

Perfect cubes: $1, 8, 27, 64, 125, 216, \ldots, x^3, x^6, x^9, \ldots$

Perfect fourths: $1, 16, 81, 256, \ldots, x^4, x^8, x^{12}, \ldots$

To simplify radicals, we use properties that were presented in Sections 8.1 and 8.2. We will assume all variables represent positive real numbers,

Property	Examples
$\sqrt[n]{a^n} = a$	$\sqrt{x^2} = x$, $\sqrt[3]{x^3} = x$, and $\sqrt[4]{(x^3)^4} = x^3$
$\sqrt[n]{a^m} = a^{m/n}$	$\sqrt[3]{x^6} = x^{6/3} = x^2$ and $\sqrt{x^4} = x^{4/2} = x^2$

There are two methods that we can use to simplify radical expressions. One method relies on listing the factors for the radicand and the other method relies on expressing the radicand in its prime factorization.

Procedure: Simplifying Radical Expressions

Method 1

Step 1: List the pairs of factors of the radicand.

Step 2: Select the pair of factors for which one factor is a *perfect power whose exponent is the index.*

Step 3: Rewrite the radicand as the product of numbers found in Step 2. Write the perfect power first.

Step 4: Apply the product rule and extract any roots.

Method 2

Step 1: Rewrite the radicand as a product of its prime factors.

Step 2: Apply the product rule and extract any roots.

Note: *For a radical expression to be completely simplified, the radicand cannot contain any exponents greater than the index.*

Objective 1 Examples **Simplify each radical completely. Assume all variables represent positive real numbers.**

1a. $\sqrt{80}$ **1b.** $2\sqrt{72}$ **1c.** $\sqrt{y^5}$ **1d.** $\sqrt{24x^3}$ **1e.** $\sqrt[3]{54}$ **1f.** $\sqrt[4]{80a^7b^8}$

Solutions

1a.

Method 1: Factors of 80

1, 80
2, 40
4, 20
5, 16
8, 10

4 and 16 are both perfect square factors. We will use 16 since it is larger.

$\sqrt{80} = \sqrt{16 \cdot 5}$ Rewrite 80 as $16 \cdot 5$.

$= \sqrt{16}\sqrt{5}$ Apply the product rule.

$= 4\sqrt{5}$ Write $\sqrt{16}$ as 4.

Method 2: Prime factorization of 80

$80 = 4 \cdot 20$
$= 2 \cdot 2 \cdot 4 \cdot 5$
$= 2 \cdot 2 \cdot 2 \cdot 2 \cdot 5$
$= 2^4 \cdot 5$

$\sqrt{80} = \sqrt{2^4 \cdot 5}$ Rewrite 80 as $2^4 \cdot 5$.

$= \sqrt{2^4}\sqrt{5}$ Apply the product rule.

$= 2^2\sqrt{5}$ Write $\sqrt{2^4}$ as 2^2.

$= 4\sqrt{5}$ Simplify 2^2.

1b.

Method 1: Factors of 72

1, 72
2, 36
3, 24
4, 18
6, 12
8, 9

36, 4, and 9 are perfect square factors. We will use 36 since it is the largest.

$2\sqrt{72} = 2\sqrt{36 \cdot 2}$ Rewrite 72 as $36 \cdot 2$.

$= 2\sqrt{36}\sqrt{2}$ Apply the product rule.

$= 2(6)\sqrt{2}$ Write $\sqrt{36}$ as 6.

$= 12\sqrt{2}$ Multiply the coefficients.

Method 2: Prime factorization of 72

$72 = 2 \cdot 36$
$= 2 \cdot 6 \cdot 6$
$= 6^2 \cdot 2$

$2\sqrt{72} = 2\sqrt{6^2 \cdot 2}$ Rewrite 72 as $6^2 \cdot 2$.

$= 2\sqrt{6^2}\sqrt{2}$ Apply the product rule.

$= 2(6)\sqrt{2}$ Write $\sqrt{6^2}$ as 6.

$= 12\sqrt{2}$ Multiply the coefficients.

1c. We apply method 1 to simplify this radical. The factors of y^5 are

y^4, y
y^3, y^2

INSTRUCTOR NOTE: Remind students that $b^m \cdot b^n = b^{m+n}$.

Both y^4 and y^2 are perfect squares. We use y^4 since it is the largest perfect square factor.

$\sqrt{y^5} = \sqrt{y^4 \cdot y}$ Rewrite y^5 as $y^4 \cdot y$.

$= \sqrt{y^4}\sqrt{y}$ Apply the product rule for radicals.

$= y^2\sqrt{y}$ Write $\sqrt{y^4}$ as y^2.

1d. We apply method 1 to simplify this radical. The factors of 24 and x^3 are

1, 24
2, 12 $\quad x^2, x$
3, 8
4, 6

4 and x^2 are perfect square factors.

$\sqrt{24x^3} = \sqrt{4x^2 \cdot 6x}$ Rewrite $24x^3$ as $4x^2 \cdot 6x$.

$= \sqrt{4x^2}\sqrt{6x}$ Apply the product rule.

$= 2x\sqrt{6x}$ Write $\sqrt{4x^2}$ as $2x$.

1e. We need to find perfect cube factors of 54 since the index is 3.

Method 1: Factors of 54

1, 54
2, 27 — 27 is a perfect cube.
3, 18
6, 9

$$\sqrt[3]{54} = \sqrt[3]{27 \cdot 2} \quad \text{Rewrite 54 as } 27 \cdot 2.$$
$$= \sqrt[3]{27}\sqrt[3]{2} \quad \text{Apply the product rule.}$$
$$= 3\sqrt[3]{2} \quad \text{Write } \sqrt[3]{27} \text{ as 3.}$$

Method 2: Prime factorization of 54

$$54 = 6 \cdot 9$$
$$= 2 \cdot 3 \cdot 3 \cdot 3$$
$$= 2 \cdot 3^3$$

$$\sqrt[3]{54} = \sqrt[3]{3^3 \cdot 2} \quad \text{Rewrite 54 as } 3^3 \cdot 2.$$
$$= \sqrt[3]{3^3}\sqrt[3]{2} \quad \text{Apply the product rule.}$$
$$= 3\sqrt[3]{2} \quad \text{Write } \sqrt[3]{3^3} \text{ as 3.}$$

1f. We apply method 2 to simplify this radical. The prime factorization of 80 is

$$80 = 8 \cdot 10$$
$$= 2 \cdot 2 \cdot 2 \cdot 2 \cdot 5$$
$$= 2^4 \cdot 5$$

A perfect fourth factor of a^7 is a^4, so $a^7 = a^4 \cdot a^3$. Note that b^8 is a perfect fourth since $b^8 = (b^2)^4$.

$$\sqrt[4]{80a^7b^8} = \sqrt[4]{2^4 \cdot 5a^4a^3(b^2)^4}$$
$$= \sqrt[4]{2^4a^4(b^2)^4}\sqrt[4]{5a^3}$$
$$= 2ab^2\sqrt[4]{5a^3}$$

Student Check 1 Simplify each radical completely. Assume all variables represent positive real numbers.

a. $\sqrt{18}$ **b.** $-3\sqrt{76}$ **c.** $\sqrt{a^9}$ **d.** $\sqrt{32y^{10}}$ **e.** $\sqrt[3]{128}$ **f.** $\sqrt[4]{162x^6y^4}$

Note: *In Example 1b, we would have obtained the same result if we used a different pair of factors. We would just have more steps of simplification. For instance,*

$$2\sqrt{72} = 2\sqrt{9 \cdot 8} \quad \text{Rewrite 72 as } 9 \cdot 8.$$
$$= 2\sqrt{9}\sqrt{8} \quad \text{Apply the product rule.}$$
$$= 2(3)\sqrt{8} \quad \text{Write } \sqrt{9} \text{ as 3.}$$
$$= 6\sqrt{8} \quad \text{Multiply the coefficients.}$$
$$= 6\sqrt{4 \cdot 2} \quad \text{Rewrite 8 as } 4 \cdot 2.$$
$$= 6\sqrt{4}\sqrt{2} \quad \text{Apply the product rule.}$$
$$= 6(2)\sqrt{2} \quad \text{Write } \sqrt{4} \text{ as 2.}$$
$$= 12\sqrt{2} \quad \text{Multiply the coefficients.}$$

If we do not use the largest perfect square factor, the radicand will contain another perfect square factor. Therefore, the steps to simplify have to be repeated.

Objective 2 Use the quotient rule to simplify radicals.

Simplifying Radical Expressions Using the Quotient Rule

When the radicand is a quotient, we can simplify it by applying the root to the numerator and denominator. Consider the following example.

$$\sqrt{\frac{36}{4}} = \sqrt{9} = 3 \quad \Big| \quad \frac{\sqrt{36}}{\sqrt{4}} = \frac{6}{2} = 3 \longrightarrow \sqrt{\frac{36}{4}} = \frac{\sqrt{36}}{\sqrt{4}} = 3$$

This example illustrates another important property of radicals. The nth root of a quotient is the same as the quotient of the nth roots.

Property: Quotient Rule for Radicals

If $\sqrt[n]{a}$ and $\sqrt[n]{b}$ are real numbers with $b \neq 0$ and n is a positive integer, then

$$\sqrt[n]{\frac{a}{b}} = \frac{\sqrt[n]{a}}{\sqrt[n]{b}}$$

For a radical to be in its simplest form,

- There cannot be any fractions in the radical.
- There cannot be any radicals in the denominator of a fraction. In Section 8.5, we will learn how to remove a radical from a denominator.

Procedure: Simplifying the nth Root of a Quotient

Step 1: If the radicand is a quotient, apply the root to the numerator and denominator.
Step 2: Rewrite each radicand as a product of a perfect power whose exponent is the index and another factor.
Step 3: Apply the product rule.

Objective 2 Examples Simplify each radical completely. Assume the variables represent positive real numbers.

2a. $\sqrt{\frac{4}{9}}$ **2b.** $\sqrt[3]{\frac{x}{8}}$ **2c.** $\sqrt{\frac{20}{9y^4}}$ **2d.** $\sqrt[3]{\frac{192x^5}{27}}$

Solutions

2a. $\sqrt{\frac{4}{9}} = \frac{\sqrt{4}}{\sqrt{9}} = \frac{2}{3}$

2b. $\sqrt[3]{\frac{x}{8}} = \frac{\sqrt[3]{x}}{\sqrt[3]{8}} = \frac{\sqrt[3]{x}}{2}$ or $\frac{1}{2}\sqrt[3]{x}$

2c. $\sqrt{\frac{20}{9y^4}} = \frac{\sqrt{20}}{\sqrt{9y^4}}$ Apply the quotient rule for radicals.

$= \frac{\sqrt{4 \cdot 5}}{3y^2}$ Write the numerator with a perfect square factor and simplify the denominator.

$= \frac{\sqrt{4}\sqrt{5}}{3y^2}$ Apply the product rule for radicals.

$= \frac{2\sqrt{5}}{3y^2}$ Simplify.

2d. We apply the quotient rule for radicals and then find the factors of the radicand that are perfect cubes. The factors of 192 and x^5 are

1, 192	x^4, x
2, 96	x^3, x^2
3, 64	
4, 48	
6, 32	
8, 24	
12, 16	

$$\sqrt[3]{\frac{192x^5}{27}} = \frac{\sqrt[3]{192x^5}}{\sqrt[3]{27}}$$ Apply the quotient rule for radicals.

$$= \frac{\sqrt[3]{64 \cdot 3x^3x^2}}{\sqrt[3]{27}}$$ Write the appropriate factors of the radicand.

$$= \frac{\sqrt[3]{64x^3}\sqrt[3]{3x^2}}{3}$$ Apply the product rule for radicals and simplify the denominator.

$$= \frac{4x\sqrt[3]{3x^2}}{3}$$ Simplify.

Student Check 2 Simplify each radical completely. Assume the variables represent positive real numbers.

a. $\sqrt{\frac{81}{49}}$ **b.** $\sqrt[3]{\frac{a^2}{64}}$ **c.** $\sqrt{\frac{44}{9x^2}}$ **d.** $\sqrt[3]{\frac{40y^8}{343}}$

The Distance Formula

Use the distance formula.

The distance formula provides a way for us to measure the distance between two ordered pairs on a coordinate system. The distance formula is based on the Pythagorean theorem, $a^2 + b^2 = c^2$.

If we let the distance between (x_1, y_1) and (x_2, y_2) be represented by d, then we have by the Pythagorean theorem

$$(x_2 - x_1)^2 + (y_2 - y_1)^2 = d^2$$

So, $d = \sqrt{(x_2 - x_1)^2 + (y_2 - y_1)^2}$.

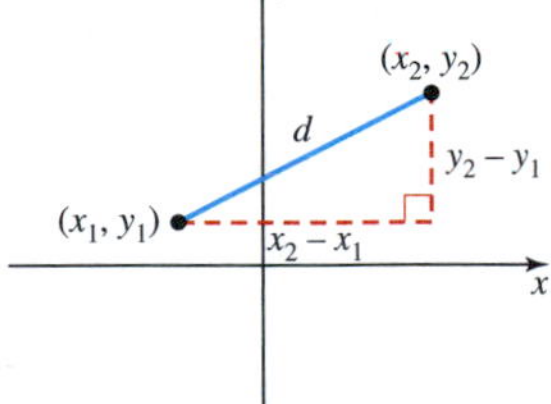

Property: Distance Formula

If (x_1, y_1) and (x_2, y_2) are two points, then the distance between them is

$$d = \sqrt{(x_2 - x_1)^2 + (y_2 - y_1)^2}$$

Procedure: Finding the Distance Between Two Points

Step 1: Label one point (x_1, y_1) and the other point (x_2, y_2).
Step 2: Substitute the values in the formula.
Step 3: Simplify the radicand.
Step 4: Simplify the radical by applying the product rule.

Objective 3 Examples

Find the distance between each pair of points. State the exact distance by expressing answers with simplified radicals and approximate them to two decimal places.

3a. $(-3, 5)$ and $(3, 8)$

Solution

3a. $d = \sqrt{(x_2 - x_1)^2 + (y_2 - y_1)^2}$ State the distance formula.

$= \sqrt{[3 - (-3)]^2 + (8 - 5)^2}$ Let $(x_1, y_1) = (-3, 5)$ and $(x_2, y_2) = (3, 8)$.

$= \sqrt{(6)^2 + (3)^2}$ Simplify inside each set of parentheses.

$= \sqrt{36 + 9}$ Simplify the exponents.

$= \sqrt{45}$ Add.

$= \sqrt{9 \cdot 5}$ Write 45 with a perfect square factor.

$= 3\sqrt{5}$ Apply the product rule for radicals and simplify.

The exact distance between $(-3, 5)$ and $(3, 8)$ is $3\sqrt{5}$ units, or approximately 6.71 units.

3b. $(4, \sqrt{7})$ and $(-7, 0)$

Solution

3b. Let $(x_1, y_1) = (4, \sqrt{7})$ and $(x_2, y_2) = (-7, 0)$.

$d = \sqrt{(x_2 - x_1)^2 + (y_2 - y_1)^2}$ State the distance formula.

$= \sqrt{(-7 - 4)^2 + (0 - \sqrt{7})^2}$ Let $(x_1, y_1) = (4, \sqrt{7})$ and $(x_2, y_2) = (-7, 0)$.

$= \sqrt{(-11)^2 + (-\sqrt{7})^2}$ Simplify inside each set of parentheses.

$= \sqrt{121 + 7}$ Simplify the exponents.

$= \sqrt{128}$ Add.

$= \sqrt{64 \cdot 2}$ Write 128 with a perfect square factor.

$= 8\sqrt{2}$ Apply the product rule for radicals and simplify.

The exact distance between $(4, \sqrt{7})$ and $(-7, 0)$ is $8\sqrt{2}$ units, or approximately 11.31 units.

3c. Use the map to approximate the distance between Key West, Florida, and Pensacola, Florida. Each tick mark represents 60 mi. (Source: http://www.sitesatlas.com/Maps/index.htm)

Solution **3c.** We first identify coordinates of the two cities. Key West is approximately located at (60, −180) and Pensacola is approximately located at (−240, 180).

$d = \sqrt{(x_2 - x_1)^2 + (y_2 - y_1)^2}$	State the distance formula.
$= \sqrt{[180 - (-180)]^2 + (-240 - 60)^2}$	Let $(x_1, y_1) = (60, -180)$ and $(x_2, y_2) = (-240, 180)$.
$= \sqrt{(360)^2 + (-300)^2}$	Simplify inside both sets of parentheses.
$= \sqrt{129{,}600 + 90{,}000}$	Simplify the exponents.
$= \sqrt{219{,}600}$	Add.
$= \sqrt{3600 \cdot 61}$	Write 219,600 with a perfect square factor.
$= 60\sqrt{61}$	Apply the product rule for radicals and simplify.

So, the distance from Key West to Pensacola is $60\sqrt{61}$ mi or approximately 468.61 mi.

Student Check 3 Find the distance between each pair of points. State the exact distances by expressing answers with simplified radicals and approximate them to two decimal places.

a. $(-4, 2)$ and $(4, 6)$

b. $(5, \sqrt{3})$ and $(-2, 0)$

c. Use the map to approximate the distance between Orlando (90, 90) and Boca Raton (180, −60). Each tick mark represents 60 mi. (Source: http://www.sitesatlas.com/Maps/index.htm)

Objective 4 ▶ Troubleshoot common errors.

Troubleshooting Common Errors

Some common errors associated with simplifying radical expressions are shown.

Objective 4 Examples **A problem and an incorrect solution are given. Provide the correct solution and an explanation of the error.**

4a. Simplify $\sqrt[3]{24}$.

Incorrect Solution	Correct Solution and Explanation
$\sqrt[3]{24} = \sqrt[3]{4 \cdot 6} = 2\sqrt[3]{6}$	The index is 3 so we should remove perfect cube factors not perfect square factors. $\sqrt[3]{24} = \sqrt[3]{8 \cdot 3} = 2\sqrt[3]{3}$

4b. Simplify $\sqrt{162x^4}$.

Incorrect Solution	Correct Solution and Explanation
$\sqrt{162x^4} = \sqrt{9 \cdot 2x^2} = 3x\sqrt{2}$	The factors inside the radical should be factors of the original radicand. $\sqrt{162x^4} = \sqrt{81 \cdot 2x^4} = \sqrt{81x^4}\sqrt{2} = 9x^2\sqrt{2}$

4c. Find the distance between $(6, -1)$ and $(4, 3)$.

Incorrect Solution

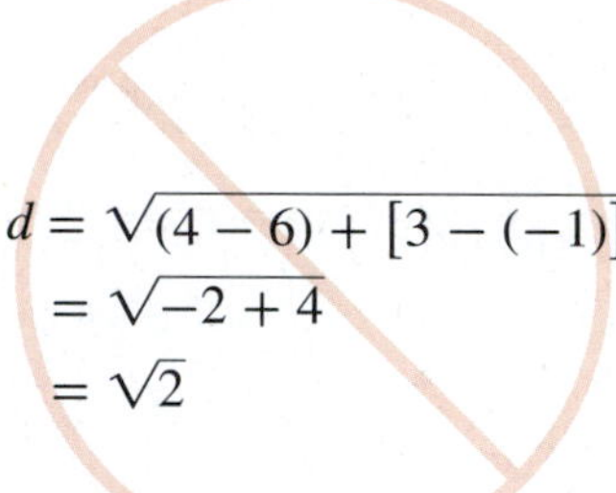

$$d = \sqrt{(4-6) + [3-(-1)]} = \sqrt{-2+4} = \sqrt{2}$$

Correct Solution and Explanation

The distance formula has a square on each of the differences inside the radical.

$$\begin{aligned} d &= \sqrt{(4-6)^2 + [3-(-1)]^2} \\ &= \sqrt{(-2)^2 + (4)^2} \\ &= \sqrt{4+16} \\ &= \sqrt{20} \\ &= \sqrt{4 \cdot 5} \\ &= 2\sqrt{5} \end{aligned}$$

ANSWERS TO STUDENT CHECKS

Student Check 1 **a.** $3\sqrt{2}$ **b.** $-6\sqrt{19}$ **c.** $a^4\sqrt{a}$ **d.** $4y^5\sqrt{2}$ **e.** $4\sqrt[3]{2}$ **f.** $3xy\sqrt[4]{2x^2}$

Student Check 2 **a.** $\frac{9}{7}$ **b.** $\frac{1}{4}\sqrt[3]{a^2}$ **c.** $\frac{2\sqrt{11}}{3x}$ **d.** $\frac{2y^2\sqrt[3]{5y^2}}{7}$

Student Check 3 **a.** $4\sqrt{5}$ units or 8.94 units **b.** $2\sqrt{13}$ units or 7.21 units **c.** $30\sqrt{34}$ or 174.93 mi

SUMMARY OF KEY CONCEPTS

1. The product rule for radicals tells us that the nth root of a product is the product of nth roots. This enables us to simplify radicals by writing the radicand as a product involving a power whose exponent is the index.
2. The quotient rule for radicals tells us that the nth root of a quotient is the quotient of nth roots.
3. The distance between two points (x_1, y_1) and (x_2, y_2) is $d = \sqrt{(x_2 - x_1)^2 + (y_2 - y_1)^2}$.

GRAPHING CALCULATOR SKILLS

The graphing calculator can help us learn powers, find factors of numbers, and check answers.

Example 1: Find perfect squares, cubes, and fourths.

Solution: Enter x raised to the appropriate exponent and view the table. The values in the Y_1, Y_2, and Y_3 column are perfect squares, cubes, and fourths, respectively.

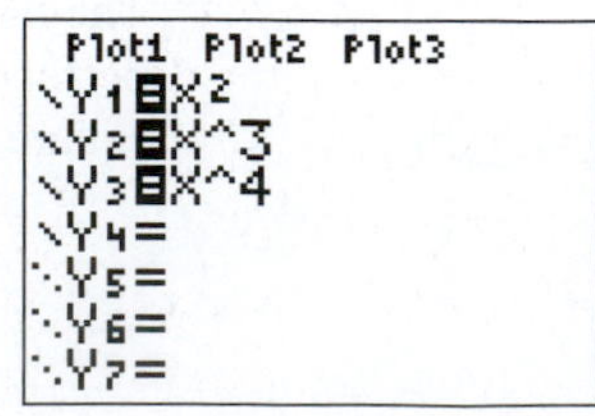

X	Y1	Y2
1	1	1
2	4	8
3	9	27
4	16	64
5	25	125
6	36	216
7	49	343

X=1

X	Y2	Y3
1	1	1
2	8	16
3	27	81
4	64	256
5	125	625
6	216	1296
7	343	2401

Y3=1

Example 2: Find factors of 54.

Solution: In the equation editor, , enter $54/x$ for Y_1. View the TABLE.

X	Y1	
1	54	
2	27	
3	18	
4	13.5	
5	10.8	
6	9	
7	7.7143	

X=1

2nd GRAPH

From the table, we see that the factors of 54 are 1 and 54, 2 and 27, 3 and 18, 6 and 9.

Example 3: Show that $\sqrt{54} = 3\sqrt{6}$ and that $\sqrt[3]{54} = 3\sqrt[3]{2}$.

Solution:

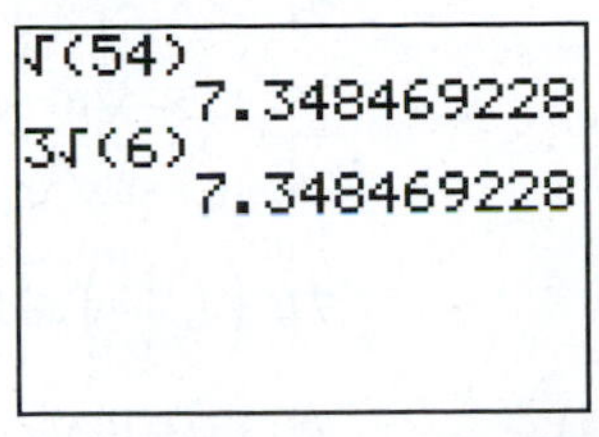

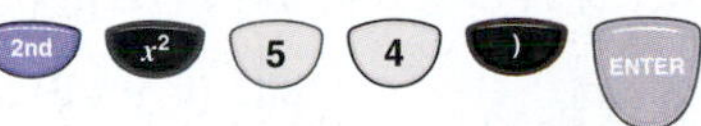

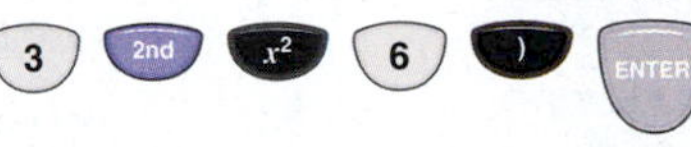

Since the decimal values are the same, $\sqrt{54} = 3\sqrt{6}$.

```
³√(54)
         3.77976315
3³√(2)
         3.77976315
```

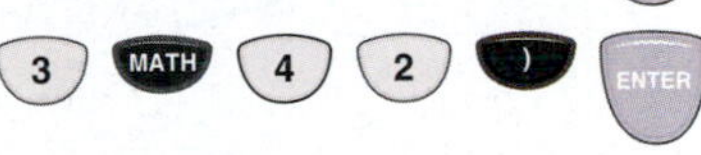

Since the decimal values are the same, $\sqrt[3]{54} = 3\sqrt[3]{2}$.

SECTION 8.3 EXERCISE SET

Write About It!

Use complete sentences in your answer to each exercise.

1. What does it mean for a square root expression to be written in simplified form?
2. What does it mean for a cube root expression to be written in simplified form?
3. How can the product rule for radicals be used to simplify a radical expression?
4. How can the quotient rule for radicals be used to simplify a radical expression?
5. Explain how to simplify $\sqrt{72x^2y^4}$.
6. Explain how to simplify $\sqrt[3]{\dfrac{54x^5}{125y^{15}}}$.

Practice Makes Perfect!

Simplify each radical completely. Assume all variables represent positive real numbers. (*See Objective 1.*)

7. $\sqrt{72}$ $6\sqrt{2}$
8. $3\sqrt{50}$ $15\sqrt{2}$
9. $-2\sqrt{90}$ $-6\sqrt{10}$
10. $-3\sqrt{75}$ $-15\sqrt{3}$
11. $\sqrt{-68}$ not a real number
12. $\sqrt{-18}$ not a real number
13. $5\sqrt{288}$ $60\sqrt{2}$
14. $3\sqrt{500}$ $30\sqrt{5}$
15. $\sqrt{48a^8}$ $4a^4\sqrt{3}$
16. $\sqrt{80b^4}$ $4b^2\sqrt{5}$
17. $\sqrt{300x^5}$ $10x^2\sqrt{3x}$
18. $\sqrt{98y^3}$ $7y\sqrt{2y}$
19. $\sqrt{54a^7}$ $3a^3\sqrt{6a}$
20. $\sqrt{112b^{13}}$ $4b^6\sqrt{7b}$
21. $\sqrt{125r^9t^{11}}$ $5r^4t^5\sqrt{5rt}$
22. $\sqrt{76x^3y^5}$ $2xy^2\sqrt{19xy}$
23. $\sqrt{144x^5y^6}$ $12x^2y^3\sqrt{x}$
24. $\sqrt{49r^4t^{13}}$ $7r^2t^6\sqrt{t}$
25. $\sqrt[3]{250}$ $5\sqrt[3]{2}$
26. $\sqrt[3]{72}$ $2\sqrt[3]{9}$
27. $\sqrt[3]{-2000}$ $-10\sqrt[3]{2}$
28. $\sqrt[3]{-40}$ $-2\sqrt[3]{5}$

Additional answers can be found in the Instructor Answer Appendix.

29. $\sqrt[3]{192a^6}$ $4a^2\sqrt[3]{3}$
30. $\sqrt[3]{80b^9}$ $2b^3\sqrt[3]{10}$
31. $\sqrt[3]{56x^8y^3}$ $2x^2y\sqrt[3]{7x^2}$
32. $\sqrt[4]{80x^7y^{10}}$ $2xy^2\sqrt[4]{5x^3y^2}$
33. $\sqrt[5]{64x^7y^{12}}$ $2xy^2\sqrt[5]{2x^2y^2}$
34. $\sqrt[5]{160a^{17}b^9}$ $2a^3b\sqrt[5]{5a^2b^4}$

Simplify each radical completely. Assume all variables represent positive real numbers. (*See Objective 2.*)

35. $\sqrt{\dfrac{45}{49}}$ $\dfrac{3\sqrt{5}}{7}$
36. $\sqrt{\dfrac{48}{121}}$ $\dfrac{4\sqrt{3}}{11}$
37. $\sqrt{\dfrac{32x^5}{9}}$ $\dfrac{4x^2\sqrt{2x}}{3}$
38. $\sqrt{\dfrac{125y^9}{4}}$ $\dfrac{5y^4\sqrt{5y}}{2}$
39. $\sqrt{\dfrac{375x^7}{27}}$ $\dfrac{5x^3\sqrt{5x}}{3}$
40. $\sqrt{\dfrac{196y^5}{18}}$ $\dfrac{7y^2\sqrt{2y}}{3}$
41. $\sqrt[3]{\dfrac{16r^8}{125}}$ $\dfrac{2r^2\sqrt[3]{2r^2}}{5}$
42. $\sqrt[3]{\dfrac{375t^5}{8}}$ $\dfrac{5t\sqrt[3]{3t^2}}{2}$

Find the distance between the points. Express each answer with simplified radicals. (*See Objective 3.*)

43. $(4, -1)$ and $(-5, -13)$ 15
44. $(10, 0)$ and $(30, -15)$ 25
45. $(3, -11)$ and $(-12, -3)$ 17
46. $(-5, 13)$ and $(-17, 29)$ 20
47. $(6, -4\sqrt{5})$ and $(-1, -5\sqrt{5})$ $3\sqrt{6}$
48. $(9, -3\sqrt{2})$ and $(12, -6\sqrt{2})$ $3\sqrt{3}$
49. $(-9, 5\sqrt{3})$ and $(-13, 3\sqrt{3})$ $2\sqrt{7}$
50. $(-1, 9\sqrt{15})$ and $(4, 10\sqrt{15})$ $2\sqrt{10}$
51. $\left(\dfrac{4}{5}, -\dfrac{1}{5}\right)$ and $\left(\dfrac{12}{5}, \dfrac{14}{5}\right)$ $\dfrac{17}{5}$
52. $(0, 2)$ and $\left(\dfrac{1}{2}, \dfrac{8}{3}\right)$ $\dfrac{5}{6}$
53. $(-0.5, 0.1)$ and $(-1.7, 0.6)$ 1.3
54. $(0.7, -0.6)$ and $(6.7, 0.5)$ 6.1

55. Use the map to approximate the distance between Missoula, Montana $(-180, 180)$, and Billings, Montana $(90, 90)$. Each tick mark represents 90 mi. Round the answer to two decimal places. (Source: http://www.sitesatlas.com/Maps/index.htm) 284.60 mi

56. Use the map to approximate the distance between Ciudad Guayana, Venezuela $(400, 400)$, and Oranjestad, Aruba $(-400, 800)$. Each tick mark represents 400 km. Round the answer to two decimal places. (Source: http://www.sitesatlas.com/Maps/index.htm) 894.43 km

Mix 'Em Up!

Simplify each radical. Assume all variables represent positive real numbers.

57. $-\sqrt{48a^6b^4}$ $-4a^3b^2\sqrt{3}$

58. $\sqrt{128x^8y^2}$ $8x^4y\sqrt{2}$

59. $\sqrt{450h^7k^{11}}$ $15h^3k^5\sqrt{2hk}$

60. $\sqrt{162m^3n^5}$ $9mn^2\sqrt{2mn}$

61. $\sqrt{\dfrac{75r^5}{16s^4}}$ $\dfrac{5r^2\sqrt{3r}}{4s^2}$

62. $\sqrt{\dfrac{98a^3}{25b^2}}$ $\dfrac{7a\sqrt{2a}}{5b}$

63. $\sqrt[3]{160a^5b^7}$ $2ab^2\sqrt[3]{20a^2b}$

64. $\sqrt[3]{108c^4d^8}$ $3cd^2\sqrt[3]{4cd^2}$

65. $\sqrt[5]{-486r^7t^{13}}$ $-3rt^2\sqrt[5]{2r^2t^3}$

66. $\sqrt[5]{-96x^6y^{11}}$ $-2xy^2\sqrt[5]{3xy}$

67. $\sqrt[3]{\dfrac{128x^8}{27y^6}}$ $\dfrac{4x^2\sqrt[3]{2x^2}}{3y^2}$

68. $\sqrt[3]{-\dfrac{375r^{11}}{216t^9}}$ $-\dfrac{5r^3\sqrt[3]{3r^2}}{6t^3}$

69. $\dfrac{\sqrt{192a^7b^2}}{\sqrt{75}}$ $\dfrac{8a^3b\sqrt{a}}{5}$

70. $\dfrac{\sqrt{98y^5}}{\sqrt{18x^4}}$ $\dfrac{7y^2\sqrt{y}}{3x^2}$

71. $\dfrac{\sqrt{24d^7}}{\sqrt{75c^4}}$ $\dfrac{2d^3\sqrt{2d}}{5c^2}$

72. $\dfrac{\sqrt{56y^7}}{\sqrt{63x^6}}$ $\dfrac{2y^3\sqrt{2y}}{3x^3}$

Find the distance between the points. Express each answer with simplified radicals.

73. $(-10, 0)$ and $(14, -10)$ 26

74. $(6, -2)$ and $(46, 7)$ 41

75. $(5, -\sqrt{6})$ and $(13, -3\sqrt{6})$ $2\sqrt{22}$

76. $(8, -8\sqrt{5})$ and $(1, -7\sqrt{5})$ $3\sqrt{6}$

77. $\left(1, \dfrac{11}{6}\right)$ and $\left(\dfrac{1}{2}, \dfrac{5}{2}\right)$ $\dfrac{5}{6}$

78. $\left(\dfrac{1}{4}, \dfrac{1}{12}\right)$ and $\left(-\dfrac{37}{12}, -\dfrac{2}{3}\right)$ $\dfrac{41}{12}$

79. $(3, -10\sqrt{2})$ and $(7, -8\sqrt{2})$ $2\sqrt{6}$

80. $(9, 9\sqrt{3})$ and $(18, 10\sqrt{3})$ $2\sqrt{21}$

81. $(-0.04, 0.02)$ and $(0.05, -0.38)$ 0.41

82. $(3.5, -7)$ and $(5.5, -8.5)$ 2.5

83. $\left(\dfrac{1}{2}, \dfrac{2}{3}\right)$ and $\left(-\dfrac{1}{4}, \dfrac{5}{3}\right)$ $\dfrac{5}{4}$

84. $\left(\dfrac{1}{3}, \dfrac{7}{6}\right)$ and $\left(-\dfrac{1}{3}, \dfrac{29}{12}\right)$ $\dfrac{17}{12}$

You Be the Teacher!

Correct each student's errors, if any.

85. Simplify the expression $\sqrt{63x^7}$.

Vincent's work:

$$\sqrt{63x^7} = \sqrt{9 \cdot 7x^2x^5}$$
$$= 3x\sqrt{7x^5}$$

86. Simplify the expression $\sqrt[3]{64a^{10}b^4}$.

Emily's work:

$$\sqrt[3]{64a^{10}b^4} = \sqrt{8^2a^{10}b^4} = 8a^5b^2$$

87. Find the distance between the points $(3, -1)$ and $(11, -16)$.

Doris's work:

$$\sqrt{(11-3)^2 + (-16+1)^2} = \sqrt{(8)^2 + (-15)^2}$$
$$= \sqrt{64 + 225}$$
$$= 8 + 15$$
$$= 23$$

88. Find the distance between the points $(-5, 3)$ and $(-9, 6)$.

Nancy's work:

$$\sqrt{(-9-5)^2 + (6+3)^2} = \sqrt{(-14)^2 + (9)^2}$$
$$= \sqrt{196 + 81}$$
$$= \sqrt{277}$$

Calculate It!

Use a calculator to verify the simplification of each radical expression.

89. $\sqrt{54x^5} = 3x^2\sqrt{6x}$

90. $\sqrt{\dfrac{98}{25x^4}} = \dfrac{7\sqrt{2}}{5x^2}$

91. $\sqrt[5]{96x^{12}} = 2x^2\sqrt[5]{3x^2}$

92. $\sqrt[3]{250x^7} = 5x^2\sqrt[3]{3x}$

SECTION 8.4 Adding, Subtracting, and Multiplying Radical Expressions

OBJECTIVES

As a result of completing this section, you will be able to

1. Add or subtract radicals.
2. Multiply radical expressions.
3. Solve applications.
4. Troubleshoot common errors.

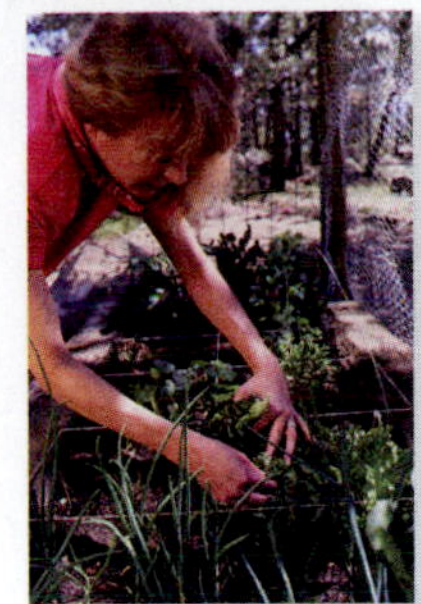

A homeowner wants to fence in her rectangular garden to keep animals out. The dimensions of the fenced area are $5\sqrt{2}$ ft by $10\sqrt{2}$ ft. How many linear feet of fencing will the homeowner need? To answer this question, we need to find the perimeter of the fenced area, which is

$$P = 5\sqrt{2} + 5\sqrt{2} + 10\sqrt{2} + 10\sqrt{2}$$

To simplify this expression, we must add radical expressions. In this section we will learn this and other operations of radicals.

Adding and Subtracting Radical Expressions

Objective 1 Add or subtract radicals.

To add radical expressions together, they must be of the same form. Recall we can only add variable expressions that have like terms. For instance,

$$2x + 4x = (2 + 4)x = 6x$$

This same rule applies to radicals. We can only add them if they are *like radicals.*

$$2\sqrt{5} + 4\sqrt{5} = (2 + 4)\sqrt{5} = 6\sqrt{5}$$

Definition: **Like radicals** are radicals with the same index and the same radicand.

Some examples of like and unlike radicals are shown.

Like Radicals	Unlike Radicals
$2\sqrt{5}, 4\sqrt{5}$	$2\sqrt{3}, 3\sqrt{5}$
$-\sqrt{6}, \sqrt{6}$	$\sqrt{6}, \sqrt[3]{6}$

Just because radicals are unlike does not mean that we cannot combine them. Consider, for example, $\sqrt{8}$ and $\sqrt{50}$. These radicals are not simplified since the radicands contain perfect square factors. When simplified, $\sqrt{8} = \sqrt{4 \cdot 2} = 2\sqrt{2}$ and $\sqrt{50} = \sqrt{25 \cdot 2} = 5\sqrt{2}$. So, after simplification, the radicals are alike and can therefore be added or subtracted.

Procedure: Adding or Subtracting Radical Expressions

Step 1: Use the product rule or quotient rule for radicals to simplify each radical, if necessary.

Step 2: Add or subtract like radicals by adding their coefficients and keeping the radical the same.

Objective 1 Examples Perform the indicated operation.

1a. $\sqrt{6} + \sqrt{6}$ **1b.** $\sqrt{15} - 3\sqrt{15}$ **1c.** $4\sqrt[3]{5} + 3\sqrt{5}$

1d. $\sqrt{12} + 2\sqrt{48} - \sqrt{75}$ **1e.** $\sqrt[3]{16} + 5\sqrt[3]{54}$ **1f.** $-3x\sqrt{4x} + 2\sqrt{x^3}$

1g. $\sqrt[3]{\dfrac{16a^4}{27}} - 4a\sqrt[3]{54a}$

Solutions

1a.

$$\begin{aligned} \sqrt{6} + \sqrt{6} &= 1\sqrt{6} + 1\sqrt{6} && \text{Recall } \sqrt{6} = 1\sqrt{6}. \\ &= (1 + 1)\sqrt{6} && \text{Add the coefficients.} \\ &= 2\sqrt{6} && \text{Simplify.} \end{aligned}$$

1b. $\sqrt{15} - 3\sqrt{15} = 1\sqrt{15} - 3\sqrt{15}$ — Recall $\sqrt{15} = 1\sqrt{15}$.
$= (1 - 3)\sqrt{15}$ — Subtract the coefficients.
$= -2\sqrt{15}$ — Simplify.

1c. The radicals, $\sqrt[3]{5}$ and $\sqrt{5}$, are not like radicals since their indices are different. So, this expression cannot be simplified any further.

1d. $\sqrt{12} + 2\sqrt{48} - \sqrt{75}$
$= \sqrt{4 \cdot 3} + 2\sqrt{16 \cdot 3} - \sqrt{25 \cdot 3}$ — Find perfect square factors.
$= \sqrt{4}\sqrt{3} + 2\sqrt{16}\sqrt{3} - \sqrt{25}\sqrt{3}$ — Apply the product rule.
$= 2\sqrt{3} + 2(4)\sqrt{3} - 5\sqrt{3}$ — Simplify each square root.
$= 2\sqrt{3} + 8\sqrt{3} - 5\sqrt{3}$ — Multiply the coefficients.
$= (2 + 8 - 5)\sqrt{3}$ — Add like radicals.
$= 5\sqrt{3}$ — Simplify.

1e. $\sqrt[3]{16} + 5\sqrt[3]{54} = \sqrt[3]{8 \cdot 2} + 5\sqrt[3]{27 \cdot 2}$ — Find perfect cube factors.
$= \sqrt[3]{8}\sqrt[3]{2} + 5\sqrt[3]{27}\sqrt[3]{2}$ — Apply the product rule.
$= 2\sqrt[3]{2} + 5(3)\sqrt[3]{2}$ — Simplify each cube root.
$= 2\sqrt[3]{2} + 15\sqrt[3]{2}$ — Multiply the coefficients.
$= (2 + 15)\sqrt[3]{2}$ — Add like radicals.
$= 17\sqrt[3]{2}$ — Simplify.

1f. $-3x\sqrt{4x} + 2\sqrt{x^3} = -3x\sqrt{4 \cdot x} + 2\sqrt{x^2 \cdot x}$ — Find perfect square factors.
$= -3x\sqrt{4}\sqrt{x} + 2\sqrt{x^2}\sqrt{x}$ — Apply the product rule.
$= -3x(2)\sqrt{x} + 2x\sqrt{x}$ — Simplify each square root.
$= -6x\sqrt{x} + 2x\sqrt{x}$ — Multiply the coefficients.
$= (-6x + 2x)\sqrt{x}$ — Add like radicals.
$= -4x\sqrt{x}$ — Simplify.

1g. $\sqrt[3]{\dfrac{16a^4}{27}} - 4a\sqrt[3]{54a}$

$= \dfrac{\sqrt[3]{16a^4}}{\sqrt[3]{27}} - 4a\sqrt[3]{27 \cdot 2a}$ — Apply the quotient rule for radicals and find a perfect cube factor of $54a$.

$= \dfrac{\sqrt[3]{8a^3 \cdot 2a}}{3} - 4a\sqrt[3]{27}\sqrt[3]{2a}$ — Find a perfect cube factor of $16a^4$ and apply the product rule for radicals.

$= \dfrac{\sqrt[3]{8a^3}\sqrt[3]{2a}}{3} - 4a(3)\sqrt[3]{2a}$ — Apply the product rule for radicals and simplify $\sqrt[3]{27}$.

$= \dfrac{2a\sqrt[3]{2a}}{3} - 12a\sqrt[3]{2a}$ — Simplify $\sqrt[3]{8a^3}$ and multiply the coefficients.

$= \dfrac{2a\sqrt[3]{2a}}{3} - \dfrac{36a\sqrt[3]{2a}}{3}$ — Convert the second expression to a fraction with the common denominator, 3.

$= \dfrac{2a\sqrt[3]{2a} - 36a\sqrt[3]{2a}}{3}$ — Combine the fractions.

$$= \frac{(2a - 36a)\sqrt[3]{2a}}{3}$$ Subtract the like radicals.

$$= -\frac{34a\sqrt[3]{2a}}{3}$$ Simplify.

Student Check 1 Perform the indicated operation.

a. $\sqrt{10} + \sqrt{10}$ **b.** $\sqrt{7} - 5\sqrt{7}$ **c.** $6\sqrt[3]{2} + 4\sqrt{2}$

d. $\sqrt{24} + 2\sqrt{54} - \sqrt{6}$ **e.** $\sqrt[3]{24} + 9\sqrt[3]{81}$ **f.** $-7y\sqrt{16y} + 4\sqrt{y^3}$

g. $\sqrt{\frac{20b^5}{49}} - 2b^2\sqrt{45b}$

Multiplying Radical Expressions

Objective 2 Multiply radical expressions.

In Section 8.3, the product rule for radicals was introduced to enable us to simplify radical expressions. The rule stated that, for $\sqrt[n]{a}$ and $\sqrt[n]{b}$ real numbers,

$$\sqrt[n]{a \cdot b} = \sqrt[n]{a} \cdot \sqrt[n]{b}$$

Rewriting this property provides us with a method to multiply radicals.

Property: Product Rule for Radicals

For $\sqrt[n]{a}$ and $\sqrt[n]{b}$ real numbers,

$$\sqrt[n]{a} \cdot \sqrt[n]{b} = \sqrt[n]{a \cdot b}$$

The product rule states that the product of nth roots is the nth root of the product of the radicands. Note that the radicals must have the same index to apply this rule. This property, along with the distributive property, enables us to multiply radical expressions.

Procedure: Multiplying Radical Expressions

Step 1: Use the distributive property to eliminate parentheses, if needed.
Step 2: Multiply two radicals with the same index by forming the root of the product of the radicands.
Step 3: Simplify the product.
Step 4: Simplify the radical, if necessary.

Objective 2 Examples Simplify each product and write each answer in simplest radical form.

2a. $\sqrt{2} \cdot \sqrt{6}$ **2b.** $(3\sqrt{7})(2\sqrt{8})$ **2c.** $\sqrt[3]{4x^2} \cdot \sqrt[3]{2x}$ **2d.** $\sqrt{6}(2 + \sqrt{3})$

2e. $(2\sqrt{3} + \sqrt{5})(\sqrt{3} - 2\sqrt{5})$ **2f.** $(2 + \sqrt{7})^2$ **2g.** $(2 + \sqrt{7})(2 - \sqrt{7})$

Solutions

2a.

$\sqrt{2} \cdot \sqrt{6} = \sqrt{2 \cdot 6}$ Apply the product rule.
$= \sqrt{12}$ Simplify the radicand.
$= \sqrt{4 \cdot 3}$ Find a perfect square factor of 12.
$= \sqrt{4} \cdot \sqrt{3}$ Apply the product rule.
$= 2\sqrt{3}$ Simplify.

2b.

$(3\sqrt{7})(2\sqrt{8}) = 3 \cdot 2 \cdot \sqrt{7}\sqrt{8}$ Apply the commutative property.
$= 6\sqrt{7 \cdot 8}$ Multiply the coefficients and apply the product rule.
$= 6\sqrt{56}$ Simplify the radicand.
$= 6\sqrt{4 \cdot 14}$ Find a perfect square factor of 56.

$= 6\sqrt{4}\sqrt{14}$ Apply the product rule.

$= 6(2)\sqrt{14}$ Simplify the square root.

$= 12\sqrt{14}$ Multiply the coefficients.

2c. $\sqrt[3]{4x^2} \cdot \sqrt[3]{2x} = \sqrt[3]{4x^2 \cdot 2x}$ Apply the product rule.

$= \sqrt[3]{8x^3}$ Simplify the radicand.

$= 2x$ Simplify.

2d. $\sqrt{6}(2 + \sqrt{3}) = \sqrt{6}(2) + \sqrt{6}(\sqrt{3})$ Apply the distributive property.

$= 2\sqrt{6} + \sqrt{6 \cdot 3}$ Apply the product rule.

$= 2\sqrt{6} + \sqrt{18}$ Simplify the radicand.

$= 2\sqrt{6} + \sqrt{9 \cdot 2}$ Find a perfect square factor of 18.

$= 2\sqrt{6} + \sqrt{9}\sqrt{2}$ Apply the product rule.

$= 2\sqrt{6} + 3\sqrt{2}$ Simplify the square root.

2e. Apply the FOIL method to multiply the two radical expressions.

$$\begin{aligned}(2\sqrt{3} + \sqrt{5})(\sqrt{3} - 2\sqrt{5}) &= 2\sqrt{3}\sqrt{3} - 2\sqrt{3}(2\sqrt{5}) + \sqrt{5}\sqrt{3} - \sqrt{5}(2\sqrt{5}) \\ &= 2\sqrt{9} - 4\sqrt{15} + \sqrt{15} - 2\sqrt{25} \\ &= 2(3) + (-4 + 1)\sqrt{15} - 2(5) \\ &= 6 - 3\sqrt{15} - 10 \\ &= -4 - 3\sqrt{15}\end{aligned}$$

2f. A binomial can be squared by applying the FOIL method or by applying the rule for squaring a binomial.

Method 1

$$\begin{aligned}(2 + \sqrt{7})^2 &= (2 + \sqrt{7})(2 + \sqrt{7}) \\ &= 2 \cdot 2 + 2\sqrt{7} + 2\sqrt{7} + \sqrt{7}\sqrt{7} \\ &= 4 + 4\sqrt{7} + \sqrt{49} \\ &= 4 + 4\sqrt{7} + 7 \\ &= 11 + 4\sqrt{7}\end{aligned}$$

Method 2

Recall $(a + b)^2 = a^2 + 2ab + b^2$.

$$\begin{aligned}(2 + \sqrt{7})^2 &= 2^2 + 2 \cdot 2\sqrt{7} + (\sqrt{7})^2 \\ &= 4 + 4\sqrt{7} + 7 \\ &= 11 + 4\sqrt{7}\end{aligned}$$

2g. Conjugates can be multiplied by applying the FOIL method or by applying the product of conjugates property.

Method 1

$$\begin{aligned}(2 + \sqrt{7})(2 - \sqrt{7}) &= 2 \cdot 2 - 2\sqrt{7} + 2\sqrt{7} - \sqrt{7}\sqrt{7} \\ &= 4 + 0\sqrt{7} - \sqrt{49} \\ &= 4 - 7 \\ &= -3\end{aligned}$$

Method 2

Recall $(a + b)(a - b) = a^2 - b^2$.

$$\begin{aligned}(2 + \sqrt{7})(2 - \sqrt{7}) &= 2^2 - (\sqrt{7})^2 \\ &= 4 - 7 \\ &= -3\end{aligned}$$

Student Check 2 Simplify each product and write each answer in simplest radical form.

a. $\sqrt{8} \cdot \sqrt{3}$ **b.** $(2\sqrt{6})(5\sqrt{15})$ **c.** $\sqrt[4]{9x^3} \cdot \sqrt[4]{9x}$ **d.** $\sqrt{3}(4 - \sqrt{15})$

e. $(7\sqrt{2} + \sqrt{6})(2\sqrt{2} - 3\sqrt{6})$ **f.** $(3 + \sqrt{5})^2$ **g.** $(3 + \sqrt{5})(3 - \sqrt{5})$

Applications

Objective 3 Solve applications.

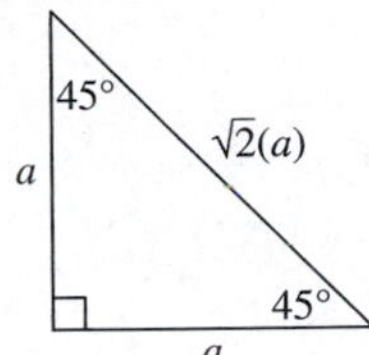

The applications for adding, subtracting, and multiplying radicals that we will solve involve perimeter and area formulas and finding sides of special right triangles. We will examine two special triangles. One is a 45°-45°-90° triangle and the other is a 30°-60°-90° triangle.

The **45°-45°-90° triangle** is a triangle with angles of 45°, 45°, and 90°. The legs of this type of triangle are equal and the length of the hypotenuse of the triangle is $\sqrt{2}$ times the length of a leg.

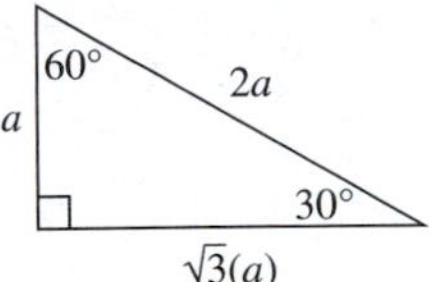

The **30°-60°-90° triangle** is a triangle with angles of 30°, 60°, and 90°. If a is the length of the side opposite the 30° angle, then the length of the other leg is $\sqrt{3}$ times the value of a, and the length of the hypotenuse is twice the value of a.

Objective 3 Examples Solve each problem. Write each answer in simplest radical form and approximate to two decimal places.

3a. Find the perimeter and area of the triangle in Figure 8.1.

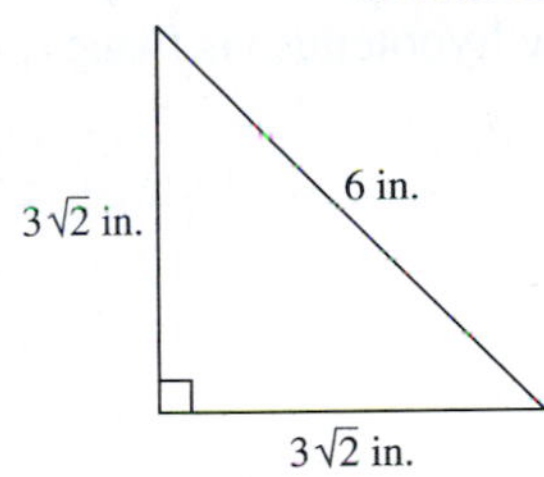

Figure 8.1

Solution 3a.

$P = 3\sqrt{2} + 3\sqrt{2} + 6$ Add the lengths of each side.

$P = (3 + 3)\sqrt{2} + 6$ Add like radicals.

$P = 6\sqrt{2} + 6$ Simplify.

So, the perimeter of the triangle is $6\sqrt{2} + 6$ in. or approximately 14.49 in.

$A = \frac{1}{2}bh$ State the area formula.

$A = \frac{1}{2}(3\sqrt{2})(3\sqrt{2})$ Substitute $3\sqrt{2}$ for b and h.

$A = \frac{1}{2}(3)(3)\sqrt{2}\sqrt{2}$ Apply the commutative property.

$A = \frac{1}{2}(9)\sqrt{2 \cdot 2}$ Multiply the coefficients and apply the product rule for radicals.

$A = \frac{1}{2}(9)\sqrt{4}$ Simplify the radicand.

$A = \frac{1}{2}(9)(2)$ Simplify the square root.

$A = \frac{1}{2}(18)$ Multiply.

$A = 9$ Simplify.

So, the area of the triangle is 9 in^2.

3b. Find the length of a room's diagonal if the room measures (i) 12 ft by 12 ft and (ii) $12\sqrt{3}$ ft by $12\sqrt{3}$ ft.

Solution 3b. (i) A square cut by a diagonal produces two 45°-45°-90° right triangles. So, a square room cut in half by the diagonal is an example of a 45°-45°-90° triangle.

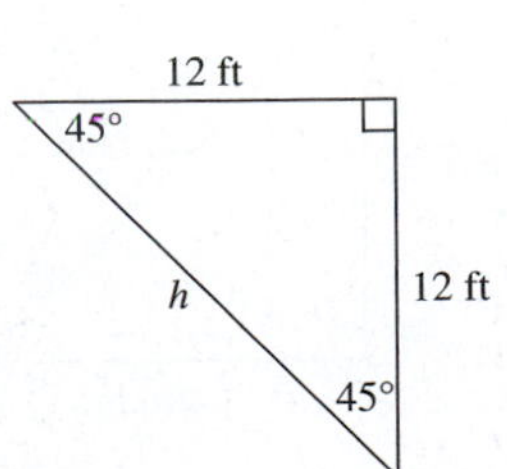

The hypotenuse of the right triangle is $\sqrt{2}$ times the length of the leg. Therefore, the length of the diagonal is

$$h = \sqrt{2}(12) = 12\sqrt{2}$$

So, the room's diagonal is $12\sqrt{2}$ ft or approximately 16.97 ft.

(ii) If the room is $12\sqrt{3}$ ft by $12\sqrt{3}$ ft, the diagonal of the room is $\sqrt{2}$ times the length of the leg. Therefore, the length of the diagonal is

$$h = \sqrt{2}(12\sqrt{3}) = 12\sqrt{2}\sqrt{3} = 12\sqrt{6}$$

So, the room's diagonal is $12\sqrt{6}$ ft or approximately 29.39 ft.

Note: *Since we are working with right triangles, we can check our work by using the Pythagorean theorem,* $a^2 + b^2 = c^2$.

$$a^2 + b^2 = c^2 \rightarrow (12\sqrt{3})^2 + (12\sqrt{3})^2 = (12\sqrt{6})^2$$
$$144(3) + 144(3) = 144(6)$$
$$432 + 432 = 864$$
$$864 = 864$$

Since this is a true statement, our work is correct.

3c. The triangle shown in Figure 8.2 is a 30°-60°-90° triangle. Find the length of the other leg and the hypotenuse.

Solution

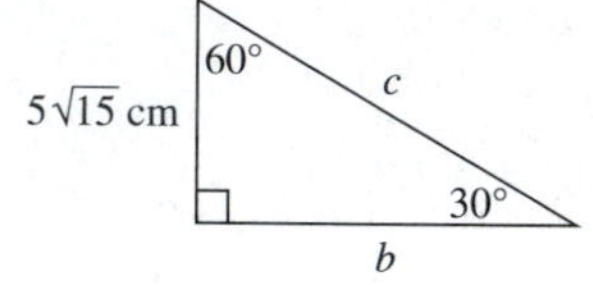

Figure 8.2

3c. The side opposite the 30° angle is given, $a = 5\sqrt{15}$. The hypotenuse is twice the value of a. So,

$$c = 2a$$
$$c = 2(5\sqrt{15})$$
$$c = 10\sqrt{15}$$

The leg opposite the 60° angle is $b = \sqrt{3}(a)$. So,

$b = \sqrt{3}(a)$	State the relationship between a and b.
$b = \sqrt{3}(5\sqrt{15})$	Replace a with $5\sqrt{15}$.
$b = 5\sqrt{45}$	Multiply the radicals.
$b = 5\sqrt{9 \cdot 5}$	Find a perfect square factor of 45.
$b = 5(3)\sqrt{5}$	Apply the product rule.
$b = 15\sqrt{5}$	Simplify.

Check: $a^2 + b^2 = c^2 \rightarrow (5\sqrt{15})^2 + (15\sqrt{5})^2 = (10\sqrt{15})^2$

$$25(15) + 225(5) = 100(15)$$
$$375 + 1125 = 1500$$
$$1500 = 1500$$

So, the other leg is $15\sqrt{5}$ cm, or approximately 33.54 cm, and the hypotenuse is $10\sqrt{15}$ cm, or approximately 38.73 cm.

Student Check 3 Solve each problem. Write each radical in simplest radical form and approximate to two decimal places.

a. Find the perimeter and area of the triangle shown

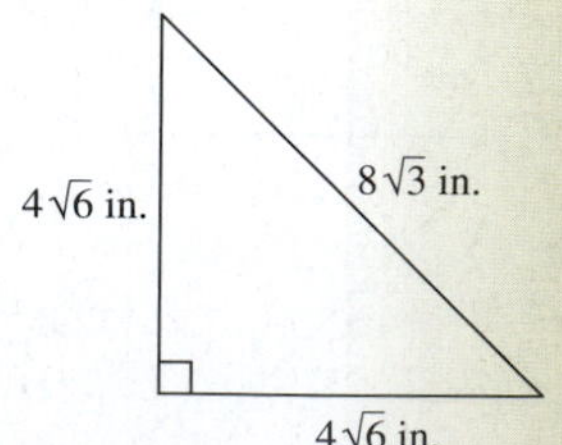

b. Find the length of the hypotenuse in a 45°-45°-90° triangle if the length of each leg is $2\sqrt{10}$ in.

c. In a 30°-60°-90° triangle, the leg opposite the 30° angle measures $2\sqrt{6}$ ft. Find the length of the other leg and the hypotenuse.

Objective 4

Troubleshoot common errors.

Troubleshooting Common Errors

Some common errors associated with adding, subtracting, and multiplying radicals are shown.

Objective 4 Examples

A problem and an incorrect solution are given. Provide the correct solution and an explanation of the error.

4a. Add: $\sqrt{3} + \sqrt{3}$.

Incorrect Solution	Correct Solution and Explanation
$\sqrt{3} + \sqrt{3} = \sqrt{6}$	To add like radicals, we add the coefficients of the radicals, not the radicands. $\sqrt{3} + \sqrt{3} = 1\sqrt{3} + 1\sqrt{3} = 2\sqrt{3}$

4b. Simplify: $(4\sqrt{7})^2$.

Incorrect Solution	Correct Solution and Explanation
$(4\sqrt{7})^2 = 4(7) = 28$	Because we have a product squared, we must square each factor. $(4\sqrt{7})^2 = 4^2(\sqrt{7})^2 = 16(7) = 112$

4c. Simplify: $(\sqrt{3} + \sqrt{5})^2$.

Incorrect Solution	Correct Solution and Explanation
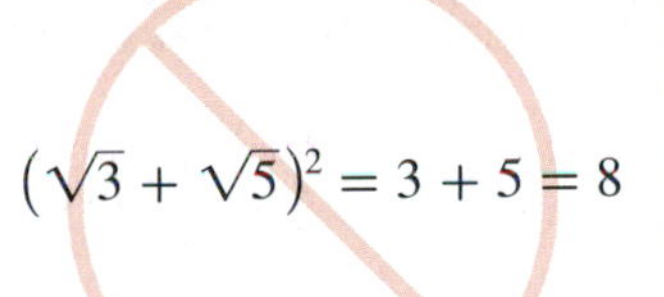 $(\sqrt{3} + \sqrt{5})^2 = 3 + 5 = 8$	We cannot square each term in the binomial. We must use the FOIL method or the squaring pattern. $(\sqrt{3} + \sqrt{5})^2 = (\sqrt{3})^2 + 2(\sqrt{3})(\sqrt{5}) + (\sqrt{5})^2 = 3 + 2\sqrt{15} + 5 = 8 + 2\sqrt{15}$

ANSWERS TO STUDENT CHECKS

Student Check 1 **a.** $2\sqrt{10}$ **b.** $-4\sqrt{7}$ **c.** $3\sqrt[3]{5}$ **d.** can't be combined **e.** $7\sqrt{6}$ **f.** $29\sqrt[3]{3}$ **g.** $-24y\sqrt{y}$ **h.** $-\dfrac{40b^2\sqrt{5b}}{7}$

Student Check 2 **a.** $2\sqrt{6}$ **b.** $30\sqrt{10}$ **c.** $3x$ **d.** $4\sqrt{3} - 3\sqrt{5}$ **e.** $10 - 38\sqrt{3}$ **f.** $14 + 6\sqrt{5}$ **g.** 4

Student Check 3 **a.** $8\sqrt{6} + 8\sqrt{3}$ in. or 33.45 in.; 48 in.^{2} **b.** $4\sqrt{5}$ in. **c.** $6\sqrt{2}$ ft; $4\sqrt{6}$ ft

SUMMARY OF KEY CONCEPTS

1. Like radicals are radicals with the same index and same radicand. Like radicals can be added by adding their coefficients and keeping the radical the same. If the radicals are unlike, try simplifying them according to the methods shown in Section 8.3. After simplifying the radicals, combine them, if possible.
2. Multiply radicals using the product rule for radicals. If the indices are the same, multiply the radicands.
3. Special cases of right triangles can be solved using the relationships to the right.

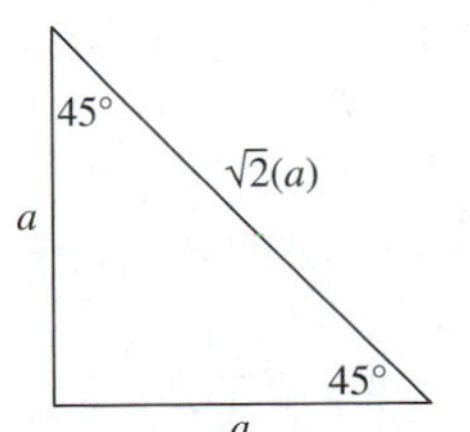

45°-45°-90° triangle

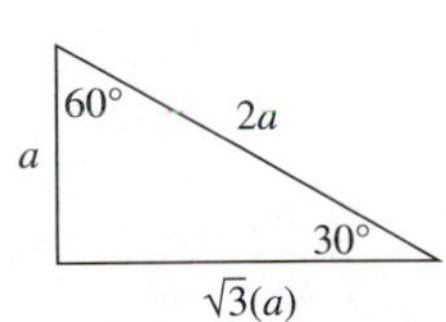

30°-60°-90° triangle

GRAPHING CALCULATOR SKILLS

We can use the graphing calculator to verify that our answers are correct.

Example 1: Verify that $\sqrt{12} + 2\sqrt{48} - \sqrt{75} = 5\sqrt{3}$.

Solution:

```
√(12)+2√(48)-√(7
5)
       8.660254038
5√(3)
       8.660254038
```

The two radical expressions have the same decimal value and are, therefore, equivalent.

Example 2: Verify that $(2 + \sqrt{7})^2 = 11 + 4\sqrt{7}$.

Solution:

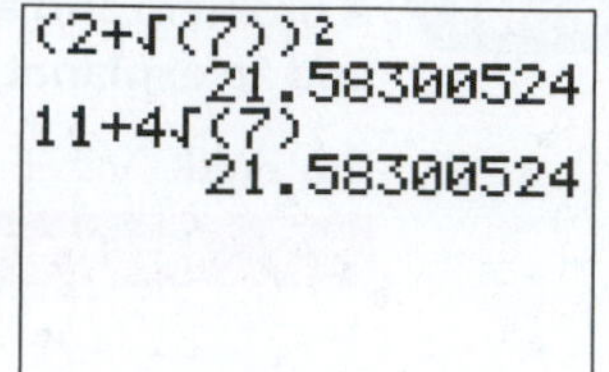

The two radical expressions have the same decimal value and are, therefore, equivalent.

SECTION 8.4 EXERCISE SET

Write About It!

Use complete sentences in your answer to each exercise.

1. Define like radicals.
2. Explain how to add and subtract radical expressions.
3. Explain how to multiply radical expressions.
4. Can you multiply like radicals? Give an example.
5. Explain why $\sqrt{18x} + \sqrt{12x} \neq \sqrt{30x^2} = x\sqrt{30}$.
6. Explain why $\sqrt[3]{10x^2} - \sqrt[3]{2x^2} \neq \sqrt[3]{8x^2}$.
7. Explain how to simplify $(\sqrt{x} + \sqrt{y})^2$.
8. Explain why $\sqrt{a} + \sqrt{a} \neq \sqrt{2a}$.

Practice Makes Perfect!

Add or subtract. (*See Objective 1.*)

9. $3\sqrt{6} + 2\sqrt{6}$ $5\sqrt{6}$
10. $11\sqrt{5} + \sqrt{5}$ $12\sqrt{5}$
11. $-2\sqrt{7} - 12\sqrt{7}$ $-14\sqrt{7}$
12. $3\sqrt{5} - 15\sqrt{5}$ $-12\sqrt{5}$
13. $2\sqrt{54} + 7\sqrt{216}$ $48\sqrt{6}$
14. $3\sqrt{125} + 5\sqrt{20}$ $25\sqrt{5}$
15. $10\sqrt{8x^3} - 4\sqrt{18x^3}$ $8x\sqrt{2x}$
16. $7y^2\sqrt{27y} + 2\sqrt{75y^5}$ $31y^2\sqrt{3y}$
17. $-7\sqrt[3]{5} + 6\sqrt[3]{5} - 8\sqrt[3]{5}$ $-9\sqrt[3]{5}$
18. $16\sqrt[3]{11} - 9\sqrt[3]{11} - 2\sqrt[3]{11}$ $5\sqrt[3]{11}$
19. $8\sqrt[3]{16} - \sqrt[3]{54} + 18\sqrt[3]{2}$ $31\sqrt[3]{2}$
20. $6\sqrt[3]{40} - 2\sqrt[3]{135} + 8\sqrt[3]{5}$ $14\sqrt[3]{5}$
21. $-15\sqrt[3]{24a^3b} + 9\sqrt[3]{81a^3b} + 2\sqrt[3]{6a^3b}$ $-3a\sqrt[3]{3b} + 2a\sqrt[3]{6b}$
22. $4\sqrt[3]{250xy^3} + 8\sqrt[3]{54xy^3} - 7\sqrt[3]{2y^3}$ $44y\sqrt[3]{2x} - 7y\sqrt[3]{2}$
23. $\dfrac{3\sqrt{5}}{4} + \sqrt{\dfrac{20}{9}}$ $\dfrac{17\sqrt{5}}{12}$
24. $\dfrac{3\sqrt{7}}{2} - \sqrt{\dfrac{28}{25}}$ $\dfrac{11\sqrt{7}}{10}$
25. $\dfrac{7\sqrt{20}}{3} - \sqrt{\dfrac{125}{9}}$ $3\sqrt{5}$
26. $\dfrac{\sqrt{45}}{4} - \sqrt{\dfrac{80}{9}}$ $-\dfrac{7\sqrt{5}}{12}$
27. $\dfrac{7\sqrt[3]{24}}{4} - \sqrt[3]{\dfrac{375}{64}}$ $\dfrac{9\sqrt[3]{3}}{4}$
28. $\dfrac{\sqrt[3]{80}}{5} - \sqrt[3]{\dfrac{270}{343}}$ $-\dfrac{\sqrt[3]{10}}{35}$
29. $\dfrac{\sqrt{50r^3}}{9} + \sqrt{\dfrac{98r^3}{9}}$ $\dfrac{26r\sqrt{2r}}{9}$
30. $\dfrac{s^2\sqrt{300s}}{7} - \sqrt{\dfrac{48s^5}{49}}$ $\dfrac{6s^2\sqrt{3s}}{7}$

Simplify each product and write each answer in simplest radical form. (*See Objective 2.*)

31. $\sqrt{32} \cdot \sqrt{7}$ $4\sqrt{14}$
32. $\sqrt{18} \cdot \sqrt{3}$ $3\sqrt{6}$
33. $(6\sqrt{20})(2\sqrt{8})$ $48\sqrt{10}$
34. $(5\sqrt{12})(6\sqrt{28})$ $120\sqrt{21}$
35. $(2\sqrt{12x^2})(\sqrt{8x^3})$ $8x^2\sqrt{6x}$
36. $(\sqrt{98y})(2\sqrt{27y^2})$ $42y\sqrt{6y}$
37. $\sqrt[3]{32x}\sqrt[3]{18x^4}$ $4x\sqrt[3]{9x^2}$
38. $\sqrt[3]{243y^2}\sqrt[3]{15y^8}$ $9y^3\sqrt[3]{5y}$
39. $\sqrt{7}(5 + \sqrt{21})$ $5\sqrt{7} + 7\sqrt{3}$
40. $\sqrt{5}(16 + \sqrt{15})$ $16\sqrt{5} + 5\sqrt{3}$
41. $(\sqrt{5} - 2\sqrt{30})(3\sqrt{5} + \sqrt{30})$ $-45 - 25\sqrt{6}$
42. $(4\sqrt{7} - \sqrt{35})(2\sqrt{7} + \sqrt{35})$ $21 + 14\sqrt{5}$
43. $(3\sqrt{2} - \sqrt{6})^2$ $24 - 12\sqrt{3}$
44. $(2\sqrt{5} + \sqrt{10})^2$ $30 + 20\sqrt{2}$
45. $(1 - 2\sqrt{10})^2$ $41 - 4\sqrt{10}$
46. $(5 + 3\sqrt{7})^2$ $88 + 30\sqrt{7}$
47. $(2 - \sqrt{6x})(2 + \sqrt{6x})$ $4 - 6x$
48. $(5 - \sqrt{2y})(5 + \sqrt{2y})$ $25 - 2y$
49. $(2\sqrt{3x} - 5\sqrt{2y})(\sqrt{3x} + \sqrt{2y})$ $6x - 3\sqrt{6xy} - 10y$
50. $(3\sqrt{7a} + 2\sqrt{5b})(\sqrt{7a} - 6\sqrt{5b})$ $21a - 16\sqrt{35ab} - 60b$

Solve each problem. (*See Objective 3.*)

In Exercises 51–54, find the perimeter and area of the triangle shown. Write each answer in simplest radical form and approximate answers to two decimal places.

51. $a = 18\sqrt{2}$ cm and $c = 36$ cm
52. $a = 11\sqrt{2}$ cm and $c = 22$ cm
53. $a = 10\sqrt{14}$ ft and $c = 20\sqrt{7}$ ft
54. $a = 3\sqrt{6}$ ft and $c = 6\sqrt{3}$ ft

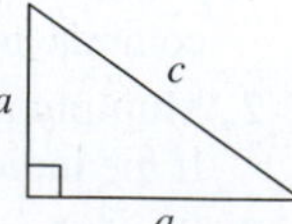

Additional answers can be found in the Instructor Answer Appendix.

55. In a 30°-60°-90° triangle, the leg opposite the 30° angle measures $4\sqrt{21}$ cm. Find the length of the other leg and the hypotenuse.

56. In a 30°-60°-90° triangle, the leg opposite the 30° angle measures $14\sqrt{6}$ m. Find the length of the other leg and the hypotenuse.

Mix 'Em Up!

Perform each operation and write each answer in simplest radical form.

57. $11\sqrt{6} - 3\sqrt{6} + \sqrt{6}$ $9\sqrt{6}$

58. $9\sqrt{5} + 4\sqrt{5} - 12\sqrt{5}$ $\sqrt{5}$

59. $21\sqrt[3]{24r^3t} - 5\sqrt[3]{81r^3t}$ $27r\sqrt[3]{3t}$

60. $12\sqrt[3]{40xy^6} - 6\sqrt[3]{135xy^6}$ $6y^2\sqrt[3]{5x}$

61. $-8\sqrt[3]{16a^3b} + 4\sqrt[3]{54a^3b} + 10\sqrt[3]{3a^3b}$ $-4a\sqrt[3]{2b} + 10a\sqrt[3]{3b}$

62. $7\sqrt{108a^2b} - 6\sqrt{75a^2b}$ $12a\sqrt{3b}$

63. $10\sqrt{98cd^3} - 16\sqrt{50cd^3}$ $-10d\sqrt{2cd}$

64. $6\sqrt[3]{80xy^3} + 11\sqrt[3]{270xy^3} - 9\sqrt[3]{40xy^3}$ $45y\sqrt[3]{10x} - 18y\sqrt[3]{5x}$

65. $5\sqrt[4]{16x^4y^7} - x\sqrt[4]{81y^7}$ $7xy\sqrt[4]{y^3}$

66. $12\sqrt[4]{5a^4b^9} - 2a\sqrt[4]{48b^9}$ $12ab^2\sqrt[4]{5b} - 4ab^2\sqrt[4]{3b}$

67. $(5 - \sqrt{3a})(5 + \sqrt{3a})$ $25 - 3a$

68. $(3 + \sqrt{5b})(3 - \sqrt{5b})$ $9 - 5b$

69. $-5\sqrt{2x^2} + 8x\sqrt{2} - \sqrt{18x^2}$ 0

70. $7\sqrt{5y^3} - 8\sqrt{20y^3} + 3y\sqrt{45y}$ 0

71. $\dfrac{3\sqrt{75r^2t}}{2} - \sqrt{\dfrac{363r^2t}{49}}$ $\dfrac{83r\sqrt{3t}}{14}$

72. $\dfrac{\sqrt{45ab^3}}{2} - \sqrt{\dfrac{80ab^3}{9}}$ $\dfrac{b\sqrt{5ab}}{6}$

73. $\sqrt[3]{\dfrac{375a}{64}} - \dfrac{\sqrt[3]{24a}}{5}$ $\dfrac{17\sqrt[3]{3a}}{20}$

74. $\sqrt[3]{\dfrac{135}{8b^3}} + \dfrac{\sqrt[3]{40}}{3b}$ $\dfrac{13\sqrt[3]{5}}{6b}$

75. $\sqrt[5]{16x^2y}\sqrt[5]{10x^6y^4}$ $2xy\sqrt[5]{5x^3}$

76. $\sqrt[5]{81a^3b^2}\sqrt[5]{6a^8b^4}$ $3a^2b\sqrt[5]{2ab}$

77. $(8\sqrt{2} - \sqrt{6})(\sqrt{2} + 3\sqrt{6})$ $-2 + 46\sqrt{3}$

78. $(\sqrt{20} - \sqrt{3})(2\sqrt{20} + 9\sqrt{3})$ $13 + 14\sqrt{15}$

79. $(4\sqrt{3} - \sqrt{15})^2$ $63 - 24\sqrt{5}$

80. $(3\sqrt{6} + \sqrt{2})^2$ $56 + 12\sqrt{3}$

81. $\sqrt{6}(2 - \sqrt{12})$ $2\sqrt{6} - 6\sqrt{2}$

82. $\sqrt{5}(3 + \sqrt{30})$ $3\sqrt{5} + 5\sqrt{6}$

83. $(\sqrt{5x} - 2\sqrt{3y})(\sqrt{5x} + 4\sqrt{3y})$ $5x + 2\sqrt{15xy} - 24y$

84. $(7\sqrt{6a} + 4\sqrt{2b})(\sqrt{6a} - \sqrt{2b})$ $42a - 6\sqrt{3ab} - 8b$

In Exercises 85–88, find the perimeter and area of the triangle shown. Write each answer in simplest radical form and approximate answers to two decimal places.

85. $a = 21\sqrt{2}$ m and $c = 42$ m

86. $a = 16\sqrt{2}$ m and $c = 32$ m

87. $a = 18\sqrt{6}$ in. and $c = 36\sqrt{3}$ in.

88. $a = 7\sqrt{10}$ in. and $c = 14\sqrt{5}$ in.

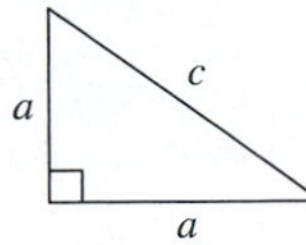

89. In a 30°-60°-90° triangle, the leg opposite the 30° angle measures $5\sqrt{21}$ cm. Find the length of the other leg and the hypotenuse. $15\sqrt{7}$ cm; $10\sqrt{21}$ cm

90. In a 30°-60°-90° triangle, the leg opposite the 30° angle measures $18\sqrt{6}$ m. Find the length of the other leg and the hypotenuse. $54\sqrt{2}$ m; $36\sqrt{6}$ m

You Be the Teacher!

Correct each student's errors, if any.

91. Simplify $\sqrt{28x^2y} + \sqrt{63x^2y}$.

Vincent's work:

$$\begin{aligned}\sqrt{28x^2y} + \sqrt{63x^2y} &= \sqrt{4 \cdot 7x^2y} + \sqrt{9 \cdot 7x^2y}\\ &= 2x\sqrt{7y} + 3x\sqrt{7y}\\ &= 5x^2\sqrt{49y^2}\\ &= 35x^2y\end{aligned}$$

92. Simplify $3\sqrt{75a^4b} - a^2\sqrt{98b}$.

Daron's work:

$$\begin{aligned}3\sqrt{75a^4b} - a^2\sqrt{98b} &= 3\sqrt{25 \cdot 3a^4b} - a^2\sqrt{49 \cdot 2b}\\ &= 15a^2\sqrt{3b} - 7a^2\sqrt{2b}\\ &= 8a^2\sqrt{b}\end{aligned}$$

93. Simplify $(2\sqrt{x} - y)^2$.

Doris's work:

$$\begin{aligned}(2\sqrt{x} - y)^2 &= (2\sqrt{x})^2 + (-y)^2\\ &= 4x + y^2\end{aligned}$$

94. Simplify $(\sqrt{7} - \sqrt{3})^2$.

Nancy's work:

$$\begin{aligned}(\sqrt{7} - \sqrt{3})^2 &= (\sqrt{7})^2 - (\sqrt{3})^2\\ &= 7 - 3\\ &= 4\end{aligned}$$

Calculate It!

Use a calculator to verify the simplification of each radical expression.

95. $\sqrt{72} + 3\sqrt{50} - 6\sqrt{18} = 3\sqrt{2}$

96. $10\sqrt{48} - 3\sqrt{192} + 6\sqrt{12} = 28\sqrt{3}$

97. $7\sqrt[3]{40} - 2\sqrt[3]{135} = 8\sqrt[3]{5}$

98. $(3\sqrt{12} - 2\sqrt{5})^2 = 128 - 24\sqrt{15}$

PIECE IT TOGETHER SECTIONS 8.1–8.4

Simplify each expression. Assume all variables represent positive real numbers. (*Section 8.1, Objectives 1–3*)

1. $\sqrt{49a^{12}}$ $7a^6$

2. $\sqrt[3]{\dfrac{125x^9}{8y^6}}$ $\dfrac{5x^3}{2y^2}$

3. $\sqrt[5]{-243a^{10}b^{25}}$ $-3a^2b^5$

4. $\sqrt[7]{128a^7b^{21}}$ $2ab^3$

Simplify each expression. (*Section 8.1, Objective 5*)

5. $\sqrt{(2x + 1)^2}$ $|2x + 1|$

6. $\sqrt[7]{(5b)^7}$ $5b$

Graph the radical function using the accompanying table. (*Section 8.1, Objective 7*)

7. $f(x) = \sqrt[3]{x} - 2$

x	-8	-1	0	1	8	12
$f(x)$						

Simplify each expression and write each answer with positive exponents. Assume all variables represent positive real numbers. (*Section 8.2, Objectives 1–4*)

8. $(-16)^{1/2}$ not a real number

9. $\left(\frac{1}{27}\right)^{2/3}$ $\frac{1}{9}$

10. $(64x^9y^6)^{2/3}$ $16x^6y^4$

11. $(32x^{10})^{-2/5}$ $\frac{1}{4x^4}$

12. $(a^{-8}b^{12})^{-3/4}$ $\frac{a^6}{b^9}$

13. $\frac{(x^{5/6}y^{1/5})^3}{y}$ $\frac{x^{5/2}}{y^{2/5}}$

14. $\frac{(a^{2/7}b^{1/3})^7}{b^3}$ $\frac{a^2}{b^{2/3}}$

15. $(12x^{2/5} - 5y^{1/2})(x^{2/5} + y^{1/2})$ $12x^{4/5} + 7x^{2/5}y^{1/2} - 5y$

16. $(7r^{1/2} - s^{2/3})(7r^{1/2} + s^{2/3})$ $49r - s^{4/3}$

Simplify each radical. Assume all variables represent positive real numbers. (*Section 8.3, Objectives 1 and 2*)

17. $3\sqrt{75}$ $15\sqrt{3}$

18. $\sqrt{144wx^5y^6}$ $12x^2y^3\sqrt{wx}$

19. $\sqrt[3]{40a^7b^{12}}$ $2a^2b^4\sqrt[3]{5a}$

20. $\sqrt[4]{48x^5y^{11}}$ $2xy^2\sqrt[4]{3xy^3}$

21. $\sqrt{\frac{80x^5}{9y^2}}$ $\frac{4x^2\sqrt{5x}}{3y}$

Find the distance between the points. Write the answer with simplified radicals. (*Section 8.3, Objective 3*)

22. $(-9, 5\sqrt{3})$ and $(-13, 3\sqrt{3})$ $2\sqrt{7}$

Perform each operation and write each answer in simplest radical form. (*Section 8.4, Objectives 1 and 2*)

23. $14\sqrt{98} - 6\sqrt{8}$ $86\sqrt{2}$

24. $8\sqrt[3]{16} - \sqrt[3]{54} + 19\sqrt[3]{2}$ $32\sqrt[3]{2}$

25. $(2\sqrt{12x^4})(\sqrt{8x^3})$ $8x^3\sqrt{6x}$

26. $(\sqrt{5} - 2\sqrt{30})(3\sqrt{5} + \sqrt{30})$ $-45 - 25\sqrt{6}$

27. $(3\sqrt{2} - \sqrt{6})^2$ $24 - 12\sqrt{3}$

SECTION 8.5 Dividing Radical Expressions and Rationalizing

OBJECTIVES

As a result of completing this section, you will be able to

1. Divide radical expressions.
2. Rationalize the denominator.
3. Rationalize the denominator using conjugates.
4. Troubleshoot common errors.

In Section 8.4, we learned how to add, subtract, and multiply radicals. We will now learn how to divide radicals using the quotient rule for radicals. We will also learn how to rewrite radical expressions by removing the radical from the denominator.

Dividing Radical Expressions

Objective 1

Divide radical expressions.

In Section 8.3, the quotient rule for radicals was introduced. The rule stated that if $\sqrt[n]{a}$ and $\sqrt[n]{b}$ are real numbers and $b \neq 0$,

$$\sqrt[n]{\frac{a}{b}} = \frac{\sqrt[n]{a}}{\sqrt[n]{b}}$$

This property enables us to divide radical expressions.

Property: Quotient Rule for Radicals

For $\sqrt[n]{a}$ and $\sqrt[n]{b}$ real numbers with $b \neq 0$,

$$\frac{\sqrt[n]{a}}{\sqrt[n]{b}} = \sqrt[n]{\frac{a}{b}}$$

In this section, we will continue to assume that all variables represent positive real numbers.

Procedure: Dividing Radical Expressions

Step 1: Rewrite the problem as the nth root of the quotient of the radicands in the numerator and denominator.
Step 2: Simplify the quotient.
Step 3: Simplify the radical, if necessary.

Objective 1 Examples **Simplify each quotient and write each answer in simplest radical form.**

1a. $\dfrac{\sqrt{48}}{\sqrt{3}}$ **1b.** $\dfrac{\sqrt{21x^3}}{\sqrt{3x}}$ **1c.** $\dfrac{\sqrt{40}}{5\sqrt{2}}$ **1d.** $\dfrac{7\sqrt[3]{16x^5}}{\sqrt[3]{2x}}$ **1e.** $\dfrac{\sqrt[4]{64a^5b}}{\sqrt[4]{ab^{-3}}}$

Solutions

1a.

$\dfrac{\sqrt{48}}{\sqrt{3}} = \sqrt{\dfrac{48}{3}}$	Apply the quotient rule for radicals.
$= \sqrt{16}$	Simplify the radicand.
$= 4$	Simplify the square root.

1b.

$\dfrac{\sqrt{21x^3}}{\sqrt{3x}} = \sqrt{\dfrac{21x^3}{3x}}$	Apply the quotient rule for radicals.
$= \sqrt{7x^2}$	Simplify the radicand.
$= x\sqrt{7}$	Simplify: $\sqrt{x^2} = x$.

1c.

$\dfrac{1\sqrt{40}}{5\sqrt{2}} = \dfrac{1}{5}\sqrt{\dfrac{40}{2}}$	Apply the quotient rule for radicals.
$= \dfrac{1}{5}\sqrt{20}$	Simplify the radicand.
$= \dfrac{1}{5}\sqrt{4 \cdot 5}$	Find a perfect square factor of 20.
$= \dfrac{1}{5}(2)\sqrt{5}$	Apply the product rule for radicals and simplify.
$= \dfrac{2}{5}\sqrt{5}$	Multiply the coefficients.

1d.

$\dfrac{7\sqrt[3]{16x^5}}{\sqrt[3]{2x}} = 7\sqrt[3]{\dfrac{16x^5}{2x}}$	Apply the quotient rule for radicals.
$= 7\sqrt[3]{8x^4}$	Simplify the radicand.
$= 7\sqrt[3]{8x^3}\sqrt[3]{x}$	Find a perfect square factor of $8x^4$.
$= 7(2x)\sqrt[3]{x}$	Simplify.
$= 14x\sqrt[3]{x}$	Multiply the coefficients.

1e.

$\dfrac{\sqrt[4]{64a^5b}}{\sqrt[4]{ab^{-3}}} = \sqrt[4]{\dfrac{64a^5b}{ab^{-3}}}$	Apply the quotient rule for radicals.
$= \sqrt[4]{64a^{5-1}b^{1-(-3)}}$	Apply the quotient of like bases rule.
$= \sqrt[4]{64a^4b^4}$	Simplify.
$= \sqrt[4]{16a^4b^4}\sqrt[4]{4}$	Find a perfect fourth factor of $64a^4b^4$.
$= 2ab\sqrt[4]{4}$	Simplify.

Student Check 1 Simplify each quotient and write each answer in simplest radical form.

a. $\dfrac{\sqrt{50}}{\sqrt{2}}$ **b.** $\dfrac{\sqrt{30y^5}}{\sqrt{10y}}$ **c.** $\dfrac{\sqrt{24}}{4\sqrt{3}}$ **d.** $\dfrac{5\sqrt[3]{54y^7}}{\sqrt[3]{2y}}$ **e.** $\dfrac{\sqrt[4]{48a^9b}}{\sqrt[4]{ab^{-3}}}$

Rationalizing the Denominator

Objective 2 ▶

Rationalize the denominator.

Now we examine quotients, such as $\dfrac{\sqrt{2}}{\sqrt{3}}$, in which the radicand in the denominator does not divide evenly into the radicand in the numerator. Such a quotient is not in simplest radical form since there is either a radical in the denominator or a fraction in the radical. Our goal is to rewrite the quotient as an equivalent expression that doesn't contain a radical in the denominator.

The process of removing the radical from the denominator of a fraction is called **rationalizing the denominator**. It is called rationalizing because the original denominator is an irrational number but after we go through this process, the denominator is a rational number. We will use the following facts.

- Multiplying a fraction by a form of 1 produces an equivalent fraction.

$$\frac{a}{b} \cdot \frac{c}{c} = \frac{a \cdot c}{b \cdot c}$$

- An nth root raised to an exponent of n removes the radical.

$$(\sqrt[n]{x})^n = x, \text{ provided } \sqrt[n]{x} \text{ is a real number}$$

Procedure: Rationalizing the Denominator

Step 1: Multiply the fraction by a form of 1.

a. If the radical in the denominator is a square root, multiply the numerator and denominator by the square root expression that makes the radicand in the denominator a perfect square.

b. If the radical in the denominator is a cube root, multiply the numerator and denominator by the cube root expression that makes the radicand in the denominator a perfect cube.

c. If the radical in the denominator is an nth root, multiply the numerator and denominator by the nth root expression that makes the radicand in the denominator a perfect nth.

Step 2: Simplify the products in the numerator and denominator.

Objective 2 Examples Simplify each expression by rationalizing the denominator.

2a. $\frac{\sqrt{3}}{\sqrt{7}}$ **2b.** $\frac{1}{\sqrt{2x}}$ **2c.** $\frac{2}{\sqrt[3]{9}}$ **2d.** $\frac{6\sqrt[4]{5}}{\sqrt[4]{2}}$

Solutions

2a. $\frac{\sqrt{3}}{\sqrt{7}} = \frac{\sqrt{3}}{\sqrt{7}} \cdot \frac{\sqrt{7}}{\sqrt{7}}$ Multiply by a form of 1, $\frac{\sqrt{7}}{\sqrt{7}}$.

$= \frac{\sqrt{3} \cdot \sqrt{7}}{\sqrt{7} \cdot \sqrt{7}}$ Multiply the numerators and denominators.

$= \frac{\sqrt{21}}{\sqrt{49}}$ Apply the product rule for radicals.

$= \frac{\sqrt{21}}{7}$ Simplify the denominator.

2b. $\frac{1}{\sqrt{2x}} = \frac{1}{\sqrt{2x}} \cdot \frac{\sqrt{2x}}{\sqrt{2x}}$ Multiply by a form of 1, $\frac{\sqrt{2x}}{\sqrt{2x}}$.

$= \frac{1 \cdot \sqrt{2x}}{\sqrt{2x} \cdot \sqrt{2x}}$ Multiply the numerators and denominators.

$= \frac{\sqrt{2x}}{\sqrt{4x^2}}$ Apply the product rule for radicals.

$= \frac{\sqrt{2x}}{2x}$ Simplify the denominator.

2c. We must make the radicand in the denominator, 9, a perfect cube. Since 27 is a perfect cube, we use the factor $\sqrt[3]{3}$.

$$\frac{2}{\sqrt[3]{9}} = \frac{2}{\sqrt[3]{9}} \cdot \frac{\sqrt[3]{3}}{\sqrt[3]{3}}$$ Multiply by a form of 1, $\frac{\sqrt[3]{3}}{\sqrt[3]{3}}$.

$$= \frac{2 \cdot \sqrt[3]{3}}{\sqrt[3]{9} \cdot \sqrt[3]{3}}$$ Multiply the numerators and denominators.

$$= \frac{2\sqrt[3]{3}}{\sqrt[3]{27}}$$ Apply the product rule for radicals.

$$= \frac{2\sqrt[3]{3}}{3}$$ Simplify the denominator.

2d. We must make the radicand in the denominator, 2, a perfect fourth. Since 16 is a perfect fourth, we use the factor $\sqrt[4]{8}$.

$$\frac{6\sqrt[4]{5}}{\sqrt[4]{2}} = \frac{6\sqrt[4]{5}}{\sqrt[4]{2}} \cdot \frac{\sqrt[4]{8}}{\sqrt[4]{8}}$$ Multiply by a form of 1, $\frac{\sqrt[4]{8}}{\sqrt[4]{8}}$.

$$= \frac{6\sqrt[4]{5} \cdot \sqrt[4]{8}}{\sqrt[4]{2} \cdot \sqrt[4]{8}}$$ Multiply the numerators and denominators.

$$= \frac{6\sqrt[4]{40}}{\sqrt[4]{16}}$$ Apply the product rule for radicals.

$$= \frac{6\sqrt[4]{40}}{2}$$ Simplify the denominator.

$$= 3\sqrt[4]{40}$$ Simplify the fraction.

Student Check 2 Simplify each expression by rationalizing the denominator.

a. $\frac{\sqrt{5}}{\sqrt{6}}$ **b.** $\frac{\sqrt{10}}{\sqrt{3x}}$ **c.** $\frac{5}{\sqrt[3]{2}}$ **d.** $\frac{7\sqrt[5]{3}}{\sqrt[5]{2}}$

Rationalizing the Denominator Using Conjugates

Objective 3 Rationalize the denominator using conjugates.

If the denominator of a fraction contains a sum or difference of square roots, then the form of 1 that must be used to rationalize the denominator involves the *conjugate* of the denominator. Recall that **conjugates** are binomial expressions that only differ in the sign of a term. Two important facts are

- $a + b$ and $a - b$ are conjugates.
- $(a + b)(a - b) = a^2 - b^2$

In general, the product of conjugates will not contain any radical terms.

Procedure: Rationalizing the Denominator Using Conjugates

Step 1: Multiply the fraction by a form of 1. The form of 1 is the conjugate of the denominator divided by itself.
Step 2: Multiply the numerators and denominators.
Step 3: Simplify the expressions in the numerators and denominators.

Objective 3 Examples Simplify each expression. Rationalize the denominators using conjugates.

3a. $\frac{6}{\sqrt{2} + \sqrt{3}}$ **3b.** $\frac{4 + \sqrt{5}}{3 - \sqrt{2}}$

Solutions **3a.** The conjugate of the denominator, $\sqrt{2}+\sqrt{3}$, is $\sqrt{2}-\sqrt{3}$.

$$\frac{6}{\sqrt{2}+\sqrt{3}} = \frac{6}{\sqrt{2}+\sqrt{3}} \cdot \frac{\sqrt{2}-\sqrt{3}}{\sqrt{2}-\sqrt{3}}$$ Multiply by a form of 1, $\frac{\sqrt{2}-\sqrt{3}}{\sqrt{2}-\sqrt{3}}$.

$$= \frac{6(\sqrt{2}-\sqrt{3})}{(\sqrt{2}+\sqrt{3})(\sqrt{2}-\sqrt{3})}$$ Multiply the numerators and denominators.

$$= \frac{6\sqrt{2}-6\sqrt{3}}{(\sqrt{2})^2-(\sqrt{3})^2}$$ Apply the distributive property and the product of conjugates property.

$$= \frac{6\sqrt{2}-6\sqrt{3}}{2-3}$$ Simplify.

$$= \frac{6\sqrt{2}-6\sqrt{3}}{-1}$$ Combine the terms in the denominator.

$$= -6\sqrt{2}+6\sqrt{3}$$ Simplify the fraction.

3b. The conjugate of $3-\sqrt{2}$ is $3+\sqrt{2}$.

$$\frac{4+\sqrt{5}}{3-\sqrt{2}} = \frac{4+\sqrt{5}}{3-\sqrt{2}} \cdot \frac{3+\sqrt{2}}{3+\sqrt{2}}$$ Multiply by a form of 1, $\frac{3+\sqrt{2}}{3+\sqrt{2}}$.

$$= \frac{(4+\sqrt{5})(3+\sqrt{2})}{(3-\sqrt{2})(3+\sqrt{2})}$$ Multiply the numerators and denominators.

$$= \frac{12+4\sqrt{2}+3\sqrt{5}+\sqrt{10}}{(3)^2-(\sqrt{2})^2}$$ Apply the FOIL method and the product of conjugates property.

$$= \frac{12+4\sqrt{2}+3\sqrt{5}+\sqrt{10}}{9-2}$$ Simplify.

$$= \frac{12+4\sqrt{2}+3\sqrt{5}+\sqrt{10}}{7}$$ Combine the terms in the denominator.

Student Check 3 Simplify each expression. Rationalize the denominators using conjugates.

a. $\frac{5}{\sqrt{6}-\sqrt{2}}$ **b.** $\frac{7+\sqrt{3}}{5-\sqrt{7}}$

Objective 4 ▶

Troubleshoot common errors.

Troubleshooting Common Errors

Some common errors with dividing radicals and rationalizing denominators are shown.

Objective 4 Examples A problem and an incorrect solution are given. Provide the correct solution and an explanation of the error.

4a. Simplify $\frac{\sqrt{6}}{3}$.

Incorrect Solution	Correct Solution and Explanation
$\frac{\sqrt{6}}{3} = \sqrt{2}$	The denominator does not have a radical, so we cannot simplify the quotient. $\frac{\sqrt{6}}{3} = \frac{1}{3}\sqrt{6}$

4b. Simplify by rationalizing the denominator. $\dfrac{\sqrt[3]{5}}{\sqrt[3]{2}}$.

Incorrect Solution	Correct Solution and Explanation
$\dfrac{\sqrt[3]{5}}{\sqrt[3]{2}} = \dfrac{\sqrt[3]{5}}{\sqrt[3]{2}} \cdot \dfrac{\sqrt[3]{2}}{\sqrt[3]{2}} = \dfrac{\sqrt[3]{10}}{2}$	We need to multiply by an expression that will make the denominator a perfect cube. The radicand of the denominator has a single factor of 2. So, we need to get two more factors of 2, or $2^2 = 4$. $$\dfrac{\sqrt[3]{5}}{\sqrt[3]{2}} = \dfrac{\sqrt[3]{5}}{\sqrt[3]{2}} \cdot \dfrac{\sqrt[3]{4}}{\sqrt[3]{4}} = \dfrac{\sqrt[3]{20}}{\sqrt[3]{8}} = \dfrac{\sqrt[3]{20}}{2}$$

ANSWERS TO STUDENT CHECKS

Student Check 1 **a.** 5 **b.** $y^2\sqrt{3}$ **c.** $\dfrac{\sqrt{2}}{2}$

d. $15y^2$ **e.** $2a^2b\sqrt[4]{3}$

Student Check 2 **a.** $\dfrac{\sqrt{30}}{6}$ **b.** $\dfrac{\sqrt{30x}}{3x}$ **c.** $\dfrac{5\sqrt[3]{4}}{2}$ **d.** $\dfrac{7\sqrt[5]{48}}{2}$

Student Check 3 **a.** $\dfrac{5\sqrt{6} + 5\sqrt{2}}{4}$

b. $\dfrac{35 + 7\sqrt{7} + 5\sqrt{3} + \sqrt{21}}{18}$

SUMMARY OF KEY CONCEPTS

1. We can divide radical expressions with the same index by dividing the radicands and then taking the root of the quotient.
2. Rationalizing the denominator is the process of removing the radical from the denominator.
   a. If there is a single term in the denominator, multiply the fraction by a form of 1 that makes the radicand in the denominator a perfect square if the index is 2, a perfect cube if the index is 3, and so on.
   b. If the denominator is a binomial, multiply the numerator and denominator of the fraction by the conjugate of the denominator.

GRAPHING CALCULATOR SKILLS

The calculator can be used to check our answers when rationalizing expressions. Enter the given expression and the result to determine if their values are the same. If they are, then the expressions are equivalent.

Example: Verify that $\dfrac{1}{\sqrt{2}} = \dfrac{\sqrt{2}}{2}$.

Solution:

```
1/√(2)
          .7071067812
√(2)/2
          .7071067812
```

Since the decimal values of the two expressions are equal, the expressions are equivalent.

SECTION 8.5 EXERCISE SET

Write About It!

Use complete sentences in your answer to each exercise.

1. Explain how to divide radical expressions.
2. Explain why we need to rationalize denominators.
3. Define the conjugate of $a + \sqrt{b}$, where a and b are real numbers and $b > 0$.
4. Explain how we multiply $a + \sqrt{b}$ and its conjugate, where a and b are real numbers and $b > 0$.
5. Explain how to rationalize the denominator of the expression $\dfrac{1}{\sqrt[5]{6}}$.
6. Explain why $\dfrac{5}{\sqrt{11} - \sqrt{10}} \neq 5$.

Practice Makes Perfect!

Simplify each quotient and write each answer in simplest radical form. (*See Objective 1.*)

7. $\dfrac{\sqrt{48}}{\sqrt{3}}$ 4
8. $\dfrac{\sqrt{180}}{\sqrt{5}}$ 6
9. $\dfrac{\sqrt{300s^{15}}}{\sqrt{3s^7}}$ $10s^4$
10. $\dfrac{\sqrt{216t^{11}}}{\sqrt{6t^5}}$ $6t^3$
11. $\dfrac{\sqrt{150a^3}}{\sqrt{6a^{13}}}$ $\dfrac{5}{a^5}$
12. $\dfrac{\sqrt{128b^3}}{\sqrt{2b^7}}$ $\dfrac{8}{b^2}$
13. $\dfrac{\sqrt{24c^{13}}}{\sqrt{2c^{10}}}$ $2c\sqrt{3c}$
14. $\dfrac{\sqrt{96d^{15}}}{\sqrt{2d^4}}$ $4d^5\sqrt{3d}$
15. $\dfrac{\sqrt{294}}{21\sqrt{3}}$ $\dfrac{\sqrt{2}}{3}$
16. $\dfrac{\sqrt{224}}{14\sqrt{7}}$ $\dfrac{2\sqrt{2}}{7}$
17. $\dfrac{\sqrt{54}}{12\sqrt{3}}$ $\dfrac{\sqrt{2}}{4}$
18. $\dfrac{\sqrt{90}}{21\sqrt{5}}$ $\dfrac{\sqrt{2}}{7}$
19. $\dfrac{\sqrt{24a^6}}{18\sqrt{2a}}$ $\dfrac{a^2\sqrt{3a}}{9}$
20. $\dfrac{\sqrt{126b^9}}{6\sqrt{2b^4}}$ $\dfrac{b^2\sqrt{7b}}{2}$
21. $\dfrac{6\sqrt[3]{112t^{11}}}{\sqrt[3]{7t^2}}$ $12t^3\sqrt[3]{2}$
22. $\dfrac{4\sqrt[3]{120s^{18}}}{\sqrt[3]{3s^9}}$ $8s^3\sqrt[3]{5}$
23. $\dfrac{9\sqrt[3]{810x^{18}}}{\sqrt[3]{10x^3}}$ $27x^5\sqrt[3]{3}$
24. $\dfrac{3\sqrt[3]{2500y^{11}}}{\sqrt[3]{5y^5}}$ $15y^2\sqrt[3]{4}$
25. $\dfrac{7\sqrt[5]{192r^{13}}}{\sqrt[5]{3r^2}}$ $14r^2\sqrt[5]{2r}$
26. $\dfrac{5\sqrt[5]{486y^{13}}}{\sqrt[5]{2y^7}}$ $15y\sqrt[5]{y}$
27. $\dfrac{\sqrt{192a^5b^3}}{\sqrt{75a^{-2}b^9}}$ $\dfrac{8a^3\sqrt{a}}{5b^3}$
28. $\dfrac{\sqrt{98x^{-1}y^7}}{\sqrt{18x^3y^2}}$ $\dfrac{7y^2\sqrt{y}}{3x^2}$
29. $\dfrac{9\sqrt[3]{810x^{18}y^6}}{\sqrt[3]{10x^3y^2}}$ $27x^5y\sqrt[3]{3y}$
30. $\dfrac{6\sqrt[4]{144u^{14}v^{-1}}}{\sqrt[4]{3u^6v^{-10}}}$ $12u^2v^2\sqrt[4]{3v}$
31. $\dfrac{10\sqrt[3]{189a^{20}b^{12}}}{\sqrt[3]{7ab^{-4}}}$ $30a^6b^5\sqrt[3]{ab}$
32. $\dfrac{4\sqrt[4]{1215s^{17}t^8}}{\sqrt[4]{3s^2t^{-9}}}$ $12s^3t^4\sqrt[4]{5s^3t}$

Additional answers can be found in the Instructor Answer Appendix.

Simplify each expression by rationalizing the denominators. (*See Objective 2.*)

33. $\dfrac{\sqrt{44}}{\sqrt{6}}$ $\dfrac{\sqrt{66}}{3}$
34. $\dfrac{\sqrt{21}}{\sqrt{15}}$ $\dfrac{\sqrt{35}}{5}$
35. $\dfrac{\sqrt{18}}{\sqrt{3a}}$ $\dfrac{\sqrt{6a}}{a}$
36. $\dfrac{\sqrt{2}}{\sqrt{6x}}$ $\dfrac{\sqrt{3x}}{3x}$
37. $\dfrac{\sqrt{15}}{\sqrt{12y}}$ $\dfrac{\sqrt{5y}}{2y}$
38. $\dfrac{\sqrt{49}}{\sqrt{21x}}$ $\dfrac{\sqrt{21x}}{3x}$
39. $\dfrac{2}{\sqrt[3]{25}}$ $\dfrac{2\sqrt[3]{5}}{5}$
40. $\dfrac{7}{\sqrt[3]{2}}$ $\dfrac{7\sqrt[3]{4}}{2}$
41. $\dfrac{6}{\sqrt[5]{9a^2}}$ $\dfrac{2\sqrt[5]{27a^3}}{a}$
42. $\dfrac{12}{\sqrt[5]{4b}}$ $\dfrac{6\sqrt[5]{8b^4}}{b}$
43. $\dfrac{6}{\sqrt[6]{8x}}$ $\dfrac{3\sqrt[6]{8x^5}}{x}$
44. $\dfrac{15}{\sqrt[6]{25y}}$ $\dfrac{3\sqrt[6]{625y^5}}{y}$
45. $\dfrac{18}{\sqrt[4]{3x^3}}$ $\dfrac{6\sqrt[4]{27x}}{x}$
46. $\dfrac{24}{\sqrt[4]{2y}}$ $\dfrac{12\sqrt[4]{8y^3}}{y}$

Simplify each expression. Rationalize the denominator using its conjugate. (*See Objective 3.*)

47. $\dfrac{12}{\sqrt{6} + \sqrt{3}}$ $4(\sqrt{6} - \sqrt{3})$
48. $\dfrac{8}{\sqrt{11} - \sqrt{7}}$ $2(\sqrt{11} + \sqrt{7})$
49. $\dfrac{10}{\sqrt{11} - \sqrt{3}}$ $\dfrac{5(\sqrt{11} + \sqrt{3})}{4}$
50. $\dfrac{6}{\sqrt{10} + \sqrt{6}}$ $\dfrac{3(\sqrt{10} - \sqrt{6})}{2}$
51. $\dfrac{4 - \sqrt{6}}{2 - \sqrt{5}}$ $-8 - 4\sqrt{5} + 2\sqrt{6} + \sqrt{30}$
52. $\dfrac{2 - \sqrt{2}}{5 + \sqrt{3}}$ $\dfrac{10 - 2\sqrt{3} - 5\sqrt{2} + \sqrt{6}}{22}$
53. $\dfrac{2 - \sqrt{3}}{2 + \sqrt{3}}$ $7 - 4\sqrt{3}$
54. $\dfrac{10 - \sqrt{7}}{4 + \sqrt{7}}$ $\dfrac{47 - 14\sqrt{7}}{9}$
55. $\dfrac{6 + \sqrt{15}}{5 - \sqrt{15}}$ $\dfrac{45 + 11\sqrt{15}}{10}$
56. $\dfrac{10 - \sqrt{10}}{2 - \sqrt{10}}$ $-\dfrac{5 + 4\sqrt{10}}{3}$
57. $\dfrac{h}{\sqrt{h + 100} - 10}$ $\sqrt{h + 100} + 10$
58. $\dfrac{k}{\sqrt{k + 16} - 4}$ $\sqrt{k + 16} + 4$

Mix 'Em Up!

Simplify each expression. Assume all variables represent positive real numbers.

59. $\dfrac{\sqrt{640x^3}}{\sqrt{10x}}$ $8x$
60. $\dfrac{\sqrt{300y^5}}{\sqrt{3y}}$ $10y^2$
61. $\dfrac{\sqrt{245a^4}}{\sqrt{5a^8}}$ $\dfrac{7}{a^2}$
62. $\dfrac{\sqrt{68b^{10}}}{\sqrt{17b^2}}$ $2b^4$
63. $\dfrac{10\sqrt[3]{540x^{13}}}{\sqrt[3]{10x^2}}$ $30x^3\sqrt[3]{2x^2}$
64. $\dfrac{8\sqrt[3]{112y^{14}}}{\sqrt[3]{7y^7}}$ $16y^2\sqrt[3]{2y}$

65. $\dfrac{4\sqrt[3]{405a^{12}}}{\sqrt[3]{3a^3}}$ $12a^3\sqrt[3]{5}$

66. $\dfrac{3\sqrt[3]{1500b^{15}}}{\sqrt[3]{6b^9}}$ $15b^2\sqrt[3]{2}$

67. $\dfrac{\sqrt{600s^{21}}}{\sqrt{2s^4}}$ $10s^8\sqrt{3s}$

68. $\dfrac{\sqrt{96d^{15}}}{\sqrt{2d^4}}$ $4d^5\sqrt{3d}$

69. $\dfrac{\sqrt{150x^4}}{20\sqrt{3x}}$ $\dfrac{x\sqrt{2x}}{4}$

70. $\dfrac{3\sqrt[4]{112a^9}}{\sqrt[4]{7a^2}}$ $6a\sqrt[4]{a^3}$

71. $\dfrac{2\sqrt[4]{486b^{13}}}{\sqrt[4]{6b^2}}$ $6b^2\sqrt[4]{b^3}$

72. $\dfrac{\sqrt{96y^6}}{18\sqrt{2y^3}}$ $\dfrac{2y\sqrt{3y}}{9}$

73. $\dfrac{7\sqrt[5]{320r^8}}{\sqrt[5]{2r^2}}$ $14r\sqrt[5]{5r}$

74. $\dfrac{6\sqrt[5]{3402t^{14}}}{\sqrt[5]{2t^3}}$ $18t^2\sqrt[5]{7t}$

75. $\dfrac{\sqrt{96a^7b^{-3}}}{2\sqrt{2a^{-2}b}}$ $\dfrac{2a^4\sqrt{3a}}{b^2}$

76. $\dfrac{\sqrt{1080x^{-1}y^7}}{8\sqrt{6x^3y^2}}$ $\dfrac{3y^2\sqrt{5y}}{4x^2}$

77. $\dfrac{5\sqrt[3]{54x^{12}y}}{\sqrt[3]{2x^5y^{-5}}}$ $15x^2y^2\sqrt[3]{x}$

78. $\dfrac{3\sqrt[3]{48x^{10}y^{-11}}}{\sqrt[3]{2x^3y^{-2}}}$ $\dfrac{6x^2\sqrt[3]{3x}}{y^3}$

79. $\dfrac{\sqrt{216a^7}}{15\sqrt{2b^4}}$ $\dfrac{2a^3\sqrt{3a}}{5b^2}$

80. $\dfrac{\sqrt{120x^5}}{8\sqrt{6y^2}}$ $\dfrac{x^2\sqrt{5x}}{4y}$

81. $\dfrac{6\sqrt[5]{800r^9s^{-16}}}{\sqrt[5]{5r^2s^{-6}}}$ $\dfrac{12r\sqrt[5]{5r^2}}{s^2}$

82. $\dfrac{3\sqrt[5]{576r^{16}s^{-12}}}{\sqrt[5]{2r^4s^{-2}}}$ $\dfrac{6r^2\sqrt[5]{9r^2}}{s^2}$

Simplify each expression by rationalizing the denominator.

83. $\dfrac{\sqrt{28}}{\sqrt{6}}$ $\dfrac{\sqrt{42}}{3}$

84. $\dfrac{\sqrt{40}}{\sqrt{14}}$ $\dfrac{2\sqrt{35}}{7}$

85. $\dfrac{3}{\sqrt[4]{18a^3}}$ $\dfrac{\sqrt[4]{72a}}{2a}$

86. $\dfrac{15}{\sqrt[4]{125b}}$ $\dfrac{3\sqrt[4]{5b^3}}{b}$

87. $\dfrac{8}{\sqrt{21}+\sqrt{5}}$ $\dfrac{\sqrt{21}-\sqrt{5}}{2}$

88. $\dfrac{6}{\sqrt{11}-\sqrt{3}}$ $\dfrac{3(\sqrt{11}+\sqrt{3})}{4}$

89. $\dfrac{\sqrt{42}}{\sqrt{7x}}$ $\dfrac{\sqrt{6x}}{x}$

90. $\dfrac{\sqrt{75}}{\sqrt{15y}}$ $\dfrac{\sqrt{5y}}{y}$

91. $\dfrac{9}{\sqrt[3]{12}}$ $\dfrac{3\sqrt[3]{18}}{2}$

92. $\dfrac{12}{\sqrt[3]{45}}$ $\dfrac{4\sqrt[3]{75}}{5}$

93. $\dfrac{1+\sqrt{2}}{-3+\sqrt{5}}$

94. $\dfrac{7-\sqrt{5}}{3-\sqrt{3}}$

95. $\dfrac{-3-\sqrt{21}}{5-\sqrt{21}}$ $-9-2\sqrt{21}$

96. $\dfrac{-9+\sqrt{14}}{4+\sqrt{14}}$ $\dfrac{-50+13\sqrt{14}}{2}$

97. $\dfrac{5+\sqrt{13}}{4-\sqrt{13}}$ $11+3\sqrt{13}$

98. $\dfrac{-4-\sqrt{6}}{2+\sqrt{6}}$ $1-\sqrt{6}$

99. $\dfrac{6}{\sqrt[5]{4x^2}}$ $\dfrac{3\sqrt[5]{8x^3}}{x}$

100. $\dfrac{8}{\sqrt[5]{48y^3}}$ $\dfrac{4\sqrt[5]{162y^2}}{3y}$

101. $\dfrac{2-\sqrt{6}}{1+\sqrt{3}}$

102. $\dfrac{1-\sqrt{15}}{3-\sqrt{5}}$

103. $\dfrac{a}{\sqrt{a+36}+6}$ $\sqrt{a+36}-6$

104. $\dfrac{b}{\sqrt{b+49}-7}$ $\sqrt{b+49}+7$

You Be the Teacher!

Correct each student's errors, if any.

105. Simplify $\dfrac{\sqrt{28}}{14}$.

Andrew's work:

$$\frac{\sqrt{28}}{14} = \frac{\sqrt{\cancel{28}^{\,2}}}{\cancel{14}_{\,1}} = \sqrt{2}$$

$$\frac{\sqrt{28}}{14} = \frac{\sqrt{4\cdot 7}}{14} = \frac{2\sqrt{7}}{14} = \frac{\sqrt{7}}{7}$$

106. Simplify by rationalizing the denominator: $\dfrac{6}{\sqrt[5]{4x^2}}$.

Rebecca's work:

$$\frac{6}{\sqrt[5]{4x^2}} = \frac{6\sqrt[5]{4^4x^3}}{\sqrt[5]{4x^2}\sqrt[5]{4^4x^3}} = \frac{6\sqrt[5]{256x^3}}{4x} = \frac{3\sqrt[5]{256x^3}}{2x}$$

$$\frac{6}{\sqrt[5]{4x^2}} = \frac{6\sqrt[5]{2^3x^3}}{\sqrt[5]{2^2x^2}\sqrt[5]{2^3x^3}} = \frac{6\sqrt[5]{8x^3}}{2x} = \frac{3\sqrt[5]{8x^3}}{x}$$

107. Simplify by rationalizing the denominator: $\dfrac{\sqrt{7}-\sqrt{3}}{\sqrt{7}+\sqrt{3}}$.

Roy's work:

$$\frac{\sqrt{7}-\sqrt{3}}{\sqrt{7}+\sqrt{3}} = \frac{(\sqrt{7}-\sqrt{3})}{(\sqrt{7}+\sqrt{3})}\cdot\frac{(\sqrt{7}+\sqrt{3})}{(\sqrt{7}+\sqrt{3})} = \frac{7-3}{7+2\sqrt{21}+3} = \frac{4}{10+2\sqrt{21}} = \frac{\cancel{4}^{\,2}}{{}^{5}\cancel{10}+{}^{1}\cancel{2}\sqrt{21}} = \frac{2}{5+\sqrt{21}}$$

108. Simplify by rationalizing the denominator: $\dfrac{\sqrt{6}}{\sqrt{12}-\sqrt{3}}$.

Rita's work:

$$\frac{\sqrt{6}}{\sqrt{12}-\sqrt{3}} = \frac{\sqrt{6}}{\sqrt{12-3}} = \frac{\sqrt{6}}{\sqrt{9}} = \frac{\sqrt{6}}{3} = \frac{\sqrt{\cancel{6}^{\,2}}}{\cancel{3}^{\,1}} = \sqrt{2}$$

$$\frac{\sqrt{6}}{\sqrt{12}-\sqrt{3}} = \frac{\sqrt{6}}{\sqrt{4\cdot 3}-\sqrt{3}} = \frac{\sqrt{6}}{2\sqrt{3}-\sqrt{3}} = \frac{\sqrt{6}}{\sqrt{3}} = \sqrt{2}$$

Calculate It!

Use a calculator to verify the simplification of each radical expression.

109. $\dfrac{12}{\sqrt{11}-\sqrt{2}} = \dfrac{4(\sqrt{11}+\sqrt{2})}{3}$

110. $\dfrac{-6+\sqrt{3}}{-11-\sqrt{5}} = \dfrac{66-6\sqrt{5}-11\sqrt{3}+\sqrt{15}}{116}$

111. $\dfrac{12}{\sqrt[6]{3}} = 4\sqrt[6]{243}$

112. $\dfrac{16}{\sqrt[5]{20}} = \dfrac{8\sqrt[5]{5000}}{5}$

SECTION 8.6 Radical Equations and Their Applications

▶ OBJECTIVES

As a result of completing this section, you will be able to

1. Solve equations that contain radical expressions.
2. Solve problems involving the Pythagorean theorem and other formulas.
3. Troubleshoot common errors.

Mosteller's formula is used to calculate a person's body surface area (BSA). This value is often used in calculating drug dosages. A person's BSA in square meters is given by $\text{BSA} = \sqrt{\frac{hw}{3131}}$, where h is a person's height in inches and w is a person's weight in pounds. The average BSA for women is 1.6 m^2. If a woman is 5 ft tall, how much should she weigh to have an average BSA?

In this section, we will learn how square roots can be used to solve this problem.

Radical Equations

Objective 1 ▶

Solve equations that contain radical expressions.

Now that we know how to work with radical expressions, we will solve equations that contain these expressions. This type of equation is called a **radical equation**. Some examples of radical equations are

$$\sqrt{x-1} = 3 \qquad \sqrt{4x-1} = \sqrt{3x} \qquad \sqrt[3]{2x+3} = 6$$

To solve these equations, our goal is to produce an equivalent equation that doesn't contain radicals. We use the property that we learned in Section 8.1 to accomplish this goal. Recall that for $a \geq 0$,

$$\left(\sqrt[n]{a}\right)^n = a$$

Here are some illustrations of this property. We will assume the radicands are nonnegative.

Radical Expression	Value of n	$\left(\sqrt[n]{a}\right)^n$
$\sqrt{x-1}$	$n = 2$	$\left(\sqrt{x-1}\right)^2 = x - 1$
$\sqrt[3]{2x+3}$	$n = 3$	$\left(\sqrt[3]{2x+3}\right)^3 = 2x + 3$

In addition to this property, we must apply a fundamental principle of solving equations; what we do to one side of an equation, we must do to the other side. This means that if we have to square one side of the equation to remove a radical, we must square the other side. If we have to cube one side of an equation to remove a radical, we must cube the other side. This leads to the following property.

Property: Power Property

If $a = b$, then $a^n = b^n$.

This property tells us that if two quantities are equal, then the quantities raised to the nth are equal as well. The equation that results from raising each side to the nth does not necessarily have the same solutions as the original equation. Consider the following illustration.

Original equation	$x = -4$	Solution set: $\{-4\}$
Squared equation	$(x)^2 = (-4)^2$ $x^2 = 16$ $x^2 - 16 = 0$ $(x-4)(x+4) = 0$ $x = 4$ or $x = -4$	Solution set: $\{-4, 4\}$

Notice that the solution set of the squared equation contains an extra solution, 4, that is *not* a solution of the original equation. In this case, 4 is an **extraneous solution** that results from squaring the original equation.

When we apply the power property, we find solutions of the original equation, but we can also get solutions that do not satisfy the original equation. That is, we can obtain extraneous solutions when we apply the power property. For this reason, we must *always* check the proposed solutions in the original equation.

Procedure: Solving Radical Equations

Step 1: Isolate the radical expression on one side of the equation.
Step 2: Raise each side of the equation to the exponent equal to the index and simplify each side.
Step 3: If the equation still contains a radical expression, repeat Steps 1 and 2.
Step 4: Solve the resulting equation.
Step 5: Check the proposed solutions in the original equation. Discard any extraneous solutions.
Step 6: Write the solution set of the equation.

Objective 1 Examples **Solve each radical equation.**

1a. $\sqrt{x-1}=5$ **1b.** $\sqrt[3]{2x+3}=6$ **1c.** $\sqrt{y-3}+4=2$

1d. $\sqrt{5x-2}=\sqrt{3x}$ **1e.** $\sqrt{y-4}=y-10$ **1f.** $\sqrt{x+1}-\sqrt{x}=1$

Solutions **1a.**

$$\sqrt{x-1}=5$$
$$(\sqrt{x-1})^2=(5)^2 \quad \text{Apply the power property.}$$
$$x-1=25 \quad \text{Simplify each side.}$$
$$x-1+1=25+1 \quad \text{Add 1 to each side.}$$
$$x=26 \quad \text{Simplify.}$$

Check:

$$\sqrt{x-1}=5 \quad \text{Original equation}$$
$$\sqrt{26-1}=5 \quad \text{Replace } x \text{ with 26.}$$
$$\sqrt{25}=5 \quad \text{Simplify.}$$
$$5=5 \quad \text{Simplify.}$$

The solution 26 makes the original equation true. So, the solution set is $\{26\}$.

1b.

$$\sqrt[3]{2x+3}=6$$
$$(\sqrt[3]{2x+3})^3=(6)^3 \quad \text{Apply the power property.}$$
$$2x+3=216 \quad \text{Simplify each side.}$$
$$2x+3-3=216-3 \quad \text{Subtract 3 from each side.}$$
$$2x=213 \quad \text{Simplify.}$$
$$\frac{2x}{2}=\frac{213}{2} \quad \text{Divide each side by 2.}$$
$$x=\frac{213}{2} \quad \text{Simplify.}$$

Check:

$$\sqrt[3]{2x+3}=6 \quad \text{Original equation}$$
$$\sqrt[3]{2\left(\frac{213}{2}\right)+3}=6 \quad \text{Replace } x \text{ with } \frac{213}{2}.$$
$$\sqrt[3]{213+3}=6 \quad \text{Multiply 2 and } \frac{213}{2}.$$
$$\sqrt[3]{216}=6 \quad \text{Add.}$$
$$6=6 \quad \text{Simplify.}$$

The solution $\frac{213}{2}$ makes the original equation true, so the solution set is $\left\{\frac{213}{2}\right\}$.

1c. The square root expression is not isolated. Before squaring, we must subtract 4 from each side of the equation.

$$\sqrt{y-3}+4=2$$
$$\sqrt{y-3}+4-4=2-4 \quad \text{Subtract 4 from each side.}$$

$$\sqrt{y-3} = -2 \quad \text{Simplify.}$$
$$(\sqrt{y-3})^2 = (-2)^2 \quad \text{Apply the power property.}$$
$$y - 3 = 4 \quad \text{Simplify each side.}$$
$$y - 3 + 3 = 4 + 3 \quad \text{Add 3 to each side.}$$
$$y = 7 \quad \text{Simplify.}$$

Check:
$$\sqrt{y-3} + 4 = 2 \quad \text{Original equation}$$
$$\sqrt{7-3} + 4 = 2 \quad \text{Replace } y \text{ with 7.}$$
$$\sqrt{4} + 4 = 2 \quad \text{Simplify the radicand.}$$
$$2 + 4 = 2 \quad \text{Simplify the square root.}$$
$$6 = 2 \quad \text{Simplify.}$$

The solution 7 makes the original equation false and, therefore, must be excluded from the solution set. It is an extraneous solution. So, the solution set is the empty set, or $\varnothing$.

Note: *We can conclude that the equation has no solution without going through all of the steps to solve and check. After we isolate the radical, we obtain the equation, $\sqrt{y-3} = -2$. This statement is always false. Because $\sqrt{y-3}$ represents the principal, or positive, square root of $y - 3$, it will never equal a negative number.*

1d. Each square root expression is isolated on one side of the equation. So, we can apply the power property and square each side.

$$\sqrt{5x-2} = \sqrt{3x}$$
$$(\sqrt{5x-2})^2 = (\sqrt{3x})^2 \quad \text{Square each side.}$$
$$5x - 2 = 3x \quad \text{Simplify each side.}$$
$$5x - 2 + 2 = 3x + 2 \quad \text{Add 2 to each side.}$$
$$5x = 3x + 2 \quad \text{Simplify.}$$
$$5x - 3x = 3x + 2 - 3x \quad \text{Subtract } 3x \text{ from each side.}$$
$$2x = 2 \quad \text{Simplify.}$$
$$\frac{2x}{2} = \frac{2}{2} \quad \text{Divide each side by 2.}$$
$$x = 1 \quad \text{Simplify.}$$

Check:
$$\sqrt{5x-2} = \sqrt{3x} \quad \text{Original equation}$$
$$\sqrt{5(1)-2} = \sqrt{3(1)} \quad \text{Replace } x \text{ with 1.}$$
$$\sqrt{5-2} = \sqrt{3} \quad \text{Multiply.}$$
$$\sqrt{3} = \sqrt{3} \quad \text{Simplify.}$$

The solution 1 makes the original equation true. So, the solution set is $\{1\}$.

INSTRUCTOR NOTE:
Remind students that squaring a binomial creates a trinomial.

1e.
$$\sqrt{y-4} = y - 10$$
$$(\sqrt{y-4})^2 = (y-10)^2 \quad \text{Apply the power property.}$$
$$y - 4 = y^2 - 20y + 100 \quad \text{Simplify. Recall } (a-b)^2 = a^2 - 2ab + b^2.$$

To solve the resulting quadratic equation, we must write the equation in standard form, that is, get one side of the equation equal to 0. Then we solve the equation by factoring.

$$y - 4 - y = y^2 - 20y + 100 - y \quad \text{Subtract } y \text{ from each side.}$$
$$-4 = y^2 - 21y + 100 \quad \text{Simplify.}$$
$$-4 + 4 = y^2 - 21y + 100 + 4 \quad \text{Add 4 to each side.}$$

$$0 = y^2 - 21y + 104 \quad \text{Simplify.}$$
$$0 = (y - 13)(y - 8) \quad \text{Factor.}$$
$$y - 13 = 0 \text{ or } y - 8 = 0 \quad \text{Set each factor equal to 0.}$$
$$y = 13 \qquad y = 8 \quad \text{Solve the resulting equations.}$$

Check: Replace y with 13.

$$\sqrt{y-4} = y - 10$$
$$\sqrt{13-4} = 13 - 10$$
$$\sqrt{9} = 3$$
$$3 = 3 \quad \text{True}$$

Replace y with 8.

$$\sqrt{y-4} = y - 10$$
$$\sqrt{8-4} = 8 - 10$$
$$\sqrt{4} = -2$$
$$2 = -2 \quad \text{False}$$

The only solution that makes the original equation true is 13. So, the solution set is $\{13\}$.

1f. We must isolate one of the radicals on one side of the equation before squaring.

$$\sqrt{x+1} - \sqrt{x} = 1$$
$$\sqrt{x+1} - \sqrt{x} + \sqrt{x} = 1 + \sqrt{x} \quad \text{Add } \sqrt{x} \text{ to each side.}$$
$$\sqrt{x+1} = 1 + \sqrt{x} \quad \text{Simplify.}$$
$$(\sqrt{x+1})^2 = (1 + \sqrt{x})^2 \quad \text{Apply the power property.}$$
$$x + 1 = 1 + 2\sqrt{x} + x \quad \text{Simplify each side.}$$

Since the equation still has a radical expression, we must isolate the radical $2\sqrt{x}$ on one side of the equation.

$$x + 1 - 1 = 1 + 2\sqrt{x} + x - 1 \quad \text{Subtract 1 from each side.}$$
$$x = 2\sqrt{x} + x \quad \text{Simplify.}$$
$$x - x = 2\sqrt{x} + x - x \quad \text{Subtract } x \text{ from each side.}$$
$$0 = 2\sqrt{x} \quad \text{Simplify.}$$
$$(0)^2 = (2\sqrt{x})^2 \quad \text{Apply the power property.}$$
$$0 = 4x \quad \text{Simplify: } (2\sqrt{x})^2 = (2)^2(\sqrt{x})^2 = 4x.$$
$$\frac{0}{4} = \frac{4x}{4} \quad \text{Divide each side by 4.}$$
$$0 = x \quad \text{Simplify.}$$

The solution 0 makes the original equation true. So, the solution set is $\{0\}$.

Student Check 1 Solve each radical equation.

a. $\sqrt{x+3} = 4$ **b.** $\sqrt[3]{4x-3} = 5$ **c.** $\sqrt{3a-2} + 5 = 1$

d. $\sqrt{3x-1} = \sqrt{2x}$ **e.** $\sqrt{z+2} + 4 = z$ **f.** $\sqrt{3x-2} - \sqrt{x} = 2$

Objective 2

Solve problems involving the Pythagorean theorem and other formulas.

Applications

Some applications of radical equations deal with right triangles. Recall the *Pythagorean theorem* shows the relationship between the sides of a right triangle.

Property: Pythagorean Theorem

For a right triangle with legs a and b and hypotenuse c,

$$a^2 + b^2 = c^2$$

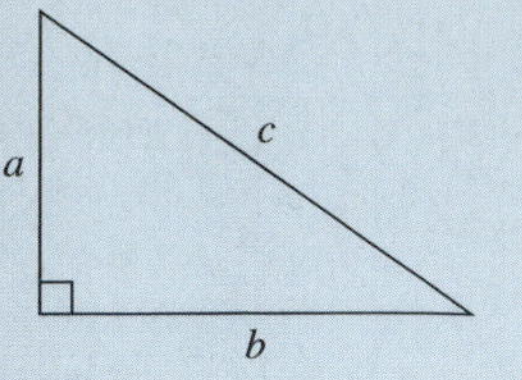

We will also solve some other applications that are based on formulas.

Objective 2 Examples

Solve each problem.

2a. Find the exact length of the unknown side of the right triangle in Figure 8.4.

5 ft
15 ft
b

Figure 8.4

Solution 2a.

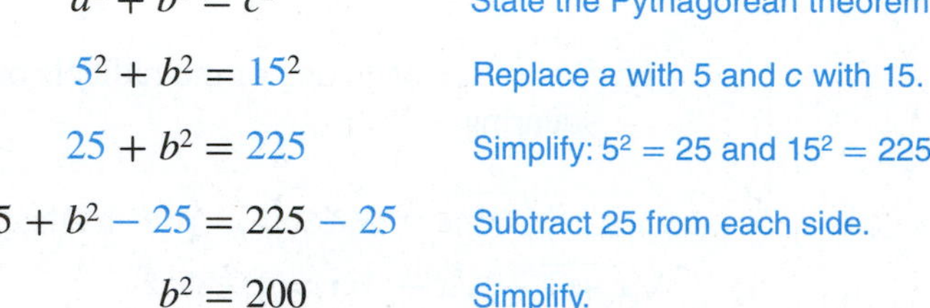

$a^2 + b^2 = c^2$	State the Pythagorean theorem.
$5^2 + b^2 = 15^2$	Replace a with 5 and c with 15.
$25 + b^2 = 225$	Simplify: $5^2 = 25$ and $15^2 = 225$.
$25 + b^2 - 25 = 225 - 25$	Subtract 25 from each side.
$b^2 = 200$	Simplify.
$b = \sqrt{200}$	Solve for b.
$b = \sqrt{100 \cdot 2}$	Factor 200 as $100 \cdot 2$.
$b = 10\sqrt{2}$	Apply the product rule.

The value of the unknown leg of the right triangle is $10\sqrt{2}$ ft, or approximately 14.14 ft.

2b. Mosteller's formula is used to calculate a person's body surface area (BSA). This value is often used in calculating drug dosages. A person's BSA in square meters is given by $\text{BSA} = \sqrt{\frac{hw}{3131}}$, where h is a person's height in inches and w is a person's weight in pounds. (i) Find the BSA for a female who is 5 ft 6 in. tall and weighs 165 lb. (ii) The average BSA for women is 1.6 m^2. If a woman is 5 ft tall, how much should she weigh to have an average BSA?

Solution 2b. i.

$\text{BSA} = \sqrt{\frac{hw}{3131}}$	State the BSA formula.
$\text{BSA} = \sqrt{\frac{(66)(165)}{3131}}$	Replace h with 66 and w with 165.
$\text{BSA} = \sqrt{\frac{10{,}890}{3131}}$	Multiply 66(165).
$\text{BSA} \approx 1.86$	Approximate the value.

So, the BSA for a female who is 5 ft 6 in. tall and weighs approximately 165 lb is approximately 1.86 m^2.

ii. Let BSA = 1.6, $h = 60$ in., and solve for w.

$$\text{BSA} = \sqrt{\frac{hw}{3131}} \quad \text{State the BSA formula.}$$

$$1.6 = \sqrt{\frac{60w}{3131}} \quad \text{Replace BSA with 1.6 and } h \text{ with 60.}$$

$$(1.6)^2 = \left(\sqrt{\frac{60w}{3131}}\right)^2 \quad \text{Square each side.}$$

$$2.56 = \frac{60w}{3131} \quad \text{Simplify each side.}$$

$$2.56(3131) = \left(\frac{60w}{3131}\right)(3131) \quad \text{Multiply each side by 3131.}$$

$$8015.36 = 60w \quad \text{Simplify.}$$

$$\frac{8015.36}{60} = \frac{60w}{60} \quad \text{Divide each side by 60.}$$

$$133.6 \approx w \quad \text{Approximate.}$$

A woman 5 ft tall should weigh approximately 134 lb to have an average BSA.

Student Check 2 Solve each problem.

a. One of the ropes that is used to tie down a tent needs to be replaced. Determine the distance from the tent to the tie-down stake if the tent is 6 ft tall and the tie-down stake is placed 4 ft from the tent.

b. **i.** Find the BSA of a person who is 5 ft 4 in. and weighs 140 lb.

ii. The average BSA of a man is 1.9 m^2. If a man is 5 ft 8 in. tall, how much should he weigh to have an average BSA?

Objective 3 Troubleshoot common errors.

Troubleshooting Common Errors

Some common errors associated with radical equations are shown.

Objective 3 Examples **A problem and an incorrect solution are given. Provide the correct solution and an explanation of the error.**

3a. Solve $\sqrt{3x - 5} + 2 = 7$.

Incorrect Solution	Correct Solution and Explanation
$\sqrt{3x-5}+2=7$ $(\sqrt{3x-5}+2)^2=7^2$ $3x-5+4=49$ $3x-1=49$ $3x=50$ $x=\frac{50}{3}$ $\left\{\frac{50}{3}\right\}$	We must isolate the radical before squaring. $\sqrt{3x-5}+2=7$ $\sqrt{3x-5}+2-2=7-2$ $\sqrt{3x-5}=5$ $(\sqrt{3x-5})^2=5^2$ $3x-5=25$ $3x=30$ $x=10$ The solution set is $\{10\}$.

3b. Solve $\sqrt{z+2}+4=z$.

Incorrect Solution	Correct Solution and Explanation
$\sqrt{z+2}+4=z$ $\sqrt{z+2}=z-4$ $(\sqrt{z+2})^2=(z-4)^2$ $z+2=z^2-8z+16$ $0=z^2-9z+14$ $0=(z-2)(z-7)$ $z=2$ or $z=7$ $\{2, 7\}$	The equation was solved correctly but the solutions were not checked to determine if the solutions are extraneous. **Check:** $z=2$: $\sqrt{z+2}+4=z$ $\sqrt{2+2}+4=2$ $\sqrt{4}+4=2$ $2+4=2$ $6=2$ False **Check:** $z=7$: $\sqrt{z+2}+4=z$ $\sqrt{7+2}+4=7$ $\sqrt{9}+4=7$ $3+4=7$ $7=7$ True So, the solution set is $\{7\}$.

ANSWERS TO STUDENT CHECKS

Student Check 1 **a.** $\{13\}$ **b.** $\{32\}$ **c.** $\varnothing$ **d.** $\{1\}$ **e.** $\{7\}$ **f.** $\{9\}$

Student Check 2 **a.** $2\sqrt{13}$ ft **b. i.** 1.69 m^2 **ii.** 166.2 lb

SUMMARY OF KEY CONCEPTS

1. There are five main steps that we must follow to solve equations with radicals.
   - First, isolate the radical on one side of the equation.
   - Second, raise both sides of the equation to the exponent equal to the index and simplify. Use the FOIL method when appropriate.
   - Third, if the equation contains a radical, repeat the first two steps.
   - Fourth, solve the resulting equation.
   - Fifth, check the proposed answers in the original equation. It is possible that extraneous solutions may result, so do not skip the checking step.
2. When solving applications involving the Pythagorean theorem, it is helpful to draw a diagram of the right triangle. Be sure to identify which sides are legs and which side is the hypotenuse. Then substitute the known values into $a^2+b^2=c^2$ and solve.

GRAPHING CALCULATOR SKILLS

As we have seen in previous sections, we can use a graphing calculator to check the answers found when solving square root equations.

Example: In Example 1 part (c), we found that $\sqrt{y-3}+4=2$ has no solution. Use a graphing calculator to verify this result.

Solution: We can determine where the graphs of the expressions on the left and right side of the equation intersect. Enter the left side of the original equation as Y_1 and the right side of the original equation as Y_2. Note: We use the variable x in the calculator, rather than y like we have in our equation.

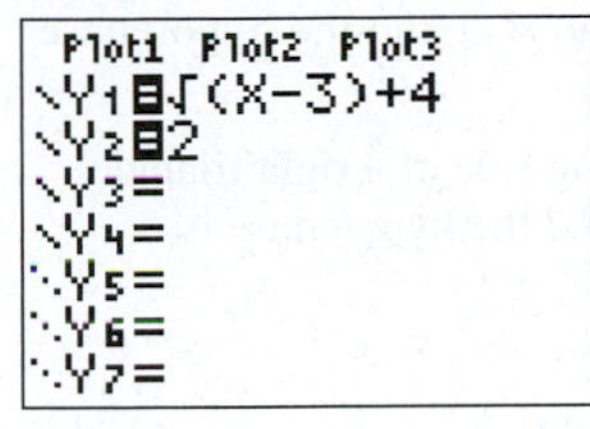

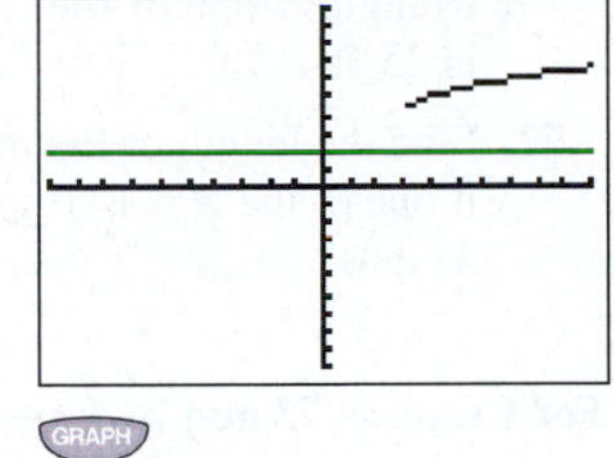

GRAPH

Since the two graphs do not intersect, the equation has no solution.

We can also use the table to check the solution. The proposed solution for this problem is 7. When we examine the table, we notice that when x is 7, the values of Y_1 and Y_2 are not equal. Therefore, 7 is not a solution of the equation.

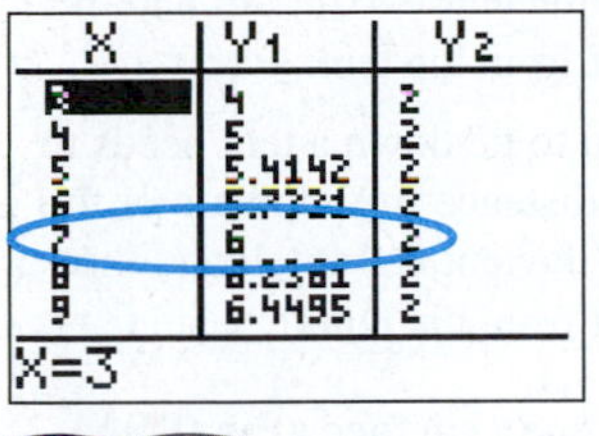

X	Y1	Y2
3	4	2
4	5	2
5	5.4142	2
6	5.7321	2
7	6	2
8	6.2361	2
9	6.4495	2

X=3

2nd GRAPH

SECTION 8.6 EXERCISE SET

Write About It!

Use complete sentences in your answer to each exercise.

1. Explain how to apply the power property to solve a radical equation of the form $\sqrt[n]{x} = a$, a is a real number.
2. Explain why it is important to check the solutions when solving radical equations. It is possible to get extraneous solutions.
3. Explain how to solve an equation with two radicals, such as $\sqrt{x-5} + 1 = \sqrt{x}$.
4. Explain how to solve an equation with one radical, such as $\sqrt{x+5} + 1 = x$.
5. Explain how to determine whether 655 is a solution of the equation $\sqrt[3]{2x+21} - 9 = 2$.
6. Explain why 3 and -2 are not solutions of the equation $\sqrt{70-15x} = x - 8$.

Practice Makes Perfect!

Solve each radical equation. (*See Objective 1.*)

7. $\sqrt{2x-19} = 1$ $\{10\}$
8. $\sqrt{3y-11} = 8$ $\{25\}$
9. $\sqrt{4a+1} = 9$ $\{20\}$
10. $\sqrt{9b+22} = 7$ $\{3\}$
11. $\sqrt{7r+15} - 6 = 2$ $\{7\}$
12. $\sqrt{5r-6} + 7 = 19$ $\{30\}$
13. $\sqrt{5r-1} + 7 = 3$ $\varnothing$
14. $\sqrt{6s+5} - 1 = -10$ $\varnothing$
15. $\sqrt[3]{2x-9} = 1$ $\{5\}$
16. $\sqrt[3]{3y+11} = 5$ $\{38\}$
17. $\sqrt[3]{3x+1} = -2$ $\{-3\}$
18. $\sqrt[3]{6x+2} = -4$ $\{-11\}$
19. $\sqrt{6x+37} = x + 5$ $\{2\}$
20. $\sqrt{2y-5} = y - 4$ $\{7\}$
21. $\sqrt{32-2a} = a - 4$ $\{8\}$
22. $\sqrt{-4b-7} = b + 3$ $\{-2\}$
23. $\sqrt{2x+27} = x + 6$ $\{-1\}$
24. $\sqrt{-25y+19} = y - 7$ $\varnothing$
25. $\sqrt{7y-3} - 1 = y$ $\{1, 4\}$
26. $\sqrt{14r-62} + 1 = r$ $\{7, 9\}$
27. $\sqrt{2x+7} = \sqrt{x} + 2$ $\{1, 9\}$
28. $\sqrt{3y+97} = \sqrt{y} + 9$ $\{1, 64\}$
29. $\sqrt{3z+61} = \sqrt{z} - 9$ $\varnothing$
30. $\sqrt{3w+121} = \sqrt{w} - 9$ $\varnothing$
31. $\sqrt{6x-5} = 1 + \sqrt{4x-4}$ $\{1, 5\}$
32. $\sqrt{25x+19} = 4 + \sqrt{17x-21}$ $\{5, 6\}$
33. $\sqrt{4r+29} = 1 + \sqrt{2r+26}$ $\{5\}$
34. $\sqrt{6x+31} = 1 + \sqrt{4x+24}$ $\{3\}$
35. $\sqrt{31-5x} = -4 + \sqrt{3x-9}$ $\varnothing$
36. $\sqrt{-13x+29} = -1 + \sqrt{-11x+20}$ $\varnothing$

Solve each problem. (*See Objective 2.*)

37. Find the length of the missing side of a right triangle if one of the legs is 12 ft and the hypotenuse is 37 ft. 35 ft
38. Find the length of the missing side of a right triangle if one of the legs is 30 cm and the hypotenuse is 34 cm. 16 cm
39. Find the length of the hypotenuse of a right triangle if the lengths of the two legs are 11 cm and 60 cm. 61 cm
40. Find the length of the hypotenuse of a right triangle if the lengths of the two legs are 15 m and 8 m. 17 m
41. Find the length of the hypotenuse of a right triangle if the lengths of the two legs are 12 ft and 16 ft. 20 ft

Additional answers can be found in the Instructor Answer Appendix.

42. Find the length of the hypotenuse of a right triangle if the lengths of the two legs are 15 in. and 20 in. 25 in.

43. One of the ropes that is used to tie down a tent needs to be replaced. Determine the distance from the top of the tent to the tie-down stake if the tent is 10 ft tall and the tie-down stake is placed 5 ft from the tent. $5\sqrt{5}$ ft

44. One of the ropes that is used to tie down a tent needs to be replaced. Determine the distance from the top of the tent to the tie-down stake if the tent is 9 ft tall and the tie-down stake is placed 3 ft from the tent. $3\sqrt{10}$ ft

For Exercises 45–48, a person's body surface area (BSA) in square meters is given by BSA $= \sqrt{\frac{hw}{3131}}$, where *h* is a person's height in inches and *w* is a person's weight in pounds. Round each answer to two decimal places.

45. (a) Find the BSA of a person who is 5 ft 11 in. and weighs 233 lb. (b) The average BSA of a man is 1.9 m^2. If a man is 6 ft 5 in. tall, how much should he weigh to have an average BSA? (a) 2.30 (b) 146.79 lb

46. (a) Find the BSA of a person who is 6 ft 5 in. and weighs 180 lb. (b) The average BSA of a man is 1.9 m^2. If a man is 5 ft 5 in. tall, how much should he weigh to have an average BSA? (a) 2.10 (b) 173.90 lb

47. (a) Find the BSA of a person who is 5 ft 4 in. and weighs 136 lb. (b) The average BSA of a woman is 1.6 m^2. If a woman is 4 ft 6 in. tall, how much should she weigh to have an average BSA? (a) 1.67 (b) 148.43 lb

48. (a) Find the BSA of a person who is 4 ft 8 in. and weighs 140 lb. (b) The average BSA of a woman is 1.6 m^2. If a woman is 5 ft 7 in. tall, how much should she weigh to have an average BSA? (a) 1.58 (b) 119.63 lb

Mix 'Em Up!

Solve each radical equation.

49. $\sqrt{12x+21}-7=2$ $\{5\}$

50. $\sqrt{11y-13}+2=5$ $\{2\}$

51. $\sqrt{6a+7}-1=-10$ $\varnothing$

52. $\sqrt{3b-5}+7=1$ $\varnothing$

53. $\sqrt[4]{5x-7}+1=3$ $\left\{\frac{23}{5}\right\}$

54. $\sqrt[4]{6y-1}-8=-5$ $\left\{\frac{41}{3}\right\}$

55. $\sqrt[3]{7a-13}=2$ $\{3\}$

56. $\sqrt[3]{4y+19}-6=-1$ $\left\{\frac{53}{2}\right\}$

57. $\sqrt{16x+20}-5=x$ $\{1, 5\}$

58. $\sqrt{21y+15}=y+5$ $\{1, 10\}$

59. $\sqrt{7x+39}=x+3$ $\{6\}$

60. $\sqrt{5b-24}+4=b$ $\{5, 8\}$

61. $\sqrt{3x-11}+7=\sqrt{x}$ $\varnothing$

62. $\sqrt{2y+7}+2=\sqrt{y}$ $\varnothing$

63. $\sqrt{3r-23}-1=\sqrt{r}$ $\{16\}$

64. $\sqrt{2s-63}+3=\sqrt{s}$ $\{36\}$

65. $\sqrt{2x-3}=-1+\sqrt{4x-8}$ $\{6\}$

66. $\sqrt{3x+6}=1+\sqrt{x+3}$ $\{1\}$

67. $\sqrt{26-5x}=-4+\sqrt{3x-6}$ $\varnothing$

68. $\sqrt{11-7y}=-4+\sqrt{y+3}$ $\varnothing$

69. $\sqrt{13x+3}=1+\sqrt{11x-2}$ $\{1, 6\}$

70. $\sqrt{5y-1}=-3+\sqrt{11y+14}$ $\{1, 2\}$

Solve each problem. Round each answer to two decimal places.

71. Find the length of the missing side of a right triangle if one of the legs is 24 ft and the hypotenuse is 25 ft. 7 ft

72. Find the length of the missing side of a right triangle if one of the legs is 16 cm and the hypotenuse is 34 cm. 30 cm

For Exercises 73 and 74, a person's body surface area (BSA) in square meters is given by BSA $= \sqrt{\frac{hw}{3131}}$, where *h* is a person's height in inches and *w* is a person's weight in pounds.

73. (a) Find the BSA of a person who is 6 ft 4 in. tall and weighs 236 lb. (b) The average BSA of a man is 1.9 m^2. If a man is 5 ft 9 in. tall, how much should he weigh to have an average BSA? (a) 2.39 (b) 163.81 lb

74. (a) Find the BSA of a person who is 4 ft 4 in. and weighs 103 lb. (b) The average BSA of a woman is 1.6 m^2. If a woman is 5 ft 9 in. tall, how much should she weigh to have an average BSA? (a) 1.31 (b) 116.16 lb

For Exercises 75 and 76, the period *T*, in seconds, of a simple pendulum is the time it takes for it to make a complete swing and can be approximated by

$$T = 2\pi\sqrt{\frac{L}{9.8}}$$

where *L* is the length of the pendulum in meters.

75. Find the exact value of the period of a simple pendulum if its length is 2.8 m. $\frac{2\pi\sqrt{14}}{7}$ sec

76. Find the exact value of the period of a simple pendulum if its length is 12.25 m. $\sqrt{5}\pi$ sec

For Exercises 77 and 78, the speed *s*, in miles per hour, of a car that skids to a stop can be calculated by measuring the distance *d*, in feet, of the skid marks. The formula is

$$s = \sqrt{30kd}$$

where *k* is the coefficient of friction which depends on the road surface and condition. Round answers to one decimal place.

Road Surface and Condition	*k*
Dry tar	1.0
Wet tar	0.5
Dry concrete	0.8
Wet concrete	0.4

77. Find the speed of a car if it leaves skid marks 190 ft long on a dry concrete road. 67.5 mph

78. Find the speed of a car if it leaves skid marks 90 ft long on a wet tar road. 36.7 mph

You Be the Teacher!

Correct each student's errors, if any.

79. Solve the equation $\sqrt{4x + 25} = 9$.

Mary Beth's work:

$\sqrt{4x + 25} = 9$ $\sqrt{4x+25} = 9$
$2x + 5 = 9$ $(\sqrt{4x+25})^2 = 9^2$
$2x = 4$ $4x + 25 = 81$
$x = 2$ $4x = 56$
  $x = 14$ The solution checks.

80. Solve the equation $\sqrt[3]{2x - 5} + 2 = 3$.

Rebecca's work:

$\sqrt[3]{2x - 5} + 2 = 3$ $\sqrt[3]{2x-5} + 2 = 3$
$2x - 5 + 8 = 27$ $\sqrt[3]{2x-5} = 1$
$2x + 3 = 27$ $2x - 5 = 1^3$
$2x = 24$ $2x = 6$
$x = 12$ $x = 3$ The solution checks.

81. Solve the equation $\sqrt{13 - 3x} = x - 3$.

Darius's work:

$\sqrt{13 - 3x} = x - 3$ $\sqrt{13-3x} = x - 3$
$13 - 3x = x^2 + 9$ $(\sqrt{13-3x})^2 = (x-3)^2$
$0 = x^2 + 3x - 4$ $13 - 3x = x^2 - 6x + 9$
$0 = (x + 4)(x - 1)$ $0 = x^2 - 3x - 4$
$x = -4, 1$ $0 = (x-4)(x+1)$
  $x = 4, -1$
Since -1 is an extraneous solution, the solution set is $\{4\}$.

82. Solve the equation $\sqrt{12x - 11} = 3 + \sqrt{6x - 14}$.

Carol's work:

$\sqrt{12x - 11} = 3 + \sqrt{6x - 14}$
$12x - 11 = 9 + 6x - 14$
$6x = 6$
$x = 1$

83. Solve the equation $\sqrt{2x - 23} = \sqrt{x} - 1$.

James's work:

$\sqrt{2x - 23} = \sqrt{x} - 1$
$2x - 23 = x + 1$
$x = 24$

84. Solve the equation $\sqrt{x + 7} = -2 + \sqrt{5x + 15}$.

Randy's work:

$\sqrt{x + 7} = -2 + \sqrt{5x + 15}$
$x + 7 = 4 + 5x + 15$
$-4x = 12$
$x = -3$

Calculate It!

Solve each equation and then use a calculator to check the solutions.

85. $\sqrt{4x + 57} = x + 3$

86. $\sqrt[3]{4x + 19} - 10 = -7$

87. $\sqrt{8x + 49} = 3 + \sqrt{2x + 10}$

88. $\sqrt{5x + 15} = 4 + \sqrt{-3x + 7}$

SECTION 8.7 Complex Numbers

OBJECTIVES

As a result of completing this section, you will be able to

1. Write square roots of negative numbers as complex numbers.
2. Add and subtract complex numbers.
3. Multiply complex numbers.
4. Divide complex numbers.
5. Find powers of *i*.
6. Troubleshoot common errors.

In Section 8.1, we discussed square roots such as $\sqrt{-16}$, $\sqrt{-8}$, and $\sqrt{-7}$. We stated that these are not real numbers since there is no real number that, when squared, equals a negative number. While it is true that they are not real numbers, we now introduce a new set of numbers, called *complex numbers*, that are used to represent these expressions. Complex numbers are used in many aspects of science, such as engineering. They are used in particular in electrical engineering.

Complex Numbers

Objective 1 Write square roots of negative numbers as complex numbers.

Centuries ago, mathematicians realized that the square root of negative numbers arose when solving certain types of equations. They actually referred to these types of solutions as "imaginary" or "ghostly." Mathematicians devised a new number system to include numbers like $\sqrt{-16}$ and $\sqrt{-8}$. This new number system is called the *complex number* system. The *imaginary unit*, represented by the symbol *i*, is included in this system. The imaginary unit is a number that can be squared to obtain a negative number.

Definition: Imaginary Unit i

The number i is the number whose square is -1.

$$i = \sqrt{-1} \quad \text{and} \quad i^2 = -1$$

We use i to write numbers like $\sqrt{-16}$ as the product of a real number and i. Since $i = \sqrt{-1}$,

$$\sqrt{-16} = \sqrt{-1 \cdot 16} = \sqrt{-1} \cdot \sqrt{16} = i(4) \text{ or } 4i.$$

Definition: Any number of the form $a + bi$, where a and b are real numbers, is a **complex number**. This form is the standard form of a complex number, where a is the real part and b is the imaginary part of the complex number.

If $a = 0$, then $a + bi = 0 + bi = bi$. This number is purely imaginary.
If $b = 0$, then $a + bi = a + 0(i) = a$. This number is purely real.

Some examples of complex numbers are $3 + 2i$, $-1 - 5i$, $0 + i\sqrt{7}$, and $-2 + 0i$.

Note: *The number* $3 + 2i$ *represents* $3 + \sqrt{-4}$ *since* $\sqrt{-4} = \sqrt{-1 \cdot 4} = \sqrt{-1} \cdot \sqrt{4} = 2i$.

The set of complex numbers includes all of the real numbers since any real number a can be written as $a + 0i$. We can visualize the complex number system as shown.

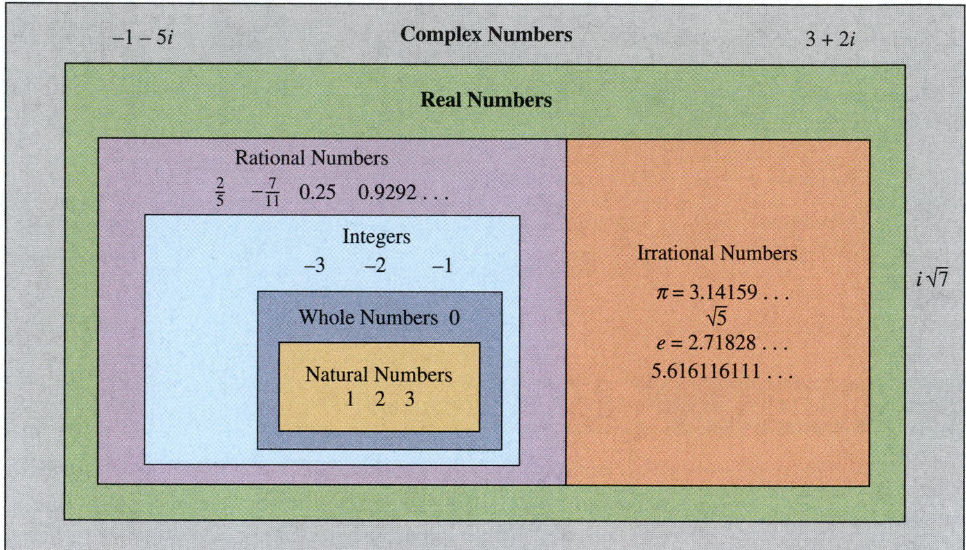

Property: Square Root of a Negative Number

For c a real number and $c > 0$,

$$\sqrt{-c} = \sqrt{-1 \cdot c} = \sqrt{-1}\sqrt{c} = i\sqrt{c}$$

Objective 1 Examples Write each expression in terms of i.

Problems	Solutions
1a. $\sqrt{-16}$	$\sqrt{-16} = \sqrt{16 \cdot -1} = \sqrt{16}\sqrt{-1} = 4i$
1b. $\sqrt{-7}$	$\sqrt{-7} = \sqrt{-1 \cdot 7} = \sqrt{-1}\sqrt{7} = i\sqrt{7}$
1c. $\sqrt{-8}$	$\sqrt{-8} = \sqrt{-1 \cdot 4 \cdot 2} = \sqrt{4 \cdot -1}\sqrt{2} = 2i\sqrt{2}$

Problems	Solutions
1d. $4 + \sqrt{-25}$	$4 + \sqrt{-25} = 4 + \sqrt{25 \cdot -1} = 4 + 5i$
1e. $\dfrac{8 + \sqrt{-36}}{2}$	$\dfrac{8 + \sqrt{-36}}{2} = \dfrac{8 + \sqrt{-1 \cdot 36}}{2} = \dfrac{8 + \sqrt{-1}\sqrt{36}}{2} = \dfrac{8 + 6i}{2} = \dfrac{8}{2} + \dfrac{6i}{2} = 4 + 3i$
1f. $\dfrac{3 + \sqrt{-12}}{4}$	$\dfrac{3 + \sqrt{-12}}{4} = \dfrac{3 + \sqrt{-1 \cdot 4 \cdot 3}}{4} = \dfrac{3 + \sqrt{4 \cdot -1}\sqrt{3}}{4} = \dfrac{3 + 2i\sqrt{3}}{4} = \dfrac{3}{4} + \dfrac{2\sqrt{3}}{4}i = \dfrac{3}{4} + \dfrac{\sqrt{3}}{2}i$

Student Check 1 Write each expression in terms of i.

a. $\sqrt{-49}$ **b.** $\sqrt{-5}$ **c.** $\sqrt{-20}$

d. $6 + \sqrt{-4}$ **e.** $\dfrac{10 + \sqrt{-25}}{5}$ **f.** $\dfrac{2 + \sqrt{-18}}{6}$

Adding and Subtracting Complex Numbers

Objective 2 Add and subtract complex numbers.

Any operation that can be performed on real numbers can also be performed on complex numbers. That is, we can add, subtract, multiply, and divide complex numbers.

Operations with complex numbers are done the same way we perform operations on polynomials. To add or subtract complex numbers, combine like terms. Like terms are the real parts and the imaginary parts.

Procedure: Adding or Subtracting Complex Numbers

Step 1: If adding, remove parentheses and combine the real parts and the imaginary parts.

Step 2: If subtracting, remove the parentheses by distributing -1 to the number being subtracted. Combine the real parts and imaginary parts.

Objective 2 Examples Perform the indicated operation on the complex numbers.

2a. $(4 - 5i) + (7 - 9i)$ **2b.** $(4 - 5i) - (7 - 9i)$ **2c.** $2i + (7 - i) - (3 + i)$

Solutions

2a.

$(4 - 5i) + (7 - 9i) = 4 - 5i + 7 - 9i$	Clear parentheses.
$= 4 + 7 - 5i - 9i$	Group real parts and imaginary parts.
$= 11 - 14i$	Add.

2b. We first find the opposite of $7 - 9i$.

$$\begin{aligned}(4 - 5i) - 1(7 - 9i) &= 4 - 5i - 7 + 9i && \text{Clear parentheses.}\\ &= 4 - 7 - 5i + 9i && \text{Group real parts and imaginary parts.}\\ &= -3 + 4i && \text{Add.}\end{aligned}$$

2c. We first find the opposite of $3 + i$.

$$\begin{aligned}2i + (7 - i) - 1(3 + i) &= 2i + 7 - i - 3 - i && \text{Clear parentheses.}\\ &= 7 - 3 + 2i - i - i && \text{Group real parts and imaginary parts.}\\ &= 4 + 0i && \text{Add.}\\ &= 4\end{aligned}$$

Student Check 2 Perform the indicated operation on the complex numbers.

a. $(2 + 3i) + (8 - 5i)$ **b.** $(2 + 3i) - (8 - 5i)$ **c.** $4 + (8 + 7i) - (4 + 6i)$

Multiplying Complex Numbers

Objective 3 Multiply complex numbers.

We multiply complex numbers just like we multiply polynomials. We will use the distributive property and the FOIL method. The key is to express the answers in the form $a + bi$. After multiplying, we must rewrite all instances of i^2 as -1.

Procedure: Multiplying Complex Numbers

Step 1: Rewrite any square roots of negative numbers as complex numbers.
Step 2: If a single term is multiplied by a binomial, then distribute.
Step 3: If the complex numbers being multiplied are binomials, then use the FOIL method.
Step 4: If necessary, rewrite i^2 as -1 and combine like terms.
Step 5: Write the answer in the form $a + bi$.

Objective 3 Examples Simplify each product of complex numbers and write each answer in standard form.

3a. $\sqrt{-25} \cdot \sqrt{-4}$ **3b.** $3i(2 + 6i)$ **3c.** $(4 - 5i)(7 - 9i)$
3d. $(4 - 5i)(4 + 5i)$ **3e.** $(1 - 3i)^2$

Solutions

3a.
$$\begin{aligned}\sqrt{-25} \cdot \sqrt{-4} &= (5i)(2i) && \text{Write each radical as a complex number.}\\ &= 10i^2 && \text{Multiply.}\\ &= 10(-1) && \text{Rewrite } i^2 \text{ as } -1.\\ &= -10 && \text{Multiply.}\end{aligned}$$

3b.
$$\begin{aligned}3i(2 + 6i) &= (3i)(2) + (3i)(6i) && \text{Apply the distributive property.}\\ &= 6i + 18i^2 && \text{Multiply.}\\ &= 6i + 18(-1) && \text{Rewrite } i^2 \text{ as } -1.\\ &= 6i - 18 && \text{Simplify.}\\ &= -18 + 6i && \text{Write in standard form.}\end{aligned}$$

3c. $(4 - 5i)(7 - 9i)$
$$\begin{aligned}&= 4(7) - 4(9i) - 5i(7) - 5i(-9i) && \text{Apply the FOIL method.}\\ &= 28 - 36i - 35i + 45i^2 && \text{Multiply.}\\ &= 28 - 71i + 45(-1) && \text{Rewrite } i^2 \text{ as } -1 \text{ and combine like terms.}\\ &= 28 - 71i - 45 && \text{Simplify.}\\ &= -17 - 71i && \text{Combine like terms.}\end{aligned}$$

3d. $(4 - 5i)(4 + 5i)$

$= 4(4) + 4(5i) - 5i(4) - 5i(5i)$ Apply the FOIL method.

$= 16 + 20i - 20i - 25i^2$ Simplify.

$= 16 - 25(-1)$ Rewrite i^2 as -1 and combine like terms.

$= 16 + 25$ Simplify.

$= 41$ Combine like terms.

3e. We can square the complex number by either repeating the base or applying the property to square a binomial.

Method 1

$$\begin{aligned}(1 - 3i)^2 &= (1 - 3i)(1 - 3i)\\ &= 1 - 3i - 3i + 9i^2\\ &= 1 - 6i + 9(-1)\\ &= 1 - 6i - 9\\ &= -8 - 6i\end{aligned}$$

Method 2

$$\begin{aligned}(1 - 3i)^2 &= (1)^2 - 2(1)(3i) + (3i)^2\\ &= 1 - 6i + 9i^2\\ &= 1 - 6i + 9(-1)\\ &= 1 - 6i - 9\\ &= -8 - 6i\end{aligned}$$

Student Check 3 Simplify each product of complex numbers and write each answer in standard form.

a. $\sqrt{-25} \cdot \sqrt{-16}$ **b.** $7i(3 + 2i)$ **c.** $(2 - 3i)(4 - i)$

d. $(2 - 3i)(2 + 3i)$ **e.** $(4 - 5i)^2$

Dividing Complex Numbers

Objective 4 Divide complex numbers.

The goal in dividing complex numbers is to write the result as a complex number in standard form $a + bi$. That is, we must remove the complex number from the denominator. This will involve the product of conjugates. Complex numbers also have conjugates. For example, the numbers $4 + 5i$ and $4 - 5i$ are *complex conjugates*.

Definition: The **complex conjugate** of $a + bi$ is $a - bi$.

Some other examples of complex conjugates are shown in the table.

Complex Number	Complex Conjugate
$-2 - 7i$	$-2 + 7i$
$5i = 0 + 5i$	$-5i = 0 - 5i$
$1 + \sqrt{11}i$	$1 - \sqrt{11}i$
$\frac{3}{5} - \frac{\sqrt{2}}{5}i$	$\frac{3}{5} + \frac{\sqrt{2}}{5}i$
$4 = 4 + 0i$	$4 = 4 - 0i$

Note from Example 3 part (d) that the product of complex conjugates is a real number, that is,

$$(4 - 5i)(4 + 5i) = 16 + 25 = 41$$

We generalize this relationship as follows.

Property: Product of Complex Conjugates

$$(a + bi)(a - bi) = a^2 + b^2$$

We use this fact to divide complex numbers.

Procedure: Dividing Complex Numbers

Step 1: If necessary, write the square root of a negative number as an imaginary number.

Step 2: Multiply the numerator and the denominator by the complex conjugate of the denominator.
Step 3: Simplify the result.

Objective 4 Examples Simplify each quotient of complex numbers and write answers in standard form.

4a. $\dfrac{\sqrt{-36}}{\sqrt{-4}}$ **4b.** $\dfrac{5-3i}{4i}$ **4c.** $\dfrac{9}{1+i}$ **4d.** $\dfrac{-2+i}{4+3i}$

Solutions

4a. $\dfrac{\sqrt{-36}}{\sqrt{-4}}=\dfrac{6i}{2i}$ Rewrite the numerator and denominator as complex numbers.

$=\dfrac{6i(-2i)}{2i(-2i)}$ Multiply the numerator and denominator by the complex conjugate of $2i$, $-2i$.

$=\dfrac{-12i^2}{-4i^2}$ Simplify each product.

$=\dfrac{-12(-1)}{-4(-1)}$ Rewrite i^2 as -1.

$=\dfrac{12}{4}$ Simplify the numerator and denominator.

$=3$ Divide.

Note that the standard form of 3 is $3+0i$. When the number is purely real, there is no need to write it in this form.

4b. $\dfrac{5-3i}{4i}=\dfrac{(5-3i)(-4i)}{(4i)(-4i)}$ Multiply the numerator and denominator by $-4i$.

$=\dfrac{-20i+12i^2}{-16i^2}$ Simplify each product.

$=\dfrac{-20i+12(-1)}{-16(-1)}$ Rewrite i^2 as -1.

$=\dfrac{-20i-12}{16}$ Simplify.

$=\dfrac{-20i}{16}-\dfrac{12}{16}$ Divide each term in the numerator by 16.

$=-\dfrac{12}{16}-\dfrac{20}{16}i$ Write in the standard form of a complex number.

$=-\dfrac{3}{4}-\dfrac{5}{4}i$ Simplify the real and imaginary parts.

4c. $\dfrac{9}{1+i}=\dfrac{(9)(1-i)}{(1+i)(1-i)}$ Multiply the numerator and denominator by $1-i$.

$=\dfrac{9-9i}{1-i^2}$ Simplify each product.

$=\dfrac{9-9i}{1-(-1)}$ Rewrite i^2 as -1.

$=\dfrac{9-9i}{2}$ Simplify.

$=\dfrac{9}{2}-\dfrac{9}{2}i$ Divide each term in the numerator by 2.

4d. $\dfrac{-2+i}{4+3i} = \dfrac{(-2+i)(4-3i)}{(4+3i)(4-3i)}$ Multiply the numerator and denominator by $4-3i$.

$= \dfrac{-8+6i+4i-3i^2}{16-9i^2}$ Simplify each product.

$= \dfrac{-8+10i-3(-1)}{16-9(-1)}$ Rewrite i^2 as -1.

$= \dfrac{-8+10i+3}{16+9}$ Simplify.

$= \dfrac{-5+10i}{25}$ Combine like terms.

$= \dfrac{-5}{25} + \dfrac{10}{25}i$ Write in the standard form of a complex number.

$= -\dfrac{1}{5} + \dfrac{2}{5}i$ Simplify the real and imaginary parts.

Student Check 4 Simplify each quotient of complex numbers. Write answers in standard form.

a. $\dfrac{\sqrt{-64}}{\sqrt{-16}}$ **b.** $\dfrac{4+3i}{5i}$ **c.** $\dfrac{8}{3-i}$ **d.** $\dfrac{1+2i}{4+i}$

Powers of *i*

Objective 5 Find powers of *i*.

We can raise complex numbers to exponents just as we can raise real numbers to exponents. So, we will examine what happens when we find powers of i. Note from the following examples that all powers of i simplify to one of four quantities: 1, -1, i, or $-i$. We find the powers of i by using the fact that $i^2 = -1$.

$i^1 = i$ $\qquad$ $i^5 = i^4 \cdot i = 1 \cdot i = i$

$i^2 = -1$ $\qquad$ $i^6 = i^4 \cdot i^2 = (1)(-1) = -1$

$i^3 = i^2(i) = -1i = -i$ $\qquad$ $i^7 = i^4 \cdot i^3 = (1)(-i) = -i$

$i^4 = i^2 \cdot i^2 = (-1)(-1) = 1$ $\qquad$ $i^8 = i^4 \cdot i^4 = (i^4)^2 = (1)^2 = 1$

We could continue finding the powers of i in this manner, but it is not necessary since there is a pattern to the powers of i. Observe the following.

- i^5 has the same value as i^1 and 5 divided by 4 has remainder 1.
- i^6 has the same value as i^2 and 6 divided by 4 has remainder 2.
- i^7 has the same value as i^3 and 7 divided by 4 has remainder 3.
- i^8 has the same value as i^4 and 8 divided by 4 has remainder 0. (Recall $i^0 = 1$.)

So, if we know the remainder of the exponent of i divided by 4, then we can find the value of i raised to that exponent.

Property: The Power of *i*

Let n be an integer. Then

$$i^n = i^r$$

where r is the remainder of n divided by 4.

Procedure: Finding a Power of i

Step 1: Divide the exponent of i by 4 and find its remainder (0, 1, 2, or 3).
Step 2: The result of i to this exponent is as follows.
- **a.** 1 if the remainder is 0.
- **b.** i if the remainder is 1.
- **c.** -1 if the remainder is 2.
- **d.** $-i$ if the remainder is 3.

Objective 5 Examples Find each power of i.

5a. i^{14} **5b.** i^{49} **5c.** i^{23} **5d.** i^{-16}

Solutions

5a. $\frac{14}{4} = 7R2$, so $i^{14} = i^2 = -1$.

5b. $\frac{49}{4} = 12R1$, so $i^{49} = i^1 = i$.

5c. $\frac{23}{4} = 5R3$, so $i^{23} = i^3 = -i$.

5d. $-\frac{16}{4} = -4R0$, so $i^{-16} = i^0 = 1$.

Student Check 5 Find each power of i.

a. i^{13} **b.** i^{35} **c.** i^{30} **d.** i^{-24}

Objective 6

Troubleshoot common errors.

Troubleshooting Common Errors

Some common errors related to complex numbers are shown.

Objective 6 Examples A problem and an incorrect solution are given. Provide the correct solution and an explanation of the error.

6a. Simplify $\sqrt[3]{-8}$.

Incorrect Solution	Correct Solution and Explanation
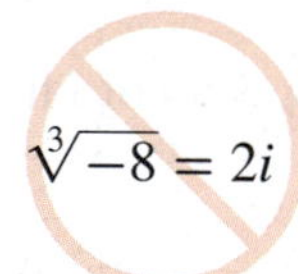 $\sqrt[3]{-8} = 2i$	The cube root of a negative number is a negative number. The square root of a negative number is imaginary. So, $\sqrt[3]{-8} = -2$ but $\sqrt{-8} = 2i\sqrt{2}$.

6b. Perform the operation: $(2 + 3i) + (-7 + 6i)$.

Incorrect Solution	Correct Solution and Explanation
$(2 + 3i) + (-7 + 6i)$ $= 2 - 7 + 3i + 6i$ $= -5 + 9i$ $= -5 + 9(-1)$ $= -5 - 9$ $= -14$	The value of i is not -1. Recall $i^2 = -1$. $(2 + 3i) + (-7 + 6i) = 2 - 7 + 3i + 6i$ $= -5 + 9i$

ANSWERS TO STUDENT CHECKS

Student Check 1 **a.** $7i$ **b.** $i\sqrt{5}$ **c.** $2i\sqrt{5}$ **d.** $6 + 2i$ **e.** $2 + i$ **f.** $\frac{1}{3} + \frac{\sqrt{2}}{2}i$

Student Check 2 **a.** $10 - 2i$ **b.** $-6 + 8i$ **c.** $8 + i$

Student Check 3 **a.** -20 **b.** $-14 + 21i$ **c.** $5 - 14i$ **d.** 13 **e.** $-9 - 40i$

Student Check 4 **a.** 2 **b.** $\frac{3}{5} - \frac{4}{5}i$ **c.** $\frac{12}{5} + \frac{4}{5}i$ **d.** $\frac{6}{17} + \frac{7}{17}i$

Student Check 5 **a.** i **b.** $-i$ **c.** -1 **d.** 1

SUMMARY OF KEY CONCEPTS

1. A complex number is a number that can be written in the form $a + bi$, where $i^2 = -1$ and $i = \sqrt{-1}$.
2. Add and subtract complex numbers by combining like terms.
3. Multiply complex numbers by applying the distributive property and the FOIL method. Rewrite i^2 as -1 and write the answer in the form $a + bi$.
4. Divide complex numbers by multiplying the numerator and denominator by the complex conjugate of the denominator. Write the answer in the form $a + bi$.
5. Any power of i is $1, -1, i$, or $-i$. The result of i to an exponent is the same as i raised to the remainder of dividing the given exponent by 4.

GRAPHING CALCULATOR SKILLS

The calculator can express square roots of negative numbers in terms of i as well as perform operations with complex numbers.

Example 1: Simplify $4 + \sqrt{-25}$.

Solution: Change the calculator MODE from REAL to $a + bi$.

```
NORMAL SCI ENG
FLOAT 0123456789
RADIAN DEGREE
FUNC PAR POL SEQ
CONNECTED DOT
SEQUENTIAL SIMUL
REAL a+bi re^θi
FULL HORIZ G-T
SET CLOCK 01/01/02 20:55
```

```
4+√(-25)
                4+5i
```

Example 2: Perform the operations (a) $(4 - 5i) - (7 - 9i)$ and (b) $(4 - 5i)(7 - 9i)$.

Solution: To enter the imaginary unit i, press 2nd .

```
(4-5i)-(7-9i)
              -3+4i
(4-5i)(7-9i)
             -17-71i
```

SECTION 8.7 EXERCISE SET

Write About It!

Use complete sentences in your answer to each exercise.

1. Define an imaginary number.
2. Define a complex number.
3. Define the conjugate of $a + bi$, where a and b are real numbers.
4. Explain how to add and subtract complex numbers.
5. Explain how to multiply complex numbers.
6. Explain the role of the conjugate in dividing complex numbers.

Additional answers can be found in the Instructor Answer Appendix.

7. Show why $(5 - i)^2 \neq 24$.
8. Show why $\sqrt{-4} \cdot \sqrt{-4} \neq 4$.

Practice Makes Perfect!

Write each expression in terms of *i*. (See *Objective 1*.)

9. $\sqrt{-36}$ $6i$
10. $\sqrt{-25}$ $5i$
11. $\sqrt{-75}$ $5i\sqrt{3}$
12. $\sqrt{-60}$ $2i\sqrt{15}$
13. $1 - \sqrt{-4}$ $1 - 2i$
14. $8 - \sqrt{-225}$ $8 - 15i$
15. $10 + \sqrt{-16}$ $10 + 4i$
16. $5 + \sqrt{-400}$ $5 + 20i$

17. $\dfrac{2+\sqrt{-81}}{2}$ $1+\frac{9}{2}i$ **18.** $\dfrac{8+\sqrt{-196}}{6}$ $\frac{4}{3}+\frac{7}{3}i$

19. $\dfrac{7-\sqrt{-9}}{15}$ $\frac{7}{15}-\frac{1}{5}i$ **20.** $\dfrac{12-\sqrt{-36}}{10}$ $\frac{6}{5}-\frac{3}{5}i$

21. $\dfrac{10-\sqrt{-363}}{14}$ **22.** $\dfrac{9-\sqrt{-52}}{17}$

23. $\dfrac{5+\sqrt{-32}}{11}$ $\frac{5}{11}+\frac{4\sqrt{2}}{11}i$ **24.** $\dfrac{6+\sqrt{-405}}{15}$ $\frac{2}{5}+\frac{3\sqrt{5}}{5}i$

Perform the indicated operation on the complex numbers. (*See Objective 2.*)

25. $(3-6i)+(7+10i)$ $10+4i$ **26.** $(2+5i)+(5-7i)$ $7-2i$

27. $(-3-4i)-(1-9i)$ $-4+5i$ **28.** $(-5+9i)-(-2-8i)$ $-3+17i$

29. $(-1+7i)+(2+12i)+(-11+5i)$ $-10+24i$

30. $(-2+8i)-(5-2i)+(-9-6i)$ $-16+4i$

31. $(5-8i)-(1+9i)-(-3+2i)$ $7-19i$

32. $(6-i)+(6+3i)-(-6+5i)$ $18-3i$

Simplify each product of complex numbers and write each answer in standard form. (*See Objective 3.*)

33. $\sqrt{-4}\cdot\sqrt{-49}$ -14 **34.** $\sqrt{-64}\cdot\sqrt{-9}$ -24

35. $\sqrt{9}\cdot\sqrt{-25}$ $15i$ **36.** $\sqrt{16}\cdot\sqrt{-225}$ $60i$

37. $\sqrt{-9}(2+\sqrt{-100})$ $-30+6i$ **38.** $\sqrt{-25}(-3-\sqrt{-16})$ $20-15i$

39. $3i(7+4i)$ $-12+21i$ **40.** $2i(9-8i)$ $16+18i$

41. $(1+\sqrt{-25})(-3-\sqrt{-16})$ $17-19i$

42. $(2-\sqrt{-49})(1+\sqrt{-4})$ $16-3i$

43. $(9+2i)(-2+3i)$ $-24+23i$ **44.** $(3-7i)(11+2i)$ $47-71i$

45. $(3-4i)(6+7i)$ $46-3i$ **46.** $(1-4i)(9+3i)$ $21-33i$

47. $(\sqrt{5}+3i)(2\sqrt{5}-i)$ $13+5i\sqrt{5}$ **48.** $(3\sqrt{6}-5i)(\sqrt{6}+2i)$ $28+i\sqrt{6}$

49. $(3-\sqrt{-49})^2$ $-40-42i$ **50.** $(1+\sqrt{-25})^2$ $-24+10i$

51. $(6-7i)^2$ $-13-84i$ **52.** $(5-4i)^2$ $9-40i$

53. $(8+i)^2$ $63+16i$ **54.** $(2+3i)^2$ $-5+12i$

55. $(\sqrt{7}-2i)^2$ $3-4i\sqrt{7}$ **56.** $(\sqrt{10}+3i)^2$ $1+6i\sqrt{10}$

57. $(1-\sqrt{-25})(1+\sqrt{-25})$ 26

58. $(3-\sqrt{-1})(3+\sqrt{-1})$ 10

59. $(2+5i)(2-5i)$ 29 **60.** $(3-5i)(3+5i)$ 34

Simplify each quotient of complex numbers and write each answer in standard form. (*See Objective 4.*)

61. $\dfrac{\sqrt{-4}}{\sqrt{-25}}$ $\frac{2}{5}$ **62.** $\dfrac{\sqrt{-81}}{\sqrt{-49}}$ $\frac{9}{7}$

63. $\dfrac{3}{5i}$ $-\frac{3i}{5}$ **64.** $\dfrac{7}{3i}$ $-\frac{7i}{3}$

65. $-\dfrac{12}{8i}$ $\frac{3i}{2}$ **66.** $-\dfrac{25}{15i}$ $\frac{5i}{3}$

67. $\dfrac{-3+5i}{2i}$ $\frac{5}{2}+\frac{3}{2}i$ **68.** $\dfrac{1-3i}{6i}$ $-\frac{1}{2}-\frac{1}{6}i$

69. $\dfrac{-3+i}{-6+7i}$ $\frac{5}{17}+\frac{3}{17}i$ **70.** $\dfrac{1+10i}{-2+5i}$ $\frac{48}{29}-\frac{25}{29}i$

71. $\dfrac{7+9i}{1-3i}$ $-2+3i$ **72.** $\dfrac{11-3i}{8-i}$ $\frac{7}{5}-\frac{1}{5}i$

73. $\dfrac{-2-4i}{7-i}$ $-\frac{1}{5}-\frac{3}{5}i$ **74.** $\dfrac{13+4i}{7+9i}$ $\frac{127}{130}-\frac{89}{130}i$

Find each power of *i*. (*See Objective 5.*)

75. i^{30} -1 **76.** i^{52} 1 **77.** i^{21} i **78.** i^{51} $-i$

79. i^{82} -1 **80.** i^{47} $-i$ **81.** i^{-12} 1 **82.** i^{-7} i

Mix 'Em Up!

Simplify each complex number and write each answer in standard form.

83. $\dfrac{4-\sqrt{-196}}{8}$ $\frac{1}{2}-\frac{7}{4}i$ **84.** $\dfrac{10-\sqrt{-25}}{25}$ $\frac{2}{5}-\frac{1}{5}i$

85. $\dfrac{8+\sqrt{-225}}{10}$ $\frac{4}{5}+\frac{3}{2}i$ **86.** $\dfrac{1+\sqrt{-400}}{4}$ $\frac{1}{4}+5i$

87. $(7-8i)+(-4+11i)$ $3+3i$ **88.** $(-2+10i)+(3+2i)$ $1+12i$

89. $(9+2i)^2$ $77+36i$ **90.** $(-8+2i)^2$ $60-32i$

91. $(3-i)+(-1-6i)$ $2-7i$ **92.** $(-5-7i)-(8-2i)$ $-13-5i$

93. $(-1-3i)-(6+4i)-(13+4i)$ $-20-11i$

94. $(-3+9i)-(5+5i)-(-12+3i)$ $4+i$

95. $\dfrac{\sqrt{32}}{\sqrt{-36}}$ $-\frac{2\sqrt{2}}{3}i$ **96.** $\dfrac{\sqrt{48}}{\sqrt{-64}}$ $-\frac{\sqrt{3}}{2}i$

97. $\sqrt{-9}(\sqrt{49}+\sqrt{-25})$ $-15+21i$ **98.** $\sqrt{-4}(\sqrt{64}-\sqrt{-9})$ $6+16i$

99. $\dfrac{-6+15i}{3i}$ $5+2i$ **100.** $\dfrac{12-3i}{6i}$ $-\frac{1}{2}-2i$

101. $(7-\sqrt{-16})^2$ $33-56i$ **102.** $(1+\sqrt{-36})^2$ $-35+12i$

103. $9i(8-3i)$ $27+72i$ **104.** $4i(-6+4i)$ $-16-24i$

105. $(5+\sqrt{-1})(-4-\sqrt{-9})$ $-17-19i$

106. $(1-\sqrt{-81})(3+\sqrt{-49})$ $66-20i$

107. $(-5+i)(15-3i)$ $-72+30i$ **108.** $(-6+4i)(-2+2i)$ $4-20i$

109. $(2\sqrt{7}+4i)(\sqrt{7}-5i)$ $34-6i\sqrt{7}$ **110.** $(3-2i\sqrt{5})(1+i\sqrt{5})$ $13+i\sqrt{5}$

111. $\dfrac{7+9i}{1-i}$ $-1+8i$ **112.** $\dfrac{-13+4i}{1-6i}$ $-1-2i$

113. $(\sqrt{7}-6i)(\sqrt{7}+6i)$ 43 **114.** $(5-i\sqrt{3})(5+i\sqrt{3})$ 28

115. $(8-6i)(8+6i)$ 100 **116.** $(12+5i)(12-5i)$ 169

117. $\dfrac{12}{18i}$ $-\frac{2}{3}i$ **118.** $-\dfrac{15}{36i}$ $\frac{5}{12}i$

119. $\dfrac{13+5i}{7+3i}$ $\frac{53}{29}-\frac{2}{29}i$ **120.** $\dfrac{5+i}{5+3i}$ $\frac{14}{17}-\frac{5}{17}i$

Find each power of *i*.

121. i^{41} i

122. i^{71} $-i$

123. i^{192} 1

124. i^{130} -1

125. $(2i)^7$ $-128i$

126. $(3i)^5$ $243i$

127. $(-2i)^{-3}$ $-\frac{1}{8}i$

128. $(-5i)^{-1}$ $\frac{1}{5}i$

You Be the Teacher!

Correct each student's errors, if any.

129. Multiply: $\sqrt{-9} \cdot \sqrt{-16}$.

Andrew's work:

$\sqrt{-9} \cdot \sqrt{-16} = \sqrt{-9 \cdot -16} = \sqrt{144} = 12$

$\sqrt{-9} \cdot \sqrt{-16} = 3i \cdot 4i = 12i^2 = -12$

130. Subtract: $(6 - \sqrt{-4}) - (11 + \sqrt{-9})$.

Monica's work:

$(6 - \sqrt{-4}) - (11 + \sqrt{-9}) = (6 - 2i) - (11 + 3i)$
$= 6 - 2i - 11 - 3i$
$= -5 - 5i$

$(6 - \sqrt{-4}) - (11 + \sqrt{-9}) = (6 + \sqrt{4}) - (11 + 3i)$
$= 6 + 2 - 11 + 3i$
$= -3 + 3i$

131. Simplify $(4 - i\sqrt{3})^2$.

Jeannette's work:

$(4 - i\sqrt{3})^2 = 16 - 9i^2 = 16 + 9 = 25$

$(4 - i\sqrt{3})^2 = 16 - 8i\sqrt{3} + 3i^2 = 13 - 8i\sqrt{3}$

132. Simplify $\dfrac{3 - 4i}{6 + 2i}$.

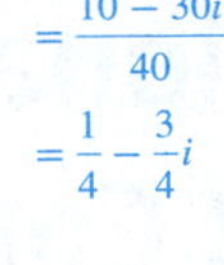

Mark's work:

$$\frac{3-4i}{6+2i} = \frac{3-4i}{6+2i} \cdot \frac{6+2i}{6+2i} = \frac{18 - 24i + 6i - 8i^2}{36 + 4i^2} = \frac{26 - 18i}{32} = \frac{13}{16} - \frac{9}{16}i$$

$$\frac{3-4i}{6+2i} = \frac{3-4i}{6+2i} \cdot \frac{6-2i}{6-2i} = \frac{18 - 6i - 24i + 8i^2}{36 - 4i^2} = \frac{10 - 30i}{40} = \frac{1}{4} - \frac{3}{4}i$$

Calculate It!

Use a calculator to verify each simplification.

133. $(3 - 5i)(7 + 2i) = 31 - 29i$

134. $(6 - \sqrt{-48})^2 = -12 - 48i\sqrt{3}$

135. $\dfrac{-1 - 3i}{3 + 4i} = -\dfrac{3}{5} - \dfrac{1}{5}i$

136. $(-5i)^{-3} = -\dfrac{1}{125}i$

GROUP ACTIVITY Solutions of Quadratic Equations

In Chapter 9, we will solve quadratic equations that do not factor. The solutions of these equations will involve radical expressions. We must use radical operations to verify that the number is a solution of the given equation. Radical operations are also used to simplify the solutions.

1. Write each radical expression in simplified radical form.
   a. $-2 \pm \sqrt{50}$ $-2 \pm 5\sqrt{2}$
   b. $\dfrac{6 \pm \sqrt{20}}{2}$ $3 \pm \sqrt{5}$
   c. $-1 \pm \sqrt{-4}$ $-1 \pm 2i$
2. Verify that the values in part 1 are solutions of the corresponding equations. (1a are solutions of 2a, 1b are solutions of 2b, and 1c are solutions of 2c.)
   a. $(x + 2)^2 = 50$
   b. $x^2 - 6x + 4 = 0$
   c. $3(x + 1)^2 + 12 = 0$
3. Find the decimal approximation of the radicals in 1a and 1b. Round to the nearest hundredth.
4. Use the decimal approximations to show that the values are solutions of the corresponding equations given in 2a and 2b.

Chapter 8 / REVIEW

Rational Exponents, Radicals, and Complex Numbers

What's the big idea? Now that you have completed Chapter 8, you should be able to see a connection between radical expressions and exponents. You should also be able to simplify radicals, perform operations with radicals, and solve equations containing radical expressions. Lastly, you should be able to express square roots of negative numbers as complex numbers.

The Tools

Listed below are the key terms, skills, formulas, and properties you should know for this chapter.

The page reference is provided if you need additional help with the given topic. The Study Tips will assist in your preparation for an exam.

Study Tips

1. Learn all of the terms, formulas, and properties. Make flash cards and have someone quiz you.
2. Rework problems from the exercises and also the ones you worked in class. Work additional problems from the review exercises.
3. Review the summaries of key concepts.
4. Work the chapter test.
5. Be sure to review the online resources for additional study materials.

Terms

Conjugates 601
Complex conjugates 619
Complex number 616
Cube root 553
Extraneous solution 606
Imaginary unit i 616
Index 555
Like radicals 589
Negative square root 552
nth root 555
Perfect cube 554
Perfect square 552
Principal square root 552
Radical 552
Radical equation 606
Radical expression 553
Radical function 558
Radical sign 553
Radicand 553
Rational exponent 568
Rationalizing the denominator 600
Square root 552

Formulas and Properties

- $b^{1/n}$ 568
- $b^{m/n}$ 569
- $b^{-m/n}$ 570
- $\sqrt[n]{a^n}$ 558
- $(\sqrt[n]{a})^n$ 557
- Cube root 553
- Distance formula 583
- nth root 555
- Power of i 621
- Power property 606
- Product of complex conjugates 619
- Product rule for radicals 578 & 591
- Pythagorean theorem 610
- Quotient rule for radicals 582 & 598
- Rules of exponents 571
- Special triangles, 30°-60°-90° and 45°-45°-90° 593
- Square root of a negative number 616

CHAPTER 8 / SUMMARY

How well do you know this chapter? Complete the following questions to find out. Take a look back at the section if you need help.

SECTION 8.1 Radicals and Radical Functions

1. The square root of a real number A is a real number b if $b^2 = A$.
2. The symbol $\sqrt{\ }$ denotes the principal or positive square root.
3. The number inside the radical is called the radicand.
4. If $A < 0$, then the $\sqrt{A}$ is not a(n) real number.
5. The cube root of a real number A is a real number b if $b^3 = A$.
6. The nth root of a real number A is a real number b if $b^n = A$.
7. If n is an even number, $A \geq 0$ for the nth root of A to be a real number.
8. If n is an odd number, A can be any real number for the nth root of A to be a real number.
9. If n is even, then $\sqrt[n]{a^n} = a$ for $a \geq 0$ and $\sqrt[n]{a^n} = |a|$ for $a < 0$.
10. If n is odd, then $\sqrt[n]{a^n} = a$ for all values of a.
11. As long as $\sqrt[n]{a}$ is a real number, $(\sqrt[n]{a})^n = a$.
12. If n is even, the domain of $f(x) = \sqrt[n]{x}$ is $[0, \infty)$.
13. If n is odd, the domain of $f(x) = \sqrt[n]{x}$ is $(-\infty, \infty)$.

SECTION 8.2 Rational Exponents

14. An exponent with a fraction is called a(n) rational exponent.
15. For n a positive integer greater than 1 and $\sqrt[n]{b}$ a real number, $b^{1/n} = \sqrt[n]{b}$.
16. If m and n are positive integers greater than 1 and $\sqrt[n]{b}$ is a real number, then $b^{m/n} = \sqrt[n]{b^m}$.
17. If m and n are positive integers greater than 1 and $\sqrt[n]{b}$ is a real number, then for $b \neq 0$, $b^{-m/n} = \dfrac{1}{b^{m/n}}$.
18. The exponent rules work for integer exponents but also for rational exponents.

SECTION 8.3 Simplifying Radical Expressions and the Distance Formula

19. The product rule for radicals states that for a and b real numbers, $\sqrt[n]{ab} = \sqrt[n]{a}\,\sqrt[n]{b}$, where $a, b \geq 0$.
20. For a radical to be in its simplest form, the radicand cannot contain any perfect power factors or any exponents greater than or equal to the index.
21. The quotient rule for radicals states that for a and b real numbers and $b \neq 0$, $\sqrt[n]{\dfrac{a}{b}} = \dfrac{\sqrt[n]{a}}{\sqrt[n]{b}}$.
22. The distance between two points (x_1, y_1) and (x_2, y_2) is $d = \sqrt{(x_2 - x_1)^2 + (y_2 - y_1)^2}$.

SECTION 8.4 Adding, Subtracting, and Multiplying Radical Expressions

23. Like radicals are radicals with the same radicand and the same index.
24. To add like radicals, combine the coefficients of the like radical expressions then keep the radicand the same.
25. Radicals should be simplified before adding.
26. We can multiply radicals with the same index. In this case, we multiply the radicands and simplify the product.
27. We can multiply radical expressions by applying the distributive property and the FOIL method.
28. The product of conjugates doesn't contain a radical.

SECTION 8.5 Dividing Radical Expressions and Rationalizing

29. We can divide radicals with the same index. In this case, divide the radicands and simplify the quotient.
30. The process of removing a radical from the denominator of a fraction is called rationalizing the denominator.
31. To remove a radical from the denominator, we must multiply the fraction by a form of one that makes the denominator a perfect power whose exponent is the index.
32. To remove a radical from the denominator if the denominator is a binomial, we must multiply by a form of one that involves the conjugate of the denominator.

SECTION 8.6 Radical Equations and Their Applications

33. The power property states that if $a = b$, then $a^n = b^n$.
34. When we apply the power property, we obtain the solutions of the original equation but we can also obtain extraneous solutions. Therefore, we must check the proposed solutions in the original equation.
35. To apply the power property, the radical expression must be isolated to one side of the equation.
36. If an equation contains more than one radical expression, we may have to apply the power property twice.
37. The Pythagorean theorem states that if a and b are legs of a right triangle and c is a hypotenuse, then $a^2 + b^2 = c^2$.

SECTION 8.7 Complex Numbers

38. The imaginary unit can be defined as the number whose square is equal to -1. We write it as i. $i = \sqrt{-1}$ and $i^2 = -1$.
39. A complex number is a number of the form $a + bi$, where a and b are real numbers.
40. We add and subtract complex numbers by adding like terms.
41. We multiply complex numbers by applying the distributive property and the FOIL method.

42. To divide complex numbers, we multiply the numerator and denominator by the conjugate of the denominator. If the denominator is $a + bi$, then we will multiply by $a - bi$.

43. To find i^n find the remainder of n divided by 4. Then $i^n = i^r$.

CHAPTER 8 / REVIEW EXERCISES

SECTION 8.1

Simplify each radical. Assume all radicands with even indices are positive real numbers. (*See Objectives 1–3.*)

1. $-\sqrt{196t^{20}}$ $-14t^{10}$ **2.** $\sqrt{4y^{12}}$ $2y^6$ **3.** $\sqrt{100h^8k^2}$ $10h^4k$

4. $\sqrt{625m^{10}n^6}$ $25m^5n^3$ **5.** $\sqrt{\dfrac{81s^{16}}{25}}$ $\dfrac{9s^8}{5}$ **6.** $\sqrt{\dfrac{144t^6}{49}}$ $\dfrac{12t^3}{7}$

7. $\sqrt{3.61a^6}$ $1.9a^3$ **8.** $\sqrt{2.25c^{14}}$ $1.5c^7$ **9.** $\sqrt{(3x-1)^2}$ $3x - 1$

10. $(\sqrt{5x+3})^2$ for $x \ge -\dfrac{3}{5}$ $5x + 3$ **11.** $\sqrt[3]{-1000b^3}$ $-10b$

12. $\sqrt[5]{-32a^{10}b^5}$ $-2a^2b$ **13.** $\sqrt[4]{64x^8y^4}$ $2x^2y\sqrt[4]{4}$

14. $-\sqrt[4]{81a^4b^{16}}$ $-3ab^4$ **15.** $\sqrt[6]{4096a^{12}b^6}$ $4a^2b$

16. $\sqrt[7]{x^7y^{28}}$ xy^4

Find the domain of each radical function. Write answers in interval notation. Evaluate the radical functions for the indicated values. Round answers to two decimal places as needed. (*See Objective 6.*)

17. $f(x) = \sqrt{8x-5}$; $x = -2, x = 0, x = 2$

18. $f(x) = \sqrt{7x-9}$; $x = -4, x = 0, x = 6$

19. $f(x) = \sqrt[3]{3-8x}$; $x = -3, x = 0, x = 2$

20. $f(x) = \sqrt[3]{4-x}$; $x = -4, x = 0, x = 4$

Graph each radical function using the accompanying tables. (*See Objective 7.*)

21. $f(x) = \sqrt{16-3x}$

x	$f(x)$
-3	
0	
1	
4	

22. $f(x) = 2 - \sqrt[3]{x}$

x	$f(x)$
-8	
-1	
0	
1	
8	

SECTION 8.2

Rewrite each expression as a radical expression and simplify, if possible. Assume all variables represent positive real numbers. (*See Objectives 1 and 2.*)

23. $(-64x^{12})^{1/3}$ $-4x^4$ **24.** $(-27y^9)^{2/3}$ $9y^6$

25. $(1000a^6b^3)^{2/3}$ $100a^4b^2$ **26.** $(125c^3d^6)^{4/3}$ $625c^4d^8$

27. $(32x^{-15}y^5)^{3/5}$ $\dfrac{8y^3}{x^9}$ **28.** $(y+5)^{2/3}$ $\sqrt[3]{(y+5)^2}$

29. $(2x-3)^{2/5}$ $\sqrt[5]{(2x-3)^2}$ **30.** $\left(\dfrac{a^{12}}{81b^8}\right)^{3/4}$ $\dfrac{a^9}{27b^6}$

Simplify each expression and write each answer with positive exponents. Assume all variables represent positive real numbers. (*See Objectives 3 and 4.*)

31. $(-125a^3)^{-1/3}$ $-\dfrac{1}{5a}$ **32.** $(27y^{-6})^{-2/3}$ $\dfrac{y^4}{9}$

33. $(x^{-3}y^{12})^{-4/3}$ $\dfrac{x^4}{y^{16}}$ **34.** $(-r^{10}s^{-5})^{-2/5}$ $-\dfrac{s^2}{r^4}$

35. $2^{5/4} \cdot 2^{3/4}$ 4 **36.** $3^{4/7} \cdot 3^{3/7}$ 3

37. $\dfrac{x^{4/7}}{x^{2/3}}$ $\dfrac{1}{x^{2/21}}$ **38.** $\dfrac{y^{3/5}}{y^{1/2}}$ $y^{1/10}$

39. $(-32x^{15}y^{20})^{3/5}$ $-8x^9y^{12}$ **40.** $(-64a^6b^3)^{2/3}$ $16a^4b^2$

41. $(5x^{1/2} - 3y^{1/3})(5x^{1/2} + 3y^{1/3})$ $25x - 9y^{2/3}$

42. $(a^{4/5} - 2b^{1/2})(a^{4/5} + 6b^{1/2})$ $a^{8/5} + 4a^{4/5}b^{1/2} - 12b$

43. $x^{1/2}(2x^{1/2} + 3x^{2/3})$ $2x + 3x^{7/6}$ **44.** $2s^{2/3}(3s^{2/3} - 4s^{1/2})$ $6s^{4/3} - 8s^{7/6}$

Use rational exponents to simplify each expression. Assume all variables represent positive real numbers. (*See Objective 5.*)

45. $\sqrt[3]{125x^9}$ $5x^3$ **46.** $\sqrt[5]{32y^{25}}$ $2y^5$

47. $\sqrt[6]{u^3v^6}$ $v\sqrt{u}$ **48.** $\sqrt[6]{x^6y^9}$ $xy\sqrt{y}$

49. $\sqrt[4]{\dfrac{36}{c^6}}$ $\sqrt{\dfrac{6}{c^3}}$ **50.** $\sqrt[4]{\dfrac{9}{y^2}}$ $\sqrt{\dfrac{3}{y}}$

51. $\sqrt{8} \cdot \sqrt[3]{4}$ $\sqrt[6]{8192}$ **52.** $\sqrt[3]{9} \cdot \sqrt[6]{3}$ $\sqrt[6]{243}$

Solve each problem. (*See Objective 6.*)

53. How much money will be in an account in 6 months if you invest \$1200 at an annual interest rate of 1.2% compounded quarterly ($n = 4$)? \$1207.21

54. What expression would you need to enter on a scientific calculator to evaluate $\sqrt[5]{650}$? Use the calculator to approximate the value and round your answer to two decimal places. $(650)^{(1/5)}$, 3.65

SECTION 8.3

Simplify each radical. Assume all variables represent positive real numbers. (*See Objectives 1 and 2.*)

55. $\sqrt{360a^5b^{12}}$ $6a^2b^6\sqrt{10a}$ **56.** $\sqrt{\dfrac{50r^9}{49s^2}}$ $\dfrac{5r^4\sqrt{2r}}{7s}$

57. $\sqrt[3]{250x^8y^{16}}$ $5x^2y^5\sqrt[3]{2x^2y}$ **58.** $\sqrt[5]{-64r^9t^{18}}$ $-2rt^3\sqrt[5]{2r^4t^3}$

59. $\sqrt[3]{\dfrac{2000x^5}{27y^{12}}}$ $\dfrac{10x\sqrt[3]{2x^2}}{3y^4}$ **60.** $\sqrt[4]{\dfrac{32r^{15}}{81t^{12}}}$ $\dfrac{2r^3\sqrt[4]{2r^3}}{3t^3}$

61. $\dfrac{\sqrt{54d^{11}}}{\sqrt{18c^6}}$ $\dfrac{d^5\sqrt{3d}}{c^3}$ **62.** $\dfrac{\sqrt[3]{14y^7}}{\sqrt[3]{56x^6}}$ $\dfrac{y^2\sqrt[3]{2y}}{2x^2}$

Find the distance between each pair of points. Write each answer in simplest radical form. (*See Objective 3.*)

63. $(3\sqrt{2}, -\sqrt{6})$ and $(5\sqrt{2}, -3\sqrt{6})$ $4\sqrt{2}$

64. $(-2, 10\sqrt{5})$ and $(1, 7\sqrt{5})$ $3\sqrt{6}$

65. $\left(\frac{\sqrt{3}}{2}, -4\sqrt{2}\right)$ and $\left(\frac{5\sqrt{3}}{2}, -2\sqrt{2}\right)$ $2\sqrt{5}$

66. $\left(\frac{\sqrt{5}}{3}, \frac{3}{4}\right)$ and $\left(-\frac{\sqrt{5}}{6}, \frac{1}{4}\right)$ $\sqrt{\frac{3}{2}}$

SECTION 8.4

Perform each operation and write each answer in simplest radical form. (*See Objectives 1 and 2.*)

67. $-9\sqrt{6} + 14\sqrt{6} + \sqrt{6}$ $6\sqrt{6}$

68. $4\sqrt{14x^3} - 8\sqrt{14x^3} - 9x\sqrt{14x}$ $-13x\sqrt{14x}$

69. $-7a\sqrt{216ab^2} + 15b\sqrt{294a^3}$ $63ab\sqrt{6a}$

70. $13\sqrt{45cd^5} - 2d^2\sqrt{245cd}$ $25d^2\sqrt{5cd}$

71. $9\sqrt[3]{3xy^3} - 5y\sqrt[3]{27x} + \sqrt[3]{24xy^3}$ $11y\sqrt[3]{3x} - 15y\sqrt[3]{x}$

72. $14t\sqrt[5]{729r^5t} - 10r\sqrt[5]{96t^6}$ $22rt\sqrt[5]{3t}$

73. $\frac{7\sqrt{9a^2b}}{2} + \sqrt{\frac{16a^2b}{25}}$ $\frac{113a\sqrt{b}}{10}$

74. $\frac{\sqrt{80ab^3}}{3} - \sqrt{\frac{45ab^3}{4}}$ $-\frac{b\sqrt{5ab}}{6}$

75. $\sqrt[3]{\frac{24x^2y^3}{125}} - \frac{\sqrt[3]{375x^2y^3}}{5}$ $-\frac{3y\sqrt[3]{3x^2}}{5}$

76. $\sqrt[3]{\frac{162a}{125b^3}} + \frac{\sqrt[3]{48a}}{3b}$ $\frac{19\sqrt[3]{6a}}{15b}$

77. $\sqrt[5]{10x^4y^{-10}}\sqrt[5]{16x^8y^{15}}$ $2x^2y\sqrt[5]{5x^2}$

78. $\sqrt[3]{96a^5b^6}\sqrt[3]{6a^7b^{-4}}$ $4a^4\sqrt[3]{9b^2}$

79. $\sqrt{7}(5 - \sqrt{14})$ $5\sqrt{7} - 7\sqrt{2}$

80. $\sqrt{5}(6 + \sqrt{15})$ $6\sqrt{5} + 5\sqrt{3}$

81. $(\sqrt{3} - 2\sqrt{6})(\sqrt{3} + 5\sqrt{6})$ $9\sqrt{2} - 57$

82. $(3\sqrt{2} - \sqrt{10})^2$ $28 - 12\sqrt{5}$

83. $(7 - \sqrt{2a})(7 + \sqrt{2a})$ $49 - 2a$

84. $(2\sqrt{12a} + 5\sqrt{b})(\sqrt{3a} - \sqrt{16b})$ $12a - 11\sqrt{3ab} - 20b$

In Exercises 85 and 86, find the perimeter and area of the triangle. Approximate each answer to one decimal place. (*See Objective 3.*)

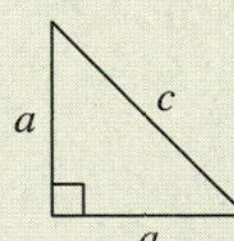

85. $a = 13\sqrt{2}$ and $c = 26$ 62.8, 169

86. $a = 6\sqrt{2}$ and $c = 12$ 29, 36

87. In a 30°-60°-90° triangle, the leg opposite the 30° angle measures $12\sqrt{6}$ cm. Find the length of the other leg and the hypotenuse. $36\sqrt{2}$ cm, $24\sqrt{6}$ cm

88. In a 30°-60°-90° triangle, the leg opposite the 30° angle measures $8\sqrt{2}$ m. Find the length of the other leg and the hypotenuse. $8\sqrt{6}$ m, $16\sqrt{2}$ m

SECTION 8.5

Perform each operation and write each answer in simplest radical form. Assume all variables represent positive real numbers. (*See Objective 1.*)

89. $\frac{\sqrt{150x^{15}}}{\sqrt{2x^6}}$ $5x^4\sqrt{3x}$

90. $\frac{\sqrt[3]{48t^9}}{\sqrt[3]{3t^8}}$ $2\sqrt[3]{2t}$

91. $\frac{\sqrt{126y^7}}{2\sqrt{7y^2}}$ $\frac{3y^2\sqrt{2y}}{2}$

92. $\frac{5\sqrt[3]{72a^{10}}}{\sqrt[3]{3a^2}}$ $10a^2\sqrt[3]{3a^2}$

93. $\frac{6\sqrt[5]{640c^{12}}}{\sqrt[5]{5c^4}}$ $12c\sqrt[5]{2^2c^3}$

94. $\frac{7\sqrt[4]{320a^{15}}}{\sqrt[4]{5a^5}}$ $14a^2\sqrt[4]{2^2a^2}$

95. $\frac{\sqrt[4]{64a^{11}b^{11}}}{2\sqrt[4]{2a^6b^{-3}}}$ $ab^3\sqrt[4]{2ab^2}$

96. $\frac{3\sqrt[3]{945a^8b^{-4}}}{\sqrt[3]{7a^2b^{-9}}}$ $9a^2b\sqrt[3]{5b^2}$

Simplify by rationalizing each denominator. (*See Objectives 2 and 3.*)

97. $\frac{\sqrt{40}}{\sqrt{15x}}$ $\frac{2\sqrt{6x}}{3x}$

98. $\frac{6}{\sqrt[5]{16x^3}}$ $\frac{3\sqrt[5]{2x^2}}{x}$

99. $\frac{12}{\sqrt[4]{27a}}$ $\frac{4\sqrt[4]{3a^3}}{a}$

100. $\frac{18}{\sqrt[4]{4b^3}}$ $\frac{9\sqrt[4]{2^3b^2}}{b}$

101. $\frac{2}{\sqrt{7} + \sqrt{5}}$ $\sqrt{7} - \sqrt{5}$

102. $\frac{2 + \sqrt{7}}{5 - \sqrt{7}}$ $\frac{17 + 7\sqrt{7}}{18}$

103. $\frac{5 - \sqrt{6}}{\sqrt{3} - 2}$ $3\sqrt{2} + 2\sqrt{6} - 5\sqrt{3} - 10$

104. $\frac{\sqrt{2}}{\sqrt{6} - \sqrt{2}}$ $\frac{\sqrt{3} + 1}{2}$

105. $\frac{a}{\sqrt{a + 4} - 2}$ $\sqrt{a + 4} + 2$

106. $\frac{b}{\sqrt{b + 16} - 4}$ $\sqrt{b + 16} + 4$

SECTION 8.6

Solve each radical equation. (*See Objective 1.*)

107. $\sqrt{2x - 5} + 2 = 1$ $\varnothing$

108. $\sqrt{12b - 23} + 2 = 7$ $\{4\}$

109. $\sqrt[4]{3x - 5} - 6 = -4$ $\{7\}$

110. $\sqrt[3]{7y + 6} + 9 = 12$ $\{3\}$

111. $\sqrt{2x - 17} + 8 = x$ $\{9\}$

112. $\sqrt{72 - 4y} + 3 = y$ $\{9\}$

113. $\sqrt{2y + 124} - 8 = \sqrt{y}$ $\{36, 100\}$

114. $\sqrt{2z + 41} + 5 = \sqrt{z}$ $\varnothing$

115. $\sqrt{6x + 31} = 2 + \sqrt{2x + 11}$ $\{-1\}$

116. $\sqrt{3x + 10} + 2 = \sqrt{7x + 22}$ $\{2\}$

Solve each problem. (*See Objective 2.*)

117. Find the length of the missing side of a right triangle if one of the legs is 35 ft and the hypotenuse is 37 ft. 12 ft

118. Find the length of the hypotenuse of a right triangle if the lengths of the two legs are 40 in. and 9 in. 41 in.

119. One of the ropes that is used to tie down a tent needs to be replaced. Determine the distance from the top of the tent to the tie-down stake if the tent is 9 ft tall and the tie-down stake is placed 3 ft from the tent. $3\sqrt{10}$ ft or 9.4868 ft

120. A person's body surface area (BSA) in square meters is given by BSA $= \sqrt{\dfrac{hw}{3131}}$, where h is a person's height in inches and w is a person's weight in pounds.

a. Find the BSA of a person who is 4 ft 9 in. and weighs 116 lb. 1.453198 m^2

b. The average BSA of a woman is 1.9 m^2. If a woman is 5 ft 11 in. tall, how much should she weigh to have an average BSA? 159.1959 lb

SECTION 8.7

Perform each operation and write each answer in standard form. (*See Objectives 1–5.*)

121. $(3 - 5i) + (-11 + 7i)$ $-8 + 2i$

122. $(-7 + 9i) - (13 - 2i)$ $-20 + 11i$

123. $(-4 - 2i) - (9 + 5i) + (13 + 14i)$ $7i$

124. $(-9 + 4i) + (1 + 3i) - (-15 + 7i)$ 7

125. $\sqrt{-9} \cdot \sqrt{-25}$ -15

126. $6i(-8 - 5i)$ $30 - 48i$

127. $(-2 + 3i)(5 - 4i)$ $2 + 23i$

128. $(6 - 3i)(-4 + 8i)$ $60i$

129. $(1 - \sqrt{6}i)(4 + 3\sqrt{6}i)$ $22 - \sqrt{6}i$

130. $(3 + \sqrt{5}i)^2$ $4 + 6\sqrt{5}i$

131. $(\sqrt{11} - 2i)(\sqrt{11} + 2i)$ 15

132. $\dfrac{-8 + 12i}{4i}$ $3 + 2i$

133. $\dfrac{9 + i}{1 + 3i}$ $\dfrac{6 - 13i}{5}$

134. $\dfrac{1 + 5i}{4 - 3i}$ $\dfrac{23i - 11}{25}$

135. $(2i)^6$ -64

136. $(-3i)^5$ $-243i$

137. i^{-7} i

138. $(-i)^{-9}$ i

CHAPTER 8 TEST / RATIONAL EXPONENTS, RADICALS, AND COMPLEX NUMBERS

1. $-\sqrt{25a^{16}} =$

a. $5a^4$ **b.** $-5a^4$ **c.** $5a^8$ **(d.)** $-5a^8$

2. $\sqrt{-256} =$

a. -16 **b.** 16 **c.** -128 **(d.)** not a real number

3. $\sqrt[3]{-729} =$

a. 9 **(b.)** -9 **c.** 243 **d.** not a real number

4. $-\sqrt[4]{16x^{12}} =$

(a.) $-2x^3$ **b.** $2x^3$ **c.** $-4x^3$ **d.** $4x^3$

5. The best approximation for $\sqrt[7]{12{,}516}$ is

(a.) 3.849 **b.** 1788 **c.** 111.875 **d.** 783.125

6. $\sqrt{75} - \sqrt{12} + \sqrt{27} =$

a. $\sqrt{60}$ **(b.)** $6\sqrt{3}$ **c.** $2\sqrt{15}$ **d.** $10\sqrt{3}$

7. $(\sqrt{5} - 1)(2\sqrt{5} + 7) =$

a. $-7 + 7\sqrt{5}$ **b.** $3 - 5\sqrt{5}$ **(c.)** $3 + 5\sqrt{5}$ **d.** $-7 - 5\sqrt{5}$

8. $(4\sqrt{3} - 6\sqrt{2})^2 =$

(a.) $120 - 48\sqrt{6}$ **b.** $-24 - 48\sqrt{6}$ **c.** 120 **d.** -24

9. $\dfrac{5\sqrt{12}}{10\sqrt{2}} =$

a. $\sqrt{3}$ **(b.)** $\dfrac{\sqrt{6}}{2}$ **c.** 3 **d.** 12

10. The solution set of $\sqrt{m^2 + 5m - 8} = m + 1$ is

(a.) $\{3\}$ **b.** $\left\{\dfrac{9}{5}\right\}$ **c.** $\{-3\}$ **d.** $\{2\}$

11. The solution set of $\sqrt{x + 6} = x$ is

a. $\{-2, 3\}$ **b.** $\{-2\}$ **(c.)** $\{3\}$ **d.** $\varnothing$

12. $625^{3/4} =$

a. 15 **b.** 25 **(c.)** 125 **d.** 5

13. $4^{3/5} \cdot 4^{7/5} =$

a. 8 **(b.)** 16 **c.** 256 **d.** 4

14. $\dfrac{16^{-5/4}}{16^{-3/4}}$

a. 8 **b.** 4 **c.** $\dfrac{1}{8}$ **(d.)** $\dfrac{1}{4}$

15. $(32k^{10})^{1/5} =$

(a.) $2k^2$ **b.** $32k^2$ **c.** $32k^{50}$ **d.** $2k^{50}$

16. When simplified, $3i(5 + 2i) - 4i$ is

(a.) $-6 + 11i$ **b.** $-6 + 4i$ **c.** $5i$ **d.** $17i$

Simplify each radical. Assume all variables represent positive real numbers.

17. $\sqrt{50x^4y^3}$ $5x^2y\sqrt{2y}$

18. $\sqrt{\dfrac{49a^6}{16}}$ $\dfrac{7a^3}{4}$

19. $\sqrt[3]{40}$ $2\sqrt[3]{5}$

20. $\sqrt[3]{\dfrac{8}{27b^6}}$ $\dfrac{2}{3b^2}$

21. $\sqrt[4]{162a^7}$ $3a\sqrt[4]{2a^3}$

Perform each operation and write each answer in simplest radical form. Rationalize the denominators when necessary.

22. $\sqrt{6x} \cdot \sqrt{2x}$ $2x\sqrt{3}$

23. $\sqrt[3]{4y^2} \cdot \sqrt[3]{6y^2}$ $2y\sqrt[3]{3y}$

24. $(\sqrt{4x} - 9)^2$ $4x - 9$

25. $\dfrac{\sqrt{18b^3}}{\sqrt{2b}}$ $3b$

26. $\dfrac{1}{\sqrt{3}}$ $\dfrac{\sqrt{3}}{3}$

27. $\dfrac{6\sqrt[3]{2}}{\sqrt[3]{5}}$ $\dfrac{6\sqrt[3]{50}}{5}$

28. $\dfrac{\sqrt{3}}{\sqrt{5} + \sqrt{2}}$ $\dfrac{\sqrt{15} - \sqrt{6}}{3}$

29. $\sqrt{48} - 5\sqrt{300}$ $-46\sqrt{3}$

30. $2\sqrt[3]{16} - 4\sqrt[3]{54}$ $-8\sqrt[3]{2}$

31. $(\sqrt{5} + 6)^2$ $41 + 12\sqrt{5}$

32. $(\sqrt{5} + 6)(\sqrt{5} - 6)$ -31

Solve each radical equation.

33. $\sqrt{x + 2} = 3$ $\{7\}$

34. $\sqrt{5x + 2} + 4 = 3$ $\varnothing$

35. $\sqrt{2x + 1} = x - 1$ $\{4\}$

36. A right triangle has legs that are each $4\sqrt{5}$ units. Find the simplest radical expression that represents the length of the hypotenuse. Then find the perimeter and area of the triangle. The hypotenuse is $4\sqrt{10}$ units. The perimeter is $8\sqrt{5} + 4\sqrt{10}$ units. The area is 40 units^2.

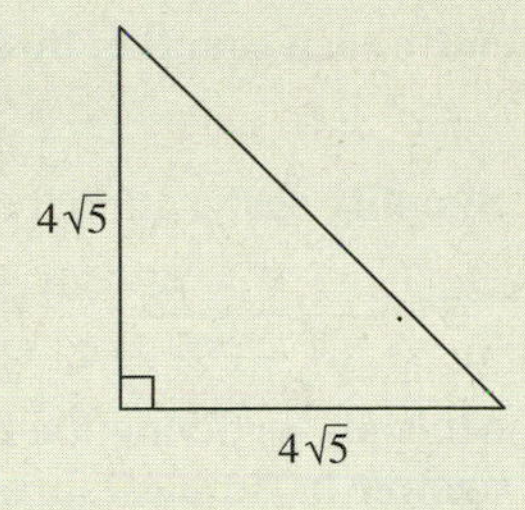

Write each complex number in terms of *i*.

37. $\sqrt{-64}$ $8i$

38. $4 + \sqrt{-9}$ $4 + 3i$

Perform each operation with the complex numbers.

39. $(-2 + 10i) + (-5 - 5i) + (12 + 2i)$ $5 + 7i$

40. $3i(-7 + 4i)$ $-12 - 21i$

41. $(9 + 4i)(9 - 4i)$ 97

42. $(9 + 4i)^2$ $65 + 72i$

43. $\dfrac{4}{3i}$ $-\dfrac{4}{3}i$

44. $\dfrac{13}{1 + 5i}$ $\dfrac{1}{2} - \dfrac{5}{2}i$

CUMULATIVE REVIEW EXERCISES / CHAPTERS 1–8

Translate each problem into a linear equation and solve it to find the information. (*Section 2.2, Objective 1*)

1. The sum of two consecutive even integers is 66. Find the numbers. 32, 34
2. The sum of two consecutive integers is 43. Find the numbers. 21, 22
3. In 2011, the annual salary of the Secretary of State, Hillary Clinton, was $199,700. The annual salary of the Speaker of the House of Representatives, John Boehner, was $23,800 more than the salary of Secretary Clinton. What was the annual salary of Speaker Boehner? (Source: http://usgovinfo.about.com) $223,500
4. In 2011, the annual salary of the rank-and-file members of the House and Senate was $174,000. The annual salary of the Majority Leader was $19,400 more than the salary of the rank-and-file members of the House and Senate. What was the annual salary for the Majority Leader? (Source: http://usgovinfo.about.com) $193,400

Solve each formula for the specified variable. (*Section 2.3, Objective 4*)

5. $S = 2\pi rh + \pi r^2$ for h $h = \dfrac{S - \pi r^2}{2\pi r}$
6. $A = P(1 + rt)^n$ for P $P = \dfrac{A}{(1 + rt)^n}$

Solve each compound inequality. Write the solution in interval notation. (*Section 2.5, Objectives 3 and 4*)

7. $4(1 - 3x) < -5$ and $8x - 2 < 14$ $\left(\dfrac{3}{4}, 2\right)$
8. $6 - 2x \le 0$ or $5x - 2 \le 3$ $(-\infty, 1] \cup [3, \infty)$

Solve each absolute value inequality. (*Section 2.7, Objectives 1 and 2*)

9. $|4 - 3x| + 5 \le 12$ $\left[-1, \dfrac{11}{3}\right]$
10. $|5x + 1| - 3 > 16$ $(-\infty, -4) \cup \left(\dfrac{18}{5}, \infty\right)$

Find the midpoint of each line segment formed by the ordered pairs. (*Section 3.1, Objective 5*)

11. (3.6, 4.8) and (–9.4, 12.2) (–2.9, 8.5)
12. (–7.1, 1.6) and (–14.3, 10.8) (–10.7, 6.2)

Find $f(a + 2)$ for each function. (*Section 3.3, Objective 3*)

13. $f(x) = x^2 - 3x - 5$ $a^2 + a - 7$
14. $f(x) = -3x + 8$ $-3a + 2$

Rewrite the equation using function notation, if possible. Identify *m* and *b*. (*Section 4.1, Objective 2*)

15. $6x - 5y = 10$ $f(x) = \dfrac{6}{5}x - 2, \dfrac{6}{5}, -2$
16. $7y + 21 = 0$ $f(x) = -3, 0, -3$

Write the equation of the line that satisfies the given information. Write each answer in slope-intercept form. (*Section 4.4, Objective 2*)

17. (–3, –14) and (2, 11) $y = 5x + 1$
18. passes through (3, –8) and parallel to $y = 5$ $y = -8$

Solve each system of linear equations graphically. (*Section 5.1, Objectives 2 and 3*)

19. $\begin{cases} -5x + y = 24 \\ 10x - 2y = 16 \end{cases}$ $\varnothing$

20. $\begin{cases} x + 3y = 9 \\ x + 2y = 0 \end{cases}$ $\{(-18, 9)\}$

Solve each system of linear equations using substitution or elimination. (*Section 5.2, Objectives 1 and 2*)

21. $\begin{cases} x + 7y = 16 \\ -2x + 4y = 22 \end{cases}$ $\{(-5, 3)\}$

22. $\begin{cases} 0.6x - 0.1y = 0.5 \\ -0.5x + 0.3y = 1.1 \end{cases}$ $\{(2, 7)\}$

23. $\begin{cases} \dfrac{1}{3}x - y = 1 \\ -\dfrac{1}{2}x + y = \dfrac{3}{4} \end{cases}$ $\left\{\left(-\dfrac{21}{2}, -\dfrac{9}{2}\right)\right\}$

24. $\begin{cases} \dfrac{4}{9}x - \dfrac{1}{3}y = 4 \\ \dfrac{1}{6}x + \dfrac{3}{2}y = -5 \end{cases}$ $\{(6, -4)\}$

Solve each problem. (*Section 5.3, Objectives 2 and 3*)

25. A chemist has a 48% acid solution. She needs 80 L of an 18% acid solution. How many liters of water and how many liters of the 48% acid solution must she mix together to obtain 80 L of an 18% acid solution? 50 L of water and 30 L of 48% acid solution

26. A chemist has a 64% alcohol solution. He needs 30 gal of a 24% alcohol solution. How many gallons of water and how many gallons of the 64% alcohol solution must he mix together to obtain 30 gal of a 24% alcohol solution? 18.75 gal of water and 11.25 gal of 64% alcohol solution

27. Jennifer invests \$8000 in two different accounts. She invests part of her money in a savings account that earns 1.25% simple interest and the rest of her money in a money market fund that earns 0.9% simple interest. If she earns a total of \$94.75 interest in 1 yr, how much did she invest in each account? \$6500 at 1.25% in the savings account and \$1500 at 0.9% in the money market fund

28. Nancy invests \$15,000 in two different accounts. She invests part of her money in a savings account that earns 1.5% simple interest and the rest of her money in a CD that earns 4% simple interest. If she earns a total of \$512.50 interest in 1 yr, how much did she invest in each account? \$3500 at 1.5% in the savings account and \$11,500 at 4% in the CD account

Simplify each expression. Write each answer with positive exponents. (*Section 6.1, Objectives 3 and 4*)

29. $-9(7a)^0$ -9

30. $-2(-3x)^0$ -2

31. $\left(\frac{3}{5}\right)^{-2}$ $\frac{25}{9}$

32. $\left(-\frac{2}{7}\right)^{-3}$ $-\frac{343}{8}$

Solve each problem. (*Section 6.1, Objective 6*)

33. Write an expression for the surface area of a can if the height is twice its radius. $6\pi r^2$

34. Write an expression for the surface area of a can if the radius is twice its height. $12\pi h^2$

Simplify each expression using the rules of exponents. Write each answer with positive exponents. (*Section 6.2, Objectives 1 and 2*)

35. $(-2x^2y^8)^4$ $16x^8y^{32}$

36. $(-3a^{-6})^{-2}$ $\frac{a^{12}}{9}$

37. $\left(\frac{3p^3}{5q^{-4}}\right)^{-3}$ $\frac{125}{27p^9q^{12}}$

38. $\left(\frac{2r^3}{7s^{-4}}\right)^{-2}$ $\frac{49}{4r^6s^8}$

Simplify each expression. (*Section 6.4, Objectives 1, 2, and 4*)

39. $(2x^2 + 6x + 16) - (-3x^2 + 8x - 10)$ $5x^2 - 2x + 26$

40. $(-7y^2 + 8y - 4) - (8y^2 - 6y + 13)$ $-15y^2 + 14y - 17$

41. $(a + 5)(a^2 - 5a + 25)$ $a^3 + 125$

42. $(2b - 1)(4b^2 + 2b + 1)$ $8a^3 - 1$

43. $(a + 2b)^2$ $a^2 + 4ab + 4b^2$

44. $(7c - d)^2$ $49c^2 - 14cd + d^2$

Use grouping to factor each polynomial completely, if possible. (*Section 6.5, Objective 3*)

45. $6x^3 + 21x^2 - 10x - 35$ $(2x + 7)(3x^2 - 5)$

46. $15y^3 - 3y^2 + 10y + 2$ can't be factored by grouping

Factor each trinomial. (*Section 6.6, Objectives 1 and 4*)

47. $a^6 - 2a^3 - 35$ $(a^3 + 5)(a^3 - 7)$

48. $d^8 + 9d^4 - 36$ $(d^4 - 3)(d^4 + 12)$

Factor completely. (*Section 6.7, Objectives 1 and 2*)

49. $x^4 - 81$ $(x^2 + 9)(x + 3)(x - 3)$

50. $625y^4 - 1$ $(25y^2 + 1)(5y + 1)(5y - 1)$

51. $a^3 + 1000$ $(a + 10)(a^2 - 10a + 100)$

52. $64b^3 - 1$ $(4b - 1)(16b^2 + 4b + 1)$

Solve each equation by factoring. (*Section 6.8, Objective 1*)

53. $y(4y + 1) = 3$ $\left\{-1, \frac{3}{4}\right\}$

54. $z(2z + 3) = 14$ $\left\{-\frac{7}{2}, 2\right\}$

Simplify each rational expression. (*Section 7.1, Objective 3*)

55. $\frac{x^2 - 49}{5x^2 - 29x - 42}$ $\frac{x + 7}{5x + 6}$

56. $\frac{16x^2 - 1}{4x^2 - 11x - 3}$ $\frac{4x - 1}{x - 3}$

Multiply or divide the rational expressions. Write each answer in simplest form. (*Section 7.1, Objectives 4 and 5*)

57. $\frac{x^2 - 5x}{x^2 - 2x - 15} \cdot \frac{2x^2 - x - 21}{4x^2}$ $\frac{2x - 7}{4x}$

58. $\frac{x^3 - 8}{x^2 + 3x - 10} \div \frac{2x^2 + 4x + 8}{x^2 + 4x - 5}$ $\frac{x - 1}{2}$

Perform each operation using long or synthetic division. (*Section 7.2, Objectives 2 and 3*)

59. $\frac{2x^3 - 7x^2 - 16}{x - 4}$ $2x^2 + x + 4$

60. $\frac{3x^3 - 79x - 20}{x + 5}$ $3x^2 - 15x - 4$

Add or subtract the rational expressions. Write each answer in lowest terms. (*Section 7.3, Objective 3*)

61. $\frac{2}{x^2 - 4} + \frac{3}{x^2 - 3x - 10}$ $\frac{5x - 16}{(x + 2)(x - 2)(x - 5)}$

62. $\frac{x + 1}{9x^2 - 1} - \frac{1}{12x - 4}$ $\frac{x + 3}{4(3x + 1)(3x - 1)}$

Simplify each complex fraction. (*Section 7.4, Objectives 1 and 2*)

63. $\frac{\frac{3}{a} + \frac{5}{b}}{5a + 3b}$ $\frac{1}{ab}$

64. $\frac{\frac{7}{a} - \frac{3}{b}}{3a - 7b}$ $-\frac{1}{ab}$

Solve each rational equation. (*Section 7.5, Objective 1*)

65. $\frac{5}{x - 5} - \frac{4}{x + 1} = \frac{7}{x^2 - 4x - 5}$ $\{-18\}$

66. $\frac{2}{3x - 2} - \frac{3}{x - 4} = \frac{12}{3x^2 - 14x + 8}$ $\{-2\}$

Boyle's law states that if the temperature remains the same, the volume *V* of a gas is inversely proportional to the pressure *P* applied to it. (*Section 7.7, Objective 2*)

67. A given mass of a gas occupies 252 mL at 640 mm Hg. If the temperature remains constant, what volume will the gas occupy if the pressure is increased to 720 mm Hg? 224 mL

68. A given mass of a gas occupies 320 mL at 540 mm Hg. If the temperature remains constant, what volume will the gas occupy if the pressure is increased to 900 mm Hg? 192 mL

Find the constant of variation and the equation of variation for each problem. (*Section 7.7, Objectives 2 and 3*)

69. x varies jointly as the cube of y and the sum of u and v; $x = 125$ when $y = 2.5$, $u = 2.3$, and $v = 1.7$ $x = 2y^3(u + v)$

70. F varies inversely as the square root of G; $F = 1.2$ when $G = 6.25$. $F = \frac{3}{\sqrt{G}}$

Simplify each expression. Assume all variables represent positive real numbers. (*Section 8.1, Objectives 1–3*)

71. $\sqrt{\dfrac{49x^6}{4}}$ $\dfrac{7x^3}{2}$

72. $\sqrt{\dfrac{64y^{12}}{25}}$ $\dfrac{8y^6}{5}$

73. $\sqrt[3]{\dfrac{125x^6}{64y^3}}$ $\dfrac{5x^2}{4y}$

74. $\sqrt[3]{\dfrac{a^9}{8b^{12}}}$ $\dfrac{a^3}{2b^4}$

75. $\sqrt[7]{128x^{14}y^7}$ $2x^2y$

76. $\sqrt[7]{-a^{14}b^{21}}$ $-a^2b^3$

Find the domain of each radical function and write in interval notation. Evaluate each function for the indicated values. (*Section 8.1, Objective 6*)

77. $f(x) = \sqrt{5 - x};\ x = -4, x = 1, x = 6$

78. $f(x) = \sqrt{x - 4};\ x = 5, x = 4, x = 3$

Rewrite each expression as a radical expression and simplify, if possible. Assume all variables represent positive real numbers. (*Section 8.2, Objectives 1–3*)

79. $(64x^9)^{1/3}$ $4x^3$

80. $(27y^{15})^{1/3}$ $3y^5$

81. $(-64)^{2/3}$ 16

82. $(-36)^{3/2}$ not a real number

83. $(64a^6b^{12})^{-5/6}$ $\dfrac{1}{32a^5b^{10}}$

84. $(a^{-8}b^{12})^{-5/4}$ $\dfrac{a^{10}}{b^{15}}$

Simplify each expression. Assume all variables represent positive real numbers. Write each answer with positive exponents. (*Section 8.2, Objective 4*)

85. $\dfrac{x^{3/4}}{x^{1/3}}$ $x^{5/12}$

86. $\dfrac{y^{1/3}}{y^{-1/6}}$ $y^{1/2}$

87. $(6x^{2/5} - y^{1/2})(x^{2/5} - 3y^{1/2})$ $6x^{4/5} - 19x^{2/5}y^{1/2} + 3y$

88. $(r^{1/2} - 2s^{2/3})(r^{1/2} - 2s^{2/3})$ $r - 4r^{1/2}s^{2/3} + 4s^{4/3}$

Use rational exponents to simplify each expression. (*Section 8.2, Objective 5*)

89. $\sqrt[12]{a^3}$ $\sqrt[4]{a}$

90. $\sqrt[6]{b^3}$ $\sqrt{b}$

91. $\sqrt[10]{32}$ $\sqrt{2}$

92. $\sqrt[6]{27}$ $\sqrt{3}$

Simplify each radical expression. Assume all variables represent positive real numbers. (*Section 8.3, Objectives 1 and 2*)

93. $\sqrt{72x^4}$ $6x^2\sqrt{2}$

94. $\sqrt{50y^6}$ $5y^3\sqrt{2}$

95. $\sqrt{98a^5b^4}$ $7a^2b^2\sqrt{2a}$

96. $\sqrt{48r^2t^7}$ $4rt^3\sqrt{3t}$

97. $\sqrt[3]{\dfrac{8x^4}{y^3}}$ $\dfrac{2x\sqrt[3]{x}}{y}$

98. $\sqrt[3]{\dfrac{125a^7}{b^6}}$ $\dfrac{5a^2\sqrt[3]{a}}{b^2}$

Find the distance between each pair of points. Write answers with simplified radicals. (*Section 8.3, Objective 3*)

99. $(3, -6)$ and $(1, 2)$ $2\sqrt{17}$

100. $(-1, 9)$ and $(6, 10)$ $5\sqrt{2}$

Perform the indicated operation and write answers in simplest radical form. (*Section 8.4, Objectives 1 and 2*)

101. $2\sqrt{50x^2} - \sqrt{18x^2} + 3x\sqrt{98}$ $28x\sqrt{2}$

102. $5\sqrt[3]{40} + 3\sqrt[3]{135} - 4\sqrt[3]{5}$ $15\sqrt[3]{5}$

103. $\sqrt{3}(2 + \sqrt{6})$ $2\sqrt{3} + 3\sqrt{2}$

104. $(\sqrt{5} + \sqrt{10})(3\sqrt{5} - \sqrt{40})$ $-5 + 5\sqrt{2}$

Simplify each quotient of radical expressions and write answers in simplest radical form. (*Section 8.5, Objective 1*)

105. $\dfrac{\sqrt{150x^{14}}}{\sqrt{3x^{11}}}$ $5x\sqrt{2x}$

106. $\dfrac{\sqrt{54y^{11}}}{\sqrt{2y^8}}$ $3y\sqrt{3y}$

107. $\dfrac{\sqrt{75c^4d^9}}{\sqrt{3cd^3}}$ $5cd^3\sqrt{c}$

108. $\dfrac{\sqrt{20x^8y^3}}{\sqrt{5x^5y^{-1}}}$ $2xy^2\sqrt{x}$

Simplify each expression by rationalizing the denominator. (*Section 8.5, Objectives 2 and 3*)

109. $\dfrac{10}{\sqrt{5a}}$ $\dfrac{2\sqrt{5a}}{a}$

110. $\dfrac{6}{\sqrt{2b}}$ $\dfrac{3\sqrt{2b}}{b}$

111. $\dfrac{10}{\sqrt{7} + \sqrt{3}}$ $\dfrac{5(\sqrt{7} - \sqrt{3})}{2}$

112. $\dfrac{6}{\sqrt{11} - \sqrt{2}}$ $\dfrac{2(\sqrt{11} + \sqrt{2})}{3}$

Solve each radical equation. (*Section 8.6, Objective 1*)

113. $\sqrt{3r + 4} - 1 = 6$ $\{15\}$

114. $\sqrt{2t - 1} + 5 = 8$ $\{5\}$

115. $\sqrt{14 - 2x} + 3 = x$ $\{5\}$

116. $\sqrt{2x + 12} - 2 = x$ $\{2\}$

Perform each operation on the complex numbers and write each answer in standard form. (*Section 8.7, Objectives 2–4*)

117. $(1 - 4i) + (-3 + 11i)$ $-2 + 7i$

118. $(-8 + 9i) - (10 + 4i)$ $-18 + 5i$

119. $(3 - 2i)^2$ $5 - 12i$

120. $(1 - 2i)(1 + 2i)$ 5

121. $\dfrac{2 + 12i}{4i}$ $3 - \dfrac{1}{2}i$

122. $\dfrac{15 + 2i}{-3i}$ $-\dfrac{2}{3} + 5i$

123. $\dfrac{3 + 5i}{2 - i}$ $\dfrac{1}{5} + \dfrac{13}{5}i$

124. $\dfrac{8 + 5i}{1 + 3i}$ $\dfrac{23}{10} - \dfrac{19}{10}i$

Find each power of *i*. (*Section 8.7, Objective 5*)

125. i^{16} 1

126. i^6 -1

127. i^{-3} i

128. i^{-5} $-i$

CHAPTER 9

Quadratic Equations and Functions and Nonlinear Inequalities

Reflection

As you approach the end of this book and perhaps the end of the course, take a few moments to reflect on the past weeks. Chances are this is not the last math course required for your program of study (talk to your instructor or advisor to be sure). It is our hope that you will take not only math skills from this course but also skills that can help you succeed in other classes. It is important to understand the methods that helped you successfully understand and retain the material as well as those things that did not work as well for you. You should also leave the course with information about the concepts you understand well and the ones that you do not.

Question For Thought: What is working in this course that you want to continue doing as you take other classes? What are things you did not do in this course that you want to start doing as you take other classes? What are some things that did not work that you need to stop doing as you take other classes?

Chapter Outline

Section 9.1 Quadratic Functions and Their Graphs 636

Section 9.2 Solving Quadratic Equations Using the Square Root Property and Completing the Square 651

Section 9.3 Solving Quadratic Equations Using the Quadratic Formula 665

Section 9.4 Solving Equations Using Quadratic Methods 677

Section 9.5 More on Graphing Quadratic Functions 690

Section 9.6 Solving Polynomial and Rational Inequalities in One Variable 700

Coming Up...

In Section 9.5, we will learn how to find the maximum height of a ball that is tossed upward with an initial velocity of 30 ft/sec from a height of 6 ft, and whose height is modeled by the function $f(t) = -16t^2 + 30t + 6$.

"Everyone thinks of changing the world, but no one thinks of changing himself."

—Leo Tolstoy (Novelist)

SECTION 9.1 Quadratic Functions and Their Graphs

In Chapter 6, we learned how to solve quadratic equations by factoring. Factoring, however, does not solve every quadratic equation. In this chapter, we will present several additional methods for solving quadratic equations. We will also explore the graphs of quadratic functions and solve polynomial and rational inequalities.

OBJECTIVES

As a result of completing this section, you will be able to

1. Graph a function of the form $f(x) = x^2 + k$.
2. Graph a function of the form $f(x) = (x - h)^2$.
3. Graph a function of the form $f(x) = ax^2$.
4. Graph a function of the form $f(x) = a(x - h)^2 + k$.
5. Solve equations of the form $a(x - h)^2 + k = 0$ graphically.
6. Troubleshoot common errors.

In Chapter 3, we learned how to graph an equation in two variables by plotting points. In fact, we graphed a basic quadratic equation of the form $y = x^2$ and found its shape to be a parabola. Many structures in the real world are parabolic-shaped. In this section, we will examine the graphs of quadratic functions more closely.

Objective 1

Graph a function of the form $f(x) = x^2 + k$.

Graphs of Functions of the Form $f(x) = x^2 + k$

Definition: A **quadratic function** is a function of the form $f(x) = ax^2 + bx + c$, where a, b, and c are real numbers with $a \neq 0$. The graph of a quadratic function is a **parabola**.

Before we examine graphs of the form $f(x) = x^2 + k$, we will review the graph of $y = x^2$. A table of values is shown for reference.

x	$y = x^2$	(x, y)
-3	$(-3)^2 = 9$	$(-3, 9)$
-2	$(-2)^2 = 4$	$(-2, 4)$
-1	$(-1)^2 = 1$	$(-1, 1)$
$-\frac{1}{2}$	$\left(-\frac{1}{2}\right)^2 = \frac{1}{4}$	$\left(-\frac{1}{2}, \frac{1}{4}\right)$
0	$(0)^2 = 0$	$(0, 0)$
$\frac{1}{2}$	$\left(\frac{1}{2}\right)^2 = \frac{1}{4}$	$\left(\frac{1}{2}, \frac{1}{4}\right)$
1	$(1)^2 = 1$	$(1, 1)$
2	$(2)^2 = 4$	$(2, 4)$
3	$(3)^2 = 9$	$(3, 9)$

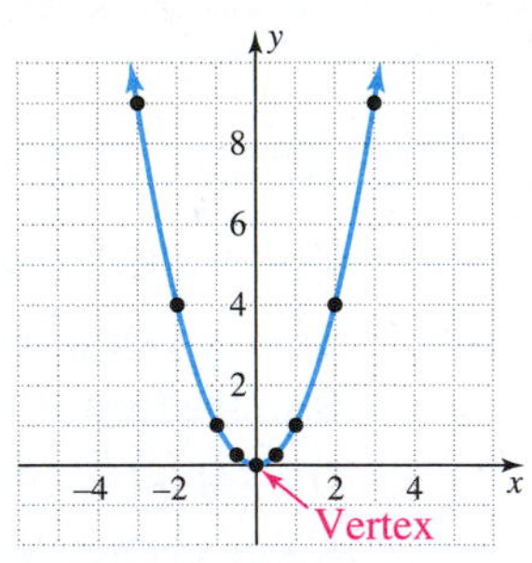

This graph represents a function because it passes the vertical line test. The domain of the function is $(-\infty, \infty)$ because the graph extends indefinitely in the left and right directions. The range of the function is $[0, \infty)$ because the graph begins from (0, 0) in the bottom-most direction and extends upward indefinitely. The point (0, 0) is the lowest point on the graph. The x-intercept and y-intercept of the graph are also (0, 0).

Parabolas have some important characteristics.

1. Parabolas are smooth curves.
2. Parabolas have a highest or lowest point, called the **vertex**.
3. Parabolas are symmetric about their vertex. If a parabola is folded vertically in half through its vertex, the left side and the right side of the parabola are mirror images of one another. This means that x-values with the same distance from the vertex have the same y-values.
4. The vertical line through the vertex is called the **axis of symmetry**. The axis of symmetry is of the form $x = h$, where h is the x-value of the vertex.
5. The direction that a parabola opens is determined by a, the coefficient of the x^2 term. If $a > 0$, the parabola opens up and if $a < 0$, the parabola opens down.

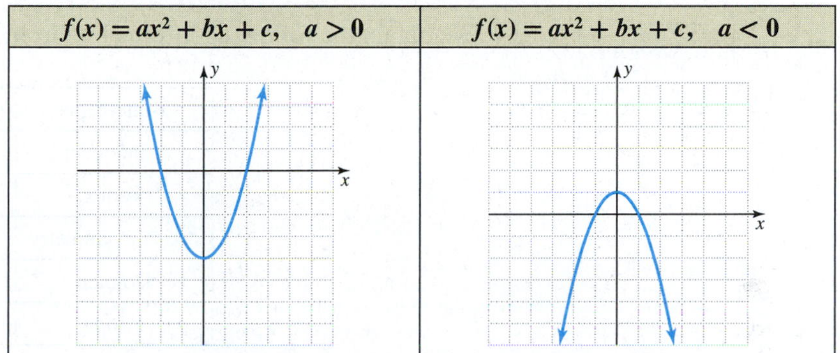

Key points on the graph of a parabola are

1. the x-intercepts, points where the graph crosses the x-axis and of the form $(x, 0)$,
2. the y-intercept, point where the graph crosses the y-axis and of the form $(0, y)$,
3. and the vertex.

The following graph is a parabola with its key points identified.

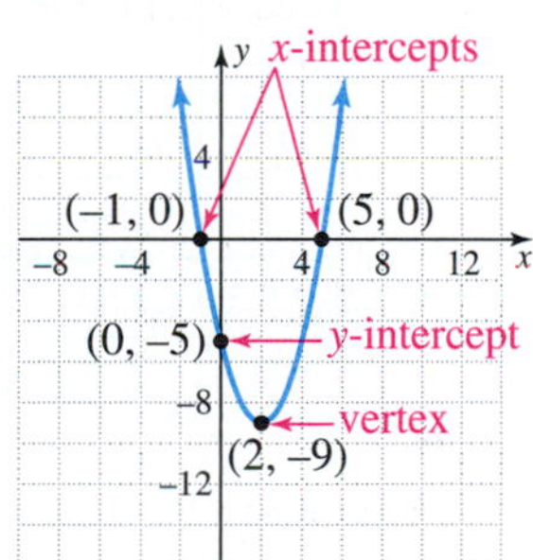

Now that we have an understanding of the basic shape and characteristics of a parabola, we will graph transformations of the basic parabola formed by adding or subtracting a number from the term x^2; that is, we will graph equations of the form $y = x^2 + k$.

Procedure: Graphing a Function of the Form $f(x) = x^2 + k$

Step 1: Make a table of values.
Step 2: Plot the points.
Step 3: Connect the points with a smooth curve.

Objective 1 Examples **Create a table of values to graph each function. Identify the x-intercepts, y-intercept, vertex, the axis of symmetry, domain, and range. Explain how the graph of the function relates to the graph of $f(x) = x^2$.**

1a. $f(x) = x^2 - 4$ **1b.** $f(x) = x^2 + 1$

Solutions **1a.** A table of values and the graph of $f(x) = x^2 - 4$ are shown. The graph of $f(x) = x^2$ is shown for comparison.

x	$f(x) = x^2 - 4$	(x, y)
-2	$(-2)^2 - 4 = 4 - 4 = 0$	$(-2, 0)$
-1	$(-1)^2 - 4 = 1 - 4 = -3$	$(-1, -3)$
0	$(0)^2 - 4 = 0 - 4 = -4$	$(0, -4)$
1	$(1)^2 - 4 = 1 - 4 = -3$	$(1, -3)$
2	$(2)^2 - 4 = 4 - 4 = 0$	$(2, 0)$

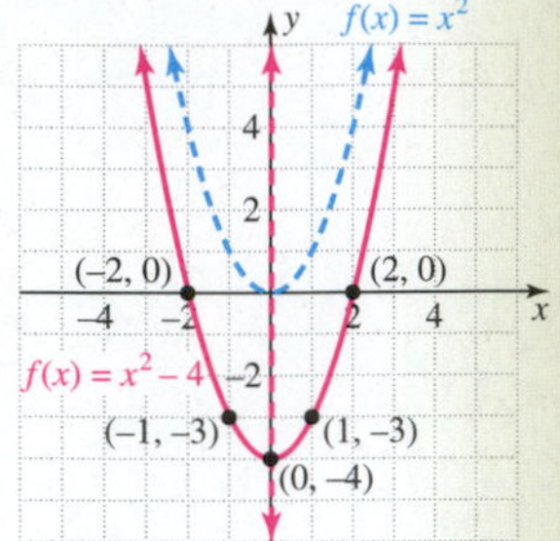

We obtain the following information from the graph.

x-intercepts	$(-2, 0)$ and $(2, 0)$
y-intercept	$(0, -4)$
Vertex	$(0, -4)$
Axis of symmetry	$x = 0$
Domain	$(-\infty, \infty)$
Range	$[-4, \infty)$

The graph of $f(x) = x^2 - 4$ is the graph of $f(x) = x^2$ shifted *down* 4 units.

1b. A table of values and the graph of $f(x) = x^2 + 1$ are shown. The graph of $f(x) = x^2$ is shown for comparison.

x	$f(x) = x^2 + 1$	(x, y)
-2	$(-2)^2 + 1 = 4 + 1 = 5$	$(-2, 5)$
-1	$(-1)^2 + 1 = 1 + 1 = 2$	$(-1, 2)$
0	$(0)^2 + 1 = 0 + 1 = 1$	$(0, 1)$
1	$(1)^2 + 1 = 1 + 1 = 2$	$(1, 2)$
2	$(2)^2 + 1 = 4 + 1 = 5$	$(2, 5)$

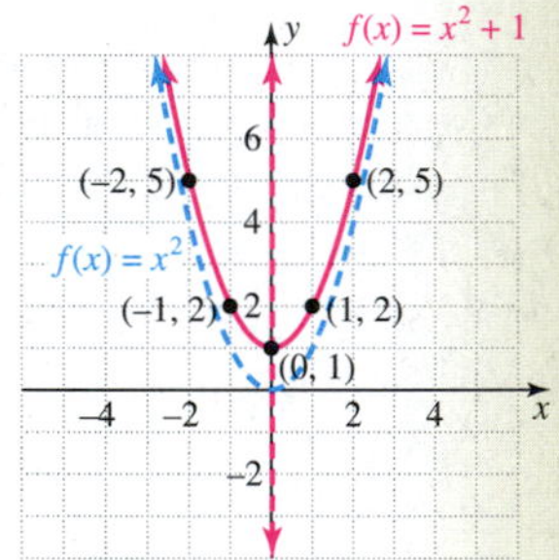

We obtain the following information from the graph.

x-intercepts	None
y-intercept	$(0, 1)$
Vertex	$(0, 1)$
Axis of symmetry	$x = 0$
Domain	$(-\infty, \infty)$
Range	$[1, \infty)$

The graph of $f(x) = x^2 + 1$ is the graph of $f(x) = x^2$ shifted *up* 1 unit.

Student Check 1 Create a table of values to graph each function. Identify the x-intercepts, y-intercept, vertex, the axis of symmetry, domain, and range. Explain how the graph of the function relates to the graph of $f(x) = x^2$.

a. $f(x) = x^2 + 2$ **b.** $f(x) = x^2 - 1$

Fact: Important Features of a Parabola of the Form $f(x) = x^2 + k$

The graph of $f(x) = x^2 + k$ is the graph of $f(x) = x^2$ shifted up or down.

- If $k > 0$, the graph of $f(x) = x^2 + k$ is the graph of $f(x) = x^2$ shifted up k units.
- If $k < 0$, the graph of $f(x) = x^2 + k$ is the graph of $f(x) = x^2$ shifted down $|k|$ units.
- The vertex is $(0, k)$ and the axis of symmetry is the vertical line $x = 0$, or the y-axis.

Graphs of Functions of the Form $f(x) = (x - h)^2$

Objective 2 Graph a function of the form $f(x) = (x - h)^2$.

Now we examine what happens to the graph of $f(x) = x^2$ when we add or subtract a number from x before squaring.

Procedure: Graphing a Function of the Form $f(x) = (x - h)^2$

Step 1: Make a table of values.
Step 2: Plot the points.
Step 3: Connect the points with a smooth curve.

Objective 2 Examples **Create a table of values to graph each function. Identify the x-intercepts, y-intercept, vertex, the axis of symmetry, domain, and range. Explain how the graph of the function relates to the graph of $f(x) = x^2$.**

2a. $f(x) = (x + 1)^2$ **2b.** $f(x) = (x - 2)^2$

Solutions **2a.** A table of values and the graph of $f(x) = (x + 1)^2$ are shown. The graph of $f(x) = x^2$ is shown for comparison.

x	$f(x) = (x + 1)^2$	(x, y)
-3	$(-3 + 1)^2 = (-2)^2 = 4$	$(-3, 4)$
-2	$(-2 + 1)^2 = (-1)^2 = 1$	$(-2, 1)$
-1	$(-1 + 1)^2 = (0)^2 = 0$	$(-1, 0)$
0	$(0 + 1)^2 = (1)^2 = 1$	$(0, 1)$
1	$(1 + 1)^2 = (2)^2 = 4$	$(1, 4)$

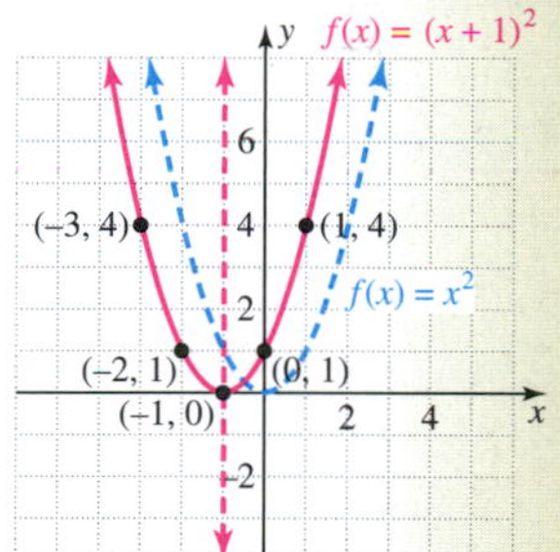

We obtain the following information from the graph.

x-intercept	$(-1, 0)$
y-intercept	$(0, 1)$
Vertex	$(-1, 0)$
Axis of symmetry	$x = -1$
Domain	$(-\infty, \infty)$
Range	$[0, \infty)$

The graph of $f(x) = (x + 1)^2$ is the graph of $f(x) = x^2$ shifted *left* 1 unit.

2b. A table of values and the graph of $f(x) = (x - 2)^2$ are shown. The graph of $f(x) = x^2$ is shown for comparison.

x	$f(x) = (x - 2)^2$	(x, y)
-1	$(-1 - 2)^2 = (-3)^2 = 9$	$(-1, 9)$
0	$(0 - 2)^2 = (-2)^2 = 4$	$(0, 4)$
1	$(1 - 2)^2 = (-1)^2 = 1$	$(1, 1)$
2	$(2 - 2)^2 = (0)^2 = 0$	$(2, 0)$
3	$(3 - 2)^2 = (1)^2 = 1$	$(3, 1)$
4	$(4 - 2)^2 = (2)^2 = 4$	$(4, 4)$

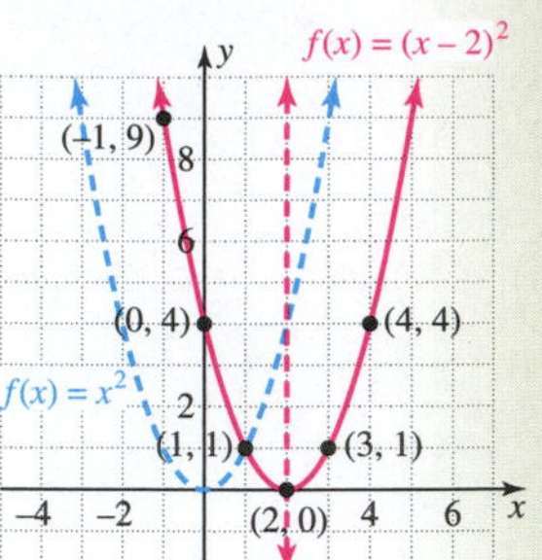

We obtain the following information from the graph.

x-intercept	$(2, 0)$
y-intercept	$(0, 4)$
Vertex	$(2, 0)$
Axis of symmetry	$x = 2$
Domain	$(-\infty, \infty)$
Range	$[0, \infty)$

The graph of $f(x) = (x - 2)^2$ is the graph of $f(x) = x^2$ shifted *right* 2 units.

Student Check 2 Create a table of values to graph each function. Identify the x-intercepts, y-intercept, vertex, the axis of symmetry, domain, and range. Explain how the graph of the function relates to the graph of $f(x) = x^2$.

a. $f(x) = (x + 2)^2$ **b.** $f(x) = (x - 1)^2$

Fact: Important Features of Parabolas of the Form $f(x) = (x - h)^2$

The graph of $f(x) = (x - h)^2$ is the graph of $f(x) = x^2$ shifted left or right.

- If $h > 0$, the graph of $f(x) = (x - h)^2$ is the graph of $f(x) = x^2$ shifted right h units.
- If $h < 0$, the graph of $f(x) = (x - h)^2$ is the graph of $f(x) = x^2$ shifted left $|h|$ units.
- The vertex is $(h, 0)$ and the axis of symmetry is the vertical line $x = h$.

Graphs of Functions of the Form $f(x) = ax^2$

Objective 3 Graph a function of the form $f(x) = ax^2$.

The previous two objectives illustrate that the values h and k, from the functions $f(x) = (x - h)^2$ and $f(x) = x^2 + k$, shift the graph of $f(x) = x^2$ left or right h units and up or down k units. We now investigate the effects of multiplying the function, $f(x) = x^2$, by a constant.

Objective 3 Examples **Create a table of values to graph each function. Identify the x-intercepts, y-intercept, vertex, the axis of symmetry, domain, and range. Explain how the graph of the function relates to the graph of $f(x) = x^2$.**

3a. $f(x) = 2x^2$ **3b.** $f(x) = \frac{1}{2}x^2$ **3c.** $f(x) = -x^2$

Solutions **3a.** A table of values and the graph of $f(x) = 2x^2$ are shown. The graph of $f(x) = x^2$ is shown for comparison.

x	$f(x) = 2x^2$	(x, y)
-2	$2(-2)^2 = 2(4) = 8$	$(-2, 8)$
-1	$2(-1)^2 = 2(1) = 2$	$(-1, 2)$
0	$2(0)^2 = 2(0) = 0$	$(0, 0)$
1	$2(1)^2 = 2(1) = 2$	$(1, 2)$
2	$2(2)^2 = 2(4) = 8$	$(2, 8)$

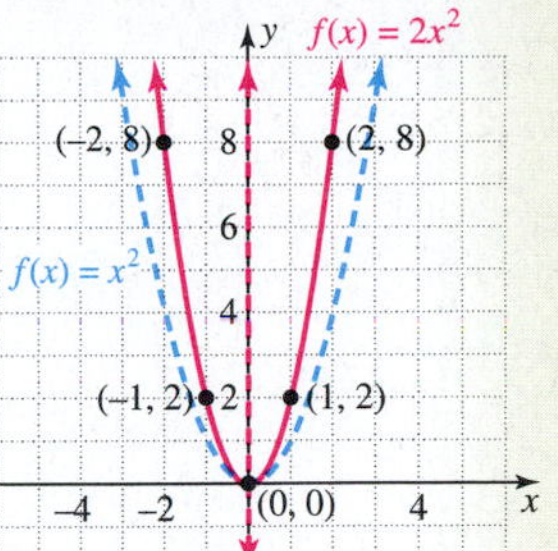

We obtain the following information from the graph.

x-intercept	(0, 0)
y-intercept	(0, 0)
Vertex	(0, 0)
Axis of symmetry	$x = 0$
Domain	$(-\infty, \infty)$
Range	$[0, \infty)$

The graph of $f(x) = 2x^2$ is more *narrow* than the graph of $f(x) = x^2$. We say that $f(x) = 2x^2$ is the graph of $f(x) = x^2$ *vertically stretched* by a factor of 2. The y-values of $f(x) = 2x^2$ are twice as large as the y-values of $f(x) = x^2$, which has the effect of pulling the graph of $f(x) = x^2$ upward; that is, stretching it vertically.

3b. A table of values and the graph of $f(x) = \frac{1}{2}x^2$ are shown. The graph of $f(x) = x^2$ is shown for comparison.

x	$f(x) = \frac{1}{2}x^2$	(x, y)
-2	$\frac{1}{2}(-2)^2 = \frac{1}{2}(4) = 2$	$(-2, 2)$
-1	$\frac{1}{2}(-1)^2 = \frac{1}{2}(1) = \frac{1}{2}$	$\left(-1, \frac{1}{2}\right)$
0	$\frac{1}{2}(0)^2 = \frac{1}{2}(0) = 0$	$(0, 0)$
1	$\frac{1}{2}(1)^2 = \frac{1}{2}(1) = \frac{1}{2}$	$\left(1, \frac{1}{2}\right)$
2	$\frac{1}{2}(2)^2 = \frac{1}{2}(4) = 2$	$(2, 2)$

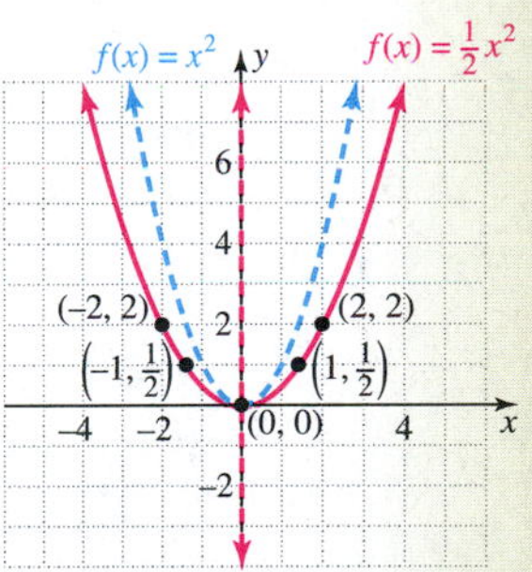

We obtain the following information from the graph.

x-intercept	(0, 0)
y-intercept	(0, 0)
Vertex	(0, 0)
Axis of symmetry	$x = 0$
Domain	$(-\infty, \infty)$
Range	$[0, \infty)$

The graph of $f(x) = \frac{1}{2}x^2$ is *wider* than the graph of $f(x) = x^2$. The graph of $f(x) = \frac{1}{2}x^2$ is the graph of $f(x) = x^2$ *vertically shrunk* by a factor of $\frac{1}{2}$. The y-values of $f(x) = \frac{1}{2}x^2$ are half as large as the y-values of $f(x) = x^2$. The effect is that the graph is being pushed downward, that is, vertically shrunk.

3c. A table of values and the graph of $f(x) = -x^2$ are shown. The graph of $f(x) = x^2$ is shown for comparison.

x	$f(x) = -x^2$	(x, y)
-2	$-(-2)^2 = -(4) = -4$	$(-2, -4)$
-1	$-(-1)^2 = -(1) = -1$	$(-1, -1)$
0	$-(0)^2 = -(0) = 0$	$(0, 0)$
1	$-(1)^2 = -(1) = -1$	$(1, -1)$
2	$-(2)^2 = -(4) = -4$	$(2, -4)$

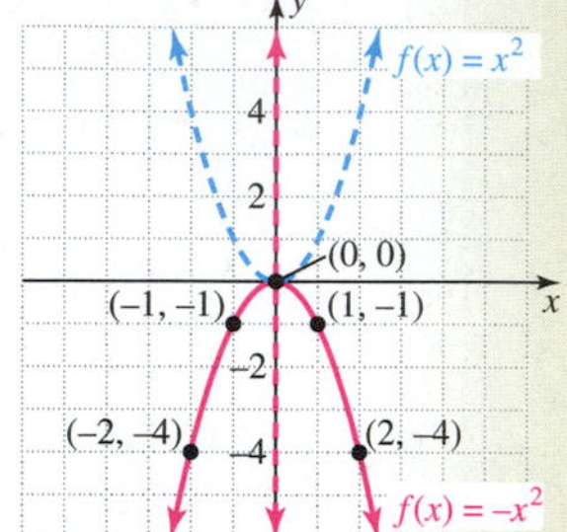

We obtain the following information from the graph.

x-intercept	$(0, 0)$
y-intercept	$(0, 0)$
Vertex	$(0, 0)$
Axis of symmetry	$x = 0$
Domain	$(-\infty, \infty)$
Range	$(-\infty, 0]$

The graph of $f(x) = -x^2$ is the graph of $y = x^2$ *reflected over the x-axis.*

Student Check 3 Create a table of values to graph each function. Identify the x-intercepts, y-intercept, vertex, the axis of symmetry, domain, and range. Explain how the graph of the function relates to the graph of $f(x) = x^2$.

a. $y = 3x^2$ **b.** $y = \frac{1}{3}x^2$ **c.** $y = -2x^2$

Fact: Important Features of a Parabola of the Form $f(x) = ax^2$

The graph of $f(x) = ax^2$ is the graph of $f(x) = x^2$ stretched, shrunk, and/or reflected over the x-axis.

- *If $a > 1$, the graph of $f(x) = ax^2$ is the graph of $f(x) = x^2$ stretched vertically by a factor of a.*
- *If $0 < a < 1$, the graph of $f(x) = ax^2$ is the graph of $f(x) = x^2$ shrunk vertically by a factor of a.*
- *If $a < 0$, the graph of $y = ax^2$ is also reflected over the x-axis.*
- *The vertex is $(0, 0)$ and the axis of symmetry is the vertical line $x = 0$.*
- *If $a > 0$, the parabola opens up. If $a < 0$, the parabola opens down.*

Graphs of Functions of the Form $f(x) = a(x - h)^2 + k$

Objective 4

Graph a function of the form $f(x) = a(x - h)^2 + k$.

We now incorporate what we know about the values of a, h, and k and the graphs of functions of the form $f(x) = x^2 + k$, $f(x) = (x - h)^2$, and $f(x) = ax^2$ to graph functions of the form $f(x) = a(x - h)^2 + k$. The following table summarizes functions from Examples 1–3 and the effects of a, h, and k.

Function	Vertex	Direction	Function in the form $f(x) = a(x - h)^2 + k$	Values of a, h, and k
$f(x) = x^2 - 4$	$(0, -4)$	opens up	$f(x) = 1(x - 0)^2 - 4$	$a = 1, (h, k) = (0, -4)$
$f(x) = x^2 + 1$	$(0, 1)$	opens up	$f(x) = 1(x - 0)^2 + 1$	$a = 1, (h, k) = (0, 1)$
$f(x) = (x + 1)^2$	$(-1, 0)$	opens up	$f(x) = 1[x - (-1)]^2 + 0$	$a = 1, (h, k) = (-1, 0)$
$f(x) = (x - 2)^2$	$(2, 0)$	opens up	$f(x) = 1(x - 2)^2 + 0$	$a = 1, (h, k) = (2, 0)$
$f(x) = 2x^2$	$(0, 0)$	opens up	$f(x) = 2(x - 0)^2 + 0$	$a = 2, (h, k) = (0, 0)$
$f(x) = \frac{1}{2}x^2$	$(0, 0)$	opens up	$f(x) = \frac{1}{2}(x - 0)^2 + 0$	$a = \frac{1}{2}, (h, k) = (0, 0)$
$f(x) = -x^2$	$(0, 0)$	opens down	$f(x) = -1(x - 0)^2 + 0$	$a = -1, (h, k) = (0, 0)$

Property: Vertex Form of a Quadratic Function $f(x) = a(x - h)^2 + k$

The graph of $f(x) = a(x - h)^2 + k$ is the graph of $f(x) = x^2$ where

- The vertex of the graph is the point (h, k).
- The axis of symmetry is the vertical line through the vertex, $x = h$.
- The parabola opens up if $a > 0$ and opens down if $a < 0$.

$f(x) = a(x - h)^2 + k \quad (a > 0)$

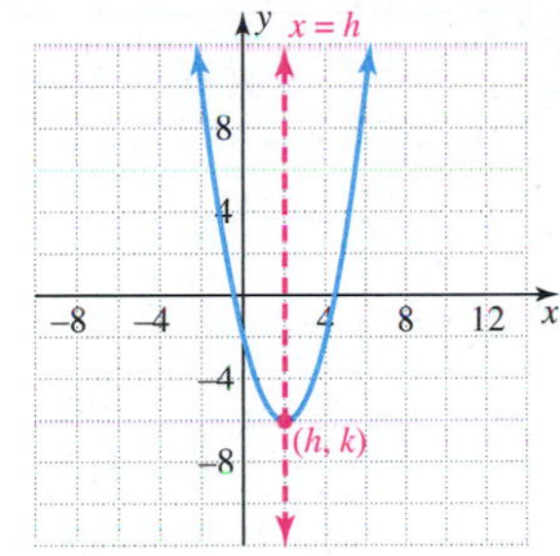

$f(x) = a(x - h)^2 + k \quad (a < 0)$

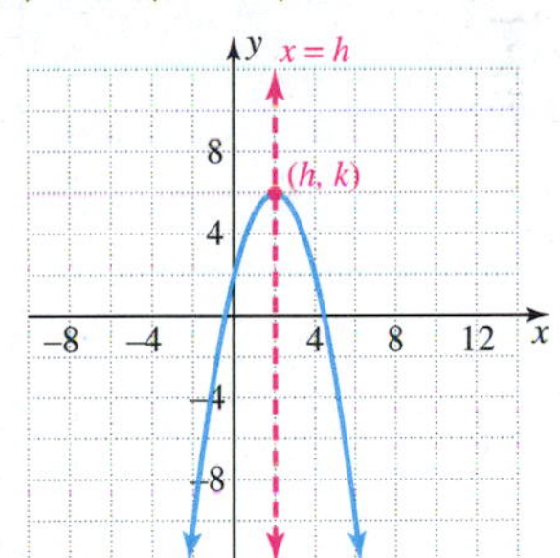

Procedure: Graphing a Function of the Form $f(x) = a(x - h)^2 + k$

Step 1: Identify the vertex (h, k).

Step 2: Plot two additional points, one on each side of the vertex. Find the y-intercept if one of the two additional points does not provide this information.

Step 3: Connect the points with a parabola.

Objective 4 Examples

Graph each function by plotting the vertex and two additional points. Identify the x-intercepts, y-intercept, vertex, the axis of symmetry, domain, and range. Explain how the graph of the function relates to the graph of $y = x^2$.

4a. $f(x) = (x - 1)^2 - 4$ **4b.** $f(x) = (x + 3)^2 + 1$ **4c.** $f(x) = -2(x - 2)^2 + 1$

Solutions

4a. The vertex of $f(x) = 1(x - 1)^2 - 4$ is $(h, k) = (1, -4)$. The graph opens up since $a = 1 > 0$.

	x	$y = (x - 1)^2 - 4$	(x, y)
x-value to the left of the vertex	-1	$(-1 - 1)^2 - 4 = 0$	$(-1, 0)$
vertex	1	-4	$(1, -4)$
x-value to the right of the vertex	3	$(3 - 1)^2 - 4 = 0$	$(3, 0)$
y-intercept	0	$(0 - 1)^2 - 4 = -3$	$(0, -3)$

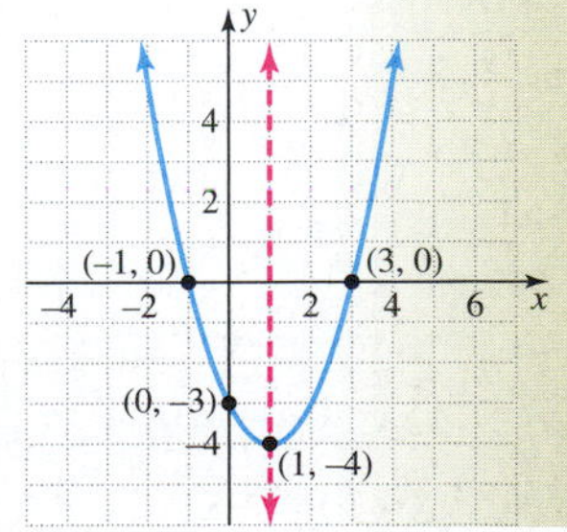

We obtain the following information from the graph.

x-intercepts	$(-1, 0)$ and $(3, 0)$
y-intercept	$(0, -3)$
Vertex	$(1, -4)$
Axis of symmetry	$x = 1$
Domain	$(-\infty, \infty)$
Range	$[-4, \infty)$

The graph of $f(x) = (x - 1)^2 - 4$ is the graph of $f(x) = x^2$ shifted right 1 unit and down 4 units.

4b. The vertex of $f(x) = 1[x - (-3)]^2 + 1$ is $(h, k) = (-3, 1)$. The graph opens up since $a = 1 > 0$.

	x	$y = (x + 3)^2 + 1$	(x, y)
x-value to the left of the vertex	-4	$(-4 + 3)^2 + 1 = 2$	$(-4, 2)$
vertex	-3	1	$(-3, 1)$
x-value to the right of the vertex	-2	$(-2 + 3)^2 + 1 = 2$	$(-2, 2)$
y-intercept	0	$(0 + 3)^2 + 1 = 10$	$(0, 10)$

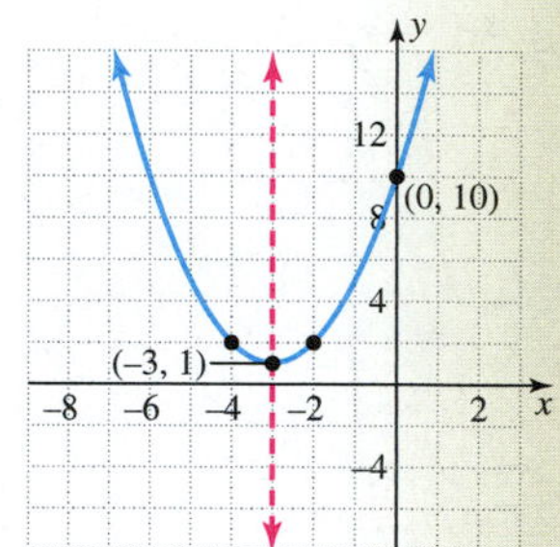

We obtain the following information from the graph.

x-intercepts	none
y-intercept	$(0, 10)$
Vertex	$(-3, 1)$
Axis of symmetry	$x = -3$
Domain	$(-\infty, \infty)$
Range	$[1, \infty)$

The graph of $f(x) = (x + 3)^2 + 1$ is the graph of $f(x) = x^2$ shifted left 3 units and up 1 unit.

4c. The vertex of $f(x) = -2(x - 2)^2 + 1$ is $(h, k) = (2, 1)$. The graph opens down since $a = -2 < 0$.

	x	$y = -2(x - 2)^2 + 1$	(x, y)
x-value to the left of the vertex	0	$-2(0 - 2)^2 + 1 = -7$	$(0, -7)$
vertex	2	1	$(2, 1)$
x-value to the right of the vertex	4	$-2(4 - 2)^2 + 1 = -7$	$(4, -7)$

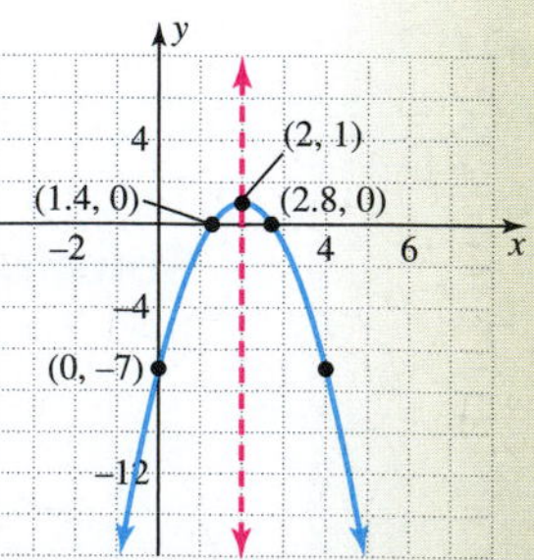

We find the following information from the graph. The x-intercepts are approximated from the graph. We will learn how to find these precisely later in this chapter.

x-intercepts	Approximately (1.4, 0) and (2.8, 0)
y-intercept	$(0, -7)$
Vertex	$(2, 1)$
Axis of symmetry	$x = 2$
Domain	$(-\infty, \infty)$
Range	$(-\infty, 1]$

The graph of $f(x) = -2(x-2)^2 + 1$ is the graph of $f(x) = x^2$ shifted right 2 units, reflected over the x-axis and stretched by a factor of 2, and shifted up 1 unit.

Student Check 4 Graph each function by plotting the vertex and two additional points. Identify the x-intercepts, y-intercept, vertex, the axis of symmetry, domain, and range. Explain how the graph of the function relates to the graph of $f(x) = x^2$.

a. $f(x) = (x-2)^2 - 9$ **b.** $f(x) = (x+4)^2 + 3$ **c.** $f(x) = -(x-3)^2 + 4$

Note: *While we obtain the graphs using a table of values, it is important to note that there is an order of the transformations. We must first translate horizontally, reflect over the x-axis, stretch or shrink vertically, and finally translate vertically.*

Objective 5 Solve equations of the form $a(x-h)^2 + k = 0$ graphically.

Solve Quadratic Equations Using a Graph

In Sections 9.2 and 9.3, we will learn several methods to solve quadratic equations algebraically. We can also solve quadratic equations graphically. Being able to visualize the solutions will assist us in later sections.

Property: The Graph of $f(x) = a(x-h)^2 + k$ and Solutions of $a(x-h)^2 + k = 0$

If $(r, 0)$ and/or $(s, 0)$ are points on the graph of $y = a(x-h)^2 + k$, then r and s are solutions of the equation $a(x-h)^2 + k = 0$.

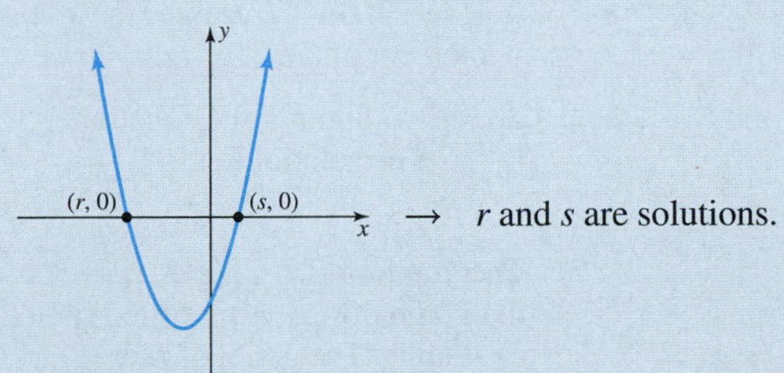

So, the solutions of the equation $a(x-h)^2 + k = 0$ correspond to the x-values whose y-values are zero. Graphically, these are the x-intercepts of the function $f(x) = a(x-h)^2 + k$. There are three possible cases.

- If the graph of the parabola crosses the x-axis two times, $a(x-h)^2 + k = 0$ has two real solutions.
- If the graph of the parabola crosses the x-axis one time, $a(x-h)^2 + k = 0$ has one real solution.

INSTRUCTOR NOTE: Remind students that in Chapter 3, we learned that if $f(r) = 0$ and $f(s) = 0$, then r and s are zeros of the function $f(x)$.

- If the graph of the parabola doesn't cross the x-axis, $a(x - h)^2 + k = 0$ has no real solutions.

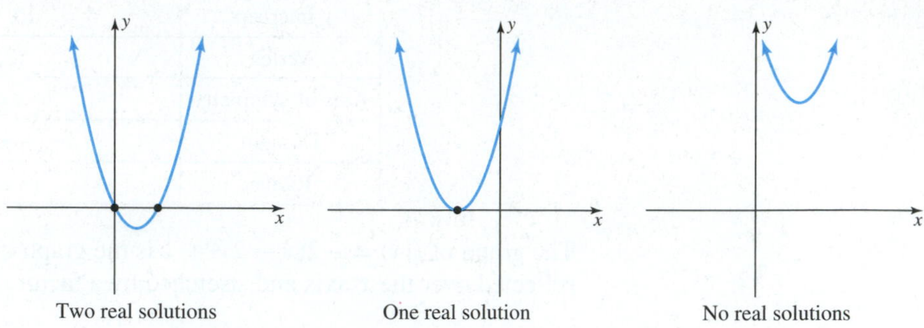

Procedure: Solving an Equation of the Form $a(x - h)^2 + k = 0$ Graphically

Step 1: Graph the function $f(x) = a(x - h)^2 + k$.
Step 2: State the x-intercepts of the function.
Step 3: The x-value(s) of the x-intercepts are the solutions of the equation.

Objective 5 Examples

Graph each function and then use the graph to solve the equation $f(x) = 0$.

5a. $f(x) = x^2 - 4$ **5b.** $f(x) = -3(x + 4)^2$ **5c.** $f(x) = (x - 2)^2 + 4$

Solutions

5a. The graph of $f(x) = x^2 - 4$ is the same as the graph of $f(x) = x^2$ shifted down 4 units. The graph of $f(x) = x^2 - 4$ crosses the x-axis twice at the x-intercepts of $(-2, 0)$ and $(2, 0)$.

So, the solutions of the equation $x^2 - 4 = 0$ are -2 and 2. The solution set is $\{\pm 2\}$.

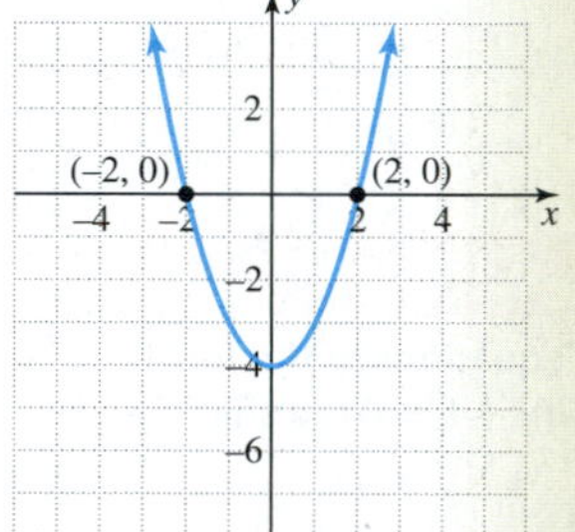

5b. The graph of $f(x) = -3(x + 4)^2$ is the same as the graph of $f(x) = x^2$ shifted left 4 units but opens down since $a = -3$. It is also more narrow than the graph of $f(x) = x^2$. The graph of $f(x) = -3(x + 4)^2$ crosses the x-axis once at the x-intercept of $(-4, 0)$.

So, the solution of the equation $-3(x + 4)^2 = 0$ is -4. The solution set is $\{-4\}$.

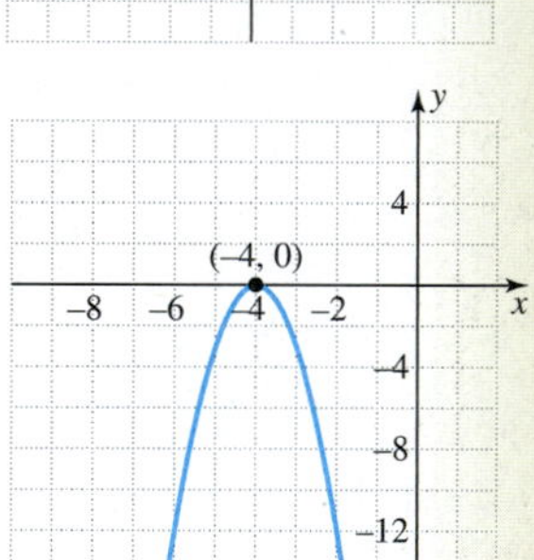

5c. The graph of $f(x) = (x - 2)^2 + 4$ is the same as the graph of $f(x) = x^2$ shifted right 2 units and up 4 units. The graph of $f(x) = (x - 2)^2 + 4$ does not cross the x-axis.

There are no values of x for which $y = 0$. So, there are no real solutions of this equation.

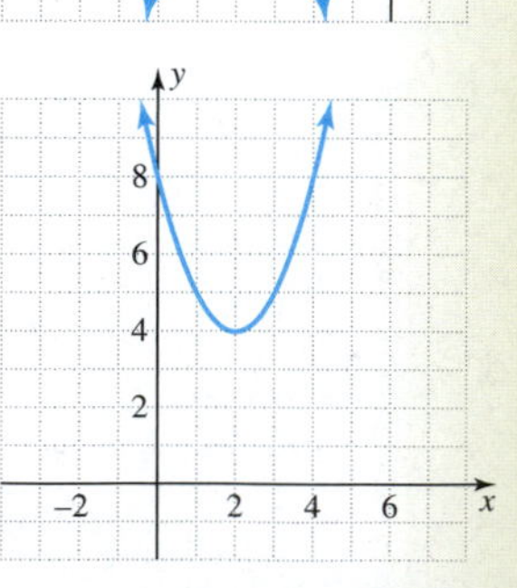

Student Check 5 Graph each function and then use the graph to solve the equation $f(x) = 0$.

a. $f(x) = (x - 3)^2 - 4$ **b.** $f(x) = 2(x + 1)^2$ **c.** $f(x) = -x^2 - 2$

Objective 6 ▶

Troubleshoot common errors.

Troubleshooting Common Errors

A common error associated with graphing quadratic functions is shown.

Objective 6 Example

A problem and an incorrect solution are given. Provide the correct solution and an explanation of the error.

Find the vertex of $y = (x + 3)^2$.

Incorrect Solution	Correct Solution and Explanation
The vertex is (3, 0).	The vertex is the point (h, k) when the function is written in the form $y = a(x - h)^2 + k$. $y = (x + 3)^2 = [x - (-3)]^2 + 0$ So, the vertex is $(-3, 0)$.

ANSWERS TO STUDENT CHECKS

Student Check 1 **a.** none; (0, 2); (0, 2); $x = 0$; $(-\infty, \infty)$; $[2, \infty)$; It is the graph of $f(x) = x^2$ shifted up 2 units.

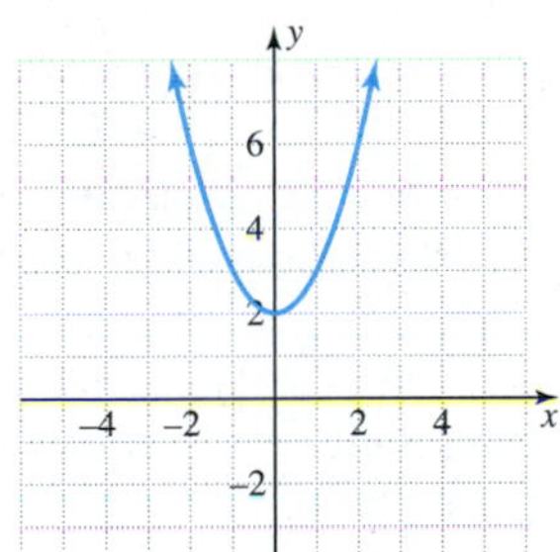

b. $(-1, 0)$ and $(1, 0)$; $(0, -1)$; $(0, -1)$; $x = 0$; $(-\infty, \infty)$; $[-1, \infty)$; It is the graph of $f(x) = x^2$ shifted down 1 unit.

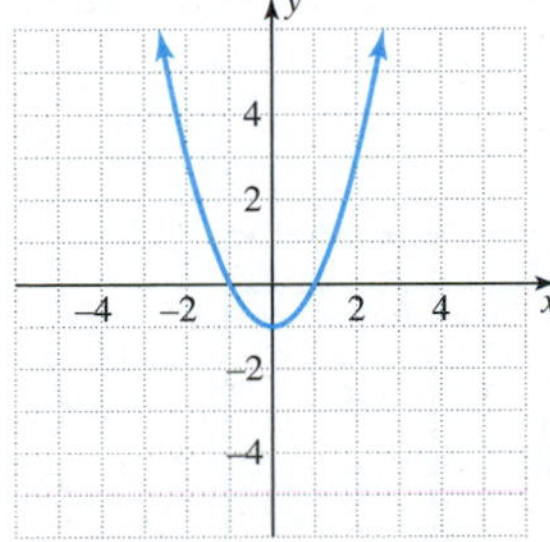

Student Check 2 **a.** $(-2, 0)$; $(0, 4)$; $(-2, 0)$; $x = -2$; $(-\infty, \infty)$; $[0, \infty)$; It is the graph of $f(x) = x^2$ shifted left 2 units.

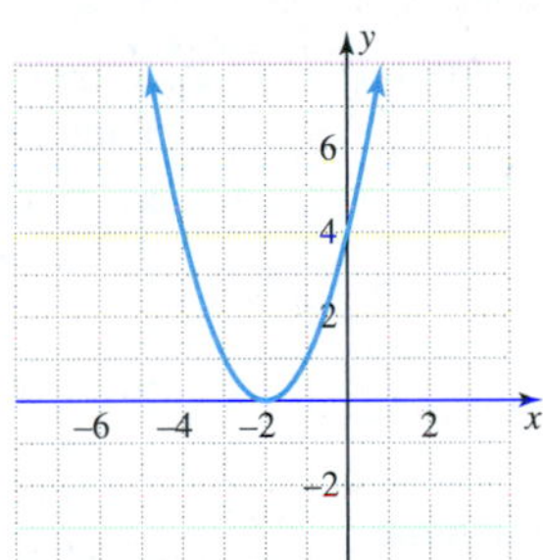

b. $(1, 0)$; $(0, 1)$; $(1, 0)$; $x = 1$; $(-\infty, \infty)$; $[0, \infty)$; It is the graph of $f(x) = x^2$ shifted right 1 unit.

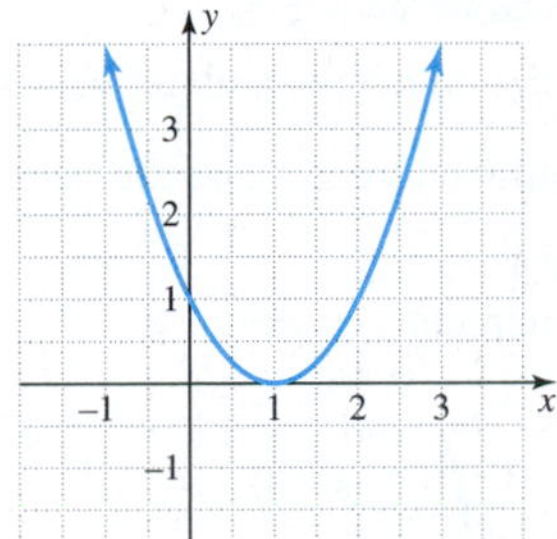

Student Check 3 **a.** (0, 0); (0, 0); (0, 0); $x = 0$; $(-\infty, \infty)$; $[0, \infty)$; It is the graph of $f(x) = x^2$ vertically stretched by a factor of 3.

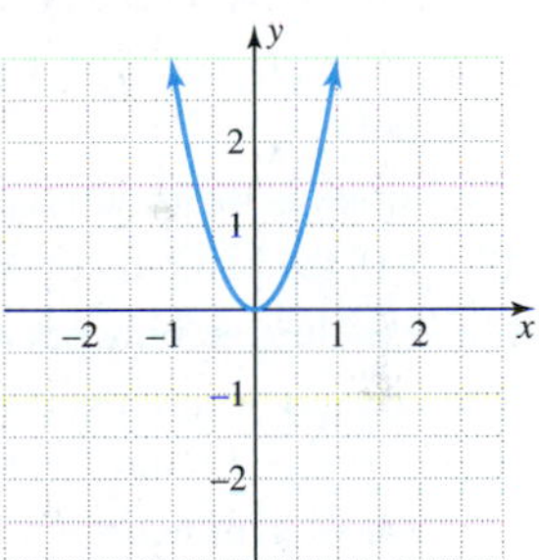

b. (0, 0); (0, 0); (0, 0); $x = 0$; $(-\infty, \infty)$; $[0, \infty)$; It is the graph of $f(x) = x^2$ vertically shrunk by a factor of $\frac{1}{3}$.

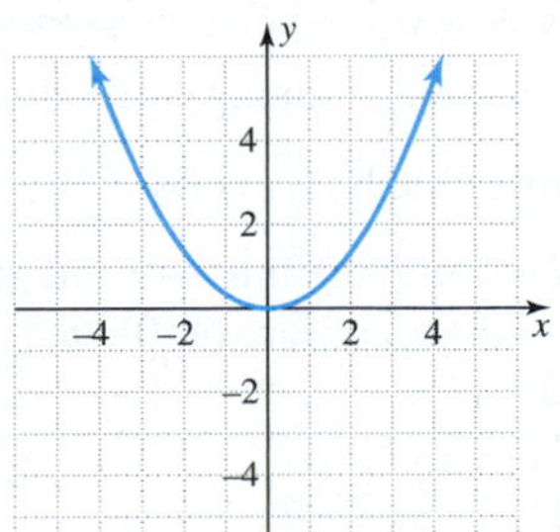

c. (0, 0); (0, 0); (0, 0), $x = 0$; $(-\infty, \infty)$; $(-\infty, 0]$; It is the graph of $f(x) = x^2$ vertically stretched by a factor of 2 and reflected over the x-axis.

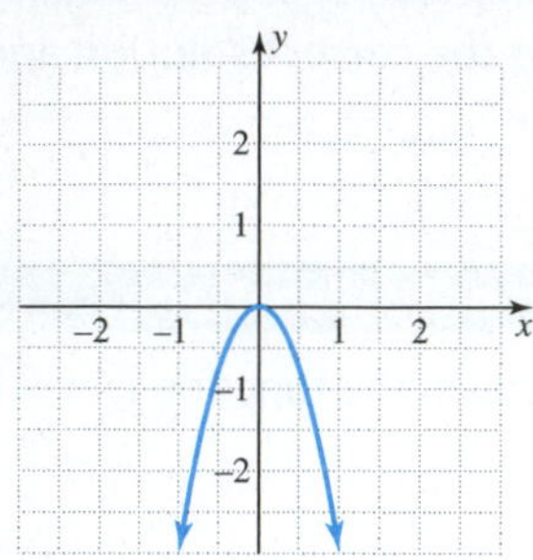

Student Check 4 **a.** (−1, 0) and (5, 0); (0, −5); (2, −9); $x = 2$; $(-\infty, \infty)$; $[-9, \infty)$; It is the graph of $f(x) = x^2$ shifted right 2 units and down 9 units.

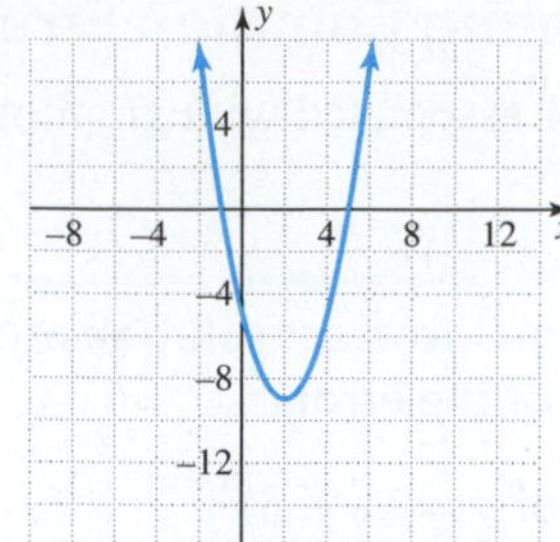

b. None; (0, 19); (−4, 3); $x = -4$; $(-\infty, \infty)$; $[-4, \infty)$; It is the graph of $f(x) = x^2$ shifted left 4 units and up 3 units.

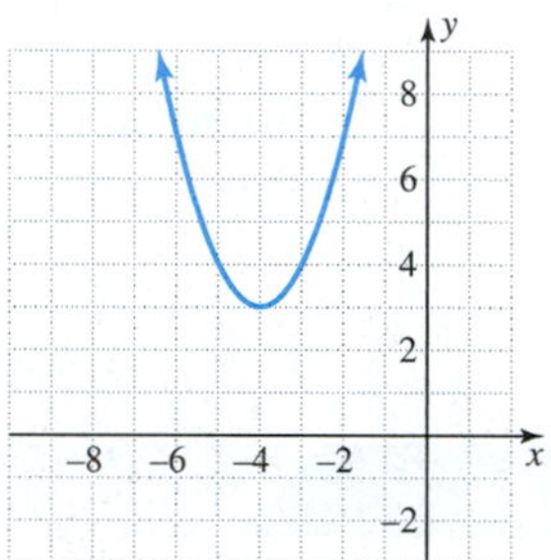

c. (1, 0) and (5, 0); (0, −5); (3, 4); $x = 3$; $(-\infty, \infty)$; $(-\infty, 4]$; It is the graph of $y = x^2$ shifted right 3 units up 4 units, and reflected over the x-axis.

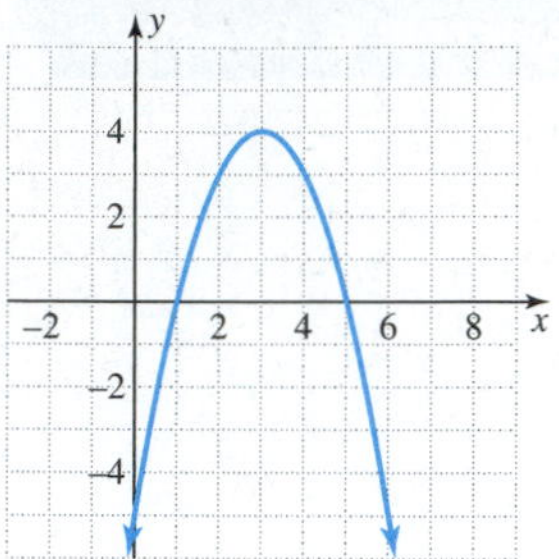

Student Check 5 **a.** {1, 5} **b.** {−1} **c.** no real solutions

SUMMARY OF KEY CONCEPTS

1. The graph of a function of the form $f(x) = a(x - h)^2 + k$ is a parabola.
   a. The vertex is the ordered pair (h, k).
   b. If $a > 0$, the parabola opens upward.
   c. If $a < 0$, the parabola opens downward.
   d. The domain of the function is $(-\infty, \infty)$ and the range is $[k, \infty)$ if $a > 0$ and $(-\infty, k]$ if $a < 0$.
   e. The axis of symmetry is the vertical line through the vertex, $x = h$.
2. Solutions of the equation $a(x - h)^2 + k = 0$ are the x-intercepts of the graph of $f(x) = a(x - h)^2 + k$.
   a. If the graph of $f(x) = a(x - h)^2 + k$ crosses the x-axis two times, there are two real solutions of the equation $a(x - h)^2 + k = 0$.
   b. If the graph of $f(x) = a(x - h)^2 + k$ crosses the x-axis one time, there is one real solution of the equation $a(x - h)^2 + k = 0$.
   c. If the graph of $f(x) = a(x - h)^2 + k$ does not cross the x-axis, there are no real solutions of the equation $a(x - h)^2 + k = 0$.

GRAPHING CALCULATOR SKILLS

We can use the graphing calculator to find the x- and y-intercepts and the vertex of a parabola.

Example: Graph $f(x) = (x - 3)^2 - 4$. Find the x- and y-intercepts and the vertex.

Solution: Enter the function and graph. From the table, we see the x-intercepts are (1, 0), and (5, 0) and the y-intercept is (0, 5).

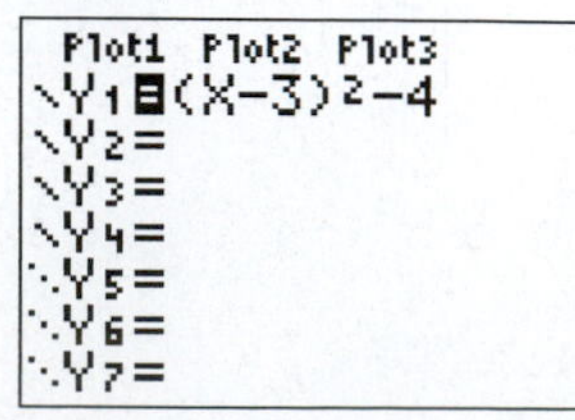

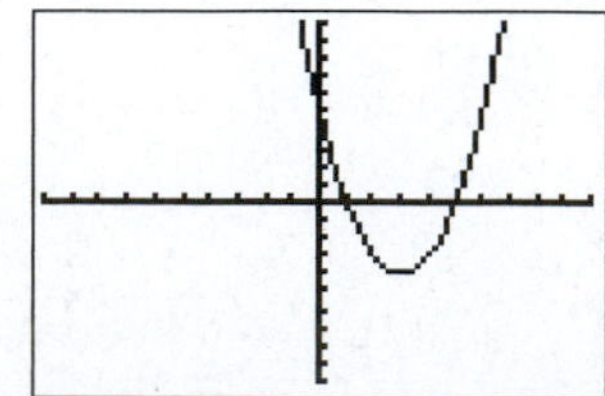

X	Y1
0	5
1	0
2	−3
3	−4
4	−3
5	0
6	5

X=6

To find the vertex, press 2nd TRACE and choose either minimum (if the parabola opens up) or maximum (if the parabola opens down). Once the selection is made, we are prompted to enter a left bound, a right bound, and a guess. The left bound is a point on the left side of the vertex. The right bound is a point on the right side of the vertex. So, move to the appropriate point and press ENTER. To enter a guess, move the cursor between the markings made from the left and right bound and press ENTER.

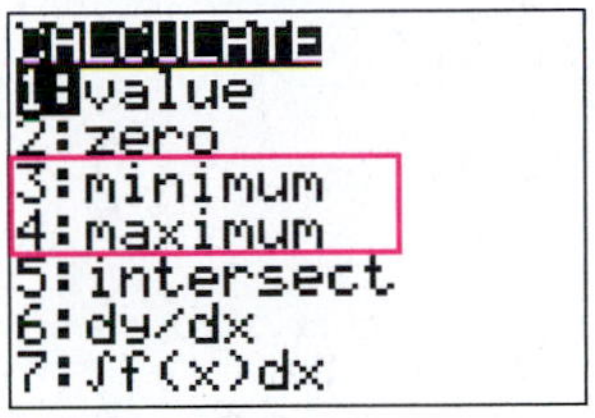

CALCULATE
1:value
2:zero
3:minimum
4:maximum
5:intersect
6:dy/dx
7:∫f(x)dx

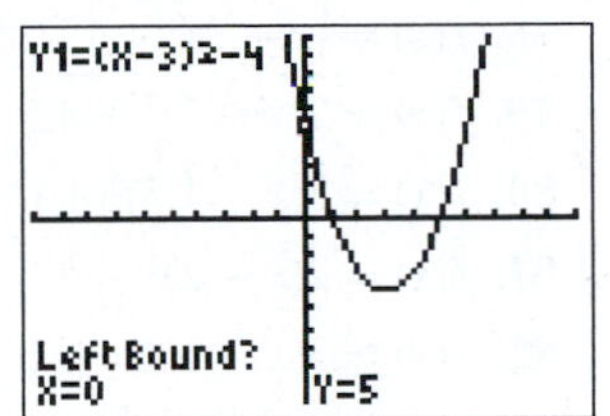

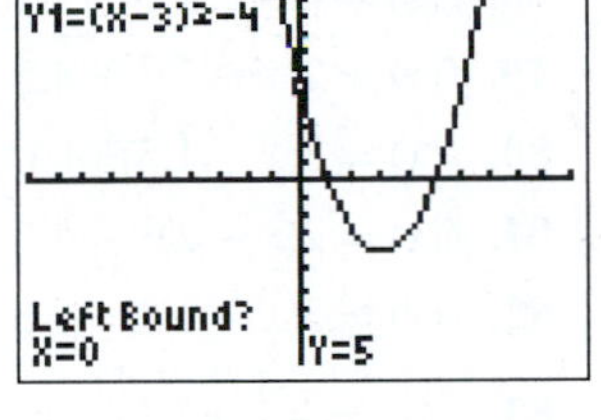

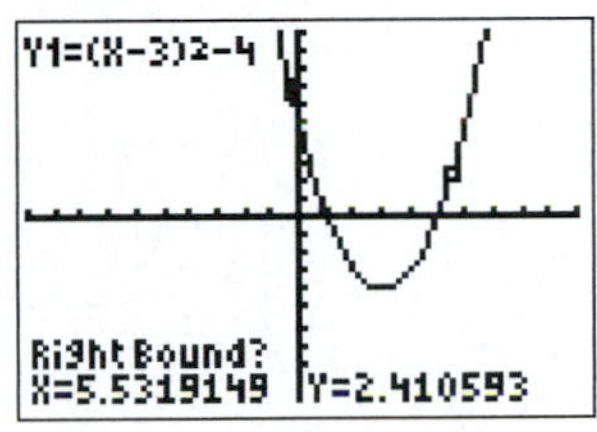

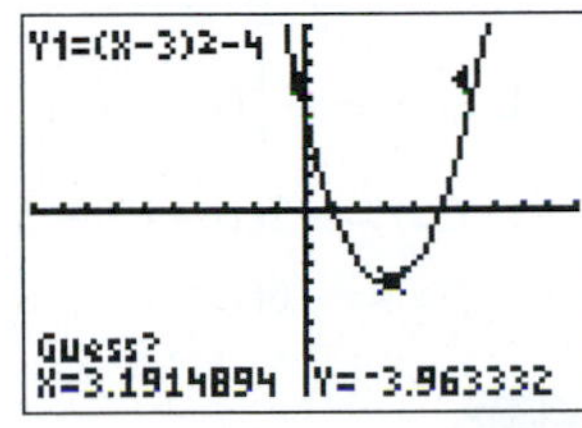

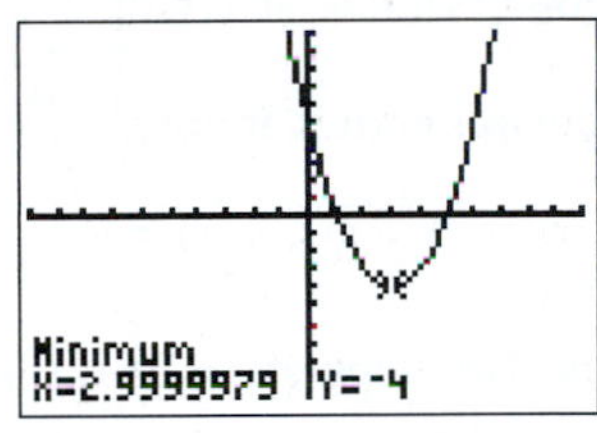

The x-intercepts are found using the zero function. Press 2nd TRACE and select zero. We are prompted to enter a left and right bound of the zero as well. The left bound is a point on the left side of the x-intercept and the right bound is a point on the right side of the x-intercept. To enter a guess, move between the markings placed by the bounds and press ENTER.

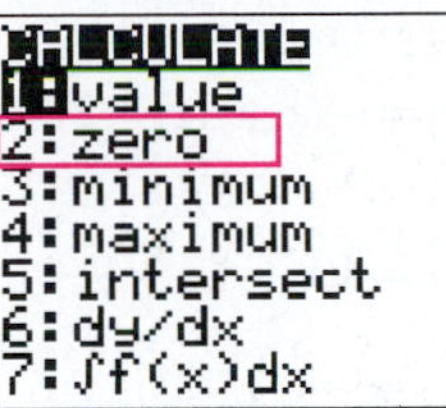

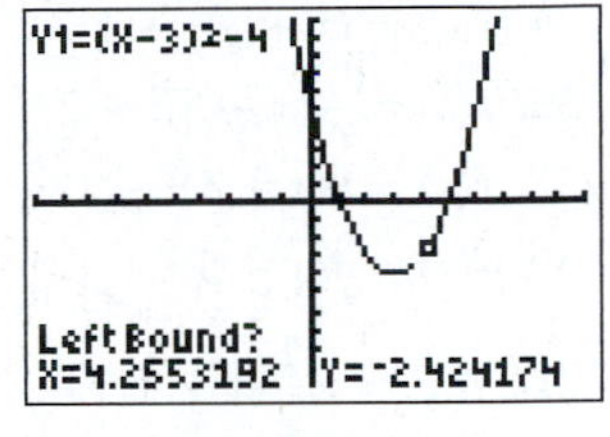

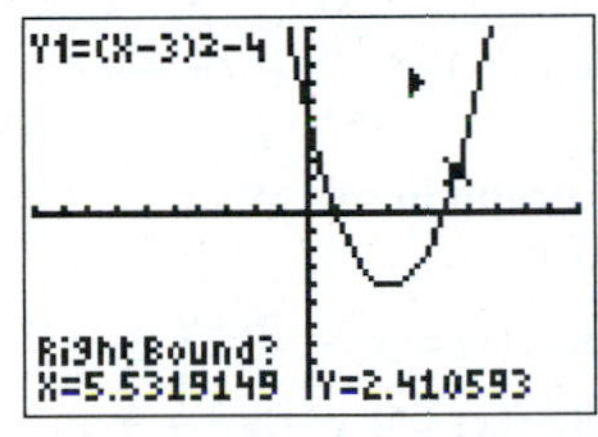

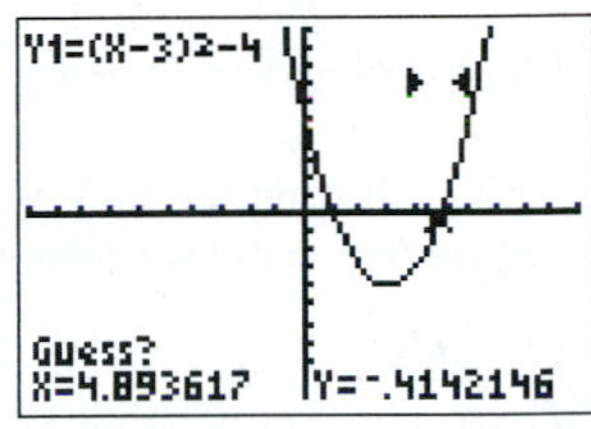

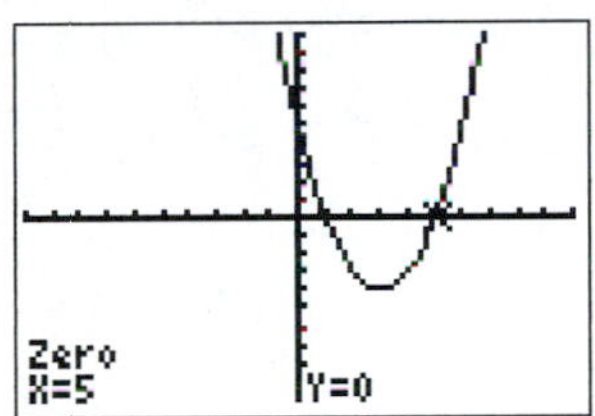

SECTION 9.1 EXERCISE SET

Write About It!

Use complete sentences in your answer to each exercise.

1. Describe the graph of the function $f(x) = x^2 + k, k > 0$.
2. Describe the graph of the function $f(x) = (x - h)^2$.
3. Describe the graph of the function $f(x) = -(x - h)^2 + k$.
4. Describe the graph of the function $f(x) = a(x - h)^2 + k$.

Practice Makes Perfect!

Create a table of values to graph each function. Identify the x-intercepts, y-intercept, vertex, the axis of symmetry, domain, and range. Explain how the graph of the function relates to the graph of $f(x) = x^2$. (*See Objective 1.*)

5. $f(x) = x^2 + 5$
6. $f(x) = x^2 + 7$
7. $f(x) = x^2 - 4$
8. $f(x) = x^2 - 25$
9. $f(x) = x^2 - 2.25$
10. $f(x) = x^2 - 12.96$
11. $f(x) = x^2 + 5.76$
12. $f(x) = x^2 + 6.25$

Additional answers can be found in the Instructor Answer Appendix.

(*See Objective 2.*)

13. $f(x) = (x + 3)^2$
14. $f(x) = (x + 5)^2$
15. $f(x) = (x - 4)^2$
16. $f(x) = (x - 6)^2$
17. $f(x) = \left(x + \frac{3}{2}\right)^2$
18. $f(x) = \left(x + \frac{3}{4}\right)^2$
19. $f(x) = \left(x - \frac{5}{8}\right)^2$
20. $f(x) = \left(x - \frac{5}{2}\right)^2$

(*See Objective 3.*)

21. $f(x) = 4x^2$
22. $f(x) = 5x^2$
23. $f(x) = \frac{1}{4}x^2$
24. $f(x) = \frac{1}{5}x^2$
25. $f(x) = -3x^2$
26. $f(x) = -6x^2$
27. $f(x) = -\frac{1}{5}x^2$
28. $f(x) = -\frac{1}{10}x^2$
29. $f(x) = -2.5x^2$
30. $f(x) = -1.6x^2$

Graph each function by plotting the vertex and two additional points. Identify the x-intercepts, y-intercept, the axis of symmetry, domain, and range. Explain how the graph of the function relates to the graph of $y = x^2$. (*See Objective 4.*)

31. $f(x) = (x - 1)^2 - 9$
32. $f(x) = (x - 3)^2 - 4$
33 $f(x) = (x + 3)^2 + 4$
34. $f(x) = (x + 2)^2 + 5$
35. $f(x) = -(x - 5)^2 - 2$
36. $f(x) = -(x + 4)^2 - 2$
37. $f(x) = 2(x + 4)^2 - 18$
38. $f(x) = 3(x - 2)^2 - 12$
39. $f(x) = 4(x - 5)^2 + 1$
40. $f(x) = 4(x + 1)^2 + 9$
41. $f(x) = -\frac{1}{2}(x + 4)^2 + 18$
42. $f(x) = -\frac{2}{3}(x + 3)^2 + 6$
43. $f(x) = \frac{2}{3}(x - 2)^2 - 6$
44. $f(x) = \frac{3}{4}(x + 2)^2 - 12$

Graph each function and use the graph to solve the equation $f(x) = 0$. (*See Objective 5.*)

45. $f(x) = x^2 - 16$
46. $f(x) = x^2 - 12.25$
47. $f(x) = 12(x + 2)^2 - 3$
48. $f(x) = 4(x - 1)^2 - 9$
49. $f(x) = -2(x + 6)^2 - 3$
50. $f(x) = -3(x - 1)^2 - 6$
51. $f(x) = \frac{1}{2}(x - 1)^2 - 2$
52. $f(x) = \frac{1}{3}(x - 2)^2 + 3$
53. $f(x) = -\frac{1}{2}(x - 1)^2 + 2$
54. $f(x) = -\frac{1}{3}(x + 2)^2 + 3$

Mix 'Em Up!

Graph each function by plotting the vertex and two additional points. Identify the x-intercepts, y-intercept, axis of symmetry, domain, and range. Explain how the graph of the function relates to the graph of $f(x) = x^2$.

55. $f(x) = x^2 - \frac{25}{9}$
56. $f(x) = x^2 - \frac{9}{4}$
57. $f(x) = \left(x - \frac{7}{2}\right)^2$
58. $f(x) = \left(x + \frac{5}{2}\right)^2$
59. $f(x) = -x^2 + 1$
60. $f(x) = -x^2 + 4$
61. $f(x) = 2(x + 1)^2 - 8$
62. $f(x) = 3(x - 2)^2 - 12$
63. $f(x) = -8(x - 5)^2 + 2$
64. $f(x) = -18(x + 1)^2 - 2$
65. $f(x) = (x - 2.5)^2 - 12.96$
66. $f(x) = (x - 1.5)^2 - 1.96$
67. $f(x) = -\frac{1}{2}(x + 1)^2 + 2$
68. $f(x) = -\frac{3}{2}(x - 3)^2 + 6$
69. $f(x) = x^2 + \frac{1}{4}$
70. $f(x) = x^2 + \frac{1}{9}$
71. $f(x) = \frac{1}{4}(x - 1)^2 + 1$
72. $f(x) = \frac{1}{2}(x - 1)^2 + 1$
73. $f(x) = -(x + 1.8)^2 + 20.25$
74. $f(x) = -(x + 3.2)^2 + 6.25$
75. $f(x) = 2(x - 1.2)^2 - 6.48$
76. $f(x) = 3(x + 1.5)^2 - 21.87$

Graph each function and use the graph to solve the equation $f(x) = 0$.

77. $f(x) = (x - 2)^2 - 2.25$
78. $f(x) = (x + 2)^2 - 12.25$
79. $f(x) = 2(x + 2)^2 + 4.5$
80. $f(x) = 3(x - 1.5)^2 + 0.75$
81. $f(x) = 2(x - 2)^2 - 4.5$
82. $f(x) = 3(x + 1.5)^2 - 0.75$
83. $f(x) = -\frac{1}{2}(x - 3)^2 + 8$
84. $f(x) = -\frac{4}{3}(x + 1)^2 + 3$
85. $f(x) = -2(x - 1)^2 - 5$
86. $f(x) = -6(x + 1)^2 - 2$

You Be the Teacher!

Correct each student's errors, if any.

87. Describe how the graph of $f(x) = (x - 3)^2 + 25$ relates to the graph of $f(x) = x^2$.

Patel's work: The graph of $y = x^2$ is shifted left 3 units and down 25 units.

The graph of $y = x^2$ is shifted right 3 units and up 25 units.

88. Describe how the graph of $f(x) = -(x + 9)^2 - 4$ relates to the graph of $f(x) = x^2$.

Blake's work: The graph of $y = x^2$ is shifted left 9 units, down 4 units, and reflected over the x-axis.

89. Describe how the graph of $f(x) = -\frac{1}{2}(x - 5)^2 + 4$ relates to the graph of $f(x) = x^2$.

Mary's work: The graph of $y = x^2$ is shifted right 5 units, up 4 units, reflected over the x-axis, and shrunk by a factor of 2.

90. Describe how the graph of $f(x) = 2(x + 7)^2 + 3$ relates to the graph of $f(x) = x^2$.

Jeannette's work: The graph of $y = x^2$ is shifted left 7 units, up 3 units, and stretched by a factor of 2.

Calculate It!

Use a graphing calculator to find the x-intercepts and the vertex of each quadratic function. Round each answer to two decimal places.

91. $f(x) = 2.7x^2 - 5.2x - 4.8$ $(-0.68, 0), (2.61, 0); (0.96, -7.30)$
92. $f(x) = -1.6x^2 + 7.3x + 8.1$ $(-0.92, 0), (5.49, 0); (2.28, 16.43)$

SECTION 9.2 Solving Quadratic Equations Using the Square Root Property and Completing the Square

OBJECTIVES

As a result of completing this section, you will be able to

1. Solve quadratic equations by applying the square root property.
2. Create a perfect square trinomial and express it in factored form.
3. Solve a quadratic equation of the form $x^2 + bx + c = 0$ by completing the square.
4. Solve a quadratic equation of the form $ax^2 + bx + c = 0$ $(a \neq 1)$ by completing the square.
5. Solve applications of quadratic equations.
6. Find zeros of quadratic functions of the form $f(x) = a(x - h)^2 + k$.
7. Troubleshoot common errors.

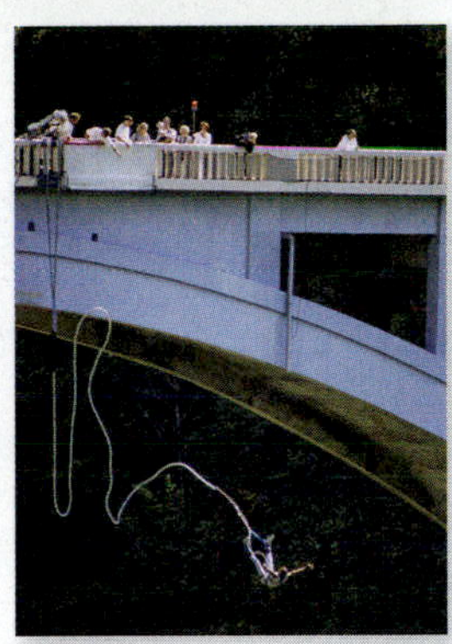

One of the world's highest commercial bungee jumps is off of the Bloukrans River Bridge in South Africa. The jump takes place from a platform below the roadway of the bridge and is 216 m (710 ft) above the ground. A jumper is in free fall until the elasticity of the cord affects the rate of fall. While the jumper is free-falling, his height, in feet, above the valley floor is $h = -16t^2 + 710$, where t is in seconds. How many seconds will it take for the jumper to be 410 ft above the valley floor? (Source: http://www.faceadrenalin.com/)

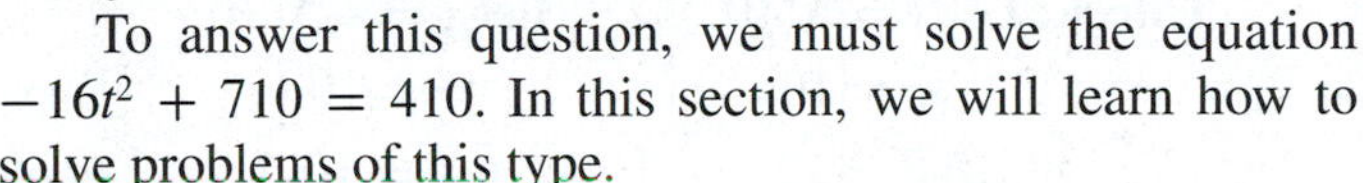

To answer this question, we must solve the equation $-16t^2 + 710 = 410$. In this section, we will learn how to solve problems of this type.

Using the Square Root Property to Solve Quadratic Equations

Objective 1 Solve quadratic equations by applying the square root property.

In Chapter 6, we learned that a **quadratic equation** is an equation of the form $ax^2 + bx + c = 0$, where a, b, and c are real numbers with $a \neq 0$. We also learned how to solve these equations by applying the zero products property. We will review this property for reference.

To solve the following equations using the zero products property, we first write the equation in standard form, $ax^2 + bx + c = 0$. Then we factor and set each factor equal to zero.

$x^2 = 36$	$a^2 = 25$	$y^2 = 49$
$x^2 - 36 = 0$	$a^2 - 25 = 0$	$y^2 - 49 = 0$
$(x - 6)(x + 6) = 0$	$(a - 5)(a + 5) = 0$	$(y - 7)(y + 7) = 0$
$x - 6 = 0$ or $x + 6 = 0$	$a - 5 = 0$ or $a + 5 = 0$	$y - 7 = 0$ or $y + 7 = 0$
$x = 6$ or $x = -6$	$a = 5$ or $a = -5$	$y = 7$ or $y = -7$
$\{-6, 6\}$	$\{-5, 5\}$	$\{-7, 7\}$

These equations illustrate that when an equation is of the form $x^2 = k$, where k is a perfect square, there are two solutions of the equation, the positive and negative square root of the value k. Thus, we can also solve the equation $x^2 = 36$ as shown.

$$x^2 = 36$$
$$x = \pm\sqrt{36}$$
$$x = \pm 6$$

This leads to another important property for solving quadratic equations.

Property: The Square Root Property

If $a^2 = k$, where k is a real number, then $a = \sqrt{k}$ or $a = -\sqrt{k}$, or $a = \pm\sqrt{k}$.

Note: *If the radicand is a negative number, then the equation has two complex, nonreal solutions since the square root of a negative number is imaginary. Recall* $\sqrt{-k} = i\sqrt{k}$, *for* $k > 0$.

Procedure: Solving a Quadratic Equation Using the Square Root Property

Step 1: Isolate the squared term to one side of the equation.
Step 2: Remove the square on the variable by applying the square root property.
Step 3: Solve the resulting equations.
Step 4: Check solutions by substituting the values into the original equation.

Objective 1 Examples Solve each equation by applying the square root property. Write each radical in simplest form.

1a. $y^2 = 10$ **1b.** $a^2 + 3 = 7$ **1c.** $2y^2 - 5 = 9$

1d. $(x + 3)^2 = 25$ **1e.** $(3a - 4)^2 - 1 = 7$ **1f.** $(5y + 2)^2 = -49$

Solutions **1a.** $y^2 = 10$

$y = \pm\sqrt{10}$ Apply the square root property.

The solutions are $-\sqrt{10}$ or $\sqrt{10}$. The solution set is $\{-\sqrt{10}, \sqrt{10}\}$ or $\{\pm\sqrt{10}\}$.

Check: Let $y = \sqrt{10}$.	Let $y = -\sqrt{10}$.	
$y^2 = 10$	$y^2 = 10$	Original equation.
$(\sqrt{10})^2 = 10$	$(-\sqrt{10})^2 = 10$	Replace y with proposed solution.
$10 = 10$	$10 = 10$	Apply $(\sqrt[n]{x})^n = x$.

1b.

$a^2 + 3 = 7$

$a^2 + 3 - 3 = 7 - 3$ Subtract 3 from each side.

$a^2 = 4$ Simplify.

$a = \pm\sqrt{4}$ Apply the square root property.

$a = \pm 2$ Simplify.

The solutions are -2 or 2, and the solution set is $\{-2, 2\}$ or $\{\pm 2\}$. The check is left for the reader.

1c.

$2y^2 - 5 = 9$

$2y^2 - 5 + 5 = 9 + 5$ Add 5 to each side.

$2y^2 = 14$ Simplify.

$\frac{2y^2}{2} = \frac{14}{2}$ Divide each side by 2.

$y^2 = 7$ Simplify.

$y = \pm\sqrt{7}$ Apply the square root property.

The solutions are $-\sqrt{7}$ or $\sqrt{7}$, and the solution set is $\{-\sqrt{7}, \sqrt{7}\}$ or $\{\pm\sqrt{7}\}$. The check is left for the reader.

1d.

$(x + 3)^2 = 25$

$x + 3 = \pm\sqrt{25}$ Apply the square root property.

$x + 3 = \pm 5$ Simplify.

$x + 3 - 3 = \pm 5 - 3$ Subtract 3 from each side.

$x = -3 \pm 5$ Rewrite using the commutative property.

$x = -3 + 5$ or $x = -3 - 5$ Separate the two solutions.

$x = 2$ $\quad$ $x = -8$ Solve the resulting equations.

The solutions are -8 or 2, and the solution set is $\{-8, 2\}$. The check is left for the reader.

1e.

$$(3a-4)^2-1=7$$

$$(3a-4)^2-1+1=7+1 \quad \text{Add 1 to each side.}$$

$$(3a-4)^2=8 \quad \text{Simplify.}$$

$$3a-4=\pm\sqrt{8} \quad \text{Apply the square root property.}$$

$$3a-4=\pm 2\sqrt{2} \quad \text{Simplify: } \sqrt{8}=\sqrt{4\cdot 2}=2\sqrt{2}.$$

$$3a-4+4=\pm 2\sqrt{2}+4 \quad \text{Add 4 to each side.}$$

$$3a=4\pm 2\sqrt{2} \quad \text{Rewrite using the commutative property.}$$

$$a=\frac{4\pm 2\sqrt{2}}{3} \quad \text{Divide each side by 3.}$$

The exact solutions are $\dfrac{4+2\sqrt{2}}{3}$ or $\dfrac{4-2\sqrt{2}}{3}$, and the solution set is $\left\{\dfrac{4+2\sqrt{2}}{3}, \dfrac{4-2\sqrt{2}}{3}\right\}$ or $\left\{\dfrac{4\pm 2\sqrt{2}}{3}\right\}$. The approximate solutions are $\dfrac{4+2\sqrt{2}}{3}\approx 2.28$ and $\dfrac{4-2\sqrt{2}}{3}\approx 0.39$.

1f.

$$(5y+2)^2=-49$$

$$5y+2=\pm\sqrt{-49} \quad \text{Apply the square root property.}$$

$$5y+2=\pm 7i \quad \text{Simplify: } \sqrt{-49}=7i.$$

$$5y+2-2=\pm 7i-2 \quad \text{Subtract 2 from each side.}$$

$$5y=-2\pm 7i \quad \text{Rewrite using the commutative property.}$$

$$\frac{5y}{5}=\frac{-2\pm 7i}{5} \quad \text{Divide each side by 5.}$$

$$y=\frac{-2\pm 7i}{5} \quad \text{Simplify.}$$

The solutions are $\dfrac{-2+7i}{5}$ or $\dfrac{-2-7i}{5}$, and the solution set is $\left\{\dfrac{-2+7i}{5}, \dfrac{-2-7i}{5}\right\}$ or $\left\{\dfrac{-2\pm 7i}{5}\right\}$.

Student Check 1 Solve each equation by applying the square root property. Write each radical in simplest form.

a. $y^2=21$ **b.** $a^2+7=43$ **c.** $3y^2-8=22$

d. $(x+7)^2=16$ **e.** $(5a-7)^2=12$ **f.** $(7y+6)^2=-1$

Perfect Square Trinomials

Objective 2 ▶

Create a perfect square trinomial and express it in factored form.

We can use the square root property to solve a quadratic equation as long as one side of the equation can be expressed as a binomial squared. Squared binomials result from factoring perfect square trinomials. We will learn how to create perfect square trinomials from binomials and then use this in Objective 3 to solve a quadratic equation.

Recall that a **perfect square trinomial** is a trinomial that results from squaring a binomial. Some examples of these trinomials are shown.

$$(x+3)^2=x^2+6x+9$$

$$(x+5)^2=x^2+10x+25$$

$$(x-6)^2=x^2-12x+36$$

$$(x-10)^2=x^2-20x+100$$

In a perfect square trinomial with a leading coefficient of 1, the constant term relates to the coefficient of x in an interesting way.

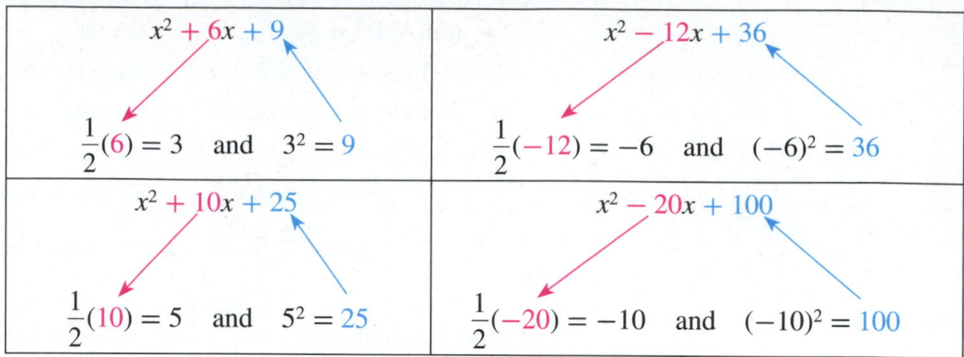

So, in a perfect square trinomial whose leading coefficient is 1, the constant term is half of the coefficient of x, squared. For $x^2 + bx + c$ to be a perfect square trinomial, $c = \left(\frac{b}{2}\right)^2$. We can use this fact to create a perfect square trinomial from a binomial of the form, $x^2 + bx$. This process is called **completing the square**.

> **Procedure: Creating a Perfect Square Trinomial from $x^2 + bx$**
>
> **Step 1:** Find half of the value of b and square it.
> **Step 2:** Add this number to the expression $x^2 + bx$.
> **Step 3:** Factor the trinomial as $\left(x + \frac{b}{2}\right)^2$.

Objective 2 Examples

Create a perfect square trinomial from each binomial by completing the square and then write it in factored form.

2a. $x^2 + 4x$ **2b.** $y^2 - 8y$ **2c.** $x^2 - 3x$

Solutions

2a. In $x^2 + 4x$, the value of $b = 4$, so

$$c = \left(\frac{4}{2}\right)^2 = (2)^2 = 4$$

Therefore, $x^2 + 4x + 4$ is a perfect square trinomial, and it factors as

$$x^2 + 4x + 4 = (x + 2)(x + 2) = (x + 2)^2$$

2b. In $y^2 - 8y$, the value of $b = -8$, so

$$c = \left(\frac{-8}{2}\right)^2 = (-4)^2 = 16$$

Therefore, $y^2 - 8y + 16$ is a perfect square trinomial, and it factors as

$$y^2 - 8y + 16 = (y - 4)(y - 4) = (y - 4)^2$$

2c. In $x^2 - 3x$, the value of $b = -3$, so

$$c = \left(\frac{-3}{2}\right)^2 = \frac{9}{4}$$

Therefore, $x^2 - 3x + \frac{9}{4}$ is a perfect square trinomial, and it factors as

$$x^2 - 3x + \frac{9}{4} = \left(x - \frac{3}{2}\right)\left(x - \frac{3}{2}\right) = \left(x - \frac{3}{2}\right)^2$$

Student Check 2 Create a perfect square trinomial from each binomial by completing the square and then write it in factored form.

a. $x^2 + 12x$ **b.** $y^2 - 2y$ **c.** $x^2 - 5x$

Completing the Square to Solve $x^2 + bx + c = 0$

Objective 3 Solve a quadratic equation of the form $x^2 + bx + c = 0$ by completing the square.

We now apply the process of *completing the square* to solving a quadratic equation. Our goal is to make one side of the equation a perfect square trinomial, which we will factor and then apply the square root property to solve the equation.

Consider the equation $x^2 - 6x + 2 = 0$. This equation cannot be solved by factoring. So, we want to isolate the binomial $x^2 - 6x$ on one side of the equation and then complete the square.

$x^2 - 6x = -2$ Move the constant term to the right side of the equation.

$x^2 - 6x + 9 = -2 + 9$ Make $x^2 - 6x$ a perfect square trinomial by completing the square. Take half of -6 and square it. Add this number to each side of the equation.

$(x - 3)^2 = 7$ Factor the perfect square trinomial and simplify the right side of the equation.

$x - 3 = \pm\sqrt{7}$ Apply the square root property.

$x = 3 \pm \sqrt{7}$ Solve for the variable by adding 3 to each side.

Procedure: Solving an Equation of the Form $x^2 + bx + c = 0$ by Completing the Square

Step 1: Move the constant term to the right side.
Step 2: Complete the square by finding half of the coefficient of x and squaring it. Add this number to each side of the equation.
Step 3: Factor the perfect square trinomial and simplify the right side.
Step 4: Apply the square root property to solve the equation.

Objective 3 Examples Solve each equation by completing the square.

3a. $x^2 + 8x - 5 = 0$ **3b.** $y^2 - y - 4 = 0$

Solutions **3a.** $x^2 + 8x - 5 = 0$

$x^2 + 8x - 5 + 5 = 0 + 5$ Add 5 to each side.

$x^2 + 8x = 5$ Simplify.

$x^2 + 8x + 16 = 5 + 16$ Add $\left(\frac{8}{2}\right)^2 = 4^2 = 16$ to each side to complete the square.

$(x + 4)^2 = 21$ Factor the trinomial and simplify the right side.

$x + 4 = \pm\sqrt{21}$ Apply the square root property.

$x = -4 \pm \sqrt{21}$ Subtract 4 from each side.

The solution set is $\{-4 - \sqrt{21}, -4 + \sqrt{21}\}$ or $\{-4 \pm \sqrt{21}\}$.

3b. $y^2 - y - 4 = 0$

$y^2 - y - 4 + 4 = 0 + 4$ — Add 4 to each side.

$y^2 - y = 4$ — Simplify.

$y^2 - 1y + \frac{1}{4} = 4 + \frac{1}{4}$ — Add $\left(\frac{-1}{2}\right)^2 = \frac{1}{4}$ to each side to complete the square.

$\left(y - \frac{1}{2}\right)^2 = \frac{17}{4}$ — Factor the trinomial and simplify the right side.

$y - \frac{1}{2} = \pm\sqrt{\frac{17}{4}}$ — Apply the square root property.

$y - \frac{1}{2} = \pm\frac{\sqrt{17}}{2}$ — Apply the quotient rule for radicals.

$y = \frac{1}{2} \pm \frac{\sqrt{17}}{2}$ — Add $\frac{1}{2}$ to each side.

The solution set is $\left\{\frac{1}{2} + \frac{\sqrt{17}}{2}, \frac{1}{2} - \frac{\sqrt{17}}{2}\right\}$ or $\left\{\frac{1}{2} \pm \frac{\sqrt{17}}{2}\right\}$.

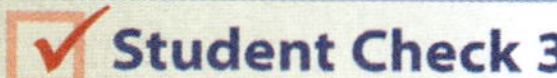

Student Check 3 Solve each equation by completing the square.

a. $a^2 + 10a + 20 = 0$ **b.** $x^2 - 3x - 11 = 0$

Objective 4

Solve a quadratic equation of the form $ax^2 + bx + c = 0$ ($a \neq 1$) by completing the square.

Completing the Square to Solve $ax^2 + bx + c = 0, a \neq 1$

In the equations in Example 3, the coefficient of the squared term is 1. This must be the case in order to complete the square. Consider the following trinomial.

$$(2x - 3)^2 = 4x^2 - 12x + 9$$

This is a perfect square trinomial, but the relationship between the constant term and the middle term that we discussed earlier does not exist; that is,

$$\frac{1}{2}(-12) = -6 \quad \text{but} \quad (-6)^2 = 36 \neq 9$$

So, to solve a quadratic equation with $a \neq 1$ by completing the square, we must write an equivalent equation with $a = 1$. We do this by dividing each side of the equation by a.

Procedure: Solving $ax^2 + bx + c = 0, a \neq 1$, by Completing the Square

Step 1: Divide each side of the equation by a, the coefficient of the squared term.
Step 2: Move the constant term to the right side of the equation.
Step 3: Complete the square to make one side a perfect square trinomial. Add this number to each side of the equation.
Step 4: Factor the trinomial and simplify the other side.
Step 5: Apply the square root property and solve the equation.

Objective 4 Examples Solve each equation by completing the square.

4a. $2x^2 - 24x + 30 = 0$ **4b.** $3x^2 - 2x - 5 = 0$

Solutions **4a.**

$$2x^2 - 12x + 30 = 0$$

$$\frac{2x^2 - 12x + 30}{2} = \frac{0}{2}$$ Divide each side by 2.

$$x^2 - 6x + 15 = 0$$ Simplify.

$$x^2 - 6x + 15 - 15 = 0 - 15$$ Subtract 15 from each side.

$$x^2 - 6x = -15$$ Simplify.

$$x^2 - 6x + 9 = -15 + 9$$ Add $\left(\frac{-6}{2}\right)^2 = (-3)^2 = 9$ to each side.

$$(x - 3)^2 = -6$$ Factor and simplify.

$$x - 3 = \pm\sqrt{-6}$$ Apply the square root property.

$$x = 3 \pm i\sqrt{6}$$ Add 3 to each side and simplify $\sqrt{-6}$.

The solution set is $\{3 + i\sqrt{6}, 3 - i\sqrt{6}\}$ or $\{3 \pm i\sqrt{6}\}$.

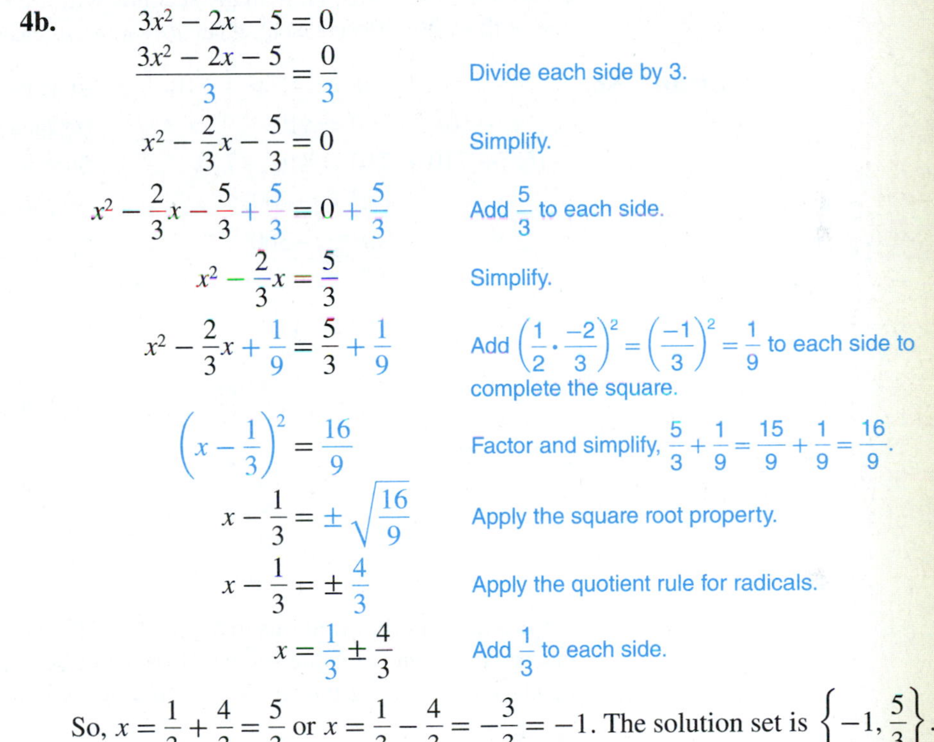

4b.

$$3x^2 - 2x - 5 = 0$$

$$\frac{3x^2 - 2x - 5}{3} = \frac{0}{3}$$ Divide each side by 3.

$$x^2 - \frac{2}{3}x - \frac{5}{3} = 0$$ Simplify.

$$x^2 - \frac{2}{3}x - \frac{5}{3} + \frac{5}{3} = 0 + \frac{5}{3}$$ Add $\frac{5}{3}$ to each side.

$$x^2 - \frac{2}{3}x = \frac{5}{3}$$ Simplify.

$$x^2 - \frac{2}{3}x + \frac{1}{9} = \frac{5}{3} + \frac{1}{9}$$ Add $\left(\frac{1}{2} \cdot \frac{-2}{3}\right)^2 = \left(\frac{-1}{3}\right)^2 = \frac{1}{9}$ to each side to complete the square.

$$\left(x - \frac{1}{3}\right)^2 = \frac{16}{9}$$ Factor and simplify, $\frac{5}{3} + \frac{1}{9} = \frac{15}{9} + \frac{1}{9} = \frac{16}{9}$.

$$x - \frac{1}{3} = \pm\sqrt{\frac{16}{9}}$$ Apply the square root property.

$$x - \frac{1}{3} = \pm\frac{4}{3}$$ Apply the quotient rule for radicals.

$$x = \frac{1}{3} \pm \frac{4}{3}$$ Add $\frac{1}{3}$ to each side.

So, $x = \frac{1}{3} + \frac{4}{3} = \frac{5}{3}$ or $x = \frac{1}{3} - \frac{4}{3} = -\frac{3}{3} = -1$. The solution set is $\left\{-1, \frac{5}{3}\right\}$.

Student Check 4 Solve each equation by completing the square.

a. $4x^2 - 8x + 20 = 0$ **b.** $2y^2 - y - 3 = 0$

Applications

Objective 5 Solve applications of quadratic equations.

Quadratic equations model many real-life situations. Some of these include the area of geometric figures, the height of an object, and investment-related problems. For these types of problems, there are formulas we use to write equations that solve them.

Objective 5 Examples **Solve each problem.**

5a. The base of the Great Pyramid of Giza, Egypt, is a square. If the area of the base of the pyramid is 570,780.25 ft^2, how long is each side of the base of the pyramid? Recall the area of a square is $A = s^2$. (Source: http://www.plim.org/greatpyramid.html)

Solution 5a.

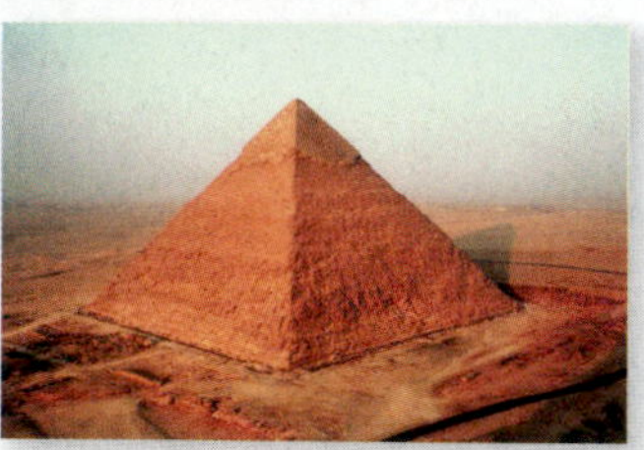

$$s^2 = A \quad \text{State the area formula.}$$
$$s^2 = 570{,}785.25 \quad \text{Replace } A \text{ with } 570{,}785.25.$$
$$s = \pm\sqrt{570{,}785.25} \quad \text{Apply the square root property.}$$
$$s = \pm 755.5 \quad \text{Simplify.}$$

Because the variable s represents the length of the side of a square, the negative value must be discarded. Therefore, the length of each side of the base of the pyramid is 755.5 ft.

5b. One of the world's highest commercial bungee jumps is off of the Bloukrans River Bridge in South Africa. The jump takes place from a platform below the roadway of the bridge and is 710 ft above the valley floor. A jumper is in free fall until the elasticity of the cord affects the rate of fall. While the jumper is free-falling, his height, in feet, above the valley floor is $h = -16t^2 + 710$, where t is in seconds. How many seconds will it take for the jumper to be 410 ft above the valley floor? (Source: http://www.faceadrenalin.com/)

Solution 5b.

$$h = -16t^2 + 710 \quad \text{State the given model.}$$
$$-16t^2 + 710 = 410 \quad \text{Replace } h \text{ with 410.}$$
$$-16t^2 + 710 - 710 = 410 - 710 \quad \text{Subtract 710 from each side.}$$
$$-16t^2 = -300 \quad \text{Simplify.}$$
$$\frac{-16t^2}{-16} = \frac{-300}{-16} \quad \text{Divide each side by } -16.$$
$$t^2 = \frac{300}{16} \quad \text{Simplify.}$$
$$t = \pm\sqrt{\frac{300}{16}} \quad \text{Apply the square root property.}$$
$$t = \pm\frac{\sqrt{300}}{\sqrt{16}} \quad \text{Apply the quotient rule for radicals.}$$
$$t = \pm\frac{10\sqrt{3}}{4} \quad \text{Simplify.}$$

Approximating the solutions gives us $t \approx 4.33$ sec or $t \approx -4.33$ sec. We must discard the negative value because time can't be negative. So, the jumper reaches 410 ft above the valley floor in approximately 4.33 sec.

5c. At what rate would you have to invest \$1000 into a 2-yr CD to have \$1060.90 once the CD matures? Use the formula $A = P(1 + r)^2$, where P is the amount invested and r is the annual interest rate.

Solution 5c.

$$P(1 + r)^2 = A \quad \text{State the given model.}$$
$$1000(1 + r)^2 = 1060.90 \quad \text{Replace } P \text{ with 1000 and } A \text{ with 1060.90.}$$
$$\frac{1000(1 + r)^2}{1000} = \frac{1060.90}{1000} \quad \text{Divide each side by 1000.}$$
$$(1 + r)^2 = 1.0609 \quad \text{Simplify.}$$
$$1 + r = \pm\sqrt{1.0609} \quad \text{Apply the square root property.}$$
$$1 + r = \pm 1.03 \quad \text{Simplify the radical.}$$
$$r = -1 \pm 1.03 \quad \text{Subtract 1 from each side.}$$
$$r = -1 + 1.03 \quad \text{or} \quad r = -1 - 1.03 \quad \text{Write two equivalent expressions for } r.$$
$$r = 0.03 \qquad r = -2.03 \quad \text{Simplify.}$$

The negative value doesn't make sense, so the money needs to be invested at a 3% annual interest rate.

Student Check 5 Solve each problem.

a. A square city block covers an area of 435,600 ft^2. What is the length of the side of the block?

b. In 2010, the Burj Khalifa in Dubai officially opened as the tallest building in the world at 2717 ft. If a penny is dropped from the top of this building, its height, in feet, t sec after it is dropped is given by $h = -16t^2 + 2717$. How many seconds will it take for the penny to reach the ground? (Source: http://www.burjkhalifa.ae/language/en-us/home.aspx)

c. At what rate would you have to invest \$25,000 in a 2-yr CD to have \$28,090 once the CD matures? Use the formula $A = P(1 + r)^2$, where P is the amount invested and r is the annual interest rate.

Zeros of Quadratic Functions

Objective 6

Find zeros of quadratic functions of the form $f(x) = a(x - h)^2 + k$.

In Section 9.1, we learned how to graph functions of the form $f(x) = a(x - h)^2 + k$ by plotting points and shifting and reflecting (if necessary) the graph of $f(x) = x^2$. We will apply the square root property and complete the square to find zeros of quadratic functions more precisely. The **zeros of a quadratic function** are the values of x for which $y = f(x) = 0$. The zeros correspond to the x-intercepts of the graph.

Procedure: Finding Zeros of Quadratic Functions

Step 1: Set $f(x) = 0$.

Step 2: Complete the square or apply the square root property to solve the equation.

Step 3: Confirm by graphing the function.

Objective 6 Examples **Find the zeros of each function. State the x-intercepts of the graph of the function. Confirm the solutions by graphing.**

6a. $f(x) = (x - 1)^2 - 4$ **6b.** $f(x) = (x + 2)^2 - 3$ **6c.** $f(x) = 2(x - 3)^2 + 8$

Solutions **6a.**

$(x - 1)^2 - 4 = 0$	Set $f(x) = 0$.
$(x - 1)^2 = 4$	Add 4 to each side.
$x - 1 = \pm\sqrt{4}$	Apply the square root property.
$x = 1 \pm 2$	Add 1 to each side and simplify the square root.
$x = 1 + 2$ or $x = 1 - 2$	Find the two values for x.
$x = 3$ $\quad$ $x = -1$	Simplify.

The zeros of the function are $x = 3$ and $x = -1$. So, the x-intercepts are (3, 0) and (−1, 0). The graph of this function confirms our work is correct.

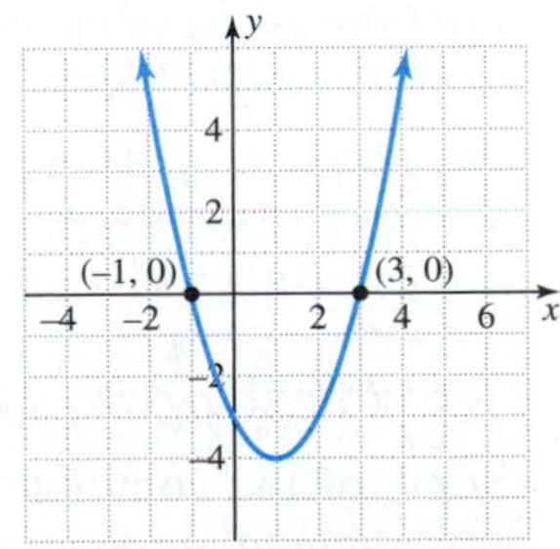

6b.

$$(x + 2)^2 - 3 = 0 \qquad \text{Set } f(x) = 0.$$
$$(x + 2)^2 = 3 \qquad \text{Add 3 to each side.}$$
$$x + 2 = \pm\sqrt{3} \qquad \text{Apply the square root property.}$$
$$x = -2 \pm \sqrt{3} \qquad \text{Subtract 2 from each side.}$$
$$x = -2 + \sqrt{3} \quad \text{or} \quad x = -2 - \sqrt{3} \qquad \text{Find the two values for } x.$$
$$x \approx -0.27 \qquad x \approx -3.73 \qquad \text{Approximate the solutions.}$$

The zeros of the function are approximately $x \approx -3.73$ and $x \approx -0.27$. So, the x-intercepts are $(-3.73, 0)$ and $(-0.27, 0)$. The graph of this function confirms our work is correct.

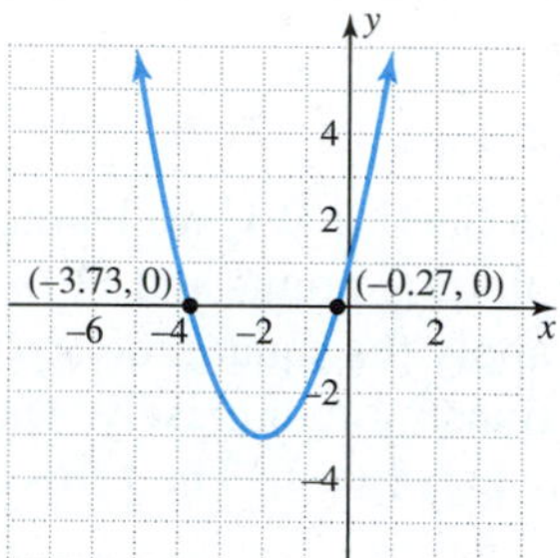

6c.

$$2(x - 3)^2 + 8 = 0 \qquad \text{Set } f(x) = 0.$$
$$2(x - 3)^2 = -8 \qquad \text{Subtract 8 from each side.}$$
$$\frac{2(x - 3)^2}{2} = \frac{-8}{2} \qquad \text{Divide each side by 2.}$$
$$(x - 3)^2 = -4 \qquad \text{Simplify.}$$
$$x - 3 = \pm\sqrt{-4} \qquad \text{Apply the square root property.}$$
$$x = 3 \pm 2i \qquad \text{Add 3 to each side and simplify the square root.}$$
$$x = 3 + 2i \quad \text{or} \quad x = 3 - 2i \qquad \text{Find the two values for } x.$$

Because the solutions are not real numbers, there are no zeros of this function. The graph confirms this since it does not cross the x-axis.

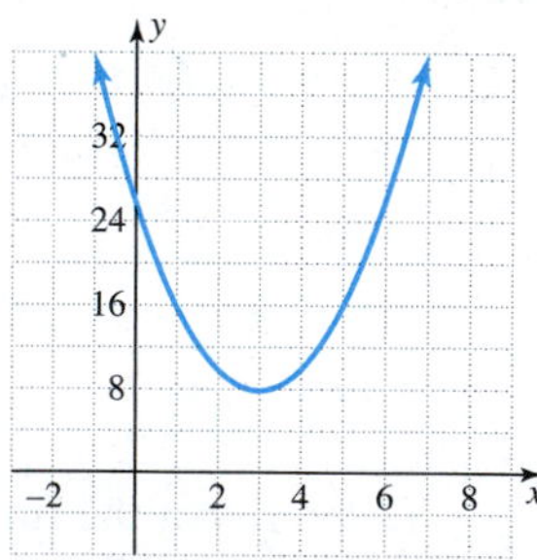

Student Check 6 Find the zeros of each function. State the x-intercepts of the graph of the function. Confirm the solutions by graphing.

a. $f(x) = (x + 4)^2 - 1$ **b.** $f(x) = (x - 5)^2 - 3$ **c.** $f(x) = -3(x + 1)^2 - 3$

Objective 7 ▶ Troubleshoot common errors.

Troubleshooting Common Errors

Some of the common errors associated with applying the square root property and completing the square are shown.

Objective 7 Examples **A problem and an incorrect solution are given. Provide the correct solution and an explanation of the error.**

7a. Solve $(x - 1)^2 = 9$.

Incorrect Solution	Correct Solution and Explanation
$(x - 1)^2 = 9$ $x - 1 = 3$ $x = 4$ The solution set is $\{4\}$.	When we apply the square root property, we take both the positive and negative square root. $(x - 1)^2 = 9$ $x - 1 = \pm 3$ $x = 1 \pm 3$ $x = 1 + 3$ or $x = 1 - 3$ $x = 4$ or $x = -2$ The solution set is $\{-2, 4\}$.

7b. Solve $x^2 - 8x - 1 = 0$ by completing the square.

Incorrect Solution	Correct Solution and Explanation
$x^2 - 8x - 1 = 0$ $x^2 - 8x = 1$ $x^2 - 8x + 16 = 1$ $(x - 4)^2 = 1$ $x - 4 = \pm 1$ $x = 4 \pm 1$ $x = 5$ or $x = -3$ The solution set is $\{-3, 5\}$.	When we complete the square, we add the number to both sides of the equation. $x^2 - 8x - 1 = 0$ $x^2 - 8x = 1$ $x^2 - 8x + 16 = 1 + 16$ $(x - 4)^2 = 17$ $x - 4 = \pm\sqrt{17}$ $x = 4 \pm \sqrt{17}$ $x = 4 + \sqrt{17}$ or $x = 4 - \sqrt{17}$ The solution set is $\{4 \pm \sqrt{17}\}$.

ANSWERS TO STUDENT CHECKS

Student Check 1 **a.** $\{\pm\sqrt{21}\}$ **b.** $\{\pm 6\}$ **c.** $\{\pm\sqrt{10}\}$ **d.** $\{-11, -3\}$ **e.** $\left\{\frac{7 \pm 2\sqrt{3}}{5}\right\}$ **f.** $\left\{\frac{-6 \pm i}{7}\right\}$

Student Check 2 **a.** $x^2 + 12x + 36 = (x + 6)^2$ **b.** $y^2 - 2y + 1 = (y - 1)^2$ **c.** $x^2 - 5x + \frac{25}{4} = \left(x - \frac{5}{2}\right)^2$

Student Check 3 **a.** $\{-5 \pm \sqrt{5}\}$ **b.** $\left\{\frac{3}{2} \pm \frac{\sqrt{53}}{2}\right\}$

Student Check 4 **a.** $\{1 \pm 2i\}$ **b.** $\left\{-1, \frac{3}{2}\right\}$

Student Check 5 **a.** 660 ft **b.** approximately 13.03 sec **c.** 6%

Student Check 6 **a.** The zeros are $x = -5, -3$. The x-intercepts are $(-5, 0)$ and $(-3, 0)$. **b.** The zeros are $x = 5 \pm \sqrt{3}$. The x-intercepts are approximately $(6.73, 0)$ and $(3.27, 0)$. **c.** The zeros are $x = -1 \pm i$, so there are no x-intercepts of the graph.

SUMMARY OF KEY CONCEPTS

1. When an equation is of the form $a^2 = k$, where k is a real number, the square root property can be applied. The property enables us to remove the square from the expression on the left side of the equation and set that expression equal to the positive and negative square root of the number on the right side of the equation. If a squared expression is equal to a negative number, the equation has complex, nonreal solutions.
2. A perfect square trinomial results from squaring a binomial. If given a binomial of the form $x^2 + bx$, we can

make it a perfect square trinomial by adding the number $c = \left(\frac{b}{2}\right)^2$. That is, adding half of the coefficient of x, squared, makes the binomial a perfect square trinomial.

3. Completing the square can be used to solve any quadratic equation. To apply this method, we must
   - **a.** Make the leading coefficient 1. If it is not 1, divide each side by the given coefficient.
   - **b.** Isolate the constant term on one side of the equation.
   - **c.** Find the number that completes the square and add it to both sides of the equation.
   - **d.** Factor the trinomial as a binomial squared and simplify the constant side.
   - **e.** Apply the square root property and solve.
4. To solve applications of quadratic equations use the appropriate mathematical formula. Substitute the known values and apply the square root property to solve the resulting equation.
5. To find zeros of a quadratic function of the form $f(x) = a(x - h)^2 + k$, set $f(x) = 0$ and apply the square root property to solve the equation. The zeros correspond to the x-intercepts on the graph.

GRAPHING CALCULATOR SKILLS

The calculator can be used to approximate the solutions of quadratic equations and to check solutions. The solutions of a quadratic equation correspond to the x-intercepts of the graph.

Example 1: Show that $\frac{4 + 2\sqrt{2}}{3} \approx 2.28$ and $\frac{4 - 2\sqrt{2}}{3} \approx 0.39$.

Solution: Enter parentheses around the numerator of the fraction. Close the parentheses after the radicand.

```
(4+2√(2))/3
          2.276142375
(4-2√(2))/3
          .3905242918
```

Example 2: Show that $\frac{4 + 2\sqrt{2}}{3}$ is a solution of $(3a - 4)^2 = 8$.

Method 1 Store the solution as the value of x. Enter the expression on the left side of the equation. If the result agrees with the right side of the equation, then the value entered is a solution of the equation.

```
(4+2√(2))/3→X
          2.276142375
(3X-4)²
                    8
```

Method 2 Graph the function $f(x) = (3a - 4)^2 - 8$ and determine if the given value corresponds to an x-intercept of the graph. After the graph is displayed, press TRACE and enter the given value. If it is a solution, the y-value is 0.

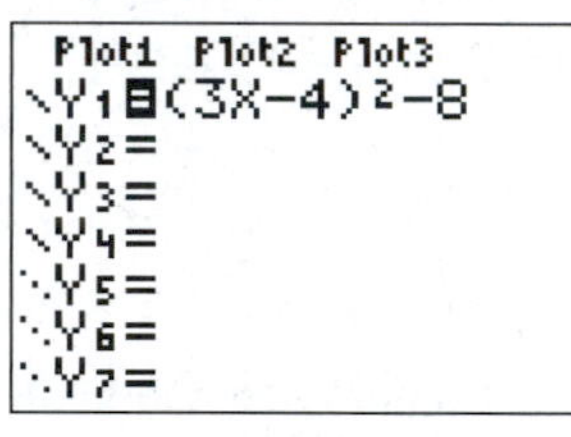


```
WINDOW
 Xmin=-5
 Xmax=5
 Xscl=1
 Ymin=-10
 Ymax=10
 Yscl=1
 Xres=1
```

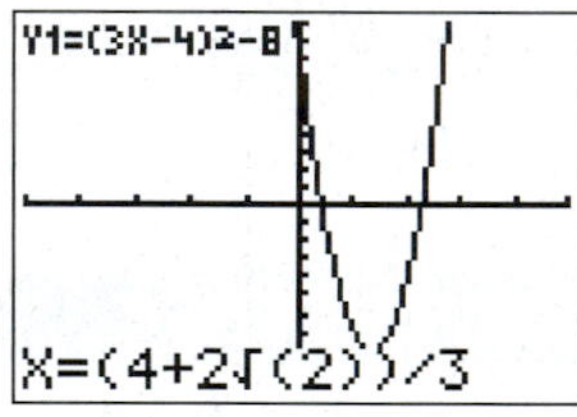

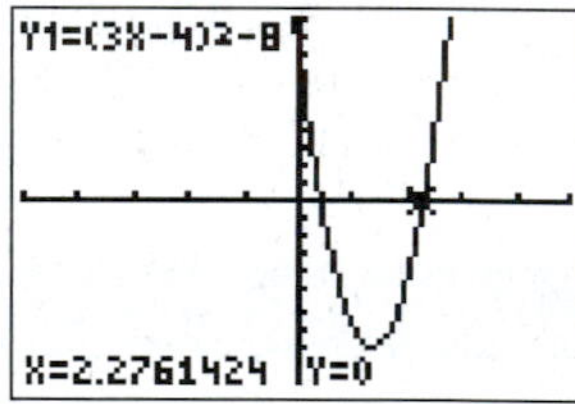

SECTION 9.2 EXERCISE SET

Write About It!

Use complete sentences in your answer to each exercise.

1. Explain the difference in solving the quadratic equation $x^2 - a^2 = 0$ using factoring and applying the square root property.
2. Use an example to explain how to complete the square for an expression of the form $x^2 + bx$, $b > 0$.
3. Explain how to complete the square for the expression $2x^2 - 3x$.
4. What is the connection between the solutions of a quadratic equation and the zeros of the corresponding quadratic function? They are the same.
5. If the solutions of a quadratic equation are complex, what does it mean about the graph of the corresponding quadratic function?
6. If the solutions of a quadratic equation are two distinct real numbers, what does it mean about the graph of the corresponding quadratic function?

Additional answers can be found in the Instructor Answer Appendix.

Practice Makes Perfect!

Solve each equation by applying the square root property. (*See Objective 1.*)

7. $x^2 + 5 = 54$ $\{-7, 7\}$

8. $y^2 + 10 = 46$ $\{-6, 6\}$

9. $r^2 - 3 = 61$ $\{-8, 8\}$

10. $s^2 - 8 = 17$ $\{-5, 5\}$

11. $49a^2 - 9 = -8$

12. $25b^2 - 7 = 74$

13. $81x^2 + 9 = 13$

14. $9y^2 + 5 = 69$

15. $z^2 - 2100 = 0$ $\{\pm 10\sqrt{21}\}$

16. $z^2 - 175 = 0$ $\{\pm 5\sqrt{7}\}$

17. $a^2 + 180 = 0$ $\{\pm 6i\sqrt{5}\}$

18. $b^2 + 24 = 0$ $\{\pm 2i\sqrt{6}\}$

19. $(4x - 1)^2 - 5 = 20$

20. $(5y + 7)^2 - 8 = -4$

21. $(7z - 10)^2 + 9 = 45$

22. $(4w + 8)^2 - 7 = 18$

23. $(3a - 1)^2 = 48$

24. $(4b - 7)^2 - 6 = 0$

25. $(6z + 5)^2 = -150$

26. $(9w - 2)^2 + 3 = 0$

Create a perfect square trinomial from each binomial by completing the square and then express it in factored form. (*See Objective 2.*)

27. $x^2 + 18x$ $x^2 + 18x + 81 = (x + 9)^2$

28. $y^2 + 16y$ $y^2 + 16y + 64 = (y + 8)^2$

29. $a^2 - 14a$ $a^2 - 14a + 49 = (a - 7)^2$

30. $b^2 - 24b$ $b^2 - 24b + 144 = (b - 12)^2$

31. $r^2 + r$

32. $s^2 + 7s$

33. $a^2 - 9a$

34. $b^2 - 15b$

Solve each equation by completing the square. (*See Objective 3.*)

35. $x^2 - 6x - 7 = 0$ $\{-1, 7\}$

36. $y^2 - 8y + 16 = 0$ $\{4\}$

37. $r^2 + 5r - 50 = 0$ $\{-10, 5\}$

38. $s^2 - 3s - 10 = 0$ $\{-2, 5\}$

39. $x^2 - 10x + 21 = 0$ $\{3, 7\}$

40. $y^2 + 6y + 5 = 0$ $\{-5, -1\}$

Solve each equation by completing the square. (*See Objective 4.*)

41. $8x^2 - 2x - 3 = 0$

42. $3y^2 - 11y - 42 = 0$

43. $2a^2 + 17a + 21 = 0$

44. $4b^2 + 27b + 18 = 0$

45. $4x^2 + 16x + 5 = 0$

46. $3y^2 + 12y + 7 = 0$

47. $2r^2 + 8r + 13 = 0$

48. $2r^2 + 3r + 3 = 0$

Solve each problem. (*See Objective 5.*)

49. A downtown square city block in Portland, Oregon, is 68,000 ft^2. What is the length of each side of the city block? Round the answer to two decimal places. (Source: http://www.land4ever.com/block.htm) 260.77 ft

50. A downtown square city block in Chicago, Illinois, is 217,000 ft^2. What is the length of each side of the city block? Round the answer to two decimal places. (Source: http://www.land4ever.com/block.htm) 465.83 ft

51. If a penny is dropped from the top of the Central Plaza in Hong Kong, its height, in meters, t sec after it is dropped is given by $s = -9.8t^2 + 374$. How many seconds will it take for the penny to reach the ground? Round the answer to two decimal places. (Source: http://www.centralplaza.com.hk) 6.18 sec

52. If a penny is dropped from the top of the Willis Tower in Chicago, its height, in meters, t sec after it is dropped is given by $s = -9.8t^2 + 442$. How many seconds will it take for the penny to reach the ground? Round the answer to two decimal places. (Source: http://www.willistower.com) 6.72 sec

53. If a penny is dropped from the top of the Chrysler Building in New York, its height, in feet, t sec after it is dropped is given by $s = -16t^2 + 1046$. How many seconds will it take for the penny to reach the ground? Round the answer to two decimal places. (Source: http://www.nyc.com) 8.09 sec

54. If a penny is dropped from the top of the Taipei Tower in Taiwan, its height, in feet, t sec after it is dropped is given by $s = -16t^2 + 1667$. How many seconds will it take for the penny to reach the ground? Round the answer to two decimal places. (Source: http://www.taipei-101.com.tw) 10.21 sec

55. At what rate would you have to invest \$5100 in a 2-yr CD to have \$5421.10 once the CD matures? Use the formula $A = P(1 + r)^2$, where P is the amount invested and r is the annual interest rate. 3.1%

56. At what rate would you have to invest \$14,200 in a 2-yr CD to have \$14,773.68 once the CD matures? Use the formula $A = P(1 + r)^2$, where P is the amount invested and r is the annual interest rate. 2%

57. At what rate would you have to invest \$25,600 in a 2-yr CD to have \$27,370.39 once the CD matures? Use the formula $A = P(1 + r)^2$, where P is the amount invested and r is the annual interest rate. 3.4%

58. At what rate would you have to invest \$4800 in a 2-yr CD to have \$5052.84 once the CD matures? Use the formula $A = P(1 + r)^2$, where P is the amount invested and r is the annual interest rate. 2.6%

Find the zeros of each function. State the x-intercepts of the graph of the function. (*See Objective 6.*)

59. $f(x) = (x + 7)^2 - 36$

60. $f(x) = (x - 8)^2 - 81$

61. $f(x) = (3x + 2)^2 - 1$

62. $f(x) = (3x - 1)^2 - 4$

63. $f(x) = (x + 1)^2 + 24$ zeros: $x = -1 \pm 2i\sqrt{6}$; no x-intercepts

64. $f(x) = (x - 11)^2 + 20$

65. $f(x) = (x - 3)^2 - 50$

66. $f(x) = (x + 11)^2 - 48$

67. $f(x) = (10x + 1)^2 - 3$

68. $f(x) = (5x - 1)^2 - 50$

69. $f(x) = (3x + 5)^2 + 84$

70. $f(x) = (x - 3)^2 + 45$ zeros: $x = 3 \pm 3i\sqrt{5}$, no x-intercepts

Mix 'Em Up!

Solve each equation by applying the square root property or by completing the square.

71. $64x^2 - 1 = 24$

72. $49y^2 - 6 = 3$

73. $r^2 + 99 = -1$ $\{\pm 10i\}$

74. $49s^2 + 41 = 5$ $\left\{\pm \frac{6}{7}i\right\}$

75. $z^2 = 252$ $\{\pm 6\sqrt{7}\}$

76. $w^2 = -243$ $\{\pm 9i\sqrt{3}\}$

77. $(2x + 10)^2 + 8 = 57$

78. $(4y - 8)^2 + 5 = 14$

79. $(9z + 2)^2 + 75 = 0$

80. $(7w + 3)^2 + 50 = 5$

81. $(4a - 3)^2 - 21 = 0$

82. $(3b - 6)^2 - 7 = 0$

83. $4(4x-3)^2+8=0$
84. $2(3w+8)^2+20=0$
85. $3(6x-1)^2-900=0$
86. $2(5w-4)^2-270=0$
87. $x^2+5x-36=0$ $\{-9,4\}$
88. $y^2+2y-63=0$ $\{-9,7\}$
89. $2r^2-r-6=0$
90. $8s^2-2s-3=0$
91. $2x^2+7x=0$
92. $5x^2-3x=0$
93. $2r^2-r+2=0$
94. $2s^2+s+1=0$
95. $4x(x-3)+3=0$
96. $8y(2y+1)=4$
97. $3a(3a-8)+19=0$
98. $4y(y+5)+39=0$

Create a perfect square trinomial from each binomial by completing the square and then express it in factored form.

99. x^2+8x
100. y^2-26y
101. $a^2-1.8a$
102. $b^2-0.6b$
103. $r^2+0.9a$
104. $s^2+0.5s$
105. $r^2+\frac{1}{2}r$
106. $s^2+\frac{5}{3}s$
107. $a^2-\frac{3}{2}a$
108. $b^2-\frac{7}{4}b$

Solve each problem.

109. The largest church clock face in Europe is on the steeple of St. Peter's Church in Zurich, Switzerland. It has an area of approximately 639.73 ft^2. What is the diameter of the clock face? Round the answer to two decimal places. (Source: http://en.wikipedia.org) 28.54 ft

110. The Busch Stadium in St. Louis, Missouri, is almost a perfect circle. It has an area of approximately 502,654.82 ft^2. What is the diameter of the stadium? (Source: http://www.baseball-statistics.com/) 800 ft

111. BASE jumping is an extreme sport in which people parachute off of fixed structures such as buildings, antennas, spans (bridges), or earth (cliffs). The Perrine Bridge in Twin Falls, Idaho, is 486 ft above the Snake River. The height of a base jumper t sec after the start of his fall can be represented by $s=-16t^2+486$. How many seconds does the jumper have before he reaches the Snake River without opening his parachute? Round the answer to two decimal places. (Source: http://en.wikipedia.org/wiki/BASE_jumping) 5.51 sec

112. The mountain Kjerag in Lysefjorden, Norway, is a BASE jumping site with a drop of 984 m. The height of a base jumper t sec after the start of his fall can be represented by $h=-9.8t^2+984$. How many seconds does the jumper have before reaching the ground without opening his parachute? Round the answer to two decimal places. (Source: http://en.wikipedia.org/wiki/Kjerag) 10.02 sec

113. At what rate would you have to invest \$12,600 in a 2-yr CD to have \$13,263.72 once the CD matures? Use the formula $A=P(1+r)^2$, where P is the amount invested and r is the annual interest rate. 2.6%

114. At what rate would you have to invest \$21,200 in a 2-yr CD to have \$21,711.85 once the CD matures? Use the formula $A=P(1+r)^2$, where P is the amount invested and r is the annual interest rate. 1.2%

You Be the Teacher!

Correct each student's errors, if any.

115. Solve $4x^2-9=0$ using the square root property.

Josh's work:

$$4x^2-9=0$$
$$2x+3i=0$$
$$2x=-3i$$
$$x=-\frac{3}{2}i$$

$4x^2-9=0$; $4x^2=9$; $x^2=\frac{9}{4}$; $x=\pm\frac{3}{2}$

116. Solve $(3x+1)^2+9=25$ using the square root property.

Matt's work:

$$(3x+1)^2+9=25$$
$$3x+1+3=5$$
$$3x=1$$
$$x=\frac{1}{3}$$

$(3x+1)^2+9=25$; $(3x+1)^2=16$; $3x+1=\pm4$; $3x=-1\pm4$; $x=\frac{-1\pm4}{3}$

117. Complete the square for x^2+14x and factor the trinomial.

Allison's work:

$$x^2+14x+196=(x+14)^2$$

$x^2+14x+\left(\frac{14}{2}\right)^2=x^2+14x+49$; $=(x+7)^2$

118. Solve $3x^2+8x+6=0$ by completing the square.

Michelle's work:

$$3x^2+8x+6=0$$
$$x^2+\frac{8}{3}x=-6$$
$$x^2+\frac{8}{3}x+\frac{16}{9}=-6$$
$$\left(x+\frac{4}{3}\right)^2=-6$$
$$x+\frac{4}{3}=\pm i\sqrt{6}$$
$$x=-\frac{4}{3}\pm i\sqrt{6}$$

$3x^2+8x+6=0$; $x^2+\frac{8}{3}x+2=0$; $x^2+\frac{8}{3}x=-2$; $x^2+\frac{8}{3}x+\frac{16}{9}=-2+\frac{16}{9}$; $\left(x+\frac{4}{3}\right)^2=-\frac{18}{9}+\frac{6}{9}$; $\left(x+\frac{4}{3}\right)^2=-\frac{12}{9}$; $x+\frac{4}{3}=\sqrt{-\frac{12}{9}}$; $x=-\frac{4}{3}\pm\frac{2i\sqrt{3}}{3}$; $x=\frac{-4\pm2i\sqrt{3}}{3}$

Calculate It!

Use a graphing calculator to determine if the two expressions are the same after completing the square.

119. $x^2-15x+\frac{225}{4}$ and $\left(x-\frac{15}{2}\right)^2$

120. $x^2-\frac{7}{10}x+\frac{49}{400}$ and $\left(x-\frac{7}{20}\right)^2$

121. $x^2-2.3x+1.3225$ and $(x-1.15)^2$

122. $x^2-\frac{5}{7}x+\frac{25}{196}$ and $\left(x-\frac{5}{14}\right)^2$

SECTION 9.3 Solving Quadratic Equations Using the Quadratic Formula

OBJECTIVES

As a result of completing this section, you will be able to

1. Solve quadratic equations using the quadratic formula.
2. Use the discriminant to determine the number and types of solutions of a quadratic equation.
3. Solve applications of quadratic equations.
4. Troubleshoot common errors.

The height of a basketball, in feet, thrown upward by a player at a free throw line is given by $h = -16t^2 + 22t + 7$. When will the ball reach the height of the basketball hoop, which is 10 ft from the floor? To answer this question, we must solve the equation $-16t^2 + 22t + 7 = 10$.

In this section, we will learn another method for solving quadratic equations that enables us to solve this problem.

Objective 1

Solve quadratic equations using the quadratic formula.

The Quadratic Formula

Thus far, we have learned three methods to solve quadratic equations. These include applying the zero products property (factoring), applying the square root property, and completing the square.

- Factoring is an efficient method but it does not solve every equation.
- The square root property only applies to equations of the form $a^2 = k$, where k is a real number.
- Completing the square does solve every quadratic equation but can be somewhat tedious, especially if the leading coefficient is not 1 and the coefficient of the x term is not even.

Another method for solving quadratic equations involves using a formula. This formula, the *quadratic formula*, is derived from completing the square on the standard form of the quadratic equation, $ax^2 + bx + c = 0$, where $a \neq 0$.

Derivation of the Quadratic Formula

Begin with $ax^2 + bx + c = 0$ and complete the square.

$$\frac{ax^2 + bx + c}{a} = \frac{0}{a}$$ Divide each side by a to get a coefficient of 1 on x^2.

$$x^2 + \frac{b}{a}x + \frac{c}{a} = 0$$ Simplify.

$$x^2 + \frac{b}{a}x = -\frac{c}{a}$$ Subtract $\frac{c}{a}$ from each side.

$$x^2 + \frac{b}{a}x + \frac{b^2}{4a^2} = -\frac{c}{a} + \frac{b^2}{4a^2}$$ Complete the square $\left(\frac{1}{2} \cdot \frac{b}{a}\right)^2 = \frac{b^2}{4a^2}$ and add to each side.

$$\left(x + \frac{b}{2a}\right)^2 = \frac{b^2 - 4ac}{4a^2}$$ Factor the left and simplify the right. $-\frac{c}{a} + \frac{b^2}{4a^2} = -\frac{4ac}{4a \cdot a} + \frac{b^2}{4a^2} = \frac{b^2 - 4ac}{4a^2}$

$$x + \frac{b}{2a} = \pm\sqrt{\frac{b^2 - 4ac}{4a^2}}$$ Apply the square root property.

$$x + \frac{b}{2a} = \pm\frac{\sqrt{b^2 - 4ac}}{\sqrt{4a^2}}$$ Apply the quotient rule for radicals.

$$x = -\frac{b}{2a} \pm \frac{\sqrt{b^2 - 4ac}}{2a}$$ Simplify the radical and subtract $\frac{b}{2a}$ from each side.

$$x = \frac{-b \pm \sqrt{b^2 - 4ac}}{2a}$$ Add the fractions.

So, this formula depends on the values of a, b, and c from the standard form of the quadratic equation.

Property: The Quadratic Formula

If $ax^2 + bx + c = 0$ $(a \neq 0)$, then

$$x = \frac{-b \pm \sqrt{b^2 - 4ac}}{2a}$$

Procedure: Solving an Equation Using the Quadratic Formula

Step 1: Write the equation in standard form, $ax^2 + bx + c = 0$.
Step 2: Identify the values of a, b, and c from the standard form.
Step 3: Substitute the values of a, b, and c in the quadratic formula.
Step 4: Use the order of operations to simplify the resulting expression.
Step 5: Simplify the quotient as much as possible.
Step 6: Check solutions by substituting the values into the original equation.

Objective 1 Examples Solve each equation using the quadratic formula.

1a. $x^2 - 6x + 9 = 0$ **1b.** $y^2 + 4y = 7$ **1c.** $6x^2 + x = 0$ **1d.** $\frac{2y^2}{3} + \frac{1}{6}y + 4 = 0$

Solutions **1a.** The equation is in standard form, so

$$1x^2 - 6x + 9 = 0 \rightarrow a = 1, b = -6, c = 9$$

$$x = \frac{-b \pm \sqrt{b^2 - 4ac}}{2a}$$ State the quadratic formula.

$$x = \frac{-(-6) \pm \sqrt{(-6)^2 - 4(1)(9)}}{2(1)}$$ Replace a, b, and c with the appropriate values.

$$x = \frac{6 \pm \sqrt{36 - 36}}{2}$$ Simplify the expression.

$$x = \frac{6 \pm \sqrt{0}}{2}$$ Simplify the radicand.

$$x = \frac{6 \pm 0}{2}$$ Simplify the radical expression.

$$x = \frac{6 + 0}{2} = \frac{6}{2} = 3 \quad \text{or} \quad x = \frac{6 - 0}{2} = \frac{6}{2} = 3$$ Write the two values for x.

So, the solution set is $\{3\}$. This quadratic equation has *one repeating rational* solution.

1b. We first write the equation in standard form and identify a, b, and c.

$$y^2 + 4y = 7$$

$$1y^2 + 4y - 7 = 0$$ Subtract 7 from each side.

$$a = 1, b = 4, c = -7$$ Identify a, b, and c.

$$y = \frac{-b \pm \sqrt{b^2 - 4ac}}{2a}$$ State the quadratic formula.

$$y = \frac{-(4) \pm \sqrt{(4)^2 - 4(1)(-7)}}{2(1)}$$ Replace a, b, and c with the appropriate values.

$$y = \frac{-4 \pm \sqrt{16 + 28}}{2}$$ Simplify the expression.

$$y = \frac{-4 \pm \sqrt{44}}{2}$$ Simplify the radicand.

$y = \dfrac{-4 \pm \sqrt{4 \cdot 11}}{2}$ Find a perfect square factor of 44.

$y = \dfrac{-4 \pm 2\sqrt{11}}{2}$ Apply the product rule for radicals.

$y = \dfrac{2(-2 \pm \sqrt{11})}{2}$ Factor the numerator.

$y = -2 \pm \sqrt{11}$ Divide out the common factor, 2.

$y = -2 + \sqrt{11}$ or $y = -2 - \sqrt{11}$ Write the two values for x.

The solution set is $\{-2 + \sqrt{11}, -2 - \sqrt{11}\}$ or $\{-2 \pm \sqrt{11}\}$. This quadratic equation has *two irrational* solutions.

1c. The equation is in standard form, so

$$6x^2 + 1x + 0 = 0 \rightarrow a = 6, b = 1, c = 0$$

$x = \dfrac{-b \pm \sqrt{b^2 - 4ac}}{2a}$ State the quadratic formula.

$x = \dfrac{-(1) \pm \sqrt{(1)^2 - 4(6)(0)}}{2(6)}$ Replace a, b, and c with the appropriate values.

$x = \dfrac{-1 \pm \sqrt{1 - 0}}{12}$ Simplify the expression.

$x = \dfrac{-1 \pm \sqrt{1}}{12}$ Simplify the radicand.

$x = \dfrac{-1 \pm 1}{12}$ Simplify the radical expression.

$x = \dfrac{-1 + 1}{12} = \dfrac{0}{12} = 0$ or Write the two values for x.

$x = \dfrac{-1 - 1}{12} = \dfrac{-2}{12} = -\dfrac{1}{6}$

The solution set is $\left\{-\dfrac{1}{6}, 0\right\}$. This quadratic equation has *two rational* solutions.

1d. We first clear the fractions before identifying a, b, and c.

$6\left(\dfrac{2y^2}{3} + \dfrac{1}{6}y + 4\right) = 6(0)$ Multiply each side of the equation by the LCD, 6.

$4y^2 + 1y + 24 = 0$ Simplify.

$a = 4, b = 1, c = 24$ Identify a, b, and c.

$y = \dfrac{-b \pm \sqrt{b^2 - 4ac}}{2a}$ State the quadratic formula.

$y = \dfrac{-(1) \pm \sqrt{(1)^2 - 4(4)(24)}}{2(4)}$ Replace a, b, and c with the appropriate values.

$y = \dfrac{-1 \pm \sqrt{1 - 384}}{8}$ Simplify the expression.

$y = \dfrac{-1 \pm \sqrt{-383}}{8}$ Simplify the radicand.

$y = \dfrac{-1 \pm i\sqrt{383}}{8}$ Simplify the radical expression.

The solution set is $\left\{\dfrac{-1 \pm i\sqrt{383}}{8}\right\}$. This quadratic equation has *no real* solutions but two *complex, nonreal* solutions.

Student Check 1 Solve each equation using the quadratic formula.

a. $x^2 - 12x + 36 = 0$ **b.** $5h^2 = 4h - 1$

c. $7x^2 + 2x = 0$ **d.** $\frac{3}{2}x^2 + \frac{1}{4}x - 2 = 0$

The Discriminant of a Quadratic Equation

Objective 2

Use the discriminant to determine the number and types of solutions of a quadratic equation.

Example 1 illustrates that a quadratic equation will have either two real solutions, one real solution, or no real solutions. In Section 9.1, we discovered these same possibilities by graphing. Recall that the solutions of a quadratic equation $f(x) = 0$ correspond to the x-intercepts of the graph of $y = f(x)$.

We can also determine the number and types of solutions of a quadratic equation by the number inside the radical of the quadratic formula, $b^2 - 4ac$. This number is called the *discriminant*.

INSTRUCTOR NOTE: Remind students that since the solutions of $f(x) = 0$ are x-intercepts of $f(x)$, the discriminant also tells us the number of x-intercepts.

Definition: Discriminant

For $ax^2 + bx + c = 0$ $(a \neq 0)$, the discriminant is $b^2 - 4ac$.

Description of $b^2 - 4ac$	Types of Solutions
Positive and a perfect square	Two rational solutions
Positive and not a perfect square	Two irrational solutions
Zero	One repeating rational solution
Negative	Two complex, nonreal solutions

The discriminant can also tell us whether or not a quadratic equation can be solved by factoring. If the discriminant is a positive perfect square, then the equation can also be solved by factoring. In Example 1 part (c), the discriminant is $b^2 - 4ac = 1$. So, the equation can also be solved by factoring.

$$6x^2 + x = 0$$
$$x(6x + 1) = 0$$
$$x = 0 \quad \text{or} \quad 6x + 1 = 0$$
$$x = 0 \qquad\qquad x = -\frac{1}{6}$$

So, there may be more than one method that can be used to solve a quadratic equation.

Procedure: Using the Discriminant to Determine the Number and Types of Solutions

Step 1: Write the equation in standard form.
Step 2: Identify the values of a, b, and c.
Step 3: Substitute the values into the expression $b^2 - 4ac$.
Step 4: Use the chart in the definition box to determine the number and types of solutions.

Objective 2 Examples **Find the discriminant of each equation and use it to determine the number and types of solutions.**

2a. $3x^2 - 5x + 6 = 0$ **2b.** $7x^2 + x = 6$ **2c.** $4x(x + 2) = 7$

Solutions

2a. The equation is in standard form, so $a = 3$, $b = -5$, $c = 6$. The discriminant is

$$\begin{aligned} b^2 - 4ac &= (-5)^2 - 4(3)(6) \\ &= 25 - 72 \\ &= -47 \end{aligned}$$

A negative discriminant means that the equation has two complex, nonreal solutions.

2b. First write the equation in standard form.

$$7x^2 + x = 6$$

$$7x^2 + 1x - 6 = 0 \qquad \text{Subtract 6 from each side.}$$

So, $a = 7$, $b = 1$, $c = -6$.

$$\begin{aligned} b^2 - 4ac &= (1)^2 - 4(7)(-6) \\ &= 1 + 168 \\ &= 169 \end{aligned}$$

The discriminant is a positive perfect square, so the equation $7x^2 + x - 6 = 0$ has two rational solutions. This equation is also factorable.

2c. First write the equation in standard form.

$$4x(x + 2) = 7$$

$$4x^2 + 8x = 7 \qquad \text{Apply the distributive property.}$$

$$4x^2 + 8x - 7 = 0 \qquad \text{Subtract 7 from each side.}$$

So, $a = 4$, $b = 8$, and $c = -7$.

$$\begin{aligned} b^2 - 4ac &= (8)^2 - 4(4)(-7) \\ &= 64 + 112 \\ &= 176 \end{aligned}$$

The discriminant is positive but not a perfect square, so the equation $4x(x + 2) = 7$ has two irrational solutions.

Student Check 2 Find the discriminant of each equation and use it to determine the number and types of solutions.

a. $x^2 + x + 1 = 0$ **b.** $2x^2 - 3x = 5$ **c.** $5x(x - 2) = 6$

Objective 3

Solve applications of quadratic equations.

Applications

In Chapter 6, we solved applications of quadratic equations by factoring. We will solve similar types of applications but will use the quadratic formula to solve the problems.

Objective 3 Examples **Solve each problem.**

3a. The height, in feet, of a basketball thrown upward by a player at a free throw line is $h = -16t^2 + 22t + 7$ after t seconds. To the nearest hundredth of a second, when will the ball reach the height of the basketball hoop, which is 10 ft from the floor?

Solution **3a.**

$$-16t^2 + 22t + 7 = h \qquad \text{State the given model.}$$

$$-16t^2 + 22t + 7 = 10 \qquad \text{Replace the value of } h \text{ with 10.}$$

$$-16t^2 + 22t + 7 - 10 = 10 - 10 \qquad \text{Subtract 10 from each side.}$$

$$-16t^2 + 22t - 3 = 0 \qquad \text{Simplify.}$$

$$16t^2 - 22t + 3 = 0 \qquad \text{Multiply each side by } -1.$$

The equation is in standard form, so $a = 16$, $b = -22$, and $c = 3$.

$$t = \frac{-b \pm \sqrt{b^2 - 4ac}}{2a}$$ State the quadratic formula.

$$t = \frac{-(-22) \pm \sqrt{(-22)^2 - 4(16)(3)}}{2(16)}$$ Replace a, b, and c with the appropriate values.

$$t = \frac{22 \pm \sqrt{484 - 192}}{32}$$ Simplify the expression.

$$t = \frac{22 \pm \sqrt{292}}{32}$$ Simplify the radicand.

$$t = \frac{22 + \sqrt{292}}{32} \approx 1.22 \quad \text{or}$$ Write the two solutions and approximate their value.

$$t = \frac{22 - \sqrt{292}}{32} \approx 0.15$$

So, the ball reaches a height of 10 ft in approximately 0.15 sec after the ball is thrown and in approximately 1.22 sec after the ball is thrown.

3b. A company knows that the cost, in dollars, to produce x items is given by the cost function $C(x) = 5x^2 + 800x$. It also knows that the revenue from selling x items is given by the revenue function $R(x) = 1000x + 200$. How many items does the company need to produce to break even, that is, when does revenue = cost?

Solution 3b.

$$C(x) = R(x)$$

$$5x^2 + 800x = 1000x + 200$$ Write the model.

$$5x^2 + 800x - 1000x - 200 = 0$$ Subtract $1000x$ and 200 from each side.

$$5x^2 - 200x - 200 = 0$$ Simplify.

$$\frac{5x^2 - 200x - 200}{5} = \frac{0}{5}$$ Divide each side by 5.

$$x^2 - 40x - 40 = 0$$ Simplify.

The equation is in standard form, so $a = 1$, $b = -40$, and $c = -40$.

$$x = \frac{-b \pm \sqrt{b^2 - 4ac}}{2a}$$ State the quadratic formula.

$$x = \frac{-(-40) \pm \sqrt{(-40)^2 - 4(1)(-40)}}{2(1)}$$ Replace a, b, and c with the appropriate values.

$$x = \frac{40 \pm \sqrt{1600 + 160}}{2}$$ Simplify the expression.

$$x = \frac{40 \pm \sqrt{1760}}{2}$$ Simplify the radicand.

$$x = \frac{40 + \sqrt{1760}}{2} \approx 41 \quad \text{or}$$ Write the two solutions and find their values.

$$x = \frac{40 - \sqrt{1760}}{2} \approx -1$$

Because it doesn't make sense to produce -1 items, the company needs to produce 41 items to break even.

3c. The owners of a child care center want to construct a playground at the back of their building, using the building as one of its borders. They can afford 200 ft of fencing. What dimensions should they construct the playground to have an enclosed area of 5000 ft^2?

Solution

3c. Construct a diagram of the situation. Let x represent the length of the playground and let y represent the width.

The total length of the fence is 200 ft, so $2x + y = 200$. The area of the playground is 5000 ft^2, so $xy = 5000$. Since we have two variables and two equations, we must solve one of the equations for one of the variables and substitute the expression in the other equation.

When we solve $2x + y = 200$ for y, we get $y = 200 - 2x$. Now we substitute $200 - 2x$ in $xy = 5000$ in place of y.

$xy = 5000$	Begin with the second equation.
$x(200 - 2x) = 5000$	Replace y with $200 - 2x$.
$200x - 2x^2 = 5000$	Apply the distributive property.
$-2x^2 + 200x - 5000 = 0$	Subtract 5000 from each side.
$x^2 - 100x + 2500 = 0$	Divide each side by -2.

The equation is in standard form, so $a = 1$, $b = -100$, and $c = 2500$.

$x = \dfrac{-b \pm \sqrt{b^2 - 4ac}}{2a}$	State the quadratic formula.
$x = \dfrac{-(-100) \pm \sqrt{(-100)^2 - 4(1)(2500)}}{2(1)}$	Replace a, b, and c with the appropriate values.
$x = \dfrac{100 \pm \sqrt{10{,}000 - 10{,}000}}{2}$	Simplify the expression.
$x = \dfrac{100 \pm \sqrt{0}}{2}$	Simplify the radicand.
$x = \dfrac{100 \pm 0}{2}$	Simplify the radical.
$x = 50$	Simplify.

So, the child care center should construct the playground with a length of 50 ft. The width of the playground is

$$\begin{aligned} y &= 200 - 2x \\ &= 200 - 2(50) \\ &= 200 - 100 \\ &= 100 \text{ ft} \end{aligned}$$

3d. At noon, Nguyen left on his bike and traveled due north with an average speed of 5 mph. One hour later, Tranh left from the same point and traveled due east with an average speed of 7 mph. At what time will the boys be 50 mi apart?

Solution

3d. Construct a diagram and label the distances traveled. Let x represent the time that Nguyen has traveled. Since Nguyen travels at 5 mph, his total distance traveled is $5x$ mi. Tranh left an hour later, so his time traveled is $x - 1$. Tranh's total distance traveled is $7(x - 1)$ mi. Since Nguyen traveled due north and Tranh traveled due east, a right triangle is formed.

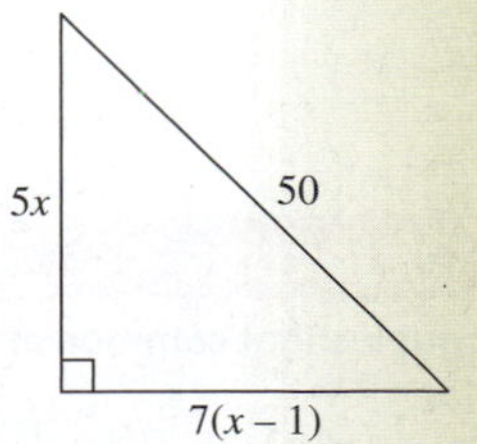

So, we use the Pythagorean theorem to solve the problem.

$a^2 + b^2 = c^2$	State the Pythagorean theorem.
$(5x)^2 + [7(x - 1)]^2 = (50)^2$	Replace *a*, *b*, and *c* with the legs and hypotenuse.
$25x^2 + 49(x^2 - 2x + 1) = 2500$	Square each expression.
$25x^2 + 49x^2 - 98x + 49 = 2500$	Apply the distributive property.
$74x^2 - 98x + 49 - 2500 = 2500 - 2500$	Combine like terms. Subtract 2500 from each side.
$74x^2 - 98x - 2451 = 0$	Simplify.

The equation is in standard form, so $a = 74$, $b = -98$, and $c = -2451$.

$x = \dfrac{-b \pm \sqrt{b^2 - 4ac}}{2a}$	State the quadratic formula.
$x = \dfrac{-(-98) \pm \sqrt{(-98)^2 - 4(74)(-2451)}}{2(74)}$	Replace *a*, *b*, and *c* with the appropriate values.
$x = \dfrac{98 \pm \sqrt{735{,}100}}{148}$	Simplify the expression.
$x = \dfrac{98 + \sqrt{735{,}100}}{148} \approx 6.46$	Write the two solutions and approximate their values.
$x = \dfrac{98 - \sqrt{735{,}100}}{148} \approx -5.13$	

Because time is not negative, we can omit -5.13 as a solution. So, Nguyen and Tranh will be 50 mi apart in 6.46 hr after noon, or at approximately 6:28 P.M.

Student Check 3 Solve each problem.

a. The height, in feet, of a ball thrown upward from a height of 10 ft is $h = -16t^2 + 32t + 10$ after t seconds. How many seconds will it take for the ball to reach a height of 20 ft? Round the answer to the nearest hundredth.

b. A company knows that the cost, in thousands of dollars, to produce x hundred items is given by the cost function $C(x) = -0.1x^2 + 2x + 5$. It also knows that the revenue, in thousands of dollars, from selling x hundred items is given by the revenue function $R(x) = 0.35x$. How many items does the company need to produce to break even, that is, when will revenue = cost? Round the answer to the nearest integer.

c. A farmer wants to enclose a pen at the back of his barn, using the barn as one of its borders. The farmer has 400 ft of fencing. What dimensions should the pen have to enclose an area of 16,000 ft^2?

d. Susan and her husband, Jim, both have to travel out of town. Susan leaves at 7 A.M. and travels west at 60 mph. Jim leaves 2 hr later and travels south at 75 mph. At what time will Susan and Jim be 200 mi apart?

Objective 4 Troubleshoot common errors.

Troubleshooting Common Errors

Some common errors associated with the quadratic formula are shown.

Objective 4 Examples **A problem and an incorrect solution are given. Provide the correct solution and an explanation of the error.**

4a. Solve $x^2 - 7x = 9$.

Incorrect Solution	Correct Solution and Explanation
$x^2 - 7x = 9$ So, $a = 1$, $b = -7$, and $c = 9$. $x = \frac{-b \pm \sqrt{b^2 - 4ac}}{2a}$ $x = \frac{-(-7) \pm \sqrt{(-7)^2 - 4(1)(9)}}{2(1)}$ $x = \frac{7 \pm \sqrt{13}}{2}$ The solution set is $\left\{\frac{7 \pm \sqrt{13}}{2}\right\}$.	The equation must be written in standard form before using the quadratic formula. $x^2 - 7x - 9 = 0$ So, $a = 1$, $b = -7$, and $c = -9$. $x = \frac{-b \pm \sqrt{b^2 - 4ac}}{2a}$ $x = \frac{-(-7) \pm \sqrt{(-7)^2 - 4(1)(-9)}}{2(1)}$ $x = \frac{7 \pm \sqrt{85}}{2}$ The solution set is $\left\{\frac{7 \pm \sqrt{85}}{2}\right\}$.

4b. Solve $x^2 + 4x + 2 = 0$.

Incorrect Solution	Correct Solution and Explanation
$x^2 + 4x + 2 = 0$ So, $a = 1$, $b = 4$, and $c = 2$. $x = \frac{-b \pm \sqrt{b^2 - 4ac}}{2a}$ $x = \frac{-(4) \pm \sqrt{(4)^2 - 4(1)(2)}}{2(1)}$ $x = \frac{-4 \pm \sqrt{8}}{2}$ $x = -2 \pm \sqrt{8}$ $x = -2 \pm 2\sqrt{2}$ The solution set is $\{-2 \pm 2\sqrt{2}\}$.	The denominator of 2 must divide into *each* term in the numerator. Simplify the radical first and then divide out the common factor. $x = \frac{-4 \pm \sqrt{8}}{2}$ $x = \frac{-4 \pm 2\sqrt{2}}{2}$ $x = \frac{2(-2 \pm \sqrt{2})}{2}$ $x = -2 \pm \sqrt{2}$ The solution set is $\{-2 \pm \sqrt{2}\}$.

ANSWERS TO STUDENT CHECKS

Student Check 1 **a.** $\{6\}$ **b.** $\left\{\frac{2 \pm i}{5}\right\}$ **c.** $\left\{-\frac{2}{7}, 0\right\}$ **d.** $\left\{\frac{-1 \pm \sqrt{193}}{12}\right\}$

Student Check 2 **a.** -3; two complex, nonreal solutions **b.** 49; two rational solutions **c.** 220; two irrational solutions

Student Check 3 **a.** The ball will reach 10 ft in 0.39 sec and 1.61 sec. **b.** The company needs to produce 1912 items to break even. **c.** The pen can either be 55.3 ft by 289.4 ft or 144.7 ft by 110.6 ft. **d.** They will be 200 mi apart in approximately 3 hr or at 10 A.M.

SUMMARY OF KEY CONCEPTS

1. We now have four different methods for solving quadratic equations.
   a. The zero products property (factoring)—doesn't solve all equations.
   b. The square root property—solves equations of the form $a^2 = k$.
   c. Completing the square—solves all equations but can be very tedious.
   d. The quadratic formula—solves all equations that are in standard form.
2. Quadratic equations have at most two real solutions. The discriminant, $b^2 - 4ac$, can be used to determine the number and type of solutions of a quadratic equation. If the discriminant is positive, there are two real solutions. If it is zero, there is one real solution. If it is negative, there are no real solutions.

SECTION 9.3 EXERCISE SET

Write About It!

Use complete sentences in your answer to each exercise. For Exercises 1–4, explain the method that you would use to solve the given equation.

1. $(x-2)(x-3) = 0$
2. $(x+4)^2 = 24$
3. $x^2 - 4x + 2 = 0$
4. $x^2 - 3x + 4 = 0$
5. Explain the advantage of using the quadratic formula to solve a quadratic equation.
6. Explain what you must do to an equation before using the quadratic formula to solve it. Write the equation in standard form and identify the coefficients.
7. How can you use the discriminant to determine the type and number of solutions of a quadratic equation?
8. How can you use the discriminant to determine if the equation can be solved by factoring? If the discriminant is a positive perfect square, then the equation can be solved by factoring.

Practice Makes Perfect!

Solve each equation using the quadratic formula. (*See Objective 1.*)

9. $x^2 - 7x + 12 = 0$ $\{3, 4\}$
10. $y^2 + 4y - 12 = 0$ $\{-6, 2\}$
11. $x^2 - 9x + 13 = 0$
12. $9y^2 - 6y = 1$ $\left\{\frac{1 \pm \sqrt{2}}{3}\right\}$
13. $x(x+8) + 18 = 0$ $\{-4 \pm i\sqrt{2}\}$
14. $y(y+6) + 12 = 0$ $\{-3 \pm i\sqrt{3}\}$
15. $2a^2 + 11a = 0$
16. $3b^2 - 2b = 0$
17. $81z^2 = 18z - 11$
18. $6w^2 = 24w - 25$ $\left\{\frac{12 \pm i\sqrt{6}}{6}\right\}$
19. $a^2 = -20a - 77$ $\{-10 \pm \sqrt{23}\}$
20. $4b^2 = 8b + 9$ $\left\{\frac{2 \pm \sqrt{13}}{2}\right\}$
21. $x^2 - \frac{1}{3}x - 8 = 0$ $\left\{-\frac{8}{3}, 3\right\}$
22. $b^2 + \frac{8}{15}b + \frac{1}{15} = 0$ $\left\{-\frac{1}{3}, -\frac{1}{5}\right\}$
23. $z^2 - \frac{5}{3}z + \frac{1}{18} = 0$
24. $w^2 - \frac{1}{4}w - \frac{5}{16} = 0$ $\left\{\frac{1 \pm \sqrt{21}}{8}\right\}$

Find the discriminant and use it to determine the number and types of solutions of each equation. (*See Objective 2.*)

25. $8x^2 - 5x - 2 = 0$ two irrational solutions
26. $10y^2 - 4y - 3 = 0$ two irrational solutions
27. $3a^2 - 4a + 7 = 0$ two complex, nonreal solutions
28. $8b^2 + 4b + 1 = 0$ two complex, nonreal solutions
29. $8r^2 - 7r - 1 = 0$ two rational solutions
30. $2s^2 - s - 3 = 0$ two rational solutions
31. $4a^2 - 12a + 9 = 0$ one repeating rational solution
32. $25b^2 + 20b + 4 = 0$ one repeating rational solution

Solve each problem. (*See Objective 3.*)

33. The height, in feet, of a ball thrown upward from a height of 9 ft is $h = -16t^2 + 45t + 9$ after t seconds. How many seconds will it take for the ball to reach a height of 24 ft? Round the answer to two decimal places. 0.39 sec and 2.43 sec
34. The height, in feet, of a ball thrown upward from a height of 11 ft is $h = -16t^2 + 50t + 11$ after t seconds. How many seconds will it take for the ball to reach a height of 20 ft? Round the answer to two decimal places. 0.19 sec and 2.93 sec
35. A company knows that the cost, in thousands of dollars, to produce x hundred items is given by the cost function $C(x) = -0.2x^2 + 10x + 5$. It also knows that the revenue, in thousands of dollars, from selling x hundred items is given by the revenue function $R(x) = 0.1x$. How many items does the company need to produce to break even, that is, when does revenue = cost? 5000 items
36. A company knows that the cost, in thousands of dollars, to produce x hundred items is given by the cost function $C(x) = -0.3x^2 + 5x + 10$. It also knows that the revenue, in thousands of dollars, from selling x hundred items is given by the revenue function $R(x) = 3.85x$. How many items does the company need to produce to break even, that is, when does revenue = cost? 800 items
37. A farmer wants to enclose a pen at the back of his barn, using the barn as one of its borders. The farmer has 320 ft of fencing. What dimensions should the pen have to enclose an area of 12,800 ft²? 80 ft by 160 ft

38. A farmer wants to enclose a pen at the back of his barn, using the barn as one of its borders. The farmer has 460 ft of fencing. What dimensions should the pen have to enclose an area of 26,450 ft²? 115 ft by 230 ft

Additional answers can be found in the Instructor Answer Appendix.

39. Josh and his wife, Christy, each have to travel out of town for meetings. The towns where they are meeting are 250 mi apart and are due north and due east from their home. Josh leaves at 9 A.M. and travels at 50 mph, and Christy leaves 2 hr later and travels at 75 mph. When will they reach their destinations and be 250 mi apart? 1 P.M.

40. Mike and Brian each have to travel out of town for meetings. The towns where they are meeting are 230 mi apart and are due east and due south from their home. Mike leaves at 9 A.M. and travels at 60 mph, and Brian leaves 1 hr later and travels at 70 mph. When will they reach their destinations and be 230 mi apart? 12:01 P.M.

Mix 'Em Up!

Solve each equation using the quadratic formula.

41. $x^2 + 6x - 16 = 0$ $\{-8, 2\}$

42. $y^2 - 4y - 32 = 0$ $\{-4, 8\}$

43. $3a^2 + 4a - 20 = 0$

44. $2b^2 + 13b + 20 = 0$ $\left\{-4, -\frac{5}{2}\right\}$

45. $2x^2 - 16x + 27 = 0$

46. $4y^2 + 8y - 11 = 0$ $\left\{\frac{-2 \pm \sqrt{15}}{2}\right\}$

47. $x(x - 3) + 6 = 0$

48. $2y(y + 4) + 15 = 0$

49. $5a^2 + 9a = 0$

50. $3b^2 + b = 0$ $\left\{-\frac{1}{3}, 0\right\}$

51. $2z^2 = 10z - 5$

52. $w^2 = 2w + 21$ $\{1 \pm \sqrt{22}\}$

53. $a^2 = \frac{1}{2}a + \frac{9}{16}$

54. $\frac{1}{7}b^2 = b - \frac{23}{28}$ $\left\{\frac{7 \pm \sqrt{26}}{2}\right\}$

55. $x^2 - \frac{7}{6}x - 4 = 0$

56. $b^2 + \frac{3}{4}b + \frac{1}{8} = 0$ $\left\{-\frac{1}{2}, -\frac{1}{4}\right\}$

57. $z^2 - \frac{2}{3}z + \frac{1}{3} = 0$

58. $w^2 + \frac{1}{2}w + \frac{7}{4} = 0$ $\left\{\frac{-1 \pm 3i\sqrt{3}}{4}\right\}$

Find the discriminant and use it to determine the number and types of solutions of each equation.

59. $x^2 + 2x - 6 = 0$ two irrational solutions

60. $3y^2 - 5y - 10 = 0$ two irrational solutions

61. $9a^2 + a + 2 = 0$ two complex, nonreal solutions

62. $2b^2 + 3b + 11 = 0$ two complex, nonreal solutions

63. $2r^2 - 9r - 18 = 0$ two rational solutions

64. $3s^2 - 11s - 42 = 0$ two rational solutions

65. $9a^2 + 30a + 25 = 0$ one repeating rational solution

66. $4b^2 - 28b + 49 = 0$ one repeating rational solution

Solve each problem.

67. A company knows that the cost, in thousands of dollars, to produce x hundred items is given by the cost function $C(x) = -0.2x^2 + x + 5$. It also knows that the revenue, in thousands of dollars, from selling x hundred items is given by the revenue function $R(x) = 1.45x$. How many items does the company need to produce to break even, that is, when does revenue = cost? 400 items

68. A company knows that the cost, in thousands of dollars, to produce x hundred items is given by the cost function $C(x) = -0.5x^2 + 2x + 3$. It also knows that the revenue, in thousands of dollars, from selling x hundred items is given by the revenue function $R(x) = 3.25x$. How many items does the company need to produce to break even, that is, when does revenue = cost? 150 items

69. A farmer wants to enclose a pen at the back of his barn, using the barn as one of its borders. The farmer has 280 ft of fencing. What dimensions should the pen have to enclose an area of 9800 ft^2? 70 ft by 140 ft

70. A farmer wants to enclose a pen at the back of his barn, using the barn as one of its borders. The farmer has 640 ft of fencing. What dimensions should the pen have to enclose an area of 51,200 ft^2? 160 ft by 320 ft

71. Mark and Nina each have to travel out of town for meetings. The towns where they are meeting are 220 mi apart and are due west and due north from their home. Mark leaves at 7 A.M. and travels at 60 mph, and Nina leaves 2 hr later and travels at 77 mph. When will they reach their destinations and be 220 mi apart? 10:17 A.M.

72. Pat and Nancy each have to travel out of town for meetings. The towns where they are meeting are 170 mi apart and are due east and due south from their home. Pat leaves at 6 A.M. and travels at 65 mph, and Nancy leaves 3 hr later and travels at 77 mph. When will they reach their destinations and be 170 mi apart? 8:34 A.M.

You Be the Teacher!

Correct each student's errors, if any.

73. Solve the equation $x^2 - 3x - 4 = 0$ using the quadratic formula.

Josh's work:

$$x^2 - 3x - 4 = 0$$
$$x = \frac{3 \pm \sqrt{-3^2 - 4(-4)}}{2}$$
$$x = \frac{3 \pm \sqrt{-9 + 16}}{2}$$
$$x = \frac{3 \pm \sqrt{7}}{2}$$

74. Solve the equation $4x^2 + 16x + 13 = 0$ using the quadratic formula.

Matt's work:

$$4x^2 + 16x + 13 = 0$$
$$x = \frac{-16 \pm \sqrt{16^2 - 4(4)(13)}}{4}$$
$$x = \frac{-16 \pm \sqrt{256 - 208}}{4}$$
$$x = \frac{-16 \pm \sqrt{48}}{4}$$
$$x = -4 \pm \sqrt{12}$$
$$x = -4 \pm 2\sqrt{3}$$

75. Solve the equation $2x^2 - 5x - 8 = 0$ using the quadratic formula and approximate the solutions to two decimal places.

Bruce's work:

$2x^2 - 5x - 8 = 0$

$$x = \frac{5 \pm \sqrt{5^2 - 4(2)(-8)}}{4}$$

$$x = \frac{5 \pm \sqrt{89}}{4}$$

```
5+√(89)/4
          7.358495283
5-√(89)/4
          2.641504717
```

The approximate solutions are 7.36 and 2.64.

76. Solve the equation $3x^2 + 5x - 6 = 0$ using the quadratic formula and approximate the solutions to two decimal places.

Beth's work:

$3x^2 + 5x - 6 = 0$

$$x = \frac{-5 \pm \sqrt{-5^2 - 4(3)(-6)}}{6}$$

```
-5+√(-5²-4*3*-6)
/6
          -3.8573909
-5-√(-5²-4*3*-6)
/6
          -6.1426091
```

So, the approximate solutions are −3.86 and −6.14.

PIECE IT TOGETHER SECTIONS 9.1–9.3 Review

Graph each function by plotting the vertex and two additional points. Identify the x-intercepts, y-intercept, the vertex, the axis of symmetry, domain, and range. Explain how the graph of the function relates to the graph of $y = x^2$. (*Section 9.1, Objective 4*)

1. $f(x) = (x - 2)^2 + 6$ **2.** $f(x) = (x + 4)^2 - 9$

Graph each function and use the graph to solve the equation $f(x) = 0$. State the zeros of $f(x)$ and the x-intercepts of the graph of $f(x)$. (*Section 9.1, Objective 5; Section 9.2, Objective 6*)

3. $f(x) = (x - 5)^2 - 9$ **4.** $f(x) = -(x + 3)^2 + 1$

Solve each equation by applying the square root property, completing the square, or using the quadratic formula. (*Section 9.2, Objectives 1 and 4; Section 9.3, Objective 1*)

5. $(7a + 1)^2 + 2 = 6$ **6.** $18x^2 - 8 = 90$

7. $3y^2 - y - 14 = 0$ **8.** $2a^2 + 10a + 13 = 0$

Create a perfect square trinomial from each binomial by completing the square and then write it in factored form. (*Section 9.2, Objective 2*)

9. $y^2 - 22y$
$y^2 - 22y + 121 = (y - 11)^2$

10. $b^2 - 7b$
$b^2 - 7b + \frac{49}{4} = \left(b - \frac{7}{2}\right)^2$

Find the discriminant and use it to determine the number and types of solutions of each equation. (*Section 9.3, Objective 2*)

11. $2x^2 + 2x - 3 = 0$
28; two irrational solutions

12. $x^2 - 6x + 16 = 0$
−28; two complex solutions

13. $3y^2 - 14y - 5 = 0$
256; two rational solutions

14. $25y^2 - 20y + 4 = 0$
0; one repeating rational solution

Solve each problem. (*Section 9.2, Objective 5; Section 9.3, Objective 3*)

15. If a penny is dropped from the top of the John Hancock Center in Chicago, its height, in meters, t sec after it is dropped is given by $s = -9.8t^2 + 344$. How many seconds will it take for the penny to reach the ground? Round the answer to two decimal places. (Source: http://www.emporis.com) 5.92 sec

16. A company knows that the cost, in thousands of dollars, to produce x hundred items is given by the cost function $C(x) = -0.3x^2 + 2x + 6$. It also knows that the revenue, in thousands of dollars, from selling x hundred items is given by the revenue function $R(x) = 2.3x$. How many items does the company need to produce to break even, that is, when does revenue = cost? 400 items

SECTION 9.4 Solving Equations Using Quadratic Methods

OBJECTIVES

As a result of completing this section, you will be able to

1. Solve rational equations.
2. Solve radical equations.
3. Solve higher degree polynomial equations.
4. Solve equations by substitution.
5. Solve applications of equations that use quadratic methods.
6. Troubleshoot common errors.

A local YMCA is holding a special biathlon event that consists of 1 mi of running and $\frac{1}{2}$ mi of swimming. Taylor finished the event in 25 min. He swam an average of 4 mph less than he ran. What was Taylor's speed running and swimming?

To solve this problem, we must solve a rational equation that can be written as a quadratic equation. We then will use one of the methods we have learned for solving quadratic equations: factoring, applying the square root property, completing the square, or using the quadratic formula. In this section, we will use these methods to solve rational, radical, and higher degree polynomial equations.

Rational Equations

Objective 1 Solve rational equations.

Recall that a **rational equation** is an equation that contains a rational expression. Generally speaking, these are equations that contain variables in the denominator. When we solve a rational equation, we clear the fractions and then solve the resulting equation. The key to working with these equations is that we must check our solutions to make sure we do not have an extraneous solution; that is, one that makes one of the rational expressions undefined.

Procedure: Solving a Rational Equation

Step 1: Multiply each side of the equation by the LCD to clear fractions.
Step 2: Solve the resulting equation. (In this section, the resulting equation will be quadratic.)
Step 3: Exclude from the solution set any possible solutions that make any of the denominators zero.
Step 4: Check by substituting the possible solutions in the original equation.

Objective 1 Examples Solve each equation.

1a. $\frac{14}{x} = x - 5$ **1b.** $\frac{7y}{y-3} - \frac{y+2}{y+3} = \frac{4}{y^2-9}$

Solutions **1a.**

$$\frac{14}{x} = x - 5$$

$$x\left(\frac{14}{x}\right) = x(x-5) \quad \text{Multiply each side by the LCD, } x.$$

$$14 = x^2 - 5x \quad \text{Simplify.}$$

$$0 = x^2 - 5x - 14 \quad \text{Subtract 14 from each side.}$$

$$0 = (x-7)(x+2) \quad \text{Factor.}$$

$$x - 7 = 0 \quad \text{or} \quad x + 2 = 0 \quad \text{Set each equation equal to zero.}$$

$$x = 7 \qquad x = -2 \quad \text{Solve each equation.}$$

We can check the solutions in the original equation.

Let $x = 7$:	Let $x = -2$:
$\frac{14}{x} = x - 5$	$\frac{14}{x} = x - 5$
$\frac{14}{7} = 7 - 5$	$\frac{14}{-2} = -2 - 5$
$2 = 2$	$-7 = -7$
True	True

Since both solutions make the original equation true, the solution set is $\{-2, 7\}$.

1b. Factoring the last denominator gives us $y^2 - 9 = (y - 3)(y + 3)$. So, the LCD is $(y - 3)(y + 3)$. Multiply each side by the LCD and simplify.

$$\frac{7y}{y-3} - \frac{y+2}{y+3} = \frac{4}{y^2-9}$$

$$(y-3)(y+3)\frac{7y}{y-3} - (y-3)(y+3)\frac{y+2}{y+3} = (y-3)(y+3)\frac{4}{(y-3)(y+3)}$$

$$7y(y+3) - (y+2)(y-3) = 4$$

$$7y^2 + 21y - (y^2 - y - 6) = 4$$

$$7y^2 + 21y - y^2 + y + 6 = 4$$

$$6y^2 + 22y + 2 = 0$$

$$3y^2 + 11y + 1 = 0$$

This equation does not factor, so we will use the quadratic formula to solve it with $a = 3$, $b = 11$, $c = 1$.

$$y = \frac{-b \pm \sqrt{b^2 - 4ac}}{2a}$$ State the quadratic formula.

$$y = \frac{-(11) \pm \sqrt{(11)^2 - 4(3)(1)}}{2(3)}$$ Replace *a*, *b*, and *c* with the appropriate values.

$$y = \frac{-11 \pm \sqrt{121 - 12}}{6}$$ Simplify the expression.

$$y = \frac{-11 \pm \sqrt{109}}{6}$$ Simplify the radicand.

Since neither solution makes the denominators zero, the solution set is $\left\{\frac{-11 + \sqrt{109}}{6}, \frac{-11 - \sqrt{109}}{6}\right\}$ or $\left\{\frac{-11 \pm \sqrt{109}}{6}\right\}$.

Student Check 1 Solve each equation.

a. $\frac{20}{x} = x + 8$ **b.** $\frac{3a}{a-1} + \frac{a+2}{2a} = \frac{1}{2a^2 - 2a}$

Radical Equations

Objective 2 Solve radical equations.

Radical equations are equations containing radicals. In this section, we will use only square roots, not higher roots, in the equations. To solve these equations, recall that we first isolate the square root on one side of the equation and then square both sides to eliminate the radical. These equations are like rational equations in that we must check the possible solutions to ensure they are not extraneous solutions.

Procedure: Solving a Radical Equation

Step 1: Isolate the radical expression to one side of the equation.
Step 2: Square each side of the equation to eliminate the radical.
Step 3: Solve the resulting quadratic equation.
Step 4: Check the possible solutions and exclude any extraneous solutions.

Objective 2 Examples Solve each equation.

2a. $2x = \sqrt{7x + 2}$ **2b.** $p - 2\sqrt{p} = 8$

Solutions **2a.**

$$2x = \sqrt{7x + 2}$$

$$(2x)^2 = \left(\sqrt{7x + 2}\right)^2 \quad \text{Square each side.}$$

$$4x^2 = 7x + 2 \quad \text{Simplify.}$$

$$4x^2 - 7x - 2 = 0 \quad \text{Subtract } 7x \text{ and 2 from each side.}$$

$$(4x + 1)(x - 2) = 0 \quad \text{Factor the resulting equation.}$$

$$4x + 1 = 0 \quad \text{or} \quad x - 2 = 0 \quad \text{Set each factor equal to 0.}$$

$$x = -\frac{1}{4} \qquad x = 2 \quad \text{Solve each equation.}$$

We check the solutions in the original equation.

$x = -\frac{1}{4}$:

$$2x = \sqrt{7x + 2}$$

$$2\left(-\frac{1}{4}\right) = \sqrt{7\left(-\frac{1}{4}\right) + 2}$$

$$-\frac{1}{2} = \sqrt{\frac{1}{4}}$$

$$-\frac{1}{2} = \frac{1}{2}$$

False

$x = 2$:

$$2x = \sqrt{7x + 2}$$

$$2(2) = \sqrt{7(2) + 2}$$

$$4 = \sqrt{16}$$

$$4 = 4$$

True

The only solution that checks is 2. So, the solution set is $\{2\}$.

2b.

$$p - 2\sqrt{p} = 8$$

$$p - 8 = 2\sqrt{p} \quad \text{Subtract 8 and add } 2\sqrt{p} \text{ to each side.}$$

$$(p - 8)^2 = \left(2\sqrt{p}\right)^2 \quad \text{Square each side.}$$

$$p^2 - 16p + 64 = 4p \quad \text{Simplify.}$$

$$p^2 - 20p + 64 = 0 \quad \text{Subtract } 4p \text{ from each side.}$$

$$(p - 16)(p - 4) = 0 \quad \text{Factor the trinomial.}$$

$$p - 16 = 0 \quad \text{or} \quad p - 4 = 0 \quad \text{Set each factor equal to zero.}$$

$$p = 16 \qquad p = 4 \quad \text{Solve each equation.}$$

We check the solutions in the original equation.

$p = 16$:	$p = 4$:
$p - 2\sqrt{p} = 8$	$p - 2\sqrt{p} = 8$
$16 - 2\sqrt{16} = 8$	$4 - 2\sqrt{4} = 8$
$16 - 2(4) = 8$	$4 - 2(2) = 8$
$16 - 8 = 8$	$4 - 4 = 8$
$8 = 8$	$0 = 8$
True	False

The only solution that checks is 16. So, the solution set is $\{16\}$.

Student Check 2 Solve each equation.

a. $3x = \sqrt{5x + 4}$ **b.** $a - \sqrt{a} = 12$

Higher Degree Polynomial Equations

Objective 3 Solve higher degree polynomial equations.

Higher degree polynomial equations are polynomial equations with a degree greater than 2. We will solve these types of equations by initially applying the zero products property. That is, we set the equation equal to zero and factor. In Section 6.8, we solved these types of equations. An example is shown.

$$x^3 = 64x$$

$$x^3 - 64x = 0 \quad \text{Write the equation in standard form.}$$

$$x(x^2 - 64) = 0 \quad \text{Factor out the GCF.}$$

$$x(x - 8)(x + 8) = 0 \quad \text{Factor the difference of squares.}$$

$$x = 0 \quad \text{or} \quad x - 8 = 0 \quad \text{or} \quad x + 8 = 0 \quad \text{Set each factor equal to zero.}$$

$$x = 8 \qquad x = -8 \quad \text{Solve.}$$

The solution set is $\{-8, 0, 8\}$. Note that the degree of the polynomial is 3 and there are three solutions. Notice that we were able to factor the polynomial until it consisted of three linear factors.

In this section, we will solve polynomial equations that we can initially factor. But we will find that some of the factors can't be factored any further. So, we may have to apply the square root property, complete the square, or use the quadratic formula to solve the equation. Consider the following example.

$$x^3 = -64x$$

$$x^3 + 64x = 0 \quad \text{Write the equation in standard form.}$$

$$x(x^2 + 64) = 0 \quad \text{Factor out the GCF.}$$

$$x = 0 \quad \text{or} \quad x^2 + 64 = 0 \quad \text{Set each factor equal to zero.}$$

$$x^2 = -64 \quad \text{Write the equation in the form } a^2 = k.$$

$$x = \pm\sqrt{-64} \quad \text{Apply the square root property.}$$

$$x = \pm 8i \quad \text{Simplify the radical.}$$

So, the solution set is $\{0, -8i, 8i\}$. Note that we were able to factor after we wrote the equation in standard form, but then we couldn't factor $x^2 + 64$ since it is the sum of squares. So, we had to set that factor equal to zero and apply another technique of solving quadratic equations, the square root property. Also note the degree of the equation is 3 and there are three solutions (one real solution and two imaginary solutions).

Procedure: Solving a Higher Degree Polynomial Equation

Step 1: Write the equation in standard form (one side is equal to zero).
Step 2: Factor the polynomial.
Step 3: Set each factor equal to zero and solve.

Objective 3 Examples

Solve each equation.

3a. $x^4 - 5x^2 - 36 = 0$ **3b.** $4y^4 - 5y^2 = -1$ **3c.** $x^3 = 8$

Solutions

3a.

$$x^4 - 5x^2 - 36 = 0$$

$$(x^2 - 9)(x^2 + 4) = 0 \quad \text{Factor the polynomial.}$$

$$x^2 - 9 = 0 \quad \text{or} \quad x^2 + 4 = 0 \quad \text{Set each factor equal to zero.}$$

$$x^2 = 9 \qquad x^2 = -4 \quad \text{Write each equation in the form } a^2 = k.$$

$$x = \pm\sqrt{9} \qquad x = \pm\sqrt{-4} \quad \text{Apply the square root property.}$$

$$x = \pm 3 \qquad x = \pm 2i \quad \text{Simplify each radical.}$$

So, the solution set is $\{-3, 3, -2i, 2i\}$ or $\{\pm 3, \pm 2i\}$. We can check the solutions.

Let $x = -3$:

$$x^4 - 5x^2 - 36 = 0$$
$$(-3)^4 - 5(-3)^2 - 36 = 0$$
$$81 - 5(9) - 36 = 0$$
$$81 - 45 - 36 = 0$$
$$36 - 36 = 0$$
$$0 = 0$$

Let $x = 3$:

$$x^4 - 5x^2 - 36 = 0$$
$$(3)^4 - 5(3)^2 - 36 = 0$$
$$81 - 5(9) - 36 = 0$$
$$81 - 45 - 36 = 0$$
$$36 - 36 = 0$$
$$0 = 0$$

Let $x = -2i$:

$$x^4 - 5x^2 - 36 = 0$$
$$(-2i)^4 - 5(-2i)^2 - 36 = 0$$
$$16i^4 - 5(4i^2) - 36 = 0$$
$$16 - 5(-4) - 36 = 0$$
$$16 + 20 - 36 = 0$$
$$0 = 0$$

Let $x = 2i$:

$$x^4 - 5x^2 - 36 = 0$$
$$(2i)^4 - 5(2i)^2 - 36 = 0$$
$$16i^4 - 5(4i^2) - 36 = 0$$
$$16 - 5(-4) - 36 = 0$$
$$16 + 20 - 36 = 0$$
$$0 = 0$$

3b.

$$4y^4 - 5y^2 = -1$$

$$4y^4 - 5y^2 + 1 = 0 \quad \text{Add 1 to each side.}$$

$$(4y^2 - 1)(y^2 - 1) = 0 \quad \text{Factor.}$$

$$4y^2 - 1 = 0 \quad \text{or} \quad y^2 - 1 = 0 \quad \text{Set each factor equal to zero.}$$

$$y^2 = \frac{1}{4} \qquad y^2 = 1 \quad \text{Write each equation in the form } a^2 = k.$$

$$y = \pm\sqrt{\frac{1}{4}} \qquad y = \pm\sqrt{1} \quad \text{Apply the square root property.}$$

$$y = \pm\frac{1}{2} \qquad y = \pm 1 \quad \text{Simplify each radical.}$$

The solution set is $\left\{-\frac{1}{2}, -1, \frac{1}{2}, 1\right\}$ or $\left\{\pm\frac{1}{2}, \pm 1\right\}$.

3c.

$$x^3 = 8$$

$$x^3 - 8 = 0 \quad \text{Write the equation in standard form.}$$

$$(x - 2)(x^2 + 2x + 4) = 0 \quad \text{Factor.}$$

$$x - 2 = 0 \quad \text{or} \quad x^2 + 2x + 4 = 0 \quad \text{Set each factor equal to zero.}$$

$$x = 2 \qquad x = \frac{-b \pm \sqrt{b^2 - 4ac}}{2a} \quad \text{Solve. Apply the quadratic formula.}$$

$$x = \frac{-(2) \pm \sqrt{(2)^2 - 4(1)(4)}}{2(1)}$$ Let $a = 1$, $b = 2$, and $c = 4$.

$$x = \frac{-2 \pm \sqrt{-12}}{2}$$ Simplify the radicand.

$$x = \frac{-2 \pm \sqrt{-4 \cdot 3}}{2}$$ Find a perfect square factor of -12.

$$x = \frac{-2 \pm 2i\sqrt{3}}{2}$$ Apply the product rule for radicals.

$$x = -1 \pm i\sqrt{3}$$ Simplify.

The solution set is $\{2, -1 + i\sqrt{3}, -1 - i\sqrt{3}\}$ or $\{2, -1 \pm i\sqrt{3}\}$.

Student Check 3 Solve each equation.

a. $x^4 - 15x^2 - 16 = 0$ **b.** $9x^4 - 10x^2 = -1$ **c.** $x^3 = 27$

Note: *Example 3 illustrates that the degree of the equation relates to the number of solutions of the equation. In parts 3a and 3b, the degree is 4 and there are 4 solutions. In part 3c, the degree is 3 and there are 3 solutions. It turns out that the degree determines the maximum number of solutions since solutions can repeat.*

Solving Equations Using Substitution

Objective 4 Solve equations by substitution.

To solve an equation by substitution requires us to replace an expression by a variable so that a quadratic equation results. The following table shows some examples of how this is done.

Original Equation	Expression to be Replaced	New Equation
$x^4 - 9x^2 - 10 = 0$ $(x^2)^2 - 9x^2 - 10 = 0$	$u = x^2$	$(u)^2 - 9u - 10 = 0$
$(x - 1)^2 + 5(x - 1) + 4 = 0$	$u = (x - 1)$	$(u)^2 + 5(u) + 4 = 0$
$x^{2/3} + 8x^{1/3} + 16 = 0$ $(x^{1/3})^2 + 8x^{1/3} + 16 = 0$	$u = x^{1/3}$	$(u)^2 + 8u + 16 = 0$

Note that the expression that we replaced with u

- is the base of the squared term.
- is also found in the middle term of the original equation.

After we replace the expression with u, the resulting equation is a quadratic equation that we can solve using the methods of this chapter. When we solve this new equation, we find the values of u that satisfy it. Our goal is to solve for the variable in the original equation. So, after we solve the new equation, we must use these solutions to solve for the original variable.

Procedure: Solving an Equation Using Substitution

Step 1: Let "u" equal the base of the squared term.
Step 2: Rewrite the equation using u.
Step 3: Solve the resulting quadratic equation.
Step 4: Back-substitute to solve for the original variable.

Objective 4 Examples **Solve each equation by substitution.**

4a. $(x+4)^2 - 5(x+4) + 6 = 0$ **4b.** $x^{2/5} - 3x^{1/5} - 4 = 0$

Solutions **4a.** The expression $(x+4)$ is the base of the squared term and it also repeats in the middle term of the equation. So, we let $u = x + 4$.

$(x+4)^2 - 5(x+4) + 6 = 0$	
$u^2 - 5u + 6 = 0$	Substitute u for $x + 4$.
$(u-3)(u-2) = 0$	Factor.
$u - 3 = 0$ or $u - 2 = 0$	Apply the zero products property.
$u = 3$ $\quad u = 2$	Solve.

Now we solve for x using the equation $u = x + 4$.

$u = 3$ $\quad u = 2$	State the solutions for u.
$x + 4 = 3$ $\quad x + 4 = 2$	Substitute $x + 4$ for u.
$x = -1$ $\quad x = -2$	Subtract 4 from each side of each equation.

Both solutions check, so the solution set is $\{-2, -1\}$.

4b. The expression $x^{2/5} = (x^{1/5})^2$, so let $u = x^{1/5}$.

$x^{2/5} - 3x^{1/5} - 4 = 0$	
$(x^{1/5})^2 - 3x^{1/5} - 4 = 0$	Write $x^{2/5}$ as $(x^{1/5})^2$.
$u^2 - 3u - 4 = 0$	Substitute u for $x^{1/5}$.
$(u-4)(u+1) = 0$	Factor.
$u - 4 = 0$ or $u + 1 = 0$	Apply the zero products property.
$u = 4$ $\quad u = -1$	Solve.

Now we solve for x using the equation $u = x^{1/5}$.

$u = 4$ $\quad u = -1$	State the solutions for u.
$x^{1/5} = 4$ $\quad x^{1/5} = -1$	Substitute $x^{1/5}$ for u.
$(x^{1/5})^5 = 4^5$ $\quad (x^{1/5})^5 = (-1)^5$	Apply the power property.
$x = 1024$ $\quad x = -1$	Simplify.

Both solutions check, so the solution set is $\{-1, 1024\}$.

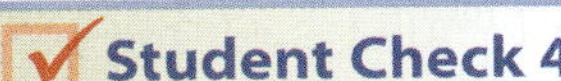

Student Check 4 Solve each equation by substitution.

a. $(x-2)^2 - 4(x-2) + 3 = 0$ **b.** $x^{2/3} + 2x^{1/3} - 8 = 0$

Applications

Objective 5

Solve applications of equations that use quadratic methods.

Distance and work problems are often modeled by equations that require us to use quadratic methods to solve them. Example 5 illustrates these concepts.

Objective 5 Examples **Solve each problem.**

5a. A local YMCA is holding a special biathlon event that consists of 1 mi of running and $\frac{1}{2}$ mi of swimming. Taylor finished the event in 25 min. He swam an average of 4 mph less than he ran. What was his speed running and swimming?

Solution **5a.** We let x represent the speed running and $x - 4$ represent the speed swimming. We organize the information in a table. Recall $d = rt$ or $t = \frac{d}{r}$.

	Distance	Rate	Time
Running	1 mi	x	$\frac{1}{x}$
Swimming	$\frac{1}{2}$ mi	$x - 4$	$\frac{\frac{1}{2}}{x-4} = \frac{1}{2(x-4)}$

The total time running and swimming is 25 min, which is $\frac{25}{60} = \frac{5}{12}$ hr.

$$\frac{1}{x} + \frac{1}{2(x-4)} = \frac{5}{12}$$ Sum of times is 25 min.

$$12x(x-4)\left(\frac{1}{x}\right) + 12x(x-4)\left[\frac{1}{2(x-4)}\right] = 12x(x-4)\left(\frac{5}{12}\right)$$ Multiply each side by the LCD.

$$12(x-4) + 6x = 5x(x-4)$$ Simplify each product.

$$12x - 48 + 6x = 5x^2 - 20x$$ Apply the distributive property.

$$0 = 5x^2 - 38x + 48$$ Write equation in standard form.

$$0 = (5x - 8)(x - 6)$$ Factor.

$$5x - 8 = 0 \quad \text{or} \quad x - 6 = 0$$ Apply the zero products property.

$$x = \frac{8}{5} \qquad x = 6$$ Solve each equation.

The only solution that makes sense is 6 since $\frac{8}{5}$ makes the swimming speed negative. So, Taylor ran at a speed of 6 mph and swam at a speed of 2 mph.

5b. Peter and Jason own a lawn-mowing service. Together, they can mow a lawn in 24 min. Peter mows 20 min faster than John. How long will it take each man working alone to mow the lawn?

Solution **5b.** The unknowns are the times it takes each man to mow the lawn alone. Since Peter's time is 20 min faster than John's time, we let

$$x = \text{time for Jason to mow the lawn}$$
$$x - 20 = \text{time for Peter to mow the lawn}$$

Recall that we need to find the portion of the job completed in 1 min.

	Time to Complete the Job	Portion Completed in 1 Min
Peter	$x - 20$	$\frac{1}{x-20}$
Jason	x	$\frac{1}{x}$
Working together	24	$\frac{1}{24}$

So, the sum of the portions of the job completed in 1 min working alone is equal to the portion of the job completed in 1 min working together.

$$\frac{1}{x-20} + \frac{1}{x} = \frac{1}{24}$$ Write the model.

$24x(x-20)\dfrac{1}{x-20} + 24x(x-20)\dfrac{1}{x} = 24x(x-20)\dfrac{1}{24}$	Multiply each side by the LCD.
$24x + 24(x-20) = x(x-20)$	Simplify each product.
$24x + 24x - 480 = x^2 - 20x$	Apply the distributive property.
$48x - 480 = x^2 - 20x$	Combine like terms.
$0 = x^2 - 68x + 480$	Write the equation in standard form.
$0 = (x-60)(x-8)$	Factor.
$x - 60 = 0 \quad \text{or} \quad x - 8 = 0$	Apply the zero products property.
$x = 60 \qquad x = 8$	Solve.

The only solution that makes sense is 60 since 8 makes Peter's time negative. Therefore, it takes Jason 60 min to mow the lawn alone and Peter 60 − 20 = 40 min to mow the lawn alone.

Student Check 5 Solve each problem.

a. Maria and Juan travel out of town for work. Juan travels 245 mi and Maria travels 420 mi. Maria's speed is 14 mph less than Juan's. If their total time traveling was 11 hr, what was each of their speeds?

b. Tom and Barbara can clean their house in 3 hr working together. Working alone, Tom takes 2 hr longer than Barbara to clean their house. How long will it take each of them working alone to clean their house?

Objective 6 Troubleshoot common errors.

Troubleshooting Common Errors

Some common errors associated with solving equations using quadratic methods are shown.

Objective 6 Examples **A problem and an incorrect solution are given. Provide the correct solution and an explanation of the error.**

6a. Solve $3x = \sqrt{2-7x}$.

Incorrect Solution	Correct Solution and Explanation
$3x = \sqrt{2-7x}$ $(3x)^2 = (\sqrt{2-7x})^2$ $9x^2 = 2 - 7x$ $9x^2 + 7x - 2 = 0$ $(9x-2)(x+1) = 0$ $9x - 2 = 0 \quad \text{or} \quad x + 1 = 0$ $x = \dfrac{2}{9} \qquad x = -1$ The solution set is $\left\{-1, \dfrac{2}{9}\right\}$.	When solving a radical equation, we must check for extraneous solutions. $x = \dfrac{2}{9}$: $3\left(\dfrac{2}{9}\right) = \sqrt{2 - 7\left(\dfrac{2}{9}\right)}$ $\dfrac{2}{3} = \sqrt{2 - \dfrac{14}{9}}$ $\dfrac{2}{3} = \sqrt{\dfrac{4}{9}}$ $\dfrac{2}{3} = \dfrac{2}{3}$ True $x = -1$: $3(-1) = \sqrt{2 - 7(-1)}$ $-3 = \sqrt{2+7}$ $-3 = \sqrt{9}$ $-3 = 3$ False So, the solution set is $\left\{\dfrac{2}{9}\right\}$.

6b. $x^4 - 13x^2 + 36 = 0.$

Incorrect Solution	Correct Solution and Explanation
$x^4 - 13x^2 + 36 = 0$ $(x^2 - 4)(x^2 - 9) = 0$ $x^2 = 4 \quad \text{or} \quad x^2 = 9$ $x = 2 \qquad x = 3$ The solution set is $\{2, 3\}$.	The degree of the equation is 4, which means we should have four solutions. When we solve the two quadratic equations, we get $x = \pm 2 \quad \text{or} \quad x = \pm 3$ The solution set is $\{-3, -2, 2, 3\}$.

6c. $(y + 3)^2 - 4(y + 3) - 32 = 0.$

Incorrect Solution	Correct Solution and Explanation
Let $u = (y + 3)$: $(y + 3)^2 - 4(y + 3) - 32 = 0$ $u^2 - 4u - 32 = 0$ $(u - 8)(u + 4) = 0$ $u = 8 \quad \text{or} \quad u = -4$ The solution set is $\{-4, 8\}$.	We need to back-substitute and find the solutions for y. $u = 8 \quad \text{or} \quad u = -4$ $y + 3 = 8 \qquad y + 3 = -4$ $y = 5 \qquad y = -7$ The solution set is $\{-7, 5\}$.

ANSWERS TO STUDENT CHECKS

Student Check 1 **a.** $\{-10, 2\}$ **b.** $\left\{\frac{-1 \pm \sqrt{85}}{14}\right\}$

Student Check 2 **a.** $\{1\}$ **b.** $\{16\}$

Student Check 3 **a.** $\{-4, 4, -i, i\}$ **b.** $\left\{-1, 1, -\frac{1}{3}, \frac{1}{3}\right\}$ **c.** $\left\{3, \frac{-3 \pm 3i\sqrt{3}}{2}\right\}$

Student Check 4 **a.** $\{3, 5\}$ **b.** $\{-64, 8\}$

Student Check 5 **a.** Juan's speed was 70 mph and Maria's speed was 56 mph. **b.** It will take Barbara 5.16 hr and Tom 7.16 hr to clean their house working alone.

SUMMARY OF KEY CONCEPTS

1. We must apply quadratic techniques to solve rational equations if, after clearing the fractions, a quadratic equation results. The solution set includes only those solutions that do not make any of the rational expressions undefined.
2. After eliminating the radicals from a square root equation, a quadratic equation may result. We solve the resulting equation using any of the methods we have learned in the previous sections. The possible solutions must be checked in the original equation. Exclude any extraneous solutions from the solution set.
3. We solve higher degree polynomial equations by initially factoring the polynomial. Once we obtain either linear or quadratic factors, we set the factors equal to zero and solve. The degree indicates the number of solutions including repetitions.
4. Other equations can be solved by substituting a variable for an appropriate expression that creates a quadratic equation. The key is to back-substitute to find the solutions of the original equation.
5. Work and distance problems can result in rational equations. We solve these by clearing the fractions by multiplying by the LCD.

GRAPHING CALCULATOR SKILLS

The graphing calculator can be used to verify solutions or to determine which solutions might be extraneous.

Example 1: Determine if $-\frac{1}{4}$ and 2 are solutions of $2x = \sqrt{7x + 2}$.

Solution: Enter the left side and right side of the equation into Y_1 and Y_2, respectively.

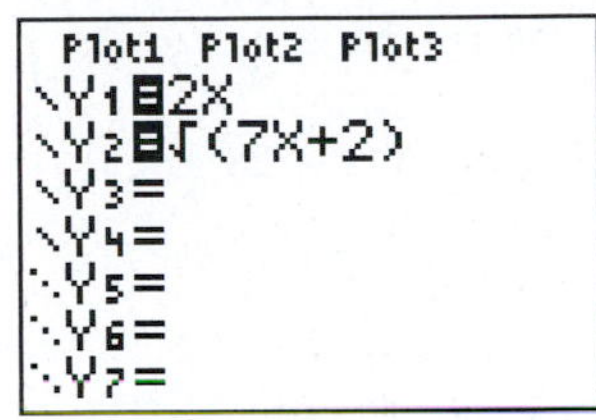

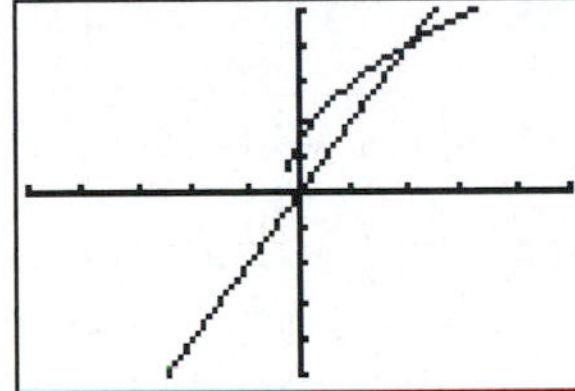

Notice that the graphs of the two functions intersect at only one point. This means there is only one solution of the equation. From the graph we see that the two functions intersect at $x = 2$.

Example 2: Find the solutions of $x^4 - 13x^2 + 36 = 0$ graphically.

Solution: Let $Y_1 = x^4 - 13x^2 + 36$. The solutions of $Y_1 = 0$ correspond to the x-intercepts of the graph.

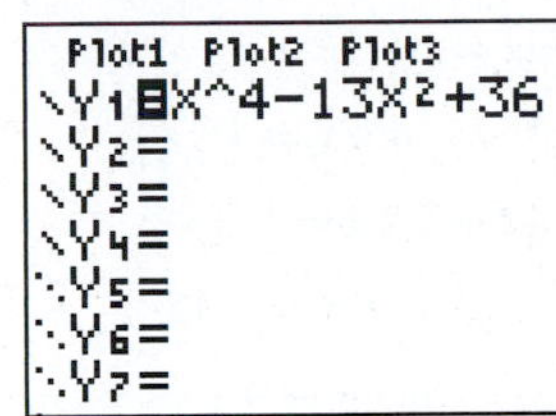

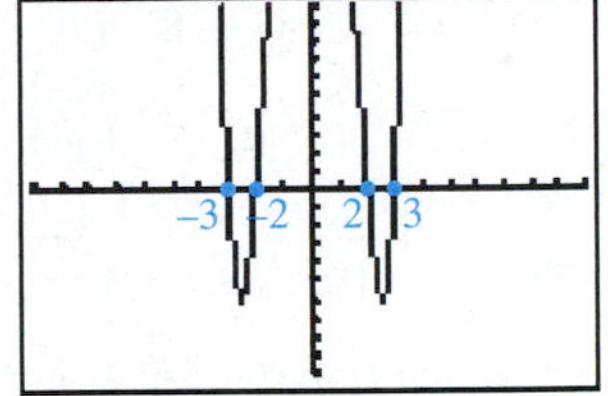

The graph crosses the x-axis at $x = -3, -2, 2, 3$. So, these values are in the solution set. Recall that if an equation has complex solutions, the graph will not intersect the x-axis at these values.

SECTION 9.4 EXERCISE SET

Write About It!

Use complete sentences in your answer to each exercise. For Exercises 1 and 2, explain how to change the equation to a quadratic form. Do not solve the equation.

1. $21x = 8 + \frac{4}{x}$

2. $3x = \sqrt{8 - 6x}$
Square both sides of the equation.

3. Explain how to use the substitution method to solve the equation $(2x - 1)^2 - 2(2x - 1) - 8 = 0$.
Let $u = 2x - 1$. Rewrite the equation as $u^2 - 2u - 8 = 0$.

4. If you solve the equation in Exercise 3 by simplifying the equation and writing it in standard form, $ax^2 + bx + c = 0$, do we get the same answer?
Yes, we will get the same answer.

5. After solving the radical equation using substitution, explain the next step required to finish the problem.

$$x^{1/2} - 2x^{1/4} - 15 = 0$$
$$w = x^{1/4}$$
$$w^2 - 2w - 15 = 0$$
$$(w - 5)(w + 3) = 0$$
$$w = 5 \quad \text{or} \quad w = -3$$

We back-substitute to solve for x.
$w = 5$ or -3
$x^{1/4} = 5$ or -3
$x = 5^4$ or $(-3)^4$
$x = 625$ or 81

6. Can we use the substitution method to solve the following equation? Explain.
$3(x + 2)^2 + 10(x - 2) - 8 = 0$
No, because $x + 2$ and $x - 2$ are not the same.

Practice Makes Perfect!

Solve each rational equation. (*See Objective 1.*)

7. $6x = 1 + \frac{15}{x}$ $\left\{-\frac{3}{2}, \frac{5}{3}\right\}$

8. $7x = 31 - \frac{12}{x}$ $\left\{\frac{3}{7}, 4\right\}$

9. $\frac{2x}{x - 1} + \frac{14x + 52}{x + 1} = -\frac{76}{x^2 - 1}$ $\left\{-\frac{3}{2}\right\}$

10. $\frac{3x}{x + 4} + \frac{6x + 12}{x - 4} = \frac{33}{x^2 - 16}$ $\left\{-\frac{5}{3}, -1\right\}$

11. $\frac{5x}{x - 2} - \frac{4x + 28}{x + 2} = \frac{23}{x^2 - 4}$ $\{5 \pm 2i\sqrt{2}\}$

12. $\frac{3x}{x - 3} - \frac{2x + 21}{x + 3} = \frac{41}{x^2 - 9}$ $\{3 \pm i\sqrt{13}\}$

13. $\frac{7x}{2x - 3} + \frac{x - 6}{x} = \frac{14}{2x^2 - 3x}$ $\left\{\frac{1}{3}, \frac{4}{3}\right\}$

14. $\frac{3x}{3x + 5} + \frac{x - 11}{2x} = -\frac{29}{3x^2 + 5x}$ $\left\{\frac{1}{9}, 3\right\}$

15. $\frac{10x}{2x + 1} + \frac{x - 3}{x} = -\frac{3}{2x^2 + x}$ $\left\{\frac{5}{12}\right\}$

16. $\frac{x}{3x - 2} + \frac{x - 2}{x} = \frac{4}{3x^2 - 2x}$ $\{2\}$

17. $\frac{6x}{x - 2} + \frac{3x - 12}{x} = \frac{21}{x^2 - 2x}$ $\left\{\frac{3 \pm \sqrt{6}}{3}\right\}$

18. $\frac{9x}{x - 2} - \frac{9x + 6}{5x} = \frac{26}{5x^2 - 10x}$ $\left\{\frac{-1 \pm \sqrt{15}}{6}\right\}$

Additional answers can be found in the Instructor Answer Appendix.

19. $\frac{x}{x+6} + \frac{x-12}{x} = -\frac{78}{x^2+6x}$ $\left\{\frac{3 \pm i\sqrt{3}}{2}\right\}$

20. $\frac{x}{x-5} - \frac{4x+24}{5x} = \frac{96}{5x^2-25x}$ $\{2 \pm 2i\sqrt{5}\}$

Solve each radical equation. (*See Objective 2.*)

21. $x = \sqrt{15x - 56}$ $\{7, 8\}$ **22.** $x = \sqrt{-5x + 66}$ $\{6\}$

23. $x + 2 = \sqrt{8x + 16}$ $\{-2, 6\}$ **24.** $x + 6 = \sqrt{17x + 60}$ $\{-3, 8\}$

25. $3x = \sqrt{14x + 8}$ $\{2\}$ **26.** $2x = \sqrt{13x + 12}$ $\{4\}$

27. $14x + 3\sqrt{x} = 27$ $\left\{\frac{81}{49}\right\}$ **28.** $15x - 4\sqrt{x} = 4$ $\left\{\frac{4}{9}\right\}$

29. $x + 14\sqrt{x} = -45$ $\varnothing$ **30.** $4x + 7\sqrt{x} = -3$ $\varnothing$

31. $4x - 19\sqrt{x} = -12$ **32.** $3x - 14\sqrt{x} = -15$ $\left\{\frac{25}{9}, 9\right\}$

Solve each polynomial equation. (*See Objective 3.*)

33. $x^4 - 24x^2 - 25 = 0$ $\{\pm 5, \pm i\}$ **34.** $9x^4 + 80x^2 - 9 = 0$ $\left\{\pm\frac{1}{3}, \pm 3i\right\}$

35. $9x^4 - 13x^2 + 4 = 0$ **36.** $4x^4 - 17x^2 + 4 = 0$

37. $x^4 + 10x^2 = -9$ $\{\pm 3i, \pm i\}$ **38.** $4x^4 + 37x^2 = -9$

39. $x^4 - 16 = 0$ $\{\pm 2, \pm 2i\}$ **40.** $81x^4 - 625 = 0$ $\left\{\pm\frac{5}{3}, \pm\frac{5}{3}i\right\}$

41. $x^3 = 64$ $\{4, -2 \pm 2i\sqrt{3}\}$ **42.** $x^3 = -125$ $\left\{-5, \frac{5 \pm 5i\sqrt{3}}{2}\right\}$

Solve each equation using substitution. (*See Objective 4.*)

43. $5(x-3)^2 + 23(x-3) - 10 = 0$ $\left\{-2, \frac{17}{5}\right\}$

44. $15(3x-1)^2 - 31(3x-1) + 14 = 0$ $\left\{\frac{5}{9}, \frac{4}{5}\right\}$

45. $9(x+1)^2 + 12(x+1) + 4 = 0$ $\left\{-\frac{5}{3}\right\}$

46. $4(x-2)^2 - 20(x-2) + 25 = 0$ $\left\{\frac{9}{2}\right\}$

47. $3x^{2/3} - 8x^{1/3} + 4 = 0$ **48.** $2x^{2/3} + x^{1/3} - 3 = 0$ $\left\{-\frac{27}{8}, 1\right\}$

49. $2x^{2/5} + 3x^{1/5} + 1 = 0$ **50.** $x^{2/5} + x^{1/5} - 6 = 0$ $\{-243, 32\}$

Solve each problem. Round each answer to two decimal places. (*See Objective 5.*)

51. Carolina and Marilyn own a lawn service. Together they can mow the lawns in one development in 4 hr. Carolina does all the mowing 2 hr faster than Marilyn. How long will it take each working alone to do the job? Marilyn: 9.12 hr; Carolina: 7.12 hr

52. Melanie and Jim own a house cleaning company. Together they can clean the houses in one development in 5 hr. Melanie does all the house cleaning 4 hr slower than Jim. How long will it take each working alone to do the job? Jim: 8.39 hr; Melanie: 12.39 hr

53. Tony drove a distance of 500 mi to visit his friend Johnson in Palmyra, Pennsylvania. During the return trip, he was caught in a rainstorm and had to decrease his speed by 10 mph. If the return trip took 1 hr longer than the trip to Palmyra, find his original speed. 75.89 mph

54. Allison drove a distance of 240 mi to visit her friend Ricki in Northridge, California. During the return trip, she tried a new toll road and was able to increase her speed by 14 mph. If her return trip took 1 hr less time than the trip to Northridge, find her original speed. 51.39 mph

Mix 'Em Up!

Solve each equation.

55. $x = -3 + \frac{18}{x}$ $\{-6, 3\}$ **56.** $5x = -36 + \frac{32}{x}$ $\left\{-8, \frac{4}{5}\right\}$

57 $x = \sqrt{16x - 55}$ $\{5, 11\}$ **58.** $x = \sqrt{11x - 28}$ $\{4, 7\}$

59. $\frac{5x}{x-3} + \frac{20x+5}{x+3} = -\frac{7}{x^2-9}$ $\left\{\frac{4 \pm 2\sqrt{6}}{5}\right\}$

60. $\frac{3x}{x-1} + \frac{x-14}{x+1} = \frac{10}{x^2-1}$ $\left\{\frac{3 \pm \sqrt{5}}{2}\right\}$

61. $8x - 15\sqrt{x} = 27$ $\{9\}$ **62.** $2x + 17\sqrt{x} = 30$ $\left\{\frac{9}{4}\right\}$

63. $\frac{2x}{x+2} + \frac{2x+12}{x-2} = \frac{1}{x^2-4}$ $\left\{\frac{-3 \pm i\sqrt{14}}{2}\right\}$

64. $\frac{2x}{x+4} + \frac{2x+4}{x-4} = \frac{12}{x^2-16}$ $\left\{\frac{-1 \pm i\sqrt{3}}{2}\right\}$

65. $2x - 11\sqrt{x} = -15$ **66.** $7x - 37\sqrt{x} = -10$ $\left\{\frac{4}{49}, 25\right\}$

67. $4x = -25 - \frac{25}{x}$ **68.** $3x = 29 - \frac{56}{x}$ $\left\{\frac{8}{3}, 7\right\}$

69. $\frac{5x}{x+3} + \frac{20x+15}{x-3} = \frac{13}{x^2-9}$ $\left\{-\frac{8}{5}, -\frac{4}{5}\right\}$

70. $\frac{5x}{x+4} + \frac{11x+16}{x-4} = \frac{48}{x^2-16}$ $\left\{-2, -\frac{1}{2}\right\}$

71. $4x = \sqrt{-8x + 15}$ $\left\{\frac{3}{4}\right\}$ **72.** $3x = \sqrt{20x - 4}$ $\left\{\frac{2}{9}, 2\right\}$

73. $12(5x+6)^2 + 7(5x+6) - 10 = 0$ $\left\{-\frac{29}{20}, -\frac{16}{15}\right\}$

74. $4(x-1)^2 + 4(x-1) + 1 = 0$ $\left\{\frac{1}{2}\right\}$

75. $\frac{x}{3x+5} - \frac{x-4}{x} = \frac{20}{3x^2+5x}$ $\left\{\frac{7}{2}\right\}$

76. $\frac{4x}{x-1} - \frac{9x+5}{5x} = \frac{1}{x^2-x}$ $\left\{-\frac{4}{11}\right\}$

77. $4x^4 - 29x^2 = -25$ **78.** $9x^4 - 16x^2 = 25$ $\left\{\pm\frac{5}{3}, \pm i\right\}$

79. $x^{2/3} - 3x^{1/3} - 4 = 0$ $\{-1, 64\}$ **80.** $x^{2/5} + x^{1/5} - 2 = 0$ $\{-32, 1\}$

81. $x + 5 = \sqrt{15x + 49}$ $\{-3, 8\}$ **82.** $x - 1 = \sqrt{-3x + 13}$ $\{3\}$

83. $(x-2)^2 - 6(x-2) + 9 = 0$ $\{5\}$

84. $(x+4)^2 - 10(x+4) + 25 = 0$ $\{1\}$

85. $25x + 35\sqrt{x} = -6$ $\varnothing$ **86.** $7x + 9\sqrt{x} = -2$ $\varnothing$

87. $x^4 - 625 = 0$ $\{\pm 5, \pm 5i\}$ **88.** $16x^4 - 81 = 0$

89. $3(4x-3)^2 + 16(4x-3) + 5 = 0$ $\left\{-\frac{1}{2}, \frac{2}{3}\right\}$

90. $2(2x+1)^2 - 9(2x+1) + 10 = 0$ $\left\{\frac{1}{2}, \frac{3}{4}\right\}$

91. $8x^3 - 125 = 0$ **92.** $x^3 + 1 = 0$ $\left\{-1, \frac{1 \pm i\sqrt{3}}{2}\right\}$

Solve each problem. Round each answer to two decimal places.

93. Ron and Ted own a printing service. Together they can complete all print jobs in 3 hr. Ron does all the print jobs 1 hr faster than Ted. How long will it take each working alone to do the job? Ted: 6.54 hr; Ron: 5.54 hr

94. Albert and Alex own a lawn service. Together they can mow the lawns in one development in 3 hr. Albert does all the mowing 2 hr faster than Alex. How long will it take each working alone to do the job? Alex: 7.16 hr; Albert: 5.16 hr

95. Ruth drove a distance of 480 mi to visit her friend, Brian, in Irvine, California. During her trip home, she had to decrease her speed by 12 mph because of a traffic accident. If her trip back home took 2 hr longer than the trip to Irvine, find her original speed. 60 mph

96. Seung drove a distance of 280 mi to visit her parents in Atlanta, Georgia. During her trip home, she was caught in a rainstorm and had to decrease her speed by 15 mph. If her trip back home took 1.5 hr longer than her trip to Atlanta, find her original speed. 60.94 mph

You Be the Teacher!

Correct each student's errors, if any.

97. Solve the equation $\frac{2x}{x+4} - \frac{x-6}{x-4} = \frac{32}{x^2-16}$.

Jeung's work:

$$\frac{2x}{x+4} - \frac{x-6}{x-4} = \frac{32}{x^2-16}$$
$$\frac{2x}{x+4} - \frac{x-6}{x-4} = \frac{32}{(x+4)(x-4)}$$
$$2x(x-4) - (x-6)(x+4) = 32$$
$$2x^2 - 8x - x^2 - 2x - 24 = 32$$
$$x^2 - 10x - 56 = 0$$
$$(x-14)(x+4) = 0$$
$$x = 14 \quad \text{or} \quad x = -4$$

98. Solve the equation $x = \sqrt{5x+6}$.

Mark's work:

$$x = \sqrt{5x+6}$$
$$x^2 = 5x + 6$$
$$x^2 - 5x - 6 = 0$$
$$(x-6)(x+1) = 0$$
$$x = 6 \quad \text{or} \quad x = -1$$

99. Solve the equation $4x^4 + 7x^2 - 36 = 0$.

Ken's work:

$$4x^4 + 7x^2 - 36 = 0$$
$$(4x^2 - 9)(x^2 + 4) = 0$$
$$4x^2 - 9 = 0 \quad \text{or} \quad x^2 + 4 = 0$$
$$x^2 = \frac{9}{4} \qquad x^2 = -4$$
$$x = \frac{3}{2} \qquad x = 2i$$

100. Solve the equation $x^{2/3} - 3x^{1/3} - 4 = 0$.

Beth's work: Let $u = x^{1/3}$.

$$x^{2/3} - 3x^{1/3} - 4 = 0$$
$$u^2 - 3u - 4 = 0$$
$$(u-4)(u+1) = 0$$
$$u - 4 = 0 \quad \text{or} \quad u + 1 = 0$$
$$u = 4 \qquad u = -1$$

The solution set is $\{-1, 4\}$.

Calculate It!

Use a graphing calculator to verify or find solutions of each equation.

101. Determine if 2 and 8 are solutions of $x + 3 = \sqrt{16x - 7}$. Both are solutions to the equation.

102. Determine if 6 and -4 are solutions of $x - 2 = \sqrt{28 - 2x}$. 6 is the solution to the equation.

103. Find the solutions of $3x = \frac{8}{x} - 10$ graphically. The solution set is $\left\{-4, \frac{2}{3}\right\}$.

104. Find the solutions of $25x^4 - 29x^2 + 4 = 0$ graphically. The solution set is $\left\{\pm\frac{2}{5}, \pm 1\right\}$.

SECTION 9.5 More on Graphing Quadratic Functions

OBJECTIVES

As a result of completing this section, you will be able to

1. Graph a quadratic function of the form $f(x) = ax^2 + bx + c$ by converting it to the form $f(x) = a(x - h)^2 + k$.
2. Graph a quadratic function of the form $f(x) = ax^2 + bx + c$ using the vertex formula.
3. Solve applications of finding maximum or minimum values.
4. Troubleshoot common errors.

A ball is tossed upward with an initial velocity of 30 ft/sec from a height of 6 ft. The height of the ball t sec after it is tossed is given by $f(t) = -16t^2 + 30t + 6$. What is the maximum height the ball will reach and when will it reach this height?

Since the height of the ball is modeled by a quadratic function, its graph is a parabola. We know from Section 9.1 that the parabola has its highest or lowest point at its vertex. To answer this question, we must find the vertex of this function. We will learn methods that enable us to do that in this section.

Graphing Quadratic Functions

Objective 1

Graph a quadratic function of the form $f(x) = ax^2 + bx + c$ by converting it to the form $f(x) = a(x - h)^2 + k$.

Recall from Section 9.1 that the vertex of the parabola given by $y = a(x - h)^2 + k$ is the point (h, k). The vertex, along with a few other key points, is used to determine the shape of a parabola. In this section, the quadratic functions have the form $f(x) = ax^2 + bx + c$. If we can convert this form to the vertex form, $f(x) = a(x - h)^2 + k$, we will know the vertex of the given function. We complete the square to make the conversion.

Recall, we can make $x^2 + 10x$ a perfect square by adding $\left(\frac{10}{2}\right)^2 = (5)^2 = 25$; that is, $x^2 + 10x + 25 = (x + 5)^2$.

Procedure: Converting $f(x) = ax^2 + bx + c$ to the Form $f(x) = a(x - h)^2 + k$

Step 1: Write $f(x) = ax^2 + bx + c$ as $y = ax^2 + bx + c$.
Step 2: Move the constant term c to the opposite side of the equation.
Step 3: Divide each side by a if $a \neq 1$.
Step 4: Add the number that completes the square to each side of the equation.
Step 5: Factor the perfect square trinomial as $(\text{binomial})^2$ and simplify the other side.
Step 6: Isolate y by moving the constant back to the other side.
Step 7: Replace y with $f(x)$.

Objective 1 Examples

Convert $f(x) = ax^2 + bx + c$ to the vertex form, $f(x) = a(x - h)^2 + k$. Identify the vertex and intercepts, and graph each function.

1a. $f(x) = x^2 - 4x + 3$ **1b.** $f(x) = -x^2 + 6x - 5$ **1c.** $f(x) = 2x^2 + 4x - 1$

Solutions **1a.**

$f(x) = x^2 - 4x + 3$

$y = x^2 - 4x + 3$ Let $f(x) = y$.

$y - 3 = x^2 - 4x$ Subtract 3 from each side.

$y - 3 + 4 = x^2 - 4x + 4$ Add $\left(\frac{-4}{2}\right)^2 = (-2)^2 = 4$ to each side.

$y + 1 = (x - 2)^2$ Simplify and factor $x^2 - 4x + 4 = (x - 2)^2$.

$y = (x - 2)^2 - 1$ Subtract 1 from each side.

$f(x) = (x - 2)^2 - 1$ Replace y with $f(x)$.

So, the vertex of $f(x) = (x - 2)^2 - 1$ is $(2, -1)$.

x-intercepts: Let $y = 0$ or $f(x) = 0$.

$$y = x^2 - 4x + 3$$
$$0 = x^2 - 4x + 3$$
$$0 = (x - 1)(x - 3)$$
$$x - 1 = 0 \quad \text{or} \quad x - 3 = 0$$
$$x = 1 \qquad\qquad x = 3$$

The x-intercepts are $(3, 0)$ and $(1, 0)$.

y-intercept: Let $x = 0$ or find $f(0)$.

$$f(x) = x^2 - 4x + 3$$
$$f(0) = (0)^2 - 4(0) + 3$$
$$f(0) = 3$$

The y-intercept is $(0, 3)$.

Note that $a = 1$. Since $a > 0$, the parabola opens up. So, the graph is

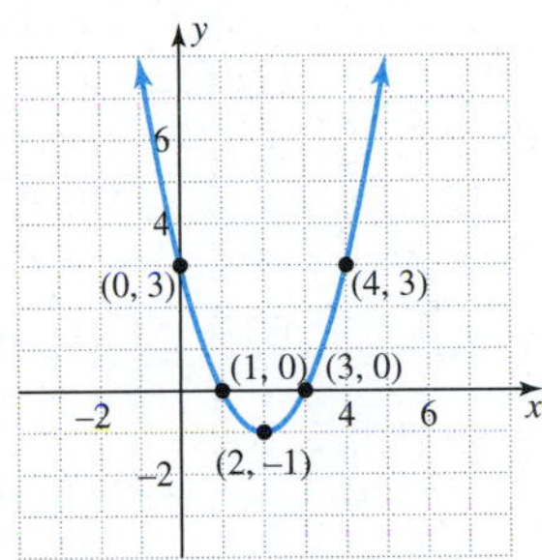

1b.

$f(x) = -x^2 + 6x - 5$	
$y = -x^2 + 6x - 5$	Let $f(x) = y$.
$y + 5 = -x^2 + 6x$	Add 5 to each side.
$\dfrac{y + 5}{-1} = \dfrac{-x^2 + 6x}{-1}$	Divide each side by the coefficient of x^2, -1.
$-y - 5 = x^2 - 6x$	Simplify.
$-y - 5 + 9 = x^2 - 6x + 9$	Add $\left(\dfrac{-6}{2}\right)^2 = (-3)^2 = 9$ to each side.
$-y + 4 = (x - 3)^2$	Simplify and factor $x^2 - 6x + 9 = (x - 3)^2$.
$-y = (x - 3)^2 - 4$	Subtract 4 from each side.
$y = -(x - 3)^2 + 4$	Multiply each side by -1.
$f(x) = -(x - 3)^2 + 4$	Replace y with $f(x)$.

So, the vertex of $f(x) = -(x - 3)^2 + 4$ is $(3, 4)$.

x- intercepts: Let $y = 0$ or $f(x) = 0$.

$$y = -x^2 + 6x - 5$$
$$0 = -x^2 + 6x - 5$$
$$-1(0) = -1(-x^2 + 6x - 5)$$
$$0 = x^2 - 6x + 5$$
$$0 = (x - 1)(x - 5)$$
$$x - 1 = 0 \quad \text{or} \quad x - 5 = 0$$
$$x = 1 \qquad\qquad x = 5$$

The x-intercepts are $(1, 0)$ and $(5, 0)$.

y-intercept: Let $x = 0$ or find $f(0)$.

$$f(x) = -x^2 + 6x - 5$$
$$f(0) = -(0)^2 + 6(0) - 5$$
$$f(0) = -5$$

The y-intercept is $(0, -5)$.

Note that $a = -1$. Since $a < 0$, the parabola opens down. So, the graph is

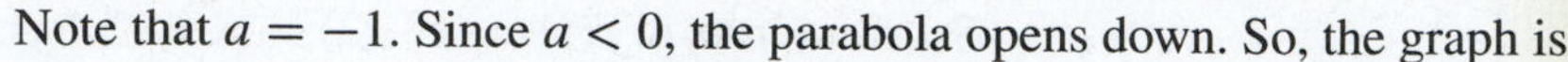

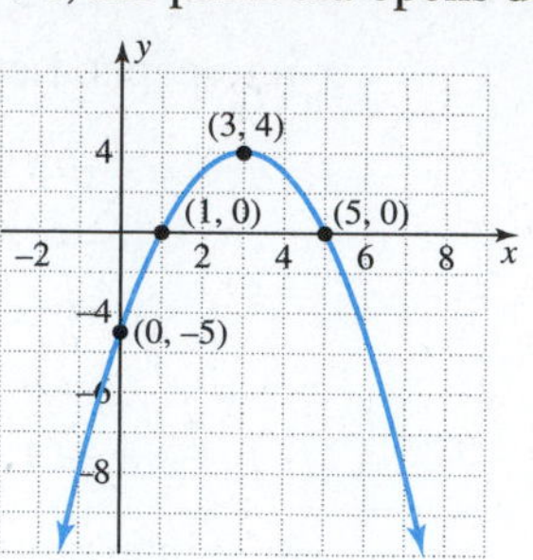

1c.

$$f(x) = 2x^2 + 4x - 1$$

$$y = 2x^2 + 4x - 1 \quad \text{Let } f(x) = y.$$

$$y + 1 = 2x^2 + 4x \quad \text{Add 1 to each side.}$$

$$\frac{y+1}{2} = \frac{2x^2 + 4x}{2} \quad \text{Divide each side by the coefficient of } x^2,\ 2.$$

$$\frac{y}{2} + \frac{1}{2} = x^2 + 2x \quad \text{Simplify.}$$

$$\frac{y}{2} + \frac{1}{2} + 1 = x^2 + 2x + 1 \quad \text{Add } \left(\frac{2}{2}\right)^2 = (1)^2 = 1 \text{ to each side.}$$

$$\frac{y}{2} + \frac{3}{2} = (x + 1)^2 \quad \text{Simplify and factor } x^2 + 2x + 1 = (x + 1)^2.$$

$$\frac{y}{2} = (x + 1)^2 - \frac{3}{2} \quad \text{Subtract } \frac{3}{2} \text{ from each side.}$$

$$2\left(\frac{y}{2}\right) = 2\left[(x + 1)^2 - \frac{3}{2}\right] \quad \text{Multiply each side by 2.}$$

$$y = 2(x + 1)^2 - 3 \quad \text{Simplify.}$$

$$f(x) = 2(x + 1)^2 - 3 \quad \text{Replace } y \text{ with } f(x).$$

So, the vertex of $f(x) = 2[x - (-1)]^2 - 3$ is $(-1, -3)$.

x-intercepts: Let $y = 0$ or $f(x) = 0$.

$$y = 2x^2 + 4x - 1$$

$$0 = 2x^2 + 4x - 1$$

$$x = \frac{-(4) \pm \sqrt{(4)^2 - 4(2)(-1)}}{2(2)}$$

$$x = \frac{-4 \pm \sqrt{24}}{4}$$

$$x = \frac{-4 \pm 2\sqrt{6}}{4}$$

$$x = \frac{2(-2 \pm \sqrt{6})}{2(2)}$$

$$x = \frac{-2 \pm \sqrt{6}}{2}$$

$$x = \frac{-2 + \sqrt{6}}{2} \quad \text{or} \quad x = \frac{-2 - \sqrt{6}}{2}$$

$$x \approx 0.22 \qquad x \approx -2.22$$

The x-intercepts are $(0.22, 0)$ and $(-2.22, 0)$.

y-intercept: Let $x = 0$ or find $f(0)$.

$$f(x) = 2x^2 + 4x - 1$$

$$f(0) = 2(0)^2 + 4(0) - 1$$

$$f(0) = -1$$

The y-intercept is $(0, -1)$.

Note that $a = 2$. Since $a > 0$, the parabola opens up. So, the graph is

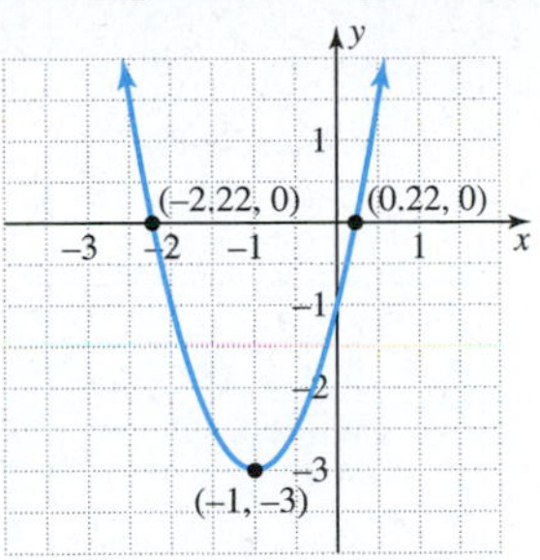

Student Check 1 Convert $f(x) = ax^2 + bx + c$ to the vertex form, $f(x) = a(x - h)^2 + k$. Identify the vertex and intercepts, and graph each function.

a. $f(x) = x^2 - 10x + 9$ **b.** $f(x) = -x^2 - 8x + 3$ **c.** $f(x) = 3x^2 + 12x - 2$

The Vertex Formula

Objective 2 Graph a quadratic function of the form $f(x) = ax^2 + bx + c$ using the vertex formula.

Completing the square to obtain the form $f(x) = a(x - h)^2 + k$ can often be tedious. There is a formula that we can use to find the vertex of a parabola. In Example 1, notice that the vertex is always located halfway between the two x-intercepts. The following table shows this result.

Example	x-Intercepts	x-Value of Vertex	Average of the x-Values of the x-Intercepts
1a	$(1, 0)$ and $(3, 0)$	$x = 2$	$\frac{1+3}{2} = \frac{4}{2} = 2$
1b	$(1, 0)$ and $(5, 0)$	$x = 3$	$\frac{1+5}{2} = \frac{6}{2} = 3$
1c	$(-2.22, 0)$ and $(0.22, 0)$	$x = -1$	$\frac{-2.22+0.22}{2} = \frac{-2}{2} = -1$

So, if the x-intercepts are $x = \dfrac{-b + \sqrt{b^2 - 4ac}}{2a}$ and $x = \dfrac{-b - \sqrt{b^2 - 4ac}}{2a}$, then the x-value of the vertex can be found by averaging these x-values. This gives us

$$x = \frac{\dfrac{-b + \sqrt{b^2 - 4ac}}{2a} + \dfrac{-b - \sqrt{b^2 - 4ac}}{2a}}{2}$$

$$x = \frac{\dfrac{-b + \sqrt{b^2 - 4ac} - b - \sqrt{b^2 - 4ac}}{2a}}{2}$$

$$x = \frac{-2b}{2a} \cdot \frac{1}{2}$$

$$x = -\frac{b}{2a}$$

Note that the y-value of the vertex is found by substituting the x-value in the function.

Property: The Vertex Formula

The x-value of the vertex of the graph of $f(x) = ax^2 + bx + c$ is $x = -\dfrac{b}{2a}$.

The y-value is found by substituting this value of x in the function, or $f\left(-\dfrac{b}{2a}\right)$.

The Key Points on a Parabola	
x-intercepts	Recall *x-intercepts* are solutions of the equation $f(x) = 0$ or $$ax^2 + bx + c = 0$$ Since this type of equation has either 2, 1, or 0 real solutions, there will be 2 x-intercepts, 1 x-intercept, or no x-intercepts.
y-intercept	The *y-intercept* is found by replacing x with 0, that is, $f(0)$. $$y = a(0)^2 + b(0) + c = c$$ This shows that the y-intercept of $f(x) = ax^2 + bx + c$ is the point $(0, c)$. A quadratic function has only one y-intercept.
Vertex	The *vertex* is of the form $\left[-\frac{b}{2a}, f\left(-\frac{b}{2a}\right)\right]$.

Objective 2 Examples

Graph each function by finding the intercepts, the vertex, and additional points, if needed.

2a. $f(x) = -x^2 + 1$ **2b.** $f(x) = x^2 + 4x - 5$

Solutions

2a. Note that $f(x) = -1x^2 + 0x + 1$, so $a = -1$, $b = 0$, and $c = 1$.

x-intercepts	$-x^2 + 1 = 0$ $-x^2 = -1$ $x^2 = 1$ $x = \pm\sqrt{1}$ $x = \pm 1$ $(1, 0)$ and $(-1, 0)$
y-intercept	$f(0) = -(0)^2 + 1$ $f(0) = 1$ $(0, 1)$
Vertex	$x = -\frac{b}{2a} = -\frac{0}{2(-1)} = -\frac{0}{-2} = 0$ $f(0) = -(0)^2 + 1 = 0 + 1 = 1$ $(0, 1)$

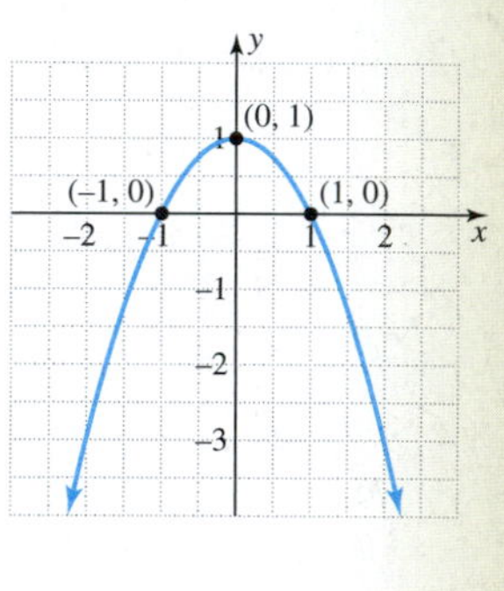

2b. Note that $a = 1$, $b = 4$, and $c = -5$.

x-intercepts	$x^2 + 4x - 5 = 0$ $(x + 5)(x - 1) = 0$ $x + 5 = 0$ or $x - 1 = 0$ $x = -5$ or $x = 1$ $(-5, 0)$ and $(1, 0)$
y-intercept	$f(0) = (0)^2 + 4(0) - 5$ $f(0) = -5$ $(0, -5)$
Vertex	$x = -\frac{b}{2a} = -\frac{4}{2(1)} = -\frac{4}{2} = -2$ $f(-2) = (-2)^2 + 4(-2) - 5$ $f(-2) = 4 - 8 - 5$ $f(-2) = -9$ $(-2, -9)$

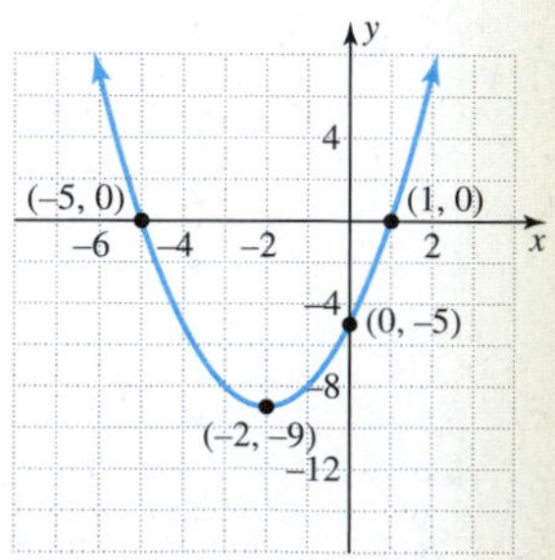

Student Check 2 Graph each function by finding the intercepts, the vertex, and additional points, if needed.

a. $y = -x^2 + 4$ **b.** $y = x^2 + 6x$

Maximum and Minimum Values of Quadratic Functions

Objective 3 Solve applications of finding the maximum and minimum values.

We have learned that the vertex of a parabola corresponds to either the highest or lowest point on the graph of a quadratic function. When a parabola opens up, the vertex is the lowest point on the graph. When a parabola opens down, the vertex is the highest point on the graph.

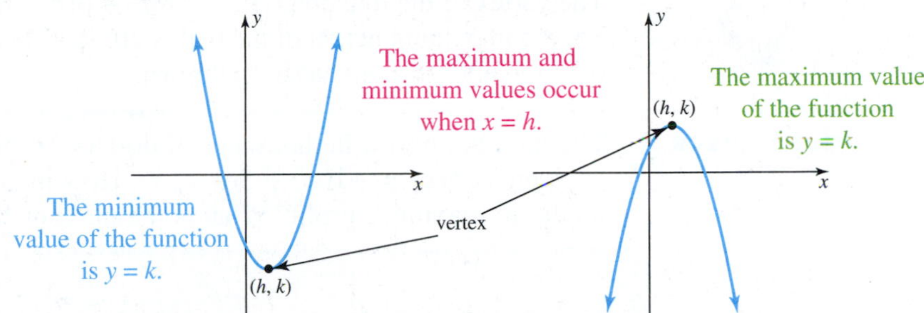

We can use this information to solve applications that ask us to determine where the maximum or minimum values of a function occur. If the given model is a quadratic function, we find the vertex of the parabola to solve the problem. For example, the graph of $f(x) = -x^2 + 6x - 5$ is shown.

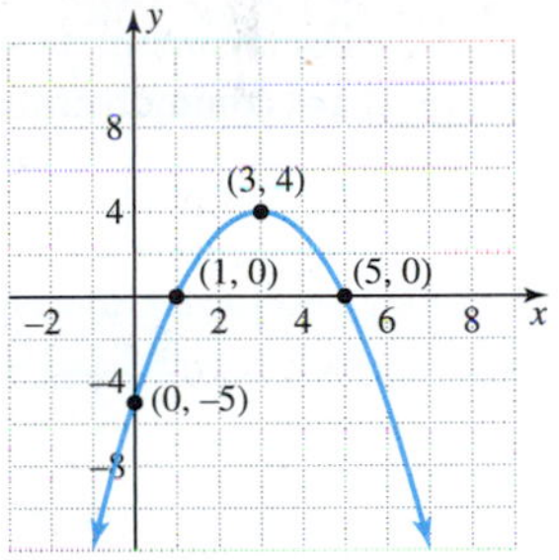

Since the vertex of the parabola is (3, 4), we say that the function has a maximum value of 4 when $x = 3$. Note that the parabola extends indefinitely downward, so there is no minimum value.

Procedure: Solving Applications by Finding Maximum or Minimum Values

Step 1: Identify the value of a and determine if the parabola opens up ($a > 0$) or down ($a < 0$).

Step 2: Determine if the vertex is the maximum or minimum point of the graph of the function.

a. If the parabola opens up, the vertex is the minimum point.

b. If the parabola opens down, the vertex is the maximum point.

Step 3: Find the vertex of the quadratic function.

Step 4: The x-value of the vertex is where the maximum or minimum occurs.

Step 5: The y-value of the vertex is the maximum or minimum value of the function.

Objective 3 Example A ball is tossed upward with an initial velocity of 30 ft/sec from a height of 6 ft. The height of the ball t sec after it is tossed is given by $f(t) = -16t^2 + 30t + 6$. What is the maximum height the ball will reach and when will it reach this height?

Solution The quadratic function opens down since the value of $a = -16 < 0$. Therefore, the maximum value of the function occurs at the vertex. We find the vertex by using the vertex formula.

$$f(t) = -16t^2 + 30t + 6 \rightarrow a = -16, b = 30, c = 6$$

We substitute these values into the vertex formula, $t = -\dfrac{b}{2a}$.

$$t = -\frac{b}{2a} = -\frac{30}{2(-16)} = -\frac{30}{-32} = \frac{15}{16} \approx 0.94$$

$$
\begin{aligned}
f\left(\frac{15}{16}\right) &= -16\left(\frac{15}{16}\right)^2 + 30\left(\frac{15}{16}\right) + 6\\
&= -16\left(\frac{225}{256}\right) + \frac{225}{8} + 6\\
&= -\frac{225}{16} + \frac{450}{16} + \frac{96}{16}\\
&= \frac{321}{16} \approx 20.06
\end{aligned}
$$

The vertex of the function $f(t) = -16t^2 + 30t + 6$ is approximately (0.94, 20.06). So, the maximum height of the ball is 20.06 ft. It will reach this height about 0.94 sec after the ball has been thrown.

Student Check 3 The monthly profit, in thousands of dollars, from selling x hundred computers is given by $f(x) = -10x^2 + 56x + 37$. How many computers must be sold in a month to maximize profit? What is the maximum monthly profit?

Objective 4 Troubleshoot common errors.

Troubleshooting Common Errors

A common error associated with converting a quadratic function from standard form to vertex form is shown.

Objective 4 Example **A problem and an incorrect solution are given. Provide the correct solution and an explanation of the error.**

Write $y = x^2 - 6x + 5$ in the form $y = a(x - h)^2 + k$.

Incorrect Solution	Correct Solution and Explanation
$y = x^2 - 6x + 5$ $y - 5 = x^2 - 6x + 9$ $y - 5 = (x - 3)^2$ $y = (x - 3)^2 + 5$	When we complete the square, we must add that number to both sides of the equation. $y - 5 + 9 = x^2 - 6x + 9$ $y + 4 = (x - 3)^2$ $y = (x - 3)^2 - 4$

ANSWERS TO STUDENT CHECKS

Student Check 1 **a.** $f(x) = (x - 5)^2 - 16$; vertex: $(5, -16)$; x-intercepts: $(1, 0)$ and $(9, 0)$; y-intercept: $(0, 9)$

b. $f(x) = -(x + 4)^2 + 19$; vertex: $(-4, 19)$; x-intercepts: $(-4 + \sqrt{19}, 0)$; and $(-4 - \sqrt{19}, 0)$; y-intercept: $(0, 3)$

c. $f(x) = 3(x + 2)^2 - 14$; vertex: $(-2, -14)$; x-intercepts:

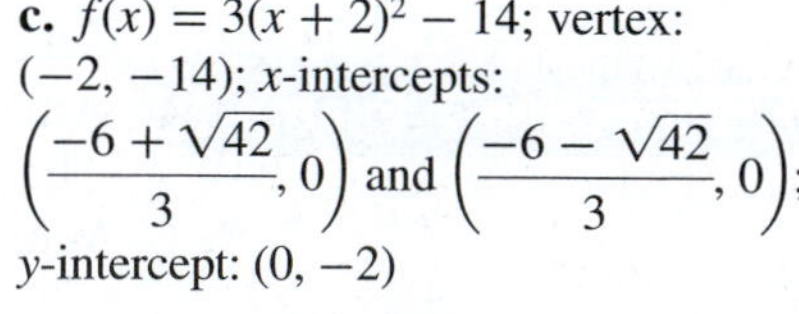

$\left(\frac{-6 + \sqrt{42}}{3}, 0\right)$ and $\left(\frac{-6 - \sqrt{42}}{3}, 0\right)$; y-intercept: $(0, -2)$

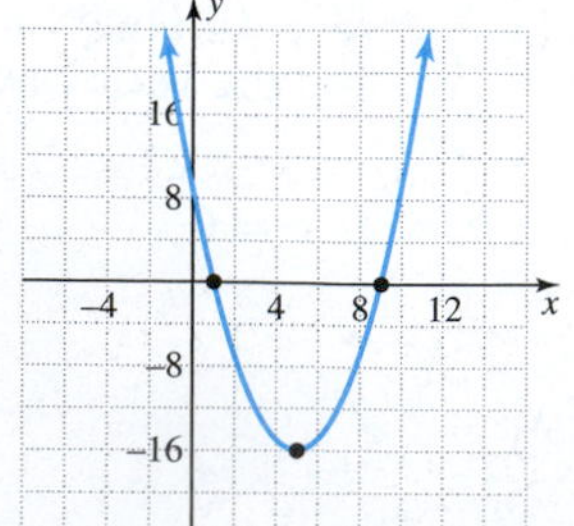

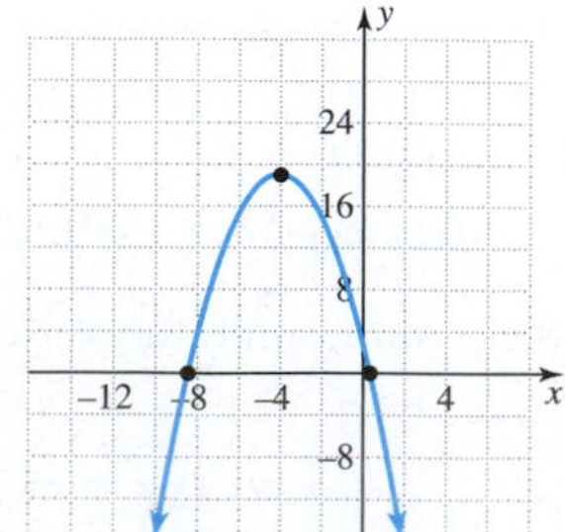

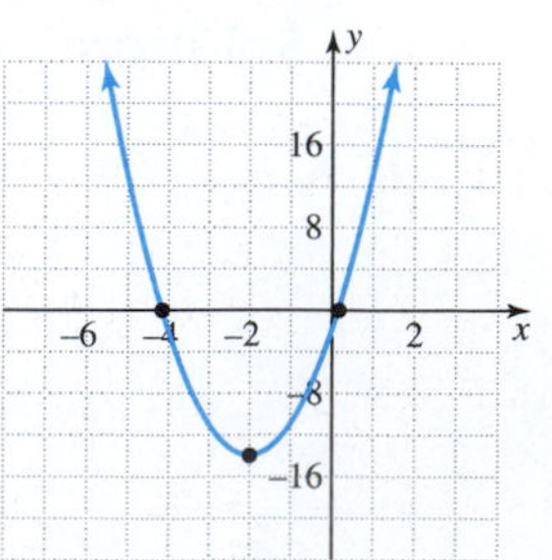

Student Check 2 **a.** vertex: (0, 4); x-intercepts: (−2, 0) and (2, 0); y-intercept: (0, 4)

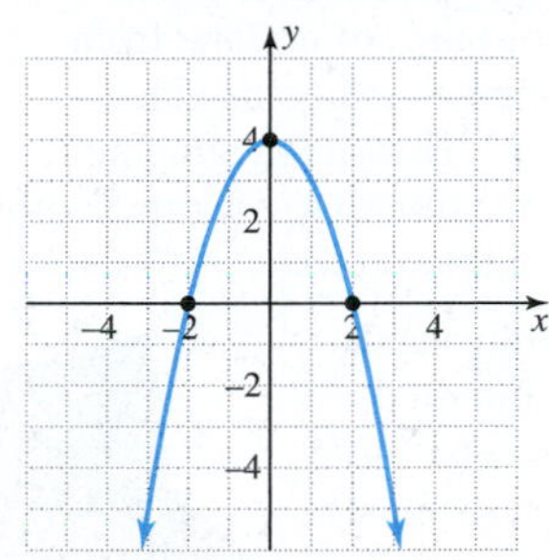

b. vertex: (−3, −9); x-intercepts: (−6, 0) and (0, 0); y-intercept: (0, 0)

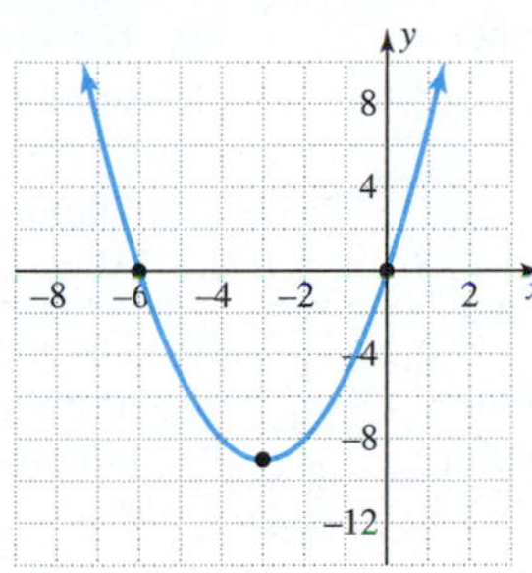

Student Check 3 The company must sell 2.8 hundred or 280 computers in a month to maximize profit. The maximum monthly profit is 115.4 thousand or \$115,400.

SUMMARY OF KEY CONCEPTS

1. Complete the square to convert a quadratic function from standard form, $f(x) = ax^2 + bx + c$, to vertex form, $f(x) = a(x - h)^2 + k$. The steps that are used to solve an equation by completing the square are also used in this process. We can then readily identify the vertex of the parabola when written in the vertex form.
2. The vertex formula can be used to find the vertex of $f(x) = ax^2 + bx + c$. The x-value of the vertex is $x = -\dfrac{b}{2a}$ and the corresponding y-value is found by substituting this x-value into the function and solving for y.
3. For quadratic models, the maximum or minimum value will occur at its vertex. The y-value of the vertex is the maximum or minimum value of the function. The x-value of the vertex is where this maximum or minimum occurs.

GRAPHING CALCULATOR SKILLS

The graphing calculator can be used to graph quadratic functions.

Example: Verify that (2, −1) is the vertex of $y = x^2 - 4x + 3$. Also verify that (1, 0) and (3, 0) are the x-intercepts and (0, 3) is the y-intercept.

Solution: Enter the function into the calculator. View the table and graph.

```
Plot1 Plot2 Plot3
\Y1=X²-4X+3
\Y2=
\Y3=
\Y4=
\Y5=
\Y6=
\Y7=
```

X	Y_1
-2	15
-1	8
0	3
1	0
2	-1
3	0
4	3

X= -2

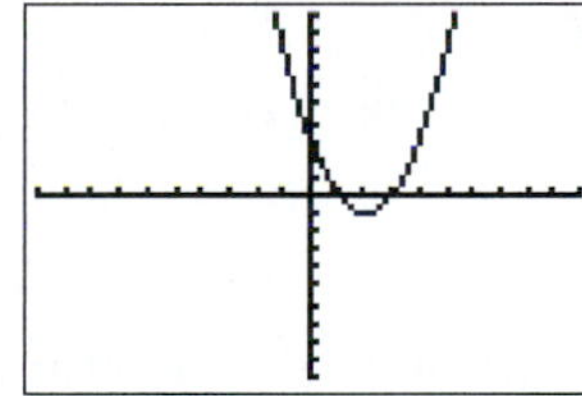

From the table, we see the points (0, 3), (1, 0), and (3, 0) are solutions and the graph confirms these are intercepts. The point (2, −1) is also the lowest point on the graph and is the vertex of the parabola.

SECTION 9.5 EXERCISE SET

Write About It!

Use complete sentences in your answer to each exercise.

1. Explain how to convert a quadratic function of the form $y = x^2 + bx + c$ to the vertex form $y = (x - h)^2 + k$.
2. Explain how to find the x-intercepts of a quadratic function in vertex form, $y = (x - h)^2 + k$.
3. Explain which form, standard or vertex, makes it easier to find the y-intercept. The standard form, $y = ax^2 + bx + c$, will make it easier to find the y-intercept, $(0, c)$.
4. Explain which form, standard or vertex, makes it easier to find the x-intercepts. The vertex form, $y = a(x - h)^2 + k$ will make it easier to find the x-intercepts.
5. What can you say about the location of the vertex and the two x-intercepts? The x-coordinate of the vertex is between the two x-intercepts.

Additional answers can be found in the Instructor Answer Appendix.

6. Explain how the solutions of the quadratic equation, $ax^2 + bx + c = 0$, are related to the x-intercepts of the quadratic function, $f(x) = ax^2 + bx + c$. They are the same.

7. Explain how the quadratic formula is related to the vertex formula.

$$ax^2 + bx + c = 0$$

$$x = \frac{-b \pm \sqrt{b^2 - 4ac}}{2a}$$

8. Explain how to apply the vertex formula to find the maximum or minimum value of a quadratic function.

Practice Makes Perfect!

Convert $f(x) = ax^2 + bx + c$ to the vertex form, $f(x) = a(x - h)^2 + k$. Identify the vertex and intercepts, and graph each function. (*See Objective 1.*)

9. $f(x) = x^2 + 2x - 15$
10. $f(x) = x^2 - 4x - 5$
11. $f(x) = x^2 + 4x - 60$
12. $f(x) = x^2 + 12x + 35$
13. $f(x) = x^2 + 3x - 18$
14. $f(x) = x^2 - 7x + 12$
15. $f(x) = -x^2 - 2x + 8$
16. $f(x) = -x^2 - 6x + 7$
17. $f(x) = -x^2 - 5x + 36$
18. $f(x) = -x^2 + x + 2$
19. $f(x) = 2x^2 + 12x + 2$
20. $f(x) = 3x^2 + 6x - 3$
21. $f(x) = -5x^2 - 10x + 7$
22. $f(x) = -3x^2 + 12x - 5$
23. $f(x) = -2x^2 + 12x - 27$
24. $f(x) = -2x^2 + 8x - 25$
25. $f(x) = 2x^2 + 8x + 11$
26. $f(x) = 3x^2 - 6x + 9$

Use the vertex formula to determine the vertex of each function. (*See Objective 2.*)

27. $f(x) = x^2 - 9x + 2$
28. $f(x) = x^2 + 5x + 7$
29. $f(x) = -2x^2 + 6x + 1$
30. $f(x) = -3x^2 - 5x + 6$
31. $f(x) = 5x^2 + 3x + 1$
32. $f(x) = 5x^2 + 2x - 7$

Graph each quadratic function by finding the intercepts, the vertex, and additional points, if needed. (*See Objective 2.*)

33. $f(x) = x^2 - 10$
34. $f(x) = x^2 - 7$
35. $f(x) = -x^2 - 5$
36. $f(x) = x^2 + 3$
37. $f(x) = x^2 + 6x$
38. $f(x) = x^2 - 5x$
39. $f(x) = -x^2 + 7x$
40. $f(x) = -x^2 - 3x$
41. $f(x) = (x - 2)^2 - 1$
42. $f(x) = (x + 4)^2 - 9$
43. $f(x) = -(x + 3)^2 + 1$
44. $f(x) = -(x - 2)^2 + 9$
45. $f(x) = 2x^2 - 9x - 5$
46. $f(x) = 3x^2 + 2x - 5$

Solve each problem. (*See Objective 3.*)

47. The monthly revenue, in thousands of dollars, from selling x hundred computers is given by $f(x) = -2x^2 + 40x + 235$. How many computers must be sold in a month to maximize revenue? What is the maximum monthly revenue? 1000 computers with a maximum monthly revenue of $435,000

48. The monthly revenue, in thousands of dollars, from selling x hundred computers is given by $f(x) = -17x^2 + 85x - 35$. How many computers must be sold in a month to maximize revenue? What is the maximum monthly revenue? 250 computers with a maximum monthly revenue of $71,250

49. The monthly profit, in thousands of dollars, from selling x thousand graphing calculators is given by $f(x) = -3x^2 + 63x + 17$. How many graphing calculators must be sold in a month to maximize profit? What is the maximum monthly profit? 10,500 graphing calculators with a maximum monthly profit of $347,750

50. The monthly revenue, in thousands of dollars, from selling x hundred laptops is given by $f(x) = -6x^2 + 75x - 23$. How many laptops must be sold in a month to maximize revenue? What is the maximum monthly revenue? 625 laptops with a maximum monthly revenue of $211,375

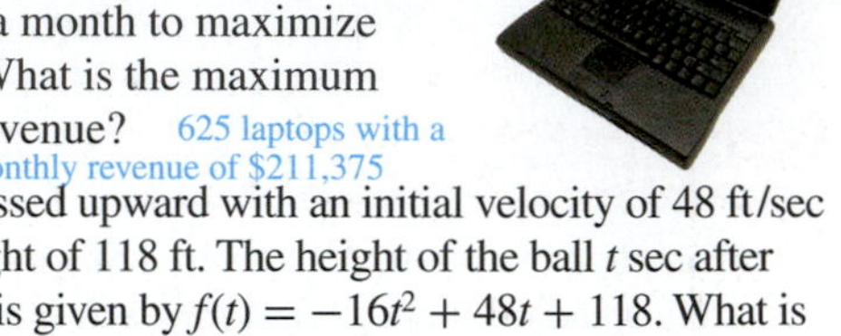

51. A ball is tossed upward with an initial velocity of 48 ft/sec from a height of 118 ft. The height of the ball t sec after it is tossed is given by $f(t) = -16t^2 + 48t + 118$. What is the maximum height the ball will reach and when will it reach this height? 154 ft in 1.5 sec

52. A ball is tossed upward with an initial velocity of 67.2 ft/sec from a height of 21 ft. The height of the ball t sec after it is tossed is given by $f(t) = -16t^2 + 67.2t + 21$. What is the maximum height the ball will reach and when will it reach this height? 91.56 ft in 2.1 sec

53. A ball is tossed upward with an initial velocity of 102.4 ft/sec from a height of 15 ft. The height of the ball t sec after it is tossed is given by $f(t) = -16t^2 + 102.4t + 15$. What is the maximum height the ball will reach and when will it reach this height? 178.84 ft in 3.2 sec

54. A ball is tossed upward with an initial velocity of 54.4 ft/sec from a height of 99 ft. The height of the ball t sec after it is tossed is given by $f(t) = -16t^2 + 54.4t + 99$. What is the maximum height the ball will reach and when will it reach this height? 145.24 ft in 1.7 sec

Mix 'Em Up!

Graph each quadratic function by finding the intercepts, the vertex, and additional points, if needed.

55. $f(x) = x^2 + 11x + 24$
56. $f(x) = x^2 - 3x + 2$
57. $f(x) = -x^2 - 3x + 10$
58. $f(x) = -x^2 - 3x - 2$
59. $f(x) = -(x - 4)^2 + 20$
60. $f(x) = 4x^2 - 8x - 1$
61. $f(x) = 2x^2 - 5x$
62. $f(x) = -3x^2 + 12x$
63. $f(x) = 4\left(x + \frac{1}{2}\right)^2 + 3$
64. $f(x) = -3(x - 2)^2 + 12$
65. $f(x) = 4(x + 1.4)^2 - 6.25$
66. $f(x) = -2(x - 2.1)^2 + 11.52$
67. $f(x) = x^2 + 5.6x - 21.32$
68. $f(x) = -x^2 + 6.4x - 7.68$

Solve each problem.

69. The unemployment rate among individuals aged 25 and older who are not high school graduates can be modeled by $f(x) = 0.0051x^2 - 0.0177x + 0.0542$, where x is the number of years after 2004. In what year did the unemployment rate reach its minimum value? What was the minimum unemployment rate? (Source: http://www.trends.collegeboard.org) In about 1.74 years after 2004, or 2005, the unemployment rate reached a minimum of about 3.88%.

70. The unemployment rate among individuals aged 25 and older who have a bachelor's degree or higher can be modeled by $f(x) = 0.0028x^2 - 0.011x + 0.0289$, where x is the number of years after 2004. In what year did the unemployment rate reach its minimum value? What was the minimum unemployment rate? (Source: http://www.trends.collegeboard.org) In about 1.96 years after 2004, or 2005, the unemployment rate reached a minimum of about 1.81%.

71. The mean maximum temperature for Topeka, Kansas can be modeled by $f(x) = -2.1585x^2 + 30.564x - 20.28$, where $x = 1$ corresponds to January, $x = 2$ corresponds to February and so on. In what month does this temperature reach a maximum? What is the maximum temperature? (Source: http://www.esrl.noaa.gov/psd/data/usstations/) In month 7, or July, the maximum mean temperature in Topeka reaches a maximum of about 88°F.

72. The mean maximum temperature for Bismark, North Dakota can be modeled by $f(x) = -2.5503x^2 + 36.542x - 49.625$, where $x = 1$ corresponds to January, $x = 2$ corresponds to February and so on. In what month does this temperature reach a maximum? What is the maximum temperature? (Source: http://www.esrl.noaa.gov/psd/data/usstations/) In month 7, or July, the maximum mean temperature in Bismark reaches about 81°F.

You Be the Teacher!

Correct each student's errors, if any.

73. Convert $y = -x^2 + 6x - 1$ to vertex form.

Elaine's work:

$$\begin{aligned} y &= -x^2 + 6x - 1 \\ y + 1 &= -x^2 + 6x \\ -y + 1 &= x^2 + 6x \\ -y + 1 + 9 &= x^2 + 6x + 9 \\ -y + 10 &= (x + 3)^2 \\ -y &= (x + 3)^2 - 10 \\ y &= -(x + 3)^2 + 10 \end{aligned}$$

$$\begin{aligned} y &= -x^2 + 6x - 1 \\ y + 1 &= -x^2 + 6x \\ -y - 1 &= x^2 - 6x \\ -y - 1 + 9 &= x^2 - 6x + 9 \\ -y + 8 &= (x - 3)^2 \\ -y &= (x - 3)^2 - 8 \\ y &= -(x - 3)^2 + 8 \end{aligned}$$

74. Use the vertex formula to find the coordinates of the vertex of the function, $f(x) = -3x^2 + 12x + 8$.

Mark's work:

$$x = -\frac{12}{3(2)} = -2$$

$$\begin{aligned} f(-2) &= -3(-2)^2 + 12(-2) + 8 \\ &= 12 - 14 + 8 \\ &= 6 \end{aligned}$$

$$x = -\frac{12}{2(-3)} = 2$$

$$\begin{aligned} f(2) &= -3(2)^2 + 12(2) + 8 \\ &= -12 + 24 + 8 \\ &= 20 \end{aligned}$$

The vertex is (2, 20).

75. Find the x-intercepts of $f(x) = -x^2 + 5x + 6$.

Loretta's work:

$$\begin{aligned} -x^2 + 5x + 6 &= 0 \\ -(x^2 - 5x + 6) &= 0 \\ -(x - 2)(x - 3) &= 0 \\ x = -2, x &= 3 \end{aligned}$$

$$\begin{aligned} -x^2 + 5x + 6 &= 0 \\ -(x^2 - 5x - 6) &= 0 \\ -(x - 6)(x + 1) &= 0 \\ x = 6, x &= -1 \end{aligned}$$

The x-intercepts are (6, 0) and (−1, 0).

76. Graph $f(x) = -2(x - 1)^2 + 8$.

Allison's work:

x-intercepts: $-2(x - 1)^2 + 8 = 0$

$$\begin{aligned} (-2x + 2)^2 &= -8 \\ -2x + 2 &= 2i\sqrt{2} \end{aligned}$$

no x-intercepts

y-intercept: $f(0) = -2(0 - 1)^2 + 8 = 4 + 8 = 12$

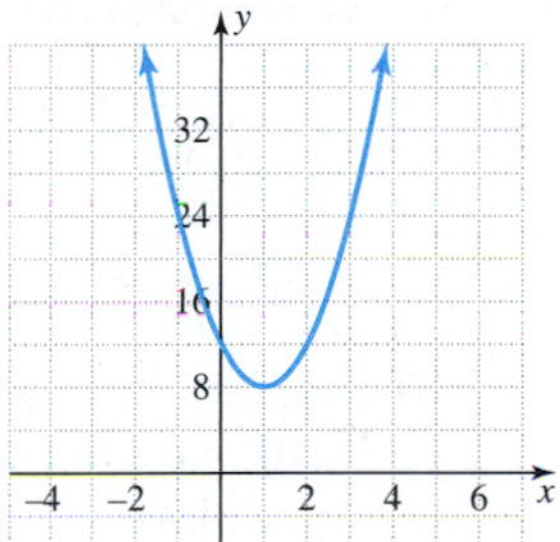

Calculate It!

Use a graphing calculator to verify the vertex form of each quadratic function.

77. $f(x) = x^2 - 8x + 17 = (x - 4)^2 + 1$

78. $f(x) = -x^2 - 4x = -(x + 2)^2 + 4$

SECTION 9.6 Solving Polynomial and Rational Inequalities in One Variable

OBJECTIVES

As a result of completing this section, you will be able to

1. Solve quadratic inequalities using the boundary number method.
2. Solve higher degree polynomial inequalities using the boundary number method.
3. Solve rational inequalities using the boundary number method.
4. Find solutions of nonlinear inequalities from a graph.
5. Solve applications of nonlinear inequalities.
6. Troubleshoot common errors.

The monthly profit, in thousands of dollars, from selling x thousand digital cameras is given by $f(x) = -4x^2 + 80x + 14$. How many cameras must be sold to have a profit greater than \$350,000? To solve this problem, we can solve the inequality $-4x^2 + 80x + 14 > 350$.

In this section, we use our knowledge of solving polynomial and rational equations to solve polynomial and rational inequalities, or *nonlinear inequalities*.

The Boundary Number Method

Objective 1 Solve quadratic inequalities using the boundary number method.

In Sections 9.2 and 9.3, we learned how to solve quadratic equations of the form $ax^2 + bx + c = 0$. We now learn how to solve inequalities where one side is a quadratic expression and the other side is zero. These inequalities are called **quadratic inequalities**. Some examples of quadratic inequalities are

$$x^2 + x - 6 < 0 \qquad x^2 - 8x > 9$$

Consider the inequality $x^2 + x - 6 < 0$. Solving this inequality requires us to find the values of x that make the expression $x^2 + x - 6$ negative. It is impossible to substitute every value of x in $x^2 + x - 6$ to determine which ones make the expression negative. So, we use a method, called the *boundary number method*, to solve this problem. **Boundary numbers** are solutions of the equation formed by replacing the inequality with an equals sign. These values form boundaries of regions on the number line.

To use this method, we first find the solutions of the equation $x^2 + x - 6 = 0$.

$$\begin{aligned} x^2 + x - 6 &= 0 \\ (x+3)(x-2) &= 0 \\ x + 3 = 0 \quad &\text{or} \quad x - 2 = 0 \\ x = -3 \quad &\qquad\quad x = 2 \end{aligned}$$

The solutions tell us that $x^2 + x - 6$ is zero for only $x = -3$ and $x = 2$. So, all other values of x will make the expression $x^2 + x - 6$ either positive or negative. To determine which x-values make the expression positive or negative, we test one point from each of the regions formed by the solutions of the equation. Marking the solutions on a number line enables us to select our test points. It is sufficient to test only one point from each region.

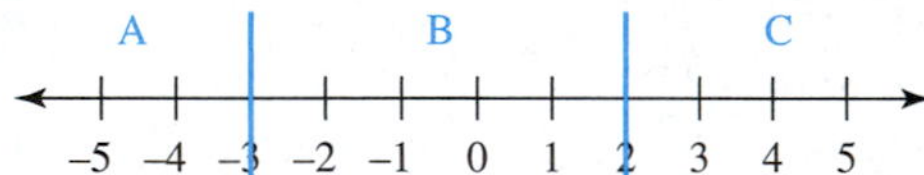

Region	Intervals of x	Test Value	Test	True/False
A	Values less than −3	−4	$(-4)^2 + (-4) - 6 < 0$ $64 - 4 - 6 < 0$ $54 < 0$	False
B	Values between −3 and 2	0	$(0)^2 + (0) - 6 < 0$ $0 + 0 - 6 < 0$ $-6 < 0$	True
C	Values greater than 2	3	$(3)^2 + (3) - 6 < 0$ $9 + 3 - 6 < 0$ $6 < 0$	False
	Boundary number	−3	$(-3)^2 + (-3) - 6 < 0$ $9 - 3 - 6 < 0$ $0 < 0$	False
	Boundary number	2	$(2)^2 + (2) - 6 < 0$ $4 + 2 - 6 < 0$ $0 < 0$	False

The test point in region B satisfies the inequality, so all of the solutions of the inequality lie in region B. The boundary numbers -3 and 2 are not included in the solution set since they do not satisfy the inequality. Therefore, we will use a parenthesis on these numbers. Recall that the inequality symbol $<$ tells us that the boundary numbers are not included in the solution set.

We shade region B and place the appropriate symbols on the boundary numbers. The graph and solution set are

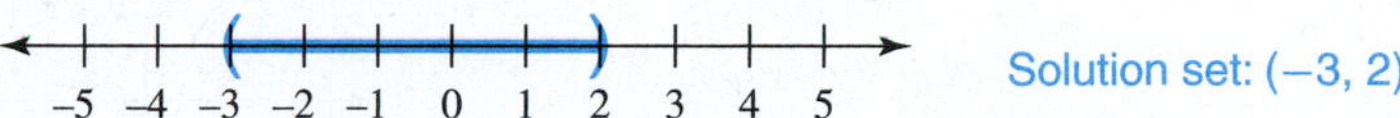

Solution set: $(-3, 2)$

> **Procedure: Solving a Quadratic Inequality Using the Boundary Number Method**
>
> **Step 1:** Find the boundary numbers by solving the associated equation.
> **Step 2:** Mark the boundary numbers on a number line.
> **Step 3:** Test one point from each region formed by the boundary numbers in the original inequality.
> **Step 4:** Shade the region that contains the test point that satisfies the inequality.
> **Step 5:** Determine the appropriate symbols to use on the boundary numbers.
> **a.** If the inequality is $<$ or $>$, then use parentheses on the boundary numbers.
> **b.** If the inequality is $\le$ or $\ge$, use brackets on the boundary numbers.
> **Step 6:** Write the solution in interval notation.

Objective 1 Examples

Solve each inequality using the boundary number method. Write each solution in interval notation.

1a. $(x-2)(x+4) \le 0$ **1b.** $x^2 - 8x > 9$ **1c.** $(x-1)^2 \le 0$ **1d.** $(x+3)^2 < -25$

Solutions

1a.

$$(x-2)(x+4) = 0 \quad \text{Write the associated equation.}$$
$$x - 2 = 0 \quad \text{or} \quad x + 4 = 0 \quad \text{Set each factor equal to zero.}$$
$$x = 2 \qquad x = -4 \quad \text{Solve.}$$

Mark the boundary numbers, -4 and 2, on the number line.

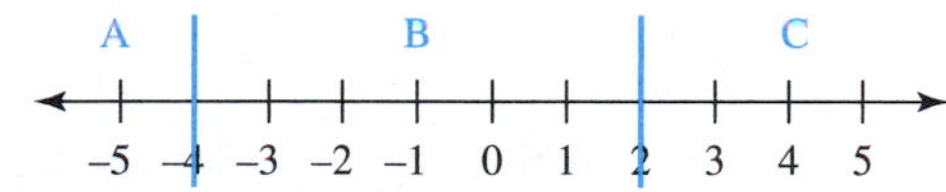

Region	Interval	Test Value	Test	True/False
A	Values less than -4	-5	$(-5-2)(-5+4) \le 0$ $(-7)(-1) \le 0$ $7 \le 0$	False
B	Values between -4 and 2	0	$(0-2)(0+4) \le 0$ $(-2)(4) \le 0$ $-8 \le 0$	True
C	Values greater than 2	3	$(3-2)(3+4) \le 0$ $(1)(7) \le 0$ $7 \le 0$	False

We shade region B and place brackets on the boundary numbers since they satisfy inequality. The graph and solution set are

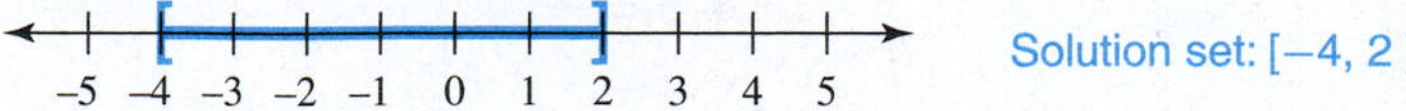

Solution set: $[-4, 2]$

1b.

$x^2 - 8x = 9$	Write the associated equation.
$x^2 - 8x - 9 = 0$	Write the equation in standard form.
$(x - 9)(x + 1) = 0$	Factor.
$x - 9 = 0$ or $x + 1 = 0$	Set each factor equal to zero.
$x = 9$ $\quad x = -1$	Solve.

Mark the boundary numbers -1 and 9 on the number line.

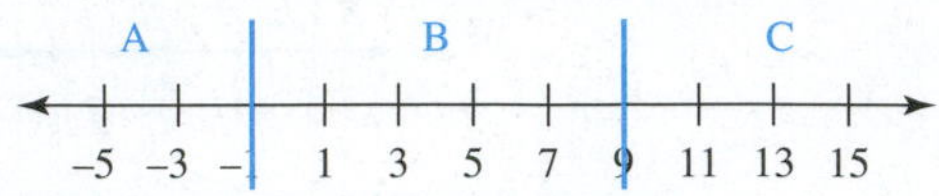

Region	Intervals of x	Test Value	Test	True/False
A	Values less than -1	-2	$(-2)^2 - 8(-2) > 9$ $4 + 16 > 9$ $20 > 9$	True
B	Values between -1 and 9	0	$(0)^2 - 8(0) > 9$ $0 - 0 > 9$ $0 > 9$	False
C	Values greater than 9	10	$(10)^2 - 8(10) > 9$ $100 - 80 > 9$ $20 > 9$	True

We shade regions A and C and place parentheses on the boundary numbers since the inequality symbol is $>$. The graph and solution set are

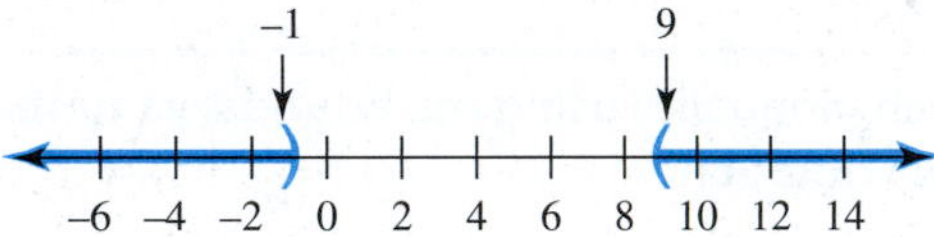

Solution set: $(-\infty, -1) \cup (9, \infty)$

1c.

$(x - 1)^2 = 0$	Write the associated equation.
$x - 1 = \pm 0$	Apply the square root property.
$x = 1$	Add 1 to each side.

Mark the boundary number 1 on the number line.

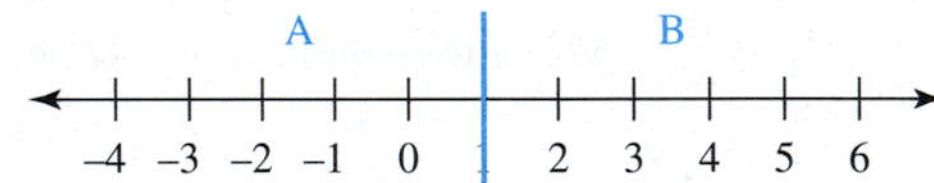

Region	Intervals of x	Test Value	Test	True/False
A	Values less than 1	0	$(0 - 1)^2 \leq 0$ $(-1)^2 \leq 0$ $1 \leq 0$	False
B	Values greater than 1	2	$(2 - 1)^2 \leq 0$ $(1)^2 \leq 0$ $1 \leq 0$	False

The two regions do not satisfy the inequality but the boundary number does satisfy it. So, the solution set consists of a single value, 1. The graph and solution set are

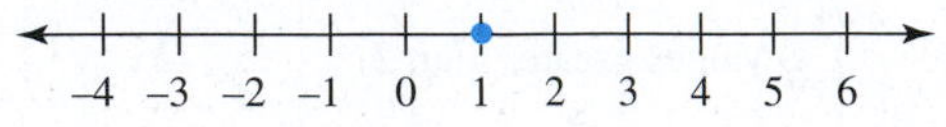

Solution set: $\{1\}$

1d.

$(x + 3)^2 = -25$	Write the associated equation.
$x + 3 = \pm\sqrt{-25}$	Apply the square root property.
$x + 3 = \pm 5i$	Simplify the radical.
$x = -3 \pm 5i$	Subtract 3 from each side.

Because the solutions are not real numbers, there are no boundary numbers to mark on the number line. So, the entire number line is a region itself.

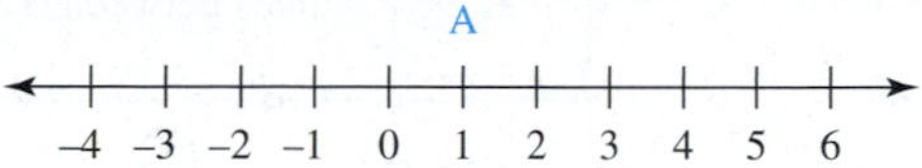

Region	Intervals of x	Test Value	Test	True/False
A	Any real number	0	$(0+3)^2 < -25$ $(3)^2 < -25$ $9 < -25$	False

Since this region doesn't contain values that satisfy the inequality, there are no solutions of the inequality. So, the solution set is the empty set, $\varnothing$.

Note: *The fact that there is no solution could have also been obtained from the inequality. The expression on the left is always positive. A positive number will never be less than a negative number. So, this is a contradiction and has no solution.*

Student Check 1 Solve each inequality using the boundary number method. Write each solution in interval notation.

a. $(x-6)(x-3) > 0$ **b.** $x^2 + 4x \geq 21$ **c.** $(x-5)^2 \leq 0$ **d.** $(x+8)^2 < -16$

Higher Degree Polynomial Inequalities

Objective 2 Solve higher degree polynomial inequalities using the boundary number method.

We solve higher degree polynomial inequalities in exactly the same way that we solve quadratic inequalities. The only difference is that we solve a higher degree equation instead of a quadratic equation to find the boundary numbers. We mark the numbers on a number line and test points from each of the corresponding regions.

Objective 2 Examples **Solve each inequality using the boundary number method. Write each solution set in interval notation.**

2a. $x(x-3)(x+2) > 0$ **2b.** $x^4 - 5x^2 + 4 \leq 0$

Solutions **2a.**

$x(x-3)(x+2) = 0$ Write the associated equation.

$x = 0 \quad x-3=0 \quad x+2=0$ Set each factor equal to zero.

$x = 3 \quad x = -2$ Solve.

Mark the boundary numbers -2, 0, and 3 on a number line.

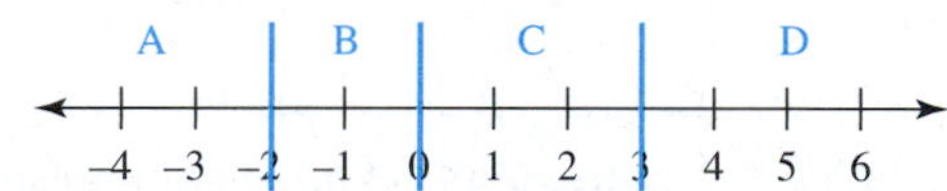

Region	Intervals of x	Test Value	Test	True/False
A	Values less than -2	-4	$-4(-4-3)(-4+2) > 0$ $-4(-7)(-2) > 0$ $-56 > 0$	False
B	Values between -2 and 0	-1	$-1(-1-3)(-1+2) > 0$ $-1(-4)(1) > 0$ $4 > 0$	True

Region	Intervals of x	Test Value	Test	True/False
C	Values between 0 and 3	1	$1(1-3)(1+2)>0$ $1(-2)(3)>0$ $-6>0$	False
D	Values greater than 3	4	$4(4-3)(4+2)>0$ $4(1)(6)>0$ $24>0$	True

Regions B and D contain solutions of the inequality. The boundary numbers are not part of the solution set. The graph and solution set are

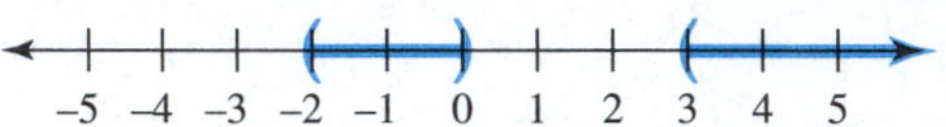

Solution set: $(-2, 0) \cup (3, \infty)$

2b.

$$x^4 - 5x^2 + 4 = 0$$ Write the associated equation.

$$(x^2 - 4)(x^2 - 1) = 0$$ Factor.

$$(x + 2)(x - 2)(x + 1)(x - 1) = 0$$ Apply the difference of squares.

$$x + 2 = 0,\ x - 2 = 0,\ x + 1 = 0,\ x - 1 = 0$$ Set each factor equal to zero.

$$x = -2, \quad x = 2, \quad x = -1, \quad x = 1$$ Solve.

Mark the boundary numbers -2, -1, 1, and 2 on a number line.

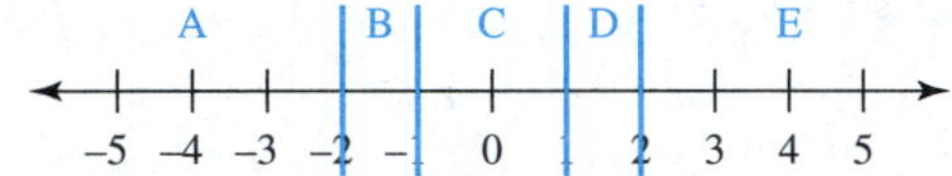

Region	Intervals of x	Test Value	Test	True/False
A	Values less than -2	-3	$x^4-5x^2+4\le 0$ $(-3)^4-5(-3)^2+4\le 0$ $81-5(9)+4\le 0$ $40\le 0$	False
B	Values between -2 and -1	-1.5	$x^4-5x^2+4\le 0$ $(-1.5)^4-5(-1.5)^2+4\le 0$ $5.0625-5(2.25)+4\le 0$ $-2.1875\le 0$	True
C	Values between -1 and 1	0	$x^4-5x^2+4\le 0$ $(0)^4-5(0)^2+4\le 0$ $0-0+4\le 0$ $4\le 0$	False
D	Values between 1 and 2	1.5	$x^4-5x^2+4\le 0$ $(1.5)^4-5(1.5)^2+4\le 0$ $5.0625-5(2.25)+4\le 0$ $-2.1875\le 0$	True
E	Values greater than 2	3	$x^4-5x^2+4\le 0$ $(3)^4-5(3)^2+4\le 0$ $81-5(9)+4\le 0$ $40\le 0$	False

Regions B and D contain solutions of the inequality. The boundary numbers are part of the solution set since the inequality is $\le$. The graph and solution set are

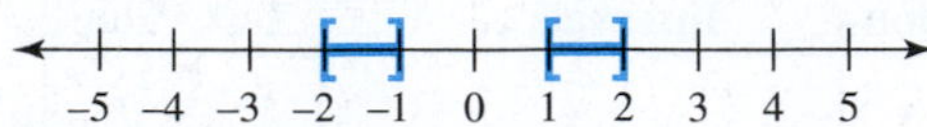

Solution set: $[-2, -1] \cup [1, 2]$

Student Check 2 Solve each inequality using the boundary number method. Write each solution set in interval notation.

a. $(x - 4)(x - 7)(x + 3) < 0$ **b.** $x^4 - 10x^2 + 9 \ge 0$

Rational Inequalities

Objective 3 Solve rational inequalities using the boundary number method.

Rational inequalities are inequalities that involve rational expressions. They are solved in much the same way as the previous types of inequalities. Boundary numbers, however, come not only from the solutions of the associated equation, but also from the values that make the denominators zero; that is, that make the rational expression undefined.

Procedure: Solving a Rational Inequality

Step 1: Find solutions of the associated equation.
Step 2: Find the values that make each rational expression undefined.
Step 3: Mark the numbers from steps 1 and 2 on a number line.
Step 4: Test points from each of the regions formed by the boundary numbers.
Step 5: Shade the regions that tested true.
Step 6: Determine if the boundary numbers are included. Boundary numbers that come from the solutions of the equation are only included if the inequality symbol is $\leq$ or $\geq$. Boundary numbers that make the denominator zero are *never* included in the solution set.
Step 7: Write the solution in interval notation.

Objective 3 Examples Solve each inequality using the boundary number method. Write each solution set in interval notation.

3a. $\dfrac{x-1}{x+2} \geq 0$ **3b.** $\dfrac{7}{x-3} < -1$

Solutions **3a.** We find the boundary numbers.

Associated Equation	Values That Make the Expression Undefined
$\dfrac{x-1}{x+2} = 0$	
$(x+2)\dfrac{x-1}{x+2} = 0(x+2)$	$x+2=0$
$x-1=0$	$x=-2$
$x=1$	

Mark the boundary numbers -2 and 1 on a number line.

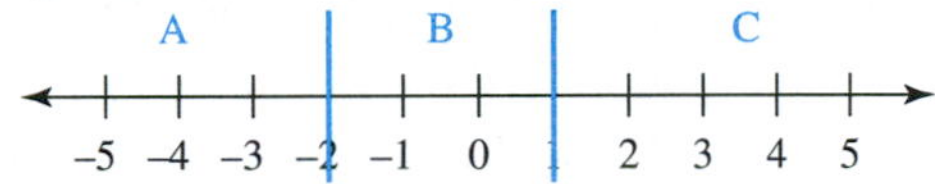

Region	Intervals of x	Test Value	Test	True/False
A	Values less than -2	-5	$\dfrac{-5-1}{-5+2} \geq 0$ $\dfrac{-6}{-3} \geq 0$ $2 \geq 0$	True
B	Values between -2 and 1	0	$\dfrac{0-1}{0+2} \geq 0$ $\dfrac{-1}{2} \geq 0$ $-\dfrac{1}{2} \geq 0$	False
C	Values greater than 1	3	$\dfrac{3-1}{3+2} \geq 0$ $\dfrac{2}{5} \geq 0$	True

Regions A and C contain solutions of the inequality. The boundary number 1 is included in the solution set but −2 is *not* included in the solution. The graph and solution set are

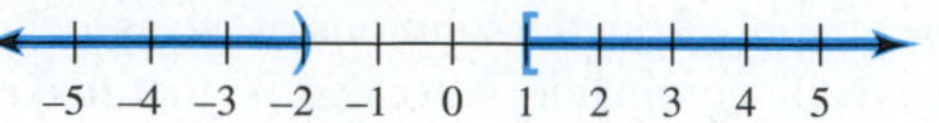

Solution set: $(-\infty, -2) \cup [1, \infty)$

3b. We find the boundary numbers.

Associated Equation	**Values That Make the Expression Undefined**
$\dfrac{7}{x-3} = -1$ $(x-3)\dfrac{7}{x-3} = -1(x-3)$ $7 = -x + 3$ $4 = -x$ $-4 = x$	$x - 3 = 0$ $x = 3$

Mark the boundary numbers −4 and 3 on a number line.

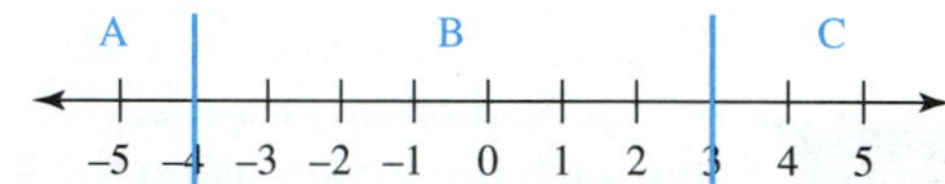

Region	Intervals of x	Test Value	Test	True/False
A	Values less than −4	−5	$\dfrac{7}{-5-3} < -1$ $\dfrac{7}{-8} < -1$ $-0.875 < -1$	False
B	Values between −4 and 3	0	$\dfrac{7}{0-3} < -1$ $\dfrac{7}{-3} < -1$ $-2.33 < -1$	True
C	Values greater than 3	4	$\dfrac{7}{4-3} < -1$ $\dfrac{7}{1} < -1$ $7 < -1$	False

Region B contains solutions of the inequality. Neither of the boundary numbers is included in the solution. The graph and solution set are

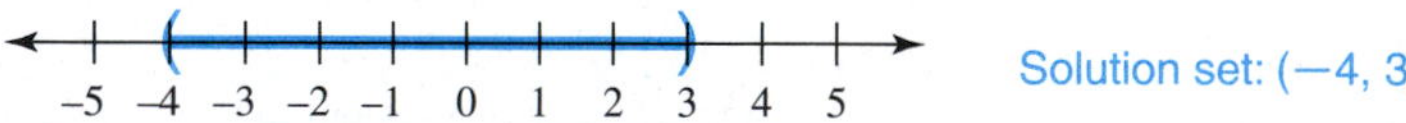

Solution set: $(-4, 3)$

Student Check 3 Solve each inequality using the boundary number method. Write each solution set in interval notation.

a. $\dfrac{x-5}{x+1} \le 0$ **b.** $\dfrac{4}{x-8} \ge -2$

Solving Nonlinear Inequalities Graphically

Objective 4 Find solutions of nonlinear inequalities from a graph.

We can also find solutions of inequalities from a graph. When we solve an inequality of the form $f(x) < 0$, we find the values of x whose corresponding y-values are negative. Points with negative y-values lie in Quadrants III and IV. So, our solution will consist of the intervals of x whose points lie *below* the x-axis.

When we solve an inequality of the form $f(x) > 0$, we find the values of x whose corresponding y-values are positive. Points with positive y-values lie in Quadrants I and II. So, our solution will consist of the intervals of x whose points lie *above* the x-axis.

To solve an inequality graphically, we must find the solutions of the associated equation. These solutions correspond to the x-intercepts and are our boundary numbers. The solutions divide the graph into regions where the graph of $f(x)$ is either above or below the x-axis. Consider the graph of $f(x) = x^2 - 4$. The boundary numbers and x-intercepts are $x = -2$ and $x = 2$.

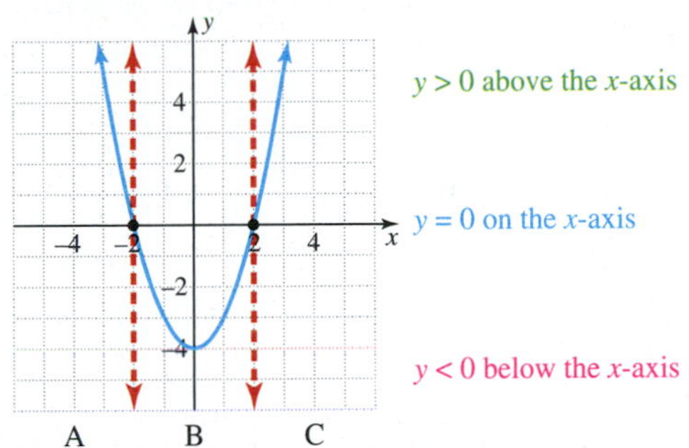

Note that we have three regions.

Region A consists of x values less than -2.

Region B consists of x values between -2 and 2.

Region C consists of x values greater than 2.

In regions A and C, the graph lies *above* the x-axis which means the y-values are *positive*. In region B, the graph lies *below* the x-axis which means the y-values are *negative*.

To solve the inequality $f(x) > 0$ or $x^2 - 4 > 0$ using the graph of $f(x)$, we make a table as shown.

Region	Intervals of x	Graph Above/Below x-axis	Is $f(x) > 0$?	Solution?
A	Values less than -2	Above	Yes	Yes
B	Values between -2 and 2	Below	No	No
C	Values greater than 2	Above	Yes	Yes
Boundary numbers: -2, 2		On x-axis	No	No

So, the solution set of $f(x) > 0$ or $x^2 - 4 > 0$ consists of regions A and C. We write the solution set as $(-\infty, -2) \cup (2, \infty)$.

Procedure: Solving a Nonlinear Inequality Graphically

Step 1: Rewrite the inequality so that one side of it is zero, that is, put it in the form $f(x) > 0, f(x) \geq 0, f(x) < 0$, or $f(x) \leq 0$.

Step 2: Draw the graph of the function $f(x)$. Find the x-intercepts, or boundary numbers.

Step 3: Draw a dashed vertical line through the x-intercepts to determine the regions in the coordinate plane.

Step 4: Complete a chart to determine the solution of the inequality.

a. If the graph is below the x-axis, $f(x) < 0$.

b. If the graph is above the x-axis, $f(x) < 0$.

Step 5: Write the solution set in interval notation.

a. The boundary numbers will be included in the solution set if the inequality is $\leq$ or $\geq$.

b. The boundary values will not be included in the solution set if the inequality is $<$ or $>$.

c. If the problem involves a rational inequality, do not include a boundary number in the solution set that makes the expression undefined.

Objective 4 Examples **Use a graph to solve each associated inequality. Write each solution set in interval notation.**

4a. The graph of $f(x) = \frac{1}{4}x^2 - \frac{1}{4}x - 3$ is given. Use this graph to solve the inequality $\frac{1}{4}x^2 - \frac{1}{4}x - 3 \leq 0$.

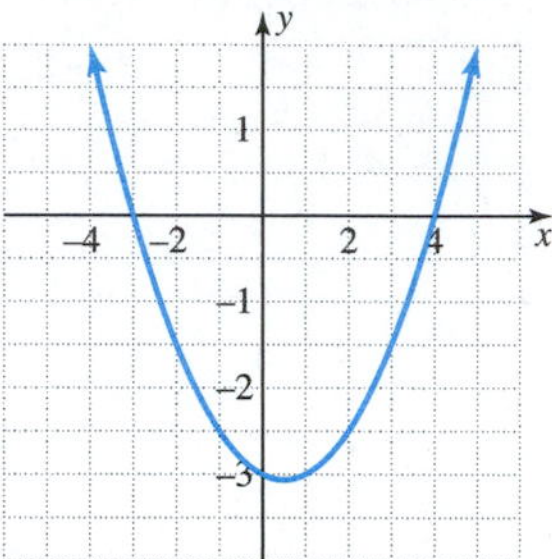

Solution **4a.** The boundary numbers are -3 and 4 since these are the x-intercepts of the graph. We can show the regions by drawing a vertical line through the boundary numbers.

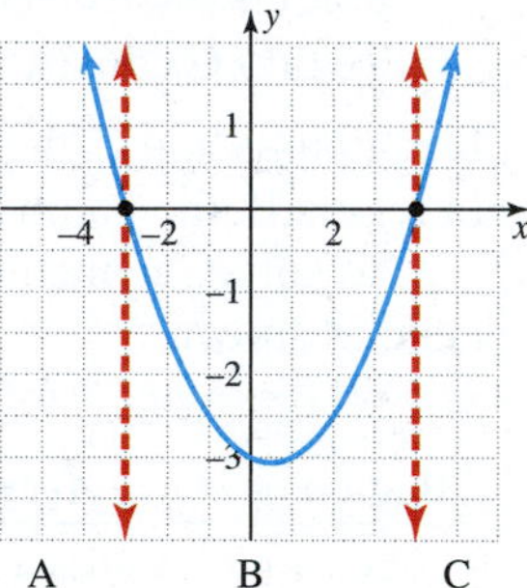

Region	Intervals of x	Graph Above/Below x-axis	Is $f(x) \leq 0$?	Solution?
A	Values less than -3	Above	No	No
B	Values between -3 and 4	Below	Yes	Yes
C	Values greater than 4	Above	No	No
Boundary numbers: -3, 4		On the x-axis	Yes	Yes

So, the solution set consists of region B and the boundary numbers. The solution set is $[-3, 4]$.

4b. The graph of $f(x) = x(x - 2)(x + 2)$ is shown. Use this to solve $x(x - 2)(x + 2) > 0$.

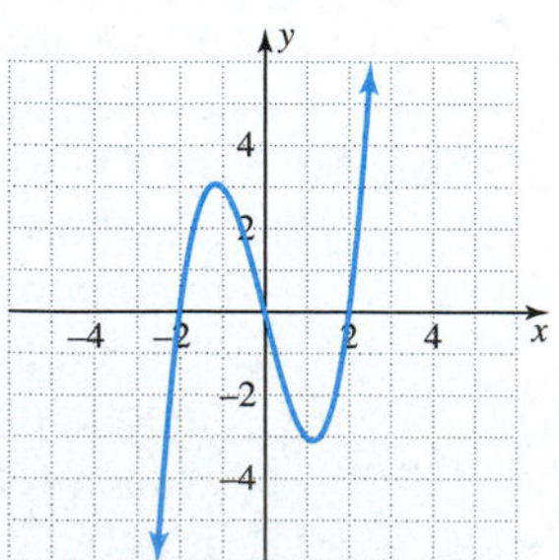

Solution **4b.** The boundary numbers are -2, 0, and 2 since these are the x-intercepts of the graph. We can show the regions on the graph by drawing a vertical line through these boundary numbers.

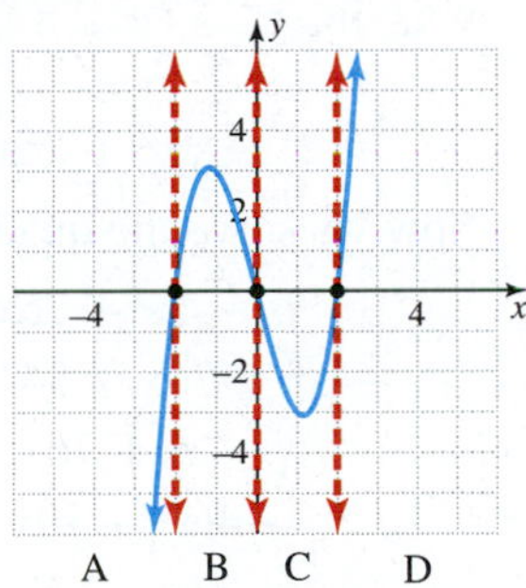

Region	Intervals of x	Graph Above/Below x-axis	Is $f(x) > 0$?	Solution?
A	Values less than -2	Below	No	No
B	Values between -2 and 0	Above	Yes	Yes
C	Values between 0 and 2	Below	No	No
D	Values greater than 2	Above	Yes	Yes
Boundary numbers: $-2, 0, 2$		On the x-axis	No	No

The solution set consists of regions B and D. The solution set is $(-2, 0) \cup (2, \infty)$.

Student Check 4 Use the graph to solve each associated inequality. Write each solution set in interval notation.

a. The graph of $f(x) = -x^2 + 1$ is given below. Use this graph to solve $-x^2 + 1 > 0$.

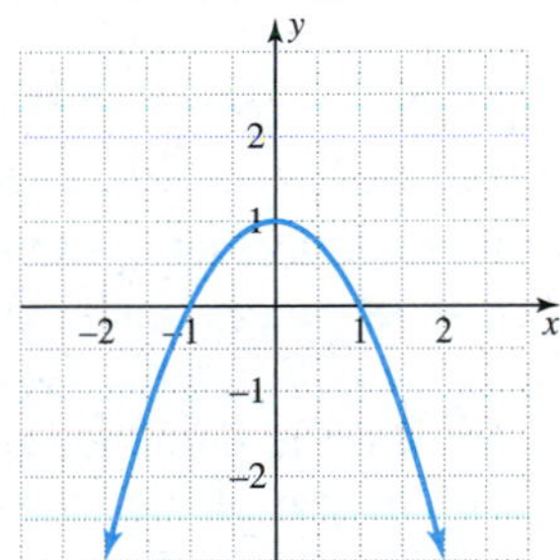

b. The graph of $f(x) = (x - 1)(x + 2)(x - 3)$ is given. Use the graph to solve $(x - 1)(x + 2)(x - 3) \leq 0$.

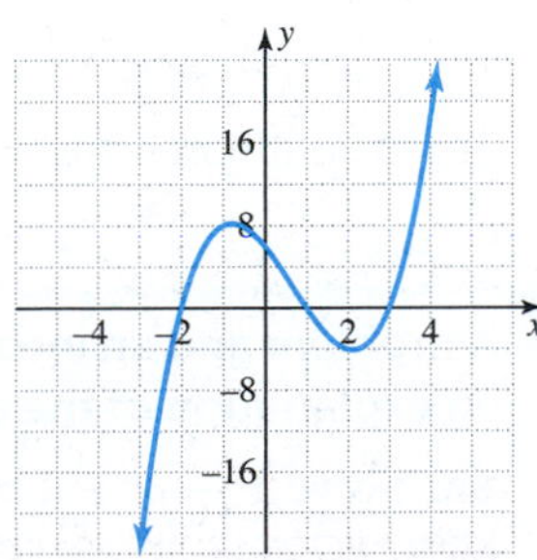

Objective 5

Solve applications of nonlinear inequalities.

Applications

Nonlinear inequalities are used to represent situations in which a quantity modeled by a nonlinear function is less than or greater than another quantity. Key phrases to denote less than or more than will determine the appropriate symbol to use.

Objective 5 Example The monthly profit, in thousands of dollars, from selling x thousand digital cameras is given by $P(x) = -4x^2 + 76x + 14$. How many cameras must be sold to have a profit greater than \$350,000?

Solution

$$\text{Profit} > 350 \quad \text{Profit is greater than \$350,000.}$$
$$-4x^2 + 76x + 14 > 350 \quad \text{Replace profit with the given function.}$$

Now we solve the quadratic inequality using the boundary number method.

$$-4x^2 + 76x + 14 = 350 \quad \text{Write the associated equation.}$$
$$-4x^2 + 76x + 14 - 350 = 350 - 350 \quad \text{Subtract 350 from each side.}$$
$$-4x^2 + 76x - 336 = 0 \quad \text{Simplify.}$$
$$-4(x^2 - 19x + 84) = 0 \quad \text{Factor out the GCF, } -4.$$
$$-4(x - 12)(x - 7) = 0 \quad \text{Factor the trinomial.}$$
$$x - 12 = 0 \quad \text{or} \quad x - 7 = 0 \quad \text{Set each factor equal to 0.}$$
$$x = 12 \qquad x = 7 \quad \text{Solve the resulting equations.}$$

Mark the boundary numbers 7 and 12 on the number line.

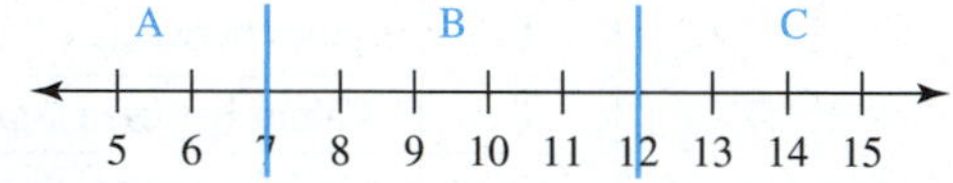

Region	Intervals of x	Test Value	Test	True/False
A	Values less than 7	0	$-4x^2 + 76x + 14 > 350$ $-4(0)^2 + 76(0) + 14 > 350$ $14 > 350$	False
B	Values between 7 and 12	8	$-4x^2 + 76x + 14 > 350$ $-4(8)^2 + 76(8) + 14 > 350$ $366 > 350$	**True**
C	Values greater than 12	13	$-4x^2 + 76x + 14 > 350$ $-4(13)^2 + 76(13) + 14 > 350$ $326 > 350$	False

The solution set of the inequality is the interval (7, 12), which means that the profit will be greater than \$350,000 if the company sells between 7000 and 12,000 cameras.

Student Check 5 A rocket is launched upward from the ground at an initial velocity of 64 ft/sec. The height of the rocket t sec after it is launched is $h(t) = -16t^2 + 64t$. When will the rocket be greater than 48 ft above the ground?

Objective 6 Troubleshoot common errors.

Troubleshooting Common Errors

Some common errors associated with nonlinear inequalities are shown.

Objective 6 Examples **A problem and an incorrect solution are given. Provide the correct solution and an explanation of the error.**

6a. Solve $x^2 - x - 2 > 0$.

Incorrect Solution	Correct Solution and Explanation
$x^2 - x - 2 = 0$ $(x - 2)(x + 1) = 0$ $x = 2, x = -1$ Regions A and C test true. The solution set is $(-\infty, -1] \cup [2, \infty)$.	The boundary numbers are correct and regions A and C do test true. The error is that the symbol on the boundary numbers should be ( ) and not [ ]. The solution set should be $(-\infty, -1) \cup (2, \infty)$.

6b. Solve $(x + 3)^2 > -4$.

Incorrect Solution	Correct Solution and Explanation
$(x + 3)^2 = -4$ $x + 3 = \pm\sqrt{-4}$ $x + 3 = \pm 2i$ $x = -3 \pm 2i$ There are no boundary numbers, so the solution of the inequality is the empty set, ∅.	While there are no boundary numbers, we cannot automatically conclude that there is no solution until we test a point in the region. The only region to test is the entire number line. We test 0: $(0 + 3)^2 > -4$ $(3)^2 > -4$ $9 > -4$ True Since the test point makes the inequality true, the entire region contains solutions. Therefore, the solution set is $(-\infty, \infty)$. Note: We could have noticed this from the inequality. Any number squared is always greater than a negative number.

ANSWERS TO STUDENT CHECKS

Student Check 1 **a.** $(-\infty, 3) \cup (6, \infty)$ **b.** $(-\infty, -7] \cup [3, \infty)$ **c.** $\{5\}$ **d.** ∅

Student Check 2 **a.** $(-\infty, -3) \cup (4, 7)$ **b.** $(-\infty, -3] \cup [-1, 1] \cup [3, \infty)$

Student Check 3 **a.** $(-1, 5]$ **b.** $(-\infty, 6] \cup (8, \infty)$

Student Check 4 **a.** $(-1, 1)$ **b.** $(-\infty, -2] \cup [1, 3]$

Student Check 5 between 1 and 3 sec

SUMMARY OF KEY CONCEPTS

1. The boundary number method can be used to solve any inequality. We use it to solve quadratic inequalities, higher degree polynomial inequalities, and rational inequalities.
   a. Find the boundary numbers by finding solutions of the associated equation and any values of x that make the function undefined.
   b. Mark the boundary numbers on the number line.
   c. Test a point from each region formed by the boundary numbers in the *original* inequality.
   d. Shade the regions that contain solutions. Determine the appropriate symbol on the endpoints.
   e. Write the solution set in interval notation.
2. We can also use a graph to solve an inequality. Draw the graph of $f(x)$. Intervals of x whose y-values lie above the x-axis are solutions of the inequality $f(x) > 0$. Intervals of x whose y-values lie below the x-axis are solutions of the inequality $f(x) < 0$.

GRAPHING CALCULATOR SKILLS

A graphing calculator can help us test points in an inequality. The calculator can also graph the solution set of an inequality.

Example 1: Use the calculator to test points to solve $x^2 - 8x > 9$.

Solution: After solving the associated equation, we know the boundary numbers are -1 and 9. Enter the left and right side of the inequality in Y_1 and Y_2, respectively. Now we find the values of x for which $Y_1 > Y_2$.

```
Plot1 Plot2 Plot3
\Y1=X²-8X
\Y2=9
\Y3=
\Y4=
\Y5=
\Y6=
\Y7=
```

To test points in the regions formed by the boundary numbers of -1 and 9, we can view the table. We choose test points of -2, 2, and 10.

X	Y1	Y2
-3	33	9
-2	20 >	9
-1	9	9
0	0	9
1	-7	9
2	-12 >	9
3	-15	9

X=-3

X	Y1	Y2
6	-12	9
7	-7	9
8	0	9
9	9	9
10	20 >	9
11	33	9
12	48	9

X=12

For $x = -2$, we have $20 > 9$ which is true.
For $x = 2$, we have $-12 > 9$ which is false
For $x = 10$, we have $20 > 9$ which is true.

So, regions A and C are included in the solution set.

Example 2: Graph the solution set of $x^2 - 8x > 9$.

Solution: Enter the complete inequality in Y_1. Use 2nd MATH to select the inequality symbol and then graph.

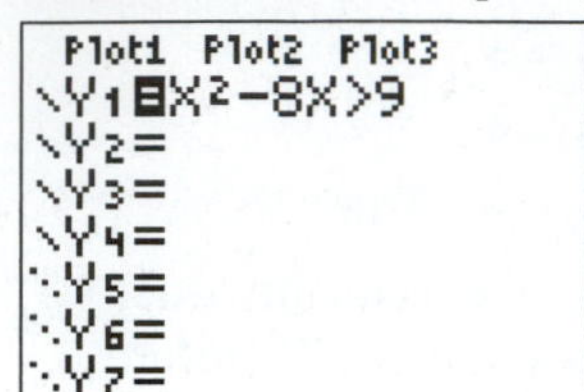

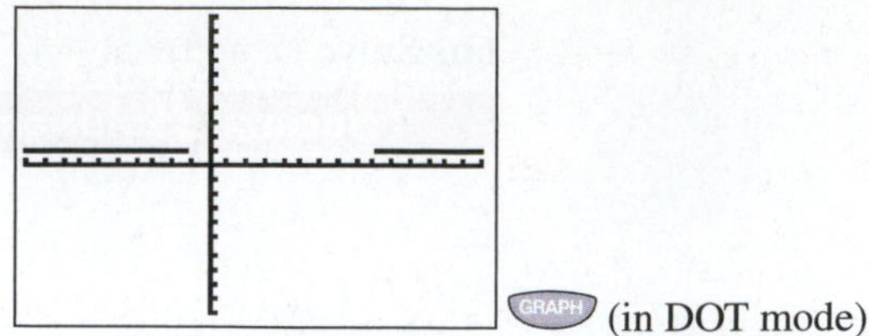

GRAPH (in DOT mode).

The solution of the inequality is shaded slightly above the x-axis. We must determine whether or not the endpoints are included, but the graph shows us that regions A and C contain solutions of the inequality.

SECTION 9.6 EXERCISE SET

Write About It!

Use complete sentences in your answer to each exercise.

1. Explain how to find the boundary numbers of a polynomial or rational inequality.
2. Explain how to determine the appropriate symbols, parentheses or brackets, on the boundary numbers when solving polynomial inequalities.
3. When solving the rational inequality, what can you say about the boundary numbers that make the rational expression undefined?
4. If there is no solution of a quadratic inequality, $ax^2 + bx + c \le 0$, what can you say about the graph of the corresponding quadratic function $f(x) = ax^2 + bx + c$? The graph of the quadratic function lies above the x-axis for all values of x.
5. If there is no solution of a quadratic inequality, $ax^2 + bx + c \ge 0$, what can you say about the graph of the corresponding quadratic function $f(x) = ax^2 + bx + c$? The graph of the quadratic function lies below the x-axis for all values of x.
6. Consider a quadratic inequality of the form $(x + b)^2 \le c$ and explain how to determine the solution if $c < 0$ and $c = 0$. If $c < 0$, there is no solution because $(x + b)^2$ is always nonnegative. If $c = 0$, then $x = -b$.
7. Consider a quadratic inequality of the form $(x + b)^2 \ge c$ and explain how to determine the solution if $c > 0$ and $c = 0$. If $c > 0$, then $x = \sqrt{c} - b$. If $c = 0$, then $x = -b$.

Practice Makes Perfect!

Solve each inequality using the boundary number method. Write each solution in interval notation. (*See Objective 1.*)

8. $(x + 2)(x - 5) > 0$ $(-\infty, -2) \cup (5, \infty)$
9. $(x - 1)(x + 12) > 0$ $(-\infty, -12) \cup (1, \infty)$
10. $(x - 3)(x - 8) \ge 0$ $(-\infty, 3] \cup [8, \infty)$
11. $(x + 5)(x + 2) \ge 0$ $(-\infty, -5] \cup [-2, \infty)$
12. $(x + 4)(x - 6) < 0$ $(-4, 6)$
13. $(x + 10)(x + 2) < 0$ $(-10, -2)$
14. $x^2 + 6x \le 40$ $[-10, 4]$
15. $x^2 - 2x \le 35$ $[-5, 7]$
16. $x^2 - 5x > 24$ $(-\infty, -3) \cup (8, \infty)$
17. $x^2 + 5x > 36$ $(-\infty, -9) \cup (4, \infty)$
18. $(x + 4)^2 > 0$ $(-\infty, -4) \cup (-4, \infty)$
19. $(x - 2)^2 > 0$ $(-\infty, 2) \cup (2, \infty)$
20. $(x + 5)^2 \ge 0$ $(-\infty, \infty)$
21. $(x + 6)^2 \ge 0$ $(-\infty, \infty)$
22. $3x^2 + 5x + 2 \le 0$
23. $2x^2 - 13x + 20 \le 0$ $\left[\frac{5}{2}, 4\right]$
24. $5x^2 + 7x - 6 > 0$
25. $6x^2 - x - 1 > 0$
26. $(2x - 1)^2 \le 0$ $\left\{\frac{1}{2}\right\}$
27. $(3x + 5)^2 \le 0$ $\left\{-\frac{5}{3}\right\}$
28. $(3x + 4)^2 < 0$ $\varnothing$
29. $(4x - 1)^2 < 0$ $\varnothing$
30. $(x + 7)^2 + 9 < 0$ $\varnothing$
31. $(x - 1)^2 + 4 < 0$ $\varnothing$

Solve each inequality using the boundary number method. Write each solution in interval notation. (*See Objective 2.*)

32. $x(x + 3)(x - 6) \ge 0$ $[-3, 0] \cup [6, \infty)$
33. $x(x - 3)(x + 4) \ge 0$ $[-4, 0] \cup [3, \infty)$
34. $(x + 8)(x + 1)(x - 5) < 0$ $(-\infty, -8) \cup (-1, 5)$
35. $(x - 9)(x + 2)(x - 2) < 0$ $(-\infty, -2) \cup (2, 9)$
36. $(3x - 5)(x + 2)(2x - 9) > 0$ $\left(-2, \frac{5}{3}\right) \cup \left(\frac{9}{2}, \infty\right)$
37. $(7x + 2)(x - 1)(7x - 3) > 0$
38. $x^4 - 13x^2 + 36 \ge 0$ $(-\infty, -3] \cup [-2, 2] \cup [3, \infty)$
39. $9x^4 - 10x^2 + 1 > 0$
40. $x^4 - 26x^2 + 25 \le 0$ $[-5, -1] \cup [1, 5]$
41. $x^4 - 3x^2 - 4 \le 0$ $[-2, 2]$
42. $x^4 + 5x^2 + 6 < 0$ $\varnothing$
43. $x^4 + 8x^2 + 7 > 0$ $(-\infty, \infty)$

Solve each rational inequality using the boundary number method. Write each solution in interval notation. (*See Objective 3.*)

44. $\dfrac{x + 3}{x - 4} \ge 0$ $(-\infty, -3] \cup (4, \infty)$
45. $\dfrac{x - 5}{x + 1} \ge 0$ $(-\infty, -1) \cup [5, \infty)$
46. $\dfrac{2x + 5}{x - 3} \le 0$ $\left[-\frac{5}{2}, 3\right)$
47. $\dfrac{3x - 8}{x + 6} \le 0$
48. $\dfrac{x + 3}{2x - 5} > 0$
49. $\dfrac{x + 4}{x - 9} > 0$ $(-\infty, -4) \cup (9, \infty)$
50. $\dfrac{3}{x - 5} \ge 2$
51. $\dfrac{7}{x + 3} \ge 1$ $(-3, 4]$
52. $\dfrac{5}{x - 6} \le -2$ $\left[\frac{7}{2}, 6\right)$
53. $\dfrac{4}{x + 8} \le -2$ $[-10, -8)$

Additional answers can be found in the Instructor Answer Appendix.

Use the graph of $y = f(x)$ to solve each inequality. Write each solution in interval notation. (*See Objective 4.*)

54. $(x + 6)(x - 2) \le 0$

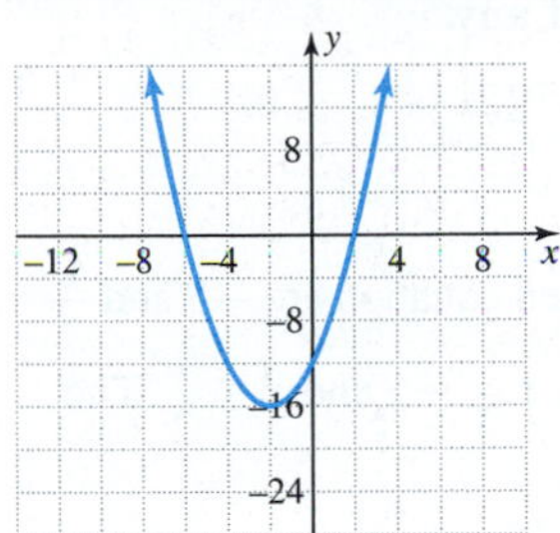

$[-6, 2]$

55. $(2x + 5)(x - 3) \le 0$

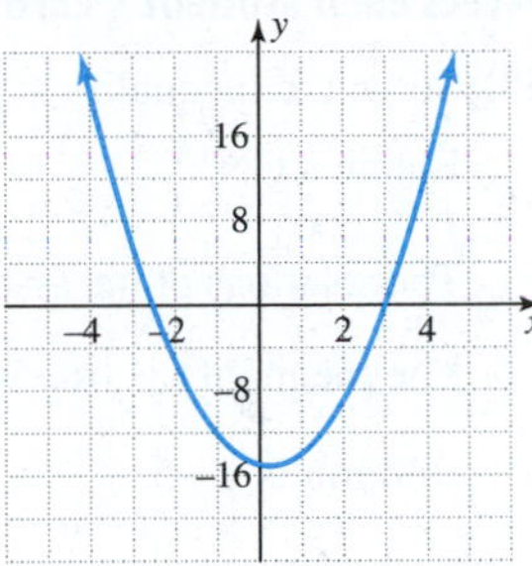

56. $(x + 4)(x - 1) > 0$

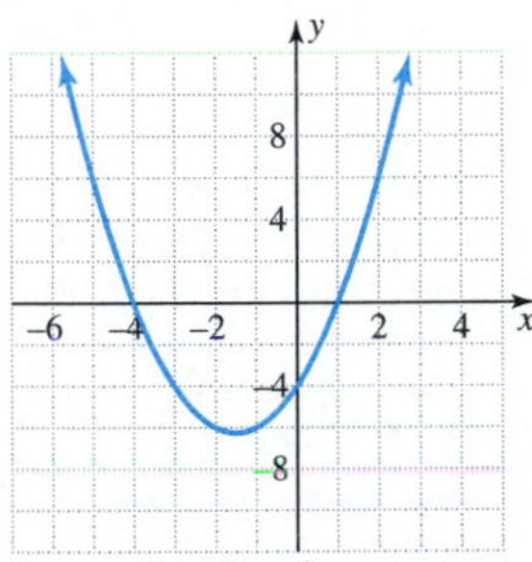

$(-\infty, -4) \cup (1, \infty)$

57. $(x - 7)(x - 2) > 0$

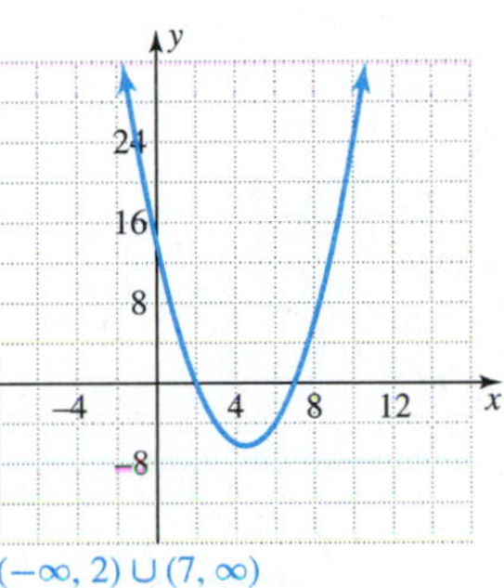

$(-\infty, 2) \cup (7, \infty)$

58. $x(x + 5)(x - 4) \ge 0$

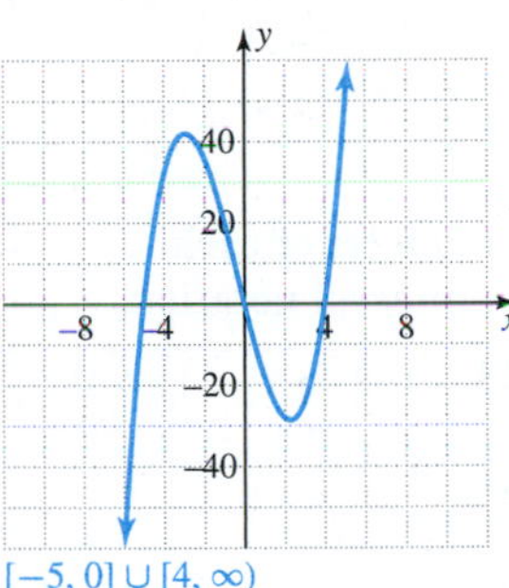

$[-5, 0] \cup [4, \infty)$

59. $x(x - 6)(x - 3) \ge 0$

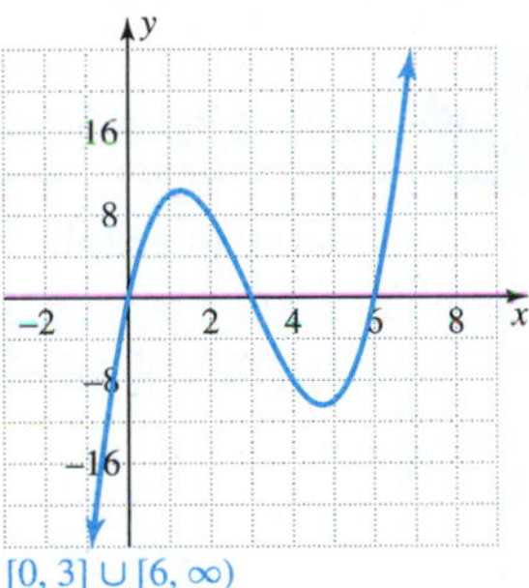

$[0, 3] \cup [6, \infty)$

60. $2x^2 + 5x - 3 \le 0$

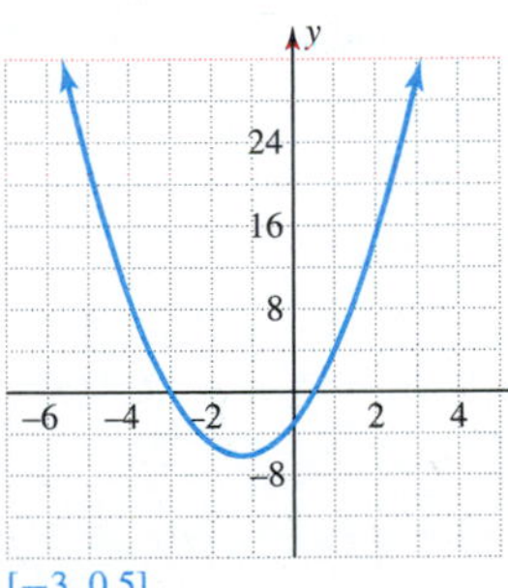

$[-3, 0.5]$

61. $2x^2 + 3x - 20 \le 0$

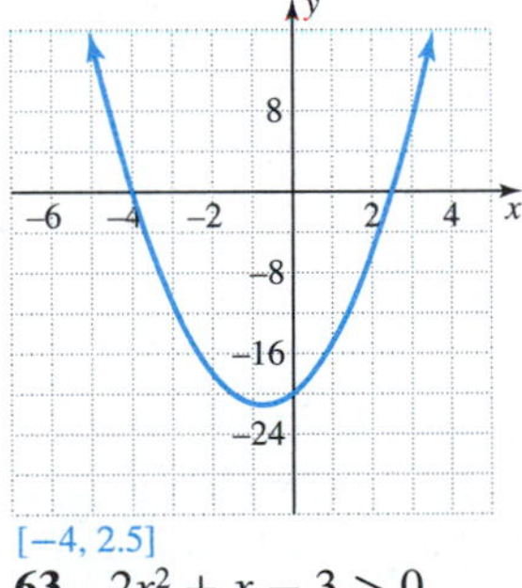

$[-4, 2.5]$

62. $5x^2 - 13x + 6 > 0$

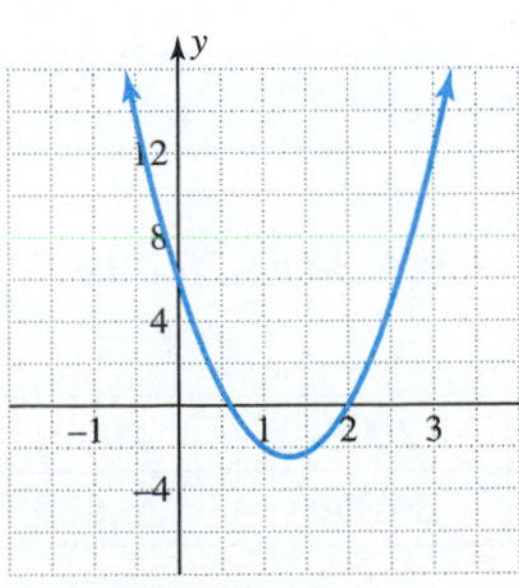

$(-\infty, 0.6) \cup (2, \infty)$

63. $2x^2 + x - 3 > 0$

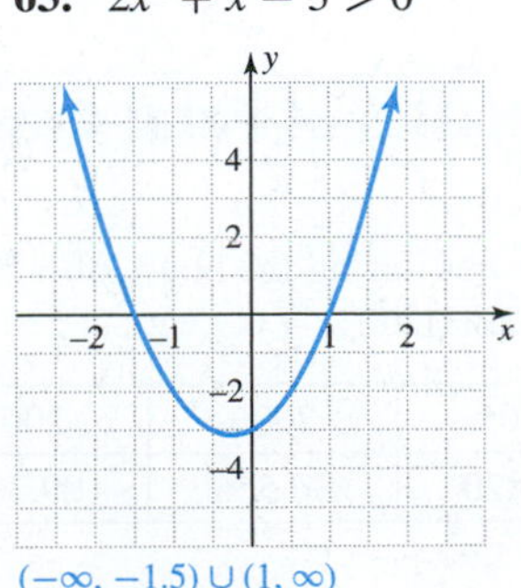

$(-\infty, -1.5) \cup (1, \infty)$

Solve each inequality using its graph. Write each solution in interval notation. (*See Objective 4.*)

64. $x^2 - 2x - 15 \ge 0$ $(-\infty, -3] \cup [5, \infty)$

65. $x^2 - 13x + 40 > 0$ $(-\infty, 5) \cup (8, \infty)$

66. $x^2 - 12x + 20 < 0$ $(2, 10)$

67. $x^2 + 6x - 40 \le 0$ $[-10, 4]$

68. $3x^2 + 4x - 4 \ge 0$

69. $9x^2 + 12x + 4 \ge 0$ $(-\infty, \infty)$

70. $x^2 + 6x + 9 < 0$ $\varnothing$

71. $2x^2 + x - 15 \le 0$ $\left[-3, \frac{5}{2}\right]$

Solve each problem. (*See Objective 5.*)

72. The weekly profit, in hundreds of dollars, from selling x hundred shirts is given by $P(x) = -x^2 + 8x - 12$. How many shirts must be sold for the company to make a profit? between 200 and 600 shirts each week

73. The profit, in dollars, for selling x units is given by $P(x) = -x^2 + 20x$. How many units must be sold to have a profit greater than \$75? between 5 and 15 units

74. A projectile is shot upward from ground level at an initial velocity of 256 ft/sec. Its height in t sec is given by $h(t) = -16t^2 + 256t$. When will the height of the projectile be greater than 768 ft above the ground? between 4 and 12 sec

75. A cannon ball is shot into the air from a cannon at an initial velocity of 320 ft/sec from a height of 10 ft. Its height in t sec is given by $h(t) = -16t^2 + 320t + 10$. When will the height of the cannon ball be more than 586 ft above the ground? between 2 and 18 sec

Mix 'Em Up!

Solve each inequality using the boundary number method or graphing. Write each solution in interval notation.

76. $(x - 3)(x + 9) \le 0$ $[-9, 3]$

77. $(x - 5)(x - 2) \le 0$ $[2, 5]$

78. $(x - 4)(x + 6) > 0$ $(-\infty, -6) \cup (4, \infty)$

79. $(x - 8)(x + 3) > 0$ $(-\infty, -3) \cup (8, \infty)$

80. $x^2 - 3x \le 28$ $[-4, 7]$

81. $x^2 - 7x > 60$ $(-\infty, -5) \cup (12, \infty)$

82. $(x - 11)^2 > 0$ $(-\infty, 11) \cup (11, \infty)$

83. $(x - 1)^2 \le 0$ $\{1\}$

84. $4x^2 + 11x + 6 > 0$

85. $2x^2 - x - 1 < 0$ $\left(-\frac{1}{2}, 1\right)$

86. $\dfrac{1}{x - 3} > -1$ $(-\infty, 2) \cup (3, \infty)$

87. $\dfrac{4}{x + 5} > 1$ $(-5, -1)$

88. $(x + 7)(x + 2)(x - 1) \ge 0$ $[-7, -2] \cup [1, \infty)$

89. $(x - 10)(x + 5)(x - 4) \ge 0$ $[-5, 4] \cup [10, \infty)$

90. $\dfrac{x - 4}{5x - 2} \ge 0$

91. $\dfrac{2x + 5}{x - 3} \ge 0$ $\left(-\infty, -\frac{5}{2}\right] \cup (3, \infty)$

92. $x(5x - 2)(2x + 7) < 0$

93. $x(3x - 2)(5x - 6) < 0$

94. $\dfrac{x + 7}{x - 1} \le 0$ $[-7, 1)$

95. $\dfrac{x - 11}{x + 6} \le 0$ $(-6, 11]$

96. $x^4 + 5x^2 - 36 < 0$ $(-2, 2)$

97. $x^4 - 26x^2 + 25 > 0$ $(-\infty, -5) \cup (-1, 1) \cup (5, \infty)$

98. $x^4 + 7x^2 + 10 > 0$ $(-\infty, \infty)$

99. $x^4 + 4x^2 + 3 < 0$ $\varnothing$

100. $\dfrac{2}{x - 5} \le -1$ $[3, 5)$

101. $\dfrac{3}{x - 6} \le 1$ $(-\infty, 6) \cup [9, \infty)$

Use the graph of $y = f(x)$ to solve each associated inequality. Write each solution in interval notation.

102. $(x + 5)(x - 4) \le 0$

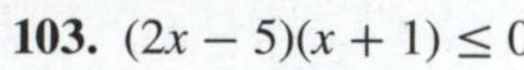

103. $(2x - 5)(x + 1) \le 0$

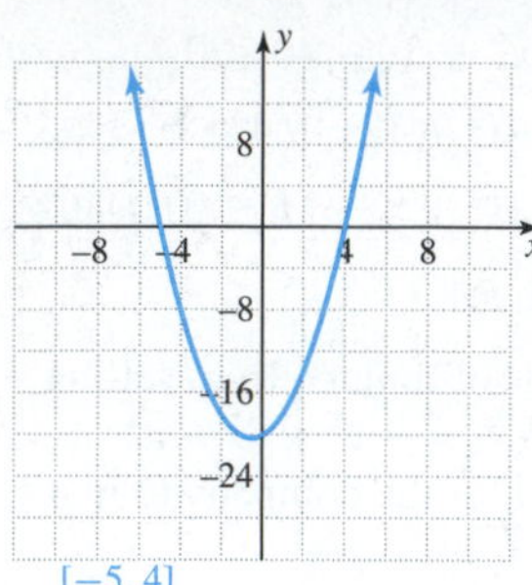

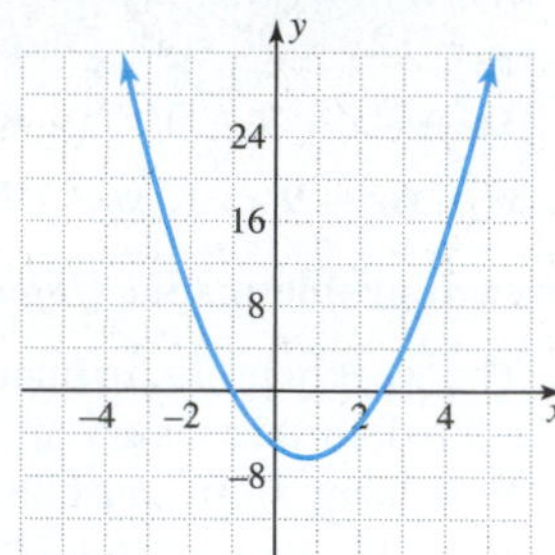

$[-5, 4]$

104. $(x + 7)(x - 2) > 0$

105. $(2x + 7)(x - 3) > 0$

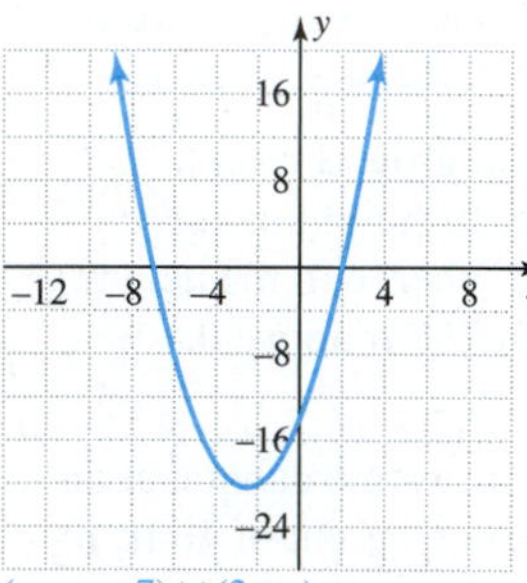

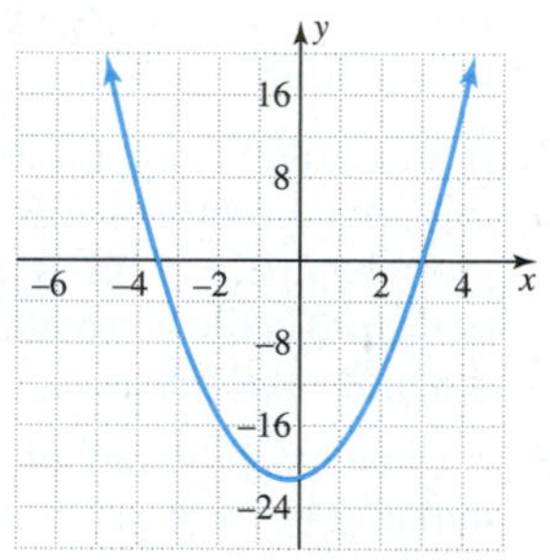

$(-\infty, -7) \cup (2, \infty)$

106. $x(2x + 7)(x - 3) \ge 0$

107. $(x - 3)(x - 6)(x + 3) \ge 0$

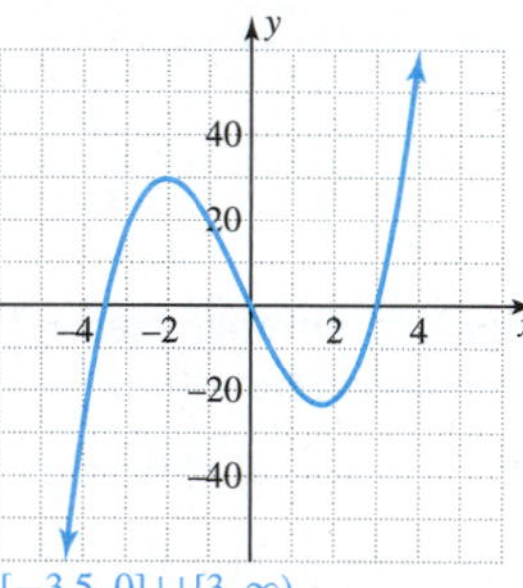

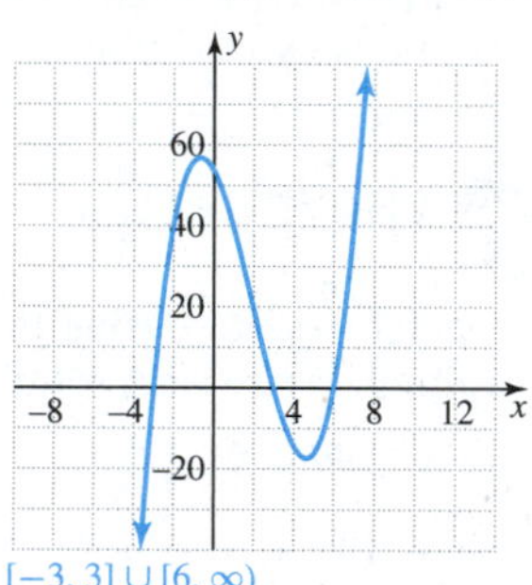

$[-3.5, 0] \cup [3, \infty)$

$[-3, 3] \cup [6, \infty)$

108. $x^4 - 9x^2 < 0$

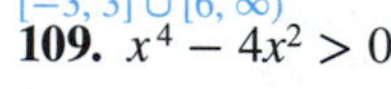

109. $x^4 - 4x^2 > 0$

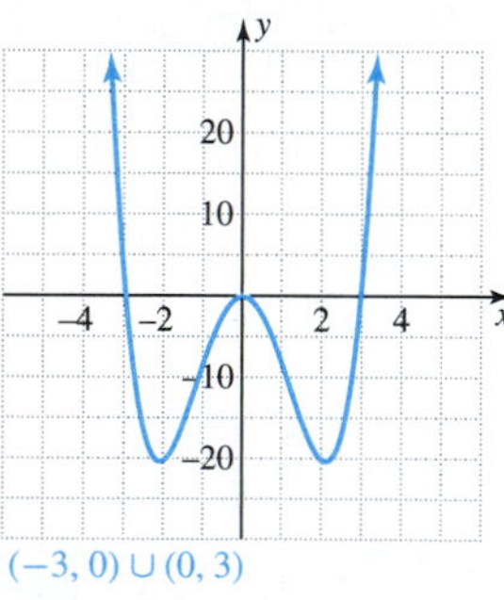

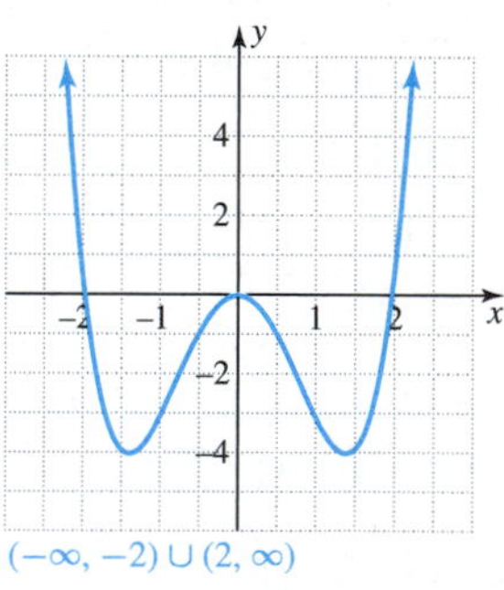

$(-3, 0) \cup (0, 3)$

$(-\infty, -2) \cup (2, \infty)$

You Be the Teacher!

Correct each student's errors, if any.

110. Solve the inequality $x(x + 5)(2x - 1) \le 0$.

Elaine's work:

The solutions of the associated equation are -5 and $\frac{1}{2}$. The inequality is true when $x \ge -5$ and $x \le \frac{1}{2}$. The solution is $\left[-5, \frac{1}{2}\right]$

111. Solve the inequality $\frac{5}{x + 3} \ge 2$.

Mark's work:

$$\frac{5}{x + 3} \ge 2$$

$$5 \ge 2x + 6$$

$$2x + 1 \le 0$$

$$x \le -\frac{1}{2}$$

$$\left(-\infty, -\frac{1}{2}\right]$$

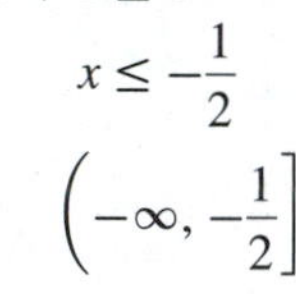

$\frac{5}{x+3} = 2$ $\quad x + 3 = 0$

$5 = 2x + 6$ $\quad x = -3$

$2x + 1 = 0$

$x = -\frac{1}{2}$

The boundary numbers are -3 and $-\frac{1}{2}$. The inequality is true when $x > -3$ and $x \le -\frac{1}{2}$. The solution is $\left(-3, -\frac{1}{2}\right]$.

Calculate It!

Use a calculator to graph the solution set of each inequality. Write each solution in interval notation.

112. $(x + 5)(x - 7) \le 0$

113. $(x + 4)(x - 1) > 0$

114. $\frac{2x - 1}{x + 3} \ge 0$

115. $\frac{x + 1}{x - 3} \le 0$

GROUP ACTIVITY The Mathematics of Fatal Crashes

The following table shows the number of drivers involved in fatal crashes during the hours of 6:00 A.M. and 5:59 P.M. for the years 2002 to 2009. (Source: http://www-nrd.nhtsa.dot.gov/Pubs/811402EE.pdf)

Year	2002	2003	2004	2005	2006	2007	2008	2009
Number of Drivers	31,135	31,863	31,686	31,820	30,566	29,307	26,377	23,625

1. Write the ordered pairs that correspond to the information given in the table, where x is the years after 2002.
2. Plot the ordered pairs by hand, on a graphing calculator, or in a spreadsheet.
3. Do the data look quadratic? If so, what can you say about the value of a for the quadratic function that models the data? Will the function that models the data have a maximum or minimum value?
4. Determine the quadratic function, $f(x) = ax^2 + bx + c$, that represents the data.
   a. What is the y-intercept of the data? This is the value of c in the function. (0, 31,135), so $c = 31{,}135$
   b. Choose two other points from the table. Write two equations by substituting these ordered pairs into the function $f(x) = ax^2 + bx + c$.
   c. You should now have two equations with two unknowns, a and b. Use either substitution or elimination to solve for a and b. Round answers to two decimal places. Answers vary. $a = -300.14$ and $b = 1028.14$
   d. Write the quadratic function that models the data, using the values of a, b, and c.
5. Find the vertex of your model and interpret its meaning in the context of the problem.
6. Use your model to approximate when the number of drivers involved in fatal crashes will be 10,000.
7. Now use a graphing calculator or spreadsheet to determine the quadratic function that models these data. $f(x) = -317.55x^2 + 1170.6x + 31{,}007$

Chapter 9 / REVIEW

Quadratic Equations and Functions and Nonlinear Inequalities

What's the big idea? Now that we have completed Chapter 9, we should be able to solve a quadratic equation using various methods and see how the solutions of the equation relate to the graph of the corresponding function. We should also be able to use the properties of parabolas to solve problems that relate to maximizing and minimizing functions. We should be able to apply our skills for solving polynomial and rational equations to solving polynomial and rational inequalities.

The Tools

Listed below are the key terms, skills, formulas, and properties you should know for this chapter.

The page reference is provided if you need additional help with the given topic. The Study Tips will assist in your preparation for an exam.

Study Tips

1. Learn all of the terms, formulas, and properties. Make flash cards and have someone quiz you.
2. Rework problems from the exercises and also the ones you worked in class. Work additional problems from the review exercises.
3. Review the summaries of key concepts.
4. Work the chapter test.
5. Be sure to review the online resources for additional study materials.

Terms

Axis of symmetry 637
Boundary number 700
Completing the square 654
Discriminant 668
Higher degree polynomial equation 680
Parabola 636
Perfect square trinomial 653
Quadratic equation 651
Quadratic function 636
Quadratic inequalities 700
Radical equation 678
Rational equation 677
Rational inequalities 705
Vertex 637
Zeros of a quadratic function 659

Formulas and Properties

- Discriminant 668
- Quadratic formula 666
- Quadratic function 636
- Square root property 651
- Vertex formula 693
- Vertex form of a quadratic function 643

CHAPTER 9 / SUMMARY

How well do you know this chapter? Complete the following questions to find out. Take a look back at the section if you need help.

SECTION 9.1 Quadratic Functions and Their Graphs

1. The graph of the basic quadratic function $f(x) = x^2$ is a(n) parabola.
2. The highest or lowest point on a parabola is called the vertex.
3. Parabolas are symmetric about their vertex. The vertical line through the vertex is called the axis of symmetry.
4. The graph of $f(x) = x^2 + k$ is the graph of $f(x) = x^2$ shifted up or down. If $k > 0$, the graph goes up k units. If $k < 0$, the graph goes down $|k|$ units. Its vertex is $(0, k)$.
5. The graph of $f(x) = (x - h)^2$ is the graph of $f(x) = x^2$ shifted left or right. If $h > 0$, the graph goes right h units. If $h < 0$, the graph goes left $|h|$ units. Its vertex is $(h, 0)$.

6. The graph of $f(x) = ax^2$ is the graph of $f(x) = x^2$ __stretched__, __shrunk__, or __reflected__ over the x-axis. If $a > 1$, the parabola opens __up__ and is more __narrow__ than the graph of $f(x) = x^2$. If $0 < a < 1$, the parabola opens __up__ and is __wider__ than the graph of $f(x) = x^2$. If $a < 0$, the parabola opens __down__.

7. The vertex form of a quadratic function is __$a(x - h)^2 + k$__. The vertex of the parabola is __(h, k)__. The x-intercepts are found by solving the equation __$f(x) = 0$__ and the y-intercept is __$f(0)$__.

8. If the point $(r, 0)$ is on the graph of a quadratic function $f(x) = a(x - h)^2 + k$, then __$x = r$__ is a solution of the equation $a(x - h)^2 + k = 0$. The solution is also called a(n) __zero__ of the function.

SECTION 9.2 Solving Quadratic Equations Using the Square Root Property and Completing the Square

9. The square root property states that if $x^2 = k$, where k is a real number, then $x =$ __$\sqrt{k}$__ or $x =$ __$-\sqrt{k}$__.
   a. If $k > 0$, the equation has __two__ real solution(s).
   b. If $k = 0$, the equation has __one__ real solution(s).
   c. If $k < 0$, the equation has __two__ complex, nonreal solution(s).

10. To apply the square root property, the squared expression must be __isolated__ to one side and equal to a(n) __constant__.

11. A(n) __perfect__ __square__ __trinomial__ results from squaring a binomial.

12. In a perfect square trinomial with a leading coefficient of 1, the __middle__ term is half of the __coefficient__ of x, __squared__. In symbols, $c =$ __$\left(\frac{b}{2}\right)^2$__.

13. When we add the number that makes a binomial a perfect square trinomial, we are __completing__ __the__ __square__.

14. To solve an equation by completing the square,
   - The coefficient of the squared term must be __one__.
   - If it is not, we __divide__ both sides of the equation by the coefficient.
   - Move the __constant__ term to the side opposite the variable terms.
   - Find the number that __completes__ the __square__ and __add__ it to both sides of the equation.
   - __Factor__ the perfect square trinomial and simplify the other side.
   - Apply the __square__ __root__ __property__ and solve the equation.

SECTION 9.3 Solving Quadratic Equations Using the Quadratic Formula

15. The __quadratic__ __formula__ is derived by completing the square. If $ax^2 + bx + c = 0$, then $x =$ __$\dfrac{-b \pm \sqrt{b^2 - 4ac}}{2a}$__.

16. The __discriminant__ of a quadratic equation determines the number and types of solutions. Its value is __$b^2 - 4ac$__.
   a. If this value is a positive perfect square, there is/are __two__ __rational__ solution(s).
   b. If this value is positive but not a perfect square, there is/are __two__ __irrational__ solution(s).
   c. If this value is zero, there is/are __one__ __rational__ solution(s).
   d. If this value is negative, there is/are __two__ __complex__, __nonreal__ solution(s).

SECTION 9.4 Solving Equations Using Quadratic Methods

17. A(n) __rational__ equation is an equation that contains a rational expression. Multiplying both sides by the __LCD__ clears fractions. We must exclude any solutions that make the expression __undefined__.

18. A(n) __radical__ equation is an equation that contains radicals. We must isolate the __radical__ and __square__ both sides. We must check for __extraneous__ solutions.

19. We can solve some other equations by __substitution__. We replace the base of the squared term with __u__ to create a(n) __quadratic__ equation. We must back __substitute__ to solve for the original variable.

SECTION 9.5 More on Graphing Quadratic Functions

20. We can convert $f(x) = ax^2 + bx + c$ to the vertex form $f(x) = a(x - h)^2 + k$ by __completing__ the __square__.

21. The vertex of $f(x) = ax^2 + bx + c$ is (__$-\dfrac{b}{2a}$__, __$f\left(-\dfrac{b}{2a}\right)$__).

22. The key points on the graph of a parabola are the __x-intercepts__, the __y-intercepts__, and the __vertex__.

23. If a parabola opens up, the vertex is the __lowest__ point. If a parabola opens down, the vertex is the __highest__ point. If (h, k) is the vertex, then the maximum or minimum value of a quadratic function is __k__ and occurs at __$x = h$__.

SECTION 9.6 Solving Polynomial and Rational Inequalities in One Variable

24. One method for solving polynomial inequalities is the __boundary__ __number__ __method__. It requires us to solve the __associated__ equation which is found by replacing the __inequality__ symbol with a(n) __equal__ sign.

25. The __boundary__ __numbers__ are marked on a number line and one point from each region is __tested__ in the original inequality. It is sufficient to test only __one__ point from each region.

26. If the inequality is $<$ or $>$, the boundary numbers are __not__ __included__ in the solution set. We use __parentheses__ for them in the interval notation.

27. If the inequality is $\leq$ or $\geq$, the boundary numbers are __included__ in the solution set. We use __brackets__ for them in the interval notation.

28. The boundary numbers for a rational inequality come from the solutions of the __associated__ equation as well as values that make the expression __undefined__.

29. The boundary numbers of $f(x) = 0$ are also the __x-intercepts__ of the graph of $f(x)$. To solve $f(x) > 0$, we find the regions where the graph of $f(x)$ is __above__ the x-axis. To solve $f(x) < 0$, we find the regions where the graph of $f(x)$ is __below__ the x-axis.

CHAPTER 9 / REVIEW EXERCISES

SECTION 9.1

Graph each function. Identify the x-intercepts, y-intercept, vertex, axis of symmetry, domain, and range. Explain how the graph of the function relates to the graph of $y = x^2$. (*See Objectives 1–4.*)

1. $f(x) = x^2 + 9$ **2.** $f(x) = \left(x - \frac{3}{2}\right)^2$

3. $f(x) = -\frac{1}{2}(x - 5)^2 + 2$ **4.** $f(x) = 4(x + 1)^2 - 16$

Graph each function and use its graph to solve $f(x) = 0$. (*See Objective 5.*)

5. $f(x) = -3(x + 8)^2 + 12$ $\{-10, -6\}$ **6.** $f(x) = -\frac{5}{4}(x - 1)^2 + 5$ $\{-1, 3\}$

7. $f(x) = 4.5(x - 2)^2 - 18$ $\{0, 4\}$ **8.** $f(x) = 3(x + 1)^2 + 10$ no real solutions

SECTION 9.2

Solve each equation by applying the square root property. (*See Objective 1.*)

9. $z^2 - 15 = 3$ $\{-3\sqrt{2}, 3\sqrt{2}\}$ **10.** $25x^2 + 10 = 131$ $\left\{-\frac{11}{5}, \frac{11}{5}\right\}$

11. $9(y - 5)^2 - 2 = 47$ **12.** $25(2b + 3)^2 + 49 = 13$

13. $25(y - 3)^2 - 11 = 38$ **14.** $9(z - 6)^2 + 2 = 22$

15. $(3b - 2)^2 + 73 = 9$ **16.** $(3w - 1)^2 + 33 = 8$

Create a perfect square trinomial from each binomial by completing the square and then write it in factored form. (*See Objective 2.*)

17. $x^2 + 12x$ **18.** $s^2 + 15s$ **19.** $r^2 - \frac{3}{2}r$ **20.** $s^2 + \frac{5}{2}s$

Solve each equation by completing the square. (*See Objectives 3 and 4.*)

21. $x^2 + 6x - 9 = 0$ $\{-3\sqrt{2} - 3, 3\sqrt{2} - 3\}$ **22.** $y^2 - 10y + 5 = 0$ $\{5 - 2\sqrt{5}, 5 + 2\sqrt{5}\}$

23. $9r^2 - 18r - 7 = 0$ **24.** $4r^2 + 24r + 45 = 0$

25. $2x(x - 1) - 7 = 0$ **26.** $2x(x + 1) = 1$

Solve each problem. (*See Objective 5.*)

27. A square city block covers an area of 67,600 ft². What is the length of the side of the block? 260 ft

28. If a ball is dropped from the top of the Q1 Tower in Gold Coast, its height, in feet, t sec after it is dropped is given by $s = -16t^2 + 1058$. How many seconds will it take for the ball to reach the ground? Round the answer to two decimal places. (Source: http://www.skypoint.com.au) 8.13 sec

29. At what rate would you have to invest \$25,500 in a 2-yr CD to have \$26,270.74 once the CD matures? Use the formula $A = P(1 + r)^2$, where P is the amount invested and r is the annual interest rate. 1.5%

SECTION 9.3

Solve each equation using the quadratic formula. (*See Objective 1.*)

30. $3x^2 - 14x + 16 = 0$ **31.** $7y^2 - 34y - 5 = 0$ $\left\{-\frac{1}{7}, 5\right\}$

32. $3a^2 + 17a + 22 = 0$ **33.** $20b^2 - b - 12 = 0$

34. $5x(5x + 18) + 76 = 0$ **35.** $3x(3x + 4) = 13$

36. $x^2 = 10x - 31$ $\{5 \pm i\sqrt{6}\}$ **37.** $z^2 - 6z = -16$ $\{3 \pm i\sqrt{7}\}$

Use the discriminant to determine the number and types of solutions of each equation. (*See Objective 2.*)

38. $5x^2 - 9x - 2 = 0$ two rational solutions **39.** $6y^2 + 4y + 5 = 0$ two complex, nonreal solutions

40. $11a^2 - 7a + 7 = 0$ two complex, nonreal solutions **41.** $3b^2 + 4b - 8 = 0$ two irrational solutions

42. $4r^2 - 20r + 25 = 0$ one rational solution **43.** $9s^2 - 24s - 16 = 0$ two irrational solutions

Solve each problem. (*See Objective 3.*)

44. A company knows that the cost, in thousands of dollars, to produce x hundred items is given by the cost function $C(x) = -0.35x^2 + 2x + 4$. It also knows that the revenue, in thousands of dollars, from selling x hundred items is given by the revenue function $R(x) = 0.55x$. How many items does the company need to produce to break even, that is, when does revenue = cost? Round the answer to the nearest integer. 604 items

45. A farmer wants to enclose a pen at the back of his barn, using the barn as one of its borders. The farmer has 320 ft of fencing. What dimensions should the pen have to enclose an area of 12,800 ft²? 80 ft by 160 ft

46. The height, in feet, of a ball thrown upward from a height of 6 ft is $h = -16t^2 + 56t + 6$, where t is in seconds. How many seconds will it take for the ball to reach a height of 30 ft? 0.5 sec or 3 sec

47. Josh and his wife, Christy, each have to travel out of town for meetings. The towns where they are meeting are 210 mi apart and are due east and due south from their home. Josh leaves at 9 A.M. and travels at 54 mph, and Christy leaves 1 hr later and travels at 72 mph. When will they reach their destinations and be 210 mi apart? 11:55 A.M.

SECTION 9.4

Solve each equation. (*See Objectives 1 and 2.*)

48. $3x = 22 + \frac{80}{x}$ $\left\{-\frac{8}{3}, 10\right\}$ **49.** $8x = -16 - \frac{11}{x}$ $\left\{\frac{-4 \pm i\sqrt{6}}{4}\right\}$

50. $\frac{4x}{x + 1} + \frac{5x - 31}{x - 1} = \frac{52}{x^2 - 1}$ $\left\{\frac{5 \pm 6\sqrt{3}}{3}\right\}$

51. $\frac{4x}{x - 2} + \frac{12x + 8}{x + 2} = \frac{7}{x^2 - 4}$ $\left\{\frac{1 \pm 2\sqrt{6}}{4}\right\}$

52. $\sqrt{18 - 2x} = x - 5$ $\{7\}$ **53.** $\sqrt{10x + 24} = x + 4$ $\{-2, 4\}$

54. $x + 5 = 4\sqrt{x + 1}$ $\{3\}$ **55.** $x - 3 = \sqrt{5 - 2x}$ $\varnothing$

56. $x + 5\sqrt{x} + 4 = 0$ $\varnothing$ **57.** $20x + \sqrt{x} = 12$ $\left\{\frac{9}{16}\right\}$

Solve each equation using substitution, if needed. (*See Objectives 3 and 4.*)

58. $100x^4 - 21x^2 - 1 = 0$ **59.** $x^4 - 20x^2 + 64 = 0$ $\{\pm 2, \pm 4\}$

60. $16x^4 = 32x^2 + 9$ **61.** $4x^4 = 9x^2 + 100$ $\{\pm 2i, \pm 2.5\}$

62. $25(x - 4)^2 + 20(x - 4) + 3 = 0$ $\left\{\frac{17}{5}, \frac{19}{5}\right\}$

63. $4(2x + 3)^2 + 5(2x + 3) - 6 = 0$ $\left\{-\frac{5}{2}, -\frac{9}{8}\right\}$

64. $(5x + 2)^2 + 6(5x + 2) - 16 = 0$ $\{-2, 0\}$

65. $\frac{3}{(x+2)^2} + \frac{10}{x+2} - 8 = 0$ $\left\{-\frac{9}{4}, -\frac{1}{2}\right\}$

66. $\frac{3}{(4x+1)^2} - \frac{4}{4x+1} - 4 = 0$ $\left\{-\frac{5}{8}, -\frac{1}{8}\right\}$

67. $x^{2/3} + x^{1/3} - 20 = 0$ $\{-125, 64\}$

68. $3x^{1/2} - 23x^{1/4} + 40 = 0$

69. $2x^{2/5} + 3x^{1/5} - 2 = 0$

70. $x^{4/3} - 13x^{2/3} + 36 = 0$ $\{8, 27\}$

Solve each problem. (*See Objective 5.*)

71. Mel and Jay own a lawn service. Together they can mow lawns in one development in 4 hr. Mel does all the mowing 3 hr faster than Jay. How long will it take each working alone to do the job? Round answers to two decimal places. Jay takes 9.77 hr and Mel takes 6.77 hr.

72. Monica drove a distance of 250 mi to visit her parents in Hershey, Pennsylvania. During the return trip she was caught in a rainstorm and had to decrease her speed by 12 mph. If her return trip took 1 hr longer than her trip to Hershey, find her original speed. Round the answer to two decimal places. 61.10 mph

SECTION 9.5

Graph each quadratic function by finding the intercepts, the vertex, and additional points, if needed. (*See Objectives 1 and 2.*)

73. $f(x) = x^2 + 2x - 24$

74. $f(x) = -x^2 + 4x + 5$

Solve each problem. (*See Objective 3.*)

75. The monthly profit, in thousands of dollars, from selling x thousand digital cameras is given by $f(x) = -3x^2 + 60x + 50$. How many digital cameras must be sold in a month to maximize profit? What is the maximum monthly profit? 10,000 cameras sold with a maximum profit of $350,000

76. A ball is tossed upward with an initial velocity of 89.6 ft/sec from a height of 60 ft. The height of the ball t sec after it is tossed is given by $f(t) = -16t^2 + 89.6t + 60$. What is the maximum height the ball will reach and when will it reach this height? 185.44 ft in 2.8 sec

SECTION 9.6

Solve each inequality using the boundary number method. Write each solution set in interval notation. (*See Objectives 1–3.*)

77. $(x + 9)(x - 4) \le 0$ $[-9, 4]$

78. $(3x + 8)(x - 3) \ge 0$

79. $x^2 - 4x > 32$ $(-\infty, -4) \cup (8, \infty)$

80. $3x^2 + 15 < 14x$ $\left(\frac{5}{3}, 3\right)$

81. $x^2 + 12x + 36 > 0$ $(-\infty, -6) \cup (-6, \infty)$

82. $25x^2 + 10x + 1 \le 0$ $\left\{-\frac{1}{5}\right\}$

83. $(3x - 5)^2 + 1 < 0$ $\varnothing$

84. $(2x - 3)^2 - 16 > 0$

85. $(2x + 3)(x - 4)(x + 5) \le 0$ $(-\infty, -5) \cup \left[-\frac{3}{2}, 4\right]$

86. $(2x - 5)(x + 2)(x - 6) \ge 0$ $\left[-2, \frac{5}{2}\right] \cup [6, \infty)$

87. $x(7x - 3)(5x + 12) > 0$ $\left(-\frac{12}{5}, 0\right) \cup \left(\frac{3}{7}, \infty\right)$

88. $x(x + 2)(4x + 15) < 0$

89. $25x^4 - 29x^2 + 4 > 0$

90. $x^4 - 16 \le 0$ $[-2, 2]$

91. $\frac{6}{2x - 5} \ge -3$ $\left(-\infty, \frac{3}{2}\right] \cup \left(\frac{5}{2}, \infty\right)$

92. $\frac{6}{x + 5} \le 2$ $(-\infty, -5) \cup [-2, \infty)$

93. $\frac{2x - 7}{x + 4} \le 0$ $\left(-4, \frac{7}{2}\right]$

94. $\frac{5x + 12}{x - 2} \ge 0$ $\left(-\infty, -\frac{12}{5}\right] \cup (2, \infty)$

Use the graph of $y = f(x)$ to solve each associated inequality. Write each solution set in interval notation. (*See Objective 4.*)

95. $(x + 5)(2x - 3) \ge 0$ $(-\infty, -5] \cup \left[\frac{3}{2}, \infty\right)$

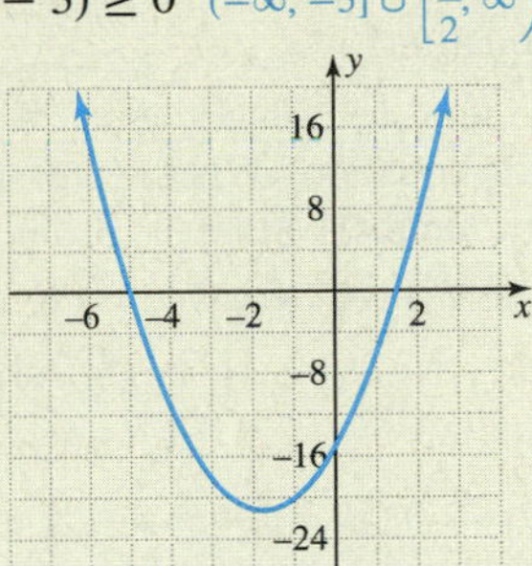

96. $(2x + 7)^2 \le 0$ $\left\{-\frac{7}{2}\right\}$

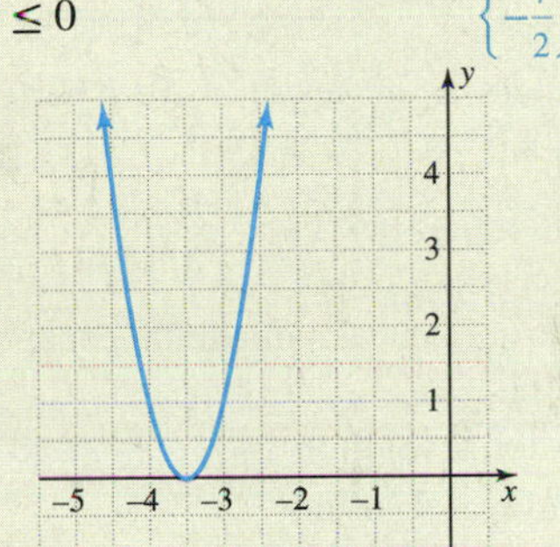

97. $(x - 3)(x + 6)(x + 1) \le 0$ $(-\infty, -6] \cup [-1, 3]$

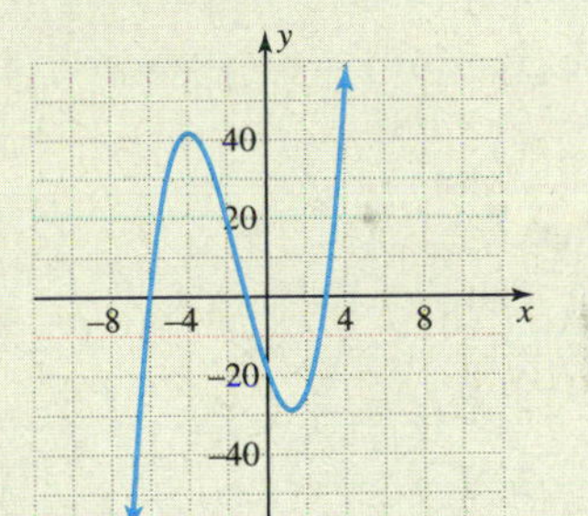

98. $x(2x - 7)(x + 4) > 0$ $(-4, 0) \cup \left(\frac{7}{2}, \infty\right)$

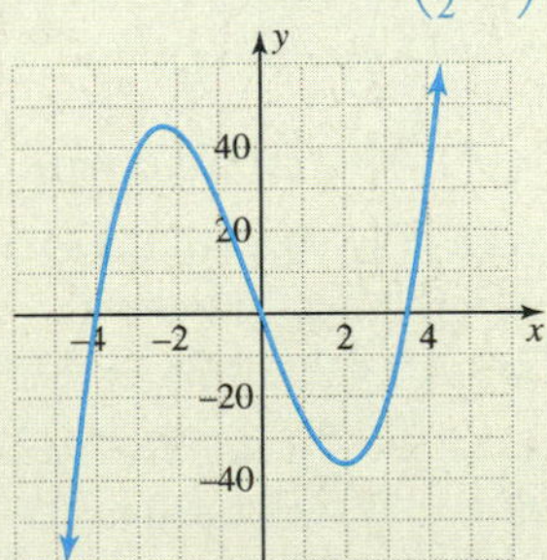

Solve each problem. (*See Objective 5.*)

99. A toy rocket is launched from the ground with an initial velocity of 160 ft/sec. Its height in t sec is given by $h(t) = -16t^2 + 160t$. When will the rocket be more than 384 ft above the ground? between 4 and 6 sec

100. The profit, in thousands of dollars, of selling x hundred units of an item is given by $P(x) = -2x^2 + 16x - 14$. How many items must be sold for the company to have a profit? between 100 and 700 units

CHAPTER 9 TEST / QUADRATIC EQUATIONS AND FUNCTIONS AND NONLINEAR INEQUALITIES

1. What are the four methods for solving a quadratic equation? Which method(s) can be used to solve any quadratic equation?

2. What determines the number and types of solutions of a quadratic equation? What expression is used to calculate this number? State the different cases.

3. The vertex of $f(x) = 2(x + 5)^2 - 3$ is

a. $(5, 3)$ **b.** $(5, -3)$
c. $(-5, 3)$ **d.** $(-5, -3)$

4. The solution set of $(x + 5)^2 + 2 = 9$ is

a. $\{-4\}$ **b.** $\{-10, -4\}$
c. $\{-5 + \sqrt{7}\}$ **d.** $\{-5 - \sqrt{7}, -5 + \sqrt{7}\}$

5. The number that makes $x^2 - 20x$ a perfect square trinomial is

a. 10 **b.** 20
c. 100 **d.** 400

6. If the discriminant of a quadratic equation is 33, the equation has

a. one rational solution
b. two rational solutions
c. two irrational solutions
d. two complex nonreal solutions

7. Using the graph of $y = ax^2 + bx + c$, the solution set of $ax^2 + bx + c = 0$ is

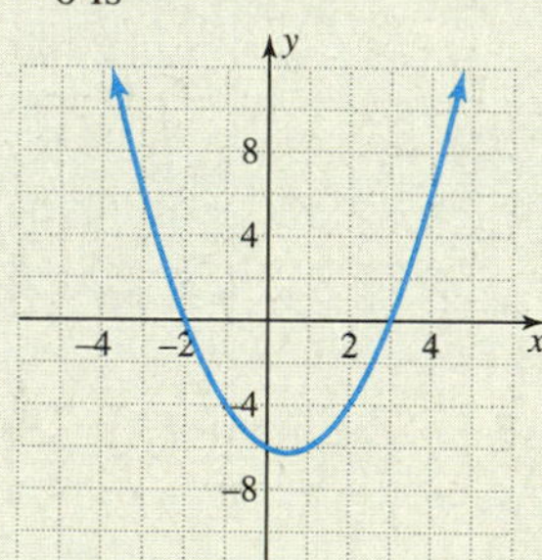

a. $\{-3, 2\}$ **b.** $\{-2, 3\}$
c. $\{6\}$ **d.** $\{-6\}$

8. Using the graph in Exercise 7, the solution set of $ax^2 + bx + c < 0$ is

a. $(-\infty, -2) \cup (3, \infty)$ **b.** $(-\infty, -2] \cup [3, \infty)$
c. $(-2, 3)$ **d.** $[-2, 3]$

9. The vertex of $y = 2x^2 - 8x + 3$ is

a. $(2, -5)$ **b.** $(2, 11)$
c. $(-2, 27)$ **d.** $(-2, 11)$

10. The boundary number(s) used to solve $\dfrac{x+5}{x-2} \geq 0$ is (are)

a. $x = -5$ **b.** $x = 2$
c. $x = -5, x = 2$ **d.** no boundary numbers

Solve each equation.

11. $y^2 + 16 = 8y$ $\{4\}$

12. $2a^2 - 3a = 4$

13. $b^2 + b + 5 = 0$

14. $5x = 19 - \dfrac{12}{x}$ $\{0.8, 3\}$

15. $8 - x = \sqrt{1 - 32x}$ $\{-9, -7\}$

16. $\dfrac{5x}{x+1} - \dfrac{4x-19}{x-1} = \dfrac{4}{x^2-1}$ $\{-5 \pm \sqrt{10}\}$

17. $x^4 - 8x^2 + 12 = 0$ $\{\pm\sqrt{2}, \pm\sqrt{6}\}$

18. $(x - 1)^{2/3} + 5(x - 1)^{1/3} + 4 = 0$ $\{-63, 0\}$

Graph each quadratic function by finding the intercepts, vertex, and additional points, if necessary. State the domain and range.

19. $f(x) = x^2 - 4x - 12$

20. $f(x) = -(x + 3)^2 + 4$

21. $f(x) = 2x^2 - 6x + 5$

22. If \$5000 is invested in a savings account for 2 yr, the amount in the account may be represented by $A = 5000(1 + r)^2$, where r is the annual interest rate. Find the interest rate that is required for an investment of \$5000 to grow to \$6050 in 2 yr. 10%

23. The Devil's Pool is a swimming area located atop the Victoria Falls along the Zambezi River, which is located on the border of Zambia and Zimbabwe. Swimmers can literally look over the edge of the falls. One swimmer was trying to take a picture over the edge and dropped his camera. The height, in meters, of his camera t sec after he dropped it is represented by $h = -9.8t^2 + 108$. How many seconds (to the nearest tenth) will it take for the camera to reach the bottom of the falls? (Source: http://goafrica.about.com/od/zambia/p/devilspool.htm) 3.3 sec

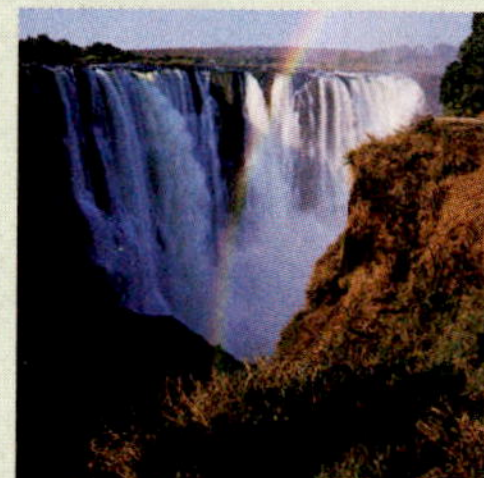

24. The national unemployment rate (not seasonally adjusted) in January for the years 2004–2010 is shown in the table. The unemployment rate is the percent of the U.S. labor force who is unemployed and can be modeled by the equation $y = 0.39x^2 - 1.69x + 6.61$, where x is the years after 2004. Assuming the unemployment rate continues to grow in this manner, in what year will the unemployment rate be 25? (Source: http://www.google.com/publicdata/home)
In approximately 9.4 yr after 2004, or the during the year 2013

Year	Unemployment Rate
2004	6.3
2005	5.7
2006	5.1
2007	5
2008	5.4
2009	8.5
2010	10.6

Solve each inequality using the boundary number method or graphing. Write each solution set in interval notation.

25. $x^2 - 3x - 10 \geq 0$ $(-\infty, -2] \cup [5, \infty)$

26. $(x + 5)(x - 3)(x - 6) < 0$ $(-\infty, -5) \cup (3, 6)$

27. $\dfrac{2x}{x-2} \geq 0$ $(-\infty, 0] \cup (2, \infty)$

28. $(2x - 5)^2 < 0$ $\varnothing$

CUMULATIVE REVIEW EXERCISES / CHAPTERS 1–9

Evaluate each expression for the given values. (*Section 1.2, Objective 5*)

1. $b^2 - 4ac$ for $a = -2, b = 5, c = 3$ 49
2. $b^2 - 4ac$ for $a = 4, b = -7, c = -1$ 65

Translate each sentence into an algebraic inequality. Use the variable x to represent the unknown number. (*Section 1.3, Objective 4*)

3. Four less twice a number is greater than the number plus six. $4 - 2x > x + 6$
4. Two less than a number is less than two more than three times the number. $x - 2 < 2 + 3x$

Solve each linear equation and check the answer. (*Section 2.1, Objectives 4 and 5*)

5. $15 + 3x = 6(x - 4)$ {13}
6. $26 - 6x = -4(x - 9)$ {-5}

Translate each problem into a linear equation and solve the problem. (*Section 2.2, Objective 1*)

7. Twice the sum of the two consecutive odd integers is 96. Find the two numbers. {23, 25}
8. Six less than three times a number yields 27. Find the number. {11}

Solve each formula for the specified variable. (*Section 2.3, Objective 4*)

9. $p = \frac{3}{4}q - rs$ for r
10. $A = P(1 + rt)$ for t $t = \frac{A - P}{Pr}$

Solve each compound inequality. Write each solution set in interval notation. (*Section 2.5, Objectives 3 and 4*)

11. $3x - 6 \le 0$ and $2 - 5x \le 12$ $[-2, 2]$
12. $2x + 7 > 1$ or $4 + x < -1$ $(-\infty, -5) \cup (-3, \infty)$

Find $f(x + 1)$ for each function. (*Section 3.3, Objective 3*)

13. $f(x) = 4 - x^2$ $-x^2 - 2x + 3$
14. $f(x) = 8x + 5$ $8x + 13$

Rewrite each linear equation using function notation, if possible. Identify m and b. (*Section 4.1, Objective 2*)

15. $6x + 7 = 3$ can't be written in function notation
16. $2x + 3y = 0$

Find the slope and y-intercept of each line from its equation. Write the y-intercept as an ordered pair. (*Section 4.3, Objective 1*)

17. $f(x) = 4x - 1$ $m = 4, b = -1; (0, -1)$
18. $5x + 2y = 3$ $m = -\frac{5}{2}, b = \frac{3}{2}; \left(0, \frac{3}{2}\right)$

Write the equation of each line that passes through the given point and is either parallel or perpendicular to the given line. Write each answer in slope-intercept form. (*Section 4.4, Objective 4*)

19. $(2, -5)$, parallel to $2x + y = 15$ $y = -2x - 1$
20. $(-8, 3)$, perpendicular to $x - 3y = 1$ $y = -3x - 21$

Solve each system of linear equations using substitution or elimination. (*Section 5.2, Objectives 1 and 2*)

21. $\begin{cases} 12x - y = 43 \\ -5x + 2y = -29 \end{cases}$ $\{(3, -7)\}$
22. $\begin{cases} 6x + 7y = 50 \\ 8x - 3y = 42 \end{cases}$ $\{(6, 2)\}$

Solve each problem. (*Section 5.3, Objective 4*)

23. A plane can travel 1560 mi with the wind in 4 hr. The return trip against the wind takes 5 hr. What are the speed of the plane in still air and the speed of the wind? Speed of the plane is 351 mph and speed of the wind is 39 mph.
24. A plane can travel 3240 mi with the wind in 6 hr. The return trip against the wind takes 7.5 hr. What are the speed of the plane in still air and the speed of the wind? Speed of the plane is 486 mph and speed of the wind is 54 mph.

Simplify each expression. Write each answer with positive exponents. (*Section 6.1, Objectives 1 and 2*)

25. $(-5x^2y^3)(13x^5y^2)$ $-65x^7y^5$
26. $(7a^4b)(-4a^2b^3)$ $-28a^6b^4$
27. $\dfrac{28p^{12}q^5}{21p^3q}$ $\frac{4}{3}p^9q^4$
28. $\dfrac{36r^{16}s^5}{60r^8s^2}$ $\frac{3r^8s^3}{5}$

Simplify each expression using the rules of exponents. Write each answer with positive exponents. (*Section 6.2, Objectives 1 and 2*)

29. $(a^4b^{-5})^3 \cdot a^{-10}b^8$ $\frac{a^2}{b^7}$
30. $(x^{-6}y^3)^3 \cdot x^{20}y^{-20}$ $\frac{x^2}{y^{11}}$
31. $\left(\dfrac{2x^{-2}}{y^{-3}}\right)^3\left(\dfrac{2y^{-2}}{x^{-1}}\right)^{-5}$ $\frac{y^{19}}{4x^{11}}$
32. $\left(\dfrac{5a^{-1}}{b^{-4}}\right)^2\left(\dfrac{5b^{-3}}{a^{-2}}\right)^{-3}$ $\frac{b^{17}}{5a^8}$

Simplify each expression. (*Section 6.4, Objectives 1–6*)

33. $(2x^2 - 5)(3x^2 - 10)$ $6x^4 - 35x^2 + 50$
34. $(7y^2 - 4)(2y^2 + 3)$ $14y^4 + 13y^2 - 12$
35. $(a + 5)(a - 5)(a + 2)$ $a^3 + 2a^2 - 25a - 50$
36. $(b - 3)(b + 3)(b + 1)$ $b^3 + b^2 - 9b - 9$

Factor each polynomial completely. (*Section 6.5, Objectives 1–3*)

37. $4x^4 - 12x^3 + 8x^2 - 24x$ $4x(x - 3)(x^2 + 2)$
38. $6y^4 + 6y^3 + 15y^2 + 15y$ $3y(y + 1)(2y^2 + 5)$

Factor using subsitution. (*Section 6.6, Objectives 1–4*)

39. $(2x + 1)^2 - (2x + 1) - 12$ $2(x + 2)(2x - 3)$
40. $(a + 3)^2 + 7(a + 3) - 18$ $(a + 1)(a + 12)$

Factor completely. (*Section 6.7, Objectives 1–3*)

41. $8x^3 + 125$ $(2x + 5)(4x^2 - 10x + 25)$
42. $a^4 - 16b^4$ $(a^2 + 4b^2)(a + 2b)(a - 2b)$

Solve each equation by factoring. (*Section 6.8, Objective 1*)

43. $(x + 1)^2 + (x + 2)^2 = (x + 3)^2$ $\{-2, 2\}$
44. $(y + 2)^2 + (y + 4)^2 = (y + 6)^2$ $\{-4, 4\}$

Multiply or divide the rational expressions. Write each answer in simplest form. (*Section 7.1, Objectives 3–5*)

45. $\dfrac{x^2 - 4}{x^2 - x - 6} \cdot \dfrac{2x^2 - x - 15}{4x^2 + 10x}$ $\frac{x - 2}{2x}$
46. $\dfrac{x^2 + 8x + 16}{x^3 + 64} \div \dfrac{5x^2 + 20x}{5x}$ $\frac{1}{x^2 - 4x + 16}$

Use the remainder theorem and synthetic division to evaluate $P(x)$ at the specified x-value. (*Section 7.2, Objective 4*)

47. $P(x) = 3x^2 - 7x + 4, P(3)$ 10
48. $P(x) = x^3 - 3x + 9, P(1)$ 7

Add or subtract the rational expressions. Write each answer in lowest terms. (*Section 7.3, Objectives 1–3*)

49. $\dfrac{4}{x^2 - 4x + 4} + \dfrac{3}{x^2 + 3x - 10}$ $\frac{7x + 14}{(x - 2)^2(x + 5)}$

50. $\dfrac{x}{3x^2 - 5x - 2} - \dfrac{2x}{9x^2 + 6x + 1}$ $\dfrac{x^2 + 5x}{(3x + 1)^2(x - 2)}$

Solve each problem. (*Section 7.6, Objective 2*)

51. A number minus eight times its reciprocal is 2. Find the number. 4 and −2

52. A number plus four times its reciprocal is 4. Find the number. 2

Solve each problem. (*Section 7.7, Objective 3*)

53. The cost of gas for a trip is directly proportional to the number of miles traveled and inversely proportional to the gas mileage of the car. If the cost of gas for a 150-mi trip in a car with gas mileage of 20 mpg is \$26.25, what is the cost of gas for a 240-mi trip in a car with gas mileage of 16 mpg? \$52.50

54. The cost of gas for a trip is directly proportional to the number of miles traveled and inversely proportional to the gas mileage of the car. If the cost of gas for a 390-mi trip in a car with gas mileage of 19.5 mpg is \$83.00, what is the cost of gas for a 240-mi trip in a car with gas mileage of 48 mpg? \$20.75

Simplify each expression. Assume that all variables represent positive real numbers. (*Section 8.1, Objectives 1 and 2*)

55. $\sqrt{\dfrac{75x^8}{3}}$ $5x^4$ 56. $\sqrt{\dfrac{98y^{10}}{2}}$ $7y^5$ 57. $\sqrt[3]{\dfrac{64a^9}{b^6}}$ $\dfrac{4a^3}{b^2}$ 58. $\sqrt[3]{\dfrac{c^3}{27d^6}}$

Rewrite each expression as a radical expression and simplify, if possible. Assume all variables represent positive real numbers. (*Section 8.2, Objective 3*)

59. $(a^{-15}b^6)^{-2/3}$ $\dfrac{a^{10}}{b^4}$ 60. $(x^8y^{-12})^{-3/4}$ $\dfrac{y^9}{x^6}$

Simplify each expression. Assume all variables represent positive real numbers. Write each answer with positive exponents. (*Section 8.2, Objective 4*)

61. $(x^{1/3} - y^{1/4})(x^{1/3} + y^{1/4})$ $x^{2/3} - y^{1/2}$ 62. $(r^{1/2} - s^{1/3})^2$ $r - 2r^{1/2}s^{1/3} + s^{2/3}$

Simplify each radical. Assume all variables represent positive real numbers. (*Section 8.3, Objectives 1 and 2*)

63. $\sqrt{80x^3y^6}$ $4xy^3\sqrt{5x}$ 64. $\sqrt{18r^4t}$ $3r^2\sqrt{2t}$ 65. $\sqrt[3]{\dfrac{a^5}{b^3}}$ 66. $\sqrt[3]{\dfrac{c^4}{b^6}}$

Perform the indicated operation and write each answer in simplest radical form. (*Section 8.4, Objectives 1 and 2*)

67. $(3 + \sqrt{5})^2$ $14 + 6\sqrt{5}$ 68. $(\sqrt{10} - 4)^2$ $26 - 8\sqrt{10}$

Divide the radical expressions and write each answer in simplest radical form. (*Section 8.5, Objective 1*)

69. $\dfrac{\sqrt{80c^7d^{14}}}{\sqrt{2c^3d^3}}$ $2c^2d^5\sqrt{10d}$ 70. $\dfrac{\sqrt{36x^8y^3}}{\sqrt{3x^5y^{-2}}}$ $2xy^2\sqrt{3xy}$

Solve each radical equation. (*Section 8.6, Objective 1*)

71. $\sqrt{3x + 4} - x = 2$ $\{-1, 0\}$ 72. $\sqrt{2t - 1} + t = 8$ $\{5\}$

Perform the operation on the complex numbers and write each answer in standard form. (*Section 8.7, Objectives 2–4*)

73. $\dfrac{5i}{3 - i}$ $-\dfrac{1}{2} + \dfrac{3}{2}i$ 74. $\dfrac{10i}{1 + 7i}$ $\dfrac{7}{5} + \dfrac{1}{5}i$

Create a table of values to graph each function. Identify the x-intercepts, y-intercept, vertex, and the axis of symmetry. Explain how the graph of the function relates to the graph of $y = x^2$. State the domain and range of the function. (*Section 9.1, Objective 4*)

75. $f(x) = (x - 1)^2 + 3$ 76. $f(x) = (x + 2)^2 - 1$

77. $f(x) = -(x - 2)^2 + 1$ 78. $f(x) = -(x + 3)^2 + 1$

Graph each function and use the graph to solve the equation $f(x) = 0$. (*Section 9.1, Objective 5*)

79. $f(x) = \dfrac{1}{2}(x + 2)^2 - 8$ 80. $f(x) = -(x - 5)^2 + 1$

Solve each equation by completing the square. (*Section 9.2, Objectives 3 and 4*)

81. $x^2 + 8x + 9 = 0$ $\{-4 \pm \sqrt{7}\}$ 82. $y^2 + 12y + 26 = 0$ $\{-6 \pm \sqrt{10}\}$

83. $4r^2 - 8r + 29 = 0$ 84. $4x^2 + 12x + 13 = 0$

Find the zeros of each function. State the x-intercepts of the graph of the function. (*Section 9.2, Objective 6*)

85. $f(x) = 3(x - 1)^2 - 12$ −1, 3; (−1, 0), (3, 0) 86. $f(x) = \dfrac{1}{2}(x + 4)^2 - 18$ −10, 2; (−10, 0), (2, 0)

Solve each equation using the quadratic formula. (*Section 9.3, Objective 1*)

87. $b^2 - 2b + 82 = 0$ $\{1 \pm 9i\}$ 88. $a^2 + 4a + 53 = 0$ $\{-2 \pm 7i\}$

89. $x^2 + 6x - 8 = 0$ $\{-3 \pm \sqrt{17}\}$ 90. $y^2 - 4y - 10 = 0$ $\{2 \pm \sqrt{14}\}$

Use the discriminant to determine the number and types of solutions of each equation. (*Section 9.3, Objective 2*)

91. $4x^2 - 5x + 6 = 0$ two complex solutions 92. $3y^2 - 7y - 5 = 0$ two irrational solutions

93. $x^2 - 6x + 9 = 0$ one rational solution 94. $4y^2 + 7y + 3 = 0$ two rational solutions

Solve each problem. (*Section 9.3, Objective 3*)

95. John wants to enclose a pen at the back of his barn, using the barn as one of its borders. John has 440 ft of fencing. What dimensions should the pen have to enclose an area of 24,000 ft^2? 120 ft by 200 ft

96. Ivan wants to enclose a pen at the back of his barn, using the barn as one of its borders. Ivan has 380 ft of fencing. What dimensions should the pen have to enclose an area of 12,000 ft^2? 80 ft by 150 ft

Solve each equation. (*Section 9.4, Objectives 1–4*)

97. $\dfrac{x}{2x + 1} + \dfrac{2x}{2x - 1} = \dfrac{5}{4x^2 - 1}$ $\left\{-1, \dfrac{5}{6}\right\}$

98. $\dfrac{5x}{x - 6} - \dfrac{2x}{x + 5} = \dfrac{26}{x^2 - x - 30}$ $\left\{-13, \dfrac{2}{3}\right\}$

99. $\sqrt{5x + 6} = x$ $\{6\}$ 100. $\sqrt{5x + 14} = x$ $\{7\}$

101. $x^{2/3} - 4x^{1/3} - 5 = 0$ $\{-1, 125\}$ 102. $x^{1/2} - 6x^{1/4} + 9 = 0$ $\{81\}$

Convert $y = ax^2 + bx + c$ to the vertex form, $y = a(x - h)^2 + k$. Identify the vertex and intercepts, and graph the function. (*Section 9.5, Objective 1*)

103. $f(x) = 2x^2 - 12x + 19$

104. $f(x) = x^2 + x - 2$

Solve each inequality using the boundary number method. Write each solution set in interval notation. (*Section 9.6, Objective 1*)

105. $3x^2 - 7x \geq 20$

106. $2x^2 + x \leq 28$ $\left[-4, \frac{7}{2}\right]$

Use the given graph of $y = f(x)$ to solve each associated inequality. Write each solution set in interval notation. (*Section 9.6, Objective 4*)

107. $x(2x + 5)(x - 3) > 0$ $\left(-\frac{5}{2}, 0\right) \cup (3, \infty)$

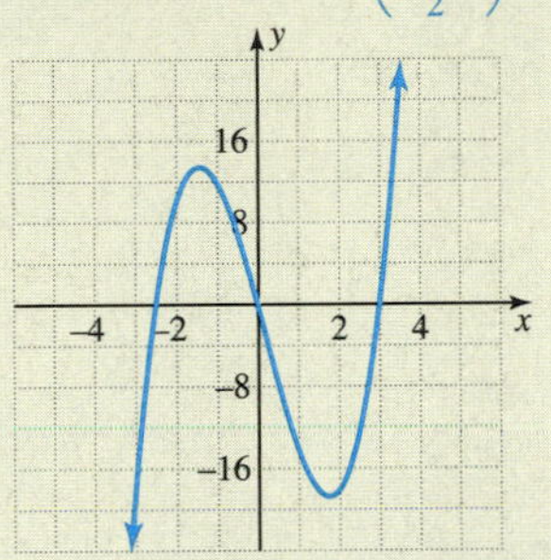

108. $x^4 - 10x^2 + 9 < 0$ $(-3, -1) \cup (1, 3)$

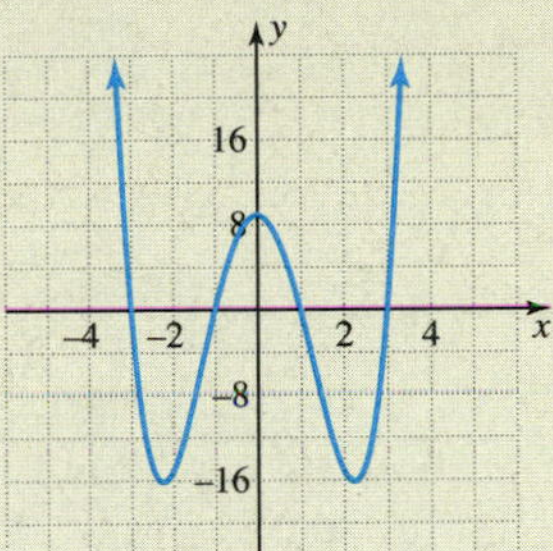

CHAPTER 10

Exponential and Logarithmic Functions

Greatness

When we think of greatness, perhaps we think of star athletes such as Peyton Manning or Cal Ripken. *Fortune Magazine* published an article entitled "Secrets of Greatness." The findings of this article are very pertinent to college students. In response to the question, what does it take to be great? the article stated

> Well, folks, it's not so simple. For one thing, you do not possess a natural gift for a certain job, because targeted natural gifts don't exist . . . You are not a born CEO or investor or chess grandmaster. You will achieve greatness only through an enormous amount of hard work over many years. . . . The good news is that your lack of a natural gift is irrelevant—talent has little or nothing to do with greatness. You can make yourself into any number of things, and you can even make yourself great. The best people in any field are those who devote the most hours to what the researchers call "deliberate practice." It is activity that's explicitly intended to improve performance, that reaches for objectives just beyond one's level of competence, provides feedback on results and involves high levels of repetition.

The researchers questioned why some people are more motivated than others to devote their lives to this deliberate practice to become great. Do you have the motivation to become great? You do not have to have some special gift or talent. You are the only thing you need to make yourself great! (Source: http://money.cnn.com/magazines/fortune/fortune_archive/2006/10/30/8391794/index.htm)

Question For Thought: What do you desire to be great in? What will it take for you to achieve this? What can you do today that will enable you to accomplish greatness?

Chapter Outline

Section 10.1 Operations and Composition of Functions 726

Section 10.2 Inverse Functions 739

Section 10.3 Exponential Functions 754

Section 10.4 Logarithmic Functions 762

Section 10.5 Properties of Logarithms 769

Section 10.6 The Common Log, Natural Log, and Change-of-Base Formula 778

Section 10.7 Exponential and Logarithmic Equations and Applications 787

Coming Up...

In Section 10.6, we will learn how to use logarithms to determine how proficient we are at a task after we no longer practice it by using the model $P = 90 - 12.5 \ln d$, where P is the proficiency rate and d is the number of days after practicing.

"I long to accomplish a great and noble task, but it is my chief duty to accomplish small tasks as if they were great and noble."

—Helen Keller (Author, Teacher)

SECTION 10.1 Operations and Composition of Functions

We have covered a few concepts related to functions throughout this text. In this chapter, we will study how to combine functions in different ways and how to find the inverse of a function, if it exists. Lastly, we will cover two other families of functions, exponential and logarithmic.

OBJECTIVES

As a result of completing this section, you will be able to

1. Add, subtract, multiply, and divide functions.
2. Evaluate combined functions.
3. Find the composition of two functions.
4. Solve applications.
5. Troubleshoot common errors.

Josh makes \$25 an hour, so his income can be represented by the function $f(x) = 25x$, where x is the number of hours he works. He pays 15% of his income in taxes. The amount of tax he owes can be represented by the function $g(x) = 0.15x$, where x is his income. Write a function that represents the total taxes Josh owes as a function of the number of hours he works.

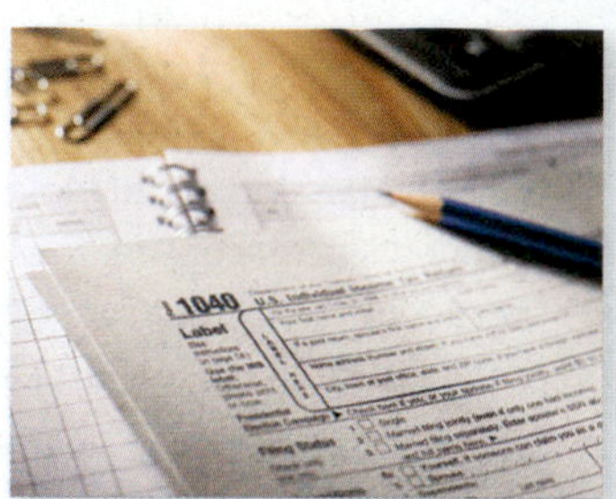

To complete this exercise, we need to know how to combine two functions.

Function Operations

Objective 1 Add, subtract, multiply, and divide functions.

Recall a function represents a real number. So, we can perform operations on them just as we do with real numbers. We can add, subtract, multiply, and divide functions. In fact, we have already seen some examples of this in earlier sections. When we perform these types of operations on functions, we get another function as its result. For example, suppose $f(x) = 2x + 1$ and $g(x) = 3x$, we can add these functions together to obtain

$$f(x) + g(x) = 2x + 1 + 3x = 5x + 1$$

The *domain* of this combined function is the intersection of the domains of the individual functions. Adding, subtracting, multiplying, and dividing functions to generate a new function is called the *algebra of functions*.

INSTRUCTOR NOTE: Inform students that the product of two functions f and g can also be denoted as $(fg)(x)$. If the multiplication dot is not written, multiplication is the implied operation.

Property: The Algebra of Functions

If we let $f(x)$ and $g(x)$ be two functions, we have

Sum	$(f + g)(x) = f(x) + g(x)$
Difference	$(f - g)(x) = f(x) - g(x)$
Product	$(f \cdot g)(x) = f(x) \cdot g(x)$
Quotient	$\left(\dfrac{f}{g}\right)(x) = \dfrac{f(x)}{g(x)},\ g(x) \neq 0$

Property: The Domain of the Combined Functions

Let the domain of $f(x)$ be set A and the domain of $g(x)$ be set B. Then the domain of $(f + g)(x)$, $(f - g)(x)$, $(f \cdot g)(x)$ is $A \cap B$. The domain of $\left(\dfrac{f}{g}\right)(x)$ is $A \cap B$, excluding the values that make the denominator zero.

Objective 1 Examples Find the sum, difference, product, and quotient of the functions. State the domain of each combined function.

1a. $f(x) = 3x + 2,\ g(x) = x - 5$ **1b.** $f(x) = \sqrt{x},\ g(x) = x^2 + x$

Solutions

1a. Both $f(x)$ and $g(x)$ are linear functions since they are of the form $y = mx + b$. The domain of each function is all real numbers or $(-\infty, \infty)$, and the intersection of their domains is $(-\infty, \infty)$.

$$\begin{aligned}(f+g)(x) = f(x) + g(x) &= (3x+2) + (x-5) \\ &= 3x + 2 + x - 5 \\ &= 4x - 3\end{aligned}$$

Domain of $f + g$ is $(-\infty, \infty)$.

$$\begin{aligned}(f-g)(x) = f(x) - g(x) &= (3x+2) - (x-5) \\ &= 3x + 2 - x + 5 \\ &= 2x + 7\end{aligned}$$

Domain of $f - g$ is $(-\infty, \infty)$.

$$\begin{aligned}(f \cdot g)(x) = f(x) \cdot g(x) &= (3x+2)(x-5) \\ &= 3x^2 - 15x + 2x - 10 \\ &= 3x^2 - 13x - 10\end{aligned}$$

Domain of $f \cdot g$ is $(-\infty, \infty)$.

$$\begin{aligned}\left(\frac{f}{g}\right)(x) = \frac{f(x)}{g(x)} &= \frac{3x+2}{x-5} \\ &= \frac{3x+2}{x-5}\end{aligned}$$

$g(x) \neq 0$
$x - 5 \neq 0$
$x \neq 5$
Domain of $\frac{f}{g}$ is $(-\infty, 5) \cup (5, \infty)$.

1b. The function $f(x)$ is a square root function and is defined for $x \geq 0$ or $[0, \infty)$. The function $g(x)$ is a quadratic function and has domain $(-\infty, \infty)$. The domain of f intersected with the domain of g is $[0, \infty)$.

$$\begin{aligned}(f+g)(x) = f(x) + g(x) &= (\sqrt{x}) + (x^2 + x) \\ &= \sqrt{x} + x^2 + x\end{aligned}$$

Domain of $f + g$ is $[0, \infty)$.

$$\begin{aligned}(f-g)(x) = f(x) - g(x) &= (\sqrt{x}) - (x^2 + x) \\ &= \sqrt{x} - x^2 - x\end{aligned}$$

Domain of $f - g$ is $[0, \infty)$.

$$\begin{aligned}(f \cdot g)(x) = f(x) \cdot g(x) &= (\sqrt{x})(x^2 + x) \\ &= x^2\sqrt{x} + x\sqrt{x}\end{aligned}$$

Domain of $f \cdot g$ is $[0, \infty)$.

$$\begin{aligned}\left(\frac{f}{g}\right)(x) = \frac{f(x)}{g(x)} &= \frac{\sqrt{x}}{x^2 + x} \\ &= \frac{\sqrt{x}}{x^2 + x}\end{aligned}$$

$g(x) \neq 0$
$x^2 + x \neq 0$
$x(x + 1) \neq 0$
$x \neq 0$ or $x \neq -1$
Domain of $\frac{f}{g}$ is $(0, \infty)$.

Student Check 1 Find the sum, difference, product, and quotient of the functions. State the domain of each combined function.

a. $f(x) = x^2 - x + 3$, $g(x) = 2x + 1$

b. $f(x) = \sqrt{2x}$, $g(x) = x^2 - 4$

Evaluating Combined Functions

Objective 2 ▶ Evaluate combined functions.

Recall we have evaluated functions in several sections. When we *evaluate a function*, we find the y-value that corresponds to the given x-value. We can also evaluate the sum, difference, product, or quotient of functions. When we evaluate a combined function, we add, subtract, multiply, or divide the y-values of the two functions. For example, let $f(x) = 3x + 2$ and $g(x) = x - 5$, to find $(f - g)(6)$, we can either find $f(6) - g(6)$ or we can find $(f - g)(x)$ and evaluate it when $x = 6$.

Method 1	Method 2
$f(6) = 3(6) + 2 = 18 + 2 = 20$ $g(6) = 6 - 5 = 1$ $f(6) - g(6) = 20 - 1 = 19$	$f(x) - g(x) = 3x + 2 - (x - 5)$ $= 3x + 2 - x + 5$ $= 2x + 7$ So, $(f - g)(6) = 2(6) + 7$ $= 12 + 7$ $= 19$

Procedure: Evaluating Combined Functions

Method 1 Evaluate each function at the given x-value. Then find the sum, difference, product, or quotient of the results, as required.

Method 2 Find the combined function first and then evaluate this function at the given x-value.

Note: *If a function is given graphically or numerically, then we need to apply method 1.*

Objective 2 Examples

Evaluate each function.

2a. Let $f(x) = x^2 - x + 3$ and $g(x) = 2x + 1$. Find $(f + g)(4)$.

Solution 2a.

Method 1	Method 2
$(f + g)(4) = f(4) + g(4)$ $f(4) = (4)^2 - (4) + 3 = 16 - 4 + 3 = 15$ $g(4) = 2(4) + 1 = 8 + 1 = 9$ $(f + g)(4) = f(4) + g(4)$ $= 15 + 9$ $= 24$	$(f + g)(x) = f(x) + g(x)$ $= x^2 - x + 3 + (2x + 1)$ $= x^2 + x + 4$ $(f + g)(4) = (4)^2 + (4) + 4$ $= 16 + 4 + 4$ $= 24$

2b. Use the graphs to find $(f + g)(3)$, $(f - g)(0)$, $(f \cdot g)(-2)$, and $\left(\frac{f}{g}\right)(1)$.

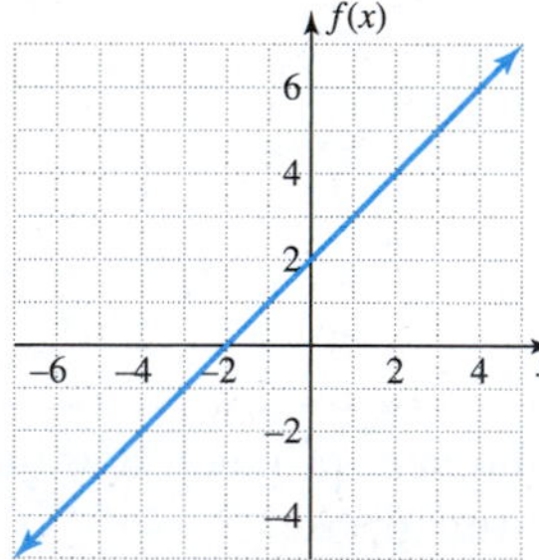

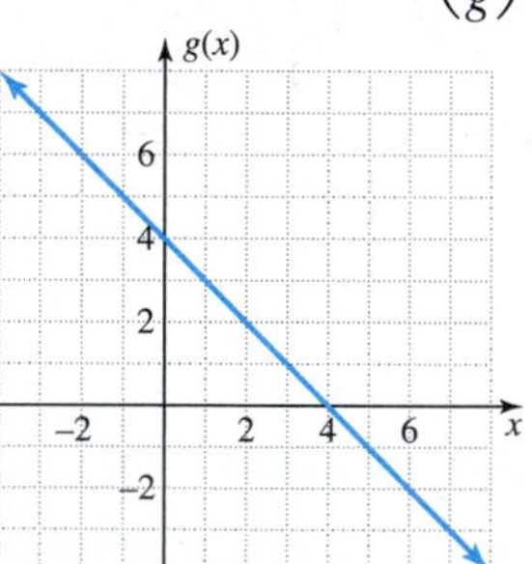

Solution 2b. We first identify the points on each graph that correspond to the x-values of 3, 0, −2, and 1.

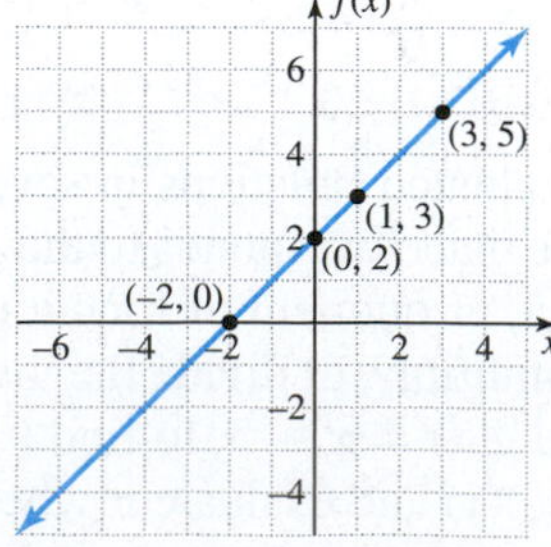

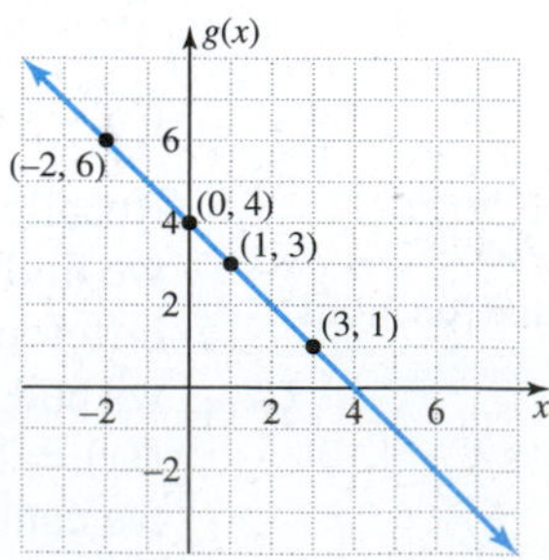

(3, 5) is on the graph of $f(x)$, so $f(3) = 5$. → $(f + g)(3) = f(3) + g(3)$
(3, 1) is on the graph of $g(x)$, so $g(3) = 1$. $= 5 + 1$
$= 6$

(0, 2) is on the graph of $f(x)$, so $f(0) = 2$. → $(f - g)(0) = f(0) - g(0)$
(0, 4) is on the graph of $g(x)$, so $g(0) = 4$. $= 2 - 4$
$= -2$

(−2, 0) is on the graph of $f(x)$, so $f(-2) = 0$. → $(f \cdot g)(-2) = f(-2) \cdot g(-2)$
(−2, 6) is on the graph of $g(x)$, so $g(-2) = 6$. $= (0)(6)$
$= 0$

(1, 3) is on the graph of $f(x)$, so $f(1) = 3$. → $\left(\frac{f}{g}\right)(1) = \frac{f(1)}{g(1)}$
(1, 3) is on the graph of $g(x)$, so $g(1) = 3$. $= \frac{3}{3}$
$= 1$

2c. Use the table to find $(f + g)(-3)$, $(f - g)(-2)$, $(f \cdot g)(0)$, and $\left(\frac{f}{g}\right)(3)$.

x	$f(x)$	$g(x)$
−3	8.5	8
−2	7	3
−1	5.5	0
0	4	−1
1	2.5	0
2	1	3
3	−0.5	8

Solution **2c.** $(f + g)(-3) = f(-3) + g(-3) = 8.5 + 8 = 16.5$

$(f - g)(-2) = f(-2) - g(-2) = 7 - 3 = 4$

$(f \cdot g)(0) = f(0) \cdot g(0) = (4)(-1) = -4$

$$\left(\frac{f}{g}\right)(3) = \frac{f(3)}{g(3)} = \frac{-0.5}{8} = -\frac{1}{2} \cdot \frac{1}{8} = -\frac{1}{16}$$

Student Check 2 Evaluate each function.

a. Let $f(x) = 3x^2 - 2x + 1$ and $g(x) = 5x + 4$. Find $(f + g)(2)$.

b. Use the graphs to find $(f + g)(0)$, $(f - g)(-1)$, $(f \cdot g)(3)$, and $\left(\frac{f}{g}\right)(4)$.

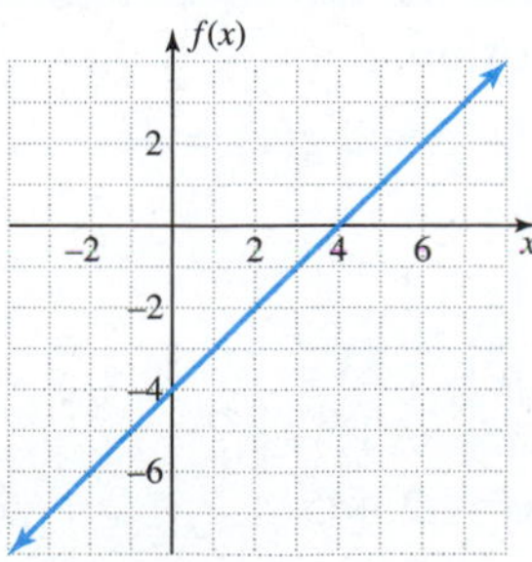

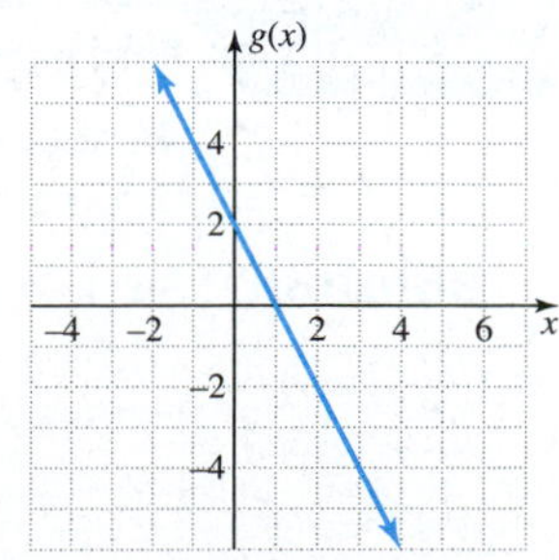

c. Use the table to find $(f+g)(1)$, $(f-g)(2)$, $(f \cdot g)(-3)$, and $\left(\frac{f}{g}\right)(0)$.

x	$f(x)$	$g(x)$
−4	7	−13
−3	6	−11
−2	5	−9
−1	4	−7
0	3	−5
1	2	−3
2	1	−1

Function Composition

Objective 3 Find the composition of two functions.

Functions can also be combined in another way. The *composition* of functions takes one function's output and makes it the input of another function. So, we are inputting a function into a function. In more mathematical terms, we are evaluating a function at a function.

Consider the two functions, $f(x) = x^2$ and $g(x) = x + 1$. If we begin with $x = 3$ and find $f(3)$, we get $f(3) = 3^2 = 9$. Now if we evaluate $g(x)$ at $x = 9$, we get $g(9) = 9 + 1 = 10$. This can be represented by the following diagram.

$x = 3$ → $f(x) = x^2$ → $f(3) = 9$ → $g(x) = x + 1$ → $g(9) = 10$

So, $g(9) = 10$, but since $f(3) = 9$, it follows that $g(9) = g(f(3))$. The notation $g(f(3))$ means "g composed with f." It is more commonly denoted as $(g \circ f)(3)$ which is read as "g of f of 3." We generalize this as follows.

INSTRUCTOR NOTE: Show students the difference between the product of two functions and the composition of two functions.

Definition: The Composition of Functions

If f and g are functions, then the function f composed with the function g can be represented as $(f \circ g)(x)$ and

$$(f \circ g)(x) = f(g(x))$$

The function g composed with the function f can be represented as $(g \circ f)(x)$ and

$$(g \circ f)(x) = g(f(x))$$

The domain of a composed function is the domain of the final function, excluding any values that make the inner function undefined.

Procedure: Finding f Composed with g

Step 1: Substitute the function g for the variable in the function f.
Step 2: Simplify.

Objective 3 Examples Find each composition.

3a. Let $f(x) = 2x + 5$ and $g(x) = 3x - 1$. Find $(f \circ g)(x)$ and $(g \circ f)(x)$ and state their domains.

Solution

3a. $(f \circ g)(x) = f(g(x))$ — Apply the definition of composition.

$= f(3x - 1)$ — Replace $g(x)$ with $3x - 1$.

$= 2(3x - 1) + 5$ — Substitute $3x - 1$ for x in the function f.

$= 6x - 2 + 5$ — Apply the distributive property.

$= 6x + 3$ — Simplify.

$$\begin{aligned}(g \circ f)(x) &= g(f(x)) && \text{Apply the definition of composition.}\\ &= g(2x + 5) && \text{Replace } f(x) \text{ with } 2x + 5.\\ &= 3(2x + 5) - 1 && \text{Substitute } 2x + 5 \text{ for } x \text{ in the function } g.\\ &= 6x + 15 - 1 && \text{Apply the distributive property.}\\ &= 6x + 14 && \text{Simplify.}\end{aligned}$$

The functions $f \circ g$ and $g \circ f$ are both linear functions. The functions f and g are linear as well. Therefore, the domain of $f \circ g$ and $g \circ f$ is $(-\infty, \infty)$.

3b. Let $f(x) = x^2 - 2x + 3$ and $g(x) = \sqrt{x + 1}$. Find $(f \circ g)(3)$.

Solution **3b.**

$$\begin{aligned}(f \circ g)(3) &= f(g(3)) && \text{Apply the definition of composition.}\\ &= f(2) && \text{Evaluate } g(3) = \sqrt{3+1} = \sqrt{4} = 2.\\ &= (2)^2 - 2(2) + 3 && \text{Evaluate } f(2).\\ &= 4 - 4 + 3 && \text{Simplify.}\\ &= 3 && \text{Simplify.}\end{aligned}$$

We can also find $(f \circ g)(x)$ and then evaluate this function at $x = 3$.

$$\begin{aligned}(f \circ g)(x) &= f(g(x)) && \text{Apply the definition of composition.}\\ &= f(\sqrt{x+1}) && \text{Replace } g(x) \text{ with } \sqrt{x+1}.\\ &= (\sqrt{x+1})^2 - 2(\sqrt{x+1}) + 3 && \text{Substitute } \sqrt{x+1} \text{ for } x \text{ in } f(x).\\ &= x + 1 - 2\sqrt{x+1} + 3 && \text{Apply } (\sqrt[n]{x})^n = x.\\ (f \circ g)(3) &= 3 + 1 - 2\sqrt{3+1} + 3 && \text{Replace } x \text{ with } 3.\\ &= 7 - 2\sqrt{4} && \text{Combine like terms and simplify the radicand.}\\ &= 7 - 2(2) && \text{Simplify the radical.}\\ &= 7 - 4 && \text{Multiply.}\\ &= 3 && \text{Subtract.}\end{aligned}$$

3c. Use the graphs of f and g to find $(g \circ f)(4)$.

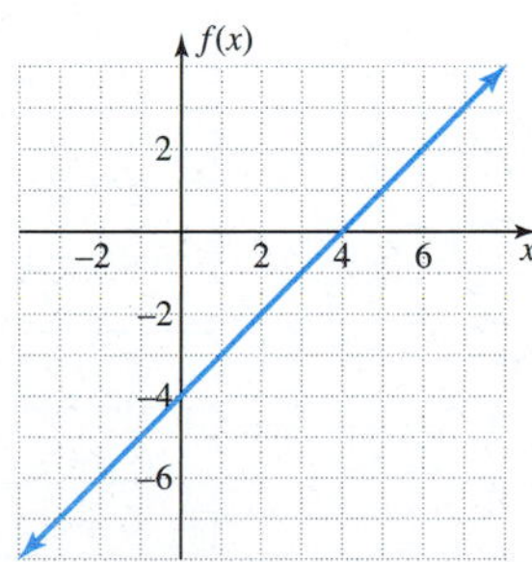

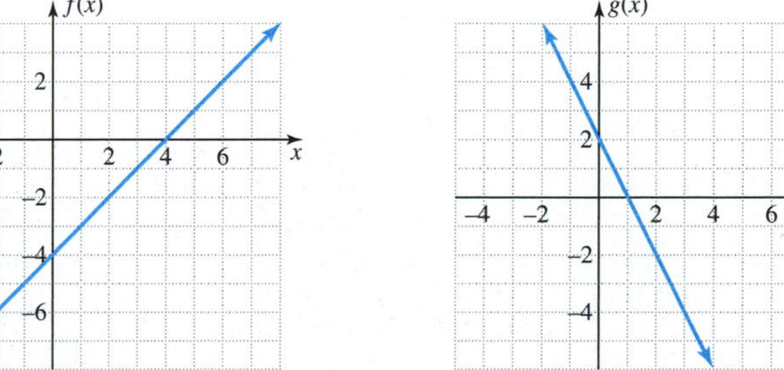

Solution **3c.**

$$\begin{aligned}(g \circ f)(4) &= g(f(4)) && \text{Apply the definition of composition.}\\ &= g(0) && \text{Since } (4, 0) \text{ is on the graph of } f(x),\ f(4) = 0.\\ &= 2 && \text{Since } (0, 2) \text{ is on the graph of } g(x),\ g(0) = 2.\end{aligned}$$

3d. Use the table to find $(g \circ f)(0)$ and $(f \circ g)(3)$.

x	$f(x)$	$g(x)$
0	1	−1
1	0	1
2	−1	3
3	−2	5
4	−3	7
5	−4	9
6	−5	11

Solution **3d.**

$(g \circ f)(0) = g(f(0))$	$(f \circ g)(3) = f(g(3))$	Apply the definition of composition.
$= g(1)$	$= f(5)$	The table shows that $f(0) = 1$ and $g(3) = 5$.
$= 1$	$= -4$	The table shows that $g(1) = 1$ and $f(5) = -4$.

Student Check 3 Find each composition.

a. Let $f(x) = 5x + 7$ and $g(x) = 4x - 8$. Find $(f \circ g)(x)$ and $(g \circ f)(x)$ and state their domains.

b. Let $f(x) = x^2 - 4x + 1$ and $g(x) = |x + 3|$. Find $(g \circ f)(3)$.

c. Use the graphs of f and g to find $(f \circ g)(5)$.

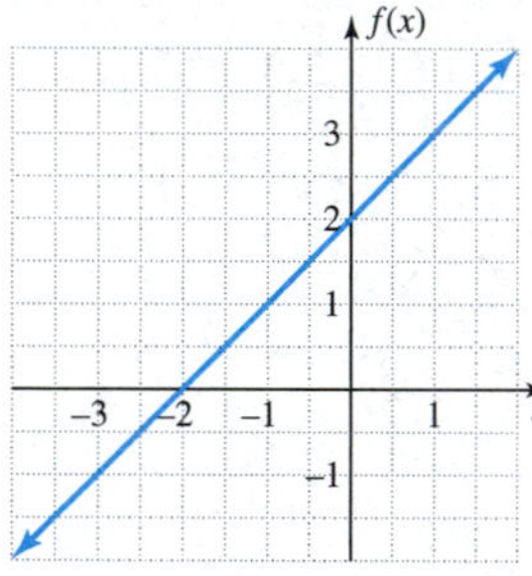

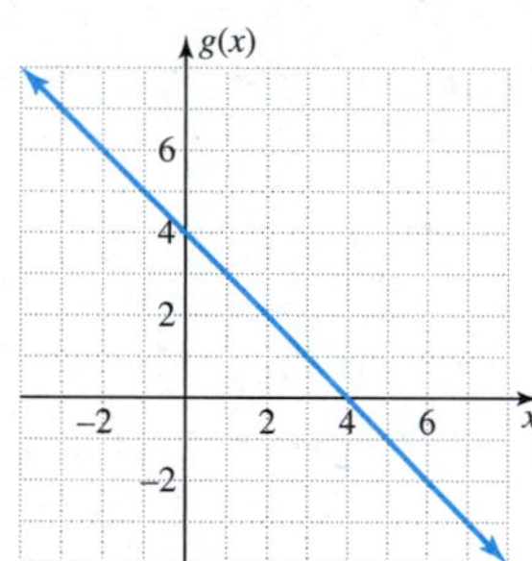

d. Use the table to find $(g \circ f)(1)$ and $(f \circ g)(2)$.

x	$f(x)$	$g(x)$
0	1	−1
1	0	1
2	−1	3
3	−2	5
4	−3	7
5	−4	9
6	−5	11

Applications

Objective 4 Solve applications.

Applications of combining functions occur in many areas of our lives though we may not think of them as such. There are applications in business, physics, biology, and so on. Examples 4a and 4b illustrate some of these applications.

Objective 4 Examples **Solve each problem using an appropriate combination of functions.**

4a. A bookstore buys math textbooks from a publishing company for \$80 each and has other fixed costs of \$200. So, the bookstore's total cost of the math books can be represented by the function $C(x) = 80x + 200$, where x is the number of math books purchased. The bookstore sells the math books at a 40% markup, or at \$112 each. So, the bookstore's total revenue from selling x math books can be represented by the function $R(x) = 112x$. Write a function that represents the profit from selling x math books. (Recall: Profit = Revenue − Cost.)

Solution **4a.**

$P(x) = R(x) - C(x)$	Write the definition of profit.
$= 112x - (80x + 200)$	Replace $R(x)$ and $C(x)$ with the given functions.
$= 112x - 80x - 200$	Apply the distributive property.
$= 32x - 200$	Combine like terms.

So, the function $P(x) = 32x - 200$ represents the profit from selling x math books.

4b. Josh makes \$25 an hour, so his income can be represented by the function $f(x) = 25x$, where x is the number of hours he works. He pays 15% of his income in taxes. The amount of tax he owes can be represented by the function $g(x) = 0.15x$, where x is his income. Write a function that represents the total taxes Josh owes as a function of the number of hours he works.

Solution **4b.** To find the total taxes Josh owes, we have to determine his total income and then find 15% of that amount. This is the composition of functions because the output of the income function becomes the input of the tax function. We will let $t(x)$ represent the total taxes as a function of the number of hours Josh works.

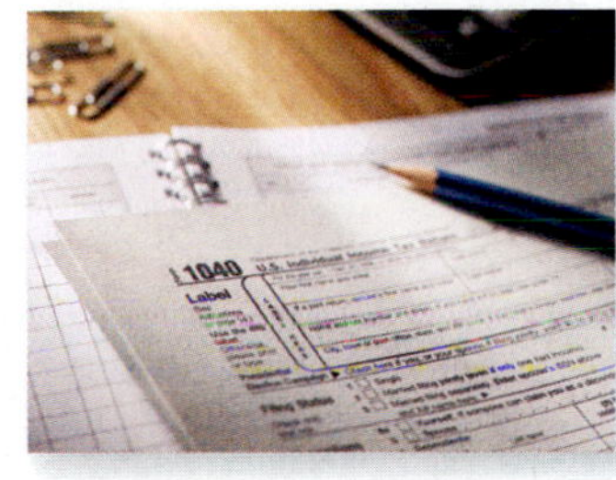

$t(x) = (g \circ f)(x)$	Write $t(x)$ as a composition of g and f.
$= g(f(x))$	Apply the definition of the composition.
$= g(25x)$	Replace $f(x)$ with $25x$.
$= 0.15(25x)$	Substitute $25x$ for x in $g(x)$.
$= 3.75x$	Simplify.

The function $t(x) = 3.75x$ represents the amount of taxes Josh owes as a function of the number of hours he works. So, Josh pays \$3.75 in taxes for each hour he works.

Student Check 4 Solve each problem using an appropriate combination of functions.

a. An online company sells only one product each day. They pay \$60 for each item they sell and have other fixed costs of \$500 a day. So, the company's total daily cost can be represented by the function $C(x) = 60x + 500$, where x is the number of items sold. The company sells the items for \$85 each. So, the total revenue from selling x items can be represented by the function $R(x) = 85x$. Write a function that represents the daily profit from selling x items.

b. Sarah makes \$20 an hour, so her income can be represented by the function $f(x) = 20x$, where x is the number of hours she works. She pays 12% of her income in taxes. The amount of tax she owes can be represented by the function $g(x) = 0.12x$, where x is her income. Write a function that represents the total taxes as a function of the number of hours Sarah works.

Objective 5 ▶

Troubleshoot common errors.

Troubleshooting Common Errors

Some common errors associated with combining functions are shown.

Objective 5 Examples **A problem and an incorrect solution are given. Provide the correct solution and an explanation of the error.**

5a. Let $f(x) = 4x - 1$ and $g(x) = 7x + 9$. Find $(f - g)(x)$.

Incorrect Solution	Correct Solution and Explanation
$(f - g)(x) = f(x) - g(x)$ $= 4x - 1 - 7x + 9$ $= -3x + 8$	We must distribute -1 to each term of $g(x)$. $(f - g)(x) = f(x) - g(x)$ $= 4x - 1 - (7x + 9)$ $= 4x - 1 - 7x - 9$ $= -3x - 10$

5b. Let $f(x) = 4x - 1$ and $g(x) = 7x + 9$. Find $f\big(g(x)\big)$.

Incorrect Solution	Correct Solution and Explanation
$f\big(g(x)\big) = (4x - 1)(7x + 9)$ $= 28x^2 + 36x - 7x - 9$ $= 28x^2 + 29x - 9$	The product of the functions is written as $(f \cdot g)(x)$ or $f(x) \cdot g(x)$ but $f\big(g(x)\big)$ represents the composition. $f\big(g(x)\big) = f(7x + 9)$ $= 4(7x + 9) - 1$ $= 28x + 36 - 1$ $= 28x + 35$

ANSWERS TO STUDENT CHECKS

Student Check 1 **a.** $(f + g)(x) = x^2 + x + 4$, domain $= (-\infty, \infty)$; $(f - g)(x) = x^2 - 3x + 2$, domain $= (-\infty, \infty)$; $(f \cdot g)(x) = 2x^3 - x^2 + 5x + 3$; domain $= (-\infty, \infty)$; $\left(\frac{f}{g}\right)(x) = \frac{x^2 - x + 3}{2x + 1}$; domain $= \left(-\infty, -\frac{1}{2}\right) \cup \left(-\frac{1}{2}, \infty\right)$

b. $(f + g)(x) = x^2 + \sqrt{2x} - 4$, domain $= [0, \infty)$; $(f - g(x) = -x^2 + \sqrt{2x} + 4$, domain $= [0, \infty)$; $(f \cdot g)(x) = x^2\sqrt{2x} - 4\sqrt{2x}$; domain $= [0, \infty)$; $\left(\frac{f}{g}\right)(x) = \frac{\sqrt{2x}}{x^2 - 4}$; domain $= [0, 2) \cup (2, \infty)$

Student Check 2 **a.** 23 **b.** $-2, -9, 4, 0$ **c.** $-1, 2, -66, -\frac{3}{5}$

Student Check 3 **a.** $f \circ g = 20x - 33$, $g \circ f = 20x + 20$; domain of both is $(-\infty, \infty)$ **b.** 1 **c.** 1 **d.** $-1, -2$

Student Check 4 **a.** $R(x) = 25x - 500$ **b.** $t(x) = 2.4x$

SUMMARY OF KEY CONCEPTS

1. We can add, subtract, multiply, and divide functions by performing these operations on the expressions that represent the functions.

$$(f + g)(x) = f(x) + g(x)$$
$$(f - g)(x) = f(x) - g(x)$$
$$(f \cdot g)(x) = f(x) \cdot g(x)$$
$$\left(\frac{f}{g}\right)(x) = \frac{f(x)}{g(x)}$$

2. When a function's output value is input into another function, we are evaluating the composition of functions.
3. To compose two functions together, replace the variable x in the outer function with the expression that represents the inner function.

$$(f \circ g)(x) = f\big(g(x)\big)$$
$$(g \circ f)(x) = g\big(f(x)\big)$$

GRAPHING CALCULATOR SKILLS

A graphing calculator can perform operations on functions and evaluate combined functions.

Example: Let $f(x) = 4x - 1$ and $g(x) = 7x + 9$. Find the following:

a. $(f + g)(-2)$ **b.** $(f \cdot g)(-2)$ **c.** $(f \circ g)(-2)$

Solution: Enter the functions f and g into Y_1 and Y_2. Then let Y_3 equal the operation to be performed. We can then view the table. The Y_3 value for $x = -2$ is the value of the combined function.

a. $(f + g)(-2)$

```
Plot1 Plot2 Plot3
\Y1=4X-1
\Y2=7X+9
\Y3=Y1+Y2
\Y4=
\Y5=
\Y6=
\Y7=
```

X	Y1	Y2	Y3
-2	-9	-5	-14
-1	-5	2	-3
0	-1	9	8
1	3	16	19
2	7	23	30
3	11	30	41
4	15	37	52

X=-2

So, $(f + g)(-2) = -14$.

b. $(f \cdot g)(-2)$

```
Plot1 Plot2 Plot3
\Y1=4X-1
\Y2=7X+9
\Y3=Y1Y2
\Y4=
\Y5=
\Y6=
\Y7=
```

X	Y1	Y2	Y3
-2	-9	-5	45
-1	-5	2	-10
0	-1	9	-9
1	3	16	48
2	7	23	161
3	11	30	330
4	15	37	555

X=-2

So, $(f \cdot g)(-2) = 45$.

c. $(f \circ g)(-2) = f(g(-2))$

```
Plot1 Plot2 Plot3
\Y1=4X-1
\Y2=7X+9
\Y3=Y1(Y2)
\Y4=
\Y5=
\Y6=
\Y7=
```

X	Y1	Y2	Y3
-2	-9	-5	-21
-1	-5	2	7
0	-1	9	35
1	3	16	63
2	7	23	91
3	11	30	119
4	15	37	147

X=-2

So, $(f \circ g)(-2) = f(g(-2)) = -21$.

Note: To access Y_1 and Y_2, press VARS, select Y-VARS, 1, 1 (for Y_1) or 2 (for Y_2).

SECTION 10.1 EXERCISE SET

Write About It!

Use complete sentences in your answer to each exercise.

1. Explain how to add and subtract functions, $f(x)$ and $g(x)$. Give an example of each.
2. Explain how to multiply binomial functions, $f(x)$ and $g(x)$. Give an example.
3. Explain how to determine the restriction on the domain when finding the quotient of two functions, $f(x)$ and $g(x)$.
4. Explain how to evaluate the sum, $(f + g)(2)$.
5. Explain how to evaluate the difference, $(f - g)(2)$.
6. Explain how to find the composite function, $f(g(x))$, if $f(x) = 2x^2 - x$ and $g(x) = 5x - 1$.
7. Explain how to find the composite function, $g(f(x))$, if $f(x) = 2x^2 - x$ and $g(x) = 5x - 1$.
8. If $f(x) = |7x - 2|$ and $g(x) = 2x - 6$, explain how to evaluate $f(g(2))$ and $g(f(2))$.

Practice Makes Perfect!

Find the sum, difference, product, and quotient of the functions. State the domain of each combined function. (*See Objective 1.*)

9. $f(x) = 4x + 11$ and $g(x) = x + 8$
10. $f(x) = 7x - 3$ and $g(x) = 2x + 1$
11. $f(x) = \sqrt{x}$ and $g(x) = 11x - 2$
12. $f(x) = 5\sqrt{2x}$ and $g(x) = 3x - 12$
13. $f(x) = \sqrt{5x}$ and $g(x) = x^2 + 9$
14. $f(x) = 2\sqrt{x}$ and $g(x) = 4x^2 + 1$
15. $f(x) = 4x - 3$ and $g(x) = -x^2 + 5x$
16. $f(x) = 6x + 1$ and $g(x) = x^2 - 3x$

Let $f(x) = x^2 - 5$, $g(x) = -3x^2 + x - 4$, $h(x) = 2x + 3$. Evaluate each function. (*See Objective 2.*)

17. $(f + g)(-3)$ -30
18. $(f + h)(-2)$ -2
19. $(f - g)(5)$ 94
20. $(h - g)(2)$ 21
21. $(f \cdot g)(-2)$ 18
22. $(f \cdot h)(3)$ 36

Additional answers can be found in the Instructor Answer Appendix.

23. $\left(\frac{f}{g}\right)(4)$ $-\frac{11}{48}$

24. $\left(\frac{g}{h}\right)(-1)$ -8

25. $(g + h)\left(\frac{1}{2}\right)$ $-\frac{1}{4}$

26. $(g - h)\left(-\frac{1}{3}\right)$ -7

Use the graphs of $f(x)$, $g(x)$, and $h(x)$ to find the value of each combined function. (*See Objective 2.*)

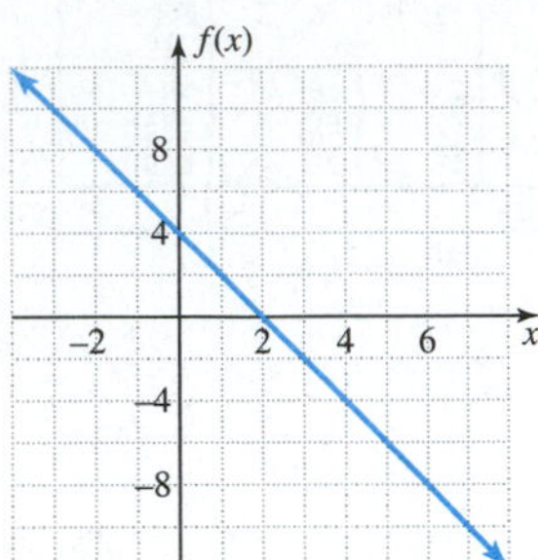

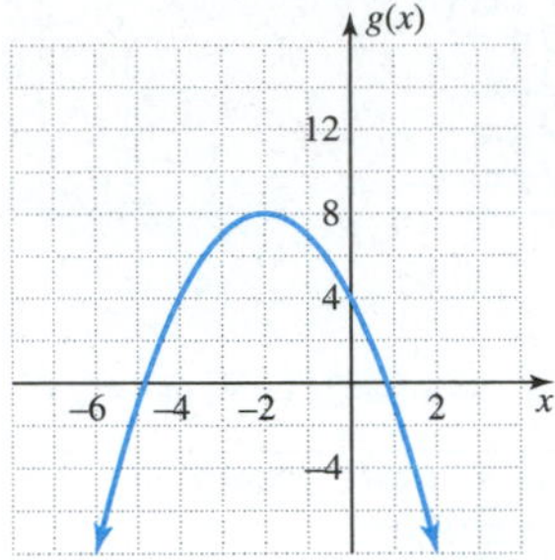

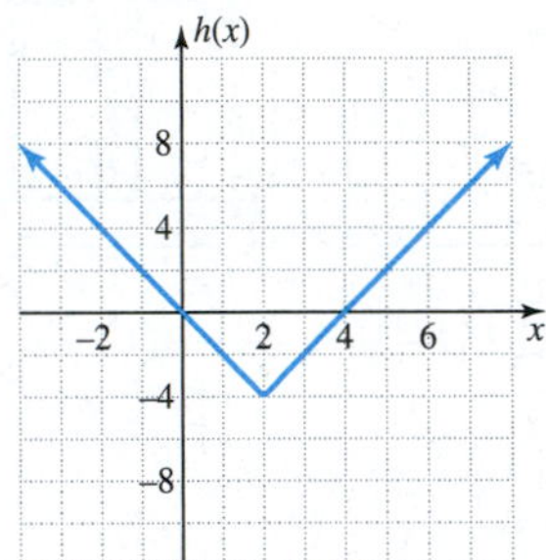

27. $(f + g)(0)$ 8
28. $(g + h)(-2)$ 12
29. $(f - h)(4)$ -4
30. $(g - h)(-4)$ -4
31. $(f \cdot g)(0)$ 16
32. $(g \cdot h)(-2)$ 32
33. $(f \circ h)(5)$ -12
34. $\left(\frac{f}{g}\right)(2)$ 0
35. $\left(\frac{f}{h}\right)(4)$ undefined
36. $\left(\frac{g}{f}\right)(-4)$ $\frac{1}{3}$

Use the table to find the value of each combined function. (*See Objective 2.*)

37. $(g + h)(0)$ -6
38. $(f + g)(-3)$ -8
39. $(f - g)(2)$ 8
40. $(g - h)(-2)$ -30
41. $(f \cdot g)(-1)$ 33
42. $(g \cdot h)(-2)$ -221
43. $\left(\frac{f}{g}\right)(1)$ 3
44. $\left(\frac{g}{h}\right)(-1)$ $-\frac{11}{3}$

x	$f(x)$	$g(x)$	$h(x)$
-3	15	-23	29
-2	3	-17	13
-1	-3	-11	3
0	-3	-5	-1
1	3	1	1
2	15	7	9
3	33	13	23

For the given functions $f(x)$ and $g(x)$, find the composition functions $(f \circ g)(x)$ and $(g \circ f)(x)$ and state their domain. (*See Objective 3.*)

45. $f(x) = x - 2$ and $g(x) = 12x - 11$
46. $f(x) = 5x + 7$ and $g(x) = 6x + 1$
47. $f(x) = 5x + 9$ and $g(x) = x - 4$
48. $f(x) = 3x - 15$ and $g(x) = 2x + 1$

For the given functions $f(x)$ and $g(x)$, evaluate each composition function. (*See Objective 3.*)

49. Let $f(x) = 2x^2 + 7x - 5$ and $g(x) = 3x - 4$. Find $(f \circ g)(2)$ and $(g \circ f)(-3)$.
$(f \circ g)(2) = 17$; $(g \circ f)(-3) = -28$
50. Let $f(x) = x^2 - 3x$ and $g(x) = 4 - 2x$. Find $(f \circ g)(4)$ and $(g \circ f)(-1)$.
$(f \circ g)(4) = 28$; $(g \circ f)(-1) = -4$
51. Let $f(x) = 3x^2 - 4x + 1$ and $g(x) = \sqrt{x + 18}$. Find $(f \circ g)(-2)$ and $(g \circ f)(3)$.
$(f \circ g)(-2) = 33$; $(g \circ f)(3) = \sqrt{34}$
52. Let $f(x) = 2x^2 - 3x - 9$ and $g(x) = \sqrt{x - 3}$. Find $(f \circ g)(7)$ and $(g \circ f)(4)$.
$(f \circ g)(7) = -7$; $(g \circ f)(4) = 2\sqrt{2}$
53. Let $f(x) = x^2 - 5x + 9$ and $g(x) = |x + 7|$. Find $(f \circ g)(-6)$ and $(g \circ f)(-1)$.
$(f \circ g)(-6) = 5$; $(g \circ f)(-1) = 22$
54. Let $f(x) = 3x^2 - 3x + 10$ and $g(x) = |x + 6|$. Find $(f \circ g)(-5)$ and $(g \circ f)(3)$.
$(f \circ g)(-5) = 10$; $(g \circ f)(3) = 34$

Use the graphs to find each value. (*See Objective 3.*)

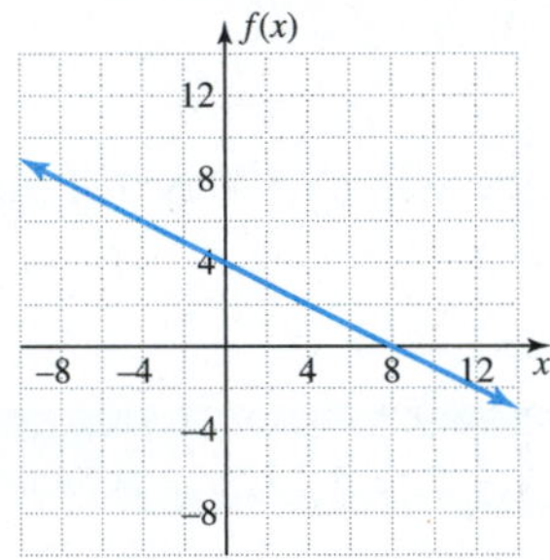

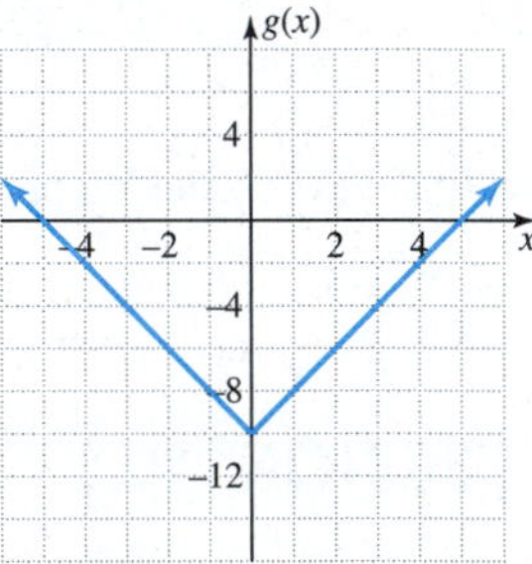

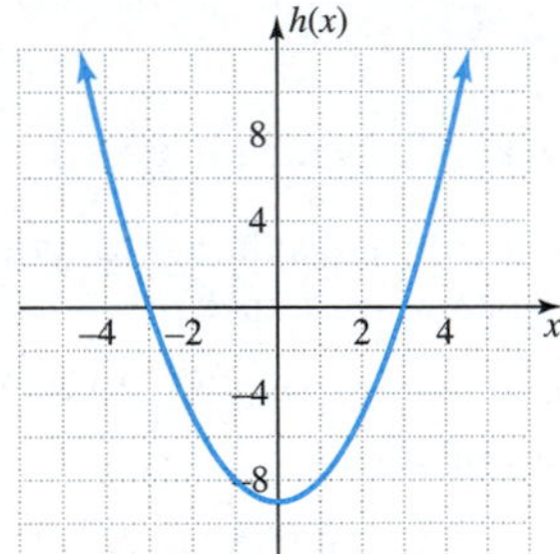

55. $(g \circ f)(0)$ -2
56. $(h \circ g)(5)$ -9
57. $(g \circ f)(2)$ -4
58. $(f \circ g)(-3)$ 6
59. $(g \circ h)(-2)$ 0
60. $(h \circ f)(2)$ 0

Use the table to find each value. (*See Objective 3.*)

61. Find $(f \circ g)(-2)$ and $(g \circ f)(-2)$.

x	$f(x)$	$g(x)$
-2	7	-19
7	16	17
-19	-10	-87

$(f \circ g)(-2) = -10$
$(g \circ f)(-2) = 17$

62. Find $(f \circ g)(0)$ and $(g \circ f)(0)$.

x	$f(x)$	$g(x)$
0	−5	3
−5	−10	13
3	−2	−3

$(f \circ g)(0) = -2$
$(g \circ f)(0) = 13$

63. Find $(f \circ g)(-3)$ and $(g \circ f)(-3)$.

x	$f(x)$	$g(x)$
−3	−11	1
−11	−19	−31
1	−7	17

$(f \circ g)(-3) = -7$
$(g \circ f)(-3) = -31$

64. Find $(f \circ g)(2)$ and $(g \circ f)(2)$.

x	$f(x)$	$g(x)$
2	−5	9
−5	−12	−5
9	2	23

$(f \circ g)(2) = 2$
$(g \circ f)(2) = -5$

Solve each problem. (*See Objective 4.*)

65. Amiee makes $18 an hour, so her income can be represented by the function $f(x) = 18x$, where x is the number of hours she works. She pays 15% of her income in taxes. The amount of tax she owes can be represented by the function $g(x) = 0.15x$, where x is her income. Write a function that represents the total taxes Amiee owes as a function of the number of hours she works. $t(x) = 2.7x$

66. Blake makes $36 an hour, so his income can be represented by the function $f(x) = 36x$, where x is the number of hours he works. He pays 30% of his income in taxes. The amount of tax he owes can be represented by the function $g(x) = 0.30x$, where x is his income. Write a function that represents the total taxes Blake owes as a function of the number of hours he works. $t(x) = 10.8x$

67. A bookstore buys math textbooks from a publishing company for $105 each and has other fixed costs of $421. So, the bookstore's total cost of the math books can be represented by the function $C(x) = 105x + 421$, where x is the number of math books purchased. The bookstore sells the math books at a 36% markup or at $142.80 each. So, the bookstore's total revenue from selling x math books can be represented by the function $R(x) = 142.80x$. Write a function that represents the profit from selling x math books. $P(x) = 37.8x - 421$

68. A bookstore buys biology textbooks from a publishing company for $60 each and has other fixed costs of $474. So, the bookstore's total cost of the biology books can be represented by the function $C(x) = 60x + 474$, where x is the number of biology books purchased. The bookstore sells the biology books at a 41% markup or at $84.60 each. So, the bookstore's total revenue from selling x biology books can be represented by the function $R(x) = 84.60x$. Write a function that represents the profit from selling x biology books. $P(x) = 24.6x - 474$

69. An organization is planning a bingo fund-raiser. The cost of using the hall is $27 per person. The cost of food is $20 per person and the cost of prizes is $16 for every four players. Let x represent the number of players.

a. Write a function that represents the cost of the hall, the cost of food, and the cost of prizes. $h(x) = 27x, f(x) = 20x, p(x) = 4x$

b. Write a function that represents the total cost of the bingo fund-raiser as a function of the number of players. $c(x) = 51x$

70. An organization is planning a golf outing. The cost of using the golf course is $39 per person. The cost of food is $23 per person and the cost of prizes is $60 for every six golfers. Let x represent the number of golfers.

a. Write a function that represents the cost of the golf course, the cost of food, and the cost of prizes. $g(x) = 39x, f(x) = 23x, p(x) = 10x$

b. Write a function that represents the total cost of the golf outing as a function of the number of golfers. $c(x) = 72x$

Mix 'Em Up!

Find each function. State the restriction for each quotient function.

For Exercises 71–76, let $f(x) = 2x + 1$, $g(x) = 3x^2 - 2x$, and $h(x) = 4 - x$.

71. $(f + g)(x)$ and $(f \cdot g)(x)$ **72.** $(f + h)(x)$ and $(f \cdot h)(x)$

73. $(g \cdot h)(x)$ and $\left(\frac{h}{g}\right)(x)$ **74.** $(g - h)(x)$ and $\left(\frac{g}{h}\right)(x)$

75. $(f \circ g)(x)$ and $(g \circ f)(x)$ **76.** $(f \circ h)(x)$ and $(h \circ f)(x)$

For Exercises 77–82, let $f(x) = 2x^2 + 3$, $g(x) = x^2 - 7x + 6$, and $h(x) = 5 - 3x$.

77. $(f + g)(x)$ and $(f - g)(x)$

78. $(g + h)(x)$ and $(g - h)(x)$

79. $(f \cdot g)(x)$ and $\left(\frac{f}{g}\right)(x)$ **80.** $(g \cdot h)(x)$ and $\left(\frac{g}{h}\right)(x)$

81. $(f \circ h)(x)$ and $(h \circ f)(x)$ **82.** $(g \circ h)(x)$ and $(h \circ g)(x)$

Let $f(x) = x^2 + 2$, $g(x) = x^2 - 4x - 5$, $h(x) = 2x - 1$. Evaluate each combined function.

83. $(f + g)(4)$ 13 **84.** $(f - h)(-3)$ 18

85. $(f \cdot g)(1)$ −24 **86.** $(h \cdot g)(-2)$ −35

87. $\left(\frac{f}{g}\right)(2)$ $-\frac{2}{3}$ **88.** $\left(\frac{g}{h}\right)(3)$ $-\frac{8}{5}$

89. $(f \circ g)(2)$ 83 **90.** $(g \circ f)(-1)$ −8

91. $(g \circ h)\left(\frac{1}{2}\right)$ −5 **92.** $(h \circ f)\left(-\frac{1}{2}\right)$ $\frac{7}{2}$

Use the graphs of $f(x)$, $g(x)$, and $h(x)$ to find the value of each combined function.

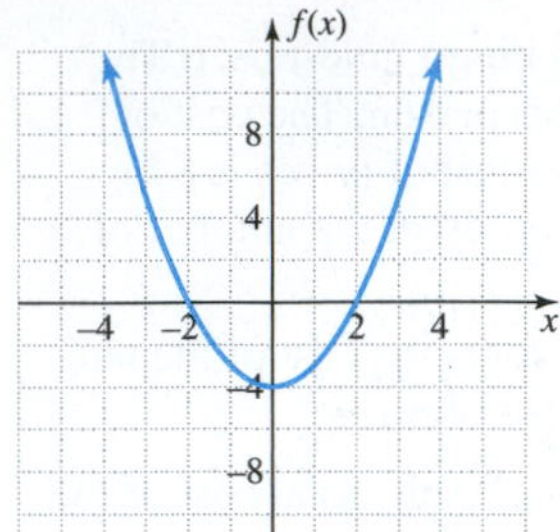

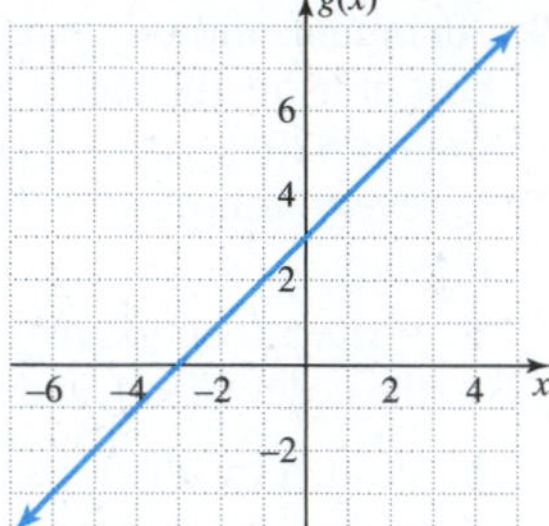

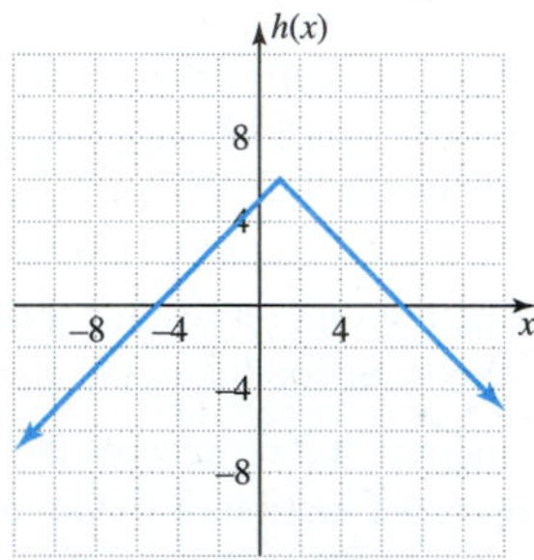

93. $(f + g)(0)$ -1 **94.** $(g + h)(-3)$ 2 **95.** $(f - h)(3)$ 1

96. $(g - h)(-5)$ -2 **97.** $(f \cdot g)(1)$ -12 **98.** $(g \cdot h)(-2)$ 3

99. $\left(\frac{f}{g}\right)(2)$ 0 **100.** $\left(\frac{f}{h}\right)(7)$ undefined **101.** $(f \circ g)(-2)$ -3

102. $(h \circ g)(-2)$ 6 **103.** $(h \circ f)(3)$ 1 **104.** $(f \circ h)(-5)$ -4

Use the table to find the value of each combined function.

x	$f(x)$	$g(x)$	$h(x)$
−3	23	−8	2
−2	16	−6	1
−1	11	−4	0
0	8	−2	1
1	5	0	2
2	8	2	3
3	11	4	4
4	16	6	5
5	23	8	6

105. $(g + h)(0)$ -1

106. $(f + g)(5)$ 31

107. $(f - g)(-1)$ 15

108. $(h - g)(1)$ 2

109. $(f \cdot g)(2)$ 16

110. $(g \cdot h)(4)$ 30

111. $\left(\frac{f}{g}\right)(3)$ $\frac{11}{4}$ **112.** $\left(\frac{g}{h}\right)(5)$ $\frac{4}{3}$ **113.** $(f \circ g)(1)$ 8

114. $(g \circ h)(-3)$ 2 **115.** $(h \circ h)(2)$ 4 **116.** $(g \circ g)(0)$ -6

Solve each problem.

117. Vicki makes $10.25 an hour, so her income can be represented by the function $f(x) = 10.25x$, where x is the number of hours she works. She pays 15% of her income in taxes. The amount of tax she owes can be represented by the function $g(x) = 0.15x$, where x is her income. Write a function that represents the total taxes Vicki owes as a function of the number of hours she works. $t(x) = (g \circ f)(x) = 1.54x$

118. Emilie makes $38 an hour, so her income can be represented the function $f(x) = 38x$, where x is the number of hours she works. She pays 30% of her income in taxes. The amount of tax she owes can be represented by the function $g(x) = 0.30x$, where x is her income. Write a function that represents the total taxes Emilie owes as a function of the number of hours she works. $t(x) = (g \circ f)(x) = 11.40x$

119. A bookstore buys math textbooks from a publishing company for $110 each and has other fixed costs of $214. So, the bookstore's total cost of the math books can be represented by the function $C(x) = 110x + 214$, where x is the number of math books purchased. The bookstore sells the math books at a 24% markup or at $136.40 each. So, the bookstore's total revenue from selling x math books can be represented by the function $R(x) = 136.40x$. Write a function that represents the profit from selling x math books. $P(x) = 26.4x - 214$

120. A bookstore buys physics textbooks from a publishing company for $90 each and has other fixed costs of $167. So, the bookstore's total cost of the physics books can be represented by the function $C(x) = 90x + 167$, where x is the number of physics books purchased. The bookstore sells the physics books at a 27% markup or at $114.30 each. So, the bookstore's total revenue from selling x physics books can be represented by the function $R(x) = 114.30x$. Write a function that represents the profit from selling x physics books. $P(x) = 24.3x - 167$

121. An organization is planning a bingo fund-raiser. The cost of using the hall is $29 per person. The cost of food is $15 per person and the cost of prizes is $48 for every four players. Let x represent the number of players.

a. Write a function that represents the cost of the hall, the cost of food, and the cost of prizes. $h(x) = 29x$, $f(x) = 15x$, $p(x) = 12x$

b. Write a function that represents the total cost of the bingo fund-raiser as a function of the number of players. $c(x) = 56x$

122. An organization is planning a golf outing. The cost of using the golf course is $45 per person. The cost of food is $26 per person and the cost of prizes is $40 for every eight players. Let x represent the number of golfers.

a. Write a function that represents the cost of the golf course, the cost of food, and the cost of prizes.

b. Write a function that represents the total cost of the golf outing as a function of the number of golfers.

You Be the Teacher!

Correct each student's errors, if any.

123. Find the difference function $(f - g)(x)$ for $f(x) = 2x^2 + 4x - 10$ and $g(x) = 3x^2 - 2x + 9$.

Melanie's work:

$$(f - g)(x) = (2x^2 + 4x - 10) - (3x^2 - 2x + 9)$$
$$= 2x^2 + 4x - 10 - 3x^2 - 2x + 9$$
$$= -x^2 + 2x - 1$$

124. Find the quotient function $\left(\dfrac{f}{g}\right)(x)$ for $f(x) = 9x - 4$ and $g(x) = x^2 - 4x$.

Don's work:

$$\left(\frac{f}{g}\right)(x) = \frac{9x - 4}{x^2 - 4x}, x \neq -2, 2$$

$$\left(\frac{f}{g}\right)(x) = \frac{9x - 4}{x^2 - 4x} = \frac{9x - 4}{x(x - 4)}, x \neq 0, 4$$

125. Find the composition function $(f \circ g)(x)$ for $f(x) = 4x - 9$ and $g(x) = 5 - 2x$.

Fred's work:

$$(f \circ g)(x) = 5 - 2(4x - 9)$$
$$= 5 - 8x + 18$$
$$= -8x + 23$$

$$(f \circ g)(x) = 4(5 - 2x) - 9 = 20 - 8x - 9 = -8x + 11$$

126. Find the composition function $(g \circ f)(x)$ for $f(x) = x^2 - 3x$ and $g(x) = 2x + 1$.

Deb's work:

$$(g \circ f)(x) = 2(x^2 - 3x)^2 + 1$$
$$= (2x^2 - 6x)^2 + 1$$
$$= 4x^4 + 36x^2 + 1$$

$$(g \circ f)(x) = 2(x^2 - 3x) + 1 = 2x^2 - 6x + 1$$

Calculate It!

Use a calculator to find each value.

127. Let $f(x) = 2x + 5$ and $g(x) = 3 - 7x$. Find $(f \circ g)(-4)$, $(f \circ g)(-2)$, and $(f \circ g)(2)$.

128. Let $f(x) = x^2 + 1$ and $g(x) = 2 - x$. Find $(g \circ f)(-2)$, $(g \circ f)(1)$, and $(g \circ f)(3)$.

Think About It!

Let $f(x) = x^2$, $g(x) = 2x + 3$, and $h(x) = \sqrt{x}$. Write each function as a composition of two of these functions.

129. $c(x) = 2x^2 + 3$ $c(x) = (g \circ f)(x)$

130. $c(x) = (2x + 3)^2$ $c(x) = (f \circ g)(x)$

131. $c(x) = \sqrt{2x + 3}$ $c(x) = (h \circ g)(x)$

132. $c(x) = 2\sqrt{x} + 3$ $c(x) = (g \circ h)(x)$

SECTION 10.2 Inverse Functions

OBJECTIVES

As a result of completing this section, you will be able to

1. Determine if a function is one-to-one.
2. Use the horizontal line test.
3. Find the inverse of a function.
4. Find the equation of the inverse of a function.
5. Evaluate inverse functions.
6. Graph a function and its inverse.
7. Verify two functions are inverses.
8. Troubleshoot common errors.

We have studied many functions throughout this course, linear, quadratic, polynomial, radical, rational, and so on. Our goal in this section is to find a function that enables us to "undo" the effects of that function; that is, to find the *inverse* of a function.

Objective 1 Determine if a function is one-to-one.

One-to-One Functions

Often, it is necessary for us to represent a formula, or a function, for "undoing" the operations that have been performed on a number. This formula or function is called an inverse function. The *inverse function* is a function that takes us back to the place where we started before the original function was applied. The following are some examples of inverses in real life.

Original "function"	Inverse "function"
Sit down	Stand up
Open a door	Close a door
Turn a light on	Turn a light off

We see from these examples that the inverse function is what must be performed to get us back to where we started. Can every function be inverted? That is, are there operations that cannot be undone? Consider the following examples.

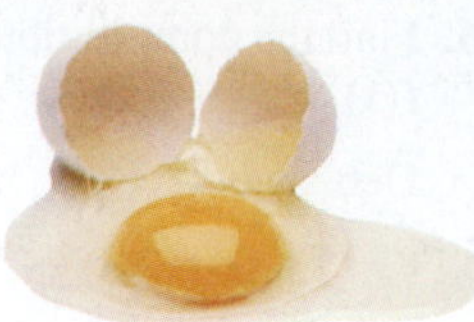

Original "function"	Inverse "function"
Crack an egg	Cannot be undone
Cut your hair	Cannot be undone
Spill a drink	Cannot be undone

These examples are operations that cannot be undone. We cannot get back to the egg before it was cracked. We cannot get back to our hairstyle after our hair has been cut. We cannot get back to the original drink after the drink has spilled.

We will use this idea to consider the inverse of a mathematical function. The following table shows examples of performing an operation on a given number, and then performing another operation that takes us back to that original number.

Original Number	Function	Inverse Function	Original Number?
3	Add 2 $3 + 2 = 5$	Subtract 2 $5 - 2 = 3$	Yes
5	Multiply by 4 $5 \cdot 4 = 20$	Divide by 4 $\frac{20}{4} = 5$	Yes
-3	Square $(-3)^2 = 9$	Square root $\sqrt{9} = 3$	No

We see from the table that squaring is a mathematical function that does *not* have an inverse function. This is because if we square a negative number, we get a positive number. When we take the square root of a positive number, the result is positive. We will never get back to the original negative number. Only special types of functions have inverse functions. These functions are called *one-to-one functions*.

Definition: A function is a **one-to-one function** if each y-value corresponds to only one x-value. If $f(a) = f(b)$ implies that $a = b$, then f is one-to-one.

Procedure: Determining if a Function Is One-to-One

Step 1: Examine the y-values in the given points.
Step 2: If any y-value corresponds to more than one x-value, then the function is *not* one-to-one.

Objective 1 Examples

Determine if each function is one-to-one.

INSTRUCTOR NOTE: Use a mapping to show how a one-to-one function has one arrow coming from each x-value and only one arrow going to each y-value.

1a. $f = \{(1, 2), (3, 4), (5, 6), (0, 1)\}$

1b. $f = \{(-2, 4), (-1, 1), (0, 0), (1, 1), (2, 4)\}$

1c.

x	y
−4	2
2	2
4	2

1d.

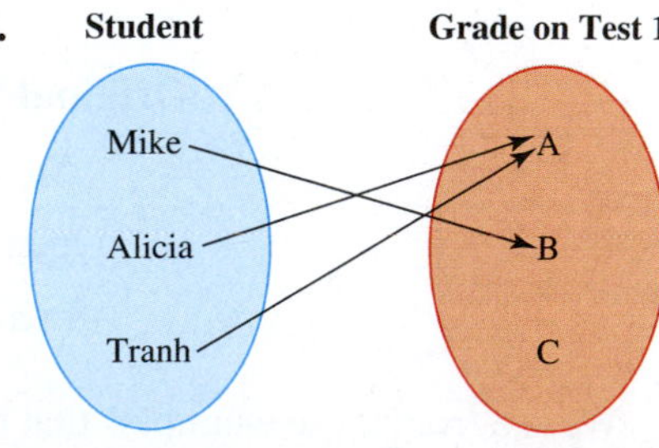

1e.

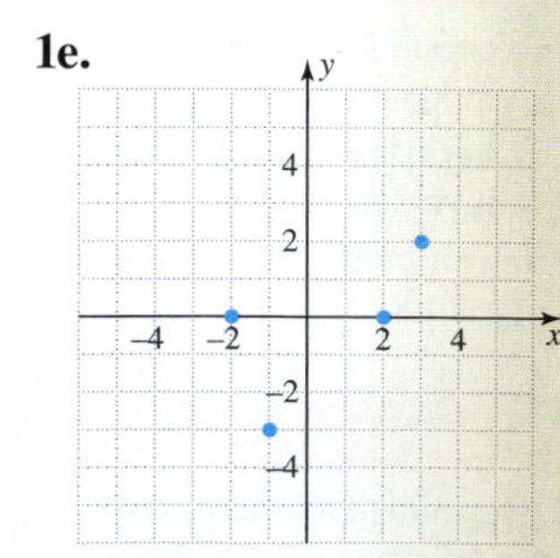

Solutions **1a.** This function is one-to-one because every value of y corresponds to only one value of x.

1b. This function is *not* one-to-one because the y-values of 1 and 4 correspond to more than one x-value.

1c. This function is *not* one-to-one because the y-value of 2 corresponds to more than one x-value.

1d. This function is *not* one-to-one since the y-value of "A" corresponds to more than one x-value, Alicia and Tranh.

1e. The points plotted on the graph are $(-1, -3)$, $(-2, 0)$, $(2, 0)$, and $(3, 2)$. Since the y-value of 0 corresponds to more than one x-value, -2 and 2, this function is *not* one-to-one.

Student Check 1 Determine if each function is one-to-one.

a. $f = \{(-3, 8), (-1, 4), (0, 2), (1, 0), (2, -2)\}$

b. $f = \{(-4, 0), (4, 0), (0, 4)\}$

c.

x	y
-1	7
0	7
3	7

d.

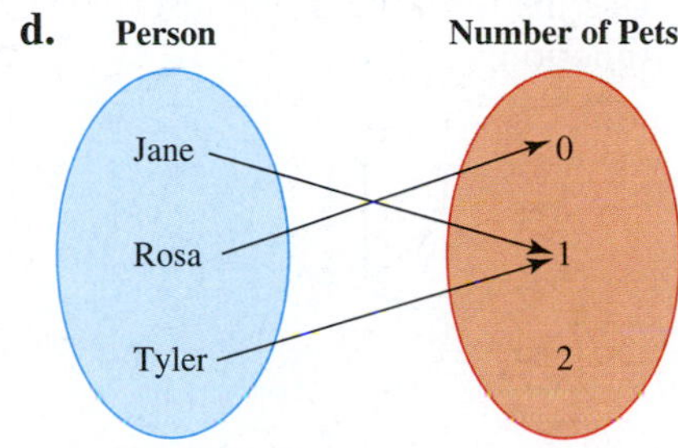

e.

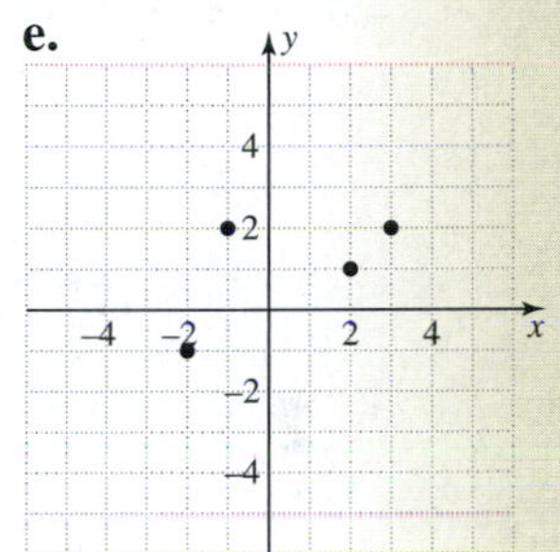

The Horizontal Line Test

Objective 2 Use the horizontal line test.

Recall that we use a vertical line test to determine if a graph represents a function. If every vertical line drawn through a graph intersects the graph at no more than one point, then the graph represents a function.

We use a *horizontal line test* to determine if a function is one-to-one. Consider the graph in Figure 10.1. If we draw a horizontal line through $y = 3$, we see that it touches two points on the graph of the function, $(-1, 3)$ and $(3, 3)$. This means that the y-value of 3 corresponds to two x-values, -1 and 3; therefore, this function is *not* one-to-one.

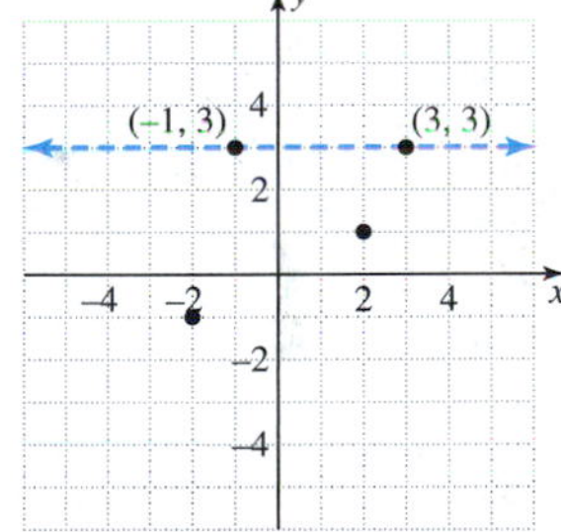

Figure 10.1

Property: Horizontal Line Test

If every horizontal line intersects the graph of a function at no more than one point, the function is one-to-one.

Objective 2 Examples **Use the horizontal line test to determine if each graph represents a one-to-one function.**

2a.

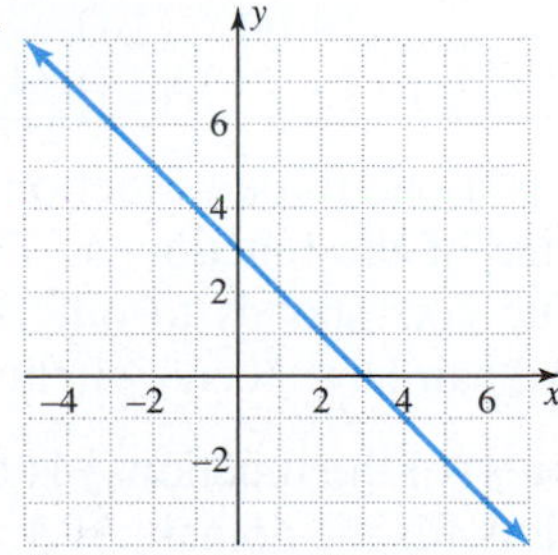

2b.

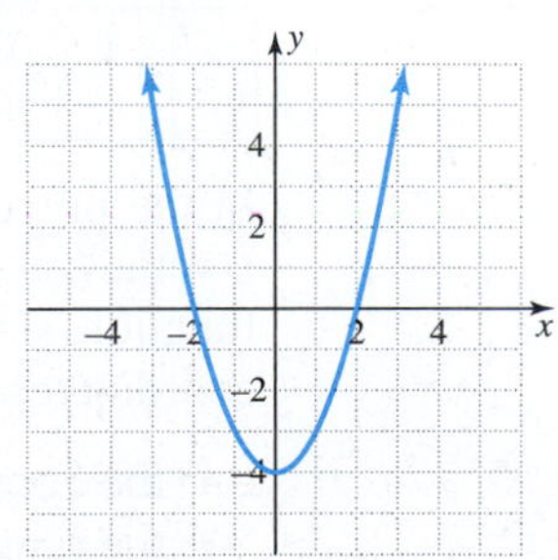

Solutions **2a.**

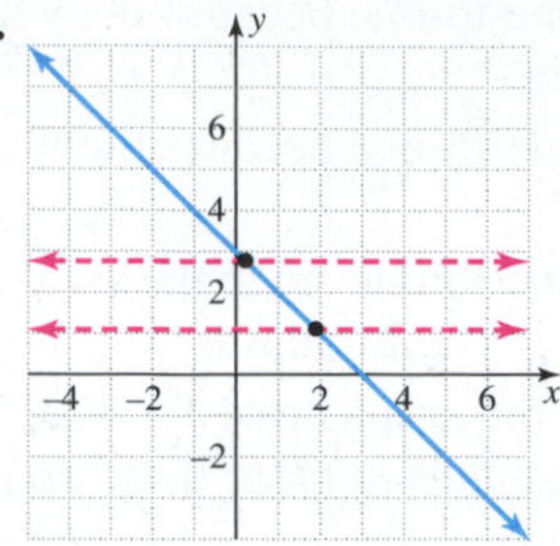

The function is one-to-one since every horizontal line intersects the graph at no more than one point.

2b.

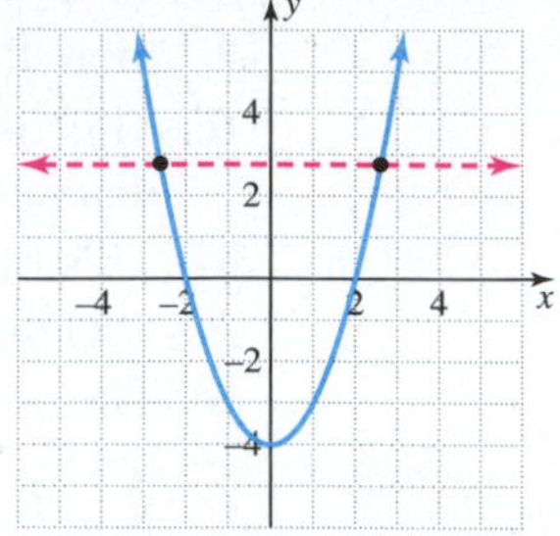

The function is *not* one-to-one since there is a horizontal line that intersects the graph at two points.

Student Check 2 Use the horizontal line test to determine if each graph represents a one-to-one function.

a.

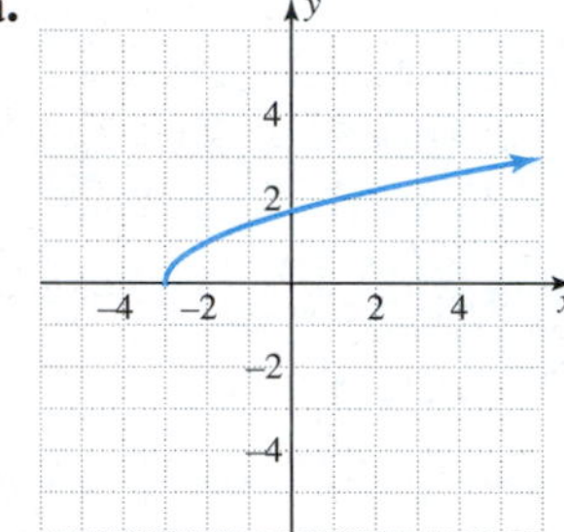

b.

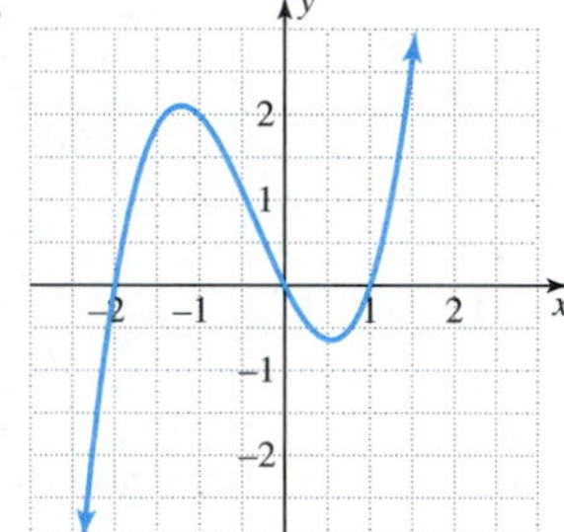

Note: *All linear equations are one-to-one functions except for vertical and horizontal lines. Recall that vertical lines are not functions since they fail the vertical line test. Horizontal lines are functions, but they are not one-to-one functions since they do not pass the horizontal line test.*

The Inverse of a Function

Objective 3 Find the inverse of a function.

INSTRUCTOR NOTE: Caution students to not confuse the notation for the inverse function with the negative exponent.

Before we defined a one-to-one function, we briefly introduced the idea of the inverse of a function. As we know, one-to-one functions pass both the vertical and horizontal line tests. One-to-one functions are also special because for each one-to-one function, we can find its *inverse function*.

To find the inverse of a function, we interchange the x- and y-values. In other words, if a function $f(x)$ takes a value x to a value y, then the inverse of the function $f(x)$, denoted $f^{-1}(x)$ (read "f inverse of x"), takes the value y back to the value x. Consider the two following functions denoted by A and B.

	Function	1-1	Inverse—interchange x and y	Is the inverse a function?
A	$\{(1, 3), (2, 5), (3, 7), (4, 9)\}$	Yes	$\{(3, 1), (5, 2), (7, 3), (9, 4)\}$	Yes
B	$\{(-2, 4), (-1, 1), (0, 0), (1, 1), (2, 4)\}$	No	$\{(4, -2), (1, -1), (0, 0), (1, 1), (4, 2)$	No, the x-value of 4 corresponds to 2 y-values.

In the preceding table, A is a one-to-one function. When we interchange the x- and y-values, we get the inverse of the function A. The inverse of A, denoted A^{-1}, is a function since each x-value corresponds to only one y-value. Notice that since the coordinates of each ordered pair have been switched,

- the domain (set of inputs) of the function A is the range (set of outputs) of A^{-1}.
- the range of the function A is the domain of A^{-1}.

We also note that the function B is not one-to-one. When we interchange the x- and y-values, we get an inverse relation but *not* an inverse function. This is because $x = 4$ corresponds to the y-values of -1 and 1.

Definition: If a function f is one-to-one, then the inverse of f, denoted f^{-1} exists. The **inverse function** f^{-1} consists of all ordered pairs (y, x) such that (x, y) belongs to the function f. The domain of f is the range of f^{-1}. The range of f is the domain of f^{-1}.

Procedure: Finding the Inverse of a Function

Step 1: Determine if the function is one-to-one. If the function is not one-to-one, then it doesn't have an inverse function.

Step 2: Interchange the x- and y-values of the ordered pairs.

Objective 3 Examples

Determine if each function is one-to-one and if so, state its inverse.

3a. $f = \{(0, 1), (1, 2), (2, 3), (3, 4)\}$

3b.

State (x)	Governor in 2011 (y)
Alabama	Robert Bentley
Florida	Rick Scott
Georgia	Nathan Deal
North Carolina	Beverly Perdue
South Carolina	Nikki Haley
Tennessee	Bill Haslam

(Source: http://www.multistate.com/site.nsf/G_L2011?OpenPage)

3c.

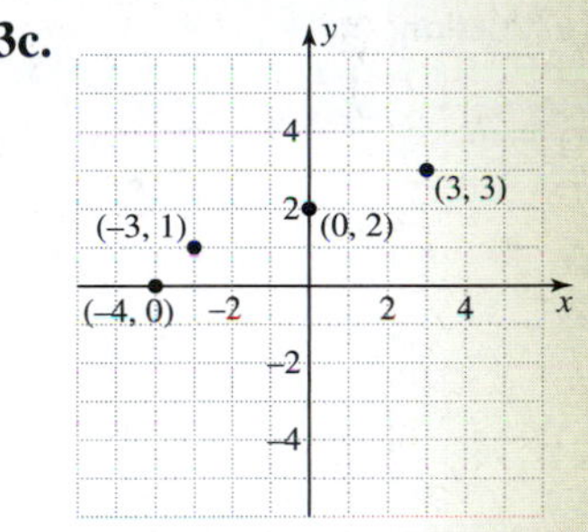

Solutions

3a. The function f is one-to-one since each y-value corresponds to only one x-value. So, we interchange the x- and y-values to find the inverse function.

$$f^{-1} = \{(1, 0), (2, 1), (3, 2), (4, 3)\}$$

3b. The table represents a one-to-one function since each y-value corresponds to only one x-value. The inverse function interchanges the x- and y-values. So, the table is the inverse function.

Governor in 2011 (x)	State (y)
Robert Bentley	Alabama
Rick Scott	Florida
Nathan Deal	Georgia
Beverly Perdue	North Carolina
Nikki Haley	South Carolina
Bill Haslam	Tennessee

3c. The function is one-to-one since it passes the horizontal line test. We find the inverse function by graphing the ordered pairs that result from interchanging the x- and y-coordinates. So, the graph is the inverse function.

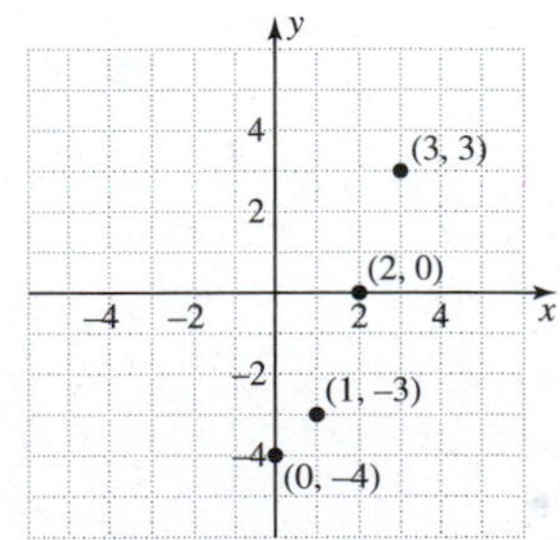

Student Check 3 Determine if each function is one-to-one and if so, state its inverse.

a. $f = \{(-3, -27), (-2, -8), (-1, -1), (0, 0), (1, 1), (2, 8), (3, 27)\}$

b.

Year Term Began (input)	President of the United States (output)
2009	Barack Hussein Obama
2001	George Walker Bush
1993	William Jefferson Clinton
1989	George Herbert Walker Bush
1981	Ronald Wilson Reagan
1977	James Earl Carter, Jr.
1974	Gerald Rudolph Ford
1969	Richard Milhous Nixon

c.

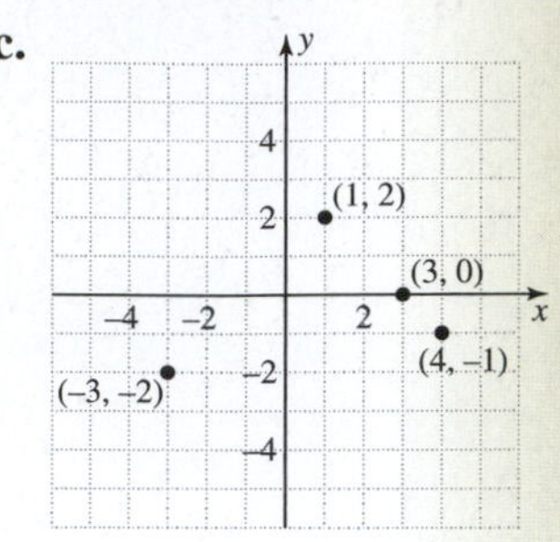

Finding the Equation of the Inverse of a Function

Objective 4 ▶

Find the equation of the inverse of a function.

We can apply the concept from the last objective to find the equation of the inverse of a function. The functions we work with in this objective are one-to-one functions, so that we know the inverse exists.

Procedure: Finding the Equation of the Inverse of a One-to-One Function

Step 1: Replace $f(x)$ with y.
Step 2: Interchange x and y.
Step 3: Solve the equation for y.
Step 4: Replace y with the notation $f^{-1}(x)$.

Objective 4 Examples Find the equation of the inverse of each function.

4a. $f(x) = 2x - 6$ **4b.** $f(x) = (x + 1)^3$ **4c.** $f(x) = \dfrac{4}{x - 3}$

Solutions **4a.**

$f(x) = 2x - 6$

$y = 2x - 6$ Replace $f(x)$ with y.

$x = 2y - 6$ Interchange x and y.

$x + 6 = 2y$ Add 6 to each side.

$\dfrac{x + 6}{2} = y$ Divide each side by 2.

$f^{-1}(x) = \dfrac{x + 6}{2}$ Replace y with $f^{-1}(x)$.

$f^{-1}(x) = \dfrac{1}{2}x + 3$ Simplify.

4b.

$f(x) = (x + 1)^3$

$y = (x + 1)^3$ Replace $f(x)$ with y.

$x = (y + 1)^3$ Interchange x and y.

$\sqrt[3]{x} = y + 1$ Take the cube root of each side.

$\sqrt[3]{x} - 1 = y$ Subtract 1 from each side.

$f^{-1}(x) = \sqrt[3]{x} - 1$ Replace y with $f^{-1}(x)$.

4c.

$$f(x) = \frac{4}{x-3}$$

$$y = \frac{4}{x-3}$$ Replace $f(x)$ with y.

$$x = \frac{4}{y-3}$$ Interchange x and y.

$$x(y-3) = 4$$ Multiply each side by $(y - 3)$.

$$xy - 3x = 4$$ Apply the distributive property.

$$xy = 4 + 3x$$ Add $3x$ to each side.

$$y = \frac{4+3x}{x}$$ Divide each side by x.

$$f^{-1}(x) = \frac{4+3x}{x}$$ Replace y with $f^{-1}(x)$.

Student Check 4 Find the equation of the inverse of each function.

a. $f(x) = 4x - 8$ **b.** $f(x) = (x-2)^3$ **c.** $f(x) = \dfrac{1}{x+6}$

Evaluating Inverse Functions

Objective 5

Evaluate inverse functions.

We will now learn how to use function notation with inverse functions. Recall that if the point (x, y) is on the graph of f, then we write $f(x) = y$. For instance, if the point $(2, -3)$ is on the graph of f, then $f(2) = -3$.

Knowing one point on the graph of f provides us with information about a point on the graph of f^{-1}. Recall that if the point (x, y) is on the graph of f, then the point (y, x) is on the graph of f^{-1}. Therefore, $f^{-1}(y) = x$. So, knowing that $(2, -3)$ is on the graph of f tells us that $(-3, 2)$ is on the graph of f^{-1}. Therefore, $f^{-1}(-3) = 2$.

Objective 5 Examples **Evaluate each function using the given information.**

5a. If $f(6) = 10$, what is $f^{-1}(10)$? **5b.** If $f^{-1}(3) = -4$, what is $f(-4)$?

Solutions

5a. $f(6) = 10$ means that $(6, 10)$ is on the graph of f. Therefore, $(10, 6)$ is on the graph of f^{-1}. So, $f^{-1}(10) = 6$.

5b. $f^{-1}(3) = -4$ means that $(3, -4)$ is on the graph of f^{-1}. Therefore, $(-4, 3)$ is on the graph of f. So, $f(-4) = 3$.

Student Check 5 Evaluate each function using the given information.

a. If $f(-1) = 5$, what is $f^{-1}(5)$? **b.** If $f^{-1}\left(\frac{1}{2}\right) = 7$, what is $f(7)$?

The Graph of a Function and the Graph of Its Inverse

Objective 6

Graph a function and its inverse.

The graph of a function and the graph of its inverse relate to each other in a special way. Consider the graphs of the functions from Example 4a, $f(x) = 2x - 6$ and $f^{-1}(x) = \frac{1}{2}x + 3$. The points on the graph of the function and the graph of its inverse are mirror images of each other across the line $y = x$. We say that the graph of $f(x)$ and the graph of $f^{-1}(x)$ are *symmetric* about the line $y = x$.

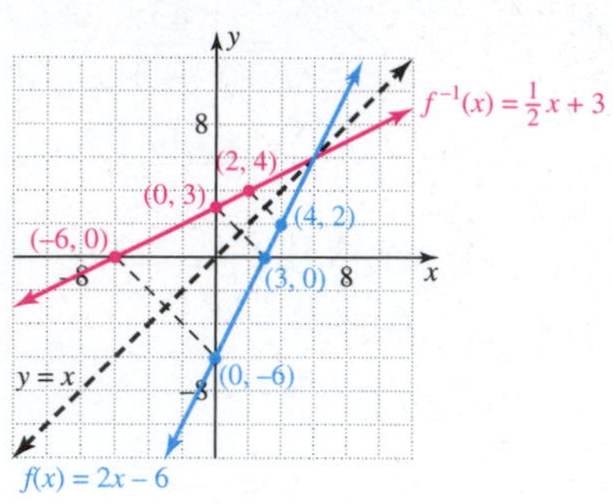

Procedure: Graphing the Inverse of a One-to-One Function

Step 1: Draw the line of symmetry, $y = x$.

Step 2: Interchange the coordinates of ordered pairs on the graph of the function f.

Step 3: Use the line of symmetry to assist in drawing the graph of the inverse function.

Objective 6 Examples

Graph the inverse of each function.

6a.

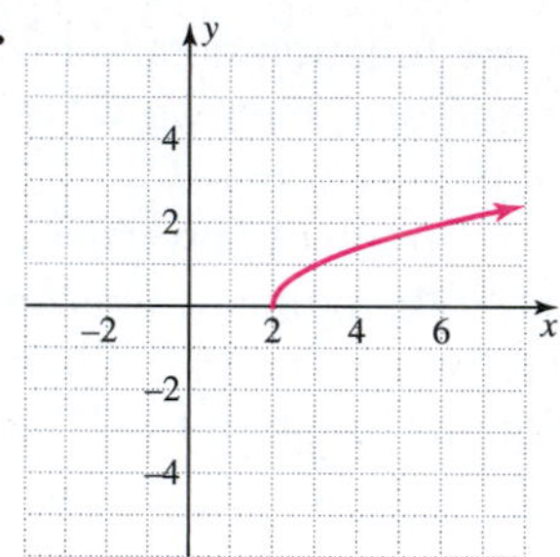

6b.

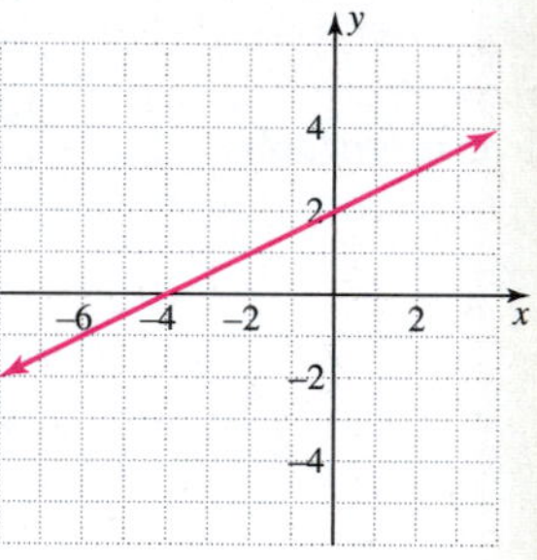

Solutions

6a. The inverse is shown in blue.

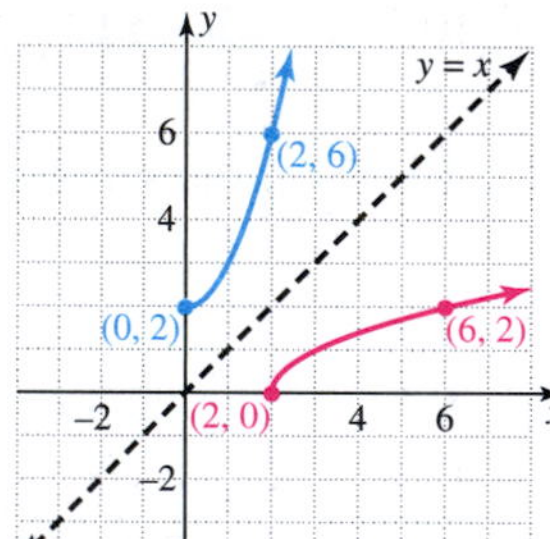

6b. The inverse is shown in blue.

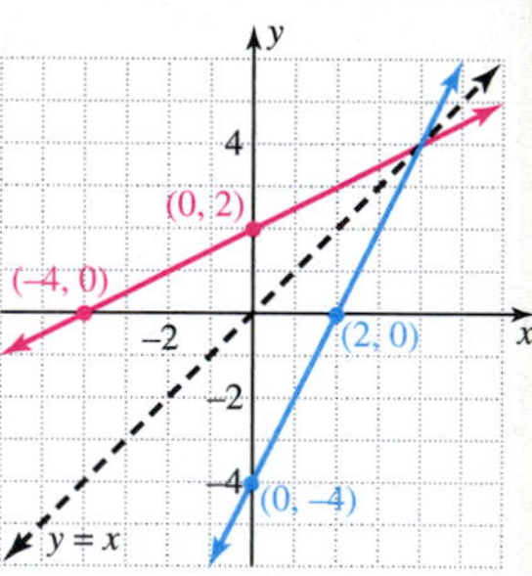

Student Check 6

Graph the inverse of each function.

a.

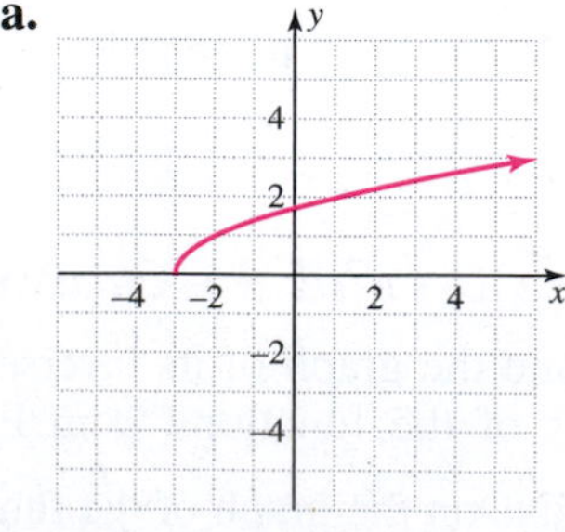

b.

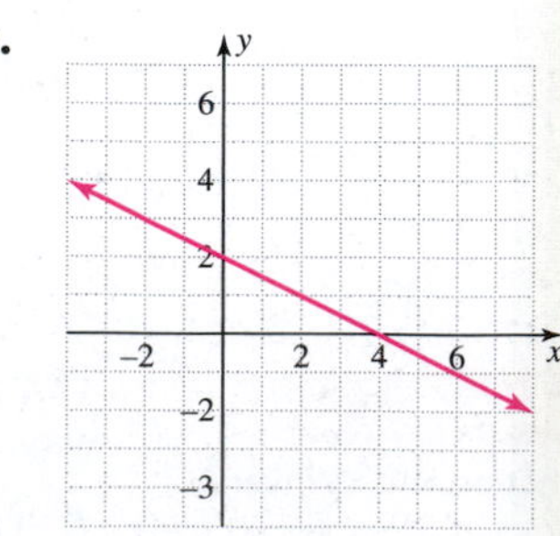

Verifying Functions are Inverses

Objective 7 ▶

Verify two functions are inverses.

Two functions are inverses of one another if, when we apply a function and then apply its inverse, we end up with the original input value. For instance, we know that if $f(x) = 2x - 6$, then its inverse function is $f^{-1}(x) = \frac{1}{2}x + 3$. We evaluate the function f at 0 and then apply the inverse function to that output value.

$$f(x) = 2x - 6$$
$$f(0) = 2(0) - 6$$
$$f(0) = -6$$

Now, we evaluate the inverse function, f^{-1}, at $x = -6$.

$$f^{-1}(x) = \frac{1}{2}x + 3$$
$$f^{-1}(-6) = \frac{1}{2}(-6) + 3$$
$$f^{-1}(-6) = -3 + 3$$
$$f^{-1}(-6) = 0$$

So, the function f takes the value 0 to -6 and the function f^{-1} takes the value -6 back to 0. When we compose a function with its inverse, we always get back to the starting value.

Property: The Composition of Inverse Functions

The inverse of a one-to-one function f is the function f^{-1} such that

$$(f \circ f^{-1})(x) = x \text{ and } (f^{-1} \circ f)(x) = x$$

Procedure: Verifying Two One-to-One Functions, f and f^{-1}, Are Inverses

Step 1: Find $(f \circ f^{-1})(x)$ and $(f^{-1} \circ f)(x)$.

Step 2: If both of the compositions produce x, then the functions f and f^{-1} are inverses.

Objective 7 Examples Verify that the functions are inverses of each other.

7a. $f(x) = 4x + 12$ and $f^{-1}(x) = \frac{1}{4}x - 3$ **7b.** $f(x) = x^3 - 1$ and $f^{-1}(x) = \sqrt[3]{x + 1}$

Solutions

7a.

$$\begin{aligned}(f \circ f^{-1})(x) &= f\left(f^{-1}(x)\right)\\ &= f\left(\frac{1}{4}x - 3\right)\\ &= 4\left(\frac{1}{4}x - 3\right) + 12\\ &= x - 12 + 12\\ &= x\end{aligned}$$

$$\begin{aligned}(f^{-1} \circ f)(x) &= f^{-1}\left(f(x)\right)\\ &= f^{-1}(4x + 12)\\ &= \frac{1}{4}(4x + 12) - 3\\ &= x + 3 - 3\\ &= x\end{aligned}$$

Since both compositions produce x, the two functions are inverses.

7b.

$$\begin{aligned}(f \circ f^{-1})(x) &= f\left(f^{-1}(x)\right)\\ &= f\left(\sqrt[3]{x+1}\right)\\ &= \left(\sqrt[3]{x+1}\right)^3 - 1\\ &= x + 1 - 1\\ &= x\end{aligned}$$

$$\begin{aligned}(f^{-1} \circ f)(x) &= f^{-1}\left(f(x)\right)\\ &= f^{-1}(x^3 - 1)\\ &= \sqrt[3]{x^3 - 1 + 1}\\ &= \sqrt[3]{x^3}\\ &= x\end{aligned}$$

Since both compositions produce x, the two functions are inverses.

Student Check 7 Verify that the functions are inverses of each other.

a. $f(x) = 3x - 6$ and $f^{-1}(x) = \frac{1}{3}x + 2$ **b.** $f(x) = (x + 5)^3$ and $f^{-1}(x) = \sqrt[3]{x} - 5$

Objective 8 ▶ Troubleshoot common errors.

Troubleshooting Common Errors

Some common errors associated with inverse functions are shown.

Objective 8 Examples A problem and an incorrect solution are given. Provide the correct solution and an explanation of the error.

8a. Determine if the function $f(x) = x^2 + 4$ is one-to-one and if so, find the equation of its inverse.

Incorrect Solution	Correct Solution and Explanation
$f(x) = x^2 + 4$ $y = x^2 + 4$ $x = y^2 + 4$ $x - 4 = y^2$ $\sqrt{x - 4} = y$	The function $f(x) = x^2 + 4$ is a parabola and is not one-to-one. Therefore, it does not have an inverse function.

8b. Find $f^{-1}(x)$ if $f(x) = 4x - 5$.

Incorrect Solution	Correct Solution and Explanation
$f^{-1}(x) = \frac{1}{4x - 5}$	The notation $f^{-1}(x)$ represents the inverse function. It is not the same as a negative exponent. So, to find $f^{-1}(x)$, $y = 4x - 5$ $x = 4y - 5$ $x + 5 = 4y$ $\frac{x + 5}{4} = y$ So, $f^{-1}(x) = \frac{1}{4}x + \frac{5}{4}$.

8c. Show that if $f(x) = \frac{1 + x}{x}$, then $\frac{1}{x - 1}$ is its inverse.

Incorrect Solution	Correct Solution and Explanation
$(f \circ f^{-1})(x) = f(f^{-1}(x))$ $= f\left(\frac{1}{x - 1}\right)$ $= 1 + \frac{1}{x - 1}$ $= \frac{x - 1 + 1}{x - 1}$ $= \frac{x}{x - 1}$ Since $f \circ f^{-1} \neq x$, the functions are not inverses.	The expression $\frac{1}{x - 1}$ should replace both occurrences of x in the function f. $f\left(\frac{1}{x - 1}\right) = \frac{1 + \frac{1}{x - 1}}{\frac{1}{x - 1}}$ $= \frac{\left(1 + \frac{1}{x - 1}\right)(x - 1)}{\left(\frac{1}{x - 1}\right)(x - 1)}$ $= \frac{x - 1 + 1}{1}$ $= x$ So, $f \circ f^{-1} = x$. It can be shown that $f^{-1} \circ f$ is also x.

ANSWERS TO STUDENT CHECKS

Student Check 1 **a.** yes **b.** no **c.** no **d.** no **e.** no

Student Check 2 **a.** one-to-one **b.** not one-to-one

Student Check 3 **a.** one-to-one, $\{(-27, -3), (-8, -2), (-1, -1), (0, 0), (1, 1), (8, 2), (27, 3)\}$

b. one-to-one,

President of the United States (input)	Year Term Began (output)
Barack Hussein Obama	2009
George Walker Bush	2001
William Jefferson Clinton	1993
George Herbert Walker Bush	1989
Ronald Wilson Reagan	1981
James Earl Carter, Jr.	1977
Gerald Rudolph Ford	1974
Richard Milhous Nixon	1969

c. one-to-one,

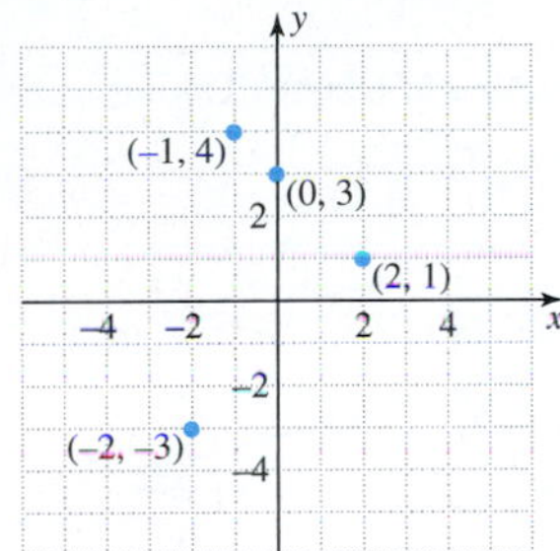

Student Check 4 **a.** $f^{-1}(x) = \frac{1}{4}x + 2$ **b.** $f^{-1}(x) = \sqrt[3]{x} + 2$ **c.** $f^{-1}(x) = \frac{1 - 6x}{x}$

Student Check 5 **a.** -1 **b.** $\frac{1}{2}$

Student Check 6 **a.**

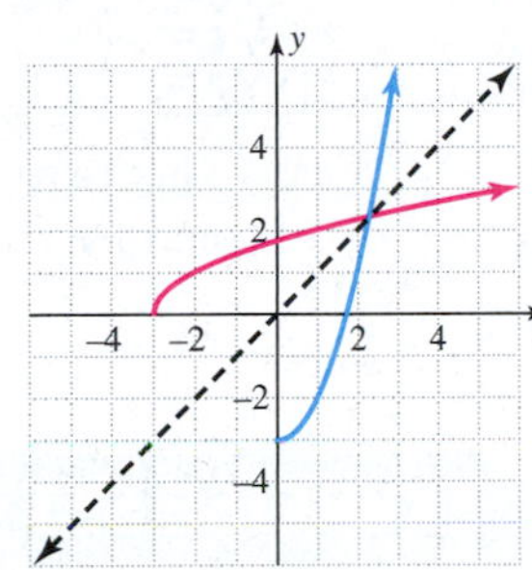

b. 

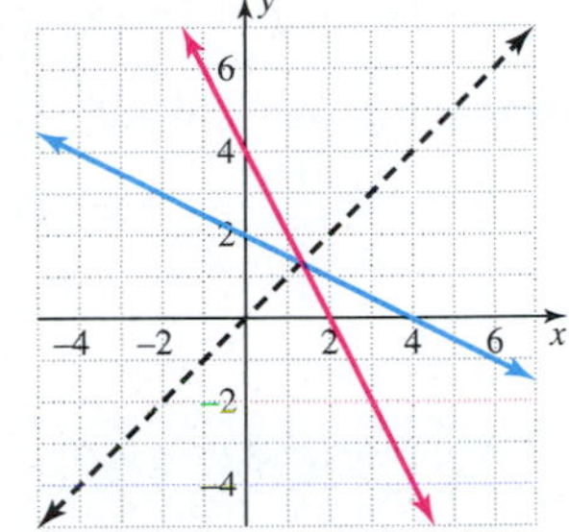

Student Check 7 **a.** f and f^{-1} are inverses. **b.** f and f^{-1} are inverses.

SUMMARY OF KEY CONCEPTS

1. A function is one-to-one if every y-value corresponds to only one x-value.
2. The horizontal line test is used to determine if the graph of a function is one-to-one. If every horizontal line crosses the graph at no more than one point, then the graph is one-to-one.
3. To find the inverse of a one-to-one function, interchange the ordered pairs of the function f.
4. To find the equation of the inverse of a one-to-one function, replace $f(x)$ with y and interchange the x- and y-variables. Then solve the new equation for y. This result is $f^{-1}(x)$.
5. If (x, y) is on the graph of a one-to-one function f, then $f^{-1}(y) = x$ since the point (y, x) is on the graph of f^{-1}.
6. The graphs of a one-to-one function and its inverse are symmetric about the line $y = x$.
7. We can verify two functions f and f^{-1} are inverses by showing that $(f \circ f^{-1})(x) = x$ and $(f^{-1} \circ f)(x) = x$.

GRAPHING CALCULATOR SKILLS

A graphing calculator can draw the graph of an inverse function. It can also assist us in verifying that two functions are inverses.

Example 1: Draw the inverse of the function $f(x) = 2x - 6$.

Solution: Input the function into the equation editor and graph the function.

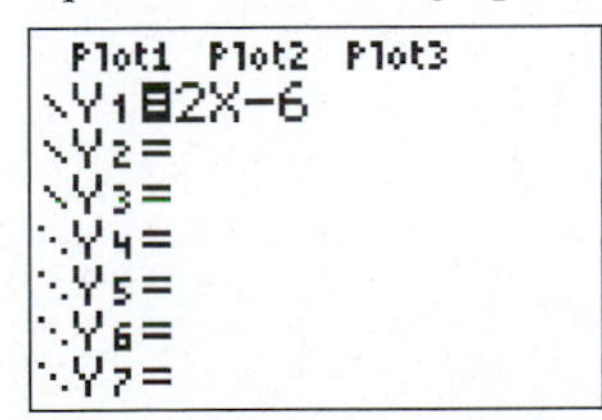

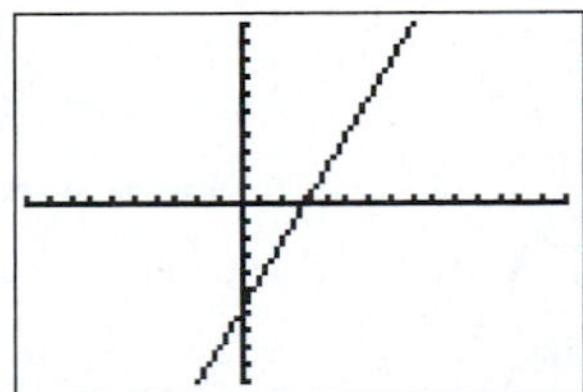

To draw the inverse of the function, press 2nd PRGM 8. Then enter the function for which you want the inverse drawn.

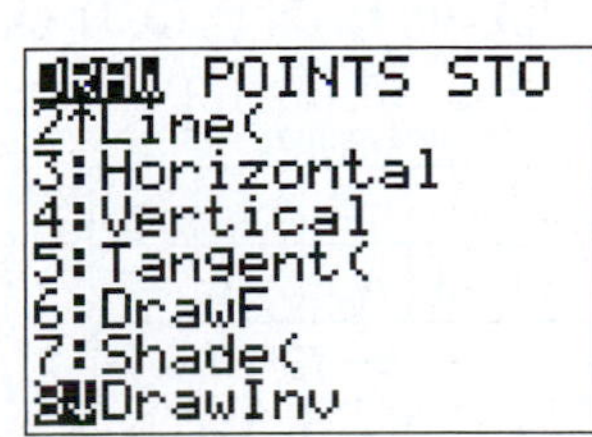

Press 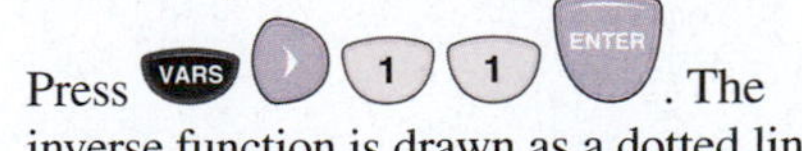 . The inverse function is drawn as a dotted line.

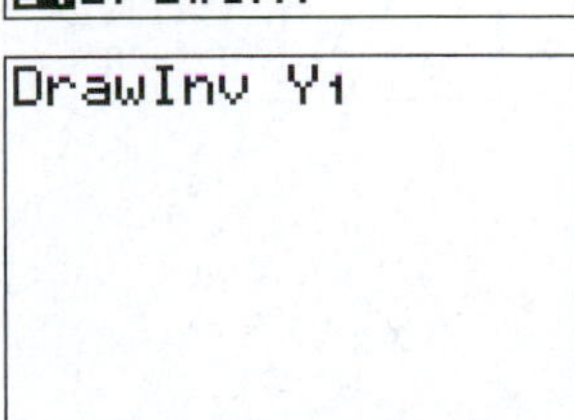

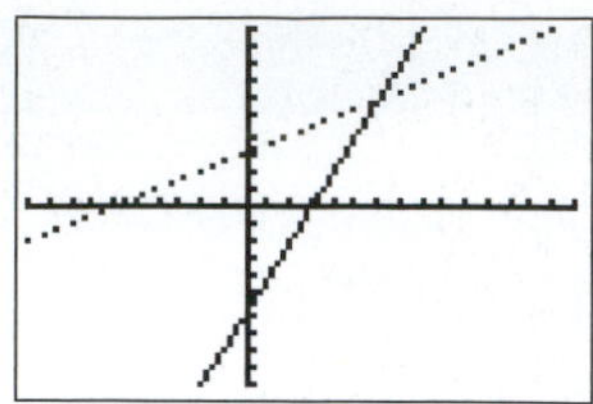

Note that the graphs might not look symmetric about the line $y = x$ unless we use the square viewing window.

Example 2: Verify that $f(x) = 2x - 6$ and $f^{-1}(x) = \frac{1}{2}x + 3$ are inverses.

Solution: Enter the two functions into Y_1 and Y_2. In Y_3 and Y_4, enter the expressions for the compositions of the two functions.

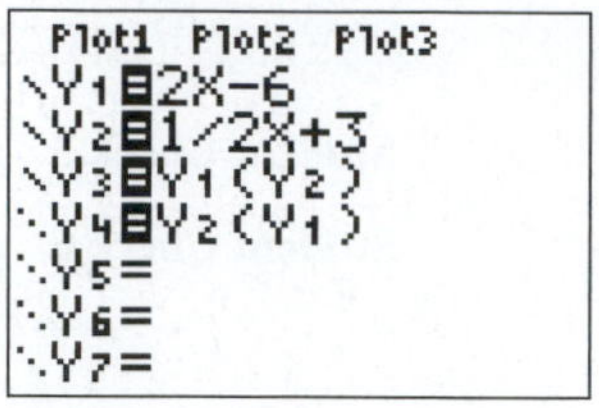

```
Plot1 Plot2 Plot3
\Y1=2X-6
\Y2=1/2X+3
\Y3=Y1(Y2)
\Y4=Y2(Y1)
\Y5=
\Y6=
\Y7=
```

View the TABLE and examine the columns for Y_3 and Y_4.

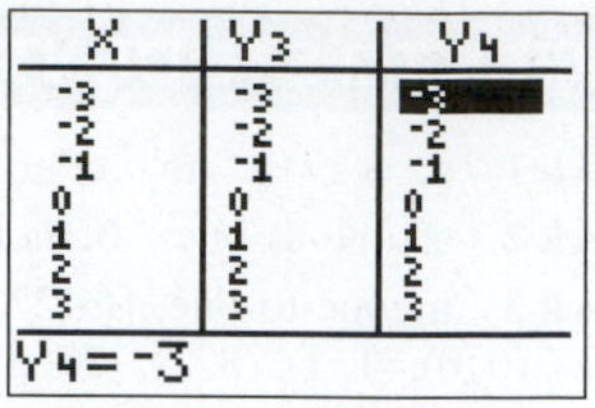

X	Y3	Y4
-3	-3	-3
-2	-2	-2
-1	-1	-1
0	0	0
1	1	1
2	2	2
3	3	3

Y4=-3

As we can see from the TABLE, the values in Y_3 and Y_4 equal the corresponding x-value; therefore, the two functions are inverses.

SECTION 10.2 EXERCISE SET

Write About It!

Use complete sentences in your answer to each exercise.

1. Define a one-to-one function. Answers vary. A function is one-to-one if each y-value corresponds to only one x-value.
2. Explain how to determine if a function is one-to-one.
3. Define the inverse of a one-to-one function f.
4. Explain how to find the equation of the inverse of a one-to-one function.
5. Explain how to graph the inverse of a one-to-one function from the graph of the function.
6. Explain how to determine if two functions are inverses of each other.

Practice Makes Perfect!

Determine if each function is one-to-one. (*See Objective 1.*)

7. $f = \{(1, 4), (2, 7), (3, 10), (4, 13), (5, 16)\}$ yes, a one-to-one function
8. $f = \{(-3, 9), (-2, 7), (-1, 5), (0, 3), (1, 1)\}$ yes, a one-to-one function
9. $f = \{(-2, 7), (-1, 4), (0, 3), (1, 4), (2, 7)\}$ not a one-to-one function
10. $f = \{(-1, 0), (0, 3), (1, 4), (2, 3), (3, 0)\}$ not a one-to-one function
11. $f = \{(-2, 1), (0, 1), (2, 1), (3, 1), (5, 1)\}$ not a one-to-one function
12. $f = \{(-2, -8), (0, 0), (1, 1), (2, 8), (3, 27)\}$ yes, a one-to-one function

13.

x	-2	0	2	4	6
y	-3	-3	-3	-3	-3

not a one-to-one function

14.

x	-1	0	1	2	3
y	-2	-1	2	9	28

yes, a one-to-one function

15. **Number of Classes Taking** not a one-to-one function

Blake → 2; Sue → 3; James → 3 (range values: 1, 2, 3)

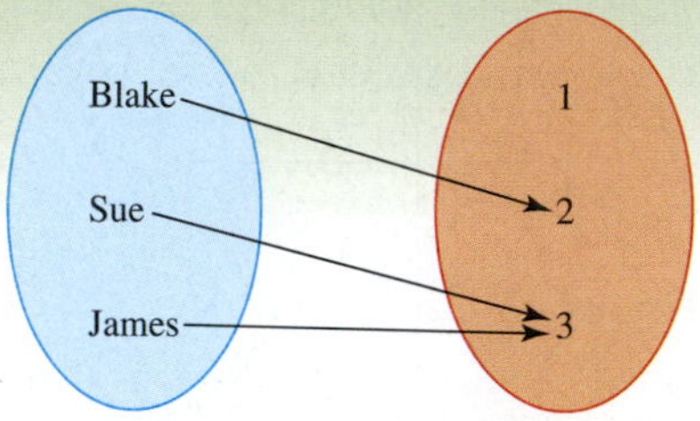

16. **Number of TV Sets** yes, a one-to-one function

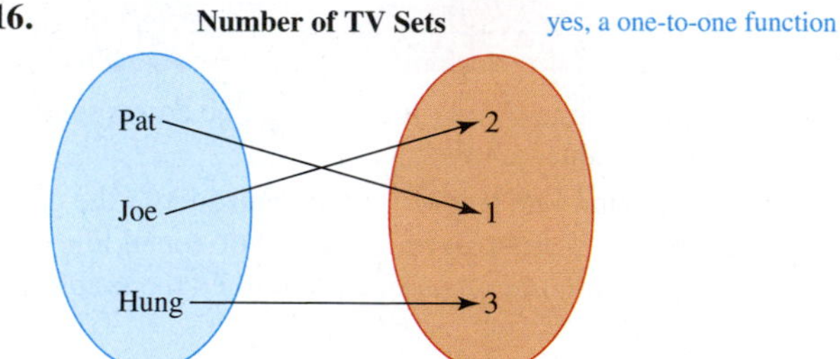

17. not a one-to-one function

18. not a one-to-one function

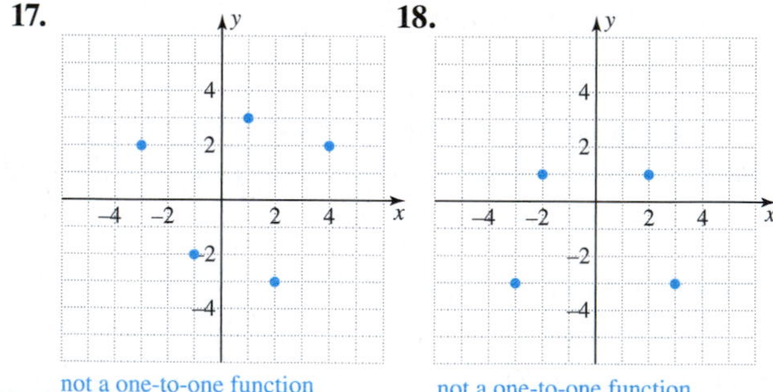

Additional answers can be found in the Instructor Answer Appendix.

Use the horizontal line test to determine if each graph is a one-to-one function. (*See Objective 2.*)

19.

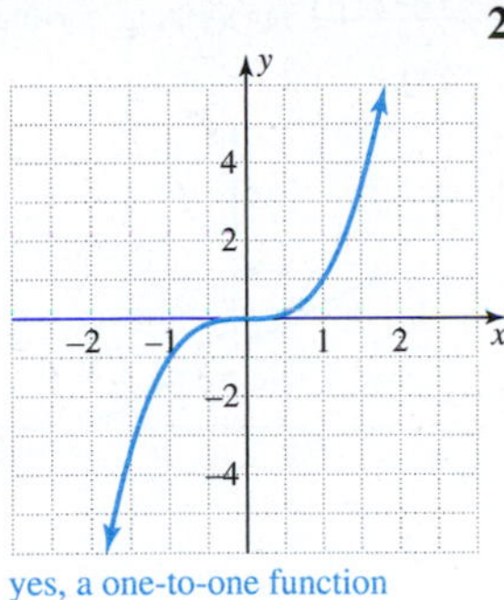

yes, a one-to-one function

20.

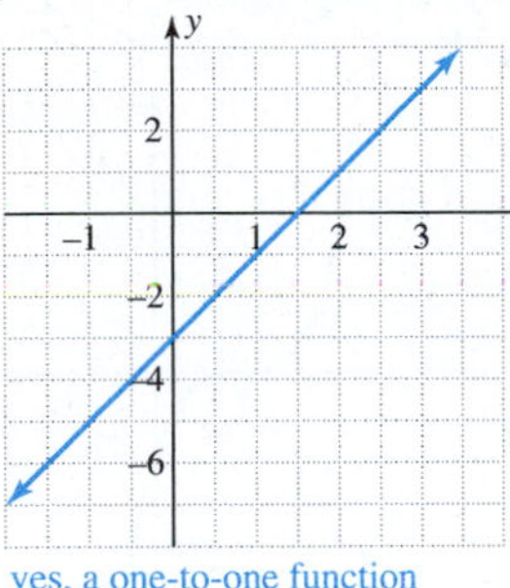

yes, a one-to-one function

21.

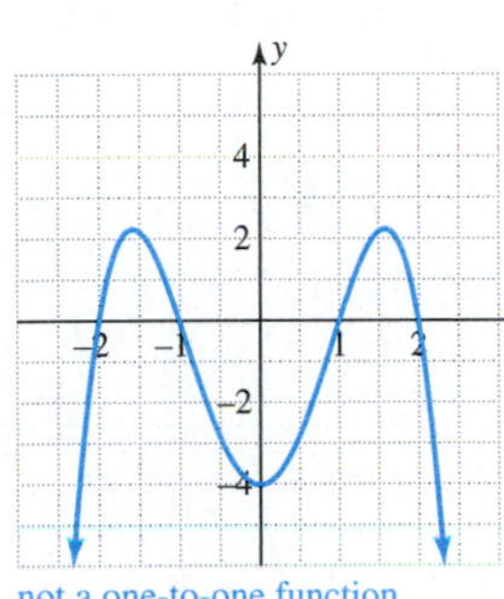

not a one-to-one function

22.

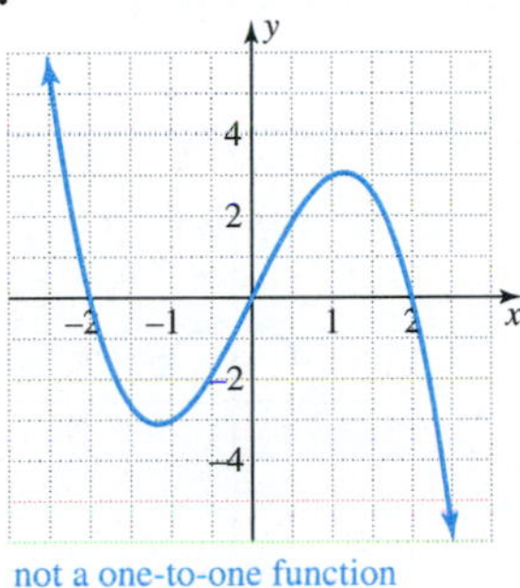

not a one-to-one function

23.

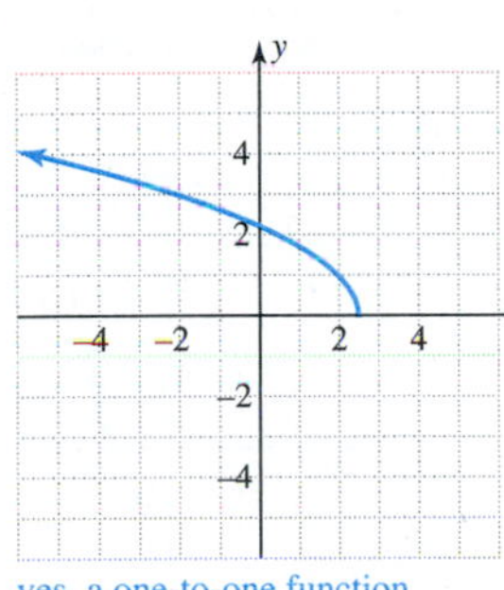

yes, a one-to-one function

24.

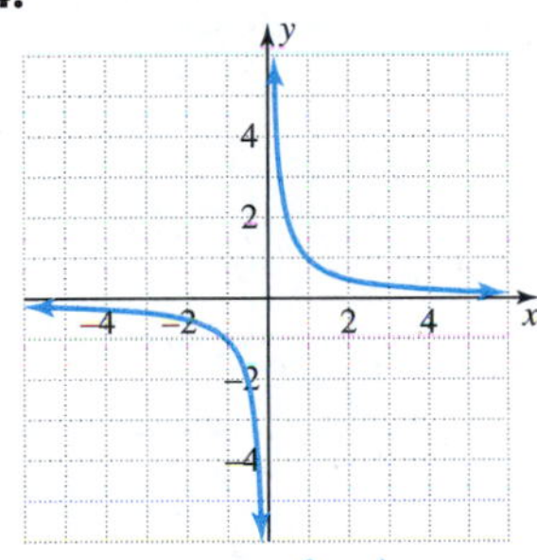

yes, a one-to-one function

Determine if each function is one-to-one and if so, state its inverse. (*See Objective 3.*)

25. $f(x) = \{(-2, 3), (0, 4), (1, 6), (2, -1), (3, 1)\}$
one-to-one; $f^{-1}(x) = \{(3, -2), (4, 0), (6, 1), (-1, 2), (1, 3)\}$

26. $f(x) = \{(-3, 1), (1, 2), (3, 3), (4, 5), (5, 7)\}$
one-to-one; $f^{-1}(x) = \{(1, -3), (2, 1), (3, 3), (5, 4), (7, 5)\}$

27. $f(x) = \{(-5, 4), (1, 3), (4, 2), (6, 3), (7, 1)\}$
not a one-to-one function

28. $f(x) = \{(-1, 5), (0, 9), (2, 6), (5, -4), (7, 9)\}$
not a one-to-one function

29.

City (x)	Mayor in 2011 (y)
New York City	Michael Bloomberg
Los Angeles	Antonio Villaraigosa
Chicago	Rahm Emanuel
Houston	Annise Parker
Phoenix	Phil Gordon

(Source: http://www.citymayors.com)

30.

City(x)	Number of States with the same city name (y)
Riverside	46
Fairview	43
Franklin	42
Midway	40
Fairfield	39
Pleasant Valley	39

(Source: http://geography.about.com/od/lists/a/placename50.htm)
not a one-to-one function

31. **32.**

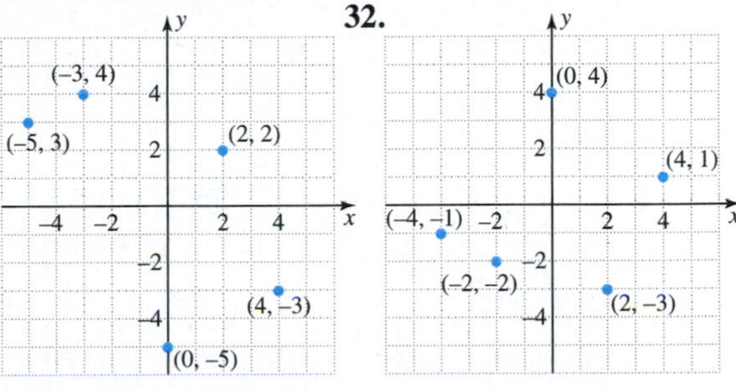

Find the equation of the inverse of each function. (*See Objective 4.*)

33. $f(x) = x - 8$ $f^{-1}(x) = x + 8$

34. $f(x) = x + 5$ $f^{-1}(x) = x - 5$

35. $f(x) = 11x - 2$

36. $f(x) = 5x - 9$

37. $f(x) = \dfrac{3x - 5}{2}$

38. $f(x) = \dfrac{3x + 7}{6}$

39. $f(x) = (2x + 9)^3$

40. $f(x) = (4x - 7)^3$

41. $f(x) = 8x^3 + 1$

42. $f(x) = 27x^3 + 5$

43. $f(x) = \dfrac{15}{x + 3}$

44. $f(x) = \dfrac{10}{x - 1}$

Evaluate each function using the given information. (*See Objective 5.*)

45. If $f(-10) = 6$, what is $f^{-1}(6)$? -10

46. If $f(4) = 9$, what is $f^{-1}(9)$? 4

47. If $f\left(\dfrac{3}{2}\right) = \dfrac{1}{5}$, what is $f^{-1}\left(\dfrac{1}{5}\right)$? $\dfrac{3}{2}$

48. If $f\left(-\dfrac{7}{4}\right) = \dfrac{1}{6}$, what is $f^{-1}\left(\dfrac{1}{6}\right)$? $-\dfrac{7}{4}$

49. If $f^{-1}(3) = -5$, what is $f(-5)$? 3

50. If $f^{-1}(-4) = 12$, what is $f(12)$? -4

51. If $f^{-1}\left(-\dfrac{1}{6}\right) = \dfrac{5}{4}$, what is $f\left(\dfrac{5}{4}\right)$? $-\dfrac{1}{6}$

52. If $f^{-1}\left(\dfrac{7}{10}\right) = -\dfrac{1}{4}$, what is $f\left(-\dfrac{1}{4}\right)$? $\dfrac{7}{10}$

Graph the inverse of the given function on the same set of axes. (*See Objective 6.*)

53. **54.**

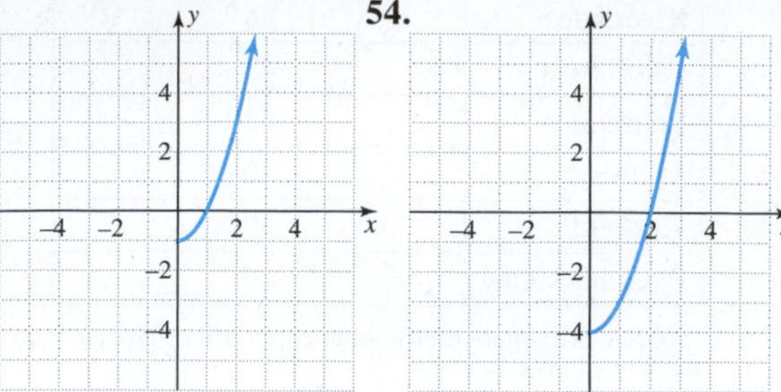

55. **56.**

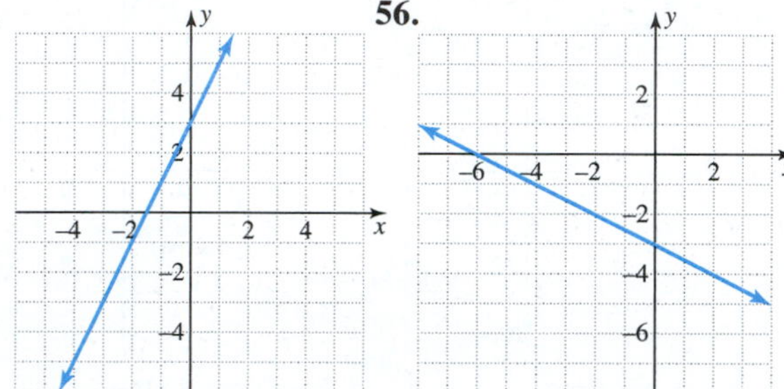

Determine if the functions are inverses of one other. (*See Objective 7.*)

57. $f(x) = 3x - 4,\ g(x) = \frac{1}{3}x + \frac{4}{3}$ yes

58. $f(x) = 4x - 1,\ g(x) = \frac{1}{4}x + \frac{1}{4}$ yes

59. $f(x) = (2x + 1)^3,\ g(x) = \frac{\sqrt[3]{x} - 1}{2}$ yes

60. $f(x) = (5x - 9)^3,\ g(x) = \frac{\sqrt[3]{x} + 9}{5}$ yes

61. $f(x) = \frac{2x + 1}{4},\ g(x) = 2x + \frac{1}{2}$ no

62. $f(x) = \frac{3x - 5}{2},\ g(x) = \frac{2}{3}x - \frac{5}{3}$ no

63. $f(x) = x^3 - 2,\ g(x) = \sqrt[3]{x + 2}$ yes

64. $f(x) = 8x^3 + 9,\ g(x) = \frac{\sqrt[3]{x - 9}}{2}$ yes

65. $f(x) = x^3 + 8,\ g(x) = \sqrt[3]{x} - 2$ no

66. $f(x) = x^3 - 27,\ g(x) = \sqrt[3]{x} + 3$ no

Mix 'Em Up!

Determine if each function is one-to-one.

67. $f = \{(2, 4), (1, 7), (-2, 4), (-1, 7), (0, 5)\}$
not a one-to-one function

68. $f = \{(-1, -1.5), (0, -1), (3, 0.5), (5, 1.5), (6, 2)\}$
yes, a one-to-one function

69.

x	-5	-3	-1	1	3
y	-3	-1	1	5	7

yes, a one-to-one function

70.

x	-4	-2	0	2	4
y	1	4	9	4	25

not a one-to-one function

71.

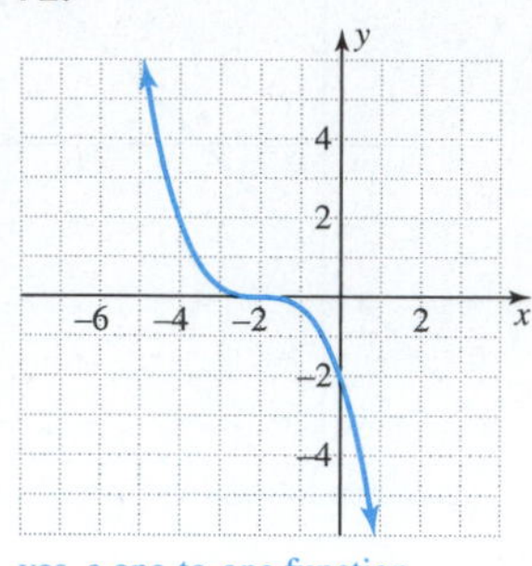

yes, a one-to-one function

72.

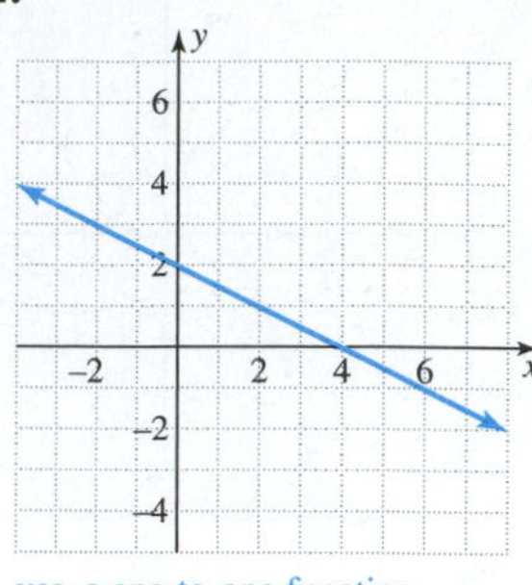

yes, a one-to-one function

73.

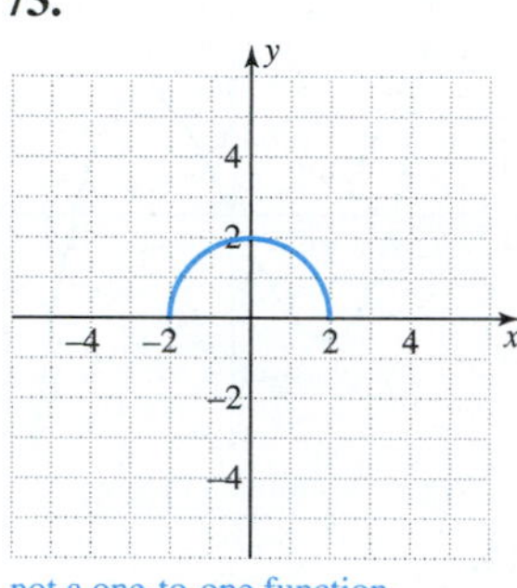

not a one-to-one function

74.

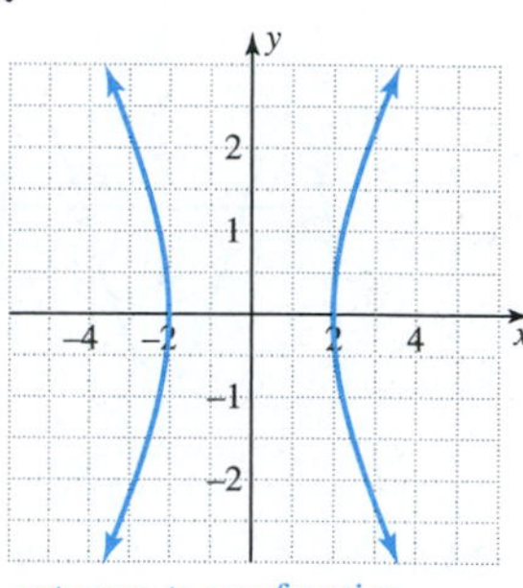

not a one-to-one function

75.

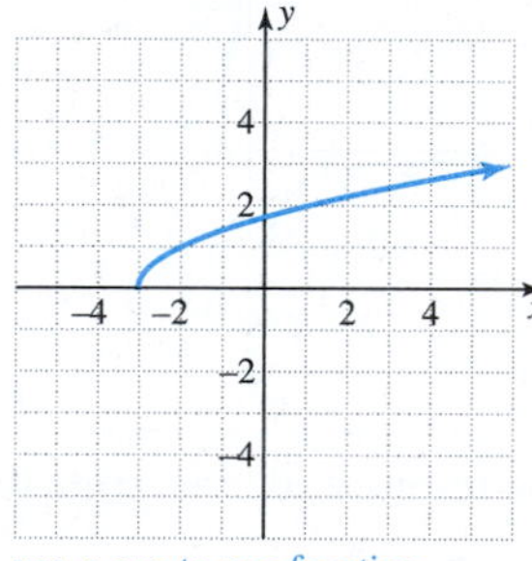

yes, a one-to-one function

76.

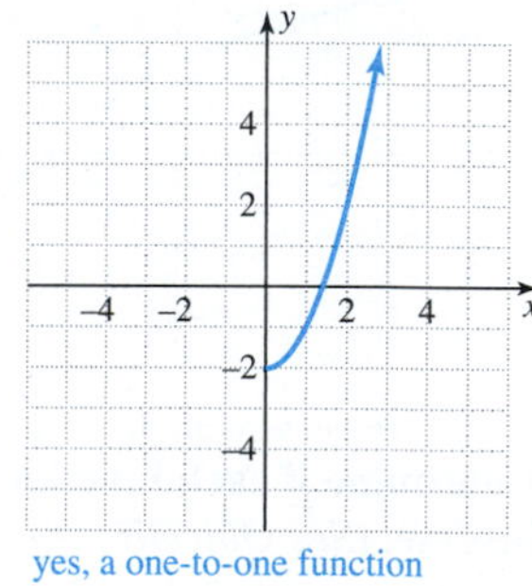

yes, a one-to-one function

Determine if each function is one-to-one and if so, state its inverse.

77. $f(x) = \{(-2, -8), (0, 2), (1, 7), (2, 12), (4, 22)\}$
one-to-one; $f^{-1}(x) = \{(-8, -2), (2, 0), (7, 1), (12, 2), (22, 4)\}$

78. $f(x) = \{(-3, -2.3), (-1, -1.1), (2, 0.7), (5, 2.5), (9, 4.9)\}$
one-to-one; $f^{-1}(x) = \{(-2.3, -3), (-1.1, -1), (0.7, 2), (2.5, 5), (4.9, 9)\}$

79. $f(x) = \{(-2.5, 11.5), (0, -1), (-1.5, 3.5), (2.5, 11.5), (1.5, 3.5)\}$ not a one-to-one function

80.

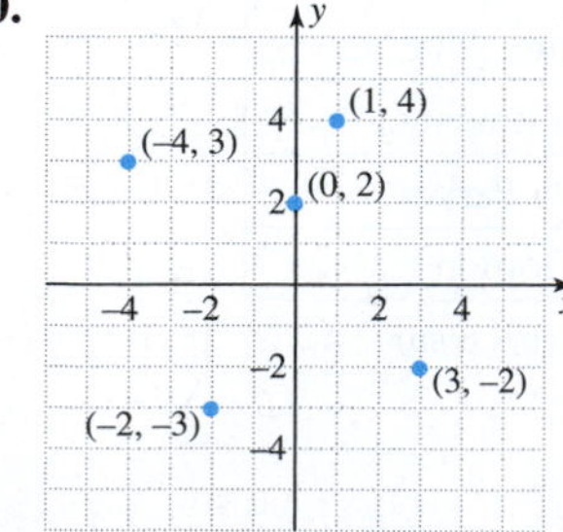

one-to-one;

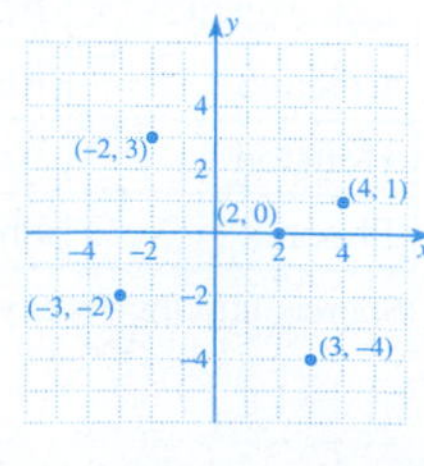

81.

State (x)	Senator in 2011 (y)
Connecticut	Joseph Lieberman
South Carolina	Lindsey Graham
Hawaii	Daniel Inouye
Georgia	Johnny Isakson
Louisiana	Mary Landrieu

(Source: http://www.senate.gov/index.htm)

82.

Car Model (x)	Average Highway MPG (y)
Volkswagen Jetta	42
Toyota Yaris	36
Honda Fit	35
Hyundai Accent	36
Chevrolet Aveo	35

(Source: http://www.fueleconomy.gov) not a one-to-one function

Find the equation of the inverse of each function.

83. $f(x) = 7x + 3$

84. $f(x) = 8x + 12$

85. $f(x) = \dfrac{5}{8x - 3}$ $f^{-1}(x) = \dfrac{3x + 5}{8x}$

86. $f(x) = \dfrac{13}{6x + 1}$

87. $f(x) = 4x^3 + 2$

88. $f(x) = 3x^3 - 5$

89. $f(x) = \dfrac{6x - 2}{5}$

90. $f(x) = \dfrac{3x + 10}{6}$

91. $f(x) = (5x + 4)^3$

92. $f(x) = (3x + 1)^3$

Evaluate each function given the information.

93. If $f(-7) = -2$, what is $f^{-1}(-2)$? -7

94. If $f(4.5) = -1.5$, what is $f^{-1}(-1.5)$? 4.5

95. If $f\left(-\dfrac{4}{5}\right) = \dfrac{3}{8}$, what is $f^{-1}\left(\dfrac{3}{8}\right)$? $-\dfrac{4}{5}$

96. If $f^{-1}\left(\dfrac{2}{7}\right) = -\dfrac{1}{4}$, what is $f\left(-\dfrac{1}{4}\right)$? $\dfrac{2}{7}$

Graph the inverse of each function.

97.

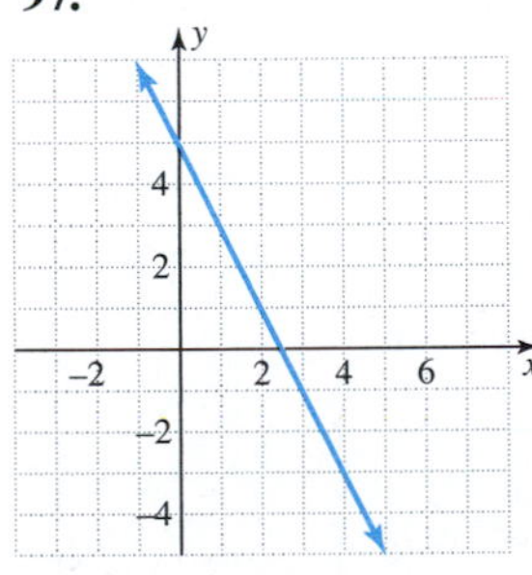

98.

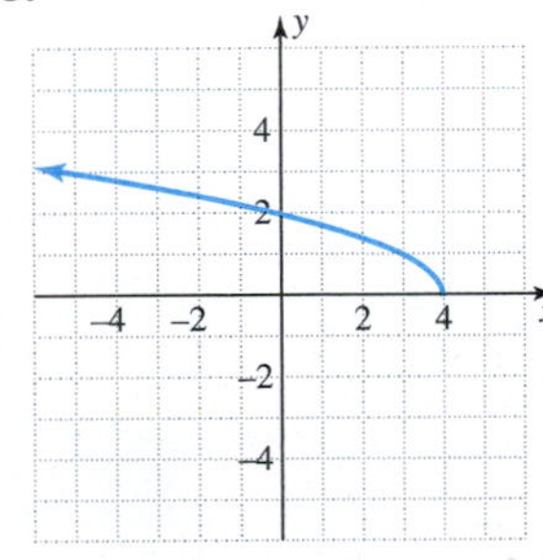

99.

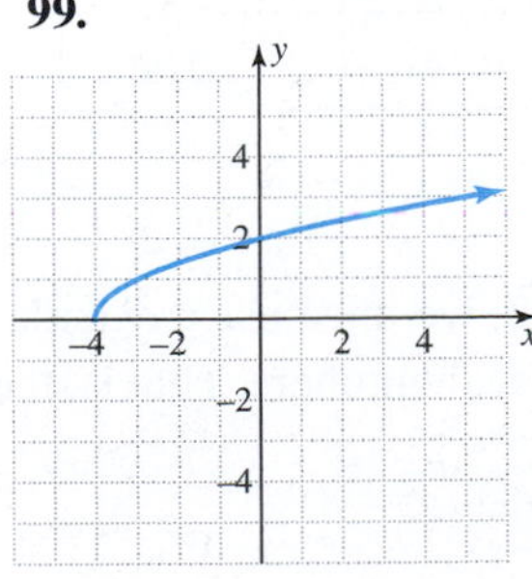

100.

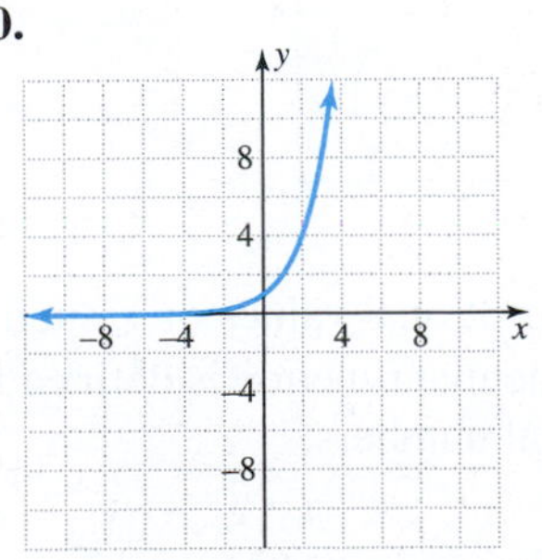

You Be the Teacher!

Correct each student's errors, if any.

101. Find the equation of the inverse of the function $f(x) = \dfrac{3x - 7}{8}$.

Eric's work:

$$f(x) = \frac{3x - 7}{8}$$

$$f^{-1}(x) = \frac{1}{f(x)} = \frac{8}{3x - 7}$$

102. Find the equation of the inverse of the function $f(x) = 8x^3 + 27$.

Mona's work:

$$y = 8x^3 + 27$$

$$x = 8y^3 + 27$$

$$\sqrt[3]{x} = 2y + 3$$

$$\sqrt[3]{x} - 3 = 2y$$

$$y = f^{-1}(x) = \frac{\sqrt[3]{x} - 3}{2}$$

Calculate It!

Use both the TABLE and graphing features of a graphing calculator to determine if the functions are inverses of one another.

103. $f(x) = 2x + 4$ and $g(x) = \dfrac{1}{2}x - 2$

104. $f(x) = (x - 1)^3$ and $g(x) = \sqrt[3]{x} + 1$

Think About It!

105. Find the inverse function, $f^{-1}(x)$, if $f(x)$ is a linear function and $f(3) = -6$ and $f(0) = 2$. $f^{-1}(x) = -\dfrac{3}{8}x + \dfrac{3}{4}$

106. Find the inverse function, $f^{-1}(x)$, if $f(x)$ is a linear function and $f(-2) = 5$ and $f(6) = 13$. $f^{-1}(x) = x - 7$

107. Find the inverse function, $f^{-1}(x)$, if $f(x)$ is a linear function and $f(4) = -1$ and $f(3) = 0$. $f^{-1}(x) = -x + 3$

108. Find the inverse function, $f^{-1}(x)$, if $f(x)$ is a linear function and $f(7) = -2$ and $f(-1) = 0$. $f^{-1}(x) = -4x - 1$

SECTION 10.3 Exponential Functions

OBJECTIVES

As a result of completing this section, you will be able to

1. Graph exponential functions.
2. Solve exponential equations of the form $b^x = b^y$.
3. Solve applications of exponential functions.
4. Troubleshoot common errors.

The number of Facebook accounts (in millions) can be modeled by $f(x) = 108.8449(1.0658)^x$, where x is the months after August 2008. Use this model to estimate the number of Facebook accounts in December 2013. (Source: http://www.benphoster.com/facebook-user-growth-chart-2004-2010/)

To solve this problem, we must evaluate the function for $x = 64$. This type of function is called an exponential function. In this section, we will learn to work with functions and equations of this type.

Exponential Functions

Objective 1 Graph exponential functions.

Recall from previous chapters that expressions that contain exponents, such as 3^4 and x^3, are called exponential expressions. In this section, we will look at functions that involve exponential expressions. In these functions, the base is constant and the exponent changes. This type of function is called an *exponential function*.

Definition: An **exponential function** is a function of the form $f(x) = b^x$, where $b > 0$ and $b \neq 1$, and x is a real number.

Note that in the definition of an exponential function, $b > 0$ and $b \neq 1$.

- If the base of an exponential function is 1, then $1^x = 1$ for all x, which is a horizontal line and not an exponential function.
- If the base is a negative value, then the function would not be defined for all values. For instance, $(-4)^{1/2}$ is not a real number.

Consider the function $f(x) = 4^x$. Recall $x^{m/n} = \left(\sqrt[n]{x}\right)^m$ and $x^{-n} = \dfrac{1}{x^n}$.

x	$y = 4^x$	(x, y)
-2	$4^{-2} = \frac{1}{16}$	$\left(-2, \frac{1}{16}\right)$
$-\frac{3}{2}$	$4^{-3/2} = \frac{1}{8}$	$\left(-\frac{3}{2}, \frac{1}{8}\right)$
-1	$4^{-1} = \frac{1}{4}$	$\left(-1, \frac{1}{4}\right)$
$-\frac{1}{2}$	$4^{-1/2} = \frac{1}{2}$	$\left(-\frac{1}{2}, \frac{1}{2}\right)$

x	$y = 4^x$	(x, y)
0	$4^0 = 1$	$(0, 1)$
$\frac{1}{2}$	$4^{1/2} = 2$	$\left(\frac{1}{2}, 2\right)$
1	$4^1 = 4$	$(1, 4)$
$\frac{3}{2}$	$4^{3/2} = 8$	$\left(\frac{3}{2}, 8\right)$
2	$4^2 = 16$	$(2, 16)$

To evaluate the function for irrational values of x, such as $\sqrt{2}$, we must use a calculator; $4^{\sqrt{2}} \approx 7.10$. So, an exponential function is defined for all real numbers. That is, the domain of $f(x) = b^x$ is all real numbers.

Using the points obtained in the previous tables, we can graph the function $f(x) = 4^x$. (See Figure 10.2.)

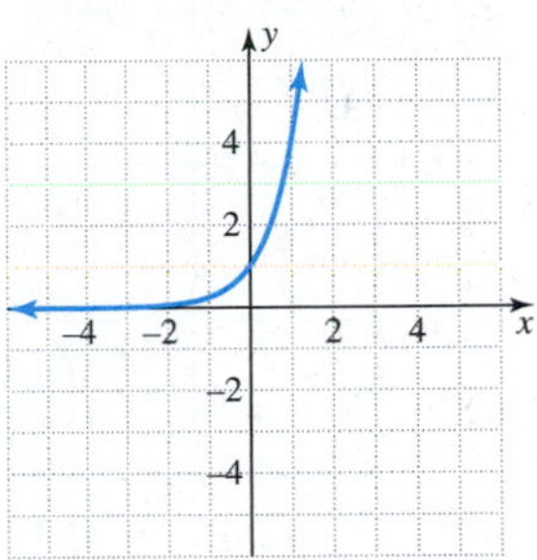

Figure 10.2

Property: Characteristics of the Graph of $y = b^x, b > 0, b \neq 1$

- The domain of the function is $(-\infty, \infty)$.
- The range is $(0, \infty)$.
- The y-intercept is $(0, 1)$.
- There is no x-intercept. The graph gets very close to the x-axis but never touches it.
- The point $(1, b)$ lies on the graph.
- The function is one-to-one.
- If $b > 1$, the function is increasing.
- If $0 < b < 1$, the function is decreasing.

$f(x) = b^x$ for $b > 1$

(0, 1)

$f(x) = b^x$ for $0 < b < 1$

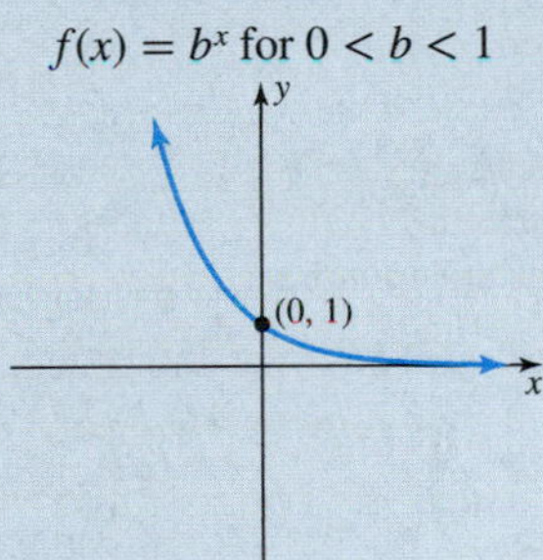

Objective 1 Examples **Graph each exponential function. Plot and label at least three points on the graph. State the domain and range of each function.**

1a. $f(x) = 2^x$ **1b.** $f(x) = \left(\frac{1}{2}\right)^x$

Solutions **1a.**

x	$y = 2^x$	(x, y)
-2	$2^{-2} = \frac{1}{4}$	$\left(-2, \frac{1}{4}\right)$
-1	$2^{-1} = \frac{1}{2}$	$\left(-1, \frac{1}{2}\right)$
0	$2^0 = 1$	$(0, 1)$
1	$2^1 = 2$	$(1, 2)$
2	$2^2 = 4$	$(2, 4)$

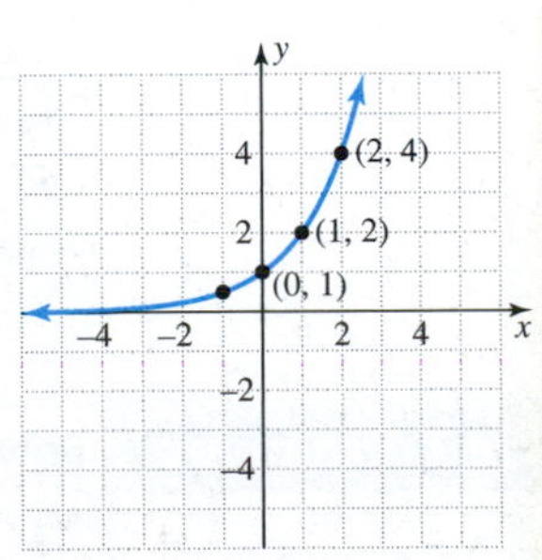

The domain is $(-\infty, \infty)$, and the range is $(0, \infty)$.

1b.

x	$y = \left(\frac{1}{2}\right)^x$	(x, y)
-2	$\left(\frac{1}{2}\right)^{-2} = 4$	$(-2, 4)$
-1	$\left(\frac{1}{2}\right)^{-1} = 2$	$(-1, 2)$
0	$\left(\frac{1}{2}\right)^{0} = 1$	$(0, 1)$
1	$\left(\frac{1}{2}\right)^{1} = \frac{1}{2}$	$\left(1, \frac{1}{2}\right)$
2	$\left(\frac{1}{2}\right)^{2} = \frac{1}{4}$	$\left(2, \frac{1}{4}\right)$

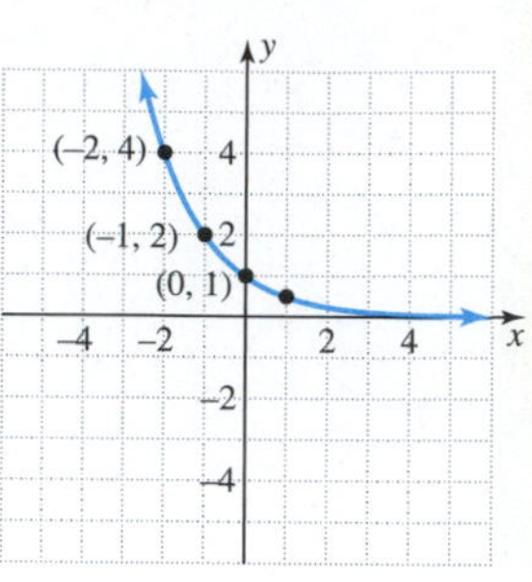

The domain is $(-\infty, \infty)$, and the range is $(0, \infty)$.

Student Check 1 Graph each exponential function. Plot and label at least three points on the graph. State the domain and range of each function.

a. $f(x) = 3^x$ **b.** $f(x) = \left(\frac{1}{3}\right)^x$

Solving Exponential Equations

Objective 2 Solve exponential equations of the form $b^x = b^y$.

We can use the one-to-one property of exponential functions to solve a special type of exponential equation. Recall a function is one-to-one if every y-value corresponds to only one x-value. So, the expression b^x is different for every value of x. The only way that exponential expressions with the same base are equal is if their exponents are equal.

Property: One-to-One Property of b^x

If $b > 0$ and $b \neq 1$, then $b^x = b^y$ if and only if $x = y$.

This property indicates that if each side of an equation can be expressed with the same base, then the exponents of the bases must be equal. Later in the chapter, we will learn how to solve exponential equations in which the sides of the equation cannot be expressed with the same base.

Procedure: Solving Exponential Equations of the Form $b^x = b^y$

Step 1: Rewrite each side of the equation as a power of the same base.
Step 2: Set the exponents equal.
Step 3: Solve the resulting equation.
Step 4: Check the solution by substitution.

Objective 2 Examples Solve each equation.

2a. $3^x = 27$ **2b.** $4^x = 32$ **2c.** $16^{x-1} = 64^x$

Solutions **2a.**

$$3^x = 27$$
$$3^x = 3^3 \quad \text{Rewrite each side with the same base.}$$
$$x = 3 \quad \text{Apply the one-to-one property of } b^x.$$

The solution set is $\{3\}$. To check the solution, replace x with 3 in the original equation.

2b.

$$4^x = 32$$
$$(2^2)^x = 2^5 \quad \text{Rewrite each side with the same base.}$$
$$2^{2x} = 2^5 \quad \text{Apply the power of a power rule, } (b^m)^n = b^{mn}.$$
$$2x = 5 \quad \text{Apply the one-to-one property of } b^x.$$
$$x = \frac{5}{2} \quad \text{Divide each side by 2.}$$

The solution set is $\left\{\frac{5}{2}\right\}$. The check is left for the reader.

2c.

$$16^{x-1} = 64^x$$
$$(4^2)^{x-1} = (4^3)^x \quad \text{Rewrite each side with the same base.}$$
$$4^{2x-2} = 4^{3x} \quad \text{Apply the power of a power rule, } (b^m)^n = b^{mn}.$$
$$2x - 2 = 3x \quad \text{Apply the one-to-one property of } b^x.$$
$$-2 = x \quad \text{Subtract } 2x \text{ from each side.}$$

The solution set is $\{-2\}$. The check is left for the reader.

Student Check 2 Solve each equation.

a. $10^x = 100$ **b.** $8^x = 4$ **c.** $9^{x+2} = 27^x$

Applications

Objective 3 Solve applications of exponential functions.

Exponential functions are used to model quantities that grow or decline very rapidly, such as the population of a country, the growth of bacteria, the amount of a chemical element, or the amount of money in an account.

Property: Compound Interest Formula

The compound interest formula determines the amount of money A that will be in an account if P dollars is invested for t years at an annual interest rate r, compounded n times a year.

$$A = P\left(1 + \frac{r}{n}\right)^{nt}$$

Objective 3 Examples Solve each problem.

3a. Sandra invests \$5000 in a savings account that pays 5% annual interest compounded quarterly. Use the compound interest formula to determine the amount of money she will have in the account at the end of 3 yr.

Solution **3a.** $A = P\left(1 + \frac{r}{n}\right)^{nt}$ State the compound interest formula.

$A = 5000\left(1 + \frac{0.05}{4}\right)^{4(3)}$ Substitute the values: $P = 5000$, $r = 0.05$, $n = 4$, and $t = 3$.

$A = 5000(1.0125)^{12}$ Simplify the expression in parentheses.

$A \approx \$5803.77$ Approximate.

So, Sandra will have \$5803.77 at the end of 3 yr.

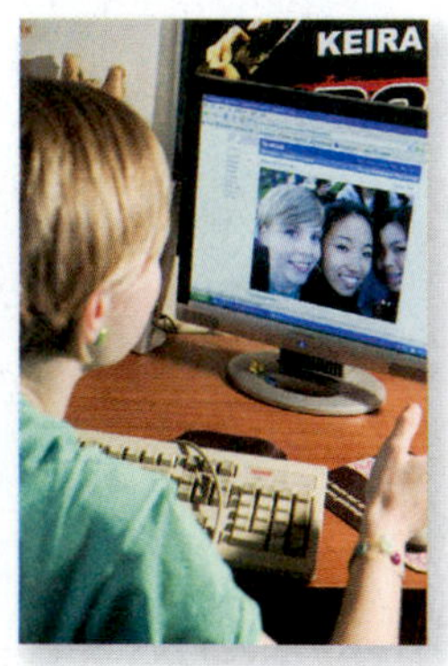

3b. The number of Facebook accounts is shown in the following table. The number of accounts (in millions) can be modeled by $f(x) = 108.8449(1.0658)^x$, where x is the months after August 2008. Assuming this growth continues, use this model to estimate the number of Facebook accounts in December 2013. (Source: http://www.benphoster.com/facebook-user-growth-chart-2004-2010/)

Month	Total Accounts (in millions)
August 2008 ($x = 0$)	100
January 2009 ($x = 5$)	150
December 2009 ($x = 16$)	350
July 2010 ($x = 23$)	500
February 2011 ($x = 30$)	650

Solution **3b.** December 2013 is 64 months after August 2008.

$f(x) = 108.8449(1.0658)^x$ State the model.

$f(64) = 108.8449(1.0658)^{64}$ Replace x with 64.

$f(64) \approx 6427.67$ Approximate.

According to the model, there will be about 6427.67 million Facebook accounts, or 6,427,670,000 accounts in December 2013.

Student Check 3 Solve each problem.

a. Use the compound interest formula to determine the amount Fred will have in a savings account at the end of 10 yr if he invests \$2000 at 4.5% annual interest compounded monthly.

b. The population of Russia (in millions) can be modeled by the function $f(x) = 146.25(0.996)^x$, where x is the number of years after 2000. What is an estimate for the population in 2000? 2015? 2050? (Source: http://www.google.com/publicdata)

Objective 4 Troubleshoot common errors.

Troubleshooting Common Errors

Some common errors associated with exponential functions are shown.

Objective 4 Examples **A problem and an incorrect solution are given. Provide the correct solution and an explanation of the error.**

4a. Graph $y = 2^x$.

Incorrect Solution	Correct Solution and Explanation
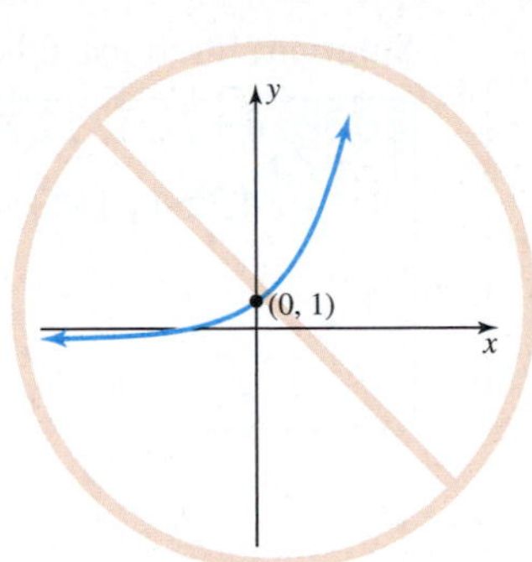	The graph is incorrect because the output of $2^x > 0$. The graph should not cross the x-axis or go below the x-axis. The correct graph is 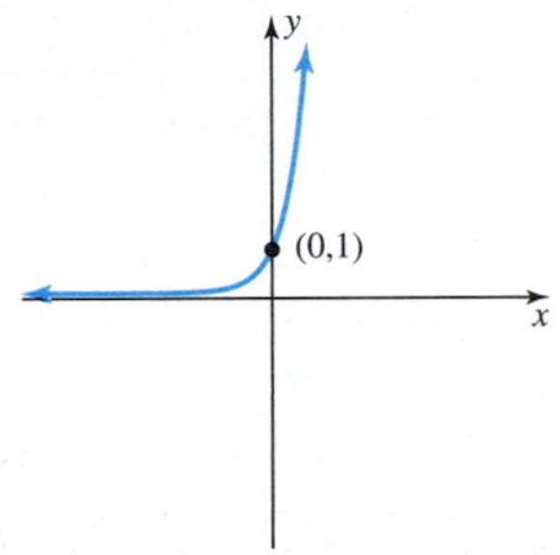

4b. Find the amount Sung will have in an account after 7 yr if he invests \$3000 at 5% annual interest compounded monthly.

Incorrect Solution	Correct Solution and Explanation
$A = P\left(1 + \frac{r}{n}\right)^{nt}$ $A = 3000\left(1 + \frac{0.05}{12}\right)^{12(7)}$ $A = 22{,}074.40$	The error occurred in entering the expression on a calculator. Be sure to put the entire exponent in parentheses. This will give us the correct amount of $A = \$4254.11$

ANSWERS TO STUDENT CHECKS

Student Check 1

a.

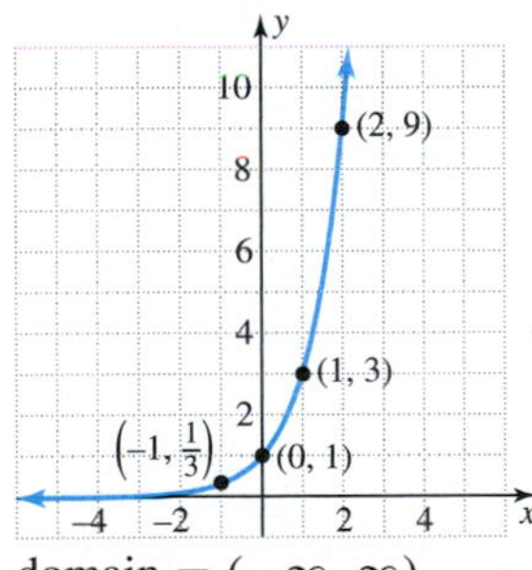

domain = $(-\infty, \infty)$

range = $(0, \infty)$

b.

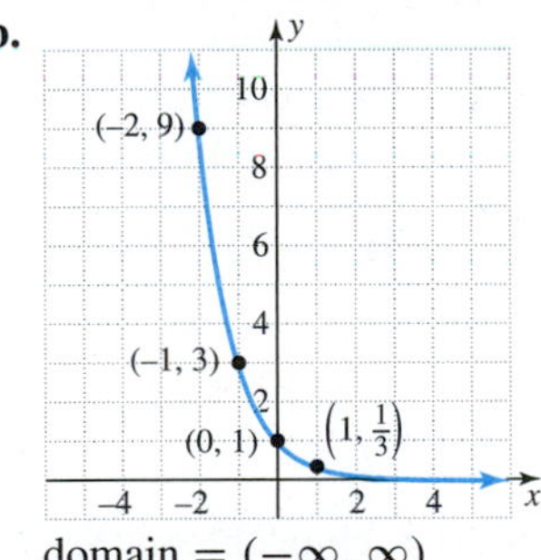

domain = $(-\infty, \infty)$

range = $(0, \infty)$

Student Check 2 **a.** $\{2\}$ **b.** $\left\{\frac{2}{3}\right\}$ **c.** $\left\{-\frac{1}{2}\right\}$

Student Check 3 **a.** \$3133.99 **b.** about 142.3 million, about 135.61 million, about 109.85 million

SUMMARY OF KEY CONCEPTS

1. An exponential function is a function of the form $y = b^x$, $b > 0, b \neq 1$.
   - The domain of the function is $(-\infty, \infty)$ and the range is $(0, \infty)$.
   - The y-intercept is $(0, 1)$ and there is no x-intercept.
   - If $b > 1$, the function is increasing. If $0 < b < 1$, the function is decreasing.
2. If an equation can be written in the form $b^x = b^y$, $b > 0$, then we can use the one-to-one property of exponential functions to conclude that $x = y$.
3. Exponential functions model many real-life situations. The compound interest formula is such a model. It is given by $A = P\left(1 + \frac{r}{n}\right)^{nt}$.

GRAPHING CALCULATOR SKILLS

A graphing calculator can be used to graph exponential functions as well as solve application problems.

Example 1: Graph $y = 2^x$.

Solution: Enter the function in the equation editor. Graph and review the TABLE.

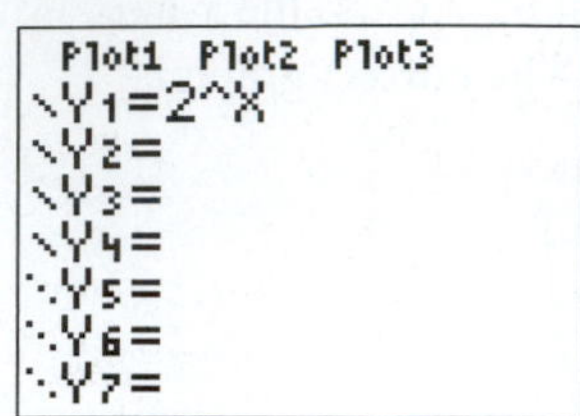

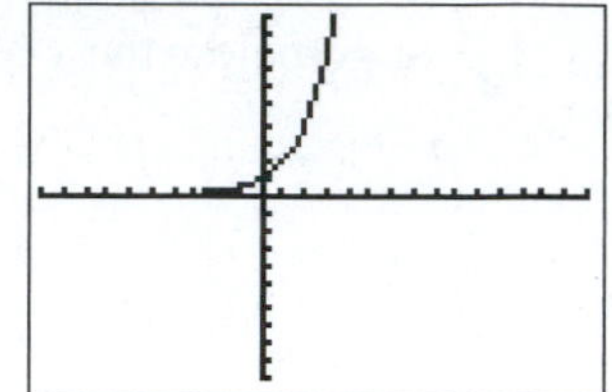

X	Y1
-3	.125
-2	.25
-1	.5
0	1
1	2
2	4
3	8

X=-3

Example 2: Simplify $A = 3000\left(1 + \frac{0.05}{12}\right)^{12(7)}$.

Solution: Insert parentheses around the entire exponent.

```
3000(1+.05/12)^(
12*7)
              4254.108157
```

SECTION 10.3 EXERCISE SET

Write About It!

Use complete sentences in your answer to each exercise.

1. Describe the behavior of the graph of $f(x) = 6^x$ at each end of the x-axis.
2. Describe the behavior of the graph of $f(x) = \left(\frac{1}{8}\right)^x$ at each end of the x-axis.
3. Explain how to solve the equation $5^{2x-1} = 625$.
4. Explain how to solve the equation $25^{2x-1} = 625$.
5. Explain how to solve the equation $4^{x+1} = \frac{1}{64}$ by using the base 4.
6. Explain how to solve the equation, $4^{x+1} = \frac{1}{64}$ by using the base 2.

Practice Makes Perfect!

Graph each exponential function. Plot and label at least three points on the graph. State the domain and range of each function. (*See Objective 1.*)

7. $y = 5^x$
8. $y = 6^x$
9. $y = \left(\frac{1}{5}\right)^x$
10. $y = \left(\frac{2}{3}\right)^x$
11. $y = \left(\frac{2}{5}\right)^x$
12. $y = \left(\frac{1}{6}\right)^x$

Solve each exponential equation. (*See Objective 2.*)

13. $5^x = 125$ $\{3\}$
14. $6^x = 36$ $\{2\}$
15. $\left(\frac{1}{2}\right)^x = 8$ $\{-3\}$
16. $\left(\frac{1}{3}\right)^x = 81$ $\{-4\}$
17. $\left(\frac{3}{2}\right)^x = \frac{16}{81}$ $\{-4\}$
18. $\left(\frac{5}{3}\right)^x = \frac{27}{125}$ $\{-3\}$
19. $27^x = 9$ $\left\{\frac{2}{3}\right\}$
20. $216^x = 36$ $\left\{\frac{2}{3}\right\}$
21. $125^x = \frac{1}{25}$ $\left\{-\frac{2}{3}\right\}$
22. $4^{x-2} = 64^x$ $\{-1\}$
23. $2^{3x-1} = 128$ $\left\{\frac{8}{3}\right\}$
24. $25^{2x+3} = 125$ $\left\{-\frac{3}{4}\right\}$
25. $\left(\frac{1}{2}\right)^{3x+2} = 32^x$ $\left\{-\frac{1}{4}\right\}$
26. $\left(\frac{1}{3}\right)^{2x-1} = 27^x$ $\left\{\frac{1}{5}\right\}$

Solve each problem. (*See Objective 3.*)

For Exercises 27 and 28, use the formula $A = P\left(1 + \frac{r}{n}\right)^{nt}$.

27. Hixon invests \$6000 in a savings account that pays 4% annual interest compounded quarterly. Determine the amount of money he will have in the account at the end of 3 yr. \$6760.95
28. Ruth invests \$7000 in a savings account that pays 1.3% annual interest compounded quarterly. Determine the amount of money she will have in the account at the end of 6 yr. \$7566.90

Additional answers can be found in the Instructor Answer Appendix.

29. The population of the United Kingdom (in millions) can be modeled by the function $f(x) = 62.06(1.005)^x$, where x is the number of years after 2009. What is an estimate for the population in 2015? 2025? 2050? Round answers to two decimal places. (Source: http://www.census.gov/population/international/data/idb/informationGateway.php) 63.95 million, 67.22 million, 76.14 million

30. The population of the United States (in millions) can be modeled by the function $f(x) = 281.4(1.0093)^x$, where x is the number of years after 2000. What is an estimate for the population in 2010? 2015? 2025? Round answers to two decimal places. (Source: http://www.census.gov/prod/cen2010/briefs/c2010br-01.pdf) 308.70 million, 323.32 million, 354.68 million

31. The radioactive isotope sodium-24 is used to detect circulatory problems in patients. The function $A(t) = 20(0.9548)^t$ determines the amount of sodium-24, in grams, that a hospital has in storage after t hr. How many grams does the hospital have initially? How many grams will remain after 24 hr? after 48 hr? (Sources: http://www.chemistryexplained.com/elements/P-T/Sodium.html and http://www.3rd1000.com/nuclear/halflife.htm) 20 g, 6.59 g, 2.17 g

32. Rachel has been on a medication for several months and has reached a steady level of 150 mg in her body. It is time for Rachel to discontinue taking the medication. The function $A(t) = 150(0.7071)^t$ determines the amount of drug, in milligrams, that remains in Rachel's body t days after she discontinues the drug. How much is in her body 1 week after she stops taking the drug? 2 weeks after? 13.26 mg, 1.17 mg

Mix 'Em Up!

Graph each exponential function. State the domain and range of the function.

33. $f(x) = 8^x$
34. $f(x) = 9^x$
35. $f(x) = 4.5^x$
36. $f(x) = 1.6^x$
37. $f(x) = \left(\frac{1}{10}\right)^x$
38. $f(x) = \left(\frac{3}{5}\right)^x$

Solve each exponential equation.

39. $\left(\frac{2}{5}\right)^x = \frac{625}{16}$ $\{-4\}$
40. $\left(\frac{2}{3}\right)^x = \frac{243}{32}$ $\{-5\}$
41. $9^{x+2} = 243$ $\left\{\frac{1}{2}\right\}$
42. $3^{4x+7} = 3$ $\left\{-\frac{3}{2}\right\}$
43. $16^{5x+2} = 64$ $\left\{-\frac{1}{10}\right\}$
44. $2^{3x+8} = 4$ $\{-2\}$
45. $27^{5x+1} = \frac{1}{9}$ $\left\{-\frac{1}{3}\right\}$
46. $64^{x-1} = \frac{1}{256}$ $\left\{-\frac{1}{3}\right\}$
47. $3^{2x+3} = 27$ $\{0\}$
48. $2^{x+5} = 32$ $\{0\}$
49. $10^{3x+2} = 1000$ $\left\{\frac{1}{3}\right\}$
50. $10^{2x-1} = 100$ $\left\{\frac{3}{2}\right\}$
51. $3^{2x-5} = 27^x$ $\{-5\}$
52. $2^{5x-5} = 4^x$ $\left\{\frac{5}{3}\right\}$
53. $3^{3x-6} = 81^x$ $\{-6\}$
54. $8^{4x+8} = 16^x$ $\{-3\}$
55. $27^{2x-2} = \frac{1}{3}$ $\left\{\frac{5}{6}\right\}$
56. $2^{2x+4} = \frac{1}{8}$ $\left\{-\frac{7}{2}\right\}$

Solve each problem.

57. Seyfried invests \$9500 in a savings account that pays 2.6% annual interest compounded monthly. Use the compound interest formula to determine the amount of money she will have in the account at the end of 5 yr. \$10,817.35

58. Morgan invests \$7200 in a savings account that pays 2.1% annual interest compounded monthly. Use the compound interest formula to determine the amount of money he will have in the account at the end of 6 yr. \$8165.93

59. The population of China (in billions) can be modeled by the function $f(x) = 1.33(1.003)^x$, where x is the number of years after 2009. What is an estimate for the population in 2012? 2020? 2030? Round answers to two decimal places. (Source: http://www.census.gov/population/international/data/idb/informationGateway.php) 1.34 billion, 1.37 billion, 1.42 billion

60. The population of France (in millions) can be modeled by the function $f(x) = 64.57(1.004)^x$, where x is the number of years after 2009. What is an estimate for the population in 2010? 2025? 2040? Round answers to two decimal places. (Source: http://www.census.gov/population/international/data/idb/informationGateway.php) 64.83 million, 68.83 million, 73.08 million

61. Iodine-131 is a radioactive isotope that is used in treating thyroid cancer and in imaging the thyroid and can also be used in nuclear reactors. The function $A(t) = 30(0.917)^t$ determines the amount, in milligrams, of the isotope that a hospital has in storage after t days. How much does the hospital have initially? How much of the substance will remain after 8 days? after 30 days? (Source: http://www.ead.anl.gov) 30 mg, 15 mg, 2.23 mg

62. Gallium-67 is a radioactive isotope that is used in imaging various inflammatory conditions and tumors such as Hodgkin's disease. The function $A(t) = 10(0.8085)^t$ determines the amount, in milliliters, of the substance that an imaging center has in storage after t days. How much of the substance will remain after 2 days? after 1 week? (Source: http://www.nordion.com) 6.54 mL, 2.26 mL

You Be the Teacher!

Correct each student's errors, if any.

63. Solve the equation, $16^{3x-2} = 8$.

Ellen's work:

$$16^{3x-2} = 8$$
$$8^{2(3x-2)} = 8$$
$$2(3x-2) = 0$$
$$6x - 4 = 0$$
$$6x = 4$$
$$x = \frac{4}{6} = \frac{2}{3}$$

$16^{3x-2} = 8$
$2^{4(3x-2)} = 2^3$
$4(3x-2) = 3$
$12x - 8 = 3$
$12x = 11$
$x = \frac{11}{12}$

64. Solve the equation, $125^{x-7} = \frac{1}{25}$.

Fourlas's work:

$$125^{x-7} = \frac{1}{25}$$
$$5^{3x-7} = 5^{-2}$$
$$3x - 7 = -2$$
$$3x = 5$$
$$x = \frac{5}{3}$$

$$125^{x-7} = \frac{1}{25}$$
$$5^{3(x-7)} = 5^{-2}$$
$$3(x-7) = -2$$
$$3x - 21 = -2$$
$$3x = 19$$
$$x = \frac{19}{3}$$

Use a graphing calculator to graph or evaluate each function. Round each answer to two decimal places if appropriate.

65. $y = \left(\frac{1}{5}\right)^x$

66. $y = 7^x$

67. Evaluate $114.5(5.14)^{-0.72x}$ at $x = 5$.

68. Evaluate $2500\left(1 + \frac{0.023}{12}\right)^{12t}$ at $t = 8$.

SECTION 10.4 Logarithmic Functions

OBJECTIVES

As a result of completing this section, you will be able to

1. Use logarithmic notation to write an exponential equation and vice versa.
2. Simplify logarithmic expressions.
3. Solve logarithmic equations using exponential notation.
4. Graph basic logarithmic functions.
5. Troubleshoot common errors.

The graph shows the annual carbon dioxide (CO_2) emissions for a car with the given gas mileage. The annual tons of CO_2 emitted can be modeled by $f(x) = 13.272(0.8369)^x$, where x is the mileage of a car in miles per gallon. Determine the gas mileage of a car that emits 1 ton of CO_2 annually. To solve this problem, we must solve

$$13.272(0.8369)^x = 1$$

Solving this equation requires us to apply a property that enables us to move the variable from the exponent. In this section, we will introduce the concept of logarithms, which we will apply in Section 10.7 to solve this problem.

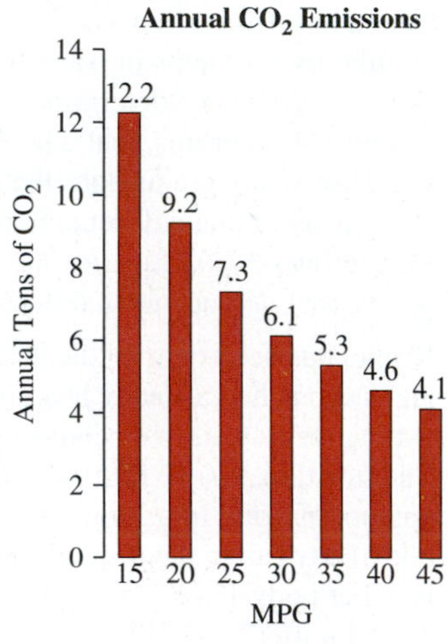

(Source: http://www.fueleconomy.gov/)

Logarithms

Objective 1

Use logarithmic notation to write an exponential equation and vice versa.

Consider the exponential function $f(x) = 2^x$. Because $f(x) = 2^x$ is a one-to-one function, we know it has an inverse. To find the inverse of $f(x) = 2^x$ algebraically, we must interchange the variables and solve for y. This gives us

$$f(x) = 2^x$$
$$y = 2^x$$
$$x = 2^y$$

We have not yet learned a technique that enables us to solve for the variable y. So, we need a new symbol, called a *logarithm*, to denote an equivalent form of y. The symbol $\log_b x$ is read "log base b of x" and means the exponent of b that produces x. So, in this example, we say that

$$y = \log_2 x$$

This notation states that y is the exponent of 2 that produces x. So, a logarithm is simply an exponent. As we can see, logarithmic functions arise as we find the inverse of an exponential function. So, the inverse of $f(x) = 2^x$ is $f^{-1}(x) = \log_2 x$.

INSTRUCTOR NOTE: x is called the argument of the logarithm.

Definition: Logarithm

For $b > 0, b \neq 1$,

Logarithmic form

$$y = \log_b x \text{ is equivalent to } b^y = x$$

Exponential form

for every $x > 0$ and every real number y. The value x is called the **argument** of the logarithmic expression.

Some examples of this definition are as follows.

$$2^3 = 8 \text{ is equivalent to } \log_2 8 = 3.$$
$$5^2 = 25 \text{ is equivalent to } \log_5 25 = 2.$$
$$3^{-1} = \frac{1}{3} \text{ is equivalent to } \log_3\left(\frac{1}{3}\right) = -1.$$

Objective 1 Examples **Rewrite each logarithmic equation in exponential form and each exponential equation in logarithmic form.**

Problems	Solutions
1a. $\log_2 16 = 4$	$\log_2 16 = 4$ is equivalent to $2^4 = 16$.
1b. $\log_5 1 = 0$	$\log_5 1 = 0$ is equivalent to $5^0 = 1$.
1c. $\log_3 3 = 1$	$\log_3 3 = 1$ is equivalent to $3^1 = 3$.
1d. $\log_4 \frac{1}{16} = -2$	$\log_4 \frac{1}{16} = -2$ is equivalent to $4^{-2} = \frac{1}{16}$.
1e. $5^3 = 125$	$5^3 = 125$ is equivalent to $\log_5 125 = 3$.
1f. $3^{-1} = \frac{1}{3}$	$3^{-1} = \frac{1}{3}$ is equivalent to $\log_3 \frac{1}{3} = -1$.
1g. $8^{2/3} = 4$	$8^{2/3} = 4$ is equivalent to $\log_8 4 = \frac{2}{3}$.
1h. $\sqrt{4} = 2$	$\sqrt{4} = 4^{1/2} = 2$ is equivalent to $\log_4 2 = \frac{1}{2}$.

Student Check 1 Rewrite each logarithmic equation in exponential form and each exponential equation in logarithmic form.

a. $\log_3 81 = 4$ **b.** $\log_2 1 = 0$ **c.** $\log_6 6 = 1$ **d.** $\log_5 \frac{1}{125} = -3$

e. $6^2 = 36$ **f.** $4^{-1} = \frac{1}{4}$ **g.** $9^{3/2} = 27$ **h.** $\sqrt{9} = 3$

Evaluating Logarithmic Expressions

Objective 2 Simplify logarithmic expressions.

Now that we know that $\log_b x$ means to find the exponent of b that produces x, we can simplify logarithmic expressions. It is helpful to be able to express in words what the logarithm denotes.

$\log_4 16$ means the exponent of 4 that produces 16. Since $4^2 = 16$, $\log_4 16 = 2$.

$\log_{10} 10$ means the exponent of 10 that produces 10. Since $10^1 = 10$, $\log_{10} 10 = 1$.

$\log_2 1$ means the exponent of 2 produces that 1. Since $2^0 = 1$, $\log_2 1 = 0$.

It is also helpful to recall the following properties of exponents.

$$b^0 = 1 \qquad b^1 = b \qquad b^{-n} = \frac{1}{b^n} \qquad b^{1/n} = \sqrt[n]{b} \qquad b^{m/n} = \sqrt[n]{b^m}$$

Recall also that when we raise a base b, $b > 0$ and $b \neq 1$, to any number, the result is always positive. So, the logarithm of a negative number is undefined. For instance,

$\log_2(-4)$ means the exponent of 2 that produces -4.

There is no exponent of 2 that produces a negative value, so $\log_2(-4)$ is undefined.

Procedure: Simplifying a Logarithmic Expression of the Form $\log_b x$

Step 1: Find the exponent of b that produces x by using the exponential form.
Step 2: The value of the logarithmic expression is this exponent.

Objective 2 Example Simplify each logarithmic expression.

Problems	Solutions
2a. $\log_6 1$	$6^0 = 1 \Rightarrow \log_6 1 = 0$
2b. $\log_4 4$	$4^1 = 4 \Rightarrow \log_4 4 = 1$
2c. $\log_3 \frac{1}{27}$	$3^{-3} = \frac{1}{27} \Rightarrow \log_3 \frac{1}{27} = -3$
2d. $\log_2 \sqrt{2}$	$2^{1/2} = \sqrt{2} \Rightarrow \log_2 \sqrt{2} = \frac{1}{2}$
2e. $\log_{1/2} 4$	$\left(\frac{1}{2}\right)^{-2} = 4 \Rightarrow \log_{1/2} 4 = -2$
2f. $\log_5(-25)$	$\log_5(-25)$ is undefined since $5^y \neq -25$ for any real number y.

Student Check 2 Simplify each logarithmic expression.

a. $\log_9 1$ **b.** $\log_5 5$ **c.** $\log_2 \frac{1}{4}$ **d.** $\log_6 \sqrt[3]{6}$

e. $\log_{1/3} 3$ **f.** $\log_7(-49)$

Note: *Examples 2a and 2b show us two important properties of logarithms. For $b > 0$, $b \neq 1$, we have*

$$\log_b 1 = 0$$
$$\log_b b = 1$$

Solving Logarithmic Equations

Objective 3 Solve logarithmic equations using exponential notation.

We can use the exponential form of a logarithmic equation to solve logarithmic equations.

Procedure: Solving a Logarithmic Equation

Step 1: Rewrite the equation in exponential form.
Step 2: Solve the resulting equation.

Objective 3 Examples Solve each equation.

3a. $\log_2 32 = x$ **3b.** $\log_6 x = 3$ **3c.** $\log_x 49 = 2$ **3d.** $\log_b b = x$

Solutions

3a. $\log_2 32 = x$

$2^x = 32$ Rewrite as an exponential equation.

$2^x = 2^5$ Rewrite 32 with a base of 2.

$x = 5$ Apply the one-to-one property of b^x.

The solution set is $\{5\}$. To check, note that $\log_2 32 = 5$ since $2^5 = 32$.

3b. $\log_6 x = 3$

$6^3 = x$ — Rewrite as an exponential equation.

$216 = x$ — Simplify the exponent.

The solution set is $\{216\}$.

3c. $\log_x 49 = 2$

$x^2 = 49$ — Rewrite as an exponential equation.

$x = \pm\sqrt{49}$ — Apply the square root property.

$x = \pm 7$ — Simplify the square root.

Because the base of a logarithm must be greater than zero, the only solution is $x = 7$. So, the solution set is $\{7\}$.

3d. $\log_b b = x$

$b^x = b$ — Rewrite as an exponential equation.

$b^x = b^1$ — Write b as b^1.

$x = 1$ — Apply the one-to-one property of b^x.

The solution set is $\{1\}$.

Student Check 3 Solve each equation.

a. $\log_8 64 = x$ **b.** $\log_{10} x = 2$ **c.** $\log_x 8 = 3$ **d.** $\log_b b^2 = x$

Graphing Basic Logarithmic Functions

Objective 4 Graph basic logarithmic functions.

Recall in Objective 1, we found that the inverse of $f(x) = 2^x$ is $f^{-1}(x) = \log_2 x$. The inverse function, $f^{-1}(x) = \log_2 x$, is called a *logarithmic function*.

Definition: For $x > 0$ and $b > 0$ $(b \neq 1)$, a function of the form $f(x) = \log_b x$ is called a **logarithmic function**. The domain of $f(x)$ is $(0, \infty)$ and the range is $(-\infty, \infty)$.

Tables of values for $f(x) = 2^x$ and $f^{-1}(x) = \log_2 x$ are shown. Recall we interchange the x- and y-values to obtain the inverse function.

x	$f(x) = 2^x$
-2	$\frac{1}{4}$
-1	$\frac{1}{2}$
0	1
1	2
2	4

x	$f^{-1}(x) = \log_2 x$
$\frac{1}{4}$	-2
$\frac{1}{2}$	-1
1	0
2	1
4	2

When we plot the points for f and f^{-1}, we see that their graphs are symmetric about the line $y = x$.

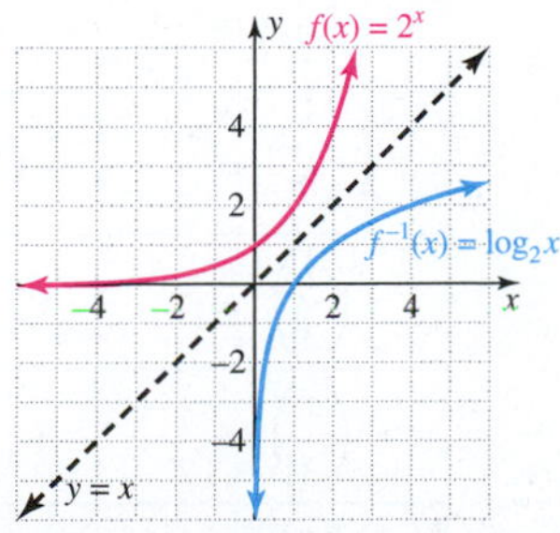

Because $y = b^x$ and $y = \log_b x$ are inverse functions, we know some characteristics about the graph of the logarithmic function.

Characteristics of the Graph of $f(x) = b^x, b > 0, b \neq 1$	Characteristics of the Graph of $f(x) = \log_b x, b > 0, b \neq 1$
The domain of the function is $(-\infty, \infty)$.	The domain of the function is $(0, \infty)$.
The range is $(0, \infty)$.	The range is $(-\infty, \infty)$.
The y-intercept is $(0, 1)$.	The x-intercept is $(1, 0)$.
There is no x-intercept. The graph gets very close to the x-axis but never touches it.	There is no y-intercept. The graph gets very close to the y-axis but never touches it.
The point $(1, b)$ lies on the graph.	The point $(b, 1)$ lies on the graph.
The function is one-to-one.	The function is one-to-one.
If $b > 1$, the function is increasing.	If $b > 1$, the function is increasing.
If $0 < b < 1$, the function is decreasing.	If $0 < b < 1$, the function is decreasing.

Procedure: Graphing a Function of the Form $y = \log_b x$

Step 1: Convert the equation to its exponential form $b^y = x$.
Step 2: Substitute values for y and solve for x.
Step 3: Plot the ordered pairs (x, y) and connect the points with a smooth curve.

Objective 4 Examples Graph each logarithmic function. State the domain and range of each function.

4a. $f(x) = \log_3 x$ **4b.** $f(x) = \log_{1/3} x$

Solutions **4a.** The function $f(x) = \log_3 x$ is equivalent to $3^y = x$. So, we make a table, assign values to y and solve for x.

$x = 3^y$	y	(x, y)
$3^{-2} = \frac{1}{9}$	-2	$\left(\frac{1}{9}, -2\right)$
$3^{-1} = \frac{1}{3}$	-1	$\left(\frac{1}{3}, -1\right)$
$3^0 = 1$	0	$(1, 0)$
$3^1 = 3$	1	$(3, 1)$
$3^2 = 9$	2	$(9, 2)$

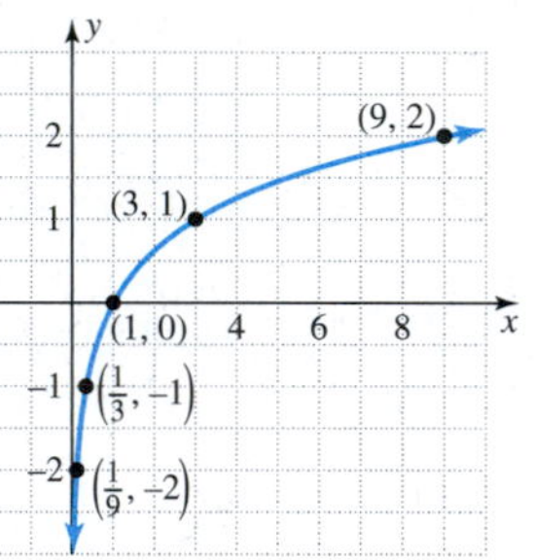

The domain is $(0, \infty)$ and the range is $(-\infty, \infty)$. Notice that the x-intercept is $(1, 0)$ and there is no y-intercept.

4b. The equation $y = \log_{1/3} x$ is equivalent to $\left(\frac{1}{3}\right)^y = x$. So, we make a table, assigning values to y and solving for x. We get

$x = \left(\frac{1}{3}\right)^y$	y	(x, y)
$\left(\frac{1}{3}\right)^{-2} = 9$	-2	$(9, -2)$
$\left(\frac{1}{3}\right)^{-1} = 3$	-1	$(3, -1)$
$\left(\frac{1}{3}\right)^{0} = 1$	0	$(1, 0)$
$\left(\frac{1}{3}\right)^{1} = \frac{1}{3}$	1	$\left(\frac{1}{3}, 1\right)$
$\left(\frac{1}{3}\right)^{2} = \frac{1}{9}$	2	$\left(\frac{1}{9}, 2\right)$

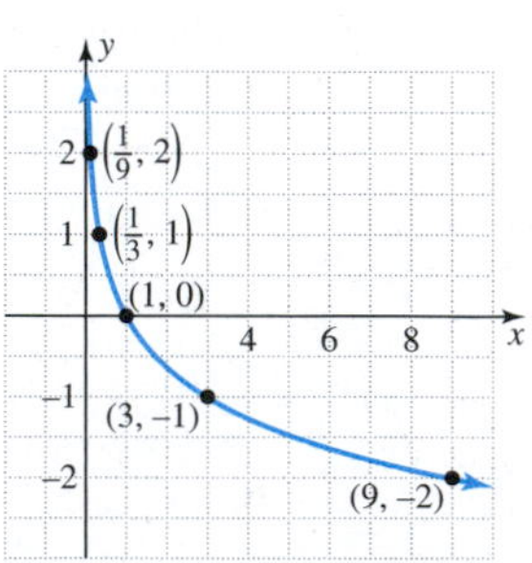

The domain is $(0, \infty)$ and the range is $(-\infty, \infty)$. Notice that the x-intercept is $(1, 0)$ and there is no y-intercept.

Student Check 4 Graph each logarithmic function. State the domain and range of each function.

a. $y = \log_4 x$ **b.** $y = \log_{1/4} x$

Objective 5 Troubleshoot common errors.

Troubleshooting Common Errors

Some common errors associated with logarithms are shown.

Objective 5 Examples

A problem and an incorrect solution are given. Provide the correct solution and an explanation of the error.

5a. Write $3^{-2} = \frac{1}{9}$ in logarithmic form.

Incorrect Solution	Correct Solution and Explanation
$3^{-2} = \frac{1}{9}$ is equivalent to $\log_3(-2) = \frac{1}{9}$.	The logarithm is always equal to the exponent of the base. So, $3^{-2} = \frac{1}{9}$ is equivalent to $\log_3 \frac{1}{9} = -2$.

5b. Simplify $\log_2(-4)$.

Incorrect Solution	Correct Solution and Explanation
$\log_2(-4) = -2$	The argument of a logarithm must be greater than zero. So, $\log_2(-4)$ is not defined. There is no exponent of 2 that will produce -4. Note that $2^{-2} = \frac{1}{4}$ not -4.

ANSWERS TO STUDENT CHECKS

Student Check 1 **a.** $3^4 = 81$ **b.** $2^0 = 1$ **c.** $6^1 = 6$ **d.** $5^{-3} = \frac{1}{125}$ **e.** $\log_6 36 = 2$ **f.** $\log_4 \frac{1}{4} = -1$ **g.** $\log_9 27 = \frac{3}{2}$ **h.** $\log_9 3 = \frac{1}{2}$

Student Check 2 **a.** 0 **b.** 1 **c.** -2 **d.** $\frac{1}{3}$ **e.** -1 **f.** undefined

Student Check 3 **a.** $\{2\}$ **b.** $\{100\}$ **c.** $\{2\}$ **d.** $\{2\}$

Student Check 4

a.

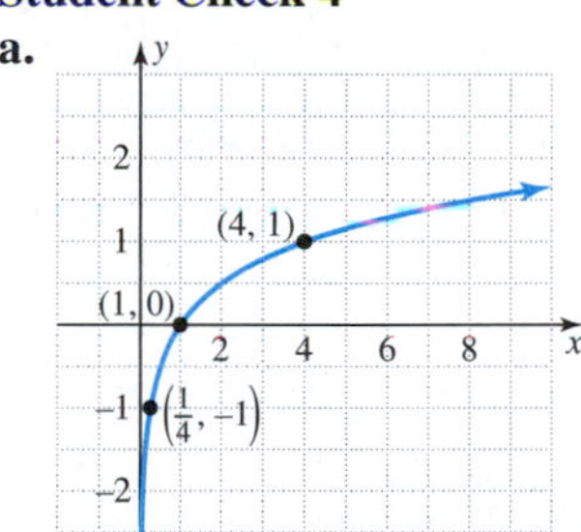

b.

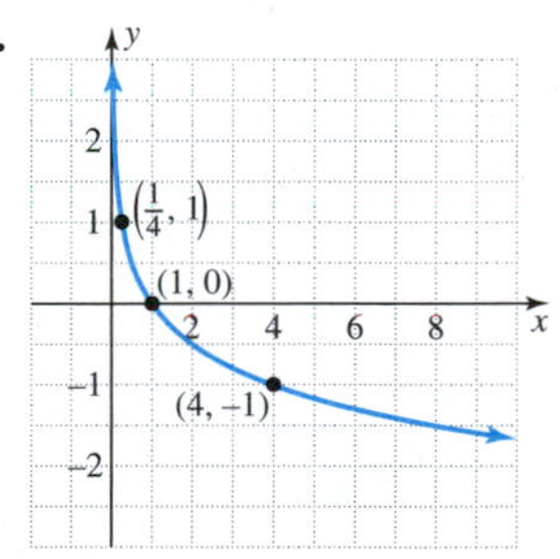

SUMMARY OF KEY CONCEPTS

1. For $b > 0$ and $b \neq 1$, $y = \log_b x$ is equivalent to $b^y = x$. The log base b of x means to find the exponent of b that produces x.
2. We can simplify logarithmic expressions by finding the exponent of the base that produces the argument. Some key logarithms for $b > 0, b \neq 1$ are as follows.
   - **a.** $\log_b 1 = 0$ since $b^0 = 1$.
   - **b.** $\log_b b = 1$ since $b^1 = b$.
   - **c.** $\log_b b^x = x$ since $b^x = b^x$.
3. We can solve logarithmic equations by converting the equation to its equivalent exponential form.
4. We can graph logarithmic functions by converting them to their equivalent exponential form. Some of the key properties are
   - The domain of the function is $(0, \infty)$ and the range is $(-\infty, \infty)$.
   - The x-intercept is $(1, 0)$. There is no y-intercept. The graph gets very close to the y-axis but never touches it.
   - The point $(b, 1)$ lies on the graph.
   - The function is one-to-one.
   - If $b > 1$, the function is increasing and if $0 < b < 1$, the function is decreasing.

SECTION 10.4 EXERCISE SET

Write About It!

Use complete sentences in your answer to each exercise.

1. Define a logarithm. Answers vary. A logarithm is an exponent.
2. Explain how to write the exponential equation, $x^a = b$, in logarithmic form.
3. Explain how to write the logarithmic equation, $\log_w y = z$, in exponential form.
4. Explain why $\log_a 1 = 0$, $a > 0$. Answers vary. Write it in exponential form: $a^0 = 1$.
5. Explain why $\log_a a = 1$, $a > 0$. Answers vary. Write it in exponential form: $a^1 = a$.
6. Explain how to set up a table to graph a logarithmic function.
7. Explain how to simplify the expression $\log_b b^5$.
8. Explain how to simplify the expression $a^{\log_a 2}$. Answers vary. Set the expression equal to x, convert to logarithmic form, and solve the resulting equation for x.

Practice Makes Perfect!

Rewrite each logarithmic equation in exponential form and each exponential equation in logarithmic form. (*See Objective 1.*)

9. $\log_2 32 = 5$ $2^5 = 32$
10. $\log_{10} 100 = 2$ $10^2 = 100$
11. $\log_8 8 = 1$ $8^1 = 8$
12. $\log_5 5 = 1$ $5^1 = 5$
13. $\log_7 1 = 0$ $7^0 = 1$
14. $\log_{12} 1 = 0$ $12^0 = 1$
15. $\log_2 \frac{1}{16} = -4$ $2^{-4} = \frac{1}{16}$
16. $\log_4 \frac{1}{64} = -3$ $4^{-3} = \frac{1}{64}$
17. $6^3 = 216$ $\log_6 216 = 3$
18. $12^2 = 144$ $\log_{12} 144 = 2$
19. $11^1 = 11$ $\log_{11} 11 = 1$
20. $10^1 = 10$ $\log_{10} 10 = 1$
21. $15^0 = 1$ $\log_{15} 1 = 0$
22. $2^0 = 1$ $\log_2 1 = 0$
23. $7^{-2} = \frac{1}{49}$ $\log_7 \frac{1}{49} = -2$
24. $10^{-3} = \frac{1}{1000}$ $\log_{10} \frac{1}{1000} = -3$
25. $4^{3/2} = 8$ $\log_4 8 = \frac{3}{2}$
26. $125^{2/3} = 25$ $\log_{125} 25 = \frac{2}{3}$
27. $\sqrt{144} = 12$ $\log_{144} 12 = \frac{1}{2}$
28. $\sqrt[3]{64} = 4$ $\log_{64} 4 = \frac{1}{3}$
29. $\sqrt[4]{81} = 3$ $\log_{81} 3 = \frac{1}{4}$
30. $\sqrt[3]{1000} = 10$ $\log_{1000} 10 = \frac{1}{3}$

Simplify each expression. (*See Objective 2.*)

31. $\log_5 1$ 0
32. $\log_{11} 1$ 0
33. $\log_{13} 13$ 1
34. $\log_7 7$ 1
35. $\log_{1/2} \frac{1}{2}$ 1
36. $\log_{1/5} \frac{1}{5}$ 1
37. $\log_2 \frac{1}{32}$ -5
38. $\log_3 \frac{1}{243}$ -5
39. $\log_{1/2} 4$ -2
40. $\log_{1/3} 81$ -4
41. $\log_3(-9)$ undefined
42. $\log_{10}(-10)$ undefined

Solve each logarithmic equation. (*See Objective 3.*)

43. $\log_5 625 = x$ $\{4\}$
44. $\log_3 243 = x$ $\{5\}$
45. $\log_6 216 = x$ $\{3\}$
46. $\log_{10} 1000 = x$ $\{3\}$
47. $\log_8 x = 3$ $\{512\}$
48. $\log_3 x = 4$ $\{81\}$
49. $\log_x 343 = 3$ $\{7\}$
50. $\log_x 100 = 2$ $\{10\}$
51. $\log_{10} 1 = x$ $\{0\}$
52. $\log_7 1 = x$ $\{0\}$
53. $\log_a a^3 = x$ $\{3\}$
54. $\log_a a^5 = x$ $\{5\}$
55. $\log_b b^{-2} = x$ $\{-2\}$
56. $\log_b b^{-6} = x$ $\{-6\}$

Graph each logarithmic function. Plot and label at least three points. State the domain and range of each function. (*See Objective 4.*)

57. $y = \log_5 x$
58. $y = \log_{10} x$
59. $y = \log_{1.5} x$
60. $y = \log_{2.5} x$
61. $y = \log_{1/2} x$
62. $y = \log_{1/10} x$

Mix 'Em Up!

Solve each logarithmic equation.

63. $\log_5 125 = x$ $\{3\}$
64. $\log_2 128 = x$ $\{7\}$
65. $\log_c c^4 = x$ $\{4\}$
66. $\log_c c^8 = x$ $\{8\}$
67. $\log_5 x = 4$ $\{625\}$
68. $\log_3 x = 5$ $\{243\}$
69. $\log_x \frac{1}{125} = 3$ $\left\{\frac{1}{5}\right\}$
70. $\log_x \frac{1}{100} = 2$
71. $\log_x 3 = \frac{1}{3}$ $\{27\}$
72. $\log_x 2 = \frac{1}{4}$ $\{16\}$
73. $\log_{1.2} 1 = x$ $\{0\}$
74. $\log_{5.4} 1 = x$ $\{0\}$
75. $\log_{1/4} 4 = x$ $\{-1\}$
76. $\log_{1/5} 25 = x$ $\{-2\}$
77. $\log_d d^{-3} = x$ $\{-3\}$
78. $\log_d d^{-7} = x$ $\{-7\}$
79. $\log_6 \sqrt{6} = x$
80. $\log_{10} \sqrt{10} = x$

Graph each logarithmic function. Plot and label at least three points. State the domain and range of each function.

81. $y = \log_8 x$
82. $y = \log_{1.6} x$
83. $y = \log_{0.4} x$
84. $y = \log_{0.8} x$

You Be the Teacher!

Correct each student's errors, if any.

85. Solve the equation $\log_{1/4} 16 = x$.

Ahmed's work:

$$\log_{1/4} 16 = x$$
$$\frac{1}{4}x = 16$$
$$x = 64$$

$\log_{1/4} 16 = x$
$\left(\frac{1}{4}\right)^x = 16$
$4^{-x} = 4^2$
$-x = 2$
$x = -2$

86. Solve the equation $\log_x(-25) = 2$.

Celina's work:

$$\log_x(-25) = 2$$
$$x^2 = \frac{1}{25}$$
$$x = \frac{1}{5}$$

$\log_x(-25) = 2$
$x^2 = -25$
There is no solution.

Additional answers can be found in the Instructor Answer Appendix.

PIECE IT TOGETHER SECTIONS 10.1–10.4

1. Let $f(x) = 4x + 15$ and $g(x) = 3 - 2x$. Find $f + g, f - g, f \cdot g, \frac{f}{g}, f \circ g$, and $g \circ f$. State the domain of each combined function. (*Section 10.1, Objectives 1 and 3*)

Let $f(x) = 5x^2 + 1$ and $g(x) = 6 - x^2$. Evaluate each function. (*Section 10.1, Objective 2*)

2. $(f + g)(-3)$ 43
3. $(f - g)(4)$ 91
4. $(f \cdot g)(3)$ -138
5. $\left(\frac{f}{g}\right)(-2)$ $\frac{21}{2}$

Determine if each function is one-to-one. If so, find its inverse. Verify the functions are inverses. (*Section 10.2, Objectives 1–4*)

6. $f(x) = \{(-4, 1), (-2, 5), (0, 5), (2, 6)\}$ not a one-to-one function
7. $f(x) = \frac{x}{x - 4}$ one-to-one; $f^{-1}(x) = \frac{4x}{x - 1}$

Graph each function. Plot and label at least three points on the graph. State the domain and range of each function. (*Section 10.3, Objective 1, and Section 10.4, Objective 4*)

8. $f(x) = 1.5^x$
9. $f(x) = \log_{1/5} x$

Solve each exponential equation. (*Section 10.3, Objective 2*)

10. $3^{2x-5} = 27^x$ $\{-5\}$
11. $2^{5x-5} = 4^x$ $\left\{\frac{5}{3}\right\}$
12. The population of Mexico (in millions), can be modeled by the function $f(x) = 111.21(1.0112)^x$, where x is the number of years after 2009. What is an estimate for the population in 2012? 2020? 2030? Round answers to the nearest million. 115 million, 126 million, 141 million

Simplify each expression. (*Section 10.4, Objective 2*)

13. $\log_8 \frac{1}{512}$ -3
14. $\log_{1/5} 125$ -3
15. $\log_{3.5} 3.5$ 1
16. $\log_4(-16)$ undefined

Solve each logarithmic equation. (*Section 10.4, Objective 3*)

17. $\log_6 x = 4$ $\{1296\}$
18. $\log_x 32 = 5$ $\{2\}$

SECTION 10.5 Properties of Logarithms

OBJECTIVES

As a result of completing this section, you will be able to

1. Apply the product rule for logarithms.
2. Apply the quotient rule for logarithms.
3. Apply the power rule for logarithms.
4. Rewrite logarithmic expressions using two or more rules.
5. Apply inverse properties of logarithms.
6. Troubleshoot common errors.

In Section 10.4, we introduced the concept of a logarithm and discovered that a logarithm is actually an exponent. In this section, we will apply properties of exponents and inverse properties to develop properties of logarithms.

Objective 1 Apply the product rule for logarithms.

The Product Rule for Logarithms

The first rule that we will develop is the product rule for logarithms, or logs for short. That is, we want to discover a rule for simplifying the log of a product. Consider the following expressions.

Based on Definition	Observed Rule
$\log_2(4 \cdot 8) = \log_2 32 = 5$ since $2^5 = 32$	The log of a product can be obtained by adding the log of each factor.
$\log_2 4 = 2$ and $\log_2 8 = 3$ since $2^2 = 4$ and $2^3 = 8$	$\log_2(4 \cdot 8) = \log_2 4 + \log_2 8$ $5 = 2 + 3$

Property: Product Rule for Logarithms

For x, y, and $b > 0$ with $b \neq 1$,

$$\log_b(xy) = \log_b x + \log_b y$$

This property states that the log of a product is the sum of the log of the factors. To verify this property, we use the product of like bases rule for exponents, $b^m \cdot b^n = b^{m+n}$. We assign variables to each exponential expression and write them in their logarithmic form.

$$x = b^m \text{ is equivalent to } \log_b x = m$$
$$y = b^n \text{ is equivalent to } \log_b y = n$$

Next multiply the left sides and right sides of the exponential equations to obtain $x \cdot y = b^m \cdot b^n$. So, $x \cdot y = b^{m+n}$. If we write this equation in its logarithmic form, we have

$$\log_b(x \cdot y) = m + n$$

Since $m = \log_b x$ and $n = \log_b y$, we have

$$\log_b(xy) = \log_b x + \log_b y$$

Procedure: Applying the Product Rule for Logarithms

Step 1: If the expression involves the sum of two logarithms with the same base, then the expression can be combined into one logarithm. It is equal to the logarithm of the product of the arguments of the two logarithms.

Step 2: If the expression involves the logarithm of an expression with factors, then the expression can be written as the sum of the logarithms of each factor.

Objective 1 Examples **Use the product rule for logarithms to write each expression as a single logarithm. Simplify, if possible.**

1a. $\log_2 3 + \log_2 5$ **1b.** $\log_4 \frac{1}{3} + \log_4 12$ **1c.** $\log_3 x + \log_3(x + 1)$

Solutions

1a.
$$\log_2 3 + \log_2 5 = \log_2(3 \cdot 5)$$
$$= \log_2 15$$

1b.
$$\log_4 \frac{1}{3} + \log_4 12 = \log_4\left(\frac{1}{3} \cdot 12\right)$$
$$= \log_4 4$$
$$= 1$$

1c.
$$\log_3 x + \log_3(x + 1) = \log_3[x(x + 1)]$$
$$= \log_3(x^2 + x)$$

Student Check 1 Use the product rule for logarithms to write each expression as a single logarithm. Simplify, if possible.

1a. $\log_6 2 + \log_6 7$ **1b.** $\log_5 \frac{1}{2} + \log_5 10$ **1c.** $\log_2 x + \log_2(x - 3)$

The Quotient Rule for Logarithms

Objective 2

Apply the quotient rule for logarithms.

Now we investigate what happens when we simplify the log of a quotient.

Based on Definition

$$\log_2 \frac{4}{8} = \log_2 \frac{1}{2} = -1$$

$$\text{since } 2^{-1} = \frac{1}{2}$$

Observed Rule

The log of a quotient can be obtained by subtracting the log of the numerator and the log of the denominator.

$$\log_2 \frac{4}{8} = \log_2 4 - \log_2 8$$
$$-1 = 2 - 3$$

Property: Quotient Rule for Logarithms

For x, y, and $b > 0$ and $b \neq 1$,

$$\log_b \frac{x}{y} = \log_b x - \log_b y$$

This property states that the log of a quotient is the difference of the log of the numerator and denominator. To verify this property, we use the quotient of like bases rule for exponents, $\frac{b^m}{b^n} = b^{m-n}$. We assign variables to each exponential expression and write them in their logarithmic form.

$$x = b^m \text{ is equivalent to } \log_b x = m$$
$$y = b^n \text{ is equivalent to } \log_b y = n$$

Next divide the left sides and right sides of the exponential equations to obtain $\frac{x}{y} = \frac{b^m}{b^n}$. So, $\frac{x}{y} = b^{m-n}$. If we write this equation in its logarithmic form, we have

$$\log_b \frac{x}{y} = m - n$$

Since $m = \log_b x$ and $n = \log_b y$, we have

$$\log_b \frac{x}{y} = \log_b x - \log_b y$$

Procedure: Applying the Quotient Rule for Logarithms

Step 1: If the expression involves the difference of two logarithms with the same base, then the expression can be combined into one logarithm. It is equal to the logarithm of the quotient of the arguments of the two logarithms.

Step 2: If the expression involves the logarithm of an expression with a quotient, then the expression can be written as the difference of the logarithm of the numerator and the logarithm of the denominator.

Objective 2 Examples Use the quotient rule for logarithms to write each expression as a single logarithm. Simplify, if possible.

2a. $\log_2 20 - \log_2 5$ **2b.** $\log_3 x - \log_3 7$ **2c.** $\log_{10}(x + 2) - \log_{10}(x - 3)$

Solutions

2a. $\log_2 20 - \log_2 5 = \log_2 \frac{20}{5}$

$= \log_2 4$

$= 2$

2b. $\log_3 x - \log_3 7 = \log_3 \frac{x}{7}$

2c. $\log_{10}(x + 2) - \log_{10}(x - 3) = \log_{10}\left(\frac{x + 2}{x - 3}\right)$

Student Check 2 Use the quotient rule for logarithms to write each expression as a single logarithm. Simplify, if possible.

a. $\log_3 54 - \log_3 2$ **b.** $\log_2 y - \log_2 5$ **c.** $\log_5(x + 1) - \log_5(x - 2)$

Power Rule for Logarithms

Objective 3

Apply the power rule for logarithms.

The last rule of logarithms we develop is the power rule.

Based on Definition	**Observed Rule**
$\log_2 4^3 = \log_2 64 = 6$ since $2^6 = 64$	The log of a power can be obtained by multiplying the exponent by the log of the base. $\log_2 4^3 = 3\log_2 4$, $6 = 3(2)$

Property: Power Rule for Logarithms

For x and $b > 0$ with $b \neq 1$,

$$\log_b x^n = n\log_b x$$

This property states that the log of an exponential expression is the exponent times the log of the base of the exponential expression. To verify this property, we must use the power of a power rule for exponents, $(b^m)^n = b^{mn}$. So, we assign a variable to b^m and convert it to its logarithmic form.

$$x = b^m \text{ is equivalent to } \log_b x = m$$

So, if we raise each side of the exponential equation to an exponent of n, we get

$$x^n = (b^m)^n$$
$$x^n = b^{mn}$$

If we write this in logarithmic form, we have

$$\log_b x^n = mn$$

Since $\log_b x = m$, we have

$$\log_b x^n = [\log_b x]n$$
$$\log_b x^n = n\log_b x$$

This rule enables us to solve exponential equations in Section 10.6.

Procedure: Applying the Power Rule for Logarithms

Step 1: If the expression involves a coefficient of a logarithm, then the expression can be written with a coefficient of one. The coefficient of the logarithm becomes the exponent of the argument of the logarithm.

Step 2: If the expression involves the logarithm of an exponential expression, then the exponent of the exponential expression becomes the coefficient of the logarithm.

Objective 3 Examples Use the power rule for logarithms to rewrite each expression.

3a. $\log_2 3^5$ **3b.** $\log_4 x^2$ **3c.** $\log_3 \sqrt{6}$ **3d.** $\log_{10} 10^{x-1}$

Solutions

3a. $\log_2 3^5 = 5\log_2 3$

3b. $\log_4 x^2 = 2\log_4 x$

3c. $\log_3 \sqrt{6} = \log_3 6^{1/2}$

$$= \frac{1}{2}\log_3 6$$

3d. $\log_{10}10^{x-1} = (x-1)\log_{10}10$

$= (x-1)(1)$

$= x - 1$

Student Check 3 Use the power rule for logarithms to rewrite each expression.

a. $\log_4 2^5$ **b.** $\log_3 x^4$ **c.** $\log_2\sqrt{7}$ **d.** $\log_{10}10^{x+2}$

Applying Two or More Logarithmic Properties

Objective 4 Rewrite logarithmic expressions using two or more rules.

To rewrite some expressions, we must apply more than one rule of logarithms. A summary of the rules is provided.

Property: Summary of Logarithmic Properties

Product rule: $\log_b xy = \log_b x + \log_b y$

Quotient rule: $\log_b \frac{x}{y} = \log_b x - \log_b y$

Power rule: $\log_b x^n = n\log_b x$

Procedure: Rewriting Logarithmic Expressions as a Single Logarithm

Step 1: Apply the power rule by making the coefficients of the logarithm the exponent of the argument of the logarithm.

Step 2: Work from left to right, applying the product or quotient rules as necessary.

Procedure: Expanding a Single Logarithm as a Combination of Logarithmic Expressions

1. If the argument of the logarithm is a quotient, apply the quotient rule.
   - **a.** If either the numerator or denominator is a product, apply the product rule.
   - **b.** If any of the arguments of the logarithms have a power, apply the power rule.
2. If the argument of the logarithm is a product, apply the product rule. If any of the factors has an exponent, apply the power rule.

Objective 4 Examples Rewrite each logarithmic expression. For parts (a)–(c), write the expression as a single logarithm. For parts (d) and (e), expand the logarithmic expression.

4a. $3\log_9 2 + 4\log_9 3$ **4b.** $2\log_3 x + \frac{1}{2}\log_3(x+1)$

4c. $\log_4 x - 2\log_4 y$ **4d.** $\log_2\left(\frac{3x}{y}\right)$ **4e.** $\log_5\left(\frac{x^3}{y^4}\right)$

Solutions

4a. $3\log_9 2 + 4\log_9 3 = \log_9 2^3 + \log_9 3^4$ — Apply the power rule for logarithms.

$= \log_9 8 + \log_9 81$ — Simplify the exponents.

$= \log_9(8 \cdot 81)$ — Apply the product rule for logarithms.

$= \log_9 648$ — Simplify.

4b. $2\log_3 x + \frac{1}{2}\log_3(x+1)$

$= \log_3 x^2 + \log_3(x+1)^{1/2}$ Apply the power rule for logarithms.

$= \log_3 x^2 + \log_3\sqrt{x+1}$ Apply $b^{\frac{1}{n}} = \sqrt[n]{b}$.

$= \log_3\left(x^2\sqrt{x+1}\right)$ Apply the product rule for logarithms.

4c. $\log_4 x - 2\log_4 y = \log_4 x - \log_4 y^2$ Apply the power rule for logarithms.

$= \log_4\left(\frac{x}{y^2}\right)$ Apply the quotient rule for logarithms.

4d. $\log_2\left(\frac{3x}{y}\right) = \log_2 3x - \log_2 y$ Apply the quotient rule for logarithms.

$= \log_2 3 + \log_2 x - \log_2 y$ Apply the product rule for logarithms.

4e. $\log_5\left(\frac{x^3}{y^4}\right) = \log_5 x^3 - \log_5 y^4$ Apply the quotient rule for logarithms.

$= 3\log_5 x - 4\log_5 y$ Apply the power rule for logarithms.

Student Check 4 Rewrite each logarithmic expression. For parts (a)–(c), write the expression as a single logarithm. For parts (d) and (e), expand the logarithmic expression.

a. $2\log_6 3 + 3\log_6 2$ **b.** $3\log_2 y + \frac{1}{3}\log_2(y+2)$

c. $\log_3 x - 4\log_3 y$ **d.** $\log_4\left(\frac{3a}{b}\right)$ **e.** $\log_7\left(\frac{x^2}{y^3}\right)$

Inverse Properties of Logarithms

Objective 5 Apply inverse properties of logarithms.

In Section 10.4 we learned that the functions $f(x) = b^x$ and $f(x) = \log_b x$ ($b > 0, b \neq 1$) are inverse functions, and in Section 10.2 we learned that when we compose a function and its inverse, we get the value x. That is, $(f \circ f^{-1})(x) = x$ and $(f^{-1} \circ f)(x) = x$. Applying these facts to $f(x) = b^x$ and $f^{-1}(x) = \log_b x$, we get

$$(f \circ f^{-1})(x) = f(\log_b x) = b^{\log_b x} = x$$

$$(f^{-1} \circ f)(x) = f^{-1}(b^x) = \log_b b^x = x$$

Property: Inverse Properties of Logarithms

For $b > 0, b \neq 1$,

$$\log_b b^x = x$$

$$b^{\log_b x} = x, x > 0$$

Objective 5 Examples Simplify each expression.

Problems	Solutions
5a. $\log_{10} 10^5$	$\log_{10} 10^5 = 5$
5b. $\log_2 2^{x+1}$	$\log_2 2^{x+1} = x + 1$
5c. $3^{\log_3 9}$	$3^{\log_3 9} = 9$
5d. $6^{\log_6(x+2)}$	$6^{\log_6(x+2)} = x + 2$

Student Check 5 Simplify each expression.

a. $\log_5 5^4$ **b.** $\log_3 3^{x-5}$ **c.** $2^{\log_2 8}$ **d.** $10^{\log_{10}(x-1)}$

Note: *The problems in Example 5 can be simplified in another way. For instance, we can apply the power rule to simplify* $\log_{10}10^5$. *We get* $\log_{10}10^5 = 5\log_{10}10 = 5(1) = 5$. *We can also simplify* $3^{\log_3 9}$ *by first evaluating* $\log_3 9$. *Since* $\log_3 9 = 2$, *we have* $3^{\log_3 9} = 3^2 = 9$.

Objective 6 ▶ Troubleshooting Common Errors

Troubleshoot common errors.

Some common errors associated with logarithmic properties are shown.

Objective 6 Examples **A problem and an incorrect solution are given. Provide the correct solution and an explanation of the error.**

6a. Rewrite $\log_5(x^2 + 3)$, if possible.

Incorrect Solution	Correct Solution and Explanation
$\log_5(x^2+3) = \log_5 x^2 + \log_5 3$ $= 2\log_5 x + \log_5 3$	The sum of logarithms comes from the logarithm of a product, not the logarithm of a sum. The original expression cannot be rewritten. The expression $\log_5(3x^2)$ is written as $\log_5 3 + 2\log_5 x$.

6b. Rewrite $\dfrac{\log_5 3}{\log_5 7}$, if possible.

Incorrect Solution	Correct Solution and Explanation
$\dfrac{\log_5 3}{\log_5 7} = \log_5 3 - \log_5 7$	The difference of logarithms comes from the logarithm of a quotient, not the quotient of logarithms. The original expression cannot be rewritten. The expression $\log_5 \dfrac{3}{7}$ is written as $\log_5 3 - \log_5 7$.

6c. Rewrite $(\log_2 x)^4$, if possible.

Incorrect Solution	Correct Solution and Explanation
$(\log_2 x)^4 = 4\log_2 x$	The power rule can only be applied if the exponent is on the expression inside the logarithm, not when the exponent is on the entire logarithmic expression. The original expression cannot be rewritten. The expression $\log_2 x^4$ is written as $4\log_2 x$.

ANSWERS TO STUDENT CHECKS

Student Check 1 **a.** $\log_6 14$ **b.** 1 **c.** $\log_2(x^2 - 3x)$

Student Check 2 **a.** 3 **b.** $\log_2 \dfrac{y}{5}$ **c.** $\log_5\left(\dfrac{x+1}{x-2}\right)$

Student Check 3 **a.** $5\log_4 2 = \dfrac{5}{2}$ **b.** $4\log_3 x$ **c.** $\dfrac{1}{2}\log_2 7$ **d.** $x + 2$

Student Check 4 **a.** $\log_6 72$ **b.** $\log_2\left(y^3\sqrt[3]{y+2}\right)$ **c.** $\log_3 \dfrac{x}{y^4}$ **d.** $\log_4 3 + \log_4 a - \log_4 b$ **e.** $2\log_7 x - 3\log_7 y$

Student Check 5 **a.** 4 **b.** $x - 5$ **c.** 8 **d.** $x - 1$

SUMMARY OF KEY CONCEPTS

1. The product, quotient, and power rules for logarithms come from the properties of exponents. These rules enable us to rewrite a logarithmic expression as a single logarithm or as a combination of logarithms. The rules are, for x, y, and $b > 0$ with $b \neq 1$,

   Product rule: $\log_b xy = \log_b x + \log_b y$

   Quotient rule: $\log_b \frac{x}{y} = \log_b x - \log_b y$

   Power rule: $\log_b x^n = n\log_b x$

2. The inverse properties of logarithms are, for $b > 0$, $b \neq 1$,

   $\log_b b^x = x$

   $b^{\log_b x} = x, x > 0$

SECTION 10.5 EXERCISE SET

Write About It!

Use complete sentences in your answer to each exercise.

1. Explain how to apply the product rule for logarithms.
2. Explain how to apply the quotient rule for logarithms.
3. Explain how to apply the power rule for logarithms.
4. Explain the difference between $\log_a x^m$ and $(\log_a x)^m$.
5. Explain why $\log_a(x + y) \neq \log_a x + \log_a y$. Give an example.
6. Explain why $\log_a(x - y) \neq \log_a x - \log_a y$. Give an example.

Practice Makes Perfect!

Use the product rule for logarithms to write each expression as a single logarithm. Simplify, if possible. Assume all logarithmic arguments are positive numbers. (*See Objective 1.*)

7. $\log_2 5 + \log_2 3$ $\log_2 15$
8. $\log_5 4 + \log_5 2$ $\log_5 8$
9. $\log_3 5 + \log_3 6$ $\log_3 30$
10. $\log_9 7 + \log_9 6$ $\log_9 42$
11. $\log_3 \frac{2}{5} + \log_3 20$ $\log_3 8$
12. $\log_7 \frac{1}{3} + \log_7 12$ $\log_7 4$
13. $\log_2 \frac{1}{5} + \log_2 10$ 1
14. $\log_6 \frac{2}{7} + \log_6 21$ 1
15. $\log_4 7 + \log_4 x$ $\log_4 7x$
16. $\log_{10} 6 + \log_{10} y$ $\log_{10} 6y$
17. $\log_2 x + \log_2(x - 4)$ $\log_2(x^2 - 4x)$
18. $\log_3 x + \log_3(x + 5)$ $\log_3(x^2 + 5x)$
19. $\log_a(x + 2) + \log_a(x - 3)$ $\log_a(x^2 - x - 6)$
20. $\log_b(x + 3) + \log_b(x + 4)$ $\log_b(x^2 + 7x + 12)$

Use the quotient rule for logarithms to write each expression as a single logarithm. Simplify, if possible. Assume all logarithmic arguments are positive numbers. (*See Objective 2.*)

21. $\log_2 15 - \log_2 3$ $\log_2 5$
22. $\log_4 36 - \log_4 2$ $\log_4 18$
23. $\log_5 45 - \log_5 9$ 1
24. $\log_3 12 - \log_3 4$ 1
25. $\log_3 x - \log_3 8$ $\log_3 \frac{x}{8}$
26. $\log_2 x - \log_2 5$ $\log_2 \frac{x}{5}$
27. $\log_6 24x - \log_6 2x$ $\log_6 12$
28. $\log_7 42y - \log_7 3y$ $\log_7 14$
29. $\log_2(x - 2) - \log_2(x + 3)$ $\log_2\left(\frac{x-2}{x+3}\right)$
30. $\log_3 x - \log_3(x + 6)$ $\log_3\left(\frac{x}{x+6}\right)$
31. $\log_9(x^2 - 25) - \log_9(x - 5)$ $\log_9(x + 5)$
32. $\log_{10}(x^2 - 4) - \log_{10}(x + 2)$ $\log_{10}(x - 2)$

Use the power rule for logarithms to rewrite each expression. Assume all logarithmic arguments are positive numbers. (*See Objective 3.*)

33. $\log_3 5^2$ $2\log_3 5$
34. $\log_2 7^3$ $3\log_2 7$
35. $\log_5 x^4$ $4\log_5 x$
36. $\log_{11} y^6$ $6\log_{11} y$
37. $\log_3(x - 2)^4$ $4\log_3(x - 2)$
38. $\log_5(x + 3)^6$ $6\log_5(x + 3)$
39. $\log_3 \sqrt[5]{x}$ $\frac{1}{5}\log_3 x$
40. $\log_6 \sqrt[3]{y}$ $\frac{1}{3}\log_6 y$
41. $\log_{10} \sqrt{10}$ $\frac{1}{2}$
42. $\log_{12} \sqrt[5]{12}$ $\frac{1}{5}$
43. $\log_{10} 10^{2x-3}$ $2x - 3$
44. $\log_3 3^{4x+5}$ $4x + 5$

Write each expression as a single logarithm. Assume all logarithmic arguments are positive numbers. (*See Objective 4.*)

45. $2\log_3 5 + 3\log_3 2$ $\log_3 200$
46. $5\log_9 2 + \log_9 6$ $\log_9 192$
47. $3\log_6 3 + 2\log_6 5$ $\log_6 675$
48. $3\log_{10} 5 + 2\log_{10} 4$ $\log_{10} 2000$
49. $4\log_5 x + \frac{1}{5}\log_5(x + 6)$ $\log_5 x^4\sqrt[5]{x+6}$
50. $2\log_4 y + \frac{1}{3}\log_4(y - 2)$ $\log_4 y^2\sqrt[3]{y-2}$
51. $\log_{10} x - 3\log_{10} y$ $\log_{10}\frac{x}{y^3}$
52. $3\log_2 x - 5\log_2 y$ $\log_2\frac{x^3}{y^5}$
53. $2\log_5 a - 4\log_5 b$ $\log_5\frac{a^2}{b^4}$
54. $8\log_{10} a - 2\log_{10} b$ $\log_{10}\frac{a^8}{b^2}$

Expand each logarithm. Assume all logarithmic arguments are positive numbers. (*See Objective 4.*)

55. $\log_{10} a^5\sqrt{a - 3}$
56. $\log_6 b^7\sqrt[3]{b + 5}$
57. $\log_2 \frac{x^{15}}{y^3}$ $15\log_2 x - 3\log_2 y$
58. $\log_7 \frac{a^6}{b^8}$ $6\log_7 a - 8\log_7 b$
59. $\log_4 \frac{3x}{y^2}$ $\log_4 3 + \log_4 x - 2\log_4 y$
60. $\log_6 \frac{10a^2}{b}$ $\log_6 10 + 2\log_6 a - \log_6 b$
61. $\log_3 \frac{x^2}{27y}$ $2\log_3 x - 3 - \log_3 y$
62. $\log_7 \frac{z^3}{49w^2}$ $3\log_7 z - 2 - 2\log_7 w$

Additional answers can be found in the Instructor Answer Appendix.

Simplify each expression. Assume all logarithmic arguments are positive numbers. (*See Objective 5.*)

63. $\log_{10}10^6$ 6 **64.** $\log_9 9^5$ 5

65. $\log_3 3^{4x-5}$ $4x-5$ **66.** $\log_{10}10^{3x+7}$ $3x+7$

67. $3^{\log_3 6}$ 6 **68.** $2^{\log_2 x}$ x

69. $4^{\log_4(x+5)}$ $x+5$ **70.** $5^{\log_5(2x+3)}$ $2x+3$

71. $2^{\log_2(2x)}$ $2x$ **72.** $3^{\log_3(4y+5)}$ $4y+5$

Mix 'Em Up!

Use the rules for logarithms to write each expression as a single logarithm. Simplify, if possible. Assume all logarithmic arguments are positive numbers.

73. $\log_3 6+\log_3 14-\log_3 7$ $\log_3 12$

74. $\log_6 4-\log_6 2+\log_6 5$ $\log_6 10$

75. $\log_5 x-\log_5 y+\log_5 z$ $\log_5\left(\frac{xz}{y}\right)$

76. $\log_9 a+\log_9 b-\log_9 c$ $\log_9\left(\frac{ab}{c}\right)$

77. $\log_{10}\frac{4}{3}+\log_{10}18-\log_{10}8$ $\log_{10}3$

78. $\log_7\frac{3}{2}-\log_7 21+\log_7 28$ $\log_7 2$

79. $\log_3 20-\log_3\frac{15}{4}+\log_3 6$ $\log_3 32$

80. $\log_5 8-\log_5\frac{6}{7}+\log_5 3$ $\log_5 28$

81. $2\log_9 6+\log_9 5-\frac{1}{3}\log_9 64$ $\log_9 45$

82. $2\log_2 3+\log_2 10-\frac{1}{3}\log_2 125$ $\log_2 18$

83. $\log_3 x+\log_3(x-1)-\log_3(x^2+4)$ $\log_3\frac{x^2-x}{x^2+4}$

84. $\log_4(y-3)-\log_4(y+1)+\log_4(y+3)$ $\log_4\frac{y^2-9}{y+1}$

85. $2\log_{10}x-\frac{1}{2}\log_{10}x-3\log_{10}(x+1)$ $\log_{10}\frac{x^{3/2}}{(x+1)^3}$

86. $2\log_5 y-\frac{3}{2}\log_5 y-2\log_5(y+4)$ $\log_5\frac{y^{1/2}}{(y+4)^2}$

87. $3\log_6 a+2\log_6 b-5\log_6 c$ $\log_6\left(\frac{a^3b^2}{c^5}\right)$

88. $2\log_7 x-4\log_7 y+\log_7 z$ $\log_7\left(\frac{x^2z}{y^4}\right)$

89. $\log_{10}a-2\log_{10}b-\frac{1}{3}\log_{10}c$ $\log_{10}\left(\frac{a}{b^2\sqrt[3]{c}}\right)$

90. $\log_9 r-\frac{1}{2}\log_9 s+3\log_9 t$ $\log_9\left(\frac{rt^3}{\sqrt{s}}\right)$

91. $\log_6(x^3-8)-\log_6(x-2)$ $\log_6(x^2+2x+4)$

92. $\log_2(x^3+8)-\log_2(x+2)$ $\log_2(x^2-2x+4)$

Write each expression as a combination of logarithms. Assume all logarithmic arguments are positive numbers.

93. $\log_3\frac{5x^2}{y}$ **94.** $\log_6\frac{7a}{b^2}$ $\log_6 7+\log_6 a-2\log_6 b$

95. $\log_7\sqrt{\frac{6x}{y}}$ **96.** $\log_4\sqrt[3]{\frac{a}{5b}}$

97. $\log_2\left(x^5y^3\sqrt{x+3}\right)$ **98.** $\log_4\left(a^6b^3\sqrt[5]{b+1}\right)$

99. $\log_{10}\left(\frac{100a^3}{b^5}\right)$ **100.** $\log_5\left(\frac{25x^2}{y^6}\right)$

101. $\log_4\frac{r^2(r+3)}{s^3}$ **102.** $\log_{10}\frac{x(x^2+1)}{y^4}$

Simplify each expression. Assume all logarithmic arguments are positive numbers.

103. $\log_{10}10^{6x+5}$ $6x+5$ **104.** $\log_3 3^{x+5}$ $x+5$

105. $\log_a a^x$ x **106.** $\log_b b^y$ y

107. $5^{\log_5 16x}$ $16x$ **108.** $2^{\log_2 7y}$ $7y$

109. $a^{\log_a(2x+1)}$ $2x+1$ **110.** $3^{4\log_3 y}$ y^4

111. $6^{2\log_6(3a)}$ $9a^2$ **112.** $5^{4\log_5 2y}$ $16y^4$

You Be the Teacher!

Correct each student's errors, if any.

113. Write the expression $\log_3 4+\log_3 x$ as a single logarithm.

Karen's work:

$\log_3(4+x)$ $\log_3 4+\log_3 x=\log_3(4x)$

114. Write the expression $\log_{10}15-\log_{10}5$.

Mona's work: $\log_{10}15-\log_{10}5=\log_{10}(15-5)=\log_{10}10=1$ $\log_{10}15-\log_{10}5=\log_{10}\left(\frac{15}{5}\right)=\log_{10}3$

115. Simplify the expression $(\log_2 8)^2$.

Myra's work:

$(\log_2 8)^2=(\log_2 2^3)^2=\log_2 2^6=6$

$(\log_2 8)^2=(\log_2 2^3)^2=(3\log_2 2)^2=3^2=9$

116. Simplify the expression $3^{2\log_3 5}$.

Cara's work:

$3^{2\log_3 5}=3^{\log_3 10}=10$ $3^{2\log_3 5}=3^{\log_3 5^2}=5^2=25$

SECTION 10.6 The Common Log, Natural Log, and Change-of-Base Formula

OBJECTIVES

As a result of completing this section, you will be able to

1. Simplify common logarithms.
2. Simplify natural logarithms.
3. Use a calculator to approximate common and natural logarithms.
4. Solve simple logarithmic equations.
5. Use the change-of-base formula.
6. Solve applications of logarithms.
7. Troubleshoot common errors.

Suppose you learn a new task over a one-week period and you become about 90% proficient at it. However, as time goes by, you don't practice and you begin forgetting how to do the task though some of your ability to complete the task is retained. This situation can be modeled with the equation: $P = 90 - 12.5 \ln d$, where P is the percent proficiency and d is the number of days without practicing. What is your proficiency level if 5 days go by without practicing? 10 days? 30 days?

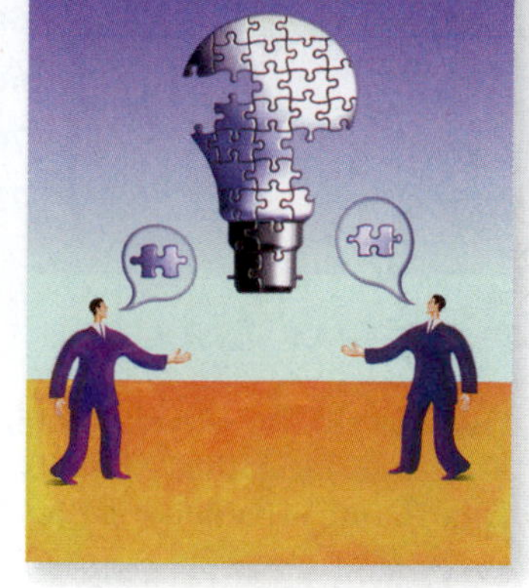

The expression $\ln d$ represents a special type of logarithm. In this section, we will learn about this logarithm along with another special logarithm. These are logarithms with the bases of 10 and the number e.

Common Logarithms

Objective 1 Simplify common logarithms.

On a calculator, we find a special command, the LOG command. This represents a logarithm with an understood base. If we use this command to simplify some logarithms, we can determine what it means. Using a calculator, we find the following.

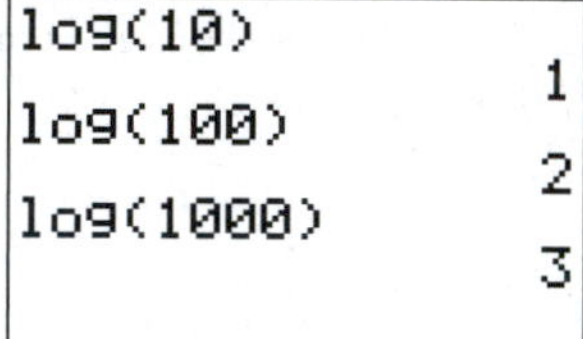

Note: *On a calculator, we must enter the argument of a log in parentheses.*

To determine the base of the LOG function, we let x represent the base. So, we have the following logarithmic statements and their equivalent exponential form.

$$\log_x 10 = 1 \rightarrow x^1 = 10$$
$$\log_x 100 = 2 \rightarrow x^2 = 100$$
$$\log_x 1000 = 3 \rightarrow x^3 = 1000$$

Solving each of these equations for x gives us $x = 10$. Therefore, the LOG command is used to simplify logarithms with a base of 10. So, the LOG function tells us the exponent of 10 that will produce the argument of the logarithm. A logarithm with a base of 10 is called a common logarithm and is written as $\log x$.

Definition: For $x > 0$,

$$y = \log x \text{ is equivalent to } y = \log_{10} x$$

The equivalent exponential form of the **common logarithm** is $10^x = y$.

So, the common logarithm means the exponent of 10 that produces the argument value.

Procedure: Simplifying Common Logarithms

Step 1: If possible, rewrite the argument of the logarithm as a power of 10; that is, as 10^x.

Step 2: Apply the inverse property to simplify the logarithm. Recall $\log 10^x = x$ since the base of the logarithm and the base of the exponent are the same.

Objective 1 Examples Simplify each common logarithm.

Problems	Solutions
1a. $\log 100$	$\log 100 = \log 10^2 = 2$
1b. $\log \frac{1}{100}$	$\log \frac{1}{100} = \log \frac{1}{10^2} = \log 10^{-2} = -2$
1c. $\log 1$	$\log 1 = \log 10^0 = 0$
1d. $\log \sqrt{10}$	$\log \sqrt{10} = \log 10^{1/2} = \frac{1}{2}$

Student Check 1 Simplify each common logarithm.

a. $\log 10$ **b.** $\log \frac{1}{10}$ **c.** $\log 1000$ **d.** $\log \sqrt[3]{10}$

Natural Logarithms

Objective 2 Simplify natural logarithms.

Natural logarithms are another type of logarithm that has a special notation. Natural logarithms are denoted by $\log_e x$ but are usually written as $\ln x$. We read this as "el en of x." On a calculator, the function LN denotes the natural logarithm. The natural logarithm is a logarithm with a base of e. Recall that e is an irrational number and is approximately equal to 2.71828. This number arises in various real-life situations. It is used in financing, exponential growth and decay, and other applications.

Definition: Natural Logarithm

For $x > 0$,

$$y = \ln x \text{ is equivalent to } y = \log_e x$$

The equivalent exponential form of $y = \ln x$ is $e^y = x$.

So, the natural logarithm means to find the exponent of e that produces the argument value.

Procedure: Simplifying Natural Logarithms

Step 1: If possible, rewrite the argument as a power of e; that is, as e^x.

Step 2: Apply the inverse property to simplify the logarithm. $\ln(e^x) = x$ since the base of the logarithm and the base of the exponent are the same.

Objective 2 Examples Simplify each natural logarithm.

Problems	Solutions
2a. $\ln e$	$\ln e = \ln e^1 = 1$
2b. $\ln e^3$	$\ln e^3 = 3$
2c. $\ln 1$	$\ln 1 = \ln e^0 = 0$
2d. $\ln \frac{1}{e^2}$	$\ln \frac{1}{e^2} = \ln e^{-2} = -2$

Student Check 2 Simplify each natural logarithm.

a. $\ln e^2$ **b.** $\ln \sqrt{e}$ **c.** $\ln \frac{1}{e}$

Simplifying Common and Natural Logarithms on a Calculator

Objective 3 ▶

Use a calculator to approximate common and natural logarithms.

We are not restricted to using only powers of 10 and powers of e as arguments of common logarithms and natural logarithms, respectively. We can use a calculator to simplify common and natural logarithms for any positive real number.

Objective 3 Examples **Use a calculator to approximate each logarithm to two decimal places.**

3a. $\log 5$ **3b.** $\ln 10$

Solutions **3a.** Press LOG 5) to get $\log 5 \approx 0.70$.

Check: $10^{0.70} \approx 5.01$. Because of rounding, the value will not be exactly 5 but it is very close, so our work is correct.

3b. Press LN 1 0) to get $\ln 10 \approx 2.30$.

Check: $e^{2.30} \approx 9.97$.

Student Check 3 Use a calculator to approximate each logarithm to two decimal places.

a. $\log 80$ **b.** $\ln 17$

Solving Logarithmic Equations

Objective 4 ▶

Solve simple logarithmic equations.

We can solve simple logarithmic equations by converting them to their exponential form.

Procedure: Solving a Logarithmic Equation Using LOG and LN

Step 1: Write the equation in exponential form.

- **a.** If the equation is of the form $y = \log x$, rewrite it as $10^y = x$.
- **b.** If the equation is of the form $y = \ln x$, rewrite it as $e^y = x$.

Step 2: Solve the resulting equation.

Objective 4 Examples **Solve each equation for *x*. Give an exact solution and approximate the solution to two decimal places, if necessary.**

4a. $\log x = 4$ **4b.** $\log(x + 1) = 2$ **4c.** $\log(2x - 3) = -0.5$

4d. $\ln x = 2$ **4e.** $\ln(2x) = 7$

Solutions **4a.**

$\log x = 4$

$10^4 = x$ Write in exponential form.

$x = 10{,}000$ Simplify.

So, the solution set is $\{10{,}000\}$.

4b. $\log(x + 1) = 2$

$10^2 = x + 1$ Write in exponential form.

$100 = x + 1$ Simplify.

$99 = x$ Subtract 1 from each side.

So, the solution set is $\{99\}$.

4c. $\log(2x - 3) = -0.5$

$$10^{-0.5} = 2x - 3 \quad \text{Write in exponential form.}$$
$$10^{-0.5} + 3 = 2x \quad \text{Add 3 to each side.}$$
$$\frac{10^{-0.5} + 3}{2} = x \quad \text{Divide each side by 2.}$$
$$x \approx 1.66 \quad \text{Simplify.}$$

So, the solution set is $\left\{\frac{10^{-0.5} + 3}{2}\right\}$ or $\{1.66\}$.

4d. $\ln x = 2$

$$e^2 = x \quad \text{Write in exponential form.}$$
$$x \approx 7.39 \quad \text{Simplify.}$$

So, the solution set is $\{e^2\}$ or $\{7.39\}$.

4e. $\ln(2x) = 7$

$$e^7 = 2x \quad \text{Write in exponential form.}$$
$$\frac{e^7}{2} = x \quad \text{Divide each side by 2.}$$
$$x \approx 548.32 \quad \text{Simplify.}$$

So, the solution set is $\left\{\frac{e^7}{2}\right\}$ or $\{548.32\}$.

Student Check 4 Solve each equation for x. Give an exact solution and approximate the solution to two decimal places, if necessary.

a. $\log x = 3$ **b.** $\log(x - 4) = 1$ **c.** $\log(5x + 2) = -0.25$

d. $\ln x = 5$ **e.** $\ln(6x) = 3$

The Change-of-Base Formula

Objective 5 Use the change-of-base formula.

A calculator has built-in functions for only the common logarithm and the natural logarithm; that is, logs with bases of 10 and e. If we need to simplify $\log_3 7$, we need a way to express it in terms of a common logarithm or natural logarithm. So, let $y = \log_3 7$, which is equivalent to $3^y = 7$. This is an exponential equation in which the left side and right side cannot be expressed with the same base. So, we have to use a property of logarithms to solve for y. Recall the power rule states $\log x^n = n \log x$. It is also true that $\ln x^n = n \ln x$. So, we apply the common log to each side, which we can do since logarithms are one-to-one functions.

$$3^y = 7 \quad \text{Write the exponential form of } y = \log_3 7.$$
$$\log 3^y = \log 7 \quad \text{Apply the common log to each side.}$$
$$y \log 3 = \log 7 \quad \text{Apply the power rule for logarithms.}$$
$$y = \frac{\log 7}{\log 3} \quad \text{Divide each side by } \log 3.$$

So, $\log_3 7 = \frac{\log 7}{\log 3}$. Now we can use our calculator to approximate the value of this logarithm. We could have also taken the natural log of both sides to get $\log_3 7 = \frac{\ln 7}{\ln 3}$.

Definition: The Change-of-Base Formula

For $a, b > 0, b \neq 1$,

$$\log_b a = \frac{\log a}{\log b} \quad \text{or} \quad \frac{\ln a}{\ln b}$$

Procedure: Using the Change-of-Base Formula

Step 1: Rewrite the logarithm using either the common log or the natural log.
 a. The log of the argument goes in the numerator.
 b. The log of the base goes in the denominator. (Think "basement.")

Step 2: Use a calculator to simplify the resulting expression.

Objective 5 Examples Use the change-of-base formula to rewrite each logarithm. Approximate its value to two decimal places.

5a. $\log_2 20$ **5b.** $\log_3 0.25$ **5c.** $f(x) = \log_4 x$

Solutions

5a. $\log_2 20 = \frac{\log 20}{\log 2} \approx 4.32$ | $\log_2 20 = \frac{\ln 20}{\ln 2} \approx 4.32$

Check: $2^{4.32} \approx 19.97$

5b. $\log_3 0.25 = \frac{\log 0.25}{\log 3} \approx -1.26$ | $\log_3 0.25 = \frac{\ln 0.25}{\ln 3} \approx -1.26$

Check: $3^{-1.26} \approx 0.25$

5c. $f(x) = \log_4 x = \frac{\log x}{\log 4}$ | $f(x) = \log_4 x = \frac{\ln x}{\ln 4}$

Student Check 5 Use the change-of-base formula to rewrite each logarithm. Approximate its value to two decimal places.

a. $\log_4 20$ **b.** $\log_5 0.1$ **c.** $f(x) = \log_6 x$

Applications

Objective 6

Solve applications of logarithms.

Logarithms are used to find the magnitude of an earthquake, the acidity of a solution, the brightness of a star, the color ratio between two stars, and the intensity level of a sound. Logarithms are also used to solve equations that have variable exponents. These types of equations come up when dealing with compound interest or exponential growth and decay. Section 10.7 will develop some of these ideas further. Example 6 provides examples of applications that can be modeled using logarithms.

Objective 6 Examples Solve each problem.

6a. Suppose you learn a new task over a 1-week period and you become about 90% proficient at it. However, as time goes by, you don't practice and begin to forget how to do the task, though some of your ability to complete the task is retained. This situation can be modeled by the equation $P = 90 - 12.5 \ln d$, where P is the percent proficiency and d is the number of days without practicing. What is the proficiency level if 5 days go by without practicing? 10 days? 30 days? Round answers to two decimal places.

Solution **6a.**

Let $d = 5$:	Let $d = 10$:	Let $d = 30$:
$P = 90 - 12.5 \ln d$	$P = 90 - 12.5 \ln d$	$P = 90 - 12.5 \ln d$
$P = 90 - 12.5 \ln 5$	$P = 90 - 12.5 \ln 10$	$P = 90 - 12.5 \ln 30$
$P \approx 69.88$	$P \approx 61.22$	$P \approx 47.49$

The proficiency level will be 69.88% after 5 days without practicing, 61.22% after 10 days without practicing, and 47.49% after 30 days without practicing.

6b. The moment magnitude scale, denoted M_W, is a scale that measures the energy released by an earthquake and is now preferred over the Richter scale for medium to large earthquakes. The formula is $M_W = \frac{2}{3}\log_{10} M_0 - 10.7$, where M_0 is the magnitude of the seismic moment, or the energy released, in dyne centimeters. As of July 2011, the largest reported moment is 2.5×10^{30} dyn • cm for the 1960 Chile earthquake. Compute the moment magnitude for this earthquake. (Source: http://earthquake.usgs.gov/learn/topics/measure.php)

Solution **6b.**

$M_W = \frac{2}{3}\log_{10} M_0 - 10.7$ State the moment magnitude formula.

$M_W = \frac{2}{3}\log_{10}(2.5 \times 10^{30}) - 10.7$ Replace M_0 with 2.5×10^{30}.

$M_W \approx \frac{2}{3}(30.39794) - 10.7$ Simplify the logarithm.

$M_W \approx 9.6$ Simplify.

Student Check 6 Solve each problem.

a. Suppose you learn a new task over a 1-week period and you become about 85% proficient at it. However, as time goes by, you don't practice and begin to forget how to do the task. Some of your ability to complete the task is retained. This situation can be modeled with the equation: $P = 85 - 12.5 \ln d$, where P is the percent proficiency and d is the number of days without practicing. What is your proficiency level if 5 days go by without practicing? 10 days? 30 days? Round answers to two decimal places.

b. As of July 2011, the second largest reported moment is 7.5×10^{29} dyn•cm for the 1964 Alaska earthquake. Use the moment magnitude scale to compute the moment magnitude for this earthquake. (Source: http://earthquake.usgs.gov/learn/topics/measure.php)

Objective 7 Troubleshoot common errors.

Troubleshooting Common Errors

Some common errors associated with common and natural logs are shown.

Objective 7 Examples **A problem and an incorrect solution are given. Provide the correct solution and an explanation of the error.**

7a. Use the change-of-base formula to approximate $\log_2 3$ to two decimal places.

Incorrect Solution	Correct Solution and Explanation
$\log_2 3 = \dfrac{\log 2}{\log 3} \approx 0.63$	The log of the base goes in the denominator, not the numerator. $\log_2 3 = \dfrac{\log 3}{\log 2} \approx 1.58$

7b. Solve $\log 4x = 2$.

Incorrect Solution	Correct Solution and Explanation
$\log 4x = 2$ $e^2 = 4x$ $\dfrac{e^2}{4} = x$	The notation log represents the common logarithm, which has base 10, not e. $\log 4x = 2$ $10^2 = 4x$ $100 = 4x$ $\dfrac{100}{4} = x$ $25 = x$

ANSWERS TO STUDENT CHECKS

Student Check 1 **a.** 1 **b.** −1 **c.** 3 **d.** $\frac{1}{3}$

Student Check 2 **a.** 2 **b.** $\frac{1}{2}$ **c.** −1

Student Check 3 **a.** 1.90 **b.** 2.83

Student Check 4 **a.** $\{1000\}$ **b.** $\{14\}$
c. $\left\{\frac{10^{-0.25} - 2}{5}\right\}$ or $\{-0.29\}$
d. $\{e^5\}$ or $\{148.41\}$ **e.** $\left\{\frac{e^3}{6}\right\}$ or $\{3.35\}$

Student Check 5 **a.** 2.16 **b.** −1.43
c. $f(x) = \frac{\log x}{\log 6}$ or $\frac{\ln x}{\ln 6}$

Student Check 6 **a.** 64.88%, 56.22%, 42.49% **b.** 9.22

SUMMARY OF KEY CONCEPTS

1. A common logarithm is $y = \log x$, which is equivalent to $y = \log_{10} x$. It represents the exponent of 10 that produces x; that is, $10^y = x$. If the argument of the logarithm is a power of 10, we can use the inverse property of logarithms, $\log 10^x = x$, to simplify the expression.
2. A natural logarithm is $y = \ln x$, which is equivalent to $y = \log_e x$. It represents the exponent of e that produces x; that is, $e^y = x$. If the argument of the logarithm is a power of e, we can use the inverse property of logarithms, $\ln e^x = x$, to simplify the expression.
3. A calculator can be used to simplify common and natural logarithms when the argument is not a power of the base. Use the LOG and LN keys.
4. A simple logarithmic equation can be solved by converting the equation to its exponential form.
5. Any logarithm can be converted to a common log or natural log using the change-of-base formula. For $a, b > 0, b \neq 1$,

$$\log_b a = \frac{\log a}{\log b} \text{ or } \frac{\ln a}{\ln b}$$

GRAPHING CALCULATOR SKILLS

A graphing calculator can be used to simplify common and natural logarithms. We can also graph basic logarithmic functions.

Example 1: Find $\log 50$ and $\ln 50$.

Solution:

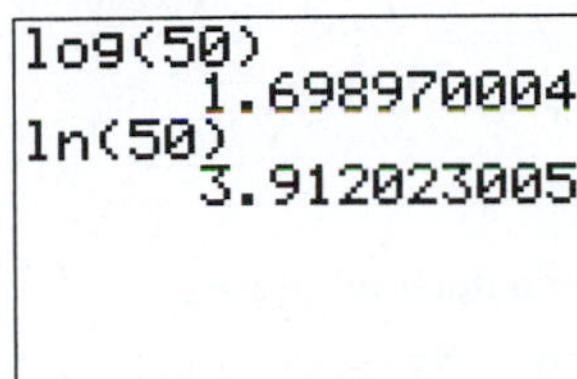

Example 2: Simplify $\log_4 9$.

Solution: $\log_4 9 = \dfrac{\log 9}{\log 4}$

log(9)/log(4)
1.584962501

Example 3: Graph $y = \log x$.

Solution:

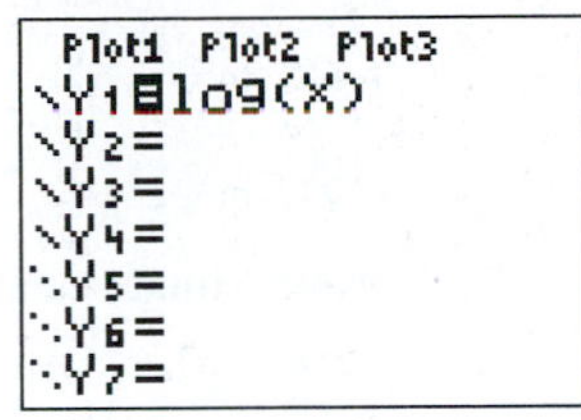

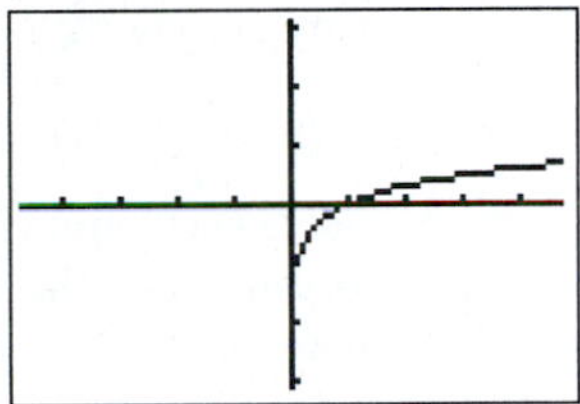

Example 4: Graph $y = \log_2 x$.

Solution: By the change-of-base formula, $y = \log_2 x = \dfrac{\log x}{\log 2}$.

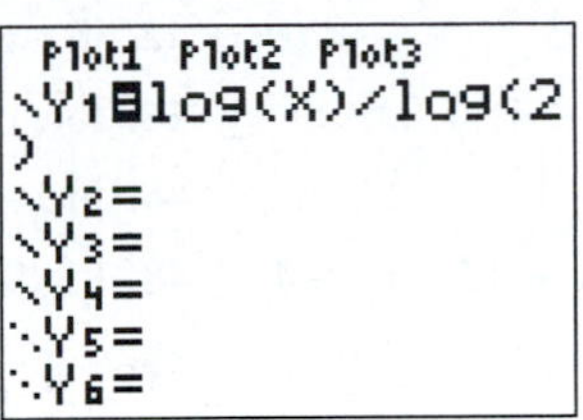

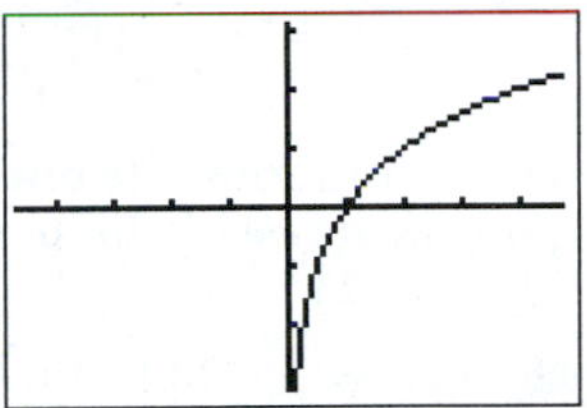

SECTION 10.6 EXERCISE SET

Write About It!

Use complete sentences in your answer to each exercise.

1. What is the base of $y = \log x$? Write the equivalent exponential form. Answers vary. Base is 10. The equivalent exponential form is $x = 10^y$.
2. What is the base of $y = \ln x$? Write the equivalent exponential form. Answers vary. Base is e. The equivalent exponential form is $x = e^y$.
3. Explain how to apply the change-of-base formula. Give an example.
4. Explain why $\dfrac{\log 5}{\log 3}$, $\dfrac{\ln 5}{\ln 3}$, and $\log_3 5$ are the same.
5. Explain why $\ln 1$ and $\log 1$ are both equal to zero. Answers vary. The equivalent exponential form is $10^0 = 1$ and $e^0 = 1$.
6. Explain how to evaluate each expression exactly.
   a. $\log_2 64$ b. $\log_4 64$
   c. $\log_8 64$ d. $\log_{1/2} 64$
7. Explain how to apply the change-of-base formula to evaluate each expression.
   a. $\log_3 64$ b. $\log_5 64$ c. $\log_{3.75} 64$
8. Explain the connection between the parts of Exercises 7 and 8.

Practice Makes Perfect!

Simplify each common logarithm without a calculator. (*See Objective 1.*)

9. $\log 10^5$ 5
10. $\log 10^7$ 7
11. $\log 0.001$ -3
12. $\log 0.0001$ -4
13. $\log \dfrac{1}{1000}$ -3
14. $\log \dfrac{1}{10{,}000}$ -5
15. $\log \sqrt[3]{10}$ $\frac{1}{3}$
16. $\log \sqrt[3]{100}$ $\frac{2}{3}$
17. $\log \sqrt[5]{100}$ $\frac{2}{5}$
18. $\log \sqrt[5]{1000}$ $\frac{3}{5}$

Simplify each natural logarithm without a calculator. (*See Objective 2.*)

19. $\ln e^4$ 4
20. $\ln e^{-3}$ -3
21. $\ln e^8$ 8
22. $\ln e^{-1}$ -1
23. $\ln \dfrac{1}{e^4}$ -4
24. $\ln \dfrac{1}{e^6}$ -6
25. $\ln \sqrt[3]{e}$ $\frac{1}{3}$
26. $\ln \sqrt[3]{e^2}$ $\frac{2}{3}$
27. $\ln \sqrt[5]{e^3}$ $\frac{3}{5}$
28. $\ln \sqrt[4]{e^3}$ $\frac{3}{4}$

Approximate each logarithm to two decimal places. (*See Objective 3.*)

29. $\log 20$ 1.30
30. $\log 45$ 1.65
31. $\log 0.085$ -1.07

Additional answers can be found in the Instructor Answer Appendix.

32. $\log 0.0054$ -2.27 **33.** $\log e$ 0.43 **34.** $\log\sqrt{e}$ 0.22
35. $\ln 30$ 3.40 **36.** $\ln 100$ 4.61 **37.** $\ln\sqrt{10}$ 1.15
38. $\ln\sqrt[3]{100}$ 1.54

Solve each equation for x. Give an exact solution and approximate the solution to two decimal places, if necessary. (*See Objective 4.*)

39. $\log x = 6$ $\{1{,}000{,}000\}$ **40.** $\log x = -1.5$ $\{0.03\}$
41. $\ln x = 3$ $\{e^3\}$ or $\{20.09\}$ **42.** $\ln x = -4$ $\{e^{-4}\}$ or $\{0.02\}$
43. $\log(x - 4) = 0.6$ $\{10^{0.6} + 4\}$ or $\{7.98\}$ **44.** $\log(x + 10) = 1.5$ $\{10^{1.5} - 10\}$ or $\{21.62\}$
45. $\ln(x - 4) = -2$ $\{e^{-2} + 4\}$ or $\{4.14\}$ **46.** $\ln(x - 8) = -2.2$ $\{e^{-2.2} + 8\}$ or $\{8.11\}$
47. $\log(5x + 12) = -4.4$ **48.** $\log(9x + 3) = 3.1$
49. $\ln(2x - 10) = 1.9$ **50.** $\log(4x - 7) = -5.9$
51. $\ln(3x + 5) = -0.9$ **52.** $\ln(12x - 3) = -3.4$
53. $\log(2x) = -1$ **54.** $\log(7x) = 2$ $\left\{\frac{10^2}{7}\right\}$ or $\{14.29\}$

Use the change-of-base formula to rewrite each logarithm. Approximate each value to two decimal places. (*See Objective 5.*)

55. $\log_{12}39.5$ 1.48 **56.** $\log_7 90.1$ 2.31 **57.** $\log_4 2.4$ 0.63
58. $\log_2 0.00518$ -7.59 **59.** $\log_5 0.395$ -0.58 **60.** $\log_6 0.75$ -0.16

Solve each problem. (*See Objective 6.*)

61. Samuel becomes about 95% proficient at a new task over a 1-week period. The equation $P = 95 - 28.782\log d$ models his percent proficiency P after d days of not practicing the new task. What is Samuel's proficiency level after 2 days without practicing? 7 days? 14 days? Round answers to two decimal places. 86.34% after 2 days, 70.68% after 7 days, 62.01% after 14 days

62. Bethany becomes about 90% proficient at a new task over a 1-week period. The equation $P = 90 - 28.782\log d$ models her percent proficiency P after d days of not practicing the new task. What is Bethany's proficiency level after 4 days without practicing? 8 days? 20 days? Round answers to two decimal places. 72.67% after 4 days, 64.01% after 8 days, 52.55% after 20 days

For Exercises 63 and 64, compute the moment magnitude given by $M_w = \frac{2}{3}\log_{10}M_0 - 10.7$, where M_0 is the magnitude of the seismic moment, or the energy released, in dyne centimeters for each earthquake. Round each answer to two decimal places.

63. The reported magnitude of the seismic moment for the earthquake in Tohoku off the coast of Honshu, Japan in March 2011 was 4.04×10^{29} dyn • cm. (Sources: http://earthquake.usgs.gov/learn/topics/measure.php and http://warships1discussionboards.yuku.com/reply/212609/Re-USGS-Press-Release-9-0-Earthquake-no-reactors-string-) 9.04

64. The reported magnitude of the seismic moment for the earthquake in Haiti in January 2010 was 4.4×10^{26} dyn • cm. (Source: http://earthquake.usgs.gov/learn/topics/measure.php) 7.06

Mix 'Em Up!

Simplify each logarithm without a calculator.

65. $\log 10^{5/2}$ $\frac{5}{2}$ **66.** $\log\frac{1}{10^7}$ -7 **67.** $\ln e^{5/3}$ $\frac{5}{3}$
68. $\ln\sqrt[4]{e}$ $\frac{1}{4}$ **69.** $\log\sqrt[5]{10}$ $\frac{1}{5}$ **70.** $\log\frac{1}{\sqrt[4]{10}}$ $-\frac{1}{4}$
71. $\ln\sqrt[7]{e^4}$ $\frac{4}{7}$ **72.** $\ln\frac{1}{e^6}$ -6

Approximate each logarithm to two decimal places.

73. $\log 54$ 1.73 **74.** $\log 0.0279$ -1.55 **75.** $\ln 67$ 4.20
76. $\ln 238$ 5.47 **77.** $\ln\sqrt[3]{10}$ 0.77 **78.** $\ln\sqrt{120}$ 2.39
79. $\ln 0.00812$ -4.81 **80.** $\ln 0.0572$ -2.86

Solve each equation for x. Give an exact solution and approximate the solution to two decimal places, if necessary.

81. $\ln(x - 8) = 5.5$ $\{e^{5.5} + 8\}$ or $\{252.69\}$ **82.** $\ln(x + 13) = 7.1$ $\{e^{7.1} - 13\}$ or $\{1198.97\}$
83. $5\log(6x - 1) = 8.5$ **84.** $6\ln(2x - 3) = 9.6$
85. $\log(10x - 9) = 1.2$ **86.** $\log(5x - 6) = 1.9$
87. $2\ln(4x + 9) = 3$ **88.** $3\ln(7x - 11) = -6.9$
89. $\log(5x) = 1$ $\{2\}$ **90.** $\ln(10x) = 2.7$
91. $\log(4x + 3) = 0$ **92.** $\ln(4x - 5) = 0$

Use the change-of-base formula to rewrite each logarithm. Approximate each value to two decimal places.

93. $\log_{2.3}638.7$ 7.76 **94.** $\log_{8.5}50.9$ 1.84
95. $\log_{0.24}3.1$ -0.79 **96.** $\log_{1/3}0.85$ 0.15

Solve each problem.

97. Harry has become about 88% proficient at a new task over a 1-week period. The equation $P = 88 - 12.5\ln d$ models his percent proficiency P after d days of not practicing the new task. What is Harold's proficiency level after 6 days without practicing? 12 days? 18 days? Round answers to two decimal places. 65.60% after 6 days, 56.94% after 12 days, 51.87% after 18 days

98. Alyse has become about 92% proficient at a new task over a 1-week period. The equation $P = 92 - 12.5\ln d$ models her percent proficiency P after d days of not practicing the new task. What is Alyse's proficiency level after 5 days without practicing? 10 days? 15 days? Round answers to two decimal places. 71.88% after 5 days, 63.22% after 10 days, 58.15% after 15 days

For Exercises 99 and 100, the formula $D(x) = 10\log(10^{12}x)$, represents the intensity of sound, in decibels (dB) where x is the intensity of sound in watts per square meter (W/m²). Compute the intensity of each sound in decibels.

99. The intensity of a rocket launching is 10^6 W/m². 180 dB

100. The intensity of a soft whisper is 10^{-9} W/m². 30 dB

You Be the Teacher!

Correct each student's errors, if any.

101. Solve $\log(3x - 5) = 2$.

Brad's work:

$$\log(3x - 5) = 2$$
$$3x - 5 = 10$$
$$3x = 15$$
$$x = 5$$

$$\log(3x - 5) = 2$$
$$3x - 5 = 10^2 = 100$$
$$3x = 105$$
$$x = 35$$

102. Solve $2\ln(x - 1) = 4.8$.

Mona's work:

$$2\ln(x - 3) = 4.8$$
$$\ln(2x - 6) = 4.8$$
$$2x - 6 = e^{4.8}$$
$$2x = e^{4.8} + 6$$
$$x = \frac{e^{4.8} + \cancel{6}^3}{\cancel{2}}$$
$$x = e^{4.8} + 3$$
$$x \approx 124.510$$

$$2\ln(x - 3) = 4.8$$
$$\ln(x - 3) = 2.4$$
$$x - 3 = e^{2.4}$$
$$x = e^{2.4} + 3 \approx 14.02$$

103. Solve $\log(4x - 15) = 1.7$.

Tracy's work:

$$\log(4x - 15) = 1.7$$
$$4x - 15 = 10^{1.7}$$
$$4x = 10^{1.7} + 15$$
$$x = \frac{10^{1.7} + 15}{4}$$

The screen shot of Tracy's calculator is shown.

```
10^1.7+15/4
         53.86872336
log(4*Ans-15)
         2.302059991
```

104. Solve $\ln(5x + 3) = -1.5$.

Brian's work:

$$\ln(5x + 3) = -1.5$$
$$5x + 3 = e^{-1.5}$$
$$5x = e^{-1.5} - 3$$
$$x = \frac{e^{-1.5} - 3}{5}$$

The screen shot of Brian's calculator is shown.

```
e^(-1.5)-3/5
        -.3768698399
ln(5*-Ans+3)
         .1094379122
```

Calculate It!

Apply the change-of-base formula and simplify each logarithm exactly.

105. $\log_2 32$ **106.** $\log_5 625$

SECTION 10.7 Exponential and Logarithmic Equations and Applications

OBJECTIVES

As a result of completing this section, you will be able to

1. Solve exponential equations.
2. Solve logarithmic equations.
3. Solve applications that involve exponential and logarithmic models.
4. Troubleshoot common errors.

Maurice invests \$10,000 in a money market account that earns 4% annual interest compounded quarterly. How long will it take the account to grow to \$20,000? To answer this question, we need to solve the equation

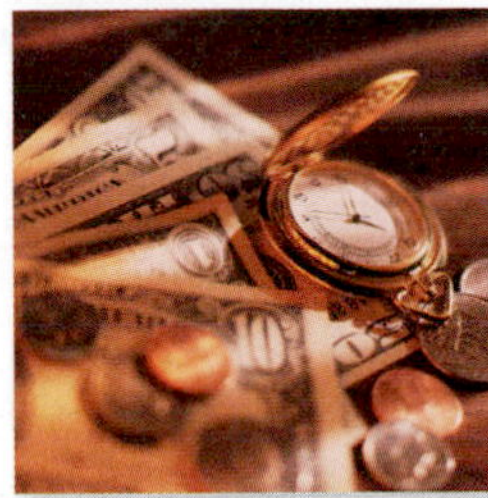

$$10{,}000\left(1 + \frac{0.04}{4}\right)^{4t} = 20{,}000$$

In this section, we will learn how to solve exponential and logarithmic equations.

Exponential Equations

Objective 1 Solve exponential equations.

In Section 10.3, we learned how to solve exponential equations using the one-to-one property of exponential functions, which stated that for $b > 0$, $b \neq 1$, $b^x = b^y$ if and only if $x = y$. For example,

$$2^x = 8$$
$$2^x = 2^3$$
$$x = 3$$

If the left and right sides of an equation cannot be expressed as a power of the same base, we must use a property of logarithms to solve the exponential equation. In the equation $2^x = 10$, we cannot express 10 as a power of 2, so we must apply a property that enables us to remove the variable from the exponent. The power rule of logarithms provides us with this ability. Recall the power rule states that $\log_b a^n = n\log_b a$.

To apply this property, we must first take the logarithm of each side of the equation. Remember that what we do to one side of an equation, we must do to the other. Because common logarithms and natural logarithms can be approximated on a calculator, we will use them in solving the equations.

$$2^x = 10$$

$$\log(2^x) = \log(10) \quad \text{Take the common log of each side.}$$

$$x\log 2 = 1 \quad \text{Apply the power rule for logarithms.}$$

$$x = \frac{1}{\log 2} \quad \text{Divide each side by log 2.}$$

$$x \approx 3.32 \quad \text{Approximate.}$$

This leads us to an important property for solving equations.

Property: One-to-One Property of Logarithms

For $x, y, b > 0, b \neq 1$, if $x = y$, then

$$\log_b x = \log_b y$$

Procedure: Solving an Exponential Equation Using Logarithms

Step 1: Isolate the exponential expression to one side of the equation, if necessary.
Step 2: Apply the one-to-one property of logs using common logs or natural logs.
Step 3: Apply the power rule for logarithms to make the exponent the coefficient of the logarithm.
Step 4: Solve the resulting equation.
Step 5: Check the proposed solution by substituting it into the original equation.

Objective 1 Examples **Solve each equation. Give an exact solution and an approximation of the solution to two decimal places.**

1a. $5^x = 30$ **1b.** $e^x = 12$ **1c.** $2^{x+1} = 6$ **1d.** $1.05^{12x} = 2$ **1e.** $4(0.6^x) = 12$

Solutions **1a.**

$$5^x = 30$$

$$\log 5^x = \log 30 \quad \text{Apply the one-to-one property of logs.}$$

$$x\log 5 = \log 30 \quad \text{Apply the power rule for logarithms.}$$

$$x = \frac{\log 30}{\log 5} \quad \text{Divide each side by log 5.}$$

$$x \approx 2.11 \quad \text{Approximate.}$$

The solution set is $\left\{\frac{\log 30}{\log 5}\right\}$ or $\{2.11\}$. The solution can be checked in the original equation.

1b. Since the base is e, we take the natural logarithm of each side.

$$e^x = 12$$

$$\ln e^x = \ln 12 \quad \text{Apply the one-to-one property of logs.}$$

$$x = \ln 12 \quad \text{Apply the inverse property, } \ln e^x = x.$$

$$x \approx 2.48 \quad \text{Approximate.}$$

The solution set is $\{\ln 12\}$ or $\{2.48\}$. The solution can be checked in the original equation.

1c.

$$2^{x+1} = 6$$

$$\log 2^{x+1} = \log 6 \quad \text{Apply the one-to-one property of logs.}$$

$$(x+1)\log 2 = \log 6 \quad \text{Apply the power rule for logs.}$$

$$x + 1 = \frac{\log 6}{\log 2} \quad \text{Divide each side by log 2.}$$

$$x = \frac{\log 6}{\log 2} - 1 \quad \text{Subtract 1 from each side.}$$

$$x \approx 1.58 \quad \text{Approximate.}$$

The solution set is $\left\{\frac{\log 6}{\log 2} - 1\right\}$ or $\{1.58\}$. The solution can be checked in the original equation.

1d.

$$1.05^{12x} = 2$$

$$\ln 1.05^{12x} = \ln 2 \quad \text{Apply the one-to-one property of logs.}$$

$$12x \ln 1.05 = \ln 2 \quad \text{Apply the power rule for logs.}$$

$$x = \frac{\ln 2}{12 \ln 1.05} \quad \text{Divide each side by } 12 \ln 1.05.$$

$$x \approx 1.18 \quad \text{Approximate.}$$

The solution set is $\left\{\frac{\ln 2}{12 \ln 1.05}\right\}$ or $\{1.18\}$. The solution can be checked in the original equation.

1e.

$$4(0.6^x) = 12$$

$$0.6^x = \frac{12}{4} \quad \text{Divide each side by 4.}$$

$$0.6^x = 3 \quad \text{Simplify.}$$

$$\log 0.6^x = \log 3 \quad \text{Apply the one-to-one property of logs.}$$

$$x \log 0.6 = \log 3 \quad \text{Apply the power rule for logs.}$$

$$x = \frac{\log 3}{\log 0.6} \quad \text{Divide each side by log 0.6.}$$

$$x \approx -2.15 \quad \text{Approximate.}$$

The solution set is $\left\{\frac{\log 3}{\log 0.6}\right\}$ or $\{-2.15\}$. The solution can be checked in the original equation.

Student Check 1 Solve each equation. Give an exact solution and an approximation of the solution to two decimal places.

a. $7^x = 53$ **b.** $e^x = 25$ **c.** $3^{x-2} = 80$ **d.** $1.02^{4x} = 3$ **e.** $5(0.2)^x = 2$

Logarithmic Equations

Objective 2 ▶ Solve logarithmic equations.

To solve a logarithmic equation, we may have to apply some of the properties we learned in Section 10.5 that enable us to combine two or more logarithms into a single logarithm. Then we can convert the logarithmic equation to its exponential form to solve it.

INSTRUCTOR NOTE:
Point out that we use logs to solve exponential equations and exponents to solve logarithmic equations.

Procedure: Solving a Logarithmic Equation

Step 1: Write each side of the equation as a single logarithm, if necessary.

a. $\log_b x + \log_b y = \log_b(xy)$

b. $\log_b x - \log_b y = \log_b \frac{x}{y}$

Step 2: If the equation is in the form $\log_b x = y$, convert it to $b^y = x$ to solve the equation.

Step 3: Exclude any values that make the arguments of the logarithms negative or zero.

Objective 2 Examples

Solve each equation.

2a. $\log_2(x + 1) = 4$

2b. $\log_3 x + \log_3(x - 8) = 2$

2c. $\log_4(x + 3) - \log_4(2x) = 1$

2d. $2 \log x - \log(x + 6) = 0$

Solutions

2a. Note that $x + 1 > 0$ for the log to be defined, so $x > -1$.

$$\log_2(x + 1) = 4$$

$2^4 = x + 1$	Rewrite in exponential form.
$16 = x + 1$	Simplify the exponent.
$15 = x$	Subtract 1 from each side.

Check:

$\log_2(15 + 1) = 4$	Replace x with 15.
$\log_2 16 = 4$	Simplify.
$4 = 4$	Simplify the logarithm.

Since 15 makes the equation true, the solution set is $\{15\}$.

2b. Note that $x > 0$ and $x - 8 > 0$, or $x > 8$, for the log to be defined. So, $x > 8$ satisfies both of these conditions.

$$\log_3 x + \log_3(x - 8) = 2$$

$\log_3 x(x - 8) = 2$	Apply the product rule.
$3^2 = x(x - 8)$	Rewrite in exponential form.
$9 = x^2 - 8x$	Simplify and apply the distributive property.
$0 = x^2 - 8x - 9$	Subtract 9 from each side.
$0 = (x - 9)(x + 1)$	Factor.
$x - 9 = 0$ or $x + 1 = 0$	Apply the zero products property.
$x = 9$ $\quad$ $x = -1$	Solve each equation.

Because $x > 8$, we must exclude -1 from the solution set since it makes the argument of the logarithm negative. So, the solution set is $\{9\}$.

2c. Note that $x + 3 > 0$, or $x > -3$, and $x > 0$ for the log to be defined. So, $x > 0$ satisfies both of these conditions.

$$\log_4(x + 3) - \log_4(2x) = 1$$

$\log_4 \frac{x + 3}{2x} = 1$	Apply the quotient rule for logs.
$4^1 = \frac{x + 3}{2x}$	Rewrite in exponential form.

$$4 = \frac{x+3}{2x}$$ Simplify the exponent.

$$8x = x + 3$$ Multiply each side by $2x$ and simplify.

$$7x = 3$$ Subtract x from each side and simplify.

$$x = \frac{3}{7}$$ Divide each side by 7.

The solution set is $\left\{\frac{3}{7}\right\}$. We can check the solution in the original equation.

2d. Note that $x > 0$ and $x + 6 > 0$, or $x > -6$, for the log to be defined. So, $x > 0$ satisfies both of these conditions.

$$2 \log x - \log(x + 6) = 0$$

$$\log x^2 - \log(x + 6) = 0$$ Apply the power rule for logs.

$$\log \frac{x^2}{x+6} = 0$$ Apply the quotient rule for logs.

$$10^0 = \frac{x^2}{x+6}$$ Rewrite in exponential form.

$$(x + 6)1 = \frac{x^2}{x+6}(x + 6)$$ Simplify and multiply by the LCD, $x + 6$.

$$x + 6 = x^2$$ Simplify.

$$0 = x^2 - x - 6$$ Subtract x and 6 from each side.

$$0 = (x - 3)(x + 2)$$ Factor.

$$x - 3 = 0 \quad \text{or} \quad x + 2 = 0$$ Apply the zero products property.

$$x = 3 \qquad x = -2$$ Solve each equation.

We must discard -2 from the solution set since it makes the argument of the logarithm negative. So, the solution set is $\{3\}$.

Student Check 2 Solve each equation.

a. $\log_3(x - 1) = 1$ **b.** $\log_2(x) + \log_2(x - 2) = 3$

c. $\log_5(2x - 5) - \log_5(x) = 2$ **d.** $2\log(x) - \log(2x + 15) = 0$

Applications

Objective 3 ▶

Solve applications that involve exponential and logarithmic models.

Exponential and logarithmic models arise in many different areas. We have already seen a few of these in this chapter. In Example 6, we will use the properties of exponents to solve for the variable. One of the formulas that we will use is the compound interest formula. We will also use a formula for continuous compound interest.

$$\text{Compound interest formula: } A = P\left(1 + \frac{r}{n}\right)^{nt}$$

Property: Continuous Compound Interest Formula

If P dollars are invested in an account that pays $r\%$ compounded continuously, then the amount in the account after t years is

$$A = Pe^{rt}$$

Objective 3 Examples **Solve each problem.**

3a. David invests \$200 in an account that pays 5% compounded continuously. How much will he have in the account after 2 yr? How long will it take his account to grow to \$1000?

Solution **3a.** To find the amount David will have after two years, we use the compound interest formula with $P = 200$, $r = 0.05$, and $t = 2$.

$$A = Pe^{rt} \quad \text{State the compound interest formula.}$$
$$A = 200e^{(0.05)(2)} \quad \text{Let } P = 200,\ r = 0.05, \text{ and } t = 2.$$
$$A = 200e^{0.1} \quad \text{Simplify the exponent.}$$
$$A \approx 221.03 \quad \text{Approximate.}$$

After 2 yr, David will have \$221.03 in his account.

To determine how long it will take for his account to grow to \$1000, we must solve the equation $A = Pe^{rt}$ for t.

$$A = Pe^{rt} \quad \text{State the compound interest formula.}$$
$$1000 = 200e^{0.05t} \quad \text{Let } A = 1000,\ P = 200, \text{ and } r = 0.05.$$
$$\frac{1000}{200} = e^{0.05t} \quad \text{Divide each side by 200.}$$
$$5 = e^{0.05t} \quad \text{Simplify.}$$
$$\ln 5 = \ln e^{0.05t} \quad \text{Apply the one-to-one property of logs.}$$
$$\ln 5 = 0.05t \quad \text{Apply the inverse properties of logs.}$$
$$\frac{\ln 5}{0.05} = t \quad \text{Divide each side by 0.05.}$$
$$32.19 \approx t \quad \text{Approximate.}$$

It will take about 32 yr for David's account to grow to \$1000.

3b. Maurice invests \$10,000 in a money market account that earns 4% annual interest compounded quarterly. How long will it take his account to grow to \$20,000?

Solution **3b.** We need to use the compound interest formula and solve for t.

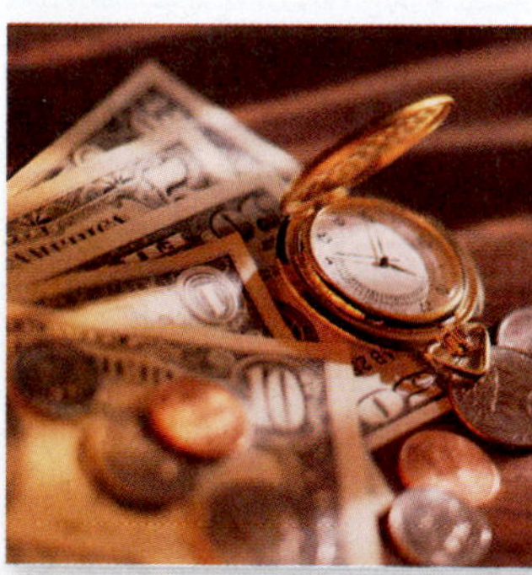

$$P\left(1 + \frac{r}{n}\right)^{nt} = A \quad \text{State the compound interest formula.}$$
$$10{,}000\left(1 + \frac{0.04}{4}\right)^{4t} = 20{,}000 \quad \text{Let } P = 10{,}000,\ r = 0.04,\ n = 4, \text{ and } A = 20{,}000.$$
$$(1.01)^{4t} = \frac{20{,}000}{10{,}000} \quad \text{Divide each side by 10,000 and simplify in parentheses.}$$
$$(1.01)^{4t} = 2 \quad \text{Simplify.}$$
$$\log(1.01)^{4t} = \log 2 \quad \text{Apply the one-to-one property of logs.}$$
$$4t \log 1.01 = \log 2 \quad \text{Apply the power rule for logs.}$$
$$t = \frac{\log 2}{4 \log 1.01} \quad \text{Divide each side by } 4 \log 1.01.$$
$$t \approx 17.42 \quad \text{Approximate.}$$

So, it will take about 17 yr for Maurice's account to grow to \$20,000.

3c. In 2009, the country of India, the second most populous country in the world, had a population of approximately 1.166 billion. Its population is growing at a rate of 1.548% per year. Its population (in billions) can be modeled by $f(t) = 1.166e^{0.01548t}$, where t is the number of years after 2009. What is an estimate for the population of India in 2015 if it continues to grow at this rate? How long will it take the population to reach 1.5 billion? (Source: https://www.cia.gov)

Solution **3c.** The population in 2015 is found by replacing t with $2015 - 2009 = 6$.

$$f(t) = 1.166e^{0.01548t} \quad \text{State the given model.}$$
$$f(6) = 1.166e^{0.01548(6)} \quad \text{Replace } t \text{ with 6.}$$
$$f(6) = 1.166e^{0.09288} \quad \text{Simplify the exponent.}$$
$$f(6) \approx 1.28 \quad \text{Approximate.}$$

In 2015, the population of India will be approximately 1.28 billion.

To determine when the population will reach 1.5 billion, we replace y with 1.5 and solve for t.

$$y = 1.166e^{0.01548t}$$
$$1.5 = 1.166e^{0.01548t} \quad \text{Replace } y \text{ with 1.5.}$$
$$\frac{1.5}{1.166} = e^{0.01548t} \quad \text{Divide each side by 1.166.}$$
$$1.28645 \approx e^{0.01548t} \quad \text{Approximate.}$$
$$\ln 1.28645 \approx \ln e^{0.01548t} \quad \text{Apply the one-to-one property of logs.}$$
$$\ln 1.28645 \approx 0.01548t \quad \text{Apply the inverse property of logs.}$$
$$\frac{\ln 1.28645}{0.01548} \approx t \quad \text{Divide each side by 0.01548.}$$
$$16.27 \approx t \quad \text{Approximate.}$$

In approximately 16 yr after 2009, or 2025, the population of India will reach 1.5 billion.

Student Check 3 Solve each problem.

a. Jeanette invests \$500 in an account that pays 6% compounded continuously. How much will she have in the account after 4 yr? How long will it take her account to grow to \$2000?

b. Suppose Cantrell invests \$30,000 in a money market account that earns 5% annual interest compounded monthly. How long will it take his account to grow to \$60,000?

c. In 2009, the country of China, the most populous country in the world, had a population of approximately 1.339 billion. Its population is growing at a rate of 0.655% per year. Its population (in billions) can be modeled by $y = 1.339e^{0.00655t}$, where t is the number of years after 2009. What is an estimate for the population of China in 2015 if it continues to grow at this rate? How long will it take the population to reach 1.5 billion? (Source: https://www.cia.gov)

Objective 4 Troubleshoot common errors.

Troubleshooting Common Errors

Some common errors associated with solving exponential and logarithmic equations are shown.

Objective 4 Examples **A problem and an incorrect solution are given. Provide the correct solution and an explanation of the error.**

4a. Solve $4^x = 12$.

Incorrect Solution	Correct Solution and Explanation
$4^x = 12$ $\log 4^x = \log 12$ $x \log 4 = \log 12$ $x = \dfrac{\log 12}{\log 4}$ $x = 3$	The logarithm is a function; we cannot divide out logs from the numerator and denominator of a fraction. We must divide the log values. $x = \dfrac{\log 12}{\log 4} \approx 1.792$

4b. Solve $\log x + \log(x - 3) = 1$.

Incorrect Solution	Correct Solution and Explanation
$\log x + \log(x-3) = 1$ $\log x(x-3) = 1$ $10^1 = x(x-3)$ $10 = x^2 - 3x$ $x^2 - 3x - 10 = 0$ $(x-5)(x+2) = 0$ $x = 5, x = -2$ The solution set is $\{-2, 5\}$.	The logarithm of a negative number is undefined. Therefore, we must exclude -2 from the solution set. So, the solution set is $\{5\}$.

ANSWERS TO STUDENT CHECKS

Student Check 1 **a.** $\left\{\dfrac{\ln 53}{\ln 7}\right\}$ or $\{2.04\}$ **b.** $\{\ln 25\}$ or $\{3.22\}$ **c.** $\left\{\dfrac{\log 80 + 2\log 3}{\log 3}\right\}$ or $\{5.99\}$ **d.** $\left\{\dfrac{\log 3}{4\log 1.02}\right\}$ or $\{13.87\}$ **e.** $\left\{\dfrac{\log 0.4}{\log 0.2}\right\}$ or $\{0.57\}$

Student Check 2 **a.** $\{4\}$ **b.** $\{4\}$ **c.** $\varnothing$ **d.** $\{5\}$

Student Check 3 **a.** \$635.62; 23.10 yr **b.** 13.89 yr **c.** 1.39 billion; approximately 17 yr or in 2026

SUMMARY OF KEY CONCEPTS

1. Exponential equations are solved by applying the one-to-one property of logarithms, which enables us to take the logarithm of both sides. Any logarithm will work but we generally use the common logarithm or the natural logarithm since we can use our calculator to simplify those expressions. The exponential expression should be isolated before taking the log of each side.

2. Logarithmic equations are solved by converting the equation to exponential form. Before doing this, we must combine any logarithms. We apply the product rule and quotient rule for logarithms to combine the logarithmic expressions. After solving the equation, we must exclude any values that make the argument of the logarithm negative or zero.

GRAPHING CALCULATOR SKILLS

Most of the skills in this section have been presented in earlier sections. The one new skill is using the number e.

Example: Simplify $1.166e^{0.01548(6)}$.

Solution: To enter e raised to an exponent, use 2nd LN or 2nd ÷ ^.

```
1.166e^(.01548*6
)
      1.279486836
```

SECTION 10.7 EXERCISE SET

Write About It!

Use complete sentences in your answer to each exercise.

1. Explain why you can take either the common or natural log of each side when solving the exponential equation $7^x = 15$.
2. Explain how to solve the exponential equation $3e^{x-1} = 48$ using an appropriate logarithm.
3. Explain how to solve the exponential equation $6(10^{x+2}) = 102$ using an appropriate logarithm.
4. Explain how to solve the logarithmic equation $\log_4(2x - 1) = 3$.
5. Explain how to solve the equation $\log_6 x + \log_6(x - 16) = 2$.
6. Explain how to solve the equation $\log_2(6x) - \log_2(2x - 10) = 3$.

Practice Makes Perfect!

Solve each equation. Give an exact solution and an approximation of the solution to two decimal places. (*See Objective 1.*)

7. $10^x = 29$ $\{\log 29\}$ or $\{1.46\}$
8. $10^x = 40$ $\{\log 40\}$ or $\{1.60\}$
9. $e^x = 6$ $\{\ln 6\}$ or $\{1.79\}$
10. $10^x = 9$ $\{\log 9\}$ or $\{0.95\}$
11. $10^{7x} = 90$
12. $10^{5x} = 88$
13. $e^{4x} = 13$
14. $e^{6x} = 31$
15. $7^{x-2} = 38$
16. $4^{x+9} = 12$
17. $(5.75)^{4x} = 19$
18. $(6.71)^{2x} = 11$
19. $9(0.53)^{5x} = 94$
20. $3(1.26)^{4x} = 92$

Solve each logarithmic equation. (*See Objective 2.*)

21. $\log_5(x - 18) = 1$ $\{23\}$
22. $\log_2(x - 3) = 5$ $\{35\}$
23. $\log_2(2x - 9) = 2$ $\left\{\frac{13}{2}\right\}$
24. $\log_3(9x - 7) = 3$ $\left\{\frac{34}{9}\right\}$
25. $\log_2 x + \log_2(x + 2) = 3$ $\{2\}$
26. $\log_6 x + \log_6(x + 35) = 2$ $\{1\}$
27. $\log_2 x + \log_2(x - 3) = 2$ $\{4\}$
28. $\log_4 x + \log_4(x - 6) = 2$ $\{8\}$
29. $\log_4 4x - \log_4(4x - 5) = 2$ $\left\{\frac{4}{3}\right\}$
30. $\log_4 6x - \log_4(3x - 12) = 1$ $\{8\}$
31. $\log 2x - \log(x - 8) = 1$ $\{10\}$
32. $\log_5 5x - \log_5(x - 11) = 2$ $\left\{\frac{55}{4}\right\}$
33. $\log_2 6x - \log_2(5x + 11) = 2$ $\varnothing$
34. $\log 2x - \log(x + 6) = 2$ $\varnothing$
35. $2\log_3 x - \log_3(x + 6) = 1$ $\{6\}$
36. $2\log_4 x - \log_4(x + 3) = 1$ $\{6\}$
37. $2\log x - \log(7x - 12) = 0$ $\{3, 4\}$
38. $2\log_7 x - \log_7(6x + 16) = 0$ $\{8\}$

Solve each problem. Round each answer to two decimal places. (*See Objective 3.*)

39. Hunter invests \$990 in an account that pays 2.8% compounded continuously. How much will he have in the account after 4 yr? How long will it take his account to grow to \$1500? \$1107.33 after 4 yr; It will take about 15 yr for his account to grow to \$1500.
40. Camille invests \$1320 in an account that pays 2.5% compounded continuously. How much will she have in the account after 2 yr? How long will it take her account to grow to \$2500? \$1387.68 after 2 yr; It will take about 26 yr for her account to grow to \$2500.
41. In 2011, Mexico had a population of approximately 113.724 million. Its population is growing at a rate of 1.102% per year. Its population (in millions) can be modeled by $y = 113.724e^{0.01102t}$, where t is the number of years after 2011. What is an estimate for the population of Mexico in 2021 if it continues to grow at this rate? How long will it take the population to reach 120 million? (Source: http://www.cia.gov) The population of Mexico in 2021 will be 126.97 million. It will take about 4.87 yr to reach 120 million.

Additional answers can be found in the Instructor Answer Appendix.

42. The annual tons of Carbon Dioxide (CO_2) emitted can be modeled by $f(x) = 13.272(0.8369)^x$, where x is the mileage of a car in miles per gallon. What is an estimate for the annual tons of CO_2 emitted for a car that gets 30 mpg? Determine the gas mileage of a car that emits 1 ton of CO_2 annually. A car that gets 30 mpg emits about 0.06 tons of CO_2 annually. A car that gets about 14.52 mpg will emit 1 ton of CO_2 annually.

43. The amount of the blood thinner Warfarin (in milligrams) in a person's body t hr after ingestion is $y = a_0e^{-0.01733t}$, where a_0 is the initial amount taken. If Wing takes 10 mg of Warfarin, how much will be in her body after 4 hr? (Source: http://www.aafp.org/afp/990201ap/635.html) 9.33 mg

44. The amount of ibuprofen (in milligrams) in a person's body t hr after ingestion is $y = a_0e^{-0.347t}$, where a_0 is the initial amount taken. If Lisa takes 600 mg of ibuprofen, how much ibuprofen will be in her body after 5 hr? (Source: http://www.rxlist.com/ibuprofen-drug.htm) 105.84 mg

Mix 'Em Up!

Solve each equation. For solutions of the exponential equations, give an exact solution and an approximation of the solution to two decimal places.

45. $\log_4(5x - 4) = 2$ $\{4\}$
46. $\log_2(4x - 12) = 3$ $\{5\}$
47. $e^{4.6x} = 37.5$
48. $e^{1.7x} = 109$ $\left\{\frac{\ln 109}{1.7}\right\}$ or $\{2.76\}$
49. $\log_6 x + \log_6(x - 9) = 2$ $\{12\}$
50. $\log_5 x + \log_5(x - 24) = 2$ $\{25\}$
51. $12^{6x-8} = 2$
52. $6^{5x-2} = 30$ $\left\{\frac{\ln 30 + 2\ln 6}{5\ln 6}\right\}$ or $\{0.78\}$
53. $\log_6 4x - \log_6(x - 2) = 2$ $\left\{\frac{9}{4}\right\}$
54. $\log_2 3x - \log_2(x - 6) = 2$ $\{24\}$
55. $2\ln x - \ln(x + 42) = 0$ $\{7\}$
56. $2\log x - \log(7x - 6) = 0$ $\{1, 6\}$
57. $7(4.08)^{4x} = 1$
58. $5(3.49)^{4x} = 1$
59. $\log_3(2x + 1) = 4$ $\{40\}$
60. $\log_5(12x + 5) = 3$ $\{10\}$
61. $\log_4 x + \log_4(x + 6) = 2$ $\{2\}$
62. $\log_6 x + \log_6(x - 5) = 2$ $\{9\}$
63. $5^{7x+1} = 21$
64. $7^{2x+8} = 5$ $\left\{\frac{\ln 5 - 8\ln 7}{2\ln 7}\right\}$ or $\{-3.59\}$
65. $\log_2 6x - \log_2(x - 5) = 1$ $\varnothing$
66. $\log 5x - \log(x + 7) = 1$ $\varnothing$
67. $\log_2 6x - \log_2(5x + 11) = 2$ $\varnothing$
68. $\log 2x - \log(x + 6) = 2$ $\varnothing$
69. $2(6.25)^x = 4$
70. $9(3.67)^{6x} = 57$ $\left\{\frac{\ln 57 - \ln 9}{6\ln 3.67}\right\}$ or $\{0.24\}$
71. $2\log_5 x - \log_5(x - 4) = 2$ $\{5, 20\}$
72. $2\log_2 x - \log_2(2x - 6) = 3$ $\{4, 12\}$
73. $10^{4.7x} = 17.5$
74. $10^{2.1x} = 29.5$ $\left\{\frac{\log 29.5}{2.1}\right\}$ or $\{0.70\}$
75. $2\ln x - \ln(5x + 36) = 0$ $\{9\}$
76. $2\log_7 x - \log_7(13x - 40) = 0$ $\{5, 8\}$

Solve each problem. Round each answer to two decimal places.

77. Kristen invests \$1870 in an account that pays 4.6% compounded continuously. How much will she have in the account after 4 yr? How long will it take her account to grow to \$2400? \$2247.77 after 4 yr; It will take about 5 yr for her account to grow to \$2400.

78. Michael invests \$1650 in an account that pays 3.8% compounded continuously. How much will he have in the account after 6 yr? How long will it take his account to grow to \$2800? \$2072.54 after 6 yr; It will take about 14 yr for his account to grow to \$2800.

79. Lance invests \$10,500 into a money market account that earns 5% annual interest compounded quarterly. How long will it take his account to grow to \$23,000? It will take about 16 yr for his account to grow to \$23,000.

80. Lawton invests \$12,000 in a money market account that earns 1.2% annual interest compounded semiannually. How long will it take his account to grow to \$14,500? It will take about 16 yr for his account to grow to \$14,500.

81. In 2011, Japan had a population of approximately 126.476 million. Its population is decreasing at a rate of 0.278% per year. Its population (in millions) can be modeled by $y = 126.476e^{-0.00278t}$, where t is the number of years after 2011. What is an estimate of the population of Japan in 2029 if it continues to grow at this rate? How long will it take the population to reach 115 million? (Source: http://www.cia.gov)

82. In 2011, the United States had a population of approximately 313.232 million. Its population is growing at a rate of 0.963% per year. Its population (in millions) can be modeled by $y = 313.232e^{0.00963t}$, where t is the number of years after 2011. What is an estimate of the population of the United States in 2026 if it continues to grow at this rate? How long will it take the population to reach 340 million? (Source: http://www.cia.gov)

83. The amount of acetaminophen (in milligrams) in a person's body t hr after ingestion is $y = a_0e^{-0.231t}$, where a_0 is the initial amount taken. If Isabel takes 600 mg of acetaminophen, how much will be in her body after 3 hr? (Source: http://www.rxlist.com/tylenol-drug.htm) 300.04 mg

84. The amount of blood pressure medication Lisinopril in a person's body t hr after ingestion is $y = a_0e^{-0.0578t}$, where a_0 is the initial amount taken. If Jason takes 10 mg of Lisinopril, how much Lisinopril will be in his body after 12 hr? (Source: http://www.drugs.com/pro/lisinopril.html) 5 mg

You Be the Teacher!

Correct each student's errors, if any.

85. Solve $8(2)^{3x} = 48$.

Lois's work:

$$8(2)^{3x} = 48$$
$$16^{3x} = 48$$
$$3x\ln 16 = \ln 48$$
$$x = \frac{\ln 48}{3\ln 16}$$
$$x \approx 0.465$$

$$8(2)^{3x} = 48$$
$$2^{3x} = 6$$
$$3x\ln 2 = \ln 6$$
$$x = \frac{\ln 6}{3\ln 2} \approx 0.86$$

86. Solve $8^{5x-6} = 13$.

Chuck's work:

$$\begin{aligned} 8^{5x-6} &= 13 \\ 5x - 6\ln 8 &= \ln 13 \\ 5x &= 6\ln 8 + \ln 13 \\ x &= \frac{6\ln 8 + \ln 13}{5} \\ x &\approx 3.008 \end{aligned}$$

$$\begin{aligned} 8^{5x-6} &= 13 \\ (5x-6)\ln 8 &= \ln 13 \\ 5x\ln 8 - 6\ln 8 &= \ln 13 \\ 5x\ln 8 &= \ln 13 + 6\ln 8 \\ x &= \frac{\ln 13 + 6\ln 8}{5\ln 8} \approx 1.45 \end{aligned}$$

87. Solve $\log_6 x + \log_6(x - 35) = 2$.

Ulrey's work:

$$\begin{aligned} \log_6 x + \log_6(x-35) &= 2 \\ \log_6(2x - 35) &= 2 \\ 2x - 35 &= 6^2 \\ 2x &= 71 \\ x &= \frac{71}{2} \end{aligned}$$

$$\begin{aligned} \log_6 x + \log_6(x-35) &= 2 \\ \log_6 x(x-35) &= 2 \\ x^2 - 35x &= 6^2 \\ x^2 - 35x - 36 &= 0 \\ (x-36)(x+1) &= 0 \\ x &= 36, -1 \end{aligned}$$

Since -1 doesn't check, the solution set is $\{36\}$.

88. Solve $2\log_2 x - \log_2(8x - 12) = 0$.

Soner's work:

$$\begin{aligned} 2\log_2 x - \log_2(8x-12) &= 0 \\ \log_2(2x - 8x + 12) &= 0 \\ -6x + 12 &= 0 \\ x &= 2 \end{aligned}$$

Calculate It!

Solve each equation and use a calculator to check each solution.

89. $4.02^{8x} = 64$

90. $4^{7x-2} = 18$

91. $\log_6 x + \log_6(x + 5) = 2$

92. $\log_3 3x - \log_3(3x - 10) = 2$

GROUP ACTIVITY The Mathematics of Financing a College Education

The average in-state cost (tuition and fees) for attending a public 4-yr institution averaged \$7605 in the 2010–2011 academic year. On average, college costs tend to increase about 8% each year. The function $f(x) = 7605(1.08)^x$ models the average cost of a college education x yr after 2010. (Sources: http://trends.collegeboard.org/college_pricing/ and http://www.finaid.org/savings/tuition-inflation.phtml).

1. Planning for the future is important to you, so you want to determine the amount you need to save to pay for your child's college education. Use the given function to determine the average cost of college each year your child attends college. Assume your child was born in 2010 and that he or she will enter college at age 18 for 4 yr. first year: \$30,389.73; second year: \$32,820.91; third year: \$35,446.58; fourth year: \$38,282.31
2. What is the total cost of your child's 4-yr college education? \$136,939.53
3. The amount in step 2 is the amount that you need to save. The formula

$$FV = PMT\frac{\left(1 + \frac{i}{12}\right)^n - 1}{\frac{i}{12}}$$

gives the future value (FV) of an ordinary annuity, where i is the annual interest rate of a savings account, n is the number of payments, and PMT is the payment amount. Use the formula to determine the payment necessary to have the funds for your child's college education in 18 yr if you pay this amount each month into an account that earns 6% annual interest. (Note: An annuity is any sequence of equal periodic payments. If payments are made at the end of each time interval, then the annuity is called an ordinary annuity.) \$353.53 each month

Chapter 10 / REVIEW

Exponential and Logarithmic Functions

What's the big idea? Functions can be combined in several ways to produce other functions. Functions that are one-to-one can be inverted to find the function that "un-does" the effect of the function. Exponential and logarithmic functions are inverses of one another. They are very important in many fields. The properties of exponents enable us to develop properties of logarithms.

The Tools

Listed below are the key terms, skills, formulas, and properties you should know for this chapter.

The page reference is provided if you need additional help with the given topic. The Study Tips will assist in your preparation for an exam.

Study Tips

1. Learn all of the terms, formulas, and properties. Make flash cards and have someone quiz you.
2. Rework problems from the exercises and also the ones you worked in class. Work additional problems from the review exercises.
3. Review the summaries of key concepts.
4. Work the chapter test.
5. Be sure to review the online resources for additional study materials.

Terms

Argument 762
Common logarithm 778
Exponential function 754
Inverse function 743
Logarithm 762
Logarithmic function 765
Natural logarithm 779
One-to-one function 740

Formulas and Properties

- Change-of-base formula 782
- Composition of functions 730
- Compound interest formula 757
- Continuous compound interest formula 791
- Difference of functions 726
- Horizontal line test 741
- Inverse properties of logarithms 774
- One-to-one property of b^x 756
- One-to-one property of logarithms 788
- Product of functions 726
- Product rule for logarithms 769
- Power rule for logarithms 772
- Quotient of functions 726
- Quotient rule for logarithms 771
- Sum of functions 726

CHAPTER 10 / SUMMARY

How well do you know this chapter? Complete the following questions to find out. Take a look back at the section if you need help.

SECTION 10.1 Operations and Composition of Functions

1. We can perform _operations_ on functions just like we do real numbers. We can _add_ them, _subtract_ them, _multiply_ them, and _divide_ them. The domain of the combined function is the _intersection_ of the domains of the individual functions. For the quotient function, we must also exclude values that make the denominator _zero_.
2. To evaluate a combined function, we can either evaluate the functions _individually_ and then perform the operations on the results or we can find the _combined_ function first and then evaluate this function.
3. The composition of functions takes one function's _output_ and makes it the other function's _input_. The notation $(f \circ g)(x) =$ _$f(g(x))$_ and $(g \circ f)(x) =$ _$g(f(x))$_.

SECTION 10.2 Inverse Functions

4. The inverse function is a function that takes us back to the value we started with before the original function was applied.

5. Only one-to-one functions have inverse functions. These are functions such that every input value corresponds to a(n) different output value.

6. The horizontal line test is used to determine if a function is one-to-one. If each horizontal line touches the graph at no more than one point, the function is one-to-one.

7. To find the inverse of a one-to-one function, we interchange the x- and y-variables. The inverse of $f(x)$ is denoted by $f^{-1}(x)$. If (x, y) belongs to $f(x)$, then (y, x) belongs to $f^{-1}(x)$.

8. To find the equation of the inverse function, we replace $f(x)$ with y, interchange x and y, solve the equation for y, and replace y with $f^{-1}(x)$.

9. If $f(x) = y$, then $f^{-1}(y) = x$.

10. The graph of a function and its inverse are symmetric about the line $y = x$.

11. To verify that $f(x)$ and $f^{-1}(x)$ are inverses, $(f \circ f^{-1})(x) = x$ and $(f^{-1} \circ f)(x) = x$.

SECTION 10.3 Exponential Functions

12. Exponential functions are functions of the form $f(x) = b^x$, where $b > 0$ and $b \neq 1$. The domain is $(-\infty, \infty)$ and the range is $(0, \infty)$. The points $(0, 1)$ and $(1, b)$ are on the graph of this function. If $b > 1$, the function is increasing and if $0 < b < 1$, the function is decreasing. The graph of a basic exponential function gets very close to the x-axis but never touches it.

13. We can use the one-to-one property of b^x to solve exponential equations. It states that if $b^x = b^y$, then $x = y$.

14. The compound interest formula is an application of exponential functions and is $A = P\left(1 + \frac{r}{n}\right)^{nt}$.

SECTION 10.4 Logarithmic Functions

15. The expression $\log_b x$ represents the exponent of b that produces x. Two of the basic logarithms are $\log_b(1) = 0$, $\log_b b = 1$

16. $y = \log_b x$ written in exponential form is $b^y = x$.

17. A logarithmic function is a function of the form $f(x) = \log_b x$, where $b > 0$, $b \neq 1$. The domain of the function is $(0, \infty)$ and the range is $(-\infty, \infty)$. The x-intercept is $(1, 0)$ and there is no y-intercept.

18. To graph an equation of the form $f(x) = \log_b x$, we should convert the equation to its exponential form and substitute values for y and solve for x. The graph of a basic logarithmic functions gets very close to the y-axis but never touches it.

SECTION 10.5 Properties of Logarithms

19. The log of a product is the sum of the logs of the factors. The product rule for logs is $\log_b(xy) = \log_b x + \log_b y$, for $x, y, b > 0$, and $b \neq 1$.

20. The log of a quotient is the difference of the logs of the numerator and denominator. The quotient rule for logs is $\log_b \frac{x}{y} = \log_b x - \log_b y$, for $x, y, b > 0$, and $b \neq 1$.

21. The log of an exponential expression is the exponent times the log of the base of the exponential expression. The power rule for logs is $\log_b x^n = n \log_b x$, for $x, y, b > 0$ and $b \neq 1$.

22. The inverse properties of logs state that $\log_b b^x$ and $b^{\log_b x}$ both equal x.

SECTION 10.6 The Common Log, Natural Log, and Change-of-Base-Formula

23. The common log is written as $\log x$ and it represents the exponent of 10 that produces x.

24. The natural log is written as $\ln x$ and it represents the exponent of e that produces x.

25. The change-of-base formula is $\log_b a = \frac{\log a}{\log b}$ or $\frac{\ln a}{\ln b}$ and it enables us to calculate logs with bases other than 10 or e.

SECTION 10.7 Exponential and Logarithmic Equations and Applications

26. To solve an exponential equation, we must isolate the exponent and then take the log of each side.

27. To solve a logarithmic equation, we must combine the logs into a single log and then convert the equation to its exponential form.

CHAPTER 10 / REVIEW EXERCISES

SECTION 10.1

Find each combined function and state its domain. (*See Objectives 1–3.*)

Let $f(x) = 2x - 7$, $g(x) = x^2 - x - 2$, and $h(x) = 4 - 2x$.

1. $(f + g)(x)$ and $(f - g)(x)$
2. $(f \cdot h)(x)$ and $\left(\frac{f}{h}\right)(x)$
3. $(f \circ h)(x)$ and $(h \circ f)(x)$
4. $(f \cdot g)(x)$ and $\left(\frac{f}{g}\right)(x)$

Let $f(x) = 2x^2 - x$, $g(x) = x^2 + 3x - 7$, $h(x) = 5x + 3$. Evaluate each function. (*See Objectives 1–3.*)

5. $(f + g)(-2)$ 1
6. $(f - h)(3)$ -3
7. $(g \cdot h)(0)$ -21
8. $\left(\frac{g}{h}\right)(-1)$ 4.5
9. $(f \circ g)(2)$ 15
10. $(h \circ f)\left(-\frac{1}{2}\right)$ 8

Use the graphs of $f(x)$, $g(x)$, and $h(x)$ to find each value. (*See Objectives 1–3.*)

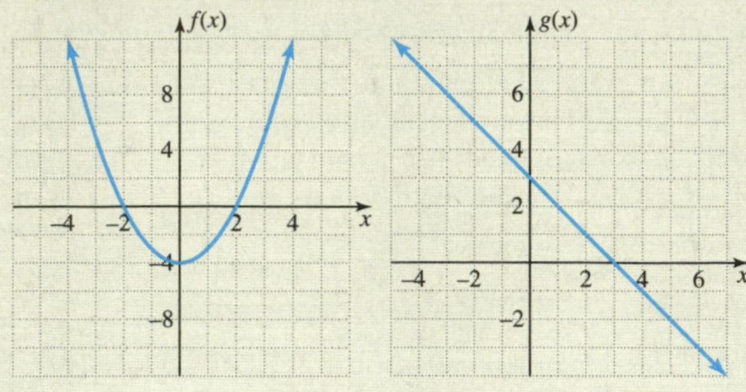

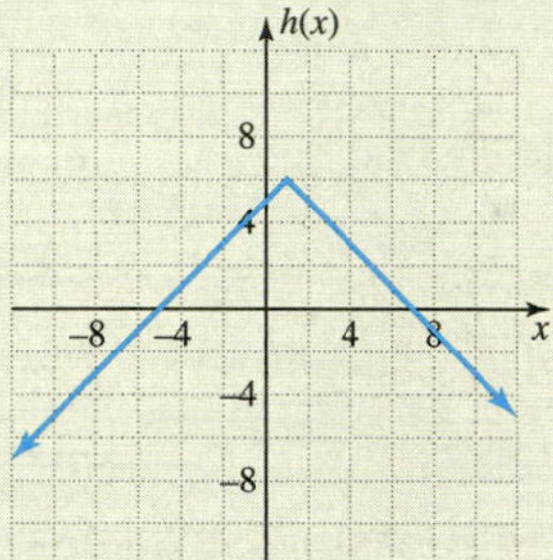

11. $(f + g)(0)$ -1
12. $(f - h)(3)$ 1
13. $(f \cdot g)(0)$ -12
14. $(g \cdot h)(-2)$ 15
15. $\left(\frac{f}{g}\right)(2)$ 0
16. $(f \circ g)(5)$ 0
17. $(h \circ g)(2)$ 6
18. $(h \circ f)(0)$ 1

Use the table to find each value. (*See Objectives 1–3.*)

x	$f(x)$	$g(x)$	$h(x)$
−3	15	13	0
−2	9	10	−1
−1	5	7	−2
0	3	4	−3
1	3	1	−4
2	5	−2	−5
3	9	−5	−4
4	15	−8	−3
5	23	−11	−2

19. $(g + h)(0)$ 1
20. $(f - g)(-3)$ 2
21. $(f \cdot g)(2)$ -10
22. $\left(\frac{f}{h}\right)(4)$ -5
23. $(f \circ g)(2)$ 9
24. $(g \circ g)(2)$ 10

Solve each problem. (*See Objective 4.*)

25. Loretta makes $12.50 an hour, so her income can be represented by the function $f(x) = 12.50x$, where x is the number of hours she works. She pays 18% of her income in taxes. The amount of tax she owes can be represented by the function $g(x) = 0.18x$, where x is her income. Write a function that represents the total taxes Loretta owes as a function of the number of hours she works? $g(x) = 2.25x$

26. An organization is planning a bingo fund-raiser. There are several costs involved for this event. The cost of using the hall is $25 per person. The cost of food is $18 per person and the cost of prizes is $52 for every four players. Let x represent the number of players. Write a function that represents the cost of the hall, the cost of food, and the cost of prizes. Then write a function that represents the total cost of the fund-raiser as a function of players. $h(x) = 25x, f(x) = 18x, p(x) = 13x, c(x) = 56x$

SECTION 10.2

Determine if each function is one-to-one. (*See Objectives 1 and 2.*)

27. $f = \{(-2, 1), (1, 5), (-1, 2), (1, 0), (0, 4)\}$ yes, a one-to-one function

28. $f = \{(0, -5), (0, 2), (-\sqrt{3}, 4), (\sqrt{3}, 4)\}$ not a one-to-one function

29.

x	−2	−1	0	1	2
y	−4	−3	0	3	4

yes, a one-to-one function

30.

x	−4	−2	0	2	4
y	5	1	3	4	6

yes, a one-to-one function

31. 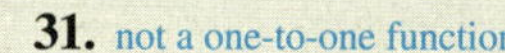 not a one-to-one function

32. 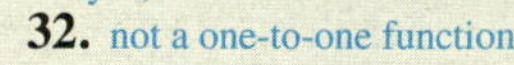 not a one-to-one function

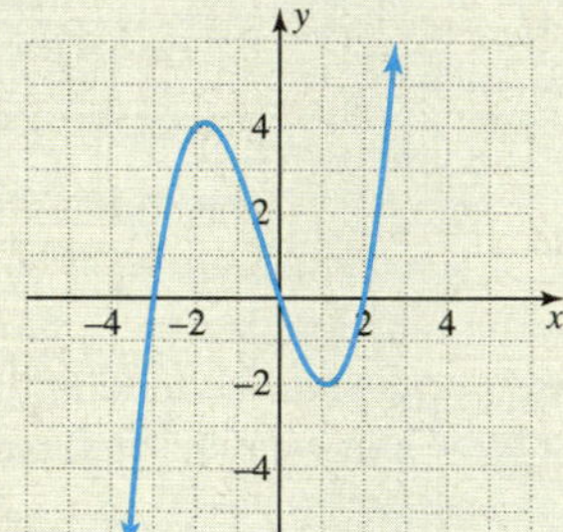

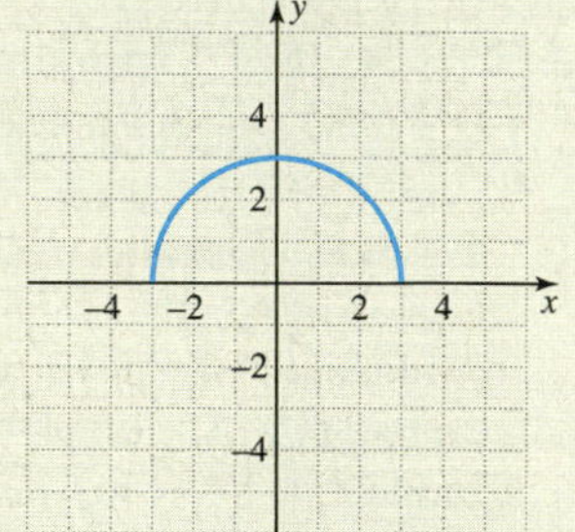

33. yes, a one-to-one function

34. not a one-to-one function

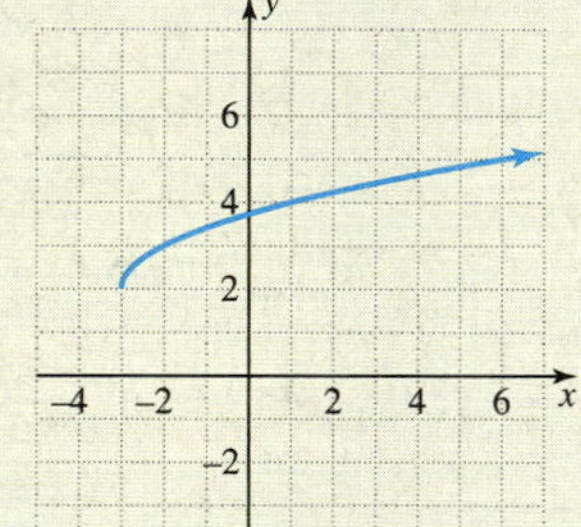

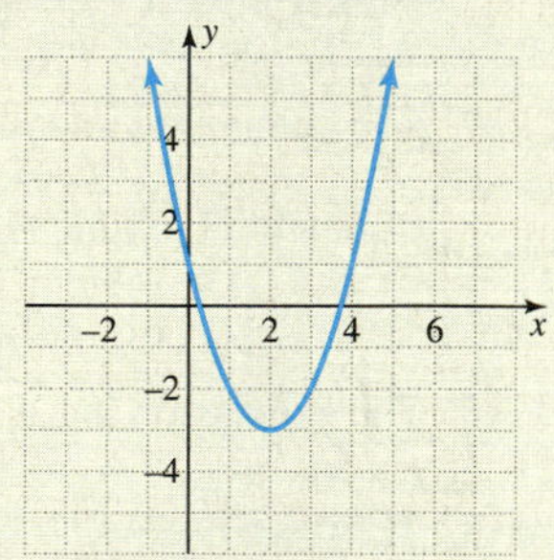

Find the inverse of each function. (*See Objectives 3 and 4.*)

35. $f(x) = \{(-2, -5), (0, 1), (3, 5), (2, 3), (4, 7)\}$

36. $f(x) = \{(-3, -3.2), (-1, -2.5), (2, 0.9), (5, 2.1), (9, 6.9)\}$

37.

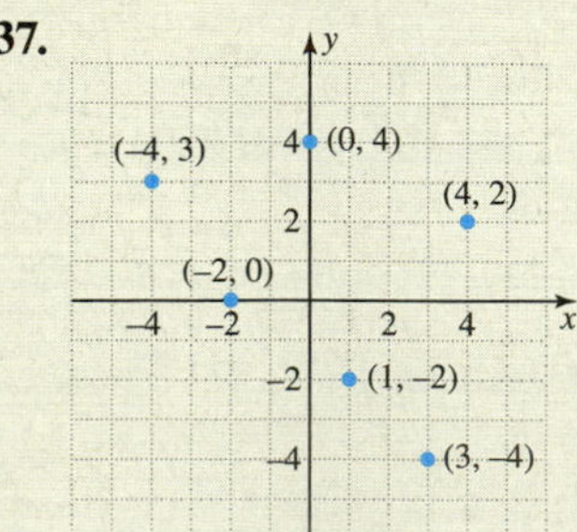

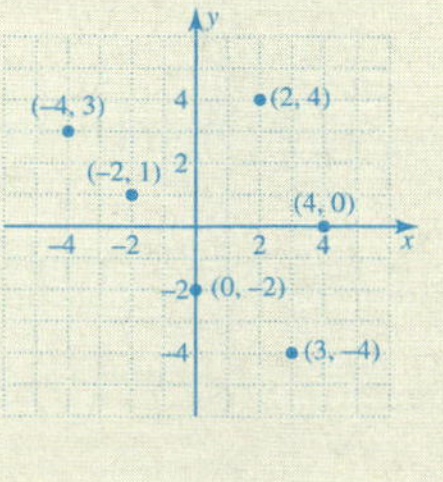

38.

State (x)	Congressmen in 2009 (y)
Arizona	Harry Mitchell
Kansas	Lynn Jenkins
Florida	Connie Mack
Illinois	Debbie Halvorson
Ohio	Marcy Kaptur

39. $f(x) = 5x - 4$ **40.** $f(x) = \frac{2x - 11}{5}$

41. $f(x) = 2x^3 + 9$ **42.** $f(x) = \frac{6}{3 - 5x}$

Evaluate each function using the given information. (*See Objective 5.*)

43. If $f(-5) = -1$, what is $f^{-1}(-1)$? -5

44. If $f(7.2) = -2.4$, what is $f^{-1}(-2.4)$? 7.2

45. If $f\left(-\frac{1}{6}\right) = \frac{9}{5}$, what is $f^{-1}\left(\frac{9}{5}\right)$? $-\frac{1}{6}$

46. If $f(0) = 8.4$, what is $f^{-1}(8.4)$? 0

Graph the inverse of each function. (*See Objective 6.*)

47.

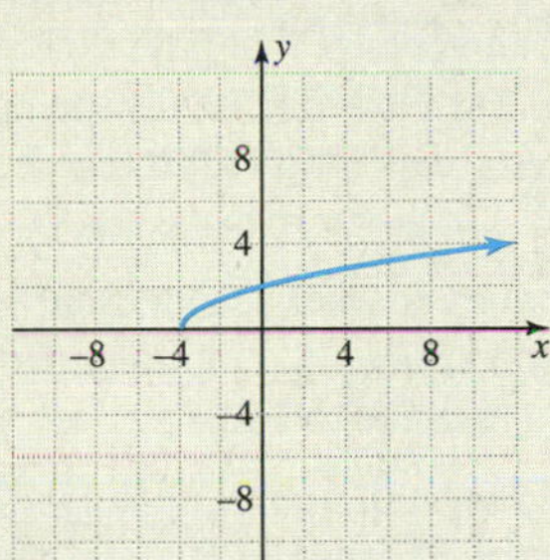

48.

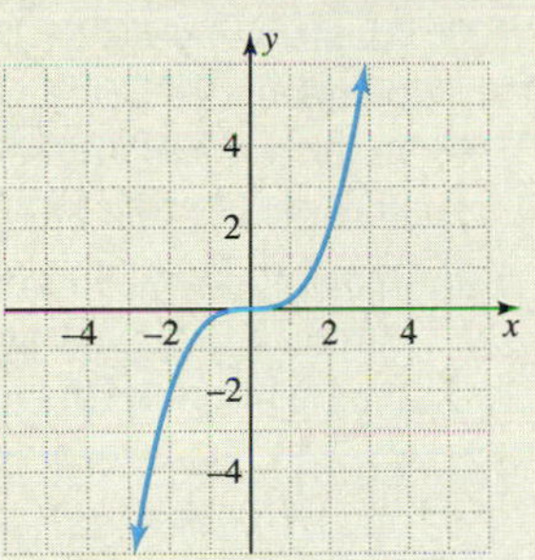

Determine if the functions are inverses of one other. (*See Objective 7.*)

49. $f(x) = 5x - 12,\ g(x) = \frac{1}{5}x + \frac{12}{5}$ yes

50. $f(x) = (4x - 11)^3,\ g(x) = \frac{\sqrt[3]{x} + 11}{4}$ yes

51. $f(x) = 6x^5 + 9,\ g(x) = \frac{\sqrt[5]{x - 9}}{6}$ no

52. $f(x) = \frac{2}{1 - 3x},\ g(x) = \frac{1}{2} - \frac{3x}{2}$ no

SECTION 10.3

Graph each exponential function. Plot and label at least three points on the graph. State the domain and range of each function. (*See Objective 1.*)

53. $y = 5^x$ **54.** $y = 0.3^x$

Solve each exponential equation. (*See Objective 2.*)

55. $\left(\frac{3}{7}\right)^x = \frac{49}{9}$ $\{-2\}$ **56.** $5^{3x-2} = 625$ $\{2\}$

57. $2^{x+3} = 1$ $\{-3\}$ **58.** $\left(\frac{1}{2}\right)^{3x-1} = 8$ $\left\{-\frac{2}{3}\right\}$

59. $25^{3-2x} = \frac{1}{5}$ $\left\{\frac{7}{4}\right\}$ **60.** $3^{5x-7} = 9^x$ $\left\{\frac{7}{3}\right\}$

Solve each problem. Round each answer to two decimal places. (*See Objective 3.*)

61. Elaine invests \$3500 in a saving account that pays 2.2% annual interest compounded quarterly. Use the compound interest formula to determine the amount of money she will have in the account at the end of 2 yr. \$3657.00

62. The population of China (in billions) can be modeled by $y = 1.34(1.005)^x$, where x is the number of years after 2011. What is an estimate for the population in 2015? 2025? 2035? (Source: http://www.cia.gov) about 1.37 billion, about 1.44 billion, about 1.51 billion

SECTION 10.4

Rewrite each logarithmic equation in exponential form and each exponential equation in logarithmic form. (*See Objective 1.*)

63. $\log_2 128 = 7$ $128 = 2^7$ **64.** $\log_{1/10} \frac{1}{100} = 2$ $\frac{1}{100} = \left(\frac{1}{10}\right)^2$

65. $9^{3/2} = 27$ $\log_9 27 = \frac{3}{2}$ **66.** $\left(\frac{1}{81}\right)^{-3/4} = 27$ $\log_{1/81} 27 = -\frac{3}{4}$

Simplify each expression. (*See Objective 2.*)

67. $\log_{1/5} 1$ 0 **68.** $\log_{0.25} 0.25$ 1 **69.** $\log_3 \frac{1}{27}$ -3

70. $\log 0.01$ -2 **71.** $\log_{1/2} 32$ -5 **72.** $\log_2(-4)$ undefined

Solve each equation for *x*. (*See Objective 3.*)

73. $\log_6 216 = x$ $\{3\}$ **74.** $\log_a \sqrt{a} = x, a > 0$ $\left\{\frac{1}{2}\right\}$

75. $\log_4 x = 3$ $\{64\}$ **76.** $\log_x \frac{1}{64} = 3$ $\left\{\frac{1}{4}\right\}$

77. $\log_{3.6} 1 = x$ $\{0\}$ **78.** $\log_c c^{-4} = x, c > 0$ $\{-4\}$

Graph each logarithmic function. Plot and label at least three points. State the domain and range of each function. (*See Objective 4.*)

79. $y = \log_5 x$ **80.** $y = \log_{0.6} x$

SECTION 10.5

Use the rules for logarithms to write each expression as a single logarithm. Simplify, if possible. Assume all logarithmic arguments are positive numbers. (*See Objectives 1–4.*)

81. $\log_3 8 + \log_3 7 - \log_3 28$ $\log_3 2$

82. $\log_4 24 - \log_4 3 + \log_4 8$ 3

83. $2\log_5 x - 3\log_5 y + 5\log_5 z$ $\log_5 \frac{x^2 z^5}{y^3}$

84. $\frac{1}{2}\log_7 a + 3\log_7 b - \frac{1}{3}\log_7 c$ $\log_7 \frac{a^{1/2} b^3}{c^{1/3}}$

Write each expression as a combination of logarithms. (*See Objectives 1–4.*)

85. $\log_4\frac{7x^3}{y^2}$ $\log_4 7 + 3\log_4 x - 2\log_4 y$

86. $\log_2\sqrt[3]{\frac{a^2}{2b}}$ $\frac{2}{3}\log_2 a - \frac{1}{3} - \frac{1}{3}\log_2 b$

87. $\log_{10}\left(x^2y^5\sqrt{x-1}\right)$

88. $\log_{10}\frac{100(x^2+4)}{y^4}$

Simplify each expression. Assume all logarithmic arguments are positive numbers. (*See Objective 5.*)

89. $\log_{10}10^{5x+7}$ $5x+7$

90. $\log_2 2^{x-7}$ $x-7$

91. $\log_b b^x$ x

92. $6^{\log_6 15x}$ $15x$

93. $a^{\log_a(3x-5)}$ $3x-5$

94. $e^{2\log_e(x)}$ x^2

SECTION 10.6

Simplify each logarithm without a calculator. (*See Objectives 1 and 2.*)

95. $\log 10^{5/3}$ $\frac{5}{3}$

96. $\ln\frac{1}{e^4}$ -4

97. $\log\sqrt[3]{100}$ $\frac{2}{3}$

98. $\ln\sqrt[3]{e^4}$ $\frac{4}{3}$

Approximate each logarithm to two decimal places. (*See Objective 3.*)

99. $\log 6.589$ 0.82

100. $\log 0.0158$ -1.80

101. $\ln 58.3$ 4.07

102. $\ln\sqrt[3]{7.5}$ 0.67

Solve each equation for *x*. Give an exact solution and approximate the solution to two decimal places, if necessary. (*See Objective 4.*)

103. $\ln(2x-5) = 4$

104. $2\log(7x+3) = 6$

105. $\ln(x-2.5) = 0$ $\{3.5\}$

106. $3\ln(2x+17) = 4.5$

107. $\ln(5x) = 1$ $\left\{\frac{e}{5}\right\}$ or $\{0.54\}$

108. $\log(0.25x) = 1$ $\{40\}$

Use the change-of-base formula to rewrite each logarithm. Approximate its value to two decimal places. (*See Objective 5.*)

109. $\log_{3.24}8.56$ 1.83

110. $\log_{0.075}0.469$ 0.29

111. $\log_{3/2}16.43$ 6.90

112. $\log_{1/3}0.986$ 0.01

Solve each problem. Round each answer to two decimal places. (*See Objective 6.*)

113. Carrie becomes about 92% proficient at a new task after a 1-week period. The equation $P = 92 - 28.782\log d$ models her percent proficiency P after d days without practicing the new skill. What is Carrie's proficiency level after 5 days without practicing? 10 days? 71.88% after 5 days, 63.22% after 10 days

114. Mark becomes about 96% proficient at a new task after a 1-week period. The equation $P = 96 - 12.5\ln d$ models his percent proficiency P after d days without practicing the new skill. What is Mark's proficiency level after 7 days without practicing? 12 days? 71.68% after 7 days, 64.94% after 12 days

SECTION 10.7

Solve each equation. For solutions of exponential equations, give an exact solution and an approximation of the solution to two decimal places. (*See Objectives 1 and 2.*)

115. $10^{5.2x} = 32.6$

116. $e^{1.2x} = 45$ $\left\{\frac{\ln 45}{1.2}\right\}$ or $\{3.17\}$

117. $7^{3x-2} = 28$

118. $10^{5x-2} = 3.5$

119. $\ln x + \ln(2x+1) = \ln 15$ $\left\{\frac{5}{2}\right\}$

120. $\log 2x - \log(x-4) = 1$ $\{5\}$

121. $\log_3(5x-2) = 4$

122. $\log_5(1-3x) = 2$ $\{-8\}$

123. $\log_2 x + \log_2(3x-22) = 4$ $\{8\}$

124. $6(2.17)^{3x} = 84$

125. $12(0.54)^{8x} = 3$ $\left\{\frac{\ln 0.25}{8\ln 0.54}\right\}$ or $\{0.28\}$

126. $\log_3(5x+2) - \log_3(x-1) = 2$ $\left\{\frac{11}{4}\right\}$

127. $2\log_2 x - \log_2(3x+8) = 1$ $\{8\}$

128. $2\ln x - \ln(56-x) = 0$ $\{7\}$

Solve each problem. (*See Objective 3.*)

129. Madison invests \$1270 in an account that pays 2.6% compounded continuously. (a) How much will she have in the account after 5 yr? (b) How long will it take her account to grow to \$2750? (a) \$1446.31 (b) about 30 yr

130. Troy invests \$2500 in a money market account that earns 2.2% annual interest compounded monthly. How long will it take his money to grow to \$4200? about 24 yr

131. In 2011, Brazil had a population of approximately 203.43 million. Its population is growing at a rate of 1.134%. The population (in millions) can be modeled by $y = 203.43e^{0.01134t}$, where t is the number of years after 2011. (a) What is an estimate for the population of Brazil in 2023 if it continues to grow at this rate? (b) How long will it take the population to reach 210 million? (Source: https://www.cia.gov/library/publications/the-world-factbook/geos/br.html) (a) 233.08 million (b) about 2.80 yr

132. The amount of ibuprofen (in milligrams) in a person's body t hour after ingestion is $y = a_0e^{-0.347t}$, where a_0 is the initial amount taken. If Lynn takes 1000 mg of ibuprofen, how much ibuprofen will be in his body after 4 hr? Round to the nearest integer. (Source: http://www.rxlist.com/ibuprofen-drug.htm) 250 mg

CHAPTER 10 TEST / EXPONENTIAL AND LOGARITHMIC FUNCTIONS

1. If $f(x) = 5x + 3$ and $g(x) = x^2 - 1$, then $(f \circ g)(x)$ is

a. $25x^2 + 2$ **b.** $25x^2 + 30x + 8$

(c.) $5x^2 - 2$ **d.** $5x^3 + 3x^2 - 5x - 3$

2. The function that is one-to-one is

(a.)

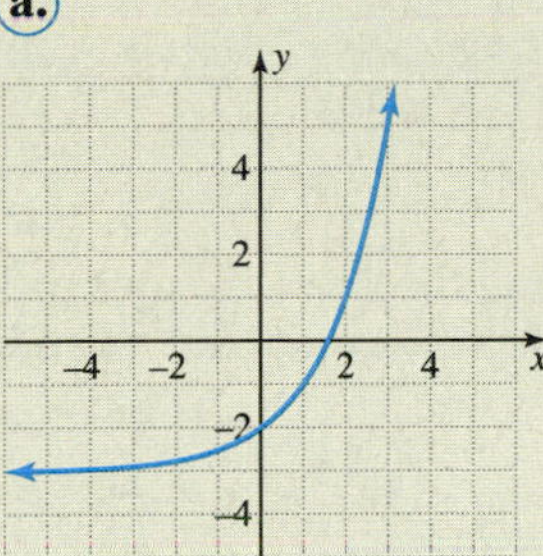

b.

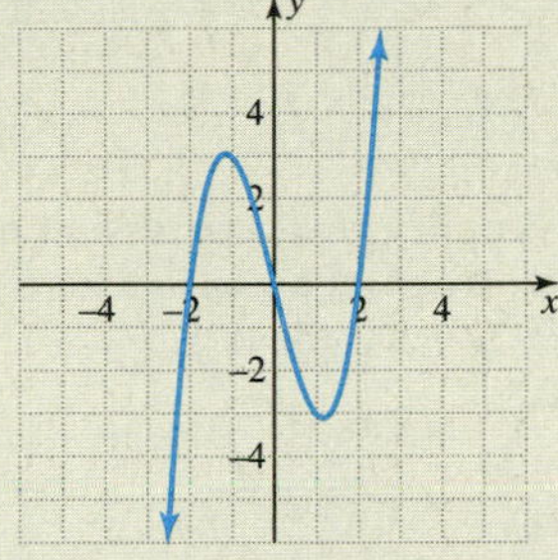

c.

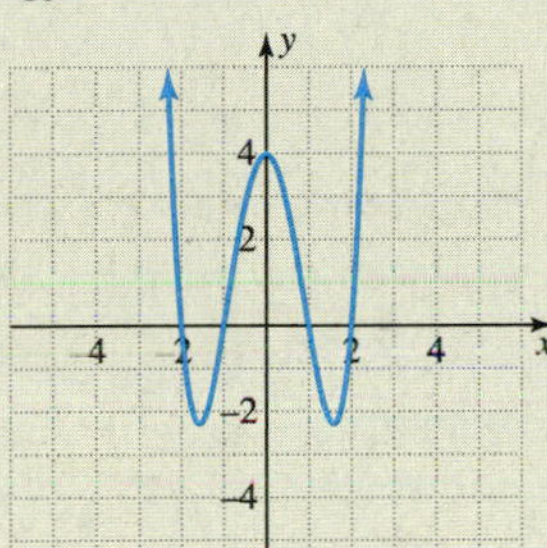

d.

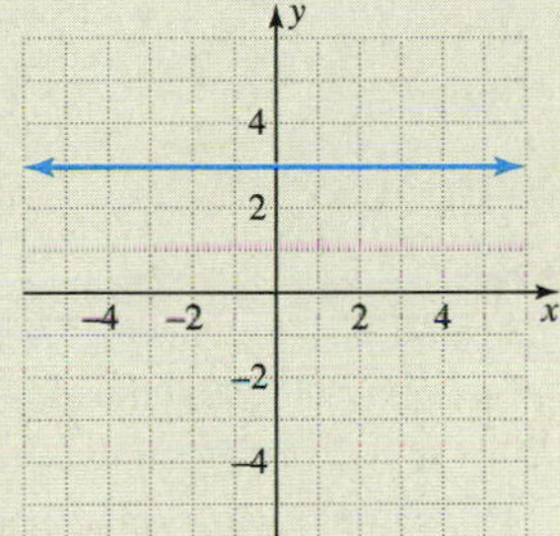

3. The statement that is false about the function $f(x) = 5^x$ is

(a.) The graph goes through the point (1, 0).

b. The domain of the function is $(-\infty, \infty)$.

c. The range of the function is $(0, \infty)$.

d. The graph is increasing.

4. The equation $4^3 = 64$ is equivalent to

a. $\log_3 64 = 4$ **(b.)** $\log_4 64 = 3$

c. $\log_3 4 = 64$ **d.** $\log_4 3 = 64$

5. The expression $2\log_b 5 + \log_b 3$ is equivalent to

a. $\log_b 13$ **b.** $\log_b 28$

c. $\log_b 30$ **(d.)** $\log_b 75$

6. When simplified, $\ln\frac{1}{e} + \log\frac{1}{1000} + e^{\ln 5} - 10^{\log 7}$ is

a. 2 **b.** 5

(c.) -6 **d.** -2

7. The solution set of $6^x = 12$ is

a. $\{2\}$ **(b.)** $\left\{\frac{\ln 12}{\ln 6}\right\}$

c. $\{\ln 6\}$ **d.** $\{\ln 2\}$

8. A conference committee is planning a reception. The cost of using a banquet hall is \$10 per person. The cost of food is \$25 per person and the cost of prizes is \$40 for every three attendees. Let x represent the number of attendees. Write a function that represents the cost of the reception hall, the cost of food, and the cost of the prizes. Finally write a function that represents the total cost of the reception as a function of attendees.

9. Let $f(x) = \frac{9}{x - 2}$ and $g(x) = 4x + 1$. Find the following.

a. $(f + g)(1)$ -4

b. $(f \cdot g)(x)$ and state the domain.

c. $(f \circ g)(7)$ $\frac{1}{3}$ **d.** $f^{-1}(x)$ $f^{-1}(x) = \frac{9 + 2x}{x}$

10. Graph each function. State the domain and range. Then sketch the graph of the inverse of the function on the same set of axes.

a. $f(x) = 3^x$ **b.** $f(x) = \log_4 x$

11. Solve each equation.

a. $2^{x+1} = \frac{1}{4}$ $\{-3\}$ **b.** $9^x = 27$ $\left\{\frac{3}{2}\right\}$

c. $5^{x-2} = 25^x$ $\{-2\}$ **d.** $\log_4 \frac{1}{16} = x$ $\{-2\}$

e. $\log x = 0$ $\{1\}$ **f.** $\log_x 81 = 2$ $\{9\}$

g. $\ln(x + 1) = 3$ $\{e^3 - 1\}$ **h.** $2\log(5x) = 4$ $\{20\}$

i. $3e^{x+1} = 12$ $\{\ln 4 - 1\}$

12. Use the properties of logarithms to rewrite the expression $4\log_2 x - \log_2 y + \frac{1}{3}\log_2 z$ as a single logarithms. $\log_2 \frac{x^4\sqrt[3]{z}}{y}$

13. Juan invests \$2000 in an account that earns 4% annual interest compounded quarterly.

a. How much will Juan have in the account at the end of 5 yr. \$2440.38

b. How long will it take the account to grow to \$10,000? about 40 yr

14. The human memory model given by $f(t) = 75 - 6\ln(t + 1)$ determines the percentage of information the average person retains t months after the information was presented.

a. What percentage of information is retained 6 months after its presentation? 63.32%

b. For how long does the average person retain 60% of presented information? 11.18 months

CUMULATIVE REVIEW EXERCISES / CHAPTERS 1–10

Use the order of operations to simplify each expression. (*Section 1.2, Objectives 3 and 4*)

1. $25 - [(12 - 7) - (6 - 10)] - 28 \div 7 \cdot 4$ 0

2. $\frac{17 - 3\sqrt{13 - 4}}{|4 - 5|^2}$ 8

Solve each problem. (*Section 1.3, Objective 5*)

3. Jonathan has collected 165 dimes and nickels. If x represents the number of dimes he collected, write an expression for the number of nickels. $165 - x$

4. Referring to Exercise 3, write an algebraic expression that represents the total value of Jonathan's coins. $0.10x + 0.05(165 - x)$

Solve each linear equation. (*Section 2.1, Objectives 2–6*)

5. $6(x - 4) - x = 5(x + 2)$ $\varnothing$ **6.** $3(2 - x) - 2x = x - 12$ $\{3\}$

Solve each equation. (*Section 2.6, Objectives 1 and 2*)

7. $|4x + 1| = |2x - 7|$ $\{-4, 1\}$ **8.** $|3x - 5| = 4$ $\left\{\frac{1}{3}, 3\right\}$

Find the midpoint of the line segment formed by the ordered pairs. (*Section 3.1, Objective 5*)

9. $(10, 9)$ and $(-4, -7)$ $(3, 1)$ **10.** $(-2.4, 3.1)$ and $(6.8, -5.5)$ $(2.2, -1.2)$

Find the domain of each function. (*Section 3.4, Objective 3*)

11. $f(x) = 4x^3 - 5$ $(-\infty, \infty)$ **12.** $f(x) = \sqrt{12 - 4x}$ $(-\infty, 3]$

Rewrite each equation using function notion, if possible. Identify *m* and *b*. (*Section 4.1, Objective 2*)

13. $4x - 7y = 28$ **14.** $x = -2$ can't be written in function notation

Determine if the two lines are parallel, perpendicular, or neither. (*Section 4.3, Objective 4*)

15. $y = -\frac{2}{5}x + 2$ and $y = \frac{2}{5}x$ neither

16. $y = 2x + 1$ and $4x - 2y = 3$ parallel

Solve each system of linear equations graphically. (*Section 5.1, Objectives 2 and 3*)

17. $\begin{cases} 2x - y = 4 \\ y = x \end{cases}$ $\{(4, 4)\}$ **18.** $\begin{cases} 2y + 3x = 12 \\ x + y = 2 \end{cases}$ $\{(8, -6)\}$

Solve each system of linear equations using substitution or elimination. (*Section 5.2, Objectives 1 and 2*)

19. $\begin{cases} y = 5x + 11 \\ x + 4y = 2 \end{cases}$ $\{(-2, 1)\}$ **20.** $\begin{cases} 7x - 15y = 51 \\ 3x + 4y = 1 \end{cases}$ $\{(3, -2)\}$

21. $\begin{cases} -0.1x + 0.6y = 7.8 \\ 0.4x + 0.5y = 12.3 \end{cases}$ $\{(12, 15)\}$ **22.** $\begin{cases} x + 6y = -3 \\ y = \frac{2}{3}x - 3 \end{cases}$ $\{(3, -1)\}$

Solve each problem. (*Section 5.3, Objective 5*)

23. Two angles are complementary. The measure of one angle is 15° less than twice the measure of the other angle. Find the measure of each angle. 35°, 55°

24. Two angles are supplementary. The measure of one angle is 12° more than three times the measure of the other angle. Find the measure of each angle. 42°, 138°

Solve each system of linear equations in three variables using elimination. (*Section 5.4, Objective 1*)

25. $\begin{cases} 3x - y - 2z = -1 \\ -x + y + 2z = 3 \\ 4x - 3y + z = 13 \end{cases}$ $\{(1, -2, 3)\}$ **26.** $\begin{cases} x + y - z = 9 \\ x - y - 2z = 3 \\ 2x - 3y + z = -8 \end{cases}$ $\{(3, 4, -2)\}$

Solve each system of linear inequalities in two variables. (*Section 5.5, Objective 1*)

27. $\begin{cases} 2x - y \geq -4 \\ y \leq x \end{cases}$ **28.** $\begin{cases} x - 3 < 0 \\ y > 2x \end{cases}$

Simplify each expression by applying the product or quotient rule. Express each answer with positive exponents. (*Section 6.1, Objectives 1–4*)

29. $(-15x^2) \cdot (2x^{13})$ $-30x^{15}$ **30.** $(4rs^3)(-16r^7s^9)$ $-64r^8s^{12}$

31. $\dfrac{a^{12}b}{a^5b^7}$ $\dfrac{a^7}{b^6}$ **32.** $\dfrac{p^2q^9}{p^{13}q^4}$ $\dfrac{q^5}{p^{11}}$

Simplify each expression using the rules of exponents. Express each answer with positive exponents. (*Section 6.2, Objectives 1 and 2*)

33. $\dfrac{(5x^3y^{-1})^{-3}}{x^{-7}y^{-10}}$ $\dfrac{y^{13}}{125x^2}$ **34.** $\dfrac{(r^{-4}s^3)^{-2}}{(3rs)^{-3}}$ $\dfrac{27r^{11}}{s^3}$

Simplify each expression. (*Section 6.4, Objectives 1–5*)

35. $(13a - 23) + (-15a + 17)$ $-2a - 6$

36. $(12p - 10) - (-8p + 1)$ $20p - 11$

37. $(7x - 4)(2x + 5)$ $14x^2 + 27x - 20$ **38.** $(6y - 1)^2$ $36y^2 - 12y + 1$

39. $(3x - 2)^3$ $27x^3 - 54x^2 + 36x - 8$ **40.** $(a + 4)^3$ $a^3 + 12a^2 + 48a + 64$

41. Write a polynomial function that represents the perimeter of a triangle with sides of lengths: $8y - 5$, $3y^2 - 12y - 7$, and $2y^2 + 16y + 9$. (*Section 6.4, Objective 1*) $5y^2 + 12y - 3$

42. Write a polynomial function that represents the perimeter of a trapezoid with sides of lengths: $4x - 1$, $10x$, $8x^2 + 3x$, and $2x^2 - 5x + 9$. (*See Section 6.4, Objective 1*) $10x^2 + 12x + 8$

Use grouping to factor each polynomial completely. (*Section 6.5, Objectives 1–3*)

43. $w^3 + 3w - 2w^2 - 6$ $(w - 2)(w^2 + 3)$ **44.** $4x^3 - 7x^2 + 4x - 7$ $(4x - 7)(x^2 + 1)$

Factor each trinomial. (*Section 6.6, Objectives 1–4*)

45. $25x^2 + 40xy + 16y^2$ $(5x + 4y)^2$ **46.** $12x^2 - 4xy - 5y^2$ $(6x - 5y)(2x + y)$

Factor completely. (*Section 6.7, Objectives 1–3*)

47. $8x^3 + 125y^3$ $(2x + 5y)(4x^2 - 10xy + 25y^2)$ **48.** $a^3 - b^3$ $(a - b)(a^2 + ab + b^2)$

49. $16x^2 - 8x + 1 - 9y^2$ $(4x - 1 + 3y)(4x - 1 - 3y)$ **50.** $y^2 + 10y + 25 - 49z^2$ $(y + 5 + 7z)(y + 5 - 7z)$

Solving each equation by factoring. (*Section 6.8, Objective 1*)

51. $10x^2 + 33x - 7 = 0$ **52.** $3y^2 - 10y - 8 = 0$ $\left\{-\frac{2}{3}, 4\right\}$

53. $y(2y - 5) = 52$ $\left\{-4, \frac{13}{2}\right\}$ **54.** $a(2a - 15) = 27$ $\left\{-\frac{3}{2}, 9\right\}$

Simplify each rational expression. (*Section 7.1, Objective 3*)

55. $\dfrac{x^2 + 2x - 8}{x^3 - 8}$ $\dfrac{x + 4}{x^2 + 2x + 4}$ **56.** $\dfrac{6x^2 + 2x}{9x^2 - 1}$ $\dfrac{2x}{3x - 1}$

Multiply or divide the rational expressions. Write each answer in simplest form. (*Section 7.1, Objectives 4 and 5*)

57. $\dfrac{4x^2 - 16}{15x + 10} \cdot \dfrac{3x^2 - x - 2}{x^2 + x - 2}$ $\frac{4}{5}(x - 2)$

58. $\dfrac{x^3 + 125}{(x + 5)^2} \div \dfrac{2x^2 - 10x + 50}{3x^2 + 6x - 45}$ $\frac{3}{2}(x - 3)$

Perform each operation. (*Section 7.2, Objectives 1–3*)

59. $\dfrac{-2x^5 + 3x^4 - 4x^3}{2x^3}$

60. $\dfrac{3x^3 - 6x^2 - 4x}{6x}$

61. $\dfrac{7x^3 + 10x^2 - 14}{x + 1}$

62. $\dfrac{12x^3 + 5x + 2}{2x - 1}$

Add or subtract the rational expressions. (*Section 7.3, Objective 1–3*)

63. $\dfrac{3}{x^2 - 16} + \dfrac{2}{x^2 - 3x - 4}$ $\dfrac{5x + 11}{(x + 4)(x - 4)(x + 1)}$

64. $\dfrac{4}{x^2 - 9} - \dfrac{5}{x^2 + x - 12}$ $\dfrac{-x + 1}{(x - 3)(x + 3)(x + 4)}$

65. $\dfrac{3}{x^2 + 2x + 1} - \dfrac{2}{x^2 - 1}$ $\dfrac{x - 5}{(x + 1)^2(x - 1)}$

66. $\dfrac{2}{4x^2 + 12x + 9} + \dfrac{1}{4x^2 - 9}$ $\dfrac{6x - 3}{(2x + 3)^2(2x - 3)}$

Simplify each complex fraction using method 1 or method 2. (*Section 7.4, Objective 1 and 2*)

67. $\dfrac{\dfrac{6c}{3ab^2}}{\dfrac{4c^2}{2a^2b}}$ $\dfrac{a}{bc}$

68. $\dfrac{\dfrac{2z^2}{5x^2y}}{\dfrac{z}{10xy^2}}$ $\dfrac{4yz}{x}$

69. $\dfrac{\dfrac{1}{x} - \dfrac{2}{y}}{2x - y}$ $-\dfrac{1}{xy}$

70. $\dfrac{\dfrac{3}{r} + \dfrac{1}{s}}{r + 3s}$ $\dfrac{1}{rs}$

Find the slope of the line that goes through the two given points. (*Section 7.4, Objective 3*)

71. $\left(-\dfrac{5}{6}, \dfrac{1}{4}\right)$ and $\left(\dfrac{1}{3}, -\dfrac{11}{8}\right)$ $-\dfrac{39}{28}$

72. $\left(\dfrac{1}{2}, -\dfrac{1}{3}\right)$ and $\left(\dfrac{1}{10}, -\dfrac{1}{3}\right)$ 0

Simplify each rational expression. (*Section 7.4, Objective 3*)

73. $\dfrac{3^{-2} + 2^{-4}}{2^{-2}}$ $\dfrac{25}{36}$

74. $\dfrac{1 + 5^{-2}}{2^{-2}}$ $\dfrac{104}{25}$

Solve each rational equation. (*Section 7.5, Objective 1*)

75. $\dfrac{5}{y} + \dfrac{2}{y - 5} = \dfrac{3}{y^2 - 5y}$ {4}

76. $\dfrac{2}{x} - \dfrac{1}{x + 3} = \dfrac{3}{x^2 + 3x}$ ∅

Solve each rational equation for the specified variable. (*Section 7.6, Objective 1*)

77. $\dfrac{1}{a} = \dfrac{2}{b} - \dfrac{3}{c}$ for b $b = \dfrac{2ac}{3a + c}$

78. $\dfrac{4}{x} + \dfrac{1}{y} = \dfrac{5}{z}$ for z $z = \dfrac{5xy}{x + 4y}$

Simplify each expression. Assume all variables represent positive real numbers. Write each answer with positive exponents. (*Section 8.2, Objectives 1–4*)

79. $(x^{-12}y^8)^{3/4}$ $\dfrac{y^6}{x^9}$

80. $(x^{-6}y^9)^{-2/3}$ $\dfrac{x^4}{y^6}$

81. $(a^{1/2} - b^{1/3})(a^{1/2} + 2b^{1/3})$

82. $(2x^{2/5} - y^{1/4})^2$

Simplify each radical expression. Assume all variables represent positive real numbers. (*Section 8.3, Objectives 1 and 2*)

83. $\sqrt{120x^5y^3}$

84. $\sqrt{90rt^7}$

85. $\sqrt[3]{\dfrac{x^4}{8y^3}}$

86. $\sqrt[5]{\dfrac{32a^7}{b^5}}$

Perform the indicated operation and write each answer in simplest radical form. (*Section 8.4, Objectives 1 and 2*)

87. $(2 - \sqrt{3})^2$ $7 - 4\sqrt{3}$

88. $(\sqrt{7} + 1)^2$ $8 + 2\sqrt{7}$

Solve each radical equation. (*Section 8.6, Objective 1*)

89. $\sqrt{6x + 28} + 2 = x$ {12}

90. $\sqrt{4t + 60} = t$ {10}

Graph each function $y = f(x)$ by plotting points. Identify the x-intercepts, y-intercept, vertex, and the axis of symmetry. Explain how the graph of the function relates to the graph of $y = x^2$. State the domain and range of the function. (*Section 9.1, Objectives 1–4*)

91. $f(x) = (x + 3)^2 - 1$

92. $f(x) = -(x + 2)^2 + 4$

Solve each equation by completing the square. (*Section 9.2, Objectives 1–4*)

93. $x^2 - 4x + 1 = 0$ $\{2 \pm \sqrt{3}\}$

94. $x^2 + 6x + 14 = 0$ $\{-3 \pm i\sqrt{5}\}$

Solve each equation using the quadratic formula. (*Section 9.3, Objective 1*)

95. $x^2 - 2x + 50 = 0$ $\{1 \pm 7i\}$

96. $x^2 + 6x + 4 = 0$ $\{-3 \pm \sqrt{5}\}$

Solve each equation. (*Section 9.4, Objectives 1–4*)

97. $6x^{2/3} - 7x^{1/3} - 3 = 0$

98. $x - 5x^{1/2} + 6 = 0$ {4, 9}

Use the graph of $y = f(x)$ to solve each associated inequality. Write each solution set in interval notation. (*Section 9.6, Objective 4*)

99. $x(2x + 5)(x - 3) < 0$

100. $x^4 - 10x^2 + 9 > 0$

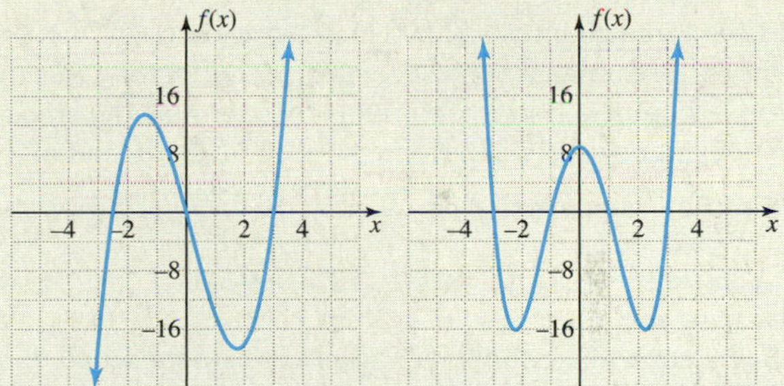

Find the sum, difference, product, and quotients of the functions. State the domain of the combined functions. Also, state the restriction for each quotient function. (*Section 10.1, Objective 1*)

101. $f(x) = 5x + 4$ and $g(x) = 3x - 2$

102. $f(x) = \sqrt{x + 2}$ and $g(x) = 3x - 6$

Let $f(x) = x^2 - 3x + 5$ and $g(x) = 4x + 3$. Evaluate each function. (*Section 10.1, Objectives 2 and 3*)

103. $(f + g)(-2)$ 10

104. $(f \cdot g)(2)$ 33

105. $(f \circ g)(3)$ 185

106. $\left(\dfrac{f}{g}\right)(-1)$ −9

For the given functions, $f(x)$ and $g(x)$, find the composition functions, $(f \circ g)(x)$ and $(g \circ f)(x)$. State the domain. (*Section 10.1, Objective 3*)

107. $f(x) = 3x - 5$ and $g(x) = 4 - x$

108. $f(x) = 7x - 10$ and $g(x) = 2x$

Determine if each function is one-to-one. (*Section 10.2, Objectives 1 and 2*)

109. $f = \{(-9, 3), (-7, 2), (-5, 1), (3, 0), (2, -1)\}$ yes, a one-to-one function

110. $f = \{(-3, 1), (-2, 4), (0, 5), (3, 2), (2, 2)\}$ not a one-to-one function

Find the equation of the inverse of each function. (*Section 10.2, Objective 4*)

111. $f(x) = (7x - 2)^3$ **112.** $f(x) = x^3 + 125$ $f^{-1}(x) = \sqrt[3]{x - 125}$

Evaluate each function given the information. (*Section 10.2, Objective 5*)

113. If $f^{-1}(6.9) = 1.5$, what is $f(1.5)$? 6.9

114. If $f(0) = 2.1$, what is $f^{-1}(2.1)$? 0

Graph each exponential function. Plot and label at least three points on the graph. State the domain and range of the function. (*Section 10.3, Objective 1*)

115. $y = 0.2^x$ **116.** $y = 3^x$

Solve each exponential equation. (*Section 10.3, Objective 2*)

117. $4^{2x-5} = 8^x$ 10 **118.** $27^{3x-4} = 9^{4x}$ 12

For Exercises 121–123, use the formula $A = P\left(1 + \frac{r}{n}\right)^{nt}$.

119. Joshua invests \$1850 in a savings account that pays 2.5% annual interest compounded quarterly. Use the compound interest formula to determine the amount of money he will have in the account at the end of 2 yr. \$1944.55

120. Candice invests \$5200 in a savings account that pays 1.8% annual interest compounded monthly. Use the compound interest formula to determine the amount of money she will have in the account at the end of 3 yr. \$5488.30

Simplify each expression. (*Section 10.4, Objectives 1 and 2*)

121. $\log_3 \frac{1}{81}$ -4 **122.** $\log_{1/2} 1$ 0

Solve each logarithmic equation for x. (*Section 10.4, Objective 3*)

123. $\log_3 x = 5$ 243 **124.** $\log_x 144 = 2$ 12

Graph each logarithmic function. Plot and label at least three points. State the domain and range of the function. (*Section 10.4, Objective 4*)

125. $y = \log_{1/6} x$ **126.** $y = \log_4 x$

Use the properties of logarithms to write each expression as a single logarithm. Simplify, if possible. (*Section 10.5, Objectives 1–3*)

127. $\log_3(x + 5) + \log_3(x - 4)$ $\log_3(x^2 + x - 20)$

128. $\log_6 x + \log_6(2x + 1)$ $\log_6(2x^2 + x)$ **129.** $\log_4 x - \log_4 (x + 6)$

130. $\log_2 x - \log_2 10$ $\log_2 \frac{x}{10}$ **131.** $\log(x + 2)^5$ $5\log(x + 2)$

132. $\ln(x - 3)^7$ $7\ln(x - 3)$

Rewrite each logarithmic expression. If the expression is a single logarithm, write the expression as a combination of logarithms. If the expression is a combination of logarithms, write the expression as a single logarithm. (*Section 10.5, Objective 4*)

133. $\ln x^3\sqrt{x - 1}$ **134.** $\log_6 \frac{y^3}{\sqrt{y + 12}}$

135. $5\log_{10} x - 2\log_{10} y$ **136.** $\frac{1}{2}\log_2 x + 2\log_2 y$

Simplify each the expression. (*Section 10.5, Objective 5*)

137. $\log_6 6^{3x+2}$ $3x + 2$ **138.** $\log_{10} 10^{5x-4}$ $5x - 4$

139. $5^{\log_5(2x+7)}$ $2x + 7$ **140.** $3^{\log_3(2x)}$ $2x$

Simplify each common or natural logarithm without a calculator. (*Section 10.6, Objectives 1 and 2*)

141. $\log 10^9$ 9 **142.** $\log \frac{1}{1000}$ -3 **143.** $\ln \frac{1}{e^8}$ -8 **144.** $\ln \sqrt[5]{e^3}$ $\frac{3}{5}$

Solve each equation for x. Give an exact solution and approximate the solution to two decimal places, if necessary. (*Section 10.6, Objective 4*)

145. $\log(5x - 4) = 1.5$ **146.** $\ln(2x + 21) = 4.5$

Solve each equation. Give an exact solution and an approximation of the solution to two decimal places. (*Section 10.7, Objectives 1 and 2*)

147. $(0.75)^{8x} = 72$ **148.** $(0.44)^{5x} = 91$

149. $\log_5(2x - 7) = 1$ $\{6\}$ **150.** $\log(3x + 46) = 2$ $\{18\}$

CHAPTER

11

Conic Sections and Nonlinear Systems

Responsibility

A responsible person is someone who is both trustworthy and dependable. It's a trait that people put their confidence in and it may make the difference in you beating out the competition for a job or completing a college degree.

As a student, you must be responsible for your education. Remember, college is a privilege and a magnificent way for you to grow yourself while at the same time making yourself extremely marketable and competitive in today's workplace. Some ways to demonstrate responsibility as a student are to

- make sure you've done the work your instructor has assigned.
- be an active learner by reading ahead in the book, asking questions, and seeking help.
- own up to mistakes, fix them, and move on toward your goal.

If you model yourself around the attributes given in this book, you're well on your way to success in both college and career.

Question For Thought: Would your current employer or teacher describe you as a responsible person? How do you define responsibility?

Chapter Outline

Section 11.1 The Parabola and the Circle 808

Section 11.2 The Ellipse and the Hyperbola 817

Section 11.3 Solving Nonlinear Systems of Equations 824

Section 11.4 Solving Nonlinear Inequalities and Systems of Inequalities 831

Coming Up...

In Section 11.2, we will learn how to graph an equation that can be used to model the orbit of the planets around the sun.

"The willingness to accept responsibility for one's own life is the source from which self-respect springs."

—Joan Didion (Author)

SECTION 11.1 The Parabola and the Circle

In **Chapter 9,** we studied quadratic functions that are transformations of the function $f(x) = x^2$. The graphs of these functions are vertical parabolas. In this chapter, we study other types of equations that produce horizontal parabolas, circles, ellipses, and hyperbolas. To conclude the chapter, we will solve nonlinear systems of equations and inequalities.

OBJECTIVES

As a result of completing this section, you will be able to

1. Graph parabolas of the form $x = a(y - k)^2 + h$.
2. Graph circles of the form $(x - h)^2 + (y - k)^2 = r^2$.
3. Write the equation of a circle given its center and radius.
4. Complete the square to write the equation of a circle in standard form.
5. Troubleshoot common errors.

Centuries ago, mathematicians studied the graphs of parabolas and circles as simply mathematical objects. But these graphs have been found to have significant value in the study of science and in real-life architectural structures. The parabola, for example, is found in the design of satellite dishes and is also used to model projectile motion.

The parabola is an example of a conic section. **Conic sections** are curves that are obtained by the intersection of a cone and a plane. In this chapter, we will study the four types of curves that result from this intersection: the parabola, the circle, the ellipse, and the hyperbola.

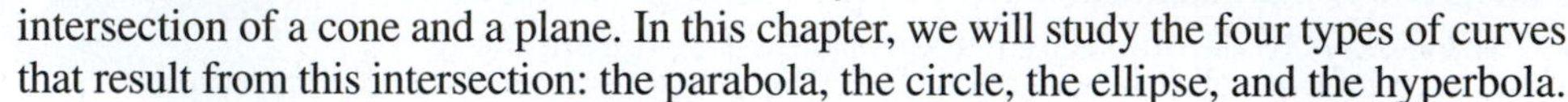

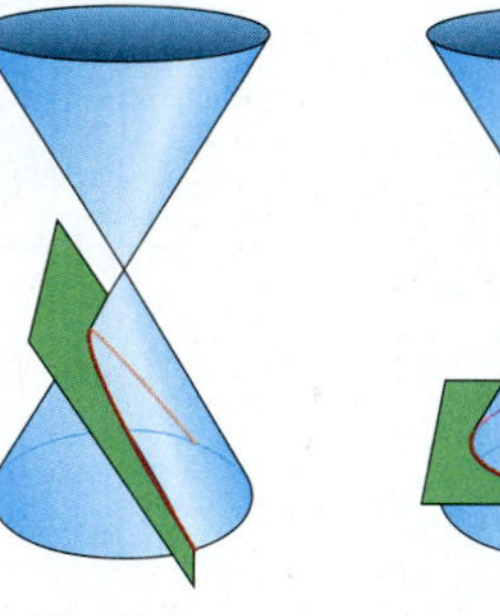

Parabola

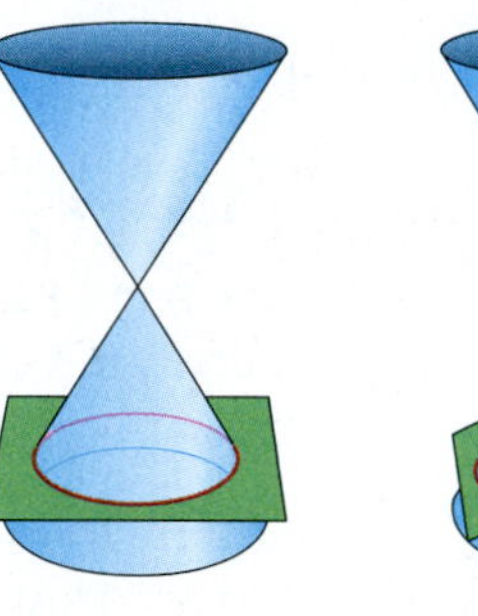

Circle

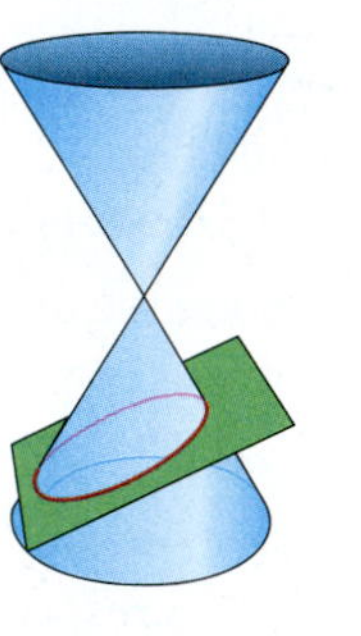

Ellipse

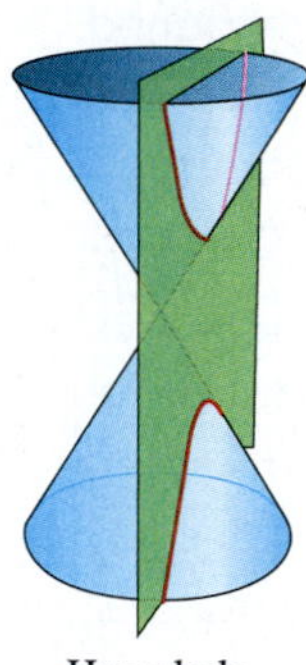

Hyperbola

Parabolas of the Form $x = a(y - k)^2 + h$

Objective 1

Graph parabolas of the form $x = a(y - k)^2 + h$.

Recall in Chapter 9, we studied quadratic functions of the form $f(x) = a(x - h)^2 + k$. This function is the graph of a *vertical* parabola whose vertex is (h, k) and opens upward if $a > 0$ and opens downward if $a < 0$. Now we will turn our attention to equations of the form $x = a(y - k)^2 + h$.

We will first examine the equations $x = y^2$ and $x = -y^2$. To determine their graphs, we plot ordered pairs that satisfy the equations. Since the variable x is isolated, we assign values to y and solve for x.

Graph of $x = y^2$

$x = y^2$	y	(x, y)
$(-3)^2 = 9$	-3	$(9, -3)$
$(-2)^2 = 4$	-2	$(4, -2)$
$(-1)^2 = 1$	-1	$(1, -1)$
$(0)^2 = 0$	0	$(0, 0)$
$(1)^2 = 1$	1	$(1, 1)$
$(2)^2 = 4$	2	$(4, 2)$
$(3)^2 = 9$	3	$(9, 3)$

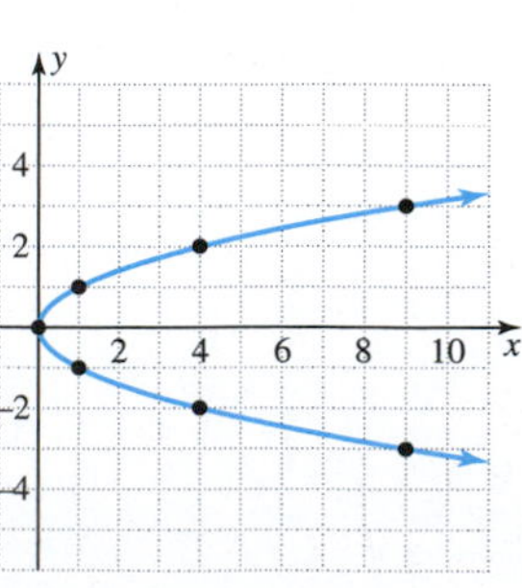

Plotting these points gives us the graph of a *horizontal* parabola. This equation is a relation but not a function since it fails the vertical line test. The vertex of the parabola is (0, 0) and the axis of symmetry is the horizontal line $y = 0$. The parabola opens to the right.

Graph of $x = -y^2$

$x = -y^2$	y	(x, y)
$-(-3)^2 = -9$	-3	$(-9, -3)$
$-(-2)^2 = -4$	-2	$(-4, -2)$
$(-1)^2 = -1$	-1	$(-1, -1)$
$-(0)^2 = 0$	0	(0, 0)
$-(1)^2 = -1$	1	$(-1, 1)$
$-(2)^2 = -4$	2	$(-4, 2)$
$-(3)^2 = -9$	3	$(-9, 3)$

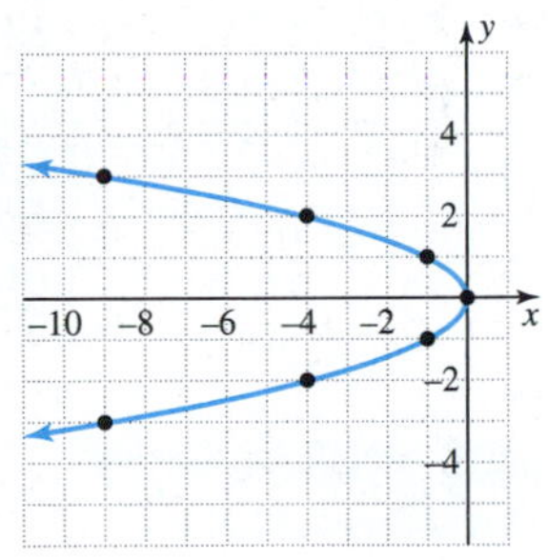

These two graphs illustrate that $x = a(y - k)^2 + h$ is the equation of a horizontal parabola that opens right if $a > 0$ and opens left if $a < 0$. The vertex of the parabola is (h, k) and the axis of symmetry is $y = k$. Just as the value of a stretches or shrinks a vertical parabola, it does the same to a horizontal parabola. If $|a| > 1$, then the parabola is horizontally stretched (made more narrow). If $|a| < 1$, then the parabola is horizontally shrunk (made wider).

Equation	$x = a(y - k)^2 + h$	Vertex	Direction
$x = y^2 - 4$	$x = 1(y - 0)^2 - 4$	$(-4, 0)$	opens right
$x = (y + 1)^2$	$x = 1[y - (-1)]^2 + 0$	$(0, -1)$	opens right
$x = -(y - 2)^2 + 3$	$x = -1(y - 2)^2 + 3$	(3, 2)	opens left

Property: Standard Form of a Parabola

The graph of $x = a(y - k)^2 + h$ is the graph of $x = y^2$ where

- The vertex of the graph is the point (h, k).
- The axis of symmetry is the horizontal line through the vertex, $y = k$.
- The parabola opens right if $a > 0$ and opens left if $a < 0$.

It is helpful to note the differences between the function $f(x) = a(x - h)^2 + k$ and the equation $x = a(y - k)^2 + h$, as shown. These two forms are the *standard forms* of a **parabola**.

$f(x) = a(x - h)^2 + k,$ $a > 0$

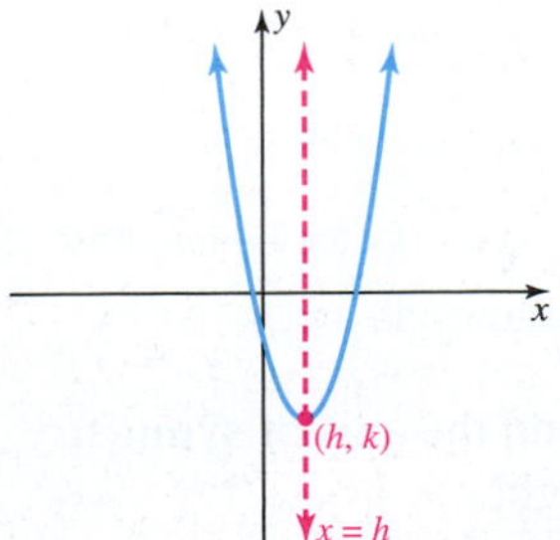

$f(x) = a(x - h)^2 + k,$ $a < 0$

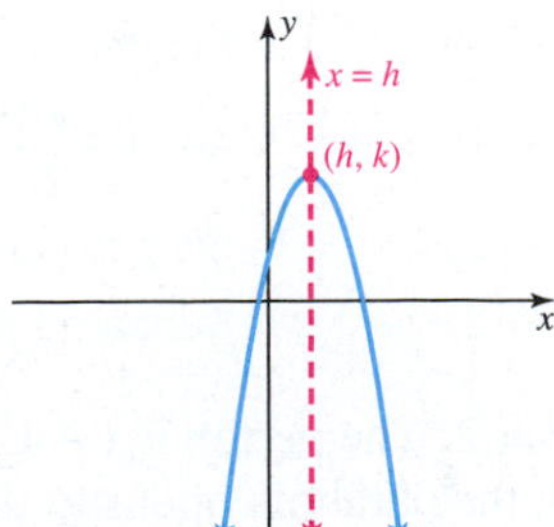

$x = a(y - k)^2 + h,$ $a > 0$

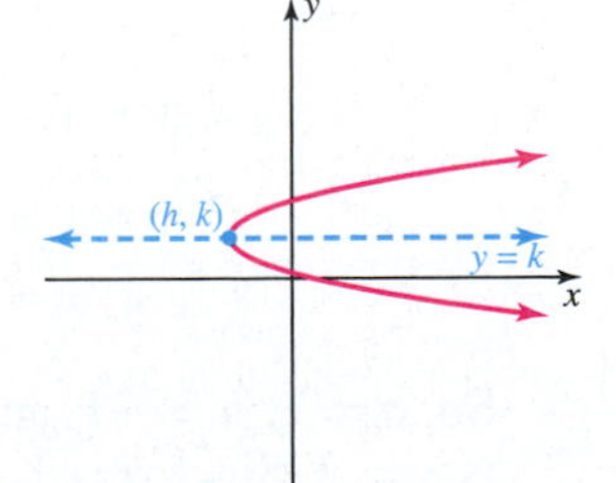

$x = a(y - k)^2 + h,$ $a < 0$

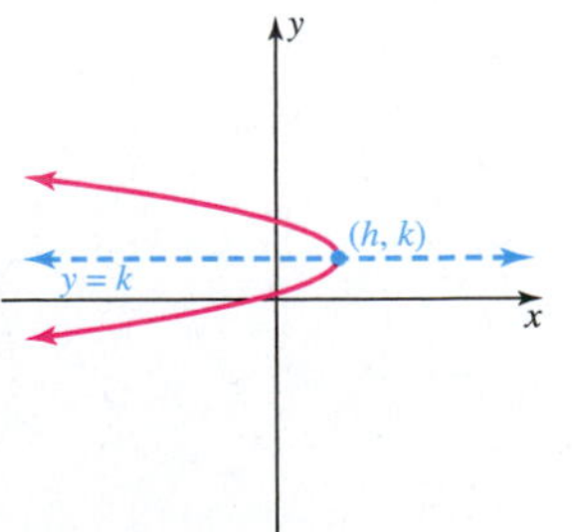

Procedure: Graphing a Parabola of the Form $x = a(y - k)^2 + h$

Step 1: If the equation is not in standard form, complete the square to write it in this form.
Step 2: Identify and plot the vertex, (h, k).
Step 3: Determine the axis of symmetry, $y = h$.
Step 4: Determine the direction the parabola opens.
  a. If $a > 0$, the parabola opens to the right.
  b. If $a < 0$, the parabola opens to the left.
Step 5: Plot additional points by finding the x- and y-intercepts or by using symmetry.
Step 6: Connect the points with a smooth curve to form a parabola.

Objective 1 Examples

Graph each equation by plotting the vertex and at least two additional points. State the axis of symmetry.

1a. $x = (y - 1)^2 - 4$ **1b.** $x = -2y^2 + 2$ **1c.** $x = y^2 - 4y + 3$

Solutions

INSTRUCTOR NOTE: Show students how they can obtain additional points using symmetry.

1a. The equation is in standard form with $a = 1$, $h = -4$, and $k = 1$, so the vertex is $(-4, 1)$ and the axis of symmetry is the horizontal line $y = 1$. Since $a = 1 > 0$, the parabola opens to the right.

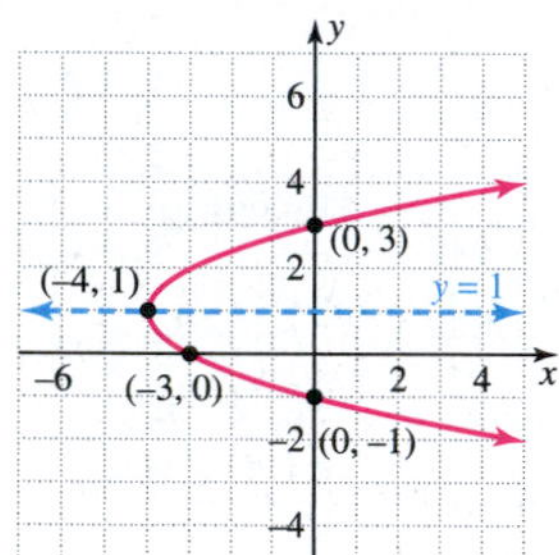

x-intercept ($y = 0$):

$$x = (y - 1)^2 - 4$$
$$x = (0 - 1)^2 - 4$$
$$x = (-1)^2 - 4$$
$$x = 1 - 4$$
$$x = -3$$

The x-intercept is $(-3, 0)$.

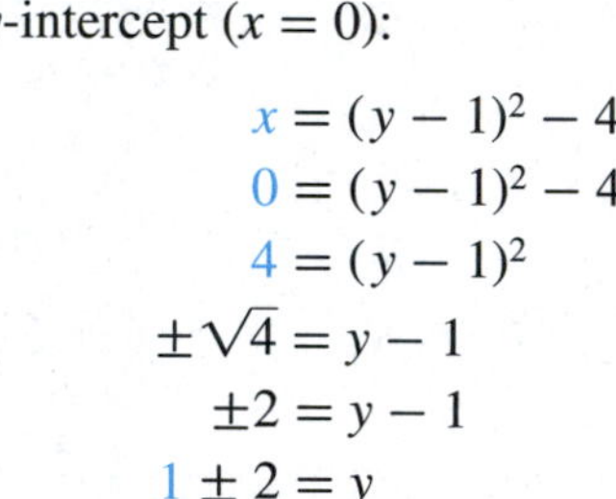

y-intercept ($x = 0$):

$$x = (y - 1)^2 - 4$$
$$0 = (y - 1)^2 - 4$$
$$4 = (y - 1)^2$$
$$\pm\sqrt{4} = y - 1$$
$$\pm 2 = y - 1$$
$$1 \pm 2 = y$$

The y-intercepts are $(0, 3)$ and $(0, -1)$.

1b. The equation, written in standard form, is $x = -2(y - 0)^2 + 2$, so the vertex is $(2, 0)$ and the axis of symmetry is $y = 0$. Since $a = -2 < 0$, the parabola opens to the left.

x-intercept ($y = 0$):

$$x = -2y^2 + 2$$
$$x = -2(0)^2 + 2$$
$$x = 2$$

The x-intercept is $(2, 0)$.

y-intercept ($x = 0$):

$$x = -2y^2 + 2$$
$$0 = -2y^2 + 2$$
$$2y^2 = 2$$
$$y^2 = 1$$
$$y = \pm 1$$

The y-intercepts are $(0, 1)$ and $(0, -1)$.

1c. We complete the square to write the equation in the form $x = a(y - k)^2 + h$.

$x = y^2 - 4y + 3$	
$x - 3 = y^2 - 4y$	Subtract 3 from each side.
$x - 3 + 4 = y^2 - 4y + 4$	Add $\left(-\frac{4}{2}\right)^2 = 4$ to each side.
$x + 1 = (y - 2)^2$	Simplify the left side and factor the trinomial.
$x = (y - 2)^2 - 1$	Subtract 1 from each side.

So, $a = 1$, $h = -1$, and $k = 2$. The vertex is $(-1, 2)$ and the axis of symmetry is $y = 2$. Since $a = 1 > 0$, the parabola opens to the right.

Note that we can use the vertex formula, $-\dfrac{b}{2a}$, to find the vertex as well. Since the equation is of the form $x = ay^2 + by + c$, the vertex is

$$y = -\frac{b}{2a} = -\frac{-4}{2(1)} = \frac{4}{2} = 2$$

$$x = (2)^2 - 4(2) + 3 = 4 - 8 + 3 = -1$$

So, the vertex formula also gives us the vertex of $(-1, 2)$.

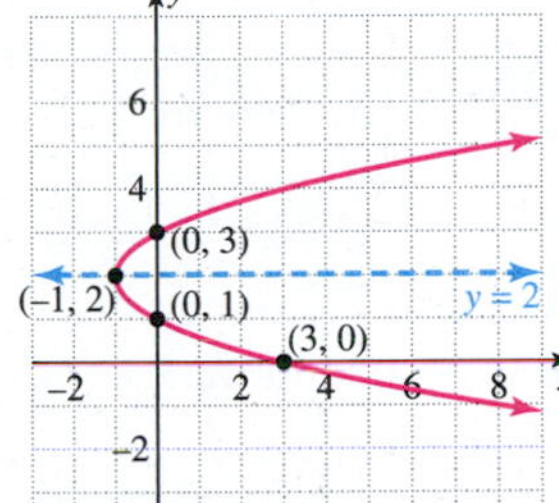

x-intercept ($y = 0$):

$$x = y^2 - 4y + 3$$
$$x = (0)^2 - 4(0) + 3$$
$$x = 3$$

The x-intercept is $(3, 0)$.

y-intercept ($x = 0$):

$$x = y^2 - 4y + 3$$
$$0 = y^2 - 4y + 3$$
$$0 = (y - 3)(y - 1)$$
$$y - 3 = 0 \quad \text{or} \quad y - 1 = 0$$
$$y = 3 \qquad\qquad y = 1$$

The y-intercepts are $(0, 3)$ and $(0, 1)$.

Student Check 1 Graph each equation by plotting the vertex and at least two additional points. State the axis of symmetry.

a. $x = (y - 3)^2 - 1$ **b.** $x = -y^2 + 4$ **c.** $x = y^2 - 6y + 5$

Circles

Objective 2 Graph circles of the form $(x - h)^2 + (y - k)^2 = r^2$.

A **circle** is defined to be the set of all points (x, y) that have the same distance, called the **radius**, from a fixed point, called the **center**. The following figure shows the graph of a circle whose center is at the origin, $(0, 0)$, and whose radius is r. The Pythagorean theorem provides the equation of the circle, which is $x^2 + y^2 = r^2$.

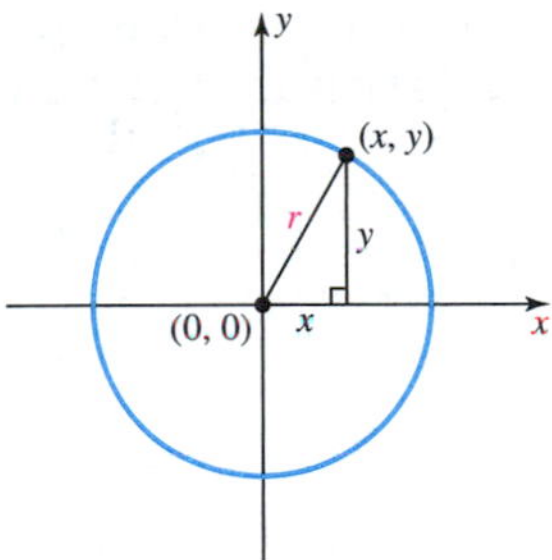

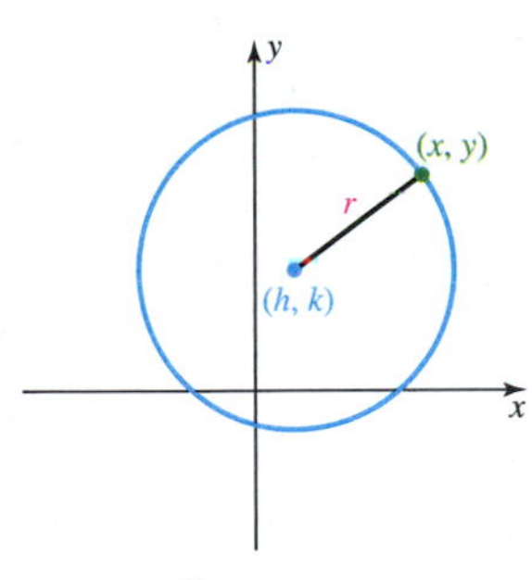

Figure 11.1

Figure 11.1 shows a circle whose center is not at the origin but at a point (h, k). The distance formula gives us $r = \sqrt{(x - h)^2 + (y - k)^2}$, or after squaring each side,

$$r^2 = (x - h)^2 + (y - k)^2$$

Property: Standard Form of a Circle

A circle with center (h, k) and radius r is given by the equation

$$(x - h)^2 + (y - k)^2 = r^2$$

If the center is $(0, 0)$, then the equation is $x^2 + y^2 = r^2$.

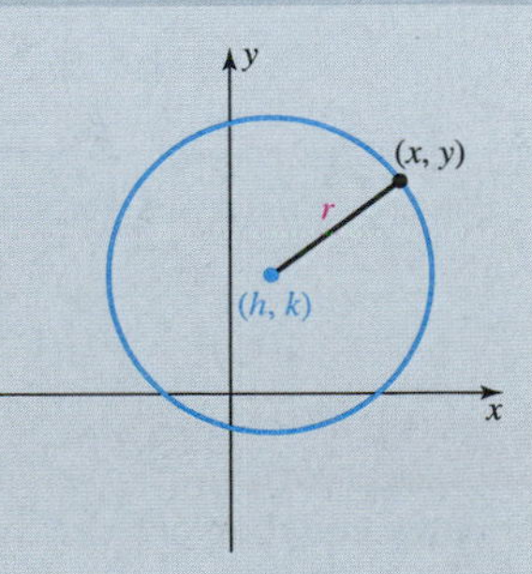

Note that in the equation of a circle, the coefficients of x^2 and y^2 are the same and the terms are connected by addition. Some examples are shown in the table.

Equation	Center	Radius
$x^2 + y^2 = 4$	$(0, 0)$	$r = \sqrt{4} = 2$
$x^2 + (y - 5)^2 = 9$	$(0, 5)$	$r = \sqrt{9} = 3$
$(x + 1)^2 + y^2 = 25$	$(-1, 0)$	$r = \sqrt{25} = 5$
$(x - 2)^2 + (y + 3)^2 = 2$	$(2, -3)$	$r = \sqrt{2}$

Procedure: Graphing a Circle with the Equation $(x - h)^2 + (y - k)^2 = r^2$

Step 1: Identify the center (h, k).
Step 2: Identify the radius r.
Step 3: Plot the center and then plot four key points that are r units away from the center. These four points are $(h + r, k)$, $(h - r, k)$, $(h, k + r)$, and $(h, k - r)$.
Step 4: Draw a circle through these four points.

Objective 2 Examples Graph each circle. Identify the center, the radius, and four key points.

2a. $x^2 + y^2 = 4$ **2b.** $(x - 1)^2 + (y + 3)^2 = 9$ **2c.** $(x + 2)^2 + y^2 = 5$

Solutions

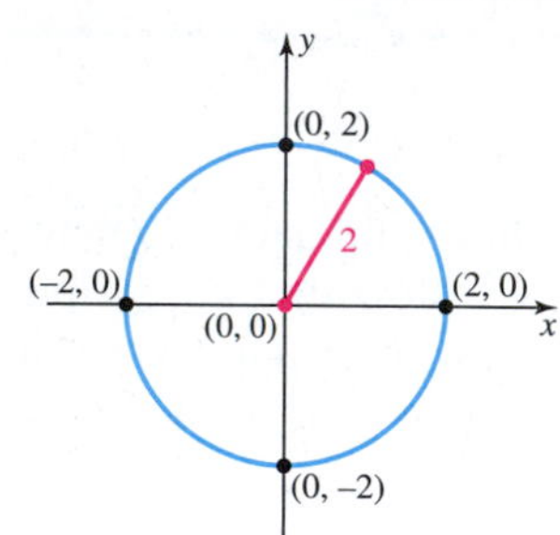

2a. The equation is equivalent to $(x - 0)^2 + (y - 0)^2 = 2^2$. So, the center is $(0, 0)$ and the radius is 2. The four key points that are 2 units from the center are

$$(0 + 2, 0) = (2, 0)$$
$$(0 - 2, 0) = (-2, 0)$$
$$(0, 0 + 2) = (0, 2)$$
$$(0, 0 - 2) = (0, -2)$$

2b. The equation is equivalent to $(x - 1)^2 + [y - (-3)]^2 = 3^2$. So, the center is $(1, -3)$ and the radius is 3. The four key points that are 3 units from the center are

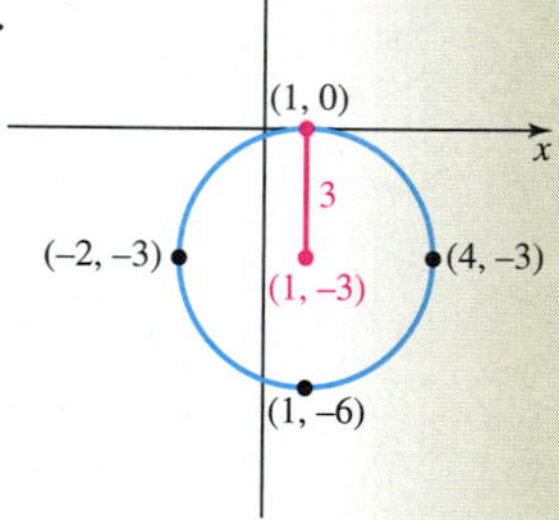

$$(1 + 3, -3) = (4, -3)$$
$$(1 - 3, -3) = (-2, -3)$$
$$(1, -3 + 3) = (1, 0)$$
$$(1, -3 - 3) = (1, -6)$$

2c. The equation is equivalent to $[x - (-2)]^2 + (y - 0)^2 = (\sqrt{5})^2$. So, the center is $(-2, 0)$ and the radius is $\sqrt{5}$. The four key points that are $\sqrt{5}$ units from the center are

$$(-2 + \sqrt{5}, 0) \approx (0.24, 0)$$
$$(-2 - \sqrt{5}, 0) \approx (-4.24, 0)$$
$$(-2, 0 + \sqrt{5}) \approx (-2, 2.24)$$
$$(-2, 0 - \sqrt{5}) \approx (-2, -2.24)$$

Student Check 2 Graph each circle. Identify the center, the radius, and four key points.

a. $x^2 + y^2 = 16$ **b.** $(x + 4)^2 + (y - 2)^2 = 4$ **c.** $x^2 + (y - 2)^2 = 15$

Objective 3 ▶

Write the equation of a circle given its center and radius.

Write the Standard Form of a Circle

If we know the center and radius of a circle, we can write its equation by substituting these values into the standard form of a circle.

Procedure: Writing the Equation of a Circle

Step 1: Let the center be (h, k) and the radius be r.
Step 2: Substitute the values into the standard form $(x - h)^2 + (y - k)^2 = r^2$ and simplify.

Objective 3 Examples **Write the equation of each circle given its center and radius.**

3a. center $= (-5, 4)$ and radius $= 7$ **3b.** center $= (0, 0)$ and radius $= 5$

Solutions

3a.

$(x - h)^2 + (y - k)^2 = r^2$ State the standard form.

$[x - (-5)]^2 + (y - 4)^2 = 7^2$ Let $(h, k) = (-5, 4)$, $r = 7$.

$(x + 5)^2 + (y - 4)^2 = 49$ Simplify.

3b.

$(x - h)^2 + (y - k)^2 = r^2$ State the standard form.

$(x - 0)^2 + (y - 0)^2 = 5^2$ Let $(h, k) = (0, 0)$, $r = 5$.

$x^2 + y^2 = 25$ Simplify.

Student Check 3 Write the equation of each circle given its center and radius.

a. center $= (5, -2)$ and radius $= 10$ **b.** center $= (0, 0)$ and radius $= 6$

Complete the Square to Write the Equation of a Circle

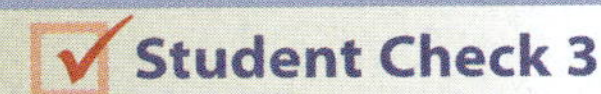

Objective 4

Complete the square to write the equation of a circle in standard form.

Just as with parabolas, we sometimes have to complete the square to write the equation of a circle in standard form. For a circle, we have to complete the square twice, once with the x-terms and once with the y-terms.

Procedure: Completing the Square to Write the Equation of a Circle in Standard Form

Step 1: Isolate the constant on one side of the equation.
Step 2: Group the x-terms and y-terms together.
Step 3: If the coefficient of x^2 and/or y^2 is not 1, factor out the coefficient from the x-terms and y-terms, as necessary.
Step 4: Find the number that makes the x-terms a perfect square trinomial and add this to each side of the equation. Find the number that makes the y-terms a perfect square trinomial and add this to each side of the equation.
Step 5: Factor the two trinomials and simplify the right side.
Step 6: Identify the center and radius.

Objective 4 Example **Write the equation $x^2 + y^2 - 4x + 6y = 3$ as a circle in standard form and identify its center and radius.**

Solution

$x^2 + y^2 - 4x + 6y = 3$ The constant is already isolated on one side.

$(x^2 - 4x) + (y^2 + 6y) = 3$ Group the x-terms and y-terms together.

$(x^2 - 4x + 4) + (y^2 + 6y + 9) = 3 + 4 + 9$ Complete the squares and add to each side.

$(x - 2)^2 + (y + 3)^2 = 16$ Factor the left side and simplify the right side.

So, the center of the circle is $(2, -3)$ and the radius is $\sqrt{16} = 4$.

INSTRUCTOR NOTE: Point out that we know this equation represents a circle since the coefficients of x^2 and y^2 are the same and because x^2 and y^2 are connected by addition.

Student Check 4 Write the equation $x^2 + y^2 - 2x + 10y = 10$ as a circle in standard form and identify its center and radius.

Objective 5 Troubleshooting Common Errors

Troubleshoot common errors.

Some common errors associated with the parabola and circle are shown.

Objective 5 Examples

A problem and an incorrect solution are given. Provide the correct solution and an explanation of the error.

5a. Find the vertex of $x = (y - 4)^2 + 3$.

Incorrect Solution	Correct Solution and Explanation
The vertex of $x = (y - 4)^2 + 3$ is $(4, 3)$.	The vertex of a horizontal parabola $x = a(y - k)^2 + h$ is (h, k). So, the vertex of $x = (y - 4)^2 + 3$ is $(3, 4)$.

5b. Find the center and radius of $(x + 3)^2 + (y - 4)^2 = 3$.

Incorrect Solution	Correct Solution and Explanation
The center is $(3, -4)$ and the radius is 3.	The equation in standard form is $(x + 3)^2 + (y - 4)^2 = 3$ $[x - (-3)]^2 + (y - 4)^2 = (\sqrt{3})^2$ So, the center is $(-3, 4)$ and the radius is $\sqrt{3}$.

ANSWERS TO STUDENT CHECKS

Student Check 1 **a.**

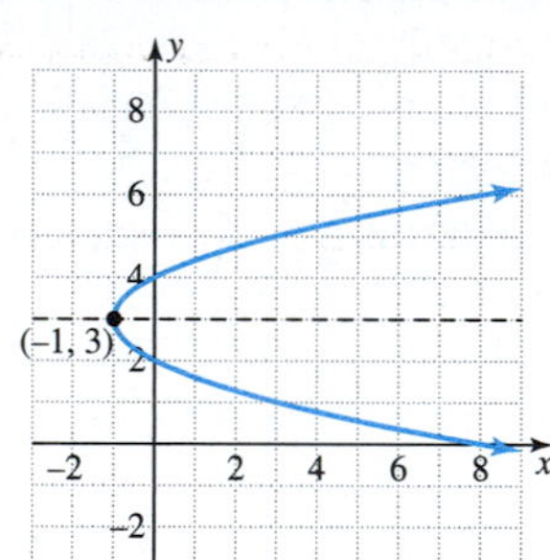

b.

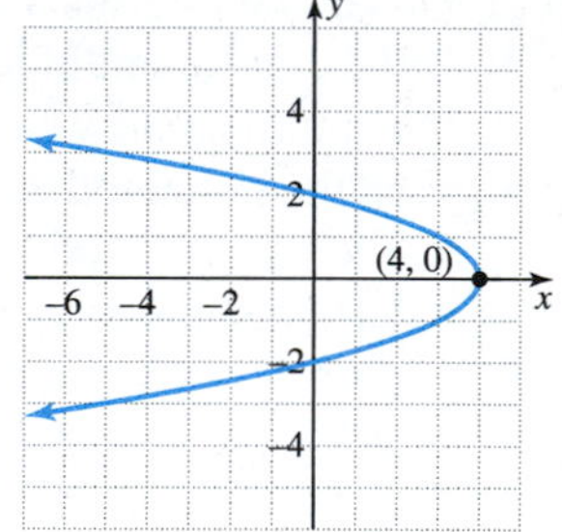

c.

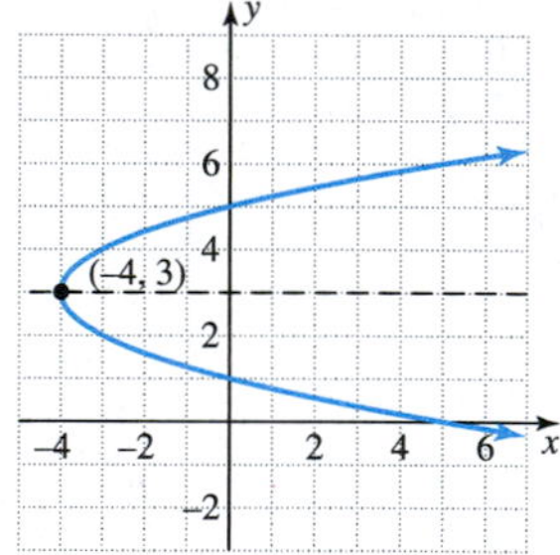

Student Check 2 **a.**

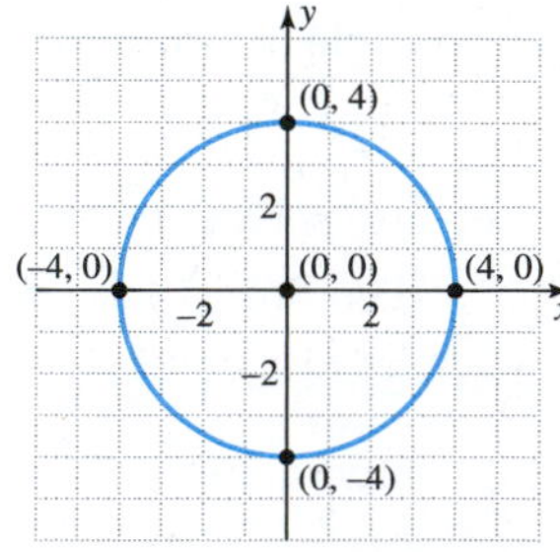

b.

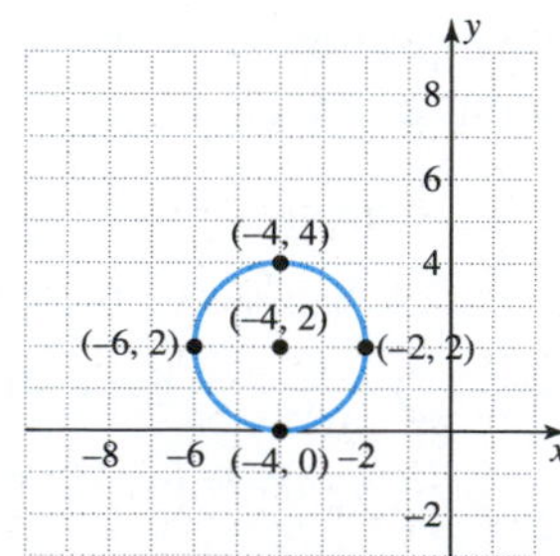

c.

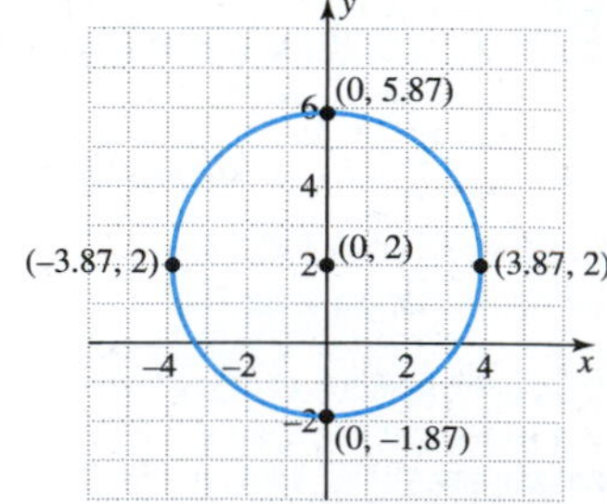

Student Check 3 **a.** $(x - 5)^2 + (y + 2)^2 = 100$ **b.** $x^2 + y^2 = 36$

Student Check 4 $(x - 1)^2 + (y + 5)^2 = 36$, center $= (1, -5)$ and radius $= 6$

SUMMARY OF KEY CONCEPTS

1. The equation $x = a(y - k)^2 + h$ is a horizontal parabola with vertex (h, k). The parabola opens to the right if $a > 0$ and to the left if $a < 0$.
2. The equation $(x - h)^2 + (y - k)^2 = r^2$ represents a circle with center (h, k) and radius r.
   a. We can identify the center and radius from the standard form of a circle.
   b. If we know the center and radius, we can write the standard form of a circle.
   c. We may have to complete the square to write the standard form of a circle.

GRAPHING CALCULATOR SKILLS

The graphing calculator can graph horizontal parabolas and circles; however, we must solve the equations for y. Because the y-variable is squared, we apply the square root property to solve for y and will, therefore, have two expressions to enter in the calculator to graph the equation.

Example 1: Graph $x = (y - 1)^2 - 4$.

Solution: Solve the equation for y.

$$x = (y - 1)^2 - 4$$

$$x + 4 = (y - 1)^2 \quad \text{Add 4 to each side.}$$

$$\pm\sqrt{x + 4} = y - 1 \quad \text{Apply the square root property.}$$

$$1 \pm \sqrt{x + 4} = y \quad \text{Add 1 to each side.}$$

Enter the two expressions, $1 + \sqrt{x + 4}$ and $1 - \sqrt{x + 4}$, to graph the parabola. Y_1 is the top half of the parabola and Y_2 is the bottom half of the parabola.

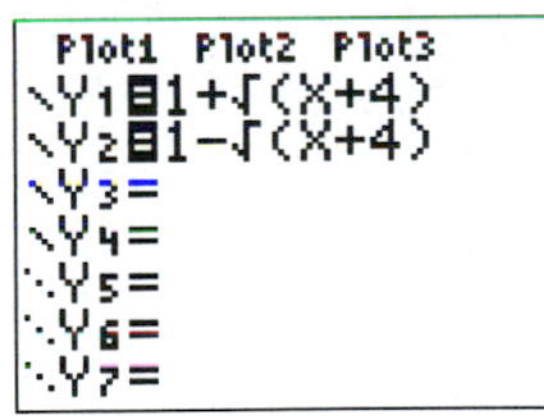

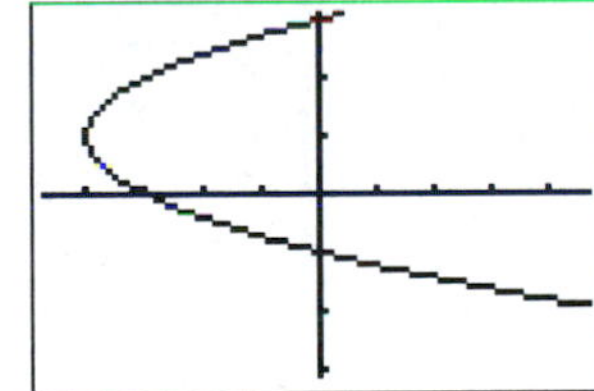

Example 2: Graph $x^2 + y^2 = 4$.

Solution: Solve the equation for y.

$$x^2 + y^2 = 4$$

$$y^2 = 4 - x^2 \quad \text{Subtract } x^2 \text{ from each side.}$$

$$y = \pm\sqrt{4 - x^2} \quad \text{Apply the square root property.}$$

Enter the two expressions, $\sqrt{4 - x^2}$ and $-\sqrt{4 - x^2}$, to graph the circle. It is best to graph a circle in the ZSquare window so that the graph will not be distorted. Y_1 is the top half of the circle and Y_2 is the bottom half of the circle.

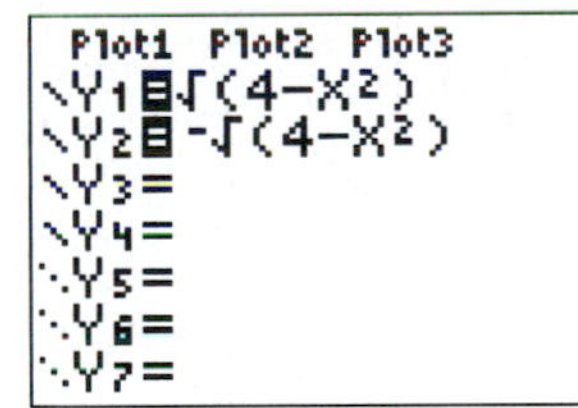

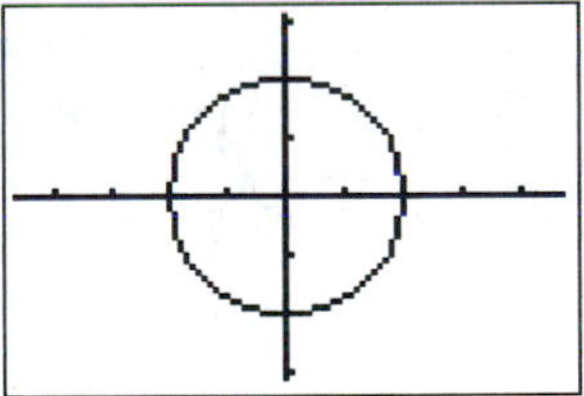

SECTION 11.1 EXERCISE SET

Write About It!

Use complete sentences in your answer to each exercise.

1. Describe the graph of $x = \frac{1}{2}(y - 3)^2 + 1$.
2. Describe the graph of $x = -2(y + 1)^2 - 3$.
3. Describe how to graph the equation $x = -y^2 + 25$.
4. Explain how to write the equation, $x = y^2 + 8y - 12$ in standard form.
5. State the advantages of writing the equation of a circle in standard form.

Additional answers can be found in the Instructor Answer Appendix.

6. Explain how to write the equation of the circle when the center (h, k) and radius r are given.
7. Explain how to write the equation of a circle, $x^2 + y^2 - 6x + 10y = 15$, in standard form.
8. Describe the graph of the equation $x^2 + y^2 + 4x + 5 = 0$.

Practice Makes Perfect!

Graph each equation by plotting the vertex and at least two additional points. State the axis of symmetry. (*See Objective 1.*)

9. $x = (y + 2)^2 + 3$
10. $x = (y + 1)^2 + 5$

11. $x = (y - 2)^2 + 1$
12. $x = (y - 4)^2 + 2$
13. $x = (y - 3)^2 - 4$
14. $x = (y - 2)^2 - 6$
15. $x = -(y + 2)^2 + 4$
16. $x = -(y - 4)^2 + 5$
17. $x = -(y - 2)^2 - 1$
18. $x = -(y + 2)^2 - 2$
19. $x = -y^2 + 9$
20. $x = -y^2 - 4$
21. $x = y^2 - 9$
22. $x = y^2 - 6$
23. $x = y^2 + 2y - 8$
24. $x = -y^2 + 6y - 8$
25. $x = y^2 + 3y + 2$
26. $x = y^2 + y - 2$

Graph each circle. Identify the center, the radius, and four key points. (*See Objective 2.*)

27. $x^2 + y^2 = 25$
28. $x^2 + y^2 = 9$
29. $(x - 2)^2 + (y - 1)^2 = 9$
30. $(x + 1)^2 + (y - 3)^2 = 4$
31. $x^2 + (y + 3)^2 = 25$
32. $(x - 1)^2 + y^2 = 4$

Write the equation of each circle given its center and radius. (*See Objective 3.*)

33. center = (−4, 5) and radius = 1 $(x + 4)^2 + (y - 5)^2 = 1$
34. center = (−2, 3) and radius = 3 $(x + 2)^2 + (y - 3)^2 = 9$
35. center = (4, −2) and radius = $\sqrt{3}$ $(x - 4)^2 + (y + 2)^2 = 3$
36. center = (1, −6) and radius = $\sqrt{7}$ $(x - 1)^2 + (y + 6)^2 = 7$
37. center = (−2, −7) and radius = $\sqrt{6}$ $(x + 2)^2 + (y + 7)^2 = 6$
38. center = (−3, −1) and radius = $\sqrt{2}$ $(x + 3)^2 + (y + 1)^2 = 2$

Write the equation of each circle in standard form and identify the center and radius. (*See Objective 4.*)

39. $x^2 + y^2 - 12x - 8y = -51$
$(x - 6)^2 + (y - 4)^2 = 1$; center: (6, 4); radius = 1
40. $x^2 + y^2 - 4x - 16y = -52$
$(x - 2)^2 + (y - 8)^2 = 16$; center: (2, 8); radius = 4
41. $x^2 + y^2 + 2x + 2y = 2$
$(x + 1)^2 + (y + 1)^2 = 4$; center: (−1, −1); radius = 2
42. $x^2 + y^2 + 10x + 4y = -13$
$(x + 5)^2 + (y + 2)^2 = 16$; center: (−5, −2); radius = 4
43. $x^2 + y^2 - 6x = 16$
$(x - 3)^2 + y^2 = 25$; center: (3, 0); radius = 5
44. $x^2 + y^2 + 12x = 64$
$(x + 6)^2 + y^2 = 100$; center: (−6, 0); radius = 10
45. $x^2 + y^2 - x + y = \frac{1}{2}$
46. $x^2 + y^2 + 3x - y = \frac{3}{2}$

Mix 'Em Up!

Graph each equation. If it is a parabola, plot the vertex and at least two additional points, and state the axis of symmetry. If it is a circle, identify the center, the radius, and the four key points.

47. $x = (y + 1)^2 - 5$
48. $x = (y - 1)^2 - 4$
49. $(x + 1)^2 + (y - 1)^2 = 1$
50. $(x - 2)^2 + (y + 3)^2 = 9$
51. $x = 2(y - 3)^2 + 4$
52. $x = 2(y + 2)^2 - 3$
53. $x = -3(y - 1)^2 + 1$
54. $x = -4(y + 2)^2 + 5$
55. $(x - 1)^2 + y^2 = 16$
56. $(x + 2)^2 + y^2 = 4$
57. $x = \frac{1}{2}(y - 1)^2 + 3$
58. $x = \frac{1}{3}(y + 1)^2 - 2$
59. $x = -2y^2 + 4$
60. $x = -3y^2 + 1$
61. $x = 2y^2 + 3$
62. $x = 2y^2 - 6$
63. $x = y^2 + 2y - 3$
64. $x = y^2 + 6y + 8$
65. $x = -y^2 + 2y + 24$
66. $x = -y^2 + 6y - 4$
67. $x = y^2 - 5y + 5$
68. $x = y^2 - 7y + 13$

Write the equation of each circle given its center and radius.

69. center = (−1, −2) and radius = 4
$(x + 1)^2 + (y + 2)^2 = 16$
70. center = (−5, −6) and radius = 12
$(x + 5)^2 + (y + 6)^2 = 144$
71. center = (−4, 6) and radius = $\sqrt{6}$
$(x + 4)^2 + (y - 6)^2 = 6$
72. center = (−1, 4) and radius = $\sqrt{15}$
$(x + 1)^2 + (y - 4)^2 = 15$
73. center = (−7, 0) and radius = $\sqrt{5}$ $(x + 7)^2 + y^2 = 5$
74. center = (−4, 0) and radius = $\sqrt{3}$ $(x + 4)^2 + y^2 = 3$
75. center = $\left(0, -\frac{3}{4}\right)$ and radius = $\frac{1}{2}$ $x^2 + \left(y + \frac{3}{4}\right)^2 = \frac{1}{4}$
76. center = $\left(0, \frac{2}{3}\right)$ and radius = $\frac{1}{3}$ $x^2 + \left(y - \frac{2}{3}\right)^2 = \frac{1}{9}$
77. center = (1.2, −2.5) and radius = 4.5
$(x - 1.2)^2 + (y + 2.5)^2 = 20.25$
78. center = (−1.6, 3.2) and radius = 1.5
$(x + 1.6)^2 + (y - 3.2)^2 = 2.25$

79. $(x - 2)^2 + y^2 = 9$

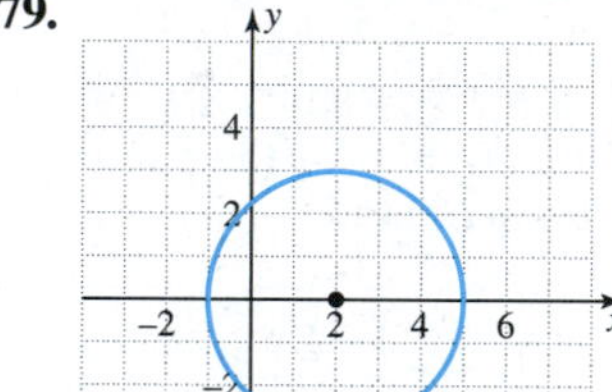

80. $x^2 + (y + 1)^2 = 4$

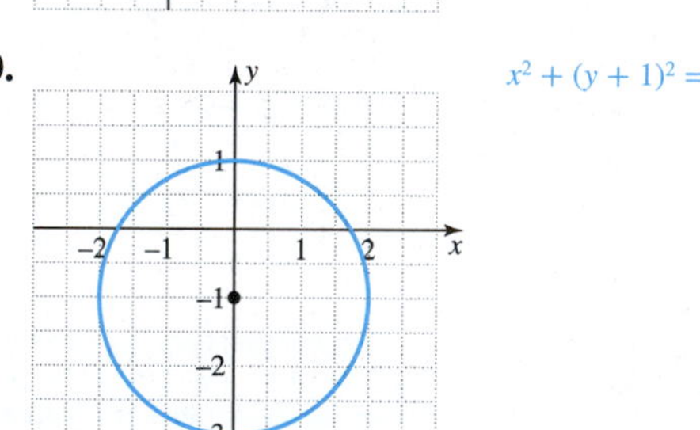

Write the equation of each circle in standard form and identify its center and radius.

81. $x^2 + y^2 + 14x + 6y = -22$
$(x + 7)^2 + (y + 3)^2 = 36$; center: (−7, −3); radius = 6
82. $x^2 + y^2 + 2x - 6y = -9$
$(x + 1)^2 + (y - 3)^2 = 1$; center: (−1, 3); radius = 1
83. $x^2 + y^2 + 6x - 12y = 0$
$(x + 3)^2 + (y - 6)^2 = 45$; center: (−3, 6); radius = $3\sqrt{5}$
84. $x^2 + y^2 - 4x + 2y = 0$
$(x - 2)^2 + (y + 1)^2 = 5$; center: (2, −1); radius = $\sqrt{5}$
85. $x^2 + y^2 + 8x - 6y = 0$
$(x + 4)^2 + (y - 3)^2 = 25$; center: (−4, 3); radius = 5
86. $x^2 + y^2 - 10x - 24y = 0$
$(x - 5)^2 + (y - 12)^2 = 169$; center: (5, 12); radius = 13

87. $x^2 + y^2 - 2y = 8$
$x^2 + (y - 1)^2 = 9$; center: $(0, 1)$; radius $= 3$

88. $x^2 + y^2 - 12y = 45$
$x^2 + (y - 6)^2 = 81$; center: $(0, 6)$; radius $= 9$

89. $x^2 + y^2 - 3x - 5y = \frac{1}{2}$

90. $x^2 + y^2 + 7x - 3y = \frac{3}{2}$

91. $x^2 + y^2 - 1.6x + 4.2y = 0.71$
$(x - 0.8)^2 + (y + 2.1)^2 = 5.76$; center: $(0.8, -2.1)$; radius $= 2.4$

92. $x^2 + y^2 + 3.6x - 5.6y = 29.88$
$(x + 1.8)^2 + (y - 2.8)^2 = 40.96$; center: $(-1.8, 2.8)$; radius $= 6.4$

You Be the Teacher!

Correct each student's errors, if any.

93. Find the vertex from the equation $x = (y + 1)^2 - 2$.

Eric's work:

$(-1, -2)$ $(-2, -1)$

94. Write the equation of the circle in standard form: $x^2 + y^2 - 10x + 8y = 9$.

Monica's work:

$(x - 5)^2 + (y + 4)^2 = 9$

$x^2 + y^2 - 10x + 8y = 9$
$x^2 - 10x + 25 + y^2 + 8y + 16 = 9 + 25 + 16$
$(x - 5)^2 + (y + 4)^2 = 50$

Calculate It!

Use a graphing calculator to graph each function and find the vertex.

95. $x = (y - 1.2)^2 - 1.5$

96. $x = -2(y + 2.4)^2 + 3.2$

Use a graphing calculator to graph each circle.

97. $(x - 2)^2 + (y + 3)^2 = 9$

98. $(x + 3)^2 + (y - 1)^2 = 25$

SECTION 11.2 The Ellipse and the Hyperbola

OBJECTIVES

As a result of completing this section, you will be able to

1. Identify the standard form of an ellipse and graph an ellipse.
2. Identify the standard form of a hyperbola and graph a hyperbola.
3. Troubleshoot common errors.

The other two conic sections that we will discuss are the ellipse and the hyperbola. Both of these shapes occur in real-life situations. The orbits of the planets, moons, and satellites are elliptical. Ellipses are also used in microscopes, telescopes, and many architectural designs. The hyperbola is used to model the path of a comet, telescope lenses, and gears for some machines.

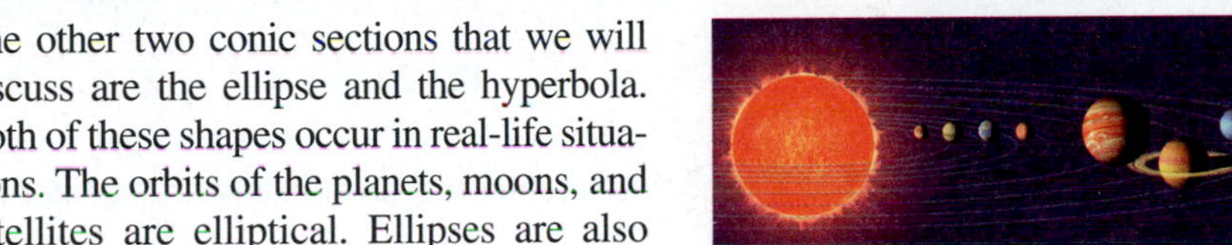

Objective 1

Identify the standard form of an ellipse and graph an ellipse.

Ellipses

An ellipse basically looks like an oval. It is a circle that has been flattened or stretched. Mathematically, we define an **ellipse** as the set of all points such that the sum of their distances from two fixed points is constant. The two fixed points are called **foci**. The midpoint of the foci is the **center** of the ellipse.

To draw an ellipse by hand, we can tie the ends of a piece of string to two pins that are stuck in a sheet of paper. Then, keeping the string tight with the point of a pencil, allow the pencil to trace a path around the pins. The resulting curve is an ellipse, with the two pins, or fixed points, representing its foci. (See Figure 11.2.)

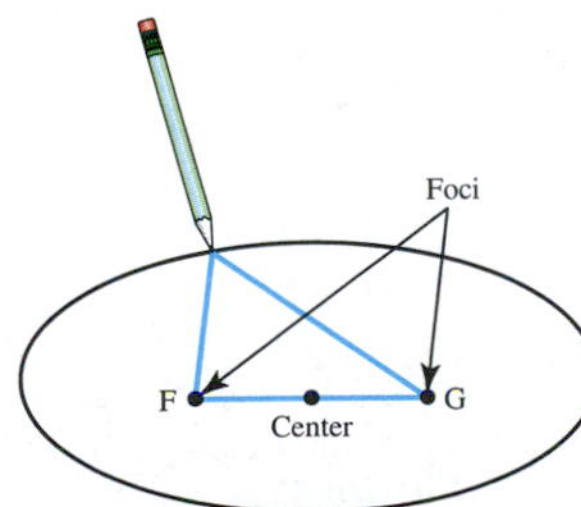

Figure 11.2

Property: Standard Form of an Ellipse

The equation

$$\frac{x^2}{a^2} + \frac{y^2}{b^2} = 1$$

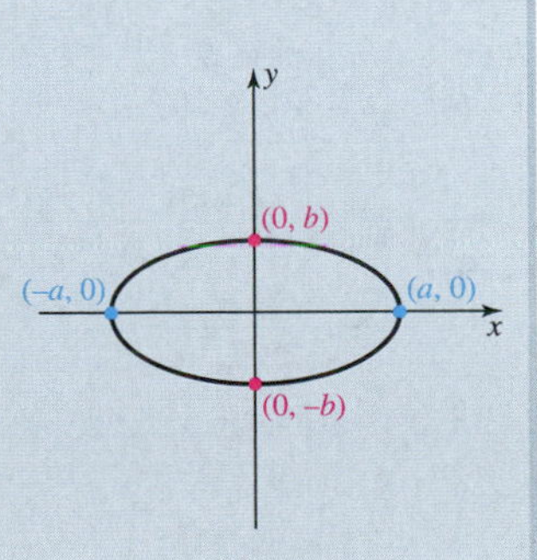

is an ellipse with center $(0, 0)$. The x-intercepts are $(a, 0)$ and $(-a, 0)$, and the y-intercepts are $(0, b)$ and $(0, -b)$.

The equation

$$\frac{(x-h)^2}{a^2} + \frac{(y-k)^2}{b^2} = 1$$

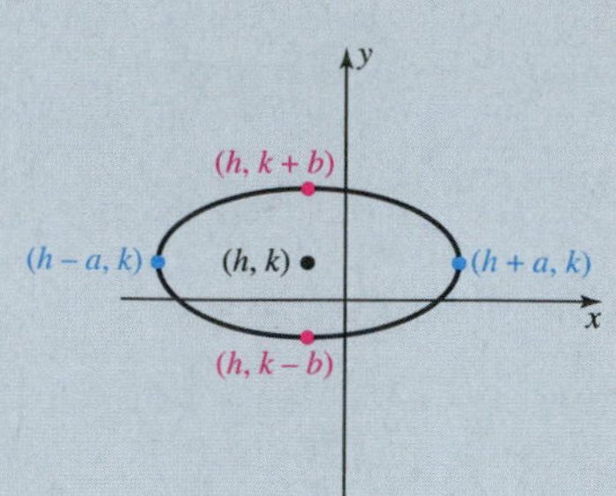

is an ellipse with center (h, k).

Procedure: Graphing an Ellipse

Step 1: Identify the center and the values of a and b.
Step 2: Plot the center.
Step 3: From the center, move a units right and left to find two key points on the ellipse.
Step 4: From the center, move b units up and down to find two more key points on the ellipse.
Step 5: Connect the points with a smooth curve.

Objective 1 Examples

Graph each ellipse. Label four key points and the center on the graph.

1a. $\frac{x^2}{16} + \frac{y^2}{9} = 1$ **1b.** $25x^2 + 4y^2 = 100$ **1c.** $\frac{(x-2)^2}{36} + \frac{(y+4)^2}{9} = 1$

Solutions

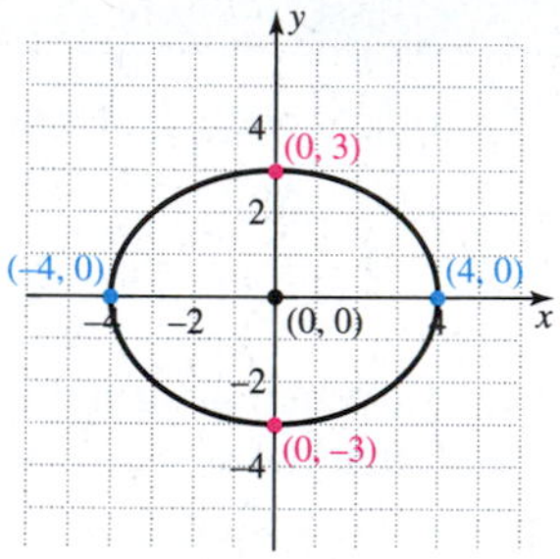

1a. The equation represents an ellipse whose center is at (0, 0) and can be written as

$$\frac{x^2}{4^2} + \frac{y^2}{3^2} = 1$$

Plot (0, 0) and move 4 units to the right and left, which takes us to the x-intercepts of (−4, 0) and (4, 0). Then move 3 units up and down, which takes us to the y-intercepts of (0, 3) and (0, −3). (See Figure 11.3.)

Figure 11.3

1b. The equation is not in standard form. To be in standard form, the constant on one side must be 1.

$$25x^2 + 4y^2 = 100$$

$$\frac{25x^2}{100} + \frac{4y^2}{100} = \frac{100}{100} \quad \text{Divide each side by 100.}$$

$$\frac{x^2}{4} + \frac{y^2}{25} = 1 \quad \text{Simplify.}$$

$$\frac{x^2}{2^2} + \frac{y^2}{5^2} = 1 \quad \text{Identify } a \text{ and } b.$$

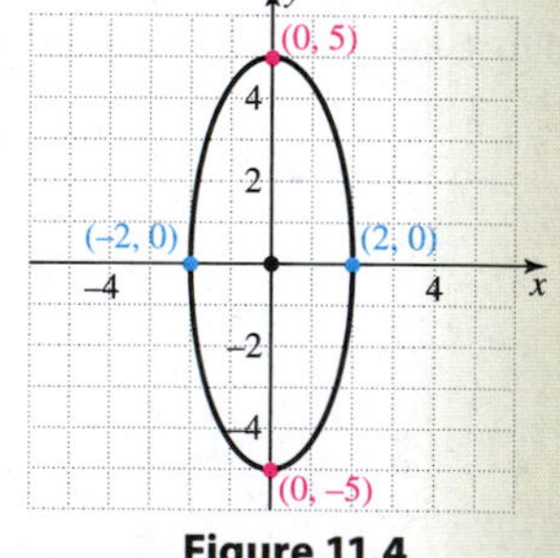

Figure 11.4

So, this is an ellipse with center (0, 0). The x-intercepts are (2, 0) and (−2, 0), and the y-intercepts are (0, 5) and (0, −5). (See Figure 11.4.)

1c. The equation represents an ellipse whose center is (2, −4) and can be written as

$$\frac{(x-2)^2}{6^2} + \frac{[y-(-4)]^2}{3^2} = 1$$

Since $a = 6$, we move 6 units right and left from the center (2, −4) to (2 + 6, −4) = (8, −4) and (2 − 6, −4) = (−4, −4). Since $b = 3$, we move 3 units up and down from the center (2, −4) to (2, −4 + 3) = (2, −1) and (2, −4 − 3) = (2, −7). (See Figure 11.5.)

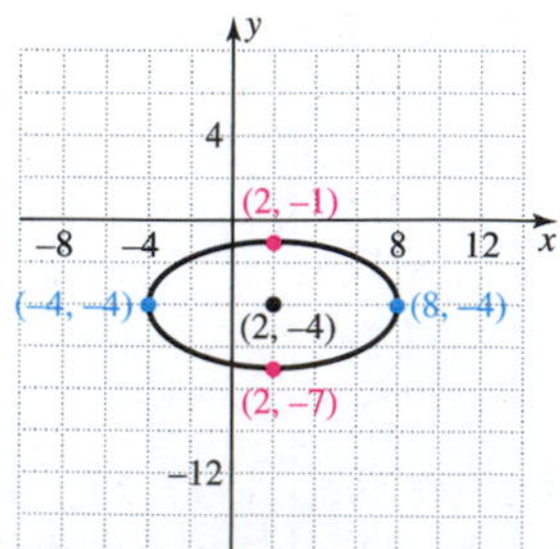

Figure 11.5

Student Check 1 Graph each ellipse. Label four key points and the center on the graph.

a. $\dfrac{x^2}{36} + \dfrac{y^2}{25} = 1$ **b.** $9x^2 + 4y^2 = 36$ **c.** $\dfrac{(x + 1)^2}{4} + \dfrac{(y - 3)^2}{16} = 1$

Note: *From Example 1, we observe the following.*

1. *In the equation of an ellipse, the coefficients of x^2 and y^2 are not the same. If they were the same, the equation would represent a circle.*
2. *If $a > b$, then the ellipse is horizontal.*
3. *If $a < b$, then the ellipse is vertical.*

Hyperbolas

Objective 2

Identify the standard form of a hyperbola and graph a hyperbola.

The last conic section we will discuss is the hyperbola. The graph of a **hyperbola** consists of two branches and is defined as the set of points such that the absolute value of the difference of their distances from two fixed points is constant. The fixed points are the **foci** and the midpoint of the foci is the **center** of the hyperbola.

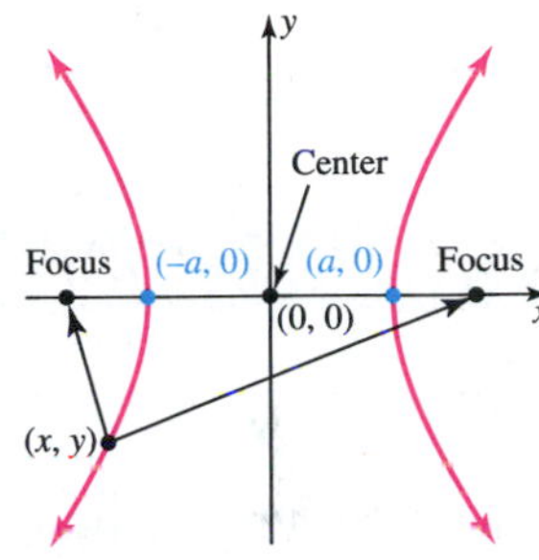

INSTRUCTOR NOTE:
Be sure to point out the difference between the equation of an ellipse and the equation of a hyperbola.

Property: Standard Form of a Hyperbola

The equation

$$\frac{x^2}{a^2} - \frac{y^2}{b^2} = 1$$

is a hyperbola with center (0, 0). The x-intercepts are $(a, 0)$ and $(-a, 0)$.

The equation

$$\frac{y^2}{b^2} - \frac{x^2}{a^2} = 1$$

is a hyperbola with center (0, 0). The y-intercepts are $(0, b)$ and $(0, -b)$.

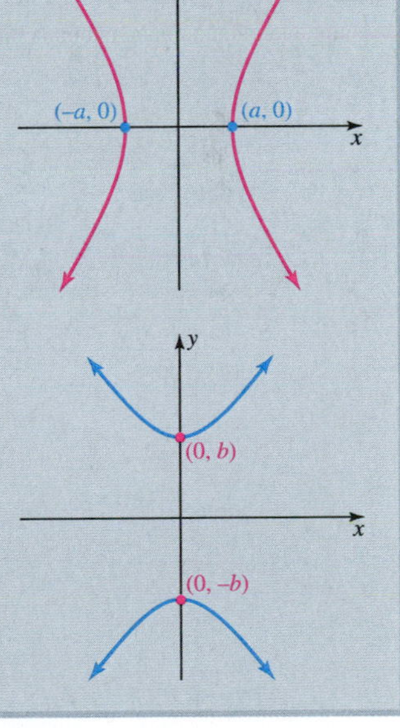

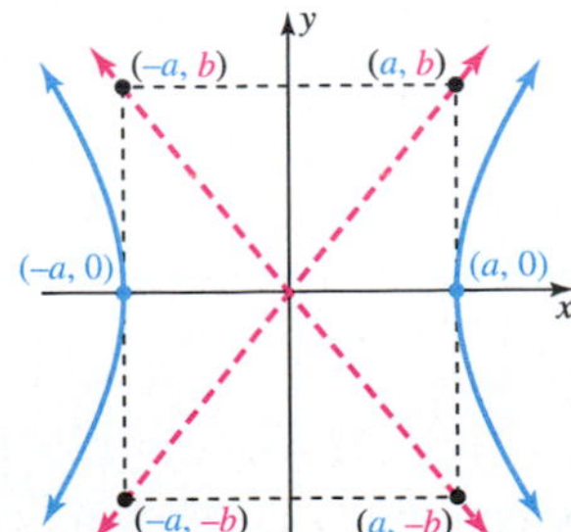

We will construct graphs of hyperbolas whose center is only at the origin. To draw the graph of a hyperbola $\dfrac{x^2}{a^2} - \dfrac{y^2}{b^2} = 1$, we will make use of a rectangle that has vertices at

$$(a, b), (-a, b), (a, -b), \text{ and } (-a, -b)$$

The diagonals of this rectangle form lines, called **asymptotes**, that the branches of the hyperbola approach. These lines are not part of the graph but enable us to draw the graph with more accuracy as shown in the figure to the left.

Procedure: Graphing a Hyperbola

Step 1: Plot the intercepts of the graph.

a. The graph of $\dfrac{x^2}{a^2} - \dfrac{y^2}{b^2} = 1$ is a horizontal hyperbola and has x-intercepts at $(a, 0)$ and $(-a, 0)$.

b. The graph of $\dfrac{y^2}{b^2} - \dfrac{x^2}{a^2} = 1$ is a vertical hyperbola and has y-intercepts at $(0, b)$ and $(0, -b)$.

Step 2: Plot the four vertices of the rectangle: (a, b), $(-a, b)$, $(a, -b)$, and $(-a, -b)$.

Step 3: Draw the two diagonals of the rectangle with dashed lines, extending beyond the rectangle. These are the asymptotes.

Step 4: Draw the two branches of the hyperbola going through the intercepts and approaching the asymptotes.

Objective 2 Examples

Graph each hyperbola.

2a. $\dfrac{x^2}{16} - \dfrac{y^2}{9} = 1$ **2b.** $16y^2 - 4x^2 = 64$

Solutions

2a. The equation represents a horizontal hyperbola with the center at the origin and is equivalent to

$$\frac{x^2}{4^2} - \frac{y^2}{3^2} = 1$$

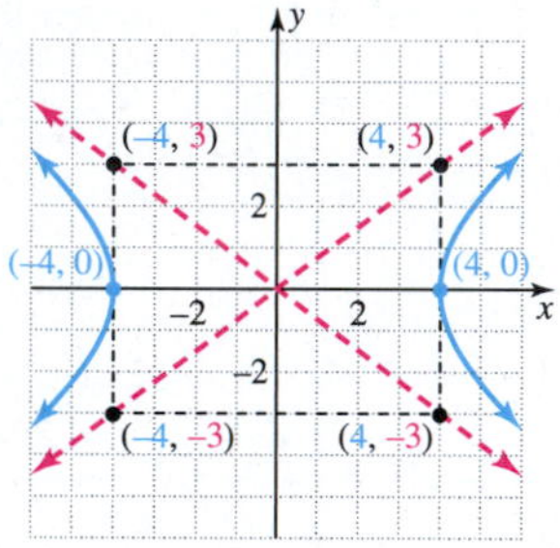

Because $a = 4$, the x-intercepts are $(4, 0)$ and $(-4, 0)$. Since $b = 3$, the four vertices of the rectangle are $(4, 3)$, $(-4, 3)$, $(4, -3)$, and $(-4, -3)$. The diagonals of the rectangle are drawn and extended. (See Figure 11.6.)

Figure 11.6

2b. The equation is not in the standard form of a hyperbola. The constant on one side of the equation must be 1.

$$16y^2 - 4x^2 = 64$$

$$\frac{16y^2}{64} - \frac{4x^2}{64} = \frac{64}{64} \qquad \text{Divide each side by 64.}$$

$$\frac{y^2}{4} - \frac{x^2}{16} = 1 \qquad \text{Simplify.}$$

$$\frac{y^2}{2^2} - \frac{x^2}{4^2} = 1 \qquad \text{Identify } a \text{ and } b.$$

So, this equation represents a vertical hyperbola with center at the origin. Since $b = 2$, the y-intercepts are $(0, 2)$ and $(0, -2)$. Since $a = 4$, the four vertices of the rectangle are $(4, 2)$, $(-4, 2)$, $(4, -2)$, and $(-4, -2)$. The diagonals of the rectangle are drawn and extended. (See Figure 11.7.)

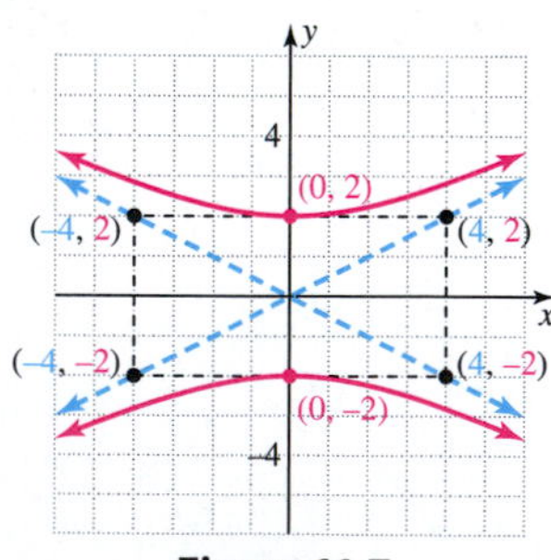

Figure 11.7

Student Check 2

Graph each hyperbola.

a. $\dfrac{x^2}{25} - \dfrac{y^2}{16} = 1$ **b.** $49y^2 - 4x^2 = 196$

Objective 3

Troubleshoot common errors.

Troubleshooting Common Errors

Some common errors associated with graphing ellipses and hyperbolas are shown.

Objective 3 Examples

A problem and an incorrect solution are given. Provide the correct solution and an explanation of the error.

3a. Graph $\frac{x^2}{9} + \frac{y^2}{25} = 1$.

Incorrect Solution	Correct Solution and Explanation
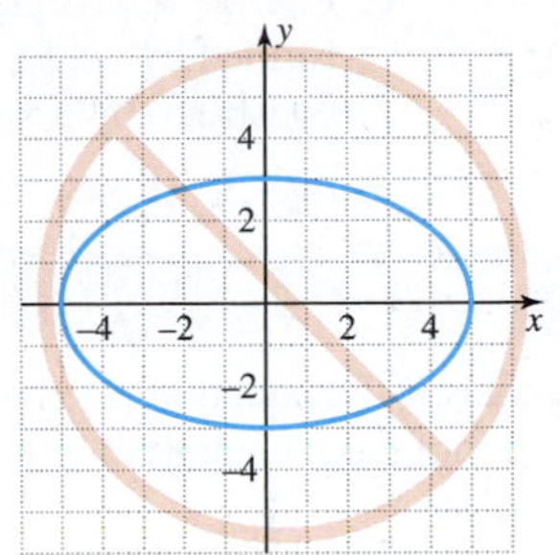	Since $a = 3$ and $b = 5$, the ellipse is vertical. The intercepts are $(0, 5)$, $(0, -5)$, $(3, 0)$, and $(-3, 0)$. 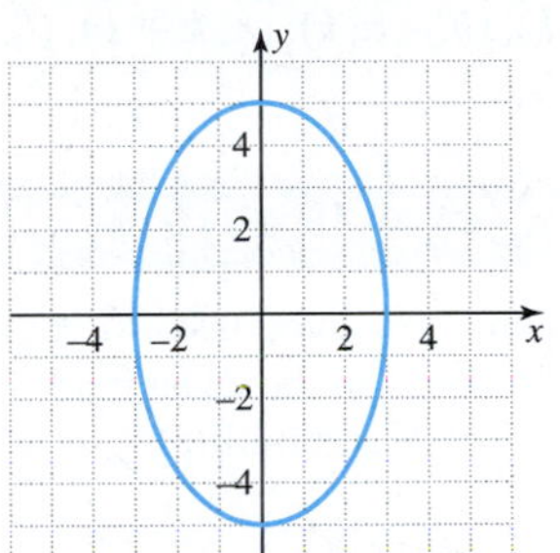

3b. Graph $\frac{x^2}{9} - \frac{y^2}{25} = 1$.

Incorrect Solution	Correct Solution and Explanation
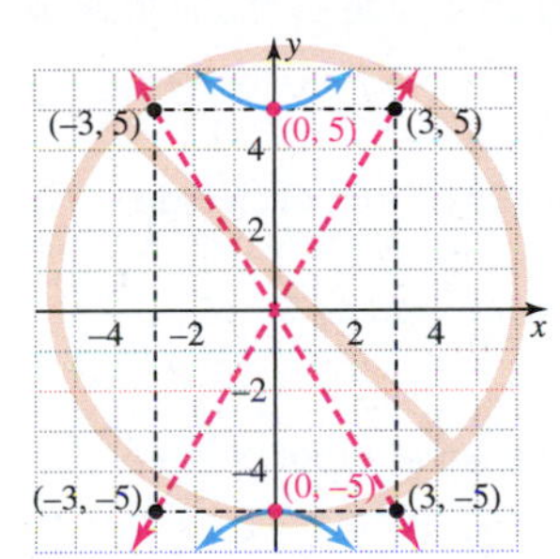	The hyperbola is horizontal since the x^2-term is positive. The x-intercepts are $(3, 0)$ and $(-3, 0)$. 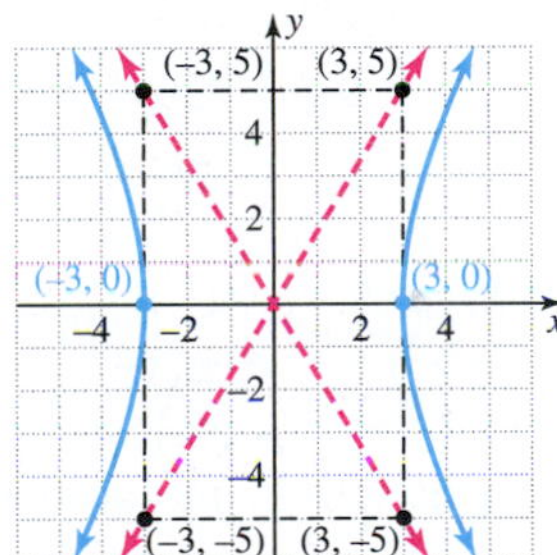

ANSWERS TO STUDENT CHECKS

Student Check 1 **a.**

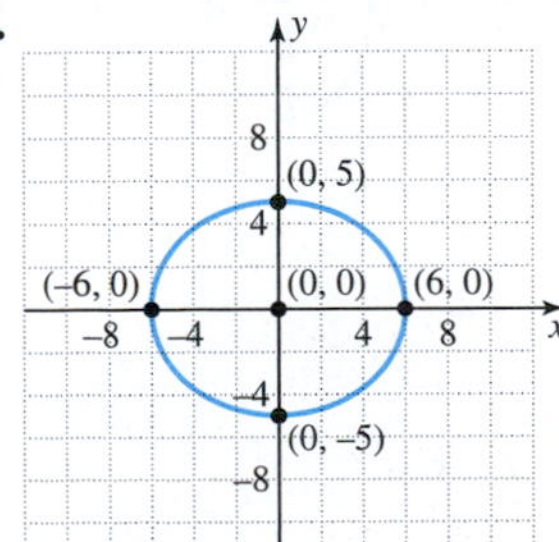

b.

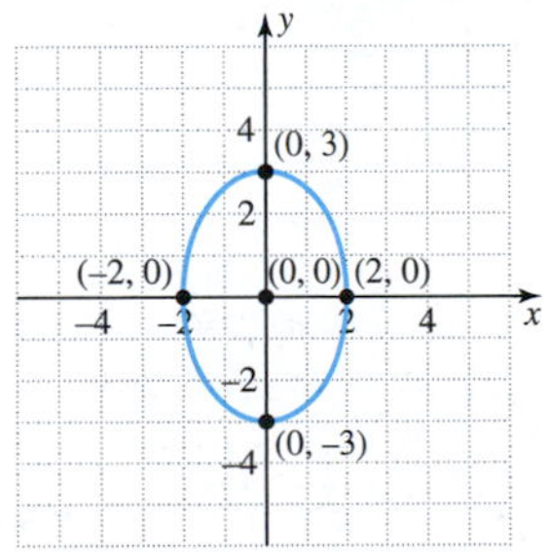

c.

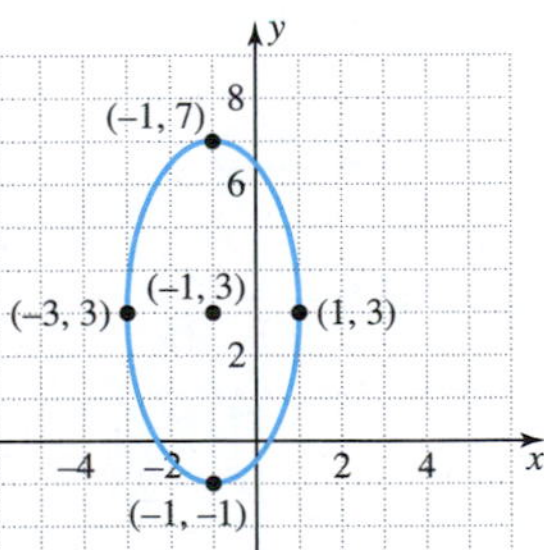

Student Check 2 **a.**

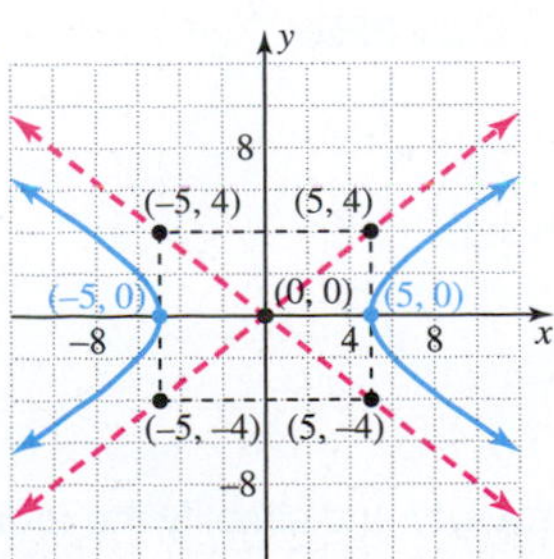

b.

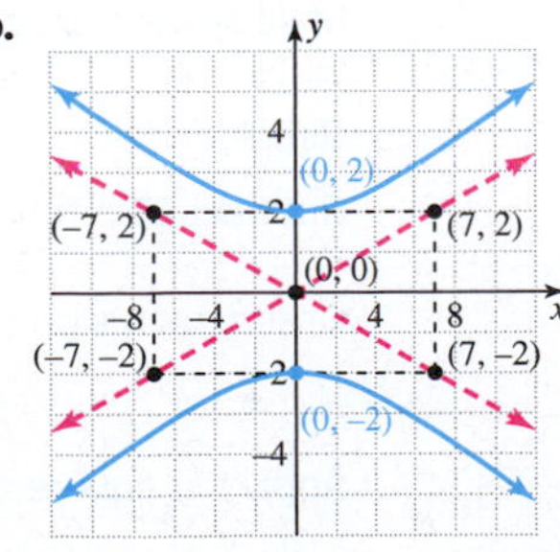

SUMMARY OF KEY CONCEPTS

1. An ellipse is defined by the equation $\dfrac{(x-h)^2}{a^2}+\dfrac{(y-k)^2}{b^2}=1.$
   a. The center is (h, k).
   b. If $a > b$, the ellipse is horizontal. If $a < b$, the ellipse is vertical.
   c. The four key points on the graph of an ellipse are $(h + a, k)$, $(h - a, k)$, $(h, k + b)$, $(h, k - b)$.

2. A hyperbola whose center is (0, 0) is given by the equation
   $$\frac{x^2}{a^2}-\frac{y^2}{b^2}=1 \quad \text{or} \quad \frac{y^2}{b^2}-\frac{x^2}{a^2}=1$$
   a. If the x^2-term is positive, the hyperbola opens horizontally.
   b. If the y^2-term is positive, the hyperbola opens vertically.
   c. Use the intercepts and the diagonals of the rectangle formed by (a, b), $(-a, b)$, $(a, -b)$, and $(-a, -b)$ to construct the graph.

GRAPHING CALCULATOR SKILLS

The graphing calculator can be used to graph ellipses and hyperbolas if the equation is solved for y. When graphing ellipses and hyperbolas, it is best to use the ZDecimal Setting. If the window is not large enough to view the graph, then use multiples of 4.7 for the Xmin and Xmax and multiples of 3.1 for the Ymin and Ymax.

Example 1: Graph $\dfrac{x^2}{9}+\dfrac{y^2}{25}=1.$

Solution:

$$\frac{x^2}{9}+\frac{y^2}{25}=1$$

$$225\left(\frac{x^2}{9}+\frac{y^2}{25}\right)=225(1) \qquad \text{Multiply each side by the LCD, 225.}$$

$$25x^2+9y^2=225 \qquad \text{Simplify.}$$

$$9y^2=225-25x^2 \qquad \text{Subtract } 25x^2 \text{ from each side.}$$

$$y^2=\frac{225-25x^2}{9} \qquad \text{Divide each side by 9.}$$

$$y=\pm\sqrt{\frac{225-25x^2}{9}} \qquad \text{Apply the square root property.}$$

Now enter the two expressions in the equation editor. Change the window to view the intercepts and graph.

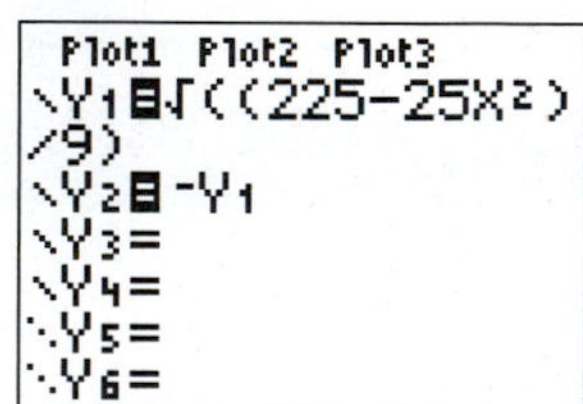


```
WINDOW
 Xmin=-9.4
 Xmax=9.4
 Xscl=1
 Ymin=-6.2
 Ymax=6.2
 Yscl=1
 Xres=1
```

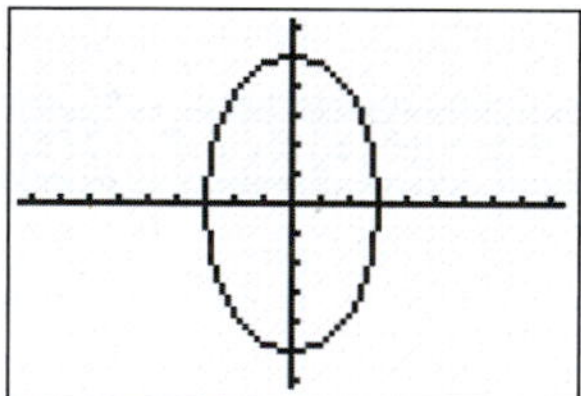

Example 2: Graph $\dfrac{x^2}{9}-\dfrac{y^2}{25}=1.$

Solution:

$$\frac{x^2}{9}-\frac{y^2}{25}=1$$

$$225\left(\frac{x^2}{9}-\frac{y^2}{25}\right)=225(1) \qquad \text{Multiply each side by the LCD, 225.}$$

$$25x^2-9y^2=225 \qquad \text{Simplify.}$$

$$-9y^2=225-25x^2 \qquad \text{Subtract } 25x^2 \text{ from each side.}$$

$$y^2=\frac{225-25x^2}{-9} \qquad \text{Divide each side by } -9.$$

$$y=\pm\sqrt{\frac{25x^2-225}{9}} \qquad \text{Apply the square root property and simplify the radicand.}$$

Now enter the two expressions in the equation editor. Change the window to view the intercepts and graph.

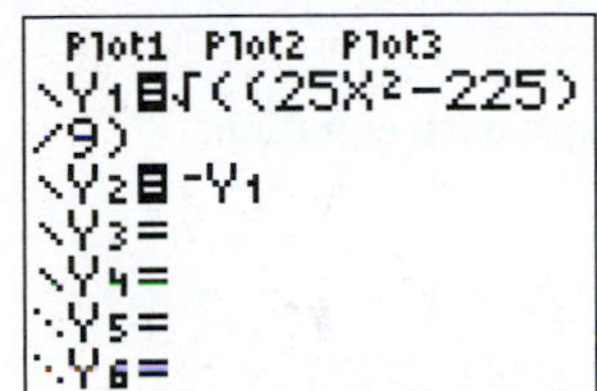

WINDOW
Xmin=-9.4
Xmax=9.4
Xscl=1
Ymin=-6.2
Ymax=6.2
Yscl=1
Xres=1

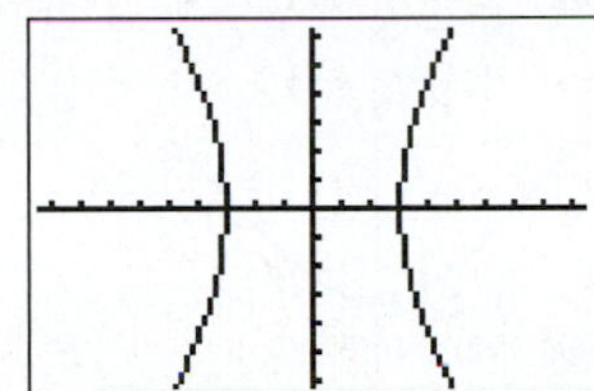

SECTION 11.2 EXERCISE SET

Write About It!

Use complete sentences in your answer to each exercise.

1. Describe how to find the four key points of the graph of the ellipse $\frac{x^2}{9} + \frac{y^2}{4} = 1$.
2. Describe how to graph the ellipse $\frac{x^2}{9} + \frac{y^2}{4} = 1$.
3. Describe how to find the x-intercepts of the graph of the hyperbola $\frac{x^2}{25} - \frac{y^2}{4} = 1$.
4. Describe how to find the two asymptotes for the graph of the hyperbola $\frac{x^2}{25} - \frac{y^2}{4} = 1$.
5. Describe how to graph the hyperbola $\frac{x^2}{25} - \frac{y^2}{4} = 1$.
6. Describe the difference between the two hyperbolas $\frac{x^2}{9} - \frac{y^2}{4} = 1$ and $\frac{y^2}{9} - \frac{x^2}{4} = 1$.
7. Explain how to use a graphing calculator to graph the ellipse $\frac{x^2}{144} + \frac{y^2}{225} = 1$. Describe what will happen if the viewing window is set to $[-10, 10]$ by $[-10, 10]$.
8. Explain how to use graphing calculator to graph the hyperbola $\frac{y^2}{144} - \frac{x^2}{225} = 1$. Describe what will happen if the viewing window is set to $[-10, 10]$ by $[-10, 10]$.

Practice Makes Perfect!

Graph each ellipse. Label four key points and the center on the graph. (*See Objective 1.*)

9. $\frac{x^2}{4} + \frac{y^2}{36} = 1$
10. $\frac{x^2}{49} + \frac{y^2}{9} = 1$
11. $\frac{x^2}{25} + y^2 = 1$
12. $x^2 + \frac{y^2}{9} = 1$
13. $49x^2 + 4y^2 = 196$
14. $x^2 + 64y^2 = 64$
15. $4x^2 + 25y^2 = 1$
16. $9x^2 + 64y^2 = 1$
17. $\frac{(x-1)^2}{36} + \frac{(y+3)^2}{25} = 1$
18. $\frac{(x+5)^2}{4} + \frac{(y-1)^2}{64} = 1$
19. $\frac{(x+3)^2}{16} + \frac{(y-2)^2}{81} = 1$
20. $\frac{(x-2)^2}{25} + \frac{(y-1)^2}{4} = 1$

Graph each hyperbola and its asymptotes. Label the intercepts. (*See Objective 2.*)

21. $\frac{y^2}{9} - \frac{x^2}{25} = 1$
22. $\frac{y^2}{16} - \frac{x^2}{4} = 1$
23. $\frac{x^2}{25} - \frac{y^2}{64} = 1$
24. $\frac{x^2}{49} - \frac{y^2}{4} = 1$
25. $x^2 - 4y^2 = 4$
26. $25y^2 - x^2 = 25$

Mix 'Em Up!

Identify each equation as a parabola, circle, ellipse, or hyperbola. Sketch the graph of each equation.

27. $x^2 + 36y^2 = 36$
28. $x^2 + \frac{y^2}{64} = 1$
29. $x = y^2 - 6y + 8$
30. $x = -y^2 - 3y + 4$
31. $2x^2 = 8 - 2y^2$
32. $x^2 + y^2 = 6y + 16$
33. $x^2 = 9y^2 - 36$
34. $y^2 = 4x^2 - 16$
35. $(x-1)^2 + \frac{(y+3)^2}{49} = 1$
36. $\frac{(x+2)^2}{4} + (y+1)^2 = 1$
37. $\frac{x^2}{16} - y^2 = 4$
38. $y^2 = 25x^2 + 1$
39. $\frac{x^2}{2.25} - y^2 = 1$
40. $\frac{y^2}{6.25} - x^2 = 1$
41. $x = y^2 + 7y - 8$
42. $x = -y^2 + 5y - 3.69$

You Be the Teacher!

Correct each student's errors, if any.

43. Graph the ellipse $\frac{x^2}{25} + \frac{y^2}{4} = 1$.

Randy's work:

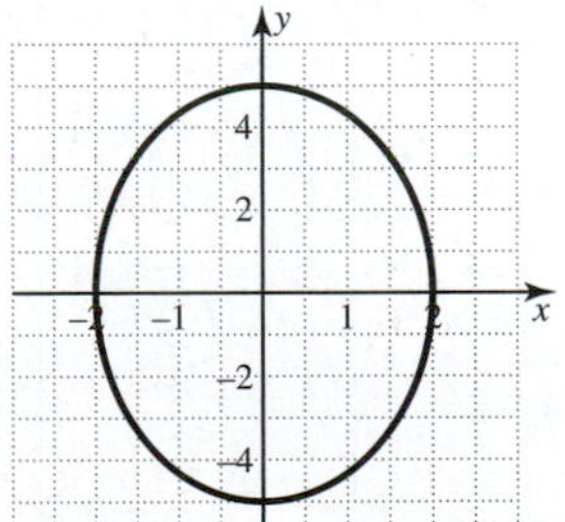

Additional answers can be found in the Instructor Answer Appendix.

44. Graph the hyperbola $\dfrac{x^2}{49} - \dfrac{y^2}{4} = 1$.

Monica's work:

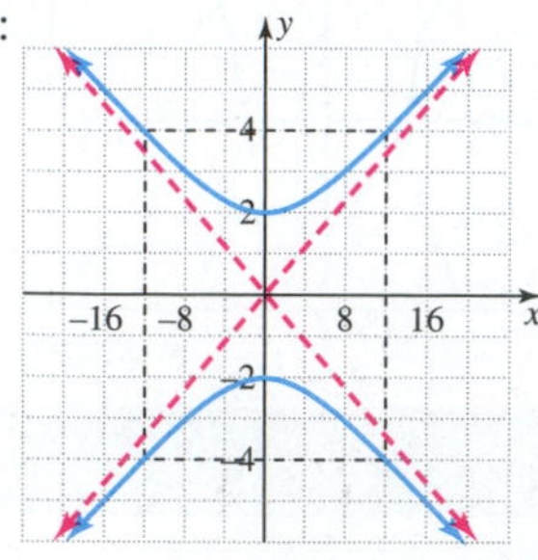

Calculate It!

Use a graphing calculator to graph each equation.

45. $\dfrac{(x - 0.6)^2}{2.25} + (y + 1.6)^2 = 1$

46. $4y^2 - 25x^2 = 49$

PIECE IT TOGETHER SECTIONS 11.1–11.2

Graph each equation. For parabolas, plot the vertex and two additional points, and state the axis of symmetry. For circles, identify the center, the radius, and four key points. For ellipses, label four key points and the center on the graph. For hyperbolas, use a rectangle and asymptotes to graph. (*Sections 11.1 and 11.2, Objectives 1 and 2*)

1. $x = y^2 + 6y - 27$
2. $x = -y^2 - 4y + 21$
3. $x^2 + y^2 = 100$
4. $x^2 + y^2 = 36$
5. $x^2 + (y + 1)^2 = 4$
6. $(x + 2)^2 + y^2 = 1$
7. $25x^2 + 9y^2 = 225$
8. $81x^2 + 4y^2 = 324$
9. $9x^2 - y^2 = 9$
10. $y^2 - 16x^2 = 16$

Write the equation of each circle in standard form and identify the center and radius. (*Section 11.1, Objective 4*)

11. $x^2 + y^2 + 10x - 14y + 58 = 0$
12. $x^2 + y^2 - 2x - 12y + 36 = 0$

SECTION 11.3 Solving Nonlinear Systems of Equations

OBJECTIVES

As a result of completing this section, you will be able to

1. Solve a nonlinear system by substitution.
2. Solve a nonlinear system by elimination.
3. Troubleshoot common errors.

In Chapter 5, we learned how to solve systems of linear equations by graphing, substitution, and elimination. While graphing is a good method to confirm answers and to determine the number of solutions of a system, it is not very precise. So, we will focus on the methods of substitution and elimination.

Objective 1

Solve a nonlinear system by substitution.

Use Substitution to Solve Nonlinear Systems of Equations

A **nonlinear system of equations** is a system of equations in which at least one of the equations is not linear. A **solution of a nonlinear system** of equations in two variables is an ordered pair that satisfies each equation in the system. Because at least one of the equations is not linear, the number of solutions will vary. A nonlinear system can have no solution, one solution, two solutions, three solutions, four solutions, and so on. Here are some illustrations of the different possibilities.

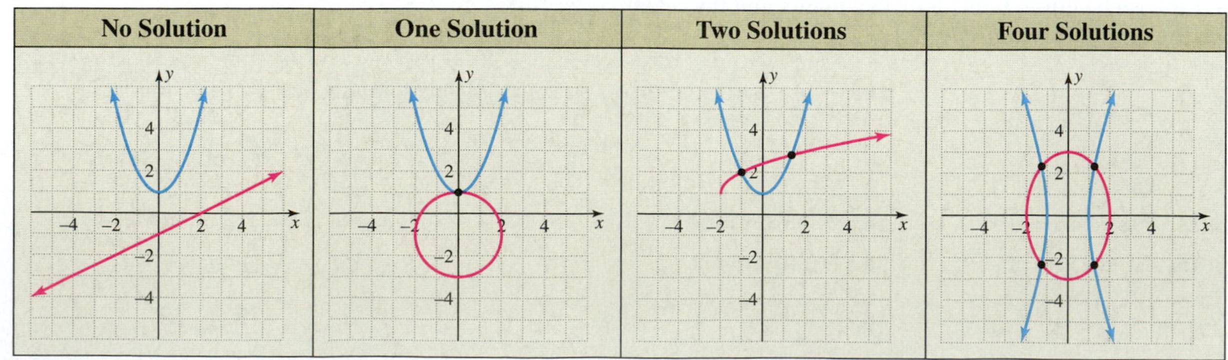

Procedure: Solving a Nonlinear System by Substitution

Step 1: Solve one of the equations for one of the variables.
Step 2: Substitute this expression into the other equation and solve the equation.
Step 3: Back substitute to find the other coordinate of each solution.
Step 4: Check by substituting the solutions into the original equations in the system.

Objective 1 Examples

Solve each system by substitution.

1a. $\begin{cases} y = (x-1)^2 \\ x - 3y = -1 \end{cases}$ **1b.** $\begin{cases} x = \frac{1}{2}y^2 + 4 \\ y = \sqrt{x-1} \end{cases}$ **1c.** $\begin{cases} x^2 + y^2 = 1 \\ x - y = 2 \end{cases}$

Solutions

1a. The first equation is solved for y so we substitute it into the second equation and solve for x.

$$x - 3y = -1 \qquad \text{Begin with the second equation.}$$
$$x - 3(x-1)^2 = -1 \qquad \text{Replace } y \text{ with } (x-1)^2.$$
$$x - 3(x^2 - 2x + 1) = -1 \qquad \text{Square the binomial.}$$
$$x - 3x^2 + 6x - 3 = -1 \qquad \text{Apply the distributive property.}$$
$$-3x^2 + 7x - 2 = 0 \qquad \text{Add 1 to each side and combine like terms.}$$
$$3x^2 - 7x + 2 = 0 \qquad \text{Multiply each side by } -1.$$
$$(3x-1)(x-2) = 0 \qquad \text{Factor.}$$
$$3x - 1 = 0 \quad \text{or} \quad x - 2 = 0 \qquad \text{Apply the zero products property.}$$
$$x = \frac{1}{3} \qquad x = 2 \qquad \text{Solve each equation.}$$

Now solve for the other variable by replacing x with the values $\frac{1}{3}$ and 2 in the first equation.

Let $x = \frac{1}{3}$.

$$y = (x-1)^2$$
$$y = \left(\frac{1}{3} - 1\right)^2$$
$$y = \left(-\frac{2}{3}\right)^2$$
$$y = \frac{4}{9}$$

Let $x = 2$.

$$y = (x-1)^2$$
$$y = (2-1)^2$$
$$y = (1)^2$$
$$y = 1$$

Now check the proposed solutions of $\left(\frac{1}{3}, \frac{4}{9}\right)$ and (2, 1) in the original system.

Let $(x, y) = \left(\frac{1}{3}, \frac{4}{9}\right)$.

$$\begin{cases} y = (x-1)^2 \\ x - 3y = -1 \end{cases}$$

Let $(x, y) = (2, 1)$.

$$\begin{cases} y = (x-1)^2 \\ x - 3y = -1 \end{cases}$$

$$\begin{cases} \frac{4}{9} = \left(\frac{1}{3} - 1\right)^2 \\ \frac{1}{3} - 3\left(\frac{4}{9}\right) = -1 \end{cases} \rightarrow \begin{cases} \frac{4}{9} = \left(-\frac{2}{3}\right)^2 \\ \frac{1}{3} - \frac{4}{3} = -1 \end{cases} \rightarrow \begin{cases} \frac{4}{9} = \frac{4}{9} \\ -\frac{3}{3} = -1 \end{cases}$$

$$\begin{cases} 1 = (2-1)^2 \\ 2 - 3(1) = -1 \end{cases} \rightarrow \begin{cases} 1 = (1)^2 \\ 2 - 3 = -1 \end{cases} \rightarrow \begin{cases} 1 = 1 \\ -1 = -1 \end{cases}$$

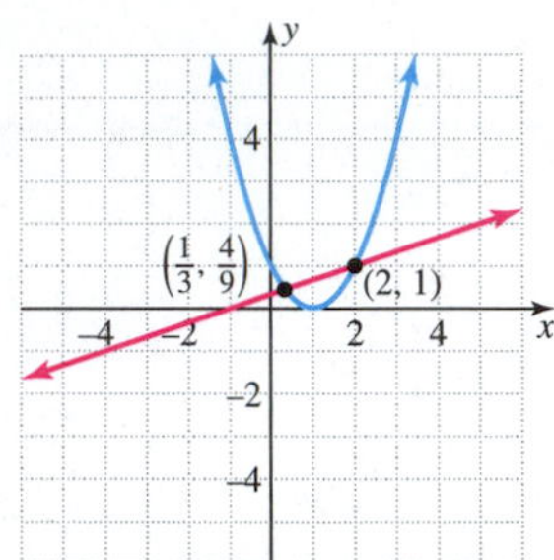

Figure 11.8

Since both ordered pairs make each equation in the system true, the solutions are $\left(\frac{1}{3}, \frac{4}{9}\right)$ and (2, 1). So, the solution set is written as $\left\{\left(\frac{1}{3}, \frac{4}{9}\right), (2, 1)\right\}$. The graph of the system is shown in Figure 11.8. Notice that the graphs intersect at the points $\left(\frac{1}{3}, \frac{4}{9}\right)$ and (2, 1).

1b. The second equation is solved for y so we substitute it into the first equation and solve for x.

$x = \frac{1}{2}y^2 + 4$	Begin with the first equation.
$x = \frac{1}{2}(\sqrt{x-1})^2 + 4$	Replace y with $\sqrt{x-1}$.
$x = \frac{1}{2}(x - 1) + 4$	Simplify.
$x = \frac{1}{2}x - \frac{1}{2} + 4$	Apply the distributive property.
$2(x) = 2\left(\frac{1}{2}x - \frac{1}{2} + 4\right)$	Multiply each side by the LCD, 2.
$2x = x - 1 + 8$	Simplify.
$2x = x + 7$	Combine like terms.
$x = 7$	Subtract x from each side.

Now solve for the other variable by replacing x with 7 in the second equation.

$$y = \sqrt{x-1}$$
$$y = \sqrt{7-1}$$
$$y = \sqrt{6}$$

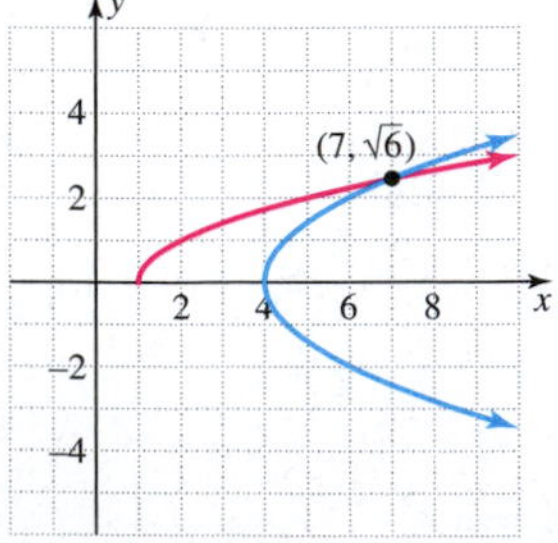

Figure 11.9

Now check the proposed solution $(7, \sqrt{6})$ in the original system.

$$\begin{cases} x = \frac{1}{2}y^2 + 4 \\ y = \sqrt{x-1} \end{cases} \rightarrow \begin{cases} 7 = \frac{1}{2}(\sqrt{6})^2 + 4 \\ \sqrt{6} = \sqrt{7-1} \end{cases} \rightarrow \begin{cases} 7 = \frac{1}{2}(6) + 4 \\ \sqrt{6} = \sqrt{6} \end{cases} \rightarrow \begin{cases} 7 = 3 + 4 \\ \sqrt{6} = \sqrt{6} \end{cases}$$

Since the ordered pair makes each equation in the system true, the solution set is $\{(7, \sqrt{6})\}$. The graph of the system is shown in Figure 11.9. Notice that the graphs intersect at the point $(7, \sqrt{6})$.

Note: After we found that $x = 7$, we could have solved for y using the other equation.

$$x = \frac{1}{2}y^2 + 4 \quad \text{Begin with the first equation.}$$

$$7 = \frac{1}{2}y^2 + 4 \quad \text{Substitute 7 for } x.$$

$$7 - 4 = \frac{1}{2}y^2 + 4 - 4 \quad \text{Subtract 4 from each side.}$$

$$3 = \frac{1}{2}y^2 \quad \text{Simplify.}$$

$$2(3) = 2\left(\frac{1}{2}y^2\right) \quad \text{Multiply each side by the LCD, 2.}$$

$$6 = y^2 \quad \text{Simplify.}$$

$$\pm\sqrt{6} = y \quad \text{Apply the square root property.}$$

So, this would give us two proposed solutions $(7, \sqrt{6})$ and $(7, -\sqrt{6})$. When we check $(7, -\sqrt{6})$, we find that it doesn't check in the second equation.

$$y = \sqrt{x - 1}$$

$$-\sqrt{6} = \sqrt{7 - 1} \quad \text{Replace } x \text{ with 7 and } y \text{ with } -\sqrt{6}.$$

$$-\sqrt{6} = \sqrt{6} \quad \text{Simplify.}$$

Since the resulting equation is false, we discard the solution $(7, -\sqrt{6})$.

1c. Solve the second equation for x. $x - y = 2$

$$x = y + 2$$

Now substitute this expression in the first equation and solve for y.

$$x^2 + y^2 = 1 \quad \text{Begin with the first equation.}$$

$$(y + 2)^2 + y^2 = 1 \quad \text{Replace } x \text{ with } y + 2.$$

$$y^2 + 4y + 4 + y^2 = 1 \quad \text{Square the binomial.}$$

$$2y^2 + 4y + 3 = 0 \quad \text{Subtract 1 from each side and combine like terms.}$$

To solve the equation, we apply the quadratic formula with $a = 2$, $b = 4$, and $c = 3$.

$$y = \frac{-b \pm \sqrt{b^2 - 4ac}}{2a} \quad \text{State the quadratic formula.}$$

$$y = \frac{-4 \pm \sqrt{4^2 - 4(2)(3)}}{2(2)} \quad \text{Replace } a \text{ with 2, } b \text{ with 4, and } c \text{ with 3.}$$

$$y = \frac{-4 \pm \sqrt{16 - 24}}{4} \quad \text{Simplify each term in the radical.}$$

$$y = \frac{-4 \pm \sqrt{-8}}{4} \quad \text{Simplify the radical.}$$

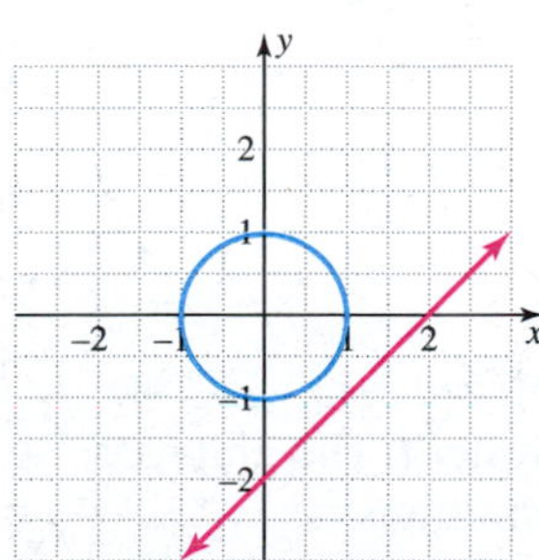

Figure 11.10

Since $\sqrt{-8}$ is not a real number, the system of equations has no solution. We confirm this by graphing as shown in Figure 11.10. So, the solution set is $\varnothing$.

Student Check 1 Solve each system by substitution.

a. $\begin{cases} y = (x + 2)^2 \\ x + y = 4 \end{cases}$ **b.** $\begin{cases} x = \frac{1}{2}y^2 + 2 \\ y = \sqrt{x + 4} \end{cases}$ **c.** $\begin{cases} x^2 + y^2 = 4 \\ x - y = 3 \end{cases}$

Use Elimination to Solve Nonlinear Systems of Equations

Objective 2

Solve a nonlinear system by elimination.

Recall that elimination is a method that requires us to add the equations in the system so that one variable is eliminated. The coefficients of the eliminated variable must be opposites. If they are not opposites, we have to multiply one or both of the equations by a nonzero number.

Procedure: Solving a Nonlinear System by Elimination

Step 1: Write the equations in the same form.
Step 2: Determine the variable to be isolated. If its coefficients are not opposites, multiply one or both of the equations by the number that will make them opposites.
Step 3: Add the equations.
Step 4: Solve the resulting equation.
Step 5: Back substitute to determine the other coordinate of each solution.

Objective 2 Example

Solve the system $\begin{cases} x^2 + y^2 = 10 \\ 2x^2 - y^2 = 17 \end{cases}$ by elimination.

Solution We can add the equations in the system since the coefficients of the y^2-terms are opposites and the equations are already in the same form.

$$\begin{cases} x^2 + y^2 = 10 \\ 2x^2 - y^2 = 17 \end{cases}$$

$$3x^2 = 27 \quad \text{Add the equations.}$$

$$x^2 = 9 \quad \text{Divide each side by 3.}$$

$$x = \pm\sqrt{9} \quad \text{Apply the square root property.}$$

$$x = \pm 3 \quad \text{Simplify.}$$

Now we find the y-coordinates for each of the solutions.

Let $x = 3$.

$$x^2 + y^2 = 10$$
$$(3)^2 + y^2 = 10$$
$$9 + y^2 = 10$$
$$y^2 = 1$$
$$y = \pm 1$$

Let $x = -3$.

$$x^2 + y^2 = 10$$
$$(-3)^2 + y^2 = 10$$
$$9 + y^2 = 10$$
$$y^2 = 1$$
$$y = \pm 1$$

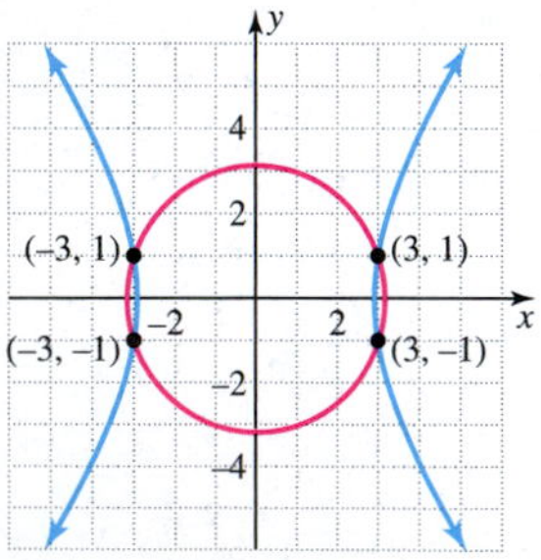

Figure 11.11

So, the four solutions of the system are (3, 1), (3, −1), (−3, 1), and (−3, −1). Be sure to check each solution in the original equations in the system. The graph of the system is shown in Figure 11.11 and confirms the four solutions.

Student Check 2

Solve the system $\begin{cases} x^2 + y^2 = 16 \\ 4x^2 - y^2 = 4 \end{cases}$ by elimination.

Objective 3

Troubleshoot common errors.

Troubleshooting Common Errors

Some common errors associated with solving nonlinear systems of equations are shown.

Objective 3 Example **A problem and an incorrect solution are given. Provide the correct solution and an explanation of the error.**

Solve the system $\begin{cases} x^2 + y^2 = 13 \\ 4x^2 - y^2 = 7 \end{cases}$ by elimination.

Incorrect Solution	Correct Solution and Explanation
Adding the equations, we get $5x^2 = 20$ $x^2 = 4$ $x = 2$ Solving for y: $(2)^2 + y^2 = 13$ $4 + y^2 = 13$ $y^2 = 9$ $y = \pm 3$ The solutions are $(2, -3)$ and $(2, 3)$.	When the square root property is applied, there are two solutions. So, $x = \pm 2$. When we solve for y, we get $x = 2$: $(2)^2 + y^2 = 13$ $4 + y^2 = 13$ $y^2 = 9$ $y = \pm 3$ $x = -2$: $(-2)^2 + y^2 = 13$ $4 + y^2 = 13$ $y^2 = 9$ $y = \pm 3$ The solutions are $(2, -3)$, $(2, 3)$, $(-2, -3)$, and $(-2, 3)$.

ANSWERS TO STUDENT CHECKS

Student Check 1 **a.** $\{(-5, 9), (0, 4)\}$ **b.** $\{(8, 2\sqrt{3})\}$ **c.** $\varnothing$

Student Check 2 **a.** $\{(2, 2\sqrt{3}), (2, -2\sqrt{3}), (-2, 2\sqrt{3}), (-2, -2\sqrt{3})\}$

SUMMARY OF KEY CONCEPTS

1. Nonlinear systems can be solved by substitution. We solve one equation for one variable and substitute this expression into the other equation in the system. We then back substitute to solve for the other variable. Solutions can be checked algebraically or graphically.
2. Nonlinear systems can also be solved by elimination. We transform the equations so that the coefficients of one of the variables are opposites, and then add the equations. We solve the resulting equation and back substitute to solve for the other variable.

SECTION 11.3 EXERCISE SET

Write About It!

Use complete sentences in your answer to each exercise.

1. What is the number of possible solutions of a system of equations consisting of a parabola and a line?
2. What is the number of possible solutions of a system of equations consisting of a circle and either a circle or an ellipse?
3. Explain how to apply the first two steps of the substitution method to solve the following system. Do not solve.

   $\begin{cases} x^2 + y^2 = 100 \\ 2x - y = 20 \end{cases}$
4. Explain how to apply the first two steps of the substitution method to solve the following system. Do not solve.

   $\begin{cases} y = x^2 + 8 \\ 2x - y = 23 \end{cases}$

Additional answers can be found in the Instructor Answer Appendix.

5. Explain how to apply the first two steps of the elimination method to solve the following system. Do not solve.
$$\begin{cases} 2x^2 + 4y^2 = 36 \\ 2y^2 - 5x^2 = -78 \end{cases}$$

6. Explain how to apply the first two steps of the elimination method to solve the following system. Do not solve.
$$\begin{cases} 2x^2 + 2y^2 = 40 \\ y^2 + 3x^2 = 52 \end{cases}$$

7. Explain how to choose the method, elimination or substitution, to solve the following system. Do not solve.
$$\begin{cases} x^2 + y^2 = 25 \\ 4x + 5y = 8 \end{cases}$$

8. Explain how to choose the method, elimination or substitution, to solve the following system. Do not solve.
$$\begin{cases} 4x^2 + 3y^2 = 79 \\ x^2 + 3y^2 = 76 \end{cases}$$

Practice Makes Perfect!

Solve each system by substitution. (*See Objective 1.*)

9. $\begin{cases} y = (x-5)^2 \\ 7x + y = 25 \end{cases}$ $\{(0, 25), (3, 4)\}$

10. $\begin{cases} y = (x-3)^2 \\ -x + y = 9 \end{cases}$ $\{(0, 9), (7, 16)\}$

11. $\begin{cases} y = (x+1)^2 \\ -5x + y = 1 \end{cases}$ $\{(0, 1), (3, 16)\}$

12. $\begin{cases} y = (x+6)^2 \\ -8x + y = 36 \end{cases}$ $\{(0, 36), (-4, 4)\}$

13. $\begin{cases} x = \frac{1}{4}y^2 + 3 \\ y = \sqrt{3x - 8} \end{cases}$ $\{(4, 2)\}$

14. $\begin{cases} x = \frac{3}{8}y^2 - 4 \\ y = \sqrt{10x - 4} \end{cases}$ $\{(2, 4)\}$

15. $\begin{cases} x = \frac{2}{3}y^2 - 4 \\ y = \sqrt{3x + 9} \end{cases}$ $\{(-2, \sqrt{3})\}$

16. $\begin{cases} x = \frac{1}{4}y^2 + 8 \\ y = \sqrt{2x - 7} \end{cases}$ $\left\{\left(\frac{25}{2}, 3\sqrt{2}\right)\right\}$

17. $\begin{cases} y = x^2 - 5 \\ -x + y = -3 \end{cases}$ $\{(-1, -4), (2, -1)\}$

18. $\begin{cases} y = x^2 + 2 \\ 4x - y = 1 \end{cases}$ $\{(3, 11), (1, 3)\}$

19. $\begin{cases} x^2 + y^2 = 100 \\ x - y = 2 \end{cases}$ $\{(-6, -8), (8, 6)\}$

20. $\begin{cases} x^2 + y^2 = 25 \\ x + y = 1 \end{cases}$ $\{(-3, 4), (4, -3)\}$

Solve each system by elimination. (*See Objective 2.*)

21. $\begin{cases} x^2 + y^2 = 10 \\ 5x^2 + 8y^2 = 53 \end{cases}$

22. $\begin{cases} x^2 + y^2 = 25 \\ x^2 + 2y^2 = 41 \end{cases}$

23. $\begin{cases} x^2 + y^2 = 29 \\ x^2 - 4y^2 = 9 \end{cases}$

24. $\begin{cases} x^2 + y^2 = 40 \\ x^2 - y^2 = 32 \end{cases}$

25. $\begin{cases} 3x^2 + y^2 = 28 \\ y^2 - x^2 = 12 \end{cases}$

26. $\begin{cases} x^2 + 4y^2 = 52 \\ x^2 - 2y^2 = 28 \end{cases}$

27. $\begin{cases} x^2 + 2y^2 = 54 \\ 6x^2 + y^2 = 49 \end{cases}$

28. $\begin{cases} x^2 + y^2 = 20 \\ x^2 + 2y^2 = 24 \end{cases}$

29. $\begin{cases} x^2 + 4y^2 = 65 \\ x^2 + y^2 = 17 \end{cases}$

30. $\begin{cases} 5x^2 + 2y^2 = 38 \\ y^2 - x^2 = 5 \end{cases}$

31. $\begin{cases} y = 2x^2 - 4 \\ y = x^2 + 3x \end{cases}$ $\{(4, 28), (-1, -2)\}$

32. $\begin{cases} y = 4x^2 + 2 \\ y = 3x^2 + 3x \end{cases}$ $\{(2, 18), (1, 6)\}$

Mix 'Em Up!

Solve each system by substitution or elimination.

33. $\begin{cases} y = (x-8)^2 \\ x + y = 64 \end{cases}$ $\{(0, 64), (15, 49)\}$

34. $\begin{cases} y = (x+4)^2 \\ -3x + y = 16 \end{cases}$ $\{(0, 16), (-5, 1)\}$

35. $\begin{cases} x = \frac{3}{8}y^2 - 4 \\ y = \sqrt{10x - 4} \end{cases}$ $\{(2, 4)\}$

36. $\begin{cases} x = \frac{1}{7}y^2 + 11 \\ y = \sqrt{3x - 5} \end{cases}$ $\{(18, 7)\}$

37. $\begin{cases} x = \frac{2}{3}y^2 - 4 \\ y = \sqrt{x + 5} \end{cases}$ $\{(-2, \sqrt{3})\}$

38. $\begin{cases} x = \frac{3}{4}y^2 - 3 \\ y = \sqrt{2x - 4} \end{cases}$ $\{(12, 2\sqrt{5})\}$

39. $\begin{cases} x^2 + y^2 = 25 \\ 2x + y = 5 \end{cases}$ $\{(4, -3), (0, 5)\}$

40. $\begin{cases} x^2 + y^2 = 100 \\ 3x - y = 10 \end{cases}$ $\{(6, 8), (0, -10)\}$

41. $\begin{cases} y = x^2 - 9 \\ x + y = -3 \end{cases}$ $\{(-3, 0), (2, -5)\}$

42. $\begin{cases} y = x^2 + 2 \\ 5x + y = -4 \end{cases}$ $\{(-3, 11), (-2, 6)\}$

43. $\begin{cases} y = x^2 - 2 \\ -x + y = 4 \end{cases}$ $\{(3, 7), (-2, 2)\}$

44. $\begin{cases} y = x^2 + 4 \\ 7x + y = -8 \end{cases}$ $\{(-3, 13), (-4, 20)\}$

45. $\begin{cases} x^2 + y^2 = 100 \\ y^2 - x^2 = 28 \end{cases}$

46. $\begin{cases} x^2 + y^2 = 10 \\ 5x^2 - 2y^2 = 43 \end{cases}$

47. $\begin{cases} 5x^2 + 4y^2 = 21 \\ 7y^2 - 2x^2 = 26 \end{cases}$

48. $\begin{cases} 4x^2 + y^2 = 32 \\ x^2 + y^2 = 20 \end{cases}$

49. $\begin{cases} y = 9x^2 - 25 \\ y = 3x^2 - 5x \end{cases}$ $\left\{\left(\frac{5}{3}, 0\right), \left(-\frac{5}{2}, \frac{125}{4}\right)\right\}$

50. $\begin{cases} y = 2x^2 - 8 \\ y = x^2 - 2x \end{cases}$ $\{(-4, 24), (2, 0)\}$

You Be the Teacher!

Correct each student's errors, if any.

51. Solve $\begin{cases} y = (x-1)^2 \\ 8x + y = 1 \end{cases}$ by substitution.

Rosa's work:

Solve the second equation for y and substitute into the first equation.

$$y = 1 - 8x$$
$$1 - 8x = (x-1)^2$$
$$1 - 8x = x^2 + 1$$
$$0 = x^2 + 8x$$
$$0 = x(x+8)$$
$$x = 0 \quad \text{or} \quad x = -8$$

Substitute the values into the first equation to solve for y.

$x = 0$: $y = (0-1)^2 = 1$

$x = -8$: $y = (-8-1)^2 = 81$

$\{(0, 1), (-8, 81)\}$

52. Solve $\begin{cases} x = \frac{1}{4}y^2 - 2 \\ y = \sqrt{6x+4} \end{cases}$ by substitution.

Ricki's work:

Substitute the second equation into the first equation.

$$x = \frac{1}{4}(6x+4)^2 - 2$$

$$4x = (6x+4)^2 - 8$$

$$4x = 36x^2 + 48x + 16 - 8$$

$$0 = 36x^2 + 44x + 8$$

$$0 = 4(9x^2 + 11x + 2)$$

$$0 = 4(9x+2)(x+1)$$

$$x = -1 \quad \text{or} \quad x = -\frac{2}{9}$$

Substitute the values into the second equation to solve for y.

$$x = -1\text{: } y = \sqrt{-6+4} = \sqrt{-2}$$

$$x = -\frac{2}{9}\text{: } y = \sqrt{-\frac{4}{3} + 4} = \sqrt{\frac{8}{3}}$$

$$\left\{\left(-\frac{2}{9}, \sqrt{\frac{8}{3}}\right)\right\}$$

53. Solve $\begin{cases} 3x^2 + 2y^2 = 12 \\ 3x^2 - y^2 = -3 \end{cases}$ by elimination.

Allison's work:

Multiply the second equation by -1 and add to the first equation.

$$\begin{aligned} 3x^2 + 2y^2 &= 12 \\ -3x^2 - y^2 &= -3 \\ \hline y^2 &= 9 \\ y &= 3 \end{aligned}$$

Substitute the value into the second equation to solve for y.

$$y = 3\text{: } 3x^2 - 9 = -3$$

$$3x^2 = 6$$

$$x^2 = 2$$

$$x = \sqrt{2}$$

$$\{(\sqrt{2}, 3)\}$$

54. Solve $\begin{cases} 4x^2 + y^2 = 28 \\ x^2 + y^2 = 19 \end{cases}$ by elimination.

Sally's work:

Multiply the second equation by -1 and add to the first equation.

$$\begin{aligned} 4x^2 + y^2 &= 28 \\ -x^2 - y^2 &= -19 \\ \hline 4x^2 &= 9 \end{aligned}$$

$$x^2 = \frac{9}{4}$$

$$x = \frac{3}{2}$$

Substitute the value into the first equation to solve for y.

$$x = \frac{3}{2}\text{: } \frac{9}{4} + y^2 = 19$$

$$y^2 = \frac{67}{4}$$

$$y = \frac{\sqrt{67}}{2} \qquad \left\{\left(\frac{3}{2}, \frac{\sqrt{67}}{2}\right)\right\}$$

Calculate It!

Use a graphing calculator to solve each nonlinear system of equations.

55. $\begin{cases} y = x^2 - 2 \\ x + y = 10 \end{cases}$

56. $\begin{cases} 2x^2 + y^2 = 11 \\ y^2 - 3x^2 = 6 \end{cases}$

SECTION 11.4 Solving Nonlinear Inequalities and Systems of Inequalities

OBJECTIVES

As a result of completing this section, you will be able to

1. **Graph a nonlinear inequality.**
2. **Solve a nonlinear system of inequalities.**
3. **Troubleshoot common errors.**

In Chapter 2, we solved linear inequalities in two variables and then applied this to solving systems of linear inequalities in Chapter 5. The method that we learned to solve linear inequalities in two variables will be used to solve nonlinear inequalities. This will be the foundation for solving nonlinear systems of inequalities, as well.

Objective 1

Graph a nonlinear inequality.

Nonlinear Inequalities

A **nonlinear inequality** is an inequality that contains an expression that is not linear. Some examples of nonlinear inequalities are

$$x^2 + y^2 > 9 \qquad x \le y^2 + 1$$

When we graph a nonlinear inequality in two variables, our goal is to find the region of the coordinate plane that contains solutions of the inequality. We begin by finding the boundary curve of the region of solutions and then test points to determine where the solutions are located. This is exactly the process that we used with linear inequalities in two variables.

Procedure: Graphing a Nonlinear Inequality in Two Variables

Step 1: Graph the associated nonlinear equation.

- **a.** If the inequality symbol is $\le$ or $\ge$, draw the graph as a solid curve.
- **b.** If the inequality symbol is $<$ or $>$, draw the graph as a dashed curve.

Step 2: Test a point from each region to determine if it makes the inequality true or false.

- **a.** If the test point makes the inequality true, shade the region that contains the point.
- **b.** If the test point makes the inequality false, do not shade the region containing the test point.

Step 3: The solution set consists of the shaded regions.

Objective 1 Examples

Graph each inequality.

1a. $y > (x - 1)^2 + 3$ **1b.** $\dfrac{x^2}{16} + \dfrac{y^2}{9} < 1$ **1c.** $\dfrac{x^2}{4} - \dfrac{y^2}{9} \le 1$

Solutions **1a.** Begin by graphing the parabola $y = (x - 1)^2 + 3$ with a dashed curve since the inequality is $>$. The parabola opens up because $a > 0$ and it has a vertex of (1, 3).

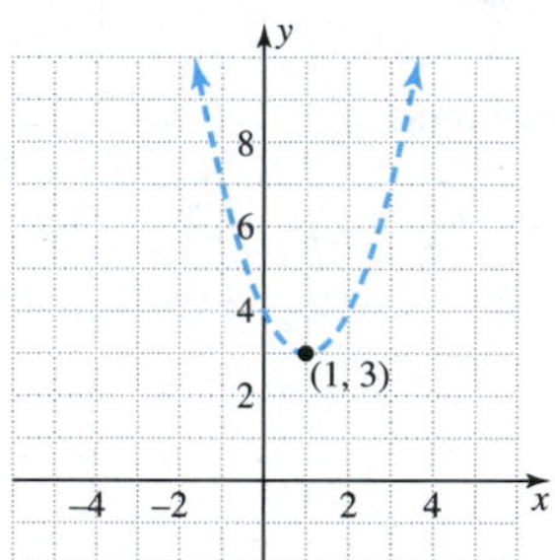

The boundary curve divides the plane into two regions, the region inside the parabola and the region outside of the parabola. We test a point from each region.

Test point	$y > (x - 1)^2 + 3$	True/False
(0, 0)	$0 > (0 - 1)^2 + 3$ $0 > 4$	False
(1, 4)	$4 > (1 - 1)^2 + 3$ $4 > 3$	True

Since the point (1, 4) satisfies the inequality, the region containing this point is the solution set of the inequality. Only the points inside the parabola are solutions of $y > (x - 1)^2 + 3$. This is shown on the graph.

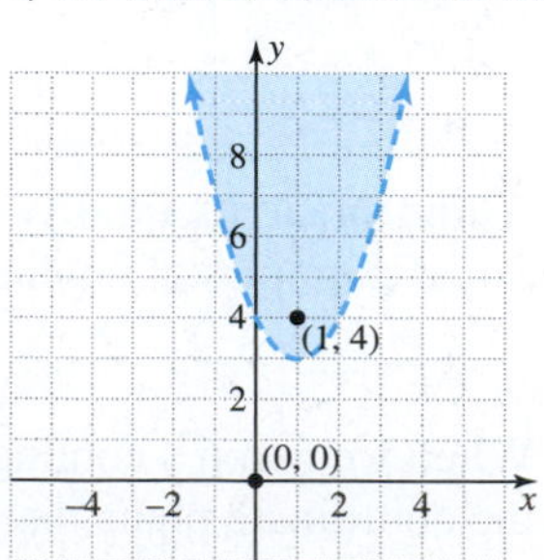

1b. We graph the ellipse $\frac{x^2}{16} + \frac{y^2}{9} = 1$ with a dashed curve since the inequality is <. The ellipse is centered at the origin and contains the key points (4, 0), (−4, 0), (0, 3), and (0, −3).

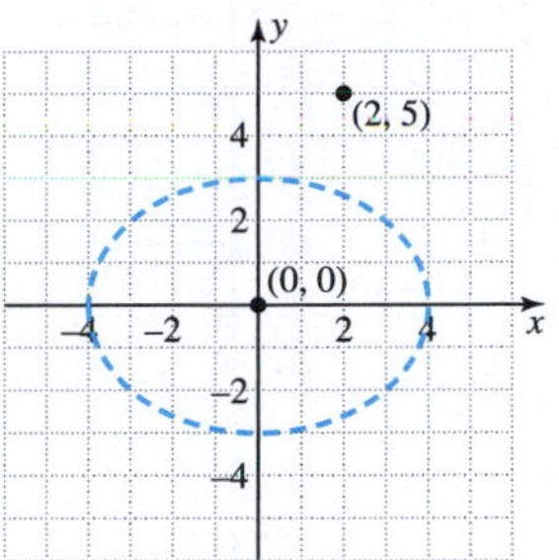

The boundary curve divides the plane into two regions, the region inside the ellipse and the region outside the ellipse. We test a point from each region.

Test point	$\frac{x^2}{16} + \frac{y^2}{9} < 1$	**True/False**
(0, 0)	$\frac{0^2}{16} + \frac{0^2}{9} < 1$ $0 < 1$	True
(2, 5)	$\frac{2^2}{16} + \frac{5^2}{9} < 1$ $\frac{1}{4} + \frac{25}{9} < 1$	False

Because the point (0, 0) makes the inequality true, the solution set consists of the points inside the ellipse.

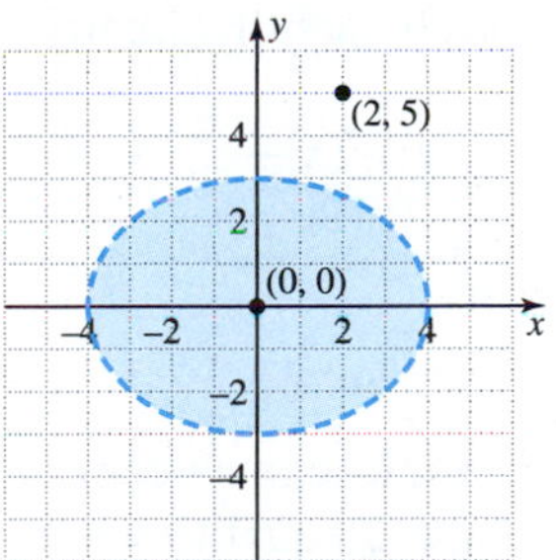

1c. Begin by graphing the hyperbola $\frac{x^2}{4} - \frac{y^2}{9} = 1$ with a solid curve since the inequality symbol is ≤. The hyperbola is horizontal and has x-intercepts (2, 0) and (−2, 0).

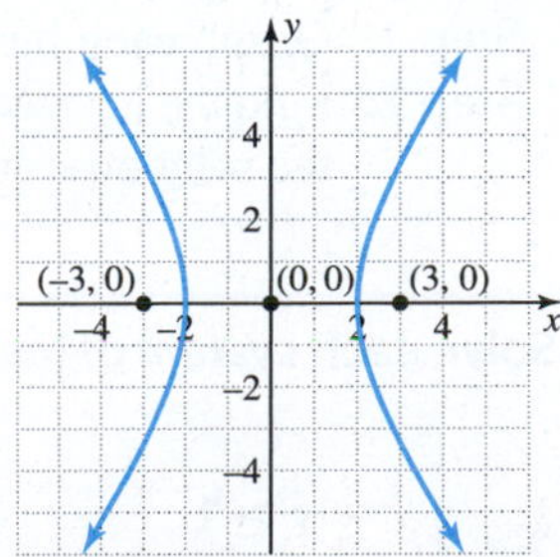

The boundary curves divide the plane into three regions, the region to the left of the left branch of the hyperbola, the region between the branches of the hyperbola, and the region to the right of the right branch of the hyperbola. We test a point in each region.

Test point	$\frac{x^2}{4} - \frac{y^2}{9} \le 1$	True/False
(0, 0)	$\frac{0^2}{4} - \frac{0^2}{9} \le 1$ $0 \le 1$	True
(3, 0)	$\frac{3^2}{4} - \frac{0^2}{9} \le 1$ $\frac{9}{4} - 0 \le 1$	False
(−3, 0)	$\frac{(-3)^2}{4} - \frac{0^2}{9} \le 1$ $\frac{9}{4} - 0 \le 1$	False

Since the point (0, 0) satisfies the inequality, the solution set consists of the graph of the hyperbola and all points between the branches.

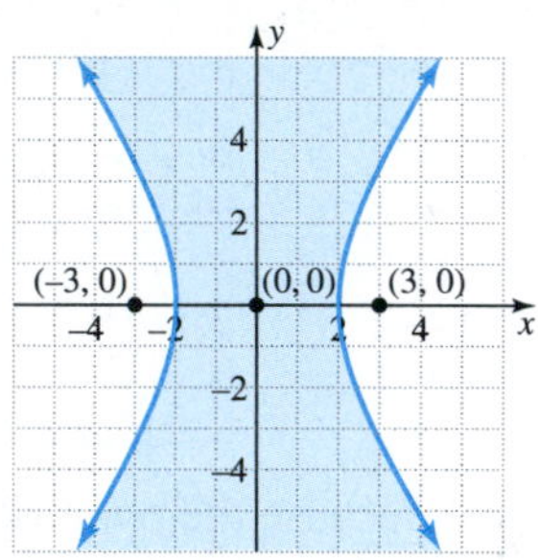

Student Check 1 Graph each inequality.

a. $y > (x - 2)^2 - 4$ **b.** $\frac{x^2}{4} + \frac{y^2}{16} < 1$ **c.** $\frac{y^2}{16} - \frac{x^2}{9} \le 1$

Nonlinear Systems of Inequalities

Objective 2

Solve a nonlinear system of inequalities.

A **nonlinear system of inequalities** is a system of inequalities in which at least one of the inequalities is not linear. When we studied systems of linear inequalities, we learned that the solution set is the intersection of the solution sets of each inequality in the system. This is exactly the same for a nonlinear system of inequalities.

Procedure: Solving a Nonlinear System of Inequalities

Step 1: Graph each inequality in the system.

Step 2: Find the intersection of the graphs of these inequalities. The intersection is the solution set of the system.

Objective 2 Examples Solve each system of inequalities.

2a. $\begin{cases} x - y \ge 1 \\ x^2 + y^2 \le 4 \end{cases}$ **2b.** $\begin{cases} \frac{x^2}{9} + \frac{y^2}{25} < 1 \\ \frac{y^2}{16} - \frac{x^2}{4} < 1 \end{cases}$

Solutions **2a.** Graph the inequalities and find their intersection.

$$x - y \geq 1$$

The boundary line is $x - y = 1$, which goes through the points $(1, 0)$ and $(0, -1)$. The points on and below the line satisfy the inequality.

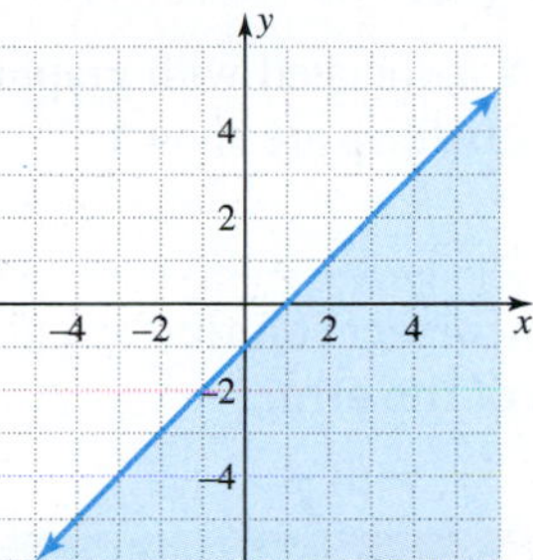

$$x^2 + y^2 \leq 4$$

The boundary curve is $x^2 + y^2 = 4$, which is a circle centered at the origin with radius 2. The points on and inside the circle satisfy the inequality.

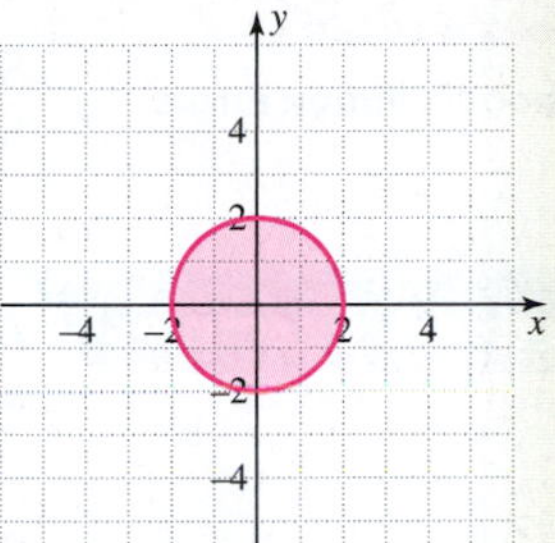

Graphing the two inequalities on the same coordinate system enables us to find the solution. The region where the two graphs intersect is the solution set of the system as shown in purple.

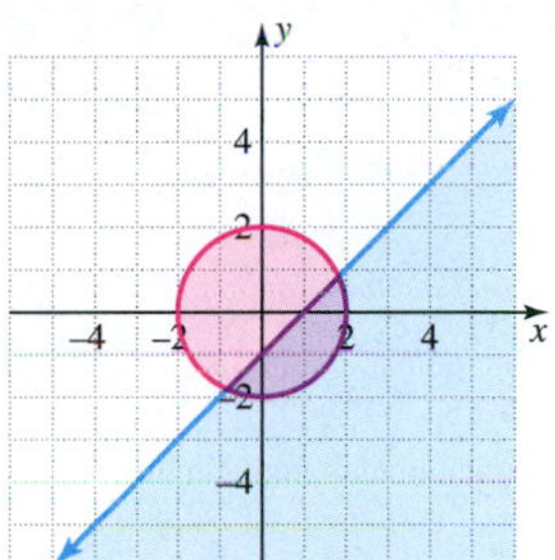

2b. Graph the inequalities and find their intersection.

$$\frac{x^2}{9} + \frac{y^2}{25} < 1$$

The boundary curve is the ellipse $\frac{x^2}{9} + \frac{y^2}{25} = 1$, which goes through the key points $(3, 0)$, $(-3, 0)$, $(0, 5)$, and $(0, -5)$. Only the points inside the ellipse satisfy the inequality.

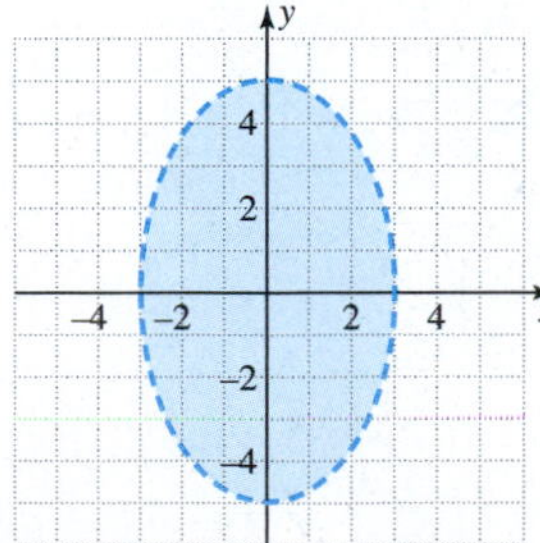

$$\frac{y^2}{16} - \frac{x^2}{4} < 1$$

The boundary curve is the hyperbola $\frac{y^2}{16} - \frac{x^2}{4} < 1$, which has y-intercepts at $(0, 4)$ and $(0, -4)$. The points between the branches satisfy the inequality.

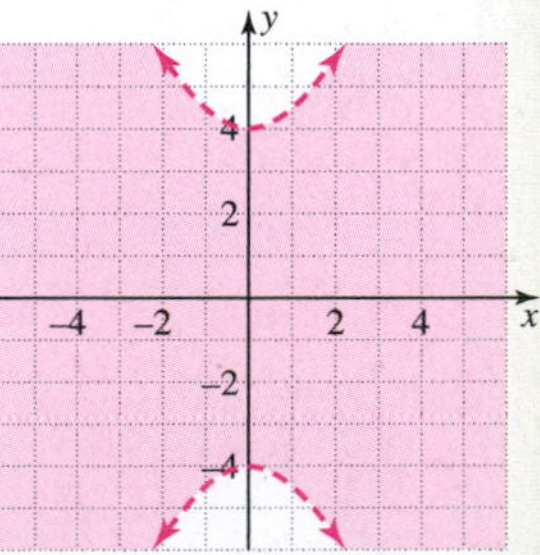

Graphing the two inequalities on the same coordinate system enables us to find the solution. The region where the two graphs intersect is the solution set of the system as shown in purple in Figure 11.12.

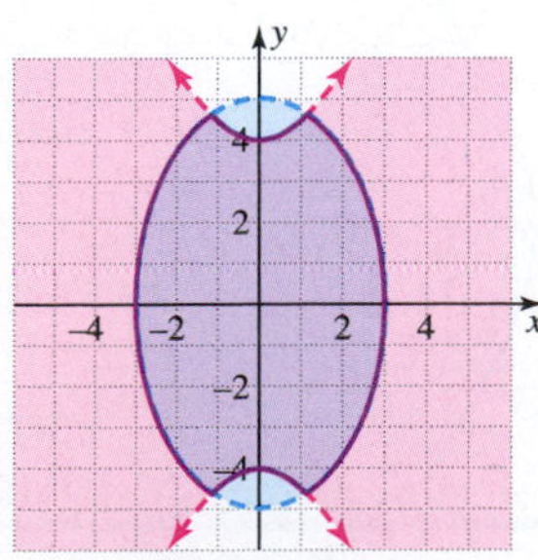

Figure 11.12

Student Check 2 Solve each system of inequalities.

a. $\begin{cases} x + y \geq 2 \\ x^2 + y^2 \leq 9 \end{cases}$ **b.** $\begin{cases} \dfrac{x^2}{4} + \dfrac{y^2}{16} < 1 \\ \dfrac{y^2}{9} - \dfrac{x^2}{25} < 1 \end{cases}$

Objective 3

Troubleshoot common errors.

Troubleshooting Common Errors

Some common errors associated with graphing nonlinear inequalities and solving nonlinear systems of inequalities are shown.

Objective 3 Examples

A problem and an incorrect solution are given. Provide the correct solution and an explanation of the error.

3a. Graph $x > (y - 3)^2 - 4$.

Incorrect Solution

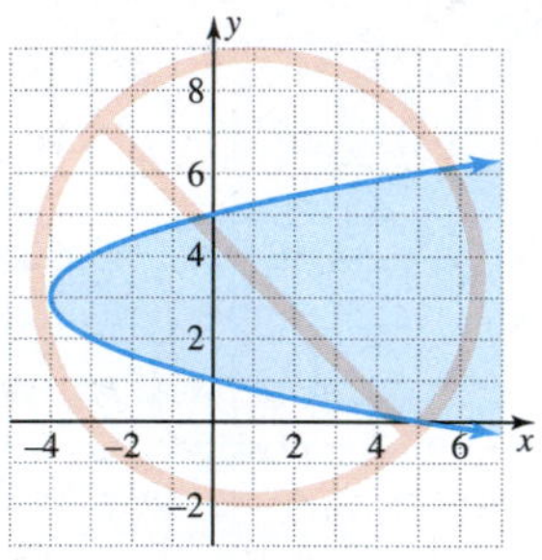

Correct Solution and Explanation

Because the inequality symbol is $>$, the boundary curve is not included in the solution set.

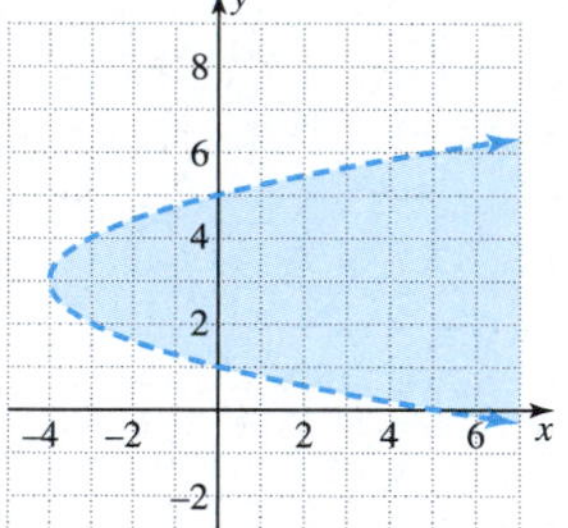

3c. Solve $\begin{cases} y \leq 3x \\ \dfrac{x^2}{4} - \dfrac{y^2}{9} \geq 1 \end{cases}$.

Incorrect Solution

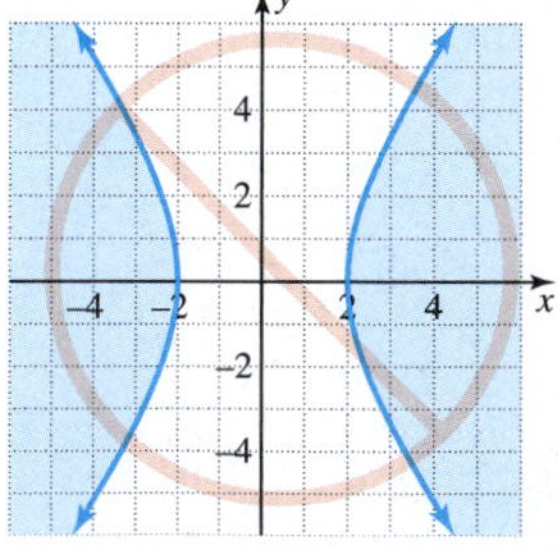

Correct Solution and Explanation

The solution of the linear inequality includes all points on and below the line $y = 3x$. Since the line $y = 3x$ does not intersect the left branch of the hyperbola, none of those points can be included in the solution of the system.

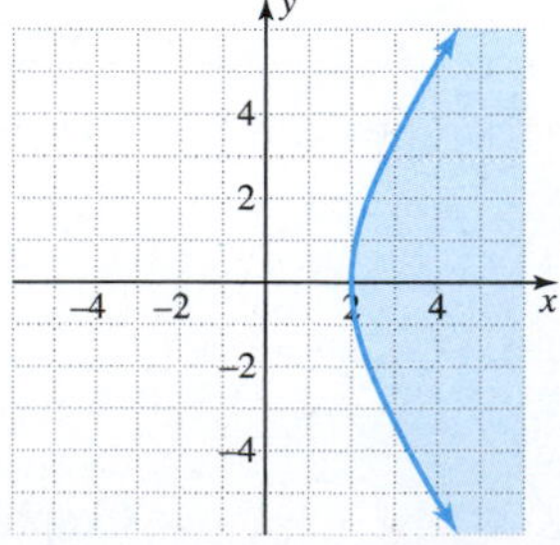

ANSWERS TO STUDENT CHECKS

Student Check 1 **a.**

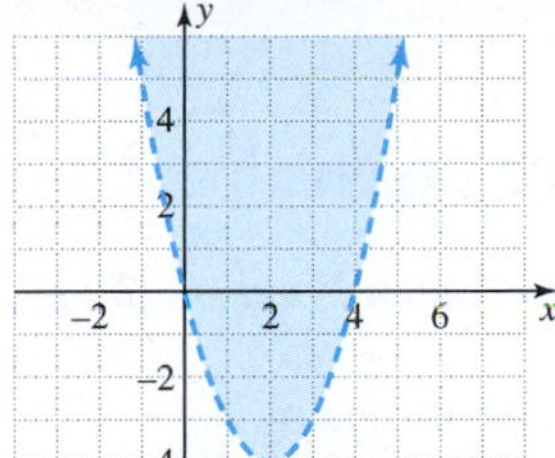

b.

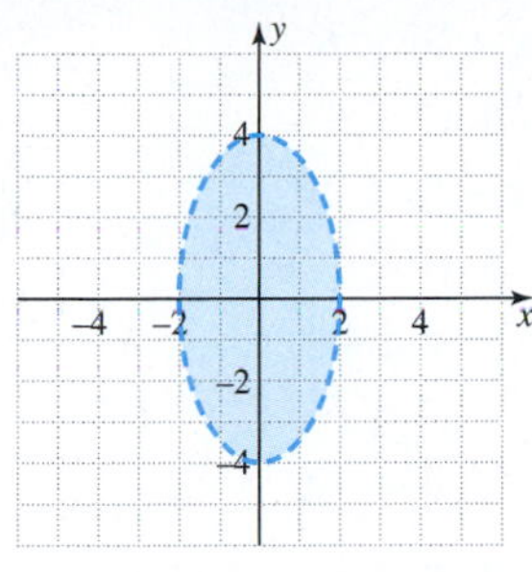

c.

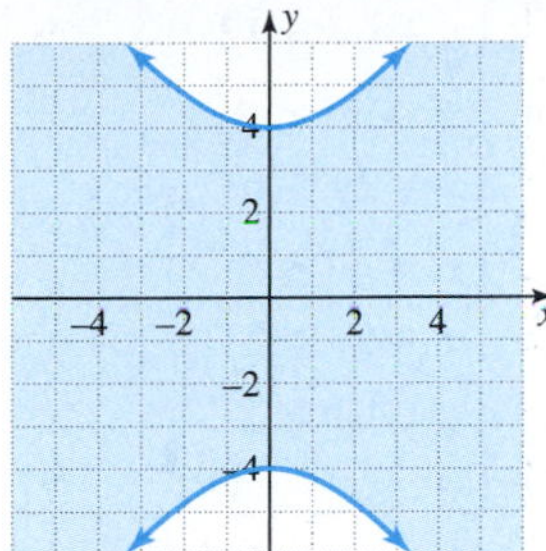

Student Check 2 **a.**

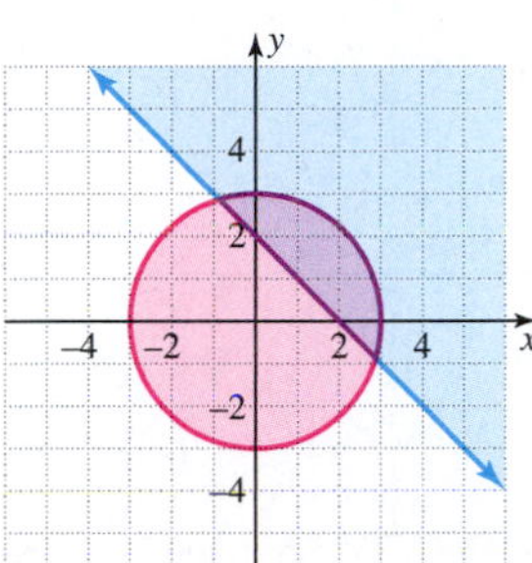

b.

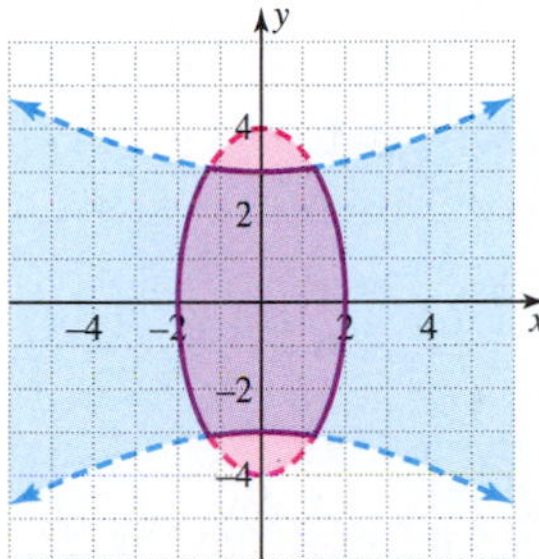

SUMMARY OF KEY CONCEPTS

1. Graphing a nonlinear inequality is similar to graphing a quadratic inequality. We graph the boundary curve, dashed or solid according to the inequality symbol. We test a point in each of the regions, and then shade the region containing the test point that satisfies the inequality. The most difficult part of this process is graphing the boundary curve. Be sure that you are familiar with the basic shapes of all the conic sections.
2. Solving a nonlinear system of inequalities is similar to solving a system of linear inequalities. We graph each inequality and then find the intersection of the graphs to determine the solution set of the system.

SECTION 11.4 EXERCISE SET

Write About It!

Use complete sentences in your answer to each exercise.

1. Explain how to graph the boundary curve when graphing a nonlinear inequality.
2. Explain how to use test points to determine the graph of a nonlinear inequality.
3. Explain how to draw the boundary curve for the inequality $x^2 - 4y^2 \le 16$.
4. Explain how to use test points to determine the graph of the solution set of $x^2 - 4y^2 \le 16$.
5. Explain how to draw the boundary curves for the inequalities in the system $\begin{cases} y > x^2 - 4 \\ \dfrac{x^2}{9} + \dfrac{y^2}{16} < 1 \end{cases}$.
6. Explain how to solve the system $\begin{cases} y > x^2 - 4 \\ \dfrac{x^2}{9} + \dfrac{y^2}{16} < 1 \end{cases}$.

Additional answers can be found in the Instructor Answer Appendix.

Practice Makes Perfect!

Graph each inequality. (*See Objective 1.*)

7. $y \le (x - 2)^2 + 6$
8. $y \le (x - 1)^2 - 4$
9. $y \ge (x + 3)^2 - 4$
10. $y \ge (x - 1)^2 + 4$
11. $y < -2(x - 4)^2 + 1$
12. $y < -(x + 2)^2 + 9$
13. $y > -2(x + 3)^2 + 8$
14. $y > -2(x - 1)^2 + 10$
15. $x^2 + y^2 \le 36$
16. $x^2 + y^2 \le 100$
17. $x^2 + y^2 \ge 16$
18. $x^2 + y^2 \ge 81$
19. $x^2 + y^2 < 100$
20. $x^2 + y^2 < 49$
21. $x^2 + y^2 > 1$
22. $x^2 + y^2 > 144$
23. $\dfrac{x^2}{4} + y^2 \le 1$
24. $x^2 + \dfrac{y^2}{25} \le 1$
25. $\dfrac{x^2}{4} - \dfrac{y^2}{9} < 1$
26. $\dfrac{y^2}{49} - \dfrac{x^2}{4} < 1$
27. $4x^2 - y^2 \ge 64$
28. $y^2 - 4x^2 \ge 100$

Solve each system. (*See Objective 2.*)

29. $\begin{cases} x^2 + y^2 < 16 \\ x + 6y > 12 \end{cases}$

30. $\begin{cases} x^2 + y^2 < 25 \\ 2x + 3y < 6 \end{cases}$

31. $\begin{cases} x^2 + y^2 < 4 \\ x - 2y > 6 \end{cases}$ ∅

32. $\begin{cases} x^2 + y^2 < 1 \\ 2x - 5y > 10 \end{cases}$ ∅

33. $\begin{cases} x^2 + y^2 \le 9 \\ x + 5y \le 8 \end{cases}$

34. $\begin{cases} x^2 + y^2 \le 9 \\ 2x - y \ge 4 \end{cases}$

35. $\begin{cases} x^2 + y^2 \le 4 \\ x - y \le -3 \end{cases}$ ∅

36. $\begin{cases} x^2 + y^2 \le 1 \\ 2x + y \le -4 \end{cases}$ ∅

37. $\begin{cases} x^2 + y^2 \ge 100 \\ 3x - 4y \le 12 \end{cases}$

38. $\begin{cases} x^2 + y^2 \ge 4 \\ x + 2y \le 2 \end{cases}$

39. $\begin{cases} \dfrac{x^2}{4} + \dfrac{y^2}{49} \le 1 \\ \dfrac{y^2}{4} - \dfrac{x^2}{64} \le 1 \end{cases}$

40. $\begin{cases} \dfrac{x^2}{4} + \dfrac{y^2}{49} \le 1 \\ \dfrac{y^2}{4} - \dfrac{x^2}{64} \ge 1 \end{cases}$

41. $\begin{cases} x^2 + \dfrac{y^2}{81} < 1 \\ \dfrac{x^2}{4} - \dfrac{y^2}{144} > 1 \end{cases}$ ∅

42. $\begin{cases} x^2 + \dfrac{y^2}{81} \le 1 \\ \dfrac{x^2}{4} - \dfrac{y^2}{64} \ge 1 \end{cases}$ ∅

43. $\begin{cases} \dfrac{x^2}{16} - \dfrac{y^2}{9} \le 1 \\ \dfrac{x^2}{4} + \dfrac{y^2}{64} \ge 1 \end{cases}$

44. $\begin{cases} \dfrac{y^2}{16} - \dfrac{x^2}{4} \le 1 \\ \dfrac{x^2}{25} + \dfrac{y^2}{81} \ge 1 \end{cases}$

45. $\begin{cases} \dfrac{x^2}{4} - \dfrac{y^2}{25} \le 1 \\ \dfrac{y^2}{25} - \dfrac{x^2}{4} \le 1 \end{cases}$

46. $\begin{cases} \dfrac{x^2}{16} - \dfrac{y^2}{9} \le 1 \\ \dfrac{y^2}{9} - \dfrac{x^2}{16} \le 1 \end{cases}$

Mix 'Em Up!

Graph each inequality or solve each system.

47. $y \ge x^2 - 9$

48. $y \le \dfrac{1}{2}x^2 - 8$

49. $\dfrac{x^2}{5} + \dfrac{y^2}{5} \ge 1$

50. $\dfrac{x^2}{4} + \dfrac{y^2}{2.25} \le 1$

51. $\dfrac{x^2}{2.25} - \dfrac{y^2}{6.25} \ge 1$

52. $\dfrac{y^2}{12.25} - \dfrac{x^2}{1.44} \ge 1$

53. $\begin{cases} y \le (x - 3)^2 + 1 \\ x^2 + y^2 \le 36 \end{cases}$

54. $\begin{cases} y \ge (x - 1)^2 + 4 \\ x^2 + y^2 \le 10 \end{cases}$ ∅

55. $\begin{cases} y < -(x + 2)^2 + 5 \\ x^2 + y^2 > 9 \end{cases}$

56. $\begin{cases} y < -(x - 1)^2 - 1 \\ x^2 + y^2 > 4 \end{cases}$

57. $\begin{cases} \dfrac{x^2}{4} + y^2 \ge 1 \\ x^2 + \dfrac{y^2}{25} \ge 1 \end{cases}$

58. $\begin{cases} \dfrac{y^2}{49} + \dfrac{x^2}{4} < 1 \\ \dfrac{x^2}{25} + \dfrac{y^2}{4} > 1 \end{cases}$

59. $\begin{cases} 4x^2 - y^2 \le 64 \\ 9x^2 + y^2 \ge 9 \end{cases}$

60. $\begin{cases} y^2 - 4x^2 \ge 16 \\ \dfrac{x^2}{25} + \dfrac{y^2}{81} \le 1 \end{cases}$

61. $\begin{cases} x^2 + y^2 < 16 \\ x + y \le 3 \end{cases}$

62. $\begin{cases} x^2 + y^2 \le 9 \\ 3x - 2y \ge 6 \end{cases}$

You Be the Teacher!

Correct each student's errors, if any.

63. Graph the inequality $\dfrac{x^2}{16} + y^2 \le 1$.

Josh's work:

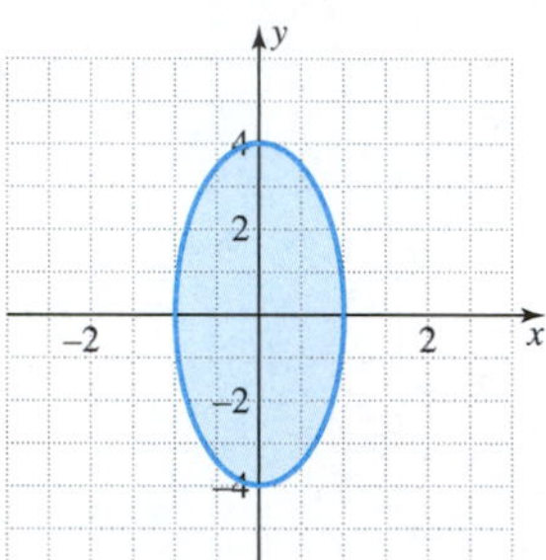

64. Solve the system $\begin{cases} y < -(x - 1)^2 + 1 \\ x^2 + 4y^2 \le 36 \end{cases}$.

Deb's work:

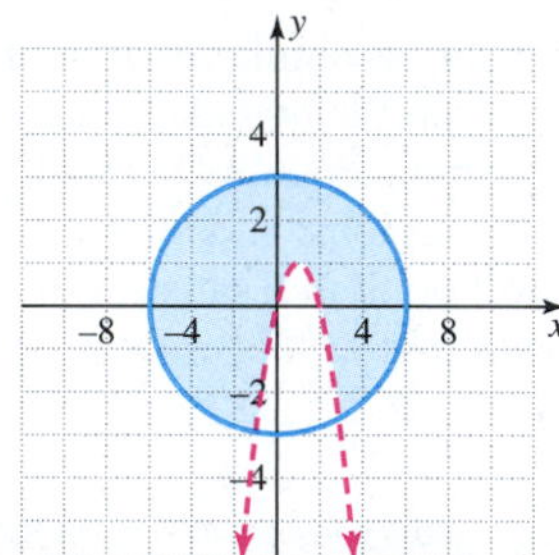

Calculate It!

Use a graphing calculator to solve each inequality.

65. $y > (x + 1)^2 - 4$

66. $\begin{cases} x^2 + 4y^2 \le 4 \\ 2x - 3y \ge 6 \end{cases}$

Think About It!

Give an example of a system of nonlinear inequalities that satisfies the given conditions.

67. The solution set is inside two ellipses, including the boundaries.

68. The solution set is outside a circle and inside an ellipse, including the boundaries.

69. The solution set is outside an ellipse and between two branches of a vertical hyperbola, excluding the boundaries.

GROUP ACTIVITY The Mathematics of Orbits

In the early 1600s, Johannes Kepler discovered that planets orbit the sun in a path that resembles an ellipse with the sun being one of the foci. This is Kepler's first law of planetary motion and is sometimes called the law of ellipses.

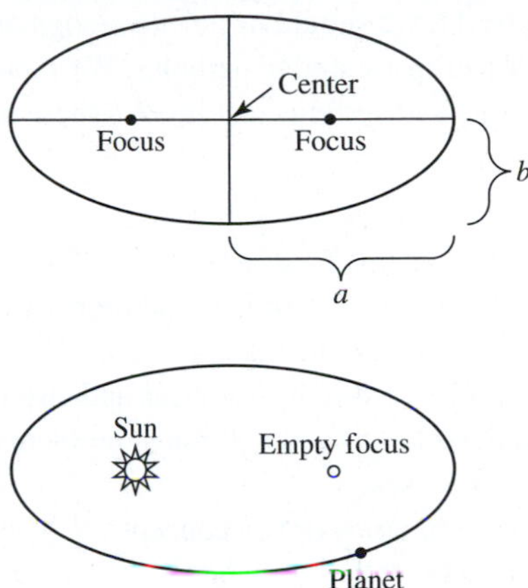

We know that the standard form of an ellipse is $\frac{x^2}{a^2} + \frac{y^2}{b^2} = 1$, where a is the horizontal distance from the origin, or center, to the ellipse, and b is the vertical distance from the origin, or center, to the ellipse. We can also describe ellipses by their eccentricity. The eccentricity, e, is a measure of how circular the ellipse is. If $e = 0$, then the ellipse is actually a circle. As e gets closer to 1, the ellipse is long and narrow. Eccentricity is defined as the ratio of the distance from the origin to a foci and the value of a or b, whichever is larger. We will assume $a > b$ for this activity. So, we define e with the formula

$$e = \frac{\sqrt{a^2 - b^2}}{a}$$

The eccentricity and value of a, in astronomical units (AU), for each planet and the dwarf planet, Pluto, are given in the table. (Source: http://www.windows2universe.org/our_solar_system/planets_table.html)

Planet or Dwarf Planet	Value of a (AU)	Orbital Eccentricity (e)
Mercury	0.3871	0.2056
Venus	0.7233	0.0068
Earth	1.000	0.0167
Mars	1.5273	0.0934
Jupiter	5.2028	0.0483
Saturn	9.5388	0.0560
Uranus	19.1914	0.0461
Neptune	30.0611	0.0097
Pluto	39.5294	0.2482

1. Which orbit is least eccentric, that is, most circular? Venus
2. Which orbit is most eccentric, that is, longest and narrowest? Pluto
3. Solve the eccentricity formula for b. Then use the information in the table to find the value of b for each planet and dwarf planet. Round to four decimal places.
4. Write the equation that models the path of each orbit. Leave the equation in exponential form.
5. Perihelion and aphelion are the nearest and farthest points on the orbit, as shown in the following figure. The perihelion distance is $a(1 - e)$ and the aphelion distance is $a(1 + e)$. Find the perihelion and aphelion distances, in astronomical units and in miles, for three orbits. (Note: 1 AU $\approx$ 92,955,807 mi.)

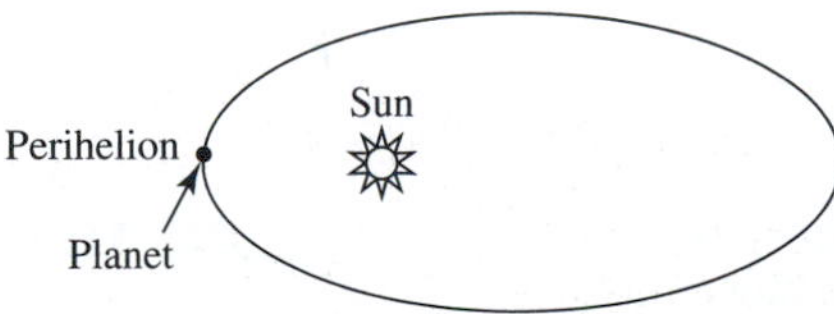

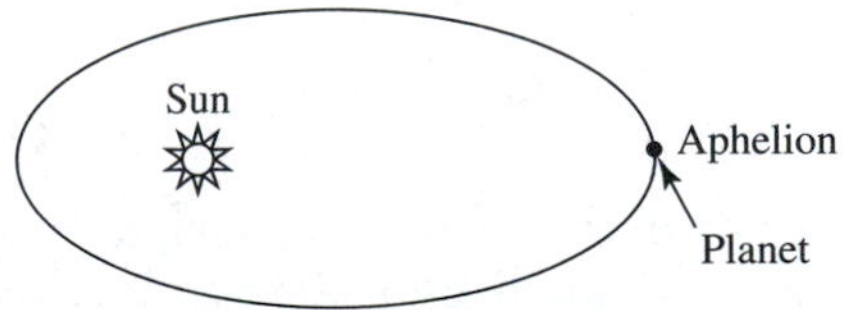

Chapter 11 / REVIEW

Conic Sections and Nonlinear Systems

What's the big idea? Now that we have completed Chapter 11, we should be able to recognize the equations and graphs of the conic sections, which are parabolas, circles, ellipses, and hyperbolas. We should also be able to solve inequalities involving the equations represented by conic sections and systems that involve these nonlinear equations.

The Tools

Listed below are the key terms, skills, formulas, and properties you should know for this chapter.

The page reference is provided if you need additional help with the given topic. The Study Tips will assist in your preparation for an exam.

Study Tips

1. Learn all of the terms, formulas, and properties. Make flash cards and have someone quiz you.
2. Rework problems from the exercises and also the ones you worked in class. Work additional problems from the review exercises.
3. Review the summaries of key concepts.
4. Work the chapter test.
5. Be sure to review the online resources for additional study materials.

Terms

Asymptotes 819
Center of a circle 811
Center of a hyperbola 819
Center of an ellipse 817
Circle 811
Conic sections 808
Ellipse 817
Foci of a hyperbola 819
Foci of an ellipse 817
Hyperbola 819
Nonlinear inequality 831
Nonlinear system of equations 824
Nonlinear system of inequalities 834
Parabola 809
Radius 811
Solution of a nonlinear system 824

Formulas and Properties

- Standard form of a circle 811
- Standard form of a hyperbola 819
- Standard form of a parabola 809
- Standard form of an ellipse 817

CHAPTER 11 / SUMMARY

How well do you know this chapter? Complete the following questions to find out. Take a look back at the section if you need help.

SECTION 11.1 The Parabola and the Circle

1. The equation $x = a(y - k)^2 + h$ is a(n) horizontal parabola with vertex (h, k). The parabola opens right if $a > 0$ and opens left if $a < 0$.
2. The equation $(x - h)^2 + (y - k)^2 = r^2$ is the standard form of a(n) circle with center (h, k) and radius r.

SECTION 11.2 The Ellipse and the Hyperbola

3. The equation $\dfrac{(x-h)^2}{a^2} + \dfrac{(y-k)^2}{b^2} = 1$ is the standard form of a(n) ellipse with center (h, k). If $a > b$, the ellipse is horizontal. If $a < b$, the ellipse is vertical.
4. The x- and y-intercepts of an ellipse with center (0, 0) are $(a, 0)$, $(-a, 0)$, $(0, b)$, and $(0, -b)$.
5. A hyperbola centered at the origin is represented by $\frac{x^2}{a^2} - \frac{y^2}{b^2} = 1$ or $\frac{y^2}{b^2} - \frac{x^2}{a^2} = 1$. If the x^2-term is positive, the

hyperbola opens horizontally. If the y^2-term is positive, the hyperbola opens vertically.

6. A(n) rectangle is constructed to graph the hyperbola. The vertices are (a, b), $(a, -b)$, $(-a, b)$, and $(-a, -b)$. The diagonals through this rectangle are called asymptotes.

SECTION 11.3 Solving Nonlinear Systems of Equations

7. A nonlinear system can be solved by substitution or elimination.

8. Solutions of nonlinear systems can be checked algebraically or graphically.

9. The solution set of a nonlinear system may consist of 0 points, 1 point, 2 points, or more points.

SECTION 11.4 Solving Nonlinear Inequalities and Systems of Inequalities

10. A nonlinear inequality can be graphed using test points. We first graph the boundary curve and then test a point in each region.

11. To solve a nonlinear system of inequalities, we graph each inequality and then find the intersection of the graphs of the inequalities in the system.

CHAPTER 11 / REVIEW EXERCISES

SECTION 11.1

Graph each parabola by plotting the vertex and two additional points. State the axis of symmetry. (*See Objective 1.*)

1. $x = (y + 3)^2 - 4$
2. $x = -(y - 3)^2 + 9$
3. $x = 2(y - 1)^2 - 10$
4. $x = -3(y + 4)^2 + 1$
5. $x = \frac{1}{2}(y - 2)^2$
6. $x = -\frac{1}{5}(y + 3)^2$
7. $x = 2y^2 + 7$
8. $x = 2y^2 - 10$
9. $x = y^2 - 4y - 5$
10. $x = -y^2 + y + 1$
11. $x = -y^2 + 6y + 7$
12. $x = y^2 - 2y - 24$

Graph each circle. Identify the center, the radius, and four key points. (*See Objective 2.*)

13. $x^2 + y^2 = 49$
14. $(x + 2)^2 + (y - 8)^2 = 25$
15. $(x - 4)^2 + (y + 5)^2 = 9$
16. $(x + 3)^2 + y^2 = 16$

Write the equation of each circle given its center and radius. (*See Objective 3.*)

17. center $= (-1, -3)$ and radius $= 1$
18. center $= (5, -2)$ and radius $= 4$
19. center $= (-6, 7)$ and radius $= \sqrt{10}$
20. center $= \left(0, \frac{3}{5}\right)$ and radius $= \frac{3}{2}$
21. center $= (3.5, 1.5)$ and radius $= 4.5$
22. center $= (-2.5, 0.5)$ and radius $= 1.5$

Write the equation of each circle in standard form and identify its center and radius. (*See Objective 4.*)

23. $x^2 + y^2 - 2x - 14y = 150$
24. $x^2 + y^2 - 2x - 4y = 54$
25. $x^2 + y^2 - 20y = 116$
26. $x^2 + y^2 - 2x = 26$
27. $x^2 + y^2 - x + 3y = \frac{1}{2}$
28. $x^2 + y^2 - 0.8x + 1.2y = 0.29$

SECTION 11.2

Identify each equation as a parabola, circle, ellipse, or hyperbola. Sketch the graph of each equation. (*See Objectives 1 and 2.*)

29. $x^2 + 49y^2 = 49$
30. $x = -y^2 + 5y + 6$
31. $x^2 + y^2 = 10y + 11$
32. $x^2 = 4y^2 + 25$
33. $x^2 - \frac{y^2}{9} = 1$
34. $x^2 = 4y^2 - 25$
35. $y^2 = 4x^2 + 1$
36. $\frac{x^2}{3.24} - y^2 = 1$

SECTION 11.3

Solve each system by substitution or elimination. (*See Objectives 1 and 2.*)

37. $\begin{cases} y = (x - 5)^2 \\ 3x + y = 25 \end{cases}$
38. $\begin{cases} x = \frac{2}{3}y^2 + 2 \\ y = \sqrt{2x - 5} \end{cases}$
39. $\begin{cases} x^2 + y^2 = 25 \\ 4x - 5y = 31 \end{cases}$
40. $\begin{cases} y = x^2 + 6 \\ 5x + y = 2 \end{cases}$
41. $\begin{cases} y = x^2 - 12 \\ 2x + y = -9 \end{cases}$
42. $\begin{cases} 3x^2 + 2y^2 = 315 \\ 3y^2 - x^2 = 27 \end{cases}$
43. $\begin{cases} x^2 + y^2 = 29 \\ 5x^2 + 4y^2 = 120 \end{cases}$
44. $\begin{cases} 2x^2 + 3y^2 = 59 \\ x^2 - y^2 = 7 \end{cases}$
45. $\begin{cases} y = 11x^2 - 3 \\ y = 3x^2 + 2x \end{cases}$
46. $\begin{cases} y = 6x^2 - 3 \\ y = x^2 + 2x \end{cases}$
47. $\begin{cases} xy = 1 \\ 4x - 6y = 5 \end{cases}$
48. $\begin{cases} xy = 1 \\ 3x - 14y = 1 \end{cases}$

SECTION 11.4

Graph each inequality. (*See Objective 1.*)

49. $y \geq (x + 3)^2 - 7$
50. $y \leq (x - 6)^2 + 2$
51. $\frac{x^2}{25} + \frac{y^2}{25} > 1$
52. $\frac{x^2}{16} + \frac{y^2}{49} \leq 1$
53. $\frac{x^2}{9} - \frac{y^2}{16} > 1$
54. $\frac{y^2}{9} - x^2 < 1$

Solve each system. (*See Objective 2.*)

55. $\begin{cases} y \geq x^2 \\ x^2 + y^2 \leq 2 \end{cases}$

56. $\begin{cases} y < x^2 - 4 \\ x^2 + y^2 > 10 \end{cases}$

57. $\begin{cases} \dfrac{x^2}{16} + \dfrac{y^2}{9} \geq 1 \\ \dfrac{x^2}{9} + \dfrac{y^2}{4} \leq 1 \end{cases}$

58. $\begin{cases} 4x^2 - y^2 \leq 8 \\ x^2 + 3y^2 \geq 28 \end{cases}$

CHAPTER 11 TEST / CONIC SECTIONS AND NONLINEAR SYSTEMS

1. The vertex of $x = -2(y + 4)^2 - 5$ is
   a. $(-4, -5)$ b. $(4, -5)$
   c. $(-5, 4)$ (d.) $(-5, -4)$
2. The radius of $x^2 + 2x + y^2 - 4y = 8$ is
   a. 13 (b.) $\sqrt{13}$
   c. 8 d. $\sqrt{8}$
3. The equation $x^2 + 4y^2 = 4$ represents the graph of a(n)
   a. parabola b. circle
   (c.) ellipse d. hyperbola
4. The graph of $\dfrac{y^2}{9} - \dfrac{x^2}{4} = 1$ has
   a. x-intercepts of $(-2, 0)$ and $(2, 0)$.
   b. x-intercepts of $(-3, 0)$ and $(3, 0)$.
   c. y-intercepts of $(0, -2)$ and $(0, 2)$.
   (d.) y-intercepts of $(0, -3)$ and $(0, 3)$.

Graph each equation and label any key points of the graph.

5. $x = 2(y + 1)^2 - 8$
6. $x^2 + y^2 = 25$
7. $x^2 + y^2 - 4x + 6y = 3$
8. $9x^2 + y^2 = 9$
9. $\dfrac{(x + 2)^2}{25} + \dfrac{(y - 1)^2}{9} = 1$
10. $9x^2 - y^2 = 9$
11. $\dfrac{y^2}{4} - \dfrac{x^2}{9} = 1$

Graph each inequality.

12. $x > y^2 - 9$
13. $x^2 - 4y^2 \leq 16$

Solve each system of nonlinear equations.

14. $\begin{cases} x^2 + y^2 = 5 \\ x + y = 3 \end{cases}$ $\{(2, 1), (1, 2)\}$

15. $\begin{cases} y = \sqrt{x + 6} \\ x + 2y^2 = 6 \end{cases}$ $\{(-2, 2)\}$

16. $\begin{cases} x^2 + 3y^2 = 19 \\ x^2 - 3y^2 = 13 \end{cases}$ $\{(-4, -1), (-4, 1), (4, -1), (4, 1)\}$

Solve each system of nonlinear inequalities.

17. $\begin{cases} x + 3y \leq 6 \\ x^2 + \dfrac{y^2}{9} \leq 1 \end{cases}$

18. $\begin{cases} x^2 + y^2 < 16 \\ \dfrac{x^2}{4} - \dfrac{y^2}{9} > 1 \end{cases}$

CUMULATIVE REVIEW EXERCISES / CHAPTERS 1–11

Evaluate each expression for the given values. (*Section 1.2, Objective 5*)

1. $-x^3 + 2x^2 - x + 4$ for $x = 3$ -8
2. $12x - 21y$ for $x = 2, y = -1$ 45

Simply each expression. (*Section 1.3, Objective 3*)

3. $-2x(8 - 3x) - 9x - 2x(x - 6)$ $4x^2 - 13x$
4. $\dfrac{3}{5}x + \dfrac{1}{4} - \dfrac{1}{2}x + \dfrac{1}{6}$ $\dfrac{1}{10}x + \dfrac{5}{12}$

Find the measure of the angles in each figure. (*Section 2.3, Objective 1*)

5. 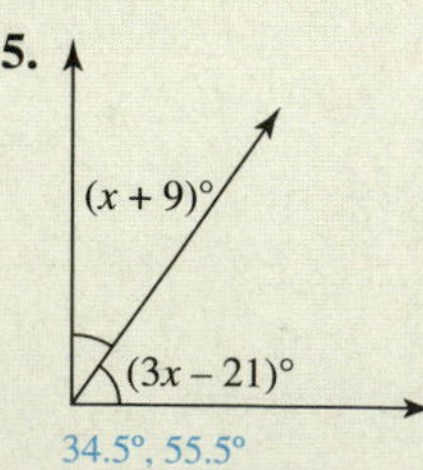

34.5°, 55.5°

6. 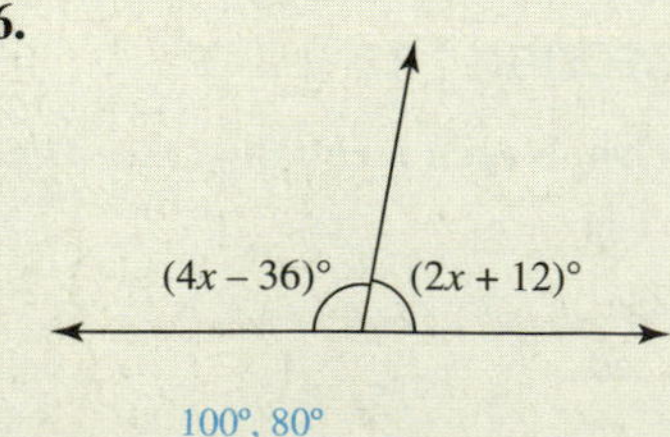

100°, 80°

Solve each problem. (*Section 2.6, Objective 3*)

7. Suppose the digital scale reflects a weight (in pounds) with an absolute error of 0.3 lb. If someone weighs 104.6 lb according to the scale, find the exact weight of the person. The exact weight of the person is between 104.3 lb and 104.9 lb.
8. Suppose a countertop food scale reflects a weight (in ounces) with an absolute error of 0.1 oz. If a piece of cheese weighs 4.5 oz according to the scale, find the exact weight of the piece of cheese. The exact weight of the piece of cheese is between 4.4 oz and 4.6 oz.

Find the midpoint of each line segment formed by the ordered pairs. (*Section 3.1, Objective 5*)

9. $(-12, 7)$ and $(-8, -7)$ $(-10, 0)$
10. $(-3.7, 2.1)$ and $(9.3, -5.7)$ $(2.8, -1.8)$

Express each relation as a graph. (*Section 3.2, Objective 1*)

11. The price of Apple Inc. stock at the end of March of the specified year is given in the table. (Source: http://www.dailyfinance.com/)

Year	2007	2008	2009	2010	2011
Price per share (in dollars)	92.91	143.50	105.12	235.00	348.51

12. The price of Exxon Mobil Corp. stock at the end of March of the specified year is given in the table.
(Source: http://www.dailyfinance.com)

Year	2007	2008	2009	2010	2011
Price per share (in dollars)	75.45	84.58	68.10	67.77	84.13

Find the requested information. (*Section 3.3, Objective 3*)

13. Find $f(x-3)$ if $f(x) = 2x^2 - x$. $2x^2 - 13x + 21$

14. Find $f(x+1)$ if $f(x) = 4 - x^2$. $-x^2 - 2x + 3$

Find the slope and y-intercept of each line from its equation. Write the y-intercept as an ordered pair. Graph the line using its slope and y-intercept. Label at least two points on the graph. (*Section 4.3, Objectives 1 and 2*)

15. $2x + 5y = 20$

16. $y + 3 = 0$

Write the equation of each line that satisfies the given conditions. Express your answer in slope-intercept form. (*Section 4.4, Objective 4*)

17. passes through $(2, -6)$, parallel to $y = 2$ $y = -6$

18. passes through $(-3, 1)$, perpendicular to $x + 2y = 1$ $y = 2x + 7$

Graph each linear inequality in two variables. (*Section 4.5, Objective 2*)

19. $x + 4y \le 12$

20. $3x - y \le 6$

Determine how the lines relate, the number of solutions of the system, and the type of system without graphing. (*Section 5.1, Objective 4*)

21. $\begin{cases} y = 5x + 11 \\ x + 4y = 2 \end{cases}$

22. $\begin{cases} 6x + 8y = 5 \\ 3x + 4y = 1 \end{cases}$

Solve each problem. (*Section 5.3, Objective 1*)

23. 550 tickets were sold at a benefit concert at a college. Adult tickets cost \$12 and student tickets cost \$5. If a total of \$4150 was collected for the tickets, how many adult tickets and student tickets were sold?
200 adult tickets and 350 student tickets

24. 1500 tickets were sold at a movie theater. Adult tickets cost \$9.50 and student tickets cost \$5.50. If a total of \$12,170 was collected for the tickets, how many adult tickets and student tickets were sold? 980 adult tickets and 520 student tickets

Solve each system of linear equations in three variables using elimination. (*Section 5.4, Objective 1*)

25. $\begin{cases} x - y - 3z = -15 \\ -x + 2y + z = 8 \\ 2x + y + 3z = 9 \end{cases}$
$\{(-2, 1, 4)\}$

26. $\begin{cases} 2x + 5y - z = 17 \\ x - y + 8z = -7 \\ x - 3y + 6z = -9 \end{cases}$
$\{(3, 2, -1)\}$

Evaluate and simplify each expression. Express all answers with positive exponents. (*Section 6.1, Objectives 1–4*)

27. $4x^0 - 5$ -1

28. $-7(5x)^0$ -7

29. $(3a^4b^{-7})(-5a^{-3}b^7)$ $-15a$

30. $(12x^8y^5)(2x^{-6}y^{-5})$ $24x^2$

Simplify each expression. (*Section 6.4, Objectives 1–5*)

31. $(2x^2 - 3x + 5) - (4x^2 + 7x + 1)$ $-2x^2 - 10x + 4$

32. $2(5a - 9)(2a + 3)$
$20a^2 - 6a - 54$

33. $3(5c - d)^2$
$75c^2 - 30cd + 3d^2$

34. $(x - 2y)(x^2 + 2xy + 4y^2)$ $x^3 - 8y^3$

Factor each trinomial. (*Section 6.6, Objectives 1–4*)

35. $(x - 2)^2 - 5(x - 2) - 14$ $x(x - 9)$

36. $3(y + 4)^2 + 4(y + 4) - 4$ $(3y + 10)(y + 6)$

Factor completely. (*Section 6.7, Objectives 1–3*)

37. $3x^4 + 2x^3 - 3x - 2$
$(3x + 2)(x - 1)(x^2 + x + 1)$

38. $4y^4 - 5y^3 + 4y - 5$
$(4y - 5)(y + 1)(y^2 - y + 1)$

Solve each polynomial equation. (*Section 6.8, Objective 1*)

39. $18x^3 - 8x = 0$
$\left\{-\frac{2}{3}, 0, \frac{2}{3}\right\}$

40. $12x^3 - 12x^2 + 3x = 0$
$\left\{0, \frac{1}{2}\right\}$

Simplify each rational expression. (*Section 7.1, Objective 3*)

41. $\dfrac{2x^2 + 8x + 32}{x^3 - 64}$ $\dfrac{2}{x - 4}$

42. $\dfrac{6x^2 - 18x + 54}{x^3 + 27}$ $\dfrac{6}{x + 3}$

Add or subtract the rational expressions. (*Section 7.3, Objectives 1–3*)

43. $\dfrac{5x}{9x^2 - 6x + 1} - \dfrac{x}{3x^2 + 5x - 2}$ $\dfrac{2x^2 + 11x}{(3x - 1)^2(x + 2)}$

44. $\dfrac{6}{x^2 + 4x - 5} - \dfrac{2}{x^2 - 25}$ $\dfrac{4(x - 7)}{(x + 5)(x - 5)(x - 1)}$

Evaluate each rational function at the given value. (*Section 7.4, Objective 3*)

45. $f(x) = \dfrac{x + 1}{x - 2}, f\left(\dfrac{7}{5}\right)$ -4

46. $f(x) = \dfrac{2x + 3}{x + 1}, f\left(-\dfrac{5}{3}\right)$ $\dfrac{1}{2}$

Solve each rational equation. (*Section 7.5, Objective 1*)

47. $\dfrac{4}{x^2 - 2x} - \dfrac{7}{2 - x} = \dfrac{5}{x}$ $\{-7\}$

48. $\dfrac{6}{y^2 - 3y} + \dfrac{5}{3 - y} = \dfrac{4}{y}$ $\{2\}$

Simplify each radical expression. Assume all variables represent positive real numbers. (*Section 8.3, Objectives 1 and 2*)

49. $\dfrac{\sqrt{56x^3y^9}}{\sqrt{2x^4y^{-3}}}$ $\dfrac{2y^6\sqrt{7x}}{x}$

50. $\dfrac{\sqrt{100x^{-3}y^6}}{\sqrt{5x^2y^{-2}}}$ $\dfrac{2y^4\sqrt{5x}}{x^3}$

Perform the indicated operation and write each answer in simplest radical form. (*Section 8.4, Objectives 1 and 2*)

51. $(7 - 2\sqrt{3})(5 + 6\sqrt{3})$ $-1 + 32\sqrt{3}$

52. $(5 - 3\sqrt{2})^2$ $43 - 30\sqrt{2}$

Perform the indicated operation and write each answer in standard form. (*Section 8.7, Objectives 2–4*)

53. $\dfrac{2 + i}{3 - i}$ $\dfrac{1}{2} + \dfrac{1}{2}i$

54. $\dfrac{1 + 2i}{4 + 3i}$ $\dfrac{2}{5} + \dfrac{1}{5}i$

Graph each function and then use the graph to solve the equation $f(x) = 0$. (*Section 9.1, Objective 5*)

55. $f(x) = \dfrac{1}{3}(x - 1)^2 - 3$

56. $f(x) = -\dfrac{1}{2}(x + 1)^2 + 2$

Solve each equation by completing the square. (*Section 9.2, Objectives 1–4*)

57. $2x^2 + 2x - 1 = 0$ $\left\{\dfrac{-1 \pm \sqrt{3}}{2}\right\}$

58. $3y^2 - 2y + 1 = 0$ $\left\{\dfrac{1 \pm i\sqrt{2}}{3}\right\}$

Use the discriminant to determine the number and type of solutions of each equation. (*Section 9.3, Objective 2*)

59. $x^2 - 3x + 6 = 0$ Two complex, nonreal solutions

60. $2y^2 - 7y - 5 = 0$ Two irrational solutions

61. $x^2 - 10x + 25 = 0$ One rational solution

62. $4y^2 + 9y + 2 = 0$ Two rational solutions

Solve each equation. (*Section 9.4, Objectives 1–4*)

63. $x^4 + 5x^2 - 36 = 0$ $\{\pm 3i, \pm 2\}$

64. $x - 7\sqrt{x} + 10 = 0$ $\{4, 25\}$

65. $x^{2/5} - 2x^{1/5} - 3 = 0$ $\{-1, 243\}$

66. $x^{1/2} + 3x^{1/4} - 4 = 0$ $\{1\}$

Solve each inequality using the boundary number method. Write the solution set in interval notation. (*Section 9.6, Objectives 1–3*)

67. $x^2 - 4 < 0$ $(-2, 2)$

68. $x^2 + x > 20$ $(-\infty, -5) \cup (4, \infty)$

Find the sum, difference, product, and quotient of the functions. State the domain of each combined function. Also, state the restriction for the quotient function. (*Section 10.1, Objective 1*)

69. $f(x) = x^2 + 4$ and $g(x) = x^2 - 2x - 3$

70. $f(x) = \sqrt{x + 1}$ and $g(x) = \sqrt{x - 3}$

Use the graphs of $f(x)$ and $g(x)$ to find the values of each combined function. (*Section 10.1, Objectives 1–3*)

$f(x)$ $g(x)$

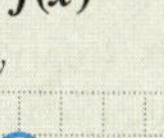

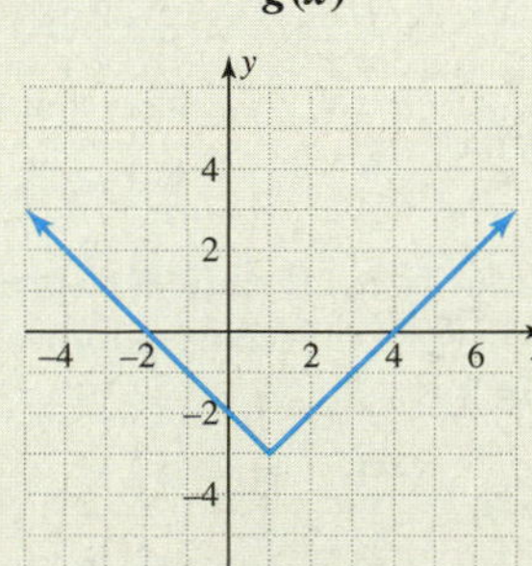

71. $(f + g)(1)$ 1

72. $(f \cdot g)(2)$ −6

73. $\left(\dfrac{f}{g}\right)(-1)$ 0

74. $(f \circ g)(3)$ 0

For the given functions $f(x)$ and $g(x)$, find the composition functions $(f \circ g)(x)$ and $(g \circ f)(x)$. State the domain of each function. (*Section 10.1, Objective 3*)

75. $f(x) = 5x - 1$ and $g(x) = 2 - 3x$

76. $f(x) = 4x - 10$ and $g(x) = 6x$

Find the equation of the inverse of each function. (*Section 10.2, Objectives 3 and 4*)

77. $f(x) = 8x^3 - 5$

78. $f(x) = x^5 + 32$

Evaluate each function given the information. (*Section 10.2, Objective 5*)

79. If $f^{-1}(-1) = 5$, what is $f(5)$? −1

80. If $f(2.5) = 0$, what is $f^{-1}(0)$? 2.5

Solve each exponential equation. (*Section 10.3, Objective 2*)

81. $\left(\dfrac{1}{2}\right)^{2x-5} = 8^x$ $\{1\}$

82. $\left(\dfrac{1}{3}\right)^{2x-5} = 9^{4x}$ $\left\{\dfrac{1}{2}\right\}$

Simplify each expression. (*Section 10.4, Objectives 1 and 2*)

83. $\log_5 \dfrac{1}{625}$ −4

84. $\log_{1/2} 4$ −2

Solve each logarithmic equation for x. (*Section 10.4, Objective 3*)

85. $\log_{0.5} x = 2$ $\{0.25\}$

86. $\log_x 121 = 2$ $\{11\}$

Rewrite each logarithmic expression. If the expression is a single logarithm, write the expression as a combination of logarithms. If the expression is a combination of logarithms, write the expression as a single logarithm. (*Section 10.5, Objectives 1–4*)

87. $\ln[\sqrt{x}(x + 2)^3]$

88. $\log_7\left(\dfrac{y^2}{\sqrt{y - 3}}\right)$

89. $2\log_{10} x + 3\log_{10} y - \log_{10} z$

90. $\log_2 x - 5\log_2 y - \log_2 z$

Evaluate each expression. (*Section 10.5, Objective 5*)

91. $\log_5 5^{2x+6}$ $2x + 6$

92. $\log_{10} 10^{x-3}$ $x - 3$

93. $e^{\ln(x+1)}$ $x + 1$

94. $1.5^{\log_{1.5}(4x)}$ $4x$

Solve each equation for x. Give an exact solution and approximate the solution to two decimal places, if necessary. (*Section 10.6, Objectives 3 and 4*)

95. $\log(2x + 9) = 1.8$

96. $\ln(3x - 5) = 4.2$

Solve each equation. Give an exact solution and an approximation of the solution to two decimal places. (*Section 10.7, Objectives 1 and 2*)

97. $6(1.035)^{4x} = 42$

98. $12(2.44)^{3x} = 96$

99. $\log_4(3x - 11) = 3$ $\{25\}$

100. $\log(3x + 52) = 2$ $\{16\}$

101. $\log x + \log(3x + 2) = 0$ $\left\{\dfrac{1}{3}\right\}$

102. $\log x - \log(3x + 2) = 0$ $\varnothing$

Graph each parabola by plotting the vertex and two additional points. State the axis of symmetry. (*Section 11.1, Objective 1*)

103. $x = y^2 + 6y - 27$

104. $x = -y^2 - 4y + 21$

Graph each circle. Identify the center, the radius, and four key points. (*Section 11.1, Objective 2*)

105. $x^2 + y^2 = 100$

106. $x^2 + y^2 = 36$

107. $x^2 + (y + 1)^2 = 4$

108. $(x + 2)^2 + y^2 = 1$

Write the equation of each circle given its center and radius. (*Section 10.1, Objective 3*)

109. center = $(5, -6)$ and radius = 9 $(x - 5)^2 + (y + 6)^2 = 81$

110. center = $(3, -1)$ and radius = $\sqrt{7}$ $(x - 3)^2 + (y + 1)^2 = 7$

Write the equation of each circle in standard form and identify the center and radius. (*Section 10.1, Objective 4*)

111. $x^2 + y^2 - 4x + 2y = 4$

112. $x^2 + y^2 + 10x - 4y + 28 = 0$

Graph each ellipse. Label four key points and the center on the graph. (*Section 11.2, Objective 1*)

113. $25x^2 + 9y^2 = 225$

114. $81x^2 + 4y^2 = 324$

Graph each hyperbola. (*Section 11.2, Objective 2*)

115. $9x^2 - y^2 = 9$

116. $y^2 - 16x^2 = 16$

Solve each system by substitution or elimination. (*Section 11.3, Objectives 1 and 2*)

117. $\begin{cases} y = \dfrac{2}{3}x^2 - 4 \\ x = \sqrt{3y + 9} \end{cases}$

$\{(-\sqrt{3}, -2), (\sqrt{3}, -2)\}$

118. $\begin{cases} y = x^2 - 8 \\ x = \sqrt{2y + 7} \end{cases}$

$\{(-3, 1), (3, 1)\}$

119. $\begin{cases} y = (x - 1)^2 \\ -4x + y = 1 \end{cases}$

$\{(0, 1), (6, 25)\}$

120. $\begin{cases} y = (x - 6)^2 \\ 8x + y = 36 \end{cases}$

$\{(0, 36), (4, 4)\}$

121. $\begin{cases} x^2 + 2y^2 = 32 \\ x^2 - 2y^2 = 8 \end{cases}$

122. $\begin{cases} x^2 + y^2 = 12 \\ 4x^2 + y^2 = 36 \end{cases}$

123. $\begin{cases} y = 4x^2 - x \\ y = 2x^2 + 3 \end{cases}$

$\left\{\left(\dfrac{3}{2}, \dfrac{15}{2}\right), (-1, 5)\right\}$

124. $\begin{cases} y = 2x^2 + x - 7 \\ y = x^2 - x + 8 \end{cases}$

$\{(-5, 38), (3, 14)\}$

Graph each inequality. (*Section 11.4, Objective 1*)

125. $\dfrac{x^2}{64} + \dfrac{y^2}{9} \geq 1$

126. $\dfrac{x^2}{4} + \dfrac{y^2}{81} \geq 1$

127. $\dfrac{x^2}{81} - \dfrac{y^2}{4} > 1$

128. $\dfrac{y^2}{9} - x^2 > 1$

Solve each system. (*Section 11.4, Objective 2*)

129. $\begin{cases} x^2 + y^2 < 64 \\ x - 4y < 12 \end{cases}$

130. $\begin{cases} x^2 + y^2 > 81 \\ 3x + 2y > 18 \end{cases}$

131. $\begin{cases} x^2 + \dfrac{y^2}{2.25} \leq 1 \\ \dfrac{y^2}{0.25} - x^2 \leq 1 \end{cases}$

132. $\begin{cases} x^2 - \dfrac{y^2}{0.81} \geq 1 \\ \dfrac{x^2}{6.25} + \dfrac{y^2}{1.96} \geq 1 \end{cases}$

Instructor Answer Appendix

CHAPTER 1

Section 1.1

1. Answers vary. Yes, the set of natural numbers is a subset of the set of whole numbers.
2. Answers vary. No, 0 is a whole number but not a natural number.
3. Answers vary. A rational number is a number that can be written as the quotient, or ratio, of two integers, providing the integer in the denominator is not equal to zero.
4. Answers vary. An irrational number is a number whose decimal value does not terminate or repeat in a pattern.
5. Answers vary. Two numbers are opposites if they have the same distance from 0 on a number line but lie on opposite sides of 0.
6. Answers vary. Since absolute value refers to distance, the absolute value of opposite numbers is the same.

49.–57.

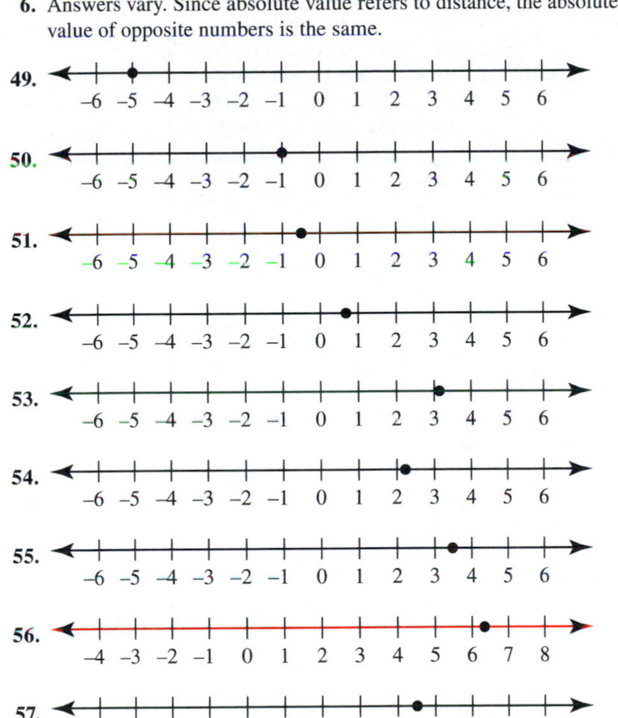

58.–62.

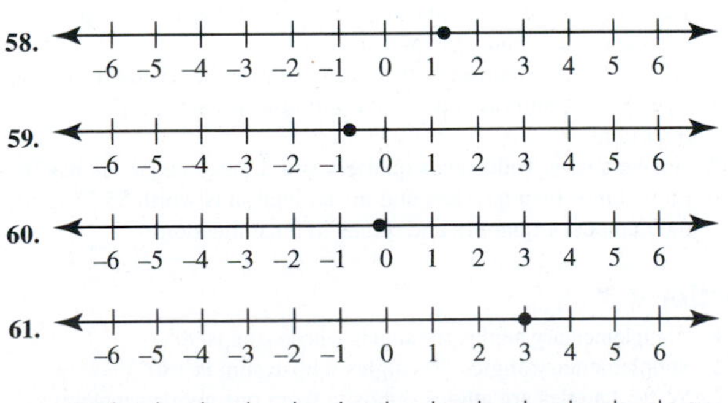

63.

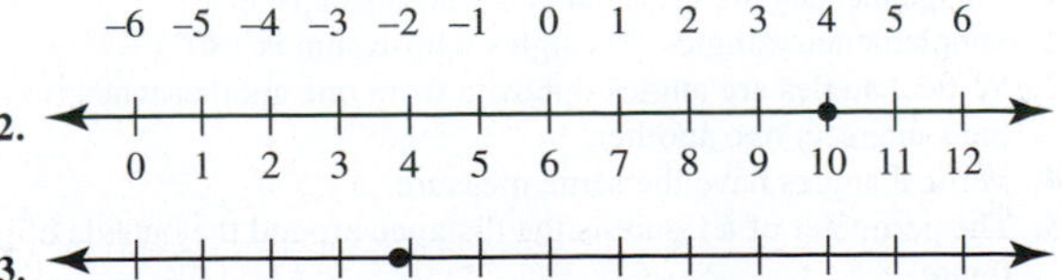

64.–143.

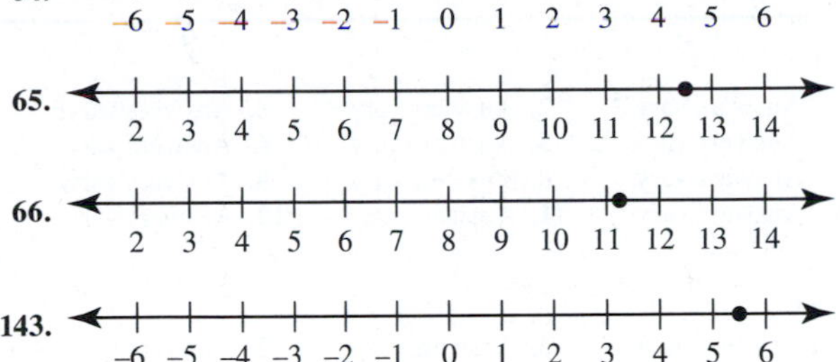

144.–145.

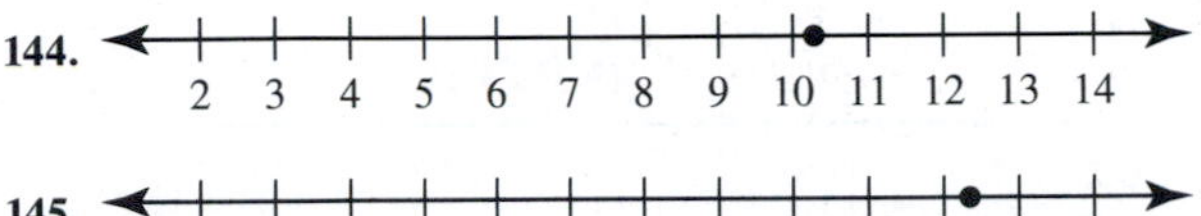

146.

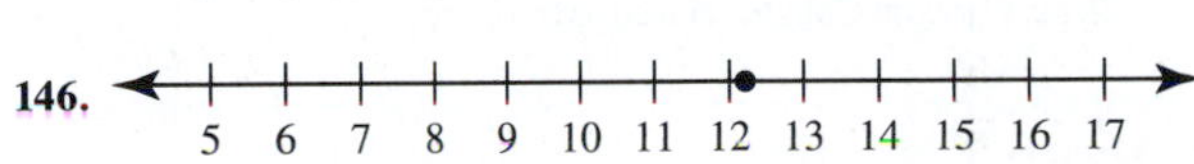

147.

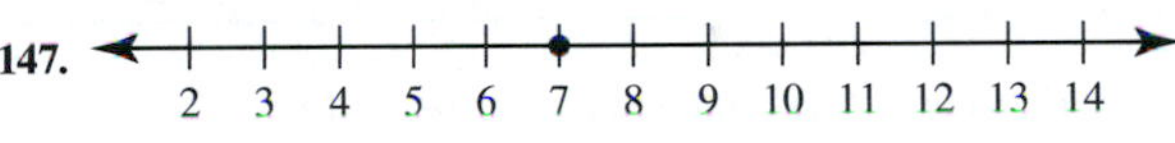

148.

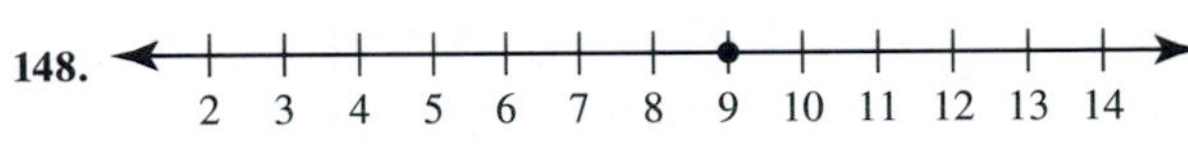

149. Since x is negative, and the absolute value of a number is always positive, the opposite of a negative number will be positive.
150. Yes, every integer is a rational number because an integer can be written as a ratio of the integer and 1.
151. No, not every rational number is an integer. A rational number is an integer if the numerator is divisible by the denominator.
152. Yes, you can use a larger scale on the real number line.

Section 1.2

1. Answers vary. To add two real numbers with the same sign, we add their absolute values and keep the sign.
2. Answers vary. To add two real numbers with the different signs, we subtract their absolute values. The sign of the result is the sign of the number with the larger absolute value.
3. Answers vary. If a number is raised to the nth, we multiply the number by itself n times.
4. Answers vary. First perform the multiplication and division in order from left to right. Then perform addition and subtraction in order from left to right.
5. The expression 7^3 represents 7 raised to the third, or 7 cubed.
6. Answers vary. The expression 12^2 represents 12 raised to the second, or 12 squared.

137. Answer vary. I turn my clock 10 min forward and then turn my clock 15 min backward.
138. Answer vary. Last week, John borrowed $2.25 from me. Yesterday, John borrowed $1.33 from me.

Section 1.3

1. Answers vary. An identity element is a number which leaves another number unchanged when an operation is performed on it.
2. Answers vary. An inverse is a number which produces the identity element when an operation is performed on it.

3. Answers vary. The commutative property of real numbers states that the order of adding or multiplying real numbers does not change the result.
4. Answers vary. The associative property of real numbers states that the way numbers are grouped when they are added or multiplied does not change the result.

GROUP ACTIVITY

Part 1

1. Answers vary. 2. Answers vary. 3. Answers vary.
4. Answers vary. 5. Answers vary. 6. Answers vary.
7. Answers vary. 8. Answers vary. 9. Answers vary.
10. Answers vary. 11. Answers vary. 12. Answers vary.

Part 2

1. Answers vary. 2. Answers vary. 3. Answers vary.

CHAPTER 1 REVIEW EXERCISES

Section 1.1

5. {Williams College, Amherst College, Middlebury College, Bowdoin College, Carleton College, Haverford College}

CHAPTER 1 TEST

4. **a.** 10; natural, whole, integer, rational; -10; 10
   b. 2π; irrational; -2π, 2π **c.** $-\frac{2}{3}$; rational; $\frac{2}{3}, \frac{2}{3}$
   d. 0; whole, integer, rational; 0, 0

CHAPTER 2

Section 2.1

1. A linear equation in one variable is an equation that can be written in the form $ax + b = c$, where a, b, and c are real numbers and $a \neq 0$.
2. A conditional equation is an equation that is true for some values of the variable and not true for other values, e.g., $3x - 5 = 10$.
3. A linear equation with no solution is called a contradiction, e.g., $4(x - 5) = 4x + 7$.
4. A linear equation with infinitely many solutions is called an identity. e.g., $2x - 2(x - 3) = 6$.
7. You use the addition and multiplication properties of equality to go through the process of solving the equation. If a false equation results, then the linear equation is a contradiction and has no solution.
8. You use the addition and multiplication properties of equality to solve the equation. If a true statement results, then the linear equation is an identity and has all real numbers as its solution set.

117.
$$\begin{aligned} \frac{4d}{3} - 4 &= \frac{2}{5} \\ 15\left(\frac{4d}{3} - 4\right) &= 15\left(\frac{2}{5}\right) \\ 15\left(\frac{4d}{3}\right) - 15(4) &= 15\left(\frac{2}{5}\right) \\ 20d - 60 &= 6 \\ 20d &= 66 \\ d &= \frac{66}{20} \\ d &= \frac{33}{10} \end{aligned}$$

118.
$$\begin{aligned} \frac{5m}{3} - 1 &= \frac{4}{7} \\ 21\left(\frac{5m}{3} - 1\right) &= 21\left(\frac{4}{7}\right) \\ 21\left(\frac{5m}{3}\right) - 21(1) &= 21\left(\frac{4}{7}\right) \\ 35m - 21 &= 12 \\ 35m &= 33 \\ m &= \frac{33}{35} \end{aligned}$$

119.
$$\begin{aligned} 0.75(2x - 4) + 0.3x &= 7.8 \\ 100[0.75(2x - 4) + 0.3x] &= 100(7.8) \\ 100[0.75(2x - 4)] + 100(0.3x) &= 100(7.8) \\ 75(2x - 4) + 30x &= 780 \\ 150x - 300 + 30x &= 780 \\ 180x &= 1080 \\ x &= \frac{1080}{180} \\ x &= 6 \end{aligned}$$

120.
$$\begin{aligned} 0.45(6 - 2x) - 0.2x &= 1249 \\ 100[0.45(6 - 2x) - 0.2x] &= 100(1249) \\ 100[0.45(6 - 2x)] - 100(0.2x) &= 124{,}900 \\ 45(6 - 2x) - 20x &= 124{,}900 \\ 270 - 90x - 20x &= 124{,}900 \\ -110x &= 124{,}630 \\ x &= \frac{124{,}630}{-110} \\ x &= -1133 \end{aligned}$$

121.
$$\begin{aligned} 3x - 2(x + 4) &= x + 5 \\ 3x - 2x - 8 &= x + 5 \\ x - 8 &= x + 5 \\ -8 &= 5 \end{aligned}$$
Since $-8 = 5$ is false, there is no solution.

122.
$$\begin{aligned} 8 + 5(x - 1) &= 3(2x + 1) - x \\ 8 + 5x - 5 &= 6x + 3 - x \\ 5x + 3 &= 5x + 3 \\ 3 &= 3 \end{aligned}$$
This is a true statement. So x is any real number. The solution is $\{x|x \text{ is real number}\}$.

Section 2.2

1. Consecutive integers are integers that follow one another. They are x and $x + 1$.
2. Consecutive even integers are even integers that follow one another. They are x and $x + 2$.
3. Read the problem carefully. Assign a variable to the unknown and express other unknowns in terms of the initially assigned variable. Use the given information and your knowledge to translate phrases into expressions and write the equation.
4. After solving the equation, you need to check the proposed solution.
35. $n - 5 = 15$; Subtraction is not commutative so the order is important.
37. Answers vary. Todd saves quarters and dimes in a jar. He has five more dimes than quarters and his collection is worth $5.75. Find the number of quarters and dimes in his collection.

Section 2.3

1. Complementary angles are angles whose sum is 90°.
2. Supplementary angles are angles whose sum is 180°.
3. Vertical angles are angles opposite from one another when two lines intersect one another.
4. Vertical angles have the same measure.
5. The perimeter of a figure is the distance around the outside of the figure.
6. The area of a figure is the number of square units it takes to cover the inside of the figure.

7. Substitute the known values of the perimeter or area and one of the dimensions into the appropriate formula and solve the resulting equation for the missing dimensions of the figure.
8. Use the addition and multiplication properties of equaliy to isolate the specific variable.
93. Let x be the amount invested in the first account and $2x - 300$ be the amount invested in the second account.

$$\begin{aligned} x + 2x - 300 &= 4500 \\ 3x - 300 &= 4500 \\ 3x &= 4800 \\ x &= \frac{4800}{3} \\ x &= 1600 \end{aligned}$$

Agnes invested \$1600 in the first account.

94.
$$\begin{aligned} S &= \frac{a}{1-r} \\ S(1-r) &= a \\ S - Sr &= a \\ -Sr &= a - S \\ r &= \frac{a-S}{-S} \\ r &= \frac{S-a}{S} \end{aligned}$$

Section 2.4

3. Start with the smallest endpoint in the set. Use a parenthesis if this endpoint is not included and a bracket if it is. End the interval with the largest endpoint in the set. Use a parenthesis if this endpoint is not included in the set and a bracket if it is.
4. When an endpoint is not included, a parenthesis is used in both the solution sets on a number line and in interval notation. When an endpoint is included, a bracket is used in both the solution sets on a number line and in interval notation.
5. Parentheses should be used with ∞ in interval notation because it means that the solutions of the inequality increase indefinitely.
6. Multiplying or dividing by a negative number cause the inequality symbol to reverse.

7.
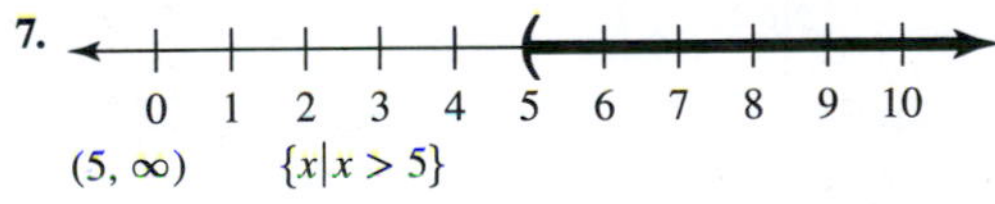

$(5, \infty)$ $\{x \mid x > 5\}$

8. 2 3 4 5 6 7 8 9 10 11 12

$(7, \infty)$ $\{x \mid x > 7\}$

9. −9 −8 −7 −6 −5 −4 −3 −2 −1 0 1

$(-\infty, -4)$ $\{x \mid x < -4\}$

10. −13 −12 −11 −10 −9 −8 −7 −6 −5 −4 −3

$(-\infty, -8)$ $\{x \mid x < -8\}$

11. 5 6 7 8 9 10 11 12 13 14 15

$[10, \infty)$ $\{x \mid x \geq 10\}$

12. −17 −16 −15 −14 −13 −12 −11 −10 −9 −8 −7

$[-12, \infty)$ $\{x \mid x \geq -12\}$

13.
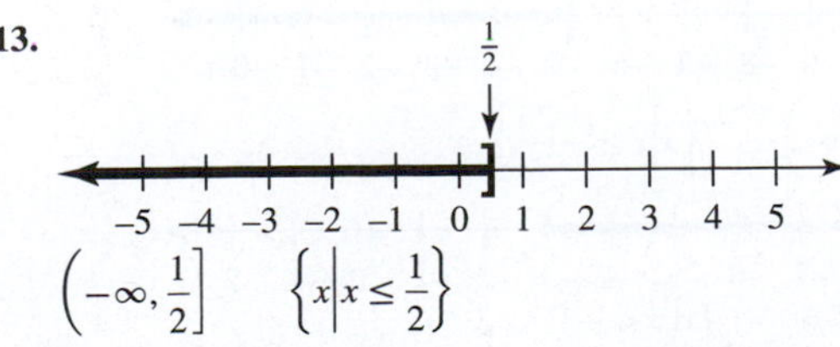

$\left(-\infty, \frac{1}{2}\right]$ $\left\{x \mid x \leq \frac{1}{2}\right\}$

14.
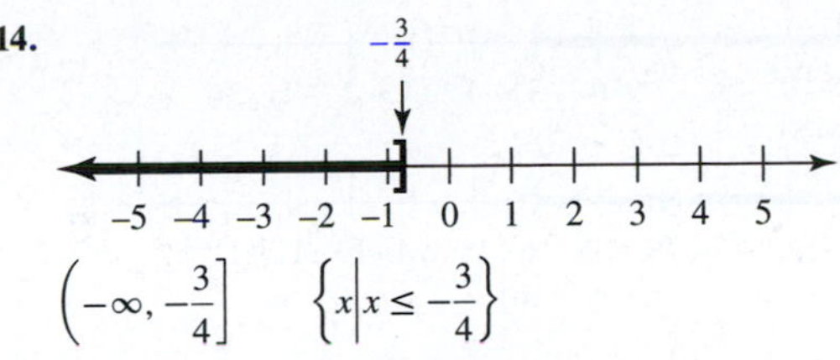

$\left(-\infty, -\frac{3}{4}\right]$ $\left\{x \mid x \leq -\frac{3}{4}\right\}$

15. −12 −11 −10 −9 −8 −7 −6 −5 −4 −3 −2

$(-9, -4]$ $\{x \mid -9 < x \leq -4\}$

16. −4 −3 −2 −1 0 1 2 3 4 5 6

$(0, 2]$ $\{x \mid 0 < x \leq 2\}$

17. −7 −6 −5 −4 −3 −2 −1 0 1 2 3

$[-4, 0]$ $\{x \mid -4 \leq x \leq 0\}$

18. 0 1 2 3 4 5 6 7 8 9 10

$[5, 6]$ $\{x \mid 5 \leq x \leq 6\}$

19. −2 −1 0 1 2 3 4 5 6 7 8

$(3, \infty)$ $\{x \mid x > 3\}$

20. −1 0 1 2 3 4 5 6 7 8 9

$[4, \infty)$ $\{x \mid x \geq 4\}$

21. −11 −10 −9 −8 −7 −6 −5 −4 −3 −2 −1

$[-10, -2)$ $\{x \mid -10 \leq x < -2\}$

22. −7 −6 −5 −4 −3 −2 −1 0 1 2 3

$(-6, 3]$ $\{x \mid -6 < x \leq 3\}$

23. −15 −14 −13 −12 −11 −10 −9 −8 −7 −6 −5

$(-\infty, -10]$ $\{x \mid x \leq -10\}$

24. −14 −13 −12 −11 −10 −9 −8 −7 −6 −5 −4

$(-\infty, -9]$ $\{x \mid x \leq -9\}$

25. −5 −4 −3 −2 −1 0 1 2 3 4 5

$(0, \infty)$ $\{x \mid x > 0\}$

26. 9 10 11 12 13 14 15 16 17 18 19

$(14, \infty)$ $\{x \mid x > 14\}$

27.
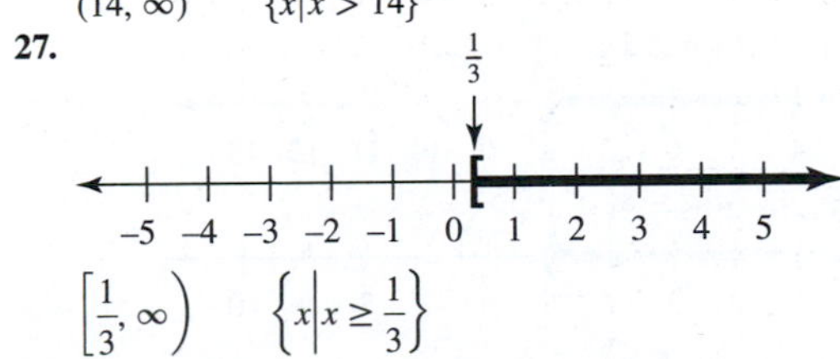

$\left[\frac{1}{3}, \infty\right)$ $\left\{x \mid x \geq \frac{1}{3}\right\}$

28.
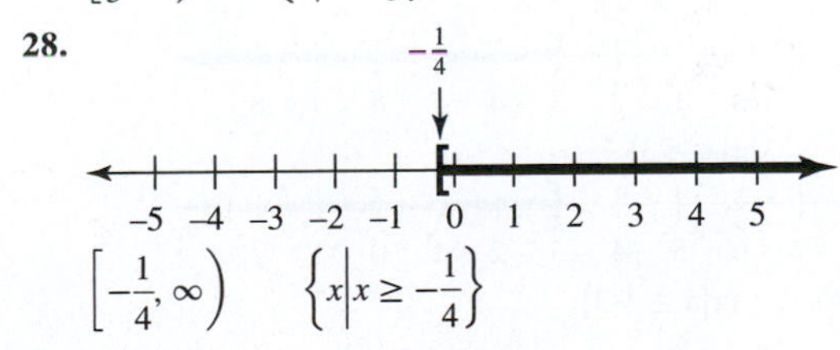

$\left[-\frac{1}{4}, \infty\right)$ $\left\{x \mid x \geq -\frac{1}{4}\right\}$

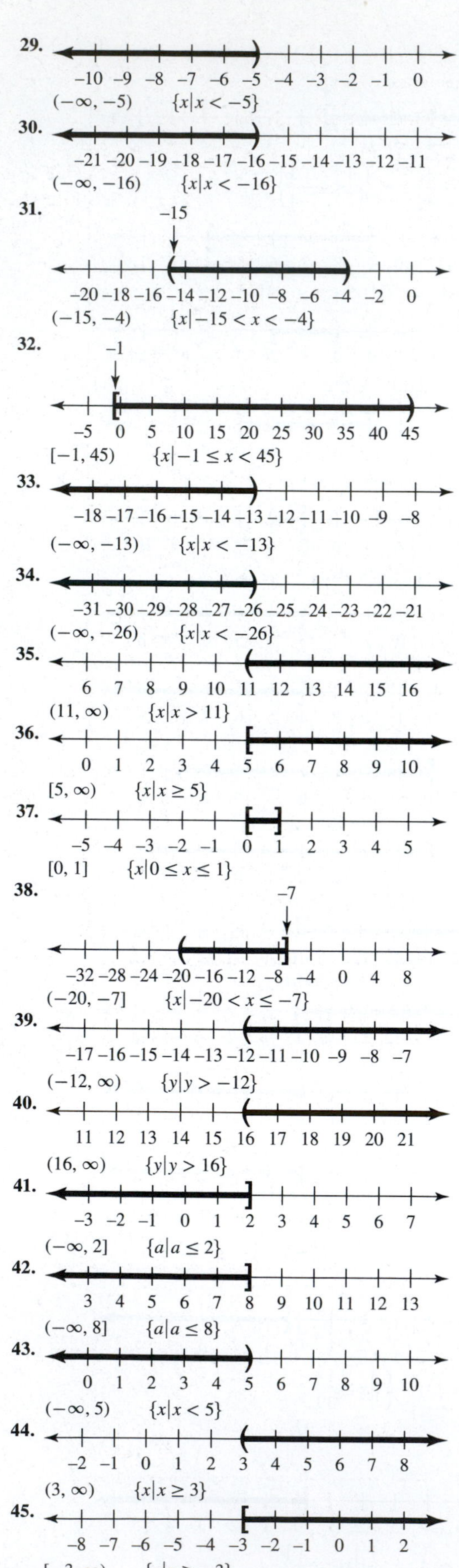
29. $(-\infty, -5)$ $\{x \mid x < -5\}$
30. $(-\infty, -16)$ $\{x \mid x < -16\}$
31. $(-15, -4)$ $\{x \mid -15 < x < -4\}$
32. $[-1, 45)$ $\{x \mid -1 \le x < 45\}$
33. $(-\infty, -13)$ $\{x \mid x < -13\}$
34. $(-\infty, -26)$ $\{x \mid x < -26\}$
35. $(11, \infty)$ $\{x \mid x > 11\}$
36. $[5, \infty)$ $\{x \mid x \ge 5\}$
37. $[0, 1]$ $\{x \mid 0 \le x \le 1\}$
38. $(-20, -7]$ $\{x \mid -20 < x \le -7\}$
39. $(-12, \infty)$ $\{y \mid y > -12\}$
40. $(16, \infty)$ $\{y \mid y > 16\}$
41. $(-\infty, 2]$ $\{a \mid a \le 2\}$
42. $(-\infty, 8]$ $\{a \mid a \le 8\}$
43. $(-\infty, 5)$ $\{x \mid x < 5\}$
44. $(3, \infty)$ $\{x \mid x \ge 3\}$
45. $[-3, \infty)$ $\{x \mid x \ge -3\}$

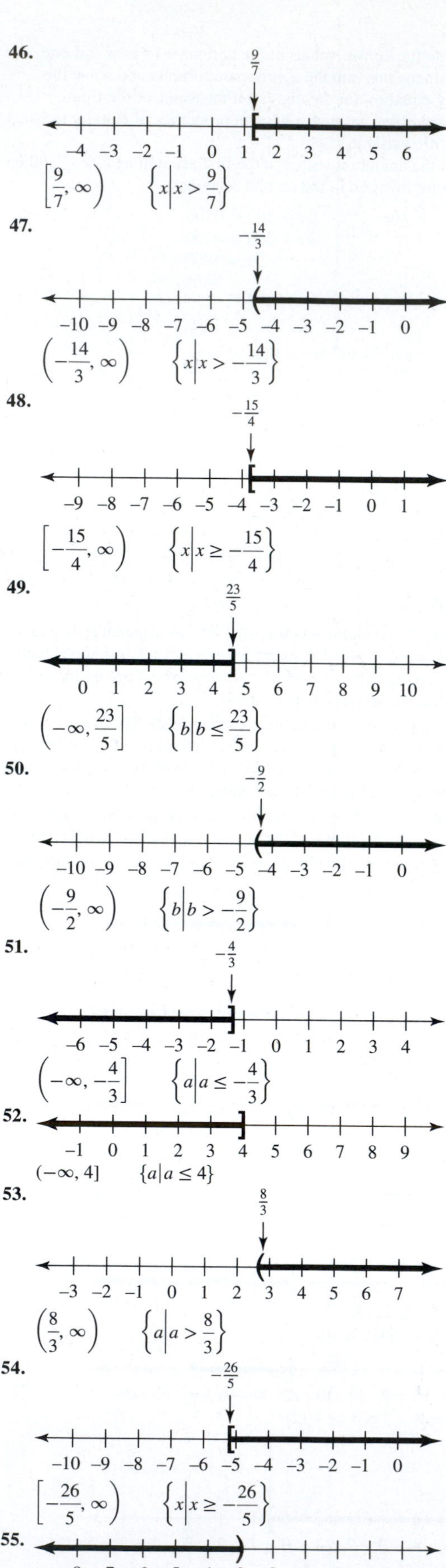
46. $\left[\frac{9}{7}, \infty\right)$ $\left\{x \middle| x > \frac{9}{7}\right\}$
47. $\left(-\frac{14}{3}, \infty\right)$ $\left\{x \middle| x > -\frac{14}{3}\right\}$
48. $\left[-\frac{15}{4}, \infty\right)$ $\left\{x \middle| x \ge -\frac{15}{4}\right\}$
49. $\left(-\infty, \frac{23}{5}\right]$ $\left\{b \middle| b \le \frac{23}{5}\right\}$
50. $\left(-\frac{9}{2}, \infty\right)$ $\left\{b \middle| b > -\frac{9}{2}\right\}$
51. $\left(-\infty, -\frac{4}{3}\right]$ $\left\{a \middle| a \le -\frac{4}{3}\right\}$
52. $(-\infty, 4]$ $\{a \mid a \le 4\}$
53. $\left(\frac{8}{3}, \infty\right)$ $\left\{a \middle| a > \frac{8}{3}\right\}$
54. $\left[-\frac{26}{5}, \infty\right)$ $\left\{x \middle| x \ge -\frac{26}{5}\right\}$
55. $(-\infty, -3)$ $\{x \mid x < -3\}$

56. 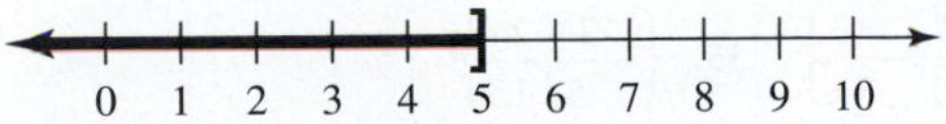
0 1 2 3 4 5 6 7 8 9 10
$(-\infty, 5]$ $\{x \mid x \le 5\}$

57. 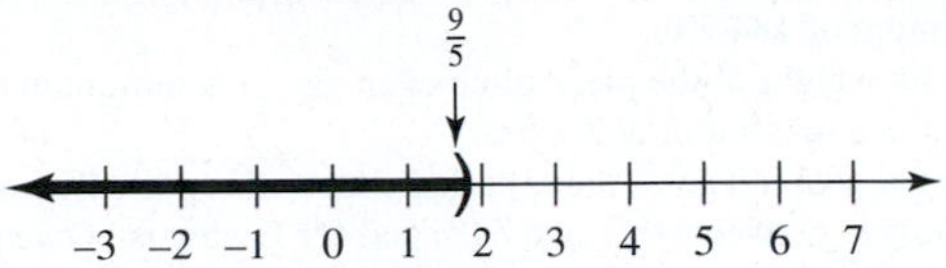
$\frac{9}{5}$
−3 −2 −1 0 1 2 3 4 5 6 7
$\left(-\infty, \frac{9}{5}\right)$ $\left\{x \,\middle|\, x < \frac{9}{5}\right\}$

58. 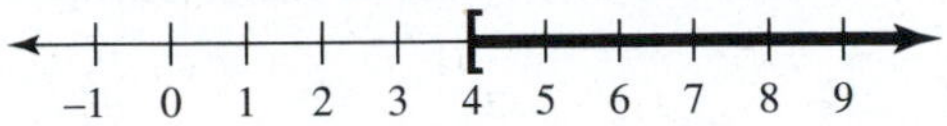
−1 0 1 2 3 4 5 6 7 8 9
$[4, \infty)$ $\{x \mid x \ge 4\}$

59. 
1 2 3 4 5 6 7 8 9 10 11
$[6, \infty)$ $\{x \mid x \ge 6\}$

60. 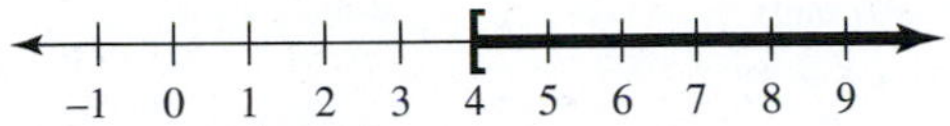
−1 0 1 2 3 4 5 6 7 8 9
$[4, \infty)$ $\{x \mid x \ge 4\}$

61. $\frac{85 + 78 + 100 + 87 + x}{5} \ge 90$; Lucinda must earn 100 on her fifth quiz.

62. $\frac{66 + 79 + 60 + 65 + x}{5} \ge 70$; Aaron must earn 80 or higher on his fifth quiz.

63. $0.15(80) + 0.25(60) + 0.40(68) + 0.20x \ge 70$; Aleksandr must score 79 or higher on the final exam.

64. $0.15(100) + 0.25(75) + 0.40(80) + 0.20x \ge 80$; Christopher must score 71.25 or higher on the final exam.

65. $39.99 + 0.45x \le 60$; Joyce can afford to use at most 44 additional minutes.

66. $29.99 + 0.30x \le 45$; Sandra can afford to use at most 50 additional minutes.

67. $1500 + 15x \le 3000$; Rebecca can invite at most 100 people to her son's party.

68. $700 + 30x \le 2000$; At most, 43 people can attend the Christmas party.

73. 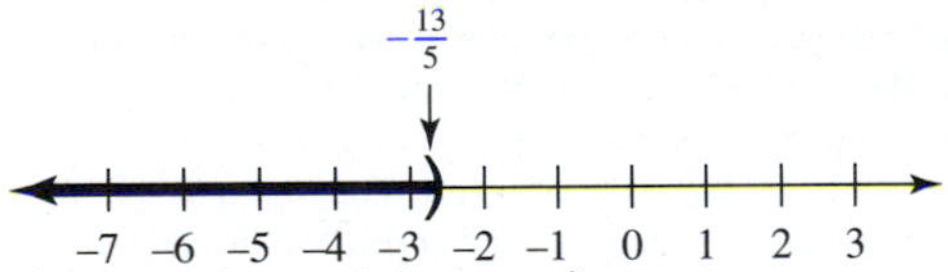
$-\frac{13}{5}$
−7 −6 −5 −4 −3 −2 −1 0 1 2 3
$\left(-\infty, -\frac{13}{5}\right)$ $\left\{a \,\middle|\, a < -\frac{13}{5}\right\}$

74. 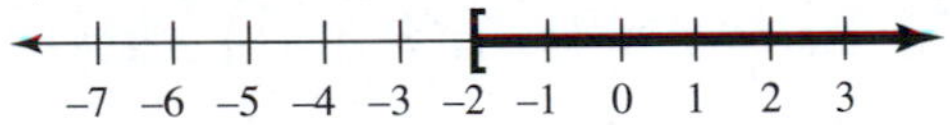
−7 −6 −5 −4 −3 −2 −1 0 1 2 3
$[-2, \infty)$ $\{a \mid a \ge -2\}$

75. 
−2 −1 0 1 2 3 4 5 6 7 8
$(3, \infty)$ $\{x \mid x > 3\}$

76. 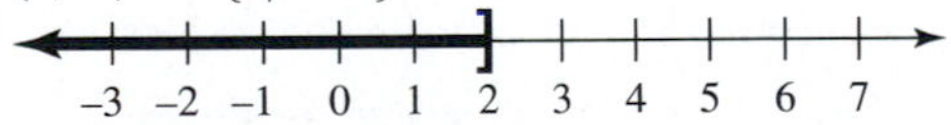
−3 −2 −1 0 1 2 3 4 5 6 7
$(-\infty, 2]$ $\{x \mid x \le 2\}$

77. 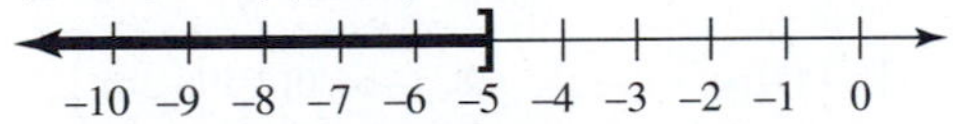
−10 −9 −8 −7 −6 −5 −4 −3 −2 −1 0
$(-\infty, -5]$ $\{x \mid x \le -5\}$

78. 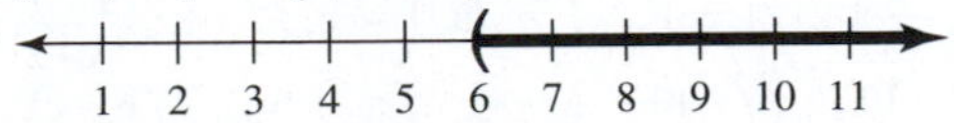
1 2 3 4 5 6 7 8 9 10 11
$(6, \infty)$ $\{x \mid x > 6\}$

79. 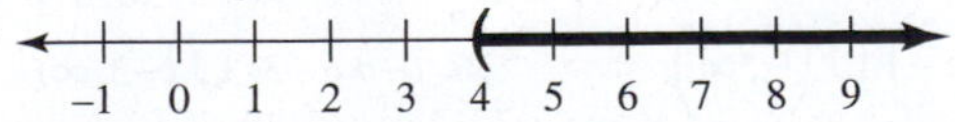
−1 0 1 2 3 4 5 6 7 8 9
$(4, \infty)$ $\{x \mid x > 4\}$

80. 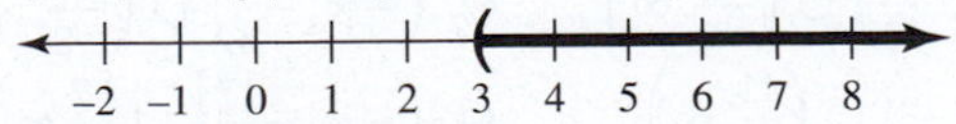
−2 −1 0 1 2 3 4 5 6 7 8
$(3, \infty)$ $\{x \mid x > 3\}$

81. 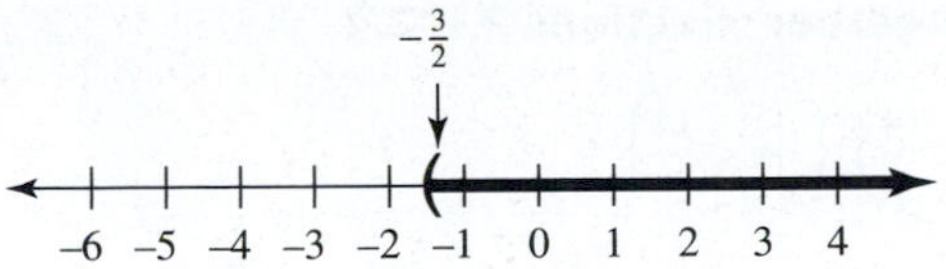
$-\frac{3}{2}$
−6 −5 −4 −3 −2 −1 0 1 2 3 4
$\left(-\frac{3}{2}, \infty\right)$ $\left\{y \,\middle|\, y > -\frac{3}{2}\right\}$

82. 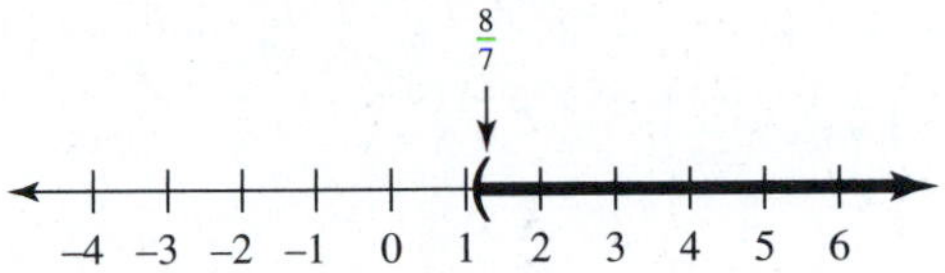
$\frac{8}{7}$
−4 −3 −2 −1 0 1 2 3 4 5 6
$\left(\frac{8}{7}, \infty\right)$ $\left\{y \,\middle|\, y > \frac{8}{7}\right\}$

83. 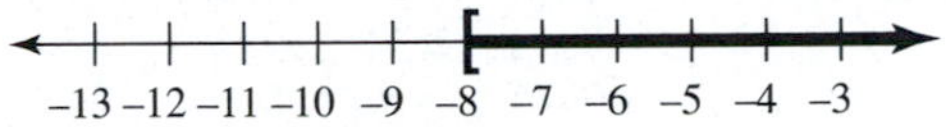
−13 −12 −11 −10 −9 −8 −7 −6 −5 −4 −3
$[-8, \infty)$ $\{x \mid x \ge -8\}$

84. 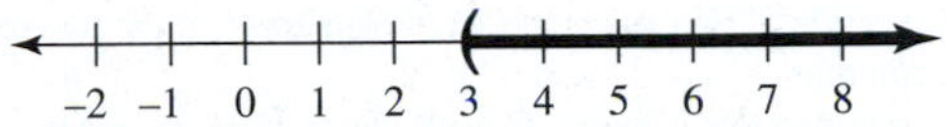
−2 −1 0 1 2 3 4 5 6 7 8
$(3, \infty)$ $\{x \mid x > 3\}$

85. 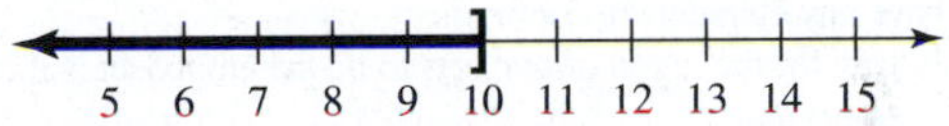
5 6 7 8 9 10 11 12 13 14 15
$(-\infty, 10]$ $\{x \mid x \le 10\}$

86. 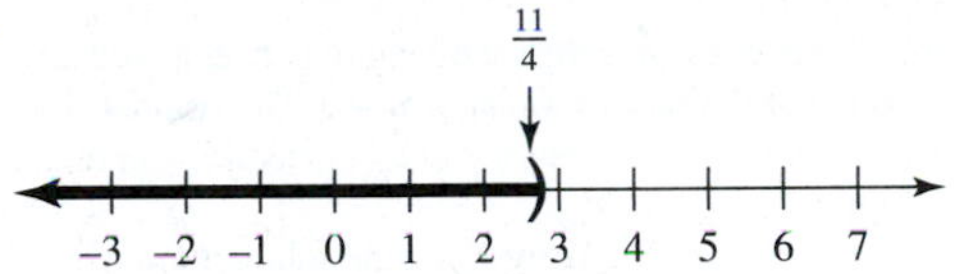
$\frac{11}{4}$
−3 −2 −1 0 1 2 3 4 5 6 7
$\left(-\infty, \frac{11}{4}\right)$ $\left\{x \,\middle|\, x < \frac{11}{4}\right\}$

87. 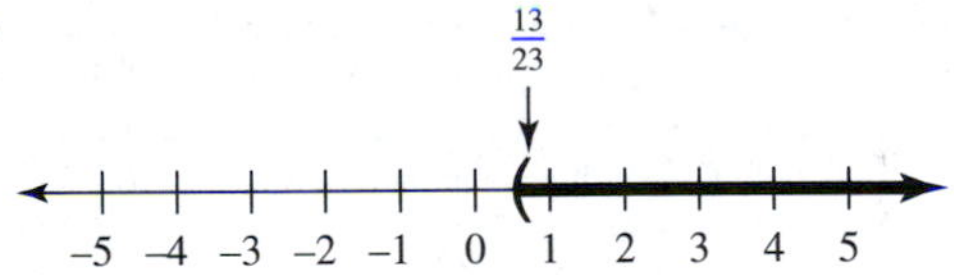
$\frac{13}{23}$
−5 −4 −3 −2 −1 0 1 2 3 4 5
$\left(\frac{13}{23}, \infty\right)$ $\left\{x \,\middle|\, x > \frac{13}{23}\right\}$

88. 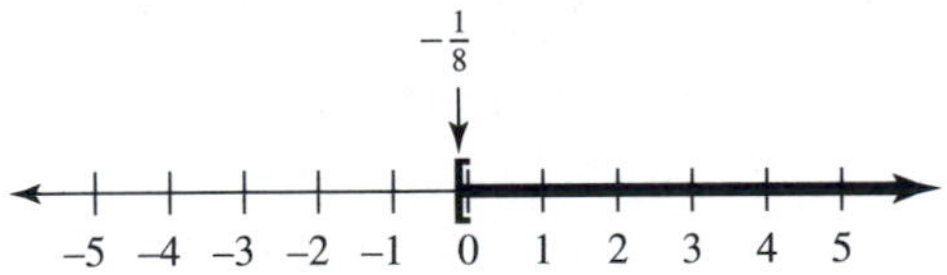
$-\frac{1}{8}$
−5 −4 −3 −2 −1 0 1 2 3 4 5
$\left[-\frac{1}{8}, \infty\right)$ $\left\{x \,\middle|\, x \ge -\frac{1}{8}\right\}$

89. 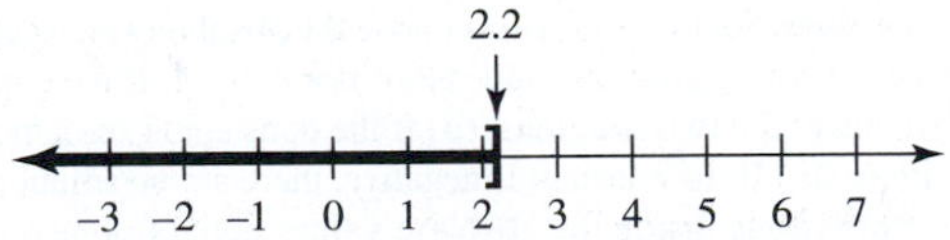
2.2
−3 −2 −1 0 1 2 3 4 5 6 7
$(-\infty, 2.2]$ $\{x \mid x \le 2.2\}$

90. 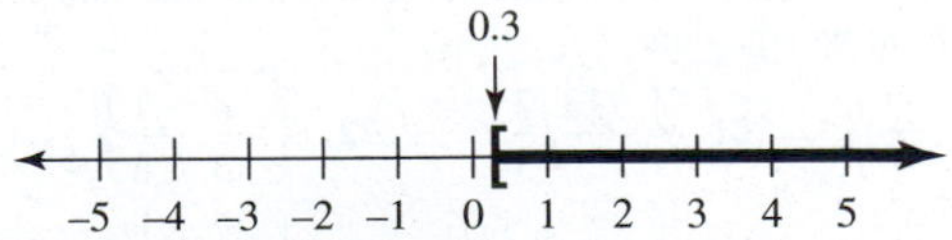
0.3
−5 −4 −3 −2 −1 0 1 2 3 4 5
$[0.3, \infty)$ $\{x \mid x \ge 0.3\}$

Piece It Together Sections 2.1–2.4

14. $\left(-\infty, \frac{14}{3}\right), \left\{x \middle| x < \frac{14}{3}\right\}$

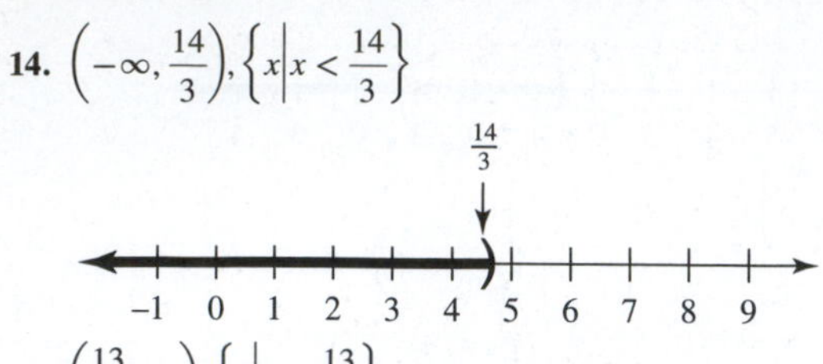

15. $\left(\frac{13}{23}, \infty\right), \left\{x \middle| x > \frac{13}{23}\right\}$

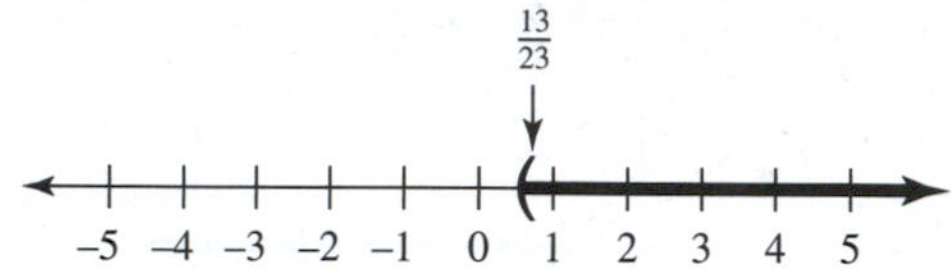

Section 2.5

1. The intersection of two sets is the set of elements that the two sets have in common.

2. The union of two sets is the set of elements that belong to either set.

3. The intersection of sets A and B can be the empty set if the two sets do not have any elements in common.

4. The only way for the union of two sets to be the empty set is if each set is the empty set.

5. A compound inequality consists of two inequalities joined by the terms "and" or "or."

6. The graph of $-a < x < a$ is the set of values between $-a$ and a; it is the intersection of the sets $x < a$ and $x > -a$. The graph of $x < -a$ or $x > a$ is the union of two sets, the numbers to the right of a and the numbers to the left of $-a$.

7. Find the solution set of inequality 1 and the solution set of inequality 2. Find the intersection of these two solution sets and write the final solution in interval notation and provide its graph.

8. Find the solution set of inequality 1 and the solution set of inequality 2. Find the union of the solution sets of inequalities 1 and 2 and write the final solution in interval notation and provide its graph.

108. Since the empty set does not have a number, the intersection of the empty set and another set will have no number in common. Therefore the intersection is also an empty set.

109.
$$\begin{aligned} 10 - 3x &< 4 &\quad \text{and} \quad 5x + 23 &> 3 \\ -3x &< -6 &\quad \text{and} \quad 5x &> -20 \\ x &> 2 &\quad \text{and} \quad x &> -4 \end{aligned}$$
The intersection is $x > 2$. The solution set is $(2, \infty)$.

110.
$$\begin{aligned} 28 - 4x &> 4 &\quad \text{or} \quad 28 - 4x &< -4 \\ -4x &> -24 &\quad \text{or} \quad -4x &< -32 \\ x &< 6 &\quad \text{or} \quad x &> 8 \end{aligned}$$
The solution set is $(-\infty, 6) \cup (8, \infty)$.

Section 2.6

3. Isolate the absolute value on one side of the equation and the constant on the other side. Set the expression inside the absolute value equal to the number(s) whose absolute value is the constant. (i) If the constant is positive, there are two solutions. (ii) If the constant is zero, there is one solution. (iii) If the constant is negative, there are no solutions.

4. Set the expressions inside the absolute values equal to one another. Set the expression inside the first absolute value equal to the opposite of the expression inside the second absolute value. Solve these resulting equations.

15. $\left\{-\frac{17}{4}, \frac{11}{4}\right\}$ **21.** $\left\{-\frac{13}{5}, \frac{7}{5}\right\}$ **22.** $\left\{-\frac{4}{3}, -\frac{1}{3}\right\}$

23. $\left\{-\frac{2}{9}, 4\right\}$ **24.** $\left\{-7, -\frac{1}{13}\right\}$ **25.** $\left\{-\frac{7}{2}, \frac{3}{4}\right\}$

29. $\left\{-15, \frac{1}{5}\right\}$ **30.** $\left\{-\frac{5}{3}, \frac{15}{17}\right\}$ **31.** $\left\{-7, -\frac{1}{3}\right\}$

33. $\left\{-\frac{17}{16}, -\frac{13}{14}\right\}$ **35.** $\left\{-\frac{24}{5}, \frac{24}{13}\right\}$

41. The exact weight of the person can be a minimum of 183.9 lb and a maximum of 184.7 lb.

42. The exact weight of the piece of chicken can be a minimum of 2.9 oz and a maximum of 3.1 oz.

43. The actual percentage of those polled who will reportedly vote for Lipsey is between 48% and 54% and for Easley is between 42% and 48%.

44. The actual percentage of those polled who will reportedly vote for Johnson is between 40% and 44% and for Hunter is between 46% and 50%.

49. $\left\{-\frac{25}{41}, -\frac{25}{29}\right\}$ **51.** $\left\{-4, \frac{32}{5}\right\}$

55. $\left\{\frac{5}{11}, 9\right\}$ **63.** $\left\{-\frac{85}{9}, \frac{65}{9}\right\}$

68. The actual measurement of the object is between 2558.77 units and 2570.07 units.

76.
$$\begin{aligned} |2x - 3| + 5 &= 8 \\ |2x - 3| &= 3 \\ 2x - 3 = 3 \quad &\text{or} \quad 2x - 3 = -3 \\ 2x = 6 \quad &\quad\quad 2x = 0 \\ x = 3 \quad &\quad\quad x = 0 \end{aligned}$$
So, the solution set is $\{0, 3\}$.

77. $\left\{-\frac{28}{17}, \frac{38}{17}\right\}$ **78.** $\left\{-\frac{36}{13}, \frac{6}{13}\right\}$

79. $\left\{-\frac{13}{2}, \frac{17}{6}\right\}$ **80.** $\left\{0, \frac{24}{5}\right\}$

Section 2.7

1. Isolate the absolute value expression to one side of the inequality. Remove the absolute value sign by writing $X < k$ and $X > -k$. Solve the resulting compound inequality.

2. Isolate the absolute value expression to one side of the inequality. Remove the absolute value sign by writing $X > k$ or $X < -k$. Solve the resulting compound inequality.

3. After isolating the absolute value expression to one side of the inequality, you have a negative number on the other side. If the inequality sign is $<$ or $\leq$, then there is no solution. Otherwise, all real numbers.

19. $\left(-2, \frac{8}{3}\right)$ **20.** $\left(-1, -\frac{1}{5}\right)$ **21.** $\left(-3, -\frac{1}{2}\right)$ **22.** $\left(-\frac{5}{2}, \frac{17}{2}\right)$

29. $\left(-\infty, -\frac{22}{5}\right) \cup \left(\frac{12}{5}, \infty\right)$ **30.** $\left(-\infty, -\frac{5}{2}\right) \cup \left(\frac{1}{6}, \infty\right)$

31. $\left(-\infty, -\frac{6}{5}\right) \cup (2, \infty)$ **32.** $\left(-\infty, -\frac{5}{4}\right) \cup (2, \infty)$

33. $\left(-\infty, \frac{1}{2}\right) \cup \left(\frac{7}{2}, \infty\right)$ **34.** $\left(-\infty, \frac{2}{3}\right) \cup (4, \infty)$

35. $(-\infty, -8] \cup [1, \infty)$ **36.** $\left(-\infty, -\frac{5}{3}\right] \cup [5, \infty)$

37. $(-\infty, -1] \cup [7, \infty)$ **38.** $(-\infty, 0] \cup \left[\frac{7}{2}, \infty\right)$

39. $\left(-\infty, -\frac{2}{3}\right) \cup (1, \infty)$ **40.** $\left(-\infty, -\frac{17}{2}\right) \cup \left(-\frac{1}{2}, \infty\right)$

51. $\left(-\infty, -\frac{10}{4}\right) \cup \left(-\frac{10}{4}, \infty\right)$ **52.** $\left(-\infty, \frac{6}{5}\right) \cup \left(\frac{6}{5}, \infty\right)$

55. $\left(-\infty, \frac{3}{4}\right) \cup \left(\frac{3}{4}, \infty\right)$ **56.** $(-\infty, -3) \cup (-3, \infty)$

61. $\left(-\infty, -\frac{3}{2}\right) \cup \left(-\frac{3}{2}, \infty\right)$ **62.** $\left(-\infty, -\frac{5}{2}\right) \cup \left(-\frac{5}{2}, \infty\right)$

63. $(-\infty, -3) \cup \left(\frac{11}{3}, \infty\right)$ **64.** $\left(-\infty, -\frac{17}{2}\right] \cup \left[\frac{7}{2}, \infty\right)$

65. $\left[-\frac{11}{5}, 1\right]$

67. According to the poll, 70.9% to 77.1% of Americans think that STEM education and training is very important to U.S. competitiveness and prosperity.

68. According to the poll, 84% to 92% of Americans with private health insurance thought that the quality of their health care was excellent or good.

75. $(-\infty, -1] \cup \left[\frac{3}{2}, \infty\right)$ **76.** $(-\infty, 0] \cup [1, \infty)$

79. $(-\infty, -4] \cup \left[\frac{12}{5}, \infty\right)$ **80.** $\left(-\infty, -\frac{13}{7}\right] \cup [1, \infty)$

91. $\left[-\frac{7}{2}, \frac{17}{2}\right]$ **92.** $\left[-\frac{10}{3}, 4\right]$ **93.** $(-19, 23)$ **94.** $\left(-\frac{27}{2}, \frac{33}{2}\right)$

95. $\left[-\frac{13}{3}, \frac{4}{3}\right]$ **96.** $[-3, -2]$ **99.** $\left(-\frac{5}{2}, \frac{7}{2}\right)$ **100.** $\left(-\frac{5}{4}, \frac{11}{4}\right)$

102. $(-\infty, -10) \cup (10, \infty)$ **103.** $(-\infty, -3] \cup \left[\frac{3}{5}, \infty\right)$

104. $\left(-\infty, -\frac{3}{4}\right] \cup \left[\frac{3}{2}, \infty\right)$ **105.** $(-\infty, 6) \cup (6, \infty)$

106. $(-\infty, -3) \cup (-3, \infty)$ **109.** $\left[\frac{1}{2}, \frac{11}{2}\right]$

110. $\left(-\infty, -\frac{10}{3}\right] \cup \left[\frac{14}{3}, \infty\right)$ **123.** $\left(-\infty, -\frac{4}{3}\right] \cup \left[\frac{1}{3}, \infty\right)$

124. $\left(-\infty, -\frac{11}{5}\right] \cup [-1, \infty)$

132.
$$\begin{aligned} |4x + 1| + 2 &\geq 15 \\ |4x + 1| &\geq 13 \\ 4x + 1 \leq -13 \quad &\text{or} \quad 4x + 1 \geq 13 \\ 4x \leq -14 \quad &\text{or} \quad 4x \geq 12 \\ x \leq -\frac{7}{2} \quad &\text{or} \quad x \geq 3 \end{aligned}$$

The solution set is $\left(-\infty, -\frac{7}{2}\right] \cup [3, \infty)$

135. $(-\infty, 0] \cup \left[\frac{28}{5}, \infty\right)$ **136.** $(-\infty, 0] \cup \left[\frac{9}{2}, \infty\right)$

CHAPTER 2 REVIEW EXERCISES

Section 2.4

57. −6 −5 −4 −3 −2 −1 0 1 2 3 4 5 6

$[-6, \infty)$, $\{x \mid x \geq -6\}$

58. −6 −5 −4 −3 −2 −1 0 1 2 3 4 5 6

$\left[-\frac{4}{9}, \infty\right)$, $\left\{x \mid x \geq -\frac{4}{9}\right\}$

59. −1 0 1 2 3 4 5 6 7 8 9 10 11

$(-\infty, 10)$, $\{x \mid x < 10\}$

60. −7 −6 −5 −4 −3 −2 −1 0 1 2 3 4 5

$(-2, 0)$, $\{x \mid -2 < x < 0\}$

61. −27 −26 −25 −24 −23 −22 −21 −20 −19 −18 −17

$(-25, -20)$, $\{x \mid -25 < x < -20\}$

62. −3 −2 −1 0 1 2 3 4 5 6 7 8 9

$(-3, 9]$, $\{x \mid -3 < x \leq 9\}$

CHAPTER 2 TEST

16.

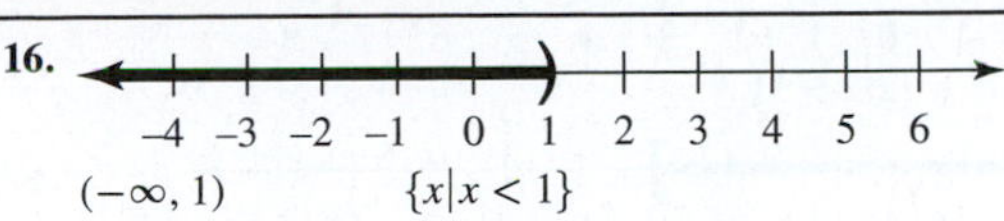

$(-\infty, 1)$ $\{x \mid x < 1\}$

17.

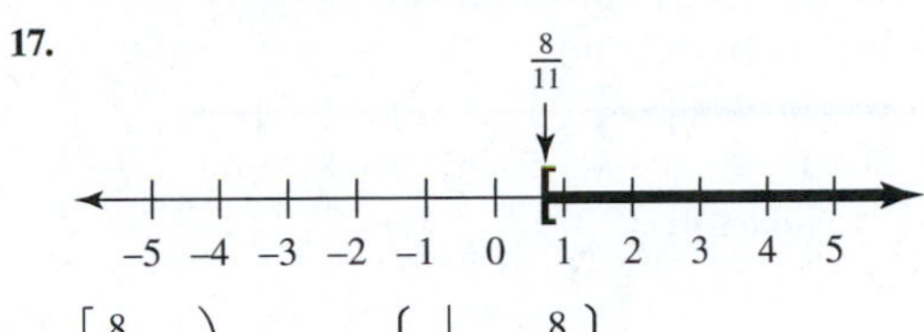

$\left[\frac{8}{11}, \infty\right)$ $\left\{m \mid m \geq \frac{8}{11}\right\}$

18.

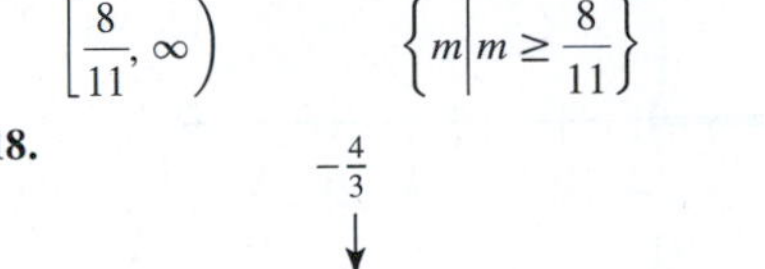

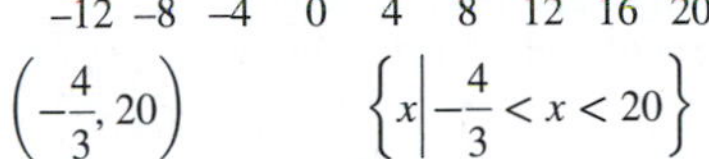

$\left(-\frac{4}{3}, 20\right)$ $\left\{x \mid -\frac{4}{3} < x < 20\right\}$

19.

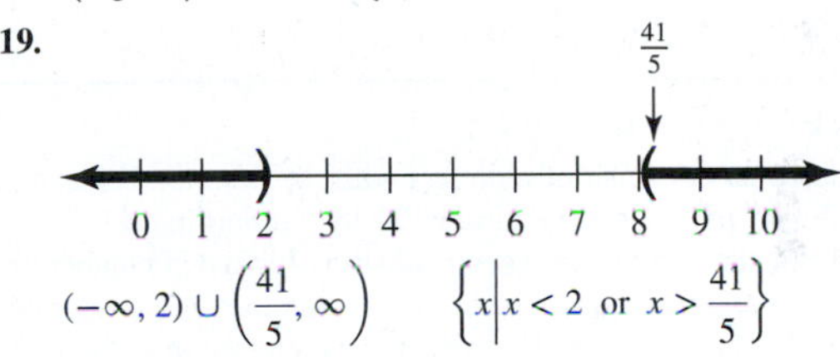

$(-\infty, 2) \cup \left(\frac{41}{5}, \infty\right)$ $\left\{x \mid x < 2 \text{ or } x > \frac{41}{5}\right\}$

20.

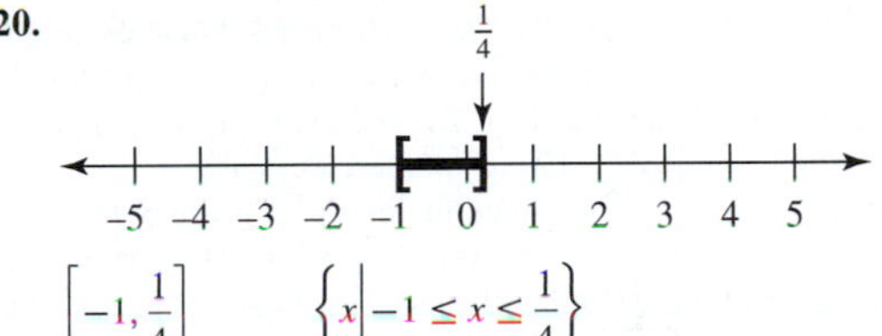

$\left[-1, \frac{1}{4}\right]$ $\left\{x \mid -1 \leq x \leq \frac{1}{4}\right\}$

21.

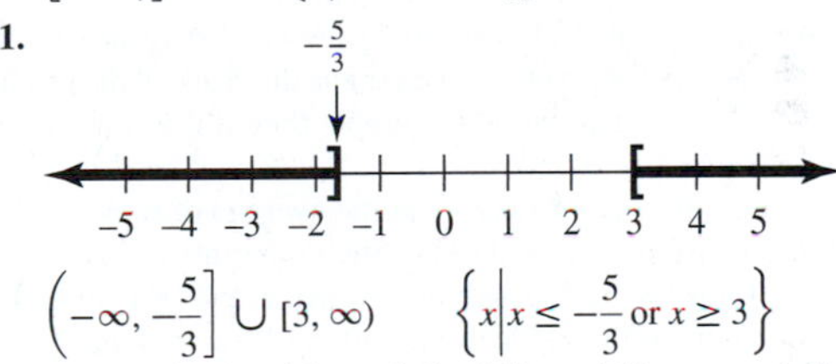

$\left(-\infty, -\frac{5}{3}\right] \cup [3, \infty)$ $\left\{x \mid x \leq -\frac{5}{3} \text{ or } x \geq 3\right\}$

22. The cashier has 16 ten-dollar bills and 12 twenty-dollar bills.

23. The length of the rectangle is 20 ft and the width is 85 ft.

24. Jan's RMR is 1357.12, which means she burns approximately 1357 Cal a day while at rest.

25. The Smartphone Company needs to sell 7 phones to make a profit. Note: The solution is approximately 6.17. Since a partial phone cannot be produced, we must round up to the next integer.

26. Juan can make between, and including, 34.2 and 70.2 to have a final grade of a C.

27. $|x - 100| = 40$; The easement begins at 60 ft from the entrance of the subdivision and ends at 140 ft from the entrance of the subdivision.

28. $|s - 65| > 15$; Officers stop vehicles that travel at speeds greater than 80 mph or less than 50 mph.

CHAPTERS 1 AND 2 CUMULATIVE REVIEW EXERCISES

7. **8.**

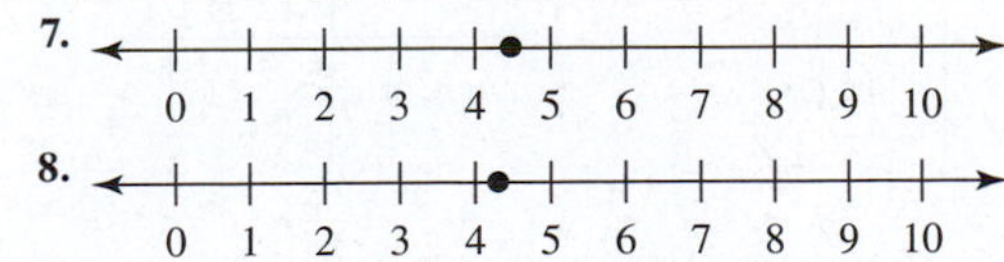

68. −2 −1 0 1 2 3 4 5 6 7 8

$(3, \infty)$ $\{x \mid x > 3\}$

69. −15 −14 −13 −12 −11 −10 −9 −8 −7 −6 −5

$(-\infty, -10]$ $\{x \mid x \leq -10\}$

70. −1 0 1 2 3 4 5 6 7 8 9

$(-\infty, 4]$ $\{a \mid a \leq 4\}$

71. $\frac{9}{5}$

−3 −2 −1 0 1 2 3 4 5 6 7

$\left(-\infty, \frac{9}{5}\right)$ $\left\{x \mid x < \frac{9}{5}\right\}$

72. You must earn between 77 and 100 on the final exam.

73. You must earn between 65 and 94 on the final exam.

CHAPTER 3

Section 3.1

1. An ordered pair is a pair of numbers with the first number as the x-coordinate and the second number as the y-coordinate.
2. The rectangular coordinate system consists of two real number lines intersecting at right angles. The horizontal number line is referred to as the x-axis and the vertical number line is referred to as the y-axis. The point where the two number lines intersect is called the origin.
3. The two axes divide the plane into four regions called quadrants. The quadrants are labeled as I, II, III, and IV, beginning in the upper right quadrant and rotating counterclockwise.
4. Replace the values of x and y with the numbers given in the ordered pair. Simplify each side of the equation. If the resulting equation is true, then the ordered pair is a solution of the equation. Otherwise, it is not a solution of the equation.
5. To graph an equation, find a few ordered pairs that satisfy the equation. Plot the ordered pairs as points and determine the shape of the graph.
6. If the ordered pair is a point on the graph, then it is a solution of the graphed equation.
7. The average of the x-coordinates and the average of the y-coordinates give the coordinates of the midpoint.
8. If the input values of both points are the same, then the midpoint lies in the middle of the vertical line joining the two points.

Answers for 9–18

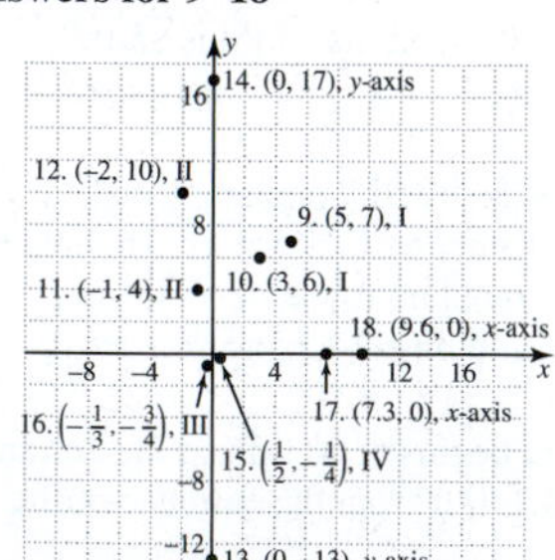

19.

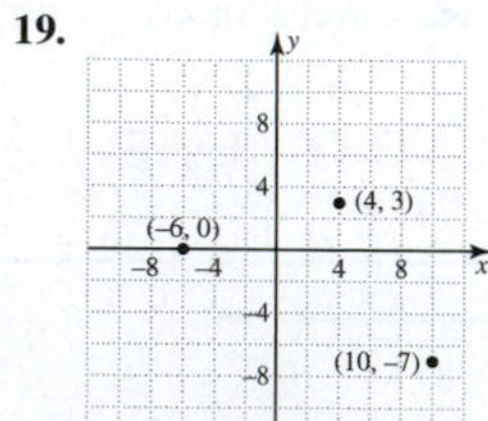

20.

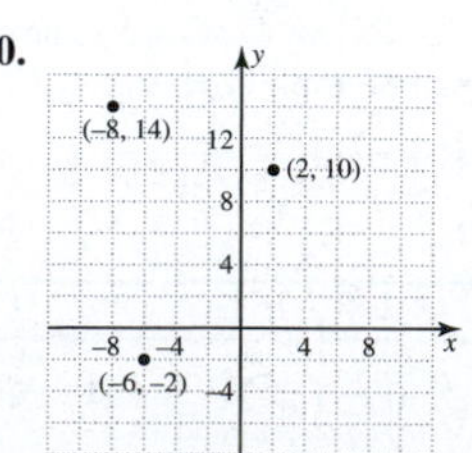

21.

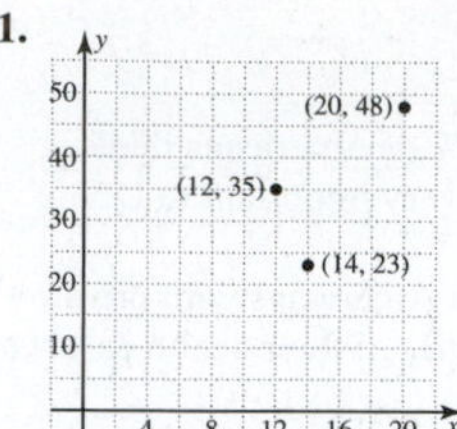

22. **23.** **24.**

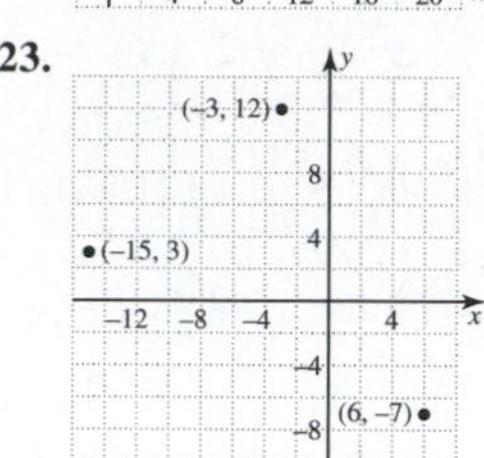
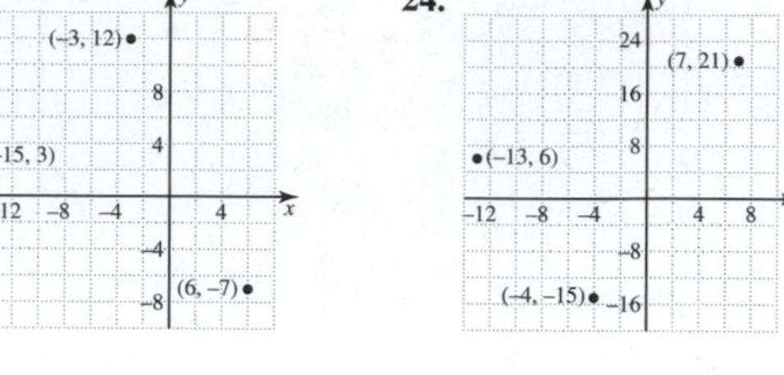

39.

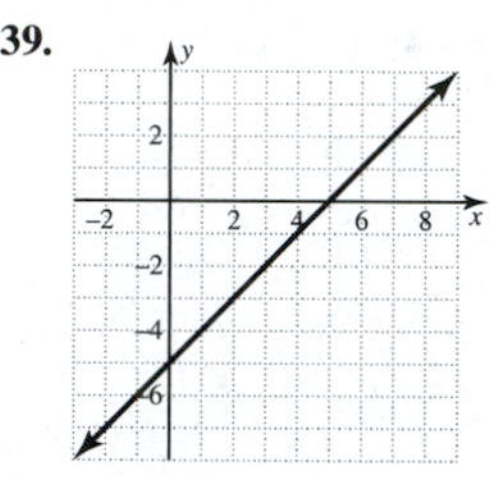

40.

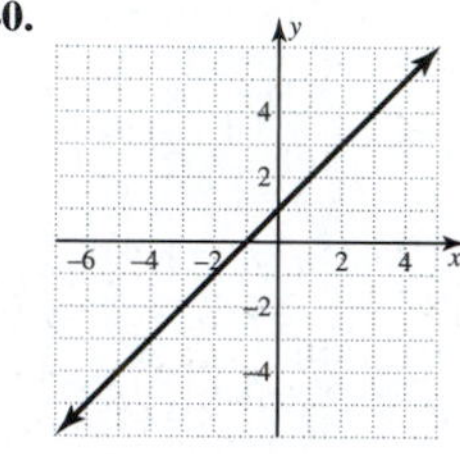

41.

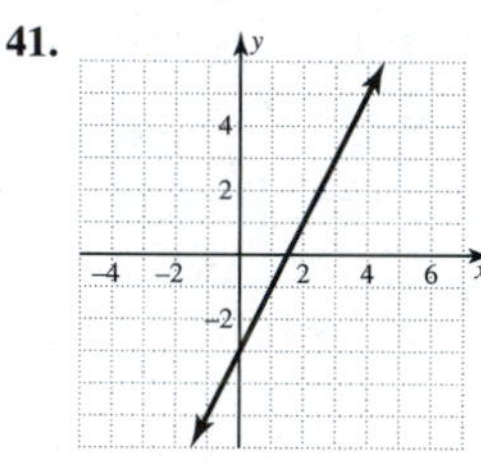

42.

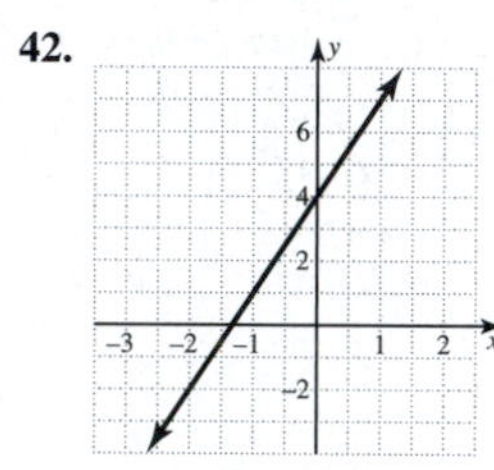

43.

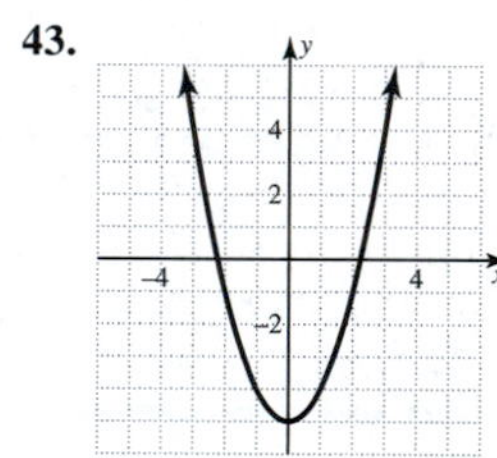

44.

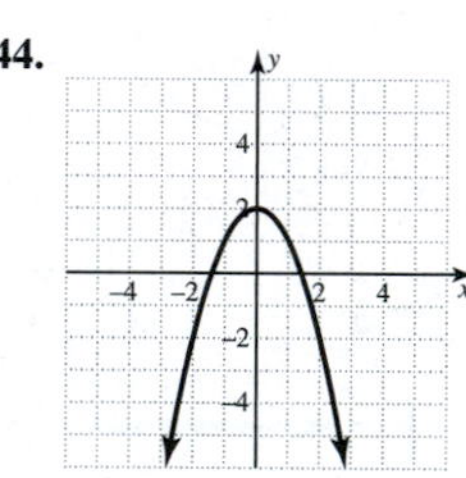

45.

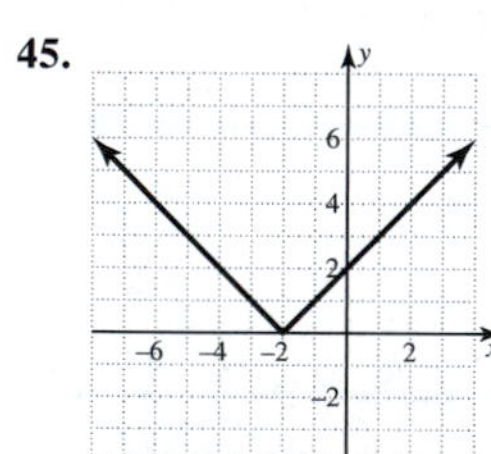

46.

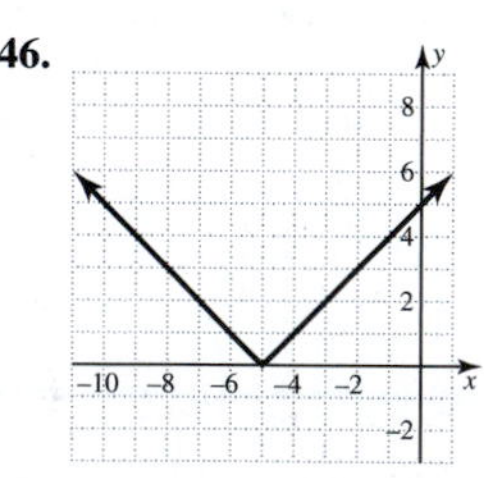

47.

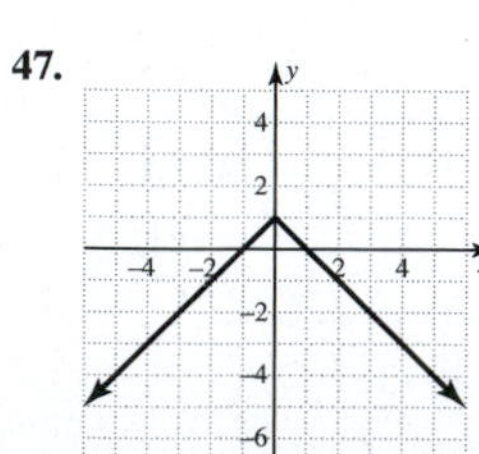

48.

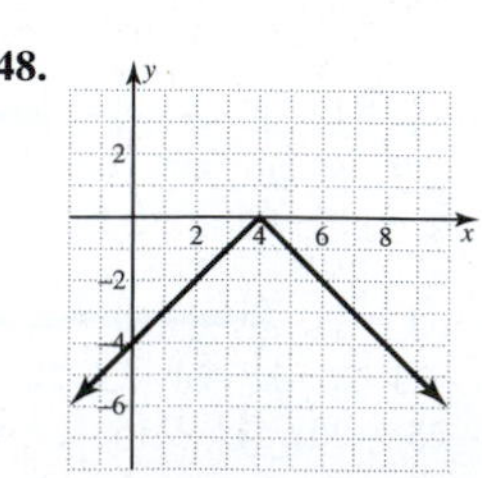

65. a.

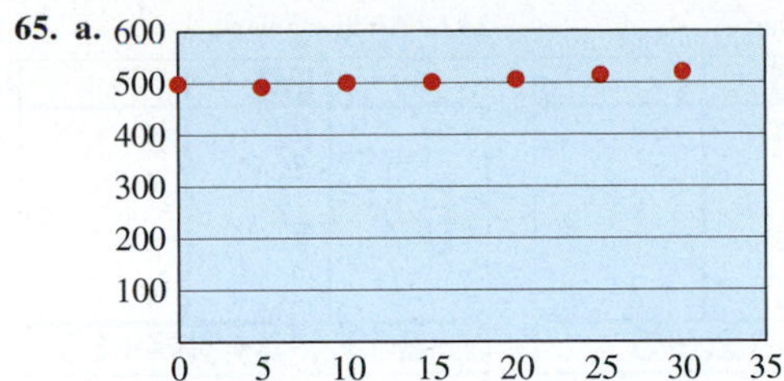

b. The mean SAT mathematics score was 498 in 1975 and 520 in 2005.

c. The mean SAT mathematics score decreased from 1975 to 1980 but slowly increased from 1980 to 2005.

66. a.

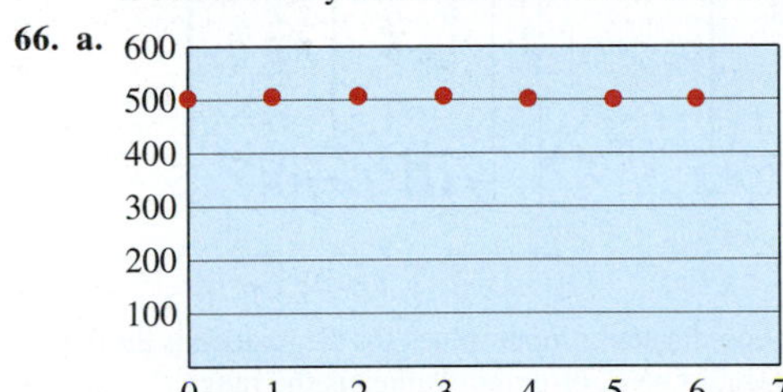

b. The mean SAT critical reading score for college-bound Georgia seniors was 504 in 2002 and 502 in 2008

c. The mean SAT critical reading score for college-bound Georgia seniors was increasing from year 2002 to 2004, remained steady for a year, was slowly decreasing from 2006 to 2007, then remained constant the following year.

67. a. (0, 3391), (1, 3652), (2, 3925), (3, 3517), (4, 3392)

b.

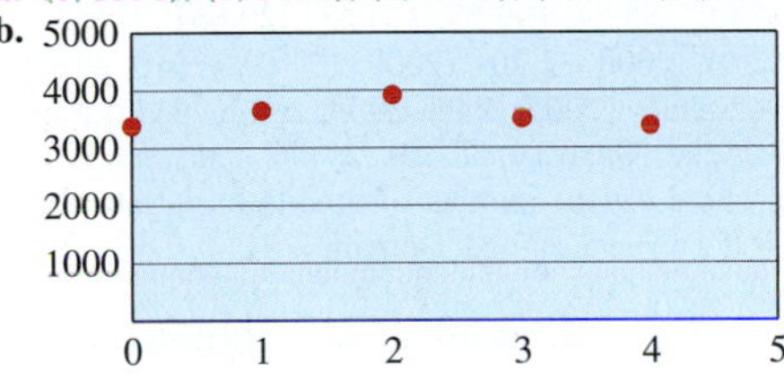

c. Honda automobiles sales were increasing from 2006 to 2008 and decreasing from 2008 to 2010.

68. a. (a) (0, 7), (1, 6.2), (2, 5.3), (3, 5.4), (4, 6.4) (5, 10.3), (6, 13.2), (7, 12.4)

b.

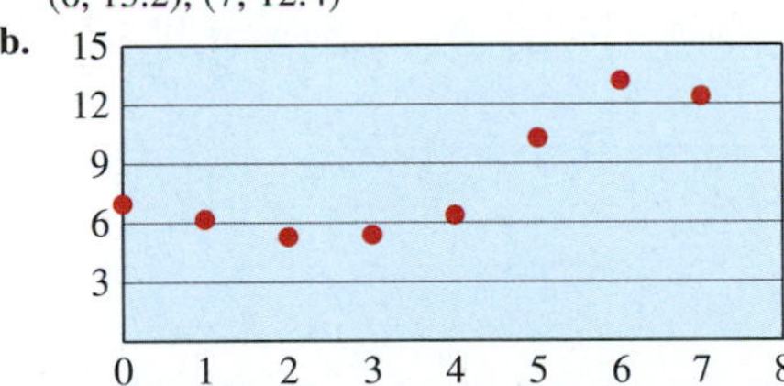

c. The percent of the labor force that is unemployed in California was decreasing from 2004 to 2007, increasing sharply from 2007 to 2010, and deceasing from 2010 to 2011.

69. II, II, IV

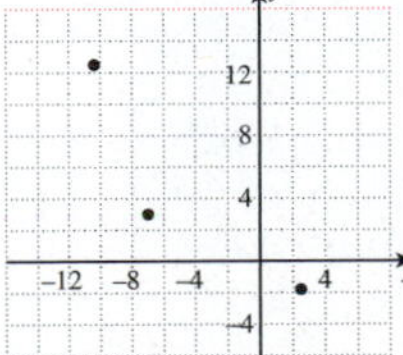

70. IV, I, II

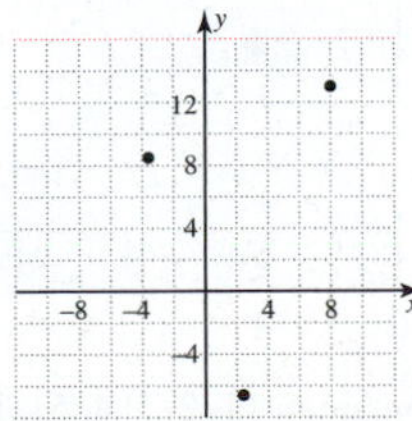

71. III, IV, III

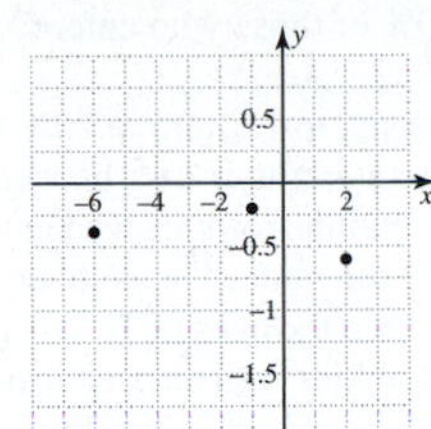

72. II, IV, III

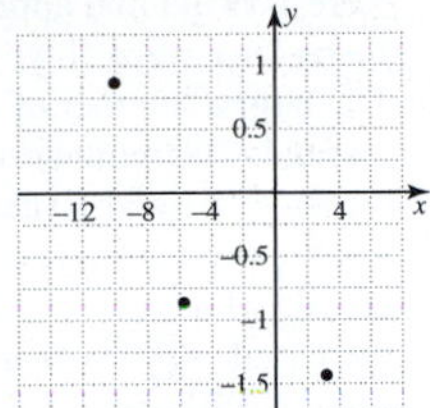

73. III, IV, I

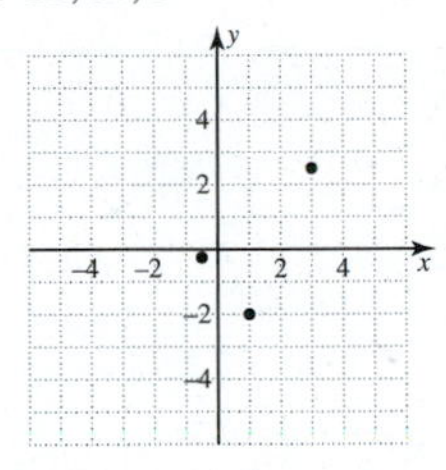

74. IV, II, I

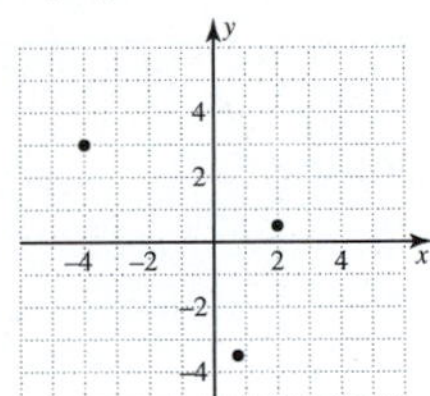

83.

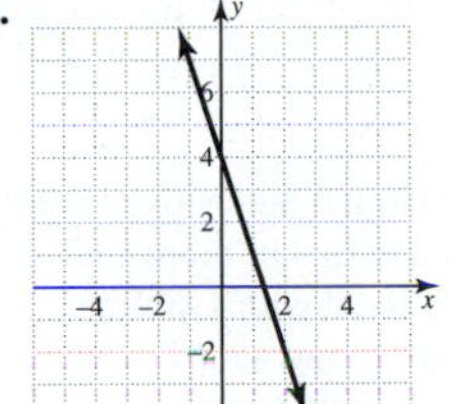

84.

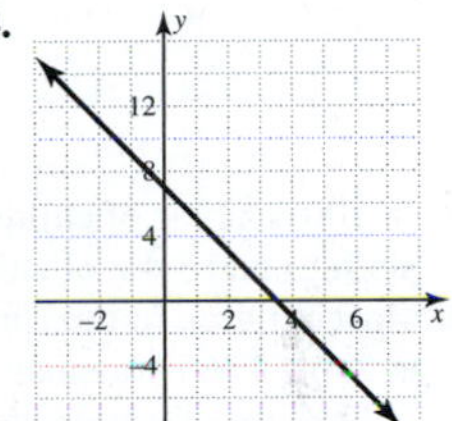

85.

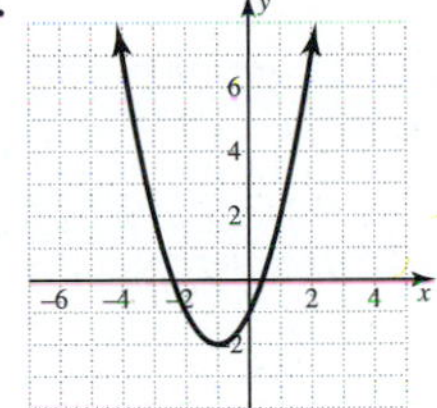

86.

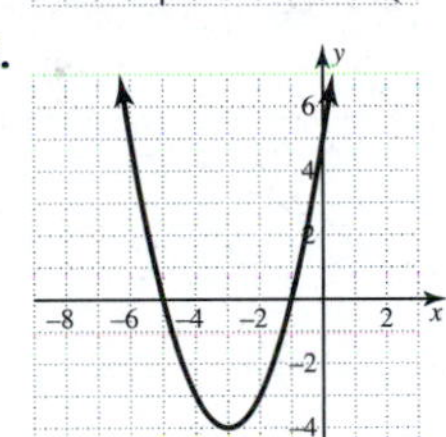

87.

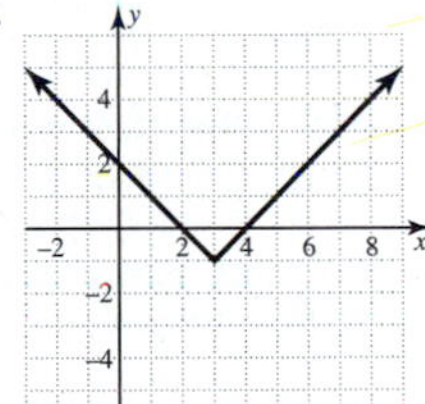

88.

89.

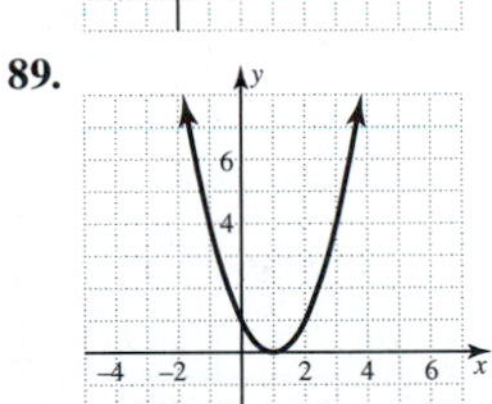

90.

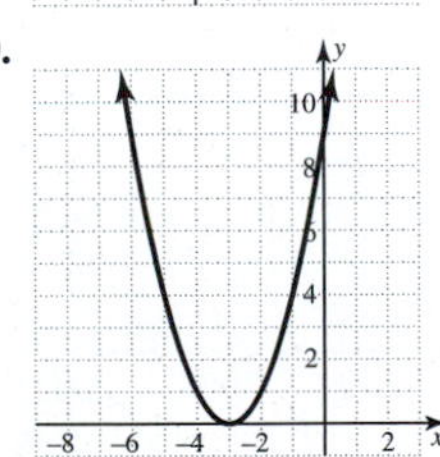

103. a.

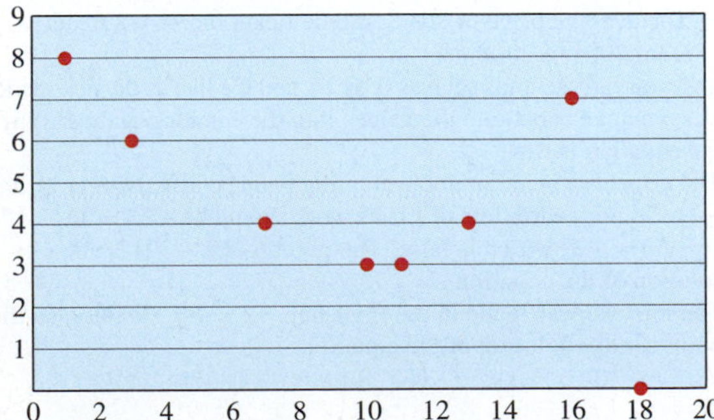

b. In 2010, approximately 8% of those who entered foster care were 1 yr old and approximately 0% of those who entered foster care were 18 yr old.

c. The largest number of children entering foster care in 2010 were 1 yr old. The number of kids who entered foster care between the ages of 1 and 10 decreased. The number of kids who entered foster care at the age of 11 remained the same. The number of kids who entered foster care between the ages of 11 and 16 increased. Approximately none of the kids who entered foster care in 2010 were 18 yr old.

104. a.

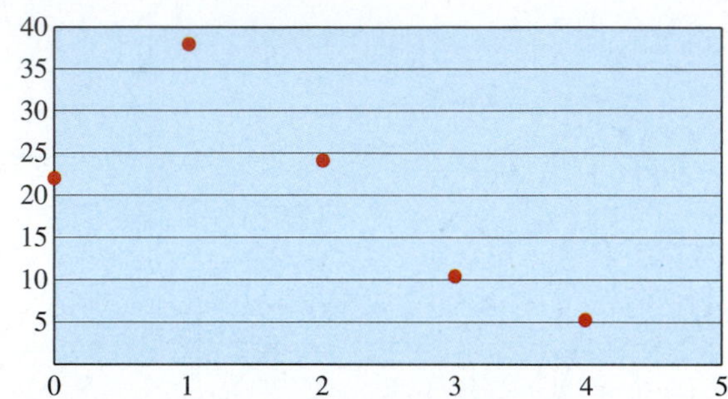

b. In 2009, 22.1% of children in the United States lived with 0 siblings and 5.4% of children lived with 4 or more siblings.

c. Most children in the United States lived with 1 sibling in 2009. The percentage of children living with 2 or more siblings decreases as the number of siblings increases.

105. a.

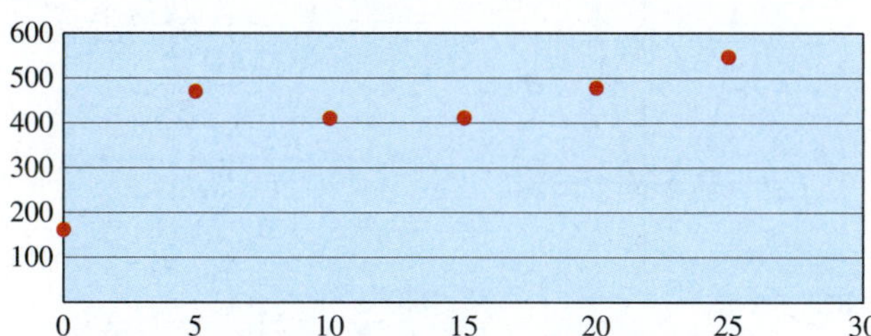

b. There were 161 domestic movies released 0 yr after 1980, or 1980, and 547 domestic movies released 25 yr after 1980, or 2005.

c. The number of domestic movies released increased sharply from 1980 to 1985, decreased from 1985 to 1990, increased slowly from 1990 to 1995, but increased sharply from 1995 to 2005.

106. a.

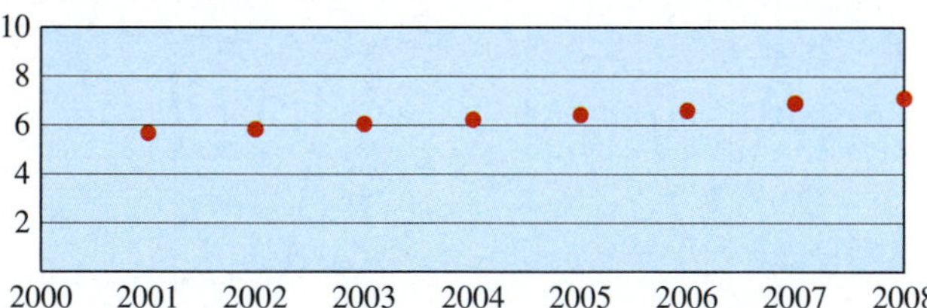

b. The average price of a U.S. movie ticket was \$5.66 in 2001 and \$7.08 in 2008.

c. The average price of a U.S. movie ticket increased steadily from 2001 to 2008.

107. Yes, you can use units such as 0, 5, 10, and the like as the tick marks.

108. Yes, you can substitute the values into the equation and verify if the equation is true.

109. The error was made in not evaluating $|-11|$ correctly. It is 11 not -11. So, the fourth line of Lisa's work should be $-33 = 11 - 12$ so $-33 = -1$, which is false. The point $(-25, -33)$ is not a solution of the equation.

110. The mistake was made in the final step. $12 = -12$ is false, so the point is not a solution of the equation.

111. Answers vary.

X	Y1
-2	201
-1	60
0	5
1	36
2	153

Y1=43X²-12X+5

112. Answers vary.

X	Y1
-2	89
-1	27
0	13
1	17
2	9

Y1=-5X^3+9X²+13

113. Answers vary.

X	Y1
-2	-6.5
-1	-5
0	-3.5
1	-2
2	-.5

Y1=(3X-7)/2

114. Answers vary.

X	Y1
-2	2.8
-1	2
0	1.2
1	.4
2	-.4

Y1=(6-4X)/5

Section 3.2

3. The set of x-coordinates or input values for the relation is the domain.

4. The set of y-coordinates or output values is the range.

11. $y = 8.10x$, where x is the number of hours worked and y is the amount of money earned. Domain $= \{0, 1, 2, 3, \ldots\}$, Range $= \{0, 8.10, 16.20, 24.30, \ldots\}$

12. $y = 4.50x$, where x is the number of gallons Calista puts in her car and y is the cost of the sale. Domain $= \{0, 1, 2, 3, \ldots\}$, Range $= \{0, 4.50, 9.00, 13.50, \ldots\}$

15. $\{(2007, 84.05), (2008, 199.27), (2009, 85.84), (2010, 213.43), (2011, 325.65)\}$; Domain $= \{2007, 2008, 2009, 2010, 2011\}$, Range $= \{84.05, 85.84, 199.27, 213.43, 325.65\}$

16. $\{(2007, 30.19), (2008, 32.20), (2009, 17.10), (2010, 28.18), (2011, 27.75)\}$; Domain $= \{2007, 2008, 2009, 2010, 2011\}$, Range $= \{17.10, 27.75, 28.18, 30.19, 32.20\}$

18. $y = 2.50x$, where x is the number of double-chocolate chip cookies sold and y is the revenue earned; Domain $= \{0, 1, 2, 3, \ldots\}$, Range $= \{0, 2.50, 5.00, 7.5, \ldots\}$

20. $y = 2.00x$, where x is the number of salted pretzels sold and y is the revenue earned; Domain $= \{0, 1, 2, 3, \ldots\}$, Range $= \{0, 2.00, 4.00, 6.00, \ldots\}$

43. $y = 0.50x$, where x is the number of days late and y is the late fee. Domain $= \{0, 1, 2, 3, \ldots\}$, Range $= \{0, 0.50, 1.00, 1.50, \ldots\}$

44. $y = 1.25x$, where x is the number of shirts washed and y is the total cost; Domain $= \{0, 1, 2, 3, \ldots\}$, Range $= \{0, 1.25, 2.25, 3.75, \ldots\}$

47.

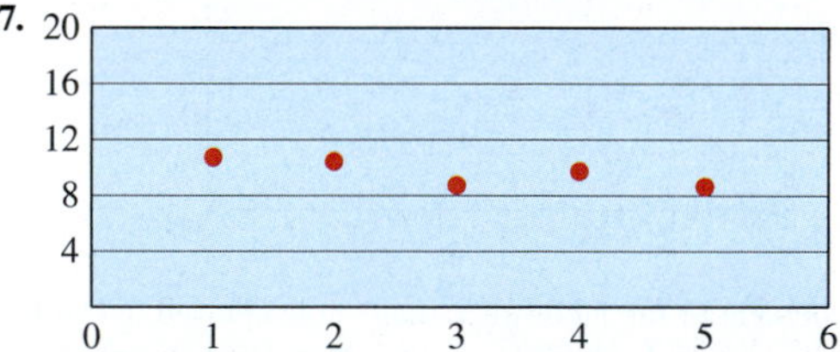

Domain $= \{1, 2, 3, 4, 5\}$, Range $= \{8.6, 8.7, 9.7, 10.4, 10.7\}$

48.

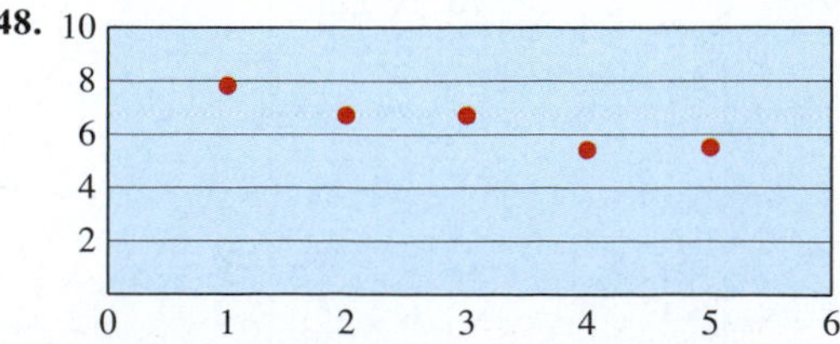

Domain $= \{1, 2, 3, 4, 5\}$, Range $= \{5.4, 5.5, 6.7, 7.8\}$

49. Domain $= \{-90, -71, -15, 20, 53\}$, Range $= \{-34, 4, 9, 34, 91\}$

50. Domain $= \{-100, -93, 33, 52\}$, Range $= \{23, 42, 83\}$

51. Domain $= \{-17, -16, 0, 18\}$, Range $= \{-17, 4, 18\}$

52. Domain = $\{-9, -6, 1, 24\}$, Range = $\{3, 7, 24\}$

75. Answers vary. Let x be a mom and let y be the name of her children.

76. Answers vary. Let x be a person and let y be their Social Security number.

77. Answers vary. The graph will be a set of discrete points.

78. Answers vary. The graph will be connected on this interval.

Piece It Together Sections 3.1 and 3.2

1.

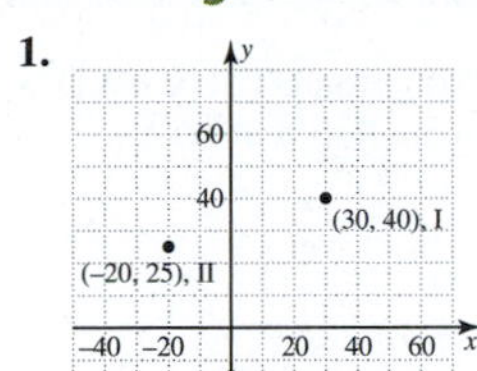

5.

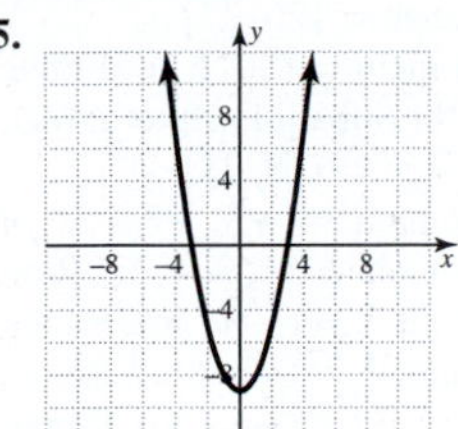

6.

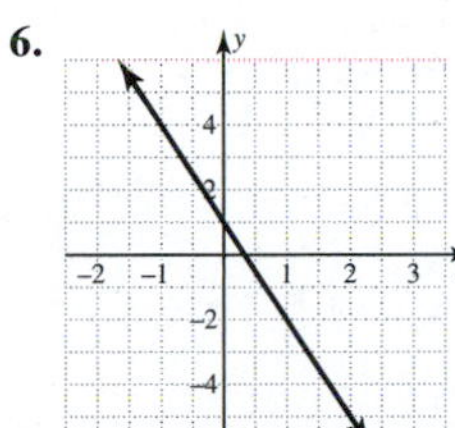

7.

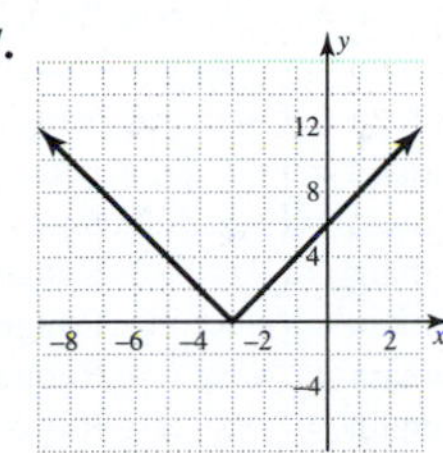

Section 3.3

1. No, not every relation is a function. A relation is a set of ordered pairs, a correspondence between a set of input values and output values.

2. Yes, a function is a relation in which each input value can have only one output value.

3. The vertical line test is used to determine if a graph represent a function. If any vertical line intersects a graph in more than one point, the graph does not represent a function.

4. Function notation, $f(x)$, is another name for the output value y. The notation $f(a) = b$ means that the input value a corresponds to the output value b.

6. not a function; -1 corresponds to 6 and 3.

7. not a function; -6 corresponds to -9 and 10.

17. not a function; For example, $x = 2$ corresponds to $y = 1$ and $y = -1$.

18. not a function; For example, $x = 0$ corresponds to $y = 1$ and $y = -1$.

19. not a function; For example, $x = 1$ corresponds to $y = -2$ and $y = -4$.

20. not a function; For example, $x = 1$ corresponds to $y = 3$ and $y = 1$.

61. $f(30) = 77$; In 2000, the life expectancy for all sexes and races was about 77 yr.

62. $f(15) = 75$; In 1985, the life expectancy for all sexes and races was about 75 yr.

63. $f(39) = 173$; In 2009, there were about 173 thousand twin births in the United States.

64. $f(20) = 114$; In 1990, there were about 114 thousand twin births in the United States.

65. $m(10) = 6$; In 2010, about 6% of middle school students used a tobacco product.

66. $m(12) = 4$; In 2012, about 4% of middle school students used a tobacco product.

67. $h(4) = 27$; In 2004, about 27% of high school students used a tobacco product.

68. $h(8) = 14$; In 2008, about 14% of high school students used a tobacco product.

89. $d(10) = 76{,}485$; In 2010, about 76,485 AIDS cases were diagnosed.

90. $d(13) = 184{,}527$; In 2013, about 184,527 AIDS cases were diagnosed.

91. $x = 7$; In 2007, about 20% of adults age 18 and over were smokers.

92. $x = 16$; In 2016, about 16% of adults age 18 and over were smokers.

93. $g(3)$ means to evaluate the function at $x = 3$ not multiply the function by 3.

$g(x) = 7 - x^2$

$g(3) = 7 - (3)^2 = 7 - 9 = -2$

94. The order of operations tells us to evaluate the exponent and then multiply.

$g(x) = 2x^2 + 5$

$g(-4) = 2(-4)^2 + 5 = 2(16) + 5 = 32 + 5 = 37$

99. a. $\{-6, 2\}$

b.

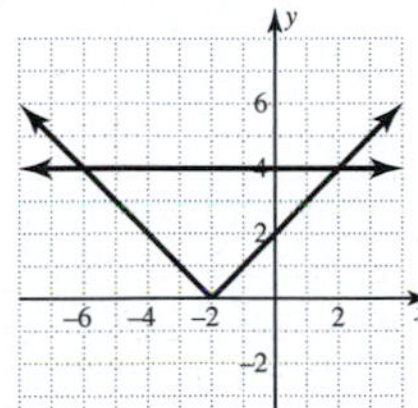

c. $(2, 4)$ and $(-6, 4)$; The x-values of the points of intersection are the solutions of the equation $|x + 2| = 4$.

d. $(-6, 2)$

e.

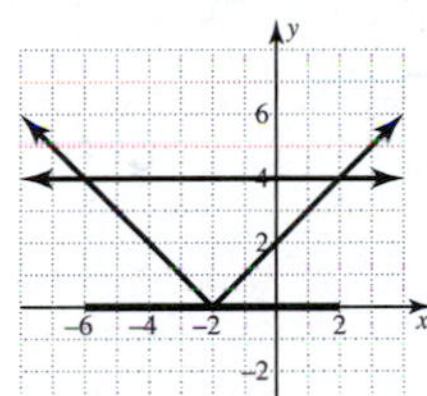

f. To solve $f(x) < 4$ using the graph, we can find the values of x for which the graph of $f(x)$ is below the graph of $y = 4$.

100. a. $\{-1, 3\}$

b.

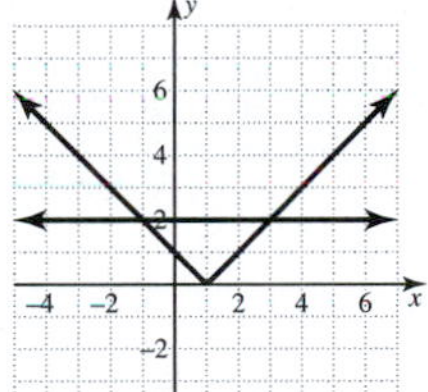

c. $(-1, 2)$ and $(3, 2)$; The x-values of the points of intersection are the solutions of the equation $|x - 1| = 2$.

d. $(-\infty, -1) \cup (3, \infty)$

e.

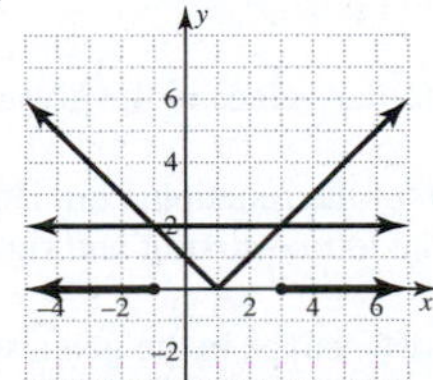

f. To solve $f(x) > 2$ using the graph, we can find the values of x for which the graph of $f(x)$ is above the line through $y = 2$.

101. a. ∅

b.

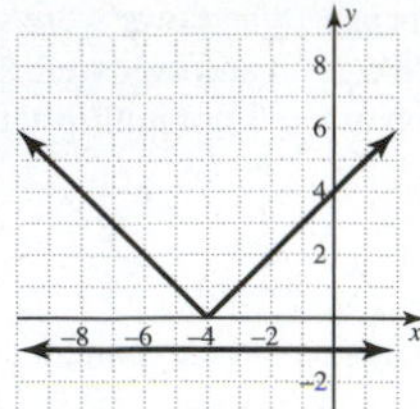

c. There are no points where the two graphs intersect, and there are no solutions of the equation $|x + 4| = -1$.

d. $(-\infty, \infty)$

e.

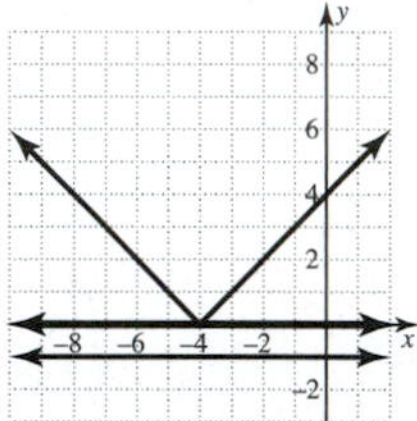

f. To solve $f(x) > -1$, we can find the values of x such that the graph of $f(x)$ is above the graph of $y = -1$.

102. a. ∅

b.

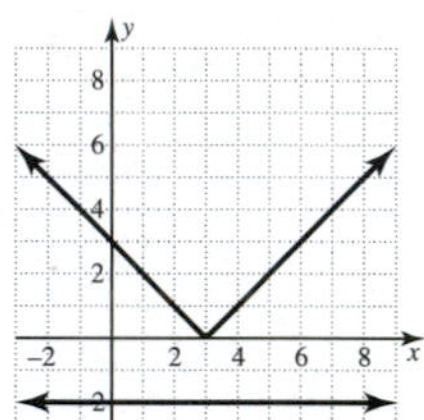

c. There are no points where the two graphs intersect, and there are no solutions of the equation $|x - 3| = -2$.

d. ∅

e.

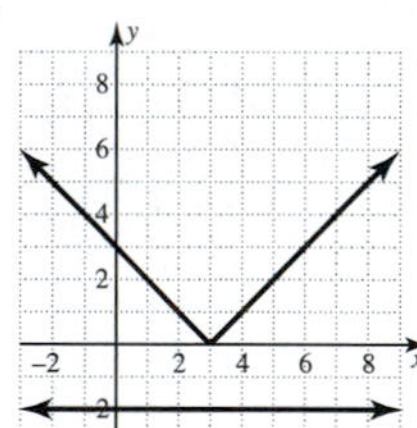

f. To solve $f(x) < -2$ using the graph, we can find the values of x for which the graph of $f(x)$ is below the line through $y = -2$.

Section 3.4

1. The domain of a function is the set of x-values in the given set of ordered pairs.
2. To find the domain, read the graph of the function from left to right to determine the x-value of the leftmost point and rightmost point on the graph.
3. The range of a function is the set of y-values in the given set of ordered pairs.
4. To find the range, read the graph from bottom to top to determine the y-value of the lowest point and the highest point on the graph.
5. To find the domain of an algebraic function with a fraction, exclude any number that makes the denominator of the fraction equal to zero or the function undefined.
6. To find the domain of an algebraic function with a square root, set the expression inside the square root greater than or equal to zero and solve the resulting inequality.
7. Domain = $\{-3, -2, -1, 0\}$, Range = $\{-4, 0, 1, 5\}$
8. Domain = $\{-1, 0, 1, 2\}$, Range = $\{-10, -7, 4, 6\}$
9. Domain = $\{-10, 0, 2, 5\}$, Range = $\{-7, 3, 4\}$
10. Domain = $\{-9, -8, 6, 7\}$, Range = $\{-11, -9, 11\}$
11. Domain = {2011 Ford Edge AWD, 2011 Honda CR-V 4WD, 2011 Subaru Forester AWD, 2011 Toyota RAV4 4WD}, Range = {26, 27}
12. Domain = {2011 Ford Edge AWD, 2011 Honda CR-V 4WD, 2011 Subaru Forester AWD, 2011 Toyota RAV4 4WD}, Range = {18, 21}

19. Domain = $\{-3, -2, 1, 2\}$, Range = $\{0, 1\}$
20. Domain = $\{1, 2, 4, 5\}$, Range = $\{0, 3\}$
21. Domain = $(-\infty, \infty)$, Range = $(-\infty, \infty)$
22. Domain = $(-\infty, \infty)$, Range = $(-\infty, \infty)$
23. Domain = $(-\infty, \infty)$, Range = $\{-3\}$
24. Domain = $(-\infty, \infty)$, Range = $\{1\}$
25. Domain = $(-\infty, \infty)$, Range = $(-\infty, 4]$
26. Domain = $(-\infty, \infty)$, Range = $(-\infty, 3]$
27. Domain = $[-2, 2]$, Range = $[0, 2]$
28. Domain = $[-3, 3]$, Range = $[-3, 0]$
29. Domain = $(-\infty, \infty)$, Range = $(-\infty, \infty)$
30. Domain = $(-\infty, \infty)$, Range = $(-\infty, \infty)$
31. Domain = $(-\infty, \infty)$, Range = $[1, \infty)$
32. Domain = $(-\infty, \infty)$, Range = $[1, \infty)$

39. $(-\infty, -5) \cup (-5, \infty)$
40. $(-\infty, 7) \cup (7, \infty)$

42. $[-2, \infty)$
43. $\left(-\infty, \frac{3}{4}\right]$
44. $\left(-\infty, \frac{5}{3}\right]$

57. Domain = $\{0, 4, 13, 20\}$, Range = $\{-3, -2, 44\}$
58. Domain = $\{-4, -2, 0, 4\}$, Range = $\{-14, 13, 21\}$

74. A fraction is undefined when the denominator is equal to zero, not the numerator. Since $x - 1 = 0$ when $x = 1$, the domain is $(-\infty, 1) \cup (1, \infty)$.

76.

X	Y1
-4	ERROR
-3	0
0	3
2	2.2361
5	ERROR

Y1=√(9−X²)

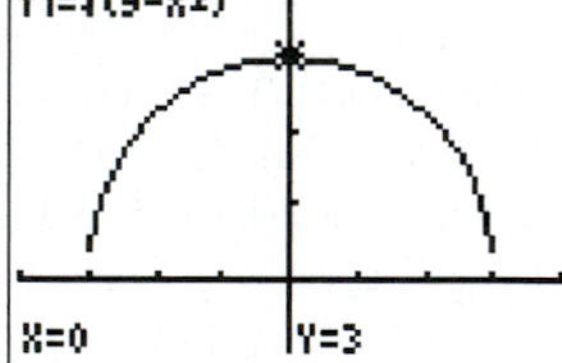

77.

X	Y1
-4	2.6458
-3	0
0	ERROR
2	ERROR
5	4

Y1=√(X²−9)

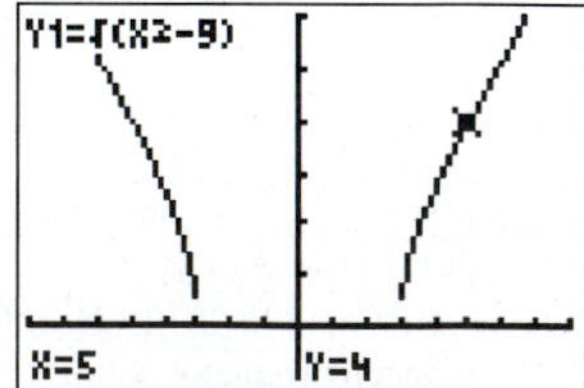

78.

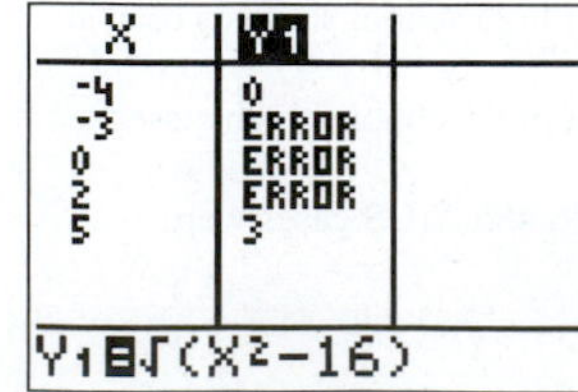

X	Y1
-4	0
-3	ERROR
0	ERROR
2	ERROR
5	3

Y1=√(X²−16)

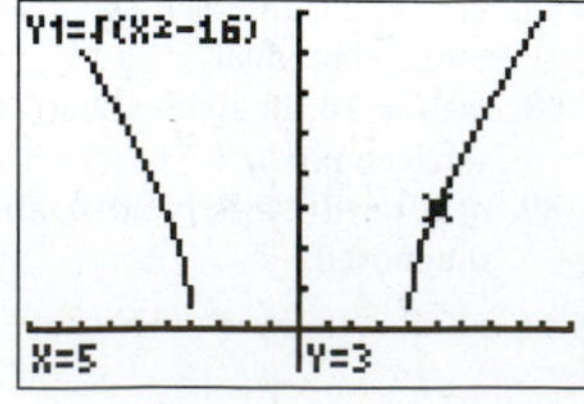

79.

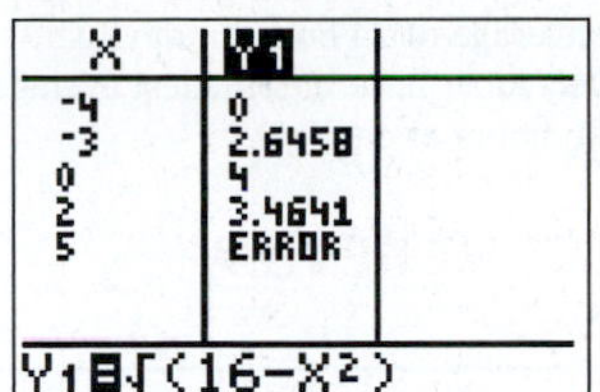

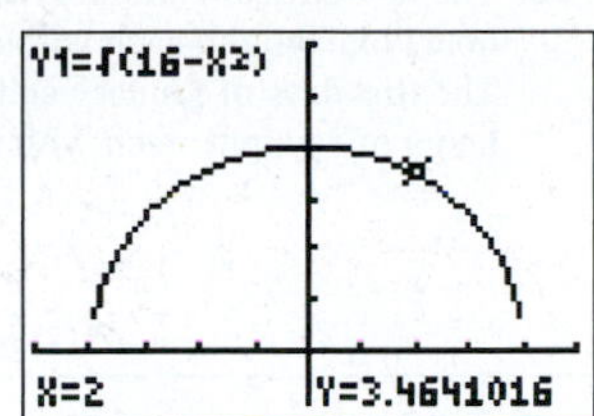

80. Answers vary.

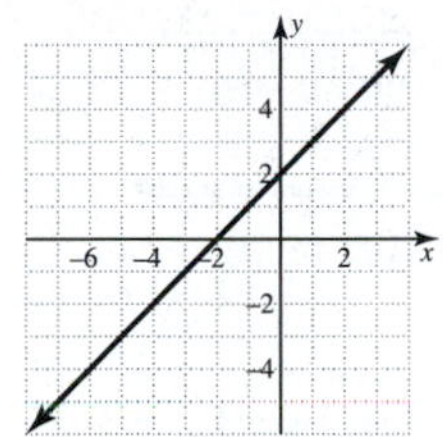

81. Answers vary.

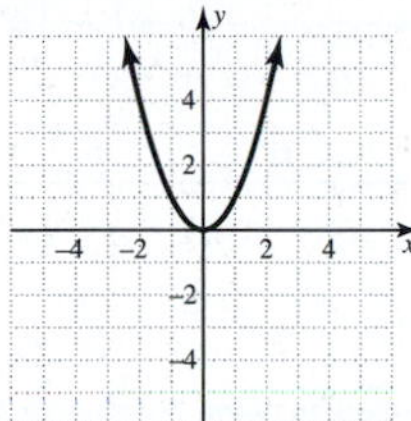

82. Answers vary.

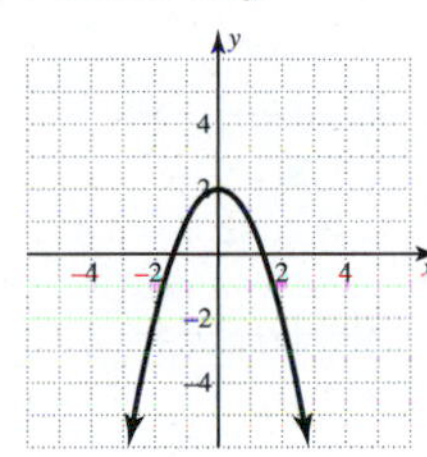

83. Answers vary.

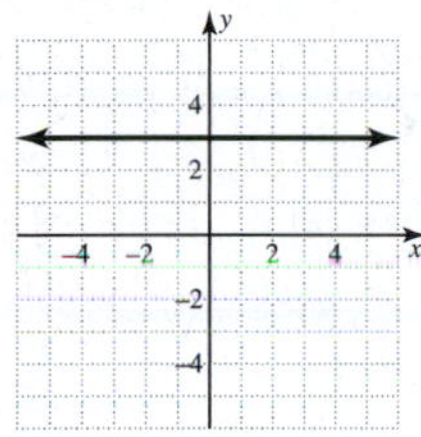

84. Answers vary.

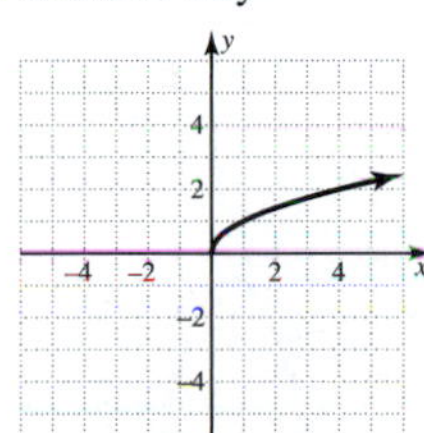

85. Answers vary.

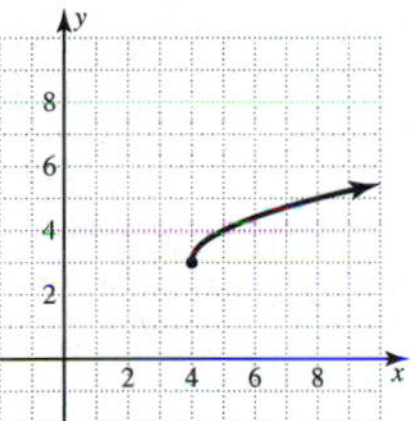

86. Answers vary.

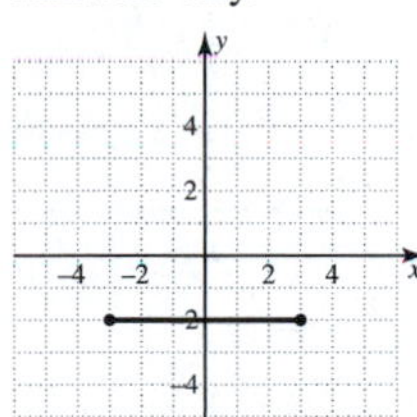

87. Answers vary.

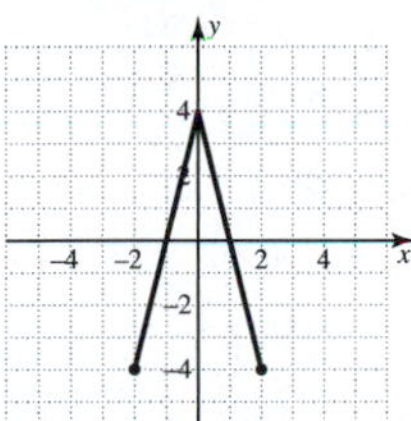

GROUP ACTIVITY

1. Answers vary. **2.** Answers vary.
3. Answers vary. **4.** Answers vary.

CHAPTER 3 REVIEW EXERCISES

Section 3.1

1.

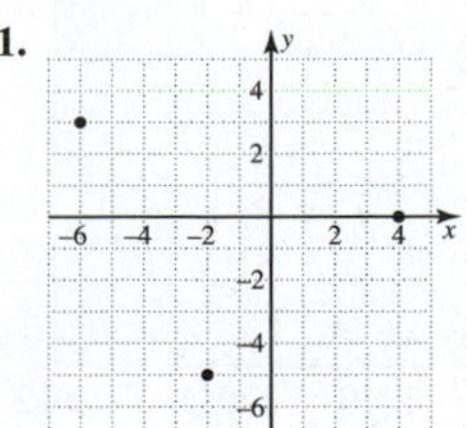

2.

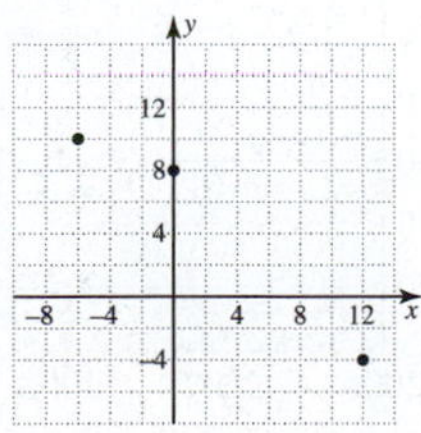

7.

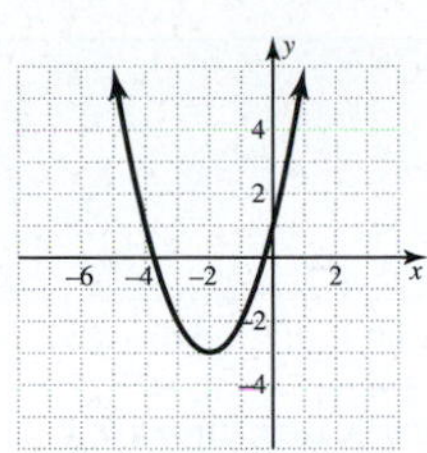

8.

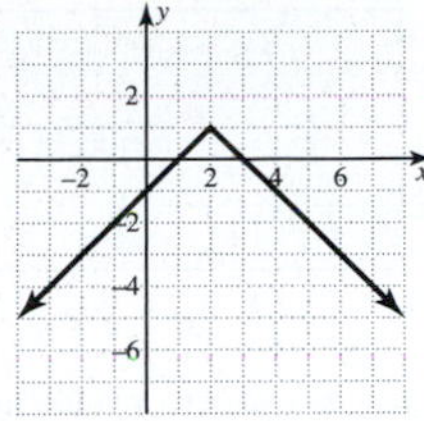

11. a.

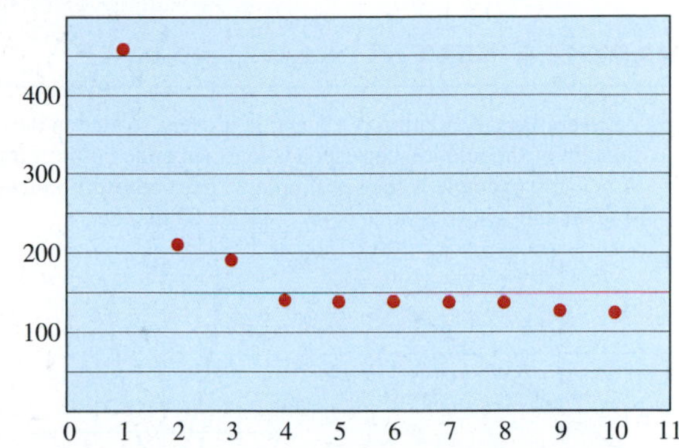

b. There were 461 deaths by lightning in the state of Florida between 1959 and 2010. There were 126 deaths by lightning in the state of Maryland between 1959 and 2010.

12. a.

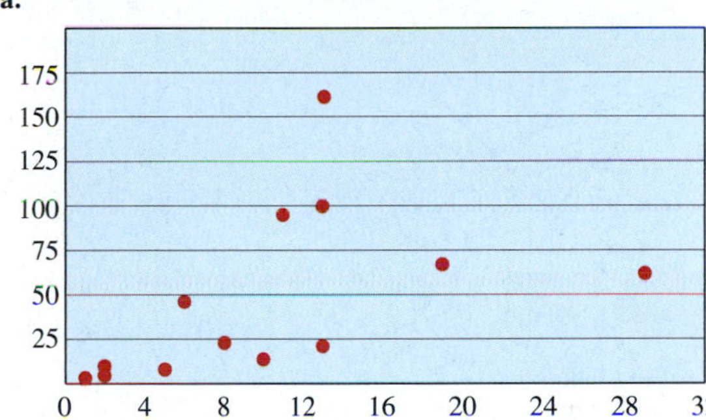

b. Deleware has 1 U.S. Representative and 3 counties. South Carolina has 6 U.S. Representatives and 46 counties.

c. The plot is a scatter plot with no particular trend. Georgia has 13 U.S. Representatives but has the most counties, 159. New York has the most U.S. Representatives, 29, but only 62 counties. Delaware has the least U.S. Representative, 1, and the least counties, 3.

Section 3.2

15.

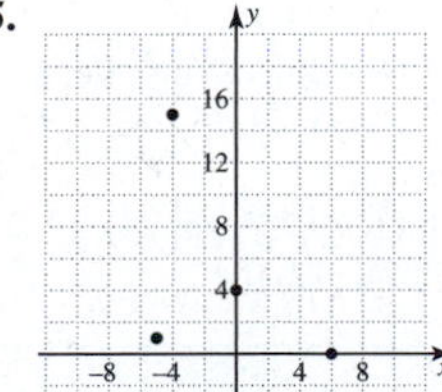

16.

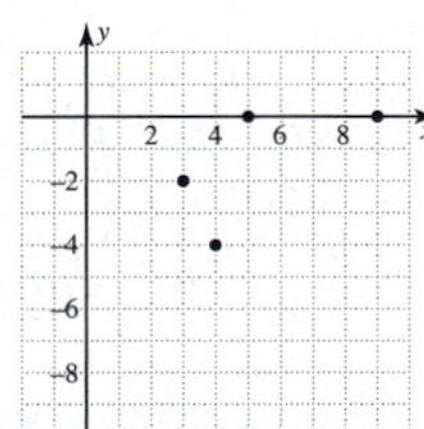

17.

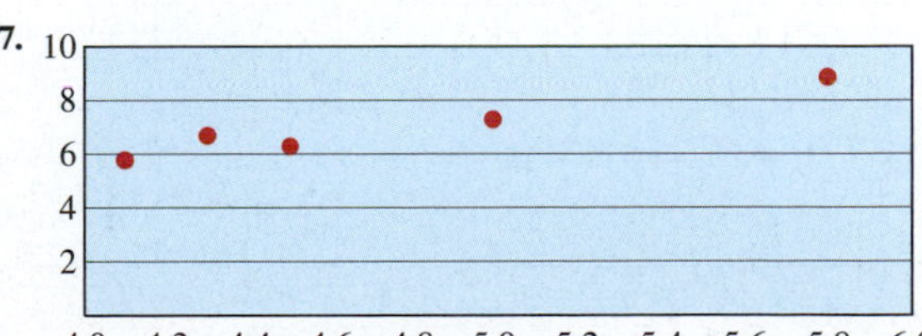

18.

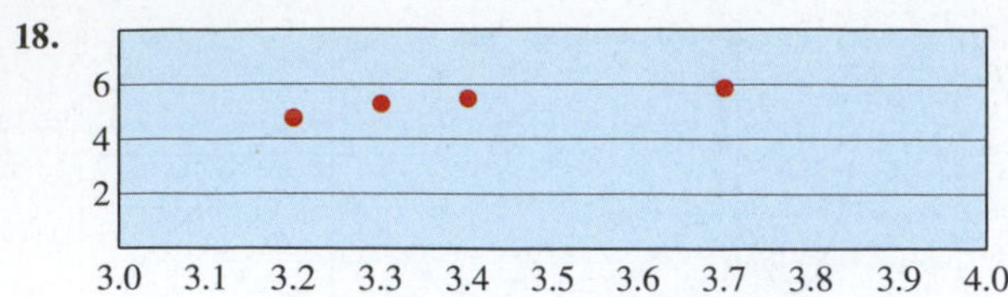

19. $y = 5.25x$, where x is the number of games, Domain $= \{0, 1, 2, \ldots\}$, Range $= \{0, 5.25, 10.50, \ldots\}$

CHAPTER 3 TEST

8. Answers vary. A relation is a function if every x value in the domain of the relation corresponds to exactly one output value. A real-life example is the set of ordered pairs whose x-value is a student and whose y-value is the students ID number.

9.

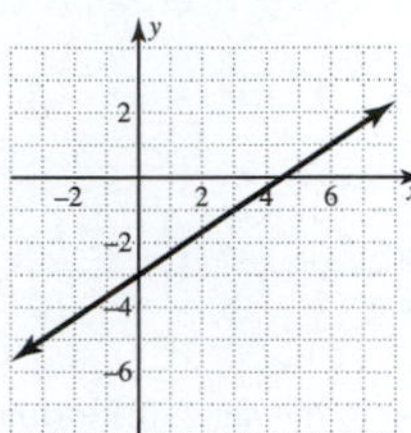

It is a function.
Domain $= (-\infty, \infty)$,
Range $= (-\infty, \infty)$

10.

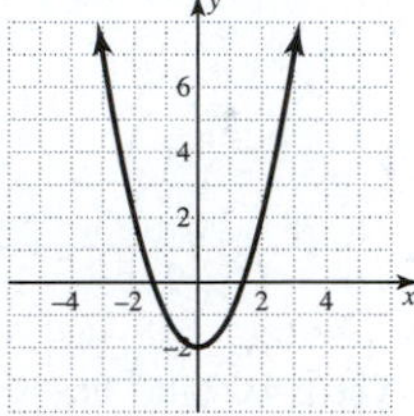

It is a function.
Domain $= (-\infty, \infty)$,
Range $= (-2, \infty)$

11.

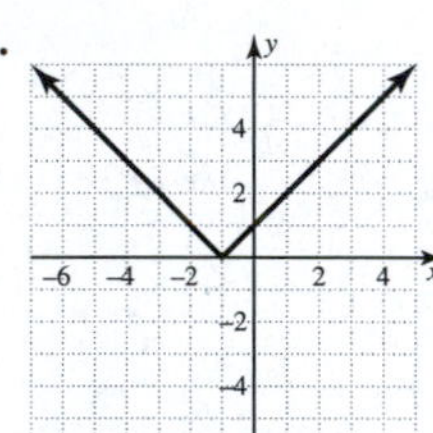

It is a function.
Domain $= (\infty, \infty)$,
Range $= [0, \infty)$

12. Answers vary. The vertical line test is used to determine if the graph of a relation represents a function. If every vertical line drawn through a graph touches at most one point on the graph, then the relation is a function. If at least one vertical line drawn through a graph touches two or more points on the graph, then the relation is not a function.

19. The number of hours of daylight is a function of the month, since each month has exactly one average number of hours that corresponds to it.

20. Domain: {Jan., Feb., Mar., Apr., May, June, July, Aug., Sept., Oct., Nov., Dec.}, Range: {0, 4.08, 5.87, 9.33, 11.05, 14.22, 14.75, 19.73, 24.00}

21. {(1, 0), (2, 4.08), (3, 9.33), (4, 14.22), (5, 19.73), (6, 24), (7, 24), (8, 24), (9, 14.75), (10, 11.05), (11, 5.87), (12, 0)}

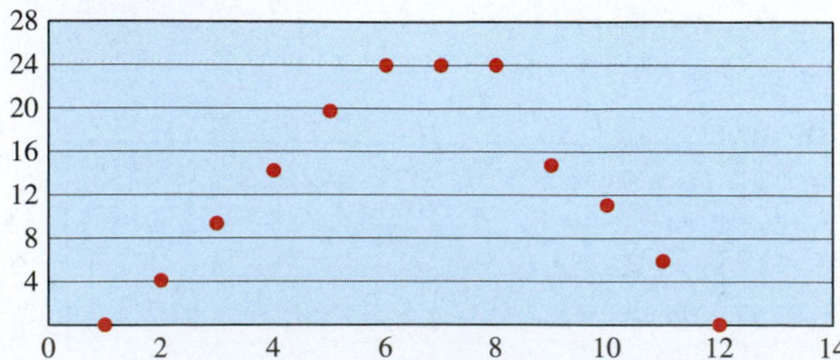

22. The first days of June, July, and August have the largest average hours of daylight, each with an average of 24 hours of daylight. The first days of January and December have the smallest average hours of daylight, each with zero hours of daylight.

CHAPTERS 1–3 CUMULATIVE REVIEW EXERCISES

7.

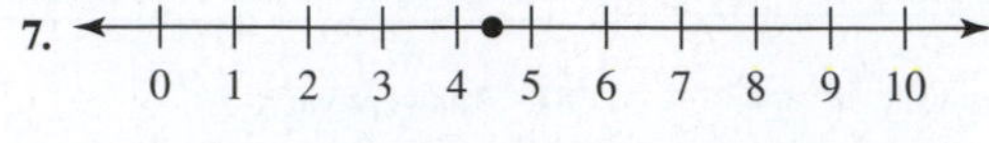

8.

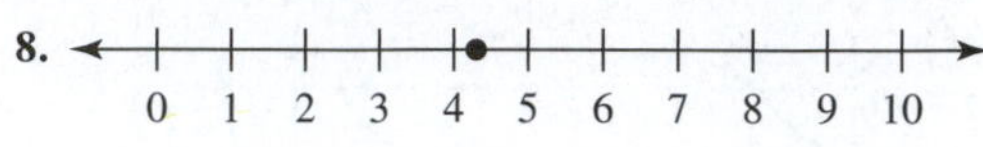

67.

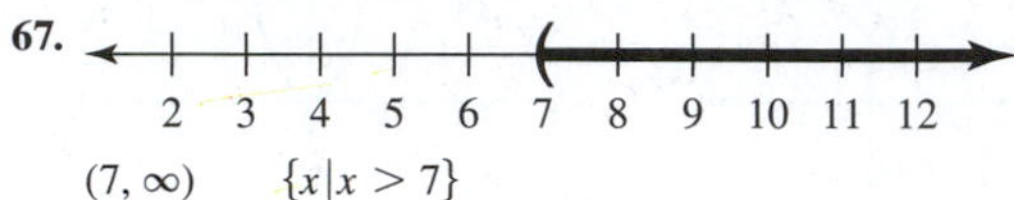

$(7, \infty)$ $\{x \mid x > 7\}$

68.

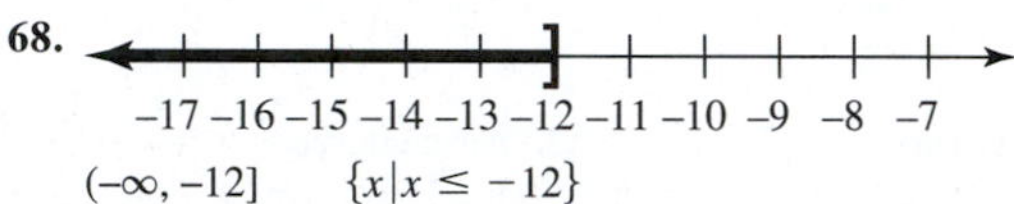

$(-\infty, -12]$ $\{x \mid x \leq -12\}$

69.

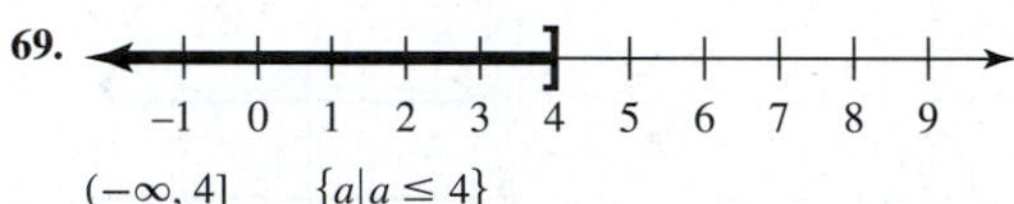

$(-\infty, 4]$ $\{a \mid a \leq 4\}$

70.

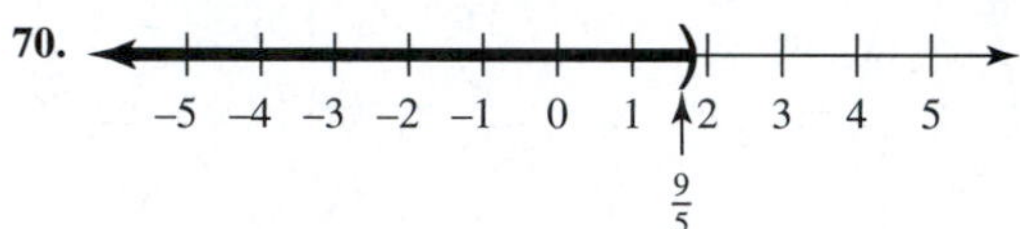

$\left(-\infty, \frac{9}{5}\right), \left\{x \,\middle|\, x < \frac{9}{5}\right\}$

85.

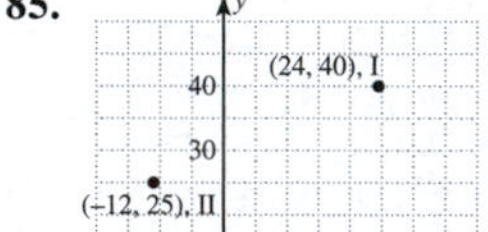

86.

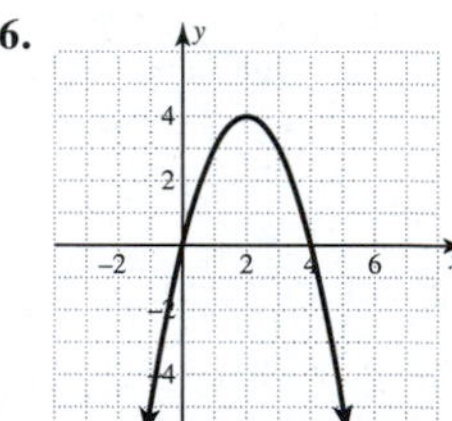

88. a. {(0, 10,271), (1, 10,369), (2, 9320), (3, 10,114), (4, 9639)}

b.

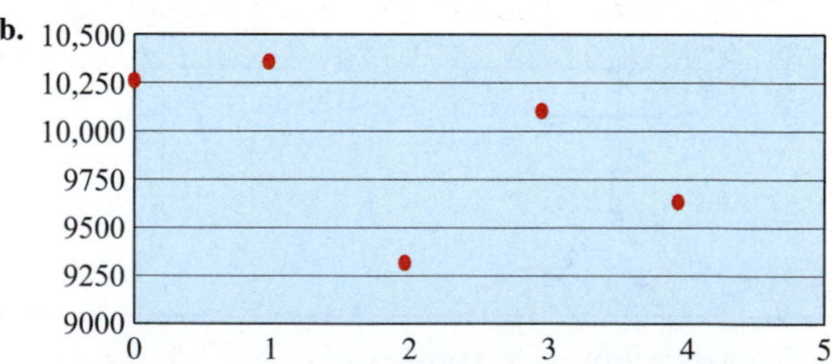

c. The number of units of Honda motorcycles increased from 2006 to 2007, decreased from 2007 to 2008, increased from 2008 to 2009, and decreased from 2009 to 2010.

89.

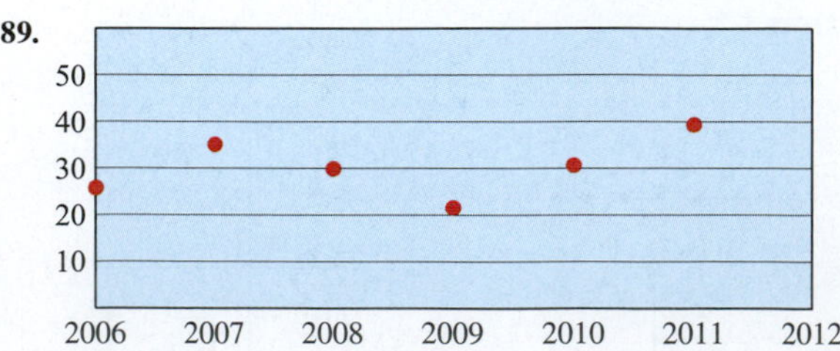

90.

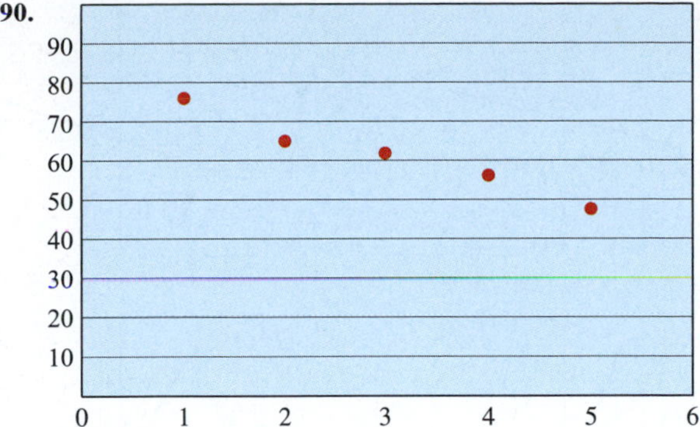

91.

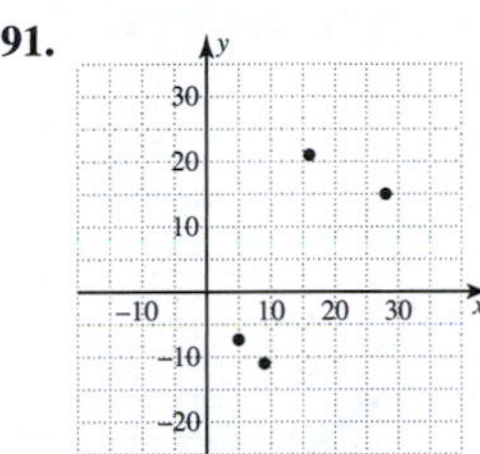

93. $y = 12x$, where x is the number of one-way tickets sold; Domain $= \{0, 1, 2, \ldots\}$, Range $= \{0, 12, 24, \ldots\}$

CHAPTER 4

Section 4.1

2. To evaluate a linear function, you substitute the given x value into the function in place of the variable and simplify.

4. You evaluate $f(0)$ by substituting 0 into the function in place of the variable x and simplify. You solve the equation $f(x) = 0$ for x.

5. To write a linear equation in two variables in function notation, solve the equation for y and replace y with $f(x)$; No, not all linear equations in two variables represent functions. An equation of the form $x = a$, where a is a real number, does not represent a function.

15. linear, $m = \frac{3}{2}, b = 0$ **16.** linear, $m = \frac{2}{5}, b = 0$

17. linear, $m = -\frac{1}{4}, b = -8$ **18.** linear, $m = -\frac{1}{6}, b = 12$

27. $f(x) = \frac{2}{3}x - 5, m = \frac{2}{3}, b = -5$

28. $f(x) = \frac{5}{4}x + 1, m = \frac{5}{4}, b = 1$

33. $f(x) = -\frac{1}{6}x - 3, m = -\frac{1}{6}, b = -3$

34. $f(x) = -\frac{1}{9}x - 1, m = -\frac{1}{9}, b = -1$

35. $f(x) = -\frac{4}{3}x + 4, m = -\frac{4}{3}, b = 4$

36. $f(x) = -\frac{6}{5}x + 6, m = -\frac{6}{5}, b = 6$

37. $f(x) = \frac{9}{2}x + 36, m = \frac{9}{2}, b = 36$

38. $f(x) = \frac{8}{3}x + 8, m = \frac{8}{3}, b = 8$ **39.** $f(x) = \frac{1}{7}x, m = \frac{1}{7}, b = 0$

40. $f(x) = -\frac{1}{4}x, m = -\frac{1}{4}, b = 0$ **41.** $f(x) = \frac{2}{7}x, m = \frac{2}{7}, b = 0$

42. $f(x) = \frac{3}{8}x, m = \frac{3}{8}, b = 0$ **63.** $x = \frac{3}{5}$ **64.** $x = \frac{1}{7}$

65. $x = -\frac{14}{3}$

69. b. $f(300) = 174.99$; The total monthly cost for an additional 300 min is \$174.99.
c. $x = 100$; The total monthly cost for an additional 100 min in the first month is \$124.99.

70. a. $f(x) = 94.99 + 0.40x$ for the first month
b. $f(100) = 134.99$; The total charge for an additional 100 min in the first month is \$134.99.
c. $x = 300$; The total charge for an additional 300 min in the first month is \$214.99.

71. b. $f(19) = 11.65$; The cost of 19 additional one-fifth miles, or 4 mi, is \$11.65.
c. $x = 29$; The cost for 29 additional one-fifth miles, or 6 mi, is \$16.15.

72. b. $f(5) = 12.50$; The cost of a 5-mi cab ride is \$12.50.

73. $f(x) = \frac{5}{2}x + 5, m = \frac{5}{2}, b = 5, f(0) = 5, x = -2$

74. $f(x) = -\frac{7}{3}x + 7, m = -\frac{7}{3}, b = 7, f(0) = 7, x = 3$

75. $f(x) = -1.1x + 4.4, m = -1.1, b = 4.4, f(0) = 4.4, x = 4$

76. $f(x) = -3.7x - 11.1, m = -3.7, b = -11.1, f(0) = -11.1, x = -3$

77. $f(x) = 3x - 5, m = 3, b = -5, f(0) = -5, x = \frac{5}{3}$

78. $f(x) = -\frac{4}{3}x + 5, m = -\frac{4}{3}, b = 5, f(0) = 5, x = \frac{15}{4}$

79. $f(x) = 4.8x - 3.6, m = 4.8, b = -3.6, f(0) = -3.6, x = 0.75$

80. $f(x) = 0.3x - 2.1, m = 0.3, b = -2.1, f(0) = -2.1, x = 7$

81. $f(x) = -2, m = 0, b = -2, f(0) = -2$, no solution

82. $f(x) = 4, m = 0, b = 4, f(0) = 4$, no solution

85. b. $f(5) = 625$; The cost of a 5-hr repair on an electric window is \$625.
c. $x = 2$; The cost of a 2-hr repair on an electric window is \$400.

86. a. $f(x) = 45x + 50$
b. $f(3) = 185$; The cost of a 3-hr house call is \$185.
c. $x = 1.5$; The cost of a 1.5-hr house call is \$117.50.

87. b. $f(24) = 235$; The cost of joining the fitness club for 24 months is \$235.
c. $x = 36$; The cost of joining the fitness club for 36 months is \$343.

88. a. $f(x) = 19.95x + 99$
b. $f(12) = 338.40$; The cost of joining the fitness club for 12 months is \$338.40.
c. $x = 60$; The cost of joining the fitness club for 60 months is \$1296.

89. b. $f(41{,}000) = 15{,}000$; A single person making \$34,500 + \$41,000 = \$75,500 must pay \$15,000 in federal income tax.
c. $x = 17{,}000$; A single person making \$34,500 + \$17,000 = \$51,500 must pay \$9000 in federal income tax.

90. b. $f(1475) = 13{,}956.75$; A married person filing separate who makes \$69,675 + \$1475 = \$71,150 must pay \$13,956.75 in federal income tax.

97.
$$
\begin{aligned}
f(x) &= 9x - 2 \\
f(x) &= 0 \\
9x - 2 &= 0 \\
9x - 2 + 2 &= 0 + 2 \\
9x &= 2 \\
\frac{9x}{9} &= \frac{2}{9} \\
x &= \frac{2}{9}
\end{aligned}
$$

99.

```
5/6(0)-1
                  -1
5/6(-12)-1
                 -11
```

100.

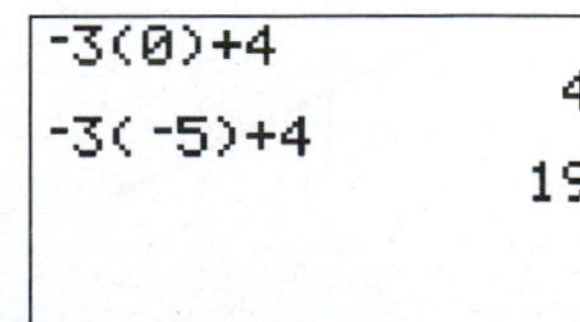

101.

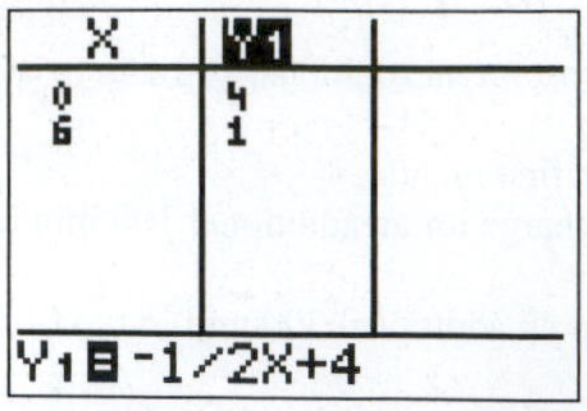

X	Y1
0	4
6	1

Y1 = −1/2X+4

102.

X	Y1
0	−2
−3	−1

Y1 = −1/3X−2

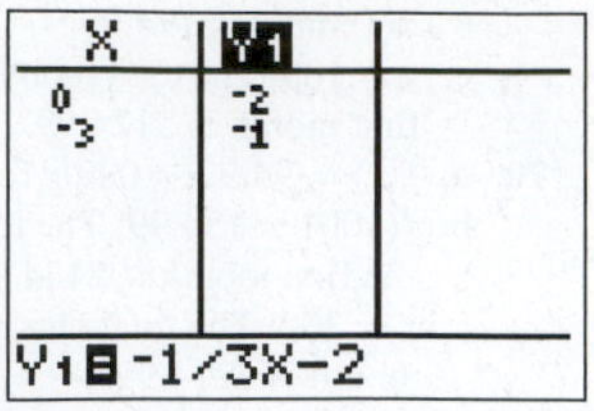

Section 4.2

1. An equation is a linear equation in two variables if it is of the form $f(x) = mx + b$, or $y = mx + b$, where m and b are real numbers. A linear equation can also be in standard form, $Ax + By = C$, where A, B, and C are real numbers and $B \neq 0$.
2. To graph a linear equation, make a table of at least three ordered pairs that satisfy the equation. Plot the ordered pairs and connect the points to draw the line.
3. To find the x-intercept, replace y with 0 and solve for x. Write the x-intercept in the form $(x, 0)$.
4. To find the y-intercept, replace x with 0 and solve for y. Write the y-intercept in the form $(0, y)$.
5. We must find another solution of the equation to have enough information to graph the line.
6. The equation of the form $x = h$ represents a vertical line and the equation of the form $y = k$ represents a horizontal line.

7.
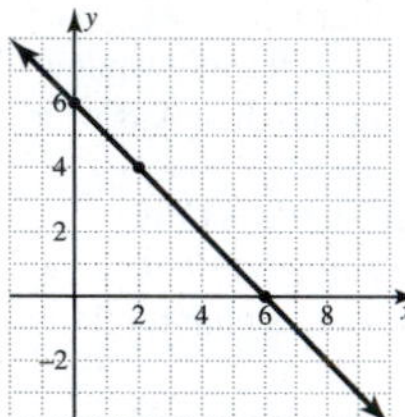
Domain: $(-\infty, \infty)$, Range: $(-\infty, \infty)$

8.
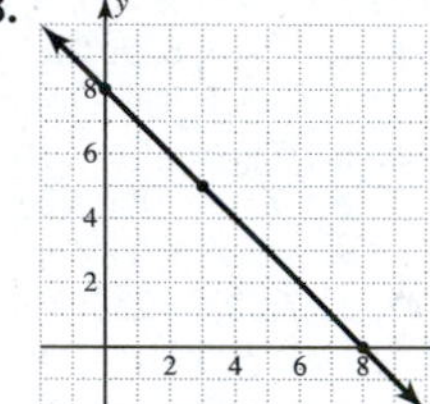
Domain: $(-\infty, \infty)$, Range: $(-\infty, \infty)$

9.
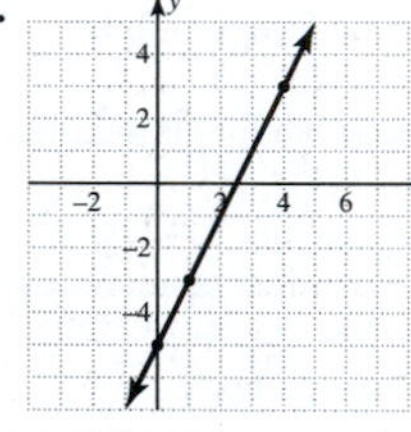
Domain: $(-\infty, \infty)$, Range: $(-\infty, \infty)$

10.
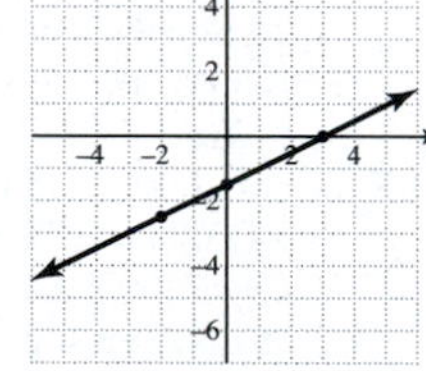
Domain: $(-\infty, \infty)$, Range: $(-\infty, \infty)$

11.
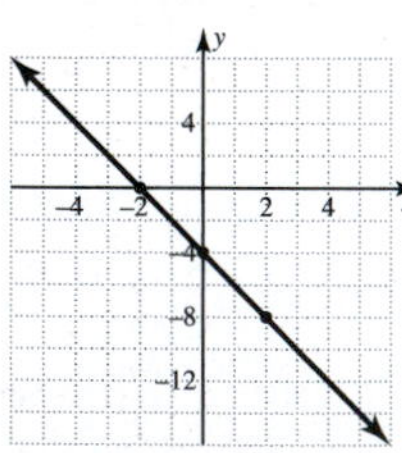
Domain: $(-\infty, \infty)$, Range: $(-\infty, \infty)$

12. 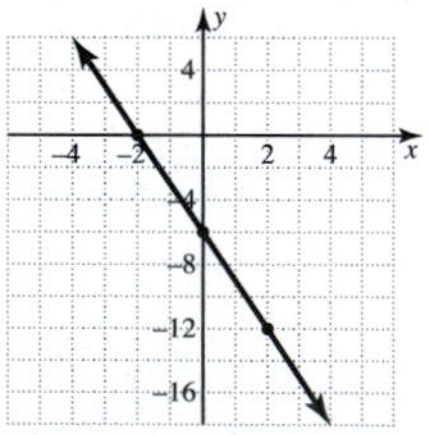
Domain: $(-\infty, \infty)$, Range: $(-\infty, \infty)$

13.
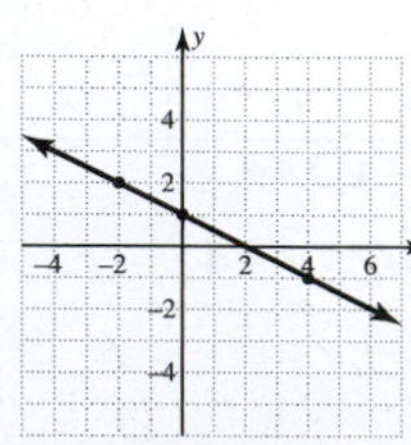
Domain: $(-\infty, \infty)$, Range: $(-\infty, \infty)$

14.
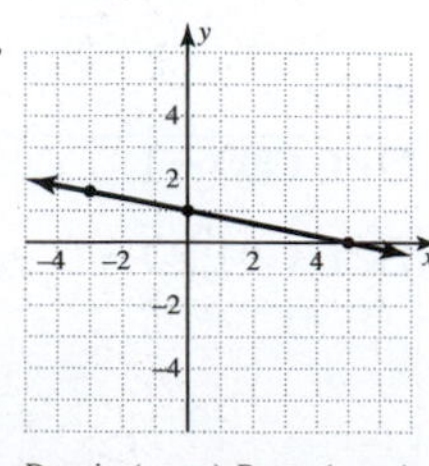
Domain: $(-\infty, \infty)$, Range: $(-\infty, \infty)$

15.
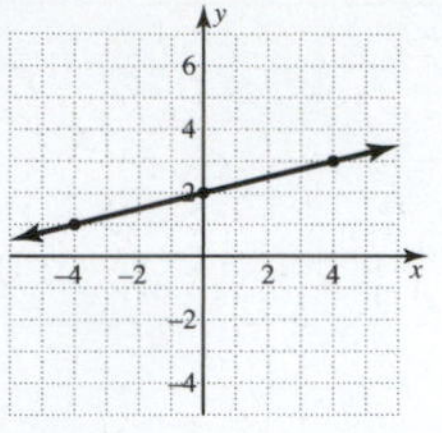
Domain: $(-\infty, \infty)$, Range: $(-\infty, \infty)$

16.
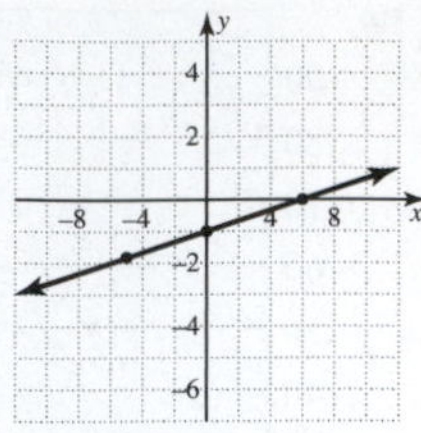
Domain: $(-\infty, \infty)$, Range: $(-\infty, \infty)$

17.
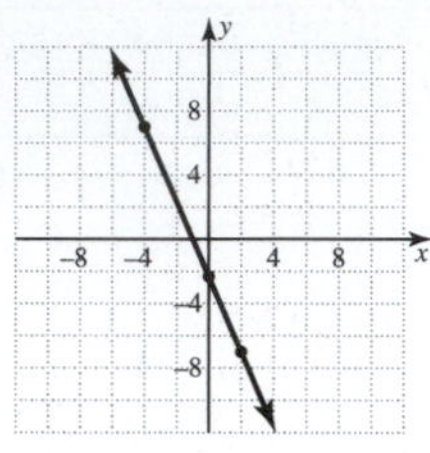
Domain: $(-\infty, \infty)$, Range: $(-\infty, \infty)$

18.
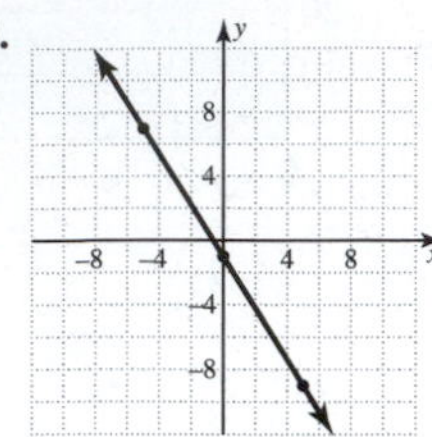
Domain: $(-\infty, \infty)$, Range: $(-\infty, \infty)$

19.
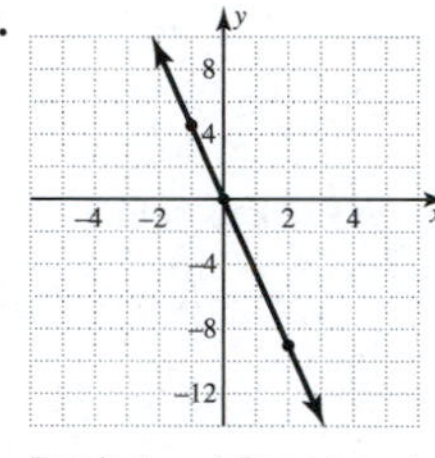
Domain: $(-\infty, \infty)$, Range: $(-\infty, \infty)$

20.
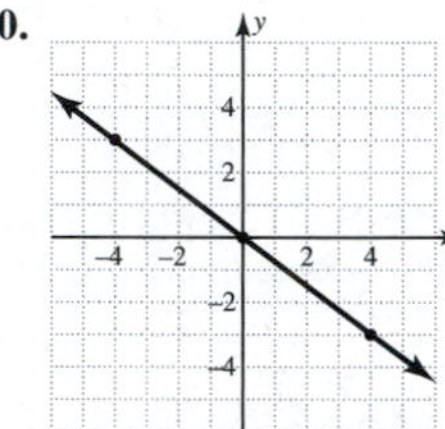
Domain: $(-\infty, \infty)$, Range: $(-\infty, \infty)$

21.
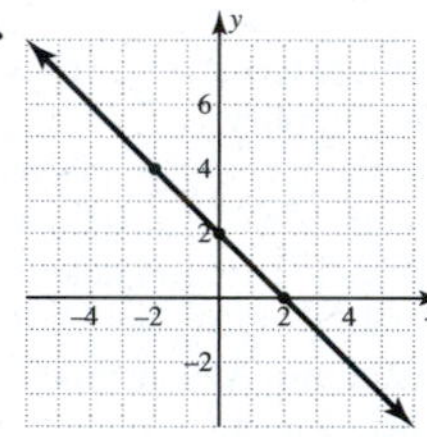
Domain: $(-\infty, \infty)$, Range: $(-\infty, \infty)$

22.
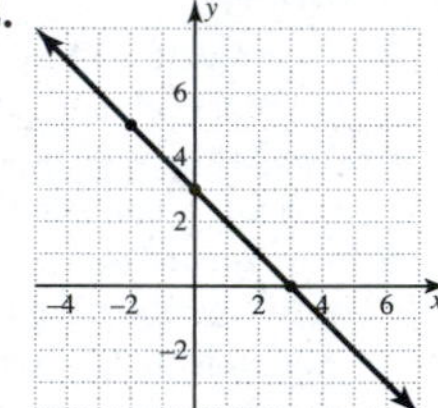
Domain: $(-\infty, \infty)$, Range: $(-\infty, \infty)$

23.
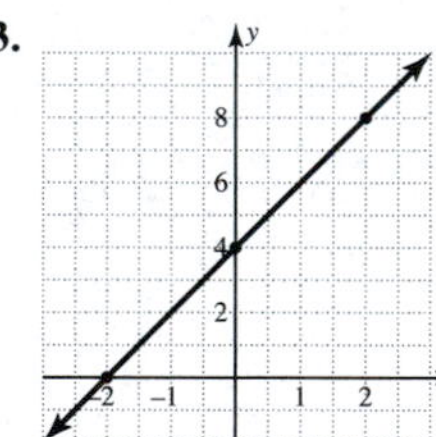
Domain: $(-\infty, \infty)$, Range: $(-\infty, \infty)$

24.
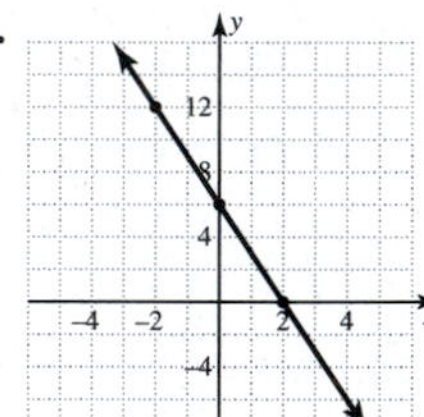
Domain: $(-\infty, \infty)$, Range: $(-\infty, \infty)$

25.
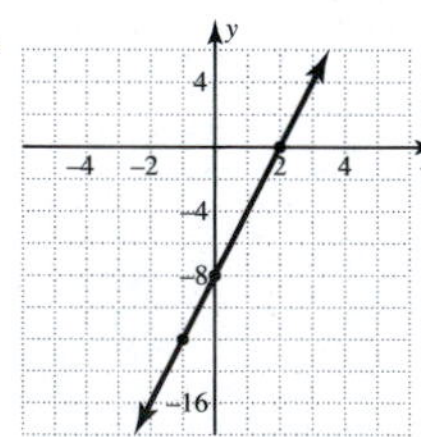
Domain: $(-\infty, \infty)$, Range: $(-\infty, \infty)$

26.
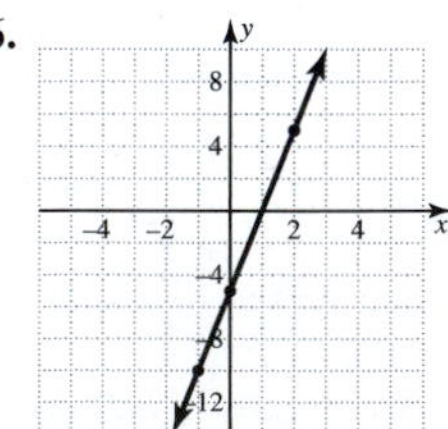
Domain: $(-\infty, \infty)$, Range: $(-\infty, \infty)$

27.
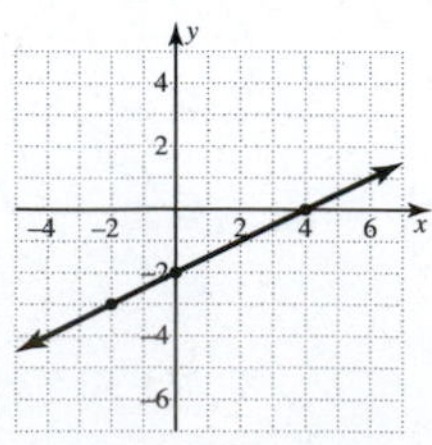
Domain: $(-\infty, \infty)$, Range: $(-\infty, \infty)$

28.
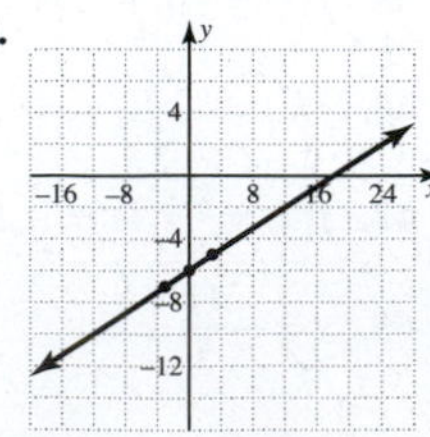
Domain: $(-\infty, \infty)$, Range: $(-\infty, \infty)$

29.

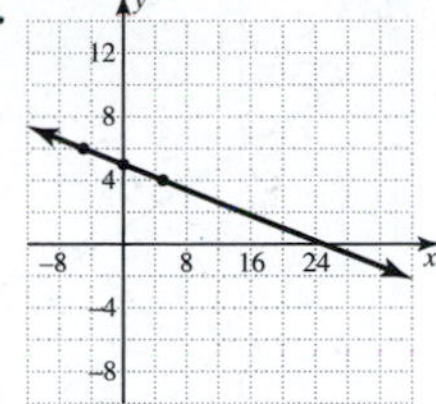

Domain: $(-\infty, \infty)$, Range: $(-\infty, \infty)$

30.

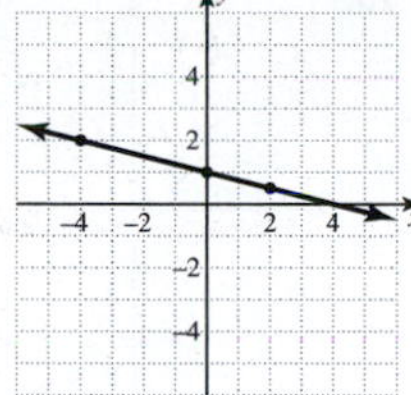

Domain: $(-\infty, \infty)$, Range: $(-\infty, \infty)$

31.

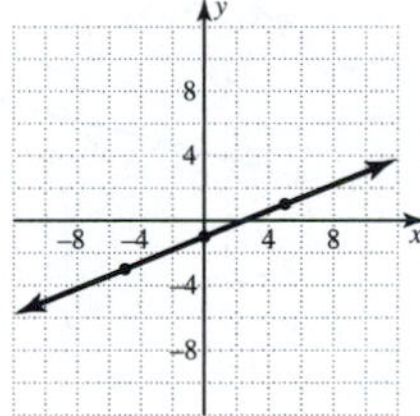

Domain: $(-\infty, \infty)$, Range: $(-\infty, \infty)$

32.

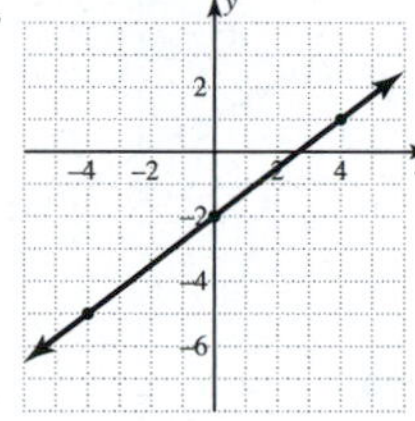

Domain: $(-\infty, \infty)$, Range: $(-\infty, \infty)$

33.

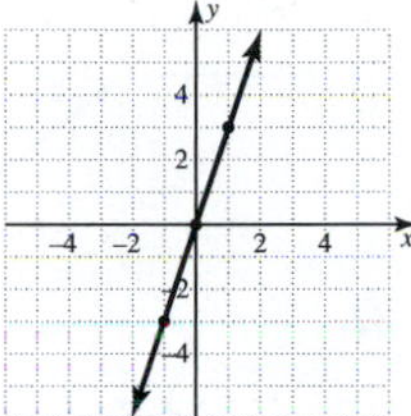

Domain: $(-\infty, \infty)$, Range: $(-\infty, \infty)$

34.

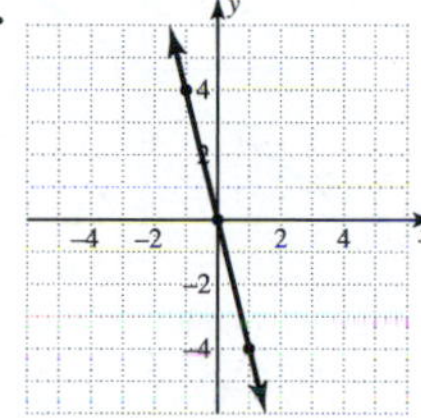

Domain: $(-\infty, \infty)$, Range: $(-\infty, \infty)$

35.

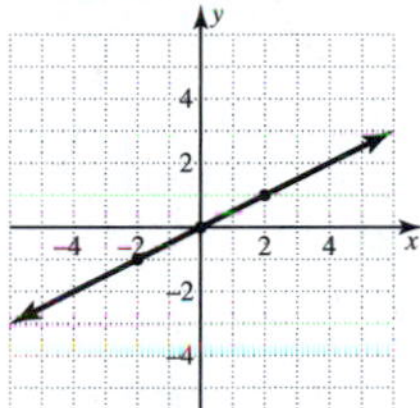

Domain: $(-\infty, \infty)$, Range: $(-\infty, \infty)$

36.

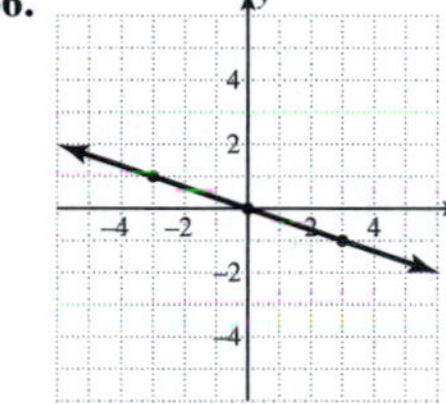

Domain: $(-\infty, \infty)$, Range: $(-\infty, \infty)$

37.

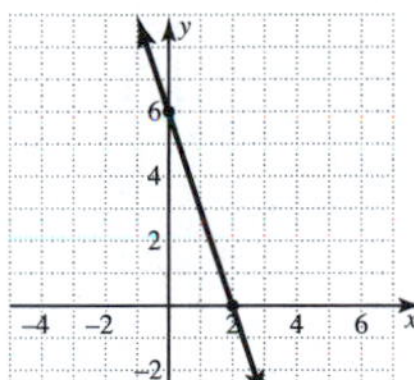

38.

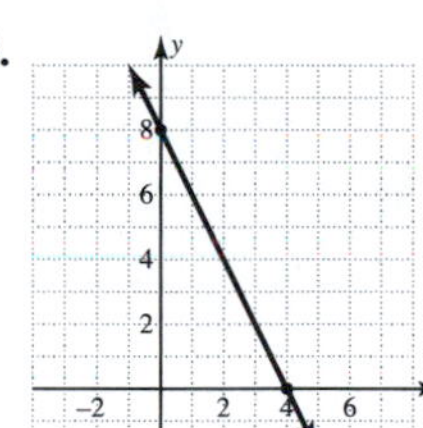

39.

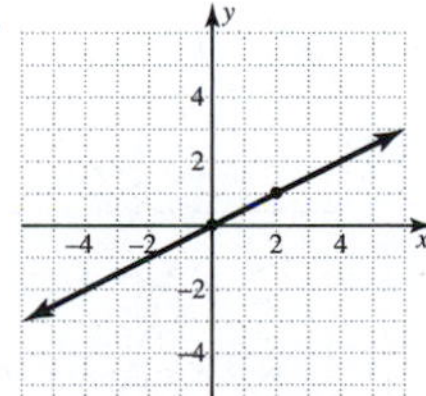

40.

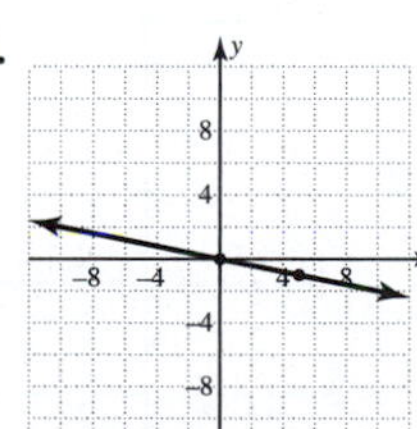

41.

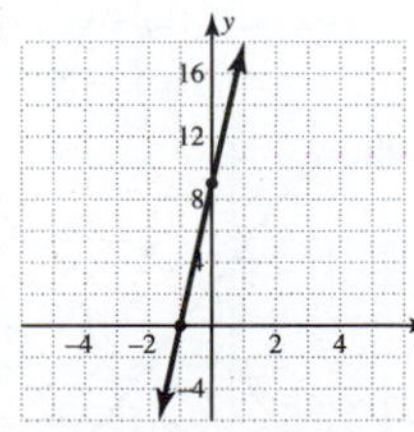

42.

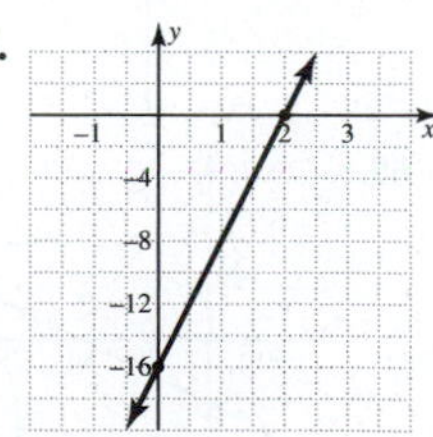

43.

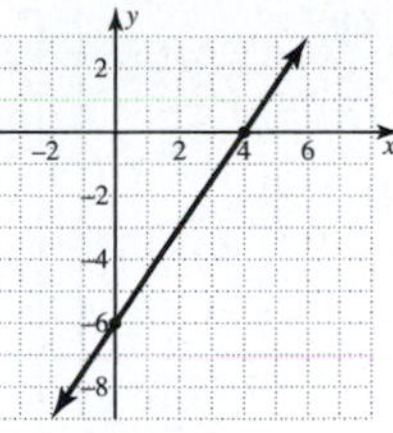

44.

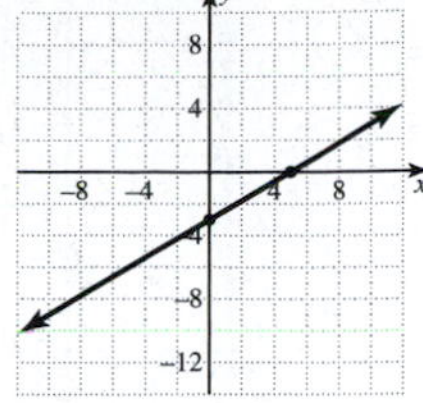

45.

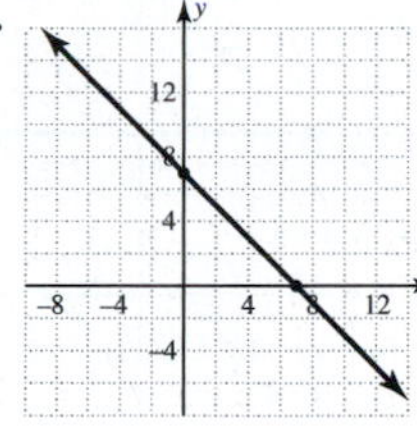

46.

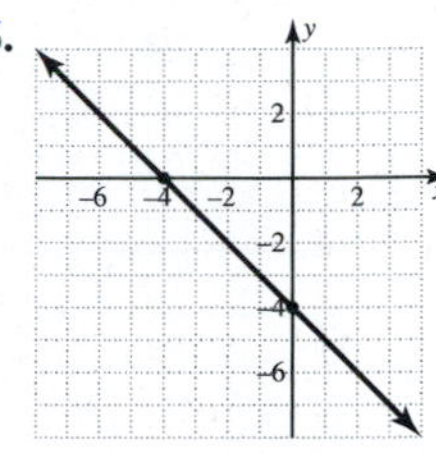

47.

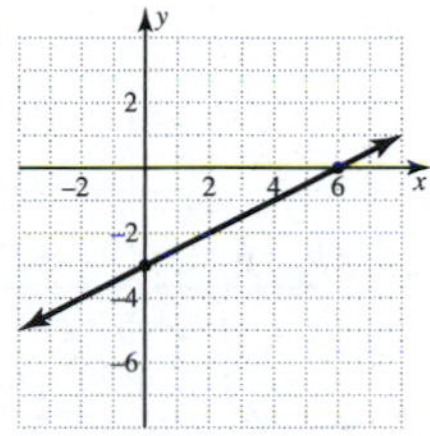

48.

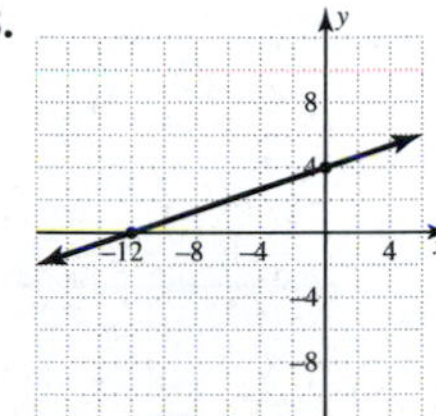

49.

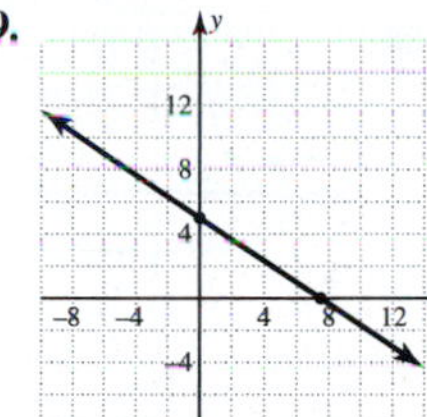

50.

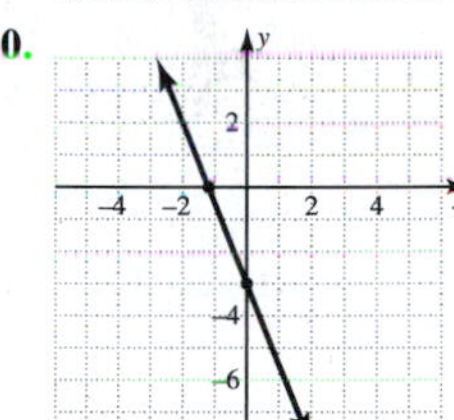

51.

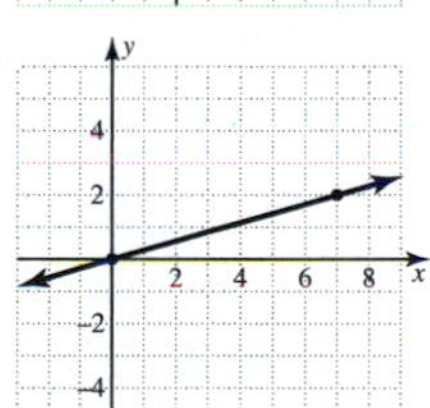

52.

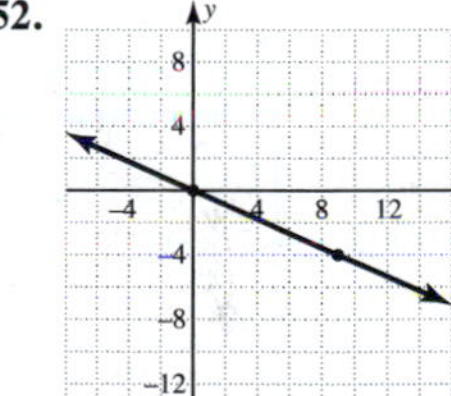

53.

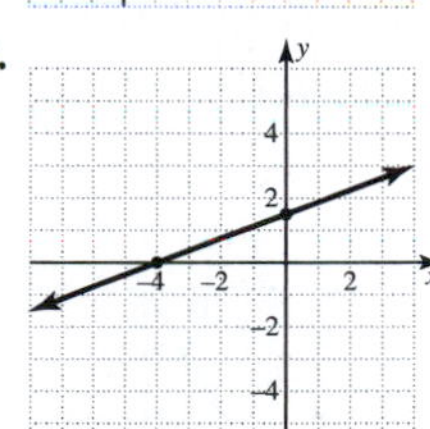

54.

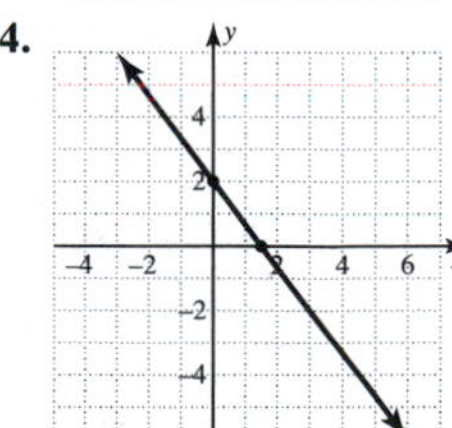

55.

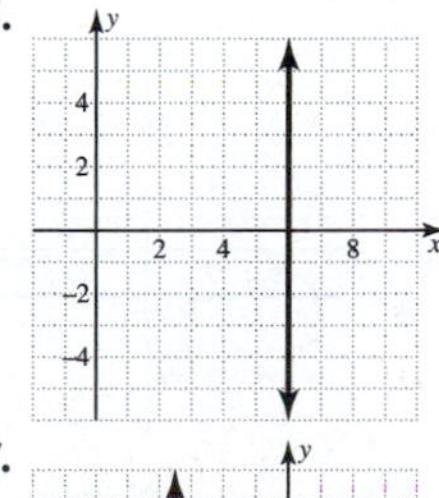

56.

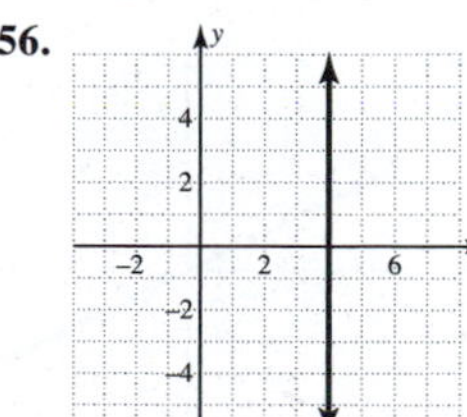

57.

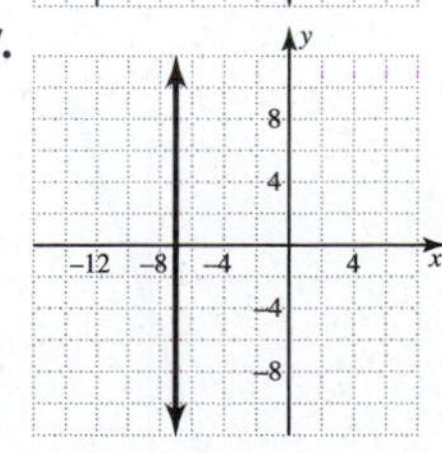

58.

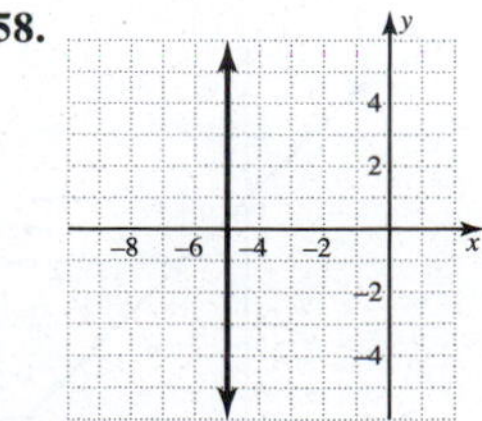

59.

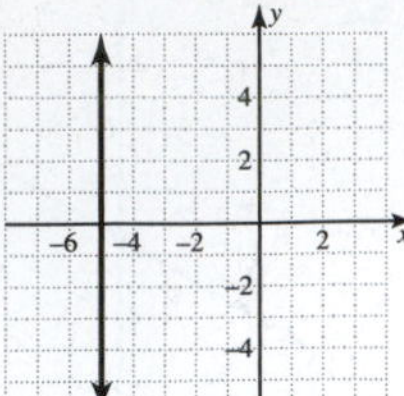

60.

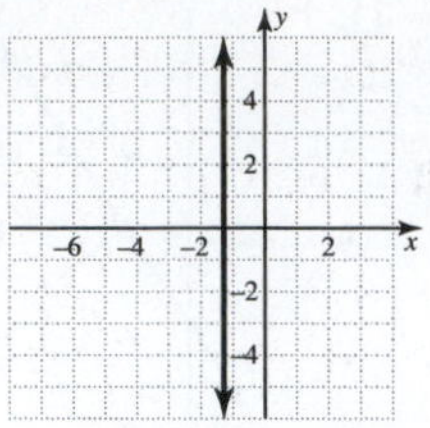

61.

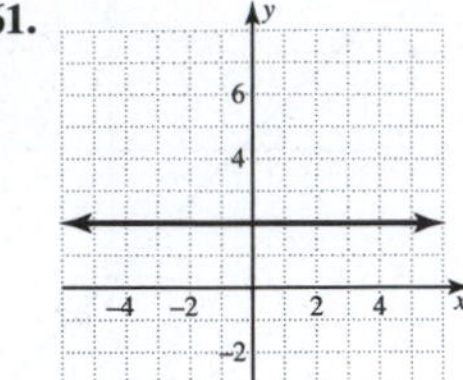

62.

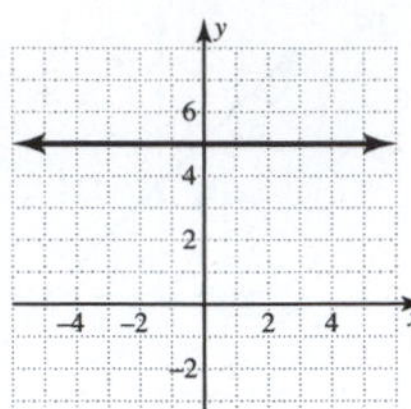

63.

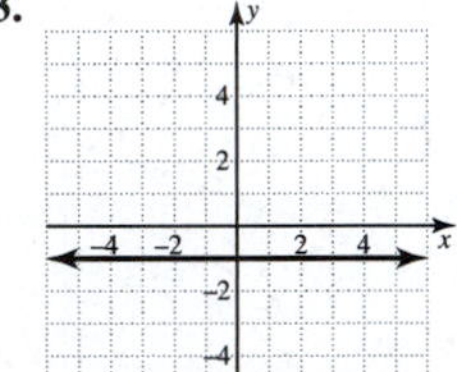

64.

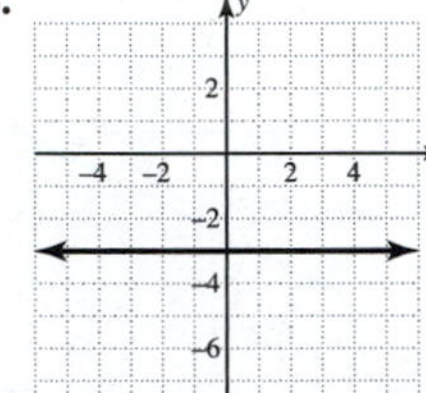

65.

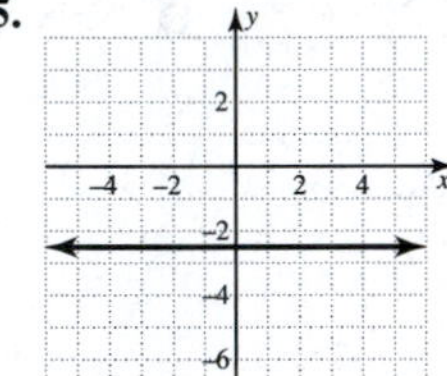

66.

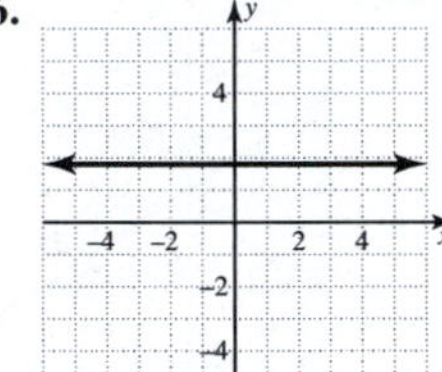

67.

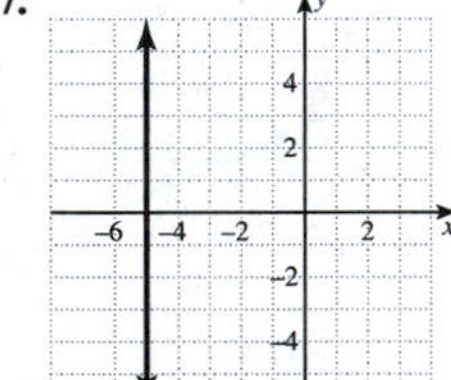

68.

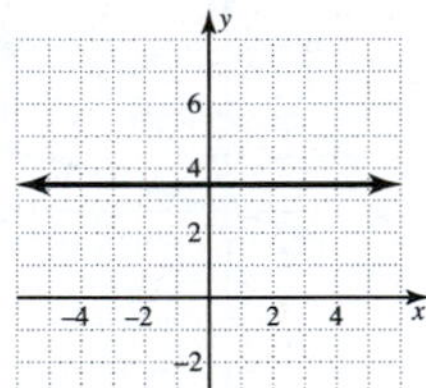

69.

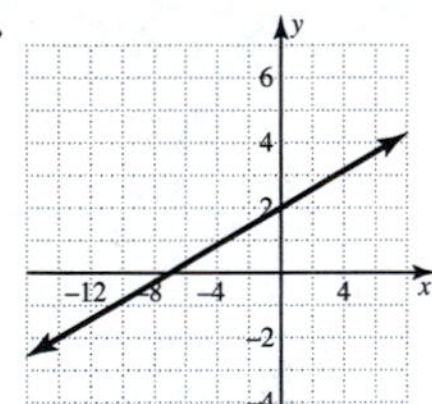

70.

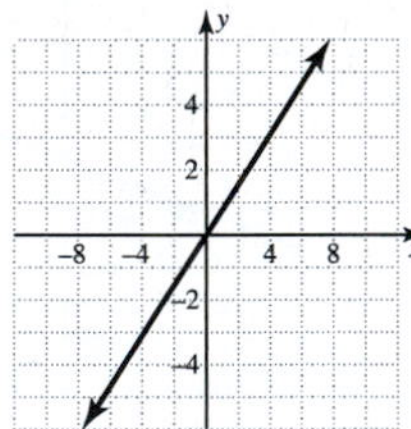

71.

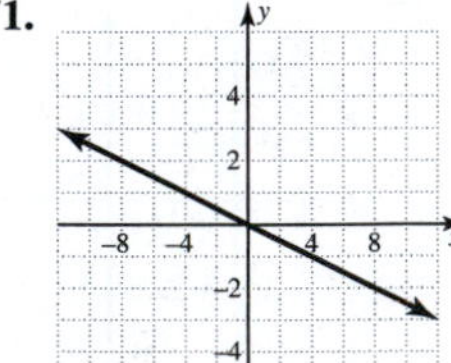

72.

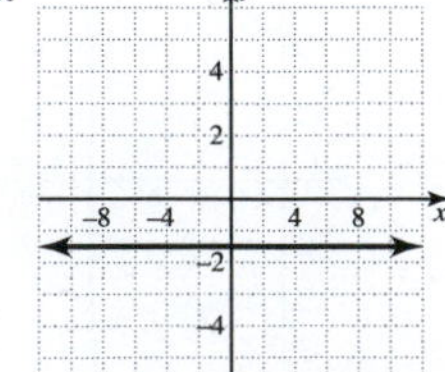

73.

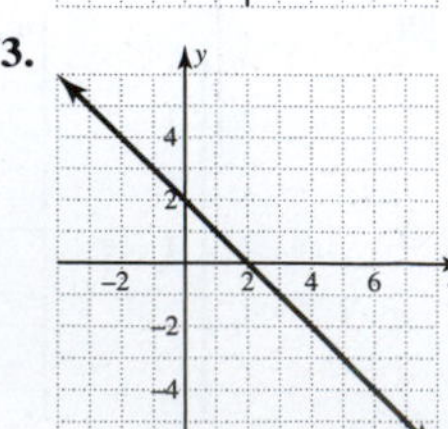

74.

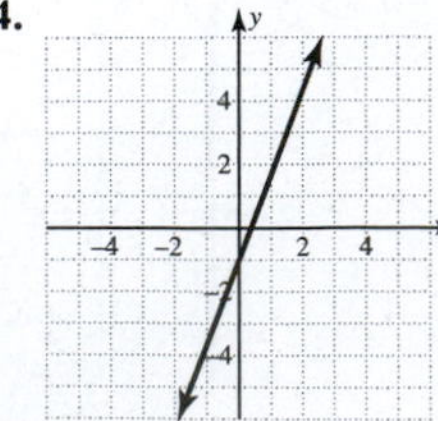

75.

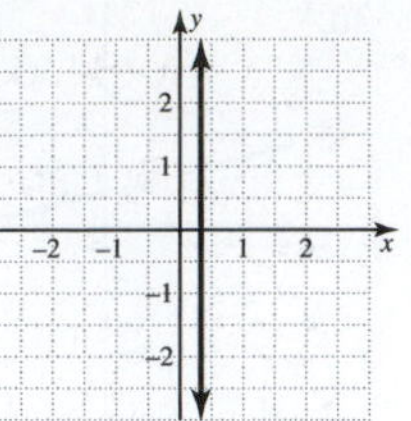

76.

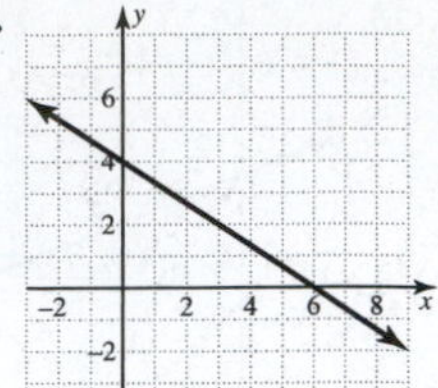

77.

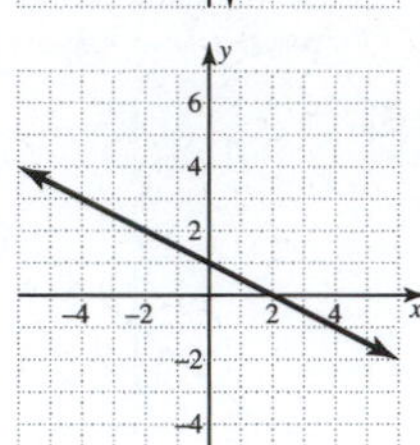

78.

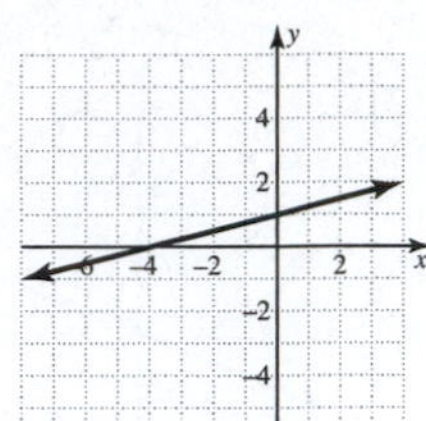

79.

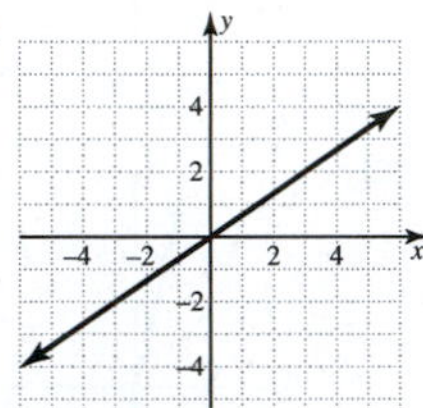

80.

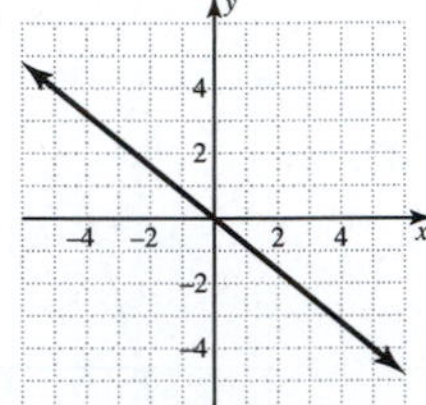

81.

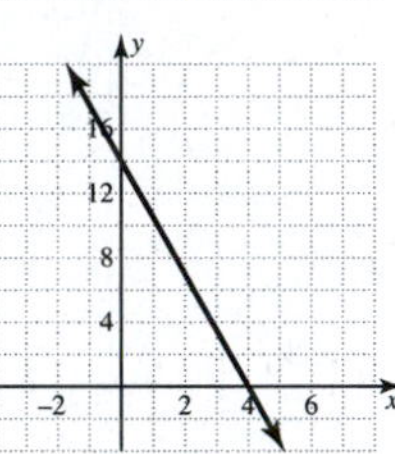

82.

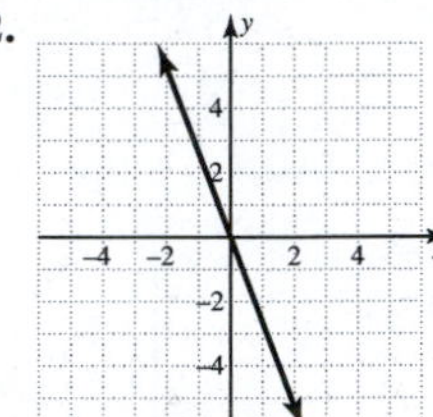

83.

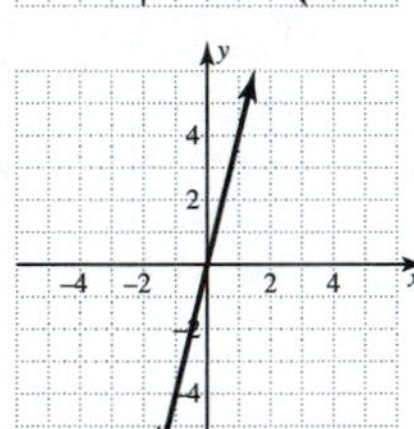

84.

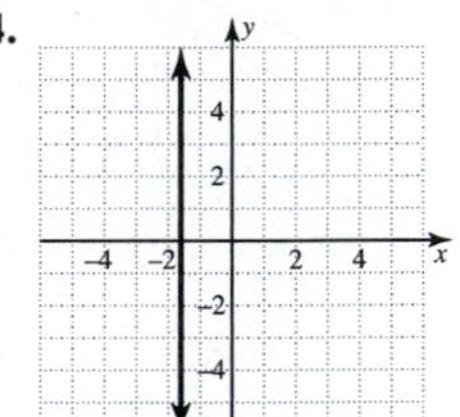

85.

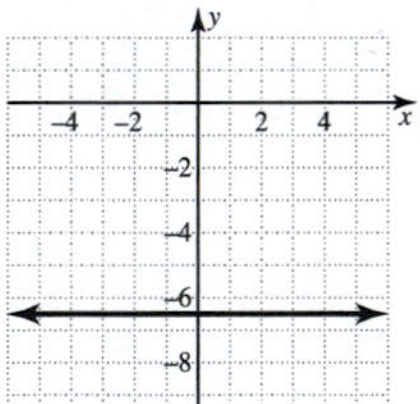

86.

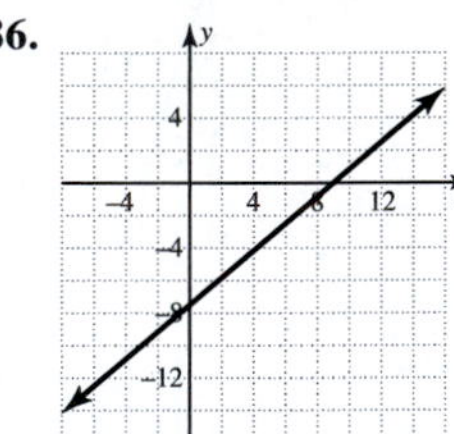

87.

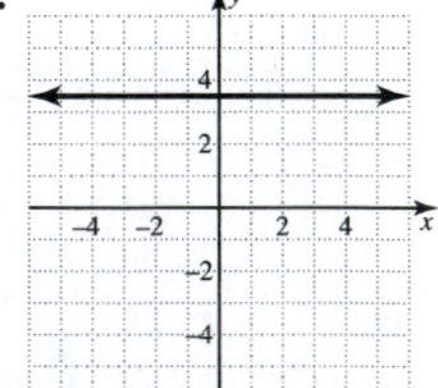

88.

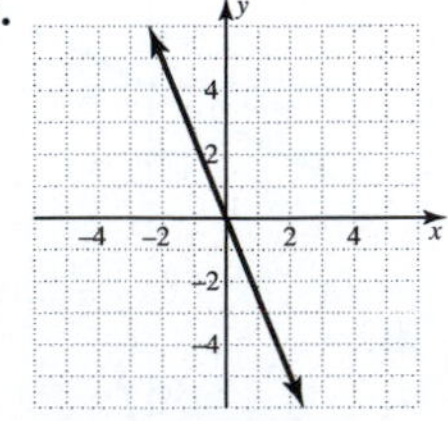

89.

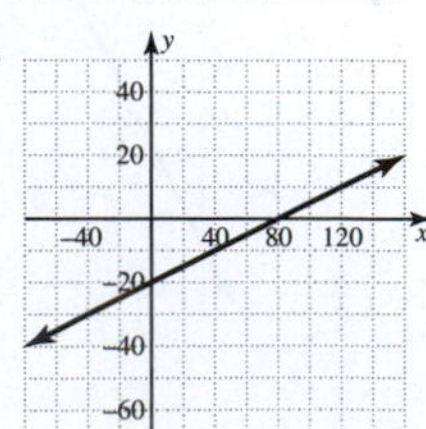

90.

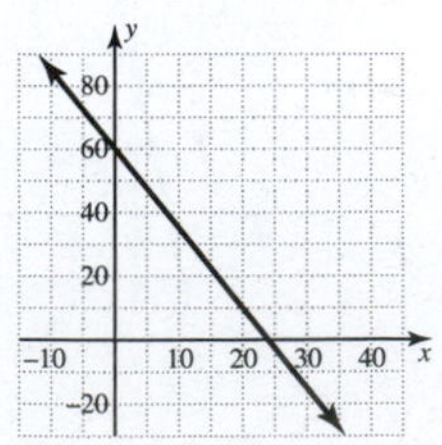

93. e.

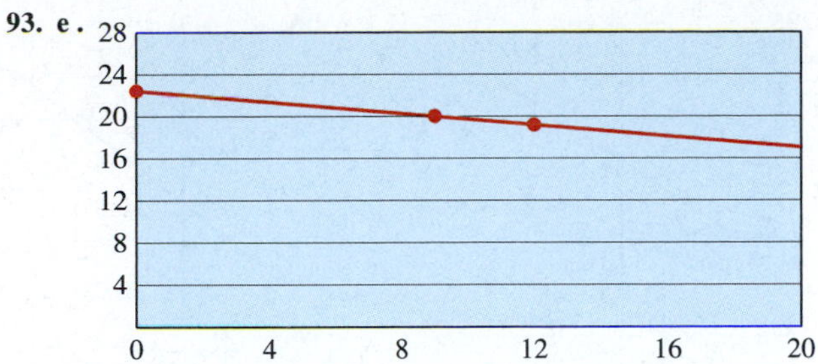

94. c. 25.43% of adults aged 20 and over will be at a healthy weight in 2020.

d.

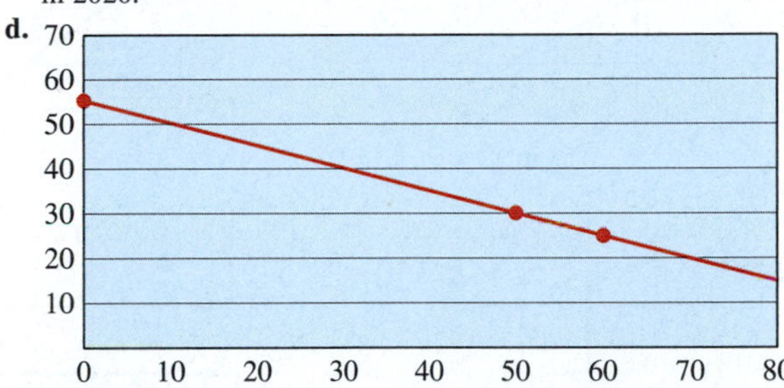

e. The percent of adults who are at a healthy weight is declining.

f. 33.47% of adults would be at a healthy weight if the Healthy People Campaign is successful.

95. a. The x-intercept is (10, 0), which means that when the equipment is 10 yr old, its value will be \$0. The y-intercept is (0, 50,000), which means that the equipment has an initial value of \$50,000.

b. After 4 yr, the value of the equipment is \$30,000.

c.

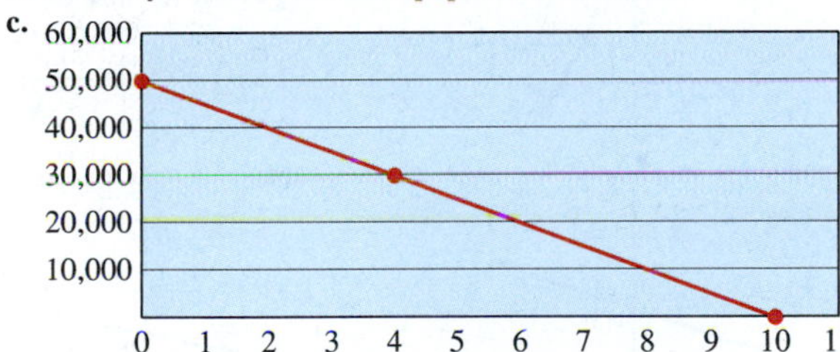

96. a. The x-intercept is (3, 0), which means that when the computer is 3 yr old, its value will be \$0. The y-intercept is (0, 2400), which means that the initial value of the computer is \$2400.

b. After 2 yr, the computer's value is \$800.

c.

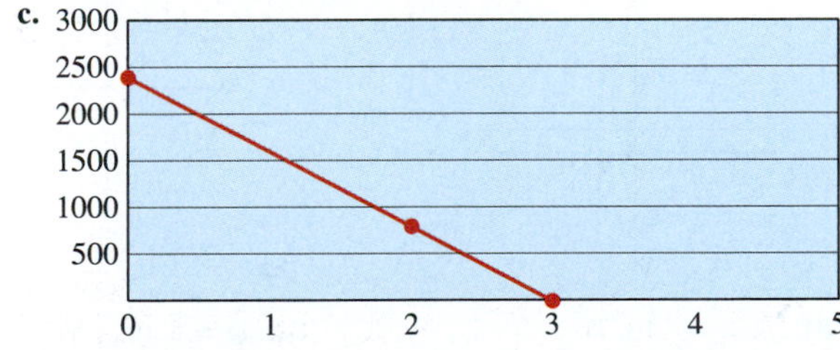

97. The equation doesn't have a y-variable, so it is equivalent to $x = 4$, which is the equation of a vertical line. The graph is

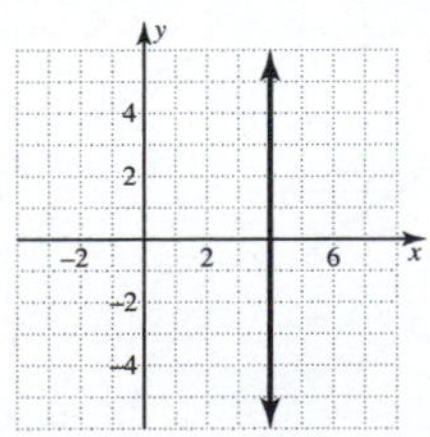

98. The x-intercept is $(-3, 0)$ and the y-intercept is $(0, 4)$. The graph is

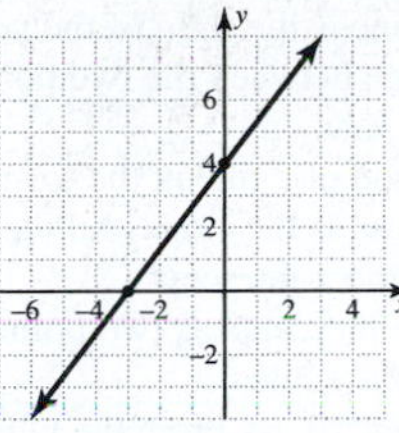

99. The function is $f(x) = 5x$, not $f(x) = 5$.

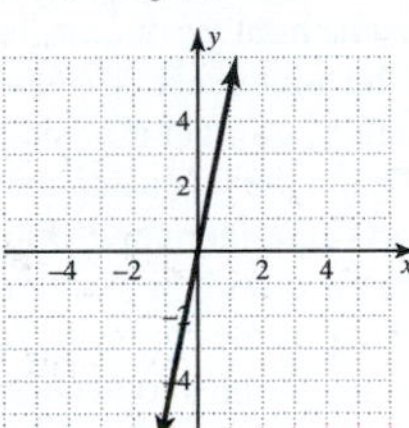

101.

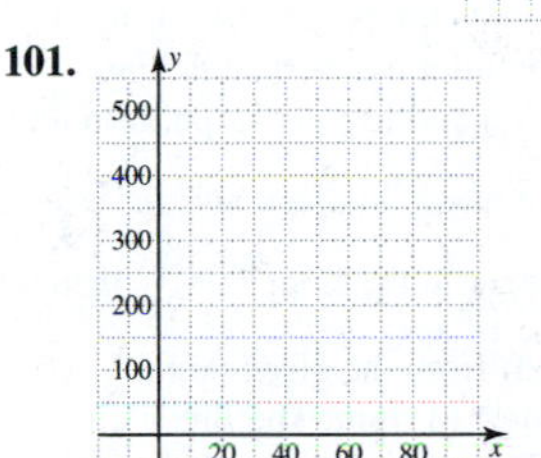

102.

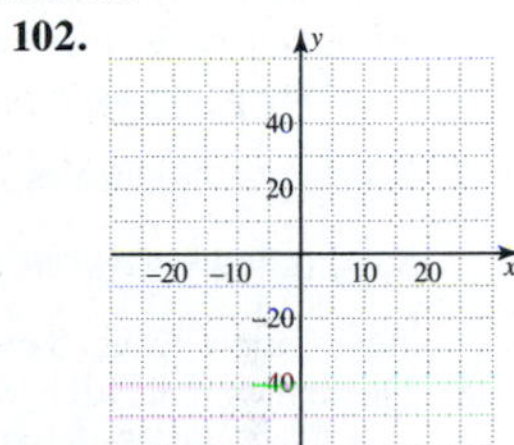

105.

a.

d.

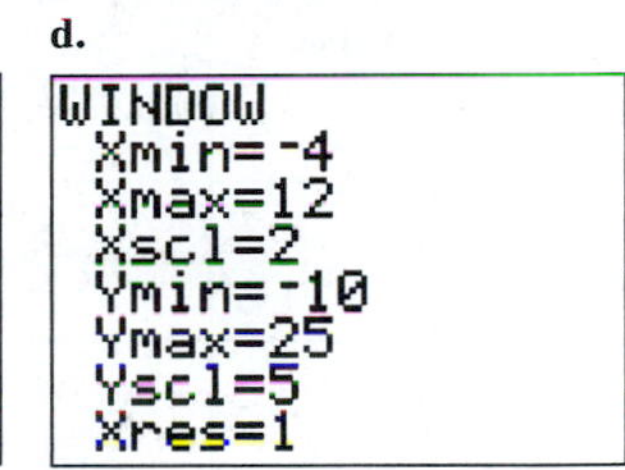

e.

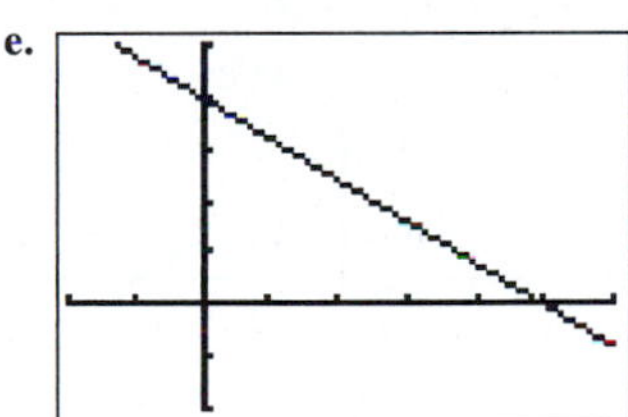

106.

a.

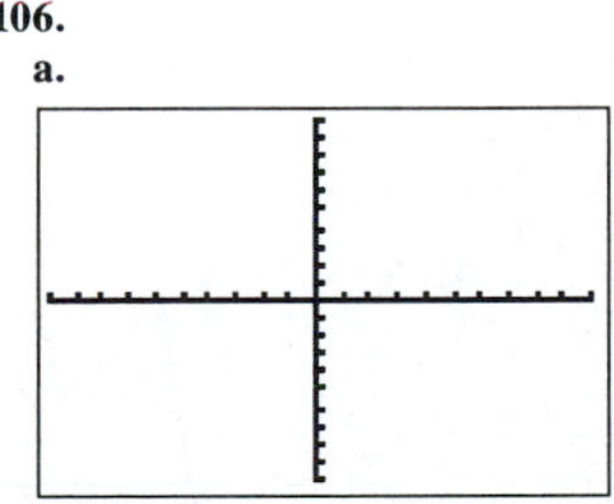

d.

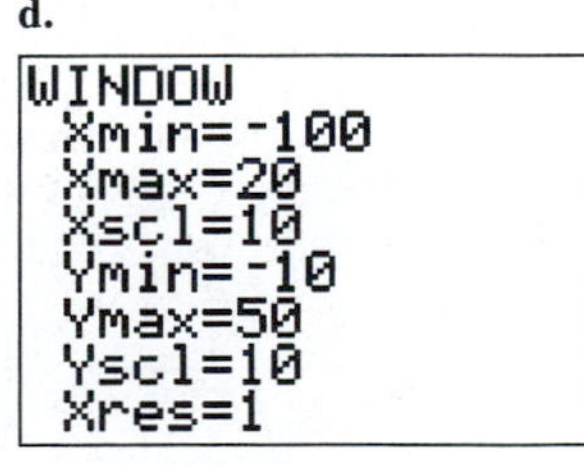

e.

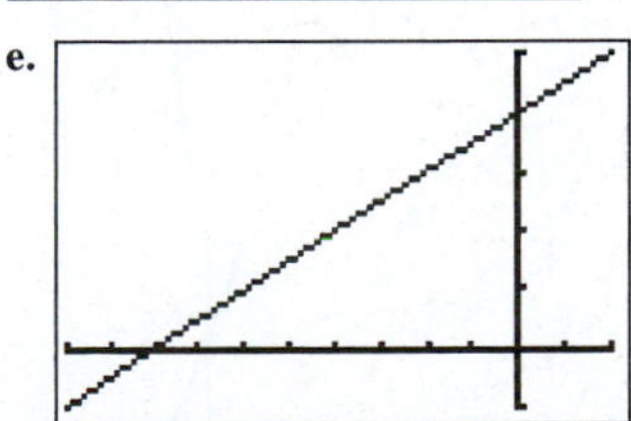

Section 4.3

1. The slope of a line is the ratio of the vertical change to the horizontal change between the two points. It tells the direction and steepness of the line.
2. Write the equation of a line in the form $y = mx + b$. The coefficient of x, m, is the slope and the constant term, b, is the y-coordinate of the y-intercept.
3. Use the x-and y-intercepts as two points and the slope formula to find the slope of the line.
4. Plot the y-intercept $(0, b)$. From the y-intercept, move to another point by using the information based on the slope. The numerator of the slope gives the number of units to go up if positive and down if negative from the y-intercept. The denominator of the slope gives the number of units to go right if positive and left if negative from the y-intercept. Graph the line using these two points.
5. Write each line in slope-intercept form and find its slope. If the slopes are the same and the y-intercepts are different, then the lines are parallel. If the slopes of two nonvertical lines are negative reciprocals of one another, then they are perpendicular to each other. Vertical and horizontal lines are perpendicular to each other. Otherwise, they are neither parallel nor perpendicular.
6. Label the points as (x_1, y_1) and (x_2, y_2) and use the slope formula.
7. If we write the slope as $\dfrac{-3}{5}$, then we move 3 units down and 5 units to the right from a given point. If we write the slope as $\dfrac{3}{-5}$, then we move 3 units up and 5 units to the left from the given point.
8. The line with $m > 0$ is rising from left to right. The line with $m < 0$ is falling from left to right. The line with $m = 0$ is horizontal and the line with m undefined is vertical. For example,

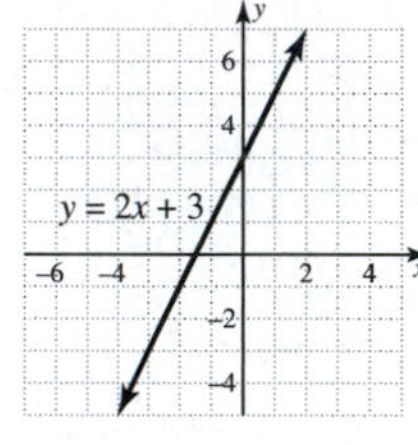

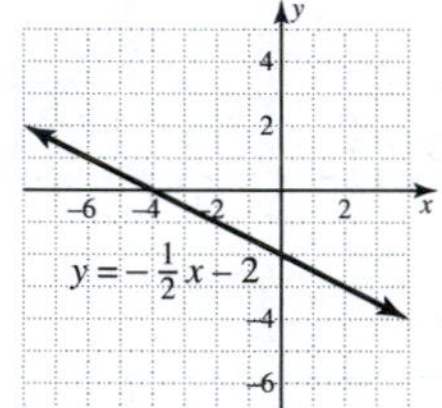

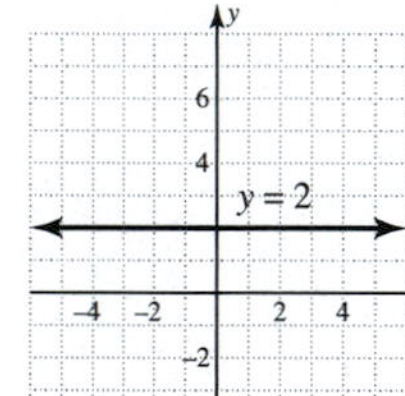

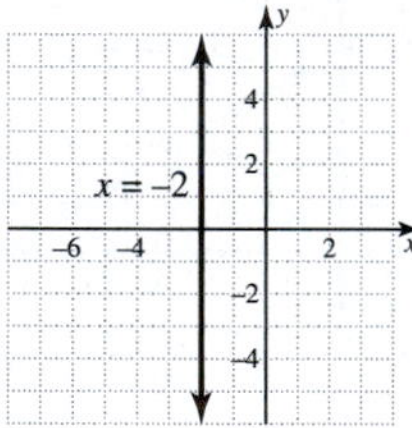

18. $m = -\dfrac{5}{4}$, $(0, 5)$ **22.** $m = \dfrac{7}{3}$, $\left(0, -\dfrac{4}{3}\right)$

25. $m = -\dfrac{8}{5}$, $(0, -8)$ **26.** $m = -\dfrac{2}{9}$, $(0, -2)$

31.

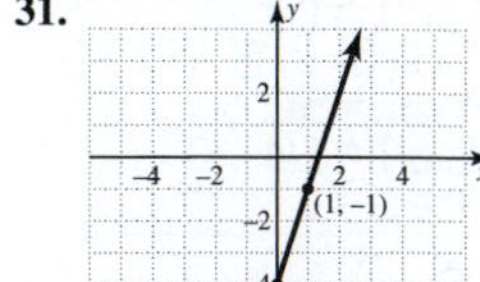

32.

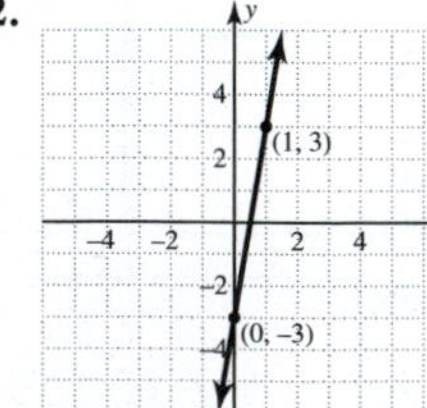

33.

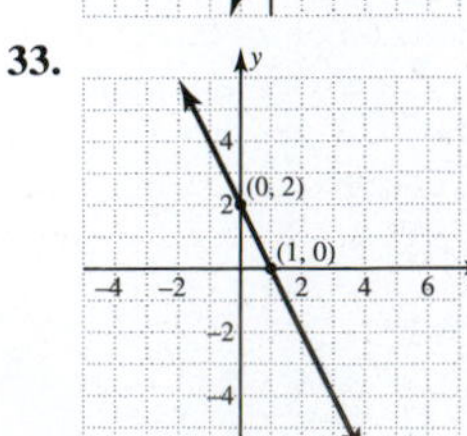

34.

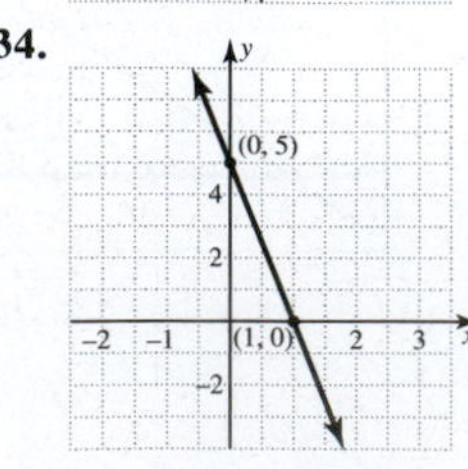

35.

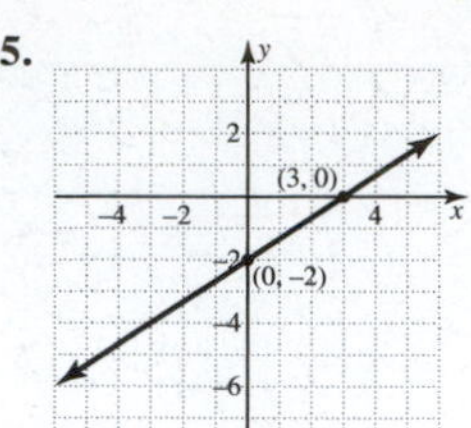

36.

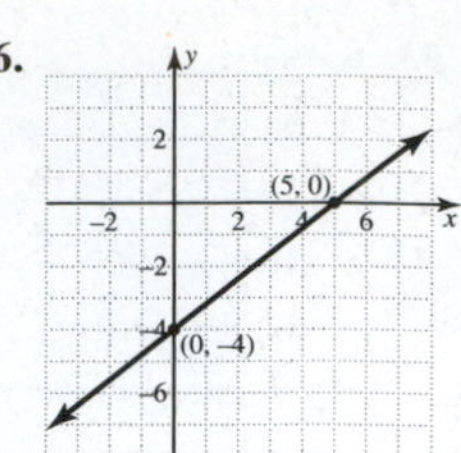

37.

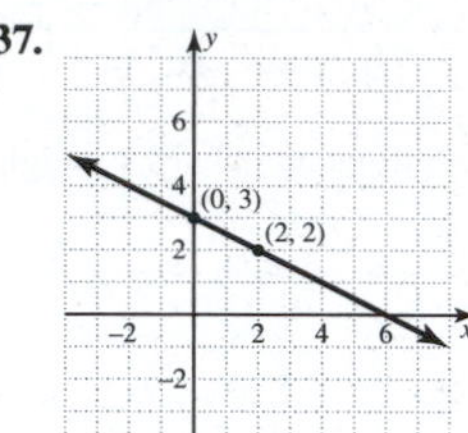

38.

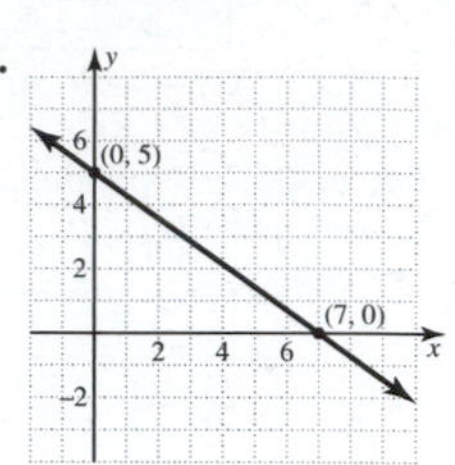

39.

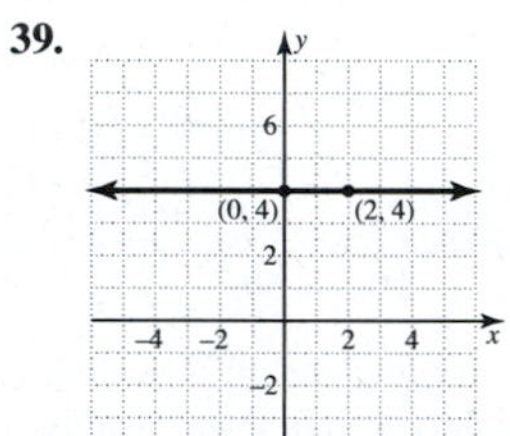

40.

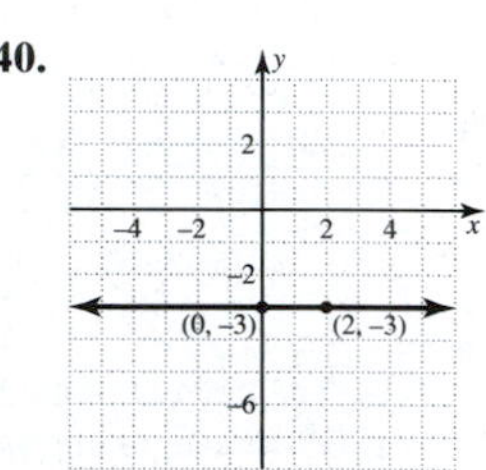

41.

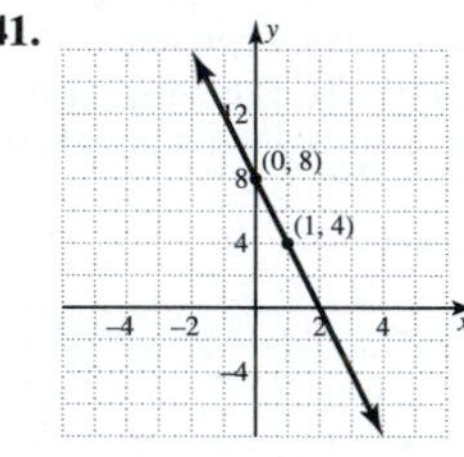

42.

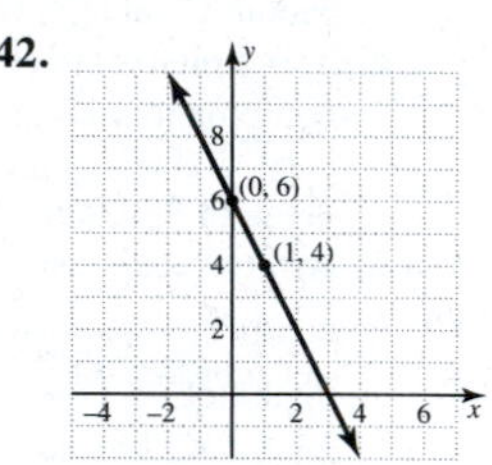

43.

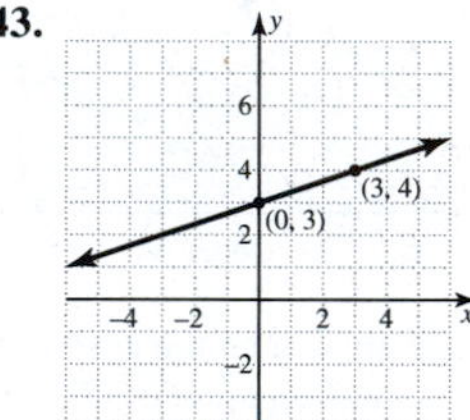

44.

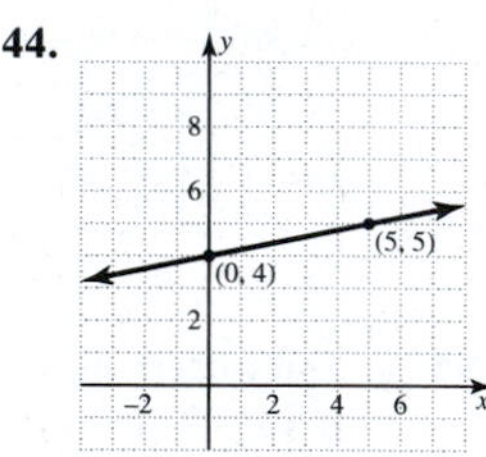

45.

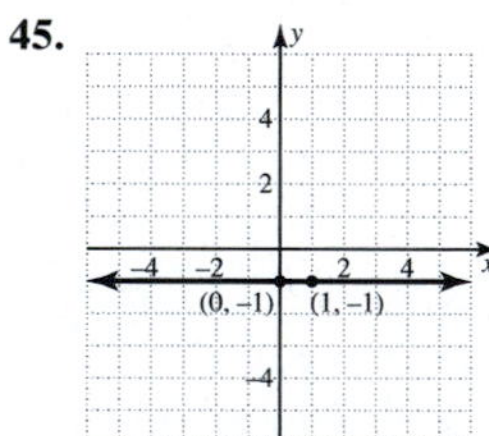

46.

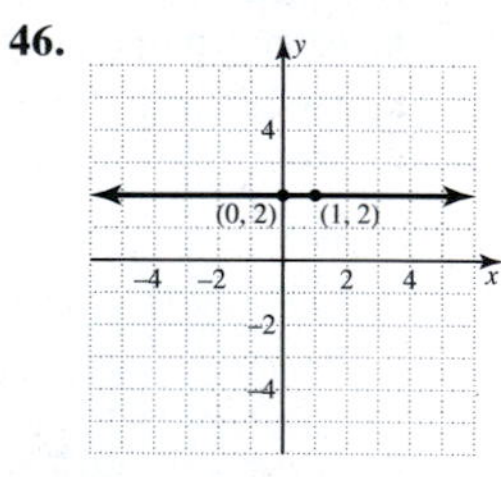

81. $m = -\dfrac{1}{6}$, $(0, 1)$

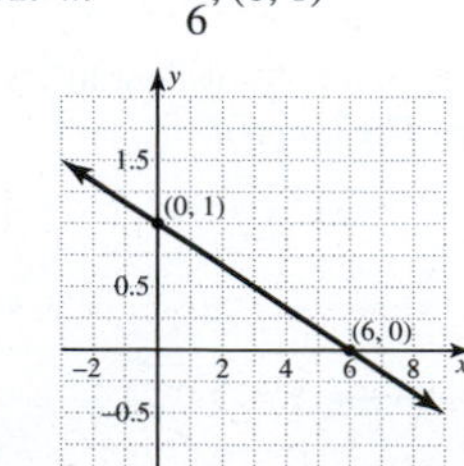

82. $m = 5$, $(0, -5)$

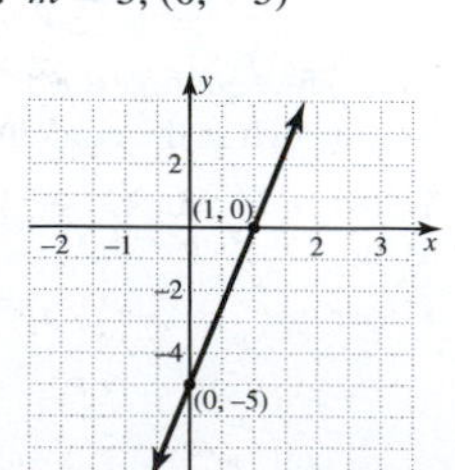

83. $m = 0, \left(0, \dfrac{5}{2}\right)$

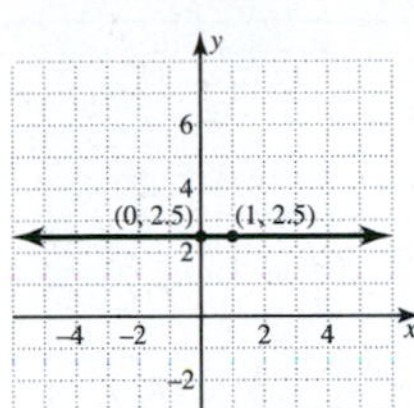

84. not defined, no y-intercept

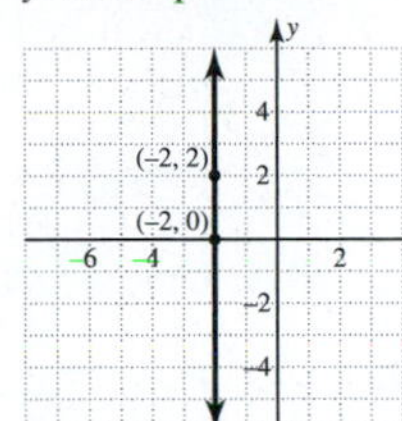

85. $m = 7, (0, -5)$

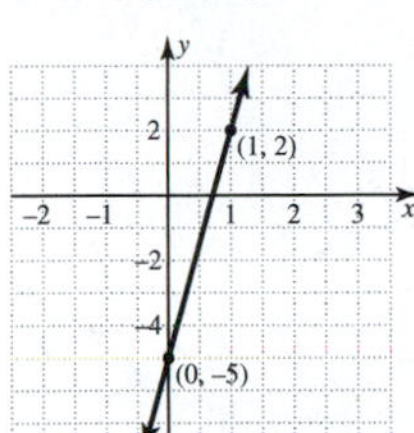

86. $m = \dfrac{1}{4}, (0, 2)$

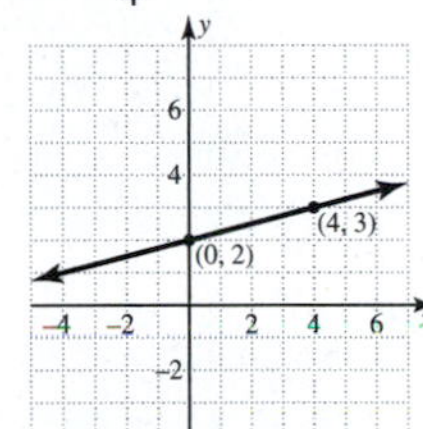

87. $m = 0, (0, -2)$

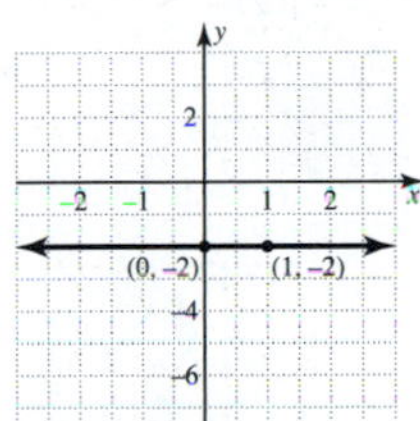

88. $m = \dfrac{2}{3}, (0, -4)$

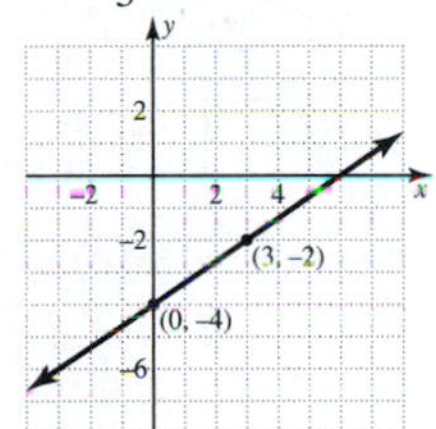

89. $m = -\dfrac{1}{2}, (0, 0)$

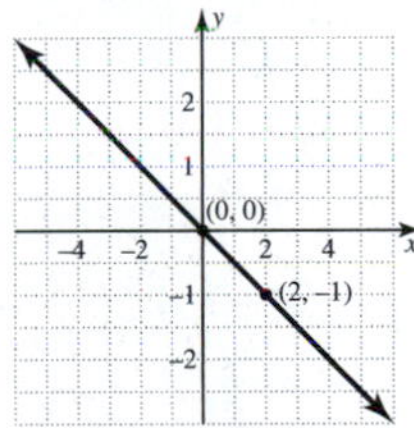

90. $m = 3, (0, 9)$

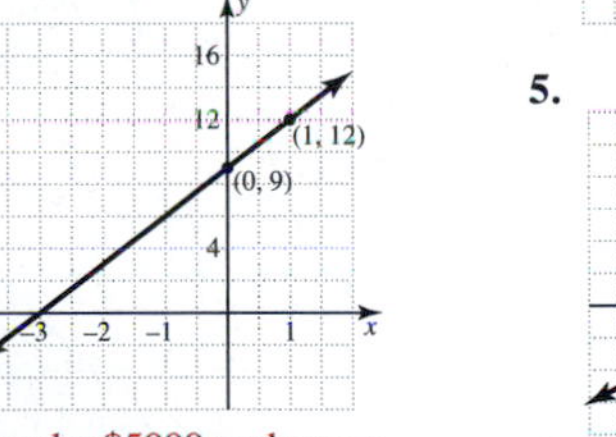

103. $m = 5000$, The slope means the salary increases by \$5000 each year.; (0, 72,000), The y-intercept means the starting salary is \$72,000.

104. $m = 6000$, The slope means the salary increases by \$6000 each year.; (0, 61,500), The y-intercept means the starting salary is \$61,500.

105. $m = 0.5$, The slope means a person will burn 0.5 cal each mile per pound over 100.; (0, 53), The y-intercept means a 100-lb person will burn 53 cal each mile.

106. $m = 0.7$, The slope means a person will burn 0.7 cal each mile per pound over 100.; (0, 73), The y-intercept means a 100-lb person will burn 73 cal each mile.

107. $m = -2000$, The slope means the car's value decreases by \$2000 every year.; (0, 25,000), The y-intercept means the initial value of the car. That is, the car was originally worth \$25,000.

108. $m = -3500$, The slope means the truck's value decreases by 3500 every year.; (0, 38,000), The y-intercept means the initial value of the truck. That is, the truck was originally worth \$38,000.

109. $m = \dfrac{5}{7}$, The slope means the staircase rises 5 in. for every 7 in. across.

110. $m = \dfrac{1}{18}$; The ramp with a 3 ft rise and length of 12 ft will have a slope of $\dfrac{1}{4}$, which is more than the suggested guideline.

111. $m = \dfrac{1}{100} = 1\%$

112. $m = \dfrac{1}{4}$, The slope means for the road rises 1 mi for every 4 mi.

114. You need to plot the point (0, 3), go down 3 units and right 5 units to get to the point (5, 0).

116. Slopes of perpendicular lines are negative reciprocals, so the slope is $\dfrac{9}{4}$.

123. As the slope ($m > 0$) of the line increases, the graph of the line is getting closer to the y-axis.

124. As the slope ($m > 0$) of the line decreases, the graph of the line is getting closer to the x-axis.

125. As the slope ($m < 0$) of the line decreases, the graph of the line is getting closer to the y-axis.

126. As the slope ($m < 0$) of the line increases, the graph of the line is getting closer to the x-axis.

Piece It Together Sections 4.1–4.3

1. Domain $= (-\infty, \infty)$, Range $= (-\infty, \infty)$

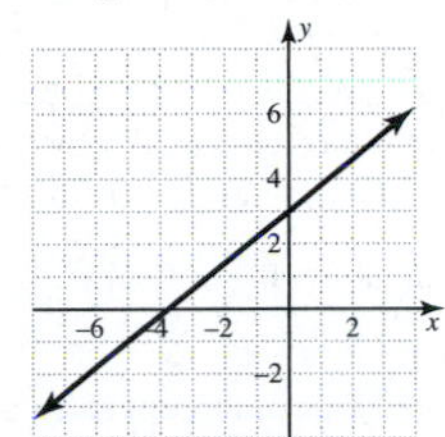

2. Domain $= (-\infty, \infty)$, Range $= \{6\}$

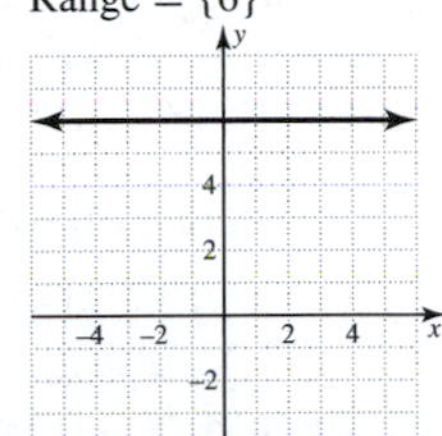

3.

4.

5.

6.

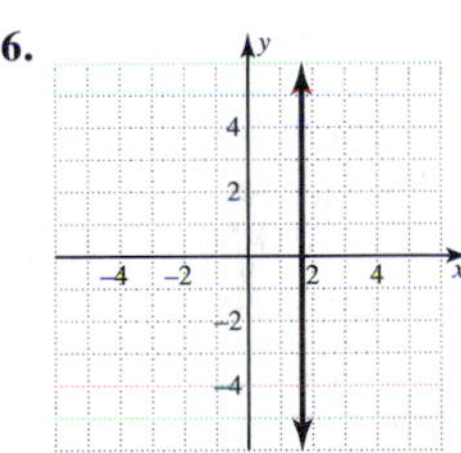

7. $m = \dfrac{4}{7}, (0, -4)$

8. $m = -1, (0, 3)$

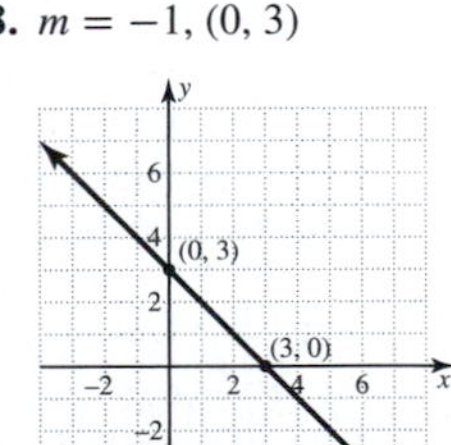

Section 4.4

1. Use the slope-intercept form of a line $y = mx + b$ by substituting the slope for m and the y-coordinate of the y-intercept for b.

2. Find the slope of the line using the slope formula. Use the slope and one of the two points to find the equation of the line using either the slope-intercept form or point-slope form of a line.

3. If the slope and y-intercept of an equation are known, then use the slope-intercept form. If a point and a slope, two points, or a point and a line parallel or perpendicular are given, then use the point-slope form.

4. False. The equation should be $y = 7x - 35$.

5. False. The equation is $y = \dfrac{4}{3}x + 4$.

6. False. The equation is $y = -\frac{1}{3}x + \frac{20}{3}$.

27. $y = \frac{2}{3}x - 12$ **28.** $y = -\frac{1}{4}x - \frac{7}{2}$ **37.** $y = \frac{1}{2}x - 6$

38. $y = \frac{3}{4}x - 9$ **59.** $y = -\frac{1}{2}x + 3$ **60.** $y = -\frac{2}{3}x - 1$

62. $y = -\frac{10}{11}x - \frac{50}{11}$

99. The slope $m = \frac{4}{3}$ is correct. The error is that the y-value of the given point is not the value of b since x is not 0.

The slope $m = \frac{4}{3}$ is correct.

$$y - 3 = \frac{4}{3}(x - 8)$$
$$y - 3 = \frac{4}{3}x - \frac{32}{3}$$
$$y = \frac{4}{3}x - \frac{32}{3} + 3$$
$$y = \frac{4}{3}x - \frac{32}{3} + \frac{9}{3}$$
$$y = \frac{4}{3}x - \frac{23}{3}$$

101. Substitute $(5, -1)$ and $m = -2$ into the slope-intercept form and solve for b.

$$y = mx + b$$
$$-1 = -2(5) + b$$
$$-1 = -10 + b$$
$$-1 + 10 = b$$
$$9 = b$$
$$y = -2x + 9$$

102. (0, 2) does not lie on the given line; (−1, −3) lies on the given line.
103. (0, 2) lies on the given line; (−1, −3) does not lie on the given line.
104. (0, 2) does not lie on the given line; (−1, −3) lies on the given line.
105. (0, 2) lies on the given line; (−1, −3) lies on the given line.

Section 4.5

1. Replace the variables with the corresponding values of x and y in the ordered pair. Simplify the resulting inequality. If the inequality is true, then the ordered pair is a solution. If the inequality is false, then it is not a solution.

3. If the inequality symbol is $\leq$ or $\geq$, the boundary line is a solid line. If the inequality symbol is $<$ or $>$, then the boundary line is a dashed line.

4. One point in a half-plane must be tested. If the point makes the original inequality true, then the half-plane containing the test point includes solutions of the linear inequality. If the point makes the original inequality false, then the half-plane not containing the test point includes solutions of the linear inequality.

19.
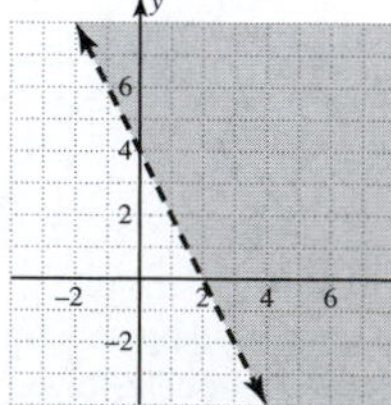

20.
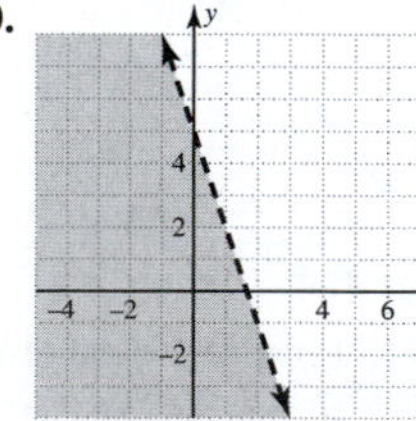

21.
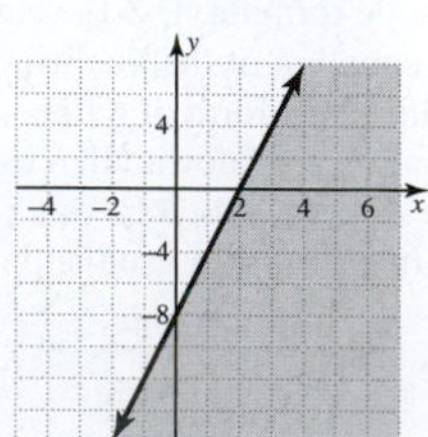

22.
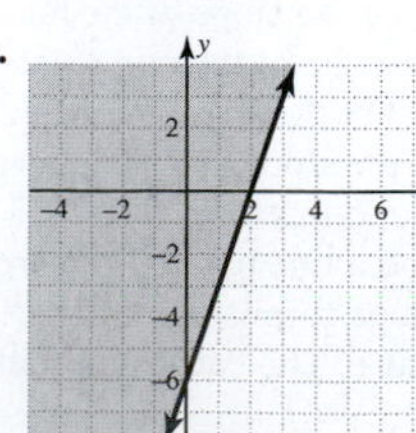

23.
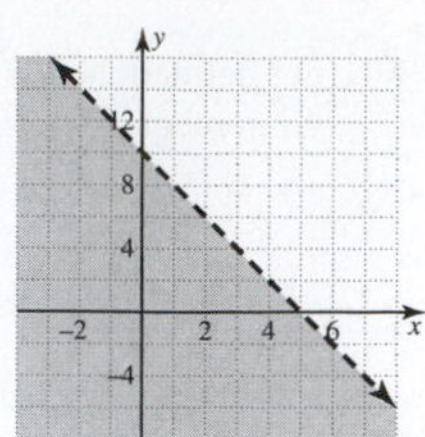

24.
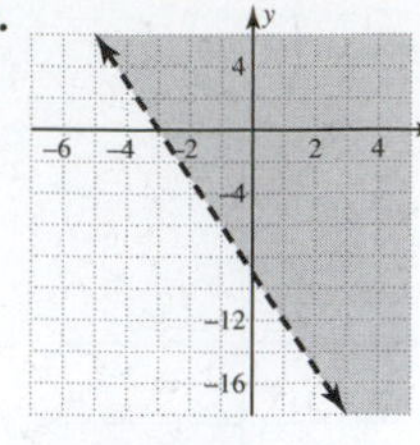

25.
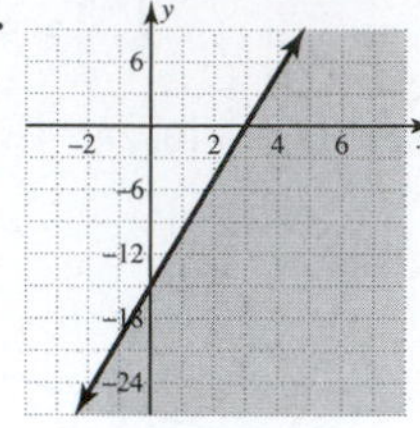

26.
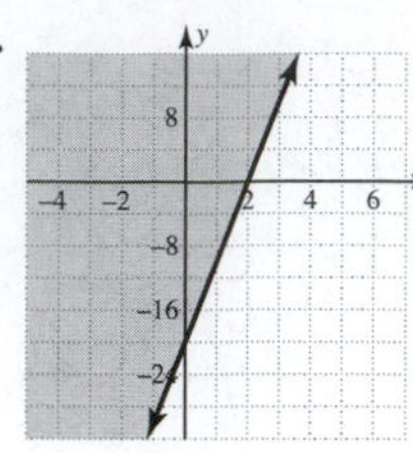

27.
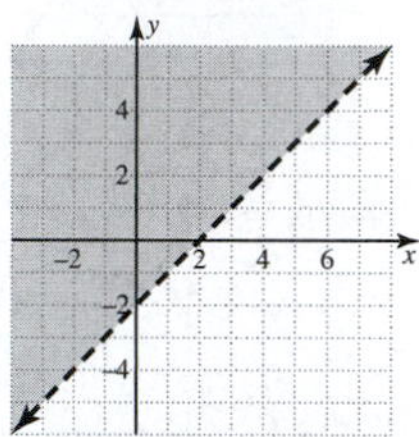

28.
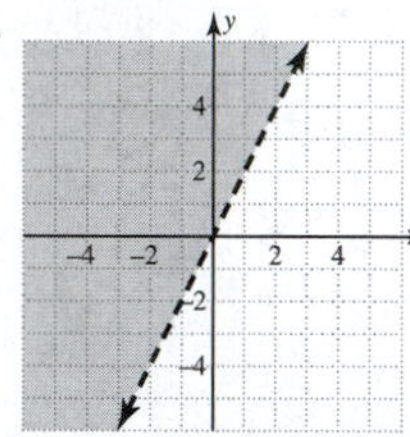

29.
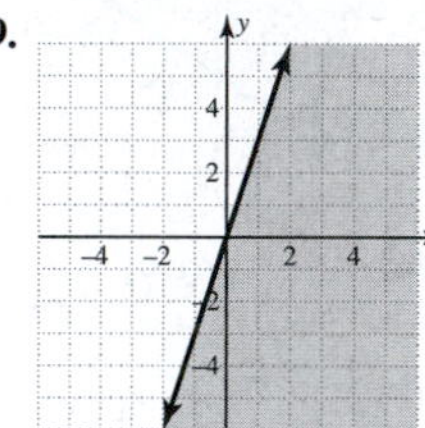

30.
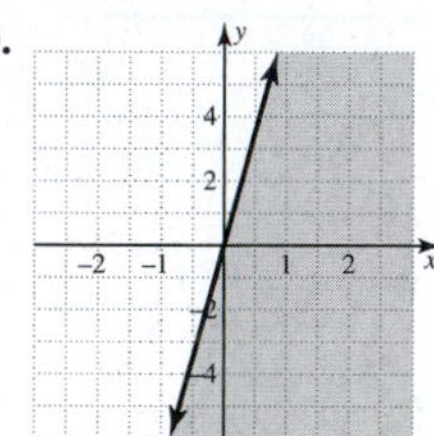

31.
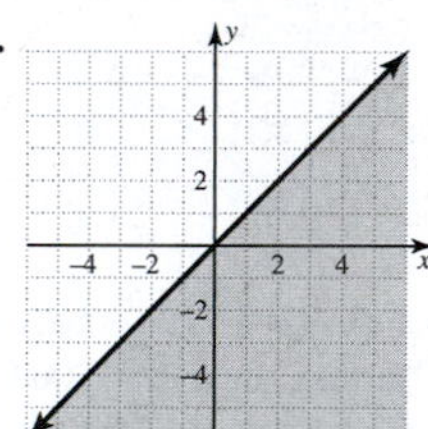

32.
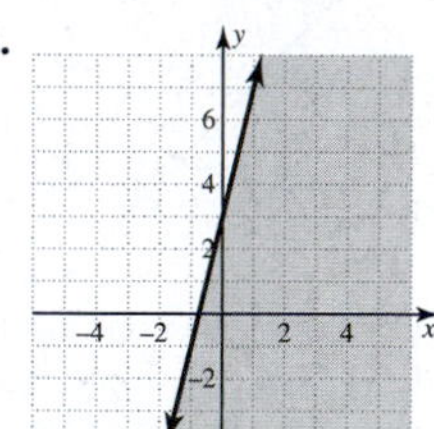

33.
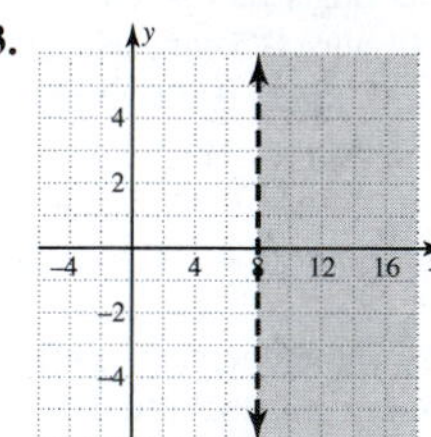

34.
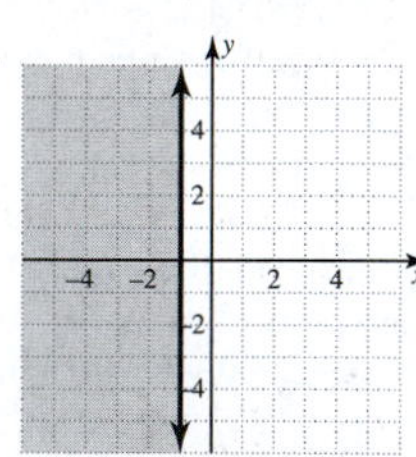

35.
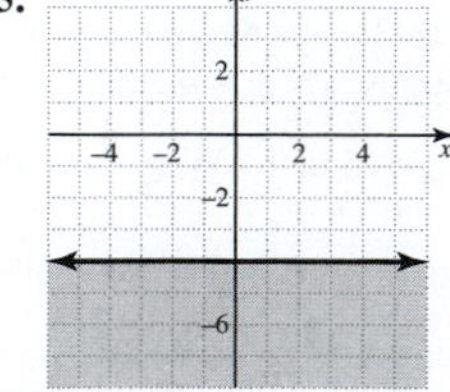

36.
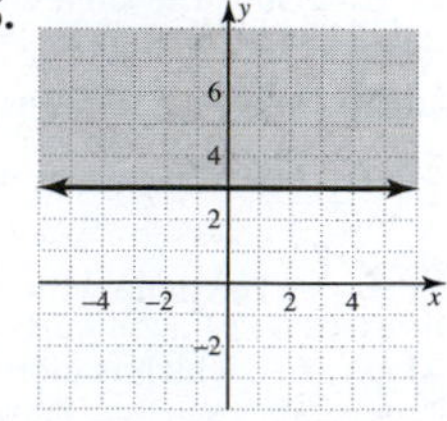

37. b.

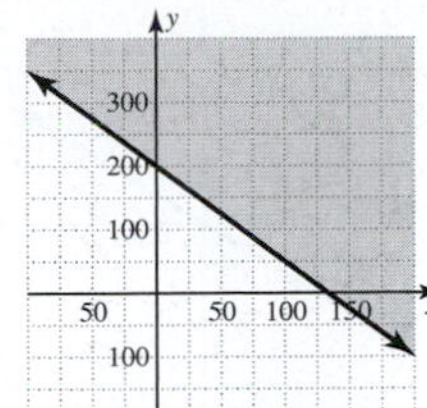

c. Answers vary; test average = 74 and final exam average = 89; test average = 90 and final exam average = 65; test average = 80 and final exam average = 80

38. b.

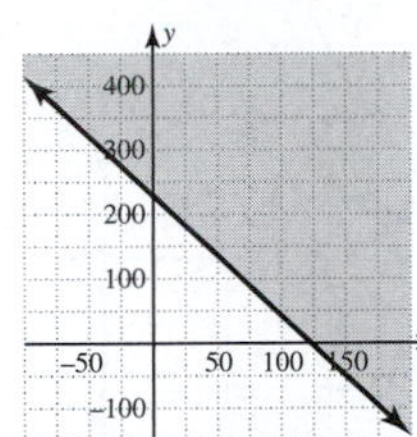

c. Answers vary; test average = 90 and final average = 62; test average = 76 and final average = 88; test average = 72 and final average = 95

39. b.

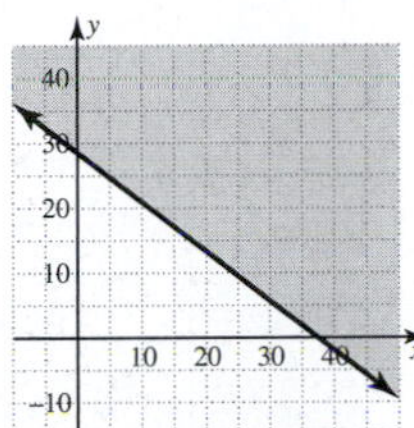

c. Answers vary; 30 hr at the bookstore and 6 hr at the school library; 25 hr at the bookstore and 10 hr at the school library; 10 hr at the bookstore and 21 hr at the school library

40. b.

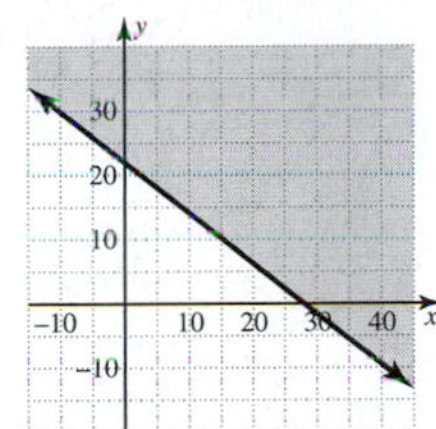

c. Answers vary; 10 hr at the food factory and 15 hr at the company; 16 hr at the food factory and 10 hr at the company; 8 hr at the food factory and 16 hr at the company

41. b.

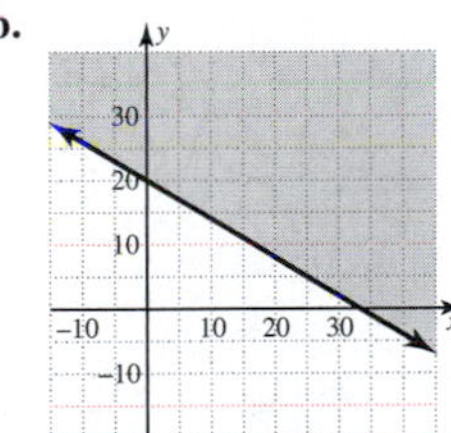

c. Answers vary; 10 pairs of pants and 14 suit jackets; 25 pairs of pants and 5 suit jackets; 7 pairs of pants and 16 suit jackets

42. b.

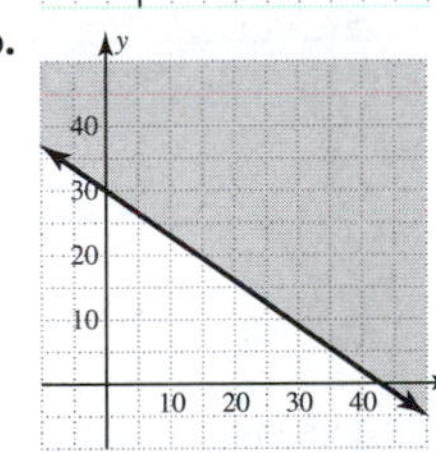

c. Answers vary; 30 lawns less than one-half an acre and 9 lawns from one-half to an acre; 10 lawns less than one-half an acre and 23 lawns from one-half to an acre; 20 lawns less than one-half an acre and 16 lawns from one-half to an acre

43. The intercepts should be $(0, -5)$ and $(5, 0)$. The point $(0, 0)$ does satisfy the inequality.

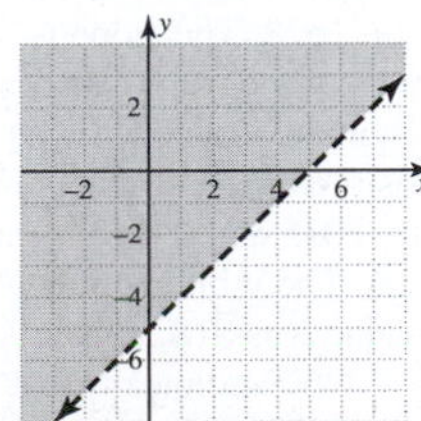

44. Since the boundary line goes through $(0, 0)$, we must use another test point, such as $(-5, 2)$.

$3(-5) + 2 > 0$

$-13 > 0$ False

So, we shade the half-plane that does not contain $(-5, 2)$.

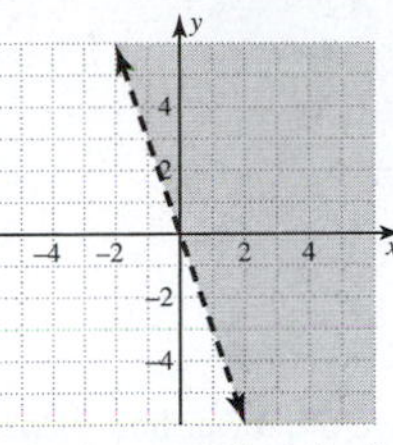

45. All of Lily's work is correct except that the boundary line should be solid since the inequality symbol is $\leq$.

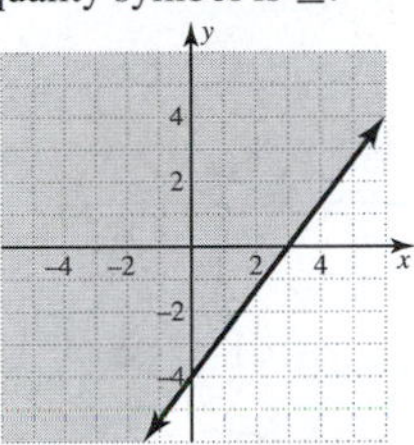

46.

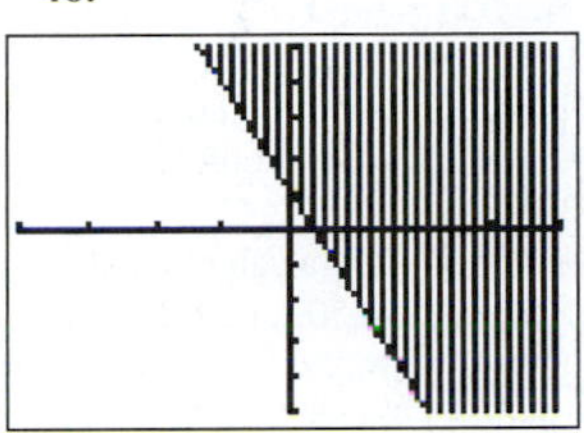

47.

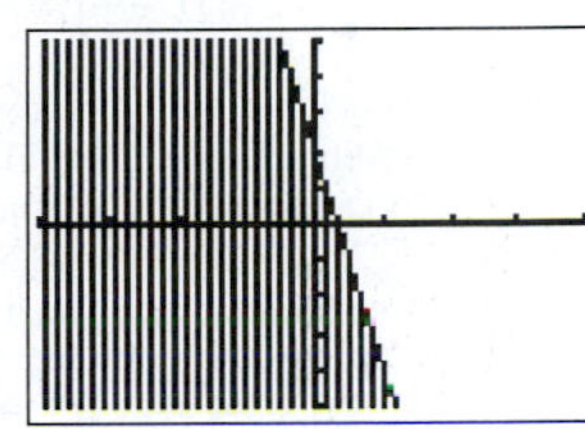

48.

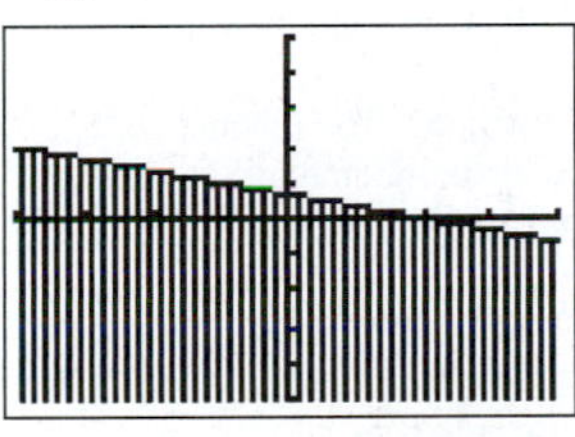

49.

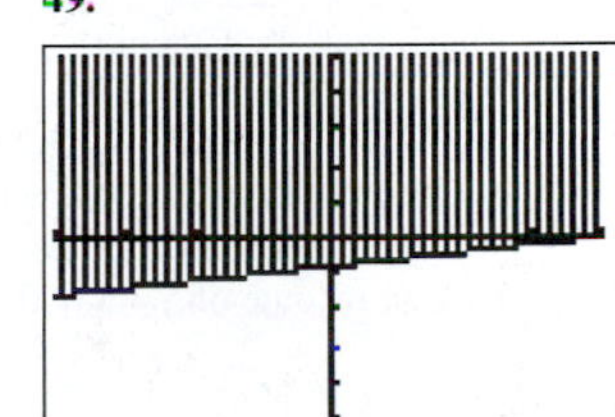

50. a.

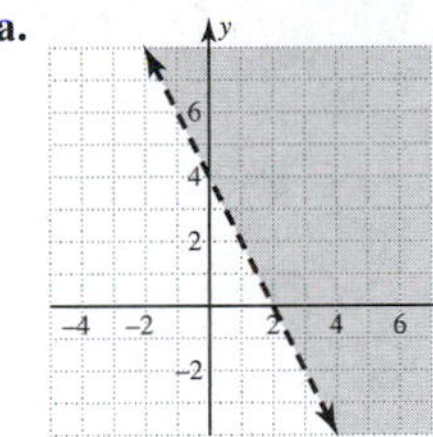

b.

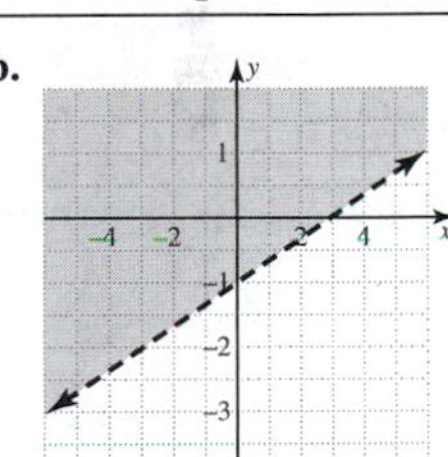

c.

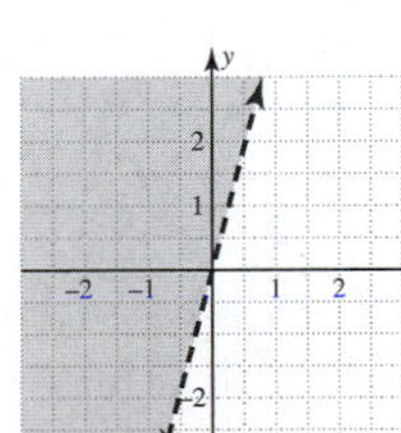

d.

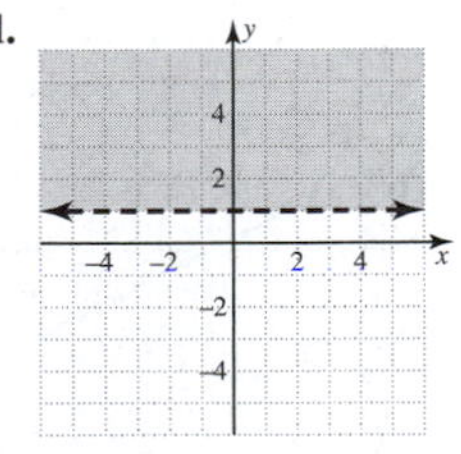

g. Both will have the upper-half plane as solutions, but $y \geq mx + b$ will have a solid boundary line and $y > mx + b$ will have a dashed boundary line.

51. a.

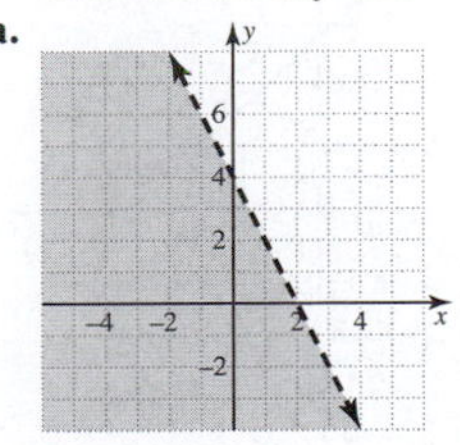

b.

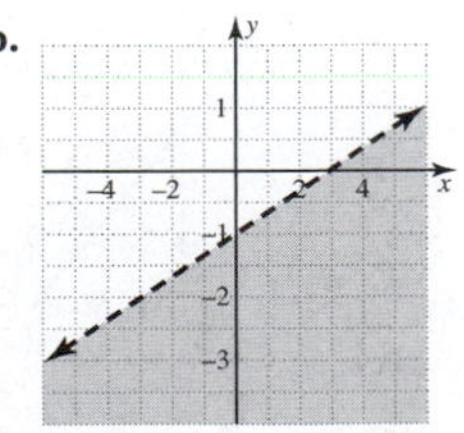

c.

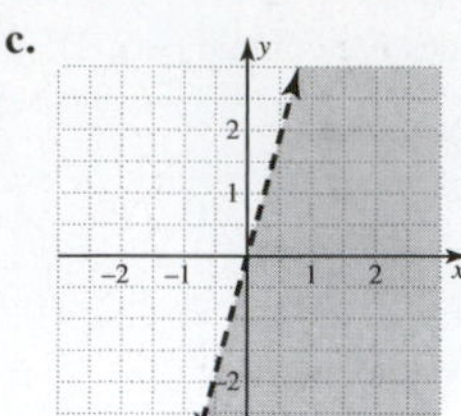

d.

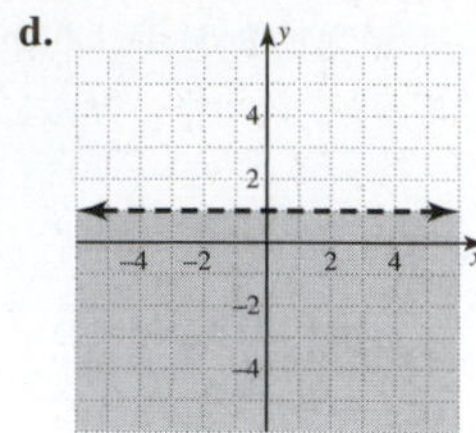

g. Both will have the lower half-plane as solutions but $y \leq mx + b$ will have a solid boundary line and $y < mx + b$ will have a dashed boundary line.

CHAPTER 4 REVIEW EXERCISES

Section 4.1

2. yes, $m = -7$, $b = 10$, $f\left(\frac{1}{7}\right) = 9$

5. $f(x) = \frac{8}{3}x - \frac{1}{3}; f\left(-\frac{7}{4}\right) = -5$ **6.** $f(x) = -\frac{2}{7}x + \frac{19}{7}; f\left(\frac{5}{2}\right) = 2$

13. $f(x) = 0.45x + 69.99; f(100) = 114.99$, The monthly cost for 100 additional minutes is \$114.99.; $x = 250$, The monthly cost for 250 additional minutes is \$182.49.

14. $f(x) = 2x + 2.5, f(8) = 18.5$, It costs \$18.50 for a cab ride of 9 mi in no traffic.; $x = 16$, The cab fare is \$34.50 for a ride of 16 additional miles, or 17 mi.

15. $f(x) = 30x + 89, f(2) = 149$, It costs \$149 to clean 2 additional rooms.; $x = 5$, It costs \$239 to clean 5 additional rooms.

16. $f(x) = 29x + 20.99, f(12) = 368.99$, It costs \$368.99 to join a fitness club for 12 months.; $x = 18$, It costs \$542.99 to join a fitness club for 18 months.

17. $f(x) = 0.28x + 16{,}750, f(85{,}400) = 40{,}662$, The income tax is \$40,662 for a person filing single with an income of \$167,650.; $x = 26{,}350$, The income tax is \$24,128 for a person filing single with an income of \$108,600.

Section 4.2

21.

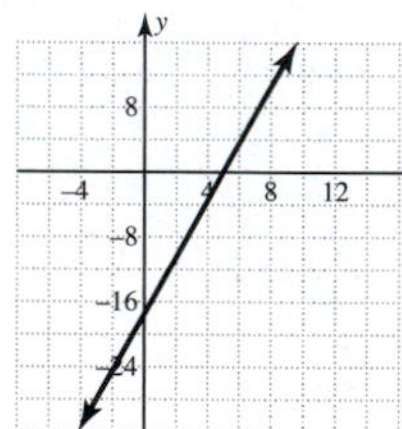

22.

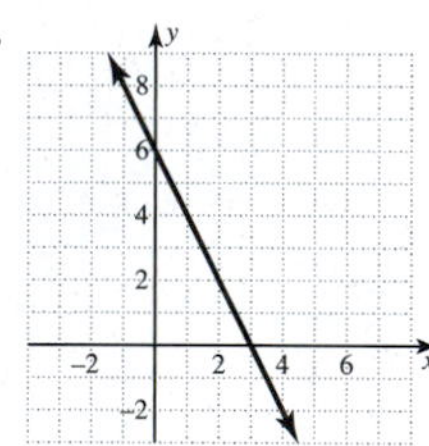

23.

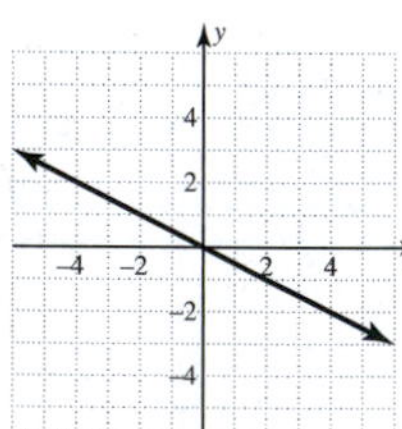

24.

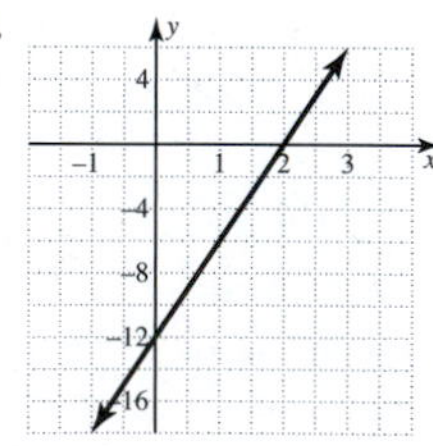

25.

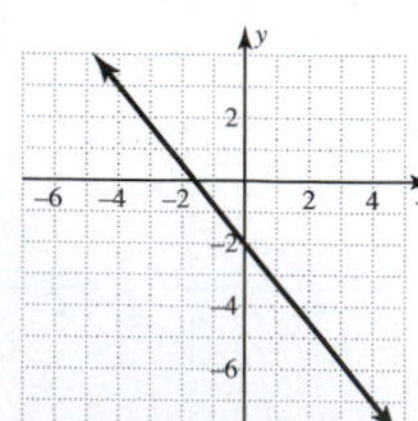

26.

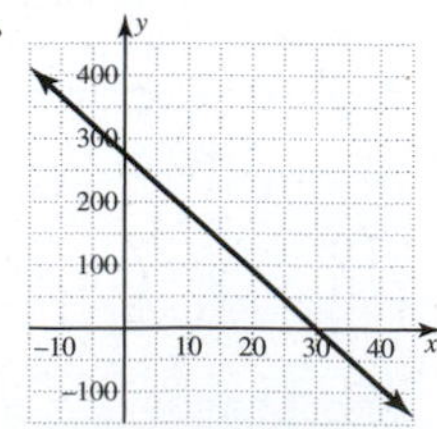

27.

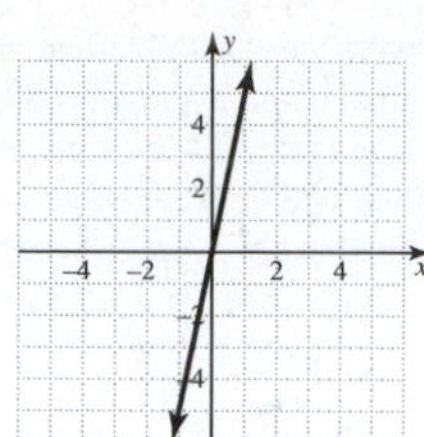

28.

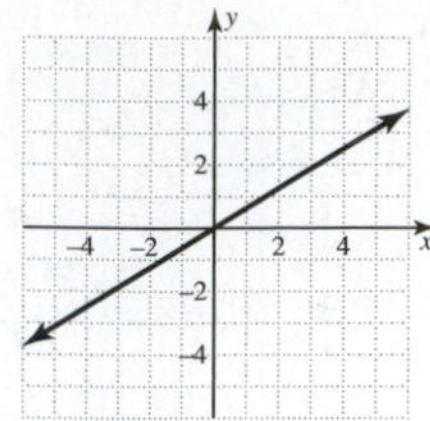

29.

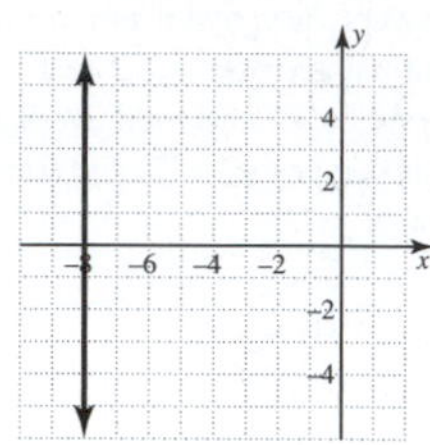

30.

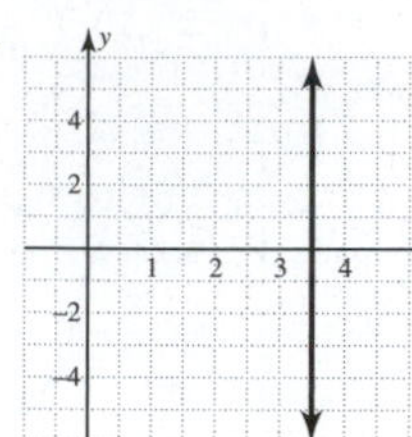

31.

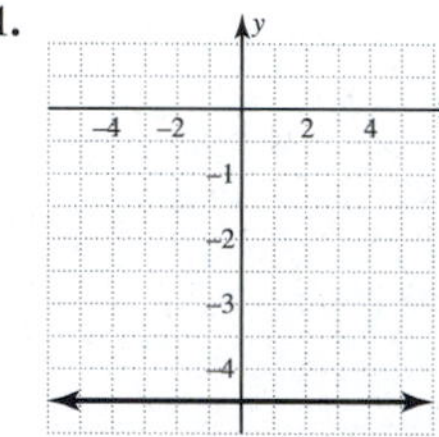

32.

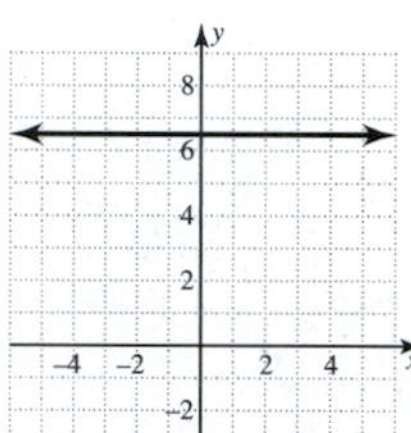

33. a. (0, 4889.50); In 2000, the cost of higher education at 2-yr public institutions was about \$4889.50. **b.** \$6567.50 **c.** \$9923.50

34. a. (0, 5969.80), In 1994, the expenditure per pupil in public elementary and secondary schools was about \$5969.80. **b.** \$8562.60 **c.** \$13,377.80

Section 4.3

35. $m = -\frac{1}{9}$, y-intercept $= \left(0, \frac{4}{3}\right)$

36. $m = \frac{1}{3}$, y-intercept $= \left(0, \frac{11}{3}\right)$

39.

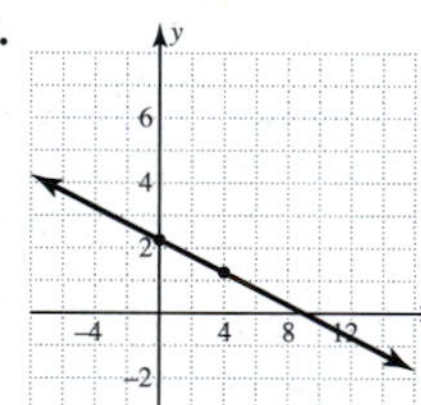

40.

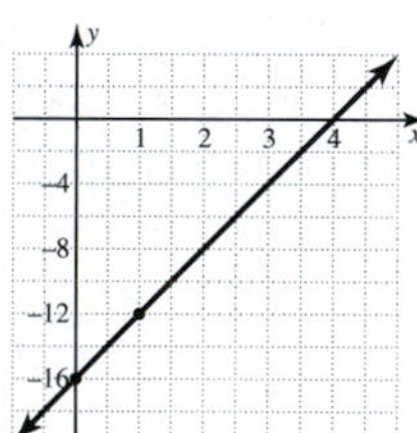

47. $f(x) = 70.2x + 2394$; The slope is 70.2 and means that every year after 2004, the number of registered nurses increases by 70.2 thousand. The y-intercept is (0, 2394) and means that there were 2394 thousand registered nurses in 2004.

48. $f(x) = 52.5x + 1628$; The slope is 52.5 and means that every year after 2004, the number of postsecondary teachers increases by 52,500. The y-intercept is (0, 1628) and means that there were 1,628,000 postsecondary teachers in 2004.

49. The slope is −2035 and it means that the car's value decreases \$2035 per year. The y-intercept is (0, 18,500) and means that the car's initial value was \$18,500.

50. The slope is −3960 and means that the truck's value decreases \$3960 every year. The y-intercept is (0, 36,000) and means that the truck's initial value was \$36,000.

51. The slope of the incline is 0.709. So, the incline raises 0.709 units for every unit the vehicle travels.

52. The slope of the incline is 0.585. So, the incline raises 0.585 units for every unit the vehicle travels or runs.

Section 4.4

60. $y = \frac{3}{2}x - \frac{33}{40}$ or $y = \frac{3}{2}x - 0.825$

63. $y = -\frac{5}{6}x + \frac{3}{2}$

70. $y = -\frac{2}{3}x - \frac{4}{5}$

Section 4.5

81.

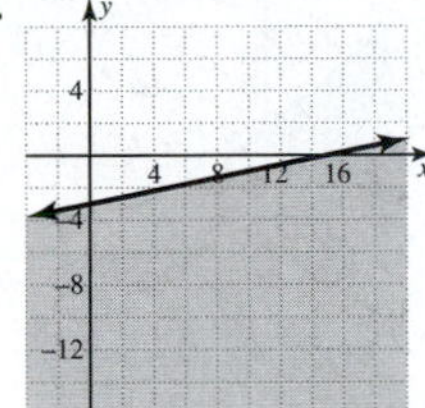

82.

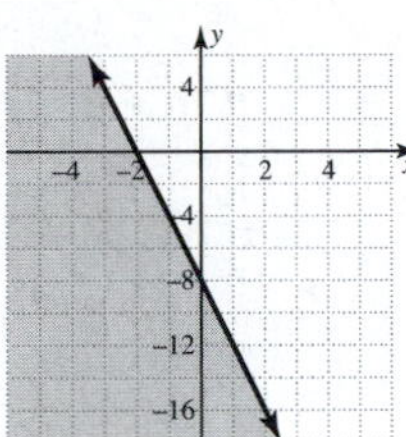

83.

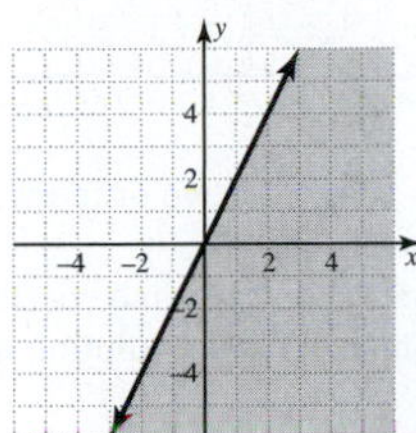

84.

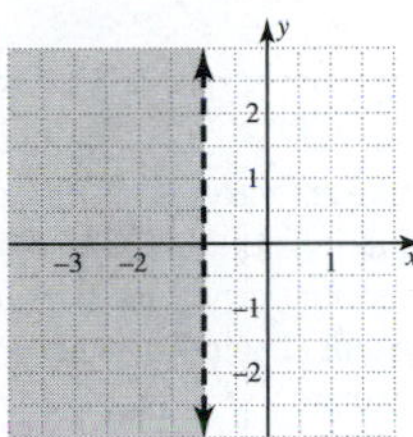

85. a. $15.25x + 10.5y \geq 400$;

b.

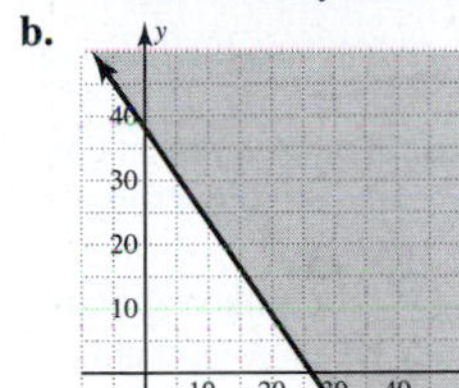

c. Answers vary; 20 hr as manager and 10 hr as teller; 21 hr as manager and 8 hr as teller; 22 hr as manager and 7 hr as teller

CHAPTER 4 TEST

9. a. (0, −6); (3, 0)

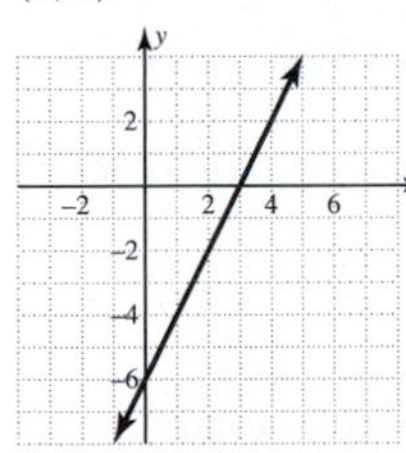

b. (0, 1); (−5, 0)

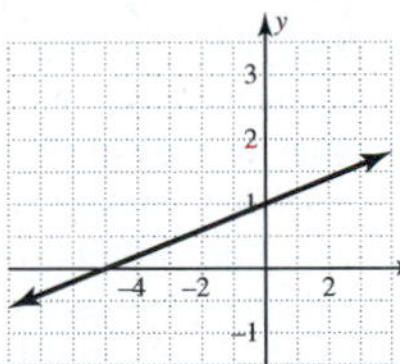

c. (0, 0)

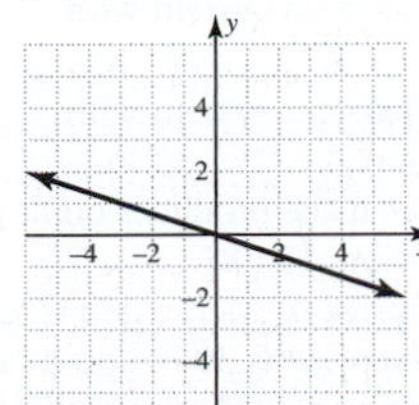

10. b. The x-intercept is (16, 0). This means that it will take 16 min for the plane to reach an altitude of 0 ft, or to reach the ground.

c. The y-intercept is (0, 8000). This means that the initial altitude of the plane was 8000 ft above the ground.

11. a. $m = \frac{4}{5}$ **b.** $m = -\frac{3}{5}$

12.

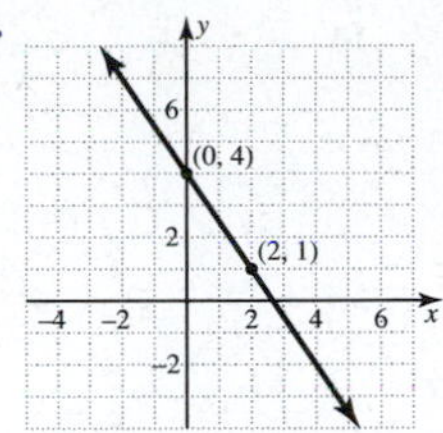

14. a. $m = 138.41$, y-intercept $= (0, 2961.8)$

b. The average undergraduate cost of attending a public 2-yr college was \$2961.80 in 1986. The cost is increasing by \$138.41 each year after 1986.

15. a. $y = \frac{3}{4}x - 6$ **e.** $y = \frac{1}{3}x - 1$

18. a.

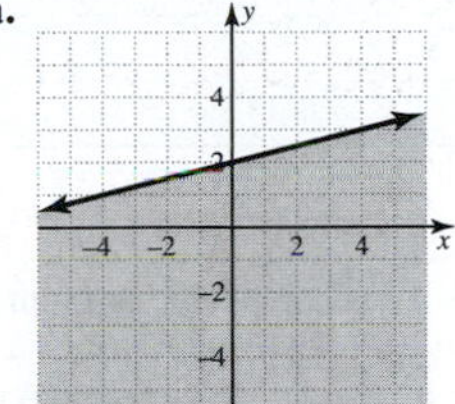

b.

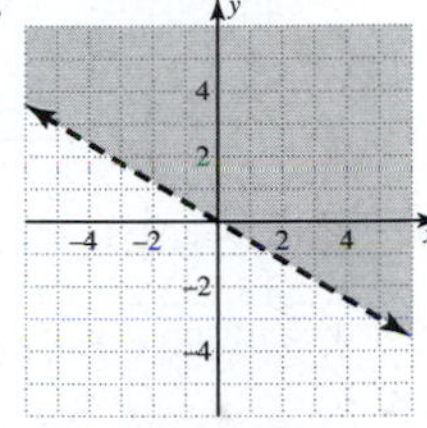

19. a. $6x + 12y \geq 12{,}000$

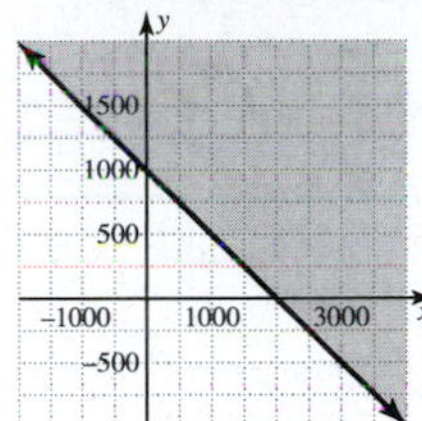

Three combinations are 500 student tickets and 1000 general admission tickets; 0 student tickets and 1000 general admission tickets; and 600 student tickets and 1500 general admission tickets.

CHAPTERS 1–4 CUMULATIVE REVIEW EXERCISES

12. $\frac{1}{15}x + 4$ **13.** $-\frac{1}{4}x + \frac{4}{3}$ **29.** $d = \frac{a - 3b}{8c}$

30. $z = \frac{3y - 5x}{10w}$

31. 6 7 8 9 10 11 12 13 14 15 16

$(11, \infty)$; $\{x \mid x > 11\}$

32. 18 19 20 21 22 23 24 25 26 27 28

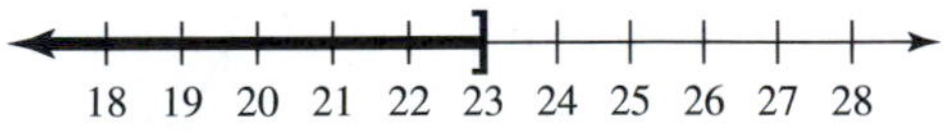

$(-\infty, 23]$

$\{y \mid y \leq 23\}$

40. $\left\{\frac{16}{7}, 38\right\}$

43.

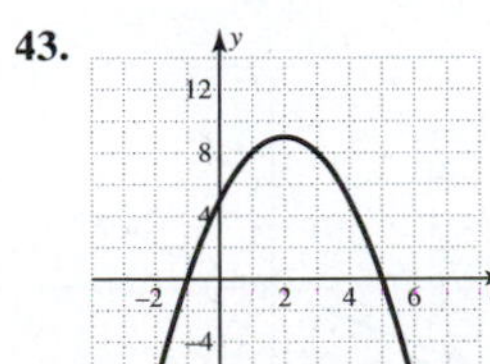

44.

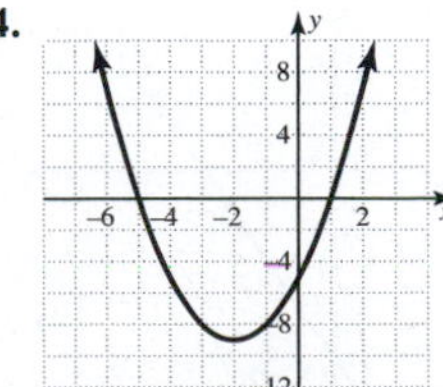

48.

55. 81; In 2015, the life expectancy for a person of any sex and race is approximately 81 yr.

73. e. 120,000; 100,000; 80,000; 60,000; 40,000; 20,000; 0 1 2 3 4 5 6

74. a. $f(x) = 0.25x + 6227.25$

b. \$13,952.25; The tax owed is \$13,952.25 for the amount of \$76,400 over \$45,500.

75.

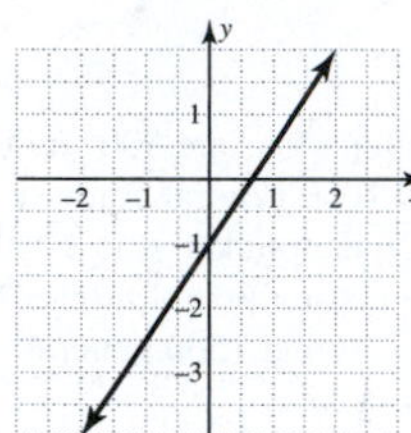

Domain = $(-\infty, \infty)$, Range = $(-\infty, \infty)$

76.

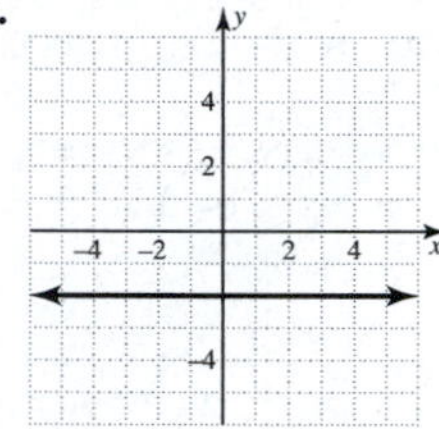

Domain = $(-\infty, \infty)$, Range = $\{-2\}$

77.

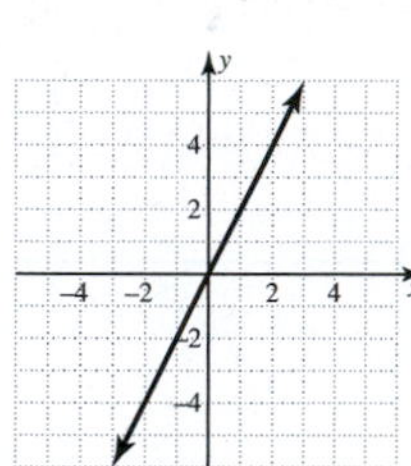

Domain = $(-\infty, \infty)$, Range = $(-\infty, \infty)$

78.

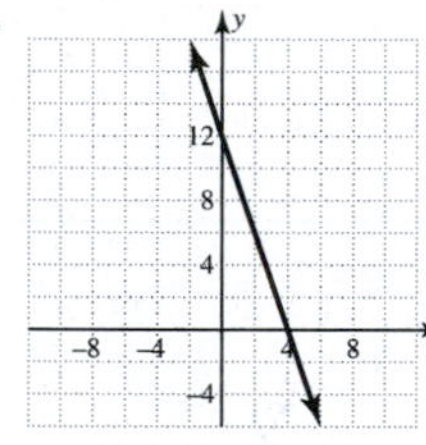

Domain = $(-\infty, \infty)$, Range = $(-\infty, \infty)$

79. $m = -9$, $(0, 6)$

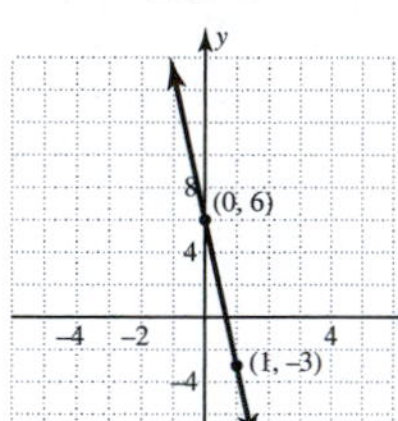

80. $m = -2$, $(0, -5)$

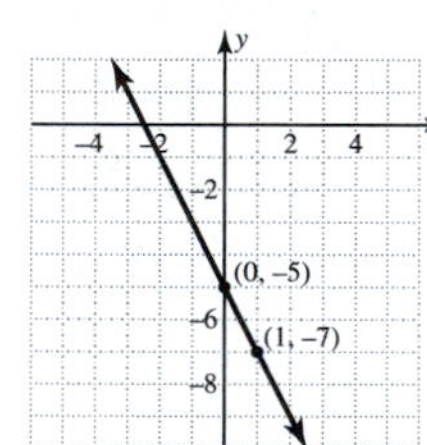

81. $m = \frac{3}{2}$, $(0, -6)$

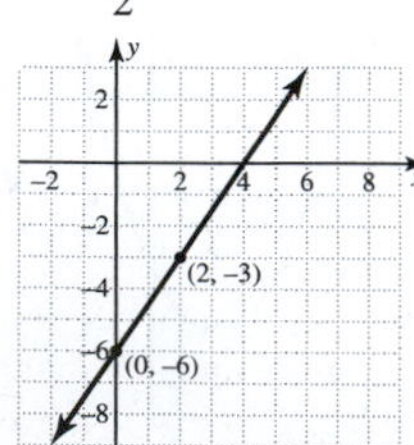

82. $m = \frac{1}{5}$, $(0, 0)$

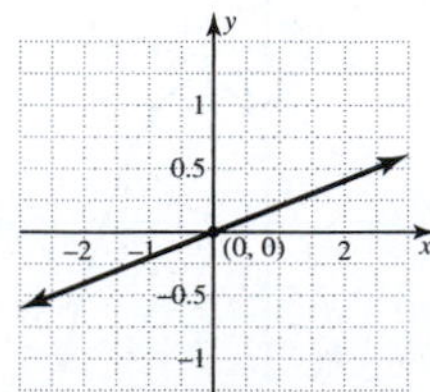

99.

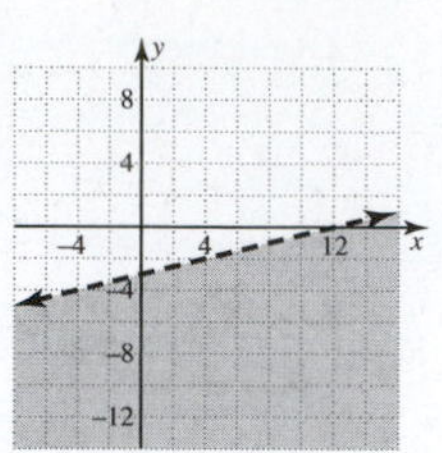

100.

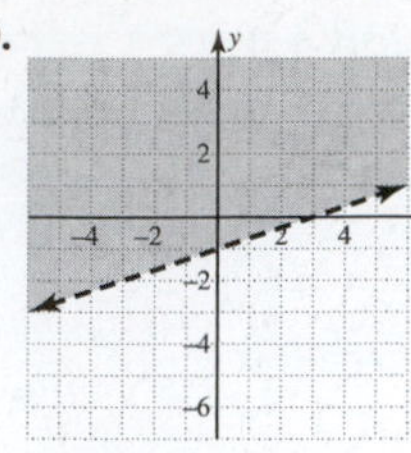

101.

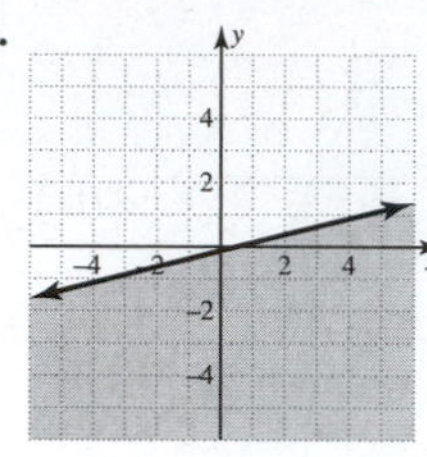

102.

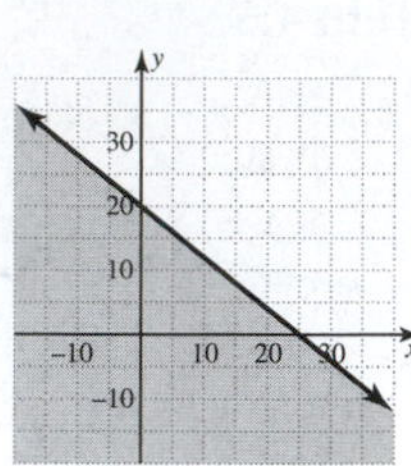

CHAPTER 5

Section 5.1

1. Answers vary. A system of equations is a set of two or more equations that must be solved simultaneously.

2. Answers vary. I must graph each equation on the same coordinate system. The point where the graphs intersect is the solution of the system.

3. Answers vary. A system of dependent equations is a set of two equations that are the same line.

4. Answers vary. An inconsistent system contains lines that are parallel.

5. Answers vary. Demand is the amount of a product that consumers wish to buy. Supply is the amount of product that a supplier has to sell to the consumer.

6. Answers vary. The break-even point is the number of units sold and produced that makes the company break even, that is, make no profit. It is where cost is equal to revenue.

36. $\{(x, y)|2x - 3y = 6\}$ **37.** $\{(x, y)|3x + y = -9\}$

39. $\{(x, y)|y = 2x + 5\}$ **40.** $\left\{(x, y)\middle|y = \frac{3}{4}x + 1\right\}$

41. $\left\{(x, y)\middle|y = \frac{5}{4}x - 2\right\}$ **42.** $\left\{(x, y)\middle|y = \frac{3}{4}x - 4\right\}$

45. $\left\{(x, y)\middle|y = \frac{3}{5}x + 2\right\}$

49. same line, infinitely many solutions, consistent system with dependent equations

50. same line, infinitely many solutions, consistent system with dependent equations

51. same line, infinitely many solutions, consistent system with dependent equations

52. parallel lines, no solution, inconsistent system

53. parallel lines, no solution, inconsistent system

54. parallel lines, no solution, inconsistent system

55. intersecting lines, one solution, consistent system with independent equations

56. intersecting lines, one solution, consistent system with independent equations

58. The break-even point is reached when she produces eight pairs of earrings. The cost and revenue are both \$240.

59. The equilibrium point is when the price of each video is \$3. The number of videos demanded and supplied is 30 hundred or 3000.

60. The equilibrium point is when the price of each video is \$4.50. The number of videos demanded and supplied is 54 hundred or 5400.

61. The point of intersection is approximately (3.1, 32), which means that in 3.1 yr after March 2008, or in April 2011, there were 32 million visitors to Twitter and to MySpace. Before this time, MySpace had more visitors than Twitter. After this time, Twitter had more visitors than MySpace.

62. The point of intersection is approximately (13, 83), which means that in 13 yr after 2006, or in 2019, the revenue for Target Corporation and Amazon will each be \$83 billion. Before 2019, Target's revenue is greater than Amazon revenue. After 2019, Amazon revenue will be greater than Target's revenue.
81. parallel lines, no solution, inconsistent system
82. intersecting lines, one solution, consistent system with independent equations
83. intersecting lines, one solution, consistent system with independent equations
84. same line, infinitely many solutions, consistent system with dependent equations
85. same line, infinitely many solutions, consistent system with dependent equations
86. parallel lines, no solution, inconsistent system
87. The equilibrium point is when the price of each item is \$8. The number of items demanded and supplied is 72 hundred or 7200.
88. The equilibrium point is when the price of each item is \$4. The number of items demanded and supplied is 32 hundred or 3200.
89. The equilibrium point is when the price of each item is \$3. The number of items demanded and supplied is 63 hundred or 6300.
91. The mistake was made in solving the first equation for y. The first equation should be

$$-2y = -x - 8$$
$$y = \frac{1}{2}x + 4$$

The graph of each equation is shown.

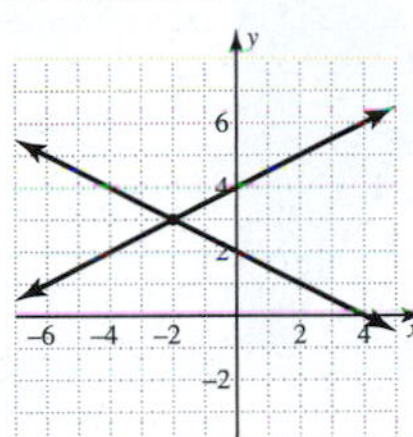

The solution set is $\{(-2, 3)\}$.

92. The intercepts of $x - y = 3$ are (3, 0) and (0, −3). The points on $y = -3x + 1$ are (0, 1) but then we move down 3 units and right 1 unit to get to the point (1, −2).

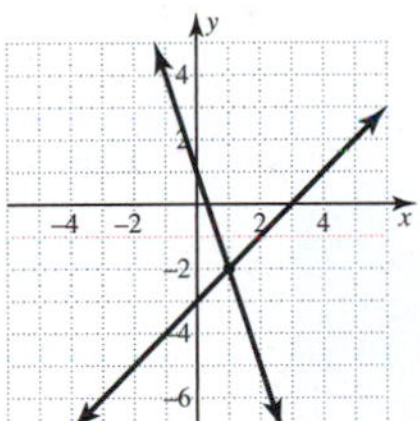

The solution set is $\{(1, -2)\}$.

95. $\{(2.5, 3)\}$

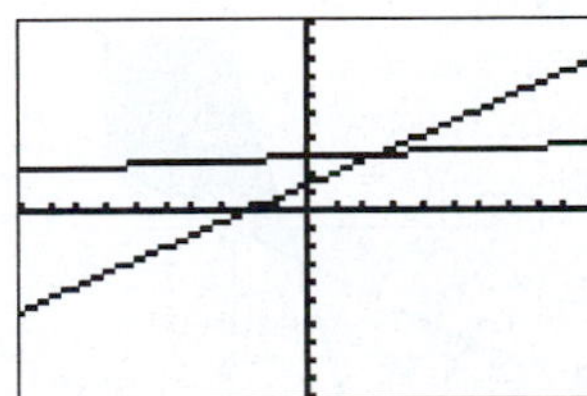

96. $\{(-1.5, 0.5)\}$

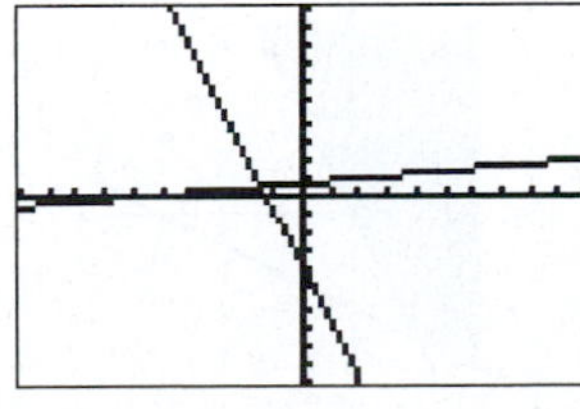

97. Answers vary; $\begin{cases} x + y = 1 \\ x - y = 5 \end{cases}$
98. Answers vary; $\begin{cases} x + 4y = 0 \\ 2x - y = -9 \end{cases}$
99. Answers vary; $\begin{cases} y = 2x \\ x - 5y = 0 \end{cases}$
100. Answers vary; $\begin{cases} 3x - 2y = -1 \\ x + y = 3 \end{cases}$
101. Answers vary; $y = \frac{1}{4}x + 3$
102. Answers vary; $y = -\frac{3}{5}x - 4$

Section 5.2

1. Answers vary. In substitution, solve for the variable that has a coefficient of 1 or −1, if possible.
2. Answers vary. One equation in the system is used to solve for the variable. The other equation is where this expression is substituted into.
3. Answers vary. The coefficients of one variable must be opposite so that one of the variables is eliminated when the equations are added. If the coefficients of neither variable are opposite, then we must multiply one or both equations by some nonzero number.
4. Answers vary. After one coordinate is found, we must substitute this value into one of the original equations in the system to solve for the value of the other variable.
24. $\left\{\left(-5, \frac{3}{5}\right)\right\}$
94. Answers vary; $m = \frac{1}{7}$ and $b \neq \frac{1}{7}$, one example is $y = \frac{1}{7}x + 2$.

Piece It Together Sections 5.1 and 5.2

8. intersecting lines, one solution, consistent system with independent equations
9. same line, infinitely many solutions, consistent system with dependent equations
10. intersecting lines, one solution, consistent system with independent equations

Section 5.3

1. Answers vary. The first step in solving an application problem is determining what is unknown.
5. Answers vary. A plane's speed flying with the wind is $x + y$ mph and against the wind is $x - y$ mph.
6. Answers vary. After the system is set up, we can use either substitution or elimination to solve the system. We must check the reasonableness of the solution.
15. Amie's hourly wage as an office assistant is \$10.50 and her hourly wage as a store clerk is \$8.50.
71.

$$x + y = 180$$
$$x = 3y - 34$$

$$3y - 34 + y = 180$$
$$4y - 34 = 180$$
$$4y = 214$$
$$y = \frac{214}{4} = 53.5$$
$$x = 3y - 34$$
$$x = 3(53.5) - 34 = 126.5$$

The measures of the angles are 126.5° and 53.5°.

72.

$$x + y = 9000$$
$$0.055x + 0.07y = 615$$

$$55(x + y = 9000)$$
$$-1000(0.055x + 0.07y = 615)$$
$$55x + 55y = 49{,}5000$$
$$\underline{-55x - 70y = -615{,}000}$$
$$-15y = -120{,}000$$
$$y = 8000$$
$$x + y = 9000$$
$$x + 8000 = 9000$$
$$x = 1000$$

Wang invested \$1000 at 5.5% and \$8000 at 7%.

Section 5.4

1. Answers vary. A solution of a system of three variables will be an ordered triple (x, y, z) that satisfies all three equations in the system.
2. Answers vary. To create a system of two linear equations from a system of three linear equations, we must pair together two equations and eliminate one variable and then pair together two other equations and eliminate the same variable.
3. Answers vary. A false statement means that there is no solution of the system of equations.

4. Answers vary. A true statement means that there are infinitely many solutions of the system.
5. Answers vary. We know we must use a system of three linear equations to solve a word problem if there are three unknowns.
6. Answers vary. To solve this problem, we would get one equation from the total amount to invest, that is, $x + y + z = A$. The other equation comes from the interest earned. It would be $(r/100)x + (s/100)y + (t/100)z = I$.

24. $\left\{\left(-\frac{5}{2}z - 1, -\frac{3}{2}z - 4, z\right) \middle| z \text{ is a real number}\right\}$

45. $\left\{\left(\frac{1}{3}z + \frac{4}{3}, \frac{1}{3}z + \frac{16}{3}, z\right) \middle| z \text{ is a real number}\right\}$

46. $\left\{\left(\frac{8}{5}z + \frac{23}{5}, \frac{14}{5}z + \frac{4}{5}, z\right) \middle| z \text{ is a real number}\right\}$

61. $\begin{cases} 3x - 4y + 6z = 3 \\ -3(x - y - 2z = 8) \end{cases} \rightarrow \begin{cases} 3x - 4y + 6z = 3 \\ -3x + 3y + 6z = -24 \end{cases}$

$$-y + 12z = -21$$

$\begin{cases} 2(x - y - 2z = 8) \\ -2x + 7y - 3z = -17 \end{cases} \rightarrow \begin{cases} 2x - 2y - 4z = 16 \\ -2x + 7y - 3z = -17 \end{cases}$

$$5y - 7z = -1$$

$\begin{cases} 5(-y + 12z = -21) \\ 5y - 7z = -1 \end{cases} \rightarrow \begin{cases} -5y + 60z = -105 \\ 5y - 7z = -1 \end{cases}$

$$53z = -106$$
$$z = -2$$

$5y - 7z = -1$	$x - y - 2z = 8$
$5y - 7(-2) = -1$	$x - (-3) - 2(-2) = 8$
$5y + 14 = -1$	$x + 3 + 4 = 8$
$5y = -15$	$x + 7 = 8$
$y = -3$	$x = 1$

The solution set is $\{(1, -3, -2)\}$.

62. $\begin{cases} -3(x + 2y + 3z = -1) \\ 3x - 2y + z = 7 \end{cases} \rightarrow \begin{cases} -3x - 6y - 9z = 3 \\ 3x - 2y + z = 7 \end{cases}$

$$-8y - 8z = 10$$

$\begin{cases} 2(4y + 4z = -5) \\ -8y - 8z = 10 \end{cases} \rightarrow \begin{cases} 8y + 8z = -10 \\ -8y - 8z = 10 \end{cases}$

$$0 = 0$$

$4y + 4z = -5$	$x + 2y + 3z = -1$
$4y = -4z - 5$	$x + 2\left(-z - \frac{5}{4}\right) + 3z = -1$
$y = -z - \frac{5}{4}$	$x - 2z - \frac{5}{2} + 3z = -1$
	$x + z - \frac{5}{2} = -1$
	$x = -z + \frac{3}{2}$

The solution set is $\left\{\left(-z + \frac{3}{2}, -z - \frac{5}{4}, z\right) \middle| z \text{ is a real number}\right\}$.

63. Answers vary. One example is $x + y + z = 8$
$x - y - z = 2$
$2x + y + z = 13$

64. Answers vary. One example is $x + y + z = 8$
$x - y - z = -12$
$3x + 2y - z = -7$

Section 5.5

1. Answers vary. A system of linear inequalities is a set of two or more linear inequalities that must be solved simultaneously.
2. Answers vary. A solution of a system of linear inequalities consists of the intersection of half-planes.
3. Answers vary. Inequalities will be needed in a system if there are phrases such as "at most," "at least," "more than," "less than," and so on.
4. Answers vary. To solve a system of linear inequalities, we graph each inequality in the system and find the intersection of the half-planes.
5. Answers vary. A boundary line is solid if the inequality symbol is ≥ or ≤. A boundary line is dashed if the inequality symbol is < or >.

6.

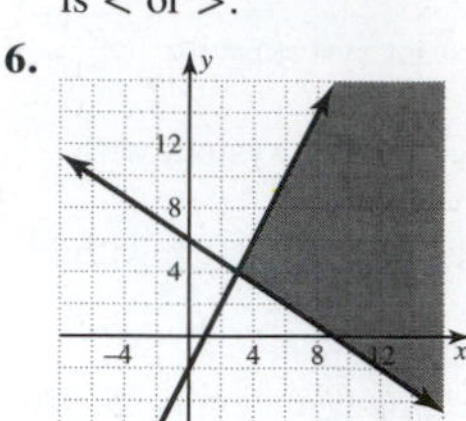

7.

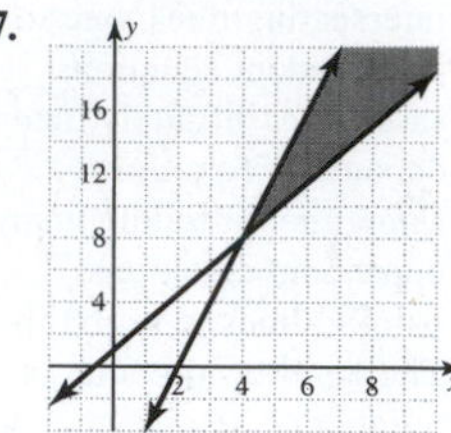

8.

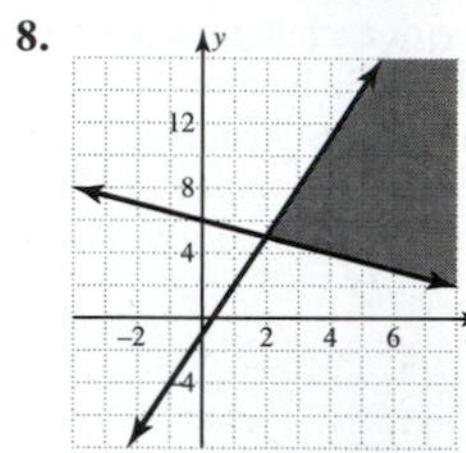

9.

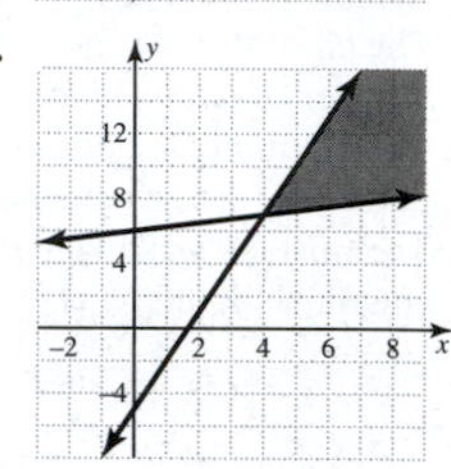

10.

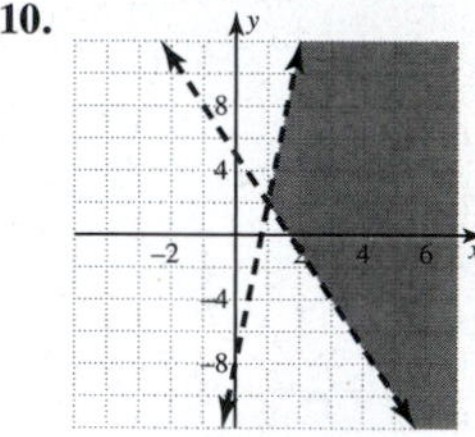

11.

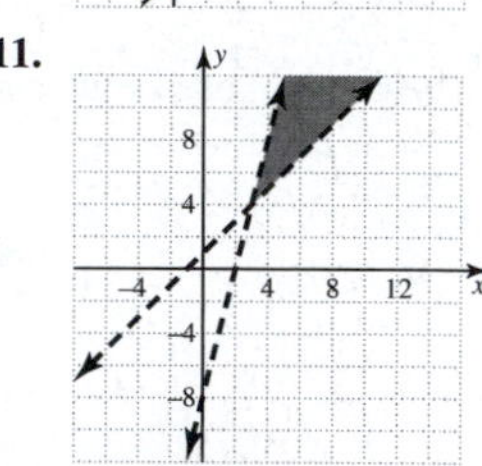

12.

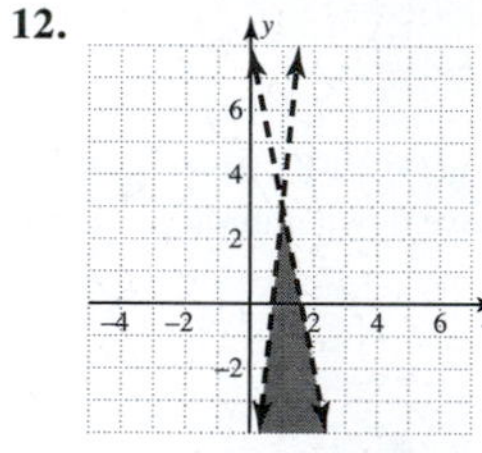

13.

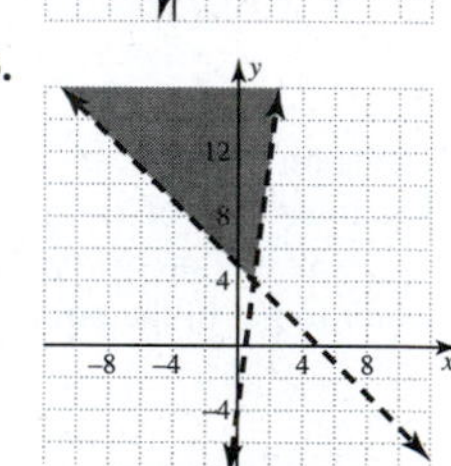

14.

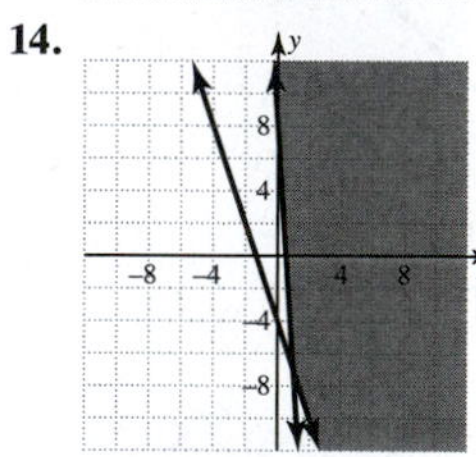

15.

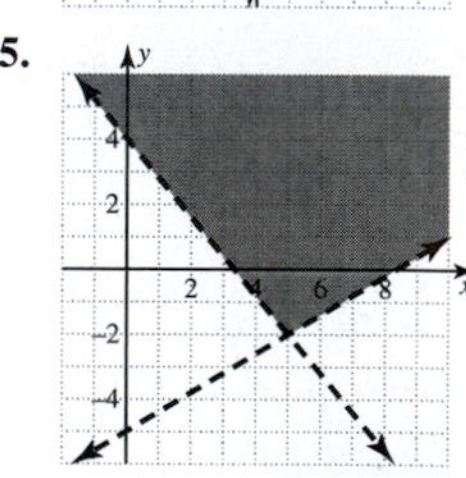

16.

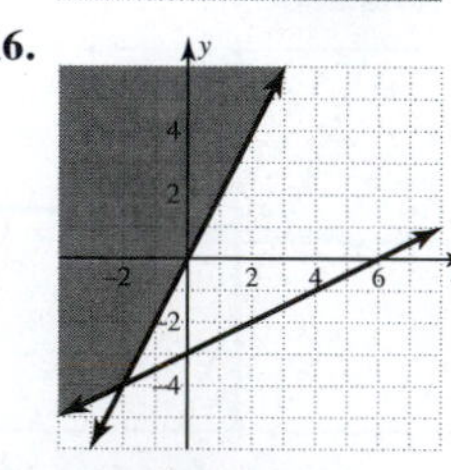

17.

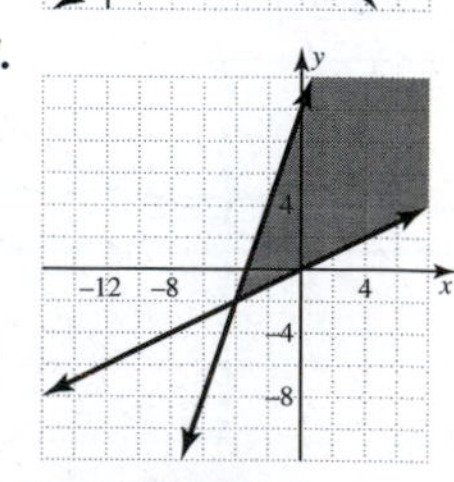

18.

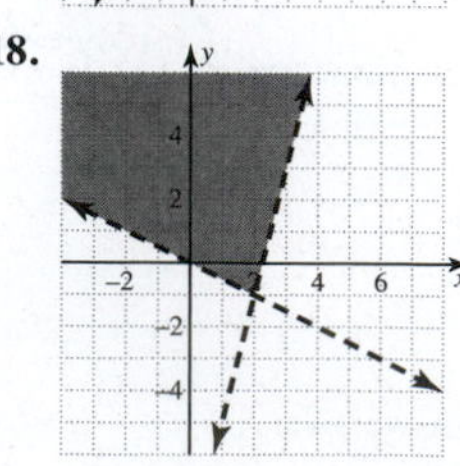

19.

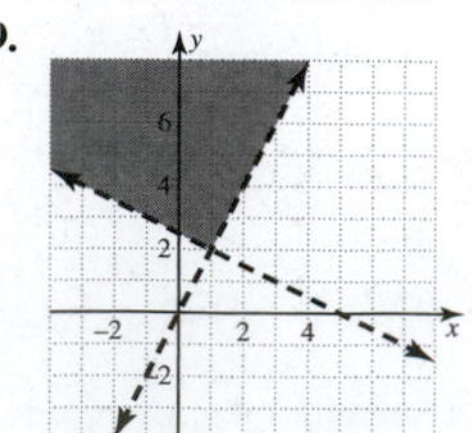

20.

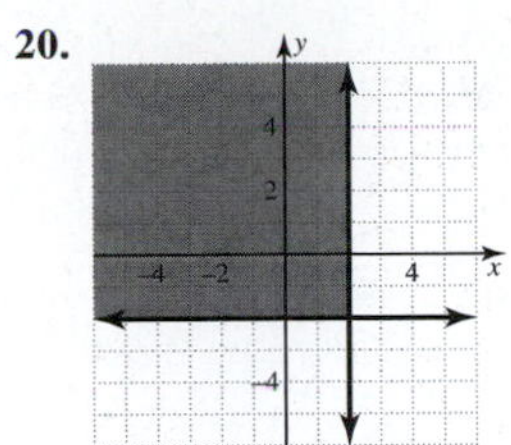

21.

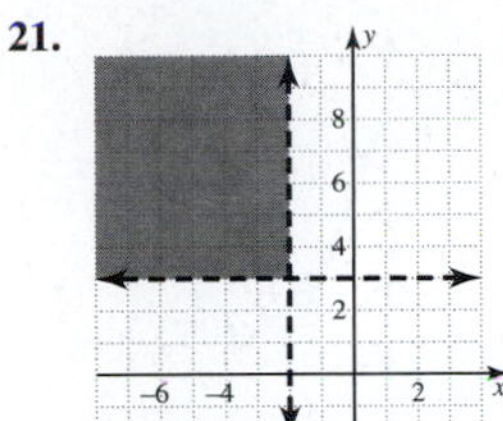

22.

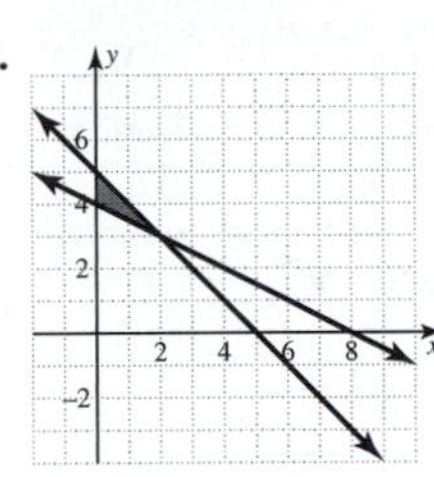

23.

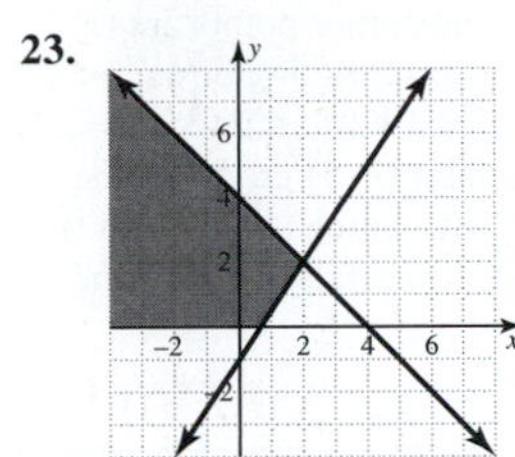

24.

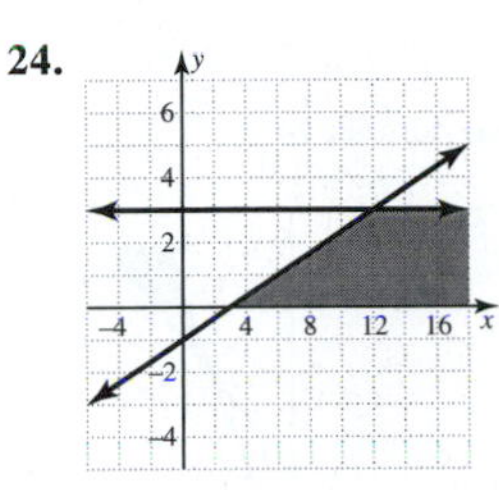

25.

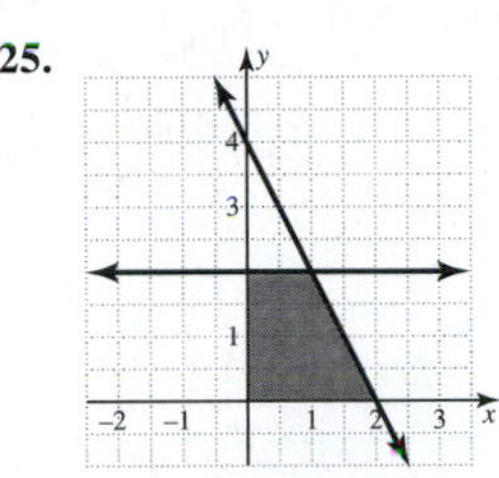

26.

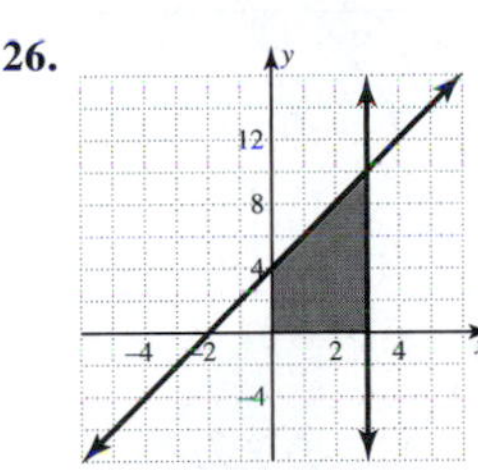

27.

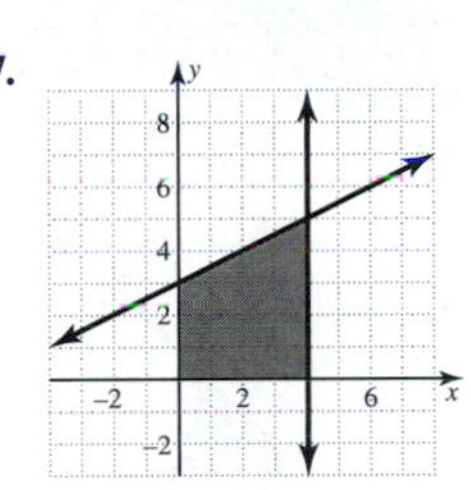

32. Answers vary; 400 rolls and 800 boxes, 0 rolls and 1050 boxes, 240 rolls and 960 boxes

33. Answers vary; 100 rolls and 600 boxes, 0 rolls and 750 boxes, 150 rolls and 500 boxes

34. Answers vary; 270 drinks and 540 hot dogs, 20 drinks and 790 hot dogs, 0 drinks and 800 hot dogs

35. Answers vary; 240 drinks and 480 hot dogs, 0 drinks and 600 hot dogs, 200 drinks and 400 hot dogs

36. Answers vary; 160 double and 80 king-size rooms, 240 double and 120 king-size rooms, 180 double and 60 king-size rooms

37. Answers vary; 375 double and 125 king-size rooms, 200 double and 60 king-size rooms, 350 double and 100 king-size rooms

38. a.

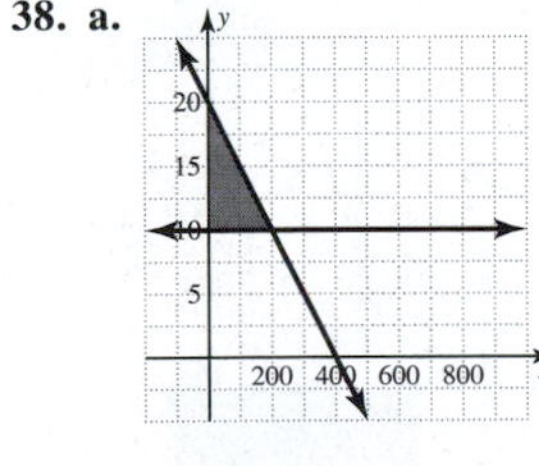

39. a.

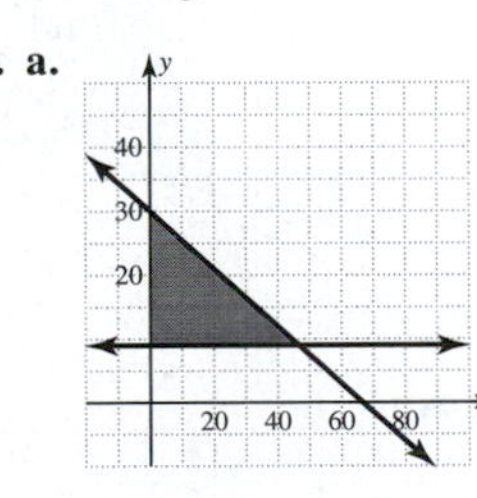

40. a.

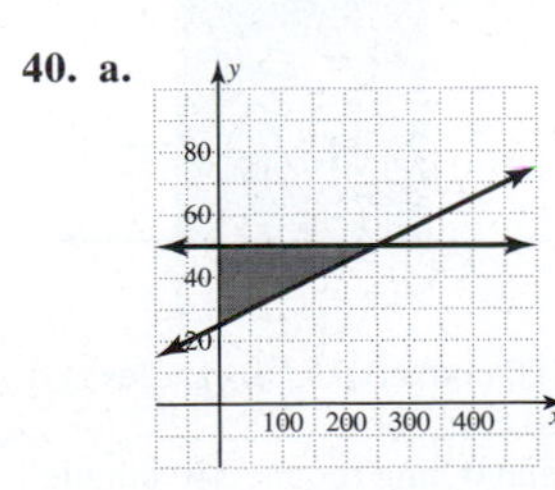

41. a.

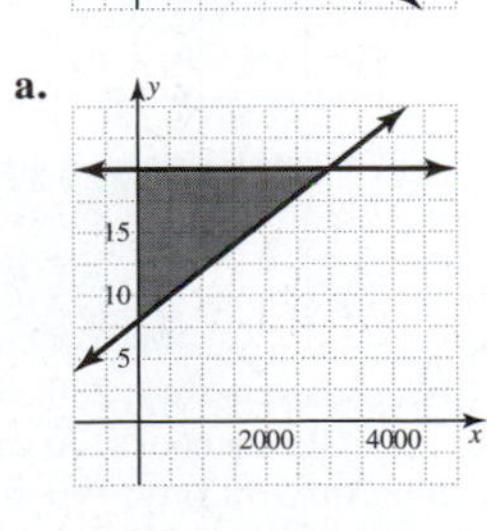

42.

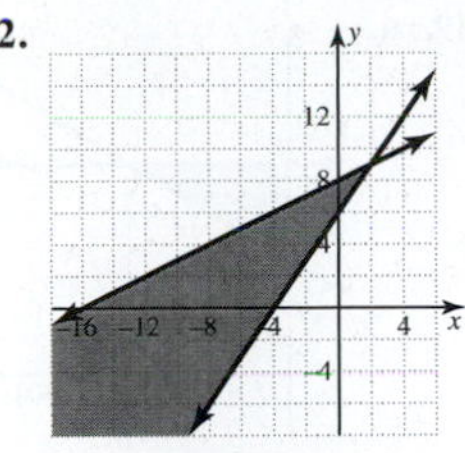

43.

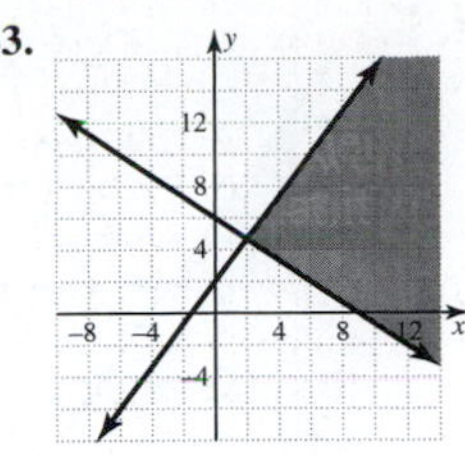

44.

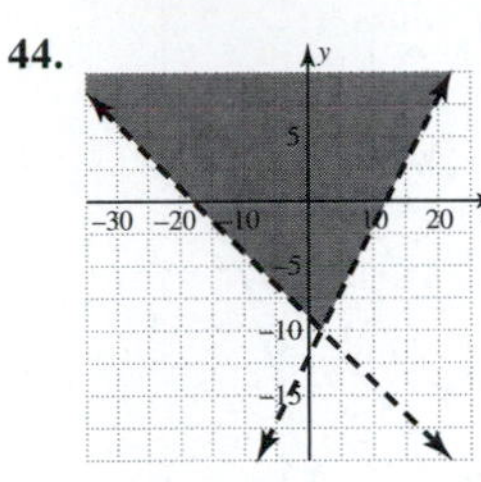

45.

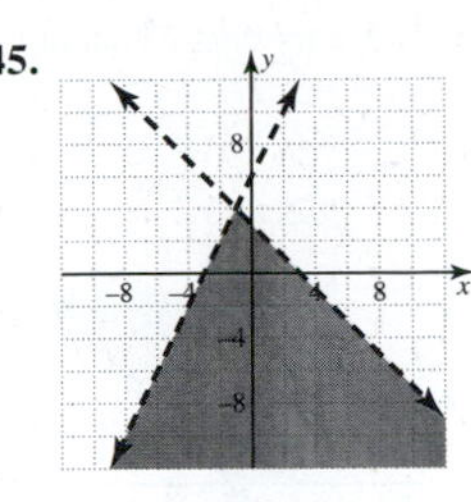

46.

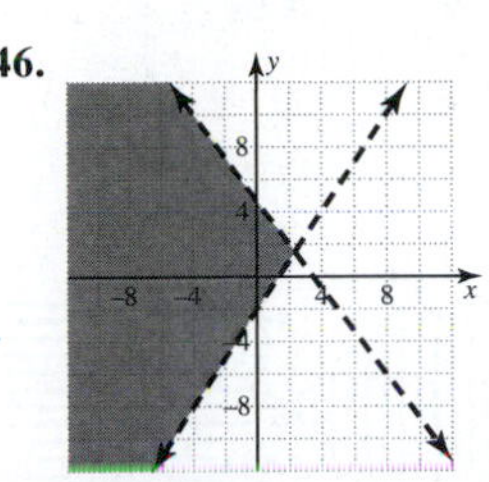

47.

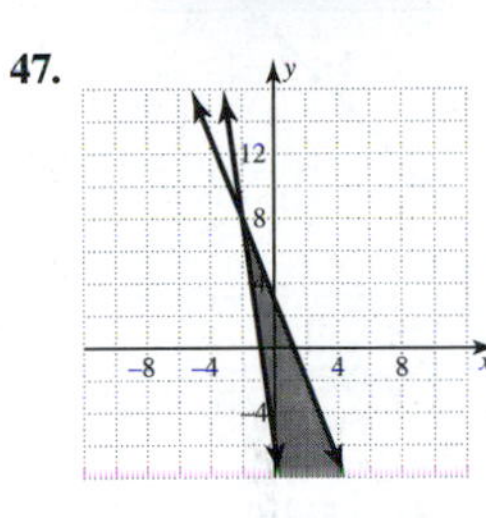

48.

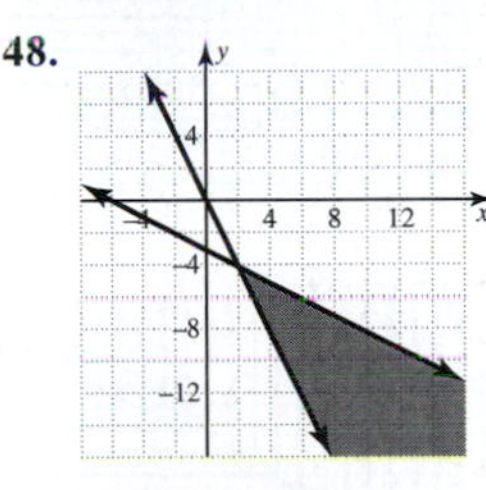

49.

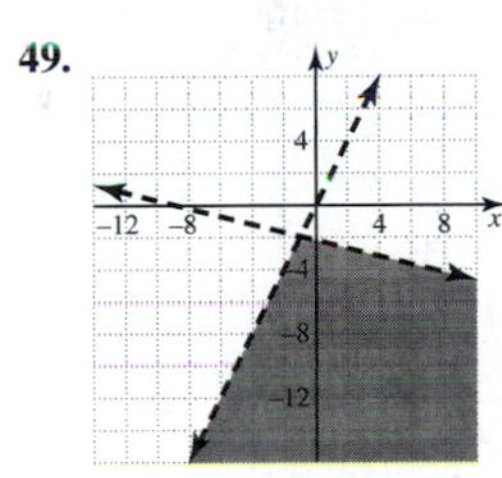

50.

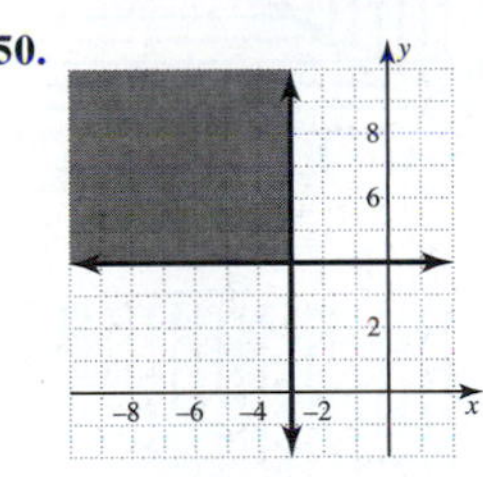

51.

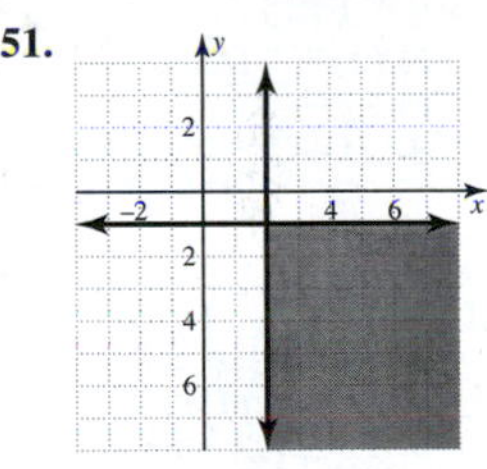

52.

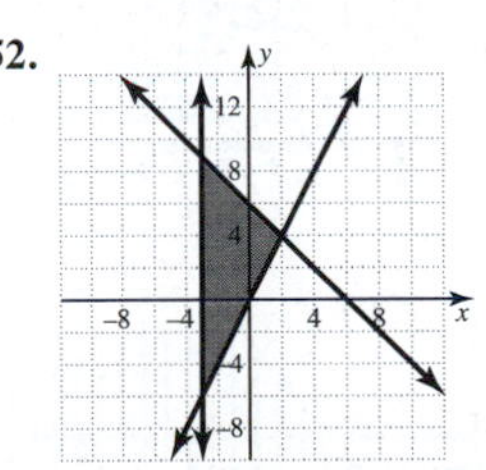

53.

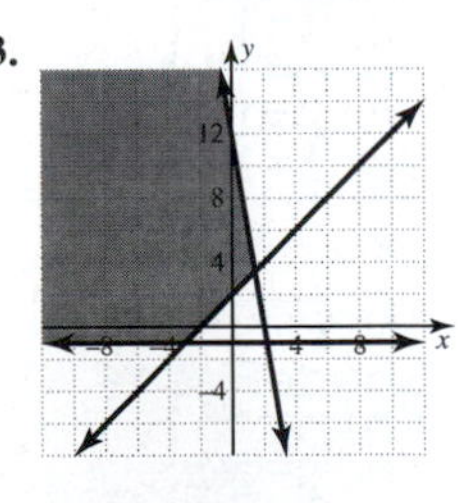

54.

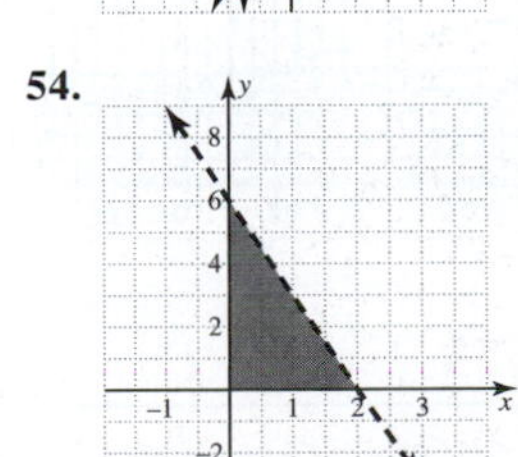

55.

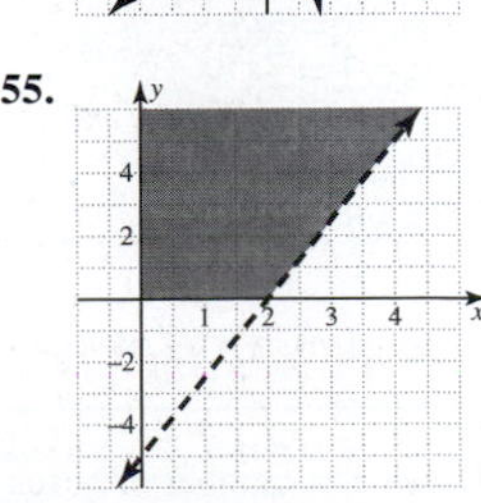

56. Answers vary; 300 drinks and 780 hot dogs, 360 drinks and 720 hot dogs, 0 drinks and 960 hot dogs

57. Answers vary; 240 double and 80 king-size rooms, 60 double and 20 king-size rooms, 200 double and 50 king-size rooms

58. a.

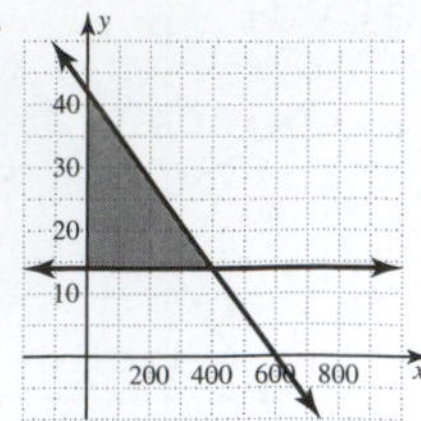

59. a.

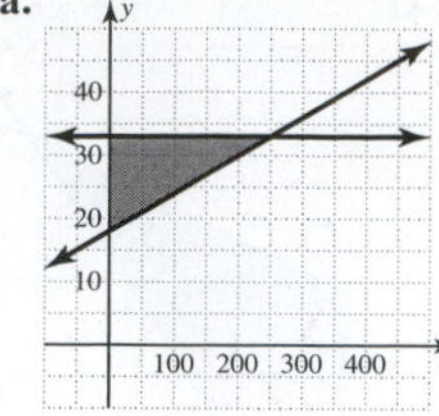

60. Solve the first inequality for y first.

$x - y \le 2$

$-y \le -x + 2$

$y \ge x - 2$

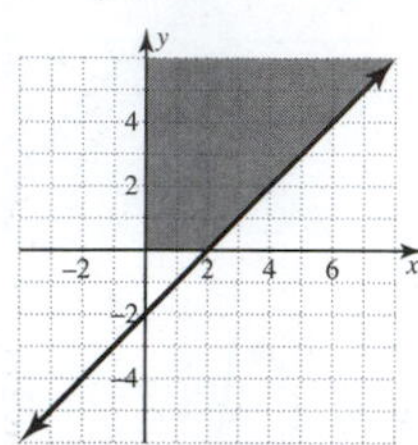

61.

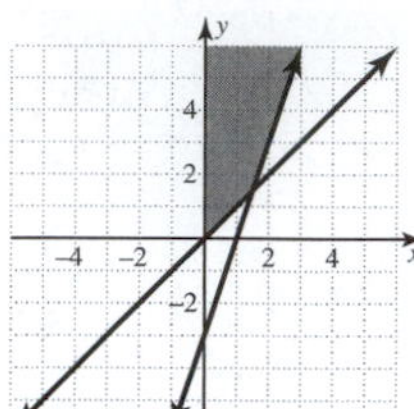

62.

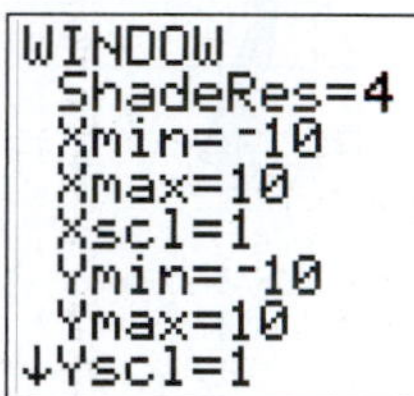

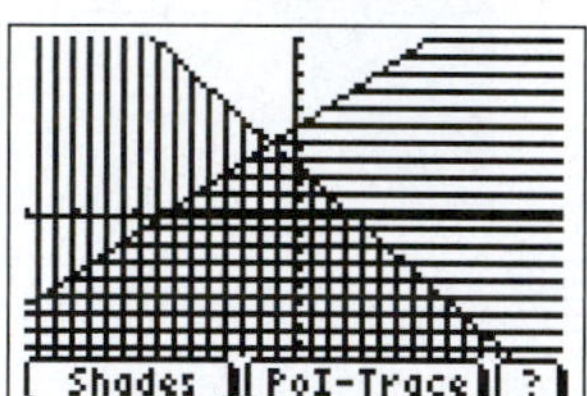

63.

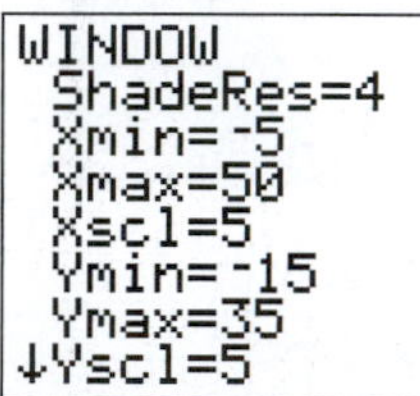

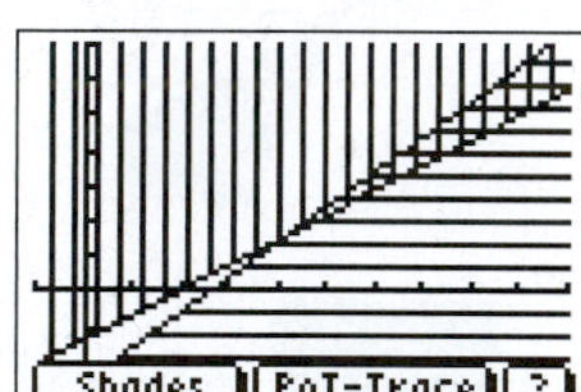

64.

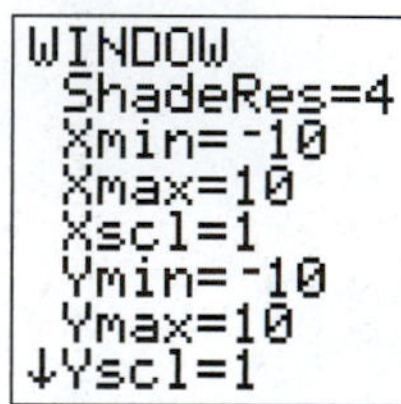

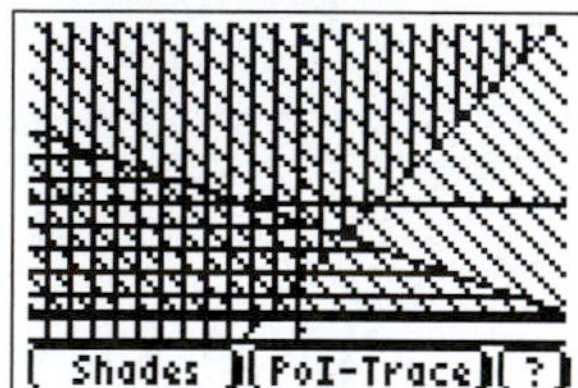

65.

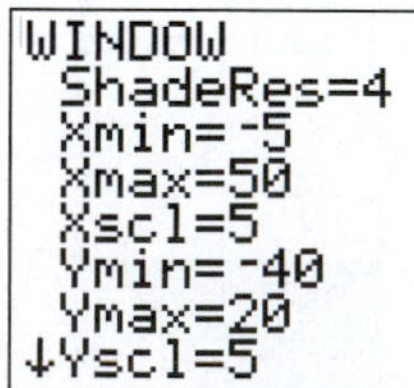

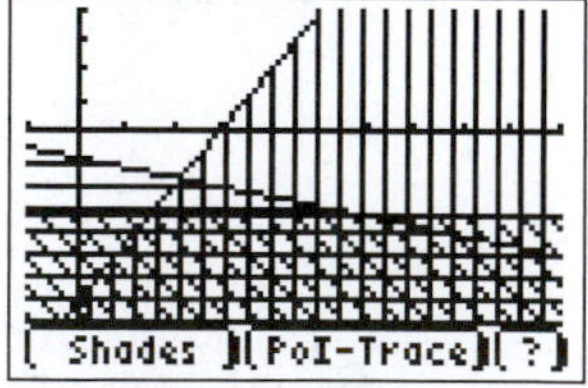

CHAPTER 5 GROUP ACTIVITY

1. Let x = number of buses and y = number of vans.

2. $\begin{cases} 30x + 6y \ge 90 \\ 2x + y \le 9 \\ x \ge 0 \\ y \ge 0 \end{cases}$

3.

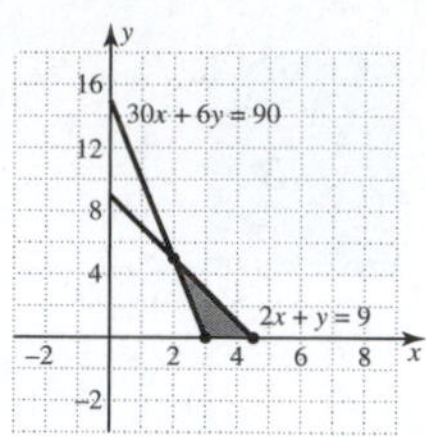

4. The corner points are (2, 5), (3, 0), and (4.5, 0).

5. Transportation costs; $c = 700x + 100y$

6. $c = 1900$ for (2, 5); $c = 2100$ for (3, 0); $c = 3150$ for (4.5, 0); The point (2, 5) produces the minimum cost.

7. The church should rent two buses and five vans to minimize transportation costs. The costs will be $1900.

CHAPTER 5 REVIEW EXERCISES

Section 5.4

53. $\left\{\left(5 + \frac{2}{3}z, -15 - \frac{13}{3}z, z\right) \middle| z \text{ is a real number}\right\}$

58.

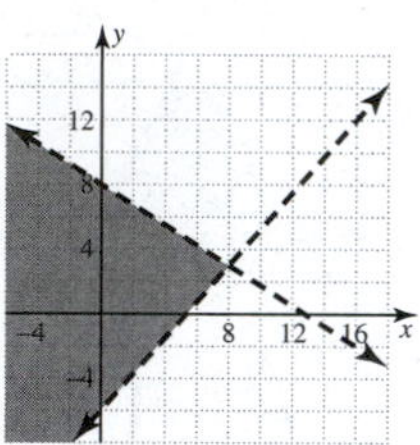

59.

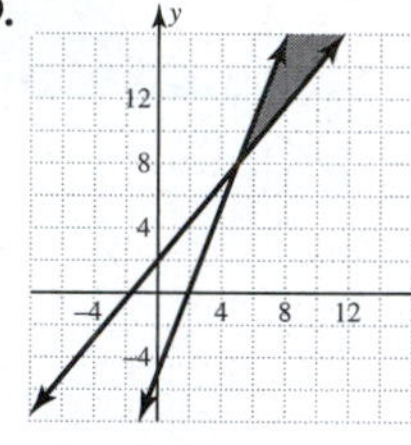

60.

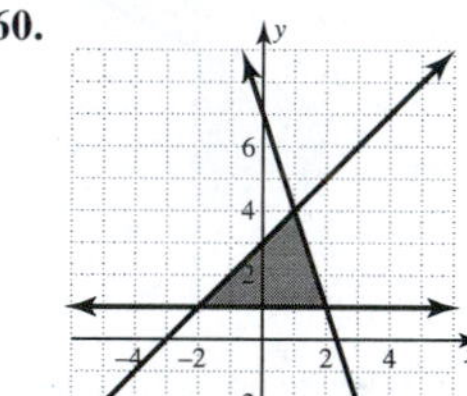

61.

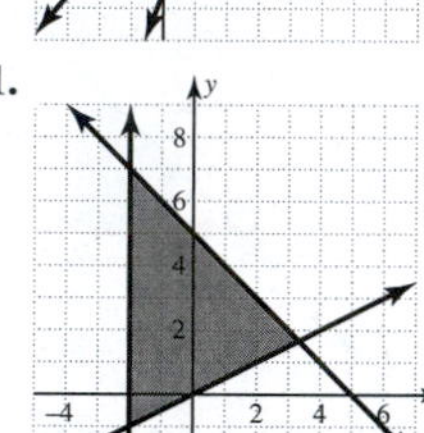

62.

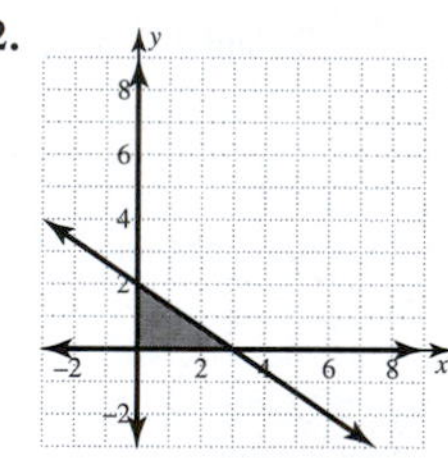

63.

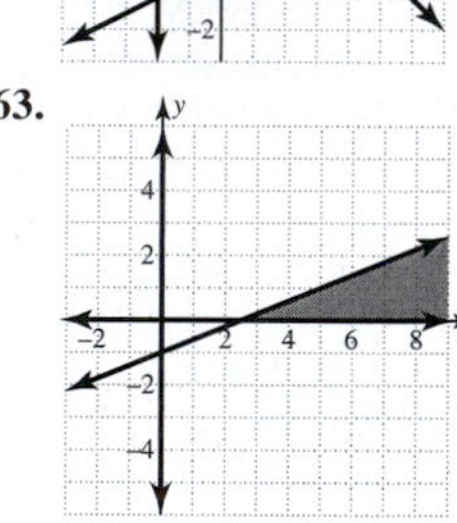

64.

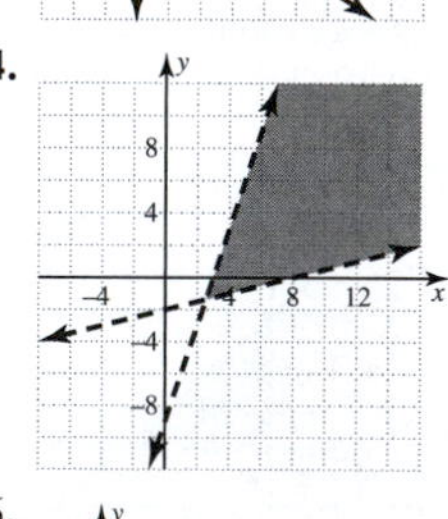

65.

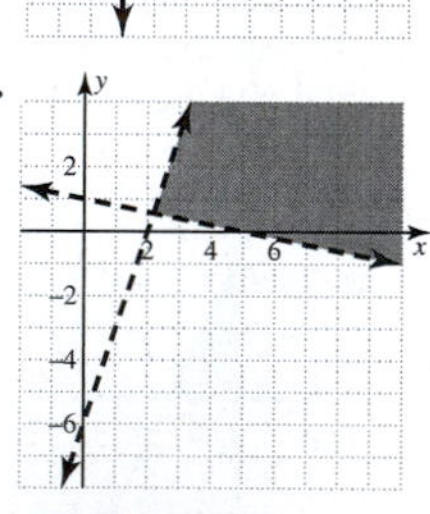

66.

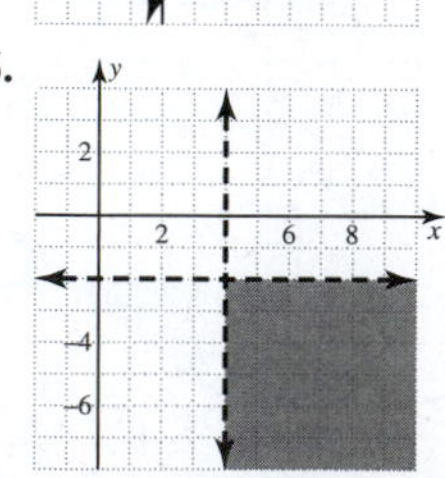

67.

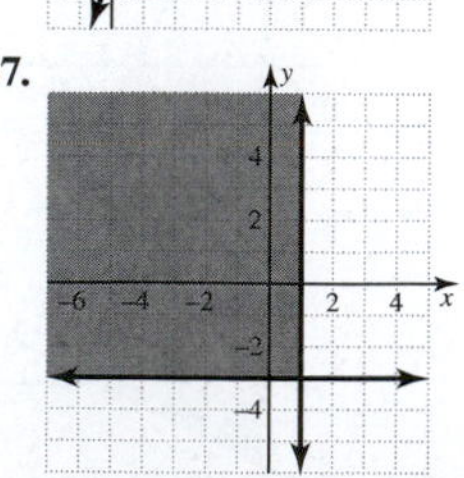

68. Answers vary; 280 candles and 1120 cookbooks, 90 candles and 360 cookbooks, 30 candles and 120 cookbooks

69. Answers vary; 790 double rooms and 0 king rooms, 500 double rooms and 150 king rooms; 590 double rooms and 195 king rooms

CHAPTER 5 TEST

2. The solution is $(-3, -1)$.

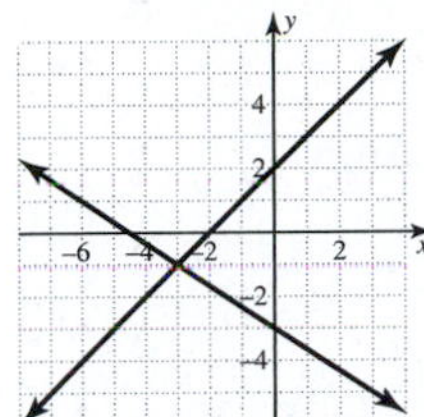

4. b. In 2008 (5 yr after 2003), about 32% of Americans used the newspaper to obtain their news and 32% of Americans used the Internet to obtain their news.

5. Answers vary. An example is one of the three methods of graphing, substitution, or elimination. Graphing is the least precise since it is impossible to determine exact fractional solutions when I graph by hand. I would use substitution if one of the equations in the system is already solved for one of the variables. I would use elimination for almost every other system as long as I first put the equations in standard form.

13.

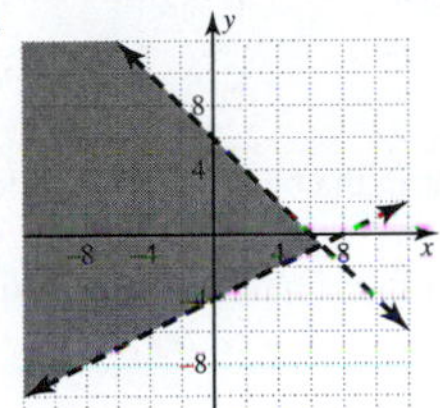

CHAPTERS 1–5 CUMULATIVE REVIEW EXERCISES

17.

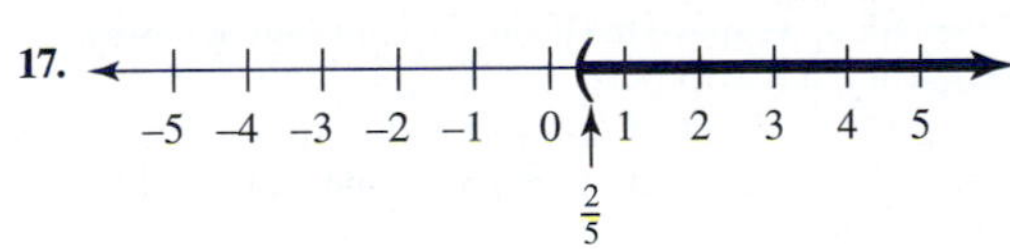

$\left(\frac{2}{5}, \infty\right)$; $\left\{x \,\middle|\, x > \frac{2}{5}\right\}$

18.

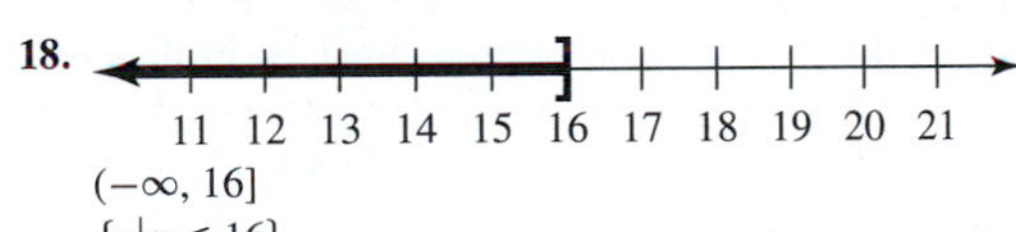

$(-\infty, 16]$

$\{y \mid y \le 16\}$

23. $\left\{-\frac{3}{11}, -\frac{7}{9}\right\}$ **24.** $\left\{\frac{22}{7}, \frac{26}{5}\right\}$ **25.** $\left\{\frac{1}{5}, 15\right\}$

27. a. $\{(0, 0.45), (1, 2), (3, 3.27), (5, 7.5), (9, 14.8), (11, 19)\}$

b.

c. The number of iPads sold steadily increased from April 2010 to March 2011.

33. $f(x) = \frac{3}{5}x - 2, m = \frac{3}{5}, b = -2$

34. $f(x) = -\frac{6}{5}x, m = -\frac{6}{5}, b = 0$

36. $f(x) = -\frac{1}{2}x + 2, m = -\frac{1}{2}, b = 2$

43. Domain $= (-\infty, \infty)$, Range $= (-\infty, \infty)$

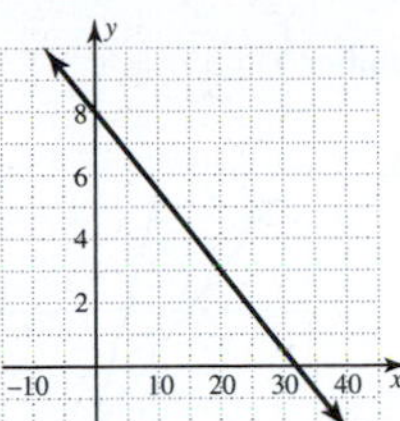

44. Domain $= (-\infty, \infty)$, Range $= \{-4\}$

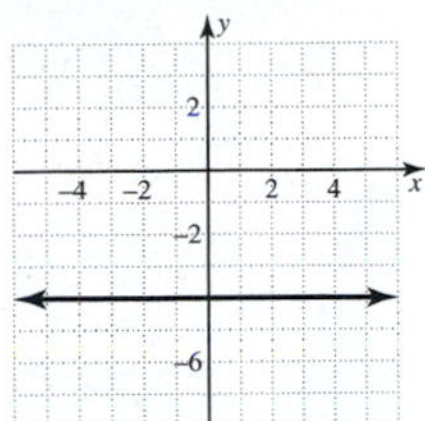

45. Domain $= (-\infty, \infty)$, Range $= (-\infty, \infty)$

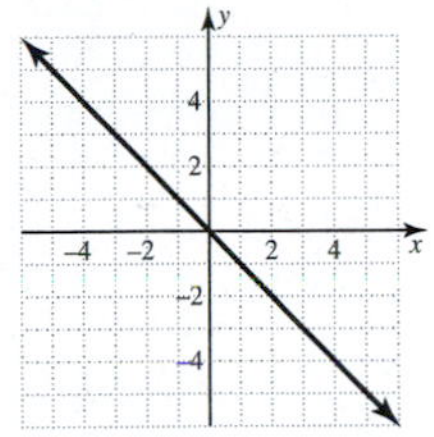

46. Domain $= (-\infty, \infty)$, Range $= (-\infty, \infty)$

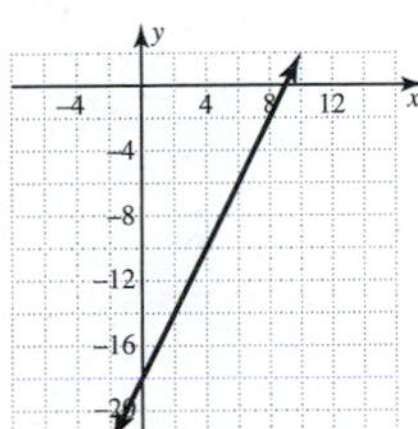

47. $m = -9$, $(0, 6)$

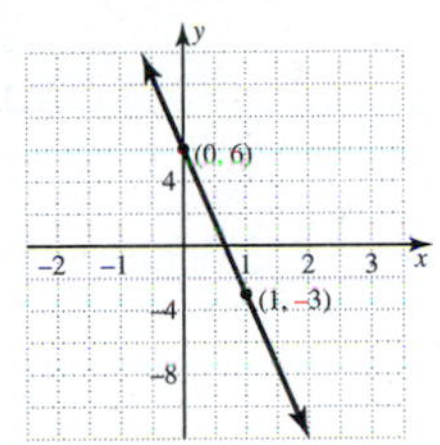

48. $m = -2$, $(0, -5)$

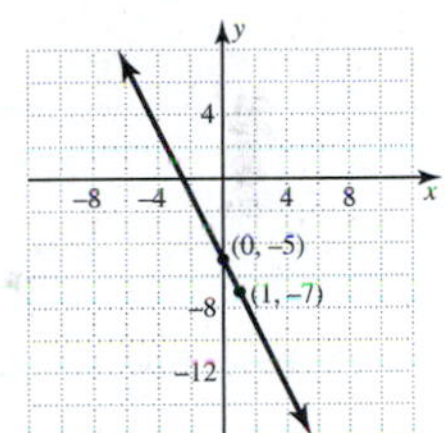

49. $m = \frac{3}{2}$, $(0, -6)$

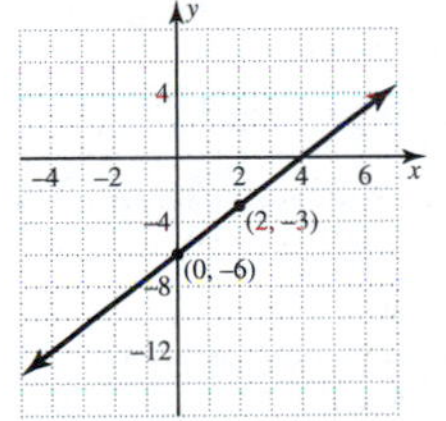

50. $m = \frac{1}{5}$, $(0, 0)$

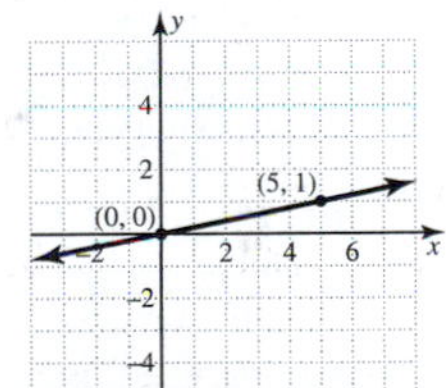

63.

64.

65.

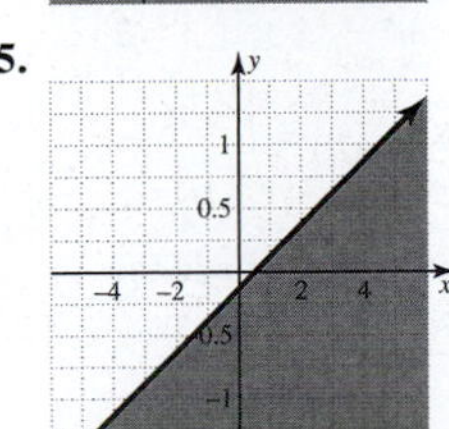

66.

73. same lines, infinitely many solutions, consistent system with dependent equations

74. intersecting lines, one solution, consistent system with independent equations

75. parallel lines, no solution, inconsistent system

76. intersecting lines, one solution, consistent system with independent equations

83. Derek's hourly wage as a tour guide is \$8 and hourly wage as a night shift manager is \$13.50.

89. 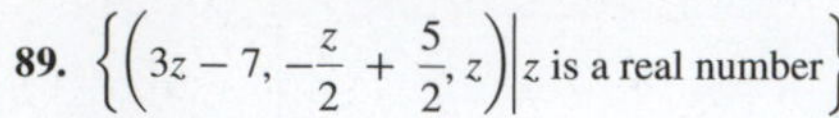
$\left\{\left(3z-7,-\frac{z}{2}+\frac{5}{2},z\right)\middle|\, z \text{ is a real number}\right\}$

93.

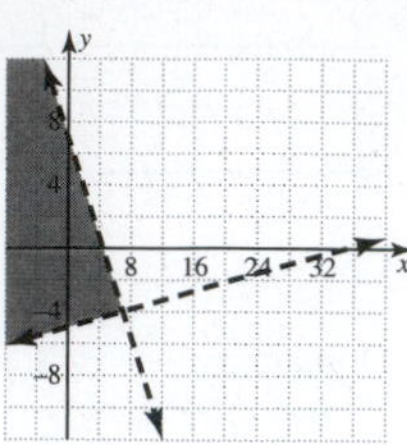

94.

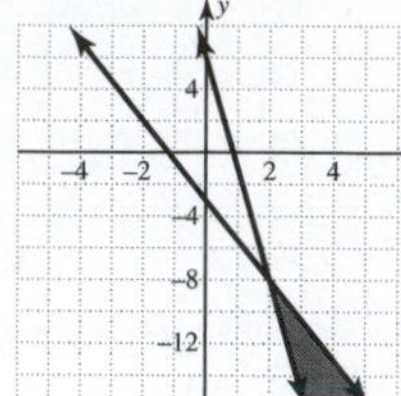

95.

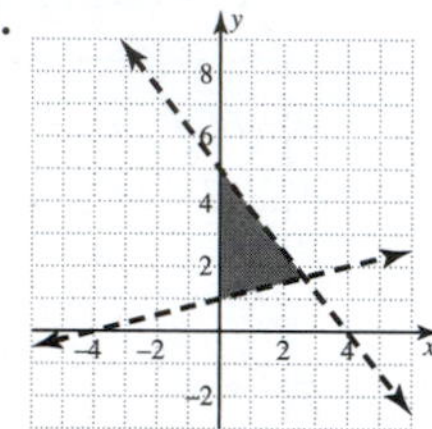

147. Enter the expression in the denominator with a pair of parentheses.

```
165000(0.045/12)
/(1-(1+0.045/12)
^(-360))
         836.0307612
```

The correct answer is \$836.03.

148. The interest rate should be entered as 0.029.

```
3200(0.029/12)/(
1-(1+0.029/12)^(
-60))
         57.35771762
```

The correct answer is \$57.36.

149. The rate should be entered as 0.0135.

```
10800(1+0.0135/5
2)^(52*2)
         11095.53339
```

The correct answer is \$11,095.53.

150. Enter the exponents in parentheses.

```
7600(1+0.021/52)
^(52*3)
         8094.101039
```

The correct answer is \$8094.10.

CHAPTER 6

Section 6.1

1. Consider the expression $\frac{x^3}{x^3}$. It simplifies to 1. Applying the quotient of like bases rule, we have $\frac{x^3}{x^3} = x^{3-3} = x^0$. So a nonzero real number raised to the zero exponent is 1.

2. With parentheses in $(-5)^4$, the base is -5. So, $(-5)^4 = 625$. Without the parentheses in -5^4, the exponent 4 applies only to the base 5. The expression is the opposite of 5^4, which is 625. So, $-5^4 = -625$.

3. In 3^{-2}, the exponent -2 applies only to the base 3. So, $3^{-2} = \frac{1}{3^2} = \frac{1}{9}$. In -3^2, the expression is the opposite of 3^2, which is 9. So, $-3^2 = -9$. Therefore, $3^{-2} \neq -3^2$.

4. Answers vary; 2^{-7} means to find the reciprocal of 2^7. So $2^{-7} = \frac{1}{2^7} = \frac{1}{128}$.

5. Yes, we can apply the product of like bases rule of exponents because the bases are the same.

6. No, we cannot apply the product of like bases rule of exponents because the bases are not the same.

77. $-\frac{4a^{17}}{b^{10}}$ **79.** $\frac{6}{r^3t^{10}}$ **80.** $\frac{3}{p^{17}q^5}$ **83.** $-\frac{15}{hp^4}$

84. $\frac{27x^3}{y^4}$ **85.** $-\frac{2}{n^{18}}$ **86.** $-\frac{4}{m^6}$

144.

```
3500(1+0.0225/12
)^(12*4)
         3829.287287
```

145. When applying the product of like bases rule, we add the exponents but keep the base the same.

```
2^(5+8)
                8192
```

The correct answer is 8192.

146. When applying the product of like bases rule, we add the exponents and keep the base the same.

```
(-3)^(5+4)
              -19683
```

The correct answer is $-19{,}683$.

Section 6.2

1. We apply the exponent m to each factor in the base that is a product. So, $(ab)^m = a^m b^m$.

2. We apply the exponent m to each factor in the base that is a quotient. So, $\left(\frac{a}{b}\right)^m = \frac{a^m}{b^m}$.

3. We apply the exponent n to each factor in the base. The factors in the base are -2 and x^m. So, $(-2x^m)^n = (-2)^n(x^m)^n = (-2)^n x^{mn} \neq -2^n x^{mn}$.

4. The scientific notation of a number is a number written in the form $a \times 10^n$, where $1 \leq a < 10$ and n is an integer.

5. Answers vary; For example, to convert 0.000234 to a number in scientific notation, we move the decimal point right 4 places. So, $0.000234 = 2.34 \times 10^{-4}$

6. Answers vary; For example, to convert 3.76×10^5 to a number in standard form, we move the decimal point right 5 places. So, $3.76 \times 10^5 = 376000$.

7. Answers vary; For example, to convert 9800 to a number in scientific notation, we move the decimal point left 3 places. So, $9800 = 9.8 \times 10^3$.

8. We can apply the rules of exponents to perform calculations of very large or very small numbers in scientific notation. We apply the product of like bases rule to multiply and the quotient of like bases rule to divide.

27. $\frac{a^4}{b^{18}}$ **29.** $\frac{16r^2}{s^{13}}$

123.
$$\begin{aligned}(-4a^3b^2)^3 &= (-4)^3(a^3)^3(b^2)^3\\ &= (-4)(-4)(-4)a^9b^6\\ &= -64a^9b^6\end{aligned}$$

124.
$$\begin{aligned}\left(\frac{5x^{-2}}{2x^5}\right)^{-3} &= \frac{5^{-3}x^6}{2^{-3}x^{-15}}\\ &= \frac{5^{-3}x^{6-(-15)}}{2^{-3}}\\ &= \frac{2^3x^{21}}{5^3}\\ &= \frac{8x^{21}}{125}\end{aligned}$$

125.
$$\begin{aligned}\frac{6.3 \times 10^{-21}}{1.4 \times 10^{-6}} &= \frac{6.3}{1.4} \times 10^{-21-(-6)}\\ &= 4.5 \times 10^{-15}\end{aligned}$$

126.
$$\begin{aligned}(4.1 \times 10^{-8})(2.8 \times 10^{-5}) &= (4.1)(2.8) \times 10^{-8+(-5)}\\ &= 11.48 \times 10^{-13}\\ &= 1.148 \times 10 \times 10^{-13}\\ &= 1.148 \times 10^{-12}\end{aligned}$$

127.

```
1.512*10^-12/(7.
2*10^-4)
             2.1E-9
1.512E-12/7.2E-4
             2.1E-9
```

The correct answer is 2.1×10^{-9}.

128.

```
2.47E2/1.3E-7
         1900000000
```

The correct answer is 1.9×10^9.

129. $7ab^5$; $(7ab^5)^2 = 49a^2b^{10}$ ($-7ab^5$ is also acceptable.)

130. $12x^2y^3$; $(12x^2y^3)^2 = 144x^4y^6$ ($-12x^2y^3$ is also acceptable.)

131. $-\dfrac{3r}{s^2}$; $\left(-\dfrac{3r}{s^2}\right)^3 = -\dfrac{27r^3}{s^6}$

132. $\dfrac{5x^4}{y^3}$; $\left(\dfrac{5x^4}{y^3}\right)^3 = \dfrac{125x^{12}}{y^9}$ **133.** x^{5a+6} **134.** 2^{12x-4} or $\dfrac{2^{12x}}{16}$

Piece It Together Sections 6.1 and 6.2

14. $-\dfrac{28k^{15}}{j^5}$ **15.** $\dfrac{14x^2}{y^{10}}$ **18.** $\dfrac{81}{a^6b^{30}}$

Section 6.3

1. A second-degree polynomial is of the form $ax^2 + bx + c$, where $a \neq 0$, a, b, and c are real numbers.
2. The degree of a term is the sum of the exponents on the variables in the term. The degree of the polynomial is the largest degree of the terms in the polynomial.
3. The coefficient of a term is the real number factor. The leading coefficient is the coefficient of the term with the largest degree.
4. To evaluate a polynomial function at $x = c$, replace each x with the value c and use the order of operations to simplify the expression.

17. $3a^3$, $2a^2$, $-5a$, 7; polynomial; 3, Cubic; 3
18. $-2p^3$, $-5p^2$, $4p$, 11; polynomial; 3, Cubic; -2
19. $-11r^4$, $13r$, 2; trinomial; 4, quartic; -11
20. $\dfrac{3}{2}s^2$, $-7s$, 10; trinomial; 2; quadratic; $\dfrac{3}{2}$
21. $32abc^2$; monomial; 4, quartic; n/a
22. $9rs^2t^2$; monomial; 5, fifth-degree; n/a
23. 7, $-16s$; binomial; 1; linear; -16
24. $2t$, -8; binomial; 1, linear; 2
43. 1358, 318; The penny will be 1358 ft above the ground 4 sec after it is dropped. The penny will be 318 ft above the ground 9 sec after it is dropped.
44. 264, 28.8; The penny will be 264 m above the ground 5 sec after it is dropped. The penny will be 28.8 m above the ground 7 sec after it is dropped.
45. 17.7, 28.4; Video game revenue was approximately \$17.7 billion in 2007 and \$28.4 billion in 2009.
46. 4.7, 12.8; The unemployment rate for men over 20 yr old was approximately 4.7% in 2007 and approximately 12.8% in 2010.
47. Domain = $(-\infty, \infty)$, Range = $(-\infty, 0]$

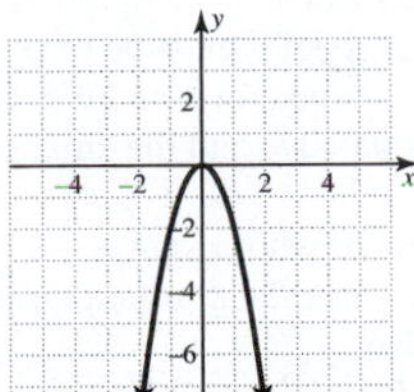

48. Domain = $(-\infty, \infty)$, Range = $(-\infty, \infty)$

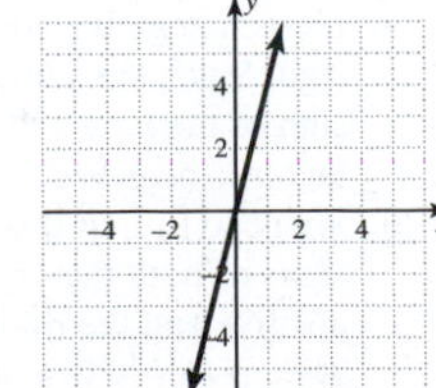

49. Domain = $(-\infty, \infty)$, Range = $[0, \infty)$

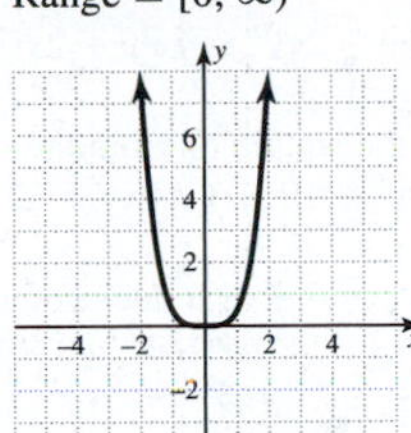

50. Domain = $(-\infty, \infty)$, Range = $(-\infty, \infty)$

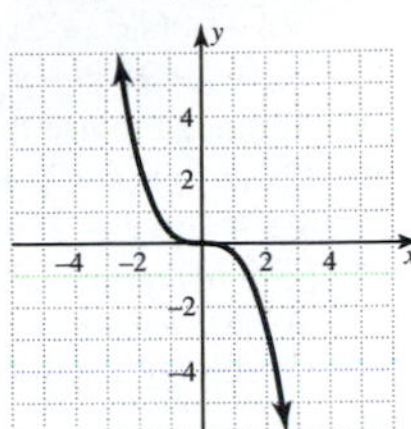

51. $6.25b^2$, -0.64; binomial; 2, quadratic; 6.25
52. $\dfrac{1}{27}z^3$, -8; binomial; 1, linear; $\dfrac{1}{27}$
53. $5a^3b^2c$; monomial; 6, sixth-degree; n/a
54. $-6p^2q^4r$; monomial; 7, seventh-degree; n/a
55. $0.1x^3$, $1.6x^2$, $-2.9x$, 1.8; polynomial; 3, cubic; 0.1
56. $-0.1y^3$, $-0.9y^2$, $1.2y$, -2.1; polynomial; 3, cubic; -0.1
57. $-\dfrac{x^2}{5}$; monomial; 2, quadratic; $-\dfrac{1}{5}$
58. $\dfrac{v}{7}$; monomial; 1, linear; $\dfrac{1}{7}$
67. $C(7) = 58.16$ and $C(10) = 65.38$; The average cost of electricity was about \$58.16 per 500 kWh in 2006 and \$65.38 per 500 kWh in 2009.
68. $Q(6) = 4329$ and $Q(9) = 3700$; The average quarterly domestic airline fuel consumption was about 4329 million gallons in 2006 and about 3700 million gallons in 2009.
69. Domain = $(-\infty, \infty)$, Range = $[1, \infty)$

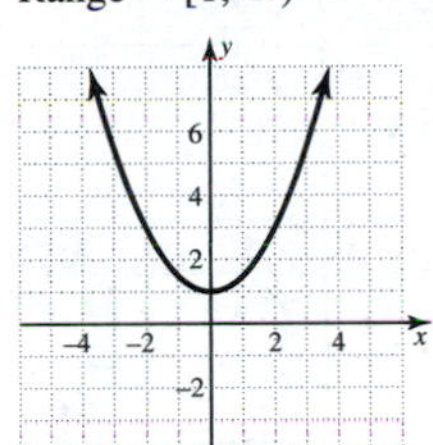

70. Domain = $(-\infty, \infty)$, Range = $(-\infty, \infty)$

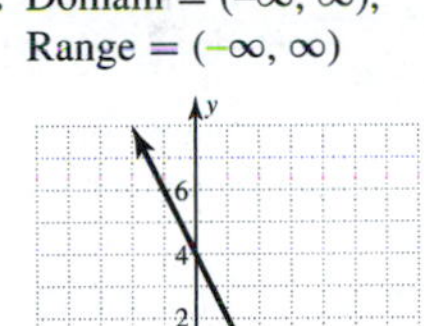

71. Domain = $(-\infty, \infty)$, Range = $(-\infty, \infty)$

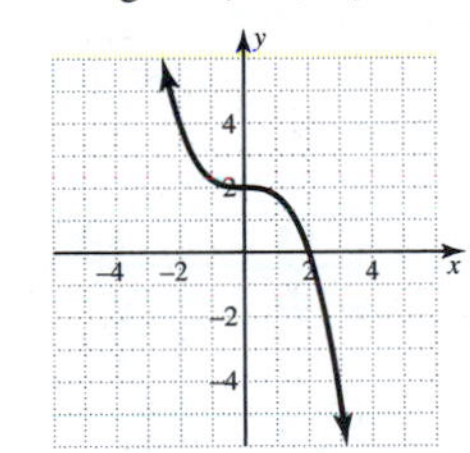

72. Domain = $(-\infty, \infty)$, Range = $[-4, \infty)$

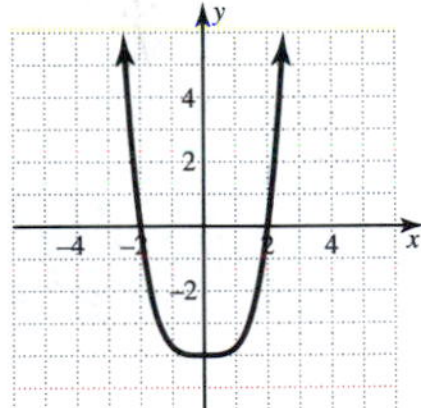

75.

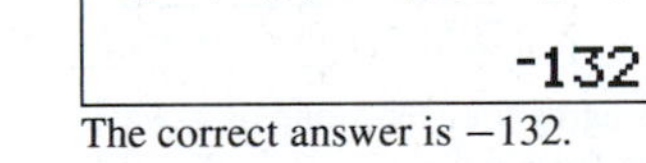

```
-3*(-5)²+10*-5-7
               -132
```

The correct answer is -132.

76.

```
-(-1)²+6*-1-4
               -11
```

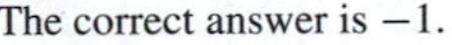

The correct answer is -1.

Section 6.4

1. To add two polynomials, remove the parentheses, combine like terms by adding the coefficients and keeping the variable component the same, and write the answer in standard form.
2. To subtract two polynomials add the first polynomial to the opposite of the second polynomial.
3. To subtract two polynomials vertically, align like terms in the same column, put the polynomial being subtracted at the bottom and find its opposite, add like terms, and write the answer in standard form.
4. Answers vary; A common mistake made by students when subtracting two polynomials is not multiplying each term in the polynomial being subtracted by -1. For example, $(-5x + 11) - (-3x + 18) = -5x + 11 + 3x + 18 = -2x + 29$.

5. $(2x + y)(4x^2 - 2xy + y^2) = 8x^3 - 4x^2y + 2xy^2 + 4x^2y - 2xy^2 + y^3 = 8x^3 + y^3$ is the sum of two cubes.
$(2x + y)(4x^2 + 2xy - y^2) = 8x^3 + 4x^2y - 2xy^2 + 4x^2y + 2xy^2 - y^3 = 8x^3 + 8x^2y - y^3$ is not.
6. You need to use the FOIL method on the square of a binomial: $(2x + 5y)^2 = (2x + 5y)(2x + 5y) = 4x^2 + 10xy + 10xy + 25y^2 = 4x^2 + 20xy + 25y^2$
7. $(a + b)^3$ means to multiply three factors of $a + b$. It does not mean to cube each term in parentheses. $(a + b)^3 = (a + b)^2 (a + b) = (a^2 + 2ab + b^2)(a + b) = a^3 + 3a^2b + 3ab^2 + b^3$
8. The expression $(x + y)^2$ is $(x + y)(x + y)$. The base is repeated as a factor twice. The left side is the product of conjugates and does not repeat the factor $x + y$.

117. $25m^4 - \frac{4}{9}$ **118.** $9n^4 - \frac{16}{49}$

128. $(a^2 + 23a - 18) + (a^2 + 4a - 33)$
$= a^2 + 23a - 18 + a^2 + 4a - 33$
$= 2a^2 + 27a - 51$

129. $(9b^2 - 15b + 2) - (-2b^2 - 6b + 7)$
$= 9b^2 - 15b + 2 + 2b^2 + 6b - 7$
$= 11b^2 - 9b - 5$

134.

```
Plot1 Plot2 Plot3
\Y1■(-2X²+5X-4)+
(5X²-7X+3)
\Y2■3X²-2X-1
\Y3=
\Y4=
\Y5=
\Y6=
```

X	Y1	Y2
-1	4	4
0	-1	-1
1	0	0
2	7	7
3	20	20
4	39	39
5	64	64

Y2■3X²-2X-1

Since the Y1 and Y2 columns agree for each value of x, we can conclude that the expressions are equal.

135.

```
Plot1 Plot2 Plot3
\Y1■(5X-2)²
\Y2■25X²+4
\Y3=
\Y4=
\Y5=
\Y6=
\Y7=
```

X	Y1	Y2
-1	49	29
0	4	4
1	9	29
2	64	104
3	169	229
4	324	404
5	529	629

Y2■25X²+4

Since the Y1 and Y2 columns do not agree for each value of x, we can conclude that the expressions are not equal.

136.

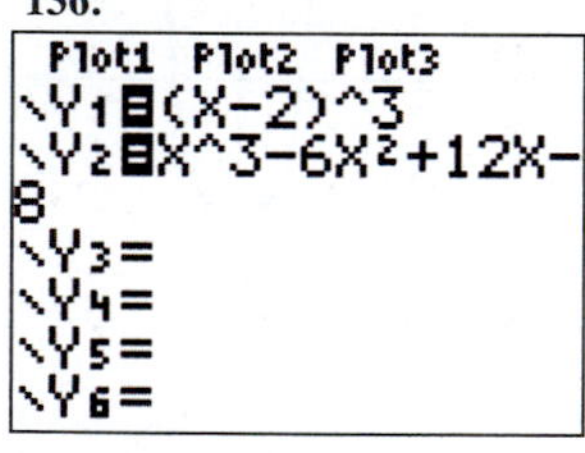

X	Y1	Y2
-1	-27	-27
0	-8	-8
1	-1	-1
2	0	0
3	1	1
4	8	8
5	27	27

Y2■X^3-6X²+12X-8

Since the Y1 and Y2 columns agree for each value of x, we can conclude that the expressions are equal.

137.

```
Plot1 Plot2 Plot3
\Y1■(X+2)^4
\Y2■X^4+16
\Y3=
\Y4=
\Y5=
\Y6=
\Y7=
```

X	Y1	Y2
-1	1	17
0	16	16
1	81	17
2	256	32
3	625	97
4	1296	272
5	2401	641

Y2■X^4+16

Since the Y1 and Y2 columns do not agree for each value of x, we can conclude that the expressions are not equal.

Piece It Together Sections 6.3 and 6.4

1. $-5.4c^6$; monomial; 6, sixth-degree; -5.4
2. $-x$, 3.5; binomial; 1, linear; -1
3. $6x^2$, $-4.5x$, -9.4; trinomial; 2, quadratic; 6
4. $4s^2t$; monomial; 3, cubic; n/a

9. Domain $= (-\infty, \infty)$, Range $= (-\infty, \infty)$

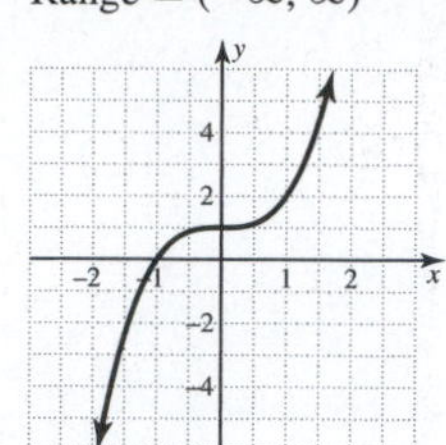

10. Domain $= (-\infty, \infty)$, Range $= (-\infty, \infty)$

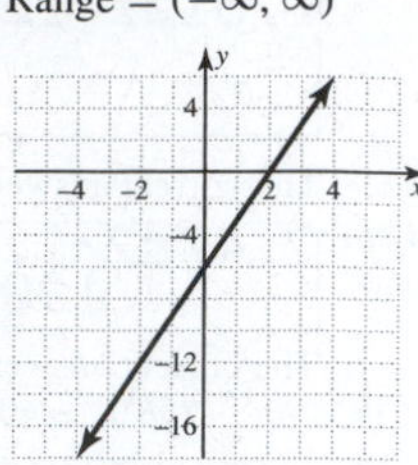

Section 6.5

1. The GCF of two integers is the largest integer that divides evenly into both integers.
2. The GCF of two variable terms with the same base is the base raised to the smallest power of the terms.
3. To factor a polynomial, find the GCF of the terms in the polynomial, rewrite each term as a product of the GCF and the remaining factor, and apply the distributive property to factor out the GCF.
4. If the GCF is a negative term, rewrite each term of the polynomial as a product of the GCF and the opposite of the remaining factor, then apply the distributive property to factor out the GCF.
5. Group together the first two terms and last two terms and factor out the GCF from each pair of terms.
6. If there is no common factor after the initial pairing, rearrange the grouping of terms and repeat the steps described in Exercise 5.

76. $8c^3 - 2c^2 + 12c - 3 = (8c^3 - 2c^2) + (12c - 3)$
$= 2c^2(4c - 1) + 3(4c - 1)$
$= (4c - 1)(2c^2 + 3)$

77. The expression cannot be factored using grouping because $3w^3 + 2w^2 - 6w + 4 = w^2(3w + 2) - 2(3w - 2)$.

78. $30y^3 + 25y^2 - 6y - 5 = (30y^3 + 25y^2) - (6y + 5)$
$= 5y^2(6y + 5) - 1(6y + 5)$
$= (6y + 5)(5y^2 - 1)$

79.

```
gcd(3225,3975)
                75
```

80.

```
gcd(1344,936)
                24
```

81.

```
gcd(gcd(228,336)
,552)
                12
```

82.

```
gcd(gcd(480,1152
),2400)
                96
```

Section 6.6

1. If c is a positive constant, then the factors of c are either both positive or both negative. Since the sum of the factors of c is equal to b, the signs of the factors of c are the same as the sign of b.
2. If c is a negative constant, then the factors of c must be of opposite sign. Since the sum of the factors of c is equal to b, the sign of the larger of the two factors is the same as the sign of b and the other factor will be of the opposite sign.
3. If c is a positive constant, then the factors of c are of the same sign. Since the sum of the factors of c is equal to b and $b < 0$, the signs of both factors of c are also negative.
4. If c is a negative constant, then the factors of c must be of opposite sign. Since the sum of the factors of c is equal to b and $b > 0$, the sign of the larger of the two factors is positive and the smaller factor is negative.
5. We need to find the factors of $5 \cdot 14 = 70$ whose sum is -19. The signs of both factors must be negative since the constant term is positive and the middle term is negative.
$70 = (-1)(-70) = (-2)(-35) = (-5)(-14) = (-7)(-10)$
Now replace $-19x = -5x - 14x$
$5x^2 - 19x + 14 = 5x^2 - 5x - 14x + 14 = 5x(x - 1) - 14(x - 1)$
$= (x - 1)(5x - 14)$

6. Let $u = a^4$ and $u^2 = a^8$. Rewrite the polynomial in terms of u.
$a^8 - 4a^4 - 21 = (a^4)^2 - 4(a^4) - 21 = u^2 - 4u - 21 =$
$(u - 7)(u + 3) = (a^4 - 7)(a^4 + 3)$

7. Let $u = (y + 5)$ and $u^2 = (y + 5)^2$. Rewrite the polynomial in terms of u.
$10(y + 5)^2 - 13(y + 5) - 9 = 10u^2 - 13u - 9$
$= (5u - 9)(2u + 1) = [5(y + 5) - 9)][2(y + 5) + 1]$
$= [5y + 25 - 9][2y + 10 + 1]$
$= (5y + 16)(2y + 11)$

8. First we use FOIL to multiply out the square, combine like terms, and factor.
$10(y + 5)^2 - 13(y + 5) - 9$
$= 10(y^2 + 10y + 25) - 13y - 65 - 9$
$= 10y^2 + 100y + 250 - 13y - 74$
$= 10y^2 + 87y + 176$
$= (5y + 16)(2y + 11)$

110. $-6a^2 - 30a + 36 = -6(a^2 + 5a - 6)$
$= -6(a^2 + 6a - a - 6)$
$= -6[a(a + 6) - 1(a + 6)]$
$= -6(a + 6)(a - 1)$

111. $12(a - 4)^2 + 4(a - 4) - 5$
$= 12u^2 + 4u - 5$
$= (6u + 5)(2u - 1)$
$= [6(a - 4) + 5][2(a - 4) - 1]$
$= (6a - 24 + 5)(2a - 8 - 1)$
$= (6a - 19)(2a - 9)$

112. $2(b - 6)^2 + 23(b - 6) + 45$
$= 2u^2 + 23u + 45$
$= (2u + 5)(u + 9)$
$= [2(b - 6) + 5)][(b - 6) + 9)]$
$= (2b - 12 + 5)(b - 6 + 9)$
$= (2b - 7)(b + 3)$

117.

```
Plot1 Plot2 Plot3
\Y1■X²-12X+35
\Y2■(X-5)(X-7)
\Y3=
\Y4=
\Y5=
\Y6=
\Y7=
```

X	Y1	Y2
-3	80	80
-2	63	63
-1	48	48
0	35	35
1	24	24
2	15	15
3	8	8

X=3

118.

```
Plot1 Plot2 Plot3
\Y1■X²+5X-104
\Y2■(X-8)(X+13)
\Y3=
\Y4=
\Y5=
\Y6=
\Y7=
```

X	Y1	Y2
-3	-110	-110
-2	-110	-110
-1	-108	-108
0	-104	-104
1	-98	-98
2	-90	-90
3	-80	-80

X= -3

119.

```
Plot1 Plot2 Plot3
\Y1■X²-9X-90
\Y2■(X-15)(X+6)
\Y3=
\Y4=
\Y5=
\Y6=
\Y7=
```

X	Y1	Y2
-3	-54	-54
-2	-68	-68
-1	-80	-80
0	-90	-90
1	-98	-98
2	-104	-104
3	-108	-108

X= -3

120.

```
Plot1 Plot2 Plot3
\Y1■X²-12X-45
\Y2■(X-15)(X+3)
\Y3=
\Y4=
\Y5=
\Y6=
\Y7=
```

X	Y1	Y2
-3	0	0
-2	-17	-17
-1	-32	-32
0	-45	-45
1	-56	-56
2	-65	-65
3	-72	-72

X= -3

Section 6.7

1. Answers vary; For example, to factor $121x^2 - 49y^2$, rewrite $121x^2$ as $(11x)^2$ and $49y^2$ as $(7y)^2$, and write the factorization as $(11x + 7y)(11x - 7y)$.

2. The expression $(2x + y)^3$ is the sum, $2x + y$, raised to the third power. You need to multiply using the distributive property. The expression $8x^3 + y^3$ is the sum of two cubes.

3. The expression $8x^3 + y^3$ is the sum of two cubes and is factored as $(2x + y)(4x^2 - 2xy + y^2)$. It is not the same as $(2x + y)$ $(2x^2 - 2xy + y^2)$.

4. Answers vary; For example, to factor $125x^3 + 729y^3$, rewrite $125x^3$ as $(5x)^3$ and $729y^3$ as $(9y)^3$, and write the factorization as
$(5x + 9y)[(5x)^2 - (5x)(9y) + (9y)^2]$
$= (5x + 9y)(25x^2 - 45xy + 81y^2)$.

5. You need to factor out the GCF, 2, and then apply the difference of two squares method.
$18x^2 - 50y^2 = 2(9x^2 - 25y^2) = 2[(3x)^2 - (5y)^2]$
$= 2(3x + 5y)(3x - 5y)$

6. You need to rewrite w^6 as $(w^2)^3$ and $64z^6$ as $(4z^2)^3$ and factor as the difference of two cubes.
$w^6 - 64z^6 = (w^2)^3 - (4z^2)^3$
$= (w^2 - 4z^2[(w^2)^2 + (w^2)(4z^2) + (4z^2)^2]$
$= (w^2 - 4z^2)(w^4 + 4w^2z^2 + 16z^4)$
$= (w + 2z)(w - 2z)(w^4 + 4w^2z^2 + 16z^4)$

7. You need to rewrite w^6 as $(w^3)^2$ and $64z^6$ as $(8z^3)^2$ and factor as the difference of two squares.
$w^6 - 64z^6 = (w^3)^2 - (8z^3)^2$
$= (w^3 - 8z^3)(w^3 + 8z^3)$
$= [w^3 - (2z)^3][w^3 + (2z)^3]$
$= (w - 2z)[w^2 + w(2z) + (2z)^2](w + 2z)][w^2 - w(2z) + (2z)^2]$
$= (w - 2z)(w^2 + 2wz + 4z^2)(w + 2z)(w^2 - 2wz + 4z^2)$

8. Solutions to Exercises 6 and 7 should be the same because $w^4 + 4w^2z^2 + 16z^4$ and $(w^2 + 2wz + 4z^2)(w^2 - 2wz + 4z^2)$ are the same.

61. $\left(7x - \dfrac{5}{3}\right)\left(7x + \dfrac{5}{3}\right)$ **62.** $\left(5y - \dfrac{1}{8}\right)\left(5y + \dfrac{1}{8}\right)$

84. $(x - y)(x + y)(x^2 - xy + y^2)(x^2 + xy + y^2)$

96. $5ab + 3b - 10a - 6 = b(5a + 3) - 2(5a + 3)$
$= (5a + 3)(b - 2)$

97. $x^2 + 6x + 9 - z^2 = (x^2 + 6x + 9) - z^2$
$= (x + 3)^2 - z^2$
$= [(x + 3) + z][(x + 3) - z]$
$= (x + 3 + z)(x + 3 - z)$

98. $x^2 - 4x + 4 - z^2 = (x^2 - 4x + 4) - z^2$
$= (x - 2)^2 - z^2$
$=[(x - 2) + z][(x - 2) - z]$
$= (x - 2 + z)(x - 2 - z)$

99.

```
Plot1 Plot2 Plot3
\Y1■X^3-1
\Y2■(X-1)(X²+X+1
)
\Y3=
\Y4=
\Y5=
\Y6=
```

X	Y1	Y2
-1	-2	-2
0	-1	-1
1	0	0
2	7	7
3	26	26
4	63	63
5	124	124

Y2■(X-1)(X²+X+1)

Since the Y_1 and Y_2 columns agree for each value of x, we can conclude that the factoring is correct.

100.

```
Plot1 Plot2 Plot3
\Y1■9X²-16
\Y2■(3X+4)(3X-4)

\Y3=
\Y4=
\Y5=
\Y6=
```

X	Y1	Y2
-1	-7	-7
0	-16	-16
1	-7	-7
2	20	20
3	65	65
4	128	128
5	209	209

Y2■(3X+4)(3X-4)

Since the Y_1 and Y_2 columns agree for each value of x, we can conclude that the factoring is correct.

101.

```
Plot1 Plot2 Plot3
\Y1■16X^4-81
\Y2■(4X²+9)(2X+3
)(2X-3)
\Y3=
\Y4=
\Y5=
\Y6=
```

X	Y1	Y2
-1	-65	-65
0	-81	-81
1	-65	-65
2	175	175
3	1215	1215
4	4015	4015
5	9919	9919

Y2■(4X²+9)(2X+3...

Since the Y_1 and Y_2 columns agree for each value of x, we can conclude that the factoring is correct.

102.

```
Plot1 Plot2 Plot3
\Y1■X^3+27
\Y2■(X+3)(X²-3X+
9)
\Y3=
\Y4=
\Y5=
\Y6=
```

X	Y1	Y2
-1	26	26
0	27	27
1	28	28
2	35	35
3	54	54
4	91	91
5	152	152

Y2■(X+3)(X²-3X+...

Since the Y_1 and Y_2 columns agree for each value of x, we can conclude that the factoring is correct.

Section 6.8

1. Answers vary; For example, $3x^3 - 4x^2 + 7x - 10 = 0$ is a third-degree polynomial equation in standard form.
2. Answers vary; For example, using the zero products property to solve $(2x - 3)(x - 7) = 0$, we set each factor equal to 0.
   $2x - 3 = 0$ or $x - 7 = 0$
   $2x = 3$ or $x = 7$
   $x = \frac{3}{2}$ or $x = 7$
3. Answers vary; For example, we need to rewrite the equation $x^2 = 5x + 24$ in standard form.
   $x^2 - 5x - 24 = 0$
   $(x - 8)(x + 3) = 0$
   $x - 8 = 0$ or $x + 3 = 0$
   $x = 8$ or $x = -3$
4. The equation $(x - 3)(x + 4) = 0$ is a second-degree polynomial equation. When solving, we have two solutions.
   $(x - 3)(x + 4) = 0 \rightarrow x - 3 = 0$ or $x + 4 = 0 \rightarrow x = 3$ or $x = -4$
   The equation $x(x - 3)(x + 4) = 0$ is a third-degree polynomial equation. When solving we have three solutions.
   $x(x - 3)(x + 4) = 0 \rightarrow x = 0$ or $x - 3 = 0$ or $x + 4 = 0 \rightarrow x = 0$ or $x = 3$ or $x = -4$
5. The equation $(x - 3)(x + 2) = 0$ is in standard form and we can apply the zero products property to solve it. We need to rewrite $(x - 3)(x + 2) = 6$ in standard form before applying the zero products property.
6. We need to rewrite $x(x - 2) = 15$ in standard form and apply the zero products property.
   $x(x - 2) = 15$
   $x^2 - 2x = 15$
   $x^2 - 2x - 15 = 0$
   $(x - 5)(x + 3) = 0$
   $x - 5 = 0$ or $x + 3 = 0$
   $x = 5$ or $x = -3$
7. The zeros of a polynomial function are the solutions of the equation $f(x) = 0$.
8. We set $f(x) = 0$ and solve the equation.
   $f(x) = x^3 - 3x^2 - 10x = 0$
   $x(x^2 - 3x - 10) = 0$
   $x(x - 5)(x + 2) = 0$
   $x = 0$ or $x - 5 = 0$ or $x + 2 = 0$
   $x = 0$ or $x = 5$ or $x = -2$
   The zeros of the function are (0, 0), (5, 0), and (−2, 0).

9. $\left\{\frac{5}{2}, 7\right\}$ **10.** $\left\{-3, \frac{9}{5}\right\}$ **11.** $\left\{\frac{7}{3}, \frac{7}{2}\right\}$

12. $\left\{\frac{7}{5}, 4\right\}$ **13.** $\left\{\frac{1}{4}, \frac{4}{3}\right\}$ **14.** $\left\{-2, -\frac{4}{5}\right\}$

15. $\left\{-9, \frac{1}{4}\right\}$ **16.** $\left\{\frac{11}{4}, 4\right\}$ **17.** $\left\{-\frac{2}{3}, \frac{11}{3}\right\}$

18. $\left\{-5, \frac{5}{4}\right\}$ **19.** $\left\{\frac{3}{10}, \frac{3}{4}\right\}$ **20.** $\left\{-\frac{3}{8}, \frac{5}{6}\right\}$

47. $x = 0, x = -\frac{2}{3}, x = \frac{2}{3}; \left(-\frac{2}{3}, 0\right), (0, 0), \left(\frac{2}{3}, 0\right)$

48. $x = 0, x = -\frac{9}{5}, x = \frac{9}{5}; \left(-\frac{9}{5}, 0\right), (0, 0), \left(\frac{9}{5}, 0\right)$

49. $x = 0, x = -\frac{1}{3}, x = \frac{1}{3}; \left(-\frac{1}{3}, 0\right), (0, 0), \left(\frac{1}{3}, 0\right)$

50. $x = 0, x = -\frac{5}{3}, x = \frac{5}{3}; \left(-\frac{5}{3}, 0\right), (0, 0), \left(\frac{5}{3}, 0\right)$

51. $x = 0, x = 3, x = -\frac{3}{2}; \left(-\frac{3}{2}, 0\right), (0, 0), (3, 0)$

52. $x = 0, x = 5, x = -\frac{5}{2}; \left(-\frac{5}{2}, 0\right), (0, 0), (5, 0)$

55. $\left\{-\frac{7}{4}, 0, \frac{9}{5}\right\}$ **56.** $\left\{-\frac{13}{5}, 0, 2\right\}$ **57.** $\left\{\frac{5}{2}, 13\right\}$

58. $\left\{-15, -\frac{7}{3}\right\}$ **59.** $\left\{1, \frac{6}{5}\right\}$ **60.** $\left\{-\frac{3}{2}, \frac{4}{5}\right\}$

61. $\left\{-\frac{3}{5}, 0, 1\right\}$ **62.** $\left\{0, \frac{1}{2}, \frac{3}{5}\right\}$ **63.** $\{-6, -5\}$

64. $\left\{-\frac{2}{5}, \frac{11}{5}\right\}$ **65.** $\left\{-2, -\frac{5}{3}\right\}$ **66.** $\left\{\frac{1}{9}, 2\right\}$

71. $\left\{-\frac{10}{3}, \frac{5}{2}\right\}$ **72.** $\left\{4, \frac{23}{2}\right\}$ **73.** $\left\{-\frac{4}{3}, \frac{2}{5}\right\}$

74. $\left\{\frac{3}{2}, 5\right\}$ **75.** $\left\{-\frac{2}{3}, 0, \frac{2}{3}\right\}$ **76.** $\left\{-\frac{5}{2}, 0, \frac{5}{2}\right\}$

77. $\left\{-\frac{11}{6}, 0, \frac{11}{6}\right\}$ **78.** $\left\{-\frac{1}{3}, 0, \frac{1}{3}\right\}$

109.

```
Plot1 Plot2 Plot3
\Y1■5X²-43X+24
\Y2■0
\Y3=
\Y4=
\Y5=
\Y6=
\Y7=
```

X	Y1	Y2
.6	0	0
8	0	0
1	-14	0

Y1■5X²-43X+24

110.

```
Plot1 Plot2 Plot3
\Y1■49X^3-9X
\Y2■0
\Y3=
\Y4=
\Y5=
\Y6=
\Y7=
```

X	Y1	Y2
0	0	0
-.4286	0	0
.42857	0	0
3	1296	0

Y1■49X^3-9X

GROUP ACTIVITY

1. $A = \dfrac{P\left(\frac{r}{12}\right)}{1 - \left(1 + \frac{r}{12}\right)^{-n}} = \dfrac{100{,}000\left(\frac{0.06}{12}\right)}{1 - \left(1 + \frac{0.06}{12}\right)^{-360}} = \$599.55$
2. 360(299.55)=\$215,838 is the total paid to the lending institution. The total interest paid is \$215,838 - 100,000 = \$115,838

3.

Payment Number	Beginning Balance	Scheduled Payment	Interest	Principal	Ending Balance
1	100,000	599.55	500.00	99.55	99,900.45
2	99,900.45	599.55	499.50	100.05	99,800.40
3	99,800.40	599.55	499.00	100.55	99,699.85
4	99,699.85	599.55	498.50	101.05	99,598.80
5	99,598.80	599.55	497.99	101.56	99,497.24
6	99,497.24	599.55	497.49	102.06	99,395.18
7	99,395.18	599.55	496.98	102.57	99,292.61
8	99,292.61	599.55	496.46	103.09	99,189.52
9	99,189.52	599.55	495.95	103.60	99,085.92
10	99,085.92	599.55	495.43	104.12	98,981.80
11	98,981.80	599.55	494.91	104.64	98,877.16
12	98,877.16	599.55	494.39	105.16	98,772.00

4. Early in the schedule, the majority of each payment is interest.

5.

Payment Number	Beginning Balance	Scheduled Payment	Extra Payment	Total Payment	Interest	Principal	Ending Balance
1	100,000.00	599.55	500.00	1099.55	500.00	599.55	99,400.45
2	99,400.45	599.55	500.00	1099.55	497.00	602.55	98,797.90
3	98,797.90	599.55	500.00	1099.55	493.99	605.56	98,192.34
4	98,192.34	599.55	500.00	1099.55	490.96	608.59	97,583.75
5	97,583.75	599.55	500.00	1099.55	487.92	611.63	96,972.12
6	96,972.12	599.55	500.00	1099.55	484.86	614.69	96,357.43
7	96,357.43	599.55	500.00	1099.55	481.79	617.76	95,739.67
8	95,739.67	599.55	500.00	1099.55	478.70	620.85	95,118.82
9	95,118.82	599.55	500.00	1099.55	475.59	623.96	94,494.86
10	94,494.86	599.55	500.00	1099.55	472.47	627.08	93,867.78
11	93,867.78	599.55	500.00	1099.55	469.34	630.21	93,237.57
12	93,237.57	599.55	500.00	1099.55	466.19	633.36	92,604.21

CHAPTER 6 REVIEW EXERCISES

Section 6.3

51. $5.76b^3, -1.24b, 2.1$; trinomial; 3, cubic; 576

52. $\frac{1}{8}x^3, -27$; binomial; 3, cubic; $\frac{1}{8}$

53. a^2b^4c; monomial; 7, seventh-degree; n/a

54. $-pq^4r^3$; monomial; 8, eighth-degree; n/a

55. $-1.2x^3, 0.4x^2, 3.1x, -1.6$; polynomial; 3, cubic; -1.2

56. $-0.6y^3, -0.5y^2, -2.8y, 1.1$; polynomial; 3, cubic; -0.6

63. 199.22, about 199,220,000 tickets were sold in 2002; 132.50, about 132,500,000 tickets were sold in 2010.

64. 7.08, about 7.08% of market share for the movie distributor Paramount Pictures; 13.97, about 13.97% of market share for the movie distributor Paramount Pictures in 2010

Section 6.4

73. $-6x^6 + 15x^5 - 3x^4$

74. $-2y^4 - 6y^3 + 5y^2$

75. $-20x^3y^2 + 45x^2y^2 - 10xy^4$

76. $-12x^4y + 6x^3y^2 - 24x^2y^3$

77. $4q^2 - 17q + 15$

78. $28v^2 - 31v - 5$

81. $7w^2 + w - 30$

82. $6z^2 - 19z - 272$

83. $5p^4 + 7p^2 - 6$

84. $9x^2 - 30x + 25$

85. $9m^4 - \frac{4}{25}$

86. $16n^4 - \frac{9}{4}$

87. $a^3 + 3a^2 - 4a - 12$

88. $x^3 - 12x^2y + 48xy^2 - 64y^3$

89. $27a^3 - 27a^2b + 9ab^2 - b^3$

90. $u^4 - 8u^3v + 24u^2v^2 - 32uv^3 + 16v^4$

Section 6.7

141. $\left(3x - \frac{1}{8}\right)\left(3x + \frac{1}{8}\right)$

142. $\left(7y - \frac{1}{4}\right)\left(7y + \frac{1}{4}\right)$

Section 6.8

165. $\left\{-\frac{7}{4}, -\frac{1}{6}\right\}$

166. $\left\{\frac{3}{4}, 1\right\}$

169. $\{-3, 13\}$

170. $\left\{-\frac{2}{5}, 2\right\}$

171. $\left\{-\frac{4}{3}, 0, \frac{4}{3}\right\}$

172. $\{-6, 0, 6\}$

183. $\{-2, 0, 2\}$

184. $\left\{-\frac{1}{5}, 0, \frac{1}{5}\right\}$

CHAPTER 6 TEST

14. $-y^2 - 4y + 5$; trinomial; 2; -1

15. $-27x^3 + 1$; binomial; 3; -27

16. $9a^6$; monomial; 6; 9

CHAPTERS 1–6 CUMULATIVE REVIEW EXERCISES

15.

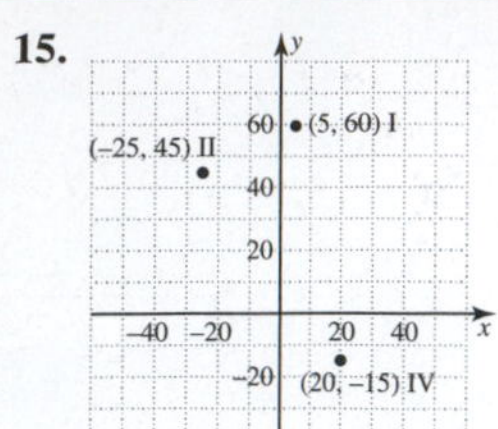

27. Domain = $(-\infty, \infty)$, Range = $(-\infty, \infty)$

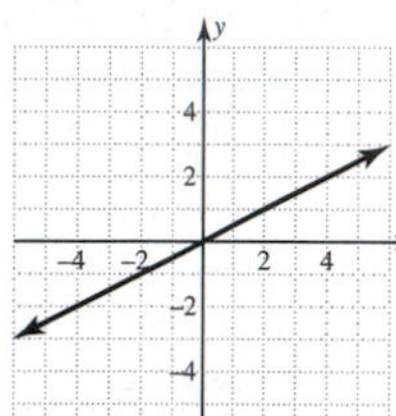

28. Domain = $(-\infty, \infty)$, Range = $\{2\}$

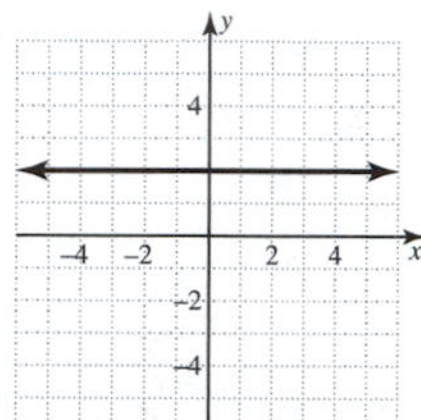

29. $m = 2$, $(0, -3)$

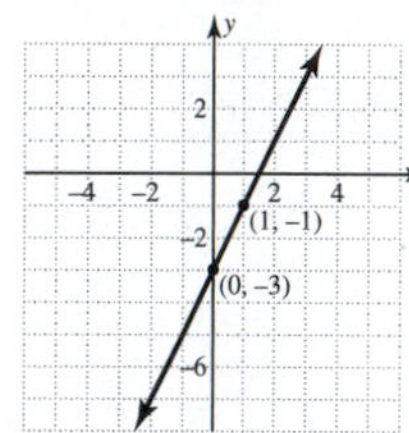

30. $m = -3$, $(0, 6)$

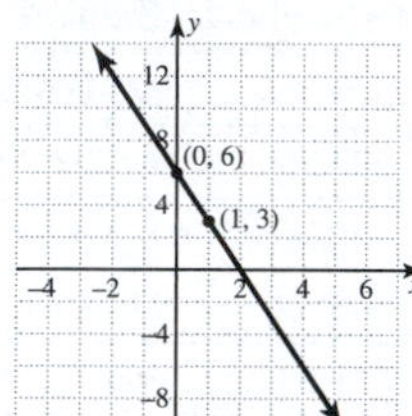

37.

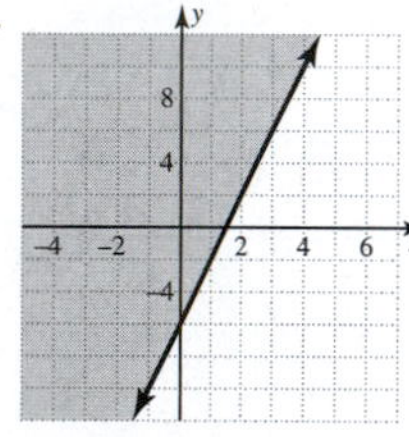

38.

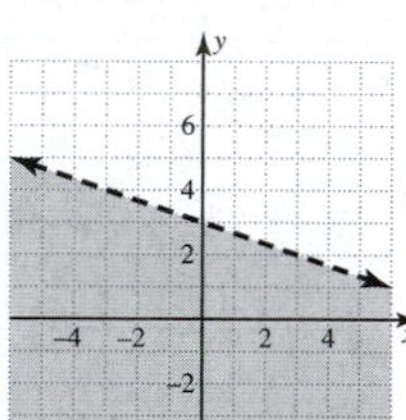

49. $\{(-2, 3, 1)\}$
50. $\{(3, -2, 4)\}$
51. $\{(-6z - 8, 4z + 5, z) \mid z \text{ is a real number}\}$
52. no solution

53.

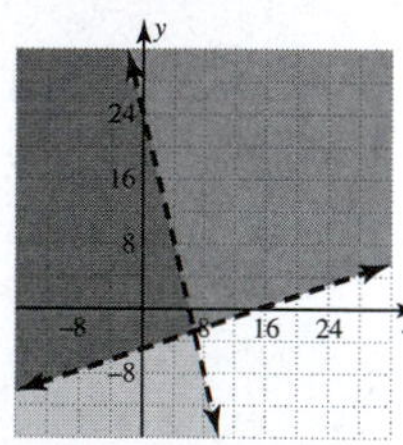

54.

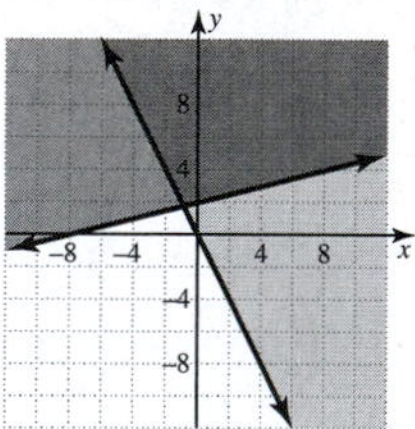

89. $2x^3, -5x, 17$; trinomial; 3, cubic; 2
90. $2y^2, -10y, -12$; trinomial; 2, quadratic; 2
91. $-ab^2c$; monomial; 4, quartic; n/a
92. $-5t, 9$; binomial; 1, linear; -5
102. $45x^3y - 20x^2y^2 + 30xy^3$
103. $w^3 + 6w^2 - 27$
105. $4r^2 + r - 3$
106. $6y^2 - 29y - 5$
107. $4a^2 + 36a + 81$
108. $49b^2 - 42b + 9$
115. $(5p + 2)(2p + 7)$
116. $(8q - 1)(8q - 7)$
117. $-7(2x + 1)(7x - 10)$
118. $-(5y + 4)(8y + 1)$
121. $r(3r + 8)(3r - 8)$
122. $-5s(s + 4)(s - 4)$
123. $(5a + 4)(5a - 6)$
124. $8(2b + 3)(b - 3)$
126. prime
127. $(4z^2 + 9)(2z + 3)(2z - 3)$
128. $(4w^2 + 1)(2w + 1)(2w - 1)$
129. $(3a + 1)(9a^2 - 3a + 1)$
130. $(2b - 5)(4b^2 + 10b + 25)$
131. $(4c - 1)(3c + 4)(3c - 4)$
132. $(4d - 1)(d + 3)(d - 3)$
133. $(5u - 1 - 3v)(5u - 1 + 3v)$
134. $(3r + s - 7)(3r + s + 7)$
135. $\left\{\frac{7}{3}, \frac{7}{2}\right\}$
136. $\left\{\frac{7}{5}, 4\right\}$
137. $\left\{-5, \frac{5}{4}\right\}$
138. $\left\{\frac{3}{10}, \frac{3}{4}\right\}$
139. $\left\{-\frac{9}{7}, 0, \frac{9}{7}\right\}$
143. $x = -5, x = 2$
144. $x = -\frac{8}{3}, x = 1$
145. $x = -2, x = 0, x = \frac{5}{2}$
146. $x = 0, x = \frac{1}{2}, x = 1$

CHAPTER 7

Section 7.1

1. The difference between a rational expression and a rational function is that only a rational function uses function notation to represent the quotient of polynomials. Both have the restriction that the denominator cannot equal zero.
2. To find the domain of a rational function, find the values that make the denominator equal to 0 and exclude these values from the set of all real numbers.
3. To simplify a rational expression, factor the numerator and denominator completely and divide out their common factors.
4. A common mistake made by students when multiplying two rational expressions is to divide out terms instead of common factors of the numerator and denominator. For example, $\frac{\cancel{x^2} - \overset{4}{\cancel{12}}x}{\cancel{x^2} + \cancel{3}} = 4x$. This expression can't be simplified since the numerator and denominator do not have a common factor other than 1.
5. To divide two rational expressions, convert the division to multiplication by the reciprocal of the divisor. Rewrite each numerator and denominator in factored form. Divide out the common factors and multiply the remaining factors.
6. The division is incorrect because x^2 and 6 are terms in the numerator and x^2 and -3 are terms in the denominator and they are not factors.
7. The division is incorrect in the first rational expression because x^2 is a term in the numerator and x^2 is a term in the denominator and they are not factors. The division is incorrect in the second rational expression because x is a term and not a factor in the denominator.
8. We have to exclude $x = -3$ because it makes the denominator equal to zero.

15. $\{x \mid x \text{ is a real number and } x \neq 3\}$
16. $\{x \mid x \text{ is a real number and } x \neq -9\}$
17. $\{x \mid x \text{ is a real number and } x \neq -7, x \neq 7\}$
18. $\{x \mid x \text{ is a real number and } x \neq -6, x \neq 6\}$
19. $\{x \mid x \text{ is a real number and } x \neq 0, x \neq 2\}$
20. $\{x \mid x \text{ is a real number and } x \neq -3, x \neq 0\}$
21. $\left\{x \,\middle|\, x \text{ is a real number and } x \neq -\frac{7}{2}\right\}$
22. $\left\{x \,\middle|\, x \text{ is a real number and } x \neq \frac{2}{5}\right\}$
25. $\left(-\infty, -\frac{1}{2}\right) \cup \left(-\frac{1}{2}, \frac{1}{2}\right) \cup \left(\frac{1}{2}, \infty\right)$
26. $\left(-\infty, -\frac{2}{5}\right) \cup \left(-\frac{2}{5}, \frac{2}{5}\right) \cup \left(\frac{2}{5}, \infty\right)$
27. $(-\infty, 0) \cup \left(0, \frac{2}{3}\right) \cup \left(\frac{2}{3}, \infty\right)$
28. $\left(-\infty, -\frac{5}{2}\right) \cup \left(-\frac{5}{2}, 0\right) \cup (0, \infty)$
29. $\left(-\infty, -\frac{1}{2}\right) \cup \left(-\frac{1}{2}, \infty\right)$
30. $(-\infty, 0) \cup (0, \infty)$

31. $\left(-\infty, \frac{5}{2}\right) \cup \left(\frac{5}{2}, \infty\right)$ **32.** $\left(-\infty, \frac{2}{3}\right) \cup \left(\frac{2}{3}, \infty\right)$

48. $\dfrac{5x-4}{x^2-3x+9}$ **50.** $\dfrac{3x+7}{9x^2+6x+4}$ **51.** $\dfrac{x+1}{4x^2+3}$

52. $\dfrac{7x+4}{2x^2+7}$ **53.** $\dfrac{2(x+4)}{4x^2+13}$ **54.** $\dfrac{3(x-3)}{x^2+4}$

71. a. The reunion will cost \$105 per person if 50 people attend.
72. a. The reunion will cost \$80 per person if 100 people attend.
73. a. The cost to remove 80% of the pollutants is \$330,000.
74. a. The cost to remove 75% of the pollutants is \$262,500.

81. $\left\{x \middle| x \text{ is a real number and } x \neq -\frac{4}{3}\right\}$; $\left(-\infty, -\frac{4}{3}\right) \cup \left(-\frac{4}{3}, \infty\right)$

82. $\left\{x \middle| x \text{ is a real number and } x \neq \frac{12}{5}\right\}$; $\left(-\infty, \frac{12}{5}\right) \cup \left(\frac{12}{5}, \infty\right)$

83. $\left\{x \middle| x \text{ is a real number and } x \neq 0, x \neq \frac{3}{4}\right\}$;
$(-\infty, 0) \cup \left(0, \frac{3}{4}\right) \cup \left(\frac{3}{4}, \infty\right)$

84. $\{x | x \text{ is a real number and } x \neq -7, x \neq 0\}$;
$(-\infty, -7) \cup (-7, 0) \cup (0, \infty)$

85. $\left\{x \middle| x \text{ is a real number and } x \neq -\frac{3}{2}, x \neq \frac{3}{2}\right\}$;
$\left(-\infty, -\frac{3}{2}\right) \cup \left(-\frac{3}{2}, \frac{3}{2}\right) \cup \left(\frac{3}{2}, \infty\right)$

86. $\{x | x \text{ is a real number and } x \neq -4\}$; $(-\infty, -4) \cup (-4, \infty)$
87. $\{x | x \text{ is a real number and } x \neq 3\}$; $(-\infty, 3) \cup (3, \infty)$

88. $\left\{x \middle| x \text{ is a real number and } x \neq -\frac{5}{3}, x \neq \frac{5}{3}\right\}$;
$\left(-\infty, -\frac{5}{3}\right) \cup \left(-\frac{5}{3}, \frac{5}{3}\right) \cup \left(\frac{5}{3}, \infty\right)$

111. a. The reunion will cost \$70 per person if 100 people attend.
112. a. The reunion will cost \$140 per person if 75 people attend.
113. a. The cost to remove 85% of the pollutants is \$382,500.
114. a. The cost to remove 82% of the pollutants is \$328,000.

117. $2x \neq 0$
$x \neq 0$
The domain of the function is $(-\infty, 0) \cup (0, \infty)$.

118. $x + 1 \neq 0$
$x \neq -1$
The domain of the function is $(-\infty, -1) \cup (-1, \infty)$.

119.
$$\frac{9x^2-4}{3x^2+x-2} = \frac{(3x-2)(3x+2)}{(x+1)(3x-2)} = \frac{(3x+2)\overset{1}{\cancel{(3x-2)}}}{(x+1)\underset{1}{\cancel{(3x-2)}}} = \frac{3x+2}{x+1}$$

120.
$$\frac{4x^2-x-3}{2x^2+x-3} = \frac{(4x+3)(x-1)}{(2x+3)(x-1)} = \frac{(4x+3)\overset{1}{\cancel{(x-1)}}}{(2x+3)\underset{1}{\cancel{(x-1)}}} = \frac{4x+3}{2x+3}$$

121.

```
Plot1 Plot2 Plot3
\Y1■(5X+4)/(2X-1
)
\Y2=
\Y3=
\Y4=
\Y5=
\Y6=
```

```
2→X
                2
Y1▸Frac
             14/3
```

122.

```
Plot1 Plot2 Plot3
\Y1■(3X-1)/(X+2)
\Y2=
\Y3=
\Y4=
\Y5=
\Y6=
```

```
-6→X
               -6
Y1▸Frac
             19/4
```

123.

```
Plot1 Plot2 Plot3
\Y1■(4X²+12X+9)/
(X-2)/((2X+3)^4/
(X²-X-2))
\Y2■(X+1)/(2X+3)
2
\Y3=
\Y4=
```

X	Y1	Y2
-3	-.2222	-.2222
-2	-1	-1
-1	ERROR	0
0	.11111	.11111
1	.08	.08
2	ERROR	.06122
3	.04938	.04938

Y2■(X+1)/(2X+3)²

Since the Y_1 and Y_2 columns agree for each value of x except where the denominator of Y_1 is zero, we can conclude that the two expressions are the same.

124.

```
Plot1 Plot2 Plot3
\Y1■(X^3+1)/(X²-
1)/((X+2)/(X²+X-
2))
\Y2■X²-X+1
\Y3=
\Y4=
\Y5=
```

X	Y1	Y2
-4	21	21
-3	13	13
-2	ERROR	7
-1	ERROR	3
0	1	1
1	ERROR	1
2	3	3

X=-2

Since the Y_1 and Y_2 columns agree for each value of x except where the denominator of Y_1 is zero, we can conclude that the two expressions are the same.

Section 7.2

1. To divide a polynomial by a monomial, divide each term in the numerator by the monomial and apply the rules of exponents to simplify each term. For example,
$$\frac{12x^3+10x^2-8x}{4x^2} = \frac{12x^3}{4x^2} + \frac{10x^2}{4x^2} - \frac{8x}{4x^2} = 3x + \frac{5}{2} - \frac{2}{x}$$

5. In synthetic division, the first step is to write down the coefficients of the dividend in standard form, inserting zero for any missing terms.

6. After completing the synthetic division, the numbers on the bottom row are the coefficients of the quotient whose degree is one less than the degree of the dividend. The last number in the row is the remainder.

7. Let c be the given x-value. Using the synthetic division, we can write the polynomial function as $P(x) = (x - c)\,Q(x) + R$, where $Q(x)$ is the quotient and R is the remainder. To evaluate the polynomial function a $x = c$, we substitute the value c and get $P(c) = (c - c)\,Q(c) + R = R$

8. We need to factor the numerator completely before division.
$$\frac{x^3+64}{x+4} = \frac{(x+4)(x^2-4x+16)}{x+4} = x^2 - 4x + 16 \neq x^2 + 16$$

17. $-\frac{3}{2}x^2 + \frac{11}{6}x + 1$ **23.** $4x - 1 + \dfrac{1}{x+2}$

31. $x^2 - 4x + 16 - \dfrac{128}{x+4}$ **32.** $25x^2 + 10x + 4 + \dfrac{16}{5x-2}$

37. $x + 4 - \dfrac{13}{x+10}$ **43.** $2x^2 - 9x - 2 + \dfrac{2}{x-3}$

44. $6x^2 - x + 2 - \dfrac{23}{x+8}$ **65.** $2x + 8 - \dfrac{1}{x+3}$

66. $x + 3 + \dfrac{17}{x-1}$ **78.** $4x^2 - 2x + 4 + \dfrac{5}{x+2}$

79. $8x^2 - 6x + 2 + \dfrac{1}{3x+1}$ **80.** $4x^2 - 5x - 2 - \dfrac{16}{5x-2}$

93. $\dfrac{8x^3+12x^2+4x}{4x} = \dfrac{\overset{2x^2}{\cancel{8x^3}}}{\cancel{4x}} + \dfrac{\overset{3x}{\cancel{12x^2}}}{\cancel{4x}} + \dfrac{\overset{1}{\cancel{4x}}}{\cancel{4x}} = 2x^2+3x+1$

95.

$$\begin{array}{r} 3x^2 - 6x + 14 \\ 2x+4\overline{\smash{)}6x^3 + 0x^2 + 4x - 16} \\ \underline{6x^3 + 12x^2} \qquad\qquad\quad \\ -12x^2 + 4x \qquad\quad \\ \underline{-12x^2 - 24x} \qquad\quad \\ 28x - 16 \\ \underline{28x + 56} \\ -72 \end{array}$$

So, $\dfrac{6x^3+4x-16}{2x+4} = 3x^2 - 6x + 14 - \dfrac{72}{2x+4}$.

96.

$$\begin{array}{c|cccc} 2 & 5 & 0 & -2 & -16 \\ & & 10 & 20 & 36 \\ \hline & 5 & 10 & 18 & 20 \end{array}$$

So, $(5x^3 - 2x - 16) \div (x-2) = 5x^2 + 10x + 18 + \dfrac{20}{x-2}$.

97.

```
Plot1 Plot2 Plot3
\Y1■(4X^2+13X+3)
/(X+3)
\Y2■4X+1
\Y3=
\Y4=
\Y5=
\Y6=
```

X	Y1	Y2
-4	-15	-15
-3.5	-13	-13
-3	ERROR	-11
-2.5	-9	-9
-2	-7	-7
-1.5	-5	-5
-1	-3	-3

X=-4

98.

```
Plot1 Plot2 Plot3
\Y1■(X^3+7X²+13X
-5)/(X+2)
\Y2■X²+5X+3-11/(
X+2)
\Y3=
\Y4=
\Y5=
```

X	Y1	Y2
-3.5	5.0833	5.0833
-3	8	8
-2.5	18.75	18.75
-2	ERROR	ERROR
-1.5	-24.25	-24.25
-1	-12	-12
-.5	-6.583	-6.583

X=-2.5

99.

```
Plot1 Plot2 Plot3
\Y1■(64X^3-125)/
(4X-5)
\Y2■16X²+20X+25
\Y3=
\Y4=
\Y5=
\Y6=
```

X	Y1	Y2
-1.5	31	31
-1	21	21
-.5	19	19
0	25	25
.5	39	39
1	61	61
1.5	91	91

X=-1.5

100.

```
Plot1 Plot2 Plot3
\Y1■(64X^3+125)/
(4X-5)
\Y2■16X²+20X+25+
250/(4X-5)
\Y3=
\Y4=
\Y5=
```

X	Y1	Y2
-1.5	8.2727	8.2727
-1	-6.778	-6.778
-.5	-16.71	-16.71
0	-25	-25
.5	-44.33	-44.33
1	-189	-189
1.5	341	341

X=0

Section 7.3

1. To find the LCD, factor each denominator. Find the product of all the factors from the first denominator and include any additional factors from the second denominator.

$12x^3y = 2^2 \cdot 3x^3y$

$18x^2y^5 = 2 \cdot 3^2 x^2y^5$

$\text{LCD} = 2^2 \cdot 3x^3y \cdot 3 \cdot y^4 = 36x^3y^5$

2. You need to subtract the second numerator $x - 2$ from the first numerator $4x$.

$\dfrac{4x}{3x-2} - \dfrac{x-2}{3x-2} = \dfrac{4x-(x-2)}{3x-2} = \dfrac{4x-x+2}{3x-2}$

$= \dfrac{3x+2}{3x-2} \neq \dfrac{3x-2}{3x-2}$

3. Since $2x - 5$ and $5 - 2x$ are opposites, we write $5 - 2x$ as $-1(2x - 5)$ and use $2x - 5$ as the LCD. We build the numerator for the second fraction as: $\dfrac{1}{5-2x} = \dfrac{1}{-1(2x-5)} = \dfrac{-1}{2x-5}$

4. To find the LCD, we find the product of all the factors from the first denominator and include any additional factor from the second denominator. The LCD $= (x-1)(2x+1)(x+3)$. Now we build the numerators for each rational expression.

$\dfrac{x}{(x-1)(2x+1)} = \dfrac{5(x+3)}{(x-1)(2x+1)(x+3)}$

$\dfrac{5x}{(x+3)(2x+1)} = \dfrac{5x(x-1)}{(x-1)(2x+1)(x+3)}$

5. You cannot add numerators and add denominators without finding the LCD and building the numerators.

6. You need to factor each denominator completely and find the LCD.

$x^2 - x - 6 = (x-3)(x+2)$, $x^2 - 2x - 8 = (x-4)(x+2)$

$\text{LCD} = (x-3)(x+2)(x-4)$

Next build each numerator and subtract.

$\dfrac{4x+1}{x^2-x-6} = \dfrac{4x+1}{(x-3)(x+2)} = \dfrac{(4x+1)(x-4)}{(x-3)(x+2)(x-4)}$

$\dfrac{3x-1}{x^2-2x-8} = \dfrac{3x-1}{(x+2)(x-4)} = \dfrac{(x-3)(3x-1)}{(x-3)(x+2)(x-4)}$

$\dfrac{(4x+1)(x-4)}{(x-3)(x+2)(x-4)} - \dfrac{(x-3)(3x-1)}{(x-3)(x+2)(x-4)}$

$= \dfrac{x^2-5x-7}{(x-3)(x+2)(x-4)}$

7. Factor each denominator completely and find the product of the factors from the first denominator and include any additional factors from the second and third denominators.

$(5x-2)^2$, $2 \cdot 3(5x-2)$, $2^3(5x-2)$

$\text{LCD} = (5x-2)^2 \cdot 2 \cdot 3 \cdot 2^2 = 2^3 \cdot 3(5x-2)^2 = 24(5x-2)^2$

8. Use the LCD found in Exercise 7 to build each numerator and simplify.

$\dfrac{3}{(5x-2)^2} + \dfrac{1}{30x-12} - \dfrac{5}{40x-16}$

$= \dfrac{3}{(5x-2)^2} + \dfrac{1}{2 \cdot 3(5x-2)} - \dfrac{5}{2^3(5x-2)}$

$= \dfrac{3 \cdot 2^3 \cdot 3}{2^3 \cdot 3(5x-2)^2} + \dfrac{2^2(5x-2)}{2^3 \cdot 3(5x-2)^2} - \dfrac{5 \cdot 3(5x-2)}{2^3 \cdot 3(5x-2)^2}$

$= \dfrac{72 + 20x - 8 - 75x + 30}{24(5x-2)^2}$

$= \dfrac{-55x + 94}{24(5x-2)^2}$

49. $\dfrac{x-3}{(x+1)(x-1)}$ **50.** $\dfrac{9x-6}{(x+2)(x-2)}$

51. $\dfrac{19x+10}{2(3x+4)(3x-4)}$ **52.** $\dfrac{13x-21}{3(2x+3)(2x-3)}$

63. $\dfrac{1}{3x+5}$ **64.** $\dfrac{1}{8x+3}$ **65.** $\dfrac{1}{7x-5}$ **66.** $\dfrac{1}{x-10}$

77. $\dfrac{29x+7}{3(2x+1)(2x-1)}$ **78.** $\dfrac{17x-16}{2(3x+2)(3x-2)}$

79. $\dfrac{18r-35s^2}{84r^5s^3}$ **85.** $\dfrac{4x+6}{x(x-2)(x+2)}$ **86.** $\dfrac{9x+15}{x(x+3)(x-3)}$

87. $\dfrac{3x-43}{5(x+9)(x-9)}$ **88.** $\dfrac{51x-18}{5(6x+7)(6x-7)}$

91. $\dfrac{11x-22}{(x+1)(x-1)(2x-3)}$ **92.** $\dfrac{2x+16}{(x+3)(x-3)(x-5)}$

93. $\dfrac{3x^3-2x^2+4x}{(x+2)(x-2)(x^2-2x+4)}$ **94.** $\dfrac{3x^3+13x^2+39x}{(x+3)(x-3)(x^2+3x+9)}$

97. Jane used a common denominator, not the LCD. Her answer is not entirely incorrect; it hasn't been simplified.

$$\frac{5}{4x-3}+\frac{3}{3-4x}=\frac{5}{4x-3}+\frac{-1(3)}{-1(3-4x)}$$
$$=\frac{5-3}{4x-3}$$
$$=\frac{2}{4x-3}$$

98. $$\frac{3}{x^3+64}+\frac{2x}{x^2-16}$$
$$=\frac{3}{(x+4)(x^2-4x+16)}+\frac{2x}{(x-4)(x+4)}$$
$$=\frac{3(x-4)}{(x+4)(x-4)(x^2-4x+16)}+\frac{2x(x^2-4x+16)}{(x-4)(x+4)(x^2-4x+16)}$$
$$=\frac{3x-12+2x^3-8x^2+32x}{(x+4)(x-4)(x^2-4x+16)}$$
$$=\frac{2x^3-8x^2+35x-12}{(x+4)(x-4)(x^2-4x+16)}$$

100. $$\frac{x+2}{x(x-1)}+\frac{x-2}{x(x+1)}-\frac{x}{x^2-1}$$
$$=\frac{x+2}{x(x-1)}+\frac{x-2}{x(x+1)}-\frac{x}{(x-1)(x+1)}$$
$$=\frac{(x+2)(x+1)}{x(x-1)(x+1)}+\frac{(x-2)(x-1)}{x(x-1)(x+1)}-\frac{x(x)}{x(x-1)(x+1)}$$
$$=\frac{x^2+3x+2+x^2-3x+2-x^2}{x(x-1)(x+1)}$$
$$=\frac{x^2+4}{x(x-1)(x+1)}$$

101.

```
Plot1 Plot2 Plot3
\Y1■7/(2X+6)-5/(7X)
\Y2■(39X-30)/(14X(X+3))
\Y3=
\Y4=
\Y5=
```

X	Y1	Y2
-2	-.0909	-.0909
-1	-.1429	-.1429
0	-.3333	-.3333
1	1	1
2	.2	.2
3	.11111	.11111
4	.07692	.07692

Y2■1/(4X-3)

102.

```
Plot1 Plot2 Plot3
\Y1■14/(4X-3)+13/(3-4X)
\Y2■1/(4X-3)
\Y3=
\Y4=
\Y5=
\Y6=
```

X	Y1	Y2
-2	-.0909	-.0909
-1	-.1429	-.1429
0	-.3333	-.3333
1	1	1
2	.2	.2
3	.11111	.11111
4	.07692	.07692

Y2■1/(4X-3)

103.

```
Plot1 Plot2 Plot3
\Y1■(6X+2)/(9X²-49)-5/(6X-14)
\Y2■(-3X-31)/(2(3X+7)(3X-7))
\Y3=
\Y4=
\Y5=
```

X	Y1	Y2
-2	.96154	.96154
-1	.35	.35
0	.31633	.31633
1	.425	.425
2	1.4231	1.4231
3	-.625	-.625
4	-.2263	-.2263

Y2■(-3X-31)/(2(...

104.

```
Plot1 Plot2 Plot3
\Y1■(3X-5)/(X²+X-12)+(3X-5)/(5X²-7X-24)
\Y2■(18X²+6X-60)/((X-3)(X+4)(5X+8))
\Y3=
```

X	Y1	Y2
-1	1.3333	1.3333
0	.625	.625
1	.27692	.27692
2	-.2222	-.2222
3	ERROR	ERROR
4	1.125	1.125
5	.70707	.70707

X= -1

Section 7.4

2. Find the LCD of the denominators in the complex fraction. Multiply the numerator and denominator of the complex fraction by the LCD and simplify. Multiply the remaining factors in the numerator and denominator. Simplify by dividing common factors.

3. Find the LCD of the fractions in the numerator of the complex fraction. Multiply the numerator and denominator of the complex fraction by the LCD and simplify. Factor the denominator completely. Divide out the common factors.

87. $$\frac{\frac{2}{x}+\frac{5}{y}}{10x+4y}=\frac{\frac{2y+5x}{xy}}{2(5x+2y)}$$
$$=\frac{2y+5x}{xy}\div\frac{2(2y+5x)}{1}$$
$$=\frac{2y+5x}{xy}\cdot\frac{1}{2(2y+5x)}$$
$$=\frac{\overset{1}{\cancel{2y+5y}}}{xy}\cdot\frac{1}{2\underset{1}{\cancel{(2y+5x)}}}=\frac{1}{2xy}$$

88. $$\frac{\frac{3}{x}-\frac{2}{y}}{\frac{1}{x}+\frac{1}{y}}=\left(\frac{3}{x}-\frac{2}{y}\right)\div\left(\frac{1}{x}+\frac{1}{y}\right)$$
$$=\frac{3y-2x}{xy}\div\frac{y+x}{xy}$$
$$=\frac{3y-2x}{\underset{1}{\cancel{xy}}}\cdot\frac{\overset{1}{\cancel{xy}}}{y+x}$$
$$=\frac{3y-2x}{y+x}$$

89. $$\frac{4x^{-2}-y^{-2}}{2x^{-1}-y^{-1}}=\frac{\frac{4}{x^2}-\frac{1}{y^2}}{\frac{2}{x}-\frac{1}{y}}$$
$$=\frac{\left(\frac{4}{x^2}-\frac{1}{y^2}\right)x^2y^2}{\left(\frac{2}{x}-\frac{1}{y}\right)x^2y^2}$$
$$=\frac{4y^2-x^2}{2xy^2-x^2y}$$
$$=\frac{(2y+x)\overset{1}{\cancel{(2y-x)}}}{xy\underset{1}{\cancel{(2y-x)}}}$$
$$=\frac{2y+x}{xy}$$

90. $$\frac{7^{-2}+8^{-1}}{2^{-3}}=\frac{\frac{1}{49}+\frac{1}{8}}{\frac{1}{8}}=\frac{\left(\frac{1}{49}+\frac{1}{8}\right)\cdot 392}{\frac{1}{8}\cdot 392}=\frac{8+49}{49}=\frac{57}{49}$$

91. The numerator and denominator of the complex fraction must be entered in parentheses.

```
(-5/8-2)/(4(-5/8)-3)
              .4772727273
Ans►Frac
                    21/44
```

92. The numerator and denominator of the complex fraction must be entered in parentheses.

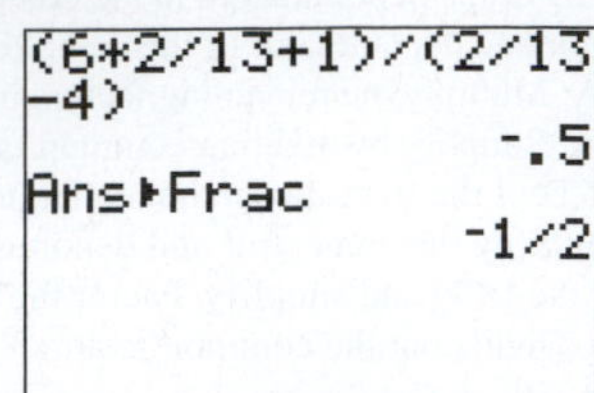

93. **94.** **95.** **96.**

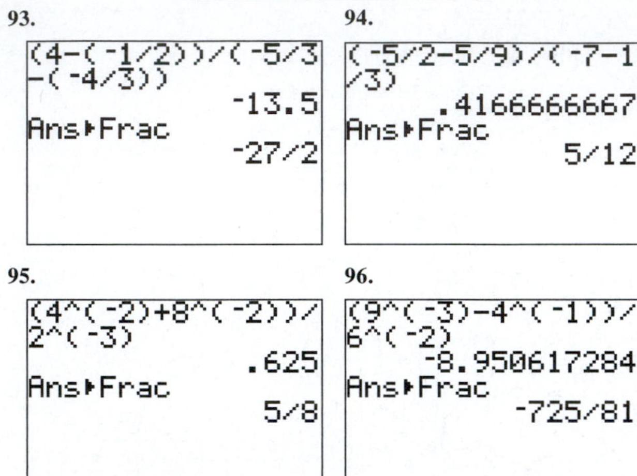

Section 7.5

3. When solving a rational equation, factor each denominator completely and find the LCD of all the denominators in the equation. Multiply both sides of the equation by the LCD which will eliminate all of the fractions. Solve the resulting equation.

4. There are answers that may work in the new equation after we multiply both sides of the rational equation by the LCD but they may not work in the original equation.

5. We need to factor each denominator completely. $x^2 - 4x + 4 = (x - 2)^2$, $x - 2$, and $x^2 - x - 2 = (x - 2)(x + 1)$. We find the LCD, $(x - 2)^2(x + 1)$. Multiply both sides of the equation by the LCD and solve the resulting equation.

30. $\left\{\frac{3}{7}\right\}$ **32.** $\left\{\frac{9}{10}\right\}$ **33.** $\left\{-\frac{3}{4}\right\}$ **34.** $\left\{\frac{40}{33}\right\}$

35. $\left\{\frac{12}{23}\right\}$ **38.** $\left\{-\frac{5}{4}, -\frac{4}{5}\right\}$ **39.** $\left\{-\frac{5}{3}, \frac{5}{8}\right\}$ **56.** $\left\{-\frac{8}{3}\right\}$

58. $\left\{-\frac{2}{3}, \frac{8}{3}\right\}$ **62.** $\left\{-\frac{2}{3}\right\}$

83.
$$\frac{4x + 28}{x^2 - x} + \frac{5}{x - 1} = \frac{10}{x^2 - x}$$
$$\frac{4x + 28}{x(x - 1)} + \frac{5}{x - 1} = \frac{10}{x(x - 1)}$$
$$(4x + 28) + 5(x) = 10$$
$$9x + 28 = 10$$
$$9x = -18$$
$$x = -2$$

The solution set is $\{-2\}$.

84. $\{10\}$

```
Plot1 Plot2 Plot3
\Y1=X/(X-9)-4
\Y2=6/(X-9)
\Y3=
\Y4=
\Y5=
\Y6=
\Y7=
```

X	Y1	Y2
6	-6	-2
7	-7.5	-3
8	-12	-6
9	ERROR	ERROR
10	6	6
11	1.5	3
12	0	2

X=10

85. $\{2\}$

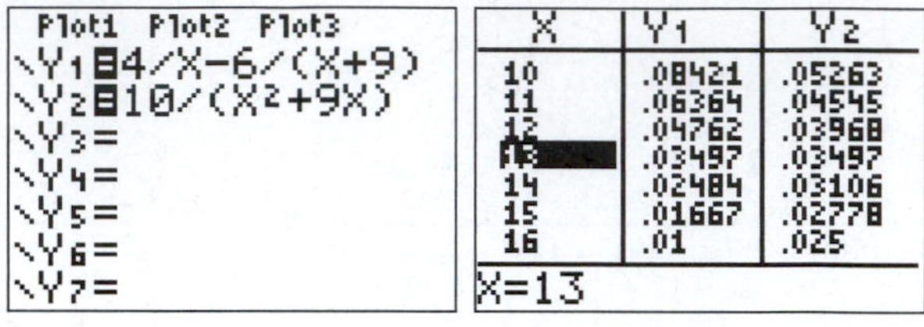

86. $\{13\}$

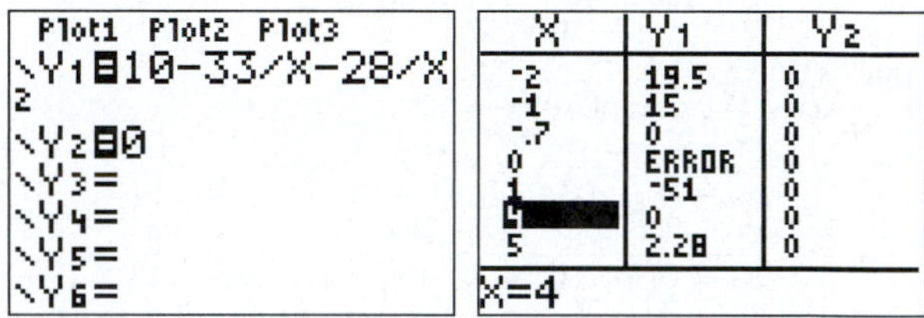

87. $\left\{4, -\frac{7}{10}\right\}$

```
Plot1 Plot2 Plot3
\Y1=10-33/X-28/X2
\Y2=0
\Y3=
\Y4=
\Y5=
\Y6=
```

X	Y1	Y2
-2	19.5	0
-1	15	0
-.7	0	0
0	ERROR	0
1	-51	0
4	0	0
5	2.28	0

X=4

Section 7.6

1. Find the LCD of all the denominators in the rational equation. Multiply each side of the equation by the LCD and simplify. Solve the resulting equation for the specific variable.

4. The sum means the portion of the job completed by Melissa in 1 hr and the portion of the job completed by Jenny in 1 hr.

6. If Mark drives y mph slower than Nicole, then his speed is y mph subtracted from Nicole's speed. That is, Mark's speed is $(x - y)$ mph.

11. $y = \frac{3xz}{10z - 5x}$ **12.** $b = \frac{12ac}{9c - 2a}$ **13.** $s = \frac{3rt}{5t + 2r}$

14. $c = \frac{7ab}{10a - 6b}$ **16.** $x_1 = \frac{mx_2 - y_2 + y_1}{m}$ **21.** $V_1 = \frac{P_2R_1V_2}{P_1R_2}$

22. $R_2 = \frac{P_2R_1V_2}{P_1V_1}$ **39.** $s = \frac{14rt}{8t + r}$ **42.** $x_2 = \frac{mx_1 + y_2 - y_1}{m}$

43. $a = \frac{2bc}{3c - b}$ **45.** $b = \frac{2A - ah}{h}$

57. Let x be the number of hours to complete the job when working together.
$$\frac{1}{10} + \frac{1}{6} = \frac{1}{x}$$
$$30x \cdot \frac{1}{10} + \frac{1}{6} \cdot 30x = \frac{1}{x} \cdot 30x$$
$$3x + 5x = 30$$
$$8x = 30$$
$$x = \frac{30}{8} = 3.75$$

It takes 3 hr and 45 min to complete the job when working together.

Section 7.7

1. Direct variation describes the situation where y varies directly as x or y is directly proportional to x. The equation is $y = kx$, where k is the constant of variation.

2. Inverse variation describes the situation where y varies inversely as x or y is inversely proportional to x. The equation is $y = \frac{k}{x}$, where k is the constant of variation.

3. Joint variation describes the situation where z varies jointly as x and y or z is jointly proportional to x and y. The equation is $z = kxy$, where k is the constant of variation.

4. A varies directly as the square of r.
5. V varies jointly as the square of r and h.
6. V varies directly as r^3.
7. G varies jointly as m_1 and m_2 and inversely as the square of r.
8. y varies inversely as the cube root of x.

11. $k = \frac{2}{3}; y = \frac{2}{3}x$ **12.** $k = \frac{6}{5}; y = \frac{6}{5}x$ **25.** $k = \frac{3}{2}; y = \frac{3}{2x}$

26. $k = \frac{3}{5}; y = \frac{3}{5x}$ **27.** $k = 1.6; y = \frac{1.6}{x}$ **28.** $k = 2.1; y = \frac{2.1}{x}$

45. $k = 6.4; w = \frac{6.4u^2}{v^3}$ **46.** $k = 0.00288; z = \frac{0.00288\sqrt{x}}{y^4}$

50. $b = \frac{5}{2}\sqrt{a}$ **51.** $y = \frac{3.6}{x^2}$

CHAPTER 7 REVIEW EXERCISES

Section 7.1

5. $\{x|x$ is a real number and $x \neq 10\}$ or $(-\infty, 10) \cup (10, \infty)$

6. $\left\{x \middle| x \text{ is a real number and } x \neq \frac{2}{3}\right\}$ or $\left(-\infty, \frac{2}{3}\right) \cup \left(\frac{2}{3}, \infty\right)$

7. $\{x|x$ is a real number and $x \neq 6, x \neq -6\}$ or $(-\infty, -6) \cup (-6, 6) \cup (6, \infty)$

8. $\{x|x$ is a real number and $x \neq 7, x \neq -7\}$ or $(-\infty, -7) \cup (-7, 7) \cup (7, \infty)$

9. $\{x|x$ is a real number and $x \neq -2, x \neq 0\}$ or $(-\infty, -2) \cup (-2, 0) \cup (0, \infty)$

10. $\{x|x$ is a real number and $x \neq 0, x \neq 3\}$ or $(-\infty, 0) \cup (0, 3) \cup (3, \infty)$

20. $\frac{x-1}{x^2+3}$ **35.** $\frac{x(x^2+2x+4)}{(x+2)(5x-2)^2}$

Section 7.2

55. $x^2 - 2x - 12 - \frac{92}{x-6}$ **56.** $x^2 - 3x + 13 - \frac{51}{x+3}$

Section 7.3

73. $-\frac{14x}{2x+9}$ **75.** $\frac{2a-20}{5a+7}$ **83.** $\frac{39x-46}{5(x+6)(x-6)}$

84. $\frac{14x-1}{(3x+2)(3x-2)}$ **87.** $\frac{15}{5x-6}$ **93.** $\frac{x+2}{(x+1)(x^2+x+1)}$

94. $\frac{7x^2-14x+8}{(x+2)(x-2)(x^2-2x+4)}$

Section 7.4

109. $\frac{7x+4y}{xy}$ **112.** $\frac{3b-2a}{ab}$ **115.** $\frac{12}{23}$ **116.** $-\frac{3}{28}$

117. $\frac{15}{2}$ **118.** $-\frac{22}{3}$

Section 7.7

151. $k = 50; w = \frac{50\sqrt[3]{u}}{v^2}$

CHAPTERS 1–7 CUMULATIVE REVIEW EXERCISES

5.
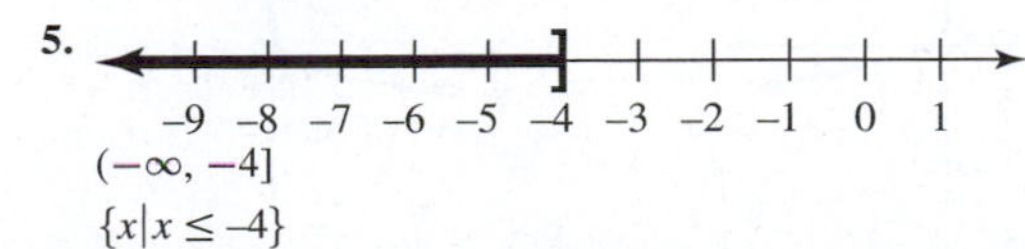

$(-\infty, -4]$
$\{x|x \leq -4\}$

6.
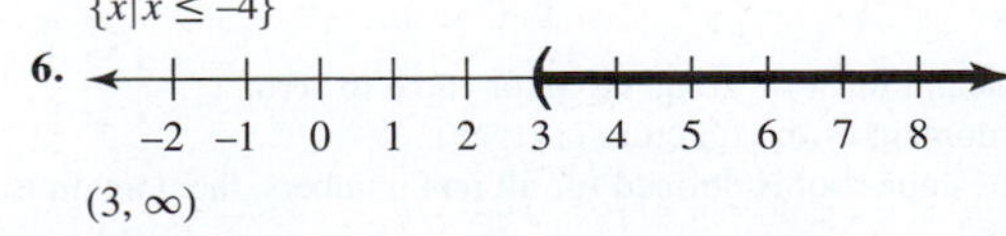

$(3, \infty)$
$\{y|y > 3\}$

11. Domain $= (-\infty, \infty)$, Range $= (-\infty, \infty)$

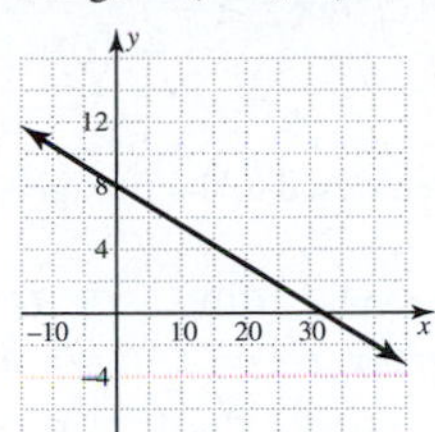

12. Domain $= (-\infty, \infty)$, Range $= \{-4\}$

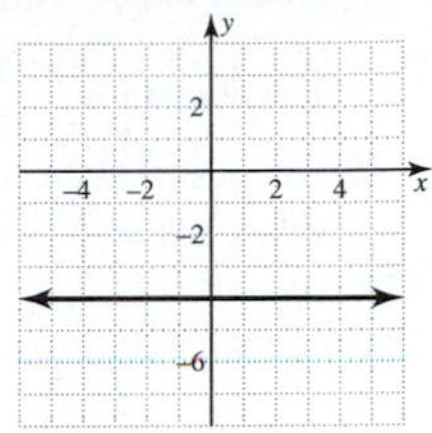

13. $m = -2$, $(0, -5)$

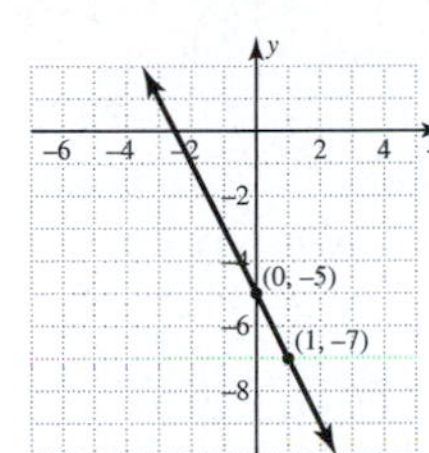

14. $m = \frac{3}{2}$, $(0, -6)$

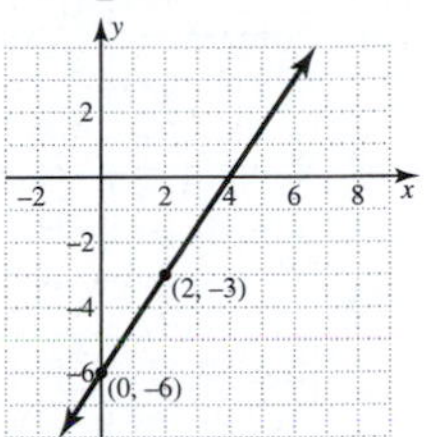

19.

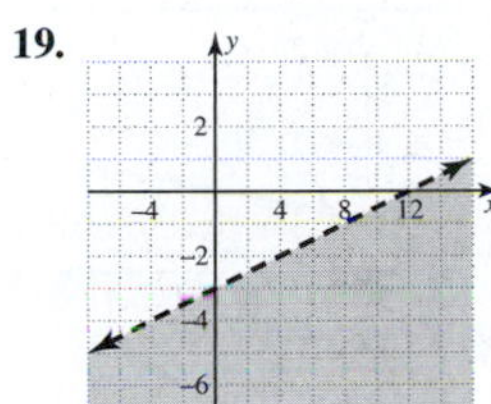

20.

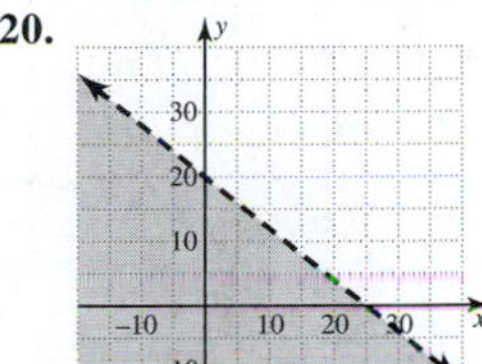

23. same line, infinitely many solutions, consistent system with dependent equations
24. intersecting lines, one solution, consistent system with independent equations

35.

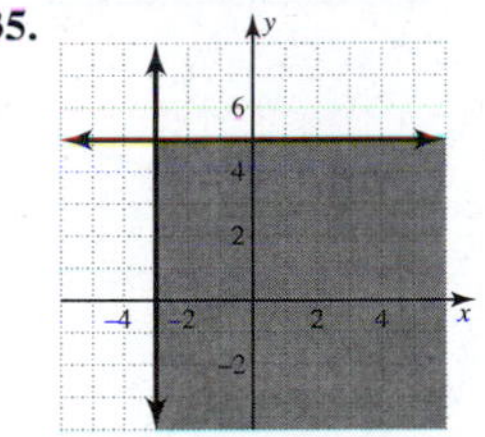

36.

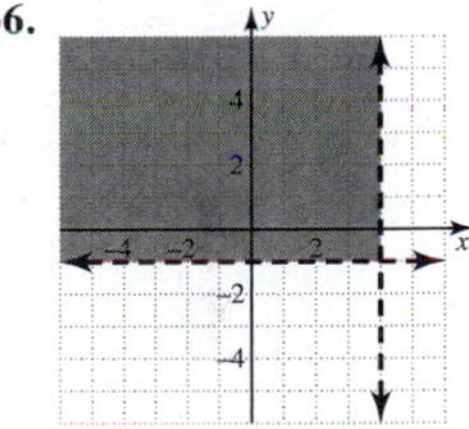

79. $\left\{\frac{7}{3}, \frac{7}{2}\right\}$ **80.** $\left\{\frac{7}{5}, 4\right\}$ **91.** $\frac{6(x-2)}{x^2+6}$ **92.** $\frac{x+1}{4x^2+3}$

97. $-x^2 + \frac{9}{5}x - 2$ **98.** $-\frac{1}{3}x^2 + 2x + \frac{3}{2}$

99. $3x^2 - 10x + 7 - \frac{35}{2x+1}$ **100.** $3x^2 + 5x + 8 + \frac{17}{5x-7}$

101. $x^2 - 6x + 36 - \frac{432}{x+6}$ **102.** $2x^2 + 6x - 17 - \frac{27}{x-3}$

107. $\frac{7}{x-7}$ **109.** $\frac{3x-2}{5(x+1)^2}$ **110.** $\frac{2x-5}{2(4x+5)^2}$

121. $\left\{-\frac{3}{2}, -\frac{6}{5}\right\}$ **122.** $\left\{-\frac{5}{8}, \frac{2}{5}\right\}$ **127.** $y = \frac{xz}{2x-12z}$

128. $b = \frac{3ac}{4a-c}$

CHAPTER 8

Section 8.1

4. $\sqrt[4]{-256}$ is not a real number because there is no real number that can be raised to the fourth to get -256.
5. Answers vary. To find the cube root of 125, find b such that $b^3 = 125$. So, $b = 5$.

6. To find the domain of a square root function, the radicand must be greater than or equal to 0. So, $2x + 9 \geq 0$ or $x \geq -\frac{9}{2}$.
The domain is $\left[-\frac{9}{2}, \infty\right)$.

8. We must find the expression that when raised to the fifth gives us x^{10}. $\sqrt[5]{x^{10}} = \sqrt[5]{(x^2)^5} = x^2$.

83. domain $= [-5, \infty)$; $f(-6)$ not a real number, $f(0) = \sqrt{5}, f(4) = 3$

84. domain $= [2, \infty)$; $f(0)$ not a real number, $f(3) = 1,\ f(7) = \sqrt{5}$

85. domain $= \left[-\frac{1}{6}, \infty\right)$; $f(-1)$ not a real number, $f\left(-\frac{1}{6}\right) = 0$, $f(4) = 5$

86. domain $= \left[\frac{3}{5}, \infty\right)$; $f(0)$ not a real number, $f\left(\frac{3}{5}\right) = 0$, $f(2) = \sqrt{7}$

87. domain $= (-\infty, \infty)$; $f(-5) = -2, f(0) = -\sqrt[3]{3}, f(3) = 0$

88. domain $= (-\infty, \infty)$; $f(-7) = 0, f(0) = \sqrt[3]{7}, f(1) = 2$

93.

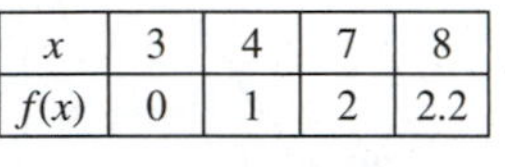

x	3	4	7	8
$f(x)$	0	1	2	2.2

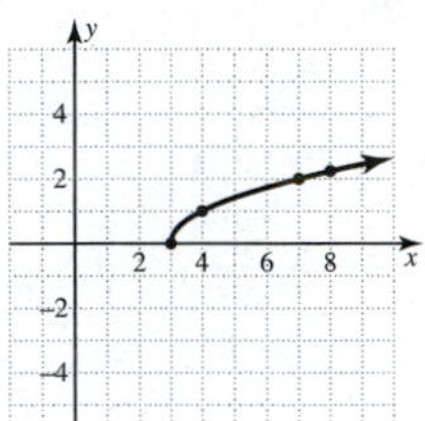

94.

x	−2	−1	2	6
$f(x)$	0	1	2	2.8

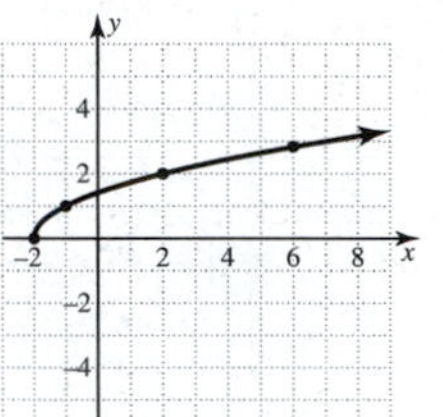

95.

x	−2.5	−1	0	2
$f(x)$	0	1.7	2.2	3

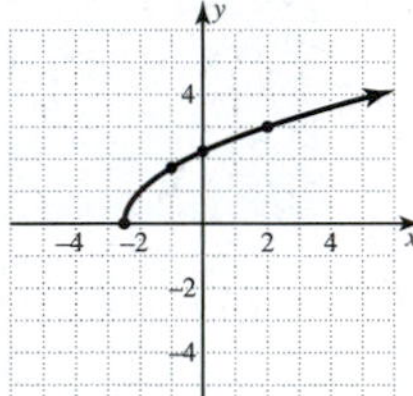

96.

x	1.5	2	4	6
$f(x)$	0	1	2.2	3

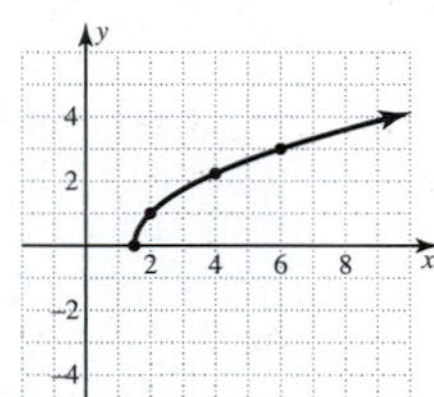

97.

x	−6	0	1	2	3	10
$f(x)$	−2	−1.3	−1	0	1	2

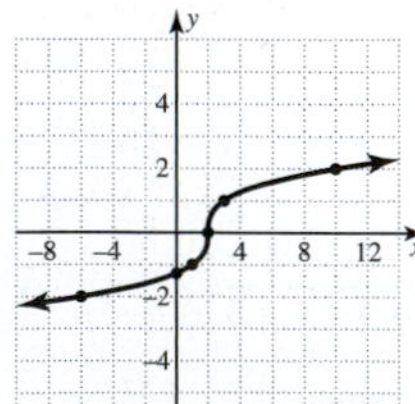

98.

x	−10	−2	−1	0	6
$f(x)$	−2	0	1	1.3	2

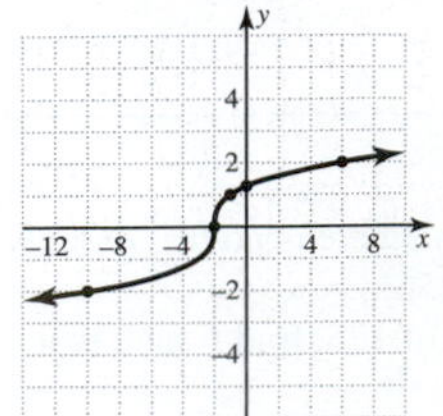

99.

x	−8	−1	0	1	8	12
$f(x)$	−4	−3	−2	−1	0	0.3

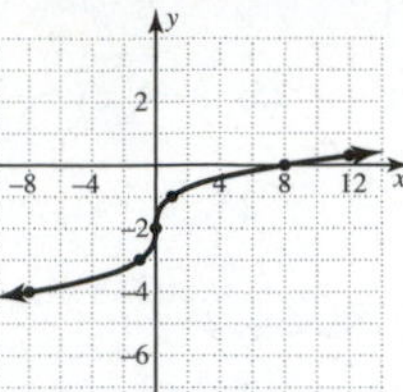

100.

x	−8	−1	0	1	8
$f(x)$	0	1	2	3	4

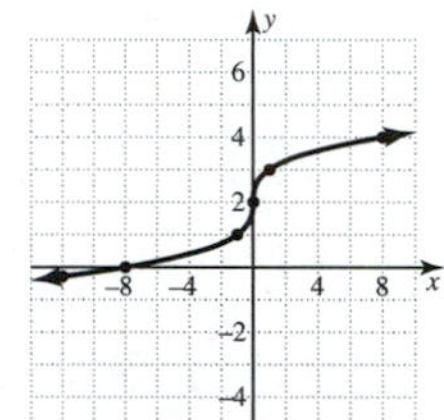

123. domain $= \left[-\frac{4}{3}, \infty\right)$; $f\left(-\frac{4}{3}\right) = 0, f(0) = 2, f(2) = \sqrt{10}$

124. domain $= \left[\frac{7}{2}, \infty\right)$; $f(0)$ not a real number, $f\left(\frac{7}{2}\right) = 0$, $f(5) = \sqrt{3}$

125. domain $= (-\infty, 3]$; $f(-6) = 3, f(0) = \sqrt{3}, f(4)$ not a real number

126. domain $= (-\infty, 5]$; $f(-4) = 3, f(0) = \sqrt{5}, f(5) = 0$

127. domain $= (-\infty, \infty)$; $f(-4) = 0, f(0) = \sqrt[3]{4}, f(4) = 2$

128. domain $= (-\infty, \infty)$; $f(-3) = -2, f(0) = -\sqrt[3]{5}, f(6) = 1$

131.

x	−3	0	1	1.5
$f(x)$	3	1.7	1	0

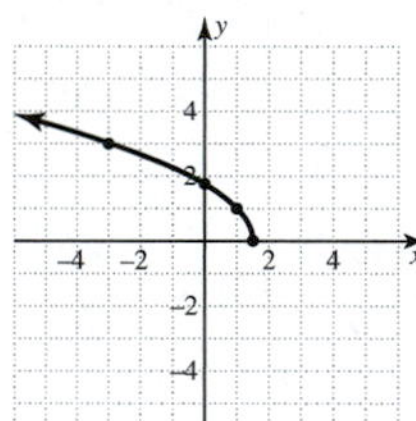

132.

x	−3	−1	0	0.5
$f(x)$	2.6	1.7	1	0

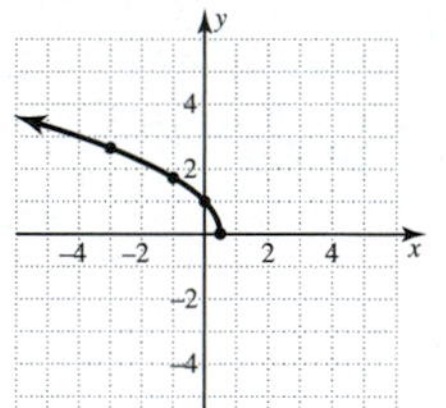

133.

x	−8	−1	0	1	8
$f(x)$	3	2	1	0	−1

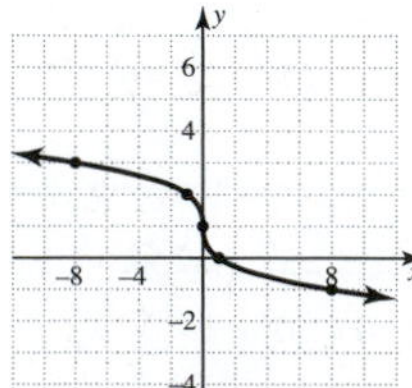

134.

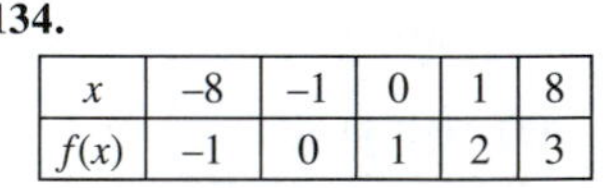

x	−8	−1	0	1	8
$f(x)$	−1	0	1	2	3

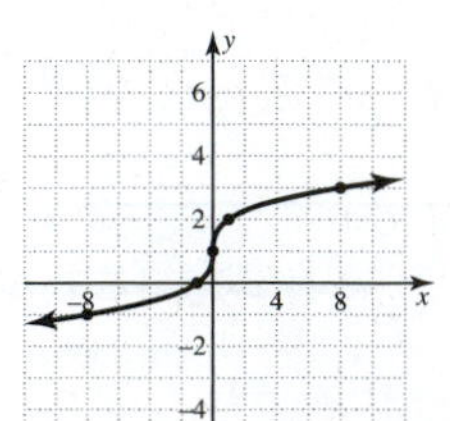

135. The radicand must be greater than or equal to zero: $x - 1 \geq 0$, $x \geq 1$
So, the domain of the function is $[1, \infty)$.

136. Since the cube root is defined for all real numbers, the domain is $(-\infty, \infty)$.

137. Since this expression is the square root of a negative value, it is not a real number.

138. Since this expression is the fourth root of a negative value, it is not a real number.

139. 42.83

```
√(1834)
         42.82522621
```

140. 7.56

```
³√(432)
         7.559526299
```

141. 4.31

```
5ˣ√(1492)
         4.312744844
```

142. 2.52

```
7ˣ√(654)
         2.524790763
```

Section 8.2

1. Answers vary; An exponent that is a fraction is called a rational exponent. An example is $x^{2/3}$.

2. Answers vary; An expression with a rational exponent $x^{1/n}$ is the same as the nth root of x, $\sqrt[n]{x}$. An expression of the form $x^{m/n}$ is the same as the nth root of x^m. An example is $a^{3/4} = \sqrt[4]{a^3}$.

3. Answers vary; Add the exponents: $x^{2/5} \cdot x^{1/3} = x^{2/5+1/3} = x^{6/15+5/15} = x^{11/15}$.

4. Answers vary; Subtract the exponent in the denominator from the exponent in the numerator, $\dfrac{x^{1/2}}{x^{1/3}} = x^{1/2-1/3} = x^{3/6-2/6} = x^{1/6}$.

5. Answers vary; Write the rational exponent as 1 divided by the fifth root of 32 raised to the third and then simplify the radical.

$$32^{-3/5} = \frac{1}{\sqrt[5]{32^3}} = \frac{1}{(\sqrt[5]{32})^3} = \frac{1}{2^3} = \frac{1}{8}$$

6. The bases are not the same.

163. When like bases are multiplied, the exponents are added but the bases stay the same. $5^{1/3} \cdot 5^{2/3} = 5^{1/3+2/3} = 5^{3/3} = 5^1 = 5$

164. The exponent of 3/4 is applied to -16 not just 16. $(-16x^8)^{3/4} = (-16)^{3/4}(x^8)^{3/4}$ is not a real number since the 4th root of -16 is not a real number.

165. We can't take the opposite of the base until after we apply the exponent $-(-x^3y^6)^{2/3} = -(-x^3)^{2/3}(y^6)^{2/3} = -x^2y^4$.

166. $-3x$ is the square root of $9x^2$. To find the 4th root, convert the radical to exponential form. $-\sqrt[4]{9x^2} = -(9x^2)^{1/4} = -[(3x)^2]^{1/4} = -(3x)^{1/2} = -\sqrt{3x}$

167. 392.50

```
36^(5/3)
         392.4980481
```

168. 0.46

```
7^(-2/5)
           .45915655
```

169. $\dfrac{1}{16}$

```
32^(-4/5)
               .0625
Ans▸Frac
                1/16
```

170. $\dfrac{1}{625}$

```
125^(-4/3)
               .0016
Ans▸Frac
               1/625
```

Section 8.3

1. The radicand must not contain any perfect square factors or fractions. There also can't be radicals in the denominators of a fraction.

2. The radicand must not contain any perfect cube factors or fractions. There also can't be radicals in the denominators of a fraction.

3. We must rewrite the radical as a product in which one of the factors is a power with an exponent equal to the index. Then we apply the product rule and simplify by extracting any roots.

4. The quotient rule for radicals enables us to rewrite a fraction in a radical. We apply the nth root to the numerator and denominator and simplify.

5. x^2 and y^4 are perfect squares since their exponents are even. Factor 72 as $36 \cdot 2$. Apply the product rule. Group the perfect square factors, x^2, y^4, and 36 in the first square root and the remaining factor 2 in the other square root. Simplify.
$\sqrt{72x^2y^4} = \sqrt{36x^2y^4} \cdot \sqrt{2} = 6xy^2\sqrt{2}$

6. Rewrite each term as a product of a perfect cube and other factors. Factor 54 as $27 \cdot 2$, and x^5 as $x^3 \cdot x^2$. Apply the quotient rule for radicals. Group the perfect cube factors, x^3 and 27 in the first cube root and the remaining factors in the other cube root. Simplify.

$$\sqrt[3]{\frac{54x^5}{125y^{15}}} = \frac{\sqrt[3]{27x^3} \cdot \sqrt[3]{2x^2}}{\sqrt[3]{125y^{15}}} = \frac{3x}{5y^5}\sqrt[3]{2x^2}$$

85. The variable in the radicand has an exponent greater than 2, so it is not simplified.
$\sqrt{63x^7} = \sqrt{9 \cdot 7x^6x} = 3x^3\sqrt{7x}$

86. The index is 3, so we should find the cube root not the square root.
$\sqrt[3]{64a^{10}b^4} = \sqrt[3]{64a^9b^3}\sqrt[3]{ab} = 4a^3b\sqrt[3]{ab}$

87. The order of operations requires us to simplify the expression inside the radical before we apply the square root.

$$\begin{aligned}\sqrt{(11-3)^2 + (-16+1)^2} &= \sqrt{(8)^2 + (-15)^2} \\ &= \sqrt{64+225} \\ &= \sqrt{289} \\ &= 17\end{aligned}$$

88. The difference in the x-values is $-9 - (-5) = -9 + 5 = -4$ and the difference in the y-values is $6 - 3 = 3$.

$$\begin{aligned}\sqrt{(-9+5)^2 + (6-3)^2} &= \sqrt{(-4)^2 + (3)^2} \\ &= \sqrt{16+9} \\ &= \sqrt{25} \\ &= 5\end{aligned}$$

89.

```
Plot1 Plot2 Plot3
\Y1■√(54X^5)
\Y2■3X²√(6X)
\Y3=
\Y4=
\Y5=
\Y6=
\Y7=
```

X	Y1	Y2
-1	ERROR	ERROR
0	0	0
1	7.3485	7.3485
2	41.569	41.569
3	114.55	114.55
4	235.15	235.15
5	410.79	410.79

Y2■3X²√(6X)

Since the Y_1 and Y_2 columns agree for each value of x, we can conclude that the two expressions are the same.

90.

```
Plot1 Plot2 Plot3
\Y1■√(98/(25X^4)
)
\Y2■7√(2)/(5X²)
\Y3=
\Y4=
\Y5=
\Y6=
```

X	Y1	Y2
-1	1.9799	1.9799
0	ERROR	ERROR
1	1.9799	1.9799
2	.49497	.49497
3	.21999	.21999
4	.12374	.12374
5	.0792	.0792

Y2■7√(2)/(5X²)

Since the Y_1 and Y_2 columns agree for each value of x, we can conclude that the two expressions are the same.

91.

```
Plot1 Plot2 Plot3
\Y1■5*√(96X^12)
\Y2■2X²*5*√(3X²)

\Y3=
\Y4=
\Y5=
\Y6=
```

X	Y1	Y2
-1	2.4915	2.4915
0	0	0
1	2.4915	2.4915
2	13.15	13.15
3	34.797	34.797
4	69.406	69.406
5	118.57	118.57

Y2■2X²*5*√(3X²)

Since the Y_1 and Y_2 columns agree for each value of x, we can conclude that the two expressions are the same.

92.

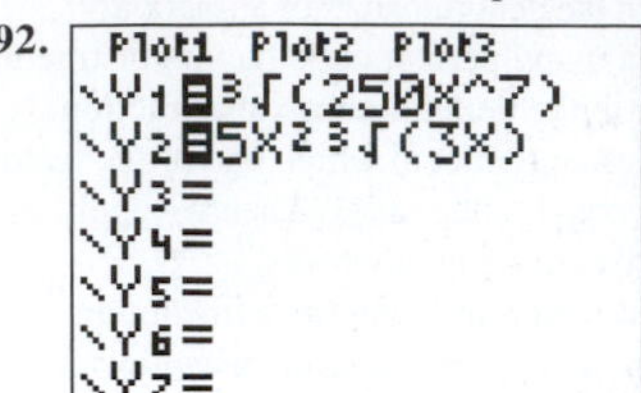

X	Y1	Y2
-1	-6.3	-7.211
0	0	0
1	6.2996	7.2112
2	31.748	36.342
3	81.77	93.604
4	160	183.15
5	269.3	308.28

X=0

Since the Y_1 and Y_2 columns do not agree for all values of x, we can conclude that the two expressions are not the same. Thus, the simplification is incorrect.

95.

```
√(72)+3√(50)-6√(
18)
        4.242640687
3√(2)
        4.242640687
```

The two radical expressions have the same decimal value. Therefore, they are equivalent.

96.

```
10√(48)-3√(192)+
6√(12)
        48.49742261
28√(3)
        48.49742261
```

The two radical expressions have the same decimal value. Therefore, they are equivalent.

97.

```
7³√(40)-2³√(135)
        13.67980757
8³√(5)
        13.67980757
```

The two radical expressions have the same decimal value. Therefore, they are equivalent.

98.

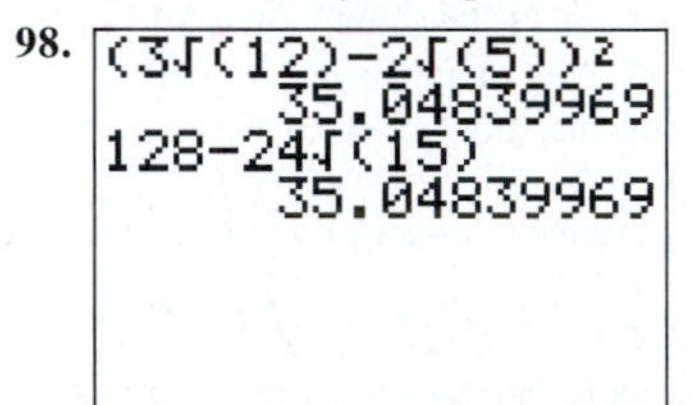

```
(3√(12)-2√(5))²
        35.04839969
128-24√(15)
        35.04839969
```

The two radical expressions have the same decimal value. Therefore, they are equivalent.

Section 8.4

1. Like radicals are radicals with the same index and same radicand.
2. We add or subtract radical expressions by adding or subtracting the coefficients of the radicals and keeping the radicals the same.
3. If the indices are the same, we multiply the radicands of the radicals using the product rule for radicals.
4. Yes, like radicals are radicals with the same indices and same radicands. Answers vary. For example, $(3\sqrt{2})(4\sqrt{2}) = 12\sqrt{4} = 12(2) = 24$.
5. We must first simplify each radical. $\sqrt{18x} + \sqrt{12x} = 3\sqrt{2x} + 2\sqrt{3x}$; The radicals are not alike, so we cannot add the expressions.
6. The radicals cannot be simplified any further. They are not like radicals, so we can't subtract them.
7. To multiply, we use the FOIL method.
$$\begin{aligned}(\sqrt{x}+\sqrt{y})^2 &= (\sqrt{x}+\sqrt{y})(\sqrt{x}+\sqrt{y})\\ &= \sqrt{x^2}+\sqrt{xy}+\sqrt{xy}+\sqrt{y^2}\\ &= x+2\sqrt{xy}+y\end{aligned}$$
8. These are like radicals so we add the coefficients and keep the radical the same. We get $\sqrt{a} + \sqrt{a} = 1\sqrt{a} + 1\sqrt{a} = 2\sqrt{a}$.

51. $36\sqrt{2} + 36$ or 86.9 cm; 324 cm^2
52. $22\sqrt{2} + 22$ or 53.1 cm; 121 cm^2
53. $20\sqrt{14} + 20\sqrt{7}$ or 127.7 ft; 700 ft^2
54. $6\sqrt{6} + 6\sqrt{3}$ or 25.1 ft; 27 ft^2
55. $12\sqrt{7}$ cm; $8\sqrt{21}$ cm
56. $42\sqrt{2}$ m; $28\sqrt{6}$ m
85. $42\sqrt{2} + 42$ or 101.40 m; 441 m^2
86. $32\sqrt{2} + 32$ or 77.25 m; 256 m^2
87. $36\sqrt{6} + 36\sqrt{3}$ or 150.54 in.; 972 in.^{2}
88. $14\sqrt{10} + 14\sqrt{5}$ or 75.58 in.; 245 in.^{2}
91. $$\begin{aligned}\sqrt{28x^2y} + \sqrt{63x^2y} &= \sqrt{4\cdot 7x^2y} + \sqrt{9\cdot 7x^2y}\\ &= 2x\sqrt{7y} + 3x\sqrt{7y}\\ &= 5x\sqrt{7y}\end{aligned}$$
92. $$\begin{aligned}3\sqrt{75a^4b} - a^2\sqrt{98b} &= 3\sqrt{25\cdot 3a^4b} - a^2\sqrt{49\cdot 2b}\\ &= 15a^2\sqrt{3b} - 7a^2\sqrt{2b}\end{aligned}$$
93. $$\begin{aligned}(2\sqrt{x}-y)^2 &= (2\sqrt{x}-y)(2\sqrt{x}-y)\\ &= (2\sqrt{x})^2 - 2(2\sqrt{xy}) + (y)^2\\ &= 4x - 4y\sqrt{x} + y^2\end{aligned}$$
94. $$\begin{aligned}(\sqrt{7}-\sqrt{3})^2 &= (\sqrt{7}-\sqrt{3})(\sqrt{7}-\sqrt{3})\\ &= (\sqrt{7})^2 - \sqrt{21} - \sqrt{21} + (\sqrt{3})^2\\ &= 7 - 2\sqrt{21} + 3\\ &= 10 - 2\sqrt{21}\end{aligned}$$

Piece It Together Sections 8.1–8.4

7.

x	−8	−1	0	1	8	12
$f(x)$	−4	−3	−2	−1	0	0.3

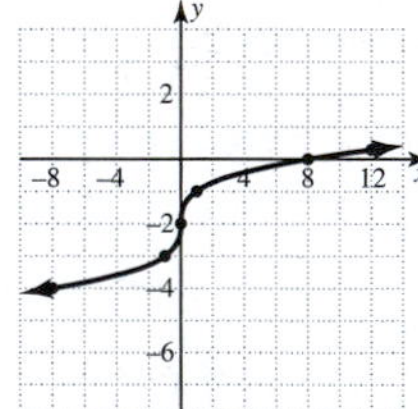

Section 8.5

1. To divide radical expressions with the same index, you divide the radicands and keep the root of the quotient.
2. We need to rationalize the denominator because we need to remove the radical from the denominator.
3. The conjugate of $a + \sqrt{b}$ is $a - \sqrt{b}$.
4. When we multiply $a + \sqrt{b}$ and its conjugate $a - \sqrt{b}$, we obtain an expression without the square root of b.
$$\begin{aligned}(a+\sqrt{b})(a-\sqrt{b}) &= a^2 - a\sqrt{b} + a\sqrt{b} - (\sqrt{b})^2\\ &= a^2 - b\end{aligned}$$

5. To rationalize the denominator of $\frac{1}{\sqrt[5]{6}}$, you need to multiply the numerator and the denominator by a term that makes the radicand in the denominator a power of the index, that is, $\sqrt[5]{6^4}$:

$$\frac{1}{\sqrt[5]{6}} \cdot \frac{\sqrt[5]{6^4}}{\sqrt[5]{6^4}} = \frac{\sqrt[5]{6^4}}{\sqrt[5]{6^5}} = \frac{\sqrt[5]{6^4}}{6}.$$

6. You need to rationalize the denominator and compare the result.

$$\frac{5}{\sqrt{11}-\sqrt{10}} \cdot \frac{\sqrt{11}+\sqrt{10}}{\sqrt{11}+\sqrt{10}} = \frac{5(\sqrt{11}+\sqrt{10})}{11-10} = \frac{5(\sqrt{11}+\sqrt{10})}{1} \neq 5$$

93. $\frac{-3-\sqrt{5}-3\sqrt{2}-\sqrt{10}}{4}$

94. $\frac{21-3\sqrt{5}+7\sqrt{3}-\sqrt{15}}{6}$

101. $\frac{-2+\sqrt{6}+2\sqrt{3}-3\sqrt{2}}{2}$

102. $\frac{3-3\sqrt{15}+\sqrt{5}-5\sqrt{3}}{4}$

107.
$$\frac{\sqrt{7}-\sqrt{3}}{\sqrt{7}+\sqrt{3}} = \frac{(\sqrt{7}-\sqrt{3})}{(\sqrt{7}+\sqrt{3})} \cdot \frac{(\sqrt{7}-\sqrt{3})}{(\sqrt{7}-\sqrt{3})} = \frac{7-2\sqrt{21}+3}{7-3} = \frac{10-2\sqrt{21}}{4} = \frac{{}^{5}\cancel{10}-{}^{1}\cancel{2}\sqrt{21}}{\cancel{4}_2} = \frac{5-\sqrt{21}}{2}$$

109.
```
12/(√(11)-√(2))
       6.30778447
4(√(11)+√(2))/3
       6.30778447
```
The two radical expressions have the same decimal value. Therefore, they are equivalent.

110.
```
(-6+√(3))/(-11-√
(5))
      .3224484189
(66-6√(5)-11√(3)
+√(15))/116
      .3224484189
```
The two radical expressions have the same decimal value. Therefore, they are equivalent.

111.
```
12/6 ×√(3)
       9.992198132
4(6 ×√(243))
       9.992198132
```
The two radical expressions have the same decimal value. Therefore, they are equivalent.

112.
```
16/5 ×√(20)
       8.788484346
8(5 ×√(5000))/5
       8.788484346
```
The two radical expressions have the same decimal value. Therefore, they are equivalent.

Section 8.6

1. You raise both sides of the equation to an exponent equal to the index n and solve the resulting equation.

3. With one radical on one side of the equation, you square each side.

$(\sqrt{x-5}+1)^2 = (\sqrt{x})^2$

Next you use the FOIL method to multiply.

$$(\sqrt{x-5}+1)(\sqrt{x-5}+1) = (\sqrt{x})^2$$
$$x-5+2\sqrt{x-5}+1 = x$$
$$2\sqrt{x-5} = 4$$
$$\sqrt{x-5} = 2$$
$$(\sqrt{x-5})^2 = (2)^2$$
$$x-5 = 4$$
$$x = 9$$

4. Isolate the radical on one side of the equation and square each side.

$\sqrt{x+5} = x-1$

$(\sqrt{x+5})^2 = (x-1)^2$

Next you use the FOIL method to multiply.

$$x+5 = x^2-2x+1$$
$$x^2-3x-4 = 0$$
$$(x-4)(x+1) = 0$$
$$x = 4, x = -1$$

Check: $x = 4$; $x = -1$

$x = 4$	$x = -1$
$\sqrt{(4)+5}+1 = 4$	$\sqrt{(-1)+5}+1 = -1$
$\sqrt{9}+1 = 4$	$\sqrt{4}+1 = -1$
$3+1 = 4$	$2+1 = -1$
$4 = 4$	$3 = -1$
	not true

5. You substitute 655 for x and check the equation.

$$\sqrt[3]{2(655)+21}-9 \stackrel{?}{=} 2$$
$$\sqrt[3]{1331}-9 \stackrel{?}{=} 2$$
$$11-9 = 2$$
$$2 = 2$$

6. You substitute 3 and –2 for x and check.

$$x = 3: \sqrt{70-15(3)} \stackrel{?}{=} (3)-8$$
$$\sqrt{25} \stackrel{?}{=} -5$$
$$5 \neq -5$$
$$x = -2: \sqrt{70-15(-2)} \stackrel{?}{=} (-2)-8$$
$$\sqrt{100} \stackrel{?}{=} -10$$
$$10 \neq -10$$

82.
$$\sqrt{12x-11} = 3+\sqrt{6x-14}$$
$$(\sqrt{12x-11})^2 = (3+\sqrt{6x-14})^2$$
$$12x-11 = 9+6\sqrt{6x-14}+6x-14$$
$$6x-6 = 6\sqrt{6x-14}$$
$$x-1 = \sqrt{6x-14}$$
$$(x-1)^2 = (\sqrt{6x-14})^2$$
$$x^2-2x+1 = 6x-14$$
$$x^2-8x+15 = 0$$
$$(x-3)(x-5) = 0$$
$$x = 3, x = 5$$

The solution set is $\{3, 5\}$. The solutions check.

83.
$$\sqrt{2x-23} = \sqrt{x}-1$$
$$(\sqrt{2x-23})^2 = (\sqrt{x}-1)^2$$
$$2x-23 = x-2\sqrt{x}+1$$
$$x-24 = -2\sqrt{x}$$
$$(x-24)^2 = (-2\sqrt{x})^2$$
$$x^2-48x+576 = 4x$$
$$x^2-52x+576 = 0$$
$$(x-16)(x-36) = 0$$
$$x = 16, x = 36$$

Since 36 is an extraneous solution, the solution set is $\{16\}$.

84.
$$\sqrt{x+7} = -2+\sqrt{5x+15}$$
$$(\sqrt{x+7})^2 = (-2+\sqrt{5x+15})^2$$
$$x+7 = 4-4\sqrt{5x+15}+5x+15$$
$$4\sqrt{5x+15} = 4x+12$$

$$\sqrt{5x+15} = x+3$$
$$5x + 15 = x^2 + 6x + 9$$
$$0 = x^2 + x - 6$$
$$0 = (x+3)(x-2)$$
$$x = -3, x = 2$$

Since -3 is an extraneous solution, the solution set is $\{2\}$.

85.

X	Y1	Y2
-9	4.5826	-6
-8	5	-5
-4	6.4031	-1
0	7.5498	3
4	8.544	7
	9	9
-8	5	-5

X=6

The two radical expressions have the same value at 6 but not at -8. Therefore, the solution set is $\{6\}$.

86.

X	Y1	Y2
-1	-7.534	-7
0	-7.332	-7
1	-7.156	-7
	-7	-7
3	-6.859	-7
4	-6.729	-7
5	-6.609	-7

X=2

The two radical expressions have the same value at 2. Therefore, the solution set is $\{2\}$.

87.

X	Y1	Y2
-6	1	ERROR
	3	3
-4	4.1231	4.4142
-3	5	5
-2	5.7446	5.4495
-1	6.4031	5.8284
0	7	6.1623

X=-5

The two radical expressions have the same value at -5 and -3. Therefore, the solution set is $\{-5, -3\}$.

88.

X	Y1	Y2
-4	ERROR	8.3589
-3	0	8
-2	2.2361	7.6056
-1	3.1623	7.1623
0	3.873	6.6458
1	4.4721	6
	5	5

X=2

The two radical expressions have the same value at 2 but not at -3. Therefore, the solution set is $\{2\}$.

Section 8.7

1. An imaginary number is a number whose value, when squared, is a negative number. We write the imaginary number in the form of $\sqrt{-c} = i\sqrt{c}$, where $c > 0$.
2. A complex number is a number in the form of $a + bi$, where $i = \sqrt{-1}$.
3. The conjugate of $a + bi$ is $a - bi$.
4. You add and subtract complex numbers by combining like terms, that is, combining real parts and imaginary parts.
5. You multiply complex numbers by applying the distributive property and the FOIL method.
6. To divide complex numbers, you multiply the denominator by its conjugate. The resulting denominator is a real number.
7. $(5-i)^2 = 25 - 2(5)(i) + i^2 = 25 - 10i - 1 = 24 - 10i \neq 24$
8. $\sqrt{-4} \cdot \sqrt{-4} = 2i \cdot 2i = 4i^2 = -4 \neq 4$

21. $\frac{5}{7} - \frac{11\sqrt{3}}{14}i$ **22.** $\frac{9}{17} - \frac{2\sqrt{13}}{17}i$

133.
```
(3-5i)(7+2i)
              31-29i
```

134.
```
(6-√(-48))²
-12-83.13843876i
-12-48i√(3)
-12-83.13843876i
```

135.
```
(-1-3i)/(3+4i)
             -.6-.2i
-3/5-1/5i
             -.6-.2i
```

136.
```
(-5i)^(-3)
              -.008i
Ans▸Frac
             -1/125i
```

GROUP ACTIVITY

2. a.

$$(-2 + 5\sqrt{2} + 2)^2 = 50$$
$$(5\sqrt{2})^2 = 50$$
$$25(2) = 50$$
$$50 = 50$$

$$(-2 - 5\sqrt{2} + 2)^2 = 50$$
$$(5\sqrt{2})^2 = 50$$
$$25(2) = 50$$
$$50 = 50$$

b.

$$(3+\sqrt{5})^2 - 6(3+\sqrt{5}) + 4 = 0$$
$$9 + 6\sqrt{5} + 5 - 18 - 6\sqrt{5} + 4 = 0$$
$$14 - 18 + 4 = 0$$
$$0 = 0$$

$$(3-\sqrt{5})^2 - 6(3-\sqrt{5}) + 4 = 0$$
$$9 - 6\sqrt{5} + 5 - 18 + 6\sqrt{5} + 4 = 0$$
$$14 - 18 + 4 = 0$$
$$0 = 0$$

c.

$$3(-1 + 2i + 1)^2 + 12 = 0$$
$$3(2i)^2 + 12 = 0$$
$$3(4i^2) + 12 = 0$$
$$12(-1) + 12 = 0$$
$$-12 + 12 = 0$$
$$0 = 0$$

$$3(-1 - 2i + 1)^2 + 12 = 0$$
$$3(-2i)^2 + 12 = 0$$
$$3(4i^2) + 12 = 0$$
$$12(-1) + 12 = 0$$
$$-12 + 12 = 0$$
$$0 = 0$$

3. $-2 + \sqrt{50} \approx 5.07,\ -2 - \sqrt{50} \approx -9.07,\ \frac{6+\sqrt{20}}{2} \approx 5.24,\ \frac{6-\sqrt{20}}{2} \approx 0.76$

4.

$$(5.07 + 2)^2 \approx 50$$
$$(7.07)^2 \approx 50$$
$$49.98 \approx 50$$

$$(-9.07 + 2)^2 \approx 50$$
$$(-7.07)^2 \approx 50$$
$$49.98 \approx 50$$

$$(5.24)^2 - 6(5.24) + 4 \approx 0$$
$$27.4576 - 31.44 + 4 \approx 0$$
$$0.0176 \approx 0$$

$$(0.76)^2 - 6(0.76) + 4 \approx 0$$
$$0.5776 - 4.56 + 4 \approx 0$$
$$0.0176 \approx 0$$

CHAPTER 8 REVIEW EXERCISES

Section 8.1

17. domain $= \left[\frac{5}{8}, \infty\right)$ or $\left\{x \middle| x \geq \frac{5}{8}\right\}$; not a real number, 3.32

18. domain $= \left[\frac{9}{7}, \infty\right)$ or $\left\{x \middle| x \geq \frac{9}{7}\right\}$; not a real number, 5.74

19. domain $= (-\infty, \infty)$ or $\{x \mid x \in \mathbb{R}\}$; 3, 1.44, -2.35

20. domain $= (-\infty, \infty)$ or $\{x \mid x \in \mathbb{R}\}$; 2, 1.59, 0

21.

x	$f(x)$
-3	5
0	4
1	3.61
4	2

22.

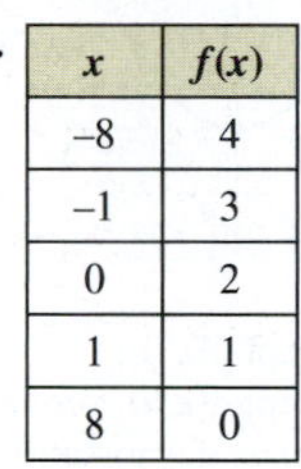

x	$f(x)$
-8	4
-1	3
0	2
1	1
8	0

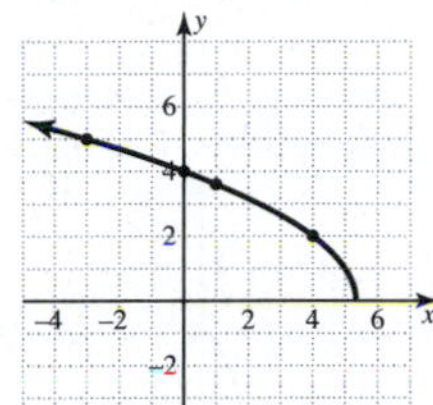

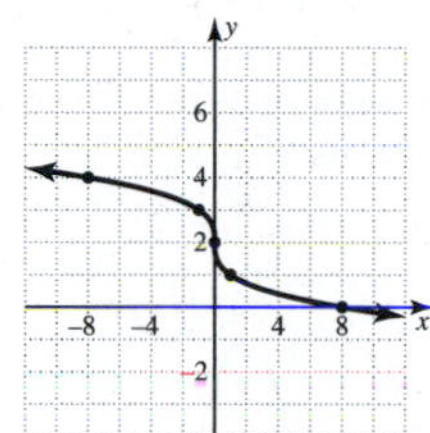

CHAPTERS 1–8 CUMULATIVE REVIEW EXERCISES

77. domain $= (-\infty, 5]$; $f(-4) = 3$, $f(1) = 2$, $f(6)$ undefined

78. domain $= [4, \infty)$; $f(5) = 1$, $f(4) = 0$, $f(3)$ undefined

CHAPTER 9

Section 9.1

1. The vertex is the ordered pair $(0, k)$, the parabola opens upward, and the range is $[k, \infty)$.
2. The vertex is the ordered pair $(h, 0)$, the parabola opens upward, and the range is $[0, \infty)$.
3. The vertex is the ordered pair (h, k), the parabola opens downward, and the range is $(-\infty, k]$.
4. The vertex is the ordered pair (h, k), the parabola opens upward if $a > 0$ and downward if $a < 0$, and the range is $[k, \infty)$ if $a > 0$ and $(-\infty, k]$ if $a < 0$.
5. no x-intercept; y-intercept: $(0, 5)$; vertex: $(0, 5)$; axis of symmetry: $x = 0$; domain: $(-\infty, \infty)$; range: $[5, \infty)$; shifts up 5 units

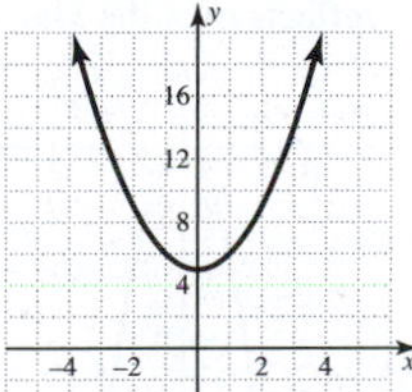

6. no x-intercept; y-intercept: $(0, 7)$; vertex: $(0, 7)$; axis of symmetry: $x = 0$; domain: $(-\infty, \infty)$; range: $[7, \infty)$; shifts up 7 units

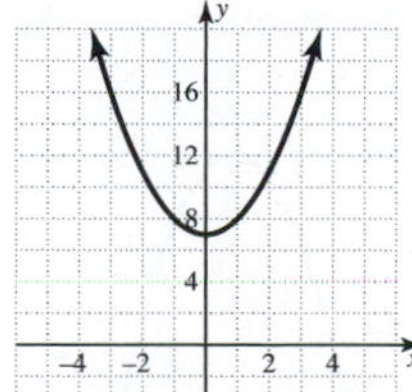

7. x-intercepts: $(-2, 0)$ and $(2, 0)$; y-intercept: $(0, -4)$; vertex: $(0, -4)$; axis of symmetry: $x = 0$; domain: $(-\infty, \infty)$; range: $[-4, \infty)$; shifts down 4 units

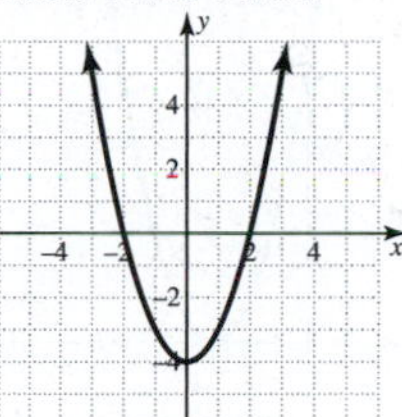

8. x-intercepts: $(-5, 0)$ and $(5, 0)$; y-intercept: $(0, -25)$; vertex: $(0, -25)$; axis of symmetry: $x = 0$; domain: $(-\infty, \infty)$; range: $[-25, \infty)$; shifts down 25 units

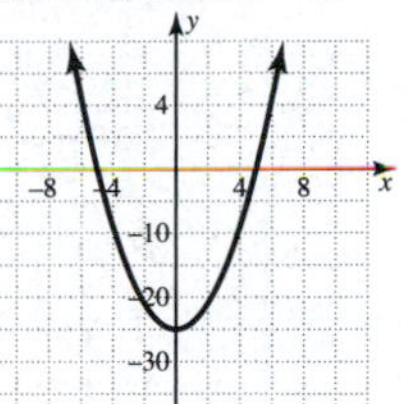

9. x-intercepts: $(-1.5, 0)$ and $(1.5, 0)$; y-intercept: $(0, -2.25)$; vertex: $(0, -2.25)$; axis of symmetry: $x = 0$; domain: $(-\infty, \infty)$; range: $[-2.25, \infty)$; shifts down 2.25 units

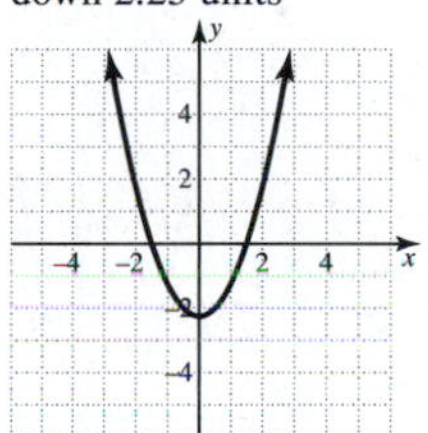

10. x-intercepts: $(-3.6, 0)$ and $(3.6, 0)$; y-intercept: $(0, -12.96)$; vertex: $(0, -12.96)$; axis of symmetry: $x = 0$; domain: $(-\infty, \infty)$; range: $[-12.96, \infty)$; shifts down 12.96 units

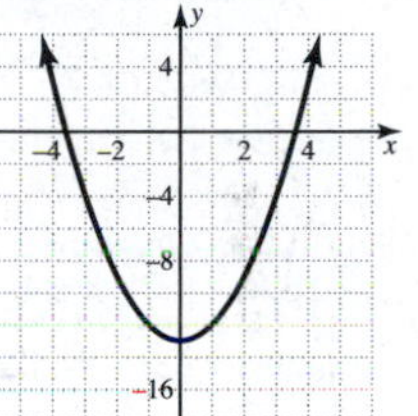

11. no x-intercept; y-intercept: $(0, 5.76)$; vertex: $(0, 5.76)$; axis of symmetry: $x = 0$; domain: $(-\infty, \infty)$; range: $[5.76, \infty)$; shifts up 5.76 units

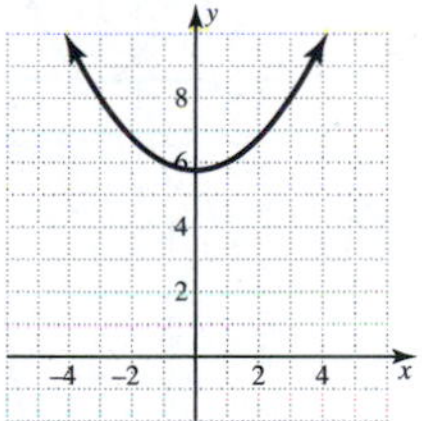

12. no x-intercept: y-intercept: $(0, 6.25)$; vertex: $(0, 6.25)$; axis of symmetry: $x = 0$; domain: $(-\infty, \infty)$; range: $[6.25, \infty)$; shifts up 6.25 units

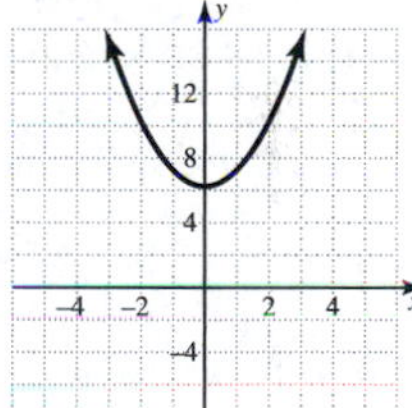

13. x-intercept: $(-3, 0)$; y-intercept: $(0, 9)$; vertex: $(-3, 0)$; axis of symmetry: $x = -3$; domain: $(-\infty, \infty)$; range: $[0, \infty)$; shifts left 3 units

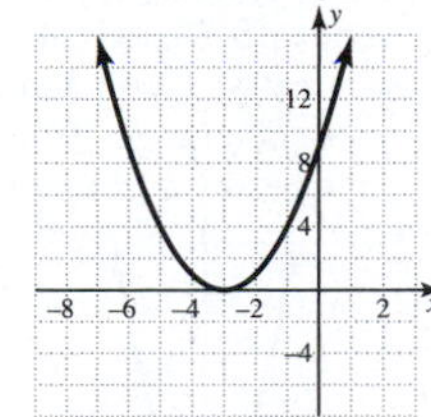

14. x-intercept: $(-5, 0)$; y-intercept: $(0, 25)$; vertex: $(-5, 0)$; axis of symmetry: $x = -5$; domain: $(-\infty, \infty)$; range: $[0, \infty)$; shifts left 5 units

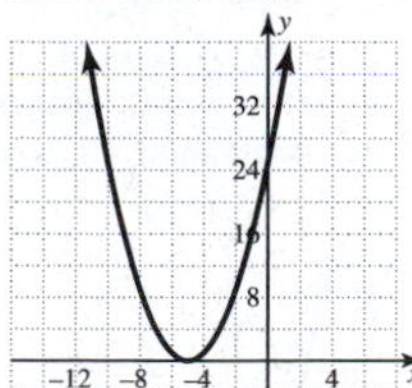

15. x-intercept: (4, 0); y-intercept: (0, 16); vertex: (4, 0); axis of symmetry: $x = 4$; domain: $(-\infty, \infty)$; range: $[0, \infty)$; shifts right 4 units

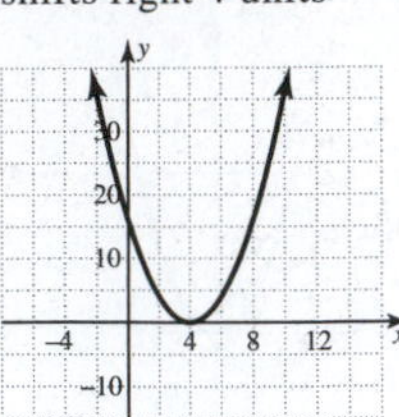

16. x-intercept: (6, 0); y-intercept: (0, 36); vertex: (6, 0); axis of symmetry: $x = 6$; domain: $(-\infty, \infty)$; range: $[0, \infty)$; shifts right 6 units

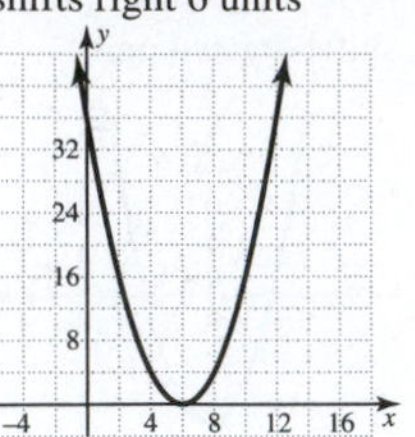

17. x-intercept: $\left(-\frac{3}{2}, 0\right)$; y-intercept: $\left(0, \frac{9}{4}\right)$; vertex: $\left(-\frac{3}{2}, 0\right)$; axis of symmetry: $x = -\frac{3}{2}$; domain: $(-\infty, \infty)$; range: $[0, \infty)$; shifts left $\frac{3}{2}$ units

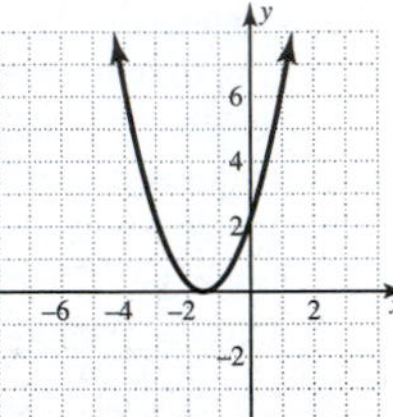

18. x-intercept: $\left(-\frac{3}{4}, 0\right)$; y-intercept: $\left(0, \frac{9}{16}\right)$; vertex: $\left(-\frac{3}{4}, 0\right)$; axis of symmetry: $x = -\frac{3}{4}$; domain: $(-\infty, \infty)$; range: $[0, \infty)$; shifts left $\frac{3}{4}$ units

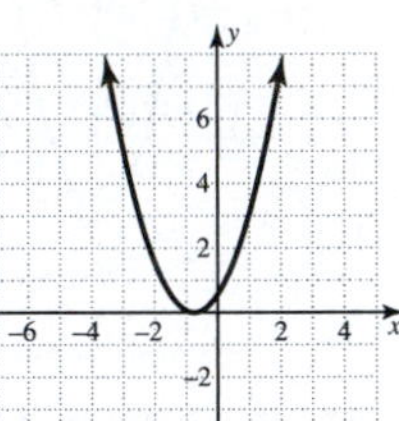

19. x-intercept: $\left(\frac{5}{8}, 0\right)$; y-intercept: $\left(0, \frac{25}{64}\right)$; vertex: $\left(\frac{5}{8}, 0\right)$; axis of symmetry: $x = \frac{5}{8}$; domain: $(-\infty, \infty)$; range: $[0, \infty)$; shifts right $\frac{5}{8}$ units

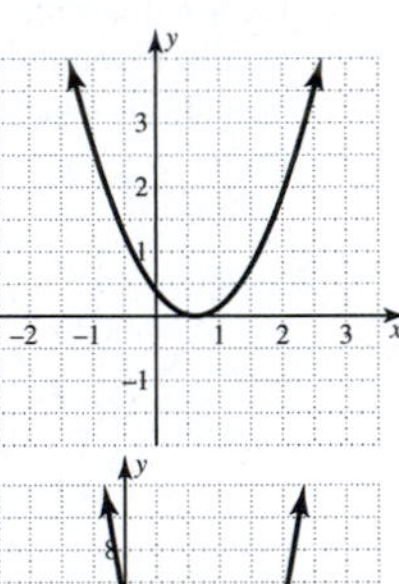

20. x-intercept: $\left(\frac{5}{2}, 0\right)$; y-intercept: $\left(0, \frac{25}{4}\right)$; vertex: $\left(\frac{5}{2}, 0\right)$; axis of symmetry: $x = \frac{5}{2}$; domain: $(-\infty, \infty)$; range: $[0, \infty)$; shifts right $\frac{5}{2}$ units

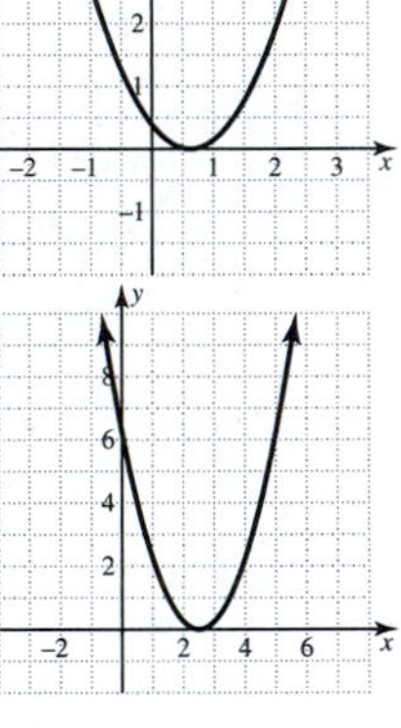

21. x-intercept: (0, 0); y-intercept: (0, 0); vertex: (0, 0); axis of symmetry: $x = 0$; domain: $(-\infty, \infty)$; range: $[0, \infty)$; stretches by a factor of 4

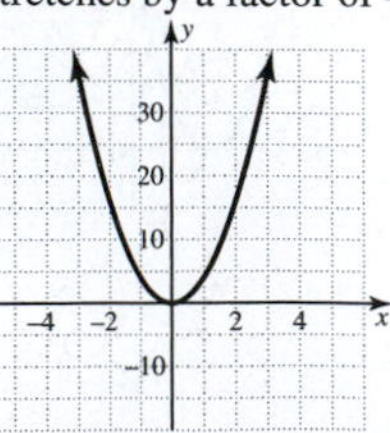

22. x-intercept: (0, 0); y-intercept: (0, 0); vertex: (0, 0); axis of symmetry: $x = 0$; domain: $(-\infty, \infty)$; range: $[0, \infty)$; stretches by a factor of 5

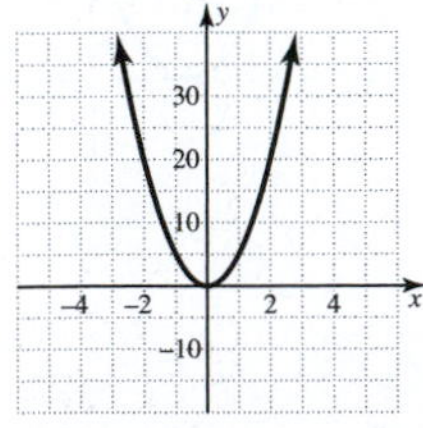

23. x-intercept: (0, 0); y-intercept: (0, 0); vertex: (0, 0); axis of symmetry: $x = 0$; domain: $(-\infty, \infty)$; range: $[0, \infty)$; shrinks by a factor of $\frac{1}{4}$

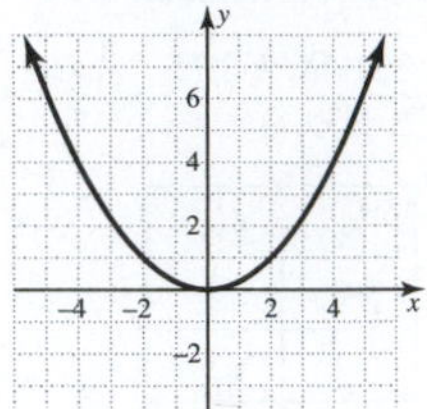

24. x-intercept: (0, 0); y-intercept: (0, 0); vertex: (0, 0); axis of symmetry: $x = 0$; domain: $(-\infty, \infty)$; range: $[0, \infty)$; shrinks by a factor of $\frac{1}{5}$

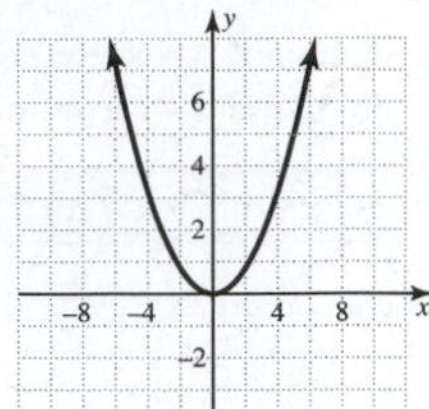

25. x-intercept: (0, 0); y-intercept: (0, 0); vertex: (0, 0); axis of symmetry: $x = 0$; domain: $(-\infty, \infty)$; range: $(-\infty, 0]$; stretches by a factor of 3 and reflects over the x-axis

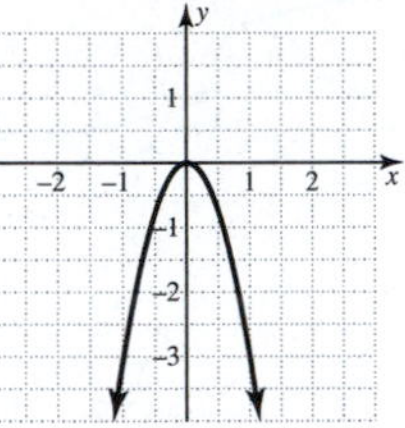

26. x-intercept: (0, 0); y-intercept: (0, 0); vertex: (0, 0); axis of symmetry: $x = 0$; domain: $(-\infty, \infty)$; range: $(-\infty, 0]$; stretches by a factor of 6 and reflects over the x-axis

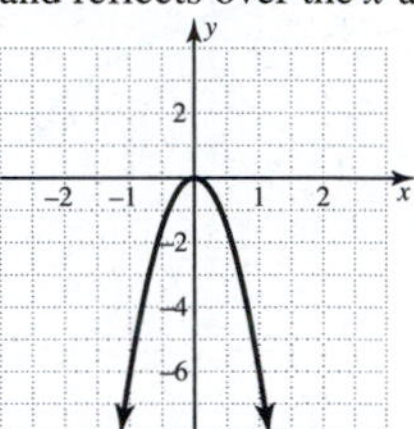

27. x-intercept: (0, 0); y-intercept: (0, 0); vertex: (0, 0); axis of symmetry: $x = 0$; domain: $(-\infty, \infty)$; range: $(-\infty, 0]$; shrinks by a factor of $\frac{1}{5}$ and reflects over the x-axis

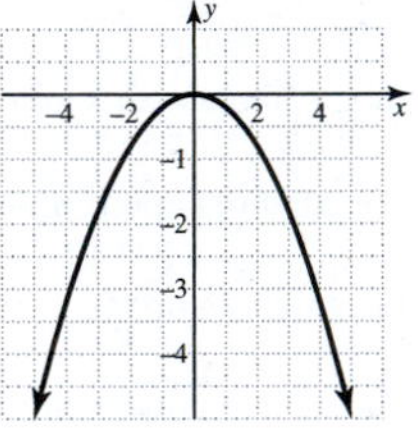

28. x-intercept: (0, 0); y-intercept: (0, 0); vertex: (0, 0); axis of symmetry: $x = 0$; domain: $(-\infty, \infty)$; range: $(-\infty, 0]$; shrinks by a factor of $\frac{1}{10}$ and reflects over the x-axis

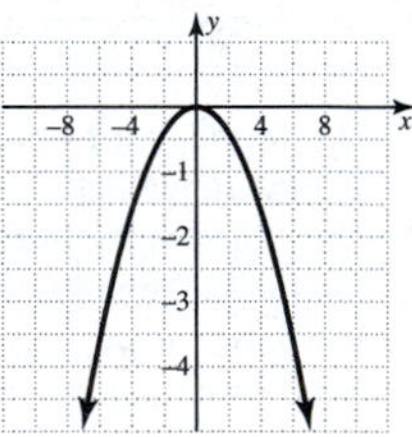

29. x-intercept: (0, 0); y-intercept: (0, 0); vertex: (0, 0), axis of symmetry: $x = 0$; domain: $(-\infty, \infty)$; range: $(-\infty, 0]$; stretches by a factor of 2.5 and reflects over the x-axis

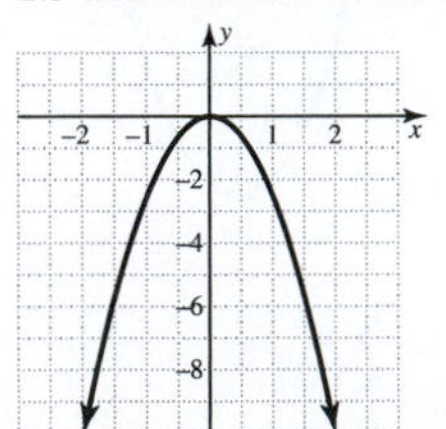

30. x-intercept: (0, 0); y-intercept: (0, 0); vertex: (0, 0); axis of symmetry: $x = 0$; domain: $(-\infty, \infty)$; range: $(-\infty, 0]$; stretches by a factor of 1.6 and reflects over the x-axis

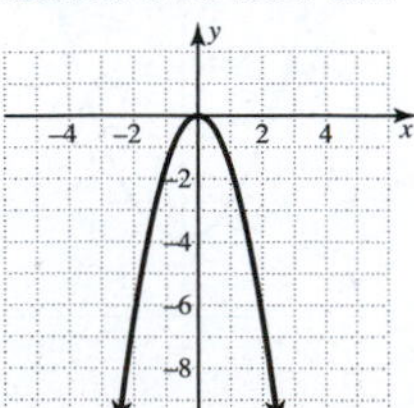

31. x-intercepts: $(-2, 0)$ and $(4, 0)$; y-intercept: $(0, -8)$; vertex: $(1, -9)$; axis of symmetry: $x = 1$; domain: $(-\infty, \infty)$; range: $[-9, \infty)$; shifts right 1 unit and down 9 units

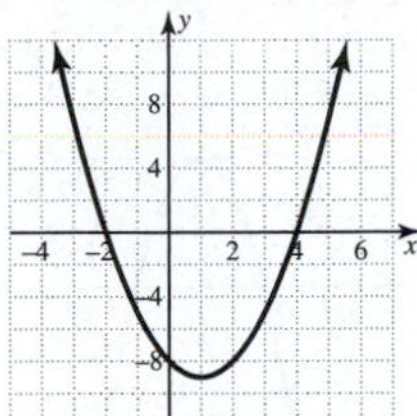

32. x-intercepts: $(1, 0)$ and $(5, 0)$; y-intercept: $(0, 5)$; vertex: $(3, -4)$; axis of symmetry: $x = 3$; domain: $(-\infty, \infty)$; range: $[-4, \infty)$; shifts right 3 units and down 4 units

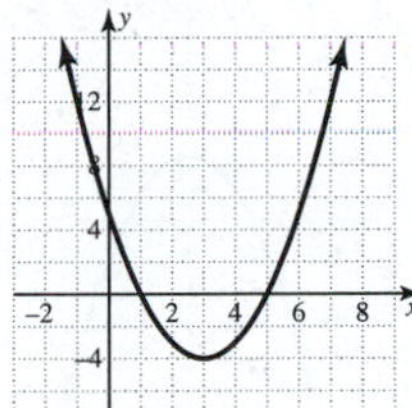

33. no x-intercept; y-intercept: $(0, 13)$; vertex: $(-3, 4)$; axis of symmetry: $x = -3$; domain: $(-\infty, \infty)$; range: $[4, \infty)$; shifts left 3 units and up 4 units

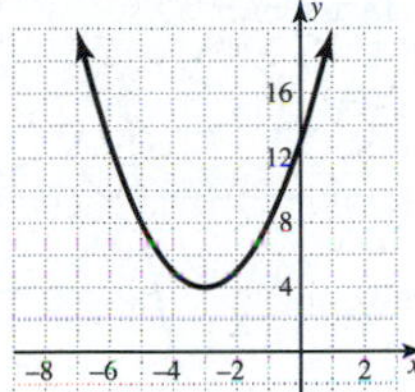

34. no x-intercept; y-intercept: $(0, 9)$; vertex: $(-2, 5)$; axis of symmetry: $x = -2$; domain: $(-\infty, \infty)$; range: $[5, \infty)$; shifts left 2 units and up 5 units

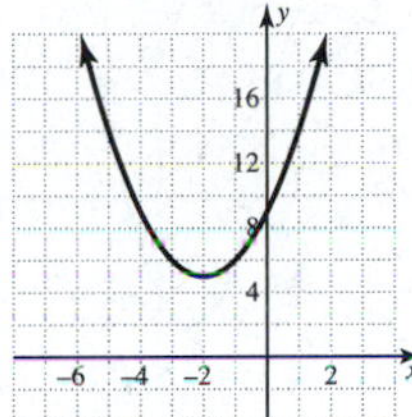

35. no x-intercepts; y-intercept: $(0, -27)$; vertex: $(5, -2)$; axis of symmetry: $x = 5$; domain: $(-\infty, \infty)$; range: $(-\infty, 2]$; reflects over the the x-axis, and shifts right 5 units and down 2 units

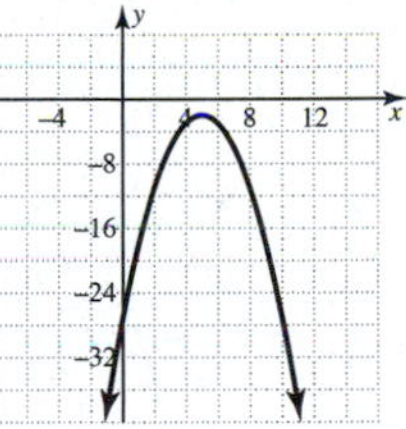

36. no x-intercepts; y-intercept: $(0, -18)$; vertex: $(-4, -2)$; axis of symmetry: $x = -4$; domain: $(-\infty, \infty)$; range: $(-\infty, -2]$; reflects over the x-axis, and shifts left 4 units and down 2 units

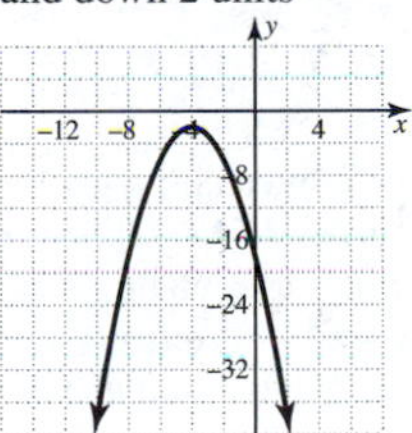

37. x-intercepts: $(-7, 0)$ and $(-1, 0)$; y-intercept: $(0, 14)$; vertex: $(-4, -18)$; axis of symmetry: $x = -4$; domain: $(-\infty, \infty)$; range: $[-18, \infty)$; stretches by a factor of 2, and shifts left 4 units and down 18 units

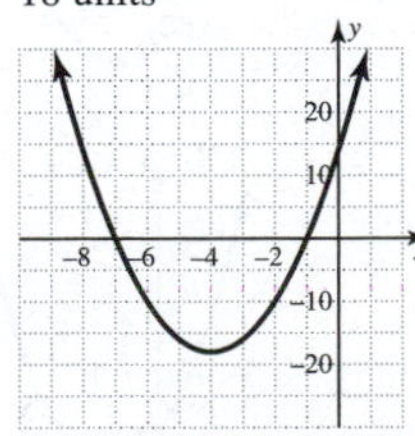

38. x-intercepts: $(0, 0)$ and $(4, 0)$; y-intercept: $(0, 0)$; vertex: $(2, -12)$; axis of symmetry: $x = 2$; domain: $(-\infty, \infty)$; range: $[-12, \infty)$; stretches by a factor of 3, and shifts right 2 units and down 12 units

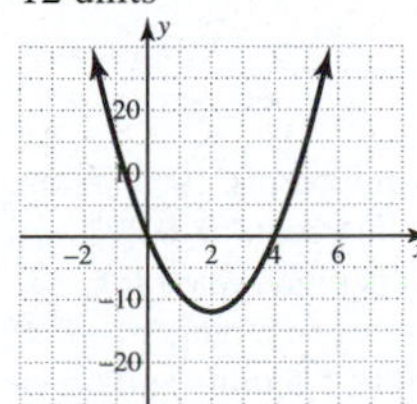

39. no x-intercepts; y-intercept: $(0, 101)$; vertex: $(5, 1)$; axis of symmetry: $x = 5$; domain: $(-\infty, \infty)$; range: $[1, \infty)$; stretches by a factor of 5, and shifts up 1 unit and right 5 units

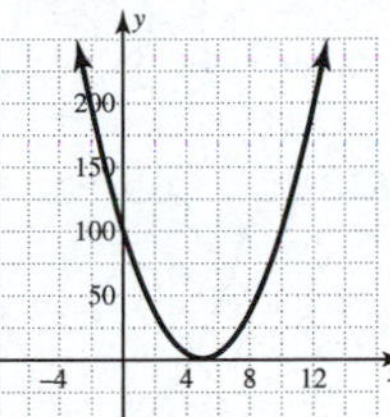

40. no x-intercept; y-intercept: $(0, 13)$; vertex: $(-1, 9)$; axis of symmetry: $x = -1$; domain: $(-\infty, \infty)$; range: $[9, \infty)$; stretches by a factor of 4, and shifts left 1 unit and up 9 units

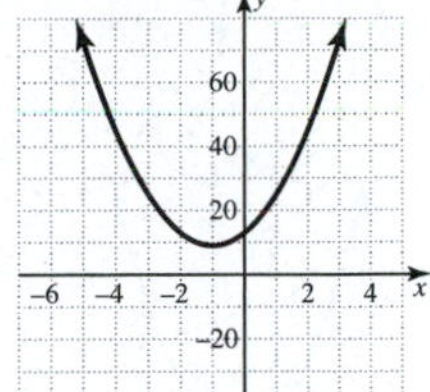

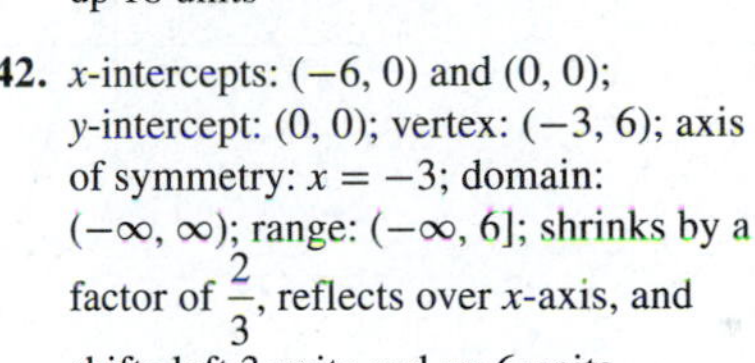

41. x-intercepts: $(-10, 0)$ and $(2, 0)$; y-intercept: $(0, 10)$; vertex: $(-4, 18)$; axis of symmetry: $x = -4$; domain: $(-\infty, \infty)$; range: $(-\infty, 18]$; shrinks by a factor of $\frac{1}{2}$, reflects over the x-axis, and shifts left 4 units and up 18 units

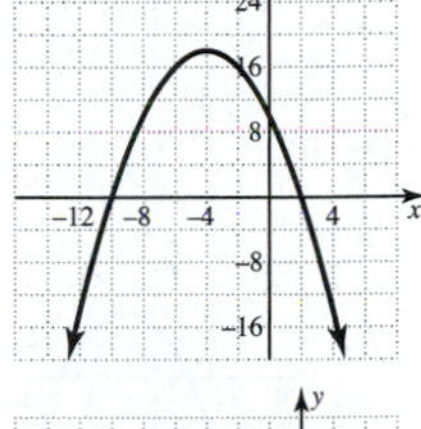

42. x-intercepts: $(-6, 0)$ and $(0, 0)$; y-intercept: $(0, 0)$; vertex: $(-3, 6)$; axis of symmetry: $x = -3$; domain: $(-\infty, \infty)$; range: $(-\infty, 6]$; shrinks by a factor of $\frac{2}{3}$, reflects over x-axis, and shifts left 3 units and up 6 units

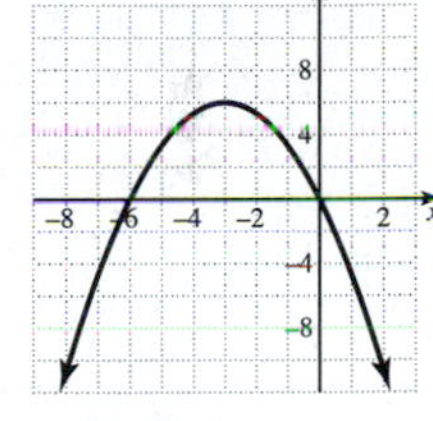

43. x-intercepts: $(-1, 0)$ and $(5, 0)$; y-intercept: $\left(0, -\frac{10}{3}\right)$; vertex: $(2, -6)$; axis of symmetry: $x = 2$; domain: $(-\infty, \infty)$; range: $[-6, \infty)$; shrinks by a factor of $\frac{2}{3}$, and shifts right 2 units and down 6 units

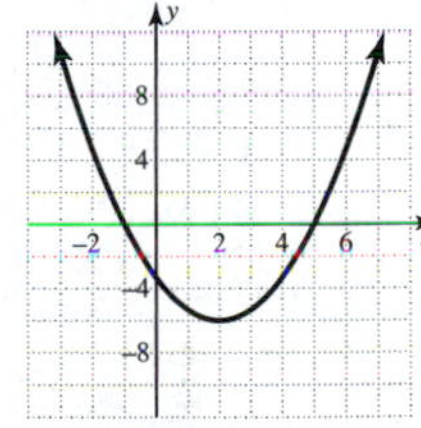

44. x-intercepts: $(-6, 0)$ and $(2, 0)$; y-intercept: $(0, -9)$; vertex: $(-2, -12)$; axis of symmetry: $x = -2$; domain: $(-\infty, \infty)$; range: $[-12, \infty)$; shrinks by a factor of $\frac{3}{4}$, and shifts left 2 units and down 12 units

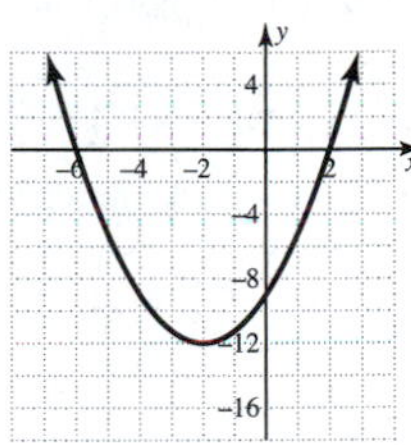

45.

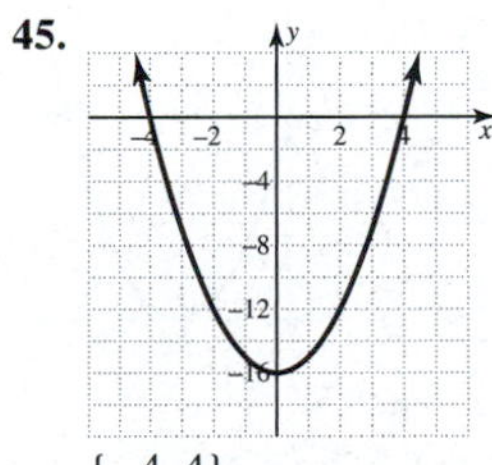

$\{-4, 4\}$

46.

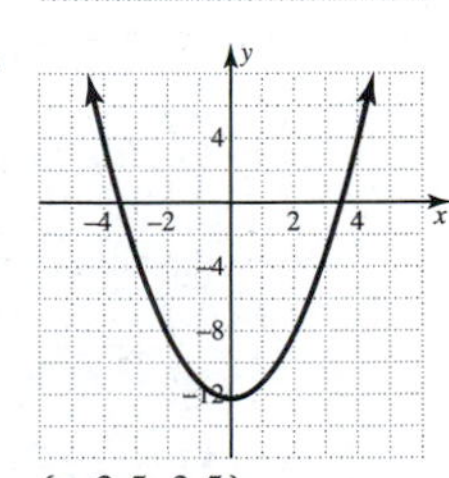

$\{-3.5, 3.5\}$

47.

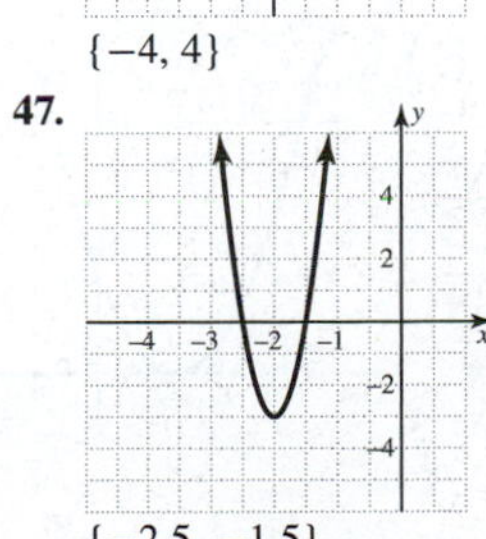

$\{-2.5, -1.5\}$

48.

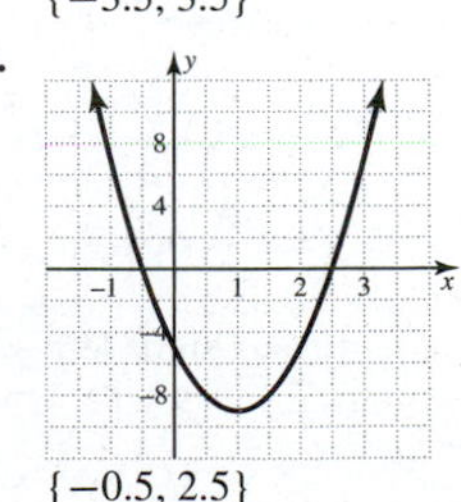

$\{-0.5, 2.5\}$

49.

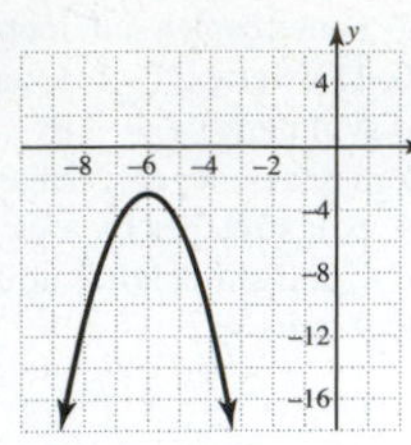

$\varnothing$

50.

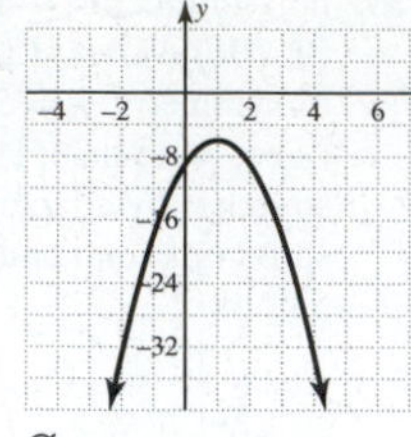

$\varnothing$

51.

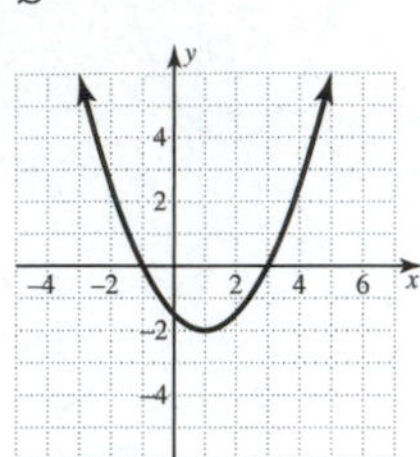

$\{-1, 3\}$

52.

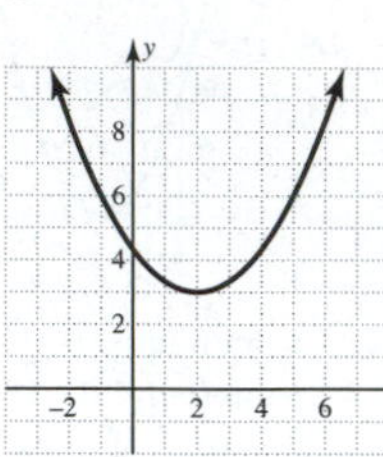

$\varnothing$

53.

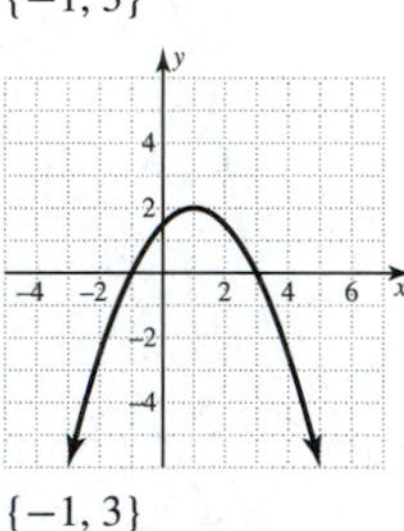

$\{-1, 3\}$

54.

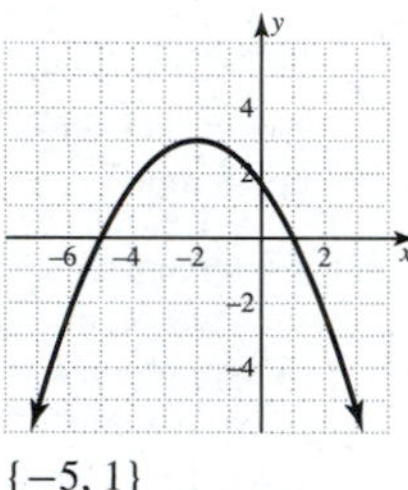

$\{-5, 1\}$

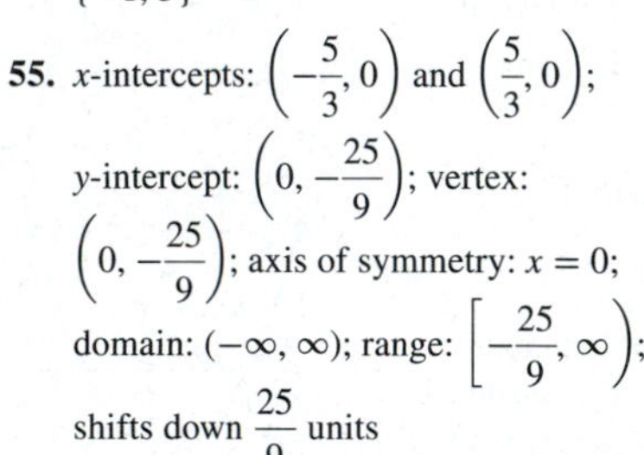

55. x-intercepts: $\left(-\frac{5}{3}, 0\right)$ and $\left(\frac{5}{3}, 0\right)$; y-intercept: $\left(0, -\frac{25}{9}\right)$; vertex: $\left(0, -\frac{25}{9}\right)$; axis of symmetry: $x = 0$; domain: $(-\infty, \infty)$; range: $\left[-\frac{25}{9}, \infty\right)$; shifts down $\frac{25}{9}$ units

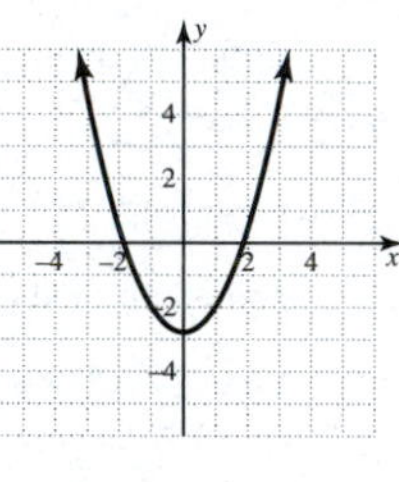

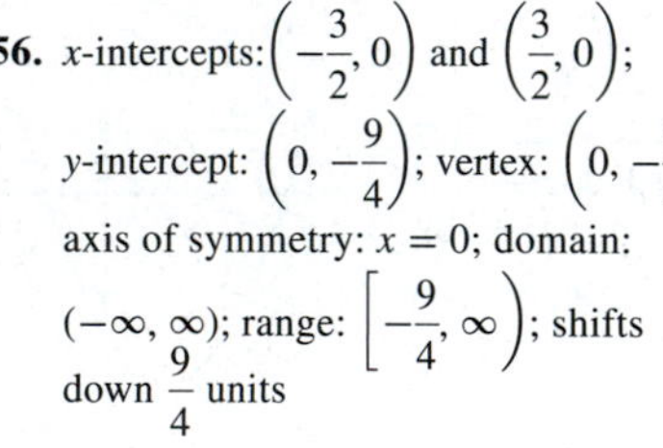

56. x-intercepts: $\left(-\frac{3}{2}, 0\right)$ and $\left(\frac{3}{2}, 0\right)$; y-intercept: $\left(0, -\frac{9}{4}\right)$; vertex: $\left(0, -\frac{9}{4}\right)$; axis of symmetry: $x = 0$; domain: $(-\infty, \infty)$; range: $\left[-\frac{9}{4}, \infty\right)$; shifts down $\frac{9}{4}$ units

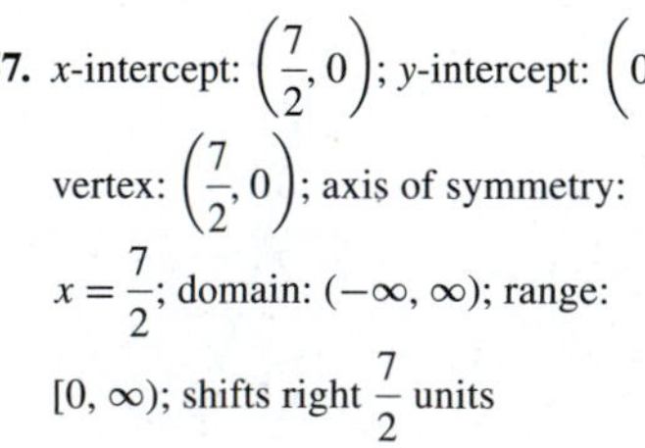

57. x-intercept: $\left(\frac{7}{2}, 0\right)$; y-intercept: $\left(0, \frac{49}{4}\right)$; vertex: $\left(\frac{7}{2}, 0\right)$; axis of symmetry: $x = \frac{7}{2}$; domain: $(-\infty, \infty)$; range: $[0, \infty)$; shifts right $\frac{7}{2}$ units

58. x-intercept: $\left(-\frac{5}{2}, 0\right)$; y-intercept: $\left(0, \frac{25}{4}\right)$; vertex: $\left(-\frac{5}{2}, 0\right)$; axis of symmetry: $x = -\frac{5}{2}$; domain: $(-\infty, \infty)$; range: $[0, \infty)$; shifts left $\frac{5}{2}$ units

59. x-intercepts: $(-1, 0)$ and $(1, 0)$; y-intercept: $(0, 1)$; vertex: $(0, 1)$; axis of symmetry: $x = 0$; domain: $(-\infty, \infty)$; range: $(-\infty, 1]$; reflects over the x-axis and shifts up 1 unit

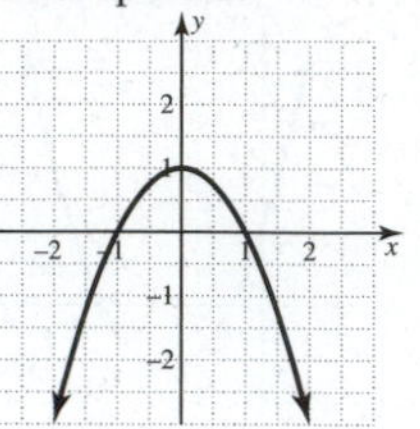

60. x-intercepts: $(-2, 0)$ and $(2, 0)$; y-intercept: $(0, 4)$; vertex: $(0, 4)$; axis of symmetry: $x = 0$; domain: $(-\infty, \infty)$; range: $[-\infty, 4)$; reflects over the x-axis and shifts up 4 units

61. x-intercepts: $(-3, 0)$ and $(1, 0)$; y-intercept: $(0, -6)$, vertex: $(-1, -8)$; axis of symmetry: $x = -1$; domain: $(-\infty, \infty)$; range: $[-8, \infty)$; shifts down 8 units and left 1 unit, and stretches by a factor of 2

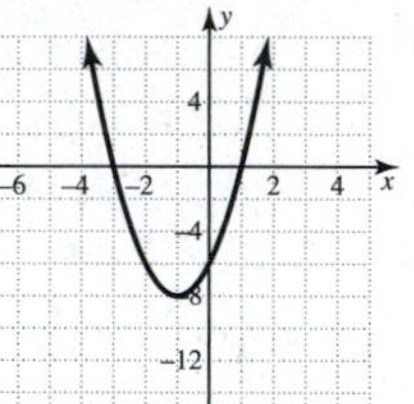

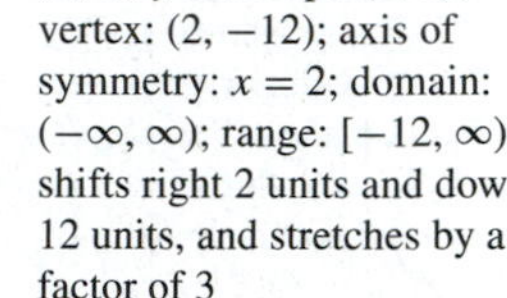

62. x-intercepts: $(0, 0)$ and $(4, 0)$; y-intercept: $(0, 0)$; vertex: $(2, -12)$; axis of symmetry: $x = 2$; domain: $(-\infty, \infty)$; range: $[-12, \infty)$; shifts right 2 units and down 12 units, and stretches by a factor of 3

63. x-intercepts: $(4.5, 0)$ and $(5.5, 0)$; y-intercept: $(0, -198)$; vertex: $(5, 2)$; axis of symmetry: $x = 5$; domain: $(-\infty, \infty)$; range: $(-\infty, 2]$; reflects over the x-axis, shifts right 5 units and up 2 units, and stretches by a factor of 8

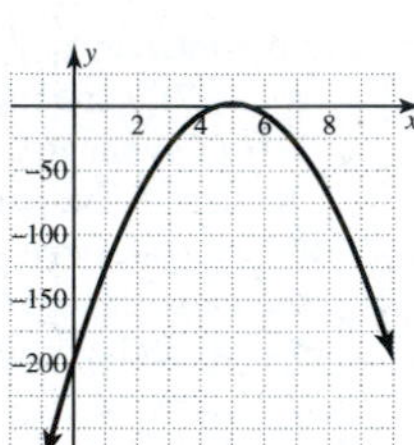

64. no x-intercept; y-intercept: $(0, -20)$; vertex: $(-1, -2)$; axis of symmetry: $x = -1$; domain: $(-\infty, \infty)$; range: $(-\infty, -2]$; reflects over the x-axis, shifts left 1 unit and down 2 units, and stretches by a factor of 18

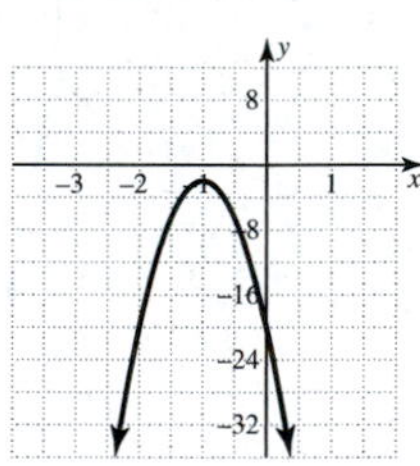

65. x-intercepts: $(-1.1, 0)$ and $(6.1, 0)$; y-intercept: $(0, -6.71)$; vertex: $(2.5, -12.96)$; axis of symmetry: $x = 2.5$; domain: $(-\infty, \infty)$; range: $[-12.96, \infty)$; shifts right 2.5 units and down 12.96 units

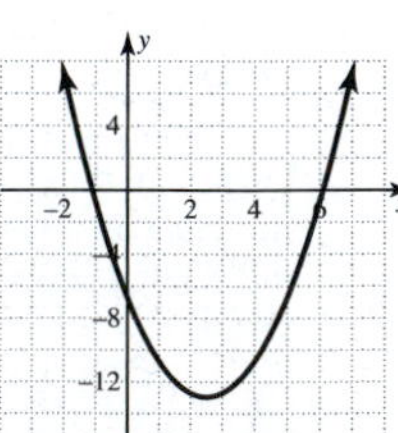

66. x-intercepts: $(0.1, 0)$ and $(2.9, 0)$; y-intercept: $(0, 0.29)$ vertex: $(1.5, -1.96)$; axis of symmetry: $x = 1.5$; domain: $(-\infty, \infty)$; range: $[-1.96, \infty)$; shifts right 1.5 units and down 1.96 units

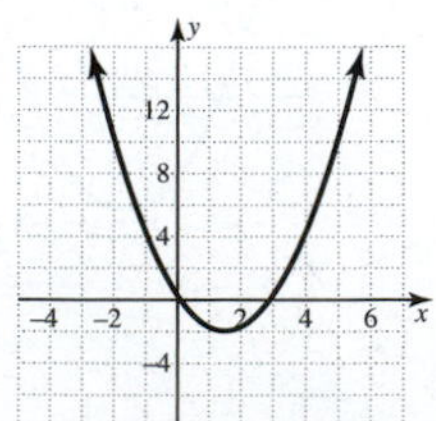

67. x-intercepts: $(-3, 0)$ and $(1, 0)$; y-intercept: $\left(0, \frac{3}{2}\right)$; vertex: $(-1, 2)$; axis of symmetry: $x = -1$; domain: $(-\infty, \infty)$; range: $(-\infty, 2]$; reflects over the x-axis, shifts left 1 unit and up 2 units, and shrinks by a factor of $\frac{1}{2}$

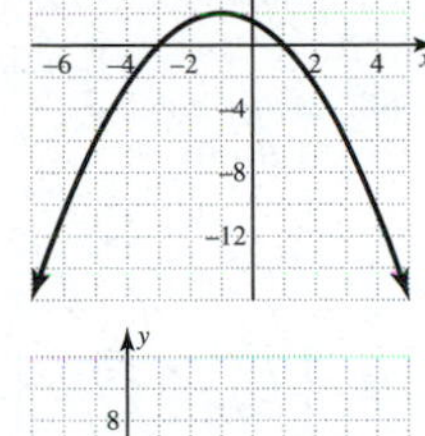

68. x-intercepts: $(1, 0)$ and $(5, 0)$; y-intercept: $(0, -7.5)$; vertex: $(3, 6)$; axis of symmetry: $x = 3$; domain: $(-\infty, \infty)$; range: $(-\infty, 6]$; reflects over the x-axis, shifts right 3 units and up 6 units, and stretches by a factor of 1.5

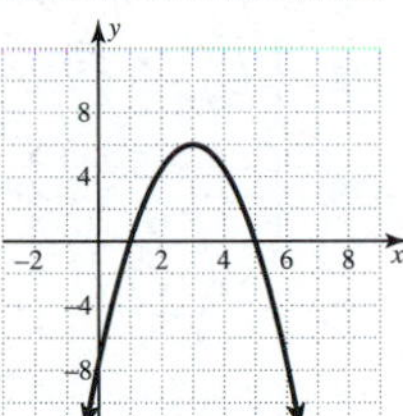

69. no x-intercepts; y-intercept: $\left(0, \frac{1}{4}\right)$; vertex: $\left(0, \frac{1}{4}\right)$; axis of symmetry: $x = 0$; domain: $(-\infty, \infty)$; range: $\left[\frac{1}{4}, \infty\right)$; shifts up $\frac{1}{4}$ unit

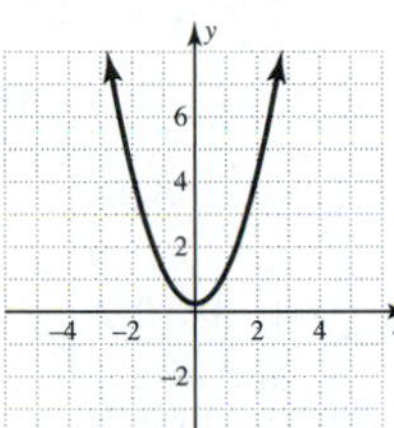

70. no x-intercept; y-intercept: $\left(0, \frac{1}{9}\right)$; vertex: $\left(0, \frac{1}{9}\right)$; axis of symmetry: $x = 0$; domain: $(-\infty, \infty)$; range: $\left[\frac{1}{9}, \infty\right)$; shifts up $\frac{1}{9}$ unit

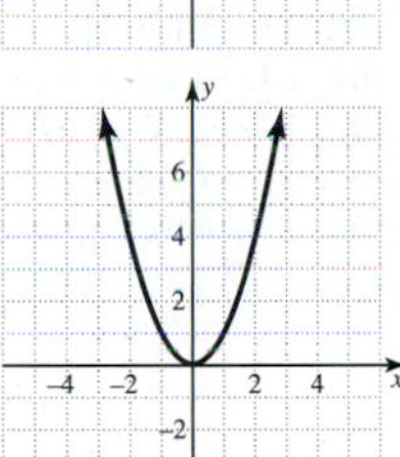

71. no x-intercept; y-intercept: $\left(0, \frac{5}{4}\right)$; vertex: $(1, 1)$; axis of symmetry: $x = 1$; domain: $(-\infty, \infty)$; range: $[1, \infty)$; shifts right 1 unit and up 1 unit, and shrinks by a factor of $\frac{1}{4}$

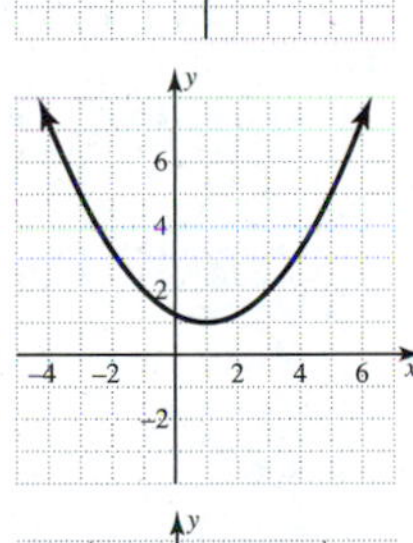

72. no x-intercept; y-intercept: $\left(0, \frac{3}{2}\right)$; vertex: $(1, 1)$; axis of symmetry: $x = 1$; domain: $(-\infty, \infty)$; range: $[1, \infty)$; shifts right 1 unit and up 1 unit, and shrinks by a factor of $\frac{1}{2}$

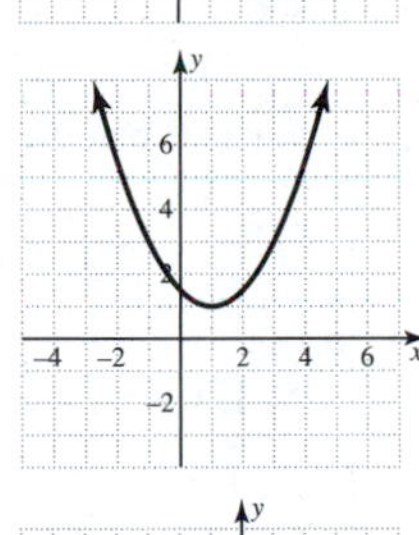

73. x-intercepts: $(-6.3, 0)$ and $(2.7, 0)$; y-intercept: $(0, 17.01)$; vertex: $(-1.8, 20.25)$; axis of symmetry: $x = -1.8$; domain: $(-\infty, \infty)$; range: $(-\infty, 20.25]$; reflects over the x-axis, and shifts left 1.8 units and up 20.25 units

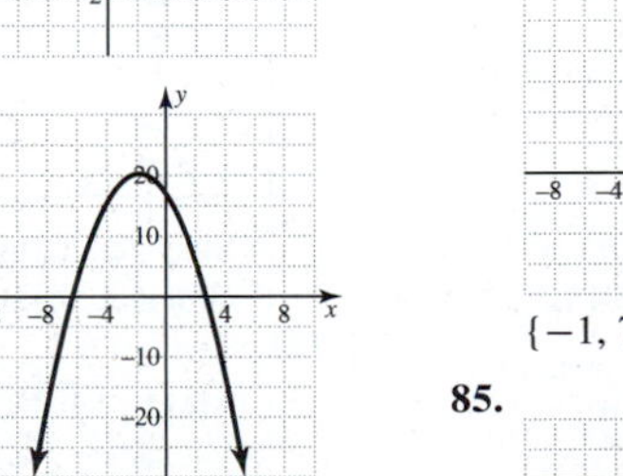

74. x-intercepts: $(-5.7, 0)$ and $(-0.7, 0)$; y-intercept: $(0, -3.99)$; vertex: $(-3.2, 6.25)$; axis of symmetry: $x = -3.2$; domain: $(-\infty, \infty)$; range: $(-\infty, 6.25]$; reflects over the x-axis, and shifts left 3.2 units and up 6.25 units

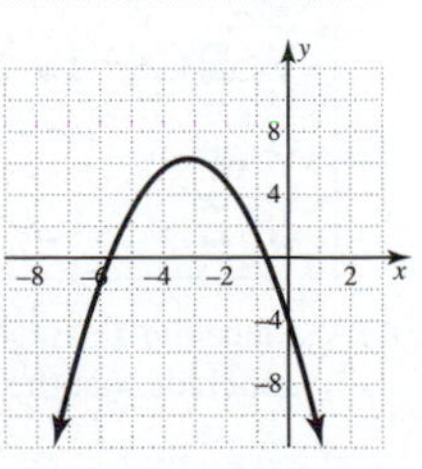

75. x-intercepts: $(-0.6, 0)$ and $(3, 0)$; y-intercept: $(0, -3.6)$; vertex: $(1.2, -6.48)$; axis of symmetry: $x = 1.2$; domain: $(-\infty, \infty)$; range: $[-6.48, \infty)$; shifts right 1.2 units and down 6.48 units, and stretches by a factor of 2

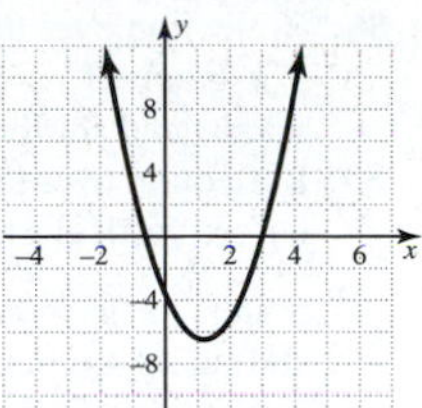

76. x-intercepts: $(-4.2, 0)$ and $(1.2, 0)$; y-intercept: $(0, -15.12)$; vertex: $(-1.5, -21.87)$; axis of symmetry: $x = -1.5$; domain: $(-\infty, \infty)$; range: $[-21.87, \infty)$; shifts left 1.5 units and down 21.87 units, and stretches by a factor of 3

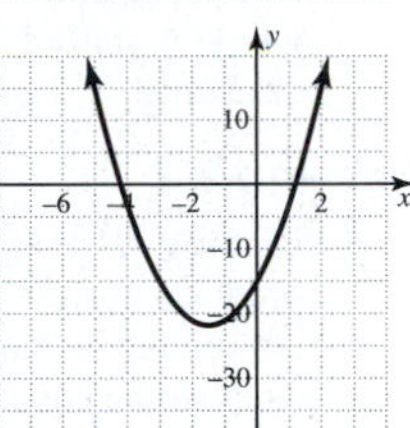

77.

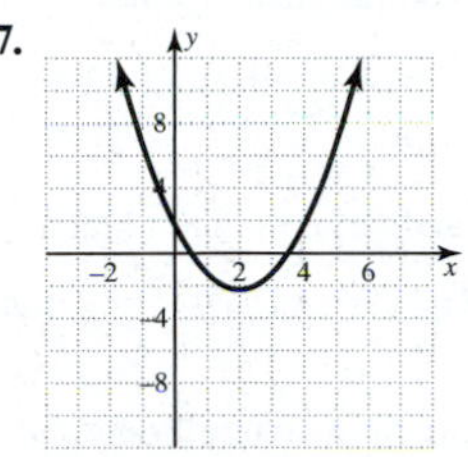

$\{0.5, 3.5\}$

78.

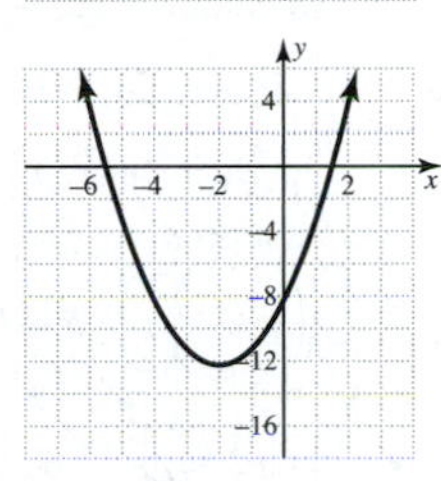

$\{-5.5, 1.5\}$

79.

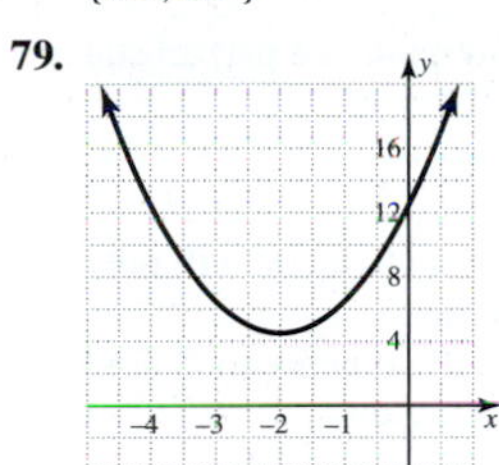

$\varnothing$

80.

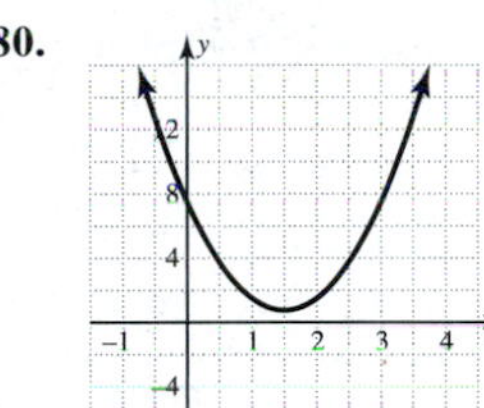

$\varnothing$

81.

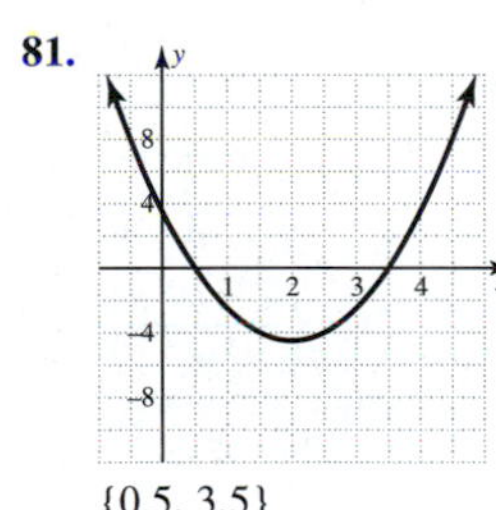

$\{0.5, 3.5\}$

82.

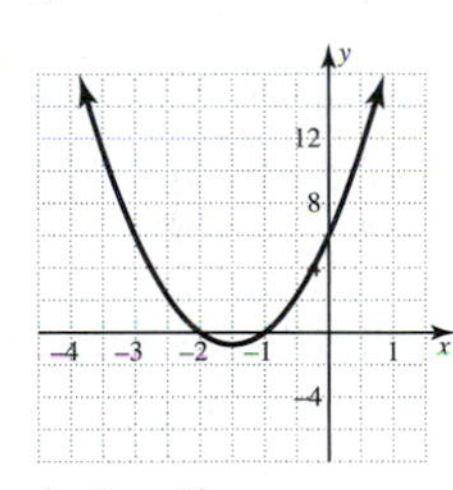

$\{-2, -1\}$

83.

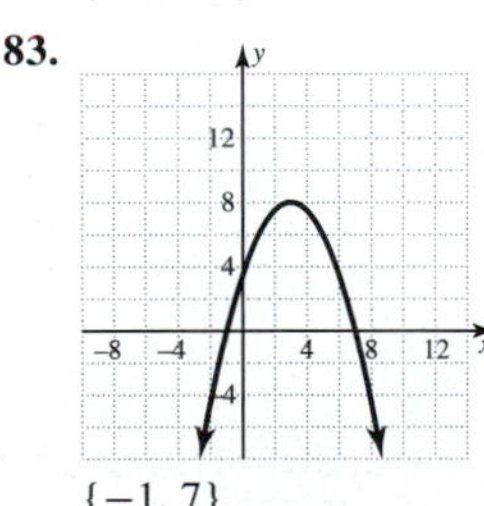

$\{-1, 7\}$

84.

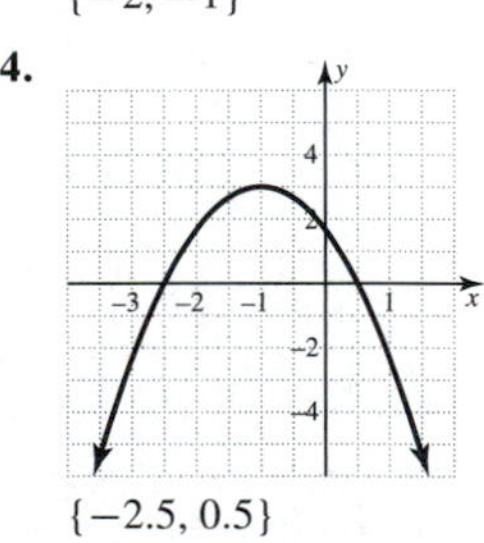

$\{-2.5, 0.5\}$

85.

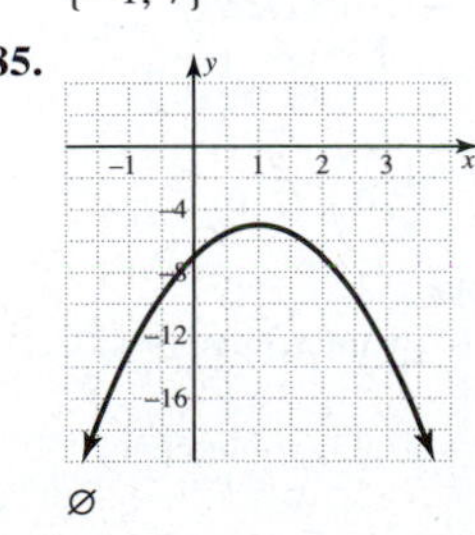

$\varnothing$

86.

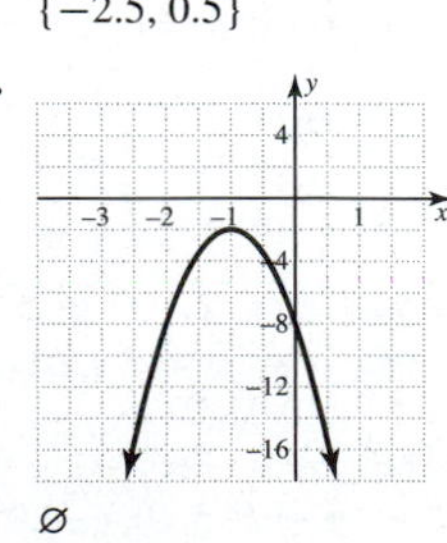

$\varnothing$

88. Reflection over the x-axis must occur before vertical translation. The graph of $y = x^2$ is shifted left 9 units, reflected over the x-axis, and shifted down 4 units.

89. Reflection over the x-axis and stretching/shrinking must occur before vertical translation. The graph of $y = x^2$ is shifted right 5 units, reflected over the x-axis, shrunk by a factor of $\frac{1}{2}$, and shifted up 4 units.

90. Stretching must occur before vertical translations. The graph of $y = x^2$ is shifted left 7 units, stretched by a factor of 2, and shifted up 3 units.

Section 9.2

1. Factor $x^2 - a^2$ as a difference of two squares and use the zero products property to solve. To apply the square root property, rewrite the equation as $x^2 = a^2$ and take the square root of both sides of the equation.

2. To complete the square for $x^2 + 6x$, add the square of half of the coefficient of x, that is, $\left(\frac{6}{2}\right)^2 = 9$, to the binomial and it becomes a perfect square trinomial.

3. First take out 2 as a common factor and write the binomial as $2\left(x^2 - \frac{3}{2}x\right)$. Add the square of half of the coefficient of x, that is, $\left(\frac{1}{2} \cdot \frac{3}{2}\right)^2$ or $\frac{9}{16}$, to the binomial inside the parentheses and get $2\left(x^2 - \frac{3}{2}x + \frac{9}{16}\right)$. Write the expression as a perfect square trinomial $2\left(x - \frac{3}{4}\right)^2$.

5. The graph of the corresponding quadratic function will not cross the x-axis; that is, there are no x-intercepts.

6. The graph of the corresponding quadratic function will cross the x-axis twice; that is, there are two x-intercepts.

11. $\left\{-\frac{1}{7}, \frac{1}{7}\right\}$ **12.** $\left\{-\frac{9}{5}, \frac{9}{5}\right\}$ **13.** $\left\{-\frac{2}{9}, \frac{2}{9}\right\}$

14. $\left\{-\frac{8}{3}, \frac{8}{3}\right\}$ **19.** $\left\{-1, \frac{3}{2}\right\}$ **20.** $\left\{-\frac{9}{5}, -1\right\}$

21. $\left\{\frac{4}{7}, \frac{16}{7}\right\}$ **22.** $\left\{-\frac{13}{4}, -\frac{3}{4}\right\}$ **23.** $\left\{\frac{1 \pm 4\sqrt{3}}{3}\right\}$

24. $\left\{\frac{7 \pm \sqrt{6}}{4}\right\}$ **25.** $\left\{\frac{-5 \pm 5i\sqrt{6}}{6}\right\}$ **26.** $\left\{\frac{2 \pm i\sqrt{3}}{9}\right\}$

31. $r^2 + r + \frac{1}{4} = \left(r + \frac{1}{2}\right)^2$ **32.** $s^2 + 7s + \frac{49}{4} = \left(s + \frac{7}{2}\right)^2$

33. $a^2 - 9a + \frac{81}{4} = \left(a - \frac{9}{2}\right)^2$ **34.** $b^2 - 15b + \frac{225}{4} = \left(b - \frac{15}{2}\right)^2$

41. $\left\{-\frac{1}{2}, \frac{3}{4}\right\}$ **42.** $\left\{-\frac{7}{3}, 6\right\}$ **43.** $\left\{-7, -\frac{3}{2}\right\}$

44. $\left\{-6, -\frac{3}{4}\right\}$ **45.** $\left\{\frac{-4 \pm \sqrt{11}}{2}\right\}$ **46.** $\left\{\frac{-6 \pm \sqrt{15}}{3}\right\}$

47. $\left\{\frac{-4 \pm i\sqrt{10}}{2}\right\}$ **48.** $\left\{\frac{-3 \pm i\sqrt{15}}{4}\right\}$

59. zeros: $x = -13, x = -1$; x-intercepts: $(-13, 0), (-1, 0)$

60. zeros: $x = -1, x = 17$; x-intercepts: $(-1, 0), (17, 0)$

61. zeros: $x = -1, x = -\frac{1}{3}$; x-intercepts: $(-1, 0), \left(-\frac{1}{3}, 0\right)$

62. zeros: $x = -\frac{1}{3}, x = 1$; x-intercepts: $\left(-\frac{1}{3}, 0\right), (1, 0)$

64. zeros: $x = 11 \pm 2i\sqrt{5}$; no x-intercepts

65. zeros: $x = 3 \pm 5\sqrt{2} \approx 10.07$ or -4.07; x-intercepts: $(-4.07, 0)$, $(10.07, 0)$

66. zeros: $x = -11 \pm 4\sqrt{3} \approx -4.07$ or -17.93; x-intercepts: $(-17.93, 0), (-4.07, 0)$

67. zeros: $x = \frac{-1 \pm \sqrt{3}}{10} \approx 0.07$ or -0.27; x-intercepts: $(-0.27, 0)$, $(0.07, 0)$

68. zeros: $x = \frac{1 \pm 5\sqrt{2}}{5} \approx 1.61$ or -1.21; x-intercepts: $(-1.21, 0)$, $(1.61, 0)$

69. zeros: $x = \frac{-5 \pm 2i\sqrt{21}}{3}$, no x-intercepts **71.** $\left\{\pm\frac{5}{8}\right\}$

72. $\left\{\pm\frac{3}{7}\right\}$ **77.** $\left\{-\frac{17}{2}, -\frac{3}{2}\right\}$ **78.** $\left\{\frac{5}{4}, \frac{11}{4}\right\}$

79. $\left\{\frac{-2 \pm 5i\sqrt{3}}{9}\right\}$ **80.** $\left\{\frac{-3 \pm 3i\sqrt{5}}{7}\right\}$ **81.** $\left\{\frac{3 \pm \sqrt{21}}{4}\right\}$

82. $\left\{\frac{6 \pm \sqrt{7}}{3}\right\}$ **83.** $\left\{\frac{3 \pm i\sqrt{2}}{4}\right\}$ **84.** $\left\{\frac{-8 \pm i\sqrt{10}}{3}\right\}$

85. $\left\{\frac{1 \pm 10\sqrt{3}}{6}\right\}$ **86.** $\left\{\frac{4 \pm 3\sqrt{15}}{5}\right\}$ **89.** $\left\{-\frac{3}{2}, 2\right\}$

90. $\left\{-\frac{1}{2}, \frac{3}{4}\right\}$ **91.** $\left\{-\frac{7}{2}, 0\right\}$ **92.** $\left\{0, \frac{3}{5}\right\}$

93. $\left\{\frac{1 \pm i\sqrt{15}}{4}\right\}$ **94.** $\left\{\frac{-1 \pm i\sqrt{7}}{4}\right\}$ **95.** $\left\{\frac{3 \pm \sqrt{6}}{2}\right\}$

96. $\left\{\frac{-1 \pm \sqrt{5}}{4}\right\}$ **97.** $\left\{\frac{4 \pm i\sqrt{3}}{3}\right\}$ **98.** $\left\{\frac{-5 \pm i\sqrt{14}}{2}\right\}$

99. $x^2 + 8x + 16 = (x + 4)^2$ **100.** $y^2 - 26y + 169 = (y - 13)^2$

101. $a^2 - 1.8a + 0.81 = (a - 0.9)^2$

102. $b^2 - 0.6b + 0.09 = (b - 0.3)^2$

103. $a^2 + 0.9a + 0.2025 = (a + 0.45)^2$

104. $s^2 + 0.5s + 0.0625 = (s + 0.25)^2$

105. $r^2 + \frac{1}{2}r + \frac{1}{16} = \left(r + \frac{1}{4}\right)^2$ **106.** $s^2 + \frac{5}{3}s + \frac{25}{36} = \left(s + \frac{5}{6}\right)^2$

107. $a^2 - \frac{3}{2}a + \frac{9}{16} = \left(a - \frac{3}{4}\right)^2$ **108.** $b^2 - \frac{7}{4}b + \frac{49}{64} = \left(b - \frac{7}{8}\right)^2$

119.

```
Plot1 Plot2 Plot3
\Y1=X²-15X+225/4
\Y2=(X-15/2)²
\Y3=
\Y4=
\Y5=
\Y6=
```

X	Y1	Y2
-4	132.25	132.25
-3	110.25	110.25
-2	90.25	90.25
-1	72.25	72.25
0	56.25	56.25
1	42.25	42.25
2	30.25	30.25

Y2=(X-15/2)²

Since the Y_1 and Y_2 columns agree for each value of x, we can conclude that the two expressions are the same.

120.

```
Plot1 Plot2 Plot3
\Y1=X²-7/10X+49/
400
\Y2=(X-7/20)²
\Y3=
\Y4=
\Y5=
\Y6=
```

X	Y1	Y2
-4	18.923	18.923
-3	11.223	11.223
-2	5.5225	5.5225
-1	1.8225	1.8225
0	.1225	.1225
1	.4225	.4225
2	2.7225	2.7225

Y2=(X-7/20)²

Since the Y_1 and Y_2 columns agree for each value of x, we can conclude that the two expressions are the same.

121.

```
Plot1 Plot2 Plot3
\Y1=X²-2.3X+1.32
25
\Y2=(X-1.15)²
\Y3=
\Y4=
\Y5=
\Y6=
```

X	Y1	Y2
0	1.3225	1.3225
1	.0225	.0225
2	.7225	.7225
3	3.4225	3.4225
4	8.1225	8.1225
5	14.823	14.823
6	23.523	23.523

Y2=(X-1.15)²

Since the Y_1 and Y_2 columns agree for each value of x, we can conclude that the two expressions are the same.

122.

```
Plot1 Plot2 Plot3
\Y1■X²-5/7X+25/1
96
\Y2■(X-5/14)²
\Y3=
\Y4=
\Y5=
\Y6=
```

X	Y1	Y2
0	.12755	.12755
1	.41327	.41327
2	2.699	2.699
3	6.9847	6.9847
4	13.27	13.27
5	21.556	21.556
6	31.842	31.842

Y2■(X-5/14)²

Since the Y_1 and Y_2 columns agree for each value of x, we can conclude that the two expressions are the same.

Section 9.3

1. I would use the zero products property to solve the equation since the equation is factored and set equal to zero.
2. I would use the square root property since the equation is of the form $a^2 = k$.
3. I would use completing the square since the coefficient of x^2 is 1 and the coefficient of x is even.
4. I would use the quadratic formula since the equation cannot be factored and the coefficient of x is odd.
5. The advantage is that we can just substitute values into a formula and not have to complete the square when the coefficient of x^2 is not 1 and the coefficient of x is not even.
7. If the discriminant is positive, there are two real solutions. If the discriminant is zero, there is one real solution. If the discriminant is negative, there are no real solutions.

11. $\left\{\dfrac{9 \pm \sqrt{29}}{2}\right\}$ **15.** $\left\{-\dfrac{11}{2}, 0\right\}$ **16.** $\left\{0, \dfrac{2}{3}\right\}$

17. $\left\{\dfrac{1 \pm i\sqrt{10}}{9}\right\}$ **23.** $\left\{\dfrac{5 \pm \sqrt{23}}{6}\right\}$ **43.** $\left\{-\dfrac{10}{3}, 2\right\}$

45. $\left\{\dfrac{8 \pm \sqrt{10}}{2}\right\}$ **47.** $\left\{\dfrac{3 \pm i\sqrt{15}}{2}\right\}$ **48.** $\left\{\dfrac{-4 \pm i\sqrt{14}}{2}\right\}$

49. $\left\{-\dfrac{9}{5}, 0\right\}$ **51.** $\left\{\dfrac{5 \pm \sqrt{15}}{2}\right\}$ **53.** $\left\{\dfrac{1 \pm \sqrt{10}}{4}\right\}$

55. $\left\{-\dfrac{3}{2}, \dfrac{8}{3}\right\}$ **57.** $\left\{\dfrac{1 \pm i\sqrt{2}}{3}\right\}$

73. $x^2 - 3x - 4 = 0$

$$\begin{aligned} x &= \frac{-(-3) \pm \sqrt{(-3)^2 - 4(1)(-4)}}{2(1)} \\ &= \frac{3 \pm \sqrt{9 + 16}}{2} \\ &= \frac{3 \pm \sqrt{25}}{2} \\ &= \frac{3+5}{2} \text{ or } \frac{3-5}{2} \\ &= 4 \text{ or } -1 \end{aligned}$$

74. $4x^2 + 16x + 13 = 0$

$$\begin{aligned} x &= \frac{-(16) \pm \sqrt{(16)^2 - 4(4)(13)}}{2(4)} \\ &= \frac{-16 \pm \sqrt{256 - 208}}{8} \\ &= \frac{-16 \pm \sqrt{48}}{8} \\ &= \frac{-16 \pm 4\sqrt{3}}{8} \\ &= \frac{-4 \pm \sqrt{3}}{2} \end{aligned}$$

75. The quadratic formula was used and simplified correctly. The error was in approximating the solutions.

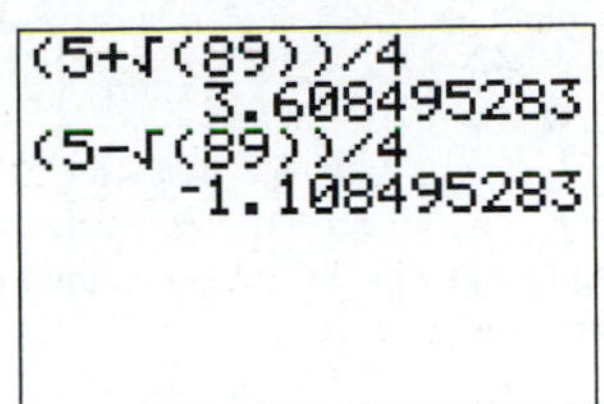

So, the approximate solutions are 3.61 and −1.11.

76. $3x^2 + 5x - 6 = 0$

$$\begin{aligned} x &= \frac{-5 \pm \sqrt{(5)^2 - 4(3)(-6)}}{2(3)} \\ &= \frac{-5 \pm \sqrt{25 + 72}}{6} \\ &= \frac{-5 \pm \sqrt{97}}{6} \end{aligned}$$

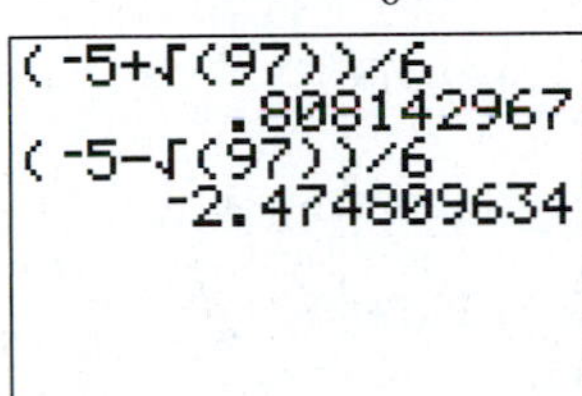

So, the approximate solutions are 0.81 and −2.47.

Piece It Together Sections 9.1–9.3

1. none; (0, 10); (2, 6); $x = 2$; $(-\infty, \infty)$; $[6, \infty)$; It is the graph of $f(x) = x^2$ shifted right 2 units and up 6 units.

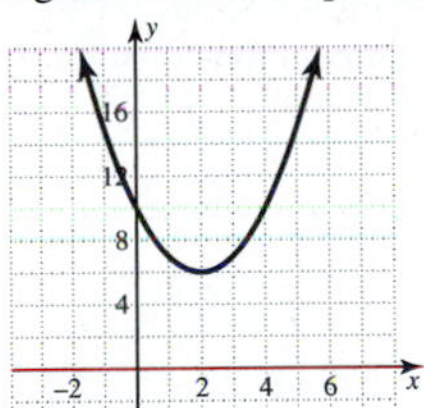

2. (−7, 0) and (−1, 0); (0, 7); (−4, −9); $x = -4$; $(-\infty, \infty)$; $[-9, \infty)$; It is the graph of $f(x) = x^2$ shifted left 4 units and down 9 units.

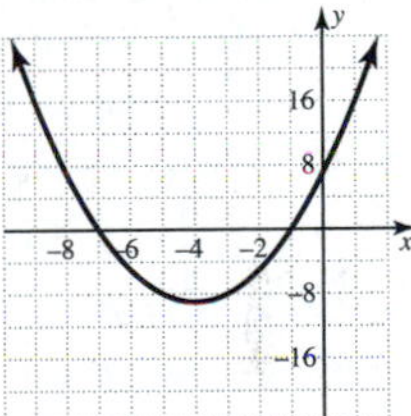

3. The solution set of $f(x) = 0$ is {2, 8}. The zeros of $f(x)$ are $x = 2$ and $x = 8$. The x-intercepts of the graph of $f(x)$ are (2, 0) and (8, 0).

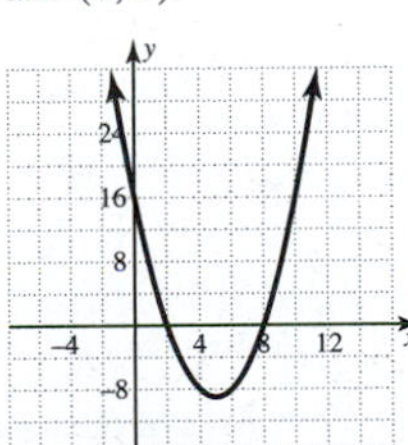

4. The solution set of $f(x) = 0$ is {−4, −2}. The zeros of $f(x)$ are $x = -4$ and $x = -2$. The x-intercepts of the graph of $f(x)$ are (−4, 0) and (−2, 0).

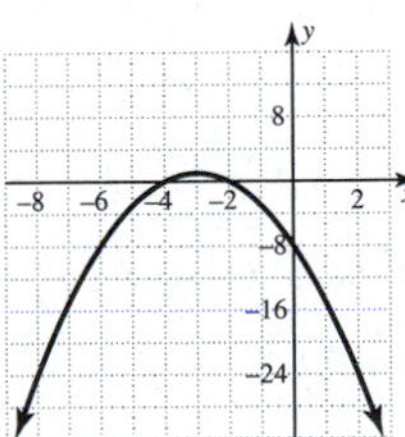

5. $\left\{-\dfrac{3}{7}, \dfrac{1}{7}\right\}$ **6.** $\left\{-\dfrac{7}{3}, \dfrac{7}{3}\right\}$ **7.** $\left\{-2, \dfrac{7}{3}\right\}$ **8.** $\left\{\dfrac{-5 \pm i}{2}\right\}$

Section 9.4

1. Multiply both sides of the equation by the least common denominator, x.

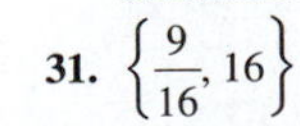

31. $\left\{\dfrac{9}{16}, 16\right\}$ **35.** $\left\{\pm 1, \pm\dfrac{2}{3}\right\}$ **36.** $\left\{\pm 2, \pm\dfrac{1}{2}\right\}$

38. $\left\{\pm\dfrac{1}{2}i, \pm 3i\right\}$ **47.** $\left\{\dfrac{8}{27}, 8\right\}$ **49.** $\left\{-1, -\dfrac{1}{32}\right\}$

65. $\left\{\frac{25}{4}, 9\right\}$ **67.** $\left\{-5, -\frac{5}{4}\right\}$ **77.** $\left\{\pm 1, \pm\frac{5}{2}\right\}$

88. $\left\{\pm\frac{3}{2}, \pm\frac{3}{2}i\right\}$ **91.** $\left\{\frac{5}{2}, \frac{-5 \pm 5i\sqrt{3}}{4}\right\}$

97. After multiplying by the LCD, Jeung didn't distribute the negative sign to the middle product.

$$\frac{2x}{x+4} - \frac{x-6}{x-4} = \frac{32}{x^2-16}$$
$$\frac{2x}{x+4} - \frac{x-6}{x-4} = \frac{32}{(x+4)(x-4)}$$
$$2x(x-4) - (x-6)(x+4) = 32$$
$$2x^2 - 8x - (x^2 - 2x - 24) = 32$$
$$2x^2 - 8x - x^2 + 2x + 24 = 32$$
$$x^2 - 6x - 8 = 0$$
$$x = \frac{-(-6) \pm \sqrt{(-6)^2 - 4(1)(-8)}}{2(1)}$$
$$x = \frac{6 \pm \sqrt{68}}{2}$$
$$x = \frac{6 \pm 2\sqrt{17}}{2}$$
$$x = 3 \pm \sqrt{17}$$

The solution set is $\{3 \pm \sqrt{17}\}$.

98. Mark solved the equation correctly but didn't check his solutions.

$x = \sqrt{5x+6}$	$x = \sqrt{5x+6}$
$6 = \sqrt{5(6)+6}$	$-1 = \sqrt{5(-1)+6}$
$6 = \sqrt{36}$	$-1 = \sqrt{1}$
$6 = 6$	$-1 = 1$
True	False

Since $x = -1$ is an extraneous solution, the solution set is $\{6\}$.

99. Ken didn't apply the square root property correctly. When $x^2 = k$, $x = \pm\sqrt{k}$. There should be four solutions of the equation.

$$4x^4 + 7x^2 - 36 = 0$$
$$(4x^2 - 9)(x^2 + 4) = 0$$
$$4x^2 - 9 = 0 \quad \text{or} \quad x^2 + 4 = 0$$
$$x^2 = \frac{9}{4} \quad \text{or} \quad x^2 = -4$$
$$x = \pm\sqrt{\frac{9}{4}} \quad \text{or} \quad x = \pm\sqrt{-4}$$
$$x = \pm\frac{3}{2} \quad \text{or} \quad x = \pm 2i$$

The solution set is $\left\{\pm\frac{3}{2}, \pm 2i\right\}$.

100. Beth didn't find the solutions for x. We need to back-substitute to find x.

$u = x^{1/3}$	$u = x^{1/3}$
$-1 = x^{1/3}$	$4 = x^{1/3}$
$(-1)^3 = (x^{1/3})^3$	$(4)^3 = (x^{1/3})^3$
$-1 = x$	$64 = x$

The solution set is $\{64, -1\}$.

Section 9.5

1. Subtract c from each side, add $\left(\frac{1}{2}b\right)^2$ to both sides of the equation, and complete the square on the right side of the equation.

$$y = x^2 + bx + c$$
$$y - c = x^2 + bx$$
$$y - c + \frac{b^2}{4} = x^2 + bx + \frac{b^2}{4}$$
$$y - c + \frac{b^2}{4} = \left(x + \frac{b}{2}\right)^2$$
$$y = \left(x + \frac{b}{2}\right)^2 + c - \frac{b^2}{4}$$

The vertex form is $y = \left(x + \frac{b}{2}\right)^2 + c - \frac{b^2}{4}$.

2. Subtract k from each side of the equation. You solve the equation by applying the square root property.

7. The x-coordinate of the vertex, $x = \frac{-b}{2a}$, is midway between the solutions of the quadratic equation;

that is, solutions from the quadratic formula,
$x = \frac{-b - \sqrt{b^2 - 4ac}}{2a}$ and $x = \frac{-b + \sqrt{b^2 - 4ac}}{2a}$

8. The maximum or minimum of a quadratic function occurs at the vertex. The maximum or minimum value of a quadratic function is the y-value of the vertex; that is, $f\left(\frac{-b}{2a}\right)$.

9. $f(x) = (x + 1)^2 - 16$
x-intercepts: (3, 0) and (−5, 0); y-intercept: (0, −15); vertex: (−1, −16)

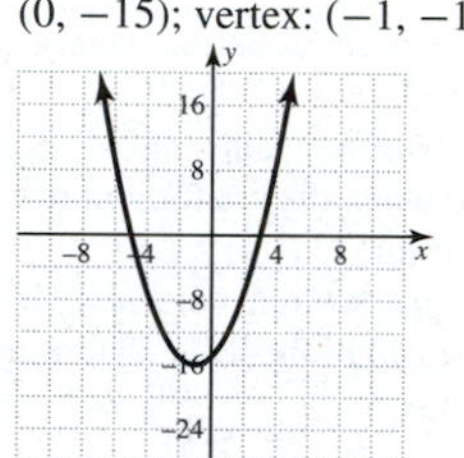

10. $f(x) = (x - 2)^2 - 9$
x-intercepts: (5, 0) and (−1, 0); y-intercept: (0, −5); vertex: (2, −9)

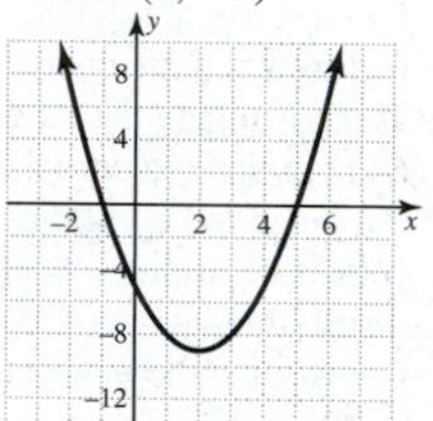

11. $f(x) = (x + 2)^2 - 64$
x-intercepts: (6, 0) and (−10, 0); y-intercept: (0, −60); vertex: (−2, −64)

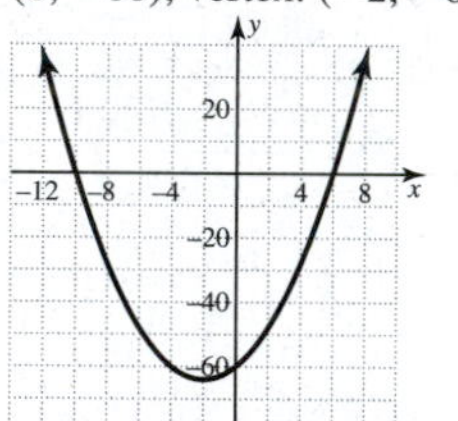

12. $f(x) = (x + 6)^2 - 1$
x-intercepts: (−5, 0) and (−7, 0); y-intercept: (0, 35); vertex: (−6, −1)

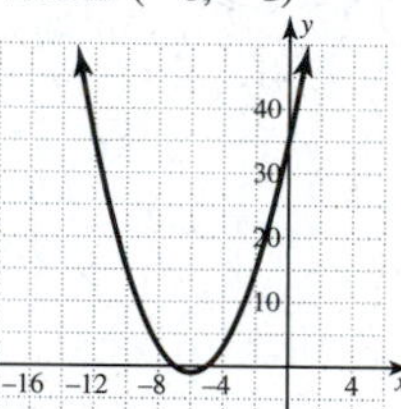

13. $f(x) = \left(x + \frac{3}{2}\right)^2 - \frac{81}{4}$
x-intercepts: (3, 0) and (−6, 0); y-intercept: (0, −18); vertex: $\left(-\frac{3}{2}, \frac{81}{4}\right)$

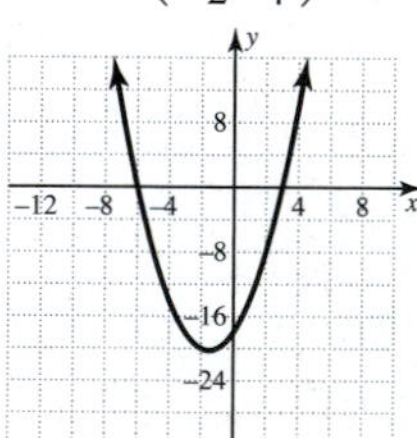

14. $f(x) = \left(x - \frac{7}{2}\right)^2 - \frac{1}{4}$
x-intercepts: (3, 0) and (4, 0); y-intercept: (0, 12); vertex: $\left(\frac{7}{2}, -\frac{1}{4}\right)$

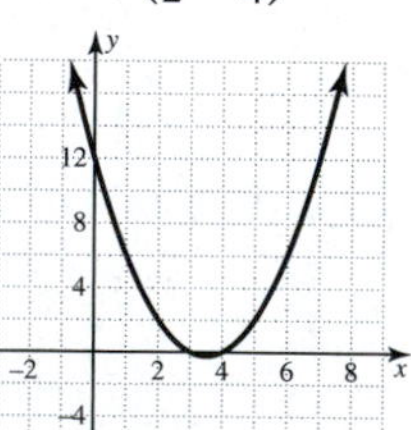

15. $f(x) = -(x + 1)^2 + 9$
x-intercepts: (2, 0) and (−4, 0); y-intercept: (0, 8); vertex: (−1, 9)

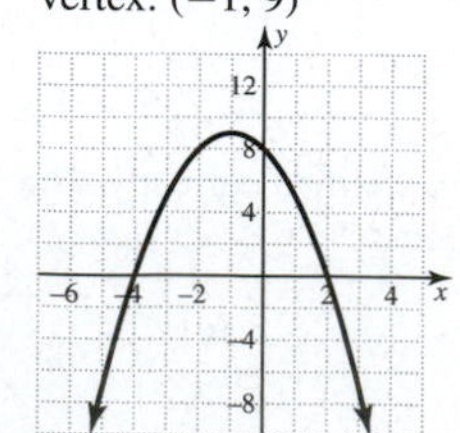

16. $f(x) = -(x + 3)^2 + 16$
x-intercepts: (1, 0) and (−7, 0); y-intercept: (0, 7); vertex: (−3, 16)

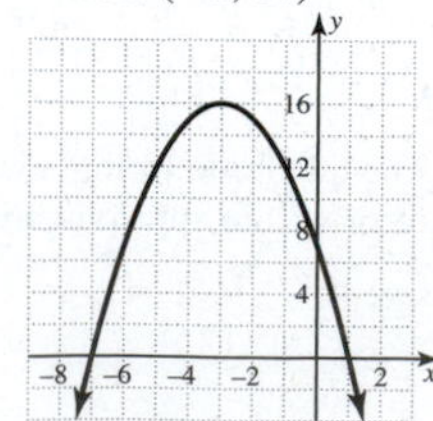

17. $f(x) = -\left(x + \frac{5}{2}\right)^2 + \frac{169}{4}$
x-intercepts: (4, 0) and (−9, 0); y-intercept: (0, 36); vertex: $\left(-\frac{5}{2}, \frac{169}{4}\right)$

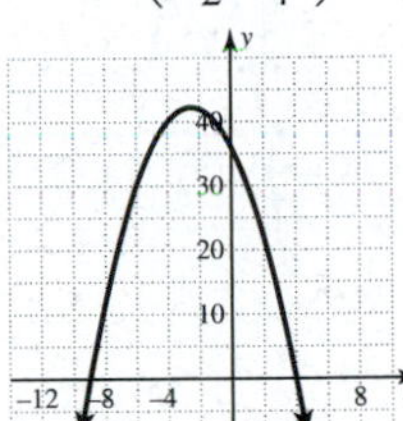

18. $f(x) = -\left(x - \frac{1}{2}\right)^2 + \frac{9}{4}$
x-intercepts: (2, 0) and (−1, 0); y-intercept: (0, 2); vertex: $\left(\frac{1}{2}, \frac{9}{4}\right)$

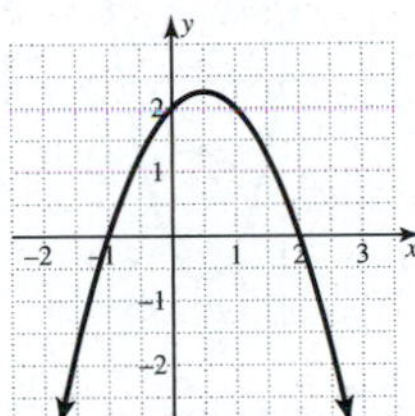

19. $f(x) = 2(x + 3)^2 - 16$
x-intercepts: $(-3 \pm 2\sqrt{2}, 0) \approx$ (−0.17, 0) and (−5.83, 0); y-intercept: (0, 2); vertex: (−3, −16)

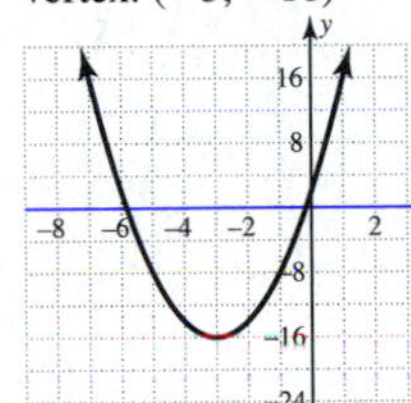

20. $f(x) = 3(x + 1)^2 - 6$
x-intercepts: $(-1 \pm \sqrt{2}, 0) \approx$ (0.41, 0) and (−2.41, 0); y-intercept: (0, −3); vertex: (−1, −6)

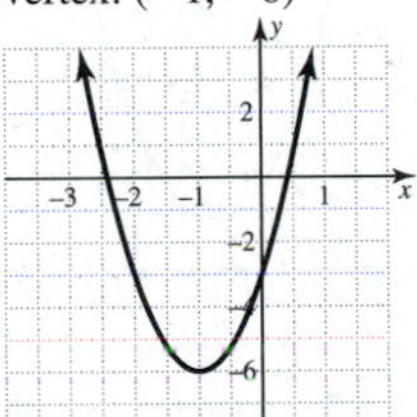

21. $f(x) = -5(x + 1)^2 + 12$
x-intercepts: $\left(-1 \pm \frac{2\sqrt{15}}{5}, 0\right) \approx$ (0.55, 0) and (−2.55, 0); y-intercept: (0, 7); vertex: (−1, 12)

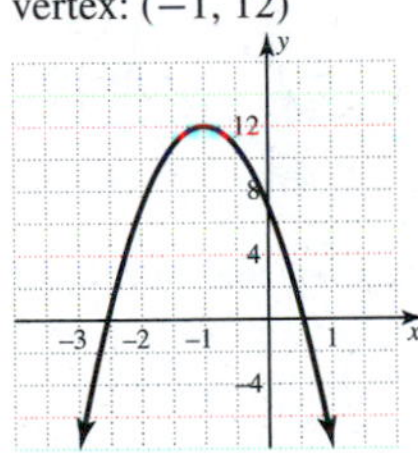

22. $f(x) = -3(x - 2)^2 + 7$
x-intercepts: $\left(2 \pm \frac{\sqrt{21}}{3}, 0\right) \approx$ (3.53, 0) and (0.47, 0); y-intercept: (0, −5); vertex: (2, 7)

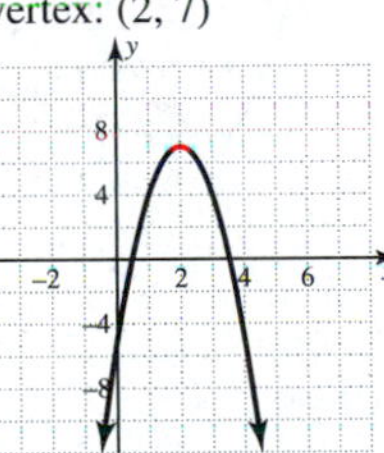

23. $f(x) = -2(x - 3)^2 - 9$
no x-intercepts; y-intercept: (0, −27); vertex: (3, −9)

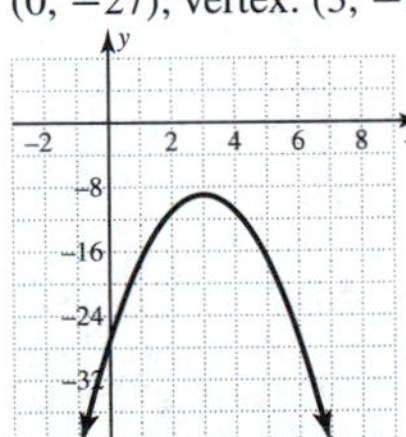

24. $f(x) = -2(x - 2)^2 - 17$
no x-intercepts; y-intercept: (0, −25); vertex: (2, −17)

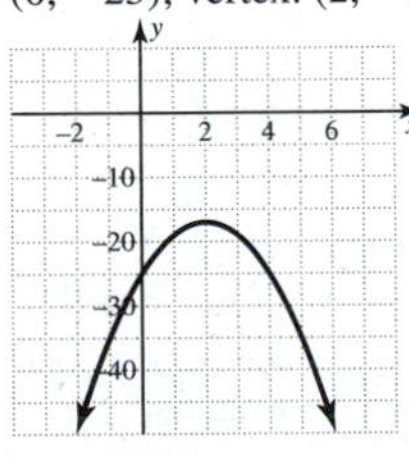

25. $f(x) = 2(x + 2)^2 + 3$
no x-intercepts; y-intercept: (0, 11); vertex: (−2, 3)

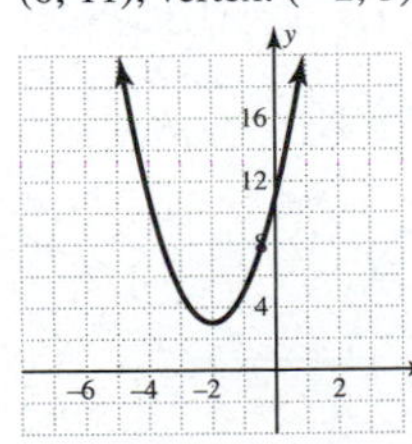

26. $f(x) = 3(x - 1)^2 + 6$
x-intercepts: none; y-intercept: (0, 9); vertex: (1, 6)

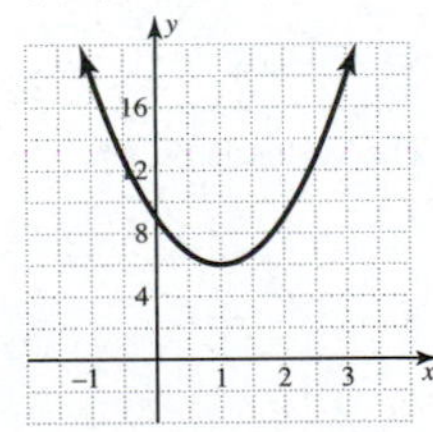

27. $\left(\frac{9}{2}, -\frac{73}{4}\right)$ **28.** $\left(-\frac{5}{2}, \frac{3}{4}\right)$ **29.** $\left(\frac{3}{2}, \frac{11}{2}\right)$

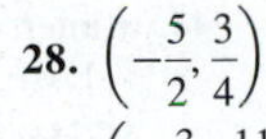

30. $\left(-\frac{5}{6}, \frac{97}{12}\right)$ **31.** $\left(-\frac{3}{10}, \frac{11}{20}\right)$ **32.** $\left(-\frac{1}{5}, -\frac{36}{5}\right)$

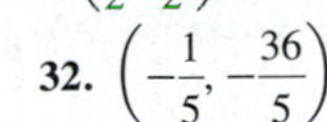

33. x-intercepts: (±3.16, 0); y-intercept: (0, −10); vertex: (0, −10)

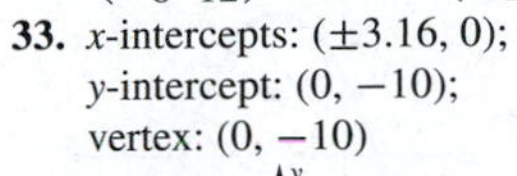

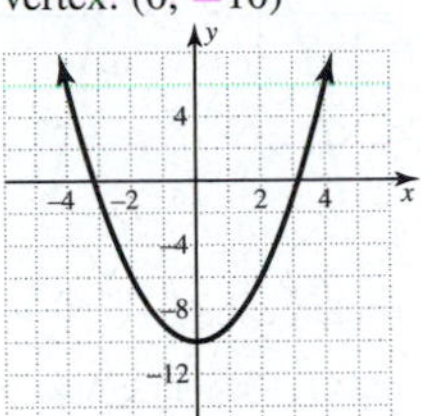

34. x-intercepts: (±2.65, 0); y-intercept: (0, −7); vertex: (0, −7)

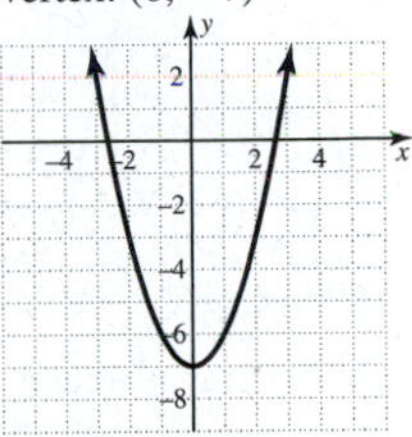

35. no x-intercepts; y-intercept: (0, −5); vertex: (0, −5)

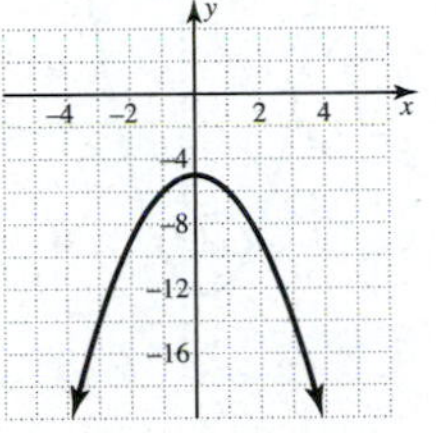

36. no x-intercepts; y-intercept: (0, 3); vertex: (0, 3)

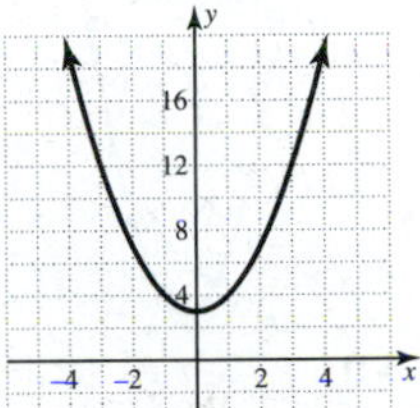

37. x-intercepts: (0, 0) and (−6, 0); y-intercept: (0, 0); vertex: (−3, −9)

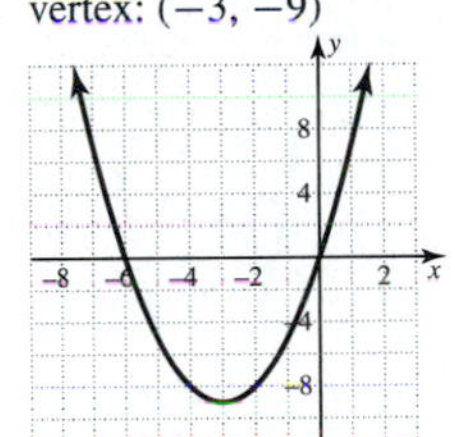

38. x-intercepts: (0, 0) and (5, 0); y-intercept: (0, 0); vertex: $\left(\frac{5}{2}, -\frac{25}{4}\right)$

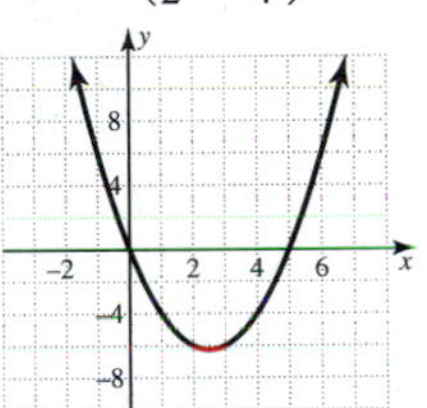

39. x-intercepts: (0, 0) and (7, 0); y-intercept: (0, 0); vertex: $\left(\frac{7}{2}, \frac{49}{4}\right)$

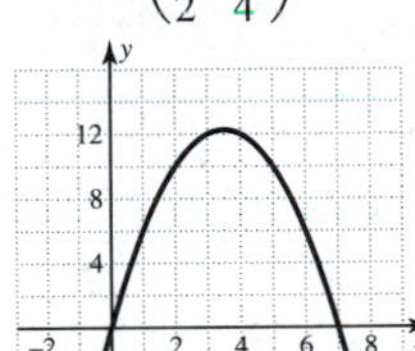

40. x-intercepts: (0, 0) and (−3, 0); y-intercept: (0, 0); vertex: $\left(-\frac{3}{2}, \frac{9}{4}\right)$

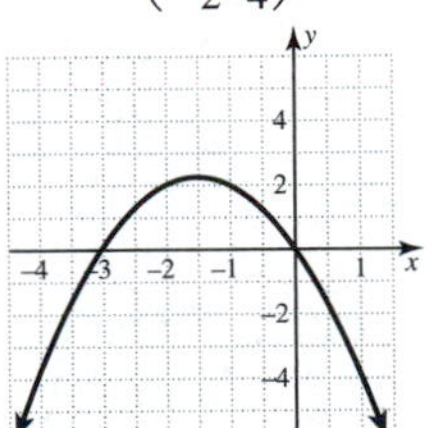

41. x-intercepts: (1, 0) and (3, 0); y-intercept: (0, 3); vertex: (2, −1)

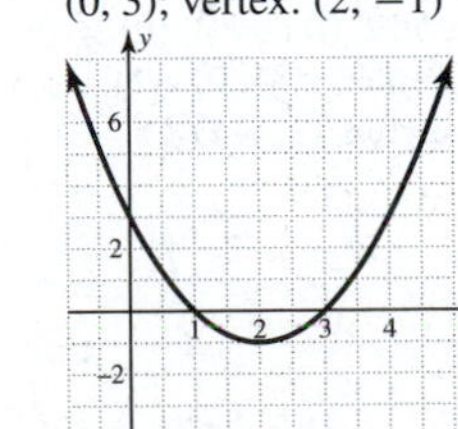

42. x-intercepts: (−1, 0) and (−7, 0); y-intercept: (0, 7); vertex: (−4, −9)

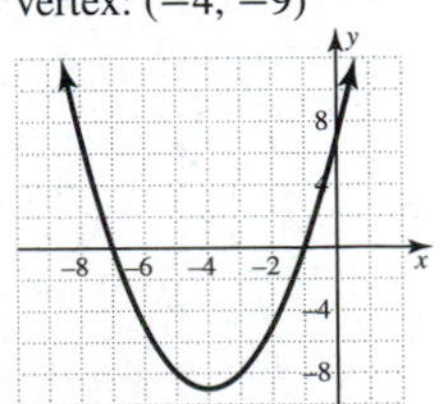

43. x-intercepts: $(-2, 0)$ and $(-4, 0)$; y-intercept: $(0, -8)$; vertex: $(-3, 1)$

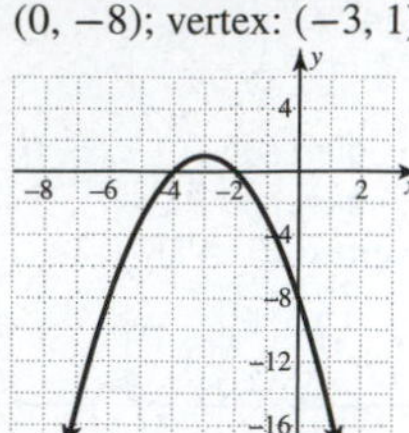

44. x-intercepts: $(5, 0)$ and $(-1, 0)$; y-intercept: $(0, 5)$; vertex: $(2, 9)$

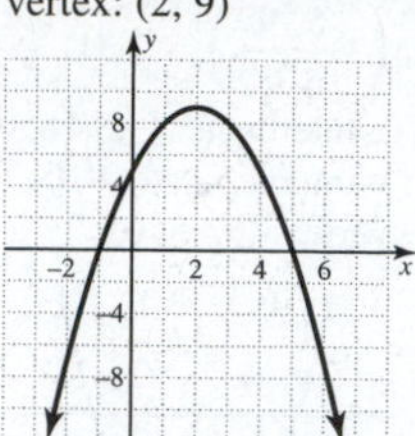

45. x-intercepts: $\left(-\frac{1}{2}, 0\right)$ and $(5, 0)$; y-intercept: $(0, -5)$; vertex: $\left(\frac{9}{4}, -\frac{121}{8}\right)$

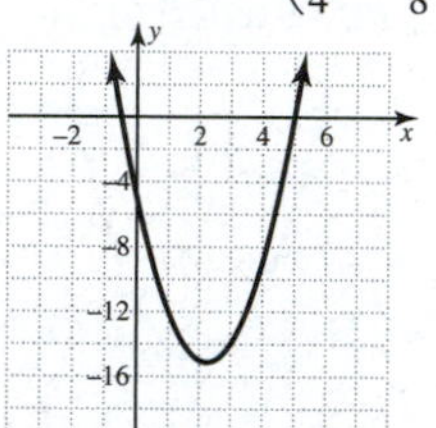

46. x-intercepts: $\left(-\frac{5}{3}, 0\right)$ and $(1, 0)$; y-intercept: $(0, -5)$; vertex: $\left(-\frac{1}{3}, -\frac{16}{3}\right)$

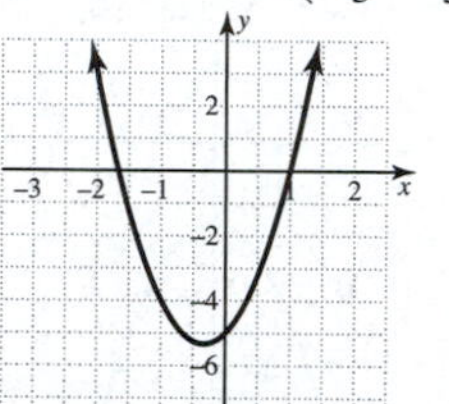

55. x-intercepts: $(-3, 0)$ and $(-8, 0)$; y-intercept: $(0, 24)$; vertex: $\left(-\frac{11}{2}, -\frac{25}{4}\right)$

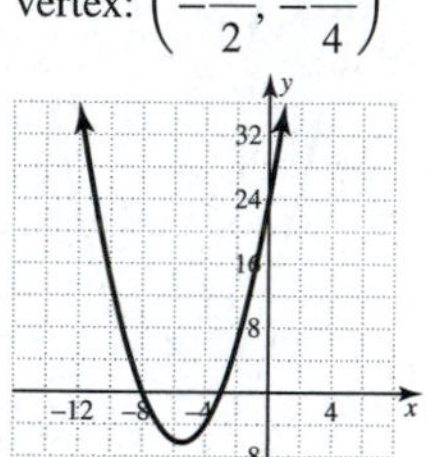

56. x-intercepts: $(1, 0)$ and $(2, 0)$; y-intercept: $(0, 2)$; vertex: $\left(\frac{3}{2}, -\frac{1}{4}\right)$

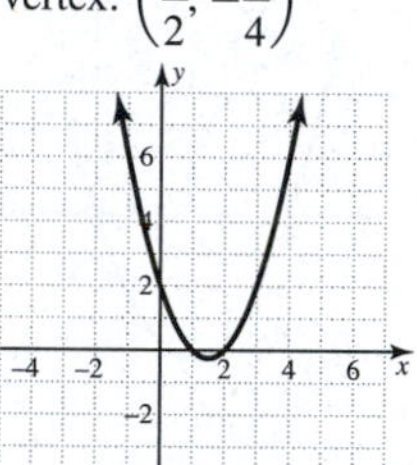

57. x-intercepts: $(-5, 0)$ and $(2, 0)$; y-intercept: $(0, 10)$; vertex: $\left(-\frac{3}{2}, \frac{49}{4}\right)$

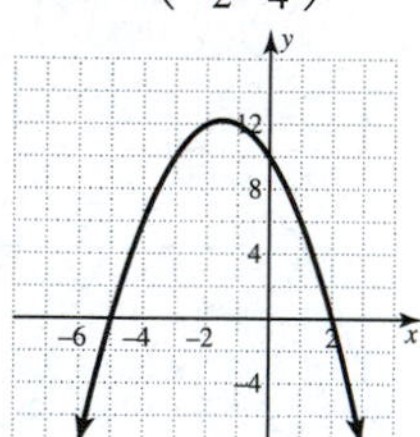

58. x-intercepts: $(-2, 0)$ and $(-1, 0)$; y-intercept: $(0, -2)$; vertex: $\left(-\frac{3}{2}, \frac{1}{4}\right)$

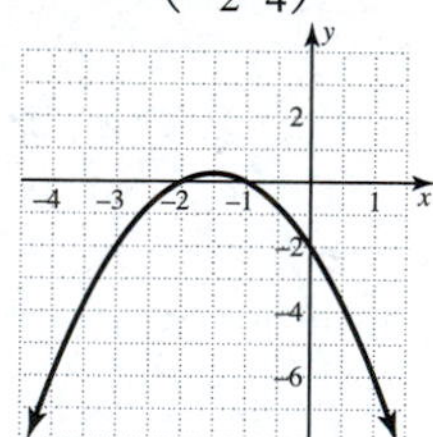

59. x-intercepts: $(4 \pm 2\sqrt{5}, 0) \approx (8.47, 0)$ and $(-0.47, 0)$; y-intercept: $(0, 4)$; vertex: $(4, 20)$

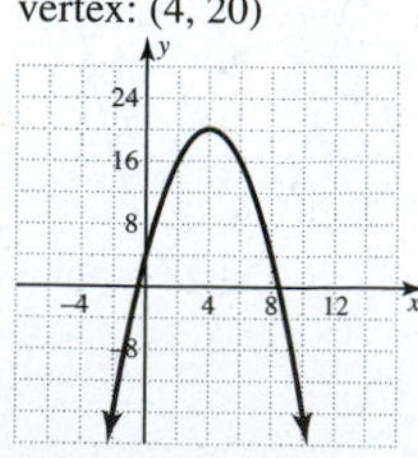

60. x-intercepts: $\left(1 \pm \frac{\sqrt{5}}{2}, 0\right) \approx (2.12, 0)$ and $(-0.12, 0)$; y-intercept: $(0, -1)$; vertex: $(1, -5)$

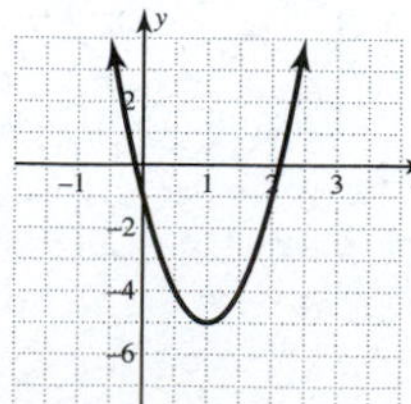

61. x-intercepts: $(0, 0)$ and $\left(\frac{5}{2}, 0\right)$; y-intercept: $(0, 0)$; vertex: $\left(\frac{5}{4}, -\frac{25}{8}\right)$

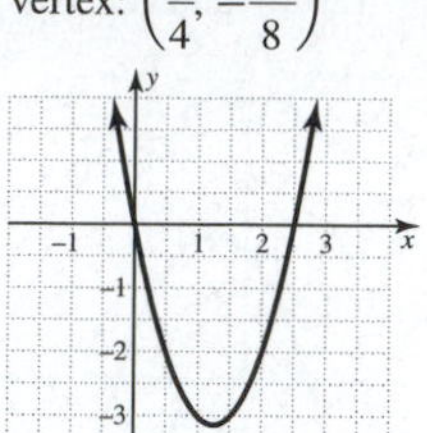

62. x-intercepts: $(0, 0)$ and $(4, 0)$; y-intercept: $(0, 0)$; vertex: $(2, 12)$

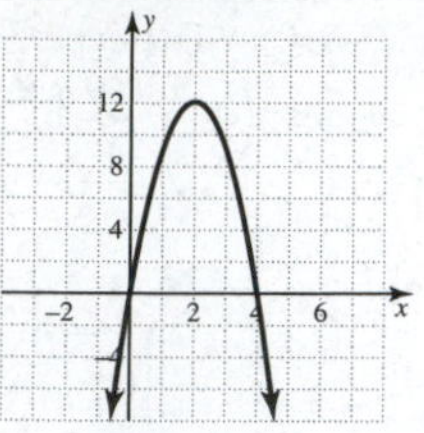

63. no x-intercepts; y-intercept: $(0, 4)$; vertex: $\left(-\frac{1}{2}, 3\right)$

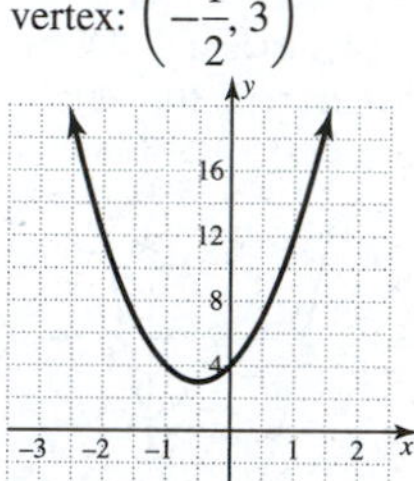

64. x-intercepts: $(4, 0)$ and $(0, 0)$; y-intercept: $(0, 0)$; vertex: $(2, 12)$

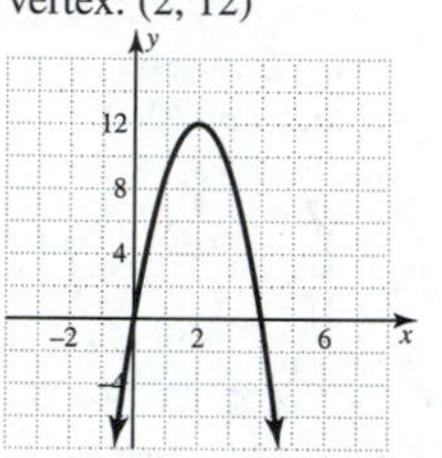

65. x-intercepts: $(-0.15, 0)$ and $(-2.65, 0)$; y-intercept: $(0, 1.59)$; vertex: $(-1.4, -6.25)$

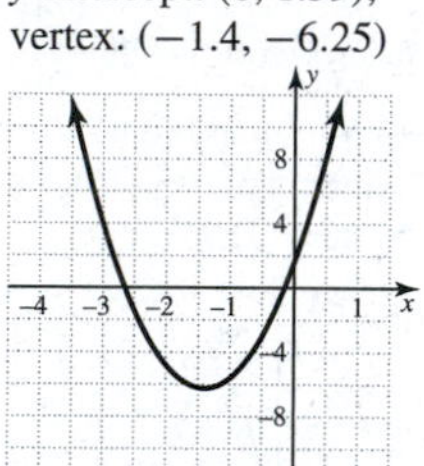

66. x-intercepts: $(4.5, 0)$ and $(-0.3, 0)$; y-intercept: $(0, 2.7)$; vertex: $(2.1, 11.52)$

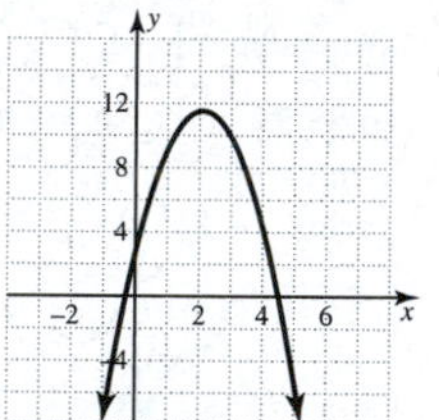

67. x-intercepts: $(-8.2, 0)$ and $(2.6, 0)$; y-intercept: $(0, -21.32)$; vertex: $(-2.8, -29.16)$

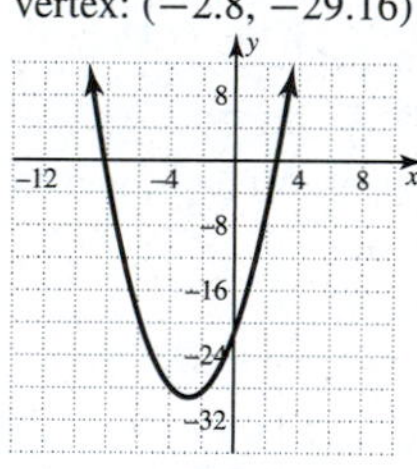

68. x-intercepts: $(4.8, 0)$ and $(1.6, 0)$; y-intercept: $(0, -7.68)$; vertex: $(3.2, 2.56)$

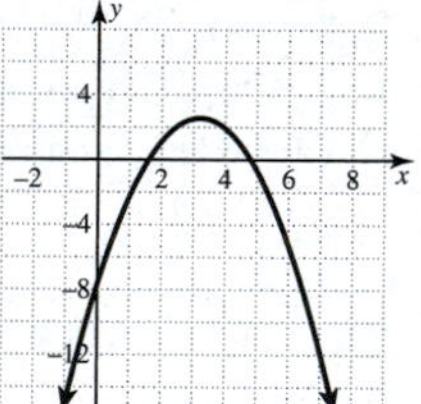

76. x-intercepts:

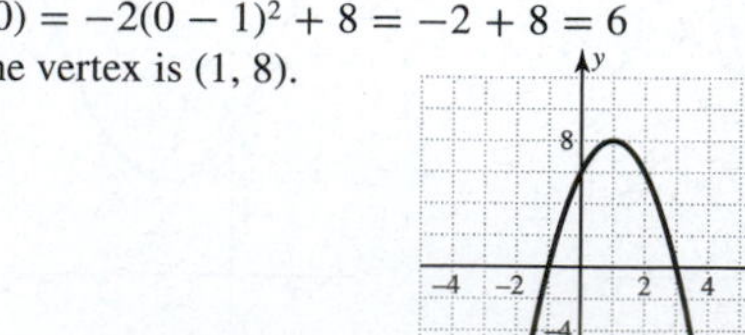

$$-2(x-1)^2 + 8 = 0$$
$$-2(x-1)^2 = -8$$
$$(x-1)^2 = 4$$
$$x - 1 = \pm\sqrt{4}$$
$$x = 1 \pm 2 = 3 \quad \text{or} \quad -1$$

The x-intercepts are $(3, 0)$ and $(-1, 0)$.

y-intercept:
$f(0) = -2(0-1)^2 + 8 = -2 + 8 = 6$
The vertex is $(1, 8)$.

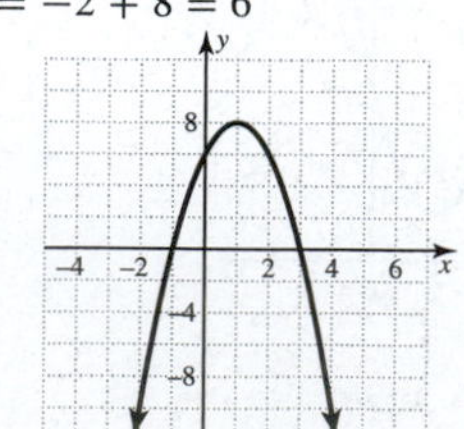

77.

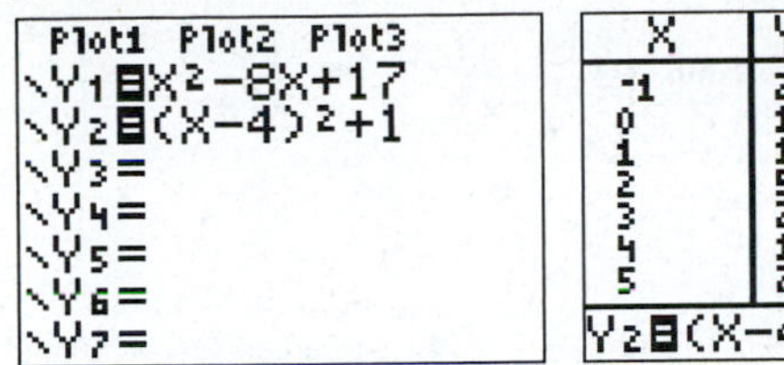

X	Y1	Y2
-1	26	26
0	17	17
1	10	10
2	5	5
3	2	2
4	1	1
5	2	2

Y2■(X-4)²+1

Y1=X²-8X+17

X=4 Y=1

Since the Y_1 and Y_2 columns agree for each value of x, we can conclude that the two forms are the same.

78.

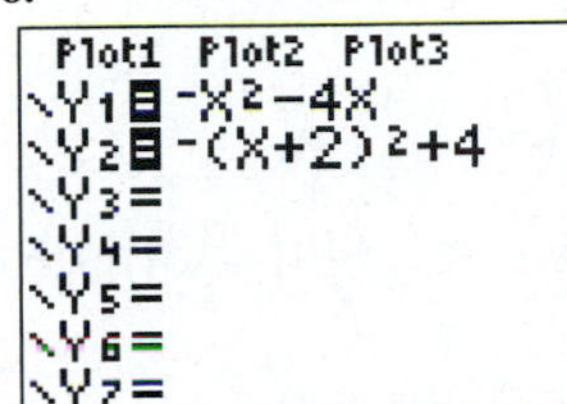

X	Y1	Y2
-4	0	0
-3	3	3
-2	4	4
-1	3	3
0	0	0
1	-5	-5
2	-12	-12

Y2■-(X+2)²+4

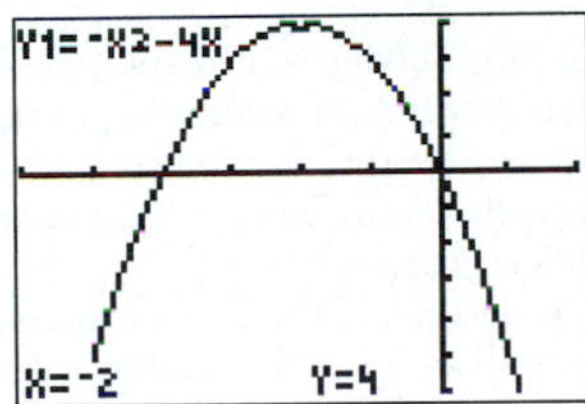

Since the Y_1 and Y_2 columns agree for each value of x, we can conclude that the two forms are the same.

Section 9.6

1. To find the boundary numbers, first find solutions of the associated equation. For rational inequalities, we also find the values that make the function undefined.
2. If the boundary number makes the inequality true, use a bracket on the boundary number. Otherwise, use a parenthesis on the boundary number. We can also determine the symbol by the inequality sign. If the symbol is $<$ or $>$, then we use a parenthesis. If the symbol is $\leq$ or $\geq$, we use a bracket.
3. We always use parentheses on the boundary numbers that make the expression undefined since they are not in the domain of the function.

22. $\left[-1, -\frac{2}{3}\right]$

24. $(-\infty, -2) \cup \left(\frac{3}{5}, \infty\right)$ **25.** $\left(-\infty, -\frac{1}{3}\right) \cup \left(\frac{1}{2}, \infty\right)$

37. $\left(-\frac{2}{7}, \frac{3}{7}\right) \cup (1, \infty)$ **39.** $(-\infty, -1) \cup \left(-\frac{1}{3}, \frac{1}{3}\right) \cup (1, \infty)$

47. $\left(-6, \frac{8}{3}\right]$ **48.** $(-\infty, -3) \cup \left(\frac{5}{2}, \infty\right)$ **50.** $\left(5, \frac{13}{2}\right]$

55. $\left[-\frac{5}{2}, 3\right]$ **68.** $(-\infty, -2] \cup \left[\frac{2}{3}, \infty\right)$

84. $(-\infty, -2) \cup \left(-\frac{3}{4}, \infty\right)$ **90.** $\left(-\infty, \frac{2}{5}\right) \cup [4, \infty)$

92. $\left(-\infty, -\frac{7}{2}\right) \cup \left(0, \frac{2}{5}\right)$ **93.** $(-\infty, 0) \cup \left(\frac{2}{3}, \frac{6}{5}\right)$

103. $\left[-1, \frac{5}{2}\right]$ **105.** $\left(-\infty, -\frac{7}{2}\right) \cup (3, \infty)$

110. The solutions of the associated equation are 0, -5, and, $\frac{1}{2}$. The inequality is true when $x \leq -5$ or $x \geq 0$ and $x \leq \frac{1}{2}$. The solution is $(-\infty, -5] \cup \left[0, \frac{1}{2}\right]$.

112. $[-5, 7]$

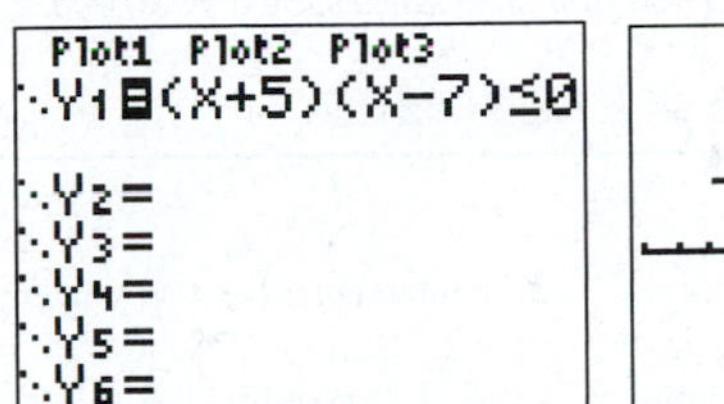

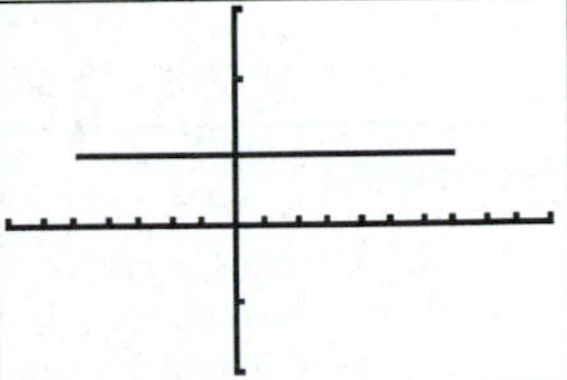

113. $(-\infty, -4) \cup (1, \infty)$

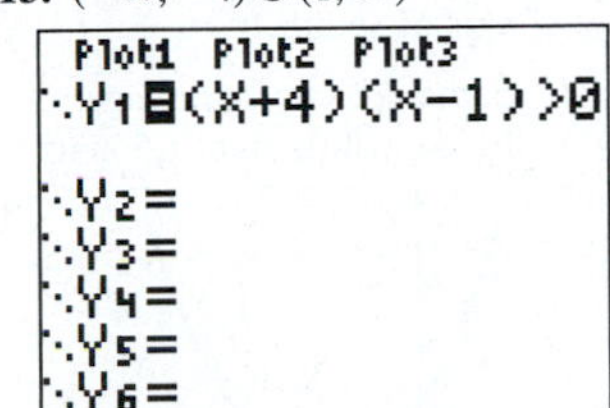

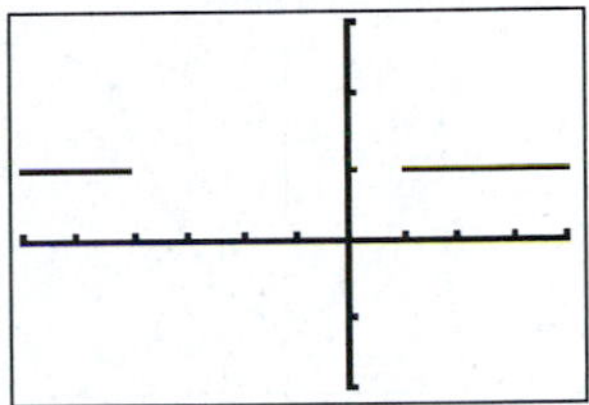

114. $(-\infty, -3) \cup \left[\frac{1}{2}, \infty\right)$

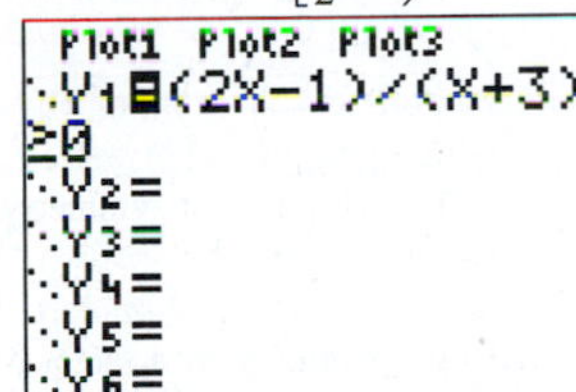

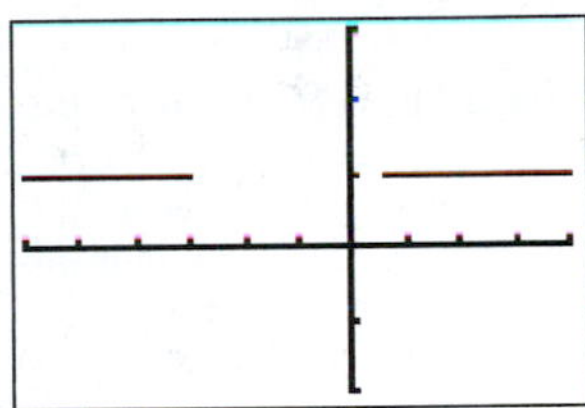

115. $[-1, 3)$

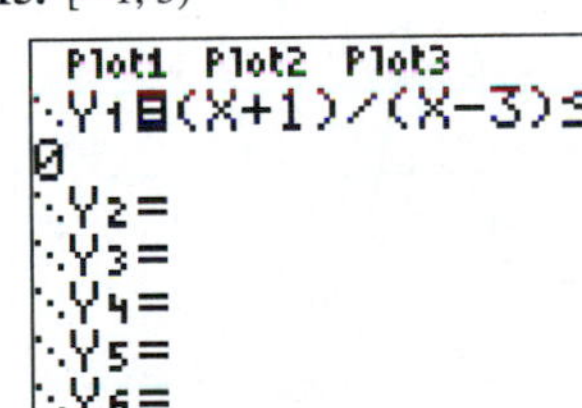

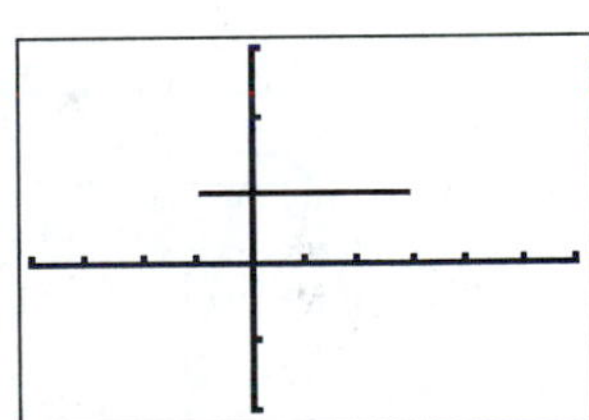

CHAPTER 9 GROUP ACTIVITY

1. (0, 31,135), (1, 31,863), (2, 31,686), (3, 31,820), (4, 30,566), (5, 29,307), (6, 26,377), (7, 23,625)
2. 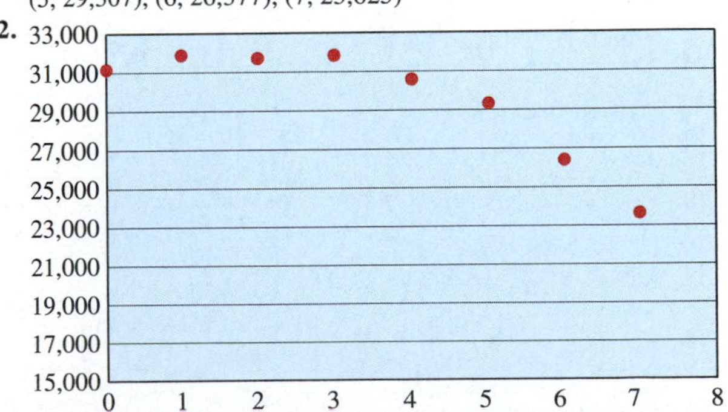

3. Yes, it looks like a parabola. The parabola opens down, so $a < 0$ in the quadratic function. The function has a maximum value.
4. **b.** Answers vary. Using (1, 31,863) and (7, 23,625), the two equations are

$$a + b + 31{,}135 = 31{,}863$$
$$49a + 7b + 31{,}135 = 23{,}625$$

d. Answers vary. $f(x) = -300.14x^2 + 1028.14x + 31{,}135$

5. Answers vary. The vertex is (1.71, 32,015.48). In approximately 2 years after 2002, or in 2004, the number of drivers involved in fatal crashes during the day whose BAC was greater than 0.01 reached a maximum of 32,015.
6. Answers vary. The solutions of $f(x) = 10{,}000$ are $x \approx 10.28$ or $x \approx -6.85$. So, in approximately 10 yr after 2002, or in 2012, the number of drivers involved in fatal crashes should be 10,000.

CHAPTER 9 REVIEW EXERCISES

Section 9.1

1. no x-intercepts; y-intercept: (0, 9); vertex: (0, 9); axis of symmetry: $x = 0$; domain: $(-\infty, \infty)$; range: $[9, \infty)$; shifts up 9 units

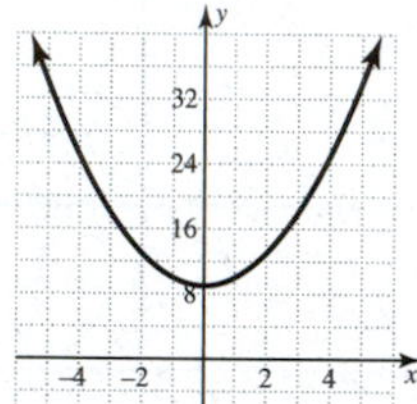

2. x-intercepts: $\left(\frac{3}{2}, 0\right)$; y-intercept: (0, 2.25); vertex: $\left(\frac{3}{2}, 0\right)$; axis of symmetry: $x = \frac{3}{2}$; domain: $(-\infty, \infty)$; range: $[0, \infty)$; shifts right 1.5 units

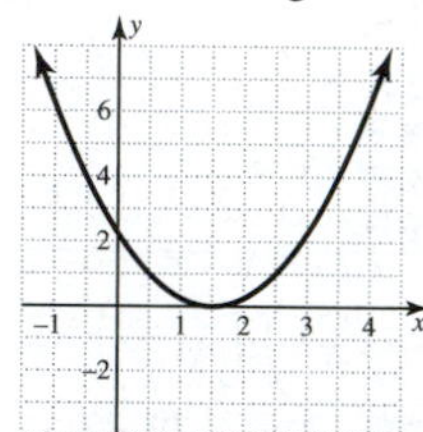

3. x-intercepts: (3, 0), (7, 0); y-intercept: (0, −10.5); vertex: (5, 2); axis of symmetry: $x = 5$; domain: $(-\infty, \infty)$; range: $(-\infty, 2]$; shifts right 5 units, reflects over the x-axis, shrinks by a factor of $\frac{1}{2}$, and shifts up 2 units

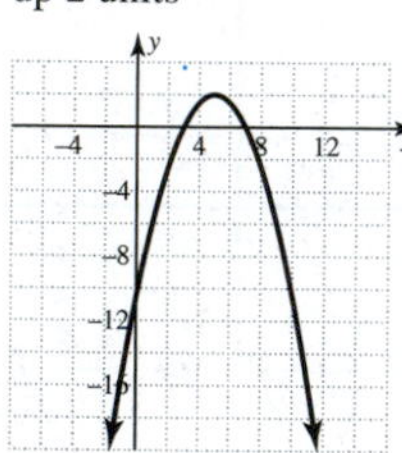

4. x-intercepts: (−3, 0), (1, 0); y-intercept: (0, −12); vertex: (−1, −16); axis of symmetry: $x = -1$; domain: $(-\infty, \infty)$ range: $[-16, \infty)$; shifts left 1 unit, stretches by a factor of 4, and shifts down 16 units

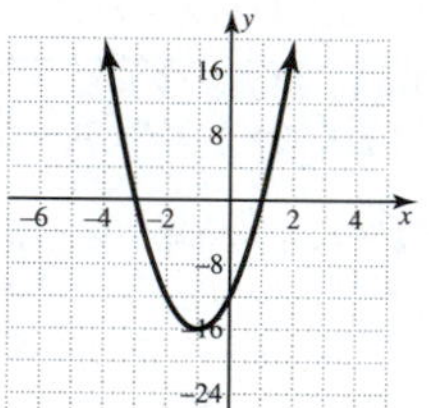

Section 9.2

11. $\left\{\frac{8}{3}, \frac{22}{3}\right\}$ 12. $\left\{\frac{-15-6i}{10}, \frac{-15+6i}{10}\right\}$ 13. $\left\{\frac{8}{5}, \frac{22}{5}\right\}$
14. $\left\{\frac{18-2\sqrt{5}}{3}, \frac{18+2\sqrt{5}}{3}\right\}$ 15. $\left\{\frac{2-8i}{3}, \frac{2+8i}{3}\right\}$
16. $\left\{\frac{1-5i}{3}, \frac{1+5i}{3}\right\}$ 17. $x^2 + 12x + 36 = (x+6)^2$
18. $s^2 + 15s + 56.25 = (s+7.5)^2$ 19. $r^2 - \frac{3}{2}r + \frac{9}{16} = \left(r - \frac{3}{4}\right)^2$
20. $s^2 + \frac{5}{2}s + \frac{25}{16} = \left(s + \frac{5}{4}\right)^2$ 23. $\left\{-\frac{1}{3}, \frac{7}{3}\right\}$
24. $\left\{-3 - \frac{3i}{2}, -3 + \frac{3i}{2}\right\}$ 25. $\left\{\frac{1-\sqrt{15}}{2}, \frac{1+\sqrt{15}}{2}\right\}$
26. $\left\{\frac{-1-\sqrt{3}}{2}, \frac{-1+\sqrt{3}}{2}\right\}$

Section 9.3

30. $\left\{2, \frac{8}{3}\right\}$ 32. $\left\{-\frac{11}{3}, -2\right\}$ 33. $\left\{-\frac{3}{4}, \frac{4}{5}\right\}$
34. $\left\{\frac{-9 \pm \sqrt{5}}{5}\right\}$ 35. $\left\{\frac{-2 \pm \sqrt{17}}{3}\right\}$

Section 9.4

58. $\left\{\pm\frac{1}{2}, \pm\frac{1}{5}i\right\}$ 60. $\left\{\pm\frac{1}{2}i, \pm\frac{3}{2}\right\}$ 68. $\left\{\frac{4096}{81}, 625\right\}$
69. $\left\{-32, \frac{1}{32}\right\}$

Section 9.5

73. x-intercepts: (−6, 0) and (4, 0); y-intercept: (0, −24); vertex: (−1, −25)

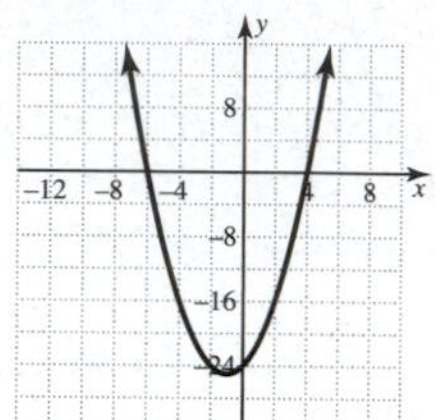

74. x-intercepts: (−1, 0) and (5, 0); y-intercept: (0, 5); vertex: (2, 9)

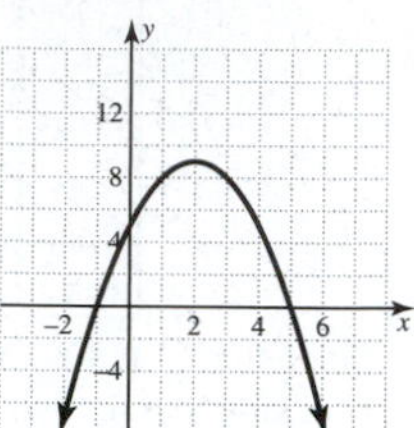

Section 9.6

78. $\left(-\infty, -\frac{8}{3}\right] \cup [3, \infty)$ 84. $\left(-\infty, -\frac{1}{2}\right) \cup \left(\frac{7}{2}, \infty\right)$
88. $\left(-\infty, -\frac{15}{4}\right) \cup (-2, 0)$ 89. $(-\infty, -1) \cup \left(-\frac{2}{5}, \frac{2}{5}\right) \cup (1, \infty)$

CHAPTER 9 TEST

1. The four methods for solving a quadratic equation are applying the zero products property (factoring), applying the square root property, completing the square, and using the quadratic formula. You can complete the square and use the quadratic formula to solve any quadratic equation.
2. The discriminant determines the number and types of solutions of a quadratic equation. It can be found by evaluating $b^2 - 4ac$, where $ax^2 + bx + c = 0$, $a \neq 0$. If the discriminant is a positive perfect square, the equation has two rational solutions. If the discriminant is positive but not a perfect square, the equation has two irrational solutions. If the discriminant is zero, the equation has one rational solution. If the discriminant is negative, the equation has two complex, nonreal solutions.

12. $\left\{\frac{3-\sqrt{41}}{4}, \frac{3+\sqrt{41}}{4}\right\}$
13. $\left\{\frac{-1-i\sqrt{19}}{2}, \frac{-1+i\sqrt{19}}{2}\right\}$ or $\left\{-\frac{1}{2} - \frac{\sqrt{19}}{2}i, -\frac{1}{2} + \frac{\sqrt{19}}{2}i\right\}$
19. x-intercepts: (−2, 0) and (6, 0); y-intercept: (0, −12); vertex: (2, −16); domain: $(-\infty, \infty)$; range: $[-16, \infty)$

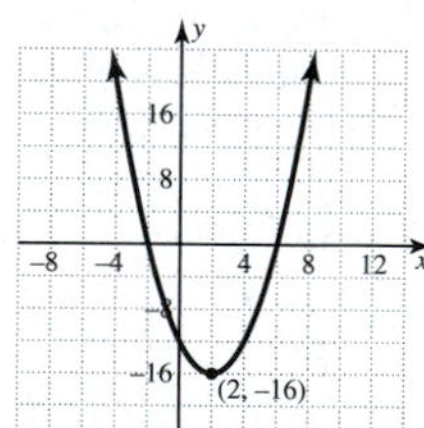

20. x-intercepts: (−5, 0), (−1, 0); y-intercept: (0, −5); vertex: (−3, 4); domain: $(-\infty, \infty)$; range: $(-\infty, 4]$

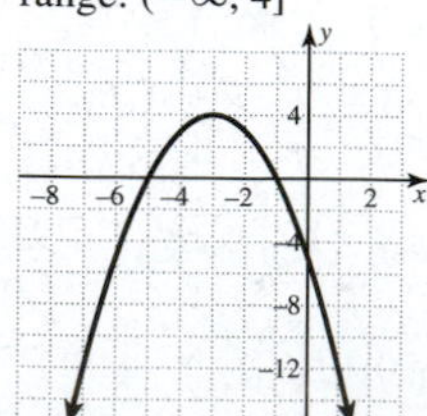

21. no x-intercepts; y-intercept: (0, 5); vertex: $\left(\frac{3}{2}, \frac{1}{2}\right)$; domain: $(-\infty, \infty)$; range: $\left[\frac{1}{2}, \infty\right)$

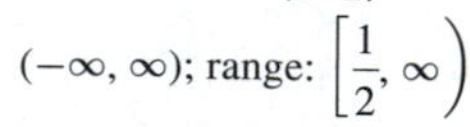

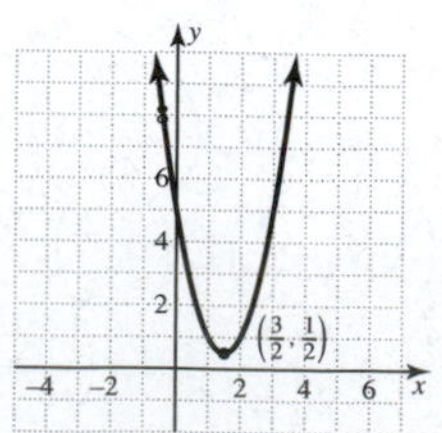

CHAPTERS 1–9 CUMULATIVE REVIEW EXERCISES

9. $r = \dfrac{3q - 4p}{4s}$ **16.** $f(x) = -\dfrac{2}{3}x;\ m = -\dfrac{2}{3};\ b = 0$

58. $\dfrac{c}{3d^2}$ **65.** $\dfrac{a\sqrt[3]{a^2}}{b}$ **66.** $\dfrac{c\sqrt[3]{c}}{b^2}$

75. no x-intercept; y-intercept: $(0, 4)$; vertex: $(1, 3)$; axis of symmetry: $x = 1$; domain: $(-\infty, \infty)$; range: $[3, \infty)$; shifts right 1 unit and up 3 units

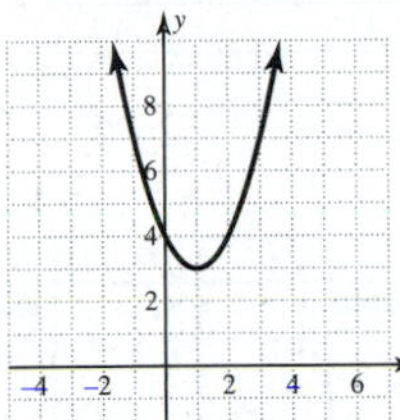

76. x-intercepts: $(-1, 0)$, $(-3, 0)$; y-intercept: $(0, 3)$; vertex: $(-2, -1)$; axis of symmetry: $x = -2$; domain: $(-\infty, \infty)$; range: $[-1, \infty)$; shifts left 2 units and down 1 unit

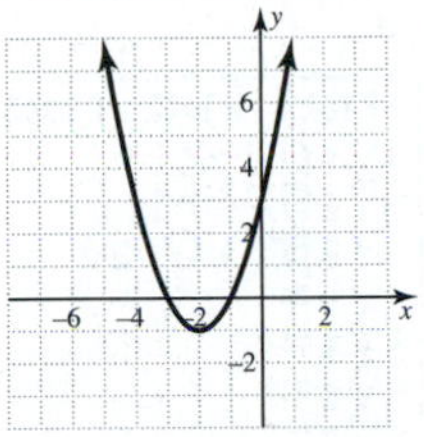

77. x-intercepts: $(1, 0)$, $(3, 0)$; y-intercept: $(0, -3)$; vertex: $(2, 1)$; axis of symmetry: $x = 2$; domain: $(-\infty, \infty)$; range: $(-\infty, 1]$; shifts right 2 units, reflects over the x-axis, and shifts up 1 unit

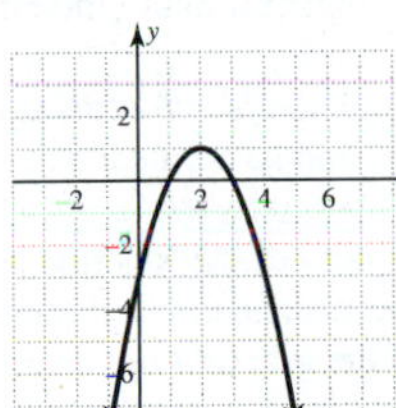

78. x-intercepts: $(-4, 0)$, $(-2, 0)$; y-intercept: $(0, -8)$; vertex: $(-3, 1)$; axis of symmetry: $x = -3$; domain: $(-\infty, \infty)$; range: $(-\infty, 1]$ shifts left 3 units, reflects over the x-axis, and shifts up 1 unit

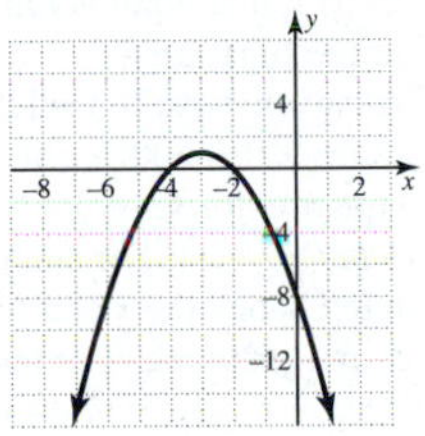

79.

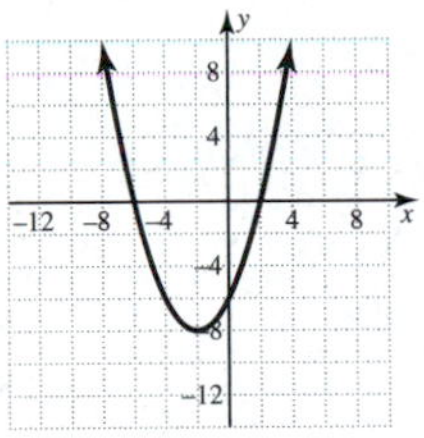

$\{-6, 2\}$

80.

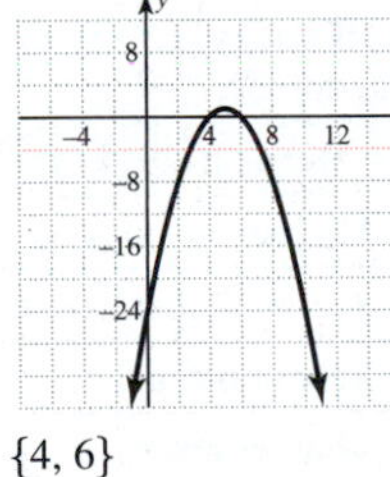

$\{4, 6\}$

83. $\left\{\dfrac{2 \pm 5i}{2}\right\}$ **84.** $\left\{\dfrac{-3 \pm 2i}{2}\right\}$

103. $f(x) = 2(x - 3)^2 + 1$; no x-intercepts; y-intercept: $(0, 19)$; vertex: $(3, 1)$

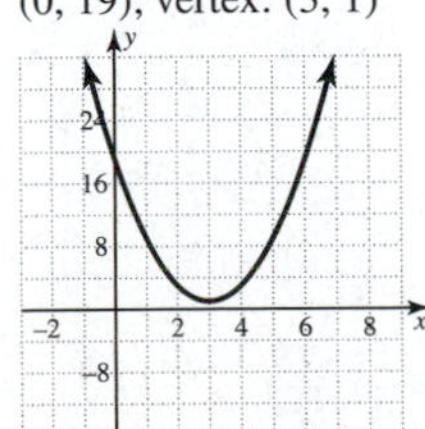

104. $f(x) = (x + 0.5)^2 - 2.25$; x-intercepts: $(-2, 0)$, $(1, 0)$; y-intercept: $(0, -2)$; vertex: $\left(-\dfrac{1}{2}, -\dfrac{9}{4}\right)$

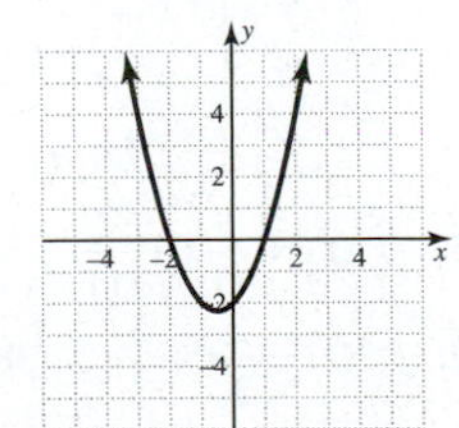

105. $\left(-\infty, -\dfrac{5}{3}\right] \cup [4, \infty)$

CHAPTER 10

Section 10.1

1. Answers vary. We add and subtract functions by combining like terms. For example, let $f(x) = 13x + 15$ and $g(x) = 7x - 8$.
$f(x) + g(x) = 13x + 15 + 7x - 8 = 20x + 7$
$f(x) - g(x) = 13x + 15 - 7x + 8 = 6x + 23$

2. Answers vary. We multiply functions by using FOIL method. For example, let $f(x) = 3x + 1$ and $g(x) = 2x - 5$.
$$f(x) \cdot g(x) = (3x + 1)(2x - 5) = 6x^2 - 15x + 2x - 5 = 6x^2 - 13x - 5$$

3. Answers vary. When finding the quotient of two functions, the restriction on the domain is the values of x that make the denominator zero. For example, let $f(x) = 2x + 5$ and $g(x) = 3x - 15$.
$$\frac{f(x)}{g(x)} = \frac{2x + 5}{3x - 15}$$
So, $g(x) \neq 0$.
$3x - 15 \neq 0$
$3x \neq 15$
$x \neq 5$
The restriction is $x \neq 5$.

4. Answers vary. Find the sum of the two functions f and g first and evaluate the resulting function at $x = 2$, or find $f(2)$ and $g(2)$ and add these values.

5. Answers vary. Find the difference of the two functions f and g first and evaluate the resulting function at $x = 2$, or find $f(2)$ and $g(2)$ and subtract these values.

6. Answers vary. To find the composite function, $f(g(x))$, substitute the function g for x in the function f. That is, substitute $5x - 1$ in place of x in the function f.
$f(g(x)) = f(5x - 1) = 2(5x - 1)^2 - (5x - 1)$

7. Answers vary. To find the composite function, $g(f(x))$, substitute the function f for x in the function of g. That is, substitute $2x^2 - x$ in place of x in the function g.
$g(f(x)) = g(2x^2 - x) = 5(2x^2 - x) - 1$

8. Answers vary. To evaluate $f(g(2))$, evaluate $g(2)$ and substitute the result into the function f. To evaluate $g(f(2))$, evaluate $f(2)$ and substitute the result into the function g.
$f(g(2)) = f(2(2) - 6) = f(-2) = |7(-2) - 2| = |-16| = 16$
$g(f(2)) = g(|7(2) - 2|) = g(12) = 2(12) - 6 = 18$

9. $(f + g)(x) = 5x + 19$, $(f - g)(x) = 3x + 3$, $(f \cdot g)(x) = 4x^2 + 43x + 88$
domain for $f + g,\ f - g,\ f \cdot g$: $(-\infty, \infty)$
$\left(\dfrac{f}{g}\right)(x) = \dfrac{4x + 11}{x + 8},\ x \neq -8$
domain for $\dfrac{f}{g}$: $(-\infty, -8) \cup (-8, \infty)$

10. $(f + g)(x) = 9x - 2$, $(f - g)(x) = 5x - 4$, $(f \cdot g)(x) = 14x^2 + x - 3$
domain for $f + g, f - g, f \cdot g$: $(-\infty, \infty)$
$\left(\dfrac{f}{g}\right)(x) = \dfrac{7x - 3}{2x + 1},\ x \neq -\dfrac{1}{2}$
domain for $\dfrac{f}{g}$: $\left(-\infty, -\dfrac{1}{2}\right) \cup \left(-\dfrac{1}{2}, \infty\right)$

11. $(f + g)(x) = \sqrt{x} + 11x - 2$, $(f - g)(x) = \sqrt{x} - 11x + 2$, $(f \cdot g)(x) = 11x\sqrt{x} - 2\sqrt{x}$
domain for $f + g,\ f - g,\ f \cdot g$: $[0, \infty)$
$\left(\dfrac{f}{g}\right)(x) = \dfrac{\sqrt{x}}{11x - 2},\ x \neq \dfrac{2}{11}$
domain for $\dfrac{f}{g}$: $\left[0, \dfrac{2}{11}\right) \cup \left(\dfrac{2}{11}, \infty\right)$

12. $(f+g)(x) = 5\sqrt{2x} + 3x - 12$, $(f-g)(x) = 5\sqrt{2x} - 3x + 12$,
$(f \cdot g)(x) = 15x\sqrt{2x} - 60\sqrt{2x}$
domain for $f+g$, $f-g$, $f \cdot g$: $[0, \infty)$
$\left(\frac{f}{g}\right)(x) = \frac{5\sqrt{2x}}{3x-12}, x \neq 4$
domain for $\frac{f}{g}$: $[0, 4) \cup (4, \infty)$

13. $(f+g)(x) = \sqrt{5x} + x^2 + 9$, $(f-g)(x) = \sqrt{5x} - x^2 - 9$,
$(f \cdot g)(x) = x^2\sqrt{5x} + 9\sqrt{5x}$
domain for $f+g$, $f-g$, $f \cdot g$: $[0, \infty)$
$\left(\frac{f}{g}\right)(x) = \frac{\sqrt{5x}}{x^2+9}$
domain for $\frac{f}{g}$: $[0, \infty)$

14. $(f+g)(x) = 2\sqrt{x} + 4x^2 + 1$, $(f-g)(x) = 2\sqrt{x} - 4x^2 - 1$,
$(f \cdot g)(x) = 8x^2\sqrt{x} + 2\sqrt{x}$
domain for $f+g$, $f-g$, $f \cdot g$: $[0, \infty)$
$\left(\frac{f}{g}\right)(x) = \frac{2\sqrt{x}}{4x^2+1}$
domain for $\frac{f}{g}$: $[0, \infty)$

15. $(f+g)(x) = -x^2 + 9x - 3$, $(f-g)(x) = x^2 - x - 3$,
$(f \cdot g)(x) = -4x^3 + 23x^2 - 15x$
domain for $f+g$, $f-g$, $f \cdot g$: $(-\infty, \infty)$
$\left(\frac{f}{g}\right)(x) = \frac{4x-3}{-x^2+5x}, x \neq 0, 5$
domain for $\frac{f}{g}$: $(-\infty, 0) \cup (0, 5) \cup (5, \infty)$

16. $(f+g)(x) = x^2 + 3x + 1$, $(f-g)(x) = -x^2 + 9x + 1$,
$(f \cdot g)(x) = 6x^3 - 17x^2 - 3x$
domain for $f+g$, $f-g$, $f \cdot g$: $(-\infty, \infty)$
$\left(\frac{f}{g}\right)(x) = \frac{6x+1}{x^2-3x}, x \neq 0, 3$
domain for $\frac{f}{g}$: $(-\infty, 0) \cup (0, 3) \cup (3, \infty)$

45. $(f \circ g)(x) = 12x - 13$
$(g \circ f)(x) = 12x - 35$
domain: $(-\infty, \infty)$

46. $(f \circ g)(x) = 30x + 12$
$(g \circ f)(x) = 30x + 43$
domain: $(-\infty, \infty)$

47. $(f \circ g)(x) = 5x - 11$
$(g \circ f)(x) = 5x + 5$
domain: $(-\infty, \infty)$

48. $(f \circ g)(x) = 6x - 12$
$(g \circ f)(x) = 6x - 29$
domain: $(-\infty, \infty)$

71. $(f+g)(x) = 3x^2 + 1$
$(f \cdot g)(x) = 6x^3 - x^2 - 2x$

72. $(f+h)(x) = x + 5$
$(f \cdot h)(x) = -2x^2 + 7x + 4$

73. $(g \cdot h)(x) = -3x^3 + 14x^2 - 8x$
$\left(\frac{h}{g}\right)(x) = \frac{4-x}{3x^2-2x}, x \neq 0, \frac{2}{3}$

74. $(g-h)(x) = 3x^2 - x - 4$
$\left(\frac{g}{h}\right)(x) = \frac{3x^2-2x}{4-x}, x \neq 4$

75. $(f \circ g)(x) = 6x^2 - 4x + 1$
$(g \circ f)(x) = 12x^2 + 8x + 1$

76. $(f \circ h)(x) = -2x + 9$
$(h \circ f)(x) = -2x + 3$

77. $(f+g)(x) = 3x^2 - 7x + 9$
$(f-g)(x) = x^2 + 7x - 3$

78. $(g+h)(x) = x^2 - 10x + 11$
$(g-h)(x) = x^2 - 4x + 1$

79. $(f \cdot g)(x) = 2x^4 - 14x^3 + 15x^2 - 21x + 18$
$\left(\frac{f}{g}\right)(x) = \frac{2x^2+3}{x^2-7x+6}, x \neq 1, 6$

80. $(g \cdot h)(x) = -3x^3 + 26x^2 - 53x + 30$
$\left(\frac{g}{h}\right)(x) = \frac{x^2-7x+6}{5-3x}, x \neq \frac{5}{3}$

81. $(f \circ h)(x) = 18x^2 - 60x + 53$
$(h \circ f)(x) = -6x^2 - 4$

82. $(g \circ h)(x) = 9x^2 - 9x - 4$
$(h \circ g)(x) = -3x^2 + 21x - 13$

122. **a.** cost of the golf course = \$45x, cost of food = \$26x, cost of prizes = \$5x, **b.** total cost = \$76x

123. $(f-g)(x) = (2x^2 + 4x - 10) - (3x^2 - 2x + 9)$
$= 2x^2 + 4x - 10 - 3x^2 + 2x - 9$
$= -x^2 + 6x - 19$

127.

```
Plot1 Plot2 Plot3
\Y1=2X+5
\Y2=3-7X
\Y3=Y1(Y2)
\Y4=
\Y5=
\Y6=
\Y7=
```

X	Y1	Y2	Y3
[illegible]	-3	31	[illegible]
-2	1	17	39
2	9	-11	-17

X= -4

128.

```
Plot1 Plot2 Plot3
\Y1=2-X
\Y2=X²+1
\Y3=Y1(Y2)
\Y4=
\Y5=
\Y6=
\Y7=
```

X	Y1	Y2	Y3
[illegible]	4	5	[illegible]
1	1	2	0
3	-1	10	-8

X= -2

Section 10.2

2. Answers vary. Use a horizontal line test to determine if a function is one-to-one. If the horizontal line intersects the graph at no more than one point, then the graph is one-to-one.

3. Answers vary. The inverse of a one-to-one function f is a function that when applied to the output of the function f takes us back to the original input value. It is a function that "un-does" the effects of the function f.

4. Answers vary. To find the equation of the inverse of a function, replace $f(x)$ with y, interchange x and y, and solve the resulting equation for y.

5. Answers vary. For each of the ordered pairs of the function f, interchange the x and y. Plot the new points and connect them with a smooth graph. The resulting graph is the graph of the inverse.

6. Answers vary. If two functions are inverses of each other, their composite functions give x. That is, $f(f^{-1}(x)) = x$ and $f^{-1}(f(x)) = x$.

29. one-to-one;

Mayor in 2011 (x)	City (y)
Michael Bloomberg	New York City
Antonio Villaraigosa	Los Angeles
Rahm Emanuel	Chicago
Annise Parker	Houston
Phil Gordon	Phoenix

31. one-to-one;

32. one-to-one;

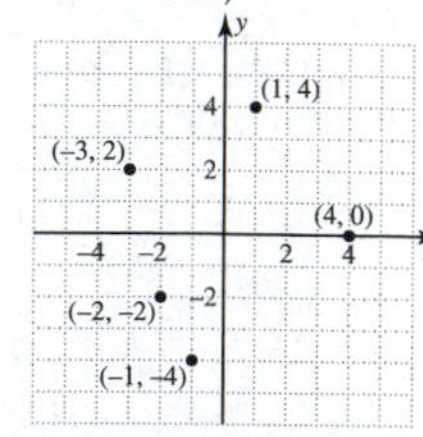

35. $f^{-1}(x) = \frac{1}{11}x + \frac{2}{11}$

36. $f^{-1}(x) = \frac{1}{5}x + \frac{9}{5}$

37. $f^{-1}(x) = \frac{2}{3}x + \frac{5}{3}$

38. $f^{-1}(x) = 2x - \frac{7}{3}$

39. $f^{-1}(x) = \dfrac{\sqrt[3]{x} - 9}{2}$ **40.** $f^{-1}(x) = \dfrac{\sqrt[3]{x} + 7}{4}$

41. $f^{-1}(x) = \dfrac{\sqrt[3]{x} - 1}{2}$ **42.** $f^{-1}(x) = \dfrac{\sqrt[3]{x} - 5}{3}$

43. $f^{-1}(x) = \dfrac{15 - 3x}{x}$ **44.** $f^{-1}(x) = \dfrac{10 + x}{x}$

53.

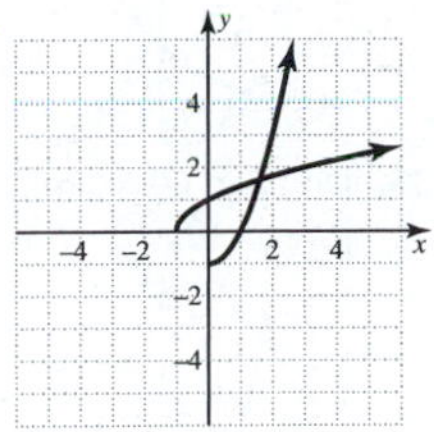

54.

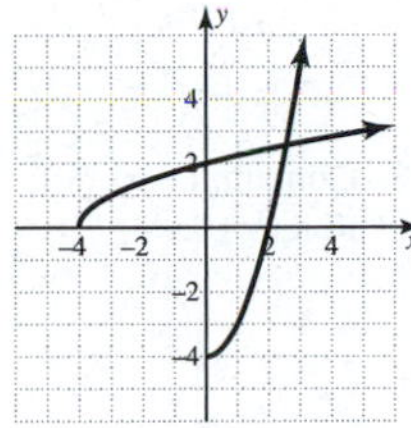

55.

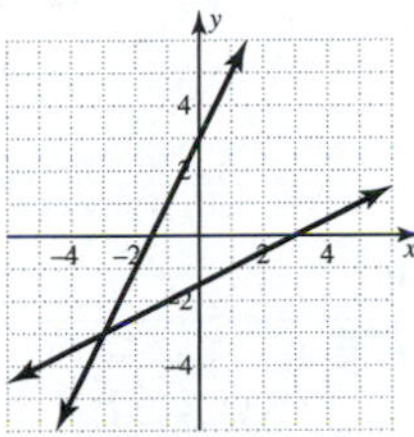

56.

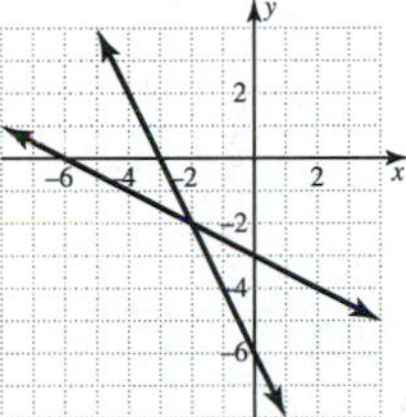

81. one-to-one;

Senator in 2011 (y)	State (x)
Joseph Lieberman	Connecticut
Lindsey Graham	South Carolina
Daniel Inouye	Hawaii
Johnny Isakson	Georgia
Mary Landrieu	Louisiana

83. $f^{-1}(x) = \dfrac{1}{7}x - \dfrac{3}{7}$ **84.** $f^{-1}(x) = \dfrac{1}{8}x - \dfrac{3}{2}$

86. $f^{-1}(x) = \dfrac{13 - x}{6x}$ **87.** $f^{-1}(x) = \dfrac{\sqrt[3]{2x - 4}}{2}$

88. $f^{-1}(x) = \dfrac{\sqrt[3]{9x + 45}}{3}$ **89.** $f^{-1}(x) = \dfrac{5}{6}x + \dfrac{1}{3}$

90. $f^{-1}(x) = 2x - \dfrac{10}{3}$ **91.** $f^{-1}(x) = \dfrac{\sqrt[3]{x} - 4}{5}$

92. $f^{-1}(x) = \dfrac{\sqrt[3]{x} - 1}{3}$

97.

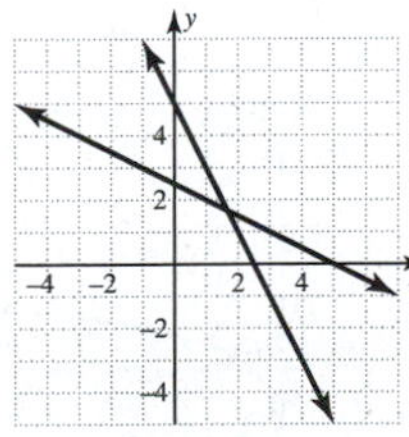

98.

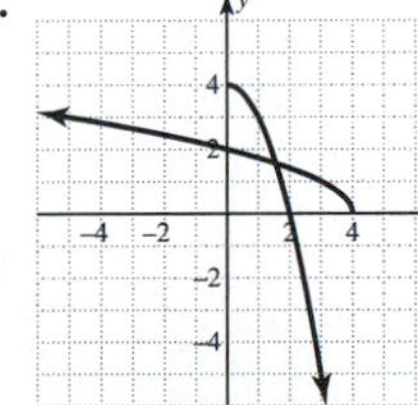

99.

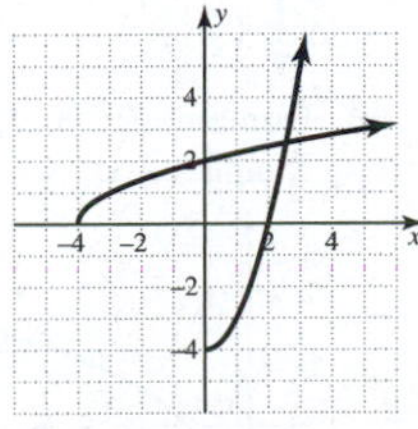

100.

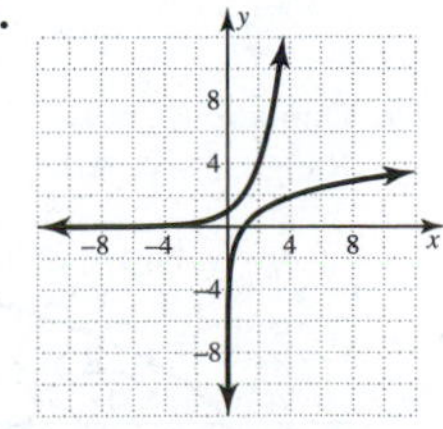

101.

$$f(x) = \frac{3x - 7}{8}$$
$$x = \frac{3y - 7}{8}$$
$$8x = 3y - 7$$
$$8x + 7 = 3y$$
$$y = \frac{8x + 7}{3}$$
$$f^{-1}(x) = \frac{8x + 7}{3}$$

102.

$$y = 8x^3 + 27$$
$$x = 8y^3 + 27$$
$$x - 27 = 8y^3$$
$$y^3 = \frac{x - 27}{8}$$
$$y = \sqrt[3]{\frac{x - 27}{8}}$$
$$f^{-1}(x) = \frac{\sqrt[3]{x - 27}}{2}$$

103.

```
Plot1 Plot2 Plot3
\Y1=2X+4
\Y2=1/2X-2
\Y3=Y1(Y2)
\Y4=Y2(Y1)
\Y5=
\Y6=
\Y7=
```

X	Y1	Y2
-5	-6	-4.5
-3	-2	-3.5
-1	2	-2.5
1	6	-1.5
3	10	-.5
5	14	.5
7	18	1.5

Y2=1/2X−2

X	Y3	Y4
-5	-5	-5
-3	-3	-3
-1	-1	-1
1	1	1
3	3	3
5	5	5
7	7	7

Y4=Y2(Y1)

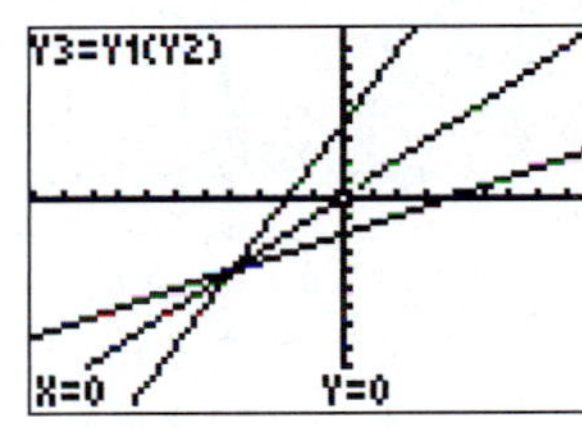

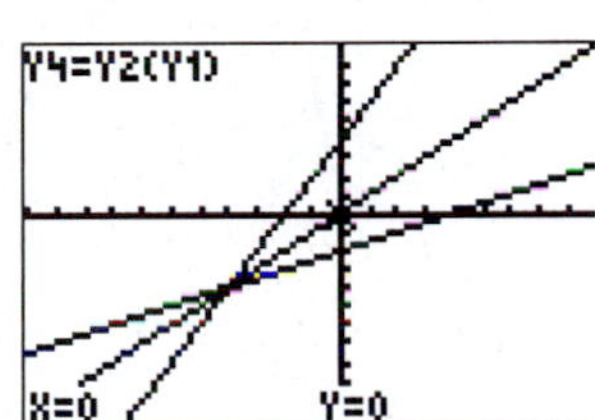

104.

```
Plot1 Plot2 Plot3
\Y1=(X-1)^3
\Y2=³√(X)+1
\Y3=Y1(Y2)
\Y4=Y2(Y1)
\Y5=
\Y6=
\Y7=
```

X	Y1	Y2
-27	-21952	-2
-8	-729	-1
-1	-8	0
0	-1	1
1	0	2
8	343	3
27	17576	4

Y2=³√(X)+1

X	Y3	Y4
-27	-27	-27
-8	-8	-8
-1	-1	-1
0	0	0
1	1	1
8	8	8
27	27	27

Y4=Y2(Y1)

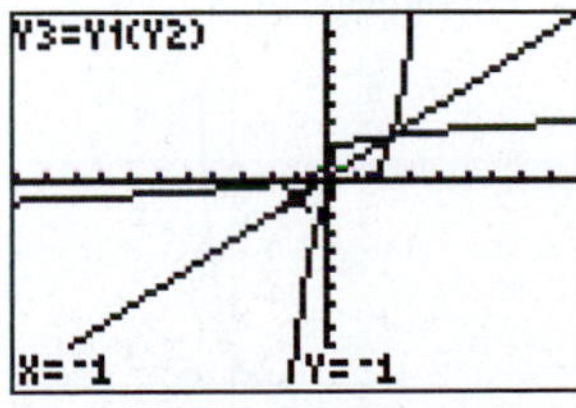

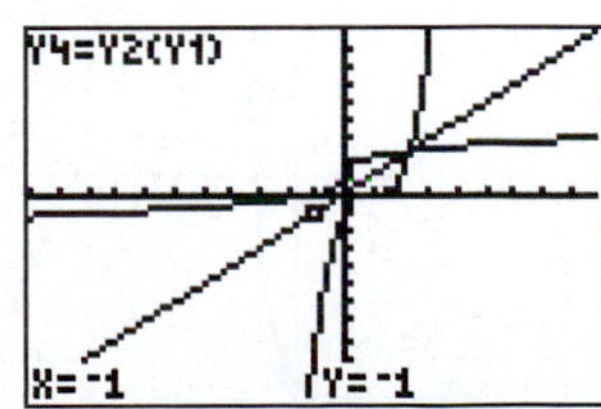

Section 10.3

1. Answers vary. Since the base $6 > 1$, the function is increasing. At the left end of the x-axis, the graph gets very close to the x-axis but never touches it. At the right end of the x-axis, the graph is increasing to ∞.
2. Answers vary. Since the base $0 < \frac{1}{8} < 1$, the function is decreasing. At the left end of the x-axis, the graph is decreasing from ∞. At the right end of the x-axis, the graph gets very close to the x-axis but never touches it.
3. Answers vary. We must express 625 as a power of 5; that is, $625 = 5^4$. Set the exponents equal and solve the resulting equation for x.
4. Answers vary. We must express 625 as a power of 25, i.e. $625 = 25^2$. Set the exponents equal and solve the resulting equation for x.
5. Answers vary. We must express $\frac{1}{64}$ as a power of 4; that is, $\frac{1}{64} = 4^{-3}$. Set the exponents equal and solve the resulting equation for x.
6. Answers vary. We must express both 4 and $\frac{1}{64}$ as a power of 2; that is, $4 = 2^2$ and $\frac{1}{64} = 2^{-6}$. We then apply the power of a power rule on the left side of the equation. We set the exponents equal and solve the resulting equation for x.

7. Domain: $(-\infty, \infty)$, Range: $(0, \infty)$

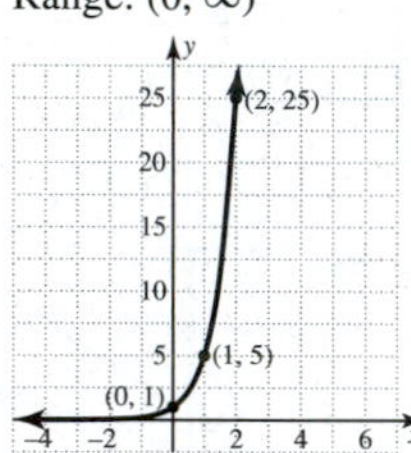

8. Domain: $(-\infty, \infty)$, Range: $(0, \infty)$

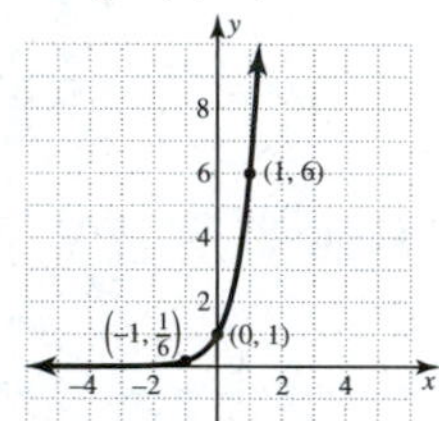

9. Domain: $(-\infty, \infty)$, Range: $(0, \infty)$

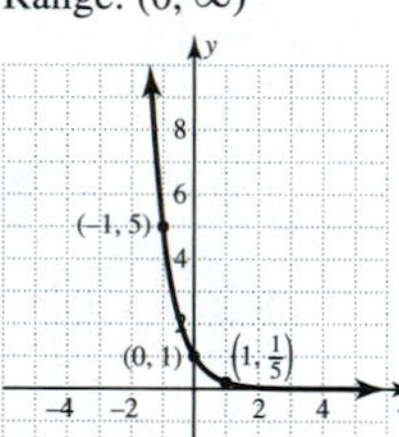

10. Domain: $(-\infty, \infty)$, Range: $(0, \infty)$

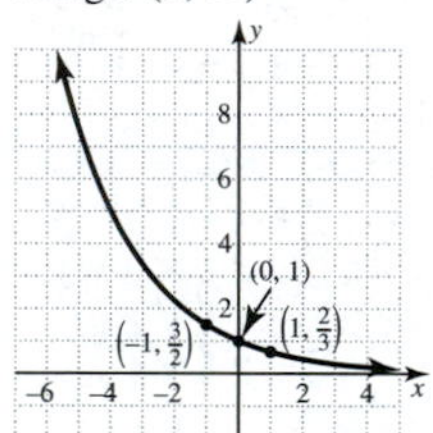

11. Domain: $(-\infty, \infty)$, Range: $(0, \infty)$

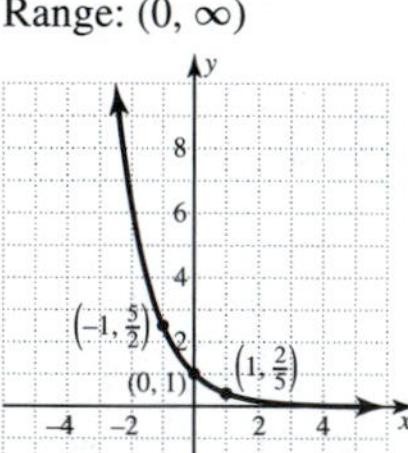

12. Domain: $(-\infty, \infty)$, Range: $(0, \infty)$

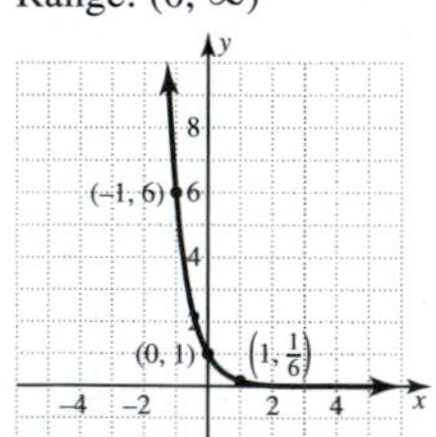

33. Domain: $(-\infty, \infty)$, Range: $(0, \infty)$

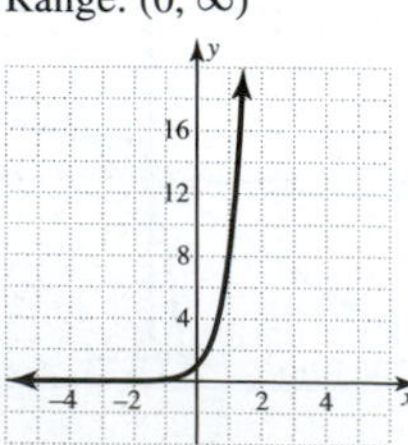

34. Domain: $(-\infty, \infty)$, Range: $(0, \infty)$

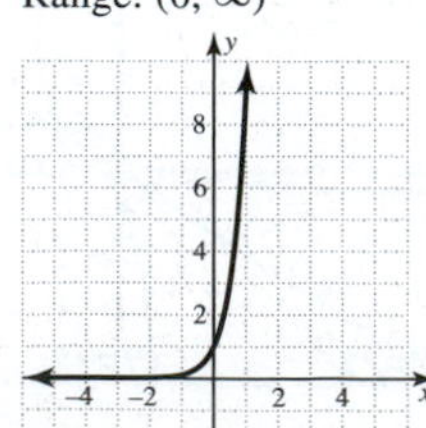

35. Domain: $(-\infty, \infty)$, Range: $(0, \infty)$

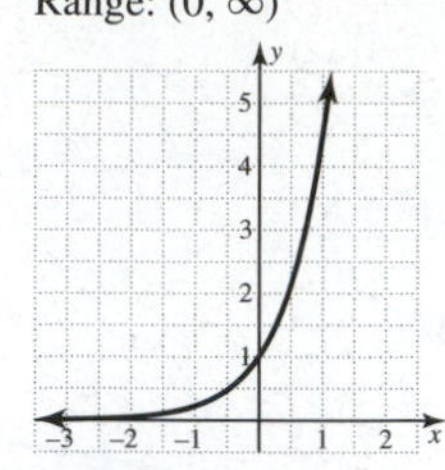

36. Domain: $(-\infty, \infty)$, Range: $(0, \infty)$

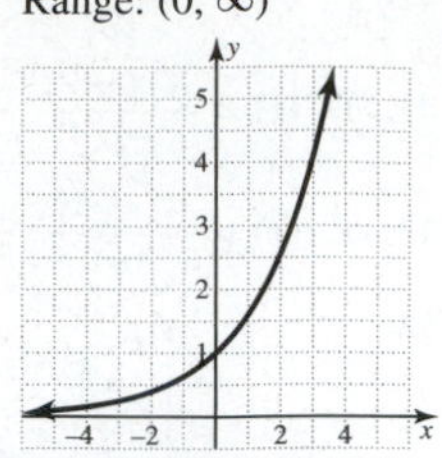

37. Domain: $(-\infty, \infty)$, Range: $(0, \infty)$

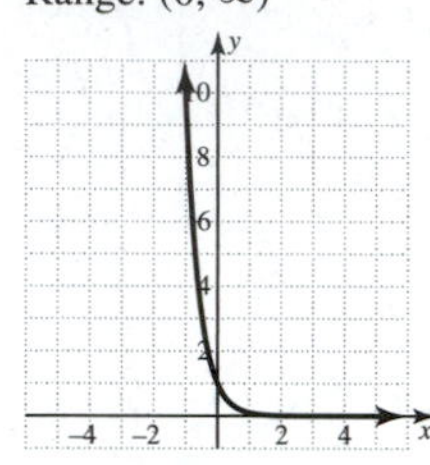

38. Domain: $(-\infty, \infty)$, Range: $(0, \infty)$

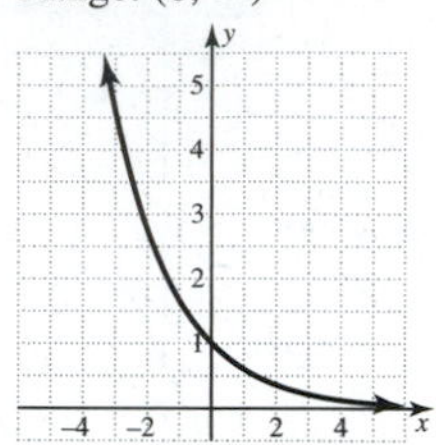

65.

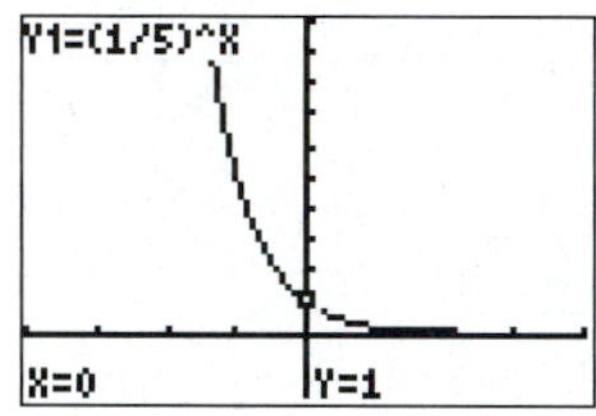

66.

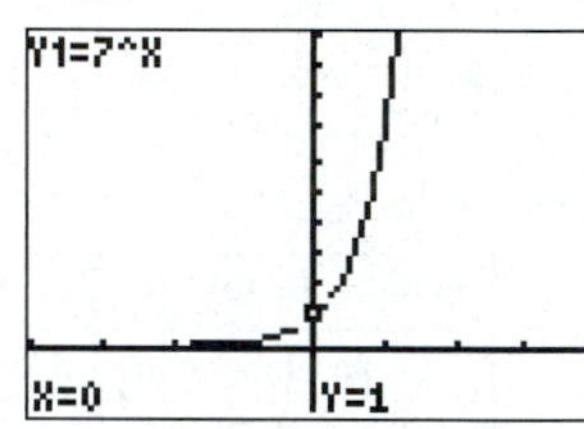

67. 0.32

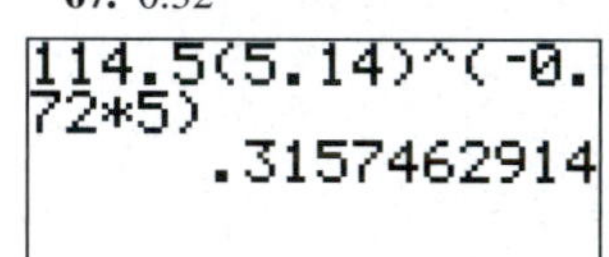

68. 3004.51

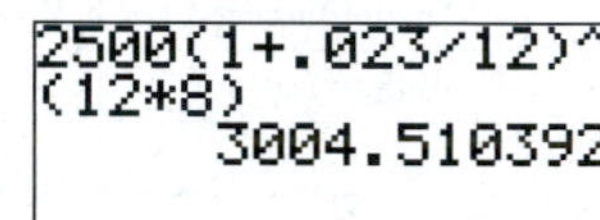

Section 10.4

2. Answers vary. Since a is the exponent of x that produces b, we say a is the logarithm of b with base x. We write $a = \log_x b$.
3. Answers vary. Since z is the log of y with base w, we say z is the exponent of w that produces y. We write $w^z = y$.
6. Answers vary. To set up a table to graph a logarithmic function, convert the equation to its exponential form and substitute values for y and solve for x.
7. Answers vary. Set the expression equal to x, convert to exponential form, and solve the resulting equation for x.

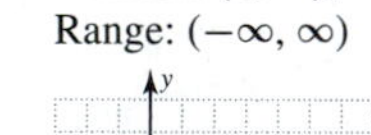

57. Domain: $(0, \infty)$, Range: $(-\infty, \infty)$

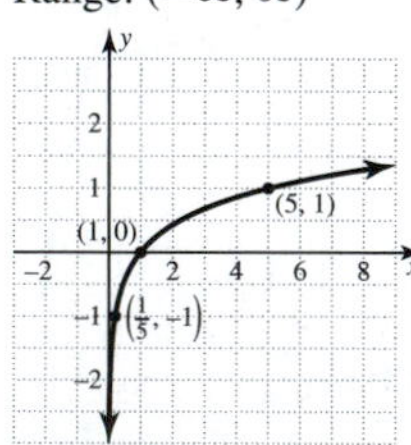

58. Domain: $(0, \infty)$, Range: $(-\infty, \infty)$

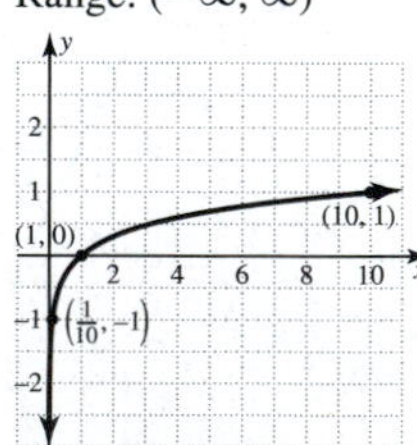

59. Domain: $(0, \infty)$, Range: $(-\infty, \infty)$

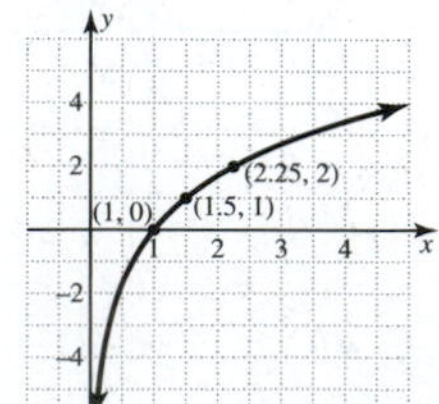

60. Domain: $(0, \infty)$, Range: $(-\infty, \infty)$

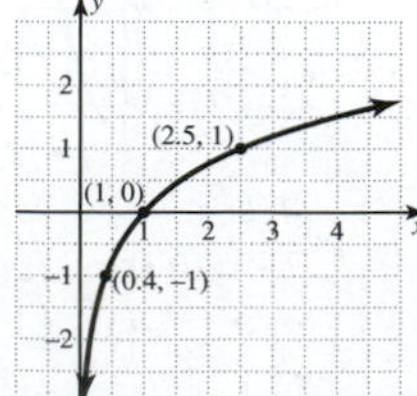

61. Domain: $(0, \infty)$, Range: $(-\infty, \infty)$

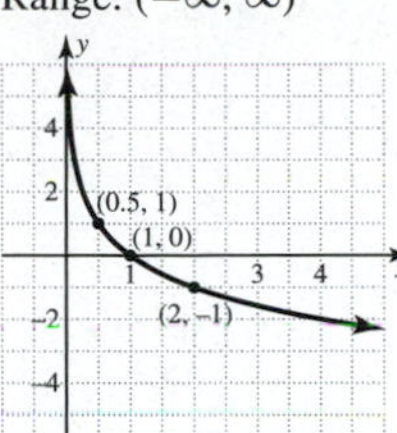

62. Domain: $(0, \infty)$, Range: $(-\infty, \infty)$

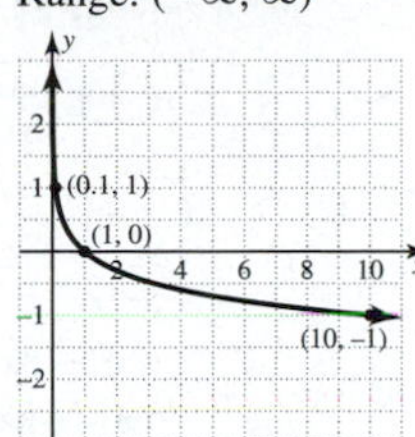

70. $\left\{\frac{1}{10}\right\}$ **79.** $\left\{\frac{1}{2}\right\}$ **80.** $\left\{\frac{1}{2}\right\}$

81. Domain: $(0, \infty)$, Range: $(-\infty, \infty)$

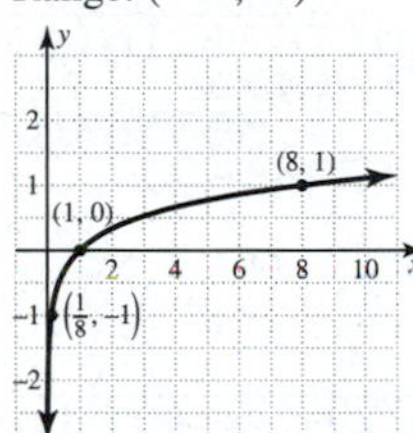

82. Domain: $(0, \infty)$, Range: $(-\infty, \infty)$

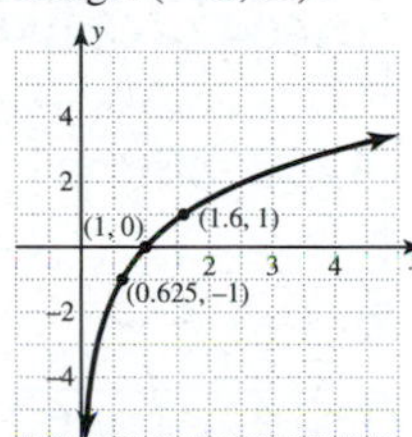

83. Domain: $(0, \infty)$, Range: $(-\infty, \infty)$

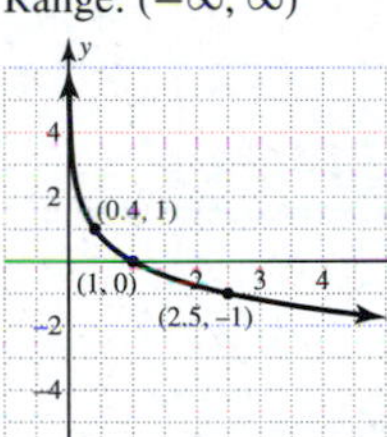

84. Domain: $(0, \infty)$, Range: $(-\infty, \infty)$

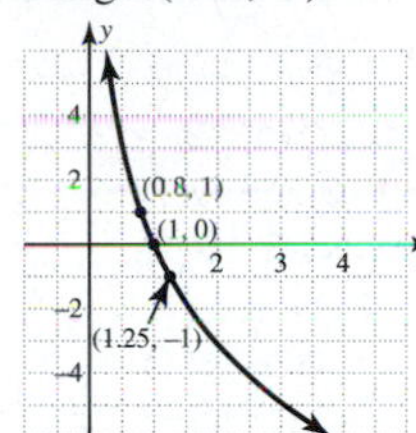

Piece It Together Sections 10.1–10.4

1. $(f + g)(x) = 2x + 18$, domain is $(-\infty, \infty)$; $(f - g)(x) = 6x + 12$, domain is $(-\infty, \infty)$; $(f \cdot g)(x) = -8x^2 - 18x + 45$, domain is $(-\infty, \infty)$; $\left(\frac{f}{g}\right)(x) = \frac{4x + 15}{3 - 2x}$, domain is $\left(-\infty, \frac{3}{2}\right) \cup \left(\frac{3}{2}, \infty\right)$; $(f \circ g)(x) = -8x + 27$, domain is $(-\infty, \infty)$; $(g \circ f)(x) = -8x - 27$, domain is $(-\infty, \infty)$.

8. Domain: $(-\infty, \infty)$, Range: $(0, \infty)$

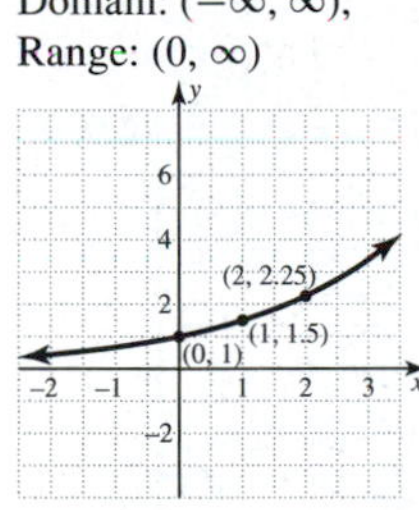

9. Domain: $(0, \infty)$, Range: $(-\infty, \infty)$

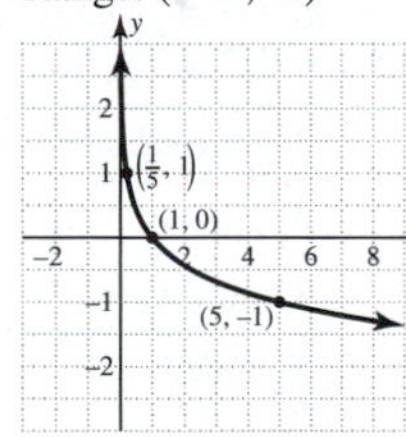

Section 10.5

1. Answers vary. We apply the product rule for logarithms to combine a sum of logarithms with the same base into one logarithm, which is written as the logarithm of the product of the arguments of the logarithms.

2. Answers vary. We apply the quotient rule for logarithms to combine a difference of logarithms with the same base into one logarithm, which is written as the logarithm of the quotient of the arguments.

3. Answers vary. We apply the power rule for logarithms to an expression with a coefficient of a logarithm that becomes the exponent of the argument of the logarithm.

4. Answers vary. In $\log_a x^m$, the argument x is raised to the exponent m, whereas in $(\log_a x)^m$, the logarithm of x with base a is raised to the exponent m. The first one, $\log_a x^m$, can be simplified as $m \log_a x$, while the other expression cannot be simplified.

5. Answers vary. Applying the product rule for logarithms, the sum of logarithms with the same base is equal to the logarithm of the product of the arguments. For example,

$$\log_2(2 + 4) = \log_2 6$$
$$\log_2 2 + \log_2 4 = 1 + 2 = 3$$
$$\log_2 6 \neq 3$$

6. Answers vary. Applying the quotient rule for logarithms, the difference of logarithms with the same base is equal to the logarithm of the quotient of the arguments. For example,

$$\log_2(8 - 2) = \log_2 6$$
$$\log_2 8 - \log_2 2 = 3 - 1 = 2$$
$$\log_2 6 \neq 2$$

55. $5 \log_{10} a + \frac{1}{2}\log_{10}(a - 3)$ **56.** $7 \log_6 b + \frac{1}{3}\log_6(b + 5)$

93. $\log_3 5 + 2\log_3 x - \log_3 y$ **95.** $\frac{1}{2}\log_7 6 + \frac{1}{2}\log_7 x - \frac{1}{2}\log_7 y$

96. $\frac{1}{3}\log_4 a - \frac{1}{3}\log_4 5 - \frac{1}{3}\log_4 b$

97. $5\log_2 x + 3\log_2 y + \frac{1}{2}\log_2(x + 3)$

98. $6 \log_4 a + 3\log_4 b + \frac{1}{5}\log_4(b + 1)$

99. $2 + 3\log_{10} a - 5\log_{10} b$ **100.** $2 + 2\log_5 x - 6\log_5 y$

101. $2\log_4 r + \log_4(r + 3) - 3\log_4 s$

102. $\log_{10} x + \log_{10}(x^2 + 1) - 4\log_{10} y$

Section 10.6

3. Answers vary. You rewrite any logarithm by applying the change-of-base formula and using either the common or natural logarithm. For example, $\log_6 20 = \frac{\log 20}{\log 6} \approx 1.67$ and $\log_6 20 \approx \frac{\ln 20}{\ln 6} \approx 1.67$.

4. Answers vary. When rewriting in the equivalent exponential form, they all represent the exponent of 3 that gives 5. That is, $\log_3 5 = \frac{\log 5}{\log 3} = \frac{\ln 5}{\ln 3} \approx 1.465$ and $3^{1.465} \approx 5$.

6. Answers vary. Use each base to write 64 in the equivalent exponential form.
a. $\log_2 64 = \log_2(2^6) = 6$ **b.** $\log_4 64 = \log_4(4^3) = 3$
c. $\log_8 64 = \log_8(8^2) = 2$ **d.** $\log_{1/2} 64 = \log_{1/2}\left(\frac{1}{2}\right)^{-6} = -6$

7. Answers vary. Use the common or natural logarithm to rewrite 64 in the equivalent exponential form.
a. $\log_3 64 = \frac{\log 64}{\log 3}$ **b.** $\log_5 64 = \frac{\log 64}{\log 5}$ **c.** $\log_{3.75} 64 = \frac{\log 64}{\log 3.75}$

8. Answers vary. When rewriting in the equivalent exponential form, they all represent the exponent of different bases that gives 64.

47. $\left\{\frac{10^{-4.4} - 12}{5}\right\}$ or $\{-2.40\}$ **48.** $\left\{\frac{10^{3.1} - 3}{9}\right\}$ or $\{139.55\}$

49. $\left\{\frac{e^{1.9} + 10}{2}\right\}$ or $\{8.34\}$ **50.** $\left\{\frac{10^{-5.9} + 7}{4}\right\}$ or $\{1.75\}$

51. $\left\{\frac{e^{-0.9} - 5}{3}\right\}$ or $\{-1.53\}$ **52.** $\left\{\frac{e^{-3.4} + 3}{12}\right\}$ or $\{0.25\}$

53. $\left\{\frac{10^{-1}}{2}\right\}$ or $\{0.05\}$ **83.** $\left\{\frac{10^{1.7} + 1}{6}\right\}$ or $\{8.52\}$

84. $\left\{\frac{e^{1.6} + 3}{2}\right\}$ or $\{3.98\}$ **85.** $\left\{\frac{10^{1.2} + 9}{10}\right\}$ or $\{2.48\}$

86. $\left\{\frac{10^{1.9} + 6}{5}\right\}$ or $\{17.09\}$ **87.** $\left\{\frac{e^{1.5} - 9}{4}\right\}$ or $\{-1.13\}$

88. $\left\{\frac{e^{-2.3} + 11}{7}\right\}$ or $\{1.59\}$ **90.** $\left\{\frac{e^{2.7}}{10}\right\}$ or $\{1.49\}$

91. $\left\{\frac{10^0 - 3}{4}\right\}$ or $\{-0.5\}$ **92.** $\left\{\frac{e^0 + 5}{4}\right\}$ or $\{1.5\}$

103.
```
(10^1.7+15)/4
      16.27968084
log(4*Ans-15)
              1.7
```

104.
```
(e^(-1.5)-3)/5
      -.555373968
ln(5Ans+3)
             -1.5
```

105. 5
```
log(32)/log(2)
                5
ln(32)/ln(2)
                5
```

106. 4
```
log(625)/log(5)
                4
ln(625)/ln(5)
                4
```

Section 10.7

1. Answers vary. When solving for x, apply the change-of-base formula to evaluate $x = \log_7 15$ using either the common or natural logarithm.
2. Answers vary. Since the given base is e, you should use the natural logarithm to solve for x.
3. Answers vary. Since the given base is 10, you should use the common logarithm to solve for x.
4. Answers vary. You rewrite the equation in exponential form and solve the resulting equation for x.
5. Answers vary. You apply the product rule of logarithms to combine the sum of logarithms on the left side into one logarithm. Then write the resulting equation in exponential form and solve for x.
6. Answers vary. You apply the quotient rule of logarithms to combine the difference of logarithms on the left side into one logarithm. Then write the resulting equation in exponential form and solve for x.

11. $\left\{\frac{\log 90}{7}\right\}$ or $\{0.28\}$ **12.** $\left\{\frac{\log 88}{5}\right\}$ or $\{0.39\}$

13. $\left\{\frac{\ln 13}{4}\right\}$ or $\{0.64\}$ **14.** $\left\{\frac{\ln 31}{6}\right\}$ or $\{0.57\}$

15. $\left\{\frac{\ln 38}{\ln 7} + 2\right\}$ or $\{3.87\}$ **16.** $\left\{\frac{\ln 12}{\ln 4} - 9\right\}$ or $\{-7.21\}$

17. $\left\{\frac{\ln 19}{4\ln 5.75}\right\}$ or $\{0.42\}$ **18.** $\left\{\frac{\ln 11}{2\ln 6.71}\right\}$ or $\{0.63\}$

19. $\left\{\frac{\ln 94 - \ln 9}{5\ln 0.53}\right\}$ or $\{-0.74\}$ **20.** $\left\{\frac{\ln 92 - \ln 3}{4\ln 1.26}\right\}$ or $\{3.70\}$

47. $\left\{\frac{\ln 37.5}{4.6}\right\}$ or $\{0.79\}$ **51.** $\left\{\frac{\ln 2 + 8\ln 12}{6\ln 12}\right\}$ or $\{1.38\}$

57. $\left\{\frac{-\ln 7}{4\ln 4.08}\right\}$ or $\{-0.35\}$ **58.** $\left\{\frac{-\ln 5}{4\ln 3.49}\right\}$ or $\{-0.32\}$

63. $\left\{\frac{\ln 21 - \ln 5}{7\ln 5}\right\}$ or $\{0.13\}$ **69.** $\left\{\frac{\ln 2}{\ln 6.25}\right\}$ or $\{0.38\}$

73. $\left\{\frac{\log 17.5}{4.7}\right\}$ or $\{0.26\}$

81. The population of Japan in 2029 will be 120.30 million. It will take about 34.22 yr to reach 115 million.

82. The population of the United States in 2026 will be 361.91 million. It will take about 8.52 yr to reach 340 million.

88.
$$\begin{aligned}
2\log_2 x - \log_2(8x - 12) &= 0\\
\log_2 x^2 - \log_2(8x - 12) &= 0\\
\log_2 \frac{x^2}{8x - 12} &= 0\\
\frac{x^2}{8x - 12} &= 2^0 = 1\\
x^2 &= 8x - 12\\
x^2 - 8x + 12 &= 0\\
(x - 6)(x - 2) &= 0\\
x &= 6, 2
\end{aligned}$$

89. $x = \frac{\ln 64}{8\ln 4.02} \approx 0.374$

```
ln(64)/(8ln(4.02
))
        .37365568
4.02^(8Ans)
               64
```

90. $x = \frac{\ln 18 + 2\ln 4}{7\ln 4} \approx 0.584$

```
(ln(18)+2ln(4))/
(7ln(4))
      .5835660715
4^(7Ans-2)
               18
```

91. $x = 4$

```
ln(4)/ln(6)+ln(4
+5)/ln(6)
                2
```

92. $x = \frac{15}{4}$

```
ln(3*15/4)/ln(3)
-ln(3*15/4-10)/l
n(3)
                2
```

CHAPTER 10 REVIEW EXERCISES

Section 10.1

1. $(f + g)(x) = x^2 + x - 9$, $(f - g)(x) = -x^2 + 3x - 5$; domain: $(-\infty, \infty)$
2. $(f \cdot h)(x) = -4x^2 + 22x - 28$; domain: $(-\infty, \infty)$
   $\left(\frac{f}{h}\right)(x) = \frac{2x - 7}{4 - 2x}, x \neq 2$
   domain: $(-\infty, 2) \cup (2, \infty)$
3. $(f \circ h)(x) = -4x + 1$, $(h \circ f)(x) = -4x + 18$; domain: $(-\infty, \infty)$
4. $(f \cdot g)(x) = 2x^3 - 9x^2 + 3x + 14$; domain: $(-\infty, \infty)$
   $\left(\frac{f}{g}\right)(x) = \frac{2x - 7}{(x - 2)(x + 1)}, x \neq -1, 2$
   domain: $(-\infty, -1) \cup (-1, 2) \cup (2, \infty)$

Section 10.2

35. $f^{-1}(x) = \{(-5, -2), (1, 0), (5, 3), (3, 2), (7, 4)\}$

36. $f^{-1}(x) = \{(-3.2, -3), (-2.5, -1), (0.9, 2), (2.1, 5), (6.9, 9)\}$

38.

Congressmen in 2009 (x)	State (y)
Harry Mitchell	Arizona
Lynn Jenkins	Kansas
Connie Mack	Florida
Debbie Halvorson	Illinois
Marcy Kaptur	Ohio

39. $f^{-1}(x) = \frac{x + 4}{5}$ **40.** $f^{-1}(x) = \frac{5x + 11}{2}$

41. $f^{-1}(x) = \left(\frac{x - 9}{2}\right)^{1/3}$ **42.** $f^{-1}(x) = \frac{3x - 6}{5x}$

47.

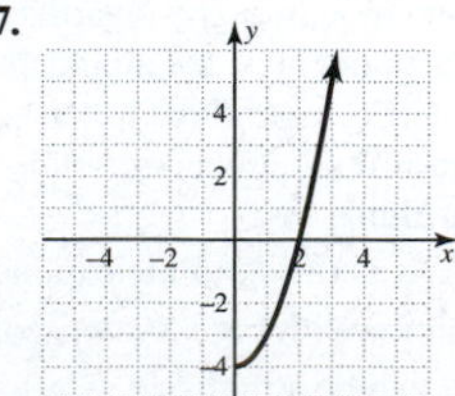

48.

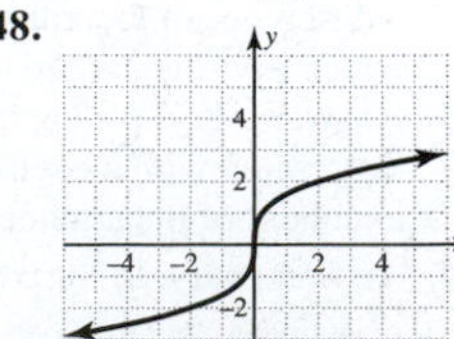

Section 10.3

53.

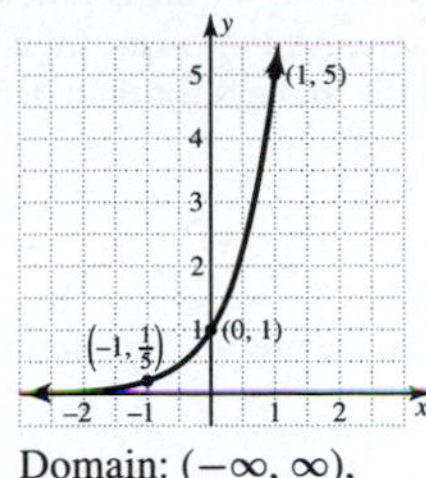

Domain: $(-\infty, \infty)$,
Range: $(0, \infty)$

54.

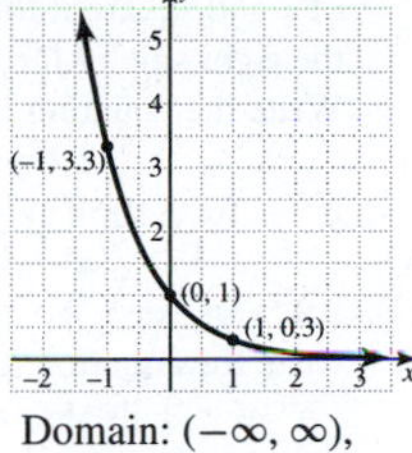

Domain: $(-\infty, \infty)$,
Range: $(0, \infty)$

Section 10.4

79.

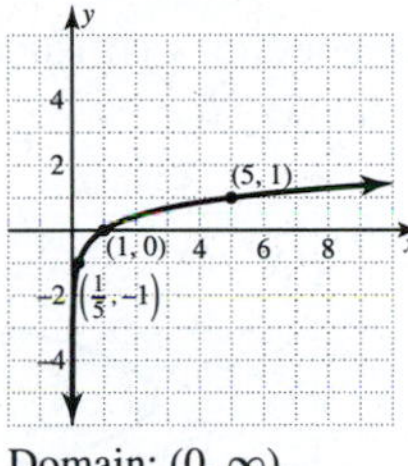

Domain: $(0, \infty)$,
Range: $(-\infty, \infty)$

80.

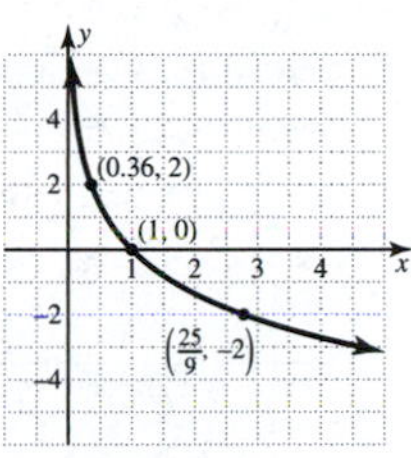

Domain: $(0, \infty)$,
Range: $(-\infty, \infty)$

Section 10.5

87. $2\log_{10}x + 5\log_{10}y + \frac{1}{2}\log_{10}(x-1)$

88. $2 + \log_{10}(x^2+4) - 4\log_{10}y$

Section 10.6

103. $\left\{\frac{e^4+5}{2}\right\}$ or $\{29.80\}$ **104.** $\left\{\frac{997}{7}\right\}$ or $\{142.43\}$

106. $\left\{\frac{e^{1.5}-17}{2}\right\}$ or $\{-6.26\}$

Section 10.7

115. $\left\{\frac{\log 32.6}{5.2}\right\}$ or $\{0.29\}$ **117.** $\left\{\frac{\ln 28 + 2\ln 7}{3\ln 7}\right\}$ or $\{1.24\}$

118. $\left\{\frac{2+\log 3.5}{5}\right\}$ or $\{0.51\}$ **121.** $\left\{\frac{83}{5}\right\}$ or $\{16.6\}$

124. $\left\{\frac{\ln 14}{3\ln 2.17}\right\}$ or $\{1.14\}$

CHAPTER 10 TEST

8. $r(x) = 10x$, $f(x) = 25x$, $p(x) = \frac{40}{3}x$, $T(x) = \frac{145}{3}x$

9. b. $(f \cdot g)(x) = \frac{36x+9}{x-2}$; domain: $(-\infty, 2) \cup (2, \infty)$

10. a.

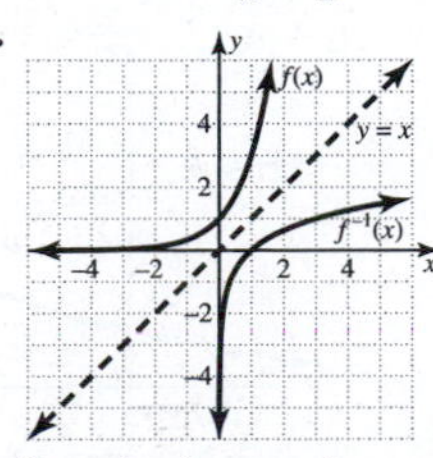

Domain: $(-\infty, \infty)$,
Range: $(0, \infty)$

b.

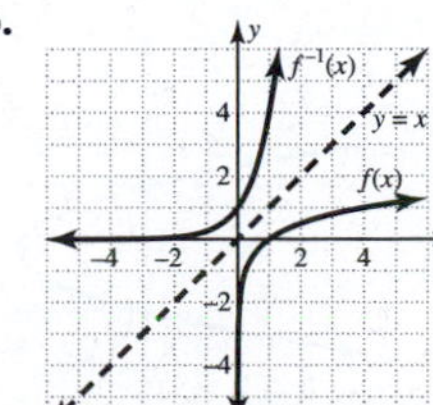

Domain: $(0, \infty)$,
Range: $(-\infty, \infty)$

CHAPTERS 1–10 CUMULATIVE REVIEW EXERCISES

13. $f(x) = \frac{4}{7}x - 4$, $m = \frac{4}{7}$, $b = -4$

27.

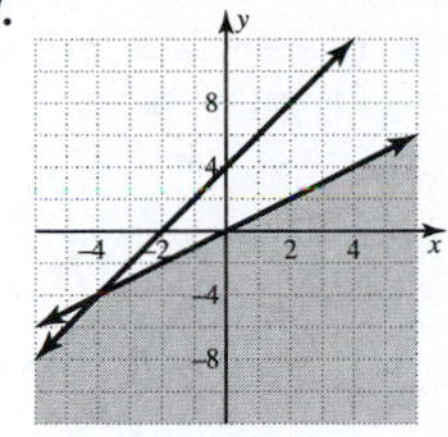

28.

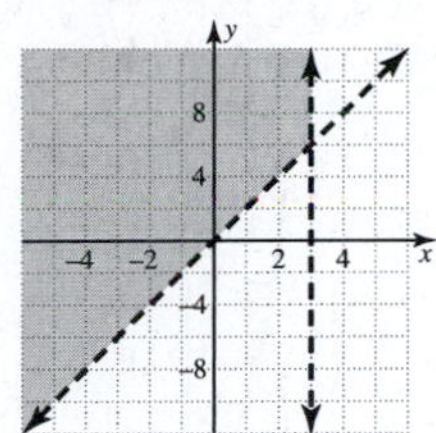

51. $\left\{\frac{1}{5}, -\frac{7}{2}\right\}$ **59.** $-x^2 + \frac{3}{2}x - 2$ **60.** $\frac{1}{2}x^2 - x - \frac{2}{3}$

61. $7x^2 + 3x - 3 - \frac{11}{x+1}$ **62.** $6x^2 + 3x + 4 + \frac{6}{2x-1}$

81. $a + a^{1/2}b^{1/3} - 2b^{2/3}$ **82.** $4x^{4/5} - 4x^{2/5}y^{1/4} + y^{1/2}$

83. $2x^2y\sqrt{30xy}$ **84.** $3t^3\sqrt{10rt}$ **85.** $\frac{x\sqrt[3]{x}}{2y}$ **86.** $\frac{2a\sqrt[5]{a^2}}{b}$

91. x-intercepts: $(-4, 0)$, $(-2, 0)$; y-intercept: $(0, 8)$; vertex: $(-3, -1)$; axis of symmetry: $x = -3$; shift graph of $y = x^2$ left 3 units and down 1 unit; domain: $(-\infty, \infty)$; range: $[-1, \infty)$

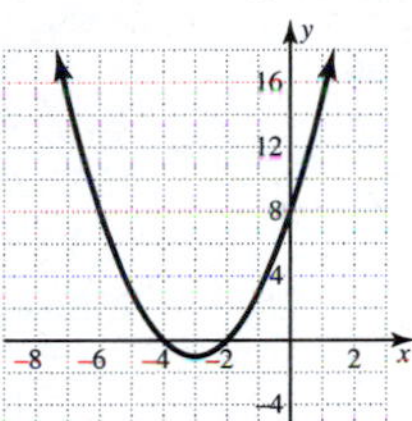

92. x-intercepts: $(-4, 0)$, $(0, 0)$; y-intercept: $(0, 0)$; vertex: $(-2, 4)$; axis of symmetry: $x = -2$; shift graph of $y = x^2$ left 2 units, reflect over the x-axis, and shift up 4 units; domain: $(-\infty, \infty)$; range: $(-\infty, 4]$

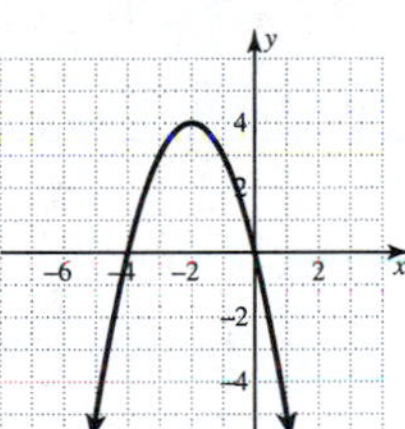

97. $\left\{-\frac{1}{27}, \frac{27}{8}\right\}$ **99.** $\left(-\infty, -\frac{5}{2}\right) \cup (0, 3)$

100. $(-\infty, -3) \cup (-1, 1) \cup (3, \infty)$

101. $(f+g)(x) = 8x + 2$
$(f-g)(x) = 2x + 6$
$(f \cdot g)(x) = 15x^2 + 2x - 8$
domain: $(-\infty, \infty)$
$\left(\frac{f}{g}\right)(x) = \frac{5x+4}{3x-2}, x \neq \frac{2}{3}$
domain: $\left(-\infty, \frac{2}{3}\right) \cup \left(\frac{2}{3}, \infty\right)$

102. $(f+g)(x) = \sqrt{x+2} + 3x - 6$
$(f-g)(x) = \sqrt{x+2} - 3x + 6$
$(f \cdot g)(x) = 3x\sqrt{x+2} - 6\sqrt{x+2}$
domain: $[-2, \infty)$
$\left(\frac{f}{g}\right)(x) = \frac{\sqrt{x+2}}{3x-6}, x \neq 2$
domain: $[-2, 2) \cup (2, \infty)$

107. $(f \circ g)(x) = -3x + 7$
$(g \circ f)(x) = -3x + 9$
domain: $(-\infty, \infty)$

108. $(f \circ g)(x) = 14x - 10$
$(g \circ f)(x) = 14x - 20$
domain: $(-\infty, \infty)$

111. $f^{-1}(x) = \dfrac{\sqrt[3]{x} + 2}{7}$

115. Domain: $(-\infty, \infty)$,
Range: $(0, \infty)$

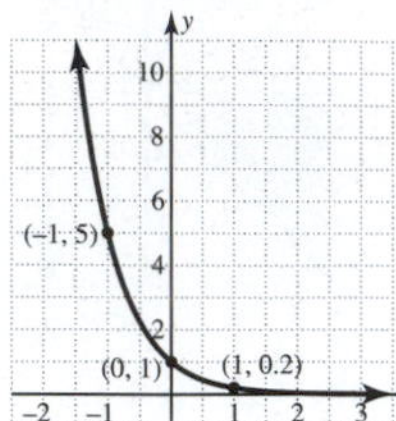

116. Domain: $(-\infty, \infty)$,
Range: $(0, \infty)$

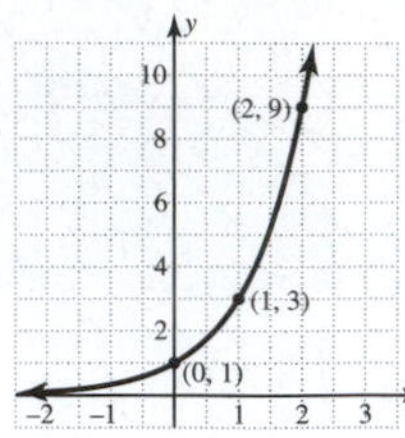

125. Domain: $(0, \infty)$,
Range: $(-\infty, \infty)$

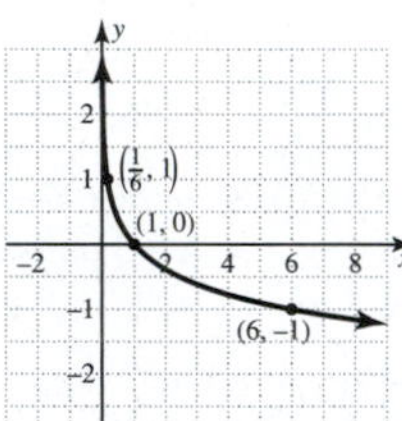

126. Domain: $(0, \infty)$,
Range: $(-\infty, \infty)$

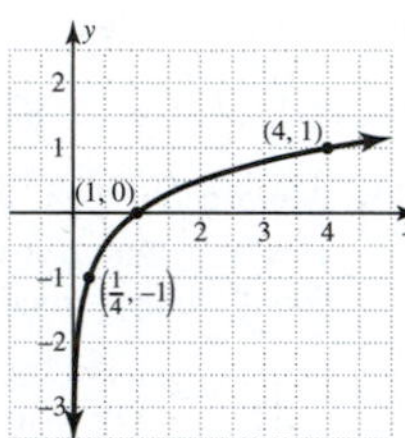

129. $\log_4\left(\dfrac{x}{x+6}\right)$

133. $3\ln x + \dfrac{1}{2}\ln(x-1)$

134. $3\log_6 y - \dfrac{1}{2}\log_6(y+12)$

135. $\log_{10}\dfrac{x^5}{y^2}$

136. $\log_2 \sqrt{x}y^2$

145. $\left\{\dfrac{10^{1.5}+4}{5}\right\}$ or $\{7.12\}$

146. $\left\{\dfrac{e^{4.5}-21}{2}\right\}$ or $\{34.51\}$

147. $\left\{\dfrac{\ln 72}{8\ln 0.75}\right\}$ or $\{-1.86\}$

148. $\left\{\dfrac{\ln 91}{5\ln 0.44}\right\}$ or $\{-1.10\}$

CHAPTER 11

Section 11.1

1. Answers vary. The vertex is $(1, 3)$. The axis of symmetry is $y = 3$. There is no y-intercept. The x-intercept is $\left(\dfrac{11}{2}, 0\right)$. Since $a = \dfrac{1}{2} > 0$, the parabola opens to the right.

2. Answers vary. The vertex is $(-3, -1)$. The axis of symmetry is $y = -1$. There is no y-intercept. The x-intercept is $(-5, 0)$. The parabola opens to the left since $a = -2$.

3. Answers vary. We rewrite the equation as $x = -1(y - 0)^2 + 25$. The vertex is $(25, 0)$. The y-intercepts are $(0, 5)$ and $(0, -5)$. Plot the points and connect them to draw a parabola that opens to the left since $a = -1$.

4. Answers vary. We need to write the equation in standard form by completing the square for the y-terms. First, isolate the y-terms to one side of the equation. Then find one-half of the coefficient of the y-term and square it. Add the number to each side of the equation. Then isolate the x-variable.

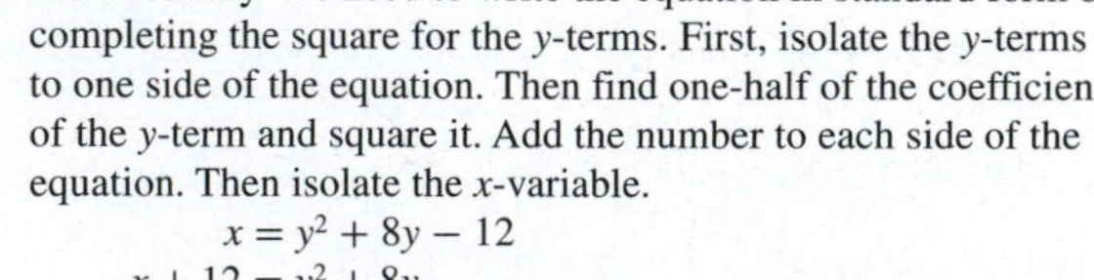

$$x = y^2 + 8y - 12$$
$$x + 12 = y^2 + 8y$$
$$x + 12 + 16 = y^2 + 8y + 16$$
$$x + 28 = (y + 4)^2$$
$$x = (y + 4)^2 - 28$$

5. Answers vary. You can easily identify the center and radius of the circle.

6. Answers vary. Since the distance between a point (x, y) on the circle and the center (h, k) is always the radius of the circle, we can use the distance formula to write the equation of the circle.

7. Answers vary. We need to complete the squares for the x-terms and the y-terms. For the x-terms, take one-half of the coefficient of x and square it to complete the square. Add the same number to the right side of the equation. Repeat the process for the y-terms. Write the equation in standard form.

$$x^2 + y^2 - 6x + 10y = 15$$
$$(x^2 - 6x + 9) + (y^2 + 10y + 25) = 15 + 9 + 25$$
$$(x - 3)^2 + (y + 5)^2 = 49$$

8. Answers vary. We need to complete the square for the x-terms and write the equation in standard form.

$$x^2 + y^2 + 4x + 5 = 0$$
$$(x^2 + 4x + 4) + y^2 = -5 + 4$$
$$(x + 2)^2 + y^2 = -1$$

Since $r^2 \neq -1$ for any real number r, this equation does not give an equation of a circle.

9. vertex: $(3, -2)$; axis of symmetry: $y = -2$

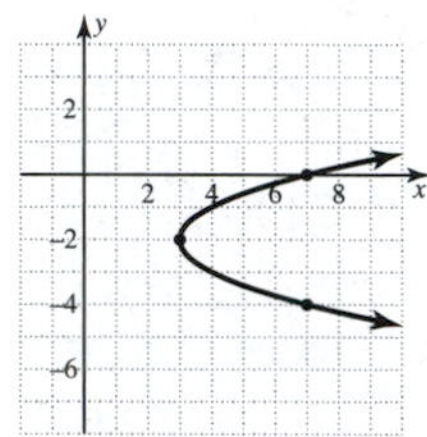

10. vertex: $(5, -1)$; axis of symmetry: $y = -1$

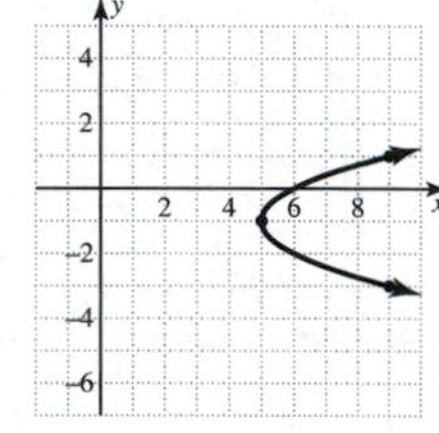

11. vertex: $(1, 2)$; axis of symmetry: $y = 2$

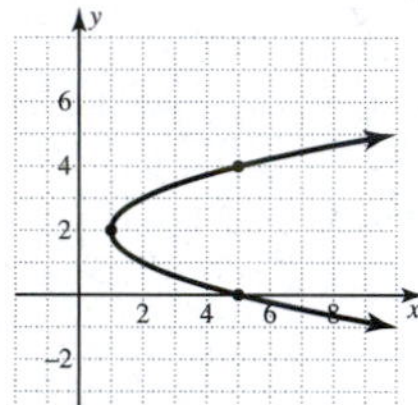

12. vertex: $(2, 4)$; axis of symmetry: $y = 4$

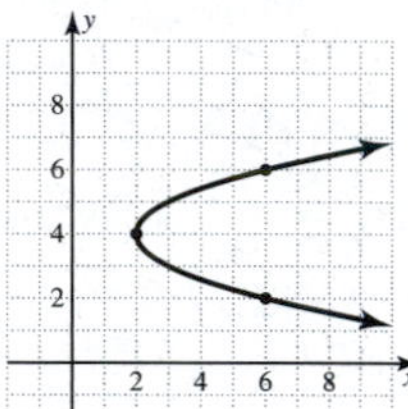

13. vertex: $(-4, 3)$; axis of symmetry: $y = 3$

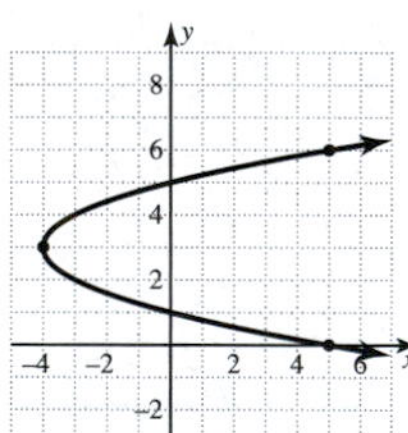

14. vertex: $(-6, 2)$; axis of symmetry: $y = 2$

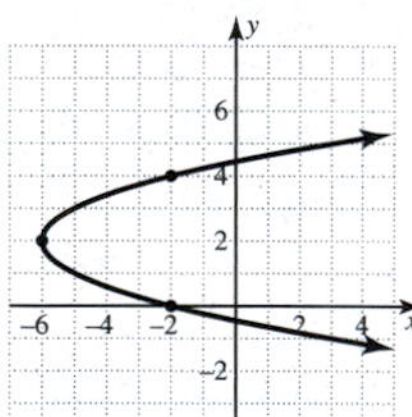

15. vertex: $(4, -2)$; axis of symmetry: $y = -2$

16. vertex: $(5, 4)$; axis of symmetry: $y = 4$

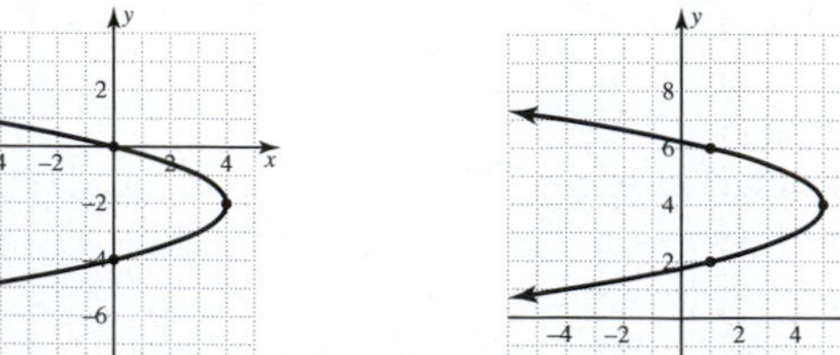

17. vertex: $(-1, 2)$; axis of symmetry: $y = 2$

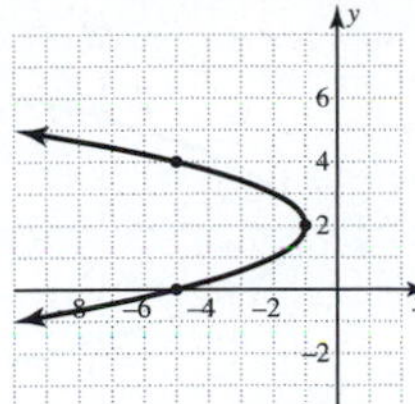

18. vertex: $(-2, -2)$; axis of symmetry: $y = -2$

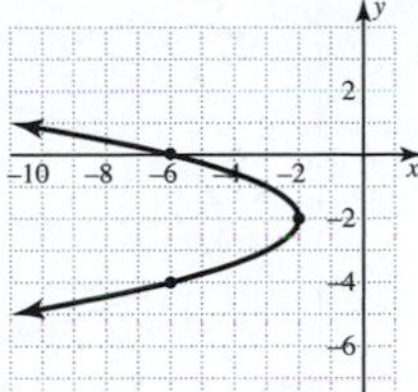

19. vertex: $(9, 0)$; axis of symmetry: $y = 0$

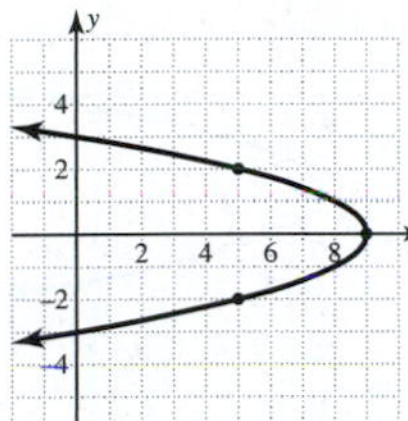

20. vertex: $(-4, 0)$; axis of symmetry: $y = 0$

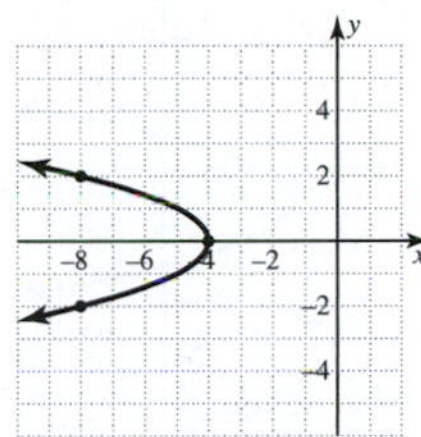

21. vertex: $(-9, 0)$; axis of symmetry: $y = 0$

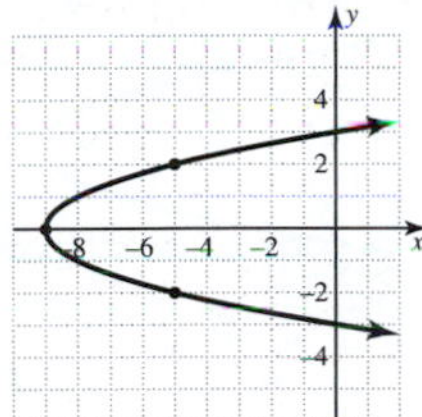

22. vertex: $(-6, 0)$; axis of symmetry: $y = 0$

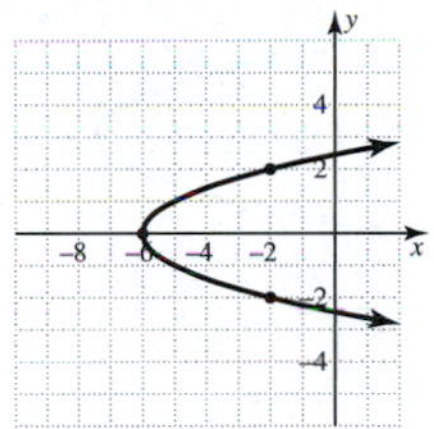

23. vertex: $(-9, -1)$; axis of symmetry: $y = -1$

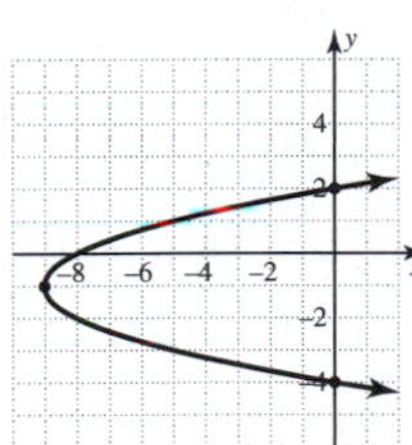

24. vertex: $(1, 3)$; axis of symmetry: $y = 3$

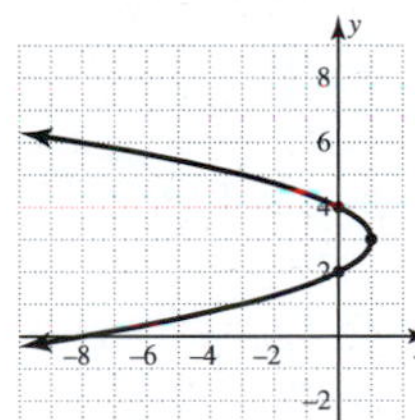

25. vertex: $\left(-\frac{1}{4}, -\frac{3}{2}\right)$; axis of symmetry: $y = -\frac{3}{2}$

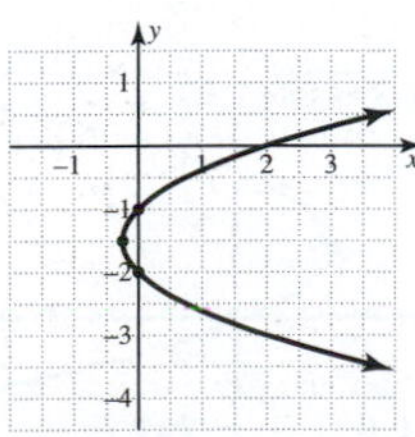

26. vertex: $\left(-\frac{9}{4}, -\frac{1}{2}\right)$; axis of symmetry: $y = -\frac{1}{2}$

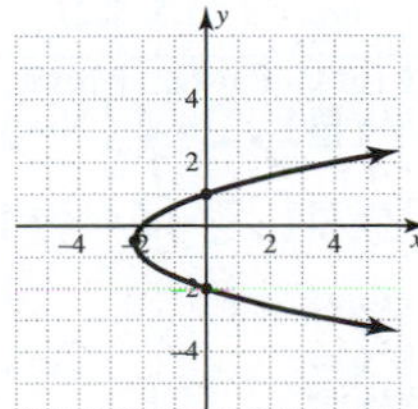

27. center: $(0, 0)$; radius: 5; points: $(5, 0)$, $(0, 5)$, $(-5, 0)$, $(0, -5)$

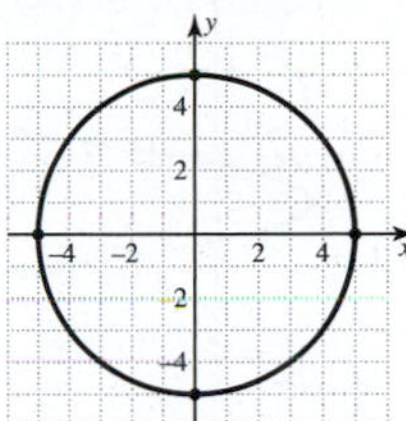

28. center: $(0, 0)$; radius: 3; points: $(3, 0)$, $(0, 3)$, $(-3, 0)$, $(0, -3)$

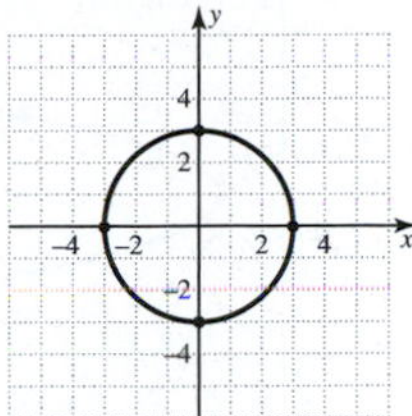

29. center: $(2, 1)$; radius: 3; points: $(2, 4)$, $(2, -2)$, $(5, 1)$, $(-1, 1)$

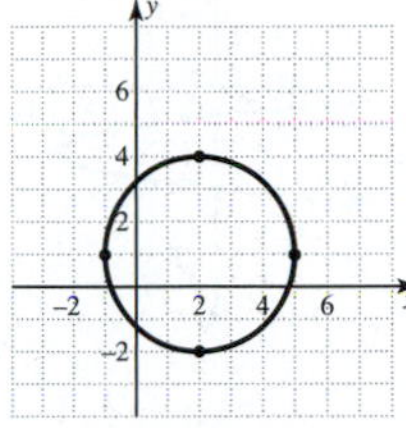

30. center: $(-1, 3)$; radius: 2; points: $(-1, 5)$, $(-1, 1)$, $(1, 3)$, $(-3, 3)$

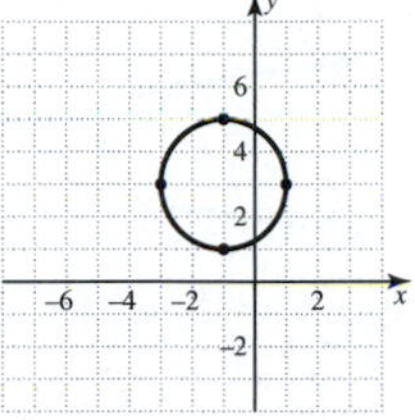

31. center: $(0, -3)$; radius: 5; points: $(0, 2)$, $(0, -8)$, $(5, -3)$, $(-5, -3)$

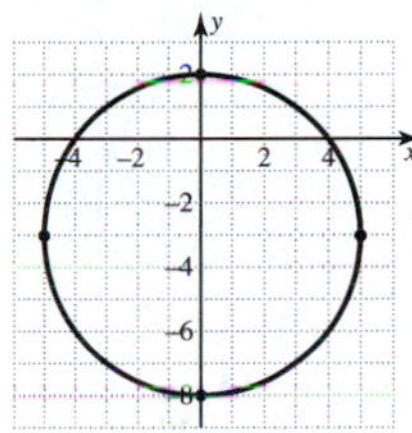

32. center: $(1, 0)$; radius: 2; points: $(1, 2)$, $(1, -2)$, $(-1, 0)$, $(3, 0)$

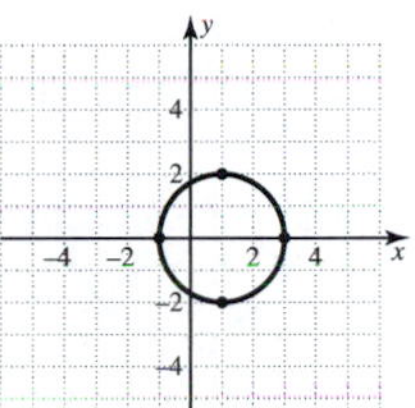

45. $\left(x - \frac{1}{2}\right)^2 + \left(y + \frac{1}{2}\right)^2 = 1$; center: $\left(\frac{1}{2}, -\frac{1}{2}\right)$; radius $= 1$

46. $\left(x + \frac{3}{2}\right)^2 + \left(y - \frac{1}{2}\right)^2 = 4$; center: $\left(-\frac{3}{2}, \frac{1}{2}\right)$; radius $= 2$

47. vertex: $(-5, -1)$; axis of symmetry: $y = -1$

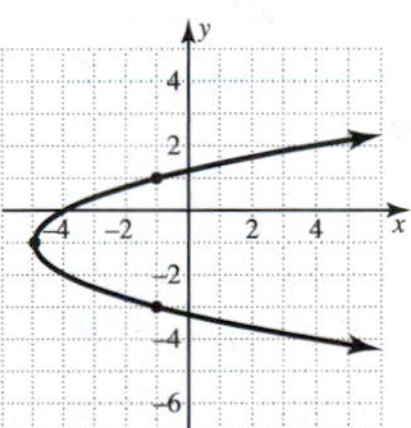

48. vertex: $(-4, 1)$; axis of symmetry: $y = 1$

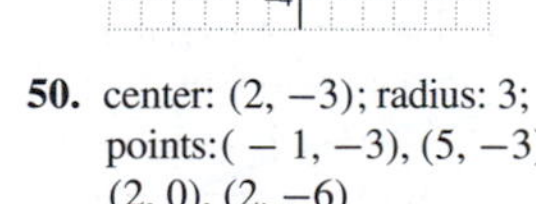

49. center: $(-1, 1)$; radius: 1; points: $(-1, 0)$, $(-1, 2)$, $(0, 1)$, $(-2, 1)$

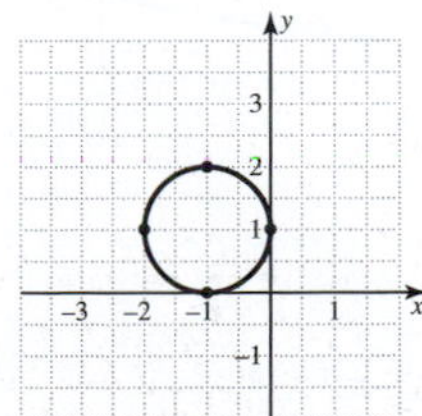

50. center: $(2, -3)$; radius: 3; points: $(-1, -3)$, $(5, -3)$, $(2, 0)$, $(2, -6)$

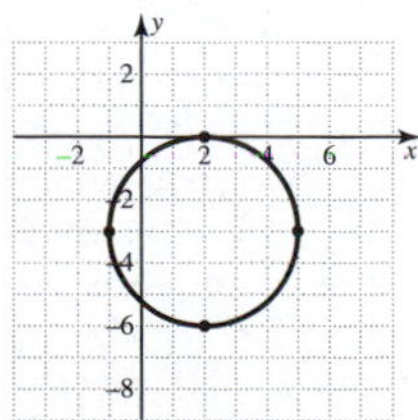

51. vertex: (4, 3); axis of symmetry: $y = 3$

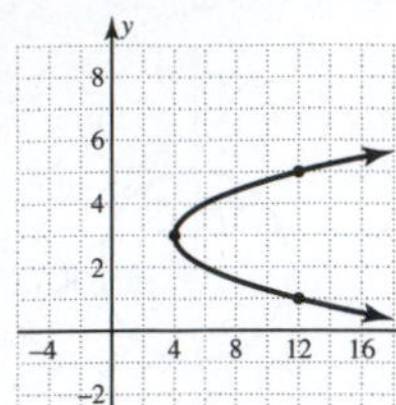

52. vertex: (−3, −2); axis of symmetry: $y = -2$

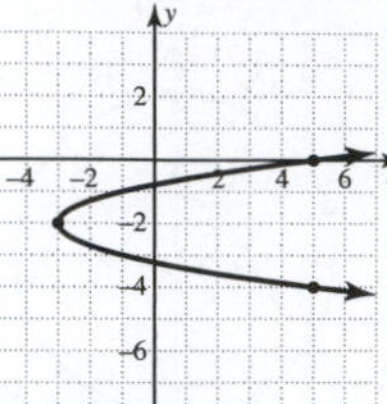

53. vertex: (1, 1); axis of symmetry: $y = 1$

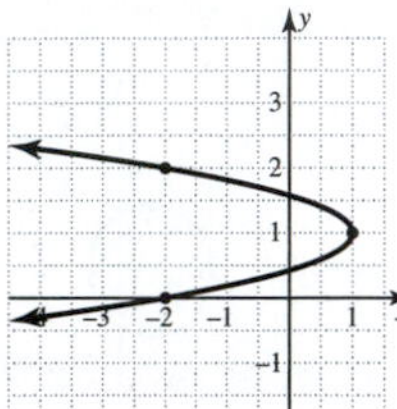

54. vertex: (5, −2); axis of symmetry: $y = -2$

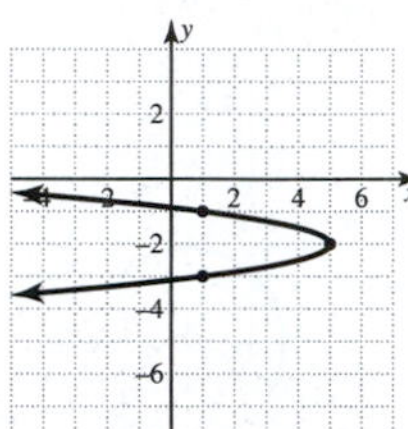

55. center: (1, 0); radius: 4; points: (1, 4), (1, −4), (5, 0), (−3, 0)

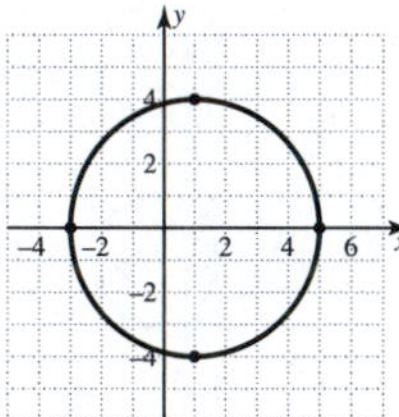

56. center: (−2, 0); radius: 2; points: (−4, 0), (0, 0), (−2, −2), (−2, 2)

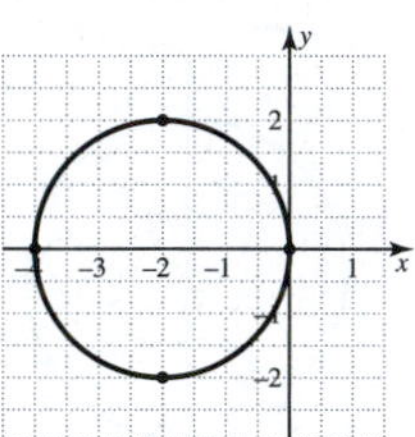

57. vertex: (3, 1); axis of symmetry: $y = 1$

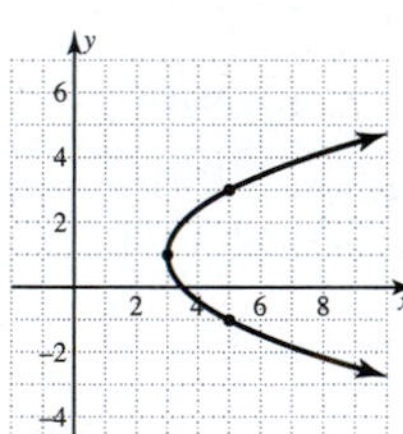

58. vertex: (−2, −1); axis of symmetry: $y = -1$

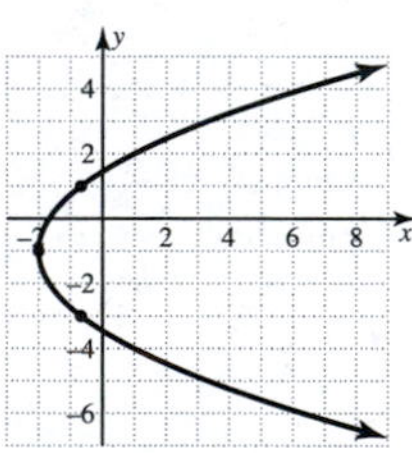

59. vertex: (4, 0); axis of symmetry: $y = 0$

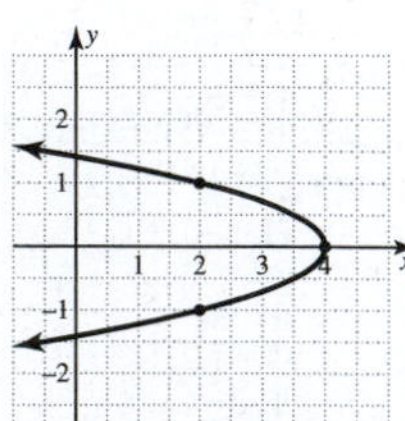

60. vertex: (1, 0); axis of symmetry: $y = 0$

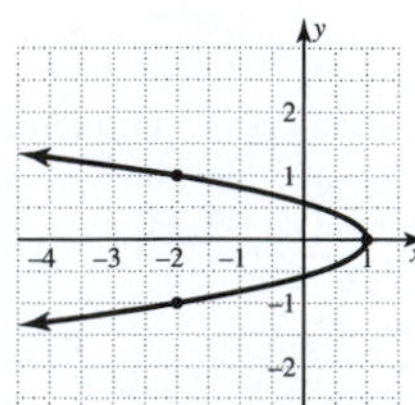

61. vertex: (3, 0); axis of symmetry: $y = 0$

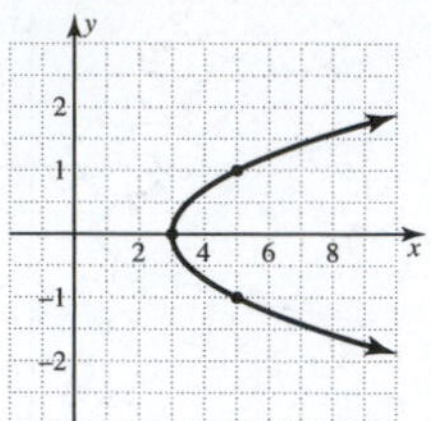

62. vertex: (−6, 0); axis of symmetry: $y = 0$

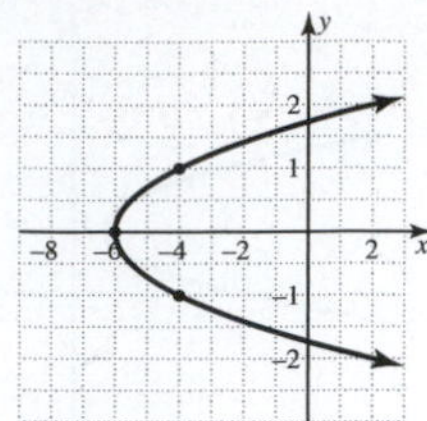

63. vertex: (−4, −1); axis of symmetry: $y = -1$

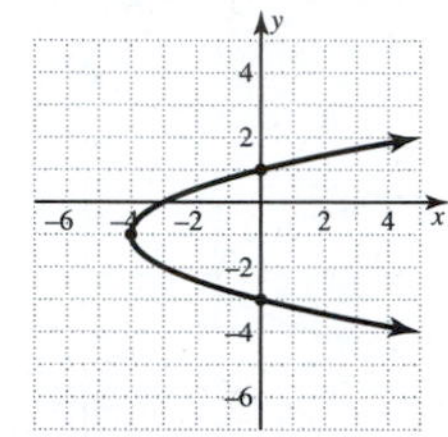

64. vertex: (−1, −3); axis of symmetry: $y = -3$

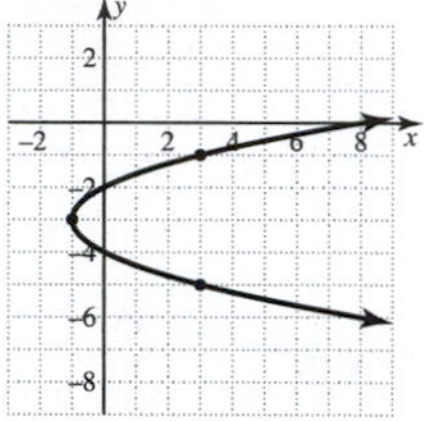

65. vertex: (25, 1); axis of symmetry: $y = 1$

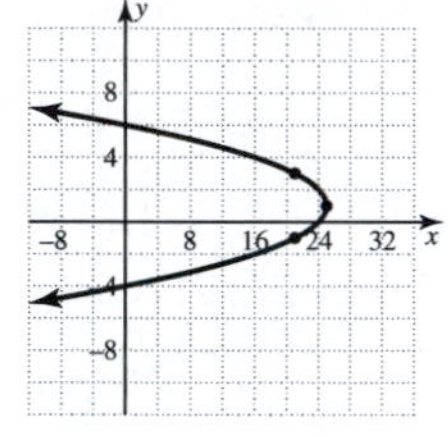

66. vertex: (5, 3); axis of symmetry: $y = 3$

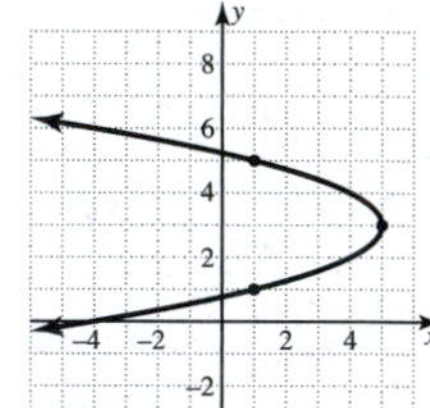

67. vertex: $\left(-\frac{5}{4}, \frac{5}{2}\right)$; axis of symmetry: $y = \frac{5}{2}$

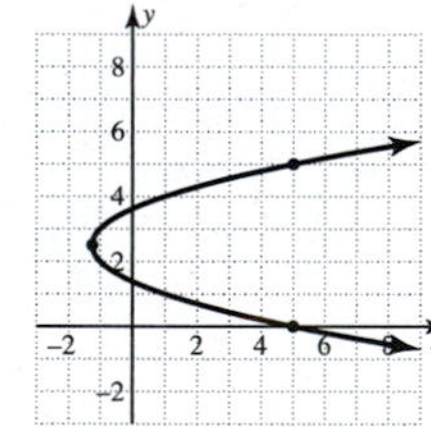

68. vertex: $\left(\frac{3}{4}, \frac{7}{2}\right)$; axis of symmetry: $y = \frac{7}{2}$

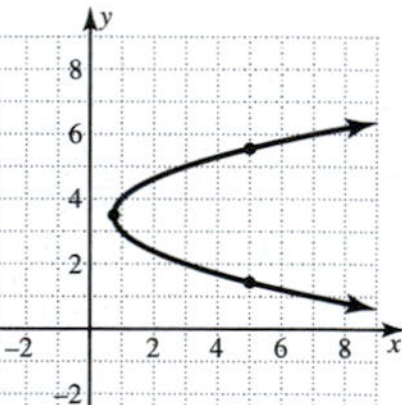

89. $\left(x - \frac{3}{2}\right)^2 + \left(y - \frac{5}{2}\right)^2 = 9$; center: $\left(\frac{3}{2}, \frac{5}{2}\right)$; radius = 3

90. $\left(x + \frac{7}{2}\right)^2 + \left(y - \frac{3}{2}\right)^2 = 16$; center: $\left(-\frac{7}{2}, \frac{3}{2}\right)$; radius = 4

95. vertex: (−1.5, 1.2)

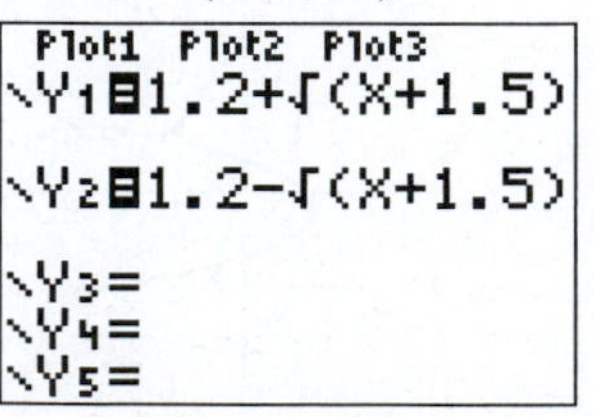

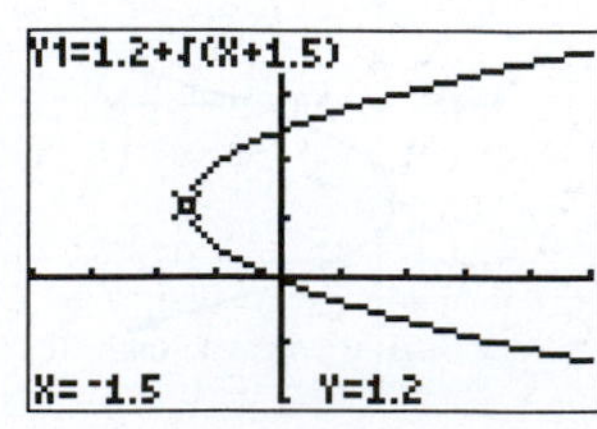

96. vertex: (3.2, −2.4)

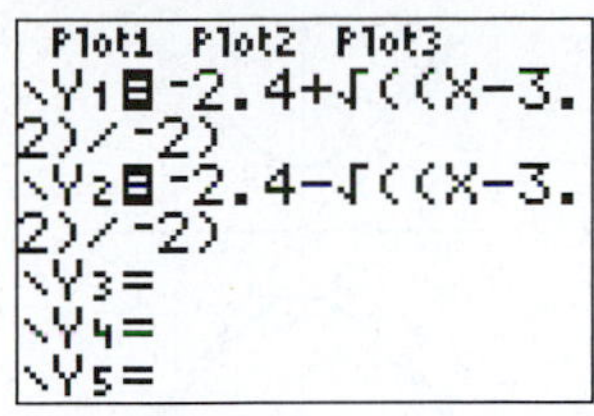

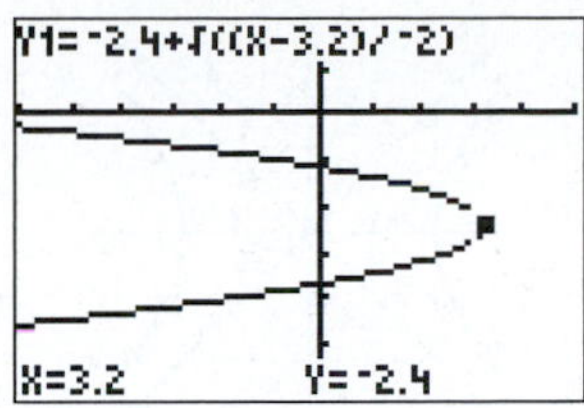

97.

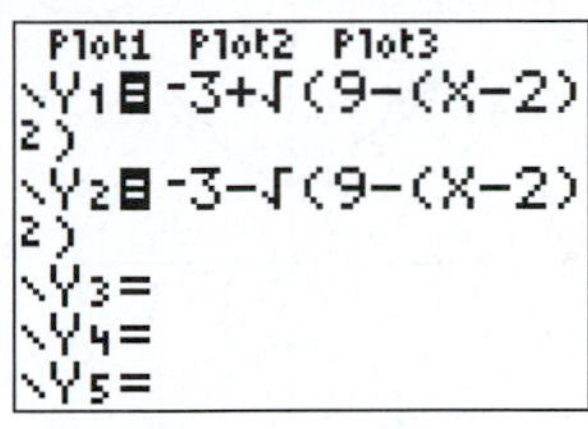

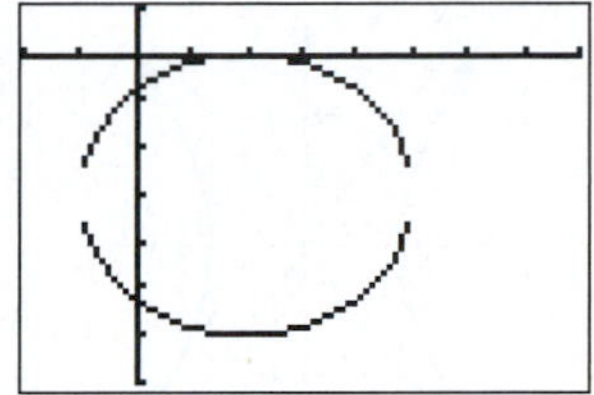

98.

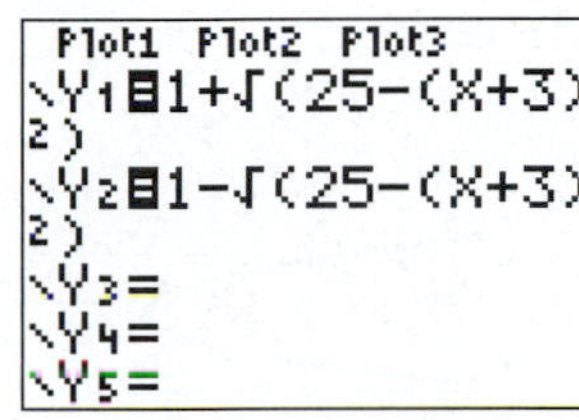

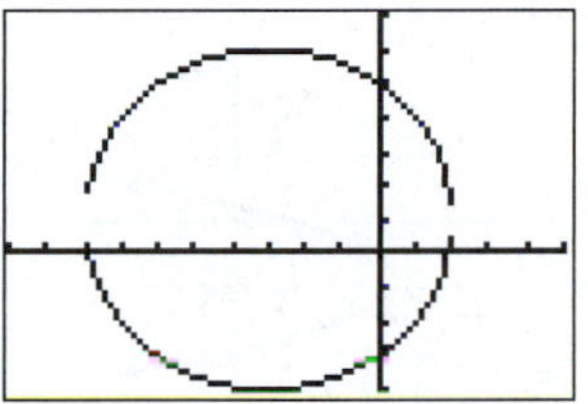

Section 11.2

1. Answers vary. Since the center of the ellipse is (0, 0), the key points will be the two x-intercepts and two y-intercepts. The two x-intercepts are (3, 0) and (−3, 0). The two y-intercepts are (0, 2) and (0, −2).
2. Answers vary. From the center of the ellipse, move 3 units right and left to find the two x-intercepts. From the center, move 2 units up and down to find the two y-intercepts. Connect the points with a smooth curve.
3. Answers vary. To find the x-intercepts, set y equal to zero and find the corresponding x-values. The x-intercepts are (5, 0) and (−5, 0).
4. Answers vary. Plot the four vertices of the rectangle (5, 2), (5, −2), (−5, 2), and (−5, −2). Draw two diagonals with dashed lines, through the vertices of the rectangle. The diagonals are the two asymptotes of the hyperbola.
5. Answers vary. The hyperbola is horizontal with x-intercepts of (5, 0) and (−5, 0). Draw the rectangle and asymptotes as described in Exercise 4. Draw two branches of the hyperbola passing through the x-intercepts and approaching the asymptotes.
6. Answers vary. The equation $\frac{x^2}{9} - \frac{y^2}{4} = 1$ represents a hyperbola that is horizontal with x-intercepts. The equation $\frac{y^2}{9} - \frac{x^2}{4} = 1$ represents a hyperbola that is vertical with y-intercepts.
7. Answers vary. To graph the ellipse, you need to solve for y first and enter the resulting equations as Y_1 and Y_2. The intercepts of the ellipse are (12, 0), (−12, 0), (0, 15), and (0, −15). If the viewing window is set to [−10, 10] by [−10, 10], you will not see the graph.
8. Answers vary. To graph the hyperbola, you need to solve for y first and enter the resulting equations as Y_1 and Y_2. The hyperbola is vertical with intercepts (0, 12) and (0, −12). If the viewing window is set to [−10, 10] by [−10, 10], you will not see the graph.

9.

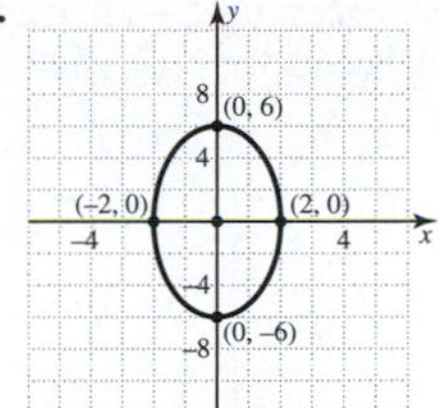

10.

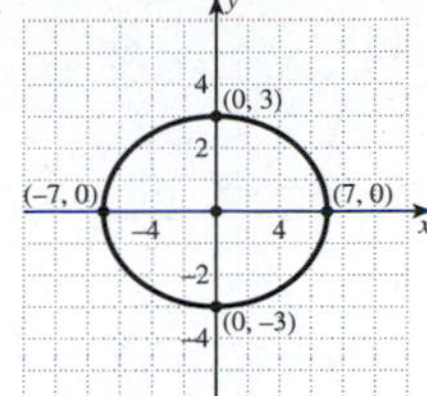

11.

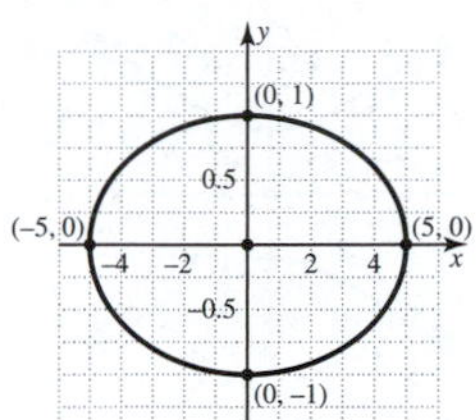

12.

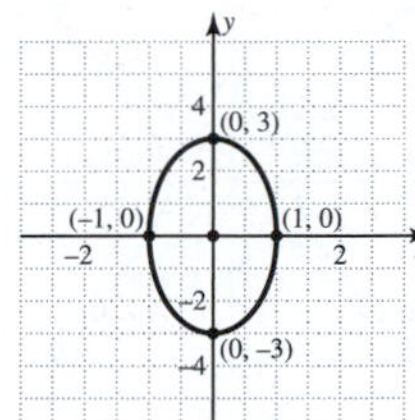

13.

14.

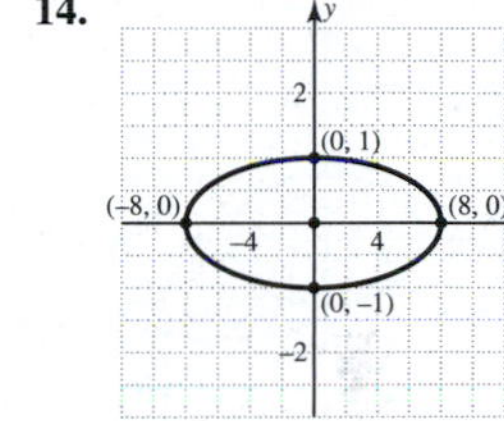

15.

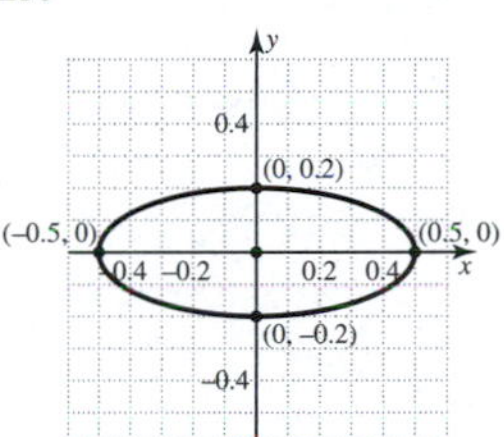

16.

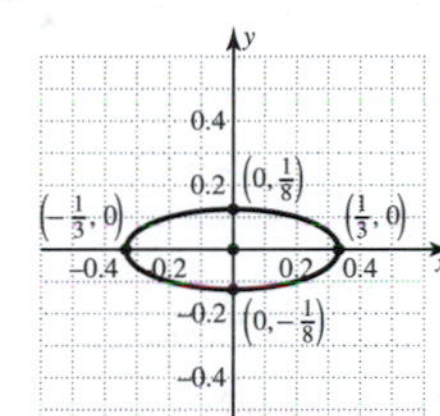

17.

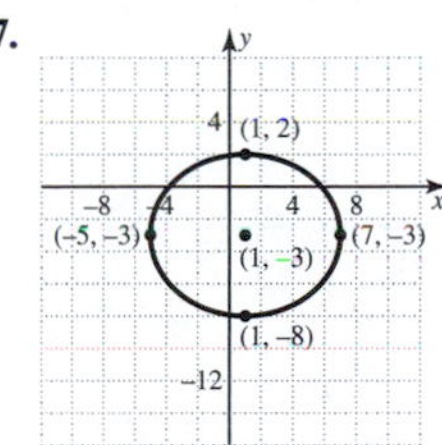

18.

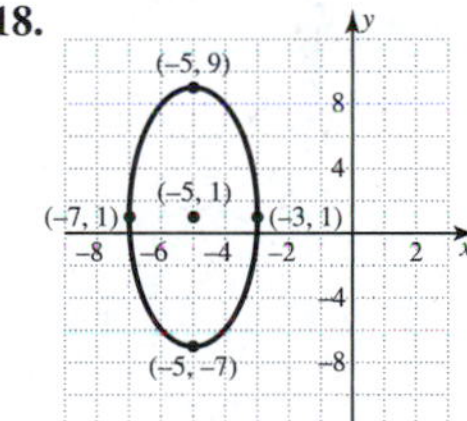

19.

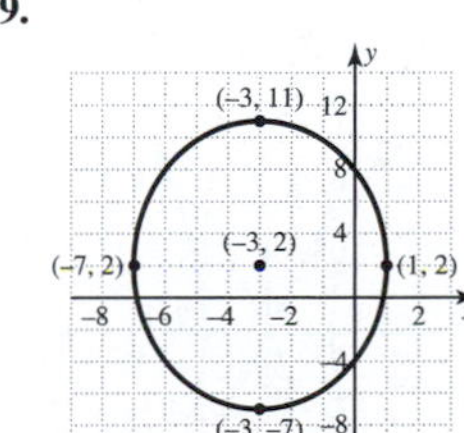

20.

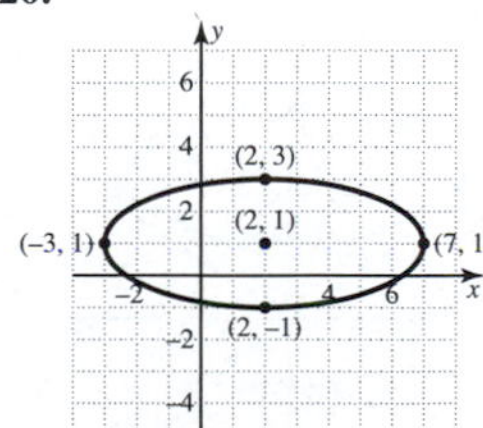

21.

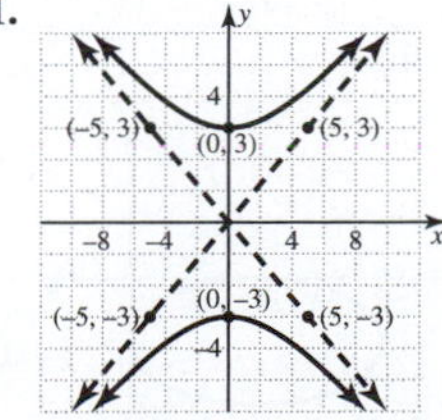

22.

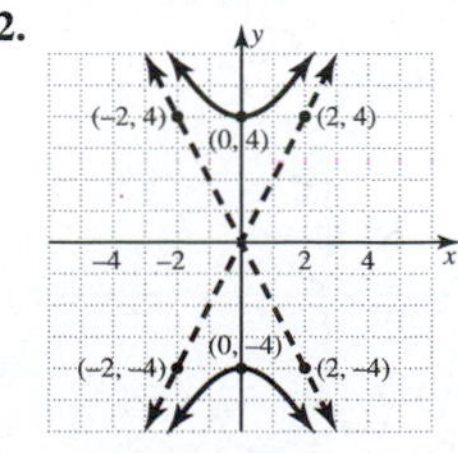

23.

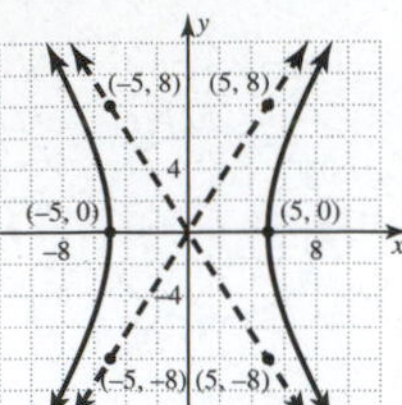

24.

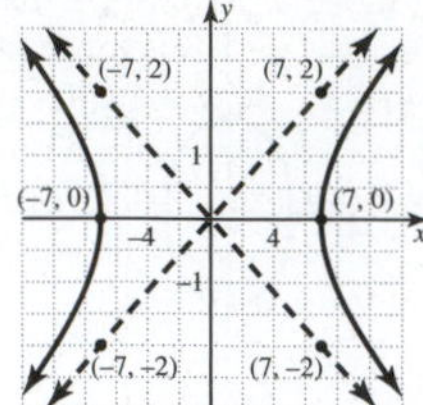

25.

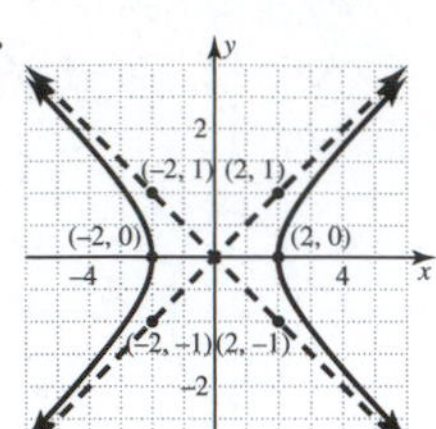

26.

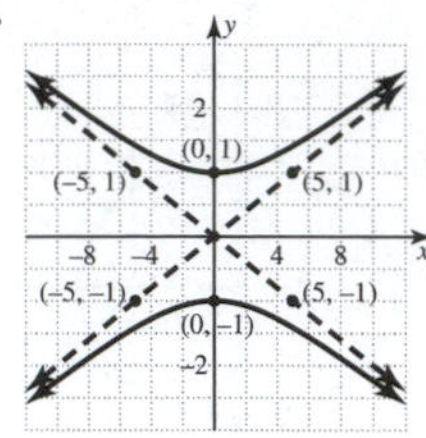

27. ellipse

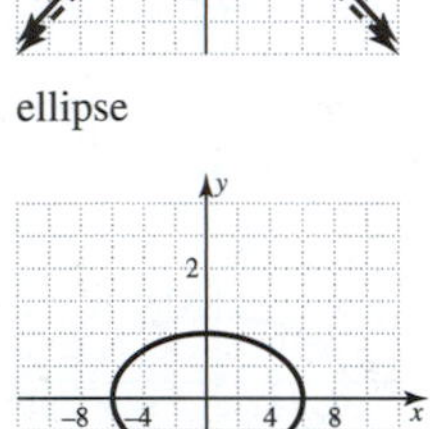

28. ellipse

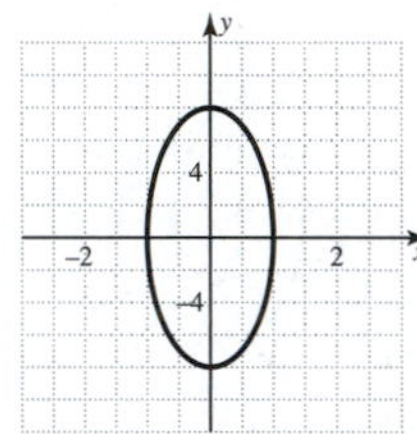

29. parabola

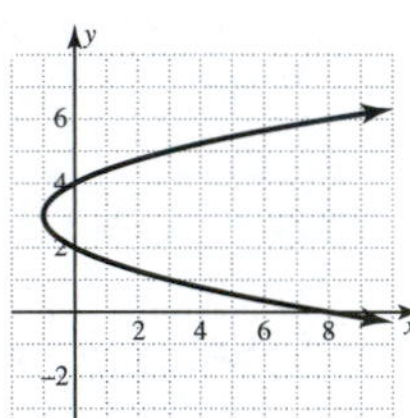

30. parabola

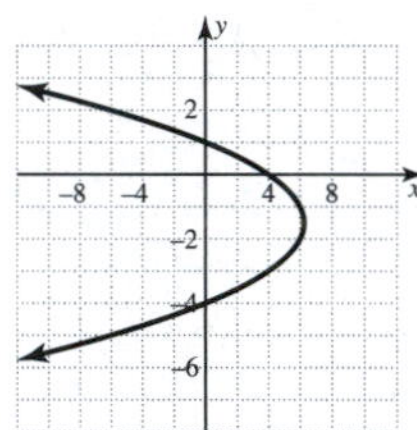

31. circle

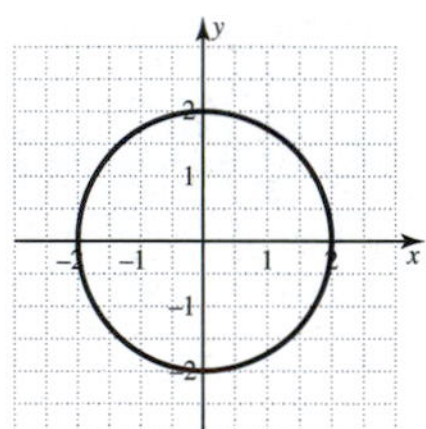

32. circle

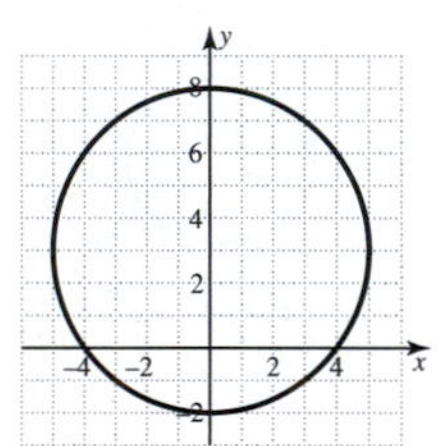

33. hyperbola

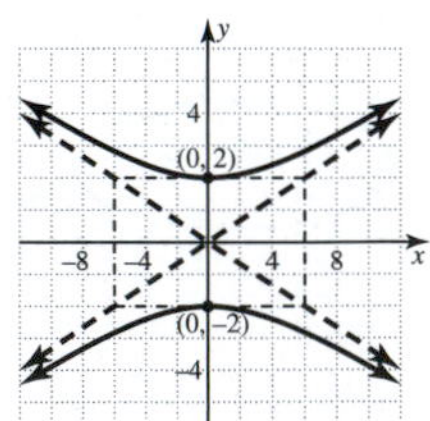

34. hyperbola

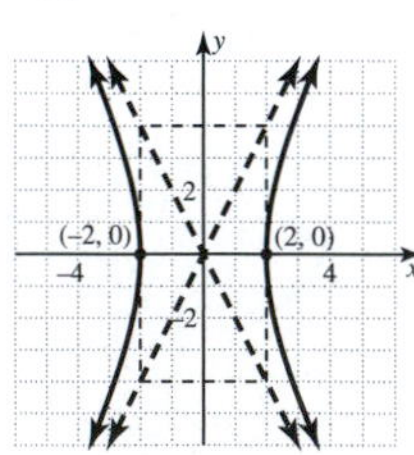

35. ellipse

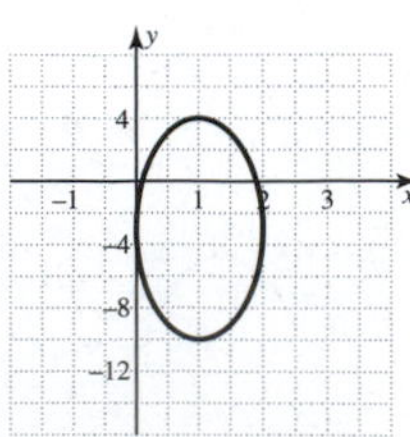

36. ellipse

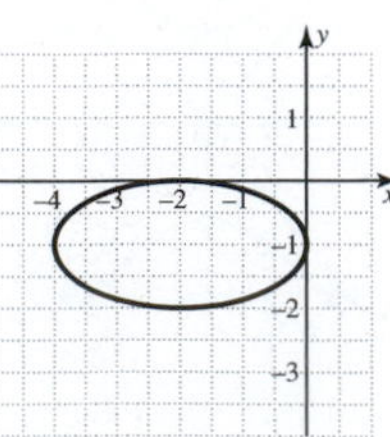

37. hyperbola

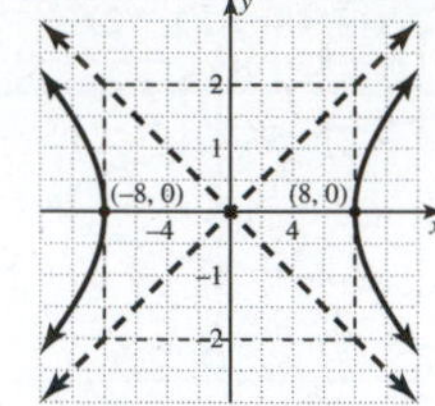

38. hyperbola

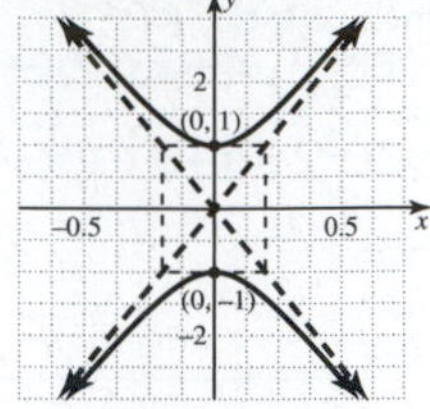

39. hyperbola

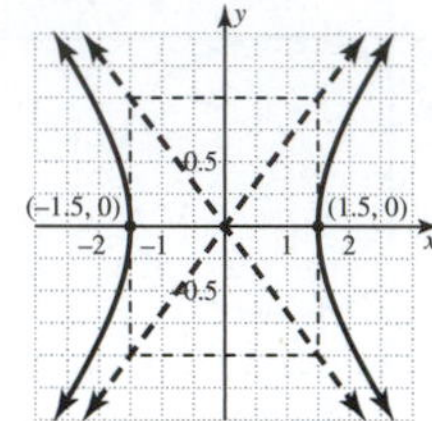

40. hyperbola

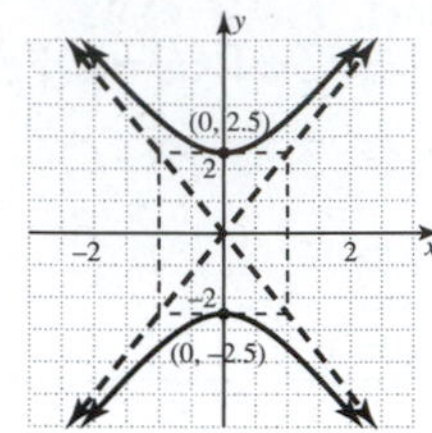

41. parabola

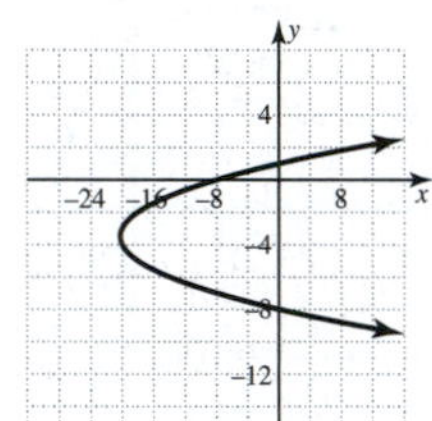

42. parabola

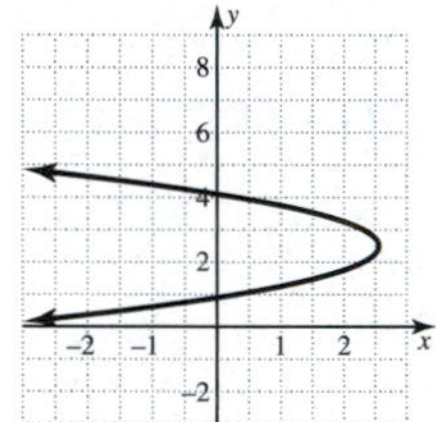

43.

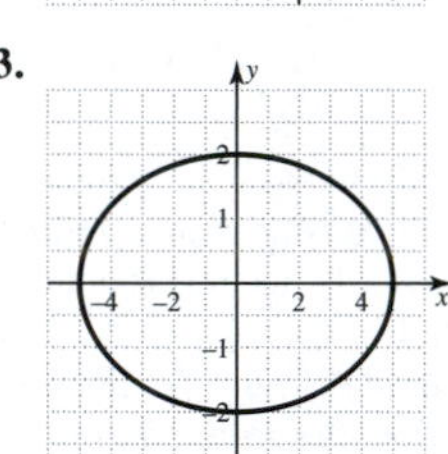

44.

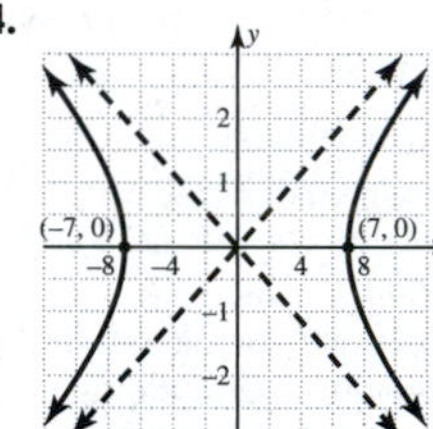

45.

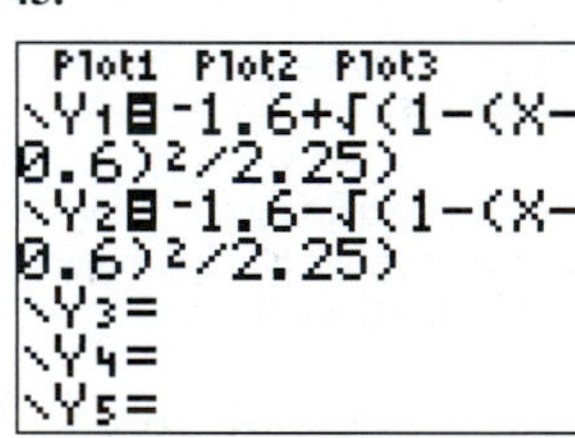

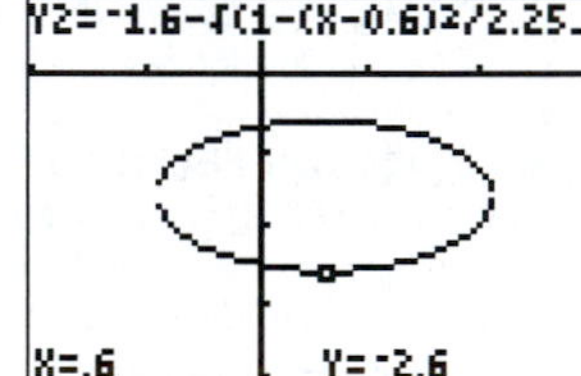

46.

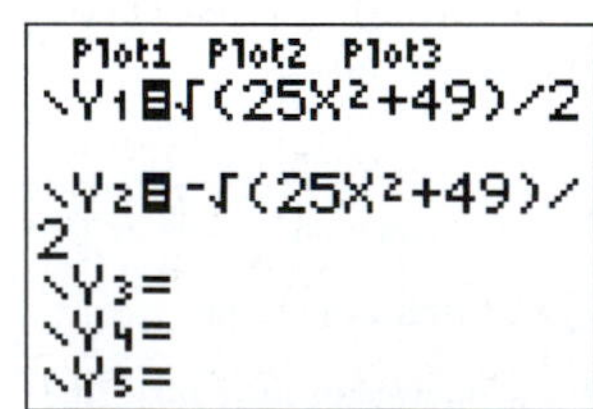

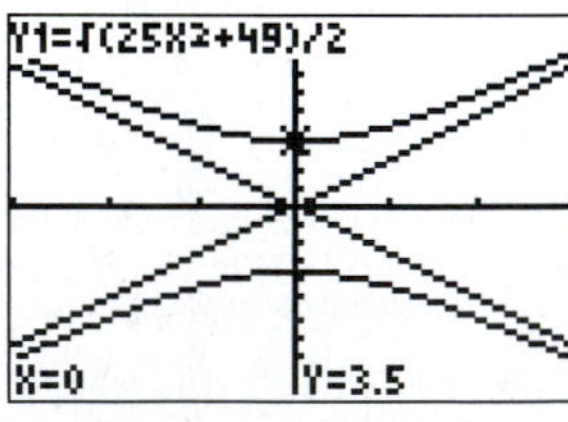

Piece It Together Sections 11.1–11.2

1. vertex: $(-36, -3)$; axis of symmetry: $y = -3$

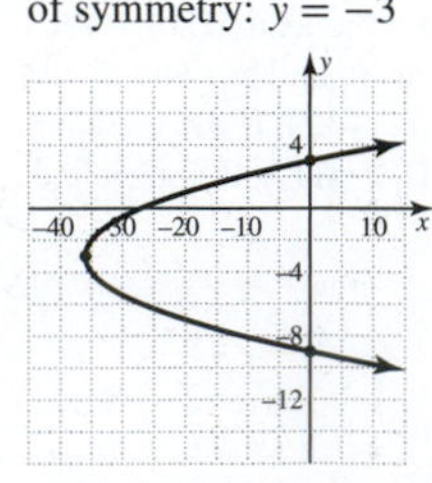

2. vertex: $(25, -2)$; axis of symmetry: $y = -2$

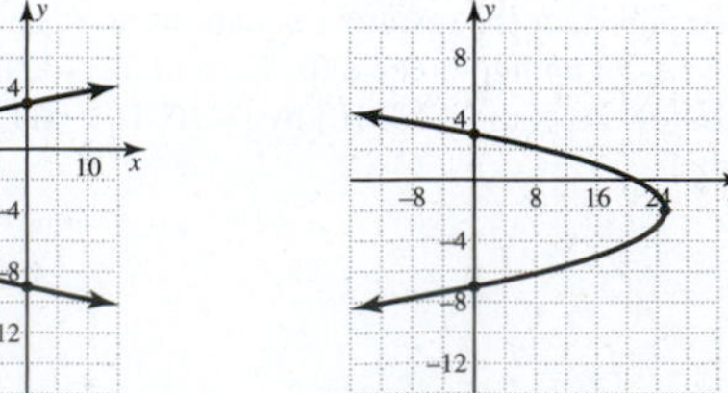

3. center: (0, 0); radius: 10; points: (10, 0), (0, 10), (−10, 0), (0, −10)

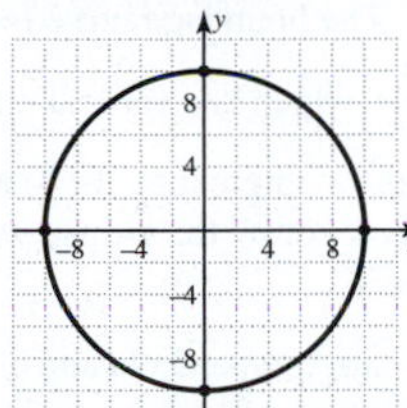

4. center: (0, 0); radius: 6; points: (6, 0), (0, 6), (−6, 0), (0, −6)

5. center: (0, −1); radius: 2; points: (2, −1), (−2, −1), (0, 1), (0, −3)

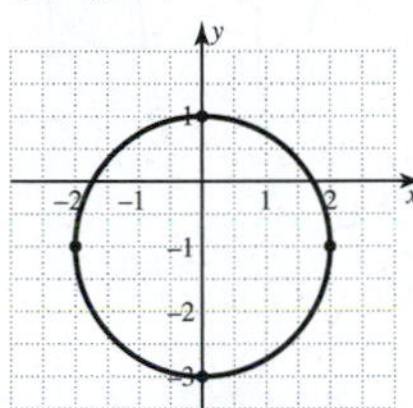

6. center: (−2, 0); radius: 1; points: (−2, 1), (−2, −1), (−1, 0), (−3, 0)

7.

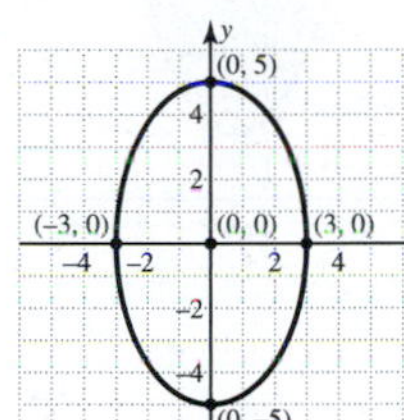

8.

9.

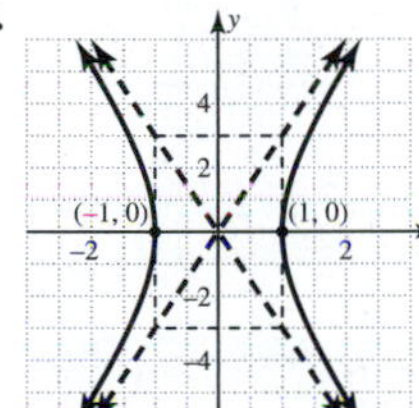
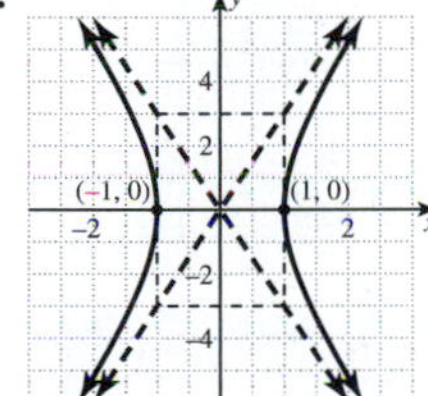

10.

11. $(x + 5)^2 + (y - 7)^2 = 16$; center: $(-5, 7)$; radius $= 4$
12. $(x - 1)^2 + (y - 6)^2 = 1$; center: $(1, 6)$; radius $= 1$

Section 11.3

1. Answers vary. The number of possible solutions can be two if the line intersects the parabola at two points, one if the line touches the parabola at one point, and none if the line does not intersect the parabola.
2. Answers vary. There is no solution if they do not intersect. The number of possible solutions can be one if the circle is touching the other circle or ellipse and two or four if the circle intersects the other graph at two or four points. There can be three points as well. For example: $x^2 + y^2 = 4$ and $(x - 1)^2 + 9y^2 = 9$. The number of solutions is infinite if the two circles are identical.
3. Answers vary. Solve the second equation for y and substitute the expression into the first equation.
4. Answers vary. Replace y with $x^2 + 8$ in the second equation and solve for x.
5. Answers vary. Rewrite the second equation in the same form as in the first equation. Multiply the second equation by −2 and add to the first equation.
6. Answers vary. Rewrite the second equation in the same form as in the first equation. Multiply the second equation by −2 and add to the first equation.
7. Answers vary. Since the second equation is linear, use the substitution method. Solve the second equation for one of the two variables and substitute the expression into the first equation.
8. Answers vary. Since both equations are quadratic, use the elimination method. Multiply the second equation by −1 and add to the first equation.

21. $\{(3, 1), (-3, 1), (3, -1), (-3, -1)\}$
22. $\{(3, 4), (-3, 4), (3, -4), (-3, -4)\}$
23. $\{(5, 2), (-5, 2), (5, -2), (-5, -2)\}$
24. $\{(6, 2), (-6, 2), (6, -2), (-6, -2)\}$
25. $\{(2, 4), (-2, 4), (2, -4), (-2, -4)\}$
26. $\{(6, 2), (-6, 2), (6, -2), (-6, -2)\}$
27. $\{(2, 5), (-2, 5), (2, -5), (-2, -5)\}$
28. $\{(4, 2), (-4, 2), (4, -2), (-4, -2)\}$
29. $\{(1, 4), (-1, 4), (1, -4), (-1, -4)\}$
30. $\{(2, 3), (-2, 3), (2, -3), (-2, -3)\}$
45. $\{(6, 8), (-6, 8), (6, -8), (-6, -8)\}$
46. $\{(3, 1), (-3, 1), (3, -1), (-3, -1)\}$
47. $\{(1, 2), (-1, 2), (1, -2), (-1, -2)\}$
48. $\{(2, 4), (-2, 4), (2, -4), (-2, -4)\}$

51.
$$y = 1 - 8x$$
$$1 - 8x = (x - 1)^2$$
$$1 - 8x = x^2 - 2x + 1$$
$$0 = x^2 + 6x$$
$$0 = x(x + 6)$$
$$x = 0 \quad \text{or} \quad x = -6$$
Substitute the values into the first equation to solve for y.
$x = 0$: $y = (0 - 1)^2 = 1$
$x = -6$: $y = (-6 - 1)^2 = 49$
$\{(0, 1), (-6, 49)\}$

52.
$$x = \frac{1}{4}(\sqrt{6x + 4})^2 - 2$$
$$4x = (6x + 4) - 8$$
$$4x = 6x - 4$$
$$4 = 2x$$
$$x = 2$$
Substitute the value into the second equation to solve for y.
$x = 2$: $y = \sqrt{12 + 4} = \sqrt{16} = 4$
$\{(2, 4)\}$

53.
$$3x^2 + 2y^2 = 12$$
$$\underline{-3x^2 + y^2 = 3}$$
$$3y^2 = 15$$
$$y^2 = 5$$
$$y = \sqrt{5} \quad \text{or} \quad y = -\sqrt{5}$$
$y = \sqrt{5}$ or $y = -\sqrt{5}$:
$$3x^2 + 10 = 12$$
$$3x^2 = 2$$
$$x^2 = \frac{2}{3}$$
$$x = \sqrt{\frac{2}{3}} \quad \text{or} \quad x = -\sqrt{\frac{2}{3}}$$
$$\left\{\left(\sqrt{\frac{2}{3}}, \sqrt{5}\right), \left(-\sqrt{\frac{2}{3}}, \sqrt{5}\right), \left(\sqrt{\frac{2}{3}}, -\sqrt{5}\right), \left(-\sqrt{\frac{2}{3}}, -\sqrt{5}\right)\right\}$$

54.
$$4x^2 + y^2 = 28$$
$$\underline{-x^2 - y^2 = -19}$$
$$3x^2 = 9$$
$$x^2 = 3$$
$$x = \sqrt{3} \quad \text{or} \quad x = -\sqrt{3}$$
$x = \sqrt{3}$ or $x = -\sqrt{3}$:
$$12 + y^2 = 28$$
$$y^2 = 16$$
$$y = \pm 4$$
$\{(\sqrt{3}, 4), (\sqrt{3}, -4), (-\sqrt{3}, 4), (-\sqrt{3}, -4)\}$

55. $\{(-4, 14), (3, 7)\}$

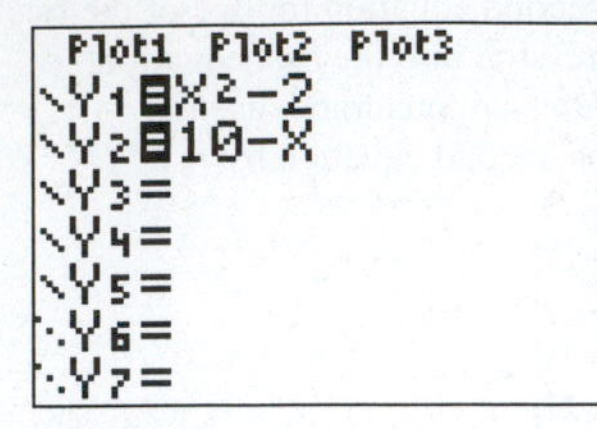

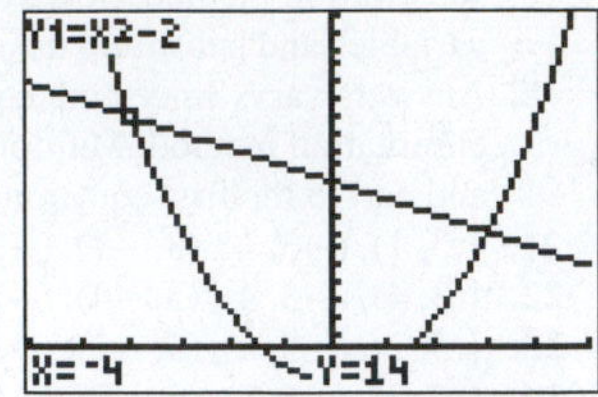

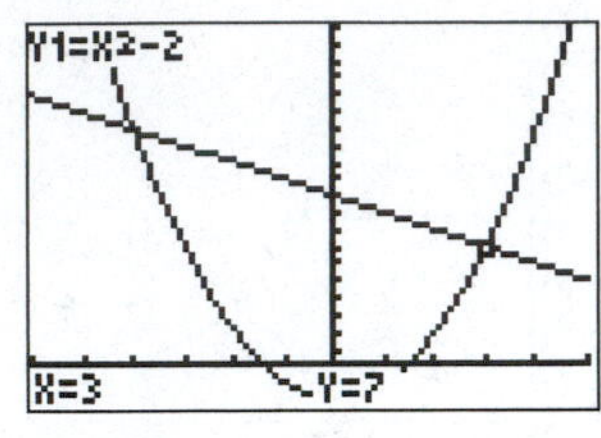

X	Y1	Y2
-5	23	15
-4	14	14
-1	-1	11
0	-2	10
1	-1	9
3	7	7
4	14	6

Y2=10-X

56. $\{(1, 3), (-1, 3), (1, -3), (-1, -3)\}$

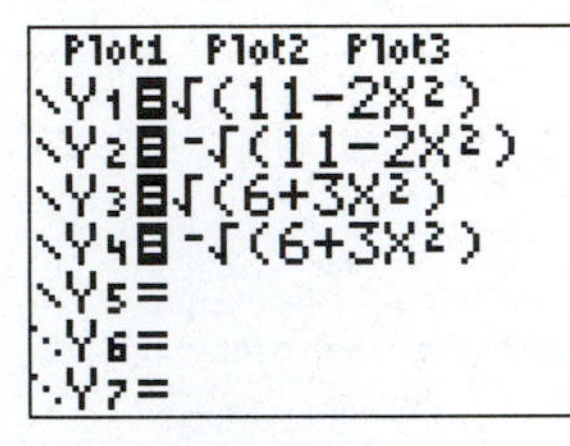

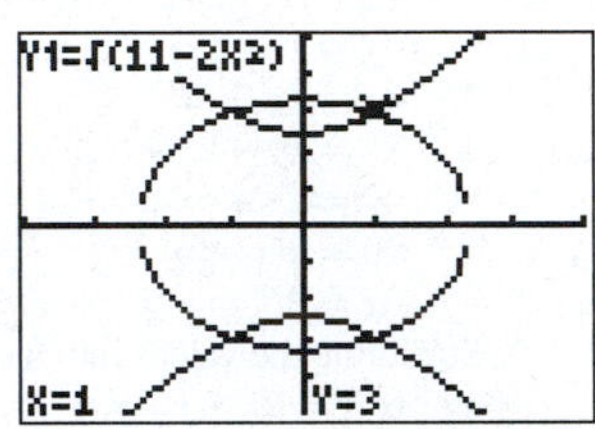

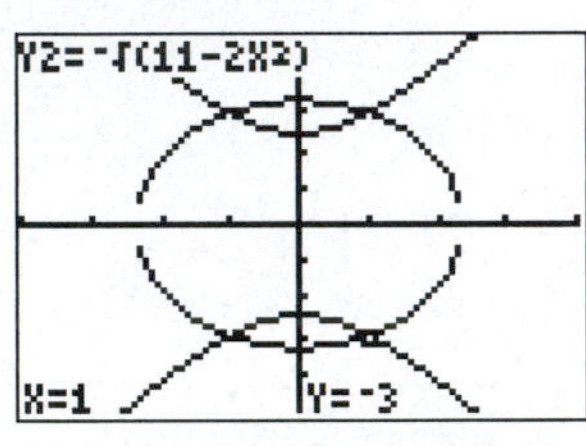

X	Y1	Y2
-2	1.7321	-1.732
-1	3	-3
0	3.3166	-3.317
1	3	-3
2	1.7321	-1.732

Y2=-√(11-2X²)

X	Y3	Y4
-2	4.2426	-4.243
-1	3	-3
0	2.4495	-2.449
1	3	-3
2	4.2426	-4.243

Y4=-√(6+3X²)

Section 11.4

1. Answers vary. The boundary curve is the graph of the equation formed by replacing the inequality symbol with an equals sign. It is dashed if the inequality symbol is $<$ or $>$. It is solid if the inequality symbol is $\le$ or $\ge$.
2. Answers vary. Choose a test point from each region formed by the boundary curve. If the test point makes the inequality true, shade the region that contains the point.
3. Answers vary. The boundary curve for the inequality is $x^2 - 4y^2 = 16$, which is a horizontal hyperbola with x-intercepts, $(4, 0)$ and $(-4, 0)$. Since the inequality symbol is $\le$, draw the hyperbola with a solid curve.
4. Answers vary. The boundary curve, the solid horizontal hyperbola, divides the plane into three regions, the region to the left of the left branch, the region between the branches, and the region to the right of the right branch of the hyperbola. Use a test point in each region, e.g. $(-5, 0)$, $(0, 0)$, and $(5, 0)$. If the test point makes the inequality true, shade the region that contains the point. Since $(0, 0)$ satisfies the inequality, we shade the region between the branches of the hyperbola.
5. Answers vary. The boundary curve for the inequality, $y > x^2 - 4$, is $y = x^2 - 4$ which is a parabola with x-intercepts, $(2, 0)$ and $(-2, 0)$, and y-intercept $(0, -4)$. Since the inequality symbol is $>$, draw the parabola with a dashed curve. The boundary curve for the inequality, $\frac{x^2}{9} + \frac{y^2}{16} < 1$, is $\frac{x^2}{9} + \frac{y^2}{16} = 1$ which is a vertical ellipse with x-intercepts, $(3, 0)$ and $(-3, 0)$, and y-intercepts, $(0, 4)$ and $(0, -4)$. Since the inequality symbol is $<$, draw the ellipse with a dashed curve.
6. Answers vary. We graph each boundary curve, a parabola and ellipse. Then we use test points to determine the graph of each inequality. The intersection of the two graphs is the solution of the system.

7.

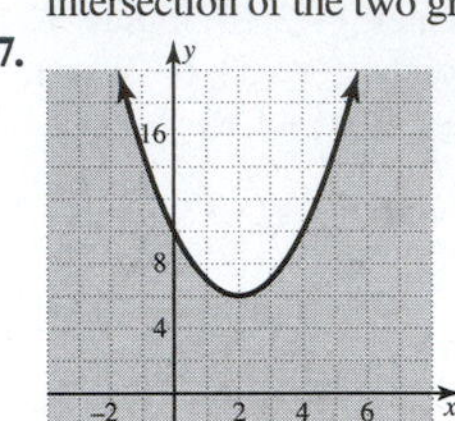

8.

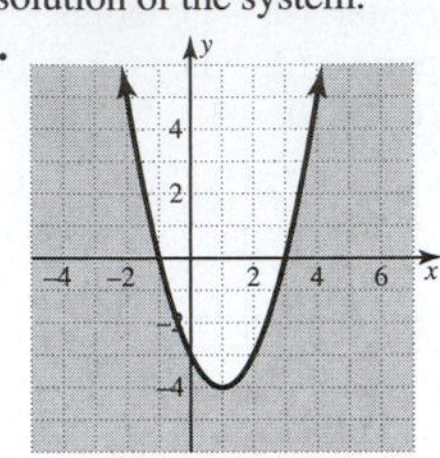

9.

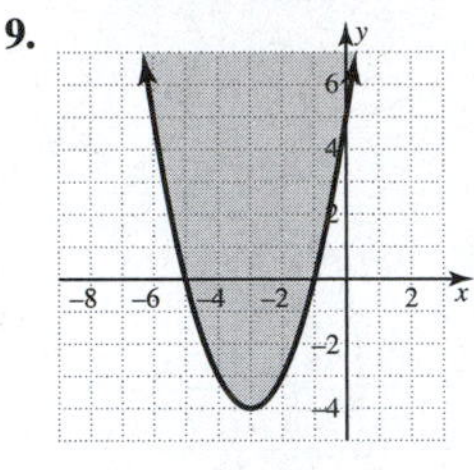

10.

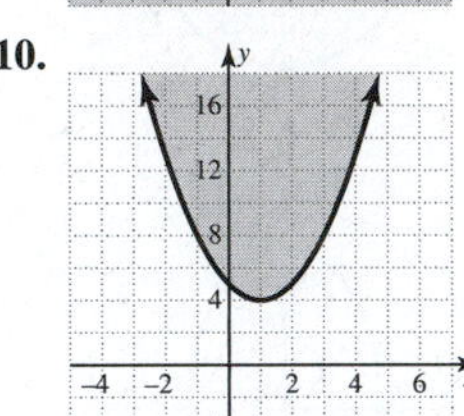

11.

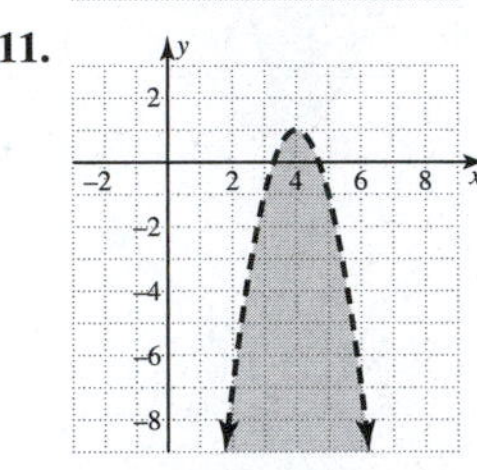

12.

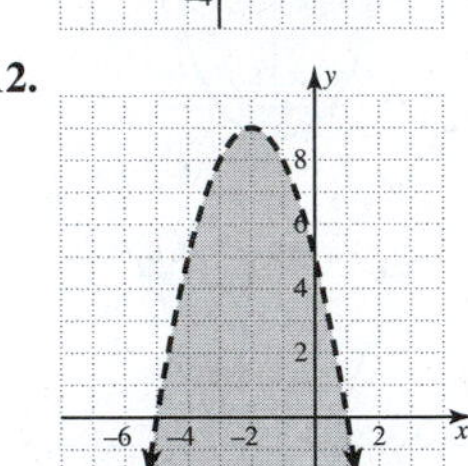

13.

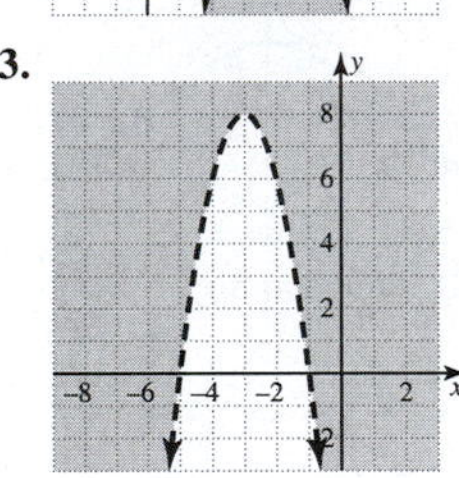

14.

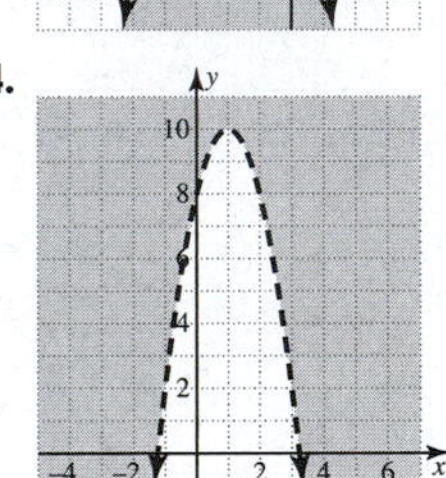

15.

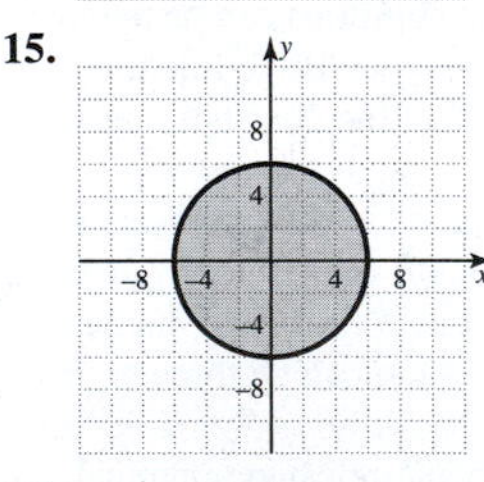

16.

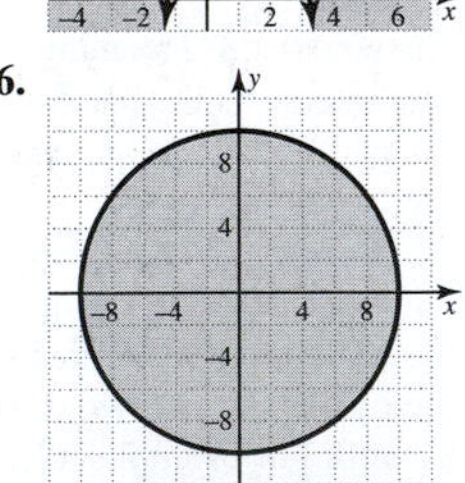

17.

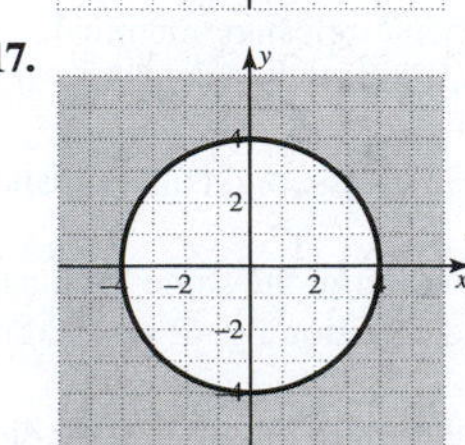

18.

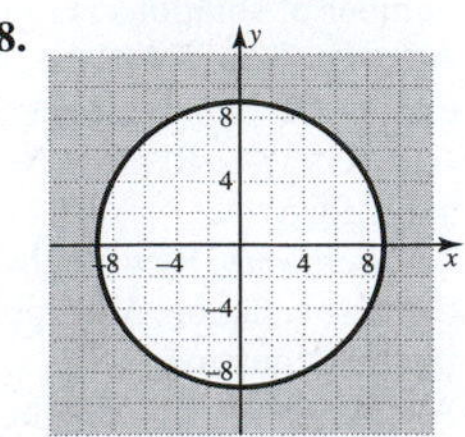

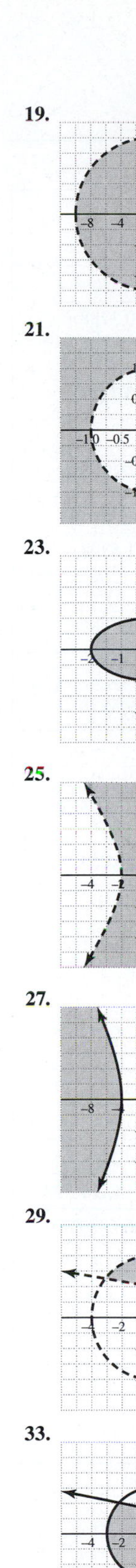

19. **20.**

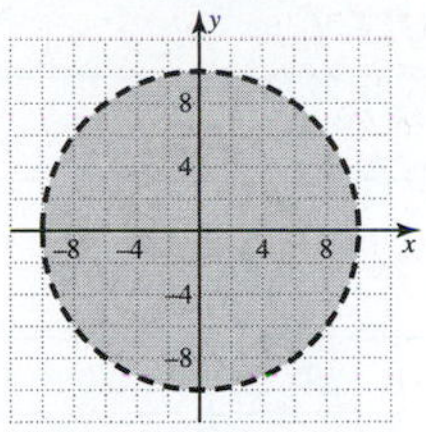

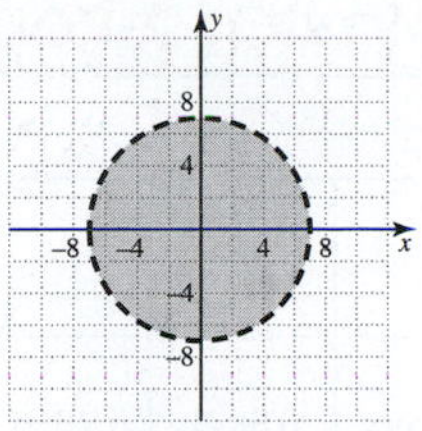

21. 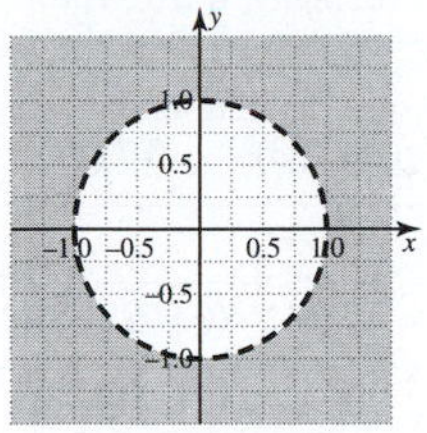 **22.**

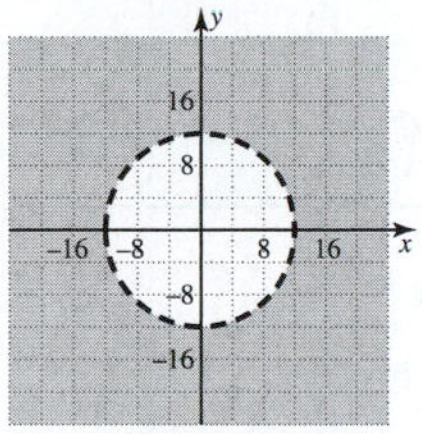

23. 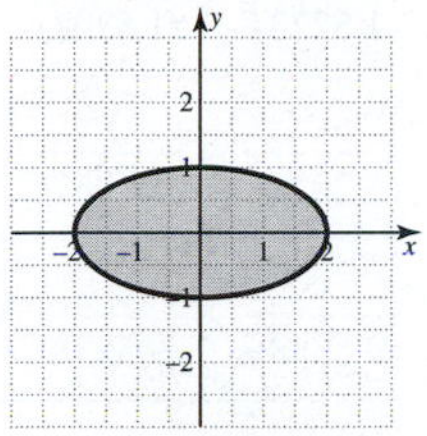 **24.**

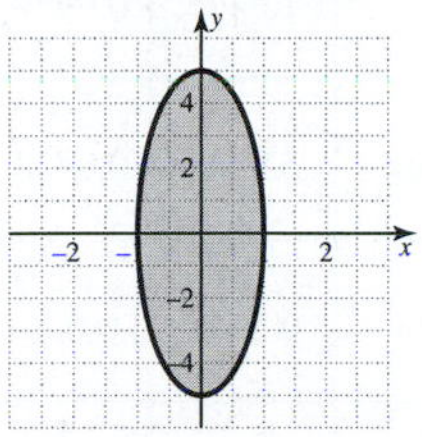

25. **26.**

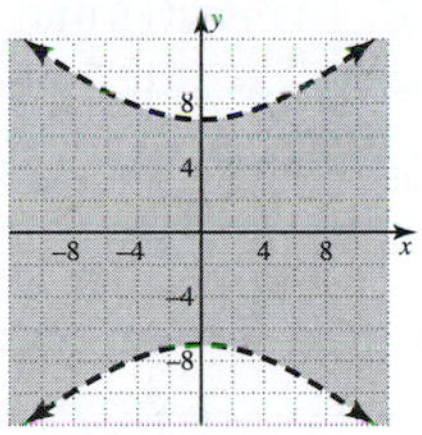

27. **28.**

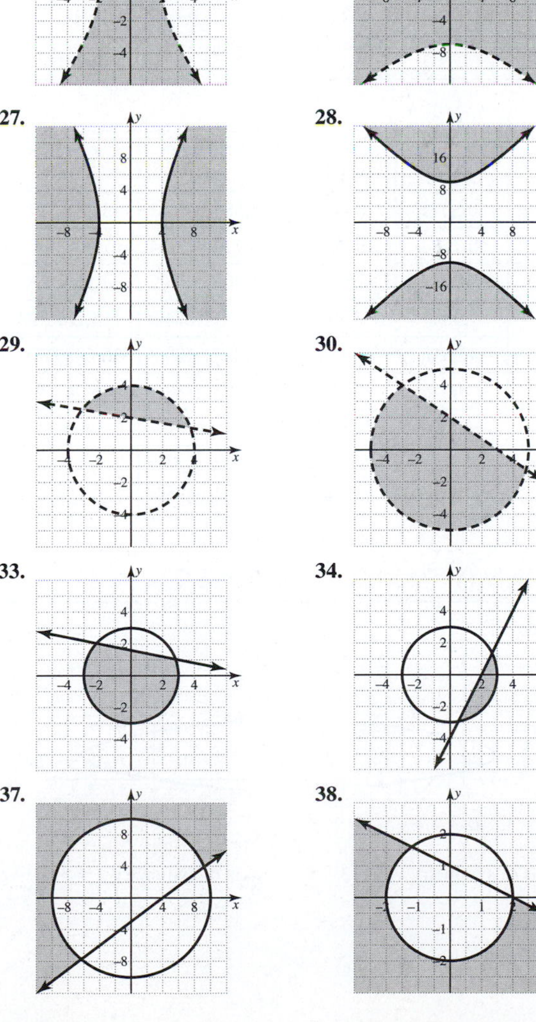

29. **30.**

33. **34.**

37. **38.**

39. 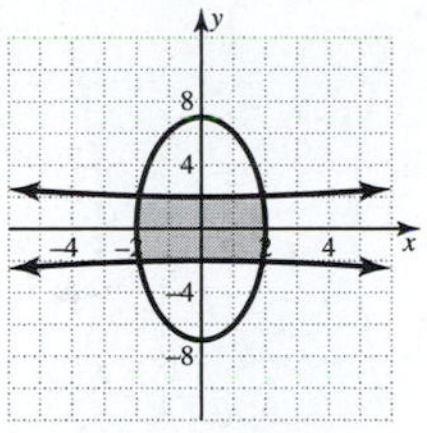 **40.**

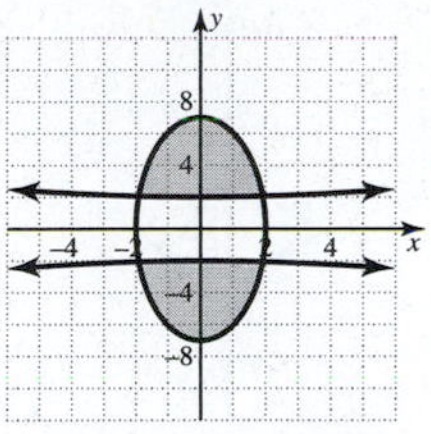

43. 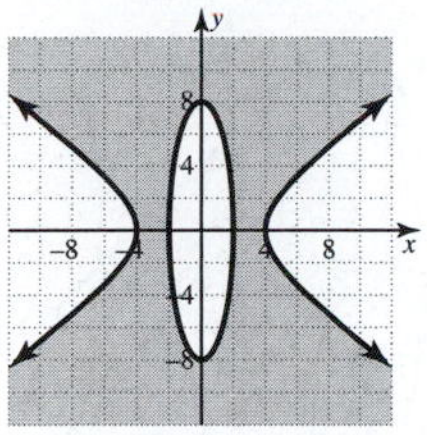 **44.**

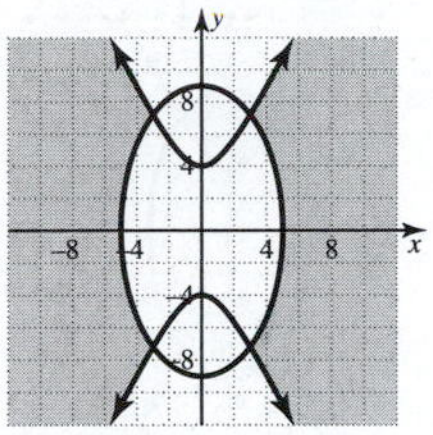

45. 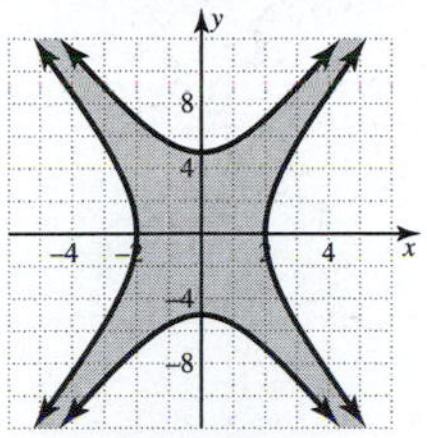 **46.**

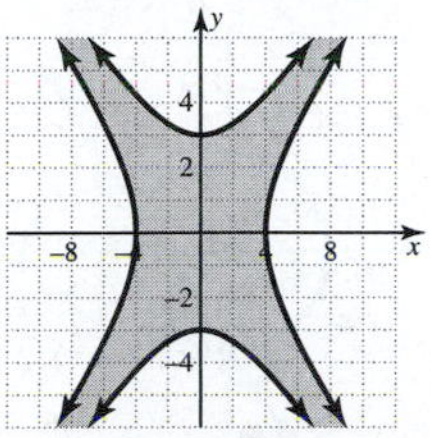

47. 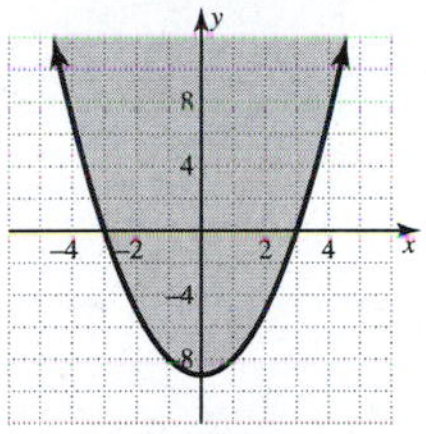 **48.**

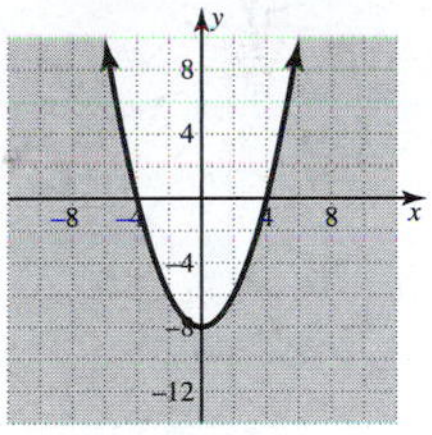

49. 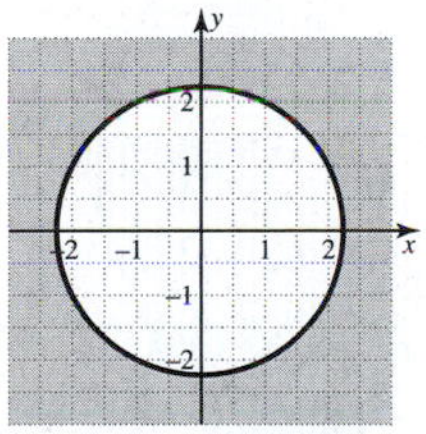 **50.**

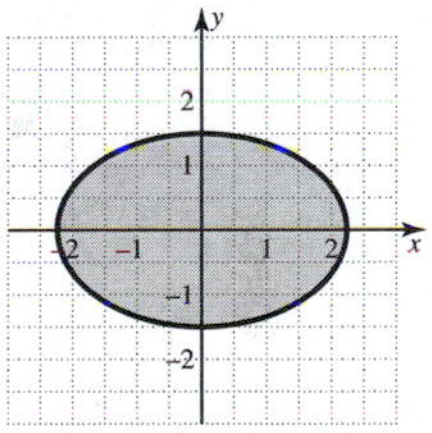

51. 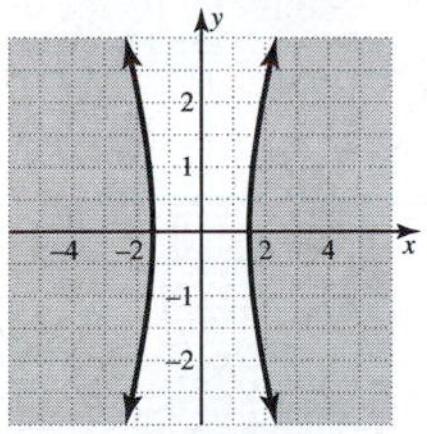 **52.**

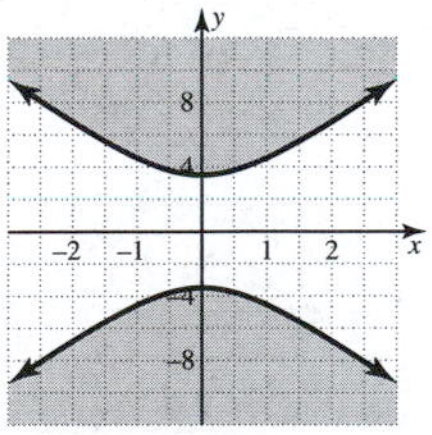

53. 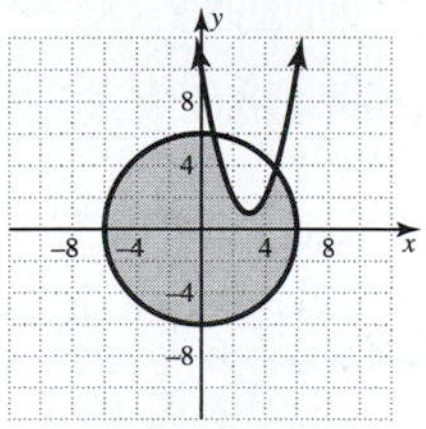 **55.**

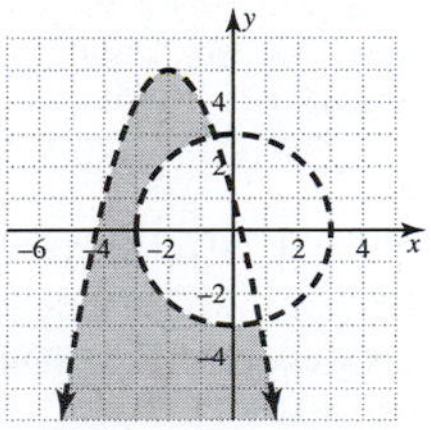

56. 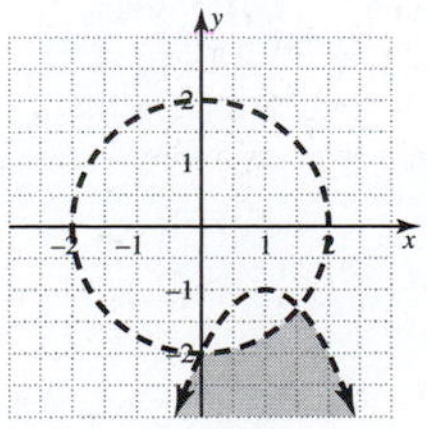 **57.**

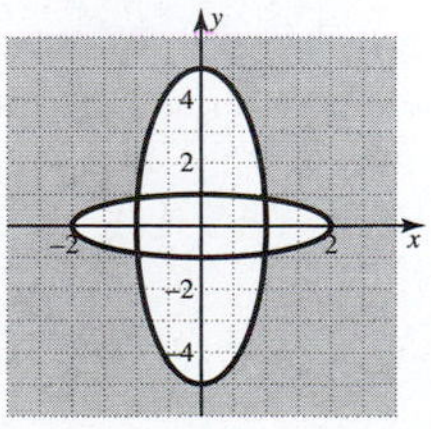

58. 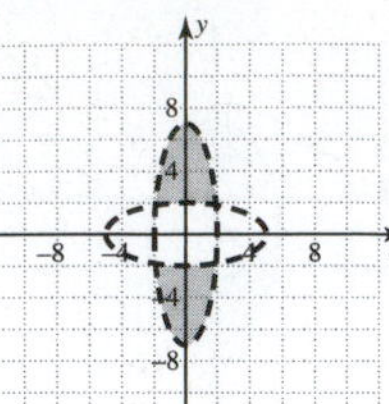

59. 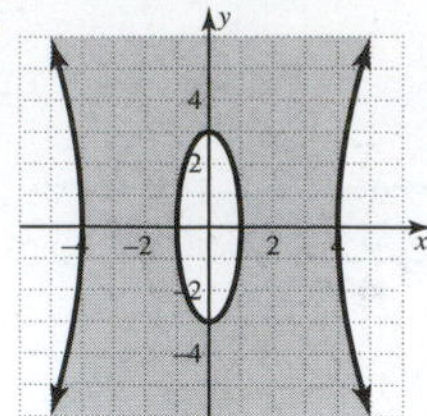

60. 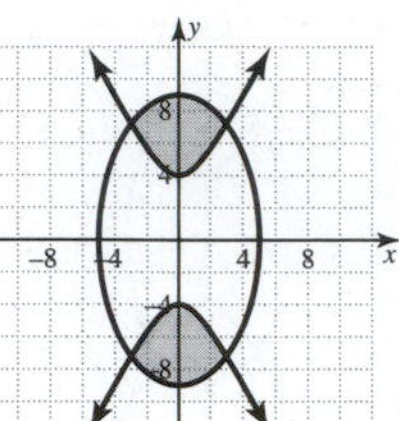

61. 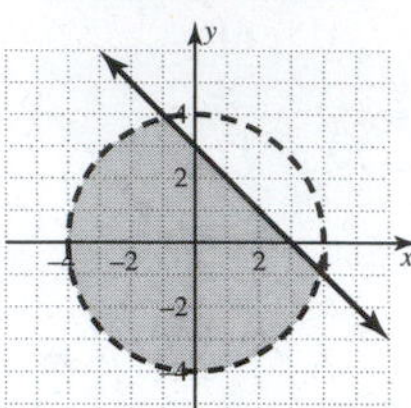

62. 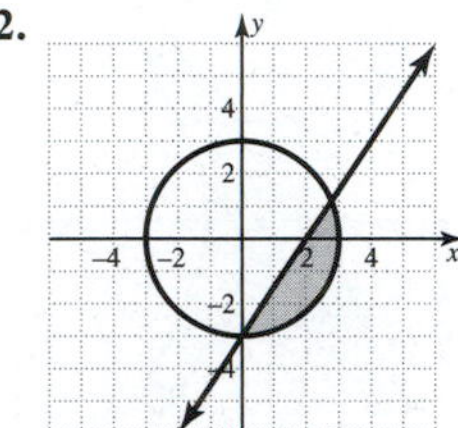

63. 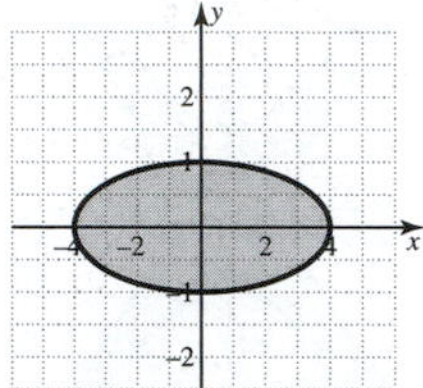

64. 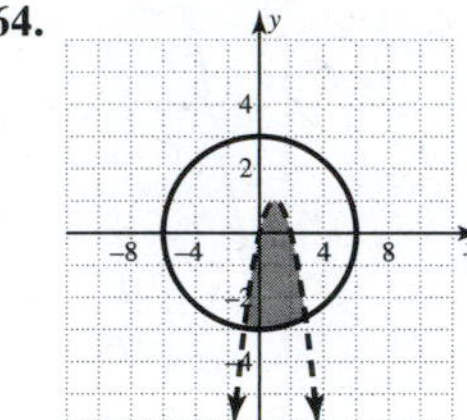

65. 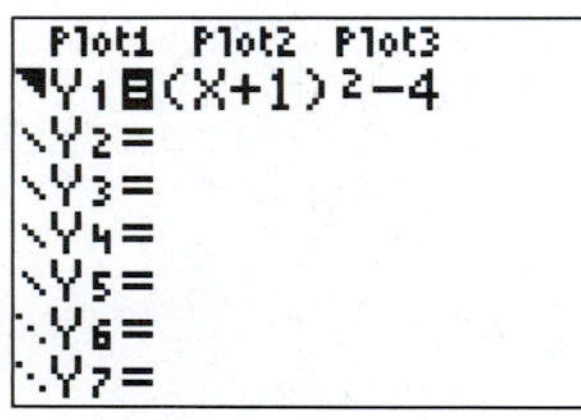

66. 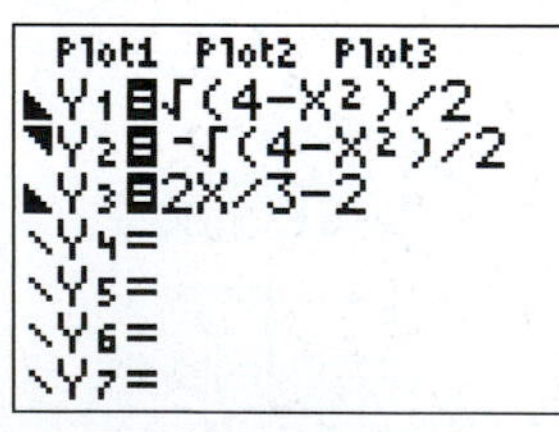

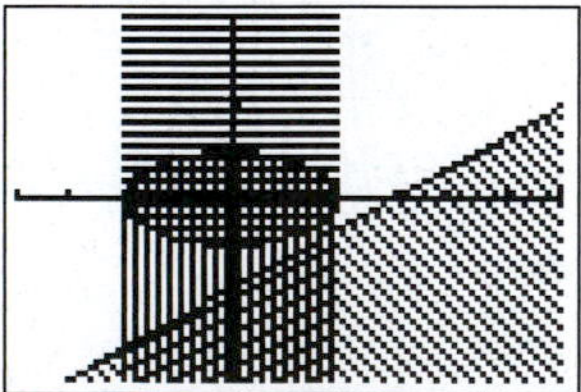

67. Answers vary. For example, $\frac{x^2}{4} + y^2 \le 1$ and $x^2 + \frac{y^2}{9} \le 1$.

68. Answers vary. For example, $x^2 + y^2 \ge 1$ and $\frac{x^2}{4} + \frac{y^2}{9} \le 1$.

69. Answers vary. For example, $\frac{y^2}{9} - x^2 < 1$ and $\frac{x^2}{9} + \frac{y^2}{4} > 1$.

CHAPTER 11 GROUP ACTIVITY

3., 4.

	a	e	b	$\frac{x^2}{a^2} + \frac{y^2}{b^2} = 1$
Mercury	0.3871	0.2056	0.3788	$\frac{x^2}{(0.3871)^2} + \frac{y^2}{(0.3788)^2} = 1$
Venus	0.7233	0.0068	0.7233	$\frac{x^2}{(0.7233)^2} + \frac{y^2}{(0.7233)^2} = 1$
Earth	1	0.0167	0.9999	$\frac{x^2}{(1)^2} + \frac{y^2}{(0.9999)^2} = 1$
Mars	1.5273	0.0934	1.5206	$\frac{x^2}{(1.5273)^2} + \frac{y^2}{(1.5206)^2} = 1$
Jupiter	5.2028	0.0483	5.1967	$\frac{x^2}{(5.2028)^2} + \frac{y^2}{(5.1967)^2} = 1$
Saturn	9.5388	0.056	9.5238	$\frac{x^2}{(9.5388)^2} + \frac{y^2}{(9.5238)^2} = 1$
Uranus	19.1914	0.0461	19.171	$\frac{x^2}{(19.1914)^2} + \frac{y^2}{(19.171)^2} = 1$
Neptune	30.0611	0.0097	30.0597	$\frac{x^2}{(30.0611)^2} + \frac{y^2}{(30.0597)^2} = 1$
Pluto	39.5294	0.2482	38.2059	$\frac{x^2}{(39.5294)^2} + \frac{y^2}{(38.2059)^2} = 1$

5.

	Perihelion (AU)	Aphelion (AU)	Perihelion (mi)	Aphelion (mi)
Mercury	0.3075	0.4667	28,585,048.43	43,381,337.35
Venus	0.7184	0.7282	66,777,737.64	67,692,132.76
Earth	0.9833	1.0167	91,403,445.02	94,508,168.98
Mars	1.3847	1.6699	128,711,274.9	155,231,533.2
Jupiter	4.9515	5.4541	46,0271,120.8	506,989,824.5
Saturn	9.0046	10.0730	837,032,388.1	936,341,315.5
Uranus	18.3067	20.0761	1,701,711,884	1,866,192,265
Neptune	29.7695	30.3527	2,767,248,578	2,821,459,042
Pluto	29.7182	49.3406	2,762,479,535	4,586,495,019

CHAPTER 11 REVIEW EXERCISES

Section 11.1

1. 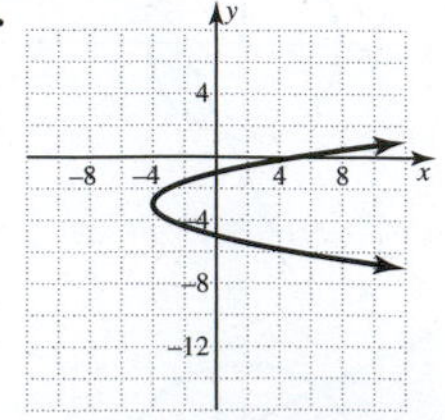

axis of symmetry $y = -3$

2. 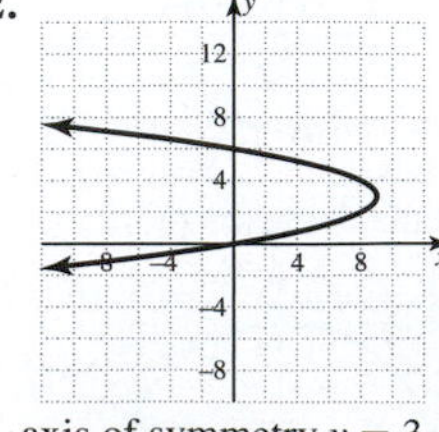

axis of symmetry $y = 3$

3.

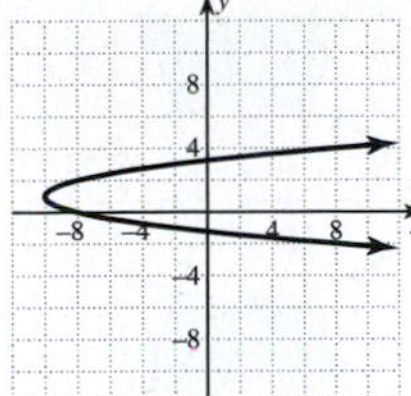

axis of symmetry $y = 1$

4.

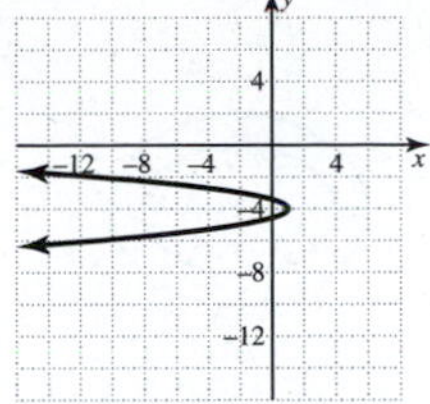

axis of symmetry $y = -4$

5.

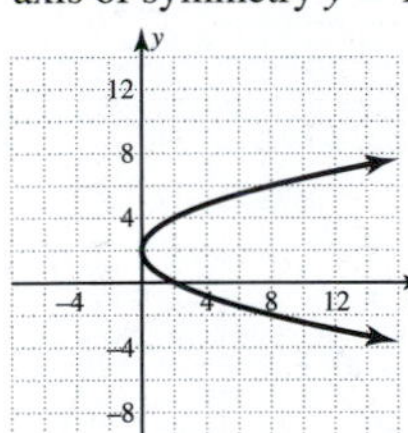

axis of symmetry $y = 2$

6.

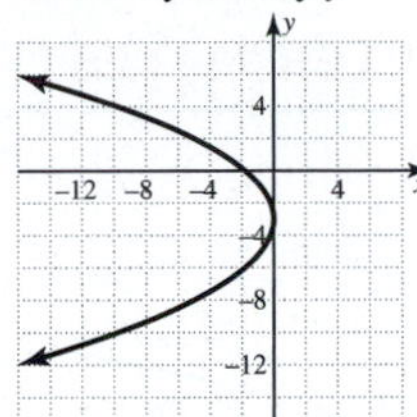

axis of symmetry $y = -3$

7.

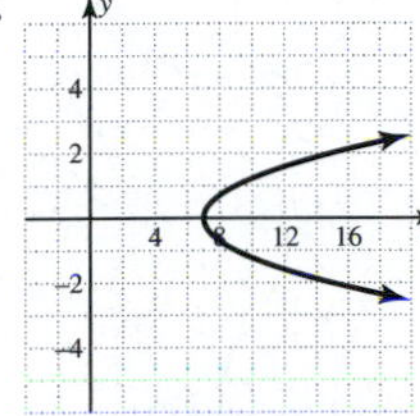

axis of symmetry $y = 0$

8.

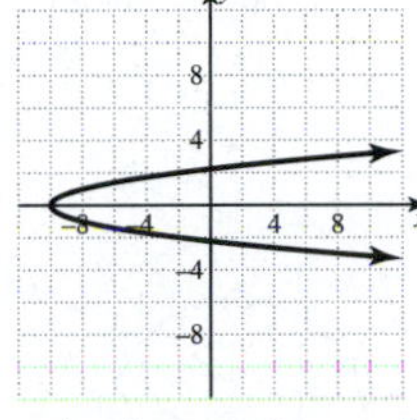

axis of symmetry $y = 0$

9.

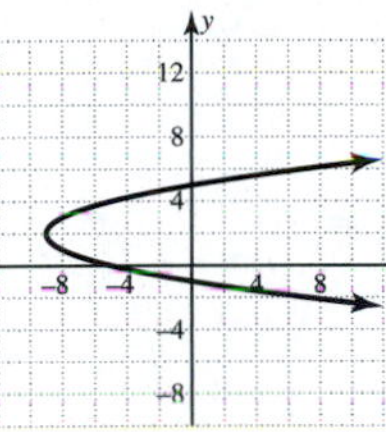

axis of symmetry $y = 2$

10.

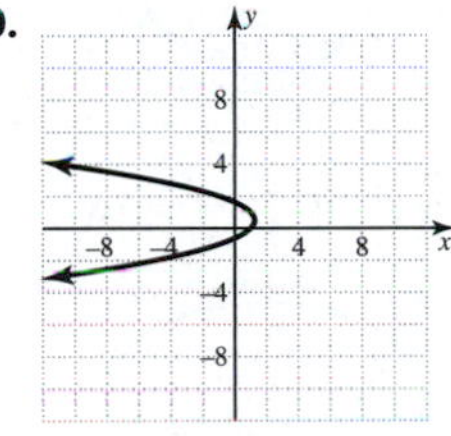

axis of symmetry $y = \frac{1}{2}$

11.

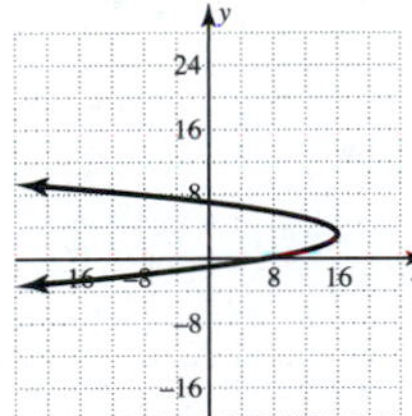

axis of symmetry $y = 3$

12.

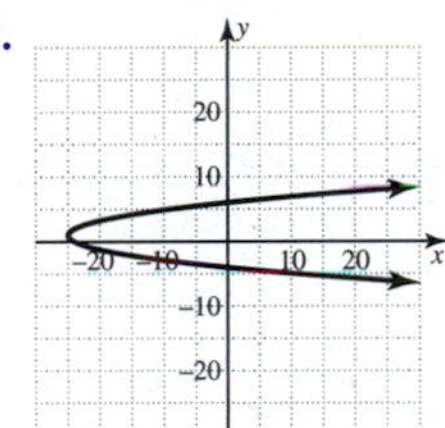

axis of symmetry $y = 1$

13.

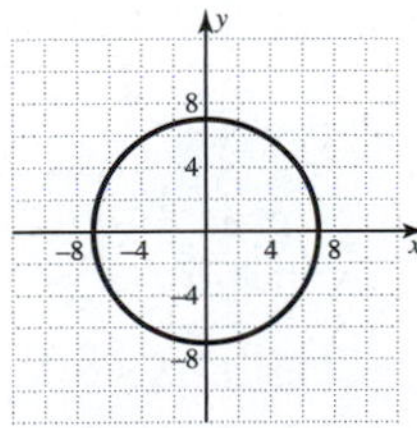

The center is (0, 0) and the radius is $r = 7$. Four key points on the circle are (7, 0), (0, 7), (−7, 0), and (0, −7).

14.

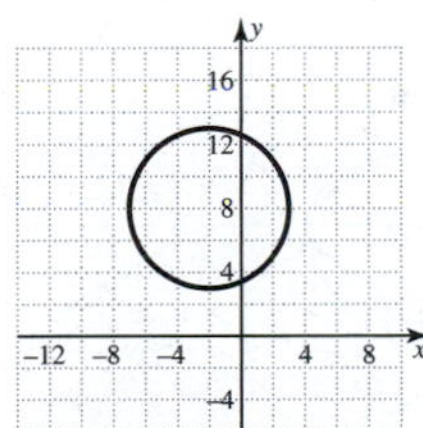

The center is (−2, 8) and the radius is $r = 5$. Four key points on the circle are (3, 8), (−2, 13), (−7, 8), and (−2, 3).

15.

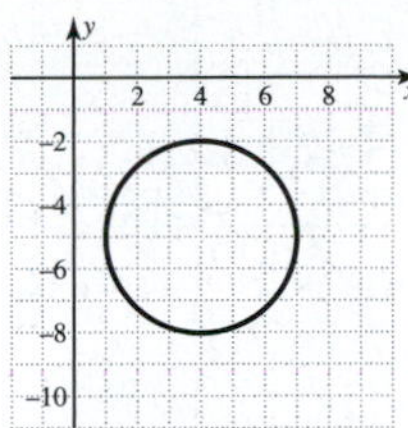

The center is (4, −5) and the radius is $r = 3$. Four key points on the circle are (7, −5), (4, −2), (1, −5), and (4, −8).

16. The center is (−3 , 0) and the radius is $r = 4$. Four key points on the circle are (1, 0), (−3, 4), (−7, 0), and (−3, −4).

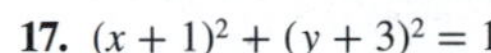

17. $(x + 1)^2 + (y + 3)^2 = 1$

18. $(x - 5)^2 + (y + 2)^2 = 16$

19. $(x + 6)^2 + (y - 7)^2 = 10$

20. $x^2 + \left(y - \frac{3}{5}\right)^2 = \frac{9}{4}$

21. $(x - 3.5)^2 + (y - 1.5)^2 = 20.25$

22. $(x + 2.5)^2 + (y - 0.5)^2 = 2.25$

23. $(x - 1)^2 + (y - 7)^2 = 200$
center: (1, 7); radius $= 10\sqrt{2}$

24. $(x - 1)^2 + (y - 2)^2 = 59$
center: (1, 2); radius $= \sqrt{59}$

25. $x^2 + (y - 10)^2 = 216$
center: (0, 10); radius $= 6\sqrt{6}$

26. $(x - 1)^2 + y^2 = 27$
center: (1, 0); radius $= 3\sqrt{3}$

27. $\left(x - \frac{1}{2}\right)^2 + \left(y + \frac{3}{2}\right)^2 = 3$
center: $\left(\frac{1}{2}, -\frac{3}{2}\right)$; radius $= \sqrt{3}$

28. $(x - 0.4)^2 + (y + 0.6)^2 = 0.81$
center: (0.4, −0.6); radius = 0.9

Section 11.2

29.

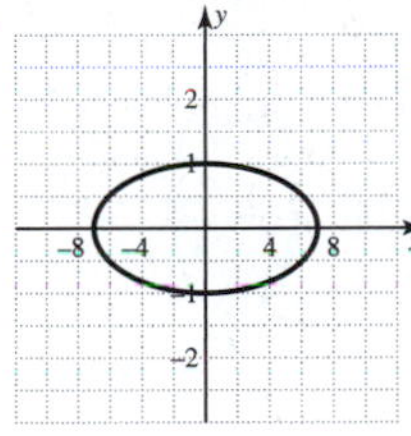

ellipse

30.

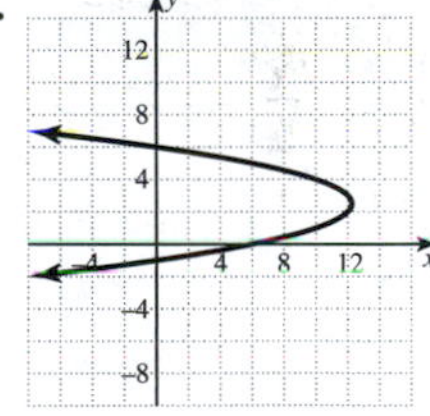

parabola

31.

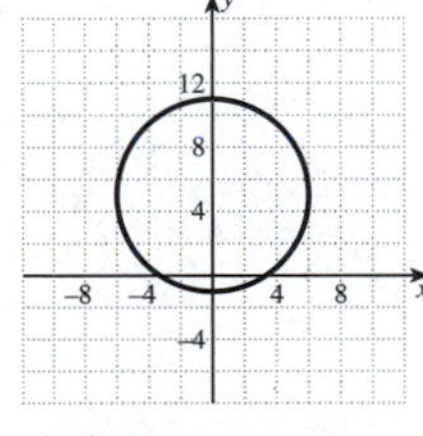

circle

32.

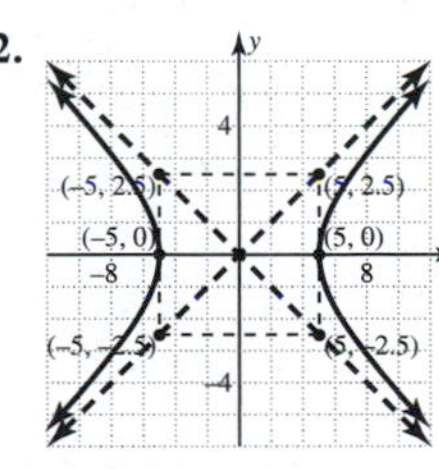

hyperbola

33.

hyperbola

34.

hyperbola

35.

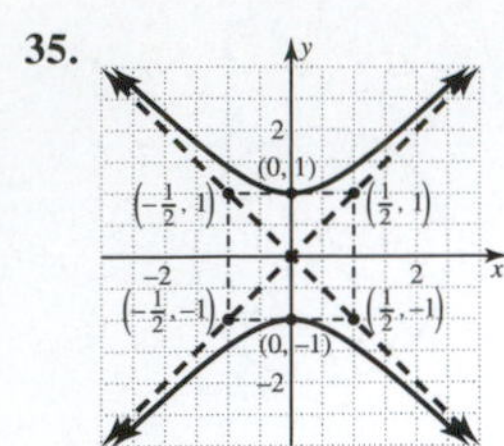

hyperbola

36.

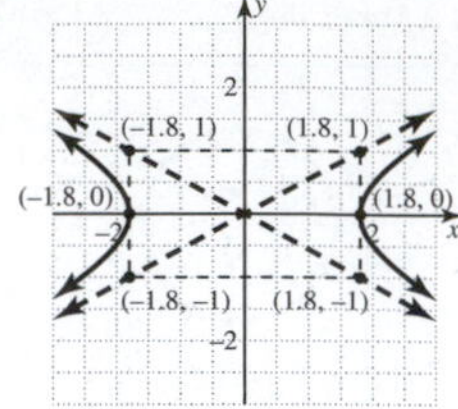

hyperbola

Section 11.3

37. $\{(0, 25), (7, 4)\}$

38. $\{(4, \sqrt{3})\}$

39. $\left\{(4, -3), \left(\frac{84}{41}, -\frac{187}{41}\right)\right\}$

40. $\{(-4, 22), (-1, 7)\}$

41. $\{(-3, -3), (1, -11)\}$

42. $\{(-9, -6), (-9, 6), (9, -6), (9, 6)\}$

43. $\{(-2, -5), (-2, 5), (2, -5), (2, 5)\}$

44. $\{(-4, -3), (-4, 3), (4, -3), (4, 3)\}$

45. $\left\{\left(\frac{3}{4}, \frac{51}{16}\right), \left(-\frac{1}{2}, -\frac{1}{4}\right)\right\}$

46. $\left\{\left(-\frac{3}{5}, -\frac{21}{25}\right), (1, 3)\right\}$

47. $\left\{\left(-\frac{3}{4}, -\frac{4}{3}\right), \left(2, \frac{1}{2}\right)\right\}$

48. $\left\{\left(-2, -\frac{1}{2}\right), \left(\frac{7}{3}, \frac{3}{7}\right)\right\}$

Section 11.4

49.

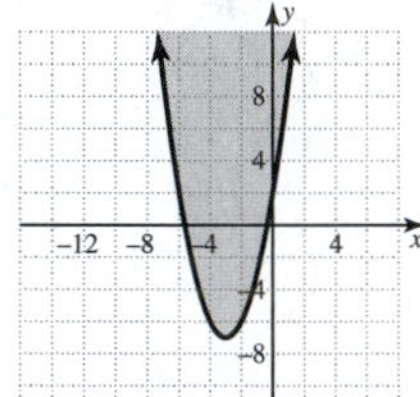

50.

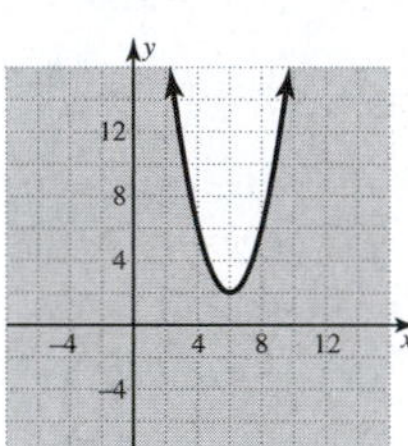

51.

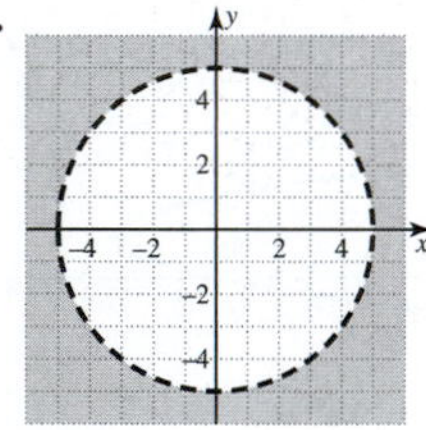

52.

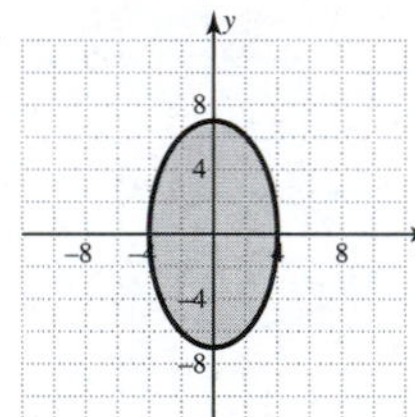

53.

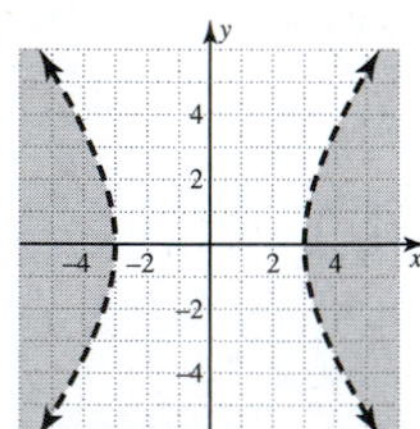

54.

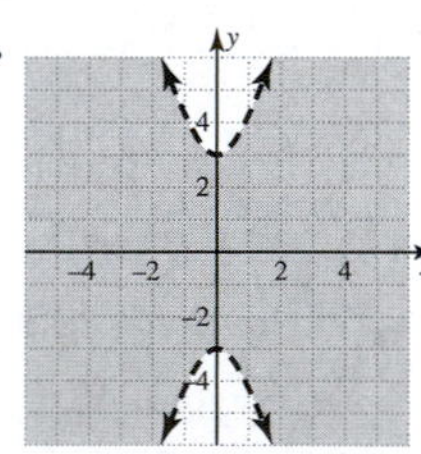

55.

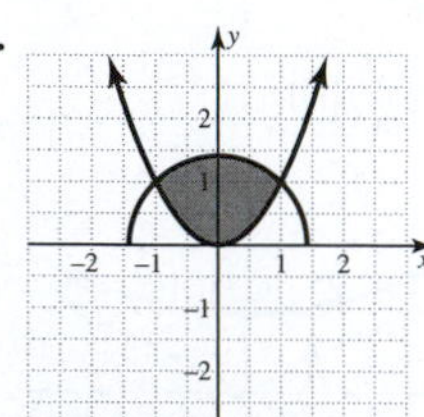

56.

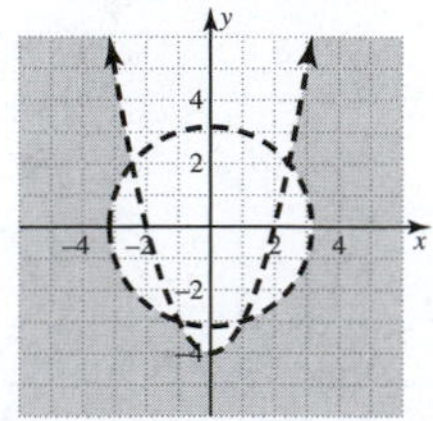

57.

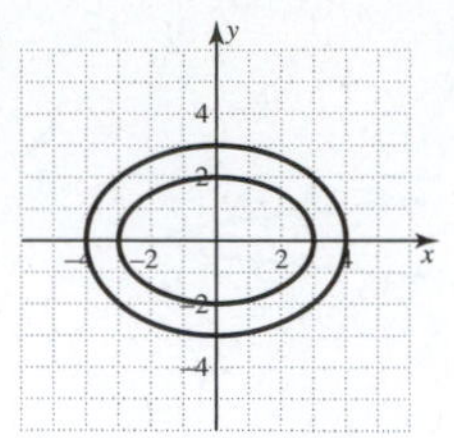

∅

58.

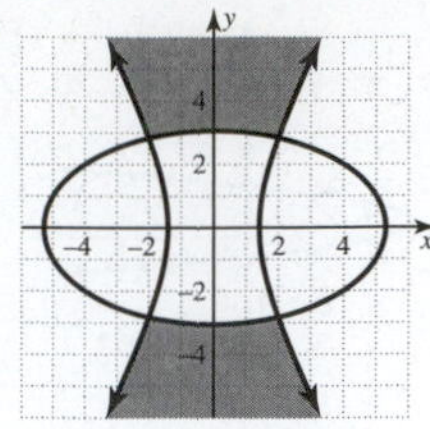

CHAPTER 11 TEST

5.

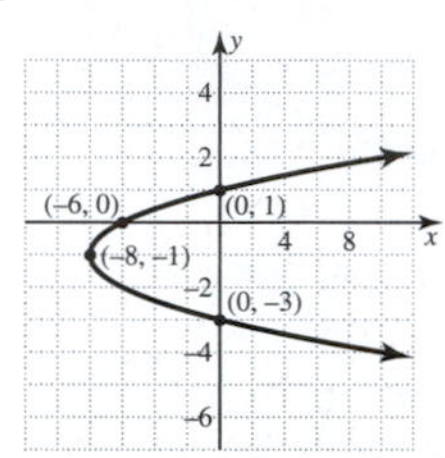

6.

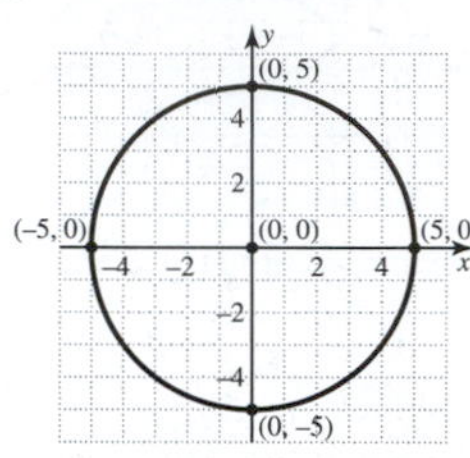

7.

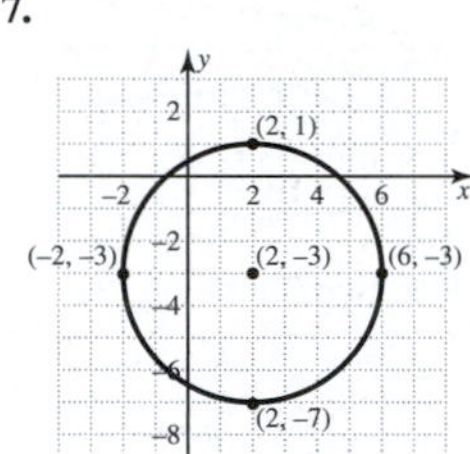

8.

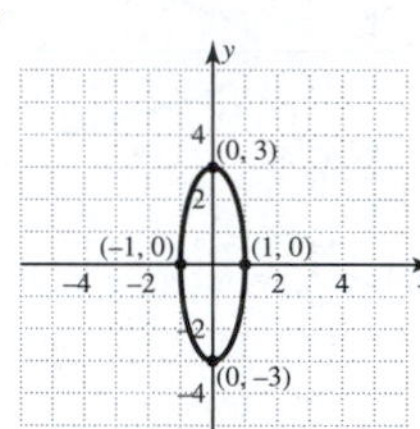

9.

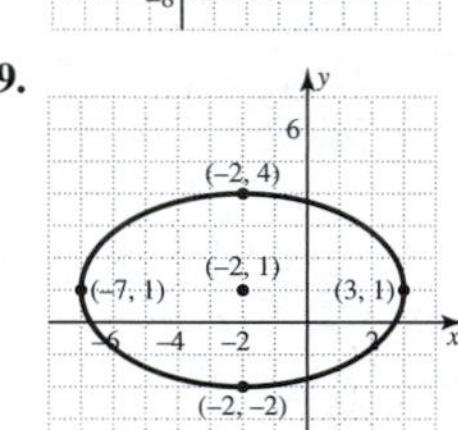

10.

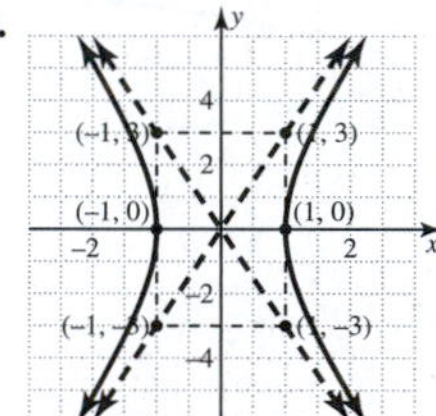

11.

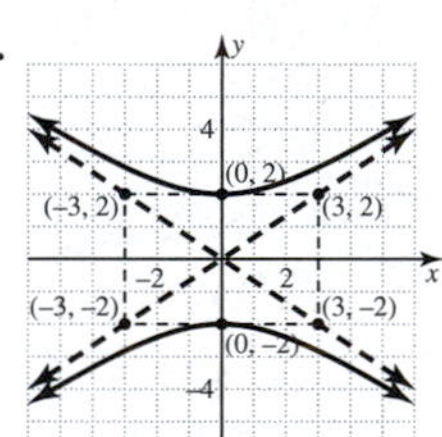

12.

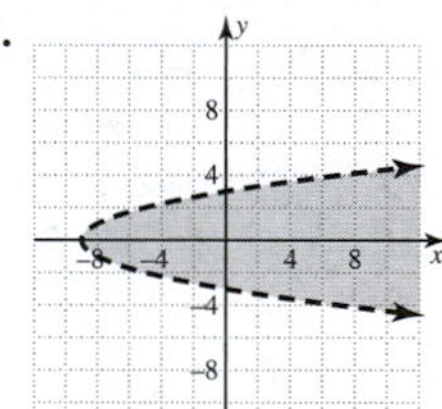

13.

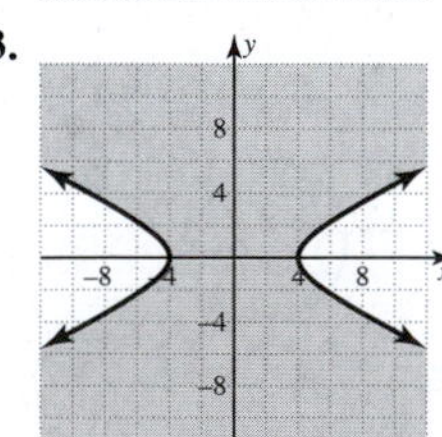

17.

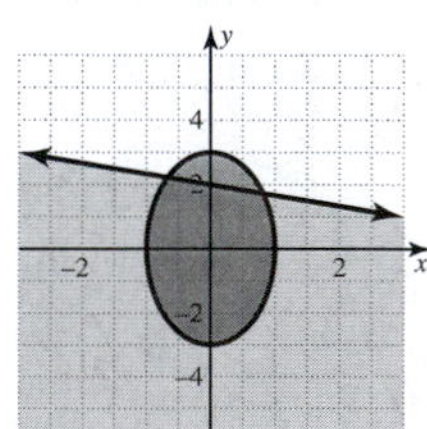

18.

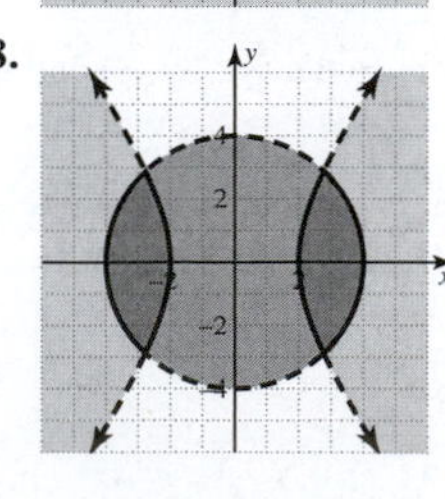

CHAPTERS 1–11 CUMULATIVE REVIEW EXERCISES

11.

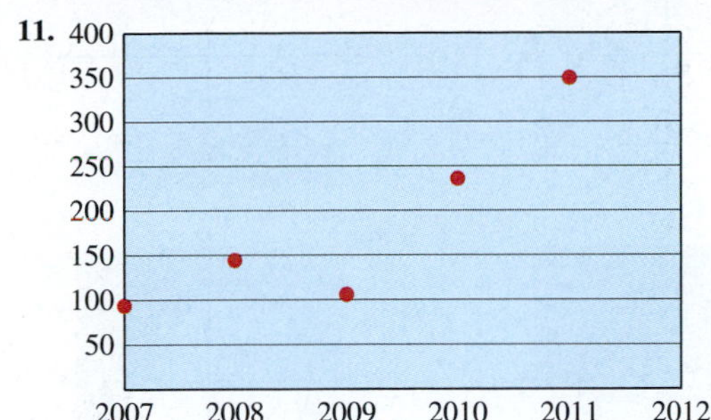

12.

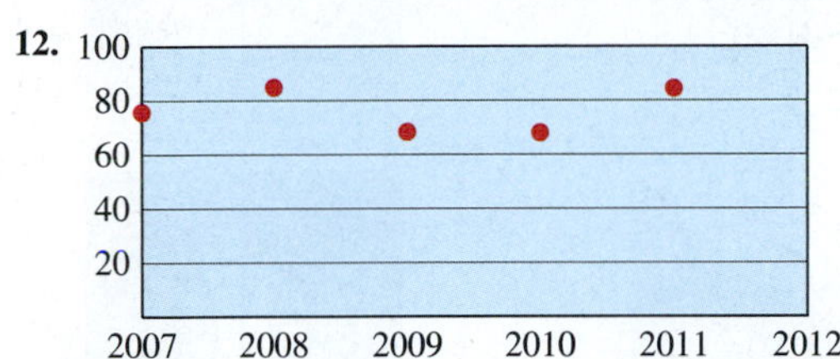

15. $m = -\frac{2}{5}$, (0, 4) **16.** $m = 0$, (0, −3)

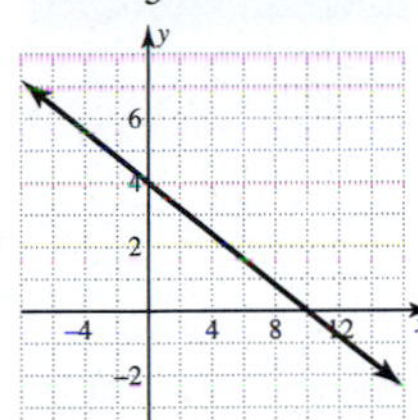

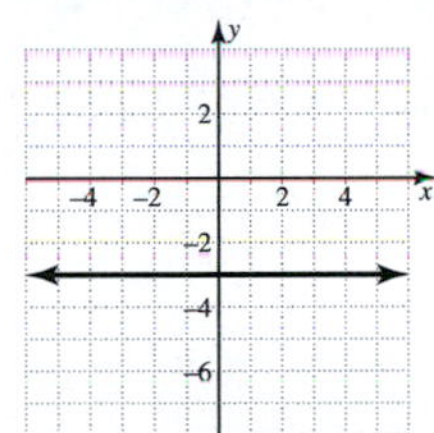

19.

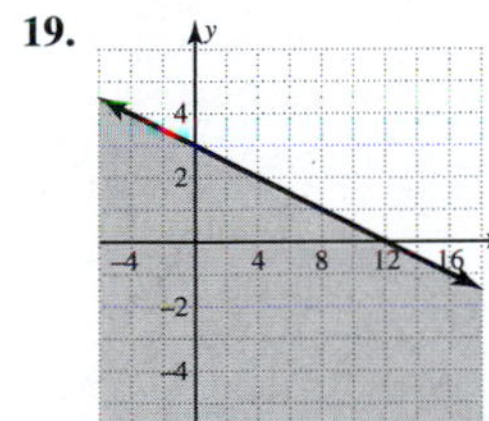

20.

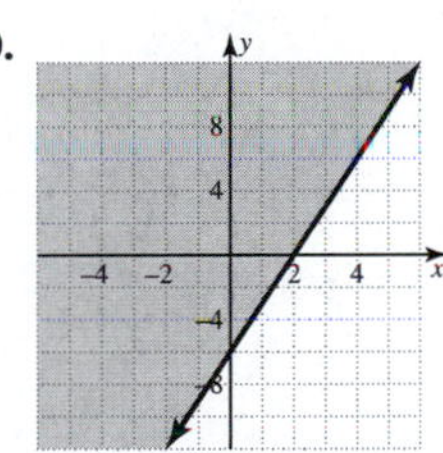

21. intersecting lines, one solution, consistent system with independent equations

22. parallel lines, no solution, inconsistent system

55.

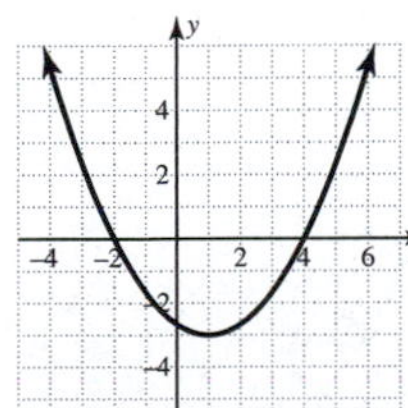

The solution set is $\{-2, 4\}$.

56.

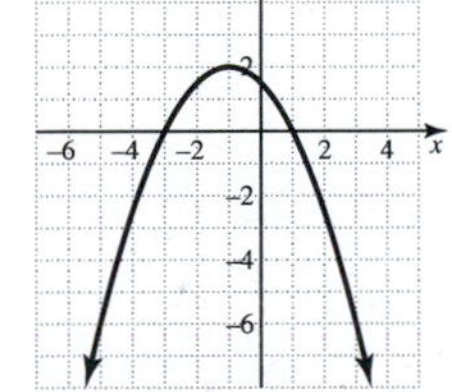

The solution set is $\{-3, 1\}$.

69. $(f+g)(x) = 2x^2 - 2x + 1$

$(f-g)(x) = 2x + 7$

$(fg)(x) = x^4 - 2x^3 + x^2 - 8x - 12$

domain is $(-\infty, \infty)$

$\left(\frac{f}{g}\right)(x) = \frac{x^2+4}{(x-3)(x+1)}, x \neq 3, -1$

domain is $(-\infty, -1) \cup (-1, 3) \cup (3, \infty)$

70. $(f+g)(x) = \sqrt{x+1} + \sqrt{x-3}$

$(f-g)(x) = \sqrt{x+1} - \sqrt{x-3}$

$(fg)(x) = \sqrt{x^2 - 2x - 3}$

domain is $[3, \infty)$

$\left(\frac{f}{g}\right)(x) = \frac{\sqrt{x+1}}{\sqrt{x-3}} = \frac{\sqrt{x^2-2x-3}}{x-3}, x \neq 3$

domain is $(3, \infty)$

75. $(f \circ g)(x) = -15x + 9$

$(g \circ f)(x) = -15x + 5$

domain is $(-\infty, \infty)$

76. $(f \circ g)(x) = 24x - 10$

$(g \circ f)(x) = 24x - 60$

domain is $(-\infty, \infty)$

77. $f^{-1}(x) = \frac{\sqrt[3]{x+5}}{2}$ **78.** $f^{-1}(x) = \sqrt[5]{x-32}$

87. $\frac{1}{2}\ln x + 3\ln(x+2)$ **88.** $2\log_7 y - \frac{1}{2}\log_7(y-3)$

89. $\log_{10}\frac{x^2y^3}{z}$ **90.** $\log_2\frac{x}{y^5z}$

95. $\left\{\frac{10^{1.8}-9}{2}\right\}$ or $\{27.05\}$ **96.** $\left\{\frac{e^{4.2}+5}{3}\right\}$ or $\{23.90\}$

97. $\left\{\frac{\ln 7}{4\ln 1.035}\right\}$ or $\{14.14\}$ **98.** $\left\{\frac{\ln 8}{3\ln 2.44}\right\}$ or $\{0.78\}$

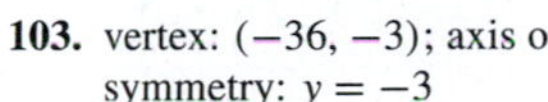

103. vertex: (−36, −3); axis of symmetry: $y = -3$

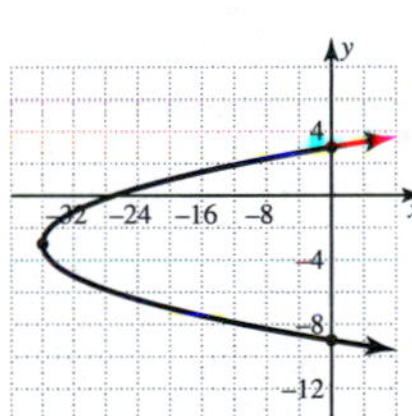

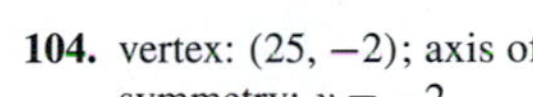

104. vertex: (25, −2); axis of symmetry: $y = -2$

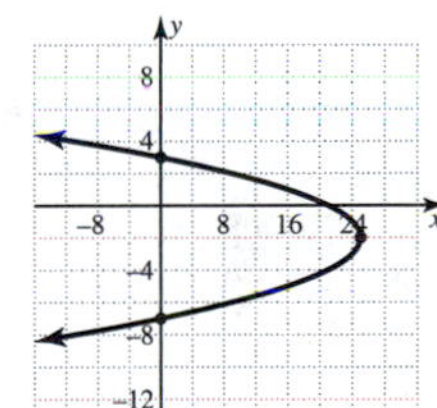

105. center: (0, 0); radius: 10; points: (10, 0), (0, 10), (−10, 0), (0, −10)

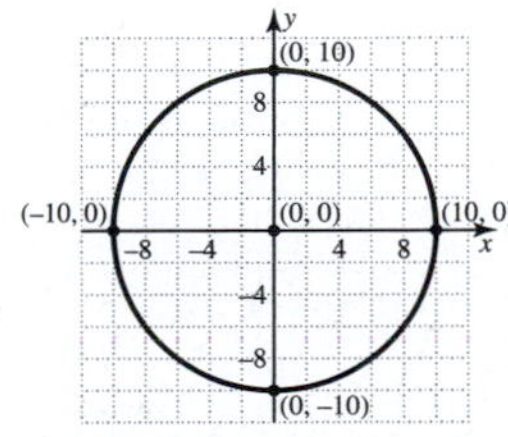

106. center: (0, 0); radius: 6; points: (6,0), (0, 6), (−6, 0), (0, −6)

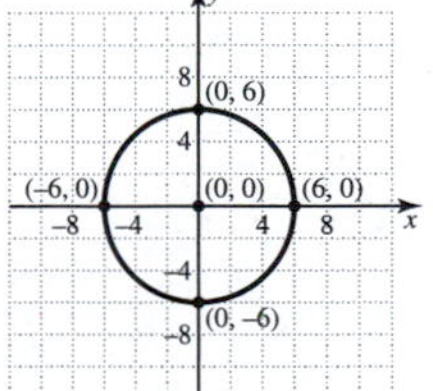

107. center: (0, −1); radius: 2; points: (−2, −1), (0, −3), (2, −1), (0, 1)

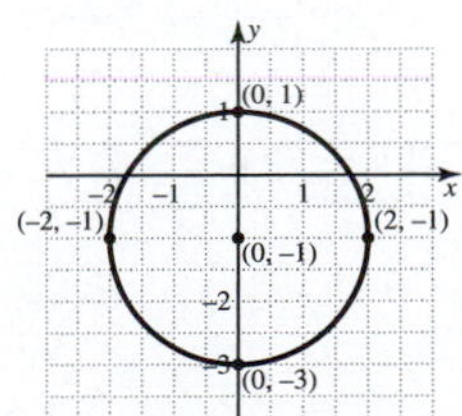

108. center: (−2, 0); radius: 1; points: (−2, 1), (−2, −1), (−1, 0), (−3, 0)

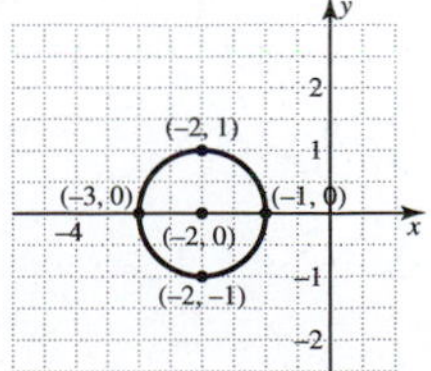

111. $(x - 2)^2 + (y + 1)^2 = 9$; center: $(2, -1)$; radius $= 3$

112. $(x + 5)^2 + (y - 2)^2 = 1$; center: $(-5, 2)$; radius $= 1$

113.

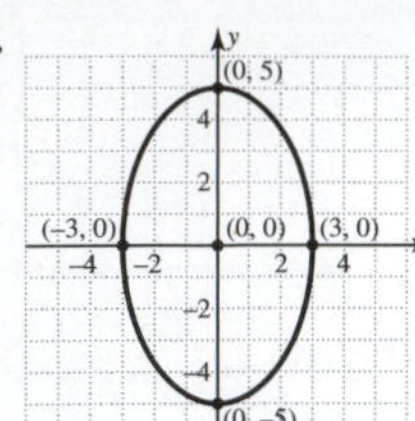

114.

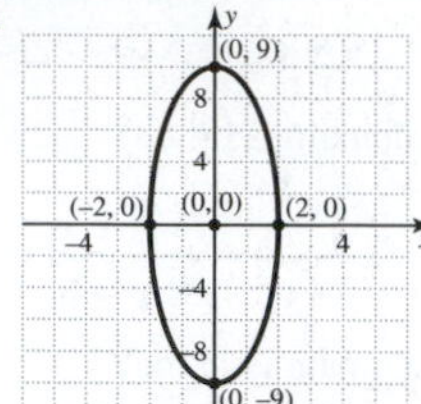

115.

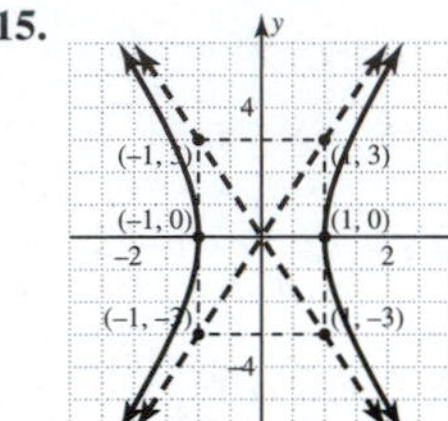

116.

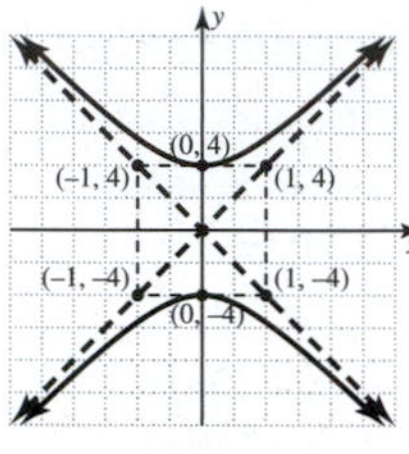

121. $\{(-2\sqrt{5}, -\sqrt{6}), (-2\sqrt{5}, \sqrt{6}), (2\sqrt{5}, -\sqrt{6}), (2\sqrt{5}, \sqrt{6})\}$

122. $\{(-2\sqrt{2}, -2), (-2\sqrt{2}, 2), (2\sqrt{2}, -2), (2\sqrt{2}, 2)\}$

125.

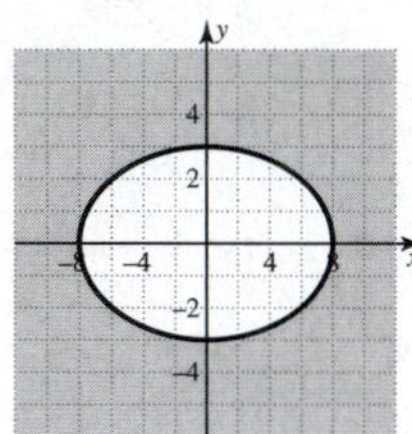

126.

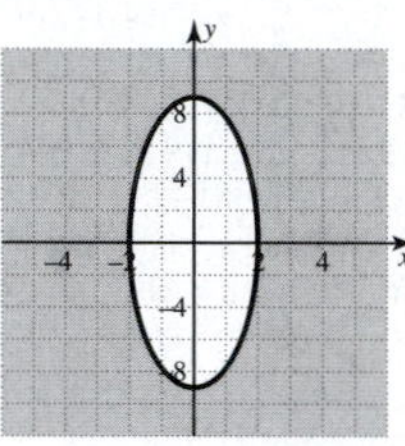

127.

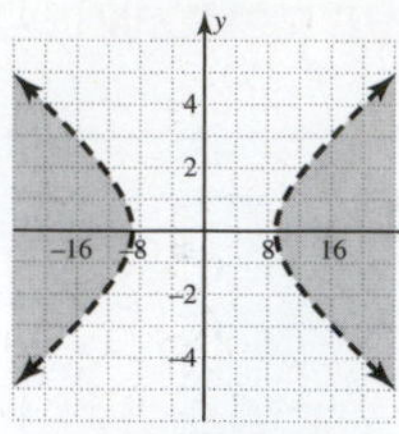

128.

129.

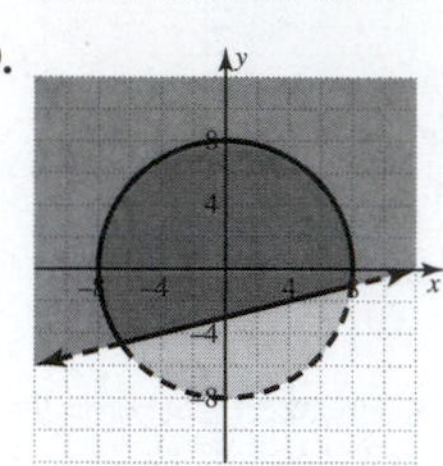

130.

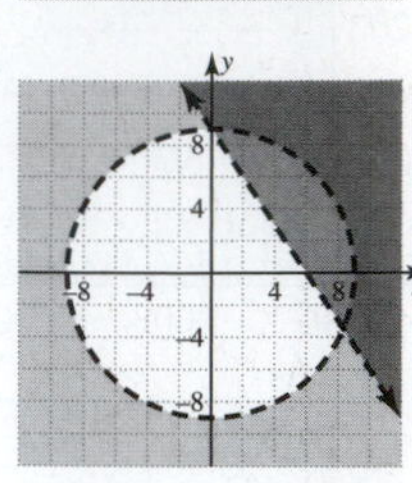

131.

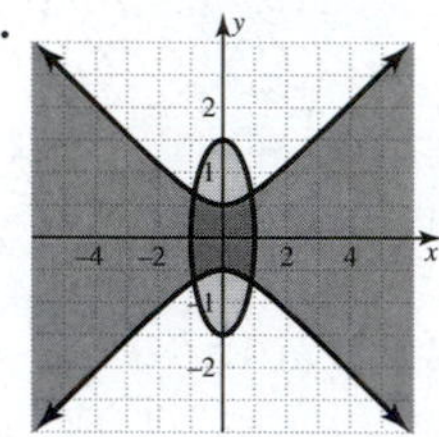

132.

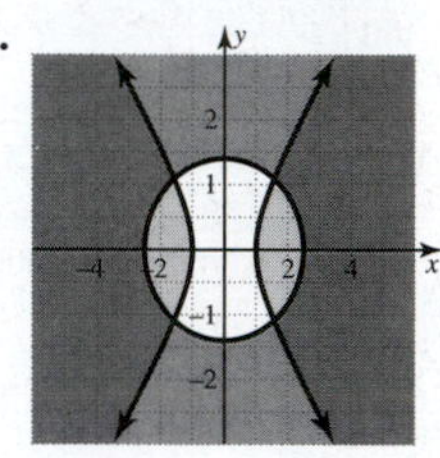

Credits

Photo Credits

Design Icons

Write about it: © Blend Images/Getty RF; **You be the teacher:** © Rubberball/Getty RF; **Practice makes perfect, Think about it, Calculate it, Mix' em up:** © Getty RF; **Piece it together** © PhotoDisc/Getty RF; **Group Activity:** © Banana Stock/PictureQuest RF.

Chapter S

Opener: © Purestock/Punchstock RF; **p. S2(top)** Library of Congress Prints and Photographs Division[LC-USW33-019093-C; **p. S2(right):** © Getty RF; **P. S3:** © Ingram Publishing/Superstock RF; **p. S6(top):** © Janice Ortiz; **p. S6(right), p. S7, p. S8(both), p. S9:** © Getty RF; **p. S10:** © Alamy RF; **p. S11(top):** © Todd Hendricks; **p. S11(right):** © Punchstock RF; **p. S11(bottom):** © Getty RF; **p. S12:** © Ingram Publishing RF; **p. S13:** © Corbis RF; **p. S14:** © Getty RF; **p. S15:** © Ingram Publishing RF; **p. S16, p. S18:** © Getty RF; **p. S19(top):** © Todd Hendricks; **p. S19(right):** © Veer RF; **p. S20:** © Corbis RF; **p. S23, p. S24:** © Getty RF; **p. S25(top):** Courtesy of Grace Shyu; **p. S25(bottom):** © Veer RF; **p. S25(right)** © Getty RF; **p. S26:** © Veer RF; **p. S27, p. S 28:** © Getty RF.

Chapter 1

Opener: © Getty RF; **p. 1(bottom):** © Punchstock RF; **p. 2(bottom):** © Getty RF; **p. 2(top):** Library of Congress, Prints and Photographs Division [LC-USZ62-115064]; **p. 3:** © McGraw-Hill Companies, Inc.; **p. 4, p. 7:** © Getty RF; **p. 15:** © The McGraw-Hill Companies Inc. John Flournoy, photographer; **p. 18, p. 26:** © Corbis RF; **p. 33, 39(top):** © Punchstock RF; **p. 39(bottom):** © Getty RF; **p. 44(both):** © The McGraw-Hill Companies Inc. Ken Cavanagh Photographer.

Chapter 2

Opener: © Ashley Zellmer; **p. 52(bottom):** © Getty RF; **p. 52(top):** Library of Congress Prints and Photographs Division [LC-USZ62-25564]; **p. 67:** © Veer RF; **p. 70:** © Corbis RF; **p. 71(both):** © The McGraw-Hill Companies, Inc.; **p. 72:** © PictureQuest RF; **p. 77:** © AGE Fotostock RF; **p. 78, p.79:** © Getty RF; **p. 83:** © Stephen Reynolds; **p. 84:** © Getty RF; **p. 85:** © Blend Images RF; **p. 93:** © Veer RF; **p. 94:** © Stockdisc/PunchStock RF; **p. 98(left):** © Alamy RF; **p. 98(right), 101(both):** © Getty RF; **p. 114:** © Creatas/PunchStock RF; **p. 115, 121:** © Corbis RF; **p. 125(both):** © The McGraw-Hill Companies, Inc.; **p. 137:** © Ingram Publishing RF.

Chapter 3

Opener: © Getty RF; **p. 147(bottom):** © Getty RF; **p. 148(top):** © Photo Courtesy Minding International, Ltd.; **p. 164:** © Brand X Pictures/PunchStock RF; **p. 165:** © BananaStock RF; **p. 166, p. 178:** © Corbis RF; **p. 190:** © The McGraw-Hill Companies, Inc. Jill Braaten, photographer; **p. 200(baseball, football):** © Getty RF; **p. 200(hockey):** © Photolink/Getty RF.

Chapter 4

Opener: © Getty RF; **p. 211(bottom):** © Getty RF; **p. 212(top):** Library of Congress Prints and Photographs Division [LC-USZ62-104276]; **p. 212(middle), p. 217:** © Lars A. Niki; **p. 218:** © Getty RF; **p. 219, p. 222(left):** © Corbis RF; **p. 222(right):** © Getty RF; **p. 224:** © Stockbyte/PunchStock RF; **p. 231:** © Artville/Veer RF; **p. 238:** © Corbis RF; **p. 247:** © Corbis RF; **p. 262:** © Comstock/Alamy RF; **p. 269:** © Alamy RF.

Chapter 5

Opener: © Getty RF; **p. 289(bottom):** © Ashley Zellmer; **p. 290(top):** Library of Congress, Prints and Photographs Division (LC-USZ62-95719); **p. 297:** © Alamy RF; **p.298:** © Ashley Zellmer; **p. 305, p. 313:** © The McGraw-Hill Companies, Inc. Jill Braaten, photographer; **p. 317(bottom), p. 319:** © Corbis RF; **p. 320:** © Ingram Publishing/SuperStock RF; **p. 321, p. 322, p. 323, p. 324:** © Getty RF; **p. 325:** © Alamy RF; **p.327, p. 329:** © Corbis RF; **p. 330:** © Getty RF; **p. 331:** © Alamy RF; **p. 332(tiger):** © Comstock/PunchStock RF; **p.332(chemist):** © Getty RF; **p. 333(biker):** © Corbis RF; **p. 333(squash):** © Getty RF; **p. 335, 340:** © Corbis RF; **p. 341:** © TongRo Image/Alamy RF; **p. 342:** © Getty RF; **p. 347, p. 351:** Brand X Pictures/PunchStock RF; **p. 362:** © Ingram Publishing RF.

Chapter 6

Opener: © Getty RF; **p. 365(bottom):** © Pixtal/age fotostock RF; **p. 366(top):** Library of Congress, Prints and Photographs Division (LC-J601-302); **p. 366(right):** © Jupiter RF; **p. 376:** © Mark Steinmetz RF; **p. 377(left):** © Getty RF; **p. 377(right):** © Ingram Publishing RF; **p. 379, p. 383:** © Getty RF; **p. 385(top):** © Comstock/Alamy RF; **p. 385(bottom), p. 386:** © Getty RF; **p. 389(left):** © Ingram Publishing RF; **p. 389(right):** © Getty RF; **p. 391, p.395(top):** © Alasdair Drysdale; **p. 395(middle):** © Corbis RF; **p. 395(bottom):** © Goodshoot/PunchStock RF; **p. 400(right):** © Getty RF; **p. 402:** © Corbis RF; **p. 411, p. 415(top):** © Getty RF; **p. 415(middle):** © Ingram Publishing/SuperStock RF; **p. 416(bottom):** © Alamy RF; **p. 416(top):** © Getty RF; **p. 418:** © Pixtal/age fotostock RF; **p. 425:** © Corbis RF; **p. 437:** © Brand X/Jupiter RF; **p. 445, p. 449:** © Pixtal/age fotostock RF; **p. 450:** © SuperStock RF.

Chapter 7

Opener: © Getty RF; **p. 468, p. 477(top):** © Getty RF; **p. 477(bottom):** © Corbis RF; **p. 478:** © Jupiterimages/Getty RF; **p. 505, p. 509, p. 523, p. 525, p. 526:** © Corbis RF; **p. 530, p. 531:** © Brand X/Fotosearch RF; **p. 532:** © The McGraw-Hill Companies, Inc. Ken Cavanagh photographer; **p. 534:** © Ingram Publishing/SuperStock; **p. 535:** © Getty RF; **p. 536:** © Stockbyte/PunchStock RF; **p. 537:** © Rubberball/SuperStock RF.

Chapter 8

Opener: © Getty RF; **p. 551(bottom):** © Comstock Images/Jupiterimages RF; **p. 552(top):** Used with permission from Jim Rohn International © 2011; **p. 552(bottom):** © Comstock Images/Alamy RF; **p. 565:** © Rubberball/SuperStock RF; **p. 566:** © David R. Frazier Photolibrary RF; **p. 567:** © Corbis RF; **p. 578:** © Pixtal/SuperStock RF; **p. 589:** © Getty RF; **p. 615:** © Comstock Images/Jupiterimages RF.

Chapter 9

Opener: © Getty RF; **p. 635(bottom):** © Getty RF; **p. 636(top):** Library of Congress, Prints and Photographs Division [LC-USZ62-128302]; **p. 636(bridge):** © Corbis RF; **p. 636(monument):** © Design Pics/PunchStock RF; **p. 636(satellite, fountain):** © Getty RF; **p. 651:** © Brand X Pictures/Jupiter RF; **p. 658(top):** © Alamy RF; **p. 658(bottom):** © imageshop/PunchStock RF; **p. 663, p. 664:** © Corbis RF; **p. 665:** © The McGraw-Hill Companies, Inc. Gerald Wofford, photographer; **p. 670:** © Getty RF; **p. 671(top):** © Valerie Martin RF; **p. 671(bottom):** © Corbis RF; **p. 674, p. 675, p. 677, p. 684:** © Corbis RF; **p. 685:** © Ingram Publishing/Alamy RF; **p. 688, p. 690, p.696:** ©Getty RF; **p. 698:** © Comstock Images/Alamy RF; **p. 700:** ©Alamy RF; **p. 720:** © Corbis RF.

Chapter 10

Opener: © Corbis RF; **p. 726(top):** Library of Congress Prints and Photographs Division [LC-USZ62-112513]; **p. 726(bottom), p. 733, p. 740:** © Getty RF; **p. 754, p. 758:** © The McGraw-Hill Companies Inc. John Flournoy, photographer; **p. 761(top):** © Getty RF; **p. 761(middle):** © Imagestate Media RF; **p. 761(bottom):** © Corbis RF; **p. 778:** © Getty RF; **p. 783,** © Corbis RF; **p. 784,** © Getty RF; **p. 786:** USGS photo by Walter D. Mooney; **p. 787, p. 792:** © Getty RF; **p. 793:** ©Alasdair Drysdale RF.

Chapter 11

Opener: © The McGraw-Hill Companies, Inc. Mark Dierker, photogapher; **p. 807(bottom):** © EyeWire, Inc. RF; **p. 808(top):** David Shankbone(http://shankbone.org); http://en.wikipedia.org/wiki/File:Joan_Didion_at_the_Brooklyn_Book_Festival.jpg; **p. 808(bottom):** © EyeWire, Inc. RF; **p. 817:** © The McGraw-Hill Companies, Inc.

Text Credits

Chapter S

Page S-2: Reproduced with permission of Curtis Brown, London on behalf of the Estate of Sir Winston Churchill. Copyright © Winston S. Churchill.

Chapter 1

Page 2: Source: Carl Sandburg, Carl Sandburg Family Trust, Asheville, NC.

Chapter 2

Page 52: Source: Benjamin Franklin, (Scientist and statesman).

Chapter 3

Page 148: Source: Edward de Bono, The World Centre for New Thinking, Villa Bighi Kalkara, Malta.

Chapter 4

Page 212: Source: Alexander Graham Bell, (Scientist and inventor).

Chapter 5

Page 290: Source: Abraham Lincoln, United States President.

Chapter 6

Page 366: Copyright Tuskegee University Archives.

Chapter 7

Page 468: Source: Buddhist Proverb, International Buddhism Center.

Chapter 8

Page 552: SOURCE: Article by Jim Rohn, America's Foremost Business Philosopher, reprinted with permission from Jim Rohn International © 2010. As a world-renowned author and success expert, Jim Rohn touched millions of lives during his 46-year career as a motivational speaker and messenger of positive life change. For more information on Jim and his popular personal achievement resources or to subscribe to the weekly Jim Rohn Newsletter, visit www.JimRohn.com.

Chapter 9

Page 636: Source: Leo Tolstoy, Leo Tolstoy Museum-Estate.

Chapter 10:

Page 726: Source: Helen Keller, "Words of Life," © 1966 Harper.

Chapter 11:

Page 808: Excerpt from SLOUCHING TOWARDS BETHLEHEM by Joan Didion. Copyright © 1966, 1968, renewed 1996 by Joan Didion. Reprinted by permission of Farrar, Straus and Giroux, LLC.

Index

A

absolute error, 120
absolute value, of real numbers, 9–10
absolute value equations
  applications of, 120–121
  definition of, 116
  properties of, 116, 119
  solving, 116–118, 121–122
  with two absolute values, 119–120
absolute value inequalities
  applications of, 130–131
  properties of, 125, 127, 128, 129
  solving
    of form $|X| < k$ or $|X| \leq k$, 125–126, 133
    of form $|X| > k$ or $|X| \geq k$, 127–128, 133
    using test points, 131–133
  special cases of, 128–130
addition
  associative property of, 34–35
  commutative property of, 34–35
  of complex numbers, 617–618, 622
  of functions, 726–727
  identity property of, 33
  inverse property of, 33–34
  of polynomials, 402–404, 412
  of radical expressions, 589–591, 595
  of rational expressions
    with like denominators, 495–496
    with unlike denominators, 498–500, 501
  of real numbers, 15–16
    with different signs, 16
    with same signs, 15
  translating phrases into expressions, 37–38
addition method. *See* elimination method
addition property of equality
  definition of, 54
  in elimination method, 309
  linear equations in one variable solved by, 54–55, 61
addition property of inequality
  definition of, 88
  linear inequalities in one variable solved by, 89–92
additive identity, 33
additive inverse, 33–34
algebraic expressions
  coefficients in, 35
  definition of, 23
  evaluating, 23–25
    applications of, 25–27
  simplifying, 35–37, 40
  terms in, 35
  translating phrases into, 37–40, 41
"and," compound inequalities with, 104–107
angles, 72–75
  complementary, 329
    definition of, 72
    finding, 72–75
  finding unknown, 72–75
  right, 72
  straight, 72
  supplementary, 329
    definition of, 72
    finding, 72–75, 81
  vertical, 72
applications
  of absolute value equations, 120–121
  of absolute value inequalities, 130–131
  of algebraic expressions, evaluating, 25–27
  of combined functions, 732–733
  of combined variation, 535–537
  of complex fractions, 509–511
  of compound inequalities, 109–111
  of direct variation, 530–532
  of distance, 527
  of exponential equations, 791–793
  of exponential functions, 757–758, 759
  of exponents, 372–374
  of functions, 184
    domain of, 194–195
  of inverse variation, 533–535
  of joint variation, 535–537
  of linear equations in one variable, 65–71
  of linear equations in two variables, 230–232, 261–263
  of linear functions, 217–219, 230–232
  of linear inequalities in one variable, 92–95
  of linear inequalities in two variables, 272–274
  of logarithmic equations, 791–793
  of logarithms, 782–783
  of nonlinear inequalities, 709–710
  of ordered pairs, 156–158
  of order of operations, 25–27
  of polynomial equations, 448–450
  of polynomials, 410–412
    division of, 490–491
  of proportions, 525–526
  of quadratic equations, 657–659, 669–672
  of quadratic functions, 695–696
  of quadratic methods, 683–685
  of radical equations, 609–611
  of radical expressions, 593–594
  of rational equations
    distance, 527
    with number relationships, 524–525
    proportions, 525–526
    work, 526
  of rational exponents, 573–574
  of rational expressions, 476–478
  setting up equations for, 65
  of simple interest, 323–325
  of slope of a line, 247–248
  of systems of linear equations in three variables, 339–342
  of systems of linear equations in two variables, 297–299, 313–314
    distance, 326–328
    geometry, 329–330
    investment, 323–325
    mixtures, 325–326
    money, 320–323
  of systems of linear inequalities in two variables, 350–352
  translating phrases into expressions in, 38–40
  of work, 526
area formula, 75–77, 80
argument of logarithms, 762
associative property of addition, 34–35
associative property of multiplication, 34–35
asymptotes, of hyperbolas, 819
$ax^2 + bx + c$, factoring
  by grouping, 430–432
  by substitution, 432–433
  by trial and error, 428–430
axis of symmetry, of parabola, 637

B

base
  in exponential expressions, 20, 366
  of logarithms, 762
    changing, 781–782, 784
    common logarithms, 779
    natural logarithms, 779
  negative
    raised to even power, 21
    raised to odd power, 21
base fee, 217

binomials
  completing square of, 653–655
  conjugates
    definition of, 409
    multiplication of, 409–410
  definition of, 393
  division of polynomials by, 488–489
  factoring
    difference of two cubes, 439–442
    difference of two squares, 437–438, 440–442
    sum of two cubes, 439–442
    sum of two squares, 442
  multiplication of
    FOIL method for, 406–407
    square of binomial, 407–409, 413
  perfect square trinomials from, 408
$b^{-m/n}$, evaluating, 570–571
$b^{m/n}$, evaluating, 569–570
$b^{1/n}$, evaluating, 568
boundary line, 270
boundary number, 700
boundary number method
  higher degree polynomial inequalities solved by, 703–704
  quadratic inequalities solved by, 700–703, 710–711
  rational inequalities solved by, 705–706
break-even point, 297

C

calculator. *See* graphing calculator
Cartesian coordinate system. *See* rectangular coordinate system
center
  of circle, 811
  of ellipse, 817
  of hyperbolas, 819
change-of-base formula, 781–782, 784
circles, 811–812, 814
  circumference of, 75–77
  equation for, 812–813
  graphing, 812
  perimeter of, 75–77
  standard form of, 811
circumference formula, 75–77
coefficient
  definition of, 35, 391
  in polynomials, 393
  in scientific notation, 383
combined functions
  applications of, 732–733
  domain of, 726–727
  evaluating, 727–730, 734
combined variation, 535–537
commitment, 211
common logarithms, 778–779, 784
commutative property of addition, 34–35
commutative property of multiplication, 34–35
complementary angles
  definition of, 72, 329
  finding, 72–75
completing the square
  for equation of a circle, 813
  perfect square trinomials from, 653–655
  quadratic equations solved by
    of the form $ax^2 + bx + c$, 656–657
    of the form $x^2 + bx + c$, 655–656, 661
complex conjugate, 619
complex fractions
  applications of, 509–511
  definition of, 505
  simplification of
    by division, 506–507, 511
    by multiplication by the LCD, 507–509
complex numbers
  addition of, 617–618, 622
  conjugates of, 619
  definition of, 615–616
  division of, 619–621
  multiplication of, 618–619
  set of, 616
  subtraction of, 617–618
composition of functions, 730–732
composition of inverse functions, 747, 748
compound inequalities
  applications of, 109–111
  definition of, 104
  joined by "and," 104–107
  joined by "or," 107–109, 111
compound interest formula, 78–79, 757
conditional equations, 59–60
conic sections, 808. *See also* circle; ellipses; hyperbolas; parabola
conjugates
  definition of, 409
  multiplication of, 409–410
  product of. *See* difference of two squares
  rationalizing the denominator using, 601–602
consecutive integers, applications of, 66
consistent systems of linear equations in three variables, 336
consistent systems with dependent equations, 295–297
consistent systems with independent equations, 295–297
constant of variation (proportionality), 530–532
constant term, 391
continuous compound interest formula, 791
contradiction, 59–60
creativity, 147
cube(s)
  difference of two, factoring, 439–442
  perfect, 554
  sum of two, factoring, 439–442
cube root(s)
  approximating, 556–557
  definition of, 553
  of negative number, 622
  simplifying, 554, 562
cube root functions, graphing, 561

D

degree
  of polynomial equations, 445
  of polynomials, 393
  of term, 391–392
denominator
  of radical expressions
    rationalizing, 599–601, 603
    rationalizing using conjugates, 601–602
  zero as, 468
dependent variable, 182
difference of functions, 726, 734
difference of two cubes, factoring, 439–442
difference of two squares
  definition of, 409, 437
  factoring, 437–438, 440–442
direct variation, 530–532, 537
distance applications, 326–328, 527
distance formula, 80, 326, 583–585, 586
distance-rate-time formula, 527
distributive property
  definition of, 34–35
  in factoring of polynomials, 420
  in multiplication of polynomials, 404–405
dividend, zero as, 19
division
  complex fractions simplified by, 506–507, 511
  of complex numbers, 619–621
  of exponential expressions, with like bases, 368
  of functions, 726–727
  of opposites, 472
  of polynomial functions, 489–490
  of polynomials
    applications of, 490–491
    by binomials, 488–489
    long division, 485–487, 491
    by monomials, 483–485, 491
    synthetic division, 487–489, 492
  property of, for fractions, 483–484
  of radical expressions, 598–599, 602
  of rational expressions, 475–476
  of real numbers, 18–20

translating phrases into expressions, 37–38
by zero, 19
domain
of exponential functions, 755
of functions
applications of, 194–195
definition of, 193
equation in finding, 193–194
finding, 190–191
graph in finding, 191–193, 195
of linear equations in two variables, 225–226
of linear functions, 225–226
of logarithmic functions, 766–767
of polynomial functions, 396–398
of radical functions, 559–560
of rational functions, 470–471, 478
of relations, 165, 167–170, 171–172

E

element of set, 2
elimination method
definition of, 308
solving nonlinear systems of equations by, 828–829
solving systems of linear equations in three variables by, 336–339
solving systems of linear equations in two variables by, 308–311
special cases of, 311–313
ellipses
definition of, 817
graphing, 818–819, 821
standard form for, 817–818
ellipsis (…), 2
empty set, 2–3
equality
addition property of
definition of, 54
in elimination method, 309
linear equations in one variable solved by, 54–55
multiplication property of
definition of, 55
linear equations in one variable solved by, 56, 61
equations. *See also* algebraic equations; linear equations in one variable; linear equations in two variables; quadratic equations; radical equations; rational equations
conditional, 59–60
definition of, 37
equivalent, 54
relations as, 165
solving, by substitution, 682–683, 686
equilibrium point, 297
equivalent equations, 54
even integers, consecutive, 66
exponent(s)
applications of, 372–374
definition of, 366
even, 21
in exponential expressions, 20
i raised to, 621–622
negative, 369–371, 374, 381
fractions with, 381
odd, 21
power of a power rule for, 379–381, 382
power of a product rule for, 379–381, 382
power of a quotient rule for, 380–381, 382
product of like bases rule for, 366–367, 374, 381
quotient of like bases rule for, 367–368, 374, 381
in scientific notation, 383
zero as, 369, 381
exponential equations, 787–789
applications of, 791–793
logarithms as, 762–763, 767
one-to-one property of, 756
solving, 756–757
with logarithms, 788–789, 794
exponential expressions, 20–21
definition of, 366
division of, with like bases, 368
evaluating, 27
multiplication of, with like bases, 367
with negative exponents, 370–371, 374, 381
product as base of, raising to a power, 379–381, 382
product of like bases rule for, 366–367, 374, 381
quotient as base of, raising to a power, 380–381, 382
quotient of like bases rule for, 367–368, 374, 381
raising to a power
even power, 21
with negative exponents, 370–371, 374, 381
odd power, 21
power of a power rule for, 379–381, 382
power of a product rule for, 379–381, 382
power of a quotient rule for, 380–381, 382
with product as base of, 379–381, 382
with quotient as base of, 380–381, 382
with rational exponents
of the form $b^{-m/n}$, 570–571
of the form $b^{m/n}$, 569–570, 574
of the form $b^{1/n}$, 568
simplifying, 571–572
simplifying, 371–372, 386–387
exponential functions
applications of, 757–758, 759
definition of, 754
graphing, 755–756, 759
exponential notation, 20
expressions. *See* algebraic expressions; radical expressions; rational expressions
extraneous solutions
to radical equations, 606
to rational equations, 517

F

factoring
of binomials
difference of two cubes, 439–442
difference of two squares, 437–438, 440–442
sum of two cubes, 439–442
sum of two squares, 442
of perfect square trinomials, 441
polynomial equations solved by, 446–448
of polynomials
with greatest common factor, 419–421, 422–423
by grouping, 421–422
strategy for, 440–442
of trinomials
$ax^2 + bx + c$, 428–432
by grouping, 430–432, 441–442
perfect square trinomials, 441
by substitution, 432–433
by trial and error, 428–430, 441–442
$x^2 + bx + c$, 425–428
finite set, 2
first coordinate. *See* *x*-coordinate
first-degree equations. *See* linear equations in one variable
foci
of ellipses, 817
of hyperbolas, 819
FOIL method, 406–407, 428, 618
formulas
for area, 75–77, 80
for change-of-base, 781–782
for circumference, 75–77
for compound interest, 78–79, 757
for continuous compound interest, 791
definition of, 75–77
for distance, 80, 326, 583–585, 586
for distance-rate-time, 527
for midpoint, 155–156
for perimeter, 75–77, 80
for simple interest, 323
for slope of a line, 243–245
solving for specified variable, 79–81
for temperature conversion, 78–79
for vertex, 693–695
for volume, 81

fraction(s)
  as exponents. *See* rational exponents
  in lowest terms, 471
  negative powers of, 381
  property of division for, 483–484
fractional exponents. *See* rational exponents
function(s)
  algebra of, 726–727, 734
  applications of, 184
  combined
    applications of, 732–733
    domain of, 726–727
    evaluating, 727–730, 734
  composition of, 730–732
  cube root, graphing, 561
  definition of, 178
  dependent variable in, 182
  domain of
    applications of, 194–195
    equation in finding, 193–194
    finding, 190–191
    graph in finding, 191–193, 195
  evaluating, 182
  exponential. *See* exponential functions
  graph of, 745–746
    domain and range from, 191–193, 195
  independent variable in, 182
  inverse. *See* inverse functions
  linear. *See* linear functions
  one-to-one. *See* one-to-one functions
  operations on, 726–727, 734
  polynomial. *See* polynomial functions
  quadratic. *See* quadratic function
  radical. *See* radical functions
  range of
    finding, 190–191
    graph in finding, 191–193, 195
  rational. *See* rational function
  relations as
    determining, 178–179, 185
    vertical line test for, 180–181
  solving, 182, 185
    of form $f(x) = c$, 216–217
  square root, graphing, 561
  zero of, 216, 450
function notation, 181–184, 219
  for linear equations in, 213–214
fundamental property of rational expressions, 471–474

G

geometry. *See also* angles; triangles
  applications of, 329–330
  area formula, 75–77, 80
  circumference, 25, 75–77
goal setting, 365
graphing
  of circle, 812
  of cube root functions, 561
  of ellipses, 818–819, 821
  of exponential functions, 755–756, 759
  of functions, 745–746
    finding domain and range from, 191–193, 195
  of hyperbolas, 820, 821
  of inverse functions, 745–746
  of linear equations in two variables, 152–154, 224–226
    finding domain and range from, 225–226
    horizontal lines, 228–230, 233
    using intercepts, 226–228, 232–233
    using slope and y-intercept, 240–243, 249
    vertical lines, 228–230
  of linear functions, 224–226
    finding domain and range from, 225–226
    using slope and y-intercept, 240–243
  of linear inequalities in one variable, solution sets of, 87–88, 95, 271–272
  of linear inequalities in two variables, 274
  of nonlinear inequalities, 706–709, 832–834, 836
  of nonlinear systems of inequalities, 834–836
  of parabolas, of form $x = a(y - k)^2 + h$, 810–811
  of polynomial functions, 395–398
    zeros of, 450–451
  of quadratic equations, solving by, 645–646
  of quadratic functions, 636–637, 647
    $f(x) = (x - h)^2$, 639–640
    $f(x) = ax^2$, 640–642
    $f(x) = ax^2 + bx + c$, 690–693
    $f(x) = a(x - h)^2 + k$, 642–645
    $f(x) = x^2 + k$, 637–639
    vertex formula for, 693–695
  on real number line, 8
  of relations, 165
    determining domain and range from, 167–170, 171–172
  of square root functions, 561
  of systems of linear equations in two variables, solving by, 291–293
  of systems of linear inequalities in two variables, 347–350, 352
graphing calculators
  absolute value equations on, 122
  absolute value inequalities on, 134–135
  absolute values on, 12
  algebraic expressions on, 28–29
  circles on, 815
  combined functions on, 735
  common logarithms on, 780, 785
  complex fractions on, 512
  complex numbers on, 623
  compound inequalities on, 112
  ellipses on, 822
  e on, 795
  evaluating expressions on, 28
  exponential expressions on, 387
  exponential functions on, 760
  exponents on, 28–29
  factoring binomials on, 443
  factoring trinomials on, 434
  formulas on, 82
  functions on, 186
    domain and range of, 196
  greatest common factor on, 423
  hyperbolas on, 822–823
  inverse functions on, 749–750
  irrational numbers on, 11
  linear equations in one variable on, 62
  linear equations in two variables on, 234–235, 265
  linear functions on, 220–221
  linear inequalities in one variable on, 97
  linear inequalities in two variables on, 275–276
  logarithms on, 780, 785
  natural logarithms on, 780, 785
  negative exponents on, 375
  nonlinear inequalities on, 711–712
  opposites on, 12
  ordered pairs on, plotting, 160–161
  parabolas on, 815
  polynomial equations on, 453
  polynomial functions on, 399
  polynomials on
    division of, 492
    subtraction of, 414
  quadratic equations solved on, 662
  quadratic functions on, 648–649, 697
  radical equations on, 612–613, 686–687
  radical expressions on, 564, 586–587
  rational equations on, 521
  rational exponents on, 575
  rational expressions on
    division of, 603
    evaluating, 479
    least common denominator of, 502
    subtraction of, 502
  relations on, 172–173
  slope of a line on, 250–251
  systems of linear equations in two variables on, 301, 316
  systems of linear inequalities in two variables on, 353
greatest common factor (GCF)
  definition of, 418
  factoring from polynomial, 419–421, 422–423
  finding, 418–419
greatness, 725

grouping
factoring polynomials by, 421–422
factoring trinomials by, 430–432, 441–442

H

half-planes, 270
higher degree polynomial equations, solving using quadratic methods, 680–682, 686
higher degree polynomial inequalities. *See also* nonlinear inequalities
solving using boundary number method, 703–704
horizontal lines
equation for, 228
graph of, 228–230, 233
slope of, 244–245
horizontal line test, 741–742
hyperbolas
definition of, 819
graphing, 820, 821
standard form for, 819
hypotenuse, 448

I

i. *See* imaginary unit
identity
additive, 33
definition of, 60
multiplicative, 33
identity element, 33
imaginary unit (i)
definition of, 615–616
powers of, 621–622
inconsistent systems of linear equations in three variables, 336
inconsistent systems of linear equations in two variables, 295–297
independent variable, 182
index, of nth roots, 555
inequalities, 37. *See also* compound inequalities; linear inequalities in one variable; linear inequalities in two variables; quadratic inequalities; rational inequalities
inequality
addition property of
definition of, 88
linear inequalities in one variable solved by, 89–92
multiplication property of
definition of, 89
linear inequalities in one variable solved by, 89–92
infinite set, 2
initial value of y-intercept, 247
input value, 165, 170–171
integers
consecutive, 66
set of, 4, 6–7
intercepts. *See* x-intercepts; y-intercepts
interest
compound, 78–79, 757
continuous compound, 791
simple
applications of, 323–325
formula for, 323
intersecting lines, as systems of linear equations, 311–313
intersection of sets, 3, 101–102
interval, intersection of, 101–102
interval notation, for linear inequalities in one variable, 86–87, 88, 95
inverse
additive, 33–34
definition of, 33
multiplicative, 33–34
inverse functions
composition of, 747, 748
definition of, 739, 742–743
equations of, 744–745, 748
evaluating, 745, 748
finding, 743–744
graphing, 745–746
verifying, 747–748
inverse properties of logarithms, 774–775
inverse variation, 533–535, 538
investment applications, 323–325
irrational numbers
set of, 4, 5–7
square roots as, 5, 10

J

joint variation, 535–537

L

leading coefficient of polynomials, 393
learning strategies, 467
least common denominator (LCD)
of complex fractions, 507–509
of rational expressions, 496–498, 501
legs, of triangle, 448
like radicals
addition of, 589–591
definition of, 589
subtraction of, 589–591
like terms
definition of, 35
in polynomials, 402
line(s)
equation of. *See* linear equations in one variable; linear equations in two variables
graph of, from linear equations in two variables, 152
horizontal. *See* horizontal lines
midpoint formula for, 155–156
parallel. *See* parallel lines
perpendicular. *See* perpendicular lines
slope of. *See* slope of a line
as systems of linear equations, 294–295
vertical. *See* vertical lines
linear equations in one variable
applications of, 65–71
conditional, 59–60
contradictions, 59–60
definition of, 52
determining an equation is linear, 53
formulas. *See* formulas
with infinitely many solutions, 60
with no solution, 59–60
solutions of, 52–53
solving
using addition property of equality, 54–55, 61
using multiplication property of equality, 56, 61
strategy for, 57–59
translating statements into, 65–69
linear equations in two variables. *See also* systems of linear equations in two variables
applications of, 230–232
definition of, 150
in function notation, 213–214, 219
functions. *See* linear functions
graphing, 152–154, 224–226
determining solutions using, 154
finding domain and range from, 225–226
using intercepts, 226–228, 232–233
using slope and y-intercept, 240–243, 249
horizontal lines, 228–230, 233
point from
two points, writing equation from, 258–260, 264
writing equation from, with slope, 256–258
in point-slope form, 256–258
in slope-intercept form, 239–240
slope of. *See* slope of a line
finding, 240, 248–249
with point, writing equation from, 256–258
with y-intercept, writing equation from, 255, 264
solutions of
determining graphically, 154
ordered pair as, 150–152, 159
in standard form, 213–214
vertical lines, 228–230
writing
applications of, 261–263
for parallel lines, 260–261
for perpendicular lines, 261–263, 264
from point and slope, 256–258
from slope and y-intercept, 255, 264
from two points, 258–260, 264
x-intercepts of, graphing using, 226–228, 232–233

linear equations in two variables. *See also* systems of linear equations in two variables (*Continued*)
y-intercepts of
finding, 240
graphing equation from, with slope, 240–243, 249
graphing using, 226–228, 232–233
initial value of, 247
writing equation from, with slope, 255, 264
linear functions, 212–213
applications of, 217–219, 230–232
definition of, 212
determining, 212–213, 219
evaluating, 214–216
graphing, 224–226
finding domain and range from, 225–226
using slope and y-intercept, 240–243
in slope-intercept form, 239–240
slope of, finding, 240
solving, 216–217
y-intercepts of, finding, 240
linear inequalities in one variable
applications of, 92–95
definition of, 86
solution sets of, 86–88
graphing, 87–88, 95
interval notation for, 86–87, 88, 95
set-builder notation for, 88
solving
using addition property of inequality, 89–92
using multiplication property of inequality, 89–92
linear inequalities in two variables. *See also* systems of linear inequalities in two variables
applications of, 272–274
definition of, 269
graphing, 270–272, 274
solution sets of, 271–272, 274
solutions of, 269–270
linear programming, 272–273
logarithm(s)
applications of, 782–783
change-of-base formula for, 781–782, 784
as combination of logarithmic expressions, 773
common, 778–779, 784
definition of, 762–763
exponential equations solved with, 788–789
inverse properties of, 774–775
natural, 779
one-to-one property of, 788
power rule for, 772–773, 775
product rule for, 769–770, 775
quotient rule for, 770–771, 775
logarithmic equations
applications of, 791–793
solving
converting to single logarithm, 789–791
using exponential form, 764–765, 780–781
logarithmic expressions
as exponential expressions, 762–763, 767
rewriting as single logarithm, 773
simplifying, 763–764, 767
logarithmic functions
definition of, 765
graphing, 765–767
long division, of polynomials, 485–487, 491
lower half-plane, 270
lowest terms, 471

M

mapping, relations expressed as, 165
maximum values, of quadratic function, 695–696
member of a set, 2
midpoint formula, 155–156
minimum values, of quadratic function, 695–696
mixture applications, 325–326
money applications, of systems of linear equations in two variables, 320–323
monomials
definition of, 393
division of polynomials by, 483–485, 491
multiplication
associative property of, 34–35
of binomials, 406–407
commutative property of, 34–35
complex fractions simplified by, 507–509
of complex numbers, 618–619
of conjugates, 409–410
of exponential expressions, with like bases, 367
of functions, 726–727, 734
identity property of, 33
inverse property of, 33–34
of polynomials, 404–406
of conjugates, 409–410
square of binomial, 407–409, 413
of three or more polynomials, 410
of radical expressions, 591–592, 595
of rational expressions, 474–475
of real numbers, 18–20
translating phrases into expressions, 37–38
multiplication property of equality
definition of, 55
linear equations in one variable solved by, 56, 61
multiplication property of inequality
definition of, 89
linear inequalities in one variable solved by, 89–92
multiplicative identity, 33
multiplicative inverse, 33–34

N

natural logarithms, 779
natural numbers, set of, 4, 6–7
negative exponents, 369–371, 374
fractions with, 381
negative numbers
cube roots of, 622
roots of, 559
square root of, 616
negative rational expressions, 469
negative slope, 243, 245
negative square roots, 552
net worth, 25
nonlinear inequalities. *See also* quadratic inequalities; rational inequalities
applications of, 709–710
definition of, 831
graphing, 832–834
solving
using boundary number method, 700–706
by graphing, 706–709
nonlinear systems of equations
solutions of, 824
solving
by elimination, 828–829
by substitution, 824–828
nonlinear systems of inequalities, solving, 834–836
nth roots
approximating, 556–557
definition of, 555
of a quotient, 582
simplifying, 555–557
$\sqrt[n]{a^n}$, 557
$(\sqrt[n]{a})^n$, 556–557
numerator, zero as, 468

O

odd integers, consecutive, 66
one-to-one functions
definition of, 740
determining, 740–741
horizontal line test for, 741–742
inverse of
definition of, 739, 742–743
equations of, 744–745, 748
evaluating, 745, 748
finding, 743–744
graphing, 745–746
verifying, 747–748

one-to-one property of b^x, 756
one-to-one property of logarithms, 788
opposite(s)
division of, 472
of polynomials, 402
of real numbers, 8–9, 10
"or," compound inequalities with, 107–109, 111
ordered pairs
applications of, 156–158
definition of, 148
plotting in rectangular coordinate system, 149–150, 159
as solution to linear equations in two variables, 150–152, 290
as solution to linear inequalities in two variables, 269–270
ordered triples, 336
order of operations, 22–23, 27
applications of, 25–27
organization skills, 51
origin, definition of, 7, 148
output value, 165, 170–171

P

paired data, 156
parabolas
of form $x = a(y - k)^2 + h$, 808–811, 814
as graph of linear equations in two variables, 152
as graph of quadratic function, 636
in standard form, 809
parallel lines. *See also* inconsistent systems
definition of, 245
determining, 246–247
as systems of linear equations, 294–295, 311–313
writing equations of, 260–261
perfect cube, 554
perfect square, 552
perfect square trinomials
from completing the square, 653–655
factoring, 441
from squaring binomials, 408
perimeter
formula for, 75–77, 80
of rectangle, 329
perpendicular lines
definition of, 245–246
determining, 246–247
writing equations of, 260–261, 264
perseverance, 211
phrases
translating into algebraic expressions, 37–40, 41
translating into linear inequalities, 93
pi (π), 5
point(s)
distance between two, 583–585, 586
identifying location in rectangular coordinate system, 149
from linear equations in two variables
with two points, writing equation from, 258–260, 264
writing equation from, with slope, 256–258
point-slope form, definition of, 256
polynomial(s). *See also* binomial; monomial; trinomial
addition of, 402–404, 412
applications of, 410–412
division of, 490–491
definition of, 392–393
degree of, 393
division of
applications of, 410–412
by binomials, 488–489
long division, 485–487, 491
by monomials, 483–485, 491
synthetic division, 487–489, 492
factoring
with greatest common factor, 419–421, 422–423
strategy for, 440–442
leading coefficient of, 393
like terms in, 402
multiplication of, 404–406
square of binomial, 407–409
of three or more polynomials, 410
opposite of, 402
prime, 426
in standard form, 392
subtraction of, 402–404, 413
polynomial equations. *See also* quadratic equations
applications of, 448–450
definition of, 445
degree of, 445
solutions to, 445
solving, 445–448, 451–452
by factoring, 446–448
using quadratic methods, 680–682, 686
in standard form, 445
polynomial functions
definition of, 394
division of, 489–490
domain of, 396–398
evaluating, 394–395, 398
graphing, 395–398, 450–451
range of, 396–398
x-intercepts of, 450–451
zeros of, 450–451
polynomial inequalities, higher degree, solving by boundary number method, 703–704
positive slope, 243, 245
power. *See* exponent; exponent(s)
power of a power rule for exponents, 379–381, 382
power of a product rule for exponents, 379–381, 382
power of a quotient rule for exponents, 380–381, 382
power property, 606
power rule for logarithms, 772–773, 775
prime numbers, definition of, 418
prime polynomials, 426
principal square root, 552
product of complex conjugates, 619
product of functions, 726, 734
product of like bases rule for exponents, 366–367, 374, 381
product property of zero, 18
product rule for logarithms, 769–770, 775
product rule for radicals
definition of, 578
in multiplying radical expressions, 591
in simplifying radical expressions, 578–581, 585–586
property of division for fractions, 483–484
property of equality
addition
definition of, 54
in elimination method, 309
linear equations in one variable solved by, 54–55, 61
multiplication
definition of, 55
linear equations in one variable solved by, 56
property of inequality
addition
definition of, 89
linear inequalities in one variable solved by, 89–92
multiplication
definition of, 89
linear inequalities in one variable solved by, 89–92
proportions
applications of, 525–526
definition of, 525
Pythagorean theorem, 448, 610

Q

quadrants, of rectangular coordinate system, 149
quadratic equations
applications of, 657–659, 669–672
definition of, 446
discriminant of
definition of, 668
using, 668–669

quadratic equations (*Continued*)
 solving
  by completing the square, 655–657, 661
  determining solutions to, 668–669
  of the form $ax^2 + bx + c$, 656–657
  of the form $x^2 + bx + c$, 655–656, 661
  by graphing, 645–646
  using quadratic formula, 666–668, 673
  using square root property, 651–653, 661
quadratic formula
 definition of, 666
 derivation of, 665
 solving quadratic equations using, 666–668, 673
quadratic function
 definition of, 636
 $f(x) = ax^2 + bx + c$, converting, 690–693, 696
 graph of, 636–637, 647
  $f(x) = (x - h)^2$, 639–640
  $f(x) = ax^2$, 640–642
  $f(x) = ax^2 + bx + c$, 690–693
  $f(x) = a(x - h)^2 + k$, 642–645
  $f(x) = x^2 + k$, 637–639
  vertex formula for, 693–695
 maximum values of, 695–696
 minimum values of, 695–696
 zeros of, 659–660
quadratic inequalities. *See also* nonlinear inequalities
 definition of, 700
 solving, using boundary number method, 700–703, 710–711
quotient, *n*th root of, 582
quotient of functions, 726
quotient of like bases rule for exponents, 367–368, 374
quotient rule for logarithms, 770–771, 775
quotient rule for radicals
 definition of, 582
 in dividing radical expressions, 598
 in simplifying radical expressions, 582–583

R

radical(s)
 definition of, 552–553
 product rule for
  in multiplying radical expressions, 591
  in simplifying radical expressions, 578–581
 quotient rule for
  in dividing radical expressions, 598
  in simplifying radical expressions, 582–583
radical equations
 applications of, 609–611
 definition of, 606
 extraneous solution to, 606
 solving, 607–609, 611–612
  using quadratic methods, 678–680, 685
radical expressions. *See also* cube root(s); square root(s)
 addition of, 589–591, 595
 applications of, 593–594
 definition of, 553
 denominator of
  rationalizing, 599–601, 603
  rationalizing using conjugates, 601–602
 division of, 598–599, 602
 multiplication of, 591–592, 595
 as rational exponents, 567–568
 simplifying
  in form $(\sqrt[n]{a})^n$, 556–557
  in form $\sqrt[n]{a^n}$, 557
  using product rule for radicals, 578–581, 585–586
  using quotient rule for radicals, 582–583
  using rational exponents, 572–573, 575
 subtraction of, 589–591
radical functions
 definition of, 558
 domain of, 559–560
 evaluating, 559–560
radical sign, 553
radicand, 553
radius, 811
range
 of exponential functions, 755
 of function
  finding, 190–191
  graph in finding, 191–193, 195
 of linear equations in two variables, finding from graph, 225–226
 of linear functions, finding from graph, 225–226
 of logarithmic functions, 766–767
 of polynomial functions, 396–398
 of relation, 165, 167–170, 171–172
rate, 527
rate of change, slope as, 247
ratio, 525
rational equations
 applications of
  distance, 527
  with number relationships, 524–525
  proportions, 525–526
  work, 526
 definition of, 516
 extraneous solutions to, 517
 rewriting, 523–524
 solving, 516–520
  using quadratic methods, 677–678
  for specific variable, 523–524
rational exponents
 applications of, 573–574
 definition of, 567
 exponential expressions with
  of the form $b^{-m/n}$, 570–571
  of the form $b^{m/n}$, 569–570, 574
  of the form $b^{1/n}$, 568
  simplifying, 571–572, 575
rational expressions
 addition of
  with like denominators, 495–496
  with unlike denominators, 498–500, 501
 applications of, 476–478
 definition of, 468
 division of, 475–476
 evaluating, 469–470
 fundamental property of, 471–474
 least common denominator of, 496–498, 501
 multiplication of, 474–475
 negative, 469
 reciprocals of, 475–476
 simplifying, 472–474, 478
 subtraction of
  with like denominators, 495–496, 500–501
  with unlike denominators, 498–500
rational functions
 definition of, 468
 domain of, 470–471, 478
 evaluating, 469–470
rational inequalities. *See also* nonlinear inequalities
 solving, using boundary number method, 705–706
rationalizing the denominator, 599–601, 603
 using conjugates, 601–602
rational numbers
 as exponents. *See* rational exponents
 set of, 4, 5–7
real number(s), 4–7
 addition of, 15–16
  with different signs, 16
  with same signs, 15
 definition of, 5
 division of, 18–20
 multiplication of, 18–20
 opposites of, 8–9, 10
 set of, 4–7
 subtraction of, 16–17
real number line, 7–8
reciprocals
 definition of, 19, 475
 of rational expressions, 475–476
rectangle
 area of, 75–77
 perimeter of, 75–77

rectangular coordinate system
definition of, 148
plotting ordered pairs in, 149–150, 159
reflection, 635
refocusing, 551
relations, 165–167. *See also* functions
definition of, 165
domain of, 165, 167–170, 171–172
expressions of, 165
as function
determining, 178–179, 185
vertical line test for, 180–181
input values for, 165, 170–171
output values for, 165, 170–171
range of, 165, 167–170, 171–172
remainder theorem, 489–490
responsibility, 807
right angle, 72
right triangle, 448
roots. *See also* cube root(s); square root(s)
approximating, 556–557
roster method, 2

S

scatter plot, 156
scientific notation, 383–385
conversion to standard notation, 384
operation with numbers in, 385–386
second coordinate. *See* y-coordinate
set
definition of, 2
intersection of, 3, 101–102
symbols for, 2–3
union of, 3, 102–104
set-builder notation
definition of, 2
for linear inequalities in one variable, 88
simple interest applications, 323–325
simple interest formula, 323
simplification
of algebraic expressions, 35–37, 40
of common logarithms, 778–779
of complex fractions
by division, 506–507, 511
by multiplication by the LCD, 507–509
of cube roots, 554, 562
of exponential expressions, 371–372, 386–387
with rational exponents, 571–572
of logarithmic expressions, 763–764
of natural logarithms, 779
of nth roots, 555–557
of radical expressions
in form $(\sqrt[n]{a})^n$, 556–557
in form $\sqrt[n]{a^n}$, 557
using product rule for radicals, 578–581, 585–586
using quotient rule for radicals, 582–583
using rational exponents, 572–573, 575
of rational expressions, 472–474, 478
of square roots, 553, 562
slope-intercept form
definition of, 239
for linear equations in two variables, 239–240
for linear functions, 239–240
slope of a line
applications of, 247–248
definition of, 238–239
finding, 240, 248–249
formula for, 243–245
of horizontal lines, 244–245
of linear equations in two variables
with point, writing equation from, 256–258
with y-intercept
graphing equation from, 240–243, 249, 264
writing equation from, 255
negative, 243, 245
positive, 243, 245
as rate of change, 247
of vertical lines, 244–245
solutions
to linear equations in one variable, 52–53
to linear equations in two variables, 290
determining graphically, 154
ordered pair as, 150–152, 159
to linear inequalities in one variable, 86
of nonlinear systems of equations, 824
to polynomial equations, 445
to radical equations, 606
to rational equations, extraneous, 517
to systems of linear equations in three variables, 336
to systems of linear equations in two variables, 295–297
solution sets
to linear equations in one variable, 52–53
to linear inequalities in one variable, 86–88
graphing, 87–88
to linear inequalities in two variables, graphing, 271–272
to systems of linear inequalities in two variables, 347
square(s)
of binomials, 407–409, 413
difference of two
definition of, 409, 437
factoring, 437–438, 440–442
perfect, 552
sum of two, 442
square (geometry)
area of, 75–77
perimeter of, 75–77
square root(s)
approximating, 556–557
definition of, 5, 552
irrational, 5, 10
negative, 552
of negative numbers, 616
principal, 552
simplifying, 553, 562
square root functions, graphing, 561
square root property
definition of, 651
solving quadratic equations using, 651–653, 661
standard form
for ellipses, 817–818
for hyperbolas, 819
for linear equations in two variables, 213–214
for parabola, 809
for polynomial equations, 445
for polynomials, 392
standard notation, conversion from scientific notation, 384
straight angles, 72
study skills, 289
subscripts, 24
substitution method
definition of, 305
factoring trinomials with, 432–433
solving equations by, 682–683, 686
solving nonlinear systems of equations by, 824–828
solving systems of linear equations in two variables by, 305–308, 314–315
special cases of, 311–313
subtraction
of complex numbers, 617–618
of functions, 726–727, 734
of polynomials, 402–404, 413
of radical expressions, 589–591
of rational expressions
with like denominators, 495–496, 500–501
with unlike denominators, 498–500
of real numbers, 16–17
translating phrases into expressions, 37–38
sum of functions, 726
sum of two cubes, factoring, 439–442
sum of two squares, factoring, 442
supplementary angles, 329
definition of, 72
finding, 72–75, 81
synthetic division, 487–489, 492

systems of linear equations in three variables
applications of, 339–342
consistent, 336
definition of, 336
solutions to, 336
solving, 336–339, 343
systems of linear equations in two variables
applications of, 297–299, 313–314
distance, 326–328
geometry, 329–330
investment, 323–325
mixtures, 325–326
money, 320–323
consistent with dependent equations, 295–297
consistent with independent equations, 295–297
definition of, 290
inconsistent, 295–297
solutions to, types of, 295–297
solving
by elimination, 308–311
by graphing, 291–293, 299–300
by substitution, 305–308, 314–315
special cases of
solving by elimination, 311–313
solving by graphing, 294–295
solving by substitution, 311–313
systems of linear inequalities in two variables
applications of, 350–352
definition of, 347
graphing, 347–350, 352
solution set of, 347
systems of nonlinear equations. *See* nonlinear systems of equations
systems of nonlinear inequalities. *See* nonlinear systems of inequalities

T

tables, relations expressed as, 165
temperature conversion formula, 78–79
terms
in algebraic expressions, 35
constant, 391
definition of, 391
degree of, 391–392
greatest common factor of a set of, 418–419
like, 35
test points, solving absolute value inequalities with, 131–133
time management skills, 1
trial-and-error method, factoring trinomials by, 428–430, 441–442
triangles
30°-60°-90°, 593
45°-45°-90°, 593
area of, 75–77
length of legs of, 448
perimeter of, 75–77
right, 448
trinomials
definition of, 393
factoring
$ax^2 + bx + c$, 428–432
by grouping, 430–432, 441–442
perfect square trinomials, 441
by substitution, 432–433
by trial and error, 441–442
$x^2 + bx + c$, 425–428, 433
perfect square
from completing the square, 653–655
factoring, 441
from squaring binomials, 408

U

undefined slope, 244–245
union of sets, 3, 102–104
unlike radicals, 589
upper half-plane, 270

V

variable fee, 217
variables
in algebraic expressions, 23–24
definition of, 2
dependent, 182
independent, 182
solving formula for, 79–81
variation
combined, 535–537
direct, 530–532, 537
inverse, 533–535, 538
joint, 535–537
vertex formula, 693–695
vertex of parabola, in graph of quadratic function, 637, 647
$f(x) = a(x - h)^2 + k$, 643
vertical angles, 72
vertical lines
equation for, 228
graph of, 228–230
slope of, 244–245
vertical line test, 180–181
volume formula, 81
V-shaped graph, from linear equations in two variables, 152

W

whole numbers, set of, 4, 6–7
work applications, 526
work equation, 526

X

x-axis, definition of, 148
$x^2 + bx + c$, factoring, 425–428, 433
x-coordinate, 148
x-intercepts
of linear equations in two variables, graphing using, 226–228, 232–233
of parabola, 694
of polynomial functions, 450–451

Y

y-axis, definition of, 148
y-coordinate, 148
y-intercepts
of linear equations in two variables
graphing equation from, with slope, 240–243, 249
graphing using, 226–228, 232–233
initial value of, 247
writing equation from, with slope, 255, 264
of parabola, 694

Z

zero
as denominator, 469
as dividend, 19
as exponent, 369, 381
as numerator, 468
product property of, 18
as slope of a line, 244–245
zero of a function, 216, 450
zero of a polynomial function, 450–451
zero of a quadratic function, 659–660
zero products property, 445–446, 452